Third Reference Catalogue
of Bright Galaxies

Gérard de Vaucouleurs   Antoinette de Vaucouleurs
Harold G. Corwin, Jr.   Ronald J. Buta
Georges Paturel   Pascal Fouqué

Third Reference Catalogue of Bright Galaxies

Volume II

Springer Science+Business Media, LLC

Gérard de Vaucouleurs
Department of Astronomy
University of Texas
Austin, TX 78712-1083
USA

Antoinette de Vaucouleurs (deceased)
Department of Astronomy
University of Texas
Austin, TX 78712-1083
USA

Harold G. Corwin, Jr.
Department of Astronomy
University of Texas
Austin, TX 78712-1083
USA

Ronald J. Buta
Department of Astronomy
University of Texas
Austin, TX 78712-1083
USA

Georges Paturel
Observatoire de Lyon
69230 Saint-Genis Laval
France

Pascal Fouqué
Observatoire d'Astrophysique
92195 Meudon
France

Library of Congress Cataloging-in-Publication Data
Third reference catalogue of bright galaxies / Gérard de Vaucouleurs . . . [et al.].
      p.   cm.
      Includes bibliographical references.
      Contents: v. 1. Introduction, references, notes, and appendices —
v. 2. Data for galaxies between 0h and 12h — v. 3. Data for
galaxies between 12h and 24h.
      ISBN 978-1-4757-4362-3   ISBN 978-1-4757-4360-9 (eBook)
      DOI 10.1007/978-1-4757-4360-9
      1. Galaxies — Catalogs.   I. Vaucouleurs, Gérard Henri de. 1918–.
      QB857.T47   1991
      523.1′12′0216 — dc20   91-9186

Printed on acid-free paper.

© 1991 Springer Science+Business Media New York
Originally published by Springer-Verlag New York, Inc. in 1991
Softcover reprint of the hardcover 1st edition 1991

Camera-ready copy provided by the authors.

9 8 7 6 5 4 3 2 1

ISBN 978-1-4757-4362-3

1. Introduction

This second volume of RC3 includes data for galaxies between 0^h and 12^h. For convenience, we repeat the explanation of the catalogue entries below. Details of the reduction procedures, and Notes, References, and Appendices are in Volume 1.

2. The Catalogue

The data for each galaxy are found on four successive lines on a single page. The entries are as follows:

Column 1: Positions

> Line 1: **RA** and **DEC** = right ascension and declination for the equinox 2000.0, precessed from the 1950.0 position in Column 1, Line 4, given to 0.1 second of time and 1 arcsec when available, and to 0.1 minute of time and 1 arcmin otherwise (Section 3.1.a).
>
> Line 2: l and **b** = galactic longitude and latitude in the IAU 1958 system (Blaauw *et al.* 1960); both to $0°01$.
>
> Line 3: **SGL** and **SGB** = supergalactic longitude and latitude in the RC2 system (Section 3.1.b), both to $0°01$.
>
> Line 4: **RA** and **DEC** = right ascension and declination for the equinox 1950.0 (Section 3.1.a).

Column 2: Names = commonly used designations for the galaxies (Section 3.2).

> Line 1: Names (*e.g.*, LMC, SMC) or NGC and IC designations.
>
> Line 2: UGC (Nilson 1973), ESO (Lauberts 1982), MCG (Vorontsov-Velyaminov *et al.* 1962–1974), UGCA (Nilson 1974), and CGCG (Zwicky *et al.* 1961–1968) designations, given in that order of preference. MCG designations not listed here are given in UGC and ESO.
>
> Line 3: Other common designations (see Table 1 for a complete list).
>
> Line 4: PGC (Paturel *et al.* 1989a,b) designation. For cross identifications of various catalogues with PGC, see Appendix 10.

Column 3: Types and Luminosity Classes

> Line 1: **Type** = mean revised morphological type in the RC2 system, coded as in RC2 (see Section 3.3.a).
>
> Line 2: S_T and n_L = sources of revised type estimates and number of luminosity class estimates.
>
> Line 3: **T** = mean numerical index of stage along the Hubble sequence in RC2 system, coded as explained in Section 3.3.c, and its mean error.

V

Line 4: L = mean numerical luminosity class in RC2 system, coded as explained in Section 3.3.d, and its mean error.

Column 4: Optical Diameters and Axis Ratios

Line 1: $\log D_{25}$ = mean decimal logarithm of the apparent major *isophotal* diameter measured at or reduced to surface brightness level $\mu_B = 25.0$ B-m/ss, and its mean error, as explained in Section 3.4.a. Unit of D is 0.1 arcmin to avoid negative entries.

Line 2: $\log R_{25}$ = mean decimal logarithm of the ratio of the major isophotal diameter, D_{25}, to the minor isophotal diameter, d_{25}, measured at or reduced to the surface brightness level $\mu_B = 25.0$ B-m/ss, and its mean error, as explained in Section 3.4.b.

Line 3: $\log A_e$ = decimal logarithm of the apparent diameter (in 0.1 arcmin) of the "effective aperture," the circle centered on the nucleus within which one-half of the total B-band flux is emitted, and its mean error, both derived as explained in Section 3.4.c.

Line 4: $\log D_o$ = decimal logarithm of the isophotal major diameter corrected to "face-on" (i = 0°), and corrected for galactic extinction to $A_g = 0$, but not for redshift, as explained in Section 3.4.d.

Column 5: Major Axis Position Angle, Galactic and Internal Extinctions

Line 1: p. a. = position angle, measured in degrees from north through east (all < 180°), taken when available from UGC, ESO, and ESGC (and in a few cases from H I data) (Section 3.5.a).

Line 2: A_g = Galactic extinction in B-band magnitudes, calculated following Burstein and Heiles (1978a,b, 1982, 1984), as explained in Section 3.5.b.

Line 3: A_i = internal extinction in B-band magnitudes (for correction to face-on), calculated from $\log R$ and T, as explained in Section 3.5.c.

Line 4: A_{21} = H I line self-absorption in magnitudes (for correction to face-on), calculated from $\log R$ and $T \geq 1$, as explained in section 3.5.d.

Column 6: Optical and Infrared Magnitudes

Line 1: B_T = total (asymptotic) magnitude in the B system, and its mean error, derived by extrapolation from photoelectric aperture-magnitude data, B_T^A, and from surface photometry with photoelectric zero point, B_T^S, as explained in Section 3.6.a. The magnitude is followed by an "M" when it is the weighted mean of B_T^A and B_T^S, by a "V" when it is a V-band magnitude rather than a B-band magnitude, and by a "v" when the nucleus of the galaxy is variable. The magnitude is replaced by an asterisk (*) when deriving B_T^A would have required an extrapolation in excess of 0.75 mag.

Line 2: m_B = photographic magnitude and its mean error from Ames (1930), Shapley and Ames (1932), CGCG, Buta and Corwin (1986), and/or Lauberts and Valentijn (1989) reduced to the B_T system, as explained in Section 3.6.b.

Line 3: m_{FIR} = far-infrared magnitude calculated from $m_{FIR} = -20.0 - 2.5 \log$ FIR, where FIR is the far infrared continuum flux measured at 60 and 100 microns as listed in the *IRAS Point Source Catalog* (1987). For galaxies larger than 8′ in RC2 and for the Virgo cluster area, resolved by the IRAS beam, integrated fluxes are taken from Rice *et al.* (1988) or Helou *et al.* (1988). See Section 3.6.c for details.

Line 4: B_T^o = total "face-on" magnitude corrected for Galactic and internal extinction, and for redshift, as explained in Section 3.6.d.

Column 7: Total Color Indices

Line 1: $(B-V)_T$ = total (asymptotic) color index in the Johnson B−V system, and its mean error, derived by extrapolation from photoelectric color-aperture data, and/or from surface photometry with photoelectric zero point, as explained in Section 3.7.a.

Line 2: $(U-B)_T$ = total (asymptotic) color index in the Johnson U−B system, and its mean error, derived by extrapolation from photoelectric color-aperture data, and/or from surface photometry with photoelectric zero point, as explained in Section 3.7.a.

Line 3: $(B-V)_T^o$ = total B−V color index corrected for Galactic and internal extinction, and for redshift, as explained in Section 3.7.b.

Line 4: $(U-B)_T^o$ = total U−B color index corrected for Galactic and internal extinction, and for redshift, as explained in Section 3.7.b.

Column 8: Effective Color Indices and B-band Surface Brightness

Line 1: $(B-V)_e$ = mean B−V color index, and its mean error, within the effective aperture A_e, derived by interpolation from photoelectric color-aperture data, as explained in Section 3.7.a.

Line 2: $(U-B)_e$ = mean U−B color index, and its mean error, within the effective aperture A_e, derived by interpolation from photoelectric color-aperture data, as explained in Section 3.7.a.

Line 3: m_e' = mean B-band surface brightness in magnitudes per square arcmin (B-m/sm) within the effective aperture A_e, and its mean error, as given by the relation $m_e' = B_T + 0.75 + 5 \log A_e - 5.26$. m_e' is statistically related to the effective mean surface brightness, μ_e' (RC2, p. 31; Olson and de Vaucouleurs 1981), with which it coincides when $\log R = 0$ ($i = 0°$) (Section 3.8.a).

Line 4: m_{25}' = the mean surface brightness in magnitudes per square arcmin (B-m/sm) within the $\mu_B = 25.0$ B-m/ss elliptical isophote of major axis $\log D_{25}$ and axis ratio $\log R_{25}$, defined as in RC2 (Equation 21) by:

$$m_{25}' = B_T + \Delta m_{25} + 5 \log D_{25} - 2.5 \log R_{25} - 5.26,$$

where $\Delta m_{25} = 2.5 \log L_T/L_{D_{25}} = B_{25} - B_T$ is the magnitude increment contributed by the outer regions of a galaxy fainter than $\mu_B = 25.0$ B-m/ss. For details, see Section 3.8.b.

Column 9: 21-cm Magnitude and Linewidths, Hydrogen Index

> **Line 1:** m_{21} = 21-cm emission line magnitude, and its mean error, defined by m_{21} $= 21.6 - 2.5 \log S_H$, where S_H is the measured neutral hydrogen flux density in units of $10^{-24}\,\mathrm{W\,m^{-2}}$. For details, see Section 3.9.a.

> **Line 2:** W_{20} = neutral hydrogen line full width (in $\mathrm{km\,s^{-1}}$) measured at the 20% level (I_{20}/I_{max}), and its mean error, as explained in Section 3.9.b.

> **Line 3:** W_{50} = neutral hydrogen line full width (in $\mathrm{km\,s^{-1}}$) measured at the 50% level (I_{50}/I_{max}), and its mean error, as explained in Section 3.9.b.

> **Line 4:** *HI* = corrected neutral hydrogen index, which is the difference $m^o_{21} - B^o_T$ between the corrected (face-on) 21-cm emission line magnitude and the similarly corrected magnitude in the B_T system. Details are given in Section 3.9.c.[1]

Column 10: Radial Velocities

> **Line 1:** $V_{21} = cz$ is the mean heliocentric radial velocity, and its mean error, in $\mathrm{km\,s^{-1}}$ derived from neutral hydrogen observations, as explained in Section 3.10.a.

> **Line 2:** $V_{opt} = cz$ is the mean heliocentric radial velocity, and its mean error, in $\mathrm{km\,s^{-1}}$ derived from optical observations, as explained in Section 3.10.b.

> **Line 3:** V_{GSR} = the weighted mean of the neutral hydrogen and optical velocities, corrected to the "Galactic standard of rest," as explained in Section 3.10.c.

> **Line 4:** V_{3K} = the weighted mean velocity corrected to the reference frame defined by the 3°K microwave background radiation, as explained in Section 3.10.d.

---

[1]Since m_{21} and B_T are listed separately in columns 6 and 9, line 1, there is no need to print the uncorrected index.

R.A. 2000 DEC. / l b / SGL SGB / R.A. 1950 DEC.	Names / PGC	Type / S_T n_L / T / L	$\log D_{25}$ / $\log R_{25}$ / $\log A_e$ / $\log D_o$	p.a. / A_g / A_i / A_{21}	B_T / m_B / m_{FIR} / B_T^o	$(B-V)_T$ / $(U-B)_T$ / $(B-V)_T^o$ / $(U-B)_T^o$	$(B-V)_e$ / $(U-B)_e$ / m'_e / m'_{25}	m_{21} / W_{20} / W_{50} / HI	V_{21} / V_{opt} / V_{GSR} / V_{3K}
000001.8+471628	A 2357+47	.SBT3..	1.33± .04	165				15.19±.1	5017± 9
113.96 -14.70	UGC 12889	U	.07± .05	.52	13.4 ±.3			455± 11	5217
341.85 20.74		3.0± .7		.10				464± 8	
235728.5+465946	PGC 2		1.38	.04	12.79			2.37	4751
000008.7-062229		.S..1P?	1.18± .05	168					
90.19 -65.93	MCG -1- 1- 16	E (1)	.74± .05	.09					6493± 60
286.61 11.35		1.0±1.8		.75					6581
235734.9-063911	PGC 12	7.0±1.3	1.19	.37					6147
000021.3-023638		.E+..*.	1.15± .06	120					
94.25 -62.60	MCG -1- 1- 20	E	.25± .05	.12					
290.34 12.41		-4.0±1.2		.00					
235747.5-025320	PGC 23		1.09						
000022.5-804730		PSXT4P.	1.19± .04						7900
305.44 -36.09	ESO 12- 12	S (1)	.07± .04	.36	13.94 ±.14				7747
215.32 -13.35	IRAS23577-8104	4.0± .8		.10	13.25				7880
235747.0-810412	PGC 30	2.2± .8	1.23	.04	13.42				
0000.4 +3930		.I..9..	.96± .09					15.51±.1	329± 6
112.33 -22.31	UGC 12894	U	.00± .06	.47				55± 4	
333.64 20.33		10.0± .9		.00				34± 5	522
2357.9 +3914	PGC 35		1.00	.00					35
0000.4 +1713		.SA.8..	1.23± .06						1108
105.91 -43.95	UGC 12893	U	.01± .06	.07	14.6 ±.3				
310.37 17.28		8.0± .8		.01					1262
2357.9 +1657	PGC 38		1.24	.00	14.48				761
0000.4 +0750		.SB.2..	.91± .07						
101.69 -52.91	UGC 12892	U	.24± .05	.19	15.40 ±.12				
300.81 15.18		2.0± .9		.30					
2357.9 +0734	PGC 39		.93	.12					
000029.0-402900		.SBR4*.	1.28± .05	153					3164± 34
337.93 -73.00	ESO 293- 27	S (1)	.55± .05	.03	14.40 ±.14				3135
253.93 -.18		4.0±1.0		.81					2921
235755.0-404542	PGC 43	5.2± .7	1.29	.28	13.54				
000038.0+200333		.S..7..	.90± .07	100					6752
106.99 -41.23	UGC 12895	U	.09± .08	.10					6912
313.30 17.78	KUG 2358+197	7.0± .9		.13					6409
235804.3+194651	PGC 53		.91	.05					
000038.3+282306		.S..2..	1.06± .06	11	14.84S±.15			17.52±.3	8705± 10
109.57 -33.17	UGC 12897	U	.48± .05	.16	14.87 ±.12				8705± 41
321.95 19.13	HICK 99A	2.0± .9		.59				607± 7	8882
235804.6+280624	PGC 54		1.07	.24	14.02		13.77± .35	3.26	8378
0000.6 +3336		.S..6*.	.96± .17	10					4778± 10
110.96 -28.08	UGC 12898	U	.39± .12	.18					
327.43 19.77	KAZ 237	6.0±1.3		.58					4963
2358.1 +3320	PGC 55		.98	.20					4465
000044.3+282405		.S?....	.89± .11		15.71S±.15				
109.60 -33.15	MCG 5- 1- 21		.24± .07	.16	15.37 ±.12				8216± 41
321.97 19.11	HICK 99C			.35					8393
235810.6+280723	PGC 58		.90	.12	14.96		14.40± .62		7889
000046.8+282410		.L?....	.98± .09		14.68S±.15				
109.61 -33.16	UGC 12899		.01± .03	.16	14.22 ±.10				8870± 31
321.97 19.11	HICK 99B			.00					9047
235813.1+280728	PGC 63		1.00		14.07		14.43± .46		8543
000052.9-472118		RLBR+*/	1.07± .05	156					
325.70 -67.48	ESO 193- 9	Sr	.55± .03	.04	15.18 ±.14				5912± 34
247.46 -2.67	FAIR 623	-.9± .6		.00					5859
235819.0-473800	PGC 65		.99		15.05				5701
000055.9-404242		.SBT5*.	1.02± .05	17					
337.20 -72.88	ESO 293- 29	Sr (1)	.26± .05	.03	14.95 ±.14				14710
253.74 -.34		5.4± .7		.39					14679
235822.0-405924	PGC 69	2.2±1.3	1.02	.13	14.45				14467
0000.9 +2020		.S..6*.	1.26± .04	111					6803
107.19 -40.98	UGC 12900	U	1.09± .05	.27	15.50 ±.12				6964
313.61 17.76		6.0±1.4		1.47					6460
2358.4 +2004	PGC 70		1.29	.50	13.73				
000057.9-333642		.S?....	1.08± .05	120					6916± 35
359.07 -77.34	ESO 349- 17		.10± .05	.02	14.46 ±.14				6910
260.46 2.15				.13					6643
235824.1-335324	PGC 73		1.08	.05	14.24				

R.A. 2000 DEC. / l b / SGL SGB / R.A. 1950 DEC.	Names / PGC	Type / S_T n_L / T / L	$\log D_{25}$ / $\log R_{25}$ / $\log A_e$ / $\log D_o$	p.a. / A_g / A_i / A_{21}	B_T / m_B / m_{FIR} / B_T^o	$(B-V)_T$ / $(U-B)_T$ / $(B-V)_T^o$ / $(U-B)_T^o$	$(B-V)_e$ / $(U-B)_e$ / m' / m'_{25}	m_{21} / W_{20} / W_{50} / HI	V_{21} / V_{opt} / V_{GSR} / V_{3K}
000102.9-431948		.E+2...	1.16± .05	172					
331.87 -70.86	ESO 241- 10	S	.10± .04	.03	14.09 ±.14				
251.27 -1.28		-4.0± .6		.00					
235829.0-433630	PGC 75		1.13						
000102.9+285442		.SB.3..	1.18± .05	48				15.59±.3	6899± 10
109.82 -32.67	UGC 12901	U	.42± .05	.22	14.60 ±.12				
322.51 19.12		3.0± .8		.59				405± 7	7076
235829.1+283800	PGC 76		1.20		.21	13.74			1.64	6573
0001.0 +0427	IC 5374	.S?....	.82± .13	178					
99.96 -56.15	MCG 1- 1- 10		.08± .07	.12	15.19 ±.12				8932± 57
297.44 14.18				.09					9053
2358.5 +0411	PGC 79		.83		.04 14.89				8579
0001.0 +0430	IC 5375	.S?....	.99± .10						
99.99 -56.10	MCG 1- 1- 9		.62± .07	.12	15.24 ±.13				9073± 57
297.49 14.19				.76					9194
2358.5 +0414	PGC 80		1.00		.31 14.35				8720
0001.0 +0613	NGC 7802	.L.....	1.05± .08	52					
101.02 -54.49	UGC 12902	U	.30± .03	.16	14.48 ±.10				
299.21 14.63		-2.0± .9		.00					
2358.5 +0557	PGC 81		1.02						
000113.6+130838		.S?....	.94± .11		15.87S±.15				
104.52 -47.92	MCG 2- 1- 9		.28± .07	.16					5461± 41
306.23 16.23	HICK 100C			.42					5605
235839.8+125156	PGC 89		.95		.14 15.27		14.69± .58		5110
000114.3+344033		.SB.2..	1.02± .06	40					
111.36 -27.06	UGC 12904	U	.14± .05	.34	15.08 ±.14				
328.56 19.76		2.0± .9		.17					
235840.6+342351	PGC 91		1.05		.07				
0001.2 +8039		.S..6*.	1.04± .08	55					
120.81 18.00	UGC 12905	U	.98± .06	.99					
16.72 17.67		6.0±1.4		1.44					
2358.7 +8023	PGC 94		1.13		.49				
0001.2 +0619		.S..4..	1.02± .06	155					14742
101.16 -54.41	UGC 12903	U	.70± .05	.16	15.56 ±.12				
299.33 14.61		4.0±1.0		1.03					14868
2358.7 +0603	PGC 96		1.03		.35 14.28				14388
000117.8-530030		.S..3*/	1.13± .05	37					
319.36 -62.49	ESO 149- 12	S	.80± .04	.00	16.06 ±.14				
242.12 -4.69		3.0±1.3		1.10					
235844.0-531712	PGC 99		1.13		.40				
000118.8-272506		.LX.-..	1.08± .05	114					
28.57 -78.86	ESO 409- 1	S	.21± .03	.04	14.92 ±.14				
266.38 4.23		-3.0± .8		.00					
235845.1-274148	PGC 100		1.05						
000119.8+130642	NGC 7803	.S...0..	1.00± .06	85	14.06S±.15			15.91±.1	5367± 6
104.54 -47.96	UGC 12906	U	.21± .05	.16	13.85 ±.15			321± 11	5260± 27
306.20 16.20	IRAS23587+1249	.0± .9		.15	12.28			232± 7	5505
235846.0+125000	PGC 101		1.00			13.57		13.38± .37		5011
000120.0+343133	IC 5376	.S..2..	1.30± .04	4					5029
111.34 -27.22	UGC 12909	U	.76± .05	.22	14.59 ±.12				
328.41 19.73	KAZ 239	2.0± .9		.94					5215
235846.2+341451	PGC 102		1.32		.38 13.38				4718
000126.2+130646		.S?....	.89± .11		15.26S±.15				
104.58 -47.97	MCG 2- 1- 12		.14± .07	.16	15.07 ±.12				5253± 41
306.21 16.18	HICK 100B			.21					5397
235852.4+125004	PGC 108		.90		.07 14.75		14.19± .62		4902
000126.6+312602	NGC 7805	.LX.0*P	1.07± .07		14.2 ±.3	.91± .04	1.00± .02		
110.59 -30.24	UGC 12908	R	.11± .03	.17	14.13 ±.10	.47± .06	.56± .03		4850± 31
325.16 19.36	MK 333	-2.0± .6	.51± .12	.00		.81	12.20± .39		5032
235852.7+310920	PGC 109		1.07		13.88	.45	14.09± .47		4531
0001.4 +0521		.SBR9..	1.00± .16						3949
100.68 -55.34	UGC 12910	U	.00± .12	.12					
298.37 14.31		9.0± .8		.00					4072
2358.9 +0505	PGC 110		1.01		.00				3596
000130.2+312633	NGC 7806	.SAT4$P	1.04± .06	20				16.89±.2	4768± 7
110.61 -30.23	UGC 12911	R	.16± .05	.17	14.31 ±.12			349± 14	4761± 30
325.17 19.35		4.0± .5		.23				273± 7	4949
235856.4+310951	PGC 112		1.05		.08 13.88			2.93	4448

R.A. 2000 DEC. / l b / SGL SGB / R.A. 1950 DEC.	Names / PGC	Type / ST nL / T / L	logD25 / logR25 / logA. / logD0	p.a. / Ag / Al / A21	BT / mB / mFIR / BT°	(B-V)T / (U-B)T / (B-V)T° / (U-B)T°	(B-V). / (U-B). / m'. / m'25	m21 / W20 / W50 / HI	V21 / Vopt / VGSR / V3K
0001.5 +0900		.S?....	.96± .17						9249
102.72 -51.89	UGC 12912		.10± .12	.21	15.18 ±.14				
302.05 15.20				.15					9382
2359.0 +0844	PGC 116		.98	.05	14.77				8896
000136.1+333345			.62± .11					16.54±.3	7431± 10
111.17 -28.17	CGCG 498- 66		.11± .12	.19	15.25 ±.18				
327.40 19.57	KAZ 241							222± 10	7616
235902.3+331703	PGC 119		.64						7118
000137.9+232905 A 2359+23A		RS.R6*P	1.36± .04	160				14.66±.2	4371± 10
108.40 -37.98	UGC 12914	R	.25± .05	.16	13.07 ±.12			628± 8	4425± 25
316.88 18.16	KAZ 240			.38				487± 10	4545
235904.1+231223	PGC 120		1.37	.13	12.51			2.02	4041
000138.7+265522									7679± 10
109.42 -34.64	CGCG 477- 39			.12	15.6 ±.4				
320.46 18.70									7853
235904.9+263840	PGC 121								7349
0001.6 +0330		.S..6*.	1.11± .05	5					6340
99.61 -57.09	UGC 12913	U	.84± .05	.07					
296.53 13.78		6.0±1.4		1.24					6458
2359.1 +0314	PGC 124		1.12	.42					5987
000142.5+232941 A 2359+23B		.S?....	1.19± .05	137				14.77±.3	4336± 10
108.43 -37.97	UGC 12915		.48± .05	.16	13.95 ±.13			571± 13	4418± 30
316.90 18.15				.71				544± 10	4511
235908.6+231259	PGC 129		1.20	.24	13.06			1.48	4007
000152.7-365100		PSXT2..	1.06± .05						15117± 81
347.00 -75.71	ESO 349- 19	Sr	.11± .04	.03	15.32 ±.14				15100
257.46 .84		1.7± .5		.14					14858
235919.0-370742	PGC 138		1.06	.06	15.00				
000155.7-273736		PSAS3*.	1.13± .04		14.66 ±.16	.94± .02	.95± .02		
27.49 -79.00	ESO 409- 3	Sr (1)	.07± .04	.04	14.16 ±.14	.19± .04	.20± .03		8507± 52
266.23 4.03		2.7± .5	.43± .04	.10		.86	12.30± .10		8522
235922.0-275418	PGC 140	2.2± .6	1.14	.03	14.18	.13	15.00± .28		8213
000156.9-152701 WLM		.IBS9..	2.06± .02	4	11.03M±.08	.44± .04	.39± .01	11.24±.1	-120± 4
75.87 -73.61	MCG -3- 1- 15	PUE (2)	.46± .03	.10			-.23± .02	79± 4	-75± 21
277.91 8.09	DDO 221	10.0± .3	1.75± .02	.35	14.07	.31	15.24± .04	57± 4	-61
235923.2-154343	PGC 143	9.0± .5	2.07	.23	10.58		15.03± .14	.42	-447
0001.9 +1733		.S..7*.	1.00± .08	170					6352
106.52 -43.73	UGC 12916	U	.19± .06	.07					
310.79 16.99		7.0±1.3		.26					6506
2359.4 +1717	PGC 146		1.01	.09					6006
0001.9 +4019		.SBR3..	1.04± .08	95					
112.82 -21.57	UGC 12917	U	.13± .06	.44					
334.53 20.10		3.0± .9		.18					
2359.4 +4003	PGC 148		1.08	.06					
000203.7-332806		PSBS1..	.95± .06	8					
359.17 -77.61	ESO 349- 20	r	.23± .05	.02	14.71 ±.14				8758± 81
260.68 1.98	IRAS23594-3344	1.0± .9		.23	13.15				8752
235930.0-334448	PGC 151		.95	.11	14.34				8485
0002.0 +1635 IC 5377		.I..9*.	1.02± .06						1050
106.19 -44.67	UGC 12918	U	.25± .05	.08	15.20 ±.14				
309.80 16.77		10.0±1.2		.19					1202
2359.5 +1619	PGC 156		1.02	.12	14.93				703
000209.5+025626 NGC 7809		.I..9?.	.67± .11	95					
99.45 -57.67	MCG 0- 1- 19	E	.07± .05	.09	15.14 ±.15				
295.99 13.51	3ZW 126	10.0±2.0		.05					
235935.7+023944	PGC 158		.68	.04					
000210.5-435831		.LX.-..	1.18± .09						
330.23 -70.45	MCG -7- 1- 8	S	.09± .08	.03					
250.73 -1.70		-3.0±1.6		.00					
235937.0-441513	PGC 160		1.17						
000219.4+125819 NGC 7810		.L.....	1.09± .07	80					
104.83 -48.17	UGC 12919	U	.15± .03	.16	14.03 ±.10				5532± 50
306.12 15.94	IRAS23597+1241	-2.0± .9		.00	11.77				5675
235945.5+124137	PGC 163		1.08		13.79				5181
0002.3 +2712		.S..4..	1.09± .06	47					7613
109.69 -34.40	UGC 12920	U	.82± .05	.13	15.48 ±.12				
320.78 18.59		4.0±1.0		1.20					7787
2359.8 +2656	PGC 165		1.10	.41	14.10				7284

R.A. 2000 DEC. / l b / SGL SGB / R.A. 1950 DEC.	Names / PGC	Type / S_T n_L / T / L	$\log D_{25}$ / $\log R_{25}$ / $\log A_e$ / $\log D_o$	p.a. / A_g / A_i / A_{21}	B_T / m_B / m_{FIR} / B_T^o	$(B-V)_T$ / $(U-B)_T$ / $(B-V)_T^o$ / $(U-B)_T^o$	$(B-V)_e$ / $(U-B)_e$ / m'_e / m'_{25}	m_{21} / W_{20} / W_{50} / HI	V_{21} / V_{opt} / V_{GSR} / V_{3K}
000226.7+032108	NGC 7811	CI..9??	.61± .12	15				16.89±.3	7650± 12
99.85 -57.31	MCG 0- 1- 20	E	.04± .05	.09	15.06 ±.18			247± 21	7641±173
296.42 13.55	MK 543							226± 15	7767
235953.0+030426	PGC 168		.62						7297
0002.5 +4455		.S..8*.	1.00± .08	120					
113.91 -17.09	UGC 2	U	.73± .06	.48					
339.39 20.23		8.0±1.4		.89					
0000.0 +4439	PGC 175		1.04	.36					
000234.8-034238		.SBS4?.	1.09± .06	0					
94.32 -63.83	MCG -1- 1- 24	E (1)	.20± .05	.14					
289.42 11.56		4.0± .9		.29	12.87				
000001.0-035920	PGC 176	1.9±1.2	1.10	.10					
000240.4+084415		.S?....	.82± .13						5538
103.00 -52.24	MCG 1- 1- 13		.74± .07	.23	15.83 ±.17				
301.85 14.87	IRAS00001+0827			.91	13.56				5670
000006.5+082733	PGC 179		.84	.37	14.68				5185
000244.3-802048		.SBS9..	1.36± .03	18					1953
305.45 -36.54	ESO 12- 14	S (1)	.30± .05	.35	14.88 ±.14				
215.78 -13.32		9.0± .7		.30					1801
000013.0-803730	PGC 181	10.0± .8	1.40	.15	14.22				1930
0002.7 +1853		.SB.1..	1.20± .05	90					7882
107.24 -42.49	UGC 3	U	.25± .05	.08	14.37 ±.15				
312.20 17.07		1.0± .8		.25					8039
0000.2 +1837	PGC 186		1.21	.12	13.93				7538
000246.2-524606		.IBS9..	1.18± .05	100					
319.13 -62.80	ESO 149- 13	S (1)	.41± .04	.00	15.45 ±.14				
242.43 -4.81		9.5± .6		.31					
000013.0-530248	PGC 187	8.9± .7	1.18	.21					
000251.1+312915		.L?....	.90± .17					16.92±.3	4949± 10
110.94 -30.25	MCG 5- 1- 27		.00± .07	.18	15.24 ±.11				
325.26 19.07				.00				108± 7	5130
000017.0+311233	PGC 190		.92		14.99				4630
000254.5-341406	NGC 7812	.SXT3*.	1.04± .05	125					6842± 81
355.68 -77.39	ESO 349- 21	BSr	.19± .04	.02	13.94 ±.14				
260.02 1.55	IRAS00003-3431	2.8± .6		.26	13.08				6833
000021.0-343048	PGC 195		1.04	.09	13.60				6572
0003.0 +0412		.S..4..	1.01± .06	155					8630
100.65 -56.55	UGC 4	U	.19± .05	.08	15.11 ±.14				
297.32 13.63		4.0± .9		.28					8749
0000.5 +0356	PGC 201		1.01	.10	14.69				8277
000305.3-421042		.SAR2*.	1.02± .05	128					
333.08 -72.00	ESO 293- 31	Sr (1)	.34± .04	.03	15.34 ±.14				
252.49 -1.23		1.7± .7		.42					
000032.0-422724	PGC 204	3.3±1.3	1.02	.17					
000305.8-015447		.SX.4..	1.25± .04	45					7323± 50
96.22 -62.25	UGC 5	U	.29± .04	.16	13.97 ±.13				
291.24 11.95	IRAS00005-0211	4.0± .9		.43	13.03				7424
000032.0-021129	PGC 205		1.26	.15	13.32				6973
000308.1+310222								18.27±.3	4797± 10
110.90 -30.70	CGCG 498- 68			.20	15.0 ±.4				
324.80 18.95								404± 7	4977
000034.0+304540	PGC 206								4477
000309.5+215735	A 0001+21	.P.....	.99± .05	105	14.38 ±.17	.37± .03	.45± .02	17.11±.3	6600± 9
108.37 -39.54	UGC 6	R	.14± .04	.13	14.36 ±.12	.09± .05	.17± .04	407± 10	6608± 40
315.37 17.55	MK 334	99.0	.43± .03	.20		.27	12.02± .08		6763
000035.5+214053	PGC 207		1.00	.07	14.01	.02	13.82± .31	3.03	6261
000310.1+251042								18.54±.3	7753± 10
109.33 -36.42	CGCG 477- 44			.11	15.50 ±.12				
318.70 18.10	4ZW 2							76± 7	7923
000036.0+245400	PGC 208								7420
000310.2-544458		.LXS-..	1.22± .08						
317.35 -61.01		S	.13± .08	.00					
240.56 -5.54		-3.0± .8		.00					
000037.2-550140	PGC 211		1.20						
000311.6+155756	IC 5381	.S..2$/	1.16± .05	54					11230
106.32 -45.34	UGC 7	P	.58± .04	.10	14.71 ±.12				
309.22 16.38		2.0±1.8		.71					11380
000037.6+154114	PGC 212		1.17	.29	13.78				10882

R.A. 2000 DEC. / l b / SGL SGB / R.A. 1950 DEC.	Names / PGC	Type $S_T\ n_L$ / T / L	$\log D_{25}$ / $\log R_{25}$ / $\log A_e$ / $\log D_o$	p.a. / A_g / A_I / A_{21}	B_T / m_B / m_{FIR} / B_T^o	$(B-V)_T$ / $(U-B)_T$ / $(B-V)_T^o$ / $(U-B)_T^o$	$(B-V)_e$ / $(U-B)_e$ / m'_e / m'_{25}	m_{21} / W_{20} / W_{50} / HI	V_{21} / V_{opt} / V_{GSR} / V_{3K}
000312.4-355612		.L?....	1.07± .07		14.66 ±.14				14897±115
349.30 -76.48	ESO 349- 22		.06± .05	.03					14882
258.42 .91				.00					14634
000039.0-361254	PGC 213		1.07		14.41				
000315.1+160845	NGC 7814	.SAS2*/	1.74± .02	135	11.56 ±.13	.99± .02	1.07± .01	14.10±.2	1054± 5
106.41 -45.17	UGC 8	R	.38± .02	.10	11.75 ±.10	.51± .03	.57± .01	460± 6	1042± 23
309.40 16.40		2.0± .3	1.17± .02	.47		.89	12.90± .06	464± 5	1204
000041.2+155203	PGC 218		1.75	.19	11.09	.41	14.14± .17	2.82	706
0003.2 -1045		.S?....	.94± .11		14.04V±.13	1.02± .04	1.04± .04		
85.93 -70.08	MCG -2- 1- 12		.57± .07	.06					8982± 62
282.57 9.28			.28± .02	.84		.85			9054
0000.7 -1102	PGC 219		.94	.28	14.11		13.17± .58		8644
000320.3+294742		.S..6*.	.90± .06	100				16.11±.2	6984± 7
110.63 -31.93	UGC 12	U	.10± .04	.16	15.42 ±.13			249± 13	
323.51 18.74		6.0±1.2		.15				239± 7	7162
000046.2+293100	PGC 223		.92	.05	15.08			.98	6661
0003.3 +0836		.S..6*.	1.18± .05						11944
103.20 -52.41	UGC 10	U	.03± .05	.25					
301.76 14.67	IRAS00007+0820	6.0±1.1		.04	13.70				12075
0000.7 +0820	PGC 226		1.20	.01					11591
000321.7+220610		.S?....	.97± .05	55				16.69±.3	4447± 7
108.47 -39.42	UGC 11		.13± .05	.13	15.35 ±.10				
315.53 17.53				.19				152± 7	4610
000047.6+214928	PGC 227		.98	.06	15.02			1.60	4108
000324.1-494730		.L..0*P	1.19± .05	148	14.32 ±.14				10800±190
321.86 -65.53	ESO 193- 11	S	.32± .04	.04					10737
245.29 -3.90	FAIR 361	-2.0± .8		.00					10602
000051.1-500412	PGC 233		1.15		14.12				
000326.0-531948		.SXR6..	1.01± .05	15					
318.44 -62.33	ESO 149- 15	S (1)	.10± .05	.00	15.35 ±.14				11119
241.93 -5.10		6.0± .8		.15					11044
000053.0-533630	PGC 235	5.6± .8	1.01	.05	15.15				10939
0003.4 +2721		PSBS0..	1.04± .06						7721
110.02 -34.30	UGC 13	U	.04± .05	.13	14.69 ±.14				
320.98 18.37		.0± .9		.03					7895
0000.9 +2705	PGC 240		1.05		14.41				7392
000332.3-104442	NGC 7808	PLA.0*.	1.10± .06		13.48V±.13	.85± .03			
86.11 -70.11	MCG -2- 1- 13	E	.00± .05	.06					8923± 62
282.60 9.22	IRAS00009-1101	-2.0±1.2		.00		.75			8995
000058.6-110124	PGC 243		1.11		14.14		14.71± .35		8585
000335.3+231201		.S?....	1.27± .04	32					
108.87 -38.36	UGC 14		.26± .05	.19	13.79 ±.13				7290± 46
316.67 17.67	IRAS00010+2255			.39	12.89				7455
000101.2+225519	PGC 250		1.28	.13	13.17				6953
0003.7 +1513	A 0001+14	.S..9*.	1.39± .04		14.8 ±.2	.61± .09	.69± .04	15.35±.1	878± 6
106.21 -46.10	UGC 17	PU (1)	.15± .05	.13		-.02± .11	-.03± .06	109± 7	
308.48 16.10	DDO 222	9.3± .6	1.14± .05	.15		.54	16.03± .12	99± 12	1026
0001.1 +1456	PGC 255	9.0±1.0	1.40	.07	14.56	-.07	16.29± .30	.72	530
0003.7 +0416		.S..2..	1.10± .05	63					11562
100.99 -56.55	UGC 15	U	.73± .05	.07					
297.44 13.48		2.0± .9		.89					11681
0001.2 +0400	PGC 258		1.11	.36					11209
000349.1+072845	NGC 7816	.S..4..	1.23± .05					14.86±.2	5239± 7
102.81 -53.52	UGC 16	U	.06± .05	.19	13.61 ±.14			184± 6	5141± 60
300.65 14.28	IRAS00012+0712	4.0± .8		.09	13.21			172± 6	5365
000115.2+071203	PGC 263		1.25	.03	13.30			1.53	4885
000359.0+204500	NGC 7817	.SA.4*/	1.55± .02	45				14.37±.1	2309± 4
108.23 -40.76	UGC 19	PU	.58± .03	.14	12.56 ±.12			423± 5	2157± 42
314.16 17.14	IRAS00014+2028	4.0± .5		.85	11.10			404± 5	2468
000124.9+202818	PGC 279		1.57	.29	11.56			2.53	1967
000401.0-111037		.S..1?.	1.04± .09	10					
85.71 -70.54	MCG -2- 1- 14	E	.09± .07	.08					
282.22 8.97		1.0±1.7		.10					
000127.4-112719	PGC 281		1.04	.05					
000401.7-111027		.SBT5*.	1.02± .08	90					
85.72 -70.54	MCG -2- 1- 15	E (1)	.10± .05	.08					
282.22 8.97		5.0±1.0		.15					
000128.1-112709	PGC 282	4.2±1.2	1.03	.05					

R.A. 2000 DEC. / l b / SGL SGB / R.A. 1950 DEC.	Names / PGC	Type / S_T n_L / T / L	$\log D_{25}$ / $\log R_{25}$ / $\log A_e$ / $\log D_o$	p.a. / A_g / A_i / A_{21}	B_T / m_B / m_{FIR} / B_T^o	$(B-V)_T$ / $(U-B)_T$ / $(B-V)_T^o$ / $(U-B)_T^o$	$(B-V)_e$ / $(U-B)_e$ / m'_e / m'_{25}	m_{21} / W_{20} / W_{50} / HI	V_{21} / V_{opt} / V_{GSR} / V_{3K}
0004.1 +0721	NGC 7818	.S..6*.	1.02± .06						
102.89 -53.66	UGC 21	U	.00± .05	.19	14.67 ±.15				
300.56 14.17		6.0±1.2		.00					
0001.6 +0705	PGC 288		1.03	.00					
000414.0-525458		.LA.-*.	1.13± .09						
318.56 -62.76		S	.13± .08	.00					
242.36 -5.07		-3.0±1.2		.00					
000141.3-531140	PGC 290		1.11						
000414.7-503848		.S?....	.91± .06	27					
320.68 -64.82	ESO 193- 14		.16± .05	.04	15.13 ±.14				11896± 62
244.52 -4.31				.22					11830
000142.0-505530	PGC 292		.92	.08	14.78				11702
0004.2 +1018		.L.....	1.14± .07	25					
104.36 -50.85	UGC 22	U	.17± .03	.21					
303.53 14.86		-2.0± .8		.00					
0001.7 +1002	PGC 295		1.14						
0004.2 +1046		.SBT3..	1.13± .05	20					7976
104.57 -50.40	UGC 23	U	.19± .05	.20	14.73 ±.16				
304.01 14.97		3.0± .9		.26					8112
0001.7 +1030	PGC 296		1.15	.10	14.20				7625
0004.2 +2235		.SB.6*.	.97± .09	50					4442
108.88 -38.99	UGC 24	U	.22± .06	.22	15.18 ±.13				
316.07 17.41		6.0±1.2		.33					4606
0001.7 +2219	PGC 298		.99	.11	14.62				4104
0004.3 +0609		.S..6*.	1.02± .06	123					5074
102.33 -54.82	UGC 25	U	.65± .05	.18					
299.36 13.82		6.0±1.3		.95					5198
0001.8 +0553	PGC 301		1.03	.32					4721
000424.8+312825	NGC 7819	.SBS3..	1.19± .03		14.18 ±.14	.66± .02	.69± .02	14.95±.1	4958± 6
111.32 -30.34	UGC 26	U (1)	.12± .05	.17	14.02 ±.13	-.03± .04		251± 9	4858± 59
325.29 18.73	IRAS00018+3111	3.0± .8	.83± .02	.16	12.97	.57	13.82± .04	245± 6	5137
000150.5+311143	PGC 303	2.7± .8	1.20	.06	13.73	-.10	14.68± .24	1.16	4639
0004.4 +0550		.S..6*.	1.24± .05	135					3111
102.19 -55.12	UGC 27	U	.22± .05	.18	14.7 ±.2				
299.05 13.72		6.0±1.1		.32					3234
0001.9 +0534	PGC 305		1.26	.11	14.16				2758
000430.6+051155	NGC 7820	.S..0..	1.12± .05	165					
101.85 -55.74	UGC 28	U	.36± .05	.11	13.87 ±.13				3064± 50
298.41 13.54		.0± .9		.27					3185
000156.8+045513	PGC 307		1.11		13.44				2711
000432.8+333332		.S?....	1.08± .05	126				16.40±.3	4760± 10
111.85 -28.30	UGC 30		.84± .05	.16					
327.47 18.96	KUG 0001+332			1.26				257± 7	4943
000158.4+331650	PGC 309		1.09	.42					4447
0004.5 -0704	A 0002-07	.SBR3*.	1.38± .04	75					3812
91.78 -67.06	MCG -1- 1- 28	E (1)	.32± .05	.23					
286.27 10.10		3.0± .9		.45					3896
0002.0 -0721	PGC 312	3.6± .8	1.40	.16					3469
0004.5 +2818		.E...*.	1.11± .16						
110.56 -33.43	UGC 29	U	.03± .08	.14	14.63 ±.14				
322.00 18.27		-5.0±1.1		.00					
0002.0 +2802	PGC 313		1.12						
000439.8-453742		PSXT1..	.94± .06	120					
326.56 -69.29	ESO 241- 12	r	.16± .05	.03	15.21 ±.14				
249.32 -2.68		1.0± .9		.16					
000207.0-455424	PGC 319		.94	.08					
000439.9-080602		.LAR-?.	1.15± .10						
90.57 -67.98	MCG -1- 1- 30	E	.00± .07	.11					
285.27 9.76		-3.0± .8		.00					
000206.3-082244	PGC 320		1.17						
000442.2-302900		.E?....	1.09± .06	154	13.95 ±.13	.94± .01	.98± .01		
12.03 -79.23	ESO 409- 12		.14± .04	.02	14.08 ±.14				8004± 25
263.71 2.48			.70± .02	.00		.86	12.94± .07		8008
000209.0-304542	PGC 322		1.05		13.87		14.04± .34		7720
000444.8+264955								17.32±.3	7570± 10
110.22 -34.88	CGCG 477- 49			.15	14.6 ±.4				
320.47 18.01								190± 7	7742
000210.5+263313	PGC 323								7241

R.A. 2000 DEC. / l b / SGL SGB / R.A. 1950 DEC.	Names / PGC	Type / S_T n_L / T / L	$\log D_{25}$ / $\log R_{25}$ / $\log A_e$ / $\log D_o$	p.a. / A_g / A_l / A_{21}	B_T / m_B / m_{FIR} / B_T^o	$(B-V)_T$ / $(U-B)_T$ / $(B-V)_T^o$ / $(U-B)_T^o$	$(B-V)_e$ / $(U-B)_e$ / m'_e / m'_{25}	m_{21} / W_{20} / W_{50} / HI	V_{21} / V_{opt} / V_{GSR} / V_{3K}
000445.8-160155		.SBR5..	1.17± .05	100					
76.27 -74.50	MCG -3- 1- 18	E (1)	.06± .05	.06					
277.57 7.27		5.0± .8		.09					
000212.3-161837	PGC 325	3.1±1.2	1.17	.03					
000446.9-620342	NGC 7823	.SBS4?.	1.06± .05						
312.25 -54.21	ESO 111- 12	S (1)	.05± .04		.00 13.42 ±.14				
233.63 -8.11	IRAS00022-6220	3.8± .6		.07 11.70				
000215.0-622024	PGC 328	4.4±1.2	1.06	.02					
0004.7 -0134		.S?....	.82± .13						
97.35 -62.10	MCG 0- 1- 25		.28± .07	.13 15.10 ±.14				7240± 53
291.70 11.64	IRAS00022-0150			.43 13.28				7341
0002.2 -0150	PGC 329		.83	.14 14.50				6891
000448.8-012953		.S?....	.64± .17						
97.42 -62.04	MCG 0- 1- 24		.11± .07	.13 15.14 ±.18				7110± 51
291.77 11.66	MK 544			.14				7211
000215.0-014635	PGC 330		.65	.06 14.85				6760
000451.3-542906		PLXS+?.	1.07± .05	10					
317.09 -61.34	ESO 149- 16	S	.29± .03	.00 14.74 ±.14				
240.89 -5.68		-1.0± .9		.00				
000219.0-544548	PGC 331		1.02						
0004.8 +1711		.IA.9..	1.06± .06	55					1034
107.32 -44.26	UGC 31	U	.08± .05	.08 14.82 ±.16				
310.56 16.24		10.0± .8		.06				1186
0002.3 +1655	PGC 332		1.07	.04 14.68				688
000452.7-452912		PSXS1*.	1.01± .05	164					
326.68 -69.43	ESO 241- 13	Sr	.27± .04	.03 15.53 ±.14				11900±190
249.47 -2.67	FAIR 629	.5± .6		.28				11851
000220.0-454554	PGC 336		1.02	.14 15.07				11681
000500.2-303024		RLB.0..	.96± .06	118					
11.81 -79.29	ESO 409- 13	r	.16± .03	.02 15.00 ±.14				
263.71 2.41		-2.4± .9		.00				
000227.0-304706	PGC 348		.94						
0005.0 +0507		.SB.0..	1.00± .06						
102.04 -55.85	UGC 33	U	.12± .05	.10 14.99 ±.14				
298.38 13.39		.0± .9		.09				
0002.5 +0451	PGC 353		1.01						
000506.1+065515	NGC 7824	.S..2..	1.20± .05	145					
103.03 -54.15	UGC 34	U	.13± .05	.18 14.07 ±.14				6134± 56
300.17 13.83		2.0± .8		.16				6260
000232.2+063833	PGC 354		1.22	.06 13.68				5782
000506.7+221125					.15 15.7 ±.4			17.37±.3	14451± 10
109.01 -39.43	CGCG 478- 23								
315.69 17.15								613± 7	14614
000232.4+215443	PGC 355								14113
000507.4-504918		.L?....	.95± .07	171					
320.21 -64.72	ESO 193- 17		.10± .05	.04 14.41 ±.14				11169± 62
244.40 -4.50				.00				11102
000235.1-510600	PGC 358		.94	14.20				10976
000513.3-113012	IC 1529	PLAR0P*	1.22± .05						
85.91 -70.97	MCG -2- 1- 19	E	.02± .05	.08				6903
282.00 8.59		-2.0±1.1		.00				6972
000239.7-114654	PGC 364		1.22					6567
0005.2 +0615		.S..9..	1.19± .12						3113
102.75 -54.79	UGC 35	U	.05± .12	.16				
299.52 13.63		9.0± .8		.05				3237
0002.7 +0559	PGC 365		1.21	.02				2761
0005.2 +0646		.SAR1*.	1.12± .06	18					
103.02 -54.30	UGC 36	U	.34± .05	.18 14.45 ±.12				6250± 56
300.04 13.76		1.0± .8		.34				6375
0002.7 +0630	PGC 366		1.14	.17 13.86				5898
000516.5-162842	NGC 7821	.S..6P/	1.14± .06	111					7426
75.46 -74.91	MCG -3- 1- 19	E	.48± .05	.05				
277.18 7.01	IRAS00027-1645	6.0±1.3		.71 11.94				7478
000243.1-164524	PGC 367		1.14	.24				7100
0005.4 +0510	NGC 7825	.SBS3..	1.03± .06	27					5629
102.24 -55.84	UGC 37	U	.34± .05	.10 14.49 ±.12				
298.45 13.30		3.0± .9		.46				5749
0002.9 +0454	PGC 377		1.04	.17 13.88				5277

R.A. 2000 DEC.	Names	Type	logD$_{25}$	p.a.	B$_T$	(B-V)$_T$	(B-V)$_e$	m$_{21}$	V$_{21}$
l b		S$_T$ n$_L$	logR$_{25}$	A$_g$	m$_B$	(U-B)$_T$	(U-B)$_e$	W$_{20}$	V$_{opt}$
SGL SGB		T	logA$_e$	A$_i$	m$_{FIR}$	(B-V)$_T^o$	m'$_e$	W$_{50}$	V$_{GSR}$
R.A. 1950 DEC.	PGC	L	logD$_o$	A$_{21}$	B$_T^o$	(U-B)$_T^o$	m'$_{25}$	HI	V$_{3K}$
0005.4 +0513	NGC 7827	.LB....	1.08± .10						
102.27 -55.79	UGC 38	U	.12± .05	.10	14.93 ±.14				
298.50 13.32		-2.0± .8		.00					
0002.9 +0457	PGC 378		1.07						
000529.0+323152		.S?....	.89± .11					17.20±.3	10422± 10
111.82 -29.35	MCG 5- 1- 31		.00± .07	.16	14.85 ±.12				
326.42 18.64				.00				142± 7	10603
000254.4+321510	PGC 381		.90	.00	14.63			2.56	10107
000529.3-501612		.SXT5?P	1.29± .04	78	14.33 ±.14	.78± .02	.91± .01		
320.63 -65.25	ESO 193- 19	S (1)	.24± .05	.04	14.24 ±.14				10381± 33
244.95 -4.38	FAIR 631	5.0± .8	.78± .02	.36		.66	13.72± .05		10316
000257.1-503254	PGC 382	1.1± .8	1.30	.12	13.83		15.04± .28		10185
000529.4-493712		RSAR3*.	1.14± .04	84					11400±190
321.30 -65.84	ESO 193- 18	S (1)	.38± .04	.04	14.74 ±.14				11337
245.57 -4.16	FAIR 362	3.0± .8		.53					11201
000257.1-495354	PGC 383	4.4± .8	1.14	.19	14.09				
0005.5 +5338		.S..6*.	1.07± .07	7					
116.09 -8.62	UGC 39	U	.32± .06	.94					
348.58 19.89		6.0±1.3		.48					
0003.0 +5322	PGC 393		1.16	.16					
000539.5-534859		.LAR-*.	1.25± .08						
317.39 -62.01		S	.21± .08	.00					
241.57 -5.57		-3.0± .8		.00					
000307.5-540541	PGC 399		1.22						
0005.7 +2726		.SB?...	1.06± .06	115					7531
110.65 -34.34	UGC 40		.21± .05	.13	14.92 ±.13				
321.15 17.87				.31					7704
0003.2 +2710	PGC 415		1.07	.10	14.44				7203
0005.8 -1359		.SB?...	1.04± .09						5921
81.68 -73.10	MCG -2- 1- 22		.49± .07	.07					
279.63 7.67				.73					5981
0003.3 -1416	PGC 425		1.04	.25					5590
000556.0+222928		.SB.6*.	1.01± .05	30					6597
109.33 -39.18	UGC 41	U	.49± .04	.14					
316.04 17.01	KUG 0003+222	6.0±1.3		.72					6760
000321.6+221246	PGC 431		1.03	.24					6259
000558.8-360812		.SXT2P.	.94± .05						8411± 69
347.11 -76.80	ESO 349- 27	Sr	.07± .04	.02	14.66 ±.14				
258.43 .31		1.5± .6		.08					8394
000326.0-362454	PGC 437		.94	.03	14.47				8150
000558.8-360654		.L?....	1.05± .06						8683± 69
347.18 -76.81	ESO 349- 26		.64± .05	.02	14.52 ±.14				
258.45 .32				.00					8666
000326.0-362336	PGC 438		.96		14.36				8422
000604.0-521306		.LA.-*.	1.02± .06	10					
318.60 -63.51	ESO 193- 22	S	.29± .04	.00	14.36 ±.14				
243.12 -5.10		-3.0±1.2		.00					
000332.0-522948	PGC 449		.97						
0006.0 +1307		.SB.7*.	.96± .17						5481
106.21 -48.26	UGC 42	U	.00± .12	.19					
306.49 15.08		7.0±1.2		.00					5622
0003.5 +1251	PGC 450		.98	.00					5132
0006.0 +1425		.S..8*.	1.00± .10	175					5291
106.71 -47.01	UGC 43	U	.49± .04	.22					
307.81 15.37		8.0±1.4		.61					5436
0003.5 +1409	PGC 451		1.02	.25					4943
000609.1-503848		.L?....	.81± .07	155					10166± 62
320.03 -64.95	ESO 193- 23		.38± .05	.04	15.10 ±.14				
244.62 -4.60				.00					10099
000337.0-505530	PGC 456		.76		14.91				9972
0006.1 +1944		.SX.8*.	.96± .09						5936
108.57 -41.86	UGC 44	U	.05± .06	.11					
313.23 16.45		8.0± .9		.06					6093
0003.6 +1928	PGC 458		.97	.02					5594
0006.2 +5432		.S..8..	.96± .09	150					
116.34 -7.75	UGC 45	U	.39± .06	1.05	15.38 ±.12				
349.52 19.79		8.0± .9		.48					
0003.6 +5416	PGC 461		1.06	.20					

R.A. 2000 DEC. / l b / SGL SGB / R.A. 1950 DEC.	Names / PGC	Type / S_T n_L / T / L	$\log D_{25}$ / $\log R_{25}$ / $\log A_e$ / $\log D_o$	p.a. / A_g / A_l / A_{21}	B_T / m_B / m_{FIR} / B_T^o	$(B-V)_T$ / $(U-B)_T$ / $(B-V)_T^o$ / $(U-B)_T^o$	$(B-V)_e$ / $(U-B)_e$ / m'_e / m'_{25}	m_{21} / W_{20} / W_{50} / HI	V_{21} / V_{opt} / V_{GSR} / V_{3K}
000614.2+085315		.S?....	1.04± .09						11542
104.45 -52.35	MCG 1- 1- 29		.49± .07		15.40 ±.12				
302.22 14.05				.73					11672
000340.2+083633	PGC 465		1.06	.25	14.31				11190
000615.4-524639		PLA.0*.	1.24± .08						
318.06 -63.00		S	.18± .08	.00					
242.59 -5.32		-2.0± .8		.00					
000343.5-530321	PGC 468		1.21						
000619.5+201209	A 0003+19		.50± .14						7688± 31
108.76 -41.42			.00± .12	.11					
313.71 16.50	MK 335								7846
000345.1+195527	PGC 473		.51						7347
000620.5-412936	A 0003-41	.SBS6P/	1.50± .04	20	13.70S±.15			13.70±.2	1542± 11
332.83 -72.92	ESO 293- 34	PS	.51± .05	.03	12.79 ±.14			307± 16	1562± 62
253.35 -1.57		6.1± .6		.75				297± 12	1507
000348.0-414618	PGC 474		1.51	.25	12.43		14.80± .27	1.01	1305
000621.6+172543			.82± .07	78					5506± 46
107.88 -44.12	UGC 46		.21± .06	.06	14.60 ±.16				
310.87 15.94	KUG 0003+171								5658
000347.3+170901	PGC 477		.83						5161
0006.4 -4130		.S?....	.78?		14.58V±.15	.47± .04	.55± .04		
332.76 -72.92	MCG -7- 1- 10		.38?	.03					1428± 62
253.35 -1.59			.67± .03	.57		.38			1392
0003.9 -4147	PGC 482		.78	.19	14.44		12.84±1.59		1190
000627.0-132454	NGC 7828	.RING.B	.94± .11	140	14.4 ±.2	.51± .03			
83.24 -72.72	MCG -2- 1- 25	R	.28± .07	.07		-.21± .05			5660±115
280.24 7.71	IRAS00038-1341			.21	12.36	.39			5722
000353.6-134136	PGC 483		.94	.14	14.12	-.29	13.22± .60		5328
000628.5-034258	NGC 7832	.E+....	1.28± .09	25					
96.34 -64.24	MCG -1- 1- 33	E	.26± .07	.14					
289.71 10.63		-4.0± .8		.00					
000354.8-035939	PGC 485		1.23						
000629.1-132514	NGC 7829	.RING.A	.82± .19	121	14.6 ±.2	.66± .02			
83.25 -72.73	MCG -2- 1- 24	R	.00± .07	.07		.24± .05			5792±115
280.23 7.70		-1.0± .9		.00		.59			5854
000355.7-134156	PGC 488		.82		14.45	.26	13.55± .98		5460
0006.5 +1716		.S..8*.	1.11± .07	64					870± 10
107.90 -44.28	UGC 47	U	.83± .06	.06					
310.73 15.86		8.0±1.4		1.02					1021
0004.0 +1700	PGC 496		1.12	.42					525
0006.5 +4752		.S..7..	1.18± .07					14.64±.1	4322± 11
115.20 -14.32	UGC 48	U	.07± .06	.62					
342.52 19.65		7.0± .8		.09				81± 8	4520
0004.0 +4736	PGC 499		1.24	.03					4060
0006.6 +0821	NGC 7834	.S..6*.	1.06± .08						5226
104.37 -52.89	UGC 49	U	.17± .06	.28	14.95 ±.15				
301.72 13.82		6.0±1.2		.24					5355
0004.1 +0805	PGC 504		1.09	.08	14.40				4874
0006.6 +2609		.S..2..	.97± .07	10					7561
110.56 -35.64	UGC 50	U	.48± .05	.14	14.97 ±.14				
319.84 17.47	IRAS00041+2552	2.0±1.0		.59	13.54				7731
0004.1 +2552	PGC 507		.98	.24	14.16				7231
0006.7 -0638	A 0004-06B	.RING.B	.85?	45					22615± 63
93.53 -66.92		R	.07?	.16					
286.86 9.70		10.0± .3		.05					22699
0004.2 -0655	PGC 509		.87	.04					22272
0006.7 -0638	A 0004-06A	.RING.A	.85?	40					
93.53 -66.92		R	.07?	.16					
286.86 9.70		-2.0± .3		.00					
0004.2 -0655	PGC 510		.86						
0006.7 +0506		.SB.4..	.96± .07	30					5372
102.75 -56.00	UGC 51	U	.21± .05	.18	15.02 ±.12				
298.47 12.97		4.0± .9		.31					5492
0004.2 +0450	PGC 511		.98	.10	14.50				5020
0006.8 +0837		.SA.5..	1.26± .04						5204
104.56 -52.64	UGC 52	U	.03± .05	.29	14.3 ±.2				
302.00 13.84	IRAS00042+0821	5.0± .8		.05	13.14				5333
0004.2 +0821	PGC 515		1.28	.02	13.95				4853

R.A. 2000 DEC. / l b / SGL SGB / R.A. 1950 DEC.	Names / PGC	Type / S_T n_L / T / L	$\log D_{25}$ / $\log R_{25}$ / $\log A_e$ / $\log D_o$	p.a. / A_g / A_i / A_{21}	B_T / m_B / m_{FIR} / B_T^o	$(B-V)_T$ / $(U-B)_T$ / $(B-V)_T^o$ / $(U-B)_T^o$	$(B-V)_e$ / $(U-B)_e$ / m'_e / m'_{25}	m_{21} / W_{20} / W_{50} / HI	V_{21} / V_{opt} / V_{GSR} / V_{3K}
000650.4+191916		.SX.4..	.95± .06						7894
108.65 -42.31	UGC 53	U	.10± .05	.10	15.12 ±.14				
312.83 16.21	KUG 0004+190	4.0± .9		.15					8050
000416.0+190234	PGC 517		.96	.05	14.82				7552
000650.7-521206		.SBR5*.	1.04± .05	5					
318.37 -63.57	ESO 193- 25	S (1)	.29± .05	.00	15.48 ±.14				
243.17 -5.21		5.0± .9		.43					
000419.0-522848	PGC 520	2.2±1.3	1.04	.14					
000652.8-502454		.LA.-..	1.15± .05	89					
320.01 -65.21	ESO 193- 26	S	.20± .04	.04	14.27 ±.14				10037± 62
244.88 -4.63		-3.0± .8		.00					9971
000421.0-504136	PGC 526		1.13		14.08				9842
0006.8 +4144		.S..6*.	1.04± .08	28					
114.09 -20.36	UGC 54	U	1.06± .06	.40					
336.08 19.27		6.0±1.5		1.47					
0004.3 +4128	PGC 527		1.08	.50					
000659.2-065714		.SXS8*.	1.07± .05	50					
93.30 -67.23	MCG -1- 1- 34	UE (1)	.10± .05	.20					
286.57 9.56		8.3± .7		.12					
000425.6-071356	PGC 538	7.5± .8	1.09	.05					
0007.0 +4639		.SB.8..	1.04± .08	115					
115.04 -15.53	UGC 55	U	.07± .06	.44					
341.24 19.53		8.0± .9		.08					
0004.4 +4623	PGC 540		1.08	.03					
000703.7-442205		.E+1...	1.10± .10						
327.39 -70.59	S	.24± .08	.03						
250.66 -2.66		-4.0± .9		.00					
000431.6-443847	PGC 546		1.03						
0007.0 +1404		.S..6*.	.96± .09	30					5431
106.93 -47.41	UGC 56	U	.10± .06	.24	15.18 ±.14				
307.51 15.06		6.0±1.3		.14					5574
0004.5 +1348	PGC 548		.98	.05	14.77				5083
000706.3-412124		.S?....	1.00± .06	50					13868
332.71 -73.11	ESO 293- 37		.07± .05	.03	14.23 ±.14				13833
253.53 -1.66				.10					13630
000434.0-413806	PGC 551		1.01	.04	14.00				
000707.6-801830		.S..5*/	1.17± .04	120					
305.24 -36.62	ESO 12- 15	S	.97± .05	.35	16.09 ±.14				
215.87 -13.49		5.0±1.3		1.46					
000444.1-803512	PGC 553		1.20	.49					
000710.1-214712		PSBS2..	1.27± .04	161					10138
58.48 -78.60	ESO 538- 22	r	1.07± .05	.07	15.32 ±.14				10171
272.22 4.86		2.2± .9		1.23					9827
000437.0-220354	PGC 558		1.27	.50	13.92				
000714.5-523712		.SBS9..	1.29± .06	5					
317.89 -63.21	ESO 149- 18	S (1)	.11± .05	.00	15.85 ±.14				
242.79 -5.41		9.0± .8		.11					
000443.0-525354	PGC 562	11.1± .8	1.29	.06					
000716.0+274232	NGC 1	.SAS3*.	1.21± .03	120	13.65 ±.13	.76± .02	.85± .02	14.78±.1	4545± 5
111.11 -34.15	UGC 57	PU (1)	.12± .04	.18	13.33 ±.12	.12± .03	.21± .03	338± 6	4545± 59
321.47 17.59	IRAS00047+2725	3.0± .6	.74± .02	.17	12.03	.67	12.84± .06	319± 5	4718
000441.3+272550	PGC 564	3.5±1.1	1.22	.06	13.09	.05	14.22± .22	1.63	4218
000716.8+081805	NGC 3	.L...?.	1.03± .08	113	14.2 ±.3	.86± .05			
104.58 -52.99	UGC 58	U	.23± .03	.33	14.37 ±.10	.38± .07			3900± 49
301.70 13.65	ARAK 1	-2.0±1.7		.00	12.98	.73			4028
000442.7+080123	PGC 565		1.04		13.97	.31	13.66± .50		3549
000717.2+274046	NGC 2	.S..2..	1.02± .03	115	14.96 ±.13	.77± .02	.82± .02	16.50±.3	7550± 10
111.11 -34.18	UGC 59	PU	.22± .04	.18	14.67 ±.12	.08± .04	.17± .04		7366± 85
321.44 17.58		1.5± .6	.53± .01	.27		.63	13.10± .04	362± 7	7720
000442.5+272404	PGC 567		1.04	.11	14.27	.00	14.38± .23	2.11	7221
000719.4+323635	NGC 7831	.S..3*/	1.24± .04	38				15.07±.3	5072± 10
112.28 -29.35	UGC 60	PU (1)	.61± .04	.15	13.60 ±.16				5028±118
326.55 18.27	ARAK 2	3.0± .7		.84	12.81			476± 7	5252
000444.5+321953	PGC 569	4.5±1.2	1.25	.30	12.57			2.20	4757
000724.5+470228		.L.....	1.20± .14	145					
115.18 -15.17	UGC 61	U	.05± .08	.44	13.8 ±.3				
341.64 19.48		-2.0± .8		.00					
000448.8+464546	PGC 574		1.24						

R.A. 2000 DEC. / l b / SGL SGB / R.A. 1950 DEC.	Names / PGC	Type / S_T n_L / T / L	$\log D_{25}$ / $\log R_{25}$ / $\log A_e$ / $\log D_o$	p.a. / A_g / A_i / A_{21}	B_T / m_B / m_{FIR} / B_T^o	$(B-V)_T$ / $(U-B)_T$ / $(B-V)_T^o$ / $(U-B)_T^o$	$(B-V)_e$ / $(U-B)_e$ / m'_e / m'_{25}	m_{21} / W_{20} / W_{50} / HI	V_{21} / V_{opt} / V_{GSR} / V_{3K}
000731.1-204354 63.11 -78.11 273.26 5.13 000458.0-210036	ESO 538- 23 PGC 583	.SAT4P. r 4.5± .8	1.03± .05 .13± .05 1.03	87 .00 .19 .06	15.48 ±.14 15.18				19603± 60 19640 19289
0007.5 -1150 86.85 -71.57 281.85 7.94 0005.0 -1207	MCG -2- 1- 27 PGC 585	.S?....	1.12± .08 .28± .07 1.12	 .07 .43 .14					6749 6816 6414
000740.0+253519 110.68 -36.24 319.30 17.16 000505.4+251837	CGCG 477- 56 PGC 590			 .07	 15.5 ±.4			18.48±.3 121± 7	7054± 10 7222 6723
000748.9+352145 112.99 -26.67 329.43 18.50 000513.7+350503	NGC 5 UGC 62 4ZW 7 PGC 595	.E...*. U -5.0±1.2	1.08± .17 .22± .08 1.06	115 .28 .00	14.33 ±.10				
000750.8+355759 113.12 -26.07 330.06 18.56 000515.6+354118	UGC 63 PGC 598	.I.9.. U 10.0± .9	.94± .09 .18± .06 .96	55 .22 .13 .09	15.34 ±.13 14.99				439 624 134
000801.5+330412 112.55 -28.92 327.05 18.18 000526.5+324730	NGC 7836 UGC 65 MK 336 PGC 608	.I?....	.93± .06 .21± .05 .94	133 .13 .16 .11	14.4 ±.2 14.03 ±.17 12.73 13.90	.71± .02 .20± .04 .60 .11	 13.37± .37	16.87±.3 313± 10 2.87	4904± 10 5100±110 5087 4593
0008.1 +0943 105.55 -51.68 303.17 13.80 0005.5 +0926	UGC 66 IRAS00055+0926 PGC 613	.SX.3.. U 3.0± .8	1.13± .05 .13± .05 1.15	85 .25 .18 .07	14.46 ±.15 13.49 13.99				6421 6553 6070
0008.1 +0746 104.69 -53.55 301.23 13.31 0005.6 +0730	UGC 67 PGC 616	.S..2.. U 2.0± .9	1.10± .05 .43± .05 1.13	24 .25 .53 .22	15.06 ±.13 14.15				11834 11960 11483
000810.1+270013 111.17 -34.88 320.78 17.28 000535.3+264331	UGC 68 IRAS00055+2643 PGC 617	.SB?...	.97± .09 .09± .06 .98	145 .15 .13 .04	14.36 ±.12 12.67 14.03				8761± 46 8932 8433
0008.1 +2731 111.30 -34.36 321.32 17.36 0005.6 +2715	UGC 69 PGC 619	.S..6*. U 6.0±1.2	1.07± .06 .15± .05 1.09	42 .16 .23 .08	14.49 ±.13 14.09				4638 4810 4311
000813.4-343442 351.48 -78.12 260.07 .40 000541.0-345124	Scl DIG ESO 349- 31 PGC 621	.IB.9.. S 10.0± .9	1.04± .05 .07± .05 1.04	 .02 .06 .04	 15.54 ±.14 15.46			15.55±.3 30± 12 20± 9 .05	207± 9 195 -59
000818.3-593054 312.91 -56.77 236.22 -7.73 000548.0-594736	ESO 111- 14 PGC 624	.SXT6.. Sr (1) 5.8± .6 3.3± .9	1.34± .03 .28± .04 1.34	160 .00 .42 .14	14.43 ±.14 13.98				7782 7686 7637
000820.6-295454 14.00 -80.14 264.52 1.93 000548.0-301136	NGC 7 ESO 409- 22 PGC 627	.SBS5?/ PS (1) 4.5± .6 5.6±1.2	1.35± .03 .61± .03 1.35	29 .02 .92 .31	14.4 ±.2 13.88 ±.14 13.10	.51± .04 -.18± .07 .38 -.27	 14.47± .27		1497± 39 1501 1213
000834.4-335130 354.21 -78.59 260.78 .58 000602.1-340812	NGC 10 ESO 349- 32 IRAS00060-3408 PGC 634	.SXT4.. PBSr (1) 3.6± .4 2.2± .8	1.38± .04 .30± .04 1.38	25 .02 .44 .15	13.3 ±.4 13.13 ±.14 13.08 12.64	.78± .05 .68	 14.30± .45		6818± 32 6808 6548
000834.4-105658 88.95 -70.94 282.79 7.98 000601.1-111340	MCG -2- 1- 28 PGC 635	.SBR1.. E 1.0± .9	1.07± .06 .03± .05 1.08	10 .10 .03 .01					
000835.8-051303 96.06 -65.83 288.39 9.68 000602.3-052945	MCG -1- 1- 36 IRAS00059-0529 PGC 637	PS..1P* E 1.0±1.2	1.09± .06 .23± .05 1.10	155 .10 .23 .11	 13.39				
000836.0+440547 114.86 -18.10 338.57 19.12 000600.1+434906	UGC 71 PGC 638	.S..4.. U 4.0± .9	1.04± .08 .23± .06 1.08	100 .39 .33 .11	15.33 ±.15				

R.A. 2000 DEC. / l b / SGL SGB / R.A. 1950 DEC.	Names / PGC	Type / ST nL / T / L	logD25 / logR25 / logAe / logDo	p.a. / Ag / Ai / A21	BT / mB / mFIR / BT°	(B-V)T / (U-B)T / (B-V)T° / (U-B)T°	(B-V)e / (U-B)e / m'e / m'25	m21 / W20 / W50 / HI	V21 / Vopt / VGSR / V3K
0008.6 -0044		.S..1..	.99± .06	18					11924
99.89 -61.68	UGC 72	U	.29± .05	.07	15.18 ±.13				12026
292.80 10.95	IRAS00060-0100	1.0± .9		.30	13.40				11575
0006.0 -0100	PGC 639		1.00	.15	14.66				
000842.4+372702	NGC 11	.S..1..	1.17± .05	111					
113.61 -24.65	UGC 73	U	.76± .05	.32	14.59 ±.14				
331.63 18.54	IRAS00061+3710	1.0± .9		.77	13.85				
000607.0+371021	PGC 642		1.20	.38					
000844.8+243234		.S?....	1.04± .06	136				16.17±.3	4581± 10
110.70 -37.31	UGC 76		.54± .05	.16					4747
318.27 16.74				.81				204± 7	4248
000610.0+241553	PGC 644		1.05	.27					
000844.9+043646	NGC 12	.SXT5*.	1.23± .04	125	13.8 ±.4	.66± .06		15.45±.1	3940± 10
103.33 -56.62	UGC 74	PU (1)	.05± .04	.11	13.99 ±.17			267± 15	
298.11 12.37		4.5± .6		.08		.60			4057
000611.0+042005	PGC 645	4.0± .8	1.24	.03	13.76		14.65± .46	1.67	3589
000846.4+154854	NGC 14	RIBS9P.	1.45± .04	25	12.71 ±.16	.58± .04	.56± .03	14.22±.1	865± 4
108.13 -45.83	UGC 75	R	.13± .06	.13	12.87 ±.15	.10± .05	.03± .04	117± 6	
309.36 15.04	ARP 235	10.0± .3	1.17± .05	.09		.52	14.05± .19	88± 8	1012
000612.0+153213	PGC 647		1.46	.06	12.57	.05	14.47± .31	1.58	519
000848.1+332602	NGC 13	RS..2*.	1.39± .04	53				15.45±.3	4808± 10
112.81 -28.60	UGC 77	PU (1)	.62± .04	.15	14.01 ±.12				4748±118
327.45 18.06		2.3± .7		.76				467± 7	4989
000612.9+330921	PGC 650	4.5±1.1	1.40	.31	13.05			2.09	4496
000849.2-495406		.L?....	.83± .06	75					10198± 62
319.82 -65.81	ESO 193- 31		.42± .03	.04	15.50 ±.14				10133
245.48 -4.76				.00					10001
000618.0-501048	PGC 651		.77		15.31				
000854.9+234905	NGC 9	.S..3*P	1.10± .05	155				16.14±.3	4528± 10
110.56 -38.03	UGC 78	P	.24± .04	.22	14.35 ±.12			184± 13	4512± 46
317.53 16.58	IRAS00063+2332	3.0±1.2		.33	13.40				4692
000620.1+233224	PGC 652		1.13	.12	13.77			2.25	4193
0008.9 +2537		.S..6*.	1.20± .05	85				15.93±.3	4345± 7
111.05 -36.26	UGC 79	U	.14± .05	.11	15.1 ±.3			196± 14	
319.39 16.87		6.0±1.1		.21					4513
0006.4 +2521	PGC 654		1.21	.07	14.73			1.13	4014
0009.0 +1053		.S..3..	1.15± .05	47					6674
106.41 -50.61	UGC 81	U (1)	.53± .05	.20	15.29 ±.14				6808
304.41 13.85		3.0± .9		.73					6324
0006.5 +1037	PGC 658	5.5±1.2	1.17	.27	14.31				
000904.3-325454		.S?....	1.17± .05	20					6940± 81
358.11 -79.17	ESO 349- 33		.77± .05	.02	14.76 ±.14				6933
261.71 .79	IRAS00064-3311			1.13					6667
000632.0-331136	PGC 659		1.18	.38	13.56				
000904.7+274349	NGC 16	.LX.-./	1.25± .05	16	13.00 ±.14	1.00± .03			3056± 23
111.59 -34.21	UGC 80	R	.27± .05	.19	12.80 ±.11	.41± .04			3228
321.56 17.20		-3.0± .4		.00		.92			2730
000629.7+272708	PGC 660		1.24		12.64	.37	13.46± .33		
0009.0 +2137	NGC 15	.S..1..	1.00± .06	30					6332
110.01 -40.18	UGC 82	U	.20± .05	.12	14.67 ±.12				6492
315.29 16.14		1.0± .9		.21					5994
0006.5 +2121	PGC 661		1.01	.10	14.26				
000907.0-372118		.S?....	.99± .07	171					8695± 81
341.40 -76.43	ESO 293- 43		.21± .06	.03	14.98 ±.14				8673
257.48 -.69				.31					8440
000635.0-373800	PGC 662		.99	.10	14.59				
0009.3 +0036		.S..3..	1.00± .08	39					
101.19 -60.47	UGC 83	U (1)	.94± .06	.09					
294.18 11.15		3.0±1.0		1.30					
0006.8 +0020	PGC 670	4.5±1.4	1.01	.47					
000925.0-364306		.S?....	.98± .06	57					6567± 81
343.17 -76.92	ESO 349- 34		.71± .05	.02	15.48 ±.14				6547
258.11 -.54				.73					6309
000653.0-365948	PGC 671		.98	.36	14.64				
000929.1+472126		.S..4..	1.09± .06	87				15.28±.1	5154± 11
115.60 -14.92	UGC 85	U	.35± .05	.49	14.3 ±.3				5351
341.99 19.14	IRAS00068+4704	4.0± .9		.51				316± 8	5351
000652.7+470445	PGC 676		1.14	.17	13.27			1.84	4891

R.A. 2000 DEC.	Names	Type	logD$_{25}$	p.a.	B$_T$	(B-V)$_T$	(B-V)$_e$	m$_{21}$	V$_{21}$
l b		S$_T$ n$_L$	logR$_{25}$	A$_g$	m$_B$	(U-B)$_T$	(U-B)$_e$	W$_{20}$	V$_{opt}$
SGL SGB		T	logA$_\bullet$	A$_i$	m$_{FIR}$	(B-V)$_T^o$	m'$_e$	W$_{50}$	V$_{GSR}$
R.A. 1950 DEC.	PGC	L	logD$_o$	A$_{21}$	B$_T^o$	(U-B)$_T^o$	m'$_{25}$	HI	V$_{3K}$

000932.9+331834 NGC 20		.L..-*.	1.19± .07	140					
112.95 -28.75 UGC 84		PU	.02± .05	.15	14.04 ±.13				5058±118
327.34 17.89		-3.0± .8		.00					5239
000657.5+330153 PGC 679			1.21		13.82				4746
000936.3-321636 IC 1531		.LX.-P*	1.25± .04	138					
.87 -79.57 ESO 349- 35		BS	.11± .04	.02	13.48 ±.14				7684± 39
262.36 .90		-3.0± .6		.00					7679
000704.0-323318 PGC 684			1.24		13.34				7408
0009.7 +2821		.E...?.	1.04± .11						
111.92 -33.62 UGC 87		U	.00± .05	.10	14.62 ±.12				
322.23 17.14		-5.0±1.6		.00					
0007.2 +2805 PGC 687			1.06						
000947.6+274947 NGC 22		.S..3..	1.22± .12	160				15.51±.3	8312± 10
111.80 -34.14 UGC 86		U	.11± .12	.13	14.43 ±.18				8484
321.69 17.06		3.0± .8		.15				390± 7	7986
000712.5+273306 PGC 690			1.23		.05 14.08				1.37
000951.5+264724								16.79±.3	17225± 10
111.57 -35.16 CGCG 478- 35				.15	15.5 ±.4				17395
320.62 16.87								268± 7	16897
000716.4+263043 PGC 693									
000951.6-642218 IC 1532		.S..3./	1.21± .05	74					
310.24 -52.16 ESO 78- 17		S (1)	.56± .04	.00	14.83 ±.14				1871
231.57 -9.36		4.3± .7		.82					1760
000723.0-643900 PGC 695		5.6±1.3	1.21	.28	14.00				1753
0009.8 +0441		.S..0..	1.04± .15	25					
103.85 -56.63 UGC 88		U	.13± .12	.12					
298.26 12.12		.0± .9		.10					
0007.3 +0425 PGC 696			1.04						
000953.6+255523 NGC 23		.SBS1..	1.32± .03	8	12.85 ±.17	.82± .05	.86± .05	15.04±.1	4565± 4
111.37 -36.01 UGC 89		R	.19± .04	.16	12.64 ±.11			414± 5	4635± 44
319.73 16.72 MK 545		1.0± .4	.68± .04	.19	10.75	.71	11.74± .13	359± 6	4733
000718.6+253842 PGC 698			1.33	.09	12.30		13.82± .26	2.65	4236
000956.6-245743 NGC 24		.SAS5..	1.76± .01	46	12.19 ±.13	.58± .02	.72± .02	13.07±.1	554± 4
43.70 -80.43 ESO 472- 16		R (3)	.63± .02	.06	12.07 ±.11	-.07± .03	.01± .04	229± 5	595± 50
269.38 3.23 IRAS00073-2514		5.0± .3	1.12± .01	.95	12.64	.44	13.28± .03	210± 5	575
000724.0-251424 PGC 701		4.7± .5	1.77	.32	11.10	-.17	14.27± .16	1.65	253
0009.9 +1656		.I..9*.	1.00± .08	88					908
108.90 -44.79 UGC 90		U	.55± .06	.13					
310.57 14.99		10.0±1.3		.41					1057
0007.4 +1640 PGC 703			1.01	.27					564
000959.9-570112 NGC 25		.LB.-?P	1.14± .05	88	13.0 V±.2		1.11± .03		
313.96 -59.21 ESO 149- 19		S	.23± .04	.00	13.99 ±.14				9429± 40
238.69 -7.17 FAIR 1		-3.3± .7	.72± .07	.00					9340
000730.0-571754 PGC 706			1.11		13.85				9270
001001.7+281234		.S..3..	1.02± .08					15.66±.3	8168± 10
111.95 -33.78 UGC 92		U (1)	.05± .06	.13	14.50 ±.12				8340
322.09 17.06		3.0± .8		.07				97± 7	7843
000726.6+275553 PGC 707		2.5±1.1	1.03	.03	14.23			1.40	
001003.3+280901		.S?....	1.04± .06	115				15.52±.3	8216± 10
111.94 -33.84 UGC 91			.61± .05	.13	15.57 ±.12				8388
322.03 17.05				.92				378± 7	7891
000728.1+275220 PGC 709			1.05	.31	14.47			.74	
001006.9-061919		.SXS9..	1.17± .05	35					3851
95.80 -66.99 MCG -1- 1- 42		E (1)	.21± .05	.15					
287.42 9.00		9.0± .8		.21					3934
000733.4-063600 PGC 714		7.5± .8	1.18	.10					3508
001017.9-181554		.IBS9..	1.15± .03	85				15.07±.1	1546± 5
73.88 -77.03 ESO 538- 24		SUE (1)	.07± .03	.05	15.13 ±.14			35± 5	1589
275.85 5.30		10.0± .5		.05				24± 5	1226
000745.1-183236 PGC 721		10.0± .8	1.16	.03	15.03			.01	
001020.0-462507		PSBS3*.	1.31± .04	55					
323.07 -69.07 ESO 241- 21		Sr (1)	.63± .04	.03	14.33 ±.14				6087
248.89 -3.87		2.8± .7		.86					6033
000749.0-464148 PGC 725		2.2±1.2	1.32	.31	13.39				5873
001022.4+305057		.SA.8..	1.32± .05	60				14.72±.3	4947± 7
112.63 -31.20 UGC 93		U	.13± .06	.17	14.8 ±.3				5124
324.82 17.39		8.0± .7		.16				271± 10	5124
000747.0+303416 PGC 726			1.34	.07	14.50			.16	4629

R.A. 2000 DEC. / l b / SGL SGB / R.A. 1950 DEC.	Names / PGC	Type / S_T n_L / T / L	$logD_{25}$ / $logR_{25}$ / $logA_\bullet$ / $logD_\circ$	p.a. / A_g / A_i / A_{21}	B_T / m_B / m_{FIR} / B_T^o	$(B-V)_T$ / $(U-B)_T$ / $(B-V)_T^o$ / $(U-B)_T^o$	$(B-V)_\bullet$ / $(U-B)_\bullet$ / $m'_\bullet$ / m'_{25}	m_{21} / W_{20} / W_{50} / HI	V_{21} / V_{opt} / V_{GSR} / V_{3K}
001025.0-462930		.SBS4..	1.10± .05	69					
322.94 -69.01	ESO 241- 22	S (1)	.38± .04	.03	14.20 ±.14				
248.83 -3.91		4.0± .9		.55					
000754.1-464612	PGC 729	3.3± .9	1.10	.19					
001026.3+285921		.S..6*.	1.28± .04	7				15.14±.3	7851± 10
112.23 -33.03	UGC 95	U	1.02± .05	.14	15.52 ±.12				8024
322.90 17.10		6.0±1.4		1.47				459± 7	7528
000751.0+284240	PGC 731		1.29	.50	13.88			.76	
001026.4+254957	NGC 26	.SAT2..	1.29± .03	100				14.69±.1	4589± 5
111.50 -36.13	UGC 94	R (1)	.15± .03	.13	13.62 ±.13			322± 6	
319.66 16.59	IRAS00078+2533	2.0± .4		.18	13.32			306± 5	4757
000751.4+253316	PGC 732	3.5±1.1	1.30	.07	13.26			1.36	4259
001031.0+105829								18.01±.2	26150± 14
106.98 -50.63				.21				563± 26	26809± 28
304.58 13.53									26419
000756.8+104148	PGC 737								25936
001033.0+285950	NGC 27	.S?....	1.09± .06	117				15.83±.3	7033± 10
112.26 -33.03	UGC 96		.37± .05	.14	14.45 ±.12				7054± 38
322.92 17.07	IRAS00079+2843			.55	13.10			453± 7	7208
000757.7+284309	PGC 742		1.11	.18	13.72			1.93	6712
0010.5 +2840		.L.....	1.03± .11	70					
112.20 -33.34	UGC 97	U	.21± .05	.10	15.21 ±.12				
322.59 17.02		-2.0± .8		.00					
0008.0 +2824	PGC 748		1.01						
001038.6-565906	NGC 28	.E.1...	1.03± .05	5					
313.82 -59.27	ESO 149- 20	S	.24± .04	.00	14.63 ±.14				
238.76 -7.24		-4.0± .7		.00					
000809.0-571548	PGC 751		.96						
001039.6-564807		PSBT1..	1.02± .06	97					
313.92 -59.45	ESO 149- 21	r	.26± .05	.00	15.07 ±.14				
238.93 -7.19		1.0± .9		.27					
000810.0-570448	PGC 753		1.02	.13					
0010.6 +1343		.S..9..	1.40± .04					15.04±.1	1741± 5
108.06 -47.96	UGC 99	U	.00± .05	.16				92± 4	
307.36 14.13		9.0± .7		.00				82± 4	1882
0008.1 +1327	PGC 757		1.42	.00					1394
001041.5+325859	NGC 21	.SBR4..	1.07± .05	42					
113.16 -29.11	UGC 98	PU	.31± .04	.16	13.99 ±.14				4859±118
327.04 17.61	IRAS00080+3242	3.5± .6		.45	13.05				5039
000805.9+324218	PGC 759		1.08	.15	13.35				4547
001043.5+283351		.SB?...	.74± .14					17.67±.3	8651± 10
112.21 -33.46	MCG 5- 1- 47		.00± .07	.10					8823
322.48 16.97				.00				168± 7	
000808.2+281710	PGC 762		.75	.00					8327
001044.8-470430		PSXR2..	1.26± .04	163	13.92 ±.15	.85± .02	.93± .01		
322.10 -68.51	ESO 241- 23	Sr	.32± .04	.04	14.17 ±.14				9353± 33
248.29 -4.15		2.1± .5	.91± .05	.39		.71	13.96± .16		9297
000814.0-472112	PGC 764		1.26	.16	13.54		14.28± .27		9142
001047.1+332105	NGC 29	.SXS4*.	1.23± .03	154					
113.25 -28.76	UGC 100	PU	.34± .03	.15	13.51 ±.13				4589±118
327.42 17.64	IRAS00082+3304	3.5± .6		.50	13.12				4769
000811.6+330424	PGC 767		1.25	.17	12.83				4278
0010.8 +2533		.SBS1..	1.08± .06	155					10231
111.55 -36.41	UGC 101	U	.33± .05	.13	15.11 ±.14				
319.40 16.44		1.0± .9		.34					10398
0008.3 +2517	PGC 772		1.10	.16	14.52				9901
001054.7-210407		RL..+..	1.06± .05	103	15.1 ±.2	.93± .03	.99± .03		
64.01 -78.95	ESO 538- 25	r	.57± .04	.03	15.26 ±.14	.01± .05	.10± .04		7768± 22
273.19 4.27		-1.3± .9	.53± .08	.00		.77	13.22± .28		7802
000822.0-212048	PGC 774		.97		15.07	-.04	13.80± .34		7456
001057.8-351237		PSX.0..	.94± .06						
347.37 -78.17	ESO 349- 37	r	.04± .05	.02	15.04 ±.14				14910± 81
259.65 -.34		-.1± .9		.03					14895
000826.1-352918	PGC 775		.94		14.77				14647
001101.5+300321		.SX.1..	1.17± .05	85				16.61±.3	6791± 10
112.62 -32.01	UGC 102	U	.07± .05	.14	14.39 ±.17				
324.02 17.13		1.0± .8		.07				377± 7	6966
000826.1+294640	PGC 779		1.18	.03	14.10			2.48	6471

R.A. 2000 DEC. / l b / SGL SGB / R.A. 1950 DEC.	Names / PGC	Type / S_T n_L / T / L	$\log D_{25}$ / $\log R_{25}$ / $\log A_o$ / $\log D_o$	p.a. / A_g / A_i / A_{21}	B_T / m_B / m_{FIR} / B_T^o	$(B-V)_T$ / $(U-B)_T$ / $(B-V)_T^o$ / $(U-B)_T^o$	$(B-V)_e$ / $(U-B)_e$ / m'_e / m'_{25}	m_{21} / W_{20} / W_{50} / HI	V_{21} / V_{opt} / V_{GSR} / V_{3K}
001106.7-120627	NGC 34	.P.....	1.34± .06	30				14.87±.1	5931± 11
88.78 -72.25	MCG -2- 1- 32	E	.43± .07	.07				502± 16	5772± 25
281.86 7.03	MK 938	99.0		.59	10.27				5970
000833.5-122308	PGC 781		1.34	.21					5573
001111.4+290507								16.89±.3	8153± 10
112.45 -32.97	CGCG 499- 70			.14	15.7 ±.4				
323.03 16.95								295± 7	8326
000836.0+284826	PGC 786								7831
001112.9-333449		PSXT3*.	1.16± .04	57					7847± 35
353.84 -79.20	ESO 349- 38	Sr (1)	.38± .04	.02	13.98 ±.14				7837
261.23 .15		3.3± .6		.52					7577
000841.0-335130	PGC 789	2.8± .6	1.16	.19	13.38				
001115.1+285414	UGC 105	PLXS+..	1.11± .07					16.86±.3	8046± 10
112.42 -33.15		Ur	.03± .03	.10	14.65 ±.14				
322.85 16.91		-.6± .6		.00				243± 7	8219
000839.7+283733	PGC 791		1.11		14.42				7723
0011.2 +0240		.S?....	.79± .08						12752± 60
103.36 -58.66	MCG 0- 1- 35		.08± .06	.05	15.38 ±.12				12862
296.36 11.25		.		.13					12402
0008.7 +0224	PGC 793		.80	.04	15.13				
001121.2-285119		.E+4...	1.15± .05	23	14.1 ±.2	1.20± .05	1.16± .03		18475± 52
19.56 -80.99	ESO 409- 25	S	.20± .04	.04	14.30 ±.14				18482
265.76 1.66		-4.0± .8	1.18± .07	.00		1.02	15.47± .21		18188
000849.0-290800	PGC 796		1.09		13.93		14.32± .34		
001122.5+062323	NGC 36	.SXT3..	1.34± .04	21	13.95 ±.18				6106± 60
105.35 -55.11	UGC 106	U	.21± .05	.18	13.28				6227
300.05 12.19	IRAS00088+0606	3.0± .8		.28					5756
000848.4+060642	PGC 798		1.36	.10	13.44				
001124.2-412349		.SBS8?.	1.27± .04	137					1462± 60
330.32 -73.53	ESO 293- 45	S	.59± .05	.03	15.13 ±.14				1425
253.76 -2.43		8.0± .8		.73					1225
000853.1-414030	PGC 800		1.27	.30	14.37				
001124.2-565725	NGC 37	PLXS+*.	1.04± .06	35					
313.64 -59.33	ESO 149- 22	Sr	.22± .04	.00	14.70 ±.14				
238.81 -7.34		-1.5± .5		.00					
000855.0-571406	PGC 801		1.01						
001137.5-365619		.SBS7P.	1.20± .07	163					
341.08 -77.08	ESO 349- 39	S (1)	.40± .07	.02	15.71 ±.14				
258.04 -1.03		6.8± .6		.55					
000906.0-371300	PGC 809	6.7± .6	1.20	.20					
001137.9+275654		.S?....	1.00± .08	18				17.02±.3	7446± 10
112.31 -34.10	UGC 107		.73± .06	.13					7617
321.88 16.67	IRAS00090+2740			1.09	14.07			376± 7	7121
000902.5+274013	PGC 810		1.01	.36					
001142.9+205828			.62± .11						13900
110.62 -40.95	CGCG 456- 37		.00± .12	.11	15.09 ±.16				14057
314.75 15.41	MK 337								13562
000907.9+204147	PGC 814		.63						
001145.5+283001		.SB.3*.	.95± .07	51				16.66±.3	8036± 10
112.46 -33.56	UGC 108	U	.29± .05	.10	14.80 ±.13				8208
322.45 16.73		3.0± .9		.40				456± 7	7712
000910.0+281320	PGC 816		.96	.15	14.24			2.28	
001147.1-053513	NGC 38	RSA.1*.	1.13± .06	80					
97.50 -66.48	MCG -1- 1- 47	E	.02± .05	.12					
288.26 8.82		1.0±1.2		.02					
000913.7-055154	PGC 818		1.14	.01					
0011.8 -0105		.S..8*.	.89± .10						6227
101.26 -62.28	UGC 109	U	.01± .06	.10					6326
292.68 10.08		8.0±1.2		.01					5880
0009.3 -0122	PGC 825		.90	.00					
001155.6+230128								16.41±.3	7574± 10
111.22 -38.95	CGCG 478- 38			.21	15.7 ±.4				7735
316.85 15.75								255± 7	7239
000920.5+224447	PGC 828								
0011.9 +2623		.SAS6..	1.07± .07	145					4510
112.05 -35.64	UGC 110	U	.21± .06	.13	14.98 ±.14				4678
320.30 16.34		6.0± .9		.30					4182
0009.4 +2607	PGC 830		1.08	.10	14.52				

R.A. 2000 DEC. / l b / SGL SGB / R.A. 1950 DEC.	Names / PGC	Type / S_T n_L / T / L	$\log D_{25}$ / $\log R_{25}$ / $\log A_e$ / $\log D_o$	p.a. / A_g / A_i / A_{21}	B_T / m_B / m_{FIR} / B_T^o	$(B-V)_T$ / $(U-B)_T$ / $(B-V)_T^o$ / $(U-B)_T^o$	$(B-V)_e$ / $(U-B)_e$ / m'_e / m'_{25}	m_{21} / W_{20} / W_{50} / HI	V_{21} / V_{opt} / V_{GSR} / V_{3K}
001204.4-471743		.SBS3*.	1.05± .05	178					13250±190
321.29 -68.41	ESO 193- 35	S (1)	.19± .05	.04	15.15 ±.14				13193
248.15 -4.44	FAIR 364	3.0±1.2		.26					13041
000934.0-473424	PGC 831	4.4± .9	1.05	.10	14.75				
001205.8+245914								15.97±.3	10749± 10
111.75 -37.03	CGCG 478- 40			.13	15.7 ±.4				
318.86 16.07								329± 7	10914
000930.5+244233	PGC 832								10418
001208.0+120241		.S..6*.	1.07± .05	76				15.37±.3	6342± 10
107.98 -49.68	UGC 111	U	.50± .04	.21	15.14 ±.12			265± 13	
305.75 13.39		6.0±1.3		.73					6478
000933.6+114600	PGC 833		1.09	.25	14.17			.95	5994
001209.2+275440		.S?....	.87± .08					16.36±.3	7491± 10
112.44 -34.16	CGCG 499- 74		.48± .12	.13	15.40 ±.14				
321.86 16.55	KUG 0009+276A			.70				475± 7	7662
000933.7+273759	PGC 835		.88	.24	14.54			1.58	7166
001209.8+291910		.S?....	.62± .11					17.65±.3	7389± 10
112.75 -32.77	CGCG 499- 75		.26± .12	.15	15.80 ±.16				
323.31 16.77	KUG 0009+290			.39				129± 7	7562
000934.2+290229	PGC 837		.64	.13	15.23			2.29	7068
0012.2 +4144		.S..9..	1.26± .06					16.06±.1	5029± 11
115.13 -20.53	UGC 112	U	.00± .06	.29					5220
336.17 18.28		9.0± .8		.00				62± 8	
0009.6 +4128	PGC 841		1.29	.00					4745
001215.8+221919		.S..1..	1.06± .04	105					7629
111.14 -39.65	UGC 113	U	.45± .04	.11	14.85 ±.12				
316.15 15.55	IRAS00096+2202	1.0± .9		.46	13.54				7789
000940.7+220238	PGC 847		1.07	.23	14.18				7293
001219.3+310344	NGC 39	.SAT5..	1.06± .06					16.13±.3	4857± 11
113.15 -31.07	UGC 114	U	.04± .05	.21	14.21 ±.12			198± 7	4855± 46
325.10 17.00	IRAS00096+3046	5.0± .8		.06				186± 7	5033
000943.6+304703	PGC 852		1.08	.02	13.92			2.20	4540
001236.9+053014		.S?....	.96± .05	68				15.61±.3	8474± 10
105.45 -56.04	UGC 116	U	.46± .04	.17	15.25 ±.12			355± 13	
299.25 11.67	IRAS00100+0513			.69					8592
001002.8+051333	PGC 859		.98	.23	14.34			1.04	8124
001248.1+220126	NGC 41	.S?....	.93± .06					16.82±.3	5949± 10
111.22 -39.97	MCG 4- 1- 39		.20± .06	.13	14.63 ±.13				
315.87 15.37	IRAS00101+2144			.30	13.26			331± 7	6108
001012.9+214445	PGC 865		.94	.10	14.17			2.55	5613
0012.8 +2206	NGC 42	.L..-*.	1.03± .08	115					
111.27 -39.89	UGC 118	U	.25± .03	.13	14.76 ±.10				
315.96 15.36		-3.0±1.2		.00					
0010.3 +2150	PGC 867		1.01						
001254.9+332134		.SXS6..	1.16± .07	110				15.03±.3	4756± 10
113.75 -28.82	UGC 117	U	.19± .06	.16	14.78 ±.17				
327.49 17.20		6.0± .8		.27				216± 7	4935
001019.0+330453	PGC 869		1.18	.09	14.32			.62	4445
001300.6+391448		.SBS5..	1.05± .05	105					4940±125
114.86 -23.02	UGC 121	U	.22± .04	.31	15.09 ±.14				5128
333.59 17.89		5.0± .9		.34					4647
001024.0+385807	PGC 874		1.08	.11	14.41				
001301.0+305457	NGC 43	.LB....	1.21± .06						4785± 10
113.29 -31.24	UGC 120	U	.04± .03	.21	13.60 ±.11				4782± 46
324.98 16.83		-2.0± .8		.00					4960
001025.1+303816	PGC 875		1.23		13.32				4468
001303.0+142439		.S?....	1.00± .06	78					2025
109.12 -47.43	UGC 119		.21± .05	.22	14.16 ±.13				2166
308.18 13.72				.32					1679
001028.3+140758	PGC 878		1.02	.11	13.60				
001303.3-241255		.SAR6..	1.05± .05	153					10269
49.42 -80.88	ESO 472- 20	r	.15± .04	.00	15.61 ±.14				10291
270.33 2.80	IRAS00105-2429	5.6± .9		.22	13.37				9967
001031.1-242936	PGC 879		1.05	.08	15.34				
001315.7-492031		.L?....	.95± .07	2					10300±190
318.72 -66.60	ESO 193- 36		.22± .05	.04	14.98 ±.14				10235
246.25 -5.26	FAIR 637			.00					10101
001046.0-493712	PGC 887		.92		14.79				

R.A. 2000 DEC.	Names	Type	logD$_{25}$	p.a.	B$_T$	(B-V)$_T$	(B-V)$_\bullet$	m$_{21}$	V$_{21}$
l　　b		S$_T$　n$_L$	logR$_{25}$	A$_g$	m$_B$	(U-B)$_T$	(U-B)$_\bullet$	W$_{20}$	V$_{opt}$
SGL　SGB		T	logA$_\bullet$	A$_l$	m$_{FIR}$	(B-V)$_T^o$	m'$_\bullet$	W$_{50}$	V$_{GSR}$
R.A. 1950 DEC.	PGC	L	logD$_o$	A$_{21}$	B$_T^o$	(U-B)$_T^o$	m'$_{25}$	HI	V$_{3K}$

0013.2　+1701		.I..9..	1.35± .04	109				14.49±.1	853±　6
110.01 -44.88	UGC　122	U	.81± .05	.12				106±　7	
310.83　14.24		10.0± .9		.61				93± 12	1000
0010.7　+1645	PGC　889		1.36	.41					509
001320.2-373049		.LX.-..	1.19± .09						
338.25 -76.88	MCG -6- 1- 35	S	.21± .08	.02					15236± 81
257.60　-1.53		-3.0± .8		.00					15212
001049.2-374730	PGC　892		1.16						14983
0013.4　+2822		.S..1..	1.00± .08	32					
112.86 -33.75	UGC　124	U	.73± .06	.09	15.54 ±.13				
322.38　16.36		1.0±1.0		.74					
0010.8　+2806	PGC　896		1.01	.36					
001327.3+172906	IC　　4	.S?....	1.02± .06	10					
110.20 -44.45	UGC　123		.08± .05	.09	14.04 ±.12				4992± 50
311.30　14.30	IRAS00108+1712			.11	12.59				5140
001052.4+171226	PGC　897		1.03	.04	13.81				4649
001334.3-050537		.SAT3P?	1.08± .06	170					
98.99 -66.19	MCG -1- 1- 52	E	.24± .05	.13					
288.88　 8.54	IRAS00109-0522	3.0±1.7		.34	13.27				
001100.9-052218	PGC　902		1.09	.12					
001343.5+301141		.S?....	.94± .11					15.87±.3	7148± 10
113.32 -31.97	MCG 5- 1- 55		.28± .07	.16	14.74 ±.13				
324.26　16.58				.35				458±　7	7322
001107.6+295500	PGC　908		.95	.14	14.16			1.58	6829
001345.4+480904	IC　1534	.L.....	1.02± .08	72					
116.46 -14.25	UGC　125	U	.32± .03	.63	14.8 ±.3				
342.86　18.46	5ZW　 6	-2.0± .9		.00					
001107.6+475224	PGC　910		1.04						
0013.8　+2658		.SB.6*.	1.15± .07	135					4685
112.67 -35.14	UGC　127	U	.57± .06	.10					
320.96　16.04		6.0±1.2		.84					4853
0011.2　+2642	PGC　912		1.16	.29					4359
0013.8　+3559		.S..8..	1.32± .05	65					4533± 10
114.45 -26.26	UGC　128	U	.09± .06	.20					
330.24　17.36		8.0± .7		.11					4716
0011.2　+3543	PGC　913		1.34	.04					4230
0013.8　-0428		.S?....	1.08± .09						10913
99.70 -65.63	MCG -1- 1- 53		.53± .07	.13					11000
289.51　 8.65				.65					10570
0011.3　-0445	PGC　916		1.09	.27					
0013.8　+0342		PSBS2..	.97± .07						
105.11 -57.85	UGC　129	U	.07± .05	.04	15.07 ±.15				
297.56　10.90		2.0± .9		.08					
0011.3　+0326	PGC　917		.97	.03					
001355.8+284546		.S?....	.92± .11					17.25±.3	6945± 10
113.08 -33.39	MCG 5- 1- 56		.03± .07	.07	15.36 ±.15				
322.80　16.31				.04				91±　7	7116
001120.0+282906	PGC　919		.92	.01	15.18			2.06	6623
001356.9+305258		.S?....	.61± .14	160					
113.52 -31.30	UGC　130		.11± .06	.19	14.7 ±.3				4792± 46
324.97　16.63				.15					4967
001120.9+303618	PGC　921		.63	.05	14.29				4475
0013.9　+1258		.SX.8..	1.20± .05	14					1680
108.98 -48.88	UGC　132	U	.45± .05	.16	14.88 ±.14				
306.79　13.17		8.0± .9		.56					1817
0011.4　+1242	PGC　924		1.21	.23	14.16				1333
001359.5-700119		.SBS6..	1.24± .04						4135
307.59 -46.77	ESO　50- 6	S　　(1)	.10± .05	.03	14.52 ±.14				4008
226.15 -11.30		6.0± .8		.15					4051
001136.0-701800	PGC　926	4.4± .8	1.24	.05	14.32				
001402.2+481406	NGC　48	.SX.4P*	1.15± .04	15				14.24±.1	1776± 14
116.52 -14.17	UGC　133	PU	.19± .04	.63	14.4 ±.3			428± 18	
342.95　18.42	IRAS00113+4757	4.0± .6		.28	13.38			278± 19	1972
001124.3+475726	PGC　929		1.21	.10	13.51			.63	1517
001403.2-231101	NGC　45	.SAS8..	1.93± .01	142	11.32M±.08	.71± .03	.67± .02	11.43±.1	468±　3
55.89 -80.67	ESO 473- 1	R　　(3)	.16± .02	.06	11.30 ±.12	-.05± .04	-.02± .03	189±　4	
271.39　 2.91	DDO 223	8.0± .3	1.57± .03	.20	12.34	.66	14.60± .07	167±　5	493
001131.0-232742	PGC　930	7.3± .3	1.93	.08	11.05	-.09	15.39± .11	.30	163

R.A. 2000 DEC.	Names	Type	logD$_{25}$	p.a.	B$_T$	(B-V)$_T$	(B-V)$_e$	m$_{21}$	V$_{21}$
l b		S$_T$ n$_L$	logR$_{25}$	A$_g$	m$_B$	(U-B)$_T$	(U-B)$_e$	W$_{20}$	V$_{opt}$
SGL SGB		T	logA$_e$	A$_i$	m$_{FIR}$	(B-V)^o_T	m'$_e$	W$_{50}$	V$_{GSR}$
R.A. 1950 DEC.	PGC	L	logD$_o$	A$_{21}$	B^o_T	(U-B)^o_T	m'$_{25}$	HI	V$_{3K}$
0014.1 +0725		.S..7*.	1.04± .08	27					5843
106.97 -54.28	UGC 135	U	1.06± .06	.19					
301.26 11.78		7.0±1.5		1.38					5965
0011.6 +0709	PGC 937		1.06	.50					5494
001422.5+481447	NGC 49	.L...?.	1.05± .08	165					
116.58 -14.17	UGC 136	U	.04± .03	.63	14.7 ±.3				
342.96 18.37		-2.0±1.7		.00					
001144.5+475807	PGC 952		1.11						
001428.4+484014	NGC 137	.S..4..	1.08± .06	77					
116.66 -13.75	UGC 137	U	.58± .05	.61	15.3 ±.3				
343.41 18.37		4.0± .9		.86					
001150.3+482334	PGC 962		1.13	.29					
001428.9-004443	NGC 139	.SXS5?.	1.33± .02	82					3953
102.82 -62.17	UGC 139	UE (1)	.34± .04	.13	14.26 ±.18				
293.21 9.55	KUG 0011-010	4.7± .6		.51					4052
001155.1-010123	PGC 963	3.1±1.2	1.35	.17	13.60				3606
001431.0-071006	NGC 47	.SBT4..	1.35± .04						5702
97.55 -68.20	MCG -1- 1- 55	E (1)	.03± .05	.14					
286.92 7.71	IRAS00119-0726	4.0± .7		.05	13.06				5780
001157.7-072646	PGC 967	3.1± .8	1.37	.02					5362
0014.5 +2826		.SB.0..	1.04± .08	145					
113.18 -33.72	UGC 141	U	.29± .06	.07	15.37 ±.14				6814
322.50 16.12	IRAS00119+2810	.0± .9		.22	12.11				6985
0011.9 +2810	PGC 970		1.03		14.98				6491
001434.9+481522	NGC 51	.L..0P*	1.10± .06						
116.62 -14.16	UGC 138	PU	.11± .05	.63	14.1 ±.3				
342.98 18.33		-2.0± .6		.00					
001156.8+475842	PGC 974		1.16						
0014.6 +1834	NGC 52	.S?....	1.33± .05	127					5392
110.89 -43.43	UGC 140		.70± .06	.11	14.34 ±.12				
312.47 14.25	IRAS00120+1818			1.05	12.83				5542
0012.0 +1818	PGC 978		1.34	.35	13.15				5051
001439.6-393655			.85± .06	12					12525
332.08 -75.36	ESO 293- 48		.12± .06	.03	15.23 ±.14				12493
255.67 -2.45									
001209.0-395336	PGC 979		.85						12281
001440.8-373919		.SBS3..	1.05± .05	60					
336.95 -76.94	ESO 293- 49	S (1)	.25± .04	.02	14.71 ±.14				8335± 81
257.55 -1.83		3.0± .9		.34					8310
001210.1-375600	PGC 981	3.3± .9	1.05	.12	14.28				8083
001441.9-601944	NGC 53	PSBR3..	1.30± .04	160	13.33 ±.16	.72± .04	.83± .01		
311.10 -56.21	ESO 111- 20	Sr	.16± .04	.00	13.70 ±.14				
235.67 -8.73	FAIR 3	2.6± .6	1.06± .03	.22		.66	14.12± .09		4568± 45
001215.1-603624	PGC 982		1.30	.08	13.29		14.28± .26		4469
									4429
001444.7-072044	NGC 50	.L..-P.	1.36± .06	155					
97.51 -68.39	MCG -1- 1- 58	E	.13± .08	.13					
286.77 7.60		-3.0± .8		.00					
001211.4-073725	PGC' 983		1.36						
0014.7 +0513		.SB.3..	1.07± .07	133					12600
106.25 -56.45	UGC 142	U	.50± .06	.15	15.29 ±.13				
299.12 11.08		3.0± .9		.69					12716
0012.2 +0457	PGC 986		1.08	.25	14.36				12251
001455.0-240525		.SBT2..	.87± .05						
51.17 -81.23	ESO 473- 2	r	.00± .04	.03	15.01 ±.14				
270.59 2.43	IRAS00124-2421	2.2± .9		.00	13.42				
001223.0-242206	PGC 991		.88						
001455.5-865938		.LA.-*.	1.11± .07	19					
303.48 -30.10	ESO 2- 6	S	.48± .04	.53					
209.22 -15.16		-3.7± .7		.00					
001345.1-871618	PGC 993		1.04						
001455.5+261950		.S?....	.82± .13					16.02±.3	7350± 10
112.83 -35.82	MCG 4- 1- 44		.08± .07	.15	14.84 ±.13				
320.35 15.68				.11				157± 7	7516
001219.7+260310	PGC 994		.83	.04	14.53			1.45	7023
001506.0-350838			.79± .07	0					
344.79 -78.84	ESO 350- 3		.11± .06	.02	15.41 ±.14				7946± 87
259.99 -1.11									7929
001235.0-352518	PGC 1009		.79						7684

R.A. 2000 DEC. l　b SGL　SGB R.A. 1950 DEC.	Names PGC	Type S_T　n_L T L	$\log D_{25}$ $\log R_{25}$ $\log A_o$ $\log D_o$	p.a. A_g A_i A_{21}	B_T m_B m_{FIR} B_T^o	$(B-V)_T$ $(U-B)_T$ $(B-V)_T^o$ $(U-B)_T^o$	$(B-V)_o$ $(U-B)_o$ m'_o m'_{25}	m_{21} W_{20} W_{50} HI	V_{21} V_{opt} V_{GSR} V_{3K}
001507.7-070626 97.99 -68.20 287.03　7.58 001234.5-072307	NGC　54 MCG -1- 1- 60 PGC　1011	.SBR1?. E 1.0±1.3 	1.10± .06 .45± .05 1.11	93 .14 .46 .22					
001508.5-391313 332.67 -75.74 256.08　-2.41 001238.1-392954	NGC　55 ESO 293- 50 PGC　1014	.SBS9*/ R　　(1) 9.0± .7 5.6± .6	2.51± .01 .76± .02 2.51	108 .03 .77 .38	8.42S±.05 8.29 ±.12 8.32 7.60	.55± .03 .12± .03 .38 .00	 13.92± .09	9.17±.1 169± 5 172± 4 1.19	125± 4 121± 13 94 -120
0015.2 +0553 106.77 -55.84 299.81 11.13 0012.7 +0537	 UGC　143 PGC　1021	.SBR1.. U 1.0± .8 	1.04± .06 .07± .05 1.05	 .08 .08 .04					13314 13431 12965
001517.4-571438 312.49 -59.21 238.70 -7.92 001250.0-573118	 ESO 111- 22 FAIR　4 PGC　1024	RSAR3*. Sr 2.8± .5 	1.10± .04 .26± .04 1.10	8 .00 .35 .13	 15.15 ±.14 14.72				9790 9700 9633
0015.3 -0810 97.00 -69.21 286.01　7.22 0012.8 -0827	 PGC　1027	.SBS8.. E　　(1) 8.0± .9 7.5±1.2	1.12± .08 .19± .08 1.14	40 .14 .23 .09					
001525.2-212638 65.71 -80.02 273.17　3.16 001253.0-214318	NGC　59 ESO 539- 4 PGC　1034	.LAT-*. S -2.5± .5 	1.42± .04 .30± .04 .86± .02 1.38	127 .06 .00 	13.12 ±.14 13.09 ±.14 13.04	.69± .02 .01± .02 .66 -.01	.62± .01 -.01± .02 12.91± .07 14.33± .28		 382± 60 412 73
001531.0+171936 110.83 -44.70 311.25 13.78 001255.9+170256	NGC　57 UGC　145 PGC　1037	.E..... U -5.0± .7 	1.35± .07 .07± .05 1.15± .05 1.35	40 .15 .00 	12.67 ±.15 13.14 ±.12 12.73	1.05± .03 .56± .04 .96 .55	1.09± .01 .63± .02 13.91± .16 14.21± .40		 5440± 22 5587 5098
001531.2-321056 357.84 -80.71 262.86　-.26 001300.1-322736	 ESO 410- 5 PGC　1038	DE.3.*. S　　(1) -.3± .5 10.0± .8	1.12± .05 .10± .05 1.12	54 .02 .07 	 14.90 ±.14 				
001541.2+250650 112.78 -37.05 319.14 15.29 001305.5+245010	 CGCG 478- 46 PGC　1042			 .11 	 15.7 ±.4 			18.25±.3 122± 7 	20363± 10 20527 20033
0015.7 +2726 113.27 -34.76 321.52 15.70 0013.1 +2710	 UGC　146 PGC　1044	.L..... U -2.0± .9 	1.10± .07 .34± .03 1.06	65 .06 .00 	 15.05 ±.11 				
0015.7 +2939 113.72 -32.57 323.78 16.07 0013.1 +2923	 UGC　147 PGC　1046	.S..0.. U .0± .9 	1.21± .06 .70± .06 1.19	156 .13 .53 	 14.59 ±.12 				
001544.1-330150 353.37 -80.29 262.06　-.57 001313.0-331830	 ESO 350- 4 PGC　1047	.SXT0.. r -.1± .9 	.97± .06 .16± .05 .96	16 .02 .12 	 14.51 ±.14 14.26				 7363± 81 7353 7092
001551.3+160525 110.60 -45.93 310.02 13.43 001316.3+154845	 UGC　148 IRAS00132+1548 PGC　1051	.S?.... 	1.11± .05 .58± .05 1.12	98 .11 .87 .29	 14.09 ±.16 12.16 13.09			15.24±.3 294± 10 1.86	4203± 10 4156± 50 4345 3858
001552.6+194628 111.59 -42.31 313.74 14.21 001317.3+192948	 MCG 3- 1- 33 ARAK　3 PGC　1052	.L?.... 	.72± .22 .18± .07 .70	 .07 .00 	 15.04 ±.14 14.86				 7554± 54 7706 7215
001556.6+140420 110.03 -47.91 308.00 12.96 001321.7+134740	 UGC　149 PGC　1056	.S?.... 	1.12± .04 .65± .04 1.14	60 .16 .98 .33	 15.32 ±.12 14.15			16.09±.3 310± 13 1.61	5474± 10 5613 5129
001558.4-001815 103.86 -61.86 293.74　9.31 001324.7-003455	NGC　60 UGC　150 PGC　1058	.SAR6P. E　　(1) 6.0± .8 4.2± .8	1.11± .05 .03± .04 1.11	155 .04 .04 .01	 14.80 ±.20 				
001613.6+101957 108.91 -51.58 304.28 12.01 001339.0+100317	 UGC　151 PGC　1074	.L...?. U -2.0±1.6 	1.08± .06 .06± .05 1.10	 .25 .00 	 14.29 ±.11 				

R.A. 2000 DEC. / l b / SGL SGB / R.A. 1950 DEC.	Names / PGC	Type / S_T n_L / T / L	$\log D_{25}$ / $\log R_{25}$ / $\log A_e$ / $\log D_o$	p.a. / A_g / A_i / A_{21}	B_T / m_B / m_{FIR} / B_T^o	$(B-V)_T$ / $(U-B)_T$ / $(B-V)_T^o$ / $(U-B)_T^o$	$(B-V)_e$ / $(U-B)_e$ / m'_e / m'_{25}	m_{21} / W_{20} / W_{50} / HI	V_{21} / V_{opt} / V_{GSR} / V_{3K}
001624.3-061906 / 99.55 -67.59 / 287.89 7.51 / 001351.0-063546	NGC 61A / MCG -1- 1- 62 / / PGC 1083	.L...P? / E / -2.0±1.7 /	1.04± .09 / .21± .07 / / 1.02	/ .12 / .00 /					
001630.4+295513 / 113.97 -32.34 / 324.08 15.94 / 001354.1+293833	/ UGC 152 / / PGC 1089	.S..6*. / U / 6.0±1.3 /	1.18± .05 / .58± .05 / / 1.19	73 / .14 / .86 / .29	15.38 ±.14 / / 14.36			15.60±.3 / 227± 7 / .95 /	4863± 10 / 5035 / 4544
0016.6 +4756 / 116.93 -14.52 / 342.67 17.97 / 0014.0 +4740	/ UGC 153 / / PGC 1099	.S..8*. / U / 8.0±1.3 /	1.04± .08 / .37± .06 / / 1.09	178 / .50 / .46 / .19					
0016.7 +0227 / 105.80 -59.25 / 296.52 9.87 / 0014.2 +0211	/ UGC 154 / / PGC 1105	.S..9*. / U / 9.0±1.2 /	1.04± .15 / .00± .12 / / 1.05	/ .10 / .00 / .00					4109 / 4216 / 3762
0016.7 +0703 / 107.90 -54.80 / 301.07 11.07 / 0014.2 +0647	/ UGC 155 / / PGC 1106	.S?.... / / /	1.17± .05 / .61± .05 / / 1.18	3 / .12 / .91 / .30	14.48 ±.12 / / 13.43				3967 / 4087 / 3619
0016.7 +1220 / 109.80 -49.64 / 306.32 12.36 / 0014.2 +1204	/ UGC 156 / / PGC 1107	.I..9.. / U / 10.0± .7 /	1.40± .04 / .19± .05 / / 1.42	5 / .23 / .14 / .10	14.5 ±.3 / / 14.08			14.74±.1 / 135± 5 / 121± 4 / .56	1133± 6 / 1267 / 787
001651.0+293652 / 114.00 -32.66 / 323.78 15.81 / 001414.6+292013	/ CGCG 499- 95 / / PGC 1108			/ .13 / /	14.8 ±.4 / /			18.79±.3 / 136± 7 / /	6895± 10 / 7067 / 6576
001651.0-051607 / 100.75 -66.64 / 288.95 7.70 / 001417.6-053247	/ MCG -1- 1- 64· / IRAS00142-0532 / PGC 1109	.SBS1P* / E / 1.0±1.3 /	1.14± .06 / .47± .05 / / 1.15	45 / .13 / .48 / .24	/ 12.17 /				3943 / 4026 / 3602
0016.9 +2751 / 113.68 -34.39 / 321.99 15.51 / 0014.3 +2735	/ UGC 157 / / PGC 1113	.SB.3*. / U / 3.0±1.2 /	.98± .07 / .05± .05 / / .98	15 / .04 / .07 / .03	14.85 ±.13 / /				
001702.6+295627 / 114.11 -32.34 / 324.12 15.83 / 001426.2+293947	/ CGCG 499- 98 / / PGC 1119			/ .15 / /	14.8 ±.4 / /			17.44±.3 / 339± 7 / /	9690± 10 / 9862 / 9372
001705.0+420940 / 116.16 -20.26 / 336.69 17.42 / 001427.1+415301	/ UGC 158 / / PGC 1123	.S..3.. / U (1) / 3.0± .9 / 4.5±1.2	1.20± .05 / .60± .05 / / 1.24	173 / .42 / .82 / .30	15.21 ±.14 / / 13.93			14.90±.1 / 336± 8 / .67 /	5071± 11 / 5260 / 4790
0017.0 +1731 / 111.40 -44.57 / 311.53 13.46 / 0014.5 +1715	/ UGC 159 / / PGC 1124	.I..9*. / U / 10.0±1.2 /	1.11± .14 / .00± .12 / / 1.12	/ .16 / .00 / .00					1003 / 1150 / 662
001705.5-132912 / 90.77 -74.17 / 280.98 5.22 / 001432.9-134552	NGC 62 / MCG -2- 1- 43 / IRAS00145-1345 / PGC 1125	RSBR1*. / E / 1.0±1.3 /	1.02± .07 / .11± .05 / / 1.02	130 / .06 / .11 / .05	/ 13.49 /				
001705.7-062227 / 99.92 -67.70 / 287.89 7.33 / 001432.5-063906	/ MCG -1- 1- 65 / VV 721 / PGC 1126·	.SXS6.. / E (1) / 6.0± .8 / 4.2± .8	1.07± .06 / .01± .05 / / 1.09	/ .13 / .02 / .01					
001706.3-043639 / 101.44 -66.04 / 289.61 7.83 / 001432.9-045319	/ MCG -1- 1- 66 / / PGC 1127	.L..+*/ / E / -1.0±1.3 /	1.13± .06 / .56± .05 / / 1.06	123 / .13 / .00 /					
0017.1 +3428 / 114.94 -27.86 / 328.77 16.49 / 0014.5 +3412	/ UGC 160 / / PGC 1128	.SA.8.. / U / 8.0± .8 /	1.10± .07 / .00± .06 / / 1.12	/ .14 / .01 / .00					4670 / 4849 / 4364
001709.6+234507 / 112.90 -38.45 / 317.82 14.71 / 001433.8+232827	/ CGCG 478- 49 / / PGC 1129			/ .09 / /	15.7 ±.4 / /			16.54±.3 / 625± 7 / /	18043± 10 / 18203 / 17711

R.A. 2000 DEC.	Names	Type	logD₂₅	p.a.	B_T	(B-V)_T	(B-V)_e	m₂₁	V₂₁
l b		S_T n_L	logR₂₅	A_g	m_B	(U-B)_T	(U-B)_e	W₂₀	V_opt
SGL SGB		T	logA_∘	A_i	m_FIR	(B-V)_T^∘	m'_e	W₅₀	V_GSR
R.A. 1950 DEC.	PGC	L	logD_∘	A₂₁	B_T^∘	(U-B)_T^∘	m'₂₅	HI	V_3K

001710.2-191802		.SXT5?.	1.22± .03	38					
75.99 -78.91	ESO 539- 5	SE (1)	.18± .03	.06	13.53 ±.14				3190± 52
275.36 3.43	IRAS00146-1934	4.5± .8		.28	12.93				3227
001438.0-193442	PGC 1130	4.4±1.1	1.22	.09	13.18				2876
0017.1 +0643		.I..9*.	1.00± .16						5659
107.93 -55.14	UGC 161	U	.09± .12	.12					
300.76 10.89		10.0±1.2		.07					5778
0014.6 +0627	PGC 1131		1.01	.04					5311
001717.0+301249		.S?....	.74± .14					18.07±.3	6292± 10
114.22 -32.08	MCG 5- 1- 61		.09± .07	.15	15.32 ±.13				6464
324.40 15.82				.12				247± 7	6464
001440.5+295610	PGC 1138		.75	.05	15.04			2.98	5974
001719.3+064324		.SBS6*.	.98± .07	103					5659
107.99 -55.16	UGC 163	U	.20± .05	.12	15.07 ±.13				
300.77 10.85		6.0±1.3		.29					5778
001445.0+062644	PGC 1139		.99	.10	14.64				5311
0017.4 +1805		.SB.4..	1.17± .05	65					5443
111.65 -44.04	UGC 164	U	.35± .05	.08	14.77 ±.15				
312.11 13.51	IRAS00148+1748	4.0± .8		.52	13.12				5591
0014.8 +1748	PGC 1144		1.18	.18	14.14				5103
001729.6+261539								17.15±.3	7456± 10
113.52 -35.99	CGCG 478- 50			.14	15.4 ±.4				7621
320.38 15.10								163± 7	
001453.5+255900	PGC 1147								7129
001730.4-064932	NGC 64	.SBS4..	1.19± .05	30					
99.75 -68.16	MCG -1- 1- 68	E (1)	.15± .05	.13					
287.48 7.10	IRAS00149-0706	4.0± .8		.22	13.08				
001457.2-070612	PGC 1149	1.9± .8	1.20	.08					
0017.6 +2229	LGS 1							17.27±.3	3695± 9
112.75 -39.71				.13				106± 10	
316.57 14.37								72± 7	3852
0015.0 +2213	PGC 1153								3361
001736.1+301219		.S..6*.	1.27± .04	18				14.83±.3	4834± 10
114.30 -32.10	UGC 166	U	.31± .05	.17	14.9 ±.2				
324.41 15.75		6.0±1.1		.45				258± 7	5006
001459.5+295540	PGC 1154		1.29	.15	14.30			.38	4517
0017.7 +2440		.S..2..	1.00± .06	37					6047
113.25 -37.57	UGC 165	U	.68± .05	.03	15.44 ±.13				
318.77 14.76	IRAS00150+2423	2.0±1.0		.83	13.05				6209
0015.1 +2423	PGC 1158		1.00	.34	14.51				5717
001745.6+112658	NGC 63	.S...P.	1.24± .05	108				15.75±.3	1172± 10
109.87 -50.57	UGC 167	R	.18± .05	.23	12.63 ±.14				
305.49 11.91	IRAS00151+1110			.27	12.05			160± 10	1303
001510.8+111018	PGC 1160		1.26	.09	12.12			3.53	826
001748.5-682132		.S..5*/	1.07± .05	157					
307.60 -48.46	ESO 50- 7	S	.77± .04	.00	15.72 ±.14				
227.88 -11.20		5.0±1.4		1.15					
001527.0-683812	PGC 1162		1.07	.38					
001755.4-475544		.SXT4..	.95± .07	59					11450±190
318.17 -68.20	ESO 194- 1	r	.15± .06	.04	15.15 ±.14				11389
247.85 -5.57	FAIR 367	4.5± .9		.23					
001527.0-481224	PGC 1166		.95	.08	14.82				11246
001759.9+243342		.S?....	.94± .11					16.17±.3	5794± 10
113.31 -37.68	MCG 4- 1- 48		.11± .07	.03	14.71 ±.12				5956
318.68 14.68				.17				255± 7	
001523.9+241703	PGC 1170		.94	.06	14.48			1.63	5464
001801.6-590508		.S..2*/	1.22± .04	158					8948
310.92 -57.52	ESO 111- 24	S	.69± .04	.00	15.59 ±.14				8851
237.00 -8.79	IRAS00156-5921	2.3± .7		.85	13.56				
001536.0-592148	PGC 1172		1.22	.34	14.66				8802
001804.5-335227		RSBT0..	1.16± .04	90					7569± 39
347.69 -80.15	ESO 350- 7	Sr	.11± .04	.02	14.17 ±.14				7555
261.40 -1.30		.3± .5		.09					
001534.0-340906	PGC 1173		1.16		13.95				7302
0018.0 +1817		.S..1..	1.06± .06	162					5521
111.92 -43.86	UGC 168	U	.65± .05	.11					
312.36 13.39		1.0± .9		.67					5669
0015.5 +1801	PGC 1175		1.07	.33					5181

R.A. 2000 DEC. l b SGL SGB R.A. 1950 DEC.	Names PGC	Type ST nL T L	logD25 logR25 logAe logDo	p.a. Ag Ai A21	BT mB mFIR BT^0	(B-V)T (U-B)T (B-V)T^0 (U-B)T^0	(B-V)e (U-B)e m'e m'25	m21 W20 W50 HI	V21 Vopt VGSR V3K
001806.0-374238		.SBT3?.	1.09± .05	166	15.31 ±.14				
334.39 -77.31	ESO 294- 2	S (1)	.58± .05	.02					
257.71 -2.49	IRAS00156-3759	3.0±1.7		.80					
001536.0-375918	PGC 1176	4.4± .9	1.09	.29					
001815.0+300345	NGC 67	.E...*.	1.02± .10						
114.44 -32.26	MCG 5- 1- 64	RC	.17± .06	.17	15.21 ±.12				6844± 72
324.28 15.59	ARAK 4	-5.0± .8		.00					7016
001538.3+294706	PGC 1185		.99		14.94				6526
0018.2 +1923		.SB.6?.	1.00± .08						5422
112.25 -42.79	UGC 169	U	.09± .06	.13	14.52 ±.12				
313.47 13.58	IRAS00156+1906	6.0±1.2		.13	13.33				5572
0015.7 +1906	PGC 1186		1.01	.04	14.24				5084
001818.3+300419	NGC 68	.LA.-..	1.31± .05		*				
114.45 -32.25	UGC 170	R (1)	.05± .04	.17	13.91 ±.16				5803± 40
324.30 15.58	IRAS00157+2947	-3.0± .4		.00					5975
001541.7+294740	PGC 1187	4.5±1.1	1.32		13.66				5486
001820.6+300222	NGC 69	.LBS-..	.95± .10		15.8 ±.2	1.05± .08			
114.46 -32.29	MCG 5- 1- 66	RC	.06± .06	.17	15.27 ±.12				6754± 44
324.26 15.56	ARAK 5	-2.8± .6		.00			.94		6926
001543.9+294543	PGC 1191		.96		15.14		15.26± .57		6436
001821.1-730908		RL..+?.	1.05± .05	136	14.95 ±.14				
306.24 -43.76	ESO 28- 12	r	.41± .05	.08	12.87				
223.14 -12.40	IRAS00160-7325	-1.3±1.8		.00					
001604.0-732548	PGC 1193		1.00						
001822.5+300450	NGC 70	.SAT5..	1.15± .05	0				16.59±.3	7167± 10
114.47 -32.25	UGC 174	R (1)	.06± .05	.17	14.18 ±.14				7251± 57
324.31 15.56		5.0± .4		.09				345± 7	7341
001545.8+294811	PGC 1194	4.5±1.1	1.17	.03	13.89			2.67	6852
0018.3 +1750		.S..6*.	.96± .09	110					11209
111.91 -44.32	UGC 172	U	.90± .06	.11					
311.92 13.23		6.0±1.5		1.33					11356
0015.8 +1734	PGC 1195		.97	.45					10869
001823.7+300346	NGC 71	.LA.-P*	1.09± .09		14.2 ±.3	1.02± .03	1.04± .02		
114.48 -32.27	UGC 173	R (1)	.06± .06	.17	14.42 ±.11				6716± 44
324.29 15.56		-3.0± .4	.77± .12	.00		.92	13.50± .41		6888
001547.1+294707	PGC 1197	4.5±1.1	1.10		14.12		14.37± .56		6399
001823.8+484350		.I..9..	1.05± .06	30	14.0 ±.3				
117.33 -13.78	UGC 171	U	.14± .05	.62	11.81				
343.51 17.73	IRAS00157+4827	10.0± .9		.11					
001544.4+482711	PGC 1198		1.11	.07					
0018.4 +4959		.SX.6*.	1.15± .05	50				15.34±.1	5283± 11
117.51 -12.53	UGC 175	U	.16± .05	.66	14.6 ±.3				
344.82 17.80		6.0±1.2		.24				130± 8	5478
0015.8 +4943	PGC 1202		1.21	.08	13.69			1.57	5033
001828.5+300224	NGC 72	.SBT2..	1.04± .05	15	14.5 ±.3	1.04± .05	1.10± .04	19.21±.3	7259± 10
114.49 -32.29	UGC 176	R	.07± .04	.17	14.70 ±.14				7050± 44
324.27 15.54		2.0± .4	.76± .08	.09		.93	13.80± .24	119± 7	7421
001551.8+294545	PGC 1204		1.06	.04	14.34		14.39± .41	4.84	6932
001834.4+300209	NGC 72A	.E.3.*.	.52?		15.7 ±.2	1.03± .08			
114.52 -32.30	MCG 5- 1- 70	P	.09± .07	.17	15.33 ±.20				6807± 56
324.27 15.51	ARAK 6	-5.0±1.5		.00		.93			6979
001557.7+294530	PGC 1208		.52		15.24		13.04±1.50		6489
001839.1-151920	NGC 73	.SXT4*.	1.21± .04	145					7714
88.44 -75.95	MCG -3- 1- 26	E (1)	.15± .04	.04					
279.32 4.31		4.0± .6		.22					7764
001606.7-153600	PGC 1211	3.1± .8	1.21	.07					7391
001848.0-190027	A 0016-19	.SXS9..	1.17± .03	165	14.83 ±.16	.52± .12	.48± .05	14.80±.3	2060± 17
78.50 -78.95	ESO 539- 7	SUE (3)	.06± .03	.05	14.91 ±.14	-.37± .12	-.28± .07	208± 34	
275.77 3.16	DDO 1	8.7± .4	1.06± .03	.06		.48	15.62± .09	187± 25	2097
001616.0-191706	PGC 1218	8.5± .4	1.18	.03	14.76	-.40	15.38± .23	.01	1745
001850.1-102140		.SBS5P.	1.05± .05	65	13.6 V±.3		.46± .03		
96.77 -71.56	MCG -2- 1- 52	E	.17± .04	.12					8281± 45
284.14 5.75	IRAS00163-1039	4.5± .6	.92± .07	.26	11.17				8347
001617.3-103819	PGC 1221		1.06	.09					7948
001850.8-082735		PSBR0?.	1.07± .05	175					
98.98 -69.80	MCG -2- 1- 50	E	.09± .04	.13					
285.98 6.31		.0± .6		.07					
001617.8-084414	PGC 1223·		1.08						

R.A. 2000 DEC.	Names	Type	$logD_{25}$	p.a.	B_T	$(B-V)_T$	$(B-V)_o$	m_{21}	V_{21}
l b		S_T n_L	$logR_{25}$	A_g	m_B	$(U-B)_T$	$(U-B)_o$	W_{20}	V_{opt}
SGL SGB		T	$logA_o$	A_i	m_{FIR}	$(B-V)_T^o$	m'_o	W_{50}	V_{GSR}
R.A. 1950 DEC.	PGC	L	$logD_o$	A_{21}	B_T^o	$(U-B)_T^o$	m'_{25}	HI	V_{3K}
001850.9-102238		.SBS3P?	1.04± .07	73	14.81 ±.13	.48± .02			
96.76 -71.57	MCG -2- 1- 51	E	.24± .04	.12					8123± 52
284.12 5.74		3.0±1.3		.33		.36			8188
001618.1-103917	PGC 1224		1.05	.12	14.31		14.29± .40		7789
0018.8 -5738		RSASOP*	1.35± .06		15.36 ±.15	.63± .03		13.52±.3	1745± 9
311.39 -58.95		S	.11± .08	.00		-.07± .06		261± 10	
238.45 -8.50		.4± .5		.08		.60			1652
0016.4 -5755	PGC 1225		1.34		15.25	-.07	16.65± .39		1591
0019.0 +2328		.S..6*.	1.07± .06	3					4464
113.38 -38.79	UGC 179	U	.22± .05	.13	14.98 ±.14				
317.63 14.24		6.0±1.2		.33					4623
0016.4 +2312	PGC 1231		1.08	.11	14.51				4133
001904.6-225615	NGC 66	.SBR3P.	1.07± .05	32					7568
60.93 -81.56	ESO 473- 10	S (1)	.22± .04	.00	14.21 ±.14				
272.00 1.89	IRAS00165-2312	3.0± .8		.30	12.27				7591
001633.0-231254	PGC 1236	2.2± .8	1.07	.11	13.86				7264
0019.0 +1544		.S..2..	1.04± .06	170					10991
111.60 -46.41	UGC 180	U	.43± .05	.11	15.06 ±.12				
309.85 12.60		2.0± .9		.53					11132
0016.5 +1528	PGC 1237		1.05	.22	14.31				10649
0019.1 +1923		.S..1..	1.04± .06	14					5354
112.51 -42.82	UGC 181	U	.58± .05	.13	15.23 ±.12				
313.51 13.39		1.0± .9		.59					5504
0016.5 +1907	PGC 1238		1.05	.29	14.45				5016
001909.4-243121		.SB?...	1.02± .05	149					9831
50.87 -82.28	ESO 473- 11		.28± .04	.09	15.09 ±.14				
270.48 1.38				.42					9849
001638.0-244800	PGC 1241		1.03	.14	14.52				9532
001911.8-224014		.I..9*P	1.18± .06						669
62.55 -81.44	MCG -4- 2- 3	S	.10± .05	.07					
272.26 1.94		10.0± .9		.07					694
001640.2-225653	PGC 1242		1.19	.05					365
0019.3 +0626	NGC 75	.L.....	1.16± .09						
108.75 -55.54	UGC 182	U	.00± .05	.07	14.22 ±.14				
300.62 10.29		-2.0± .7		.00					
0016.8 +0610	PGC 1255		1.17						
001935.0+471429		.S..2..	1.20± .05	50					
117.34 -15.28	UGC 183	U	.37± .05	.44	13.7 ±.3				
341.98 17.42		2.0± .9		.45					
001655.5+465750	PGC 1264		1.24	.18					
0019.5 +1311		.I..9*.	.96± .09						5253
111.08 -48.94	UGC 184	U	.00± .06	.15					
307.33 11.90		10.0±1.2		.00					5388
0017.0 +1255	PGC 1265		.97	.00					4909
001937.8+295607	NGC 76	.S?....	1.00± .08	80					
114.77 -32.43	UGC 185	U	.05± .06	.19	13.99 ±.12				7389± 47
324.21 15.27				.06					7560
001700.9+293928	PGC 1267		1.02	.03	13.66				7072
001940.1-511603		.LBR+P*	1.19± .05	29					
314.79 -65.12	ESO 194- 4	Sr	.47± .03	.04	14.52 ±.14				6586± 24
244.69 -6.82		-.8± .6		.00					6513
001713.0-513242	PGC 1270		1.13		14.38				6399
001948.1-342821		RSBR1..	1.13± .04	121					
343.74 -80.00	ESO 350- 9	Sr	.14± .04	.02	14.61 ±.14				7609± 39
260.94 -1.82		1.0± .6		.15					7592
001718.0-344500	PGC 1275		1.14	.07	14.35				7345
0019.8 +2346	IC 1540	.SB.3..	.98± .07	30					5834
113.67 -38.52	UGC 186	U	.39± .05	.10	14.91 ±.13				
317.97 14.12		3.0± .9		.54					5993
0017.2 +2330	PGC 1276		.99	.20	14.22				5503
001958.5-770527		.SBS7..	1.17± .04					14.26±.3	1810± 9
305.22 -39.90	ESO 28- 14	S (1)	.25± .04	.23	15.90 ±.14				
219.23 -13.36		7.0± .5		.35				124± 7	1664
001749.0-772206	PGC 1284	8.6± .5	1.19	.13	15.31			-1.18	1769
0019.9 +0908		.S?....	1.00± .08	66					5461
109.98 -52.93	UGC 187		.55± .06	.22					
303.33 10.82				.82					5585
0017.4 +0852	PGC 1285		1.02	.27					5115

R.A. 2000 DEC. / l b / SGL SGB / R.A. 1950 DEC.	Names / PGC	Type / S_T n_L / T / L	$\log D_{25}$ / $\log R_{25}$ / $\log A_o$ / $\log D_o$	p.a. / A_g / A_i / A_{21}	B_T / m_B / m_{FIR} / B_T^o	$(B-V)_T$ / $(U-B)_T$ / $(B-V)_T^o$ / $(U-B)_T^o$	$(B-V)_o$ / $(U-B)_o$ / m' / m'_{25}	m_{21} / W_{20} / W_{50} / HI	V_{21} / V_{opt} / V_{GSR} / V_{3K}
0020.0 +1959		.SB?...	1.10± .05	13					7755± 57
112.93 -42.27	UGC 188		.54± .05	.08	15.36 ±.13				7906
314.16 13.31				.80					7418
0017.4 +1943	PGC 1286		1.10	.27	14.43				
002000.1-062007		.SBS2P?	1.18± .04	35					
101.75 -67.90	MCG -1- 2- 1	E (1)	.32± .04	.12					
288.13 6.65		1.5±1.2		.39					
001727.0-063645	PGC 1288	3.1±1.6	1.19	.16					
0020.0 +1052	A 0017+10	.S..9..	1.21± .05	150	13.87 ±.16	.44± .04	.42± .03	14.23±.1	1144± 6
110.58 -51.24	UGC 191	U (1)	.13± .05	.22	14.89 ±.18	-.21± .06	-.11± .04	137± 8	1273
305.06 11.22	DDO 2	9.0± .7	1.08± .03	.13		.35	14.76± .07	125± 12	798
0017.5 +1036	PGC 1292	9.0±1.0	1.23	.07	13.97	-.27	14.45± .31	.19	
0020.0 +1506		.S..7*.	1.04± .15	120					7618
111.78 -47.08	UGC 189	U	.13± .12	.12					
309.28 12.22		7.0±1.2		.18					7757
0017.5 +1450	PGC 1293		1.05	.06					7275
002006.0-543127		.SXS5..	1.04± .05	15	14.79 ±.17	.64± .04	.72± .02		
312.64 -62.01	ESO 150- 1	S (1)	.17± .04	.00	15.04 ±.14				
241.54 -7.80		4.5± .6	.78± .03	.26			14.18± .07		
001740.0-544806	PGC 1295	1.7± .6	1.04	.09			14.40± .31		
0020.3 +0741	IC 13	.S..4..	1.16± .05	163					5708
109.64 -54.37	UGC 195	U	.50± .05	.16	14.73 ±.12				
301.92 10.36		4.0± .9		.73					5828
0017.8 +0725	PGC 1301		1.18	.25	13.81				5361
002022.9-140207		.SBS9..	1.11± .04	33					
92.43 -75.04	MCG -2- 2- 2	E (1)	.10± .04	.03					
280.69 4.30		9.3± .5		.10					
001750.6-141845	PGC 1303	9.4± .5	1.11	.05					
002024.6+591730	IC 10	.IB.9$.	1.80± .02		11.8 S±.2			10.53±.1	-344± 3
118.97 -3.34	UGC 192	R	.09± .05	3.5				85± 3	-342± 22
354.48 17.86	IRAS00177+5900	10.0± .3		.07	9.29			63± 2	-146
001741.5+590052	PGC 1305		2.13	.05	8.18		15.36± .26	2.30	-552
002025.5+004936	NGC 78A	.SBR0*.	1.05± .08	80	13.68V±.14	.86± .05	.88± .05		
106.74 -61.07	UGC 193	UE	.20± .06	.06					5052± 42
295.16 8.55	IRAS00178+0032	-.3± .7	.79± .03	.15	13.66	.77			5152
001751.6+003258	PGC 1306		1.04		14.26		14.15± .44		4707
002027.4+004959	NGC 78B	.L..0P?	.84± .07	43	14.4 ±.2	.90± .04	.99± .02		
106.76 -61.06	UGC 194	UE	.10± .05	.06					5443± 49
295.17 8.54	MK 547	-2.3±1.0	.64± .08	.00		.82	13.04± .28		5543
001753.5+003320	PGC 1309		.83		14.26		13.22± .44		5098
002034.3+472604		.SBS5..	1.18± .05	140	13.9 ±.3			15.71±.1	5136± 9
117.53 -15.11	UGC 196	U	.15± .05	.44				249± 11	
342.20 17.27	IRAS00179+4709	5.0± .8		.23	13.39				5329
001754.4+470926	PGC 1315		1.22	.08	13.18			2.46	4876
002041.4-231657		PSBR2..	1.20± .04	80	14.44 ±.14				7788± 60
60.19 -82.06	ESO 473- 16	Sr	.34± .03	.05					7810
271.78 1.43		1.6± .6		.42					
001810.0-233336	PGC 1325		1.20	.17	13.89				7486
002042.0-042600		.LA.0?.	1.10± .06	36					
103.68 -66.15	MCG -1- 2- 5	E	.37± .05	.12					
290.04 7.02		-2.0±1.2		.00					
001808.7-044239	PGC 1326		1.06						
002044.8+223558	IC 1542	.S?....	.85± .12					16.42±.3	7352± 10
113.71 -39.72	MCG 4- 2- 1		.10± .07	.14	14.97 ±.12				7508
316.82 13.68				.08				212± 7	7020
001808.6+221920	PGC 1328		.85		14.64				
0020.8 +2001		.SX.1..	1.00± .06						7690
113.19 -42.26	UGC 197	U	.02± .05	.08	14.70 ±.14				
314.24 13.14		1.0± .8		.02					7840
0018.2 +1945	PGC 1330		1.01	.01	14.50				7354
002055.9+215156	IC 1543	.S?....	.87± .08	30	14.30 ±.14				5622± 37
113.62 -40.45	UGC 198		.04± .05	.10					5776
316.09 13.49	IRAS00183+2135			.07	13.48				5288
001819.8+213518	PGC 1333		.88	.02	14.11				
002057.0-635127		.S..3*/	1.31± .04	172					1775
308.50 -52.94	ESO 78- 22	S (1)	.77± .04	.00	14.56 ±.14				1663
232.41 -10.38		4.3± .6		1.14					1656
001835.0-640806	PGC 1335	5.6± .9	1.31	.39	13.41				

R.A. 2000 DEC. / l b / SGL SGB / R.A. 1950 DEC.	Names / PGC	Type S_T n_L / T / L	$\log D_{25}$ / $\log R_{25}$ / $\log A_e$ / $\log D_o$	p.a. / A_g / A_i / A_{21}	B_T / m_B / m_{FIR} / B_T^o	$(B-V)_T$ / $(U-B)_T$ / $(B-V)_T^o$ / $(U-B)_T^o$	$(B-V)_e$ / $(U-B)_e$ / m'_e / m'_{25}	m_{21} / W_{20} / W_{50} / HI	V_{21} / V_{opt} / V_{GSR} / V_{3K}
0020.9 +1251		.I..9*.	.96± .09						1803
111.50 -49.33	UGC 199	U	.00± .06	.16					
307.09 11.49		10.0±1.2		.00					1936
0018.4 +1235	PGC 1336		.98	.00					1459
0021.0 +1741		.I..9..	1.00± .16						5227± 10
112.73 -44.57	UGC 200	U	.04± .12	.16					
311.91 12.59		10.0± .9		.03					5372
0018.4 +1725	PGC 1337		1.01	.02					4888
0021.0 +0736		.S..6*.	1.04± .08	18					11831
109.90 -54.49	UGC 201	U	.76± .06	.15					
301.89 10.17		6.0±1.4		1.12					11950
0018.5 +0720	PGC 1346		1.05	.38					11485
002106.7+263051		.S?....	.77± .11	.14	15.73 ±.17			17.41±.3	11665± 10
114.55 -35.86	CGCG 479- 4	(1)	.58± .10	.86					11586
320.79 14.35				.29	14.66			375± 7	11829
001830.0+261413	PGC 1347	1.0±1.0	.79					2.46	11340
002108.7+271242		RLB.0..	1.15± .08	25	14.25 ±.10				9246± 46
114.69 -35.17	UGC 202	U	.24± .04	.10					
321.50 14.47		-2.0± .8		.00	14.02				9411
001831.9+265604	PGC 1349		1.13						8923
002108.8-502927		.S?....	.93± .07	.04	15.19 ±.14				10107
314.80 -65.93	ESO 194- 7		.04± .06	.05					
245.51 -6.82				.02	14.98				10036
001842.0-504606	PGC 1350		.94						9916
002111.3+222129	NGC 80	.LA.-*.	1.26± .09		13.1 ±.2	1.03± .02	1.07± .01		5672± 27
113.79 -39.98	UGC 203	PU	.03± .07	.14	13.36 ±.11	.64± .06	.65± .03		5827
316.60 13.53		-2.5± .5	.99± .04	.00		.94	13.54± .15		5339
001835.1+220451	PGC 1351		1.27		13.08	.63	14.19± .52		
0021.2 +3028		.L?....	.82± .19	109	15.15 ±.11				4679± 57
115.26 -31.94	MCG 5- 2- 1		.20± .07	.22					
324.82 15.02			.81	.00	14.85				4850
0018.6 +3012	PGC 1353								4364
002115.2-483746	NGC 87	.IB.9P.	.94± .03		14.67 ±.14	.35± .03	.30± .03		3491± 44
316.16 -67.72	ESO 194- 8	S	.09± .04	.04	14.70 ±.14	-.39± .05	-.34± .05		3426
247.33 -6.30		10.0± .6	.53± .02	.07		.30	12.81± .05		3291
001848.0-485424	PGC 1357		.94	.04	14.58	-.43	14.01± .23		
002116.9+303025		.SB?...	.94± .11		15.01 ±.14			17.37±.3	4840± 10
115.28 -31.92	MCG 5- 2- 2		.00± .07	.22					4713± 57
324.85 15.02				.00	14.74			252± 7	5007
001839.8+301347	PGC 1359		.96	.00				2.63	4521
002118.1+295131								17.45±.3	6791± 10
115.18 -32.56	CGCG 500- 3			.20	15.6 ±.4				
324.19 14.90								142± 7	6961
001841.0+293453	PGC 1361								6474
0021.3 +2306	IC 1544	.SXS5..	1.07± .06	150	14.42 ±.12			15.69±.3	5714± 10
113.98 -39.24	UGC 204	U	.15± .05	.08				260± 13	
317.36 13.65		5.0± .8		.23				257± 10	5871
0018.7 +2250	PGC 1362		1.08	.08	14.12				5383
002122.1-483827	NGC 88	.SBT0*P	.90± .05	145	14.98 ±.14	.57± .04			3433± 44
316.11 -67.71	ESO 194- 10	S	.21± .04	.04	15.11 ±.14	.28± .07			3368
247.33 -6.32		-.3± .7		.16		.49			3233
001855.1-485506	PGC 1370		.90		14.80	.27	13.84± .31		
002122.8+222608	NGC 83	.E.....	1.19± .09		13.58 ±.14	1.08± .02	1.12± .01		6359± 27
113.87 -39.91	UGC 206	PU	.02± .06	.14	13.85 ±.12				6515
316.69 13.50		-5.0± .6	.90± .03	.00		.99	13.57± .12		6027
001846.6+220930	PGC 1371		1.21		13.50		14.48± .48		
0021.4 +1719		.SB?...	.89± .11	10	15.36 ±.12				5273
112.78 -44.95	UGC 205		.24± .07	.16					
311.56 12.41				.36					5417
0018.8 +1703	PGC 1372		.90	.12	14.82				4933
002124.1-483957	NGC 89	.SBS0?P	1.08± .03	148	14.18 ±.13	.69± .02	.71± .03		3283± 54
316.07 -67.69	ESO 194- 11	S	.28± .04	.04	14.28 ±.14	.14± .03	.17± .04		3218
247.30 -6.34		.3± .7	.50± .02	.21		.60	12.17± .06		3083
001857.0-485636	PGC 1374		1.07		13.93	.11	13.73± .22		
002129.2+223002	IC 1546	.S?....	.94± .11		15.59 ±.09			17.05±.3	5804± 10
113.91 -39.85	MCG 4- 2- 8		.57± .07	.14					5960
316.76 13.49				.84				340± 7	
001853.0+221324	PGC 1382		.95	.28	14.58			2.18	5472

R.A. 2000 DEC. / l b / SGL SGB / R.A. 1950 DEC.	Names / PGC	Type: ST nL / T / L	logD25 / logR25 / logAe / logDo	p.a. / Ag / Ai / A21	BT / mB / mFIR / BT°	(B-V)T / (U-B)T / (B-V)T° / (U-B)T°	(B-V)e / (U-B)e / m'e / m'25	m21 / W20 / W50 / HI	V21 / Vopt / VGSR / V3K
002131.1+293938		.S..3..	.97± .07	17				17.07±.3	7041± 10
115.20 -32.76	UGC 207	U (1)	.51± .05	.21	15.52 ±.12				
324.00 14.82		3.0±1.0		.71				303± 7	7211
001854.0+292300	PGC 1387	4.5±1.3	.99	.26	14.55			2.27	6724
002132.1-483734	NGC 92	.SA.1*P	1.29± .03	148	13.81 ±.13	.74± .02	.82± .02		
316.05 -67.73	ESO 194- 12	S (1)	.32± .05	.04	13.71 ±.14	.10± .03	.08± .03		3375± 27
247.35 -6.35		1.0± .7	.58± .02	.32		.64	12.20± .07		3310
001905.1-485412	PGC 1388	1.29	.16	13.36	.05	14.31± .22		3175
002151.7+222401	NGC 91	.SXS5P.	1.29± .03	132	14.54 ±.13	.84± .02	.92± .02	15.03±.2	5353± 7
114.00 -39.96	UGC 208	R (1)	.40± .05	.12	14.23 ±.13	.21± .05	.48± .04	453± 16	5173± 38
316.68 13.38		5.0± .4	.80± .02	.60		.71	14.03± .04	428± 12	5502
001915.5+220723	PGC 1405	1.1± .8	1.30	.20	13.64	.09	14.82± .22	1.19	5014
0022.0 +2224	NGC 93	.S?....	1.14± .05	48	14.34 ±.13	1.06± .02		15.00±.3	5380± 10
114.05 -39.95	UGC 209		.39± .05	.12	14.55 ±.12	.58± .03		600± 6	
316.69 13.35				.59		.93			5535
0019.4 +2208	PGC 1412	1.15	.20	13.72	.44	13.93± .31	1.08	5048
0022.0 +2344		.S..3..	1.06± .06	19					4481
114.30 -38.63	UGC 210	U (1)	.40± .05	.10	14.89 ±.12				
318.03 13.62		3.0± .9		.55					4639
0019.4 +2328	PGC 1413	4.5±1.2	1.07	.20	14.20				4151
002210.2-042050		.SAR4?.	1.11± .06	95					
104.61 -66.17	MCG -1- 2- 6	E (1)	.10± .05	.11					
290.23 6.69		4.0±1.2		.15					
001936.9-043727	PGC 1419	4.2±1.2	1.12	.05					
0022.2 +2036		.S..8*.	1.00± .08	165					5637
113.75 -41.74	UGC 211	U	.73± .06	.12					
314.89 12.94		8.0±1.4		.89					5788
0019.6 +2020	PGC 1422	1.01	.36					5302
002214.2+102938	NGC 95	.SXT5P.	1.29± .03		13.19 ±.14	.67± .04	.84± .04	14.34±.2	5401± 8
111.30 -51.72	UGC 214	R (1)	.24± .03	.20	13.21 ±.10	.04± .03	.27± .03	345± 12	4886± 50
304.81 10.62	IRAS00196+1012	5.0± .4	.86± .02	.36	12.12	.54	12.98± .05	307± 9	5513
001939.3+101300	PGC 1426	1.31	.12	12.61	-.05	13.89± .21	1.60	5042
0022.3 +0626		.L...*.	1.11± .16	150					
110.05 -55.70	UGC 213	U	.20± .08	.08	15.08 ±.15				
300.82 9.57		-2.0±1.2		.00					
0019.8 +0610	PGC 1433	1.09						
002226.3+293011		.SB.2..	1.07± .06	135				16.67±.3	7091± 10
115.41 -32.94	UGC 215	U	.17± .05	.21	14.65 ±.13				
323.87 14.60		2.0± .9		.22				376± 7	7260
001949.0+291333	PGC 1439	1.09	.09	14.16			2.42	6774
002226.4-533846		.SXS8..	1.43± .04	15				14.71±.3	1438± 9
312.40 -62.95	ESO 150- 5	S (1)	.16± .04	.00	14.02 ±.14				
242.49 -7.89		8.0± .5		.19				121± 7	1357
002001.0-535524	PGC 1440	7.8± .5	1.43	.08	13.82			.81	1264
002230.0+294443	NGC 97	.E...*.	1.18± .15		13.29 ±.10				4780± 21
115.47 -32.71	UGC 216	U	.07± .08	.21					4949
324.12 14.63		-5.0±1.1		.00					4463
001952.8+292806	PGC 1442	1.19		13.01				
002233.3+280717		.S..9*.	1.04± .15	10				16.74±.3	4756± 10
115.22 -34.32	UGC 217	U	.18± .12	.10					
322.48 14.33		9.0±1.2		.18				126± 7	4922
001956.2+275040	PGC 1450	1.05	.09					4435
002234.0-082914		RL..+?/	1.20± .05	118					5692
101.47 -70.15	MCG -2- 2- 7	E	.56± .05	.10					
286.23 5.42		-1.0±1.2		.00					5762
002001.2-084552	PGC 1451	1.12						5357
002238.7-483452		.SXR6..	1.10± .05	47	13.9 ±.3	.29± .04			
315.62 -67.83	ESO 194- 13	S (1)	.09± .05	.04	14.00 ±.14	-.11± .06			3400±190
247.44 -6.51	FAIR 650	6.0± .8		.13		.24			3335
002012.1-485130	PGC 1452	5.6± .8	1.10	.04	13.80	-.14	14.01± .41		3200
002245.8+295625		.S?....	.82± .13					16.43±.3	9182± 10
115.56 -32.52	MCG 5- 2- 8		.28± .07	.21	15.49 ±.12				
324.33 14.60				.35				294± 7	9351
002008.5+293947	PGC 1460	.84	.14	14.84			1.45	8866
002248.6+240905								16.98±.3	12556± 10
114.61 -38.26	CGCG 479- 18			.06	15.4 ±.4				
318.48 13.52								247± 7	12714
002012.1+235227	PGC 1462							12227

R.A. 2000 DEC. l b SGL SGB R.A. 1950 DEC.	Names PGC	Type S_T n_L T L	$\log D_{25}$ $\log R_{25}$ $\log A_e$ $\log D_o$	p.a. A_g A_i A_{21}	B_T m_B m_{FIR} B_T^o	$(B-V)_T$ $(U-B)_T$ $(B-V)_T^o$ $(U-B)_T^o$	$(B-V)_e$ $(U-B)_e$ m'_e m'_{25}	m_{21} W_{20} W_{50} HI	V_{21} V_{opt} V_{GSR} V_{3K}
002249.4-451610 318.53 -70.98 250.67 -5.59 002022.0-453248	NGC 98 ESO 242- 5 PGC 1463	.SBT4.. Sr (1) 4.2± .6 2.2± .8	1.22± .03 .11± .05 .82± .01 1.22	0 .03 .16 .05	13.41 ±.13 13.59 ±.14 13.27	.73± .01 .66	.82± .01 13.00± .04 14.09± .23		6148± 26 6093 5932
0022.8 +0658 110.45 -55.20 301.38 9.58 0020.3 +0642	IC 1549 UGC 218 PGC 1464	.S?....	.92± .11 .00± .06 .93	.12 .00 .00	14.64 ±.12 14.46				12156 12273 11810
002257.1+251247 114.84 -37.21 319.56 13.70 002020.3+245610	CGCG 479- 19 PGC 1472			.04	15.6 ±.4			16.96±.3 436± 7	12189± 10 12349 11862
0023.0 +1614 113.07 -46.08 310.57 11.80 0020.4 +1558	 UGC 219 PGC 1478	.S..8*. U 8.0±1.3	.99± .06 .36± .05 1.00	148 .08 .45 .18	 15.38 ±.13 14.84				5241 5382 4901
002301.4-565009 310.68 -59.86 239.39 -8.82 002037.4-570647	 PGC 1481	.LAS-.. S -3.0± .6	1.24± .08 .17± .08 1.21	 .00 .00					
0023.0 +0226 108.80 -59.65 296.93 8.35 0020.5 +0210	 UGC 220 PGC 1485	.L..... U -2.0± .9	1.15± .15 .43± .08 1.09	177 .07 .00					
002311.2+272558 115.27 -35.02 321.81 14.06 002034.2+270921	 UGC 221 PGC 1491	.S?....	.81± .21 .15± .12 .82	155 .14 .20 .07	14.66 ±.15 14.29				3874± 46 4038 3552
002316.3+245717 114.88 -37.48 319.31 13.58 002039.6+244040	CGCG 479- 21 PGC 1495			.03	15.5 ±.4			17.09±.3 135± 7	17974± 10 18134 17647
002316.3-082933 101.95 -70.21 286.27 5.25 002043.5-084611	MCG -2- 2- 8 PGC 1496	.SBT6.. E (1) 6.0± .9 5.3± .8	1.01± .07 .08± .05 1.02	100 .11 .11 .04					
0023.3 +3508 116.47 -27.37 329.64 15.33 0020.7 +3452	 UGC 222 PGC 1498	.I?....	1.04± .15 1.06± .06 1.06	138 .22 .75 .50					6109 6287 5807
0023.5 +2014 114.08 -42.15 314.59 12.56 0020.9 +1958	 UGC 223 PGC 1502	.I..9*. U 10.0±1.1	1.16± .13 .00± .12 1.17	 .12 .00 .00					5721 5871 5386
002348.0+144114 112.98 -47.65 309.07 11.25 002112.5+142437	A 0021+14 UGC 226 MK 338 PGC 1511	.S?....	1.04± .06 .26± .05 1.05	4 .14 .39 .13	 14.81 ±.12 14.25				5400±110 5536 5059
002353.8+281954 115.61 -34.15 322.75 14.08 002116.5+280317	 UGC 229 PGC 1516	.S..3.. U (1) 3.0± .9 4.5±1.2	.95± .06 .27± .05 .97	35 .13 .37 .14	 15.16 ±.12 14.60			16.01±.2 274± 13 256± 7 1.27	7208± 7 7374 6888
002354.2-621616 308.45 -54.56 234.06 -10.31 002133.0-623254	 ESO 78- 23 IRAS00215-6232 PGC 1517		.82± .07 .08± .06 .82	102 .00	15.05 ±.14 12.99				10232± 69 10124 10105
002354.6-323211 349.09 -81.95 263.07 -2.05 002125.0-324848	NGC 101 ESO 350- 14 IRAS00214-3248 PGC 1518	.SXT6*. BSr (1) 5.7± .4 5.6± .5	1.34± .03 .03± .04 .99± .03 1.34	 .02 .04 .01	13.37 ±.17 13.36 ±.14 13.46 13.29	.53± .02 -.08± .03 .50 -.10	.61± .02 -.03± .03 13.81± .07 14.83± .26		3400± 34 3388 3130
0023.9 +2418 114.96 -38.14 318.69 13.30 0021.3 +2401	 UGC 228 IRAS00213+2401 PGC 1520	.S..4.. U 4.0± .9	1.01± .08 .22± .06 1.02	43 .06 .32 .11	 14.68 ±.12 14.27				5683 5841 5355
002359.7+154610 113.30 -46.59 310.15 11.46 002124.1+152933	NGC 99 UGC 230 IRAS00214+1529 PGC 1523	.S..6*. U 6.0±1.1	1.15± .03 .04± .05 .76± .01 1.15	 .10 .06 .02	13.99 ±.13 13.71 ±.12 13.22 13.66	.34± .02 -.21± .03 .27 -.26	.40± .02 -.12± .03 13.28± .03 14.46± .23	14.86±.1 182± 10 132± 9 1.18	5322± 7 5184± 60 5459 4980

R.A. 2000 DEC. l b SGL SGB R.A. 1950 DEC.	Names PGC	Type S_T n_L T L	$\log D_{25}$ $\log R_{25}$ $\log A_e$ $\log D_0$	p.a. A_g A_i A_{21}	B_T m_B m_{FIR} B_T^0	$(B-V)_T$ $(U-B)_T$ $(B-V)_T^0$ $(U-B)_T^0$	$(B-V)_e$ $(U-B)_e$ m'_e m'_{25}	m_{21} W_{20} W_{50} HI	V_{21} V_{opt} V_{GSR} V_{3K}
002402.3+253944 115.22 -36.80 320.06 13.54 002125.3+252307	 MCG 4- 2- 16 PGC 1524	.S?.... 	.74± .14 .21± .07 .74	 .07 .26 .10	 	 	 	16.64±.3 252± 7 	7378± 10 7539 7053
002402.6+162909 113.48 -45.88 310.87 11.61 002126.9+161232	NGC 100 UGC 231 PGC 1525	.S..6*/ PU 6.0± .7 	1.74± .02 .90± .04 .97± .01 1.75	56 .10 1.32 .45	13.91 ±.13 14.0 ±.2 12.50	.65± .03 .02± .04 .46 -.13	.74± .02 .11± .04 14.25± .03 15.22± .19	13.18±.1 224± 5 210± 7 .23	842± 5 982 502
002428.5-504105 313.42 -65.89 245.48 -7.38 002203.0-505742	 ESO 194- 15 PGC 1535	.S?.... 	1.04± .07 .14± .06 1.05	 .04 .20 .07	 14.87 ±.14 14.58	 	 	 	 7300 7227 7111
0024.5 -1129 99.64 -73.13 283.46 4.08 0022.0 -1146	 MCG -2- 2- 12 PGC 1538	.S?.... 	1.22± .07 .66± .07 1.23	 .11 .92 .33	 	 	 	 	 5210 5269 4881
0024.6 +2916 115.94 -33.23 323.73 14.09 0022.0 +2900	A 0022+29A CGCG 500- 15 PGC 1544	 	 	 .14 	 14.8 ±.4 	 	 	 	 4762± 72 4929 4445
002438.8+331510 116.51 -29.28 327.76 14.76 002200.6+325833	 UGC 232 IRAS00220+3258 PGC 1546	.SBR1.. U 1.0± .8 	1.02± .08 .14± .06 1.04	 .18 .14 .07	 14.55 ±.12 12.95 14.17	 	 	15.55±.3 259± 7 1.31	4842± 10 5016 4535
002443.0+144920 113.34 -47.55 309.25 11.07 002207.4+143243	A 0022+14 UGC 233 MK 339 PGC 1550	.P..... R 99.0 	.96± .09 .00± .06 .97	 .15 .00 .00	 14.37 ±.12 14.18	 	 	15.80±.3 139± 10 1.62	5279± 9 5400±110 5416 4939
002446.6+293337 116.02 -32.95 324.02 14.11 002208.9+291700	 UGC 234 PGC 1552	.S..6*. U 6.0±1.3 	1.04± .08 .37± .06 1.06	72 .24 .55 .19	 15.34 ±.13 14.54	 	 	16.75±.3 218± 7 2.03	4668± 10 4836 4352
0024.8 +4509 118.03 -17.46 339.93 16.32 0022.2 +4453	 UGC 235 PGC 1559	.L...?. U -2.0±1.8 	1.00± .08 .33± .06 .99	135 .37 .00 	 	 	 	 	
0024.9 +0639 111.24 -55.61 301.20 9.00 0022.4 +0623	 UGC 237 PGC 1566	.SB.1.. U 1.0± .9 	1.00± .06 .38± .05 1.00	70 .07 .39 .19	 15.16 ±.12 	 	 	 	
002458.8+433945 117.88 -18.95 338.39 16.14 002218.4+432309	 UGC 236 PGC 1567	.E..... U -5.0± .8 	.85± .10 .01± .06 .90	 .33 .00 	 14.80 ±.10 14.39	 	 	 	 5104±125 5292 4831
002503.6+312041 116.35 -31.19 325.84 14.36 002225.6+310404	 UGC 238 IRAS00224+3104 PGC 1572	.S?.... 	1.28± .04 .56± .05 1.29	178 .19 .85 .28	 14.27 ±.12 13.25 13.19	 	 	 	 6848± 50 7019 6536
002509.7-143218 95.69 -76.00 280.55 3.04 002237.7-144855	 MCG -3- 2- 13 PGC 1575	.SBS7*/ E (1) 7.0±1.3 4.2± .8	1.23± .05 .80± .05 1.23	81 .08 1.11 .40	 	 	 	 	
0025.1 +4203 117.73 -20.55 336.75 15.91 0022.5 +4147	 UGC 239 PGC 1576	.I..9.. U 10.0± .9 	1.00± .16 .14± .12 1.02	 .25 .10 .07	 	 	 	 	
0025.2 +0629 111.28 -55.78 301.05 8.90 0022.6 +0613	 UGC 240 IRAS00226+0612 PGC 1577	.SXT3.. U 3.0± .9 	1.39± .04 .36± .05 1.39	68 .07 .50 .18	 13.20 	 	 	 	8661 8775 8316
002513.7+300210 116.21 -32.49 324.52 14.10 002235.9+294533	A 0022+29B MCG 5- 2- 11 PGC 1578	.SB?... 	.94± .11 .00± .07 .96	 .22 .00 .00	 15.16 ±.15 14.90	 	 	16.45±.3 102± 7 1.55	4855± 10 4782± 72 5022 4539
002517.0+125306 113.09 -49.49 307.37 10.47 002241.6+123630	NGC 105 UGC 241 IRAS00226+1236 PGC 1583	.SA.2*. PU 1.5± .6 	1.05± .06 .19± .05 1.06	167 .18 .23 .09	13.9 ±.2 12.91 13.43	.71± .09 .59 	 13.52± .38	 	5290± 60 5421 4948

R.A. 2000 DEC. l　　b SGL　　SGB R.A. 1950 DEC.	Names PGC	Type S_T　n_L T L	$\log D_{25}$ $\log R_{25}$ $\log A_e$ $\log D_o$	p.a. A_g A_i A_{21}	B_T m_B m_{FIR} B_T^o	$(B-V)_T$ $(U-B)_T$ $(B-V)_T^o$ $(U-B)_T^o$	$(B-V)_e$ $(U-B)_e$ m'_e m'_{25}	m_{21} W_{20} W_{50} HI	V_{21} V_{opt} V_{GSR} V_{3K}
002529.0+201415 114.70 -42.22 314.69 12.11 002252.6+195739	UGC　242 IRAS00228+1957 PGC　1591	.SX.7.. U 7.0± .8 	1.24± .04 .10± .05 1.25	 .12 .13 .05	13.85 ±.15 13.37 13.58				4300± 50 4449 3966
002529.9+455518 118.23 -16.71 340.72 16.29 002248.9+453842	UGC　243 IRAS00228+4538 PGC　1592	.S..3.. U　　(1) 3.0± .9 5.5±1.2	1.27± .04 .61± .05 1.30	2 .34 .84 .30	14.68 ±.13 12.55				
0025.5 +2449 115.50 -37.67 319.29 13.05 0022.9 +2433	UGC　244 PGC　1594	.SB.4?. U 4.0± .9 	.99± .06 .33± .05 1.00	154 .04 .49 .17	14.85 ±.12 14.30				4544 4702 4218
002531.2-330247 344.80 -81.85 262.68 -2.53 002302.0-331924	ESO　350- 15 PGC　1595	.E+3... S -5.0± .7 	1.33± .04 .23± .04 1.27	23 .02 .00	14.27 ±.14 14.03				14933± 30 14918 14666
002547.9-021705 107.96 -64.41 292.50 6.40 002314.4-023341	MCG -1- 2- 11 IRAS00232-0233 PGC　1609	.S..4P/ E 4.2± .8 	1.30± .04 .90± .04 1.32	15 .14 1.32 .45	13.03				5338 5426 4997
002549.1-621948 308.06 -54.54 234.06 -10.54 002329.1-623624	ESO　79- 1 PGC　1610	.SXT5*. Sr　　(1) 4.7± .6 2.2± .9	1.06± .05 .29± .04 1.06	126 .00 .43 .14	14.83 ±.14				
0025.8 +0524 111.23 -56.88 300.03 8.46 0023.3 +0508	UGC　245 PGC　1611	.SB.2.. U 2.0± .9 	1.13± .05 .46± .05 1.14	3 .08 .57 .23	15.27 ±.15 14.50				11754 11865 11410
002554.5+254324 115.75 -36.79 320.21 13.14 002317.3+252648	HICK　1C PGC　1614	.S?....	 	 .07	15.89S±.15				10056± 41 10216 9732
002559.0+291241 116.28 -33.33 323.72 13.79 002321.2+285605	NGC　108 UGC　246 PGC　1619	RLBR+.. U -1.0± .7 	1.31± .07 .10± .05 1.31	 .15 .00	13.09 ±.10 12.88				4749± 36 4916 4432
0026.0 +1420 113.69 -48.07 308.85 10.65 0023.4 +1404	UGC　247 PGC　1621	.S..6*. U 6.0±1.4 	1.00± .08 .73± .06 1.01	19 .11 1.07 .36					11177 11311 10836
002603.0-325044 345.26 -82.07 262.91 -2.57 002333.9-330720	PGC　1624	.LX.-.. S -3.0± .8 	1.28± .08 .30± .08 1.24	 .02 .00					
002606.1+254308 115.80 -36.80 320.21 13.10 002328.8+252632	HICK　1B PGC　1625	.L?....	 	 .07	15.54S±.15				10261± 38 10421 9937
002607.2+254330 115.81 -36.80 320.22 13.10 002329.9+252654	UGC　248 VV　622 PGC　1627	.S?.... 	1.22± .07 .17± .07 1.22	 .07 .26 .09	14.88S±.15 14.47 ±.17 14.32		15.39± .43		10237± 41 10397 9913
002608.2+253936 115.80 -36.86 320.15 13.08 002331.0+252300	CGCG 479- 30 PGC　1628	 	.50± .17 .07± .06 .51	 .07	15.90 ±.12			18.26±.3 279± 7	10086± 10 10246 9762
0026.1 +1338 113.60 -48.77 308.17 10.44 0023.6 +1322	UGC　249 PGC　1631	.S..8*. U 8.0±1.2 	1.16± .05 .16± .05 1.18	90 .15 .20 .08					5242± 41 5374 4901
0026.2 +1624 114.20 -46.03 310.91 11.09 0023.6 +1608	UGC　250 PGC　1632	.SB.6*. U 6.0±1.2 	1.09± .07 .16± .06 1.10	155 .17 .24 .08	15.06 ±.18 14.60				12732 12871 12393
0026.2 +2148 115.20 -40.68 316.30 12.28 0023.6 +2132	NGC　109 UGC　251 PGC　1633	.SBR1.. U 1.0± .8 	1.06± .06 .04± .05 1.07	77 .11 .04 .02	14.7 ±.4 14.66 ±.15 14.45	.99± .02 .54± .04 .91 .52	14.74± .51		5512± 57 5664 5181

R.A. 2000 DEC. / l b / SGL SGB / R.A. 1950 DEC.	Names / PGC	Type / S_T n_L / T / L	$\log D_{25}$ / $\log R_{25}$ / $\log A_o$ / $\log D_o$	p.a. / A_g / A_i / A_{21}	B_T / m_B / m_{FIR} / B_T^o	$(B-V)_T$ / $(U-B)_T$ / $(B-V)_T^o$ / $(U-B)_T^o$	$(B-V)_e$ / $(U-B)_e$ / m' / m'_{25}	m_{21} / W_{20} / W_{50} / HI	V_{21} / V_{opt} / V_{GSR} / V_{3K}
002617.4-042933		.SBS1P*	1.13± .06	55					
106.99 -66.58	MCG -1- 2- 14	E	.48± .05	.15					
290.38 5.67	IRAS00236-0446	1.0±1.3		.48	13.51				
002344.3-044609	PGC 1634		1.15	.24					
0026.4 +0617		.SXS3..	1.08± .06	30					11423
111.78 -56.04	UGC 253	U	.23± .05	.16	14.88 ±.14				
300.94 8.54		3.0± .9		.32					11536
0023.9 +0601	PGC 1638		1.10	.11	14.32				11079
002633.2-415112		DE.5P*	1.04± .05	6					
320.42 -74.42	ESO 294- 10	S	.19± .04	.03	15.66 ±.14				
254.19 -5.27		-5.0±1.0		.00					
002406.0-420748	PGC 1641		.98						
002646.8-334036	NGC 115	.SBS4*.	1.29± .04	127					
340.49 -81.57	ESO 350- 17	BS (1)	.33± .04	.02	13.73 ±.14				1779± 39
262.15 -2.96		4.0± .6		.49					1762
002418.0-335712	PGC 1651	4.4± .8	1.29	.17	13.21				1515
0026.7 +1133		.S?....	.94± .11	41					
113.34 -50.85	MCG 2- 2- 10		.40± .07	.17	15.47 ±.12				2074± 57
306.15 9.80				.60					2201
0024.2 +1117	PGC 1652		.95	.20	14.69				1732
002648.9+314211	NGC 112	.S?....	1.03± .06	108					
116.83 -30.88	UGC 255		.31± .05	.24	14.49 ±.12				6650± 46
326.27 14.05	IRAS00241+3125			.46	13.25				6820
002410.6+312535	PGC 1654		1.05	.15	13.76				6340
002651.3+490701			.64± .17						
118.80 -13.56	MCG 8- 2- 2		.11± .07	.69	15.0 ±.3				5774± 82
344.02 16.38	5ZW 20								5966
002409.0+485025	PGC 1655		.70						5522
002654.7-023005	NGC 113	.LA.-*.	1.15± .10	45					
108.48 -64.68	MCG -1- 2- 16	E	.13± .07	.17					
292.36 6.07		-3.0± .9		.00					
002421.3-024640	PGC 1656		1.15						
002656.4+500151		.S..4..	1.20± .05	171				17.21±.1	5170± 11
118.90 -12.65	UGC 256	U	.77± .05	.81	15.0 ±.3				
344.96 16.44	IRAS00242+4945	4.0±1.0		1.13	12.08			160± 8	5362
002413.8+494515	PGC 1658		1.28	.39	13.04			3.79	4922
002657.2-565842	NGC 119	.LA.-P.	1.02± .06						7340± 51
309.57 -59.83	ESO 150- 8	S	.02± .04	.00	14.14 ±.14				7247
239.39 -9.38		-2.5± .6		.00					7185
002435.0-571518	PGC 1659		1.02		14.03				
002658.3-014709	NGC 114	.LBT.*.	.93± .06	165					
108.88 -63.99	UGC 259	UE	.10± .04	.11	14.73 ±.10				
293.07 6.26	MK 946	-2.0± .7		.00					
002424.7-020345	PGC 1660		.92						
002702.9+113503		.S..6*.	1.48± .03	21				13.52±.1	2134± 4
113.45 -50.84	UGC 260	U	.72± .05	.16	13.71 ±.12			273± 6	2184± 38
306.19 9.74	IRAS00244+1118	6.0±1.1		1.05	12.70			261± 8	2261
002427.6+111827	PGC 1665		1.50	.36	12.50			.67	1792
0027.1 +5105		.S..2..	1.00± .08	167					
119.04 -11.59	UGC 258	U	.33± .06	.99					
346.06 16.51		2.0± .9		.41					
0024.4 +5049	PGC 1672		1.09	.17					
002709.2-405900		PSBT2*.	1.08± .05	135					
321.23 -75.27	ESO 294- 16	Sr	.27± .04	.03	15.06 ±.14				7994
255.07 -5.13		1.6± .6		.33					7952
002442.1-411536	PGC 1673		1.08	.13	14.63				7760
002710.9+012001	NGC 117	.L..+*/	.87± .08	100					
110.37 -60.95	MCG 0- 2- 29	E	.25± .05	.05	15.29 ±.10				
296.13 7.06		-1.0±1.4		.00					
002437.0+010325	PGC 1674		.84						
0027.2 +2410		.S?....	1.08± .07						7348
115.88 -38.36	UGC 261		.11± .06	.06	14.53 ±.13				
318.71 12.54				.16					7504
0024.6 +2354	PGC 1675		1.08	.05	14.27				7021
0027.2 +2004		.S?....	.96± .09	117					5593
115.22 -42.44	UGC 265		.17± .06	.12	14.93 ±.12				
314.62 11.67	IRAS00246+1947			.25					5740
0024.6 +1947	PGC 1676		.97	.08	14.52				5259

R.A. 2000 DEC.	Names	Type	$\log D_{25}$	p.a.	B_T	$(B-V)_T$	$(B-V)_e$	m_{21}	V_{21}
l b		S_T n_L	$\log R_{25}$	A_g	m_B	$(U-B)_T$	$(U-B)_e$	W_{20}	V_{opt}
SGL SGB		T	$\log A_e$	A_i	m_{FIR}	$(B-V)_T^o$	m'_e	W_{50}	V_{GSR}
R.A. 1950 DEC.	PGC	L	$\log D_o$	A_{21}	B_T^o	$(U-B)_T^o$	m'_{25}	HI	V_{3K}

002716.3-014648	NGC 118	.I.0..?	.85± .06	40					11250± 61
109.05 -64.00	UGC 264	E	.15± .04	.11	14.83 ±.13				11339
293.09 6.19	MK 947	90.0		.22	11.99				10910
002442.7-020324	PGC 1678		.86	.08	14.42				
002716.3+394732	A 0024+39	.P.....	1.04± .08	67					10748± 72
117.90 -22.84	UGC 262	R	.29± .06	.28	14.92 ±.12				10930
334.49 15.23	4ZW 20	99.0		.41					10462
002436.3+393057	PGC 1679		1.07	.15	14.20				
0027.2 +4402		.S..8..	1.19± .05	171					
118.36 -18.61	UGC 263	U	.81± .05	.31					
338.83 15.77		8.0± .9		.99					
0024.6 +4346	PGC 1681		1.22	.40					
002728.1-323848		.LA.-..	1.25± .08						
344.75 -82.42	MCG -6- 2- 7	S	.10± .08	.02					
263.19 -2.80		-3.0± .8		.00					
002459.2-325524	PGC 1691		1.24						
002730.2-013050	NGC 120	.LB.0*.	1.19± .05	73					
109.30 -63.75	UGC 267	PUE	.39± .03	.05	14.42 ±.11				
293.37 6.21		-1.8± .5		.00					
002456.6-014725	PGC 1693		1.14						
002731.6+305942								15.95±.3	9385± 10
116.92 -31.60	CGCG 500- 22			.23	15.6 ±.4				
325.58 13.78								237± 10	9554
002453.2+304307	PGC 1695								9073
002734.3+085229	IC 1551	.S?....	1.40± .03	15				18.05±.3	13040± 10
112.98 -53.54	UGC 268		.31± .05	.13	14.3 ±.2			146± 13	13170± 76
303.55 8.94	IRAS00250+0836			.46	12.87				13161
002459.3+083553	PGC 1700		1.41	.15	13.65			4.24	12699
002734.6-341148	ESO 350- 19	.S..2*/	1.17± .04	60					
337.43 -81.27		S	.88± .04	.02	15.65 ±.14				
261.69 -3.27		2.0±1.3		1.08					
002506.0-342824	PGC 1701		1.17	.44					
002749.9-011200		.SXS7*.	1.16± .03	130					3897
109.63 -63.46	UGC 272	UE (1)	.37± .04	.04	15.03 ±.16				3987
293.70 6.21	KUG 0025-014	6.5± .8		.52					3556
002516.3-012835	PGC 1713	5.3± .8	1.16	.19	14.46				
002752.6-014837	NGC 124	.SAS5..	1.15± .03	168					4044± 50
109.37 -64.06	UGC 271	R (1)	.22± .03	.08	13.67 ±.12				4132
293.11 6.03	IRAS00253-0205	5.0± .4		.32	13.05				3704
002519.1-020512	PGC 1715	1.9± .8	1.15	.11	13.24				
0027.9 +2600		.S..8*.	1.04± .15	115					5599
116.35 -36.56	UGC 273	U	.18± .12	.09					5759
320.58 12.75		8.0±1.2		.22					5276
0025.3 +2544	PGC 1717		1.05	.09					
0027.9 +3036		.SB.0..	.97± .07	25					
116.97 -31.99	UGC 274	U	.13± .05	.22	15.28 ±.14				
325.21 13.62		.0± .9		.10					
0025.3 +3020	PGC 1720		.99						
002757.7+324635		.S?....	1.04± .15					16.45±.3	4923± 10
117.24 -29.84	UGC 270		.04± .12	.24					5095
327.40 14.00				.05				169± 7	4616
002519.0+323000	PGC 1721		1.06	.02					
002758.5+023025	IC 17	.SXT3P?	1.25± .04	128					4391
111.21 -59.83	UGC 275	E (1)	.28± .04	.08	14.40 ±.16				4492
297.33 7.18		3.0± .8		.38					4048
002524.4+021350	PGC 1723	3.1±1.2	1.26	.14	13.91				
002759.8+304555		.S..2..	.98± .07	33				16.85±.3	6331± 10
117.00 -31.84	UGC 276	U	.66± .05	.22	15.60 ±.13				6499
325.37 13.64	IRAS00254+3030	2.0±1.0		.81	13.03			306± 10	6019
002521.4+302920	PGC 1724		1.00	.33	14.51			2.01	
002804.2-580613		.S.R6*.	1.12± .05	29					
308.88 -58.75	ESO 112- 4	r	.85± .05	.00	15.93 ±.14				
238.31 -9.80		5.6±1.4		1.25					
002543.0-582248	PGC 1729		1.12	.43					
002812.2+232722								17.66±.3	18612± 10
116.06 -39.11	CGCG 479- 33			.05	15.7 ±.4				
318.04 12.17								450± 7	18766
002535.0+231047	PGC 1733								18284

R.A. 2000 DEC.	Names	Type	logD$_{25}$	p.a.	B$_T$	(B-V)$_T$	(B-V)$_e$	m$_{21}$	V$_{21}$
l b		S$_T$ n$_L$	logR$_{25}$	A$_g$	m$_B$	(U-B)$_T$	(U-B)$_e$	W$_{20}$	V$_{opt}$
SGL SGB		T	logA$_{\bullet}$	A$_i$	m$_{FIR}$	(B-V)$_T^o$	m'	W$_{50}$	V$_{GSR}$
R.A. 1950 DEC.	PGC	L	logD$_o$	A$_{21}$	B$_T^o$	(U-B)$_T^o$	m'$_{25}$	HI	V$_{3K}$

002815.0+304808 A 0025+30A		.SB..$.	1.23± .05	118					
117.07 -31.81	UGC 279	R	.52± .05	.22	14.22 ±.12				6357± 41
325.42 13.59				.78					6526
002536.6+303133	PGC 1736		1.25	.26	13.20				6045

0028.2 +0313		.S..6*.	1.09± .07	160					4362
111.61 -59.13	UGC 277	U	.09± .06	.06	15.1 ±.2				
298.05 7.31		6.0±1.1		.13					4465
0025.7 +0257	PGC 1737		1.10	.05	14.87				4019

002816.9+311158								16.70±.3	14719± 10
117.13 -31.41	CGCG 500- 27			.22	15.1 ±.4				
325.82 13.66								292± 7	14888
002538.4+305523	PGC 1739								14408

002818.7+032427		.SXS6*.	1.06± .08	168					4035
111.69 -58.96	UGC 283	UE (1)	.14± .06	.06					
298.23 7.34		5.7± .7		.21					4138
002544.5+030752	PGC 1741	4.2±1.2	1.07	.07					3692

0028.3 +2723		.SB.4..	1.18± .05	5					9619
116.65 -35.20	UGC 278	U	.64± .05	.13	15.12 ±.12				
321.99 12.94		4.0± .9		.94					9781
0025.7 +2707	PGC 1743		1.19	.32	13.99				9299

002820.7-490437		.SB?...	.96± .07	15					
312.89 -67.60	ESO 194- 20		.22± .05	.04	15.45 ±.14				2900±190
247.22 -7.55	FAIR 654			.23					2831
002556.0-492112	PGC 1745		.97	.11	15.15				2704

002821.6+032316		.SXT7..	1.24± .04	45					4090
111.70 -58.98	UGC 282	UE (1)	.03± .05	.06					
298.21 7.33		7.0± .5		.04					4193
002547.4+030641	PGC 1746	4.2± .8	1.25	.01					3747

0028.3 -0013		.SB.3..	1.08± .06	55					
110.35 -62.53	UGC 280	U	.06± .05	.05					
294.69 6.35		3.0± .8		.09					
0025.8 -0030	PGC 1747		1.09	.03					

002822.2+023135		.S..6*.	1.04± .04	22					4298
111.41 -59.83	UGC 281	U	.36± .04	.07	15.32 ±.14				
297.37 7.09	KUG 0025+022	6.0±1.2		.53					4399
002548.2+021500	PGC 1748		1.05	.18	14.69				3956

002827.7-072330		.S..3*/	1.04± .07	115					
106.45 -69.52	MCG -1- 2- 21	E	.72± .05	.11					
287.72 4.33		3.0±1.4		.99					
002554.9-074005	PGC 1751		1.05	.36					

002831.1+331618		.SXS5..	1.13± .05	23				15.51±.3	4731± 10
117.44 -29.36	UGC 284	U	.15± .05	.24	14.71 ±.16				
327.92 13.97		5.0± .8		.22				229± 7	4903
002552.1+325943	PGC 1754		1.15	.07	14.23			1.21	4426

002834.7+310915 A 0025+30B			.95?		15.6 ±.2	.51± .04	.59± .03	15.77±.3	6070± 10
117.20 -31.46	CGCG 500- 28		.48?	.22	15.29 ±.12	-.12± .07	-.15± .05		6212± 60
325.78 13.59	MK 340		.28± .03				12.50± .07	233± 7	6242
002556.1+305240	PGC 1758		.97						5763

002838.1+025715 A 0026+02		.S..4*/	.94± .08	55				14.93±.3	4469± 17
111.69 -59.42	MCG 0- 2- 45	PE	.64± .05	.06				422± 34	4460± 56
297.81 7.14		3.5±1.0		.95				198± 25	4570
002604.0+024040	PGC 1760		.95	.32					4126

002841.9-595649		.SB?...	1.04± .06						
308.15 -56.95	ESO 112- 5		.08± .06	.00	14.54 ±.14				4545
236.51 -10.32				.12					4443
002622.0-601324	PGC 1765		1.04	.04	14.40				4406

002845.7-322419		.SBT4..	1.13± .04						
344.56 -82.79	ESO 350- 20	Sr (1)	.08± .04	.02	14.52 ±.14				13774± 39
263.50 -2.99		4.2± .6		.11					13760
002617.1-324054	PGC 1767	1.1± .8	1.13	.04	14.30				13506

002848.3+032552			.64± .17						
111.93 -58.96	MCG 0- 2- 47		.19± .07	.06	15.19 ±.19				4500± 49
298.29 7.23	ARAK 11								4603
002614.0+030917	PGC 1770		.64						4157

0028.8 +2856		.S..1..	1.00± .06	107					
116.99 -33.67	UGC 285	U	.53± .05	.12	15.05 ±.13				
323.57 13.12		1.0±1.0		.55					
0026.2 +2840	PGC 1771		1.01	.27					

R.A. 2000 DEC. / l b / SGL SGB / R.A. 1950 DEC.	Names / PGC	Type / S_T n_L / T / L	$\log D_{25}$ / $\log R_{25}$ / $\log A_e$ / $\log D_o$	p.a. / A_g / A_i / A_{21}	B_T / m_B / m_{FIR} / B_T^o	$(B-V)_T$ / $(U-B)_T$ / $(B-V)_T^o$ / $(U-B)_T^o$	$(B-V)_e$ / $(U-B)_e$ / m'_e / m'_{25}	m_{21} / W_{20} / W_{50} / HI	V_{21} / V_{opt} / V_{GSR} / V_{3K}
002850.4+025023	NGC 125	RLA.+P*	1.22± .04	85	13.11 ±.15	.97± .05	.93± .02	17.42±.3	5306 9
111.75 -59.54	UGC 286	PUE	.05± .04	.07	13.75 ±.12			363± 10	5289± 56
297.71 7.06		-1.0± .4	.98± .04	.00		.90	13.50± .13	263± 7	5407
002616.2+023348	PGC 1772		1.22		13.35		13.98± .26		4963
0028.8 -3225		.S?....	1.04± .09		14.6 ±.3	.68± .07			
344.32 -82.79	MCG -5- 2- 6		.09± .07	.02					
263.49 -3.02				.14					
0026.4 -3242	PGC 1775		1.04	.05			14.40± .58		
0028.8 +1008		.SA.7..	1.01± .06						5560
113.83 -52.33	UGC 287	U	.07± .05	.14	15.11 ±.16				
304.88 8.95		7.0± .8		.10					5682
0026.3 +0952	PGC 1776		1.02	.04	14.84				5218
0028.9 +4325		.S?....	1.11± .07	135				15.60±.1	188± 11
118.62 -19.26	UGC 288		.20± .06	.36					
338.24 15.39				.29				45± 8	374
0026.3 +4309	PGC 1777		1.14	.10					-84
0029.1 +1553		.I..9?.	1.28± .11	136				15.07±.1	758± 10
115.10 -46.64	UGC 290	U	1.00± .12	.16				121± 6	
310.57 10.29		10.0±1.9		.75				100± 5	895
0026.5 +1537	PGC 1781		1.30	.50					420
002907.4+310015								17.65±.3	6098± 10
117.32 -31.62	CGCG 500- 32			.22	15.0 ±.4				
325.65 13.44								307± 7	6266
002628.8+304340	PGC 1782								5787
002908.2+024838	NGC 126	.LB.0?.	.93± .11	110					
111.88 -59.58	MCG 0- 2- 49	PE	.28± .05	.07	15.22 ±.10				4252± 60
297.70 6.99		-2.0± .7		.00					4353
002634.0+023203	PGC 1784		.89		15.08				3910
002912.2+025221	NGC 127	.LA.0*.	.91± .10	70					
111.94 -59.52	MCG 0- 2- 50	R	.16± .05	.06					4050± 18
297.77 6.99	IRAS00266+0235	-2.0± .5		.00	13.28				4151
002638.0+023546	PGC 1787		.89						3708
002912.8+330607		.S..9*.	.96± .09					16.59±.3	4821± 10
117.58 -29.54	UGC 291	U	.05± .06	.26					
327.77 13.80		9.0±1.2		.05				165± 7	4993
002633.7+324933	PGC 1788		.98	.02					4515
002915.0+025155	NGC 128	.L...P/	1.47± .02	1	12.77 ±.13	1.02± .01	1.04± .01		
111.96 -59.53	UGC 292	R	.52± .02	.06	12.80 ±.14	.58± .01	.63± .01		4241± 16
297.76 6.97		-2.0± .4	.76± .01	.00		.92	12.06± .03		4342
002640.9+023520	PGC 1791		1.40		12.66	.53	13.71± .18		3899
002918.4+025211	NGC 130	.LA.-*.	.87± .12	40					
111.98 -59.53	MCG 0- 2- 52	R	.31± .05	.06					4433± 18
297.77 6.96		-3.0± .5		.00					4535
002644.3+023537	PGC 1794		.83						4091
002923.1+312335		.S..6*.	1.04± .15	121				16.12±.3	6335± 10
117.43 -31.24	UGC 294	U	1.06± .06	.20					
326.06 13.46		6.0±1.5		1.47				257± 7	6504
002644.3+310700	PGC 1797		1.06	.50					6025
0029.5 -0105		.SXS3..	1.02± .06	154					
110.61 -63.45	UGC 295	U	.27± .05	.05					
293.91 5.83	IRAS00269-0122	3.0± .9		.37	13.42				
0026.9 -0122	PGC 1805		1.02	.13					
002938.4-331531	NGC 131	.SBS3*/	1.27± .03	63	13.78 ±.13	.56± .02	.64± .02	13.37±.3	1410± 17
339.13 -82.27	ESO 350- 21	R	.51± .03	.02	13.69 ±.14	.02± .05	.11± .05	151± 34	1419± 44
262.73 -3.41	IRAS00271-3332	3.0± .4	.73± .01	.70	13.34	.45	12.92± .02	100± 25	1394
002710.1-333206	PGC 1813		1.27	.25	13.01	-.06	13.70± .20	.11	1147
002942.3-513107		.LA.-..	1.17± .05	112	12.15V±.15	1.03± .02			
311.03 -65.27	ESO 194- 21	S	.06± .05	.04	13.35 ±.14				3312± 56
244.88 -8.39	FAIR 655	-3.0± .8	.72± .05	.00					3235
002719.0-514742	PGC 1816		1.16		13.26				3129
002942.8+212821	IC 1552	.S?....	.97± .05	127				16.13±.3	5600± 10
116.22 -41.12	UGC 297		.60± .04	.06	15.33 ±.13			390± 13	
316.14 11.41	IRAS00271+2111			.90	14.03				5749
002705.8+211147	PGC 1817		.97	.30	14.34			1.49	5270
002953.5+312334		.S..5..	1.10± .06	74				15.68±.3	6305± 10
117.55 -31.25	UGC 299	U	.23± .05	.20	14.96 ±.15				
326.08 13.35		5.0± .8		.35				344± 7	6474
002714.7+310700	PGC 1828		1.12	.12	14.37			1.19	5995

R.A. 2000 DEC.	Names	Type	logD$_{25}$	p.a.	B$_T$	(B-V)$_T$	(B-V)$_e$	m$_{21}$	V$_{21}$
l b		S$_T$ n$_L$	logR$_{25}$	A$_g$	m$_B$	(U-B)$_T$	(U-B)$_e$	W$_{20}$	V$_{opt}$
SGL SGB		T	logA$_e$	A$_I$	m$_{FIR}$	(B-V)$_T^o$	m'$_e$	W$_{50}$	V$_{GSR}$
R.A. 1950 DEC.	PGC	L	logD$_o$	A$_{21}$	B$_T^o$	(U-B)$_T^o$	m'$_{25}$	HI	V$_{3K}$
0029.9 +0331			1.04± .09	.04				15.83±.1	1345± 8
112.52 -58.91	MCG 0- 2- 62		.09± .07					33± 9	
298.46 6.98								22± 7	1448
0027.4 +0315	PGC 1835		1.04						1003
0030.0 +0331		.I..9*.	1.11± .14	.04				15.82±.1	1346± 6
112.56 -58.91	UGC 300	U	.00± .12	.00				36± 5	
298.47 6.95		10.0±1.2		.00				33± 5	1449
0027.5 +0315	PGC 1838		1.11						1004
0030.1 -1106	A 0027-11	RSBR1..	1.29± .05	170					3523
104.55 -73.21	MCG -2- 2- 30	E	.37± .05	.11					
284.23 2.88		1.0± .8		.38					3580
0027.6 -1123	PGC 1841		1.30	.18					3196
003008.5-365250		.SXR5*.	1.02± .05	42					
326.40 -79.26	ESO 350- 22	Sr (1)	.27± .04	.02	14.95 ±.14				
259.23 -4.54		5.3± .6		.40					
002741.0-370924	PGC 1842	4.1± .7	1.03	.13					
003010.7+020537	NGC 132	.SXS4..	1.28± .03	40				14.34±.3	5361± 10
112.15 -60.33	UGC 301	R (1)	.13± .03	.03	13.45 ±.13			426± 13	5316± 50
297.07 6.54	IRAS00276+0149	4.0± .4		.20	11.80			362± 10	5458
002736.7+014903	PGC 1844	1.9± .8	1.28	.07	13.19			1.09	5018
003012.4-410602		.S..5*/	1.29± .05	104					8983
318.94 -75.36	ESO 294- 17	S	1.05± .04	.03	15.87 ±.14				8939
255.12 -5.72		5.0±1.3		1.50					8751
002746.0-412236	PGC 1845		1.29	.50	14.29				
0030.3 +2505		.S..9..	.96± .17						5410
116.90 -37.53	UGC 302	U	.05± .12	.06					
319.78 12.04		9.0± .9		.05					5567
0027.7 +2449	PGC 1848		.97	.02					5086
003022.2-331444	NGC 134	.SXS4..	1.93± .02	50	11.23 ±.13	.84± .01	.93± .01	11.80±.1	1579± 5
338.30 -82.38	ESO 350- 23	R (2)	.62± .03	.02	11.10 ±.11	.23± .04	.32± .02	492± 5	1602± 33
262.78 -3.56	IRAS00278-3331	4.0± .3	1.24± .02	.91	9.69	.71	12.92± .04	448± 5	1562
002754.0-333118	PGC 1851	3.7± .6	1.93	.31	10.21	.11	14.16± .18	1.28	1315
003024.1-484102		.S..2*/	1.20± .04	114					
312.25 -68.06	ESO 194- 22	S	.60± .04	.04	14.76 ±.14				
247.70 -7.77		2.0±1.2		.74					
002800.1-485736	PGC 1855		1.21	.30					
0030.4 +4205		.I..9..	1.26± .04	175				15.58±.1	5623± 11
118.78 -20.61	UGC 303	U	.15± .05	.26					
336.92 14.94		10.0± .8		.11				223± 8	5807
0027.8 +4149	PGC 1861		1.28	.08					5346
003029.8-084702		.SBS3?.	1.09± .09	148					
106.85 -71.00	MCG -2- 2- 33	E (1)	.54± .08	.14					
286.51 3.45	KAZ 358	3.0±1.4		.75					
002757.4-090336	PGC 1862	4.2±1.2	1.11	.27					
003029.9+230657				.05	15.7 ±.4			18.58±.3	9101± 10
116.69 -39.50	CGCG 479- 36								9253
317.82 11.59								314± 7	8774
002752.5+225023	PGC 1863								
0030.6 +1240		.S..0..	.89± .10	125					
115.03 -49.88	UGC 304	U	.17± .06	.17	15.39 ±.13				
307.48 9.17		.0± .9		.13					
0028.0 +1224	PGC 1868		.90						
0030.6 +1321		.SA.5..	1.11± .05						
115.16 -49.20	UGC 305	U	.07± .05	.15	14.75 ±.18				9934± 60
308.15 9.33		5.0± .8		.11					10063
0028.0 +1305	PGC 1869		1.13	.04	14.44				9595
003038.5-314956		PSBS1*.	1.01± .05	16					
345.58 -83.47	ESO 410- 14	Sr	.30± .04	.02	15.72 ±.14				14980± 20
264.17 -3.21	FAIR 1067	1.5± .7		.37					14967
002810.0-320630	PGC 1874		1.01	.15	15.18				14711
003040.1-284244		.S?....	1.05± .05	95					
12.68 -85.15	ESO 410- 15		.75± .04	.09	15.19 ±.14				7304±106
267.20 -2.32				1.13					7302
002811.0-285918	PGC 1876		1.06	.38	13.93				7023
003046.0-292038		.S?....	.91± .05	168					
5.78 -84.94	ESO 410- 16		.17± .04	.02	15.17 ±.14				7216±106
266.60 -2.52				.26					7212
002817.0-293712	PGC 1879		.91	.09	14.86				6938

R.A. 2000 DEC.	Names	Type	$\log D_{25}$	p.a.	B_T	$(B-V)_T$	$(B-V)_o$	m_{21}	V_{21}
l　b		S_T　n_L	$\log R_{25}$	A_g	m_B	$(U-B)_T$	$(U-B)_o$	W_{20}	V_{opt}
SGL　SGB		T	$\log A_o$	A_i	m_{FIR}	$(B-V)_T^o$	m'_o	W_{50}	V_{GSR}
R.A. 1950 DEC.	PGC	L	$\log D_o$	A_{21}	B_T^o	$(U-B)_T^o$	m'_{25}	HI	V_{3K}
0030.8 +1039		.L.....	1.04± .08	100					
114.74 -51.89	UGC 307	U	.26± .03	.19					
305.51　8.60		-2.0± .9		.00					
0028.3 +1023	PGC 1882		1.02						
003053.7-562614		.L...P/	1.00± .05	168					
308.69 -60.46	ESO 150- 9	S	.21± .03	.00	15.22 ±.14				
240.06 -9.77		-2.0±1.3		.00					
002833.0-564248	PGC 1883		.97						
003056.5+421133		.S..6*.	1.00± .06	105					
118.88 -20.52	UGC 306	U	.25± .05	.22	15.39 ±.14				
337.03 14.87		6.0±1.3		.36					
002815.0+415459	PGC 1885		1.02	.12					
0030.9 -0713		.S?....	1.04± .09						7639
108.27 -69.51	MCG -1- 2- 23		.28± .07	.13					7709
288.05　3.79				.42					7306
0028.4 -0730	PGC 1886		1.05	.14					
003058.1+101232	NGC 137	.L.....	1.11± .03	100	13.74 ±.15	.96± .03	1.06± .03		
114.67 -52.33	UGC 309	U	.00± .05	.15	13.83 ±.10	.57± .04	.58± .03		5276± 31
305.07　8.47		-2.0± .8	.78± .04	.00		.88	13.13± .13		5397
002822.8+095558	PGC 1888		1.13		13.57	.56	14.16± .25		4935
0030.9 +0510	NGC 138	.S..1*.	1.11± .05	175					12015
113.47 -57.32	UGC 308	U	.36± .05	.06	14.55 ±.12				
300.14　7.17		1.0±1.2		.37					12122
0028.4 +0454	PGC 1889		1.12	.18	13.98				11673
003102.3-365320		PLAT0*.	1.07± .05	34					
325.51 -79.33	ESO 350- 27	S	.20± .04	.02	15.03 ±.14				7290± 39
259.27 -4.71		-2.0± .6		.00					7260
002835.0-370954	PGC 1896		1.04		14.90				7041
003106.6+050444	NGC 139	.SB?...	.91± .09						13725
113.50 -57.42	CGCG 409- 22		.25± .06	.06	15.26 ±.09				
300.05　7.11				.31					13832
002832.1+044810	PGC 1900		.92	.12	14.75				13383
003109.2-223708	NGC 142	.SBS3?P	1.04± .04	101					
76.26 -83.56	ESO 473- 21	S　(1)	.27± .03	.09	14.59 ±.14				
273.14　-.67	IRAS00286-2253	3.4± .6		.37	11.77				
002839.0-225342	PGC 1901	1.1± .9	1.05	.14					
003110.5+043020		.S?....	.82± .13						12615
113.37 -57.99	MCG 1- 2- 17		.36± .07	.01	15.52 ±.13				
299.50　6.95				.44					12720
002836.0+041347	PGC 1903		.82	.18	15.04				12273
003112.1-002425	IC 25	.L?....	.86± .07						5820± 61
111.79 -62.84	MCG 0- 2- 64		.10± .06	.04	15.24 ±.10				5911
294.70　5.62	MK 952			.00	12.59				5480
002838.4-004058	PGC 1905		.85		15.11				
003114.1-075029		.SAR3*.	1.11± .06	70					
108.07 -70.13	MCG -1- 2- 25	E　(1)	.27± .05	.17					
287.48　3.55		3.0± .9		.37					
002841.6-080703	PGC 1908	3.6± .8	1.13	.14					
0031.2 -1028	A 0028-10	.SBR4?.	1.15± .06	170					
106.03 -72.68	MCG -2- 2- 38	E　(1)	.12± .05	.11					3480± 96
284.92　2.80	KAZ 1	4.0±1.0		.18					3539
0028.7 -1045	PGC 1909	1.9± .8	1.16	.06					3153
003114.2-365114		.LBR0?.	1.07± .05	39					
325.40 -79.38	ESO 350- 28	S	.37± .03	.02	14.69 ±.14				
259.32 -4.74		-2.0± .9		.00					
002847.0-370748	PGC 1910		1.01						
003116.1-223339	NGC 143	.SBR3?/	.99± .05	20					
76.79 -83.54	ESO 473- 22	S	.64± .03	.09	15.26 ±.14				
273.21　-.68		3.0± .8		.89					
002846.0-225012	PGC 1911		1.00	.32					
003118.4+285933		.S..6*.	1.07± .06	30				15.66±.3	4663± 10
117.64 -33.67	UGC 310	U	.55± .05	.15	14.81 ±.13				4827
323.72 12.60		6.0±1.3		.80				218± 7	
002839.8+284300	PGC 1913		1.09	.27	13.84			1.55	4348
003119.0+082832	A 0028+08	.S?....	.82± .13		14.48S±.15				
114.43 -54.06	MCG 1- 2- 18		.08± .07	.15	14.89 ±.12				4357± 27
303.39　7.94	MK 552			.06	11.78				4473
002843.9+081159	PGC 1914		.83		14.45		13.23± .68		4015

R.A. 2000 DEC. / l b / SGL SGB / R.A. 1950 DEC.	Names / PGC	Type / S_T n_L / T / L	$\log D_{25}$ / $\log R_{25}$ / $\log A_e$ / $\log D_o$	p.a. / A_g / A_i / A_{21}	B_T / m_B / m_{FIR} / B_T^o	$(B-V)_T$ / $(U-B)_T$ / $(B-V)_T^o$ / $(U-B)_T^o$	$(B-V)_e$ / $(U-B)_e$ / m'_e / m'_{25}	m_{21} / W_{20} / W_{50} / HI	V_{21} / V_{opt} / V_{GSR} / V_{3K}
003120.5+304724	NGC 140	.S..6*.	1.18± .05	45				15.00±.3	6433± 10
117.85 -31.88	UGC 311	U	.08± .05	.22	13.94 ±.13				6455± 50
325.53 12.94	IRAS00287+3031	6.0±1.2		.12	13.15			126± 7	6601
002841.5+303051	PGC 1916		1.20	.04	13.57			1.39	6123
003121.1-223845	NGC 144	.S..5*P	.91± .05						
76.39 -83.61	ESO 473- 23	S (1)	.03± .03	.09	14.42 ±.14				
273.13 -.73		4.7± .7		.05					
002851.0-225518	PGC 1917	2.2±1.2	.92	.02					
003122.9-482150		.SBS3?/	1.05± .05	156					
312.01 -68.41	ESO 194- 23	S	.63± .04	.04	15.99 ±.14				
248.05 -7.85		3.0± .9		.87					
002859.1-483824	PGC 1919		1.05	.31					
003123.1-224602		.IBS9..	1.05± .04	29					541
75.66 -83.70	ESO 473- 24	S (1)	.24± .04	.09	16.11 ±.14				
273.02 -.77		10.0± .5		.18					559
002853.1-230236	PGC 1920	10.0± .6	1.06	.12	15.84				242
003124.0+082805		.SB?...	1.16± .05	7	13.88S±.15				
114.47 -54.07	UGC 312		.34± .05	.15	14.33 ±.12				4333± 28
303.39 7.92	HICK 2A			.51					4449
002848.9+081132	PGC 1921		1.18	.17	13.46		13.70± .31		3992
003125.7+061226		.S?....	1.08± .06	10					
113.94 -56.31	UGC 313		.23± .05	.17	14.13 ±.12				2055± 50
301.18 7.33				.34					2165
002851.0+055553	PGC 1924		1.10	.11	13.60				1713
003128.6-194551		.S..7?/	1.23± .04	0					
90.03 -81.33	ESO 539- 14	E	.98± .03	.07	15.66 ±.14				
275.93 .08		7.0±1.9		1.36					
002858.0-200224	PGC 1926		1.24	.49					
003129.4+082402		.S?....	1.04± .06	165	15.00S±.15				
114.49 -54.14	UGC 314		.11± .05	.13	14.94 ±.15				4274± 32
303.33 7.88	HICK 2C			.17					4389
002854.3+080729	PGC 1927		1.05	.06	14.65		14.76± .35		3932
003135.7-103022		.SBR4?/	1.15± .06	48					
106.29 -72.73	MCG -2- 2- 40	E	.55± .05	.10					
284.92 2.70	IRAS00290-1046	4.0± .9		.81	12.29				
002903.6-104656	PGC 1932		1.16	.28					
0031.6 +1436		.S..6*.	1.11± .14	40					11433
115.75 -47.99	UGC 316	U	.83± .12	.15	15.53 ±.12				
309.44 9.40		6.0±1.4		1.22					11565
0029.0 +1420	PGC 1933		1.12	.42	14.10				11095
003138.4+082326		.S..4..	.96± .09						4274
114.55 -54.16	UGC 315	U	.00± .06	.13					
303.33 7.84	HICK 2D	4.0± .9		.00					4390
002903.3+080653	PGC 1934		.97	.00					3933
003138.7-115127		.SBT8..	1.08± .06	100					
105.05 -74.04	MCG -2- 2- 39	E (1)	.17± .05	.09					
283.61 2.31		8.0± .9		.21					
002906.8-120801	PGC 1935	7.5± .8	1.09	.09					
0031.6 +0055		.S..8*.	.96± .09						5364
112.51 -61.55	UGC 317	U	.05± .06	.06					
296.03 5.87		8.0±1.2		.06					5458
0029.1 +0039	PGC 1936		.97	.02					5024
0031.7 -0230		.SBR7*.	1.09± .06	0					
111.26 -64.93	MCG -1- 2- 28	E (1)	.12± .05	.16					
292.70 4.91		7.0±1.2		.16					
0029.2 -0247	PGC 1940	8.7± .8	1.10	.06					
003145.7-050909	NGC 145	.SBS8..	1.25± .03	115	13.17 ±.13	.43± .03		14.70±.3	4140± 17
110.02 -67.53	MCG -1- 2- 27	R	.14± .04	.16		-.21± .04		140± 34	4146± 34
290.12 4.18	ARP 19	8.0± .4		.18	11.91	.34		140± 25	4217
002912.7-052542	PGC 1941		1.27	.07	12.82	-.28	13.92± .24	1.81	3806
003150.3-264315		.S..5./	1.41± .03	83					
37.15 -85.61	ESO 473- 25	S	.90± .03	.09	14.68 ±.14				7235
269.21 -2.00	VV 582	4.7± .7		1.35					7239
002921.1-265948	PGC 1942		1.42	.45	13.21				6948
003201.0-642327		.SBS7*P	1.28± .04	112	13.66V±.15	.42± .06	.44± .06		
306.39 -52.61	ESO 79- 2	S	.95± .05	.00	14.57 ±.14				2772± 62
232.16 -11.68	IRAS00297-6440	7.1± .6	.98± .03	1.31		.22			2656
002946.0-644000	PGC 1951		1.28	.47	13.02		12.98± .28		2659

R.A. 2000 DEC. / l b / SGL SGB / R.A. 1950 DEC.	Names / PGC	Type / S_T n_L / T / L	$\log D_{25}$ / $\log R_{25}$ / $\log A_e$ / $\log D_o$	p.a. / A_g / A_i / A_{21}	B_T / m_B / m_{FIR} / B_T^o	$(B-V)_T$ / $(U-B)_T$ / $(B-V)_T^o$ / $(U-B)_T^o$	$(B-V)_e$ / $(U-B)_e$ / m'_e / m'_{25}	m_{21} / W_{20} / W_{50} / HI	V_{21} / V_{opt} / V_{GSR} / V_{3K}
003202.1-641509 / 306.41 -52.74 / 232.30 -11.65 / 002947.0-643142	ESO 79- 3 / IRAS00297-6431 / PGC 1952	.SB.3?/ / S / 3.0± .5	1.43± .04 / .81± .05 / .96± .02 / 1.43	134 / .00 / 1.11 / .40	12.59V±.14 / 13.78 ±.14 / 10.89 / 12.45	.80± .03 / .63	.92± .02 / / / 13.38± .27		2592± 62 / 2477 / 2478
0032.0 -0819 / 108.31 -70.64 / 287.06 3.22 / 0029.5 -0836	MCG -2- 2- 43 / VV 570 / PGC 1953	.SBS7*. / E (1) / 7.3± .7 / 6.4± .7	1.12± .05 / .28± .04 / / 1.13	175 / .12 / .39 / .14					4993 / / 5058 / 4663
003209.2-401557 / 318.61 -76.27 / 256.04 -5.84 / 002943.1-403230	ESO 294- 20 / / PGC 1956	.SBS8.. / S (1) / 8.0± .8 / 8.9± .8	1.23± .05 / .13± .05 / / 1.24	5 / .03 / .16 / .07	14.49 ±.14				
003212.2+314053 / 118.16 -31.01 / 326.46 12.92 / 002932.9+312420	A 0029+31 / UGC 319 / PGC 1957	.S..4.. / U (1) / 4.0± .8 / 3.0±1.3	1.14± .05 / .19± .05 / / 1.16	135 / .23 / .27 / .09	14.72 ±.16 / / / 14.18			15.16±.3 / 279± 7 / / .89	6343± 10 / 6511 / 6035
003212.3+312500 / 118.13 -31.28 / 326.19 12.87 / 002933.0+310827	CGCG 500- 40 / / PGC 1958			.23	15.7 ±.4			18.22±.3 / 272± 7	6136± 10 / 6304 / 5827
003214.8-412257 / 317.16 -75.21 / 254.95 -6.16 / 002949.0-413930	ESO 294- 21 / IRAS00298-4139 / PGC 1961	.SXT3.. / Sr (1) / 3.2± .6 / 3.3±1.2	1.20± .05 / .44± .04 / / 1.20	176 / .03 / .60 / .22	14.29 ±.14 / 12.68 / 13.60				7825 / 7780 / 7595
003223.5-421021 / 316.21 -74.46 / 254.18 -6.40 / 002958.0-422654	ESO 294- 22 / / PGC 1968	PSAT3.. / r / 3.3± .8	.98± .07 / .11± .06 / / .98	/ .03 / .15 / .06	15.02 ±.14 / / / 14.77				9825 / 9777 / 9598
0032.4 +0234 / 113.44 -59.95 / 297.70 6.12 / 0029.9 +0218	UGC 320 / / PGC 1970	.S..6*. / U / 6.0±1.4	1.09± .06 / .77± .05 / / 1.10	172 / .05 / 1.13 / .39					2376 / / 2475 / 2036
0032.5 +2324 / 117.33 -39.26 / 318.21 11.19 / 0029.9 +2308	UGC 321 / / PGC 1971	.SB.6?. / U / 6.0±1.3	1.13± .05 / .70± .05 / / 1.13	149 / .04 / 1.04 / .35	15.29 ±.12 / / / 14.19				4668 / / 4820 / 4342
0032.5 +0129 / 113.16 -61.02 / 296.65 5.81 / 0030.0 +0113	UGC 322 / / PGC 1973	.S..6*. / U / 6.0±1.3	1.06± .06 / .47± .05 / / 1.06	68 / .03 / .69 / .23					
003242.7-111907 / 106.48 -73.59 / 284.21 2.21 / 003010.7-113540	MCG -2- 2- 49 / IRAS00301-1135 / PGC 1979	.S..5*/ / E / 5.0±1.3	1.23± .05 / .80± .05 / / 1.24	30 / .08 / 1.20 / .40	/ / 12.78				
003243.8+260706 / 117.70 -36.56 / 320.92 11.72 / 003005.5+255033	CGCG 479- 38 / / PGC 1981			.10	15.7 ±.4			18.68±.3 / 61± 7	20361± 10 / 20518 / 20040
003244.4+304500 / 118.20 -31.95 / 325.55 12.63 / 003005.2+302827	MCG 5- 2- 23 / / PGC 1982	.S?.... / /	1.08± .09 / .37± .07 / / 1.09	/ .19 / .46 / .19	15.00 ±.12 / / / 14.29			15.73±.3 / 317± 7 / 1.25	6120± 10 / 6286 / 5810
0032.8 +1144 / 115.70 -50.88 / 306.69 8.40 / 0030.2 +1127	UGC 323 / IRAS00302+1127 / PGC 1986	.S?.... / /	.95± .07 / .49± .05 / / .97	110 / .20 / .74 / .25	15.21 ±.13 / 13.24 / 14.24				4924 / / 5048 / 4585
003303.6-080652 / 109.19 -70.49 / 287.34 3.04 / 003031.2-082324	MCG -1- 2- 30 / / PGC 1996	.SBS8*. / E (1) / 8.0± .7 / 6.1± .7	1.10± .05 / .37± .04 / / 1.12	140 / .13 / .46 / .19					
003303.9+221155 / 117.34 -40.47 / 317.03 10.81 / 003026.4+215523	UGC 325 / / PGC 1997	.S?.... / /	.91± .10 / .02± .06 / / .92	10 / .06 / .04 / .01				16.59±.3 / 130± 7	7758± 10 / 7907 / 7430
003307.9-321533 / 339.59 -83.51 / 263.91 -3.84 / 003040.0-323206	IC 1554 / ESO 350- 33 / IRAS00306-3232 / PGC 2000	.LBT+P* / PSr / -.5± .5	1.15± .05 / .23± .04 / / 1.12	24 / .02 / .00	13.61 ±.14 / 12.15 / 13.56				1776± 87 / 1761 / 1509

R.A. 2000 DEC. l b SGL SGB R.A. 1950 DEC.	Names PGC	Type S_T n_L T L	$\log D_{25}$ $\log R_{25}$ $\log A_e$ $\log D_0$	p.a. A_g A_i A_{21}	B_T m_B m_{FIR} B_T^o	$(B-V)_T$ $(U-B)_T$ $(B-V)_T^o$ $(U-B)_T^o$	$(B-V)_e$ $(U-B)_e$ m'_e m'_{25}	m_{21} W_{20} W_{50} HI	V_{21} V_{opt} V_{GSR} V_{3K}
003310.3-130847 105.04 -75.38 282.47 1.58 003038.7-132520	 MCG -2- 2- 51 IRAS00306-1325 PGC 2001	PSBT4.. E (1) 4.0± .9 1.9±1.2	1.09± .06 .29± .05 1.10	115 .08 .43 .15	 13.86 				6314 6363 5992
0033.1 +0754 115.10 -54.68 302.96 7.35 0030.6 +0738	 UGC 327 PGC 2003	.S..4.. U 4.0± .9 	1.04± .06 .42± .05 1.05	121 .14 .62 .21	 15.36 ±.13 14.51				 13527± 57 13641 13186
003311.6+483028 119.82 -14.25 343.51 15.27 003027.4+481356	NGC 147 UGC 326 DDO 3 PGC 2004	.E.5.P. R -5.0± .3 	2.12± .02 .23± .04 1.71± .02 2.17	25 .76 .00 	10.47M±.05 10.43 ±.15 9.71	.95± .05 .78 	.90± .01 14.61± .05 15.49± .14		 -160± 18 28 -412
0033.3 -0106 112.72 -63.62 294.17 4.91 0030.8 -0123	 UGC 328 PGC 2016	.SBT8.. UE (1) 8.0± .6 7.5± .8	1.09± .05 .09± .04 1.10	125 .07 .11 .05				14.35±.1 155± 8 132± 12 	1986± 6 2073 1648
003326.3+075249 115.20 -54.72 302.95 7.28 003051.2+073617	 MCG 1- 2- 26 PGC 2017	.S?.... 	.94± .11 .19± .07 .95	163 .15 .28 .09	 15.26 ±.13 14.75				 13497± 57 13610 13156
003330.3-564455 307.88 -60.21 239.83 -10.19 003110.9-570127	 PGC 2019	.LA.+*P S -1.3± .7 	1.15± .09 .34± .08 1.10	 .00 .00 					
003331.7+225415 117.56 -39.78 317.76 10.86 003054.0+223743	 MCG 4- 2- 29 PGC 2021	.S?.... 	.91± .07 .54± .06 .91	 .06 .80 .27	 15.69 ±.12 14.80			17.41±.3 221± 7 2.34	4599± 10 4749 4273
0033.5 +0240 114.01 -59.89 297.87 5.88 0031.0 +0224	 UGC 329 PGC 2022	.SXS6.. UE (1) 5.5± .5 5.3± .8	1.29± .04 .18± .04 1.30	170 .01 .26 .09	 14.9 ±.3 14.59				4391 4489 4051
003336.5+232405 117.64 -39.29 318.26 10.95 003058.7+230733	 MCG 4- 2- 30 PGC 2023	.S?.... 	.94± .11 .40± .07 .94	 .04 .60 .20	 15.34 ±.12 14.68			16.41±.3 255± 7 1.53	5179± 10 5330 4854
003341.9+393241 119.21 -23.20 334.43 13.97 003100.4+391609	 UGC 330 PGC 2026	.L..... U -2.0± .9 	1.15± .07 .42± .03 1.11	140 .19 .00 	 14.74 ±.10 				
0033.7 +0714 115.21 -55.36 302.35 7.04 0031.2 +0658	 UGC 331 PGC 2027	.SB.1.. U 1.0± .9 	.95± .07 .25± .05 .96	170 .08 .25 .12	 14.99 ±.12 				
0033.8 +3043 118.48 -31.99 325.57 12.39 0031.2 +3027	NGC 149 UGC 332 PGC 2028	.L...*. U -2.0±1.2 	1.09± .07 .22± .03 1.07	155 .19 .00 	 14.69 ±.11 				
003356.2+312705 118.56 -31.27 326.30 12.52 003116.6+311033	A 0031+31 UGC 334 DDO 4 PGC 2031	.S..9.. U (1) 9.0± .8 9.0±1.4	1.24± .06 .01± .06 1.02± .04 1.27	 .23 .01 .01	15.07 ±.19 14.4 ±.2 14.56	.45± .06 -.14± .08 .36 -.20	.53± .04 -.05± .06 15.66± .09 16.10± .38	15.47±.3 131± 7 .91	4633± 10 4800 4325
003358.1+485703 119.98 -13.82 343.98 15.20 003113.6+484031	 UGC 333 PGC 2032	.SB?... 	1.06± .06 .04± .05 1.13	 .80 .06 .02	 14.9 ±.3 14.06			15.71±.1 128± 8 1.63	5214± 11 5403 4964
003401.4-031000 112.33 -65.68 292.21 4.18 003128.2-032632	 MCG -1- 2- 31 PGC 2034	.L..0*. E -2.0±1.3 	1.07± .06 .50± .05 1.02	20 .16 .00 					
003402.8-094220 108.86 -72.10 285.87 2.35 003130.7-095852	NGC 151 MCG -2- 2- 54 IRAS00315-0958 PGC 2035	.SBR4.. R (3) 4.0± .3 2.7± .5	1.57± .02 .34± .03 1.07± .01 1.58	75 .12 .50 .17	12.31 ±.13 12.29 ±.20 12.24 11.66	.72± .01 .13± .02 .61 .04	.83± .01 .28± .02 13.15± .03 14.16± .17	13.93±.1 463± 6 452± 5 2.10	3747± 5 3654± 50 3806 3419
003403.3-341728 329.40 -81.92 261.98 -4.58 003136.0-343400	 ESO 350- 34A PGC 2036	.L...P. S -2.0±1.1 	.85? .85? .72	 .02 .00 	 15.18 ±.14 				

R.A. 2000 DEC.	Names	Type	logD$_{25}$	p.a.	B$_T$	(B-V)$_T$	(B-V)$_\bullet$	m$_{21}$	V$_{21}$
l　　b		S$_T$　n$_L$	logR$_{25}$	A$_g$	m$_B$	(U-B)$_T$	(U-B)$_\bullet$	W$_{20}$	V$_{opt}$
SGL　SGB		T	logA$_\bullet$	A$_i$	m$_{FIR}$	(B-V)$_T^o$	m'$_\bullet$	W$_{50}$	V$_{GSR}$
R.A. 1950 DEC.	PGC	L	logD$_o$	A$_{21}$	B$_T^o$	(U-B)$_T^o$	m'$_{25}$	HI	V$_{3K}$
003407.6+440844		.SA.7..	1.10± .04					16.10±.1	5646± 11
119.66 -18.62	UGC　　336	U	.03± .04	.40	15.1 ±.2				
339.10　14.56		7.0± .8		.04				233± 8	5831
003124.6+435212	PGC　2039		1.14	.01	14.60			1.49	5378
003408.4-212609		.S?....			16.46S±.15				
87.31 -83.08				.07					
274.49 -1.00	HICK　　4E								
003138.4-214241	PGC　2040								
003409.8-073608		.E?....			15.81S±.15				
110.30 -70.05				.16					7804± 41
287.91　2.92	HICK　　3D								7871
003137.4-075240	PGC　2043								7474
003411.1-304628	A　0031-31	.SXS9*.	1.27± .03					14.31±.2	1585± 7
347.90 -84.75	ESO　410- 18	SU　(2)	.04± .04	.02	14.04 ±.14			72± 16	1619±106
265.42 -3.64	DDO　224	9.2± .5		.04				55± 6	1575
003143.0-310300	PGC　2044	8.8± .5	1.27	.02	13.98			.31	1313
003413.2-073354		.S?....			15.46S±.15				
110.36 -70.02				.16					7302± 41
287.95　2.91	HICK　　3A								7369
003140.7-075025	PGC　2045								6972
003414.0-212816		PSBR3*P	1.05± .04	122	15.56S±.15				
87.30 -83.12	ESO　540- 2	S	.43± .03	.07	15.45 ±.14				7065± 41
274.46 -1.03	HICK　　4B	3.0± .6		.60					7086
003144.0-214448	PGC　2046		1.06	.22	14.78		14.59± .28		6764
003414.0-212628		PSBR4..	1.15± .04	21	13.66S±.15				
87.44 -83.09	ESO　540- 1	r	.17± .04	.07	13.66 ±.14				8076± 34
274.49 -1.02	IRAS00317-2142	4.5± .8		.25	11.59				8097
003144.0-214300	PGC　2047		1.15	.08	13.30		13.83± .29		7775
0034.2 +2438		.SX.8..	1.04± .06	150					5283
117.97 -38.06	UGC　　337	U	.02± .05	.05					
319.52　11.08		8.0± .8		.03					5437
0031.6 +2422	PGC　2048		1.04	.01					4960
003415.6-212504		.L?....			15.79S±.15				
87.59 -83.08				.07					8863± 41
274.52 -1.02	HICK　　4C								8884
003145.7-214136	PGC　2051								8562
003415.8-274818	NGC　　150	.SBT3*.	1.59± .02	118	12.00 ±.13	.64± .01	.69± .01	13.11±.1	1584± 5
21.87 -86.13	ESO　410- 19	R　　(3)	.32± .03	.09	12.04 ±.11	.02± .02	.04± .02	336± 8	1523± 29
268.31 -2.83	IRAS00317-2804	3.0± .4	1.00± .01	.45	10.67	.55	12.49± .03	308± 6	1581
003147.1-280450	PGC　2052	2.5± .4	1.59	.16	11.48	-.05	13.97± .17	1.48	1300
003415.8-314710	NGC　　148	.L..0*/	1.30± .03	90	13.13 ±.13	.95± .01	.96± .01	14.83±.3	1516± 10
340.65 -84.03	ESO　410- 20	PS	.40± .03	.02	13.11 ±.10	.50± .01	.48± .01	507± 6	1644± 30
264.44 -3.93		-1.5± .6	.51± .01	.00		.89	11.17± .03		1515
003148.0-320342	PGC　2053		1.24		13.08	.46	13.47± .20		1261
003416.8-212840		.E?....			15.80S±.15				
87.35 -83.13				.07					8215± 41
274.46 -1.04	HICK　　4D								8236
003146.8-214512	PGC　2057								7914
0034.3 +1011		.S?....	.94± .11						5409
116.04 -52.45	UGC　　339		.28± .07	.18	15.32 ±.12				
305.27　7.65				.35					5528
0031.8 +0955	PGC　2061		.95	.14	14.78				5070
0034.4 +1216	IC　　31	.S..1..	1.21± .05	89					9516
116.40 -50.38	UGC　　340	U	.81± .05	.20	15.23 ±.12				
307.32　8.17		1.0± .9		.83					9641
0031.8 +1200	PGC　2062		1.23	.40	14.09				9178
003425.1-073559		.L?....			15.62S±.15				
110.48 -70.06				.16					7860± 41
287.93　2.86	HICK　　3B								7926
003152.6-075231	PGC　2064								7530
0034.4 +3023	A　0031+30								6087
118.60 -32.34				.16					6252
325.26　12.20									5777
0031.8 +3007	PGC　2065								
0034.4 +0334		.SX.5..	1.02± .06	65					6434
114.69 -59.02	UGC　　342	U	.09± .05	.04	15.18 ±.17				
298.81　5.91		5.0± .9		.14					6535
0031.9 +0318	PGC　2068		1.02	.05	14.97				6094

R.A. 2000 DEC. l　b SGL　SGB R.A. 1950 DEC.	Names PGC	Type S_T　n_L T L	$\log D_{25}$ $\log R_{25}$ $\log A_e$ $\log D_o$	p.a. A_g A_i A_{21}	B_T m_B m_{FIR} B_T^o	$(B-V)_T$ $(U-B)_T$ $(B-V)_T^o$ $(U-B)_T^o$	$(B-V)_e$ $(U-B)_e$ m'_e m'_{25}	m_{21} W_{20} W_{50} HI	V_{21} V_{opt} V_{GSR} V_{3K}
003432.5-433922 313.51 -73.12 252.83　-7.17 003208.1-435554	 ESO 242- 14 PGC 2070	RLB.0.. r -2.4± .8 	1.05± .06 .13± .05 .58± .04 1.03	75 .03 .00 	14.21 ±.15 14.19 ±.14 14.08	.97± .02 .46± .03 .90 .47	1.02± .01 .49± .02 12.60± .14 14.01± .38	 	5937 5883 5718
003433.2-300110 353.99 -85.30 266.18　-3.50 003205.0-301742	IC 1555 ESO 410- 21 PGC 2071	.SAS7.. S　　(1) 6.5± .6 7.2± .6	1.11± .04 .17± .03 1.11	136 .02 .23 .08	 14.39 ±.14 14.14	 	 	 	1575±106 1567 1301
003436.5-554722 307.84 -61.18 240.82 -10.12 003217.1-560354	NGC 159 ESO 150- 11 PGC 2073	RSBR0.. Sr .0± .6 	1.14± .05 .50± .03 1.11	95 .00 .37 	 14.86 ±.14 	 	 	 	
003440.1-104601 108.57 -73.17 284.88　1.91 003208.2-110233	NGC 155 MCG -2- 2- 55 PGC 2076	.L..0P. E -2.0± .8 	1.24± .09 .13± .07 1.23	175 .05 .00 	 	 	 	 	
003446.5-082348 110.27 -70.86 287.19　2.55 003214.2-084019	NGC 157 MCG -2- 2- 56 PGC 2081	.SXT4.. R　　(3) 4.0± .3 1.8± .4	1.62± .02 .19± .02 1.25± .01 1.63	40 .13 .27 .09	11.00M±.12 11.1 ±.2 10.60	.59± .01 -.02± .01 .51 -.07	.66± .01 .05± .01 12.71± .02 13.48± .16	12.81±.2 317± 9 214± 7 2.11	1668± 6 1730± 27 1734 1342
0034.7　+5326 120.42　-9.35 348.57 15.55 0032.0　+5310	 UGC 343 PGC 2083	.S..8*. U 8.0±1.2 	1.11± .14 .15± .12 1.19	70 .85 .18 .07	 	 	 	16.07±.1 210± 8 	5185± 11 5377 4954
003448.5+282431 118.50 -34.32 323.30 11.73 003209.5+280800	 UGC 345 PGC 2085	.SB.6?. U 6.0±1.4 	1.02± .06 .67± .05 1.03	51 .12 .99 .34	 15.56 ±.12 14.43	 	 	15.63±.2 196± 13 185± 7 .87	4177± 7 4338 3862
003451.2+393242 119.45 -23.21 334.46 13.75 003209.4+391611	 UGC 344 PGC 2088	.SXS6.. U 6.0± .9 	1.02± .06 .17± .05 1.04	110 .17 .25 .09	 15.09 ±.14 14.64	 	 	15.50±.1 222± 8 .78	5780± 11 5959 5496
003454.9-122956 107.30 -74.87 283.22　1.36 003223.4-124627	 MCG -2- 2- 58 PGC 2092	.SXR5.. E　　(1) 5.0± .8 5.3± .8	1.13± .06 .08± .05 1.13	5 .03 .12 .04	 	 	 	 	
003456.6+315651 118.86 -30.79 326.84 12.40 003216.7+314020	 UGC 346 PGC 2094	.S..4.. U 4.0± .9 	.98± .07 .31± .05 1.00	128 .22 .46 .16	 14.78 ±.12 14.06	 	 	17.25±.3 287± 7 3.03	5871± 10 6038 5565
003501.9-501228 309.58 -66.71 246.38　-8.88 003240.0-502900	 ESO 194- 27 FAIR 658 PGC 2099	.S..9?/ S 9.0±1.9 	1.13± .05 .76± .04 1.14	36 .04 .78 .38	 15.75 ±.14 	 	 	 	
003506.5+231855 118.08 -39.40 318.25 10.59 003228.5+230224	 CGCG 479- 41 PGC 2106	 	 	 .04 	 15.1 ±.4 	 	 	16.86±.3 198± 7 	11334± 10 11484 11009
003513.2-373329 320.06 -79.03 258.85　-5.69 003247.0-375000	 ESO 294- 23 PGC 2112	RSBR1.. Sr .5± .6 	1.05± .05 .10± .05 1.06	 .02 .11 .05	 14.79 ±.14 14.57	 	 	 	7232 7198 6987
003513.4+294758 118.74 -32.94 324.70 11.92 003234.1+293127	 CGCG 500- 48 PGC 2113	 	 	 .19 	 15.0 ±.4 	 	 	17.78±.3 270± 7 	6972± 10 7135 6661
003518.2-440453 312.74 -72.74 252.45　-7.41 003254.0-442124	 ESO 242- 17 PGC 2115	RSXT2*. r 2.2± .9 	.95± .07 .03± .06 .95	 .03 .04 .02	 14.71 ±.14 14.57	 	 	 	7502± 34 7447 7285
0035.3　+0255 114.96 -59.70 298.24　5.52 0032.8　+0239	 UGC 348 PGC 2118	.S..8*. U 8.0±1.2 	.98± .07 .00± .05 .98	70 .03 .00 .00	 15.17 ±.17 15.13	 	 	 	4213 4311 3874
0035.3　+0323 115.08 -59.23 298.69　5.64 0032.8　+0307	 UGC 349 PGC 2119	.SXS4.. U 4.0± .9 	1.00± .09 .28± .06 1.00	136 .04 .41 .14	 15.24 ±.13 14.76	 	 	 	6395 6495 6056

R.A. 2000 DEC. / l b / SGL SGB / R.A. 1950 DEC.	Names / PGC	Type / S_T n_L / T / L	$\log D_{25}$ / $\log R_{25}$ / $\log A_e$ / $\log D_o$	p.a. / A_g / A_i / A_{21}	B_T / m_B / m_{FIR} / B_T^o	$(B-V)_T$ / $(U-B)_T$ / $(B-V)_T^o$ / $(U-B)_T^o$	$(B-V)_e$ / $(U-B)_e$ / m'_e / m'_{25}	m_{21} / W_{20} / W_{50} / HI	V_{21} / V_{opt} / V_{GSR} / V_{3K}
0035.3 +0452		.S..0*.	1.04± .15	115					
115.42 -57.76	UGC 350	U	.29± .12	.00	15.28 ±.14				
300.14 6.03		.0±1.3		.22					
0032.8 +0436	PGC 2120		1.03						
0035.3 +3630	And III	.E?....							
119.34 -26.25									
331.42 13.14				.21					
0032.7 +3614	PGC 2121								
003523.5-232235	NGC 167	.SBR5?.	.99± .04	171					
78.02 -84.78	ESO 473- 29	S (1)	.13± .03	.07	14.37 ±.14				
272.69 -1.83	IRAS00328-2339	5.0± .9		.19	13.27				
003254.0-233906	PGC 2122	3.3± .9	1.00	.06					
003527.2-553117		RSXT1..	.94± .07						
307.67 -61.46	ESO 150- 13	r	.05± .06	.00	14.81 ±.14				8340
241.12 -10.18		1.0± .9		.05					8249
003308.0-554748	PGC 2125		.94	.02	14.66				8179
003534.1-025056	NGC 161	.L..0*.	1.13± .11	150					
113.37 -65.43	MCG -1- 2- 36	E	.22± .07	.16					
292.63 3.90		-2.0±1.3		.00					
003300.8-030727	PGC 2131		1.12						
003536.1-044400		.S..4P?	1.05± .06	85					
112.67 -67.29	MCG -1- 2- 38	E (1)	.43± .05	.17					
290.80 3.37	IRAS00330-0500	4.0±1.8		.63	13.31				
003303.2-050031	PGC 2133	5.3±1.2	1.06	.21					
003536.2+090722	IC 34	.SBR1..	1.45± .03	156					
116.34 -53.55	UGC 351	U	.44± .05	.16	13.50 ±.14				5299± 50
304.30 7.09		1.0± .8		.44					5415
003300.8+085051	PGC 2134		1.46	.22	12.84				4960
0035.6 +2416		.I?....	1.11± .14	140					5146
118.33 -38.46	UGC 352		.20± .12	.05					
319.23 10.69				.15					5298
0033.0 +2400	PGC 2136		1.11	.10					4823
003540.1-200735		PSXR3..	1.31± .03	153					
94.83 -82.12	ESO 540- 3	SEr (1)	.40± .03	.12	13.92 ±.14				3351± 60
275.87 -.97		2.8± .4		.56					3376
003310.0-202406	PGC 2138	3.3± .6	1.33	.20	13.22				3047
0035.6 +0859		.S..6*.	1.08± .06	124					5072
116.36 -53.68	UGC 353	U	.46± .05	.16	15.25 ±.13				
304.18 7.03		6.0±1.3		.68					5187
0033.1 +0843	PGC 2139		1.10	.23	14.39				4733
003548.0-252223	IC 1558	.SXS9..	1.53± .02	150	12.61 ±.15	.41± .06	.46± .02	13.63±.2	1558± 7
58.02 -86.08	ESO 474- 2	SU (2)	.14± .04	.05	13.21 ±.14	-.16± .06	-.13± .03	138± 16	
270.78 -2.47	DDO 225	8.7± .4	1.44± .02	.14		.36	15.30± .06	130± 6	1564
003319.0-253854	PGC 2142	8.8± .5	1.53	.07	12.74	-.20	14.74± .21	.82	1269
003555.9+315216		.SB.2..	1.04± .06	123				16.79±.3	6356± 10
119.10 -30.88	UGC 355	U	.49± .05	.22	14.59 ±.14				6455± 50
326.80 12.18		2.0± .9		.60				425± 7	6526
003316.0+313546	PGC 2147		1.06	.24	13.70			2.84	6054
0035.9 +2358		.S..4..	.95± .07	119				17.35±.3	5618± 9
118.39 -38.76	UGC 354	U	.58± .05	.10				244± 10	
318.94 10.55		4.0±1.0		.86				221± 7	5769
0033.3 +2342	PGC 2148		.96	.29					5295
003559.8-100719	NGC 163	.E.0...	1.19± .08	85	13.64 ±.14	.94± .02	1.02± .02		
110.12 -72.61	MCG -2- 2- 66	R	.10± .06	.10		.49± .04	.49± .03		
285.60 1.77		-5.0± .4	.85± .03	.00			13.38± .11		
003327.9-102350	PGC 2149		1.18				14.35± .46		
0036.0 +1237		.S?....	.82± .13						
117.06 -50.07	MCG 2- 2- 22		.17± .07	.21	15.18 ±.12				10064± 60
307.76 7.88				.21					10189
0033.4 +1221	PGC 2150		.84	.09	14.66				9727
0036.0 -0953	A 0033-10	.SBS5P*	1.09± .06	20					
110.28 -72.38	MCG -2- 2- 64	E (1)	.07± .05	.10					4882± 72
285.82 1.84	VV 548	5.0±1.2		.10					4940
0033.4 -1010	PGC 2151	4.2±1.6	1.10	.03					4556
003604.3+235730	NGC 160	RLA.+P.	1.47± .03	45	13.6 ±.2	.95± .09	1.03± .06	15.66±.2	5254± 5
118.43 -38.78	UGC 356	R	.24± .03	.10	13.27 ±.13			556± 5	5295± 37
318.93 10.52		-1.0± .3	.78± .07	.00		.85	13.03± .23	521± 5	5406
003326.0+234100	PGC 2154		1.44		13.20		15.23± .27		4931

R.A. 2000 DEC. / l b / SGL SGB / R.A. 1950 DEC.	Names / PGC	Type / S_T n_L / T / L	logD_25 / logR_25 / logA_e / logD_o	p.a. / A_g / A_i / A_21	B_T / m_B / m_FIR / B_T^o	(B-V)_T / (U-B)_T / (B-V)_T^o / (U-B)_T^o	(B-V)_e / (U-B)_e / m' / m'_25	m_21 / W_20 / W_50 / HI	V_21 / V_opt / V_GSR / V_3K
003607.3-323617 / 333.23 -83.60 / 263.75 -4.54 / 003340.0-325248	ESO 350- 37 / IRAS00336-3252 / PGC 2157	.S..1.. / S / 1.0± .9	.95± .05 / .20± .04 / .95	58 / .02 / .21 / .10	14.53 ±.14 / 13.47				
0036.1 +2128 / 118.19 -41.26 / 316.48 9.95 / 0033.5 +2112	UGC 357 / PGC 2159	.S..4.. / U / 4.0± .9	1.07± .07 / .32± .06 / 1.08	177 / .07 / .48 / .16	15.18 ±.14 / 14.58				9265 / 9411 / 8938
003608.7-073213 / 111.75 -70.08 / 288.12 2.46 / 003336.3-074844	MCG -1- 2- 39 / PGC 2160	.SXS8?. / E (1) / 8.0± .8 / 7.5± .8	1.15± .06 / .20± .05 / 1.17	45 / .14 / .25 / .10					5417 / 5483 / 5087
0036.1 +0142 / 115.06 -60.93 / 297.10 5.00 / 0033.6 +0126	UGC 358 / PGC 2162	.S?....	1.06± .06 / .00± .05 / 1.06	/ .03 / .00 / .00	14.77 ±.17 / 14.71				5451 / 5545 / 5113
003612.8-121532 / 108.69 -74.71 / 283.54 1.12 / 003341.3-123202	MCG -2- 2- 68 / PGC 2166	.SXR1*. / E / 1.0±1.2	1.14± .06 / .30± .05 / 1.14	135 / .02 / .30 / .15					6738 / 6788 / 6416
0036.2 +2548 / 118.65 -36.94 / 320.78 10.88 / 0033.6 +2532	UGC 360 / PGC 2168	.I?....	1.19± .06 / .52± .06 / 1.20	/ .09 / .39 / .26					9760± 35 / 9915 / 9440
003619.3+325410 / 119.27 -29.86 / 327.85 12.29 / 003339.0+323740	MCG 5- 2- 28 / PGC 2169	.SB?...	.92± .11 / .27± .07 / .94	/ .28 / .37 / .13	15.13 ±.12 / 14.44			17.35±.3 / 198± 7 / 2.77	4877± 10 / 5045 / 4574
003620.9+453953 / 120.17 -17.13 / 340.69 14.38 / 003336.8+452323	CGCG 535- 12 / PGC 2172			.32	15.46 ±.16				14524± 82 / 14709 / 14262
003622.4-274712 / 20.92 -86.59 / 268.46 -3.27 / 003354.0-280342	ESO 410- 24 / PGC 2173	.L?....	1.04± .06 / .35± .04 / 1.00	107 / .09 / .00	14.61 ±.14 / 14.36				10426±106 / 10424 / 10145
003629.0-100625 / 110.52 -72.62 / 285.65 1.66 / 003357.0-102255	NGC 165 / MCG -2- 2- 69 / IRAS00339-1022 / PGC 2182	.SBT4.. / R (2) / 4.0± .4 / 1.2± .6	1.19± .03 / .08± .05 / .94± .02 / 1.20	50 / .10 / .12 / .04	13.88 ±.14 / 13.18 / 13.62	.80± .02 / .24± .06 / .72 / .18	.75± .02 / .22± .06 / 14.07± .04 / 14.49± .23	15.67±.1 / 308± 11 / 298± 11 / 2.00	5874± 10 / 5944± 59 / 5933 / 5551
003634.1+324450 / 119.32 -30.02 / 327.70 12.21 / 003353.7+322820	UGC 362 / PGC 2183	.S..9.. / U / 9.0± .9	1.04± .15 / .13± .12 / 1.07	/ .28 / .13 / .06				16.41±.3 / 184± 7	6066± 7 / 6234 / 5762
003637.4-274724 / 20.70 -86.65 / 268.47 -3.32 / 003409.1-280354	ESO 410- 25 / PGC 2184	.L?....	.92± .06 / .15± .04 / .91	48 / .06 / .00	15.16 ±.14 / 15.00				6905±106 / 6903 / 6624
003637.9-565424 / 306.99 -60.10 / 239.77 -10.64 / 003420.0-571054	ESO 150- 14 / PGC 2190	.L..+*/ / S / -.7±1.1	1.28± .04 / .99± .03 / 1.13	106 / .00 / .00	14.95 ±.14 / 14.83				8257 / 8161 / 8104
003639.5-223536 / 85.66 -84.36 / 273.54 -1.89 / 003410.0-225206	NGC 168 / ESO 474- 4 / IRAS00341-2251 / PGC 2192	.L..-?/ / S / -1.4± .8	1.07± .05 / .68± .03 / .97	26 / .09 / .00	14.87 ±.14 / 13.20				
003645.6+213405 / 118.40 -41.17 / 316.60 9.83 / 003407.8+211735	UGC 364 / IRAS00341+2117 / PGC 2194	.S?....	1.00± .06 / .36± .05 / 1.01	58 / .07 / .53 / .18	14.47 ±.13 / 12.58 / 13.82				8925± 50 / 9071 / 8598
003645.8+015311 / 115.41 -60.77 / 297.32 4.90 / 003411.7+013641	NGC 170 / MCG 0- 2- 91 / PGC 2195	.L..-?. / E / -3.0±2.0	.64± .11 / .19± .05 / .61	85 / .00 / .00	15.43 ±.10				
003651.2-282206 / 11.07 -86.54 / 267.93 -3.53 / 003423.1-283836	ESO 410- 27 / PGC 2199	.SBS5*. / S (1) / 5.3± .7 / 3.3± .6	1.25± .03 / .35± .03 / 1.26	44 / .07 / .52 / .17	14.52 ±.14 / 13.89				7141± 47 / 7138 / 6862

R.A. 2000 DEC. / l b / SGL SGB / R.A. 1950 DEC.	Names / PGC	Type / S_T n_L / T / L	$\log D_{25}$ / $\log R_{25}$ / $\log A_e$ / $\log D_o$	p.a. / A_g / A_i / A_{21}	B_T / m_B / m_{FIR} / B_T^o	$(B-V)_T$ / $(U-B)_T$ / $(B-V)_T^o$ / $(U-B)_T^o$	$(B-V)_e$ / $(U-B)_e$ / m'_e / m'_{25}	m_{21} / W_{20} / W_{50} / HI	V_{21} / V_{opt} / V_{GSR} / V_{3K}
003651.9+235904	IC 1559	.LX..P*	.89± .11	94	14.7 ±.3	.74± .05		14.99±.2	4595± 17
118.66 -38.77	MCG 4- 2- 34	R (1)	.24± .07	.06		.12± .08		623± 18	4694± 24
319.00 10.35	MK 341	-2.0± .6		.00		.66		626± 16	4779
003413.5+234234	PGC 2201	3.5±1.2	.86		14.57	.12	13.41± .67		4305
003652.1+235930	NGC 169	.SAS2*/	1.42± .03	88				14.98±.3	4627± 9
118.66 -38.76	UGC 365	R (1)	.60± .04	.06				611± 10	4507± 29
319.01 10.35	IRAS00342+2342	2.0± .4		.74	12.84			555± 7	4768
003413.7+234300	PGC 2202	3.5±1.1	1.43	.30					4294
003652.9-333324			.69± .07	95					
328.07 -82.85	ESO 350- 38		.12± .06	.02	14.31 ±.14				6154± 67
262.86 -4.95	IRAS00344-3349				11.38				6132
003426.0-334954	PGC 2204		.69						5893
003658.9-292842	NGC 174	.SBTO*.	1.15± .04	152					
355.64 -86.04	ESO 411- 1	Sr	.38± .03	.02	13.82 ±.14				3470±106
266.85 -3.86	IRAS00345-2945	.0± .6		.29	10.52				3462
003431.0-294512	PGC 2206		1.13		13.46				3195
0037.0 +2541	A 0034+25	.E?....	1.23± .14						9620±101
118.87 -37.07	UGC 367		1.17± .04	.07	14.89 ±.15				9774
320.70 10.68				.00					9301
0034.4 +2525	PGC 2210		.89		14.68				
003706.7-463836		.SBT6*.	1.27± .04	83					3555
310.23 -70.28	ESO 242- 18	Sr (1)	.30± .05	.03	14.00 ±.14				3491
250.00 -8.35	IRAS00347-4655	5.9± .5		.44	13.57				3351
003444.1-465506	PGC 2215	5.0± .6	1.28	.15	13.52				
003708.0+285007					.15 15.0 ±.4				15.91±.3	1986± 10
119.16 -33.93	CGCG 500- 52								
323.83 11.32								109± 7	2147
003428.5+283337	PGC 2216								1673
003710.7+255023					.07 15.29 ±.12				17.51±.3	9014± 10
118.92 -36.92	CGCG 479- 46								
320.85 10.68								297± 7	9169
003432.0+253353	PGC 2221								8695
0037.2 +1303		.SA.8..	1.01± .06						4515
117.57 -49.67	UGC 370	U	.02± .05	.17					
308.26 7.70		8.0± .8		.02					4640
0034.6 +1247	PGC 2222		1.02	.01					4179
003712.4+015633	NGC 173	.SAT5..	1.50± .03	90	*			13.94±.1	4366± 5
115.64 -60.73	UGC 369	UE (2)	.08± .04	.00	13.7 ±.3			310± 6	4358± 42
297.40 4.81	IRAS00346+0140	4.5± .5		.13				298± 5	4460
003438.4+014003	PGC 2223	2.5± .5	1.50	.04	13.52			.38	4028
003714.4-223506	NGC 172	.SBR4?/	1.31± .04	12	14.0 ±.2	.57± .06	.61± .03		
86.84 -84.43	ESO 474- 5	S	.79± .03	.09	14.66 ±.14	-.01± .08	.08± .04		2928
273.58 -2.01	IRAS00348-2251	3.8± .6	1.07± .06	1.16		.38	14.85± .16		2944
003445.1-225136	PGC 2228		1.32	.39	13.18	-.14	13.43± .28		2631
003715.7-531536		.SBS5*P	1.02± .05	30					
307.73 -63.74	ESO 150- 15	S (1)	.04± .05	.00	14.73 ±.14				8375
243.43 -9.93		5.0± .8		.05					8290
003456.0-533206	PGC 2229	2.2± .8	1.02	.02	14.63				8203
003721.6+290856		.S..6*.	1.22± .05	45				15.40±.3	5276± 10
119.24 -33.62	UGC 371	U	1.01± .05	.15	15.07 ±.13				
324.15 11.34		6.0±1.4		1.47				330± 7	5437
003442.0+285226	PGC 2231		1.23	.50	13.43			1.47	4964
003722.0-195612	NGC 175	.SBR2..	1.33± .02		12.90 ±.13	.73± .01	.83± .01		
98.05 -82.12	ESO 540- 6	R (3)	.06± .03	.11	12.87 ±.12	.12± .03	.23± .03		3824± 37
276.16 -1.30	VV 791A	2.0± .3	.98± .01	.07	13.26	.67	13.29± .02		3849
003452.0-201242	PGC 2232	2.4± .5	1.34	.03	12.67	.09	14.24± .18		3521
0037.5 +4254		.SA.9..	1.17± .07					16.10±.1	5418± 11
120.22 -19.89	UGC 372	U	.01± .06	.28					
337.94 13.77		9.0± .7		.01				122± 8	5600
0034.8 +4238	PGC 2240		1.19	.00					5146
003735.4-223254	NGC 177	.SAR2..	1.35± .03	9					
87.74 -84.45	ESO 474- 6	Sr	.65± .03	.09	14.14 ±.14				3892± 60
273.64 -2.08		1.8± .4		.79					3908
003506.0-224924	PGC 2241		1.36	.32	13.22				3595
003735.9+001650		.S?....	.82± .13						10435± 30
115.45 -62.39	MCG 0- 2- 94		.17± .07	.04	15.25 ±.12				10524
295.81 4.27	MK 955			.25	13.09				10098
003502.2+000021	PGC 2243		.82	.09	14.89				

R.A. 2000 DEC. l b SGL SGB R.A. 1950 DEC.	Names PGC	Type S_T n_L T L	logD_25 logR_25 logA_e logD_o	p.a. A_g A_i A_21	B_T m_B m_FIR B_T^o	(B-V)_T (U-B)_T (B-V)_T^o (U-B)_T^o	(B-V)_e (U-B)_e m'_e m'_25	m_21 W_20 W_50 HI	V_21 V_opt V_GSR V_3K
003739.4+102116 117.38 -52.37 305.64 6.91 003503.7+100447	IC 35 UGC 374 PGC 2246	.S..6*. U 6.0±1.2 	1.00± .06 .08± .05 1.02	 .22 .12 .04	 14.67 ±.13 14.30			15.95±.2 223± 6 207± 7 1.60	4587± 7 4704 4249
003741.7-334301 326.29 -82.78 262.75 -5.15 003515.0-335930	A 0035-34 ESO 350- 40 VV 784 PGC 2248	.RING.. S 	1.06± .06 .12± .05 1.05	128 .02 .00 	 13.46 			14.93±.3 349± 24 248± 18 	8934± 14 9072± 26 8942 8704
003746.4-175101 103.46 -80.20 278.22 -.81 003516.0-180730	NGC 179 ESO 540- 7 PGC 2253	.LX.-.. SE -3.0± .6 	.97± .03 .08± .02 .97	113 .09 .00 	14.19 ±.13 14.01 ±.14 13.93	.94± .01 .86 	 13.69± .22 		 6013± 27 6044 5704
0037.7 +0401 116.37 -58.67 299.47 5.23 0035.2 +0345	 UGC 377 PGC 2254	.S..8*. U 8.0±1.2 	1.07± .07 .25± .06 1.07	157 .04 .31 .13					5332 5432 4994
003749.5-263854 41.14 -86.93 269.66 -3.26 003521.0-265524	 ESO 474- 7 PGC 2258	.LXR+.. r -1.3± .9 	.96± .06 .04± .03 .96	 .04 .00 	 14.81 ±.14 				
0037.8 +0508 116.63 -57.56 300.56 5.50 0035.3 +0452	 UGC 379 PGC 2260	.S..6*. U 6.0±1.3 	1.12± .05 .76± .05 1.12	120 .06 1.11 .38					5222 5325 4884
003754.5+324115 119.64 -30.10 327.70 11.93 003513.9+322446	 UGC 376 PGC 2261	.S..6*. U 6.0±1.4 	1.11± .07 1.13± .06 1.13	142 .23 1.47 .50				15.05±.3 213± 7 	4817± 10 4984 4514
003755.9-285524 1.09 -86.52 267.45 -3.91 003528.0-291154	 ESO 411- 2 IRAS00354-2911 PGC 2265	.S?.... 	.98± .05 .63± .04 .98	165 .02 .95 .32	 15.12 ±.14 13.71 14.13				3604±106 3598 3327
003756.3+045442 116.61 -57.79 300.34 5.43 003521.6+043813	 MCG 1- 2- 38 IRAS00353+0438 PGC 2266	.S?.... 	.94± .11 .68± .07 .94	 .05 1.02 .34	 15.42 ±.14 13.02 14.30				8490 8592 8152
003757.5+083802 117.25 -54.09 303.98 6.40 003522.1+082133	NGC 180 UGC 380 IRAS00353+0821 PGC 2268	.SBT4.. U 4.0± .7 	1.38± .04 .09± .05 1.40	160 .19 .13 .05	 13.7 ±.2 12.74 13.32			14.67±.2 379± 8 354± 6 1.31	5264± 6 5251± 60 5376 4925
003757.6-091514 112.20 -71.84 286.58 1.55 003525.6-093144	 MCG -2- 2- 73 IRAS00354-0931 PGC 2269	.SXS6.. E (1) 6.0± .8 3.1±1.2	1.16± .05 .08± .05 1.18	115 .17 .11 .04					
003802.5-463106 309.83 -70.43 250.16 -8.48 003540.0-464736	 ESO 242- 20 PGC 2274	.SBS8.. S (1) 8.0± .8 6.7± .9	1.21± .05 .15± .05 1.22	141 .03 .18 .07	 14.19 ±.14 13.97				3329 3265 3124
0038.2 -5538 306.83 -61.39 241.09 -10.58 0035.9 -5555	 PGC 2277	.LBT+.. S -1.0± .8 	1.15± .09 .21± .08 1.12	 .00 .00 					
003812.5+024344 116.32 -59.97 298.24 4.78 003538.2+022715	NGC 182 UGC 382 PGC 2279	PSXT1P* PUE 1.3± .4 	1.30± .03 .08± .04 .99± .06 1.30	75 .02 .08 .04	13.27 ±.18 13.39 ±.14 13.18	.85± .03 .38± .04 .79 .37	.93± .03 .43± .04 13.71± .18 14.42± .26	15.66±.3 452± 13 423± 10 2.44	5261± 10 5217± 43 5355 4921
003821.9+323812 119.75 -30.15 327.66 11.82 003541.2+322143	 UGC 384 PGC 2286	.SXT7.. U 7.0± .8 	1.18± .05 .03± .05 1.20	 .20 .04 .01	 14.43 ±.19 14.17			15.41±.3 196± 7 1.22	4702± 10 4869 4399
003823.4+292822 119.53 -33.31 324.52 11.18 003543.6+291153	NGC 181 MCG 5- 2- 32 PGC 2287	.S?.... 	.74± .14 .49± .07 .75	 .17 .74 .25				17.72±.3 422± 7 	5535± 10 5696 5224
003823.5+150222 118.25 -47.71 310.27 7.91 003546.8+144553	A 0035+14 UGC 386 MK 343 PGC 2288	.L..... U -2.0± .8 	1.16± .06 .20± .03 1.16	170 .21 .00 	 14.64 ±.14 14.35				5463± 60 5593 5129

R.A. 2000 DEC. I b SGL SGB R.A. 1950 DEC.	Names PGC	Type S_T n_L T L	$\log D_{25}$ $\log R_{25}$ $\log A_\bullet$ $\log D_0$	p.a. A_g A_i A_{21}	B_T m_B m_{FIR} B_T^0	$(B-V)_T$ $(U-B)_T$ $(B-V)_T^0$ $(U-B)_T^0$	$(B-V)_\bullet$ $(U-B)_\bullet$ $m'_\bullet$ m'_{25}	m_{21} W_{20} W_{50} HI	V_{21} V_{opt} V_{GSR} V_{3K}
0038.4 +1328 118.07 -49.27 308.74 7.52 0035.8 +1312	UGC 385 PGC 2290	.S?.... 	1.32± .10 .45± .12 1.33	64 .15 .66 .23	 15.0 ±.2 14.16				11100±110 11225 10765
003825.3+030959 116.51 -59.54 298.68 4.85 003551.0+025330	NGC 186 UGC 390 PGC 2291	.LB.+P* UE -.7± .7 	1.14± .04 .22± .04 1.11	23 .01 .00	 14.40 ±.11				
003827.9+293729 119.56 -33.16 324.67 11.20 003548.0+292100	MCG 5- 2- 31 PGC 2294	.S?.... 	.94± .11 .40± .07 .95	 .17 .49 .20	 15.19 ±.12 14.52				5657 5818 5347
003829.1+293042 119.56 -33.28 324.56 11.17 003549.2+291413	NGC 183 UGC 387 PGC 2298	.E..... U -5.0± .7 	1.32± .12 .11± .08 .67± .07 1.31	130 .17 .00	13.74 ±.19 13.37 ±.11 13.22	1.00± .01 .91	1.02± .01 12.58± .26 15.05± .66		5296± 31 5457 4985
003832.0+301719 119.63 -32.50 325.33 11.32 003551.9+300050	UGC 388 PGC 2302	.S..6*. U 6.0±1.2 	1.07± .08 .26± .06 1.09	3 .21 .38 .13				16.26±.3 259± 7	5405± 10 5568 5096
003832.9-242025 76.06 -85.98 271.96 -2.79 003604.0-243654	IC 1561 ESO 474- 8 PGC 2305	.SB?... 	1.09± .05 .36± .04 1.10	97 .07 .44 .18	 14.92 ±.14 14.37				3912± 69 3921 3621
003834.9-241631 76.79 -85.94 272.02 -2.78 003606.0-243300	IC 1562 ESO 474- 9 IRAS00360-2432 PGC 2308	.SBS5.. S (1) 5.0± .8 2.2± .8	1.21± .04 .03± .03 1.22	 .07 .04 .01	 13.56 ±.14 12.78 13.43				3659± 69 3668 3368
003835.4+292649 119.58 -33.34 324.50 11.14 003555.5+291020	NGC 184 CGCG 500- 59 PGC 2309	 	.86± .10 .59± .06 .88	 .17	 15.62 ±.10			17.61±.3 397± 7	5289± 10 5450 4978
0038.6 +1724 118.58 -45.36 312.61 8.43 0036.0 +1708	UGC 393 PGC 2312	.S..8*. U 8.0±1.2 	1.32± .05 .71± .06 1.33	170 .10 .87 .35					5434 5569 5102
003843.4+415948 120.41 -20.81 337.05 13.41 003559.9+414320	UGC 394 PGC 2314	.SX.8.. U 8.0± .8 	1.30± .03 .40± .04 1.32	5 .24 .50 .20	 15.1 ±.2 14.37			15.58±.1 258± 8 1.02	5610± 11 5590± 76 5790 5335
003847.5+254251 119.36 -37.07 320.81 10.30 003608.5+252623	CGCG 479- 50 PGC 2316	 	 	 .12	 15.6 ±.4			16.67±.3 240± 7	7245± 10 7399 6926
003847.9+322502 119.84 -30.38 327.46 11.69 003607.2+320833	CGCG 500- 60 PGC 2317	 	 	 .20	 15.3 ±.4			17.50±.3 80± 7	11263± 10 11429 10960
003854.6+070327 117.41 -55.68 302.50 5.76 003619.5+064658	UGC 397 HICK 5A PGC 2324	.S?.... 	.96± .07 .08± .05 .97	 .10 .10 .04	14.94S±.15 14.64		 14.40± .38		11195± 41 11303 10857
003854.8+070322 117.41 -55.68 302.50 5.76 003619.7+064653	MCG 1- 2- 42 HICK 5B PGC 2325	.L?.... 	.42? .00± .07 .43	 .10 .00	15.67S±.15 15.39		 12.62±1.73	17.59±.3 277± 21 266± 15	12100± 12 12233± 35 12222 11777
0038.9 +2538 119.40 -37.14 320.74 10.25 0036.3 +2522	UGC 398 PGC 2327	.SAR1.. U 1.0± .8 	1.12± .05 .10± .05 1.13	25 .10 .10 .05	 14.32 ±.13 14.07				4612 4765 4293
003858.0+482018 120.79 -14.48 343.46 14.30 003612.0+480350	NGC 185 UGC 396 IRAS00362+4803 PGC 2329	.E.3.P. R -5.0± .3 	2.07± .02 .07± .04 1.58± .02 2.19	35 .83 .00	10.10M±.05 10.14 ±.16 13.66 9.27	.92± .01 .39± .03 .73 .23	.93± .01 17.13±.3 .36± .02 13.71± .04 15.27± .15		-251± 13 -64 -502
003859.7-090010 113.14 -71.64 286.89 1.38 003627.7-091639	NGC 191 MCG -2- 2- 77 PGC 2331	.SXT5*P R 5.0± .4 	1.17± .04 .08± .04 1.19	125 .14 .12 .04					

R.A. 2000 DEC.	Names	Type	logD₂₅	p.a.	B_T	(B-V)_T	(B-V)_e	m₂₁	V₂₁
l　　b		S_T　n_L	logR₂₅	A_g	m_B	(U-B)_T	(U-B)_e	W₂₀	V_opt
SGL　SGB		T	logA_e	A_i	m_FIR	(B-V)_T°	m'	W₅₀	V_GSR
R.A. 1950 DEC.	PGC	L	logD_o	A₂₁	B_T°	(U-B)_T°	m'₂₅	HI	V_3K

003900.9-090030 IC　1563		.L...P/	.83± .09	143					
113.15 -71.64 MCG -2- 2- 76		RCE	.28± .05	.14					
286.89　1.37		-2.0± .9		.00					
003628.9-091658 PGC　2332			.81						
003904.1-183637 NGC　209		.LA.-P*	1.14± .04						
103.81 -81.02 ESO 540- 8		SE	.08± .03	.04	13.91 ±.14				3931± 60
277.57 -1.32		-3.0± .6		.00					3959
003634.0-185306 PGC　2338			1.13		13.81				3625
0039.0 +0601 IC　1564		.SX.4*.	.99± .06	83					5269
117.33 -56.71 UGC　399		U	.32± .05	.08	14.68 ±.12				
301.50　5.45		4.0±1.3		.47					⸌5374
0036.5 +0545 PGC　2342			1.00		.16 14.09				4931
003908.4-141023 NGC　178		.SBS9..	1.31± .03	175	13.1 ±.2	.47± .04		13.93±.3	1449± 17
109.84 -76.73 MCG -2- 2- 78		R　　(2)	.32± .04	.08	12.9 ±.2	-.35± .05		86± 34	1485± 46
281.88 -.10 8ZW　34		9.0± .4		.33	12.72	.37		75± 25	1496
003637.5-142651 PGC　2349		7.5± .8	1.32	.16	12.60	-.42	13.70± .27	1.17	1137
003910.2-082350		.S?....			16.02S±.15				
113.55 -71.05				.12					
287.49　1.50 HICK　6B									
003638.1-084018 PGC　2350									
003911.1-082401		.E?....			15.65S±.15				
113.56 -71.05				.12					
287.49　1.50 HICK　6C									
003639.0-084029 PGC　2351									
003913.5+005143 NGC　192		PSBR1*.	1.28± .03	167	13.42M±.10	.79± .02	.86± .02		
116.45 -61.86 UGC　401		UE	.33± .04	.00	13.65 ±.12	.35± .03	.36± .03		4210± 32
296.49　4.03 IRAS00366+0035		1.0± .6	.69± .02	.34	11.83	.69	12.41± .06		4300
003639.6+003515 PGC　2352			1.28	.17	13.12	.30	13.83± .19		3874
003913.8-082339		.L?....			15.73S±.15				11669± 41
113.60 -71.04				.12					11730
287.50　1.49 HICK　6A									11342
003641.7-084008 PGC　2353									
003915.4-430431		.S..5P.	1.19± .05	128					3988± 69
310.96 -73.86 ESO 242- 23		S	.46± .05	.03	13.91 ±.14				3934
253.62 -7.85		5.0± .9		.68					3768
003652.0-432100 PGC　2356			1.19	.23	13.18				
003917.8+005444 NGC　196		.LB..P*	1.13± .07	15	13.80M±.10	.94± .01	.95± .01		
116.50 -61.81 UGC　405		UE	.22± .04	.00		.40± .02	.49± .02		4238± 41
296.54　4.03 HICK　7B		-2.2± .7	.59± .02	.00		.88	12.26± .08		4328
003643.9+003816 PGC　2357			1.10		13.74	.40	13.78± .39		3902
003918.1-272055		.SAT3*.	1.14± .04	58					6884± 49
27.57 -87.29 ESO 411- 3		S　　(1)	.43± .03	.05	14.63 ±.14				6882
269.07 -3.77 IRAS00368-2737		3.0±1.0		.59	13.28				6603
003650.0-273724 PGC　2358		3.9± .6	1.15	.21	13.94				
003918.3+031953 NGC　193		.LXS-*.	1.16± .08	55	13.25 ±.15	1.04± .02	1.02± .01		
116.98 -59.40 UGC　408		PE	.08± .05	.07	13.87 ±.11	.52± .05	.56± .03		4327± 48
298.90　4.68		-2.5± .6	.81± .05	.00		.98	12.79± .16		4424
003643.9+030325 PGC　2359			1.16		13.53	.52	13.75± .47		3989
0039.3 +1306		.S..8*.	1.04± .08	40					10618
118.37 -49.66 UGC　404		U	.38± .06	.18					
308.44　7.22		8.0±1.3		.46					10742
0036.7 +1250 PGC　2360			1.05	.19					10283
003918.4+030210 NGC　194		.E.....	1.19± .06	30	13.15 ±.13	.98± .01	.99± .01		
116.92 -59.69 UGC　407		PE	.04± .04	.00	13.46 ±.10	.58± .06	.59± .06		5150± 37
298.61　4.60		-5.0± .5	.80± .02	.00		.93	12.64± .07		5247
003644.0+024542 PGC　2362			1.18		13.26	.61	14.02± .32		4813
003918.4-295644		.L?....	.88± .06						7109± 52
345.62 -86.12 ESO 411- 4			.04± .04	.02	15.32 ±.14				7099
266.53 -4.47				.00					6837
003651.1-301312 PGC　2363			.88		15.20				
003918.6+293928		.S?....	.88± .10					15.94±.3	5507± 10
119.79 -33.14 UGC　400			.01± .06	.16	15.02 ±.12				5668
324.74 11.03				.01				183±　7	5197
003638.6+292300 PGC　2364			.90	.00	14.81			1.13	
003918.9+005331 NGC　197		.LB.0P.	1.07± .06	5	14.80M±.11	.66± .10			
116.50 -61.83 UGC　406		E	.16± .05	.00		.05± .06			4192± 25
296.52　4.02 HICK　7D		-2.0± .9		.00		.60			4282
003644.9+003703 PGC　2365			1.04		14.74	.06	14.61± .33		3856

R.A. 2000 DEC.	Names	Type	$\log D_{25}$	p.a.	B_T	$(B-V)_T$	$(B-V)_e$	m_{21}	V_{21}
l　　b		S_T　n_L	$\log R_{25}$	A_g	m_B	$(U-B)_T$	$(U-B)_e$	W_{20}	V_{opt}
SGL　SGB		T	$\log A_e$	A_i	m_{FIR}	$(B-V)_T^o$	m'_e	W_{50}	V_{GSR}
R.A. 1950 DEC.	PGC	L	$\log D_o$	A_{21}	B_T^o	$(U-B)_T^o$	m'_{25}	HI	V_{3K}

003918.9+035729		.L.....	1.11± .10	20					
117.09 -58.78	UGC 402	U	.10± .05	.04	14.35 ±.11				
299.51 4.84	MK 554	-2.0± .8		.00	13.58				
003644.4+034101	PGC 2366		1.10						
003921.0-185508		.S?....	1.06± .05	165					3927± 60
103.60 -81.33	ESO 540- 9		.26± .04	.04	14.65 ±.14				3954
277.29 -1.47				.38					3622
003651.1-191136	PGC 2369		1.06	.13	14.20				
0039.3 +0433		.S..8*.	1.00± .06	40					5167
117.22 -58.18	UGC 409	U	.18± .05	.04	15.18 ±.14				
300.10 4.99		8.0±1.2		.22					5268
0036.8 +0417	PGC 2370		1.01	.09	14.90				4830
003923.0+024751	NGC 198	.SAR5..	1.09± .04	80					
116.92 -59.93	UGC 414	R (1)	.02± .04	.00	13.85 ±.12				5262± 50
298.38 4.52	IRAS00367+0231	5.0± .4		.04	12.69				5358
003648.7+023123	PGC 2371	3.1± .8	1.09	.01	13.45				4925
0039.3 +0643	IC 1565	.S?....	1.18± .09						11150± 59
117.57 -56.02	UGC 410		.00± .05	.06	14.31 ±.19				11257
302.21 5.56				.00					10812
0036.8 +0627	PGC 2372		1.19		14.09				
003924.8+251511								16.44±.3	4638± 10
119.50 -37.54	CGCG 479- 54			.08	15.0 ±.4				
320.38 10.06								178± 7	4790
003645.8+245843	PGC 2374								4319
003929.9+253839		.E?....	1.08± .09	100					
119.56 -37.15	UGC 411		.12± .04	.10	14.20 ±.10				4614± 50
320.77 10.13				.00					4767
003650.8+252211	PGC 2377		1.06		14.03				4296
003930.3-143923	NGC 187	.SBS6..	1.11± .06	148					3947
109.77 -77.22	MCG -3- 2- 34	E (1)	.43± .05	.04					
281.44 -.32	IRAS00369-1455	6.0± .9		.63	13.44				3988
003659.5-145551	PGC 2380	5.3±1.2	1.11	.21					3632
003930.6+294545		.S..0..	.96± .09	11				18.06±.3	5484± 10
119.84 -33.04	UGC 412	U	.69± .06	.21	15.56 ±.13				
324.86 11.01		.0±1.0		.51				391± 7	5645
003650.5+292917	PGC 2381		.95		14.76				5175
003933.2+030819	NGC 199	.LA.0P*	1.09± .06	160					
117.06 -59.60	UGC 415	UE	.23± .03	.04	14.61 ±.11				
298.73 4.57		-1.5± .6		.00					
003658.8+025151	PGC 2382		1.06						
003933.7-354820		.SBS5*/	1.11± .05	25					
318.44 -80.96	ESO 351- 1	S	.91± .04	.02	16.26 ±.14				
260.80 -6.07		5.0±1.4		1.36					
003708.0-360448	PGC 2383		1.11	.45					
0039.5 +0356		.SX.8*.	1.06± .08	10					4744
117.21 -58.80	UGC 416	U	.10± .06	.04					
299.51 4.78		8.0±1.2		.12					4843
0037.0 +0340	PGC 2384		1.06	.05					4407
003934.7-202302		.SBS6..	1.07± .04	131					
100.40 -82.73	ESO 540- 10	S (1)	.21± .03	.11	15.60 ±.14				3954
275.88 -1.92		6.0± .8		.30					3976
003705.1-203930	PGC 2386	5.6±1.2	1.08	.10	15.17				3652
003934.7+025319	NGC 200	.SBS4..	1.27± .03	161	13.48 ±.14	.85± .02	.81± .02	15.80±.2	5169± 8
117.03 -59.85	UGC 420	R (1)	.28± .03	.04	13.71 ±.12	.06± .03	.16± .02	439± 21	5140± 60
298.48 4.49	IRAS00370+0236	4.0± .4	.81± .02	.42	12.50	.76	13.02± .06	388± 8	5264
003700.4+023651	PGC 2387	.8± .8	1.27	.14	13.12	-.02	13.97± .23	2.54	4832
003935.0+005135	NGC 201	.SXR5..	1.26± .03	155	13.57M±.10	.67± .02	.68± .02		
116.64 -61.87	UGC 419	UE (1)	.10± .01	.00	14.13 ±.19	.04± .03	.13± .02		4366± 41
296.51 3.95	IRAS00370+0035	4.5± .5	.91± .01	.15	13.33	.62	13.66± .03		4456
003701.1+003507	PGC 2388	3.1± .8	1.26	.05	13.52	.01	14.47± .19		4030
0039.6 +0857		.S..3..	1.33± .04	98					
117.97 -53.80	UGC 418	U (1)	.94± .05	.19	14.94 ±.12				4545± 57
304.40 6.09		3.0± .9		1.29					4658
0037.0 +0841	PGC 2390	4.5±1.3	1.34	.47	13.43				4208
003935.9-091142	NGC 195	RSBR1*.	1.03± .05	45					
113.51 -71.85	MCG -2- 2- 79	PE	.15± .04	.13					4918
286.75 1.18	IRAS00370-0928	1.0± .7		.15	12.74				4976
003703.9-092810	PGC 2391		1.04	.07					4593

R.A. 2000 DEC.	Names	Type	logD$_{25}$	p.a.	B$_T$	(B-V)$_T$	(B-V)$_e$	m$_{21}$	V$_{21}$
l b		S$_T$ n$_L$	logR$_{25}$	A$_g$	m$_B$	(U-B)$_T$	(U-B)$_e$	W$_{20}$	V$_{opt}$
SGL SGB		T	logA$_e$	A$_i$	m$_{FIR}$	(B-V)^o_T	m'	W$_{50}$	V$_{GSR}$
R.A. 1950 DEC.	PGC	L	logD$_o$	A$_{21}$	B^o_T	(U-B)^o_T	m'$_{25}$	HI	V$_{3K}$

003939.6+032634 NGC 203		.L..0P*	.93± .08	85					
117.17 -59.30 MCG 0- 2-114		E	.49± .05	.07	14.97 ±.12				
299.03 4.62		-2.0±1.4		.00					
003705.2+031006	PGC 2393		.86						
003939.8+033213 NGC 202		.LBS+?.	.95± .06	153					
117.19 -59.20 UGC 421		E	.49± .04	.07	15.34 ±.10				
299.12 4.65		-1.0±1.3		.00					
003705.4+031545	PGC 2394		.88						
0039.6 +0900		.SB?...	1.06± .08						
118.02 -53.76 UGC 422			.09± .06	.19	14.60 ±.14				4656± 57
304.45 6.08 IRAS00370+0843				.13	13.36				4769
0037.1 +0843	PGC 2396		1.08		.04	14.25				4319
003944.3+031758 NGC 204		.LAS0P*	1.08± .06	30	13.8 ±.2	.91± .02	.99± .02		
117.18 -59.44 UGC 423		E	.02± .05	.07	14.19 ±.11	.48± .04	.58± .04		
298.89 4.57		-2.0± .8	.58± .08	.00			12.23± .24		
003709.9+030130	PGC 2397		1.08				14.05± .38		
003945.3+330928								16.60±.3	8818± 10
120.12 -29.65 CGCG 500- 65				.29	14.9 ±.4				8985
328.24 11.64								277± 7	8517
003704.3+325300	PGC 2398								
0039.8 +2112		.S?....	.69± .15						
119.32 -41.58 MCG 3- 2- 22			.05± .07	.07	15.26 ±.13				9135± 60
316.42 9.05				.07					9278
0037.2 +2056	PGC 2402		.69		.02	15.07				8809
004004.1-303550		.SBT4P.	1.15± .04	140					
337.92 -85.73 ESO 411- 6		Sr (1)	.12± .04	.02	15.21 ±.14				15102± 52
265.94 -4.81		4.2± .6		.18					15089
003737.0-305218	PGC 2411	2.2± .8	1.15	.06	14.91				14832
004011.1+224259		.SB?...	.81± .11						
119.54 -40.08 UGC 425			.15± .06	.08	14.66 ±.15				5878± 46
317.92 9.32 IRAS00375+2226				.22	13.09				6025
003732.6+222632	PGC 2416		.82	.07	14.33				5555
004012.8-560908 NGC 212		.LA.-..	1.11± .05	131					
306.15 -60.91 ESO 150- 18		S	.09± .04	.00	14.43 ±.14				
240.64 -10.97		-3.0± .8		.00					
003756.0-562536	PGC 2417		1.10						
004017.5+024522 NGC 208		.SAR1?.	.87± .08						
117.36 -59.99 MCG 0- 2-118		E	.03± .05	.04	15.17 ±.13				
298.40 4.29		1.0±1.9		.03					
003743.2+022855	PGC 2420		.87	.01					
004021.3+501957		.S..6*.	1.14± .05	64				14.81±.1	5293± 11
121.12 -12.50 UGC 427		U	.39± .05	.99	14.9 ±.3				5481
345.51 14.34		6.0±1.2		.57				300± 8	5051
003733.9+500330	PGC 2427		1.24	.20	13.32			1.29	
004022.5+414111 NGC 205		.E.5.P.	2.34± .01	170	8.92M±.05	.85± .05	.80± .01	16.22±.2	-232± 10
120.72 -21.14 UGC 426		R	.30± .02	.15	8.93 ±.10				-254± 14
336.79 13.06 IRAS00376+4124		-5.0± .3	1.75± .02	.00	12.93	.82	13.21± .06	13± 10	-60
003738.7+412444	PGC 2429		2.27		8.77		14.85± .09		-514
0040.4 +0654		.L...*.	1.15± .15	120					
118.08 -55.86 UGC 429		U	.14± .08	.10	14.92 ±.17				
302.46 5.34		-2.0±1.2		.00					
0037.9 +0638	PGC 2431		1.14						
004029.6+250134 A 0037+24									
119.79 -37.78 CGCG 479- 58				.07	15.07 ±.12				4500±110
320.21 9.77 MK 345									4651
003750.5+244507	PGC 2432								4181
004031.9-455908		PSBS0*/	1.18± .05	39					
308.76 -71.02 ESO 242- 24		Sr	.52± .03	.03	14.58 ±.14				3657
250.79 -8.77 IRAS00381-4615		-.4± .6		.39	13.77				3593
003810.0-461536	PGC 2433		1.16		14.10				3451
0040.5 +0314		.S..6*.	1.09± .07	44					
117.58 -59.52 UGC 430		U	.59± .03	.04					
298.89 4.35		6.0±1.3		.86					
0038.0 +0258	PGC 2435		1.10	.29					
004034.9-135226 NGC 210		.SXS3..	1.70± .02	160	11.60 ±.14	.71± .03	.85± .02	12.49±.2	1634± 6
111.57 -76.51 MCG -2- 2- 81		R (2)	.18± .03	.04	11.8 ±.2	.07± .04	.23± .02	302± 7	1735± 58
282.27 -.35 IRAS00380-1408		3.0± .3	1.37± .02	.25	12.25	.66	13.94± .04	280± 5	1678
003804.0-140854	PGC 2437	1.1± .6	1.70	.09	11.36	.03	14.49± .18	1.04	1319

R.A. 2000 DEC.	Names	Type	logD$_{25}$	p.a.	B$_T$	(B-V)$_T$	(B-V)$_e$	m$_{21}$	V$_{21}$
l　　b		S$_T$　n$_L$	logR$_{25}$	A$_g$	m$_B$	(U-B)$_T$	(U-B)$_e$	W$_{20}$	V$_{opt}$
SGL　　SGB		T	logA$_e$	A$_i$	m$_{FIR}$	(B-V)$_T^o$	m'$_e$	W$_{50}$	V$_{GSR}$
R.A. 1950 DEC.	PGC	L	logD$_o$	A$_{21}$	B$_T^o$	(U-B)$_T^o$	m'$_{25}$	HI	V$_{3K}$
0040.6 -0018　IC　1571		.S..3..	1.06± .06	0					5841
116.98 -63.06　UGC　432		U　　　(1)	.10± .05	.03	14.46 ±.13				
295.45　3.37		3.0± .8		.14					5927
0038.1 -0035　PGC　2440		2.5±1.1	1.06	.05	14.24				5506
004043.1-813403		.SBS4..	1.03± .05	138					
303.41 -35.55　ESO　12- 22		S　　　(1)	.30± .04	.36	15.48 ±.14				
214.84 -15.01		4.0± .8		.45					
003929.0-815030 PGC　2444		3.3± .8	1.07	.15					
004043.7-632638		.SBT7P*	1.35± .04	0					
304.95 -53.64　ESO　79- 5		Sr　　(1)	.32± .05	.00	13.74 ±.14				1771± 34
233.31 -12.44　IRAS00385-6342		6.6± .5		.44	13.98				1656
003833.0-634306 PGC　2445		6.7± .9	1.35	.16	13.29				1654
004047.9-791427			.97± .07	137					
303.56 -37.87　ESO　12- 21			.61± .05	.33	14.53 ±.14				9000± 60
217.22 -14.76　IRAS00392-7930					12.69				8847
003916.0-793054 PGC　2450			1.00						8974
004048.5-561251 NGC　215		.LA.-..	1.04± .06	120					
305.97 -60.85　ESO　150- 19		S	.10± .04	.00	14.08 ±.14				8229
240.59 -11.06		-3.0± .9		.00					8134
003832.0-562918 PGC　2451			1.02		13.96				8073
004102.3+314407		.S..6*.	1.25± .04	74					4656
120.35 -31.08　UGC　433		U	.79± .05	.24	14.59 ±.12				4820
326.88　11.09　IRAS00383+3127		6.0±1.3		1.15	12.82				4352
003821.4+312740 PGC　2458			1.27	.39	13.17				
0041.1　+5044		.S..4..	1.04± .08	103				16.03±.1	5180± 11
121.26 -12.10　UGC　434		U	.29± .06	1.06					5368
345.94　14.28		4.0± .9		.43				241±　8	4940
0038.3　+5028　PGC　2461			1.14	.15					
004112.3-210757 A　0038-21		.E+..P?	.76± .06						
101.12 -83.57　ESO　540- 14		PS	.27± .03	.07	15.05 ±.14				
275.25　-2.49		-4.4± .8		.00					
003843.0-212424 PGC　2464			.69						
004112.6-133248		.IBS9..	1.16± .06	20					
112.45 -76.21　MCG -2- 2- 83		E　　　(1)	.02± .05	.02					
282.63　　-.40		10.0± .8		.02					
003841.6-134915 PGC　2465		7.5± .8	1.16	.01					
0041.2　+1628　NGC　213		.SBT1..	1.22± .05						5448
119.38 -46.33　UGC　436		U	.06± .05	.10	14.23 ±.19				
311.85　7.60		1.0± .8		.06					5579
0038.6　+1612　PGC　2469			1.23	.03	14.01				5117
004121.7-014250 A　0038-01		PSBR1P*	1.05± .05	170	13.8　±.3	.65± .06			
117.08 -64.47　UGC　439		UE	.02± .04	.07	14.12 ±.12	-.03± .10			5246± 43
294.13　2.82		1.0± .6		.02		.59			5327
003848.3-015917 PGC　2472			1.05	.01	13.93	-.03	13.82± .40		4913
0041.3　+0329		.S..9..	.96± .09						4670
118.01 -59.28　UGC　437		U	.00± .06	.09					
299.19　4.22		9.0± .9		.00					4766
0038.8　+0313　PGC　2473			.97	.00					4334
004126.9-210245 NGC　216		.L...0?/	1.31± .04	27	13.72 ±.13	.49± .01	.50± .01	14.19±.2	1557± 12
101.86 -83.50　ESO　540- 15		RS	.47± .03	.07	13.52 ±.14	-.10± .02	-.09± .01	246± 24	1573± 50
275.35　-2.53		-2.0±1.2	.59± .01	.00		.42	12.16± .03	158± 18	1576
003857.6-211912 PGC　2478			1.25		13.54	-.13	13.98± .24		1259
004128.2+252959 NGC　214		.SXR5..	1.27± .02	35	12.97 ±.13	.72± .03	.80± .02	15.28±.1	4533± 5
120.10 -37.32　UGC　438		R　　(2)	.13± .03	.08	12.94 ±.11	.13± .03		372±　6	4495± 40
320.73　9.67　IRAS00387+2513		5.0± .4	.79± .01	.19	12.06	.65	12.41± .03	355±　6	4685
003848.9+251333 PGC　2479		1.1± .6	1.28	.06	12.65	.07	13.86± .20	2.56	4215
004133.9+293103								16.33±.3	5253± 10
120.36 -33.30　CGCG 500- 68				.18	15.4　±.4				
324.71　10.52								92±　7	5413
003853.5+291437 PGC　2481									4944
004133.9-100120 NGC　217		.S..0*/	1.42± .04	110					
114.72 -72.74　MCG -2- 2- 85		E	.63± .05	.10					4001
286.08　　.49　IRAS00390-1017		.0±1.1		.47					4056
003902.3-101747 PGC　2482			1.40						3678
004137.8+291209								17.15±.3	6978± 10
120.36 -33.62　CGCG 500- 69				.18	15.4　±.4				
324.40　10.44								261±　7	7137
003857.6+285543 PGC　2483									6668

R.A. 2000 DEC.	Names	Type	logD25	p.a.	BT	(B-V)T	(B-V)e	m21	V21
l b		ST nL	logR25	Ag	mB	(U-B)T	(U-B)e	W20	Vopt
SGL SGB		T	logAe	Ai	mFIR	(B-V)T°	m'	W50	VGSR
R.A. 1950 DEC.	PGC	L	logDo	A21	BT°	(U-B)T°	m'25	HI	V3K
0041.7 +1049		.S?....	.96± .17						11775
119.05 -51.97	UGC 441		.10± .12	.20					
306.35 6.07				.15					11892
0039.1 +1033	PGC 2488		.98	.05					11440
004144.9-165124		.SB?...	1.20± .05	172					1550
110.12 -79.49	MCG -3- 2- 40	(1)	.17± .05	.05					
279.45 -1.44				.25					1582
003914.7-170751	PGC 2492	5.3± .8	1.21	.09					1241
004145.0+362125	NGC 218	.S?....	1.05± .08						4894
120.75 -26.47	UGC 440		.01± .06	.18	15.03 ±.18				
331.51 11.85				.01					5065
003902.6+360459	PGC 2493		1.06	.00	14.82				4603
004150.5-091815		.LBR0P?	1.11± .08	155	14.71 ±.19	1.00± .04	1.05± .03		16728±113
115.24 -72.03	MCG -2- 2- 86	E	.04± .05	.10					16785
286.80 .62		-2.0±1.2	.68± .04	.00		.81	13.60± .09		16404
003918.7-093442	PGC 2501		1.12		14.36		15.05± .47		
004151.5+325926		.SAT7..	1.22± .06	175				15.60±.3	4930± 7
120.62 -29.84	UGC 442	U	.30± .06	.28	14.92 ±.19				
328.16 11.17		7.0± .8		.42				216± 7	5096
003910.2+324300	PGC 2504		1.25	.15	14.21			1.24	4630
004157.1-325816		.S?....	1.02± .05	93					9621± 39
321.71 -83.81	ESO 351- 2		.17± .05	.02	15.98 ±.14				9600
263.72 -5.82				.26					9361
003931.0-331442	PGC 2510		1.02	.09	15.65				
004201.4+253406					15.4 ±.4			16.59±.3	10151± 10
120.26 -37.25	CGCG 479- 60			.11					10303
320.83 9.56								208± 7	9834
003922.0+251740	PGC 2514								
004204.7+364815		.S?....	1.06± .06	160	13.99 ±.12				10693± 52
120.85 -26.03	UGC 444		.16± .05	.18					10865
331.97 11.87	IRAS00393+3632			.24	13.07				10403
003922.2+363149	PGC 2517		1.07	.08	13.52				
004207.7+332416		.S..6*.	1.00± .08	0	15.43 ±.12			15.92±.3	6245± 10
120.70 -29.43	UGC 443	U	.55± .06	.28					6411
328.59 11.20		6.0±1.3		.81				191± 7	5946
003926.1+330750	PGC 2520		1.03	.27	14.32			1.33	
004211.3+005414	NGC 219	CE...?.	.73± .14	60	15.32 ±.10				5492
118.02 -61.88	MCG 0- 2-128	E	.03± .05	.00					5580
296.73 3.33		-5.0±2.0		.00					5157
003937.4+003748	PGC 2522		.72		15.24				
004214.9-112820		.LAT-*.	1.13± .06	75					5911
114.64 -74.20	MCG -2- 2- 92	E	.24± .05	.15					5961
284.72 -.08		-3.0± .9		.00					5591
003943.5-114446	PGC 2525		1.11						
004214.9-180940	A 0039-18	.SBS6?/	1.43± .02	76					1555
109.17 -80.79	ESO 540- 16	SE (1)	.95± .03	.06	14.32 ±.14				1641± 60
278.21 -1.92		6.0± .6		1.39					1669
003945.0-182606	PGC 2526	5.6±1.3	1.43	.47	12.87				1335
004216.0+005041	NGC 223	PSBR0*.	1.10± .05	62	14.22 ±.12				5355± 50
118.05 -61.94	UGC 450	E	.17± .05	.03					5443
296.68 3.30		.0± .9		.13					5021
003942.1+003415	PGC 2527		1.09		13.99				
004222.0+293829	IC 43	.SX.5..	1.20± .03		13.94 ±.15	.70± .03	.78± .02	15.75±.1	4858± 6
120.58 -33.19	UGC 448	U (2)	.04± .05	.18	14.04 ±.15			187± 5	4915± 46
324.87 10.37	IRAS00396+2922	5.0± .8	.80± .02	.06	13.46	.62	13.43± .07	176± 5	5018
003941.5+292203	PGC 2536	3.7± .7	1.22	.02	13.72		14.69± .24	2.01	4550
004222.6+294150	IC 45	.S?....	.97± .07	51	15.35 ±.13			15.88±.3	5281± 10
120.58 -33.13	UGC 449		.60± .05	.18					5441
324.92 10.38				.88				242± 7	4973
003942.1+292524	PGC 2537		.99	.30	14.25			1.33	
004227.5-233746	NGC 230	.S.1?P/	1.03± .04	44	15.62 ±.14				
92.34 -85.96	ESO 474- 14	S	.75± .03	.03					
272.90 -3.45	IRAS00399-2354	1.0±1.9		.77	12.90				
003959.0-235412	PGC 2539		1.04	.38					
004227.7+331106		.I..9..	1.04± .15					15.68±.3	6263± 7
120.77 -29.65	UGC 446	U	.08± .12	.28					6429
328.38 11.09		10.0± .9		.06				158± 7	5964
003946.2+325440	PGC 2540		1.07	.04					

R.A. 2000 DEC. / l b / SGL SGB / R.A. 1950 DEC.	Names / PGC	Type / S_T n_L / T / L	$\log D_{25}$ / $\log R_{25}$ / $\log A_o$ / $\log D_o$	p.a. / A_g / A_i / A_{21}	B_T / m_B / m_{FIR} / B_T^o	$(B-V)_T$ / $(U-B)_T$ / $(B-V)_T^o$ / $(U-B)_T^o$	$(B-V)_o$ / $(U-B)_o$ / m'_o / m'_{25}	m_{21} / W_{20} / W_{50} / HI	V_{21} / V_{opt} / V_{GSR} / V_{3K}
0042.5 +4034	And IV								-373± 13
121.10 -22.27				.33					
335.75 12.47									-196
0039.8 +4018	PGC 2544								-650
004236.5-013136	NGC 227	.L..-P*	1.21± .05	155	13.11 ±.14	.96± .01	.98± .01		
117.84 -64.31	UGC 456	PUE	.11± .03	.10	13.30 ±.09	.45± .06			5305± 37
294.40 2.57		-3.4± .5	.73± .03	.00		.88	12.25± .11		5386
004003.0-014802	PGC 2547		1.20		13.07	.45	13.74± .29		4972
004240.2+295226		.S..6*.	1.07± .14	173				16.37±.3	4851± 10
120.67 -32.96	UGC 453	U	.79± .12	.22					
325.11 10.36		6.0±1.4		1.17				174± 7	5011
003959.6+293600	PGC 2553		1.09	.40					4543
004241.9+405155	NGC 221	CE.2...	1.94± .02	170	9.03M±.05	.95± .01	.96± .01		
121.15 -21.98	UGC 452	R	.13± .02	.33	9.08 ±.10	.48± .01			-205± 8
336.04 12.49	ARAK 12	-6.0± .3	1.08± .01	.00		.88	10.05± .04		-28
003957.8+403529	PGC 2555		1.96		8.71	.40	13.40± .11		-481
004244.4+411608	NGC 224	.SAS3..	3.28± .01	35	4.36M±.02	.92± .02		6.15±.1	-300± 4
121.17 -21.57	UGC 454	R (2)	.49± .02	.33	3.95 ±.10	.50± .02		536± 7	-295± 7
336.45 12.55		3.0± .3	2.62± .02	.67	5.66	.76	12.93± .06	510± 7	-121
004000.2+405943	PGC 2557	2.2± .5	3.31	.24	3.34	.34	14.37± .08	2.56	-574
004245.5-233340	NGC 232	.SBR1?P	.98± .04						
93.66 -85.93	ESO 474- 15	S	.09± .03	.06	14.43 ±.14				6673± 69
272.98 -3.50	VV 830	1.0± .9		.10	10.61				6682
004017.0-235006	PGC 2559		.98	.05	14.19				6382
004247.3+232921		.S?....	.82± .13					16.04±.3	7327± 10
120.37 -39.34	MCG 4- 2- 47		.46± .07	.05	15.75 ±.13				
318.82 8.92				.67				243± 10	7474
004008.3+231255	PGC 2562		.82	.23	14.98			.84	7006
0042.8 +2329	NGC 228	RSBR2..	1.08± .06						7314
120.38 -39.34	UGC 458	U	.05± .05	.05	14.57 ±.14				
318.82 8.91	IRAS00401+2313	2.0± .8		.07	13.19				7461
0040.1 +2313	PGC 2563		1.08	.03	14.38				6993
004252.5-233228	NGC 235	.L?....	1.12± .06	117	14.08 ±.17	.92± .02	.94± .01		
94.12 -85.92	ESO 474- 16		.28± .05	.06	14.24 ±.14	.23± .03			6664± 21
273.01 -3.52			.48± .06	.00		.82	11.97± .21		6673
004024.0-234854	PGC 2569		1.08		14.02	.23	13.83± .38		6373
004253.5-233240	NGC 235A	.L...P.	.55± .17						
94.15 -85.93	ESO 474- 17	S	.00± .04	.06	13.97 ±.14				6798± 69
273.01 -3.53		-2.0±1.2		.00					6807
004025.0-234906	PGC 2570		.55		13.81				6507
004253.7+333127		.S..4..	1.04± .08	6				15.91±.3	6245± 10
120.89 -29.31	UGC 457	U	.76± .06	.28	15.51 ±.13				5898± 52
328.74 11.07	MK 958	4.0±1.0		1.12	13.45			295± 7	6399
004012.0+331501	PGC 2571		1.07	.38	14.07			1.46	5935
004254.1+323446	NGC 226	.S?....	.96± .09					15.43±.3	4825± 10
120.85 -30.26	UGC 459		.00± .06	.27	14.31 ±.12				4907± 46
327.80 10.87	IRAS00402+3218			.00	13.20			150± 7	4993
004012.7+321821	PGC 2572		.99	.00	14.01			1.42	4528
004258.0+271502	IC 46	.L?....	.77± .11					16.02±.3	5287± 10
120.62 -35.58	CGCG 479- 63		.08± .10	.18	14.75 ±.12				
322.53 9.73				.00				230± 7	5442
004018.0+265837	PGC 2575		.78		14.49				4974
0043.0 -2215	IC 1574	.IBS9..	1.33± .04	175	14.32 ±.15	.61± .04	.59± .03	15.22±.2	358± 8
101.18 -84.76	ESO 474- 18	PSU (2)	.44± .04	.07	14.51 ±.14	.00± .05	-.01± .03	67± 11	
274.28 -3.21	DDO 226	9.7± .5	1.07± .02	.33		.49	15.16± .05	48± 8	371
0040.5 -2231	PGC 2578	9.5± .5	1.33	.22	14.03	-.09	14.70± .25	.97	63
004307.1-774017		.S..5*/	1.12± .05	9					
303.51 -39.45	ESO 12- 24	S	.84± .05	.29	16.09 ±.14				
218.84 -14.70	IRAS00413-7757	4.5±1.0		1.27					
004130.1-775642	PGC 2579		1.15	.42					
0043.2 -0912					16.0 ±.2	1.03± .05			
116.37 -71.98									
286.99 .32				.11					
0040.7 -0928	PGC 2582								
0043.3 +0000					16.85 ±.13	.55± .02			
118.51 -62.79						-.04± .05			
295.95 2.81				.00					24670± 72
0040.8 -0015	PGC 2591								24755
									24337

R.A. 2000 DEC.	Names	Type	logD$_{25}$	p.a.	B$_T$	(B-V)$_T$	(B-V)$_e$	m$_{21}$	V$_{21}$
l b	S$_T$ n$_L$		logR$_{25}$	A$_g$	m$_B$	(U-B)$_T$	(U-B)$_e$	W$_{20}$	V$_{opt}$
SGL SGB		T	logA$_e$	A$_I$	m$_{FIR}$	(B-V)$_T^o$	m'$_e$	W$_{50}$	V$_{GSR}$
R.A. 1950 DEC.	PGC	L	logD$_o$	A$_{21}$	B$_T^o$	(U-B)$_T^o$	m'$_{25}$	HI	V$_{3K}$

004322.0-551941		PSBR4*.	1.24± .05	5					7938
305.36 -61.76	ESO 150- 20	Sr (1)	.42± .04	.00	14.47 ±.14				7845
241.56 -11.24		3.7± .6		.61					7779
004106.0-553606	PGC 2592	2.2± .9	1.24	.21	13.81				
004323.7+504039		.S..6*.	1.04± .08	115				14.71±.1	5219± 11
121.63 -12.17	UGC 460	U	.18± .06	1.17	14.0 ±.3				5406
345.92 13.91	IRAS00405+5024	6.0±1.2		.26	13.03			241± 8	5406
004035.1+502414	PGC 2593		1.15	.09	12.58			2.04	4979
004325.9-384523		.LXS0P.	1.04± .05	39					11868
310.62 -78.25	ESO 295- 2	S	.13± .04	.02	15.07 ±.14				11826
258.09 -7.56		-2.0± .8		.00					11631
004102.0-390148	PGC 2594		1.02		14.87				
004326.0-501105	NGC 238	PSBR3P.	1.29± .04	93	13.14 ±.16	.64± .04	.79± .03		
306.20 -66.89	ESO 194- 31	Sr	.09± .05	.04	13.28 ±.14				8621± 34
246.72 -10.18	FAIR 663	3.1± .5	1.07± .03	.13		.56	13.98± .07		8543
004107.0-502730	PGC 2595		1.30	.05	12.99		14.22± .30		8436
004327.3+025725	NGC 236	.SXS5..	1.06± .04	160				15.92±.3	5643± 11
118.96 -59.85	UGC 462	UE (1)	.06± .04	.06	14.21 ±.12			135± 7	5648± 50
298.81 3.58	IRAS00409+0241	5.0± .5		.09	13.30			122± 7	5737
004052.9+024100	PGC 2596	3.1± .6	1.06	.03	14.03			1.85	5309
004328.2-000725	NGC 237	.SXT6..	1.20± .02	175	13.7 ±.2			15.02±.1	4175± 7
118.55 -62.93	UGC 461	R (3)	.24± .02	.02	13.46 ±.11			292± 10	4109± 66
295.82 2.75	IRAS00408-0023	6.0± .4		.35	12.35				4259
004054.5-002350	PGC 2597	2.4± .7	1.20	.12	13.13		13.94± .24	1.77	3841
004328.5-062058		.S..4./	1.26± .04	82					5831
117.37 -69.14	MCG -1- 2- 49	E	.72± .04	.20					5896
289.78 1.04		3.5± .6		1.06					5504
004056.1-063723	PGC 2598		1.28	.36					
004332.5+142035	NGC 234	.SXT5..	1.21± .05					15.47±.3	4452± 10
120.05 -48.48	UGC 463	U	.00± .05	.19	13.23 ±.12				4449± 50
309.90 6.53	IRAS00409+1404	5.0± .8		.00	11.65			203± 10	4577
004055.6+140410	PGC 2600		1.22	.00	13.02			2.44	4120
004334.4-081113	IC 48	.LXT0P?	.99± .04	25	13.98 ±.13	.85± .01			
116.95 -70.97	MCG -1- 3- 1	E	.09± .03	.16		.27± .02			6039
288.00 .52	IRAS00410-0827	-2.2± .7		.00	13.76	.75			6098
004102.5-082739	PGC 2603		1.00		13.73	.26	13.59± .27		5714
004336.6+303515	NGC 233	.E...?.	1.22± .08						
120.94 -32.25	UGC 464	U	.03± .05	.16	13.44 ±.10				5451± 36
325.86 10.31		-5.0±1.5		.00					5612
004055.6+301850	PGC 2604		1.24		13.19				5145
004345.2-242517		.SB?...	1.06± .05	43					6795
89.92 -86.79	ESO 474- 19		.23± .04	.02	15.86 ±.14				6801
272.20 -3.95				.34					6507
004117.1-244142	PGC 2611		1.07	.11	15.46				
004351.1+325111		.SB.1..	1.22± .05	178				16.39±.1	4772± 6
121.09 -29.99	UGC 465	U	.58± .05	.29	14.42 ±.12			426± 9	4758±118
328.11 10.73		1.0± .9		.59				412± 7	4937
004109.4+323446	PGC 2614		1.25	.29	13.48			2.62	4472
004351.9+004804		.SBS1P?	1.13± .04	165					5460
118.90 -62.01	UGC 466	E	.29± .04	.02	14.93 ±.15				
296.75 2.90		1.0± .9		.30					5547
004118.0+003139	PGC 2615		1.13	.15	14.55				5127
004354.3-041443		PS..5*.	1.18± .04	100					
118.11 -67.05	MCG -1- 3- 5	E (1)	.52± .04	.21					
291.85 1.52		4.5± .9		.78					
004121.5-043108	PGC 2616	4.2± .8	1.19	.26					
004356.1+015053	IC 49	.SXS5..	1.18± .04	95	14.2 ±.4	.50± .07		15.57±.1	4562± 10
119.07 -60.97	UGC 468	UE (2)	.07± .04	.03	14.06 ±.15			181± 11	4571± 59
297.77 3.17		5.0± .5		.11		.45		171± 11	4652
004121.9+013428	PGC 2617	3.9± .5	1.18	.04	13.92		14.77± .45	1.61	4228
004400.8-763523		PSBR3..	1.12± .06	85					
303.50 -40.53	ESO 29- 7	Sr	.15± .05	.24					7850± 63
219.96 -14.63	FAIR 213	2.8± .5		.21					7702
004220.1-765148	PGC 2619		1.14	.08					7809
0044.0 +2611		.S?....	1.17± .05	131					10081
120.87 -36.65	UGC 469		.67± .05	.12	15.14 ±.12				
321.54 9.25				1.01					10233
0041.4 +2555	PGC 2621		1.18	.34	13.96				9766

R.A. 2000 DEC.	Names	Type	logD$_{25}$	p.a.	B$_T$	(B-V)$_T$	(B-V)$_e$	m$_{21}$	V$_{21}$
l b	S$_T$ n$_L$	logR$_{25}$	A$_g$	m$_B$	(U-B)$_T$	(U-B)$_e$	W$_{20}$	V$_{opt}$
SGL SGB		T	logA$_e$	A$_i$	m$_{FIR}$	(B-V)$_T^o$	m'$_e$	W$_{50}$	V$_{GSR}$
R.A. 1950 DEC.	PGC	L	logD$_o$	A$_{21}$	B$_T^o$	(U-B)$_T^o$	m'$_{25}$	HI	V$_{3K}$

004404.8+285235		.S..8..	1.11± .14	140				16.67±.3	4828± 7
120.99 -33.97	UGC 467	U	.44± .12	.16					
324.19 9.84		8.0± .9		.54				117± 7	4985
004124.3+283610	PGC 2622		1.13	.22					4519
0044.2 +2650		.I..9*.	.99± .06	160					5190
120.94 -36.00	UGC 470	U	.40± .05	.19	14.82 ±.13				
322.20 9.37	IRAS00415+2634	10.0±1.3		.30	13.62				5343
0041.5 +2634	PGC 2627		1.01	.20	14.32				4876
0044.2 -1235		.S?....	1.04± .09						6493
115.95 -75.36	MCG -2- 3- 4		.66± .07	.09					6538
283.77 -.85	IRAS00417-1251			.82					6176
0041.7 -1251	PGC 2629		1.05	.33					
0044.3 -5950									
304.59 -57.26				.00					
237.03 -12.21	FAIR 7								
0042.1 -6007	PGC 2633								
004421.9-172106		.L..+P*	1.28± .07						
113.08 -80.09	ESO 540- 17A	SE	.53± .06	.03	14.93 ±.14				
279.14 -2.18		-1.0± .7		.00					
004152.1-173730	PGC 2634		1.20						
004421.9-172106		.LB?...	1.15± .05	128					9290± 22
113.08 -80.09	ESO 540- 17		.53± .03	.03	15.00 ±.14				9320
279.14 -2.18				.00					8984
004152.1-173730	PGC 2635		1.07		14.83				
004437.6-034535	NGC 239	PS..2..	1.00± .07	160					
118.65 -66.57	MCG -1- 3- 7	E	.33± .05	.23					
292.37 1.48	IRAS00420-0401	2.0± .9		.40	12.96				
004204.6-040159	PGC 2642		1.03	.16					
004441.9+452517		.SAS4..	1.19± .05	50					
121.69 -17.43	UGC 471	U	.07± .05	.49	15.0 ±.3				
340.67 12.90		4.0± .8		.10					
004155.4+450853	PGC 2645		1.23	.03					
0044.9 +1656		.I..9*.	1.14± .13	115					4666± 10
120.69 -45.90	UGC 472	U	.27± .12	.12					
312.52 6.86		10.0±1.2		.21					4797
0042.3 +1640	PGC 2649		1.15	.14					4337
004456.8+272657	A 0042+27								5400±110
121.16 -35.40				.17					5554
322.83 9.34	MK 346								5088
004216.5+271033	PGC 2650								
004459.6-562248		.LXT+..	1.15± .05	120					8400
304.76 -60.73	ESO 150- 21	S	.06± .04	.00	14.06 ±.14				8303
240.54 -11.66		-1.0± .8		.00					8247
004245.0-563912	PGC 2651		1.14		13.93				
004501.9-604536		.LA.0*/	1.17± .05	24					10500±108
304.34 -56.35	ESO 112- 9	S	.48± .03	.00	15.10 ±.14				10391
236.11 -12.46		-2.0± .8		.00					10370
004251.0-610200	PGC 2652		1.10		14.94				
0045.0 +0608	NGC 240	.S..0..	1.01± .09						
120.06 -56.70	UGC 473	U	.05± .06	.14	14.49 ±.13				
302.02 4.04		.0± .8		.04					
0042.5 +0552	PGC 2653		1.02						
004506.4+254654								17.45±.3	13724± 10
121.15 -37.07	CGCG 479- 68			.13	15.6 ±.4				
321.20 8.93								198± 7	13875
004226.5+253030	PGC 2655								13409
0045.3 +1031		.S..6*.	.96± .09	83					11973
120.46 -52.32	UGC 474	U	.98± .06	.23					
306.29 5.14		6.0±1.5		1.44					12087
0042.7 +1015	PGC 2659		.98	.49					11640
0045.5 +3802	And I	.E.3.P$							
121.65 -24.82		P		.22					
333.33 11.43		-5.0±1.8							
0042.8 +3746	PGC 2666								
004533.3-263354	IC 1579	.SBR4*.	.97± .05	6					
55.83 -88.57	ESO 474- 22	r	.18± .04	.03	14.99 ±.14				
270.21 -4.91		4.5±1.3		.28					
004306.0-265018	PGC 2667		.97	.09					

R.A. 2000 DEC.	Names	Type	logD$_{25}$	p.a.	B$_T$	(B-V)$_T$	(B-V)$_e$	m$_{21}$	V$_{21}$
l b		S$_T$ n$_L$	logR$_{25}$	A$_g$	m$_B$	(U-B)$_T$	(U-B)$_e$	W$_{20}$	V$_{opt}$
SGL SGB		T	logA$_e$	A$_i$	m$_{FIR}$	(B-V)$_T^o$	m'$_e$	W$_{50}$	V$_{GSR}$
R.A. 1950 DEC.	PGC	L	logD$_o$	A$_{21}$	B$_T^o$	(U-B)$_T^o$	m'$_{25}$	HI	V$_{3K}$
004546.3-153549 NGC 244		.L...P?	1.09± .06	50				15.51±.1	944± 7
116.13 -78.39	MCG -3- 3- 3	PE	.08± .04	.02				104± 4	957± 25
280.94 -2.03	VV 728	-1.7±1.0		.00				65± 5	979
004316.2-155213	PGC 2675		1.08						635
004551.9-311155		.SBR4..	1.09± .04	75					5799± 39
319.24 -85.75	ESO 411- 10	S (1)	.14± .03	.02 14.73 ±.14				5781
265.68 -6.16		4.0± .8		.20				5534
004326.0-312818	PGC 2679	1.1± .8	1.09	.07 14.47				
004554.8+251546		.S?....	.92± .11					16.00±.3	13714± 10
121.36 -37.59	MCG 4- 2- 53		.27± .07	.12 15.28 ±.12				
320.73 8.63				.20			419± 7	13863
004315.0+245923	PGC 2680		.92	14.75				13398
004557.9-203631		.S?....	.96± .06	98					4113±106
111.78 -83.36	ESO 540- 19		.12± .05	.02 14.83 ±.14				4131
276.07 -3.42				.17				3815
004329.0-205254	PGC 2683		.96	.06 14.61				
004600.8+295730 NGC 243		.L?....	.97± .15					17.74±.3	4787± 10
121.53 -32.90	MCG 5- 2- 43		.34± .07	.22 14.62 ±.11				4945
325.35 9.67				.00			263± 7	4481
004319.6+294107	PGC 2687		.94	14.33				
0046.0 -1130	A 0043-11	.SXS9..	1.30± .05		14.18 ±.19	.45± .06	.40± .04	14.42±.1	1618± 6
118.04 -74.33	MCG -2- 3- 9	PE (2)	.06± .05	.12	-.20± .06	-.22± .05	100± 7	1666
284.94 -.98	DDO 5	9.0± .5	1.10± .04	.07	.40	15.17± .09	81± 12	1301
0043.5 -1147	PGC 2689	8.8± .6	1.32	.03 13.99	-.24	15.39± .32	.40	
004605.6-014320 NGC 245		.SAT3P$	1.14± .04	145				15.35±.3	4074± 17
119.82 -64.56	UGC 476	R (2)	.06± .03	.07 13.00 ±.12			186± 34	4123± 57
294.45 1.68	MK 555	3.0± .8		.08 11.53			128± 25	4156
004332.2-015943	PGC 2691	3.6± .7	1.14	.03 12.82			2.50	3747
004609.7-553407		PSXT4?.	1.13± .04	89					
304.50 -61.54	ESO 150- 22	Sr (1)	.49± .04	.00 14.74 ±.14				
241.39 -11.67		3.6± .6		.73				
004355.0-555030	PGC 2693	4.4± .7	1.13	.25				
0046.2 +1929		.S..8*.	1.51± .04	167				13.76±.1	2649± 7
121.25 -43.36	UGC 477	U	.71± .06	.10 14.7 ±.2			250± 9	2662± 46
315.09 7.19		8.0±1.1		.88			232± 12	2786
0043.6 +1913	PGC 2699		1.52	.36 13.74			-.34	2325
004619.4+511304		.S..6*.	1.20± .05	105				15.48±.1	5271± 11
122.11 -11.65	UGC 475	U	.38± .05	1.19 14.1 ±.3				5457
346.54 13.53	IRAS00435+5056	6.0±1.2		.56			334± 8	5035
004329.5+505641	PGC 2704		1.31	.19 12.34			2.95	
004624.3-132634 IC 51		.L...P?	1.13± .06	30					
117.77 -76.26	MCG -2- 3- 11	E	.04± .05	.00				
283.08 -1.59	ARP 230	-2.0±1.7		.00 12.23				
004353.7-134257	PGC 2710		1.13					
004624.5+293723		.SB?...	.94± .11					17.25±.3	4969± 10
121.62 -33.24	MCG 5- 2- 45		.05± .07	.18 15.34 ±.15				5126
325.04 9.51				.07			184± 7	4663
004343.3+292100	PGC 2711		.95	.02 15.07			2.16	
004625.4+301418		.S..1..	1.13± .05	116				17.40±.3	4995± 10
121.65 -32.62	UGC 478	U	.71± .05	.22 14.62 ±.14				5008± 46
325.65 9.64		1.0± .9		.72			424± 7	5154
004344.0+295755	PGC 2712		1.15	.35 13.61			3.43	4691
004627.5+253712								18.05±.3	13549± 10
121.52 -37.24	CGCG 479- 71			.10 15.6 ±.4				
321.11 8.60							181± 7	13698
004347.5+252050	PGC 2714							13234
004628.6+314835		RSBS2..	1.04± .06	15				16.03±.3	6253± 10
121.70 -31.05	UGC 479	U	.22± .05	.25 15.03 ±.13				6414
327.20 9.97		2.0± .9		.27			359± 7	5952
004346.7+313213	PGC 2717		1.06	.11 14.44			1.48	
004632.0+361933		.S?....	1.19± .05						11173± 33
121.83 -26.54	UGC 480		.12± .05	.18 13.50 ±.12				11342
331.67 10.90	IRAS00438+3603			.18 12.57				10884
004348.7+360310	PGC 2720		1.20	.06 13.08				
0046.6 +0828		.S..5..	.96± .17						11848
120.88 -54.38	UGC 482	U	.05± .12	.19				
304.38 4.29		5.0± .9		.07				11956
0044.0 +0812	PGC 2721		.98	.02				11515

R.A. 2000 DEC. / l b / SGL SGB / R.A. 1950 DEC.	Names / PGC	Type / S_T n_L / T / L	$\log D_{25}$ / $\log R_{25}$ / $\log A_o$ / $\log D_o$	p.a. / A_g / A_i / A_{21}	B_T / m_B / m_{FIR} / B_T^o	$(B-V)_T$ / $(U-B)_T$ / $(B-V)_T^o$ / $(U-B)_T^o$	$(B-V)_o$ / $(U-B)_o$ / m'_o / m'_{25}	m_{21} / W_{20} / W_{50} / HI	V_{21} / V_{opt} / V_{GSR} / V_{3K}
004636.7+294322			.92± .11					17.08±.3	4816± 10
121.68 -33.14	MCG 5- 2- 47		.09± .07	.18	14.87 ±.12			173± 7	4974
325.15 9.49									4510
004355.5+292700	PGC 2722		.93						
004638.7+361953		.S?....	.89± .11	11					
121.85 -26.53	MCG 6- 2- 17		.24± .07	.18	15.45 ±.12				11093± 49
331.68 10.88				.29					11262
004355.3+360331	PGC 2726		.90	.12	14.87				10804
004639.6+314259								17.08±.3	5982± 10
121.74 -31.14	CGCG 501- 6			.25	15.5 ±.4			208± 7	6143
327.12 9.92									5681
004357.7+312637	PGC 2727								
004641.4-510001			.97± .06						9500±190
304.78 -66.11	ESO 194- 37		.16± .06	.04	15.46 ±.14				9419
246.01 -10.86	FAIR 665								9320
004424.0-511624	PGC 2732		.98						
004643.2-354749		.S?....	1.00± .05	122					6125± 52
309.25 -81.27	ESO 351- 5		.10± .05	.02	15.24 ±.14				6092
261.18 -7.48				.13					5878
004419.0-360412	PGC 2734		1.00	.05	15.03				
004648.8+294805								17.62±.3	5230± 10
121.73 -33.06	CGCG 500- 89			.22	15.2 ±.2			183± 7	5388
325.24 9.47									4924
004407.6+293143	PGC 2738								
004650.8+262828		.I..9?.	1.07± .07	69				16.14±.3	4956± 10
121.66 -36.39	UGC 483	U	.50± .08	.12	15.47 ±.13			250± 13	5107
321.97 8.71		10.0±1.8		.38					
004410.5+261206	PGC 2739		1.08	.25	14.96			.92	4643
004656.4+324027	A 0044+32	PSBS3..	1.38± .04	25				13.94±.3	4859± 10
121.84 -30.19	UGC 484	U	.47± .05	.20	13.86 ±.13				5086± 53
328.07 10.06	VV 441	3.0± .8		.65				426± 7	5029
004414.3+322405	PGC 2743		1.40	.24	12.97			.73	4568
004656.5-215050		.S?....	.95± .05	102					6564±106
111.73 -84.62	ESO 540- 21		.24± .04	.07	15.05 ±.14				6577
274.92 -3.97				.36					6270
004428.0-220712	PGC 2744		.95	.12	14.58				
004656.7+244529								17.09±.3	12677± 10
121.64 -38.10	CGCG 480- 3			.07	15.5 ±.4			135± 7	12824
320.29 8.29									12361
004416.8+242907	PGC 2745								
004702.4+302019		.S..6*.	1.34± .04	179				14.99±.3	5238± 10
121.81 -32.52	UGC 485	U	1.01± .05	.22	14.77 ±.12				5396
325.78 9.54		6.0±1.3		1.47				359± 7	4934
004420.9+300357	PGC 2747		1.36	.50	13.05			1.44	
004703.6-484308		.S..6*.	1.06± .05	102					14807
304.89 -68.39	ESO 194- 38	S (1)	.21± .04	.04	14.80 ±.14				14732
248.32 -10.45		6.0±1.2		.31					14616
004445.0-485930	PGC 2749	5.6±1.2	1.06	.10	14.39				
004703.7-520302		.S?....	.93± .06						8164± 71
304.53 -65.06	ESO 194- 39		.12± .06	.04	14.92 ±.14				8079
244.97 -11.13	FAIR 293			.13					7990
004447.0-521924	PGC 2750		.93	.06	14.65				
004703.7-520302		.SA.1?P	1.09± .07						8091± 76
304.53 -65.06	ESO 194- 39A	S	.44± .06	.04	15.46 ±.14				8006
244.97 -11.13		1.0± .7		.44					7916
004447.0-521924	PGC 2751		1.09	.22	14.87				
004703.8+280359								17.35±.3	5143± 10
121.76 -34.80	CGCG 500- 92			.18	15.4 ±.4			152± 7	5297
323.54 9.02									4833
004423.0+274737	PGC 2752								
004706.6-313450		.SBS9*.	1.09± .05	37					
314.63 -85.45	ESO 411- 13	S (1)	.47± .04	.02	15.90 ±.14				
265.37 -6.52		9.0±1.0		.47					
004441.0-315112	PGC 2753	8.9±1.2	1.09	.23					
004707.3-223538		.SXS9*.	1.02± .04						2850
110.51 -85.36	ESO 474- 25	SU (1)	.05± .04	.08	15.99 ±.14				2861
274.20 -4.21		8.8± .5		.05					
004439.1-225200	PGC 2754	10.0± .6	1.03	.03	15.85				2559

R.A. 2000 DEC. / l b / SGL SGB / R.A. 1950 DEC.	Names / / / PGC	Type / S_T n_L / T / L	$\log D_{25}$ / $\log R_{25}$ / $\log A_e$ / $\log D_o$	p.a. / A_g / A_i / A_{21}	B_T / m_B / m_{FIR} / B_T^o	$(B-V)_T$ / $(U-B)_T$ / $(B-V)_T^o$ / $(U-B)_T^o$	$(B-V)_e$ / $(U-B)_e$ / m'_e / m'_{25}	m_{21} / W_{20} / W_{50} / HI	V_{21} / V_{opt} / V_{GSR} / V_{3K}
004708.7-204538 / 113.95 -83.56 / 275.99 -3.73 / 004440.0-210200	NGC 247 / ESO 540- 22 / IRAS00446-2101 / PGC 2758	.SXS7.. / R (3) / 7.0± .3 / 6.8± .4	2.33± .01 / .49± .02 / 1.83± .02 / 2.34	174 / .07 / .68 / .25	9.67M±.07 / 9.66 ±.12 / 10.55 / 8.92	.56± .07 / / .45 /	.59± .02 / / 14.19± .05 / 14.95± .09	10.27±.1 / 234± 4 / 210± 3 / 1.11	159± 4 / / 176 / -136
004710.7+324129 / 121.89 -30.17 / 328.10 10.02 / 004428.4+322507	/ MCG 5- 3- 3 / / PGC 2761	.S?.... / / /	.64± .17 / .00± .07 / / .66	/ .20 / .00 / .00	/ 15.24 ±.15 / / 15.01			15.77±.3 / / 199± 7 / .76	5031± 10 / 5023± 57 / 5193 / 4732
0047.2 +0754 / 121.13 -54.95 / 303.88 3.99 / 0044.6 +0738	NGC 250 / UGC 487 / IRAS00446+0738 / PGC 2765	.S..0.. / U / .0± .9 /	1.06± .06 / .26± .05 / / 1.06	153 / .13 / .20 /	/ 14.63 ±.12 / 13.41 / 14.22				5259 / / 5365 / 4926
004718.8+274929 / 121.82 -35.04 / 323.32 8.92 / 004438.0+273307	IC 1584 / UGC 489 / / PGC 2766	.SB?... / / /	1.20± .06 / .06± .06 / / 1.21	/ .19 / .09 / .03	/ 14.50 ±.19 / / 14.20			16.36±.3 / / 319± 10 / 2.13	4764± 7 / / 4918 / 4454
004719.5+144212 / 121.44 -48.16 / 310.49 5.74 / 004442.3+142550	/ UGC 488 / MK 1146 / PGC 2768	.S..2.. / U / 2.0± .9 /	.95± .07 / .17± .05 / / .97	90 / .18 / .21 / .09	/ 14.94 ±.12 / / 14.43			16.34±.3 / 411± 12 / / 1.83	11787± 9 / 11622± 49 / 11905 / 11452
004727.6-312514 / 314.12 -85.62 / 265.55 -6.55 / 004502.1-314136	NGC 254 / ESO 411- 15 / / PGC 2778	RLXR+*. / PSr / -.7± .5 / 1.36	1.39± .04 / .21± .04 / .77± .02 /	137 / .02 / .00 /	12.62 ±.13 / 12.62 ±.11 / / 12.58	.88± .01 / .38± .01 / .83 / .36	.90± .01 / .39± .01 / 11.96± .06 / 13.94± .25	15.97±.3 / 279± 34 / 276± 25 /	1624± 17 / 1612± 39 / 1603 / 1359
004728.1+505301 / 122.29 -11.98 / 346.23 13.30 / 004438.0+503639	A 0044+50 / UGC 486 / / PGC 2781	.SXS3.. / U (1) / 3.0± .8 / 2.7± .9	1.06± .06 / .10± .05 / .85± .03 / 1.17	165 / 1.19 / .14 / .05	14.33 ±.15 / 14.1 ±.3 / / 12.93	.93± .04 / / .61 /	.89± .03 / / 14.07± .08 / 14.21± .35	15.39±.1 / 315± 11 / 307± 8 / 2.41	5170± 7 / 5206± 59 / 5356 / 4933
004728.3+313915 / 121.94 -31.21 / 327.09 9.73 / 004446.3+312253	/ MCG 5- 3- 5 / / PGC 2782	.S?.... / / /	.94± .11 / .19± .07 / / .96	/ .25 / .26 / .09	/ 15.43 ±.13 / / 14.88			16.25±.3 / / 174± 7 / 1.28	5072± 10 / / 5233 / 4771
004733.1-251718 / 97.36 -87.96 / 271.58 -5.01 / 004505.7-253340	NGC 253 / ESO 474- 29 / / PGC 2789	.SXS5.. / R (1) / 5.0± .3 / 3.3± .6	2.44± .01 / .61± .01 / 1.84± .04 / 2.45	52 / .05 / .91 / .30	8.04M±.05 / 7.72 ±.11 / 5.63 / 7.02		/ / 12.76± .08 / 13.59± .07	9.57±.1 / 418± 5 / 410± 4 / 2.24	251± 4 / 240± 15 / 251 / -32
004735.8-202538 / 115.26 -83.24 / 276.35 -3.74 / 004507.0-204200	A 0045-20A / ESO 540- 23 / / PGC 2791	.SBS3P? / PS / 3.0± .9 /	1.01± .04 / .42± .03 / / 1.02	21 / .10 / .58 / .21	/ 14.46 ±.14 / / 13.74				6130± 45 / 6148 / 5833
004737.7-203108 / 115.23 -83.33 / 276.26 -3.78 / 004509.0-204730	A 0045-20B / ESO 540- 25 / / PGC 2796	.SBS5P* / PS / 4.5± .6 /	1.01± .05 / .08± .04 / / 1.02	/ .07 / .13 / .04	/ 14.69 ±.14 / / 14.46				6351±106 / 6368 / 6054
004741.5-212926 / 114.12 -84.30 / 275.31 -4.05 / 004513.0-214548	/ ESO 540- 27 / IRAS00452-2145 / PGC 2799	.S?.... / / /	.80± .06 / .06± .04 / / .80	/ .06 / .08 / .03	/ 14.52 ±.14 / 11.85 / 14.32				6402±106 / 6416 / 6108
004746.5-095008 / 119.90 -72.69 / 286.69 -.93 / 004515.1-100630	/ MCG -2- 3- 15 / IRAS00452-1006 / PGC 2800	PSB.3*. / E (1) / 3.0±1.2 / 3.1±1.6	1.23± .05 / .57± .05 / / 1.24	15 / .14 / .79 / .29	/ / 13.61 /				
004747.4-112808 / 119.62 -74.32 / 285.10 -1.38 / 004516.4-114429	NGC 255 / MCG -2- 3- 17 / IRAS00452-1144 / PGC 2802	.SXT4.. / R (3) / 4.0± .3 / 3.5± .5	1.48± .02 / .08± .03 / 1.14± .03 / 1.49	15 / .10 / .11 / .04	12.36 ±.15 / 12.4 ±.2 / 12.24 / 12.14	.49± .04 / -.25± .08 / .44 / -.28	.58± .04 / / 13.55± .08 / 14.40± .20	13.54±.1 / 204± 5 / 144± 6 / 1.36	1600± 7 / 1888± 58 / 1650 / 1287
0047.8 -0953 / 119.93 -72.74 / 286.63 -.96 / 0045.3 -1010	A 0045-10 / MCG -2- 3- 16 / / PGC 2805	.SB.7P/ / PE / 6.7±1.0 /	1.48± .03 / .82± .05 / / 1.49	97 / .14 / 1.13 / .41				14.39±.1 / 151± 8 / 137± 12 /	1345± 6 / / 1396 / 1025
0047.8 +1934 / 121.77 -43.29 / 315.27 6.84 / 0045.2 +1918	NGC 251 / UGC 490 / / PGC 2806	.S..5.. / U / 5.0± .7 /	1.38± .04 / .10± .05 / / 1.38	105 / .06 / .15 / .05	/ 13.9 ±.2 / / 13.68				4597± 46 / 4732 / 4273
004756.3+222228 / 121.87 -40.49 / 318.01 7.50 / 004517.0+220607	IC 1586 / CGCG 480- 6 / MK 347 / PGC 2813	/ / /	/ / /	/ .07 / /	/ 14.9 ±.4 / 13.15 /			16.68±.3 / 200± 6 / /	5821± 7 / 5825± 49 / 5963 / 5501

R.A. 2000 DEC. / l b / SGL SGB / R.A. 1950 DEC.	Names / PGC	Type / S_T n_L / T / L	$\log D_{25}$ / $\log R_{25}$ / $\log A_o$ / $\log D_o$	p.a. / A_g / A_i / A_{21}	B_T / m_B / m_{FIR} / B_T^o	$(B-V)_T$ / $(U-B)_T$ / $(B-V)_T^o$ / $(U-B)_T^o$	$(B-V)_e$ / $(U-B)_e$ / m'_o / m'_{25}	m_{21} / W_{20} / W_{50} / HI	V_{21} / V_{opt} / V_{GSR} / V_{3K}
004801.6+081748	NGC 257	.S..6*.	1.27± .04	105	13.3 ±.2	.69± .03	.77± .01	14.78±.2	5268± 7
121.48 -54.57	UGC 493	U	.15± .05	.19	13.40 ±.13	.08± .04	.17± .02	414± 10	5248± 39
304.30 3.91	IRAS00454+0801	6.0±1.1	.87± .06	.22	12.11	.58	13.09± .15	393± 8	5374
004525.9+080127	PGC 2818		1.29	.08	12.93	.00	14.12± .32	1.77	4935
004801.8+273724	NGC 252	RLAR+*.	1.18± .05	80					
122.01 -35.24	UGC 491	PU	.14± .04	.19	13.35 ±.10				5023± 46
323.16 8.72	IRAS00453+2721	-1.0± .6		.00					5176
004521.0+272103	PGC 2819		1.18		13.08				4713
004803.3-024635	NGC 259	.S..4*/	1.45± .03	41				13.58±.3	3811± 8
120.88 -65.63	MCG -1- 3- 15	PE (1)	.66± .04	.18				452± 9	
293.56 .92	IRAS00455-0302	4.0± .6		.97	12.08			426± 7	3885
004530.2-030256	PGC 2820	3.1±1.2	1.47	.33					3483
004805.0-783551		.S.1?P/	1.09± .05	141					
303.14 -38.53	ESO 13- 2	S	.58± .04	.29	15.32 ±.14				
217.92 -15.05	IRAS00466-7851	1.0±1.8		.59					
004640.0-785212	PGC 2821		1.12	.29					
004805.2-013359		.E+..*/	1.08± .08	128					
120.99 -64.43	UGC 492	UE	.24± .04	.08	14.28 ±.10				
294.74 1.24	ARAK 14	-3.7± .7		.00					
004531.8-015020	PGC 2822		1.02						
004814.4+275914								16.88±.3	7025± 10
122.07 -34.88	CGCG 501- 19			.19	15.6 ±.4				7178
323.52 8.75								273± 7	
004533.5+274253	PGC 2827								6716
004815.9+262801								17.30±.3	5259± 10
122.05 -36.40	CGCG 480- 8			.12	15.6 ±.4				5409
322.03 8.40								181± 7	
004535.4+261140	PGC 2828								4947
004821.0-381403	NGC 264	.LB..*/	1.02± .05	113					
306.08 -78.88	ESO 295- 6	BS	.53± .03	.02	14.42 ±.14				
258.84 -8.37		-2.0±1.2		.00					
004558.0-383024	PGC 2831		.95						
0048.3 +0720		.S..1..	1.06± .06	175					
121.60 -55.52	UGC 495	U	.47± .05	.12	15.36 ±.13				
303.40 3.57		1.0± .9		.48					
0045.8 +0704	PGC 2833		1.07	.23					
004823.8+040532	IC 52	.S?....	.98± .07	97					1958
121.47 -58.77	UGC 494		.37± .05	.07	15.17 ±.12				
300.25 2.70				.55					2052
004549.0+034911	PGC 2834		.99	.18	14.55				1626
004834.4+102019									10763± 97
121.77 -52.53	CGCG 435- 7			.26	15.21 ±.12				10875
306.32 4.31	MK 1147								
004558.1+100358	PGC 2843								10432
004834.9+274127	NGC 260	.S..6P*	.97± .06					15.84±.1	5208± 7
122.16 -35.18	UGC 497	PU	.00± .04	.25	14.23 ±.12			197± 6	
323.25 8.61	IRAS00458+2725	6.0± .7		.00	13.42			163± 7	5361
004554.0+272506	PGC 2844		.99	.00	13.96			1.88	4898
004835.5-124301	A 0046-12	RLB.0P?	.93± .11	15				16.15±.1	6407± 10
120.14 -75.57	MCG -2- 3- 19	PE	.14± .05	.09				220± 6	6361± 51
283.94 -1.91	MK 960	-2.0±1.2		.00	12.87				6447
004604.8-125922	PGC 2845		.92						6091
004844.8-114601		.SBT3?.	1.11± .06	60					5744
120.45 -74.63	MCG -2- 3- 20	E (1)	.27± .05	.10					
284.87 -1.69		3.0±1.7		.37					5789
004613.9-120222	PGC 2853	5.3± .8	1.11	.14					5428
004846.7+315720	NGC 262	.SAS0*.	1.03± .08		13.90 ±.19	.84± .04	.91± .02	14.38±.1	4507± 4
122.27 -30.91	UGC 499	R	.00± .04	.26	14.68 ±.14	.14± .06	.23± .04	91± 7	4719± 46
327.45 9.53	MK 348	.0± .4	.91± .06	.00	12.89	.74	13.94± .21	66± 5	4669
004604.4+314100	PGC 2855		1.05		14.07	.11	13.89± .47		4209
0048.8 +1122		.SX.4..	1.06± .06						12008
121.90 -51.49	UGC 500	U	.02± .05	.23	15.01 ±.20				
307.34 4.53		4.0± .8		.02					12122
0046.2 +1106	PGC 2857		1.08	.01	14.68				11677
004853.8-022245	A 0046-02	.I..9*.	.46± .16	85	15.06 ±.13	.70± .01			
121.42 -65.24		E	.08± .06	.19		.05± .02			4001± 35
294.00 .83	MK 557	10.0±1.8		.06	13.19	.61			4076
004620.6-023905	PGC 2861		.48	.04	14.81	-.02	12.01± .80		3673

R.A. 2000 DEC.	Names	Type	logD$_{25}$	p.a.	B$_T$	(B-V)$_T$	(B-V)$_e$	m$_{21}$	V$_{21}$
l b		S$_T$ n$_L$	logR$_{25}$	A$_g$	m$_B$	(U-B)$_T$	(U-B)$_e$	W$_{20}$	V$_{opt}$
SGL SGB		T	logA$_e$	A$_i$	m$_{FIR}$	(B-V)$_T^o$	m'	W$_{50}$	V$_{GSR}$
R.A. 1950 DEC.	PGC	L	logD$_o$	A$_{21}$	B$_T^o$	(U-B)$_T^o$	m'$_{25}$	HI	V$_{3K}$
004902.2+281307		.S..6*.	1.22± .06	110				15.72±.2	5090± 7
122.29 -34.65	UGC 501	U	.94± .06	.21	15.17 ±.12			400± 13	
323.79 8.64	IRAS00463+2756	6.0±1.3		1.38	13.06			385± 6	5243
004621.0+275647	PGC 2865		1.24	.47	13.56			1.69	4782
0049.0 +0431		.I..9..	.96± .09						12199
121.81 -58.34	UGC 503	U	.05± .06	.09					
300.71 2.65		10.0± .9		.04					12294
0046.5 +0415	PGC 2867		.97	.02					11867
0049.1 +4611		.S..9*.	1.11± .14						
122.51 -16.68	UGC 502	U	.04± .12	.61					
341.57 12.27		9.0±1.2		.04					
0046.3 +4555	PGC 2869		1.17	.02					
004918.4-492840		PSAT3..	.95± .07	35					9700±190
303.84 -67.65	ESO 195- 3	r	.14± .06	.04	14.84 ±.14				9622
247.63 -10.97	FAIR 666	3.3± .9		.19					9514
004701.0-494500	PGC 2876		.95	.07	14.54				
0049.4 +1211		.S?....	1.07± .14						11299
122.15 -50.68	UGC 504		.11± .12	.19					
308.18 4.60				.16					11415
0046.8 +1155	PGC 2882		1.09	.05					10969
0049.4 -0144		.SBT4*.	1.12± .04	135					3856
121.78 -64.61	UGC 505	E (1)	.15± .04	.12	14.63 ±.15				
294.66 .87		4.0± .8		.23					3932
0046.9 -0201	PGC 2883	3.1±1.2	1.14	.08	14.26				3528
004934.3+233440		.S?....	.94± .11		15.18S±.15				16014± 41
122.38 -39.29	MCG 4- 3- 8		.11± .07	.07					16157
319.28 7.43	4ZW 32			.17					15697
004654.5+231820	PGC 2886		.94	.06	14.85		14.43± .58		
004934.8-465234		PSAT3..	.99± .06	10					8855
303.87 -70.25	ESO 243- 2	r	.03± .05	.03	14.84 ±.14				8785
250.26 -10.49		3.3± .9		.05					8657
004716.1-470854	PGC 2887		.99	.02	14.69				
004935.4+233528		.S?....	.52± .20		16.24S±.15				15966± 41
122.38 -39.28	MCG 4- 3- 7		.46± .07	.07					16109
319.29 7.43	HICK 8B			.68					15649
004655.6+231908	PGC 2888		.52	.23	15.40		12.54±1.01		
004935.5+010658		.S..6*/	1.28± .04	128					5277
121.96 -61.75	UGC 507	UE	.90± .04	.02	14.99 ±.12				
297.44 1.61	IRAS00470+0050	5.5± .9		1.33	13.02				5362
004701.5+005038	PGC 2889		1.28	.45	13.62				4947
004936.0+233502		.S?....	.34?		16.14S±.15				17087± 41
122.39 -39.29	MCG 4- 3- 9		.11± .07	.07					17230
319.28 7.42	HICK 8C			.17					16770
004656.2+231842	PGC 2890		.34	.06	15.80		12.39±1.31		
004936.9+233422		.S?....			16.24S±.15				16341± 41
122.39 -39.30				.07					16484
319.27 7.42	HICK 8D								16024
004657.0+231803	PGC 2892								
004939.2+225556		.S?....	1.15± .07						7461± 50
122.40 -39.94	UGC 506		.17± .06	.09	14.25 ±.13				7603
318.65 7.25				.17					7143
004659.5+223936	PGC 2894		1.16	.09	13.90				
004946.4+313600		.S..6*.	1.11± .07	35				16.31±.3	5136± 10
122.52 -31.27	UGC 509	U	.44± .06	.23					5295
327.15 9.24		6.0±1.3		.65				213± 7	
004704.0+311940	PGC 2899		1.13	.22					4836
004948.2+321643	NGC 266	.SBT2..	1.47± .03		12.54 ±.17	.91± .04	.99± .02	14.72±.2	4661± 6
122.53 -30.59	UGC 508	U	.01± .05	.23	12.33 ±.13	.44± .05	.53± .03	502± 13	4682± 36
327.82 9.39	IRAS00471+3200	2.0± .7	1.15± .05	.01	12.88	.82	13.78± .17	475± 6	4823
004705.6+320023	PGC 2901		1.49	.01	12.12	.38	14.71± .26	2.60	4364
004949.4-210058	A 0047-21	.IBS9*.	1.23± .04	42	15.19 ±.15	.43± .05	.38± .05	15.79±.1	301± 5
119.39 -83.88	ESO 540- 31	SU (2)	.37± .04	.07	15.40 ±.14	-.06± .07	-.08± .07	39± 7	
275.91 -4.40	DDO 6	10.0± .6	.97± .03	.28		.32	15.53± .07	23± 5	315
004721.0-211718	PGC 2902	9.6± .7	1.24	.19	14.96	-.14	15.25± .26	.65	7
004953.6+252423								18.64±.3	15158± 10
122.49 -37.46	CGCG 480- 12			.14	15.7 ±.4				
321.08 7.79								329± 7	15305
004713.2+250803	PGC 2908								14845

R.A. 2000 DEC. l b SGL SGB R.A. 1950 DEC.	Names PGC	Type S_T n_L T L	$\log D_{25}$ $\log R_{25}$ $\log A_e$ $\log D_o$	p.a. A_g A_i A_{21}	B_T m_B m_{FIR} B_T^o	$(B-V)_T$ $(U-B)_T$ $(B-V)_T^o$ $(U-B)_T^o$	$(B-V)_e$ $(U-B)_e$ m'_e m'_{25}	m_{21} W_{20} W_{50} HI	V_{21} V_{opt} V_{GSR} V_{3K}
004954.1-393816 304.30 -77.49 257.52 -8.98 004732.0-395436	ESO 295- 10 PGC 2909	RSBR1.. Sr 1.0± .6 	1.19± .04 .50± .04 1.19	9 .02 .51 .25	14.18 ±.14 13.56				7253 7206 7023
0049.9 +2142 122.47 -41.16 317.47 6.89 0047.3 +2126	UGC 510 PGC 2914	.S..6*. U 6.0±1.3 	1.10± .05 .57± .05 1.10	20 .07 .83 .28	15.08 ±.12 14.15				4536 4675 4216
005003.5-033331 122.07 -66.43 292.94 .23 004730.7-034951	MCG -1- 3- 16 PGC 2918	.SXT4.. E (1) 4.0± .9 5.3±1.2	1.05± .06 .10± .05 1.08	80 .25 .15 .05					
005003.9-663310 303.15 -50.57 230.30 -13.89 004803.0-664930	ESO 79- 7 IRAS00480-6649 PGC 2919	.LA.0P. S -2.0± .8 	1.21± .06 .09± .06 .33± .02 1.19	178 .00 .00 	15.45 ±.14 13.72 ±.14 13.53 14.56	.63± .03 .07± .04 .61 .07	.71± .03 .16± .04 12.59± .06 16.13± .37		1653 1527 1556
0050.1 +0755 122.35 -54.95 304.08 3.31 0047.5 +0739	UGC 512 PGC 2922	.S..8*. U 8.0±1.2 	1.04± .08 .23± .06 1.06	135 .20 .28 .11	14.78 ±.13 14.29				5392 5496 5061
005009.5-051139 122.08 -68.06 291.36 -.24 004737.0-052758	NGC 268 MCG -1- 3- 17 IRAS00476-0527 PGC 2927	.SBS4*. PE (3) 4.0± .6 2.4± .5	1.20± .03 .14± .04 .80± .01 1.22	95 .22 .20 .07	13.56 ±.13 13.4 ±.2 12.33 13.05	.51± .02 -.18± .04 .40 -.26	.59± .02 -.08± .04 13.05± .03 14.06± .23		5485± 66 5550 5160
005010.2+314352 122.62 -31.14 327.30 9.19 004727.8+312733	UGC 511 PGC 2928	.S..6*. U 6.0±1.3 	1.22± .06 .65± .06 1.24	105 .23 .95 .32	15.29 ±.14 14.09			15.80±.3 282± 7 1.39	4591± 10 4750 4292
005012.2+242942 122.57 -38.38 320.21 7.51 004732.0+241323	MCG 4- 3- 11 PGC 2930	.SB?... 	.78± .13 .09± .07 .79	 .11 .11 .04	15.49 ±.12 15.18			15.95±.3 260± 10 .73	10103± 10 10248 9788
005021.3-411435 303.77 -75.88 255.93 -9.43 004800.0-413054	ESO 295- 12 FAIR 1070 PGC 2932	.S?.... 	.98± .06 .14± .05 .98	162 .03 .21 .07	14.62 ±.14 14.34				7050± 20 6997 6827
005024.6-195423 121.01 -82.77 277.04 -4.24 004756.0-201042	ESO 540- 32 PGC 2933	.IXS9P* SE (1) 10.0± .6 10.9± .8	1.22± .05 .21± .04 1.23	0 .12 .16 .11	16.63 ±.14 				
0050.4 +1143 122.53 -51.15 307.79 4.24 0047.8 +1127	UGC 513 PGC 2934	.S?.... 	1.00± .08 .25± .06 1.02	8 .21 .38 .13	15.30 ±.14 14.65				11901 12015 11571
005025.6+243114 122.64 -38.35 320.24 7.46 004745.4+241455	MCG 4- 3- 12 ARAK 15 PGC 2935	 	.64± .17 .11± .07 .65	 .11 	15.23 ±.18 			16.03±.3 200± 10 	10135± 10 10200± 49 10283 9823
005027.6-055132 122.26 -68.73 290.73 -.49 004755.3-060751	MCG -1- 3- 18 PGC 2936	PLAR-?. E -3.0± .8 	1.17± .10 .00± .07 1.20	 .22 .00 					
005032.5-083908 122.23 -71.52 288.02 -1.27 004800.9-085527	NGC 270 MCG -2- 3- 27 PGC 2938	.L..+.. E -1.0± .8 	1.24± .05 .05± .05 1.25	25 .19 .00 					
0050.6 -1517 121.94 -78.16 281.56 -3.07 0048.1 -1534	MCG -3- 3- 5 PGC 2943	.SXS8?. E (1) 8.0±1.7 7.5±1.6	1.18± .05 .36± .05 1.18	80 .03 .45 .18					5709 5742 5401
0050.6 +1206 122.61 -50.77 308.17 4.29 0048.0 +1150	UGC 515 PGC 2946	.S..8*. U 8.0±1.2 	.98± .07 .10± .05 .99	 .18 .12 .05					11799 11914 11470
005041.4-521315 303.20 -64.91 244.90 -11.70 004826.3-522934	 PGC 2948	RLBR+*. S -1.0± .7 	1.05± .11 .06± .08 1.04	 .04 .00 					

R.A. 2000 DEC.	Names	Type	logD$_{25}$	p.a.	B$_T$	(B-V)$_T$	(B-V)$_e$	m$_{21}$	V$_{21}$
l b		S$_T$ n$_L$	logR$_{25}$	A$_g$	m$_B$	(U-B)$_T$	(U-B)$_e$	W$_{20}$	V$_{opt}$
SGL SGB		T	logA$_e$	A$_i$	m$_{FIR}$	(B-V)$_T^o$	m'$_e$	W$_{50}$	V$_{GSR}$
R.A. 1950 DEC.	PGC	L	logD$_o$	A$_{21}$	B$_T^o$	(U-B)$_T^o$	m'$_{25}$	HI	V$_{3K}$
005042.0-015432	NGC 271	PSBT2..	1.33± .03	130					
122.50 -64.78	UGC 519	UE	.10± .04	.17	12.91 ±.12				4098± 50
294.58 .52	IRAS00481-0210	1.5± .5		.13					4173
004808.8-021051	PGC 2949		1.35	.05	12.58				3770
0050.7 +1036	IC 53	.L..-*.	1.06± .11						
122.64 -52.27	UGC 516	U	.01± .05	.22	14.86 ±.15				
306.72 3.88		-3.0±1.1		.00					
0048.1 +1020	PGC 2951		1.08						
005047.7-553629		.S.2?P/	1.28± .04	154					
303.12 -61.52	ESO 150- 24	S	.96± .05	.00	15.43 ±.14				
241.47 -12.32		1.5±1.2		1.19					
004835.0-555248	PGC 2958		1.28	.48					
005048.5-065310	NGC 273	.L..../	1.34± .03	80	13.87 ±.14	.94± .02	1.03± .02		
122.48 -69.76	MCG -1- 3- 19	R	.52± .04	.23		.45± .03	.55± .03		
289.76 -.85		-2.0± .4	.66± .03	.00			12.66± .11		
004816.5-070929	PGC 2959		1.28				14.11± .23		
005049.6+300946		.SBR5..	.98± .07					16.44±.3	10384± 10
122.77 -32.71	UGC 518	U	.04± .05	.23	14.92 ±.14				10540
325.79 8.70		5.0± .9		.06				139± 7	10081
004807.5+295327	PGC 2960		1.00	.02	14.57			1.85	
005051.1+284159								16.23±.3	4992± 10
122.78 -34.17	CGCG 501- 24			.19	15.6 ±.4				5145
324.36 8.36								217± 7	4686
004809.5+282540	PGC 2964								
005053.8-552929		PSBR1..	.99± .06	3					
303.09 -61.64	ESO 150- 25	r	.34± .05	.00	15.24 ±.14				
241.59 -12.32		1.0± .9		.35					
004841.0-554548	PGC 2971		.99	.17					
005101.9-070327	NGC 274	.LXR-P.	1.18± .04	155				13.58±.1	1750± 11
122.64 -69.93	MCG -1- 3- 21	R	.01± .03	.23	12.93 ±.19			338± 11	1765± 24
289.61 -.95		-3.0± .4		.00				266± 12	1811
004830.0-071945	PGC 2980		1.20		12.67				1431
005104.4-070356	NGC 275	.SBT6P.	1.19± .03	40				13.75±.0	1744± 5
122.67 -69.94	MCG -1- 3- 22	R	.14± .03	.23	13.2 ±.2			298± 4	1784± 33
289.60 -.97	IRAS00485-0720	6.0± .4		.21	11.39			264± 5	1804
004832.5-072015	PGC 2984		1.21	.07	12.73			.95	1423
005104.6-322517		.S.R0..	1.07± .05	163					
303.76 -84.71	ESO 351- 11	r	.48± .03	.02	14.69 ±.14				9643
264.75 -7.54		-.1± .9		.36					9619
004840.0-324136	PGC 2985		1.05		14.17				9385
0051.2 +1201		.I..9..	.98± .07	50					659
122.84 -50.85	UGC 521	U	.15± .05	.16	15.03 ±.13				
308.13 4.13		10.0± .9		.11					774
0048.6 +1145	PGC 2992		.99	.07	14.76				330
005117.2-083550	NGC 277	.L..-..	1.14± .10	50					
122.81 -71.47	MCG -2- 3- 28	E	.06± .07	.19					4411
288.13 -1.43		-3.0± .9		.00					4465
004845.7-085208	PGC 2995		1.16						4092
0051.3 -0800		.SXS8?/	1.07± .09	80					
122.85 -70.88		E (1)	.38± .08	.16					
288.70 -1.28		8.0±1.3		.46					
0048.8 -0817	PGC 2999	7.5±1.2	1.09	.19					
005123.0-083108		.SBT7..	1.13± .06	65					4229
122.89 -71.39	MCG -2- 3- 29	E (1)	.28± .05	.26					
288.21 -1.43		7.0± .9		.39					4283
004851.4-084726	PGC 3004	6.4± .8	1.16	.14					3910
005126.9+404329		.S..3..	.99± .05	65	15.3 ±.2	.11± .10		16.89±.1	5855± 11
122.93 -22.15	UGC 522	U (1)	.13± .04	.36	14.50 ±.12				6028
336.22 10.84	VV 554	3.0± .9		.18	13.85	-.04		188± 8	5582
004840.8+402711	PGC 3011	3.5±1.2	1.03	.07	14.12		14.79± .34	2.71	
0051.4 -0307		.SB?...	1.08± .09						3980
122.94 -66.00	MCG -1- 3- 23		.42± .07	.23					
293.45 .01				.61					4051
0048.9 -0324	PGC 3012		1.10	.21					3654
005130.0-125042	IC 56	.SXR6*.	.92± .08						
122.99 -75.72	MCG -2- 3- 30	E (1)	.03± .05	.06					6090± 96
284.00 -2.62	KAZ 3	6.0± .9		.04					6130
004859.6-130700	PGC 3014	5.3± .7	.92	.01					5778

R.A. 2000 DEC. l　　b SGL　SGB R.A. 1950 DEC.	Names PGC	Type S_T　n_L T L	$\log D_{25}$ $\log R_{25}$ $\log A_e$ $\log D_o$	p.a. A_g A_i A_{21}	B_T m_B m_{FIR} B_T^o	$(B-V)_T$ $(U-B)_T$ $(B-V)_T^o$ $(U-B)_T^o$	$(B-V)_e$ $(U-B)_e$ m'_e m'_{25}	m_{21} W_{20} W_{50} HI	V_{21} V_{opt} V_{GSR} V_{3K}
005134.8+292403 122.97 -33.47 325.08　8.37 004852.9+290745	UGC　　524 PGC　3019	PSBS3.. U 3.0± .9	.95± .07 .00± .05 .97	.20 .00 .00	14.40 ±.12 14.12			16.56±.2 272± 7 255± 5 2.44	10779± 7 10781± 46 10934 10475
005135.0+294251 122.97 -33.16 325.39　8.44 004853.0+292633	UGC　　525 PGC　3020	.SB?... 	1.18± .04 .31± .03 1.20	150 .20 .46 .15	15.13S±.15 15.05 ±.18 14.41	.75± .06 .03± .09 .61 -.08	 15.11± .26	16.07±.2 237± 13 206± 6 1.51	4931± 7 5086 4628
0051.6　+5206 122.97 -10.77 347.55 12.84 0048.8　+5150	UGC　　523 PGC　3023	.S..2.. U 2.0± .9	1.04± .08 .37± .06 1.19	50 1.63 .46 .19	 14.8　±.3 				
0051.6　+0311 123.05 -59.68 299.59　1.67 0049.1　+0255	UGC　　526 PGC　3024	.S..8*. U 8.0±1.3	.96± .09 .39± .06 .97	94 .13 .48 .20					5418 5508 5088
005142.5+223108 123.01 -40.35 318.36　6.69 004902.8+221450	CGCG 480- 16 PGC　3027	 	.95? .95? .96	 .09 	 15.89 ±.16 			16.32±.3 372± 7 	7296± 10 7436 6979
0051.7　+0305 123.10 -59.78 299.50　1.61 0049.2　+0249	UGC　　527 PGC　3031	.S..9*. U 9.0±1.2	.97± .09 .17± .06 .99	15 .13 .18 .09					1951 2040 1621
005151.3-124604 123.34 -75.64 284.10 -2.69 004920.9-130221	IC　　56A KAZ　　4 PGC　3035	.SB.9?. E 9.0±1.8	.59± .18 .15± .08 .60	145 .06 .15 .08					 12570± 96 12610 12258
0051.9　-1624 123.64 -79.28 280.55 -3.69 0049.5　-1641	A　0049-16 MCG -3- 3- 6 PGC　3042	.SXR4*. E　　(1) 4.0± .8 3.0±1.0	1.00± .07 .00± .05 1.01	5 .06 .00 .00					
005159.8-002912 123.24 -63.36 296.05　.60 004926.2-004529	MCG　0- 3- 18 ARAK　18 PGC　3043	 	.64± .17 .28± .07 .29± .01 .65	 .15 	15.25 ±.13 15.3　±.2 	.58± .01 -.16± .02 	.53± .01 -.11± .02 12.19± .03 		 1725± 40 1804 1397
005202.1-425248 302.53 -74.25 254.36 -10.08 004942.0-430906	ESO 243-　7 PGC　3046	.SB.5?. S　　(1) 5.0±1.3 5.6±1.3	1.06± .05 .37± .04 1.06	95 .03 .55 .18	 15.38 ±.14 				
005204.6+473300 123.04 -15.32 343.02 12.01 004914.9+471643	NGC　　278 UGC　　528 IRAS00492+4716 PGC　3051	.SXT3.. R　　(1) 3.0± .3 3.9± .8	1.32± .03 .02± .05 .79± .01 1.39	 .78 .02 .01	11.47 ±.13 11.59 ±.18 9.68 10.70	.64± .01 -.02± .02 .45 -.15	.67± .01 10.91± .02 12.86± .22	13.58±.1 143± 5 100± 7 2.87	641± 9 622± 40 821 391
005206.6-224049 124.93 -85.55 274.42 -5.34 004939.0-225706	NGC　　276 ESO 474- 34 IRAS00496-2257 PGC　3054	.SXR3?P S 3.4± .6	1.01± .05 .40± .03 1.02	90 .04 .55 .20	15.75 ±.14 15.05				 13846± 52 13853 13558
005209.2-021304 123.36 -65.09 294.38　.09 004936.1-022921	NGC　　279 UGC　　532 MK　　558 PGC　3055	PLXR+P* UE -1.3± .5	1.21± .05 .12± .04 1.20	5 .15 .00	13.66 ±.10 12.69 13.45				 3889± 46 3962 3562
005212.3+294033 123.13 -33.20 325.38　8.30 004930.2+292416	UGC　　529 ARAK　17 PGC　3057	.S..0.. U .0±1.0	.94± .07 .61± .05 .93	93 .20 .45	14.6　±.2 13.89				5564± 46 5718 5260
005213.9+441953 123.08 -18.54 339.82 11.39 004925.9+440336	UGC　　530 PGC　3058	.S..1.. U 1.0±1.0	1.02± .06 .60± .05 1.06	0 .49 .61 .30	15.44 ±.12 				
005217.3-035804 123.47 -66.84 292.69　-.42 004944.5-041421	MCG -1- 3- 27 PGC　3062	.S..7?. E　　(1) 7.0±1.7 6.4±1.2	1.17± .05 .45± .05 1.19	165 .23 .62 .22					
005225.4-651337 302.76 -51.90 231.69 -13.96 005024.0-652954	ESO　79-　7A PGC　3070	.L?.... 	1.25± .07 .39± .06 1.19	80 .00 .00					6577 6454 6473

R.A. 2000 DEC.	Names	Type	logD$_{25}$	p.a.	B$_T$	(B-V)$_T$	(B-V)$_e$	m$_{21}$	V$_{21}$
l b		S$_T$ n$_L$	logR$_{25}$	A$_g$	m$_B$	(U-B)$_T$	(U-B)$_e$	W$_{20}$	V$_{opt}$
SGL SGB		T	logA$_e$	A$_i$	m$_{FIR}$	(B-V)$_T^o$	m'$_e$	W$_{50}$	V$_{GSR}$
R.A. 1950 DEC.	PGC	L	logD$_o$	A$_{21}$	B$_T^o$	(U-B)$_T^o$	m'$_{25}$	HI	V$_{3K}$
0052.4 +1431		.S..6*.	1.18± .05	120					5475
123.29 -48.35	UGC 533	U	.42± .05	.18	14.95 ±.15				
310.63 4.50		6.0±1.2		.62					5595
0049.8 +1415	PGC 3072		1.20	.21	14.13				5148
0052.4 +2608		.SBS3..	1.00± .06	145					14797
123.22 -36.73	UGC 535	U	.24± .05	.12	15.24 ±.13				
321.94 7.40		3.0± .9		.34					14944
0049.8 +2552	PGC 3075		1.01	.12	14.67				14486
0052.5 +2420	NGC 280	.SB?...	1.23± .06	95					10169
123.24 -38.52	UGC 534		.18± .06	.13	14.23 ±.15				
320.19 6.96	IRAS00498+2404			.27	13.64				10313
0049.8 +2404	PGC 3076		1.24	.09	13.78				9855
005235.0+291234		.S?....	.96± .09	50				17.30±.3	4863± 10
123.23 -33.66	UGC 536		.29± .06	.20					
324.94 8.11				.41				218± 7	5017
004953.0+285617	PGC 3081		.98	.15					4559
0052.5 +0601		.I..9..	.96± .09						5202
123.45 -56.85	UGC 537	U	.00± .06	.15					
302.40 2.21		10.0± .9		.00					5299
0050.0 +0545	PGC 3082		.97	.00					4872
005238.1-724801	SMC	.SBS9P.	3.50± .02		2.70S±.10	.45± .03		3.00±.3	175± 7
302.81 -44.33	ESO 29- 21	R (1)	.23± .04	.18		-.20± .07			190± 49
223.91 -14.81		9.0± .3		.24	3.48	.36			34
005053.1-730418	PGC 3085	7.0± .7	3.52	.12	2.28	-.27	14.48± .17	.60	114
0052.6 +0603		.S..6*.	1.00± .08						
123.50 -56.82	UGC 538	U	.33± .06	.15	15.35 ±.13				
302.44 2.19		6.0±1.3		.49					
0050.1 +0547	PGC 3087		1.01	.17					
005242.2-311222	NGC 289	.SBT4..	1.71± .02	130	11.72 ±.13	.73± .01	.81± .01	11.78±.1	1631± 5
299.13 -85.91	ESO 411- 25	R (3)	.15± .03	.02	11.39 ±.12	.11± .03	.20± .02	308± 6	1690± 41
266.04 -7.58	VV 484	4.0± .3	1.04± .02	.22	11.18	.69	12.41± .04	272± 6	1611
005017.5-312839	PGC 3089	2.2± .5	1.72	.07	11.29	.08	14.77± .17	.41	1371
005245.2-835126		.L..-/*	1.14± .05	110					
302.89 -33.27	ESO 2- 10	S	.24± .04	.56	14.24 ±.14				
212.51 -15.56		-3.3± .7		.00					
005236.1-840742	PGC 3094		1.17						
005245.6-314307		.SBS9..	1.09± .05	29					1610
299.43 -85.40	ESO 411- 26	S (1)	.46± .05	.02	15.90 ±.14				1580± 60
265.53 -7.72		9.0± .9		.47					1557
005021.0-315924	PGC 3095	8.9± .9	1.09	.23	15.41				1321
005252.1-271931		.S?....	1.06± .07	123					
244.83 -89.63	ESO 411- 27		.68± .05	.06	16.16 ±.14				1919±108
269.89 -6.68				1.01					1911
005026.1-273548	PGC 3100		1.07	.34	15.08				1645
005253.6+415814		.S..6*.	1.00± .06						
123.22 -20.90	UGC 539	U	.02± .05	.33	14.49 ±.12				
337.51 10.82		6.0±1.2		.03					
005006.5+414157	PGC 3103		1.04	.01					
005258.1+290157		.S?....	.91± .04	137	14.17S±.15	.49± .02		15.71±.2	4982± 7
123.33 -33.84	UGC 540		.24± .03	.19	14.30 ±.17	-.19± .03		266± 13	5096± 45
324.79 7.99	ARAK 19			.36	12.72	.37		251± 6	5138
005016.0+284540	PGC 3108		.92	.12	13.66	-.28	12.95± .27	1.94	4680
0053.1 +2155		.S..6*.	1.04± .08	102					7296
123.46 -40.95	UGC 541	U	.76± .06	.09					
317.87 6.22		6.0±1.4		1.12					7434
0050.5 +2139	PGC 3120		1.05	.38					6979
005313.3-130953	NGC 283	.S..5..	1.21± .05	160					
124.73 -76.03	MCG -2- 3- 31	E (1)	.23± .05	.02					
283.81 -3.11	IRAS00507-1326	5.0± .8		.34					
005043.1-132609	PGC 3124	1.9± .8	1.21	.11					
005324.2-560320		.E+1.*.	.95± .06	123					
302.36 -61.07	ESO 151- 3	S	.17± .04	.00	15.26 ±.14				
241.08 -12.76		-4.0± .9		.00					
005113.0-561936	PGC 3130		.90						
005326.2+291610		.S?....	1.31± .04	160	14.22S±.15			14.44±.2	4508± 6
123.45 -33.60	UGC 542		.70± .03	.19	14.61 ±.12			370± 13	
325.05 7.94				1.05				368± 6	4662
005044.0+285954	PGC 3133		1.33	.35	13.18		13.89± .25	.91	4205

R.A. 2000 DEC. / l b / SGL SGB / R.A. 1950 DEC.	Names / PGC	Type / S_T n_L / T / L	$\log D_{25}$ / $\log R_{25}$ / $\log A_\bullet$ / $\log D_0$	p.a. / A_g / A_i / A_{21}	B_T / m_B / m_{FIR} / B_T^0	$(B-V)_T$ / $(U-B)_T$ / $(B-V)_T^0$ / $(U-B)_T^0$	$(B-V)_\bullet$ / $(U-B)_\bullet$ / $m'_\bullet$ / m'_{25}	m_{21} / W_{20} / W_{50} / HI	V_{21} / V_{opt} / V_{GSR} / V_{3K}
0053.4 +0546	IC 1592	.S?....	1.00± .08	165					5189
123.87 -57.10	UGC 543		.14± .06	.12	15.00 ±.14				
302.22 1.93				.21					5285
0050.9 +0530	PGC 3139		1.01	.07	14.65				4860
005329.9-084609	NGC 291	PSBR1*.	1.05± .06	45					
124.55 -71.63	MCG -2- 3- 35	E	.31± .05	.23					
288.11 -2.00	IRAS00510-0901	1.0± .9		.31	12.13				
005058.5-090225	PGC 3140		1.08	.15					
005330.5-130646	NGC 286	.LXS0P?	1.11± .06	175					
125.01 -75.98	MCG -2- 3- 34	E	.17± .05	.02					
283.88 -3.17		-2.0±1.2		.00					
005100.3-132302	PGC 3142		1.09						
005331.3-580632	IC 1597	PSBT3P*	1.22± .04	151					
302.40 -59.02	ESO 112- 10	Sr	.70± .04	.00	15.01 ±.14				5053
238.99 -13.09		3.1± .8		.96					4949
005122.0-582248	PGC 3144		1.22	.35	14.01				4911
005332.8+025536		.S..5*/	1.19± .04	42					4907
123.98 -59.94	UGC 544	UE	.84± .04	.09	15.43 ±.12				
299.47 1.15	IRAS00509+0239	5.0± .9		1.26	13.53				4995
005058.3+023919	PGC 3147		1.20	.42	14.05				4578
005333.7+213043	A 0050+21								
123.59 -41.36				.09					6932± 60
317.49 6.03	MK 349								7069
005054.0+211427	PGC 3149								6614
005335.1+124139		.S?....	.74± .12		14.36 ±.13	.39± .01		17.25±.2	18313± 7
123.75 -50.17	UGC 545		.08± .06	.20		-.74± .03		365± 11	18205± 25
308.93 3.75	1ZW 1			.11	12.47	.21		305± 19	18420
005057.9+122523	PGC 3151		.76	.04	13.99	-.87	12.71± .63	3.22	17978
0053.5 -6153					14.30V±.14	1.02± .03			
302.49 -55.23				.00		.45± .05			
235.12 -13.65	FAIR 8								
0051.5 -6210	PGC 3152								
0053.6 +1804		.S..8*.	1.06± .06	5					4508± 10
123.67 -44.80	UGC 546	U	.35± .05	.15	15.29 ±.14				
314.15 5.13		8.0±1.3		.43					4637
0051.0 +1748	PGC 3157		1.07	.17	14.70				4186
005344.0-270302		.S?....	1.10± .05	18					
204.42 -89.48	ESO 474- 39		.71± .04	.07	14.73 ±.14				5455± 76
270.21 -6.80	IRAS00512-2719			1.07	12.57				5447
005118.0-271918	PGC 3159		1.11	.36	13.57				5180
005345.1-473850	IC 1594	.SXT3?.	1.10± .05	130					
301.82 -69.48	ESO 195- 12	Sr (1)	.53± .04	.03	15.18 ±.14				
249.62 -11.33		2.6± .6		.73					
005128.0-475506	PGC 3161	3.9± .9	1.10	.27					
005346.4-451108	IC 1595	.S..3?/	1.15± .05	12					
301.60 -71.94	ESO 243- 8	S	.87± .04	.03	15.38 ±.14				7444
252.11 -10.86		3.0±1.9		1.20					7378
005128.0-452724	PGC 3162		1.16	.43	14.10				7240
005348.8-331744		.S?....	1.08± .05	5					
298.33 -83.81	ESO 351- 18		.44± .05	.02	15.27 ±.14				5403± 39
264.02 -8.30				.65					5375
005125.0-333400	PGC 3165		1.08	.22	14.57				5150
005354.6-310544		.LX.-*.	1.07± .06						
295.34 -86.00	ESO 411- 28	S	.08± .04	.02	13.94 ±.14				9604± 39
266.21 -7.81		-3.0± .7		.00					9582
005130.0-312200	PGC 3169		1.06		13.78				9343
0053.9 +2446		.L.....	1.15± .07	28					
123.65 -38.10	UGC 547	U	.56± .03	.09	15.05 ±.10				
320.68 6.75	IRAS00512+2430	-2.0± .9		.00	13.15				
0051.2 +2430	PGC 3171		1.08						
005405.7+302048		.S?....	.95± .09					17.15±.2	6774± 7
123.61 -32.52	CGCG 501- 35		.48± .10	.28	15.30 ±.13			293± 13	
326.13 8.06				.72				293± 6	6928
005123.1+300433	PGC 3184		.98	.24	14.26			2.66	6473
005414.1-622721		.SBS4..	1.12± .04	150	14.58 ±.13	.74± .02	.82± .02		
302.37 -54.67	ESO 79- 8	Sr (1)	.21± .04	.00	14.59 ±.14				10615
234.56 -13.80		4.2± .5	.79± .02	.31		.63	14.02± .03		10499
005210.0-624336	PGC 3190	1.1± .9	1.12	.11	14.20		14.51± .27		10497

R.A. 2000 DEC. l b SGL SGB R.A. 1950 DEC.	Names PGC	Type S_T n_L T L	logD25 logR25 logAe logDo	p.a. Ag Ai A21	BT mB mFIR BT^o	(B-V)T (U-B)T (B-V)T^o (U-B)T^o	(B-V)e (U-B)e m' m'25	m21 W20 W50 HI	V21 Vopt VGSR V3K
0054.2 -0212 124.60 -65.07 294.53 -.42 0051.7 -0229	 PGC 3193	.S..7P/ E (1) 7.0±1.8 7.5±1.6	1.12± .08 .58± .08 1.14	173 .19 .80 .29					
005416.0-071410 124.99 -70.10 289.65 -1.78 005144.3-073025	NGC 293 MCG -1- 3- 30 IRAS00517-0730 PGC 3195	PSBT3.. E (1) 3.0± .9 3.6± .8	1.05± .06 .10± .05 1.07	145 .25 .13 .05	 13.38 				
005416.2-233208 133.18 -86.35 273.70 -6.04 005149.2-234823	 HICK 9B PGC 3196	.L?....	 .05 		15.78S±.15 				
005420.6-233310 133.49 -86.36 273.69 -6.06 005153.6-234925	 MCG -4- 3- 28 IRAS00518-2349 PGC 3201	.S?....	.82± .13 .17± .07 .82	 .05 .25 .09	15.27S±.15 13.67 14.88		 13.78± .68 		20155± 38 20158 19870
005423.1+313948 123.66 -31.20 327.43 8.30 005139.9+312333	 UGC 548 PGC 3203	.S..6*. U 6.0±1.2	1.10± .05 .06± .05 1.11	 .16 .08 .03	 14.47 ±.14 14.20			15.72±.3 244± 7 1.49	6606± 10 6763 6309
005426.6-480127 301.52 -69.10 249.26 -11.52 005210.0-481742	 ESO 195- 13 PGC 3205	.SAR2.. r 2.2± .9	.95± .07 .22± .05 .95	166 .03 .27 .11	 15.27 ±.14 				
005428.3-043601 124.90 -67.46 292.23 -1.11 005155.8-045216	 MCG -1- 3- 31 PGC 3207	.S..6*/ E (1) 6.0±1.3 5.3± .8	1.17± .05 .59± .05 1.19	15 .23 .87 .29					5616 5680 5293
005435.9+421632 123.56 -20.59 337.87 10.57 005148.2+420017	 MCG 7- 3- 5 5ZW 40 PGC 3212	.SB?...	.82± .13 .00± .07 .85	 .33 .00 .00	 15.39 ±.13 15.00				5813± 82 5987 5547
005441.4+245218 123.87 -37.99 320.82 6.61 005200.6+243603	 MCG 4- 3- 16 PGC 3215	.S?....	.82± .13 .08± .07 .83	 .09 .11 .04	 15.36 ±.12 15.08			16.46±.3 119± 7 1.34	13511± 10 13655 13199
005441.8+054627 124.42 -57.09 302.30 1.64 005206.5+053012	IC 1598 UGC 553 MK 962 PGC 3217	.S..1.. U 1.0± .9	.99± .06 .32± .05 1.00	2 .13 .32 .16	 14.83 ±.12 14.31				4452± 52 4547 4124
005442.0+364554 123.66 -26.10 332.45 9.39 005156.8+362939	 UGC 549 PGC 3218	.S..6*. U 6.0±1.5	1.04± .05 .97± .05 1.06	0 .20 1.43 .49	 15.79 ±.14 14.14			16.31±.3 277± 13 1.69	6035± 10 6201 5751
005443.2+213115 123.95 -41.34 317.57 5.77 005203.4+211500	IC 1596 UGC 550 PGC 3219	.S?....	1.26± .04 .47± .05 1.27	120 .10 .71 .24	 14.66 ±.15 13.84			15.37±.1 235± 8 1.30	2675± 6 2811 2358
005446.7+312153 123.77 -31.50 327.16 8.15 005203.5+310538	 UGC 557 PGC 3222	.SB?...	1.04± .04 .38± .03 1.06	37 .21 .56 .19	15.06S±.15 14.81 ±.12 14.12	.46± .04 -.21± .06 .31 -.32	 14.19± .25 	15.67±.2 296± 13 255± 6 1.37	4499± 7 4655 4201
005447.7+103205 124.28 -52.33 306.92 2.89 005211.1+101550	 UGC 558 PGC 3225	.L...*. U -2.0±1.2	1.22± .04 .43± .03 1.18	37 .19 .00 	 15.09 ±.15 14.73				11731 11840 11403
0054.8 +2843 123.82 -34.15 324.58 7.52 0052.1 +2827	 UGC 554 PGC 3226	.S..2.. U 2.0± .9	1.00± .06 .51± .05 1.02	125 .20 .63 .26	 15.47 ±.12 				
0054.8 +2851 123.82 -34.01 324.71 7.55 0052.1 +2835	 UGC 555 PGC 3227	.S..0.. U .0±1.0	.97± .07 .66± .05 .96	28 .20 .49 	 15.08 ±.16 				
0054.8 +1150 124.25 -51.03 308.18 3.23 0052.2 +1134	IC 57 UGC 559 PGC 3229	.L...?. U -2.0±1.7	1.01± .08 .03± .03 1.02	 .18 .00 	 15.11 ±.15 				

R.A. 2000 DEC. l b SGL SGB R.A. 1950 DEC.	Names PGC	Type S_T n_L T L	$\log D_{25}$ $\log R_{25}$ $\log A_e$ $\log D_o$	p.a. A_g A_i A_{21}	B_T m_B m_{FIR} B_T^o	$(B-V)_T$ $(U-B)_T$ $(B-V)_T^o$ $(U-B)_T^o$	$(B-V)_e$ $(U-B)_e$ m'_e m'_{25}	m_{21} W_{20} W_{50} HI	V_{21} V_{opt} V_{GSR} V_{3K}
0054.8 +4445 123.56 -18.11 340.33 11.02 0052.0 +4429	 UGC 551 PGC 3230	.I..9*. U 10.0±1.2 	1.11± .14 .29± .12 1.15	120 .45 .22 .15					
0054.8 +1341 124.19 -49.18 309.98 3.72 0052.2 +1325	 UGC 560 PGC 3232	.S..9*. U 9.0±1.2 	1.00± .16 .04± .12 1.01	 .11 .04 .02					5434 5551 5108
005450.4+291444 123.82 -33.62 325.09 7.64 005208.0+285829	 UGC 556 IRAS00521+2858 PGC 3235	.S?.... 	1.02± .04 .31± .03 1.04	100 .18 .46 .15	15.12S±.15 15.10 ±.12 11.25 14.44	.87± .02 .29± .04 .74 .18	 14.30± .26	15.68±.2 406± 8 380± 6 1.09	4629± 5 4782 4327
005453.8-374057 299.20 -79.42 259.69 -9.50 005232.1-375712	NGC 300 ESO 295- 20 IRAS00523-3756 PGC 3238	.SAS7.. R (3) 7.0± .3 6.2± .3	2.34± .01 .15± .02 2.35	111 .02 .21 .08	8.72S±.05 9.00 ±.12 9.43 8.53	.59± .03 .11± .03 .56 .09	 14.91± .09	9.15±.2 150± 5 149± 3 .55	142± 4 88± 63 98 -93
0054.9 +5230 123.47 -10.36 348.04 12.42 0052.0 +5214	 UGC 552 PGC 3239	.S..6*. U 6.0±1.2 	1.14± .13 .39± .12 1.29	162 1.59 .58 .20	 14.6 ±.3 12.40			16.12±.1 287± 8 3.53	5165± 11 5349 4937
005455.1-320151 294.37 -85.04 265.33 -8.24 005231.0-321806	 ESO 411- 29 PGC 3242	.L..-*P/ S -1.0± .8 	1.13± .07 .43± .05 1.07	178 .02 .00 	14.9 ±.3 14.98 ±.14 14.81	1.00± .04 .85 	 14.35± .49		9607± 22 9582 9350
005457.0-320115 294.28 -85.05 265.34 -8.24 005233.0-321730	 ESO 411- 30 IRAS00525-3217 PGC 3245	.SB.2?P S 2.0± .7 	.84± .06 .15± .04 .84	103 .02 .18 .07	 14.77 ±.14 11.59 				
005458.7-353045 298.01 -81.58 261.86 -9.04 005236.1-354700	 ESO 351- 20 PGC 3246	.S?.... 	.94± .07 .27± .06 .94	167 .02 .34 .14	 15.40 ±.14 14.91				12901 12865 12657
005459.4+304818 123.83 -32.06 326.63 7.98 005216.4+303203	 UGC 561 PGC 3247	.S?.... 	1.14± .07 .32± .06 1.16	165 .22 .48 .16				15.13±.3 379± 7 	4684± 7 4840 4385
005459.7-351915 297.86 -81.77 262.06 -9.00 005237.0-353530	 ESO 351- 21 PGC 3248	.LA.-.. S -3.0± .9 	.97± .06 .03± .04 .97	 .02 .00 	 14.66 ±.14 14.38				17254 17218 17009
005501.4-072135 125.56 -70.21 289.58 -1.99 005229.7-073749	NGC 298 MCG -1- 3- 33 PGC 3250	.S..6*/ E (1) 6.0±1.2 4.2± .8	1.23± .05 .58± .05 1.25	87 .25 .86 .29					1757 1813 1438
0055.0 -0822 125.69 -71.23 288.59 -2.26 0052.5 -0839	 PGC 3251	.SBS9.. E (1) 9.0± .9 8.7± .8	1.06± .09 .14± .08 1.09	135 .33 .14 .07					
005503.4-190021 128.96 -81.84 278.21 -5.07 005235.0-191636	A 0052-19 ESO 541- 1 IRAS00525-1916 PGC 3252	.SBT4.. ESr (2) 3.8± .4 2.1± .4	1.31± .03 .19± .03 1.32	178 .06 .28 .09	 13.96 ±.14 13.88 13.58				6295± 52 6313 5999
005507.9+313229 123.85 -31.32 327.35 8.12 005224.6+311615	NGC 295 UGC 562 IRAS00523+3116 PGC 3260	.SB.3*. U 3.0± .8 	1.34± .04 .36± .05 1.36	164 .21 .49 .18	 13.39 ±.12 11.77 12.51			15.05±.2 460± 13 395± 7 	5456± 7 5477± 36 5613 5159
005512.7-435458 300.58 -73.20 253.44 -10.86 005254.1-441112	 ESO 243- 11 PGC 3263	PSXT1.. Sr 1.0± .6 	1.02± .05 .17± .04 1.02	48 .03 .18 .09	 14.62 ±.14 14.33				7056 6993 6847
005514.1+352603 123.80 -27.43 331.17 8.99 005229.3+350948	 UGC 564 PGC 3265	.S?.... 	.96± .09 .15± .06 .97	175 .15 .22 .07	 15.38 ±.14 14.94				11042 11205 10755
0055.2 -0101 125.10 -63.88 295.75 -.34 0052.7 -0118	 UGC 568 PGC 3266	.L?.... 	1.11± .16 .15± .08 1.11	 .22 .00 	 15.08 ±.16 14.65				13600± 69 13675 13275

R.A. 2000 DEC.	Names	Type	logD$_{25}$	p.a.	B$_T$	(B-V)$_T$	(B-V)$_e$	m$_{21}$	V$_{21}$
l b		S$_T$ n$_L$	logR$_{25}$	A$_g$	m$_B$	(U-B)$_T$	(U-B)$_e$	W$_{20}$	V$_{opt}$
SGL SGB		T	logA$_e$	A$_i$	m$_{FIR}$	(B-V)^o_T	m'$_e$	W$_{50}$	V$_{GSR}$
R.A. 1950 DEC.	PGC	L	logD$_o$	A$_{21}$	B^o_T	(U-B)^o_T	m'$_{25}$	HI	V$_{3K}$

0055.3 +0927		.S..6*.	1.00± .08						
124.53 -53.41	UGC 569	U	.14± .06	.15					
305.91 2.48		6.0±1.2		.20					
0052.7 +0911	PGC 3268		1.01	.07					
005518.4+314347		.SA.8..	1.07± .08					14.96±.3	6301± 7
123.89 -31.13	UGC 566	U	.00± .06	.16	15.1 ±.2				
327.54 8.13		8.0± .8		.00				219± 7	6458
005235.0+312733	PGC 3269		1.09	.00	14.97			-.01	6004
0055.3 +3144		.L.....	1.05± .08	62					
123.90 -31.13	UGC 567	U	.42± .03	.16	14.79 ±.10				
327.55 8.12		-2.0± .9		.00					
0052.6 +3128	PGC 3271		1.00						
0055.3 -0114									13390± 69
125.17 -64.10	CGCG 384- 28			.22	15.1 ±.3				13464
295.54 -.42									13065
0052.8 -0131	PGC 3272								
005521.5+314038	NGC 296	.S..6*.	1.04± .06	148	15.7 ±.4	.74± .04		16.40±.2	5647± 7
123.91 -31.19	UGC 565	U	.55± .05	.24	15.30 ±.12	-.01± .06		261± 13	
327.50 8.10		6.0±1.3		.81		.55		228± 6	5804
005238.1+312424	PGC 3274		1.06	.27	14.27	-.16	14.37± .51	1.85	5350
005523.1+302921								17.67±.3	5149± 10
123.94 -32.37	CGCG 501- 43			.22	15.7 ±.4				5304
326.34 7.82								171± 7	
005240.2+301307	PGC 3275								4849
005533.0+302354								16.74±.2	6627± 7
123.98 -32.46	CGCG 501- 47			.22	15.7 ±.4			282± 13	
326.26 7.76								223± 6	6781
005250.1+300740	PGC 3285								6327
005534.7-240916	IC 1601	.S?....	.93± .05	118					3641± 56
140.58 -86.88	ESO 474- 44		.34± .04	.04	14.50 ±.14				3642
273.17 -6.49	IRAS00531-2425			.50					3359
005308.1-242530	PGC 3287		.94	.17	13.93				
005542.5-303228		.SXS9?P	.98± .05						1563
287.87 -86.46	ESO 411- 31	S (1)	.01± .05	.02	15.26 ±.14				
266.85 -8.05		9.0±1.6		.01					1543
005318.0-304842	PGC 3290	10.0±1.2	.99	.01	15.23				1302
005556.6-594004		.S..3*/	1.14± .04	127					
301.88 -57.45	ESO 112- 11	S	.59± .04	.00	15.16 ±.14				
237.44 -13.63		3.0±1.0		.81					
005350.0-595618	PGC 3309		1.14	.29					
0055.9 -0053		.E...*.	1.04± .18	10					13270± 69
125.49 -63.75	UGC 570	U	.18± .08	.13					13345
295.92 -.47		-5.0±1.2		.00					12945
0053.4 -0110	PGC 3312		1.01						
0056.0 +1206	NGC 305	.SBS3..	1.00± .06	135					12321
124.70 -50.75	UGC 571	U	.07± .05	.16	14.96 ±.15				
308.52 3.02		3.0± .9		.10					12433
0053.4 +1150	PGC 3313		1.02	.04	14.60				11995
0056.0 +1413		.S..6*.	1.03± .06	140					11928
124.62 -48.64	UGC 572	U	.11± .05	.15	15.10 ±.16				
310.57 3.58		6.0±1.2		.16					12046
0053.4 +1357	PGC 3314		1.04	.06	14.74				11603
0056.0 +0424		.I..9*.	1.07± .14						5277
125.14 -58.45	UGC 574	U	.03± .12	.08					
301.07 .94		10.0±1.2		.02					5368
0053.5 +0408	PGC 3322		1.08	.01					4950
005606.5+240729	NGC 304	.S?....	1.06± .08	175					4980± 46
124.30 -38.73	UGC 573		.21± .06	.17	14.01 ±.13				5121
320.18 6.11				.25					4667
005325.7+235116	PGC 3326		1.08	.10	13.54				
005607.5-531122		.L..0./	1.13± .05	161					
301.34 -63.92	ESO 151- 4	S	.56± .03	.04	14.85 ±.14				
244.07 -12.68		-2.0± .7		.00					
005355.0-532736	PGC 3328		1.05						
005608.8-524946		PSXS0..	.99± .06	173					
301.29 -64.28	ESO 151- 5	r	.19± .03	.04	15.01 ±.14				
244.43 -12.63		-.1± .9		.15					
005356.0-530600	PGC 3330		.99						

R.A. 2000 DEC. / l b / SGL SGB / R.A. 1950 DEC.	Names / PGC	Type / S_T n_L / T / L	$\log D_{25}$ / $\log R_{25}$ / $\log A_e$ / $\log D_o$	p.a. / A_g / A_i / A_{21}	B_T / m_B / m_{FIR} / B_T^o	$(B-V)_T$ / $(U-B)_T$ / $(B-V)_T^o$ / $(U-B)_T^o$	$(B-V)_e$ / $(U-B)_e$ / m'_e / m'_{25}	m_{21} / W_{20} / W_{50} / HI	V_{21} / V_{opt} / V_{GSR} / V_{3K}
005609.6+310430					.19 15.6 ±.4			16.54±.3	4655± 10
124.12 -31.79	CGCG 501- 48								
326.95 7.80								291± 7	4810
005326.3+304817	PGC 3332								4357
0056.2 +3105		.S..4..	.96± .09	161				16.83±.3	4666± 10
124.14 -31.77	UGC 575	U	.98± .06	.19				304± 13	
326.96 7.79		4.0±1.0		1.44				283± 10	4821
0053.5 +3049	PGC 3336		.98	.49					4368
005614.6-470353		RLAR+?.	.96± .06	57					6300±190
300.53 -70.04	ESO 243- 12	r	.16± .05	.03	14.95 ±.14				6227
250.29 -11.64	FAIR 669	-1.3± .9		.00					6105
005358.0-472006	PGC 3338		.94		14.82				
0056.2 -0105				.13	15.21 ±.12				12565± 69
125.67 -63.94	MCG 0- 3- 31								12639
295.75 -.60									12240
0053.7 -0122	PGC 3340								
005616.1-011520		.E.....	1.16± .07	40	14.39 ±.15	1.05± .01	1.09± .01		
125.70 -64.10	UGC 579	UE	.08± .04	.08	14.36 ±.14		.60± .10		13246± 45
295.60 -.64		-4.5± .6	.79± .03	.00		.91	13.83± .09		13319
005342.8-013134	PGC 3342		1.15		14.09		14.98± .39		12921
005616.7-524658	NGC 312	.E.2.*.	1.16± .04	62	13.42 ±.13	1.02± .01	1.03± .01		
301.24 -64.33	ESO 151- 6	S	.13± .03	.04	13.40 ±.14				7978± 17
244.48 -12.64		-4.0± .6	.88± .03	.00		.93	13.31± .09		7888
005404.0-530312	PGC 3343		1.13		13.25		13.90± .23		7811
0056.3 +1353		.L...?.	1.00± .10	5					
124.76 -48.96	UGC 582	U	.14± .04	.13					
310.28 3.41	IRAS00537+1337	-2.0±1.7		.00	12.40				
0053.7 +1337	PGC 3355		.99						
0056.3 +5046		.E...?.	1.11± .16						
123.73 -12.09	UGC 577	U	.00± .08	1.11					
346.35 11.89		-5.0±1.6		.00					
0053.5 +5030	PGC 3358		1.29						
0056.4 +1152		.S..3..	1.00± .08	170					11627
124.87 -50.98	UGC 581	U (1)	.14± .06	.16					
308.32 2.86		3.0± .9		.19					11738
0053.8 +1136	PGC 3361	4.5±1.2	1.01	.07					11301
005627.0+503720		.SX.6*.	1.04± .06	125				16.41±.1	5070± 11
123.75 -12.24	UGC 576	U	.22± .05	1.03	14.6 ±.3				
346.20 11.86	IRAS00534+5019	6.0± .9		.32				254± 8	5252
005333.9+502107	PGC 3363		1.14	.11	13.22			3.08	4835
0056.4 -0114		.E...*.	1.04± .18						
125.80 -64.09	UGC 583	U	.00± .08	.15	14.38 ±.11				11541± 48
295.62 -.68		-5.0±1.2		.00					11614
0053.9 -0131	PGC 3365		1.06		14.06				11216
005633.1-014616	NGC 307	.L...0*.	1.20± .06	85	13.75 ±.13	.97± .02			
125.91 -64.61	UGC 584	UE	.36± .04	.13	13.87 ±.10	.49± .03			4002± 28
295.12 -.85		-2.3± .7		.00		.87			4074
005359.8-020229	PGC 3367		1.16		13.63	.45	13.74± .36		3678
005641.4-525835	NGC 323	.E.0.*.	1.02± .04		13.59 ±.13	1.04± .01	1.06± .01		
301.12 -64.13	ESO 151- 9	S	.03± .03	.04	13.55 ±.14	.53± .03	.56± .02		7779± 31
244.30 -12.73		-5.0± .7	.68± .01	.00		.96	12.48± .05		7688
005429.0-531448	PGC 3374		1.02		13.41	.56	13.62± .24		7612
005642.7-095451	NGC 309	.SXR5..	1.48± .02	175	12.50 ±.13	.56± .01	.62± .01	14.12±.1	5662± 5
127.31 -72.74	MCG -2- 3- 50	R (3)	.08± .03	.17	12.2 ±.2	-.07± .03	.04± .03	221± 7	5649± 66
287.21 -3.07	IRAS00542-1010	5.0± .3	1.17± .01	.12	12.11	.47	13.84± .03	203± 5	5708
005411.8-101103	PGC 3377	1.3± .5	1.50	.04	12.10	-.13	14.57± .19	1.98	5348
005647.3-530553		.LAS0*.	1.16± .05	119	12.86V±.14	1.08± .03			
301.10 -64.01	ESO 151- 12	S	.28± .04	.04	13.92 ±.14				7425± 56
244.18 -12.77		-1.7± .7		.00		.97			7334
005435.0-532206	PGC 3387		1.12		13.78		13.89± .32		7259
005651.3-632853		.SAR5P.	.95± .09	147					9230± 63
301.91 -53.64	ESO 79- 13	Sr	.39± .05	.00	15.05 ±.14				9110
233.54 -14.22	FAIR 215	5.0± .7		.59	13.47				9118
005450.0-634506	PGC 3391		.95	.20	14.41				
005652.7-315747	NGC 314	RLBT+*.	.98± .06	168					
289.51 -85.02	ESO 411- 32	Sr	.10± .04	.02	14.23 ±.14				
265.50 -8.62	IRAS00544-3214	-1.4± .5		.00	13.54				
005429.0-321400	PGC 3395		.97						

R.A. 2000 DEC. / l b / SGL SGB / R.A. 1950 DEC.	Names / PGC	Type / S_T n_L / T / L	$logD_{25}$ / $logR_{25}$ / $logA_e$ / $logD_o$	p.a. / A_g / A_i / A_{21}	B_T / m_B / m_{FIR} / B_T^o	$(B-V)_T$ / $(U-B)_T$ / $(B-V)_T^o$ / $(U-B)_T^o$	$(B-V)_e$ / $(U-B)_e$ / m'_e / m'_{25}	m_{21} / W_{20} / W_{50} / HI	V_{21} / V_{opt} / V_{GSR} / V_{3K}
005657.3-435017	NGC 319	PSXS1*.	1.01± .05		14.25 ±.14				
299.48 -73.25	ESO 243- 13	BSr	.09± .04	.03					
253.58 -11.15		.6± .6		.09					
005439.0-440630	PGC 3398		1.01	.04					
005657.4-525523	NGC 328	.SBT1P*	1.43± .04	100					
301.02 -64.18	ESO 151- 13	Sr	.71± .04	.04	14.28 ±.14				4810±190
244.36 -12.76		1.4± .4		.72					4719
005445.0-531136	PGC 3399		1.43	.35	13.45				4643
005659.4-452441	IC 1603	.SAS5*.	1.10± .04	112					
299.83 -71.68	ESO 243- 14	S (1)	.21± .04	.03	14.62 ±.14				7962
251.99 -11.46		5.0± .9		.31					7894
005442.1-454054	PGC 3401	3.3± .9	1.11	.10	14.23				7760
005702.2-005232		.S?....	.93± .08						13535± 69
126.09 -63.71	UGC 588		.02± .05	.08	14.79 ±.13				13609
296.02 -.72	ARAK 22			.02					
005428.7-010845	PGC 3405		.94	.01	14.52				13211
0057.1 +1454		.S?....	1.00± .08	145					14885
124.99 -47.95	UGC 590		.19± .06	.14	15.26 ±.15				
311.30 3.50				.28					15004
0054.5 +1438	PGC 3409		1.01	.09	14.76				14562
005710.3-434336	NGC 322	.L..?P/	1.05± .06	153					
299.31 -73.36	ESO 243- 15	S	.25± .04	.03	14.29 ±.14				7093
253.70 -11.17		-2.0±1.7		.00					7030
005452.0-435948	PGC 3412		1.01		14.16				6884
005714.7-405729	NGC 324	.S?....	1.14± .05	95	13.99 ±.13	1.14± .01	1.15± .01		
298.36 -76.12	ESO 295- 25		.45± .05	.03	14.06 ±.14	.43± .02	.47± .02		3446± 39
256.50 -10.63			.43± .02	.65		1.03	11.63± .06		3391
005455.0-411342	PGC 3416		1.14	.22	13.32	.31	13.41± .31		3224
005719.8+235319	A 0054+23	.S?....	.95± .07	165					
124.66 -38.96	UGC 591		.16± .05	.17	15.02 ±.12				5100±110
320.02 5.78	MK 350			.23					5240
005439.0+233707	PGC 3420		.97	.08	14.59				4788
005724.0-305800		.LXS0..	1.13± .04	5					
284.54 -85.95	ESO 411- 33	S	.41± .03	.02	15.29 ±.14				
266.51 -8.51		-2.0± .8		.00					
005500.1-311412	PGC 3426		1.07						
005732.8+301647	NGC 311	.L.....	1.18± .15	120	14.00 ±.13	1.00± .02			
124.50 -32.57	UGC 592	U	.27± .08	.29	13.93 ±.10	.57± .03			5062± 26
326.24 7.32		-2.0± .8		.00		.86			5216
005449.6+300035	PGC 3434		1.17		13.59	.50	14.11± .77		4764
005735.2-050009	NGC 321	.SBT6P.	.99± .07	95					
126.99 -67.83	MCG -1- 3- 41	PE (2)	.21± .05	.23					5782± 72
292.05 -1.97		5.5± .6		.30					5843
005502.9-051621	PGC 3435	3.7± .6	1.01	.10					5462
005737.0-485406	IC 1605	PSBR4..	1.19± .04	138	13.71 ±.13	.78± .01	.87± .01		
300.20 -68.19	ESO 195- 19	Sr (1)	.09± .04	.04	13.82 ±.14				6800±190
248.47 -12.19	FAIR 671	3.7± .6	.85± .02	.13		.71	13.45± .05		6721
005522.0-491018	PGC 3436	2.2± .8	1.20	.04	13.55		14.30± .27		6614
005738.6-462648		.SBS6P.	.97± .05						
299.71 -70.64	ESO 243- 17	S (1)	.12± .05	.03	15.10 ±.14				8800±190
250.97 -11.76	FAIR 670	6.0± .9		.17					8728
005522.0-464300	PGC 3441	4.4± .9	.98	.06	14.86				8603
005739.2+434804	NGC 317A	.S?....	1.14± .13						
124.12 -19.06	UGC 593		.03± .12	.44					5293± 58
339.49 10.34	IRAS00548+4331			.04	10.80				5468
005449.9+433152	PGC 3442		1.18	.01					5034
0057.6 -0121		.S?....	.94± .10						
126.50 -64.19	UGC 595		.04± .06	.14	14.56 ±.12				13427± 43
295.59 -1.00				.04					13500
0055.1 -0138	PGC 3444		.95	.02	14.22				13104
005740.5+434731	NGC 317B	.SB?...	1.04± .06	105				15.60±.1	5334± 8
124.12 -19.07	UGC 594		.32± .05	.44				374± 16	5083± 50
339.48 10.33				.48				258± 8	5503
005451.2+433119	PGC 3445		1.08	.16					5069
005741.8+332104		.S?....	.64± .17					17.44±.3	5303± 10
124.43 -29.50	MCG 5- 3- 29		.11± .07	.20	15.23 ±.18				5400± 49
329.25 8.02	ARAK 23			.17				219± 7	5465
005457.4+330452	PGC 3446		.66	.06	14.83			2.55	5016

R.A. 2000 DEC. / l b / SGL SGB / R.A. 1950 DEC.	Names / PGC	Type / S_T n_L / T / L	$\log D_{25}$ / $\log R_{25}$ / $\log A_\bullet$ / $\log D_\circ$	p.a. / A_g / A_i / A_{21}	B_T / m_B / m_{FIR} / B_T°	$(B-V)_T$ / $(U-B)_T$ / $(B-V)_T^\circ$ / $(U-B)_T^\circ$	$(B-V)_\bullet$ / $(U-B)_\bullet$ / $m'_\bullet$ / m'_{25}	m_{21} / W_{20} / W_{50} / HI	V_{21} / V_{opt} / V_{GSR} / V_{3K}
005742.3+434210 124.13 -19.15 339.40 10.31 005453.1+432558	A 0054+43 MCG 7- 3- 11 IRAS00548+4325 PGC 3448	.S..3.. R (1) 1.1± .9	.94± .11 .19± .07 .98	 .44 .28 .09	15.09 ±.14 14.83 ±.12 14.19	.59± .03 .42	 14.15± .58	16.43±.1 96± 11 88± 11 2.14	5422± 10 5397± 59 5596 5161
005745.1+323239 124.47 -30.31 328.46 7.81 005501.0+321627	 CGCG 501- 50 PGC 3450		 	 .13 	 15.2 ±.2 			17.58±.3 136± 10 	7400± 10 7557 7107
0057.7 -0023 126.44 -63.23 296.53 -.77 0055.2 -0040	 UGC 599 PGC 3451	.E...?. U -5.0±1.7 	.66± .24 .22± .07 .60	 .09 .00 	 14.85 ±.20 				
005747.3-273006 228.13 -88.54 269.98 -7.78 005522.0-274618	 ESO 411- 34 IRAS00553-2746 PGC 3453	.SXR5*. S (1) 4.9± .4 2.9± .4	1.23± .04 .03± .03 1.24	 .07 .04 .01	 13.49 ±.14 13.10 13.35				5647± 47 5635 5376
005748.1-050645 127.15 -67.93 291.96 -2.05 005515.9-052257	 MCG -1- 3- 45 PGC 3454	.S..6./ E 6.0± .9 	1.14± .06 .83± .05 1.16	93 .23 1.22 .41					
005749.1+302110 124.56 -32.50 326.33 7.28 005505.9+300458	NGC 315 UGC 597 PGC 3455	.E+..*. PU -4.0± .5 1.49	1.51± .06 .20± .06 1.09± .06 	40 .28 .00 	12.2 ±.2 12.23 ±.10 11.87	1.04± .02 .60± .03 .93 .56	1.07± .01 .61± .02 13.10± .21 14.24± .39		 4936± 24 5089 4638
005750.0+312905 124.53 -31.37 327.43 7.55 005506.3+311253	 UGC 598 PGC 3456	.S...0.. U .0± .9 	.97± .07 .42± .05 .97	29 .21 .32 	 14.56 ±.15 13.96			18.23±.3 282± 10 	5005± 10 5027± 46 5161 4710
0057.9 -0507 127.24 -67.95 291.95 -2.09 0055.3 -0524	NGC 327 MCG -1- 3- 47 PGC 3462	.SBS4*. PE (1) 4.0± .6 3.1± .8	1.20± .04 .38± .04 1.22	3 .23 .56 .19	14.3 ±.2 	.88± .06 .16± .07 	 14.21± .32		
005801.5+304219 124.60 -32.15 326.68 7.32 005518.0+302607	 CGCG 501- 55 PGC 3466		 	 .22 	 15.4 ±.4 			17.39±.3 188± 7 	4723± 10 4877 4425
005801.5-050423 127.29 -67.89 292.01 -2.09 005529.2-052034	NGC 329 MCG -1- 3- 48 PGC 3467	.S..3*/ PE 2.5± .8 	1.20± .05 .41± .04 1.22	161 .22 .57 .20	14.3 ±.2 	1.04± .12 .27± .12 	 14.14± .32		
0058.1 -0614 127.59 -69.06 290.87 -2.43 0055.6 -0631	 MCG -1- 3- 50 PGC 3472	.S?.... 	1.08± .09 .82± .07 1.10	 .25 1.22 .41					5823 5880 5505
005811.0-081301 128.07 -71.02 288.96 -2.97 005539.7-082913	 MCG -1- 3- 49 PGC 3475	.LXS-?. E -2.5± .6 	1.15± .07 .13± .05 1.17	50 .32 .00 					
005819.4-152332 131.04 -78.16 281.96 -4.89 005550.2-153944	 MCG -3- 3- 10 PGC 3481	RSBR2*. E 2.0±1.2 	1.19± .05 .31± .05 1.19	63 .02 .39 .16					
0058.3 +2652 124.84 -35.98 322.97 6.30 0055.6 +2635	NGC 326 UGC 601 4ZW 35 PGC 3482	.E?.... 	1.16± .09 .00± .05 .84± .03 1.20	 .22 .00 	14.33 ±.15 13.89	1.09± .03 .63± .08 .91 .65	1.18± .02 .68± .06 14.02± .10 15.13± .50		 14138± 59 14284 13832
005823.2+364350 124.48 -26.12 332.58 8.66 005537.2+362739	A 0055+36 UGC 602 PGC 3485	.SXS5.. U (1) 5.0± .8 2.2± .8	1.20± .04 .06± .04 1.22	 .18 .09 .03	14.68 ±.13 14.12 ±.16 14.15	.62± .02 .53 	 15.38± .25	15.37±.1 265± 8 251± 11 1.19	6143± 7 6127± 59 6307 5861
005824.0-082428 128.28 -71.21 288.79 -3.07 005552.7-084039	 MCG -2- 3- 52 PGC 3486	PSBR1P? E 1.0±1.0 	1.15± .05 .08± .04 1.18	75 .32 .08 .04					
005824.1+483941 124.12 -14.19 344.32 11.18 005531.7+482330	A 0055+48 UGC 600 PGC 3487	.SXS3.. U 3.0± .8 	1.14± .05 .10± .05 1.23	135 .95 .14 .05	 14.0 ±.3 12.88			16.11±.1 163± 11 137± 7 3.18	6813± 7 6903± 59 6994 6573

R.A. 2000 DEC. / l b / SGL SGB / R.A. 1950 DEC.	Names / PGC	Type / S_T n_L / T / L	$\log D_{25}$ / $\log R_{25}$ / $\log A_e$ / $\log D_o$	p.a. / A_g / A_i / A_{21}	B_T / m_B / m_{FIR} / B_T^o	$(B-V)_T$ / $(U-B)_T$ / $(B-V)_T^o$ / $(U-B)_T^o$	$(B-V)_e$ / $(U-B)_e$ / m'_e / m'_{25}	m_{21} / W_{20} / W_{50} / HI	V_{21} / V_{opt} / V_{GSR} / V_{3K}
0058.4 +1134 / 125.66 -51.26 / 308.16 2.31 / 0055.8 +1118	UGC 603 / PGC 3490	.S?....	.89± .08 / .22± .05 / / .91	157 / .16 / .33 / .11					14079 / / 14188 / 13754
0058.4 -0123 / 126.97 -64.21 / 295.61 -1.21 / 0055.9 -0139	/ PGC 3492		/ / .67± .05 /	.14	14.73 ±.15 / / /	1.05± .02 / / /	1.07± .01 / / 13.57± .17 /		15510± 72 / 15582 / 15187
005827.4+233231 / 125.01 -39.30 / 319.75 5.45 / 005546.5+231620	MCG 4- 3- 26 / PGC 3493	.SB?...	.89± .11 / .07± .07 / / .90	/ .10 / .10 / .03	15.19 ±.13 / / / 14.93			16.67±.3 / 291± 7 / 1.71 /	10432± 10 / 10571 / 10120
005831.0-100928 / 128.88 -72.95 / 287.09 -3.57 / 005600.2-102539	MCG -2- 3- 53 / PGC 3496	.SBS9?/ / E / 9.0±1.8	1.11± .06 / .60± .05 / / 1.12	40 / .14 / .61 / .30					6837 / / 6882 / 6525
0058.6 +4501 / 124.27 -17.83 / 340.73 10.41 / 0055.8 +4445	UGC 604 / PGC 3503	.S..6*. / U / 6.0±1.5	1.00± .08 / 1.02± .06 / / 1.06	137 / .66 / 1.47 / .50			16.54±.1 / 203± 8 / /	5215± 11 / 5391 / 4960
005840.0-451201 / 298.84 -71.87 / 252.26 -11.71 / 005623.0-452812	ESO 243- 18 / PGC 3505	.SB.5?/ / S / 5.0±2.0	1.08± .05 / .78± .04 / / 1.08	121 / .03 / 1.17 / .39	15.53 ±.14 / / /				
0058.7 +1245 / 125.70 -50.08 / 309.32 2.55 / 0056.1 +1229	UGC 607 / PGC 3508	.S..6?. / U / 6.0±1.7	.98± .07 / .00± .05 / / .99	/ .18 / .00 / .00	14.95 ±.15 / / / 14.71				11641 / / 11753 / 11317
005846.4-205025 / 138.23 -83.49 / 276.63 -6.38 / 005619.0-210636	NGC 320 / ESO 541- 3 / IRAS00563-2106 / PGC 3510	RSBT0*. / S / .0± .9	.93± .05 / .25± .04 / / .92	159 / .09 / .19 /	14.55 ±.14 / / 12.30 /				
005848.8+003517 / 126.89 -62.23 / 297.56 -.75 / 005614.9+001906	IC 1607 / UGC 611 / PGC 3512	.S?....	.95± .07 / .00± .05 / / .96	/ .13 / .00 / .00	14.31 ±.12 / / / 14.15				5410± 50 / 5488 / 5086
0058.8 +1259 / 125.72 -49.85 / 309.56 2.59 / 0056.2 +1243	UGC 610 / PGC 3513	.E..... / U / -5.0± .8	1.26± .13 / .06± .08 / .96± .08 / 1.27	/ .15 / .00 /	14.0 ±.2 / 14.07 ±.16 / /	1.05± .04 / / /	1.10± .02 / / 14.25± .27 / 15.11± .71		
005850.0-350655 / 292.19 -81.86 / 262.44 -9.72 / 005628.0-352306	NGC 334 / ESO 351- 26 / IRAS00564-3523 / PGC 3514	PSBS3P* / BS (1) / 3.4± .6 / 3.3±1.2	1.07± .05 / .28± .04 / / 1.07	169 / .02 / .39 / .14	14.64 ±.14 / 13.38 / 14.16 /				9204± 20 / 9167 / 8961
005854.7-254613 / 174.28 -87.85 / 271.76 -7.61 / 005629.0-260224	ESO 475- 3 / PGC 3523	.L?....	.87± .06 / .07± .04 / / .87	/ .07 / .00 /	15.41 ±.14 / / 15.26 /				5583± 52 / 5577 / 5308
005858.1-184437 / 134.99 -81.44 / 278.70 -5.90 / 005630.1-190048	A 0056-19 / ESO 541- 4 / PGC 3526	.SXT4.. / SE (2) / 4.0± .5 / 3.2± .5	1.47± .03 / .32± .03 / / 1.47	22 / .01 / .47 / .16	13.34 ±.14 / / 12.84 /			14.28±.3 / 276± 16 / 269± 12 / 1.28	1987± 11 / 2009± 60 / 2004 / 1693
005901.2+235108 / 125.16 -38.99 / 320.08 5.40 / 005620.1+233458	UGC 612 / IRAS00563+2334 / PGC 3527	.S?....	.93± .07 / .35± .05 / / .95	90 / .13 / .52 / .17	14.63 ±.15 / / 13.96 /				5061± 46 / 5200 / 4751
005902.4+480105 / 124.25 -14.83 / 343.70 10.95 / 005610.2+474455	UGC 608 / PGC 3528	.SX.8.. / U / 8.0± .8	1.28± .04 / .32± .05 / / 1.35	133 / .83 / .40 / .16	14.9 ±.3 / / 13.64 /			14.49±.1 / 205± 7 / 179± 10 / .68	2756± 6 / / 2935 / 2512
005904.1+010005 / 126.97 -61.81 / 297.98 -.70 / 005630.1+004355	/ PGC 3530		/ / /	.05	16.58 ±.13 / / /	.48± .03 / -.09± .05 / /			5330± 72 / 5409 / 5006
005906.7-861710 / 302.79 -30.84 / 210.01 -15.74 / 010110.0-863318	ESO 2- 11 / PGC 3533	.SBS7.. / S (1) / 7.0± .8 / 7.8± .8	1.19± .05 / .21± .05 / / 1.23	50 / .46 / .28 / .10					

R.A. 2000 DEC. l b SGL SGB R.A. 1950 DEC.	Names PGC	Type S_T n_L T L	$\log D_{25}$ $\log R_{25}$ $\log A_e$ $\log D_o$	p.a. A_g A_i A_{21}	B_T m_B m_{FIR} B_T^o	$(B-V)_T$ $(U-B)_T$ $(B-V)_T^o$ $(U-B)_T^o$	$(B-V)_e$ $(U-B)_e$ m'_e m'_{25}	m_{21} W_{20} W_{50} HI	V_{21} V_{opt} V_{GSR} V_{3K}
005918.0+065515 126.41 -55.90 303.72 .85 005642.3+063905	CGCG 410- 22 MK 559 PGC 3541			.16	15.49 ±.12 13.67				13185± 45 13281 12860
005918.5-203444 138.70 -83.21 276.92 -6.43 005651.0-205054	ESO 541- 5 PGC 3543	.SBT8?. S (1) 8.0±1.7 7.8±1.2	1.08± .04 .27± .03 1.09	87 .13 .33 .14	15.84 ±.14 15.38				1958 1969 1669
005920.3-181408 134.91 -80.92 279.23 -5.85 005652.1-183018	NGC 335 ESO 541- 6 IRAS00568-1830 PGC 3544	.S..4?/ SE 4.0±1.1	1.05± .04 .54± .03 1.05	137 .04 .80 .27	15.12 ±.14 12.52 14.25				5518 5536 5222
005923.2+235903 125.26 -38.85 320.23 5.35 005642.0+234253	CGCG 480- 29 PGC 3546			.13	15.7 ±.4			18.03±.3 168± 7	5079± 10 5218 4769
005923.6-044815 128.14 -67.60 292.36 -2.35 005651.3-050425	MCG -1- 3- 51 PGC 3547	.S..1P/ E 1.0± .8	1.32± .04 .54± .05 1.33	170 .20 .55 .27					
005924.3-341944 290.06 -82.60 263.25 -9.67 005702.1-343554	IC 1608 ESO 351- 27 PGC 3549	RLAR+P? BS -.7± .5	1.31± .04 .36± .03 1.26	170 .02 .00	13.70 ±.14 13.63				3463± 34 3428 3217
005929.4-361114 292.74 -80.78 261.39 -10.08 005708.0-362724	ESO 351- 28 IRAS00570-3627 PGC 3554	.SBS5*P S (1) 4.7± .7 4.4±1.2	1.14± .04 .38± .04 1.14	80 .02 .58 .19	14.61 ±.14 13.99				3572± 39 3531 3333
005932.2+314800 124.94 -31.04 327.83 7.27 005648.1+313150	CGCG 501- 56 PGC 3555			.24	15.0 ±.4			17.08±.3 154± 7	4616± 10 4680± 53 4773 4324
0059.5 -1226 130.73 -75.19 284.93 -4.42 0057.0 -1242	MCG -2- 3- 59 IRAS00570-1242 PGC 3557	.S?....	1.04± .09 .78± .07 1.05	.09 1.17 .39					5735 5772 5427
0059.6 +1850 125.62 -43.99 315.27 3.97 0056.9 +1834	UGC 616 IRAS00569+1834 PGC 3558	PSBR3.. U 3.0± .8	1.10± .05 .08± .05 1.11	85 .10 .11 .04	14.91 ±.19 14.62				11279 11406 10961
005936.2+353335 124.80 -27.28 331.49 8.15 005650.5+351725	UGC 614 PGC 3559	.SBS7.. U 7.0± .8	1.17± .05 .19± .05 1.19	70 .16 .27 .10	13.87 ±.12 13.43				2540± 52 2702 2256
005939.3-181810 135.46 -80.97 279.18 -5.94 005711.1-183420	PGC 3562	.SBS1P* SE .7± .9	1.18± .07 .14± .05 1.18	10 .04 .14 .07					
005939.6+151956 125.87 -47.49 311.88 3.03 005701.2+150346	UGC 615 PGC 3563	.SX.2.. U 2.0± .9	.97± .07 .24± .05 .99	23 .16 .29 .12	14.20 ±.14 13.69				5517± 50 5635 5196
005940.5+292017 125.08 -33.50 325.44 6.64 005657.3+290407	MCG 5- 3- 32 PGC 3564	.E?....		.23	15.1 ±.4			17.03±.3 158± 7	4861± 10 5011 4561
0059.7 +1801 125.72 -44.81 314.48 3.72 0057.1 +1745	UGC 617 PGC 3565	.SB.1.. U 1.0± .9	.96± .07 .08± .05 .97	.13 .08 .04	14.86 ±.13 14.58				5478 5603 5159
0059.7 +0056 127.35 -61.86 297.96 -.89 0057.2 +0040	UGC 618 PGC 3566	.S..6*. U 6.0±1.2	.98± .07 .00± .05 .98	.04 .00 .00	14.95 ±.15				
005947.4-401956 296.01 -76.68 257.23 -10.98 005728.0-403606	IC 1609 ESO 295- 26 PGC 3567	.LA.0*. BS -1.6± .7	1.15± .05 .04± .04 1.15	.02 .00	13.59 ±.14				

R.A. 2000 DEC.	Names	Type	logD₂₅	p.a.	B_T	(B-V)_T	(B-V)_e	m₂₁	V₂₁
l　b		S_T　n_L	logR₂₅	A_g	m_B	(U-B)_T	(U-B)_e	W₂₀	V_opt
SGL　SGB		T	logA_e	A_i	m_FIR	(B-V)_T°	m'_e	W₅₀	V_GSR
R.A. 1950 DEC.	PGC	L	logD_o	A₂₁	B_T°	(U-B)_T°	m'₂₅	HI	V_3K
0059.8　+1444		.S..6*.	1.07± .07	5					12196
125.96 -48.09	UGC　619	U	1.01± .06	.15					
311.31　2.83	IRAS00571+1428	6.0±1.4		1.47					12313
0057.1　+1428	PGC　3569		1.08	.50					11874
005950.0-073441 NGC　337		.SBS7..	1.46± .02	60	12.06 ±.13	.45± .01	.49± .01	13.05±.1	1650± 5
129.13 -70.35	MCG -1- 3- 53	R　　(3)	.20± .04	.34	12.1 ±.2	-.09± .02	-.11± .02	261± 6	1690± 50
289.69　-3.20	IRAS00573-0750	7.0± .3	1.02± .01	.28	10.78	.32	12.65± .02	229± 7	1702
005718.5-075051	PGC　3572	5.1± .4	1.49	.10	11.45	-.18	13.71± .20	1.50	1335
005952.8+314936 A　0057+31		.LA....						16.32±.2	4456± 6
125.02 -31.01	CGCG 501- 58	R		.24	14.8 ±.4			242± 6	4466± 32
327.87　7.21	MK　352	-2.0± .5						229± 6	4611
005708.6+313327	PGC　3575								4163
010000.1-680739		RSAR2P?	1.13± .05						
301.72 -48.98	ESO　51- 11	Sr　　(1)	.05± .05	.00	14.08 ±.14				6996± 30
228.79 -15.01	FAIR　217	2.3± .5		.07	13.10				6864
005809.1-682348	PGC　3579	2.2±1.2	1.13	.03	13.94				6910
010004.2-110459		PSXR4..	1.17± .05						5421
130.56 -73.83	MCG -2- 3- 61	E　　(1)	.04± .05	.09					
286.29　-4.18		4.0± .8		.06					5462
005733.8-112108	PGC　3584	.8± .8	1.18	.02					5111
010009.4-334233 Sculptor		.E?....	2.60± .02	110	10.50S±.10				
287.53 -83.16	ESO　351- 30		.11± .04	.02	8.78 ±.11				148± 22
263.91　-9.68				.00					115
005747.1-335842	PGC　3589		2.57		9.70		18.22± .17		-99
0100.3　+1816		.S?....	1.01± .06	2					12524
125.88 -44.55	UGC　621		.23± .05	.13	15.12 ±.14				
314.76　3.66	IRAS00576+1800			.34	13.70				12649
0057.6　+1800	PGC　3598		1.03	.11	14.57				12206
010028.2+475943		.S..6*.	1.02± .06	160	14.2 S±.3			14.71±.1	2714± 8
124.49 -14.85	UGC　622	U	.17± .05	.83	14.4 ±.3			269± 21	
343.73 10.71	IRAS00575+4743	6.0±1.3		.25	12.30			239± 7	2892
005735.6+474334	PGC　3603		1.09	.08	13.14		13.66± .42	1.49	2471
010032.7+304749 IC　66		.S..1..	.98± .07	125	14.94 ±.15	.83± .02	.87± .02	16.89±.3	4825± 10
125.24 -32.03	UGC　623	U	.31± .05	.19	14.93 ±.12	.21± .04	.26± .04		
326.91　6.82		1.0± .9	.50± .03	.32		.69	12.93± .11	369± 7	4978
005748.9+303140	PGC　3606		.99	.16	14.36	.13	13.89± .38	2.37	4529
010034.5-574503			.76± .06						
300.54 -59.33	ESO　113- 4		.19± .05	.00	15.03 ±.14				3130± 35
239.49 -13.97									3024
005828.0-580112	PGC　3608		.76						2989
010036.2+304006 NGC　338		.S..2..	1.27± .04	109	13.67 ±.14	.86± .01	.94± .01	14.65±.2	4778± 7
125.26 -32.16	UGC　624	U	.47± .05	.19	13.92 ±.12	.24± .02	.34± .02	536± 13	4726± 36
326.78　6.77	IRAS00578+3024	2.0± .8	.67± .02	.58	12.02	.69	12.51± .07	531± 6	4928
005752.4+302358	PGC　3611		1.28	.23	13.00	.11	13.68± .28	1.42	4480
010038.2-075852		.SXT5..	1.10± .06	50					
129.85 -70.73	MCG -1- 3- 56	E　　(1)	.02± .05	.32					
289.35　-3.50		5.0± .8		.03					
005806.9-081501	PGC　3613	3.1± .8	1.13	.01					
0100.6　-0141		.SXS5..	.93± .07	97					
128.28 -64.47	UGC　626	U	.03± .05	.17	14.72 ±.12				
295.47　-1.82		5.0± .9		.04					
0058.1　-0158	PGC　3614		.95	.01					
010039.5+300736		.L?....	.72± .22					17.57±.3	6755± 10
125.30 -32.70	MCG 5- 3- 35		.00± .07	.32	15.30 ±.10				
326.26　6.63				.00				182± 7	6906
005755.8+295127	PGC　3615		.76		14.87				6458
010045.8-091108		.SB?...	1.05± .06	55					
130.37 -71.92	MCG -2- 3- 63	(1)	.03± .05	.17					4494± 40
288.19　-3.84	MK　968			.04	13.31				4541
005814.8-092717	PGC　3620	3.1± .8	1.07	.01					4182
010049.6-853124		.S..6?/	1.22± .06	49					
302.72 -31.60	ESO　2- 12	S	.96± .05	.48					4490
210.79 -15.77	IRAS01021-8547	5.7±1.1		1.41					4324
010209.0-854730	PGC　3629		1.26	.48					4502
010051.7-531439 NGC　348		.S..3..	.92± .05						
299.74 -63.83	ESO　151- 17	S　　(1)	.09± .04	.04	14.58 ±.14				
244.12 -13.39		3.0± .9		.13					
005841.1-533048	PGC　3632	3.3±1.3	.93	.05					

R.A. 2000 DEC. l b SGL SGB R.A. 1950 DEC.	Names PGC	Type S_T n_L T L	$\log D_{25}$ $\log R_{25}$ $\log A_o$ $\log D_o$	p.a. A_g A_i A_{21}	B_T m_B m_{FIR} B_T^o	$(B-V)_T$ $(U-B)_T$ $(B-V)_T^o$ $(U-B)_T^o$	$(B-V)_e$ $(U-B)_e$ m'_e m'_{25}	m_{21} W_{20} W_{50} HI	V_{21} V_{opt} V_{GSR} V_{3K}
010055.8+474051 IC 65 124.59 -15.16 UGC 625 343.43 10.57 IRAS00580+4724 005803.3+472443 PGC 3635	.SXS4.. U 4.0± .7 	1.59± .03 .53± .05 1.66	155 .78 .78 .27	13.6 S±.3 13.3 ±.3 11.94 11.90	 	.65± .05 15.07± .33	13.43±.1 343± 13 328± 10 1.26	2614± 8 2792 2370
0100.9 +1328 126.48 -49.34 UGC 627 310.16 2.23 0058.3 +1312 PGC 3636	.S?.... 	1.03± .06 .14± .05 1.04	115 .11 .21 .07	* 14.82 ±.14 14.44	 	 	 	11771 11884 11449
0100.9 +1929 126.02 -43.33 UGC 628 315.98 3.83 0058.3 +1913 PGC 3639	.S..9*. U 9.0±1.1 	1.18± .07 .16± .06 1.19	 .08 .16 .08	* 	 	 	 	5446 5574 5130
0101.0 +0943 126.86 -53.08 UGC 631 306.54 1.20 0058.4 +0927 PGC 3643	.S..6*. U 6.0±1.3 	1.19± .05 .87± .05 1.20	137 .12 1.27 .43	* 	 	 	 	6047 6150 5723
010101.1+295654 125.40 -32.88 UGC 630 326.11 6.51 005817.4+294046 PGC 3644	.SA.7*. U 7.0±1.2 	.96± .09 .08± .06 .98	100 .24 .11 .04	* 	 	 	17.48±.3 146± 7 	4883± 7 5034 4586
010102.5+293601 125.43 -33.22 UGC 629 325.77 6.42 005819.0+291953 PGC 3646	.S..6*. U 6.0±1.2 	1.09± .06 .11± .05 1.11	110 .23 .16 .06	* 14.59 ±.14 14.18	 	 	15.50±.3 190± 7 1.27	4991± 10 5141 4693
010111.7+300751 125.44 -32.69 UGC 632 326.29 6.52 005827.9+295143 PGC 3651	.SB.2.. U 2.0± .9 	1.04± .06 .34± .05 1.06	163 .24 .41 .17	14.81 ±.15 14.72 ±.12 14.04	1.00± .02 .40± .04 .83 .27	 14.02± .35	18.30±.3 430± 7 4.10	6965± 10 7116 6668
010113.2+300913 125.44 -32.67 MCG 5- 3- 39 326.32 6.52 ARAK 24 005829.4+295305 PGC 3652	.S?.... 	.64± .17 .00± .07 .66	 .24 .00 	 15.09 ±.16 14.75	 	 	18.24±.3 394± 7 	7000± 10 7151 6703
010122.0+313028 125.41 -31.32 UGC 633 327.64 6.82 005837.6+311420 PGC 3664	.S..3.. U (1) 3.0± .9 3.5±1.2	1.20± .05 .67± .05 1.22	9 .20 .93 .34	14.81 ±.15 14.58 ±.12 13.50	.76± .03 .06± .04 .55 -.11	 13.99± .30	15.11±.2 404± 13 392± 6 1.27	5573± 7 5726 5279
010122.1-065305 NGC 345 130.02 -69.62 MCG -1- 3- 64 290.47 -3.38 005850.5-070913 PGC 3665	.SAS1*. E 1.0±1.2 	1.07± .06 .17± .05 1.11	45 .34 .17 .08	 	 	 	 	
010124.1+310228 IC 69 125.44 -31.78 MCG 5- 3- 41 327.19 6.70 005839.9+304620 PGC 3666	.S?.... 	.94± .11 .00± .07 .96	 .20 .00 .00	 14.56 ±.12 14.33	 	 	18.01±.3 202± 7 3.68	5017± 10 5170 4722
0101.4 +0737 A 0058+07 127.26 -55.16 UGC 634 304.55 .54 DDO 7 0058.8 +0721 PGC 3667	.SX.9*. PU (1) 9.0± .7 9.0±1.0	1.22± .06 .19± .06 .97± .04 1.24	35 .16 .19 .09	14.98 ±.17 15.0 ±.3 14.63	.48± .06 -.18± .07 .39 -.25	.55± .05 -.13± .06 15.32± .11 15.48± .38	14.51±.1 148± 8 138± 12 -.21	2212± 6 2309 1888
010134.0-073521 NGC 337A 130.40 -70.31 MCG -1- 3- 65 289.80 -3.61 IRAS00589-0751 005902.6-075129 PGC 3671	.SXS8.. PE 8.0± .4 	1.77± .03 .12± .05 1.42± .06 1.81	10 .35 .15 .06	12.7 ±.3 12.18	.53± .08 -.13± .08 .42 -.21	.61± .04 -.04± .04 15.27± .14 16.11± .31	12.45±.1 98± 6 77± 12 .21	1076± 5 388± 79 1124 759
010134.7-594435 300.57 -57.34 ESO 113- 6 237.46 -14.34 005931.0-600042 PGC 3672	.S..5*/ S 5.0± .9 	1.22± .05 .97± .04 1.22	63 .00 1.45 .48	 16.25 ±.14 14.75	 	 	 	8443 8332 8313
010135.4-531158 299.48 -63.86 ESO 151- 18 244.19 -13.49 005925.0-532806 PGC 3675	.S..1*/ S 1.0±1.3 	1.15± .05 .71± .04 1.16	170 .04 .73 .36	 14.80 ±.14 	 	 	 	
0101.6 +2403 125.93 -38.76 UGC 636 320.43 4.86 0059.0 +2347 PGC 3680	.E..... U -5.0± .8 	1.13± .10 .08± .05 1.13	110 .14 .00 	 14.26 ±.11 	 	 	 	
010142.8-153419 IC 1610 135.12 -78.20 MCG -3- 3- 20 281.99 -5.73 005913.9-155027 PGC 3681	PLX.+*. E -1.0±1.2 	1.15± .06 .15± .05 1.13	110 .02 .00 	 	 	 	 	

R.A. 2000 DEC.	Names	Type	logD$_{25}$	p.a.	B$_T$	(B-V)$_T$	(B-V)$_\bullet$	m$_{21}$	V$_{21}$
l b		S$_T$ n$_L$	logR$_{25}$	A$_g$	m$_B$	(U-B)$_T$	(U-B)$_\bullet$	W$_{20}$	V$_{opt}$
SGL SGB		T	logA$_\bullet$	A$_i$	m$_{FIR}$	(B-V)^o_T	m'	W$_{50}$	V$_{GSR}$
R.A. 1950 DEC.	PGC	L	logD$_o$	A$_{21}$	B^o_T	(U-B)^o_T	m'$_{25}$	HI	V$_{3K}$
010144.4-515205		PSXT1*.	1.07± .05		14.26 ±.14				
299.14 -65.18	ESO 195- 24	Sr	.11± .05	.04					
245.55 -13.32		1.0± .5		.12					
005933.0-520812	PGC 3683		1.07	.06					
010147.7+311334								16.56±.3	4895± 10
125.53 -31.59	CGCG 501- 68			.20	15.7 ±.4				5048
327.39 6.66								191± 7	
005903.4+305727	PGC 3685								4601
010150.8-064801	NGC 349	.LA.-..	1.10± .11	140					
130.34 -69.52	MCG -1- 3- 68	E	.16± .07	.34					
290.58 -3.47		-3.0± .9		.00					
005919.1-070408	PGC 3687		1.12						
0101.8 +0303		.SB.6*.	1.04± .06	157					4379
128.10 -59.71	UGC 637	U	.37± .05	.06					
300.16 -.82		6.0±1.3		.55					4462
0059.3 +0247	PGC 3688		1.05	.19					4056
010157.1-205405		.SAROP*	.89± .05		15.25 ±.14				
144.56 -83.33	ESO 541- 10	r	.08± .04	.08					
276.75 -7.11		-.1±1.3		.06					
005930.0-211012	PGC 3692		.90						
010157.9-015604	NGC 351	PSBRO*.	1.16± .04	142	14.06 ±.12				4194± 50
129.09 -64.68	UGC 639	UE	.27± .04	.16					4262
295.33 -2.19		.3± .7		.21					3874
005924.7-021211	PGC 3693		1.16		13.63				
010159.3+262914								17.57±.3	10075± 10
125.86 -36.32	CGCG 480- 33			.20	15.43 ±.12				10218
322.81 5.43								106± 7	
005916.8+261307	PGC 3694								9771
010203.6-192659		.LASOP?	1.11± .04	148	15.34 ±.14				16891± 60
141.09 -81.94	ESO 541- 11	SE	.20± .03	.04					16904
278.19 -6.78		-2.0± .7		.00	15.05				16600
005936.0-194306	PGC 3695		1.09						
010209.2-041445	NGC 352	PSBT3?.	1.38± .03	10	13.54 ±.15	.95± .03	.99± .02		5282
129.78 -66.98	MCG -1- 3- 71	PE	.42± .04	.17		.38± .04	.41± .03		
293.09 -2.86	IRAS00596-0430	3.0± .6	.86± .03	.58	12.56	.80	13.33± .10		5343
005936.8-043052	PGC 3701		1.40	.21	12.75	.24	14.25± .22		4964
0102.3 +0906		.S..3..	.95± .07		15.21 ±.16				11637
127.47 -53.67	UGC 640	U (1)	.00± .05	.13					
306.04 .72		3.0± .8		.00					11737
0059.7 +0850	PGC 3709	4.5±1.1	.96	.00	14.99				11314
010220.1-541935		.SAS9*.	1.13± .05	42	14.91 ±.14				1386
299.46 -62.73	ESO 151- 19	S	.10± .05	.00					1289
243.05 -13.76		9.0± .8		.10					1229
010011.1-543542	PGC 3710	8.9± .8	1.13	.05	14.81				
010221.0-043227		.I..9P.	1.02± .07						
129.98 -67.26	MCG -1- 3- 72	E (1)	.23± .05	.15					
292.82 -2.99		10.0± .9		.17					
005948.7-044834	PGC 3711	9.8± .8	1.03	.11					
010224.6+263736								16.80±.3	4988± 10
125.97 -36.18	CGCG 480- 35			.20	15.7 ±.4				5131
322.97 5.37								246± 7	
005942.0+262130	PGC 3713								4684
010224.7-015727	NGC 353	.SB.1P*	1.11± .05	26	14.55 ±.12				4178
129.36 -64.69	UGC 641	UE	.48± .04	.15					4246
295.33 -2.31	IRAS00598-0213	1.0± .7		.49	13.25				3858
005951.6-021334	PGC 3714		1.13	.24	13.86				
0102.4 +3043	A 0059+30								10370± 67
125.72 -32.09				.21					10521
326.93 6.40	1ZW 2								10075
0059.7 +3027	PGC 3715								
010232.4-390411		PSAR6?.	.90± .07		15.38 ±.14				185
292.66 -77.84	ESO 295- 29	r	.03± .06	.02					134
258.61 -11.25		5.6±1.7		.05					-40
010013.0-392018	PGC 3721		.90	.02	15.30				
010241.7-215254		.E+3.P.	1.11± .05	25	14.7 ±.2	1.14± .05	1.16± .02		17089± 53
149.54 -84.16	ESO 541- 13	S	.17± .05	.07	14.54 ±.14		.48± .04		17094
275.83 -7.52		-4.0± .8	.83± .10	.00		.96	14.34± .35		16805
010015.1-220900	PGC 3727		1.07		14.27		14.80± .32		

R.A. 2000 DEC. / l b / SGL SGB / R.A. 1950 DEC.	Names / PGC	Type S_T n_L / T / L	$\log D_{25}$ / $\log R_{25}$ / $\log A_\bullet$ / $\log D_\bullet$	p.a. / A_g / A_i / A_{21}	B_T / m_B / m_{FIR} / B_T^o	$(B-V)_T$ / $(U-B)_T$ / $(B-V)_T^o$ / $(U-B)_T^o$	$(B-V)_\bullet$ / $(U-B)_\bullet$ / $m'_\bullet$ / m'_{25}	m_{21} / W_{20} / W_{50} / HI	V_{21} / V_{opt} / V_{GSR} / V_{3K}
010245.0-801401		.SXS7?.	1.21± .04	121					1808
302.33 -36.88	ESO 13- 9	S (1)	.52± .04	.31	14.92 ±.14				1651
216.26 -15.80		7.0± .7		.72					1790
010157.0-803006	PGC 3733	6.7± .9	1.24	.26	13.89				
010250.8-241312		.S?....	1.02± .05						5586
165.03 -86.12	ESO 475- 8		.12± .04	.10	15.25 ±.14				5582
273.51 -8.11				.17					5308
010025.0-242918	PGC 3742		1.03	.06	14.95				
010251.2-653636	NGC 360	.S..4./	1.55± .04	144				14.20±.3	2299± 10
301.04 -51.48	ESO 79- 14	S	.87± .05	.00	13.42 ±.14			379± 12	2264± 34
231.41 -15.08		3.7± .5		1.27				362± 9	2170
010057.0-655242	PGC 3743		1.55	.43	12.13			1.63	2198
0103.1 +2459		.S..6*.	1.02± .06	10					9023± 10
126.28 -37.81	UGC 643	U	.07± .05	.17	15.25 ±.13				
321.42 4.79		6.0±1.2		.10					9162
0100.4 +2443	PGC 3752		1.04	.03	14.95				8717
010307.0-061929	NGC 355	.LB.0P*	1.00± .10	125					
131.06 -69.01	MCG -1- 3- 77	E	.41± .05	.26					
291.13 -3.65		-2.0±1.3		.00					
010035.3-063534	PGC 3753		.97						
010307.1-065919	NGC 356	.SXS4P*	1.17± .05	140					5891± 54
131.30 -69.67	MCG -1- 3- 78	E	.26± .05	.33					5943
290.49 -3.83	VV 486	4.0± .8		.38	12.56				5578
010035.6-071524	PGC 3754		1.20	.13					
010310.7-033633		.L?....	1.12± .12						2490± 39
130.24 -66.31	MCG -1- 3- 79		.05± .07	.14					2553
293.78 -2.94	MK 970			.00	13.59				2173
010038.1-035239	PGC 3757		1.13						
010314.5+084929		.S?....	.94± .11						11509
127.89 -53.93	MCG 1- 3- 13		.47± .07	.13	15.44 ±.12				11608
305.83 .42				.36					11186
010038.1+083323	PGC 3761		.93		14.79				
010316.1+222032	NGC 354	.SB..P.	.91± .10	29				16.40±.3	4665± 9
126.53 -40.45	UGC 645	R	.27± .06	.14	14.39 ±.16			181± 10	4861± 60
318.88 4.06	MK 353			.41	11.71				4802
010035.0+220426	PGC 3763		.92	.14	13.81			2.45	4358
010321.9-062023	NGC 357	.SBR0*.	1.38± .04	160	13.14 ±.15	1.14± .02	1.15± .01		2541± 56
131.23 -69.02	MCG -1- 3- 81	R	.14± .04	.26	12.81 ±.18	.67± .04	.69± .01		2595
291.13 -3.71		.0± .5	.86± .03	.11		1.03	12.93± .10		2226
010050.2-063628	PGC 3768		1.40		12.59	.59	14.54± .25		
010323.9-571043		.S?....	.90± .07						11114
299.70 -59.87	ESO 151- 21		.10± .06	.00	15.48 ±.14				11009
240.13 -14.28				.14					10971
010118.1-572648	PGC 3770		.90	.05	15.27				
010325.9+321406		.SB?...	1.07± .08	105				17.36±.3	5319± 10
125.88 -30.57	UGC 646		.34± .06	.18	14.84 ±.12				5418± 80
328.45 6.58				.52				399± 7	5474
010040.9+315800	PGC 3773		1.09	.17	14.11			3.07	5030
0103.6 +0625		.S..2..	1.04± .08	172					14527
128.38 -56.32	UGC 649	U	.29± .06	.09	15.28 ±.14				
303.53 -.32		2.0± .9		.36					14619
0101.0 +0609	PGC 3779		1.05	.15	14.68				14204
0103.6 +0833		.S..0..	1.00± .08	38					
128.08 -54.20	UGC 650	U	.33± .06	.13					
305.59 .26		.0± .9		.25					
0101.0 +0817	PGC 3780		1.00						
0103.6 -0028		.S..6*.	1.02± .06	165					
129.72 -63.19	UGC 651	U	.56± .05	.10					
296.85 -2.21		6.0±1.3		.82					
0101.1 -0045	PGC 3781		1.03	.28					
010346.3-274513			.86± .06	165					5215± 76
226.51 -87.19	ESO 412- 3		.39± .05	.07	14.85 ±.14				5200
270.04 -9.13	IRAS01013-2801				13.84				4949
010122.1-280118	PGC 3783		.87						
0103.7 +2201		.SB.6?.	1.11± .07	35					5670± 10
126.71 -40.76	UGC 652	U	.29± .06	.15					
318.60 3.86		6.0±1.8		.43					5802
0101.1 +2145	PGC 3784		1.12	.15					5359

R.A. 2000 DEC. / l b / SGL SGB / R.A. 1950 DEC.	Names / PGC	Type / S_T n_L / T / L	$\log D_{25}$ / $\log R_{25}$ / $\log A_e$ / $\log D_o$	p.a. / A_g / A_i / A_{21}	B_T / m_B / m_{FIR} / B^o_T	$(B-V)_T$ / $(U-B)_T$ / $(B-V)^o_T$ / $(U-B)^o_T$	$(B-V)_e$ / $(U-B)_e$ / m' / m'_{25}	m_{21} / W_{20} / W_{50} / HI	V_{21} / V_{opt} / V_{GSR} / V_{3K}
010349.6+311158 / 126.04 -31.60 / 327.47 6.24 / 010105.1+305553	UGC 653 / PGC 3787	.S..6*. / U / 6.0±1.5 /	.96± .09 / .90± .06 / / .98	174 / .17 / 1.33 / .45				16.33±.3 / 273± 7 /	6262± 10 / 6414 / 5969
0103.8 +0217 / 129.24 -60.43 / 299.55 -1.51 / 0101.3 +0201	UGC 656 / PGC 3789	.SB?... / / /	.93± .09 / .24± .06 / / .93	115 / .02 / .36 / .12	15.31 ±.12 / / / 14.90				5518 / 5598 / 5197
0103.8 +2153 / 126.75 -40.89 / 318.47 3.81 / 0101.2 +2137	LGS 3 / PGC 3792	.I?.... / / /		.10				16.65±.1 / 36± 6 / 33± 7 /	-281± 6 / -149 / -592
010359.6+325753 / 125.97 -29.83 / 329.19 6.64 / 010114.2+324148	UGC 657 / PGC 3798	.S..9*. / U / 9.0±1.2 /	1.04± .08 / .08± .06 / / 1.06	/ .19 / .08 / .04				16.34±.3 / 132± 7 /	4978± 10 / 5133 / 4689
010401.4+415034 / 125.44 -20.97 / 337.83 8.77 / 010111.6+413429	UGC 655 / PGC 3803	.S..9.. / U / 9.0± .7 /	1.40± .04 / .00± .05 / / 1.43	/ .34 / .00 / .00	14.4 ±.4 / / / 14.01			14.07±.1 / 119± 8 / .05	829± 11 / 998 / 566
010407.2-510755 / 298.06 -65.88 / 246.37 -13.58 / 010156.0-512400	IC 1615 / ESO 195- 27 / IRAS01019-5124 / PGC 3812	.SBR4.. / Sr (1) / 4.2± .6 / 2.2± .9	1.10± .05 / .34± .04 / / 1.10	143 / .04 / .50 / .17	14.29 ±.14 / / / 13.65				
010410.6-623126 / 300.40 -54.54 / 234.62 -14.95 / 010212.1-624730	ESO 79- 15 / PGC 3814	.SBT6.. / r / 5.6± .9 /	1.01± .07 / .08± .06 / / 1.01	2 / .00 / .11 / .04	14.78 ±.14 / / / 14.63				8697 / 8578 / 8582
010417.0-004554 / 130.13 -63.46 / 296.62 -2.44 / 010143.5-010159	NGC 359 / UGC 662 / PGC 3817	.L..-*. / UE / -3.0± .8 /	1.18± .08 / .15± .05 / / 1.17	135 / .08 / .00	14.26 ±.12				
010417.2-510156 / 297.97 -65.97 / 246.47 -13.59 / 010206.0-511800	IC 1617 / ESO 195- 28 / IRAS01020-5117 / PGC 3818	PS.R0?. / Sr / -.4± .7 /	1.14± .05 / .39± .04 / / 1.12	125 / .04 / .29	14.59 ±.14				
0104.3 +0637 / 128.67 -56.11 / 303.77 -.43 / 0101.7 +0621	UGC 660 / PGC 3819	.I..9*. / U / 10.0±1.2 /	.96± .09 / .00± .06 / / .97	/ .08 / .00 / .00					12302 / 12394 / 11980
010418.4-400856 / 292.18 -76.71 / 257.58 -11.79 / 010200.0-402500	ESO 295- 31 / PGC 3820	PSBT2.. / r / 2.2± .9 /	.99± .06 / .13± .05 / / .99	109 / .02 / .16 / .06	14.73 ±.14 / / / 14.47				8099 / 8044 / 7878
010418.5+184151 / 127.18 -44.07 / 315.43 2.86 / 010138.6+182547	UGC 659 / PGC 3821	.S?.... / / /	1.00± .05 / .53± .04 / / 1.01	159 / .09 / .79 / .26	15.32 ±.12 / / / 14.41			15.98±.3 / 230± 13 / 1.31	5463± 10 / 5587 / 5148
010419.0-350713 / 284.72 -81.55 / 262.67 -10.82 / 010158.0-352318	NGC 365 / ESO 352- 1 / IRAS01019-3523 / PGC 3822	.SBR4P* / PBS (1) / 4.2± .5 / 3.3± .9	.99± .05 / .21± .04 / / .99	5 / .02 / .30 / .10	14.25 ±.14 / / / 12.51				
0104.3 +8040 / 123.48 17.81 / 16.60 15.12 / 0100.0 +8024	UGC 642 / PGC 3824	.S..3.. / U / 3.0± .9 /	1.00± .08 / .43± .06 / / 1.12	46 / 1.23 / .60 / .22					
010421.6-431631 / 294.56 -73.64 / 254.41 -12.37 / 010205.0-433236	NGC 368 / ESO 243- 23 / PGC 3826	PLXT+?. / BSr / -.7± .8 /	.86± .06 / .03± .03 / / .86	/ .03 / .00	14.66 ±.14				
010427.9-640720 / 300.57 -52.95 / 232.97 -15.12 / 010232.0-642324	ESO 79- 16 / IRAS01025-6423 / PGC 3827	.S?.... / / /	1.03± .06 / .40± .05 / / 1.03	0 / .00 / .60 / .20	14.35 ±.14 / / 12.23 / 13.72				5963± 49 / 5840 / 5856
010429.8-512732 / 298.02 -65.54 / 246.04 -13.68 / 010219.0-514336	ESO 195- 29 / PGC 3828	.SXS5.. / S (1) / 5.0± .9 / 4.4±1.3	1.08± .05 / .33± .04 / / 1.08	118 / .04 / .50 / .17	15.05 ±.14				

R.A. 2000 DEC. / l b / SGL SGB / R.A. 1950 DEC.	Names / PGC	Type / S_T n_L / T / L	logD_25 / logR_25 / logA_e / logD_o	p.a. / A_g / A_i / A_21	B_T / m_B / m_FIR / B_T^o	(B-V)_T / (U-B)_T / (B-V)_T^o / (U-B)_T^o	(B-V)_e / (U-B)_e / m'_e / m'_25	m_21 / W_20 / W_50 / HI	V_21 / V_opt / V_GSR / V_3K
010430.6-333914 280.38 -82.89 264.16 -10.56 010209.0-335518	 ESO 352- 2 IRAS01021-3355 PGC 3829	.S?.... 	1.10± .05 .35± .05 1.10	168 .02 .49 .18	 14.06 ±.14 13.53 13.55				32± 60 -2 -213
0104.5 -0009 130.14 -62.85 297.22 -2.34 0102.0 -0026	 UGC 663 PGC 3830	.E..... U -5.0± .9 	1.04± .13 .32± .06 .96	7 .11 .00 					
010442.6-004341 130.36 -63.41 296.68 -2.53 010209.1-005945	NGC 364 UGC 666 PGC 3833	RLBS0*. UE -2.0± .6 	1.15± .07 .05± .04 1.15	30 .08 .00 	 14.12 ±.08 				
010442.6-373802 288.87 -79.13 260.15 -11.39 010223.0-375406	 ESO 295- 32 PGC 3834	.SBS6?/ S (1) 6.0±1.8 4.4±1.3	1.11± .05 .51± .04 1.11	17 .02 .75 .26	 14.77 ±.14 				
0104.8 +0537 129.06 -57.10 302.84 -.82 0102.2 +0521	 UGC 667 PGC 3838	.S..7*. U 7.0±1.2 	1.04± .15 .18± .12 1.05	75 .07 .24 .09					6086 6175 5764
010452.8+221854 127.02 -40.45 318.95 3.70 010211.5+220250	 MCG 4- 3- 39 PGC 3841	.SB?... 	.89± .11 .14± .07 .90	 .16 .21 .07	 15.39 ±.13 14.93			17.10±.3 154± 7 2.09	15247± 10 15379 14937
010454.2+020800 129.79 -60.56 299.47 -1.80 010219.8+015156	IC 1613 UGC 668 DDO 8 PGC 3844	.IBS9.. PUE (2) 10.0± .3 9.5± .5	2.21± .01 .05± .03 2.11± .04 2.22	50 .02 .04 .02	9.88M±.09 9.8 ±.2 12.58 9.81	.67± .04 .65 	.54± .03 15.92± .09 15.68± .13	10.73±.1 35± 4 25± 4 .89	-230± 5 -237± 22 -152 -551
010454.7-470002 296.22 -69.95 250.62 -13.08 010240.7-471606	 PGC 3845	.E.3.*. S -5.0±1.2 	1.05± .11 .20± .08 .99	 .03 .00 					
010456.2-272544 219.43 -86.99 270.43 -9.31 010232.0-274148	IC 1616 ESO 412- 4 IRAS01025-2741 PGC 3846	.SBT4.. Sr (1) 4.3± .4 1.4± .5	1.21± .04 .06± .04 1.22	 .07 .09 .03	 13.35 ±.14 13.10 13.15				5636± 47 5621 5370
010500.0-460429 295.74 -70.86 251.57 -12.95 010245.4-462033	 PGC 3847	.LBS-*. S -3.0± .9 	1.10± .10 .24± .08 1.07	 .03 .00 					
010501.9+220513 127.08 -40.67 318.74 3.60 010220.7+214910	 MCG 4- 3- 40 PGC 3849	.SB?... 	.94± .11 .00± .07 .95	 .16 .00 .00	 15.31 ±.16 15.05			16.36±.3 83± 10 1.32	16493± 10 16625 16183
010502.0+324839 126.23 -29.97 329.10 6.39 010216.4+323236	 CGCG 501- 74 PGC 3850		 	 .20 	 15.7 ±.4 			17.25±.3 371± 7 	11394± 10 11548 11106
0105.0 -0612 132.36 -68.83 291.38 -4.09 0102.5 -0628	A 0102-06 MCG -1- 3- 85 PGC 3853	.SXT7.. R (2) 7.0± .3 7.1± .6	1.62± .02 .07± .03 1.64	70 .26 .10 .04	12.39S±.15 12.2 ±.2 11.96	.75± .05 .12± .07 .67 .06	.79± .05 .15± .07 15.16± .21	12.78±.1 182± 7 168± 7 .78	1094± 6 983± 79 1147 780
010507.2-800803 302.20 -36.97 216.37 -15.90 010422.1-802406	 ESO 13- 10 PGC 3854	.SXS7*P S (1) 7.4± .8 7.0± .7	1.09± .05 .26± .05 1.11	151 .31 .36 .13	 15.51 ±.14 				
010508.6-061651 132.43 -68.90 291.31 -4.12 010236.9-063254	 MCG -1- 3- 88 PGC 3855	.SB.9?/ E 9.0±1.8 	1.13± .06 .66± .05 1.16	105 .26 .67 .33					2425 2478 2112
010508.9-174527 142.29 -80.11 280.05 -7.07 010241.0-180130	NGC 369 ESO 541- 17 PGC 3856	.SXR3*. SrE (2) 3.1± .6 1.7±1.0	.99± .04 .09± .03 .99	52 .00 .12 .04	 14.61 ±.14 14.45				6187 6203 5894
010517.6-582615 299.45 -58.60 238.86 -14.67 010314.0-584218	 ESO 113- 10 PGC 3864	RSBR1?. r 1.0± .9 	.96± .06 .10± .05 .96	115 .00 .10 .05	 14.63 ±.14 				

R.A. 2000 DEC. l b SGL SGB R.A. 1950 DEC.	Names PGC	Type S_T n_L T L	$\log D_{25}$ $\log R_{25}$ $\log A_e$ $\log D_o$	p.a. A_g A_I A_{21}	B_T m_B m_{FIR} B_T^o	$(B-V)_T$ $(U-B)_T$ $(B-V)_T^o$ $(U-B)_T^o$	$(B-V)_e$ $(U-B)_e$ m'_e m'_{25}	m_{21} W_{20} W_{50} HI	V_{21} V_{opt} V_{GSR} V_{3K}
010519.3+314056 126.38 -31.10 328.02 6.05 010234.2+312453	 UGC 669 IRAS01025+3124 PGC 3866	.S..6*. U 6.0±1.3 	1.17± .05 .61± .05 1.19	125 .17 .89 .30	 15.40 ±.14 13.45 14.31			16.56±.3 255± 7 1.95	5865± 10 6017 5574
010520.1-471200 296.10 -69.74 250.43 -13.18 010306.3-472803	 S FAIR 675 PGC 3868	RLBT+.. S -1.0± .9 	1.01± .11 .07± .08 1.00	 .03 .00 					5000±190 4923 4810
010522.4+240150 127.01 -38.73 320.63 4.05 010240.5+234547	 CGCG 480- 40 PGC 3869			 .17 	 15.6 ±.4 			17.70±.3 448± 7 	9243± 10 9379 8936
010524.9-230115 161.30 -84.82 274.85 -8.40 010259.0-231718	 ESO 475- 9 PGC 3872	PSXS3.. r 3.3± .9 	1.06± .05 .41± .04 1.07	29 .06 .57 .20	 15.45 ±.14 				
010534.0-530003 298.08 -63.99 244.48 -14.06 010325.0-531606	 ESO 151- 22 PGC 3882	.S?.... 	 .92± .06 .03± .05 .93	 .04 .04 .02	 15.08 ±.14 14.92				9532 9438 9369
010537.8-470415 295.90 -69.86 250.57 -13.21 010324.1-472018	 ESO 243- 26 IRAS01033-4720 PGC 3888	.SB.4?/ S 3.8± .8 	1.12± .05 .80± .04 1.12	52 .03 1.17 .40	 15.49 ±.14 13.64 				
0105.9 +3225 126.48 -30.35 328.76 6.11 0103.2 +3209	IC 1618 UGC 671 PGC 3899	.L..... U -2.0± .9 	1.08± .09 .36± .04 1.05	159 .18 .00 	 15.24 ±.12 14.99				4706± 80 4859 4417
010609.6+312422 126.61 -31.36 327.80 5.81 010324.6+310820	A 0103+31 UGC 673 PGC 3903	.SA.5?. PU (1) 5.0± .8 6.0±1.3	1.11± .04 .44± .03 1.13	33 .21 .65 .22	15.70S±.15 15.22 ±.13 	.58± .07 -.08± .09 	 15.01± .25 	15.81±.2 307± 13 280± 6 	6254± 7 9322± 73
0106.1 +0047 130.76 -61.86 298.25 -2.47 0103.6 +0031	 UGC 675 PGC 3904	.I..9*. U 10.0±1.1 	1.03± .08 .00± .06 1.03	 .03 .00 .00					5201 5275 4882
0106.1 +4459 125.67 -17.81 340.98 9.11 0103.3 +4443	 UGC 672 PGC 3905	.I..9.. U 10.0± .8 	1.14± .13 .03± .12 1.19	 .58 .02 .01				16.68±.1 40± 4 28± 5 	708± 6 881 457
0106.1 +2123 127.51 -41.36 318.13 3.16 0103.5 +2107	 UGC 674 PGC 3906	.I..9*. U 10.0±1.2 	1.07± .14 .00± .12 1.08	 .13 .00 .00					5635 5764 5325
010611.9-301040 257.09 -85.55 267.73 -10.18 010349.0-302642	NGC 378 ESO 412- 5 IRAS01038-3026 PGC 3907	.SBR5*. S (1) 5.0± .6 1.1± .6	1.19± .04 .15± .04 1.19	90 .02 .23 .08	 13.83 ±.14 12.12 13.53				9620± 34 9595 9363
010613.7+253302 127.12 -37.20 322.15 4.26 010331.0+251700	 MCG 4- 3- 42 IRAS01035+2517 PGC 3908	.S?.... 	 .94± .11 .68± .07 .96	 .26 .84 .34	 15.59 ±.14 14.48			16.84±.3 307± 7 2.02	6623± 10 6762 6320
010614.1+033426 130.13 -59.09 300.96 -1.73 010339.2+031824	 UGC 678 PGC 3910	.SAT7*. UE (1) 6.5± .5 6.4± .8	1.20± .05 .08± .04 1.21	65 .08 .11 .04	 14.6 ±.2 14.35				5521 5603 5201
0106.3 +1435 128.34 -48.13 311.59 1.27 0103.7 +1419	 UGC 677 PGC 3914	.S?.... 	1.00± .06 .09± .05 1.01	 .09 .14 .05	 14.97 ±.15 14.68				12090 12203 11772
010624.1-000839 131.13 -62.78 297.37 -2.78 010350.4-002441	 CGCG 384- 73 MK 560 PGC 3922			 .12 	 15.32 ±.12 				10385± 45 10456 10067
010624.6-584710 299.25 -58.23 238.52 -14.85 010422.1-590312	 ESO 113- 11 PGC 3923	.S?.... 	.95± .06 .06± .06 .63± .02 .95	101 .00 .09 .03	14.84 ±.13 14.82 ±.14 14.67	.69± .02 .60 	.77± .02 13.48± .06 14.28± .37		12688± 40 12578 12554

R.A. 2000 DEC.	Names	Type	$\log D_{25}$	p.a.	B_T	$(B-V)_T$	$(B-V)_o$	m_{21}	V_{21}
l b		S_T n_L	$\log R_{25}$	A_g	m_B	$(U-B)_T$	$(U-B)_o$	W_{20}	V_{opt}
SGL SGB		T	$\log A_o$	A_i	m_{FIR}	$(B-V)^o_T$	m'_o	W_{50}	V_{GSR}
R.A. 1950 DEC.	PGC	L	$\log D_o$	A_{21}	B^o_T	$(U-B)^o_T$	m'_{25}	HI	V_{3K}
010630.8-021149		.SBT6..	1.12± .04	140					
131.81 -64.81	MCG 0- 3- 73	E (1)	.09± .04	.16					
295.38 -3.36		6.3± .5		.13					
010357.8-022751	PGC 3928	5.8± .5	1.14	.05					
010632.8-463834		PLXR+*.	1.01± .06	129					
295.24 -70.26	ESO 243- 29	S	.09± .04	.03	14.56 ±.14				9189
251.03 -13.30		-1.0± .9		.00					9113
010419.0-465436	PGC 3930		1.00		14.39				8997
010639.7+241302								17.01±.3	11733± 10
127.37 -38.52	CGCG 480- 43			.16	15.7 ±.4				
320.89 3.81								324± 7	11869
010357.5+235700	PGC 3936								11428
010649.7-454932		.LAT-?.	1.03± .11						
294.65 -71.05		S	.06± .08	.03					
251.88 -13.22		-3.0± .9		.00					
010435.4-460533	PGC 3939		1.02						
0106.8 +7536		.SB.3?.	1.16± .05						
123.92 12.76	UGC 670	U	.02± .05	1.64	14.4 ±.3				
11.45 14.47	IRAS01031+7520	3.0± .8		.02	13.19				
0103.1 +7520	PGC 3941		1.32	.01					
010658.7-383129		.S..3?/	1.14± .05	134					
287.98 -78.15	ESO 295- 36	S	.80± .04	.02	15.21 ±.14				
259.33 -12.00	IRAS01047-3847	3.0±1.9		1.11	13.94				
010440.1-384730	PGC 3945		1.14	.40					
010701.0-801824		.S..0*/	1.44± .04	157	13.55 ±.13	1.00± .02	1.04± .01		
302.11 -36.80	ESO 13- 12	S	.51± .05	.31	13.49 ±.14	.40± .02	.45± .01		5045± 26
216.19 -15.98	IRAS01063-8034	-.3± .6	.87± .01	.38	12.23	.80	13.39± .04		4887
010621.0-803424	PGC 3948		1.44		12.76	.27	14.36± .26		5028
010703.7+322327		.S..8?.	1.02± .04	95	16.61S±.15			16.34±.2	5096± 7
126.75 -30.36	UGC 679	U	.59± .03	.19				253± 13	
328.80 5.87		8.0±1.8		.73				182± 6	5248
010418.1+320726	PGC 3950		1.04	.30	15.69		15.09± .25	.36	4808
010705.5+324747	NGC 374	.S..0..	1.06± .06	175					
126.73 -29.96	UGC 680	U	.39± .05	.21	14.36 ±.13				5067± 80
329.19 5.97		.0± .9		.29					5220
010419.6+323146	PGC 3952		1.06		13.78				4780
010706.1+322050	NGC 375	.E.2.*.	1.15± .14						
126.77 -30.41		R	.00± .12	.19					6011± 56
328.76 5.85		-5.0± .6		.00					6163
010420.4+320449	PGC 3953		1.18						5723
010706.7-420029		.SXT3*.	1.01± .05	87					
291.78 -74.78	ESO 295- 37	S	.21± .05	.03	15.53 ±.14				16434
255.79 -12.65		3.0±1.3		.29					16372
010450.1-421630	PGC 3954		1.01	.11	15.09				16222
010710.4-422323		.SAS5?.	1.06± .05	145					
292.07 -74.40	ESO 243- 30	S (1)	.27± .04	.03	14.72 ±.14				
255.40 -12.72		5.0± .9		.40					
010454.0-423924	PGC 3957	3.3± .9	1.06	.13					
010710.7-415459		.LXS-..	1.09± .06	41					
291.65 -74.87	ESO 295- 38	S	.22± .04	.03	14.08 ±.14				
255.89 -12.64		-3.0± .9		.00					
010454.0-421100	PGC 3958		1.06						
0107.2 +1051	IC 75	.S..3..	1.10± .06	30					12325
129.21 -51.83	UGC 684	U (1)	.12± .05	.15	14.57 ±.15				12427
308.06 .04		3.0± .8		.16					12006
0104.6 +1035	PGC 3959	4.5±1.1	1.11	.06	14.17				
010714.3+135714	IC 1620	.SX.4..	1.00± .06	90					
128.75 -48.74	UGC 681	U	.15± .05	.06					11401± 46
311.05 .89	IRAS01045+1341	4.0± .9		.22	12.99				11511
010435.8+134113	PGC 3960		1.01	.07					11084
010715.8+323117	NGC 379	.L.....	1.14± .04	0	13.92 ±.13	1.05± .01	1.08± .01		
126.79 -30.23	UGC 683	R	.26± .07	.19	13.91 ±.10	.55± .03	.59± .03		5475± 37
328.94 5.87	IRAS01045+3215	-2.0± .4	.51± .01	.00		.93	11.96± .05		5627
010430.0+321516	PGC 3966		1.12		13.64	.51	13.82± .29		5187
010717.1+243038		.L?....	.90± .17					17.57±.3	9102± 10
127.52 -38.22	MCG 4- 3- 44		.00± .07	.15	15.22 ±.11				
321.21 3.75				.00				219± 7	9238
010434.8+241437	PGC 3968		.91		14.94				8798

R.A. 2000 DEC.	Names	Type	logD$_{25}$	p.a.	B$_T$	(B-V)$_T$	(B-V)$_e$	m$_{21}$	V$_{21}$
l b		S$_T$ n$_L$	logR$_{25}$	A$_g$	m$_B$	(U-B)$_T$	(U-B)$_e$	W$_{20}$	V$_{opt}$
SGL SGB		T	logA$_e$	A$_i$	m$_{FIR}$	(B-V)$_T^o$	m'$_e$	W$_{50}$	V$_{GSR}$
R.A. 1950 DEC.	PGC	L	logD$_o$	A$_{21}$	B$_T^o$	(U-B)$_T^o$	m'$_{25}$	HI	V$_{3K}$
010718.1+322902	NGC 380	.E.2...	1.14± .11		13.60 ±.13	1.06± .01	1.07± .01		
126.81 -30.27	UGC 682	R	.05± .07	.19	13.63 ±.10	.58± .03	.59± .03		4384± 35
328.90 5.85		-5.0± .4	.71± .02	.00		.98	12.64± .07		4536
010432.3+321301	PGC 3969		1.16		13.37	.56	14.17± .59		4096
010720.5-364517		PSBS2..	.90± .06	45					
284.65 -79.80	ESO 352- 6	r	.23± .05	.02	14.62 ±.14				
261.14 -11.74	IRAS01049-3701	2.2± .9		.28	12.91				
010501.0-370118	PGC 3971		.90	.12					
010722.4+164102		.SA.9..	1.08± .05	120				14.66±.3	155± 10
128.43 -46.02	UGC 685	U	.14± .05	.11	14.23 ±.12			90± 13	113± 50
313.68 1.61		9.0± .8		.15					271
010443.0+162501	PGC 3974		1.09	.07	13.97			.62	-160
010722.6+005531	NGC 391	PLA.-*.	.95± .08	45					
131.36 -61.69	UGC 693	UE	.09± .04	.03	14.37 ±.10				
298.47 -2.72		-3.2± .6		.00					
010448.6+003931	PGC 3976		.94						
010724.2-695235	NGC 406	.SAS5*.	1.52± .03	160	13.1 ±.2	.59± .02	.54± .02	13.14±.3	1509± 7
300.91 -47.19	ESO 51- 18	RS (2)	.42± .03	.05	12.93 ±.11	-.11± .03	-.10± .03	250± 8	1469± 58
227.02 -15.79	IRAS01057-7008	5.0± .5	.87± .05	.62	12.51	.49	12.91± .12		1372
010543.0-700836	PGC 3980	4.0± .6	1.53	.21	12.28	-.18	14.50± .26	.65	1434
010724.5+322413	NGC 382	.E...*.	.82± .14						
126.84 -30.35	UGC 688	RCU	.00± .04	.19	14.22 ±.13				5217± 32
328.83 5.81		-5.0± .5		.00					5369
010438.7+320813	PGC 3981		.85		13.96				4929
010725.2+322447	NGC 383	.LA.-*.	1.20± .10	30					
126.84 -30.34	UGC 689	R	.05± .07	.19	13.38 ±.10				5040± 26
328.84 5.81		-3.0± .3		.00					5192
010439.4+320846	PGC 3982		1.21		13.12				4752
010725.4+321733	NGC 384	.E.3...	1.04± .04	135	14.05 ±.15	.92± .02	1.00± .02		
126.85 -30.46	UGC 686	R	.11± .05	.19	14.08 ±.10	.48± .03	.58± .04		4398± 30
328.72 5.77	ARAK 26	-5.0± .5	.59± .04	.00		.84	12.49± .13		4550
010439.7+320133	PGC 3983		1.04		13.82	.46	13.99± .27		4110
010727.7+321916	NGC 385	.LA.-*.	1.03± .09		13.93 ±.13	.97± .01	1.05± .01		
126.86 -30.43	UGC 687	R	.04± .05	.19	14.11 ±.10	.41± .04	.45± .04		4940± 34
328.75 5.77		-3.0± .4	.74± .02	.00		.88	13.12± .08		5091
010441.9+320315	PGC 3984		1.05		13.78	.39	13.84± .49		4651
010730.6-612024		RLB.0..	.98± .07	70					
299.51 -55.68	ESO 113- 12	r	.17± .05	.00	15.21 ±.14				7983
235.88 -15.23		-2.4± .9		.00					7866
010532.0-613624	PGC 3988		.95		15.09				7863
010730.8+322144	NGC 386	.E.3.*.	.97± .14		15.33 ±.13	1.02± .02			
126.87 -30.39	MCG 5- 3- 57	R	.06± .06	.19	14.98 ±.11	.58± .07			5555± 56
328.80 5.77	ARAK 27	-5.0± .7		.00		.93			5707
010445.1+320543	PGC 3989		.98		14.86	.56	14.99± .74		5267
010732.7+392357		.S..6*.	1.35± .04	105				15.17±.1	5869± 9
126.32 -23.37	UGC 690	U	.08± .05	.22	13.42 ±.16			334± 11	
335.61 7.54		6.0±1.0		.12					6033
010443.4+390757	PGC 3990		1.37	.04	13.05			2.08	5600
0107.5 +3205		.S?....	1.11± .07	136					10644
126.90 -30.66	UGC 691		.95± .06	.19					
328.53 5.69				1.43					10795
0104.8 +3149	PGC 3992		1.13	.48					10355
010735.0-333818		PSBT1..	1.20± .04	61					
275.79 -82.62	ESO 352- 7	S	.29± .04	.02	14.81 ±.14				10127± 39
264.31 -11.18		1.0± .8		.30					10091
010514.0-335418	PGC 3993		1.20	.15	14.37				9883
010737.3+325613		.SBS6..	.98± .07					16.37±.3	5789± 10
126.84 -29.81	UGC 692	U	.02± .05	.21	14.83 ±.13				
329.36 5.90		6.0± .9		.03				141± 7	5942
010451.2+324013	PGC 3998		1.00	.01	14.56			1.80	5502
010742.3-465424	IC 1625	.LAT-P*	1.22± .03	7	12.9 ±.2	.93± .02	.98± .01		
294.80 -69.96	ESO 243- 33	S	.13± .03	.03	13.11 ±.14				6679± 23
250.79 -13.54		-3.3± .6	1.12± .04	.00		.85	14.01± .13		6601
010529.0-471024	PGC 4001		1.20		12.91		13.54± .27		6489
010746.4+321835	NGC 388	.E.3.*.	.97± .15		15.42 ±.19	1.10± .04	1.11± .04		
126.93 -30.43	MCG 5- 3- 59	R	.08± .06	.19	15.06 ±.11	.47± .08	.51± .08		5114± 56
328.76 5.71	ARAK 28	-5.0± .8	.52± .07	.00		1.01	13.51± .26		5266
010500.6+320235	PGC 4005		.98		14.88	.45	15.07± .78		4826

R.A. 2000 DEC. / l b / SGL SGB / R.A. 1950 DEC.	Names / PGC	Type S_T n_L / T / L	$\log D_{25}$ / $\log R_{25}$ / $\log A_o$ / $\log D_o$	p.a. / A_g / A_i / A_{21}	B_T / m_B / m_{FIR} / B_T^o	$(B-V)_T$ / $(U-B)_T$ / $(B-V)_T^o$ / $(U-B)_T^o$	$(B-V)_e$ / $(U-B)_e$ / m'_e / m'_{25}	m_{21} / W_{20} / W_{50} / HI	V_{21} / V_{opt} / V_{GSR} / V_{3K}
010747.7-173024 IC 1623			.65± .07						
145.21 -79.66 ESO 541- 23			.08± .06	.04	14.34 ±.14				6056± 39
280.45 -7.62 ARP 236					9.92				6071
010520.0-174624	PGC 4008		.65						5764
010748.8-580436		.S?....	1.04± .06	21					
298.74 -58.92 ESO 113- 13			1.06± .05	.00	16.70 ±.14				5914
239.27 -14.96				1.50					5805
010546.0-582036	PGC 4010		1.04	.50	15.17				5777
0107.8 +0104		.S?....	1.04± .06						624
131.57 -61.53 UGC 695			.07± .05	.04	14.91 ±.16				
298.64 -2.80				.10					698
0105.3 +0048	PGC 4013		1.04	.03	14.77				306
010757.8+331813		.S?....	.94± .11					16.81±.3	4209± 10
126.90 -29.44 MCG 5- 3- 60			.11± .07	.16	15.16 ±.13				
329.73 5.93				.16				125± 7	4362
010511.5+330213	PGC 4016		.95	.06	14.81			1.95	3924
0107.9 +0233		.S..2..	1.09± .06	86					
131.23 -60.06 UGC 698		U	.53± .05	.04	15.21 ±.12				
300.09 -2.42		2.0± .9		.65					
0105.4 +0217	PGC 4017		1.10	.27					
010804.2+210716		.L..-*.	.96± .12	150					
128.12 -41.58 UGC 696		U	.13± .05	.14	14.68 ±.10				
318.00 2.66 MK 561		-3.0±1.2		.00					
010523.0+205116	PGC 4019		.96						
010804.9+332700		.SB?...	.96± .07	85				17.76±.3	4700± 10
126.91 -29.29 UGC 697			.22± .05	.13	14.67 ±.12				4719± 80
329.88 5.94 IRAS01053+3311				.33				319± 7	4854
010518.5+331100	PGC 4020		.98	.11	14.18			3.47	4415
010809.2-582718		.S?....	.96± .06	41					
298.74 -58.54 ESO 113- 14			.13± .06	.00	14.56 ±.14				4944± 34
238.88 -15.05				.20					4834
010607.0-584318	PGC 4023		.96	.07	14.33				4809
010810.7-460536 IC 1627		.SBS3*/	1.40± .04	139					
294.10 -70.75 ESO 243- 34		S (1)	.60± .04	.03	13.71 ±.14				6107
251.64 -13.50 IRAS01059-4621		3.0± .6		.83					6032
010557.1-462136	PGC 4027	3.3±1.2	1.41	.30	12.81				5914
010810.9-470806		RLBT0..	1.03± .05	11					
294.69 -69.72 ESO 243- 35		Sr	.08± .03	.03	14.66 ±.14				
250.57 -13.65		-2.2± .6		.00					
010558.0-472406	PGC 4028		1.02						
0108.1 +2105		.S..9*.	1.00± .10						5017
128.16 -41.62 UGC 699		U	.00± .04	.14					
317.97 2.63		9.0±1.2		.00					5145
0105.5 +2049	PGC 4029		1.01	.00					4708
010817.2-464512 IC 1630		.S..3P/	1.11± .05	65					
294.43 -70.10 ESO 243- 36		S	.39± .05	.03	15.18 ±.14				6925
250.97 -13.61		3.0± .6		.54					6848
010604.0-470112	PGC 4036		1.11	.20	14.56				6735
010819.0-470418		RLBR+..	.95± .07	117					
294.59 -69.78 ESO 243- 37		r	.20± .05	.03	14.91 ±.14				6900±190
250.64 -13.67 FAIR 682		-1.3± .9		.00					6822
010606.0-472018	PGC 4038		.92		14.77				6711
010823.4+330759 NGC 392		.L..-*.	1.08± .17	50	13.68 ±.13	.97± .01	1.05± .01		
127.01 -29.60 UGC 700		U	.12± .05	.19	13.79 ±.10	.49± .07			4762± 44
329.59 5.80		-3.0±1.2	.72± .03	.00		.88	12.77± .09		4915
010537.1+325200	PGC 4042		1.09		13.48	.46	13.64± .89		4477
0108.4 +0627		.SBR5..	1.01± .06						12125
130.51 -56.16 UGC 705		U	.11± .05	.07					
303.90 -1.45		5.0± .8		.17					12214
0105.8 +0612	PGC 4045		1.01	.06					11806
0108.4 +0819		.SBR2..	1.00± .08						10910
130.14 -54.30 UGC 706		U	.00± .06	.13	14.92 ±.16				
305.71 -.94		2.0± .8		.00					11004
0105.8 +0804	PGC 4046		1.01	.00	14.68				10591
0108.4 -0518		.S?....	1.04± .09						10455
134.19 -67.80 MCG -1- 4- 2			.49± .07	.20					
292.49 -4.65				.73					10509
0105.9 -0534	PGC 4048		1.06	.25					10143

R.A. 2000 DEC. / l b / SGL SGB / R.A. 1950 DEC.	Names / PGC	Type / S_T n_L / T / L	$\log D_{25}$ / $\log R_{25}$ / $\log A_e$ / $\log D_o$	p.a. / A_g / A_i / A_{21}	B_T / m_B / m_{FIR} / B_T^o	$(B-V)_T$ / $(U-B)_T$ / $(B-V)_T^o$ / $(U-B)_T^o$	$(B-V)_e$ / $(U-B)_e$ / m'_e / m'_{25}	m_{21} / W_{20} / W_{50} / HI	V_{21} / V_{opt} / V_{GSR} / V_{3K}
010829.9+394139	NGC 389	.L.....	1.11± .07	54					
126.50 -23.06	UGC 703	U	.48± .03	.22	14.82 ±.10				
335.94 7.43		-2.0± .9		.00					
010540.2+392540	PGC 4054		1.06						
0108.5 +1652		.SB.6*.	.95± .07	48					12241
128.81 -45.79	UGC 708	U	.20± .05	.10	14.73 ±.12				
313.95 1.39		6.0±1.3		.30					12358
0105.9 +1637	PGC 4058		.96	.10	14.27				11927
0108.5 +0120		.S..8..	1.04± .06	142					5376
131.85 -61.22	UGC 709	U	.36± .05	.02					
298.96 -2.89		8.0± .9		.44					5450
0106.0 +0105	PGC 4059		1.04	.18					5058
010836.9+393835	NGC 393	.L..-*.	1.23± .14	20	13.6 ±.3	1.07± .02	1.05± .01		
126.53 -23.11	UGC 707	U	.08± .08	.22	13.13 ±.10	.68± .04	.65± .02		
335.89 7.40	52W 52	-3.0±1.1	.66± .13	.00			12.43± .45		
010547.2+392236	PGC 4061		1.25				14.47± .77		
0108.6 +0138	A 0106+01	.SBS7?/	1.56± .03	118				14.23±.1	1979± 6
131.82 -60.92	UGC 711	UE	.98± .04	.02	14.39 ±.14			223± 8	2010± 43
299.26 -2.84		6.7± .7		1.36				202± 12	2055
0106.1 +0123	PGC 4063		1.56	.49	13.00			.74	1662
010844.9+332746		.S..4..	1.14± .07	56				15.99±.3	12474± 10
127.07 -29.27	UGC 710	U	.47± .06	.13	15.28 ±.14				12624± 80
329.92 5.81		4.0± .9		.69				510± 7	12630
010558.5+331147	PGC 4067		1.15	.24	14.37			1.38	12192
010845.3-462831	IC 1631	.S..2P.	.97± .05	82					
294.04 -70.35	ESO 243-40	S	.17± .04	.03	14.22 ±.14				
251.26 -13.65	IRAS01065-4644	2.0± .7		.21	13.02				
010632.0-464430	PGC 4068		.98	.09					
0108.7 -1524		.S?....	.74± .14						15424± 41
142.76 -77.60	MCG -3- 4- 8		.00± .07	.02					15445
282.60 -7.33				.00					15128
0106.3 -1540	PGC 4071		.74						
010846.4-362037		.LX.-P.	1.07± .05	42					6641± 39
282.25 -80.08	ESO 352- 8	S	.21± .04	.02	14.61 ±.14				6595
261.62 -11.94		-3.0± .8		.00					6407
010627.0-363636	PGC 4073		1.05		14.49				
0108.7 -1520	IC 80B	.LA.-P*	.52± .20						16057
142.67 -77.54	MCG -3- 4- 7	PE	.00± .07	.02					16079
282.66 -7.31		-3.0± .9		.00					15761
0106.3 -1536	PGC 4074		.52						
010847.2-283455	IC 1628	.LAR-*P	1.11± .06		13.48 ±.13	1.01± .01	1.02± .01		
234.70 -85.90	ESO 412- 7	S	.06± .04	.06	13.50 ±.14				
269.46 -10.39		-4.0± .7	.72± .02	.00			12.57± .06		
010624.1-285054	PGC 4075		1.10				13.88± .35		
0108.7 -1237		.S?....	1.12± .08						7369
139.42 -74.93	MCG -2- 4- 3		.00± .07	.00					
285.34 -6.64				.00					7400
0106.3 -1253	PGC 4076		1.12						7068
010847.8-155042	IC 78	.SAT1P*	1.22± .05	133					
143.46 -78.02	MCG -3- 4- 10	E	.37± .05	.02					
282.16 -7.44		1.0±1.2		.38					
010619.6-160641	PGC 4079		1.22	.19					
010849.8-155656	IC 79	.LA.-..	.96± .10	155					
143.66 -78.11	MCG -3- 4- 11	PE	.11± .05	.02					12534± 58
282.06 -7.48		-2.5± .7		.00					12554
010621.6-161255	PGC 4082		.95						12240
010850.8-283558			.44± .22						
234.87 -85.88	MCG -5- 3- 28		.09± .07	.06					
269.45 -10.40									
010627.7-285157	PGC 4083		.44						
010851.7-454955		.L..0*P	1.01± .05	93					
293.59 -70.98	ESO 243- 41	S	.48± .03	.03	14.47 ±.14				
251.93 -13.57		-2.0± .9		.00					
010638.1-460554	PGC 4085		.94						
010851.9+320559		.S?....	.74± .14					16.85±.3	10423± 10
127.22 -30.63	MCG 5- 3- 66		.00± .07	.18	15.04 ±.13				
328.62 5.43				.00				97± 7	10574
010606.0+315000	PGC 4086		.75	.00	14.76			2.09	10135

R.A. 2000 DEC. l b SGL SGB R.A. 1950 DEC.	Names PGC	Type S_T n_L T L	$\log D_{25}$ $\log R_{25}$ $\log A_{\bullet}$ $\log D_{\circ}$	p.a. A_g A_i A_{21}	B_T m_B m_{FIR} B_T°	$(B-V)_T$ $(U-B)_T$ $(B-V)_T^{\circ}$ $(U-B)_T^{\circ}$	$(B-V)_{\bullet}$ $(U-B)_{\bullet}$ $m'_{\bullet}$ m'_{25}	m_{21} W_{20} W_{50} HI	V_{21} V_{opt} V_{GSR} V_{3K}
0108.9 +2152 128.31 -40.81 318.79 2.67 0106.2 +2137	 PGC 4095			.20				16.01±.1 58± 7	-324± 8 -195 -632
010859.2+323804 127.20 -30.09 329.14 5.55 010613.0+322206	NGC 399 UGC 712 PGC 4096	.SB.1*. U 1.0±1.2 	.97± .07 .13± .05 .99	40 .20 .13 .06	14.45 ±.12 14.06				5215± 43 5367 4929
010900.1-053057 134.66 -67.99 292.31 -4.84 010628.3-054656	 MCG -1- 4- 3 PGC 4099	.IBS9*. E (1) 10.0± .9 8.7± .8	1.07± .06 .25± .05 1.10	165 .24 .19 .12					
010902.9-371726 283.95 -79.18 260.67 -12.17 010644.0-373324	 ESO 296- 2 IRAS01067-3733 PGC 4101	RSBR1P* Sr 1.1± .5 	1.07± .05 .08± .05 1.07	165 .02 .09 .04	14.42 ±.14 13.29 14.23				6594 6545 6364
010903.3-743050 301.34 -42.56 222.20 -16.06 010741.1-744648	 ESO 29- 34 PGC 4102	.S?.... 	1.04± .09 .08± .07 1.05	 .13 .11 .04					10568 10421 10519
010904.7-454625 293.44 -71.03 251.99 -13.60 010651.1-460224	 ESO 243- 45 PGC 4104	.L..-.. S -3.0± .7 	1.14± .05 .07± .04 1.14	 .03 .00	13.41 ±.14 13.26				7746 7671 7552
0109.1 +3750 126.78 -24.89 334.18 6.86 0106.3 +3735	 UGC 713 PGC 4106	.S..8*. U 8.0±1.2 	1.00± .16 .04± .12 1.03	 .27 .05 .02					8341 8502 8068
010913.9+320901 127.31 -30.57 328.69 5.37 010627.9+315303	 UGC 714 PGC 4110	.SA.5.. U 5.0± .9 	1.06± .06 .13± .05 1.07	10 .18 .20 .07	14.35 ±.12 13.95			16.78±.3 237± 7 2.77	4634± 10 4617± 50 4784 4346
010914.5+324503 127.25 -29.97 329.26 5.52 010628.2+322905	NGC 403 UGC 715 PGC 4111	.S..0*. PU .3± .7 	1.28± .04 .47± .04 1.28	86 .20 .36	13.38 ±.14 12.75				5063± 44 5215 4777
010920.1+332841 127.21 -29.24 329.97 5.69 010633.4+331243	 MCG 5- 3- 70 PGC 4115	.S?.... 	.74± .14 .00± .07 .75	 .14 .00 .00	15.41 ±.12 15.21			16.21±.3 305± 7 .99	10593± 10 10746 10309
0109.3 +1419 129.46 -48.31 311.55 .50 0106.6 +1404	 UGC 717 IRAS01066+1403 PGC 4116	.SB.3.. U 3.0± .8 	1.17± .05 .19± .05 1.18	160 .09 .26 .09	14.37 ±.15 13.94				11034± 97 11144 10718
010922.4-785416 301.84 -38.19 217.64 -16.10 010830.8-791013	 PGC 4120	.SAS1P. S 1.0± .9 	1.08± .09 .16± .08 1.11	 .29 .16 .08					
0109.4 +1421 129.49 -48.28 311.59 .49 0106.8 +1406	 UGC 719 PGC 4124	.SB.3.. U 3.0± .9 	.91± .07 .03± .05 .92	 .09 .04 .01	14.89 ±.13 14.68				11330 11440 11014
010927.0+354304 127.03 -27.01 332.14 6.25 010639.3+352706	NGC 404 UGC 718 IRAS01066+3527 PGC 4126	.LAS-*. R -3.0± .3 	1.54± .02 .00± .03 .97± .02 1.57	 .24 .00	11.21 ±.13 11.23 ±.09 12.18 10.99	.94± .01 .27± .02 .89 .22	.85± .01 .31± .02 11.55± .05 13.78± .18	13.37±.3 91± 11 78± 8	-48± 9 -19± 29 111 -324
010932.0-474726 294.41 -69.04 249.93 -13.97 010720.0-480324	 ESO 195- 32 PGC 4131	.SXT8.. S (1) 8.0± .6 7.8± .6	.98± .05 .17± .05 .98	114 .03 .21 .09	15.49 ±.14				
010932.6-354820 280.10 -80.51 262.19 -12.00 010713.1-360418	NGC 409 ESO 352- 12 PGC 4132	.E...*. BS -5.0± .7 	1.11± .05 .08± .04 1.09	 .02 .00	14.03 ±.14 13.91				6550± 34 6506 6315
010941.6+231738 128.36 -39.39 320.19 2.89 010659.4+230140	 CGCG 480- 45 PGC 4138			.09	15.7 ±.4			17.23±.3 178± 7	10279± 10 10411 9974

R.A. 2000 DEC. / l b / SGL SGB / R.A. 1950 DEC.	Names / PGC	Type / S_T n_L / T / L	$logD_{25}$ / $logR_{25}$ / $logA_e$ / $logD_o$	p.a. / A_g / A_i / A_{21}	B_T / m_B / m_{FIR} / B_T^o	$(B-V)_T$ / $(U-B)_T$ / $(B-V)_T^o$ / $(U-B)_T^o$	$(B-V)_e$ / $(U-B)_e$ / m'_e / m'_{25}	m_{21} / W_{20} / W_{50} / HI	V_{21} / V_{opt} / V_{GSR} / V_{3K}
010941.9+310758		.L?....	.72± .22					17.36±.3	6497± 10
127.52 -31.58	MCG 5- 3- 73		.00± .07	.22	15.30 ±.10				
327.73 5.00				.00				184± 7	6645
010656.3+305200	PGC 4139		.74		14.98				6207
0109.7 +1317		.SBS8..	1.27± .04	150					4207± 10
129.77 -49.33	UGC 722	U	.19± .05	.10					
310.59 .12		8.0± .8		.23					4314
0107.1 +1302	PGC 4142		1.28	.09					3891
0109.7 -0214		.IBS9*.	1.14± .05	10					1872
133.70 -64.72	MCG -1- 4- 5	E (1)	.27± .04	.17					
295.56 -4.15		9.5± .6		.20					1935
0107.2 -0230	PGC 4143	8.1± .6	1.15	.14					1558
0109.8 +2045		.S..6*.	1.19± .05	165					5090
128.73 -41.90	UGC 723	U	1.06± .05	.16	15.56 ±.12				
317.77 2.16		6.0±1.4		1.47					5216
0107.2 +2030	PGC 4148		1.21	.50	13.91				4782
010955.3-455551	IC 1633	.E+1...	1.46± .04	120	12.6 ±.2		1.07± .01		
293.10 -70.84	ESO 243- 46	S	.08± .04	.03	12.23 ±.14		.54± .02		7226± 29
251.85 -13.77	FAIR 683	-4.0± .5	.95± .08	.00			12.88± .28		7150
010742.0-461148	PGC 4149		1.44		12.21		14.69± .31		7032
010955.5-465932		RLBR+..	.94± .07	157					
293.77 -69.81	ESO 243- 47	r	.18± .06	.03	15.38 ±.14				
250.76 -13.93		-1.3± .9		.00					
010743.0-471530	PGC 4150		.92						
010957.6-014457		.SBS7?.	1.20± .04	144					3849
133.63 -64.24	UGC 726	UE (1)	.24± .04	.22	14.59 ±.17				
296.05 -4.07		7.3± .7		.33					3913
010724.5-020054	PGC 4151	6.4±1.2	1.22	.12	14.03				3534
010959.7+322207	A 0107+32	.S.....	1.32± .10	15				15.12±.3	5179± 10
127.47 -30.34	UGC 724	R	.08± .12	.18	13.61 ±.16			343± 13	5228± 41
328.94 5.27				.12				321± 10	5332
010713.5+320610	PGC 4153		1.34	.04	13.29			1.79	4895
011005.6-352927	NGC 415	.SBT3..	1.16± .04	55					
278.64 -80.74	ESO 352- 14	S (1)	.26± .04	.02	14.32 ±.14				6538± 39
262.54 -12.05	IRAS01077-3545	3.0± .8		.35					6494
010746.0-354524	PGC 4161	3.3± .9	1.16	.13	13.90				6302
011011.0+430635	A 0107+42	.SB.6?.	1.35± .04	43	14.4 S±.3			15.10±.1	5047± 7
126.56 -19.63	UGC 725	U	.62± .05	.38	14.29 ±.13			334± 7	4986± 57
339.33 7.96		6.0±1.2		.91				316± 7	5214
010718.8+425038	PGC 4168		1.38	.31	12.98		14.39± .35	1.80	4790
011014.3-564551		.S..2?.	1.07± .05	166					
297.75 -60.17	ESO 151- 24	S	.51± .04	.00	15.60 ±.14				
240.67 -15.16		2.0±1.9		.62					
010811.0-570148	PGC 4169		1.07	.25					
011019.4-354409		.SXS7..	1.16± .04	129					
279.05 -80.50	ESO 352- 15	S (1)	.12± .04	.02	14.38 ±.14				3603± 39
262.30 -12.14		7.0± .8		.17					3558
010800.0-360006	PGC 4173	5.6±1.2	1.16	.06	14.18				3368
011021.3+301057								16.79±.3	6826± 10
127.78 -32.51	CGCG 501-113			.26	15.5 ±.4				
326.85 4.62								219± 7	6972
010736.1+295500	PGC 4176								6535
0110.3 +0330	NGC 396	.L.....	.83± .08	140					
132.14 -59.02	UGC 729	E	.23± .05	.07	15.23 ±.10				
301.19 -2.74		-2.0± .9		.00					
0107.8 +0315	PGC 4178		.80						
011029.1+431716		.SXS5..	1.18± .05	92	14.4 S±.3			14.97±.1	4909± 7
126.61 -19.45	UGC 728	U	.03± .05	.32	13.91 ±.14			279± 7	4922± 57
339.52 7.95	IRAS01076+4301	5.0± .8		.04				245± 7	5077
010736.7+430119	PGC 4184		1.21	.01	13.61		15.05± .38	1.35	4654
011033.6-523322		.E.4.*.	1.05± .06	127					
296.22 -64.32	ESO 151- 26	S	.25± .04	.04	14.35 ±.14				
245.03 -14.75		-5.0±1.2		.00					
010826.0-524918	PGC 4187		.99						
011036.1-301315	NGC 418	.SBS5..	1.31± .04		13.12 ±.16	.57± .02	.66± .02	14.52±.3	5709± 8
250.41 -84.78	ESO 412- 9	PS (2)	.07± .04	.02	13.19 ±.14	-.04± .03	.05± .03	261± 9	5684± 59
267.90 -11.12	IRAS01082-3029	5.0± .4	1.02± .03	.11	12.46	.52	13.71± .07	230± 7	5681
010814.0-302912	PGC 4189	1.3± .4	1.31	.04	13.00	-.08	14.33± .26	1.48	5454

R.A. 2000 DEC. / l b / SGL SGB / R.A. 1950 DEC.	Names / PGC	Type / S_T n_L / T / L	$\log D_{25}$ / $\log R_{25}$ / $\log A_e$ / $\log D_o$	p.a. / A_g / A_i / A_{21}	B_T / m_B / m_{FIR} / B_T^o	$(B-V)_T$ / $(U-B)_T$ / $(B-V)_T^o$ / $(U-B)_T^o$	$(B-V)_e$ / $(U-B)_e$ / m'_e / m'_{25}	m_{21} / W_{20} / W_{50} / HI	V_{21} / V_{opt} / V_{GSR} / V_{3K}
011036.6+330729 / 127.55 -29.58 / 329.70 5.34 / 010750.0+325132	NGC 407 / UGC 730 / / PGC 4190	.S..0*/ / PU / / .2± .8	1.23± .04 / .61± .04 / / 1.22	0 / .20 / .46 /	14.28 ±.12 / / / 13.54				5610± 80 / 5762 / 5326
011039.5-073554 / 136.94 -69.95 / 290.39 -5.79 / 010808.5-075150	/ MCG -1- 4- 7 / / PGC 4194	.LBS0*. / E / / -2.0±1.2	1.11± .08 / .05± .05 / / 1.13	.28 / .00 / /					
0110.6 -0016 / 133.47 -62.75 / 297.54 -3.84 / 0108.1 -0032	/ UGC 734 / / PGC 4195	.S?....	1.02± .06 / .60± .05 / / 1.03	175 / .11 / .90 / .30	15.42 ±.12 / / / 14.38				5491 / 5559 / 5176
0110.6 +1635 / 129.57 -46.02 / 313.82 .83 / 0108.0 +1620	/ UGC 733 / / PGC 4196	.SBS3.. / U / / 3.0± .9	.83± .11 / .10± .06 / / .84	/ .12 / .13 / .05	15.42 ±.12 / / / 15.07				12307 / / 12422 / 11994
0110.7 +4936 / 126.13 -13.15 / 345.67 9.41 / 0107.7 +4920	A 0107+49 / UGC 731 / DDO 9 / PGC 4202	.I..9*. / U (1) / 10.0±1.0 / 9.0±1.3	1.34± .10 / .04± .12 / 1.05± .10 / 1.41	* / .77 / .03 / .02			.58± .04 / -.19± .06 / 15.50± .23 /	13.41±.3 / 143± 9 / 130± 7 /	639± 8 / / 815 / 407
011048.5+333441 / 127.55 -29.12 / 330.15 5.42 / 010801.5+331845	/ UGC 732 / KUG 0108+333 / PGC 4210	.SAR7.. / U / / 7.0± .8	1.01± .05 / .26± .05 / / 1.02	80 / .16 / .36 / .13	* / 14.64 ±.12 / / 14.11			16.37±.2 / 288± 13 / 272± 6 / 2.13	5436± 7 / / 5588 / 5153
0110.8 -0145 / 134.14 -64.20 / 296.11 -4.28 / 0108.3 -0201	/ UGC 736 / / PGC 4214	.S..6*. / U / / 6.0±1.2	1.10± .05 / .14± .05 / / 1.12	125 / .23 / .21 / .07	* / 14.96 ±.18 / / 14.50				5073 / / 5136 / 4759
011052.0+242719 / 128.56 -38.20 / 321.38 2.95 / 010809.2+241123	CGCG 480- 47 / / / PGC 4219			.19	15.4 ±.4			18.20±.3 / / 84± 7 /	13902± 10 / / 14036 / 13600
011053.7-054920 / 136.06 -68.20 / 292.14 -5.38 / 010822.1-060516	/ MCG -1- 4- 8 / / PGC 4222	.SAS5*. / E (1) / 5.0± .8 / 3.6± .8	1.35± .04 / .38± .05 / / 1.38	15 / .31 / .58 / .19	*				5757 / / 5808 / 5447
011058.4+330906 / 127.63 -29.54 / 329.75 5.28 / 010811.7+325310	NGC 410 / UGC 735 / / PGC 4224	.E+..*. / PU / / -4.0± .6	1.38± .05 / .26± .05 / 1.06± .03 / 1.34	30 / .20 / .00 /	12.52 ±.13 / 12.50 ±.11 / / 12.23	1.04± .01 / .56± .02 / .94 / .54	1.06± .01 / .58± .02 / 13.31± .09 / 13.79± .32		5296± 25 / / 5447 / 5012
011101.0-302616 / 251.74 -84.58 / 267.70 -11.25 / 010839.0-304212	IC 1637 / ESO 412- 10 / IRAS01086-3042 / PGC 4227	.SBT5*. / PS (2) / '4.7± .5 / 2.0± .6	1.23± .04 / .12± .04 / / 1.23	90 / .02 / .18 / .06	13.46 ±.15 / 13.59 ±.14 / 13.06 / 13.30	.64± .04 / .05± .05 / .57 / .00	/ / / 14.16± .26		6002± 59 / / 5974 / 5749
011101.0-555210 / 297.26 -61.04 / 241.61 -15.18 / 010857.1-560806	/ ESO 151- 27 / / PGC 4228	.LA.-.. / S / / -3.0± .8	1.10± .05 / .07± .04 / / 1.09	46 / .00 / .00 /	14.08 ±.14 / / / 14.00				5304 / 5200 / 5158
011105.0+335005 / 127.59 -28.86 / 330.41 5.44 / 010817.9+333409	/ UGC 738 / / PGC 4235	.S..8.. / U / / 8.0± .8	1.08± .07 / .03± .06 / / 1.09	/ .18 / .03 / .01	15.01 ±.19 / / / 14.79				4558 / / 4711 / 4276
011106.2-180858 / 150.73 -79.94 / 280.01 -8.54 / 010839.0-182454	NGC 417 / ESO 541- 24 / / PGC 4237	.LX.-*. / SE / / -2.7± .7	.80± .05 / .10± .03 / / .78	55 / .00 / .00 /	15.18 ±.14				
0111.1 +0845 / 131.18 -53.79 / 306.31 -1.46 / 0108.5 +0830	/ UGC 741 / / PGC 4238	.E...*. / U / / -5.0±1.2	1.00± .19 / .05± .08 / / 1.01	/ .15 / .00 /	14.91 ±.12				
011112.5-073249 / 137.30 -69.87 / 290.47 -5.91 / 010841.4-074845	/ MCG -1- 4- 9 / / PGC 4249	.SA.5*. / E (1) / 5.0±1.2 / 5.3±1.2	1.06± .06 / .18± .05 / / 1.08	80 / .28 / .27 / .09					
011117.5+314425 / 127.85 -30.94 / 328.41 4.84 / 010831.3+312829	/ UGC 742 / KUG 0108+314 / PGC 4255	.S..8.. / U / / 8.0± .8	.99± .05 / .02± .04 / / 1.01	/ .22 / .02 / .01	15.03 ±.15 / / / 14.78			15.56±.3 / / 124± 7 / .78	5757± 10 / / 5906 / 5470

R.A. 2000 DEC.	Names	Type	logD25	p.a.	BT	(B-V)T	(B-V)o	m21	V21
l b		ST nL	logR25	Ag mB		(U-B)T	(U-B)o	W20	Vopt
SGL SGB		T	logA.	Ai mFIR		(B-V)oT	m'	W50	VGSR
R.A. 1950 DEC.	PGC	L	logDo	A21 BoT		(U-B)oT	m'25	HI	V3K
011118.6+315316		.S..1..	1.14± .07	8					
127.84 -30.79	UGC 743	U	.32± .06	.22	14.59 ±.13				5229± 80
328.55 4.87		1.0± .9		.33					5378
010832.4+313720	PGC 4258		1.16	.16	13.98				4942
011118.9-455558		.SA.1?.	1.00± .05	51					
292.38 -70.79	ESO 243- 51	S	.24± .04	.03	14.19 ±.14				
251.89 -14.01	IRAS01091-4611	1.0± .6		.24	13.30				
010906.0-461154	PGC 4259		1.01	.12					
011122.5-291404	NGC 423	.S.0?P/	.99± .05	114					
239.71 -85.13	ESO 412- 11	S	.44± .03	.08	14.40 ±.14				
268.93 -11.08	IRAS01090-2929	.0±1.9		.33	12.55				
010900.0-293000	PGC 4266		.98						
011126.9-455616		.L..0*/	1.11± .05	106					
292.32 -70.78	ESO 243- 52	S	.73± .03	.03	14.72 ±.14				
251.89 -14.03		-1.5±1.0		.00					
010914.0-461212	PGC 4271		1.00						
011127.9-380458	NGC 424	RSBR0*.	1.26± .04	60					
283.19 -78.27	ESO 296- 4	BSr	.35± .05	.02	13.76 ±.14				3450± 53
259.95 -12.79		.4± .5		.26					3397
010910.0-382054	PGC 4274		1.24		13.42				3225
011130.6+011916		.S..8*.	1.00± .07	136					
133.37 -61.14	UGC 749	U	.47± .05	.04	14.31 ±.17				6809± 30
299.14 -3.61	ARAK 33	8.0±1.3		.58	12.74				6882
010856.4+010321	PGC 4275		1.00	.23	13.68				6494
0111.5 +3640		.S..6*.	.96± .09	170					9623
127.41 -26.02	UGC 745	U	.39± .06	.21					
333.18 6.10		6.0±1.3		.58					9781
0108.7 +3625	PGC 4276		.98	.20					9349
011131.9+231308			.95?					15.88±.3	10308± 10
128.91 -39.42	CGCG 480- 48		.95?	.10					
320.24 2.47								359± 7	10439
010849.5+225713	PGC 4278		.96						10004
011136.1+490715		.E.....	1.20± .14	100					
126.32 -13.62	UGC 746	U	.11± .08	.82	14.0 ±.3				
345.23 9.16		-5.0± .8		.00					
010839.2+485120	PGC 4282		1.30						
011143.6+351633		.S?....	1.33± .04	79					4857
127.60 -27.41	UGC 748		.52± .05	.19	14.85 ±.18				
331.83 5.69	IRAS01089+3500			.78	13.54				5012
010855.7+350038	PGC 4286		1.35	.26	13.86				4579
011146.2-613135	NGC 432	.L..-..	1.11± .05						8080
298.66 -55.43	ESO 113- 22	S	.04± .04	.00	13.96 ±.14				
235.73 -15.75		-3.0± .8		.00					7961
010950.0-614730	PGC 4290		1.10		13.84				7962
011146.6-003948	IC 1639	.P...?.	.75± .09	100					
134.22 -63.09	UGC 750	E	.03± .05	.12	14.35 ±.18				5395± 50
297.23 -4.21	MK 562	99.0		.04					5461
010913.1-005543	PGC 4292		.76	.02	14.18				5081
011147.8-471253		.S?....	.98± .07	106					
293.02 -69.52	ESO 243- 53		.24± .05	.03	15.33 ±.14				11800±190
250.58 -14.27	FAIR 684			.36					11720
010936.0-472848	PGC 4294		.98	.12	14.88				11613
011150.1-013920	A 0109-01	.L...*/	.79± .13	15					
134.65 -64.06		E	.66± .08	.27					4960± 45
296.27 -4.49	MK 563	-2.0±1.3		.00					5023
010916.9-015515	PGC 4295		.73						4647
011150.7-555123	IC 1649	.S..4./	1.23± .05	136					
297.02 -61.04	ESO 151- 30	S	.79± .04	.00	14.71 ±.14				5340
241.63 -15.29		4.0± .9		1.16					5236
010947.1-560718	PGC 4298		1.23	.39	13.52				5194
011208.9-321430		PSAR1*.	1.05± .04						
262.85 -83.19	ESO 352- 18	Sr	.02± .03	.02	14.55 ±.14				9836± 34
265.92 -11.84		1.0± .6		.02					9802
010948.0-323024	PGC 4316		1.06	.01	14.39				9590
011210.1+320723	NGC 420	.L...*.	1.30± .12						
128.03 -30.54	UGC 752	U	.00± .08	.18	13.09 ±.11				4974± 36
328.82 4.76		-2.0±1.0		.00					5123
010923.7+315128	PGC 4320		1.32		12.84				4689

R.A. 2000 DEC. / l b / SGL SGB / R.A. 1950 DEC.	Names / PGC	Type / S_T n_L / T / L	$\log D_{25}$ / $\log R_{25}$ / $\log A_e$ / $\log D_o$	p.a. / A_g / A_i / A_{21}	B_T / m_B / m_{FIR} / B_T^o	$(B-V)_T$ / $(U-B)_T$ / $(B-V)_T^o$ / $(U-B)_T^o$	$(B-V)_e$ / $(U-B)_e$ / m'_e / m'_{25}	m_{21} / W_{20} / W_{50} / HI	V_{21} / V_{opt} / V_{GSR} / V_{3K}
011213.6-581448	NGC 434	.SXS2..	1.33± .03	6	12.79 ±.13	.82± .01	.89± .01		4924± 42
297.67 -58.67	ESO 113- 23	R (1)	.25± .03	.00	12.98 ±.11	.32± .02	.34± .01		4814
239.15 -15.56		2.0± .4	.82± .01	.30		.74	12.38± .04		
011013.0-583042	PGC 4325	2.2± .8	1.33	.12	12.54	.27	13.66± .20		4790
011216.1-542730		.SBT3?.	1.09± .05	58					
296.39 -62.40	ESO 151- 32	S (1)	.46± .04	.00	15.54 ±.14				
243.09 -15.21		3.0± .9		.63					
011011.0-544324	PGC 4329	4.4±1.3	1.09	.23					
011220.0-320342	NGC 427	RSBR1*.	1.01± .05	0					10012
261.56 -83.30	ESO 412- 14	Sr	.19± .03	.02	15.06 ±.14				9978
266.11 -11.85		.7± .5		.20					
010959.0-321936	PGC 4333		1.01	.10	14.72				9765
011220.0-502406	IC 1650	.SBS3P.	1.12± .05	62	14.40 ±.13	.76± .02			
294.60 -66.39	ESO 195- 34	Sr (1)	.28± .05	.04	14.36 ±.14				7334± 40
247.30 -14.77	IRAS01101-5039	3.4± .5		.39	12.89	.65			7245
011011.0-504000	PGC 4334	3.3± .9	1.12	.14	13.90		14.12± .30		7162
011229.6-581230	NGC 434A	.SBS0P/	1.06± .05	51				14.60±.3	4828± 8
297.59 -58.70	ESO 113- 24	RS	.57± .04	.00	15.79 ±.14			297± 9	4740± 69
239.19 -15.59		.0± .4		.43					4716
011029.0-582824	PGC 4344		1.03		15.29				4693
011232.0-040847		.LB.-?.	1.05± .06	165					
136.21 -66.47	MCG -1- 4- 14	E	.12± .05	.23					
293.89 -5.33		-3.0±1.2		.00					
010959.7-042441	PGC 4346		1.06						
0112.5 -0247	NGC 413	.SBR5*.	1.05± .06	25					5808
135.55 -65.14	MCG -1- 4- 13	E (1)	.19± .05	.19					
295.22 -4.97		5.0± .9		.28					5867
0110.0 -0303	PGC 4347	1.9± .8	1.07	.09					5496
011233.3-172742		PSBS2..	1.00± .05	165					13371± 39
150.79 -79.17	ESO 541- 25	r	.16± .04	.01	15.50 ±.14				13384
280.78 -8.71		2.2± .9		.19					
011006.0-174336	PGC 4348		1.01	.08	15.16				13083
0112.5 +5036		RLX.+..	1.07± .07	160					
126.36 -12.12	UGC 754	U	.10± .03	.93					
346.73 9.35		-1.0± .8		.00					
0109.6 +5021	PGC 4351		1.14						
011237.8-375348		.SBS3*.	1.14± .05	22					6492
281.79 -78.35	ESO 296- 6	S (1)	.42± .05	.02	15.26 ±.14				6440
260.18 -12.98		3.0± .7		.58					
011020.0-380942	PGC 4353	4.4± .9	1.15	.21	14.61				6267
011238.4-460418		.L?....	.93± .07	80					8100±190
291.81 -70.60	ESO 244- 2		.30± .05	.03	15.36 ±.14				8023
251.78 -14.26	FAIR 685			.00					
011026.0-462012	PGC 4354		.89		15.21				7909
011238.8+383020		.SXT5..	1.13± .05						6457
127.48 -24.18	UGC 755	U	.03± .05	.26	14.27 ±.15				
334.99 6.35		5.0± .8		.05					6617
010948.9+381427	PGC 4355		1.16	.02	13.92				6188
011239.1-334006		.L..-P.	1.13± .05	178	15.3 ±.2	1.06± .03	1.01± .03		
269.27 -82.02	ESO 352- 20	S	.25± .04	.02	14.70 ±.14	.47± .05	.45± .04		10018± 39
264.49 -12.22		-3.0± .8	.49± .09	.00		.95	13.19± .30		9979
011019.0-335600	PGC 4356		1.10		14.70	.51	15.16± .34		9777
011243.4+325809		.S..6*.	.94± .05	120				16.45±.3	5346± 10
128.07 -29.69	UGC 756	U	.09± .04	.20	14.62 ±.09				
329.67 4.88	KUG 0109+327B	6.0±1.3		.13				206± 7	5496
010956.4+324215	PGC 4359		.96	.04	14.26			2.15	5063
011248.3-581655	NGC 440	.SAS4P*	1.06± .03	45	13.73 ±.13	.55± .01	.65± .01		5015± 36
297.54 -58.62	ESO 113- 25	RS (1)	.23± .03	.00	13.75 ±.14	-.10± .02	.00± .02		4904
239.12 -15.64		3.8± .5	.46± .01	.33		.47	11.52± .01		4881
011048.0-583248	PGC 4361	5.6±1.3	1.06	.11	13.37	-.15	13.30± .21		
011249.0-001728	NGC 426	.E+....	1.14± .07	140	13.8 ±.4	.95± .04			5264± 31
134.64 -62.68	UGC 760	UE	.14± .04	.07	13.99 ±.10	.46± .06			5331
297.66 -4.36		-4.0± .5		.00		.89			
011015.4-003322	PGC 4363		1.11		13.83	.47	14.15± .54		4950
011255.2+005859	NGC 428	.SXS9..	1.61± .02	120	11.91M±.10	.44± .02	.51± .02	12.63±.1	1162± 7
134.21 -61.42	UGC 763	R (3)	.12± .03	.03	11.60 ±.11	-.19± .03	-.10± .03	179± 8	1045± 58
298.91 -4.04	IRAS01103+0043	9.0± .3	1.22± .01	.12	12.37	.40	13.47± .03	155± 8	1231
011021.2+004305	PGC 4367	6.0± .5	1.61	.06	11.62	-.22	14.53± .15	.95	847

R.A. 2000 DEC. / l b / SGL SGB / R.A. 1950 DEC.	Names / PGC	Type / S_T n_L / T / L	$\log D_{25}$ / $\log R_{25}$ / $\log A_\bullet$ / $\log D_o$	p.a. / A_g / A_i / A_{21}	B_T / m_B / m_{FIR} / B_T^o	$(B-V)_T$ / $(U-B)_T$ / $(B-V)_T^o$ / $(U-B)_T^o$	$(B-V)_\bullet$ / $(U-B)_\bullet$ / $m'_\bullet$ / m'_{25}	m_{21} / W_{20} / W_{50} / HI	V_{21} / V_{opt} / V_{GSR} / V_{3K}
011257.1-002048	NGC 429	.L..0*/	1.14± .06	19					
134.73 -62.72	UGC 762	UE	.67± .03	.07	14.37 ±.12				5644± 50
297.62 -4.41		-1.7± .8		.00					5710
011023.5-003641	PGC 4368		1.05		14.22				5331
011257.2-312700		RSBS0..	.96± .06						
256.88 -83.62	ESO 412- 16	r	.08± .04	.02	14.57 ±.14				5625
266.76 -11.86		-.1± .9		.06					5593
011036.1-314254	PGC 4369		.96		14.41				5377
011259.4-001508	NGC 430	.E...*.	1.12± .07	155	13.5 ±.4	.98± .03			
134.72 -62.63	UGC 765	UE	.08± .04	.07	13.36 ±.10	.44± .05			5283± 31
297.71 -4.39		-5.0± .6		.00		.91			5350
011025.8-003102	PGC 4376		1.10		13.22	.45	13.87± .55		4970
011259.7-190025		.LA.-*P	1.06± .15	65					
155.43 -80.49	MCG -3- 4- 30	S	.13± .05	.01	16.68 ±.14				
279.27 -9.18		-3.0± .8		.00					
011033.0-191618	PGC 4377		1.04						
011302.5+384607	NGC 425	.S?....	1.02± .06						6355± 11
127.54 -23.91	UGC 758		.03± .05	.27	13.58 ±.14			15.64±.1	6453± 52
335.27 6.35	IRAS01102+3830			.04	12.14			269± 12	6519
011012.3+383013	PGC 4379		1.04	.01	13.24			2.39	6091
0113.1 +0217		.L?....	.96± .13						13839
133.88 -60.12	UGC 768		.13± .05	.04	15.00 ±.11				13913
300.19 -3.74	IRAS01105+0201			.00	13.56				13524
0110.5 +0201	PGC 4386		.95		14.75				
0113.2 +3458		.S..6*.	1.00± .08	12					4736
127.96 -27.68	UGC 764	U	1.02± .06	.18					
331.63 5.32		6.0±1.5		1.47					4890
0110.4 +3443	PGC 4387		1.02	.50					4458
011314.0+085200			1.00?						10789
132.03 -53.62	CGCG 411- 7		.70?	.19	15.58 ±.12				
306.56 -1.94									10882
011037.0+083607	PGC 4388		1.02						10474
0113.2 +1245		.S..8?.	.99± .06						9977
131.18 -49.76	UGC 769	U	.09± .05	.11					
310.31 -.85		8.0±1.7		.11					10081
0110.6 +1230	PGC 4389		1.01	.05					9663
0113.2 +5040		.E...?.	1.03± .08						
126.47 -12.04	UGC 761	U	.01± .03	1.08					
346.82 9.26		-5.0±1.6		.00					
0110.3 +5025	PGC 4394		1.20						
0113.4 +4936		.S..8*.	1.04± .08	170					
126.59 -13.10	UGC 766	U	.29± .06	.81					
345.79 8.98		8.0±1.3		.36					
0110.5 +4921	PGC 4401		1.12	.15					
011330.8-521807		.SBS5..	1.03± .05	140					
295.08 -64.49	ESO 151- 34	S (1)	.15± .04	.04	15.25 ±.14				
245.35 -15.17		5.3± .5		.22					
011124.0-523400	PGC 4404		1.04	.07					
011333.6-375407	NGC 438	PSXS3*.	1.14± .04						
280.97 -78.27	ESO 296- 7	BSr (1)	.10± .04	.02	13.63 ±.14				3457
260.21 -13.16	IRAS01112-3810	2.8± .4		.14	12.62				3404
011116.0-381000	PGC 4406	3.3± .8	1.14	.05	13.44				3232
011337.0-452010		.LX.-*.	1.06± .10						
290.73 -71.26		S	.41± .08	.03					
252.56 -14.33		-3.0±1.1		.00					
011124.3-453603	PGC 4409		1.00						
0113.6 -0005		.S..2..	1.03± .06	105					
135.01 -62.43	UGC 771	U	.27± .05	.07	14.75 ±.12				5156
297.92 -4.51		2.0± .9		.33					5223
0111.1 -0021	PGC 4415		1.03	.13	14.30				4843
0113.6 +0052		.I..9*.	1.08± .07					15.39±.1	1164± 10
134.63 -61.48	UGC 772	U	.12± .06	.04				84± 6	
298.86 -4.25		10.0±1.1		.09				65± 5	1234
0111.1 +0037	PGC 4416		1.08	.06					850
011346.2-310344		RSB.1..	1.04± .05						
253.15 -83.73	ESO 412- 17	r	.05± .04	.02	14.61 ±.14				5767
267.19 -11.95		1.0± .9		.05					5736
011125.1-311936	PGC 4421		1.04	.02	14.47				5518

R.A. 2000 DEC. / l b / SGL SGB / R.A. 1950 DEC.	Names / PGC	Type / S_T n_L / T / L	$\log D_{25}$ / $\log R_{25}$ / $\log A_e$ / $\log D_o$	p.a. / A_g / A_i / A_{21}	B_T / m_B / m_{FIR} / B_T^o	$(B-V)_T$ / $(U-B)_T$ / $(B-V)_T^o$ / $(U-B)_T^o$	$(B-V)_e$ / $(U-B)_e$ / m'_e / m'_{25}	m_{21} / W_{20} / W_{50} / HI	V_{21} / V_{opt} / V_{GSR} / V_{3K}
0113.7 +0222		.S?....	1.11± .14						
134.15 -60.00	UGC 773		.15± .12	.11					14100± 60
300.32 -3.86				.18					14174
0111.2 +0207	PGC 4422		1.12	.07					13786
011347.8-314450	NGC 439	.LXT-?.	1.39± .03	156	*		1.02± .01		
257.77 -83.29	ESO 412- 18	PS	.20± .03	.02	12.49 ±.11		.45± .01		5763± 30
266.49 -12.09		-3.3± .4	1.11± .05	.00			13.57± .18		5729
011127.0-320042	PGC 4423		1.36		12.39				5516
011348.9+074706	A 0111+07								
132.55 -54.67				.16					5560± 45
305.55 -2.38	MK 564								5650
011112.3+073114	PGC 4425								5245
011351.4+131619		.S?....	.95± .07	95				17.83±.2	14880± 10
131.30 -49.24	UGC 774		.14± .05	.10	14.76 ±.12			538± 26	14806± 44
310.84 -.85	MK 975			.21				335± 19	14981
011112.7+130027	PGC 4428		.96	.07	14.37			3.39	14563
011351.8-314714	NGC 441	PSBT0*.	1.15± .05	135	*				
257.95 -83.25	ESO 412- 19	Sr	.10± .05	.02	13.76 ±.14				5659± 39
266.45 -12.11		-.3± .4		.08					5625
011131.0-320306	PGC 4429		1.14		13.58				5412
011358.5-622402		.SBS6..	1.14± .04	94	*				
298.44 -54.53	ESO 80- 1	S　(1)	.27± .04	.00	14.90 ±.14				4908
234.84 -16.07		6.0± .8		.40					4786
011205.0-623954	PGC 4432	6.7± .9	1.14	.13	14.48				4796
011359.9+020414	NGC 435	.SXS7*.	1.05± .05	20	*				
134.37 -60.30	UGC 779	UE　(1)	.40± .04	.04	14.81 ±.12				
300.04 -4.00		6.7± .7		.55					
011125.4+014822	PGC 4434	5.3±1.2	1.05	.20					
011401.2-575602		PSBS5..	1.19± .05	82	14.20 ±.15	.62± .02	.58± .02		
297.12 -58.94	ESO 113- 27	Sr　(1)	.99± .05	.00	14.45 ±.14				4905± 40
239.50 -15.77		4.8± .6	.94± .02	1.49		.40	14.39± .06		4795
011201.0-581154	PGC 4435	4.4± .9	1.19	.50	12.81		12.50± .32		4770
011404.7+334212	NGC 431	.LB....	1.15± .07	20					
128.31 -28.93	UGC 776	U	.21± .03	.19	13.86 ±.10				5786± 80
330.45 4.80		-2.0± .8		.00					5937
011117.1+332620	PGC 4437		1.14		13.58				5505
0114.0 -1303		.S?....	1.04± .09						
144.67 -74.96	MCG -2- 4- 14		.49± .07	.00					5295
285.24 -7.99				.74					5321
0111.6 -1319	PGC 4438		1.04	.25					4999
011407.3-323902	IC 1657	RSBS4*.	1.37± .04	170					
262.63 -82.61	ESO 352- 24	BSr　(1)	.63± .04	.02	13.16 ±.14				3552± 19
265.59 -12.33	IRAS01117-3254	3.7± .4		.92	11.83				3516
011147.0-325454	PGC 4440	2.2± .5	1.37	.31	12.20				3309
011408.1-330920		.SBT5..	1.03± .05	133					
265.17 -82.23	ESO 352- 25	S　(1)	.16± .04	.02	15.18 ±.14				5500
265.07 -12.43		5.0± .9		.24					5462
011148.1-332512	PGC 4441	4.4± .9	1.03	.08	14.89				5258
0114.1 -0151			.86± .10						
136.04 -64.14	CGCG 385- 36		.06± .06	.24	14.90 ±.08				5089
296.23 -5.11									5150
0111.6 -0207	PGC 4442		.89						4778
0114.1 -0144		.S..3..	1.09± .06	50					4902
135.98 -64.03	UGC 784	U　(1)	.20± .05	.24	14.63 ±.13				
296.35 -5.07		3.0± .8		.27					4963
0111.6 -0200	PGC 4443	3.5±1.1	1.11	.10	14.07				4591
011410.7+421425		.S..6*.	.95± .07	70				16.49±.1	5055± 11
127.42 -20.43	UGC 777	U	.06± .05	.37	15.11 ±.14				
338.67 7.04		6.0±1.3		.08				173± 8	5220
011118.1+415833	PGC 4446		.99	.03	14.63			1.83	4799
011414.7+375724		.S..3..	1.00± .08	110					
127.88 -24.70	UGC 780	U　(1)	.76± .06	.25	15.56 ±.13				
334.55 5.91	IRAS01113+3741	3.0±1.0		1.05	13.51				
011124.7+374132	PGC 4451	4.5±1.3	1.02	.38					
011415.5-321456		.SBS3P.	.99± .04	64					
260.25 -82.88	ESO 352- 26	S	.22± .03	.02	15.16 ±.14				
266.00 -12.28		3.0± .9		.30					
011155.0-323048	PGC 4453		.99	.11					

R.A. 2000 DEC. / l b / SGL SGB / R.A. 1950 DEC.	Names / PGC	Type / S_T n_L / T / L	$\log D_{25}$ / $\log R_{25}$ / $\log A_e$ / $\log D_o$	p.a. / A_g / A_i / A_{21}	B_T / m_B / m_{FIR} / B_T^o	$(B-V)_T$ / $(U-B)_T$ / $(B-V)_T^o$ / $(U-B)_T^o$	$(B-V)_e$ / $(U-B)_e$ / m' / m'_{25}	m_{21} / W_{20} / W_{50} / HI	V_{21} / V_{opt} / V_{GSR} / V_{3K}
0114.2 +0044	IC 87	.SB?...	.74± .14						
134.99 -61.59	MCG 0- 4- 48		.09± .07	.04	15.31 ±.13				12849
298.77 -4.43				.14					12918
0111.7 +0029	PGC 4454		.74	.05	15.05				12536
0114.2 +1554		RLX.0*.	1.07± .07	155					
130.94 -46.61	UGC 785	U	.12± .03	.11	14.71 ±.12				
313.39 -.20	IRAS01116+1538	-2.0± .8		.00	12.82				
0111.6 +1538	PGC 4456		1.06						
0114.2 +5013		.S..8*.	1.00± .08	28					
126.67 -12.47	UGC 778	U	1.02± .06	1.05					
346.42 9.00		8.0±1.5		1.23					
0111.3 +4958	PGC 4457		1.10	.50					
011421.9+055533	NGC 437	.S..0..	1.12± .07	130					
133.30 -56.48	UGC 788	U	.11± .06	.10	13.79 ±.12				5291± 50
303.80 -3.03		.0± .8		.08					5375
011146.0+053941	PGC 4464		1.12		13.53				4976
011423.2-552351	NGC 454	.P.....	1.26± .06						
296.12 -61.43	ESO 151- 36	S	.00± .06	.00	13.13 ±.14				3645± 44
242.14 -15.61		99.0							3541
011220.1-553942	PGC 4468		1.26						3497
0114.4 +5109		.S..6?.	1.11± .07	140					
126.60 -11.54	UGC 782	U	.54± .06	1.07					
347.33 9.20		6.0±1.8		.79					
0111.4 +5054	PGC 4469		1.21	.27					
011426.1+423321	A 0111+42	.SXT5..	1.15± .05	153	14.44 ±.17	.76± .04	.78± .03	15.27±.1	5914± 7
127.44 -20.11	UGC 783	U (1)	.23± .05	.37	14.36 ±.13			337± 11	5914± 59
338.99 7.07		5.0± .8	.94± .03	.34		.59	14.63± .07	323± 7	6079
011133.2+421730	PGC 4473	2.7± .8	1.19	.11	13.64		14.50± .33	1.52	5658
011430.0-311051		.S?....	1.09± .06	39					
253.14 -83.54	ESO 412- 21		.55± .04	.02	14.89 ±.14				5574± 39
267.10 -12.13	IRAS01121-3126			.82	12.60				5542
011209.0-312642	PGC 4477		1.09	.28	14.02				5326
011430.5-321545	A 0112-32	.SBS4P*	1.12± .04	88	14.1 ±.3	.57± .06			
260.03 -82.83	ESO 352- 27	PS (2)	.12± .03	.02	14.21 ±.14				5262± 59
266.00 -12.34	IRAS01121-3231	4.5± .4		.18		.51			5226
011210.1-323136	PGC 4478	2.1± .4	1.12	.06	13.96		14.26± .37		5017
011438.5-010114	NGC 442	.S..0*/	.99± .06	157					
135.93 -63.31	UGC 789	RUE	.30± .05	.16	14.45 ±.13				5620± 50
297.08 -5.00	IRAS01121-0116	-.1± .5		.22					5683
011205.2-011705	PGC 4484		.99		13.98				5308
0114.6 +0110		.S..6*.	.95± .07						
135.03 -61.14	UGC 790	U	.05± .05	.04	14.71 ±.12				4635
299.22 -4.41		6.0±1.2		.07					4705
0112.1 +0055	PGC 4485		.95	.02	14.58				4322
0114.6 +0149		.SB.2..	1.02± .06	130					
134.79 -60.50	UGC 791	U	.12± .05	.05	14.67 ±.13				
299.85 -4.23		2.0± .9		.15					
0112.1 +0134	PGC 4486		1.02	.06					
011449.9-002941			.64± .17	87					
135.80 -62.78	UGC 793		.06± .07	.07	14.7 ±.2				10167± 39
297.60 -4.90	IRAS01122-0045				12.74				10232
011216.3-004532	PGC 4490		.64						9855
011452.6+015505	NGC 445	.L...P*	.88± .08	135					
134.86 -60.41	CGCG 385- 47	E	.07± .05	.05	15.14 ±.07				
299.95 -4.26		-2.0±1.3		.00					
011218.2+013914	PGC 4493		.87						
011456.3+315654	IC 1652	.S..0..	1.03± .08	169					
128.73 -30.66	UGC 792	U	.53± .06	.16	14.48 ±.16				5216± 43
328.82 4.15		.0±1.0		.40					5363
011209.5+314103	PGC 4498		1.02		13.84				4932
0114.9 +0025		.E...*.	.95± .13						
135.48 -61.87	UGC 797	U	.02± .05	.04	14.95 ±.12				13439± 30
298.51 -4.68		-5.0±1.2		.00					13507
0112.4 +0010	PGC 4500		.95		14.71				13127
011500.4-321433		RSBT0..	1.00± .04						
259.33 -82.77	ESO 352- 28	Sr	.07± .03	.02	15.08 ±.14				5852± 34
266.04 -12.44		-.3± .4		.06					5816
011240.0-323024	PGC 4505		1.00		14.92				5608

R.A. 2000 DEC. l b SGL SGB R.A. 1950 DEC.	Names PGC	Type S_T n_L T L	$\log D_{25}$ $\log R_{25}$ $\log A_o$ $\log D_o$	p.a. A_g A_i A_{21}	B_T m_B m_{FIR} B_T^o	$(B-V)_T$ $(U-B)_T$ $(B-V)_T^o$ $(U-B)_T^o$	$(B-V)_e$ $(U-B)_e$ m'_o m'_{25}	m_{21} W_{20} W_{50} HI	V_{21} V_{opt} V_{GSR} V_{3K}
011504.3+052231 133.78 -57.00 303.31 -3.35 011228.6+050640	 MCG 1- 4- 7 PGC 4509	.S?.... 	.94± .11 .40± .07 .95	 .11 .59 .20	 15.02 ±.13 14.29				5316 5398 5002
0115.1 +0650 133.35 -55.54 304.74 -2.95 0112.5 +0635	 UGC 799 PGC 4511	.S..6*. U 6.0±1.3 	1.11± .14 .54± .12 1.12	103 .13 .79 .27					5523 5609 5209
011507.7+332237 128.60 -29.23 330.20 4.50 011220.1+330646	NGC 443 UGC 796 4ZW 42 PGC 4512	.S...*. R 	.92± .07 .05± .05 .94	 .16 .07 .02	 14.01 ±.14 13.75				 4897± 52 5047 4616
011509.9-442858 289.13 -72.00 253.49 -14.48 011257.0-444448	 ESO 244- 6 PGC 4517	.LA.-.. S -3.0± .6 	1.05± .05 .12± .04 1.04	11 .03 .00 	 14.65 ±.14 14.48				9299 9226 9102
011511.9+301143 129.01 -32.40 327.15 3.61 011225.9+295552	IC 1654 UGC 798 PGC 4520	RSBR1.. U 1.0± .8 	1.13± .07 .06± .06 1.15	60 .16 .06 .03	 14.06 ±.13 13.77			15.52±.3 211± 7 1.72	4897± 10 4844± 50 5039 4607
011515.9-013730 136.55 -63.86 296.53 -5.31 011242.7-015320	NGC 448 UGC 801 PGC 4524	.L..-./ UE -2.5± .6 	1.21± .08 .33± .05 1.20	116 .28 .00 	 13.14 ±.12 12.82				 1917± 31 1978 1606
011519.3+282924 129.27 -34.09 325.52 3.11 011234.1+281334	 UGC 800 KUG 0112+282B PGC 4531	.S..6*. U 6.0±1.2 	.95± .06 .03± .05 .96	110 .11 .05 .02	 14.94 ±.13 14.76			16.20±.3 169± 7 1.42	4812± 10 4952 4520
0115.4 +0805 133.14 -54.30 305.96 -2.67 0112.8 +0750	 UGC 803 PGC 4535	.SA.8.. U 8.0± .8 	1.18± .09 .03± .05 1.20	 .14 .04 .02					2348 2438 2034
0115.5 +0653 133.51 -55.48 304.81 -3.03 0112.9 +0638	 UGC 805 PGC 4539	.S..5.. U 5.0± .9 	1.11± .14 .66± .12 1.12	142 .12 .98 .33					5523 5609 5209
011530.7-005137 136.33 -63.10 297.29 -5.17 011257.4-010727	NGC 450 UGC 806 IRAS01129-0107 PGC 4540	.SXS6*. R (3) 6.0± .4 5.6± .5	1.49± .02 .12± .03 1.41± .09 1.50	72 .13 .17 .06	* 12.3 ±.2 12.08 11.95		.43± .03 -.17± .04 14.51± .23 	13.75±.0 194± 4 169± 8 1.74	1761± 5 1858± 33 1826 1452
011533.1-262658 207.11 -84.58 271.94 -11.40 011310.0-264248	 ESO 475- 14 PGC 4543	.SBS9P. S (1) 9.0± .8 5.6± .9	1.35± .03 .69± .03 1.36	130 .07 .71 .35	* 14.65 ±.14 13.86				 3772± 52 3754 3509
011535.6-611534 297.79 -55.63 236.04 -16.19 011341.1-613124	 ESO 113- 30 PGC 4546	PSXS6.. r 5.6± .9 	1.02± .06 .12± .05 1.02	 .00 .18 .06	* 14.71 ±.14 				
011536.8-063513 139.64 -68.67 291.71 -6.71 011305.6-065103	 MCG -1- 4- 19 PGC 4547	.SBS5P? E 5.0±1.7 	1.14± .06 .37± .05 1.17	60 .34 .56 .19	* 14.71 ±.14 				6346 6392 6041
0115.6 +0917 132.90 -53.11 307.14 -2.38 0113.0 +0902	 UGC 808 PGC 4549	.SAR5.. U 5.0± .7 	1.38± .04 .12± .05 1.40	60 .13 .18 .06	* 14.2 ±.3 13.83 				11653 11746 11340
011537.7+330403 128.76 -29.53 329.93 4.32 011250.1+324814	NGC 447 UGC 804 PGC 4550	RSBT0*. PU .0± .5 	1.35± .05 .01± .06 1.37	 .22 .01 	15.1 ±.4 13.53 ±.18 13.50	1.14± .10 .44± .17 1.04 .41	 16.67± .50	15.91±.3 212± 7 	5597± 10 5746 5316
011546.0-265034 211.30 -84.57 271.55 -11.53 011323.1-270624	 ESO 475- 15 PGC 4557	PSBS3?. S (1) 3.0± .8 3.9± .8	1.13± .04 .12± .03 1.13	 .03 .16 .06	 14.08 ±.14 				
011549.6+310450 129.06 -31.50 328.04 3.73 011303.0+304900	NGC 444 UGC 810 KUG 0113+308 PGC 4561	.S..7.. U 7.0± .9 	1.28± .03 .64± .03 .76± .04 1.30	157 .18 .89 .32	15.02M±.11 14.53 ±.12 13.72	.76± .04 -.01± .06 .57 -.17	.71± .03 .07± .04 14.30± .09 14.69± .19	15.09±.2 274± 8 262± 6 1.05	4839± 5 4889± 57 4985 4554

R.A. 2000 DEC.	Names	Type	logD$_{25}$	p.a.	B$_T$	(B-V)$_T$	(B-V)$_e$	m$_{21}$	V$_{21}$
l　　b		S$_T$　n$_L$	logR$_{25}$	A$_g$	m$_B$	(U-B)$_T$	(U-B)$_e$	W$_{20}$	V$_{opt}$
SGL　SGB		T	logA$_e$	A$_i$	m$_{FIR}$	(B-V)$_T^o$	m'$_e$	W$_{50}$	V$_{GSR}$
R.A. 1950 DEC.	PGC	L	logD$_o$	A$_{21}$	B$_T^o$	(U-B)$_T^o$	m'$_{25}$	HI	V$_{3K}$
011551.9+334837		.S..6*.	1.13± .05	23	15.4 ±.4	.71± .04		15.36±.3	4216± 10
128.72 -28.79	UGC　809	U	.81± .05	.16	14.91 ±.14	-.03± .07		341± 13	
330.65　4.47		6.0±1.4		1.18		.50		326± 10	4366
011304.0+333247	PGC　4563		1.14	.40	13.60	-.20	13.87± .49	1.36	3937
011553.1-322835	A　0113-32	.SXS6*.	1.08± .04		14.14 ±.17	.67± .04	.69± .03		
259.66 -82.47	ESO　352- 30	PS　　(2)	.03± .03	.02	14.21 ±.14				6041± 59
265.84 -12.66		5.7± .4	.85± .04	.04		.62	13.88± .09		6004
011333.0-324424	PGC　4566	2.6± .4	1.09	.01	14.09		14.34± .28		5798
011555.0-501123		RLXT+..	1.07± .06	125					
293.09 -66.47	ESO　195- 35	Sr	.12± .04	.04	14.43 ±.14				5200±190
247.59 -15.31	FAIR　294	-1.4± .5		.00	13.51				5110
011347.0-502712	PGC　4569		1.05		14.31				5029
011556.6-790248		.SXS5..	1.04± .05						
301.46 -38.01	ESO　13- 14	S　　　(1)	.07± .05	.30	15.09 ±.14				
217.49 -16.41		5.0± .8		.11					
011516.0-791836	PGC　4570	2.2± .8	1.07	.04					
011557.3+051039	NGC　455	.S?....	1.29± .06	165					
134.24 -57.15	UGC　815		.20± .06	.09	13.55 ±.13				5269± 50
303.18 -3.61	ARP　164			.27					5350
011321.7+045450	PGC　4572		1.30	.10	13.15				4956
011603.8+041740	IC　89	RLX.0..	1.31± .07					17.33±.3	5446± 9
134.58 -58.02	UGC　818	U	.10± .05	.07	13.35 ±.11			312± 10	5408± 27
302.34 -3.88	MK　565	-2.0± .8		.00	12.43			281± 7	5521
011328.5+040151	PGC　4578		1.30		13.20				5129
011604.2+373856		.S..0..	.93± .09						
128.31 -24.97	UGC　811	U	.13± .06	.28	15.41 ±.14				
334.35　5.48		.0± .9		.10					
011314.0+372307	PGC　4579		.94						
011604.8-613723		.S..3..	1.16± .04	58					
297.80 -55.26	ESO　113- 32	S　　　(1)	.53± .04	.00	14.42 ±.14				
235.66 -16.27		3.0± .9		.73					
011411.0-615312	PGC　4581	3.3±1.3	1.16	.26					
011606.1+302058	IC　1659	.E.....	1.20± .14	20					
129.22 -32.22	UGC　812	U	.15± .08	.16	14.14 ±.12				
327.35　3.47		-5.0± .8		.00					
011319.9+300509	PGC　4584		1.18						
0116.1　+0638		.S..8*.	.96± .17	76					5239
133.85 -55.70	UGC　819	U	.69± .12	.11					
304.61 -3.24		8.0±1.4		.84					5324
0113.5　+0623	PGC　4585		.97	.34					4926
0116.1　+0134		.SX.3*.	.99± .06						10191
135.61 -60.69	UGC　817	U	.22± .05	.05	14.77 ±.12				
299.70 -4.65	IRAS01135+0118	3.0±1.2		.31	13.21				10261
0113.5　+0118	PGC　4586		1.00	.11	14.34				9879
011607.2+330522	NGC　449	PS...$.	.88± .03		15.01 ±.13	.77± .02	.85± .02	16.04±.2	4824± 7
128.87 -29.50	MCG　5- 4- 9	R	.21± .06	.22	15.14 ±.12	.16± .03	.17± .03	259± 6	4795± 22
329.98　4.22	MK　1		.44± .01	.31	12.39	.65	12.70± .04		4970
011319.5+324933	PGC　4587		.90	.10	14.53	.06	13.75± .26	1.41	4541
011607.2-064324		.S..3*.	1.19± .05	117					6312
140.08 -68.77	MCG -1- 4- 22	E　　　(1)	.55± .05	.27					
291.60 -6.87		3.0±1.3		.76					6357
011336.0-065913	PGC　4588	3.1±1.6	1.22	.28					6007
011611.3+251132		.S?....	.84± .10					16.46±.3	8885± 10
129.98 -37.34	CGCG 481- 1		.14± .10	.18	15.30 ±.12				
322.42　2.00				.21				345± 7	9017
011327.6+245543	PGC　4593		.86	.07	14.86			1.53	8588
011612.5+330350	NGC　451	.S?....	.82± .07					16.43±.3	4880± 10
128.89 -29.52	MCG　5- 4- 11		.15± .06	.22	14.89 ±.14				4925± 40
329.96　4.20	MK　976			.22				231± 7	5031
011324.9+324801	PGC　4594		.84	.08	14.42			1.93	4602
011615.1+310202	NGC　452	.SB.2..	1.40± .03	43	13.64S±.15	1.09± .02		15.18±.2	4962± 5
129.17 -31.54	UGC　820	U	.49± .03	.18	13.76 ±.13	.58± .03		486± 8	5024± 29
328.02　3.63	VV　430	2.0± .8		.61	13.44	.92		471± 6	5109
011328.5+304613	PGC　4596		1.42	.25	12.88	.42	14.28± .22	2.05	4679
011616.5+464425		.S?....	1.08± .06	110					
127.35 -15.92	UGC　813		.37± .05	.52	14.8 ±.3				4994± 97
343.11　7.82				.54					5164
011320.0+462836	PGC　4598		1.13	.18	13.69				4754

R.A. 2000 DEC. / l b / SGL SGB / R.A. 1950 DEC.	Names / PGC	Type / S_T n_L / T / L	$logD_{25}$ / $logR_{25}$ / $logA_e$ / $logD_o$	p.a. / A_g / A_i / A_{21}	B_T / m_B / m_{FIR} / B_T^o	$(B-V)_T$ / $(U-B)_T$ / $(B-V)_T^o$ / $(U-B)_T^o$	$(B-V)_e$ / $(U-B)_e$ / m'_e / m'_{25}	m_{21} / W_{20} / W_{50} / HI	V_{21} / V_{opt} / V_{GSR} / V_{3K}
011620.8+464451		.S?....	1.28± .04		14.2 ±.3			14.77±.1	5188± 11
127.36 -15.91	UGC 816		.30± .05	.52					5334± 97
343.12 7.81	IRAS01133+4628			.44	11.87			357± 12	5360
011324.3+462902	PGC 4600		1.33	.15	13.20			1.43	4950
0116.4 +3326	And II	.E?....							
128.89 -29.14				.15					
330.34 4.27									
0113.6 +3311	PGC 4601								
011626.3+390310		.SB.1..	1.06± .06	1	14.87 ±.13				
128.22 -23.56	UGC 822	U	.25± .05	.21					
335.71 5.79	MK 977	1.0± .9		.25					
011335.2+384722	PGC 4604		1.08	.12					
011630.6-075839	IC 90	.E+....	1.10± .08	115					
141.31 -69.95	MCG -1- 4- 23	E	.13± .08	.25					
290.40 -7.29		-4.0± .9		.00					
011359.9-081428	PGC 4606		1.10						
011633.3+285832		.S?....	.87± .08		15.39 ±.12			16.26±.3	8170± 10
129.53 -33.57	CGCG 502- 21		.21± .12	.19					8310
326.07 2.99	KUG 0113+287			.31				217± 7	
011347.7+284244	PGC 4607		.88	.11	14.84			1.31	7880
0116.5 +0112		.S..9*.	.96± .17						5089
135.98 -61.02	UGC 823	U	.05± .12	.07					
299.38 -4.86		9.0±1.2		.05					5158
0114.0 +0057	PGC 4608		.97	.02					4777
011643.8+162354		.S?....	.89± .11		14.68 ±.14				1983
131.69 -46.04	MCG 3- 4- 13		.24± .07	.10					2094
314.04 -.63				.36					1674
011403.7+160806	PGC 4612		.90	.12	14.21				
0116.9 +1300		.S..1..	1.08± .06	73	14.88 ±.14				6155
132.50 -49.38	UGC 824	U	.70± .05	.08					
310.80 -1.64		1.0±1.0		.71					6257
0114.3 +1245	PGC 4619		1.08	.35	14.01				5844
011703.6-441054		.SXT2..	1.09± .05		14.24 ±.14				7000±190
287.77 -72.18	ESO 244- 10	Sr	.07± .05	.03					6927
253.84 -14.77	FAIR 688	1.5± .6		.08					
011451.0-442642	PGC 4623		1.10	.03	14.06				6803
011713.4-585436	NGC 466	.LAT+*.	1.25± .05	103	13.58 ±.14				
296.67 -57.90	ESO 113- 34	Sr	.08± .04	.00					5311± 34
238.51 -16.26		-.7± .6		.00					5197
011516.0-591024	PGC 4632		1.24		13.50				5182
011720.1-335024	NGC 461	.SXS5*.	1.07± .05	23	14.11 ±.14				
264.80 -81.28	ESO 352- 33	BS (1)	.12± .04	.02					5704± 39
264.50 -13.21	IRAS01150-3406	5.0± .9		.18	13.61				5662
011501.0-340612	PGC 4636	2.2± .9	1.08	.06	13.88				5467
0117.5 -0918		.SBR1P?	1.11± .08	95					
143.20 -71.14		E	.24± .08	.16					
289.16 -7.86		1.0±1.8		.24					
0115.0 -0934	PGC 4647		1.13	.12					
011733.3-521907		.S?....	.90± .06	73	14.14 ±.15	.67± .02	.75± .02		
293.70 -64.34	ESO 151- 39		.40± .05	.04	16.34 ±.14				7422± 40
245.40 -15.79			.83± .03	.56		.54	13.78± .07		7325
011528.0-523454	PGC 4649		.90	.20	14.66		12.48± .35		.7262
0117.5 +4900		.S..8*.	1.11± .07	145					
127.34 -13.63	UGC 825	U	1.05± .06	.91					
345.37 8.18		8.0±1.4		1.23					
0114.6 +4845	PGC 4650		1.20	.50					
0117.6 +3725		.I..9*.	1.00± .19						
128.68 -25.15	UGCA 16	U	.10± .09	.28					
334.22 5.12		10.0±1.3		.08					
0114.8 +3710	PGC 4653		1.03	.05					
011739.1+433853		.S..3*.	1.09± .06	165	14.58 ±.13				
127.94 -18.97	UGC 826	U	.21± .05	.33					
340.19 6.79	5ZW 61	3.0±1.2		.29					
011444.7+432306	PGC 4654		1.12	.11					
0117.7 +2124		.S..2..	1.00± .06	150	15.41 ±.12				11910
131.05 -41.04	UGC 828	U	.44± .05	.12					
318.91 .59		2.0± .9		.54					12033
0115.0 +2109	PGC 4655		1.02	.22	14.63				11608

R.A. 2000 DEC. l b SGL SGB R.A. 1950 DEC.	Names PGC	Type S_T n_L T L	$\log D_{25}$ $\log R_{25}$ $\log A_o$ $\log D_o$	p.a. A_g A_i A_{21}	B_T m_B m_{FIR} B_T^o	$(B-V)_T$ $(U-B)_T$ $(B-V)_T^o$ $(U-B)_T^o$	$(B-V)_e$ $(U-B)_e$ m' m'_{25}	m_{21} W_{20} W_{50} HI	V_{21} V_{opt} V_{GSR} V_{3K}
0117.7 +1011 133.52 -52.13 308.15 -2.63 0115.1 +0956	 UGC 829 IRAS01151+0956 PGC 4657	.S?.... 	1.11± .07 .15± .06 1.12	 .10 .15 .07	15.10 ±.20 12.58 14.72				10259 10353 9947
011751.4-015726 138.16 -64.05 296.38 -6.03 011518.4-021312	 UGC 830 PGC 4659	.SXR5?. UE (1) 5.3± .6 5.3±1.6	1.35± .03 .17± .04 1.37	120 .24 .25 .08	 14.7 ±.2 14.21				5952 6011 5644
011753.5-083719 142.84 -70.46 289.85 -7.78 011523.2-085306	 MCG -2- 4- 20 PGC 4663	.SBS9.. E (1) 9.0± .8 8.7± .8	1.21± .05 .15± .05 1.22	110 .13 .15 .08					4060 4098 3759
0118.0 +1732 131.91 -44.86 315.23 -.61 0115.4 +1717	NGC 459 UGC 832 PGC 4665	.S..4.. U 4.0± .9 	.98± .07 .04± .05 .99	 .12 .06 .02	15.20 ±.16 14.94				12705 12818 12399
011808.1-442750 287.49 -71.85 253.58 -15.00 011556.1-444336	 ESO 244- 12 FAIR 690 PGC 4671	.S..3*P S 3.0±1.2 	1.04± .04 .23± .05 1.04	177 .03 .31 .11	15.70 ±.14 10.88 15.31				 6762± 65 6687 6567
011808.5+112252 133.35 -50.95 309.32 -2.39 011530.3+110706	A 0115+11 UGC 833 PGC 4672	.SBR6.. U 6.0± .7 	1.37± .03 .20± .04 1.38	50 .13 .29 .10	13.76 ±.18 13.31			14.81±.1 315± 8 314± 25 1.40	5193± 6 5062± 60 5289 4881
011810.2+382638 128.66 -24.13 335.22 5.30 011519.0+381052	 UGC 831 PGC 4674	.S?.... 	1.06± .06 .19± .05 1.08	 .25 .28 .09	14.83 ±.13 14.27				7286 7443 7021
011819.0-370614 274.96 -78.52 261.19 -13.96 011602.0-372200	 ESO 352- 38 PGC 4682	.LAS0*P S -2.0± .8 	1.00± .06 .24± .03 .97	110 .02 .00 	14.56 ±.14 14.39				9555 9502 9330
0118.4 +0444 135.49 -57.47 302.94 -4.32 0115.8 +0429	 UGC 834 PGC 4686	.SXS4.. U 4.0± .9 	.95± .07 .08± .05 .96	 .08 .11 .04	15.25 ±.15 14.97				14122 14200 13811
011840.2+310211 129.77 -31.47 328.16 3.13 011553.2+304626	 UGC 835 PGC 4698	.S..6*. U 6.0±1.2 	1.02± .06 .13± .05 1.04	75 .21 .19 .06	14.62 ±.12 14.19			16.77±.3 254± 7 2.51	6834± 10 6978 6550
011840.6+321524 129.60 -30.26 329.33 3.48 011552.9+315939	 MCG 5- 4- 14 PGC 4699	.S?.... 	.89± .11 .07± .07 .91	 .21 .10 .03	15.34 ±.13 14.97			16.70±.3 178± 7 1.70	10456± 10 10602 10175
011841.0-584351 296.25 -58.04 238.71 -16.43 011644.0-585936	 ESO 113- 35 PGC 4700	.LBR+.. r -1.3± .9 	.99± .07 .08± .06 .98	90 .00 .00 	14.63 ±.14 14.55				5063 4949 4934
0118.7 -0726 142.36 -69.27 291.07 -7.68 0116.2 -0742	 MCG -1- 4- 25 VV 478 PGC 4701	.S?.... 	1.12± .08 .12± .07 1.13	 .19 .15 .06					5321 5362 5019
011844.8-193739 164.22 -80.23 278.97 -10.64 011619.0-195324	 ESO 542- 3 IRAS01162-1953 PGC 4703	.LAR+.. r -1.3± .9 	.96± .06 .23± .03 .93	58 .01 .00 	14.36 ±.14 13.55				
011845.9-235645 187.13 -83.06 274.62 -11.59 011622.1-241230	 ESO 475- 16 IRAS01163-2412 PGC 4704	RSBR2P. r 2.2± .9 	.98± .05 .21± .04 .99	57 .07 .26 .11	14.23 ±.14 12.12				
011846.0+145929 132.70 -47.36 312.83 -1.50 011606.4+144344	 UGC 838 IRAS01161+1443 PGC 4705	.S?.... 	.79± .08 .09± .05 .80	 .08 .14 .05	14.34 ±.17 13.33 14.08				6903± 50 7009 6595
011848.7-164810 155.86 -77.89 281.80 -10.01 011621.7-170355	IC 1670B MCG -3- 4- 41 PGC 4707	.L..+P* SE -1.5±1.0 	1.14± .06 .52± .05 1.06	94 .00 .00 					

R.A. 2000 DEC.	Names	Type S_T n_L	$\log D_{25}$ $\log R_{25}$	p.a. A_g	B_T m_B	$(B-V)_T$ $(U-B)_T$	$(B-V)_e$ $(U-B)_e$	m_{21} W_{20}	V_{21} V_{opt}
l　　b		T	$\log A_e$	A_i	m_{FIR}	$(B-V)_T^o$	m'_e	W_{50}	V_{GSR}
SGL　SGB									
R.A. 1950 DEC.	PGC	L	$\log D_o$	A_{21}	B_T^o	$(U-B)_T^o$	m'_{25}	HI	V_{3K}
011848.8-531733		.SXT5P.	1.23± .05	19					7407
293.80 -63.35	ESO 151- 40	Sr (1)	.16± .05	.00	14.14 ±.14				7307
244.40 -16.06	IRAS01167-5333	4.7± .6		.25	13.94				7252
011645.1-533318	PGC 4708	1.1± .8	1.23	.08	13.85				
011852.9-164811	IC 1670A	.S..4?/	1.27± .05	126					6006
155.93 -77.88	MCG -3- 4- 40	SE	.64± .05	.00					6017
281.80 -10.02	IRAS01163-1703	3.6±1.0		.94	13.23				5721
011625.9-170356	PGC 4711		1.27	.32					
0118.9 +4417		.E...*.	1.13± .10						
128.11 -18.30	UGC 836	U	.00± .05	.34					
340.88 6.74		-5.0±1.1		.00					
0116.0 +4402	PGC 4715		1.18						
0118.9 +1256		.S?....	.96± .09						5147
133.26 -49.37	UGC 839		.10± .06	.11					
310.88 -2.13				.15					5248
0116.3 +1241	PGC 4716		.97	.05					4838
0118.9 -0100		.L?....	1.12± .10	85					
138.26 -63.05	UGC 842		.09± .05	.14	14.78 ±.15				13471±120
297.39 -6.03				.00					13532
0116.4 -0116	PGC 4717		1.12		14.44				13163
0119.0 +0533		.S..8..	.98± .07						5274
135.47 -56.64	UGC 843	U	.02± .05	.09					
303.77 -4.23		8.0± .8		.03					5354
0116.4 +0518	PGC 4720		.99	.01					4963
011902.3-170336	IC 93	.S..3P*	1.11± .06	170					5964
156.72 -78.08	MCG -3- 4- 43	E	.37± .05	.01					
281.55 -10.12	IRAS01165-1719	3.0±1.3		.51	11.65				5974
011635.4-171921	PGC 4724		1.11	.18					5679
011903.8+041934	A 0116+04A								
135.95 -57.85	CGCG 411- 21			.02	15.17 ±.12				9671± 35
302.58 -4.59	MK 566								9748
011628.5+040349	PGC 4727								9360
011907.6-340615		.LXT0?.	1.28± .04	17					
264.15 -80.83	ESO 352- 41	S	.24± .03	.02	13.74 ±.14				5612± 34
264.30 -13.62		-2.0± .8		.00					5568
011649.1-342200	PGC 4731		1.25		13.64				5377
0119.1 -0009		.S..8*.	1.14± .05	67					5237
137.93 -62.22	UGC 847	U	1.07± .05	.12					
298.23 -5.85		8.0±1.4		1.23					5300
0116.6 -0025	PGC 4734		1.15	.50					4928
011909.9+330150		.S..4..	1.17± .04	54				16.37±.2	5572± 7
129.61 -29.48	UGC 841	U	.70± .05	.15	14.94 ±.12			292± 13	
330.10 3.59	KUG 0116+327	4.0± .9		1.04				289± 6	5719
011621.8+324605	PGC 4735		1.18	.35	13.72			2.29	5293
011910.4+031803	NGC 467	.LAS0P$	1.23± .04		12.9 ±.2	1.05± .08	1.06± .02		
136.40 -58.84	UGC 848	R	.00± .04	.06	13.01 ±.10	.55± .09	.57± .03		5495± 26
301.59 -4.91		-2.0± .3	1.04± .09	.00		.98	13.55± .28		5569
011635.4+030218	PGC 4736		1.23		12.85	.56	13.90± .30		5185
0119.2 +2144		.S..3..	1.04± .08	40					9422
131.44 -40.67	UGC 845	U (1)	.08± .06	.16	14.62 ±.14				
319.33 .35		3.0± .9		.11					9545
0116.5 +2129	PGC 4738	3.5±1.1	1.06	.04	14.28				9122
0119.2 -1152		.S?....	1.04± .09						12916
147.47 -73.40	MCG -2- 4- 22		.49± .07	.07					12943
286.73 -8.93				.68					12621
0116.8 -1208	PGC 4742		1.04	.25					
011918.1+043440	A 0116+04B								
135.97 -57.59	CGCG 411- 22			.09	14.46 ±.12				9932±113
302.84 -4.58	MK 567				12.00				10009
011642.6+041855	PGC 4743								9621
011922.9+122415		.E?....	.90± .17	178					14415± 32
133.56 -49.89	MCG 2- 4- 22		.16± .07	.10	14.89 ±.10				14514
310.39 -2.38	MK 983			.00					14105
011644.3+120830	PGC 4748		.86		14.57				
011924.0+122648		.S..8*.	1.13± .05	120				17.19±.2	14265± 8
133.55 -49.85	UGC 849	U	.33± .05	.10	14.74 ±.13			393± 26	14550± 19
310.43 -2.38	MK 984	8.0±1.3		.41					14410
011645.3+121103	PGC 4750		1.14	.17	14.20			2.83	14001

R.A. 2000 DEC. l b SGL SGB R.A. 1950 DEC.	Names PGC	Type S_T n_L T L	$\log D_{25}$ $\log R_{25}$ $\log A_e$ $\log D_o$	p.a. A_g A_i A_{21}	B_T m_B m_{FIR} B_T^o	$(B-V)_T$ $(U-B)_T$ $(B-V)_T^o$ $(U-B)_T^o$	$(B-V)_e$ $(U-B)_e$ m'_e m'_{25}	m_{21} W_{20} W_{50} HI	V_{21} V_{opt} V_{GSR} V_{3K}
0119.4 +2108		.SAR4..	.96± .09	70					17201
131.62 -41.25	UGC 850	U (1)	.10± .06	.12	15.18 ±.14				
318.76 .13		4.0± .9		.14					17322
0116.7 +2053	PGC 4751	2.5±1.2	.97	.05	14.80				16900
011928.2-514652		RSB.0..	.95± .07	54					
292.73 -64.79	ESO 196- 3	r	.10± .05	.04	14.92 ±.14				
245.99 -16.03		-.1± .9		.08					
011723.0-520236	PGC 4754		.95						
0119.4 +4911		.SX.7..	1.00± .08					16.15±.1	6736± 11
127.64 -13.42	UGC 846	U	.09± .06	.89					
345.62 7.93		7.0± .9		.12				126± 8	6908
0116.5 +4856	PGC 4756		1.08	.04					6506
011930.3-204634		.SB?...	1.04± .05	177					
169.68 -80.98	ESO 542- 4		.09± .04	.03	14.99 ±.14				5490
277.86 -11.07				.13					5488
011705.0-210218	PGC 4758		1.04	.05	14.81				5215
0119.5 +1733		.I?....	1.04± .15	74					8817
132.40 -44.79	UGC 852		.47± .12	.12					
315.35 -.94				.35					8929
0116.9 +1718	PGC 4763		1.05	.24					8512
011934.9-583128	NGC 484	.LA.-..	1.29± .05	94	12.05V±.13	1.01± .01	1.02± .01		
295.96 -58.22	ESO 113- 36	S	.14± .04	.00	13.07 ±.14	.46± .02	.47± .01		5200± 51
238.93 -16.54		-3.0± .8	.76± .03	.00		.95			5086
011738.1-584712	PGC 4764		1.27		12.99	.48	14.03± .29		5070
0119.6 +0810	A 0117+07	.SBT3..	1.00± .06					16.95±.3	9566± 11
134.86 -54.04	UGC 855	U	.04± .05	.11	14.56 ±.13			270± 7	9469± 60
306.34 -3.64		3.0± .8		.06				243± 7	9650
0117.0 +0755	PGC 4769		1.02	.02	14.32			2.61	9252
011938.0+313917		.SBR3..	1.00± .08	70					
129.92 -30.83	UGC 853	U	.25± .06	.23	15.32 ±.13				
328.81 3.11		3.0± .9		.35					
011650.6+312333	PGC 4770		1.02	.13					
0119.6 -0142		.SBS9..	1.07± .14	25					4787
139.01 -63.69	UGC 856	U	.11± .12	.19					
296.75 -6.39		9.0± .8		.11					4845
0117.1 -0158	PGC 4771		1.09	.05					4480
011945.6+032437	NGC 470	.SAT3..	1.45± .02	155	12.53 ±.13	.75± .03	.77± .01	14.33±.0	2374± 3
136.63 -58.71	UGC 858	R (2)	.21± .03	.06	12.27 ±.10	.10± .05	.12± .02	394± 3	2559± 38
301.74 -5.02	IRAS01171+0308	3.0± .3	1.02± .01	.30	11.10	.68	13.12± .04	362± 5	2449
011710.5+030853	PGC 4777		1.45	.11	12.00	.04	14.09± .18	2.23	2065
0119.7 +1943		PSBS4..	.98± .07						12909
132.02 -42.65	UGC 851	U	.07± .05	.16					
317.43 -.36	IRAS01170+1927	4.0± .9		.10					13027
0117.0 +1927	PGC 4778		.99	.04					12606
011953.5+322801	IC 1666	.S..6*.	1.04± .04					16.29±.3	4880± 10
129.86 -30.02	UGC 857	U	.00± .04	.22	14.21 ±.12				4897± 50
329.60 3.29	IRAS01170+3212	6.0±1.2		.01	13.50			161± 7	5026
011705.5+321218	PGC 4782		1.06	.00	13.96			2.33	4601
011954.8+163242	NGC 473	.SXR0*.	1.24± .03	153	13.33 ±.13	.81± .01	.82± .01	15.35±.1	2133± 7
132.74 -45.78	UGC 859	R	.20± .06	.12	13.15 ±.10	.16± .02	.18± .02	245± 6	2222± 66
314.40 -1.31	IRAS01172+1616	.0± .5	.69± .02	.15	12.93	.73	12.27± .05	234± 4	2243
011714.4+161658	PGC 4785		1.24		12.92	.12	13.90± .26		1828
011955.4-341522		.LB?...	.78± .07	144					
264.03 -80.61	ESO 352- 45		.15± .05	.02	15.19 ±.14				5647± 52
264.17 -13.81				.00					5602
011737.0-343106	PGC 4787		.76		15.09				5413
0119.9 +1255		.S..6*.	.98± .07						
133.63 -49.35	UGC 860	U	.02± .05	.10	15.11 ±.16				
310.93 -2.37		6.0±1.2		.03					
0117.3 +1240	PGC 4791		.99	.01					
011957.0-411410		.S?....	1.01± .07		14.5 ±.2	.26± .08			
282.19 -74.72	ESO 296- 11		.08± .06	.03	14.58 ±.14	.34± .09			5572± 22
256.97 -14.90	VV 578			.10	13.37	.20			5506
011743.0-412954	PGC 4792		1.02	.04	14.42	.30	14.23± .41		5364
011959.7+144703	NGC 471	.L.....	1.02± .03	85	14.18 ±.13	.85± .01	.87± .01		
133.18 -47.52	UGC 861	U	.19± .03	.09		.30± .03	.28± .03		4138± 31
312.72 -1.84	IRAS01173+1431	-2.0± .9	.34± .02	.00	12.12	.77	11.37± .07		4243
011720.0+143120	PGC 4793		1.00		14.03	.29	13.69± .23		3831

R.A. 2000 DEC. / l b / SGL SGB / R.A. 1950 DEC.	Names / PGC	Type / S_T n_L / T / L	$\log D_{25}$ / $\log R_{25}$ / $\log A_e$ / $\log D_o$	p.a. / A_g / A_i / A_{21}	B_T / m_B / m_{FIR} / B_T^o	$(B-V)_T$ / $(U-B)_T$ / $(B-V)_T^o$ / $(U-B)_T^o$	$(B-V)_e$ / $(U-B)_e$ / m'_e / m'_{25}	m_{21} / W_{20} / W_{50} / HI	V_{21} / V_{opt} / V_{GSR} / V_{3K}
012004.6-335404 / 262.38 -80.85 / 264.54 -13.78 / 011746.0-340948	NGC 491A / ESO 352- 46 / PGC 4799	.SBS8*. / PS (1) / 7.7± .6 / 6.7± .8	1.30± .04 / .35± .04 / / 1.31	102 / .02 / .43 / .18	14.3 ±.3 / 14.32 ±.14 / / 13.85	.50± .06 / -.20± .11 / .41 / -.27	/ / / 14.79± .36		3595± 39 / 3551 / 3360
012006.8+032500 / 136.80 -58.68 / 301.77 -5.10 / 011731.7+030917	NGC 474 / UGC 864 / ARP 227 / PGC 4801	PLAS0.. / R / -2.0± .3	1.85± .02 / .05± .03 / 1.05± .03 / 1.84	75 / .06 / .00	12.37 ±.13 / 11.49 ±.15 / / 11.91	.86± .02 / .38± .03 / .82 / .38	.93± .01 / .44± .02 / 13.11± .09 / 16.34± .18	15.45±.2 / 368± 6 / 364± 5	2372± 7 / 2315± 18 / 2437 / 2054
012006.9+331107 / 129.81 -29.30 / 330.30 3.45 / 011718.5+325524	IC 1669 / CGCG 502- 32 / PGC 4802		1.00? / .52? / / 1.01	/ .15	15.56 ±.12			18.31±.3 / 189± 7	5717± 10 / 5864 / 5439
0120.1 -0012 / 138.48 -62.21 / 298.25 -6.11 / 0117.6 -0028	UGC 866 / PGC 4805	.S..8*. / U / 8.0±1.3	1.10± .07 / .43± .06 / / 1.11	57 / .13 / .53 / .22	15.22 ±.14 / / / 14.56				1743 / 1805 / 1435
0120.2 -0020 / 138.60 -62.33 / 298.13 -6.17 / 0117.7 -0036	UGC 867 / PGC 4812	.S..4.. / U / 4.0± .9	.89± .08 / .02± .05 / / .91	/ .13 / .03 / .01	15.14 ±.13				
012019.7-440741 / 285.92 -72.03 / 253.98 -15.35 / 011808.0-442324	ESO 244- 17 / PGC 4822	RSBR1P* / Sr / 1.0± .6	1.07± .05 / .07± .05 / / 1.07	20 / .03 / .07 / .04	14.59 ±.14 / / / 14.40				7047± 19 / 6973 / 6852
012020.1-405753 / 281.50 -74.94 / 257.26 -14.93 / 011806.1-411336	NGC 482 / ESO 296- 13 / PGC 4823	.SA.2*/ / BS / 2.4± .7	1.35± .04 / .62± .04 / / 1.35	84 / .03 / .76 / .31	14.50 ±.14 / / / 13.64				6576 / 6511 / 6368
0120.4 +0550 / 135.99 -56.29 / 304.14 -4.49 / 0117.8 +0535	UGC 871 / PGC 4827	.I...9.. / U / 10.0± .8	1.20± .05 / .00± .05 / / 1.21	/ .16 / .00 / .00				15.64±.1 / 99± 4 / 89± 4	2166± 5 / 2246 / 1856
012028.3+380919 / 129.19 -24.36 / 335.07 4.78 / 011736.9+375336	CGCG 521- 2 / MK 985 / PGC 4832	.S?....	.87± .08 / .21± .12 / / .89	/ .25 / .31 / .11	14.94 ±.13 / 13.57 / / 14.33				9093± 52 / 9249 / 8828
012033.4+293703 / 130.46 -32.83 / 326.93 2.34 / 011746.8+292120	MCG 5- 4- 23 / PGC 4840	.S?....	.89± .11 / .29± .07 / / .91	130 / .25 / .36 / .15	14.81 ±.14 / / / 14.16				4404± 57 / 4544 / 4118
012034.3-340717 / 262.87 -80.62 / 264.33 -13.92 / 011816.0-342300	ESO 352- 47 / PGC 4841	.IBS9.. / S (1) / 10.0± .8 / 7.8± .8	1.14± .04 / .19± .04 / / 1.14	73 / .02 / .14 / .10	15.42 ±.14 / / / 15.26				3809 / 3825± 60 / 3780 / 3591
012035.6-172347 / 159.12 -78.17 / 281.31 -10.56 / 011809.0-173930	ESO 542- 8 / PGC 4842	.S..3P/ / E / 3.0±1.9	1.13± .05 / .34± .05 / / 1.13	143 / .00 / .47 / .17	14.41 ±.14 / / / 13.89				5983± 65 / 5991 / 5700
012035.7-172317 / 159.09 -78.16 / 281.31 -10.56 / 011809.0-173900	ESO 542- 7 / IRAS01181-1739 / PGC 4843	.SB.3P? / E / 3.0±1.8	1.16± .04 / .44± .03 / / 1.16	85 / .00 / .61 / .22	14.55 ±.14 / 12.89				
012035.7-172029 / 158.97 -78.12 / 281.36 -10.55 / 011809.0-173612	ESO 542- 6 / PGC 4844	.LXS-*. / SE / -3.0± .6	1.14± .04 / .20± .03 / .57± .07 / 1.11	179 / .00 / .00	14.44 ±.18 / 14.46 ±.14 / / 14.31	1.02± .02 / .42± .03 / .92 / .46	1.06± .02 / .46± .03 / 12.78± .23 / 14.50± .29		9358± 39 / 9366 / 9075
012038.3+294157 / 130.47 -32.74 / 327.01 2.34 / 011751.7+292614	IC 1672 / UGC 872 / PGC 4848	.S?....	1.13± .07 / .14± .06 / / 1.15	140 / .25 / .20 / .07	13.88 ±.12 / / / 13.39				7105± 29 / 7244 / 6819
012040.1+302338 / 130.37 -32.06 / 327.67 2.54 / 011753.1+300756	UGC 873 / KUG 0117+301 / PGC 4850	.S..4.. / U / 4.0± .9	.85± .08 / .10± .08 / / .86	/ .16 / .14 / .05	15.34 ±.12				
0120.7 +0117 / 138.05 -60.71 / 299.75 -5.84 / 0118.2 +0102	UGC 874 / PGC 4853	.L...?. / U / -2.0±1.9	1.03± .11 / .63± .05 / / .94	34 / .05 / .00	15.30 ±.10				

R.A. 2000 DEC. l　　b SGL　SGB R.A. 1950 DEC.	Names PGC	Type S_T　n_L T L	$\log D_{25}$ $\log R_{25}$ $\log A_e$ $\log D_o$	p.a. A_g A_i A_{21}	B_T m_B m_{FIR} B_T^o	$(B-V)_T$ $(U-B)_T$ $(B-V)_T^o$ $(U-B)_T^o$	$(B-V)_e$ $(U-B)_e$ m'_e m'_{25}	m_{21} W_{20} W_{50} HI	V_{21} V_{opt} V_{GSR} V_{3K}
0120.7 +0125 137.99 -60.58 299.88 -5.81 0118.2 +0110	UGC 875 PGC 4854	.SXS5.. U 5.0± .9	.93± .07 .03± .05 .94	 .03 .04 .01	14.87 ±.13 14.77				5421 5488 5113
0120.7 +0411 136.79 -57.88 302.57 -5.04 0118.2 +0356	UGC 876 PGC 4856	.S..8*. U 8.0±1.4	1.00± .06 .87± .05 1.00	117 .04 1.06 .43					5079 5154 4770
0120.8 -0339 140.86 -65.50 294.93 -7.20 0118.3 -0355	 PGC 4858	.LA.0P* E -2.0±1.3	1.12± .08 .15± .08 1.12	65 .16 .00					
0120.9 +0447 136.60 -57.29 303.16 -4.90 0118.3 +0432	UGC 881 PGC 4861	.E..... U -5.0± .8	1.11± .16 .07± .08 1.10	55 .08 .00	14.71 ±.15				
0120.9 +0633 135.95 -55.56 304.87 -4.41 0118.3 +0618	UGC 882 PGC 4862	.I..9.. U 10.0± .8	1.09± .06 .07± .05 1.10	85 .14 .05 .03	15.1 ±.2 14.88				2327 2409 2017
012058.6+291353 130.63 -33.20 326.58 2.14 011812.2+285811	UGC 877 PGC 4868	.S..4.. U 4.0±1.0	.96± .09 .80± .06 .98	71 .20 1.18 .40	15.79 ±.14 14.34			16.11±.3 399± 7 1.37	9640± 10 9779 9354
0120.9 +1704 132.98 -45.22 314.98 -1.40 0118.3 +1649	UGC 883 PGC 4872	.S..8.. U 8.0± .9	1.06± .06 .10± .05 1.07	50 .13 .12 .05					2501 2611 2196
0120.9 +0258 137.39 -59.06 301.40 -5.43 0118.4 +0243	UGC 885 PGC 4873	.S..6*. U 6.0±1.3	1.00± .08 .43± .06 1.01	138 .07 .63 .22					5343 5414 5034
0121.0 +2533 131.28 -36.83 323.09 1.06 0118.3 +2518	UGC 884 PGC 4877	.I?.... 	1.14± .07 .47± .06 1.16	175 .25 .35 .24					8985 9116 8692
012103.0+335354 129.93 -28.57 331.04 3.47 011814.0+333812	UGC 878 PGC 4879	.L...?. U -2.0±1.7	1.04± .11 .08± .05 1.05	95 .20 .00	14.44 ±.10				
012104.1-360706 269.76 -79.03 262.30 -14.35 011847.1-362248	ESO 352- 49 IRAS01188-3622 PGC 4881	.S..2*P S 2.0±1.2	1.20± .04 .41± .04 1.20	151 .02 .51 .21	14.26 ±.14 13.04 13.64				9648± 39 9597 9422
0121.1 +0508 136.56 -56.94 303.51 -4.85 0118.5 +0453	UGC 887 PGC 4884	.S..4.. U 4.0± .9	1.08± .06 .12± .05 1.09	 .12 .17 .06	15.12 ±.19				
0121.1 +0525 136.45 -56.66 303.79 -4.77 0118.5 +0510	UGC 888 PGC 4885	.I..9.. U 10.0± .8	1.02± .08 .01± .06 1.03	 .14 .01 .00					2434 2512 2125
012107.1+331300 130.04 -29.24 330.39 3.26 011818.5+325718	IC 1677 MCG 5- 4- 25 VV 600 PGC 4891	.S?.... 	.98± .06 .11± .06 .99	 .15 .16 .06	14.96 ±.13 14.62			15.37±.3 88± 7 .69	5417± 10 5563 5140
012107.2-264336 211.15 -83.37 271.92 -12.68 011845.0-265918	ESO 476- 4 PGC 4892	.LX.-?/ S -2.9± .5	1.23± .05 .37± .04 .61± .04 1.18	114 .05 .00	13.82 ±.15 13.88 ±.14 13.71	.99± .02 .50± .03 .91 .50	.96± .01 .51± .02 12.36± .14 13.91± .29		5853± 52 5831 5595
012107.6-351200 266.60 -79.74 263.24 -14.21 011850.0-352742	ESO 352- 50 PGC 4894	.S..4?/ S 4.0±1.8	1.17± .04 .76± .04 1.17	145 .02 1.11 .38	15.46 ±.14				
012108.3+012222 138.20 -60.62 299.85 -5.91 011834.0+010640	UGC 890 IRAS01185+0106 PGC 4896	.SBS4*. UE (2) 3.7± .7 3.2± .9	1.29± .04 .56± .04 1.29	177 .03 .82 .28	14.08 ±.12 13.67 13.20				4947± 50 5013 4639

R.A. 2000 DEC. l b SGL SGB R.A. 1950 DEC.	Names PGC	Type S_T n_L T L	$\log D_{25}$ $\log R_{25}$ $\log A_e$ $\log D_o$	p.a. A_g A_i A_{21}	B_T m_B m_{FIR} B_T^o	$(B-V)_T$ $(U-B)_T$ $(B-V)_T^o$ $(U-B)_T^o$	$(B-V)_e$ $(U-B)_e$ m'_e m'_{25}	m_{21} W_{20} W_{50} HI	V_{21} V_{opt} V_{GSR} V_{3K}
012108.9+154143 133.36 -46.57 313.67 -1.85 011828.8+152601	A 0118+15 MCG 2- 4- 29 PGC 4897	.SB.5*. P (1) 5.0± .9 7.0±1.3	.89± .11 .14± .07 .90	 .11 .21 .07	 15.30 ±.13 				
0121.1 +1751 132.87 -44.44 315.75 -1.22 0118.5 +1736	 UGC 889 PGC 4898	.SB.6*. U 6.0±1.3 	.96± .09 .29± .06 .97	178 .10 .43 .15					8544 8656 8240
012112.6-091242 145.76 -70.75 289.48 -8.73 011842.6-092824	NGC 481 MCG -2- 4- 30 PGC 4899	.LAR-P? E -3.0± .9 	1.24± .09 .13± .07 1.24	85 .13 .00 					
012112.8-362824 270.74 -78.72 261.93 -14.43 011856.1-364406	 ESO 352- 51 PGC 4900	.E?.... 	.96± .07 .14± .05 .92	50 .02 .00 	 14.68 ±.14 14.57				5930 5878 5705
012115.8+035143 137.15 -58.18 302.28 -5.25 011840.5+033602	NGC 479 UGC 893 PGC 4905	.SBT4*. UE (1) 3.7± .7 5.3±1.2	1.03± .05 .08± .04 1.03	 .03 .12 .04	 14.71 ±.14 14.52				5238 5312 4929
012116.5-003242 139.24 -62.47 298.00 -6.47 011843.0-004824	 UGC 892 ARP 67 PGC 4906	.SBR2.. UE (1) 2.0± .5 3.1± .8	1.21± .04 .05± .04 1.23	125 .13 .06 .03	 14.01 ±.16 13.76				5232± 46 5293 4926
0121.2 -1037 147.42 -72.06 288.09 -9.10 0118.8 -1053	 MCG -2- 4- 31 PGC 4911	.SB?... 	1.12± .08 .46± .07 1.13	 .13 .56 .23					5511 5540 5216
012118.1-224806 181.97 -81.97 275.91 -11.92 011854.0-230348	 ESO 476- 5 IRAS01189-2303 PGC 4912	.SBR4.. Sr (1) 4.1± .4 2.6± .5	1.29± .03 .52± .03 1.30	10 .06 .76 .26	 13.96 ±.14 13.32 13.10				5894 5885 5625
0121.3 +1224 134.27 -49.80 310.53 -2.83 0118.6 +1209	A 0118+12 UGC 891 DDO 10 PGC 4913	.SX.9*. PU (1) 9.3± .6 8.0±1.4	1.36± .05 .34± .06 .98± .02 1.37	47 .11 .34 .17	14.67 ±.15 14.9 ±.3 14.26	.61± .05 -.11± .06 .51 -.18	.57± .04 -.13± .06 15.06± .05 15.50± .33	14.32±.1 127± 8 114± 12 -.11	643± 6 741 335
012120.2-340348 261.93 -80.55 264.42 -14.07 011902.0-341930	NGC 491 ESO 352- 53 IRAS01190-3419 PGC 4914	.SBT3*. RBCS (2) 3.0± .6 2.6± .7	1.14± .03 .12± .03 .64± .03 1.14	93 .02 .16 .06	13.21 ±.15 13.30 ±.12 11.70 13.05	.67± .02 .01± .03 .62 -.03	 11.90± .08 13.46± .22		 3885± 44 3840 3651
012120.4+402916 129.06 -22.03 337.35 5.27 011827.3+401335	NGC 477 UGC 886 PGC 4915	.SXS5.. U 5.0± .8 	1.34± .05 .27± .06 1.38	135 .37 .40 .13	 13.70 ±.14 12.89				5859± 10 6018 5602
0121.3 +0005 138.95 -61.84 298.63 -6.31 0118.8 -0010	IC 1681 UGC 894 PGC 4916	.S?.... 	.99± .06 .33± .05 1.00	99 .07 .50 .17	 14.68 ±.12 14.09				3882±120 3945 3575
0121.4 +0910 135.28 -52.97 307.43 -3.79 0118.8 +0855	 UGC 896 PGC 4918	.S..8*. U 8.0±1.4 	1.04± .08 .76± .06 1.05	101 .10 .94 .38					
012127.8+070106 136.03 -55.09 305.35 -4.41 011851.2+064525	NGC 485 UGC 895 IRAS01188+0645 PGC 4921	.S..... R 	1.24± .04 .49± .05 1.25	3 .13 .73 .24	 14.02 ±.12 12.92 13.15				 2251± 50 2334 1942
0121.4 +1556 133.41 -46.31 313.93 -1.85 0118.8 +1541	 UGC 897 PGC 4922	.S..4.. U 4.0± .9 	.91± .07 .00± .05 .92	 .11 .00 .00					8979 9086 8674
012128.4-612231 296.61 -55.38 235.95 -16.90 011937.0-613812	 ESO 113- 41 PGC 4923	.SBS3P. S 3.0± .9 	1.14± .05 .28± .04 1.14	6 .00 .39 .14	 14.58 ±.14 14.19				8622 8501 8507
012129.7-363237 270.72 -78.63 261.87 -14.50 011913.0-364818	 ESO 352- 54 PGC 4924	.S..3*/ S 3.0±1.3 	1.16± .05 .92± .05 1.16	129 .02 1.27 .46	 15.19 ±.14 				

R.A. 2000 DEC. / l b / SGL SGB / R.A. 1950 DEC.	Names / PGC	Type / S_T n_L / T / L	$logD_{25}$ / $logR_{25}$ / $logA_e$ / $logD_o$	p.a. / A_g / A_i / A_{21}	B_T / m_B / m_{FIR} / B_T^o	$(B-V)_T$ / $(U-B)_T$ / $(B-V)_T^o$ / $(U-B)_T^o$	$(B-V)_e$ / $(U-B)_e$ / m'_e / m'_{25}	m_{21} / W_{20} / W_{50} / HI	V_{21} / V_{opt} / V_{GSR} / V_{3K}
0121.5 +1951		.S..3..	.96± .09	140					9852
132.53 -42.45	UGC 898	U (1)	.29± .06	.15	15.38 ±.13				
317.68 -.71		3.0± .9		.41					9969
0118.8 +1936	PGC 4926	4.5±1.2	.97	.15	14.75				9551
012130.5-612113		.LX.-*.	.94± .06		14.7 ±.2	.55± .09			
296.59 -55.40	ESO 113- 42	S	.04± .04	.00	14.89 ±.14	.21± .12			
235.98 -16.91		-3.0± .9		.00					
011939.0-613654	PGC 4927		.94				14.20± .36		
0121.5 +2054		.S..4..	.96± .09	63					9218
132.31 -41.41	UGC 899	U	.29± .06	.14	15.00 ±.12				
318.68 -.41		4.0± .9		.43					9337
0118.8 +2039	PGC 4928		.97	.15	14.37				8918
012132.8-330925		.LAR-*.	1.00± .05						
257.58 -81.13	ESO 352- 55	S	.11± .03	.02	14.28 ±.14				3552
265.36 -13.96		-3.0± .9		.00					3510
011914.1-332506	PGC 4934		.99		14.21				3315
012140.4-174907		.S..4*/	1.15± .04	164					
161.33 -78.37	ESO 542- 9	E	.88± .03	.02	15.61 ±.14				
280.94 -10.91	IRAS01192-1804	4.0±1.4		1.29	13.11				
011914.1-180448	PGC 4940		1.15	.44					
012143.2-114613		.SXR5..	1.08± .06	100					
149.23 -73.08	MCG -2- 4- 32	E (1)	.04± .05	.07					
286.98 -9.48	IRAS01192-1201	5.0± .9		.06	13.10				
011914.3-120154	PGC 4943	3.1± .8	1.09	.02					
0121.7 +7837		.S..6?.	1.26± .06					15.73±.1	4174± 11
124.48 15.85	UGC 863	U	.00± .06	1.30					
14.63 14.08	IRAS01173+7822	6.0±1.5		.00	13.14			74± 8	4355
0117.3 +7822	PGC 4945		1.38	.00					4077
012147.1+051517	NGC 488	.SAR3..	1.72± .02	15	11.15 ±.13	.87± .01	.96± .01	14.24±.1	2269± 5
136.82 -56.80	UGC 907	R (2)	.13± .02	.09	11.24 ±.11	.35± .02	.50± .02	453± 5	2233± 22
303.67 -4.99	IRAS01191+0459	3.0± .3	1.24± .01	.18	12.01	.81	12.84± .04	449± 7	2345
011911.2+045936	PGC 4946	1.1± .5	1.73	.06	10.92	.30	14.28± .16	3.25	1959
0121.7 +1815		PSBS1..	1.04± .06	48					8308
132.97 -44.02	UGC 904	U	.25± .05	.12	14.95 ±.13				
316.17 -1.24		1.0± .9		.26					8421
0119.1 +1800	PGC 4947		1.05	.13	14.47				8005
0121.8 +1735		.S?....	1.26± .04	52					2518
133.13 -44.68	UGC 903		.70± .05	.19	14.50 ±.12				
315.53 -1.44	IRAS01191+1719			1.05	10.89				2629
0119.1 +1719	PGC 4948		1.28	.35	13.25				2214
012149.4-440325		.SBR4..	1.07± .05	173	14.11 ±.13	.73± .01	.78± .01		
285.03 -72.00	ESO 244- 21	Sr (1)	.17± .05	.03	14.44 ±.14				7170± 40
254.09 -15.61		4.2± .6	.78± .01	.26		.64	13.50± .03		7095
011938.0-441906	PGC 4949	2.2± .9	1.07	.09	13.93		13.88± .32		6975
0121.8 +2346		.S?....	1.07± .08	115					11465
131.83 -38.57	UGC 905		.02± .06	.24					
321.44 .36				.04					11591
0119.1 +2331	PGC 4951		1.09	.01					11169
012153.9+091221	NGC 489	.S?....	1.22± .05	120					
135.46 -52.92	UGC 908		.65± .05	.09	13.55 ±.18				2526± 36
307.49 -3.89				.98					2615
011916.4+085640	PGC 4957		1.23	.33	12.47				2218
012155.3-162217	NGC 487	.SBR1*.	1.04± .07	112					
157.77 -77.13	MCG -3- 4- 56	E	.20± .05	.00					
282.41 -10.63		1.0±1.3		.20					
011928.3-163758	PGC 4958		1.04	.10					
012155.9+500250		.E...*.	1.15± .15						
127.94 -12.53	UGC 902	U	.04± .08	.98	14.4 ±.3				
346.55 7.77		-5.0±1.1		.00					
011854.8+494709	PGC 4960		1.30						
012156.3+333116	NGC 483	.S?....	.87± .08						
130.19 -28.92	UGC 906		.00± .05	.15	14.12 ±.14				4744± 52
330.73 3.18				.00					4891
011907.4+331536	PGC 4961		.89	.00	13.94				4468
0121.9 +1546		.SAS5..	1.00± .08	0					6330
133.63 -46.46	UGC 910	U	.04± .06	.12	14.64 ±.13				
313.81 -2.01		5.0± .8		.06					6436
0119.3 +1531	PGC 4965		1.01	.02	14.43				6025

R.A. 2000 DEC. / l b / SGL SGB / R.A. 1950 DEC.	Names / PGC	Type S_T n_L / T / L	$\log D_{25}$ / $\log R_{25}$ / $\log A_\bullet$ / $\log D_\circ$	p.a. / A_g / A_i / A_{21}	B_T / m_B / m_{FIR} / B_T°	$(B-V)_T$ / $(U-B)_T$ / $(B-V)_T^\circ$ / $(U-B)_T^\circ$	$(B-V)_\bullet$ / $(U-B)_\bullet$ / $m'_\bullet$ / m'_{25}	m_{21} / W_{20} / W_{50} / HI	V_{21} / V_{opt} / V_{GSR} / V_{3K}
012201.4+372402		.S..7..	1.20± .05	45					5084
129.63 -25.07	UGC 909	U	.15± .05	.28	14.10 ±.14				
334.43 4.28	IRAS01191+3708	7.0± .8		.20					5238
011910.1+370822	PGC 4971		1.23	.07	13.60				4818
012202.0-341149		.LBSOP.	1.15± .05	12					
261.88 -80.35	ESO 352- 57	S	.59± .03	.02	14.63 ±.14				
264.31 -14.23		-2.0± .9		.00					
011944.1-342730	PGC 4972		1.07						
0122.0 +0522	NGC 490	.S?....	.82± .13					15.68±.3	2225± 17
136.89 -56.67	MCG 1- 4- 35		.08± .07	.09	15.34 ±.12			481± 34	
303.80 -5.02				.10				333± 25	2303
0119.4 +0506	PGC 4973		.83	.04	15.13			.51	1916
012208.9+005650	NGC 493	.SXS6*/	1.53± .03	58	12.93 ±.17	.47± .04	.54± .02	13.47±.1	2339± 6
138.91 -60.97	UGC 914	PUE (2)	.51± .04	.07	12.77 ±.12	-.16± .05	-.06± .03	296± 9	2338± 73
299.51 -6.27	IRAS01195+0041	6.0± .5	1.10± .04	.76	12.37	.34	13.92± .09	261± 11	2403
011934.8+004110	PGC 4979	5.3± .8	1.54	.26	11.99	-.25	14.17± .24	1.22	2032
012210.3+321256		.SB.3..	.92± .05	5				16.59±.3	10567± 10
130.45 -30.21	UGC 911	U	.09± .04	.22	14.86 ±.12				
329.50 2.76	IRAS01194+3157	3.0± .9		.12	12.96			258± 7	10711
011922.1+315716	PGC 4981		.94	.04	14.44			2.10	10288
012215.0+344010		.S?....	.72± .09	53					5400± 49
130.09 -27.77	UGC 913		.24± .05	.14	14.8 ±.2				5549
331.84 3.45	ARAK 37			.35	12.93				5127
011925.3+342430	PGC 4985		.73	.12	14.24				
012217.5+284755		.S?....	.99± .10					16.37±.3	4207± 10
131.05 -33.59	MCG 5- 4- 31		.52± .07	.22	15.50 ±.12				4344
326.25 1.73				.72				185± 7	3921
011931.2+283215	PGC 4988		1.01	.26	14.53			1.58	
012223.3-005229	NGC 497	.SBT4*.	1.32± .03	132	13.84 ±.15	.83± .03	.95± .03		8114± 53
140.01 -62.72	UGC 915	R (1)	.38± .04	.17	13.79 ±.13	.24± .04	.36± .04		8173
297.75 -6.82	ARP 8	4.0± .4	.86± .02	.56		.67	13.63± .05		7809
011949.9-010809	PGC 4992	1.9± .8	1.33	.19	13.02	.10	14.31± .22		
012226.6+090313									5585± 51
135.73 -53.05	CGCG 411- 37			.08	15.2 ±.3				5673
307.38 -4.06	MK 568								5277
011949.2+084733	PGC 4995								
012233.3-295856		.S?....	1.17± .05	53					11095± 60
237.43 -82.60	ESO 412- 27		1.04± .05	.02	15.71 ±.14				11062
268.66 -13.61				1.38					10848
012013.0-301436	PGC 5000		1.17	.50	14.23				
0122.5 -0035		.L?....	.82± .19						7615±120
139.94 -62.43	MCG 0- 4-101		.00± .07	.14	15.23 ±.10				7675
298.04 -6.79				.00					7309
0120.0 -0051	PGC 5001		.83		14.97				
012236.2+015325	A 0120+01	.P...?.	.84± .12	120					9945±113
138.67 -60.03	CGCG 385- 87	E	.29± .08	.06	14.92 ±.14				10012
300.46 -6.12	MK 569	99.0			13.47				9638
012001.8+013746	PGC 5006		.85						
012238.9+342616	IC 1683	.S?....	1.12± .04	177					
130.22 -27.99	UGC 916		.34± .05	.16	14.17 ±.12				4800± 61
331.65 3.30	MK 987			.52	12.34				4948
011949.3+341037	PGC 5008		1.14	.17	13.46				4527
012240.3+231009								17.61±.2	15845± 14
132.20 -39.14				.19				243± 26	15913± 58
320.92 .00	MK 357							328± 19	15973
011956.6+225430	PGC 5010								15553
012240.5+265203	A 0119+26A								9187± 72
131.49 -35.48				.29					9320
324.44 1.09	MK 355								8897
011955.0+263623	PGC 5011								
0122.7 +1902		.L...*.	1.04± .09						
133.09 -43.21	UGC 917	U	.04± .04	.11					
316.98 -1.22		-2.0±1.2		.00					
0120.0 +1847	PGC 5012		1.05						
012242.7+265200	A 0119+26B								9047± 60
131.50 -35.48				.29					9179
324.44 1.08	MK 356				12.98				8757
011957.3+263620	PGC 5015								

R.A. 2000 DEC. / l b / SGL SGB / R.A. 1950 DEC.	Names / PGC	Type / S_T n_L / T / L	$\log D_{25}$ / $\log R_{25}$ / $\log A_e$ / $\log D_o$	p.a. / A_g / A_i / A_{21}	B_T / m_B / m_{FIR} / B_T^o	$(B-V)_T$ / $(U-B)_T$ / $(B-V)_T^o$ / $(U-B)_T^o$	$(B-V)_e$ / $(U-B)_e$ / m'_e / m'_{25}	m_{21} / W_{20} / W_{50} / HI	V_{21} / V_{opt} / V_{GSR} / V_{3K}
012247.0-012334 / 140.52 -63.19 / 297.27 -7.06 / 012013.9-013913	UGC 921 / / / PGC 5019	.L...P/ / E / -2.0±1.2 /	1.26± .04 / .51± .04 / / 1.20	108 / .16 / .00 /	/ 14.26 ±.10 / / 14.01				5651±110 / 5708 / 5346 /
012254.0-043836 IC 100 / 142.77 -66.30 / 294.10 -7.95 / 012022.2-045415	MCG -1- 4- 30 / / / PGC 5029	.LX.-*. / E / -3.0± .9 /	1.05± .06 / .16± .05 / / 1.04	80 / .18 / .00 /					
0122.8 +1917 / 133.06 -42.96 / 317.23 -1.17 / 0120.1 +1902	UGC 918 / / / PGC 5020	.S..6*. / U / 6.0±1.3 /	1.00± .06 / .44± .05 / / 1.02	70 / .11 / .65 / .22	/ 15.34 ±.12 / / 14.54				9347 / 9462 / 9046 /
0122.8 +1307 / 134.64 -49.03 / 311.33 -2.99 / 0120.2 +1252	UGC 923 / / / PGC 5024	.S..4.. / U / 4.0±1.0 /	.96± .09 / .51± .06 / / .97	97 / .12 / .75 / .25					11235 / 11334 / 10929 /
012254.8+285009 / 131.20 -33.53 / 326.33 1.62 / 012008.3+283430	MCG 5- 4- 33 / KUG 0120+285 / / PGC 5032	.I?.... / / /	.84± .07 / .43± .06 / / .86	/ .22 / .32 / .22	/ 15.33 ±.15 / / 14.78			16.54±.3 / / 208± 7 / 1.54	8380± 10 / 8517 / 8094 /
012255.4+090254 NGC 502 / 135.93 -53.03 / 307.41 -4.18 / 012018.0+084715	UGC 922 / ARAK 38 / / PGC 5034	.LAR0.. / U / -2.0± .8 /	1.06± .11 / .04± .05 / / 1.06	/ .09 / .00 /	13.74 ±.13 / 13.63 ±.10 / / 13.55	.95± .02 / .48± .03 / .90 / .47	/ / 13.79± .58 /		2501± 27 / 2589 / 2193 /
012255.4+331026 NGC 494 / 130.48 -29.24 / 330.46 2.88 / 012006.5+325447	UGC 919 / IRAS01201+3254 / / PGC 5035	.S..2.. / / 2.0± .8 /	1.30± .04 / .41± .05 / / 1.31	100 / .18 / .50 / .21	13.8 ±.4 / 13.70 ±.12 / 13.58 / 12.96	.95± .03 / .52± .05 / .79 / .39	/ / 14.11± .47 /	16.81±.3 / / 506± 7 / 3.64	5462± 10 / 5314± 60 / 5604 / 5182
0122.9 +0927 NGC 505 / 135.79 -52.62 / 307.81 -4.06 / 0120.3 +0912	UGC 924 / / / PGC 5036	.L..... / U / -2.0± .9 /	.96± .12 / .17± .05 / / .95	/ .11 / .00 /	/ 14.80 ±.10 / /				
012256.0+332817 NGC 495 / 130.44 -28.94 / 330.74 2.97 / 012007.0+331238	UGC 920 / / / PGC 5037	PSBS0P* / PU / .0± .6 /	1.11± .05 / .19± .04 / / 1.12	170 / .18 / .14 /	/ 13.93 ±.12 / / 13.55				4114± 10 / 4114± 56 / 4260 / 3839
012302.8-430739 / 283.14 -72.77 / 255.08 -15.71 / 012051.0-432318	ESO 244- 23 / / / PGC 5046	.S.1*P/ / S / 1.0±1.3 /	1.24± .06 / .84± .05 / / 1.24	150 / .03 / .85 / .42	/ 15.25 ±.14 / / 14.28				7088± 39 / 7015 / 6890 /
012305.4-345909 / 264.13 -79.63 / 263.53 -14.57 / 012048.0-351448	ESO 352- 61 / / / PGC 5049	.S?.... / / /	.96± .06 / .12± .05 / / .96	18 / .02 / .18 / .06	/ 14.85 ±.14 / / 14.61				5971 / 5922 / 5742 /
012306.5-344409 / 263.15 -79.81 / 263.79 -14.54 / 012049.0-345948	ESO 352- 62 / / / PGC 5051	.S..5*/ / S / 4.7± .7 /	1.21± .04 / 1.00± .05 / / 1.21	53 / .02 / 1.50 / .50	/ 15.29 ±.14 / / 13.72				9615 / 9567 / 9385 /
0123.1 -0038 / 140.28 -62.44 / 298.03 -6.95 / 0120.6 -0054	UGC 928 / / / PGC 5055	.L..... / U / -2.0± .9 /	1.06± .07 / .28± .03 / / 1.04	45 / .13 / .00 /	/ 14.82 ±.11 / / 14.57				8014±120 / 8073 / 7709 /
0123.1 -0023 / 140.14 -62.20 / 298.28 -6.88 / 0120.6 -0039	UGC 929 / / / PGC 5056	.S..6*. / U / 6.0±1.1 /	1.09± .06 / .02± .05 / / 1.10	/ .13 / .03 / .01	/ 14.44 ±.15 / / 14.25			15.91±.3 / 89± 7 / 59± 7 / 1.66	7465± 11 / 7610±120 / 7526 / 7161
012311.6+332736 NGC 499 / 130.50 -28.94 / 330.75 2.91 / 012022.5+331157	UGC 926 / / / PGC 5060	.L..-.. / PU / -2.5± .6 /	1.21± .11 / .10± .07 / .74± .02 / 1.22	82 / .18 / .00 /	13.17 ±.13 / 12.94 ±.11 / / 12.79	1.05± .01 / .65± .04 / .96 / .62	1.06± .01 / .67± .02 / 12.36± .07 / 13.86± .59		4375± 56 / 4521 / 4100 /
012311.7+333147 NGC 496 / 130.49 -28.88 / 330.82 2.93 / 012022.6+331608	UGC 927 / IRAS01203+3316 / / PGC 5061	.S..4.. / U / 4.0± .8 /	1.21± .03 / .27± .04 / / 1.23	28 / .22 / .40 / .13	14.09 ±.13 / 13.32 / / 13.45			15.07±.2 / 355± 34 / 306± 7 /	6006± 8 / / 6152 / 5731
012313.3+221519 / 132.55 -40.02 / 320.08 -.39 / 012030.1+215940	MCG 4- 4- 9 / / / PGC 5063	.SB?... / / /	.89± .11 / .24± .07 / / .91	/ .19 / .35 / .12				18.06±.3 / / 415± 7 /	13604± 10 / 13726 / 13307 /

R.A. 2000 DEC. / l b / SGL SGB / R.A. 1950 DEC.	Names / PGC	Type / S_T n_L / T / L	$\log D_{25}$ / $\log R_{25}$ / $\log A_o$ / $\log D_o$	p.a. / A_g / A_i / A_{21}	B_T / m_B / m_{FIR} / B_T^o	$(B-V)_T$ / $(U-B)_T$ / $(B-V)_T^o$ / $(U-B)_T^o$	$(B-V)_o$ / $(U-B)_o$ / m'_o / m'_{25}	m_{21} / W_{20} / W_{50} / HI	V_{21} / V_{opt} / V_{GSR} / V_{3K}
012314.6-325027		.S..4..	1.11± .05	126					
254.54 -81.06	ESO 352- 63	S (1)	.31± .04	.02	14.39 ±.14				9348± 39
265.75 -14.25	IRAS01209-3306	3.5± .6		.46	13.06				9306
012056.0-330606	PGC 5066	2.8± .9	1.11	.16	13.85				9111
012314.7-004202		.S?....	1.07± .06	6					
140.36 -62.49	UGC 931		.70± .05	.13	15.49 ±.12				1870± 97
297.98 -6.98	MK 1153			1.05					1929
012041.2-005741	PGC 5067		1.09	.35	14.30				1565
0123.2 -0054									
140.48 -62.69	CGCG 385- 93			.17	15.4 ±.3				7707±110
297.78 -7.04									7765
0120.7 -0110	PGC 5070								7402
012319.2+331639 IC 1687									
130.56 -29.12	MCG 5- 4- 39			.18	14.58 ±.12				4881± 60
330.58 2.83									5026
012030.2+330101	PGC 5074								4605
012320.9-015836		.SB?...	.89± .11						
141.18 -63.71	MCG 0- 4-112		.43± .07	.19	14.88 ±.16				4730±110
296.74 -7.35				.64					4785
012048.0-021415	PGC 5076		.91	.21	14.01				4426
0123.3 +1512		.SB.1..	.98± .07	45					
134.25 -46.96	UGC 933	U	.05± .05	.11	15.05 ±.15				
313.36 -2.50		1.0± .9		.05					
0120.7 +1457	PGC 5078		.99	.03					
012324.1+092602 NGC 509		.L...?.	1.21± .06	82					
135.99 -52.63	UGC 932	U	.43± .05	.12	14.35 ±.10				2274
307.82 -4.18		-2.0±1.7		.00					2362
012046.5+091023	PGC 5080		1.16		14.20				1967
0123.4 +3326 NGC 501					15.50 ±.13	1.04± .02			
130.55 -28.95	CGCG 502- 62			.18	15.00 ±.12				4887± 60
330.75 2.86									5033
0120.6 +3311	PGC 5082								4612
012328.0+331216 NGC 504		.L.....	1.24± .06	47					
130.60 -29.19	UGC 935	U	.59± .03	.18	13.99 ±.11				4090± 60
330.52 2.78		-2.0± .9		.00					4235
012039.0+325638	PGC 5084		1.17		13.75				3814
012328.5+304704		.S?....	1.24± .06	150				16.55±.3	10494± 10
131.01 -31.58	UGC 934		.46± .06	.19					10332± 50
328.22 2.07	ARP 70			.69				423± 7	10629
012040.8+303125	PGC 5085		1.26	.23					10207
012328.5+331955 NGC 503		.E?....							
130.59 -29.06	MCG 5- 4- 40			.18	15.1 ±.4				5978± 60
330.64 2.82									6123
012039.4+330417	PGC 5086								5703
012329.8-564116		.SBT6*.	1.08± .04	62					
294.15 -59.89	ESO 151- 43	Sr (1)	.22± .04	.00	15.00 ±.14				5242
240.89 -16.97		5.9± .5		.32					5132
012132.0-565654	PGC 5087	5.6± .6	1.08	.11	14.66				5105
012337.8+343412 A 0120+34		.SAS5..	.99± .06	85	15.1 ±.3	.49± .05			
130.42 -27.83	UGC 940	U	.15± .05	.16	14.73 ±.12	-.12± .09			6985± 63
331.83 3.15	IRAS01206+3418	5.0±1.1		.22	12.60	.38			7133
012047.9+341834	PGC 5088		1.01	.07	14.36	-.20	14.56± .45		6713
012329.8+343408									
130.39 -27.84	CGCG 521- 15			.16	14.82 ±.16				7045± 32
331.82 3.17	MK 988								7193
012040.0+341830	PGC 5089								6773
012334.3-345610		.L..-*/	1.09± .05	63					
263.54 -79.60	ESO 352- 64	S	.60± .03	.02	14.57 ±.14				6040
263.60 -14.66		-3.0±1.2		.00					5991
012117.0-351148	PGC 5091		1.00		14.46				5811
0123.6 +0657		.I..9..	.98± .09						2726
136.96 -55.03	UGC 941	U	.12± .06	.10	15.24 ±.15				
305.45 -4.94		10.0± .9		.09					2807
0121.0 +0642	PGC 5094		.99	.06	15.04				2419
012337.3+323748		.S?....	.86± .08						4748
130.74 -29.75	UGC 937		.12± .08	.21	14.96 ±.12				
329.98 2.58	KUG 0120+323			.17					4892
012048.6+322210	PGC 5095		.88	.06	14.56				4471

R.A. 2000 DEC. / l b / SGL SGB / R.A. 1950 DEC.	Names / PGC	Type / S_T n_L / T / L	$\log D_{25}$ / $\log R_{25}$ / $\log A_o$ / $\log D_o$	p.a. / A_g / A_i / A_{21}	B_T / m_B / m_{FIR} / B_T^o	$(B-V)_T$ / $(U-B)_T$ / $(B-V)_T^o$ / $(U-B)_T^o$	$(B-V)_e$ / $(U-B)_e$ / m'_e / m'_{25}	m_{21} / W_{20} / W_{50} / HI	V_{21} / V_{opt} / V_{GSR} / V_{3K}
0123.6 -0149 / 141.25 -63.54 / 296.91 -7.38 / 0121.1 -0205	CGCG 385- 98 / PGC 5097			.18	15.4 ±.3				5648±110 / 5703 / 5344
012340.1+331522 / 130.64 -29.13 / 330.58 2.76 / 012051.0+325944	NGC 507 / UGC 938 / PGC 5098	.LAR0.. / R / -2.0± .3	1.49± .07 / .00± .07 / 1.25± .06 / 1.51	.18 / .00	12.2 ±.2 / 12.55 ±.13 / 12.19	1.00± .02 / .52± .06 / .91 / .50	1.02± .01 / .56± .03 / 13.91± .21 / 14.53± .43		4924± 25 / 5070 / 4649
012340.6+331652 / 130.64 -29.11 / 330.61 2.76 / 012051.5+330114	NGC 508 / UGC 939 / PGC 5099	.E.0.*. / R / -5.0± .6	1.11± .16 / .00± .08 / 1.14	.18 / .00	14.08 ±.11 / 13.83				5529± 25 / 5675 / 5254
0123.7 +0642 / 137.09 -55.27 / 305.21 -5.03 / 0121.1 +0627	UGC 942 / PGC 5101	.S..9.. / U / 9.0± .8	1.07± .07 / .03± .06 / 1.08	.10 / .03 / .01					2339 / 2419 / 2032
0123.7 +1117 / 135.51 -50.79 / 309.63 -3.72 / 0121.0 +1101	NGC 511 / UGC 936 / IRAS01210+1101 / PGC 5103	.E...*. / U / -5.0±1.1	1.08± .17 / .00± .08 / 1.10	.11 / .00	14.70 ±.15				
0123.7 +1130 / 135.45 -50.57 / 309.84 -3.66 / 0121.1 +1115	UGC 943 / PGC 5105	.SXR3.. / U (1) / 3.0± .8 / 4.5±1.1	1.08± .06 / .02± .05 / 1.09	.11 / .03 / .01					14810 / 14904 / 14504
012345.6-584816 / 295.07 -57.83 / 238.66 -17.10 / 012151.1-590354	ESO 113- 45 / FAIR 9 / PGC 5106		.90± .07 / .07± .06 / .08± .04 / .90	111 / .00	13.5 v±.2 / 14.08 ±.14	.32± .03 / -.85± .03	.18± .03 / -.93± .03 / 9.34± .10		13834± 17 / 13718 / 13708
012347.9+330321 / 130.71 -29.33 / 330.40 2.67 / 012058.9+324743	IC 1689 / MCG 5- 4- 46 / PGC 5108	.L?....	.97± .15 / .23± .07 / .95	16 / .18 / .00	14.68 ±.10 / 14.43				4567± 8 / 4712 / 4291
012349.6+330923 / 130.70 -29.23 / 330.50 2.70 / 012100.5+325346	IC 1690 / CGCG 502- 71 / PGC 5110			.18	14.9 ±.4				4537± 60 / 4682 / 4261
012351.3-274722 / 220.05 -82.78 / 270.96 -13.48 / 012130.0-280300	ESO 413- 2 / PGC 5112	RLBR+.. / r / -1.3± .9	.91± .05 / .15± .04 / .90	169 / .09 / .00	14.96 ±.14				
012355.2-350404 / 263.76 -79.45 / 263.48 -14.75 / 012138.1-351942	NGC 526 / ESO 352- 66 / IRAS01216-3519 / PGC 5120	.L?....	.95± .06 / .20± .05 / .92	112 / .02 / .00	14.76 ±.14 / 14.65				5762± 52 / 5713 / 5534
012356.9-334810 / 258.55 -80.33 / 264.78 -14.56 / 012139.0-340348	ESO 352- 67 / PGC 5124		.63± .07 / .11± .05 / .63	.02	15.19 ±.14				1474±104 / 1429 / 1241
012358.1-350659 / 263.90 -79.41 / 263.43 -14.77 / 012141.0-352236	NGC 527 / ESO 352- 68 / PGC 5128	.SBR0?. / S / .0± .8	1.22± .04 / .57± .04 / 1.20	14 / .02 / .43	14.07 ±.14 / 13.53				5759± 60 / 5710 / 5531
012358.6+331848 / 130.71 -29.07 / 330.66 2.71 / 012109.4+330310	MCG 5- 4- 48 / ARAK 39 / PGC 5129	.S?....	.64± .17 / .11± .07 / .65	.18 / .17 / .06	15.23 ±.18 / 14.86				4995± 60 / 5140 / 4720
012359.9+335429 / 130.61 -28.48 / 331.22 2.88 / 012110.3+333851	NGC 512 / UGC 944 / PGC 5132	.S..2.. / U / 2.0± .9	1.20± .05 / .65± .05 / 1.22	116 / .21 / .80 / .32	14.11 ±.14 / 13.06				4783± 52 / 4929 / 4509
012404.0+125459 / 135.15 -49.18 / 311.21 -3.33 / 012124.8+123922	NGC 514 / UGC 947 / IRAS01214+1239 / PGC 5139	.SXT5.. / R (2) / 5.0± .3 / 3.1± .7	1.54± .02 / .10± .03 / 1.34± .03 / 1.55	110 / .08 / .15 / .05	12.24 ±.15 / 12.25 ±.12 / 12.73 / 12.00	.59± .03 / .54	.60± .03 / 14.43± .07 / 14.53± .19	13.72±.2 / 273± 9 / 247± 6 / 1.67	2470± 5 / 2527± 40 / 2568 / 2165
0124.1 +4313 / 129.23 -19.24 / 340.11 5.54 / 0121.2 +4258	UGC 945 / PGC 5146	.L...*. / U / -2.0±1.2	1.11± .16 / .10± .08 / 1.14	175 / .40 / .00					

R.A. 2000 DEC. / l b / SGL SGB / R.A. 1950 DEC.	Names / PGC	Type / S_T n_L / T / L	$\log D_{25}$ / $\log R_{25}$ / $\log A_e$ / $\log D_o$	p.a. / A_g / A_i / A_{21}	B_T / m_B / m_{FIR} / B_T^o	$(B-V)_T$ / $(U-B)_T$ / $(B-V)_T^o$ / $(U-B)_T^o$	$(B-V)_e$ / $(U-B)_e$ / m'_e / m'_{25}	m_{21} / W_{20} / W_{50} / HI	V_{21} / V_{opt} / V_{GSR} / V_{3K}
0124.1 +0955 136.11 -52.11 308.34 -4.21 0121.5 +0940	IC 101 UGC 949 PGC 5147	.S?.... 	1.14± .05 .36± .05 1.16	127 .14 .54 .18	 14.74 ±.13 14.04				2401 2490 2094
012408.3+093305 136.24 -52.47 307.98 -4.32 012130.6+091728	NGC 516 UGC 946 PGC 5148	.L..... U -2.0± .9 	1.16± .05 .45± .04 1.11	44 .12 .00 	 14.13 ±.10 13.97				 2432± 50 2520 2125
0124.1 +2703 131.86 -35.24 324.72 .83 0121.4 +2648	 UGC 948 PGC 5150	.I..9*. U 10.0±1.3 	1.00± .08 .43± .06 1.03	171 .28 .32 .22					4918 5050 4630
012414.3-344335 262.18 -79.65 263.84 -14.76 012157.0-345912	 ESO 352- 69 IRAS01219-3459 PGC 5154	.SBS2?P S 2.2± .7 	1.19± .04 .11± .04 1.19	 .02 .13 .05	 13.96 ±.14 12.63 13.75				 6034± 34 5986 5805
012416.6-372005 270.82 -77.67 261.14 -15.17 012201.0-373542	 ESO 296- 19 IRAS01220-3736 PGC 5158	.SBT5P. S (1) 5.0± .9 2.2± .9	1.00± .05 .20± .04 1.00	98 .02 .30 .10	 14.86 ±.14 				
012417.7+091955 136.38 -52.68 307.78 -4.42 012140.1+090418	NGC 518 UGC 952 PGC 5161	.S..1*/ PU 1.3± .7 	1.23± .04 .45± .04 1.24	98 .12 .45 .22	 14.16 ±.12 13.55				 2704± 50 2792 2397
0124.3 -0144 141.57 -63.41 297.04 -7.53 0121.8 -0200	 MCG 0- 4-116 PGC 5164	.S?.... 	.82± .13 .36± .07 .83	 .18 .54 .18	 15.00 ±.16 14.25				 5275±110 5330 4972
012421.8+321326 130.98 -30.13 329.64 2.31 012133.2+315749	 UGC 950 PGC 5165	.S..2.. U 2.0± .9 	.99± .06 .51± .05 1.01	68 .19 .63 .25	 15.19 ±.12 				
0124.4 +0953 136.24 -52.12 308.33 -4.29 0121.8 +0938	IC 102 UGC 954 PGC 5172	.S..0.. U .0±1.0 	.96± .09 .51± .06 .95	 .15 .38 	 15.44 ±.12 14.69				 14423 14512 14117
012427.0+334757 130.74 -28.57 331.15 2.76 012137.4+333220	NGC 513 UGC 953 ARAK 41 PGC 5174	.S?.... 	.83± .08 .31± .05 .85	75 .21 .43 .16	 13.9 ±.3 12.37 13.26				 5949± 30 6095 5676
0124.4 +1545 134.49 -46.37 313.97 -2.59 0121.8 +1530	 UGC 955 PGC 5181	.SB.6?. U 6.0±1.4 	1.04± .06 .83± .05 1.05	168 .12 1.22 .42	 15.62 ±.13 14.25				5050 5155 4747
012428.8-013831 141.58 -63.31 297.14 -7.53 012155.7-015408	NGC 519 CGCG 385-103 PGC 5182	.E+..?. PE -4.0±1.1 	.70± .15 .25± .08 .66	140 .19 .00 	 15.28 ±.14 15.01				 5332±110 5387 5029
012431.0-153217 158.15 -76.09 283.39 -11.04 012203.9-154754	 MCG -3- 4- 61 PGC 5186	RSBR0.. E .0± .8 	1.22± .05 .02± .05 1.22	85 .00 .01 					
012431.6-043132 143.63 -66.06 294.32 -8.31 012159.7-044709	 MCG -1- 4- 37 PGC 5187	.SBS6*. E (1) 6.0± .9 4.2± .8	1.09± .06 .10± .05 1.10	60 .19 .15 .05					
012434.2+014354 139.70 -60.06 300.44 -6.64 012159.7+012817	NGC 521 UGC 962 IRAS01219+0128 PGC 5190	.SBR4.. R (2) 4.0± .3 1.4± .6	1.50± .02 .04± .03 1.14± .04 1.51	20 .08 .06 .02	12.55 ±.17 12.43 ±.13 13.04 12.29	.82± .04 .24± .06 .76 .19	.95± .03 .49± .03 13.74± .08 14.81± .22	14.60±.2 241± 34 207± 10 2.29	5040± 9 5028± 49 5105 4734
012434.2-331023 255.12 -80.63 265.45 -14.58 012216.0-332600	 ESO 352- 71 IRAS01222-3325 PGC 5191	.SBS4?. S (1) 4.0± .6 3.3± .7	1.17± .04 .31± .04 1.17	168 .02 .46 .16	 14.30 ±.14 13.62 13.76				 9250± 34 9206 9015
0124.5 +0201 139.55 -59.77 300.73 -6.56 0122.0 +0146	IC 103 UGC 963 PGC 5192	.L..-*. U -3.0±1.2 	.93± .13 .23± .05 .91	130 .08 .00 	 15.00 ±.10 				

R.A. 2000 DEC. / l b / SGL SGB / R.A. 1950 DEC.	Names / PGC	Type / S_T n_L / T / L	$\log D_{25}$ / $\log R_{25}$ / $\log A_e$ / $\log D_o$	p.a. / A_g / A_l / A_{21}	B_T / m_B / m_{FIR} / B_T^o	$(B-V)_T$ / $(U-B)_T$ / $(B-V)_T^o$ / $(U-B)_T^o$	$(B-V)_e$ / $(U-B)_e$ / m'_e / m'_{25}	m_{21} / W_{20} / W_{50} / HI	V_{21} / V_{opt} / V_{GSR} / V_{3K}
012434.7+034749 NGC 520 138.70 -58.06 UGC 966 302.45 -6.06 ARP 157 012159.4+033213 PGC 5193		.P..... R 99.0	1.65± .02 .39± .04 1.09± .02 1.63	130 .05 .30	12.24 ±.13 12.09 ±.10 9.47 11.77	.82± .02 .17± .05 .72 .11	.77± .02 .22± .03 13.18± .08 14.34± .18	13.83±.1 218± 4 149± 4	2217± 3 2059± 29 2287 1910
0124.5 +1632 134.32 -45.59 UGC 958 314.72 -2.38 0121.9 +1617 PGC 5194		.SB.6*. U 6.0±1.3	1.22± .05 .90± .05 1.23	26 .14 1.32 .45	15.20 ±.12 13.73				2418 2525 2116
012436.8+215256 133.05 -40.34 MCG 4- 4- 10 319.82 -.81 012153.5+213720 PGC 5197		.S?.... 	.82± .13 .08± .07 .83	.16 .11 .04				16.96±.3 157± 7	10166± 10 10286 9870
0124.6 +0743 137.09 -54.23 UGC 964 306.26 -4.96 0122.0 +0728 PGC 5198		.S?.... 	1.10± .05 .58± .05 1.12	91 .12 .87 .29	15.23 ±.12 14.22				2734 2817 2427
012438.6+332823 NGC 515 130.84 -28.89 UGC 956 330.85 2.63 012149.2+331246 PGC 5201		.L..... U -2.0± .8	1.14± .09 .09± .05 1.15	.19 .00	14.02 ±.11				
012440.8+035125 138.72 -57.99 UGC 957 302.52 -6.07 012205.5+033549 PGC 5208		.I..9*. U 10.0±1.2	1.07± .14 .11± .12 1.07	.05 .08 .05				14.56±.1 189± 6 82± 5	2152± 7 2223 1846
012441.5-013512 NGC 530 141.66 -63.24 UGC 965 297.21 -7.57 MK 1154 012208.4-015049 PGC 5210		.LB.+./ PUE -.7± .5	1.17± .04 .54± .04 1.11	134 .19 .00	13.96 ±.11 13.70				5009± 46 5064 4705
0124.7 +0945 136.40 -52.24 UGC 969 308.23 -4.40 0122.1 +0930 PGC 5213		.SB.3.. U 3.0± .9	.98± .09 .25± .06 1.00	135 .15 .35 .13	15.32 ±.14 14.72				14608 14696 14302
012443.8+332547 NGC 517 130.87 -28.93 UGC 960 330.81 2.60 ARAK 43 012154.5+331010 PGC 5214		.L..... U -2.0± .8	1.30± .05 .30± .03 1.28	20 .19 .00	13.42 ±.10 13.17				4260± 40 4405 3986
012445.0-380742 NGC 534 272.58 -76.97 ESO 296- 21 260.33 -15.37 012230.0-382318 PGC 5215		.LXT0.. BS -2.0± .6	1.03± .05 .05± .04 1.02	142 .02 .00	14.42 ±.14 14.31				5826 5767 5609
012445.2+320957 131.09 -30.18 UGC 959 329.61 2.22 MK 991 012156.5+315420 PGC 5217		.S..1*. U 1.0±1.2	.85± .07 .17± .05 .87	60 .19 .17 .09	14.41 ±.17 12.71 13.91				10480± 39 10623 10203
012445.7+095939 NGC 522 136.34 -52.01 UGC 970 308.45 -4.34 IRAS01221+0944 012207.8+094403 PGC 5218		.S..4*/ PU 3.7± .7	1.43± .02 .78± .04 1.00± .03 1.44	33 .15 1.14 .39	13.94 ±.15 13.97 ±.12 13.22 12.65	1.00± .04 .39± .05 .81 .19	1.08± .04 .47± .04 14.43± .07 13.98± .22		2806± 50 2895 2500
0124.7 +0101 140.17 -60.72 UGC 971 299.77 -6.88 0122.2 +0046 PGC 5220		.I..9?. U 10.0±1.9	1.07± .07 .79± .06 1.07	148 .03 .60 .40					4735 4798 4430
012447.8+093221 NGC 524 136.51 -52.45 UGC 972 308.02 -4.48 IRAS01221+0916 012210.1+091645 PGC 5222		.LAT+.. R -1.0± .3	1.44± .03 .00± .03 1.22± .02 1.45	.14 .00	11.3 ±.1 11.50 ±.09 13.31 11.23	1.05± .02 .60± .03 1.00 .58	1.07± .01 .64± .01 12.84± .08 13.36± .19		2421± 19 2509 2115
012449.7+321405 131.09 -30.11 MCG 5- 4- 55 329.68 2.22 KUG 0122+319 012201.0+315829 PGC 5226		.S?.... 	.79± .07 .14± .06 .81	.19 .20 .07	15.13 ±.13 14.69			16.97±.3 328± 7 2.21	10542± 10 10685 10265
012450.0-314524 247.42 -81.39 ESO 413- 4 266.92 -14.40 012231.0-320100 PGC 5227		.SXT5.. S (1) 5.0± .5 1.1± .5	1.20± .04 .07± .04 1.20	32 .02 .10 .03	14.36 ±.14 14.18				10845± 39 10806 10606
0124.8 -0129 141.68 -63.14 UGC 974 297.32 -7.58 0122.3 -0145 PGC 5228		.S..0.. U .0±1.0	.95± .07 .55± .05 .94	160 .17 .41	15.03 ±.14 14.37				4788±110 4844 4485

R.A. 2000 DEC.	Names	Type	$\log D_{25}$	p.a.	B_T	$(B-V)_T$	$(B-V)_e$	m_{21}	V_{21}
l b		S_T n_L	$\log R_{25}$	A_g	m_B	$(U-B)_T$	$(U-B)_e$	W_{20}	V_{opt}
SGL SGB		T	$\log A_e$	A_i	m_{FIR}	$(B-V)_T^o$	m'_e	W_{50}	V_{GSR}
R.A. 1950 DEC.	PGC	L	$\log D_o$	A_{21}	B_T^o	$(U-B)_T^o$	m'_{25}	HI	V_{3K}
012452.5-013704	IC 1696	.E+....	.94± .10	10	14.63 ±.14	1.02± .02			
141.77 -63.26	UGC 973	PUE	.04± .04	.19	14.40 ±.10				5768±110
297.19 -7.62		-4.3± .4		.00		.92			5823
012219.4-015240	PGC 5231		.95		14.20		14.21± .51		5465
012452.5+094208	NGC 525	.L.....	1.18± .06	5					
136.48 -52.29	UGC 972	U	.35± .03	.15	14.20 ±.10				2146± 50
308.18 -4.45		-2.0± .8		.00					2234
012214.7+092632	PGC 5232		1.14		14.02				1840
012454.1-683706		.SBR0..	1.00± .06						
298.37 -48.21	ESO 52- 1	r	.06± .06	.00	14.88 ±.14				10640± 60
228.35 -17.32		-.1± .8		.04					10502
012322.0-685242	PGC 5233		1.00		14.68				10563
0125.0 +0026	IC 1697	.S?....	.97± .07	110					
140.62 -61.27	UGC 976		.22± .05	.04	14.72 ±.12				8699±120
299.22 -7.10	IRAS01224+0011			.31	13.59				8760
0122.4 +0011	PGC 5238		.97	.11	14.31				8395
012507.0-380024		PSBS1..	.72± .06						
272.01 -77.03	ESO 296- 22	r	.08± .05	.02	15.49 ±.14				
260.47 -15.43		1.0±1.0		.08					
012252.0-381600	PGC 5243		.73	.04					
0125.1 +0841	IC 1695	.S?....	1.04± .09						
136.94 -53.26	UGC 977		.01± .06	.14	14.88 ±.17				14505± 38
307.23 -4.80				.02					14590
0122.5 +0826	PGC 5245		1.05	.01	14.58				14199
0125.1 +1452	IC 107	.S?....	1.04± .06	100					6366
134.97 -47.20	UGC 978		.09± .05	.13	14.92 ±.16				
313.17 -3.01				.13					6468
0122.5 +1437	PGC 5250		1.05	.04	14.63				6063
0125.1 +0203	IC 109	.LB....	1.03± .08						
139.82 -59.70	UGC 980	U	.07± .03	.07	14.50 ±.11				
300.80 -6.69		-2.0± .8		.00					
0122.6 +0148	PGC 5251		1.03						
0125.1 +0217		.S?....	.85± .11	25					6115
139.70 -59.48	UGC 981		.25± .06	.06	15.33 ±.12				
301.03 -6.63				.37					6182
0122.6 +0202	PGC 5252		.86	.12	14.87				5810
012511.9-380536	NGC 544	.LX.-*.	1.07± .05						
272.17 -76.95	ESO 296- 24	BS	.12± .04	.02	14.41 ±.14				5992
260.38 -15.46		-3.0± .7		.00					5933
012257.0-382112	PGC 5253		1.05		14.30				5775
012512.9-380406	NGC 546	.SBS3*P	1.15± .04	35					
272.10 -76.96	ESO 296- 25	S	.42± .04	.02	14.48 ±.14				6563
260.41 -15.46	IRAS01229-3819	3.0±1.2		.58	12.78				6504
012258.0-381942	PGC 5255		1.15	.21	13.83				6346
0125.2 -0130		.L.....	1.08± .07	130					
141.90 -63.13	UGC 984	U	.47± .03	.19	14.62 ±.10				5252± 67
297.33 -7.68		-2.0± .9		.00					5308
0122.7 -0146	PGC 5258		1.03		14.35				4950
0125.2 +1450	IC 1698	.L.....	1.25± .06	120					
135.02 -47.23	UGC 983	U	.39± .03	.11	14.43 ±.12				6574± 56
313.15 -3.04		-2.0± .8		.00					6676
0122.6 +1435	PGC 5261		1.21		14.21				6271
012517.2+091554	NGC 532	.S..2?/	1.39± .03	28	13.95 ±.15	1.02± .04	1.03± .04	16.01±.3	2375± 10
136.80 -52.69	UGC 982	PU	.50± .05	.14	13.32 ±.12	.74± .05	.72± .05		2332± 50
307.79 -4.68	IRAS01226+0900	2.3± .9	.90± .02	.62	12.70	.88	13.94± .05	378± 10	2460
012239.5+090018	PGC 5264		1.40	.25	12.79	.60	14.48± .24	2.97	2068
012519.7+340128	NGC 523	.P.....	1.40± .04	108	13.5 ±.4	.79± .03		14.95±.3	4750± 10
130.91 -28.32	UGC 979	R	.53± .05	.19	13.42 ±.12	.20± .05			4857± 26
331.41 2.65	4ZW 45	99.0		.77	12.25	.62			4911
012229.9+334553	PGC 5268		1.41	.26	12.43	.06	14.02± .46	2.26	4493
012521.9-180948	NGC 539	.SBT5..	1.19± .03	145	14.2 ±.4	.67± .07			
165.78 -78.11	ESO 542- 10	PSE (3)	.07± .03	.05	14.32 ±.14				9657± 59
280.80 -11.84	IRAS01229-1825	5.0± .5		.10	13.64	.58			9660
012256.1-182524	PGC 5269		1.20	.03	14.10		14.85± .44		9380
0125.3 +1431		.S..3..	1.18± .05	19					11085
135.14 -47.54	UGC 985	U (1)	.56± .05	.12	14.89 ±.13				
312.85 -3.16		3.0± .9		.77					11186
0122.7 +1416	PGC 5270	4.5±1.2	1.19	.28	13.92				10782

R.A. 2000 DEC.	Names	Type	logD$_{25}$	p.a.	B$_T$	(B-V)$_T$	(B-V)$_e$	m$_{21}$	V$_{21}$
l b		S$_T$ n$_L$	logR$_{25}$	A$_g$	m$_B$	(U-B)$_T$	(U-B)$_e$	W$_{20}$	V$_{opt}$
SGL SGB		T	logA$_e$	A$_i$	m$_{FIR}$	(B-V)$_T^o$	m'$_e$	W$_{50}$	V$_{GSR}$
R.A. 1950 DEC.	PGC	L	logD$_o$	A$_{21}$	B$_T^o$	(U-B)$_T^o$	m'$_{25}$	HI	V$_{3K}$

```
012524.8+145148 IC   1700   .E.....  1.16± .10       13.78 ±.11                              6356± 56
135.06 -47.21   UGC   986   U         .00± .05   .11                                         6458
313.18  -3.07               -5.0± .7             .00                                         6053
012244.7+143613 PGC  5271             1.18              13.57

0125.4  +0732               .SXS9..  1.13± .07   15                                          2793± 10
137.49 -54.37   UGC   989   U         .23± .06   .11
306.14  -5.20               9.0± .8              .24                                         2875
0122.8  +0717   PGC  5273             1.14         .12                                       2487

012526.2-013305 NGC   538   .SBS2*.  1.00± .06   40    14.58 ±.12                            5494± 67
142.03 -63.16   UGC   991   PUE       .30± .05   .16                                         5549
297.30  -7.74               1.6± .7              .37                                         5191
012253.2-014840 PGC  5275             1.02         .15  13.99

0125.4  +1048               .S..9*.   .96± .09                                               2537
136.32 -51.18   UGC   990   U         .00± .06   .15
309.29  -4.26               9.0±1.2              .00                                         2628
0122.8  +1033   PGC  5276              .97         .00                                       2232

012528.8-381601 NGC   549   .LBS+*/  1.30± .04   68    14.53 ±.14                            6163
272.42 -76.77   ESO  296- 26 BS       .63± .04   .02  13.57                                  6103
260.21 -15.54   IRAS01232-3831 -.5± .9           .00  14.41                                  5947
012314.0-383136 PGC  5278             1.21

012531.2-012433 NGC   535   .L..+./  1.01± .07   58    14.83 ±.11                            4939±110
141.98 -63.01   UGC   997   PUE       .51± .03   .16                                         4994
297.44  -7.72               -.8± .6              .00                                         4636
012258.1-014008 PGC  5282              .95              14.59

012531.5+014535 NGC   533   .E.3.*.  1.58± .04   50    12.41 ±.15  1.02± .02  1.04± .01      5538± 25
140.15 -59.97   UGC   992   R         .21± .04   .06  12.36 ±.10   .56± .07   .57± .03       5603
300.54  -6.86               -5.0± .4  1.18± .04   .00              .95        13.80± .14     5233
012257.0+013000 PGC  5283             1.53              12.23                 .57  14.79± .26

012531.5+320810             .S..1..  1.34± .02   32    14.39M±.10  .99± .02   .94± .03 15.40±.1  4658±  6
131.28 -30.18   UGC   987   U         .52± .03   .14  13.86 ±.12  .25± .04            404± 10 4654± 36
329.63   2.05   MK   993               .59± .02   .53              .82  12.83± .05    386±  6 4800
012242.7+315235 PGC  5284             1.35         .26  13.45       .12  14.65± .17   1.69    4382

0125.5  -0129               .S..0..   .96± .09                     14.92 ±.17                6591±110
142.05 -63.09   UGC   996   U         .65± .06   .16                                         6646
297.37  -7.75                .0±1.0              .49                                         6288
0123.0  -0145   PGC  5289              .94              14.17

012533.7+334016 NGC   528   .L.....  1.24± .06   55    13.51 ±.10                            4792± 52
131.02 -28.66   UGC   988   U         .20± .03   .19  13.58                                  4937
331.09   2.50   IRAS01226+3324 -2.0± .8          .00  13.25                                  4519
012244.0+332441 PGC  5290             1.23

0125.5  +0007               .I..9..  1.04± .06  136                                          5107
141.07 -61.54   UGC   998   U         .64± .05   .03
298.95  -7.32               10.0± .9             .48                                         5167
0123.0  -0008   PGC  5291             1.04         .32                                       4803

012535.4-412755              .72± .06   87    15.50 ±.14                                     6250± 20
279.06 -74.04   ESO  296- 27          .23± .05   .03                                         6181
256.88 -15.98   FAIR 1075                        .00                                         6047
012323.0-414330 PGC  5293              .73

0125.6  +1803               .S..6*.   .96± .09   20                                          8290
134.26 -44.06   UGC   994   U         .10± .06   .11
316.24  -2.17               6.0±1.3              .14                                         8400
0122.9  +1748   PGC  5294              .97         .05                                       7990

012537.6-302155             .E?....   .98± .06  165    14.17 ±.14                            9320
238.31 -81.84   ESO  413-  5          .15± .04   .02                                         9284
268.39 -14.33               .00                                                             9077
012318.0-303730 PGC  5295              .93              14.01

0125.6  +1458               .S..6*.  1.00± .08   62
135.12 -47.08   UGC   999   U         .55± .06   .11
313.30  -3.10               6.0±1.3              .81
0123.0  +1443   PGC  5298             1.01         .27

012540.3+344247 NGC   529   .L..-*.  1.38± .11  160    13.14M±.10  1.00± .02
130.87 -27.63   UGC   995   U         .06± .08   .16  12.78 ±.11   .56± .03                  4799± 22
332.09   2.79   HICK  10B   -3.0±1.0             .00              .91                        4946
012250.0+342712 PGC  5299             1.39              12.73       .54      14.77± .59      4529

012541.0-345028             .LX.-?.  1.19± .09
261.49 -79.36   MCG  -6- 4- 32 S      .26± .08   .02
263.77 -15.07               -3.0±1.2             .00
012324.0-350603 PGC  5301             1.15
```

R.A. 2000 DEC. / l b / SGL SGB / R.A. 1950 DEC.	Names / / / PGC	Type / S_T n_L / T / L	$\log D_{25}$ / $\log R_{25}$ / $\log A_o$ / $\log D_o$	p.a. / A_g / A_i / A_{21}	B_T / m_B / m_{FIR} / B_T^o	$(B-V)_T$ / $(U-B)_T$ / $(B-V)_T^o$ / $(U-B)_T^o$	$(B-V)_e$ / $(U-B)_e$ / m'_e / m'_{25}	m_{21} / W_{20} / W_{50} / HI	V_{21} / V_{opt} / V_{GSR} / V_{3K}
012544.4-012242	NGC 541	.L..-*.	1.25± .08		13.03 ±.15	.95± .03	.99± .01		5415± 30
142.08 -62.97	UGC 1004	R	.01± .05	.16	13.49 ±.12	.39± .07	.48± .04		5470
297.49 -7.77	ARP 133	-3.0± .4	1.08± .04	.00		.86	13.92± .15		5113
012311.2-013817	PGC 5305		1.27		13.07	.38	14.13± .43		
0125.7 -0126		.L.....	1.00± .10						5321±110
142.13 -63.03	UGC 1003	U	.49± .04	.16	14.88 ±.11				5376
297.43 -7.79		-2.0± .9		.00					5019
0123.2 -0142	PGC 5307		.94		14.64				
012549.1-390619		PSXS0*.	1.06± .05	8					
274.19 -76.04	ESO 296- 28	Sr	.24± .03	.02	15.11 ±.14				
259.34 -15.72		-.5± .6		.00					
012335.1-392154	PGC 5310		1.03						
012550.1-011738	NGC 543	.L..-*/	.77± .13	90	14.1 ±.2	1.04± .04	1.05± .02		5238±110
142.07 -62.88	MCG 0- 4-138	PE	.35± .05	.18	15.05 ±.15				5294
297.58 -7.77		-3.0± .8	.57± .06	.00		.93	12.39± .22		4935
012317.0-013312	PGC 5311		.75		14.46		11.97± .71		
012550.2-342925		.SAT0*.	1.07± .05						14891± 60
259.98 -79.58	ESO 352- 76	S	.05± .05	.02	14.93 ±.14				14843
264.14 -15.05		.3± .6		.03					14662
012333.0-344500	PGC 5312		1.07		14.65				
0125.8 -0119		.S?....	.64± .17						5374±110
142.10 -62.91	MCG 0- 4-140		.00± .07	.16	14.1 ±.3				5429
297.55 -7.78				.00					5071
0123.3 -0135	PGC 5314		.65		13.85				
012551.3-372001		.L...P.	.97± .07	100					9411± 87
269.66 -77.48	ESO 296- 29	S	.07± .05	.02	14.59 ±.14				9354
261.19 -15.48		-2.0±1.1		.00					9192
012336.0-373536	PGC 5315		.96		14.43				
0125.8 +0129		.S..0..	1.04± .08						
140.46 -60.20	UGC 1006	U	.04± .06	.02	14.82 ±.16				
300.30 -7.01		.0± .8		.03					
0123.3 +0114	PGC 5316		1.04						
012556.4-385706		.LXS-P.	1.19± .09						
273.76 -76.16		S	.09± .08	.02					
259.51 -15.72		-3.0± .8		.00					
012342.3-391240	PGC 5320		1.18						
012556.4+163607	IC 1702	.S..6*.	1.08± .06	170					4206± 50
134.76 -45.48	UGC 1005	U	.08± .05	.15	14.19 ±.13				4312
314.88 -2.68	IRAS01232+1620	6.0±1.2		.11					3905
012315.6+162033	PGC 5321		1.10		.04 13.91				
012559.4-012022	NGC 545	.LA.-..	1.38± .04	55					5318± 24
142.18 -62.91	UGC 1007	R	.18± .04	.16	13.21 ±.12				5373
297.54 -7.82	ARAK 45	-3.0± .3		.00					5016
012326.2-013556	PGC 5323		1.37		12.97				
012600.8-012038	NGC 547	.E.1...	1.13± .05	85					5466± 23
142.20 -62.91	UGC 1009	R	.03± .04	.16	13.16 ±.10				5522
297.54 -7.82		-5.0± .4		.00					5164
012327.6-013612	PGC 5324		1.15		12.91				
012602.7-011336	NGC 548	.E+..*.	.95± .09	135	14.65 ±.15	.95± .02	.98± .02		5332±110
142.14 -62.80	UGC 1010	PUE	.05± .04	.15	14.69 ±.10				5388
297.66 -7.80		-3.7± .7	.73± .04	.00		.86	13.79± .13		5030
012329.5-012910	PGC 5326		.96		14.45		14.15± .50		
0126.0 +5037		.S..6*.	1.00± .08	81					
128.53 -11.87	UGC 1000	U	.43± .06	1.14					
347.28 7.29		6.0±1.3		.63					
0123.0 +5022	PGC 5327		1.11		.22				
012603.1+112639	IC 112	.S..8*.	.88± .08	128					5768± 39
136.34 -50.52	UGC 1008	U	.24± .05	.15	14.32 ±.17				5860
309.94 -4.22	ARAK 46	8.0±1.3		.30	13.32				5463
012324.5+111105	PGC 5328		.89		.12 13.86				
012603.4-035716		.SBT8..	1.28± .05	105					2003
144.07 -65.40	MCG -1- 4- 42	E (1)	.27± .05	.16					
294.98 -8.53		8.0± .8		.33					2050
012331.4-041251	PGC 5329	7.5± .8	1.29		.13				1703
0126.0 +0019		.S..8*.	1.00± .08	175					1923
141.20 -61.31	UGC 1011	U	.55± .06	.03					
299.18 -7.38		8.0±1.3		.68					1983
0123.5 +0004	PGC 5330		1.00		.27				1620

R.A. 2000 DEC. / l b / SGL SGB / R.A. 1950 DEC.	Names / PGC	Type / S_T n_L / T / L	$\log D_{25}$ / $\log R_{25}$ / $\log A_e$ / $\log D_o$	p.a. / A_g / A_i / A_{21}	B_T / m_B / m_{FIR} / B_T^o	$(B-V)_T$ / $(U-B)_T$ / $(B-V)_T^o$ / $(U-B)_T^o$	$(B-V)_e$ / $(U-B)_e$ / m'_e / m'_{25}	m_{21} / W_{20} / W_{50} / HI	V_{21} / V_{opt} / V_{GSR} / V_{3K}
012612.6+332418 131.22 -28.91 330.88 2.29 012322.9+330844	 CGCG 502- 84 MK 1155 PGC 5333			.18	15.0 ±.4				4971± 97 5115 4698
012617.8-523914 290.98 -63.66 245.16 -17.14 012416.0-525448	 ESO 152- 2 PGC 5338	.S?....	.85± .06 .08± .06 .13 .85	163 .00 .04	 15.32 ±.14 15.15				8949 8848 8794
012618.9+344516 131.01 -27.57 332.17 2.68 012328.4+342942	NGC 531 UGC 1012 HICK 10C PGC 5340	.SB.0*. U .0± .9	1.27± .04 .59± .05 1.25	34 .16 .44	14.84S±.15 14.67 ±.13 14.06		 14.54± .29	15.93±.3 560± 10	5208± 9 4660± 41 5331 4915
0126.3 -0604 146.00 -67.37 292.91 -9.15 0123.8 -0620	A 0123-06 MCG -1- 4- 44 PGC 5341	.S..6*. U (1) 6.0±1.2 3.1±1.2	1.46± .03 .81± .05 1.49	21 .22 1.19 .41				14.19±.1 272± 9 245± 12	1961± 7 2001 1663
0126.3 +0955 136.97 -51.99 308.50 -4.73 0123.7 +0940	IC 114 UGC 1015 PGC 5343	.L..... U -2.0± .8	1.24± .08 .40± .05 1.20	150 .18 .00	 15.05 ±.16				
012621.6+344214 131.03 -27.62 332.12 2.65 012331.2+342640	NGC 536 UGC 1013 HICK 10A PGC 5344	.SBR3.. U (1) 3.0± .7 3.0±1.0	1.47± .02 .44± .03 .72± .01 1.49	62 .16 .61 .22	13.20M±.08 13.04 ±.12 12.34	.85± .01 .29± .02 .70 .16	.93± .01 .38± .02 12.30± .03 14.31± .15	14.85±.2 545± 6 491± 6 2.29	5192± 5 5163± 34 5338 4921
012622.0+061634 138.40 -55.54 304.98 -5.79 012345.6+060100	 UGC 1014 PGC 5345	.S..9.. U 9.0± .8	1.07± .05 .04± .04 1.07	 .07 .04 .02	 14.71 ±.16 14.60			16.05±.3 62± 13 1.43	2132± 10 2210 1827
012624.0-383533 272.59 -76.40 259.89 -15.76 012409.7-385107	 MCG -7- 4- 6 PGC 5347	PLA.0P. S -2.0± .8	1.23± .08 .17± .08 1.21	 .02 .00					
0126.4 -0804 148.03 -69.22 290.94 -9.68 0123.9 -0820	 MCG -1- 4- 47 PGC 5348	.S?....	1.04± .09 .14± .07 1.05	 .09 .21 .07					14896 14930 14601
012625.2-013822 142.61 -63.17 297.28 -8.00 012352.2-015356	NGC 557 UGC 1016 PGC 5351	.LBT+P* PUE -.8± .5	1.15± .07 .24± .04 1.13	45 .17 .00	 14.48 ±.11 14.22				5688±110 5742 5386
012626.0-372726 269.59 -77.31 261.08 -15.61 012410.9-374300	 MCG -6- 4- 35 PGC 5352	.LA.-P? S -3.0± .8	1.23± .08 .10± .08 1.22	 .02 .00					
012631.0+344032 131.07 -27.64 332.11 2.61 012340.5+342458	NGC 542 MCG 6- 4- 22 HICK 10D PGC 5360	.S?....	.99± .10 .62± .07 1.00	 .16 .85 .31	 15.68S±.15 15.41 ±.13 14.47		 13.92± .55	17.04±.3 346± 10 2.26	4662± 9 4620± 41 4807 4390
012633.3-231333 188.69 -81.15 275.74 -13.19 012410.0-232906	 ESO 476- 8 IRAS01241-2329 PGC 5362	.SXT5.. Sᴛ (1) 4.7± .6 3.3± .8	1.29± .04 .04± .04 1.04± .03 1.29	 .03 .05 .02	 13.37M±.11 13.33 ±.14 13.24	.69± .03 .64	.77± .01 13.92± .07 14.56± .23		5504± 41 5490 5240
012633.7+313646 131.62 -30.67 329.20 1.69 012345.1+312113	A 0123+31 MCG 5- 4- 59 MK 358 PGC 5364	.SXT4*. R (1) 4.0± .5 1.0±1.3	.96± .06 .12± .06 .98	 .13 .18 .06	 14.83 ±.12 14.43			16.75±.2 367± 21 351± 8 2.26	13552± 8 13549± 42 13693 13275
012634.0-231556 188.93 -81.17 275.69 -13.20 012410.7-233130	 MCG -4- 4- 10 HICK 11B PGC 5365	.S?....	.64± .17 .28± .07 .64	 .00 .43 .14	 15.79S±.15 15.32		 13.11± .86		7352± 41 7338 7088
0126.5 +0033 141.32 -61.05 299.44 -7.44 0124.0 +0018	 UGC 1018 PGC 5366	.I..9*. U 10.0±1.2	1.00± .08 .04± .06 1.00	 .04 .03 .02					
012637.7+482336 128.94 -14.07 345.17 6.58 012336.8+480802	 MCG 8- 3- 22 5ZW 68 PGC 5368	.S?....	.74± .14 .00± .07 .81	 .82 .00 .00	 15.0 ±.3 14.17				10474± 82 10642 10245

R.A. 2000 DEC.	Names	Type	logD$_{25}$	p.a.	B$_T$	(B-V)$_T$	(B-V)$_e$	m$_{21}$	V$_{21}$
l　　b		S$_T$　n$_L$	logR$_{25}$	A$_g$	m$_B$	(U-B)$_T$	(U-B)$_e$	W$_{20}$	V$_{opt}$
SGL　SGB		T	logA$_e$	A$_i$	m$_{FIR}$	(B-V)$_T^o$	m'$_e$	W$_{50}$	V$_{GSR}$
R.A. 1950 DEC.	PGC	L	logD$_o$	A$_{21}$	B$_T^o$	(U-B)$_T^o$	m'$_{25}$	HI	V$_{3K}$

0126.6　+1016		.SB.9..	.97± .09						2182± 10
136.97 -51.63	UGC　1019	U	.27± .06	.19					
308.86　-4.70		9.0± .9		.28					2271
0124.0　+1001	PGC　5370		.99	.13					1877
012642.2+020118	NGC　550	.SBS1?.	1.17± .05	120					5872± 41
140.57 -59.64	UGC　1021	PE	.37± .04	.05	13.59 ±.14				
300.87　-7.07		1.0± .8		.38					5937
012407.6+014545	PGC　5374		1.18	.19	13.09				5568
012642.2-251851		.SBS9..	1.17± .04	25					1581± 52
201.99 -81.89	ESO　476- 10	S　　　(1)	.41± .03	.08	15.20 ±.14				
273.61 -13.63		9.0± .8		.42					1561
012420.1-253424	PGC　5375	6.7±1.2	1.18	.21	14.69				1323
0126.7　-0119			.74± .14						5013±120
142.57 -62.84	MCG　0- 4-147		.38± .07	.13	15.33 ±.17				5068
297.61　-8.00									
0124.2　-0135	PGC　5379		.75						4711
0126.7　-0059			.52± .20						4800± 49
142.36 -62.52	MCG　0- 4-148		.00± .07	.13	15.37 ±.19				4856
297.94　-7.91									
0124.2　-0115	PGC　5380		.53						4498
0126.7　+1716		.S?....	1.09± .07	52					2589
134.85 -44.78	UGC　1020		.49± .06	.18	14.81 ±.12				
315.58　-2.67	IRAS01240+1700			.73					2696
0124.0　+1700	PGC　5382		1.10	.24	13.89				2289
0126.8　+4646		.SBS8*.	1.00± .08	30					
129.21 -15.66	UGC　1017	U	.14± .06	.57					
343.63　6.10		8.0±1.2		.17					
0123.8　+4631	PGC　5386		1.05	.07					
0126.8　+1201		.L.....	1.09± .10	60					9444
136.45 -49.91	UGC　1023	U	.25± .05	.15	15.14 ±.14				
310.56　-4.24		-2.0± .8		.00					9537
0124.2　+1146	PGC　5392		1.07		14.84				9140
0126.9　+1912	IC　115	.L?....	.82± .19		15.2 ±.2	1.15± .09			13040±141
134.39 -42.87	MCG　3- 4- 39		.00± .07	.13	15.15 ±.10	.27± .15			13152
317.43　-2.12				.00		1.00			
0124.2　+1857	PGC　5395		.83		14.84	.31		14.15± .98	12742
0126.9　-1509		.SBS8..	1.12± .06	145					5716
159.32 -75.46	MCG -3- 4- 67	E　　(1)	.36± .05	.00					
283.91 -11.53		8.0± .9		.44					5728
0124.5　-1525	PGC　5397	7.5± .8	1.12	.18					5434
0127.1　+1835		.S..0..	.96± .07						
134.61 -43.47	UGC　1025	U	.10± .05	.13	14.73 ±.12				
316.86　-2.35		.0± .8		.08					
0124.4　+1820	PGC　5403		.97						
012706.0-355128		.S?....	1.01± .06	80					6297
264.13 -78.44	ESO　353- 2		.31± .05	.02	15.33 ±.14				6244
262.76 -15.51				.46					
012450.0-360700	PGC　5404		1.01	.15	14.81				6073
012708.8-582116		.S?....	.98± .06	61					13659± 69
294.06 -58.15	ESO　113- 48		.08± .06	.00	15.09 ±.14				13543
239.15 -17.52				.12					
012515.0-583648	PGC　5409		.98	.04	14.90				13532
012709.4+144633	IC　1704	.S...P$	1.04± .06	165					6349± 60
135.70 -47.21	UGC　1027	R	.14± .05	.11	14.06 ±.12				6450
313.22　-3.50	IRAS01244+1431			.21	13.24				
012429.2+143101	PGC　5411		1.05	.07	13.70				6047
0127.1　+1335		.S..9*.	1.26± .11						4508± 10
136.06 -48.37	UGC　1026	U	.12± .12	.11					
312.08　-3.85		9.0±1.1		.12					4605
0124.5　+1320	PGC　5415		1.27	.06					4205
012710.5-183915	NGC　563	.LAS0*.	1.03± .05	20					
168.93 -78.20	ESO　542- 13	SE	.08± .03	.02	14.33 ±.14				
280.40 -12.37	IRAS01245-1853	-2.3± .7		.00					
012445.0-185448	PGC　5417		1.02						
012711.5-224546	NGC　555	PSBR0*.	.86± .05						
186.57 -80.80	ESO　476- 12	Sr	.05± .03	.03	15.19 ±.14				
276.24 -13.24		.0± .5		.04					
012448.0-230118	PGC　5419		.86						

R.A. 2000 DEC. l b SGL SGB R.A. 1950 DEC.	Names PGC	Type S_T n_L T L	$\log D_{25}$ $\log R_{25}$ $\log A_e$ $\log D_o$	p.a. A_g A_i A_{21}	B_T m_B m_{FIR} B_T^o	$(B-V)_T$ $(U-B)_T$ $(B-V)_T^o$ $(U-B)_T^o$	$(B-V)_e$ $(U-B)_e$ m'_e m'_{25}	m_{21} W_{20} W_{50} HI	V_{21} V_{opt} V_{GSR} V_{3K}
012713.5-224152 186.24 -80.76 276.30 -13.23 012450.0-225724	NGC 556 ESO 476- 13 PGC 5420	.L..-*P S -3.0± .8	.61± .07 .12± .03 .60	.06 .00	15.41 ±.14				
012714.0-214622 181.53 -80.26 277.25 -13.04 012450.0-220154	ESO 542- 15 PGC 5422	.L?....	.91± .06 .01± .04 .91	.00 .00	14.53 ±.14 14.45				5567± 60 5558 5300
0127.2 +1159 136.61 -49.93 310.55 -4.34 0124.6 +1144	UGC 1029 PGC 5423	.S..1.. U 1.0±1.0	1.00± .08 .63± .06 1.01	24 .15 .64 .31	15.53 ±.12 14.63				9445 9538 9142
0127.2 -0115 142.79 -62.74 297.71 -8.10 0124.7 -0131	UGC 1030 PGC 5424	.E..... U -5.0± .9	1.00± .12 .21± .05 .96	177 .13 .00	14.57 ±.10 14.36				4794±110 4849 4493
012716.1-015819 143.28 -63.42 297.01 -8.29 012443.2-021351	NGC 558 CGCG 385-143 PGC 5425	.E+..?/ PE -3.5±1.0	.57± .18 .38± .08 .49	110 .19 .00	15.3 ±.3 15.03				5018±110 5071 4717
012725.6-015443 143.32 -63.35 297.08 -8.32 012452.7-021015	NGC 560 UGC 1036 PGC 5430	.L..0./ PUE -2.4± .5	1.28± .04 .52± .03 1.22	178 .19 .00	13.95 ±.15 13.79 ±.10 13.57	.98± .02 .45± .05 .84 .38	13.93± .27		5483± 50 5535 5182
012731.2+144907 135.81 -47.15 313.28 -3.57 012451.0+143335	IC 1706 CGCG 436- 57 PGC 5433			.11	15.3 ±.3				
012732.4+191039 134.60 -42.88 317.45 -2.27 012450.2+185507	A 0124+18 UGC 1032 MK 359 PGC 5435	.P..... R 99.0	.78± .09 .08± .05 .27± .03 .80	10 .13 .10 .04	14.16 ±.13 14.0 ±.2 12.98 13.84	.70± .01 -.08± .04 .62 -.12	.68± .01 -.21± .02 11.00± .09 12.72± .46		5043± 36 5155 4746
0127.5 -0105 142.84 -62.56 297.89 -8.13 0125.0 -0121	UGC 1040 PGC 5436	.S..0.. U .0±1.0	1.06± .06 .72± .05 1.04	34 .15 .54	14.85 ±.15 14.09				4661±110 4716 4360
012733.5-044058 145.51 -65.96 294.37 -9.08 012501.8-045630	MCG -1- 4- 52 HICK 12A PGC 5437	.E?....		.18	15.08S±.15				14407± 41 14451 14109
012734.8+313317 131.88 -30.69 329.21 1.46 012446.1+311745	A 0124+31 UGC 1033 KUG 0124+312 PGC 5440	.S..6*. U 6.0±1.1	1.47± .02 .77± .03 .94± .04 1.48	133 .13 1.13 .38	14.10S±.15 14.15 ±.13 12.86	.62± .03 .07± .03 .42 -.07	.73± .02 .05± .03 14.14± .10 14.38± .20	14.06±.2 343± 13 342± 6 .82	4037± 7 4177 3761
0127.6 -0107 142.91 -62.58 297.87 -8.16 0125.1 -0123	UGC 1043 PGC 5449	.E...*. U -5.0±1.2	.94± .13 .06± .05 .95	15 .00	14.69 ±.10 14.47				5174±110 5229 4873
012740.7+371059 130.90 -25.13 334.55 3.14 012448.4+365527	NGC 551 UGC 1034 IRAS01247+3655 PGC 5450	.SB.4.. U 4.0± .8	1.26± .04 .36± .05 1.28	140 .23 .53 .18	13.48 ±.12 13.48 12.69				4901± 52 5051 4638
012744.2-190304 170.75 -78.39 280.03 -12.58 012519.0-191836	ESO 542- 17 IRAS01252-1918 PGC 5452	.S?....	.98± .05 .04± .04 .98	.04 .04 .02	14.76 ±.14 14.58				9840± 52 9839 9567
012746.7+430916 129.94 -19.22 340.23 4.88 012450.0+425345	UGC 1037 PGC 5453	.S..3.. U (1) 3.0± .9 4.5±1.2	1.04± .06 .38± .05 1.08	92 .42 .52 .19	15.20 ±.12				
012748.3-015246 143.50 -63.29 297.14 -8.40 012515.4-020817	NGC 564 UGC 1044 PGC 5455	.E..... PUE -5.0± .5	1.14± .06 .05± .04 .85± .03 1.15	145 .16 .00	13.52 ±.14 13.46 ±.10 13.24	1.01± .02 .92	1.02± .01 .63± .07 13.26± .11 14.08± .35		5790± 50 5842 5490
012748.8+484910 129.08 -13.62 345.63 6.51 012447.1+483339	UGC 1035 PGC 5457	.S..3.. U (1) 3.0± .8 3.5±1.1	1.12± .05 .26± .05 1.19	20 .74 .36 .13	14.8 ±.3 13.61				6609 6777 6383

R.A. 2000 DEC. / l b / SGL SGB / R.A. 1950 DEC.	Names / PGC	Type / S_T n_L / T / L	$\log D_{25}$ / $\log R_{25}$ / $\log A_e$ / $\log D_o$	p.a. / A_g / A_i / A_{21}	B_T / m_B / m_{FIR} / B^o_T	$(B-V)_T$ / $(U-B)_T$ / $(B-V)^o_T$ / $(U-B)^o_T$	$(B-V)_e$ / $(U-B)_e$ / m'_e / m'_{25}	m_{21} / W_{20} / W_{50} / HI	V_{21} / V_{opt} / V_{GSR} / V_{3K}
0127.8 +4315 / 129.94 -19.12 / 340.33 4.90 / 0124.9 +4300	UGC 1038 / PGC 5460	.I..9*. / U / 10.0±1.2	1.11± .14 / .36± .12 / / 1.15	30 / .42 / .27 / .18					
012755.1-020229 / 143.68 -63.43 / 296.99 -8.47 / 012522.3-021801	IC 119 / UGC 1047 / PGC 5465	PSBR0*. / PUE / -.2± .5	1.10± .05 / .32± .04 / / 1.10	77 / .18 / .24	14.70 ±.13 / / / 14.19				5977± 81 / 6029 / 5676
012757.0-354305 / 263.03 -78.42 / 262.93 -15.66 / 012541.1-355836	NGC 568 / ESO 353- 3 / PGC 5468	.LA.-P* / BS / -3.0± .5	1.35± .04 / .21± .04 / .91± .06 / 1.32	137 / .02 / .00	13.58 ±.17 / 13.43 ±.14 / / 13.38	.99± .03 / .51± .04 / .92 / .52	1.00± .01 / .49± .02 / 13.62± .22 / 14.68± .30		5624± 29 / 5570 / 5400
012759.0-290511 / 228.72 -81.71 / 269.80 -14.61 / 012539.0-292042	ESO 413- 7 / PGC 5472	.S?....	.86± .06 / .16± .06 / / .86	3 / .04 / .24 / .08	15.21 ±.14 / / / 14.93				1512± 52 / 1479 / 1266
012800.8+320158 / 131.90 -30.20 / 329.69 1.52 / 012511.7+314627	UGC 1045 / KUG 0125+317 / PGC 5473	.S?....	1.07± .03 / .53± .03 / / 1.08	143 / .80 / .27	14.45S±.15 / 15.03 ±.12 / / 13.85	.67± .04 / .02± .07 / .50 / -.11	/ / 13.33± .23	15.25±.2 / 332± 13 / 318± 6 / 1.14	6416± 7 / / 6557 / 6141
012805.4+381256 / 130.81 -24.09 / 335.56 3.37 / 012512.3+375724	UGC 1046 / PGC 5475	.SA.9.. / U / 9.0± .8	1.05± .08 / .00± .06 / / 1.08	/ .26 / .00 / .00	14.79 ±.16 / / / 14.53				5224 / / 5376 / 4964
012806.1+491429 / 129.06 -13.19 / 346.05 6.58 / 012503.8+485858	UGC 1042 / PGC 5476	.L..... / U / -2.0± .8	1.18± .08 / .31± .04 / / 1.24	95 / .91 / .00	14.3 ±.3				
0128.1 +1025 / 137.49 -51.41 / 309.11 -5.01 / 0125.5 +1010	UGC 1050 / PGC 5479	.S..8*. / U / 8.0±1.3	1.04± .15 / .29± .12 / / 1.06	/ .21 / .36 / .15					2393 / / 2481 / 2090
012810.1-011823 / 143.30 -62.72 / 297.73 -8.33 / 012537.0-013354	NGC 565 / UGC 1052 / PGC 5481	.S..1*. / PUE / 1.3± .6	1.10± .05 / .49± .04 / / 1.12	36 / .12 / .50 / .25	14.42 ±.13 / / / 13.74				4498± 46 / 4552 / 4197
012813.0-015500 / 143.75 -63.29 / 297.13 -8.51 / 012540.2-021031	IC 120 / CGCG 385-152 / PGC 5484	.L..0*. / PE / -2.0± .9	.91± .11 / .64± .08 / / .84	135 / .16 / .00	15.23 ±.13 / / / 15.00				4900±110 / 4952 / 4600
0128.2 +1625 / 135.59 -45.54 / 314.88 -3.27 / 0125.6 +1610	MCG 3- 4- 43 / PGC 5486	.S?....	.64± .17 / .28± .07 / / .65	119 / .14 / .35 / .14	15.40 ±.20 / / / 14.79				11371± 46 / 11475 / 11071
012817.4+285917 / 132.57 -33.19 / 326.82 .54 / 012530.0+284346	UGC 1051 / PGC 5488	.I..9*. / U / 10.0±1.2	1.04± .15 / .13± .12 / / 1.06	/ .23 / .10 / .06				16.20±.3 / 188± 7	7439± 10 / 7573 / 7158
012818.8+341831 / 131.55 -27.95 / 331.87 2.15 / 012528.2+340300	NGC 561 / UGC 1048 / IRAS01254+3403 / PGC 5489	RSBS1.. / U / 1.0± .8	1.20± .05 / .03± .05 / / 1.22	/ .20 / .03 / .01	13.80 ±.14 / 11.66 / 13.51				4670 / / 4815 / 4400
012821.9+023048 / 141.09 -59.05 / 301.47 -7.33 / 012547.1+021517	IC 121 / UGC 1053 / IRAS01257+0215 / PGC 5492	.S..4.. / U / 4.0± .9	.96± .07 / .23± .05 / / .97	108 / .04 / .33 / .11	14.28 ±.14 / 13.42 / 13.85				9024± 50 / 9089 / 8721
012822.7-355923 / 263.66 -78.16 / 262.66 -15.79 / 012607.0-361454	ESO 353- 5 / PGC 5494	.LAR0*. / Sr / -1.8± .6	1.19± .04 / .22± .03 / / 1.16	32 / .02 / .00	14.15 ±.14 / / / 14.05				5464± 39 / 5409 / 5242
012825.8-023524 / 144.34 -63.91 / 296.48 -8.74 / 012553.2-025054	MCG -1- 4- 54 / PGC 5498	.S..6?/ / E / 6.0±1.9	1.15± .06 / .83± .05 / / 1.17	120 / .18 / 1.22 / .41					5071±120 / 5121 / 4772
012829.6+482313 / 129.26 -14.03 / 345.25 6.28 / 012528.1+480743	NGC 562 / UGC 1049 / IRAS01254+4807 / PGC 5502	.SAT5.. / U (1) / 5.0± .8 / 1.1± .9	1.10± .05 / .08± .05 / / 1.17	20 / .76 / .11 / .04	14.0 ±.3 / 12.55 / 13.11			15.26±.1 / 189± 11 / 136± 11 / 2.11	10254± 10 / 10268± 59 / 10421 / 10027

R.A. 2000 DEC. / l b / SGL SGB / R.A. 1950 DEC.	Names / PGC	Type / S_T n_L / T / L	$\log D_{25}$ / $\log R_{25}$ / $\log A_e$ / $\log D_o$	p.a. / A_g / A_i / A_{21}	B_T / m_B / m_{FIR} / B_T^o	$(B-V)_T$ / $(U-B)_T$ / $(B-V)_T^o$ / $(U-B)_T^o$	$(B-V)_e$ / $(U-B)_e$ / m'_e / m'_{25}	m_{21} / W_{20} / W_{50} / HI	V_{21} / V_{opt} / V_{GSR} / V_{3K}
0128.5 -0143 143.79 -63.08 297.34 -8.54 0126.0 -0159	UGC 1055 PGC 5506	.SB.1.. U 1.0± .9	1.10± .06 .44± .05 1.11	157 .16 .45 .22	 14.84 ±.12 14.16				6271±110 6324 5971
012836.3-391824 NGC 572 272.84 -75.57 ESO 296- 31 259.21 -16.28 012623.0-393354 PGC 5508		.LXS0*. BS -1.7± .7	.91± .05 .04± .03 .91	 .02 .00	 15.18 ±.14				
0128.7 +1640 135.69 -45.27 315.15 -3.31 0126.1 +1625	UGC 1056 PGC 5516	.I..9*. U 10.0±1.2	.94± .11 .19± .07 .95	 .14 .14 .09	 14.81 ±.12 14.53				595 700 296
012848.3+342047 131.65 -27.89 UGC 1054 331.93 2.06 KUG 0125+340 012557.7+340517 PGC 5518		.S..6*. U 6.0±1.4	1.13± .05 1.01± .08 1.15	124 .20 1.47 .50					2663 2808 2394
012851.5+022649 IC 123 141.36 -59.08 MCG 0- 4-161 301.44 -7.47 ARAK 49 012616.7+021119 PGC 5524		.S..4 	.64± .17 .19± .07 .64	 .03	 15.19 ±.19				9179 9244 8877
0128.8 +1346 136.61 -48.10 312.38 -4.19 0126.2 +1331	UGC 1057 PGC 5527	.S..4.. U 4.0± .9	1.17± .05 .50± .05 1.18	153 .12 .74 .25	 14.64 ±.12 13.74				6333 6430 6032
012857.1-513555 NGC 576 289.32 -64.53 ESO 196- 7 246.29 -17.48 FAIR 295 012655.0-515124 PGC 5535		PLXT+.. Sr -.6± .6	.99± .06 .10± .04 .98	18 .04 .00	 14.41 ±.14 12.48 14.32				3150±190 3051 2992
012858.9-005657 NGC 570 143.48 -62.31 UGC 1061 298.13 -8.43 012625.6-011227 PGC 5539		PSBT1*. PUE .6± .5	1.19± .05 .09± .05 .84± .04 1.20	175 .12 .09 .05	13.70 ±.15 13.84 ±.13 13.50	.93± .03 .41± .04 .84 .38	.97± .03 .47± .04 13.39± .13 14.26± .32		5493± 46 5548 5193
012859.3-003342 143.22 -61.95 UGC 1062 298.51 -8.33 012625.8-004912 PGC 5540		PLBS0*. E -2.0±1.2	1.13± .05 .30± .05 1.09	60 .07 .00	 13.85 ±.10 13.70				5450± 46 5506 5150
0129.0 -0224 144.54 -63.69 MCG -1- 4- 55 296.70 -8.84 0126.5 -0240 PGC 5543		.S?....	1.08± .09 .61± .07 1.09	 .18 .89 .30					5287 5337 4988
012902.8-353555 NGC 574 261.83 -78.34 ESO 353- 6 263.09 -15.87 IRAS01268-3551 012647.0-355124 PGC 5544		PSBT3.. BSr 2.5± .6	1.05± .05 .18± .04 1.05	2 .02 .25 .09	 14.17 ±.14 11.85 13.85				5709± 39 5656 5486
012903.0+321956 NGC 566 132.09 -29.87 UGC 1058 330.04 1.40 012613.5+320426 PGC 5545		.L..... U -2.0± .9	1.20± .06 .59± .03 1.13	178 .16 .00	 14.49 ±.10				
012907.4+110751 NGC 569 137.61 -50.66 UGC 1063 309.86 -5.04 MK 997 012628.7+105222 PGC 5548		.S?.... 	1.02± .06 .35± .05 1.04	163 .22 .53 .18	 14.59 ±.12 12.78 13.81				5772± 36 5862 5470
012912.4+392535 130.84 -22.86 UGC 1059 336.77 3.52 IRAS01262+3909 012618.2+391006 PGC 5550		.SB?... 	1.19± .05 .23± .05 1.21	97 .26 .34 .11	 14.39 ±.14 13.06 13.75			15.59±.1 151± 8 1.73	8263± 11 8417 8007
012913.3+334006 131.88 -28.55 CGCG 502- 93 331.32 1.77 KUG 0126+334 012623.0+332436 PGC 5552		.S?.... 	.80± .09 .15± .12 .82	 .17 .23 .08	 15.28 ±.12 14.85			16.49±.3 250± 7 1.56	6605± 10 6748 6334
0129.2 +1108 137.65 -50.65 UGC 1065 309.88 -5.06 0126.6 +1053 PGC 5555		.SXS4.. U 4.0± .9	1.09± .06 .24± .05 1.11	15 .22 .35 .12	 15.18 ±.17 14.57				5767± 50 5856 5465
012920.8+320349 132.22 -30.12 UGC 1066 329.81 1.26 012631.5+314820 PGC 5563		.S..6*. U 6.0±1.3	1.11± .07 .66± .06 1.12	44 .10 .97 .33				16.41±.2 171± 13 164± 6	5072± 7 5211 4797

R.A. 2000 DEC. / l b / SGL SGB / R.A. 1950 DEC.	Names / PGC	Type / S_T n_L / T / L	$\log D_{25}$ / $\log R_{25}$ / $\log A_e$ / $\log D_o$	p.a. / A_g / A_i / A_{21}	B_T / m_B / m_{FIR} / B_T^o	$(B-V)_T$ / $(U-B)_T$ / $(B-V)_T^o$ / $(U-B)_T^o$	$(B-V)_e$ / $(U-B)_e$ / m'_e / m'_{25}	m_{21} / W_{20} / W_{50} / HI	V_{21} / V_{opt} / V_{GSR} / V_{3K}
012927.2-612750		.L..-P.	1.16± .05	63	14.4 ±.2	1.03± .03	1.06± .02		8657
295.01 -55.08	ESO 113- 50	S	.14± .04	.00	13.90 ±.14	.42± .05	.46± .03		
235.88 -17.86		-3.0± .8	.63± .10	.00		.94	13.04± .33		8533
012740.0-614318	PGC 5565		1.14		13.93	.46	14.74± .34		8546
0129.5 +5023		.S..9*.	1.14± .13					15.42±.1	5843± 11
129.13 -12.02	UGC 1064	U	.00± .12	1.28					6012
347.21 6.69		9.0±1.1		.00				211± 8	
0126.5 +5008	PGC 5568		1.26	.00					5623
012944.0-011428		.L.....	1.09± .12	68					5129±120
144.06 -62.53	UGC 1072	U	.18± .05	.12	14.35 ±.10				5182
297.90 -8.69	ARAK 50	-2.0± .8		.00					4830
012710.8-012956	PGC 5574		1.07		14.16				
012944.4-182026	NGC 583	.LBT0..	.83± .05	40					
170.03 -77.55	ESO 542- 20	SE	.06± .02	.00	15.15 ±.14				
280.86 -12.89		-2.0± .5		.00					
012719.0-183554	PGC 5576		.82						
0129.7 -0158	IC 126	.S?....	1.14± .13						5704±120
144.59 -63.22	UGC 1071		.07± .12	.13	15.0 ±.2				5755
297.18 -8.89				.08					5405
0127.2 -0214	PGC 5577		1.15	.03	14.78				
012946.7+453556		.S..5..	1.24± .05	30				15.32±.1	5239± 9
129.92 -16.75	UGC 1068	U	.19± .05	.41	13.56 ±.12			373± 11	5180± 57
342.66 5.26	IRAS01268+4520	5.0± .8		.29	12.93			346± 12	5400
012647.4+452028	PGC 5579		1.28	.10	12.83			2.39	5001
0129.7 -0658	IC 127	.S..3*/	1.26± .04	65					
148.99 -67.89	MCG -1- 4- 57	PE	.59± .04	.12					
292.24 -10.21		3.0± .7		.82					
0127.2 -0714	PGC 5581		1.27	.30					
0129.8 +5156		.SB.2*.	1.04± .08	87					
128.93 -10.48	UGC 1067	U	.67± .06	1.90					
348.70 7.09		2.0±1.0		.82					
0126.7 +5141	PGC 5582		1.22	.33					
012950.4-024135		.SAS8..	1.06± .06						
145.18 -63.89	MCG -1- 4- 56	E (1)	.00± .05	.21					
296.48 -9.11		8.0± .9		.00					
012717.8-025703	PGC 5583	6.4± .8	1.08	.00					
012951.7-421926		.P.....	1.05± .05	44	14.6 ±.2	.65± .08			
278.12 -72.90	ESO 244- 30	S	.34± .04	.03	14.81 ±.14	.02± .09			7572± 63
256.08 -16.86	IRAS01276-4235	99.0		.51	12.88	.53			7498
012741.0-423454	PGC 5584		1.06	.17	14.16	-.07	13.87± .33		7375
012956.0+323005	NGC 571	.S?....	1.13± .07					15.25±.3	4658± 7
132.27 -29.67	UGC 1069		.03± .06	.16	14.57 ±.17				4798
330.26 1.28				.04				83± 7	4385
012706.3+321437	PGC 5587		1.14	.01	14.33			.90	
013000.1+405826		.S..6*.	1.27± .04	55				14.81±.1	2806± 6
130.73 -21.31	UGC 1070	U	.16± .05	.33	13.94 ±.15			157± 6	2962
338.28 3.84	IRAS01270+4042	6.0±1.1		.24	13.64			124± 8	2555
012704.5+404258	PGC 5589		1.31	.08	13.37			1.36	
013004.8+331834		.I...9..	1.04± .15					16.44±.3	5516± 10
132.15 -28.87	UGC 1074	U	.00± .12	.17					5658
331.03 1.49		10.0± .8		.00				123± 7	5245
012714.5+330306	PGC 5594		1.06	.00					
0130.0 +0251		.S..8*.	.96± .09	94					2100
141.70 -58.60	UGC 1075	U	.39± .06	.01					
301.93 -7.65		8.0±1.3		.48					2165
0127.5 +0236	PGC 5596		.96	.20					1799
013005.4-424108		.SAR5..	1.20± .04	87	13.65 ±.16	.64± .03	.72± .02		
278.61 -72.57	ESO 244- 31	S (1)	.08± .04	.03	14.04 ±.14				6768± 40
255.70 -16.94	FAIR 699	5.0± .5	1.00± .03	.12		.57	14.14± .06		6693
012755.0-425636	PGC 5597	1.1± .5	1.20	.04	13.69		14.31± .28		6573
0130.1 +2551	A 0127+25	.S..9*.	1.24± .05	163	14.44 ±.18	.68± .09	.69± .04		3675
133.73 -36.19	UGC 1073	U (1)	.20± .05	.28	14.8 ±.2	-.22± .10	-.14± .07		
323.99 -.80	DDO 11	9.0±1.1	1.05± .03	.20		.55	15.18± .08		3801
0127.3 +2536	PGC 5600	9.0±1.5	1.26	.10	14.09	-.31	14.98± .31		3389
0130.2 +1701		.S?....	1.02± .06	133					12921
136.09 -44.85	UGC 1076		.53± .05	.17	15.30 ±.12				
315.59 -3.55				.79					13026
0127.6 +1646	PGC 5611		1.03	.26	14.27				12624

R.A. 2000 DEC. l b SGL SGB R.A. 1950 DEC.	Names PGC	Type S_T n_L T L	$\log D_{25}$ $\log R_{25}$ $\log A_e$ $\log D_o$	p.a. A_g A_i A_{21}	B_T m_B m_{FIR} B_T^o	$(B-V)_T$ $(U-B)_T$ $(B-V)_T^o$ $(U-B)_T^o$	$(B-V)_e$ $(U-B)_e$ m'_e m'_{25}	m_{21} W_{20} W_{50} HI	V_{21} V_{opt} V_{GSR} V_{3K}
013025.2-330209 250.32 -79.71 265.80 -15.77 012808.0-331736	 ESO 353- 7 IRAS01281-3317 PGC 5615	.S..1P. S 1.0± .8 	1.03± .05 .07± .04 1.03	145 .02 .07 .04	 13.75 ±.14 12.87 				
013025.9-264651 212.86 -81.31 272.27 -14.73 012805.1-270218	 ESO 476- 16 IRAS01281-2702 PGC 5617	.SBR4.. Sr (1) 4.5± .5 3.9± .6	1.17± .04 .30± .03 1.17	143 .02 .46 .15	 14.29 ±.14 13.78 13.78				6021± 52 5994 5770
013028.2-223957 188.30 -80.09 276.49 -13.96 012805.1-225524	NGC 578 ESO 476- 15 IRAS01280-2255 PGC 5619	.SXT5.. R (3) 5.0± .3 2.4± .4	1.69± .01 .20± .02 1.30± .01 1.69	110 .02 .29 .10	11.44 ±.13 11.60 ±.12 11.47 11.20	.51± .02 .46 	.54± .01 -.06± .03 13.43± .03 14.26± .16	12.82±.1 277± 6 267± 6 1.52	1630± 4 1597± 34 1616 1368
013028.4-411745 275.91 -73.72 257.17 -16.87 012817.0-413312	 ESO 296- 34 PGC 5620	PLBT+*. Sr -.7± .5 	1.03± .05 .21± .04 1.00	72 .03 .00 	 14.77 ±.14 14.65				6530 6459 6329
0130.5 +1936 135.41 -42.32 318.07 -2.81 0127.8 +1921	 UGC 1077 PGC 5621	.L..-*. U -3.0±1.2 	1.02± .11 .07± .05 1.02	50 .12 .00 	 14.95 ±.13 				
013037.1+221607 134.74 -39.70 320.61 -2.02 012753.0+220040	 MCG 4- 4- 12 PGC 5623	.S?.... 	.85± .12 .48± .07 .87	 .26 .73 .24	 15.73 ±.13 14.69			16.65±.3 296± 7 1.72	10280± 10 10397 9989
013040.8-015933 145.09 -63.16 297.22 -9.12 012807.9-021500	NGC 577 UGC 1080 IRAS01281-0215 PGC 5628	PSBR1P. UE 1.0± .4 	1.26± .04 .10± .04 1.27	140 .14 .10 .05	 13.76 ±.15 13.44				5952± 46 6003 5654
013042.0-374315 267.44 -76.57 260.92 -16.48 012828.0-375842	 ESO 296- 35 IRAS01285-3758 PGC 5629	.SBS3P. S (1) 3.0± .9 2.2± .9	.99± .05 .09± .04 .99	 .02 .12 .04	 14.98 ±.14 				
013043.0-510827 288.34 -64.86 246.79 -17.73 012841.1-512354	 ESO 196- 11 PGC 5631	.SBR6?. S (1) 6.0± .8 4.4± .9	1.22± .05 .39± .05 1.23	12 .04 .58 .20	 14.46 ±.14 13.83				3639 3540 3480
013046.5+212623 135.00 -40.51 319.83 -2.31 012802.8+211056	NGC 575 UGC 1081 PGC 5634	.SBT5.. U 5.0± .8 	1.23± .04 .03± .04 1.25	 .19 .05 .02	 13.45 ±.13 13.20			15.34±.1 153± 6 141± 7 2.13	3149± 6 3161± 76 3264 2857
013049.4+411525 130.85 -21.01 338.60 3.78 012753.4+405959	NGC 573 UGC 1078 IRAS01278+4100 	.S?.... 	.61± .14 .05± .06 .64	 .31 .07 .02	 14.1 ±.4 12.43 			15.97±.1 134± 6 93± 8 	2788± 6 1745± 35
013049.5-272152 216.73 -81.24 271.69 -14.92 012829.0-273718	 ESO 413- 8 IRAS01284-2737 PGC 5639	PSBR0.. r -.1± .9 	.90± .06 .28± .05 .89	32 .04 .21 	 15.11 ±.14 12.62 				
013050.6+255454 133.91 -36.12 324.09 -.94 012804.5+253927	 MCG 4- 4- 13 PGC 5640	.S?.... 	.74± .14 .09± .07 .76	 .29 .12 .05	 15.55 ±.12 15.11			17.07±.3 296± 7 1.91	11203± 10 11329 10918
0130.9 +1711 136.24 -44.66 315.79 -3.64 0128.2 +1655	IC 1711 UGC 1082 IRAS01282+1655 PGC 5643	.S..3.. U (1) 3.0± .8 4.5±1.2	1.41± .04 .73± .05 1.43	43 .17 1.01 .37	 14.41 ±.14 12.50 13.21				2804 2909 2507
0130.9 +1405 137.25 -47.68 312.84 -4.59 0128.3 +1350	 UGC 1083 PGC 5645	.S..3.. U (1) 3.0± .8 4.5±1.1	1.14± .05 .29± .05 1.16	125 .14 .40 .15	 14.39 ±.13 13.82				4519 4615 4220
013112.2-012918 144.99 -62.64 297.76 -9.11 012839.2-014444	 MCG 0- 4-167 ARAK 51 PGC 5655	 	.58± .18 .23± .07 .59	 .10 	 15.1 ±.3 				5281±120 5333 4983
013114.2+095626 138.87 -51.70 308.87 -5.88 012836.0+094100	 CGCG 436- 68 PGC 5656	.95? .65? .97	 	 .21 	 15.38 ±.14 				9501 9586 9200

R.A. 2000 DEC. I b SGL SGB R.A. 1950 DEC.	Names PGC	Type S_T n_L T L	$\log D_{25}$ $\log R_{25}$ $\log A_e$ $\log D_o$	p.a. A_g A_i A_{21}	B_T m_B m_{FIR} B_T^o	$(B-V)_T$ $(U-B)_T$ $(B-V)_T^o$ $(U-B)_T^o$	$(B-V)_e$ $(U-B)_e$ m'_e m'_{25}	m_{21} W_{20} W_{50} HI	V_{21} V_{opt} V_{GSR} V_{3K}
0131.3 +0747 139.80 -53.77 306.80 -6.53 0128.7 +0732	UGC 1085 PGC 5661	.I..9.. U 10.0± .8	1.11± .14 .20± .12 1.12	0 .13 .15 .10					652 731 351
013120.9-065206 149.81 -67.63 292.46 -10.56 012850.2-070732	NGC 584 MCG -1- 4- 60 PGC 5663	.E.4... R -5.0± .3	1.62± .02 .26± .03 .92± .01 1.56	120 .14 .00	11.44 ±.13 11.15 ±.17 11.16	.96± .01 .49± .01 .91 .47	.97± .01 .54± .01 11.53± .04 13.86± .20		1864± 16 1899 1572
0131.3 +2357 134.53 -38.01 322.26 -1.66 0128.6 +2342	UGC 1084 PGC 5664	.S..9*. U 9.0±1.2	1.11± .14 .07± .12 1.14	.27 .07 .03					3416± 10 3537 3128
013124.7-594047 293.74 -56.73 237.76 -18.09 012935.1-595612	ESO 113- 52 PGC 5669	.S?.... -1.0± .6	.89± .06 .07± .06 .90	22 .00 .10 .03	15.45 ±.14 15.27				13148 13028 13029
0131.4 +1416 137.36 -47.47 313.05 -4.65 0128.8 +1401	UGC 1087 PGC 5673	.SAT5.. U 5.0± .8	1.17± .05 .03± .05 1.18	.13 .04 .01	14.50 ±.20 14.30				4483 4580 4184
013131.4-123918 157.86 -72.75 286.69 -12.02 012903.4-125443	IC 129 MCG -2- 5- 1 PGC 5675	PLXT+?. E -1.0± .6	1.13± .05 .19± .04 1.11	60 .05 .00					
013134.7+344659 132.21 -27.36 332.52 1.65 012843.3+343134	UGC 1086 KUG 0128+345 PGC 5676	.S?.... U	1.01± .05 .44± .04 1.03	38 .17 .66 .22	14.97 ±.12 14.12				4163 4307 3897
013136.9-065340 149.99 -67.63 292.45 -10.63 012906.3-070905	NGC 586 MCG -1- 5- 1 PGC 5679	.SAS1*$ R 1.0± .9	1.20± .04 .32± .04 1.21	10 .13 .32 .16	14.1 ±.2 13.62	.94± .05 .22± .07 .84 .13	14.17± .28		1990 2025 1698
0131.6 +0826 139.64 -53.12 307.45 -6.41 0129.0 +0811	UGC 1091 PGC 5680	.I..9*. U 10.0±1.2	.96± .09 .00± .06 .97	.12 ..00 .00					2483 2563 2182
013137.6+365000 131.82 -25.34 334.46 2.28 012844.7+363435	UGC 1088 PGC 5681	.S..6*. U 6.0±1.1	1.05± .08 .01± .06 1.06	.18 .01 .00	15.2 ±.2 14.97				4885 5033 4624
013142.5-005555 144.85 -62.08 298.34 -9.08 012909.2-011120	NGC 585 UGC 1092 IRAS01291-0111 PGC 5688	.S..1*/ UE 1.0± .6	1.33± .03 .62± .04 1.33	86 .08 .63 .31	13.99 ±.12 13.34 13.22				5361± 46 5414 5063
013146.7+333655 132.48 -28.51 331.43 1.25 012856.0+332130	NGC 579 UGC 1089 IRAS01289+3321 PGC 5691	.S..6*. U 6.0±1.1	1.06± .05 .06± .05 1.07	.17 .09 .03	13.9 ±.2 13.62 ±.13 12.71 13.43	.59± .02 .04± .03 .51 -.02	15.64±.1 260± 6 214± 5 13.89± .34 2.18		4994± 4 5011± 28 5136 4725
013146.8+384301 131.50 -23.48 336.25 2.83 012852.6+382736	UGC 1090 PGC 5692	.S..3.. U (1) 3.0± .8 4.5±1.1	1.00± .06 .02± .05 1.03	.22 .03 .01	14.57 ±.13				
0131.8 +1734 136.41 -44.24 316.23 -3.72 0129.1 +1719	UGC 1093 PGC 5693	.S..3.. U 3.0± .8	1.13± .07 .16± .06 1.14	10 .16 .22 .08					
013150.9-330711 249.83 -79.41 265.76 -16.08 012934.0-332236	ESO 353- 9 IRAS01295-3322 PGC 5696	.SBR4*. S (1) 4.0±1.1 3.3±1.1	1.17± .05 .05± .05 1.18	.02 .08 .03	13.69 ±.14 12.03 13.56				4958± 19 4911 4729
013158.2+332836 132.56 -28.64 331.31 1.17 012907.6+331311	NGC 582 UGC 1094 IRAS01291+3313	.SB?...	1.35± .04 .58± .05 1.36	58 .17 .80 .29	14.1 ±.2 13.66 ±.12 12.78 12.78	.93± .03 .17± .04 .76 .00	15.04±.1 470± 6 450± 5 14.23± .30 1.97		4352± 4 4491± 52 4494 4084
013158.4-851039 301.93 -31.87 211.10 -16.43 013447.0-852600	ESO 3- 1 IRAS01348-8526 PGC 5703	RSBT0.. Sr .0± .4	1.19± .05 .03± .05 1.23	.47 .02	13.85 ±.14 11.89 13.30				4043± 34 3876 4055

R.A. 2000 DEC.	Names	Type	logD$_{25}$	p.a.	B$_T$	(B-V)$_T$	(B-V)$_e$	m$_{21}$	V$_{21}$
l b		S$_T$ n$_L$	logR$_{25}$	A$_g$	m$_B$	(U-B)$_T$	(U-B)$_e$	W$_{20}$	V$_{opt}$
SGL SGB		T	logA$_e$	A$_i$	m$_{FIR}$	(B-V)^o_T	m'$_e$	W$_{50}$	V$_{GSR}$
R.A. 1950 DEC.	PGC	L	logD$_o$	A$_{21}$	B^o_T	(U-B)^o_T	m'$_{25}$	HI	V$_{3K}$

013202.3-414123		RSX.0P.	1.19± .09						
275.79 -73.23		S	.21± .08	.03					
256.79 -17.20		-2.0± .8		.00					
012951.6-415647	PGC 5709		1.16						
013202.5+331231								16.80±.3	10506± 10
132.63 -28.90	CGCG 502-104			.13	15.4 ±.4				
331.06 1.07								221± 7	10647
012911.9+325707	PGC 5710								10236
013204.4+331043		.S?....	.80± .09						
132.64 -28.93	CGCG 502-106		.26± .12	.13	15.20 ±.14				10305± 97
331.04 1.06	MK 1156			.39					10446
012913.9+325519	PGC 5711		.81	.13	14.63				10035
013208.6+320612		.S?....	1.04± .09						
132.88 -29.98	UGC 1095		.38± .07	.16					12435± 57
330.03 .71				.57					12573
012918.7+315047	PGC 5715		1.05	.19					12163
0132.1 +1228		.S?....	1.00± .08	150					10634
138.24 -49.18	UGC 1096		.55± .06	.16					
311.37 -5.35				.82					10725
0129.5 +1213	PGC 5716		1.01	.27					10335
013214.6-332954	NGC 597	.SBS4?.	1.15± .05						
251.31 -79.14	ESO 353- 11	BS (1)	.04± .05	.02	13.98 ±.14				5072± 39
265.38 -16.22	IRAS01299-3345	3.7± .7		.06	13.43				5023
012958.0-334518	PGC 5721	3.3± .8	1.15	.02	13.86				4844
013216.7-344700			.85± .07	0					
256.64 -78.38	ESO 353- 12		.38± .05	.02	15.32 ±.14				5178±104
264.04 -16.41									5125
013001.0-350224	PGC 5724		.85						4954
0132.2 +2124		.SB?...	.97± .07	90					9862
135.46 -40.47	UGC 1098		.26± .05	.19	14.67 ±.12				
319.91 -2.65	IRAS01295+2109			.38	12.83				9976
0129.5 +2109	PGC 5725		.99	.13	14.03				9571
013220.6-075134		.S?....	1.02± .07	70	15.55S±.15				
151.51 -68.43	MCG -1- 5- 2		.36± .05	.09					12469± 41
291.53 -11.05	HICK 13A			.53					12500
012950.5-080658	PGC 5732		1.03	.18	14.87		14.58± .39		12179
013220.7-122126	NGC 593	.LBR0?/	1.07± .06	12					
157.91 -72.39	MCG -2- 5- 3	E	.70± .05	.04					
287.04 -12.15		-2.0±1.3		.00					
012952.7-123650	PGC 5733		.97						
013222.5-075227		.S?....			15.65S±.15				
151.55 -68.44	MCG -1- 5- 3			.09					
291.52 -11.06	HICK 13B								
012952.4-080751	PGC 5735								
013228.0-384048		.SBS4?.	1.09± .05						
268.91 -75.61	ESO 296- 38	S (1)	.04± .04	.02	13.99 ±.14				3651
259.96 -16.95		4.0± .6		.05					3587
013015.1-385612	PGC 5742	2.6± .7	1.09	.02	13.89				3442
013229.2+043545		.S?....	1.12± .08					14.68±.1	1981± 6
141.84 -56.76	UGC 1102		.08± .07	.05				168± 8	1917± 46
303.79 -7.73				.11				174± 12	2049
012953.4+042021	PGC 5744		1.12	.04					1680
013233.4+352132	NGC 587	.SXS3..	1.34± .04	67					
132.31 -26.76	UGC 1100	U	.43± .05	.14	13.58 ±.12				4588± 52
333.13 1.64	IRAS01296+3506	3.0± .8		.59					4733
012941.3+350608	PGC 5746		1.35	.21	12.82				4324
013235.6+415912		.L.....	1.08± .17						
131.07 -20.23	UGC 1101	U	.00± .08	.37	14.94 ±.16				
339.39 3.69	5ZW 77	-2.0± .8		.00					
012938.6+414348	PGC 5753		1.12						
0132.6 +1150		.S..6*.	1.00± .08	156					10661
138.66 -49.77	UGC 1103	U	.84± .06	.16					
310.80 -5.65		6.0±1.4		1.24					10750
0130.0 +1135	PGC 5757		1.02	.42					10362
013239.9-120234	NGC 589	RSBR0..	1.04± .07						
157.59 -72.08	MCG -2- 5- 4	E	.08± .05	.07					5177± 52
287.37 -12.15	MK 999	.0± .9		.06					5195
013011.7-121758	PGC 5758		1.04						4894

R.A. 2000 DEC. / l b / SGL SGB / R.A. 1950 DEC.	Names / PGC	Type / S_T n_L / T / L	$\log D_{25}$ / $\log R_{25}$ / $\log A_e$ / $\log D_o$	p.a. / A_g / A_i / A_{21}	B_T / m_B / m_{FIR} / B_T^o	$(B-V)_T$ / $(U-B)_T$ / $(B-V)_T^o$ / $(U-B)_T^o$	$(B-V)_e$ / $(U-B)_e$ / m'_e / m'_{25}	m_{21} / W_{20} / W_{50} / HI	V_{21} / V_{opt} / V_{GSR} / V_{3K}
013240.0+043836		.I..9..	.93± .10						1993± 10
141.89 -56.70	UGC 1105	U	.15± .06	.05					1845± 46
303.85 -7.76		10.0± .8		.12					2055
013004.1+042312	PGC 5759		.93	.08					1686
0132.7 +8500		.S..2..	1.19± .05	95				15.99±.1	5281± 11
123.90 22.22	UGC 1039	U	.36± .05	.42	15.1 ±.2				5457
21.12 14.83	IRAS01254+8445	2.0± .8		.44	13.59			415± 8	5457
0125.4 +8445	PGC 5760		1.23	.18	14.16			1.65	5218
013242.9+181857		.I..9..	1.00± .06	5					669± 50
136.48 -43.47	UGC 1104	U	.20± .05	.13	14.21 ±.13				775
317.00 -3.70		10.0± .9		.15					375
013000.7+180333	PGC 5761		1.02	.10	13.93				
013248.5-792827		.SBS8..	1.35± .04	167					
300.56 -37.46	ESO 13- 16	S (1)	.21± .05	.31	13.53 ±.14				
216.99 -17.16	IRAS01326-7943	8.0± .8		.26	13.36				
013238.1-794348	PGC 5764	6.7± .8	1.38	.10					
013251.6-144855	IC 141	.SBT4*.	1.07± .06						
163.07 -74.36	MCG -3- 5- 4	E (1)	.00± .05	.00					
284.59 -12.84	IRAS01304-1504	4.0±1.2		.00					
013024.8-150418	PGC 5765	3.1±1.2	1.07	.00					
013252.1-070157	NGC 596	.E+..P*	1.51± .05	140	11.84 ±.13	.90± .01	.94± .01		1890± 13
150.89 -67.63	MCG -1- 5- 5	PE	.19± .05	.11	11.77 ±.17	.42± .01	.47± .01		1923
292.39 -10.97		-4.0± .5	.96± .01	.00		.86	12.13± .04		1599
013021.6-071720	PGC 5766		1.47		11.67	.40	13.94± .30		
013256.9-163211	NGC 594	.S..4*.	1.13± .05	32					5420
167.28 -75.70	MCG -3- 5- 5	SE (2)	.33± .04	.01					5424
282.86 -13.24	IRAS01305-1647	3.7± .7		.49	12.26				5145
013031.0-164734	PGC 5769	2.0± .8	1.13	.17					
0132.9 -0041	IC 138	.SXS5..	1.06± .06	30					4581±120
145.30 -61.75	UGC 1106	U	.14± .05	.02	14.56 ±.13				4634
298.66 -9.32		5.0± .8		.21					4284
0130.4 -0057	PGC 5771		1.06	.07	14.30				
0133.0 +0011		.S..9..	1.00± .11						4971
144.74 -60.91	UGC 1107	U	.00± .08	.04					
299.53 -9.10		9.0± .8		.00					5026
0130.5 -0004	PGC 5776		1.00	.00					4674
013305.5-071846	NGC 600	PSBT7..	1.52± .03	85	12.92 ±.15	.55± .03	.64± .02	13.90±.1	1842± 5
151.33 -67.86	MCG -1- 5- 7	PE (1)	.07± .04	.10		-.13± .04	-.03± .03	79± 6	1867± 53
292.13 -11.09	IRAS01305-0733	6.5± .5	1.20± .03	.10		.50	14.41± .06	53± 11	1874
013035.1-073409	PGC 5777	4.2± .8	1.53	.04	12.72	-.16	15.21± .23	1.14	1552
013306.7-121230	NGC 599	.LX.-P*	1.15± .06	135					
158.18 -72.17	MCG -2- 5- 5	E	.02± .04	.04					
287.23 -12.29	MK 1000	-3.0± .8		.00					
013038.6-122753	PGC 5778		1.15						
0133.3 +1320		.S..6*.	1.28± .04	174					2761
138.36 -48.27	UGC 1110	U	.61± .05	.15	14.88 ±.14				
312.29 -5.37		6.0±1.2		.89					2854
0130.7 +1305	PGC 5792		1.29	.30	13.82				2463
0133.3 +0305		.S..6*.	1.12± .05	108					
143.07 -58.13	UGC 1112	U	.36± .05	.02	14.70 ±.13				
302.39 -8.37		6.0±1.2		.53					
0130.8 +0250	PGC 5794		1.12	.18					
013328.6+445627			.81± .11						5112± 57
130.70 -17.30	CGCG 537- 12		.29± .06	.52	15.44 ±.09				5272
342.23 4.44									4876
013028.9+444105	PGC 5796		.85						
0133.5 +1725		.S..6*.	1.08± .07						7864
137.01 -44.30	UGC 1113	U	.00± .03	.14					
316.21 -4.15		6.0±1.2		.00					7968
0130.8 +1710	PGC 5797		1.09	.00					7570
0133.5 +1901		.S..6*.	1.07± .06	60					8495
136.52 -42.74	UGC 1114	U	.28± .05	.13	15.15 ±.15				
317.73 -3.66		6.0±1.2		.42					8603
0130.8 +1846	PGC 5799		1.08	.14	14.57				8202
013331.4+354007	NGC 591	PSB.0..	1.10± .05	5				16.61±.3	4554± 9
132.47 -26.42	UGC 1111	U	.08± .05	.13	13.89 ±.12			311± 12	4495± 36
333.48 1.55	MK 1157	.0± .8		.06	12.43				4695
013039.0+352445	PGC 5800		1.11		13.63				4288

R.A. 2000 DEC. / l b / SGL SGB / R.A. 1950 DEC.	Names / PGC	Type / S_T n_L / T / L	logD_25 / logR_25 / logA_e / logD_o	p.a. / A_g / A_i / A_21	B_T / m_B / m_FIR / B_T^o	(B-V)_T / (U-B)_T / (B-V)_T^o / (U-B)_T^o	(B-V)_e / (U-B)_e / m'_e / m'_25	m_21 / W_20 / W_50 / HI	V_21 / V_opt / V_GSR / V_3K
0133.5 -0105		.S..8*.	.91± .07	118					
145.89 -62.07	UGC 1116	U	.61± .05	.10	15.26 ±.15				4862±120
298.31 -9.57		8.0±1.4		.76					4913
0131.0 -0121	PGC 5803		.92	.31	14.39				4566
013334.5+123510	IC 1715	.I..9..	.86± .06	100				16.55±.3	4176± 10
138.71 -48.99	UGC 1115	U	.16± .04	.13	14.51 ±.14			208± 13	
311.59 -5.64	IRAS01309+1219	10.0± .9		.12	13.50				4267
013054.9+121948	PGC 5805		.87	.08	14.26			2.21	3878
013341.0+445545	NGC 590	.SB.1..	1.41± .05	150					
130.74 -17.30	UGC 1109	U	.29± .06	.52	13.78 ±.17				5058± 57
342.23 4.40		1.0± .7		.29					5218
013041.3+444024	PGC 5808		1.46	.14	12.90				4822
013342.1+033236		.SAS8*.	1.34± .03	66					3515
142.95 -57.68	UGC 1118	UE (1)	.61± .04	.05	14.43 ±.13				
302.85 -8.32		7.7± .7		.75					3580
013106.8+031715	PGC 5810	6.4± .8	1.34	.31	13.62				3217
013350.9+303937	NGC 598	.SAS6..	2.85± .01	23	6.27M±.03	.55± .02		7.18±.1	-179± 3
133.61 -31.33	UGC 1117	R (2)	.23± .02	.19	6.21 ±.12	-.10± .02		199± 7	-204± 17
328.78 -.09		6.0± .3	2.43± .02	.33	6.32	.46	13.86± .05	184± 4	-46
013101.7+302415	PGC 5818	4.3± .5	2.87	.11	5.74	-.16	14.79± .07	1.33	-454
0133.9 +1713		.S..8..	1.22± .06	125					7968± 10
137.21 -44.47	UGC 1119	U	.15± .06	.13					
316.05 -4.31		8.0± .8		.18					8071
0131.2 +1658	PGC 5822		1.23	.07					7674
013358.2-362933	NGC 612	.LA.+P/	1.16± .05	172	13.9 ±.3	1.03± .05	1.01± .02		8925± 29
261.66 -77.00	ESO 353- 15	PBS	.19± .03	.02	13.90 ±.14	.42± .09	.40± .04		8867
262.30 -16.97	IRAS01317-3644	-1.2± .5	.84± .10	.00	12.31	.91	13.59± .34		8709
013144.0-364454	PGC 5827		1.14		13.74	.43	14.12± .39		
013359.7-342315		.S..2*/	1.23± .04	164					3808± 39
254.01 -78.34	ESO 353- 14	S	.65± .04	.02	14.50 ±.14				3756
264.50 -16.70	IRAS01317-3438	2.0±1.2		.80	13.05				3584
013144.0-343836	PGC 5829		1.23	.32	13.64				
013402.5-010438		.SB.2P/	1.27± .04	139					4690±120
146.12 -62.02	UGC 1120	UE	.56± .04	.10	14.55 ±.13				4741
298.36 -9.68	IRAS01314-0119	2.0± .6		.69	13.38				4394
013129.2-011959	PGC 5830		1.28	.28	13.72				
013407.9-010156		.S..2*/	1.10± .05	71					4929± 46
146.13 -61.97	UGC 1123	UE	.55± .04	.04	14.38 ±.14				4980
298.41 -9.69		1.7± .7		.68					4633
013134.6-011717	PGC 5838		1.11	.28	13.62				
013413.1-253403		.S?....	1.14± .04	145					5832± 52
206.09 -80.29	ESO 476- 25		.68± .04	.05	14.59 ±.14				5807
273.68 -15.35	IRAS01318-2549			1.02	12.59				5581
013152.0-254924	PGC 5841		1.15	.34	13.49				
013413.6-383709		.LX.-..	.98± .06	175					
267.71 -75.44	ESO 297- 3	S	.15± .04	.02	14.86 ±.14				
260.07 -17.28		-3.0± .9		.00					
013201.0-385230	PGC 5842		.96						
013417.5-292458	NGC 613	.SBT4..	1.74± .02	120	10.73M±.08	.68± .01	.76± .01	13.29±.1	1475± 5
229.07 -80.30	ESO 413- 11	R (3)	.12± .02	.02	10.75 ±.12	.06± .04	.16± .01	390± 8	1510± 15
269.70 -16.02	VV 824	4.0± .3	1.31± .02	.17	9.75	.64	12.76± .04	364± 11	1442
013158.7-294019	PGC 5849	3.0± .4	1.74	.06	10.54	.03	13.98± .12	2.69	1239
0134.3 +2919		.S..6*.	1.00± .16						
134.03 -32.62	UGC 1122	U	.09± .12	.32					
327.55 -.61		6.0±1.2		.13					
0131.5 +2904	PGC 5853		1.03	.04					
013427.1-614828		PSBR1..	.91± .06	131					
294.18 -54.58	ESO 113- 53	r	.22± .03	.00	15.39 ±.14				
235.51 -18.45		1.0± .9		.23					
013243.0-620348	PGC 5859		.91	.11					
0134.6 -1530		.S?....	1.22± .07						5452
165.95 -74.64	MCG -3- 5- 8		.95± .07	.00					
284.00 -13.42	IRAS01322-1545			1.42					5458
0132.2 -1545	PGC 5866		1.22	.47					5177
013442.0-471104		.LAR+..	.90± .06						
282.83 -68.22	ESO 244- 34	r	.05± .03	.03	15.20 ±.14				4300±190
251.02 -18.14	FAIR 374	-1.3± .9		.00					4210
013237.0-472624	PGC 5868		.90		15.10				4126

R.A. 2000 DEC. / l b / SGL SGB / R.A. 1950 DEC.	Names / PGC	Type S_T n_L / T / L	$\log D_{25}$ / $\log R_{25}$ / $\log A_o$ / $\log D_o$	p.a. / A_g / A_i / A_{21}	B_T / m_B / m_{FIR} / B_T^o	$(B-V)_T$ / $(U-B)_T$ / $(B-V)_T^o$ / $(U-B)_T^o$	$(B-V)_o$ / $(U-B)_o$ / m'_o / m'_{25}	m_{21} / W_{20} / W_{50} / HI	V_{21} / V_{opt} / V_{GSR} / V_{3K}
0134.7 -0659 / 151.92 -67.40 / 292.55 -11.40 / 0132.2 -0715	UGCA 19 / PGC 5870	.IB.9*/ / UE (1) / 10.0± .7 / 9.2± .8	1.20± .07 / .30± .07 / / 1.21	2 / .11 / .22 / .15					
013450.4+212500 / 136.22 -40.33 / 320.10 -3.21 / 013206.3+210940	NGC 606 / UGC 1126 / IRAS01321+2109 / PGC 5874	.SBR5.. / U / 5.0± .8 /	1.15± .05 / .06± .05 / / 1.16	/ .17 / .08 / .03	/ 14.10 ±.14 / 13.47 / 13.78			15.57±.3 / 346± 7 / 320± 7 / 1.75	9972± 11 / 9956± 50 / 10084 / 9682
013451.3-360811 / 259.94 -77.11 / 262.69 -17.11 / 013237.0-362330	ESO 353- 20 / IRAS01326-3623 / PGC 5875	.L..+?/ / S / -1.0±1.7 /	1.16± .05 / .45± .03 / / 1.09	76 / .02 / .00 /	/ 14.16 ±.14 / 10.93 / 14.07				4797± 87 / 4739 / 4580
0134.8 +1205 / 139.37 -49.39 / 311.20 -6.09 / 0132.2 +1150	UGC 1129 / PGC 5876	.L...*. / U / -2.0±1.2 /	1.06± .07 / .19± .03 / / 1.05	95 / .13 / .00 /	/ 14.84 ±.12				
013452.1-362916 / 261.10 -76.88 / 262.33 -17.15 / 013238.1-364436	NGC 619 / ESO 353- 21 / PGC 5878	PSBR3.. / BSr / 2.8± .5 /	1.17± .04 / .15± .04 / / 1.17	130 / .02 / .21 / .08	/ 14.34 ±.14 / / 14.05				8512± 20 / 8452 / 8296
0134.9 +5525 / 129.12 -6.93 / 352.24 7.37 / 0131.7 +5510	UGC 1124 / PGC 5879	.S..6*. / U / 6.0±1.3 /	1.07± .07 / .62± .06 / / 1.18	40 / 1.18 / .91 / .31				15.52±.1 / 329± 8 /	5752± 11 / 5924 / 5554
0134.9 +0120 / 144.88 -59.67 / 300.79 -9.24 / 0132.4 +0105	UGC 1130 / PGC 5884	.S?.... / / /	.96± .07 / .69± .05 / / .97	2 / .06 / 1.03 / .34	/ 15.54 ±.13 / / 14.41				7677 / 7735 / 7381
013459.5+350223 / 132.93 -26.99 / 332.98 1.07 / 013207.2+344703	CGCG 521- 49 / MK 1158 / PGC 5885			.17	15.03 ±.16 / 13.41			17.67±.3 / 73± 15	4585± 10 / 4485± 50 / 4724 / 4318
0135.0 +3955 / 131.94 -22.19 / 337.58 2.61 / 0132.1 +3940	UGC 1127 / PGC 5889	.SA.8.. / U / 8.0± .8 /	1.04± .15 / .00± .12 / / 1.08	/ .38 / .00 / .00					
013502.4+411453 / 131.68 -20.88 / 338.83 3.02 / 013205.5+405934	NGC 605 / UGC 1128 / PGC 5891	.L..... / U / -2.0± .8 /	1.34± .12 / .28± .08 / / 1.33	145 / .32 / .00 /	/ 13.89 ±.12				
0135.0 +0422 / 143.06 -56.78 / 303.76 -8.40 / 0132.4 +0407	A 0132+04 / UGC 1133 / DDO 12 / PGC 5892	.I..9*. / PU (1) / 10.0± .6 / 9.0±1.3	1.49± .03 / .22± .05 / 1.41± .06 / 1.50	/ .08 / .16 / .11	14.2 ±.2 / / / 13.96	.35± .15 / .02± .14 / .27 / -.04	.30± .07 / .00± .08 / 16.75± .16 / 15.98± .31	14.50±.1 / 125± 7 / 107± 12 / .43	1964± 6 / / 2031 / 1667
013505.2-412611 / 273.67 -73.12 / 257.12 -17.74 / 013255.0-414130	NGC 625 / ESO 297- 5 / IRAS01329-4141 / PGC 5896	.SBS9$/ / R (1) / 9.0± .5 / 5.6± .7	1.76± .02 / .48± .03 / 1.13± .02 / 1.76	92 / .03 / .49 / .24	11.71 ±.13 / 11.64 ±.12 / 11.33 / 11.15	.56± .01 / -.07± .04 / .45 / -.15	.52± .01 / -.19± .01 / 12.85± .04 / 14.16± .19	13.54±.2 / 115± 7 / 77± 12 / 2.14	386± 5 / 383± 58 / 312 / 189
013505.8-072027 / 152.55 -67.67 / 292.23 -11.58 / 013235.5-073547	NGC 615 / MCG -1- 5- 8 / IRAS01325-0735 / PGC 5897	.SAT3.. / R (2) / 3.0± .3 / 3.0± .6	1.56± .02 / .40± .03 / .90± .02 / 1.56	25 / .10 / .56 / .20	12.47 ±.13 / 12.45 ±.18 / 12.66 / 11.79	.86± .01 / .28± .02 / .75 / 1.04	.93± .01 / .39± .01 / 12.46± .05 / 14.08± .18	14.16±.1 / 429± 5 / 368± 5 / 2.16	1848± 5 / / 1879 / 1560
013506.0-362923 / 260.97 -76.84 / 262.33 -17.20 / 013252.0-364442	NGC 623 / ESO 353- 23 / PGC 5898	.E+..*. / BS / -3.5± .5 /	1.31± .05 / .14± .04 / .90± .06 / 1.27	94 / .02 / .00 /	13.55 ±.17 / 13.60 ±.14 / / 13.42	1.04± .03 / .54± .04 / .95 / .58	1.05± .02 / .55± .03 / 13.54± .21 / 14.62± .30		8955± 34 / 8896 / 8739
013512.0-390853 / 268.50 -74.91 / 259.53 -17.53 / 013300.0-392412	NGC 626 / ESO 297- 6 / PGC 5901	.SBT5*. / Sr (1) / 5.1± .5 / 4.4± .5	1.29± .04 / .02± .05 / / 1.29	/ .02 / .03 / .01	/ 13.41 ±.14 / / 13.32				5623± 34 / 5556 / 5417
013515.6-224217 / 191.32 -79.13 / 276.68 -15.05 / 013253.1-225736	ESO 476- 27 / PGC 5903	.S?.... / / /	1.05± .05 / .31± .05 / / 1.05	96 / .01 / .47 / .16	/ 15.68 ±.14 / / 15.17				5686 / 5669 / 5428
013516.4+342822 / 133.11 -27.53 / 332.47 .83 / 013224.5+341303	UGC 1131 / PGC 5904	.SX.8*. / U / 8.0±1.2 /	1.03± .08 / .02± .06 / / 1.04	/ .11 / .02 / .01	/ 14.98 ±.17 / / 14.84				5086 / / 5227 / 4822

R.A. 2000 DEC. / l b / SGL SGB / R.A. 1950 DEC.	Names / PGC	Type S_T n_L / T / L	$\log D_{25}$ / $\log R_{25}$ / $\log A_e$ / $\log D_o$	p.a. / A_g / A_i / A_{21}	B_T / m_B / m_{FIR} / B_T^o	$(B-V)_T$ / $(U-B)_T$ / $(B-V)_T^o$ / $(U-B)_T^o$	$(B-V)_e$ / $(U-B)_e$ / m'_e / m'_{25}	m_{21} / W_{20} / W_{50} / HI	V_{21} / V_{opt} / V_{GSR} / V_{3K}
0135.3 +0529 / 142.57 -55.70 / 304.86 -8.15 / 0132.7 +0514	UGC 1137 / PGC 5907	.I..9*. / U / 10.0±1.2	1.07± .14 / .00± .12 / / 1.08	.07 / .00 / .00					3081 / / 3151 / 2783
013521.3-324634 / 246.29 -78.95 / 266.23 -16.76 / 013304.8-330153	MCG -6- 4- 53 / PGC 5909	.LAT-*. / S / -3.0± .8	1.19± .09 / .00± .08 / / 1.19	.02 / .00					
0135.3 +3619 / 132.75 -25.71 / 334.21 1.40 / 0132.5 +3604	UGC 1134 / PGC 5910	.S..9*. / U / 9.0±1.3	1.00± .16 / .33± .12 / / 1.02	17 / .16 / .34 / .17					
0135.4 +0452 / 142.94 -56.28 / 304.27 -8.35 / 0132.8 +0437	UGC 1138 / VV 590 / PGC 5911	.IX.9.. / U / 10.0± .8	1.09± .06 / .11± .05 / / 1.10	.05 / .08 / .06					5167 / / 5235 / 4870
013528.2+333924 / 133.33 -28.32 / 331.71 .54 / 013236.7+332405	NGC 608 / UGC 1135 / PGC 5913	.S?.... / / .92	.91± .09 / .17± .06 / / .92	32 / .14 / .25 / .08	14.18 ±.16 / / 13.76				5103± 39 / 5242 / 4836
013530.7-392312 / 268.91 -74.69 / 259.29 -17.61 / 013319.0-393830	ESO 297- 8 / IRAS01333-3938 / PGC 5915	PSXT1P. / Sr / 1.3± .5	1.18± .05 / .07± .05 / / 1.19	62 / .03 / .07 / .04	13.73 ±.14 / / 13.57				5348 / 5280 / 5143
013531.5+473258 / 130.60 -14.67 / 344.80 4.91 / 013228.8+471739	UGC 1132 / PGC 5916	.S..8.. / U / 8.0± .8	1.22± .06 / .41± .06 / / 1.28	13 / .55 / .50 / .20	14.7 ±.3 / / 13.61			15.24±.1 / 315± 8 / 1.42	5327± 11 / / 5490 / 5101
0135.5 +4132 / 131.73 -20.58 / 339.13 3.02 / 0132.6 +4117	UGC 1136 / PGC 5921	.I..9*. / U / 10.0±1.1	1.14± .13 / .03± .12 / / 1.17	.33 / .02 / .01					
0135.5 +1156 / 139.68 -49.49 / 311.11 -6.30 / 0132.9 +1141	UGC 1139 / PGC 5922	.SB?... / / .96	.95± .07 / .12± .05 / / .96	.15 / .18 / .06	14.76 ±.12 / / 14.39				5835 / / 5923 / 5539
013536.7-392136 / 268.79 -74.70 / 259.32 -17.63 / 013325.0-393654	NGC 630 / ESO 297- 9 / PGC 5924	.LAT-*. / S / -3.0± .6	1.21± .05 / .07± .04 / .69± .05 / 1.21	.03 / .00	12.51V±.16 / 13.57 ±.14 / 13.46		1.04± .02 / .45± .05		5924± 19 / 5856 / 5720
013551.1-100010 / 156.48 -69.93 / 289.61 -12.42 / 013322.1-101528	NGC 624 / MCG -2- 5- 10 (1) / IRAS01333-1015 / PGC 5932	PSBR3P. / E (1) / 3.0± .8 / 3.1±1.2	1.17± .05 / .20± .05 / / 1.18	100 / .08 / .28 / .10					5870 / / 5892 / 5586
013552.3+334053 / 133.42 -28.28 / 331.76 .46 / 013300.8+332535	NGC 614 / UGC 1140 / PGC 5933	.L...?. / U / -2.0±1.5	1.15± .09 / .01± .05 / / 1.17	.15 / .00	13.66 ±.10 / / 13.43				5161± 39 / 5300 / 4895
013559.9+003950 / 145.82 -60.21 / 300.20 -9.68 / 013325.9+002432	NGC 622 / UGC 1143 / MK 571 / PGC 5939	.SBT3.. / PUE (1) / 3.3± .4 / 1.9± .8	1.26± .03 / .15± .04 / / 1.26	45 / .02 / .20 / .07	13.71 ±.14 / 12.67 / 13.45			15.54±.3 / 372± 13 / 2.01	5155± 10 / 5212± 63 / 5211 / 4861
013609.0-361801 / 259.73 -76.81 / 262.56 -17.38 / 013355.1-363318	ESO 353- 25 / PGC 5944	.S?.... / / .99	.99± .06 / .08± .05 / / .99	.02 / .11 / .04	14.73 ±.14 / / 14.53				9777± 39 / 9718 / 9561
0136.1 +1141 / 140.00 -49.70 / 310.91 -6.51 / 0133.5 +1126	UGC 1144 / PGC 5945	.S..6*. / U / 6.0±1.2	1.19± .05 / .44± .05 / / 1.20	3 / .17 / .64 / .22					5837 / / 5924 / 5541
013618.6-134146 / 163.11 -72.96 / 285.92 -13.40 / 013351.5-135703	MCG -2- 5- 13 / PGC 5952	.SAT5P* / E (1) / 5.0± .8 / 3.1±1.6	1.15± .06 / .25± .05 / / 1.15	10 / .00 / .38 / .13					
0136.3 +1032 / 140.56 -50.79 / 309.82 -6.91 / 0133.7 +1017	UGC 1146 / PGC 5954	.S..8*. / U / 8.0±1.5	.96± .09 / .98± .06 / / .98	75 / .21 / 1.21 / .49					3523 / / 3606 / 3227

R.A. 2000 DEC. / l b / SGL SGB / R.A. 1950 DEC.	Names / PGC	Type / S_T n_L / T / L	$\log D_{25}$ / $\log R_{25}$ / $\log A_e$ / $\log D_o$	p.a. / A_g / A_i / A_{21}	B_T / m_B / m_{FIR} / B_T^o	$(B-V)_T$ / $(U-B)_T$ / $(B-V)_T^o$ / $(U-B)_T^o$	$(B-V)_e$ / $(U-B)_e$ / m'_e / m'_{25}	m_{21} / W_{20} / W_{50} / HI	V_{21} / V_{opt} / V_{GSR} / V_{3K}
013624.1-371911	NGC 633	.SBR3*.	1.11± .06	177					
262.80 -76.08	ESO 297- 11	S (1)	.07± .06	.02	13.50 ±.14				5160± 51
261.49 -17.56	IRAS01341-3734	2.8± .5		.10	11.19				5098
013411.0-373428	PGC 5960	3.3± .6	1.11	.04	13.34				4948
013625.9-423537		RSXT4P*	1.05± .05	135					
275.20 -72.03	ESO 244- 36	Sr (1)	.17± .04	.03	14.64 ±.14				6352
255.92 -18.10		4.1± .4		.26					6274
013417.0-425054	PGC 5962	4.4± .6	1.05	.09	14.31				6160
013627.8-362231		.SBS3*P	1.20± .05	170					
259.80 -76.71	ESO 353- 26	S (1)	.43± .05	.02	14.29 ±.14				5461± 39
262.49 -17.46	IRAS01342-3638	3.0± .9		.60	13.43				5401
013414.1-363748	PGC 5964	3.3±1.2	1.20	.22	13.63				5246
013632.9+395517		.SB.3..	1.12± .05	157					
132.25 -22.13	UGC 1145	U	.32± .05	.27	14.91 ±.14				
337.67 2.33	IRAS01336+3940	3.0± .9		.44	13.26				
013336.7+394000	PGC 5966		1.14	.16					
013633.8-802052		.S..6./	1.25± .04	32					
300.59 -36.57	ESO 13- 18	S	1.32± .05	.35	16.12 ±.14				
216.06 -17.22		6.0± .9		1.47					
013643.0-803606	PGC 5967		1.29	.50					
0136.6 -0122		.L..-*.	1.04± .18	15					
147.65 -62.06	UGC 1151	U	.09± .08	.07	14.63 ±.11				
298.24 -10.39		-3.0±1.2		.00					
0134.1 -0138	PGC 5971		1.04						
013642.1+154711	NGC 628	.SAS5..	2.02± .01	25	9.95M±.10	.56± .03	.64± .01	10.77±.0	656± 3
138.62 -45.70	UGC 1149	R (2)	.04± .02	.13	10.01 ±.13		.01± .03	78± 3	632± 23
314.88 -5.39		5.0± .3	1.68± .02	.06	9.56	.52	13.67± .05	54± 3	753
013400.7+153155	PGC 5974	1.1± .5	2.03	.02	9.79		14.79± .12	.97	363
0136.8 +0550	NGC 631	.E.....	1.23± .14						
143.01 -55.26	UGC 1153	U	.05± .08	.13	14.25 ±.17				5634± 56
305.31 -8.40		-5.0± .8		.00					5704
0134.2 +0535	PGC 5983		1.24		14.04				5338
013649.0+353045	NGC 621	.LB....	1.28± .08						
133.23 -26.45	UGC 1147	U	.02± .05	.14	13.73 ±.14				3305± 96
333.54 .87	4ZW 54	-2.0± .8		.00					3448
013356.1+351528	PGC 5984		1.30		13.53				3044
013656.2+315907		.S..3..	1.02± .06	177				15.64±.2	6618± 6
134.06 -29.90	UGC 1152	U (1)	.25± .05	.17	14.54 ±.12			343± 9	
330.24 -.29		3.0± .9		.35				315± 7	6753
013405.7+314351	PGC 5986	3.5±1.2	1.04	.13	13.97			1.55	6349
013700.9-623224		.LA.0..	1.27± .08						
294.06 -53.80		S	.14± .08	.00					8783± 69
234.73 -18.73		-2.0± .8		.00					8654
013519.8-624740	PGC 5993		1.25						8680
013707.2+045307		.S..6*.	.85± .10	166					
143.67 -56.14	UGC 1155	U	.24± .06	.05	14.58 ±.16				3158± 50
304.41 -8.75		6.0±1.3		.35					3225
013431.2+043751	PGC 5998		.85	.12	14.17				2862
013715.4-091152		.LA.-..	1.20± .10	155					
156.19 -69.06	MCG -2- 5- 20	E	.14± .07	.10					
290.51 -12.56		-3.0± .8		.00					
013446.1-092708	PGC 6004		1.19						
0137.2 +1428		.LXT0..	1.04± .08						
139.30 -46.94	UGC 1156	U	.08± .06	.14					10594
313.67 -5.93		-2.0± .8		.00					10688
0134.6 +1413	PGC 6005		1.04						10301
013717.0+285325		.S?....	.90± .07					15.73±.3	7756± 10
134.89 -32.91	UGC 1154		.06± .05	.28	14.42 ±.13				7731± 50
327.35 -1.36	IRAS01344+2838			.08	13.29			273± 7	7884
013428.4+283810	PGC 6006		.93	.03	14.01			1.69	7480
013717.6+055239	NGC 632	.L?....	1.19± .06	170				15.89±.2	3168± 6
143.19 -55.18	UGC 1157		.10± .03	.12	13.27 ±.10			244± 10	3157± 41
305.38 -8.51	MK 1002			.00	11.53			183± 10	3238
013441.1+053723	PGC 6007		1.19		13.11				2872
013722.1-645345	NGC 646	.SXS5P*	1.11± .06						
295.13 -51.53	ESO 80- 2	RS	.13± .06	.00	14.24 ±.14				8230± 60
232.23 -18.68	VV 443	5.0± .4		.19	11.86				8096
013547.0-650900	PGC 6010		1.11	.06	14.00				8139

R.A. 2000 DEC.	Names	Type	$\log D_{25}$	p.a.	B_T	$(B-V)_T$	$(B-V)_e$	m_{21}	V_{21}
l b		S_T n_L	$\log R_{25}$	A_g	m_B	$(U-B)_T$	$(U-B)_e$	W_{20}	V_{opt}
SGL SGB		T	$\log A_e$	A_i	m_{FIR}	$(B-V)^o_T$	m'_e	W_{50}	V_{GSR}
R.A. 1950 DEC.	PGC	L	$\log D_o$	A_{21}	B^o_T	$(U-B)^o_T$	m'_{25}	HI	V_{3K}
013730.2-211427		.S?....	1.01± .07						12856
186.13 -77.97	ESO 543- 1		.12± .05	.01					12842
278.29 -15.28				.15					12596
013507.1-212942	PGC 6016		1.01	.06					
013734.5-424021		RSB.0*/	1.08± .05	15					6570
274.78 -71.84	ESO 244- 39	Sr	.47± .04	.03	14.87 ±.14				6492
255.85 -18.31		.4± .6		.35					6379
013526.1-425536	PGC 6018		1.06		14.39				
013736.4-335527	IC 1719	.LAS0*.	1.21± .05	174					5778± 34
250.12 -77.97	ESO 353- 27	BS	.12± .04	.02	13.79 ±.14				5725
265.10 -17.38		-2.3± .6		.00					5555
013521.0-341042	PGC 6020		1.20		13.68				
013740.4-605152		.SXR2..	1.11± .06						5425
293.05 -55.37	ESO 114- 1	r	.43± .06	.00	14.82 ±.14				5300
236.50 -18.86	IRAS01360-6106	2.2± .9		.53					5314
013556.0-610706	PGC 6030		1.11	.21	14.23				
013741.9+323955		.S?....	.96± .17	155				16.07±.3	13362± 10
134.08 -29.20	UGC 1158		.15± .12	.12					
330.93 -.22	6ZW 4			.22				344± 7	13499
013450.7+322440	PGC 6032		.97	.07					13095
013759.6-400410		.SBS6?/	1.19± .05	38					
269.19 -73.86	ESO 297- 16	S (1)	.52± .04	.03	14.73 ±.14				
258.62 -18.16	IRAS01358-4019	6.0±1.2		.77	13.14				
013549.0-401924	PGC 6044	3.3±1.3	1.19	.26					
013803.3+322939		.S..6*.	1.19± .06	102				16.28±.2	5446± 6
134.20 -29.36	UGC 1160	U	.91± .06	.12	15.57 ±.12			336± 9	
330.79 -.35		6.0±1.3		1.34				307± 7	5582
013512.2+321425	PGC 6045		1.20	.46	14.09			1.74	5179
013809.0-325116		RSBR0..	.93± .06	53					
245.28 -78.39	ESO 353- 28	r	.16± .05	.02	15.19 ±.14				
266.24 -17.35		-.1± .9		.12					
013553.1-330630	PGC 6051		.93						
013815.6+413914		.SBR3..	1.19± .05						
132.24 -20.37	UGC 1162	U	.04± .05	.38	14.6 ±.2				
339.40 2.58		3.0± .8		.06					
013517.6+412400	PGC 6056		1.22	.02					
0138.2 +8039		.S..3..	1.14± .13	120				16.51±.1	8261± 11
124.92 17.96	UGC 1141	U	.39± .12	.94					
16.80 13.76		3.0± .9		.54				327± 8	8439
0133.0 +8024	PGC 6057		1.23	.20					8177
013818.5+352154	NGC 634	.S..1..	1.32± .04	167					4940± 52
133.60 -26.53	UGC 1164	U	.52± .05	.19	13.89 ±.12				5082
333.50 .53	IRAS01354+3507	1.0± .8		.53					4680
013525.4+350641	PGC 6059		1.33	.26	13.11				
013819.5+073201	A 0135+07	.SAT6..	1.43± .03	120				15.11±.2	4303± 8
142.74 -53.53	UGC 1167	U	.11± .05	.11	14.0 ±.3			212± 12	4265± 47
307.07 -8.27		6.0± .8		.16				139± 25	4375
013542.2+071647	PGC 6061		1.44	.05	13.72			1.34	4007
013828.4-333629		.SB.4?P	1.05± .05	35					
248.38 -77.97	ESO 353- 29	S	.23± .04	.02	14.75 ±.14				
265.45 -17.52	IRAS01361-3351	4.0±1.2		.34	13.82				
013613.0-335142	PGC 6071		1.05	.11					
013831.5+284324		.S..6*.	1.11± .05	51				16.01±.2	10900± 7
135.25 -33.02	UGC 1165	U	.83± .05	.28	15.62 ±.12				
327.28 -1.67		6.0±1.4		1.23				425± 7	11027
013542.8+282811	PGC 6074		1.14	.42	14.06			1.54	10625
013834.8+345932		.L.....	1.14± .05	69					4663± 52
133.74 -26.89	UGC 1166	U	.55± .05	.19	14.08 ±.12				4804
333.17 .36		-2.0± .9		.00					4402
013541.9+344419	PGC 6077		1.08		13.82				
013837.5-400041		.S..1*/	1.21± .04	143					
268.72 -73.83	ESO 297- 18	S	.53± .04	.03	14.28 ±.14				
258.69 -18.27	IRAS01363-4016	1.0±1.2		.54	13.30				
013627.0-401554	PGC 6078		1.21	.27					
013839.4-423135	NGC 641	.LAS-*.	1.16± .04		13.06 ±.13	.94± .01	1.00± .01		
273.99 -71.85	ESO 244- 42	BS	.03± .03	.03	13.34 ±.14				6323± 17
256.02 -18.50		-2.8± .4	.89± .02	.00		.87	13.00± .07		6244
013631.0-424648	PGC 6081		1.16		13.07		13.64± .23		6132

R.A. 2000 DEC. / l b / SGL SGB / R.A. 1950 DEC.	Names / PGC	Type / S_T n_L / T / L	$\log D_{25}$ / $\log R_{25}$ / $\log A_e$ / $\log D_o$	p.a. / A_g / A_i / A_{21}	B_T / m_B / m_{FIR} / B_T^o	$(B-V)_T$ / $(U-B)_T$ / $(B-V)_T^o$ / $(U-B)_T^o$	$(B-V)_e$ / $(U-B)_e$ / m'_e / m'_{25}	m_{21} / W_{20} / W_{50} / HI	V_{21} / V_{opt} / V_{GSR} / V_{3K}
013839.9-174959 / 174.90 -75.68 / 281.84 -14.86 / 013615.1-180512	NGC 648 / ESO 543- 6 / / PGC 6083	.LA.-P* / SE / -2.6± .6 /	1.02± .06 / .28± .03 / / .98	114 / .00 / .00 /					
013847.0+010419 / 146.83 -59.59 / 300.80 -10.23 / 013612.8+004906	/ UGC 1169 / / PGC 6090	.L..0.. / UE / -2.0± .6 /	1.12± .06 / .30± .03 / / 1.09	74 / .07 / .00 /	14.01 ±.10 / / / 13.86				4992± 50 / 5046 / / 4699
013847.6-314917 / 240.26 -78.68 / 267.34 -17.34 / 013631.0-320430	/ ESO 413- 12 / / PGC 6092	.SAR1*. / Sr / .6± .7 /	1.02± .05 / .28± .04 / / 1.02	171 / .02 / .29 / .14	14.99 ±.14 / / / 14.57				8849± 60 / 8802 / / 8620
013848.3-832151 / 301.30 -33.61 / 212.94 -16.86 / 014021.0-833700	/ ESO 3- 3 / / PGC 6093	.SBS6P. / S (1) / 6.0± .6 / 5.6± .7	1.25± .04 / .42± .05 / / 1.30	132 / .47 / .61 / .21	14.74 ±.14 / / / 13.63				4948 / 4783 / / 4951
013848.8-491901 / 283.72 -66.01 / 248.80 -18.94 / 013647.3-493413	/ FAIR 707 / / PGC 6094	.LA.-.. / S / -3.0± .9 /	1.11± .10 / .31± .08 / / 1.07	/ .04 / .00 /					9100±190 / 9003 / / 8937
013853.3-423505 / 273.99 -71.78 / 255.96 -18.55 / 013645.0-425018	NGC 644 / ESO 244- 43 / FAIR 706 / PGC 6097	.SBR4*. / BS (1) / 4.5± .5 / 2.2± .5	1.11± .05 / .33± .04 / / 1.11	155 / .03 / .50 / .17	14.79 ±.14 / / 13.76 / 14.19				11900±190 / 11821 / / 11710
013854.2-465012 / 280.69 -68.19 / 251.44 -18.83 / 013650.0-470524	/ ESO 244- 44 / IRAS01368-4705 / PGC 6099	PSBT1*. / Sr / 1.3± .5 /	1.21± .05 / .33± .05 / / 1.21	78 / .03 / .33 / .16	14.25 ±.14 / / 12.29 / 13.80				6708 / 6617 / / 6535
013855.4-074602 / 155.28 -67.61 / 292.05 -12.60 / 013625.5-080114	/ / / PGC 6101			52 / .10	16.9 S±.5			17.24±.3 / 374± 24 / 364± 18 /	5528± 14 / / 5555 / 5244
013858.4-463430 / 280.31 -68.41 / 251.72 -18.83 / 013654.0-464942	/ ESO 244- 45 / / PGC 6104	PLA.-?. / S / -3.0± .7 /	1.04± .04 / .08± .03 / .74± .02 / 1.03	/ .03 / .00 /	13.61 ±.13 / 14.01 ±.14 / / 13.66	.95± .01 / .41± .05 / .88 / .43	.97± .01 / .45± .05 / 12.80± .06 / 13.50± .24		6446± 56 / 6356 / / 6272
013858.8-295530 / 230.72 -79.20 / 269.34 -17.11 / 013641.0-301042	NGC 639 / ESO 413- 13 / IRAS01367-3010 / PGC 6105	.S..1?. / S / / .98	.98± .05 / .65± .03 / / .98	31 / .02 / .66 / .32	14.7 ±.2 / 14.67 ±.14 / 12.01 / 13.93	.57± .03 / / .39 /	/ / / 12.85± .32		5826± 62 / 5784 / / 5591
013902.5-432206 / 275.33 -71.12 / 255.13 -18.63 / 013655.0-433718	/ ESO 244- 46 / / PGC 6107	.L...P. / S / -2.0± .9 /	.60± .07 / .13± .05 / / .59	/ .03 / .00 /	15.15 ±.14				
013906.5-432142 / 275.29 -71.12 / 255.14 -18.65 / 013659.0-433654	/ ESO 244- 47 / / PGC 6109	.L...P. / S / -2.0± .9 /	1.08± .06 / .51± .03 / / 1.00	27 / .03 / .00 /	14.69 ±.14				
013906.6-073047 / 155.06 -67.36 / 292.31 -12.58 / 013636.6-074559	NGC 636 / MCG -1- 5- 13 / / PGC 6110	.E.3... / R / -5.0± .3 /	1.45± .04 / .12± .05 / .81± .02 / 1.44	140 / .13 / .00 /	12.41 ±.13 / 12.14 ±.18 / / 12.16	.95± .01 / .48± .03 / .90 / .46	.96± .01 / .52± .02 / 11.95± .07 / 14.36± .28		1847± 17 / 1875 / / 1562
013906.8-295454 / 230.63 -79.17 / 269.35 -17.13 / 013649.0-301006	NGC 642 / ESO 413- 14 / VV 419 / PGC 6112	.SBS5.. / S (1) / 5.0± .8 / 3.3± .8	1.31± .04 / .25± .04 / .82± .02 / 1.31	31 / .02 / .38 / .13	12.88V±.14 / 13.58 ±.14 / / 13.15	.67± .03			5930± 33 / 5888 / / 5695
013908.8-470742 / 280.97 -67.91 / 251.13 -18.89 / 013705.1-472254	/ ESO 244- 48 / IRAS01370-4722 / PGC 6114	.S..3./ / S / 3.0± .6 /	1.14± .05 / .82± .04 / / 1.14	151 / .03 / 1.13 / .41	15.61 ±.14 / / 12.68 /				
013911.6-750043 / 298.82 -41.73 / 221.58 -18.01 / 013823.0-751554	NGC 643 / ESO 29- 53 / IRAS01384-7515 / PGC 6117	.LB?... / / / 1.11	1.19± .05 / .65± .03 / / 1.11	113 / .18 / .00 /	14.61 ±.14 / / 10.93 / 14.37				3966 / 3813 / / 3926
013918.0+484551 / 131.01 -13.36 / 346.14 4.69 / 013612.9+483039	/ UGC 1168 / IRAS01362+4830 / PGC 6124	.S?.... / (1) / 4.5±1.1 / 1.19	1.12± .05 / .09± .05 / / 1.19	85 / .75 / .12 / .04	14.9 ±.3 / 12.32 / 13.94			15.67±.1 / / 377± 8 / 1.69	5281± 11 / 5444 / / 5061

R.A. 2000 DEC. l b SGL SGB R.A. 1950 DEC.	Names PGC	Type S_T n_L T L	$\log D_{25}$ $\log R_{25}$ $\log A_e$ $\log D_o$	p.a. A_g A_i A_{21}	B_T m_B m_{FIR} B_T^o	$(B-V)_T$ $(U-B)_T$ $(B-V)_T^o$ $(U-B)_T^o$	$(B-V)_e$ $(U-B)_e$ m' m'_{25}	m_{21} W_{20} W_{50} HI	V_{21} V_{opt} V_{GSR} V_{3K}
013925.2-492255 283.58 -65.90 248.73 -19.04 013724.0-493806	 ESO 196- 16 FAIR 375 PGC 6131	PSBS0.. r -.1± .9 	.94± .07 .15± .05 .93	5 .04 .11 	 15.18 ±.14 14.89				 9100±190 9002 8938
013930.8-465913 280.64 -68.01 251.29 -18.95 013727.0-471424	 ESO 244- 49 FAIR 708 PGC 6136	.LBR+*/ S -.6± .6 	1.21± .05 .94± .03 1.08	68 .03 .00 	 15.33 ±.14 15.19				 6900±190 6809 6728
013933.0+350933 133.92 -26.68 333.39 .23 013639.8+345422	 MCG 6- 4- 50 KUG 0136+349 PGC 6138	.S?.... 	.96± .06 .53± .06 .98	 .19 .78 .27	 15.39 ±.12 14.40				5148 5289 4888
013934.5-120436 162.14 -71.19 287.75 -13.80 013706.7-121947	 MCG -2- 5- 32 PGC 6141	PSXR3*. E (1) 3.0±1.2 3.1± .8	1.09± .06 .05± .05 1.09	25 .00 .07 .02					
013937.9+071416 143.41 -53.71 306.88 -8.66 013700.6+065905	NGC 638 UGC 1170 MK 1003 PGC 6145	.S?.... 	.90± .07 .20± .05 .91	20 .10 .24 .10	14.5 ±.2 14.41 ±.14 13.61 14.06	.68± .05 -.16± .09 .59 -.21	 13.36± .43	16.86±.3 178± 10 2.70	3650± 9 3123± 36 3694 3328
013938.4+054700 144.21 -55.08 305.47 -9.09 013701.9+053149	 UGC 1172 PGC 6147	.S?.... 	.95± .07 .61± .05 .96	27 .11 .76 .31	 15.06 ±.15 14.16				3250 3318 2956
0139.7 +1554 139.57 -45.40 315.22 -6.04 0137.0 +1539	 UGC 1171 PGC 6150	.I..9*. U 10.0±1.2 	1.11± .14 .04± .12 1.12	 .16 .03 .02				14.88±.1 48± 4 36± 4 	667± 5 762 376
0139.8 -0201 149.77 -62.35 297.82 -11.34 0137.3 -0217	 UGC 1174 PGC 6153	.S..4.. U 4.0± .9 	.98± .09 .17± .06 .99	103 .09 .25 .08	 14.82 ±.12 				
013956.2-091434 157.84 -68.76 290.63 -13.21 013727.1-092945	NGC 647 MCG -2- 5- 33 PGC 6155	.LBR0?. E -2.0±1.2 	1.18± .10 .12± .07 1.17	45 .08 .00 					
0139.9 +1106 141.64 -50.00 310.64 -7.58 0137.3 +1051	 UGC 1175 PGC 6159	.S..9*. U 9.0±1.2 	1.04± .08 .04± .06 1.06	 .18 .04 .02					729 812 436
013957.4-284149 224.08 -79.17 270.66 -17.13 013739.0-285700	 ESO 413- 16 IRAS01376-2856 PGC 6161	.S..4./ S 4.0± .9 	1.26± .03 .64± .03 1.26	169 .05 .95 .32	 14.06 ±.14 13.75 13.03				 5931± 52 5893 5693
014000.8-280207 220.55 -79.20 271.35 -17.04 013742.0-281718	 ESO 413- 18 IRAS01377-2817 PGC 6165	PSBR1P. Sr 1.0± .6 	1.26± .04 .76± .04 1.26	162 .02 .77 .38	 14.48 ±.14 13.13 13.61				 5908 5872 5668
014008.8+054340 144.45 -55.10 305.45 -9.23 013732.3+052830	NGC 645 UGC 1177 IRAS01375+0528 PGC 6172	.SB.3*. U 3.0± .8 	1.42± .04 .35± .05 1.43	125 .12 .49 .18	 13.41 ±.14 12.87 12.78			14.35±.3 309± 10 1.40	3308± 7 3375 3015
0140.1 +4634 131.59 -15.48 344.12 3.85 0137.1 +4618	 UGC 1173 IRAS01371+4618 PGC 6173	.S..8*. U 8.0±1.3 	1.00± .08 .55± .06 1.06	96 .64 .68 .27					
0140.1 +1554 139.72 -45.37 315.26 -6.14 0137.4 +1539	A 0137+15 UGC 1176 DDO 13 PGC 6174	.I..9.. U (1) 10.0± .6 9.0± .9	1.66± .04 .10± .06 1.27± .05 1.67	 .16 .07 .05	14.4 ±.2 14.14	.57± .06 -.13± .08 .51 -.18	.65± .05 -.09± .06 16.21± .10 17.28± .31	13.61±.1 50± 4 37± 6 -.58	631± 4 727 341
014021.3-285450 225.17 -79.06 270.44 -17.25 013803.1-291000	IC 1720 ESO 413- 19 IRAS01380-2909 PGC 6180	.S..4.. S (1) 4.0± .9 4.4± .9	1.07± .04 .14± .03 1.08	164 .06 .21 .07	 13.71 ±.14 12.71 				
014021.3-354900 255.80 -76.45 263.17 -18.17 013807.8-360410	 MCG -6- 4- 64 PGC 6181	.L...*P S -2.0±1.7 	1.25± .08 .34± .08 1.20	 .02 .00 					

R.A. 2000 DEC. l b SGL SGB R.A. 1950 DEC.	Names PGC	Type S_T n_L T L	$\log D_{25}$ $\log R_{25}$ $\log A_e$ $\log D_o$	p.a. A_g A_i A_{21}	B_T m_B m_{FIR} B_T^o	$(B-V)_T$ $(U-B)_T$ $(B-V)_T^o$ $(U-B)_T^o$	$(B-V)_e$ $(U-B)_e$ m'_e m'_{25}	m_{21} W_{20} W_{50} HI	V_{21} V_{opt} V_{GSR} V_{3K}
014025.5-075408 156.30 -67.54 292.01 -13.00 013755.7-080918	 MCG -1- 5- 16 PGC 6186	.S...P* E 	1.12± .08 .57± .07 1.13	120 .10 .85 .28	 	 	 	 	5554 5580 5271
014027.9+343732 134.26 -27.17 332.95 -.13 013734.9+342222	 UGC 1178 IRAS01375+3422 PGC 6189	.S..6*. U 6.0±1.3 	1.26± .04 .76± .05 1.28	55 .13 1.12 .38	 14.71 ±.12 12.64 13.43	 	 	14.45±.3 400± 9 .64	5502± 8 5641 5242
014028.1-053118 153.49 -65.44 294.39 -12.41 013757.1-054628	 MCG -1- 5- 14 PGC 6190	.SXS7*. E (1) 7.0±1.2 5.3±1.6	1.25± .05 .63± .05 1.26	10 .07 .87 .31	 	 	 	 	2130 2163 1844
0140.4 +1432 140.36 -46.67 313.97 -6.64 0137.8 +1417	 UGC 1181 PGC 6193	.S..6*. U 6.0±1.4 	1.00± .08 .73± .06 1.01	133 .15 1.07 .36	 	 	 	 	8124 8216 7833
014037.2-384057 264.36 -74.54 260.14 -18.53 013826.1-385606	 ESO 297- 19 PGC 6200	.S..4.. S (1) 4.0± .9 3.3±1.2	1.04± .05 .26± .04 1.04	104 .02 .39 .13	 15.54 ±.14 	 	 	 	
014039.3-314915 239.54 -78.32 267.40 -17.73 013823.0-320424	 ESO 413- 20 PGC 6202	.L..../ S -2.0± .9 	1.03± .05 .60± .03 .95	128 .02 .00 	 14.05 ±.14 	 	 	 	
0140.7 +0758 143.45 -52.91 307.68 -8.70 0138.1 +0743	NGC 652 UGC 1184 IRAS01380+0743 PGC 6208	.S?.... 	.99± .06 .23± .05 1.00	55 .13 .35 .12	 14.55 ±.12 12.87 14.05	 	 	 	5328 5401 5035
014048.0-411245 270.32 -72.64 257.46 -18.79 013839.0-412754	 ESO 297- 20 PGC 6213	.SBS7?. S (1) 7.0±1.7 5.6± .9	1.12± .05 .35± .05 1.13	76 .03 .48 .17	 15.59 ±.14 	 	 	 	
0140.8 +8358 124.31 21.23 20.12 14.43 0134.0 +8343	 UGC 1148 PGC 6214	.S?.... 	1.00± .06 .00± .05 1.05	 .49 .00 .00	 15.1 ±.2 14.63	 	 	16.24±.1 151± 8 1.62	4708± 11 4884 4641
014055.6+491403 131.18 -12.84 346.67 4.59 013749.4+485854	 UGC 1182 PGC 6220	.S..6*. U 6.0±1.2 	1.00± .08 .00± .06 1.07	 .73 .00 .00	 14.8 ±.3 	 	 	 	
014107.0-053408 153.89 -65.41 294.39 -12.57 013836.0-054917	 MCG -1- 5- 17 PGC 6228	.SXS9*. E (1) 9.0± .8 6.4±1.2	1.38± .04 .40± .05 1.38	110 .04 .41 .20	 	 	 	 	1503 1536 1218
014115.4+344843 134.39 -26.95 333.18 -.22 013822.1+343335	 MCG 6- 4- 55 PGC 6232	.SB?... 	.74± .14 .00± .07 .75	 .17 .00 .00	 14.81 ±.14 14.61	 	 	 	5091 5230 4832
014124.6+083132 143.43 -52.35 308.26 -8.70 013846.6+081624	IC 1721 UGC 1187 PGC 6235	.S?.... 	.98± .07 .29± .05 .99	100 .12 .43 .14	 14.30 ±.14 13.73	 	 	 	 4299± 50 4373 4007
014131.1-461253 278.78 -68.49 252.13 -19.25 013927.0-462800	 ESO 245- 1 PGC 6241	PSBS0*. Sr .3± .5 	.99± .05 .05± .04 .99	118 .03 .04 	 15.20 ±.14 	 	 	 	
014132.8-831244 301.17 -33.74 213.08 -16.97 014306.0-832748	 ESO 3- 4 PGC 6242	.S..4*/ S (1) 4.0± .7 4.4± .9	1.27± .04 .74± .04 1.31	141 .45 1.08 .37	 15.02 ±.14 13.46	 	 	 	4579 4414 4581
014136.2-160852 172.13 -74.01 283.72 -15.19 013910.7-162359	 MCG -3- 5- 14 PGC 6244	.IBS9.. E (1) 10.0± .9 8.7± .8	1.17± .05 .34± .05 1.17	70 .00 .25 .17	 	 	 	 	1637 1637 1369
014143.0-892004 302.77 -27.78 206.85 -15.84 024256.2-893412	NGC 2573 ESO 1- 1 IRAS02425-8934 PGC 6249	.SXS6*. S (1) 6.3± .7 6.1± .8	1.30± .04 .42± .04 1.35	70 .56 .62 .21	14.1 ±.2 14.15 ±.14 13.35 	.59± .06 -.16± .07 	 14.38± .28	 	

R.A. 2000 DEC. / l b / SGL SGB / R.A. 1950 DEC.	Names / / / PGC	Type S_T n_L / T / L	$\log D_{25}$ / $\log R_{25}$ / $\log A_e$ / $\log D_o$	p.a. / A_g / A_i / A_{21}	B_T / m_B / m_{FIR} / B_T^o	$(B-V)_T$ / $(U-B)_T$ / $(B-V)_T^o$ / $(U-B)_T^o$	$(B-V)_e$ / $(U-B)_e$ / m'_e / m'_{25}	m_{21} / W_{20} / W_{50} / HI	V_{21} / V_{opt} / V_{GSR} / V_{3K}
014146.7+273007 / 136.42 -34.05 / 326.36 -2.75 / 013858.2+271500	CGCG 482- 1 / / / PGC 6253			.24	15.7 ±.4			17.47±.3 / 195± 7	10827± 10 / 10950 / 10553
014149.0-751612 / 298.69 -41.44 / 221.29 -18.15 / 014105.0-753118	NGC 643C / ESO 30- 1 / / PGC 6256	.S..6*/ / RS / 5.6± .7	1.12± .04 / .85± .04 / / 1.14	150 / .22 / 1.25 / .43	15.70 ±.14				
014149.5-281453 / 221.58 -78.80 / 271.19 -17.47 / 013931.0-283000	/ ESO 413- 23 / / PGC 6257	.S?....	1.05± .05 / .16± .05 / / 1.06	113 / .06 / .22 / .08	15.49 ±.14 / 15.16				5855 / 5817 / 5617
014155.2-130455 / 165.47 -71.65 / 286.87 -14.59 / 013928.1-132002	NGC 655 / MCG -2- 5- 37 / / PGC 6262	.L...P. / E / -2.0± .8	1.04± .09 / .14± .07 / / 1.02	45 / .00 / .00					
0141.9 +0707 / 144.38 -53.64 / 306.93 -9.24 / 0139.3 +0652	/ UGC 1189 / / PGC 6263	.S?....	1.03± .06 / .40± .05 / / 1.04	19 / .12 / .60 / .20	15.21 ±.12 / 14.46				5470 / 5540 / 5178
014156.4+224120 / 137.91 -38.71 / 321.84 -4.37 / 013910.8+222613	/ UGC 1188 / / PGC 6265	.S?....	1.00± .16 / .33± .12 / / 1.04	22 / .44 / .50 / .17	15.44 ±.13 / 14.44			16.69±.3 / 483± 7 / 2.09	13308± 10 / 13420 / 13027
014157.0+331535 / 134.93 -28.43 / 331.78 -.87 / 013904.7+330028	CGCG 503- 6 / / / PGC 6266			.15	15.2 ±.2			17.38±.3 / 333± 7	13399± 10 / 13535 / 13137
0141.9 +3923 / 133.46 -22.44 / 337.51 1.18 / 0139.0 +3908	/ / / PGC 6267			.26	16.4 ±.2	.68± .04 / -.68± .08			
014159.6-310035 / 235.30 -78.30 / 268.30 -17.90 / 013943.0-311542	/ ESO 413- 24 / / PGC 6268	.S?....	1.22± .04 / .75± .04 / / 1.22	177 / .02 / 1.12 / .37	14.04 ±.14 / 12.86				5979 / 5932 / 5750
014209.9+123609 / 141.76 -48.41 / 312.25 -7.63 / 013929.8+122102	NGC 658 / UGC 1192 / IRAS01394+1220 / PGC 6275	.S..3.. / U (1) / 3.0± .7 / 3.5±1.0	1.48± .03 / .28± .05 / .89± .03 / 1.49	20 / .13 / .39 / .14	13.12 ±.14 / 13.14 ±.15 / 12.32 / 12.59	.66± .02 / -.10± .02 / .56 / -.17	.69± .01 / .05± .02 / 13.06± .11 / 14.67± .25	13.76±.1 / 322± 8 / 304± 5 / 1.02	2986± 4 / 2985± 60 / 3071 / 2696
014215.0-331536 / 245.21 -77.44 / 265.93 -18.25 / 014000.1-333042	/ ESO 353- 33 / IRAS01399-3330 / PGC 6280	.S?....	.89± .06 / .31± .05 / / .89	153 / .02 / .46 / .15	14.80 ±.14 / 13.40 / 14.28				5865± 87 / 5812 / 5643
014217.4-473148 / 280.29 -67.30 / 250.73 -19.44 / 014015.0-474654	/ ESO 196- 19 / FAIR 376 / PGC 6283	PSBT1.. / Sr / .7± .5	1.06± .05 / .22± .04 / / 1.07	62 / .04 / .22 / .11	14.70 ±.14 / 14.36				6750±190 / 6656 / 6582
0142.4 +0643 / 144.80 -53.97 / 306.58 -9.48 / 0139.8 +0628	/ UGC 1196 / / PGC 6288	.SAS3.. / U (1) / 3.0± .9 / 3.5±1.1	1.00± .08 / .04± .06 / / 1.01	/ .13 / .05 / .02					
014225.8+353819 / 134.45 -26.09 / 334.03 -.17 / 013931.8+352312	NGC 653 / UGC 1193 / / PGC 6290	.S..2.. / U / 2.0± .9	1.17± .05 / .79± .05 / / 1.19	39 / .16 / .97 / .39	14.31 ±.17				
014227.2+135836 / 141.26 -47.07 / 313.59 -7.27 / 013946.4+134330	/ UGC 1195 / / PGC 6292	.I?....	1.53± .03 / .50± .05 / / 1.55	50 / .15 / .37 / .25	13.44 ±.15 / 12.91			14.03±.1 / 150± 5 / 127± 5 / .86	774± 4 / 862 / 484
014227.6+260836 / 136.99 -35.33 / 325.13 -3.34 / 013939.9+255330	NGC 656 / UGC 1194 / / PGC 6293	.LB.... / U / -2.0± .8	1.18± .04 / .05± .03 / / 1.21	35 / .31 / .00	13.35 ±.10 / 12.98				3942± 42 / 4061 / 3666
0142.5 +1818 / 139.58 -42.91 / 317.72 -5.92 / 0139.8 +1803	/ UGC 1197 / / PGC 6294	.I..9.. / U / 10.0± .8	1.26± .06 / .51± .06 / / 1.27	56 / .15 / .38 / .25	14.84 ±.15 / 14.30			14.70±.1 / 205± 8 / 181± 12 / .15	2796± 6 / 2896 / 2510

R.A. 2000 DEC.	Names		Type S_T n_L T L	$\log D_{25}$ $\log R_{25}$ $\log A_e$ $\log D_o$	p.a. A_g A_i A_{21}	B_T m_B m_{FIR} B_T^o	$(B-V)_T$ $(U-B)_T$ $(B-V)_T^o$ $(U-B)_T^o$	$(B-V)_e$ $(U-B)_e$ m'_e m'_{25}	m_{21} W_{20} W_{50} HI	V_{21} V_{opt} V_{GSR} V_{3K}
014242.3-181342 178.60 -75.22 281.65 -15.88 014018.0-182848	ESO 543- 12 PGC 6301		RSBR1*. SEr 1.2± .5	1.16± .04 .30± .04 1.16	127 .00 .31 .15					4998± 60 4990 4735
014242.5+081029 144.11 -52.58 308.02 -9.11 014004.7+075523	UGC 1199 PGC 6302		.S..2.. U 2.0± .9	.98± .07 .45± .05 .99	78 .12 15.38 ±.12 .55 .22 14.61					9332 9405 9041
014248.5+130922 141.74 -47.83 312.83 -7.61 014008.1+125417	A 0140+12 UGC 1200 PGC 6309		.IB.9*. U 10.0± .8	1.31± .04 .17± .05 1.32	170 .13 13.81 ±.16 .13 .08 13.55				14.92±.2 141± 9 128± 7 1.28	807± 8 839± 50 894 518
0142.8 -0608 155.45 -65.72 293.92 -13.13 0140.3 -0624	MCG -1- 5- 22 PGC 6310		.SB?... 	1.04± .09 .21± .07 1.04	 .04 .31 .10					6083 6113 5800
014251.6+312842 135.60 -30.12 330.17 -1.65 014000.4+311337	MCG 5- 5- 2 PGC 6312		.S?.... 	.74± .14 .00± .07 .75	 .15 15.49 ±.12 .00 .00 15.26				16.18±.3 283± 7 .92	10447± 10 10578 10182
0143.0 +0412 146.56 -56.29 304.17 -10.35 0140.4 +0357	IC 150 UGC 1202 PGC 6316		.S..3.. U (1) 3.0± .9 4.5±1.2	1.00± .06 .31± .05 1.00	143 .08 14.74 ±.12 .42 .15 14.20					5572 5633 5282
014301.4+133837 141.60 -47.35 313.31 -7.51 014020.8+132332	NGC 660 UGC 1201 IRAS01403+1323 PGC 6318		.SBS1P. R 1.0± .3	1.92± .02 .42± .03 1.57± .08 1.93	170 12.02M±.10 .15 12.1 ±.2 .43 8.64 .21 11.44	.86± .07 .74	.94± .03 14.97± .23 15.40± .14	11.89±.0 318± 4 306± 6 .23	853± 3 823± 46 940 564
014302.2-341119 248.58 -76.87 264.97 -18.53 014048.0-342624	IC 1722 ESO 353- 34 PGC 6319		.SXS4P. BS 4.0± .6	1.17± .04 .46± .04 1.18	50 .02 14.71 ±.14 .67 .23 13.99					4083± 49 4026 3865
014308.0+274503 136.70 -33.73 326.69 -2.95 014019.2+272958	MCG 4- 5- 3 PGC 6326		.S?.... 	.99± .10 .45± .07 1.01	 .25 14.86 ±.13 .55 .22 13.96				17.31±.3 690± 7 3.13	10156± 10 10279 9884
014309.1-341431 248.73 -76.82 264.91 -18.56 014055.0-342936	IC 1724 ESO 353- 35 PGC 6328		.L..+*. S -1.0±1.2	1.11± .05 .41± .03 1.05	126 .02 14.09 ±.14 .00 14.01					3816± 87 3759 3598
0143.2 +0417 146.59 -56.19 304.27 -10.38 0140.6 +0402	UGC 1204 PGC 6329		.S..4.. U 4.0± .9	1.11± .05 .55± .05 1.12	41 .08 15.20 ±.13 .81 .28 14.27					5602 5663 5312
014314.1+085322 143.94 -51.86 308.75 -9.02 014035.9+083817	IC 1723 UGC 1205 IRAS01406+0838 PGC 6332		.S..3.. U (1) 3.0± .8 2.5±1.1	1.52± .03 .63± .05 1.53	29 .14 13.76 ±.15 .86 13.37 .31 12.71				14.34±.3 439± 9 419± 7 1.31	5532± 8 5498± 50 5605 5241
014318.1-341220 248.53 -76.81 264.95 -18.58 014104.0-342724	ESO 353- 36 IRAS01410-3427 PGC 6334		.LBT0*. S -2.0±1.2	1.03± .05 .52± .03 .95	19 .02 15.13 ±.14 .00 11.92 15.05					3596± 87 3539 3378
014318.7+222511 138.39 -38.89 321.68 -4.75 014033.1+221007	CGCG 482- 6 PGC 6335				 .38 15.7 ±.4				16.96±.3 278± 7	9919± 10 10029 9639
0143.3 +1959 139.23 -41.24 319.38 -5.55 0140.6 +1944	UGCA 20 PGC 6337		.I..9.. U 10.0± .8	1.49± .09 .61± .18 1.50	153 .13 .46 .31				15.00±.1 79± 7 61± 12	498± 6 603 215
0143.5 +0205 148.26 -58.20 302.14 -11.10 0141.0 +0150	UGC 1208 PGC 6348		.S..2.. U 2.0± .9	.96± .07 .29± .05 .96	12 .03 15.37 ±.13 .36 .15					
014337.0-341620 248.64 -76.72 264.89 -18.66 014123.0-343124	ESO 353- 39 PGC 6350		.SBS0?/ S .0±1.3	1.04± .05 .59± .04 1.01	88 .02 15.93 ±.14 .44					

R.A. 2000 DEC. / l b / SGL SGB / R.A. 1950 DEC.	Names / PGC	Type / S_T n_L / T / L	$\log D_{25}$ / $\log R_{25}$ / $\log A_e$ / $\log D_o$	p.a. / A_g / A_i / A_{21}	B_T / m_B / m_{FIR} / B_T^o	$(B-V)_T$ / $(U-B)_T$ / $(B-V)_T^o$ / $(U-B)_T^o$	$(B-V)_e$ / $(U-B)_e$ / m'_e / m'_{25}	m_{21} / W_{20} / W_{50} / HI	V_{21} / V_{opt} / V_{GSR} / V_{3K}
014337.4-334220 / 246.44 -76.98 / 265.49 -18.59 / 014123.0-335724	/ ESO 353- 38 / IRAS01413-3357 / PGC 6351	.LB.0*P / S / -2.0± .8 /	1.12± .05 / .33± .04 / / 1.07	113 / .02 / .00 /	/ 14.36 ±.14 / 13.17 / 14.21				8859± 23 / 8804 / 8640
014342.8-040008 / 153.56 -63.72 / 296.12 -12.79 / 014111.1-041512	/ MCG -1- 5- 25 / / PGC 6356	.LAS0?. / E / -2.0± .9 /	1.21± .07 / .28± .05 / / 1.18	20 / .14 / .00 /					
014344.6-360521 / 255.05 -75.73 / 262.96 -18.88 / 014132.0-362024	/ ESO 353- 40 / / PGC 6357	RLB.+P* / Sr / -1.2± .6 /	1.30± .05 / .23± .04 / / 1.27	162 / .02 / .00 /	/ 13.61 ±.14 / / 13.51				5304± 34 / 5241 / 5093
0143.7 +1209 / 142.52 -48.71 / 311.95 -8.14 / 0141.0 +1154	A 0141+11 / UGC 1209 / MK 572 / PGC 6358	.S..8*. / U / 8.0±1.2 /	1.10± .05 / .35± .05 / / 1.12	65 / .12 / .43 / .17	/ 14.70 ±.12 / 12.26 / 14.14				4900 / 5126±113 / 5209 / 4837
014346.0+041317 / 146.87 -56.20 / 304.25 -10.53 / 014110.1+035813	NGC 664 / UGC 1210 / IRAS01411+0358 / PGC 6359	.S..3*. / U (1) / 3.0±1.1 / 2.5±1.1	1.18± .05 / .08± .05 / / 1.19	65 / .07 / .11 / .04	/ 13.61 ±.12 / 12.59 / 13.39			15.78±.3 / 254± 13 / 243± 10 / 2.35	5425± 10 / 5412± 60 / 5485 / 5135
014347.0+290957 / 136.46 -32.33 / 328.07 -2.61 / 014057.2+285453	/ CGCG 503- 12 / / PGC 6360			/ .17 / /	/ 15.3 ±.4 / /			16.54±.3 / / 133± 7 /	4014± 10 / 4140 / 3745
0143.8 +1348 / 141.83 -47.14 / 313.53 -7.66 / 0141.2 +1333	/ UGC 1211 / / PGC 6364	.I..9*. / U / 10.0±1.0 /	1.37± .04 / .10± .05 / / 1.38	/ .15 / .08 / .05					2404 / / 2492 / 2116
014354.3-775735 / 299.45 -38.81 / 218.48 -17.90 / 014339.0-781236	ESO 13- 20 / / / PGC 6365	.IBS9P. / S (1) / 10.0± .8 / 10.0± .9	1.11± .07 / .10± .05 / / 1.13	/ .20 / .08 / .05	/ 16.14 ±.14 / /				
014356.5+170350 / 140.50 -44.01 / 316.65 -6.63 / 014113.9+164847	A 0141+16 / MCG 3- 5- 13 / MK 360 / PGC 6366	CI...P* / R / 11.0± .8 /	.64± .17 / .00± .07 / .37± .03 / .65	/ .15 / .00 / .00	/ 14.85 ±.15 / 14.85 ±.18 / / 14.70	.42± .02 / -.23± .04 / .33 / -.29	.43± .02 / -.15± .03 / 12.19± .07 / 12.88± .86	16.61±.2 / 224± 5 / 168± 8 / 1.91	8034± 5 / 7984± 60 / 8130 / 7748
014357.6+022057 / 148.23 -57.92 / 302.43 -11.11 / 014122.7+020554	A 0141+02 / UGC 1214 / MK 573 / PGC 6367	RLXT+*. / UE / -1.0± .5 /	1.13± .07 / .01± .04 / / 1.14	/ .03 / .00 /	/ 13.68 ±.10 / / 13.57				5106± 28 / 5161 / 4817
014408.0+342313 / 135.14 -27.23 / 332.98 -.92 / 014114.5+340810	/ UGC 1212 / / PGC 6372	.S..3.. / U / 3.0± .8 /	1.14± .07 / .15± .06 / / 1.15	95 / .14 / .20 / .07	/ 14.25 ±.13 / / 13.83				10748 / / 10885 / 10490
014409.1+311914 / 135.96 -30.21 / 330.12 -1.96 / 014117.8+310411	/ UGC 1213 / / PGC 6373	.L..... / U / -2.0± .8 /	.98± .12 / .02± .05 / / .99	120 / .13 / .00 /	/ 14.63 ±.10 / /				
014414.6+284220 / 136.71 -32.75 / 327.67 -2.86 / 014125.1+282717	NGC 661 / UGC 1215 / / PGC 6376	.E+..*. / PU / -4.0± .6 /	1.24± .03 / .09± .07 / .68± .02 / 1.25	60 / .25 / .00 /	13.18 ±.13 / 12.88 ±.10 / / 12.69	.97± .01 / .49± .02 / .88 / .45	1.01± .01 / .54± .02 / 12.07± .07 / 14.15± .27		3845± 24 / 3969 / 3575
0144.2 +1215 / 142.66 -48.59 / 312.07 -8.23 / 0141.6 +1200	/ UGC 1218 / / PGC 6377	.SB.4*. / U / 4.0± .9 /	.95± .07 / .34± .05 / / .96	140 / .14 / .50 / .17	/ 15.27 ±.12 / / 14.56				10222 / / 10305 / 9934
014420.8+172838 / 140.47 -43.59 / 317.08 -6.59 / 014137.9+171335	/ UGC 1219 / IRAS01416+1713 / PGC 6380	.SB?... / / /	1.10± .05 / .35± .05 / / 1.11	102 / .15 / .52 / .17	/ 13.66 ±.15 / 13.20 / 12.97			15.72±.3 / 404± 13 / 397± 10 / 2.58	4611± 10 / 4606± 50 / 4708 / 4326
014428.7-403952 / 267.36 -72.60 / 258.10 -19.44 / 014220.0-405454	/ ESO 297- 23 / FAIR 710 / PGC 6387	.SXR2P* / Sr / 2.5± .7 /	1.13± .04 / .43± .04 / / 1.13	4 / .03 / .59 / .21	/ 14.46 ±.14 / / 13.77				10121± 60 / 10045 / 9927
0144.5 +0440 / 146.89 -55.72 / 304.74 -10.58 / 0141.9 +0425	/ UGC 1222 / / PGC 6388	.I?.... / / /	1.04± .08 / .37± .06 / / 1.05	85 / .07 / .28 / .19					7790± 10 / / 7851 / 7501

R.A. 2000 DEC. l b SGL SGB R.A. 1950 DEC.	Names PGC	Type S_T n_L T L	$\log D_{25}$ $\log R_{25}$ $\log A_e$ $\log D_o$	p.a. A_g A_i A_{21}	B_T m_B m_{FIR} B_T^o	$(B-V)_T$ $(U-B)_T$ $(B-V)_T^o$ $(U-B)_T^o$	$(B-V)_e$ $(U-B)_e$ m'_e m'_{25}	m_{21} W_{20} W_{50} HI	V_{21} V_{opt} V_{GSR} V_{3K}
0144.5 +1706 140.67 -43.93 316.73 -6.75 0141.8 +1651	 PGC 6390			.15					8225± 9 8321 7940
014435.4+374148 134.41 -23.99 336.10 .12 014139.3+372646	NGC 662 UGC 1220 5ZW 98 PGC 6393	.S...P. R 	.92± .05 .20± .04 .94	20 .21 .29 .10	 13.88 ±.19 12.24 13.35			16.18±.2 291± 8 2.73	5662± 6 5706± 43 5806 5413
014438.1+215542 138.94 -39.28 321.32 -5.20 014152.7+214040	 CGCG 482- 7 PGC 6395			 .29	 15.3 ±.4			17.84±.3 356± 7	10464± 10 10572 10184
014438.2+381210 134.30 -23.49 336.58 .28 014141.7+375708	 UGC 1221 PGC 6397	.S..4.. U 4.0± .9 	1.03± .06 .14± .05 1.04	145 .18 .20 .07	 14.76 ±.13				
014439.1-073713 158.26 -66.77 292.56 -13.94 014209.3-075215	 MCG -1- 5- 28 PGC 6400	PLAR+*. E -1.0±1.1 	1.18± .05 .08± .05 1.18	170 .09 .00					 5743 5767 5464
014441.3+045326 146.82 -55.49 304.97 -10.55 014205.2+043824	 MCG 1- 5- 30 PGC 6402	.S?.... 	.97± .07 .57± .06 .98	 .07 .85 .28	 15.38 ±.12 14.45				1625 1687 1336
014445.0-040810 154.22 -63.72 296.06 -13.08 014213.4-042312	 MCG -1- 5- 29 PGC 6406	.LXR0*. E -2.0± .9 	1.03± .07 .15± .05 1.01	90 .07 .00					
014446.8+170628 140.75 -43.91 316.76 -6.81 014204.1+165126	A 0142+16 MK 361 PGC 6408			 .14					8124± 45 8220 7839
014449.4+215242 139.01 -39.32 321.29 -5.26 014204.0+213740	 MCG 4- 5- 4 PGC 6409	.SB?... 	.89± .11 .24± .07 .91	 .29 .29 .12	 15.15 ±.12 14.46			17.12±.3 125± 7 2.54	10508± 10 10616 10228
014456.1+102521 143.78 -50.27 310.36 -8.95 014217.0+101020	NGC 665 UGC 1223 PGC 6415	RL..0?. U -2.0± .7 	1.38± .11 .17± .08 .81± .02 1.37	125 .17 .00	13.17 ±.13 13.08 ±.11 12.87	1.04± .01 .63± .02 .93 .60	1.08± .01 .64± .01 12.71± .08 14.51± .60		 5419± 31 5497 5130
014457.4-225517 196.72 -77.18 276.89 -17.28 014236.0-231018	NGC 667 ESO 477- 2 PGC 6418	.LA.0*. S -2.0± .8 	.78± .06 .04± .03 .77	 .00 .00	 15.23 ±.14				
0144.9 -0016 150.72 -60.24 299.90 -12.09 0142.4 -0032	 UGC 1225 PGC 6419	.SB?... 	.64± .17 .11± .07 .64	20 .04 .14 .06					5397 5443 5111
0145.0 +1022 143.85 -50.31 310.31 -9.00 0142.4 +1007	 UGC 1226 PGC 6425	.S..9.. U 9.0± .9 	1.02± .06 .17± .05 1.03	80 .17 .17 .08	 15.08 ±.14 14.73				5894 5971 5606
014503.4-360705 254.54 -75.50 262.96 -19.15 014251.0-362206	 ESO 353- 41 IRAS01428-3621 PGC 6428	PS..0*. S .0±1.2 	1.05± .05 .23± .05 1.05	64 .02 .17	 14.93 ±.14 13.80 14.65				 5378± 60 5315 5168
014503.8-433553 273.08 -70.29 254.96 -19.74 014258.0-435054	A 0143-43 ESO 245- 5 PGC 6430	.IBS9.. PS (1) 10.0± .5 10.0± .7	1.56± .03 .06± .05 1.56	122 .03 .05 .03	12.7 ±.2 12.75 ±.12 12.66	.45± .04 -.25± .09 .43 -.27	 15.18± .29	12.61±.2 83± 8 61± 12 -.08	395± 6 310 213
0145.1 +3209 135.97 -29.36 330.96 -1.88 0142.3 +3154	 UGC 1224 PGC 6434	.S..6*. U 6.0±1.5 	1.00± .08 1.02± .06 1.01	125 .10 1.47 .50					
014510.4-415247 269.69 -71.61 256.81 -19.65 014303.0-420748	 ESO 297- 27 PGC 6435	.SAT3*. Sr (1) 3.3± .6 4.4± .6	1.11± .05 .24± .05 1.11	78 .03 .33 .12	 14.90 ±.14 14.49				6282 6202 6093

R.A. 2000 DEC. / l b / SGL SGB / R.A. 1950 DEC.	Names / PGC	Type / S_T n_L / T / L	$\log D_{25}$ / $\log R_{25}$ / $\log A_e$ / $\log D_o$	p.a. / A_g / A_i / A_{21}	B_T / m_B / m_{FIR} / B_T^o	$(B-V)_T$ / $(U-B)_T$ / $(B-V)_T^o$ / $(U-B)_T^o$	$(B-V)_e$ / $(U-B)_e$ / m'_e / m'_{25}	m_{21} / W_{20} / W_{50} / HI	V_{21} / V_{opt} / V_{GSR} / V_{3K}
0145.2 +1039	IC 154	.S..3..	1.16± .07	66					
143.78 -50.03	UGC 1229	U (1)	.79± .06	.17	14.76 ±.13				
310.60 -8.96		3.0±1.0		1.09					
0142.6 +1024	PGC 6439	4.5±1.3	1.18	.40					
014515.3+320342		.S..6*.	.83± .11					16.53±.2	10775± 6
136.02 -29.44	UGC 1227	U	.07± .06	.10	15.42 ±.12			122± 9	
330.89 -1.93		6.0±1.2		.10				105± 7	10906
014223.4+314841	PGC 6440		.84	.04	15.17			1.33	10513
014519.8+043706	IC 1726								
147.26 -55.69	CGCG 412- 25			.08	14.70 ±.16				5500± 60
304.75 -10.79									5561
014243.7+042205	PGC 6441								5212
014521.8+284315		.S..8*.	1.17± .05	25				15.60±.3	3959± 10
136.99 -32.67	UGC 1228	U	.28± .05	.25	14.74 ±.15				3974± 97
327.77 -3.09		8.0±1.2		.34				202± 7	4083
014232.1+282815	PGC 6443		1.20	.14	14.14			1.32	3690
014525.4-034938		.SBT4P*	1.09± .06	5					
154.23 -63.37	MCG -1- 5- 31	E	.11± .05	.06					
296.41 -13.16	IRAS01428-0404	4.0±1.2		.16	13.30				
014253.7-040439	PGC 6447		1.09	.05					
0145.4 +1033	IC 156	.S?....	1.18± .05						5241
143.91 -50.11	UGC 1231		.11± .05	.17	14.45 ±.18				
310.52 -9.03				.16					5319
0142.8 +1018	PGC 6448		1.20	.05	14.10				4953
014528.5-100538		.L..-..	1.12± .07	155					
162.28 -68.73	MCG -2- 5- 41	E	.12± .05	.05					
290.11 -14.74		-3.0± .9		.00					
014300.1-102039	PGC 6450		1.11						
0145.5 +2531		.S..9*.	1.33± .05						3839± 10
138.00 -35.77	UGC 1230	U	.08± .06	.45					
324.77 -4.20		9.0±1.0		.09					3955
0142.7 +2516	PGC 6451		1.37	.04					3565
0145.5 +0322		.S..4..	.97± .07	143					5332
148.21 -56.82	UGC 1235	U	.30± .05	.07	14.77 ±.12				
303.54 -11.21		4.0± .9		.45					5389
0143.0 +0307	PGC 6454		.98	.15	14.21				5044
014537.1+232633								17.46±.3	10275± 10
138.70 -37.76	CGCG 482- 11			.35	15.7 ±.4				
322.83 -4.92								241± 10	10386
014250.7+231133	PGC 6455								9998
014538.5-343130		PSAT3..	1.01± .06	168					
248.73 -76.23	ESO 353- 42	r	.11± .05	.02	15.10 ±.14				8506± 60
264.67 -19.10		3.3± .9		.15					8447
014325.0-344630	PGC 6458		1.01	.05	14.86				8291
014541.5+284820		.S..6*.	1.00± .06	110				15.81±.3	7842± 10
137.05 -32.57	UGC 1233	U	.15± .05	.25	14.58 ±.12				
327.87 -3.13	IRAS01428+2833	6.0±1.3		.21				311± 7	7966
014251.7+283320	PGC 6459		1.02	.07	14.08			1.66	7574
0145.7 +4113		.S..6*.	1.00± .16	10					
133.81 -20.50	UGC 1232	U	.55± .12	.25					
339.46 1.11		6.0±1.3		.81					
0142.8 +4058	PGC 6466		1.02	.27					
014551.8+350639		.SXS5..	1.05± .04	170				15.78±.3	5653± 8
135.34 -26.45	UGC 1234	U	.16± .04	.14	14.60 ±.12			268± 9	
333.78 -1.01	KUG 0142+348	5.0± .9		.24					5790
014257.5+345139	PGC 6473		1.06	.08	14.19			1.52	5398
0145.9 +0949		.S..2..	1.18± .05	45					5204
144.45 -50.76	UGC 1237	U	.44± .05	.16	15.04 ±.15				
309.86 -9.37	IRAS01432+0934	2.0± .8		.55	13.00				5279
0143.2 +0934	PGC 6475		1.20	.22	14.29				4916
014559.1-785415		.SAS5P.	1.01± .05						
299.64 -37.87	ESO 13- 22	S (1)	.06± .05	.28	14.72 ±.14				
217.49 -17.85		5.0± .8		.10					
014559.1-790912	PGC 6479	2.2± .8	1.04	.03					
014606.1+342226	NGC 666	.S?....	.87± .08	80	14.3 ±.3	.89± .02			
135.59 -27.15	UGC 1236		.17± .05	.13	13.9 ±.2	.36± .04			4811± 52
333.11 -1.30	6ZW 26			.26		.80			4947
014312.3+340727	PGC 6483		.88	.09	13.64	.28	13.08± .50		4555

R.A. 2000 DEC. / l b / SGL SGB / R.A. 1950 DEC.	Names / PGC	Type / S_T n_L / T / L	$\log D_{25}$ / $\log R_{25}$ / $\log A_o$ / $\log D_o$	p.a. / A_g / A_i / A_{21}	B_T / m_B / m_{FIR} / B_T^o	$(B-V)_T$ / $(U-B)_T$ / $(B-V)_T^o$ / $(U-B)_T^o$	$(B-V)_e$ / $(U-B)_e$ / m'_e / m'_{25}	m_{21} / W_{20} / W_{50} / HI	V_{21} / V_{opt} / V_{GSR} / V_{3K}
0146.1 +0624		.S..8*.	1.06± .06	155					5377
146.45 -53.94	UGC 1239	U	.10± .05	.16	14.68 ±.14				
306.57 -10.44		8.0±1.2		.12					5442
0143.5 +0610	PGC 6485		1.07	.05	14.39				5089
014620.7+041552		.S..8*.	1.06± .06	91					1803
147.91 -55.92	UGC 1240	U	.44± .05	.07	14.75 ±.12				
304.48 -11.13		8.0±1.3		.54					1862
014344.8+040053	PGC 6500		1.06	.22	14.13				1516
014622.7+362739	NGC 668	.S..3..	1.25± .03	30	13.74 ±.13	.68± .01	.74± .01	14.90±.3	4506± 7
135.11 -25.11	UGC 1238	U (1)	.16± .05	.19	13.37 ±.12	.08± .02	.11± .02	308± 9	4577± 52
335.07 -.64	IRAS01434+3612	3.0± .8	.72± .01	.21	13.20	.58	12.83± .04		4647
014327.2+361240	PGC 6502	3.5±1.0	1.27	.08	13.11	.01	14.46± .22	1.72	4256
0146.4 -0838		.IBS9..	1.08± .07	100					
160.60 -67.40	MCG -2- 5- 43	E (1)	.22± .05	.10					
291.64 -14.61	IRAS01439-0853	10.0± .9		.16	13.45				
0143.9 -0853	PGC 6504	8.7±1.2	1.09	.11					
014625.1-083814	IC 159	.SBT3P*	1.16± .06	35					3920
160.60 -67.39	MCG -2- 5- 42	E	.31± .05	.10					
291.65 -14.62		3.0±1.2		.43					3940
014355.9-085312	PGC 6505		1.17	.15					3644
014627.5+345539			.87± .08					16.13±.3	5548± 8
135.52 -26.60	CGCG 522- 2		.21± .12	.19	14.79 ±.14			268± 9	5438± 52
333.65 -1.18	MK 1006								5682
014333.2+344040	PGC 6507		.88						5290
014628.1-832312		.LX.-*.	1.14± .09						
301.05 -33.54		S	.07± .08	.49					4652± 15
212.88 -17.08		-2.7± .7		.00					4486
014819.6-833806	PGC 6510		1.20						4656
014629.2-131454	IC 160	.LX.-*.	1.09± .06	85					
168.48 -71.07	MCG -2- 5- 44	E	.20± .05	.00					
286.96 -15.71		-3.0± .9		.00					
014402.5-132953	PGC 6511		1.06						
014630.7-584020		.SBS9..	1.26± .04	0					2248
289.79 -57.03	ESO 114- 7	S (1)	.08± .05	.00	14.22 ±.14				
238.79 -20.05		9.0± .8		.08					2125
014446.0-585518	PGC 6513	7.8± .8	1.26	.04	14.14				2131
0146.5 +1440		.S..9*.	1.03± .06						7389
142.35 -46.10	UGC 1242	U	.12± .05	.08	15.18 ±.17				
314.58 -8.00		9.0±1.2		.12					7477
0143.9 +1426	PGC 6516		1.03	.06	14.97				7104
0146.6 -0346		.SBS8P.	1.14± .06	105					5434
154.76 -63.18	MCG -1- 5- 34	E (1)	.09± .05	.08					
296.56 -13.43		8.0± .8		.11					5469
0144.1 -0401	PGC 6518	7.5± .8	1.15	.05					5152
014639.7-252214		.SAR3*.	.92± .06	58					
207.96 -77.50	ESO 477- 3	r	.24± .04	.00	15.41 ±.14				
274.39 -18.08		3.3±1.3		.33					
014420.0-253712	PGC 6520		.92	.12					
0146.7 +1833		.L...?.	1.14± .07	135					
140.79 -42.38	UGC 1243	U	.27± .03	.13	14.56 ±.11				
318.30 -6.77		-2.0±1.7		.00					
0144.0 +1819	PGC 6524		1.12						
014649.2+242758		.S?....	1.08± .05	80				15.88±.3	3128± 10
138.70 -36.70	UGC 1244		.33± .05	.37	15.20 ±.14			207± 13	
323.88 -4.83				.50					3241
014402.0+241300	PGC 6528		1.11	.17	14.30			1.41	2854
014653.4+332322		.S?....	.77± .11					16.54±.3	11179± 10
136.04 -28.07	CGCG 503- 21		.06± .10	.12	15.18 ±.12				
332.25 -1.80				.09				231± 7	11312
014400.2+330824	PGC 6534		.78	.03	14.91			1.59	10921
014653.7-335521		.SBT2..	1.02± .06						
245.97 -76.28	ESO 353- 45	r	.12± .06	.02	14.98 ±.14				10387± 34
265.35 -19.29		2.2± .8		.14					10329
014440.0-341018	PGC 6535		1.03	.06	14.71				10171
0146.9 +4824		.S..7*.	1.07± .14	17					
132.35 -13.44	UGC 1241	U	.32± .12	.91					
346.23 3.39		7.0±1.3		.45					
0143.8 +4810	PGC 6539		1.16	.16					

R.A. 2000 DEC.	Names	Type	logD$_{25}$	p.a.	B$_T$	(B-V)$_T$	(B-V)$_e$	m$_{21}$	V$_{21}$
l b		S$_T$ n$_L$	logR$_{25}$	A$_g$	m$_B$	(U-B)$_T$	(U-B)$_e$	W$_{20}$	V$_{opt}$
SGL SGB		T	logA$_e$	A$_i$	m$_{FIR}$	(B-V)$_T^o$	m'$_e$	W$_{50}$	V$_{GSR}$
R.A. 1950 DEC.	PGC	L	logD$_o$	A$_{21}$	B$_T^o$	(U-B)$_T^o$	m'$_{25}$	HI	V$_{3K}$
014656.4+320637		.S?....	.89± .11		14.85 ±.13			17.56±.3	10505± 10
136.40 -29.31	MCG 5- 5- 10		.24± .07	.09				599± 7	10635
331.06 -2.25				.29					
014404.1+315139	PGC 6540		.90	.12	14.36			3.08	10244
014658.0+284522		.L?....	1.02± .14		14.80 ±.12			16.24±.3	10761± 10
137.38 -32.55	MCG 5- 5- 11		.05± .07	.28				332± 7	10884
327.92 -3.41				.00					10494
014408.1+283024	PGC 6544		1.04		14.36				
0146.9 +1224		.IA.9..	1.17± .05						804
143.52 -48.22	UGC 1246	U	.06± .05	.15	14.52 ±.19				886
312.44 -8.81	VV 93	10.0± .8		.05					518
0144.3 +1210	PGC 6545		1.19	.03	14.32				
014659.5+130734	NGC 671	.S?....	1.19± .06	55					5460± 50
143.19 -47.55	UGC 1247		.50± .06	.14	14.17 ±.12				5544
313.12 -8.59	IRAS01443+1252			.75	13.26				5174
014418.9+125237	PGC 6546		1.20	.25	13.25				
0147.0 +8213		.I..9*.	1.22± .12					15.28±.1	1254± 11
124.91 19.56	UGC 1207	U	.08± .12	.65					1431
18.43 13.82		10.0±1.1		.06				125± 8	1179
0141.0 +8158	PGC 6552		1.28	.04					
014713.1-403927		.S?....	.94± .06	47					10037
266.11 -72.24	ESO 297- 29		.18± .06	.03	15.20 ±.14				9960
258.15 -19.96				.24					9845
014505.0-405424	PGC 6557		.95	.09	14.85				
014716.1+353348	NGC 669	.S..2..	1.50± .03	36	13.36 ±.13	1.04± .02	1.07± .01	16.20±.3	4694± 8
135.53 -25.94	UGC 1248	U	.73± .05	.14	12.97 ±.14	.49± .03	.56± .02	876± 9	4752± 44
334.30 -1.12	IRAS01443+3519	2.0± .8	.93± .02	.90		.84	13.50± .06		4833
014421.2+351851	PGC 6560		1.51	.37	12.08	.30	13.86± .24	3.75	4443
014718.8-625816		.SBS9..	1.15± .04	65					1499
292.43 -53.00	ESO 80- 6	S (1)	.16± .04	.00	14.41 ±.14				1366
234.20 -19.89		8.7± .5		.16					1403
014544.0-631312	PGC 6562	6.9± .5	1.15	.08	14.24				
014720.1-611122		RSB.0./	1.19± .04	67					
291.32 -54.66	ESO 114- 8	Sr	.56± .04	.00	14.93 ±.14				
236.10 -20.01		-.1± .6		.42					
014541.0-612618	PGC 6564		1.16						
0147.4 +1617		.SXT8..	1.08± .06	5					4893
141.92 -44.50	UGC 1252	U	.11± .05	.13					4985
316.19 -7.67		8.0± .9		.13					4610
0144.7 +1603	PGC 6569		1.09	.05					
014725.0+275308	NGC 670	.LA....	1.31± .04	172	13.59 ±.13	.87± .01		15.76±.3	3703± 10
137.76 -33.37	UGC 1250	PU	.33± .04	.21	13.06 ±.09	.22± .02			3819± 41
327.14 -3.80	IRAS01446+2738	-2.0± .6		.00	13.24	.76		436± 10	3830
014435.5+273811	PGC 6570		1.29		12.96	.17	14.19± .27		3441
014730.3+360202		.S?....	.96± .09	44				18.02±.3	4847± 8
135.46 -25.47	UGC 1251		.29± .06	.19	14.94 ±.12			309± 9	
334.75 -1.00				.44					4985
014434.9+354705	PGC 6572		.98	.15	14.29			3.58	4596
0147.5 +1206		.S?....	1.14± .05	18					6623
143.86 -48.47	UGC 1253		.39± .05	.15	15.06 ±.15				6704
312.18 -9.03	IRAS01448+1151			.59	13.43				6337
0144.8 +1151	PGC 6573		1.16	.20	14.28				
014730.7+271951	IC 1727	.SBS9..	1.84± .02	150	12.07 ±.19	.57± .05	.55± .03	12.93±.2	338± 5
137.96 -33.90	UGC 1249	R (1)	.35± .02	.27	11.78 ±.17			145± 10	393± 32
326.63 -4.01		9.0± .3	1.49± .04	.36		.43	15.01± .11	121± 12	458
014441.6+270455	PGC 6574	6.0± .9	1.86	.18	11.28		15.23± .21	1.48	70
0147.6 +1120		.S..8..	1.00± .06	85					5217
144.29 -49.18	UGC 1255	U	.18± .05	.13	15.25 ±.15				5295
311.46 -9.30		8.0± .9		.22					4931
0145.0 +1106	PGC 6580		1.02	.09	14.89				
014743.1-524540	NGC 685	.SXR5..	1.57± .03		11.5 ±.3	.46± .03	.54± .02	13.30±.3	1356± 6
284.48 -62.30	ESO 152- 24	R (2)	.05± .04	.00	12.03 ±.12	-.18± .10	-.09± .03	182± 7	1365± 44
245.13 -20.39	IRAS01458-5300	5.0± .7	1.42± .06	.08	12.33	.44	14.11± .14		1247
014549.0-530036	PGC 6581	4.0± .6	1.57	.03	11.87	-.19	14.07± .35	1.40	1214
014743.7+350121		.S?....	.94± .11					15.65±.3	5557± 8
135.78 -26.44	MCG 6- 5- 5		.00± .07	.18	14.79 ±.13			118± 9	
333.83 -1.39				.00					5693
014449.1+344625	PGC 6582		.95	.00	14.57			1.08	5303

R.A. 2000 DEC. l b SGL SGB R.A. 1950 DEC.	Names PGC	Type S_T n_L T L	$logD_{25}$ $logR_{25}$ $logA_e$ $logD_o$	p.a. A_g A_i A_{21}	B_T m_B m_{FIR} B_T^o	$(B-V)_T$ $(U-B)_T$ $(B-V)_T^o$ $(U-B)_T^o$	$(B-V)_e$ $(U-B)_e$ m'_e m'_{25}	m_{21} W_{20} W_{50} HI	V_{21} V_{opt} V_{GSR} V_{3K}
014745.8-333611 244.43 -76.25 265.71 -19.44 014532.0-335107	IC　1728 ESO 353- 47 IRAS01455-3350 PGC 6584	PSXS4.. BSr　(1) 3.9± .4 3.3±1.2	1.11± .03 .18± .04 .71± .01 1.11	3 .02 .27 .09	14.07 ±.13 14.17 ±.14 12.81 13.77	.73± .02 .63 	.81± .02 13.11± .03 14.00± .22	 	 8748± 45 8691 8532
014748.0+253429 138.60 -35.58 325.00 -4.67 014500.0+251933	 MCG 4- 5- 10 PGC 6586	 	.82± .13 .17± .07 .86	 	 .47 15.20 ±.12 	 	 	16.12±.3 335± 7 	12257± 10 12372 11985
014748.1-164321 177.24 -73.32 283.46 -16.76 014523.4-165816	NGC　690 MCG -3- 5- 21 PGC 6587	.SXS5*. PSE　(3) 5.0± .5 1.6± .5	1.08± .05 .17± .04 1.08	145 .00 .26 .09	14.8　±.3 14.51	.57± .08 .50 	 14.63± .41	15.34±.1 257± 11 242± 11 .74	5160± 10 5231± 59 5156 4901
014753.5+272601 138.02 -33.78 326.75 -4.05 014504.3+271105	NGC　672 UGC 1256 IRAS01450+2710 PGC 6595	.SBS6.. R　(2) 6.0± .3 5.4± .6	1.86± .01 .45± .02 1.37± .02 1.88	65 .26 .66 .23	11.47M±.10 11.27 ±.11 11.67 10.45	.58± .03 -.10± .06 .43 -.21	.59± .01 -.05± .02 13.74± .04 14.49± .13	11.79±.1 269± 4 198± 5 1.12	421± 4 390± 34 541 152
014755.5-265329 215.08 -77.42 272.84 -18.60 014537.1-270824	IC　1729 ESO 477- 4 PGC 6598	.LXS-*. S -4.0± .6 	1.22± .03 .27± .03 .44± .03 1.14	150 .00 .00 	13.54 ±.13 13.03 ±.14 13.28	.94± .01 .31± .02 .92 .32	.92± .01 .41± .02 11.23± .10 13.94± .23	 	 1495± 52 1458 1259
014805.0-520253 283.64 -62.91 245.90 -20.45 014610.0-521748	 ESO 197- 1 PGC 6605	.L..-P. S -3.0± .8 	1.16± .08 .16± .04 1.13	77 .01 .00 	 15.29 ±.14 	 	 	 	
014807.1+362708 135.48 -25.03 335.18 -.97 014511.3+361213	 UGC 1257 PGC 6607	.S..2.. U 2.0± .9 	1.02± .06 .29± .05 1.03	107 .17 .36 .15	 14.87 ±.12 14.30	 	 	16.63±.3 355± 9 2.19	4662± 8 4801 4412
014822.8+113123 144.46 -48.95 311.69 -9.41 014543.0+111628	NGC　673 UGC 1259 IRAS01457+1116 PGC 6624	.SXS5.. U 5.0± .7 	1.33± .03 .11± .04 .92± .01 1.34	0 .16 .16 .05	13.20 ±.13 13.00 ±.13 11.70 12.75	.59± .01 -.11± .02 .50 -.17	.67± .01 -.02± .02 13.29± .03 14.43± .22	14.33±.2 342± 7 313± 5 1.53	5182± 5 5241± 60 5261 4898
0148.4 -1222 167.88 -70.11 287.96 -15.98 0145.9 -1237	A　0145-12 MCG -2- 5- 50 DDO　14 PGC 6626	.IXT9.. R　(2) 10.0± .3 7.7± .6	1.45± .04 .08± .05 1.18± .04 1.45	20 .00 .06 .04	13.71 ±.18 13.65	.47± .05 .02± .06 .44 .00	.49± .03 -.01± .04 15.10± .08 15.64± .29	14.18±.1 119± 9 81± 8 .48	1621± 8 1628 1353
0148.4 +1324 143.57 -47.16 313.52 -8.84 0145.8 +1310	 UGC 1261 PGC 6627	.S..6*. U 6.0±1.3 	1.15± .05 .65± .05 1.16	92 .15 .95 .32	 15.21 ±.12 14.09	 	 	 	5014 5098 4730
0148.4 +1341 143.43 -46.89 313.79 -8.75 0145.8 +1327	 UGC 1262 PGC 6628	.I..9.. U 10.0± .9 	1.11± .14 .25± .12 1.12	 .14 .18 .12	 	 	 	 	5083 5167 4799
014833.3+123645 143.98 -47.91 312.75 -9.11 014552.9+122151	A　0145+12 UGC 1260 MK　575 PGC 6633	PSBS1.. U 1.0± .9 	.91± .09 .10± .06 .93	 .18 .10 .05	 14.04 ±.15 11.93 13.69	 	 	15.96±.2 160± 8 153± 10 2.22	5488± 6 5295± 60 5568 5202
0148.5 +1033 145.02 -49.84 310.78 -9.75 0145.9 +1019	 UGC 1263 PGC 6634	.S?.... 	1.02± .06 .05± .05 1.03	140 .17 .08 .03	 	 	 	 	5267 5343 4982
0148.5 +1345 143.44 -46.82 313.86 -8.75 0145.9 +1331	 UGC 1264 PGC 6636	.S..6*. U 6.0±1.3 	1.04± .08 .59± .06 1.05	 .14 .86 .29	 	 	 	 	4790 4875 4506
014836.5-101939 164.36 -68.47 290.07 -15.54 014608.3-103433	 MCG -2- 5- 51 PGC 6637	.L..0*/ E -2.0±1.3 	1.08± .06 .59± .05 .99	163 .03 .00 	 	 	 	 	
014838.1-484912 279.70 -65.65 249.37 -20.53 014639.0-490406	 ESO 197- 2 IRAS01465-4904 PGC 6638	.SBS6?/ S 5.7± .8 	1.18± .04 .76± .04 1.18	152 .04 1.11 .38	 15.15 ±.14 13.75 	 	 	 	
0148.7 -2733 218.13 -77.28 272.17 -18.86 0146.4 -2748	A　0146-27 PGC 6641	.S..3?. S 	 	 .00 	15.5　±.2 	.26± .07 -.25± .12 	 	 	 9069 9029 8835

R.A. 2000 DEC. / l b / SGL SGB / R.A. 1950 DEC.	Names / PGC	Type / S_T n_L / T / L	$\log D_{25}$ / $\log R_{25}$ / $\log A_e$ / $\log D_o$	p.a. / A_g / A_i / A_{21}	B_T / m_B / m_{FIR} / B_T^o	$(B-V)_T$ / $(U-B)_T$ / $(B-V)_T^o$ / $(U-B)_T^o$	$(B-V)_e$ / $(U-B)_e$ / m'_e / m'_{25}	m_{21} / W_{20} / W_{50} / HI	V_{21} / V_{opt} / V_{GSR} / V_{3K}
014842.3-483854	NGC 692	PSBR4*.	1.32± .04	.04	13.06 ±.14				6256± 34
279.46 -65.79	ESO 197- 3	Sr (1)	.06± .05	.08	12.64				6156
249.56 -20.54	FAIR 712	3.8± .4		.03	12.90				6097
014643.0-485348	PGC 6642	1.4± .4	1.32	.03					
014843.8+103029	IC 162	.L.....	1.21± .09	.17	13.71 ±.12				5132± 49
145.11 -49.88	UGC 1267	U	.00± .05	.00					5207
310.74 -9.81	MK 1007	-2.0± .8							4847
014604.6+101535	PGC 6643		1.23		13.46				
014844.1+103029		.S?....	.89± .08	65					5200
145.11 -49.88	UGC 1266		.13± .05	.17	14.69 ±.12				5275
310.74 -9.81	VV 54			.20					4915
014604.8+101535	PGC 6644		.91	.07	14.29				
014846.7+201542		.SB?...	1.08± .09	170					8967± 29
140.75 -40.62	UGC 1265		.32± .07	.24					9069
320.07 -6.67	VV 535			.48	13.16				8689
014601.9+200048	PGC 6645		1.10	.16					
014847.0-285742		PSBS3..	.98± .05	60					
224.52 -77.22	ESO 414- 3	r	.40± .04	.00	15.45 ±.14				
270.68 -19.08		3.3± .9		.55					
014630.0-291236	PGC 6646		.98	.20					
0148.8 +1034		.S?....	.95± .07	46					5217
145.12 -49.80	UGC 1268		.51± .05	.17	15.22 ±.13				5293
310.82 -9.82				.76					4932
0146.2 +1020	PGC 6654		.97	.25	14.26				
014856.4-234755	NGC 686	.LA.-*.	1.25± .05	.00	13.37 ±.14				4657± 42
201.72 -76.61	ESO 477- 6	S	.11± .04	.00					4628
276.14 -18.34		-3.0±1.1							4413
014636.0-240248	PGC 6655		1.23		13.30				
014857.5+055428	NGC 676	.S..0*/	1.60± .03	172				14.55±.1	1510± 5
147.88 -54.15	UGC 1270	R	.52± .04	.15				397± 7	1572
306.28 -11.27	ARAK 57	.0± .4		.39				380± 5	1225
014620.6+053935	PGC 6656		1.59						
0148.9 +1311		.LB....	1.22± .06	95					
143.84 -47.32	UGC 1271	U	.18± .03	.16	14.11 ±.12				
313.35 -9.02		-2.0± .8		.00					
0146.3 +1257	PGC 6657		1.21						
014904.5-145830	NGC 682	.LA.-..	1.14± .10	95					
173.64 -71.91	MCG -3- 5- 22	E	.08± .07	.00					
285.33 -16.70		-3.0± .9		.00					
014638.9-151324	PGC 6663		1.13						
0149.1 +3458		.L..-*.	.96± .12	105					
136.10 -26.41	UGC 1269	U	.17± .05	.14	15.10 ±.10				
333.89 -1.67		-3.0±1.2		.00					
0146.2 +3444	PGC 6664		.96						
014908.4-035418	IC 164	.E+..*.	1.15± .10	20					
156.12 -63.00	MCG -1- 5- 37	E	.08± .07	.04					
296.59 -14.07		-4.0±1.2		.00					
014636.8-040911	PGC 6666		1.14						
014910.3-100345		.SBS7..	1.46± .03	120				13.90±.1	1991± 11
164.24 -68.17	MCG -2- 5- 53	E (1)	.10± .05	.07				213± 16	2005
290.37 -15.62	IRAS01466-1018	7.0± .7		.14	13.44			185± 12	1719
014642.0-101838	PGC 6667	5.3± .8	1.47	.05					
014910.5+053822	A 0146+05			.16	15.13 ±.16				5304±113
148.14 -54.38	CGCG 412- 29								5365
306.04 -11.40	MK 576								5019
014633.8+052329	PGC 6668								
0149.1 +1250		.S..1*.	1.16± .05	108					
144.07 -47.64	UGC 1274	U	.46± .05	.16	14.85 ±.13				
313.03 -9.18		1.0±1.2		.46					
0146.5 +1236	PGC 6670		1.18	.23					
014910.9-102540	NGC 681	.SXS2./	1.41± .02	68	12.82M±.10	.83± .01	.91± .01	14.12±.1	1757± 10
164.84 -68.46	MCG -2- 5- 52	R	.20± .03	.03	12.75 ±.19	.27± .02	.37± .02	378± 6	1754± 21
290.00 -15.70	IRAS01467-1040	2.0± .3	.92± .01	.25	11.83	.77	12.88± .03		1769
014642.8-104033	PGC 6671		1.42	.10	12.51	.23	14.23± .16	1.51	1485
014914.6+130318	NGC 677	.E.....	1.30± .12		13.2 ±.2	.99± .05	1.07± .02		
144.00 -47.44	UGC 1275	U	.00± .08	.16	13.60 ±.15	.49± .06	.58± .03		5100± 31
313.23 -9.13	IRAS01464+1249	-5.0± .7	.13± .06	.00	13.58	.90	14.35± .17		5182
014633.8+124825	PGC 6673		1.33		13.22	.48	14.69± .68		4816

R.A. 2000 DEC. l　　b SGL　SGB R.A. 1950 DEC.	Names PGC	Type S_T　n_L T L	$\log D_{25}$ $\log R_{25}$ $\log A_e$ $\log D_o$	p.a. A_g A_i A_{21}	B_T m_B m_{FIR} B_T^o	$(B-V)_T$ $(U-B)_T$ $(B-V)_T^o$ $(U-B)_T^o$	$(B-V)_e$ $(U-B)_e$ m'_e m'_{25}	m_{21} W_{20} W_{50} HI	V_{21} V_{opt} V_{GSR} V_{3K}
014915.4-034103 155.94 -62.79 296.82 -14.04 014643.7-035556	 MCG -1- 5- 36 PGC　6674	.SAT6*. E　　(1) 6.0±1.2 4.2± .8	1.15± .06 .05± .05 1.16	170 .04 .08 .03	 				12722 12755 12442
014915.5+204241 140.72 -40.15 320.53 -6.62 014630.3+202748	IC　　163 UGC　1276 IRAS01465+2027 PGC　6675	.SB.8.. U 8.0± .8 	1.26± .04 .33± .05 1.29	95 .31 13.60 +.12 .41 13.10 .17 12.88				14.54±.1 240± 5 223± 6 1.50	2744± 5 2846 2467
0149.2 +8515 124.21 22.53 21.45 14.55 0140.9 +8500	 UGC　1198 7ZW　　3 PGC　6676	.E?.... 	1.00± .10 .09± .04 1.03	70 .39 14.80 ±.16 .00 12.26 14.40					1207 1382 1147
014915.8+350423 136.11 -26.31 333.98 -1.67 014620.9+344930	 UGC　1272 ARAK　58 PGC　6677	.L...*. U -2.0±1.2 	1.06± .11 .25± .05 1.04	23 .14 14.22 ±.10 .00 14.01					4864± 80 4999 4612
0149.2 +1242 144.17 -47.76 312.91 -9.25 0146.6 +1228	 UGC　1278 PGC　6678	.S..6*. U 6.0±1.3 	1.00± .06 .33± .05 1.01	105 .16 15.35 ±.13 .49 .17 14.66					10020 10101 9736
014916.9-324431 240.57 -76.29 266.66 -19.66 014702.7-325924	IC　1734 ESO　353- 48 IRAS01470-3259 PGC　6679	.SBT5.. BSr　(1) 5.0± .3 2.6±1.1	1.20± .05 .07± .03 .94± .02 1.20	 .02 13.59 ±.14 .11 12.97 .04 13.32	13.36 ±.14 13.59 ±.14 12.97 13.32	.61± .02 .56 	.69± .01 13.55± .04 14.03± .30		 4986± 45 4930 4768
0149.3 +1321 143.89 -47.13 313.54 -9.06 0146.7 +1307	 UGC　1279 PGC　6687	.S?.... 	.91± .07 .49± .05 .93	148 .16 15.26 ±.13 .74 .25 14.33					4608 4691 4325
014924.4-264443 214.54 -77.08 273.04 -18.90 014706.0-265936	 ESO　477- 7 PGC　6689	.L..0.. S -2.0± .8 	1.12± .05 .21± .04 .93± .07 1.09	 .00 14.85 ±.14 .00 	14.7 ±.2 14.85 ±.14 	.99± .05 	1.00± .03 14.86± .22 14.66± .34		
014925.3+215950 140.29 -38.91 321.76 -6.23 014639.3+214458	NGC　　678 UGC　1280 PGC　6690	.SBS3*/ PU　(1) 3.0± .6 4.5±1.1	1.65± .02 .75± .02 .86± .02 1.04 1.68	78 13.33 ±.13 .33 13.10 ±.12 .38 11.82	13.33 ±.13 13.10 ±.12 11.82	1.12± .02 .55± .03 .89 .29	1.20± .01 .62± .02 13.12± .07 14.53± .18	15.16±.2 408± 7 241± 25 2.97	2836± 5 2811± 61 2941 2560
014926.0+352707 136.04 -25.94 334.35 -1.57 014630.7+351214	 UGC　1277 PGC　6691	.S..0.. U .0± .8 	1.23± .06 .29± .06 1.23	75 .14 14.23 ±.13 .22 13.81				15.90±.3 502± 9 	4141± 8 4238± 80 4278 3891
014928.3-560308 287.05 -59.25 241.59 -20.56 014740.0-561800	 ESO　152- 26 PGC　6692	RSBR1.. S 1.0± .8 	1.25± .05 .15± .05 1.25	29 .00 14.12 ±.14 .16 .08 13.88					6065± 34 5947 5938
014929.3-335020 244.71 -75.83 265.50 -19.82 014716.1-340512	 ESO　353- 49 PGC　6693	.SXS7.. S　　(1) 7.0± .8 5.6± .8	1.13± .05 .06± .05 1.14	 .02 15.00 ±.14 .08 .03 14.89					3931 3872 3718
014930.1+123032 144.35 -47.93 312.72 -9.36 014649.6+121540	 UGC　1282 MK　　577 PGC　6694	.S..0.. U .0± .9 	1.12± .05 .41± .05 1.11	55 .15 14.12 ±.13 .31 13.59					5179± 38 5259 4895
014930.5-345420 248.55 -75.35 264.36 -19.93 014718.0-350912	NGC　　696 ESO　353- 50 PGC　6695	.LXS+P? BS -.8± .6 	1.23± .05 .42± .04 1.16	25 .02 14.37 ±.14 .00 14.23					8027± 32 7965 7817
014930.7-274202 218.82 -77.11 272.04 -19.06 014713.0-275654	A　0147-27 ESO　414- 4 IRAS01472-2756 PGC　6696	.P..... S 99.0 	.66± .06 .52± .07 .66	53 15.10 ±.15 .00 15.31 ±.14 	15.10 ±.15 15.31 ±.14 	.52± .06 -.54± .11 			
014931.2-270450 216.04 -77.08 272.69 -18.98 014713.0-271942	 ESO　477- 8 IRAS01472-2719 PGC　6697	.S?.... 	1.10± .05 .18± .04 1.10	130 .00 14.08 ±.14 .26 12.19 .09 13.76					8630 8591 8395
014932.4+323532 136.87 -28.70 331.69 -2.60 014639.3+322040	 UGC　1281 PGC　6699	.S..8.. U 8.0± .7 	1.65± .03 .76± .05 1.67	38 .15 12.87 ±.12 .93 .38 11.79				13.41±.1 129± 5 116± 8 1.24	157± 4 168± 52 287 -99

R.A. 2000 DEC. / l b / SGL SGB / R.A. 1950 DEC.	Names / PGC	Type / S_T n_L / T / L	$\log D_{25}$ / $\log R_{25}$ / $\log A_e$ / $\log D_o$	p.a. / A_g / A_i / A_{21}	B_T / m_B / m_{FIR} / B_T^o	$(B-V)_T$ / $(U-B)_T$ / $(B-V)_T^o$ / $(U-B)_T^o$	$(B-V)_e$ / $(U-B)_e$ / m' / m'_{25}	m_{21} / W_{20} / W_{50} / HI	V_{21} / V_{opt} / V_{GSR} / V_{3K}
014937.7-133418 / 170.89 -70.80 / 286.81 -16.52 / 014711.4-134911	MCG -2- 5- 56 / PGC 6703	.SBR3?. / E (1) / 3.0±1.2 / 4.2±1.2	1.17± .05 / .45± .05 / / 1.17	63 / .00 / .63 / .23					1441 / / 1444 / 1175
0149.6 -1249 / 169.40 -70.24 / 287.58 -16.37 / 0147.2 -1304	A 0147-13 / MCG -2- 5- 57 / DDO 15 / PGC 6706	.SBS9.. / PE (2) / 9.0± .6 / 8.8± .7	1.29± .05 / .10± .05 / 1.05± .03 / 1.29	15 / .00 / .11 / .05	14.91 ±.17 / / / 14.80	.54± .07 / -.24± .10 / .51 / -.26	.59± .04 / -.15± .08 / 15.65± .07 / 15.96± .32	16.11±.1 / 111± 8 / 127± 12 / 1.27	1710± 6 / / 1715 / 1443
014943.5-344950 / 248.20 -75.35 / 264.44 -19.97 / 014731.0-350442	NGC 698 / ESO 353- 51 / / PGC 6710	PSXT2*. / BSr / 2.4± .5	.96± .05 / .05± .04 / / .96	/ .02 / .06 / .03	14.78 ±.14 / / / 14.61				8368± 69 / 8306 / 8158
014943.8+354705 / 136.01 -25.60 / 334.68 -1.51 / 014648.2+353212	NGC 679 / UGC 1283 / 5ZW 114 / PGC 6711	.L..-*. / U / -3.0±1.0	1.32± .07 / .00± .05 / .77± .03 / 1.34	/ .20 / .00	13.33 ±.14 / 12.82 ±.10 / / 12.72	1.00± .02 / .57± .02 / .91 / .55	1.05± .01 / .62± .02 / 12.67± .12 / 14.80± .41		5040± 20 / 5177 / 4790
0149.7 +2222 / 140.24 -38.52 / 322.15 -6.17 / 0147.0 +2208	/ UGC 1287 / / PGC 6716	.I..9.. / U / 10.0± .8	1.04± .15 / .00± .12 / / 1.07	/ .33 / .00 / .00				16.70±.3 / 114± 10	2953± 9 / / 3059 / 2678
0149.7 +1142 / 144.86 -48.67 / 311.97 -9.68 / 0147.1 +1127	NGC 683 / UGC 1288 / IRAS01471+1127 / PGC 6718	.S?....	1.00± .06 / .00± .05 / / 1.01	/ .15 / .00 / .00	14.47 ±.13 / 13.18 / / 14.29				5264 / / 5342 / 4980
014947.4+215814 / 140.40 -38.91 / 321.76 -6.32 / 014701.4+214322	NGC 680 / UGC 1286 / / PGC 6719	.E+..P* / PU / -4.0± .5	1.29± .03 / .08± .06 / .68± .02 / 1.30	/ .28 / .00	12.90 ±.13 / 12.82 ±.10 / / 12.52	1.00± .01 / .49± .02 / .91 / .44	1.03± .01 / / 11.79± .08 / 14.11± .25	15.61±.2 / 437± 7	2801± 6 / 2933± 24 / 2914 / 2533
014951.9-272756 / 217.78 -77.02 / 272.30 -19.11 / 014734.1-274248	NGC 689 / ESO 414- 5 / IRAS01475-2742 / PGC 6724	PSXR2P* / Sr / 1.6± .6	.99± .04 / .18± .03 / / .99	68 / .00 / .22 / .09	14.56 ±.14 / 12.75				
0149.9 +0339 / 149.79 -56.13 / 304.16 -12.15 / 0147.3 +0325	CGCG 386- 39 / PGC 6726			/ .08	15.4 ±.3				14562 / 14617 / 14278
0149.9 +1624 / 142.67 -44.20 / 316.50 -8.20 / 0147.2 +1610	/ UGC 1289 / / PGC 6728	.S..6*. / U / 6.0±1.2	1.06± .06 / .13± .05 / / 1.07	120 / .13 / .20 / .07					4974± 10 / / 5065 / 4693
0149.9 +1148 / 144.87 -48.55 / 312.09 -9.69 / 0147.3 +1134	/ UGC 1290 / / PGC 6733	.S..4.. / U / 4.0± .9	.90± .07 / .08± .05 / / .91	140 / .15 / .12 / .04	14.87 ±.12				
0149.9 +0204 / 151.02 -57.56 / 302.60 -12.63 / 0147.4 +0150	/ UGC 1293 / / PGC 6734	.S..2.. / U / 2.0± .9	1.00± .08 / .19± .06 / / 1.00	143 / .04 / .23 / .09	14.73 ±.12 / / / 14.32				14146 / 14196 / 13863
0150.0 -0629 / 159.56 -65.11 / 294.05 -14.94 / 0147.5 -0644	MCG -1- 5- 39 / PGC 6735	.S?....	1.12± .08 / .46± .07 / / 1.12	/ .03 / .67 / .23					5301 / 5325 / 5025
0150.0 -0044 / 153.42 -60.09 / 299.82 -13.44 / 0147.5 -0059	/ UGC 1296 / / PGC 6741	.S?....	.95± .07 / .66± .05 / / .96	105 / .06 / .99 / .33	15.47 ±.13 / / / 14.39				5217 / 5259 / 4936
0150.1 +2308 / 140.08 -37.76 / 322.90 -6.00 / 0147.4 +2254	/ UGC 1294 / / PGC 6750	.I..9*. / U / 10.0±1.2	1.04± .15 / .00± .12 / / 1.07	/ .31 / .00 / .00					2856 / / 2964 / 2583
015010.9+021829 / 150.92 -57.34 / 302.84 -12.61 / 014735.9+020338	/ UGC 1297 / / PGC 6751	.IBS9.. / U / 10.0± .9	1.02± .06 / .11± .05 / / 1.02	/ .04 / .09 / .06	14.81 ±.14 / / / 14.69				1697 / / 1748 / 1414
015012.7+271149 / 138.70 -33.87 / 326.71 -4.62 / 014723.4+265658	IC 1731 / UGC 1291 / / PGC 6756	.SXS5*. / PU / 5.3± .7	1.19± .04 / .20± .04 / / 1.21	140 / .23 / .30 / .10	14.00 ±.12 / / / 13.45				3426± 50 / 3544 / 3159

R.A. 2000 DEC. / l b / SGL SGB / R.A. 1950 DEC.	Names / PGC	Type / S_T n_L / T / L	$\log D_{25}$ / $\log R_{25}$ / $\log A_o$ / $\log D_o$	p.a. / A_g / A_i / A_{21}	B_T / m_B / m_{FIR} / B_T^o	$(B-V)_T$ / $(U-B)_T$ / $(B-V)_T^o$ / $(U-B)_T^o$	$(B-V)_o$ / $(U-B)_o$ / m'_o / m'_{25}	m_{21} / W_{20} / W_{50} / HI	V_{21} / V_{opt} / V_{GSR} / V_{3K}
015014.3+273852 138.56 -33.43 327.13 -4.46 014724.6+272401	NGC 684 UGC 1292 IRAS01474+2724 PGC 6759	.S..3./ PU (1) 3.0± .5 3.5±1.1	1.51± .03 .75± .04 .90± .04 1.53	90 .22 1.04 .38	13.34 ±.15 13.20 ±.13 13.32 11.96	.99± .02 .43± .04 .78 .22	1.07± .02 .52± .03 13.33± .13 13.83± .22	14.05±.2 493± 9 481± 6 1.71	3534± 6 3694± 52 3655 3270
015015.5+332944 136.78 -27.79 332.59 -2.42 014721.7+331453	 UGC 1295 PGC 6760	.L..... U -2.0± .8	.91± .13 .05± .05 .92	 .15 .00	 15.26 ±.11				
015020.6-561740 287.05 -58.98 241.32 -20.67 014833.1-563230	 ESO 152- 28 PGC 6770	.S..4?/ S 4.0±1.9	1.11± .05 .83± .04 1.11	142 .00 1.22 .42	 15.84 ±.14				
015028.5-470952 276.79 -66.85 251.16 -20.82 014828.1-472442	 ESO 245- 6 PGC 6776	RSAR2?P S 2.0±1.1	1.23± .06 .07± .05 1.23	4 .04 .08 .03	15.33 ±.13 15.28 ±.14 15.13	.98± .02 .40± .04 .91 .37	 16.15± .36		6039± 69 5942 5875
015030.9+060841 148.34 -53.79 306.63 -11.57 014754.0+055351	NGC 693 UGC 1304 IRAS01479+0553 PGC 6778	.S..0$. P .0±1.5	1.33± .03 .32± .05 .79± .02 1.33	106 .16 .24	13.24 ±.14 13.28 ±.12 11.04 12.84	.80± .02 .15± .03 .70 .08	.82± .02 .12± .03 12.68± .07 13.95± .23	14.79±.2 270± 8 239± 5	1564± 5 1593± 50 1626 1280
015033.0+352129 136.31 -25.97 334.34 -1.82 014737.6+350639	 UGC 1299 PGC 6780	.I..9.. U 10.0± .9	1.00± .06 .24± .05 1.01	85 .15 .18 .12	 15.39 ±.14 15.06			15.93±.3 195± 9 .75	5497± 8 5632 5246
015033.3+362215 136.03 -24.99 335.28 -1.46 014737.1+360725	NGC 687 UGC 1298 PGC 6782	.L..... U -2.0± .8	1.15± .15 .00± .08 .81± .02 1.18	 .22 .00	13.30 ±.13 13.20 ±.10 12.94	1.04± .01 .60± .03 .94 .57	1.05± .01 .61± .02 12.84± .06 13.92± .81		5117± 25 5254 4869
015040.6+322946 137.16 -28.73 331.69 -2.85 014747.4+321455	 CGCG 503- 31 PGC 6789			.15	15.7 ±.4			17.30±.3 461± 7	10535± 10 10664 10278
015041.1+334426 136.80 -27.53 332.85 -2.42 014746.9+332936	 CGCG 522- 19 MK 1008 PGC 6790		.62± .11 .11± .12 .64	 .14	 15.25 ±.18 13.67				 5611± 39 5743 5357
015041.7-035620 156.90 -62.83 296.66 -14.46 014810.1-041110	 MCG -1- 5- 40 IRAS01481-0411 PGC 6791	.LBR+P? E -1.0± .9	1.02± .07 .26± .05 .98	130 .04 .00	 12.51				
015041.7+214535 140.74 -39.05 321.63 -6.59 014755.8+213045	NGC 691 UGC 1305 IRAS01479+2130 PGC 6793	.SAT4.. PU 4.0± .5	1.54± .02 .12± .03 1.33± .03 1.56	95 .24 .18 .06	12.24 ±.15 12.9 ±.2 13.10 12.04	.80± .02 .19± .03 .71 .11	.89± .02 .30± .03 14.38± .08 14.49± .21	14.85±.1 328± 5 322± 5 2.75	2665± 4 2769 2390
015043.0+330456 137.00 -28.16 332.24 -2.65 014749.4+325006	IC 1733 UGC 1301 6ZW 58 PGC 6796	.E...*. U -5.0±1.1	1.18± .15 .03± .08 .86± .07 1.20	 .17 .00	14.12 ±.18 14.69 ±.18	1.14± .03	1.12± .02 13.91± .23 14.94± .78		
015044.0-120210 168.47 -69.48 288.45 -16.45 014816.9-121700	NGC 699 MCG -2- 5- 59 IRAS01482-1217 PGC 6798	.SB.4?/ E 4.0±1.3	1.18± .05 .68± .05 1.18	130 .01 .99 .34	 13.49				5502 5509 5235
015044.2+351703 136.37 -26.04 334.28 -1.88 014748.8+350213	NGC 688 UGC 1302 MK 1009 PGC 6799	PSXT3.. Ur 3.2± .5	1.39± .03 .21± .04 1.02± .03 1.40	145 .15 .29 .11	13.35 ±.15 13.08 ±.13 12.82 12.72	.68± .03 .03± .04 .58 -.05	.76± .02 .12± .03 13.94± .10 14.61± .23	14.92±.2 368± 6 349± 10 2.10	4150± 5 4111± 35 4284 3898
015047.0+323249 137.17 -28.68 331.74 -2.86 014753.7+321759	 UGC 1306 PGC 6802	RLX.0.. U -2.0± .8	1.11± .07 .05± .03 1.12	70 .15 .00	 14.56 ±.13 14.24			16.85±.3 250± 7	11334± 10 11463 11077
015048.1+355557 136.21 -25.40 334.89 -1.66 014752.2+354107	IC 1732 UGC 1307 PGC 6805	.S?.... 	1.17± .05 .58± .05 1.19	62 .20 .88 .29	 14.93 ±.12 13.83				4889± 30 5025 4640
015051.2+361631 136.12 -25.07 335.21 -1.55 014755.1+360141	 UGC 1308 PGC 6807	.E..... U -5.0± .7	1.36± .11 .00± .08 1.40	 .22 .00	 13.77 ±.19 13.47				5075± 56 5212 4827

R.A. 2000 DEC. / l b / SGL SGB / R.A. 1950 DEC.	Names / PGC	Type / S_T n_L / T / L	$\log D_{25}$ / $\log R_{25}$ / $\log A_e$ / $\log D_o$	p.a. / A_g / A_i / A_{21}	B_T / m_B / m_{FIR} / B_T^o	$(B-V)_T$ / $(U-B)_T$ / $(B-V)_T^o$ / $(U-B)_T^o$	$(B-V)_e$ / $(U-B)_e$ / m'_e / m'_{25}	m_{21} / W_{20} / W_{50} / HI	V_{21} / V_{opt} / V_{GSR} / V_{3K}
015052.3-360053		.LA.-*.	1.16± .05		14.33 ±.15	.94± .02	.99± .02		
251.74 -74.56	ESO 354- 3	S	.16± .04	.02	14.08 ±.14	.29± .04	.39± .03		5736± 34
263.19 -20.31		-3.0± .8	.60± .05	.00		.87	12.82± .16		5670
014841.0-361542	PGC 6809		1.14		14.09	.31	14.60± .32		5530
015052.6+482107		.SB.1..	1.07± .08						
133.02 -13.35	UGC 1303	U	.02± .06	.96	14.8 ±.3				
346.40 2.75		1.0± .8		.02					
014744.5+480617	PGC 6811		1.16	.01					
0150.9 +1817	IC 1736	.S..4..	1.06± .06	33					5165
142.18 -42.34	UGC 1309	U	.51± .05	.13	14.87 ±.12				
318.37 -7.80		4.0± .9		.75					5260
0148.2 +1803	PGC 6814		1.07	.26	13.95				4887
015058.1+215950	NGC 694	.L...$P	1.58± .05	160	14.28 ±.13	.56± .02		15.43±.1	2950± 4
140.73 -38.81	UGC 1310	P	.18± .03	.24	13.2 ±.2	-.15± .04		179± 6	2935± 30
321.88 -6.57	MK 363	-2.0±1.9		.00	12.24	.46		178± 10	3054
014812.0+214500	PGC 6816		1.58		13.69	-.19	16.61± .27		2675
0150.9 +1317		.S?....	1.16± .05	96					5198
144.47 -47.07	UGC 1312		.51± .05	.18	14.81 ±.12				
313.60 -9.45				.76					5280
0148.3 +1303	PGC 6817		1.18	.25	13.84				4916
015101.0+294809		.I..9*.	1.09± .07	20				15.66±.3	4349± 10
138.07 -31.31	UGC 1311	U	.06± .06	.19					
329.20 -3.87		10.0±1.2		.04				143± 7	4472
014809.7+293319	PGC 6823		1.11	.03					4087
015103.9-094214	NGC 701	.SBT5..	1.39± .02	40	12.82 ±.13	.65± .01	.72± .01	14.10±.1	1836± 8
164.66 -67.61	MCG -2- 5- 60	R (2)	.32± .03	.05	12.8 ±.2	-.01± .02	.03± .02	266± 11	1807± 23
290.85 -15.98	IRAS01485-0957	5.0± .4	.91± .01	.48	11.06	.57	12.86± .03	243± 9	1847
014835.5-095703	PGC 6826	5.5± .6	1.40	.16	12.28	-.07	13.83± .19	1.66	1563
0151.0 +1235		.SBS8..	1.12± .07	160					3303± 10
144.85 -47.72	UGC 1314	U	.26± .06	.12					
312.93 -9.70		8.0± .8		.32					3383
0148.4 +1221	PGC 6827		1.13	.13					3021
015106.4-442641	Phoenix	.IA.9..	1.69± .03	90					
272.16 -68.95	ESO 245- 7	S (1)	.08± .05	.03	13.08 ±.14				
254.10 -20.86		10.0± .6		.06					
014903.0-444130	PGC 6830	12.2± .7	1.70	.04					
015108.1-094737	IC 1738	PSXT3..	.93± .08	80				13.45±.3	1750± 17
164.84 -67.67	MCG -2- 5- 61	RE (1)	.10± .05	.03				516± 34	
290.76 -16.02		3.0± .4		.14				269± 25	1763
014839.7-100226	PGC 6832	3.1± .8	.93	.05					1480
015108.3+215450	IC 167	.SXS5..	1.46± .03	95	13.6 ±.2	.46± .06	.53± .03	13.93±.1	2935± 4
140.81 -38.88	UGC 1313	R	.18± .05	.24	13.50 ±.20	-.28± .08	-.20± .04	178± 4	2960± 53
321.81 -6.63	ARP 31	5.0± .3	1.22± .09	.27		.35	15.19± .30	147± 4	3039
014822.3+214001	PGC 6833		1.48	.09	13.02	-.36	15.30± .28	.82	2661
0151.1 -0102		.L.....	1.04± .18						
154.18 -60.23	UGC 1321	U	.00± .08	.04					
299.60 -13.79		-2.0± .8		.00					
0148.6 -0117	PGC 6835		1.04						
015111.3-032959		.SBS8..	1.13± .06	150					
156.66 -62.39	MCG -1- 5- 42	E (1)	.15± .05	.01					
297.14 -14.46		8.0± .8		.19					
014839.5-034448	PGC 6837	7.5± .8	1.13	.08					
015113.9+223459	NGC 695	.L...$P	.90± .09	40				15.58±.2	9735± 5
140.58 -38.23	UGC 1315	P	.04± .04	.32	13.84 ±.14			348± 6	9748± 37
322.45 -6.42	5ZW 123	-2.0±1.9		.00	10.90			185± 5	9841
014827.4+222010	PGC 6844		.93		13.37				9462
015117.3+222132	NGC 697	.SXR5*.	1.65± .02	105	12.84 ±.13	.82± .03	.95± .03	12.99±.2	3117± 5
140.68 -38.44	UGC 1317	PU	.48± .03	.32	12.47 ±.13	.12± .04	.18± .04	466± 7	
322.25 -6.51	IRAS01485+2206	4.7± .6	1.05± .01	.72	11.04	.64	13.58± .03	428± 7	3223
014830.9+220643	PGC 6848		1.68	.24	11.60	-.03	14.73± .18	1.15	2844
0151.3 +3450		.S?....	.93± .10	178					4703± 10
136.63 -26.43	UGC 1316		.35± .06	.13					
333.92 -2.15				.52					4837
0148.4 +3436	PGC 6851		.94	.17					4452
015119.3-040322	NGC 702	.SBS4P.	1.19± .04	110	13.9 ±.2	.78± .06			
157.32 -62.86	MCG -1- 5- 43	R	.13± .04	.03		.04± .06			10607± 63
296.59 -14.64	ARP 75	4.0± .4		.20		.68			10638
014847.8-041811	PGC 6852		1.19	.07	13.61	-.04	14.37± .29		10330

R.A. 2000 DEC.	Names	Type	logD$_{25}$	p.a.	B$_T$	(B-V)$_T$	(B-V)$_e$	m$_{21}$	V$_{21}$
l　　b		S$_T$　n$_L$	logR$_{25}$	A$_g$	m$_B$	(U-B)$_T$	(U-B)$_e$	W$_{20}$	V$_{opt}$
SGL　SGB		T	logA$_e$	A$_i$	m$_{FIR}$	(B-V)^o_T	m'$_e$	W$_{50}$	V$_{GSR}$
R.A. 1950 DEC.	PGC	L	logD$_o$	A$_{21}$	B^o_T	(U-B)^o_T	m'$_{25}$	HI	V$_{3K}$

0151.3　+1307		.SAS5..	1.16± .05						4821± 10
144.68 -47.19　UGC　1322		U	.07± .05	.17	14.8　±.2				
313.47　-9.60		5.0± .8		.10					4902
0148.7　+1253　PGC　6855			1.18	.03	14.53				4540
015123.7+330152		.E...*.	.96± .13	135					
137.17 -28.18　UGC　1318		U	.05± .05	.16	14.98 ±.11				
332.24　-2.81		-5.0±1.2		.00					
014830.0+324703 PGC　6856			.97						
015127.5-083024 NGC　707		PLXS-*.	1.11± .05	95					
163.04 -66.58　MCG -2- 5- 63		E	.19± .04	.10					
292.10 -15.79		-3.4± .7		.00					
014858.4-084512 PGC　6861			1.09						
0151.4　+1152		.I..9*.	1.16± .07	147					
145.35 -48.36　UGC　1323		U	.49± .06	.14					4838
312.27 -10.02		10.0±1.2		.37					4915
0148.8　+1138　PGC　6863			1.17	.25					4556
015128.7-063059		.SA.7P/	1.21± .04	95					2134
160.33 -64.93　MCG -1- 5- 44		E　　(1)	.43± .05	.01					
294.12 -15.31　KUG 0148-067		7.0±1.8		.60					2157
014858.5-064548 PGC　6864		5.3±1.6	1.21	.22					1860
015129.0+360356		.S?....	.95± .05	155	14.50 ±.13	.63± .02	.69± .02	16.95±.3	5312± 8
136.32 -25.24　UGC　1319			.14± .02	.26	14.49 ±.12	.03± .04	.01± .04	276± 9	5315± 58
335.06　-1.74　IRAS01485+3549			.63± .01	.21	13.00	.51	13.14± .03		5448
014832.9+354907 PGC　6865			.97	.07	14.00	-.06	13.75± .30	2.88	5064
0151.5　+4149		.S..6*.	1.04± .08	153					
134.77 -19.66　UGC　1320		U	.98± .06	.29					
340.40　　.32		6.0±1.4		1.44					
0148.5　+4135　PGC　6867			1.07	.49					
0151.5　+1906		.SB.3..	1.13± .05	90					10090
142.05 -41.53　UGC　1324		U	.17± .05	.14	14.26 ±.13				
319.19　-7.68　IRAS01488+1851		3.0± .8		.23	13.32				10187
0148.8　+1851　PGC　6872			1.14	.08	13.82				9813
015137.2+081525		.E.....	1.26± .13						5449± 49
147.44 -51.73　UGC　1325		U	.00± .08	.13	13.58 ±.13				5516
308.78 -11.19		-5.0± .8		.00					5166
014859.0+080037 PGC　6874			1.28		13.37				
0151.6　+0817		.E...*.	.93± .13	45					
147.42 -51.69　UGC　1326		U	.00± .05	.13	14.59 ±.10				5551± 56
308.82 -11.18		-5.0±1.2		.00					5618
0149.0　+0803　PGC　6876			.95		14.37				5268
015142.0-361118		PSAT3P.	1.05± .05						10033± 20
251.97 -74.33　ESO　354- 4		Sr	.09± .05	.02	14.45 ±.14				9966
263.02 -20.49		2.7± .6		.12					9829
014931.0-362606 PGC　6881			1.05	.04	14.23				
015143.4+302306		.S?....	.74± .14					16.36±.3	7569± 10
138.05 -30.71　MCG　5- 5- 22			.28± .07	.15	15.37 ±.16				
329.80　-3.81				.35				269± 7	7693
014851.5+300818 PGC　6884			.75	.14	14.85				1.37	7309
0151.7　+0015		.S..3*.	1.07± .06	1					
153.28 -59.01　UGC　1333		U	.10± .05	.08	15.12 ±.19				
300.93 -13.57		3.0±1.2		.14					
0149.2　+0001　PGC　6888			1.07	.05					
0151.7　+1811　A　0149+17		.SB.6*.	1.28± .04	128	15.61 ±.15	.55± .05	.57± .03		4931
142.49 -42.38　UGC　1329		U	.60± .05	.13	15.16 ±.18	-.08± .07	-.09± .05		
318.34　-8.03　DDO　16		6.0±1.2	.78± .03	.88		.38	15.00± .08		5025
0149.0　+1756　PGC　6889			1.29	.30	14.39	-.20	15.38± .29		4653
015148.2+303236		.L..-*.	1.15± .15	100					
138.02 -30.55　UGC　1327		U	.28± .08	.08	14.56 ±.11				
329.96　-3.77		-3.0±1.2		.00					
014856.3+301748 PGC　6892			1.12						
0151.8　+1702		.SXT5*.	1.13± .05	125					4701
142.99 -43.46　UGC　1328		U	.15± .05	.13	14.32 ±.14				
317.25　-8.42		5.0± .8		.22					4792
0149.1　+1648　PGC　6893			1.14	.07	13.95				4423
015150.3+061744 NGC　706		.S..4?.	1.27± .03		13.20 ±.13	.70± .02	.71± .01	15.59±.2	4984± 7
148.75 -53.52　UGC　1334		PU	.13± .04	.16	13.02 ±.12	.04± .02	.06± .02	313± 10	4959± 39
306.88 -11.84　IRAS01492+0602		4.0± .9	.79± .01	.19	11.65	.61	12.64± .03	260± 8	5044
014913.2+060256 PGC　6897			1.28	.07	12.71	-.03	14.06± .21	2.81	4701

R.A. 2000 DEC.	Names	Type	logD₂₅	p.a.	B_T	(B-V)_T	(B-V)_e	m₂₁	V₂₁
l b		S_T n_L	logR₂₅	A_g	m_B	(U-B)_T	(U-B)_e	W₂₀	V_opt
SGL SGB		T	logA_e	A_i	m_FIR	(B-V)°_T	m'_e	W₅₀	V_GSR
R.A. 1950 DEC.	PGC	L	logD_o	A₂₁	B°_T	(U-B)°_T	m'₂₅	HI	V_3K

015150.4-052949		.SBS8..	1.20± .04	115					1644
159.24 -64.03	MCG -1- 5- 45	E (1)	.50± .05	.04					
295.17 -15.13	KUG 0149-057	8.0± .6		.62					1670
014919.7-054437	PGC 6898	5.3± .6	1.20	.25					1369
0151.9 +1732		.SB.8..	1.00± .08						5087± 10
142.81 -42.98	UGC 1335	U	.14± .06	.15					
317.74 -8.28		8.0± .9		.17					5180
0149.2 +1718	PGC 6902		1.01	.07					4809
015156.5-314413		.SBS8*.	1.09± .05	116					4069
235.84 -76.06	ESO 414- 8	S (1)	.25± .05	.02	15.81 ±.14				
267.81 -20.10		8.0±1.2		.31					4015
014942.0-315900	PGC 6904	6.7±1.2	1.09	.12	15.47				3851
0151.9 +0849		.S..7..	1.02± .06	32					5372
147.21 -51.17	UGC 1337	U	.53± .05	.14					
309.36 -11.09		7.0± .9		.73					5441
0149.3 +0835	PGC 6905		1.03	.26					5090
015201.8-053012		.SBS9..	1.15± .06	170					
159.33 -64.01	MCG -1- 5- 46	E (1)	.19± .05	.04					
295.18 -15.18		9.0± .8		.20					
014931.1-054459	PGC 6909	7.5± .8	1.15	.10					
015203.3-200955		.SXT3*.	1.33± .03	100					9511± 39
189.50 -74.49	ESO 543- 20	SE (1)	.38± .04	.00					9492
280.08 -18.42		2.6± .7		.52					
014941.1-202442	PGC 6912	5.3±1.6	1.33	.19					9261
015206.9+352451					14.57 ±.14	.56± .04	.54± .04		4877± 80
136.64 -25.84	CGCG 522- 26			.20	15.1 ±.4	-.20± .06	-.12± .06		5011
334.50 -2.09	5ZW 131		.75± .02				13.81± .04		4628
014911.2+351004	PGC 6916								
015207.3+350214		.SAS8..	1.16± .04					16.27±.3	4383± 8
136.75 -26.20	UGC 1330	U	.00± .05	.13				122± 9	
334.16 -2.23	KUG 0149+347	8.0± .8		.00					4517
014911.9+344727	PGC 6919		1.17	.00					4133
015209.0+392254									6964± 82
135.54 -22.00	CGCG 522- 28			.22	14.9 ±.4				7106
338.18 -.68	5ZW 132								6724
014909.9+390807	PGC 6920								
015211.0+425506		.S..6*.	.97± .07	20					
134.61 -18.57	UGC 1331	U	.22± .05	.35	14.66 ±.12				
341.45 .59		6.0±1.3		.33					
014908.5+424019	PGC 6921		1.00	.11					
015212.7+360552	NGC 700	.L.....	1.08± .09	10					
136.46 -25.17	UGC 1336	U	.57± .04	.26	15.37 ±.10				4364± 80
335.15 -1.87	5ZW 133	-2.0± .9		.00					4500
014916.4+355105	PGC 6924		1.02		15.05				4116
015215.7-565444		.S..5*/	1.04± .05	52					
287.11 -58.31	ESO 152- 29	S (1)	.50± .04	.00	16.12 ±.14				
240.64 -20.90		5.0±1.3		.76					
015030.1-570930	PGC 6926	5.6± .9	1.04	.25					
015216.9+360211		.L?....	.95± .10						
136.50 -25.23	CGCG 522- 30		.10± .04	.26	15.16 ±.08				4690± 31
335.09 -1.90				.00					4826
014920.6+354724	PGC 6928		.97		14.83				4442
015218.0+480515		.E.....	1.26± .13	95					
133.32 -13.55	UGC 1332	U	.06± .08	.96	14.6 ±.3				
346.24 2.43		-5.0± .8		.00					
014909.8+475028	PGC 6929		1.40						
015221.9+354748		.S..3..	.95± .07	75					
136.58 -25.45	UGC 1338	U	.14± .05	.20	15.01 ±.12				4099± 30
334.88 -2.00		3.0± .9		.19					4234
014925.8+353301	PGC 6934	2.5±1.2	.97	.07	14.58				3851
015224.8+355123		.LBR+..	1.04± .11						
136.58 -25.39	UGC 1339	U	.08± .05	.20	14.68 ±.11				
334.94 -1.99		-1.0± .8		.00					
014928.7+353636	PGC 6938		1.05						
015227.9+173047	NGC 711	.L.....	1.21± .06	15					
142.99 -42.97	UGC 1342	U	.33± .03	.15	14.14 ±.10				4928± 50
317.75 -8.41		-2.0± .8		.00					5020
014944.3+171600	PGC 6940		1.18		13.92				4651

R.A. 2000 DEC. / l b / SGL SGB / R.A. 1950 DEC.	Names / PGC	Type / S_T n_L / T / L	$\log D_{25}$ / $\log R_{25}$ / $\log A_e$ / $\log D_o$	p.a. / A_g / A_i / A_{21}	B_T / m_B / m_{FIR} / B_T^o	$(B-V)_T$ / $(U-B)_T$ / $(B-V)_T^o$ / $(U-B)_T^o$	$(B-V)_e$ / $(U-B)_e$ / m'_e / m'_{25}	m_{21} / W_{20} / W_{50} / HI	V_{21} / V_{opt} / V_{GSR} / V_{3K}
015230.4+315909		.SXS7..	1.09± .06					16.52±.2	10757± 6
137.74 -29.12	UGC 1341	U	.03± .05	.12	15.1 ±.2			259± 9	
331.35 -3.40		7.0± .8		.05				251± 7	10884
014937.3+314423	PGC 6944		1.10	.02	14.86			1.64	10500
015234.7+363003		RSB.1..	1.21± .05	22	13.55 ±.14	.89± .03	.97± .01	18.57±.3	4155± 9
136.43 -24.76	UGC 1344	U	.29± .05	.26	13.88 ±.12	.25± .04	.35± .03	96± 12	4407± 25
335.54 -1.79		1.0± .8	.87± .03	.29		.74	13.39± .08		4323
014938.0+361517	PGC 6948		1.24	.14	13.13	.15	13.74± .30	5.29	3940
0152.6 +4805		.E.....	1.00± .19						
133.38 -13.53	UGC 1340	U	.05± .08	.96					
346.26 2.38		-5.0± .8		.00					
0149.5 +4751	PGC 6955		1.14						
015239.1-184650		.S?....	1.03± .07						
185.47 -73.64	ESO 543- 22		.13± .05	.00					14580± 52
281.56 -18.30				.20					14565
015016.0-190136	PGC 6956		1.03	.07					14328
015239.6+361018	NGC 703	.L..-*.	1.08± .17	50	14.27 ±.15	1.01± .03	1.10± .02		
136.54 -25.08	UGC 1346	U	.12± .08	.26	14.24 ±.10	.59± .04	.61± .03		4752± 82
335.25 -1.93		-3.0±1.2	.69± .05	.00		.90	13.21± .16		4887
014943.2+355532	PGC 6957		1.09		13.92	.55	14.23± .89		4505
015241.7+360838	NGC 705	.S..0..	1.08± .06	117					
136.55 -25.10	UGC 1345	U	.63± .05	.26	14.63 ±.15				4541± 27
335.22 -1.94	6ZW 90	.0±1.0		.47					4677
014945.3+355352	PGC 6958		1.07		13.83				4294
0152.7 -1313		.SBS9*.	1.15± .08						
171.82 -70.03		E (1)	.00± .08	.00					
287.34 -17.19		9.0±1.3		.00					
0150.3 -1328	PGC 6960	9.8±1.2	1.15	.00					
015245.8+363707	A 0149+36	.SXT5..	1.10± .04		13.49 ±.17	.63± .03	.71± .02	15.80±.2	5534± 6
136.43 -24.64	UGC 1347	U (1)	.06± .04	.22	13.80 ±.12	.01± .05	.07± .03	132± 7	5520± 42
335.67 -1.78	KUG 0149+363	5.0± .8	.89± .03	.09		.53	13.43± .07		5670
014949.0+362221	PGC 6961	2.0±1.0	1.12	.03	13.36	-.06	13.69± .28	2.41	5288
015246.4+360906	NGC 708	.E.....	1.48± .10	35	*		1.03± .02		
136.57 -25.09	UGC 1348	U	.08± .08	.26	13.7 ±.2		.63± .05		4813± 24
335.24 -1.95		-5.0± .7	1.41± .09	.00			15.10± .32		4948
014950.0+355420	PGC 6962		1.50		13.34				4566
015246.7-485327		.SBS4?.	1.07± .05	118	*				
278.43 -65.22	ESO 197- 9	S (1)	.36± .04	.04	14.71 ±.14				6346±108
249.30 -21.21	IRAS01507-4908	4.0± .9		.53					6244
015049.0-490812	PGC 6964	3.3±1.3	1.08	.18	14.10				6190
015249.0-032649		.SAS5?/	1.46± .04	20	*				5018
157.36 -62.14	MCG -1- 5- 47	E	.86± .05	.07					
297.30 -14.84	IRAS01503-0341	5.0± .8		1.28	12.32				5050
015017.2-034135	PGC 6966		1.46	.43					4742
015250.7+361321	NGC 709				15.23 ±.13	.97± .03	1.05± .03		
136.56 -25.02	CGCG 522- 40			.26	15.2 ±.4	.43± .05	.52± .05		3359± 69
335.31 -1.94			.40± .03				12.72± .09		3494
014954.2+355835	PGC 6969								3112
015253.8+360312	NGC 710	.S..6*.	1.11± .04		14.27 ±.19	.60± .04	.64± .02	16.47±.3	6105± 8
136.62 -25.18	UGC 1349	U	.02± .04	.26	14.06 ±.13	-.09± .06	-.03± .04	239± 9	6110± 30
335.15 -2.01	IRAS01499+3548	6.0±1.1	.78± .04	.03	12.52	.50	13.66± .10		6241
014957.5+354826	PGC 6972		1.14	.01	13.80	-.16	14.63± .29	2.66	5858
015257.6+363048		.SBR3..	1.24± .06	55	14.21 ±.15	.98± .04	.96± .03	18.20±.3	4975± 9
136.51 -24.73	UGC 1350	U	.16± .06	.26	14.14 ±.15	.34± .05	.43± .04	245± 12	5244± 30
335.35 -1.86		3.0± .8	.89± .02	.22		.86	14.15± .05		5135
015000.8+361602	PGC 6977		1.26	.08	13.66	.23	14.84± .37	4.46	4752
015259.6+124228	IC 1743	.SB.1*.	1.26± .04	57				15.27±.1	4558± 5
145.44 -47.46	UGC 1351	U	.38± .05	.14	13.77 ±.12			420± 7	4590± 39
313.19 -10.11	IRAS01503+1227	1.0± .8		.39	11.10			401± 5	4637
015018.8+122743	PGC 6982		1.28	.19	13.19			1.90	4278
015300.4-134421	NGC 720	.E.5...	1.67± .04	135	11.16M±.05	.98± .01	.99± .01		
173.02 -70.36	MCG -2- 5- 68	R	.29± .06	.00	11.09 ±.16	.47± .03	.53± .01		1716± 11
286.82 -17.36		-5.0± .3	1.08± .02	.00		.96	12.12± .07		1715
015034.4-135906	PGC 6983		1.59		11.13	.48	13.79± .27		1453
0153.0 -0105		.S..2..	.95± .07	67					
155.08 -60.06	UGC 1354	U	.47± .05	.04	15.50 ±.12				
299.68 -14.26		2.0± .9		.57					
0150.5 -0120	PGC 6986		.95	.23					

R.A. 2000 DEC. l b SGL SGB R.A. 1950 DEC.	Names PGC	Type S_T n_L T L	$\log D_{25}$ $\log R_{25}$ $\log A_e$ $\log D_o$	p.a. A_g A_i A_{21}	B_T m_B m_{FIR} B_T^o	$(B-V)_T$ $(U-B)_T$ $(B-V)_T^o$ $(U-B)_T^o$	$(B-V)_e$ $(U-B)_e$ m'_e m'_{25}	m_{21} W_{20} W_{50} HI	V_{21} V_{opt} V_{GSR} V_{3K}
015308.5+364908 NGC 712 136.46 -24.43 UGC 1352 335.88 -1.78 015011.4+363423 PGC 6988		.L..... U -2.0± .8	1.11± .16 .10± .08 1.12	85 .22 .00	13.77 ±.10 13.47				5258± 25 5395 5013
015312.5+041148 NGC 718 150.74 -55.30 UGC 1356 304.93 -12.79 IRAS01506+0357 015036.5+035703 PGC 6993		.SXS1.. R 1.0± .3	1.37± .03 .06± .04 1.38	45 .08 .06 .03	12.59 ±.13 12.37 ±.10 13.62 12.30	.89± .01 .37± .02 .85 .35	.90± .01 14.15± .21	19.10±.3 124± 13 6.77	1733± 10 1762± 38 1789 1454
015312.7-493333 279.17 -64.62 ESO 197- 10 248.57 -21.28 015116.0-494818 PGC 6994		.L...P. S -2.0± .8	1.26± .05 .18± .04 .80± .04 1.24	178 .04 .00	13.46 ±.15 13.40 ±.14 13.30	1.05± .02 .52± .03 .97 .52	1.09± .01 .59± .02 12.95± .12 14.19± .31		6172± 19 6068 6020
015314.3+224325 IC 1742 141.09 -37.96 MCG 4- 5- 23 322.75 -6.81 015027.5+222840 PGC 6996		.S?.... 	.89± .07 .15± .06 .92	 .31 .23 .08	 15.09 ±.12 14.49			16.62±.3 318± 7 2.04	9768± 10 9873 9497
015318.9-781424 298.95 -38.41 ESO 13- 24 218.10 -18.33 IRAS01533-7829 015319.0-782906 PGC 7002		.S..2.. S 2.0± .9	1.11± .05 .53± .04 1.14	139 .29 .65 .26	 14.94 ±.14 13.13				
0153.3 +1956 142.22 -40.60 UGC 1357 320.13 -7.79 0150.6 +1942 PGC 7004		.S..1.. U 1.0± .9	1.03± .06 .65± .05 1.05	46 .22 .66 .32	 15.29 ±.12 14.30				9117 9215 8843
015322.1+285943 138.90 -31.95 CGCG 503- 46 328.63 -4.64 015031.1+284459 PGC 7005				 .16	15.6 ±.4			17.34±.3 232± 7	10259± 10 10379 9998
015323.0+365720 136.47 -24.28 UGC 1353 336.02 -1.78 6ZW 93 015025.8+364235 PGC 7006		.L..-*. U -3.0±1.2	1.08± .17 .08± .08 1.10	110 .22 .00	14.14 ±.10 13.85				5055± 80 5192 4810
015329.7+361314 NGC 714 136.70 -24.99 UGC 1358 335.35 -2.06 015033.1+355830 PGC 7009		.S..0.. U .0± .9	1.19± .05 .61± .05 1.18	112 .26 .45	14.10 ±.13 14.03 ±.15 13.29	1.04± .02 .56± .05 .84 .41	 13.36± .30		4458± 25 4593 4211
015336.5+435758 134.60 -17.49 UGC 1355 342.51 .73 IRAS01505+4343 015032.6+434314 PGC 7017		.SXS3.. U 3.0± .8	1.14± .05 .06± .05 1.19	 .49 .08 .03	13.97 ±.13 11.77 13.35				6247 6396 6021
0153.6 +1949 NGC 719 142.36 -40.68 UGC 1360 320.04 -7.89 0150.9 +1935 PGC 7019		.L...?. U -2.0±1.6	1.15± .15 .10± .08 1.16	150 .22 .00	14.23 ±.12				
015341.5+312825 138.18 -29.55 CGCG 503- 48 330.96 -3.82 015048.6+311341 PGC 7021				 .15	15.3 ±.4			17.33±.3 301± 7	10077± 10 10202 9821
0153.7 +1446 144.64 -45.45 UGC 1362 315.24 -9.59 0151.0 +1432 PGC 7022		.S..9*. U 9.0±1.2	.96± .09 .00± .06 .97	 .15 .00 .00					7918 8002 7640
015342.4+295601 138.68 -31.03 UGC 1359 329.53 -4.37 IRAS01508+2941 015050.6+294117 PGC 7023		.SB?... 	.89± .06 .02± .04 .90	 .16 .04 .01	 14.26 ±.13 13.18 14.02			16.43±.2 308± 13 278± 7 2.40	7658± 7 7683± 50 7780 7399
015345.8-234528 NGC 723 202.91 -75.54 ESO 477- 13 276.37 -19.42 IRAS01514-2400 015126.0-240012 PGC 7024		.SAR4*. Sr (1) 4.0± .6 3.3±1.1	1.17± .04 .05± .04 1.17	 .00 .07 .02	13.25 ±.14 12.63 13.17				1491± 60 1460 1251
015348.9-355123 NGC 727 250.13 -74.13 ESO 354- 10 263.42 -20.89 015138.1-360606 PGC 7027		PSXS2.. BSr (1) 2.0± .5 4.4±1.3	1.03± .05 .23± .04 1.04	76 .02 .29 .12	 14.95 ±.14 14.54				10056± 39 9988 9852
015350.2+362100 136.74 -24.84 MCG 6- 5- 40 335.50 -2.08 015053.4+360616 PGC 7029		.E?.... 		 .26	15.3 ±.4				4123± 28 4258 3877

R.A. 2000 DEC. l b SGL SGB R.A. 1950 DEC.	Names PGC	Type S_T n_L T L	$logD_{25}$ $logR_{25}$ $logA_e$ $logD_o$	p.a. A_g A_i A_{21}	B_T m_B m_{FIR} B_T^o	$(B-V)_T$ $(U-B)_T$ $(B-V)_T^o$ $(U-B)_T^o$	$(B-V)_e$ $(U-B)_e$ m'_e m'_{25}	m_{21} W_{20} W_{50} HI	V_{21} V_{opt} V_{GSR} V_{3K}
015350.7+363350 136.68 -24.64 335.69 -2.01 015053.7+361907	 UGC 1361 PGC 7030	.S..6*. U 6.0±1.3 	1.04± .08 .47± .06 1.06	 .22 .69 .24				16.55±.3 234± 9 	5740± 8 5876 5495
015355.1+361345 136.79 -24.96 335.39 -2.14 015058.4+355901	NGC 717 UGC 1363 PGC 7033	.S..0.. U .0±1.0 	1.12± .05 .73± .05 .58± .04 1.10	117 .26 .55 	14.86 ±.15 14.80 ±.14 13.94	.96± .03 .46± .05 .74 .31	1.04± .02 .47± .04 13.25± .13 13.46± .33		 4968± 80 5103 4722
015357.0-380205 256.59 -72.95 261.07 -21.08 015148.0-381648	 ESO 297- 31 PGC 7034	.SAS5?. S (1) 5.0± .9 3.3± .9	1.04± .05 .25± .05 .37 1.04	84 .03 .37 .12	 15.22 ±.14 14.79				5722 5649 5526
0153.9 -0045 155.17 -59.65 300.08 -14.39 0151.4 -0059	 UGC 1365 IRAS01514-0059 PGC 7039	.S?.... 	1.00± .06 .36± .05 1.01	118 .04 .55 .18	 14.54 ±.13 13.47 13.93				4840 4879 4563
0154.0 +1455 144.66 -45.29 315.41 -9.61 0151.3 +1441	 UGC 1364 PGC 7042	.SB.6*. U 6.0±1.2 	1.14± .13 .23± .12 1.15	30 .15 .33 .11					
015403.3-141513 174.65 -70.54 286.35 -17.72 015137.6-142956	 MCG -2- 5- 72 PGC 7045	RS..0?/ E .3± .8 	1.12± .04 .61± .03 1.09	75 .00 .45 					
0154.0 -0045 155.20 -59.64 300.08 -14.41 0151.5 -0100	 UGC 1367 PGC 7046	.I..9*. U 10.0±1.3 	1.00± .08 .43± .06 1.00	 .04 .32 .22					4745 4784 4468
0154.1 +0751 148.61 -51.86 308.59 -11.90 0151.5 +0737	 UGC 1368 PGC 7053	.S..2.. U 2.0± .9 	1.14± .05 .48± .05 1.16	53 .14 .59 .24	 14.74 ±.12 13.93				7941 8005 7661
015409.3-564124 286.48 -58.39 240.86 -21.17 015224.0-565606	NGC 745 ESO 152- 32 PGC 7054	.L..+*P S -1.4± .5 	1.12± .06 .24± .05 .87± .05 1.09	 12.6 V±.2 14.04 ±.14 .00	 .00 .00 13.95				 5953± 44 5831 5831
0154.2 +1805 143.28 -42.28 318.45 -8.61 0151.5 +1751	 UGC 1369 PGC 7059	.S..6*. U 6.0±1.1 	1.12± .05 .11± .05 1.13	50 .12 .16 .05	4927 14.46 ±.15 14.16				4927 5020 4652
015416.1-374712 255.77 -73.03 261.34 -21.12 015207.0-380154	 ESO 297- 32 PGC 7061	PSBS1*. Sr 1.0± .6 	1.02± .05 .04± .04 1.02	 .03 .04 .02	 14.41 ±.14 				
0154.2 +1331 145.45 -46.58 314.09 -10.14 0151.6 +1317	 UGC 1370 PGC 7063	.S..2.. U 2.0± .9 	.95± .07 .35± .05 .97	153 .17 .43 .17	 15.35 ±.12 				
015418.1-094253 166.32 -67.13 291.04 -16.76 015149.9-095736	 MCG -2- 5- 71 PGC 7064	.LB.-*. E -2.5± .6 	1.33± .04 .25± .04 1.30	95 .05 .00 					
015419.7+363749 136.76 -24.55 335.79 -2.07 015122.6+362306	 UGC 1366 PGC 7066	.SB.6*. U 6.0±1.3 	1.24± .05 .64± .05 1.26	140 .22 .94 .32	 14.58 ±.12 13.40				5118 5254 4873
015421.2-564542 286.50 -58.31 240.79 -21.19 015236.1-570024	NGC 754 ESO 152- 33 PGC 7068	.E.0.P* S -5.0±1.0 	.73± .06 .01± .04 .73	 .00 .00 	 15.24 ±.14 				
015422.4-003741 155.22 -59.50 300.23 -14.45 015149.0-005224	 MCG 0- 5- 46 ARAK 61 PGC 7071	 	.64± .17 .11± .07 .64	 .04 	 14.9 ±.2 				4849 4888 4572
015424.0+052518 150.33 -54.07 306.22 -12.71 015147.3+051035	 UGC 1373 KUG 0151+051 PGC 7072	.S?.... 	1.01± .05 .49± .04 1.03	80 .15 .74 .25	 15.36 ±.09 14.44				4941 4998 4661

R.A. 2000 DEC. / l b / SGL SGB / R.A. 1950 DEC.	Names / PGC	Type / S_T n_L / T / L	logD25 / logR25 / logAe / logDo	p.a. / Ag / Ai / A21	BT / mB / mFIR / BT°	(B-V)T / (U-B)T / (B-V)T° / (U-B)T°	(B-V)e / (U-B)e / m'e / m'25	m21 / W20 / W50 / HI	V21 / Vopt / VGSR / V3K
0154.4 +1739 / 143.53 -42.68 / 318.05 -8.80 / 0151.7 +1725	UGC 1372 / PGC 7073	.S?.... / / /	.99± .06 / .36± .05 / / 1.00	14 / .13 / .55 / .18	/ 14.93 ±.12 / / 14.22				6890 / / 6981 / 6614
015428.1+044819 / 150.80 -54.62 / 305.62 -12.91 / 015151.7+043337	IC 1746 / UGC 1371 / / PGC 7076	.L..... / U / -2.0± .9 /	1.16± .05 / .49± .05 / / 1.10	93 / .13 / .00 /	/ 14.75 ±.10 / /			15.93±.3 / / 448± 25 /	2540± 17 / 7800± 43 / /
0154.5 +1651 / 143.92 -43.42 / 317.30 -9.09 / 0151.8 +1637	UGC 1374 / PGC 7082	.SB.6*. / U / 6.0±1.2 /	1.00± .06 / .14± .05 / / 1.02	0 / .13 / .20 / .07	/ 15.22 ±.16 / / 14.87				5115 / / 5204 / 4839
0154.5 +2003 / 142.53 -40.40 / 320.34 -8.01 / 0151.8 +1949	UGC 1375 / PGC 7085	.SB.6*. / U / 6.0±1.2 /	1.06± .06 / .02± .05 / / 1.08	/ .20 / .03 / .01	/ 14.56 ±.15 / / 14.28				9001 / / 9099 / 8728
0154.6 +1702 / 143.87 -43.24 / 317.48 -9.05 / 0151.9 +1648	UGC 1377 / PGC 7088	.SB.6*. / U / 6.0±1.4 /	1.07± .07 / .91± .06 / / 1.08	78 / .14 / 1.34 / .46					
0154.6 -0009 / 154.91 -59.04 / 300.73 -14.39 / 0152.1 -0024	UGC 1382 / PGC 7090	.E...?. / U / -5.0±1.6 /	1.03± .11 / .00± .05 / / 1.04	/ .05 / .00 /	/ 14.24 ±.11 / / 14.10				5593± 31 / / 5633 / 5316
015440.3-620619 / 290.57 -53.48 / 235.05 -20.81 / 015307.1-622100	ESO 114- 14 / / / PGC 7091	PSAT4*. / Sr (1) / 4.1± .4 / 4.0± .4	1.19± .05 / .05± .05 / / 1.19	128 / .02 / .07 / .02	/ 14.22 ±.14 / / 14.08				7138± 34 / / 7005 / 7041
0154.7 +0601 / 150.03 -53.48 / 306.84 -12.60 / 0152.1 +0547	UGC 1383 / PGC 7096	.S..6*. / U / 6.0±1.4 /	1.15± .05 / .82± .05 / / 1.16	35 / .15 / 1.21 / .41					5245 / / 5304 / 4966
015445.5+392259 / 136.06 -21.87 / 338.36 -1.14 / 015145.9+390817	NGC 721 / UGC 1376 / IRAS01517+3908 / PGC 7097	.SBT4.. / PU / 4.0± .6 /	1.23± .03 / .23± .04 / .86± .02 / 1.25	135 / .18 / .34 / .11	14.17 ±.13 / 13.67 ±.12 / / 13.35	.70± .03 / .02± .04 / .58 / -.07	.78± .02 / .10± .04 / 13.96± .04 / 14.59± .22	15.52±.1 / / 286± 8 / 2.06	5597± 11 / 5487± 52 / 5733 / 5355
0154.7 +2041 / 142.33 -39.78 / 320.95 -7.84 / 0152.0 +2027	NGC 722 / UGC 1379 / / PGC 7098	.S?.... / / /	1.22± .05 / .49± .05 / / 1.24	138 / .21 / .74 / .25	/ 14.35 ±.12 / / 13.37				4901 / / 5000 / 4629
0154.8 +1806 / 143.45 -42.22 / 318.51 -8.74 / 0152.1 +1752	UGC 1384 / PGC 7104	.SB.7*. / U / 7.0±1.2 /	1.19± .06 / .37± .06 / / 1.20	8 / .13 / .51 / .19					4926 / / 5018 / 4651
015451.3-350943 / 247.51 -74.26 / 264.19 -21.04 / 015240.1-352424	ESO 354- 12 / IRAS01526-3524 / PGC 7106	.L?.... / / /	.95± .06 / .23± .03 / / .92	108 / .02 / .00 /	/ 15.33 ±.14 / 13.16 / 15.23				5012 / 4946 / 4807
015452.7-133914 / 173.79 -69.98 / 287.01 -17.79 / 015226.7-135355	MCG -2- 5- 74 / PGC 7109	.S..9P* / E (1) / 9.0± .9 / 6.4± .8	1.04± .05 / .16± .04 / / 1.04	40 / .00 / .16 / .08					
015453.2+365502 / 136.80 -24.24 / 336.10 -2.07 / 015155.7+364020	A 0151+36 / UGC 1385 / MK 2 / PGC 7111	RSB.0.. / U / .0± .9 /	.85± .06 / .06± .04 / / .87	170 / .20 / .04 /	13.9 ±.3 / 14.33 ±.15 / / 13.92	.59± .04 / -.14± .07 / .48 / -.15	/ / / 12.87± .42	16.57±.3 / 200± 9 / /	5621± 8 / 5476± 28 / 5745 / 5366
0154.8 +1328 / 145.68 -46.57 / 314.09 -10.29 / 0152.2 +1314	UGC 1386 / PGC 7114	.S..4.. / U / 4.0±1.0 /	.95± .07 / .70± .05 / / .97	166 / .16 / 1.04 / .35					6222 / / 6302 / 5944
015454.2+004842 / 154.14 -58.16 / 301.71 -14.17 / 015220.0+003400	IC 172 / MCG 0- 5- 49 / ARAK 63 / PGC 7116	.S?.... / / /	.64± .17 / .19± .07 / / .64	/ .01 / .23 / .09	/ 14.9 ±.2 / 12.71 / 14.56				8173 / 8216 / 7896
015456.2-090039 / 165.53 -66.48 / 291.80 -16.75 / 015227.6-091521	NGC 731 / MCG -2- 5- 73 / / PGC 7118	.E+..*. / PE / -3.7± .5 /	1.24± .06 / .00± .05 / .72± .07 / 1.25	/ .09 / .00 /	13.0 ±.2 / / /	.93± .02 / .44± .04 / /	.95± .01 / .49± .02 / 12.12± .24 / 14.08± .38		

R.A. 2000 DEC. / l b / SGL SGB / R.A. 1950 DEC.	Names / PGC	Type / ST nL / T / L	logD25 / logR25 / logAe / logDo	p.a. / Ag / Ai / A21	BT / mB / mFIR / BT°	(B-V)T / (U-B)T / (B-V)T° / (U-B)T°	(B-V). / (U-B). / m'. / m'25	m21 / W20 / W50 / HI	V21 / Vopt / VGSR / V3K
015500.2-260107 212.07 -75.75 274.01 -20.04 015242.0-261548	 ESO 477- 14 PGC 7127	.LA.0.. S -2.0± .8	1.07± .05 .16± .04 1.05	 .00 .00	14.52 ±.14				
0155.1 +4901 133.56 -12.52 347.27 2.33 0152.0 +4847	 UGC 1381 PGC 7138	.S..7.. U 7.0± .9	1.04± .15 .23± .12 1.16	107 1.28 .31 .11					
015510.4+351650 137.35 -25.80 334.61 -2.72 015214.2+350209	IC 171 UGC 1388 PGC 7139	.E?....	1.40± .11 .06± .08 1.42	105 .22 .00	 13.23 ±.14 12.93				5275± 44 5407 5027
015510.8+361541 137.05 -24.86 335.51 -2.37 015213.8+360100	 UGC 1387 KUG 0152+360 PGC 7140	.S..8*. U 8.0±1.2	.98± .05 .17± .04 1.00	175 .27 .21 .08	 15.15 ±.13 14.66			16.02±.3 240± 9 1.28	4532± 8 4667 4287
015517.1+100041 147.72 -49.76 310.77 -11.50 015237.8+094600	 UGC 1391 IRAS01525+0945 PGC 7150	.S..6*. U 6.0±1.4	1.16± .04 .89± .04 1.18	177 .18 1.31 .45	 15.35 ±.12 13.57 13.83			15.86±.3 445± 13 1.59	5928± 10 5998 5649
015517.9-125435 172.49 -69.38 287.81 -17.73 015251.5-130915	 MCG -2- 5- 76 PGC 7153	.SBS8P. E (1) 8.0± .6 7.5± .7	1.08± .06 .14± .05 1.08	5 .00 .17 .07					
0155.3 +2116 142.26 -39.18 321.55 -7.77 0152.6 +2102	 UGC 1393 PGC 7163	.L..-*. U -3.0±1.3	.86± .17 .12± .05 .87	65 .22 .00	 15.25 ±.10 14.85				12320± 57 12420 12049
015522.2+063641 149.88 -52.88 307.46 -12.58 015244.8+062200	A 0152+06 UGC 1395 IRAS01527+0622 PGC 7164	.SAT3.. U (1) 3.0± .8 4.5±1.1	1.10± .04 .10± .04 1.11	 .14 .14 .05	 14.18 ±.13 13.61 13.86			16.30±.3 255± 13 2.39	5164± 10 5184± 34 5225 4887
015525.2+240840 141.15 -36.45 324.26 -6.78 015237.2+235400	 CGCG 482- 30 PGC 7170			 .41	15.7 ±.4			17.81±.3 219± 10	9883± 10 9990 9616
0155.4 +3633 137.02 -24.55 335.81 -2.31 0152.5 +3619	 UGC 1392 PGC 7173	.L...?. U -2.0±1.7	1.04± .08 .26± .03 1.03	43 .20 .00					
015530.7+475717 133.89 -13.55 346.31 1.88 015221.7+474236	 UGC 1389 PGC 7179	.E..... U -5.0± .8	1.26± .13 .06± .08 1.40	 .99 .00	 14.5 ±.3				
0155.5 +2117 142.30 -39.15 321.58 -7.80 0152.7 +2103	 UGC 1396 IRAS01527+2103 PGC 7180	.L..... U -2.0± .9	1.05± .08 .35± .03 1.02	94 .22 .00	 14.80 ±.10 13.60 14.51				4903± 57 5003 4632
015532.0-104800 168.73 -67.77 290.00 -17.31 015304.4-110240	NGC 726 MCG -2- 6- 3 KUG 0153-110 PGC 7182	.SBS8P. E (1) 7.5± .6 5.9± .6	1.07± .04 .28± .04 1.07	100 .00 .35 .14					5359 5366 5094
015541.3-295520 227.91 -75.63 269.85 -20.69 015326.1-301000	NGC 749 ESO 414- 11 IRAS01534-3009 PGC 7191	PSBS0*. Sr .4± .6	1.29± .03 .13± .03 1.29	111 .02 .10	 13.43 ±.14 12.16 13.24				4394± 34 4343 4173
015541.3+230420 141.64 -37.45 323.27 -7.22 015254.0+224940	 CGCG 482- 31 PGC 7192	.S?....	.77± .11 .00± .10 .80	 .31 .00 .00	 15.52 ±.12 15.13			17.21±.3 175± 7 2.09	13792± 10 13896 13523
015541.4+312506 138.67 -29.48 331.07 -4.23 015248.2+311026	 CGCG 503- 51 PGC 7193			 .11	15.7 ±.4			17.29±.3 232± 7	10716± 10 10840 10461
015542.0+464807 134.22 -14.66 345.26 1.43 015234.4+463327	 UGC 1394 PGC 7197	.S..1.. U 1.0± .8	1.37± .04 .28± .05 1.43	70 .71 .28 .14	 14.3 ±.3 13.19			15.68±.1 482± 8 2.35	6388± 11 6541 6172

R.A. 2000 DEC. / l b / SGL SGB / R.A. 1950 DEC.	Names / PGC	Type: S_T n_L / T / L	$\log D_{25}$ / $\log R_{25}$ / $\log A_e$ / $\log D_o$	p.a. / A_g / A_i / A_{21}	B_T / m_B / m_{FIR} / B_T^o	$(B-V)_T$ / $(U-B)_T$ / $(B-V)_T^o$ / $(U-B)_T^o$	$(B-V)_e$ / $(U-B)_e$ / m' / m'_{25}	m_{21} / W_{20} / W_{50} / HI	V_{21} / V_{opt} / V_{GSR} / V_{3K}
015549.7-220703		.S?....	1.03± .09						13403
197.53 -74.55	ESO 543- 25		.02± .07	.00					13375
278.19 -19.62				.02					
015329.1-222142	PGC 7206		1.03	.01					13161
0155.8 +1802		.L.....	1.00± .10	2					
143.78 -42.21	UGC 1399	U	.14± .04	.13	14.74 ±.10				
318.53 -8.99		-2.0± .9		.00					
0153.1 +1748	PGC 7209		.99						
015551.5-095801		.SBR5P.	1.12± .04	125					8117
167.49 -67.08	MCG -2- 6- 4	E (1)	.10± .04	.04					
290.87 -17.19	KUG 0153-102A	4.5± .6		.15					8127
015323.4-101241	PGC 7210	3.1± .6	1.13	.05					7852
015553.8+325925		.L.....	1.03± .08	36					
138.21 -27.96	UGC 1397	U	.34± .03	.16	14.83 ±.10				
332.54 -3.70		-2.0± .9		.00					
015259.4+324445	PGC 7214		1.00						
015557.2+011706	IC 173	.SBT4..	.95± .05	110	14.9 ±.2	.88± .06			13912± 42
154.17 -57.62	UGC 1402	UE (2)	.12± .04	.03	14.68 ±.12				13955
302.25 -14.29		3.7± .5		.18		.76			13635
015322.8+010227	PGC 7217	.9± .5	.95	.06	14.44		14.19± .33		
015558.7+370748		.S..6?.	1.08± .06		14.56 ±.15			17.05±.3	5218± 8
136.97 -23.98	UGC 1398	U	.04± .05	.19				157± 9	
336.37 -2.20	IRAS01530+3653	6.0±1.7		.06					5354
015300.8+365309	PGC 7220		1.10	.02	14.28			2.75	4976
015604.4+360750		.S..3..	1.40± .04	156					
137.28 -24.93	UGC 1400	U (1)	.81± .05	.27					
335.46 -2.58		3.0± .9		1.11					
015307.4+355311	PGC 7223	4.5±1.2	1.42	.40					
0156.1 +1738	IC 1748	.SX.4..	1.02± .06	130					11341
144.05 -42.56	UGC 1403	U	.17± .05	.13	14.47 ±.12				
318.17 -9.19		4.0± .9		.25					11431
0153.4 +1724	PGC 7229		1.03	.09	14.02				11067
015611.1+064442	IC 1749	.L...*.	.96± .06	155					5108
150.10 -52.68	UGC 1407	U	.14± .03	.15	14.56 ±.10				5168
307.66 -12.73	IRAS01535+0630	-2.0±1.2		.00	12.90				4830
015333.7+063003	PGC 7235		.95		14.34				
015612.1+053520		.34?							
150.91 -53.73	MCG 1- 6- 2		.00± .07	.14	15.4 ±.3				900± 64
306.53 -13.08									956
015335.2+052041	PGC 7237		.35						622
0156.2 +0438		.S..7..	1.23± .05	93					4973
151.61 -54.58	UGC 1410	U	.88± .05	.12	15.26 ±.12				
305.60 -13.38	IRAS01536+0424	7.0± .9		1.22	12.96				5027
0153.6 +0424	PGC 7243		1.24	.44	13.91				4695
015616.1-225404		.S..4*/	1.31± .03	155					
200.43 -74.73	ESO 477- 16	S	1.00± .03	.00	14.72 ±.14				1646
277.38 -19.85		3.5± .6		1.46					1616
015356.0-230842	PGC 7244		1.31	.50	13.25				1406
0156.2 +1309		.S..7..	1.00± .06	90					7669
146.31 -46.75	UGC 1408	U	.29± .05	.16	15.32 ±.14				
313.89 -10.72		7.0± .9		.40					7747
0153.6 +1255	PGC 7246		1.02	.15	14.74				7392
0156.2 +7316		RSBT1*.	1.53± .04	5				13.45±.1	2940± 11
127.63 11.00	UGC 1378	U	.16± .06	2.24	13.5 ±.3				2930± 35
9.93 10.73	IRAS01519+7302	1.0± .6		.17	12.97			496± 8	3115
0151.9 +7302	PGC 7247		1.74	.08	11.02			2.35	2825
0156.3 +0345	IC 174	.L...?.	1.13± .10	95					
152.29 -55.37	UGC 1409	U	.22± .05	.08	14.25 ±.10				
304.73 -13.65		-2.0±1.6		.00					
0153.7 +0331	PGC 7249		1.10						
015620.9+053744	NGC 741	.E.0.*.	1.47± .08		12.2 ±.2	1.05± .03	1.06± .01		
150.93 -53.68	UGC 1413	R	.01± .07	.14	12.46 ±.11				5561± 17
306.58 -13.11	3ZW 38	-5.0± .4	1.24± .04	.00		.96	13.91± .14		5618
015344.1+052306	PGC 7252		1.49		12.18		14.53± .47		5283
015621.3+372709		.S..6*.	1.04± .08	125					
136.95 -23.65	UGC 1405	U	.98± .06	.19	15.80 ±.14				
336.70 -2.15		6.0±1.4		1.44					
015323.0+371230	PGC 7254		1.06	.49					

R.A. 2000 DEC. / l b / SGL SGB / R.A. 1950 DEC.	Names / PGC	Type / S_T n_L / T / L	$\log D_{25}$ / $\log R_{25}$ / $\log A_e$ / $\log D_o$	p.a. / A_g / A_i / A_{21}	B_T / m_B / m_{FIR} / B_T^o	$(B-V)_T$ / $(U-B)_T$ / $(B-V)_T^o$ / $(U-B)_T^o$	$(B-V)_e$ / $(U-B)_e$ / m'_e / m'_{25}	m_{21} / W_{20} / W_{50} / HI	V_{21} / V_{opt} / V_{GSR} / V_{3K}
015621.9-042804 / 160.13 -62.55 / 296.51 -15.96 / 015350.7-044242	NGC 748 / MCG -1- 6- 4 / IRAS01538-0442 / PGC 7259	PSAR3?. / E (1) / 3.0± .7 / 3.1±1.2	1.36± .03 / .31± .05 / .89± .03 / 1.36	42 / .00 / .43 / .15	13.41 ±.15 / / / 12.94	.78± .05 / .18± .06 / .69 / .10	.92± .02 / .36± .03 / 13.35± .07 / 14.31± .25		5322 / / 5348 / 5050
015622.7-090343 / 166.31 -66.30 / 291.84 -17.11 / 015354.1-091821	NGC 755 / MCG -2- 6- 5 / IRAS01538-0918 / PGC 7262	.SBT3?. / PE (1) / 3.4± .6 / 4.2±1.2	1.53± .02 / .49± .04 / .90± .02 / 1.54	50 / .05 / .68 / .25	13.09 ±.14 / / 12.67 / 12.34	.51± .02 / -.12± .03 / .39 / -.20	.58± .01 / -.10± .02 / 13.08± .06 / 14.37± .20	13.62±.1 / 244± 8 / 226± 6 / 1.03	1642± 6 / / 1654 / 1375
015623.9+371258 / 137.03 -23.87 / 336.48 -2.24 / 015325.8+365820	/ UGC 1404 / / PGC 7263	.SBS3.. / U / 3.0± .9 /	1.06± .06 / .26± .05 / / 1.08	100 / .19 / .36 / .13	/ 15.25 ±.15 / / 14.67			16.02±.3 / 279± 9 / / 1.23	4458± 8 / / 4594 / 4216
0156.4 +0537 / 150.95 -53.67 / 306.58 -13.12 / 0153.7 +0522	NGC 742 / MCG 1- 6- 4 / / PGC 7264	CE.0.*. / R / -6.0±1.0 /	.30? / .00± .04 / / .33	/ .14 / .00 /	* / 15.3 ±.2 / / 15.15				/ 5409± 72 / 5465 / 5131
0156.4 +0404 / 152.09 -55.07 / 305.05 -13.58 / 0153.8 +0350	IC 1750 / UGC 1412 / / PGC 7266	.S..0.. / U / .0±1.0 /	.96± .09 / .65± .06 / / .94	64 / .12 / .49 /	* / 15.37 ±.13 / /				
015627.8+364810 / 137.16 -24.27 / 336.11 -2.41 / 015330.1+363332	NGC 732 / UGC 1406 / MK 1011 / PGC 7270	.L..... / U / -2.0± .8 /	1.15± .07 / .14± .03 / / 1.15	10 / .22 / .00 /	* / 14.49 ±.12 / / 14.18				5890± 36 / 6025 / 5647 /
015636.3-391446 / 258.82 -71.81 / 259.79 -21.67 / 015429.0-392924	/ ESO 297- 34 / / PGC 7279	.LX.0.P / S / -2.0± .8 /	1.14± .05 / .18± .04 / / 1.11	91 / .03 / .00 /	* / 14.55 ±.14 / / 14.44				5606± 15 / 5528 / 5416 /
0156.6 +1742 / 144.18 -42.46 / 318.28 -9.28 / 0153.9 +1728	/ UGC 1417 / / PGC 7281	.S..8*. / U / 8.0±1.5 /	1.07± .07 / 1.09± .06 / / 1.08	153 / .13 / 1.23 / .50	* / / /				11425 / / 11515 / 11152
015638.0+341032 / 138.01 -26.78 / 333.69 -3.41 / 015342.5+335554	NGC 735 / UGC 1411 / / PGC 7282	.S..3.. / U (2) / 3.0± .8 / 1.2± .8	1.26± .04 / .33± .05 / .79± .02 / 1.28	138 / .23 / .45 / .16	14.07 ±.13 / 13.77 ±.12 / / 13.19	.81± .03 / .16± .05 / .67 / .04	.89± .03 / .32± .05 / 13.51± .05 / 14.38± .28	14.44±.2 / 456± 8 / 430± 10 / 1.08	4635± 6 / 4739± 42 / 4767 / 4389
0156.6 +1017 / 148.03 -49.37 / 311.15 -11.73 / 0154.0 +1003	/ UGC 1419 / / PGC 7285	.SXS8.. / U / 8.0± .9 /	1.04± .15 / .18± .12 / / 1.06	5 / .17 / .22 / .09					6157± 10 / / 6227 / 5880
015641.1+330238 / 138.38 -27.86 / 332.65 -3.83 / 015346.5+324800	NGC 736 / UGC 1414 / 6ZW 111 / PGC 7289	.E+..*. / PU / -4.0± .5 /	1.19± .07 / .02± .05 / .83± .03 / 1.20	/ .16 / .00 /	13.16 ±.14 / 13.35 ±.10 / / 13.06	1.01± .01 / .55± .02 / .93 / .54	1.03± .01 / .56± .02 / 12.80± .12 / 14.04± .38		/ 4374± 38 / 4501 / 4123
0156.7 +1500 / 145.49 -44.98 / 315.71 -10.20 / 0154.0 +1446	/ UGC 1420 / IRAS01539+1446 / PGC 7292	.S?.... / / /	.95± .07 / .20± .05 / / .97	3 / .16 / .31 / .10	/ 14.73 ±.12 / 13.49 / 14.24				4607 / 4690 / 4332 /
015644.0+362304 / 137.35 -24.65 / 335.74 -2.61 / 015346.6+360826	/ UGC 1415 / / PGC 7295	.S..0.. / U / .0±1.0 /	1.04± .06 / .59± .05 / / 1.03	1 / .22 / .44 /	/ 14.66 ±.15 / / 13.92				4796± 80 / 4930 / 4553 /
015644.5-435823 / 269.20 -68.63 / 254.64 -21.86 / 015442.1-441300	/ ESO 245- 10 / IRAS01546-4413 / PGC 7298	.S..3P. / S / 3.0± .8 /	1.38± .04 / .59± .05 / / 1.38	25 / .03 / .81 / .29	/ 14.32 ±.14 / 13.34 / 13.44				5713 / 5622 / 5541 /
015646.2+365313 / 137.20 -24.17 / 336.21 -2.43 / 015348.4+363836	/ UGC 1416 / / PGC 7300	.S?.... / / /	1.07± .06 / .28± .05 / / 1.09	65 / .22 / .43 / .14	/ 14.72 ±.12 / / 14.05				5484 / 5619 / 5242 /
015647.7-034415 / 159.49 -61.88 / 297.28 -15.87 / 015416.1-035852	/ MCG -1- 6- 5 / / PGC 7301	PL..+P* / E / -1.0±1.2 /	1.17± .05 / .22± .05 / / 1.14	80 / .02 / .00 /					
015650.2+314214 / 138.85 -29.14 / 331.42 -4.36 / 015356.7+312736	/ MCG 5- 5- 29 / MK 1167 / PGC 7304	.S?.... / / /	.94± .11 / .11± .07 / / .95	/ .11 / .16 / .06	/ 14.78 ±.12 / / 14.48				5236± 97 / 5360 / 4983 /

R.A. 2000 DEC. / l b / SGL SGB / R.A. 1950 DEC.	Names / PGC	Type / S_T n_L / T / L	$\log D_{25}$ / $\log R_{25}$ / $\log A_e$ / $\log D_o$	p.a. / A_g / A_i / A_{21}	B_T / m_B / m_{FIR} / B_T^o	$(B-V)_T$ / $(U-B)_T$ / $(B-V)_T^o$ / $(U-B)_T^o$	$(B-V)_e$ / $(U-B)_e$ / m'_e / m'_{25}	m_{21} / W_{20} / W_{50} / HI	V_{21} / V_{opt} / V_{GSR} / V_{3K}
015653.6-020110 157.68 -60.40 299.02 -15.43 015421.0-021548	IC 176 UGC 1426 PGC 7306	.S..5*. UE (1) 5.0± .8 4.2±1.2	1.26± .04 .67± .04 1.27	94 .03 1.00 .33	 14.70 ±.12 13.65				4446 4479 4172
015654.8+331601 138.36 -27.64 332.87 -3.80 015359.9+330124	NGC 739 MCG 5- 5- 30 ARAK 67 PGC 7312	.L?.... 	.72± .22 .00± .07 .74	 .16 .00 	 14.85 ±.12 14.62				4411± 52 4538 4161
015655.0+330054 138.44 -27.88 332.64 -3.89 015400.4+324617	NGC 740 UGC 1421 PGC 7316	.SB.3$. P 3.0± .9 	1.21± .04 .63± .04 1.22	137 .16 .87 .31	 14.77 ±.12 13.71			16.38±.2 368± 9 350± 7 2.35	4609± 6 4735 4358
015657.3+402030 136.23 -20.83 339.40 -1.18 015356.2+400553	 UGC 1418 IRAS01539+4005 PGC 7320	.L..... U -2.0± .8 	1.17± .09 .16± .05 1.17	50 .21 .00 	 14.18 ±.11 13.58 				
0156.9 -0038 156.34 -59.19 300.41 -15.07 0154.4 -0053	 CGCG 387- 6 PGC 7321			 .03 	 15.10 ±.12 				5786 5823 5512
015657.7-052411 161.53 -63.25 295.61 -16.34 015427.0-053848	NGC 762 MCG -1- 6- 6 MK 1012 PGC 7322	PSBT1.. E 1.0± .8 	1.13± .06 .07± .05 1.13	55 .02 .07 .04					4820± 61 4843 4550
015659.2-114653 171.18 -68.27 289.07 -17.88 015432.2-120130	 MCG -2- 6- 6 PGC 7324	.SBS8*. E (1) 8.0±1.2 6.4±1.2	1.43± .03 .69± .05 1.43	87 .00 .84 .34				14.55±.1 170± 11 138± 8 	1853± 8 1856 1591
0157.0 +0602 150.90 -53.23 307.04 -13.14 0154.4 +0548	 UGC 1427 PGC 7325	.L..... U -2.0± .9 	1.01± .08 .20± .03 .99	65 .10 .00 	 14.47 ±.10 				
0157.0 -0005 155.87 -58.70 300.97 -14.94 0154.5 -0020	IC 177 CGCG 387- 7 PGC 7326			 .01 	 15.3 ±.3 				13543 13582 13268
0157.0 -0028 156.23 -59.03 300.59 -15.06 0154.5 -0042	IC 1756 UGC 1429 IRAS01545-0042 PGC 7328	.S..6*. U 6.0±1.3 	1.13± .05 .75± .05 1.13	155 .03 1.11 .38	 15.30 ±.12 13.14 14.13				6648 6686 6374
015706.8+324720 138.56 -28.08 332.45 -4.01 015412.3+323243	 UGC 1422 PGC 7333	.S?.... 	1.07± .07 .40± .06 1.08	90 .16 .59 .20	 14.35 ±.13 13.57			15.94±.2 351± 9 331± 7 2.17	4583± 6 4799± 52 4712 4335
0157.2 +1433 145.88 -45.36 315.31 -10.47 0154.5 +1419	IC 1755 UGC 1428 PGC 7341	.S..1.. U 1.0± .9 	1.14± .05 .65± .05 1.16	154 .15 .67 .33	 14.70 ±.12 13.78				7930 8011 7655
015719.4+283524 140.02 -32.09 328.57 -5.59 015428.1+282047	IC 1753 MCG 5- 5- 33 5ZW 149 PGC 7353	.E?.... 	.75± .10 .05± .06 .76	 .21 .00 	 14.95 ±.12 14.59				10188± 42 10305 9929
015721.8+053636 151.33 -53.59 306.64 -13.35 015444.9+052200	 UGC 1435 KUG 0154+053 PGC 7355	.S?.... 	1.01± .05 .25± .04 1.02	40 .14 .38 .13	 14.77 ±.12 14.22				5677 5733 5400
015725.8+171303 144.65 -42.85 317.87 -9.62 015442.1+165827	 UGC 1432 ARP 56 PGC 7359	.S..4.. U 4.0± .9 	.97± .05 .17± .04 .98	65 .14 .25 .08	 14.61 ±.12 14.17			16.31±.3 216± 13 2.06	8054± 10 8142 7781
0157.4 +1955 143.43 -40.30 320.45 -8.70 0154.7 +1941	 UGC 1433 PGC 7362	.S..6*. U 6.0±1.2 	.82± .12 .00± .06 .84	 .19 .00 .00	 15.36 ±.13 15.14				8692 8787 8421
015730.5-092745 167.48 -66.44 291.49 -17.47 015502.2-094221	NGC 747 MCG -2- 6- 7 PGC 7366	.S..3*. E (1) 3.0±1.3 3.1± .8	1.00± .07 .35± .05 1.01	175 .04 .48 .17					

R.A. 2000 DEC. / l b / SGL SGB / R.A. 1950 DEC.	Names / PGC	Type / S_T n_L / T / L	$logD_{25}$ / $logR_{25}$ / $logA_e$ / $logD_o$	p.a. / A_g / A_i / A_{21}	B_T / m_B / m_{FIR} / B_T^o	$(B-V)_T$ / $(U-B)_T$ / $(B-V)_T^o$ / $(U-B)_T^o$	$(B-V)_e$ / $(U-B)_e$ / m'_e / m'_{25}	m_{21} / W_{20} / W_{50} / HI	V_{21} / V_{opt} / V_{GSR} / V_{3K}
0157.5 -0205		.S..8..	1.05± .08						4892
158.04 -60.38	UGC 1442	U	.15± .06	.01	15.17 ±.17				4925
298.99 -15.61		8.0± .8		.18					
0155.0 -0220	PGC 7368		1.05	.08	14.96				4619
015732.8+331224	NGC 750	.E...P.	1.22± .13						
138.52 -27.66	UGC 1430	R	.10± .11	.16	12.84 ±.18				5222± 22
332.87 -3.94		-5.0± .3		.00					5348
015437.9+325748	PGC 7369		1.22		12.60				4972
015732.9+331216	NGC 751	.E...P.	1.15± .14						
138.52 -27.66	UGC 1431	R	.00± .12	.16					5163± 34
332.86 -3.94		-5.0± .4		.00					5290
015438.1+325740	PGC 7370		1.17						4913
0157.5 +1953		.S..6*.	.85± .11						8830
143.48 -40.33	UGC 1436	U	.09± .06	.19					
320.42 -8.73		6.0±1.2		.14					8925
0154.8 +1939	PGC 7371		.87	.05					8560
015736.2+164616		.S?....	.92± .07						8040
144.91 -43.26	MCG 3- 6- 8		.18± .06	.13	14.61 ±.13				
317.46 -9.81				.27					8127
015452.7+163140	PGC 7377		.93	.09	14.16				7767
015737.8-574725	NGC 782	.SBR3..	1.37± .04	15	12.48 ±.15	.63± .01	.71± .01		
286.66 -57.20	ESO 114- 15	RS (2)	.06± .04	.00	12.60 ±.12	.06± .04	.15± .02		6000± 34
239.64 -21.56	IRAS01559-5801	3.0± .4	1.08± .02	.08	12.14	.58	13.37± .05		5875
015556.0-580200	PGC 7379	2.8± .4	1.37	.03	12.43	.03	14.05± .26		5885
015738.7+361517		.S...0..	1.10± .07	164					
137.58 -24.73	UGC 1434	U	.52± .06	.28	15.21 ±.12				
335.69 -2.83	IRAS01546+3600	.0± .9		.39					
015441.2+360041	PGC 7381		1.10						
015741.4-054033		.SAS6..	1.21± .05	135					5037
162.21 -63.38	MCG -1- 6- 7	E (1)	.22± .05	.04					
295.38 -16.59		6.0± .8		.33					5059
015510.9-055508	PGC 7385	4.2±1.2	1.21	.11					4768
015742.5+355458	NGC 753	.SXT4..	1.40± .02	125	12.97 ±.13	.66± .01	.76± .01	14.14±.1	4886± 6
137.70 -25.05	UGC 1437	R (2)	.11± .02	.23	12.54 ±.10	.00± .03	.13± .02	340± 7	4879± 12
335.38 -2.97	IRAS01547+3540	4.0± .3	.93± .01	.16	11.45	.55	13.11± .04	314± 11	5017
015445.4+354022	PGC 7387	1.6± .7	1.42	.06	12.28	-.08	14.52± .17	1.80	4641
015748.4-331425	IC 1762	.SBS4..	1.26± .04	43					
240.06 -74.43	ESO 354- 17	BS (1)	.62± .04	.02	14.36 ±.14				5663± 39
266.32 -21.48	IRAS01555-3329	4.0± .7		.91					5601
015536.1-332900	PGC 7393	2.2±1.3	1.26	.31	13.39				5454
015748.9+275155	A 0154+27								
140.40 -32.74	CGCG 503- 63			.28	15.18 ±.12				8100±110
327.93 -5.95	MK 364								8215
015458.0+273720	PGC 7394								7841
015749.7+332236	NGC 761	.SB.1*.	1.18± .04	143				17.94±.3	5029± 10
138.53 -27.48	UGC 1439	PU	.48± .04	.16	14.40 ±.12				
333.05 -3.93	VV 425	1.0± .6		.48				397± 7	5156
015454.6+330800	PGC 7395		1.20	.24	13.70			4.01	4780
015750.2+372145		.S..3..	1.04± .06	134					
137.28 -23.65	UGC 1441	U (1)	.74± .05	.23	15.50 ±.13				
336.72 -2.45		3.0±1.0		1.02					
015451.7+370709	PGC 7396	4.5±1.3	1.06	.37					
015750.4+362032	NGC 759	.E.....	1.20± .08		13.84 ±.15	1.10± .02	1.05± .01		
137.60 -24.63	UGC 1440	PU	.04± .06	.22	13.40 ±.10	.55± .03	.53± .03		4879± 44
335.78 -2.83	IRAS01548+3605	-5.0± .5	.63± .04	.00	13.16	1.00	12.48± .15		5012
015452.8+360557	PGC 7397		1.22		13.24	.52	14.73± .45		4636
015751.1-611244		PSBR3..	1.23± .05	67					
289.35 -54.13	ESO 114- 16	Sr (1)	.14± .05	.03	13.89 ±.14				6999± 60
235.96 -21.27		2.6± .6		.19					6867
015617.1-612718	PGC 7398	3.3± .8	1.23	.07	13.61				6899
015751.1+445502	NGC 746	.I...9..	1.28± .04	90				14.15±.1	712± 11
135.11 -16.38	UGC 1438	U	.16± .05	.58	13.5 ±.3			120± 16	615± 52
343.66 .38	IRAS01548+4441	10.0± .8		.12	13.44			105± 12	857
015445.1+444026	PGC 7399		1.33	.08	12.85			1.22	488
015755.5-325913	IC 1759	.SAT4*.	1.19± .05						
239.13 -74.49	ESO 354- 18	Sr	.05± .05	.02	13.82 ±.14				3853± 34
266.60 -21.49		3.9± .7		.07					3792
015543.0-331348	PGC 7400		1.19	.02	13.70				3643

R.A. 2000 DEC. / l b / SGL SGB / R.A. 1950 DEC.	Names / PGC	Type S_T n_L / T / L	$\log D_{25}$ / $\log R_{25}$ / $\log A_e$ / $\log D_o$	p.a. / A_g / A_i / A_{21}	B_T / m_B / m_{FIR} / B_T^o	$(B-V)_T$ / $(U-B)_T$ / $(B-V)_T^o$ / $(U-B)_T^o$	$(B-V)_e$ / $(U-B)_e$ / m'_e / m'_{25}	m_{21} / W_{20} / W_{50} / HI	V_{21} / V_{opt} / V_{GSR} / V_{3K}
015758.2-380113		.S..5*.	1.00± .05	82					
255.07 -72.30	ESO 297- 36	S (1)	.23± .04	.03	15.04 ±.14				
261.14 -21.87		5.0±1.3		.34					
015550.1-381548	PGC 7402	3.3±1.2	1.00	.11					
0158.0 +0423		.S?....	1.29± .04	140					4764
152.48 -54.61	UGC 1444		.65± .05	.12	15.12 ±.16				
305.49 -13.87				.98					4816
0155.4 +0409	PGC 7406		1.30	.33	14.00				4488
015803.0+032210		.I..9*.	1.12± .05	51					4837
153.29 -55.53	UGC 1446	U	.57± .06	.07	14.75 ±.12				
304.48 -14.19	KUG 0155+031	10.0±1.3		.43					4886
015527.3+030735	PGC 7411		1.13	.28	14.25				4561
015805.6+373441									4686± 62
137.27 -23.43	CGCG 522- 91			.14	15.3 ±.4				
336.94 -2.42	MK 1170								4822
015506.9+372006	PGC 7416								4447
015806.6+030510	A 0155+02	.SB.9P*	1.08± .05					15.13±.1	5555± 7
153.54 -55.77	UGC 1449	RE	.23± .05	.08	13.91 ±.13			217± 7	5416± 23
304.20 -14.28	MK 582	9.1± .4		.24	11.43			111± 10	5590
015531.1+025035	PGC 7417		1.09	.12	13.59			1.42	5267
015808.6+020342		.S..6*.	1.12± .04	110				15.88±.3	6292± 10
154.40 -56.68	UGC 1448	U	.60± .04	.04	14.91 ±.12			370± 13	
303.19 -14.59		6.0±1.3		.88					6337
015533.7+014907	PGC 7420		1.13	.30	13.97			1.62	6017
0158.1 +1906		.L.....	1.05± .08	90					
144.00 -41.02	UGC 1445	U	.18± .03	.17	14.70 ±.11				
319.73 -9.14		-2.0± .8		.00					
0155.4 +1852	PGC 7421		1.04						
015812.7-393244	ESO 297- 37	.SBT4*/	1.25± .05	64					
258.99 -71.39		Sr	.70± .04	.03	14.56 ±.14				5498
259.48 -21.99	IRAS01561-3947	4.4± .6		1.03					5418
015606.0-394718	PGC 7427		1.25	.35	13.47				5310
015818.2-541302		.L..-*/	1.20± .05	22					
283.11 -60.26	ESO 153- 3	S	.30± .04	.00	13.79 ±.14				6501
243.50 -21.91		-3.5± .8		.00					6384
015630.1-542736	PGC 7430		1.11		13.69				6371
0158.4 +2208		.S?....	1.11± .14	128					4977
142.78 -38.13	UGC 1452		.66± .12	.27					
322.62 -8.15				.97					5077
0155.7 +2154	PGC 7441		1.14	.33					4710
015830.0+252134		.SB?...	1.09± .06	85				15.86±.2	4927± 6
141.51 -35.08	UGC 1451		.32± .05	.40	14.27 ±.12			356± 10	4911± 47
325.65 -7.00	IRAS01556+2507			.44	11.04			298± 7	5036
015540.8+250700	PGC 7445		1.13	.16	13.39			2.30	4664
015830.5-561457		.L..-P.	1.13± .05						
285.07 -58.49	ESO 153- 4	S	.37± .04	.03	14.65 ±.14				5936± 44
241.30 -21.80	IRAS01568-5628	-3.0± .9		.00	13.94				5814
015646.0-562930	PGC 7447		1.08		14.53				5815
015832.5-261732	NGC 775	.SAT5*.	1.22± .04	167					
213.55 -75.00	ESO 477- 18	Sr (1)	.13± .03	.00	13.40 ±.14				4553± 52
273.84 -20.86	IRAS01562-2632	5.4± .6		.19	12.63				4511
015615.0-263206	PGC 7451	2.2± .8	1.22	.06	13.18				4324
015834.5+443432		.SX.6..	1.04± .08						
135.33 -16.67	UGC 1447	U	.16± .06	.54	14.84 ±.13				
343.40 .13		6.0± .8		.23					
015528.7+441958	PGC 7456		1.09	.08					
0158.6 +0315		.SB.9..	1.18± .06						3489
153.59 -55.56	UGC 1454	U	.07± .06	.08					
304.41 -14.35		9.0± .8		.07					3537
0156.0 +0301	PGC 7458		1.19	.03					3214
015836.0-012724			.69± .15	127					
157.84 -59.70	MCG 0- 6- 15		.33± .07	.03	15.1 ±.2				4809± 57
299.71 -15.69	ARAK 69								4843
015603.1-014158	PGC 7460		.69						4537
015840.5+003146	NGC 768	.SBR4*.	1.22± .04	30					
155.97 -57.96	UGC 1457	UE (2)	.32± .04	.01	14.01 ±.12				6973± 40
301.70 -15.16	IRAS01561+0017	4.0± .6		.47	13.52				7013
015606.5+001713	PGC 7465	1.5± .7	1.22	.16	13.49				6699

R.A. 2000 DEC.	Names	Type	logD$_{25}$	p.a.	B$_T$	(B-V)$_T$	(B-V)$_\bullet$	m$_{21}$	V$_{21}$
l b	S$_T$ n$_L$	logR$_{25}$	A$_g$	m$_B$	(U-B)$_T$	(U-B)$_\bullet$	W$_{20}$	V$_{opt}$
SGL SGB	T	logA$_e$	A$_I$	m$_{FIR}$	(B-V)$_T^o$	m'$_\bullet$	W$_{50}$	V$_{GSR}$
R.A. 1950 DEC.	PGC	L	logD$_o$	A$_{21}$	B$_T^o$	(U-B)$_T^o$	m'$_{25}$	HI	V$_{3K}$

015842.1+082054	NGC 766	.E.....	1.30± .12		.20 13.68 ±.16				8104± 50
149.96 -50.96	UGC 1458	U	.00± .08		.00				8167
309.43 -12.82		-5.0± .7			13.36				7828
015603.6+080620	PGC 7468		1.33						
0158.7 +2439		.I..9*.	1.19± .12						4897
141.83 -35.73	UGC 1453	U	.08± .12	.40					
325.01 -7.30		10.0±1.1		.06					5004
0155.9 +2425	PGC 7470		1.23	.04					4633
015846.7-782353		.SBS4P.	1.20± .04						8349
298.68 -38.17	ESO 13- 26	S	.05± .05	.30 14.39 ±.14					8188
217.88 -18.57	IRAS01589-7838	3.5± .5		.07 12.75					8330
015856.0-783824	PGC 7471		1.23	.02 13.97					
015847.6+245329	NGC 765	.SXT4..	1.44± .03		.40 13.6 ±.3				5117± 50
141.76 -35.51	UGC 1455	U	.00± .05		.00				5224
325.23 -7.23	IRAS01559+2439	4.0± .7		.00					4854
015558.7+243856	PGC 7475		1.48		.00 13.12				
0158.8 +0535		.S..6*.	1.06± .06	172					
151.89 -53.45	UGC 1461	U	.79± .05	.12					
306.73 -13.70		6.0±1.4		1.16					
0156.2 +0521	PGC 7478		1.07	.39					
015850.9-093513	NGC 767	.SB.3P?	1.06± .05	165					5390
168.31 -66.32	MCG -2- 6- 10	E	.47± .05	.03					
291.45 -17.82	IRAS01563-0949	3.0±1.9		.64					5399
015622.7-094946	PGC 7483		1.06	.23					5127
0158.8 +0033	IC 1761	.L?....	.95± .10		.00 15.16 ±.07				6940± 67
156.02 -57.91	CGCG 387- 19		.24± .04		.00				6980
301.75 -15.19					15.05				6667
0156.3 +0019	PGC 7484		.92						
015852.2-080958	A 0156-08	.SBR5P.	1.08± .04	130	13.6 ±.4	.52± .04	.60± .04	15.74±.1	4754± 10
166.15 -65.21	MCG -1- 6- 12	PE (1)	.09± .05	.02				208± 11	4814± 59
292.91 -17.49	KUG 0156-084	4.6± .6	1.04± .11	.13		.47	14.32± .25	201± 11	4769
015623.3-082431	PGC 7485	2.2± .9	1.08	.04 13.42			13.61± .47	2.28	4491
015852.3-113054	NGC 773	.SXR1P.	1.13± .06	0					
171.60 -67.76	MCG -2- 6- 11	E	.26± .05	.00					5606
289.46 -18.27		1.0± .9		.26					5609
015625.3-114527	PGC 7486		1.13	.13					5346
015855.0+364032	IC 178	.S..2..	1.10± .05	170	14.06 ±.13	.78± .02	.87± .02	17.05±.3	4845± 8
137.72 -24.25	UGC 1456	U	.13± .05	.22 13.92 ±.12		.24± .03	.22± .03	288± 9	4731± 52
336.17 -2.91	IRAS01559+3625	2.0± .8	.65± .02	.16 13.10		.67	12.80± .07		4975
015557.0+362559	PGC 7488		1.12	.06 13.56		.17	14.07± .32	3.42	4601
015900.0+864031		.S..4..	1.10± .04	170				16.10±.1	4655± 11
123.99 23.94	UGC 1285	U (1)	.11± .04	.44 14.35 ±.18					4629±125
22.89 14.77	IRAS01477+8625	4.0± .8		.16 12.83				183± 8	4829
014740.0+862549	PGC 7491	2.5±1.1	1.14	.05 13.72				2.33	4602
015904.9-562616			1.01± .06	170					7853
285.12 -58.29	ESO 153- 5		.34± .06	.00 15.48 ±.14					7731
241.09 -21.87									7733
015721.0-564048	PGC 7501		1.01						
015905.1+361531		.S..1..	1.18± .05	145					
137.89 -24.64	UGC 1460	U	.24± .05	.22 14.64 ±.15					
335.80 -3.10	IRAS01561+3600	1.0± .8		.25 13.36					
015607.4+360058	PGC 7502		1.20	.12					
015906.7+360347		.S..6*.	1.17± .05	109					
137.95 -24.83	UGC 1459	U	.85± .05	.23 15.33 ±.12					
335.62 -3.18		6.0±1.3		1.25					
015609.2+354915	PGC 7504		1.19	.43					
015907.1-675211	NGC 802	.LXS+P*	.93± .06	152	13.67V±.13	.47± .02			
293.50 -48.00	ESO 52- 13	S	.18± .03	.00 14.17 ±.14	-.14± .05			1505± 24
228.85 -20.54		-.8± .6		.00		.43			1360
015755.0-680642	PGC 7505		.90		14.13		-.15	13.21± .32	1436
0159.1 +2523		.SB.6*.	1.20± .05	65					5059
141.66 -35.01	UGC 1462	U	.16± .05	.38 15.0 ±.2					
325.73 -7.12		6.0±1.1		.23					5167
0156.3 +2509	PGC 7506		1.24	.08 14.37					4797
015907.9+015316		.IXS9*.	1.14± .07	15					2951
154.95 -56.72	UGC 1464	UE (1)	.16± .07	.05					
303.09 -14.88		10.0± .6		.12					2995
015633.2+013843	PGC 7508	9.8± .8	1.15	.08					2677

R.A. 2000 DEC. l b SGL SGB R.A. 1950 DEC.	Names PGC	Type S_T n_L T L	$\log D_{25}$ $\log R_{25}$ $\log A_e$ $\log D_o$	p.a. A_g A_i A_{21}	B_T m_B m_{FIR} B_T^o	$(B-V)_T$ $(U-B)_T$ $(B-V)_T^o$ $(U-B)_T^o$	$(B-V)_e$ $(U-B)_e$ m'_e m'_{25}	m_{21} W_{20} W_{50} HI	V_{21} V_{opt} V_{GSR} V_{3K}
0159.1 +0006 156.56 -58.27 301.32 -15.39 0156.6 -0008	 CGCG 387- 20 PGC 7511			 .01 	 14.9 ±.3 				5807 5845 5534
015913.2+185718 144.38 -41.08 319.67 -9.43 015628.3+184246	NGC 770 UGC 1463 PGC 7517	.E.3.*. R -5.0± .6 L	1.07± .06 .14± .04 1.05	15 .17 .00 	* 13.91 ±.10 13.70		1.15± .05	14.67±.2 465± 5 	2440± 7 2493± 34 2534 2172
0159.2 +1801 144.82 -41.95 318.79 -9.75 0156.5 +1747	 UGC 1465 PGC 7519	.SB?... 	.96± .09 .98± .06 .98	8 .16 1.47 .49				17.73±.3 176± 14 	2012± 10 2101 1741
015915.9+242515 142.07 -35.92 324.83 -7.50 015627.3+241043	 CGCG 482- 35 PGC 7521			 .40 	 14.8 ±.4 			17.23±.3 143± 7 	3907± 10 4013 3644
015917.3-504310 278.82 -63.11 247.29 -22.23 015724.0-505742	 ESO 197- 16 PGC 7523	.L?.... 	.97± .06 .47± .03 .90	48 .01 .00 	 14.48 ±.14 14.38				6202±108 6093 6058
015920.3+190022 144.39 -41.02 319.73 -9.44 015635.3+184550	NGC 772 UGC 1466 IRAS01565+1845 PGC 7525	.SAS3.. R (3) 3.0± .3 1.2± .5	1.86± .01 .23± .02 1.41± .02 1.88	130 .17 .32 .12	11.09 ±.13 10.99 ±.11 10.90 10.53	.78± .01 .26± .02 .68 .18	.86± .01 .29± .02 13.63± .06 14.66± .16	12.52±.0 471± 3 436± 4 1.87	2458± 3 2454± 29 2550 2188
015921.8+364934 137.77 -24.08 336.34 -2.94 015623.5+363502	 MCG 6- 5- 73 PGC 7527	.E?.... 		 .21 	 15.0 ±.4 				4817± 80 4950 4577
015926.2-674711 293.41 -48.07 228.93 -20.59 015814.0-680142	 ESO 52- 14 IRAS01581-6802 PGC 7530	.SXS0?. S .0± .9 	1.04± .05 .23± .04 1.03	47 .00 .17 	 14.80 ±.14 13.03 				5993 5871 5873
0159.5 +1356 146.94 -45.73 314.91 -11.20 0156.8 +1342	 UGC 1468 PGC 7533	.S..7.. U 7.0± .8 	1.19± .05 .23± .05 1.20	145 .13 .32 .12					4623± 10 4701 4350
015933.9-562111 284.93 -58.33 241.17 -21.94 015750.0-563542	 ESO 153- 7 PGC 7535	.SXS6*. S (1) 6.0± .6 5.6± .6	1.09± .05 .14± .05 1.09	 .03 .20 .07	 15.26 ±.14 15.01				5993 5871 5873
015935.0+140030 146.93 -45.66 314.98 -11.19 015653.1+134558	NGC 774 UGC 1469 PGC 7536	.L..... U -2.0± .8 	1.18± .09 .11± .05 1.18	165 .13 .00 	 13.97 ±.11 13.77				4595± 50 4673 4322
015936.2+305435 139.76 -29.72 330.90 -5.20 015642.9+304003	NGC 769 UGC 1467 ARAK 70 PGC 7537	.S?.... 	.92± .07 .25± .05 .94	73 .18 .37 .12	 13.8 ±.2 12.27 13.18				4472± 35 4593 4220
0159.6 -0521 162.67 -62.84 295.83 -16.97 0157.1 -0536	IC 183 MCG -1- 6- 15 PGC 7538	.L?.... 	1.12± .12 .47± .07 1.05	 .01 .00 					4016 4037 3748
015938.3+272558 141.02 -33.03 327.67 -6.49 015647.5+271127	 MCG 4- 5- 27 PGC 7540	.S?.... 	.94± .11 .40± .07 .94	 .22 .30 	 15.19 ±.12 14.59			15.95±.3 212± 7 	5268± 10 5381 5009
015938.9-763552 297.85 -39.85 219.72 -18.99 015926.1-765021	 PGC 7542	PS..3?. S (1) 3.0±1.7 5.6±1.2	1.13± .08 .10± .08 1.16	 .25 .14 .05					
015942.6-055753 163.49 -63.33 295.22 -17.15 015712.4-061224	NGC 779 MCG -1- 6- 16 IRAS01571-0612 PGC 7544	.SXR3.. R (2) 3.0± .3 3.0± .6	1.60± .02 .53± .03 .94± .01 1.60	20 .00 .73 .27	11.95 ±.13 11.95 ±.17 12.16 11.20	.79± .01 .21± .02 .68 .11	.87± .01 .31± .01 12.14± .03 13.47± .17	14.10±.1 371± 4 364± 5 2.63	1391± 4 1423± 50 1411 1125
015942.8+320504 139.39 -28.60 332.00 -4.78 015648.5+315033	 UGC 1470 PGC 7545	.S..8*. U 8.0±1.3 	1.19± .06 .91± .06 1.21	138 .17 1.12 .46				16.05±.2 259± 9 236± 7 	5166± 6 5289 4915

R.A. 2000 DEC. l b SGL SGB R.A. 1950 DEC.	Names PGC	Type S_T n_L T L	$\log D_{25}$ $\log R_{25}$ $\log A_o$ $\log D_o$	p.a. A_g A_i A_{21}	B_T m_B m_{FIR} B_T^o	$(B-V)_T$ $(U-B)_T$ $(B-V)_T^o$ $(U-B)_T^o$	$(B-V)_e$ $(U-B)_e$ m'_e m'_{25}	m_{21} W_{20} W_{50} HI	V_{21} V_{opt} V_{GSR} V_{3K}
015949.8-554930	NGC 795	.L..-*.	1.08± .05	141					
284.36 -58.77	ESO 153- 8	S	.23± .04	.03	14.22 ±.14				
241.74 -22.02		-4.0± .8		.00					
015805.1-560400	PGC 7552		1.02						
015949.9-070339		.LA.0P.	1.20± .05	30	15.53S±.15				
165.01 -64.19	MCG -1- 6- 20	E	.41± .05	.04					5929± 41
294.11 -17.45	HICK 14A	-2.0± .8		.00					5945
015720.3-071810	PGC 7553		1.14		15.40		15.38± .33		5664
015951.2-065028	IC 184	.SBR1*.	1.02± .07	7	14.66 ±.13	.82± .03			
164.72 -64.01	MCG -1- 6- 21	E	.31± .05	.08		.22± .04			
294.33 -17.40		1.0± .9		.31					
015721.5-070459	PGC 7554		1.02	.15			13.82± .38		
015951.8+072442	IC 182	.SB.3..	.90± .06						4714
151.00 -51.69	UGC 1473	U	.14± .05	.17	14.53 ±.13				
308.60 -13.39	IRAS01572+0710	3.0± .8		.19					4773
015713.8+071011	PGC 7556		.91	.07	14.13				4440
015952.2-070518		.SA.3P?	1.05± .06	6	14.86S±.15				
165.07 -64.21	MCG -1- 6- 22	E	.46± .05	.04					
294.08 -17.47	IRAS01573-0719	3.0±1.3		.63	13.26				
015722.6-071949	PGC 7557		1.06	.23			13.83± .38		
015954.7+233837	NGC 776	.SXT3..	1.24± .05					15.34±.2	4921± 6
142.56 -36.60	UGC 1471	U	.01± .05	.33	13.22 ±.12			164± 6	
324.15 -7.92	IRAS01570+2323	3.0± .8		.01	12.64			124± 6	5025
015706.6+232406	PGC 7560		1.27	.00	12.84			2.50	4657
015955.6+241841			1.00?					16.81±.3	5089± 10
142.29 -35.97	CGCG 482- 38		.70?	.37	15.60 ±.12				
324.78 -7.68								293± 7	5194
015707.0+240410	PGC 7562		1.03						4826
015959.5-110445	IC 1767	PSAR0*.	1.24± .05	75					5247
171.35 -67.26	MCG -2- 6- 12	E	.42± .05	.00					
289.97 -18.44		.0± .9		.31					5251
015732.3-111916	PGC 7568		1.22						4987
0200.0 +0657		.SX.8..	1.04± .08						6032± 10
151.37 -52.08	UGC 1477	U	.08± .06	.14	15.14 ±.18				
308.17 -13.57		8.0± .9		.10					6090
0157.4 +0643	PGC 7570		1.05	.04	14.89				5758
0200.0 +3420		.I..9*.	1.07± .14					15.90±.3	4849± 8
138.71 -26.42	UGC 1472	U	.03± .12	.21				152± 9	
334.11 -4.00		10.0±1.2		.02					4977
0157.1 +3406	PGC 7571		1.09	.01					4604
0200.0 -0016		.S..2..	1.04± .05	147					
157.29 -58.49	UGC 1481	U	.59± .04	.01	15.39 ±.08				
301.00 -15.71		2.0± .9		.72					
0157.5 -0031	PGC 7574		1.04	.29					
020009.3+123918	NGC 781	.S?....	1.18± .04	13					
147.86 -46.86	UGC 1482		.63± .04	.20	14.03 ±.15				3483± 50
313.72 -11.77				.94					3557
015728.2+122448	PGC 7577		1.20	.31	12.87				3210
020011.4+373613		.SBS8..	1.12± .05	15				15.93±.3	4044± 8
137.70 -23.29	UGC 1474	U	.07± .05	.17	14.60 ±.16			180± 9	
337.12 -2.79		8.0± .8		.08					4179
015712.2+372143	PGC 7579		1.13	.03	14.34			1.55	3806
020011.6+380115	IC 179	.E.....	1.26± .13	110	13.56 ±.18	.99± .03	1.04± .01		
137.57 -22.89	UGC 1475	U	.08± .06	.18	13.12 ±.10	.55± .05	.60± .02		4188± 25
337.50 -2.64		-5.0± .8	.74± .06	.00		.91	12.75± .23		4324
015712.1+374644	PGC 7581		1.27		12.98	.53	14.66± .71		3952
020014.9-835917		.SXT5*P	1.21± .04						
300.82 -32.86	ESO 3- 7	S (1)	.04± .04	.47	13.44 ±.14				3413± 34
212.18 -17.31	IRAS02031-8413	4.5± .5		.06	11.53				3246
020309.0-841342	PGC 7583	3.3±1.2	1.26	.02	12.89				3421
020015.0+312546	NGC 777	.E.1...	1.39± .07	155	12.49 ±.14	1.04± .02	1.05± .01		
139.73 -29.19	UGC 1476	R	.08± .06	.16	12.46 ±.09	.60± .03	.63± .02		4985± 15
331.43 -5.13		-5.0± .3	1.06± .03	.00		.96	13.28± .11		5107
015721.1+311116	PGC 7584		1.39		12.24	.59	14.20± .41		4734
0200.2 +2415		.SBT5..	.95± .07						4846
142.41 -36.00	UGC 1478	U	.05± .05	.37	14.50 ±.12				
324.76 -7.78	IRAS01574+2400	5.0± .9		.07	13.45				4951
0157.4 +2400	PGC 7588		.99	.02	14.04				4583

R.A. 2000 DEC. / l b / SGL SGB / R.A. 1950 DEC.	Names / PGC	Type / ST nL / T / L	logD25 / logR25 / logAe / logDo	p.a. / Ag / Ai / A21	BT / mB / mFIR / BoT	(B-V)T / (U-B)T / (B-V)oT / (U-B)oT	(B-V)e / (U-B)e / m'e / m'25	m21 / W20 / W50 / HI	V21 / Vopt / VGSR / V3K
0200.3 +2253 / 142.97 -37.29 / 323.48 -8.27 / 0157.5 +2239	UGC 1483 / PGC 7590	.I..9.. / U / 10.0± .8 /	1.07± .14 / .00± .12 / / 1.10	/ .36 / .00 / .00					5112 / / 5213 / 4848
020018.0-341900 / 243.09 -73.58 / 265.20 -22.09 / 015807.0-343330	ESO 354- 25 / PGC 7591	.L...P? / S / -2.0±1.2 /	1.14± .05 / .28± .04 / / 1.10	103 / .02 / .00 /	13.86 ±.14 / / / 13.76				5034± 34 / 4968 / 4830
0200.3 +2428 / 142.33 -35.79 / 324.96 -7.71 / 0157.5 +2413	UGC 1479 / IRAS01574+2413 / PGC 7594	.S?....	.99± .06 / .57± .05 / / 1.02	176 / .37 / .86 / .29	14.94 ±.15 / 12.57 / / 13.68				4927 / / 5032 / 4665
0200.3 +1558 / 146.14 -43.77 / 316.93 -10.70 / 0157.6 +1544	UGC 1484 / PGC 7596	.S..8*. / U / 8.0±1.4	.96± .09 / .80± .06 / / .97	107 / .14 / .99 / .40					5077 / / 5160 / 4806
020019.5+311847 / 139.79 -29.29 / 331.33 -5.19 / 015725.6+310417	NGC 778 / UGC 1480 / PGC 7597	.L...*. / U / -2.0±1.3	1.03± .08 / .31± .03 / / 1.00	150 / .16 / .00	14.22 ±.11				
0200.3 +2106 / 143.75 -38.96 / 321.80 -8.92 / 0157.6 +2051	UGC 1485 / VV 751 / PGC 7602	.S?....	1.04± .06 / .22± .05 / / 1.06	25 / .26 / .27 / .11	14.41 ±.12 / 13.34 / / 13.83				4851 / / 4948 / 4584
020023.6+243455 / 142.31 -35.68 / 325.07 -7.68 / 015734.7+242025	IC 1764 / UGC 1486 / IRAS01575+2420 / PGC 7603	.SBR3.. / U / 3.0± .8	1.16± .05 / .02± .05 / / 1.20	/ .37 / .02 / .01	14.14 ±.15 / 13.59 / / 13.72				5029± 50 / 5134 / 4767
020027.3+250500 / 142.13 -35.20 / 325.55 -7.52 / 015738.1+245030	UGC 1487 / PGC 7606	.L...*. / U / -2.0±1.2	1.06± .07 / .09± .03 / / 1.09	60 / .37 / .00	15.17 ±.15 / / / 14.66			18.40±.3 / / 436± 7 /	9255± 10 / 9362 / 8994
020029.1-341519 / 242.83 -73.57 / 265.27 -22.12 / 015818.0-342948	ESO 354- 26 / PGC 7609	.E.4... / S / -5.0± .9	1.07± .06 / .10± .04 / / 1.04	/ .02 / .00	14.01 ±.14 / / / 13.91				4872± 39 / 4805 / 4668
020029.7+295355 / 140.33 -30.63 / 330.04 -5.75 / 015737.0+293926	CGCG 503- 70 / PGC 7611			.16	15.5 ±.4			17.51±.3 / / 247± 7 /	11516± 10 / 11634 / 11262
020032.3+211714 / 143.72 -38.78 / 321.99 -8.90 / 015745.6+210245	UGC 1490 / IRAS01577+2102 / PGC 7613	.S?....	.66± .13 / .35± .06 / / .68	85 / .26 / .53 / .18	14.6 ±.3 / 13.68 / / 13.78				3051± 50 / 3148 / 2785
020032.9+322802 / 139.45 -28.18 / 332.42 -4.80 / 015738.1+321333	MCG 5- 5- 40 / PGC 7614	.S?....	.95± .07 / .18± .06 / / .97	/ .18 / .26 / .09	15.41 ±.14 / / / 14.88			16.41±.3 / / 329± 7 / 1.43	12921± 10 / 13045 / 12672
020036.2-250307 / 209.20 -74.35 / 275.24 -21.15 / 015818.0-251736	ESO 477- 20 / PGC 7618	.LA.0.. / S / -2.0± .8	1.11± .04 / .46± .03 / / 1.04	72 / .00 / .00	14.71 ±.14				
0200.6 +2943 / 140.44 -30.78 / 329.89 -5.85 / 0157.8 +2929	UGC 1491 / PGC 7620	.S..9*. / U / 9.0±1.2	.96± .09 / .00± .06 / / .98	/ .16 / .00 / .00					12187 / / 12304 / 11933
020040.7-761046 / 297.59 -40.22 / 220.14 -19.14 / 020024.5-762513	PGC 7621	.LA.-.. / S / -3.0± .8	1.15± .09 / .20± .08 / / 1.16	/ .24 / .00					
0200.7 +4228 / 136.34 -18.59 / 341.62 -1.04 / 0157.7 +4214	UGC 1489 / PGC 7628	.S..6*. / U / 6.0±1.2	1.04± .15 / .13± .12 / / 1.08	120 / .38 / .19 / .06					
020048.7-090010 / 168.30 -65.56 / 292.17 -18.15 / 015820.3-091438	NGC 787 / MCG -2- 6- 15 / PGC 7632	RSAT3*. / E (1) / 3.0± .7 / 3.1±1.2	1.39± .04 / .11± .05 / / 1.39	90 / .01 / .16 / .06					

R.A. 2000 DEC. / l b / SGL SGB / R.A. 1950 DEC.	Names / PGC	Type / S_T n_L / T / L	$\log D_{25}$ / $\log R_{25}$ / $\log A_e$ / $\log D_o$	p.a. / A_g / A_i / A_{21}	B_T / m_B / m_{FIR} / B_T^o	$(B-V)_T$ / $(U-B)_T$ / $(B-V)_T^o$ / $(U-B)_T^o$	$(B-V)_e$ / $(U-B)_e$ / m'_e / m'_{25}	m_{21} / W_{20} / W_{50} / HI	V_{21} / V_{opt} / V_{GSR} / V_{3K}
0200.8 +1743		.S..6*.	.95± .07	25					9510
145.44 -42.09	UGC 1495	U	.18± .05	.13	15.24 ±.13				
318.64 -10.21		6.0±1.3		.26					9597
0158.1 +1729	PGC 7637		.96	.09	14.81				9241
0200.9 +1510		.L.....	1.16± .10	165					
146.74 -44.46	UGC 1496	U	.25± .05	.13	14.23 ±.11				
316.21 -11.10		-2.0± .8		.00					
0158.2 +1456	PGC 7644		1.14						
020055.0+381242		.SB.2?.	1.26± .04	87				16.90±.3	4105± 8
137.66 -22.67	UGC 1493	U	.41± .05	.18	13.89 ±.12			286± 9	4154± 44
337.73 -2.70	IRAS01579+3758	2.0± .8		.51	12.59				4242
015755.2+375813	PGC 7646		1.28	.21	13.16			3.53	3871
020058.1+460831		.S...0..	1.14± .05	3					
135.31 -15.05	UGC 1492	U	.58± .05	.62	14.7 ±.3				
344.99 .34		.0± .9		.44					
015749.8+455402	PGC 7648		1.17						
020058.7+081843	A 0158+08	.S.....	1.00± .06	26				16.36±.3	4742± 12
150.80 -50.76	UGC 1498	R	.29± .05	.18	14.35 ±.13			327± 21	4756± 60
309.57 -13.37				.44				256± 15	4804
015820.1+080415	PGC 7649		1.01	.15	13.72			2.50	4469
020103.5+232515								16.41±.3	5036± 10
142.96 -36.73	CGCG 482- 46			.36	15.2 ±.4				
324.04 -8.25								191± 7	5138
015815.4+231047	PGC 7653								4773
0201.0 -0848		.IBS9*.	1.22± .05	25					1611
168.13 -65.37	MCG -2- 6- 16	E (1)	.19± .05	.01					
292.39 -18.17		10.0±1.2		.14					1621
0158.6 -0903	PGC 7654	7.5± .8	1.22	.09					1349
020106.4-064901	NGC 788	.SAS0*.	1.28± .04	75	13.00 ±.13	.91± .01	.95± .01		
165.26 -63.81	MCG -1- 6- 25	R	.12± .04	.07	13.08 ±.18	.43± .02	.48± .02		4109± 41
294.44 -17.70		.0± .6	.77± .02	.09		.83	12.34± .05		4125
015836.7-070329	PGC 7656		1.28		12.80	.42	13.94± .24		3845
020106.6+315256	NGC 783	.S..5..	1.20± .05	35				15.15±.1	5191± 5
139.78 -28.70	UGC 1497	U	.05± .05	.17	12.84 ±.13			94± 6	5110± 46
331.92 -5.13	MK 1171	5.0± .8		.07	12.10				5312
015812.3+313828	PGC 7657		1.22	.02	12.57			2.56	4940
0201.1 +0631		.L.....	1.07± .07	100					
152.08 -52.36	UGC 1505	U	.43± .03	.13	14.53 ±.10				
307.83 -13.96		-2.0± .9		.00					
0158.5 +0617	PGC 7658		1.02						
0201.1 +1940		.S..6*.	1.02± .06	168					
144.62 -40.24	UGC 1500	U	.79± .05	.25					
320.52 -9.60		6.0±1.4		1.16					
0158.4 +1926	PGC 7663		1.04	.39					
0201.2 +3157		.S..2..	1.00± .08	84					
139.77 -28.62	UGC 1499	U	.55± .06	.17					
332.00 -5.12		2.0± .9		.68					
0158.3 +3143	PGC 7666		1.02	.27					
020113.3+302136		.I..9*.	1.26± .06					16.60±.3	3811± 10
140.34 -30.14	UGC 1502	U	.29± .06	.16					
330.52 -5.73		10.0±1.1		.22				130± 7	3930
015820.1+300708	PGC 7667		1.27	.15					3559
020114.0-314344		RSBT1*.	.91± .05	146					13890± 20
233.96 -74.14	ESO 414- 22	r	.17± .04	.02	14.80 ±.14				13831
268.04 -22.07	FAIR 1077	1.0± .9		.18	12.48				13679
015901.0-315812	PGC 7668		.91	.09	14.43				
020117.0+285037	NGC 784	.SB.8*/	1.82± .02		12.23S±.08	.49± .05		12.86±.1	198± 5
140.90 -31.58	UGC 1501	PU	.64± .03	.23	11.91 ±.12			116± 6	221± 52
329.12 -6.30	IRAS01584+2836	7.5± .5		.79	13.52	.31		96± 4	313
015824.9+283609	PGC 7671		1.84	.32	11.11		14.58± .15	1.42	-56
020119.8+331947		.E.....	.83± .10					16.74±.2	5086± 6
139.33 -27.31	UGC 1503	U	.06± .06	.21	14.39 ±.12			289± 9	
333.27 -4.63	IRAS01583+3305	-5.0± .8		.00	13.63			278± 7	5211
015824.2+330519	PGC 7674		.85		14.11				4839
020121.1-011811		.S?....							11838
158.85 -59.21	MCG 0- 6- 22								11871
300.06 -16.31	MK 1015			.04	15.4 ±.3				
015848.1-013239	PGC 7675								11568

R.A. 2000 DEC.	Names	Type	logD$_{25}$	p.a.	B$_T$	(B-V)$_T$	(B-V)$_e$	m$_{21}$	V$_{21}$
l b		S$_T$ n$_L$	logR$_{25}$	A$_g$	m$_B$	(U-B)$_T$	(U-B)$_e$	W$_{20}$	V$_{opt}$
SGL SGB		T	logA$_e$	A$_I$	m$_{FIR}$	(B-V)$_T^0$	m'$_e$	W$_{50}$	V$_{GSR}$
R.A. 1950 DEC.	PGC	L	logD$_o$	A$_{21}$	B$_T^0$	(U-B)$_T^0$	m'$_{25}$	HI	V$_{3K}$
020121.6-052216	NGC 790	.LAR0?.	1.12± .08						
163.47 -62.60	MCG -1- 6- 26	E	.00± .08	.02					
295.93 -17.39		-2.0± .8		.00					
015851.1-053644	PGC 7677		1.13						
020124.7+153851	NGC 786	.S?....	.84± .08						
146.65 -43.98	UGC 1506		.06± .05	.15	14.33 ±.14				4520± 34
316.70 -11.06	IRAS01587+1524			.09	12.95				4602
015841.6+152423	PGC 7680		.86	.03	14.06				4250
020130.7+262855	IC 187	.SB.1..	1.31± .04	70					
141.85 -33.80	UGC 1507	U	.45± .05	.26	13.78 ±.12				5102± 50
326.94 -7.23	IRAS01587+2614	1.0± .8		.46	12.24				5211
015840.3+261428	PGC 7683		1.33	.22	13.00				4844
020131.1-245527		.SBS3P*	1.09± .04	97					
208.89 -74.12	ESO 477- 22	S (1)	.28± .03	.00	14.52 ±.14				
275.41 -21.34	VV 467	2.5± .6		.39	13.12				
015913.0-250954	PGC 7684	2.2± .9	1.09	.14					
020132.9+450012		.S..0..	1.05± .08	155					
135.74 -16.12	UGC 1504	U	.32± .06	.42	14.7 ±.3				7163± 68
343.99 -.19		.0± .9		.24					7310
015825.8+444544	PGC 7686		1.07		13.90				6946
020137.1-682628	NGC 813	.SXR0*P	1.11± .05	99					
293.52 -47.38	ESO 52- 16	S	.17± .05	.00	13.84 ±.14				8160
228.21 -20.68		.0± .9		.13					8013
020029.1-684054	PGC 7692		1.10		13.58				8095
020140.0+314936	NGC 785	.L..-*.	1.17± .09	80					
139.92 -28.72	MCG 5- 5- 46	U	.14± .05	.18					5022± 52
331.91 -5.26		-3.0±1.2		.00					5144
015845.6+313509	PGC 7694		1.17						4773
0201.7 +0829	NGC 791	.E.....	1.20± .14						
150.95 -50.52	UGC 1511	U	.00± .08	.17	14.10 ±.15				
309.81 -13.49		-5.0± .8		.00					
0159.1 +0815	PGC 7702		1.23						
020144.9-582240		RSB.1..	1.04± .06	111					
286.30 -56.42	ESO 114- 19	r	.26± .05	.00	14.68 ±.14				
238.95 -22.05		1.0± .9		.27					
020006.1-583706	PGC 7703		1.04	.13					
020146.4+263242		.S?....	.78± .09	45				16.02±.2	5009± 6
141.89 -33.72	UGC 1510		.29± .05	.26	14.7 ±.2			237± 10	5100± 49
327.02 -7.26	ARAK 71			.36	12.65			192± 8	5120
015856.0+261815	PGC 7706		.80	.15	14.05			1.82	4752
020148.4+234127									4727± 10
143.05 -36.42	CGCG 482- 50			.30	15.5 ±.4				
324.36 -8.31									4829
015900.0+232700	PGC 7707								4465
0201.8 +1615		.S..0..	1.07± .06	53					
146.46 -43.37	UGC 1512	U	.42± .05	.14	14.49 ±.12				
317.32 -10.94		.0± .9		.31					
0159.1 +1601	PGC 7709		1.07						
0201.8 +0828		.SB.0..	.90± .07	110					
150.99 -50.52	UGC 1513	U	.10± .05	.17	15.33 ±.13				
309.80 -13.52		.0± .9		.07					
0159.2 +0814	PGC 7713		.91						
020151.3-102803		.SBT5*.	1.13± .06	45					4722
171.15 -66.50	MCG -2- 6- 17	E (1)	.05± .05	.01					
290.72 -18.75		5.0± .8		.07					4726
015923.9-104229	PGC 7714	3.1± .8	1.13	.02					4463
020152.8+233314	IC 189	.SB?...	.89± .11					17.06±.3	12347± 10
143.13 -36.54	MCG 4- 5- 39		.00± .07	.31	14.78 ±.08				
324.23 -8.37	IRAS01590+2318			.00	13.00			266± 7	12449
015904.5+231847	PGC 7716		.92	.00	14.39			2.66	12085
0201.9 +1143		.I..9*.	1.07± .14						4655
149.00 -47.55	UGC 1515	U	.07± .12	.19					
312.97 -12.50		10.0±1.2		.05					4725
0159.3 +1129	PGC 7725		1.09	.03					4383
0201.9 +2106		.S?....	1.14± .13						9084
144.21 -38.83	UGC 1514		.03± .12	.26	14.24 ±.15				
321.94 -9.27				.04					9180
0159.2 +2052	PGC 7726		1.16	.01	13.91				8819

R.A. 2000 DEC. l b SGL SGB R.A. 1950 DEC.	Names PGC	Type S_T n_L T L	$\log D_{25}$ $\log R_{25}$ $\log A_e$ $\log D_o$	p.a. A_g A_i A_{21}	B_T m_B m_{FIR} B_T^o	$(B-V)_T$ $(U-B)_T$ $(B-V)_T^o$ $(U-B)_T^o$	$(B-V)_e$ $(U-B)_e$ m'_e m'_{25}	m_{21} W_{20} W_{50} HI	V_{21} V_{opt} V_{GSR} V_{3K}
020159.4-583417 286.42 -56.24 238.74 -22.06 020021.0-584842	 ESO 114- 21 IRAS02003-5848 PGC 7727	.S..5.. S (1) 5.0± .8 2.2± .8	1.20± .04 .21± .04 1.20	10 .00 .31 .10	 14.19 ±.14 13.69 13.84			6396 6268 6287
020207.3+233259 143.19 -36.53 324.25 -8.43 015919.0+231833	IC 190 MCG 4- 5- 40 PGC 7731	.E?....		 .31 	 15.1 ±.4 			4769± 10 4871 4507
020209.1-794231 299.04 -36.88 216.50 -18.42 020245.0-795654	 ESO 13- 27 PGC 7736	.IBS9P. S (1) 10.0± .9 5.6± .9	.99± .05 .14± .04 1.02	42 .32 .11 .07	 15.26 ±.14 			
0202.1 +0958 150.13 -49.13 311.29 -13.12 0159.5 +0944	IC 1771 MCG 2- 6- 14 PGC 7737		.81± .11 .00± .06 .82	 .17 	 15.16 ±.09 			4866± 79 4931 4594
020210.5-310128 231.30 -74.08 268.82 -22.20 015957.0-311554	 ESO 414- 26 PGC 7739	.L?....	.90± .06 .02± .04 .90	 .02 .00 	 14.50 ±.14 14.39			5602± 39 5544 5390
020211.9-000753 158.03 -58.10 301.30 -16.19 015938.3-002219	NGC 800 UGC 1526 IRAS01596-0021 PGC 7740	.SAT5*. UE (1) 5.0± .6 3.1±1.2	1.01± .05 .07± .04 .71± .02 1.01	10 .01 .11 .04	13.70V±.15 14.05 ±.12 13.89	.67± .04		5950± 42 5985 5680
020212.3-000603 158.00 -58.07 301.33 -16.18 015938.7-002029	NGC 799 UGC 1527 PGC 7741	PSBS1*. UE .5± .5	1.30± .04 .06± .04 .82± .06 1.30	100 .01 .06 .03	12.97V±.17 14.1 ±.2 13.91	.96± .03		5831± 32 5867 5562
020212.3-283928 222.64 -74.30 271.40 -21.96 015957.0-285354	 ESO 414- 25 PGC 7742	.SBS5.. S (1) 5.0± .9 3.3±1.0	1.21± .04 .32± .03 1.21	161 .00 .47 .16	 14.16 ±.14 13.66			4992± 52 4941 4773
020213.3-060449 164.76 -63.05 295.27 -17.78 015943.2-061915	 MCG -1- 6- 31 PGC 7743	.S..5P/ E 5.0±1.3	1.07± .06 .65± .05 1.07	10 .01 .97 .32				
0202.2 +1542 146.87 -43.85 316.83 -11.22 0159.5 +1528	NGC 792 UGC 1517 IRAS01595+1528 PGC 7744	.L..... U -2.0± .8	1.22± .08 .20± .05 1.21	130 .14 .00 	 14.12 ±.12 13.91			4614 4695 4345
0202.2 -0106 159.02 -58.93 300.32 -16.47 0159.7 -0121	 UGC 1525 PGC 7748	.E..... U -5.0± .8	1.06± .11 .15± .05 1.02	 .03 .00 	 15.05 ±.14 			
0202.2 +1911 145.16 -40.60 320.15 -10.02 0159.5 +1857	 UGC 1519 PGC 7750	.S..8*. U 8.0±1.2	1.03± .08 .26± .06 1.05	40 .23 .32 .13				2346 2436 2079
0202.2 +0958 150.17 -49.12 311.30 -13.14 0159.6 +0944	IC 1770 UGC 1522 PGC 7751	.L..... U -2.0± .8	1.06± .11 .01± .05 1.07	 .17 .00 	 14.81 ±.14 14.57			4597± 79 4662 4325
020217.3-214546 198.36 -73.03 278.83 -21.04 015957.0-220012	 ESO 544- 7 VV 583 PGC 7753	.S?....	1.04± .05 .32± .04 1.04	107 .00 .47 .16	 14.87 ±.14 14.31			13187± 69 13156 12950
0202.3 +1939 144.97 -40.16 320.60 -9.87 0159.6 +1925	 UGC 1523 PGC 7756	.S..3.. U (1) 3.0± .9 3.5±1.1	.97± .07 .07± .05 1.00	150 .29 .10 .04	 15.23 ±.16 14.74			13214 13306 12948
020226.1+320420 140.01 -28.43 332.20 -5.32 015931.3+314954	NGC 789 UGC 1520 ARAK 72 PGC 7760	.S?....	.80± .08 .23± .05 .81	3 .18 .24 .12	 14.4 ±.2 12.41 13.88		16.13±.3 257± 9 2.14	5267± 8 5116± 35 5381 5011
020228.8+221918 143.81 -37.65 323.13 -8.95 015941.3+220453	 CGCG 482- 53 PGC 7762			 .30 	 15.7 ±.4 		18.27±.3 126± 7	4823± 10 4921 4560

R.A. 2000 DEC. / l b / SGL SGB / R.A. 1950 DEC.	Names / PGC	Type / S_T n_L / T / L	$\log D_{25}$ / $\log R_{25}$ / $\log A_e$ / $\log D_o$	p.a. / A_g / A_i / A_{21}	B_T / m_B / m_{FIR} / B_T^o	$(B-V)_T$ / $(U-B)_T$ / $(B-V)_T^o$ / $(U-B)_T^o$	$(B-V)_e$ / $(U-B)_e$ / m' / m'_{25}	m_{21} / W_{20} / W_{50} / HI	V_{21} / V_{opt} / V_{GSR} / V_{3K}
020229.5+182225 NGC 794	.L..-*.	1.11± .16	45	13.81 ±.15	1.08± .03	1.10± .03		8224± 31
145.62 -41.35	UGC 1528	U	.07± .08	.19	13.70 ±.10	.54± .04	.59± .04		8312
319.39 -10.36		-3.0±1.1	.76± .04	.00		.96	13.10± .13		7957
015944.6+180800	PGC 7763		1.12		13.42	.53	14.06± .85		7957
020230.6+110512 IC 193	.SAT5..	1.24± .04					15.80±.3	4649± 10
149.56 -48.08	UGC 1529	U	.07± .05	.19	14.14 ±.18			263± 13	
312.40 -12.83		5.0± .8		.11					4717
015950.3+105047	PGC 7765		1.25	.04	13.82			1.94	4378
020231.0-505554	.LAR-?.	1.15± .05	169					6315± 33
278.19 -62.65	ESO 197- 18	S	.16± .04	.00	13.47 ±.14				6204
247.03 -22.73		-4.0± .6		.00					6174
020039.0-511018	PGC 7766		1.10		13.37				
0202.5 +4458	.S..6*.	1.00± .16	36					
135.92 -16.10	UGC 1516	U	.84± .12	.39					
344.03 -.36		6.0±1.4		1.24					
0159.4 +4444	PGC 7767		1.04	.42					
0202.5 +1601 IC 192	.L.....	.97± .08						
146.80 -43.53	UGC 1530	U	.10± .03	.14	14.52 ±.10				
317.16 -11.18		-2.0± .9		.00					
0159.8 +1547	PGC 7768		.97						
0202.5 +1709	.S..3..	1.10± .05	114					8173
146.23 -42.48	UGC 1531	U (1)	.46± .05	.15	14.86 ±.12				
318.24 -10.79		3.0± .9		.63					8258
0159.8 +1655	PGC 7770	2.5±1.2	1.11	.23	14.02				7905
020233.3-794020	.S..4./	1.28± .04	3					4576
299.01 -36.91	ESO 13- 28	S	.76± .04	.32	15.28 ±.14				4413
216.53 -18.45	IRAS02030-7954	4.0± .6		1.12	12.74				4564
020309.0-795442	PGC 7773		1.31	.38	13.81				
020235.6+222245				.30	15.7 ±.4			17.04±.3	12525± 10
143.82 -37.59	CGCG 482- 54								
323.19 -8.95								401± 7	12624
015948.1+220820	PGC 7775								12262
020237.4-424748	PSBT2P*	1.11± .04	92					5587
264.84 -68.68	ESO 245- 12	Sr	.32± .04	.03	14.64 ±.14				5496
255.95 -22.91	IRAS02005-4302	2.0± .6		.39	13.50				5415
020035.0-430212	PGC 7779		1.12	.16	14.16				
020240.9-412454	.S?....	.97± .06	120					5618± 39
261.86 -69.57	ESO 298- 3		.10± .06	.03	14.90 ±.14				5531
257.46 -22.90				.14					5441
020037.1-413918	PGC 7786		.97	.05	14.69				
0202.7 +2634	.SX.6*.	.98± .07						14601
142.13 -33.63	UGC 1533	U	.05± .05	.19	15.07 ±.14				
327.13 -7.45		6.0±1.2		.07					14710
0159.9 +2620	PGC 7789		.99	.03	14.74				14344
020254.4-144026 NGC 815	.SBS9*.	1.02± .07	50					
179.71 -69.22	MCG -3- 6- 4	E (1)	.29± .05	.00					
286.38 -19.90		9.0±1.3		.29					
020029.5-145450	PGC 7798	6.4±1.2	1.02	.14					
0202.9 +0731	.S..4..	1.00± .06	35					7591
152.03 -51.26	UGC 1536	U	.34± .05	.17	15.12 ±.12				
308.96 -14.08		4.0± .9		.50					7649
0200.3 +0717	PGC 7800		1.02	.17	14.41				7320
020258.9+450124	.S..0?.	.96± .09	90					
135.99 -16.03	UGC 1532	U	.10± .06	.39	15.0 ±.3				
344.10 -.42		.0±1.8		.07					
015951.5+444700	PGC 7801		.99						
020302.2-093925	.S..7?/	1.45± .02	37					3866
170.35 -65.70	MCG -2- 6- 19	EU	1.10± .04	.07					
291.63 -18.84	KUG 0200-098	6.8± .8		1.38					3872
020034.2-095349	PGC 7806		1.45	.50					3607
0203.0 +4628	.SB.0..	1.00± .19						
135.57 -14.63	UGC 1534	U	.05± .08	.62					
345.43 .14		.0± .9		.04					
0159.9 +4614	PGC 7808		1.06						
020305.4+023648 IC 194	.S..3*/	1.16± .04	13					6389
155.88 -55.61	UGC 1542	UE (1)	.79± .04	.05	15.21 ±.12				
304.11 -15.61		3.3± .8		1.09					6432
020030.1+022224	PGC 7812	4.5±1.3	1.16	.39	14.02				6119

R.A. 2000 DEC. l b SGL SGB R.A. 1950 DEC.	Names / PGC	Type S_T n_L / T / L	$\log D_{25}$ / $\log R_{25}$ / $\log A_e$ / $\log D_o$	p.a. / A_g / A_i / A_{21}	B_T / m_B / m_{FIR} / B_T^o	$(B-V)_T$ / $(U-B)_T$ / $(B-V)_T^o$ / $(U-B)_T^o$	$(B-V)_e$ / $(U-B)_e$ / m'_e / m'_{25}	m_{21} / W_{20} / W_{50} / HI	V_{21} / V_{opt} / V_{GSR} / V_{3K}
0203.1 +3618		.I..9..	1.00± .16					16.19±.1	4322± 6
138.74 -24.35	UGC 1535	U	.09± .12	.24				80± 6	
336.16 -3.84		10.0± .9		.07					4452
0200.2 +3604	PGC 7817		1.02	.04					4083
0203.2 +2345		.I..9*.	1.11± .14	55					2823± 10
143.40 -36.25	UGC 1538	U	.20± .12	.31					
324.53 -8.58		10.0±1.2		.15					2925
0200.4 +2331	PGC 7819		1.14	.10					2562
0203.2 +0539		.S..0..	.98± .07	124					
153.48 -52.90	UGC 1545	U	.52± .05	.13	14.97 ±.13				
307.14 -14.73		.0± .9		.39					
0200.6 +0525	PGC 7820		.96						
0203.2 +1943		.E.....	1.00± .19	155					
145.20 -40.02	UGC 1543	U	.21± .08	.30	15.16 ±.11				
320.74 -10.05		-5.0± .9		.00					
0200.5 +1929	PGC 7821		.99						
020319.6+320441	NGC 798	.E.....	1.08± .17	137					
140.22 -28.37	UGC 1539	U	.36± .08	.24	14.50 ±.10				
332.28 -5.49		-5.0± .9		.00					
020024.7+315017	PGC 7823		1.01						
020319.8+333754		.S..9*.	.96± .17	175				16.17±.3	5488± 7
139.67 -26.89	UGC 1540	U	.22± .12	.24					
333.71 -4.90		9.0±1.3		.22				177± 7	5613
020023.7+332330	PGC 7824		.98	.11					5244
020320.1+220223	A 0200+21	.IB.9..	1.35± .04		14.40 ±.15	.48± .06	.44± .05	13.83±.1	2644± 5
144.17 -37.85	UGC 1547	U (1)	.02± .05	.29	14.3 ±.3	-.15± .08	-.17± .07	153± 6	2670± 76
322.93 -9.23	DDO 17	10.0± .7	1.09± .03	.02		.39	15.34± .06	138± 12	2741
020032.7+214800	PGC 7825	9.0±1.0	1.37	.01	14.07	-.22	15.93± .27	-.25	2381
0203.3 +1838	A 0200+18	.SXS5..	1.04± .08					14.94±.1	2374± 10
145.74 -41.02	UGC 1546	U (1)	.00± .06	.19	14.42 ±.14			99± 11	2377± 59
319.72 -10.45		5.0± .8		.00				87± 11	2462
0200.6 +1824	PGC 7826	2.7± .9	1.06	.00	14.21			.73	2108
020328.0+380703	NGC 797	.SXS1..	1.20± .06	65	13.59M±.13	.94± .03	.99± .02	15.11±.1	5654± 5
138.21 -22.61	UGC 1541	U	.08± .06	.21		.39± .04	.48± .03	446± 6	5600± 52
337.84 -3.19	5ZW 170	1.0± .8	.86± .04	.08		.83	13.42± .13	404± 8	5788
020027.7+375240	PGC 7832		1.22	.04	13.23	.34	14.25± .37	1.84	5420
0203.5 +4827		.S..6*.	1.11± .05	97					
135.07 -12.71	UGC 1537	U	.86± .05	1.12					
347.27 .84		6.0±1.3		1.26					
0200.3 +4813	PGC 7833		1.22	.43					
020330.4+023358	A 0200+02								
156.08 -55.60				.05					6310± 45
304.10 -15.72	MK 585				13.60				6353
020055.2+021935	PGC 7834								6040
020331.5-095557	NGC 806	.S..6P?	1.08± .04	60					3939
171.01 -65.82	MCG -2- 6- 21	E	.51± .05	.04					
291.38 -19.02	IRAS02010-1010	6.0±1.8		.75	12.91				3944
020103.8-101020	PGC 7835		1.09	.25					3681
020332.3-605108		PSXR2*.	1.07± .05	28					8974
288.02 -54.14	ESO 114- 23	Sr (1)	.26± .04	.04	14.92 ±.14				8841
236.26 -22.00		2.0± .7		.32					
020200.0-610530	PGC 7836	5.6±1.3	1.08	.13	14.47				8876
020333.5+261810		.S..0..	.94± .09	36				18.55±.3	5219± 10
142.44 -33.82	UGC 1549	U	.52± .06	.34	14.80 ±.17				
326.94 -7.72		.0±1.0		.39				396± 7	5327
020043.0+260347	PGC 7837		.95		14.00				4962
020335.5+044708		.S..7..	1.00± .05	145					
154.29 -53.63	UGC 1553	U	.24± .04	.12	15.14 ±.13				
306.31 -15.08	KUG 0200+045	7.0± .8		.33					
020058.9+043245	PGC 7839		1.01	.12					
020337.3+240426		.SB?...	1.44± .05	135				14.49±.1	2671± 6
143.37 -35.92	UGC 1551		.07± .06	.36	13.5 ±.2			134± 7	
324.87 -8.55	IRAS02008+2350			.11	13.67			118± 9	2773
020048.4+235003	PGC 7841		1.47	.04	13.05			1.40	2411
020337.3+154456			1.00?						7973
147.28 -43.68	CGCG 461- 37		.70?	.14	15.58 ±.12				
316.99 -11.52									8053
020054.0+153033	PGC 7842		1.01						7705

R.A. 2000 DEC.	Names	Type S_T　n_L	$\log D_{25}$ $\log R_{25}$	p.a. A_g	B_T m_B	$(B-V)_T$ $(U-B)_T$	$(B-V)_e$ $(U-B)_e$	m_{21} W_{20}	V_{21} V_{opt}
l　　b		T	$\log A_e$	A_i	m_{FIR}	$(B-V)_T^o$	m'_e	W_{50}	V_{GSR}
SGL　SGB									
R.A. 1950 DEC.	PGC	L	$\log D_o$	A_{21}	B_T^o	$(U-B)_T^o$	m'_{25}	HI	V_{3K}
020340.5+261640			.78± .13					17.27±.3	5022± 10
142.48 -33.84	MCG 4- 5- 46		.13± .07	.34	15.00 ±.14				5130
326.93 -7.75								224± 7	
020050.0+260217	PGC 7843		.81						4765
0203.6 +4806		.SXT7..	1.10± .07					16.21±.1	5028± 11
135.20 -13.04	UGC 1544	U	.06± .06	1.12					5179
346.96 .68		7.0± .8		.09				90± 8	
0200.5 +4752	PGC 7844		1.20	.03					4822
020344.6+144231	IC 195	.LX.0..	1.19± .06	135					3648± 56
147.88 -44.63	UGC 1555	U	.27± .03	.15	13.98 ±.10				3725
316.00 -11.91		-2.0± .8		.00					3379
020102.0+142808	PGC 7846		1.16		13.79				
020345.0+381536	NGC 801	.S..5..	1.50± .03	150	13.96M±.12	.87± .02	.95± .02	14.44±.1	5764± 6
138.22 -22.46	UGC 1550	U	.66± .05	.21	13.38 ±.12	.35± .04	.41± .03	452± 6	5748± 11
337.99 -3.19	IRAS02007+3801	5.0± .8	.90± .02	.99	12.29	.66	13.90± .06	439± 5	5895
020044.5+380114	PGC 7847		1.52	.33	12.44	.17	14.64± .23	1.67	5527
020345.2+160153	NGC 803	.SAS5/	1.48± .02	8	13.24 ±.15	.68± .02	.81± .02	13.64±.1	2100± 4
147.17 -43.41	UGC 1554	PU	.38± .04	.14	13.13 ±.13	-.03± .03	.09± .03	266± 4	
317.27 -11.45	IRAS02010+1547	5.0± .5	.97± .03	.58	13.13	.56	13.58± .07	255± 4	2181
020101.7+154731	PGC 7849		1.49	.19	12.45	-.12	14.52± .21	1.00	1833
020349.0+430008		.S..6*.	.96± .09	80					
136.74 -17.92	UGC 1548	U	.34± .06	.38	15.13 ±.09				
342.32 -1.35		6.0±1.3		.51					
020043.7+424545	PGC 7853		.99	.17					
020350.1+144422	IC 196	.SB.3*.	1.44± .03	5					3534± 41
147.89 -44.60	UGC 1556	U	.30± .05	.15	13.65 ±.18				3611
316.04 -11.92		3.0±1.1		.42					3266
020107.4+143000	PGC 7856		1.46	.15	13.06				
0203.8 +1157		.SX.9*.	1.09± .07						5295
149.49 -47.15	UGC 1558	U	.00± .06	.22					5365
313.36 -12.86		9.0± .9		.00					5025
0201.2 +1143	PGC 7859		1.11	.00					
0203.9 +1518	IC 1774	.SXS7..	1.27± .04	140					3626
147.61 -44.07	UGC 1559	U	.03± .05	.14	14.5 ±.3				3705
316.59 -11.74		7.0± .8		.04					3358
0201.2 +1504	PGC 7863		1.28	.01	14.26				
0203.9 +1940		PSBS3*.	1.09± .07	65					8491
145.42 -40.02	UGC 1560	U	.21± .06	.30	14.56 ±.13				
320.74 -10.21	IRAS02011+1925	3.0± .9		.29	12.57				8582
0201.1 +1925	PGC 7864		1.11	.10	13.92				8226
020356.0-231850	NGC 808	PSBR4*.	1.09± .04	7	14.14 ±.13	.65± .05			
203.83 -73.18	ESO 478- 1	PSr (1)	.31± .03	.00	14.12 ±.14	.04± .05			
277.23 -21.65	IRAS02015-2333	3.7± .5		.46	11.61				
020137.0-233312	PGC 7865	2.6± .7	1.09	.16			13.66± .25		
0204.0 +2412		.I..9..	1.02± .08	100				15.99±.1	606± 6
143.42 -35.76	UGC 1561	U	.08± .06	.33	14.51 ±.12			71± 7	
325.02 -8.59	5ZW 173	10.0± .8		.06				65± 12	708
0201.2 +2358	PGC 7871		1.05	.04	14.12			1.83	346
020402.3+304958	NGC 804	.L.....	1.14± .07	7					
140.83 -29.50	UGC 1557	U	.65± .03	.19	14.67 ±.11				
331.19 -6.11		-2.0± .9		.00					
020108.3+303536	PGC 7873		1.06						
020405.2+024712	IC 197	.SB.4*.	1.00± .05	55					6332± 50
156.12 -55.33	UGC 1564	U	.30± .04	.05	14.29 ±.14				6375
304.36 -15.80	KUG 0201+025	4.0± .9		.45					6063
020129.8+023250	PGC 7875		1.00	.15	13.75				
020410.6+475833		.LBR0..	1.09± .07						
135.32 -13.14	UGC 1552	U	.06± .03	.99	14.1 ±.3				
346.87 .55	5ZW 172	-2.0± .8		.00	13.34				
020059.2+474411	PGC 7881		1.19						
020418.4+283928	A 0201+28				15.2 ±.2	.79± .03			
141.71 -31.54	CGCG 503- 80			.19	14.7 ±.4	.04± .06			4591± 66
329.20 -6.99	MK 365								4703
020126.0+282507	PGC 7888								4338
020419.0-084408	NGC 809	RL..+*.	1.17± .06	170					
169.46 -64.80	MCG -2- 6- 23	E	.15± .05	.02					5418
292.67 -18.93		-1.0±1.2		.00					5426
020150.6-085829	PGC 7889		1.15						5159

R.A. 2000 DEC. / l b / SGL SGB / R.A. 1950 DEC.	Names / PGC	Type / S_T n_L / T / L	$\log D_{25}$ / $\log R_{25}$ / $\log A_e$ / $\log D_o$	p.a. / A_g / A_i / A_{21}	B_T / m_B / m_{FIR} / B_T^o	$(B-V)_T$ / $(U-B)_T$ / $(B-V)_T^o$ / $(U-B)_T^o$	$(B-V)_e$ / $(U-B)_e$ / m'_e / m'_{25}	m_{21} / W_{20} / W_{50} / HI	V_{21} / V_{opt} / V_{GSR} / V_{3K}
020429.6+284845	NGC 805	.LB....	1.05± .08	115					
141.69 -31.38	UGC 1566	U	.17± .03	.19	14.49 ±.10				
329.36 -6.97		-2.0± .9		.00					
020137.2+283424	PGC 7899		1.04						
020431.2-061158		.SBS9..	1.51± .02	73				13.87±.1	1363± 6
165.92 -62.80	MCG -1- 6- 39	UE (1)	.39± .04	.04				178± 8	
295.30 -18.36	KUG 0202-064	8.5± .5		.40				160± 12	1378
020201.2-062619	PGC 7900	6.4± .8	1.52	.20					1101
020431.9+275533		.SB.8..	1.13± .05	4				15.52±.3	4697± 10
142.05 -32.22	UGC 1565	U	.37± .05	.21	15.08 ±.14				
328.54 -7.31		8.0± .9		.46				206± 7	4808
020140.2+274112	PGC 7902		1.15	.19	14.40			.93	4444
020432.4-521022		.L..-*.	1.19± .05						6001
279.20 -61.46	ESO 197- 21	S	.03± .04	.00	13.28 ±.14				5886
245.66 -22.97		-2.6± .5		.00					5866
020243.0-522442	PGC 7903		1.19		13.18				
020434.9-100632	NGC 811	.SBS7*.	1.04± .05	35					
171.77 -65.78	MCG -2- 6- 24	E (1)	.18± .05	.01					
291.26 -19.32	KUG 0202-103	7.0±1.3		.24					
020207.3-102053	PGC 7905	5.3±1.2	1.04	.09					
020436.5+475618		.E...*.	1.11± .16						
135.40 -13.15	UGC 1563	U	.00± .08	.99	14.5 ±.3				
346.87 .47		-5.0±1.1		.00					
020125.0+474157	PGC 7909		1.27						
020444.1+083237		.S?....	1.13± .05	62					
151.96 -50.15	UGC 1572		.60± .05	.21	14.30 ±.14				3508± 50
310.10 -14.18	IRAS02020+0818			.90	12.98				3567
020205.2+081817	PGC 7917		1.15	.30	13.18				3239
0204.7 +3540		.S..8*.	1.00± .08						
139.29 -24.86	UGC 1568	U	.00± .06	.28					
335.70 -4.39		8.0±1.2		.00					
0201.8 +3526	PGC 7922		1.03	.00					
0204.7 +2126		.S..2..	1.00± .08	25					
144.84 -38.29	UGC 1570	U	.84± .06	.33	15.52 ±.14				
322.49 -9.77		2.0±1.0		1.04					
0202.0 +2112	PGC 7925		1.03	.42					
020451.0-551259		.SBS6P.	1.10± .05	0					
282.57 -58.91	ESO 153- 16	S	.29± .05	.02	15.09 ±.14				5912± 49
242.34 -22.78		6.0± .9		.43					5790
020307.0-552718	PGC 7927		1.10	.15	14.62				5790
0204.8 +3509		.S..3..	1.06± .06	146					
139.48 -25.34	UGC 1569	U (1)	.86± .05	.28					
335.23 -4.60		3.0±1.0		1.19					
0201.9 +3455	PGC 7929	4.5±1.4	1.08	.43					
020452.7+112520		.SB?...	.94± .11						7836
150.14 -47.53	MCG 2- 6- 22		.28± .07	.17	15.25 ±.12				
312.92 -13.27				.43					7904
020212.1+111100	PGC 7930		.95	.14	14.61				7567
020453.2+321648		.S?....	.94± .11					17.14±.3	6983± 10
140.50 -28.07	MCG 5- 5- 52		.19± .07	.24	15.28 ±.13				
332.59 -5.72				.26				190± 7	7104
020157.9+320228	PGC 7931		.96	.09	14.73			2.32	6738
020453.6-572905		PSBR3*.	.99± .06	30					9176
284.82 -56.99	ESO 114- 24	r	.08± .05	.00	14.92 ±.14				9049
239.87 -22.56	IRAS02032-5743	3.3± .9		.11					
020314.0-574324	PGC 7932		.99	.04	14.75				9064
0204.9 +4309		.S..7..	1.07± .07	119					
136.90 -17.71	UGC 1567	U	.79± .06	.41					
342.54 -1.47	IRAS02018+4255	7.0±1.0		1.10	13.60				
0201.8 +4255	PGC 7933		1.11	.40					
020455.9+285917	NGC 807	.E.....	1.25± .08	145					4730± 36
141.73 -31.19	UGC 1571	U	.13± .05	.15	13.47 ±.10				
329.55 -6.99	IRAS02020+2844	-5.0± .7		.00	13.66				4843
020203.2+284457	PGC 7934		1.23		13.25				4479
020503.4+155100			1.00?						9440
147.67 -43.46	CGCG 461- 43		.30?	.15	15.33 ±.13				
317.20 -11.81									9520
020219.9+153640	PGC 7939		1.01						9173

R.A. 2000 DEC. l b SGL SGB R.A. 1950 DEC.	Names PGC	Type S_T n_L T L	$\log D_{25}$ $\log R_{25}$ $\log A_e$ $\log D_o$	p.a. A_g A_i A_{21}	B_T m_B m_{FIR} B_T^o	$(B-V)_T$ $(U-B)_T$ $(B-V)_T^o$ $(U-B)_T^o$	$(B-V)_e$ $(U-B)_e$ m' m'_{25}	m_{21} W_{20} W_{50} HI	V_{21} V_{opt} V_{GSR} V_{3K}
020505.1-550641 282.41 -58.98 242.45 -22.82 020321.0-552100	 ESO 153- 17 FAIR 720 PGC 7941	.SXR5.. Sr (1) 4.7± .5 4.4± .8	1.41± .04 .14± .05 1.42	110 .02 .21 .07	 13.92 ±.14 13.84 13.66			14.48±.3 193± 12 179± 9 .75	6529± 10 5900±190 6406 6405
020505.3-063027 166.58 -62.96 295.02 -18.58 020235.5-064446	 MCG -1- 6- 41 PGC 7942	.SAR6.. E (1) 6.0± .8 5.3± .8	1.10± .06 .04± .05 1.10	100 .01 .06 .02					
0205.1 +2440 143.52 -35.24 325.55 -8.65 0202.3 +2426	 UGC 1575 PGC 7944	.I..9?. U	1.04± .15 1.06± .06 1.07	6 .29 .75 .50					4829 4932 4571
0205.2 +0955 151.21 -48.86 311.49 -13.85 0202.5 +0941	 UGC 1580 IRAS02025+0941 PGC 7951	.S?.... 	1.13± .05 .71± .05 1.14	137 .15 1.06 .35	 15.05 ±.12 13.02 13.79				7766 7829 7497
020515.2+060615 153.89 -52.27 307.75 -15.07 020237.8+055156	IC 1776 UGC 1579 PGC 7952	.SBS7.. U 7.0± .7 	1.29± .04 .01± .05 1.30	 .10 .02 .01	 13.81 ±.19 13.69				3405± 50 3457 3136
020515.3+373801 138.73 -22.96 337.53 -3.71 020215.2+372342	 UGC 1574 PGC 7953	.S..6*. U 6.0±1.3 	1.04± .08 .53± .06 1.05	145 .13 .78 .26	 15.52 ±.12				
020519.6+300026 141.44 -30.20 330.53 -6.68 020226.1+294607	 UGC 1576 PGC 7960	.S..7.. U 7.0±1.0 	1.04± .08 1.06± .06 1.06	153 .17 1.38 .50				16.99±.3 199± 7 	5235± 10 5350 4986
020519.9-062706 166.61 -62.88 295.09 -18.62 020250.1-064125	 MCG -1- 6- 42 IRAS02028-0641 PGC 7961	PSBR2?. E 2.0±1.8 	1.12± .06 .57± .05 1.12	85 .01 .71 .29	 12.25				
020520.3+250619 143.39 -34.81 325.98 -8.54 020230.6+245200	 MCG 4- 6- 1 PGC 7962	.S?.... 	.82± .13 .17± .07 .84	 .25 .26 .09	 15.42 ±.12 14.88			15.97±.3 202± 7 1.00	4810± 10 4914 4553
0205.4 +1314 149.23 -45.82 314.72 -12.78 0202.7 +1300	NGC 810 UGC 1583 PGC 7965	.E...?. U -5.0±1.6 	1.23± .14 .11± .08 1.23	25 .23 .00					
0205.4 +1322 149.15 -45.70 314.85 -12.74 0202.7 +1308	 UGC 1584 PGC 7966	.S..6*. U 6.0±1.2 	1.18± .06 .36± .06 1.20	82 .23 .53 .18					7525 7598 7257
020526.8+311033 141.03 -29.08 331.62 -6.26 020232.3+305614	 UGC 1577 IRAS02025+3056 PGC 7967	.SB?... 	1.34± .03 .17± .04 1.35	85 .17 .24 .09	 13.65 ±.15 12.96 13.20			15.02±.1 287± 7 291± 7 1.73	5276± 4 5219± 43 5393 5028
020531.2+345507 139.71 -25.53 335.07 -4.82 020233.5+344048	 CGCG 522-109 PGC 7970	 	.66± .13 .29± .06 .68	 .27	 15.70 ±.16				4542 4668 4303
020533.6+345246 139.73 -25.56 335.03 -4.84 020236.0+343827	 UGC 1581 IRAS02025+3438 PGC 7972	.S..8*. U 8.0±1.2 	1.22± .05 .40± .05 1.24	160 .27 .49 .20	 15.06 ±.17 13.08 14.29			14.45±.3 363± 9 -.04	4411± 8 4537 4172
020537.1+064601 153.53 -51.64 308.43 -14.95 020259.4+063143	 UGC 1587 IRAS02030+0632 PGC 7977	.S?.... 	1.01± .05 .46± .04 1.02	24 .13 .69 .23	 14.60 ±.14 13.78 13.74			15.61±.3 368± 13 1.64	5658± 10 5712 5389
020539.1-560419 283.27 -58.13 241.40 -22.81 020357.0-561836	 ESO 153- 19 PGC 7978	.SBS6?. S (1) 6.0±1.7 7.8± .9	1.19± .05 .34± .05 1.20	52 .03 .49 .17	 15.07 ±.14				
0205.6 -0041 159.98 -58.12 300.99 -17.18 0203.1 -0056	 UGC 1588 PGC 7979	.S..6*. U 6.0±1.4 	1.10± .05 .83± .05 1.10	150 .01 1.21 .41	 15.54 ±.12				

R.A. 2000 DEC. / l b / SGL SGB / R.A. 1950 DEC.	Names / PGC	Type S_T n_L / T / L	$\log D_{25}$ / $\log R_{25}$ / $\log A_e$ / $\log D_o$	p.a. / A_g / A_i / A_{21}	B_T / m_B / m_{FIR} / B_T^o	$(B-V)_T$ / $(U-B)_T$ / $(B-V)_T^o$ / $(U-B)_T^o$	$(B-V)_e$ / $(U-B)_e$ / m'_e / m'_{25}	m_{21} / W_{20} / W_{50} / HI	V_{21} / V_{opt} / V_{GSR} / V_{3K}
020539.6+395017 / 138.09 -20.84 / 339.57 -2.91 / 020237.2+393559	UGC 1582 / PGC 7980	.S..7.. / U / 7.0± .9 /	1.15± .05 / .48± .05 / / 1.17	37 / .23 / .66 / .24	/ 14.30 ±.13 / / 13.40			15.35±.1 / / 285± 8 / 1.72	4815± 11 / / 4951 / 4587
020542.0-524807 / 279.66 -60.84 / 244.96 -23.10 / 020354.0-530224	ESO 153- 18 / PGC 7982	.E.3.*. / S / -5.0±1.2 /	1.08± .05 / .14± .04 / / 1.03	126 / .01 / .00 /	/ 14.30 ±.14 / /				
020543.2+241358 / 143.86 -35.60 / 325.19 -8.94 / 020254.0+235940	MCG 4- 6- 2 / PGC 7984	.SB?... / / /	.94± .11 / .11± .07 / / .96	/ .29 / .14 / .06	/ 15.16 ±.13 / / 14.60			17.16±.3 / / 548± 7 / 2.49	12664± 10 / / 12765 / 12406
020544.6-710656 / 294.58 -44.79 / 225.33 -20.54 / 020453.0-712112	ESO 52- 20 / PGC 7986	.SBS4P. / S (1) / 4.0± .9 / 2.2± .9	1.09± .05 / .28± .05 / / 1.09	10 / .02 / .42 / .14	/ 14.58 ±.14 / / 14.09				8126 / / 7974 / 8075
020545.5-324036 / 236.45 -72.99 / 267.09 -23.10 / 020334.0-325454	ESO 354- 34 / PGC 7988	RLX.+.. / r / -1.3± .8 /	1.07± .06 / .08± .06 / / 1.06	/ .02 / .00 /	/ 14.61 ±.14 / / 14.50				5826± 29 / / 5762 / 5622
0205.7 +5044 / 134.76 -10.42 / 349.48 1.41 / 0202.5 +5030	UGC 1578 / PGC 7991	.S..6*. / U / 6.0±1.4 /	1.04± .08 / .98± .06 / / 1.14	152 / 1.07 / 1.44 / .49					
020549.5-094920 / 171.84 -65.36 / 291.63 -19.55 / 020321.8-100338	MCG -2- 6- 26 / KUG 0203-100 / PGC 7998	.SBS9P* / E (1) / 9.4± .6 / 5.3± .8	1.21± .04 / .30± .05 / / 1.21	120 / .01 / .30 / .15					1900 / / 1904 / 1644
0205.9 +1455 / 148.43 -44.23 / 316.39 -12.33 / 0203.2 +1441	UGC 1589 / PGC 8003	.S..2.. / U / 2.0± .9 /	.95± .07 / .20± .05 / / .96	95 / .15 / .25 / .10	/ 15.26 ±.13 / / 14.73				12505 / / 12582 / 12239
020602.0+451135 / 136.47 -15.71 / 344.47 -.85 / 020253.6+445717	UGC 1585 / PGC 8009	.SXS2.. / U / 2.0± .9 /	1.00± .06 / .09± .05 / / 1.06	3 / .54 / .12 / .05	/ 14.4 ±.3 / / 13.68				5688± 68 / / 5834 / 5475
0206.0 +0917 / 151.91 -49.34 / 310.94 -14.25 / 0203.4 +0903	IC 198 / UGC 1592 / PGC 8011	.S..... / R / /	1.04± .06 / .26± .05 / / 1.06	55 / .22 / .39 / .13	* / 14.58 ±.12 / / 13.92		.75± .04 / .10± .06 / /	15.68±.2 / 363± 15 / 348± 11 / 1.62	9245± 9 / 9414± 60 / 9309 / 8980
020603.6-551132 / 282.27 -58.84 / 242.35 -22.95 / 020420.0-552548	ESO 153- 20 / IRAS02043-5525 / PGC 8012	PSBT3P. / Sr / 2.6± .6 /	1.20± .04 / .21± .04 / .83± .02 / 1.21	14 / .04 / .28 / .10	13.80 ±.14 / 14.06 ±.14 / 13.03 / 13.56	.81± .02 / / .72 /	.88± .01 / / 13.44± .07 / 14.15± .27		/ 5931± 40 / 5809 / 5810
020604.2+294736 / 141.70 -30.34 / 330.39 -6.91 / 020310.7+293318	UGC 1590 / PGC 8013	.L..-*. / U / -3.0±1.1 /	1.19± .09 / .11± .05 / / 1.20	130 / .17 / .00 /	13.71 ±.10 / / / 13.47				/ 5002± 50 / 5116 / 4753
0206.1 +1316 / 149.43 -45.72 / 314.81 -12.93 / 0203.4 +1302	UGC 1593 / PGC 8014	.SXT5.. / U / 5.0± .8 /	1.01± .06 / .11± .05 / / 1.03	/ .27 / .17 / .06	/ 15.13 ±.16 / / 14.65				7425 / / 7497 / 7158
020608.0+495432 / 135.07 -11.19 / 348.75 1.02 / 020253.3+494015	UGC 1586 / PGC 8015	.SX.3.. / U / 3.0± .9 /	1.04± .15 / .08± .12 / / 1.16	/ 1.31 / .11 / .04	/ 14.6 ±.3 / / 13.12			17.12±.1 / / 217± 12 / 3.97	6046± 11 / / 6199 / 5847
020612.4+295800 / 141.66 -30.17 / 330.57 -6.87 / 020318.8+294343	UGC 1591 / PGC 8019	.S?.... / / /	1.14± .05 / .80± .05 / / 1.16	153 / .17 / 1.20 / .40	/ 14.66 ±.15 / / 13.26				/ 4835± 50 / 4950 / 4587
020612.8-283555 / 222.37 -73.43 / 271.56 -22.83 / 020358.0-285012	ESO 414- 28 / PGC 8020	PLAR-?. / S / -3.0± .6 /	1.06± .05 / .16± .04 / / 1.03	111 / .00 / .00 /	/ 14.44 ±.14 / /				
020614.2-371956 / 250.53 -71.23 / 261.98 -23.47 / 020407.0-373412	ESO 298- 7 / PGC 8025	.S?.... / / /	1.07± .05 / .08± .05 / / 1.08	/ .03 / .10 / .04	/ 14.80 ±.14 / / 14.62				/ 6138± 52 / 6060 / 5949

R.A. 2000 DEC.	Names	Type	logD$_{25}$	p.a.	B$_T$	(B-V)$_T$	(B-V)$_e$	m$_{21}$	V$_{21}$
l b		S$_T$ n$_L$	logR$_{25}$	A$_g$	m$_B$	(U-B)$_T$	(U-B)$_e$	W$_{20}$	V$_{opt}$
SGL SGB		T	logA$_e$	A$_i$	m$_{FIR}$	(B-V)^o_T	m'$_e$	W$_{50}$	V$_{GSR}$
R.A. 1950 DEC.	PGC	L	logD$_o$	A$_{21}$	B^o_T	(U-B)^o_T	m'$_{25}$	HI	V$_{3K}$
0206.2 +0913	IC 199	.S..2..	1.14± .05	25					9221
152.02 -49.38	UGC 1594	U	.26± .05	.22	14.90 ±.16				
310.89 -14.31		2.0± .8		.32					9281
0203.6 +0859	PGC 8026		1.16	.13	14.27				8953
020615.9-413120		.SXT6..	1.21± .05	129					5397± 60
260.92 -68.97	ESO 298- 8	Sr (1)	.44± .05	.03	14.72 ±.14				
257.35 -23.58		5.7± .6		.65					5308
020413.0-414536	PGC 8028	5.6± .9	1.21	.22	14.02				5223
020616.1-001730		.L.....	1.04± .11	0					12702± 34
159.82 -57.70	UGC 1597	U	.14± .05	.01	14.30 ±.10				
301.44 -17.21	MK 1018	-2.0± .8		.00					12734
020342.6-003147	PGC 8029		1.02		14.10				12437
020621.1-520144		.S..5*/	1.18± .04	56					
278.55 -61.41	ESO 197- 24	S	.69± .04	.00	15.42 ±.14				
245.80 -23.26		5.0± .9		1.03					
020432.1-521600	PGC 8031		1.18	.34					
020627.4+270204		.S..6*.	1.09± .04	75				15.75±.3	4962± 10
142.88 -32.92	UGC 1595	U	.35± .04	.23	15.19 ±.14			276± 13	
327.87 -8.04		6.0±1.2		.51					5070
020336.0+264747	PGC 8035		1.11	.17	14.42			1.15	4709
0206.5 +0341	IC 1779								9570
156.27 -54.25	CGCG 387- 35			.09	15.1 ±.3				
305.45 -16.10									9614
0203.9 +0327	PGC 8039								9303
020630.5+295937		.LX.0..	1.08± .07	125					
141.72 -30.12	UGC 1596	U	.16± .03	.17	14.52 ±.11				
330.62 -6.92		-2.0± .8		.00					
020336.8+294521	PGC 8040		1.07						
020637.1-314644		.SAR3..	1.09± .05	176					10726± 39
233.31 -73.01	ESO 414- 31	r	.39± .04	.02	15.27 ±.14				
268.09 -23.21		3.3± .9		.54					10663
020425.0-320100	PGC 8049		1.09	.20	14.62				10520
020639.3-410927	NGC 822	.E...*.	1.06± .06	77	14.11 ±.13	.96± .01			5366± 37
259.99 -69.13	ESO 298- 9	BS	.26± .04	.03	14.25 ±.14				
257.76 -23.64		-5.0± .7		.00		.90			5278
020436.1-412342	PGC 8055		.99		14.06		13.75± .32		5191
020640.3+013006		.S..7..	1.07± .06	111					6812
158.25 -56.12	UGC 1600	U	.88± .05	.03					
303.28 -16.79		7.0±1.0		1.21					6849
020405.7+011550	PGC 8056		1.08	.44					6546
020644.0-565609		PSXR0..	1.07± .06	2					
283.90 -57.32	ESO 153- 23	r	.20± .03	.00	14.75 ±.14				
240.44 -22.87		-.1± .9		.15					
020504.0-571024	PGC 8059		1.06						
0206.7 -0051		.SA.8*.	1.09± .06						5994
160.59 -58.12	UGC 1603	U	.08± .05	.04	15.01 ±.19				
300.90 -17.49		8.0±1.1		.10					6024
0204.2 -0106	PGC 8060		1.09	.04	14.86				5730
0206.8 +3725		.S..8*.	1.00± .08	26					
139.12 -23.07	UGC 1599	U	.94± .06	.14					
337.46 -4.07		8.0±1.4		1.16					
0203.8 +3711	PGC 8063		1.01	.47					
020649.5+310927	IC 200							16.27±.3	3846± 10
141.36 -29.00	CGCG 504- 12			.16	15.2 ±.4				
331.72 -6.54								129± 7	3963
020354.9+305512	PGC 8064								3600
020651.1+443428	NGC 812	.S...P.	1.97± .02	160				13.49±.1	5163± 11
136.81 -16.26	UGC 1598	R	.63± .05	.56	12.2 ±.3				5285± 41
343.97 -1.23	IRAS02037+4419			.95	11.98			435± 8	5315
020343.3+442012	PGC 8066		2.02	.32	10.69			2.49	4956
0206.8 -0030	IC 1781								13035
160.28 -57.81	CGCG 387- 37			.02	15.2 ±.3				
301.26 -17.41									13066
0204.3 -0045	PGC 8067								12771
020653.0-362709	NGC 824	.SBR4..	1.16± .04						5836± 52
247.88 -71.50	ESO 354- 37	BSr (1)	.07± .04	.03	14.14 ±.14				
262.95 -23.56		3.8± .5		.11					5760
020445.1-364124	PGC 8068	1.1± .8	1.16	.04	13.97				5645

R.A. 2000 DEC. l b SGL SGB R.A. 1950 DEC.	Names PGC	Type S_T n_L T L	$\log D_{25}$ $\log R_{25}$ $\log A_e$ $\log D_o$	p.a. A_g A_i A_{21}	B_T m_B m_{FIR} B_T^o	$(B-V)_T$ $(U-B)_T$ $(B-V)_T^o$ $(U-B)_T^o$	$(B-V)_{\bullet}$ $(U-B)_{\bullet}$ $m'_{\bullet}$ m'_{25}	m_{21} W_{20} W_{50} HI	V_{21} V_{opt} V_{GSR} V_{3K}
020653.0-523439 279.09 -60.92 245.19 -23.30 020505.0-524854	 ESO 153- 24 PGC 8069	.SXT2*. r 2.2±1.2 	.94± .07 .09± .06 .94	0 .00 .11 .04	 14.94 ±.14 				
020708.5+425756 137.36 -17.77 342.53 -1.92 020402.5+424341	 UGC 1601 PGC 8078	.SB.4.. U 4.0±1.0 	1.06± .06 .77± .05 1.09	169 .35 1.13 .38	 15.49 ±.12 				
0207.1 +0859 152.49 -49.49 310.74 -14.60 0204.5 +0845	 UGC 1606 PGC 8079	.S..7.. U 7.0± .8 	1.04± .08 .04± .06 1.06	 .23 .05 .02					8020 8079 7753
0207.2 +4346 137.12 -17.00 343.26 -1.61 0204.1 +4332	 UGC 1602 PGC 8082	.S..8*. U 8.0±1.3 	1.14± .13 .69± .12 1.18	105 .42 .84 .34					
020715.1+325712 140.79 -27.28 333.40 -5.92 020418.9+324257	 MCG 5- 6- 7 IRAS02044+3243 PGC 8086	.SB?... 	.82± .13 .00± .07 .84	 .29 .00 .00	 14.86 ±.12 13.13 14.50				11982± 57 12103 11740
020717.7+370548 139.33 -23.35 337.20 -4.29 020417.7+365133	 UGC 1604 6ZW 169 PGC 8087	.L...?. U -2.0±1.6 	1.04± .12 .22± .05 1.02	48 .13 .00 	 14.76 ±.10 				
020718.5+302527 141.75 -29.66 331.08 -6.92 020424.4+301112	 UGC 1608 PGC 8090	.SA.8.. U 8.0± .9 	1.00± .16 .04± .12 1.02	 .21 .05 .02				17.72±.3 181± 7 	9269± 10 9384 9022
020720.1-252634 211.56 -72.92 275.04 -22.71 020503.1-254048	NGC 823 ESO 478- 2 IRAS02050-2540 PGC 8093	.LAR-?. S -3.0± .8 	1.25± .05 .12± .04 1.23	 .00 .00 	 13.61 ±.14 13.45 13.54				4431± 52 4386 4208
020725.4+020658 157.98 -55.50 303.95 -16.79 020450.4+015244	 MCG 0- 6- 33 ARAK 74 PGC 8096	.L?.... 	.42? .00± .07 .42	 .05 .00 	15.80S±.15 15.64		 12.75±1.73		7197± 41 7236 6932
020725.5+325712 140.83 -27.26 333.42 -5.95 020429.2+324258	 MCG 5- 6- 8 PGC 8097	.S?.... 	.89± .11 .24± .07 .91	8 .29 .29 .12	 14.85 ±.13 14.15				11126± 57 11247 10884
0207.4 -0205 162.18 -59.05 299.70 -17.99 0204.9 -0220	IC 205 UGC 1613 PGC 8098	.SBS1.. U 1.0± .8 	.99± .06 .02± .05 1.00	 .03 .02 .01	 14.49 ±.13 14.34				8323 8349 8060
0207.4 +0910 152.46 -49.29 310.95 -14.61 0204.7 +0856	IC 202 UGC 1610 IRAS02047+0856 PGC 8101	.S..3.. U (1) 3.0± .9 5.5±1.3	1.14± .05 .77± .05 1.16	132 .21 1.06 .38	 15.14 ±.12 13.65 13.81				9101 9160 8834
020733.9+171205 147.71 -41.98 318.71 -11.90 020449.4+165751	NGC 817 UGC 1611 IRAS02048+1657 PGC 8109	.S?.... 	.87± .08 .35± .05 .89	27 .17 .49 .18	 14.2 ±.2 13.04 13.48			15.97±.3 253± 13 216± 10 2.31	4520± 10 4495± 50 4601 4256
020734.2+020655 158.04 -55.48 303.96 -16.83 020459.2+015241	 UGC 1617 ARAK 75 PGC 8110	.L..-*. U -3.0±1.2 	.90± .15 .05± .05 .90	 .05 .00 	15.80S±.15 14.79 ±.10 14.69		 14.22± .80		7139± 17 7178 6874
0207.6 +1521 148.72 -43.67 316.95 -12.56 0204.9 +1507	 UGC 1614 PGC 8112	.S..0.. U .0± .9 	1.16± .05 .67± .05 1.14	91 .18 .50 	 15.36 ±.13 				
020737.6+021051 158.00 -55.42 304.03 -16.82 020502.5+015637	 UGC 1618 HICK 15D PGC 8114	.L..-*. U -3.0±1.2 	.94± .14 .05± .05 .94	 .05 .00 	15.28S±.15 15.13		 14.70± .74		6274± 17 6312 6008
020739.8+020859 158.04 -55.44 304.00 -16.84 020504.8+015445	 UGC 1620 HICK 15C PGC 8117	.L..-*. U -3.0±1.2 	.90± .15 .00± .05 .90	 .05 .00 	14.59S±.15 14.75 ±.10 14.54		 13.95± .77		7256± 15 7295 6991

R.A. 2000 DEC. l b SGL SGB R.A. 1950 DEC.	Names PGC	Type S_T n_L T L	$\log D_{25}$ $\log R_{25}$ $\log A_e$ $\log D_o$	p.a. A_g A_i A_{21}	B_T m_B m_{FIR} B_T^o	$(B-V)_T$ $(U-B)_T$ $(B-V)_T^o$ $(U-B)_T^o$	$(B-V)_e$ $(U-B)_e$ m'_e m'_{25}	m_{21} W_{20} W_{50} HI	V_{21} V_{opt} V_{GSR} V_{3K}
020742.7+453719 136.63 -15.21 344.98 -.94 020433.4+452305	 UGC 1607 IRAS02045+4523 PGC 8120	.SBS2.. U 2.0± .8 	1.23± .05 .33± .05 1.28	72 .57 .41 .16	 13.9 ±.3 13.66 12.88			16.27±.1 466± 8 3.22	6261± 11 6407 6050
020749.9-351205 243.98 -71.82 264.34 -23.69 020541.0-352618	 ESO 354- 41 PGC 8126	PS.R1.. r 1.0± .9 	1.04± .06 .43± .05 1.04	158 .03 .44 .21	 14.68 ±.14 14.14				6103 6030 5909
020751.6+445036 136.90 -15.95 344.28 -1.28 020443.2+443623	 UGC 1609 PGC 8127	.S..3.. U (1) 3.0±1.0 4.5±1.3	1.09± .06 .71± .05 1.14	81 .56 .99 .36	 15.2 ±.3 				
020753.1+021003 158.11 -55.39 304.04 -16.89 020518.1+015550	 UGC 1624 HICK 15A PGC 8128	.L...*. U -2.0±1.2 	1.03± .11 .30± .05 1.00	130 .05 .00 	14.73S±.15 14.87 ±.10 14.67		 14.03± .60		7031± 15 7070 6766
0207.9 +1613 148.34 -42.85 317.80 -12.33 0205.2 +1558	 UGC 1622 IRAS02051+1558 PGC 8131	.S..6*. U 6.0±1.3 	1.04± .06 .46± .05 1.06	165 .18 .67 .23	 14.85 ±.12 13.97				7686 7765 7423
020756.1+433519 137.31 -17.14 343.15 -1.80 020449.2+432106	 UGC 1612 PGC 8132	.SA.5.. U 5.0± .8 	1.06± .06 .18± .05 1.10	55 .42 .27 .09	 15.13 ±.15 				
0207.9 +1910 146.81 -40.13 320.62 -11.28 0205.2 +1856	 UGC 1623 PGC 8135	.SB?... 	.96± .07 .17± .05 .99	5 .30 .25 .08	 15.30 ±.14 14.71				5174 5261 4913
0207.9 +4119 138.05 -19.29 341.10 -2.72 0204.9 +4105	 UGC 1615 PGC 8139	.SBS8*. U 8.0±1.2 	1.00± .08 .09± .06 1.03	 .35 .11 .04					
0208.0 +2820 142.75 -31.57 329.21 -7.88 0205.2 +2805	 UGC 1625 IRAS02051+2805 PGC 8146	.S..4.. U 4.0±1.0 	1.02± .06 .75± .05 1.04	169 .21 1.10 .37	 12.75 				
020805.0+015339 158.43 -55.60 303.78 -17.01 020530.1+013926	 UGC 1627 MK 1174 PGC 8148	.S?.... 	.93± .07 .40± .05 .94	70 .04 .60 .20	 14.88 ±.13 14.20				4786± 97 4824 4521
020807.5-283818 222.49 -73.01 271.57 -23.25 020553.1-285230	 ESO 414- 32 PGC 8151	PSXR1P? Sr 1.0± .6 	1.21± .04 .48± .03 .41± .04 1.21	71 .00 .49 .24	14.71 ±.16 14.57 ±.14 14.08	.91± .02 .31± .03 .78 .23	.93± .02 .36± .03 12.25± .13 14.43± .26		5442 5388 5228
0208.2 +0111 159.16 -56.18 303.08 -17.27 0205.7 +0057	 MCG 0- 6- 43 PGC 8157	.S?.... 	.82± .13 .08± .07 .82	 .04 .09 .04	 15.27 ±.12 15.01				12274± 79 12309 12010
020821.1+105944 151.55 -47.56 312.80 -14.22 020540.5+104532	NGC 821 UGC 1631 PGC 8160	.E.6.$. R -5.0± .6 	1.41± .04 .20± .05 1.22± .03 1.38	25 .17 .00 	11.67 ±.13 12.33 ±.09 11.92	.99± .02 .93 	1.00± .01 13.26± .10 13.20± .27		 1718± 14 1782 1452
020821.4+412846 138.07 -19.12 341.27 -2.72 020516.8+411433	 UGC 1626 ARP 74 PGC 8161	.SXT5.. U 5.0± .8 	1.19± .05 .02± .05 1.22	 .35 .02 .01	 14.11 ±.16 13.71			15.36±.1 170± 8 1.65	5543± 11 5681 5321
020824.7+145813 149.18 -43.94 316.65 -12.87 020541.6+144401	 UGC 1630 IRAS02057+1444 PGC 8163	.S?.... 	.99± .06 .32± .05 1.01	43 .17 .48 .16	 14.30 ±.14 12.57 13.63				4405± 50 4480 4141
020825.2+142057 149.53 -44.51 316.05 -13.09 020542.5+140645	NGC 820 UGC 1629 IRAS02057+1406 PGC 8165	.S..3.. U (1) 3.0± .8 3.5±1.1	1.13± .05 .25± .05 1.15	72 .19 .35 .13	 13.64 ±.13 12.26 13.07			14.73±.3 363± 13 346± 10 1.54	4422± 7 4495 4158
0208.4 +0624 154.79 -51.64 308.30 -15.72 0205.8 +0610	IC 208 UGC 1635 PGC 8167	.SA.4.. U (1) 4.0± .8 4.5±1.0	1.26± .04 .01± .05 1.27	 .14 .02 .01	 14.2 ±.2 13.98				3449± 56 3500 3183

R.A. 2000 DEC. I b SGL SGB R.A. 1950 DEC.	Names PGC	Type S_T n_L T L	$\log D_{25}$ $\log R_{25}$ $\log A_e$ $\log D_o$	p.a. A_g A_i A_{21}	B_T m_B m_{FIR} B_T^o	$(B-V)_T$ $(U-B)_T$ $(B-V)_T^o$ $(U-B)_T^o$	$(B-V)_e$ $(U-B)_e$ m'_e m'_{25}	m_{21} W_{20} W_{50} HI	V_{21} V_{opt} V_{GSR} V_{3K}
020830.6+185952 147.06 -40.24 320.50 -11.47 020544.8+184540	 CGCG 461- 50 PGC 8171	 	.95? .65? .99	 .34 	 15.61 ±.13 				5205 5291 4944
020832.4+061927 154.89 -51.70 308.23 -15.78 020554.9+060515	NGC 825 UGC 1636 PGC 8173	.S..1.. U 1.0± .8 	1.35± .04 .41± .05 1.36	53 .12 .42 .21	 14.06 ±.15 13.48				 3388± 43 3439 3123
020834.5+291403 142.51 -30.69 330.09 -7.63 020541.2+285951	NGC 819 UGC 1632 IRAS02056+2859 PGC 8174	.S?.... 	.77± .11 .16± .06 .79	10 .22 .22 .08	 14.4 ±.2 13.02 13.94			16.51±.3 287± 7 2.49	6576± 10 6566± 50 6687 6328
020842.3-074728 169.92 -63.37 293.92 -19.76 020613.5-080139	NGC 829 MCG -1- 6- 49 IRAS02062-0801 PGC 8182	.SBS5P? PE 5.0±1.0 	1.07± .04 .25± .04 1.07	105 .01 .37 .12	 12.45 			14.04±.1 183± 25 	4056± 14 4064 3801
020844.4+384636 139.05 -21.66 338.84 -3.88 020542.5+383224	NGC 818 UGC 1633 IRAS02057+3832 PGC 8185	.SX.5*. PU 4.5± .5 	1.47± .02 .36± .04 .91± .02 1.48	113 .16 .55 .18	13.20 ±.14 12.63 ±.12 12.11 12.14	.67± .02 .08± .03 .54 -.02	.81± .02 .18± .03 13.24± .05 14.47± .21	14.49±.1 462± 5 444± 5 2.16	4244± 5 4457± 52 4379 4018
020851.4+024309 157.98 -54.80 304.67 -16.96 020616.0+022858	 UGC 1643 KUG 0206+024 PGC 8188	.SB.3.. U 3.0± .9 	.90± .05 .26± .04 .91	13 .04 .36 .13	 15.20 ±.12 14.70				 12452 12492 12188
020853.9-451732 267.67 -66.17 253.18 -24.01 020656.1-453142	 ESO 246- 1 PGC 8193	.SBS7.. S (1) 7.0± .9 7.8± .9	1.04± .05 .32± .05 1.04	155 .03 .44 .16	 16.16 ±.14 				
020855.5-564414 283.24 -57.33 240.62 -23.19 020716.0-565824	NGC 852 ESO 153- 26 PGC 8195	.SBT4*. Sr (1) 4.2± .5 5.6± .6	1.12± .05 .10± .05 1.13	83 .02 .14 .05	 14.18 ±.14 13.98				6409 6282 6296
020856.4+075819 153.81 -50.20 309.89 -15.35 020617.7+074408	NGC 827 UGC 1640 IRAS02062+0744 PGC 8196	.S?.... 	1.35± .05 .44± .06 1.37	85 .19 .66 .22	 13.70 ±.13 11.50 12.83			14.31±.3 380± 13 357± 10 1.26	3458± 10 3512 3192
020857.0+260158 143.92 -33.66 327.15 -8.94 020606.0+254747	 UGC 1638 PGC 8198	.S..6*. U 6.0±1.3 	1.00± .05 .33± .04 1.01	15 .18 .48 .16	 15.22 ±.12 14.54			16.99±.3 246± 13 2.29	4874± 10 4978 4622
020857.3+471311 136.34 -13.62 346.51 -.49 020545.6+465900	 UGC 1634 IRAS02057+4658 PGC 8199	.SXR6.. U 6.0± .7 	1.38± .05 .01± .06 1.47	 .96 .01 .00	 14.0 ±.3 13.39 12.98			15.24±.1 147± 8 2.25	4978± 11 5126 4773
020858.7-070335 168.99 -62.77 294.70 -19.65 020629.4-071745	IC 209 MCG -1- 6- 51 IRAS02064-0717 PGC 8200	.SBR4*. E (1) 4.0± .8 4.2± .8	1.24± .05 .13± .05 1.24	110 .01 .19 .06	 13.56 				
020858.8-074605 170.01 -63.31 293.96 -19.82 020630.0-080015	NGC 830 MCG -1- 6- 50 MK 1020 PGC 8201	.LB.-?. E -3.0±1.3 	1.15± .10 .21± .07 1.12	110 .00 .00 					3888± 52 3896 3633
020901.2+273224 143.31 -32.24 328.56 -8.38 020609.2+271813	 CGCG 483- 5 PGC 8203			 .16 	 15.4 ±.4 			16.51±.3 120± 7 	9856± 10 9963 9606
020901.3-755622 296.86 -40.26 220.27 -19.69 020851.0-761030	 ESO 30- 8 PGC 8204	.SBS7*/ S (1) 7.0±1.2 5.6±1.3	1.35± .04 .74± .04 1.37	18 .25 1.02 .37	 14.93 ±.14 13.66				1260 1101 1232
020904.6+050648 156.04 -52.70 307.07 -16.28 020627.7+045238	 UGC 1646 KUG 0206+048 PGC 8207	.SB.4.. U 4.0± .8 	1.28± .03 .54± .04 1.29	173 .12 .79 .27	 14.61 ±.14 13.67				3380 3427 3115
0209.1 +3020 142.20 -29.61 331.15 -7.31 0206.2 +3006	 UGC 1644 PGC 8211	.I..9*. U 10.0±1.2 	.96± .09 .00± .06 .98	 .24 .00 .00					9150 9264 8905

R.A. 2000 DEC. / l b / SGL SGB / R.A. 1950 DEC.	Names / PGC	Type / S_T n_L / T / L	$\log D_{25}$ / $\log R_{25}$ / $\log A_e$ / $\log D_o$	p.a. / A_g / A_i / A_{21}	B_T / m_B / m_{FIR} / B_T^o	$(B-V)_T$ / $(U-B)_T$ / $(B-V)_T^o$ / $(U-B)_T^o$	$(B-V)_o$ / $(U-B)_o$ / m' / m'_{25}	m_{21} / W_{20} / W_{50} / HI	V_{21} / V_{opt} / V_{GSR} / V_{3K}
020907.1+441740		.S..6*.	1.00± .06	105					
137.30 -16.40	UGC 1637	U	.18± .05	.50	15.3 ±.3				
343.87 -1.71		6.0±1.3		.26					
020559.1+440330	PGC 8212		1.04	.09					
0209.1 +0637		.SAS8..	1.07± .08						3302
154.88 -51.37	UGC 1649	U	.01± .06	.14					
308.57 -15.82		8.0± .8		.01					3353
0206.5 +0623	PGC 8214		1.08	.00					3037
020910.3+315941		.SAS8..	1.13± .05					14.80±.2	5007± 6
141.58 -28.05	UGC 1641	U	.02± .05	.30	14.33 ±.15			118± 9	
332.68 -6.67		8.0± .8		.02				97± 7	5125
020614.6+314530	PGC 8215		1.16	.01	14.00			.79	4765
0209.2 +3712		.S..8*.	1.11± .07	178					
139.69 -23.13	UGC 1642	U	1.13± .06	.12					
337.45 -4.60		8.0±1.4		1.23					
0206.2 +3658	PGC 8217		1.12	.50					
020913.8+253416		.L?....	.93± .07	75				15.51±.3	4872± 10
144.19 -34.07	UGC 1648		.22± .05	.22	14.51 ±.11			376± 13	
326.75 -9.17	ARAK 76			.00					4974
020623.2+252006	PGC 8220		.92		14.22				4620
020918.4-232457		.S..4..	1.26± .03	104					5321
205.33 -72.03	ESO 478- 6	S (1)	.27± .03	.00	13.22 ±.14				5281
277.31 -22.89	IRAS02069-2339	4.0± .8		.40	11.59				5095
020700.0-233906	PGC 8223	1.1± .8	1.26	.14	12.79				
020921.1-100800	NGC 833	PS..1*P	1.17± .05	85	13.69S±.15	.99± .02			
173.85 -64.98	MCG -2- 6- 30	R	.32± .05	.00		.51± .03			3934± 28
291.52 -20.46	HICK 16B	1.0± .6		.33		.90			3934
020653.7-102210	PGC 8225		1.17	.16	13.31	.45	13.58± .32		3682
020924.9-100809	NGC 835	.SXR2*P	1.10± .04	80	12.91S±.15	.81± .02			
173.88 -64.97	MCG -2- 6- 31	R	.08± .05	.00		.18± .03			4118± 22
291.52 -20.48	MK 1021	2.0± .4		.09	11.20	.77			4118
020657.5-102218	PGC 8228		1.10	.04	12.78	.16	13.05± .28		3867
020927.3+332759	CGCG 504- 20							16.90±.3	11577± 10
141.09 -26.64				.31	15.7 ±.4				
334.06 -6.14								435± 7	11698
020630.3+331349	PGC 8231								11338
020928.4-094049	IC 210	.SB.4?.	1.36± .03	65					
173.16 -64.64	MCG -2- 6- 32	E (1)	.56± .05	.00					
292.00 -20.39	IRAS02070-0954	4.0±1.2		.82	11.40				
020700.8-095458	PGC 8232	4.2± .8	1.37	.28					
020928.7-061254		.SBS5*.	1.11± .06	135					
168.03 -62.05	MCG -1- 6- 52	E (1)	.06± .05	.03					
295.61 -19.56		5.0± .8		.09					
020658.9-062704	PGC 8234	4.2± .8	1.11	.03					
0209.5 +3715		.S..6?.	1.30± .05	28				15.47±.3	4585± 8
139.73 -23.06	UGC 1650	U	1.31± .06	.12				253± 9	
337.52 -4.63		6.0±2.0		1.47					4714
0206.5 +3701	PGC 8237		1.31	.50					4354
020934.7-473403		.SBR2?/	1.06± .05	120					
271.35 -64.49	ESO 197- 26	S	.66± .04	.04	15.78 ±.14				
250.67 -24.04		2.0±1.9		.81					
020740.0-474812	PGC 8242		1.06	.33					
0209.5 +2114		.S..7..	1.15± .05	30					5136
146.26 -38.07	UGC 1652	U	.40± .05	.34	14.68 ±.13				
322.71 -10.88		7.0± .9		.56					5227
0206.8 +2100	PGC 8245		1.19	.20	13.76				4879
020938.3+354745	A 0206+35	.E...?.	1.36± .11		14.3 ±.4	1.46± .08	1.47± .08		
140.28 -24.43	UGC 1651	U	.06± .08	.24	14.1 ±.2	.52± .14	.57± .14		11173± 64
336.20 -5.24		-5.0±1.4		.00		1.30			11298
020639.1+353336	PGC 8249		1.38		13.74	.50	15.95± .75		10939
020938.5-100847	NGC 838	.LATOP*	1.06± .04	85	13.57M±.04	.62± .01	.49± .01		
173.99 -64.94	MCG -2- 6- 33	RCE	.12± .05	.00		-.08± .02	-.20± .02		3841± 22
291.52 -20.54	MK 1022	-2.0± .9	.35± .01	.00		.57	10.80± .05		3841
020711.1-102256	PGC 8250		1.04		13.51	-.07	13.44± .25		3590
020939.5-065524	IC 207	.L..+P/	1.27± .05	75					
169.07 -62.56	MCG -1- 6- 54	E	.55± .05	.00					
294.89 -19.78	IRAS02071-0709	-1.0± .9		.00	12.59				
020710.1-070932	PGC 8251		1.18						

R.A. 2000 DEC. l　b SGL　SGB R.A. 1950 DEC.	Names PGC	Type S_T　n_L T L	logD_25 logR_25 logA_e logD_o	p.a. A_g A_i A_21	B_T m_B m_FIR B_T°	(B-V)_T (U-B)_T (B-V)_T° (U-B)_T°	(B-V)_e (U-B)_e m'_e m'_25	m_21 W_20 W_50 HI	V_21 V_opt V_GSR V_3K
020943.0-101103 174.08 -64.96 291.49 -20.56 020715.7-102511	NGC　839 MCG -2- 6- 34 IRAS02072-1025 PGC　8254	.L..*P/ R -2.0± .6 	1.16± .03 .33± .05 .50± .02 1.11	85 .00 .00 	13.93M±.10 13.87	.80± .02 .39± .03 .73 .38	.84± .01 .40± .03 11.99± .06 13.79± .23	 	 3834± 28 3834 3583
0209.7 +0719 154.56 -50.68 309.31 -15.74 0207.1 +0705	 UGC　1653 PGC　8255	.S?.... 	.96± .09 .39± .06 .97	138 .13 .59 .20	 15.36 ±.12 14.60	 	 	 	7308 7361 7044
020949.3-074657 170.37 -63.18 294.00 -20.03 020720.4-080106	NGC　842 MCG -1- 6- 55 MK　1023 PGC　8258	.LXR.?. PE -2.0± .6 	1.09± .07 .14± .05 .55± .05 1.07	35 .00 .00 	13.61 ±.16 13.55	.95± .01 .49± .03 .90 .50	.98± .01 .52± .03 11.85± .18 13.58± .40	 	 3703± 43 3710 3449
020953.4-352316 244.05 -71.36 264.16 -24.12 020745.1-353724	 ESO　354- 45 PGC　8264	.S?.... 	1.02± .05 .06± .05 .09 1.02	101 .03 .03	 15.41 ±.14 15.24	 	 	 	 10861± 39 10787 10670
0209.9 +1602 149.04 -42.82 317.80 -12.84 0207.2 +1548	 UGC　1659 PGC　8266	.SB.6*. U 6.0±1.2 	1.20± .05 .41± .05 .60 1.22	38 .19 .20	 14.55 ±.13 13.71	 	 	 	8238 8315 7977
0209.9 +1046 152.23 -47.59 312.72 -14.67 0207.3 +1032	 UGC　1662 PGC　8270	.S..2.. U 2.0± .9 	.95± .07 .43± .05 .53 .97	153 .18 .21	 15.17 ±.12 14.39	 	 	 	6837 6899 6573
021005.3-325624 236.57 -72.05 266.87 -24.03 020754.7-331032	IC　1783 ESO　354- 46 IRAS02079-3310 PGC　8279	.SAT3?. PS　(2) 2.8± .5 3.4± .5	1.30± .03 .40± .04 .71± .01 1.30	3 .03 .55 .20	13.20 ±.13 13.33 ±.11 11.92 12.67	.69± .01 .05± .03 .59 -.03	.76± .01 .13± .02 12.24± .03 13.54± .21	15.80±.3 360± 25 2.93	3350± 17 3309± 44 3277 3146
021008.7+073838 154.46 -50.35 309.67 -15.73 020730.3+072430	 UGC　1663 KUG 0207+074 PGC　8281	.S..6*. U 6.0±1.1 	1.19± .04 .29± .04 .43 1.21	13 .16 .15	 14.64 ±.15 14.03	 	 	 	3431 3484 3167
021008.8+364212 140.06 -23.54 337.07 -4.97 020708.6+362804	 UGC　1654 PGC　8282	.S..5.. U 5.0± .9 	.98± .07 .19± .05 .28 1.00	48 .22 .09	 14.93 ±.12 	 	 	 	
021009.2+391129 139.19 -21.18 339.33 -3.97 020706.6+385721	NGC　828 UGC　1655 6ZW　177 PGC　8283	.S..1*P P 1.0±1.0 	1.46± .03 .11± .05 .98± .02 1.48	 .16 .12 .06	13.15 ±.13 12.70 ±.14 10.40 12.59	.90± .01 .37± .02 .80 .32	.92± .01 .39± .02 13.54± .07 15.03± .24	15.04±.1 427± 11 2.39	5374± 9 5181± 46 5499 5141
021011.1-221922 202.19 -71.51 278.53 -22.94 020752.0-223330	NGC　849 ESO　478- 9 PGC　8286	RL...?. S -2.0±1.1 	.73± .06 .26± .03 .00 .70	117 .00 	 15.32 ±.14 	 	 	 	
0210.1 +4134 138.38 -18.93 341.49 -3.00 0207.1 +4120	 UGC　1656 PGC　8287	.S..8*. U 8.0±1.4 	1.11± .07 1.13± .06 1.23 1.14	63 .30 .50	 	 	 	 	
021014.3+254051 144.40 -33.88 326.94 -9.34 020723.5+252643	 CGCG 483- 7 PGC　8292	 	 	 	.24 15.3 ±.4 	 	 	18.00±.3 170± 7 	4641± 10 4743 4390
0210.2 +0750 154.36 -50.17 309.86 -15.70 0207.6 +0736	NGC　840 UGC　1664 PGC　8293	.SBR3.. U 3.0± .8 	1.25± .03 .24± .05 .79± .03 1.27	73 .16 .32 .12	14.27 ±.14 14.22 ±.16 13.71	.88± .03 .07± .09 .75 -.04	.96± .02 .37± .03 13.71± .09 14.80± .23	15.45±.2 467± 5 446± 5 1.63	7270± 5 7143± 60 7322 7005
021014.8+413117 138.41 -18.97 341.45 -3.03 020709.7+411709	 UGC　1661 PGC　8294	.E...*. U -5.0±1.1 	1.15± .15 .07± .08 .00 1.18	 .30 	 14.83 ±.17 	 	 	 	
021014.9-094239 173.53 -64.53 292.01 -20.58 020747.3-095647	 MCG -2- 6- 35 KUG 0207-099 PGC　8295	.SB.5P? E 5.0±1.2 	1.39± .03 .41± .06 .62 1.39	33 .00 .21	 	 	 	 	2017 2018 1766
021016.0-222547 202.53 -71.52 278.41 -22.97 020757.0-223954	NGC　837 ESO　478- 10 IRAS02079-2239 PGC　8297	.S..3?. S　(1) 2.5±1.3 3.3±1.3	.97± .04 .34± .03 .47 .97	12 .00 .17	 14.84 ±.14 	 	 	 	

R.A. 2000 DEC.	Names	Type	$\log D_{25}$	p.a.	B_T	$(B-V)_T$	$(B-V)_e$	m_{21}	V_{21}
l b		S_T n_L	$\log R_{25}$	A_g	m_B	$(U-B)_T$	$(U-B)_e$	W_{20}	V_{opt}
SGL SGB		T	$\log A_e$	A_i	m_{FIR}	$(B-V)_T^o$	m'_e	W_{50}	V_{GSR}
R.A. 1950 DEC.	PGC	L	$\log D_o$	A_{21}	B_T^o	$(U-B)_T^o$	m'_{25}	HI	V_{3K}

021017.4-101912	NGC 848	PSBS2P?	1.17± .04	0	13.6 ±.3	.56± .04			
174.55 -64.95	MCG -2- 6- 36	PE	.18± .05	.00		.08± .06			3890± 40
291.38 -20.73	MK 1026	1.8± .7		.22	12.67	.50			3890
020750.2-103319	PGC 8299		1.17	.09	13.35	.05	13.86± .37		3640
021018.5+383541		.S..7..	.95± .07	25					
139.42 -21.74	UGC 1660	U	.21± .05	.18	15.20 ±.12				
338.80 -4.24		7.0± .9		.28					
020716.5+382133	PGC 8301		.97	.10					
0210.3 +0112		.I..9*.	.96± .09						3539
159.94 -55.89	UGC 1665	U	.00± .06	.05					
303.26 -17.76		10.0±1.2		.00					3573
0207.8 +0058	PGC 8302		.96	.00					3277
021025.2-220317	NGC 836	PLAS+*.	1.10± .05	110					
201.45 -71.37	ESO 544- 17	Sr	.13± .03	.00	14.34 ±.14				
278.83 -22.95		-1.2± .5		.00					
020806.0-221724	PGC 8304		1.08						
021025.4-420517		.IBS9.	1.00± .05	123					
260.89 -68.01	ESO 298- 14	S (1)	.37± .05	.03	15.50 ±.14				4165
256.73 -24.35		10.0± .9		.28					4072
020824.0-421924	PGC 8305	7.8±1.3	1.00	.18	15.19				3996
021031.6-535012		RSBR0*.	1.20± .05	150	14.29 ±.17	.83± .03	.91± .02		
279.71 -59.58	ESO 153- 27	Sr	.38± .05	.02	14.44 ±.14	.25± .04	.34± .03		5704± 34
243.76 -23.73		.4± .6	.66± .06	.28		.71	13.08± .19		5583
020847.0-540418	PGC 8311		1.19		13.99	.22	14.21± .31		5580
021037.8+055214	A 0208+05	.S.....	.91± .05	173					
155.98 -51.85	UGC 1669	R	.14± .04	.13	14.44 ±.13				4569± 46
307.95 -16.41	MK 587			.21					4616
020800.4+053807	PGC 8318		.92	.07	14.08				4306
021037.8-154627		.LB.0P*	1.13± .06	4					1587
185.26 -68.40	MCG -3- 6- 10	E	.45± .05	.01					
285.61 -21.95	IRAS02082-1600	-2.0±1.3		.00	11.78				1569
020814.2-160033	PGC 8319		1.06						1346
021038.6-405506		.SBT6P*	1.28± .04	58					
258.24 -68.66	ESO 298- 15	Sr (1)	.43± .04	.03	14.32 ±.14				
258.03 -24.40		6.4± .7		.64					
020836.0-410912	PGC 8320	6.7± .9	1.28	.22					
021040.9-750220		.SAT5..	1.40± .04	135					
296.27 -41.05	ESO 30- 9	S (1)	.61± .04	.19	14.31 ±.14				8147
221.17 -20.01		5.0± .8		.91					7989
021023.1-751624	PGC 8326	3.3±1.2	1.42	.30	13.17				8115
0210.7 +0646	A 0208+06	.S..9*.	1.34± .05		14.8 ±.2	.51± .10	.48± .05	14.94±.1	1601± 5
155.33 -51.04	UGC 1670	U (1)	.00± .06	.15		.06± .11	.03± .07	115± 6	
308.85 -16.16	DDO 18	9.0±1.0	1.18± .06	.00		.47	16.17± .15	79± 8	1651
0208.1 +0632	PGC 8332	9.0±1.0	1.36	.00	14.63	.03	16.35± .37	.31	1338
021052.8-024826		PSBR2..	1.11± .06	75					
164.34 -59.14	MCG -1- 6- 63	E	.23± .05	.04					11183
299.22 -19.01		2.0± .9		.28					11205
020820.9-030232	PGC 8339		1.12	.11					10925
021054.2-392154		.S..1*.	1.28± .04	53					
254.49 -69.44	ESO 298- 16	S	.67± .04	.03	13.83 ±.14				5201± 39
259.75 -24.34	IRAS02088-3936	1.0±1.3		.68	13.42				5115
020850.0-393600	PGC 8341		1.28	.34	13.05				5023
021056.6-313600			.89± .06						
232.18 -72.14	ESO 415- 3		.09± .06	.03	14.99 ±.14				12542± 39
268.37 -24.11									12478
020845.0-315006	PGC 8344		.89						12340
021058.1+324159		.SB?...	.99± .06	98				16.42±.3	11797± 10
141.71 -27.26	UGC 1671		.33± .05	.29	15.30 ±.12				11728± 67
333.48 -6.74				.50				584± 7	11914
020801.5+322753	PGC 8346		1.02	.17	14.44			1.81	11557
021059.6+324350		.S?....	.96± .06	106				17.30±.3	11795± 10
141.71 -27.22	MCG 5- 6- 12		.66± .06	.29					4078± 67
333.51 -6.73	KUG 0208+324			.96				594± 7	
020802.9+322943	PGC 8350		.99	.33					
021059.9+374923		.S?....	1.02± .06	73					
139.83 -22.42	UGC 1673		.60± .05	.16	15.22 ±.13				4334
338.16 -4.67	IRAS02079+3735			.90	12.12				4463
020758.5+373516	PGC 8351		1.03	.30	14.12				4106

R.A. 2000 DEC. I b SGL SGB R.A. 1950 DEC.	Names PGC	Type S_T n_L T L	logD₂₅ logR₂₅ logA_e logD_o	p.a. A_g A_i A₂₁	B_T m_B m_FIR B_T°	(B-V)_T (U-B)_T (B-V)_T° (U-B)_T°	(B-V)_e (U-B)_e m'_e m'₂₅	m₂₁ W₂₀ W₅₀ HI	V₂₁ V_opt V_GSR V_3K
021101.3+373959 139.89 -22.57 338.02 -4.74 020800.1+372553	NGC 834 UGC 1672 ARAK 77 PGC 8352	.S?....	1.04± .06 .34± .05 1.05	20 .16 .50 .17	13.84 ±.13 13.5 ±.2 11.04 13.06	.77± .02 .07± .03 .64 -.04	13.05± .34	15.58±.1 263± 7 193± 7 2.36	4594± 6 4790± 35 4729 4371
0211.0 +0641 155.49 -51.09 308.79 -16.25 0208.4 +0627	UGC 1677 PGC 8353	.S..8*. U 8.0±1.5	1.04± .08 1.06± .06 1.05	135 .16 1.23 .50					1583 1633 1320
0211.0 -0038 162.05 -57.34 301.43 -18.46 0208.5 -0053	MCG 0- 6- 47 PGC 8356		.74± .14 .38± .07 .74	73 .05	15.0 ±.2				5809± 79 5837 5549
021108.5+035109 157.83 -53.53 305.98 -17.16 020832.4+033703	IC 211 UGC 1678 IRAS02085+0337 PGC 8360	.SXS6.. UE (2) 5.5± .5 3.4± .6	1.36± .03 .10± .04 1.17± .14 1.37	45 .10 .15 .05	* 13.9 ±.2 13.59		.54± .03 14.69± .42	14.27±.1 233± 6 208± 5 .63	3256± 5 3247± 27 3297 2994
0211.1 +2549 144.57 -33.68 327.14 -9.48 0208.3 +2535	UGC 1675 PGC 8362	.SB?...	1.00± .08 .46± .06 1.02	125 .18 .70 .23	*				5128 5230 4878
021111.0+384531 139.54 -21.53 339.02 -4.32 020808.6+383125	UGC 1674 PGC 8365	.S..7.. U 7.0± .9	.99± .06 .18± .05 1.01	60 .20 .24 .09	* 14.90 ±.12				
021112.3+034654 157.91 -53.58 305.92 -17.19 020836.2+033249	NGC 851 UGC 1680 MK 588 PGC 8368	.LB.+*. UE -1.0± .7	1.00± .05 .20± .03 .98	135 .10 .00	* 14.49 ±.10 14.34				3111± 33 3152 2849
021113.4-012910 163.00 -58.01 300.59 -18.73 020840.6-014315	NGC 850 UGC 1679 PGC 8369	.LXS+.. UE -1.0± .6	1.06± .06 .03± .07 1.06	85 .01 .00	* 13.86 ±.10 13.72				8161± 50 8186 7902
021117.2+372949 140.01 -22.71 337.88 -4.86 020816.1+371544	NGC 841 UGC 1676 5ZW 194 PGC 8372	PSXS2.. CUr 2.1± .5	1.25± .03 .25± .05 .71± .01 1.26	135 .16 .31 .12	13.42 ±.13 12.90 ±.14 13.04 12.67	.84± .01 .28± .02 .72 .20	.93± .01 .37± .02 12.46± .04 13.89± .23	15.21±.2 428± 7 403± 7 2.42	4540± 6 4463± 52 4668 4310
0211.4 +1555 149.55 -42.78 317.82 -13.22 0208.7 +1541	UGC 1683 PGC 8379	.S..4.. U 4.0±1.0	.96± .09 .98± .06 .98	.18 1.44 .49	16.41 ±.13 14.73				7813 7889 7553
0211.4 +1558 149.53 -42.74 317.87 -13.20 0208.7 +1544	UGC 1684 PGC 8381	RLX.+.. U -1.0± .8	1.21± .06 .01± .03 1.23	.18 .00	14.19 ±.16 13.89				8005 8081 7745
021130.7-355014 244.98 -70.89 263.68 -24.47 020923.0-360418	NGC 854 ESO 354- 47 IRAS02093-3604 PGC 8388	.SBT5*. BSr (1) 5.0± .5 3.3± .6	1.26± .04 .48± .04 1.27	0 .03 .72 .24	13.85 ±.14 13.01 13.07				6218± 34 6141 6029
021133.1+135459 150.76 -44.58 315.91 -13.95 020850.5+134054	A 0208+13 CGCG 438- 40 MK 366 PGC 8391			.24	14.92 ±.16 13.07			17.57±.1 110± 5	7966± 6 7800±110 8035 7704
021133.2-730946 295.17 -42.74 223.09 -20.52 021100.0-732348	ESO 30- 11 PGC 8392	PSAS2?. S 2.0± .8	1.22± .06 .16± .05 1.23	90 .07 .19 .08	15.37 ±.14 15.03				8076± 34 7920 8036
021135.2+313036 142.31 -28.33 332.44 -7.33 020839.5+311632	UGC 1682 KUG 0208+312 PGC 8393	.S..6*. U 6.0±1.3	1.04± .04 .36± .04 1.06	107 .23 .53 .18	14.66 ±.12 13.87			15.98±.2 250± 9 225± 7 1.92	4983± 6 5098 4742
021138.5+340243 141.35 -25.94 334.77 -6.32 020840.6+334839	UGC 1685 IRAS02086+3348 PGC 8396	.S..2.. U 2.0±1.0	1.02± .06 .57± .05 1.04	72 .26 .70 .29	14.97 ±.13 13.08 13.95				6175 6296 5939
021141.6-091817 173.46 -63.99 292.53 -20.84 020913.8-093221	NGC 853 MCG -2- 6- 38 IRAS02092-0932 PGC 8397	.S..9P? E (1) 9.0±1.5 5.3±1.6	1.19± .04 .07± .05 1.19	70 .01 .07 .03	12.15				1479± 33 1480 1229

R.A. 2000 DEC. l b SGL SGB R.A. 1950 DEC.	Names PGC	Type S_T n_L T L	$\log D_{25}$ $\log R_{25}$ $\log A_e$ $\log D_o$	p.a. A_g A_i A_{21}	B_T m_B m_{FIR} B_T^o	$(B-V)_T$ $(U-B)_T$ $(B-V)_T^o$ $(U-B)_T^o$	$(B-V)_e$ $(U-B)_e$ m'_e m'_{25}	m_{21} W_{20} W_{50} HI	V_{21} V_{opt} V_{GSR} V_{3K}
0211.7 +1418 150.58 -44.22 316.29 -13.86 0209.0 +1404	UGC 1687 PGC 8399	.L...*. U -2.0±1.2	1.00± .19 .09± .08 1.01	0 .25 .00	14.86 ±.11				
021144.0-062928 169.32 -61.90 295.47 -20.17 020914.4-064332	MCG -1- 6- 67 KUG 0209-067 PGC 8400	.S..4*/ E 4.0±1.3	1.15± .04 .87± .05 1.15	145 .03 1.28 .44					
021146.8-181602 191.55 -69.50 282.99 -22.67 020925.0-183006	ESO 544- 20 PGC 8404	.SBR4?. Er (1) 4.4±1.0 5.3± .8	1.19± .04 .52± .03 1.20	123 .03 .76 .26	15.32 ±.14 14.50				5078± 60 5052 4843
0211.8 +1400 150.78 -44.48 316.01 -13.98 0209.1 +1346	UGC 1689 PGC 8406	.S..9*. U 9.0±1.2	1.06± .08 .10± .06 1.09	175 .27 .10 .05					4410 4480 4149
0211.9 +1406 150.76 -44.38 316.12 -13.97 0209.2 +1352	UGC 1693 PGC 8412	.S..9*. U 9.0±1.1	1.18± .06 .01± .06 1.21	 .27 .01 .01					3821± 10 3891 3560
021156.2-391221 253.82 -69.35 259.93 -24.63 020952.0-392624	ESO 298- 19 PGC 8413	.SXT4.. Sr (1) 4.2± .6 4.4± .9	1.13± .05 .13± .05 1.13	57 .03 .19 .06	14.31 ±.14 14.06				5378± 39 5292 5200
021157.6+291851 143.28 -30.36 330.45 -8.28 020903.7+290447	UGC 1690 PGC 8417	.S?.... 	1.16± .07 1.00± .06 1.18	106 .23 1.50 .50	15.66 ±.12 13.90			16.56±.3 396± 7 2.16	4914± 10 5024 4670
0211.9 +0933 153.71 -48.45 311.69 -15.54 0209.3 +0919	UGC 1694 PGC 8418	.SXS8.. U 8.0± .8	1.16± .13 .00± .12 1.18	 .20 .00 .00					4453± 10 4511 4191
021202.1-501709 274.77 -62.21 247.64 -24.27 021012.0-503112	ESO 197- 29 IRAS02102-5031 PGC 8422	.S?.... 	1.05± .05 .45± .05 1.05	145 .02 .68 .23	15.21 ±.14 13.22 14.47				6520±108 6406 6383
0212.0 +4622 137.14 -14.27 345.96 -1.33 0208.9 +4608	UGC 1686 PGC 8424	.SX.8*. U 8.0±1.2	.99± .06 .06± .05 1.09	 1.00 .07 .03				15.69±.1 131± 8	4872± 11 5017 4666
021206.6+313514 142.40 -28.22 332.56 -7.40 020910.8+312110	MCG 5- 6- 14 PGC 8426	.S?.... 	.74± .14 .09± .07 .76	 .28 .14 .05	15.10 ±.14 14.64			16.00±.3 141± 7 1.31	5899± 10 6014 5659
021212.2+443405 137.75 -15.97 344.35 -2.10 020903.1+442002	NGC 846 UGC 1688 IRAS02090+4420 PGC 8430	.SBT2.. U 2.0± .7	1.29± .04 .06± .05 1.34	140 .56 .07 .03	13.0 ±.3 11.82 12.31			15.94±.1 400± 8 3.60	5118± 11 4894± 52 5250 4898
0212.2 +4223 138.49 -18.03 342.38 -3.01 0209.1 +4209	UGC 1692 PGC 8431	.S..4.. U 4.0± .8	1.04± .09 .06± .06 1.07	 .36 .09 .03					
021213.3+391409 139.58 -21.01 339.53 -4.31 020910.2+390006	UGC 1691 6ZW 183 PGC 8433	.L..... U -2.0± .8	1.13± .10 .02± .05 1.16	 .23 .00	14.21 ±.12				
0212.2 -0208 164.12 -58.40 299.99 -19.16 0209.7 -0223	UGC 1697 PGC 8435	.S..3.. U (1) 3.0± .9 4.5±1.2	1.08± .06 .37± .05 1.09	135 .03 .51 .18	14.74 ±.12 14.12				11217 11240 10960
021219.5+372836 140.23 -22.66 337.95 -5.05 020918.2+371433	NGC 845 UGC 1695 IRAS02093+3714 PGC 8438	.S..3.. U (1) 3.0± .9 2.5±1.2	1.22± .05 .60± .05 .76± .03 1.24	149 .19 .82 .30	14.34 ±.14 14.44 ±.12 12.37	.85± .04 .49± .06	1.00± .04 .47± .06 13.63± .12 13.79± .29		
021219.8-004845 162.71 -57.30 301.36 -18.81 020946.6-010248	UGC 1698 PGC 8439	PSBR1*. UE 1.0± .7	1.06± .05 .07± .04 1.06	105 .04 .07 .04	14.81 ±.16 14.55				12185 12212 11927

R.A. 2000 DEC. l b SGL SGB R.A. 1950 DEC.	Names PGC	Type S_T n_L T L	$\log D_{25}$ $\log R_{25}$ $\log A_•$ $\log D_o$	p.a. A_g A_i A_{21}	B_T m_B m_{FIR} B_T^o	$(B-V)_T$ $(U-B)_T$ $(B-V)_T^o$ $(U-B)_T^o$	$(B-V)_•$ $(U-B)_•$ $m'_•$ m'_{25}	m_{21} W_{20} W_{50} HI	V_{21} V_{opt} V_{GSR} V_{3K}
021228.2+295128		.S..8..	1.06± .06					16.00±.3	5899± 7
143.18 -29.81	UGC 1696	U	.04± .05	.36	15.11 ±.19			141± 7	6010
331.00 -8.16		8.0± .8		.05					
020933.8+293725	PGC 8449		1.09	.02	14.69			1.29	5656
021229.7-222816	NGC 858	.SBT5?.	1.10± .04		14.3 ±.3	.57± .05			
203.16 -71.05	ESO 478- 13	PS (2)	.06± .03	.00	14.19 ±.14				12356± 59
278.45 -23.49	IRAS02102-2242	5.0± .5		.09	12.85	.48			12317
021011.0-224218	PGC 8451	1.3± .5	1.10	.03	14.06		14.49± .37		12131
0212.5 +1651		.S..6*.	.96± .09	130					4242
149.36 -41.82	UGC 1700	U	.69± .06	.24					4320
318.81 -13.14		6.0±1.4		1.01					3984
0209.8 +1637	PGC 8452		.98	.34					
021235.1-315646	NGC 857	.LAT0*.	1.18± .05		13.45 ±.14				3493± 39
233.09 -71.74	ESO 415- 6	Sr	.08± .04	.03					3427
268.01 -24.49		-1.8± .4		.00					3293
021024.0-321048	PGC 8455		1.18		13.37				
0212.7 +1358		.SA.7?.	1.00± .06						12313
151.08 -44.42	UGC 1702	U	.03± .05	.25					
316.06 -14.20		7.0±1.7		.04					12383
0210.0 +1344	PGC 8464		1.02	.01					12053
021245.8+361807		PS..3*.	1.09± .06						
140.74 -23.74	UGC 1701	U (1)	.09± .05	.25	14.65 ±.15				
336.92 -5.62		3.0±1.2		.12					
020945.5+360405	PGC 8469	3.5±1.1	1.12	.04					
021253.5-614736		.SXR1*.	1.01± .06	59					
287.16 -52.77	ESO 114- 31	r	.44± .05	.07	15.38 ±.14				
235.07 -22.97		1.0±1.3		.45					
021128.0-620136	PGC 8477		1.02	.22					
021254.5+374855	A 0209+37								5310±125
140.22 -22.31	CGCG 522-137			.19	15.7 ±.4				5438
338.30 -5.02									5083
020952.7+373453	PGC 8479								
021255.0-191853		.S?....	1.12± .05	155					
194.57 -69.75	ESO 544- 27		.65± .04	.07	15.44 ±.14				2483± 60
281.90 -23.11	IRAS02105-1932			.98					2453
021034.0-193254	PGC 8480		1.12	.33	14.38				2251
021256.6-150823		.SBS6..	1.15± .06	102					5030
184.74 -67.58	MCG -3- 6- 15	E (1)	.24± .05	.00					
286.41 -22.37		6.0± .9		.36					5013
021032.7-152224	PGC 8483	4.2± .8	1.15	.12					4790
0212.9 +3251		.I..9*.	1.22± .12						
142.09 -26.97	UGC 1703	U	.25± .12	.35					
333.79 -7.06		10.0±1.2		.19					
0210.0 +3237	PGC 8484		1.25	.13					
021302.9-420200	NGC 862	.E...*.	.94± .04		12.78V±.13	.96± .01	.97± .01		
260.01 -67.65	ESO 298- 20	BS	.03± .03	.03	13.86 ±.14	.40± .05			5310± 26
256.78 -24.84		-5.0± .7	.54± .02	.00		.90			5216
021102.0-421600	PGC 8487		.93		13.68	.42	13.35± .24		5142
0213.0 +0930		.SXS3..	.96± .09						
154.11 -48.37	UGC 1705	U	.29± .06	.20	15.30 ±.12				
311.74 -15.81		3.0± .9		.41					
0210.4 +0916	PGC 8489		.98	.15					
021312.3-705449		.S..5./	1.19± .05	105					
293.68 -44.72	ESO 53- 2	S	1.05± .04	.02	16.29 ±.14				
225.40 -21.17		5.0± .8		1.50					
021225.0-710848	PGC 8499		1.19	.50					
021315.7-073946		PSBT1..	1.23± .04	50					
171.59 -62.53	MCG -1- 6- 70	E	.12± .05	.03					
294.35 -20.83	KUG 0210-078	1.0± .8		.12					
021046.9-075346	PGC 8502		1.23	.06					
021316.3+411431		.S..7..	1.12± .05	35					
139.08 -19.05	UGC 1704	U	.41± .05	.26	15.19 ±.14				
341.43 -3.66		7.0± .9		.56					
021010.9+410030	PGC 8503		1.14	.20					
0213.2 +5324		.S...0..	1.19± .05	76					
135.06 -7.52	UGC 1699	U	.45± .05	1.14	14.83 ±.09				
352.36 1.45	IRAS02099+5310	.0± .8		.34	12.77				
0209.9 +5310	PGC 8504		1.27						

R.A. 2000 DEC.　l b　SGL SGB　R.A. 1950 DEC.	Names　PGC	Type　S_T n_L　T　L	$\log D_{25}$　$\log R_{25}$　$\log A_e$　$\log D_o$	p.a.　A_g　A_i　A_{21}	B_T　m_B　m_{FIR}　B_T^o	$(B-V)_T$　$(U-B)_T$　$(B-V)_T^o$　$(U-B)_T^o$	$(B-V)_e$　$(U-B)_e$　m'　m'_{25}	m_{21}　W_{20}　W_{50}　HI	V_{21}　V_{opt}　V_{GSR}　V_{3K}
0213.4　+0445　157.87 −52.47　307.06 −17.42　0210.8　+0431	UGC 1707　PGC 8512	.SBS3..　U　3.0± .9	.98± .07　.20± .05　.99	0　.10　.28　.10	15.15 ±.13				
021333.6+255107　145.16 −33.45　327.39 −9.97　021042.2+253707	UGC 1706　PGC 8520	.S..6*.　U　6.0±1.3	1.02± .05　.44± .04　1.04	154　.17　.65　.22	14.73 ±.13　13.89			16.20±.3　289± 13　2.09	4794± 10　4895　4546
0213.6　+1343　151.51 −44.55　315.89 −14.49　0210.9　+1329	UGC 1710　PGC 8522	.SBR1..　U　1.0± .9	.87± .10　.01± .06　.90	.29　.01　.01					12327　12395　12068
021337.3+170500　149.54 −41.50　319.13 −13.29　021052.4+165100	CGCG 461− 63　MK 367　PGC 8525			.24	15.4 ±.3				11019± 48　11097　10762
021337.6−004303　163.10 −57.04　301.56 −19.09　021104.3−005702	NGC 856　UGC 1713　PGC 8526	PSAT0*.　UE　.3± .7	1.11± .05　.15± .04　1.11	20　.05　.11	14.13 ±.12　13.87				5995± 50　6021　5738
021338.1+163550　149.82 −41.94　318.67 −13.47　021053.5+162150	IC 212　CGCG 461− 62　PGC 8527		.95?　.18?　.98	.24	15.23 ±.13				11092　11168　10835
021338.3−394431　254.66 −68.80　259.34 −24.97　021135.1−395830	ESO 298− 21　IRAS02115-3958　PGC 8528	RLB.+?.　r　-1.3±1.8	1.17± .05　.67± .05　1.07	170　.03　.00	14.93 ±.14　13.36　14.82				5247± 60　5159　5073
0213.6　+1020　153.73 −47.56　312.61 −15.67　0211.0　+1006	UGC 1714　PGC 8530	.S..8*.　U　8.0±1.4	1.07± .07　.79± .06　1.09	77　.21　.98　.40					3596　3655　3336
021340.4+311131　142.91 −28.47　332.33 −7.87　021044.7+305731	CGCG 504− 32　PGC 8531		.95?　.65?　.98	.26	15.70 ±.13			17.47±.3　404± 7	8886± 10　8999　8647
021344.6+245320　145.64 −34.33　326.51 −10.38　021054.1+243920	UGC 1711　PGC 8535	.I..9?.　U　10.0±1.9	.96± .09　.39± .06　.98	127　.25　.29　.20				16.67±.3　158± 7	2640± 10　2738　2391
0213.7　+1706　149.57 −41.48　319.16 −13.31　0211.0　+1652	CGCG 461− 64　PGC 8536			.28	15.32 ±.12				4360± 53　4438　4103
021345.0+040607　158.54 −52.98　306.44 −17.70　021108.7+035208	A 0211+03　UGC 1716　MK 589　PGC 8537	.S?....　　.69	.68± .09　.02± .05	.10　.02　.01	14.51 ±.20　12.20　14.35			16.33±.2　186± 6　1.98	3436± 7　3464± 24　3479　3179
021348.4+333605　141.99 −26.20　334.54 −6.92　021050.5+332206	UGC 1712　PGC 8540	.SA.8*.　U　8.0±1.2	.96± .17　.00± .12　.99	.29　.00　.00				16.31±.2　139± 9　126± 7	5094± 5　5212　4859
021350.2+315201　142.68 −27.83　332.96 −7.63　021053.8+313802	CGCG 504− 34　MK 1175　PGC 8541			.31	15.4 ±.4				5728± 97　5843　5490
0213.8　+1647　149.77 −41.74　318.88 −13.44　0211.1　+1634	UGC 1717　PGC 8543	.SB.6*.　U　6.0±1.3	.98± .07　.23± .05　1.00	73　.24　.33　.11	15.31 ±.14　14.69				10796　10873　10539
0214.0　+1627　150.01 −42.02　318.58 −13.61　0211.3　+1614	IC 213　UGC 1719　PGC 8556	.SXT3..　U (1)　3.0± .8　5.5±1.0	1.28± .04　.11± .05　1.31	150　.24　.16　.06	14.6 ±.3　14.18				8221　8297　7964
021403.7+275238　144.39 −31.53　329.31 −9.27　021110.8+273839	NGC 855　UGC 1718　IRAS02111+2738　PGC 8557	.E.....　U　-5.0± .7	1.42± .04　.44± .12　.82± .03　1.32	.22　.00	13.30 ±.13　12.85 ±.10　12.89　12.79	.71± .02　.03± .06　.65　−.01	.72± .01　12.89± .10　14.30± .38		567± 24　672　322

R.A. 2000 DEC.　l　b　SGL　SGB　R.A. 1950 DEC.	Names　　PGC	Type　S_T　n_L　T　L	$\log D_{25}$　$\log R_{25}$　$\log A_o$　$\log D_o$	p.a.　A_g　A_i　A_{21}	B_T　m_B　m_{FIR}　B_T^o	$(B-V)_T$　$(U-B)_T$　$(B-V)_T^o$　$(U-B)_T^o$	$(B-V)_o$　$(U-B)_o$　m'_o　m'_{25}	m_{21}　W_{20}　W_{50}　HI	V_{21}　V_{opt}　V_{GSR}　V_{3K}
021405.8+051032 157.76 −52.02 307.54 −17.45 021128.8+045633	IC　214 UGC　1720 MK　1027 PGC　8562	.I?.... 	.92± .05 .14± .04 .93	 .12 .10 .07	14.7　±.2 14.35 ±.13 11.37 14.23	.53± .05 −.27± .09 .41 −.35	 13.81± .34	16.03±.3 447± 14 1.73	9061± 10 9007± 39 9101 8798
021412.5−310856 230.42 −71.51 268.92 −24.78 021201.0−312254	 ESO　415− 10 PGC　8568	.SBS7.. S　　(1) 7.0± .9 5.6± .9	1.16± .04 .62± .03 1.16	30 .03 .85 .31	 15.14 ±.14 14.25				3707 3642 3507
021412.7+233731 146.35 −35.45 325.37 −10.97 021123.0+232333	 CGCG 483− 10 PGC　8569			 .26	15.7　±.4			15.92±.3 184± 7	9780± 10 9875 9530
021412.7−544108 279.86 −58.58 242.76 −24.17 021231.0−545506	 ESO　153− 29 PGC　8570	PSXR4*. r 4.5± .9 	1.02± .06 .11± .05 1.02	75 .04 .17 .06	 14.67 ±.14				
0214.2　+0733 155.91 −49.93 309.93 −16.72 0211.6　+0720	 UGC　1723 PGC　8571	.SX.6.. U 6.0± .8	1.09± .06 .13± .05 1.11	50 .23 .20 .07					
021415.7−320314 233.23 −71.37 267.92 −24.85 021205.0−321712	 ESO　415− 11 PGC　8573	.SBS3P? S　　(1) 3.0±1.8 5.6±1.3	1.03± .05 .54± .04 1.03	132 .03 .74 .27	 15.99 ±.14				
0214.2　+5000 136.30 −10.69 349.38　−.13 0211.0　+4947	 UGC　1715 PGC　8574	.SX.6*. U 6.0±1.2	1.04± .08 .13± .06 1.16	148 1.29 .19 .06					
0214.3　+0750 155.73 −49.67 310.22 −16.65 0211.7　+0737	 UGC　1724 PGC　8575	.SXS5.. U 5.0± .8	1.04± .06 .00± .05 1.07	 .28 .00 .00	 15.1　±.2				
021426.8−072129 171.61 −62.10 294.75 −21.04 021157.8−073527	A　0212−07 MCG −1− 6− 77 IRAS02119−0736 PGC　8581	.S..3P/ E 3.0± .8 1.36	1.36± .03 .81± .05 1.36	116 .04 1.12 .41	 11.80				4983 4988 4733
021427.3+012831 161.19 −55.11 303.85 −18.66 021152.6+011433	 UGC　1725 KUG 0211+012 PGC　8582	.S..8*. U 8.0±1.4	1.02± .04 .80± .05 1.03	84 .10 .99 .40					9021 9053 8764
021433.7−004600 163.50 −56.94 301.58 −19.33 021200.6−005957	NGC　863 UGC　1727 MK　590 PGC　8586	.SAS1*. UE 1.3± .7	1.03± .03 .03± .04 1.04	 .06 .03 .02	 13.85 ±.12 13.61 13.66			15.62±.1 367± 15 286± 15 1.95	7910± 12 8061± 60 7942 7660
021434.2+372429 140.70 −22.58 338.07　−5.49 021132.5+371031	 UGC　1721 PGC　8587	.SBT4.. U 4.0± .7	1.30± .05 .00± .06 1.32	 .19 .00 .00	 14.0　±.2 13.74			14.80±.2 127± 6 114± 7 1.06	4640± 7 4767 4414
021448.7−245058 210.68 −71.15 275.92 −24.32 021232.0−250454	 ESO　478− 15 PGC　8598	PSBS3P. r 3.3± .9	.92± .05 .06± .04 .92	 .00 .09 .03	 14.65 ±.14 14.48				11420± 69 11373 11203
021450.7+312817 143.07 −28.12 332.69　−7.98 021154.6+311420	 UGC　1726 IRAS02118+3114 PGC　8599	.S..4.. U 4.0± .9	1.12± .08 .57± .07 1.14	 .28 .84 .28	 14.62 ±.13 13.22 13.46			14.82±.2 365± 9 342± 7 1.08	5276± 6 5389 5038
021457.2−201240 197.50 −69.70 281.01 −23.72 021237.1−202636	 ESO　544− 30 IRAS02125−2026 PGC　8602	.SBS8P* SE　　(1) 7.7± .6 5.6±1.1	1.31± .03 .20± .03 1.31	103 .00 .25 .10	 13.52 ±.14 13.27			15.31±.3 104± 16 68± 12 1.94	1608± 11 1602± 60 1574 1380
021504.4+324330 142.61 −26.93 333.85　−7.52 021207.1+322934	 UGC　1729 PGC　8609	.SXS6.. U 6.0± .8	1.12± .08 .17± .07 1.15	 .31 .25 .09	 15.12 ±.19 14.54			15.30±.2 198± 9 176± 7 .68	4443± 6 4559 4207
0215.2　+1818 149.33 −40.22 320.45 −13.20 0212.5　+1805	 UGC　1731 PGC　8617	.E...?. U −5.0±1.6	1.01± .12 .03± .05 1.06	 .40 .00	 14.51 ±.11				

R.A. 2000 DEC. / l b / SGL SGB / R.A. 1950 DEC.	Names / PGC	Type / S_T n_L / T / L	$\log D_{25}$ / $\log R_{25}$ / $\log A_e$ / $\log D_o$	p.a. / A_g / A_i / A_{21}	B_T / m_B / m_{FIR} / B_T^o	$(B-V)_T$ / $(U-B)_T$ / $(B-V)_T^o$ / $(U-B)_T^o$	$(B-V)_e$ / $(U-B)_e$ / m' / m'_{25}	m_{21} / W_{20} / W_{50} / HI	V_{21} / V_{opt} / V_{GSR} / V_{3K}
0215.2 +1840 / 149.14 -39.89 / 320.80 -13.07 / 0212.5 +1827	UGC 1732 / PGC 8618	.S..6*. / U / 6.0±1.5	1.00± .08 / 1.02± .06 / 1.04	23 / .40 / 1.47 / .50					8217 / / 8298 / 7963
0215.2 +4951 / 136.51 -10.78 / 349.32 -.34 / 0212.0 +4938	UGC 1728 / PGC 8621	.S..7.. / U / 7.0±1.0	1.22± .12 / 1.16± .12 / 1.36	37 / 1.45 / 1.38 / .50					8217 (blank)
021520.5+220002 / 147.44 -36.85 / 323.95 -11.83 / 021231.9+214606	UGC 1733 / PGC 8624	.S..6*. / U / 6.0±1.3	1.26± .04 / 1.04± .04 / 1.29	128 / .29 / 1.47 / .50	/ 15.48 ±.12 / / 13.70			15.07±.3 / 279± 13 / / .87	4415± 10 / / 4505 / 4164
0215.4 -0401 / 167.54 -59.41 / 298.29 -20.43 / 0212.9 -0415	/ PGC 8628	.SAS9*. / E (1) / 9.0±1.3 / 10.9± .8	1.26± .07 / .47± .08 / 1.27	157 / .04 / .48 / .23					
021525.9-174653 / 191.55 -68.50 / 283.68 -23.44 / 021304.1-180048	NGC 872 / ESO 544- 32 / PGC 8629	.SBS5?. / SE (2) / 5.3± .7 / 4.3± .8	1.17± .03 / .29± .03 / 1.17	174 / .02 / .44 / .15	/ 14.55 ±.14 / / 14.07				4356± 60 / 4329 / 4124
021527.4+060005 / 157.55 -51.13 / 308.48 -17.51 / 021249.8+054610	NGC 864 / UGC 1736 / IRAS02128+0546 / PGC 8631	.SXT5.. / R (2) / 5.0± .3 / 3.9± .4	1.67± .02 / .12± .02 / 1.51± .14 / 1.68	20 / .14 / .18 / .06	11.4 ±.5 / 11.63 ±.12 / 11.60 / 11.28	.55± .07 / / .48 /	.63± .03 / / 14.44± .32 / 14.27± .51	12.45±.1 / 232± 5 / 220± 5 / 1.10	1560± 4 / 1550± 58 / 1605 / 1302
0215.6 +0139 / 161.44 -54.79 / 304.13 -18.88 / 0213.0 +0125	UGC 1741 / IRAS02130+0125 / PGC 8635	.SB.4.. / U / 4.0±1.0	.93± .07 / .47± .05 / .94	178 / .08 / .70 / .24	/ 15.07 ±.09 / 13.59 / 14.23				9004 / / 9036 / 8748
021538.3+353124 / 141.64 -24.27 / 336.45 -6.47 / 021238.3+351729	UGC 1735 / PGC 8636	.L..-*. / U / -3.0±1.1	1.10± .10 / .05± .05 / 1.13	/ .25 / .00 /	/ 13.79 ±.10 / / 13.42				8008± 52 / 8130 / 7779
0215.7 +1523 / 151.13 -42.81 / 317.70 -14.37 / 0213.0 +1510	UGC 1742 / PGC 8641	.SX.6.. / U / 6.0± .9	.95± .07 / .03± .05 / .98	/ .28 / .04 / .01	/ 15.22 ±.16 /				
0215.7 +2512 / 146.00 -33.85 / 326.99 -10.67 / 0212.9 +2459	UGC 1739 / PGC 8642	.S?....	1.03± .06 / .60± .05 / 1.04	37 / .19 / .90 / .30	/ 14.98 ±.14 / / 13.86				5085 / / 5183 / 4838
021549.7-282230 / 221.67 -71.31 / 272.04 -24.91 / 021336.0-283624	ESO 415- 14 / IRAS02136-2836 / PGC 8648		.89± .07 / .26± .06 / .89	/ .00 / /	/ 15.21 ±.14 / 13.40 /				10600± 69 / 10542 / 10393
021550.3-311206 / 230.45 -71.16 / 268.89 -25.13 / 021339.0-312600	IC 1788 / ESO 415- 15 / IRAS02136-3125 / PGC 8649	.SBS4?. / PS (2) / 4.0± .6 / 3.4± .6	1.41± .03 / .36± .03 / .90± .02 / 1.41	27 / .03 / .54 / .18	12.34V±.13 / 13.12 ±.11 / 11.84 / 12.50	.70± .02 / .09± .03 / .60 / .01	.78± .01 / .18± .02 / / 14.03± .21	13.61±.3 / 480± 14 / / .92	3526± 10 / 3358± 66 / 3456 / 3323
021551.2+355448 / 141.53 -23.89 / 336.82 -6.35 / 021250.8+354053	NGC 861 / UGC 1737 / IRAS02128+3540 / PGC 8652	.S..3.. / U (1) / 3.0± .9 / 4.5±1.2	1.18± .05 / .46±.05 / 1.20	38 / .26 / .63 / .23	/ 14.63 ±.12 / 13.76 / 13.67			15.90±.3 / 524± 9 / / 2.00	8199± 8 / / 8322 / 7971
021554.0+014654 / 161.42 -54.65 / 304.27 -18.91 / 021319.2+013300	UGC 1746 / PGC 8653	.SBT7*. / UE (1) / 6.5± .8 / 5.3±1.6	1.11± .05 / .12± .04 / 1.12	95 / .10 / .16 / .06	/ 14.71 ±.16 / / 14.42				6047 / / 6079 / 5791
021554.1+424927 / 139.01 -17.40 / 343.06 -3.44 / 021246.2+423532	UGC 1738 / PGC 8654	.S..6*. / U / 6.0±1.3	1.09± .06 / .70± .05 / 1.13	105 / .44 / 1.03 / .35	/ 15.48 ±.12 / / 13.99				5734± 67 / 5871 / 5522
021558.7-004253 / 163.97 -56.69 / 301.74 -19.66 / 021325.5-005647	NGC 868 / UGC 1748 / PGC 8659	.L..-*. / E / -3.0±1.3	1.11± .10 / .10± .06 / 1.10	95 / .08 / .00 /	/ 14.93 ±.11 /				
021601.7-231819 / 206.31 -70.51 / 277.66 -24.40 / 021344.0-233212	NGC 874 / ESO 478- 18 / PGC 8663	.S..2?P / S / 2.0±1.7	.97± .05 / .29± .03 / .97	173 / .00 / .36 / .15	/ 15.12 ±.14 /				

R.A. 2000 DEC. l b SGL SGB R.A. 1950 DEC.	Names PGC	Type S_T n_L T L	$\log D_{25}$ $\log R_{25}$ $\log A_e$ $\log D_o$	p.a. A_g A_i A_{21}	B_T m_B m_{FIR} B_T^o	$(B-V)_T$ $(U-B)_T$ $(B-V)_T^o$ $(U-B)_T^o$	$(B-V)_.$ $(U-B)_.$ $m'_.$ m'_{25}	m_{21} W_{20} W_{50} HI	V_{21} V_{opt} V_{GSR} V_{3K}
021603.1-202912 198.52 -69.57 280.75 -24.02 021343.2-204306	 PGC 8666	.IXS9.. S (1) 10.0± .8 10.9± .8	1.21± .05 .11± .06 1.21	120 .00 .09 .06					1613 1578 1388
021609.8+233844 146.85 -35.26 325.56 -11.37 021319.9+232450	 MCG 4- 6- 15 PGC 8671	.SB?... 	.89± .11 .24± .07 .91	 .25 .35 .12	 15.45 ±.12 14.79			16.90±.3 310± 7 1.99	9288± 10 9381 9040
0216.1 +1818 149.59 -40.13 320.53 -13.40 0213.4 +1805	 UGC 1749 PGC 8672	.S..2.. U 2.0± .9 	1.04± .06 .54± .05 1.07	135 .39 .67 .27	 15.44 ±.12 14.30				8067 8146 7814
021610.7-115534 179.65 -64.98 290.02 -22.49 021344.8-120927	IC 217 MCG -2- 6- 46 IRAS02137-1209 PGC 8673	.S..6?/ E 6.0±1.7 	1.34± .04 .69± .05 1.34	35 .00 1.01 .34	 13.01 				1894 1884 1652
021611.9+424919 139.06 -17.38 343.08 -3.49 021303.9+423525	 UGC 1743 PGC 8674	RSB.3*. U 3.0± .8 	1.06± .06 .18± .05 1.10	18 .44 .25 .09	 15.27 ±.17 14.49				13708± 67 13845 13496
021612.9+323858 142.89 -26.92 333.88 -7.77 021315.5+322504	IC 1784 UGC 1744 IRAS02132+3225 PGC 8676	.SAT4P* PU 4.0± .6 	1.23± .05 .23± .05 1.26	88 .29 .34 .12	14.00S±.11 14.19 ±.14 13.37 13.41	.90± .07 .76 	 14.42± .29 	15.30±.2 474± 9 451± 7 1.78	4816± 6 4772± 57 4931 4581
021615.2+283601 144.60 -30.68 330.17 -9.42 021321.3+282208	NGC 865 UGC 1747 IRAS02133+2822 PGC 8678	.S?.... 	1.19± .07 .58± .07 1.21	 .26 .87 .29	 14.09 ±.14 12.46 12.94			14.68±.2 296± 13 276± 6 1.45	2995± 7 3619± 50 3112 2765
021620.9+245313 146.30 -34.10 326.74 -10.92 021330.0+243920	 UGC 1752 PGC 8681	.SAS6.. U 6.0± .8 	1.19± .12 .00± .12 1.21	 .26 .00 .00				15.75±.3 392± 7 	17836± 10 17932 17590
021623.5+315959 143.20 -27.51 333.30 -8.07 021326.7+314606	 UGC 1750 PGC 8685	.S..2.. U 2.0± .9 	1.12± .08 .46± .07 1.14	 .29 .56 .23	 14.55 ±.12 13.61			15.45±.2 426± 9 395± 7 1.61	8751± 6 8865 8516
021629.8-753150 296.09 -40.45 220.56 -20.24 021622.4-754540	 PGC 8688	.LXS+*. S -1.0±1.3 	1.11± .10 .43± .08 1.07	 .24 .00 					
021632.3+281225 144.84 -31.02 329.83 -9.64 021338.7+275832	 UGC 1753 PGC 8691	.I..9*. U 10.0±1.3 	1.04± .09 .49± .07 1.06	 .25 .37 .25	 15.05 ±.12 14.43			16.26±.3 133± 7 1.58	2993± 10 3098 2751
021632.4-112057 178.77 -64.54 290.66 -22.46 021406.2-113449	NGC 873 MCG -2- 6- 48 IRAS02140-1134 PGC 8692	.S..5P* E 5.0±1.2 	1.20± .05 .10± .05 1.20	145 .02 .15 .05	 11.15 				4014 4006 3772
021636.9+262050 145.69 -32.73 328.12 -10.40 021344.8+260657	 CGCG 483- 14 PGC 8695	 	 	 .19 	15.5 ±.4 			18.87±.3 102± 7 	8851± 10 8951 8607
021645.5-474915 269.81 -63.45 250.30 -25.23 021453.1-480306	 ESO 198- 1 PGC 8699	.E+4... S -4.0± .5 	1.17± .05 .17± .04 1.12	107 .00 .00 	 14.40 ±.14 14.12				18574±146 18464 18431
0216.8 +1147 153.75 -45.90 314.32 -15.91 0214.2 +1134	 UGC 1755 PGC 8706	.S..4.. U 4.0± .9 	.97± .09 .17± .06 1.00	162 .31 .26 .09					13346 13407 13090
021653.9+021211 161.38 -54.16 304.78 -19.03 021418.8+015819	 UGC 1756 PGC 8707	.L..0*/ E -2.0±1.3 	1.05± .06 .41± .05 1.00	52 .10 .00 	 14.92 ±.10 14.78				3012 3045 2757
021655.7+305602 143.76 -28.46 332.38 -8.61 021359.7+304210	 UGC 1754 KUG 0213+307 PGC 8708	.SB.4?. U 4.0±1.0 	.94± .07 .76± .05 .97	42 .31 1.12 .38	 15.71 ±.14 14.21			16.62±.3 278± 7 2.03	10156± 10 10267 9919

R.A. 2000 DEC. l b SGL SGB R.A. 1950 DEC.	Names PGC	Type S_T n_L T L	$\log D_{25}$ $\log R_{25}$ $\log A_e$ $\log D_o$	p.a. A_g A_i A_{21}	B_T m_B m_{FIR} B_T^o	$(B-V)_T$ $(U-B)_T$ $(B-V)_T^o$ $(U-B)_T^o$	$(B-V)_e$ $(U-B)_e$ m'_e m'_{25}	m_{21} W_{20} W_{50} HI	V_{21} V_{opt} V_{GSR} V_{3K}
021703.0+051732 158.68 −51.54 307.90 −18.11 021425.8+050341	 CGCG 413- 68 MK 1029 PGC 8714			 	.13 15.3 ±.3 12.58 				9160± 61 9202 8904
021705.0+011439 162.37 −54.93 303.82 −19.36 021430.6+010048	NGC 875 UGC 1760 PGC 8718	.L..+*. UE -.7± .7 L	1.06± .05 .02± .04 1.07	105 .12 .00 	 13.93 ±.10 13.72				6429± 38 6459 6175
021710.5+143256 152.07 −43.42 317.01 −15.00 021427.1+141905	NGC 871 UGC 1759 IRAS02144+1419 PGC 8722	.SBS5*. R 5.0± .7	1.09± .04 .43± .04 1.12	4 .31 .65 .22	14.2 ±.2 13.73 ±.17 11.64 12.95	.59± .07 .41 	 13.41± .30	14.43±.1 271± 4 247± 4 1.27	3736± 5 3717± 47 3805 3482
021711.8+014221 161.96 −54.53 304.30 −19.25 021437.0+012830	 CGCG 387- 67 MK 591 PGC 8725		.50± .14 .00± .12 .51	 .10 	 15.39 ±.19 				12459± 81 12490 12204
0217.2 −0526 170.03 −60.21 296.94 −21.22 0214.7 −0540	 MCG -1- 6- 80 PGC 8726	.SB?...	1.12± .08 .22± .07 1.12	 .00 .33 .11					5352 5361 5103
021713.4+303508 143.97 −28.76 332.08 −8.81 021417.7+302117	 CGCG 504- 47 KUG 0214+303 PGC 8729		.62± .11 .26± .12 .65	 .31 	 15.87 ±.16 			15.70±.3 318± 7	10507± 10 10617 10270
021723.1+382450 140.89 −21.44 339.21 −5.57 021419.8+381059	 UGC 1757 ARAK 79 PGC 8737	.S?....	.90± .07 .53± .05 .92	87 .21 .79 .26	 14.1 ±.3 13.09				5157± 42 5284 4936
0217.4 +1434 152.13 −43.36 317.07 −15.05 0214.7 +1421	 UGC 1761 PGC 8739	.I..9.. U 10.0± .8	1.04± .06 .00± .05 1.07	 .32 .00 .00				15.76±.2 159± 8 136± 6	3998± 6 4066 3743
021727.9-595147 284.64 −54.10 237.05 −23.85 021559.1-600536	NGC 888 ESO 115- 2 PGC 8743	.E.1.*P S -5.0± .8	1.05± .06 .08± .04 1.04	 .11 .00 	 14.47 ±.14 				
021730.9-225340 205.42 −70.07 278.16 −24.69 021513.0-230730	 ESO 478- 21 PGC 8746	.S?....	1.12± .04 .41± .04 1.12	17 .00 .60 .20	 14.85 ±.14 14.17				11605 11562 11386
0217.5 −1141 179.73 −64.57 290.35 −22.77 0215.1 −1155	 MCG -2- 6- 49 PGC 8748	.SBR7P* E (1) 7.0±1.3 6.4± .8	1.03± .07 .07± .05 1.03	55 .01 .09 .03					
021734.0+293116 144.51 −29.72 331.14 −9.31 021439.1+291726	 CGCG 504- 48 MK 1030 PGC 8750		.62± .11 .11± .12 .66	 .36 	 15.40 ±.16 12.91 				5197± 47 5304 4958
0217.6 +1230 153.50 −45.18 315.08 −15.82 0214.9 +1217	IC 1790 UGC 1762 PGC 8752	.S?....	1.02± .06 .60± .05 1.05	65 .32 .90 .30	 15.50 ±.12 14.26				3597± 67 3659 3342
0217.7 +1228 153.55 −45.20 315.06 −15.86 0215.0 +1215	IC 1791 UGC 1764 PGC 8758	.L..... U -2.0± .8	1.00± .19 .00± .08 1.04	 .32 .00 	 14.27 ±.10 13.90				3750± 67 3812 3495
021751.3+322347 143.36 −27.02 333.79 −8.19 021453.9+320957	IC 1789 UGC 1763 IRAS02149+3210 PGC 8766	.S..1*. U 1.0±1.2	1.34± .06 .78± .07 1.37	 .32 .79 .39	 14.63 ±.12 13.72 13.45			15.98±.2 457± 9 435± 7 2.13	4794± 6 4908 4561
021753.4+143116 152.30 −43.37 317.05 −15.18 021510.1+141726	NGC 876 UGC 1766 PGC 8770	.SA.5*/ RC 5.0± .4	1.32± .05 .72± .05 1.35	20 .32 1.08 .36				13.50±.3 398± 34 320± 25	3860± 17 3928 3606
021754.5-232259 206.87 −70.12 277.63 −24.84 021537.1-233648	NGC 878 ESO 478- 22 IRAS02156-2336 PGC 8771	.S..1?. S 1.0±1.8	.88± .05 .21± .03 .88	112 .00 .21 .10	 14.65 ±.14 13.15 				

R.A. 2000 DEC.	Names	Type	logD$_{25}$	p.a.	B$_T$	(B-V)$_T$	(B-V)$_e$	m$_{21}$	V$_{21}$
l b		S$_T$ n$_L$	logR$_{25}$	A$_g$	m$_B$	(U-B)$_T$	(U-B)$_e$	W$_{20}$	V$_{opt}$
SGL SGB		T	logA$_e$	A$_i$	m$_{FIR}$	(B-V)$_T^o$	m'$_e$	W$_{50}$	V$_{GSR}$
R.A. 1950 DEC.	PGC	L	logD$_o$	A$_{21}$	B$_T^o$	(U-B)$_T^o$	m'$_{25}$	HI	V$_{3K}$
021754.6-344811		.SBS3P.	1.11± .05	128					
240.79 -69.99	ESO 355- 4	S	.16± .05	.03 14.44 ±.14					6314± 52
264.88 -25.74	IRAS02157-3501	3.0± .8		.22 13.49					6237
021547.1-350200	PGC 8772		1.11	.08 14.14					6127
021756.8-760450		.SAS5..	1.21± .04	80					
296.29 -39.91	ESO 30- 14	S (1)	.13± .04	.26 14.12 ±.14					8250
219.97 -20.17	IRAS02179-7618	5.0± .8		.19 13.74					8089
021757.1-761836	PGC 8773	2.2± .8	1.24	.06 13.62					8225
021758.7+143250	NGC 877	.SXT4..	1.38± .02	140 12.58 ±.15		.67± .03	.75± .03	13.65±.1	3913± 3
152.31 -43.33	UGC 1768	R (2)	.12± .02	.32 12.35 ±.11		.02± .08	.11± .08	422± 3	3983± 58
317.09 -15.19		4.0± .4	.97± .02	.17		.55	12.92± .05	395± 3	3981
021515.3+141901	PGC 8775	1.9± .7	1.41	.06 11.91		-.07	14.03± .19	1.69	3659
021759.7+354544		.SB.8?.	1.04± .06	46					
142.03 -23.88	UGC 1765	U	.25± .05	.27 15.12 ±.14					
336.87 -6.80		8.0±1.2		.31					
021459.1+353154	PGC 8777		1.06	.13					
021800.4+315257								15.88±.3	6062± 10
143.60 -27.49	CGCG 504- 50			.28 15.4 ±.4					6174
333.34 -8.43								301± 7	6174
021503.4+313908	PGC 8778								5828
021802.5-473218		.L?....	1.05± .06	3 14.4 ±.2		1.10± .03	1.13± .03		
269.02 -63.49	ESO 198- 2		.26± .03	.00 14.51 ±.14					
250.60 -25.47			.66± .07	.00			13.15± .23		
021610.0-474606	PGC 8780		1.01				13.81± .37		
021805.2+380438		.I..9..	1.02± .06					16.46±.1	5159± 9
141.16 -21.71	UGC 1767	U	.02± .05	.22 13.95 ±.12				139± 11	5154± 52
338.96 -5.84		10.0± .8		.02					5285
021502.2+375049	PGC 8782		1.04	.01 13.71				2.74	4937
021811.4+370545		.S..4..	1.00± .06	123					
141.55 -22.62	UGC 1769	U	.21± .05	.15 14.02 ±.14					8013
338.09 -6.27	IRAS02151+3652	4.0± .9		.31 13.14					8137
021509.4+365156	PGC 8786		1.01	.11 13.51					7789
0218.2 +1312		.SB?...	1.17± .05	48					3622
153.25 -44.50	UGC 1773		.75± .05	.32 15.10 ±.12					
315.81 -15.73	IRAS02155+1258			1.12 13.12					3686
0215.5 +1258	PGC 8788		1.20	.37 13.63					3368
0218.3 +3754		.S..9*.	1.20± .06					16.00±.1	4331± 6
141.27 -21.84	UGC 1771	U	.31± .06	.22				64± 6	
338.84 -5.95		9.0±1.1		.32					4457
0215.3 +3741	PGC 8798		1.22	.16					4109
021826.3+053905		.S?....	1.18± .05						
158.85 -51.05	UGC 1775		.01± .05	.14 13.80 ±.14					9093± 50
308.38 -18.32	ARP 10			.02 13.25					9135
021548.9+052517	PGC 8802		1.19	.01 13.59					8838
021827.3+380126		.I?....	1.00± .08	143					
141.25 -21.73	UGC 1772		.33± .06	.22 13.94 ±.18					5061± 52
338.95 -5.93	IRAS02154+3747			.25 12.73					5187
021524.2+374738	PGC 8804		1.02	.17 13.47					4840
021837.1-610344		.SAT4..	.98± .06						
285.55 -53.03	ESO 115- 4	r	.05± .06	.10 15.42 ±.14					14473
235.73 -23.77		4.5± .8		.08					14334
021712.1-611730	PGC 8809		.99	.03 15.16					14383
021839.0-065416	IC 219	.E...?.	1.08± .06	175					
172.58 -61.06	MCG -1- 6- 88	E	.14± .06	.02					
295.50 -21.94		-5.0±1.0		.00					
021609.9-070804	PGC 8813		1.04						
021842.9+352746		.SBS3..	1.12± .05	20					
142.30 -24.10	UGC 1776	U	.14± .05	.26 14.87 ±.17					
336.66 -7.06		3.0± .8		.19					
021542.4+351358	PGC 8820		1.14	.07					
021844.0+303009		.SXS6..	.96± .09	100				16.01±.3	4723± 10
144.35 -28.71	UGC 1777	U	.22± .06	.33 15.42 ±.13					
332.14 -9.14		6.0± .9		.32				186± 7	4832
021548.1+301621	PGC 8821		.99	.11 14.75				1.15	4487
021845.4-063824	NGC 881	.SXR5..	1.35± .03	45 13.23 ±.15		.79± .02	.87± .02		
172.25 -60.85	MCG -1- 6- 89	E (1)	.18± .03	.02		.20± .03	.29± .03		
295.78 -21.90	IRAS02160-0650	5.0± .6	.96± .03	.27 11.88			13.52± .07		
021616.1-065211	PGC 8822	.8± .6	1.35	.09			14.36± .23		

R.A. 2000 DEC. / l b / SGL SGB / R.A. 1950 DEC.	Names / PGC	Type / S_T n_L / T / L	logD₂₅ / logR₂₅ / logA_e / logD_o	p.a. / A_g / A_i / A₂₁	B_T / m_B / m_FIR / B_T^o	(B-V)_T / (U-B)_T / (B-V)_T^o / (U-B)_T^o	(B-V)_e / (U-B)_e / m'_e / m'₂₅	m₂₁ / W₂₀ / W₅₀ / HI	V₂₁ / V_opt / V_GSR / V_3K
021851.1+334330		.SA.8*.	1.08± .06	160				15.85±.2	5033± 6
143.03 -25.71	UGC 1778	U	.10± .05	.28	14.37 ±.13			203± 9	
335.09 -7.82		8.0±1.2		.13				175± 7	5149
021552.3+332943	PGC 8829		1.11	.05	13.95			1.85	4803
0218.8 +3841		.S..8..	1.00± .08	40					
141.07 -21.08	UGC 1779	U	.09± .06	.18					
339.59 -5.71		8.0± .9		.11					
0215.8 +3828	PGC 8832		1.02	.04					
021856.0-660227		PSBT3..	1.00± .06	40					
289.66 -48.77	ESO 81- 6	r	.30± .05	.03	15.08 ±.14				
230.39 -22.79		3.3± .9		.41					
021748.0-661612	PGC 8835		1.00	.15					
021856.6+403351		.IB.9*.	1.25± .04	158				15.72±.1	5204± 11
140.39 -19.32	UGC 1780	U	.68± .05	.16	15.27 ±.14				
341.27 -4.92		10.0± .9		.51				222± 8	5335
021550.7+402004	PGC 8836		1.27	.34	14.59			.78	4989
0219.0 +0448		.S..7..	1.11± .14	150					5795
159.76 -51.69	UGC 1785	U	.29± .12	.14					
307.59 -18.73		7.0± .9		.41					5834
0216.4 +0435	PGC 8838		1.12	.15					5541
021905.3-064731	NGC 883	.LAS-*.	1.23± .06	100					
172.58 -60.91	MCG -1- 6- 90	E	.12± .05	.02					
295.64 -22.02		-3.0± .6		.00					
021636.1-070118	PGC 8841		1.22						
021906.9-414456	NGC 889	.E...*.	1.02± .06						5258
257.78 -66.87	ESO 298- 27	BS	.08± .04	.03	14.36 ±.14				
257.07 -25.97		-4.6± .6		.00					5162
021707.0-415842	PGC 8843		1.00		14.25				5095
0219.1 +4244		.SB.8*.	1.20± .06	63					
139.62 -17.27	UGC 1782	U	.56± .06	.38					
343.24 -4.01		8.0± .8		.69					
0216.0 +4231	PGC 8844		1.23	.28					
0219.1 +4407		.I..9*.	1.14± .13						
139.12 -15.97	UGC 1783	U	.14± .12	.29					
344.48 -3.41		10.0±1.2		.10					
0216.0 +4354	PGC 8846		1.17	.07					
0219.2 +3640		.I..9*.	1.04± .15						6476
141.93 -22.93	UGC 1784	U	.23± .12	.16					
337.80 -6.64		10.0±1.2		.17					6599
0216.2 +3627	PGC 8849		1.06	.11					6252
021914.7-185556		.SBS9..	1.34± .02	53					1608
195.46 -68.24	ESO 545- 2	SE (2)	.38± .03	.07	14.79 ±.14				
282.58 -24.52		9.0± .4		.39					1576
021654.1-190942	PGC 8851	8.8± .4	1.35	.19	14.32				1383
021917.3+334818			.95?						8567
143.09 -25.60	CGCG 523- 19		.48?	.28	15.52 ±.12				
335.20 -7.87									8683
021618.3+333432	PGC 8853		.98						8338
021926.9-505109		.S?....	1.09± .05	122					14663
273.74 -60.99	ESO 198- 6		.34± .05	.02	14.82 ±.14				
246.89 -25.39				.46					14545
021740.0-510454	PGC 8860		1.09	.17	14.23				14533
021930.8+370642		.E...*.	1.00± .19	13					
141.81 -22.51	UGC 1786	U	.28± .08	.14	15.23 ±.11				
338.21 -6.51		-5.0±1.3		.00					
021628.5+365256	PGC 8866		.94						
021932.7-160413	NGC 887	.SXT5..	1.29± .04	5					4310
189.02 -66.78	MCG -3- 7- 1	E (1)	.11± .04	.16					
285.73 -24.11	IRAS02171-1617	5.0± .6		.16	12.33				4286
021710.0-161758	PGC 8868	.8± .6	1.30	.05					4079
021937.2-374909		.SAT4..	1.44± .04	23					
248.48 -68.63	ESO 298- 28	S (1)	.34± .05	.03	13.65 ±.14				5061± 34
261.49 -26.13		4.0± .5		.51					4975
021733.0-380254	PGC 8871	3.3± .6	1.44	.17	13.08				4885
021938.4+375611		.S..8*.	1.13± .05	117				15.25±.3	6421± 8
141.52 -21.73	UGC 1787	U	.48± .05	.22	14.63 ±.12			424± 9	
338.97 -6.17	IRAS02165+3742	8.0±1.3		.59	13.55				6546
021635.3+374226	PGC 8873		1.15	.24	13.80			1.21	6200

R.A. 2000 DEC. l b SGL SGB R.A. 1950 DEC.	Names PGC	Type S_T n_L T L	$\log D_{25}$ $\log R_{25}$ $\log A_\bullet$ $\log D_\circ$	p.a. A_g A_i A_{21}	B_T m_B m_{FIR} B_T°	$(B-V)_T$ $(U-B)_T$ $(B-V)_T^\circ$ $(U-B)_T^\circ$	$(B-V)_\bullet$ $(U-B)_\bullet$ $m'_\bullet$ m'_{25}	m_{21} W_{20} W_{50} HI	V_{21} V_{opt} V_{GSR} V_{3K}
0219.6 +1548 152.03 -42.02 318.46 -15.10 0216.9 +1535	NGC 882 UGC 1789 PGC 8874	.L..... U -2.0± .9 	1.09± .10 .31± .05 1.09	82 .36 .00 	 14.58 ±.10 				
021941.1-001523 164.83 -55.78 302.49 -20.41 021707.6-002908	A 0217+00 UGC 1794 MK 592 PGC 8876	PSXS3P. UE 2.5± .6 	1.06± .04 .09± .04 1.06	10 .07 .12 .04	 14.30 ±.13 12.74 14.06				7451±113 7474 7200
0219.7 +3637 142.05 -22.94 337.80 -6.75 0216.7 +3624	 UGC 1788 PGC 8877	.E...*. U -5.0±1.2 	1.04± .18 .09± .08 1.04	 .16 .00 					
021943.9-431504 260.75 -65.95 255.38 -26.03 021746.0-432848	 ESO 246- 8 PGC 8878	.S..5*. S (1) 5.0±1.3 3.3±1.3	1.16± .05 .55± .04 1.17	13 .03 .83 .28	 15.58 ±.14 14.65				11551 11451 11394
021952.9+290211 145.26 -29.97 330.90 -9.97 021658.2+284827	 UGC 1792 IRAS02169+2848 PGC 8882	.SXR5.. U 5.0± .8 	1.24± .07 .15± .07 1.27	0 .33 .23 .08	 13.98 ±.14 13.40			14.68±.3 338± 7 1.21	4987± 10 5092 4750
021953.6+281450 145.62 -30.70 330.17 -10.30 021659.5+280106	 UGC 1791 PGC 8884	.I..9*. U 10.0±1.2 	.96± .09 .00± .06 .99	 .30 .00 .00				15.99±.3 101± 7	4761± 10 4864 4523
0219.9 +0156 162.71 -53.94 304.76 -19.84 0217.4 +0143	 UGC 1797 PGC 8887	.S..0.. U .0± .9 	1.00± .07 .28± .05 .99	135 .09 .21 	 14.94 ±.12 14.46				12300± 46 12330 12048
021959.1-412416 256.81 -66.91 257.45 -26.15 021759.0-413800	NGC 893 ESO 298- 29 IRAS02179-4137 PGC 8888	.SAS5P* BS 4.5± .6 	1.13± .05 .12± .05 1.13	115 .03 .18 .06	 13.57 ±.14 12.55 13.33				5018± 34 4922 4854
022005.3+375441 141.62 -21.72 338.98 -6.27 021702.1+374057	 UGC 1793 IRAS02170+3741 PGC 8894	.S..6*. U 6.0±1.4 	1.00± .08 .69± .06 1.02	163 .21 1.01 .34	 15.66 ±.12 12.78 				
022007.1-194504 197.71 -68.40 281.72 -24.85 021747.0-195848	 ESO 545- 5 IRAS02177-1958 PGC 8896	.I..9P/ SE 10.0±1.1 	1.39± .03 .51± .03 1.39	56 .02 .38 .26	 13.82 ±.14 12.67 13.41				2332± 60 2296 2109
022008.6-371917 247.10 -68.73 262.06 -26.23 021804.0-373300	 ESO 298- 30 PGC 8898	.LA.-*. S -3.0±1.2 	1.03± .05 .49± .03 .96	48 .03 .00 	 15.53 ±.14 				
0220.3 +3554 142.45 -23.57 337.20 -7.17 0217.3 +3541	 UGC 1795 PGC 8902	.SB.7.. U 7.0± .9 	.96± .17 .05± .12 .98	 .26 .07 .02					4397 4517 4173
0220.3 +0801 157.57 -48.77 310.91 -18.00 0217.7 +0748	 UGC 1801 PGC 8904	.SBS3.. U 3.0± .9 	1.00± .06 .15± .05 1.03	80 .25 .21 .07	 15.23 ±.16 14.72				7183 7231 6930
022023.1+404731 140.57 -19.01 341.59 -5.07 021716.7+403348	 UGC 1796 PGC 8906	.SXS8.. U 8.0± .8 	1.03± .08 .00± .06 1.04	 .16 .00 .00	 15.05 ±.18 14.87			15.71±.1 100± 8 .85	6983± 11 7113 6770
022028.6+314101 144.24 -27.48 333.38 -8.99 021731.4+312718	 CGCG 504- 59 KUG 0217+314 PGC 8908	.S?.... 	.87± .08 .48± .12 .89	 .25 .70 .24	 15.55 ±.13 14.57			17.25±.3 270± 7 2.44	5830± 10 5941 5598
022029.3-333447 236.96 -69.78 266.28 -26.23 021821.0-334830	 ESO 355- 6 PGC 8909	 	.79± .06 .25± .05 .80	174 .03 	 15.11 ±.14 				9556±104 9481 9368
0220.5 +0648 158.59 -49.79 309.72 -18.44 0217.9 +0635	A 0217+06 UGC 1803 PGC 8913	.SBS9*/ PU 8.5± .6 	1.38± .04 .78± .05 1.40	57 .24 .80 .39	 14.70 ±.14 13.67				1624 1668 1372

R.A. 2000 DEC. / l b / SGL SGB / R.A. 1950 DEC.	Names / PGC	Type / S_T n_L / T / L	$\log D_{25}$ / $\log R_{25}$ / $\log A_e$ / $\log D_o$	p.a. / A_g / A_i / A_{21}	B_T / m_B / m_{FIR} / B_T^o	$(B-V)_T$ / $(U-B)_T$ / $(B-V)_T^o$ / $(U-B)_T^o$	$(B-V)_e$ / $(U-B)_e$ / m'_e / m'_{25}	m_{21} / W_{20} / W_{50} / HI	V_{21} / V_{opt} / V_{GSR} / V_{3K}
0220.6 +4826		.SXS4..	1.07± .07	23					
137.84 -11.83	UGC 1799	U	.21± .06	1.28					
348.43 -1.75		4.0± .9		.30					
0217.4 +4813	PGC 8918		1.19	.10					
022054.6+003325		.S..0..	.98± .07	60					
164.42 -54.94	UGC 1809	U	.21± .04	.08	14.85 ±.12				
303.42 -20.47	KUG 0218+003	.0± .9		.16					
021820.5+001943	PGC 8929		.97						
022058.1+383927		.SA.9?.	1.12± .05	95				16.22±.1	5188± 11
141.50 -20.96	UGC 1804	U	.19± .05	.16	15.05 ±.18				
339.73 -6.10		9.0±1.7		.19				185± 8	5314
021753.9+382545	PGC 8932		1.14	.09	14.69			1.44	4970
022100.7+332142		.I..9*.	1.04± .15	15				16.99±.3	4256± 10
143.64 -25.88	UGC 1806	U	.37± .12	.33					
334.95 -8.38		10.0±1.3		.28				109± 7	4370
021801.9+330800	PGC 8934		1.07	.19					4027
022101.8-315631		.SBS2*P	1.09± .06	0					9471± 34
232.21 -69.97	ESO 415- 19	S	.44± .04	.03	14.81 ±.14				
268.13 -26.27	IRAS02188-3210	2.0± .5		.54					9400
021852.0-321012	PGC 8936		1.09	.22	14.14				9279
022106.3+233545		.S..3..	1.01± .05					16.82±.3	9447± 10
148.14 -34.85	UGC 1808	U (1)	.03± .04	.22	14.49 ±.13			157± 13	
325.97 -12.43	IRAS02182+2322	3.0± .8		.05	13.54				9537
021816.0+232203	PGC 8941	2.5±1.1	1.03	.02	14.16			2.65	9204
022106.6+485738		.E.....	1.20± .14						
137.73 -11.32	UGC 1802	U	.02± .08	1.06	14.3 ±.3				
348.92 -1.59		-5.0± .8		.00					
021749.1+484356	PGC 8942		1.36						
022107.1-334317	NGC 897	.SAT1..	1.33± .04	17	12.79 ±.16	.96± .02	1.00± .01		
237.27 -69.62	ESO 355- 7	Sr	.21± .05	.03	13.07 ±.14	.44± .03	.53± .02		4802± 31
266.12 -26.37		1.0± .6	.98± .04	.22		.87	13.18± .12		4726
021859.0-335658	PGC 8944		1.33	.11	12.64	.40	13.74± .29		4615
0221.1 +4246		.I..9*.	1.19± .07					14.85±.1	629± 6
139.97 -17.11	UGC 1807	U	.00± .06	.32				66± 7	
343.43 -4.33		10.0±1.0		.00				53± 12	763
0218.0 +4233	PGC 8947		1.22	.00					421
022112.1-420007		PSBR1*.	1.12± .05	110					4873± 52
257.80 -66.41	ESO 298- 31	Sr	.44± .05	.03	15.08 ±.14				
256.77 -26.35		1.3± .5		.45					4775
021913.0-421348	PGC 8949		1.12	.22	14.54				4713
0221.2 +1343		.S..6*.	.93± .10	145					7838
153.80 -43.70	UGC 1811	U	.18± .06	.38					
316.59 -16.21		6.0±1.2		.26					7901
0218.5 +1330	PGC 8950		.96	.09					7587
022118.5-341908		.L..0.P	1.16± .04	165	14.9 ±.2	1.06± .03			
238.90 -69.45	ESO 355- 8	S	.22± .03	.03	14.78 ±.14	.43± .06			
265.45 -26.42		-2.0± .8		.00					
021911.0-343248	PGC 8953		1.13				15.00± .31		
0221.3 +2524		.S..6*.	.98± .07	20					4532
147.30 -33.17	UGC 1812	U	.34± .05	.20	14.65 ±.13				
327.69 -11.75		6.0±1.3		.50					4627
0218.5 +2511	PGC 8954		1.00	.17	13.93				4291
022124.3+163357		.SXS4..	1.40± .04	157				15.02±.1	4104± 6
152.08 -41.16	UGC 1814	U	.32± .05	.41	13.56 ±.15			219± 8	4134± 38
319.34 -15.21	IRAS02186+1620	4.0± .8		.47	12.89			211± 10	4176
021839.2+162016	PGC 8956		1.44	.16	12.66			2.20	3856
022128.8+392231 A 0218+39A		.SAS3P.	1.34± .04	50				14.67±.1	7563± 11
141.32 -20.26	UGC 1810	R	.18± .04	.20	13.42 ±.13				7465± 28
340.42 -5.88	5ZW 223	3.0± .4		.25				523± 8	7678
021823.8+390851	PGC 8961		1.36	.09	12.92			1.66	7335
022129.2-042443		.SBS7*/	1.14± .06	127					2318
170.29 -58.74	MCG -1- 7- 1	E (1)	.61± .05	.00					
298.31 -21.99		7.0±1.3		.84					2328
021858.5-043823	PGC 8962	7.5±1.2	1.14	.30					2073
022130.7+154540	IC 1794	.L?....	.97± .15						3996
152.60 -41.86	MCG 3- 7- 3		.10± .07	.40	14.66 ±.10				
318.58 -15.54	IRAS02187+1532			.00					4065
021846.2+153200	PGC 8963		1.00		14.19				3747

R.A. 2000 DEC.	Names	Type	logD$_{25}$	p.a.	B$_T$	(B-V)$_T$	(B-V)$_●$	m$_{21}$	V$_{21}$
l b	S$_T$ n$_L$	logR$_{25}$	A$_g$	m$_B$	(U-B)$_T$	(U-B)$_●$	W$_{20}$	V$_{opt}$
SGL SGB	T	logA$_●$	A$_I$	m$_{FIR}$	(B-V)$_T^o$	m'$_●$	W$_{50}$	V$_{GSR}$
R.A. 1950 DEC.	PGC	L	logD$_●$	A$_{21}$	B$_T^o$	(U-B)$_T^o$	m'$_{25}$	HI	V$_{3K}$
0221.5 +1412		.S..7..	1.39± .04	163					3742
153.57 -43.24	UGC 1817	U	.94± .05	.38	14.57 ±.12				
317.08 -16.11		7.0± .9		1.30					3807
0218.8 +1359	PGC 8964		1.43	.47	12.88				3492
022132.3+310236		.S..9*.	1.05± .06					16.31±.2	4762± 7
144.75 -27.98	UGC 1815	U	.05± .06	.27	15.20 ±.20			75± 13	
332.89 -9.46		9.0±1.2		.05				61± 7	4870
021835.5+304856	PGC 8968		1.08	.03	14.87			1.41	4530
022132.4+323243	IC 1793	.S..2..	1.08± .09					15.81±.2	5312± 6
144.10 -26.60	UGC 1816	U	.32± .07	.30	14.74 ±.12			567± 9	
334.26 -8.83	IRAS02186+3219	2.0± .8		.40	13.41			541± 7	5424
021834.3+321902	PGC 8969		1.10	.16	13.99			1.66	5082
022132.6+392124	A 0218+39B	.SBS1P.	1.17± .05	93					7335± 46
141.34 -20.27	UGC 1813	R	.57± .04	.20	15.08 ±.13				
340.40 -5.89		1.0± .5		.58					7462
021827.5+390743	PGC 8970		1.18	.28	14.21				7119
0221.5 +1622		.I..9*.	.96± .09						3897± 10
152.23 -41.31	UGC 1819	U	.00± .06	.43					
319.17 -15.32		10.0±1.2		.00					3968
0218.8 +1609	PGC 8972		1.00	.00					3648
022136.6-053118	NGC 895	.SAS6..	1.56± .02	65	12.26 ±.17	.53± .03		13.22±.0	2289± 5
171.78 -59.55	MCG -1- 7- 2	R (3)	.15± .03	.03	11.9 ±.2	-.05± .03		274± 4	2344± 50
297.15 -22.30	IRAS02191-0544	6.0± .3	1.20± .03	.22	12.38	.48	13.75± .08	250± 5	2296
021906.6-054458	PGC 8974	1.9± .5	1.56	.07	11.87	-.09	14.55± .21	1.27	2046
022136.9-334832		.SB.5?.	1.10± .04	66					
237.45 -69.50	ESO 355- 10	S (1)	.44± .04	.03	15.58 ±.14				
266.03 -26.47		5.0±1.7		.67					
021929.0-340212	PGC 8975	5.6±1.2	1.10	.22					
0221.6 +0646		.S..9*.	1.00± .08						5763
158.98 -49.67	UGC 1821	U	.04± .06	.22					
309.78 -18.71		9.0±1.2		.04					5806
0219.0 +0633	PGC 8977		1.02	.02					5512
0221.6 +0015		.S..0..	.93± .07	30					
164.99 -55.07	UGC 1824	U	.22± .05	.09	15.29 ±.13				
303.17 -20.74		.0± .9		.17					
0219.1 +0002	PGC 8979		.93						
022140.9-271645		PSBR3..	1.13± .05	140					4909± 69
218.57 -69.96	ESO 415- 22	r	.16± .05	.00	14.12 ±.14				
273.39 -26.10		3.3± .8		.22					4850
021927.0-273024	PGC 8980		1.13	.08	13.86				4705
0221.7 +1652		.S..6*.	1.07± .06	78					3967
151.99 -40.84	UGC 1822	U	.83± .05	.46	15.57 ±.12				
319.67 -15.18		6.0±1.4		1.22					4039
0219.0 +1639	PGC 8982		1.12	.42	13.87				3718
022146.1+330119		.S..6*.	1.25± .04					15.23±.2	3960± 6
143.95 -26.14	UGC 1820	U	.93± .06	.22	15.06 ±.12			265± 9	
334.71 -8.67	KUG 0218+327	6.0±1.3		1.37				229± 7	4073
021847.5+324739	PGC 8984		1.27	.46	13.45			1.32	3731
022147.4-100124		.L..-?/	1.06± .06	45					
178.46 -62.69	MCG -2- 7- 2	E	.37± .05	.08					
292.38 -23.42	MK 1033	-3.0±1.8		.00					
021920.5-101504	PGC 8986		1.02						
022152.7-204928	NGC 899	.IBS9..	1.27± .03	116				12.87±.2	1563± 11
200.83 -68.44	ESO 545- 7	PSU (1)	.16± .03	.00	13.08 ±.14			136± 16	1800± 48
280.60 -25.41	IRAS02195-2103	10.0± .5		.12	12.39			117± 12	1534
021933.6-210307	PGC 8990	5.6± .8	1.27	.08	12.95			-.16	1355
022159.1-372721		.SXT4*.	1.17± .04	145					
247.07 -68.34	ESO 298- 36	Sr (1)	.52± .04	.03	14.76 ±.14				4943± 60
261.90 -26.60		3.7± .6		.77					4856
021955.0-374100	PGC 8995	3.3±1.2	1.17	.26	13.93				4768
022201.1+331557	NGC 890	.LXR-$.	1.40± .04		12.2 ±.2	.98± .02	.99± .01		4006± 23
143.90 -25.89	UGC 1823	R	.16± .03	.29	12.38 ±.09	.51± .04	.55± .02		
334.96 -8.61		-3.0± .3	1.06± .05	.00		.87	12.95± .18		4119
021902.2+330218	PGC 8997		1.41		12.00	.46	13.67± .28		3778
022201.1-204421	IC 223	.IBS9P?	1.07± .04	152					
200.65 -68.38	ESO 545- 8	PS (1)	.21± .03	.00	14.04 ±.14				1600± 48
280.70 -25.43	IRAS02197-2058	9.7±1.0		.16					1560
021942.0-205800	PGC 8998	6.7±1.2	1.07	.11	13.88				1381

R.A. 2000 DEC.	Names	Type	logD25	p.a.	BT	(B-V)T	(B-V)e	m21	V21
l b		ST nL	logR25	Ag	mB	(U-B)T	(U-B)e	W20	Vopt
SGL SGB		T	logAe	Ai	mFIR	(B-V)°T	m'e	W50	VGSR
R.A. 1950 DEC.	PGC	L	logDo	A21	B°T	(U-B)°T	m'25	HI	V3K

```
022203.2-583634                  .91± .07  161 15.19 ±.14  .97± .03
282.57 -54.79  ESO  115-  7      .16± .06  .03 15.11 ±.14
238.30 -24.65
022033.0-585012 PGC  9000        .91

022203.8+321406            .SAS5.. .94± .11                            16.34±.3 10135± 10
144.35 -26.84  UGC  1825   U      .11± .07  .29 14.86 ±.12
334.02  -9.06              5.0± .9           .17                       298±  7 10246
021905.9+320026 PGC  9002         .96       .06 14.34                  1.93      9905

022205.1+335647            .SB?... 1.04± .06  45                                 5008
143.63 -25.26  UGC  1826          .47± .05  .34 14.65 ±.13
335.58  -8.33  IRAS02190+3343              .70 13.63                            5123
021905.6+334307 PGC  9004         1.07      .23 13.59                            4782

022206.8-210751            .I..9*. 1.01± .04  88
201.66 -68.50  ESO  545-  9  U    .09± .04  .00 15.72 ±.14
280.27 -25.51              10.0±1.2          .06
021948.0-212130 PGC  9005         1.01      .04

022207.3-483356            .E+2... 1.17± .09
269.67 -62.29  S                  .21± .08  .01
249.39 -26.06              -4.0± .8          .00
022017.4-484734 PGC  9006         1.10

022212.6+284403            .S?.... 1.09± .05  59                      14.81±.3  4795± 10
145.94 -30.05  UGC  1828          .69± .06  .37 15.40 ±.12                      4847± 67
330.84 -10.56  IRAS02192+2830             1.04                        230±  7  4899
021917.8+283024 PGC  9011         1.13      .35 13.97                 .50       4561

0222.2  -1003              .SBS8.. 1.09± .09 105
178.68 -62.63  E        (1)       .35± .08  .08
292.38 -23.54              8.0± .9           .43
0219.8  -1017  PGC  9013    7.5± .8 1.10     .18

022215.5+433249            .S..8*. .87± .10   0
139.88 -16.32  UGC  1827   U      .09± .06  .40 15.4  ±.3
344.20  -4.17              8.0±1.3           .12
021905.3+431910 PGC  9014         .91       .05

0222.3  +1202              .SB.6*. 1.14± .07 124                                3803
155.25 -45.05  UGC  1834   U      .27± .06  .31
315.04 -17.06              6.0±1.2           .40                                3861
0219.6  +1149  PGC  9016          1.17      .14                                3553

022223.4+284255            .SBS7.. .93± .06 148                       14.80±.3  4757± 10
145.99 -30.05  UGC  1833   U      .03± .06  .37 15.27 ±.14                      4745± 67
330.83 -10.61  KUG 0219+284 7.0± .8          .04                      165±  7  4859
021928.6+282916 PGC  9022         .96       .01 14.84                 -.06      4522

022230.2-003707            .S..7*/ 1.31± .04  45                      15.40±.1  1536±  6
166.22 -55.64  UGC  1839   UE     .93± .05  .05 15.26 ±.13           158±  8
302.33 -21.19              7.3± .7          1.28                      142± 12  1556
021956.9-005045 PGC  9028         1.32      .46 13.92                 1.01      1289

022231.2+430353            .S..1.. 1.03± .08 163
140.11 -16.75  UGC  1832   U      .36± .06  .43 15.20 ±.12
343.79  -4.43              1.0± .8           .37
021921.6+425015 PGC  9029         1.07      .18

022231.5+475100            .SB.0.. 1.39± .04 105
138.35 -12.28  UGC  1830   U      .10± .05 1.42 13.7  ±.3
348.04  -2.29  5ZW  227    .0± .7           .08 12.78
021915.3+473722 PGC  9030         1.52

022233.1+422048 NGC   891  .SAS3$/ 2.13± .01  22 10.81 ±.18  .88± .03  1.03± .02 11.67±.1  528±  4
140.38 -17.42  UGC  1831   R    (1) .73± .02  .32 10.84 ±.10  .27± .04  .36± .03 471±  4
343.16  -4.75  IRAS02193+4207 3.0± .3 1.65± .05 1.01  8.36  .67      14.55± .12 448±  4  661
021924.3+420710 PGC  9031   4.5± .9 2.16     .37  9.50  .08          14.48± .20 1.81      320

022240.5+282753            .S?.... .87± .08                           16.43±.3  9530± 10
146.17 -30.26  CGCG 504- 69       .06± .12  .26 14.91 ±.12
330.63 -10.77  KUG 0219+282                .09                        96±  7  9632
021945.9+281416 PGC  9034         .89       .03 14.52                 1.88      9295

022240.8+281523 IC   221   .S..5.. 1.22± .07                         14.87±.3  5085±  7
146.27 -30.45  UGC  1835          .13± .07  .26 13.70 ±.12           364± 13
330.44 -10.85  IRAS02197+2801 5.0± .8       .20 12.79                340± 10  5186
021946.4+280145 PGC  9035         1.24      .07 13.21                 1.60      4849

0222.7  +2421              .S..6*. 1.07± .07 169
148.17 -34.00  UGC  1838   U      .62± .06  .25
326.84 -12.47              6.0±1.3           .92
0219.9  +2408  PGC  9039          1.10      .31
```

R.A. 2000 DEC. / l b / SGL SGB / R.A. 1950 DEC.	Names / PGC	Type / S_T n_L / T / L	$\log D_{25}$ / $\log R_{25}$ / $\log A_o$ / $\log D_o$	p.a. / A_g / A_i / A_{21}	B_T / m_B / m_{FIR} / B_T^o	$(B-V)_T$ / $(U-B)_T$ / $(B-V)_T^o$ / $(U-B)_T^o$	$(B-V)_e$ / $(U-B)_e$ / m'_o / m'_{25}	m_{21} / W_{20} / W_{50} / HI	V_{21} / V_{opt} / V_{GSR} / V_{3K}
022247.5-412217	IC 1796	.LXS-?.	1.05± .05	86	13.02V±.13	1.01± .02			
256.06 -66.48	ESO 298- 38	BS	.08± .03	.03	14.16 ±.14	.41± .05			5073± 56
257.47 -26.67		-2.8± .7	.56± .02	.00		.95			4976
022048.0-413554	PGC 9041		1.04		13.98	.43	13.94± .30		4912
0222.8 +3805		.I..9?.	1.00± .16	170					
142.09 -21.35	UGC 1836	U	.25± .12	.16					
339.38 -6.68		10.0±1.8		.19					
0219.8 +3752	PGC 9043		1.02	.13					
022252.5+255637				.21	15.7 ±.4			17.28±.3	14694± 10
147.41 -32.54	CGCG 483- 19								
328.32 -11.85								166± 7	14789
022000.0+254300	PGC 9045								14456
022254.5+251837				.22	15.7 ±.4			15.77±.3	4587± 10
147.73 -33.12	CGCG 483- 18								
327.74 -12.11								182± 7	4681
022002.5+250500	PGC 9047								4348
022258.4+430042		.L.....	1.07± .10	25					6385± 39
140.21 -16.77	UGC 1837	U	.14± .05	.43	14.81 ±.12				6519
343.78 -4.52		-2.0± .8		.00					6179
021948.8+424705	PGC 9051		1.10		14.29				
022300.9-401048		.IBS9..	1.08± .08						
253.37 -67.01	ESO 298- 39	S (1)	.08± .05	.03	17.63 ±.14				
258.82 -26.76		10.0± .9		.06					
022100.0-402424	PGC 9053	11.1± .8	1.09	.04					
022302.2-204248	NGC 907	.SB.8?/	1.26± .03	81	13.21 ±.13	.57± .01	.58± .01	14.90±.2	1726± 9
200.82 -68.15	ESO 545- 10	PU (1)	.49± .03	.00	13.30 ±.14	-.06± .03	-.05± .02	204± 14	1600± 48
280.77 -25.67	IRAS02207-2056	8.0± .7	.86± .02	.61	12.06	.46	13.00± .04	197± 11	1682
022043.2-205624	PGC 9054	5.6± .9	1.26	.25	12.63	-.14	13.14± .21	2.02	1504
022304.6-211400	NGC 908	.SAS5..	1.78± .01	75	10.83 ±.13	.65± .02	.74± .01	13.34±.1	1498± 5
202.14 -68.32	ESO 545- 11	R (3)	.36± .02	.00	10.99 ±.12	.00± .08	.09± .02	419± 6	1701± 58
280.19 -25.75	IRAS02207-2127	5.0± .3	1.35± .02	.54	9.99	.57	13.07± .03	379± 6	1457
022046.0-212736	PGC 9057	1.5± .4	1.78	.18	10.36	-.06	13.68± .16	2.79	1282
0223.1 +4122	A 0220+41A	.L?....						18.11±.1	5852± 11
140.86 -18.29				.18					
342.33 -5.28								132± 8	5982
0220.0 +4108	PGC 9060								5642
022308.5+412220	A 0220+41B	.I?....	1.19± .07					16.92±.3	5425± 10
140.86 -18.28	UGC 1840		.07± .07	.18	13.84 ±.13				5254± 30
342.34 -5.28				.05				231± 8	5537
022000.8+410843	PGC 9062		1.20	.03	13.61			3.27	5197
022308.9+244443				.21	15.7 ±.4			17.84±.3	11576± 10
148.07 -33.61	CGCG 483- 21								
327.23 -12.39								311± 7	11668
022017.3+243107	PGC 9064								11336
022311.6+425930	A 0220+42	.E.....	1.48± .10						
140.25 -16.77	UGC 1841	U	.00± .08	.43					6226± 33
343.78 -4.57	5ZW 230	-5.0± .7		.00					6359
022001.9+424553	PGC 9067		1.55						6020
022319.0+321119		.S?....	.61± .10	53				17.51±.2	10083± 8
144.64 -26.78	MCG 5- 6- 35		.00± .06	.26	15.35 ±.15			273± 26	10047± 47
334.09 -9.32	ARAK 81			.00	11.05			261± 7	10192
022021.0+315743	PGC 9071		.63	.00	15.04			2.47	9853
022320.3+415704	NGC 898	.S..2./	1.29± .04	170					
140.68 -17.73	UGC 1842	PU	.63± .04	.26	13.84 ±.13				5400± 41
342.87 -5.06	IRAS02201+4143	2.0± .6		.77					5532
022011.9+414327	PGC 9073		1.32	.31	12.75				5192
022322.1+321150		.S?....	.76± .08	117				17.70±.3	10121± 7
144.65 -26.77	MCG 5- 6- 36		.12± .06	.26	15.01 ±.14			136± 10	10080± 47
334.11 -9.33	ARAK 80			.18					10230
022024.0+315814	PGC 9074		.79	.06	14.52			3.12	9891
022330.9+271929				.32	15.6 ±.4			17.51±.3	10597± 10
146.90 -31.23	CGCG 483- 24								
329.66 -11.41								307± 7	10695
022037.1+270553	PGC 9076								10361
0223.5 +2630	NGC 900	.L.....	1.03± .08	30					
147.31 -31.96	UGC 1843	U	.16± .03	.23	14.73 ±.10				
328.91 -11.76		-2.0± .9		.00					
0220.7 +2617	PGC 9079		1.03						

R.A. 2000 DEC. l b SGL SGB R.A. 1950 DEC.	Names PGC	Type S_T n_L T L	logD_25 logR_25 logA_e logD_o	p.a. A_g A_i A_21	B_T m_B m_FIR B_T^o	(B-V)_T (U-B)_T (B-V)_T^o (U-B)_T^o	(B-V)_e (U-B)_e m'_e m'_25	m_21 W_20 W_50 HI	V_21 V_opt V_GSR V_3K
0223.7 +2709 147.02 -31.36 329.52 -11.51 0220.8 +2656	 UGC 1844 PGC 9086	.SB.8.. U 8.0± .8 	1.08± .07 .09± .06 1.10	 .25 .11 .05	 15.19 ±.20 14.80				10646 10744 10410
022344.1-435820 261.18 -64.94 254.52 -26.71 022148.0-441154	 ESO 246- 9 PGC 9091	.L?.... 	.86± .07 .21± .05 .84	160 .04 .00 	 15.24 ±.14 15.13				 5152 5048 5000
022351.7-210132 201.79 -68.08 280.45 -25.90 022133.0-211506	 ESO 545- 12 PGC 9098	.S?.... 	.97± .05 .07± .04 .97	 .01 .10 .03	 15.18 ±.14 15.01				10773 10731 10556
022352.3+253235 147.85 -32.82 328.04 -12.22 022100.0+251900	 MCG 4- 6- 22 PGC 9099	.S?.... 	.89± .11 .14± .07 .91	 .24 .20 .07	 14.94 ±.12 14.47			16.71±.3 277± 7 2.17	5114± 10 5208 4876
0223.9 +2729 146.91 -31.04 329.85 -11.42 0221.0 +2716	 UGC 1848 PGC 9102	.S..3.. U (1) 3.0± .9 5.5±1.2	1.04± .06 .30± .05 1.07	22 .32 .42 .15	 15.23 ±.13 14.40				10704 10802 10469
0223.9 -0642 174.24 -60.00 296.06 -23.17 0221.5 -0656	 PGC 9105	.S..6*/ E 6.0±1.3 	1.14± .08 .91± .08 1.15	165 .08 1.33 .45					
0224.0 +2706 147.12 -31.38 329.50 -11.60 0221.1 +2653	 UGC 1850 PGC 9106	.S..9*. U 9.0±1.2 	1.00± .16 .09± .12 1.02	35 .25 .09 .04					5476 5573 5240
0224.1 +2720 147.03 -31.16 329.73 -11.52 0221.2 +2707	NGC 904 UGC 1852 PGC 9112	.E..... U -5.0± .8 	1.09± .10 .14± .05 1.10	130 .30 .00 	 14.56 ±.11 				
022407.9+475810 138.56 -12.07 348.26 -2.48 022051.1+474436	 UGC 1845 IRAS02208+4744 PGC 9115	.S..2.. U 2.0± .8 	1.08± .06 .21± .05 1.21	145 1.39 .25 .10	 14.8 ±.3 10.71 				
022412.2+052217 161.00 -50.52 308.59 -19.77 022134.8+050843	 MCG 1- 7- 4 PGC 9118	.S?.... 	.94± .11 .57± .07 .95	 .09 .85 .28	 15.20 ±.14 14.21				8278 8315 8030
0224.3 +3338 144.23 -25.36 335.50 -8.88 0221.3 +3325	 UGC 1853 PGC 9122	.S..7.. U 7.0±1.0 	1.04± .08 1.06± .06 1.07	160 .27 1.38 .50					
022419.0-795531 298.02 -36.30 215.99 -19.30 022530.0-800900	 ESO 14- 1 IRAS02254-8008 PGC 9123	.SBS9*P S (1) 9.0±1.2 5.6± .9	.96± .05 .07± .04 1.00	 .34 .07 .03	 14.66 ±.14 12.88 				
022422.8-582346 281.93 -54.78 238.47 -24.98 022253.0-583718	 ESO 115- 8 PGC 9124	.E.1.*. S -5.0±1.2 	.97± .06 .09± .04 .75± .02 .95	 .04 .00 	14.18 ±.13 14.65 ±.14 14.22	1.03± .01 .93 	1.11 ±.01 13.42± .08 13.79± .33		9254± 37 9118 9157
022424.8-020941 168.61 -56.55 300.88 -22.09 022152.6-022314	 UGC 1862 PGC 9126	.SXT7P* UE (1) 7.0± .8 6.4± .8	1.22± .03 .11± .04 1.23	10 .04 .15 .05	 13.66 ±.09 13.46				1420± 50 1435 1176
0224.4 +4930 138.05 -10.61 349.64 -1.82 0221.1 +4917	 UGC 1851 PGC 9127	.S..8.. U 8.0± .9 	1.04± .15 .08± .12 1.14	5 1.07 .10 .04					
022428.3+015016 164.36 -53.38 305.02 -20.94 022153.4+013643	 UGC 1863 PGC 9128	.SXS5.. U 5.0± .8 	.92± .06 .01± .04 .92	 .06 .01 .00	 14.95 ±.13 14.84			16.86±.3 155± 13 2.01	6712± 10 6739 6465
022429.6+405212 141.31 -18.66 342.00 -5.73 022122.2+403838	 UGC 1855 IRAS02213+4038 PGC 9130	.SBS1.. U 1.0± .9 	1.08± .06 .27± .05 1.10	97 .16 .27 .13	 14.85 ±.13 13.37 				

R.A. 2000 DEC. / l b / SGL SGB / R.A. 1950 DEC.	Names / PGC	Type: S_T n_L / T / L	$\log D_{25}$ / $\log R_{25}$ / $\log A_e$ / $\log D_o$	p.a. / A_g / A_i / A_{21}	B_T / m_B / m_{FIR} / B_T^o	$(B-V)_T$ / $(U-B)_T$ / $(B-V)_T^o$ / $(U-B)_T^o$	$(B-V)_e$ / $(U-B)_e$ / m'_e / m'_{25}	m_{21} / W_{20} / W_{50} / HI	V_{21} / V_{opt} / V_{GSR} / V_{3K}
022429.7-582604 / 281.95 -54.74 / 238.42 -24.99 / 022300.0-583936	/ ESO 115- 9 / / PGC 9132	.S..2P/ / S / 2.0± .9 /	1.24± .04 / .73± .04 / / 1.24	124 / .04 / .90 / .37	/ 15.15 ±.14 / / 14.12				9345 / 9209 / 9248
022431.5+313654 / 145.16 -27.21 / 333.68 -9.80 / 022133.8+312321	/ UGC 1856 / KUG 0221+313 / PGC 9134	.S..7.. / U / 7.0± .9 /	1.37± .03 / .97± .06 / / 1.39	/ .26 / 1.34 / .48	/ 14.78 ±.12 / / 13.16			14.30±.3 / / 253± 7 / .65	4804± 10 / 4912 / 4576
022434.4+331009 / 144.49 -25.77 / 335.10 -9.14 / 022135.2+325636	/ UGC 1857 / / PGC 9136	.S..7*. / U / 7.0±1.2 /	1.08± .09 / .42± .07 / / 1.10	/ .28 / .58 / .21	/ 15.10 ±.12 / / 14.23			14.91±.3 / / 266± 7 / .47	4413± 10 / 4525 / 4187
022436.9+283629 / 146.55 -29.95 / 330.94 -11.09 / 022141.9+282257	/ CGCG 504- 76 / / PGC 9139			/ .33 / /	/ 15.7 ±.4 / /			17.51±.3 / / 292± 7 /	9964± 10 / 10065 / 9731
0224.6 -7331 / 294.25 -42.00 / 222.45 -21.33 / 0224.3 -7345	/ / / PGC 9140	DE.3... / S / -5.0± .8 /	1.37± .07 / .41± .08 / / 1.26	/ .10 / .00 /					
022440.1-190828 / 197.41 -67.17 / 282.57 -25.82 / 022220.0-192200	/ ESO 545- 13 / IRAS02223-1922 / PGC 9141	.SXS5.. / E (1) / 5.0± .9 / 1.9± .8	1.06± .04 / .13± .03 / / 1.07	2 / .10 / .19 / .06	/ 13.97 ±.14 / 12.59 / 13.62				10133± 60 / 10096 / 9913
0224.6 +2533 / 148.04 -32.73 / 328.13 -12.38 / 0221.8 +2520	/ UGC 1860 / / PGC 9143	.SBT3.. / U / 3.0± .8 /	1.14± .05 / .21± .05 / / 1.17	30 / .24 / .30 / .11	/ 14.44 ±.13 / / 13.84				9637 / 9730 / 9400
022441.9+304902 / 145.56 -27.92 / 332.97 -10.17 / 022144.9+303529	/ UGC 1861 / / PGC 9145	.I..9.. / U / 10.0± .9 /	1.00± .16 / .19± .12 / / 1.03	90 / .32 / .14 / .09				16.44±.3 / / 121± 7 /	4791± 10 / 4897 / 4562
022442.7+253153 / 148.06 -32.75 / 328.11 -12.40 / 022150.3+251820	/ MCG 4- 6- 26 / / PGC 9147	.SB?... / / /	.74± .14 / .00± .07 / / .76	/ .24 / .00 / .00	/ 15.56 ±.12 / / 15.26			17.04±.3 / / 258± 7 / 1.78	10461± 10 / 10554 / 10224
022443.4-344952 / 239.80 -68.64 / 264.88 -27.14 / 022237.0-350324	/ ESO 355- 15 / / PGC 9149	.SBT4?. / S (1) / 4.0± .9 / 4.4±1.2	1.14± .05 / .89± .05 / / 1.14	147 / .03 / 1.31 / .45	/ 16.05 ±.14 / /				
022444.4+423723 / 140.67 -17.01 / 343.58 -4.99 / 022134.8+422350	/ UGC 1859 / / PGC 9150	.E?.... / / /	1.23± .14 / .18± .08 / / 1.23	47 / .31 / .00 /	/ 13.90 ±.11 / / 13.50				6087± 68 / 6219 / 5882
022453.2+431930 / 140.43 -16.35 / 344.22 -4.70 / 022142.7+430558	/ CGCG 538- 65 / / PGC 9157			/ .36 / /	/ 15.4 ±.3 / /				5065± 68 / 5198 / 4862
022458.7+372855 / 142.76 -21.76 / 339.02 -7.32 / 022155.1+371523	/ UGC 1864 / / PGC 9164	.S..3.. / U (1) / 3.0± .9 / 5.5±1.2	1.11± .07 / .29± .06 / / 1.13	88 / .21 / .41 / .15	/ 15.28 ±.17 / /				
022459.1+273122 / 147.15 -30.91 / 329.98 -11.62 / 022205.1+271750	/ CGCG 483- 32 / / PGC 9167			/ .30 / /	/ 15.6 ±.4 / /			17.43±.3 / / 362± 7 /	9542± 10 / 9640 / 9308
0224.9 +3602 / 143.35 -23.10 / 337.73 -7.96 / 0221.9 +3548	A 0221+35 / UGC 1865 / DDO 19 / PGC 9168	.S..9*. / U (1) / 9.0±1.0 / 9.0±1.3	1.46± .04 / .12± .06 / / 1.48	80 / .22 / .12 / .06				14.59±.1 / / 85± 6 / 76± 7	580± 5 / / 698 / 360
0225.0 +1940 / 151.25 -38.00 / 322.65 -14.81 / 0222.2 +1927	/ UGC 1870 / / PGC 9169	.S?.... / / /	1.14± .13 / .86± .12 / / 1.19	24 / .48 / 1.29 / .43					10593 / 10670 / 10350
022500.4+260302 / 147.88 -32.25 / 328.62 -12.24 / 022207.5+254930	/ CGCG 483- 31 / / PGC 9170		.95? / .48? / / .98	/ .23 / /					10082 / 10176 / 9846

R.A. 2000 DEC. / l b / SGL SGB / R.A. 1950 DEC.	Names / PGC	Type / S_T n_L / T / L	$\log D_{25}$ / $\log R_{25}$ / $\log A_e$ / $\log D_o$	p.a. / A_g / A_i / A_{21}	B_T / m_B / m_{FIR} / B_T^o	$(B-V)_T$ / $(U-B)_T$ / $(B-V)_T^o$ / $(U-B)_T^o$	$(B-V)_e$ / $(U-B)_e$ / m'_e / m'_{25}	m_{21} / W_{20} / W_{50} / HI	V_{21} / V_{opt} / V_{GSR} / V_{3K}
022503.6-244723 211.86 -68.86 276.27 -26.62 022248.0-250054	NGC 922 ESO 478- 28 IRAS02228-2500 PGC 9172	.SBS6P. R (3) 6.0± .4 4.8± .5	1.29± .03 .09± .03 1.29	 .00 .13 .05	12.45 ±.13 12.51 ±.12 11.33 12.34	.33± .01 -.40± .02 .29 -.43	 13.53± .20	14.01±.1 247± 7 212± 8 1.62	3092± 6 3069± 35 3038 2885
0225.0 +2212 149.86 -35.71 325.05 -13.82 0222.2 +2159	 UGC 1871 IRAS02222+2159 PGC 9173	.S?.... 	.90± .07 .17± .05 .93	37 .35 .26 .09	 14.87 ±.12 11.50 14.20				10121 10205 9880
022507.4+415106 141.04 -17.70 342.93 -5.40 022158.7+413735	 UGC 1866 PGC 9180	.SB.1.. U 1.0± .9 	.98± .07 .27± .05 1.00	30 .26 .27 .13	 14.83 ±.12 				
022510.1+235058 149.03 -34.23 326.59 -13.18 022219.1+233727	 MCG 4- 6- 28 PGC 9182	.S?.... 	.74± .14 .21± .07 .77	 .31 .26 .10	 15.62 ±.13 15.04			16.60±.3 175± 7 1.46	6508± 10 6596 6269
022512.7+205703 150.59 -36.83 323.87 -14.36 022224.0+204332	 CGCG 462- 9 PGC 9183	 	1.00? .52? 1.04	 .38 	 15.47 ±.12 				10296 10377 10054
022515.7+452706 139.68 -14.35 346.13 -3.79 022202.3+451335	 UGC 1867 PGC 9186	.S..6.. U 6.0± .9 	1.28± .06 .82± .06 1.35	57 .77 1.21 .41	 15.1 ±.3 13.06			15.91±.1 281± 8 2.43	5195± 11 5332 4997
022515.9+241544 148.84 -33.85 326.98 -13.03 022224.5+240213	 MCG 4- 6- 29 PGC 9187	.S?.... 	.85± .12 .31± .07 .88	 .31 .39 .16	 15.10 ±.14 14.31			17.23±.3 477± 7 2.76	9711± 10 9800 9473
022516.3+420523 140.97 -17.47 343.15 -5.32 022207.3+415152	NGC 906 UGC 1868 IRAS02221+4152 PGC 9188	.SB.2.. U (1) 2.0± .8 2.9± .8	1.25± .04 .04± .05 .94± .03 1.27	 .30 .05 .02	13.76 ±.16 13.98 ±.17 13.47	.88± .03 .77 	.96± .02 13.95± .09 14.74± .30	14.68±.3 338± 34 295± 25 1.19	4680± 17 4586± 59 4803 4467
022519.2-574324 281.03 -55.23 239.17 -25.23 022348.1-575654	 ESO 115- 11 IRAS02237-5757 PGC 9191	.S..6?/ S 6.0±1.9 	1.14± .05 .88± .04 1.14	10 .05 1.29 .44	 16.23 ±.14 				
022522.9+420208 141.01 -17.51 343.11 -5.36 022213.9+414837	NGC 909 UGC 1872 PGC 9197	.E..... U -5.0± .9 	.95± .21 .00± .08 .99	 .26 .00 	 14.28 ±.10 13.95				4899± 68 5029 4693
0225.3 +1105 156.85 -45.50 314.39 -18.11 0222.7 +1052	 UGC 1879 PGC 9198	.SX.5.. U 5.0± .8 	1.11± .07 .06± .06 1.13	 .27 .10 .03	 15.0 ±.2 14.57				6176 6229 5929
0225.4 +4042 141.54 -18.74 341.94 -5.96 0222.3 +4029	 UGC 1874 PGC 9200	.SX.7*. U 7.0±1.2 	1.11± .14 .15± .12 1.13	 .24 .20 .07					
022526.8+414927 141.11 -17.71 342.93 -5.46 022218.1+413556	NGC 910 UGC 1875 PGC 9201	.E+.... PU -4.0± .5 	1.30± .12 .00± .08 .95± .04 1.34	 .26 .00 	13.18 ±.15 12.84	1.01± .02 .50± .03 .90 .47	1.04± .02 .52± .03 13.42± .15 14.67± .67		 5126± 68 5256 4919
022527.4+371027 142.98 -22.01 338.79 -7.54 022224.0+365657	A 0222+36 MCG 6- 6- 29 PGC 9202	.E?.... 		 .21 	 15.0 ±.4 				9800±101 9920 9583
022527.8-253824 214.23 -68.92 275.32 -26.80 022313.0-255154	 ESO 479- 1 PGC 9204	.SBS6*. S (1) 5.5± .5 3.3± .6	1.21± .04 .10± .03 1.21	16 .00 .14 .05	 14.15 ±.14 13.99				4848± 52 4792 4644
022528.0+202344 150.97 -37.30 323.37 -14.63 022239.7+201013	IC 1797 UGC 1880 ARAK 83 PGC 9205	.SB?... 	.84± .11 .44± .06 .87	138 .40 .66 .22	 15.32 ±.14 13.02 14.24				3900± 49 3979 3658
0225.5 +1128 156.61 -45.16 314.78 -18.00 0222.8 +1115	 UGC 1883 PGC 9206	.S..7*. U 7.0±1.2 	1.34± .04 .58± .05 1.36	157 .27 .79 .29	 15.1 ±.2 14.00				3775± 10 3829 3528

R.A. 2000 DEC. / l b / SGL SGB / R.A. 1950 DEC.	Names / PGC	Type / S_T n_L / T / L	$\log D_{25}$ / $\log R_{25}$ / $\log A_e$ / $\log D_o$	p.a. / A_g / A_i / A_{21}	B_T / m_B / m_{FIR} / B_T^o	$(B-V)_T$ / $(U-B)_T$ / $(B-V)_T^o$ / $(U-B)_T^o$	$(B-V)_e$ / $(U-B)_e$ / m'_e / m'_{25}	m_{21} / W_{20} / W_{50} / HI	V_{21} / V_{opt} / V_{GSR} / V_{3K}
0225.5 +2644		.S..3..	.94± .11						10341
147.68 -31.57	UGC 1881	U (1)	.05± .07	.24					
329.31 -12.07	IRAS02226+2630	3.0± .9		.07					10437
0222.7 +2630	PGC 9212	5.5±1.1	.96	.02					10106
022538.2+230000		.S?....	.89± .11					16.47±.3	4194± 10
149.59 -34.95	MCG 4- 6- 31		.24± .07	.33	15.45 ±.12				
325.84 -13.62				.35				242± 7	4280
022247.8+224630	PGC 9214		.92	.12	14.74			1.61	3955
022538.2+365751		.E.....	1.11± .10		14.2 ±.3	1.03± .05	1.07± .02		
143.10 -22.19	UGC 1877	U	.05± .05	.25	14.49 ±.13				
338.61 -7.67		-5.0± .8	.88± .14	.00			14.12± .51		
022235.0+364421	PGC 9215		1.13				14.60± .60		
022541.0+245803								16.90±.3	10351± 10
148.58 -33.17	CGCG 483- 40			.24	15.5 ±.4				10345
327.68 -12.83								379± 10	10442
022249.0+244433	PGC 9219								10114
022542.4+415722	NGC 911	.E.....	1.23± .14	115					
141.10 -17.56	UGC 1878	U	.27± .08	.26	13.73 ±.10				5604± 41
343.07 -5.45		-5.0± .8		.00					5734
022233.5+414352	PGC 9221		1.19		13.38				5398
022545.3+371353		.S..5..	1.12± .05						
143.01 -21.94	UGC 1882	U	.08± .05	.21	14.54 ±.15				
338.86 -7.57		5.0± .8		.11					
022241.8+370023	PGC 9225		1.14	.04					
022545.9+244637								18.09±.3	10367± 10
148.70 -33.33	CGCG 483- 42			.24	15.7 ±.4				10457
327.51 -12.92								404± 7	10130
022254.0+243307	PGC 9227								
0225.8 +2724		.SBR3..	.94± .09						10526
147.40 -30.94	UGC 1885	U	.05± .06	.30	15.26 ±.15				
329.95 -11.84		3.0± .9		.07					10623
0222.9 +2711	PGC 9233		.97	.02	14.81				10292
022550.0-530519		.LAT+?.	1.13± .05	15					
275.37 -58.69	ESO 153- 34	S	.14± .04	.00	14.12 ±.14				6478
244.26 -26.08		-1.3± .7		.00					6352
022409.0-531848	PGC 9234		1.11		14.02				6361
022550.6+182949	NGC 918	.SXT5*.	1.54± .02	158	13.05 ±.17	.86± .03	.94± .02	13.99±.1	1509± 4
152.16 -38.96	UGC 1888	PU (1)	.23± .04	.47	13.6 ±.3	.18± .06	.27± .03	266± 5	1502± 59
321.61 -15.46	IRAS02230+1816	5.3± .6	1.16± .03	.35		.70	14.34± .08	256± 5	1583
022303.7+181620	PGC 9236	2.2± .7	1.59	.12	12.37	.04	15.02± .23	1.50	1266
022550.6-111128		PLBT0?.	1.05± .07	130					
181.81 -62.69	MCG -2- 7- 6	E	.23± .05	.03					
291.36 -24.65	IRAS02234-1124	-2.0± .9		.00	12.22				
022324.8-112458	PGC 9237		1.02						
022552.8+375323		.SBR1..	.97± .09						
142.77 -21.32	UGC 1884	U	.01± .06	.15	15.27 ±.17				
339.46 -7.30		1.0± .8		.01					
022248.5+373954	PGC 9238		.99	.00					
022556.4+245129		.S?....	.78± .13					17.71±.3	10786± 10
148.70 -33.24	MCG 4- 6- 36		.09± .07	.24	15.49 ±.12				
327.60 -12.93				.11				88± 7	10877
022304.5+243800	PGC 9241		.80	.04	15.12			2.54	10549
022557.5-244249		PSBR1*.	1.00± .06						
211.77 -68.64	ESO 479- 2	r	.37± .04	.00	15.23 ±.14				10427± 69
276.38 -26.82	IRAS02237-2456	1.0± .9		.38	11.99				10373
022342.0-245618	PGC 9243		1.00	.19	14.72				10221
022600.2-212513	A 0223-21	.SXS9*.	1.26± .03		14.2 ±.2	.49± .06	.47± .03	14.53±.2	1555± 11
203.23 -67.75	ESO 545- 16	SU (2)	.08± .03	.00	14.56 ±.14	.01± .08	-.02± .05	153± 16	
280.08 -26.45	DDO 21	9.0± .7	1.19± .04	.08		.46	15.61± .10	140± 12	1511
022342.0-213842	PGC 9246	9.6± .6	1.26	.04	14.35	-.01	15.12± .27	.14	1342
022600.5+392815		.SXT4..	1.57± .02	35				14.06±.1	4865± 8
142.15 -19.85	UGC 1886	U (1)	.26± .04	.17	12.73 ±.15			485± 16	4961± 43
340.89 -6.62	IRAS02229+3914	4.0± .7		.38	13.66			478± 8	4993
022254.4+391446	PGC 9247	2.5±1.0	1.59	.13	12.15			1.78	4656
022602.7+301021		.LB....	1.02± .14						
146.15 -28.40	UGC 1889	U	.11± .07	.31	14.71 ±.11				
332.51 -10.71		-2.0± .8		.00					
022306.1+295652	PGC 9251		1.03						

R.A. 2000 DEC.	Names	Type	logD$_{25}$	p.a.	B$_T$	(B-V)$_T$	(B-V)$_e$	m$_{21}$	V$_{21}$
l b		S$_T$ n$_L$	logR$_{25}$	A$_g$	m$_B$	(U-B)$_T$	(U-B)$_e$	W$_{20}$	V$_{opt}$
SGL SGB		T	logA$_e$	A$_i$	m$_{FIR}$	(B-V)^o_T	m'	W$_{50}$	V$_{GSR}$
R.A. 1950 DEC.	PGC	L	logD$_o$	A$_{21}$	B^o_T	(U-B)^o_T	m'$_{25}$	HI	V$_{3K}$
022605.1+420838	NGC 914	.SAS5..	1.26± .04	117					5548± 10
141.10 -17.36	UGC 1887	U	.15± .05		.28 13.66 ±.13				5383± 52
343.27 -5.43	IRAS02229+4155	5.0± .8			.23 12.75				5672
022255.9+415509	PGC 9253		1.28		.08 13.12				5336
022606.3+285950			.95?					15.94±.3	9806± 10
146.71 -29.47	CGCG 504- 78		.35?		.31 15.51 ±.12				
331.44 -11.22								260± 10	9907
022310.8+284621	PGC 9254		.98						9575
022607.1-001952	NGC 926	.SBT4*.	1.26± .03	36	14.02 ±.14	.77± .03	.83± .03		
167.16 -54.86	UGC 1901	PUE (1)	.28± .04		.09 13.64 ±.12	.17± .05	.31± .05		6485± 66
302.91 -21.98	IRAS02235-0033	4.0± .4	.80± .02		.41	.66	13.51± .04		6504
022333.7-003320	PGC 9256	4.2± .8	1.27		.14 13.26	.08	14.48± .21		6242
022607.7+315440		.S..2..	1.22± .07					16.46±.3	5388± 10
145.38 -26.80	UGC 1890	U	.22± .07		.28 14.20 ±.14				
334.10 -9.98		2.0± .8			.27			593± 7	5496
022309.5+314111	PGC 9258		1.24		.11 13.60			2.75	5162
0226.1 +2603		.E.....	1.04± .18						
148.16 -32.13	UGC 1891	U	.00± .08		.26 14.92 ±.14				
328.74 -12.48		-5.0± .8			.00				
0223.3 +2550	PGC 9260		1.08						
0226.2 +2736		.SBR3..	1.05± .08						9997
147.40 -30.72	UGC 1892	U	.04± .06		.30 14.90 ±.16				
330.17 -11.83		3.0± .8			.05				10094
0223.3 +2723	PGC 9262		1.07		.02 14.47				9764
0226.2 +1227		.S..7..	.95± .07	40					8343
156.13 -44.22	UGC 1897	U	.05± .05		.33				
315.81 -17.81		7.0± .9			.07				8400
0223.5 +1214	PGC 9263		.98		.03				8097
0226.3 +2712	NGC 919	.S..2..	1.09± .06	138					10349
147.62 -31.08	UGC 1894	U	.67± .05		.31 15.39 ±.12				
329.81 -12.02		2.0±1.0			.83				10445
0223.4 +2659	PGC 9267		1.12		.34 14.15				10116
0226.3 -0132		.L.....	.96± .09						
168.57 -55.77	UGC 1905	U	.06± .03		.05 15.11 ±.12				
301.67 -22.38		-2.0± .8			.00				
0223.8 -0146	PGC 9269		.96						
022620.7+302641		.LAR0..	.93± .16					16.96±.3	10048± 10
146.09 -28.12	UGC 1896	U	.30± .07		.37 14.65 ±.12				
332.79 -10.65	IRAS02234+3013	-2.0± .8			.00 13.65			513± 7	10152
022323.8+301312	PGC 9270		.92		14.12				9820
022620.8-442645	NGC 939	.E+..*.	1.08± .05	110					5220
261.47 -64.29	ESO 246- 11	BS	.09± .04		.04 14.06 ±.14				5114
253.95 -27.15		-4.3± .6			.00				5072
022426.0-444012	PGC 9271		1.06		13.95				
0226.3 -0950	A 0223-10	.SXS8*.	1.47± .04	105	14.57 ±.15	.52± .05	.50± .04		2108
179.76 -61.73	MCG -2- 7- 7	PE (2)	.41± .05		.03	-.23± .07	-.14± .06		
292.85 -24.48	DDO 20	8.3± .6	.97± .02		.51	.42	14.91± .05		2098
0223.9 -1004	PGC 9272	7.1± .9	1.47		.21 14.03	-.30	15.72± .27		1875
022621.9-241732	A 0224-24	.SBS8..	1.43± .03	55				14.05±.2	1515± 11
210.69 -68.46	ESO 479- 4	SU (1)	.28± .04		.00 12.87 ±.14			240± 16	
276.86 -26.87	IRAS02240-2430	8.3± .4			.34 12.61			182± 12	1462
022406.0-243100	PGC 9273	5.6± .6	1.43		.14 12.53			1.38	1308
022623.1+283026		.SBS5..	1.14± .07	125				15.87±.3	10310± 10
147.01 -29.89	UGC 1895	U	.14± .06		.35 15.1 ±.2				
331.02 -11.49	IRAS02234+2817	5.0± .8			.21			274± 7	10409
022328.0+281658	PGC 9275		1.17		.07 14.53			1.27	10079
0226.4 +2259		.S..7..	1.14± .13	50					10187
149.79 -34.88	UGC 1898	U	.69± .12		.31				
325.91 -13.79	IRAS02235+2246	7.0± .9			.95				10272
0223.5 +2246	PGC 9277		1.17		.34				9949
0226.4 +2739		.S?....	.98± .09						9726
147.42 -30.66	UGC 1899		.09± .06		.27 14.79 ±.13				
330.24 -11.85					.14				9823
0223.5 +2726	PGC 9278		1.01		.05 14.32				9493
022628.4+010939	IC 225	.E.....	.83± .09						1518± 29
165.71 -53.63	UGC 1907	U	.05± .07		.07 14.52 ±.11				1541
304.48 -21.62	MK 1038	-5.0± .8			.00				
022354.0+005611	PGC 9283		.83		14.43				1274

R.A. 2000 DEC.	Names	Type	logD$_{25}$	p.a.	B$_T$	(B-V)$_T$	(B-V)$_e$	m$_{21}$	V$_{21}$
I b		S$_T$ n$_L$	logR$_{25}$	A$_g$	m$_B$	(U-B)$_T$	(U-B)$_e$	W$_{20}$	V$_{opt}$
SGL SGB		T	logA$_e$	A$_i$	m$_{FIR}$	(B-V)$_T^o$	m'$_e$	W$_{50}$	V$_{GSR}$
R.A. 1950 DEC.	PGC	L	logD$_o$	A$_{21}$	B$_T^o$	(U-B)$_T^o$	m'$_{25}$	HI	V$_{3K}$

022629.9+303502		.S?....	.87± .08						16.47±.3	5497± 10
146.06 -27.98	CGCG 504- 82		.48± .12	.37	15.63 ±.13				
332.93 -10.62	KUG 0223+303			.70				256± 7	5601
022332.9+302134	PGC 9284		.90	.24	14.53			1.70	5269
022633.1+274935		.S..6*.	.82± .13					16.15±.3	9617± 10
147.38 -30.49	UGC 1902	U	.17± .07	.27					
330.41 -11.81		6.0±1.3		.25				282± 7	9715
022338.6+273607	PGC 9285		.84	.09					9385
022633.1+294947		.S..3..	1.06± .06					16.91±.3	10475± 10
146.43 -28.67	UGC 1903	U (1)	.76± .06	.38	15.63 ±.12				10769
332.24 -10.96		3.0±1.0		1.04				501± 7	10578
022336.8+293619	PGC 9286	3.5±1.3	1.10	.38	14.13			2.40	10246
022633.6-155051	NGC 921	.SBR5?.	1.13± .06	81					
190.72 -65.23	MCG -3- 7- 15	E (1)	.35± .05	.02					
286.30 -25.73		5.0± .9		.52					
022411.1-160418	PGC 9287	4.2±1.2	1.13	.17					
0226.5 +3724		.S..2..	1.12± .05	55					
143.10 -21.71	UGC 1900	U	.53± .05	.25					
339.09 -7.64		2.0± .9		.65					
0223.5 +3711	PGC 9288		1.14	.26					
022637.1+120911	NGC 927	.SBR5..	1.09± .06					15.52±.1	8261± 6
156.46 -44.43	UGC 1908	U (1)	.00± .05	.32	14.13 ±.13			231± 6	8252± 59
315.55 -18.01	MK 593	5.0± .8		.00	13.04			211± 7	8316
022354.9+115544	PGC 9292	1.1± .8	1.12	.00	13.76			1.76	8015
0226.6 +2245		.S..3..	1.11± .07	152					9984
149.97 -35.07	UGC 1906	U (1)	.83± .06	.32					
325.71 -13.93		3.0±1.0		1.15					10069
0223.8 +2232	PGC 9293	4.5±1.3	1.14	.42					9746
022638.7+500242		.SB.0..	1.00± .09	45					
138.19 -9.99	UGC 1893	U	.08± .06	1.21	15.14 ±.15				
350.28 -1.90		.0± .8		.06					
022318.0+494914	PGC 9294		1.11						
0226.6 +3709		.S..4..	1.00± .08	117					
143.22 -21.93	UGC 1904	U	.73± .06	.25					
338.88 -7.77		4.0±1.0		1.07					
0223.6 +3656	PGC 9295		1.02	.36					
022641.7-143057	NGC 944	.L..+?/	1.03± .06	15					
188.11 -64.49	MCG -3- 7- 16	E	.49± .04	.01					
287.78 -25.52	IRAS02242-1444	-.5±1.3		.00	12.19				
022418.3-144424	PGC 9300		.96						
022646.3+415004		.S?....	.99± .10							5723± 68
141.34 -17.60	MCG 7- 6- 20		.33± .07	.23	14.93 ±.12				5852
343.05 -5.68				.41					5518
022337.3+413637	PGC 9301		1.01	.17	14.24				
022646.6+202955	NGC 924	.L.....	1.36± .11	53					4661± 50
151.25 -37.08	UGC 1912	U	.24± .08	.34	13.37 ±.11				4740
323.60 -14.87		-2.0± .8		.00					4421
022358.1+201628	PGC 9302		1.36		12.96				
022654.2+320543		.S..7..	.94± .11					16.17±.3	4995± 10
145.47 -26.57	UGC 1909	U	.19± .07	.28	15.21 ±.12				
334.34 -10.05		7.0± .9		.26				205± 7	5103
022355.8+315216	PGC 9308		.96	.09	14.65			1.42	4770
022654.7+250201		.S?....	.94± .11						9641
148.85 -32.99	MCG 4- 6- 42		.28± .07	.29	15.04 ±.12				
327.86 -13.05				.35					9731
022402.5+244834	PGC 9309		.96	.14	14.31				9406
022656.0+203206		.SB?...	.92± .11						4212
151.27 -37.03	MCG 3- 7- 13		.09± .07	.34	15.31 ±.14				
323.65 -14.89				.14					4291
022407.5+201839	PGC 9313		.95	.05	14.80				3972
022657.2+351053		.S?....	1.28± .04	175					5128
144.12 -23.73	UGC 1910		.58± .05	.22	14.65 ±.13				
337.13 -8.70	IRAS02239+3457			.88	13.81				5243
022355.7+345726	PGC 9314		1.30	.29	13.52				4908
0227.0 -0242		.S..7./	1.07± .09	110					
170.16 -56.55		E	.82± .08	.04					
300.50 -22.87		7.0± .9		1.13					
0224.5 -0256	PGC 9321		1.08	.41					

R.A. 2000 DEC. / l b / SGL SGB / R.A. 1950 DEC.	Names / PGC	Type / ST nL / T / L	logD25 / logR25 / logAe / logDo	p.a. / Ag / Ai / A21	BT / mB / mFIR / BT^o	(B-V)T / (U-B)T / (B-V)T^o / (U-B)T^o	(B-V)e / (U-B)e / m' / m'25	m21 / W20 / W50 / HI	V21 / Vopt / VGSR / V3K
022716.8-314144				.03					4530± 14
231.05 -68.68									
268.47 -27.59									4456
022507.7-315509	PGC 9331								4343
022716.8+333441	NGC 925	.SXS7..	2.02± .01	102	10.69M±.11	.57± .05	.58± .03	11.13±.0	553± 3
144.89 -25.18	UGC 1913	R (2)	.25± .02	.27	10.55 ±.12			215± 3	562± 17
335.72 -9.47	IRAS02243+3321	7.0± .3	1.74± .04	.34	10.58	.46	14.67± .09	200± 3	665
022416.9+332115	PGC 9332	4.3± .5	2.04	.12	10.00		15.00± .14	1.00	331
022717.6-035401		.SBS8..	1.04± .07	70					
171.70 -57.39	MCG -1- 7- 13	E (1)	.10± .05	.03					
299.27 -23.25		8.0± .9		.13					
022446.7-040726	PGC 9333	7.5± .8	1.05	.05					
022718.2-120512	NGC 929	.S..1P?	1.01± .10	170					
183.83 -62.96	MCG -2- 7- 9	E	.26± .05	.00					
290.47 -25.19		1.0±1.9		.27					
022453.0-121838	PGC 9334		1.01	.13					
022720.6-152517		.SB.8*/	1.17± .05	121					3766
190.09 -64.85	MCG -3- 7- 19	E (1)	.47± .05	.02					
286.81 -25.84		8.0±1.2		.58					3739
022457.9-153842	PGC 9335	6.4±1.2	1.17	.24					3542
022725.6+230542	A 0224+22	.S?....	.89± .11					17.42±.3	9545± 10
149.99 -34.68	MCG 4- 6- 44		.81± .07	.31	15.88 ±.16				9777± 72
326.10 -13.96	5ZW 242			.99				250± 7	9634
022435.0+225217	PGC 9340		.92	.40	14.48			2.54	9312
022725.7+410335		.E.....	1.08± .17						5372± 68
141.77 -18.27	UGC 1914	U	.00± .08	.19	14.65 ±.10				
342.42 -6.14		-5.0± .8		.00					5499
022417.5+405009	PGC 9341		1.11		14.38				5165
0227.4 +2415		PSBS2..	.95± .07	105					9340
149.39 -33.63	UGC 1917	U	.13± .05	.26	15.23 ±.13				
327.19 -13.49		2.0± .9		.16					9428
0224.6 +2402	PGC 9345		.98	.07	14.71				9104
0227.4 +0207		.S..8*.	.93± .07						2876
165.08 -52.70	UGC 1923	U	.29± .05	.05					
305.56 -21.57		8.0±1.2		.35					2901
0224.9 +0154	PGC 9347		.94	.14					2633
022731.9-432830		.S?....	.99± .07						8510± 20
259.32 -64.64	ESO 246- 12		.15± .06	.04	15.39 ±.14				
255.03 -27.43	FAIR 1088			.22					8405
022536.1-434154	PGC 9349		.99	.07	15.08				8360
022732.5+254005		.SB.2..	1.07± .06	118				17.37±.3	5085± 10
148.68 -32.35	UGC 1918	U	.28± .05	.24	14.57 ±.12				
328.51 -12.92	IRAS02246+2526	2.0± .9		.35	12.71			400± 7	5177
022439.7+252640	PGC 9351		1.10	.14	13.93			3.30	4851
022732.6-001444	NGC 934	.LX.-..	1.13± .07	130	14.0 ±.2	.86± .03	.95± .02		
167.55 -54.57	UGC 1926	UE	.17± .04	.04	14.04 ±.10				6353± 31
303.11 -22.29		-2.5± .5		.00		.78	14.01± .24		6371
022459.0-002809	PGC 9352		1.11		13.90		14.09± .43		6111
022732.8-100956		.S..5?/	1.18± .04	88					1985± 27
180.68 -61.71	MCG -2- 7- 10	E (1)	.45± .05	.01					
292.58 -24.84	MK 1039	5.0±1.7		.68	12.90				1974
022506.2-102321	PGC 9354	5.3±1.6	1.19	.23					1754
022734.6+415841	NGC 923	.S..3*.	.91± .07	95					5625± 68
141.43 -17.41	UGC 1915	U (1)	.17± .05	.29	14.47 ±.14				
343.25 -5.75	IRAS02244+4145	3.0±1.3		.23	12.04				5754
022425.3+414516	PGC 9355	4.5±1.2	.94	.08	13.91				5421
022736.7+420029		.S?....	.64± .17						5436± 97
141.43 -17.38	MCG 7- 6- 23		.11± .07	.29	15.23 ±.18				
343.28 -5.74	MK 1176			.17					5565
022427.3+414704	PGC 9357		.66	.06	14.74				5232
022737.6-010917	NGC 936	.LBT+..	1.67± .02	135	11.12M±.10	.97± .01	.98± .01	15.45±.3	1340± 10
168.57 -55.26	UGC 1929	R	.06± .02	.05	11.00 ±.09	.56± .01	.57± .01	500± 6	1400± 17
302.17 -22.57		-1.0± .3	1.15± .01	.00		.94	12.36± .03		1371
022504.7-012242	PGC 9359		1.67		10.98	.55	14.21± .15		1115
022738.3+360904		.SBS3..	1.13± .05	45					10749
143.84 -22.78	UGC 1919	U	.09± .05	.22	14.60 ±.16				
338.06 -8.39	IRAS02245+3555	3.0± .8		.12					10866
022435.6+355539	PGC 9364		1.15	.05	14.17				10532

R.A. 2000 DEC. / l b / SGL SGB / R.A. 1950 DEC.	Names / PGC	Type: S_T n_L / T / L	$\log D_{25}$ / $\log R_{25}$ / $\log A_e$ / $\log D_o$	p.a. / A_g / A_i / A_{21}	B_T / m_B / m_{FIR} / B_T^o	$(B-V)_T$ / $(U-B)_T$ / $(B-V)_T^o$ / $(U-B)_T^o$	$(B-V)_e$ / $(U-B)_e$ / m'_e / m'_{25}	m_{21} / W_{20} / W_{50} / HI	V_{21} / V_{opt} / V_{GSR} / V_{3K}
022741.3+261325		.S?....	.89± .11		14.82 ±.12			16.25±.3	9505± 10
148.43 −31.84	MCG 4- 6- 49		.07± .07	.28				323± 7	9598
329.04 −12.72				.05					9272
022448.0+260000	PGC 9366		.91		14.35				
0227.7 +2635		.SBS3..	.95± .07						9786
148.25 −31.51	UGC 1921	U	.07± .05	.28	14.52 ±.12				9880
329.38 −12.56	5ZW 244	3.0± .8		.10					9553
0224.8 +2622	PGC 9367		.97	.04	14.07				
022741.6+271310	NGC 928	.S?....	.85± .12					17.22±.3	5124± 10
147.94 −30.94	MCG 4- 6- 50		.31± .07	.31	14.87 ±.16			321± 7	5219
329.96 −12.30				.47				2.99	4892
022447.5+265945	PGC 9368		.88	.16	14.07				
022745.6+281235	IC 226	.S?....	.74± .14					15.40±.3	10894± 10
147.47 −30.03	UGC 1922		.00± .07	.34				673± 7	10992
330.87 −11.89				.00					10664
022450.6+275910	PGC 9373		.77	.00					
0227.7 +5029		.E.....	1.04± .18	5					
138.19 −9.51	UGC 1916	U	.25± .08	1.21					
350.75 −1.85		−5.0± .9		.00					
0224.4 +5016	PGC 9374		1.16						
022749.8+314333		.S..6*.	1.24± .04					14.76±.3	595± 10
145.83 −26.82	UGC 1924	U	.83± .06	.32	15.23 ±.12			105± 7	701
334.09 −10.38	KUG 0224+315	6.0±1.3		1.22				.66	370
022451.5+313009	PGC 9375		1.27	.42	13.69				
022751.8+003004		.S..4*/	1.16± .04	112					
166.86 −53.94	UGC 1934	UE	.70± .04	.04	15.23 ±.12				
303.91 −22.15		4.3± .7		1.03					
022517.8+001640	PGC 9376		1.16	.35					
022752.0+455649	NGC 920	PSBS2..	1.18± .05	10				15.30±.1	6196± 11
139.93 −13.72	UGC 1920	U	.14± .05	.80	14.8 ±.3			302± 8	6333
346.77 −3.97		2.0± .8		.17				1.49	6002
022437.2+454324	PGC 9377		1.26	.07	13.74				
022752.8−504243		PSBS3..	1.05± .06	178					
271.62 −60.15	ESO 198- 11	r	.45± .05	.00	14.77 ±.14				
246.86 −26.73	IRAS02261-5056	3.3± .9		.62	13.34				
022608.0−505606	PGC 9378		1.05	.22					
022754.6+201956	NGC 930	.SA.1..	1.28± .04					16.57±.3	4078± 10
151.64 −37.10	UGC 1931	U	.07± .05	.37	13.34 ±.13			210± 13	4086± 50
323.55 −15.18	IRAS02251+2006	1.0± .8		.07				195± 10	4156
022506.1+200632	PGC 9379		1.31	.03	12.85			3.69	3839
022800.9+214804		.S?....	.82± .13					16.72±.3	9067± 10
150.85 −35.78	MCG 4- 6- 52		.28± .07	.38	15.49 ±.12			303± 7	9148
324.94 −14.61				.43				1.94	8829
022511.3+213440	PGC 9382		.85	.14	14.64				
022810.9+193558	NGC 935	.S..6*.	1.24± .05	155				14.25±.3	4142± 10
152.14 −37.72	UGC 1937	U	.21± .05	.37	13.63 ±.12			408± 13	4212± 30
322.88 −15.53		6.0±1.1		.31				393± 10	4224
022523.1+192235	PGC 9388		1.27	.11	12.92			1.22	3909
022811.3−055019		.S..3P/	1.12± .08	155					6071
174.51 −58.65	MCG -1- 7- 15	E	.65± .07	.02					6072
297.27 −23.97	VV 449	3.0±1.3		.89	13.50				5835
022541.7−060342	PGC 9390		1.12	.32					
0228.2 +4336		.S..8*.	1.00± .08	78					
140.90 −15.86	UGC 1927	U	.04± .06	.42					
344.74 −5.10		8.0±1.5		.05					
0225.0 +4323	PGC 9391		1.04	.02					
0228.2 +1934	IC 1801	.SB.3*.	1.10± .05	30					
152.16 −37.75	UGC 1936	U	.35± .05	.37	14.56 ±.12				4023± 37
322.86 −15.55		3.0±1.2		.48					4098
0225.4 +1921	PGC 9392		1.13	.17	13.68				3783
022812.7−012047		.S..8*.	1.22± .03	12					
168.99 −55.32	UGC 1945	UE (1)	.41± .04	.04	14.14 ±.12				1762± 50
302.02 −22.77	KUG 0225-015	8.0± .8		.50					1776
022539.9−013410	PGC 9394	6.4±1.2	1.23	.20	13.59				1522
022814.2+235523								16.60±.3	10402± 10
149.75 −33.86	CGCG 483- 62			.35	15.2 ±.4			143± 10	10489
326.95 −13.79									10167
022522.8+234200	PGC 9396								

R.A. 2000 DEC. l b SGL SGB R.A. 1950 DEC.	Names PGC	Type S_T n_L T L	$\log D_{25}$ $\log R_{25}$ $\log A_e$ $\log D_o$	p.a. A_g A_i A_{21}	B_T m_B m_{FIR} B_T^o	$(B-V)_T$ $(U-B)_T$ $(B-V)_T^o$ $(U-B)_T^o$	$(B-V)_e$ $(U-B)_e$ m'_e m'_{25}	m_{21} W_{20} W_{50} HI	V_{21} V_{opt} V_{GSR} V_{3K}
0228.2 +2313 150.13 -34.48 326.30 -14.08 0225.4 +2300	 UGC 1938 PGC 9398	.S..4.. U 4.0± .9 	1.13± .05 .63± .05 1.16	154 .35 .93 .32	 14.94 ±.12 13.62				6395 6480 6159
022815.0+311846 146.12 -27.16 333.76 -10.64 022517.0+310523	NGC 931 UGC 1935 MK 1040 PGC 9399	.SA.4.. U (1) 4.0± .7 2.5±1.0	1.59± .03 .67± .06 .53± .05 1.62	 .32 .99 .34	14.46 ±.16 13.57 ±.14 12.05 12.63	1.00± .02 .23± .04 .77 .00	1.04± .02 .21± .03 12.60± .19 15.56± .26	13.67±.2 476± 9 440± 10 .71	4993± 7 4914± 41 5096 4766
0228.2 +2618 148.54 -31.71 329.17 -12.80 0225.4 +2605	 UGC 1939 PGC 9402	.SB.1.. U 1.0± .9 	.97± .09 .04± .06 1.00	65 .27 .05 .02	 14.70 ±.12 14.32				5230 5323 4998
022819.6+292719 147.00 -28.85 332.07 -11.47 022523.4+291356	 CGCG 504- 90 KUG 0225+292 PGC 9405	.S?.... 	.48± .11 .03± .08 .52	 .40 .04 .01	 15.75 ±.18 15.26			17.22±.3 283± 7 1.95	9322± 10 9423 9094
022819.8-003453 168.18 -54.71 302.83 -22.58 022546.5-004816	 UGC 1949 PGC 9406	.IXS9*. UE (1) 10.0± .7 9.8± .8	1.07± .06 .10± .05 1.08	55 .05 .07 .05					
022820.6-424544 257.71 -64.89 255.83 -27.63 022624.0-425906	 ESO 246- 13 PGC 9407	.E?.... 	.78± .07 .02± .05 1.06± .07 .78	 .03 .00 	14.00 ±.19 14.82 ±.14 	.95± .05 	.98± .03 14.79± .26 12.84± .41		
022820.8-315256 231.49 -68.44 268.27 -27.82 022612.0-320618	 ESO 415- 26 PGC 9408	.L..0*P S -2.0± .7 	1.12± .04 .37± .03 .45± .06 1.07	26 .03 .00 	14.70M±.13 14.58 ±.14 14.55	.97± .03 .47± .04 .89 .45	1.03± .02 .48± .03 12.52± .21 14.24± .27	16.02±.3 	4604± 14 4629± 39 4531 4422
0228.4 +2808 147.66 -30.03 330.87 -12.05 0225.5 +2755	 UGC 1940 PGC 9411	.L..... U -2.0± .9 	1.04± .09 .48± .04 1.01	101 .34 .00 					
0228.4 +1538 154.64 -41.18 319.11 -17.13 0225.7 +1525	 UGC 1946 PGC 9413	.S..7.. U 7.0± .9 	1.24± .06 .96± .06 1.29	137 .55 1.33 .48					4078 4142 3836
022827.7-010906 168.85 -55.13 302.24 -22.77 022554.8-012229	NGC 941 UGC 1954 IRAS02259-0122 PGC 9414	.SXT5.. R (3) 5.0± .3 5.6± .5	1.42± .02 .13± .03 1.06± .01 1.43	170 .04 .20 .07	12.93 ±.13 12.95 ±.12 13.07 12.69	.52± .02 -.13± .03 .47 -.16	.53± .01 -.15± .02 13.72± .03 14.57± .17	14.69±.1 163± 5 147± 12 1.94	1608± 6 1580± 66 1622 1368
022829.9+512644 137.94 -8.57 351.64 -1.50 022506.2+511321	 UGC 1930 PGC 9416	.LBR+.. U -1.0± .7 	1.27± .08 .07± .05 1.36	 1.02 .00 	14.5 ±.2 				
022832.8+295905 146.80 -28.35 332.57 -11.28 022536.1+294543	 UGC 1944 PGC 9419	.S..9*. U 9.0±1.2 	1.14± .13 .18± .12 1.18	100 .39 .18 .09				15.86±.3 192± 7 	4599± 7 4701 4372
022832.9-190226 198.15 -66.30 282.84 -26.71 022613.0-191548	NGC 947 ESO 545- 21 IRAS02262-1915 PGC 9420	.SAR5.. SEr (2) 5.2± .5 1.5± .6	1.31± .03 .28± .03 1.00± .03 1.32	50 .08 .42 .14	13.18 ±.16 13.43 ±.14 12.77 12.80	.58± .02 -.04± .03 .48 -.11	.68± .02 .06± .03 13.67± .07 13.89± .23		5012± 60 4973 4796
022833.3+201700 151.84 -37.08 323.56 -15.34 022544.8+200337	NGC 938 UGC 1947 PGC 9423	.E..... U -5.0± .8 	1.20± .14 .11± .08 1.23	100 .38 .00 	13.43 ±.10 12.98				4089± 50 4166 3850
022837.3-103222 181.65 -61.75 292.23 -25.18 022611.0-104544	NGC 945 MCG -2- 7- 13 IRAS02261-1045 PGC 9426	.SBT5.. PE (1) 4.5± .5 1.9± .8	1.38± .03 .08± .04 1.08± .04 1.38	 .03 .13 .04	12.79 ±.19 12.76 12.61	.72± .03 .08± .04 .67 .04	.81± .01 .18± .02 13.68± .11 14.34± .26		4484 4470 4254
022839.4+234752 149.92 -33.93 326.88 -13.93 022548.0+233430	 UGC 1950 IRAS02258+2334 PGC 9430	.S?.... 	.98± .07 .28± .05 1.01	145 .35 .34 .14	 15.44 ±.13 14.69			16.44±.3 270± 7 1.61	6275± 10 6361 6040
022845.5-103052 181.65 -61.71 292.27 -25.20 022619.3-104414	NGC 948 MCG -2- 7- 15 PGC 9431	.SBS5*. PE (1) 4.5± .6 4.2± .8	1.17± .04 .09± .04 1.17	170 .03 .14 .05					4511± 63 4497 4281

R.A. 2000 DEC. l b SGL SGB R.A. 1950 DEC.	Names PGC	Type S_T n_L T L	$\log D_{25}$ $\log R_{25}$ $\log A_o$ $\log D_o$	p.a. A_g A_i A_{21}	B_T m_B m_{FIR} B_T^o	$(B-V)_T$ $(U-B)_T$ $(B-V)_T^o$ $(U-B)_T^o$	$(B-V)_e$ $(U-B)_e$ m'_e m'_{25}	m_{21} W_{20} W_{50} HI	V_{21} V_{opt} V_{GSR} V_{3K}
022846.0+455813 140.07 -13.64 346.86 -4.10 022531.1+454451	IC 1799 UGC 1943 PGC 9432	.S?.... 	1.06± .06 .41± .05 1.13	34 .80 .61 .20	 14.6 ±.3 13.20			14.66±.1 414± 12 1.26	5022± 11 5158 4828
022847.1-430427 258.24 -64.66 255.47 -27.69 022651.0-431748	 ESO 246- 15 PGC 9433	RSAS2*. Sr 1.6± .7	1.08± .05 .42± .04 1.08	150 .04 .51 .21	 14.68 ±.14 14.08				5046 4942 4895
022849.1+381002 143.22 -20.84 339.97 -7.69 022544.0+375640	 UGC 1948 PGC 9434	.S..2.. U 2.0± .9	.97± .07 .41± .05 .99	114 .19 .51 .21	 15.15 ±.12				
022852.2-412409 254.77 -65.46 257.37 -27.81 022654.0-413730	NGC 954 ESO 299- 4 IRAS02268-4137 PGC 9438	.SBT5*. PBS (2) 5.0± .5 2.6± .5	1.21± .04 .30± .04 .90± .05 1.22	19 .03 .45 .15	13.5 ±.2 13.85 ±.14 13.58 13.24	.67± .04 .57 	.75± .02 13.53± .11 13.71± .31		5353± 59 5253 5197
0228.8 +3743 143.42 -21.24 339.58 -7.91 0225.8 +3730	 UGC 1951 PGC 9439	.E...*. U -5.0±1.2	1.08± .17 .04± .08 1.10	 .20 .00					
0228.8 +2520 149.17 -32.52 328.33 -13.33 0226.0 +2507	 UGC 1955 PGC 9440	.S?.... 	1.13± .05 .35± .05 1.16	165 .30 .53 .18	 15.07 ±.14 14.21				5208 5298 4975
022857.2-222103 206.11 -67.39 279.13 -27.24 022640.0-223424	NGC 951 ESO 479- 8 PGC 9442	.SBR2*. Sr 1.6± .6	1.01± .04 .21± .03 1.01	48 .02 .26 .11	 15.54 ±.14 15.20				6150± 60 6101 5942
0228.9 +0022 167.36 -53.87 303.87 -22.45 0226.4 +0009	 UGC 1962 PGC 9445	.S..6*. U 6.0±1.2	1.07± .06 .12± .05 1.07	105 .04 .18 .06	 14.31 ±.12				
022858.9+280839 147.78 -29.98 330.93 -12.16 022603.8+275517	 UGC 1958 PGC 9447	.S..6*. U 6.0±1.3	1.14± .07 .69± .06 1.17	18 .34 1.01 .34	 15.49 ±.13 14.14			15.47±.3 101± 7 .99	1016± 10 1113 787
022859.5-355609 242.03 -67.50 263.63 -28.01 022655.0-360930	 ESO 355- 18 PGC 9448	.SBS3P. S 3.0± .9	1.01± .05 .43± .04 1.01	15 .03 .60 .22	 15.71 ±.14				
0229.0 +4028 142.31 -18.70 342.04 -6.68 0225.9 +4015	 UGC 1952 PGC 9450	.S..6*. U 6.0±1.2	1.04± .15 .08± .12 1.06	0 .21 .12 .04					
0229.1 +3432 144.85 -24.14 336.75 -9.38 0226.1 +3419	 UGC 1959 PGC 9451	.I..9*. U 10.0±1.2	1.04± .15 .18± .12 1.06	45 .24 .13 .09					3552 3664 3333
022907.5-305952 229.06 -68.35 269.29 -27.96 022658.0-311312	 ESO 415- 28 PGC 9452	.SXS4.. S (1) 4.0± .8 3.3±1.2	1.06± .04 .24± .04 1.06	16 .03 .35 .12	 15.32 ±.14 14.90				4982± 39 4908 4795
0229.1 +4504 140.48 -14.44 346.11 -4.57 0225.9 +4451	 UGC 1953 PGC 9453	.S..6*. U 6.0±1.2	1.04± .15 .08± .12 1.11	 .74 .12 .04					
022909.6-104943 182.30 -61.83 291.95 -25.37 022643.6-110304	NGC 943 MCG -2- 7- 19 PGC 9457	.I.0.$P R 90.0	1.70? .32? 1.69	15 .04 .24					
022910.3-105012 182.32 -61.83 291.94 -25.37 022644.3-110332	NGC 942 MCG -2- 7- 18 PGC 9458	.L...+P* RCE -1.3± .6	1.70? .32? 1.66	35 .04 .00					
022911.7-110132 182.64 -61.95 291.74 -25.42 022645.9-111453	NGC 950 MCG -2- 7- 21 PGC 9461	.SBT3*. E (1) 3.0±1.2 4.2±1.2	1.11± .06 .23± .05 1.11	40 .02 .31 .11					4696 4681 4467

R.A. 2000 DEC.	Names	Type	logD$_{25}$	p.a.	B$_T$	(B-V)$_T$	(B-V)$_e$	m$_{21}$	V$_{21}$
l b		S$_T$ n$_L$	logR$_{25}$	A$_g$	m$_B$	(U-B)$_T$	(U-B)$_e$	W$_{20}$	V$_{opt}$
SGL SGB		T	logA$_e$	A$_i$	m$_{FIR}$	(B-V)$_T^o$	m'	W$_{50}$	V$_{GSR}$
R.A. 1950 DEC.	PGC	L	logD$_o$	A$_{21}$	B$_T^o$	(U-B)$_T^o$	m'$_{25}$	HI	V$_{3K}$

0229.2 +2306	IC 1803	.L?....	1.12± .12		14.55 ±.15	1.19± .02	1.20± .02		
150.44 -34.49	MCG 4- 6- 57		.09± .07	.30	14.55 ±.13				
326.29 -14.34			.58± .05	.00			12.94± .17		
0226.4 +2253	PGC 9462		1.14				14.77± .65		
022916.1-482923		PSAR2*.	1.24± .05	135					6200±190
267.90 -61.45	ESO 198- 13	Sr	.15± .05	.00	13.66 ±.14				6083
249.32 -27.24	FAIR 728	2.1± .6		.19					6069
022728.0-484242	PGC 9463		1.24	.08	13.41				
022917.6+455440	NGC 933	.S?....	1.11± .08	35				14.49±.1	5105± 11
140.18 -13.66	UGC 1956		.17± .06	.79	14.8 ±.3				5241
346.85 -4.21				.21				472± 8	5110
022602.5+454119	PGC 9465		1.18	.08	13.75			.66	4912
022922.8+472927		.S..6*.	1.19± .06	175				15.65±.1	5299± 11
139.58 -12.19	UGC 1957	U	.91± .06	1.05	15.1 ±.3				5438
348.25 -3.48		6.0±1.3		1.34				292± 8	5110
022605.4+471606	PGC 9469		1.29	.46	12.73			2.47	
0229.3 +1010		PSX.2..	1.14± .05	40					8469
158.73 -45.78	UGC 1966	U (1)	.24± .05	.28	14.52 ±.14				
313.85 -19.35		2.0± .8		.30					8517
0226.7 +0957	PGC 9470	4.5±1.1	1.17	.12	13.85				8226
022923.1-425905		.L..0*/	1.02± .06	88					5153
257.93 -64.61	ESO 246- 16	S	.39± .03	.04	15.24 ±.14				5049
255.56 -27.80	IRAS02274-4312	-2.0±1.3		.00	13.19				5003
022727.0-431224	PGC 9472		.96		15.13				
0229.4 +2013		.S?....	1.19± .05	92					4162
152.10 -37.04	UGC 1965		.66± .05	.39	15.32 ±.14				
323.59 -15.55				.99					4238
0226.6 +2000	PGC 9475		1.23	.33	13.92				3924
022926.4+312814	A 0226+31	.SXT2..	1.08± .09						5098± 73
146.30 -26.92	UGC 1963	U (1)	.04± .07	.36	14.26 ±.13				5203
334.01 -10.80		2.0± .8		.05					4874
022628.2+311454	PGC 9476	3.0±1.0	1.11	.02	13.80				
022927.0-430429	IC 1810	RSXT2..	1.10± .05						5424
258.10 -64.55	ESO 246- 18	BSr	.09± .04	.04	14.19 ±.14				5319
255.46 -27.81		2.4± .5		.11					5274
022731.0-431748	PGC 9477		1.11	.05	13.99				
022927.5+313824	NGC 940	.L...*.	1.07± .13						5087± 52
146.23 -26.76	UGC 1964	U	.09± .07	.36	13.44 ±.12				5192
334.17 -10.73	ARAK 85	-2.0±1.1		.00	13.24				4864
022629.1+312504	PGC 9478		1.09		13.00				
022928.2+421500	NGC 937	.SB.6?.	1.03± .06	117					
141.66 -17.03	UGC 1961	U	.32± .05	.26	14.87 ±.12				
343.65 -5.93		6.0±1.3		.46					
022618.1+420140	PGC 9480		1.05	.16					
022928.3-393459		.S..5?/	1.13± .05	137					
250.66 -66.16	ESO 299- 5	S	1.03± .04	.03	15.99 ±.14				
259.45 -28.02		5.0±2.0		1.50					
022728.0-394818	PGC 9481		1.13	.50					
022929.3-331447		PSXR3..	1.04± .05	130					
235.04 -68.01	ESO 355- 19	Sr (1)	.25± .04	.03	15.29 ±.14				
266.71 -28.10		2.6± .6		.34					
022722.0-332806	PGC 9484	3.3± .8	1.04	.12					
022931.6-031242		.SBT0P.	1.10± .06	90					2430± 61
171.61 -56.51	MCG -1- 7- 16	E	.18± .05	.03					2438
300.16 -23.60	MK 1043	.0± .9		.13	12.63				2193
022700.2-032602	PGC 9485		1.10						
022932.3-424835	IC 1812	.E+..*.	1.30± .05		13.22 ±.15	1.01± .02	1.02± .01		5169± 34
257.55 -64.67	ESO 246- 19	BS	.09± .04	.04	13.21 ±.14				5065
255.76 -27.84		-3.8± .5	.89± .04	.00		.95	13.16± .15		5018
022736.0-430154	PGC 9486		1.28		13.10		14.36± .30		
022935.8-483753		PSBT2?.	.99± .05	51					10500±190
268.05 -61.31	ESO 198- 14	Sr	.36± .04	.00	14.98 ±.14				10382
249.16 -27.28	FAIR 729	1.8± .7		.44					10369
022748.0-485112	PGC 9490		.99	.18	14.44				
0229.7 +1603		.I..9*.	.96± .09						3896
154.73 -40.67	UGC 1969	U	.00± .06	.59					
319.64 -17.26		10.0±1.2		.00					3960
0227.0 +1550	PGC 9497		1.02	.00					3656

R.A. 2000 DEC. l　　b SGL　　SGB R.A. 1950 DEC.	Names PGC	Type S_T　n_L T L	$\log D_{25}$ $\log R_{25}$ $\log A_e$ $\log D_o$	p.a. A_g A_i A_{21}	B_T m_B m_{FIR} B_T^o	$(B-V)_T$ $(U-B)_T$ $(B-V)_T^o$ $(U-B)_T^o$	$(B-V)_e$ $(U-B)_e$ m'_e m'_{25}	m_{21} W_{20} W_{50} HI	V_{21} V_{opt} V_{GSR} V_{3K}
022948.9+312713 146.39 -26.90 334.03 -10.88 022650.6+311354	CGCG 504- 97 PGC 9501	 	.81± .11 .36± .06 .84	 .36 	 15.40 ±.10 			16.69±.3 245± 7 	5214± 10 5319 4991
022949.6-495436 270.00 -60.46 247.71 -27.15 022804.1-500754	ESO 198- 15 PGC 9502	.S?.... 	.97± .07 .35± .05 .97	26 .01 .48 .17	16.01 ±.15 15.33 ±.14 	.73± .04 	 14.84± .38 		
022950.0-425742 257.79 -64.55 255.58 -27.89 022754.0-431100	ESO 246- 20 PGC 9504	.L..0*/ S -2.0±1.3 	1.08± .06 .59± .03 .99	71 .04 .00 	 15.27 ±.14 				
022950.2+294429 147.21 -28.45 332.48 -11.64 022653.5+293110	UGC 1968 PGC 9505	.I..9*. U 10.0±1.2 	1.14± .13 .18± .12 1.18	80 .40 .13 .09				15.29±.3 217± 7 	4625± 10 4725 4399
0229.8 +0951 159.12 -45.99 313.58 -19.58 0227.2 +0938	UGC 1974 PGC 9509	.S..1.. U 1.0± .9 	1.04± .06 .29± .05 1.07	42 .29 .30 .15	 15.28 ±.14 14.58				8582 8629 8340
0229.8 +2515 149.45 -32.50 328.35 -13.57 0227.0 +2502	UGC 1970 PGC 9510	.S..6*. U 6.0±1.3 	1.37± .04 .93± .05 1.40	22 .34 1.36 .46	 15.05 ±.13 13.34				1915 2004 1683
0229.9 +0110 166.86 -53.08 304.78 -22.45 0227.4 +0057	IC 231 UGC 1978 PGC 9514	.L...*. U -2.0±1.2 	1.02± .08 .13± .03 1.00	160 .01 .00 	 14.63 ±.10 				
023000.1+283800 147.78 -29.44 331.48 -12.15 022704.4+282441	UGC 1971 PGC 9516	.L?.... 	1.00± .15 .16± .07 1.02	 .34 .00 	 14.67 ±.10 14.27			15.04±.3 272± 7 	4566± 10 4664 4338
0230.0 +4510 140.59 -14.29 346.27 -4.67 0226.8 +4457	UGC 1967 PGC 9517	.S..7*. U 7.0±1.4 	1.00± .10 .66± .04 1.07	122 .74 .92 .33					
023005.3-492543 269.20 -60.74 248.25 -27.26 022819.0-493900	ESO 198- 17 PGC 9522	 	.99± .06 .58± .05 .99	90 .01 	 15.17 ±.14 				8400± 87 8280 8273
023005.5-085953 179.69 -60.48 294.00 -25.19 022738.2-091311	MCG -2- 7- 24 MK 1044 PGC 9523	.S?.... 	.83± .07 .06± .06 .83	 .05 .09 .03				15.85±.1 246± 16 	4932± 11 4887± 56 4921 4700
023006.7+380605 143.50 -20.80 340.03 -7.95 022701.4+375247	UGC 1972 PGC 9524	.E...?. U -6.0±1.5 	1.20± .14 .00± .08 1.23	 .18 .00 	 14.8 ±.2 				
023014.8+330753 145.71 -25.33 335.59 -10.22 022714.8+325435	UGC 1975 KUG 0227+329 PGC 9526	.S?.... 	1.01± .04 .70± .05 1.04	143 .27 1.06 .35	 15.36 ±.13 14.02			16.15±.2 187± 13 181± 7 1.78	3176± 7 3198±125 3284 2956
023016.2-581008 280.64 -54.47 238.54 -25.79 022848.0-582324	ESO 115- 13 PGC 9529	.SXR4.. r 4.5± .8 	1.04± .05 .01± .05 1.04	 .04 .02 .01	 14.93 ±.14 14.82				9459 9322 9364
0230.2 +0057 167.18 -53.21 304.58 -22.59 0227.7 +0044	UGC 1981 PGC 9530	.IBS9*. UE (1) 10.0± .6 9.8± .8	1.08± .06 .04± .05 1.09	 .00 .03 .02					1507 1527 1268
023021.6+351933 144.75 -23.32 337.57 -9.26 022719.3+350615	UGC 1976 PGC 9533	.S..3.. U (1) 3.0± .8 3.5±1.1	1.21± .05 .33± .05 1.23	113 .19 .45 .16	 14.65 ±.15 13.94				9117 9230 8901
0230.4 -1044 182.58 -61.53 292.12 -25.66 0228.0 -1058	A 0228-10 MCG -2- 7- 26 DDO 23 PGC 9539	.IBS9.. PE (2) 9.5± .6 7.7± .7	1.31± .04 .17± .05 1.09± .04 1.31	90 .03 .13 .08	14.31 ±.19 14.16	.44± .07 -.02± .08 .38 -.06	.49± .04 -.04± .05 15.25± .11 15.28± .32		2110 2095 1882

R.A. 2000 DEC.	Names	Type	logD25	p.a.	BT	(B-V)T	(B-V)e	m21	V21
l b		ST nL	logR25	Ag	mB	(U-B)T	(U-B)e	W20	Vopt
SGL SGB		T	logAe	Ai	mFIR	(B-V)$_T^o$	m'e	W50	VGSR
R.A. 1950 DEC.	PGC	L	logDo	A21	B_T^o	(U-B)$_T^o$	m'25	HI	V3K

0230.4 +0842		.I..9..	1.00± .06						4145± 10
160.20 -46.88	UGC 1982	U	.12± .05	.34					
312.49 -20.12		10.0± .9		.09					4188
0227.8 +0829	PGC 9543		1.04	.06					3904

023028.8-430137		PSBR3..	1.34± .04	138	13.46 ±.16	.81± .03	.89± .02		
257.78 -64.41	ESO 246- 21	Sr (1)	.22± .05	.04	13.60 ±.14				5451± 34
255.50 -28.00		3.2± .6	.99± .03	.31					5346
022833.0-431454	PGC 9544	2.2± .8	1.35	.11	13.16	.73	13.90± .10		5302
							14.46± .29		

023033.1+305226		.I?....	.80± .09					17.09±.3	5443± 10
146.82 -27.36	CGCG 504-101		.15± .12	.35	15.06 ±.13				5546
333.58 -11.28	KUG 0227+306			.12				108± 7	
022735.3+303909	PGC 9548		.83	.08	14.59			2.42	5219

023033.6-010626	NGC 955	.S..2*/	1.44± .02	19	12.93 ±.13	.96± .01	.97± .01		
169.50 -54.76	UGC 1986	PUE	.59± .03	.04	13.04 ±.10	.36± .03	.45± .02		1504± 66
302.45 -23.26	IRAS02279-0119	1.8± .6	.71± .02	.73		.83	11.97± .06		1518
022800.7-011943	PGC 9549		1.45	.30	12.21	.23	13.52± .18		1266

023033.7+321033		.S?....	1.08± .09					16.09±.3	4778± 10
146.22 -26.18	UGC 1980		.47± .07	.25	14.54 ±.13				4884
334.76 -10.70	IRAS02275+3157			.71	11.83			424± 7	
022734.6+315716	PGC 9550		1.10	.24	13.55			2.30	4557

023035.8-313549		.SBS6?/	1.21± .04	75					
230.59 -67.99	ESO 415- 31	S (1)	.71± .03	.03	14.66 ±.14				4633± 52
268.61 -28.29		6.0± .8		1.04					4557
022827.1-314906	PGC 9551	5.6±1.3	1.21	.36	13.57				4449

023038.0+270924		.S?....	.84± .10					17.54±.3	5083± 10
148.66 -30.71	CGCG 483- 72		.64± .10	.30	15.83 ±.15				5177
330.18 -12.91				.95				327± 7	
022743.5+265607	PGC 9554		.87	.32	14.55			2.67	4854

023038.1-341550	IC 1811	PSBR2..	1.12± .04	7					
237.57 -67.58	ESO 355- 20	BSr	.17± .04	.03	14.35 ±.14				4791± 87
265.55 -28.35		1.6± .5		.20					4708
022832.1-342906	PGC 9555		1.12	.08	14.07				4615

023038.5+421358	NGC 946	.L...*.	1.16± .06	65					
141.88 -16.96	UGC 1979	U	.16± .03	.26	14.19 ±.11				5655± 68
343.73 -6.13		-2.0±1.1		.00					5783
022728.2+420041	PGC 9556		1.17		13.84				5454

0230.6 +5008		.SX.4..	1.04± .08	165					
138.76 -9.66	UGC 1977	U	.13± .06	1.27					
350.66 -2.42		4.0± .9		.19					
0227.3 +4955	PGC 9557		1.16	.06					

023041.7-034844		.SBS8*.	1.54± .03	100				14.22±.1	1627± 11
172.74 -56.76	MCG -1- 7- 20	UE (1)	.11± .05	.03				93± 16	
299.61 -24.04		8.0± .7		.14				84± 12	1632
022810.7-040201	PGC 9559	9.8± .8	1.54	.06					1392

023042.9-025621	NGC 958	.SBT5*.	1.46± .02	173	12.89 ±.13	.74± .01	.82± .01	13.78±.2	5738± 6
171.68 -56.11	MCG -1- 7- 19	PE (3)	.46± .03	.02	13.02 ±.19	.18± .06		591± 34	
300.53 -23.81	IRAS02281-0309	4.5± .6	.92± .02	.69	11.09	.61	12.98± .05	568± 25	5746
022811.3-030938	PGC 9560	2.9± .5	1.46	.23	12.09	.08	13.87± .19	1.36	5502

023048.9+370809	NGC 949	.SAT3*$	1.38± .02	145	12.40 ±.14	.61± .02		14.28±.1	612± 5
144.05 -21.63	UGC 1983	R (1)	.27± .03	.18	12.41 ±.12	-.02± .02		196± 7	596± 17
339.23 -8.52	IRAS02277+3654	3.0± .7		.37	11.62	.51		175± 12	728
022744.6+365452	PGC 9566	5.7± .8	1.40	.14	11.85	-.09	13.50± .20	2.29	399

023049.1-341314	IC 1813	.LAT+P*	1.08± .05	102					
237.44 -67.55	ESO 355- 22	BS	.14± .03	.03	14.20 ±.14				4453± 87
265.60 -28.38		-.5± .6		.00					4370
022843.0-342630	PGC 9567		1.06		14.10				4277

023050.6+261146		.S?....	.82± .13					16.71±.3	13290± 10
149.20 -31.56	MCG 4- 7- 2		.28± .07	.29	15.64 ±.12				13381
329.32 -13.37				.39				506± 7	13060
022757.0+255830	PGC 9570		.84	.14	14.86			1.71	

023053.0+252416		.S?....	.74± .14					17.45±.3	5286± 10
149.62 -32.26	MCG 4- 7- 1		.38± .07	.26	15.65 ±.15				5375
328.59 -13.71				.57				243± 7	5055
022800.1+251100	PGC 9573		.76	.19	14.79			2.46	

0231.0 +4321		.S..6*.	1.04± .08	129					
141.49 -15.90	UGC 1984	U	1.06± .06	.39					
344.75 -5.67		6.0±1.5		1.47					
0227.8 +4308	PGC 9577		1.08	.50					

R.A. 2000 DEC.	Names	Type	logD$_{25}$	p.a.	B$_T$	(B-V)$_T$	(B-V)$_e$	m$_{21}$	V$_{21}$
l b		S$_T$ n$_L$	logR$_{25}$	A$_g$	m$_B$	(U-B)$_T$	(U-B)$_e$	W$_{20}$	V$_{opt}$
SGL SGB		T	logA$_e$	A$_i$	m$_{FIR}$	(B-V)$_T^o$	m'$_e$	W$_{50}$	V$_{GSR}$
R.A. 1950 DEC.	PGC	L	logD$_o$	A$_{21}$	B$_T^o$	(U-B)$_T^o$	m'$_{25}$	HI	V$_{3K}$

023100.8-442533		.SB.5..	1.04± .05	135					5127
260.39 -63.60	ESO 246- 22	S (1)	.24± .04	.00	14.84 ±.14				5018
253.89 -27.98		5.0± .9		.36					4983
022907.1-443848	PGC 9578	5.6± .9	1.04	.12	14.45				
023104.1+274156		.I?....	1.07± .14	37					4617
148.49 -30.18	UGC 1990		.91± .12	.31	15.72 ±.13				
330.72 -12.77	IRAS02281+2728			.68	13.15				4712
022809.2+272840	PGC 9579		1.10	.46	14.72				4389
023104.5-191957		.SXR4*.	1.08± .04	98					
199.40 -65.87	ESO 545- 26	Er (1)	.09± .03	.07	14.66 ±.14				9837± 60
282.61 -27.34		4.2± .6		.13					9796
022845.1-193312	PGC 9580	3.1± .8	1.08	.05	14.40				9625
023104.8-574746		.LXR+*.	1.12± .06	85					
280.08 -54.69	ESO 115- 14	S	.65± .03	.04	15.75 ±.14				
238.93 -25.97		-1.0± .9		.00					
022936.0-580100	PGC 9581		1.03						
023106.1-360203	NGC 964	.S..2..	1.31± .04	31					
241.97 -67.07	ESO 355- 24	S	.58± .04	.03	13.48 ±.14				4780± 36
263.51 -28.44	IRAS02290-3615	2.0± .8		.71	13.75				4692
022902.0-361518	PGC 9582		1.32	.29	12.69				4609
023109.5-575504		.LBR+?.	1.04± .05	53					
280.20 -54.59	ESO 115- 15	Sr	.15± .04	.04	14.41 ±.14				9590± 63
238.79 -25.95	FAIR 222	-1.0± .7		.00					9453
022941.0-580818	PGC 9585		1.02		14.23				9495
023109.9+293518	NGC 953	.E.....							
147.58 -28.47	UGC 1991	U		.41	14.5 ±.4				
332.46 -11.96		-5.0± .8							
022813.2+292202	PGC 9586								
023111.4+011557	IC 232	.E.....	1.23± .07	155					
167.16 -52.82	UGC 1994	U	.14± .04	.02	13.71 ±.11				6358± 31
304.98 -22.71		-5.0± .8		.00					6378
022836.8+010242	PGC 9588		1.19		13.59				6120
023113.5+432021		.E...?.	.88± .14						
141.53 -15.90	UGC 1987	U	.08± .05	.39	15.3 ±.3				
344.76 -5.71	IRAS02280+4308	-5.0±1.7		.00					
022801.6+430706	PGC 9589		.91						
023114.4+402326		.S..2..	.99± .06	123					
142.75 -18.61	UGC 1988	U	.47± .05	.19	14.79 ±.14				
342.16 -7.09	IRAS02281+4010	2.0± .9		.58	12.90				
022806.2+401010	PGC 9590		1.01	.23					
023121.8+432757									
141.50 -15.77	CGCG 539- 37			.39	15.0 ±.4				5859± 68
344.88 -5.68									5989
022809.7+431442	PGC 9595								5661
0231.3 +0120	A 0228+01	.SBR5P*	1.19± .04	62				15.87±.3	7302± 12
167.15 -52.73	UGC 1995	UE (1)	.30± .04	.02	14.26 ±.13			333± 21	7390± 60
305.07 -22.74		4.7± .7		.45				317± 15	7326
0228.8 +0107	PGC 9598	1.9± .8	1.19	.15	13.76			1.97	7067
0231.3 +4255		.I..9*.	1.00± .16						
141.73 -16.27	UGC 1992	U	.04± .12	.24					
344.40 -5.93		10.0±1.2		.03					
0228.2 +4242	PGC 9599		1.02	.02					
023123.8-574528		PLBT+*.	1.00± .05						
279.98 -54.69	ESO 115- 16	Sr	.03± .03	.04	15.07 ±.14				8977
238.96 -26.02		-1.1± .6		.00					8840
022955.0-575842	PGC 9600		1.00		14.89				8881
023126.8-173910		.S?....	.99± .05	44					
195.83 -65.08	ESO 545- 28		.14± .04	.05	14.68 ±.14				10015± 60
284.51 -27.18				.11					9978
022906.0-175224	PGC 9602		.99		14.37				9800
023128.6-345446		PLX.0*P	1.13± .09						
239.12 -67.27		S	.00± .08	.03					
264.80 -28.52		-2.0± .8		.00					
022923.4-350800	PGC 9604		1.14						
0231.6 +0249	IC 233								
165.74 -51.53	CGCG 388- 33			.06	14.8 ±.2				8240± 46
306.62 -22.33									8265
0229.0 +0236	PGC 9610								8002

R.A. 2000 DEC.	Names	Type	$\log D_{25}$	p.a.	B_T	$(B-V)_T$	$(B-V)_\bullet$	m_{21}	V_{21}
I b		S_T n_L	$\log R_{25}$	A_g	m_B	$(U-B)_T$	$(U-B)_\bullet$	W_{20}	V_{opt}
SGL SGB		T	$\log A_\bullet$	A_i	m_{FIR}	$(B-V)_T^o$	m'	W_{50}	V_{GSR}
R.A. 1950 DEC.	PGC	L	$\log D_0$	A_{21}	B_T^o	$(U-B)_T^o$	m'_{25}	HI	V_{3K}
023137.9-000821 IC 234			.80± .09						6015± 61
168.79 -53.84 CGCG 388- 34			.42± .12						6031
303.55 -23.24 MK 1045				.02	15.28 ±.16				
022904.4-002135 PGC 9613			.80						5778
023138.5-443129 NGC 979		RLBR0..	1.07± .05						
260.44 -63.45 ESO 246- 23		BSr	.07± .03						
253.77 -28.08		-2.1± .5		.00	13.81 ±.14				
022945.0-444442 PGC 9614			1.06	.00					
0231.6 +2254 IC 1809		PSBS2..	.98± .07	50					5576
151.15 -34.41 UGC 1996		U	.13± .05	.36	14.82 ±.12				
326.35 -14.92 3ZW 48		2.0± .9		.16	11.97				5658
0228.8 +2241 PGC 9616			1.01	.07	14.24				5343
0231.6 +0018		.S..6*.	1.03± .06	143					6331
168.32 -53.49 UGC 1998		U	.22± .05	.03					
304.01 -23.12		6.0±1.2		.32					6348
0229.1 +0005 PGC 9617			1.03	.11					6094
023140.1+392244 A 0228+39		.S..3..	1.37± .04	140				15.32±.1	8018± 11
143.25 -19.50 UGC 1993		U (1)	.60± .05	.21	14.10 ±.12				8005± 73
341.30 -7.63		3.0± .8		.83				476± 8	8139
022833.1+390930 PGC 9618		4.5±1.1	1.39	.30	13.00			2.02	7811
023141.4-091801 NGC 960		.S..3*.	1.07± .06	125					
180.68 -60.38 MCG -2- 7- 28		E	.61± .05	.03					
293.77 -25.64		3.0±1.3		.84					
022914.4-093115 PGC 9621			1.07	.31					
023147.0-195253 NGC 966		.LA.0P*	1.00± .05	112					
200.80 -65.93 ESO 545- 30		SE	.06± .03	.06	14.37 ±.14				
282.02 -27.58		-2.4± .8		.00					
022928.0-200606 PGC 9626			1.00						
0231.8 +0537		.SX.7*.	1.11± .07						6049
163.21 -49.24 UGC 2000		U	.07± .06	.18					
309.50 -21.48		7.0±1.2		.09					6082
0229.2 +0524 PGC 9631			1.13	.03					5810
023151.3-364017 IC 1816		.SBR2P?	1.16± .04						5176± 20
243.42 -66.74 ESO 355- 25		BS	.07± .04	.03	13.86 ±.14				5086
262.77 -28.58 FAIR 1090		2.4± .6		.09	12.79				5008
022948.1-365330 PGC 9634			1.16	.04	13.69				
0231.9 +1908		.S..6*.	1.49± .03	96					971
153.39 -37.72 UGC 1999		U	.83± .05	.29	14.60 ±.16				
322.80 -16.52		6.0±1.2		1.22					1043
0229.1 +1855 PGC 9638			1.51	.41	13.09				735
023157.8+011445		.LAT-*.	1.28± .07						
167.43 -52.72 UGC 2005		UE	.00± .04	.02	14.05 ±.18				
305.02 -22.90		-3.0± .6		.00					
022923.2+010132 PGC 9642			1.28						
0231.9 +0054		.SXS5..	1.08± .06	95					6538
167.79 -52.98 UGC 2004		U	.20± .05	.00	14.49 ±.13				
304.67 -23.01 IRAS02294+0041		5.0± .8		.30	13.56				6557
0229.4 +0041 PGC 9643			1.08	.10	14.15				6301
023159.7+234513		.S?....	.89± .11					18.28±.3	9064± 10
150.77 -33.63 MCG 4- 7- 5			.07± .07	.32	15.19 ±.13				9148
327.16 -14.64				.10				64± 10	
022908.1+233200 PGC 9645			.92	.03	14.72			3.52	8832
0232.0 -1319		.SBS8..	1.15± .06	45					4602
187.51 -62.76 MCG -2- 7- 29		E (1)	.09± .05	.04					
289.36 -26.56		8.0± .8		.11					4578
0229.6 -1333 PGC 9646		7.5± .8	1.15	.05					4380
023207.3-012141		.SBT4..	1.12± .05	85					
170.30 -54.69 UGC 2010		UE (1)	.24± .05	.06	14.10 ±.12				11275± 39
302.31 -23.71		4.0± .6		.36					11287
022934.6-013454 PGC 9648		.8± .8	1.12	.12	13.61				11039
0232.1 +0943		.LB....	1.04± .18						
159.91 -45.79 UGC 2007		U	.00± .08	.23	14.75 ±.13				
313.66 -20.16		-2.0± .8		.00					
0229.5 +0930 PGC 9650			1.07						
023213.0-171300 NGC 967		.LX.-..	1.20± .04	33					
195.11 -64.72 ESO 545- 31		SE	.19± .03	.03	13.46 ±.14				
285.04 -27.29		-3.0± .6		.00					
022952.0-172612 PGC 9654			1.17						

R.A. 2000 DEC. l　　b SGL　SGB R.A. 1950 DEC.	Names PGC	Type S_T　n_L T L	$\log D_{25}$ $\log R_{25}$ $\log A_e$ $\log D_o$	p.a. A_g A_i A_{21}	B_T m_B m_{FIR} B_T^o	$(B-V)_T$ $(U-B)_T$ $(B-V)_T^o$ $(U-B)_T^o$	$(B-V)_e$ $(U-B)_e$ m'_e m'_{25}	m_{21} W_{20} W_{50} HI	V_{21} V_{opt} V_{GSR} V_{3K}
023215.5+432719 141.66 -15.72 344.95 -5.82 022903.3+431406	 UGC　1997 PGC　9655	.S..3.. U　　(1) 3.0±1.0 4.5±1.3	1.09± .06 .67± .05 1.11	73 .24 .93 .34	 15.3　±.3 14.04				6162± 68 6292 5965
023218.0-350148 239.32 -67.08 264.66 -28.69 023013.0-351500	 ESO　355- 26 IRAS02302-3515 PGC　9658	.SBS4*. S　　(1) 4.0± .8 4.4±1.2	1.20± .04 .27± .04 1.20	155 .03 .39 .13	 13.80 ±.14 13.58 13.36				2018± 60 1932 1846
023222.1-523019 273.24 -58.41 244.74 -27.15 023042.0-524330	 IRAS02307-5243 PGC　9662	 	.89± .06 .05± .05 .90	 .07 	 14.84 ±.14 13.34 				6443± 76 6316 6328
023222.7+421156 142.20 -16.86 343.85 -6.43 022912.1+415843	 UGC　2001 PGC　9663	.S..2.. U 2.0± .8 	1.15± .05 .17± .05 1.19	35 .33 .21 .09	 14.33 ±.13 13.71				6989± 68 7116 6789
023223.8-580113 280.10 -54.40 238.64 -26.09 023056.1-581424	 ESO　115- 17 PGC　9664	.L..0*/ S -2.0±1.2 	1.07± .05 .21± .04 1.04	110 .04 .00 	 14.83 ±.14 14.65				9050 8912 8956
023224.0+352942 145.09 -23.00 337.91 -9.55 022921.2+351629	NGC　　959 UGC　2002 IRAS02293+3516 PGC　9665	.S..8*. U 8.0±1.1 	1.37± .03 .21± .05 .98± .02 1.39	65 .17 .26 .11	12.95 ±.14 12.50 ±.13 12.85 12.27	.57± .02 -.08± .02 .49 -.14	.58± .01 -.10± .02 13.34± .04 14.12± .23	14.75±.1 158± 8 162± 12 2.37	601±　6 664± 42 715 389
023224.9-183824 198.19 -65.30 283.44 -27.56 023005.0-185136	NGC　　965 ESO　545- 32 PGC　9666	.SBS6P? SE　　(1) 6.0± .9 4.2± .8	1.01± .04 .11± .03 1.02	10 .07 .16 .05	 14.92 ±.14 				
023232.0+313632 146.91 -26.51 334.43 -11.33 022933.2+312320	 UGC　2008 PGC　9669	.S..7.. U 7.0± .9 	.82± .13 .00± .07 .85	 .39 .00 .00				16.20±.3 131± 7 	5042± 10 5145 4822
0232.6　+2101 152.46 -35.98 324.66 -15.91 0229.8　+2048	 UGC　2012 PGC　9675	.S..4.. U 4.0± .9 	1.03± .06 .13± .05 1.07	140 .37 .18 .06	 14.73 ±.13 14.12				8862 8938 8628
023238.2+313351 146.95 -26.54 334.40 -11.37 022939.4+312039	 UGC　2011 PGC　9676	.SXR2.. U　　(1) 2.0± .8 4.5±1.1	1.08± .09 .25± .07 1.11	 .39 .30 .12	 15.22 ±.16 14.43			16.77±.3 547± 7 2.22	9527± 10 9630 9307
023239.1+003705 168.31 -53.10 304.42 -23.26 023004.9+002353	 UGC　2019 IRAS02300+0023 PGC　9680	.S?.... 	.75± .07 .08± .04 .75	65 .01 .11 .04	 14.53 ±.17 13.78 14.36				6206± 50 6223 5970
023239.8+280412 148.67 -29.69 331.22 -12.92 022944.3+275101	NGC　　962 UGC　2013 PGC　9682	.E..... U -5.0± .8 	 	 .30 	 13.93 ±.12 				
023239.9+422543 142.16 -16.63 344.08 -6.37 022928.9+421231	 UGC　2006 PGC　9683	.E..... U -5.0± .8 	.93± .13 .00± .05 .98	 .30 .00 	 15.21 ±.12 				
023240.0+001536 168.69 -53.37 304.05 -23.37 023006.1+000225	 UGC　2018 PGC　9684	.LBT+*. E -1.0± .9 	1.11± .05 .34± .04 1.06	75 .02 .00 	 14.14 ±.10 14.03				6162± 50 6178 5926
023240.6-391750 249.42 -65.70 259.74 -28.65 023040.6-393101	NGC　986A ESO　299-　6A PGC　9685	.IB.9*. PS　　(1) 9.7± .7 8.9± .9	1.25± .03 .39± .03 .84± .02 1.25	72 .03 .29 .20	13.97V±.15 14.73 ±.10 14.40		.48± .05		1406± 62 1309 1247
0232.8　+0542 163.45 -49.03 309.68 -21.69 0230.2　+0529	 UGC　2021 PGC　9697	.L..... U -2.0± .8 	1.04± .08 .03± .03 1.06	 .18 .00 	 14.46 ±.11 				
023250.6+203838 152.74 -36.29 324.33 -16.11 023001.4+202527	IC　　235 UGC　2016 MK　　368 PGC　9698	.P..... R 99.0 	.61± .14 .11± .06 .46± .04 .64	 .32 .15 .05	14.9　±.2 14.8　±.2 13.45 14.32	.54± .04 -.09± .06 .39 -.20	.62± .03 12.67± .11 12.53± .76		8720± 43 8795 8486

R.A. 2000 DEC.	Names	Type	logD$_{25}$	p.a.	B$_T$	(B-V)$_T$	(B-V)$_\bullet$	m$_{21}$	V$_{21}$
l b	S$_T$ n$_L$	logR$_{25}$	A$_g$	m$_B$	(U-B)$_T$	(U-B)$_\bullet$	W$_{20}$	V$_{opt}$
SGL SGB	T	logA$_\bullet$	A$_i$	m$_{FIR}$	(B-V)$_T^o$	m'	W$_{50}$	V$_{GSR}$
R.A. 1950 DEC.	PGC	L	logD$_o$	A$_{21}$	B$_T^o$	(U-B)$_T^o$	m'$_{25}$	HI	V$_{3K}$

023253.7+191531			.95?						3984
153.58 -37.50	CGCG 462- 23		.11?	.29	14.67 ±.12				
323.02 -16.69									4055
023005.7+190220	PGC 9699		.98						3749
0232.9 +3840	A 0229+38	.I..9*.	1.31± .05	176	15.65 ±.18	.32± .08	.28± .06	15.14±.1	565± 6
143.79 -20.05	UGC 2014	PU (1)	.53± .06	.19		-.07± .09	-.17± .08	67± 7	
340.79 -8.17	DDO 22	10.0± .6	.87± .04	.40		.15	15.49± .08	49± 12	685
0229.8 +3827	PGC 9702	9.0±1.0	1.33	.27	15.06	-.20	15.71± .36	-.19	358
023254.9+250538		.S?....	.74± .14					16.31±.3	5415± 10
150.27 -32.34	MCG 4- 7- 6	.00± .07	.26	14.96 ±.13				5502	
328.50 -14.26			.00				144± 7		5502
023002.1+245227	PGC 9703		.76	.00	14.68			1.63	5186
0232.9 +3452		.L...?.	1.11± .16						
145.48 -23.52	UGC 2015	U	.00± .08	.21					
337.41 -9.93		-2.0±1.6		.00					
0229.9 +3439	PGC 9704		1.13						
023256.3+284911		.I..9..	1.36± .09					14.49±.1	1014± 6
148.35 -28.99	UGC 2017	U	.10± .12	.31				92± 7	
331.93 -12.65		10.0± .7		.07				84± 12	1111
023000.0+283600	PGC 9705		1.39	.05					790
0232.9 +2320		.S..6*.	1.14± .05	109					5559
151.23 -33.90	UGC 2020	U	.81± .05	.45	15.49 ±.12				
326.87 -15.02		6.0±1.3		1.20					5641
0230.1 +2307	PGC 9707		1.19	.41	13.82				5328
023301.7+002516		.S..2..	1.00± .06	153					6714± 50
168.64 -53.19	UGC 2024	U	.22± .05	.02	14.38 ±.12				6730
304.25 -23.41	IRAS02304+0012	2.0± .9		.27	12.27				6478
023027.7+001206	PGC 9711		1.01	.11	14.02				
023303.5-104536	NGC 977	PSXR1*.	1.29± .05	20					4546± 60
183.43 -61.04	MCG -2- 7- 31	E	.09± .05	.01					4529
292.26 -26.29		1.0± .8		.09					4322
023037.6-105847	PGC 9713		1.29	.05					
0233.1 +0855			1.14± .07						6210
160.84 -46.33	UGC 2027		.10± .06	.35					
312.95 -20.67									6252
0230.5 +0842	PGC 9717		1.17						5972
023315.9-114444		.SBS6..	1.23± .03	85					4476
185.13 -61.59	MCG -2- 7- 32	E (1)	.21± .05	.00					
291.18 -26.54	KUG 0230-119	6.0± .8		.30					4456
023050.8-115753	PGC 9721	5.3± .8	1.23	.10					4253
023317.2+324448		.E...*.	.61± .11						
146.53 -25.42	UGC 2022	U	.00± .06	.30	15.24 ±.13				
335.53 -10.96	IRAS02302+3231	-5.0±1.2		.00	13.10				
023017.2+323138	PGC 9724		.65						
0233.2 +2530		.S..6*.	1.13± .05	32					11084
150.14 -31.93	UGC 2025	U	.92± .05	.26	15.65 ±.12				
328.92 -14.16		6.0±1.4		1.35					11172
0230.4 +2517	PGC 9725		1.15	.46	13.99				10856
023318.1+332926	A 0230+33	.I..9*.	1.22± .07		13.90 ±.17	.62± .06	.58± .03	14.34±.1	606± 6
146.19 -24.74	UGC 2023	PU (1)	.00± .07	.37	14.4 ±.2	-.19± .07	-.21± .04	49± 6	
336.20 -10.63	DDO 25	10.0± .6	1.17± .03	.00		.53	15.24± .07	39± 12	713
023017.2+331616	PGC 9726	9.0±1.0	1.25	.00	13.72	-.25	14.83± .43	.62	389
0233.3 +2718		.I..9*.	1.14± .13						5218
149.21 -30.32	UGC 2026	U	.14± .12	.26					
330.58 -13.39		10.0±1.2		.10					5310
0230.4 +2705	PGC 9727		1.16	.07					4992
0233.3 +2223		.S..6*.	1.12± .05	110					5431
151.86 -34.70	UGC 2028	U	.34± .05	.44	14.48 ±.12				
326.02 -15.50		6.0±1.2		.50					5510
0230.5 +2210	PGC 9729		1.16	.17	13.51				5200
0233.3 +2245		.I..9*.	1.00± .08	20					5502
151.66 -34.37	UGC 2029	U	.25± .06	.42					
326.36 -15.34		10.0±1.3		.19					5582
0230.5 +2232	PGC 9730		1.04	.13					5271
023323.0+093608	NGC 975	.S..0..	1.03± .06	0					6111± 50
160.36 -45.73	UGC 2030	U	.14± .05	.25	14.07 ±.12				6154
313.66 -20.48		.0± .9		.10					5873
023042.3+092258	PGC 9735		1.04		13.62				

R.A. 2000 DEC. / l b / SGL SGB / R.A. 1950 DEC.	Names / PGC	Type / S_T n_L / T / L	$\log D_{25}$ / $\log R_{25}$ / $\log A_\bullet$ / $\log D_\circ$	p.a. / A_g / A_i / A_{21}	B_T / m_B / m_{FIR} / B_T°	$(B-V)_T$ / $(U-B)_T$ / $(B-V)_T^\circ$ / $(U-B)_T^\circ$	$(B-V)_\bullet$ / $(U-B)_\bullet$ / $m'_\bullet$ / m'_{25}	m_{21} / W_{20} / W_{50} / HI	V_{21} / V_{opt} / V_{GSR} / V_{3K}
0233.4 +2016 / 153.11 -36.56 / 324.03 -16.39 / 0230.6 +2003	/ UGC 2031 / / PGC 9736	.L..... / U / -2.0± .8 /	1.08± .09 / .17± .04 / / 1.10	10 / .37 / .00 /	/ 14.95 ±.13 / /				
0233.4 +2035 / 152.92 -36.28 / 324.33 -16.26 / 0230.6 +2022	/ UGC 2032 / / PGC 9738	.S..6*. / U / 6.0±1.3 /	1.07± .07 / .70± .06 / / 1.10	54 / .32 / 1.02 / .35	/ / /				4335 / / 4410 / 4102
023325.5-433109 / 258.14 -63.70 / 254.88 -28.49 / 023131.0-434418	/ ESO 246- 25 / IRAS02315-4344 / PGC 9740	RSBR1*. / r / 1.0± .9 /	.89± .07 / .00± .06 / / .89	/ .00 / .00 / .00	/ 14.57 ±.14 / 12.44 / 14.51				5124 / 5016 / 4979
023327.8+220936 / 152.02 -34.88 / 325.82 -15.61 / 023037.3+215627	/ CGCG 484- 9 / / PGC 9741	/ / /		/ .44 / /	/ 15.7 ±.4 / /			17.37±.3 / / 238± 7 /	12286± 10 / / 12365 / 12054
023334.2-390243 / 248.69 -65.64 / 260.02 -28.83 / 023134.0-391551	NGC 986 / ESO 299- 7 / IRAS02315-3915 / PGC 9747	.SBT2.. / R (2) / 2.0± .3 / 2.8± .5	1.59± .03 / .12± .03 / 1.15± .01 / 1.59	150 / .03 / .15 / .06	11.64 ±.13 / 11.67 ±.11 / 9.65 / 11.45	.73± .01 / .08± .01 / .68 / .05	.79± .01 / .10± .01 / 12.88± .02 / 14.12± .20	14.77±.2 / 177± 9 / 96± 12 / 3.26	2005± 7 / 1994± 49 / 1907 / 1845
023337.3+323213 / 146.70 -25.58 / 335.38 -11.12 / 023037.4+321904	/ CGCG 505- 7 / / PGC 9751	/ / /	.95? / .35? / / .98	/ .33 / /	/ 15.51 ±.12 / /			17.05±.3 / / 348± 7 /	9673± 10 / / 9778 / 9455
0233.6 +0628 / 163.03 -48.29 / 310.53 -21.62 / 0231.0 +0615	/ UGC 2037 / / PGC 9753	.S?.... / / /	1.06± .06 / .64± .05 / / 1.08	103 / .22 / .96 / .32	/ 15.39 ±.12 / / 14.17				8324 / / 8358 / 8087
023342.9+373959 / 144.38 -20.91 / 339.97 -8.78 / 023037.4+372650	/ UGC 2033 / / PGC 9758	.SBS3.. / U / 3.0± .8 /	1.13± .05 / .27± .05 / / 1.15	160 / .19 / .38 / .14	/ 14.49 ±.12 / /				
023343.0+403142 / 143.15 -18.29 / 342.59 -7.44 / 023034.2+401833	A 0230+40 / UGC 2034 / DDO 24 / PGC 9759	.I..9.. / U (1) / 10.0± .7 / 8.0±1.3	1.40± .05 / .10± .06 / 1.19± .05 / 1.42	170 / .20 / .07 / .05	13.7 ±.2 / 14.3 ±.3 / / 13.56	.47± .07 / -.17± .09 / .40 / -.22	.56± .03 / -.08± .04 / 15.09± .11 / 15.25± .35	13.58±.1 / 58± 7 / 45± 12 / -.03	579± 6 / / 702 / 377
023344.4-385228 / 248.28 -65.67 / 260.22 -28.87 / 023144.0-390536	/ ESO 299- 8 / / PGC 9761	/ / /	.80± .06 / .31± .05 / / .80	14 / .03 / /	/ 15.18 ±.14 / /				5004±104 / 4907 / 4844
023347.1+301122 / 147.86 -27.68 / 333.27 -12.21 / 023049.4+295813	/ CGCG 505- 9 / / PGC 9763	/ / /	1.00? / .00? / / 1.04	/ .42 / /	/ 15.23 ±.18 / /			15.45±.3 / / 302± 7 /	10229± 10 / / 10328 / 10008
0233.8 +0055 / 168.38 -52.67 / 304.84 -23.46 / 0231.3 +0042	/ UGC 2044 / / PGC 9765	.SBS6*. / U / 6.0±1.3 /	.91± .07 / .16± .05 / / .92	30 / .02 / .23 / .08	/ 15.35 ±.13 / /				
0233.8 +0938 / 160.48 -45.64 / 313.74 -20.58 / 0231.2 +0925	/ UGC 2041 / / PGC 9767	.S..2.. / U / 2.0± .9 /	1.08± .06 / .23± .05 / / 1.11	170 / .27 / .28 / .11	/ 14.88 ±.14 / / 14.21				12156 / / 12199 / 11919
023358.5+442042 / 141.59 -14.78 / 345.87 -5.67 / 023044.6+440733	/ UGC 2035 / IRAS02307+4407 / PGC 9773	.S..3.. / U (1) / 3.0± .8 / 2.5±1.1	1.24± .05 / .18± .05 / / 1.28	125 / .47 / .25 / .09	/ 13.9 ±.3 / 12.71 / 13.10			15.25±.1 / / 302± 8 / 2.06	5077± 11 / / 5208 / 4884
0233.9 -0621 / 177.15 -57.99 / 297.11 -25.49 / 0231.5 -0635	/ / / PGC 9774	.IBS9.. / E (1) / 10.0± .9 / 7.5±1.2	1.20± .07 / .29± .08 / / 1.21	10 / .06 / .22 / .14	/ / /				
023359.6+205836 / 152.84 -35.87 / 324.76 -16.22 / 023110.1+204528	NGC 976 / UGC 2042 / IRAS02311+2045 / PGC 9776	.SAT5*. / R (2) / 5.0± .6 / 2.5± .7	1.17± .03 / .08± .04 / .71± .01 / 1.21	/ .40 / .11 / .04	13.26 ±.13 / 12.91 ±.12 / 11.69 / 12.52	.82± .01 / .16± .02 / .69 / .05	.90± .01 / .25± .02 / 12.30± .03 / 13.78± .21	14.71±.2 / 334± 6 / 318± 6 / 2.15	4298± 7 / 4332± 66 / 4373 / 4066
023405.3+012107 / 168.00 -52.30 / 305.31 -23.38 / 023130.6+010800	/ UGC 2051 / / PGC 9778	.L..... / U / -2.0± .8 /	1.20± .14 / .02± .08 / / 1.20	/ .02 / .00 /	/ 13.88 ±.13 / / 13.76				6574± 50 / 6593 / 6339

R.A. 2000 DEC.	Names	Type	logD$_{25}$	p.a.	B$_T$	(B-V)$_T$	(B-V)$_e$	m$_{21}$	V$_{21}$
l b		S$_T$ n$_L$	logR$_{25}$	A$_g$	m$_B$	(U-B)$_T$	(U-B)$_e$	W$_{20}$	V$_{opt}$
SGL SGB		T	logA$_e$	A$_i$	m$_{FIR}$	(B-V)$_T^o$	m'$_e$	W$_{50}$	V$_{GSR}$
R.A. 1950 DEC.	PGC	L	logD$_o$	A$_{21}$	B$_T^o$	(U-B)$_T^o$	m'$_{25}$	HI	V$_{3K}$
023406.3+342847 NGC 968		.E.....	1.56± .08	60					3637± 31
145.90 -23.77	UGC 2040	U	.29± .08	.21	13.18 ±.15				3747
337.17 -10.32		-5.0± .7		.00					3423
023104.3+341539	PGC 9779		1.51		12.91				
023408.0+325651 NGC 969		.L.....	1.23± .10						4520± 61
146.61 -25.16	UGC 2039	U	.03± .07	.30	13.27 ±.10				4626
335.79 -11.03		-2.0± .8		.00					4303
023107.6+324343	PGC 9781		1.26		12.91				
0234.1 +4031		.SB.8*.	1.00± .16						
143.23 -18.27	UGC 2038	U	.33± .12	.20					
342.53 -7.52		8.0±1.3		.41					
0231.0 +4018	PGC 9784		1.02	.17					
0234.2 +0237		.SX.4..	.95± .07						
166.76 -51.29	UGC 2056	U	.08± .05	.06	14.74 ±.12				
306.63 -23.01		4.0± .9		.12					
0231.6 +0224	PGC 9785		.95	.04					
023413.3+291842 NGC 972		.S..2..	1.52± .02	152	12.27M±.10	.83± .01	.88± .01	14.75±.1	1543± 6
148.39 -28.43	UGC 2045	U	.29± .02	.37	12.18 ±.10	.21± .02	.22± .02	381± 11	1550± 31
332.51 -12.68	IRAS02312+2905	2.0± .7	1.00± .01	.36	9.36	.68	12.58± .04	261± 5	1640
023116.5+290534	PGC 9788		1.55	.15	11.48	.09	13.96± .14	3.13	1321
023415.6+300017		.S..8*.	1.06± .06	65				15.35±.3	5080± 10
148.06 -27.80	UGC 2046	U	.08± .05	.46					5179
333.14 -12.38		8.0±1.2		.10				163± 7	5179
023118.1+294710	PGC 9790		1.10	.04					4859
023420.0+322545 IC 1815		.LB....	1.22± .06						
146.90 -25.61	UGC 2047	U	.02± .03	.33	13.88 ±.13				
335.35 -11.30		-2.0± .8		.00					
023120.1+321237	PGC 9794		1.25						
023420.2+323020 NGC 973		.S..3..	1.57± .03	48				14.22±.3	4851± 10
146.87 -25.54	UGC 2048	U (1)	.84± .05	.33	13.55 ±.12				4956
335.42 -11.27	IRAS02313+3217	3.0± .8		1.16	12.48			582± 7	4634
023120.2+321713	PGC 9795	4.5±1.1	1.60	.42	12.03			1.77	
023423.7+450010		.SBS5..	.96± .09						
141.39 -14.15	UGC 2043	U	.05± .06	.62	14.6 ±.3				
346.48 -5.42		5.0± .9		.07					
023108.8+444702	PGC 9799		1.02	.02					
023424.6-105040		.SB.4?.	1.46± .04	75					4780
183.98 -60.82	MCG -2- 7- 33	E (1)	.44± .05	.00					
292.24 -26.63	IRAS02319-1103	4.0±1.1		.64	12.25				4762
023158.9-110346	PGC 9800	4.2± .8	1.46	.22					4557
023425.9+325716 NGC 974		.SXT3*.	1.39± .05						4517
146.67 -25.13	UGC 2049	U	.11± .06	.30	13.47 ±.17				
335.83 -11.08	IRAS02314+3243	3.0± .9		.16	13.37				4623
023125.5+324409	PGC 9802		1.42	.06	12.98				4301
023426.0-323454		RL..0..	1.28± .08						
232.93 -67.07		S	.33± .08	.03					
267.50 -29.13		-2.0± .8		.00					
023218.9-324800	PGC 9803		1.24						
023426.2+251607		.I..9?.	1.14± .13	174				17.19±.3	5698± 10
150.54 -32.02	UGC 2052	U	.86± .12	.29					5784
328.82 -14.50		10.0±1.9		.65				157± 7	5471
023133.0+250300	PGC 9805		1.17	.43					
023429.4+294505 A 0231+29		.I..9..	1.31± .04	43	15.10 ±.16	.40± .04	.41± .03	14.33±.1	1029± 5
148.24 -28.01	UGC 2053	U (1)	.31± .05	.39	15.0 ±.3	-.04± .06	.05± .05	77± 5	
332.94 -12.54	DDO 26	10.0± .8	.98± .03	.23		.23	15.49± .07	59± 7	1127
023132.1+293158	PGC 9808	9.0±1.0	1.35	.15	14.46	-.16	15.74± .28	-.29	808
023433.0+335631		.S..8*.	1.16± .05	80					4443
146.24 -24.22	UGC 2054	U	.25± .05	.26	14.89 ±.17				
336.73 -10.65		8.0±1.2		.31					4551
023131.6+334324	PGC 9811		1.19	.13	14.31				4229
0234.5 +0121		.S..4..	.95± .07	98					
168.16 -52.22	UGC 2062	U	.29± .05	.03	14.86 ±.12				
305.35 -23.49		4.0± .9		.43					
0232.0 +0108	PGC 9813		.95	.15					
0234.5 +3709		.S..9*.	1.11± .14	15					3774
144.78 -21.31	UGC 2055	U	.15± .12	.18					
339.59 -9.17		9.0±1.2		.15					3889
0231.5 +3656	PGC 9814		1.13	.07					3565

R.A. 2000 DEC.	Names	Type	logD$_{25}$	p.a.	B$_T$	(B-V)$_T$	(B-V)$_e$	m$_{21}$	V$_{21}$
l b	S$_T$ n$_L$		logR$_{25}$	A$_g$	m$_B$	(U-B)$_T$	(U-B)$_e$	W$_{20}$	V$_{opt}$
SGL SGB		T	logA$_e$	A$_i$	m$_{FIR}$	(B-V)$_T^o$	m'$_e$	W$_{50}$	V$_{GSR}$
R.A. 1950 DEC.	PGC	L	logD$_o$	A$_{21}$	B$_T^o$	(U-B)$_T^o$	m'$_{25}$	HI	V$_{3K}$

023437.4-084708 NGC 985	.RINGP.	.98± .05		14.01 ±.13	.63± .01	.59± .02		12929± 30
180.83 -59.49 MCG -2- 7- 35	PE	.01± .05	.10		-.56± .02	-.59± .03		12916
294.52 -26.23 MK 1048	10.0± .8	.64± .02	.01 12.85	.52	12.70± .06		12703
023210.1-090014 PGC 9817		.99	.01 13.89	-.64	13.74± .30		

023438.7+445446	.SXS5..	1.00± .08	2					
141.47 -14.21 UGC 2050	U	.19± .06	.62 14.8 ±.3				
346.42 -5.51	5.0± .9		.28				
023123.9+444139 PGC 9818		1.06	.09				

023442.8+232446 NGC 984	.LA.+..	1.48± .10	120				14.76±.3	4352± 9
151.62 -33.64 UGC 2059	U	.17± .08	.38 13.8 ±.2			614± 10	4409± 47
327.12 -15.35 5ZW 257	-1.0± .7		.00				4435
023151.2+231140 PGC 9819		1.49	14.21				4125

023447.1+325045 NGC 978A	.L..-*.	1.30± .12	80					4707± 46
146.80 -25.19 UGC 2057	U	.07± .08	.34				4812
335.77 -11.20	-3.0±1.1		.00				4491
023146.7+323739 PGC 9821		1.33					

023448.3-074100	.SBS5..	1.06± .06	60				
179.27 -58.73 MCG -1- 7- 22	E (1)	.04± .05	.06				
295.73 -26.01 IRAS02323-0754	5.0± .9		.07 12.85				
023220.3-075406 PGC 9822	1.9± .8	1.06	.02				

023453.4+410707	.SX.4..	1.07± .07					
143.11 -17.67 UGC 2058	U	.03± .06	.22 15.09 ±.20				
343.13 -7.36	4.0± .8		.04				
023143.6+405401 PGC 9825		1.09	.01				

023453.9-133933	.SXT5*.	1.29± .05	135					4768
188.95 -62.36 MCG -2- 7- 36	E (1)	.32± .05	.00				
289.15 -27.31 IRAS02325-1352	5.0± .8		.49 12.93				4741
023230.3-135238 PGC 9826	3.1± .8	1.29	.16				4550

023503.4+412126	.SB.2..	1.09± .06	177				
143.03 -17.43 UGC 2060	U	.41± .05	.26 14.63 ±.12				
343.35 -7.27 IRAS02319+4108	2.0± .9		.51 13.52				
023153.3+410821 PGC 9827		1.12	.21				

0235.1 +2051	.SXS4..	1.31± .04	165					4264
153.20 -35.85 UGC 2064	U	.18± .05	.40 14.3 ±.2				
324.75 -16.51	4.0± .8		.26				4338
0232.3 +2038 PGC 9828		1.35	.09 13.65				4033

023518.8+405535 NGC 982	.L.....	1.23± .07	110				
143.27 -17.81 UGC 2063	U	.27± .04	.22 14.02 ±.11				
342.99 -7.52	-2.0± .8		.00				
023209.2+404230 PGC 9831		1.21					

023521.8-253144	.SBT4..	.96± .05	93				
214.91 -66.70 ESO 479- 15	r	.16± .04	.03 15.35 ±.14				
275.67 -29.01	4.5± .9		.23				
023308.0-254448 PGC 9832		.97	.08				

023522.7+372910	.S..9..	.85± .07					14.78±.3	3890± 9
144.79 -20.94 UGC 2065	U	.00± .05	.16 14.59 ±.12			123± 10	
339.97 -9.16 KUG 0232+372	9.0± .8		.00			100± 7	4006
023217.2+371606 PGC 9834		.87	.00 14.43			.36	3683

023522.8+125019 IC 238	.L.....	1.22± .08	30					6007± 31
158.50 -42.74 UGC 2070	U	.27± .05	.30 13.79 ±.10				6059
317.05 -19.75	-2.0± .8		.00				5772
023239.6+123715 PGC 9835		1.22	13.40				

0235.4 +1956	.S..4..	1.07± .06	130					8669
153.82 -36.62 UGC 2071	U	.32± .05	.38 15.18 ±.14				
323.91 -16.95	4.0± .9		.48				8740
0232.6 +1943 PGC 9837		1.11	.16 14.27				8438

023524.8+405211 NGC 980	.S..1..	1.18± .05	132				15.30±.1	5737± 11
143.31 -17.85 UGC 2066	U	.42± .05	.22 13.40 ±.16				5887± 52
342.95 -7.56 IRAS02322+4039	1.0± .8		.43 12.39			568± 8	5866
023215.3+403906 PGC 9838		1.20	.21 12.67			2.41	5543

023526.9-481233	RLAR+..	.98± .07	174				
266.13 -60.81 ESO 198- 19	r	.16± .05	.00 14.70 ±.14				
249.49 -28.30	-1.3± .9		.00				
023340.0-482536 PGC 9839		.95					

023529.5+373108	.S..2..	1.29± .03	158				
144.79 -20.90 UGC 2067	U	.71± .04	.16 14.64 ±.12				3839
340.01 -9.16 IRAS02323+3717	2.0± .8		.87 12.79				3955
023223.9+371804 PGC 9841		1.31	.36 13.57				3632

R.A. 2000 DEC. / l b / SGL SGB / R.A. 1950 DEC.	Names / PGC	Type / S_T n_L / T / L	$\log D_{25}$ / $\log R_{25}$ / $\log A_e$ / $\log D_o$	p.a. / A_g / A_i / A_{21}	B_T / m_B / m_{FIR} / B_T^o	$(B-V)_T$ / $(U-B)_T$ / $(B-V)_T^o$ / $(U-B)_T^o$	$(B-V)_e$ / $(U-B)_e$ / m'_e / m'_{25}	m_{21} / W_{20} / W_{50} / HI	V_{21} / V_{opt} / V_{GSR} / V_{3K}
023529.7-092135	NGC 988	.SBS6*.	1.66± .03	112	11.38 ±.15	.41± .02	.49± .02	13.59±.1	1504± 7
181.97 -59.69	MCG -2- 7- 37	PUE (1)	.27± .05	.06		-.17± .03	-.08± .03	275± 9	1455± 63
293.94 -26.57	IRAS02330-0934	5.7± .4	.71± .03	.40	11.88	.33	10.42± .10	267± 12	1489
023302.9-093439	PGC 9843	3.1±1.2	1.67	.14	10.91	-.22	13.86± .24	2.54	1280
023532.7-070920	NGC 991	.SXT5..	1.43± .03	60				14.15±.1	1534± 6
178.75 -58.24	MCG -1- 7- 23	R (3)	.05± .03	.00	12.4 ±.2			86± 6	1485± 66
296.36 -26.06	IRAS02330-0722	5.0± .3		.07	13.12			66± 11	1526
023304.3-072224	PGC 9846	4.3± .5	1.43	.02	12.30			1.82	1308
0235.5 +4055		.S..8*.	1.04± .08						
143.31 -17.80	UGC 2068	U	.76± .06	.22					
343.01 -7.57		8.0±1.4		.94					
0232.4 +4042	PGC 9849		1.06	.38					
0235.6 -1216		.L?....	1.23± .10						
186.74 -61.44	MCG -2- 7- 38		.65± .07	.00					4879
290.72 -27.21				.00					4856
0233.2 -1230	PGC 9851		1.13						4660
023537.6+373831	A 0232+37	.SXS7..	1.37± .03	65					
144.76 -20.77	UGC 2069	U (1)	.23± .04	.16	13.03 ±.12				3715± 37
340.13 -9.13	VV 96	7.0± .8		.31					3831
023231.9+372527	PGC 9852	6.0±1.3	1.39	.11	12.54				3509
0235.6 +0616		.S..0..	1.06± .08						
163.81 -48.16	UGC 2075	U	.01± .06	.20	14.55 ±.15				
310.50 -22.16		.0± .8		.01					
0233.0 +0603	PGC 9854		1.08						
0235.7 -0525		.SBT6?.	1.10± .08	150					
176.44 -57.04		E (1)	.48± .08	.06					
298.23 -25.67		6.0± .9		.70					
0233.2 -0539	PGC 9856	5.3±1.6	1.11	.24					
0235.7 +1128		.S..4..	1.04± .15	25					7851
159.60 -43.86	UGC 2076	U	.23± .12	.33					
315.73 -20.33		4.0± .9		.33					7898
0233.0 +1115	PGC 9857		1.07	.11					7616
023544.7-133920		PSBT0..	1.21± .05	115					
189.19 -62.18	MCG -2- 7- 41	E	.17± .05	.00					
289.20 -27.51		.0± .8		.13					
023321.2-135222	PGC 9861		1.20						
023601.1+002510		.SXS6..	1.34± .03	80					2612
169.59 -52.71	UGC 2081	UE (1)	.18± .04	.04	14.4 ±.3				
304.49 -24.12		6.0± .5		.26					2626
023327.1+001208	PGC 9869	5.3±1.2	1.34	.09	14.07				2380
023604.5+422514		.L...?.	1.16± .09	105					
142.76 -16.39	UGC 2073	U	.15± .05	.33	13.8 ±.3				5165± 52
344.37 -6.93		-2.0±1.6		.00					5291
023252.8+421212	PGC 9873		1.18		13.38		.		4969
0236.0 +0133		.S..6*.	1.01± .06	176					
168.42 -51.83	UGC 2085	U	.38± .05	.03	15.38 ±.13				
305.68 -23.79		6.0±1.3		.57					
0233.5 +0120	PGC 9875		1.01	.19					
023609.1-501929		.S?....	.96± .06	74					
269.32 -59.42	ESO 198- 22		.49± .05	.03	15.31 ±.14				6500±190
247.07 -28.09	FAIR 297			.73	12.98				6376
023426.0-503230	PGC 9879		.96	.24	14.52				6380
0236.1 -0041		.S..6*.	1.00± .08	78					
170.85 -53.53	UGC 2088	U	.94± .06	.06					
303.32 -24.48		6.0±1.4		1.39					
0233.6 -0055	PGC 9880		1.01	.47					
0236.1 +2354	A 0233+23	.SXS5..	1.23± .05	157	14.66 ±.15	.75± .04	.71± .02	15.00±.1	5648± 8
151.70 -33.05	UGC 2079	U (1)	.34± .05	.43	14.49 ±.14		.07± .07	265± 9	5616± 59
327.72 -15.44		5.0± .8	.81± .02	.52		.55	14.20± .05	249± 9	5729
0233.3 +2341	PGC 9881	1.1± .8	1.27	.17	13.60		14.79± .30	1.23	5420
023611.4+343540		.S..6*.	.95± .07						11759
146.27 -23.49	UGC 2077	U	.05± .05	.15	15.18 ±.14				
337.47 -10.65		6.0±1.2		.07					11868
023309.0+342238	PGC 9882		.96	.03	14.91				11547
0236.2 +1405		.SB.2*.	1.00± .08	155					
157.85 -41.57	UGC 2086	U	.14± .06	.35					
318.36 -19.46		2.0± .9		.17					
0233.5 +1352	PGC 9885		1.03	.07					

| R.A. 2000 DEC. | Names | Type ST nL | logD25 logR25 logAo logDo | p.a. Ag Ai A21 | BT mB mFIR BoT | (B-V)T (U-B)T (B-V)oT (U-B)oT | (B-V)o (U-B)o m' m'25 | m21 W20 W50 HI | V21 Vopt VGSR V3K |
l b / SGL SGB / R.A. 1950 DEC. / PGC		T L							
023616.3+252529		.S..6*.	1.73± .02	133	13.69M±.12			13.17±.0	707± 4
150.89 -31.69	UGC 2082	U	.81± .04	.40	13.56 ±.18			206± 5	716± 76
329.14 -14.80	IRAS02333+2512	6.0±1.0		1.20				197± 12	793
023322.8+251227	PGC 9888		1.77	.41	12.05		15.16± .19	.72	482
023618.4+113829	NGC 990	.E.....	1.26± .13		13.48 ±.15	1.03± .02	1.05± .02		
159.64 -43.63	UGC 2089	U	.08± .08	.33	13.40 ±.11	.53± .04	.54± .04		3508± 31
315.96 -20.40		-5.0± .8	.76± .04	.00		.92	12.77± .15		3555
023336.2+112528	PGC 9890		1.29		13.05	.47	14.58± .70		3274
023619.9-545154	NGC 1025	.S..3..	.97± .05	6	13.84V±.13	.71± .03	.77± .03		
275.62 -56.34	ESO 154- 4	S (1)	.25± .04	.10	14.62 ±.14				6347± 50
241.99 -27.29	IRAS02347-5504	3.0± .9	.42± .02	.35	12.67	.60			6213
023446.0-550454	PGC 9891	3.3±1.3	.98	.13	14.09		13.63± .32		6243
0236.3 +5939	Maffei I	.L..-P*	.78?	100	11.4 V±.5		2.39± .05		
135.83 -.57	UGCA 34	R	.08± .18	6.6					2± 72
359.28 1.46	WEIN 19	-3.0± .8		.00					157
0232.6 +5926	PGC 9892		1.62						-146
023623.7+004226		.S..5P/	1.15± .04	165	*				6584
169.40 -52.43	UGC 2091	UE	.53± .04	.04	14.84 ±.12				
304.82 -24.12	IRAS02338+0029	4.5± .9		.79	12.76				6599
023349.5+002925	PGC 9894		1.15	.26	13.97				6352
023623.9+314241			.74± .14						5100± 49
147.69 -26.07	MCG 5- 7- 18		.28± .07	.33	14.9 ±.2				5201
334.90 -12.02	5ZW 261				12.50				4884
023324.4+312939	PGC 9895		.77						
023627.9+385812	IC 239	.SXT6..	1.66± .02		11.80S±.11	.70± .07		12.36±.1	903± 5
144.33 -19.50	UGC 2080	R	.04± .03	.25	11.73 ±.15			135± 5	
341.38 -8.64	IRAS02333+3845	6.0± .3		.07	12.65	.63		120± 7	1022
023320.5+384510	PGC 9899		1.69	.02	11.46		14.85± .17	.87	700
023630.5+324257		.S..4..	1.20± .05	49				15.33±.3	4702± 10
147.22 -25.15	UGC 2083	U	.93± .05	.29	15.04 ±.13				
335.82 -11.58		4.0± .9		1.36				416± 7	4806
023330.0+322956	PGC 9901		1.23	.46	13.35			1.52	4487
023632.0+313550		.S..6*.	.99± .06	48				16.04±.3	4889± 10
147.77 -26.16	UGC 2087	U	.47± .05	.33	15.09 ±.12				
334.81 -12.10		6.0±1.3		.69				213± 7	4990
023332.6+312248	PGC 9902		1.03	.24	14.04			1.76	4673
0236.5 +0718		.S..6*.	1.50± .03	32					6120
163.17 -47.18	UGC 2092	U	1.06± .05	.27	15.11 ±.16				
311.65 -22.01	IRAS02338+0705	6.0±1.2		1.47					6155
0233.9 +0705	PGC 9904		1.52	.50	13.35				5886
023637.9+343628		.SXS3..	.93± .05						9478
146.35 -23.44	UGC 2090	U (1)	.03± .04	.15	15.11 ±.14				
337.52 -10.73	KUG 0233+343	3.0± .9		.04					9586
023335.4+342327	PGC 9906	3.5±1.1	.95	.01	14.85				9267
023638.9-545137	NGC 1031	.SBR1*.	1.29± .04	23	13.38 ±.13	.88± .01	.96± .01		
275.55 -56.32	ESO 154- 5	Sr	.26± .05	.10	13.64 ±.14		.42± .04		5613± 28
241.98 -27.33		.5± .6	.82± .02	.26		.76	12.97± .08		5479
023505.0-550436	PGC 9907		1.30	.13	13.07		14.03± .28		5510
023646.0+020259	NGC 993	.L..-P*	.97± .09	110					
168.13 -51.34	UGC 2095	UE	.09± .04	.04	14.56 ±.10				6993± 31
306.26 -23.80		-3.4± .7		.00					7012
023410.9+014958	PGC 9910		.97		14.41				6761
023649.6+331939	NGC 987	.SB.0..	1.13± .05	30					
147.00 -24.57	UGC 2093	U	.10± .05	.27	13.41 ±.13				4534± 51
336.40 -11.36	MK 1180	.0± .8		.07	12.82				4639
023348.4+330638	PGC 9911		1.15		13.00				4321
023651.5+360645		.SBT5..	1.30± .04					15.27±.3	5130± 8
145.70 -22.06	UGC 2094	U	.03± .05	.28	13.46 ±.15			272± 7	5157± 52
338.89 -10.06	IRAS02337+3553	5.0± .7		.05	13.41			260± 7	5242
023347.3+355345	PGC 9912		1.33	.02	13.10			2.15	4922
0236.9 +1104		.S..4..	1.04± .06	164					7911
160.25 -44.03	UGC 2096	U	.62± .05	.31					
315.45 -20.75		4.0± .9		.91					7956
0234.2 +1051	PGC 9915		1.07	.31					7677
0236.9 +0527		.S..6*.	1.00± .08	6					
164.93 -48.63	UGC 2097	U	1.02± .06	.19					
309.79 -22.74		6.0±1.5		1.47					
0234.3 +0514	PGC 9917		1.02	.50					

R.A. 2000 DEC. / l b / SGL SGB / R.A. 1950 DEC.	Names / PGC	Type / S_T n_L / T / L	$\log D_{25}$ / $\log R_{25}$ / $\log A_e$ / $\log D_o$	p.a. / A_g / A_i / A_{21}	B_T / m_B / m_{FIR} / B_T^o	$(B-V)_T$ / $(U-B)_T$ / $(B-V)_T^o$ / $(U-B)_T^o$	$(B-V)_e$ / $(U-B)_e$ / m'_e / m'_{25}	m_{21} / W_{20} / W_{50} / HI	V_{21} / V_{opt} / V_{GSR} / V_{3K}
023657.2-291131		.S?....	1.06± .05	16	15.22 ±.14				5013
224.23 -66.70	ESO 416- 3		.36± .04	.00					4940
271.45 -29.58				.55					
023447.0-292430	PGC 9921		1.06	.18	14.65				4829
023658.7-052058		.SAS5P*	1.10± .05	140					
176.74 -56.75	MCG -1- 7- 24	UE (1)	.13± .05	.05					
298.42 -25.96	IRAS02344-0533	5.0± .8		.20	13.09				
023428.9-053357	PGC 9923	3.1±1.6	1.10	.07					
0237.2 +1956		.S?....	1.07± .06	127					8918
154.27 -36.40	UGC 2098		.51± .05	.34					
324.11 -17.33				.77					8988
0234.4 +1944	PGC 9931		1.10	.26					8689
0237.2 +2133		.L...*.	1.09± .10	140	15.09 ±.11				
153.30 -34.99	UGC 2099	U	.41± .05	.45					
325.64 -16.66		-2.0±1.3		.00					
0234.4 +2121	PGC 9933		1.08						
023725.6+210555	NGC 992	.S?....	.94± .07	10	15.6 ±.2	.40± .04		14.38±.1	4141± 4
153.63 -35.38	UGC 2103		.12± .05	.44	13.65 ±.18	-.43± .06		365± 6	4095± 32
325.22 -16.90	ARAK 88			.18	10.66	.25		277± 6	4213
023435.7+205256	PGC 9938		.98	.06	13.86	-.54	14.84± .41	.46	3912
0237.4 +2317		.S..6*.	.95± .07	55	15.22 ±.13				8242
152.35 -33.45	UGC 2104	U	.12± .05	.40					
327.29 -15.97	IRAS02345+2304	6.0±1.3		.18					8321
0234.5 +2304	PGC 9939		.99	.06	14.61				8016
023730.5+333754		.S..8..	1.01± .05	70	15.24 ±.15			15.98±.3	4370± 10
146.99 -24.24	UGC 2100	U	.15± .05	.27					4475
336.74 -11.34	KUG 0234+334	8.0± .9		.18				163± 7	
023428.9+332455	PGC 9941		1.04	.07	14.78			1.13	4158
0237.5 +2108			.26± .24		15.5 ±.3			15.32±.3	4075± 9
153.63 -35.32	CGCG 462- 36		.00± .06	.44				151± 10	4097± 43
325.28 -16.90	MK 369								4149
0234.7 +2056	PGC 9944		.30						3848
023736.4-325533		.SBT4*.	1.30± .04	82	13.59 ±.14				4547± 39
233.58 -66.37	ESO 355- 30	Sr (1)	.38± .04	.03					
267.10 -29.80	IRAS02355-3308	4.1± .4		.56	12.87				4463
023530.1-330830	PGC 9951	3.3± .6	1.30	.19	12.97				4374
023737.5+423809		.S..3..	1.26± .04	162	15.0 ±.3				
142.94 -16.07	UGC 2101	U (1)	.89± .05	.35					
344.70 -7.08	IRAS02344+4225	3.0± .9		1.23					
023425.1+422510	PGC 9952	4.5±1.3	1.30	.45					
023740.2+342555	A 0234+34	PSBS1*.	1.07± .04		14.2 ±.4	.72± .06		16.01±.1	4915± 9
146.64 -23.50	UGC 2105	PU (1)	.19± .04	.20	13.92 ±.13			260± 11	4853± 41
337.47 -11.00	MK 1050	1.0± .6		.20	11.43	.60			5019
023437.7+341257	PGC 9958	5.0±1.3	1.08	.10	13.49		13.89± .46	2.43	4702
023740.9-154719		.SBS7*.	1.12± .06	37					4679
193.68 -62.89	MCG -3- 7- 42	E (1)	.23± .05	.04					
286.89 -28.35		7.0±1.2		.31					4644
023519.1-160017	PGC 9960	5.3±1.2	1.12	.11					4468
023741.8+015828	NGC 1004	.E+....	1.14± .07	115	13.71 ±.16	1.00± .03	1.01± .02		6479± 31
168.49 -51.25	UGC 2112	UE	.04± .04	.05	13.85 ±.11	.45± .04	.49± .03		6497
306.26 -24.04		-3.5± .5	.88± .05	.00		.93	13.60± .19		
023506.7+014530	PGC 9961		1.14		13.66	.47	14.18± .40		6248
023745.1-612028		.SB.8*/	1.87± .03	44	13.26 ±.14			12.62±.3	513± 9
282.80 -51.43	ESO 115- 21	S (1)	.87± .05	.01					
234.85 -25.96		8.0±1.0		1.06				114± 7	368
023629.1-613324	PGC 9962	7.8±1.2	1.87	.43	12.18			.00	435
0237.8 -0141		.SB.8..	1.07± .07	160					
172.50 -53.98	UGC 2113	U	.25± .06	.08					
302.42 -25.17		8.0± .9		.31					
0235.3 -0154	PGC 9965		1.08	.13					
023754.2+021938	IC 241	.S?....	1.04± .15	150	14.30 ±.12				6931± 50
168.20 -50.94	UGC 2115		.18± .12	.06					6950
306.65 -23.98				.26					
023518.8+020641	PGC 9969		1.05	.09	13.94				6700
023755.4+020445	NGC 1008	.E.....	.91± .10	85	14.63 ±.10				6593± 31
168.46 -51.13	UGC 2114	UE	.15± .04	.05					6611
306.39 -24.06		-5.0± .6		.00					
023520.2+015148	PGC 9970		.88		14.49				6362

R.A. 2000 DEC.	Names	Type	$\log D_{25}$	p.a.	B_T	$(B-V)_T$	$(B-V)_e$	m_{21}	V_{21}
l b		S_T n_L	$\log R_{25}$	A_g	m_B	$(U-B)_T$	$(U-B)_e$	W_{20}	V_{opt}
SGL SGB		T	$\log A_e$	A_i	m_{FIR}	$(B-V)_T^o$	m'_e	W_{50}	V_{GSR}
R.A. 1950 DEC.	PGC	L	$\log D_o$	A_{21}	B_T^o	$(U-B)_T^o$	m'_{25}	HI	V_{3K}

023757.2+341444		.SXS7..	1.19± .04						
146.79 -23.64	UGC 2109	U	.06± .04	.22	14.13 ±.15				
337.33 -11.14	KUG 0234+340	7.0± .8		.09					
023454.9+340147	PGC 9972		1.21	.03					

023758.8-015036	NGC 1037	.SBR3*.	1.22± .05	2	14.2 ±.3	.77± .07			
172.72 -54.08	UGC 2119	UE (1)	.29± .05	.08	14.15 ±.13	.48± .12			8663± 50
302.26 -25.25		3.3± .7		.40		.64			8669
023526.5-020333	PGC 9973	3.1±1.2	1.23	.15	13.61	.38	14.42± .40		8435

0237.9 +0140		.SXS5..	1.15± .05						6521
168.88 -51.43	UGC 2121	U	.02± .05	.04	14.44 ±.18				
305.98 -24.20		5.0± .8		.03					6538
0235.4 +0128	PGC 9974		1.16	.01	14.33				6290

023759.2-791132		.SBS8*.	1.06± .05						
296.92 -36.66	ESO 14- 3	S	.12± .05	.38	16.04 ±.14				
216.48 -20.14		8.0± .8		.14					
023911.1-792424	PGC 9975		1.09	.06					

0238.0 +4233		.SA.8..	1.11± .14						
143.04 -16.11	UGC 2108	U	.07± .12	.35					
344.67 -7.17		8.0± .8		.08					
0234.8 +4221	PGC 9977		1.14	.03					

0238.0 +3808		.SB.4..	1.00± .08						
144.99 -20.11	UGC 2110	U	.09± .06	.26					
340.80 -9.30		4.0± .9		.13					
0234.9 +3756	PGC 9978		1.02	.04					

023803.5-334222			.91± .07	104					
235.47 -66.17	ESO 355- 31		.29± .06	.03	14.79 ±.14				5029± 87
266.19 -29.89									4943
023558.0-335518	PGC 9979		.91						4858

0238.0 +0122		.SB.4*.	1.10± .06	149					
169.22 -51.64	UGC 2120	U	.54± .05	.06	14.91 ±.12				
305.67 -24.32		4.0± .9		.79					
0235.5 +0110	PGC 9981		1.10	.27					

023805.3+414723		.S..2..	1.18± .07	119					
143.39 -16.81	UGC 2111	U	.54± .06	.28	14.67 ±.12				
344.00 -7.56	IRAS02349+4134	2.0± .9		.67	13.03				
023454.0+413426	PGC 9983		1.21	.27					

023810.4-093243	NGC 1018	PSBR0..	1.01± .07	5					
183.08 -59.30	MCG -2- 7- 48	E	.24± .05	.07					
293.91 -27.25		.0± .9		.18					
023543.8-094540	PGC 9986		1.00						

023811.2-200958		.LAT0?.	1.19± .04	35	13.87 ±.15	.87± .02	.82± .01		
202.76 -64.64	ESO 545- 40	SE	.23± .03	.04	13.83 ±.14	.19± .03	.17± .02		1494± 60
281.92 -29.11	IRAS02358-2022	-1.8± .5	.59± .04	.00	14.20	.82	12.31± .13		1446
023553.1-202254	PGC 9987		1.16		13.78	.17	14.13± .26		1291

023811.6-011904	NGC 1015	.SBR1*.	1.42± .03	10				14.25±.1	2631± 11
172.19 -53.65	UGC 2124	UE	.01± .04	.08	12.98 ±.17			181± 16	
302.83 -25.15		.5± .5		.01				162± 12	2639
023538.9-013200	PGC 9988		1.42	.01	12.86			1.39	2403

023811.7+305051		.S..8..	1.15± .07					15.37±.3	8853± 10
148.50 -26.67	UGC 2116	U	.08± .06	.47					
334.30 -12.76		8.0± .8		.10				241± 7	8951
023512.8+303754	PGC 9989		1.19	.04					8637

023819.2+021834	NGC 1009	.S..3./	1.15± .04	124					5830
168.35 -50.89	UGC 2129	UE (1)	.77± .04	.06	15.21 ±.12				
306.66 -24.09		3.0± .7		1.07					5849
023543.8+020538	PGC 9995	5.5±1.3	1.16	.39	14.04				5599

0238.3 +3311		.S..8*.	1.00± .08	58					13326
147.37 -24.55	UGC 2117	U	.43± .06	.27					
336.43 -11.69		8.0±1.3		.53					13430
0235.3 +3259	PGC 9996		1.03	.22					13114

023819.7+020705	NGC 1016	.E.....	1.38± .06		12.61 ±.13	1.01± .02	1.05± .01		
168.54 -51.04	UGC 2128	UE	.00± .05	.06	12.76 ±.11	.56± .03	.59± .02		6584± 26
306.47 -24.15		-4.5± .5	1.08± .02	.00		.93	13.50± .06		6602
023544.4+015409	PGC 9997		1.38		12.54	.58	14.48± .36		6353

023822.5+075923		.S?....	1.06± .05	124				17.27±.3	6400± 10
163.14 -46.37	UGC 2130		.51± .04	.29	15.16 ±.12			403± 13	
312.51 -22.20				.76					6435
023542.9+074627	PGC 10001		1.09	.25	14.07			2.95	6168

R.A. 2000 DEC. l b SGL SGB R.A. 1950 DEC.	Names PGC	Type S_T n_L T L	$\log D_{25}$ $\log R_{25}$ $\log A_e$ $\log D_o$	p.a. A_g A_i A_{21}	B_T m_B m_{FIR} B_T^o	$(B-V)_T$ $(U-B)_T$ $(B-V)_T^o$ $(U-B)_T^o$	$(B-V)_e$ $(U-B)_e$ m'_e m'_{25}	m_{21} W_{20} W_{50} HI	V_{21} V_{opt} V_{GSR} V_{3K}
023824.0-581424 279.34 -53.71 238.20 -26.81 023659.0-582718	 ESO 115- 22 PGC 10004	.SBS9*. S (1) 9.0± .8 10.0± .8	1.21± .07 .00± .05 1.21	 .04 .00 .00	 16.98 ±.14				
023827.5+015425 168.80 -51.18 306.26 -24.25 023552.4+014129	NGC 1019 UGC 2132 PGC 10006	.SBT4.. UE (1) 3.5± .6 .8± .8	1.01± .05 .06± .04 1.01	40 .04 .08 .03	 14.34 ±.12 14.17				7251± 60 7268 7021
023827.5+294528 149.10 -27.62 333.34 -13.31 023529.7+293232	A 0235+29 UGC 2122 PGC 10007	.SXS5.. U (1) 5.0± .8 1.1± .8	.98± .07 .00± .05 1.02	 .45 .00 .00	14.40 ±.15 14.51 ±.12 13.99	.76± .03 .62	.74± .03 14.13± .38	15.53±.1 151± 6 123± 5 1.54	5080± 6 5068± 59 5176 4863
023832.0+413146 143.58 -17.01 343.81 -7.76 023521.0+411850	NGC 995 UGC 2118 PGC 10008	.L..... U -2.0± .8 	1.22± .06 .14± .03 1.23	35 .29 .00 	 14.40 ±.14				
023832.3-065410 179.33 -57.52 296.84 -26.72 023603.7-070706	IC 243 MCG -1- 7- 26 PGC 10009	PSBT0.. E .0± .9 	1.07± .06 .17± .05 1.06	150 .05 .13 					
023832.8-064041 179.02 -57.37 297.08 -26.67 023604.1-065336	NGC 1022 MCG -1- 7- 25 IRAS02360-0653 PGC 10010	PSBS1.. R 1.0± .3 	1.38± .02 .08± .03 .96± .01 1.38	 .04 .08 .04	12.09 ±.13 12.08 ±.19 10.00 11.95	.75± .01 .24± .01 .71 .22	.73± .01 .23± .01 12.38± .03 13.62± .19		1498± 50 1489 1275
023837.1+320410 147.98 -25.54 335.44 -12.27 023536.9+315114	IC 1823 UGC 2125 IRAS02356+3151 PGC 10013	.SBR5.. U 5.0± .7 	1.33± .04 .03± .05 1.36	 .39 .05 .02	 14.2 ±.3 13.55 13.70			14.88±.3 202± 7 1.16	5187± 10 5288 4973
023838.9+040244 166.75 -49.48 308.50 -23.61 023602.2+034949	 CGCG 414- 30 MK 1181 PGC 10014	.S?.... 	1.00± .08 .17± .10 1.01	 .13 .26 .09	 14.96 ±.13 14.53				8084± 97 8108 7853
023839.6+413852 143.55 -16.89 343.93 -7.72 023528.4+412556	NGC 996 UGC 2123 PGC 10015	.E..... U -5.0± .8 	1.15± .15 .00± .08 1.20	 .30 .00 	 14.03 ±.12				
023840.5+332643 147.32 -24.30 336.68 -11.64 023538.9+331348	 UGC 2131 PGC 10017	.SBR7.. U 7.0± .8 	1.12± .05 .26± .05 1.15	93 .27 .36 .13	 15.17 ±.17 14.52			15.42±.3 200± 7 .77	4615± 10 4719 4404
023844.4+021350 168.55 -50.88 306.62 -24.21 023609.1+020055	NGC 1020 CGCG 388- 81 PGC 10018	.L..../ E -2.0±1.1 	.88± .12 .68± .08 .79	20 .06 .00 	 15.14 ±.16				
0238.7 +0654 164.17 -47.18 311.45 -22.66 0236.1 +0642	 UGC 2136 PGC 10019	.S..3.. U (1) 3.0±1.0 3.5±1.3	1.00± .08 .84± .06 1.02	25 .22 1.16 .42					10859 10891 10627
023847.3+404156 143.99 -17.74 343.11 -8.21 023537.3+402901	 UGC 2126 PGC 10025	.SX.8?. U 8.0±1.6 	1.06± .06 .00± .05 1.08	 .20 .00 .00	 14.94 ±.18				
023847.6+414014 143.57 -16.86 343.96 -7.73 023536.3+412719	NGC 999 UGC 2127 PGC 10026	PSXS1.. U 1.0± .9 	.97± .07 .06± .05 1.00	 .30 .06 .03	 14.41 ±.12				
023848.1+021300 168.59 -50.88 306.61 -24.23 023612.8+020005	NGC 1021 CGCG 388- 84 PGC 10027	.SXR4*. E (1) 4.0±1.4 4.2±1.6	.80± .13 .13± .08 .81	160 .06 .19 .06	 15.02 ±.13				
023851.9+275051 150.19 -29.28 331.64 -14.25 023555.9+273756	 UGC 2134 IRAS02359+2738 PGC 10029	.S..3.. U (1) 3.0± .8 3.5±1.1	1.22± .05 .36± .05 1.25	105 .30 .50 .18	 14.21 ±.12 12.78 13.38				4565 4655 4346
023853.4+101754 161.41 -44.40 314.88 -21.49 023612.0+100500	 CGCG 439- 21 PGC 10030	 	1.04? .56? 1.07	 .28 	 15.29 ±.12				8023 8065 7792

R.A. 2000 DEC. / l b / SGL SGB / R.A. 1950 DEC.	Names / PGC	Type / S_T n_L / T / L	$\log D_{25}$ / $\log R_{25}$ / $\log A_e$ / $\log D_o$	p.a. / A_g / A_i / A_{21}	B_T / m_B / m_{FIR} / B_T^o	$(B-V)_T$ / $(U-B)_T$ / $(B-V)_T^o$ / $(U-B)_T^o$	$(B-V)_e$ / $(U-B)_e$ / m'_e / m'_{25}	m_{21} / W_{20} / W_{50} / HI	V_{21} / V_{opt} / V_{GSR} / V_{3K}
023853.9+090541	IC 1825	.S..6*.	1.09± .05	15				14.82±.1	5125± 7
162.38 -45.39	UGC 2138	U	.19± .05	.30	14.55 ±.13			306± 8	
313.67 -21.93	IRAS02362+0853	6.0±1.2		.28	13.21				5163
023613.4+085247	PGC 10031		1.12	.10	13.94			.78	4894
023856.0+343720	NGC 1002	.SBR3*.	1.08± .07	140					
146.81 -23.22	UGC 2133	U	.12± .06	.16	13.94 ±.12				4782± 52
337.76 -11.14	IRAS02358+3424	3.0±1.2		.17	12.66				4889
023553.1+342425	PGC 10034		1.10	.06	13.57				4573
023856.4-352924		.SBS3*P	1.09± .05	78					
239.68 -65.66	ESO 356- 2	S	.45± .04	.03	15.11 ±.14				
264.10 -30.04		3.0±1.2		.62					
023653.0-354218	PGC 10035		1.09	.22					
0238.9 -1419		.S?....	1.19± .07						4567
191.28 -61.87	MCG -3- 7- 47		.92± .07	.01					
288.62 -28.41				1.38					4535
0236.6 -1432	PGC 10038		1.19	.46					4355
023903.6-272643	IC 1830	.LXT+*.	1.22± .04		13.2 ±.3	.41± .05		15.25±.3	1375± 9
219.94 -66.13	ESO 416- 6	PSUr	.08± .04	.00	12.86 ±.14	-.26± .07		224± 34	1421± 46
273.51 -29.96	IRAS02368-2739	-.5± .4		.00	12.22	.38		195± 25	1307
023652.0-273936	PGC 10041		1.21		12.90	-.26	13.95± .37		1190
023904.8+182338		.S?....			15.59S±.15				
155.72 -37.53				.27					4067± 41
322.82 -18.37	HICK 18D								4132
023617.0+181044	PGC 10042								3839
023906.0+182319		.S?....			16.10S±.15				
155.73 -37.53				.27					4143± 41
322.81 -18.38	HICK 18C								4208
023618.3+181025	PGC 10043								3915
023906.3+182258	A 0236+18A	.IB?...	1.23± .05	155	15.43S±.15			14.58±.3	4082± 17
155.74 -37.54	UGC 2140		.39± .05	.27	14.27 ±.13			145± 34	4056± 33
322.81 -18.38	HICK 18B			.29				136± 25	4141
023618.5+181004	PGC 10044		1.26	.20	14.20		15.46± .30	.19	3848
023907.9-223943		.SBS5?.	1.27± .03	153					
208.52 -65.22	ESO 479- 20	S (1)	.58± .03	.04	14.90 ±.14				3044± 60
279.07 -29.61		4.6± .5		.87					2988
023652.0-225236	PGC 10045	4.4± .6	1.27	.29	13.98				2848
023909.5+182202	A 0236+18B	.IBS9P.	.94± .11		15.48S±.15				
155.76 -37.54	UGC 2140A	R	.68± .07	.27					10019± 41
322.80 -18.40	HICK 18A	10.0± .4		.51					10084
023621.7+180909	PGC 10046		.96	.34	14.70		13.31± .58		9791
023911.1+411445		.S..7..	1.14± .13	5					
143.82 -17.21	UGC 2135	U	.18± .12	.32	15.2 ±.2				
343.62 -8.01		7.0± .8		.24					
023600.3+410151	PGC 10047		1.17	.09					
023912.2+105049	NGC 1024	PSAR2..	1.59± .02	155	13.08 ±.13	1.01± .02	1.09± .01	13.96±.1	3531± 6
161.07 -43.90	UGC 2142	R (1)	.43± .05	.29	13.25 ±.18	.41± .03	.55± .02	516± 7	3521± 14
315.46 -21.36	ARP 333	2.0± .3	.97± .03	.53	13.20	.84	13.42± .09	502± 10	3573
023630.4+103756	PGC 10048	4.5±1.0	1.62	.21	12.28	.25	14.81± .21	1.46	3298
023915.1+300911	NGC 1012	.S..0?.	1.40± .04	24				13.17±.1	977± 5
149.08 -27.19	UGC 2141	U	.36± .05	.47	13.00 ±.12			231± 5	970±141
333.78 -13.28	IRAS02362+2956	.0±1.5		.27	11.38			191± 6	1073
023616.8+295617	PGC 10051		1.43		12.26				761
023916.5+405222	NGC 1003	.SAS6..	1.74± .02	97	12.00S±.08	.55± .05		11.93±.1	627± 5
144.00 -17.55	UGC 2137	R (1)	.47± .02	.25	11.90 ±.12	-.12± .05		233± 5	585± 56
343.31 -8.20	IRAS02360+4039	6.0± .3		.69	12.03	.40		201± 8	747
023606.2+403928	PGC 10052	5.0±1.3	1.77	.24	11.02	-.23	14.39± .13	.68	430
023919.2+063238	NGC 1026	.L.....	1.30± .12						
164.66 -47.40	UGC 2145	U	.04± .08	.23	13.55 ±.14				4179± 31
311.13 -22.93		-2.0± .7		.00					4210
023640.6+061945	PGC 10055		1.32		13.25				3948
023923.6+010536	NGC 1032	.S..0./	1.52± .03	68	12.64 ±.14	1.00± .02	1.02± .01		
169.92 -51.65	UGC 2147	UE	.47± .04	.06	12.91 ±.12	.47± .02	.54± .01		2651± 36
305.48 -24.72		.0± .5	.89± .03	.35		.89	12.58± .09		2665
023649.1+005243	PGC 10060		1.51		12.34	.39	13.93± .22		2423
023929.3-080801	NGC 1035	.SAS5$.	1.35± .03	37	12.89S±.15			14.44±.1	1241± 5
181.37 -58.15	MCG -1- 7- 27	R (2)	.48± .03	.03	13.17 ±.19			276± 6	1250± 24
295.55 -27.25	IRAS02370-0820	5.0± .8		.72	11.57			246± 5	1227
023701.7-082054	PGC 10065	5.2± .7	1.35	.24	12.23		13.26± .22	1.96	1021

R.A. 2000 DEC.	Names	Type	logD₂₅	p.a.	B_T	(B-V)_T	(B-V)_e	m₂₁	V₂₁
l b		S_T n_L	logR₂₅	A_g	m_B	(U-B)_T	(U-B)_e	W₂₀	V_opt
SGL SGB		T	logA_e	A_i	m_FIR	(B-V)°_T	m'	W₅₀	V_GSR
R.A. 1950 DEC.	PGC	L	logD_o	A₂₁	B°_T	(U-B)°_T	m'₂₅	HI	V_3K

023929.3-195032		.LAT0*.	1.13± .04	7					
202.32 -64.24	ESO 545- 42	S (1)	.32± .03	.07	14.86 ±.14				4692± 60
282.34 -29.37		-2.0± .7		.00					4644
023711.0-200324	PGC 10066	7.5±1.6	1.09		14.72				4490
023929.9+291530		.S..9*.	1.04± .08					16.58±.3	4779± 7
149.59 -27.96	UGC 2144	U	.04± .06	.46					4873
332.99 -13.73		9.0±1.2		.04				131± 7	4873
023632.4+290237	PGC 10069		1.08	.02					4562
0239.5 +1240		.SB?...	1.06± .06	17					3555± 10
159.77 -42.33	UGC 2148		.56± .05	.32					
317.31 -20.74				.84					3603
0236.8 +1228	PGC 10072		1.09	.28					3325
0239.6 +1047	NGC 1029	.S..0..	1.14± .05	70					
161.22 -43.89	UGC 2149	U	.52± .05	.33	14.09 ±.14				3611± 49
315.45 -21.47		.0± .9		.39					3654
0236.9 +1035	PGC 10078		1.15		13.32				3381
023936.7+360453		.I?....	.66± .08						2721± 52
146.25 -21.84	UGC 2143		.00± .05	.19	14.4 ±.2				
339.12 -10.57	5ZW 266			.00	13.18				2831
023632.0+355200	PGC 10080		.68	.00	14.19				2515
0239.6 +0953		.S..9*.	1.04± .15						6397± 10
161.96 -44.62	UGC 2150	U	.08± .12	.31					
314.55 -21.82		9.0±1.2		.08					6437
0237.0 +0941	PGC 10084		1.07	.04					6166
023946.7+013332	IC 1827	.S..1..	1.04± .06	154					
169.56 -51.23	UGC 2152	U	.70± .05	.06	14.63 ±.17				5904± 50
306.00 -24.67		1.0±1.0		.71					5919
023711.8+012040	PGC 10087		1.04	.35	13.78				5676
023950.6+180130	NGC 1030	.S?....	1.20± .05	8				15.55±.2	8552± 6
156.16 -37.76	UGC 2153		.38± .05	.26	14.22 ±.12			674± 5	8616± 50
322.54 -18.69	IRAS02370+1748			.56	12.49			666± 5	8616
023703.0+174838	PGC 10088		1.23	.19	13.36			2.00	8325
0239.9 +4305		.I..9*.	1.19± .12					16.04±.1	5346± 11
143.14 -15.48	UGC 2146	U	.00± .12	.37					5471
345.30 -7.22		10.0±1.1		.00				170± 8	5471
0236.7 +4253	PGC 10091		1.22	.00					5155
023958.0+281932		.SA.5..	1.21± .05	117				15.25±.3	10933± 10
150.18 -28.74	UGC 2151	U	.29± .05	.35	14.47 ±.14				
332.19 -14.25		5.0± .8		.44				493± 7	11024
023701.4+280640	PGC 10092		1.25	.15	13.63			1.48	10715
023959.4-342657	Fornax	.E?....	2.80± .01	60	9.04S±.14			11.87±.5	
237.10 -65.65	ESO 356- 4		.13± .03	.03	8.25 ±.11				47± 34
265.31 -30.28				.00					-41
023755.0-343948	PGC 10093		2.76		8.52		17.69± .17		-118
0240.0 +1422		.SBS3..	1.00± .08	20					13916
158.68 -40.83	UGC 2155	U	.19± .06	.43					
319.03 -20.20		3.0± .9		.26					13969
0237.3 +1410	PGC 10094		1.04	.09					13687
024005.7+013032	NGC 1038	PS..0*.	1.08± .05	61				16.04±.1	4372± 9
169.70 -51.22	UGC 2158	UE	.49± .04	.06	14.37 ±.13			142± 11	6069± 50
305.98 -24.76		.0± .6		.36					
023730.9+011741	PGC 10096		1.06						
024008.8-114404	IC 247	.SBT4P?	1.11± .06	45					8489
187.13 -60.22	MCG -2- 7- 52	E	.06± .05	.04					
291.59 -28.19		4.0±1.7		.09					8464
023744.0-115655	PGC 10100		1.11	.03					8274
0240.2 -0243		.S..5P/	1.29± .05	105					
174.45 -54.32	MCG -1- 7- 28	E	.54± .05	.06					
301.50 -26.04		5.0±1.2		.81					
0237.7 -0256	PGC 10106		1.29	.27					
024016.1-084638	NGC 1033	.SAS5*.	1.12± .06	0					
182.54 -58.41	MCG -2- 7- 53	E (1)	.06± .05	.04					7518
294.89 -27.58		5.0± .8		.09					7502
023749.0-085929	PGC 10108	5.3± .8	1.12	.03					7299
024019.4+321544		.SAR5..	1.24± .06						4488
148.24 -25.20	UGC 2156	U	.01± .06	.43	14.04 ±.18				
335.79 -12.50	IRAS02373+3202	5.0± .7		.01	13.05				4589
023718.8+320253	PGC 10112		1.28	.00	13.57				4277

R.A. 2000 DEC. / l b / SGL SGB / R.A. 1950 DEC.	Names / PGC	Type / S_T n_L / T / L	$\log D_{25}$ / $\log R_{25}$ / $\log A_e$ / $\log D_o$	p.a. / A_g / A_i / A_{21}	B_T / m_B / m_{FIR} / B_T^o	$(B-V)_T$ / $(U-B)_T$ / $(B-V)_T^o$ / $(U-B)_T^o$	$(B-V)_e$ / $(U-B)_e$ / m'_e / m'_{25}	m_{21} / W_{20} / W_{50} / HI	V_{21} / V_{opt} / V_{GSR} / V_{3K}
0240.4 +0113 / 170.09 -51.38 / 305.71 -24.92 / 0237.8 +0100	A 0237+01 / UGC 2162 / DDO 27 / PGC 10117	.IBS9.. / UE (2) / 10.0± .6 / 9.5± .7	1.19± .04 / .09± .04 / .96± .06 / 1.20	55 / .06 / .07 / .05	16.2 ±.2 / / / 16.05	.46± .13 / -.24± .19 / .42 / -.27	.41± .10 / -.15± .17 / 16.47± .13 / 16.76± .33	15.45±.1 / 58± 7 / 34± 12 / -.64	1185± 6 / / 1199 / 957
024024.0-082602 / 182.08 -58.17 / 295.28 -27.54 / 023756.7-083852	NGC 1042 / MCG -2- 7- 54 / IRAS02379-0838 / PGC 10122	.SXT6.. / R (2) / 6.0± .3 / 2.4± .4	1.67± .02 / .11± .02 / 1.50± .03 / 1.68	15 / .06 / .16 / .05	11.56M±.11 / 11.8 ±.2 / 12.29 / 11.38	.54± .03 / -.09± .11 / .50 / -.12	.62± .02 / .00± .04 / 14.38± .06 / 14.49± .15	13.23±.1 / 113± 5 / 93± 8 / 1.79	1373± 6 / 1407± 63 / 1359 / 1155
024024.1+390346 / 145.02 -19.09 / 341.83 -9.27 / 023715.9+385055	NGC 1023 / UGC 2154 / ARP 135 / PGC 10123	.LBT-.. / R / -3.0± .3 /	1.94± .01 / .47± .02 / 1.12± .02 / 1.91	87 / .27 / .00 /	10.35M±.06 / 10.40 ±.08 / / 10.09	1.00± .01 / .56± .01 / .91 / .48	1.01± .01 / .61± .01 / 11.60± .05 / 13.76± .10	13.68±.1 / 370± 5 / 229± 8 /	637± 4 / 601± 11 / 749 / 433
024025.2+383350 / 145.26 -19.54 / 341.39 -9.52 / 023717.6+382059	/ UGC 2157 / / PGC 10124	.S..8*. / U / 8.0±1.2 /	1.24± .06 / .55± .06 / / 1.26	39 / .20 / .68 / .28	14.68 ±.13 / / /	/ / /	/ / /	/ / /	/ / /
024025.4-052626 / 177.93 -56.19 / 298.56 -26.81 / 023755.7-053916	NGC 1041 / MCG -1- 7- 30 / / PGC 10125	.L..-P. / E / -3.0± .8 /	1.22± .07 / .13± .08 / / 1.21	105 / .02 / .00 /	/ / /	/ / /	/ / /	/ / /	/ / /
024027.9+300450 / 149.37 -27.13 / 333.84 -13.54 / 023729.5+295200	/ UGC 2159 / / PGC 10126	.S..7.. / U / 7.0±1.0 /	1.00± .08 / 1.02± .06 / / 1.04	129 / .45 / 1.38 / .50	/ / /	/ / /	/ / /	16.18±.3 / / 125± 7 /	5182± 10 / 5277 / 4967
024029.0+191750 / 155.51 -36.58 / 323.82 -18.30 / 023740.3+190500	NGC 1036 / UGC 2160 / MK 370 / PGC 10127	.P...$. / R / 99.0 /	1.16± .03 / .14± .06 / .61± .03 / 1.19	5 / .26 / .17 / .07	13.75 ±.14 / 13.37 ±.12 / 12.81 / 13.10	.58± .02 / -.22± .03 / .49 / -.28	.50± .02 / -.24± .03 / 12.29± .09 / 14.07± .27	15.00±.1 / 163± 6 / 128± 4 / 1.83	783± 7 / 773± 35 / 849 / 556
0240.4 -0607 / 178.85 -56.64 / 297.83 -27.00 / 0238.0 -0620	/ MCG -1- 7- 31 / / PGC 10128	.SBS9*. / UE (1) / 9.3± .6 / 8.7± .8	1.17± .05 / .08± .05 / / 1.17	/ .05 / .08 / .04	/ / /	/ / /	/ / /	14.74±.1 / 99± 7 / 90± 12 /	1327± 6 / / 1319 / 1106
024029.2-111642 / 186.49 -59.89 / 292.12 -28.18 / 023804.1-112932	NGC 1045 / MCG -2- 7- 59 / / PGC 10129	.LA.-P? / E / -3.0± .8 /	1.37± .08 / .28± .07 / / 1.34	55 / .05 / .00 /	/ / /	/ / /	/ / /	/ / /	/ / /
024032.9-080853 / 181.70 -57.96 / 295.60 -27.50 / 023805.3-082142	NGC 1047 / MCG -1- 7- 32 / / PGC 10132	.L..+*/ / PE / -.5± .9 /	1.10± .04 / .33± .04 / / 1.06	95 / .04 / .00 /	/ / /	/ / /	/ / /	15.58±.3 / 116± 34 / 46± 25 /	1340± 17 / / 1326 / 1121
024032.9+385401 / 145.13 -19.22 / 341.70 -9.37 / 023724.8+384110	/ / / PGC 10133	/ / /	.52± .05 / .30± .04 / / .55	/ .27 / /	/ / /	/ / /	/ / /	20.45±.1 / / 44± 5 /	695± 10 / / 811 / 496
024035.8-083250 / 182.30 -58.20 / 295.16 -27.61 / 023808.5-084539	NGC 1048A / MCG -2- 7- 58 / / PGC 10137	.SB.3P/ / PE / 3.4± .8 /	.82± .13 / .28± .07 / / .82	/ .05 / .39 / .14	15.5 ±.2 / / /	.97± .06 / / /	/ / 13.72± .69 /	/ / /	/ / /
024037.7+390328 / 145.07 -19.07 / 341.85 -9.31 / 023729.4+385038	NGC 1023A / / / PGC 10139	.IB?... / / /	1.11± .05 / .30± .04 / / 1.14	50 / .27 / .23 / .15	14.5 S±.4 / / / 14.00	/ / /	/ / / 14.17± .48	19.58±.1 / / 35± 5 / 5.43	743± 10 / 742± 60 / 859 / 544
024038.0-083201 / 182.29 -58.19 / 295.18 -27.61 / 023810.8-084451	NGC 1048 / MCG -2- 7- 62 / / PGC 10140	.L..+?/ / E / -1.0±1.8 /	.99± .10 / .62± .07 / / .90	5 / .05 / .00 /	15.49 ±.11 / / /	.97± .06 / / /	/ / 13.75± .54 /	/ / /	/ / /
024039.4+392247 / 144.93 -18.78 / 342.13 -9.16 / 023730.7+390957	/ / / PGC 10143	/ / /	/ / /	/ .27 / /	/ / /	/ / /	/ / /	20.14±.1 / / 25± 5 /	903± 10 / / 1020 / 705
0240.6 +1542 / 157.91 -39.63 / 320.39 -19.81 / 0237.9 +1530	/ UGC 2163 / / PGC 10145	.S?.... / (1) / 4.5±1.2 /	1.00± .06 / .21± .05 / / 1.02	15 / .30 / .29 / .11	/ / /	/ / /	/ / /	/ / /	13882 / 13938 / 13654
0240.7 -1508 / 193.25 -61.92 / 287.77 -28.97 / 0238.4 -1521	A 0238-15 / MCG -3- 7- 52 / IRAS02383-1521 / PGC 10153	.SXS5.. / PE (2) / 4.5± .6 / 2.3± .6	1.10± .06 / .08± .05 / .88± .06 / 1.10	144 / .05 / .12 / .04	14.1 ±.2 / / / 13.93	.78± .05 / / .70 /	.79± .04 / / 14.03± .17 / 14.27± .40	15.77±.1 / 198± 11 / 182± 11 / 1.80	7756± 10 / 7653± 59 / 7718 / 7544

R.A. 2000 DEC. l b SGL SGB R.A. 1950 DEC.	Names PGC	Type S_T n_L T L	$logD_{25}$ $logR_{25}$ $logA_e$ $logD_o$	p.a. A_g A_i A_{21}	B_T m_B m_{FIR} B_T^o	$(B-V)_T$ $(U-B)_T$ $(B-V)_T^o$ $(U-B)_T^o$	$(B-V)_e$ $(U-B)_e$ m' m'_{25}	m_{21} W_{20} W_{50} HI	V_{21} V_{opt} V_{GSR} V_{3K}
024047.1+434905		.LX.0*.	1.18± .09						
142.97 -14.76	UGC 2161	U	.02± .05	.50	14.6 ±.3				
346.01 -7.00		-2.0± .7		.00					
023732.5+433615	PGC 10156		1.23						
024050.7+213634		.S?....	.89± .11						7975
154.17 -34.54	MCG 3- 7- 42		.52± .07	.43	15.46 ±.13				
326.05 -17.40				.39					8048
023800.0+212345	PGC 10160		.90		14.52				7751
0240.9 +0835		.SB.6*.	1.06± .06	140					5435
163.37 -45.50	UGC 2167	U	.50± .05	.32	14.93 ±.12				
313.36 -22.58	IRAS02382+0822	6.0±1.3		.73	13.34				5470
0238.2 +0822	PGC 10166		1.09	.25	13.86				5206
0240.9 -1327		.L?....	1.02± .14						8174
190.28 -61.00	MCG -2- 7- 65		.27± .07	.00					
289.70 -28.73				.00					8144
0238.6 -1340	PGC 10168		.98						7963
024100.5+390419			.52± .05					20.98±.1	593± 10
145.13 -19.03			.00± .04	.27					
341.89 -9.37								27± 5	709
023752.1+385130	PGC 10169		.55						394
024100.5+321050		.S?....	.82± .13					15.81±.3	4434± 10
148.43 -25.21	MCG 5- 7- 30		.17± .07	.43	15.57 ±.12				
335.78 -12.67				.26				149± 7	4534
023760.0+315801	PGC 10170		.86	.09	14.86			.86	4223
024102.6-065612	NGC 1051	.SBT9P.	1.32± .04	120				14.02±.1	1300± 11
180.14 -57.08	MCG -1- 7- 33	UE (1)	.13± .05	.04				198± 16	
296.97 -27.33		8.5± .5		.13				189± 12	1289
023834.2-070900	PGC 10172	6.4± .8	1.32	.07					1080
0241.0 +0843	NGC 1044	.L..-P*	.74± .14		14.4 ±.2	1.20± .05			
163.30 -45.38	MCG 1- 7- 23	P	.00± .07	.32		.52± .10			6420±141
313.51 -22.56		-3.0±1.0		.00		1.07			6456
0238.4 +0831	PGC 10174		.78		13.98	.47	12.95± .76		6191
024104.9-081522	NGC 1052	.E.4...	1.48± .03	120	11.41 ±.13	.94± .01	1.00± .01	14.97±.1	1507± 5
182.02 -57.93	MCG -1- 7- 34	R	.16± .03	.06	11.43 ±.18	.43± .01	.53± .01	368± 8	1474± 10
295.52 -27.66	IRAS02386-0828	-5.0± .3	1.05± .02	.00	13.35	.91	12.15± .07	350± 6	1485
023837.4-082810	PGC 10175		1.44		11.33	.42	13.41± .20		1282
0241.0 +1725	UGC 2168	.I..9*.	1.12± .07	23					784
156.87 -38.11		U	.43± .06	.32					
322.10 -19.21		10.0±1.2		.32					845
0238.3 +1713	PGC 10177		1.15	.22					557
024109.4+434057	UGC 2164	.S..4..	1.07± .07	90					
143.09 -14.85		U	.50± .06	.50	15.2 ±.3				
345.92 -7.12		4.0± .9		.73					
023754.8+432808	PGC 10181		1.12	.25					
024109.5+355132	UGC 2166	.S..7..	1.02± .05	132					3139
146.66 -21.91	KUG 0238+356	U	.31± .05	.27	15.40 ±.14				
339.08 -10.95		7.0± .9		.43					3248
023804.9+353843	PGC 10182		1.05	.16	14.70				2934
0241.1 +0842	NGC 1046	.L..-*.							
163.34 -45.37	MCG 1- 7- 24	P							
313.51 -22.59		-3.0±1.3		.32	14.8 ±.3				
0238.5 +0830	PGC 10185								
024123.8-175613		.SB?...	1.00± .05	175					7853± 60
198.78 -63.07	ESO 546- 5		.07± .04	.05	15.21 ±.14				7809
284.60 -29.56				.11					7650
023904.0-180900	PGC 10195		1.01	.04	15.00				
024125.6+174841	IC 248	.S..1*.	.98± .07	145					9022± 50
156.70 -37.74	UGC 2170	U	.20± .05	.26	14.32 ±.13				
322.50 -19.12	IRAS02386+1735	1.0±1.2		.20	13.08				9084
023838.2+173553	PGC 10197		1.00	.10	13.74				8796
024126.3-130744		.L..-P.	1.08± .06	135					10424
189.84 -60.74	MCG -2- 7- 68	E	.18± .05	.04					
290.09 -28.77		-3.0± .9		.00					10394
023902.7-132031	PGC 10198		1.06						10213
024134.9+071114					.26 14.74 ±.12				8299± 52
164.76 -46.55	CGCG 414- 40								
311.99 -23.23	MK 595								8330
023855.8+065827	PGC 10201								8071

R.A. 2000 DEC. l　　b SGL　SGB R.A. 1950 DEC.	Names PGC	Type S_T　n_L T L	$\log D_{25}$ $\log R_{25}$ $\log A_\bullet$ $\log D_o$	p.a. A_g A_i A_{21}	B_T m_B m_{FIR} B_T^o	$(B-V)_T$ $(U-B)_T$ $(B-V)_T^o$ $(U-B)_T^o$	$(B-V)_\bullet$ $(U-B)_\bullet$ $m'_\bullet$ m'_{25}	m_{21} W_{20} W_{50} HI	V_{21} V_{opt} V_{GSR} V_{3K}
024138.6-281020　IC　1833 221.83 -65.62　ESO　416- 7 272.69 -30.57 023928.0-282306　PGC 10205		.LX.0.. Sr -1.7± .6 	1.18± .04 .27± .03 1.14	61 .00 .00 	14.10 ±.14 14.03				4937± 52 4864 4756
024143.5-271820 219.74 -65.53　ESO　416- 8 273.71 -30.55 023932.0-273106　PGC 10207		.SBS6.. S　　(1) 6.0± .8 5.6± .8	1.09± .04 .27± .03 1.09	138 .00 .40 .14	15.41 ±.14 				
024144.7+002631　NGC 1055 171.33 -51.75　UGC　2173 304.99 -25.48　IRAS02391+0013 023910.7+001345　PGC 10208		.SB.3*/ PUE　(4) 3.0± .4 3.9± .4	1.88± .01 .45± .03 1.41± .01 1.89	105 .07 .63 .23	11.40M±.10 11.62 ±.13 9.61 10.78	.81± .02 .19± .03 .70 .09	.89± .01 .28± .01 13.92± .03 14.53± .14	12.29±.1 406± 4 389± 5 1.28	996± 5 958± 46 1006 770
0241.8 -0755 181.77 -57.56 295.94 -27.77 0239.4 -0808　PGC 10213		.SBS9.. E　　(1) 9.0± .9 9.8± .8	1.12± .08 .00± .08 1.13	 .04 .00 .00					
0241.9 +0556 165.95 -47.48　UGC　2176 310.76 -23.73　IRAS02392+0544 0239.2 +0544　PGC 10215		.S..3.. U　　(1) 3.0± .9 3.5±1.2	1.11± .07 .44± .06 1.13	14 .20 .61 .22	14.46 ±.12 13.59 13.60				6364 6391 6136
024154.6+593611　Maffei II 136.50　-.33　UGCA　39 359.58　.83　WEIN　21 023808.0+592324　PGC 10217		.SXT4*. R 4.0± .7 	1.00? .00? .00	 8.1 .00 .00				11.15±.1 347± 7 305± 8	-1± 6 151 -147
024154.7+320513 148.66 -25.21　UGC　2171 335.79 -12.88 023854.0+315227　PGC 10218		.S..7.. U 7.0±1.0 	1.07± .14 1.09± .06 1.12	154 .50 1.38 .50				15.99±.3 224± 7	4561± 10 4660 4351
024155.2-205751 205.19 -64.09　ESO　546- 8 281.12 -30.08 023938.1-211036　PGC 10219		PSBR3.. r 3.3± .9 	1.05± .05 .07± .04 1.06	 .09 .10 .04	14.55 ±.14 14.30				7708± 60 7655 7511
024157.1-064734 180.21 -56.81　MCG -1- 7- 35 297.19 -27.52 023928.5-070019　PGC 10220		.LBR+P. E -1.0± .9 	1.06± .06 .28± .05 1.02	75 .04 .00 					
024157.9-211715 205.90 -64.19　ESO　546- 9 280.74 -30.12 023941.0-213000　PGC 10222		PSBS9*. r 9.1±1.2 	1.04± .05 .10± .04 1.04	 .09 .10 .05	15.01 ±.14 				
024205.9+322242 148.55 -24.93　UGC　2174 336.07 -12.78 023904.9+320956　PGC 10227		.SXS5.. U 5.0± .7 	1.33± .04 .01± .05 1.37	 .50 .02 .01	14.5 ±.3 13.91			14.75±.3 96± 7 .84	5127± 10 5227 4918
024210.0-053408　NGC 1063 178.62 -55.96　MCG -1- 7- 36　(1) 298.55 -27.26　IRAS02396-0546 023940.5-054653　PGC 10232		PSAR4*. PE 4.0± .7 3.1± .8	1.14± .05 .40± .05 1.14	75 .04 .58 .20	 13.45 				
0242.2 +0225 169.39 -50.18　UGC　2181 307.13 -24.97 0239.6 +0213　PGC 10236		.S?.... 	.93± .07 .02± .05 .93	 .06 .04 .01	14.87 ±.13 14.74				6738 6754 6512
024215.7+181258　NGC 1054 156.65 -37.29　MCG 3- 7- 46 322.97 -19.14　IRAS02394+1800 023927.8+180013　PGC 10242		.S?.... 	.94± .11 .28± .07 .96	 .26 .35 .14	14.57 ±.13 13.45 13.86				9760 9822 9535
024216.5+422336 143.86 -15.93　UGC　2175 344.91　-7.94　IRAS02389+4210 023903.5+421050　PGC 10243		.SXS4.. U 4.0± .8 	1.04± .06 .02± .05 1.08	 .37 .03 .01	15.1 ±.3 				
024223.4-301916 227.03 -65.53　ESO　416- 9 270.16 -30.79 024015.0-303200　PGC 10248		.L...?P S -2.0±1.6 	1.05± .05 .17± .03 1.03	76 .03 .00 	14.19 ±.14 14.06				6506± 19 6427 6331
024223.6-092148　NGC 1064 184.04 -58.36　MCG -2- 7- 71 294.37 -28.23 023957.1-093432　PGC 10249		.SBS5*. E　　(1) 5.0± .6 3.1± .6	1.03± .05 .05± .04 1.04	30 .04 .08 .03					

R.A. 2000 DEC. l b SGL SGB R.A. 1950 DEC.	Names PGC	Type S_T n_L T L	$\log D_{25}$ $\log R_{25}$ $\log A_e$ $\log D_o$	p.a. A_g A_i A_{21}	B_T m_B m_{FIR} B_T^o	$(B-V)_T$ $(U-B)_T$ $(B-V)_T^o$ $(U-B)_T^o$	$(B-V)_e$ $(U-B)_e$ m' m'_{25}	m_{21} W_{20} W_{50} HI	V_{21} V_{opt} V_{GSR} V_{3K}
024225.7+081238 164.12 -45.60 313.12 -23.06 023945.8+075953	 CGCG 414- 42 PGC 10250	1.04? .34? 1.07	1.04? .34? 1.07	 .31 	 15.08 ±.13 				8022 8055 7795
024233.6+351208 147.25 -22.37 338.63 -11.52 023929.5+345924	 UGC 2179 PGC 10256	.SBS6.. U 6.0± .8 	1.17± .05 .00± .05 1.19	 .20 .00 .00	 14.6 ±.2 14.35				3554 3660 3350
024235.6+344549 147.47 -22.76 338.25 -11.73 023932.0+343305	NGC 1050 UGC 2178 IRAS02395+3433 PGC 10257	PSBS1.. U 1.0± .8 	1.15± .03 .12± .04 1.17	110 .22 .13 .06	 13.47 ±.12 11.31 13.08			16.77±.3 279± 13 3.63	3901± 10 3757± 26 3989 3679
024238.3-122519 188.97 -60.10 290.95 -28.92 024014.2-123802	 MCG -2- 7- 73 HICK 19A PGC 10262	.L..-P* E -3.0± .8 	1.08± .07 .13± .04 1.07	40 .03 .00 	14.19S±.15 14.10		 14.14± .39 		 4288± 41 4259 4077
024239.3+393200 145.22 -18.48 342.45 -9.42 023930.0+391916	 UGC 2180 PGC 10264	.S?.... 	.91± .07 .32± .05 .94	37 .23 .48 .16	 14.72 ±.14 13.95				9332 9448 9136
024240.2-000048 172.10 -51.94 304.59 -25.84 024006.5-001332	NGC 1068 UGC 2188 ARP 37 PGC 10266	RSAT3.. R (2) 3.0± .3 2.3± .5	1.85± .01 .07± .02 1.09± .01 1.85	70 .05 .10 .04	9.61M±.10 9.64 ±.11 7.61 9.46	.74± .01 .09± .01 .71 .06	.77± .01 .07± .01 10.65± .03 13.52± .13	13.62±.1 300± 3 252± 3 4.12	1137± 3 1093± 14 1144 911
0242.7 +0304 168.90 -49.60 307.85 -24.88 0240.1 +0252	IC 1834 UGC 2189 PGC 10267	.S..3.. U (1) 3.0± .8 3.5±1.1	.96± .07 .06± .05 .97	 .07 .08 .03	 14.85 ±.13 				
024242.2-122542 188.99 -60.09 290.95 -28.94 024018.0-123826	 MCG -2- 7- 74 HICK 19B PGC 10268	RSBR1P? E 1.0± .9 	.97± .06 .49± .05 .98	100 .05 .50 .24	15.65S±.15 15.05		 14.16± .35 		 4260± 41 4231 4049
024246.4-122354 188.96 -60.06 290.98 -28.95 024022.2-123638	 MCG -2- 7- 75 IRAS02403-1238 PGC 10270	.SB.9P? E (1) 9.0±1.0 7.5±1.2	1.12± .05 .28± .04 1.12	100 .05 .29 .14	15.44S±.15 15.10		 15.18± .30 		 4160± 41 4131 3949
024246.7-212805 206.43 -64.07 280.56 -30.33 024030.0-214048	 ESO 546- 11 PGC 10271	.SB?... 	1.08± .05 .06± .04 1.09	 .09 .08 .03	 14.51 ±.14 14.31				4609 4554 4414
024248.6+283429 150.68 -28.23 332.71 -14.69 023951.4+282145	NGC 1056 UGC 2183 MK 1183 PGC 10272	.S..1*. U 1.0±1.1 	1.37± .05 .32± .06 1.41	160 .45 .32 .16	 13.32 ±.12 11.23 12.53			13.84±.3 272± 7 1.15	1545± 10 1619± 97 1636 1332
024249.4+324111 148.55 -24.59 336.42 -12.77 023948.0+322827	 UGC 2182 PGC 10274	.I..9.. U 10.0± .9 	1.11± .14 .54± .12 1.15	161 .39 .40 .27				16.38±.3 159± 7 	5262± 7 5363 5054
024252.1-252711 215.44 -65.02 275.90 -30.70 024039.0-253954	 ESO 479- 26 PGC 10276	PSBT1.. r 1.0± .9 	1.00± .05 .21± .04 1.00	84 .00 .21 .10	 14.89 ±.14 				
024252.5+073550 164.78 -46.03 312.54 -23.38 024013.0+072307	 MCG 1- 7- 25 MK 596 PGC 10277	.S?.... 	.89± .11 .07± .07 .91	 .23 .10 .03	 14.80 ±.12 13.26 14.40				 11476± 52 11507 11249
024255.5-543436 274.07 -55.85 242.07 -28.28 024123.0-544718	 ESO 154- 9 PGC 10280	.LA.-*. S -1.0± .9 	1.09± .05 .36± .04 .48± .07 1.04	82 .10 .00 	15.0 ±.2 14.89 ±.14 	1.10± .03 	1.06± .03 12.93± .23 14.45± .34		
024259.9-081722 182.63 -57.58 295.61 -28.13 024032.6-083005	NGC 1069 MCG -1- 7- 38 IRAS02405-0829 PGC 10285	.SXS5.. E (1) 5.0± .5 1.9± .5	1.14± .04 .20± .04 1.15	145 .05 .30 .10	 13.34 				
024303.0+322928 148.69 -24.74 336.27 -12.90 024001.8+321644	NGC 1057 UGC 2184 PGC 10287	.L..... U -2.0± .9 	1.06± .07 .18± .03 1.09	115 .50 .00 .00	 15.22 ±.14 				

R.A. 2000 DEC. / l b / SGL SGB / R.A. 1950 DEC.	Names / PGC	Type / S_T n_L / T / L	$\log D_{25}$ / $\log R_{25}$ / $\log A_o$ / $\log D_o$	p.a. / A_g / A_i / A_{21}	B_T / m_B / m_{FIR} / B_T^o	$(B-V)_T$ / $(U-B)_T$ / $(B-V)_T^o$ / $(U-B)_T^o$	$(B-V)_o$ / $(U-B)_o$ / m'_o / m'_{25}	m_{21} / W_{20} / W_{50} / HI	V_{21} / V_{opt} / V_{GSR} / V_{3K}
024307.4+410350		.S..0..	1.10± .05	142					
144.60 -17.06	UGC 2186	U	.28± .05	.30	14.38 ±.12				
343.83 -8.74		.0± .8		.21					
023956.1+405107	PGC 10289		1.12						
024307.9-084629	NGC 1071	.SBT1..	1.05± .05	160					
183.38 -57.86	MCG -2- 7- 77	E	.34± .04	.05					
295.07 -28.27		1.0± .6		.34					
024040.9-085911	PGC 10290		1.06	.17					
0243.1 +1635		.S?....	.64± .17						
157.97 -38.56	MCG 3- 7- 49		.00± .07	.29	15.30 ±.14				7674± 46
321.51 -20.00				.00					7731
0240.4 +1623	PGC 10294		.66	.00	14.95				7450
024311.4+402534		.S..6*.	1.47± .03	144				14.93±.1	4359± 11
144.91 -17.63	UGC 2185	U	.50± .05	.26	13.54 ±.13				3892± 52
343.28 -9.07	IRAS02400+4012	6.0±1.1		.74	13.48			433± 8	4458
024001.0+401251	PGC 10296		1.49	.25	12.52			2.16	4146
024312.6+413003	NGC 1053	.L.....	1.22± .06	40					
144.42 -16.66	UGC 2187	U	.34± .03	.33	13.86 ±.10				4813± 52
344.22 -8.54		-2.0± .8		.00					4933
024000.7+411720	PGC 10298		1.21		13.46				4621
0243.2 +3748		.L.....	1.00± .10	172					
146.13 -19.97	UGC 2190	U	.38± .04	.24					
341.00 -10.36		-2.0± .9		.00					
0240.1 +3736	PGC 10300		.97						
024315.1+322530	NGC 1060	.L..-*.	1.36± .11	75	13.00 ±.14	1.19± .01	1.20± .01		
148.77 -24.78	UGC 2191	U	.12± .08	.50	13.07 ±.11	.71± .03	.76± .02		5190± 22
336.23 -12.97		-3.0±1.0	.92± .03	.00		1.02	13.09± .10		5290
024013.9+321247	PGC 10302		1.41		12.46	.60	14.36± .62		4982
024315.9+322800	NGC 1061	.I?....	.95± .06					17.46±.3	4026± 10
148.75 -24.74	MCG 5- 7- 36		.16± .06	.50	15.02 ±.12				4125
336.27 -12.95	KUG 0240+322			.12				148± 7	4125
024014.7+321518	PGC 10303		1.00	.08	14.39			2.98	3818
0243.2 +0145		.S..0..	1.07± .14	153					
170.40 -50.51	UGC 2199	U	.16± .12	.07	14.48 ±.13				
306.52 -25.44		.0± .9		.12					
0240.7 +0133	PGC 10305		1.07						
024322.4+045803	NGC 1070	.S..3..	1.36± .05	175				14.10±.3	4088± 10
167.28 -48.03	UGC 2200	U (1)	.08± .06	.14	12.72 ±.12			368± 13	4107± 50
309.88 -24.41	IRAS02407+0445	3.0± .7		.11	12.37			338± 10	4112
024044.9+044521	PGC 10309	4.5±1.0	1.37	.04	12.43			1.63	3863
024326.0+312816		.S..6*.	1.17± .07	160				15.08±.3	5098± 10
149.29 -25.61	UGC 2197	U	.16± .06	.56	15.1 ±.2				
335.39 -13.46		6.0±1.1		.24				256± 7	5195
024025.8+311534	PGC 10310		1.22	.08	14.23			.77	4889
024326.9+412425		.SB.3*.	1.16± .05	78				15.14±.1	5479± 11
144.51 -16.73	UGC 2194	U	.34± .05	.33	14.42 ±.12				
344.16 -8.62	IRAS02402+4111	3.0± .8		.47	13.24			305± 8	5599
024015.0+411142	PGC 10312		1.19	.17	13.58			1.39	5287
024329.1-144517	NGC 1076	.SB.0P?	1.29± .04	99					2102
193.23 -61.16	MCG -3- 8- 3	E	.25± .04	.03					
288.34 -29.56	IRAS02411-1457	.3± .7		.19	12.69				2066
024106.9-145759	PGC 10313		1.28						1896
024330.0+372030	NGC 1058	.SAT5..	1.48± .02		11.82 ±.15	.62± .02		12.62±.0	518± 3
146.41 -20.37	UGC 2193	R (3)	.03± .02	.25	11.83 ±.11			38± 2	492± 40
340.62 -10.64	IRAS02403+3707	5.0± .3		.05	11.76	.55		28± 3	629
024023.2+370748	PGC 10314	5.2± .6	1.51	.02	11.53		14.01± .18	1.07	318
024330.9+001829	NGC 1072	.SBT3*.	1.17± .04	11					
172.00 -51.56	UGC 2208	PUE (1)	.45± .04	.08	14.16 ±.12				8021± 50
305.00 -25.94	IRAS02409+0005	2.7± .5		.62	13.64				8030
024057.0+000548	PGC 10315	3.1± .8	1.18	.23	13.40				7798
024331.3+415616		.S..7..	1.00± .08	95					32168± 19
144.28 -16.24	UGC 2195	U	.25± .06	.29	15.4 ±.3				32289
344.63 -8.37		7.0± .9		.35					31978
024018.7+414334	PGC 10316		1.03	.13	14.61				
024332.7+314725		.S..7..	1.04± .06	25				15.19±.3	4714± 10
149.15 -25.31	UGC 2198	U	.71± .05	.56	15.63 ±.12				
335.69 -13.33	IRAS02405+3134	7.0±1.0		.98	12.60			306± 7	4812
024032.2+313444	PGC 10319		1.09	.35	14.08			.75	4505

R.A. 2000 DEC. l b SGL SGB R.A. 1950 DEC.	Names PGC	Type S_T n_L T L	$\log D_{25}$ $\log R_{25}$ $\log A_e$ $\log D_o$	p.a. A_g A_i A_{21}	B_T m_B m_{FIR} B_T^o	$(B-V)_T$ $(U-B)_T$ $(B-V)_T^o$ $(U-B)_T^o$	$(B-V)_e$ $(U-B)_e$ m'_e m'_{25}	m_{21} W_{20} W_{50} HI	V_{21} V_{opt} V_{GSR} V_{3K}
024334.7+474923 141.68 -10.92 349.69 -5.41 024013.2+473641	 UGC 2192 PGC 10322	.S..0.. U .0±1.0	1.00± .08 .73± .06 1.06	74 1.02 .54	 14.7 ±.3 				
024336.1-161749 196.10 -61.87 286.58 -29.84 024115.1-163030	NGC 1074 MCG -3- 8- 1 PGC 10324	.SXR2P* E 2.3± .7 	1.28± .05 .19± .04 1.29	167 .02 .23 .09					
024338.5-315638 230.90 -65.20 268.25 -31.07 024132.0-320918	 ESO 416- 12 PGC 10326	.SXT5*. S (1) 5.0± .6 5.0± .7	1.20± .04 .40± .03 1.21	49 .03 .60 .20	 14.49 ±.14 13.86				 0± 60 -83 -169
024340.4+012236 170.91 -50.73 306.15 -25.65 024105.6+010955	NGC 1073 UGC 2210 IRAS02411+0109 PGC 10329	.SBT5.. R (3) 5.0± .3 3.7± .4	1.69± .02 .04± .02 1.50± .02 1.69	15 .07 .06 .02	11.47M±.10 11.67 ±.15 12.36 11.39	.50± .03 -.10± .10 .47 -.12	.58± .01 -.04± .03 14.35± .04 14.65± .14	12.74±.1 88± 7 75± 6 1.33	1211± 5 1209± 37 1223 988
024344.6-290014 223.88 -65.21 271.73 -31.06 024135.1-291254	NGC 1079 ESO 416- 13 IRAS02415-2913 PGC 10330	RSXT0P. R .0± .3 	1.54± .02 .21± .03 .89± .01 1.53	87 .00 .16 	12.38 ±.13 12.21 ±.11 13.75 12.10	.92± .01 .44± .02 .87 .41	.94± .01 .41± .01 12.32± .03 14.40± .18	13.49±.2 340± 6 325± 5 	1447± 7 2252± 56 1384 1283
024344.7+322944 148.83 -24.67 336.34 -13.03 024043.4+321702	NGC 1062 UGC 2201 PGC 10331	.S..7.. U 7.0± .9 	1.16± .05 .66± .05 1.20	101 .43 .92 .33	 15.45 ±.13 14.09			15.15±.3 332± 7 .73	4134± 10 4233 3926
024345.2-373320 243.86 -64.22 261.63 -30.91 024145.0-374600	 ESO 299- 14 PGC 10332	.SB.3?/ S 3.0±1.3 	1.08± .05 .68± .04 1.09	62 .03 .94 .34	 15.77 ±.14 				
0243.7 -0640 180.58 -56.39 297.45 -27.93 0241.3 -0653	 MCG -1- 8- 1 PGC 10334	.IBS9.. E (1) 10.0± .6 8.7± .6	1.19± .04 .25± .04 1.20	125 .04 .19 .13					
024347.3-744323 293.57 -40.30 220.76 -22.15 024355.0-745600	 ESO 30- 19 FAIR 11 PGC 10335	 	.69± .07 .02± .06 .70	 .13 	 14.47 ±.14 13.31 				 9860± 63 9696 9837
024349.1-595445 280.38 -52.01 236.17 -27.06 024231.0-600724	NGC 1096 ESO 115- 28 IRAS02425-6007 PGC 10336	.SBT4.. Sr (1) 4.2± .5 3.3± .8	1.28± .04 .02± .05 .97± .03 1.29	 .07 .03 .01	13.49 ±.15 13.75 ±.14 13.36 13.49	.73± .03 .67 	.77± .01 13.83± .09 14.71± .29		 6682± 34 6537 6602
024349.2+322321 148.90 -24.75 336.26 -13.09 024048.0+321040	 UGC 2202 IRAS02407+3210 PGC 10337	.I..9*. U 10.0±1.2 	.87± .10 .07± .06 .92	 .57 .05 .03	 13.03 			16.60±.3 109± 7 	4048± 10 4147 3840
024350.0+322830 148.86 -24.68 336.33 -13.06 024048.7+321549	NGC 1066 UGC 2203 IRAS02407+3217 PGC 10338	.E...*. U -5.0±1.1 	1.23± .14 .03± .08 1.29	 .43 .00 	 14.25 ±.16 12.48 				
024350.7+323043 148.85 -24.64 336.37 -13.04 024049.3+321802	NGC 1067 UGC 2204 PGC 10339	.SXS5.. U (1) 5.0± .8 2.2± .8	1.02± .03 .02± .05 .77± .03 1.06	 .43 .03 .01	14.55 ±.15 14.40 ±.12 13.97	.86± .04 .73 	.82± .03 13.89± .07 14.45± .25	16.29±.2 134± 7 129± 6 2.30	4533± 7 4535± 59 4632 4325
0243.8 +0638 165.89 -46.64 311.65 -23.94 0241.2 +0626	 UGC 2211 PGC 10341	.S..6*. U 6.0±1.4 	1.07± .07 .79± .06 1.09	174 .25 1.17 .40					6108 6136 5883
024352.5+332055 148.43 -23.90 337.12 -12.65 024050.3+330815	 UGC 2206 KUG 0240+331 PGC 10343	.S?.... 	.98± .05 .18± .04 1.00	105 .29 .26 .09	 14.85 ±.12 14.28			15.50±.3 216± 7 1.14	4723± 10 4824 4517
024354.9+331007 148.53 -24.05 336.96 -12.74 024052.8+325726	 UGC 2205 PGC 10346	.SX.5*. U 5.0± .9 	1.00± .06 .18± .05 1.04	30 .33 .27 .09	 15.04 ±.13 14.41			16.47±.3 246± 7 1.97	5460± 10 5561 5254
0244.0 +0525 167.04 -47.57 310.42 -24.41 0241.4 +0513	 CGCG 415- 4 PGC 10351	 	 	 .15 	 14.8 ±.3 				7162± 79 7186 6937

R.A. 2000 DEC. l　　b SGL　SGB R.A. 1950 DEC.	Names PGC	Type S_T　n_L T L	$\log D_{25}$ $\log R_{25}$ $\log A_e$ $\log D_o$	p.a. A_g A_i A_{21}	B_T m_B m_{FIR} B_T^0	$(B-V)_T$ $(U-B)_T$ $(B-V)_T^0$ $(U-B)_T^0$	$(B-V)_e$ $(U-B)_e$ m'_e m'_{25}	m_{21} W_{20} W_{50} HI	V_{21} V_{opt} V_{GSR} V_{3K}
0244.3　+0943 163.41 −44.09 314.84 −22.96 0241.7　+0931	 UGC　2214 PGC 10372	.S..8.. U 8.0± .9 	1.07± .07 .25± .06 1.11	10 .39 .31 .13					7981 8018 7756
024425.5+334406 148.35 −23.50 337.52 −12.56 024122.7+333127	 UGC　2212 KUG 0241+335 PGC 10375	.S?.... 	1.00± .05 .37± .04 1.03	91 .29 .55 .19	 15.24 ±.12 14.38				5327 5429 5122
0244.4　+0040 171.89 −51.12 305.48 −26.05 0241.9　+0028	 UGC　2216 PGC 10381	.I..9*. U 10.0±1.2 	1.11± .07 .51± .06 1.11	15 .07 .38 .25					2773 2783 2551
024433.8+375841 146.30 −19.71 341.28 −10.51 024126.1+374602	 UGC　2213 PGC 10383	.S..3.. U　　(1) 3.0± .9 4.5±1.1	.98± .07 .07± .05 1.00	 .22 .10 .04	 15.31 ±.16 				
024436.3-192240 202.36 −62.95 283.05 −30.51 024218.0-193518	 ESO　546- 12 PGC 10384	.S?.... 	.87± .07 .01± .06 .87	 .06 .02 .01	 15.37 ±.14 15.22				14594± 60 14544 14397
0244.7　+1643 158.28 −38.25 321.80 −20.28 0241.9　+1631	 UGC　2217 IRAS02419+1631 PGC 10388	.S..3.. U　　(1) 3.0± .9 2.5±1.3	1.10± .05 .72± .05 1.13	152 .36 1.00 .36	 15.40 ±.12 13.71 13.97				9464 9521 9241
0244.7　+1514 159.32 −39.49 320.36 −20.90 0242.0　+1502	IC　1839 UGC　2220 PGC 10394	.S..4.. U 4.0± .9 	1.02± .06 .35± .05 1.05	97 .33 .52 .18	 15.04 ±.12 14.15				7517 7569 7294
024447.7-243053 213.48 −64.41 277.04 −31.07 024234.0-244330	 ESO　479- 31 PGC 10398	.LBS0.. S -2.0± .8 	1.03± .05 .16± .04 1.01	150 .00 .00 	 14.76 ±.14 				
024450.7-691907 289.29 −44.69 226.20 −24.21 024414.0-693142	 ESO　53- 13 IRAS02442-6931 PGC 10399	 	.63± .07 .13± .06 .63	 .07 	15.4 ±.2 15.59 ±.14 	.41± .05 −.32± .07 			7226± 63 7068 7182
024453.8-173935 198.99 −62.20 285.06 −30.35 024234.0-175212	NGC　1098 ESO　546- 14 HICK　21C PGC 10403	.LAS-*. SE -2.7± .6 	1.25± .04 .12± .04 .82± .05 1.24	102 .06 .00 	13.55 ±.15 13.62 ±.14 13.42	.98± .02 .34± .04 .89 .36	1.01± .02 .43± .03 13.14± .16 14.38± .28		7387± 34 7341 7187
024458.0+302238 150.18 −26.42 334.57 −14.26 024158.7+301000	 UGC　2221 PGC 10407	.S..8?. U 8.0±2.0 	1.19± .06 1.21± .06 1.24	80 .54 1.23 .50				15.80±.3 108± 7 	833± 10 926 624
0245.0　+3834 146.11 −19.13 341.85 −10.30 0241.9　+3822	 UGC　2219 PGC 10409	.S..6*. U 6.0±1.2 	1.07± .07 .11± .06 1.10	0 .27 .16 .05					
024505.5-153517 195.13 −61.23 287.46 −30.08 024244.1-154754	NGC　1081 MCG -3- 8- 10 IRAS02427-1547 PGC 10411	.SBS3?. E　　(1) 3.0± .9 3.1±1.2	1.19± .05 .44± .05 1.19	22 .00 .61 .22	 12.79 				4000 3961 3797
024509.1-554425 275.20 −54.83 240.70 −28.31 024340.0-555700	 ESO　154- 10 IRAS02436-5556 PGC 10415	PSBR1*. Sr 1.5± .6 	1.41± .04 .39± .04 1.42	93 .11 .48 .19	 13.39 ±.14 11.30 12.74				5507± 34 5369 5413
0245.1　-0442 178.39 −54.83 299.71 −27.76 0242.6　-0455	NGC　1080 MCG -1- 8- 3 IRAS02426-0455 PGC 10416	.SXS5*. PE　　(2) 4.5± .6 1.8± .7	1.05± .06 .14± .05 1.06	160 .11 .20 .07	14.1 ±.3 12.97 13.74	.56± .04 .46 	 13.84± .46	16.14±.1 205± 15 2.34	7848± 10 7843± 59 7841 7631
024509.9+325923 148.87 −24.09 336.93 −13.05 024207.9+324645	 UGC　2222 PGC 10417	.E?.... 	1.08± .17 .22± .08 1.07	105 .34 .00 	 14.56 ±.10 14.15				4939± 79 5039 4734
024513.8+325840 148.89 −24.09 336.93 −13.07 024211.7+324603	 UGC　2225 PGC 10420	.SB?... 	.98± .07 .52± .05 1.02	6 .43 .78 .26	 15.21 ±.13 13.98				4965± 79 5065 4760

R.A. 2000 DEC. l b SGL SGB R.A. 1950 DEC.	Names PGC	Type S_T n_L T L	$\log D_{25}$ $\log R_{25}$ $\log A_e$ $\log D_o$	p.a. A_g A_i A_{21}	B_T m_B m_{FIR} B_T^o	$(B-V)_T$ $(U-B)_T$ $(B-V)_T^o$ $(U-B)_T^o$	$(B-V)_e$ $(U-B)_e$ m' m'_{25}	m_{21} W_{20} W_{50} HI	V_{21} V_{opt} V_{GSR} V_{3K}
024514.4+351122 147.79 -22.13 338.89 -12.00 024209.9+345845	 UGC 2223 KUG 0242+349 PGC 10421	.S..6*. U 6.0±1.3 	1.05± .05 .55± .05 1.08	47 .28 .81 .28	 14.84 ±.13 13.72				4981 5086 4779
024517.7-174230 199.17 -62.13 285.02 -30.45 024258.0-175506	NGC 1099 ESO 546- 15 IRAS02429-1754 PGC 10422	.SBT3.. SEr (2) 3.1± .5 2.7± .9	1.26± .02 .48± .03 .81± .03 1.27	10 .04 .66 .24	13.94 ±.14 13.98 ±.14 12.83 13.20	.88± .03 .26± .05 .73 .13	.86± .03 .27± .05 13.48± .08 13.92± .20		 7613± 34 7568 7414
024521.8-173154 198.85 -62.04 285.23 -30.44 024302.0-174430	NGC 1091 ESO 546- 16 HICK 21E PGC 10424	PSXT1P? SE .7± .7 	.95± .04 .16± .03 .96	77 .06 .17 .08	16.20S±.15 14.88 ±.14 	.93± .03 .36± .05 	 15.41± .27		
024524.9+025242 169.89 -49.31 307.89 -25.58 024249.0+024006	IC 1843 UGC 2228 IRAS02428+0240 PGC 10429	.SBS2?. UE (1) 1.7±1.0 5.3± .8	1.16± .05 .31± .05 1.17	70 .12 .38 .16	 14.01 ±.12 13.37 13.45				6793± 50 6808 6571
024529.8-173231 198.90 -62.02 285.22 -30.47 024310.0-174506	NGC 1092 ESO 546- 17 HICK 21D PGC 10432	.E...?. SE -5.0±1.0 	.97± .05 .07± .03 .96	170 .06 .00 	15.64S±.15 14.40 ±.14 14.79	1.02± .02 .39± .04 .92 .42	 15.32± .29		8835± 41 8790 8636
024535.7-174113 199.20 -62.06 285.05 -30.52 024316.0-175348	NGC 1100 ESO 546- 18 HICK 21B PGC 10438	.SXR1*. SEr .6± .5 	1.22± .02 .35± .03 .80± .02 1.23	58 .04 .35 .17	13.94 ±.14 14.03 ±.14 13.49	.91± .03 .36± .05 .77 .30	.96± .03 .38± .05 13.43± .08 14.05± .20		7554± 34 7509 7356
024536.3+420927 144.54 -15.87 345.01 -8.59 024223.0+415651	 UGC 2226 PGC 10440	.E..... U -5.0± .8 	1.01± .12 .11± .05 1.03	50 .34 .00 	 15.1 ±.3 14.71				4854± 79 4974 4666
024539.4-244855 214.25 -64.28 276.70 -31.28 024326.0-250130	 ESO 479- 33 PGC 10444	.L..0./ S -2.0± .8 	1.20± .04 .59± .03 1.11	158 .01 .00 	14.55 ±.13 14.49 ±.14 14.41	.94± .01 .47± .02 .82 .45	 13.94± .26		6835± 52 6769 6650
024540.7-152128 194.85 -60.99 287.75 -30.18 024319.1-153403	NGC 1083 MCG -3- 8- 15 IRAS02433-1534 PGC 10445	.S..3P/ E 3.0±1.3 	1.21± .05 .68± .05 1.21	17 .00 .94 .34	 10.95 				
024545.2-510014 268.54 -57.80 245.98 -29.46 024406.0-511248	 ESO 198- 30 FAIR 730 PGC 10451	 	.82± .07 .15± .06 .82	62 .02 	16.05 ±.15 15.48 ±.14 	1.09± .04 			18600±190 18470 18490
024554.2+424844 144.29 -15.26 345.60 -8.31 024240.0+423608	 UGC 2227 PGC 10457	.SBR3.. U 3.0± .7 	1.28± .04 .00± .05 1.32	 .35 .00 .00	 14.0 ±.3 13.63			15.50±.1 138± 8 1.86	5916± 11 6038 5730
0245.9 +4244 144.32 -15.32 345.55 -8.35 0242.6 +4232	 PGC 10460		 	.37				15.92±.1 147± 8 	5192± 11 5314 5005
024559.8-073442 182.48 -56.55 296.60 -28.68 024331.9-074716	NGC 1084 MCG -1- 8- 7 IRAS02435-0747 PGC 10464	.SAS5.. R (3) 5.0± .3 3.1± .6	1.51± .02 .25± .02 .97± .02 1.51	115 .06 .38 .13	11.31M±.10 11.5 ±.2 9.54 10.90	.58± .02 -.09± .08 .51 -.14	.66± .03 11.61± .04 13.07± .14	12.86±.1 357± 5 293± 6 1.84	1406± 7 1414± 20 1391 1194
024601.0-320914 231.31 -64.68 268.00 -31.57 024355.0-322148	 ESO 416- 18 PGC 10466	PLXT+?. Sr -1.4± .7 	1.14± .04 .25± .03 1.11	90 .03 .00 	 14.39 ±.14 14.26				6708 6622 6541
024601.3+280140 151.68 -28.37 332.55 -15.57 024304.2+274906	 MCG 5- 7- 50 PGC 10467	.S?.... 	.94± .11 .40± .07 .97	115 .39 .49 .20	 15.49 ±.12 14.53			15.61±.3 362± 7 .88	7953± 10 8040 7742
024604.1+155108 159.23 -38.81 321.09 -20.94 024317.9+153834	A 0243+15 MCG 3- 8- 7 MK 597 PGC 10469	 	 	.32					7563±113 7616 7341
0246.1 +4206 144.65 -15.87 345.02 -8.70 0242.9 +4154	 UGC 2231 PGC 10476	.SBS9.. U 9.0± .9 	1.04± .15 .18± .12 1.07	115 .32 .18 .09					

R.A. 2000 DEC.	Names	Type	logD$_{25}$	p.a.	B$_T$	(B-V)$_T$	(B-V)$_e$	m$_{21}$	V$_{21}$
l b		S$_T$ n$_L$	logR$_{25}$	A$_g$	m$_B$	(U-B)$_T$	(U-B)$_e$	W$_{20}$	V$_{opt}$
SGL SGB		T	logA$_e$	A$_i$	m$_{FIR}$	(B-V)$_T^o$	m'$_e$	W$_{50}$	V$_{GSR}$
R.A. 1950 DEC.	PGC	L	logD$_o$	A$_{21}$	B$_T^o$	(U-B)$_T^o$	m'$_{25}$	HI	V$_{3K}$
024610.0-301345 NGC 1097A		.E...P*	.91± .04	105					1337± 30
226.81 -64.71 ESO 416- 19		RS	.25± .02	.03	14.61 ±.14				1255
270.29 -31.61		-5.0± .5		.00					1165
024402.0-302618 PGC 10479			.84		13.63				
024612.9-261827		RSAR1..	.92± .06	20					
217.70 -64.41 ESO 479- 35		r	.11± .03	.00	14.79 ±.14				
274.94 -31.50		1.0± .9		.12					
024401.0-263100 PGC 10482			.92	.06					
0246.2 +2702		.S?....	1.07± .06	26					5726
152.27 -29.21 UGC 2236			.32± .05	.46	15.20 ±.14				5810
331.67 -16.07				.49					5514
0243.3 +2650	PGC 10484		1.12	.16	14.22				
024617.3+130544		.I..9?.	1.15± .07	165				15.45±.3	6436± 9
161.29 -41.07 UGC 2238		U	.05± .06	.34	14.61 ±.19			442± 10	6481
318.40 -22.10 IRAS02435+1253		10.0±1.6		.04	10.82				6214
024333.4+125310 PGC 10486			1.18	.03	14.22			1.20	6214
024617.4-552728		.S..3*.	1.16± .05	0					6419
274.65 -54.90 ESO 154- 13		S	.27± .05	.08	13.61 ±.14				6281
240.96 -28.54 IRAS02448-5539		3.0±1.2		.38	12.15				6325
024448.1-554000 PGC 10487		3.3±1.2	1.17	.14	13.11				
024618.9-301621 NGC 1097		.SBS3..	1.97± .01	130	10.23M±.07	.75± .01	.83± .01	11.89±.1	1275± 5
226.91 -64.68 ESO 416- 20		R (3)	.17± .02	.03	9.95 ±.12	.23± .02	.26± .01	397± 5	1274± 18
270.23 -31.64 ARP 77		3.0± .3	1.29± .02	.23	8.82	.70	12.32± .05	382± 5	1193
024411.0-302854 PGC 10488		2.2± .4	1.98	.08	9.88	.19	14.52± .10	1.93	1103
024620.4+445708		.S..3..	1.19± .06	108					
143.39 -13.31 UGC 2233		U (1)	.74± .06	.66	15.0 ±.3				
347.48 -7.28 IRAS02430+4444		3.0± .9		1.01	13.04				
024302.9+444434 PGC 10489		5.5±1.3	1.25	.37					
024622.4+450835		.S..6*.	1.11± .07	126					
143.31 -13.13 UGC 2234		U	.54± .06	.95	14.9 ±.3				
347.65 -7.19		6.0±1.3		.79					
024304.6+445601 PGC 10490			1.20	.27					
024624.9-002946 NGC 1087		.SXT5..	1.57± .02	5	11.46M±.12	.52± .02	.54± .02	13.62±.1	1519± 6
173.74 -51.65 UGC 2245		R (3)	.22± .02	.12	11.32 ±.11	-.06± .03		232± 5	1508± 24
304.39 -26.87 IRAS02438-0042		5.0± .3	1.16± .01	.32	10.43	.44	12.79± .03	220± 5	1523
024351.6-004219 PGC 10496		5.5± .6	1.58	.11	10.93	-.12	13.63± .15	2.58	1300
024625.8+033623 NGC 1085		.SAS4*.	1.47± .04	15				14.32±.2	6789± 5
169.47 -48.60 UGC 2241		UE (2)	.15± .05	.13	13.07 ±.17			407± 5	6980± 60
308.74 -25.58 IRAS02438+0323		3.5± .4		.23	12.94			381± 5	6807
024349.3+032350 PGC 10498		3.1± .7	1.49	.08	12.68			1.56	6569
024629.9-261457		PSXT4P*	1.19± .04	26					7052± 52
217.58 -64.34 ESO 479- 37		Sr (1)	.17± .04	.01	14.13 ±.14				6981
275.02 -31.56 IRAS02443-2627		3.8± .5		.25					6872
024418.0-262730 PGC 10502		2.8± .8	1.20	.08	13.82				
0246.5 +4512		.S..9*.	1.00± .16	125					
143.30 -13.06 UGC 2235		U	.14± .12	.95					
347.71 -7.17		9.0±1.2		.14					
0243.2 +4500	PGC 10503		1.09	.07					
024633.2-245158		.LXS0?P	1.14± .05	11					6835± 52
214.46 -64.09 ESO 479- 38		S	.15± .04	.01	14.16 ±.14				6768
276.66 -31.49		-2.0± .6		.00					6652
024420.0-250430 PGC 10506			1.12		14.04				
024633.5-001448 NGC 1090		.SBT4..	1.60± .02	102	12.51M±.12	.68± .03	.76± .02	13.62±.1	2758± 6
173.51 -51.44 UGC 2247		R (4)	.36± .02	.11	12.44 ±.11	.11± .07	.20± .04	334± 9	2703± 46
304.67 -26.83 IRAS02440-0027		4.0± .3	1.11± .01	.53	12.80	.57	13.66± .04	319± 12	2763
024400.0-002720 PGC 10507		4.2± .5	1.61	.18	11.81	.03	14.46± .15	1.63	2538
024635.8+322700		.SB.4?.	1.11± .07	13				16.12±.3	4821± 10
149.44 -24.43 UGC 2239		U	.69± .06	.47	15.46 ±.12				4918
336.60 -13.58		4.0± .9		1.01				323± 7	4616
024334.1+321427 PGC 10512			1.15	.34	13.94			1.83	
024636.8-422158		.LAR-?.	1.14± .09						
253.48 -62.07 MCG -7- 6- 18		S	.34± .08	.00					
255.92 -31.01		-3.0± .7		.00					
024443.5-423430 PGC 10514			1.09						
0246.6 +2022		.I..9*.	1.04± .08						4052
156.35 -34.91 UGC 2242		U	.00± .06	.38					4118
325.50 -19.15		10.0±1.2		.00					3834
0243.8 +2010	PGC 10516		1.08	.00					

R.A. 2000 DEC. / l b / SGL SGB / R.A. 1950 DEC.	Names / PGC	Type / S_T n_L / T / L	$\log D_{25}$ / $\log R_{25}$ / $\log A_e$ / $\log D_o$	p.a. / A_g / A_I / A_{21}	B_T / m_B / m_{FIR} / B_T^o	$(B-V)_T$ / $(U-B)_T$ / $(B-V)_T^o$ / $(U-B)_T^o$	$(B-V)_e$ / $(U-B)_e$ / m' / m'_{25}	m_{21} / W_{20} / W_{50} / HI	V_{21} / V_{opt} / V_{GSR} / V_{3K}
024640.3+213517				.51				17.84±.3	6987± 9
155.59 -33.87								155± 10	
326.65 -18.63	IRAS02438+2122				11.38				7056
024349.2+212244	PGC 10519								6770
024641.8-252052		.SBS4*.	1.18± .04	51					
215.55 -64.15	ESO 479- 40	S (1)	.45± .03	.00	14.64 ±.14				10548± 52
276.09 -31.55	IRAS02445-2533	3.5± .6		.65					10480
024429.0-253324	PGC 10520	2.8± .6	1.18	.22	13.91				10366
024653.3-223847	A 0244-22	.SBT9*P	1.07± .04	21					
209.61 -63.50	ESO 479- 42	S (1)	.32± .03	.07	14.63 ±.14				
279.29 -31.39	IRAS02446-2251	9.2± .7		.33	13.52				
024438.0-225118	PGC 10526	5.6±1.2	1.07	.16					
024654.5+233559		.L...*.	1.05± .08	145				16.22±.3	6190± 10
154.42 -32.12	UGC 2248	U	.12± .03	.51	15.21 ±.14				
328.56 -17.78	IRAS02440+2323	-2.0±1.2		.00	13.20			368± 7	6264
024401.5+232327	PGC 10528		1.09		14.60				5975
024656.8+481136		.LB....	1.11± .10						
142.03 -10.35	UGC 2240	U	.10± .05	1.05	14.2 ±.3				
350.30 -5.70		-2.0± .8		.00					
024333.9+475903	PGC 10529		1.22						
024657.7+390045		.SB.4..	1.03± .08	100					
146.25 -18.57	UGC 2243	U	.37± .06	.34	14.83 ±.12				
342.42 -10.40		4.0± .9		.55					
024348.3+384813	PGC 10532		1.06	.19					
0246.9 +1611	NGC 1088	.S..0..	1.00± .06	105					
159.23 -38.40	UGC 2253	U	.25± .05	.34					
321.52 -21.00		.0± .9		.19					
0244.2 +1559	PGC 10536		1.02						
024702.5+083901		.S..0..	.91± .09	85					
165.05 -44.56	UGC 2255	U	.34± .06	.34	14.42 ±.17				7618
314.02 -23.96		.0± .9		.26					7650
024422.1+082630	PGC 10538		.92		13.71				7396
024706.8-025823		.S..3*/	1.10± .06	173					
176.81 -53.28	MCG -1- 8- 8	E	.67± .05	.08					7616
301.76 -27.76		3.0±1.3		.93					7613
024435.4-031054	PGC 10542		1.10	.34					7400
024707.6-312905		PSB.1..	.97± .05						
229.72 -64.48	ESO 416- 21	r	.07± .04	.03	14.93 ±.14				7463± 62
268.79 -31.81		1.0± .9		.07					7378
024501.0-314136	PGC 10543		.97	.03	14.73				7296
0247.3 +4532		.LX.+..	1.30± .12						
143.28 -12.70	UGC 2249	U	.11± .08	.95					
348.07 -7.13		-1.0± .8		.00					
0244.0 +4520	PGC 10548		1.38						
0247.3 +3731		.I..9..	1.00± .16						578± 10
147.04 -19.86	UGC 2254	U	.09± .12	.25					
341.16 -11.21		10.0± .9		.07					687
0244.2 +3719	PGC 10550		1.02	.04					382
0247.3 +4620		.I..9*.	1.07± .14						
142.92 -11.98	UGC 2250	U	.00± .12	1.05					
348.76 -6.71		10.0±1.2		.00					
0244.0 +4608	PGC 10553		1.17	.00					
024727.4-001707	NGC 1094	.SXS2..	1.11± .03	85				15.30±.3	6464± 17
173.81 -51.32	UGC 2262	R (1)	.12± .03	.12	13.43 ±.13			368± 34	6344± 54
304.70 -27.06	IRAS02449-0029	2.0± .4		.15	12.70			308± 25	6458
024454.0-002937	PGC 10559	1.9± .8	1.12	.06	13.10			2.14	6235
024735.9-250855		PSBR1*.	1.10± .03	128					
215.19 -63.91	ESO 479- 43	Sr	.51± .03	.01	15.05 ±.10				6543± 52
276.34 -31.74		.5± .6		.52					6475
024523.0-252124	PGC 10565		1.10	.25	14.43				6361
024737.8+043815	NGC 1095	.SB.5..	1.11± .05	45					
168.81 -47.61	UGC 2264	U	.19± .05	.15	13.99 ±.12				6364± 50
309.93 -25.52	IRAS02450+0425	5.0± .8		.28	13.23				6383
024500.5+042545	PGC 10566		1.13	.09	13.53				6144
024740.6+402942		.L.....	1.11± .08	85					
145.68 -17.19	UGC 2256	U	.25± .04	.33	14.76 ±.11				
343.77 -9.77		-2.0± .8		.00					
024429.2+401712	PGC 10568		1.11						

R.A. 2000 DEC. / I b / SGL SGB / R.A. 1950 DEC.	Names / PGC	Type / S_T n_L / T / L	$\log D_{25}$ / $\log R_{25}$ / $\log A_e$ / $\log D_o$	p.a. / A_g / A_i / A_{21}	B_T / m_B / m_{FIR} / B_T^o	$(B-V)_T$ / $(U-B)_T$ / $(B-V)_T^o$ / $(U-B)_T^o$	$(B-V)_e$ / $(U-B)_e$ / m'_e / m'_{25}	m_{21} / W_{20} / W_{50} / HI	V_{21} / V_{opt} / V_{GSR} / V_{3K}
024745.0+002449		.SBT3*.	1.12± .05	10					
173.13 -50.76	UGC 2271	UE (1)	.31± .05	.13	15.11 ±.11				
305.48 -26.92	IRAS02451+0012	3.0± .7		.43	13.26				
024511.0+001219	PGC 10574	4.2±1.2	1.13	.16					
0247.7 +1423		.SA.8*.	1.04± .15						7717
160.72 -39.80	UGC 2266	U	.00± .12	.43					7765
319.84 -21.91		8.0± .8		.00					
0245.0 +1411	PGC 10575		1.08	.00					7497
024747.6+030958	A 0245+02	.S...P*	.67± .11	130					
170.29 -48.71	MCG 0- 8- 17	E	.15± .05	.13	15.13 ±.17				8830± 45
308.41 -26.05	MK 599			.20					8845
024511.5+025729	PGC 10579		.68	.07	14.73				8611
0247.8 +2324		RSBR3..	1.07± .08						6051
154.76 -32.17	UGC 2267	U	.04± .06	.55	15.10 ±.20				
328.49 -18.07		3.0± .7		.05					6124
0245.0 +2312	PGC 10581		1.12	.02	14.45				5837
0247.9 +2605		.I..9?.	1.00± .16						7570
153.19 -29.85	UGC 2268	U	.00± .12	.63					
330.98 -16.85		10.0±1.7		.00					7650
0245.0 +2553	PGC 10585		1.06	.00					7358
024755.6+373219	A 0244+37	.SBS8..	1.41± .04	155				14.04±.1	585± 6
147.15 -19.80	UGC 2259	U (1)	.14± .05	.25	13.8 ±.2			129± 8	654± 46
341.23 -11.31		8.0± .7		.18				121± 12	695
024447.9+371950	PGC 10586	6.0±1.0	1.44	.07	13.33			.64	392
024756.3+411446	NGC 1086	.S..6*.	1.19± .05	35					4037± 10
145.37 -16.49	UGC 2258	U	.17± .05	.34	13.5 ±.3				4202± 52
344.45 -9.43	IRAS02447+4102	6.0±1.2		.26	12.34				4161
024443.8+410217	PGC 10587		1.22	.09	12.91				3856
0247.9 +0353	A 0245+03	.S..9*.	1.74± .03		13.6 ±.2	.51± .07	.49± .04	12.77±.2	1025± 6
169.63 -48.14	UGC 2275	U (1)	.00± .06	.14	14.2 ±.9	-.08± .09	-.09± .05	100± 5	
309.17 -25.85	DDO 28	9.0± .8	1.47± .05	.00		.47	16.41± .12	86± 7	1042
0245.3 +0340	PGC 10588	9.0± .8	1.75	.00	13.46	-.11	17.12± .31	-.68	806
0248.0 +2706	A 0245+26	.S...4..	1.25± .04	65					5648
152.64 -28.95	UGC 2272	U	.72± .05	.49	14.62 ±.12				
331.92 -16.40		4.0± .9		1.06					5731
0245.1 +2654	PGC 10592		1.30	.36	13.04				5438
024804.2-135938	IC 1853	.SBT3?.	1.02± .07	91	14.22V±.13	.57± .03	.53± .03		
193.03 -59.82	MCG -2- 8- 6	E (1)	.42± .05	.05					4172± 62
289.44 -30.52	IRAS02457-1410	3.0±1.9	.61± .01	.58	12.31	.45			4135
024541.6-141206	PGC 10595	5.3±1.2	1.02	.21	14.14		13.69± .39		3970
024806.1-135736	NGC 1103	.SBS3*.	1.33± .04	40	13.6 ±.2	.73± .02			
192.98 -59.79	MCG -2- 8- 5	E (1)	.59± .05	.05					4154± 62
289.48 -30.52		3.0± .8		.81		.58			4117
024543.4-141004	PGC 10597	3.6± .8	1.34	.29	12.71		13.63± .32		3952
024807.3-685516		.SXT1*.	.95± .07	66					
288.62 -44.83	ESO 53- 16	r	.30± .05	.06	15.15 ±.14				
226.49 -24.62		1.0±1.3		.31					
024730.1-690742	PGC 10599		.95	.15					
024807.5-221226		PSBS3..	1.03± .05	35					
208.85 -63.10	ESO 546- 23	r	.49± .04	.06	15.38 ±.14				
279.84 -31.64		3.3± .9		.68					
024552.0-222454	PGC 10600		1.04	.25					
024815.3-360127			.93± .07						
239.98 -63.70	ESO 356- 11		.55± .06	.15	15.31 ±.14				6753± 34
263.38 -31.90									6656
024614.1-361354	PGC 10604		.94						6599
024816.4+342509	NGC 1093	.SX.2?.	1.25± .04	100					
148.77 -22.52	UGC 2274	U	.19± .05	.30	13.99 ±.14				
338.52 -12.92	IRAS02452+3412	2.0±1.5		.24					
024512.4+341241	PGC 10606		1.28	.10					
024817.2-175915	NGC 1119	PLB.0?.	.70± .06	0					
200.34 -61.60	ESO 546- 24	SE	.07± .03	.02	14.83 ±.14				
284.81 -31.19	IRAS02459-1811	-1.7±1.1		.00					
024558.0-181142	PGC 10607		.69						
024817.5+504800		.E.....	1.23± .14	70					
141.08 -7.91	UGC 2261	U	.14± .08	1.16	14.55 ±.16				4903± 57
352.61 -4.52		-5.0± .8		.00					5039
024449.5+503531	PGC 10608		1.37		13.31				4737

R.A. 2000 DEC. l b SGL SGB R.A. 1950 DEC.	Names PGC	Type S_T n_L T L	$\log D_{25}$ $\log R_{25}$ $\log A_e$ $\log D_o$	p.a. A_g A_i A_{21}	B_T m_B m_{FIR} B_T^o	$(B-V)_T$ $(U-B)_T$ $(B-V)_T^o$ $(U-B)_T^o$	$(B-V)_e$ $(U-B)_e$ m'_e m'_{25}	m_{21} W_{20} W_{50} HI	V_{21} V_{opt} V_{GSR} V_{3K}
0248.2 +4439 143.84 -13.42 347.41 -7.73 0245.0 +4427	UGC 2269 PGC 10609	.I..9*. U 10.0±1.1	1.19± .12 .05± .12 1.25	100 .65 .04 .02					
0248.3 +1744 158.49 -36.93 323.16 -20.64 0245.5 +1732	UGC 2276 PGC 10610	.S?....	.98± .09 .59± .06 1.01	52 .35 .88 .29					7279 7336 7061
0248.3 +0434 169.07 -47.55 309.93 -25.70 0245.7 +0422	NGC 1101 UGC 2278 PGC 10613	.L..... U -2.0± .8	1.11± .04 .09± .05 .68± .02 1.11	 .15 .00	13.96 ±.13 14.34 ±.12 13.93	.94± .02 .52± .03 .84 .51	1.02± .02 .55± .03 12.85± .07 14.15± .26		 6207± 31 6226 5988
0248.3 -1013 187.00 -57.71 293.77 -29.84 0245.9 -1026	MCG -2- 8- 7 PGC 10617	.S?....	1.08± .09 .32± .07 1.08	 .01 .40 .16					4554 4528 4347
024826.4-293539 225.35 -64.21 271.05 -32.09 024618.1-294806	ESO 416- 23 PGC 10623	.S?....	1.07± .05 .49± .04 1.07	69 .03 .60 .25	14.79 ±.14 14.08				6782 6701 6612
024827.5-403316 249.59 -62.43 258.00 -31.56 024632.1-404542	ESO 299- 18 PGC 10624	.S..6*/ S 6.0± .9	1.30± .04 .95± .04 1.30	54 .00 1.39 .47	15.63 ±.14				
024829.1+475935 142.36 -10.41 350.26 -6.02 024506.2+474707	UGC 2273 IRAS02450+4746 PGC 10625	.SB.6*. U 6.0±1.2	1.12± .05 .29± .05 1.21	120 1.02 .43 .15	14.6 ±.3 13.33				
024831.1+504515 141.13 -7.93 352.59 -4.57 024503.1+503247	UGC 2270 PGC 10627	.S..3*. U 3.0±1.0	1.29± .04 .04± .05 1.40	85 1.16 .06 .02	14.2 ±.2 12.97			14.25±.1 606± 8 1.25	4944± 11 4937± 57 5080 4778
0248.5 -0205 176.18 -52.42 302.84 -27.85 0246.0 -0218	UGC 2284 PGC 10629	.S..8.. U 8.0± .8	1.11± .14 .07± .12 1.12	 .11 .08 .03					7446 7444 7231
024833.4+063127 167.33 -46.01 311.98 -25.08 024554.7+061900	UGC 2285 PGC 10631	.S..8*. U 8.0±1.3	1.11± .04 .48± .04 1.14	85 .30 .59 .24	 15.01 ±.12 14.10			15.67±.2 299± 9 290± 10 1.33	5959± 10 5983 5739
024838.6-001617 174.14 -51.10 304.82 -27.34 024605.1-002844	NGC 1104 UGC 2287 IRAS02461-0028 PGC 10634	PSBT0.. PUE .3± .5	1.08± .04 .15± .04 1.08	70 .12 .11	 14.46 ±.13 13.89				
0248.6 +1418 161.01 -39.74 319.85 -22.14 0245.9 +1406	UGC 2282 PGC 10635	.SBT4.. U 4.0± .9	1.02± .06 .07± .05 1.05	150 .37 .10 .04	 14.49 ±.13 13.97				7268 7315 7049
024841.3-313210 229.81 -64.15 268.73 -32.15 024635.0-314436	ESO 416- 25 PGC 10637	.S..3*/ S 3.0±1.4	1.31± .03 .82± .03 1.31	31 .03 1.13 .41	 14.68 ±.14 13.48				4997± 33 4911 4832
0248.7 +1714 158.95 -37.30 322.72 -20.95 0245.9 +1701	UGC 2283 IRAS02459+1701 PGC 10641	.SB.5?. U 5.0±1.8	.94± .07 .28± .05 .97	110 .33 .42 .14	 15.09 ±.12 14.30				6464 6519 6246
024845.4-322046 231.66 -64.09 267.76 -32.15 024640.0-323312	ESO 356- 11A PGC 10642	.S?....	1.01± .07 .30± .05 1.01	99 .03 .44 .15					6730± 52 6642 6567
024847.4-364253 241.45 -63.45 262.55 -31.97 024647.0-365518	ESO 356- 13 PGC 10643	.SAS6*P S (1) 6.0± .6 4.4± .6	1.09± .05 .13± .04 1.09	45 .03 .19 .06	 14.52 ±.14 14.28				5098 4999 4947
024849.9-004604 174.74 -51.43 304.30 -27.53 024616.8-005830	IC 1856 UGC 2291 IRAS02463-0058 PGC 10647	RSBR2*. PUE 2.3± .5	1.05± .05 .28± .04 1.06	60 .13 .34 .14	 14.35 ±.12 13.34 13.82				6486± 50 6488 6270

R.A. 2000 DEC. / l b / SGL SGB / R.A. 1950 DEC.	Names / PGC	Type / S_T n_L / T / L	$\log D_{25}$ / $\log R_{25}$ / $\log A_e$ / $\log D_o$	p.a. / A_g / A_i / A_{21}	B_T / m_B / m_{FIR} / B_T^o	$(B-V)_T$ / $(U-B)_T$ / $(B-V)_T^o$ / $(U-B)_T^o$	$(B-V)_e$ / $(U-B)_e$ / m'_e / m'_{25}	m_{21} / W_{20} / W_{50} / HI	V_{21} / V_{opt} / V_{GSR} / V_{3K}
0248.8 -0031	A 0246+00	CE...*.	.27?						7340± 67
174.48 -51.25		E	.00± .08	.13					7343
304.56 -27.46	1ZW 9	-5.0±1.3		.00					7124
0246.3 -0044	PGC 10649		.30						
0248.8 +0059		.SA.1..	.96± .07						
172.84 -50.15	UGC 2292	U	.02± .05	.16	14.98 ±.14				
306.19 -27.00		1.0± .9		.02					
0246.3 +0047	PGC 10651		.98	.01					
024853.6+530147								15.37±.3	4954± 9
140.18 -5.86				1.57				299± 11	
354.54 -3.42								264± 8	5094
024521.0+524920	PGC 10652								4794
024853.6+414640		.SA.4..	1.31± .04	44				15.29±.1	8375± 11
145.28 -15.94	UGC 2280	U (1)	.48± .05	.26	14.6 ±.3				
345.00 -9.31		4.0± .8		.71				459± 8	8493
024540.2+413413	PGC 10653	3.5±1.1	1.34	.24	13.54			1.51	8189
024853.7+281623								16.10±.3	5405± 10
152.17 -27.85	CGCG 505- 54			.48	15.7 ±.4				
333.08 -16.02								211± 7	5490
024556.1+280357	PGC 10654								5197
024857.4-283235		.SBR0..	1.05± .05	146					
222.96 -64.05	ESO 416- 26	r	.55± .04	.01	14.58 ±.14				
272.31 -32.19		-.1± .9		.41					
024648.0-284500	PGC 10656		1.03						
024858.5+031004		.S..3P/	1.16± .04	85				15.54±.3	4172± 10
170.62 -48.51	UGC 2295	E	.39± .04	.17	13.92 ±.13			369± 13	4168± 50
308.52 -26.33	IRAS02463+0257	3.0±1.2		.54	12.70			356± 10	4186
024622.3+025738	PGC 10659		1.17	.20	13.18			2.17	3954
0249.0 +1313	NGC 1109	.SXS5..	1.09± .06	3					8434
161.91 -40.58	UGC 2293	U	.15± .05	.40	14.53 ±.14				
318.82 -22.66	IRAS02462+1301	5.0± .8		.23	12.95				8478
0246.2 +1301	PGC 10660		1.13	.08	13.85				8215
0249.0 +1543		.S?....	.94± .07	130					6353
160.08 -38.52	UGC 2289		.53± .05	.26	15.32 ±.13				
321.28 -21.64	IRAS02462+1531			.79	13.63				6404
0246.2 +1531	PGC 10661		.97	.26	14.23				6135
024904.1-142817	NGC 1120	.L..-..	1.11± .05	40					
194.09 -59.84	MCG -3- 8- 28	E	.22± .04	.04					
288.94 -30.84		-3.0± .6		.00					
024642.0-144043	PGC 10664		1.08						
024904.6-311023	IC 1859	.S?....	1.09± .05	35					5970± 19
228.97 -64.08	ESO 416- 28		.18± .04	.03	14.27 ±.14				
269.16 -32.23	IRAS02469-3122			.22	13.14				5885
024658.0-312248	PGC 10665		1.09	.09	13.95				5804
0249.0 +2300		.S..9*.	1.19± .12						6276± 10
155.29 -32.37	UGC 2290	U	.00± .12	.56					
328.24 -18.49		9.0±1.1		.00					6347
0246.2 +2248	PGC 10666		1.24	.00					6063
024907.1-165937	NGC 1114	.SAR5..	1.24± .04	8					3467
198.63 -61.00	MCG -3- 8- 29	SE (2)	.33± .04	.04					
286.00 -31.25	IRAS02467-1712	4.5± .6		.50	12.77				3421
024647.1-171201	PGC 10669	2.1± .6	1.25	.17					3271
0249.1 +0207	A 0246+01	PSBT9*.	1.68± .02	60				12.85±.1	1104± 5
171.73 -49.26	UGC 2302	UE (2)	.11± .04	.26	14.1 ±.6			68± 4	
307.43 -26.71	DDO 29	9.0± .5		.11				55± 7	1114
0246.5 +0155	PGC 10670	8.2± .5	1.70	.05	13.69			-.90	886
024908.5-311724	IC 1858	.LA.+*.	1.25± .04	176					6030± 32
229.24 -64.06	ESO 416- 29	S	.50± .03	.03	14.15 ±.14				
269.02 -32.24		-1.0± .9		.00					5944
024702.0-312948	PGC 10671		1.18		14.03				5865
024909.5-075019	NGC 1110	.SBS9*/	1.46± .03	18				13.68±.1	1332± 6
183.73 -56.11	MCG -1- 8- 10	UE (1)	.73± .05	.09				188± 8	
296.52 -29.50		8.7± .7		.75				166± 12	1313
024642.0-080244	PGC 10673	6.4±1.6	1.47	.37					1123
024909.6+182007	A 0246+18	.S?....	.78± .12						10010± 40
158.31 -36.32	UGC 2296		.00± .06	.34					
323.82 -20.57	ARAK 91			.00					10068
024621.1+180742	PGC 10674		.81						9793

R.A. 2000 DEC. l　b SGL　SGB R.A.1950 DEC.	Names PGC	Type S_T　n_L T L	$\log D_{25}$ $\log R_{25}$ $\log A_e$ $\log D_o$	p.a. A_g A_i A_{21}	B_T m_B m_{FIR} B_T^o	$(B-V)_T$ $(U-B)_T$ $(B-V)_T^o$ $(U-B)_T^o$	$(B-V)_o$ $(U-B)_o$ m' m'_{25}	m_{21} W_{20} W_{50} HI	V_{21} V_{opt} V_{GSR} V_{3K}
0249.2　+1106 163.62 -42.27 316.72 -23.53 0246.5　+1054	 UGC　2299 PGC 10675	.SAS8.. U 8.0± .9 	.96± .09 .05± .06 1.00	 .41 .06 .02					10250 10287 10031
024913.2+352006 148.49 -21.63 339.43 -12.64 024607.9+350740	 UGC　2288 MK　1056 PGC 10676	.S..0.. U .0± .9 	.98± .05 .26± .04 .99	175 .27 .20 	 15.28 ±.13 				
0249.3　-0238 177.05 -52.66 302.30 -28.20 0246.8　-0251	 UGCA　44 PGC 10682	.IBS9*. UE　(1) 10.0± .7 10.3± .8	1.21± .06 .11± .07 1.22	80 .12 .08 .05				14.25±.1 111± 16 96± 12 	1094± 11 1090 880
024919.9+080534 166.15 -44.66 313.67 -24.70 024639.8+075309	NGC　1107 UGC　2307 PGC 10683	.L..... U -2.0± .8 	1.26± .13 .08± .08 .92± .05 1.29	140 .32 .00 	13.46 ±.16 13.62 ±.12 13.19	1.26± .03 .81± .04 1.15 .73	1.28± .02 .82± .03 13.55± .15 14.44± .70		 3424± 31 3452 3205
024920.3+191819 157.70 -35.48 324.77 -20.19 024630.9+190554	IC　1854 CGCG 463- 20 MK　372 PGC 10684	 		 .41 	 14.9 ±.3 				9300±110 9361 9084
024921.6+304002 150.95 -25.71 335.29 -14.96 024621.4+302737	 UGC　2297 PGC 10686	.S..0.. U .0± .9 	1.00± .08 .25± .06 1.05	 .64 .19 				16.54±.3 114± 10 	8172± 10 8263 7968
024922.7+173948 158.82 -36.86 323.20 -20.90 024634.8+172723	 UGC　2303 IRAS02465+1727 PGC 10688	.SX.3.. U　(1) 3.0± .8 3.5±1.1	1.07± .06 .00± .05 1.11	 .41 .00 .00	 14.15 ±.13 13.69				6457± 50 6513 6240
024925.7-434743 255.76 -61.01 254.17 -31.33 024735.1-440006	 ESO　247-　5 PGC 10692	 	.61± .07 .16± .05 .61	 .00 	15.91 ±.14 15.78 ±.14 	.64± .04 			
0249.4　+2207 155.92 -33.08 327.46 -18.97 0246.6　+2155	 UGC　2304 PGC 10695	.S?.... 	1.11± .05 .83± .05 1.17	61 .60 1.24 .41					7258 7327 7044
024928.1-005221 175.04 -51.39 304.24 -27.71 024655.1-010445	 UGC　2311 IRAS02469-0104 PGC 10697	PSBR3.. UE 2.5± .6 	1.10± .05 .02± .04 1.11	120 .14 .02 .01	 13.68 ±.12 13.46				7158± 50 7159 6943
024930.2+304151 150.97 -25.67 335.34 -14.97 024630.0+302927	 UGC　2300 PGC 10701	.S..2.. U 2.0± .9 	1.00± .08 .19± .06 1.06	 .64 .23 .09				15.78±.3 398±　7 	8316± 10 8407 8112
024933.7-384613 245.77 -62.78 260.09 -31.95 024736.0-385836	 ESO　299- 20 IRAS02476-3858 PGC 10705	RSBR1?. S 1.3± .7 	1.21± .04 .32± .04 1.21	37 .00 .32 .16	 14.21 ±.14 11.59 13.82				 4993± 22 4889 4848
024934.6-311125 229.01 -63.97 269.14 -32.34 024728.0-312348	IC　1860 ESO　416- 31 PGC 10707	.E.4... S -5.0± .8 	1.24± .03 .16± .03 .79± .03 1.20	6 .03 .00 	13.7 ±.2 13.00 ±.14 13.09	1.03± .03 .96 	1.11± .01 13.11± .07 14.48± .28		 6958± 30 6872 6793
024936.2-303443 227.61 -63.97 269.87 -32.35 024729.0-304706	 ESO　416- 32 IRAS02474-3046 PGC 10709	.L?.... 	.94± .06 .01± .03 .94	 .03 .00 	 13.89 ±.14 13.37 13.84				 1319± 62 1235 1152
024937.7-305949 228.57 -63.96 269.37 -32.35 024731.0-311212	 ESO　416- 33 PGC 10710	.S?.... 	1.06± .05 .22± .04 1.06	63 .03 .32 .11	 15.23 ±.14 14.83				 6565± 62 6480 6399
0249.6　+2112 156.55 -33.84 326.61 -19.42 0246.8　+2100	 UGC　2309 PGC 10713	.L..... U -2.0± .8 	1.00± .08 .12± .03 1.05	160 .61 .00 	 15.10 ±.12 				
0249.6　+1437 161.04 -39.35 320.27 -22.23 0246.9　+1425	IC　1857 UGC　2312 PGC 10715	.S?.... 	1.00± .06 .33± .05 1.02	150 .31 .50 .17	 14.90 ±.12 14.05				9069 9116 8851

R.A. 2000 DEC. / l b / SGL SGB / R.A. 1950 DEC.	Names / PGC	Type / S_T n_L / T / L	$logD_{25}$ / $logR_{25}$ / $logA_e$ / $logD_o$	p.a. / A_g / A_i / A_{21}	B_T / m_B / m_{FIR} / B_T^o	$(B-V)_T$ / $(U-B)_T$ / $(B-V)_T^o$ / $(U-B)_T^o$	$(B-V)_e$ / $(U-B)_e$ / m'_e / m'_{25}	m_{21} / W_{20} / W_{50} / HI	V_{21} / V_{opt} / V_{GSR} / V_{3K}
024940.9+410319 / 145.76 -16.52 / 344.45 -9.81 / 024628.3+405055	IC 258 / CGCG 539-106 / PGC 10721	.LB.... / U / -2.0± .8	.85± .12 / .06± .04 / / .87	35 / .00 / /	15.3 ±.2 / / / 14.84				5386± 79 / 5502 / 5199
0249.7 -0031 / 174.74 -51.09 / 304.64 -27.68 / 0247.2 -0044	A 0247+00 / MCG 0- 8- 25 (1) / PGC 10726	.SBT3P? / PE / 3.3±1.0 / 4.2± .8	.99± .07 / .15± .05 / / 1.01	55 / .15 / .21 / .08	14.80 ±.13				
024945.5+380405 / 147.23 -19.16 / 341.87 -11.35 / 024636.8+375141	UGC 2305 / PGC 10727	.SX.6.. / U / 6.0± .8	1.14± .07 / .18± .06 / / 1.17	/ .31 / .26 / .09	15.02 ±.19 / / / 14.43				5408 / / 5517 / 5216
024945.9+465837 / 143.02 -11.23 / 349.51 -6.74 / 024624.4+464613	IC 257 / UGC 2298 / PGC 10729	.L..-*. / U / -3.0±1.1	1.34± .12 / .13± .08 / .80± .11 / 1.44	155 / .94 / .00 /	13.8 ±.3 / 13.4 ±.3	1.18± .02	1.20± .01 / / 13.31± .38 / 15.06± .66		6178± 79
024946.0+410308 / 145.78 -16.51 / 344.46 -9.83 / 024633.4+405044	IC 259 / UGC 2306 / PGC 10730	.LB?...	1.15± .15 / .07± .08 / / 1.18	35 / .00 / /	15.0 ±.3 / / / 14.52				6178± 79 / 6294 / 5991
024946.3-522251 / 269.87 -56.47 / 244.27 -29.77 / 024811.0-523513	ESO 154- 16 / IRAS02482-5235 / PGC 10731	.LB?...	.95± .07 / .20± .05 / / .92	.00 / .00 / /	15.35 ±.14 / 13.70 / / 15.15				13005± 63 / 12871 / 12903
024948.1+412747 / 145.59 -16.14 / 344.81 -9.62 / 024635.0+411523	MCG 7- 6- 74 / PGC 10732	.L?....	.82± .19 / .00± .07 / / .86	.32 / .00 / /	14.6 ±.3 / / / 14.22				5245± 58 / 5362 / 5059
024952.0+345918 / 148.79 -21.87 / 339.19 -12.92 / 024647.0+344654	CGCG 524- 26 / MK 1058 / PGC 10739	.S?....	.87± .08 / .32± .12 / / .89	.25 / .48 / .16 /	15.38 ±.12 / 13.47 / / 14.63				5138± 27 / 5240 / 4941
0249.8 +2243 / 155.65 -32.52 / 328.06 -18.78 / 0247.0 +2231	UGC 2315 / PGC 10740	.S..4.. / U / 4.0±1.0	1.00± .08 / .63± .06 / / 1.06	140 / .61 / .92 / .31					5550 / / 5620 / 5337
024952.7+142410 / 161.26 -39.50 / 320.07 -22.37 / 024707.6+141147	CGCG 440- 17 / PGC 10741		.95? / .48? / / .98	/ .31 / /	15.50 ±.12				8681 / / 8727 / 8464
0249.9 -0059 / 175.32 -51.39 / 304.15 -27.87 / 0247.4 -0112	UGC 2319 / PGC 10747	.S..6*. / U / 6.0±1.3	1.10± .05 / .57± .05 / / 1.11	47 / .16 / .84 / .28	15.30 ±.12				
024958.7-120952 / 190.46 -58.46 / 291.65 -30.63 / 024734.7-122215	NGC 1118 / MCG -2- 8- 11 / IRAS02475-1222 / PGC 10748	PLBR0?. / E / -2.0± .8	1.33± .04 / .48± .05 / / 1.26	90 / .02 / .00 /	13.33				
025004.5+412322 / 145.67 -16.18 / 344.78 -9.70 / 024651.4+411059	UGC 2313 / PGC 10754	.S..6*. / U / 6.0±1.3	1.14± .07 / .76± .06 / / 1.17	141 / .32 / 1.12 / .38	15.5 ±.3 / / / 14.04				5729± 58 / 5846 / 5543
0250.1 +3912 / 146.74 -18.11 / 342.90 -10.84 / 0247.0 +3900	UGC 2316 / PGC 10758	.L..... / U / -2.0± .8	1.04± .18 / .13± .08 / / 1.06	10 / .35 / .00 /					
0250.2 +1821 / 158.55 -36.17 / 323.96 -20.79 / 0247.4 +1809	UGC 2321 / PGC 10759	.S..2.. / U / 2.0± .9	1.03± .06 / .60± .05 / / 1.06	80 / .35 / .73 / .30	15.48 ±.12 / / / 14.33				6417 / / 6475 / 6202
0250.2 +0006 / 174.18 -50.55 / 305.36 -27.61 / 0247.7 -0006	UGC 2324 / PGC 10761	.SBR3.. / U / 3.0± .9	.96± .09 / .11± .06 / / .97	5 / .13 / .15 / .05	14.66 ±.12				
0250.2 +0810 / 166.34 -44.45 / 313.85 -24.88 / 0247.6 +0758	UGC 2323 / PGC 10762	.S.R5.. / U / 5.0± .9	.95± .07 / .09± .05 / / .98	.40 / .14 / .05 /	15.27 ±.15 / / / 14.69				8049 / / 8077 / 7831

R.A. 2000 DEC. l　　b SGL　SGB R.A. 1950 DEC.	Names PGC	Type S_T　n_L T L	$\log D_{25}$ $\log R_{25}$ $\log A_e$ $\log D_o$	p.a. A_g A_i A_{21}	B_T m_B m_{FIR} B_T^o	$(B-V)_T$ $(U-B)_T$ $(B-V)_T^o$ $(U-B)_T^o$	$(B-V)_e$ $(U-B)_e$ m'_e m'_{25}	m_{21} W_{20} W_{50} HI	V_{21} V_{opt} V_{GSR} V_{3K}
0250.2　+4732 142.84　-10.69 350.03　-6.52 0246.9　+4720	UGC　2314 PGC 10764	.I..9*. U 10.0±1.1	1.16± .13 .00± .12 1.25	.93 .00 .00					
025017.7-083550 185.12　-56.35 295.74　-29.95 024750.8-084811	MCG -2- 8- 12 PGC 10766	.S..5./ E 5.0± .6	1.34± .04 .83± .04 1.35	87 .12 1.25 .42					
025020.0+190649 158.07　-35.52 324.70　-20.48 024730.7+185427	MCG 3- 8- 21 PGC 10768	.S?....	.92± .11 .38± .07 .95	.34 .57 .19	14.81 ±.14 13.89				1238 1298 1023
025024.2+464152 143.24　-11.43 349.33　-6.98 024703.0+462930	UGC　2317 PGC 10770	.S..1.. U 1.0± .8	1.19± .05 .37± .05 1.28	20 .93 .37 .18	14.6 ±.3				
025026.1-313333 229.83　-63.77 268.69　-32.52 024820.1-314554	ESO　416- 34 PGC 10773	.SAR0.. r -.1± .9	1.06± .05 .16± .03 1.05	118 .00 .12	14.60 ±.14 14.38				7104± 30 7017 6941
025030.3-345152 237.24　-63.45 264.73　-32.42 024828.1-350412	ESO　356- 14 PGC 10777		1.41± .05 .21± .06 1.42	130 .08	14.75 ±.14				10894± 60 10798 10739
025031.2+433930 144.67　-14.13 346.76　-8.59 024714.8+432708	UGC　2318 PGC 10778	.S..4.. U 4.0± .8	1.02± .06 .06± .05 1.07	 .53 .09 .03	15.0 ±.3				
025032.2-312346 229.46　-63.76 268.89　-32.54 024826.0-313606	ESO　416- 35 PGC 10779	RSAT0.. r -.1± .9	.98± .05 .08± .04 .97	 .00 .06	14.48 ±.14 14.32				6675± 43 6588 6511
0250.5　+1319 162.24　-40.29 319.07　-22.96 0247.8　+1307	NGC　1116 UGC　2326 PGC 10781	.S..2.. U 2.0± .9	1.13± .05 .71± .05 1.17	27 .42 .87 .35	15.20 ±.12 13.83				7701 7744 7484
025036.1-313110 229.74　-63.74 268.74　-32.55 024830.0-314330	ESO　416- 36 IRAS02484-3143 PGC 10785	.S?....	1.00± .05 .18± .04 1.00	20 .00 .26 .09	14.49 ±.14 13.23 14.18				7151± 62 7064 6988
0250.6　+2040 157.12　-34.16 326.22　-19.86 0247.7　+2028	UGC　2327 IRAS02477+2028 PGC 10787	.S..3*. U　　(1) 3.0±1.2 3.5±1.1	.99± .06 .07± .05 1.04	65 .48 .10 .04	14.67 ±.13 13.28 14.04				6305 6369 6092
025039.2-014358 176.36　-51.79 303.40　-28.25 024806.8-015619	NGC　1121 UGC　2332 ARAK　93 PGC 10789	.L...*. E -2.0±1.3	.94± .06 .39± .04 .90	10 .15 .00	13.92 ±.19 13.73				2597± 50 2595 2384
025039.9-025916 177.84　-52.65 302.03　-28.61 024808.5-031137	MCG -1- 8- 13 PGC 10790	.SBT7P? E　　(1) 7.0±1.7 6.4± .8	1.08± .06 .08± .05 1.09	 .11 .11 .04					
025040.6+414018 145.64　-15.88 345.08　-9.65 024727.0+412757	NGC　1106 UGC　2322 IRAS02474+4127 PGC 10792	.LA.+.. U -1.0± .8	1.25± .06 .00± .03 1.28	 .32 .00	13.3 ±.3 12.96 12.96				4205± 39 4321 4020
025044.3-064447 182.66　-55.12 297.86　-29.62 024815.9-065708	MCG -1- 8- 14 IRAS02482-0657 PGC 10794	.S..3P? E 3.0±1.8	1.12± .06 .37± .05 1.13	170 .13 .51 .18	 13.55				6997 6980 6789
0250.7　+1600 160.33　-38.05 321.74　-21.91 0248.0　+1548	UGC　2329 PGC 10797	.S..2.. U 2.0± .9	.96± .09 .09± .06 .98	177 .23 .11 .04	14.73 ±.12 14.28				9984 10035 9768
025048.5+333047 149.74　-23.07 337.99　-13.83 024745.1+331826	CGCG 505- 55 PGC 10804			.31	15.7 ±.4			17.10±.3 382± 7	11985± 10 12083 11786

R.A. 2000 DEC.　　l　b　　SGL　SGB　R.A. 1950 DEC.	Names　　　PGC	Type　S_T　n_L　T　L	$\log D_{25}$　$\log R_{25}$　$\log A_e$　$\log D_o$	p.a.　A_g　A_i　A_{21}	B_T　m_B　m_{FIR}　B_T^o	$(B-V)_T$　$(U-B)_T$　$(B-V)_T^o$　$(U-B)_T^o$	$(B-V)_e$　$(U-B)_e$　m'_e　m'_{25}	m_{21}　W_{20}　W_{50}　HI	V_{21}　V_{opt}　V_{GSR}　V_{3K}
025050.9+224621　155.84 -32.36　328.22 -18.96　024758.3+223400	MCG 4- 7- 26　PGC 10806	.L?....　　1.00	1.02± .14　.55± .07　1.00	.59　.00	15.36 ±.10　14.68				6153± 10　6223　5942
025054.1-545829　273.27 -54.72　241.31 -29.30　024925.0-551048	NGC 1136　ESO 154- 19　FAIR 732　PGC 10807	.SBR1$.　R　(1)　1.0± .4　3.3± .8	1.16± .03　.08± .04　.81± .01　1.17	80　.08　.08　.04	13.75 ±.13　13.80 ±.14　13.54	.78± .01　.70	.86± .01　13.29± .03　14.20± .21		5574± 32　5435　5482
025100.3+372758　147.76 -19.58　341.48 -11.87　024752.2+371538	UGC 2328　PGC 10810	.E.....　U　-5.0± .8	1.23± .14　.08± .08　1.25	70　.27　.00	13.46 ±.10　13.11				4962± 52　5069　4770
025101.0+465716　143.22 -11.15　349.61 -6.93　024739.2+464455	IC 260　UGC 2325　PGC 10812	.E...*.　U　-5.0±1.2	1.15± .15　.19± .08　1.24	175　.93　.00	14.1 ±.3				
025104.5+042709　169.96 -47.20　310.07 -26.39　024827.3+041450	A 0248+04　MCG 1- 8- 8　MK 600　PGC 10813		.64± .17　.28± .07　.65	.19	15.40 ±.20			15.29±.2　111± 6　71± 5	1004± 7　975± 45　1020　788
025109.3+023535　171.82 -48.58　308.11 -27.03　024833.6+022316	UGC 2338　PGC 10815	.SBS7*.　UE　(1)　6.5± .8　6.4± .8	1.16± .04　.28± .04　1.18	67　.25　.39　.14	14.30 ±.13　13.65				4499　4510　4284
0251.1 -1442　195.00 -59.52　288.76 -31.38　0248.8 -1455	MCG -3- 8- 32　PGC 10816	.S?....　　1.03	1.04± .09　.21± .07	.09　.16					8697　8656　8500
0251.1 +0804　166.67 -44.40　313.84 -25.13　0248.5 +0752	UGC 2336　PGC 10818	.SBS6..　U　6.0± .8	1.20± .06　.31± .06　1.24	165　.40　.46　.16	15.1 ±.2　14.19				6623　6650　6406
0251.2 +2524　154.33 -30.06　330.71 -17.82　0248.3 +2512	UGC 2333　PGC 10819	.S..4..　U　4.0± .9	.93± .07　.32± .05　.99	32　.64　.47　.16	15.52 ±.12　14.37				7184　7260　6975
0251.2 +4134　145.78 -15.92　345.05 -9.79　0248.0 +4122	UGC 2330　PGC 10820	.E.....　U　-5.0± .8	1.11± .16　.20± .08　1.10	80　.32　.00					
025118.5-281206　222.28 -63.50　272.73 -32.70　024909.1-282424	ESO 416- 37　IRAS02491-2824　PGC 10829	.S..3*.　S　3.0±1.3	1.17± .04　.58± .03　1.17	19　.01　.81　.29	14.11 ±.14　12.31　13.25				5206± 52　5127　5035
0251.3 +4712　143.16 -10.90　349.85 -6.85　0248.0 +4700	UGC 2331　PGC 10832	.S..1..　U　1.0±1.0	1.00± .08　.69± .06　1.09	42　.93　.70　.34					
0251.4 +1601　160.49 -37.95　321.83 -22.06　0248.7 +1548	UGC 2339　IRAS02487+1548　PGC 10834	.S..8..　U　8.0± .8	1.01± .09　.13± .06　1.03	45　.23　.16　.06	13.16				16469　16519　16254
0251.5 +1334　162.31 -39.94　319.43 -23.08　0248.8 +1322	UGC 2340　PGC 10836	.SB.6*.　U　6.0±1.3	1.04± .06　.26± .05　1.08	80　.41　.38　.13					10052　10095　9836
025135.9-254201　216.77 -63.12　275.74 -32.67　024924.0-255418	NGC 1124　ESO 480- 7　PGC 10838	RLB.+*.　Sr　-.6± .6	1.02± .05　.13± .04　1.00	.00　.00	14.88 ±.14				
0251.6 -0044　175.52 -50.91　304.56 -28.20　0249.1 -0057	UGC 2343　PGC 10843	.S..1*.　U　1.0±1.2	1.00± .08　.14± .06　1.02	.18　.14　.07	14.85 ±.13				
025142.8-424501　253.44 -61.06　255.32 -31.90　024951.0-425718	ESO 247- 8　PGC 10849	.LX.-..　S　-3.0± .6	1.10± .05　.14± .04　1.08	20　.00　.00	14.31 ±.14　14.23				5470　5355　5339

R.A. 2000 DEC. l b SGL SGB R.A. 1950 DEC.	Names PGC	Type S_T n_L T L	$\log D_{25}$ $\log R_{25}$ $\log A_e$ $\log D_o$	p.a. A_g A_i A_{21}	B_T m_B m_{FIR} B_T^o	$(B-V)_T$ $(U-B)_T$ $(B-V)_T^o$ $(U-B)_T^o$	$(B-V)_e$ $(U-B)_e$ m'_e m'_{25}	m_{21} W_{20} W_{50} HI	V_{21} V_{opt} V_{GSR} V_{3K}
025143.2+424943 145.27 -14.77 346.17 -9.21 024827.8+423724	IC 262 UGC 2335 PGC 10850	.LB.... U -2.0± .7 	1.20± .09 .06± .05 .81± .03 1.24	50 .37 .00 	 14.2 ±.3 				
0251.7 -1637 198.52 -60.28 286.54 -31.82 0249.4 -1650	NGC 1125 MCG -3- 8- 35 PGC 10851	PSBR0?. SE .0± .7 	1.26± .03 .30± .05 .81± .03 1.25	 .02 .22 	13.43 ±.14 12.97± .09 13.83± .25	.82± .02 .29± .04 	.86± .02 .26± .04 		
0251.9 -0110 176.07 -51.17 304.11 -28.38 0249.3 -0122	A 0249-01 UGC 2345 DDO 30 PGC 10854	.SBT9*. UE (2) 9.0± .6 8.8± .6	1.54± .04 .06± .05 1.26± .06 1.56	160 .17 .06 .03	14.2 ±.2 13.98	.58± .06 -.23± .08 .52 -.27	.56± .05 -.24± .06 16.01± .13 16.62± .33	13.82±.1 112± 7 97± 12 -.20	1506± 6 1505 1294
0251.9 +0040 174.05 -49.85 306.12 -27.84 0249.4 +0028	 UGC 2346 PGC 10857	.S..4.. U 4.0± .9 	1.03± .06 .31± .05 1.05	107 .20 .45 .15	 15.30 ±.10 14.59				8773 8777 8560
025158.8-332020 233.77 -63.33 266.54 -32.80 024955.0-333236	IC 1862 ESO 356- 15 IRAS02499-3332 PGC 10858	.SA.4*/ BS 3.7± .7 	1.48± .04 1.01± .04 1.49	3 .03 1.47 .50	 14.49 ±.14 12.93 12.95				6383 6290 6226
0252.1 +0423 170.31 -47.08 310.10 -26.66 0249.5 +0411	 UGC 2352 PGC 10861	.I..9.. U 10.0±1.0 	1.00± .08 .73± .06 1.01	25 .15 .54 .36					1815 1830 1601
0252.1 +1208 163.57 -41.02 318.06 -23.79 0249.4 +1156	 UGC 2347 PGC 10862	.S?.... 	1.03± .06 .36± .05 1.06	3 .39 .53 .18					8221 8259 8006
0252.1 +1359 162.15 -39.52 319.91 -23.04 0249.4 +1347	 UGC 2348 PGC 10863	.L...?. U -2.0±1.6 	1.11± .16 .00± .08 1.15	 .32 .00 	 14.58 ±.15 				
025218.6-011747 176.32 -51.19 304.02 -28.52 024946.0-013003	NGC 1126 MCG 0- 8- 38 PGC 10868	.S..3*/ E 3.0±1.4 	.87± .08 .57± .05 .89	135 .17 .79 .28	 15.43 ±.15 				
025219.9+440355 144.78 -13.62 347.28 -8.66 024902.5+435139	 CGCG 539-115 MK 1060 PGC 10869	 	 	 .60 	 14.9 ±.3 				9394± 55 9515 9216
025220.8-575652 276.83 -52.61 237.96 -28.67 025100.0-580906	 ESO 116- 1 IRAS02509-5809 PGC 10870	PSBR4.. Sr (1) 3.8± .4 1.1± .5	1.15± .07 .10± .05 .78± .01 1.15	154 .00 .15 .05	14.00 ±.13 16.63 ±.14 13.72 15.01	.76± .01 .68 	.84± .01 13.39± .03 14.37± .39		8786± 40 8641 8705
025222.5-314903 230.38 -63.35 268.37 -32.93 025017.1-320118	 ESO 416- 39 PGC 10872	.L?.... 	.87± .06 .10± .03 .85	 .00 .00 	 14.94 ±.14 14.84				6793± 62 6704 6632
0252.3 +0159 172.78 -48.82 307.57 -27.52 0249.8 +0147	 UGC 2355 PGC 10874	.I..9*. U 10.0±1.2 	1.14± .13 .23± .12 1.16	 .21 .17 .11					4237 4245 4024
025223.4-083040 185.56 -55.89 295.98 -30.44 024956.5-084255	 MCG -2- 8- 14 PGC 10875	.S..1?/ E .5±1.3 	1.14± .05 .75± .04 1.16	99 .14 .76 .37					5029 5006 4825
025227.6-304639 228.06 -63.36 269.63 -32.96 025021.0-305854	 ESO 416- 40 IRAS02503-3058 PGC 10878	.S?.... 	.99± .05 .18± .04 .99	58 .00 .24 .09	 14.67 ±.14 14.37				6815± 62 6729 6652
0252.4 -1431 194.98 -59.16 289.04 -31.66 0250.1 -1444	 MCG -3- 8- 37 PGC 10879	.SB?... 	1.04± .09 .14± .07 1.05	 .09 .21 .07					4055 4014 3859
0252.6 +4119 146.15 -16.02 344.97 -10.15 0249.4 +4107	 UGC 2349 PGC 10882	.L..... U -2.0± .9 	1.07± .07 .30± .03 1.09	48 .58 .00 					4305± 88 4420 4122

R.A. 2000 DEC. l b SGL SGB R.A. 1950 DEC.	Names PGC	Type S_T n_L T L	$\log D_{25}$ $\log R_{25}$ $\log A_e$ $\log D_o$	p.a. A_g A_i A_{21}	B_T m_B m_{FIR} B_T^o	$(B-V)_T$ $(U-B)_T$ $(B-V)_T^o$ $(U-B)_T^o$	$(B-V)_e$ $(U-B)_e$ m'_e m'_{25}	m_{21} W_{20} W_{50} HI	V_{21} V_{opt} V_{GSR} V_{3K}
025240.7+412347		.S..3..	.99± .06						
146.12 -15.95	UGC 2350	U (1)	.02± .05	.58	14.6 ±.3				3924± 88
345.04 -10.12	IRAS02494+4111	3.0± .8		.03	13.58				4039
024927.1+411132	PGC 10884	5.5±1.1	1.05	.01	13.97				3741
025244.6+465615		.SBS3..	1.20± .06						
143.50 -11.03	UGC 2351	U	.06± .06	.93	13.8 ±.3			15.82±.1	8420± 11
349.75 -7.19	IRAS02493+4644	3.0± .7		.08				294± 8	8547
024922.4+464400	PGC 10886		1.29	.03	12.75			3.04	8249
025246.8-143147	NGC 1139	.SBR0P*	1.07± .07	36					
195.06 -59.09	MCG -3- 8- 38	E	.15± .04	.09					
289.05 -31.73		.0± .9		.11					
025024.9-144401	PGC 10888		1.07						
0252.8 +1315	NGC 1127	RSBR2*.	.90± .07						9775
162.89 -40.02	UGC 2356	U	.00± .05	.40	15.27 ±.15				
319.25 -23.50		2.0± .9		.00					9816
0250.1 +1303	PGC 10889		.94	.00	14.76				9561
025251.2+421220	NGC 1122	.SX.3..	1.24± .05	40				17.37±.1	3599± 11
145.76 -15.22	UGC 2353	U	.12± .05	.40	12.9 ±.3				3624± 39
345.75 -9.72	IRAS02496+4200	3.0± .8		.17	12.15			205± 8	3718
024936.4+420005	PGC 10890		1.28	.06	12.28			5.04	3419
025251.6-011633	NGC 1132	.E.....	1.40± .06	140	13.25 ±.16	.99± .03	1.03± .02		
176.45 -51.07	UGC 2359	UE	.27± .05	.18	13.35 ±.12	.50± .04	.55± .03		6951± 26
304.09 -28.64		-4.5± .5	1.05± .05	.00		.88	13.99± .19		6949
025018.9-012847	PGC 10891		1.35		13.03	.50	14.58± .37		6740
025253.7-244710			.79± .06	66					
214.92 -62.66	ESO 480- 8		.12± .04	.00	15.00 ±.14				3798±104
276.86 -32.92									3727
025041.0-245924	PGC 10893		.79						3622
025307.3+252914	IC 1861	.LA.0..	1.16± .05	150				15.85±.1	6703± 5
154.71 -29.77	UGC 2357	U	.15± .04	.62	14.33 ±.12			439± 6	
330.99 -18.16		-2.0± .8		.00					6779
025011.9+251700	PGC 10905		1.21		13.60				6497
0253.1 +1302		.I..9..	1.09± .07					15.10±.3	3625± 9
163.13 -40.15	UGC 2362	U	.04± .06	.41	14.90 ±.20			140± 10	
319.07 -23.65	VV 606	10.0± .8		.03				110± 7	3665
0250.4 +1250	PGC 10907		1.13	.02	14.46			.62	3411
025320.3-173606		.S?....	.86± .05						
200.62 -60.34	ESO 546- 28		.15± .04	.05	15.09 ±.14				11022± 60
285.46 -32.33				.18					10971
025101.1-174818	PGC 10912		.87	.07	14.75				10832
0253.3 +0632		.S?....	1.12± .05	103				15.70±.3	5430± 10
168.63 -45.24	UGC 2364		.56± .05	.30				354± 13	
312.47 -26.19				.83				344± 10	5451
0250.7 +0620	PGC 10913		1.15	.28					5216
025322.7+314117		.SA.7*.	1.07± .14					15.84±.3	10562± 10
151.23 -24.40	UGC 2360	U	.07± .12	.58					
336.64 -15.20		7.0±1.2		.09				336± 7	10654
025021.0+312904	PGC 10914		1.12	.03					10363
025329.4-305136		.SXT4*.	1.20± .04	94					
228.24 -63.14	ESO 416- 41	Sr (1)	.43± .03	.00	14.37 ±.14				6415± 39
269.53 -33.18	IRAS02513-3103	4.2± .5		.63					6328
025123.1-310348	PGC 10919	3.3± .7	1.20	.21	13.70				6253
025333.8-830833		.LAR0*.	1.12± .05						
298.79 -32.97	ESO 3- 13	S	.06± .03	.47	13.74 ±.14				
212.43 -19.01	IRAS02575-8320	-2.0± .8		.00	12.61				
025731.0-832036	PGC 10922		1.17						
025334.7+415309		.SXS3..	1.12± .05	60					
146.04 -15.44	UGC 2361	U	.07± .05	.43	14.1 ±.3				7156± 88
345.55 -10.00	IRAS02503+4141	3.0± .8		.09	12.51				7272
025020.2+414056	PGC 10923		1.16	.03	13.49				6975
025339.6-341149	IC 1864	.E?....	1.07± .04	63	12.55V±.14	.98± .02	.99± .02		
235.58 -62.90	ESO 356- 17		.21± .03	.03	13.66 ±.14	.35± .05	.40± .04		4488± 26
265.49 -33.11			.65± .03	.00		.93			4392
025137.0-342400	PGC 10925		1.02		13.50	.37	13.37± .26		4335
025340.2+414333									
146.13 -15.57	CGCG 540- 2			.44	15.5 ±.3				6149± 88
345.42 -10.10									6264
025025.9+413120	PGC 10926								5968

R.A. 2000 DEC. / l b / SGL SGB / R.A. 1950 DEC.	Names / PGC	Type / S_T n_L / T / L	$\log D_{25}$ / $\log R_{25}$ / $\log A_e$ / $\log D_0$	p.a. / A_g / A_i / A_{21}	B_T / m_B / m_{FIR} / B_T^o	$(B-V)_T$ / $(U-B)_T$ / $(B-V)_T^o$ / $(U-B)_T^o$	$(B-V)_e$ / $(U-B)_e$ / m'_e / m'_{25}	m_{21} / W_{20} / W_{50} / HI	V_{21} / V_{opt} / V_{GSR} / V_{3K}
0253.6 +0616		.L.....	1.27± .04	145	14.80S±.15			15.27±.3	7386± 10
168.97 -45.40	UGC 2367	U	.53± .03	.31	14.80 ±.12			594± 13	7406
312.22 -26.36	IRAS02510+0603	-2.0± .8		.00	13.26			581± 10	7173
0251.0 +0603	PGC 10927		1.22		14.38		14.68± .27		
025341.2+130055	NGC 1134	.S?....	1.40± .04	148				13.56±.2	3644± 5
163.29 -40.09	UGC 2365		.46± .05	.41	13.05 ±.12			446± 7	
319.11 -23.78	ARP 200			.69	10.71			395± 5	3684
025056.9+124843	PGC 10928		1.44	.23	11.93			1.40	3431
025350.2+125054	IC 267	PSBS3..	1.31± .04	15	13.87 ±.16	.84± .03	.94± .02	14.25±.3	3570± 9
163.46 -40.20	UGC 2368	U	.11± .05	.41	13.63 ±.16	.03± .05	.16± .04	389± 10	3577± 50
318.96 -23.88	IRAS02511+1238	3.0± .8	.96± .03	.15	11.51	.70	14.16± .07	343± 7	3610
025106.1+123843	PGC 10932		1.34	.06	13.16	-.08	14.97± .28	1.03	3357
025352.4+021724		.S..2*/	1.11± .05	17					
172.88 -48.34	UGC 2371	UE	.58± .04	.27	15.37 ±.13				
308.04 -27.77	IRAS02513+0205	2.3± .7		.72					
025116.9+020513	PGC 10933		1.14	.29					
025353.7-724535		PSXR3*.	1.09± .05						8053
291.31 -41.49	ESO 31- 4 (1)	Sr	.08± .04	.11	14.29 ±.14				7889
222.41 -23.58		3.4± .5		.11					8026
025349.0-725742	PGC 10934	5.6± .6	1.10	.04	14.01				
0253.9 +0559		.SXT6*.	.95± .07					16.03±.3	7910± 10
169.30 -45.57	UGC 2372	U	.05± .05	.31	15.08 ±.14			285± 13	
311.95 -26.52		6.0± .9		.07				261± 10	7929
0251.3 +0547	PGC 10935		.98	.02	14.67			1.33	7697
025400.1+362551		.L.....	1.12± .07	63					
148.84 -20.20	UGC 2366	U	.25± .03	.44	15.20 ±.15				
340.89 -12.91		-2.0± .8		.00					
025052.7+361340	PGC 10941		1.13						
025402.4+025743	NGC 1137	PSAT3..	1.33± .03	20				13.99±.3	3042± 10
172.24 -47.82	UGC 2374 (1)	UE	.21± .04	.28	13.21 ±.12			290± 13	3004± 50
308.77 -27.59	IRAS02514+0245	2.5± .5		.29	12.52			275± 10	3051
025126.4+024532	PGC 10942	4.5±1.0	1.36	.10	12.62			1.27	2829
0254.0 +0615		.S?....	1.02± .06	114	15.6 ±.4	1.17± .05		16.41±.3	7607± 10
169.08 -45.35	UGC 2375		.60± .05	.31	15.35 ±.12	.52± .08		450± 13	
312.24 -26.45				.90		.95		440± 10	7627
0251.4 +0603	PGC 10943		1.05	.30	14.12	.27	14.04± .52	1.99	7394
025411.9-525451		.LA.-..	1.05± .05	140					
269.91 -55.61	ESO 154- 22	S	.19± .04	.00	15.42 ±.14				
243.48 -30.29		-3.0± .8		.00					
025239.0-530700	PGC 10945		1.02						
025422.7+311737		.S..6*.	1.09± .08					15.64±.3	5422± 10
151.65 -24.64	UGC 2376	U	.02± .06	.60					5512
336.40 -15.59		6.0±1.1		.03				236± 7	5224
025121.2+310527	PGC 10950		1.15	.01					
0254.4 +1144		.S..6*.	1.17± .05	174					
164.48 -41.01	UGC 2378	U	.75± .05	.49					
317.90 -24.46		6.0±1.3		1.10					
0251.7 +1132	PGC 10952		1.22	.37					
025426.4+423901		.S..8?.	1.44± .05	127				15.26±.1	2162± 11
145.81 -14.69	UGC 2370	U	1.45± .06	.44	15.4 ±.3				
346.28 -9.73		8.0±1.9		1.23				194± 8	2279
025110.6+422651	PGC 10956		1.48	.50	13.77			.99	1983
025427.7+413449	NGC 1129	.E.....	1.60± .08	90	*		1.10± .01	
146.34 -15.63	UGC 2373	U	.11± .08	.44	13.5 ±.3				5202± 28
345.37 -10.30		-5.0± .6	1.37± .10	.00			14.24± .27		5317
025113.5+412239	PGC 10959		1.64		13.03				5021
025429.1-471040		.LA.-..	.99± .06	120	*				
260.96 -58.68	ESO 247- 10	S	.18± .04	.00	15.15 ±.14				
250.04 -31.63		-3.0± .9		.00					
025245.0-472248	PGC 10960		.96						
025433.2-183809	NGC 1145	.S..5*/	1.51± .02	60	13.63 ±.16	1.16± .03	1.22± .02	14.35±.1	1968± 11
202.78 -60.49	ESO 546- 29	SUE (1)	.80± .03	.07	13.55 ±.14	.14± .04	.23± .04		1914
284.27 -32.75	IRAS02522-1850	5.0± .7	1.01± .03	1.20	12.76	.99	14.17± .09	87± 12	1782
025215.0-185018	PGC 10965	3.3±1.2	1.51	.40	12.30	-.07	14.00± .21	1.65	
025433.6-100137	NGC 1140	.IB.9P*	1.22± .03	6	12.84 ±.14	.35± .01	.37± .01	13.37±.0	1509± 4
188.32 -56.34	MCG -2- 8- 19	RE	.26± .04	.11	12.9 ±.3	-.43± .03	-.46± .03	223± 4	1498± 33
294.38 -31.30	MK 1063	10.0± .4	.46± .04	.19	11.98	.25	10.63± .12	184± 5	1479
025208.0-101346	PGC 10966		1.23	.13	12.55	-.50	13.14± .22	.68	1309

R.A. 2000 DEC. / l b / SGL SGB / R.A. 1950 DEC.	Names / PGC	Type / S_T n_L / T / L	$\log D_{25}$ / $\log R_{25}$ / $\log A_e$ / $\log D_o$	p.a. / A_g / A_i / A_{21}	B_T / m_B / m_{FIR} / B_T^o	$(B-V)_T$ / $(U-B)_T$ / $(B-V)_T^o$ / $(U-B)_T^o$	$(B-V)_e$ / $(U-B)_e$ / m'_e / m'_{25}	m_{21} / W_{20} / W_{50} / HI	V_{21} / V_{opt} / V_{GSR} / V_{3K}
0254.6 +1201 164.31 -40.75 318.21 -24.39 0251.9 +1149	 UGC 2381 PGC 10968	.S?.... 	1.02± .08 .38± .06 1.06	135 .44 .57 .19					7656 7692 7444
0254.6 +0922 166.50 -42.84 315.53 -25.45 0252.0 +0910	 UGC 2382 PGC 10973	.S..6*. U 6.0±1.4 	1.00± .08 .84± .06 1.05	35 .50 1.24 .42					
025441.7-302546 227.29 -62.87 270.04 -33.45 025235.0-303754	 ESO 417- 1 PGC 10974	.LA.0P. S -2.5± .6 	1.26± .04 .26± .04 1.22	0 .00 .00 	 14.20 ±.14 14.10				6502± 39 6416 6341
025442.5-520547 268.65 -56.03 244.38 -30.57 025308.1-521754	 ESO 199- 5 PGC 10977	.S..4?/ S (1) 4.0±1.9 5.6±1.3	1.06± .05 .55± .04 1.06	176 .00 .80 .27	 15.45 ±.14 				
025445.2-250634 215.78 -62.31 276.49 -33.36 025233.0-251842	 ESO 480- 12 IRAS02525-2518 PGC 10981	.S?.... 	.80± .06 .21± .04 .80	12 .00 .26 .11	 14.85 ±.14 13.21 14.54				4791±104 4718 4617
025448.0+411854 146.52 -15.84 345.18 -10.50 025134.1+410645	 CGCG 540- 11 PGC 10984			 .44 	 15.1 ±.4 				5825± 88 5939 5644
025449.4-304046 227.84 -62.85 269.74 -33.47 025243.0-305254	 ESO 417- 2 PGC 10985	.L?.... 	1.02± .05 .37± .03 .96	1 .00 .00 	 15.37 ±.14 15.27				6463 6376 6302
025454.1-062225 183.29 -54.08 298.58 -30.53 025225.6-063433	 MCG -1- 8- 15 IRAS02524-0634 PGC 10989	PSBR1.. E 1.0± .8 	1.21± .05 .12± .05 1.23	35 .18 .12 .06	 12.97 				
025456.2-215352 209.18 -61.50 280.37 -33.18 025241.1-220600	 ESO 546- 31 PGC 10991	.SBS6*. S (1) 5.5± .6 5.0± .9	1.12± .04 .33± .03 1.12	30 .06 .48 .16	 14.86 ±.14 14.28				8716 8652 8536
025457.3-170305 199.94 -59.76 286.17 -32.64 025237.7-171512	 MCG -3- 8- 45 PGC 10994	.SBT6*. ES (1) 6.0± .6 5.6± .8	1.23± .04 .07± .04 1.24	164 .10 .11 .04					3062 3012 2873
025503.1-011120 176.96 -50.62 304.37 -29.14 025230.3-012328	 UGC 2385 KUG 0252-013 PGC 11000	.S..5.. U 5.0± .9 	.93± .05 .06± .04 .95	 .18 .09 .03	 15.12 ±.14 				
025509.8-001040 175.86 -49.89 305.48 -28.86 025236.2-002247	NGC 1143 UGC 2388 PGC 11007	.E+..P* UE -4.3± .6 	.96± .10 .06± .06 .97	110 .19 .00 	* 				8459± 30 8459 8251
0255.1 +5154 141.54 -6.43 354.11 -4.83 0251.6 +5142	 UGC 2380 IRAS02516+5142 PGC 11011	.S..3.. U 3.0± .9 	1.16± .07 .34± .06 1.30	9 1.46 .47 .17	* 13.14 			15.32±.1 457± 8 	4629± 11 4765 4471
025512.1-001100 175.87 -49.89 305.48 -28.87 025238.5-002307	NGC 1144 UGC 2389 IRAS02526-0023 PGC 11012	.RING.B R 	1.04± .08 .21± .06 .75± .02 1.06	130 .19 .16 .10	13.78 ±.13 13.35 ±.18 11.24 13.28	.83± .02 .09± .04 .68 -.03	.92± .01 .18± .02 13.02± .05 13.32± .42		8647± 31 8646 8438
025512.7+243801 155.69 -30.25 330.44 -18.98 025217.9+242553	 CGCG 484- 24 PGC 11015			 .50 	 15.6 ±.4 			17.62±.3 103± 7 	6987± 10 7059 6782
0255.2 +1213 164.31 -40.50 318.47 -24.45 0252.5 +1201	 UGC 2387 PGC 11017	.S..6*. U 6.0±1.2 	1.16± .05 .45± .05 1.20	12 .44 .66 .23	 15.22 ±.16 14.09				7751 7788 7540
025520.0+462010 144.18 -11.36 349.48 -7.89 025158.3+460802	 UGC 2383 PGC 11026	.SB.8*. U 8.0±1.2 	1.04± .15 .18± .12 1.12	0 .84 .22 .09				15.36±.3 219± 11 90± 8 	6720± 9 6845 6550

R.A. 2000 DEC. l b SGL SGB R.A. 1950 DEC.	Names PGC	Type S_T n_L T L	$\log D_{25}$ $\log R_{25}$ $\log A_e$ $\log D_o$	p.a. A_g A_i A_{21}	B_T m_B m_{FIR} B_T^o	$(B-V)_T$ $(U-B)_T$ $(B-V)_T^o$ $(U-B)_T^o$	$(B-V)_e$ $(U-B)_e$ m' m'_{25}	m_{21} W_{20} W_{50} HI	V_{21} V_{opt} V_{GSR} V_{3K}
0255.3 +0607 169.55 -45.24 312.23 -26.80 0252.7 +0555	CGCG 415- 25 PGC 11027			.29	15.6 ±.3			16.73±.3 160± 13 143± 10	7455± 10 7473 7244
0255.3 +0849 167.15 -43.16 315.04 -25.81 0252.7 +0837	IC 1865 UGC 2391 PGC 11035	.S..4.. U 4.0± .8 L	1.03± .08 .24± .06 1.08	.54 .36 .12	14.73 ±.12 13.79				8008 8034 7797
025539.8-272525 220.79 -62.47 273.69 -33.64 025330.1-273730	A 0253-27 ESO 417- 3 IRAS02535-2737 PGC 11052	.SXT5*. PS (2) 5.0± .4 1.8± .4	1.24± .04 .07± .04 1.24	.00 .10 .03	13.44 ±.14 13.31				5272± 59 5193 5105
0255.7 +7509 130.77 14.21 13.37 7.92 0250.2 +7456	 UGC 2358 IRAS02502+7456 PGC 11056	.SBR3.. U 3.0± .8	1.24± .05 .25± .05 1.38	25 1.55 .35 .13	15.0 ±.3 13.62 13.04			14.84±.1 422± 8 1.68	4300± 11 4250± 67 4465 4210
0255.7 +3641 149.03 -19.81 341.29 -13.08 0252.6 +3629	 UGC 2390 PGC 11060	.SB.3.. U 3.0± .9	1.06± .06 .14± .05 1.10	.48 .20 .07					
025544.1-141229 195.18 -58.31 289.57 -32.38 025322.1-142435	IC 270 MCG -2- 8- 28 PGC 11061	.LA.-*. E -3.0± .6	1.12± .05 .03± .04 1.13	90 .07 .00					
025546.4+334559 150.58 -22.34 338.74 -14.59 025242.0+333353	 UGC 2392 KUG 0252+335 PGC 11065	.S..6?. U 6.0±1.7	1.27± .03 .60± .04 1.30	12 .37 .88 .30	14.75 ±.13 13.50				1548 1643 1355
025548.8+010440 174.68 -48.89 306.91 -28.62 025314.2+005235	 UGC 2404 KUG 0253+008 PGC 11067	.S..6*. U 6.0±1.5	1.11± .04 1.01± .05 1.14	9 .27 1.47 .50					5038 5041 4830
0255.8 +0612 169.61 -45.10 312.37 -26.89 0253.2 +0600	 UGC 2399 PGC 11068	.SXS6.. U 6.0± .8	1.12± .05 .00± .05 1.15	.29 .00 .00	14.7 ±.2 14.36			16.07±.3 206± 13 194± 10 1.70	8006± 10 8024 7796
025553.8+322006 151.38 -23.57 337.49 -15.34 025251.0+320800	 UGC 2393 PGC 11071	.S..7.. U 7.0± .9	1.00± .08 .33± .06 1.05	87 .53 .46 .17	15.29 ±.12 14.27			16.19±.3 440± 7 1.75	11297± 10 11389 11102
0255.9 +0629 169.38 -44.87 312.68 -26.81 0253.3 +0617	 UGC 2405 PGC 11074	.S..6*. U 6.0±1.2	1.18± .03 .42± .03 1.21	167 .31 .61 .21	14.70S±.15 14.73 ±.13 13.76		14.41± .24	15.92±.3 450± 13 422± 10 1.94	7709± 10 7728 7499
025557.3+004131 175.13 -49.14 306.50 -28.78 025323.0+002926	 UGC 2403 IRAS02533+0029 PGC 11075	.SBT1P* E 1.0±1.3	1.13± .05 .40± .05 1.15	155 .28 .41 .20	14.63 ±.12 11.02 13.90				4143 4145 3935
0255.9 +3808 148.33 -18.52 342.57 -12.36 0252.8 +3756	 UGC 2394 PGC 11076	.S..6*. U 6.0±1.4	1.11± .07 .83± .06 1.14	31 .35 1.22 .42					
0256.0 -0142 177.83 -50.80 303.87 -29.53 0253.5 -0155	 UGC 2412 PGC 11079	.SBS7.. U 7.0± .8	1.00± .08 .00± .06 1.02	.19 .00 .00					4657 4651 4451
0256.0 +1547 161.81 -37.50 322.11 -23.16 0253.3 +1535	 UGC 2406 PGC 11083	PSXS1.. U 1.0± .9	1.00± .06 .15± .05 1.03	65 .28 .15 .08	15.23 ±.16 14.67				10164 10210 9955
025607.4+413752 146.60 -15.44 345.58 -10.54 025252.8+412547	 UGC 2395 PGC 11085	.S..0.. U .0±1.0	1.00± .08 .73± .06 1.00	153 .44 .54	15.2 ±.3 14.15				5989± 62 6103 5809
0256.1 +3704 148.91 -19.43 341.67 -12.95 0253.0 +3652	 UGC 2396 PGC 11087	.S..7*. U 7.0±1.5	1.08± .09 1.03± .04 1.13	.48 1.38 .50					5114± 10 5217 4926

R.A. 2000 DEC. l b SGL SGB R.A. 1950 DEC.	Names PGC	Type S_T n_L T L	$\log D_{25}$ $\log R_{25}$ $\log A_e$ $\log D_o$	p.a. A_g A_i A_{21}	B_T m_B m_{FIR} B_T^o	$(B-V)_T$ $(U-B)_T$ $(B-V)_T^o$ $(U-B)_T^o$	$(B-V)_e$ $(U-B)_e$ m'_e m'_{25}	m_{21} W_{20} W_{50} HI	V_{21} V_{opt} V_{GSR} V_{3K}
025610.8-152352 197.29 -58.77 288.19 -32.68 025349.8-153556	MCG -3- 8- 46 PGC 11090	.SBS3.. E (1) 3.0± .9 4.2± .8	1.13± .06 .24± .05 1.14	176 .07 .34 .12					
025616.2-715111 290.37 -42.10 223.23 -24.12 025605.0-720312	ESO 53- 23 IRAS02560-7203 PGC 11094	PSBT1*. Sr (1) 1.2± .5 3.3±1.4	1.16± .05 .55± .04 1.17	71 .03 .56 .28	15.18 ±.14 13.43 14.48				8317 8153 8288
0256.3 +7505 130.84 14.17 13.33 7.85 0250.9 +7453	CGCG 327- 2 PGC 11098	.L?....	.85± .12 .00± .04 1.02	 1.56 .00	15.2 ±.2 13.63				4075± 67 4241 3986
0256.3 +0432 171.31 -46.27 310.66 -27.59 0253.7 +0420	UGC 2414 PGC 11099	.S..6*. U 6.0±1.4	1.08± .06 .70± .05 1.10	159 .27 1.03 .35	15.33 ±.12 14.00			16.22±.3 478± 13 398± 10 1.86	8268± 10 8281 8059
0256.3 +3701 148.97 -19.46 341.64 -13.01 0253.2 +3649	UGC 2401 PGC 11100	.L...?. U -2.0±1.7	1.04± .18 .04± .08 1.09	 .48 .00					
0256.3 +0609 169.79 -45.06 312.37 -27.02 0253.7 +0557	UGC 2415 PGC 11102	.SBS4.. U 4.0± .9	1.00± .08 .43± .06 1.03	1 .28 .63 .22	15.65 ±.14 15.26 ±.12 14.46	.83± .03 .18± .06 .64 .02	 14.42± .46	16.46±.3 360± 13 339± 10 1.78	7771± 10 7789 7561
025621.6-321109 231.11 -62.49 267.90 -33.76 025417.0-322312	ESO 417- 6 PGC 11104	RSA.0?. r -.1±1.7	1.05± .06 .04± .04 1.05	130 .00 .03	14.29 ±.14 14.18				4884± 24 4792 4729
0256.4 +0052 175.07 -48.91 306.75 -28.85 0253.9 +0040	UGC 2418 IRAS02539+0040 PGC 11112	.SX.4.. U 4.0± .9	1.02± .06 .24± .05 1.05	97 .28 .36 .12	14.89 ±.13				
025632.6-024618 179.20 -51.43 302.74 -29.96 025401.2-025821	A 0254-02 MK 601 PGC 11114	.RINGP* E	.79± .07 .34± .07 .81	175 .20 .50 .17					7043±113 7034 6838
025634.7+484125 143.25 -9.18 351.56 -6.78 025308.6+482921	UGC 2398 PGC 11116	.E..... U -5.0± .8	1.08± .17 .08± .08 1.22	 1.01 .00	15.07 ±.16				
025635.8+430252 145.98 -14.15 346.83 -9.85 025319.0+425048	NGC 1138 UGC 2408 PGC 11118	.LB.... U -2.0± .8	1.03± .11 .09± .05 1.09	 .65 .00	13.8 ±.3				
0256.6 +1842 159.87 -35.04 325.00 -22.00 0253.8 +1830	UGC 2416 PGC 11121	.S..3*. U (1) 3.0±1.2 4.5±1.1	.96± .17 .05± .12 1.00	 .46 .07 .02					9796 9850 9589
025638.6+412001 146.83 -15.66 345.38 -10.78 025324.3+410757	UGC 2410 PGC 11123	.S?....	1.04± .08 .13± .06 1.08	70 .44 .19 .06	15.1 ±.3 14.46				4605± 88 4718 4426
0256.7 +0719 168.81 -44.10 313.64 -26.68 0254.0 +0707	UGC 2419 IRAS02540+0707 PGC 11128	PSBS1.. U 1.0± .8	1.11± .05 .07± .05 1.14	5 .37 .07 .03	14.38 ±.14 13.19 13.84				8143 8164 7934
0256.7 +1558 161.85 -37.26 322.36 -23.23 0254.0 +1546	UGC 2420 PGC 11133	.S..3.. U (1) 3.0±1.0 5.5±1.3	1.04± .08 .76± .06 1.08	150 .44 1.05 .38					9874 9920 9665
0256.8 +5038 142.36 -7.45 353.20 -5.74 0253.3 +5026	UGC 2409 PGC 11134	.S?....	1.00± .08 .55± .06 1.12	67 1.27 .82 .27				14.29±.1 295± 8	3884± 11 4016 3725
0256.8 +0458 171.03 -45.87 311.17 -27.55 0254.2 +0446	UGC 2423 IRAS02541+0446 PGC 11136	.S..6*. U 6.0±1.3	1.06± .06 .55± .05 1.08	141 .25 .81 .28	15.42 ±.12				

R.A. 2000 DEC. l b SGL SGB R.A. 1950 DEC.	Names PGC	Type S_T n_L T L	$logD_{25}$ $logR_{25}$ $logA_e$ $logD_o$	p.a. A_g A_i A_{21}	B_T m_B m_{FIR} B_T^o	$(B-V)_T$ $(U-B)_T$ $(B-V)_T^o$ $(U-B)_T^o$	$(B-V)_e$ $(U-B)_e$ m' m'_{25}	m_{21} W_{20} W_{50} HI	V_{21} V_{opt} V_{GSR} V_{3K}
025652.3-543423 271.80 -54.30 241.49 -30.23 025524.0-544624	A 0255-54 ESO 154- 23 IRAS02554-5445 PGC 11139	.SB?... (1) 10.0± .7	1.93± .02 .68± .05 1.29± .05 1.93	39 .00 .70 .34	12.8 ±.2 12.40 ±.12 13.88 11.80	.36± .05 -.08± .10 .21 -.19	.44± .03 -.11± .07 14.78± .13 15.60± .27	11.89±.3 142± 9 -.25	578± 8 437 488
0256.9 +3620 149.44 -20.00 341.11 -13.47 0253.8 +3608	 UGC 2417 PGC 11140	.I..9.. U 10.0± .8	1.04± .15 .04± .12 1.08	 .48 .03 .02					3643 3744 3455
025705.4-074118 185.63 -54.46 297.23 -31.38 025437.9-075319	NGC 1148 MCG -1- 8- 18 PGC 11148	.SBT7P? E (1) 7.0±1.7 6.4± .8	1.14± .06 .27± .05 1.16	105 .20 .37 .13					
0257.1 +0519 170.78 -45.56 311.57 -27.50 0254.5 +0507	 UGC 2426 PGC 11150	.S..3.. U (1) 3.0± .9 4.5±1.2	.95± .07 .18± .05 .98	65 .29 .24 .09	14.86 ±.12				
0257.1 +1730 160.83 -35.94 323.92 -22.64 0254.3 +1718	 UGC 2424 IRAS02543+1718 PGC 11152	.S..4.. U 4.0± .8	1.35± .04 .59± .05 1.39	158 .44 .87 .30	14.25 ±.13 13.59 12.89				6985 7036 6778
025709.4+011938 174.78 -48.47 307.30 -28.86 025434.6+010737	 UGC 2429 PGC 11153	.SBS9*. UE (1) 9.3± .7 8.7± .8	1.04± .06 .35± .05 1.07	60 .27 .36 .18					1776 1779 1569
0257.1 -0012 176.44 -49.56 305.62 -29.35 0254.6 -0025	 UGC 2428 PGC 11154	.SX.8*. U 8.0± .8	.97± .09 .07± .06 1.00	 .24 .09 .04					4732 4730 4526
025710.8+024632 173.28 -47.42 308.87 -28.39 025434.9+023431	IC 273 UGC 2425 IRAS02546+0234 PGC 11156	.SB.1*/ UE .7± .7	1.17± .04 .47± .04 1.20	31 .28 .48 .24	14.25 ±.12 13.12 13.44				3213± 50 3220 3006
0257.3 +4445 145.26 -12.59 348.34 -9.04 0254.0 +4433	 UGC 2421 PGC 11162	.S..6*. U 6.0±1.2	1.07± .14 .07± .12 1.16	 .97 .10 .03					
0257.3 +4509 145.07 -12.24 348.67 -8.83 0254.0 +4457	 UGC 2422 PGC 11163	.S..7*. U 7.0±1.3	1.11± .07 .54± .06 1.20	120 1.01 .74 .27					
025723.8-001836 176.61 -49.59 305.54 -29.43 025450.4-003036	NGC 1149 MCG 0- 8- 58 PGC 11170	.L..-*. E -3.0±1.4	.75± .10 .06± .05 .77	130 .24 .00	14.98 ±.11				
0257.4 +1029 166.27 -41.55 316.95 -25.63 0254.7 +1017	 UGC 2430 PGC 11173	.S..1.. U 1.0± .9	.95± .07 .10± .05 1.01	75 .65 .11 .05	14.97 ±.13 14.11				7588 7618 7379
025724.4-353406 238.29 -61.95 263.77 -33.78 025524.0-354606	 ESO 356- 20 PGC 11174	.SAR2*. Sr 2.3± .7	1.14± .04 .18± .04 1.15	40 .13 .23 .09	14.56 ±.14 14.13				7426 7325 7280
0257.4 +1026 166.32 -41.57 316.92 -25.66 0254.7 +1014	 UGC 2433 IRAS02547+1014 PGC 11176	.SB.2.. U 2.0± .9	.95± .09 .19± .06 1.01	140 .65 .23 .09					7569 7599 7360
0257.5 +1007 166.60 -41.82 316.59 -25.79 0254.8 +0955	 UGC 2432 PGC 11178	.S..9*. U 9.0±1.2	1.04± .15 .08± .12 1.10	 .62 .08 .04				15.84±.1 90± 6 82± 5	757± 10 786 548
0257.5 +0558 170.28 -45.00 312.29 -27.37 0254.9 +0546	 CGCG 415- 40 PGC 11179				.34 15.0 ±.3				6870± 60 6886 6662
025733.4+413059 146.90 -15.41 345.63 -10.83 025418.7+411858	 CGCG 540- 17 PGC 11181				.44 15.3 ±.4				4853± 58 4966 4675

R.A. 2000 DEC. / l b / SGL SGB / R.A. 1950 DEC.	Names / PGC	Type / S_T n_L / T / L	$\log D_{25}$ / $\log R_{25}$ / $\log A_e$ / $\log D_o$	p.a. / A_g / A_i / A_{21}	B_T / m_B / m_{FIR} / B_T^o	$(B-V)_T$ / $(U-B)_T$ / $(B-V)_T^o$ / $(U-B)_T^o$	$(B-V)_e$ / $(U-B)_e$ / m'_e / m'_{25}	m_{21} / W_{20} / W_{50} / HI	V_{21} / V_{opt} / V_{GSR} / V_{3K}
0257.5 +0807 168.34 -43.36 314.54 -26.58 0254.9 +0755	 UGC 2434 PGC 11183	.SAS6.. U 6.0± .9 	.95± .07 .03± .05 .99	 .42 .04 .01				 	8037± 10 8060 7828
0257.6 +0602 170.24 -44.93 312.37 -27.37 0255.0 +0550	A 0255+05 MCG 1- 8- 27 PGC 11188	.E.0.*. P -5.0±1.0 	.97± .15 .34± .07 .94± .04 .92	 .34 .00 	13.84 ±.15 13.40	1.16± .02 .67± .04 1.02 .62	1.20± .01 .75± .04 14.03± .14 12.82± .79		 6941± 20 6958 6733
025743.5+273500 154.46 -27.45 333.43 -18.06 025445.5+272300	 CGCG 485- 1 PGC 11190			 .65 	 15.7 ±.4 			17.16±.3 166± 7 	10762± 10 10840 10563
025749.0-364302 240.66 -61.68 262.37 -33.77 025550.1-365500	 ESO 356- 22 IRAS02558-3654 PGC 11197	.SBR5.. Sr (1) 4.5± .5 2.8± .6	1.32± .04 .20± .05 1.04± .02 1.33	152 .09 .30 .10	13.12 ±.15 13.30 ±.14 12.63 12.79	.69± .02 .59 	.76± .01 13.81± .06 14.05± .27		6208± 10 6170± 45 6102 6064
025749.7-101004 189.34 -55.75 294.42 -32.11 025524.4-102203	 MCG -2- 8- 33 IRAS02554-1021 PGC 11198	RSBR0?. E .0± .8 	1.43± .04 .55± .05 1.41	145 .15 .41 	 13.06 				 4502±106 4470 4307
025753.5-022050 179.06 -50.89 303.33 -30.16 025521.8-023249	IC 1870 MCG -1- 8- 20 IRAS02553-0232 PGC 11202	.SBS9.. UE (1) 9.0± .4 6.4± .6	1.44± .02 .24± .04 1.46	132 .21 .25 .12	 12.92 			13.27±.1 212± 8 180± 12 	1541± 6 1467± 33 1529 1335
0257.9 +1110 165.83 -40.92 317.72 -25.47 0255.2 +1059	 UGC 2437 PGC 11204	.S..7.. U 7.0± .8 	1.22± .06 .04± .06 1.28	 .67 .06 .02	 14.7 ±.3 13.93				3460 3492 3252
025803.5+351528 150.22 -20.82 340.30 -14.23 025457.0+350329	 UGC 2435 PGC 11210	.SAS6.. U 6.0± .7 	1.32± .05 .08± .06 1.36	145 .44 .12 .04	 14.1 ±.2 13.49				4850 4948 4662
025804.3+252622 155.83 -29.22 331.51 -19.17 025508.5+251423	 MCG 4- 8- 1 PGC 11212	.S?.... 	.82± .13 .28± .07 .87	 .58 .43 .14	 15.57 ±.12 14.52			16.62±.3 161± 7 1.96	6966± 10 7038 6765
025806.1-742724 292.36 -39.96 220.60 -23.13 025822.0-743918	 ESO 31- 5 IRAS02583-7439 PGC 11217	.SXS3*/ Sr (1) 3.4± .5 3.3± .6	1.32± .04 .40± .04 1.33	10 .13 .56 .20	 14.07 ±.14 12.70 13.34				 4802± 34 4636 4783
025809.4+240158 156.72 -30.40 330.22 -19.86 025515.0+235000	 MCG 4- 8- 3 PGC 11225	.S?.... 	.82± .13 .17± .07 .87	 .62 .13 	 15.42 ±.12 14.52			17.00±.3 431± 7 	10424± 10 10492 10222
025810.6+480852 143.75 -9.54 351.26 -7.30 025445.1+475653	 UGC 2431 PGC 11228	.SA.9.. U 9.0± .9 	1.04± .15 .08± .12 1.14	 1.02 .08 .04					
025810.8+032141 172.96 -46.83 309.59 -28.43 025534.4+030943	NGC 1153 UGC 2439 PGC 11230	.L..+?. UE -1.3± .9 	1.11± .07 .02± .07 1.14	45 .33 .00 	 13.35 ±.10 12.97				 3126± 50 3134 2920
025811.7-524343 269.02 -55.23 243.50 -30.92 025639.6-525540	 PGC 11232	.LX.-.. S -2.0±1.2 	1.11± .10 .24± .08 1.08	 .00 .00 					
0258.2 +3629 149.59 -19.72 341.40 -13.61 0255.1 +3618	 UGC 2436 PGC 11237	.S..8*. U 8.0±1.3 	1.11± .14 .54± .12 1.17	87 .59 .66 .27					
025817.4+252638 155.87 -29.19 331.54 -19.21 025521.6+251440	 MCG 4- 8- 4 PGC 11240	.SB?... 	.82± .13 .00± .07 .87	 .58 .00 .00	 15.46 ±.13 14.78			16.22±.3 219± 7 1.44	10508± 10 10580 10308
025818.8+322010 151.86 -23.31 337.76 -15.78 025515.7+320812	 CGCG 506- 1 PGC 11242			 .49 	 15.5 ±.4 			16.49±.3 220± 7 	3610± 10 3700 3418

R.A. 2000 DEC. l b SGL SGB R.A. 1950 DEC.	Names PGC	Type S_T n_L T L	$\log D_{25}$ $\log R_{25}$ $\log A_e$ $\log D_o$	p.a. A_g A_i A_{21}	B_T m_B m_{FIR} B_T^o	$(B-V)_T$ $(U-B)_T$ $(B-V)_T^o$ $(U-B)_T^o$	$(B-V)_e$ $(U-B)_e$ m'_e m'_{25}	m_{21} W_{20} W_{50} HI	V_{21} V_{opt} V_{GSR} V_{3K}
0258.3 -0202 178.81 -50.60 303.72 -30.17 0255.8 -0214	 UGC 2443 PGC 11245	.S..6*. U 6.0±1.2 	1.10± .05 .28± .05 1.12	163 .21 .41 .14	 14.36 ±.12 				
0258.3 +0605 170.37 -44.77 312.51 -27.50 0255.7 +0554	 CGCG 415- 46 PGC 11246			 .34 	 15.4 ±.3 				6629± 60 6645 6422
025824.1-041749 181.55 -52.08 301.19 -30.83 025554.0-042946	 MCG -1- 8- 24 KUG 0255-044 PGC 11248	.SBT7.. UE (1) 6.5± .6 6.4± .8	1.34± .03 .28± .04 1.36	122 .21 .39 .14					2389 2374 2188
0258.4 +0351 172.53 -46.42 310.15 -28.32 0255.8 +0339	 UGC 2441 IRAS02558+0339 PGC 11252	.S..7.. U 7.0± .9 	1.34± .04 .72± .05 1.37	13 .34 .99 .36	 14.64 ±.13 13.31				3052 3062 2846
0258.5 +0618 170.22 -44.59 312.75 -27.46 0255.8 +0606	 UGC 2444 IRAS02558+0606 PGC 11255	.SB?... 	.98± .04 .21± .03 1.01	45 .34 .32 .11	15.15S±.15 14.92 ±.12 12.64 14.32	.93± .02 .15± .04 .77 .01	 14.38± .26	15.75±.3 363± 13 341± 10 1.33	6708± 10 6725 6501
025836.9-184157 203.63 -59.62 284.34 -33.71 025619.0-185354	 ESO 546- 34 PGC 11261	.SBS9P* SE (2) 9.3± .6 8.6± .6	1.13± .03 .50± .03 1.13	107 .06 .51 .25	 15.54 ±.14 14.97				1568 1511 1387
0258.6 +2517 156.03 -29.27 331.44 -19.35 0255.7 +2506	 UGC 2442 PGC 11262	.S?.... 	1.03± .06 .04± .05 1.09	150 .62 .05 .02					10451 10523 10251
025841.1-154212 198.34 -58.37 287.94 -33.33 025620.5-155408	IC 276 MCG -3- 8- 54 PGC 11264	.L..0P/ E -2.0± .8 	1.29± .05 .60± .05 1.21	64 .09 .00 					 3017 2968 2831
025841.3+032602 173.02 -46.69 309.72 -28.52 025604.9+031405	 UGC 2446 PGC 11265	.S?.... 	.66± .13 .16± .06 .69	105 .33 .16 .08	 14.7 ±.2 			15.75±.1 188± 11 	7089± 9 3098± 50
025844.2+254530 155.78 -28.87 331.88 -19.15 025548.0+253333	 UGC 2445 PGC 11268	.I..9*. U 10.0±1.3 	1.04± .08 .59± .06 1.11	45 .70 .44 .29				16.32±.3 228± 7 	7210± 10 7283 7010
025845.4+252440 155.99 -29.17 331.56 -19.32 025549.6+251243	 CGCG 485- 5 5ZW 298 PGC 11269			 .58 	 15.7 ±.4 			16.69±.3 340± 10 	10383± 10 10455 10183
025847.3-320552 230.88 -61.98 267.98 -34.28 025643.0-321748	NGC 1165 ESO 417- 8 IRAS02567-3217 PGC 11270	PSBR1*. Sr .7± .5 	1.40± .03 .42± .04 1.40	115 .00 .43 .21	 13.66 ±.14 12.18 13.17				 4893± 34 4800 4740
025849.0-345658 236.91 -61.75 264.50 -34.12 025648.1-350854	 ESO 356- 23 PGC 11271	PSBR1.. r 1.0± .9 	.97± .07 .41± .05 .98	161 .05 .42 .21	 15.49 ±.14 				
025854.9-363641 240.36 -61.48 262.47 -34.00 025656.0-364836	 ESO 356- 24 IRAS02569-3648 PGC 11273	 	.94± .06 .28± .05 .95	32 .11 	 14.83 ±.14 				 6089± 63 5984 5948
025856.0-122356 192.99 -56.73 291.87 -32.83 025632.6-123552	NGC 1162 MCG -2- 8- 36 PGC 11274	.E...*. E -5.0±1.2 	1.14± .10 .00± .07 1.16	 .11 .00 					
0258.9 +0615 170.37 -44.55 312.75 -27.58 0256.3 +0604	 CGCG 415- 49 PGC 11276			 .34 	 15.45 ±.16 				6900± 42 6916 6694
025858.8+411718 147.25 -15.48 345.58 -11.17 025544.2+410522	A 0255+41 CGCG 540- 18 5ZW 297 PGC 11279	.L...*. R -2.0± .7 		 .45 	 15.2 ±.3 				4994± 45 5106 4817

R.A. 2000 DEC.	Names	Type	$\log D_{25}$	p.a.	B_T	$(B-V)_T$	$(B-V)_e$	m_{21}	V_{21}
l　　b		S_T　n_L	$\log R_{25}$	A_g	m_B	$(U-B)_T$	$(U-B)_e$	W_{20}	V_{opt}
SGL　SGB		T	$\log A_e$	A_i	m_{FIR}	$(B-V)_T^0$	m'_e	W_{50}	V_{GSR}
R.A. 1950 DEC.	PGC	L	$\log D_0$	A_{21}	B_T^0	$(U-B)_T^0$	m'_{25}	HI	V_{3K}
0258.9　+7545		.S?....	1.61± .07	12				13.47±.1	2547± 11
130.66　14.83	UGC　2411		1.15± .12	1.55				331± 16	
13.98　8.07				1.50				312± 12	2713
0253.4　+7533	PGC 11282		1.76	.50					2461
025859.5-561042		.S?....	.89± .07	62	15.60 ±.18	.34± .04	.42± .04		
273.63 -53.08	ESO　154- 24		.40± .05	.00	15.75 ±.14				
239.60 -30.06			.48± .05	.41			13.49± .16		
025736.0-562236	PGC 11283		.89	.20			13.91± .39		
025904.6+323756		.S..4..	1.04± .08	14				16.01±.3	11469± 10
151.85 -22.98	UGC　2447	U	1.06± .06	.49					11560
338.11 -15.77		4.0±1.0		1.47				529±　7	11278
025601.0+322600	PGC 11289		1.09	.50					
025905.5+335450		.S?....	1.03± .06	97					3891
151.14 -21.87	UGC　2448		.46± .05	.52					
339.24 -15.11				.69					3985
025600.4+334254	PGC 11290		1.08	.23					3702
025910.5-042820		PSBROP?	1.06± .04	85					
181.97 -52.05	MCG -1- 8- 25	E	.31± .05	.22					4227± 52
301.05 -31.06	MK　1065	.0± .9		.23					4211
025640.5-044015	PGC 11292		1.07						4027
0259.1　+1601		.SXS3..	1.03± .08						9072
162.39 -36.87	UGC　2453	U　　(1)	.06± .06	.41	15.07 ±.17				
322.70 -23.72		3.0± .8		.08					9117
0256.4　+1550	PGC 11295	3.5±1.1	1.07	.03	14.51				8866
0259.3　+0717		.S..7..	1.10± .04	118	15.97S±.15			16.34±.3	7629± 10
169.54 -43.70	UGC　2454	U	.66± .03	.41				379± 13	
313.88 -27.30		7.0± .9		.91				368± 10	7648
0256.7　+0706	PGC 11306		1.14	.33	14.62		14.68± .25	1.39	7423
025923.0+470441		.S?....	1.22± .12					16.09±.1	4433± 11
144.45 -10.38	UGC　2449		.00± .12	.89	14.1　±.3				
350.48　-8.06	IRAS02559+4652			.00	13.45			142±　8	4557
025559.1+465246	PGC 11309		1.30	.00	13.18			2.91	4268
0259.4　+3700		.S..6*.	1.07± .07	177					
149.54 -19.15	UGC　2451	U	.62± .06	.52					
341.97 -13.54		6.0±1.3		.91					
0256.3　+3649	PGC 11313		1.12	.31					
0259.6　+4431		.I..9*.	1.07± .14						
145.74 -12.59	UGC　2452	U	.07± .12	.92					
348.38　-9.50		10.0±1.2		.05					
0256.3　+4420	PGC 11321		1.16	.03					
025942.6+251415	NGC　1156	.IBS9..	1.52± .02	25	12.32 ±.13	.58± .01	.54± .01	12.72±.0	382±　3
156.31 -29.20	UGC　2455	R　　(2)	.13± .03	.71	12.00 ±.11	-.19± .02	-.21± .02	111± 4	372± 18
331.51 -19.59	VV　531	10.0± .3	1.04± .01	.10	11.28	.38	13.01± .02	67± 4	452
025646.8+250221	PGC 11329	7.7± .7	1.58	.06	11.33	-.33	14.43± .17	1.33	183
025950.1+024617	IC　277	.SBT4..	1.09± .04	90	13.87 ±.13	.75± .02		14.46±.1	2849±　4
173.99 -46.97	UGC　2460	UE　　(1)	.13± .04	.29	13.68 ±.12			263± 5	2866± 41
309.12 -29.02	MK　602	3.5± .6		.19	11.80	.64		227± 6	2855
025714.1+023424	PGC 11336	4.2±1.2	1.12	.06	13.27		13.85± .27	1.12	2646
0259.9　+0630		.S?....	.82± .13					16.29±.3	8584± 10
170.40 -44.19	MCG　1- 8- 33		.17± .07	.35	15.33 ±.12			257± 13	
313.12 -27.72				.21				227± 10	8600
0257.3　+0619	PGC 11338		.85	.09	14.68			1.52	8379
0259.9　+2413		.SA.6..	1.03± .06						10218
156.99 -30.01	UGC　2457	U	.02± .05	.71	15.06 ±.17				
330.60 -20.13	IRAS02570+2401	6.0± .8		.02					10286
0257.0　+2401	PGC 11340		1.09	.01	14.28				10018
025957.8+364911		RLBS+..	1.24± .08	90					
149.74 -19.27	UGC　2456	U	.25± .05	.55	13.64 ±.10				3605± 22
341.86 -13.73	MK　1066	-1.0± .8		.00	10.74				3705
025649.0+363718	PGC 11341		1.26		13.03				3421
030005.9+392956									
148.36 -16.94	CGCG 524- 41			.43	14.75 ±.12				5290
344.17 -12.32	5ZW　307								5397
025653.6+391803	PGC 11344								5111
030018.4-231821		PSBR2..	1.01± .06	3					
212.68 -60.68	ESO　480- 19	r	.35± .05	.00	15.67 ±.14				
278.77 -34.52		2.2± .9		.42					
025805.1-233012	PGC 11354		1.01	.17					

R.A. 2000 DEC. l b SGL SGB R.A. 1950 DEC.	Names PGC	Type S_T n_L T L	$\log D_{25}$ $\log R_{25}$ $\log A_e$ $\log D_o$	p.a. A_g A_i A_{21}	B_T m_B m_{FIR} B_T^o	$(B-V)_T$ $(U-B)_T$ $(B-V)_T^o$ $(U-B)_T^o$	$(B-V)_e$ $(U-B)_e$ m'_e m'_{25}	m_{21} W_{20} W_{50} HI	V_{21} V_{opt} V_{GSR} V_{3K}
030022.0-170910 201.17 -58.63 286.27 -33.93 025802.9-172101	NGC 1163 MCG -3- 8- 56 IRAS02580-1720 PGC 11359	.SB.4?/ SE 3.7± .7 	1.35± .03 .85± .04 .73± .02 1.35	55 .05 1.25 .43	14.72 ±.15 13.36 13.40	.92± .03 .28± .05 .74 .10	.95± .03 .29± .05 13.86± .05 14.19± .23		2286 2232 2104
030022.7+340422 151.30 -21.60 339.52 -15.25 025717.3+335230	UGC 2461 PGC 11360	.SA.7.. U 7.0± .9 	1.02± .06 .36± .05 1.07	27 .51 .49 .18	 15.43 ±.13 14.41				6807 6901 6619
0300.4 +4438 145.81 -12.42 348.55 -9.56 0257.1 +4427	UGC 2458 PGC 11363	.S..7.. U 7.0± .8 	1.04± .15 .04± .12 1.12	 .87 .05 .02					
0300.4 -1125 191.84 -55.89 293.12 -33.00 0258.0 -1136	MCG -2- 8- 39 IRAS02580-1136 PGC 11365	.SXT1P* E 1.0±1.2 	1.13± .06 .20± .05 1.15	10 .15 .20 .10	 13.88 				9046±106 9009 8856
030031.8-154410 198.77 -57.99 287.98 -33.77 025811.4-155601	MCG -3- 8- 57 IRAS02581-1555 PGC 11367	.SAS7.. E (1) 7.0± .8 5.3±1.2	1.17± .05 .05± .05 1.18	35 .11 .08 .03	 13.20 				1575 1525 1391
0300.5 +4902 143.67 -8.56 352.22 -7.14 0257.1 +4850	UGC 2459 IRAS02571+4850 PGC 11368	.S..8*. U 8.0±1.2 	1.39± .09 .72± .12 1.49	62 1.11 .88 .36	 12.92 			13.45±.1 334± 16 321± 12 	2464± 11 2592 2304
0300.6 +3538 150.49 -20.22 340.92 -14.47 0257.5 +3527	UGC 2466 PGC 11369	.IA.9.. U 10.0± .8 	1.09± .07 .28± .06 1.14	172 .50 .21 .14					4980 5077 4795
030037.5+351009 150.75 -20.63 340.50 -14.72 025730.7+345818	UGC 2465 5ZW 308 PGC 11370	.S..1*. U 1.0±1.3 	1.04± .06 .54± .05 1.09	144 .55 .55 .27	 14.77 ±.14 13.61				5078 5174 4892
030037.6+401507 148.06 -16.23 344.87 -12.00 025724.1+400315	UGC 2463 PGC 11371	.SX.9.. U 9.0± .7 	1.36± .04 .19± .05 1.41	95 .48 .19 .10	 14.4 ±.3 13.67			14.30±.1 213± 16 195± 12 .53	1899± 11 2008 1721
0300.6 +1150 165.98 -39.99 318.69 -25.81 0257.9 +1138	NGC 1166 UGC 2471 IRAS02579+1138 PGC 11372	.S..2.. U 2.0± .8 	1.08± .06 .05± .05 1.15	 .80 .07 .03	 14.86 ±.18 13.92				7693± 57 7725 7488
030039.6+323848 152.15 -22.79 338.30 -16.04 025735.8+322657	UGC 2464 PGC 11373	.S..0.. U .0± .9 	.95± .07 .39± .05 .98	30 .56 .29 				16.46±.3 503± 7	10168± 10 10258 9979
030039.9+000106 177.12 -48.77 306.19 -30.11 025806.2-001045	UGC 2479 KUG 0258-001 PGC 11375	.S..8*/ UE 8.0± .8 	1.13± .03 .45± .04 1.16	25 .31 .55 .22					2842 2839 2641
030042.5-220828 210.44 -60.29 280.21 -34.54 025828.0-222018	ESO 546- 37 PGC 11377	.SB?... 1.07	1.07± .05 .26± .04 	32 .00 .32 .13	 14.80 ±.14 14.44				4367 4299 4194
0300.7 +1146 166.05 -40.03 318.63 -25.85 0258.0 +1135	NGC 1168 UGC 2476 PGC 11378	.SXT3*. U (1) 3.0± .9 3.5±1.2	1.06± .06 .28± .05 1.13	18 .80 .39 .14	 15.02 ±.14 13.78				7627± 57 7659 7422
030045.8+231058 157.84 -30.78 329.73 -20.79 025752.0+225907	UGC 2472 PGC 11382	.S..1.. U 1.0±1.0 	.96± .09 .69± .06 1.03	67 .73 .70 .34				16.86±.3 550± 7	10157± 10 10221 9957
0300.8 +4426 145.97 -12.56 348.43 -9.73 0257.5 +4415	UGC 2468 PGC 11384	.S..0?. U .0±1.7 	1.14± .13 .18± .12 1.21	170 .87 .13 					
030059.5+430102 146.71 -13.79 347.24 -10.54 025741.9+424912	UGC 2470 PGC 11392	.S..1.. U 1.0±1.0 	1.11± .07 .83± .06 1.17	169 .67 .85 .42	 14.6 ±.3 				

R.A. 2000 DEC. / l b / SGL SGB / R.A. 1950 DEC.	Names / PGC	Type / S_T n_L / T / L	$\log D_{25}$ / $\log R_{25}$ / $\log A_e$ / $\log D_o$	p.a. / A_g / A_i / A_{21}	B_T / m_B / m_{FIR} / B_T^o	$(B-V)_T$ / $(U-B)_T$ / $(B-V)_T^o$ / $(U-B)_T^o$	$(B-V)_e$ / $(U-B)_e$ / m'_e / m'_{25}	m_{21} / W_{20} / W_{50} / HI	V_{21} / V_{opt} / V_{GSR} / V_{3K}
030103.1-004435		.S..8*.	.98± .05						2646
178.06 -49.22	UGC 2482	U	.17± .04	.28					
305.38 -30.44	KUG 0258-009	8.0±1.2		.21					2640
025830.0-005624	PGC 11397		1.00	.08					2446
030110.2+412346		.S..6?.	1.19± .12	36					
147.57 -15.19	UGC 2473	U	.62± .12	.44	15.0 ±.3				
345.90 -11.46	IRAS02578+4112	6.0±1.8		.91					
025755.0+411156	PGC 11399		1.23	.31					
030113.8+445718	NGC 1160	.S..6*.	1.29± .03	50	13.50 ±.15	.68± .03	.74± .02		
145.78 -12.08	UGC 2475	U	.32± .05	1.03	12.6 ±.3	.03± .04	.00± .04		2429± 35
348.89 -9.51	IRAS02579+4445	6.0±1.2	.99± .02	.47	11.99	.37	13.94± .05		2548
025753.1+444528	PGC 11403		1.39	.16	11.85	-.19	14.00± .25		2261
030114.5+445351	NGC 1161	.L.....	1.45± .10	23	12.05 ±.13	1.06± .01	1.09± .01		
145.81 -12.13	UGC 2474	U	.14± .08	.87	12.0 ±.3	.54± .02	.58± .01		1940± 21
348.84 -9.54	IRAS02579+4442	-2.0± .7	.96± .02	.00		.83	12.34± .08		2059
025753.9+444201	PGC 11404		1.53		11.14	.35	13.81± .55		1772
030115.1-282806		RLBR+..	1.14± .05	175	14.14 ±.16	.84± .02	.95± .02		
223.26 -61.35	ESO 417- 11	S	.20± .04	.00	14.40 ±.14	.26± .04	.36± .03		6376± 52
272.43 -34.89		-1.0± .8	.68± .05	.00		.75	13.03± .16		6290
025907.0-283954	PGC 11405		1.11		14.19	.27	14.22± .30		6217
0301.4 +1750		.SAR4..	1.00± .06						5961
161.61 -35.08	UGC 2486	U	.08± .05	.44	14.52 ±.12				
324.72 -23.40		4.0± .9		.11					6010
0258.6 +1739	PGC 11410		1.04	.04	13.93				5759
030130.4+374557	IC 278	.E...*.	1.23± .09						
149.52 -18.30	UGC 2481	U	.00± .05	.37	14.17 ±.16				
342.84 -13.48		-5.0±1.0		.00					
025820.2+373408	PGC 11414		1.29						
030136.1-145013	NGC 1172	.E+..*.	1.37± .07	25	12.70M±.06	.84± .03	.93± .01		
197.48 -57.35	MCG -3- 8- 59	PE	.11± .06	.11	12.65 ±.18	.29± .04	.37± .02		1550± 24
289.11 -33.89		-3.7± .6	1.11± .04	.00		.80	13.67± .14		1502
025914.9-150200	PGC 11420		1.35		12.56	.28	14.15± .38		1366
030136.4+314911		.S?....	1.16± .07	137				15.50±.3	6222± 10
152.80 -23.40	UGC 2483		.59± .06	.61					6309
337.67 -16.64				.88				296± 7	6032
025833.4+313723	PGC 11421		1.22	.29					
030142.3+351219	NGC 1167	.LA.-..	1.44± .06	70				15.00±.2	4954± 7
150.94 -20.49	UGC 2487	U	.07± .05	.55	13.38 ±.17			462± 9	4895± 46
340.65 -14.89		-3.0± .7		.00				450± 7	5049
025835.3+350031	PGC 11425		1.50		12.76				4768
030142.6+284408		.I..9*.	1.04± .15					15.86±.3	3136± 10
154.61 -26.00	UGC 2488	U	.00± .12	.52					
334.93 -18.24		10.0±1.2		.00				38± 7	3215
025843.0+283220	PGC 11426		1.09	.00					2943
030153.9+354400		.SBR0..	1.25± .04	0					4875
150.68 -20.01	UGC 2491	U	.07± .05	.59	14.29 ±.20				
341.13 -14.64		.0± .8		.05					4972
025846.2+353213	PGC 11437		1.30		13.58				4691
030155.6+510848								15.02±.3	4808± 9
142.84 -6.61				1.45				334± 11	
354.07 -6.13								325± 8	4939
025823.3+505700	PGC 11438								4654
030159.9+423507	NGC 1164	PSXS2..	1.13± .05	145				15.50±.1	4175± 11
147.10 -14.08	UGC 2490	U	.11± .05	.59	14.0 ±.3				4044± 81
346.98 -10.94	MK 1067	2.0± .8		.13	12.22			308± 8	4286
025842.9+422320	PGC 11441		1.19	.05	13.19			2.26	4001
030200.7-384038		.S?....	.92± .07						
244.27 -60.46	ESO 300- 4		.02± .06	.00	15.14 ±.14				6046
259.88 -34.38				.02					5935
030005.0-385224	PGC 11442		.92	.01	15.04				5913
030204.0+360557		.S?....	.99± .06	29					7894
150.52 -19.68	UGC 2494		.38± .05	.62	14.73 ±.13				
341.47 -14.47	IRAS02589+3554			.57	12.54				7992
025855.9+355410	PGC 11447		1.05	.19	13.50				7711
030206.8+413538		.L.....	1.11± .16	105					
147.63 -14.93	UGC 2495	U	.07± .08	.52	14.4 ±.3				
346.16 -11.50		-2.0± .8		.00					
025851.1+412351	PGC 11449		1.16						

R.A. 2000 DEC.	Names	Type	$logD_{25}$	p.a.	B_T	$(B-V)_T$	$(B-V)_e$	m_{21}	V_{21}
l b		S_T n_L	$logR_{25}$	A_g	m_B	$(U-B)_T$	$(U-B)_e$	W_{20}	V_{opt}
SGL SGB		T	$logA_e$	A_i	m_{FIR}	$(B-V)_T^o$	m'_e	W_{50}	V_{GSR}
R.A. 1950 DEC.	PGC	L	$logD_o$	A_{21}	B_T^o	$(U-B)_T^o$	m'_{25}	HI	V_{3K}
030207.5+290625		.S..8..	1.45± .03	70				13.94±.3	3115± 10
154.48 -25.64	UGC 2497	U	.61± .05	.66	14.38 ±.17				
335.31 -18.13	IRAS02591+2854	8.0± .8		.75	13.49			258± 7	3195
025907.4+285439	PGC 11453		1.51	.31	12.96			.67	2923
030208.1-542445		.SXS7..	1.06± .04	7					
270.81 -53.78	ESO 154- 28	S (1)	.17± .05	.00	15.69 ±.14				
241.41 -31.01		7.0± .8		.23					
030041.0-543630	PGC 11454	6.7± .8	1.06	.08					
030209.6-354115		.SAS7*.	1.08± .04						4527
238.25 -60.98	ESO 357- 1	S (1)	.07± .04	.04	15.09 ±.14				
263.53 -34.74		7.0± .8		.09					4423
030010.1-355300	PGC 11455	5.6±1.1	1.08	.03	14.94				4387
030212.1+172039		.SX.6..	1.20± .04	65				15.57±.3	6930± 10
162.15 -35.38	UGC 2498	U	.38± .04	.41	15.13 ±.19			311± 13	
324.33 -23.79		6.0± .8		.56					6977
025923.6+170853	PGC 11456		1.24	.19	14.13			1.25	6729
030221.1+515755		.S?....	.82± .13					14.32±.3	3352± 9
142.50 -5.86	MCG 9- 6- 1		.17± .07	1.64				367± 11	
354.77 -5.72				.26				355± 8	3485
025847.0+514608	PGC 11463		.97	.09					3200
0302.4 +4626		.S..8*.	1.00± .08	113					
145.23 -10.67	UGC 2496	U	1.02± .06	1.05					
350.25 -8.86		8.0±1.5		1.23					
0259.1 +4615	PGC 11471		1.10	.50					
0302.5 +3946		.I..9?.	1.11± .14						
148.64 -16.46	UGC 2499	U	.04± .12	.44					
344.67 -12.56		10.0±1.6		.03					
0259.3 +3935	PGC 11473		1.15	.02					
030232.7-575135		RSBR0..	1.12± .04	33					5377± 88
275.28 -51.65	ESO 116- 5	Sr	.17± .04	.00	14.18 ±.14				
237.57 -29.98	IRAS03012-5803	.0± .6		.12	12.58				5229
030115.0-580318	PGC 11476		1.11		13.97				5302
030234.0+021145		.SBS9P*	1.12± .04	133				15.18±.3	5181± 10
175.30 -46.91	UGC 2501	UE (1)	.33± .04	.25	13.93 ±.13			272± 13	5201± 39
308.76 -29.85	MK 1068	9.0± .8		.34	12.86			204± 10	5184
025958.5+020000	PGC 11477	5.3±1.6	1.14	.17	13.34			1.68	4982
030237.7-225204	NGC 1187	.SBR5..	1.74± .02	130	11.34M±.11	.56± .02	.64± .01	13.03±.1	1396± 5
212.09 -60.06	ESO 480- 23	R (3)	.13± .02	.04	11.26 ±.12	-.05± .07	.04± .02	278± 6	1546± 58
279.35 -35.03	IRAS03003-2303	5.0± .3	1.36± .03	.20	10.52	.52	13.51± .06	266± 6	1325
030024.1-230348	PGC 11479	2.1± .3	1.74	.07	11.06	-.08	14.55± .15	1.90	1228
030238.4-185352	NGC 1179	.SXR6..	1.69± .02	35	12.6 ±.2	.60± .03		13.24±.2	1780± 6
204.68 -58.81	ESO 547- 1	R (3)	.11± .03	.06	12.49 ±.12	-.07± .04		188± 7	
284.23 -34.68	IRAS03003-1905	6.0± .3		.16	13.50	.56		188± 11	1719
030021.0-190536	PGC 11480	3.2± .4	1.69	.05	12.30	-.10	15.62± .23	.89	1603
030244.3+413731		.SBS1..	1.16± .13	165					
147.72 -14.84	UGC 2500	U	.20± .12	.52	14.8 ±.3				
346.25 -11.58		1.0± .8		.20					
025928.5+412546	PGC 11484		1.21	.10					
030259.5-090757	NGC 1185	.SBR3?.	1.08± .06	30					
189.10 -54.12	MCG -2- 8- 41	E (1)	.49± .05	.22					4812±106
295.96 -33.13	IRAS03005-0919	3.0±1.3		.68	13.00				4780
030033.5-091940	PGC 11488	4.2±1.6	1.10	.25					4622
0303.0 -0157		.S..6*.	1.02± .06	172					
179.97 -49.67	UGC 2508	U	.75± .05	.19					
304.20 -31.29	IRAS03005-0209	6.0±1.4		1.10					
0300.5 -0209	PGC 11492		1.04	.38					
030309.3-231154		.SBT3..	.97± .05						
212.79 -60.02	ESO 480- 24	r	.02± .04	.02	14.95 ±.14				
278.96 -35.17		3.3± .9		.03					
030056.1-232336	PGC 11494		.98	.01					
030314.8+162625	A 0300+16	.E.1.$.			16.05 ±.16	1.10± .03	1.12± .02		
163.06 -35.96		P		.43		.48± .08	.57± .08		9740± 35
323.57 -24.42		-5.0±1.8	.42± .05			.91	13.64± .19		9783
030027.0+161442	PGC 11499					.43			9539
0303.3 +0416								16.94±.3	6932± 10
173.40 -45.28	CGCG 415- 55			.39	15.4 ±.3			214± 13	
311.08 -29.31								181± 10	6939
0300.7 +0405	PGC 11501								6732

R.A. 2000 DEC. l　　b SGL　SGB R.A. 1950 DEC.	Names PGC	Type S_T　n_L T L	$\log D_{25}$ $\log R_{25}$ $\log A_e$ $\log D_o$	p.a. A_g A_i A_{21}	B_T m_B m_{FIR} B_T^o	$(B-V)_T$ $(U-B)_T$ $(B-V)_T^o$ $(U-B)_T^o$	$(B-V)_e$ $(U-B)_e$ m'_e m'_{25}	m_{21} W_{20} W_{50} HI	V_{21} V_{opt} V_{GSR} V_{3K}
030324.3-153725 199.15 -57.32 288.24 -34.44 030103.9-154907	NGC 1189 MCG -3- 8- 61 HICK 22C PGC 11503	.SBS8*. E (1) 8.0± .8 7.5± .8	1.23± .05 .04± .05 1.24	 .10 .05 .02	14.41S±.15 14.25		 15.29± .32		 2728± 41 2676 2547
0303.4　+0429 173.21 -45.11 311.32 -29.26 0300.7　+0417	 UGC 2509 IRAS03007+0417 PGC 11504	.S?.... 	1.01± .04 .48± .03 1.05	54 .40 .72 .24	15.38S±.15 15.13 ±.12 13.34 14.08	.89± .04 .14± .07 .67 -.05	 14.06± .26	15.58±.3 353± 13 344± 10 1.26	6003± 10 6011 5803
030324.7-502943 264.93 -55.76 245.80 -32.30 030149.1-504124	 ESO 199- 12 IRAS03018-5041 PGC 11505	 	1.13± .06 .52± .06 1.13	4 .00 	 15.57 ±.14 13.05 				6993± 88 6857 6896
0303.4　+7439 131.47 14.02 13.24　7.23 0258.0　+7428	 UGC 2478 PGC 11506	.S..0.. U .0± .9 	1.11± .07 .29± .06 1.24	23 1.52 .22 	 15.1 ±.3 				
030326.1-153944 199.22 -57.33 288.20 -34.45 030105.8-155126	NGC 1190 MCG -3- 8- 62 HICK 22B PGC 11508	.L..0*/ E -2.0±1.3 	.96± .07 .48± .05 .90	95 .10 .00 	15.18S±.15 15.04		 13.66± .42		 2625± 41 2573 2444
030330.9-154108 199.27 -57.32 288.17 -34.47 030110.7-155250	NGC 1191 MCG -3- 8- 64 HICK 22D PGC 11514	.LB.-*. E -3.0±1.0 	.75± .14 .08± .05 .75	60 .10 .00 	15.28S±.15 15.04		 13.69± .72		 9342± 41 9290 9162
030331.1-201037 207.14 -59.06 282.69 -35.02 030115.0-202218	 ESO 547- 4 PGC 11515	.SXT5*P SE (1) 4.7± .5 5.3±1.2	1.19± .03 .35± .03 1.19	35 .04 .52 .17	 15.02 ±.14 14.44				 3322± 60 3257 3149
030334.7-154045 199.28 -57.31 288.18 -34.49 030114.4-155226	NGC 1192 MCG -3- 8- 65 HICK 22E PGC 11519	.E...?. E 	.82± .17 .29± .07 .75	102 .10 .00 	15.36S±.15 15.12		 13.74± .87		 9506± 41 9454 9326
030334.9-182207 203.89 -58.41 284.91 -34.85 030117.0-183348	 ESO 547- 5 PGC 11520	.IBS9.. SE (1) 9.7± .5 9.4± .8	1.06± .04 .17± .03 1.07	36 .04 .13 .08	 15.51 ±.14 15.35				1593 1534 1418
030335.1+462304 145.43 -10.63 350.31 -9.05 030011.5+461121	NGC 1169 UGC 2503 PGC 11521	.SXR3.. R (2) 3.0± .3 1.6± .6	1.62± .02 .17± .03 1.27± .11 1.72	28 1.05 .24 .09	12.2 ±.3 12.22 ±.17 10.91	.95± .04 .67 	1.02± .02 .63± .05 14.04± .29 14.74± .33	13.25±.1 460± 5 450± 4 2.25	2387± 5 2342± 66 2507 2224
030335.3-120435 193.55 -55.59 292.51 -33.88 030111.8-121616	NGC 1196 MCG -2- 8- 42B PGC 11522	.LBT0.. E -2.0± .8 	1.15± .06 .01± .05 1.17	 .18 .00 					
0303.5　+0156 175.82 -46.90 308.58 -30.18 0301.0　+0145	 UGC 2513 PGC 11523	.S..1*. U 1.0±1.3 	.96± .09 .15± .06 .98	165 .24 .15 .07	 14.76 ±.12 				
030338.5-153650 199.18 -57.26 288.26 -34.49 030118.2-154831	NGC 1199 MCG -3- 8- 67 HICK 22A PGC 11527	.E.3.*. R -5.0± .5 	1.38± .06 .10± .05 .97± .03 1.37	48 .10 .00 	12.37M±.05 12.35 ±.18 12.22	1.02± .01 .45± .03 .97 .44	1.03± .01 .53± .02 12.75± .11 14.01± .31		 2668± 15 2616 2488
030343.1+303712 153.91 -24.18 336.85 -17.65 030041.2+302530	 UGC 2507 PGC 11532	.SB.6?. U 6.0±1.3 	1.00± .08 .43± .06 1.06	10 .64 .63 .22				15.95±.3 458± 7 	16091± 10 16174 15902
030343.5-152905 198.98 -57.19 288.42 -34.49 030123.0-154046	NGC 1188 MCG -3- 8- 68 PGC 11533	.LXR0*. E -2.0± .8 	1.00± .07 .34± .05 .97	170 .10 .00 					
030349.3-010614 179.18 -48.96 305.23 -31.21 030116.6-011755	NGC 1194 UGC 2514 IRAS03012-0117 PGC 11537	.LA.+*. UE -.5± .6 	1.25± .05 .25± .05 .91± .04 1.23	140 .24 .00 	13.83 ±.15 14.21 ±.13 	.96± .03 .45± .05 	1.04± .02 .52± .04 13.87± .13 14.32± .30		
030350.1-251620 216.93 -60.33 276.40 -35.42 030139.0-252800	A 0301-25 ESO 480- 25 DDO 227 PGC 11538	.SBS9*. SU (2) 9.0± .6 8.9± .6	1.28± .03 .19± .04 1.28	 .00 .19 .09	15.0 ±.3 14.97 ±.14 14.78	.61± .08 -.06± .11 .56 -.10	 15.80± .36		1726 1648 1564

R.A. 2000 DEC. l b SGL SGB R.A. 1950 DEC.	Names PGC	Type S_T n_L T L	$\log D_{25}$ $\log R_{25}$ $\log A_e$ $\log D_o$	p.a. A_g A_i A_{21}	B_T m_B m_{FIR} B_T^o	$(B-V)_T$ $(U-B)_T$ $(B-V)_T^o$ $(U-B)_T^o$	$(B-V)_e$ $(U-B)_e$ m'_e m'_{25}	m_{21} W_{20} W_{50} HI	V_{21} V_{opt} V_{GSR} V_{3K}
030352.5+093648 168.65 -41.21 316.77 -27.44 030110.7+092507	IC 1873 MCG 1- 8- 39 VV 383 PGC 11541	.L?.... 	.72± .22 .11± .07 .79	 .67 .00 	 15.20 ±.11 14.39				9177 9200 8976
030353.0-520703 267.30 -54.85 243.92 -31.93 030221.0-521842	IC 1879 ESO 199- 14 FAIR 736 PGC 11542	.S?.... 	1.07± .06 .83± .05 1.07	136 .00 1.24 .41	 15.95 ±.14 14.64				 13100±190 12961 13008
0303.8 +7425 131.62 13.84 13.07 7.07 0258.5 +7414	 UGC 2485 PGC 11543	.S..6*. U 6.0±1.1 	1.16± .13 .05± .12 1.32	 1.72 .07 .02				15.51±.1 108± 8 	2014± 11 2178 1926
030354.6-115932 193.49 -55.48 292.63 -33.94 030131.0-121112	NGC 1200 MCG -2- 8- 43 PGC 11545	.LAS-.. E -3.0± .8 	1.47± .07 .30± .07 1.45	85 .18 .00 					
030355.0-141750 197.08 -56.61 289.87 -34.35 030133.5-142930	 MCG -3- 8- 69 PGC 11546	.S..1?/ E .5±1.3 	1.15± .05 .70± .04 1.17	105 .12 .71 .35					
0303.9 +3015 154.16 -24.46 336.56 -17.88 0300.9 +3004	 UGC 2512 PGC 11547	.SA.5.. U 5.0± .9 	1.00± .08 .14± .06 1.06	0 .64 .21 .07					
030356.3-392627 245.59 -59.92 258.89 -34.65 030202.0-393806	IC 1875 ESO 300- 6 PGC 11549	.LXT-*. BS -2.8± .7 	1.14± .05 .06± .04 .98± .04 1.13	 .00 .00 	13.37 ±.15 13.89 ±.14 13.56	.87± .02 .81 	.97± .02 13.76± .15 13.78± .32		 6092± 57 5978 5963
030359.0+432353 147.01 -13.19 347.87 -10.79 030040.2+431212	NGC 1171 UGC 2510 IRAS03006+4312 PGC 11552	.S..6*. U 6.0±1.1 	1.42± .04 .36± .05 1.52	147 .99 .54 .18	 13.0 ±.3 12.60 11.52			14.19±.1 288± 8 268± 6 2.49	2742± 6 2780± 52 2857 2574
0304.0 -1201 193.58 -55.45 292.61 -33.99 0301.7 -1213	IC 285 MCG -2- 8- 44 PGC 11557	.S?.... 	1.04± .09 .66± .07 1.02	 .18 .50 					4069 4027 3884
030408.3-260403 218.54 -60.40 275.41 -35.51 030158.0-261542	NGC 1201 ESO 480- 28 PGC 11559	.LAR0*. R -2.0± .3 	1.56± .02 .23± .02 .97± .03 1.52	7 .00 .00 	11.67 ±.14 11.77 ±.11 11.71	.94± .01 .52± .01 .90 .51	.98± .01 .55± .01 12.01± .11 13.74± .19		1711± 25 1630 1551
0304.2 +0145 176.19 -46.91 308.45 -30.40 0301.7 +0134	 UGC 2518 PGC 11566	.S..6*. U 6.0±1.3 	1.00± .08 .43± .06 1.03	79 .26 .63 .21	 15.33 ±.12 14.41				7010 7009 6812
0304.3 -0659 186.51 -52.60 298.56 -32.96 0301.9 -0711	 E (1) PGC 11573	.SBS9.. 9.0± .9 9.8± .8	1.10± .08 .16± .08 1.13	20 .27 .17 .08					
0304.3 +4816 144.60 -8.93 351.94 -8.09 0300.9 +4804	 UGC 2511 IRAS03009+4804 PGC 11574	.SB.2*. U 2.0± .9 	1.10± .05 .54± .05 1.20	150 1.07 .66 .27	 13.04 				
030431.8-272734 221.36 -60.52 273.68 -35.61 030223.0-273912	IC 1876 ESO 417- 13 PGC 11577	.LA.+?P S -1.5± .6 	1.05± .05 .06± .04 1.04	 .00 .00 	14.4 ±.4 14.31 ±.14 14.22	.29± .06 -.28± .10 .22 -.25	 14.37± .48		6546 6461 6389
030432.7+422022 147.65 -14.06 347.04 -11.46 030115.5+420843	NGC 1175 UGC 2515 PGC 11578	.LAR+.. R -1.0± .4 	1.29± .03 .50± .03 .63± .03 1.27	153 .60 .00 	13.89 ±.15 13.32 ±.13 12.89	1.04± .02 .59± .04 .78 .41	1.12± .03 .63± .04 12.53± .08 13.94± .22		5540± 29 5651 5370
030437.3+422146 147.65 -14.03 347.07 -11.46 030120.0+421007	NGC 1177 MCG 7- 7- 20 PGC 11581	.S?.... 	.64± .17 .00± .07 .69	 .60 .00 	 15.5 ±.3 14.78				5584± 94 5695 5414
0304.6 +0526 172.61 -44.21 312.47 -29.20 0302.0 +0515	 CGCG 415- 58 PGC 11582			 .44 	 15.6 ±.3 			17.55±.3 319± 13 306± 10 	8409± 10 8419 8210

R.A. 2000 DEC. / l b / SGL SGB / R.A. 1950 DEC.	Names / PGC	Type S_T n_L / T / L	$\log D_{25}$ / $\log R_{25}$ / $\log A_o$ / $\log D_o$	p.a. / A_g / A_i / A_{21}	B_T / m_B / m_{FIR} / B_T^o	$(B-V)_T$ / $(U-B)_T$ / $(B-V)_T^o$ / $(U-B)_T^o$	$(B-V)_o$ / $(U-B)_o$ / m'_o / m'_{25}	m_{21} / W_{20} / W_{50} / HI	V_{21} / V_{opt} / V_{GSR} / V_{3K}
030439.9-122032	NGC 1204	.S..0*.	1.05± .06	60					
194.19 -55.49	MCG -2- 8- 45	E	.52± .05	.18					
292.26 -34.19	IRAS03022-1232	.0±1.3		.39	11.03				
030216.8-123210	PGC 11583		1.04						
0304.7 +4718		.SB.8..	1.14± .13	0				17.02±.1	5095± 11
145.14 -9.73	UGC 2516	U	.14± .12	1.20					5217
351.18 -8.68		8.0± .8		.17				91± 8	
0301.3 +4707	PGC 11584		1.25	.07					4935
030449.6+540658		.S?....	.74± .14					16.65±.3	2205± 9
141.77 -3.80	MCG 9- 6- 2		.21± .07	2.44				181± 11	
356.73 -4.78				.16				163± 8	2341
030110.0+535519	PGC 11586		.96						2060
0304.9 +2123		.SB.7*.	1.11± .14						4209
159.96 -31.71	UGC 2520	U	.15± .12	.64					
328.54 -22.50		7.0± .8		.20					4266
0302.1 +2112	PGC 11592		1.17	.07					4013
030506.9-455749		.SBS7..	1.48± .04	14					1318
257.36 -57.58	ESO 248- 2	S (1)	.66± .05	.00	14.10 ±.14				1190
251.00 -33.67		7.0± .7		.91					1209
030323.0-460924	PGC 11595	6.7±1.2	1.48	.33	13.19				
030509.0+010537		.S..7*/	1.20± .03	175					6951
177.12 -47.22	UGC 2523	UE	.71± .04	.25	15.04 ±.12				
307.80 -30.83	KUG 0302+009	7.0± .7		.98					6948
030234.4+005400	PGC 11598		1.22	.35	13.78				6755
030512.0-603456		.S?....	1.03± .06	41	14.96 ±.13	.73± .03			
278.23 -49.64	ESO 116- 9		1.03± .05	.00	16.89 ±.14				
234.48 -29.35				1.50					
030404.0-604630	PGC 11601		1.03	.50			12.39± .34		
030531.3+425009	NGC 1186	.SBR4*.	1.50± .03	122					
147.55 -13.54	UGC 2521	R	.42± .03	.68	12.2 ±.3				2658± 52
347.56 -11.33	IRAS03022+4238	4.0± .3		.62	11.52				2770
030213.1+423833	PGC 11617		1.56	.21	10.89				2490
0305.5 -0235		.SX.9*.	1.12± .08	120					
181.32 -49.62		E (1)	.28± .08	.16					
303.71 -32.06		9.0±1.3		.28					
0303.0 -0247	PGC 11618	8.7±1.2	1.14	.14					
030540.4+332336		.S?....	.91± .10	105				15.06±.3	6228± 10
152.70 -21.61	UGC 2525		.45± .06	.61	15.55 ±.12				6317
339.52 -16.54				.67				384± 7	6045
030235.1+331200	PGC 11622		.97	.22	14.24			.60	
030544.0+364717		.S..3..	1.55± .04	136					
150.81 -18.71	UGC 2526	U (1)	.68± .06	.67	13.33 ±.12				4996± 52
342.47 -14.72	IRAS03025+3635	3.0± .8		.94	12.74				5093
030234.4+363542	PGC 11625	4.5±1.1	1.61	.34	11.67				4818
0305.7 +2212		.S..8*.	1.22± .12	37					4258± 10
159.59 -30.94	UGC 2530	U	.30± .12	.65					
329.40 -22.27	IRAS03028+2200	8.0±1.2		.37	12.85				4317
0302.8 +2200	PGC 11628		1.28	.15					4064
0305.8 +1606		.L...?.	1.08± .09	28					
163.91 -35.86	UGC 2532	U	.22± .04	.47					
323.54 -25.11		-2.0±1.7		.00					
0303.0 +1555	PGC 11630		1.10						
030555.1+413533		.S..0..	1.16± .05	18					6371± 94
148.26 -14.57	UGC 2528	U	.39± .05	.51	14.4 ±.3				6480
346.56 -12.09		.0± .8		.29					6201
030238.8+412358	PGC 11634		1.18		13.54				
030558.7-192332		.IB?...	1.23± .03	60					1684
206.10 -58.25	ESO 547- 9	(2)	.46± .04	.00	16.09 ±.14				
283.72 -35.52				.35					1621
030342.0-193506	PGC 11636	9.9± .5	1.23	.23	15.74				1513
030603.1-153642	NGC 1209	.E.6.*.	1.38± .05	80	12.41M±.05	.96± .01	1.00± .01		
199.64 -56.74	MCG -3- 8- 73	R	.32± .05	.10	12.28 ±.16	.51± .02	.56± .01		2614± 20
288.37 -35.07		-5.0± .5	.79± .02	.00		.91	11.83± .05		2561
030343.0-154815	PGC 11638		1.30		12.26	.50	13.51± .27		2437
0306.0 +0136		.I..9*.	.96± .09						2968
176.81 -46.69	UGC 2535	U	.00± .06	.19					
308.45 -30.88		10.0±1.2		.00					2966
0303.5 +0125	PGC 11639		.98	.00					2773

R.A. 2000 DEC. l b SGL SGB R.A. 1950 DEC.	Names PGC	Type S_T n_L T L	$\log D_{25}$ $\log R_{25}$ $\log A_e$ $\log D_o$	p.a. A_g A_i A_{21}	B_T m_B m_{FIR} B_T^o	$(B-V)_T$ $(U-B)_T$ $(B-V)_T^o$ $(U-B)_T^o$	$(B-V)_e$ $(U-B)_e$ m'_e m'_{25}	m_{21} W_{20} W_{50} HI	V_{21} V_{opt} V_{GSR} V_{3K}
030606.5-390209 244.61 -59.60 259.31 -35.12 030412.0-391342	NGC 1217 ESO 300- 10 IRAS03041-3913 PGC 11641	RSAR1*. Sr 1.0± .6 	1.25± .04 .14± .05 1.25	50 .00 .14 .07	 13.34 ±.14 12.61 13.12				6236± 34 6122 6108
030609.8+422219 147.89 -13.88 347.24 -11.69 030252.2+421045	IC 284 UGC 2531 IRAS03029+4211 PGC 11643	.SA.8.. U 8.0± .7 	1.61± .03 .29± .05 1.31± .04 1.67	13 .69 .36 .15	12.47 ±.17 13.0 ±.3 12.82 11.52	.93± .04 .30± .05 .70 .10	.86± .03 .27± .04 14.51± .12 14.64± .25	13.40±.1 343± 9 324± 11 1.74	2723± 6 2706± 94 2833 2554
030611.9-093229 190.43 -53.69 295.69 -33.99 030346.4-094402	NGC 1208 MCG -2- 8- 47 IRAS03037-0943 PGC 11647	.SAR0?. E -.3± .7 	1.25± .04 .29± .04 1.27	75 .28 .22					4607±106 4571 4422
030612.8+415102 148.18 -14.32 346.81 -11.99 030256.0+413928	NGC 1198 UGC 2533 PGC 11648	.L..-*. U -3.0±1.1 	1.28± .08 .23± .05 1.32	120 .63 .00	 13.5 ±.3 12.85				1598± 52 1707 1429
030628.5-094355 190.76 -53.74 295.48 -34.10 030403.1-095527	IC 1880 MCG -2- 8- 49 PGC 11656	.LA.-P* E -2.5± .8 	1.20± .04 .17± .04 1.21	30 .28 .00					10235±106 10199 10050
030630.2-664631 284.78 -45.30 227.94 -27.06 030550.0-665800	NGC 1244 ESO 82- 8 IRAS03058-6657 PGC 11659	.SAR2P* S 2.3± .7 	1.28± .04 .71± .04 1.28	2 .04 .87 .35	 13.91 ±.14 13.24 12.94				5385± 69 5224 5342
030632.8+412908 148.42 -14.60 346.54 -12.25 030316.5+411735	 UGC 2534 PGC 11661	.E?.... 	1.11± .16 .07± .08 1.17	 .51 .00	 14.9 ±.3 14.28				5172± 94 5280 5002
030640.4-325153 232.35 -60.29 266.92 -35.89 030438.0-330324	IC 1885 ESO 357- 3 PGC 11665	.S..5?. BS (1) 5.3±1.0 4.4±1.2	1.14± .04 .45± .04 1.14	138 .00 .68 .23	 14.87 ±.14				
030645.4-254311 218.06 -59.76 275.86 -36.09 030435.0-255442	NGC 1210 ESO 480- 31 PGC 11666	PLBT+P. S -1.5± .5 	1.30± .04 .05± .04 1.00± .07 1.30	 .00 .00	13.46 ±.19 13.70 ±.14 13.56	.81± .03 .26± .04 .77 .28	.87± .02 .35± .03 13.95± .25 14.73± .31		3928± 52 3846 3770
030652.5-004748 179.61 -48.19 305.85 -31.84 030419.5-005919	NGC 1211 UGC 2545 IRAS03043-0059 PGC 11670	RSBR0.. UE .0± .4 	1.32± .04 .06± .04 .83± .03 1.34	30 .19 .05	13.26 ±.15 13.13 ±.13 12.91	.92± .02 .44± .03 .84 .41	1.00± .02 .48± .03 12.90± .11 14.58± .25		3199± 37 3189 3006
030656.0-093237 190.60 -53.54 295.73 -34.17 030430.5-094408	NGC 1214 MCG -2- 8- 51 HICK 23A PGC 11675	.LBT+?. E -.5± .9 	1.11± .05 .65± .04 1.04	40 .28 .00	14.99S±.15 14.64		 13.77± .30		4841± 38 4805 4657
030656.7+413155 148.47 -14.52 346.62 -12.29 030340.3+412023	 UGC 2536 PGC 11676	.L..... U -2.0± .9 	1.11± .08 .60± .04 1.08	124 .51 .00	 15.3 ±.3 14.68				4784± 94 4892 4615
0306.9 -1401 197.27 -55.83 290.35 -35.04 0304.6 -1413	 MCG -2- 8- 53 PGC 11677	.SBS9*. EU (1) 9.2± .5 8.7± .6	1.17± .04 .08± .04 1.18	95 .18 .08 .04				15.34±.1 76± 7 69± 12	1526± 6 1477 1348
030700.8+361004 151.38 -19.10 342.08 -15.28 030351.8+355833	 UGC 2540 PGC 11679	.S..2.. U 2.0± .9 	1.23± .05 .87± .05 1.29	69 .68 1.07 .43	 15.49 ±.13 13.71				3929 4024 3751
030701.1-665614 284.89 -45.15 227.75 -27.03 030622.0-670742	NGC 1246 ESO 82- 9 FAIR 229 PGC 11680	.E.5.*. S -5.0± .6 	1.12± .05 .23± .04 1.06	40 .04 .00	 13.88 ±.14 13.76				5310± 63 5149 5268
030705.0+463715 145.84 -10.13 350.84 -9.41 030340.2+462543	 UGC 2537 IRAS03036+4625 PGC 11682	.SAS6.. U 6.0± .8 	1.08± .06 .02± .05 1.22	 1.43 .03 .01	 13.9 ±.3 13.32 12.45			15.81±.1 126± 11 145± 5 3.35	4997± 5 5117 4838
0307.1 +3420 152.44 -20.65 340.51 -16.29 0304.0 +3408	 UGC 2541 IRAS03040+3408 PGC 11685	.S..8.. U 8.0± .9 	1.04± .15 .18± .12 1.10	28 .59 .22 .09					11481 11572 11301

R.A. 2000 DEC. l　　b SGL　　SGB R.A. 1950 DEC.	Names PGC	Type S_T　n_L T L	$\log D_{25}$ $\log R_{25}$ $\log A_e$ $\log D_o$	p.a. A_g A_i A_{21}	B_T m_B m_{FIR} B_T^o	$(B-V)_T$ $(U-B)_T$ $(B-V)_T^o$ $(U-B)_T^o$	$(B-V)_e$ $(U-B)_e$ m'_e m'_{25}	m_{21} W_{20} W_{50} HI	V_{21} V_{opt} V_{GSR} V_{3K}
030709.1+414459 148.38 -14.32 346.82 -12.19 030352.3+413328	 UGC 2538 PGC 11686	.SB.1.. U 1.0± .8 	1.00± .10 .29± .06 1.05	137 .62 .30 .15	 15.2 ±.3 14.17				5606± 94 5715 5438
030709.6-093532 190.72 -53.52 295.69 -34.23 030444.1-094702	NGC 1215 MCG -2- 8- 55 HICK 23B PGC 11687	RSXR1P. E 1.0± .6 	1.17± .04 .14± .04 1.20	75 .28 .15 .07	15.00S±.15 14.52		 15.35± .29		4933± 38 4897 4749
030713.1-312401 229.39 -60.20 268.74 -36.10 030509.0-313530	A 0305-31 ESO 417- 18 IRAS03051-3135 PGC 11691	.SXR6.. PSr (2) 5.9± .4 1.9± .5	1.39± .03 .23± .04 1.39	178 .00 .34 .12	13.9 ±.4 13.84 ±.14 13.48	.61± .06 .54 	 15.11± .44		4847± 22 4751 4702
030718.6-093643 190.78 -53.50 295.68 -34.27 030453.2-094813	NGC 1216 MCG -2- 8- 56 HICK 23C PGC 11693	.L..+?/ E -1.0±1.4 	.88± .07 .54± .04 .82	65 .28 .00 	15.83S±.15 15.48		 13.72± .39		5016± 41 4979 4832
030723.4+375008 150.53 -17.64 343.55 -14.42 030412.1+373838	 UGC 2543 PGC 11696	.SA.8.. U 8.0± .8 	1.27± .04 .48± .05 1.33	105 .55 .59 .24	 14.59 ±.14 				
0307.4 +0231 176.19 -45.82 309.59 -30.88 0304.8 +0220	 UGC 2547 PGC 11698	.SB.2.. U 2.0± .9 	1.00± .06 .29± .05 1.01	148 .18 .36 .15	 15.10 ±.12 				
030726.7-123515 195.16 -55.03 292.11 -34.90 030503.8-124644	IC 291 MCG -2- 9- 1 IRAS03050-1246 PGC 11699	RSBR0?. E .0± .9 	1.02± .07 .21± .05 1.02	85 .17 .15 	 12.97 				
030732.8+422314 148.11 -13.73 347.40 -11.89 030415.0+421144	IC 288 UGC 2544 PGC 11702	.S?.... 	1.00± .06 .58± .05 1.07	42 .71 .87 .29	 14.8 ±.3 13.21				5094± 94 5204 4927
030738.2+391615 149.79 -16.39 344.79 -13.66 030424.9+390445	 UGC 2546 PGC 11711	RSBR3?. U 3.0± .9 	.89± .08 .00± .05 .95	 .65 .00 .00	 15.3 ±.3 				
030740.3-664010 284.55 -45.29 227.99 -27.20 030700.0-665136	 ESO 82- 10 FAIR 230 PGC 11712	.SXR5.. Sr (1) 4.8± .5 3.3± .6	1.10± .05 .07± .04 1.10	90 .04 .10 .03	 14.19 ±.14 14.02				5748 5587 5706
030740.4-393620 245.57 -59.18 258.55 -35.33 030547.0-394748	 ESO 300- 12 PGC 11713	.SBS5*. S (1) 5.0±1.3 5.6± .9	1.08± .05 .42± .04 1.08	10 .00 .63 .21	 14.79 ±.14 				
0307.8 +0309 175.65 -45.30 310.32 -30.75 0305.2 +0258	IC 1882 UGC 2551 PGC 11718	.S?.... 	.97± .07 .31± .05 1.00	20 .28 .47 .16	 14.92 ±.12 14.14				7184 7185 6990
030800.9+233854 159.12 -29.46 331.02 -22.02 030505.9+232726	 UGC 2549 PGC 11725	.S..3.. U (1) 3.0± .9 3.5±1.2	.99± .06 .36± .05 1.05	135 .56 .50 .18			 	16.56±.3 425± 7 	10355± 10 10416 10165
030811.1-225740 212.92 -58.84 279.33 -36.31 030558.0-230906	NGC 1229 ESO 480- 33 IRAS03059-2309 PGC 11734	.SB.3*P R 3.0± .4 	1.16± .03 .21± .04 1.16	 .01 .28 .10	 14.87 ±.14 12.81 14.49				10592± 57 10516 10430
030811.1-225527 212.85 -58.83 279.38 -36.31 030558.0-230654	NGC 1228 ESO 480- 32 PGC 11735	PLBT+P* SU -1.3± .4 	1.17± .04 .23± .04 1.13	77 .01 .00 	 14.21 ±.14 14.04				10385± 44 10310 10223
030815.2+382254 150.38 -17.09 344.11 -14.26 030503.1+381126	NGC 1207 UGC 2548 IRAS03050+3811 PGC 11737	.SAT3.. U (1) 3.0± .7 2.5±1.0	1.36± .04 .14± .05 1.41	123 .61 .20 .07	 13.38 ±.14 11.96 12.54				4787± 38 4887 4614
030815.7-041535 183.99 -50.16 302.03 -33.18 030545.6-042701	NGC 1221 MCG -1- 9- 2 PGC 11739	.LBR+?/ E -1.0±1.3 	1.07± .06 .44± .05 1.02	20 .14 .00 					

R.A. 2000 DEC. l b SGL SGB R.A. 1950 DEC.	Names PGC	Type S_T n_L T L	$\log D_{25}$ $\log R_{25}$ $\log A_o$ $\log D_o$	p.a. A_g A_i A_{21}	B_T m_B m_{FIR} B_T^o	$(B-V)_T$ $(U-B)_T$ $(B-V)_T^o$ $(U-B)_T^o$	$(B-V)_e$ $(U-B)_e$ m'_e m'_{25}	m_{21} W_{20} W_{50} HI	V_{21} V_{opt} V_{GSR} V_{3K}
030815.8+362657 151.45 -18.73 342.46 -15.33 030506.3+361530	 UGC 2550 PGC 11740	.S..6*. U 6.0±1.2 	1.19± .05 .28± .05 1.25	95 .41 .14	 .68 14.72 ±.16 13.62				4022 4117 3846
030819.0-230034 213.02 -58.83 279.27 -36.34 030606.0-231200	NGC 1230 ESO 480- 34 PGC 11743	.LB.0?P S -2.0± .8 	.80± .06 .21± .03 .77	 	 .02 15.43 ±.14 .00 				
030826.5+040644 174.86 -44.51 311.43 -30.57 030549.4+035518	NGC 1218 UGC 2555 PGC 11749	.S..0.. U .0± .8	1.11± .03 .09± .05 .78± .02 1.15	155 .46 .07 	13.84 ±.13 13.79 ±.12 13.16	1.15± .02 .69± .08 .95 .60	1.21± .01 13.23± .06 14.01± .24		 8644± 38 8648 8451
030827.0-230322 213.13 -58.81 279.22 -36.38 030614.0-231448	IC 1892 ESO 480- 36 PGC 11750	.SBS7P. S 7.5± .6 6.7± .6	1.29± .03 .27± .04 1.29	10 .02 .33 .13	 13.80 ±.14 13.45			14.51±.3 180± 16 169± 12 .93	2888± 11 2986± 66 2815 2729
0308.4 -0748 188.57 -52.26 297.89 -34.14 0306.0 -0800	 PGC 11751	.SBT8.. E (1) 8.0± .9 7.5± .8	1.12± .08 .06± .08 1.15	95 .28 .08 .03					
030828.2+020630 176.89 -45.92 309.24 -31.27 030552.7+015504	NGC 1219 UGC 2556 IRAS03058+0154 PGC 11752	.SAS4*. UE (1) 4.0± .6 3.1± .8	1.08± .03 .01± .04 .82± .02 1.10	 .18 .01 .00	13.82 ±.13 13.39 ±.12 12.92 13.37	.82± .03 .26± .04 .74 .20	.86± .03 .24± .04 13.41± .03 14.07± .21		6101± 50 6098 5909
0308.5 +2046 161.17 -31.73 328.38 -23.52 0305.6 +2034	 UGC 2553 IRAS03056+2034 PGC 11755	.SBT3.. U 3.0± .8 	1.03± .06 .12± .05 1.07	 .50 .17 .06	 14.73 ±.13 11.66 14.01				8225 8278 8033
030848.4-070232 187.64 -51.74 298.82 -34.04 030620.7-071357	 MCG -1- 9- 6 IRAS03063-0713 PGC 11767	.SBT6?. E (1) 6.0± .9 5.3±1.6	1.10± .06 .54± .05 1.13	160 .30 .79 .27	 13.65 				9014 8984 8829
030849.2-351406 237.01 -59.69 263.92 -36.13 030650.0-352530	 ESO 357- 5 PGC 11768	.SBT7.. S (1) 7.0± .8 5.6± .8	1.06± .05 .09± .04 1.06	88 .02 .12 .04	 15.35 ±.14 15.20				4466 4360 4332
0308.9 +7033 133.95 10.69 10.16 4.51 0304.0 +7022	 UGC 2542 IRAS03040+7022 PGC 11770	.S..3.. U 3.0±1.0 	1.16± .07 .88± .06 1.50	145 3.64 1.22 .44	 11.74 				4276 4435 4178
030854.9+412802 148.83 -14.39 346.78 -12.63 030538.2+411636	 UGC 2554 PGC 11771	.L...?. U -2.0±1.7 	1.11± .16 .15± .08 1.16	165 .61 .00 	15.0 ±.3 14.31				2840± 94 2947 2673
030857.4-025706 182.60 -49.20 303.59 -32.98 030626.3-030831	NGC 1222 MCG -1- 9- 5 MK 603 PGC 11774	.L..-P* E -3.0±1.2 	1.04± .05 .10± .05 1.04	10 .13 .00 	13.10 ±.13 10.53 12.93	.60± .01 -.14± .02 .54 -.15	 12.92± .30	14.95±.3 192± 16 167± 12 	2462± 11 2530± 62 2446 2276
030909.1-101733 192.16 -53.49 294.98 -34.86 030644.3-102857	 MCG -2- 9- 3 IRAS03067-1028 PGC 11782	PSBR1P? E 1.0±1.7 	1.13± .06 .34± .05 1.16	175 .28 .35 .17	 13.77 				3246±106 3206 3065
0309.2 +3941 149.84 -15.88 345.32 -13.68 0306.0 +3930	 UGC 2558 PGC 11787	.I..9*. U 10.0±1.2 	1.00± .16 .04± .12 1.05	 .57 .03 .02					
030917.4+383859 150.42 -16.76 344.45 -14.28 030604.8+382735	NGC 1213 UGC 2557 PGC 11789	.SAS8.. U 8.0± .7 	1.25± .06 .10± .06 1.30	60 .59 .12 .05	 15.0 ±.3 14.26			14.27±.1 149± 8 -.04	3427± 11 3527 3256
030917.9+425822 148.09 -13.07 348.07 -11.82 030558.9+424658	 UGC 2559 PGC 11790	.L..... U -2.0± .8 	1.15± .07 .32± .03 1.22	50 1.03 .00 	14.0 ±.3 12.86 				5680± 94 5790 5516
030919.1+800755 128.82 18.87 17.91 10.11 030209.9+795625	 UGC 2519 IRAS03021+7956 PGC 11793	.S..6?. U 6.0±1.7 	1.08± .09 .22± .07 1.17	 .98 .32 .11	14.30 ±.18 11.78 12.98 			15.37±.1 226± 11 2.28	2377± 9 2545 2308

R.A. 2000 DEC. / l b / SGL SGB / R.A. 1950 DEC.	Names / / / PGC	Type / S_T n_L / T / L	$\log D_{25}$ / $\log R_{25}$ / $\log A_o$ / $\log D_o$	p.a. / A_g / A_i / A_{21}	B_T / m_B / m_{FIR} / B_T^o	$(B-V)_T$ / $(U-B)_T$ / $(B-V)_T^o$ / $(U-B)_T^o$	$(B-V)_o$ / $(U-B)_o$ / m'_o / m'_{25}	m_{21} / W_{20} / W_{50} / HI	V_{21} / V_{opt} / V_{GSR} / V_{3K}
030930.4-172456		.SXS9..	1.10± .04	116					2204
203.26 -56.74	ESO 547- 11	SE (1)	.14± .03	.11	15.60 ±.14				
286.29 -36.13		9.0± .6		.15					2144
030712.0-173618	PGC 11798	8.9±1.1	1.11	.07	15.34				2034
030931.2-045444		.SBT7*.	1.17± .06	25					3130
185.10 -50.32	MCG -1- 9- 10	E (1)	.24± .05	.20					
301.37 -33.66		7.0± .8		.34					3106
030701.7-050607	PGC 11800	6.4±1.2	1.18	.12					2944
030932.0-493545			.85± .07	8					
262.70 -55.34	ESO 199- 21		.11± .06	.00	15.09 ±.14				11500±190
246.52 -33.49	FAIR 740								11363
030756.0-494706	PGC 11801		.85						11405
030936.6-251514 IC 1895		RSXT0P*	1.17± .05		14.22 ±.15	.89± .01	.92± .01		
217.39 -59.04	ESO 481- 1	Sr	.12± .04	.00	13.95 ±.14	.34± .02	.35± .02		3823± 52
276.46 -36.72		-.3± .5	.40± .04	.09		.83	11.71± .15		3741
030726.0-252636	PGC 11807		1.17		13.93	.34	14.66± .31		3667
0309.6 +1829		.S..6*.	1.16± .07	18					10730
163.03 -33.40	UGC 2563	U	.88± .06	.35	15.34 ±.12				
326.33 -24.83	IRAS03067+1818	6.0±1.3		1.30	12.87				10775
0306.7 +1818	PGC 11808		1.19	.44	13.65				10538
030938.0-410151		.SXS9..	1.70± .03	166	13.0 ±.3	.65± .10		13.90±.2	951± 7
248.06 -58.46	ESO 300- 14	S (1)	.29± .05	.00	13.12 ±.14	-.20± .17		146± 11	
256.73 -35.46		9.0± .6		.30		.58		131± 8	830
030747.0-411312	PGC 11812	10.0± .7	1.70	.15	12.80	-.25	15.61± .34	.95	832
030939.1-075049 NGC 1234		.SBR6P.	1.15± .06	50					
188.90 -52.03	MCG -1- 9- 11	E (1)	.16± .05	.27					
297.94 -34.44		6.0± .8		.23					
030712.1-080211	PGC 11813	5.3± .8	1.18	.08					
030942.8+405828 IC 290		.S..3..	1.04± .08	131					
149.22 -14.73	UGC 2561	U (1)	.80± .06	.55	15.4 ±.3				6026± 94
346.45 -13.03		3.0±1.0		1.11					6132
030626.8+404705	PGC 11817	4.5±1.3	1.09	.40	13.68				5859
030945.3-203452 NGC 1232		.SXT5..	1.87± .01	108	10.52 ±.14	.63± .02	.71± .01	12.12±.1	1682± 5
208.78 -57.81	ESO 547- 14	R (4)	.06± .02	.04	10.58 ±.12	.00± .08	.10± .02	250± 6	1750± 38
282.35 -36.51		5.0± .3	1.54± .02	.10		.60	13.71± .04	233± 6	1614
030730.0-204613	PGC 11819	2.0± .3	1.87	.03	10.41	-.02	14.56± .16	1.68	1519
030952.8-100300		PSBR1?.	1.12± .06	135					
191.97 -53.21	MCG -2- 9- 6	E	.32± .05	.28					4018±106
295.32 -34.99	IRAS03074-1014	1.0± .9		.32	12.88				3978
030727.8-101421	PGC 11824		1.14	.16					3838
030956.1+160149			1.00?						9476
164.94 -35.31	CGCG 464- 4		.52?		.44	15.39 ±.12			
323.96 -26.04									9514
030708.3+155027	PGC 11829		1.04						9284
0310.0 +4206		.L.....	1.00± .19	135					
148.66 -13.74	UGC 2562	U	.09± .08	.80					
347.42 -12.43		-2.0± .9		.00					
0306.7 +4155	PGC 11832		1.08						
031002.3-203557 NGC 1232A		.SBS9..	.97± .04	5					
208.85 -57.75	ESO 547- 16	R (1)	.07± .03	.02	15.28 ±.14				6496± 63
282.33 -36.58		9.0± .5		.07					6426
030747.0-204718	PGC 11834	6.7±1.2	.97	.03	15.18				6332
031002.6-532004 NGC 1249		.SBS6..	1.69± .02	86	12.19 ±.14	.43± .01	.50± .01	12.47±.3	1074± 6
268.21 -53.41	ESO 155- 6	R (2)	.32± .03	.00	12.08 ±.12	-.21± .03	-.12± .02	232± 7	1003± 58
242.20 -32.45	IRAS03085-5331	6.0± .7	1.15± .02	.47	12.07	.36	13.43± .04		929
030835.1-533124	PGC 11836	5.3± .4	1.69	.16	11.65	-.26	14.68± .18	.66	989
0310.1 +3351		.SBS8..	1.00± .08	65					12502
153.27 -20.72	UGC 2566	U	.19± .06	.62					
340.44 -17.07		8.0± .9		.23					12590
0307.0 +3340	PGC 11838		1.06	.09					12325
031009.5+315527		.S..8..	1.10± .07	20				15.71±.3	12161± 10
154.41 -22.34	UGC 2565	U	.11± .06	.72					
338.76 -18.13	VV 833	8.0± .8		.13				369± 7	12244
030705.3+314406	PGC 11840		1.16	.05					11981
0310.2 +4210		.L.....	1.00± .10						
148.66 -13.66	UGC 2564	U	.05± .04	.80					
347.50 -12.42		-2.0± .8		.00					
0306.9 +4159	PGC 11844		1.08						

R.A. 2000 DEC. l b SGL SGB R.A. 1950 DEC.	Names PGC	Type S_T n_L T L	$\log D_{25}$ $\log R_{25}$ $\log A_e$ $\log D_o$	p.a. A_g A_i A_{21}	B_T m_B m_{FIR} B_T^o	$(B-V)_T$ $(U-B)_T$ $(B-V)_T^o$ $(U-B)_T^o$	$(B-V)_e$ $(U-B)_e$ m'_e m'_{25}	m_{21} W_{20} W_{50} HI	V_{21} V_{opt} V_{GSR} V_{3K}
031013.4+404559 149.42 -14.86 346.33 -13.23 030657.5+403438	IC 292 UGC 2567 IRAS03069+4034 PGC 11846	.S..8*. U 8.0±1.3 	1.08± .03 .27± .05 .62± .02 1.13	75 .51 .33 .13	14.21 ±.15 14.1 ±.3 12.26 13.35	.71± .03 .03± .05 .52 -.10	.75± .02 .12± .04 12.80± .05 13.81± .25		3018± 71 3123 2851
031014.4+421316 148.64 -13.62 347.55 -12.40 030656.3+420155	 UGC 2568 PGC 11847	.L..... U -2.0± .8 	1.18± .15 .31± .08 1.22	 .80 .00 	 14.7 ±.3 13.88				4677± 94 4785 4512
031020.5-222416 212.14 -58.22 280.07 -36.78 030807.1-223536	IC 1898 ESO 481- 2 IRAS03081-2235 PGC 11851	.SBS5*/ SU (1) 5.5± .6 4.4±1.2	1.56± .02 .80± .03 1.56	73 .00 1.17 .40	 13.58 ±.14 12.96 12.40		14.28±.2 253± 16 243± 12 1.48	1328± 11 1253 1168
0310.3 +2024 161.84 -31.77 328.26 -24.08 0307.5 +2013	 UGC 2570 PGC 11853	.S?.... 	1.16± .07 .71± .06 1.21	117 .50 1.06 .35					10390 10441 10200
031024.6-330917 232.87 -59.50 266.48 -36.65 030823.0-332036	 ESO 357- 7 PGC 11856	.SBS9./ S (1) 9.0± .8 7.8± .9	1.37± .04 .79± .04 1.37	132 .00 .80 .39	 14.66 ±.14 13.85		15.85±.2 126± 11 124± 8 1.60	1119± 7 1016 981
031030.4+225407 160.16 -29.74 330.63 -22.88 030736.1+224247	 UGC 2571 PGC 11859	.S..3.. U (1) 3.0±1.0 3.5±1.3	1.00± .08 .84± .06 1.05	25 .54 1.16 .42			16.51±.3 399± 7	10544± 10 10602 10356
031032.8-011632 181.05 -47.82 305.65 -32.86 030800.2-012751	 UGC 2576 KUG 0308-014 PGC 11862	.S..4.. U 4.0±1.0 	1.03± .05 .68± .05 1.04	141 .12 1.00 .34				
0310.5 +4131 149.07 -14.18 347.00 -12.85 0307.3 +4120	 UGC 2569 PGC 11863	.L...?. U -2.0±1.7 	1.00± .19 .05± .08 1.06	 .61 .00 				
0310.8 +4411 147.69 -11.88 349.23 -11.35 0307.5 +4400	 UGC 2572 PGC 11867	.SB.4*. U 4.0± .9 	1.00± .16 .09± .12 1.10	 1.10 .13 .04				
031052.7-104452 193.18 -53.37 294.54 -35.37 030828.4-105610	NGC 1238 MCG -2- 9- 10 PGC 11868	.E+.... E -4.0± .8 	1.20± .10 .12± .07 1.21	110 .27 .00 				
031053.8-023313 182.60 -48.58 304.22 -33.33 030822.3-024431	NGC 1239 MCG -1- 9- 12 PGC 11869	.LA.0P* E -2.0± .8 	1.05± .06 .29± .06 1.03	110 .16 .00 				
031055.8+450103 147.26 -11.16 349.92 -10.88 030733.0+444944	 UGC 2573 PGC 11872	.SB.3.. U 3.0± .8 	1.12± .06 .53± .05 1.24	82 1.30 .73 .26	 14.8 ±.3 			
0311.0 -0225 182.49 -48.47 304.37 -33.32 0308.5 -0237	 PGC 11876	.S..1?/ E 1.0±1.8 	1.16± .08 .66± .08 1.18	35 .16 .67 .33				
031103.1+403720 149.64 -14.90 346.30 -13.44 030747.4+402601	IC 294 UGC 2574 PGC 11878	RSBT0*. U .0± .8 	1.14± .07 .12± .06 1.18	 .51 .09 	 14.9 ±.3 14.25			5015± 94 5119 4848
031105.5+352314 152.57 -19.32 341.88 -16.40 030756.9+351156	NGC 1226 UGC 2575 PGC 11879	.E..... U -5.0± .7 	1.32± .07 .04± .05 1.41	95 .62 .00 	 13.85 ±.16 			
031107.8+351928 152.61 -19.37 341.83 -16.44 030759.3+350810	NGC 1227 UGC 2577 PGC 11880	RLBS+.. U -1.0± .8 	1.00± .08 .04± .06 1.05	 .62 .00 	 15.24 ±.14 			
0311.2 -0414 184.68 -49.58 302.29 -33.88 0308.7 -0426	 MCG -1- 9- 13 PGC 11885	.SAS9.. E (1) 9.0± .8 8.7± .8	1.18± .05 .13± .05 1.19	130 .19 .14 .07				2244 2221 2060

R.A. 2000 DEC. / l b / SGL SGB / R.A. 1950 DEC.	Names / PGC	Type / S_T n_L / T / L	$\log D_{25}$ / $\log R_{25}$ / $\log A_e$ / $\log D_o$	p.a. / A_g / A_i / A_{21}	B_T / m_B / m_{FIR} / B_T^o	$(B-V)_T$ / $(U-B)_T$ / $(B-V)_T^o$ / $(U-B)_T^o$	$(B-V)_e$ / $(U-B)_e$ / m'_e / m'_{25}	m_{21} / W_{20} / W_{50} / HI	V_{21} / V_{opt} / V_{GSR} / V_{3K}
031113.7+412149 / 149.26 -14.25 / 346.94 -13.04 / 030756.8+411031	NGC 1224 / UGC 2578 / PGC 11886	.L..-*. / U / -3.0±1.1	1.15± .15 / .07± .08 / / 1.22	.60 / .00	14.7 ±.3 / / / 14.00				5051± 71 / 5157 / 4886
031114.7-085519 / 190.71 -52.31 / 296.76 -35.07 / 030848.7-090636	NGC 1241 / MCG -2- 9- 11 / PGC 11887	.SBT3.. / R (2) / 3.0± .3 / 2.3± .6	1.45± .03 / .22± .03 / .99± .02 / 1.47	145 / .23 / .30 / .11	11.99V±.13 / 12.75 ±.20 / / 12.25	.85± .02 / .08± .04 / .73 / -.02	.94± .01 / .20± .04 / / 14.40± .20	429± 11 / 414± 8	4030± 9 / 3939± 31 / 3986 / 3843
0311.2 +0119 / 178.40 -45.95 / 308.64 -32.20 / 0308.7 +0108	IC 298 / MCG 0- 9- 15 / 1ZW 11 / PGC 11890	.RING.. / R	.64± .17 / .11± .07 / / .66	.24					9656± 18 / 9650 / 9468
031119.4-085408 / 190.70 -52.29 / 296.79 -35.08 / 030853.4-090525	NGC 1242 / MCG -2- 9- 12 / PGC 11892	.SBT5*. / RE (1) / 5.4± .5 / 5.3±1.2	1.07± .03 / .23± .05 / .69± .02 / 1.09	130 / .16 / .35 / .12	14.32 ±.13 / / / 13.78	.61± .02 / -.05± .04 / .50 / -.13	.63± .02 / .04± .04 / 13.26± .08 / 13.94± .24		3938± 35 / 3901 / 3758
031119.6+011846 / 178.42 -45.94 / 308.64 -32.21 / 030844.8+010729	IC 298A / MCG 0- 9- 16 / IRAS03087+0107 / PGC 11893	.RING.A / R / 4.0± .9	.84± .09 / .11± .04 / / .86	5 / .24 / .16 / .06	13.13				
031127.0+352729 / 152.59 -19.23 / 341.98 -16.42 / 030818.4+351612	UGC 2579 / PGC 11896	.E...?. / U / -5.0±1.7	1.08± .17 / .17± .08 / / 1.16	.80 / .00	15.13 ±.14				
0311.6 -1037 / 193.17 -53.16 / 294.73 -35.52 / 0309.2 -1049	MCG -2- 9- 13 / PGC 11902	.S?....	1.04± .09 / .66± .07 / / 1.03	.27 / .50					4504 / 4461 / 4327
031147.5-002412 / 180.39 -47.01 / 306.75 -32.88 / 030914.2-003527	UGC 2585 / PGC 11912	.SBR3*. / UE (1) / 3.3± .6 / 4.2± .8	1.20± .04 / .11± .04 / / 1.21	165 / .16 / .15 / .05	13.99 ±.14 / / / 13.63				6847± 50 / 6835 / 6660
0311.9 -5042 / 264.09 -54.48 / 245.10 -33.54 / 0310.4 -5053	PGC 11915			.00	16.51 ±.17	1.04± .06			18564± 88 / 18424 / 18474
031159.9-010944 / 181.28 -47.47 / 305.91 -33.17 / 030927.2-012058	UGC 2587 / PGC 11919	.S..4*/ / UE (1) / 4.0± .9 / 4.5±1.3	1.17± .04 / .75± .04 / / 1.18	28 / .17 / 1.10 / .38					
031203.9+280732 / 157.09 -25.26 / 335.62 -20.51 / 030903.9+275617	UGC 2582 / PGC 11921	.S..6?. / U / 6.0±1.7	.96± .17 / .05± .12 / / 1.04	.85 / .07 / .02				16.28±.3 / 280± 7	16992± 10 / 17063 / 16810
031205.2-423311 / 250.58 -57.60 / 254.77 -35.62 / 031017.0-424424	ESO 248- 5 / PGC 11923	.L?....	.89± .06 / .37± .03 / / .83	142 / .00	15.78 ±.16 / 15.72 ±.14	.97± .04	/ / / 14.16± .35		
031208.5-250752 / 217.38 -58.46 / 276.63 -37.29 / 030958.0-251906	ESO 481- 7 / PGC 11924	PLA.0*. / S / -2.0±1.1	1.17± .04 / .04± .03 / / 1.17	.00 / .00	14.30 ±.14 / / / 14.20				6486± 52 / 6403 / 6333
0312.1 +4015 / 150.02 -15.10 / 346.12 -13.82 / 0308.9 +4004	UGC 2581 / PGC 11926	.S..6*. / U / 6.0±1.4	1.07± .07 / .79± .06 / / 1.13	90 / .60 / 1.17 / .40					4262 / 4365 / 4096
031214.2-102850 / 193.09 -52.95 / 294.95 -35.64 / 030949.6-104004	NGC 1247 / MCG -2- 9- 14 / IRAS03098-1040 / PGC 11931	.S..4./ / UE / 4.0± .6	1.53± .02 / .81± .05 / .94± .03 / 1.55	69 / .27 / 1.19 / .41	13.47 ±.14 / / 12.33 / 11.98	.99± .02 / .33± .03 / .76 / .09	1.00± .01 / .43± .02 / 13.66± .08 / 13.93± .21		3948 / 3905 / 3772
031220.1-830727 / 298.23 -32.67 / 212.15 -19.50 / 031648.0-831830	ESO 3- 14 / PGC 11936	.SBS8.. / S (1) / 7.5± .6 / 7.2± .6	1.20± .04 / .33± .04 / / 1.24	107 / .38 / .40 / .16	15.21 ±.14				
0312.5 +1224 / 168.41 -37.71 / 320.63 -28.21 / 0309.8 +1213	UGC 2589 / PGC 11954	.S..4.. / U / 4.0± .8	1.06± .08 / .11± .06 / / 1.14	40 / .83 / .16 / .05					9874 / 9900 / 9685

R.A. 2000 DEC. l b SGL SGB R.A. 1950 DEC.	Names PGC	Type S_T n_L T L	$\log D_{25}$ $\log R_{25}$ $\log A_e$ $\log D_o$	p.a. A_g A_i A_{21}	B_T m_B m_{FIR} B_T^o	$(B-V)_T$ $(U-B)_T$ $(B-V)_T^o$ $(U-B)_T^o$	$(B-V)_e$ $(U-B)_e$ m'_e m'_{25}	m_{21} W_{20} W_{50} HI	V_{21} V_{opt} V_{GSR} V_{3K}
031233.3+391904 150.61 -15.85 345.38 -14.42 030919.2+390750	NGC 1233 UGC 2586 IRAS03093+3907 PGC 11955	.S..3.. U (1) 3.0± .8 4.5±1.1	1.25± .06 .44± .06 1.31	27 .63 .61 .22	14.0 ±.3 13.8 ±.3 12.13 12.63	.84± .06 .20± .11 .59 .00	 14.00± .45		 4465± 80 4566 4298
031245.6-002002 180.54 -46.78 306.92 -33.09 031012.3-003114	 UGC 2594 PGC 11966	.SB.0*. U .0± .9	.95± .07 .08± .05 .97	 .19 .06	 14.35 ±.12 14.00				 6826± 50 6813 6641
031247.6-051607 186.31 -49.89 301.22 -34.54 031018.4-052719	 MK 604 PGC 11968	.S?....	1.02± .06 .62± .12 1.04	 .21 .90 .31					 2097 2069 1916
031248.3-312912 229.60 -59.01 268.54 -37.28 031045.0-314024	 ESO 417- 20 PGC 11969	.SB?... 	1.18± .04 .15± .04 1.18	51 .00 .22 .07	 15.14 ±.14 14.90				 3972 3872 3833
031248.5-051328 186.26 -49.86 301.28 -34.54 031019.4-052440	NGC 1248 MCG -1- 9- 16 PGC 11970	.LAS0*. R -2.0± .5	1.06± .05 .06± .05 .70± .01 1.07	80 .21 .00	13.36 ±.13 13.11	.86± .01 .33± .02 .79 .29	.92± .01 .40± .02 12.35± .05 13.37± .32		 2243± 68 2215 2062
0312.8 +1847 163.55 -32.71 327.01 -25.37 0310.0 +1836	 UGC 2592 PGC 11971	.S..3.. U (1) 3.0± .9 4.5±1.2	.97± .07 .31± .05 1.00	93 .34 .43 .16	 15.37 ±.13 14.52				9936 9980 9749
031251.7+044218 175.36 -43.31 312.54 -31.38 031013.9+043106	IC 302 UGC 2595 IRAS03102+0431 PGC 11972	.SBT4.. U (1) 4.0± .8 3.3± .8	1.27± .04 .09± .05 1.04± .03 1.31	 .44 .14 .05	13.59 ±.16 13.59 ±.14 12.93 12.97	.84± .03 .11± .04 .68 -.02	.94± .02 14.28± .06 14.54± .29	14.22±.0 276± 5 249± 6 1.20	5904± 4 5907 5717
031255.7+223119 160.94 -29.72 330.58 -23.56 031001.5+222007	 MCG 4- 8- 12 PGC 11976	.S?....	.74± .14 .09± .07 .80	 .64 .13 .05	 15.55 ±.12 14.68		16.30±.3 393± 7 1.57		12842± 10 12897 12657
031257.7-175543 204.69 -56.18 285.77 -37.01 031040.0-180654	 ESO 547- 20 PGC 11977	.IBS9.. SE (2) 9.5± .6 9.3± .6	1.14± .04 .07± .03 1.15	75 .07 .05 .04	 15.38 ±.14 15.26				1994 1930 1830
031303.2+422724 148.97 -13.14 348.04 -12.69 030944.3+421612	 UGC 2590 PGC 11982	.S?....	1.00± .16 .09± .12 1.09	 .93 .09 .04	 14.9 ±.3 13.82				 4719± 94 4826 4558
031303.9-572127 273.28 -50.83 237.55 -31.48 031148.1-573236	 ESO 116- 12 IRAS03118-5732 PGC 11984	.SBS7*. S (1) 6.5± .6 6.7± .6	1.54± .04 .51± .05 1.54	25 .00 .71 .26	 12.96 ±.14 12.84 12.25			 229± 9	1140± 8 1150± 47 989 1071
0313.2 +4246 148.82 -12.86 348.32 -12.53 0309.9 +4235	 UGC 2591 PGC 11993	.S..4.. U 4.0± .9	1.04± .08 .37± .06 1.13	22 .93 .55 .19					
031316.0-313908 229.92 -58.91 268.32 -37.37 031113.0-315018	 ESO 417- 21 PGC 11995	.LX.-.. S -3.0± .9	.97± .04 .19± .03 .94	148 .00 .00	 14.13 ±.14 14.07				 4125± 29 4024 3987
0313.3 +1759 164.25 -33.26 326.30 -25.86 0310.5 +1748	 UGC 2597 IRAS03105+1748 PGC 12000	.S..3.. U 3.0± .9	1.02± .08 .29± .06 1.05	170 .34 .41 .15					10399 10441 10212
031328.8+410430 149.79 -14.27 346.95 -13.55 031011.9+405319	 CGCG 540- 59 PGC 12004			 .59	 15.0 ±.4				5390± 94 5494 5227
031329.8+440752 148.13 -11.68 349.47 -11.77 031008.1+435641	 UGC 2596 IRAS03101+4356 PGC 12005	.SA.9*. U 9.0± .8	1.32± .04 .31± .05 1.43	17 1.16 .31 .15	 13.6 ±.3 13.44 12.14				 5343± 94 5454 5185
031332.7-254331 218.60 -58.26 275.88 -37.62 031123.0-255440	NGC 1255 ESO 481- 13 IRAS03113-2554 PGC 12007	.SXT4.. R (3) 4.0± .3 3.3± .5	1.62± .02 .20± .02 1.35± .02 1.62	117 .00 .29 .10	11.40 ±.14 11.78 ±.12 11.57 11.33	.49± .01 -.14± .03 .44 -.17	.56± .01 -.04± .02 13.64± .04 13.86± .18	13.55±.1 257± 10 247± 11 2.13	1697± 6 1714± 33 1612 1548

R.A. 2000 DEC. / l b / SGL SGB / R.A. 1950 DEC.	Names / PGC	Type / S_T n_L / T / L	$logD_{25}$ / $logR_{25}$ / $logA_e$ / $logD_o$	p.a. / A_g / A_i / A_{21}	B_T / m_B / m_{FIR} / B_T^o	$(B-V)_T$ / $(U-B)_T$ / $(B-V)_T^o$ / $(U-B)_T^o$	$(B-V)_e$ / $(U-B)_e$ / m'_e / m'_{25}	m_{21} / W_{20} / W_{50} / HI	V_{21} / V_{opt} / V_{GSR} / V_{3K}
031340.3-251121	A 0311-25	.SBS9*/	1.60± .02	163				13.55±.2	1738± 7
217.62 -58.13	ESO 481- 14	SU (1)	.56± .04	.00	13.82 ±.14			175± 16	
276.57 -37.64		9.3± .6		.57				177± 6	1654
031130.1-252230	PGC 12011	7.8± .8	1.60	.28	13.24			.02	1587
0313.7 +0246		.S..3..	1.00± .06	149					
177.49 -44.51	UGC 2599	U (1)	.68± .05	.25	15.49 ±.12				
310.50 -32.27		3.0±1.0		.94					
0311.1 +0235	PGC 12013	4.5±1.3	1.03	.34					
031345.3-001430 IC 307		RSBR1P?	1.23± .04	73					
180.68 -46.53	UGC 2600	UE	.32± .04	.19	14.24 ±.13				
307.12 -33.29	IRAS03112-0025	1.0± .7		.33	13.49				
031111.8-002539	PGC 12017		1.24	.16					
031353.9+335759			.95?						12261
153.90 -20.20	CGCG 525- 7		.48?						
340.99 -17.67				.61	15.59 ±.12				12347
031046.8+334650	PGC 12029		1.01						12088
031358.6-215910 NGC 1256		.LX.-?.	1.04± .03	108	14.57 ±.13	.96± .01	.99± .02		
211.82 -57.29	ESO 547- 23	S	.39± .04	.01	14.40 ±.14	.35± .02	.44± .03		4312± 60
280.66 -37.59		-2.8± .8	.36± .01	.00		.90	11.86± .04		4236
031145.0-221018	PGC 12032		.98		14.41	.35	13.66± .22		4155
031404.8-214629 NGC 1258		.SXS6*.	1.13± .04	17					
211.46 -57.21	ESO 547- 24	S (1)	.16± .03	.01	13.98 ±.14				1469
280.93 -37.60		5.5± .6		.23					1393
031151.0-215736	PGC 12034	3.9± .6	1.13	.08	13.73				1312
031407.7-375929		DIXS9?.	1.09± .08	171					
242.04 -58.26	ESO 300- 20	S (1)	.22± .05	.00	17.28 ±.14				
260.30 -36.83		10.0±1.7		.17					
031213.0-381036	PGC 12036	12.2± .8	1.09	.11					
031408.5+411731		.L...*.	1.18± .15	25					
149.78 -14.02	UGC 2598	U	.22± .08	.59	14.4 ±.3				4523± 48
347.20 -13.53	MK 1072	-2.0±1.2		.00					4627
031051.2+410622	PGC 12038		1.21		13.77				4360
031408.6+424456				.93	14.3 ±.3				5426± 94
148.98 -12.79	CGCG 540- 62								5534
348.40 -12.68									5266
031049.0+423347	PGC 12039								
031409.1-013810		PSBT3*.	1.14± .04	135					
182.33 -47.36	UGC 2607	UE (1)	.05± .04	.23	14.32 ±.16				
305.57 -33.83		3.0± .6		.07					
031136.9-014917	PGC 12040	4.2±1.2	1.17	.02					
031409.3-024922 NGC 1253		.SXT6..	1.72± .02	82	12.27 ±.13	.55± .04	.66± .01	12.23±.1	1710± 5
183.69 -48.12	MCG -1- 9- 18	R (1)	.35± .04	.19		-.18± .03	.01± .03	326± 7	1749± 15
304.20 -34.18	KUG 0311-030	6.0± .3	1.26± .02	.52		.43	14.06± .04	298± 8	1693
031138.1-030029	PGC 12041	5.9± .6	1.73	.18	11.56	-.27	14.83± .19	.50	1533
0314.1 +1629		.SXS5..	1.13± .05						10099
165.56 -34.32	UGC 2602	U	.07± .05	.42	14.50 ±.16				
324.92 -26.73	IRAS03113+1617	5.0± .8		.10	13.37				10136
0311.3 +1617	PGC 12042		1.17	.03	13.92				9912
031415.2+353223			1.11?						5264
153.05 -18.85	CGCG 525- 8		.81?						
342.38 -16.84				.81	15.53 ±.12				5354
031106.0+352115	PGC 12046		1.19						5093
031423.9+024041 NGC 1254		.LA.-*.	.88± .12						
177.75 -44.45	MCG 0- 9- 33	E	.05± .08	.25	15.10 ±.10				
310.47 -32.46		-3.0±1.4		.00					
031147.9+022934	PGC 12052		.91						
031424.0-024801 NGC 1253A		.SBS9..	1.23± .05	80	14.36 ±.14	.43± .03	.38± .02		
183.72 -48.06	MCG -1- 9- 19	R (1)	.23± .05	.22		-.36± .05	-.41± .04		1831± 16
304.24 -34.24	DDO 31	9.0± .4	.85± .02	.23	12.95	.32	14.10± .04		1810
031152.8-025907	PGC 12053	9.0±1.5	1.25	.11	13.91	-.44	14.79± .31		1650
0314.4 -0153		.L.....	1.04± .09						
182.69 -47.48	UGC 2611	U	.04± .04	.22	14.40 ±.11				
305.29 -33.97		-2.0± .8		.00					
0311.9 -0205	PGC 12056		1.06						
0314.5 +3947		.S..9*.	1.07± .14	102					
150.69 -15.25	UGC 2601	U	.25± .12	.63					
346.00 -14.47		9.0±1.2		.26					
0311.3 +3936	PGC 12063		1.13	.13					

R.A. 2000 DEC. / l b / SGL SGB / R.A. 1950 DEC.	Names / PGC	Type / S_T n_L / T / L	$\log D_{25}$ / $\log R_{25}$ / $\log A_e$ / $\log D_o$	p.a. / A_g / A_i / A_{21}	B_T / m_B / m_{FIR} / B_T^o	$(B-V)_T$ / $(U-B)_T$ / $(B-V)_T^o$ / $(U-B)_T^o$	$(B-V)_e$ / $(U-B)_e$ / m'_e / m'_{25}	m_{21} / W_{20} / W_{50} / HI	V_{21} / V_{opt} / V_{GSR} / V_{3K}
0314.6 -0446 / 186.14 -49.23 / 301.95 -34.86 / 0312.1 -0457	A 0312-04 / MCG -1- 9- 21 / DDO 32 / PGC 12068	.IBS9.. / PE (2) / 9.5± .5 / 8.2± .6	1.33± .04 / .11± .05 / 1.10± .04 / 1.34	95 / .17 / .09 / .06	14.11 ±.17 / / / 13.85	.47± .08 / -.21± .07 / .39 / -.27	.50± .03 / -.12± .04 / 15.10± .09 / 15.31± .31	14.43±.1 / 153± 8 / 114± 12 / .51	2215± 6 / / 2188 / 2036
031440.5+393703 / 150.80 -15.38 / 345.87 -14.58 / 031125.6+392556	/ UGC 2604 / IRAS03113+3925 / PGC 12070	.SXS5.. / U / 5.0± .8 /	1.10± .06 / .15± .05 / / 1.16	140 / .63 / .22 / .07	14.4 ±.3 / / / 13.55				4522± 94 / 4622 / 4358
0314.7 +4017 / 150.44 -14.81 / 346.44 -14.21 / 0311.5 +4006	/ UGC 2605 / / PGC 12073	.I..9*. / U / 10.0±1.2 /	1.14± .13 / .39± .12 / / 1.20	63 / .63 / .29 / .20					
031447.8+421321 / 149.38 -13.17 / 348.04 -13.08 / 031128.9+420214	IC 301 / UGC 2606 / / PGC 12074	.E..... / U / -5.0± .8 /	1.23± .14 / .03± .08 / / 1.36	/ .83 / .00 /	14.2 ±.3 / / / 13.23				6850± 94 / 6956 / 6690
031457.9+392055 / 151.00 -15.58 / 345.68 -14.78 / 031143.3+390949	/ UGC 2610 / / PGC 12078	.S..3.. / U (1) / 3.0± .9 / 2.5±1.2	1.04± .15 / .37± .12 / / 1.10	34 / .66 / .51 / .19	15.3 ±.3 / / / 14.12				5038± 94 / 5137 / 4873
031501.0-304232 / 228.13 -58.51 / 269.49 -37.81 / 031257.0-305336	IC 1904 / ESO 417- 22 / IRAS03129-3053 / PGC 12079	PSBT2*. / Sr / 2.3± .6 /	1.14± .04 / .37± .03 / / 1.14	108 / .00 / .46 / .19	14.41 ±.14 / 12.47 / / 13.90				4643± 24 / 4544 / 4505
031501.3+375252 / 151.84 -16.80 / 344.46 -15.64 / 031148.9+374146	IC 304 / UGC 2609 / IRAS03118+3741 / PGC 12080	.S..3.. / U (1) / 3.0± .9 / 3.5±1.2	1.06± .06 / .24± .05 / / 1.14	25 / .83 / .33 / .12	14.64 ±.12 / 12.47 / / 13.44				4938 / / 5033 / 4771
031501.5+420209 / 149.51 -13.31 / 347.91 -13.23 / 031142.9+415103	/ UGC 2608 / MK 1073 / PGC 12081	PSBS3.. / U / 3.0± .9 /	.95± .07 / .03± .05 / / 1.02	/ .71 / .04 / .01	13.7 ±.3 / 10.95 / / 12.94				6991± 25 / 7097 / 6831
031505.5-544903 / 269.64 -52.00 / 240.24 -32.65 / 031343.0-550006	IC 1908 / ESO 155- 13 / IRAS03137-5500 / PGC 12085	.SBT3P. / S / 3.0± .8 /	1.13± .05 / .15± .05 / .76± .07 / 1.13	/ .00 / .21 / .08	14.7 ±.2 / 14.53 ±.14 / 11.81 / 14.31	.72± .07 / / .64 /	.61± .03 / / 14.03± .22 / 14.84± .36		8201± 88 / 8053 / 8126
0315.1 +3056 / 155.96 -22.57 / 338.50 -19.56 / 0312.1 +3045	/ UGC 2615 / / PGC 12087	.SB.0.. / U / .0± .9 /	1.04± .08 / .23± .06 / / 1.10	40 / .74 / .17 /					
031514.6+415850 / 149.58 -13.33 / 347.89 -13.29 / 031156.0+414745	/ UGC 2612 / / PGC 12089	.S..6*. / U / 6.0±1.3 /	.91± .07 / .08± .05 / / .98	110 / .71 / .12 / .04	14.9 ±.3 / / / 14.00				9605± 94 / 9710 / 9445
0315.2 -0715 / 189.40 -50.56 / 299.04 -35.65 / 0312.8 -0727	/ / / PGC 12090	.S..7*/ / E / 7.0±1.3 /	1.14± .08 / .68± .08 / / 1.16	42 / .19 / .94 / .34					
031517.9+424145 / 149.20 -12.72 / 348.49 -12.88 / 031158.2+423040	/ UGC 2614 / / PGC 12092	.S...0.. / U / .0± .9 /	1.12± .05 / .42± .05 / / 1.18	90 / .83 / .32 /	14.3 ±.3 / / / 13.07				5209± 94 / 5316 / 5050
031520.7+413645 / 149.80 -13.63 / 347.60 -13.52 / 031202.7+412540	CGCG 540- 67 / / / PGC 12097			/ .65 / /	15.3 ±.4 / / /				5941± 94 / 6045 / 5780
031521.2+412118 / 149.94 -13.85 / 347.39 -13.68 / 031203.6+411013	NGC 1250 / UGC 2613 / / PGC 12098	.L..0*/ / PU / -1.7± .6 /	1.33± .05 / .40± .03 / .75± .05 / 1.35	159 / .65 / .00 /	13.96 ±.15 / 13.7 ±.3 / / 13.17	1.14± .03 / .49± .05 / .90 / .32	1.15± .03 / .54± .05 / 13.20± .16 / 14.50± .31		6156± 57 / 6260 / 5995
031533.7-155248 / 201.82 -54.79 / 288.45 -37.37 / 031314.3-160350	NGC 1262 / MCG -3- 9- 14 / IRAS03132-1604 / PGC 12107	.SXS5*. / E (1) / 5.0± .9 / 3.1±1.2	.92± .08 / .10± .05 / / .93	135 / .09 / .15 / .05	15.0 ±.2 / / / 13.49	.79± .08 /	/ / / 14.20± .45		
031537.5-032805 / 184.79 -48.23 / 303.57 -34.72 / 031306.9-033908	A 0313-03 / MK 605 / / PGC 12111	.S..1?. / E / 1.0±1.8 /	.41± .13 / .43± .07 / / .43	15 / .19 / .44 / .22					8410±113 / 8386 / 8231

R.A. 2000 DEC. / l b / SGL SGB / R.A. 1950 DEC.	Names / PGC	Type / S_T n_L / T / L	$\log D_{25}$ / $\log R_{25}$ / $\log A_e$ / $\log D_o$	p.a. / A_g / A_i / A_{21}	B_T / m_B / m_{FIR} / B_T^o	$(B-V)_T$ / $(U-B)_T$ / $(B-V)_T^o$ / $(U-B)_T^o$	$(B-V)_e$ / $(U-B)_e$ / m'_e / m'_{25}	m_{21} / W_{20} / W_{50} / HI	V_{21} / V_{opt} / V_{GSR} / V_{3K}
031542.4-333234		.SXS8?.	1.06± .04	15					
233.55 -58.38	ESO 357- 10	S (1)	.32± .04	.00	16.24 ±.14				
265.86 -37.72		8.0±1.7		.39					
031342.0-334336	PGC 12116	8.9± .9	1.06	.16					
031546.3-120127		.L..+*/	1.35± .06	26					
196.03 -52.98	MCG -2- 9- 19	E	.57± .05	.18					
293.27 -36.80	IRAS03133-1212	-1.0±1.2		.00	13.42				
031323.3-121229	PGC 12117		1.28						
0315.8 +4151		.L..-*.	1.04± .18	45					
149.74 -13.38	UGC 2616	U	.32± .08	.65					
347.85 -13.45		-3.0±1.3		.00					
0312.5 +4140	PGC 12119		1.08						
031550.2-583424		.SB.6*/	1.22± .05	56					
274.50 -49.82	ESO 116- 14	S	.65± .05	.00	15.00 ±.14				
236.06 -31.36		6.0±1.3		.95					
031439.1-584524	PGC 12121		1.22	.32					
0316.0 +0910		.SBS3..	1.08± .07	75					7858
171.98 -39.56	UGC 2622	U	.42± .06	.77					
317.68 -30.35		3.0± .8		.58					7872
0313.3 +0859	PGC 12129		1.15	.21					7673
031600.5-053001		.S..2..	1.25± .05	155					2324
187.34 -49.39	MCG -1- 9- 24	E	.58± .05	.22					
301.21 -35.38	IRAS03135-0541	2.0± .9		.71	13.24				2294
031331.6-054102	PGC 12130		1.27	.29					2147
031600.8-022538	NGC 1266	PLBT0P*	1.19± .05	115					
183.67 -47.51	MCG -1- 9- 23	E	.21± .06	.22					
304.82 -34.51	IRAS03134-0236	-2.0± .7		.00	10.60				
031329.3-023640	PGC 12131		1.18						
031600.8+405309		.SXS7..	1.40± .04	176					
150.31 -14.17	UGC 2617	U	.45± .05	.64	13.8 ±.3				4860± 94
347.08 -14.05	IRAS03127+4042	7.0± .8		.62	13.08				4962
031243.9+404206	PGC 12132		1.46	.23	12.48				4699
031600.9+420427		.S..2..	1.08± .06	167					
149.65 -13.17	UGC 2618	U	.42± .05	.73	14.5 ±.3				5385± 94
348.06 -13.35	IRAS03127+4153	2.0± .9		.51	12.92				5490
031242.1+415324	PGC 12133		1.15	.21	13.19				5226
031605.3-342135	IC 1906	PSBS4?.	1.09± .05	64					
235.09 -58.27	ESO 357- 11	BS (1)	.45± .04	.00	14.47 ±.14				
264.81 -37.71	IRAS03141-3432	4.0± .9		.66	13.49				
031406.0-343236	PGC 12138	3.3±1.2	1.09	.22					
031605.6-130207		.S..5?/	1.10± .06	23					
197.57 -53.40	MCG -2- 9- 21	E	.71± .05	.16					
292.03 -37.06		5.0±1.9		1.07					
031343.5-131308	PGC 12139		1.11	.36					
031606.2+404817	IC 309	.LAS0..	.90± .17						
150.37 -14.23	MCG 7- 7- 43	P	.00± .07	.64	14.5 ±.3				4234± 57
347.02 -14.11		-2.0± .9		.00					4336
031249.3+403715	PGC 12141		.97		13.82				4073
031608.9-104030		.SBR0?.	1.12± .06	135					
194.19 -52.23	MCG -2- 9- 20	E	.41± .05	.25					
294.96 -36.62		.0±1.3		.31					
031344.7-105131	PGC 12144		1.12						
031609.1-241159		.IBS9..	1.09± .05					15.78±.2	2077± 11
216.03 -57.37	ESO 481- 16	SU (1)	.08± .04	.00	15.42 ±.14			67± 16	
277.85 -38.18		10.0± .6		.06				40± 12	1994
031358.0-242300	PGC 12145	10.0± .8	1.09	.04	15.36			.38	1927
0316.3 +4111		.L.....	1.08± .17		15.4 ±.2	1.17± .02	1.19± .02		
150.19 -13.90	UGC 2619	U	.00± .06	.62					
347.35 -13.92		-2.0± .8	.23± .09	.00		.97	12.02± .29		6292± 71
0313.0 +4100	PGC 12152		1.15		14.67		15.65± .90		6395
									6132
031626.2+413149	NGC 1257	.S..1..	1.12± .05	68	*				
150.02 -13.59	UGC 2621	U	.77± .05	.64	14.6 ±.3				4747± 71
347.66 -13.74		1.0± .9	.53± .18	.78			12.64± .55		4850
031308.2+412048	PGC 12157		1.18	.38	13.07				4587
031627.4+350411		.SBS7..	1.35± .05		*			14.51±.3	4424± 11
153.72 -18.99	UGC 2623	U	.04± .06	.86	14.8 ±.4			116± 7	
342.24 -17.48		7.0± .7		.05				102± 7	4511
031318.6+345310	PGC 12159		1.43	.02	13.89			.60	4255

R.A. 2000 DEC.	Names	Type	logD$_{25}$	p.a.	B$_T$	(B-V)$_T$	(B-V)$_e$	m$_{21}$	V$_{21}$
l b		S$_T$ n$_L$	logR$_{25}$	A$_g$	m$_B$	(U-B)$_T$	(U-B)$_e$	W$_{20}$	V$_{opt}$
SGL SGB		T	logA$_e$	A$_I$	m$_{FIR}$	(B-V)$_T^o$	m'$_e$	W$_{50}$	V$_{GSR}$
R.A. 1950 DEC.	PGC	L	logD$_o$	A$_{21}$	B$_T^o$	(U-B)$_T^o$	m'$_{25}$	HI	V$_{3K}$
031632.1-002808		.SBS4*.	1.18± .04	127	*				
181.59 -46.15	UGC 2628	UE (1)	.48± .04	.16	14.62 ±.12				
307.13 -34.03		4.0± .6		.70					
031358.8-003908	PGC 12163	3.1±1.2	1.19	.24					
031643.0+411927	IC 310	.LARO*.	1.11± .08		13.89M±.10	1.15± .01	1.16± .01		
150.18 -13.73	UGC 2624	PU	.00± .04	.64	13.8 ±.3	.62± .08	.66± .08		5292± 57
347.52 -13.90	IRAS03135+4108	-2.0± .6	.75± .02	.00	13.48	.96	13.19± .07		5395
031325.3+410827	PGC 12171		1.18		13.16	.49	14.31± .44		5132
031644.3+804736	NGC 1184	.S..0..	1.45± .05						
128.71 19.60	UGC 2583	U	.67± .07	.76	13.44 ±.18				
18.63 10.25	IRAS03088+8035	.0± .8		.51					
030906.3+803629	PGC 12174		1.48						
0316.7 +0600		.S?....	1.07± .07	150					6999
175.03 -41.70	UGC 2631		.79± .06	.63					
314.40 -31.78				1.19					7003
0314.1 +0550	PGC 12175		1.13	.40					6816
031646.8+400013	IC 311	.S?....	1.04± .15	113					
150.94 -14.84	UGC 2625		.00± .12	.65	15.0 ±.3				4255± 94
346.43 -14.69				.00					4355
031331.1+394913	PGC 12177		1.10	.00	14.37				4093
031653.7-353226		.SBS7..	1.44± .04	125				13.77±.2	1574± 6
237.30 -58.02	ESO 357- 12	S (1)	.23± .05	.00	13.78 ±.14			151± 11	
263.29 -37.73		6.5± .5		.32				141± 6	1462
031456.0-354324	PGC 12181	7.8± .6	1.44	.11	13.46			.20	1449
031659.3+313402	A 0313+31	.SAS5..	1.23± .05		14.89 ±.15	1.09± .04	1.05± .04	14.49±.1	4224± 5
155.92 -21.82	UGC 2627	U (1)	.08± .05	.91	14.40 ±.20			266± 6	4214± 36
339.28 -19.54		5.0± .8	.81± .02	.13		.85	14.43± .05	221± 5	4301
031354.7+312303	PGC 12184	3.8± .8	1.32	.04	13.65		15.70± .30	.80	4051
031659.8+412123		.S..1?.	1.07± .14		15.50 ±.19				
150.21 -13.68	UGC 2626	U	.79± .12	.64	15.4 ±.3				6358± 71
347.58 -13.93		1.0±1.9	.12± .05	.81			11.59± .17		6461
031342.0+411024	PGC 12185		1.13	.40	13.94		13.72± .81		6199
031703.7+413802					*				
150.07 -13.44	CGCG 540- 79			.64	15.0 ±.4				7231± 71
347.81 -13.77									7334
031345.3+412703	PGC 12193								7072
031704.4-225156		.SBT1P*	1.36± .03		*				
213.74 -56.84	ESO 481- 17	Sr	.07± .04	.00	13.18 ±.14				3928± 22
279.58 -38.35	IRAS03148-2302	1.5± .5		.08	13.22				3848
031452.0-230254	PGC 12194		1.36	.03	13.05				3777
031704.7+152845		.S?....	.89± .11		*				4762
166.99 -34.65	MCG 2- 9- 2		.35± .07	.50	14.75 ±.15				
324.29 -27.82	IRAS03142+1518			.53	13.30				4794
031417.1+151747	PGC 12195		.93	.18	13.70				4579
031706.8+313519		.SB.6*.	.93± .07		*			15.86±.3	4128± 10
155.93 -21.79	UGC 2629	U	.06± .05	.91	15.28 ±.14				4132± 46
339.32 -19.55		6.0±1.3		.08				97± 7	4206
031402.2+312420	PGC 12196		1.02	.03	14.27			1.56	3956
0317.2 +3708		.S..4..	.81± .11	40	*				5116
152.65 -17.19	UGC 2630	U	.40± .06	.80	15.57 ±.13				
344.09 -16.42	IRAS03140+3657	4.0±1.0		.58					5208
0314.0 +3657	PGC 12200		.89	.20	14.15				4951
0317.2 +0335		.S..8*.	1.04± .15		*				
177.49 -43.31	UGC 2641	U	.37± .12	.30					
311.79 -32.79		8.0±1.3		.46					
0314.6 +0325	PGC 12201		1.07	.19					
031713.4-323433	NGC 1288	.SXT5..	1.36± .02		12.78 ±.13	.68± .02	.79± .01	15.30±.3	4541± 9
231.70 -58.08	ESO 357- 13	PBSr (3)	.08± .03	.00	12.76 ±.12	.05± .06	.14± .05		4473± 30
267.06 -38.13	IRAS03152-3245	4.6± .3	1.04± .01	.11	12.97	.64	13.47± .03	355± 25	4431
031512.0-324530	PGC 12204	1.6± .4	1.36	.04	12.63	.02	14.24± .20	2.63	4405
031717.6-410628	NGC 1291	RSBS0..	1.99± .02		9.39S±.04	.93± .01	.97± .01	12.86±.1	838± 3
247.52 -57.05	ESO 301- 2	R	.08± .02	.00	9.44 ±.11	.46± .01	.52± .01	68± 4	794± 19
256.29 -36.86	IRAS03154-4117	.0± .3	1.26± .03	.06	11.84	.91	11.63± .09	40± 3	712
031528.0-411724	PGC 12209		1.99		9.32	.45	13.99± .10		726
031720.0-334127	IC 1909	.SBR3..	1.07± .05	61					
233.80 -58.04	ESO 357- 14	BSr (1)	.29± .04	.00	14.48 ±.14				
265.63 -38.04		2.7± .5		.40					
031520.1-335224	PGC 12212	4.4± .9	1.07	.14					

R.A. 2000 DEC. l b SGL SGB R.A. 1950 DEC.	Names PGC	Type ST nL T L	logD25 logR25 logA0 logD0	p.a. Ag Ai A21	BT mB mFIR BT0	(B-V)T (U-B)T (B-V)T0 (U-B)T0	(B-V)e (U-B)e m'e m'25	m21 W20 W50 HI	V21 Vopt VGSR V3K
0317.3 -1504 200.87 -54.06 289.55 -37.68 0315.0 -1515	 PGC 12213	.SBS7?/ E 7.0±1.3 	1.09± .09 .96± .08 1.11	146 .16 1.32 .48	 	 	 	 	
0317.4 +3634 153.01 -17.64 343.63 -16.78 0314.2 +3623	 UGC 2633 IRAS03142+3623 PGC 12216	.S..6*. U 6.0±1.3 	1.16± .07 .88± .06 1.24	81 .90 1.30 .44	 	 	 	 	5668 5759 5502
031723.8-391538 244.18 -57.41 258.58 -37.24 031531.4-392634	 PGC 12217	.SBS3?. S (1) 3.0± .9 3.3± .7	1.07± .09 .37± .08 1.07	 .00 .51 .18	 	 	 	 	
031727.5+412418 150.26 -13.59 347.67 -13.97 031409.4+411320	NGC 1260 UGC 2634 IRAS03141+4113 PGC 12219	.S..0*/ PU -.3± .7 	1.06± .11 .34± .05 1.10	86 .64 .26 	14.32S±.18 14.0 ±.3 13.41 13.27	 	 13.62± .58	 	 5631± 57 5734 5472
0317.4 -0004 181.37 -45.72 307.69 -34.11 0314.9 -0015	 UGC 2645 PGC 12220	.S..3*. U (1) 3.0±1.4 3.5±1.3	1.04± .06 .67± .05 1.05	27 .17 .92 .33	* 15.51 ±.12 	 	 	 	
0317.5 +3702 152.75 -17.23 344.05 -16.52 0314.3 +3652	 UGC 2636 PGC 12226	.S..6*. U 6.0±1.4 	1.22± .12 1.23± .06 1.30	10 .87 1.47 .50	* 	 	 	 	
0317.5 +3802 152.17 -16.39 344.89 -15.94 0314.3 +3752	 UGC 2637 PGC 12227	.S..6*. U 6.0±1.4 	1.14± .07 .86± .06 1.22	47 .84 1.27 .43	* 	 	 	 	5126 5220 4962
031732.0-542123 268.70 -51.94 240.61 -33.15 031609.0-543218	 ESO 155- 14 PGC 12231	.L?.... 	1.02± .07 .04± .05 .85± .09 1.01	 .00 .00 	14.2 ±.2 14.37 ±.14 14.19	1.05± .06 .97 	1.06± .03 13.91± .30 14.06± .41	 	 8393± 60 8244 8319
031736.7-014906 183.34 -46.81 305.68 -34.71 031504.7-020002	 UGC 2649 PGC 12237	.L...*. U -2.0±1.2 	1.04± .09 .00± .04 1.06	 .21 .00 	 14.15 ±.10 13.82	 	 	 	 8264± 50 8243 8086
0317.6 -0141 183.20 -46.72 305.84 -34.67 0315.1 -0152	 UGC 2650 PGC 12241	.S..2.. U 2.0±1.0 	.99± .06 .68± .05 1.01	64 .21 .84 .34	 15.50 ±.12 	 	 	 	
031744.2-572648 272.83 -50.26 237.17 -32.03 031630.0-573742	 ESO 116- 15 IRAS03164-5737 PGC 12245	 	1.14± .06 .41± .06 1.14	 .00 	 14.70 ±.14 11.85 	 	 	 	 8542± 42 8389 8477
031745.5-101720 193.98 -51.69 295.54 -36.93 031521.0-102816	NGC 1284 MCG -2- 9- 22 PGC 12247	.LA.0*. E -2.0±1.1 	1.23± .05 .04± .05 1.24	90 .19 .00 	 	 	 	 	
031750.4+415803 150.00 -13.08 348.17 -13.69 031431.5+414706	 UGC 2639 PGC 12253	.S..2.. U 2.0±1.0 	1.04± .08 .88± .06 1.12	92 .81 1.08 .44	 15.3 ±.3 13.41	 	 	 	 4033± 19 4137 3876
031751.1+412704 150.30 -13.51 347.75 -14.00 031432.9+411607	 CGCG 540- 85 PGC 12254	 	 .46± .13 	 .64 	15.0 ±.3 14.9 ±.3 	1.05± .03 .65± .06 	1.13± .03 .69± .06 12.81± .45 	 	 4370± 57 4473 4212
031752.4+431815 149.27 -11.96 349.27 -12.90 031431.2+430719	 UGC 2640 IRAS03145+4307 PGC 12257	.SB.3.. U 3.0± .9 	1.03± .06 .24± .05 1.11	70 .93 .33 .12	 14.2 ±.3 12.27 12.90	 	 	 	 6081± 94 6188 5926
031753.5-071755 190.02 -50.04 299.20 -36.29 031526.3-072850	NGC 1285 MCG -1- 9- 26 IRAS03154-0728 PGC 12259	PSBR3P. E 3.0± .8 	1.18± .05 .14± .05 1.19	145 .16 .19 .07	 11.91 	 	 	 	5243 5206 5071
031757.1-001017 181.59 -45.69 307.61 -34.26 031523.6-002112	NGC 1280 UGC 2652 IRAS03154-0021 PGC 12262	.SAT5*. E (1) 5.0±1.3 1.9± .8	.97± .06 .09± .04 .98	55 .20 .14 .05	14.09 ±.13 13.89 ±.14 12.75 13.62	.68± .02 -.02± .03 .57 -.10	 13.53± .32	 	 6870± 50 6854 6692

R.A. 2000 DEC. / l b / SGL SGB / R.A. 1950 DEC.	Names / PGC	Type / S_T n_L / T / L	$\log D_{25}$ / $\log R_{25}$ / $\log A_e$ / $\log D_o$	p.a. / A_g / A_i / A_{21}	B_T / m_B / m_{FIR} / B_T^0	$(B-V)_T$ / $(U-B)_T$ / $(B-V)_T^0$ / $(U-B)_T^0$	$(B-V)_e$ / $(U-B)_e$ / m'_e / m'_{25}	m_{21} / W_{20} / W_{50} / HI	V_{21} / V_{opt} / V_{GSR} / V_{3K}
031757.3-441418		.L?....	1.05± .07	8	14.9 ±.2	1.00± .08	1.08± .03		
252.94 -56.08	ESO 248- 6		.34± .05	.00	14.85 ±.14				22710± 69
252.42 -36.28			.95± .10	.00		.75	15.11± .33		22579
031613.0-442512	PGC 12264		1.00		14.53		14.18± .40		22608
031808.4+414516	IC 312	.E...*.	.99± .12	125	14.4 ±.3				
150.17 -13.23	UGC 2644	PU	.28± .05	.81					4835± 57
348.03 -13.86		-5.0± .7		.00					4938
031449.7+413420	PGC 12279		1.03		13.53				4677
0318.1 +4021		.L...*.	1.04± .18	73					
150.96 -14.39	UGC 2646	U	.18± .08	.67					5388± 73
346.89 -14.69		-2.0±1.2		.00					5488
0314.9 +4011	PGC 12281		1.09						5228
031815.2-273640	NGC 1292	.SAS5..	1.47± .03	7	12.84S±.15			13.77±.1	1367± 5
222.42 -57.52	ESO 418- 1	R (2)	.35± .03	.00	12.75 ±.12			261± 6	1433± 41
273.44 -38.65	IRAS03161-2747	5.0± .4		.52	12.51			245± 5	1274
031608.0-274734	PGC 12285	4.0± .4	1.47	.17	12.25		14.16± .21	1.34	1227
031815.5-662951	NGC 1313	.SBS7..	1.96± .02		9.2 S±.2	.49± .01	.47± .01	10.54±.3	457± 5
283.36 -44.64	ESO 82- 11	R (3)	.12± .02	.04	9.57 ±.12	-.24± .03	-.22± .03	190± 6	446± 35
227.64 -28.22	VV 436	7.0± .3	1.45± .03	.16	9.08	.46	12.90± .06	156± 6	292
031739.1-664042	PGC 12286	7.0± .3	1.96	.06	9.27	-.26	13.52± .22	1.21	419
031815.8+415127	NGC 1265	.E+....	1.26± .13	165	13.22 ±.18	1.12± .06	1.13± .02		7536± 71
150.13 -13.13	UGC 2651	PU	.06± .08	.81		.74± .08	.79± .03		7639
348.13 -13.82		-4.0± .5	1.14± .07	.00		.87	14.41± .23		7379
031457.0+414032	PGC 12287		1.36		12.30	.59	14.37± .71		
0318.3 +4128					16.1 ±.4	1.14± .04	1.15± .04		3410± 71
150.36 -13.45				.70					3512
347.82 -14.06			.22± .22				12.71± .78		3252
0315.0 +4117	PGC 12292								
031822.5+412436					*				6365± 57
150.40 -13.49	CGCG 540- 87			.70	15.3 ±.4				6467
347.78 -14.10									6207
031504.3+411341	PGC 12295								
031833.1-255013	A 0316-26	.SBT6..	1.46± .03	41	*			14.74±.3	1802± 11
219.20 -57.17	ESO 481- 18	SU (1)	.46± .04	.00	13.56 ±.14			201± 16	
275.74 -38.74		5.5± .6		.68				195± 12	1713
031624.0-260106	PGC 12309	6.7± .8	1.46	.23	12.87			1.64	1658
0318.6 +3736		.S..7..	1.14± .07	155	*				6939
152.62 -16.64	UGC 2653	U	.41± .06	.85					
344.65 -16.38	IRAS03154+3725	7.0± .8		.57	13.04				7032
0315.4 +3725	PGC 12318		1.22	.21					6776
031843.2+421745		.S?....	1.12± .05	177					5736± 94
149.96 -12.72	UGC 2654		.45± .05	.80	14.2 ±.3				5840
348.54 -13.63	IRAS03154+4207			.67	12.41				5580
031523.6+420651	PGC 12326		1.19	.22	12.67				
031843.4-234656		.IBS9..	1.30± .05		*			14.54±.2	1535± 7
215.53 -56.70	ESO 481- 19	SU (1)	.08± .04	.01	15.72 ±.14			93± 16	
278.41 -38.76		10.0± .6		.06				78± 6	1451
031632.1-235748	PGC 12327	10.0± .8	1.30	.04	15.65			-1.15	1387
031844.9+412804	NGC 1267	.E+..*.	1.04± .18						5059± 67
150.43 -13.41	UGC 2657	PU (1)	.09± .08	.70					5161
347.86 -14.12		-4.0± .6		.00					4902
031526.5+411710	PGC 12331	3.5±1.1	1.12						
031844.9+412918	NGC 1268	.SXT3*.	.98± .07	120	*				3124± 71
150.42 -13.39	UGC 2658	PU (1)	.17± .05	.70	14.2 ±.3				3226
347.88 -14.11		3.0± .6		.24					2967
031526.5+411825	PGC 12332	3.5±1.2	1.04	.09	13.23				
031845.3+431426		.SXS7..	1.26± .04	175	*				6149± 94
149.44 -11.92	UGC 2655	U	.37± .05	.92	13.5 ±.3				6255
349.31 -13.07	IRAS03154+4303	7.0± .8		.51	13.17				5995
031524.1+430332	PGC 12333		1.34	.18	12.03				
0318.8 +4104		.E.....	1.00± .19	30	*				5487± 19
150.66 -13.72	UGC 2656	U	.14± .08	.74					5588
347.55 -14.36		-5.0± .9		.00					5329
0315.5 +4054	PGC 12338		1.08						
031849.8-130346	NGC 1296	.SBT2P.	1.04± .07	0	*				
198.14 -52.83	MCG -2- 9- 25	E (1)	.11± .05	.20					
292.15 -37.72	IRAS03164-1314	2.0± .9		.14	12.44				
031627.9-131438	PGC 12341	3.1±1.2	1.05	.06					

R.A. 2000 DEC. / l b / SGL SGB / R.A. 1950 DEC.	Names / PGC	Type / S_T n_L / T / L	$\log D_{25}$ / $\log R_{25}$ / $\log A_o$ / $\log D_o$	p.a. / A_g / A_i / A_{21}	B_T / m_B / m_{FIR} / B_T^o	$(B-V)_T$ / $(U-B)_T$ / $(B-V)_T^o$ / $(U-B)_T^o$	$(B-V)_e$ / $(U-B)_e$ / m'_e / m'_{25}	m_{21} / W_{20} / W_{50} / HI	V_{21} / V_{opt} / V_{GSR} / V_{3K}
031849.9-015822	NGC 1289	.LBT0*.	1.26± .07	100	13.48 ±.15	.92± .03	1.00± .03		2835± 31
183.80 -46.67	UGC 2666	UE	.20± .05	.21	13.47 ±.10	.50± .04	.48± .04		2813
305.61 -35.04		-1.7± .6	.83± .04	.00		.83	13.12± .14		2659
031617.9-020914	PGC 12342		1.26		13.22	.45	14.16± .42		
031853.3+403546		.S..4..	1.07± .07	65	•				6198± 94
150.95 -14.12	UGC 2659	U	.40± .06	.67	14.6 ±.3				6298
347.17 -14.66	IRAS03156+4025	4.0± .9		.59	13.13				6039
031536.3+402453	PGC 12343		1.13	.20	13.29				
0318.9 +0847		.S..6?.	1.00± .08	10					
172.98 -39.34	UGC 2663	U	.69± .06	.66					
317.63 -31.15		6.0±1.9		1.01					
0316.2 +0837	PGC 12344		1.06	.34					
0318.9 -1033		DLB.-*.	1.21± .07	85					
194.58 -51.58		E	.28± .08	.19					
295.28 -37.26		-3.0±1.3		.00					
0316.5 -1044	PGC 12345		1.20						
031858.5+412818	NGC 1270	.E...*.	1.17± .10	15	14.26M±.10	1.19± .01	1.21± .01		4871± 43
150.47 -13.38	UGC 2660	PU (1)	.09± .07	.70	13.7 ±.3	.75± .03	.76± .03		4973
347.89 -14.16		-5.0± .7	.44± .04	.00		.99	11.97± .14		4714
031540.1+411725	PGC 12350	3.5±1.2	1.26		13.45	.60	14.89± .55		
031904.5-535225					15.07 ±.19	1.11± .03	1.13± .02		16358± 60
267.84 -51.99				.00					16209
241.05 -33.53			.52± .07				13.16± .24		16283
031740.6-540315	PGC 12357								
031911.3+412111	NGC 1271	.LB..$.			15.1 ±.3		1.21± .02		5737± 71
150.57 -13.46	CGCG 540- 96	P		.70	15.4 ±.4				5839
347.82 -14.26		-2.0± .9	.30± .13				12.06± .45		5580
031553.0+411019	PGC 12367								
0319.1 -1120		.SBS8..	1.18± .05	25					3184
195.73 -51.92	MCG -2- 9- 28	E (1)	.22± .05	.22					
294.33 -37.49		8.0± .8		.27					3134
0316.8 -1131	PGC 12368	7.5± .8	1.20	.11					3018
031911.7+392636		.L.....	1.11± .08	147					5985± 94
151.66 -15.05	UGC 2661	U	.67± .04	.66	14.6 ±.3				6082
346.25 -15.39		-2.0± .9		.00					5825
031556.3+391544	PGC 12369		1.08		13.85				
031914.1-190604	NGC 1297	.LXS0P*	1.35± .04	3	*				1569± 66
207.59 -55.22	ESO 547- 30	PSE	.07± .03	.10	12.88 ±.11				1497
284.47 -38.61		-2.3± .4		.00					1414
031658.0-191654	PGC 12373		1.35		12.76				
0319.3 +4138					15.95 ±.14	1.12± .02			6241± 94
150.42 -13.20				.75		.61± .07			6343
348.07 -14.10			.08± .02				11.84± .04		6084
0315.9 +4127	PGC 12378								
031917.6-120611		.SAR5*.	1.12± .06	165					4606
196.84 -52.27	MCG -2- 9- 29	E (1)	.17± .05	.24					
293.37 -37.66	IRAS03169-1217	5.0± .9		.26	13.82				4554
031654.8-121701	PGC 12379	3.1±1.2	1.14	.09					4441
0319.3 +4108					15.3 ±.4				7975± 49
150.70 -13.62				.74					8076
347.67 -14.40			.36± .15				12.57± .49		7818
0316.0 +4058	PGC 12381								
031921.3+412932	NGC 1272	.E+....	1.31± .09		12.86M±.10	1.08± .04	1.12± .01		4021± 60
150.51 -13.32	UGC 2662	PU	.03± .07	.70	13.6 ±.3	.61± .06	.66± .02		4123
347.95 -14.20		-4.0± .5	1.28± .05	.00		.89	14.68± .17		3865
031602.8+411840	PGC 12384		1.40		12.16	.47	14.31± .49		
031924.8-493559	IC 1914	.SXS7..	1.58± .03	99	13.3 ±.4	.42± .06	.56± .03	13.33±.3	1037± 9
261.49 -53.90	ESO 200- 3	S (1)	.29± .05	.00	13.28 ±.14	-.01± .14			895
245.94 -35.01	IRAS03179-4946	7.0± .5	1.19± .12	.40		.36	14.70± .27	205± 7	951
031751.0-494648	PGC 12390	7.2± .5	1.58	.15	12.87	-.05	15.27± .46	.31	
0319.4 +8120		.I..9..	1.24± .11	55				14.68±.1	2519± 11
128.46 20.11	UGC 2603	U	.08± .12	.74					2687
19.13 10.51		10.0± .8		.06				121± 8	2456
0311.4 +8110	PGC 12391		1.31	.04					
031925.6+402845					15.4 ±.4				8148± 94
151.10 -14.16	CGCG 540- 97			.67					8247
347.13 -14.81									7990
031608.6+401754	PGC 12392								

R.A. 2000 DEC. / l b / SGL SGB / R.A. 1950 DEC.	Names / PGC	Type / S_T n_L / T / L	$\log D_{25}$ / $\log R_{25}$ / $\log A_e$ / $\log D_o$	p.a. / A_g / A_i / A_{21}	B_T / m_B / m_{FIR} / B_T^o	$(B-V)_T$ / $(U-B)_T$ / $(B-V)_T^o$ / $(U-B)_T^o$	$(B-V)_e$ / $(U-B)_e$ / m'_e / m'_{25}	m_{21} / W_{20} / W_{50} / HI	V_{21} / V_{opt} / V_{GSR} / V_{3K}
031927.2+413225	NGC 1273	.LAR0$.	1.04± .13		14.27 ±.13	1.11± .01	1.12± .01		
150.50 -13.27	MCG 7- 7- 59	P	.00± .12	.75	14.1 ±.3	.55± .04	.59± .04		5351± 48
348.00 -14.19		-2.0± .8	.56± .02	.00		.89	12.56± .06		5453
031608.6+412134	PGC 12396		1.12		13.42	.40	14.33± .75		5194
031927.4+413807		.S..6?.	1.02± .06	122					
150.45 -13.19	UGC 2665	U	.40± .05	.75	15.0 ±.3				7806± 57
348.08 -14.13		6.0±1.8		.59					7909
031608.6+412716	PGC 12397		1.09	.20	13.63				7650
0319.4 +0808		.SXS7*.	1.04± .15						7127
173.70 -39.71	UGC 2671	U	.00± .12	.77					
317.01 -31.55		7.0±1.2		.00					7135
0316.8 +0758	PGC 12398		1.11	.00					6947
0319.4 +4044		.I..9*.	1.11± .14						
150.96 -13.93	UGC 2664	U	.15± .12	.67					
347.36 -14.66		10.0±1.2		.11					
0316.2 +4034	PGC 12400		1.17	.07					
031934.3-322753	IC 1913	.SB.3?/	1.27± .04	149					
231.49 -57.59	ESO 357- 16	PS	.86± .03	.00	14.38 ±.14				1382± 31
267.14 -38.63		3.0± .6		1.19					1276
031733.0-323842	PGC 12404		1.27	.43	13.18				1253
031934.3+413450	IC 1907	.E?....			15.42S±.18	1.20± .02	1.18± .02		
150.50 -13.23	MCG 7- 7- 61			.75	15.2 ±.4				4420± 57
348.05 -14.18			.43± .01			.99	12.65± .03		4522
031615.6+412359	PGC 12405								4264
0319.6 +3825		.SXS5..	1.00± .16						
152.31 -15.84	UGC 2667	U	.00± .12	.88					
345.46 -16.06		5.0± .8		.00					
0316.4 +3815	PGC 12410		1.08	.00					
031940.8-192441	NGC 1300	.SBT4..	1.79± .01	106	11.11M±.10	.68± .02	.78± .01	13.49±.1	1568± 6
208.16 -55.22	ESO 547- 31	R (4)	.18± .02	.08	11.15 ±.12	.11± .03	.21± .02	289± 8	1592± 58
284.08 -38.74	IRAS03174-1935	4.0± .3	1.52± .01	.26	11.65	.62	14.14± .03	272± 11	1496
031725.0-193530	PGC 12412	1.1± .3	1.80	.09	10.77	.06	14.47± .13	2.63	1415
031941.0+413256	NGC 1274	.E.3...	.72± .22		15.12M±.12		1.16± .02		
150.54 -13.24	MCG 7- 7- 62	P	.11± .07	.75	14.8 ±.3				6447± 79
348.04 -14.22		-5.0±1.0	.26± .10	.00			11.87± .35		6549
031622.3+412205	PGC 12413		.80		14.24		13.42±1.12		6291
0319.7 +0944		.L...?.	1.00± .19						
172.33 -38.51	UGC 2674	U	.05± .08	.77	14.83 ±.12				
318.74 -30.94		-2.0±1.7		.00					
0317.0 +0934	PGC 12415		1.08						
0319.7 +4137					15.8 ±.5				
150.49 -13.17				.75					8514± 57
348.11 -14.17			.44± .22				13.53± .72		8616
0316.4 +4127	PGC 12417								8358
031943.1+003354		.S?....	.87± .07	13					
181.21 -44.88	MCG 0- 9- 57		.38± .06	.25	15.47 ±.12				7232± 45
308.64 -34.44	MK 1076			.56					7217
031709.0+002305	PGC 12418		.90	.19	14.62				7056
0319.7 +0033		.S?....	.94± .11						
181.22 -44.87	MCG 0- 9- 58		.05± .07	.25	14.65 ±.12				7230± 67
308.64 -34.45				.04					7215
0317.2 +0023	PGC 12425		.96		14.26				7054
0319.7 +4027		.I..9?.	1.07± .14						
151.17 -14.14	UGC 2668	U	.00± .12	.67					
347.16 -14.88		10.0±1.7		.00					
0316.5 +4017	PGC 12426		1.13	.00					
0319.8 +4116					15.3 ±.4				
150.71 -13.45				.74					5401± 71
347.83 -14.39			.43± .19				12.98± .65		5502
0316.5 +4106	PGC 12428								5244
031948.5+413045	NGC 1275	.P.....	1.34± .03	110	12.64v±.18	.76± .03	.72± .03		
150.58 -13.26	UGC 2669	R	.12± .04	.75	12.57 ±.16	.07± .03	-.01± .03		5260± 18
348.02 -14.26		99.0	.75± .06	.00		.53	11.88± .20		5362
031629.9+411955	PGC 12429		1.41		11.77	-.03	13.91± .25		5104
0319.8 +4135					15.5 ±.3				
150.53 -13.19				.75					7370± 57
348.09 -14.21			.34± .15				12.73± .51		7472
0316.5 +4125	PGC 12430								7214

R.A. 2000 DEC. / I b / SGL SGB / R.A. 1950 DEC.	Names / PGC	Type / S_T n_L / T / L	$\log D_{25}$ / $\log R_{25}$ / $\log A_e$ / $\log D_o$	p.a. / A_g / A_i / A_{21}	B_T / m_B / m_{FIR} / B_T^o	$(B-V)_T$ / $(U-B)_T$ / $(B-V)_T^o$ / $(U-B)_T^o$	$(B-V)_e$ / $(U-B)_e$ / m'_e / m'_{25}	m_{21} / W_{20} / W_{50} / HI	V_{21} / V_{opt} / V_{GSR} / V_{3K}
031950.8-260336	NGC 1302	RSBR0..	1.59± .02		11.60 ±.13	.89± .01	.90± .01	14.26±.1	1703± 5
219.70 -56.93	ESO 481- 20	V	.02± .02	.00	11.43 ±.11	.34± .01	.38± .01	103± 6	1730± 56
275.45 -39.03	IRAS03177-2614	.0± .3	1.06± .02	.02		.87	12.39± .06	92± 5	1613
031742.0-261424	PGC 12431		1.59		11.46	.35	14.36± .17		1562
031951.8+413425	NGC 1277	.L..+*P	.98± .11		14.66 ±.18	1.13± .05	1.21± .02		
150.55 -13.20	MCG 7- 7- 64	P	.41± .06	.75	14.6 ±.3				4982± 48
348.08 -14.23		-1.0±1.4	.35± .07	.00		.86	11.90± .24		5084
031633.1+412335	PGC 12434		.99		13.82		13.41± .59		4826
031954.4+413350	NGC 1278	.E...P*	1.19± .10		13.57M±.10	1.15± .03	1.16± .01		
150.56 -13.21	UGC 2670	PU	.08± .07	.75	13.6 ±.3				6047± 48
348.08 -14.24		-5.0± .6	.91± .04	.00		.93	13.62± .13		6148
031635.7+412300	PGC 12438		1.29		12.74		14.33± .56		5891
0319.9 -0335		.S..6*/	1.18± .08	0					
185.91 -47.46		E	.66± .08	.11					
303.81 -35.78		6.0±1.3		.96					
0317.4 -0346	PGC 12439		1.19	.33					
031957.3+033523		.S..2/	1.15± .05	122					6887
178.14 -42.81	UGC 2677	UE (1)	.59± .05	.28	14.96 ±.12				
312.08 -33.43		2.0± .6		.73					6881
031720.4+032435	PGC 12446	4.5±1.2	1.17	.30	13.88				6709
032001.7+411505		.L...*.	1.20± .14	135	*				4266± 57
150.76 -13.45	UGC 2673	U	.05± .08	.74	14.6 ±.3				4367
347.83 -14.45		-2.0±1.1		.00					4109
031643.4+410416	PGC 12452		1.28		13.82				
032003.7+012141		PSBT3*.	1.30± .03	120	*				6642
180.45 -44.29	UGC 2679	UE	.59± .04	.29	14.91 ±.16				
309.58 -34.25		3.0± .7		.81					6629
031728.8+011053	PGC 12454		1.33	.30	13.76				6466
032005.2+405423		.S..1?.	.96± .09	178	*				4297± 94
150.97 -13.74	UGC 2672	U	.51± .06	.78	15.4 ±.3				4397
347.56 -14.66		1.0±1.9		.52					4140
031647.5+404334	PGC 12456		1.03	.25	14.00				
032005.5-664209	NGC 1313A	.S..3*.	1.09± .04	30	*				
283.41 -44.37	ESO 83- 1	RS	.63± .04	.08	14.72 ±.14				
227.33 -28.29	IRAS03195-6652	2.9± .7		.87	12.80				
031931.0-665254	PGC 12457		1.10	.32					
032006.4+413748	NGC 1281	.E.5...	1.01± .12		14.5 ±.3	1.17± .08	1.18± .08		
150.56 -13.13	MCG 7- 7- 67	P	.17± .06	.75	14.4 ±.3				4201± 57
348.15 -14.23		-5.0±1.0	.63± .16	.00		.96	13.13± .56		4303
031647.5+412659	PGC 12458		1.08		13.63		14.13± .70		4045
032006.8-521114	NGC 1311	.SBS9?/	1.48± .03	40	13.44 ±.15	.46± .02	.42± .01		
265.30 -52.66	ESO 200- 7	S (1)	.59± .05	.00	13.23 ±.14	-.25± .03	-.27± .03		477± 88
242.90 -34.27	IRAS03186-5222	9.0± .7	.86± .02	.60		.33	13.23± .05		331
031839.0-522200	PGC 12460	7.2± .8	1.48	.30	12.72	-.34	14.22± .23		398
032009.7-061545	NGC 1299	.SBT3?.	1.06± .04	135					
189.20 -48.99	MCG -1- 9- 28	E (1)	.27± .05	.14					
300.63 -36.57	IRAS03176-0626	3.0±1.3		.37	12.51				
031741.6-062633	PGC 12466	3.1±1.6	1.07	.13					
032012.1+412205	NGC 1282	.E...*.	1.15± .15	25	13.87M±.10	.99± .02	1.03± .01		
150.72 -13.34	UGC 2675	PU	.10± .08	.74	13.7 ±.3	.54± .07	.56± .03		2192± 57
347.95 -14.40		-5.0± .7	.80± .03	.00		.80	13.22± .10		2294
031653.7+411116	PGC 12471		1.24		13.09	.40	14.35± .80		2036
032012.9-020644	NGC 1298	.E+..P*	1.20± .08	70	14.95 ±.14	.94± .03			
184.27 -46.49	UGC 2683	UE	.07± .05	.20	13.72 ±.11	.43± .05			6528± 31
305.58 -35.41		-3.5± .8		.00		.83			6505
031741.1-021732	PGC 12473		1.21		13.89	.42	15.65± .45		6354
032015.4-105148		.S?....	.74± .14		15.72S±.15				
195.28 -51.45	MCG -2- 9- 31		.09± .07	.19					9248± 41
294.98 -37.65	HICK 24A			.14					9199
031751.5-110235	PGC 12477		.75	.05	15.34		14.02± .75		9083
032015.5+412354	NGC 1283	.E.1.*.	.86± .17	70	14.73 ±.13	1.15± .02	1.18± .01		
150.71 -13.31	UGC 2676	P	.09± .05	.74	15.0 ±.3	.68± .06			6749± 57
347.98 -14.39		-5.0±1.2	.59± .02	.00		.92	13.17± .09		6851
031657.0+411306	PGC 12478		.95		13.92	.54	13.80± .86		6593
0320.2 +8014		.S..9*.	.89± .11					14.13±.1	2244± 11
129.16 19.23	UGC 2620	U	.00± .07	.88	15.30 ±.19			144± 16	
18.28 9.80	IRAS03128+8003	9.0±1.1		.00	13.59			135± 12	2411
0312.8 +8003	PGC 12480		.97	.00	14.41			-.28	2178

R.A. 2000 DEC. l b SGL SGB R.A. 1950 DEC.	Names PGC	Type S_T n_L T L	$\log D_{25}$ $\log R_{25}$ $\log A_e$ $\log D_o$	p.a. A_g A_i A_{21}	B_T m_B m_{FIR} B_T^o	$(B-V)_T$ $(U-B)_T$ $(B-V)_T^o$ $(U-B)_T^o$	$(B-V)_e$ $(U-B)_e$ m'_e m'_{25}	m_{21} W_{20} W_{50} HI	V_{21} V_{opt} V_{GSR} V_{3K}
032017.3-262749 220.46 -56.90 274.92 -39.13 031809.0-263836	 ESO 481- 21 PGC 12484	.SB.7?/ S 7.0±1.8 	1.19± .04 .73± .03 1.19	109 .00 1.00 .36	 15.63 ±.14 				
0320.3 -0205 184.27 -46.45 305.63 -35.43 0317.8 -0215	 UGC 2687 IRAS03177-0215 PGC 12491	.SXT1.. U 1.0± .9 	.85± .11 .18± .06 .87	 .20 .18 .09	 15.06 ±.12 13.63 				
0320.3 -5416 268.25 -51.62 240.52 -33.56 0319.0 -5427	 PGC 12500	 	 .65± .14	 .00 	14.7 ±.2 	.94± .04 	.95± .00 13.41± .50 		8719± 88 8569 8647
032022.9-105203 195.31 -51.43 294.98 -37.68 031759.0-110250	 MCG -2- 9- 32 HICK 24B PGC 12501	.S?.... 	.64± .17 .00± .07 .65	 .18 .00 .00	15.46S±.15 15.23 		 13.49± .86 		9137± 41 9088 8972
032023.3+040857 177.69 -42.35 312.75 -33.33 031745.9+035810	A 0317+03 MK 606 PGC 12502	 	 	 .37 	 13.27 				8985± 45 8981 8808
0320.5 +3729 153.01 -16.52 344.78 -16.75 0317.3 +3719	 UGC 2678 PGC 12510	.SB.3*. U 3.0±1.1 	1.03± .08 .01± .06 1.11	 .84 .02 .01					5591 5682 5430
0320.5 +1718 166.34 -32.71 326.55 -27.68 0317.7 +1708	 UGC 2684 PGC 12514	.I..9?. U 10.0±1.6 	1.26± .06 .29± .06 1.30	 .42 .22 .15				15.08±.1 90± 4 79± 4 	350± 5 385 172
032035.4-184256 207.16 -54.78 285.01 -38.89 031819.0-185342	NGC 1301 ESO 547- 32 IRAS03183-1853 PGC 12521	.SBT3?. SE (2) 3.5± .6 3.1± .6	1.34± .02 .69± .03 .87± .02 1.35	140 .10 1.01 .34	14.10 ±.14 14.13 ±.14 13.39 12.98	.73± .03 .20± .05 .55 .06	.81± .03 .19± .04 13.94± .04 13.94± .20		4008± 22 3937 3854
032038.3-541745 268.25 -51.58 240.48 -33.59 031916.0-542830	 ESO 155- 20 PGC 12523	.S?.... 	1.12± .06 .33± .05 .60± .04 1.12	78 .00 .34 .17	14.60 ±.17 14.50 ±.14 14.10	.78± .04 .65 	.87± .03 13.09± .12 14.23± .38		8302± 60 8152 8230
032038.9-010207 183.16 -45.73 306.88 -35.18 031806.1-011254	 CGCG 390- 67 HICK 25D PGC 12524	.S?.... 	 	 .16 	16.19S±.15 15.6 ±.3 				6285± 41 6265 6111
032040.1-225550 214.24 -56.06 279.53 -39.18 031828.1-230636	 ESO 481- 22 PGC 12526	.S?.... 	.92± .06 .04± .04 .92	 .00 .04 .02	 14.54 ±.14 14.36				10699± 60 10616 10552
032042.9-010632 183.26 -45.76 306.80 -35.22 031810.2-011718	 UGC 2690 IRAS03181-0117 PGC 12531	.SA.5P* UE (1) 4.5± .6 3.1±1.2	1.13± .04 .35± .04 1.15	143 .16 .53 .18	14.56S±.15 14.52 ±.12 		 14.19± .28 		
032043.4-010009 183.14 -45.70 306.93 -35.19 031810.6-011055	 CGCG 390- 70 HICK 25C PGC 12533	.S?.... 	 	 .16 	15.96S±.15 15.0 ±.3 				6285± 41 6265 6111
032045.6-010315 183.21 -45.72 306.87 -35.21 031812.9-011401	 HICK 25F PGC 12538	.L?.... 	 	 .16 	16.49S±.15 				
032045.6-010242 183.20 -45.71 306.88 -35.21 031812.9-011327	 UGC 2691 HICK 25B PGC 12539	.S?.... 	.99± .10 .24± .07 .99	 .16 .18 	15.01S±.15 14.71 ±.12 14.40		 14.20± .55 		6401± 41 6381 6227
0320.8 +3905 152.13 -15.16 346.16 -15.86 0317.6 +3855	 UGC 2681 PGC 12548	.S..8*. U 8.0±1.3 	1.00± .16 .55± .12 1.07	123 .71 .68 .27					
032051.9+381514 152.62 -15.86 345.46 -16.36 031738.1+380427	 UGC 2685 IRAS03176+3804 PGC 12549	.SXS3.. U 3.0± .8 	1.24± .05 .15± .05 1.33	0 .88 .21 .08	 14.3 ±.3 13.05 13.16				5044 5137 4884

R.A. 2000 DEC. l b SGL SGB R.A. 1950 DEC.	Names PGC	Type S_T n_L T L	$\log D_{25}$ $\log R_{25}$ $\log A_e$ $\log D_o$	p.a. A_g A_i A_{21}	B_T m_B m_{FIR} B_T^o	$(B-V)_T$ $(U-B)_T$ $(B-V)_T^o$ $(U-B)_T^o$	$(B-V)_e$ $(U-B)_e$ m'_e m'_{25}	m_{21} W_{20} W_{50} HI	V_{21} V_{opt} V_{GSR} V_{3K}
032057.9+415336 150.54 -12.82 348.47 -14.20 031738.5+414250	IC 313 UGC 2682 PGC 12558	.E+..*. PU -4.0± .6 	.97± .13 .07± .05 1.06	.75 .00 					4429± 71 4531 4275
032058.3-253046 218.80 -56.58 276.16 -39.29 031849.0-254130	NGC 1306 ESO 481- 23 IRAS03188-2541 PGC 12559	.S..3?. S 3.0±1.7 	1.03± .05 .10± .04 1.03	.00 .14 .05	13.62 ±.14 12.09 				
032059.1-002204 182.51 -45.24 307.69 -35.05 031825.7-003249	 UGC 2692 IRAS03184-0032 PGC 12560	.SAS5*. UE (1) 5.3± .7 1.9± .8	1.12± .04 .05± .04 1.14	35 .20 .07 .02	13.83 ±.12 12.78 13.52		15.70±.3 206± 7 187± 7 2.15	6309± 11 6285± 50 6289 6134
0321.0 +4047 151.18 -13.73 347.57 -14.86 0317.7 +4037	 UGC 2686 PGC 12561	.SX.1.. U 1.0± .8 	1.16± .07 .25± .06 1.23	43 .78 .25 .12					
032104.0-370604 240.04 -57.02 261.14 -38.33 031909.1-371648	NGC 1310 ESO 357- 19 IRAS03191-3716 PGC 12569	.SAS5*. R (1) 5.0± .5 5.2± .5	1.30± .04 .11± .04 1.09± .04 1.30	95 .00 .17 .06	12.55 ±.19 12.98 ±.14 12.87 12.65	.47± .02 -.12± .04 .44 -.14	.55± .01 -.06± .02 13.49± .09 13.61± .28		1739± 37 1621 1622
032112.8-043503 187.38 -47.80 302.74 -36.38 031843.2-044547	NGC 1304 MCG -1- 9- 30 PGC 12575	.L..-P. E -3.0± .9 	1.12± .08 .24± .08 1.10	130 .09 .00 					
0321.3 +4155 150.58 -12.76 348.54 -14.23 0318.0 +4145	 UGC 2688 IRAS03179+4145 PGC 12578	.S?.... 	1.14± .07 .27± .06 1.21	64 .75 .40 .14	 12.24 				2994± 94 3096 2840
0321.3 +1751 166.10 -32.17 327.19 -27.58 0318.5 +1741	 UGC 2693 PGC 12579	.S..0?. U .0±1.7 	1.00± .08 .04± .06 1.04	 .41 .03 					
032120.9+412736 150.85 -13.14 348.16 -14.52 031802.1+411651	 CGCG 540-113 PGC 12580	 	 	.82	15.1 ±.4				4411± 19 4512 4257
0321.3 +0725 174.79 -39.89 316.46 -32.28 0318.7 +0715	 UGC 2695 PGC 12581	.S..6?. U 6.0±1.9 	1.16± .07 .88± .06 1.23	22 .76 1.30 .44					10973 10978 10796
032123.0-021901 184.77 -46.39 305.45 -35.75 031851.4-022945	NGC 1305 UGC 2697 PGC 12582	PL..-P* PUE -2.9± .6 	1.14± .06 .17± .05 1.14	130 .19 .00 	 14.32 ±.11 				
0321.4 +4048 151.23 -13.68 347.63 -14.92 0318.1 +4038	 UGC 2689 PGC 12585	.L...?. U -2.0±1.6 	1.18± .15 .17± .08 1.24	127 .80 .00 					
032128.8-433506 251.49 -55.66 253.02 -37.05 031944.0-434548	 ESO 248- 12 IRAS03197-4345 PGC 12589	.S?.... 	1.10± .07 .52± .05 1.10	139 .00 .53 .26	 14.93 ±.14 14.28				9150± 60 9019 9050
0321.5 -4933 261.19 -53.60 245.85 -35.36 0320.0 -4944	 PGC 12594	 	 	.00	16.14 ±.15	.59± .05			19411± 88 19268 19326
032136.6+412334 150.93 -13.17 348.13 -14.60 031817.9+411250	NGC 1293 MCG 7- 7- 75 PGC 12597	.E.0... P -5.0± .9 	1.02± .09 .00± .04 .55± .03 1.17	 .90 .00 	14.50 ±.13 14.3 ±.2 13.49	1.11± .01 .87 	1.14± .01 12.74± .10 14.61± .46		4115± 57 4216 3961
032140.1+412138 150.96 -13.19 348.11 -14.63 031821.4+411054	NGC 1294 UGC 2694 IRAS03184+4111 PGC 12600	.LA.-?. PU -2.8± .7 	1.11± .11 .05± .07 .62± .20 1.21	 .90 .00 	14.3 ±.4 14.3 ±.3 13.32	1.08± .06 .82 	1.12± .06 12.90± .68 14.58± .69		6560± 57 6661 6406
032152.7-154051 202.59 -53.32 288.96 -38.85 031933.5-155132	 MCG -3- 9- 27 PGC 12608	.SBS9.. E (1) 9.0± .8 8.7± .8	1.20± .05 .22± .05 1.20	55 .08 .23 .11					2004 1940 1847

R.A. 2000 DEC. l b SGL SGB R.A. 1950 DEC.	Names PGC	Type S_T n_L T L	$\log D_{25}$ $\log R_{25}$ $\log A_e$ $\log D_o$	p.a. A_g A_i A_{21}	B_T m_B m_{FIR} B_T^o	$(B-V)_T$ $(U-B)_T$ $(B-V)_T^o$ $(U-B)_T^o$	$(B-V)_e$ $(U-B)_e$ m'_e m'_{25}	m_{21} W_{20} W_{50} HI	V_{21} V_{opt} V_{GSR} V_{3K}
032155.4-133903 199.57 -52.43 291.56 -38.56 031934.1-134945	MCG -2- 9- 35 VV 491 PGC 12611	.S?....	1.04± .09 .84± .07 1.05	.17 1.24 .42	13.60				9678± 41 9620 9518
032157.2-133855 199.57 -52.42 291.57 -38.56 031936.0-134936	HICK 26B PGC 12614	.L?....		.17	16.42S±.15				9332± 41 9274 9173
0322.0 -0103 183.50 -45.47 307.00 -35.52 0319.5 -0114	UGC 2699 PGC 12620	.S..6*. U 6.0±1.4	1.00± .08 .73± .06 1.02	135 .22 1.07 .36					
032203.1+405151 151.31 -13.56 347.75 -14.98 031845.1+404109	UGC 2698 PGC 12622	.E..... U -5.0± .8	1.00± .19 .05± .08 1.11	.80 .00	13.9 ±.3 13.02				6412± 94 6511 6257
032205.1+421017 150.56 -12.48 348.82 -14.20 031845.0+415934	UGC 2696 PGC 12624	.S?....	.96± .09 .90± .06 1.04	175 .85 1.35 .45	15.5 ±.3 13.22				5453± 94 5555 5300
032205.8-373512 240.87 -56.76 260.48 -38.45 032011.7-374553	PGC 12625	.IBS9?. S (1) 10.0±1.5 12.2± .8	1.46± .05 .20± .08 1.46	.00 .15 .10					1614 1495 1500
032206.6-152402 202.20 -53.16 289.33 -38.87 031947.0-153443	NGC 1309 MCG -3- 9- 28 IRAS03197-1534 PGC 12626	.SAS4*. R (3) 4.0± .5 3.4± .5	1.34± .02 .03± .03 .93± .01 1.34	45 .07 .04 .01	11.97 ±.04 12.0 ± .2 11.12 11.85	.44± .01 -.17± .02 .40 -.20	.52± .01 -.08± .02 12.11± .03 13.44± .15	13.79±.0 142± 4 125± 5 1.93	2135± 5 2257± 50 2073 1979
032211.6+404337 151.41 -13.66 347.66 -15.09 031853.8+403255	MCG 7- 7- 78 PGC 12627	.L?....	.90± .17 .16± .07 .97	.80 .00	15.1 ±.3 14.20				4788± 94 4887 4633
032217.6-070527 190.70 -49.02 299.79 -37.29 031950.3-071607	MCG -1- 9- 31 IRAS03198-0716 PGC 12633	.SAR2P* E 2.0±1.2	1.06± .04 .06± .05 1.07	.16 .07 .03	13.79				
0322.4 +0926 173.21 -38.27 318.75 -31.67 0319.7 +0916	UGC 2701 PGC 12639	.S..4.. U 4.0±1.0	.96± .09 .69± .06 1.03	143 .76 1.01 .34					7292 7302 7116
032228.6-024526 185.52 -46.45 305.03 -36.15 031957.4-025606	NGC 1308 MCG -1- 9- 32 PGC 12643	.SBR0.. E .0± .9	1.07± .06 .14± .05 1.07	135 .12 .10					
0322.5 +1454 168.66 -34.22 324.43 -29.26 0319.8 +1444	UGC 2703 PGC 12645	.S..3.. U (1) 3.0± .9 5.5±1.2	1.16± .05 .54± .05 1.23	23 .70 .75 .27	15.28 ±.14 13.76				10083 10109 9907
032241.2-041116 187.23 -47.27 303.34 -36.62 032011.3-042155	NGC 1314 MCG -1- 9- 33 PGC 12650	.SAT7.. E (1) 7.0± .8 7.5± .8	1.18± .05 .02± .05 1.18	90 .06 .02 .01					3801 3770 3632
032241.6-371228 240.16 -56.69 260.94 -38.63 032047.0-372306	NGC 1316 ESO 357- 22 ARP 154 PGC 12651	PLXS0P. R -2.0± .3	2.08± .02 .15± .02 1.43± .06 2.05	50 .00 .00	9.42M±.08 9.17 ±.11 11.76 9.31	.89± .01 .39± .03 .86 .39	.93± .01 .47± .01 12.03± .21 14.31± .13		1793± 12 1674 1678
032244.7-370610 239.97 -56.69 261.07 -38.66 032050.0-371648	NGC 1317 ESO 357- 23 IRAS03208-3716 PGC 12653	.SXR1.. V 1.0± .3	1.44± .02 .06± .02 .80± .03 1.44	78 .00 .07 .03	11.91M±.06 11.87 ±.11 11.55 11.81	.89± .01 .29± .02 .86 .28	.92± .01 .33± .01 11.55± .10 13.79± .14		1941± 14 1822 1826
0322.7 +0008 182.35 -44.57 308.47 -35.30 0320.2 -0001	UGC 2705 IRAS03202-0001 PGC 12655	.SX.6.. U 6.0± .8	.99± .06 .02± .05 1.02	.22 .03 .01	14.64 ±.13 14.36				6858 6840 6686
032247.7-015518 184.64 -45.87 306.05 -35.97 032015.7-020557	UGC 2704 PGC 12656	.S?....	.96± .09 .00± .06 .97	.13 .00 .00	14.14 ±.12 13.97				8227± 50 8203 8056

R.A. 2000 DEC. l　　b SGL　SGB R.A. 1950 DEC.	Names PGC	Type S_T　n_L T L	$\log D_{25}$ $\log R_{25}$ $\log A_e$ $\log D_o$	p.a. A_g A_i A_{21}	B_T m_B m_{FIR} B_T^o	$(B-V)_T$ $(U-B)_T$ $(B-V)_T^o$ $(U-B)_T^o$	$(B-V)_e$ $(U-B)_e$ m'_e m'_{25}	m_{21} W_{20} W_{50} HI	V_{21} V_{opt} V_{GSR} V_{3K}
032253.8+423310 150.47 -12.08 349.22 -14.09 031933.0+422230	 UGC 2700 PGC 12660	.SB.3?. U 3.0± .9 	1.17± .05 .60± .05 .63± .04 1.26	132 1.05 .82 .30	 14.8 ±.3 12.88				 6685± 94 6788 6534
032254.9-421117 248.97 -55.77 254.66 -37.65 032108.0-422154	 ESO 301- 9 IRAS03211-4221 PGC 12662	.L...*/ S -2.0±1.0 	1.28± .05 .77± .03 .63± .04 1.16	81 .00 .00 	14.21 ±.15 14.27 ±.14 13.13 14.22	.78± .02 .12± .04 .70 .06	.73± .02 .10± .03 12.85± .13 13.55± .30		 1159 1030 1056
032255.0-111212 196.27 -51.06 294.72 -38.36 032031.6-112251	 MCG -2- 9- 36 PGC 12664	.SBS7.. UE　　(1) 7.0± .5 6.4± .8	1.40± .04 .16± .05 1.42	165 .17 .21 .08			14.22±.1 238± 8 219± 12 		2807± 6 2755 2645
032260.0+012155 181.11 -43.73 309.89 -34.94 032025.1+011116	 MCG 0- 9- 74 PGC 12669	RSXR1P? E 1.0±1.2 	1.12± .06 .20± .05 1.15	80 .30 .20 .10	 14.59 ±.14 				
032301.2-423611 249.68 -55.65 254.14 -37.57 032115.1-424648	 ESO 248- 14 PGC 12670	.SBR2.. r 2.2± .9 	.95± .06 .16± .05 .64± .04 .95	153 .00 .19 .08	15.0 ±.2 15.07 ±.14 	.83± .04 	.91± .03 13.70± .10 14.20± .37		
032306.5-212234 211.87 -55.09 281.60 -39.68 032053.1-213312	NGC 1315 ESO 548- 3 PGC 12671	.LBT+$. R -1.0± .4 	1.20± .04 .04± .03 1.20	 .03 .00 	 13.46 ±.14 13.41				 1706± 39 1626 1560
032325.8-191710 208.48 -54.35 284.34 -39.61 032110.2-192746	 PGC 12680	.IBS9.. SE　　(1) 10.0± .5 10.3± .8	1.11± .06 .22± .06 1.12	104 .07 .16 .11					1552 1478 1403
032329.8+011927 181.27 -43.67 309.90 -35.07 032054.9+010850	NGC 1312 UGC 2711 PGC 12682	PSB.3*/ UE 2.7± .8 	1.14± .05 .78± .05 1.17	167 .31 1.08 .39	 15.45 ±.12 14.00				9903 9888 9731
0323.5 -1647 204.55 -53.42 287.60 -39.38 0321.2 -1658	 MCG -3- 9- 32 PGC 12684	.S?.... 	1.04± .09 .28± .07 1.05	 .10 .42 .14					9436 9368 9283
032337.4-354643 237.54 -56.64 262.73 -39.05 032141.0-355718	 ESO 357- 25 PGC 12691	.LX.0*. S -2.5± .6 	1.12± .05 .49± .03 1.05	25 .00 .00 	 14.98 ±.14 14.95				1823± 29 1707 1706
032338.2-111149 196.40 -50.90 294.77 -38.53 032114.7-112225	 MCG -2- 9- 38 PGC 12692	PS..0*. E .0±1.2 	1.25± .07 .48± .05 1.24	55 .17 .36 					9446 9394 9286
0323.6 +0634 176.10 -40.09 315.81 -33.14 0321.0 +0624	 UGC 2712 PGC 12695	.SB.8*. U 8.0±1.1 	1.14± .07 .00± .06 1.20	 .59 .00 .00					7113± 10 7114 6939
0323.7 +3702 153.82 -16.53 344.80 -17.53 0320.5 +3652	 UGC 2707 PGC 12697	.SBS8*. U 8.0±1.2 	1.11± .14 .29± .12 1.19	150 .80 .36 .15					3405± 10 3493 3247
032342.9+365512 153.90 -16.63 344.69 -17.60 032030.5+364435	 UGC 2706 PGC 12698	.S..8*. U 8.0±1.2 	.94± .09 .05± .06 1.02	 .80 .06 .02	 15.34 ±.15 14.46				4735 4823 4576
032347.2-194513 209.29 -54.43 283.74 -39.73 032132.0-195548	 ESO 548- 5 PGC 12701	.SXS9.. SE　　(2) 8.7± .5 7.7± .5	1.22± .03 .06± .03 1.23	20 .06 .06 .03	 13.70 ±.14 13.58		14.75±.2 118± 16 111± 12 1.14		1838± 11 1762 1690
032348.8+403328 151.77 -13.63 347.71 -15.43 032031.0+402251	 UGC 2708 PGC 12702	.L..... U -2.0± .8 	1.12± .07 .00± .03 1.22	 .92 .00 	 14.8 ±.3 13.78				5394± 94 5491 5241
032353.7+384039 152.88 -15.17 346.17 -16.58 032038.7+383003	 UGC 2709 IRAS03206+3830 PGC 12705	.SX.3.. U 3.0± .7 	1.37± .05 .48± .06 1.43	3 .67 .67 .24	 14.3 ±.3 13.41 12.92				5191± 94 5284 5035

R.A. 2000 DEC.	Names	Type	logD₂₅	p.a.	B_T	(B-V)_T	(B-V)_e	m₂₁	V₂₁
l b		S_T n_L	logR₂₅	A_g	m_B	(U-B)_T	(U-B)_e	W₂₀	V_opt
SGL SGB		T	logA_e	A_i	m_FIR	(B-V)°_T	m'_e	W₅₀	V_GSR
R.A. 1950 DEC.	PGC	L	logD_o	A₂₁	B°_T	(U-B)°_T	m'₂₅	HI	V_3K
032354.0-373032		.RING*P	1.15± .05						1403
240.65 -56.42	ESO 301- 11	S	.04± .05	.00	13.87 ±.14				1283
260.50 -38.81	IRAS03220-3741	10.0± .7		.03	13.59				1290
032200.0-374106	PGC 12706		1.15	.02	13.84				
032356.3-213137	NGC 1319	.L...P/	1.13± .04	27					4058± 39
212.22 -54.95	ESO 548- 6	RS	.30± .03	.04	13.88 ±.14				3977
281.41 -39.88		-1.8± .4		.00					3913
032143.0-214212	PGC 12708		1.09		13.78				
032356.4-362750	NGC 1326	RLBR+..	1.59± .02	77	11.41M±.10	.87± .01	.85± .01	13.78±.1	1362± 4
238.77 -56.52	ESO 357- 26	R	.13± .02	.00	11.43 ±.11	.28± .01	.26± .01	265± 6	1365± 16
261.84 -39.00	IRAS03220-3638	-1.0± .3	.92± .01	.00	10.86	.84	11.58± .04	247± 5	1244
032201.0-363824	PGC 12709		1.57		11.40	.27	13.92± .16		1247
0323.9 +3745		.L.....	1.13± .10						5540
153.44 -15.92	UGC 2710	U	.03± .05	.70	14.5 ±.3				5630
345.42 -17.15	IRAS03207+3734	-2.0± .8		.00	11.76				5383
0320.7 +3734	PGC 12713		1.20		13.69				
0324.0 -0344		.LX.-P*	1.18± .08	80					
186.99 -46.74		E	.35± .08	.06					
304.00 -36.80		-3.0±1.3		.00					
0321.5 -0355	PGC 12716		1.14						
0324.0 +1744		.S..8*.	1.20± .05	88					381
166.77 -31.85	UGC 2716	U	.25± .05	.45	14.82 ±.19				
327.43 -28.20		8.0±1.1		.30					415
0321.2 +1734	PGC 12719		1.25	.12	14.06				208
032406.6-840940		.SBS8*.	1.12± .05	80					
298.70 -31.65	ESO 3- 15	S (1)	.50± .04	.56	14.79 ±.14				
211.06 -19.19	IRAS03302-8419	8.0±1.3		.61	13.57				
033014.0-842000	PGC 12725	7.8± .9	1.17	.25					
0324.3 +0006		.S..8*.	.89± .10						6475
182.75 -44.29	UGC 2721	U	.06± .06	.27					
308.59 -35.68		8.0±1.2		.07					6456
0321.8 -0004	PGC 12733		.91	.03					6305
032425.3-213233	NGC 1325	.SAS4..	1.67± .02	56	12.22 ±.14	.69± .04	.77± .03	13.77±.2	1589± 7
212.30 -54.84	ESO 548- 7	R (1)	.47± .03	.04	12.34 ±.11	.09± .06	.18± .06	347± 8	1783± 41
281.40 -39.99	IRAS03221-2143	4.0± .3	1.25± .02	.68	12.87	.59	13.96± .05	327± 6	1512
032212.0-214306	PGC 12737	3.3± .6	1.68	.23	11.57	.01	14.27± .17	1.97	1450
032436.4+404127		.E.....	1.00± .19						
151.82 -13.44	UGC 2717	U	.05± .08	.91	14.3 ±.3				3757± 94
347.91 -15.47		-5.0± .8		.00					3854
032118.3+403053	PGC 12747		1.13		13.38				3605
0324.7 +4158		.S..6*.	1.04± .15	78					
151.10 -12.37	UGC 2718	U	.47± .12	.95					
348.96 -14.71	IRAS03213+4147	6.0±1.3		.69	12.57				
0321.4 +4147	PGC 12750		1.13	.24					
032448.4-212016	Holmberg VI	.SXT7*.	1.34± .03		13.27S±.08	.57± .05		15.52±.3	1333± 8
212.01 -54.70	ESO 548- 10	Sr (1)	.04± .03	.03	13.33 ±.14			57± 9	
281.68 -40.07		6.7± .5		.06		.55			1252
032235.0-213048	PGC 12754	6.7± .5	1.34	.02	13.20		14.70± .18	2.30	1189
032449.0-030231	NGC 1320	.S..1*/	1.28± .03	45	13.32 ±.13	.87± .01	.88± .01		
186.36 -46.16	MCG -1- 9- 36	E	.47± .05	.08		.26± .02	.28± .02		
304.91 -36.79	MK 607	1.0±1.2	.73± .02	.48	12.37		12.46± .08		
032218.0-031303	PGC 12756		1.28	.23			13.39± .22		
0324.8 +3856		.S..6*.	1.00± .08	3					
152.89 -14.85	UGC 2719	U	1.02± .06	.66					
346.51 -16.57		6.0±1.5		1.47					
0321.6 +3846	PGC 12757		1.06	.50					
0324.8 -3526					15.30S±.10				1433± 29
236.89 -56.41				.00					1317
263.12 -39.35									1317
0322.9 -3536	PGC 12758								
032453.7-604418		RLXR+*/	1.15± .04	91	15.19 ±.15	.94± .03			
276.17 -47.64	ESO 116- 18	Sr	.66± .04	.00	15.09 ±.14				5397± 88
233.18 -31.50	IRAS03238-6054	-.6± .6		.00	12.81	.80			5237
032353.1-605448	PGC 12759		1.05		15.05		14.16± .28		5347
032455.0-214705		.SBS9..	1.05± .04	145					
212.76 -54.80	ESO 548- 11	S (1)	.16± .03	.06	15.81 ±.14				
281.09 -40.12		9.0± .8		.16					
032242.0-215736	PGC 12762	10.0± .8	1.06	.08					

R.A. 2000 DEC. / I b / SGL SGB / R.A. 1950 DEC.	Names / PGC	Type / S_T n_L / T / L	$\log D_{25}$ / $\log R_{25}$ / $\log A_e$ / $\log D_o$	p.a. / A_g / A_i / A_{21}	B_T / m_B / m_{FIR} / B_T^o	$(B-V)_T$ / $(U-B)_T$ / $(B-V)_T^o$ / $(U-B)_T^o$	$(B-V)_e$ / $(U-B)_e$ / m'_e / m'_{25}	m_{21} / W_{20} / W_{50} / HI	V_{21} / V_{opt} / V_{GSR} / V_{3K}
032455.8+390744 / 152.79 -14.68 / 346.67 -16.47 / 032140.0+385711	UGC 2720 / PGC 12763	.S..2.. / U / / 2.0±1.0	.99± .06 / .68± .05 / / 1.06	90 / .66 / .84 / .34	15.4 ±.3 / / / 13.80				6342± 94 / 6435 / 6188
0324.9 +0725 / 175.61 -39.26 / 316.89 -33.09 / 0322.3 +0715	UGC 2726 / PGC 12768	.S..3*. / U / / 3.0±1.2	1.14± .07 / .32± .06 / / 1.21	165 / .73 / .45 / .16					8298 / 8300 / 8126
032458.5-370035 / 239.72 -56.26 / 261.09 -39.11 / 032304.0-371106	NGC 1316C / ESO 357- 27 / PGC 12769	PLA.0*. / PS / / -2.2± .5	1.15± .04 / .32± .03 / / 1.10	85 / .00 / .00	14.36 ±.14 / / / 14.33				1975± 25 / 1855 / 1862
032501.8-054445 / 189.61 -47.70 / 301.66 -37.61 / 032233.3-055516	NGC 1324 / MCG -1- 9- 38 / IRAS03225-0555 / PGC 12772	.S..3*. / E (1) / 3.0±1.2 / 3.1±1.6	1.32± .04 / .42± .05 / / 1.33	45 / .13 / .58 / .21	12.79				5683 / 5646 / 5519
032502.3-491524 / 260.34 -53.20 / 245.98 -35.99 / 032329.1-492554	ESO 200- 16 / FAIR 300 / PGC 12774	.LB?...	.91± .07 / .22± .05 / / .88	165 / .00 / .00	15.31 ±.14 / / / 15.14				11333± 79 / 11189 / 11251
0325.0 +0512 / 177.71 -40.78 / 314.46 -33.99 / 0322.4 +0502	UGC 2727 / PGC 12775	.S..7.. / U / / 7.0±1.0	1.07± .07 / .79± .06 / / 1.11	172 / .46 / 1.10 / .40					8826 / 8822 / 8655
032508.4-362154 / 238.56 -56.29 / 261.91 -39.25 / 032313.0-363224	NGC 1326A / ESO 357- 28 / PGC 12783	.SBS9*. / PS (1) / 8.8± .4 / 7.8± .6	1.27± .04 / .04± .04 / / 1.27	/ .00 / .04 / .02	13.80 ±.14 / / / 13.75			14.15±.2 / 88± 16 / / .38	1836± 11 / 1823± 29 / 1716 / 1720
032515.2-395348 / 244.82 -55.81 / 257.40 -38.60 / 032325.1-400418	ESO 301- 14 / PGC 12786	.SAT7.. / S (1) / 7.0± .8 / 6.7±1.2	1.06± .05 / .06± .05 / / 1.06	2 / .00 / .08 / .03	15.01 ±.14 / / / 14.92				4282 / 4156 / 4176
0325.2 +4240 / 150.77 -11.73 / 349.59 -14.35 / 0321.9 +4230	UGC 2723 / PGC 12787	.SBR5.. / U / / 5.0± .8	1.03± .08 / .01± .06 / / 1.13	165 / 1.09 / .02 / .01					
032518.4-362300 / 238.58 -56.25 / 261.88 -39.28 / 032323.0-363330	NGC 1326B / ESO 357- 29 / PGC 12788	.SBS9*/ / PS (1) / 9.2± .5 / 6.7± .6	1.57± .03 / .54± .04 / .98± .05 / 1.57	130 / .00 / .55 / .27	13.7 ±.2 / 13.08 ±.14 / / 12.71	.41± .03 / -.09± .09 / .29 / -.18	.42± .02 / -.11± .06 / 14.10± .11 / 15.06± .29	12.89±.2 / 179± 8 / / -.09	1006± 6 / 825± 40 / 883 / 888
032523.7-254048 / 219.44 -55.63 / 275.93 -40.29 / 032315.0-255118	NGC 1327 / ESO 481- 26 / IRAS03232-2551 / PGC 12795	PSBS3.. / S (1) / 3.0±1.0 / 2.2±1.0	1.01± .05 / .57± .03 / / 1.01	176 / .00 / .78 / .28	15.57 ±.14 / / 13.54				
0325.4 +4038 / 151.97 -13.39 / 347.96 -15.62 / 0322.1 +4028	UGC 2724 / PGC 12796	.S..8*. / U / / 8.0±1.2	1.00± .16 / .09± .12 / / 1.09	/ .91 / .11 / .04					
032525.0-161410 / 204.01 -52.78 / 288.39 -39.76 / 032306.4-162440	MCG -3- 9- 41 / PGC 12798	.S..8*/ / EU / / 7.5± .8	1.42± .03 / .93± .05 / / 1.44	9 / .13 / 1.14 / .46				14.25±.1 / 240± 16 / 237± 12	1878± 11 / 1810 / 1727
032525.7-511602 / 263.34 -52.33 / 243.61 -35.36 / 032357.1-512630	IC 1929 / ESO 200- 21 / PGC 12799	.S?....	.90± .06 / .11± .05 / .63± .05 / .90	146 / .00 / .11 / .05	14.81 ±.18 / 14.88 ±.14 / / 14.58	.72± .04 / / .59	.80± .03 / / 13.45± .16 / 13.91± .37		13271± 88 / 13124 / 13194
032526.0-060828 / 190.18 -47.84 / 301.21 -37.81 / 032257.9-061858	A 0322-06 / MK 609 / PGC 12801		.64± .10 / .16± .07 / / .65	90 / .13	12.09				10212± 62 / 10173 / 10048
032529.6+411429 / 151.64 -12.89 / 348.46 -15.27 / 032210.5+410358	UGC 2725 / PGC 12802	.L..... / U / / -2.0± .8	1.03± .11 / .08± .05 / / 1.13	162 / .94 / .00	14.8 ±.3 / / 13.74				6192± 94 / 6290 / 6041
032531.4-060750 / 190.18 -47.82 / 301.23 -37.83 / 032303.3-061820	A 0323-06 / MK 610 / PGC 12803	.S?....	.60± .10 / .00± .07 / / .62	/ .13 / .00 / .00					10301± 62 / 10262 / 10137

R.A. 2000 DEC. / l b / SGL SGB / R.A. 1950 DEC.	Names / PGC	Type / S_T n_L / T / L	$\log D_{25}$ / $\log R_{25}$ / $\log A_e$ / $\log D_o$	p.a. / A_g / A_i / A_{21}	B_T / m_B / m_{FIR} / B_T^o	$(B-V)_T$ / $(U-B)_T$ / $(B-V)_T^o$ / $(U-B)_T^o$	$(B-V)_e$ / $(U-B)_e$ / m'_e / m'_{25}	m_{21} / W_{20} / W_{50} / HI	V_{21} / V_{opt} / V_{GSR} / V_{3K}
032540.0-524702 265.53 -51.63 241.85 -34.85 032415.1-525730	IC 1933 ESO 155- 25 IRAS03242-5257 PGC 12807	.SXS7*. RS (2) 6.6± .4 5.9± .4	1.35± .04 .28± .04 .91± .09 1.35	55 .00 .38 .14	12.9 ±.3 12.97 ±.12 12.68 12.58	.36± .03 .30	.44± .02 12.97± .20 13.85± .40	14.14±.3 186± 12 1.43	1068± 9 1033± 66 918 995
0325.7 +4204 151.19 -12.17 349.16 -14.79 0322.4 +4154	 UGC 2728 PGC 12812	.S..7.. U 7.0±1.0	1.04± .08 .76± .06 1.13	51 .95 1.05 .38					
032552.5+404455 151.99 -13.25 348.11 -15.63 032234.1+403425	 UGC 2730 IRAS03225+4034 PGC 12816	.S..3.. U (1) 3.0± .9 4.5±1.2	1.16± .05 .70± .05 .97 1.26	127 1.06 .35	 14.9 ±.3 12.56 12.81				 3839± 94 3936 3688
032554.4-512033 263.40 -52.23 243.49 -35.40 032426.0-513100	IC 1932 ESO 200- 22 FAIR 747 PGC 12817	.L?....	.87± .03 .11± .05 .53± .03 .85	15 .00 .00	15.39 ±.15 15.38 ±.14 15.21	.85± .03 .29± .06 .73 .35	.96± .03 .26± .05 13.53± .09 14.32± .26		11966± 32 11819 11890
032559.3+404721 151.98 -13.21 348.15 -15.62 032240.8+403652	IC 320 UGC 2732 PGC 12819	.SBT2.. U 2.0± .8	1.09± .06 .06± .05 1.19	 1.06 .08 .03	 14.6 ±.3 13.43				6972± 94 7069 6821
032600.1+034045 179.43 -41.64 312.86 -34.80 032323.0+033017	IC 322 CGCG 390- 89 PGC 12820		.95? .05? .98	 .31	 15.01 ±.14				9198 9188 9028
032602.0-325345 232.30 -56.23 266.38 -39.94 032402.1-330412	IC 1919 ESO 358- 1 PGC 12825	.LAT-?. BS -3.0± .4	1.21± .05 .14± .04 1.19	84 .00 .00	13.80S±.10 13.78 ±.14 13.77		 14.39± .29		1220± 29 1109 1100
032602.2-173526 206.18 -53.17 286.64 -40.07 032345.0-174554	NGC 1329 ESO 548- 15 PGC 12826	.SAT1?. SE 1.0± .7	1.15± .03 .09± .03 .81± .02 1.16	35 .10 .09 .04	13.49 ±.13 13.38 ±.14 13.19	.84± .02 .27± .04 .76 .24	.90± .02 .37± .04 13.03± .06 13.88± .20		4277± 39 4205 4128
032602.3-212033 212.16 -54.42 281.69 -40.36 032349.0-213100	 ESO 548- 16 PGC 12827	.S?....	1.37± .05 .53± .06 1.37	115 .04 .79 .26	 15.53 ±.14 14.69				2161± 60 2079 2018
0326.0 +4202 151.26 -12.17 349.17 -14.86 0322.7 +4152	 UGC 2731 PGC 12828	.S..7.. U 7.0±1.0	1.04± .08 .76± .06 1.13	142 .99 1.05 .38					
032603.3+411509 151.72 -12.82 348.54 -15.35 032244.0+410440	 UGC 2733 PGC 12830	.E...?. U -5.0±1.7	1.03± .11 .18± .05 1.13	120 .94 .00	 14.5 ±.3 13.52				5363± 94 5461 5213
032612.8+463802 148.65 -8.37 352.84 -12.02 032244.2+462733	 UGC 2734 PGC 12832	.S..7*. U 7.0±1.2	1.07± .14 .11± .12 1.19	 1.24 .15 .05				16.32±.3 223± 11 180± 8	5623± 9 5734 5481
032613.4-500034 261.37 -52.73 245.01 -35.92 032442.1-501100	IC 1935 ESO 200- 23 PGC 12833	.SAS6?P S (1) 6.0± .9 3.3±1.2	1.04± .05 .17± .05 .63± .03 1.04	104 .00 .25 .08	14.65 ±.17 14.36 ±.14 14.20	.38± .04 .31	.40± .03 13.29± .08 14.30± .33		5546± 88 5401 5467
032614.4-001212 183.50 -44.12 308.42 -36.23 032340.9-002239	A 0323+00 MK 611 PGC 12835		.27± .16 .09± .07 .30	155 .25					8894± 43 8872 8727
032617.3-212009 212.18 -54.37 281.70 -40.41 032404.0-213036	NGC 1332 ESO 548- 18 IRAS03240-2130 PGC 12838	.L.S-*/ R -3.0± .3	1.67± .02 .51± .02 .97± .02 1.60	148 .04 .00	11.25 ±.13 11.35 ±.10 13.51 11.25	.96± .01 .59± .02 .91 .56	1.01± .01 .63± .01 11.59± .08 13.19± .17	15.58±.3 234± 6 222± 25	1524± 10 1546± 30 1444 1384
0326.2 -0443 188.64 -46.86 303.01 -37.63 0323.8 -0454	 PGC 12839	.LA.0P? E -2.0±1.8	1.12± .08 .44± .08 1.06	152 .06 .00					
032627.8+403030 152.22 -13.38 347.98 -15.86 032309.7+402002	 UGC 2736 PGC 12845	.S..2.. U 2.0± .9	1.19± .05 .73± .05 1.28	69 .88 .89 .36	 14.4 ±.3 12.58				 5673± 94 5769 5522

R.A. 2000 DEC. l b SGL SGB R.A. 1950 DEC.	Names PGC	Type S_T n_L T L	$\log D_{25}$ $\log R_{25}$ $\log A_e$ $\log D_o$	p.a. A_g A_i A_{21}	B_T m_B m_{FIR} B_T^o	$(B-V)_T$ $(U-B)_T$ $(B-V)_T^o$ $(U-B)_T^o$	$(B-V)_e$ $(U-B)_e$ m'_e m'_{25}	m_{21} W_{20} W_{50} HI	V_{21} V_{opt} V_{GSR} V_{3K}
032628.3-212122 NGC 1331 212.24 -54.33 ESO 548- 19 281.68 -40.46 032415.1-213148 PGC 12846		.E.2.*. R -5.0± .7	.97± .03 .04± .03 .49± .02 .96	 .04 .00	14.31 ±.13 14.25 ±.14 14.23	.88± .02 .37± .05 .86 .37	.89± .02 .39± .05 12.25± .05 14.06± .22		1375± 58 1293 1233
032631.1-354252 NGC 1336 237.35 -56.06 ESO 358- 2 262.69 -39.64 032435.0-355318 PGC 12848		.LA.-.. BS -3.3± .4	1.33± .05 .15± .04 1.01± .06 1.31	22 .00 .00	13.10M±.09 13.12 ±.14 13.08	.84± .02 .25± .03 .82 .25	.85± .01 .30± .02 13.65± .22 14.25± .27		1511± 22 1393 1397
0326.8 +0742 175.75 -38.73 UGC 2740 317.43 -33.39 IRAS03241+0732 0324.1 +0732 PGC 12859		.S..3*. U (1) 3.0±1.2 5.5±1.1	1.13± .05 .28± .05 1.19	55 .63 .39 .14	 14.78 ±.14 13.01				
0326.8 +0706 176.33 -39.15 UGC 2741 316.77 -33.65 0324.2 +0656 PGC 12863		.S?....	1.07± .14 .00± .12 1.14	 .71 .00 .00					11005 11005 10835
032653.3-353112 237.00 -55.99 ESO 358- 4 262.93 -39.74 032457.0-354136 PGC 12865		DE.0... S -5.0± .7	1.03± .10 .08± .04 1.01	56 .00 .00	16.29S±.10 18.07 ±.14		 16.25± .50		
0327.0 +7250 133.91 13.38 UGCA 69 12.81 4.81 0321.7 +7240 PGC 12868		.I..9?. U 10.0±1.7	1.32± .06 .65± .09 1.46	158 1.51 .49 .32				14.22±.1 256± 9 255± 12	2073± 7 2232 1989
032705.3+465800 148.58 -8.01 UGC 2737 353.20 -11.93 032335.8+464734 PGC 12869		.S..0.. U .0± .9	1.00± .08 .55± .06 1.10	143 1.31 .41	 15.58 ±.12				
032711.1-530038 IC 1938 265.69 -51.33 ESO 155- 32 241.49 -34.97 032547.1-531100 PGC 12874		.S?....	.90± .07 .12± .06 .90	 .00 .16 .06	 15.50 ±.14 15.27				9110± 82 8960 9040
0327.2 +6833 136.42 9.87 UGC 2729 9.60 2.01 0322.5 +6823 PGC 12876			1.54? .14? 1.84	130 3.19				13.56±.1 256± 9 246± 12	1938± 7 2091 1843
032716.1-332913 233.36 -55.97 ESO 358- 5 265.57 -40.12 032517.0-333936 PGC 12877		.SXS9P* S (1) 9.0± .6 9.4± .6	1.16± .04 .12± .04 1.16	 .00 .12 .06	 14.69 ±.14 14.56				1642± 29 1529 1524
032717.7-343137 235.22 -55.95 ESO 358- 6 264.20 -39.98 032520.0-344200 PGC 12878		.LB.-?. S -3.7± .7	1.09± .03 .29± .02 .75± .03 1.00	32 .00 .00	13.92 ±.14 14.15 ±.14 14.02	.84± .02 .83	.86± .02 13.16± .12 13.51± .23		1237± 57 1122 1121
0327.3 +0236 180.79 -42.11 UGC 2744 311.79 -35.50 0324.7 +0226 PGC 12879		.S..6*. U 6.0±1.2	.96± .09 .25± .06 .99	37 .32 .36 .12	 15.20 ±.12				
0327.3 -1344 200.66 -51.30 MCG -2- 9- 41 291.72 -39.87 0325.0 -1355 PGC 12880		.S?....	1.12± .08 .57± .07 1.13	 .16 .79 .28					12181 12119 12029
032721.0+495859 146.91 -5.49 UGC 2739 355.57 -10.05 IRAS03237+4948 032345.2+494834 PGC 12881		.S..6?. U 6.0±1.9	1.00± .08 .55± .06 1.18	36 1.92 .81 .27	 12.78				
032729.0-213337 IC 1928 212.69 -54.17 ESO 548- 20 281.42 -40.70 032516.0-214400 PGC 12884		.S..2?/ S 2.0±1.8	1.20± .04 .60± .03 1.20	30 .04 .73 .30	 14.09 ±.14 13.28				4266± 60 4183 4126
032729.6-364514 239.18 -55.78 ESO 358- 7 261.30 -39.65 IRAS03255-3655 032535.0-365536 PGC 12885			.84± .06 .47± .05 .84	167 .00	15.4 ±.2 15.20 ±.14 13.11	.68± .09 .38± .11			9992± 87 9872 9882
032735.4-211344 212.16 -54.05 ESO 548- 21 281.86 -40.71 032522.0-212406 PGC 12889		.SBS8?. S (1) 8.0±1.8 7.8±1.3	1.30± .04 .77± .03 1.30	68 .06 .95 .38	 14.60 ±.14 13.59				1657 1574 1516

R.A. 2000 DEC.	Names	Type	logD$_{25}$	p.a.	B$_T$	(B-V)$_T$	(B-V)$_e$	m$_{21}$	V$_{21}$
l b		S$_T$ n$_L$	logR$_{25}$	A$_g$	m$_B$	(U-B)$_T$	(U-B)$_e$	W$_{20}$	V$_{opt}$
SGL SGB		T	logA$_e$	A$_i$	m$_{FIR}$	(B-V)^o_T	m'$_e$	W$_{50}$	V$_{GSR}$
R.A. 1950 DEC.	PGC	L	logD$_o$	A$_{21}$	B^o_T	(U-B)^o_T	m'$_{25}$	HI	V$_{3K}$
032739.9+405350 152.19 -12.94 348.44 -15.80 032421.0+404326	UGC 2742 PGC 12893	.SBS4.. U 4.0± .9	.96± .07 .12± .05 1.05	110 .99 .18 .06	14.9 ±.3 13.67				4349± 94 4445 4200
032742.1-520821 IC 1940 264.38 -51.64 ESO 200- 26 242.45 -35.37 032616.0-521842	PGC 12896	PSBR2.. r 2.2± .9	.97± .06 .13± .06 .97	.00 .16 .07	14.77 ±.14 14.47				13640± 82 13491 13568
0327.8 +4002 152.71 -13.62 347.76 -16.35 0324.5 +3952	UGC 2743 PGC 12900	.S..6*. U 6.0±1.4	1.04± .08 .88± .06 1.12	84 .85 1.30 .44					
0327.9 +0233 A 0325+02 180.98 -42.03 UGC 2748 311.80 -35.66 0325.3 +0223	PGC 12909	.E...P? UE -4.7±1.0	1.21± .07 .18± .04 .78± .03 1.20	155 .31 .00	14.85 ±.15 14.73 ±.17 14.35	1.08± .03	1.11± .02 .92 14.24± .11 15.43± .38		9060± 59 9046 8894
032757.9-370903 NGC 1341 239.86 -55.65 ESO 358- 8 260.76 -39.67 IRAS03260-3719 032604.0-371924	PGC 12911	.SXS2.. RBCSr (2) 2.2± .4 4.7± .5	1.19± .04 .08± .04 .77± .01 1.19	155 .00 .09 .04	12.28V±.13 12.98 ±.11 11.79 12.78	.51± .01 -.09± .02 .48 -.10	.56± .01 -.10± .02 13.41± .26		1859± 24 1738 1750
032759.3+395417 152.83 -13.71 347.68 -16.46 032441.9+394355	MCG 7- 8- 11 PGC 12912	.S?.... 	.89± .11 .19± .07 .97	.85 .28 .09	14.6 ±.3 13.48				4282± 94 4375 4132
032805.9-082323 NGC 1337 193.54 -48.52 MCG -2- 9- 42 298.62 -39.00 032540.0-083344	PGC 12916	.SAS6.. R (3) 6.0± .3 3.8± .4	1.76± .02 .59± .03 1.18± .02 1.78	145 .16 .87 .29	12.48M±.10 12.30 ±.19 11.41	.59± .02 .10± .03 .43 -.01	.66± .01 .08± .02 14.00± .04 14.67± .14	12.54±.1 261± 4 243± 5 .83	1240± 6 1179± 66 1192 1082
032806.5-321710 NGC 1339 231.23 -55.78 ESO 418- 4 267.12 -40.44 032606.0-322730	PGC 12917	.E+..P* PS -4.4± .3	1.28± .03 .14± .03 .76± .01 1.24	172 .00 .00	12.51 ±.13 12.65 ±.11 12.57	.93± .01 .48± .03 .92 .49	.94± .01 .52± .01 11.80± .05 13.55± .21		1354± 19 1243 1234
032814.3-172504 A 0325-17 206.24 -52.61 ESO 548- 23 286.94 -40.57 IRAS03259-1735 032557.0-173524	PGC 12922	.E+..?. SU -3.7± .7	1.02± .04 .35± .03 .93	23 .12 .00	14.9 ±.2 14.38 ±.14 14.41	.52± .04 -.21± .07 .47 -.22	14.00± .30	16.16±.1 157± 4 88± 5	1866± 7 1793 1720
032819.1-310405 NGC 1344 229.07 -55.68 ESO 418- 5 268.72 -40.62 032617.1-311424	PGC 12923	.E.5... R -5.0± .3	1.78± .02 .24± .02 .95± .01 1.71	165 .00 .00	11.27M±.10 11.03 ±.11 11.14	.88± .01 .44± .01 .87 .45	.92± .01 .48± .01 11.61± .04 14.56± .16		1169± 15 1061 1048
0328.4 +3647 154.77 -16.20 345.17 -18.43 0325.2 +3637	UGC 2746 PGC 12928	.L..-*. U -3.0±1.1	1.20± .14 .15± .08 1.30	175 .93 .00					
0328.4 +3801 154.03 -15.19 346.19 -17.69 0325.2 +3751	UGC 2747 PGC 12929	.E...?. U -5.0±1.6	1.04± .18 .00± .08 1.15	.70 .00					
0328.5 +4009 152.76 -13.45 347.94 -16.39 0325.2 +3959	CGCG 541- 11 PGC 12933			.85	15.0 ±.4				4286± 94 4380 4137
0328.6 +4004 152.82 -13.51 347.89 -16.45 0325.3 +3954	UGC 2749 PGC 12937	.S..8*. U 8.0±1.3	1.07± .07 .50± .06 1.15	165 .85 .61 .25					
0328.7 +3632 154.98 -16.36 345.00 -18.63 0325.5 +3622	UGC 2750 PGC 12946	.S..4.. U 4.0±1.0	1.16± .07 1.00± .06 1.24	156 .80 1.47 .50					4859 4943 4706
032848.6-351042 NGC 1351A 236.36 -55.62 ESO 358- 9 263.29 -40.18 032652.1-352100	PGC 12952	.SBT4*/ PS 4.0± .5	1.43± .04 .70± .04 1.43	132 .00 1.03 .35	14.21 ±.14 13.18			217± 15 164± 11	1354± 10 1332± 26 1233 1238
0328.9 +4002 152.89 -13.50 347.90 -16.52 0325.6 +3952	UGC 2751 PGC 12955	.S..0?. U .0±1.7	1.04± .09 .00± .04 1.12	.85 .00					

R.A. 2000 DEC.	Names	Type	logD$_{25}$	p.a.	B$_T$	(B-V)$_T$	(B-V)$_e$	m$_{21}$	V$_{21}$
l b		S$_T$ n$_L$	logR$_{25}$	A$_g$	m$_B$	(U-B)$_T$	(U-B)$_e$	W$_{20}$	V$_{opt}$
SGL SGB		T	logA$_e$	A$_i$	m$_{FIR}$	(B-V)^o_T	m'$_e$	W$_{50}$	V$_{GSR}$
R.A. 1950 DEC.	PGC	L	logD$_o$	A$_{21}$	B^o_T	(U-B)^o_T	m'$_{25}$	HI	V$_{3K}$

032854.5-120913	NGC 1338	PSXT3P.	1.14± .06	55					2500
198.70 -50.23	MCG -2- 9- 44	E (1)	.03± .05	.11					
293.87 -39.98	IRAS03265-1219	3.0± .8		.05	12.42				2442
032632.1-121931	PGC 12956	4.2±1.2	1.15	.02					2348
032900.3-220825		PSBS1P.	1.18± .04	79					1792± 60
213.81 -53.99	ESO 548- 25	Sr	.30± .03	.05	14.89 ±.14				1706
280.66 -41.08		.7± .5		.30					1655
032648.1-221842	PGC 12961		1.18	.15	14.51				
032902.5-262643		.L?....	.80± .06						1923±104
221.03 -54.97	ESO 481- 29		.08± .03	.00	14.72 ±.14				1826
274.88 -41.09	IRAS03269-2637			.00	13.81				1793
032655.0-263700	PGC 12962		.79		14.69				
032904.1+404926		.L...*.	1.30± .07	80					4179± 94
152.45 -12.84	UGC 2752	U	.62± .05	.93	14.8 ±.3				4274
348.55 -16.06		-2.0±1.1		.00					4031
032545.1+403907	PGC 12964		1.31		13.78				
032904.2-532828					15.84 ±.14	1.02± .05	1.03± .00		17710± 60
266.16 -50.87				.00					17558
240.83 -35.05			.55± .09				14.08± .33		17642
032741.8-533844	PGC 12965								
032922.5-523709	IC 1946	.L?....	.95± .06	62					11464± 60
264.90 -51.20	ESO 155- 39		.30± .03	.00	15.30 ±.14				11313
241.78 -35.43				.00					11394
032758.0-524724	PGC 12972		.90		15.13				
032924.0+394733	A 0326+39	.S?....	.95± .11						7314± 94
153.12 -13.65	UGC 2755		.00± .06	.84					7406
347.76 -16.75				.00					7165
032606.5+393715	PGC 12975		1.03	.00					
0329.4 +3806		.S..4..	1.00± .08	78					
154.15 -15.01	UGC 2754	U	.43± .06	.62					
346.39 -17.79		4.0± .9		.63					
0326.2 +3756	PGC 12977		1.06	.22					
032931.8-174644	NGC 1345	.SBS5P*	1.18± .03	33	14.33 ±.15	.58± .02	.54± .02	14.46±.1	1529± 9
206.98 -52.46	ESO 548- 26	RSE (1)	.13± .03	.10	13.67 ±.14	-.09± .02	-.12± .02	150± 6	1543± 43
286.50 -40.91	VV 690	4.5± .6	.95± .04	.19	13.37	.52	14.57± .11	125± 5	1455
032715.1-175700	PGC 12979	5.6±1.2	1.19	.06	13.68	-.13	14.75± .23	.72	1386
032933.0-151420		.SBS9..	1.29± .04	97				13.72±.1	1890± 11
203.19 -51.46	MCG -3- 9- 47	E (1)	.06± .04	.17				129± 16	
289.87 -40.62		9.0± .5		.06				101± 12	1822
032713.7-152435	PGC 12981	8.7± .6	1.31	.03					1743
0329.6 +4318		.S..6*.	1.04± .08	143					
151.08 -10.75	UGC 2757	U	.76± .06	1.48					
350.61 -14.60		6.0±1.4		1.12					
0326.3 +4308	PGC 12988		1.18	.38					
032942.1-221645	NGC 1347	.SBS5*P	1.19± .04						1775
214.11 -53.88	ESO 548- 27	S (1)	.08± .04	.05	13.74 ±.14				1688
280.48 -41.24	ARP 39	5.3± .5		.11					1639
032730.0-222700	PGC 12989	3.3± .5	1.19	.04	13.56				
032943.7-333327		.LAS0..	1.24± .05	59	14.60S±.10				1714± 26
233.49 -55.46	ESO 358- 10	S	.38± .04	.00	14.85 ±.14				1599
265.38 -40.62		-3.0± .8		.00					1599
032745.0-334342	PGC 12990		1.18		14.66		14.71± .28		
032950.1-870941		.SAS9..	1.07± .04	95					
300.86 -29.30	ESO 4- 3	S (1)	.38± .04	.62	16.03 ±.14				
208.48 -17.47		9.0± .5		.39					
034606.0-871924	PGC 12994	9.7± .5	1.13						
032956.7-284628		.SB?...	.92± .06	148					11052± 52
225.10 -55.12	ESO 418- 7		.09± .04	.00	14.94 ±.14				10949
271.74 -41.17	IRAS03279-2856			.12					10928
032752.0-285642	PGC 12999		.92	.04	14.74				
033000.5+390336		.LB....	1.11± .16						
153.66 -14.17	UGC 2758	U	.00± .08	.71	14.8 ±.3				
347.24 -17.29		-2.0± .8		.00					
032644.1+385320	PGC 13000		1.19						
033001.5+414957	NGC 1334	.S?....	1.18± .05	115					4233± 94
152.00 -11.92	UGC 2759		.32± .05	1.17	14.1 ±.3				4330
349.48 -15.57	IRAS03266+4139			.48	11.42				4088
032640.7+413941	PGC 13001		1.29	.16	12.38				

R.A. 2000 DEC. l b SGL SGB R.A. 1950 DEC.	Names PGC	Type S_T n_L T L	$logD_{25}$ $logR_{25}$ $logA_e$ $logD_o$	p.a. A_g A_i A_{21}	B_T m_B m_{FIR} B_T^o	$(B-V)_T$ $(U-B)_T$ $(B-V)_T^o$ $(U-B)_T^o$	$(B-V)_e$ $(U-B)_e$ m'_e m'_{25}	m_{21} W_{20} W_{50} HI	V_{21} V_{opt} V_{GSR} V_{3K}
033004.2-475141 257.68 -52.92 247.28 -37.24 032829.0-480154	 ESO 200- 29 IRAS03285-4801 PGC 13002	.L?.... 	.90± .07 .12± .05 .88	103 .00 .00 	14.15 ±.14 14.22 ±.14 12.02 14.09	.64± .02 .56 	 13.20± .39	 	 6613± 88 6470 6532
033009.7-053114 190.40 -46.52 302.37 -38.78 032741.1-054128	 MCG -1- 9- 41 PGC 13005	.S..1P/ E .5±1.3 	1.07± .06 .59± .05 1.08	25 .11 .61 .30				
033010.0+382710 154.05 -14.64 346.76 -17.69 032654.5+381655	 UGC 2761 PGC 13006	.S..0.. U .0±1.0 	1.10± .05 .78± .05 1.12	45 .62 .58 	 15.2 ±.3 			
033010.7-284547 225.09 -55.07 271.75 -41.22 032806.0-285600	 ESO 418- 7A PGC 13007	.SBS3P. S 3.0± .8 	1.08± .07 .01± .05 1.08	 .00 .02 .01	 14.09 ±.14 			
0330.1 +4257 151.36 -10.99 350.39 -14.89 0326.8 +4247	 UGC 2760 PGC 13008	.S..6*. U 6.0±1.2 	1.00± .08 .00± .06 1.13	 1.35 .00 .00				
033013.3-053234 190.44 -46.52 302.35 -38.80 032744.7-054248	 MCG -1- 9- 42 IRAS03277-0542 PGC 13009	.S..3P? E 3.0±1.3 	1.00± .04 .17± .04 1.01	100 .11 .23 .08	 11.92 			
033019.2-041435 188.92 -45.76 303.97 -38.46 032749.4-042448	A 0327-04 MCG -1- 9- 43 IRAS03278-0424 PGC 13014	.SBS5.. PE (2) 4.5± .6 1.1± .5	1.06± .04 .18± .04 .69± .04 1.07	70 .08 .27 .09	13.9 ±.2 12.46 13.50	.48± .04 .37 	.56± .03 12.89± .12 13.58± .30	15.17±.1 353± 15 1.58	8395± 10 8376± 59 8358 8236
033019.4+413424 152.20 -12.09 349.31 -15.77 032659.0+412409	NGC 1335 UGC 2762 PGC 13015	.L..-*. U -3.0±1.2 	1.03± .11 .24± .05 1.14	165 1.08 .00 	 14.8 ±.3 13.67				5442± 94 5539 5297
033025.6-531613 265.72 -50.78 240.96 -35.32 032903.0-532624	 ESO 155- 41 IRAS03290-5326 PGC 13020	.SB?... 	.90± .06 .05± .05 .57± .04 .90	 .00 .05 .02	14.79 ±.15 14.76 ±.14 12.59 14.55	.72± .03 .59 	.80± .03 13.13± .11 14.04± .35		 13767± 60 13615 13701
033029.4+365652 155.03 -15.82 345.57 -18.67 032716.0+364638	 UGC 2763 PGC 13022	.S..0.. U .0±1.0 	1.14± .07 .86± .06 1.18	165 .88 .65 	 15.5 ±.3 			
033032.7-502013 261.42 -51.96 244.33 -36.45 032903.0-503024	IC 1947 ESO 200- 30 PGC 13027	 	.73± .06 .10± .05 .73	137 .00 	15.50 ±.15 15.50 ±.14 	.71± .04 			9± 88 -138 -65
033034.8-345112 235.76 -55.27 263.64 -40.59 032838.0-350124	NGC 1351 ESO 358- 12 PGC 13028	.LA.-P* RBCS -3.0± .4 	1.45± .03 .22± .04 .83± .01 1.42	140 .00 .00 	12.46 ±.13 12.35 ±.11 12.37	.88± .01 .35± .02 .86 .35	.92± .01 .41± .01 12.10± .05 14.02± .23		 1518± 19 1400 1407
033035.6-175630 207.38 -52.29 286.32 -41.18 032819.0-180642	 ESO 548- 28 PGC 13029	.LBT+P? SE -1.3±1.0 	1.10± .03 .22± .03 1.08	173 .11 .00 	14.21 ±.14 14.07				1502± 60 1426 1360
033036.7-142536 202.19 -50.88 290.99 -40.76 032816.6-143548	IC 326 MCG -3- 9- 49 PGC 13030	.LA.-?. E -2.5± .9 	1.08± .06 .14± .04 1.08	95 .16 .00 				
033040.7-030823 187.72 -45.04 305.35 -38.22 032810.0-031835	A 0328-03 MK 612 PGC 13034	PSBT0?. E .0± .6 	.91± .05 .17± .05 .91	95 .12 .13 	 13.05 				6041± 30 6008 5882
033040.8-501831 261.37 -51.95 244.35 -36.48 032911.0-502842	NGC 1356 ESO 200- 31 IRAS03291-5028 PGC 13035	.SXR4P* Sr 4.2± .6 	1.14± .05 .11± .05 .87± .02 1.14	 .00 .17 .06	13.71 ±.13 14.05 ±.14 12.78 13.62	.70± .02 .60 	.77± .01 13.55± .05 13.98± .30		 11471± 36 11323 11396
0330.7 +4201 152.00 -11.68 349.71 -15.56 0327.4 +4151	 UGC 2764 PGC 13039	.S..0.. U .0± .9 	1.11± .07 .54± .06 1.19	31 1.17 .40 				

R.A. 2000 DEC. l b SGL SGB R.A. 1950 DEC.	Names PGC	Type S_T n_L T L	$\log D_{25}$ $\log R_{25}$ $\log A_o$ $\log D_o$	p.a. A_g A_i A_{21}	B_T m_B m_{FIR} B_T^o	$(B-V)_T$ $(U-B)_T$ $(B-V)_T^o$ $(U-B)_T^o$	$(B-V)_e$ $(U-B)_e$ m'_e m'_{25}	m_{21} W_{20} W_{50} HI	V_{21} V_{opt} V_{GSR} V_{3K}
033047.3-210325		.SB?...	1.06± .05	99					1312
212.26 -53.28	ESO 548- 29		.18± .04	.04	14.23 ±.14				1228
282.14 -41.45				.28					1175
032834.0-211336	PGC 13042		1.07	.09	13.91				
033052.8-475844	IC 1949	.S?....	1.03± .06	27	13.67 ±.18	.56± .04	.64± .02		
257.79 -52.76	ESO 200- 33		.08± .05	.00	13.85 ±.14				6726± 79
247.09 -37.33	FAIR 749		.73± .04	.13	12.16	.50	12.81± .10		6582
032918.0-480854	PGC 13047		1.03	.04	13.62		13.45± .36		6645
033101.2+431450		.S?....	.89± .11						
151.32 -10.66	MCG 7- 8- 20		.14± .07	1.53	14.8 ±.3				9421± 94
350.72 -14.82	IRAS03276+4304			.21	12.35				9521
032737.8+430438	PGC 13050		1.03	.07	13.00				9279
033104.4-502557	IC 1950	.S..5*/	1.15± .05	153	15.10 ±.16	.58± .07	.80± .03		
261.52 -51.85	ESO 200- 35	S (1)	.54± .04	.00	14.93 ±.14				11422± 88
244.18 -36.49	IRAS03295-5036	5.0±1.3	.60± .03	.80	13.08	.41	13.59± .09		11274
032935.1-503606	PGC 13053	3.3± .9	1.15	.27	14.14		14.36± .31		11348
033108.2-361738		.E?....	1.02± .11		14.60S±.10				
238.27 -55.09	MCG -6- 8- 24		.21± .08	.00					1916± 29
261.72 -40.46				.00					1795
032913.5-362747	PGC 13058		.95		14.57		14.14± .60		1809
033108.4-333744	NGC 1350	PSBR2..	1.72± .02	0	11.16M±.10	.87± .01	.92± .01	13.70±.3	1890± 9
233.61 -55.17	ESO 358- 13	PBSr (1)	.27± .03	.00	11.27 ±.11	.34± .03	.44± .01	507± 34	1856± 18
265.23 -40.90	IRAS03291-3347	1.8± .3	1.20± .02	.34		.81	12.66± .07	383± 10	1768
032910.0-334754	PGC 13059	3.0± .9	1.72	.14	10.85	.29	13.91± .17	2.71	1770
0331.2 +3925		.S..4..	1.00± .08	46					
153.65 -13.73	UGC 2768	U	.84± .06	.82					
347.69 -17.26		4.0±1.0		1.24					
0328.0 +3915	PGC 13069		1.08	.42					
033118.4+352758		.SBT3..	1.21± .06	95					4479
156.08 -16.92	UGC 2770	U	.25± .06	.92	14.53 ±.15				
344.44 -19.70	IRAS03280+3517	3.0± .8		.34	12.53				4559
032807.0+351747	PGC 13073		1.30	.12	13.24				4327
033119.2-263351		.S?....	1.15± .04	46					
221.39 -54.49	ESO 482- 1		.58± .04	.00	14.83 ±.14				4374± 52
274.70 -41.59	IRAS03292-2643			.87	13.59				4275
032912.1-264400	PGC 13075		1.15	.29	13.94				4247
0331.3 +4248		.L.....	1.08± .17	137					
151.63 -10.98	UGC 2769	U	.36± .08	1.26					
350.41 -15.15		-2.0± .9		.00					
0328.0 +4238	PGC 13080		1.17						
033123.8+394431		.L.....	1.28± .08	12					
153.47 -13.46	UGC 2771	U	.33± .05	.83	14.5 ±.3				6008± 94
347.96 -17.08		-2.0± .8		.00					6099
032806.1+393420	PGC 13083		1.32		13.62				5861
033124.4-351951		.LBS0..	1.10± .10						
236.59 -55.08	MCG -6- 8- 25	S	.14± .08	.00					1403± 29
262.97 -40.68		-2.0± .9		.00					1284
032928.4-353000	PGC 13084		1.08						1294
033126.7+042243	NGC 1349	.L.....	.85± .08		14.17 ±.16	1.13± .04	1.16± .03	16.36±.2	6595± 7
179.93 -40.16	UGC 2774	U	.00± .05	.65	14.76 ±.10	.54± .06	.59± .04	388± 9	
314.29 -35.79		-2.0± .8	.96± .05	.00		.92	14.46± .16		6583
032848.9+041233	PGC 13088		.92		13.85	.42	13.28± .47		6432
033130.8-301246		.SXT8..	1.07± .05	149					
227.66 -54.93	ESO 418- 8	Sr (1)	.18± .04	.00	13.90 ±.14				1254± 25
269.77 -41.39	IRAS03294-3022	8.0± .5		.22					1146
032820.0-302254	PGC 13089	6.7± .8	1.07	.09	13.68				1134
033131.8-515416	IC 1954	.SBS3*.	1.50± .03	66	12.15 ±.15	.52± .01	.54± .01	14.03±.2	1071± 7
263.64 -51.20	ESO 200- 36	R (2)	.31± .03	.00	12.16 ±.11	-.06± .03	-.08± .01	232± 12	1031± 48
242.44 -36.01	IRAS03300-5204	3.0± .8	.98± .02	.43	11.42	.45	12.54± .05	214± 9	919
033006.1-520424	PGC 13090	3.6± .6	1.50	.16	11.72	-.11	13.73± .23	2.15	1000
033132.2-191639	NGC 1352	.LBS0*.	1.02± .04	134	14.28 ±.14	.94± .02	.96± .02		
209.56 -52.55	ESO 548- 30	SE	.18± .02	.07	14.16 ±.14	.49± .04	.53± .04		4368± 39
284.55 -41.52		-1.8± .6	.71± .04	.00		.87	13.32± .13		4288
032917.1-192648	PGC 13091		1.00		14.09	.48	13.79± .27		4229
033137.9-250028		.S?....	1.08± .05	93					
218.80 -54.11	ESO 482- 2		.20± .04	.00	14.76 ±.14				6532± 52
276.80 -41.70				.29					6437
032929.0-251036	PGC 13094		1.08	.10	14.43				6403

R.A. 2000 DEC. l b SGL SGB R.A. 1950 DEC.	Names PGC	Type S_T n_L T L	logD_25 logR_25 logA_e logD_o	p.a. A_g A_i A_21	B_T m_B m_FIR B_T^o	(B-V)_T (U-B)_T (B-V)_T^o (U-B)_T^o	(B-V)_e (U-B)_e m'_e m'_25	m_21 W_20 W_50 HI	V_21 V_opt V_GSR V_3K
033147.7-350312 236.10 -55.01 263.32 -40.81 032951.3-351319	 PGC 13097	DLXT0*. S -2.0± .8 	1.23± .08 .31± .08 1.18	 .00 .00 	15.30S±.10 	 	 15.53± .47	 	
033155.3-312011 229.62 -54.93 268.25 -41.36 032954.0-313018	 ESO 418- 9 PGC 13106	.IBS9.. S 10.0± .8 	1.12± .04 .10± .03 1.12	105 .00 .08 .05	 14.07 ±.14 13.99	 	 	 	972 862 856
033200.7-623413 277.67 -45.90 230.81 -31.39 033110.0-624418	 ESO 83- 6 PGC 13107	PSBT5.. Sr (1) 4.6± .4 4.8± .6	1.22± .04 .36± .05 1.23	100 .08 .54 .18	 14.97 ±.14 14.31	 	 	 	5430 5266 5389
033203.2-204904 212.03 -52.93 282.48 -41.73 032949.7-205911	NGC 1353 ESO 548- 31 IRAS03298-2059 PGC 13108	.SBT3*. R (3) 3.0± .5 3.5± .5	1.53± .02 .38± .03 .98± .03 1.54	138 .03 .53 .19	12.4 ±.2 12.25 ±.11 11.80 11.71	.95± .02 .37± .04 .86 .28	.93± .02 .34± .03 12.75± .09 13.97± .24	15.16±.3 404± 9 391± 7 3.25	1525± 8 1518± 14 1439 1388
0332.0 +0015 184.28 -42.70 309.57 -37.45 0329.5 +0005	IC 329 MCG 0-10- 1 PGC 13109	.S?.... 	.74± .14 .28± .07 .76	 .25 .35 .14	 15.20 ±.16 14.53	 	 	 	7150 7126 6991
0332.1 +4135 152.47 -11.88 349.54 -16.02 0328.7 +4125	 UGC 2775 IRAS03287+4125 PGC 13113	.SXS6*. U 6.0±1.1 	1.21± .05 .11± .05 1.32	 1.25 .17 .06	 	 	 	 	4405 4501 4262
0332.1 +0021 184.19 -42.62 309.70 -37.44 0329.6 +0011	IC 330 UGC 2779 PGC 13117	.S..2.. U 2.0±1.0 	.97± .07 .51± .05 1.00	78 .29 .63 .26	 14.97 ±.13 13.96	 	 	 	9129± 56 9105 8970
0332.2 +0015 184.30 -42.67 309.60 -37.48 0329.6 +0005	IC 331 MCG 0-10- 3 IRAS03296+0005 PGC 13119	.S?.... 	.82± .13 .00± .07 .84	 .28 .00 	 14.77 ±.12 13.12 14.39	 	 	 	6888± 49 6864 6729
0332.2 +6822 136.92 9.98 9.76 1.54 0327.5 +6812	 UGC 2765 PGC 13121	 	1.60? .40? 1.87	165 2.84 	 	 	 	13.37±.1 322± 16 317± 12 	1679± 11 1830 1586
033218.7-174306 207.29 -51.82 286.67 -41.57 033002.1-175312	 ESO 548- 32 PGC 13122	.SBS9.. SUE (2) 9.0± .5 9.4± .6	1.24± .03 .30± .03 1.26	147 .13 .30 .15	 14.84 ±.14 14.41	 	 	14.78±.1 153± 16 135± 12 .22	1961± 11 1885 1821
033228.4-185649 209.17 -52.23 285.02 -41.71 033013.0-190654	 ESO 548- 33 PGC 13128	.LBT0P? SE -2.0± .8 	1.15± .05 .25± .03 1.12	110 .08 .00 	 14.18 ±.14 14.08	 	 	 	1681± 39 1602 1543
033229.3-151321 203.64 -50.81 290.02 -41.32 033010.1-152327	NGC 1354 MCG -3-10- 4 IRAS03301-1523 PGC 13130	.SXT0*. E .0±1.2 	1.35± .04 .47± .05 1.34	148 .20 .35 	 	 	 	 	1804 1735 1661
033233.7-571227 270.90 -48.68 236.43 -33.95 033123.1-572230	IC 1960 ESO 155- 50 PGC 13135	.S..2?/ S 1.5±1.3 	1.04± .05 .60± .04 1.04	110 .00 .74 .30	 15.20 ±.14 	 	 	 	
033237.6+012256 183.21 -41.88 310.97 -37.18 033002.6+011250	IC 332 MCG 0-10- 4 PGC 13137	RSXT0*. E .0± .9 	1.09± .05 .28± .05 1.11	55 .32 .21 	 14.68 ±.13 	 	 	 	
0332.7 +4020 153.32 -12.83 348.60 -16.90 0329.4 +4010	 UGC 2777 PGC 13141	.S..8*. U 8.0±1.3 	1.11± .07 .54± .06 1.20	34 .91 .66 .27	 	 	 	 	
033247.8-341423 234.68 -54.82 264.35 -41.14 033050.4-342427	 PGC 13146	.LXS-*. S -3.0± .9 	1.06± .10 .31± .08 1.01	 .00 .00 	 	 	 	 	2123± 29 2005 2013
033257.1-210520 212.57 -52.81 282.13 -41.95 033044.0-211524	 ESO 548- 34 PGC 13150	.SB?... 	1.08± .05 .07± .04 1.09	 .04 .10 .03	 14.25 ±.14 14.11	 	 	 	1763 1677 1629

R.A. 2000 DEC.	Names	Type S$_T$ n$_L$	logD$_{25}$ logR$_{25}$	p.a. A$_g$	B$_T$ m$_B$	(B-V)$_T$ (U-B)$_T$	(B-V)$_e$ (U-B)$_e$	m$_{21}$ W$_{20}$	V$_{21}$ V$_{opt}$
l b		T	logA$_e$	A$_i$	m$_{FIR}$	(B-V)^o_T	m'$_e$	W$_{50}$	V$_{GSR}$
SGL SGB						(U-B)^o_T			
R.A. 1950 DEC.	PGC	L	logD$_o$	A$_{21}$	B^o_T		m'$_{25}$	HI	V$_{3K}$
033301.8-240803		.SBS8*/	1.36± .03	79					1915
217.47 -53.61	ESO 482- 5	S (1)	.86± .03	.01	15.46 ±.14				
277.99 -42.03		8.0± .7		1.06					1822
033052.0-241806	PGC 13154	7.0± .7	1.36	.43	14.38				1787
033306.6-344827		.SB.9*P	1.08± .05	175					1552± 29
235.66 -54.75	ESO 358- 15	S (1)	.24± .04	.00	15.34 ±.14				1433
263.58 -41.11		9.0±1.2		.25					
033110.0-345830	PGC 13157	8.9±1.2	1.08	.12	15.09				1444
0333.1 +3610		.S?....	1.26± .06						4513
155.95 -16.14	UGC 2780		.04± .06	.79					
345.26 -19.56				.06					4594
0329.9 +3600	PGC 13158		1.33	.02					4364
0333.1 +1552		.S..7..	1.14± .07	158					9326
170.14 -31.81	UGC 2781	U	1.16± .06	.68					
326.80 -31.02		7.0±1.0		1.38					9349
0330.3 +1542	PGC 13160		1.20	.50					9164
033310.6-563311	IC 1965	.SXS6..	1.00± .05	105	15.6 ±.3	.53± .05	.61± .04		
269.97 -48.92	ESO 155- 51	S (1)	.09± .04	.01	15.56 ±.14				16780± 88
237.10 -34.31		6.0± .8	.64± .07	.14		.40	14.30± .16		16622
033158.0-564312	PGC 13162	6.7± .8	1.00	.05	15.34		15.23± .36		16724
033312.0-502446	IC 1959	.SBS9*/	1.45± .04	147	13.2 ±.4	.36± .04	.42± .02		
261.28 -51.54	ESO 200- 39	S (1)	.63± .05	.00	13.22 ±.14				616± 88
244.04 -36.82	IRAS03317-5034	9.0±1.2	.93± .11	.64	13.75	.22	13.38± .25		467
033143.0-503448	PGC 13163	5.6± .8	1.45	.31	12.58		13.73± .46		544
033317.2-133954	NGC 1357	.SAS2..	1.45± .03	85	12.38 ±.13	.87± .01	.92± .01	15.55±.3	2009± 17
201.56 -49.97	MCG -2-10- 1	R	.16± .04	.09	12.36 ±.19	.25± .02	.37± .01		1999± 14
292.13 -41.28	IRAS03309-1349	2.0± .4	1.04± .02	.20	12.55	.80	13.07± .05	320± 25	1937
033056.5-134957	PGC 13166		1.45	.08	12.06	.20	14.05± .23	3.41	1859
033323.6-045956	NGC 1355	.L...*/	1.16± .06	105	14.25 ±.14	.97± .02			
190.43 -45.56	MCG -1-10- 2	RE	.57± .04	.11		.44± .05			
303.31 -39.40		-1.9± .4		.00					
033054.6-050958	PGC 13169		1.08				13.48± .35		
033326.2-234246	IC 1952	.SBS4?.	1.41± .03	141					1759
216.82 -53.42	ESO 482- 8	S (1)	.64± .03	.01	13.57 ±.14				1666
278.56 -42.13	IRAS03312-2352	4.0± .6		.94	12.94				1630
033116.0-235248	PGC 13171	3.3± .7	1.41	.32	12.60				
033329.2-180846		.SXT5*.	1.17± .03	120					4303± 60
208.09 -51.72	ESO 548- 35	SE (1)	.03± .03	.10	13.42 ±.14				4225
286.13 -41.88	IRAS03312-1818	5.0± .6		.05	13.31				4165
033113.0-181848	PGC 13174	2.2± .8	1.18	.02	13.25				
033333.6-333419		PLAR-*.	1.08± .10		15.00S±.10				1553± 29
233.52 -54.66		S	.03± .08	.00					1436
265.20 -41.40		-3.0± .8		.00					
033135.4-334420	PGC 13177		1.08		14.98		15.20± .55		1443
0333.6 +7211		.S..6*.	1.04± .08	36					
134.71 13.13	UGC 2772	U	1.06± .06	1.62					
12.64 4.01		6.0±1.5		1.47					
0328.3 +7201	PGC 13178		1.19	.50					
033336.6-360817	NGC 1365	.SBS3..	2.05± .01	32	10.32M±.07	.69± .01	.77± .01	11.84±.1	1662± 4
237.95 -54.60	ESO 358- 17	R (4)	.26± .02	.00	10.13 ±.11	.16± .03	.13± .03	398± 6	1675± 11
261.80 -40.98	VV 825	3.0± .3	1.44± .03	.36	8.24	.63	12.93± .07	370± 6	1541
033142.0-361818	PGC 13179	1.3± .4	2.05	.13	9.90	.11	14.77± .11	1.82	1558
0333.6 +4022		.S..8*.	1.11± .07	93					
153.44 -12.71	UGC 2782	U	.83± .06	.91					
348.74 -17.02		8.0±1.4		1.02					
0330.3 +4012	PGC 13180		1.20	.42					
033339.8-050522	NGC 1358	.SXR0..	1.41± .04	165	13.04 ±.13	.92± .02	1.00± .01		
190.59 -45.56	MCG -1-10- 3	R	.10± .04	.11	12.76 ±.18	.42± .03	.50± .02		4021± 24
303.22 -39.49		.0± .3	.98± .03	.08		.84	13.43± .09		3980
033110.8-051524	PGC 13182		1.41		12.70	.40	14.68± .24		3868
033341.7-212841	IC 1953	.SBT7..	1.44± .02	121	12.24 ±.14	.55± .02	.63± .01	14.25±.2	1867± 6
213.26 -52.76	ESO 548- 38	R (3)	.11± .03	.03	12.38 ±.12			248± 11	1886± 41
281.61 -42.14	IRAS03314-2138	7.0± .3	1.21± .02	.15	10.92	.51	13.78± .04	204± 12	1781
033129.0-213842	PGC 13184	3.5± .5	1.45	.05	12.13		14.04± .19	2.06	1735
033342.5-355119		DE.4.*.	1.10± .10						
237.46 -54.59		S	.09± .08	.00					
262.17 -41.05		-5.0± .8		.00					
033147.5-360120	PGC 13186		1.08						

R.A. 2000 DEC. l b SGL SGB R.A. 1950 DEC.	Names PGC	Type S_T n_L T L	$\log D_{25}$ $\log R_{25}$ $\log A_e$ $\log D_o$	p.a. A_g A_i A_{21}	B_T m_B m_{FIR} B_T^o	$(B-V)_T$ $(U-B)_T$ $(B-V)_T^o$ $(U-B)_T^o$	$(B-V)_e$ $(U-B)_e$ m'_e m'_{25}	m_{21} W_{20} W_{50} HI	V_{21} V_{opt} V_{GSR} V_{3K}
033347.9-192929 210.18 -52.12 284.31 -42.06 033133.1-193930	NGC 1359 ESO 548- 39 IRAS03315-1939 PGC 13190	.SBS9$P R (4) 9.0± .5 5.5± .5	1.38± .02 .14± .02 1.16± .03 1.39	139 .08 .15 .07	12.59 ±.15 12.60 ±.12 12.30 12.37	.37± .06 -.47± .06 .31 -.51	.34± .06 -.41± .06 13.88± .06 14.00± .19	13.45±.1 248± 8 202± 11 1.01	1966± 6 1959± 58 1884 1831
033350.7-212718 213.25 -52.72 281.64 -42.18 033138.0-213718	ESO 548- 40 PGC 13194	.S?....	1.16± .05 .12± .04 1.16	.03 .15 .06	14.59 ±.14 14.37				4082 3995 3950
033352.9-201653 211.41 -52.35 283.24 -42.13 033139.0-202654	NGC 1362 ESO 548- 41 PGC 13196	.L..0P* SE -2.3± .7	1.08± .04 .04± .03 1.08	.04 .00	13.53 ±.14 13.47				1233± 60 1149 1099
033353.3-311136 229.42 -54.50 268.38 -41.79 033152.0-312136	NGC 1366 ESO 418- 10 PGC 13197	.L..0./ PS -2.0± .6	1.32± .04 .35± .03 .55± .01 1.27	2 .00 .00	11.97V±.13 13.00 ±.10 12.92	.89± .01 .37± .01 .85 .34	.93± .01 .42± .01 13.47± .25		1297± 25 1186 1182
0333.9 -5955 274.27 -47.15 233.43 -32.86 0332.9 -6005	PGC 13201		.48± .04	.07	13.91 ±.16	.92± .02	.95± .02 11.80± .14		
0334.0 -0709 193.15 -46.61 300.66 -40.13 0331.6 -0719	MCG -1-10- 4 PGC 13204	.S?....	1.22± .05 .81± .05 1.23	63 .10 1.11 .40					10215 10168 10065
0334.1 +0306 181.77 -40.50 313.14 -36.89 0331.5 +0256	UGC 2786 PGC 13209	.SBS7*. U 7.0±1.2	1.11± .14 .15± .12 1.15	.40 .20 .07					
033415.8-610542 275.68 -46.49 232.18 -32.34 033319.6-611539	PGC 13216	.IBS9.. S (1) 10.0± .6 10.0± .6	1.12± .08 .11± .08 1.13	.06 .08 .06					
033416.8-304355 228.64 -54.38 268.99 -41.93 033215.0-305354	ESO 418- 11 PGC 13217	PSAR3P. Sr (1) 3.2± .5 3.3±1.1	1.24± .04 .11± .04 1.24	87 .00 .15 .05	14.45 ±.14 14.22				11475± 39 11364 11359
033417.8-061553 192.12 -46.08 301.80 -39.97 033150.0-062553	NGC 1361 MCG -1-10- 5 PGC 13218	.E+..P* E -4.0±1.2	1.20± .07 .04± .05 1.20	130 .07 .00					
033418.5+392124 154.17 -13.44 348.02 -17.76 033101.0+391123	A 0331+39 UGC 2783 PGC 13219	.E...?. U -5.0±1.6	1.11± .10 .02± .05 1.23	.80 .00	13.9 ±.4 13.6 ±.3 12.84	1.22± .07 .66± .12 .99 .49	14.37± .66		6081± 85 6170 5937
033418.8-192525 210.14 -51.98 284.42 -42.18 033204.0-193524	ESO 548- 44 PGC 13220	.LAR+*. r -1.3±1.3	1.07± .05 .34± .04 1.03	60 .08 .00	14.26 ±.14 14.16				1696± 60 1614 1561
033420.0+393246 154.06 -13.28 348.17 -17.64 033102.3+392246	UGC 2784 PGC 13224	.L...*. U -2.0±1.1	1.11± .10 .11± .05 1.18	160 .77 .00	13.8 ±.3				
0334.3 +4054 153.22 -12.18 349.27 -16.78 0331.0 +4044	UGC 2785 IRAS03310+4044 PGC 13225	.S..6*. U 6.0±1.4	1.04± .08 .76± .06 1.15	39 1.12 1.12 .38	13.20				
033429.5-353251 236.92 -54.45 262.54 -41.26 033234.2-354249	MCG -6- 8- 27 PGC 13230	.LXT0.. S -2.0± .8	1.10± .10 .09± .08 1.09	.00 .00	14.50S±.10 14.48		14.67± .54		1194± 29 1072 1089
033431.1-341750 234.77 -54.47 264.19 -41.49 033234.1-342748	ESO 358- 19 PGC 13232	.LA.-*. S -3.0± .9	1.01± .05 .36± .03 .95	155 .00 .00	15.71 ±.14 15.69				1254± 29 1135 1146
033443.2-190144 209.59 -51.76 284.96 -42.25 033228.0-191142	ESO 548- 47 PGC 13241	PLBR+?/ SE -1.3± .7	1.39± .04 .59± .03 1.31	72 .08 .00	13.69 ±.14 13.59				1606± 60 1525 1471

R.A. 2000 DEC. l　　b SGL　SGB R.A. 1950 DEC.	Names PGC	Type S_T　n_L T L	$\log D_{25}$ $\log R_{25}$ $\log A_o$ $\log D_o$	p.a. A_g A_i A_{21}	B_T m_B m_{FIR} B_T^o	$(B-V)_T$ $(U-B)_T$ $(B-V)_T^o$ $(U-B)_T^o$	$(B-V)_o$ $(U-B)_o$ m'_o m'_{25}	m_{21} W_{20} W_{50} HI	V_{21} V_{opt} V_{GSR} V_{3K}
033447.5+361235 156.20 -15.90 345.51 -19.80 033134.6+360236	 UGC　2788 PGC 13243	.E...?. U -5.0±1.7	1.01± .13 .21± .05 1.09	90 .89 .00	 15.2　±.3				
033457.3-323822 231.93 -54.35 266.39 -41.83 033258.0-324818	 ESO　358- 20 PGC 13250	.IBS9?P S　　(1) 10.0±1.6 6.7±1.1	1.08± .04 .20± .03 1.08	167 .00 .15 .10	 14.58 ±.14 14.43				1853± 29 1738 1742
033458.8-351016 236.27 -54.36 263.01 -41.43 033303.0-352012	NGC　1373 ESO　358- 21 PGC 13252	.E+..*. BS -4.3± .6	1.06± .06 .13± .04 1.03	131 .00 .00	14.12M±.08 14.03 ±.14 14.08	.86± .02 .32± .04 .85 .33	 14.11± .30		1378± 25 1257 1273
033500.7-245604 218.96 -53.35 276.88 -42.47 033252.0-250600	NGC　1371 ESO　482- 10 IRAS03327-2505 PGC 13255	.SXT1.. R 1.0± .3	1.75± .02 .16± .03 1.10± .03 1.75	135 .00 .16 .08	11.57M±.11 11.47 ±.11 11.34	.90± .01 .48± .02 .86 .45	.94± .01 .49± .01 12.73± .10 14.78± .16	12.65±.2 410±　4 389±　5 1.23	1471±　5 1454± 32 1373 1345
0335.0 -5511 267.94 -49.32 238.46 -35.14 0333.8 -5521	 PGC 13258		 .60± .11	 .00	15.7　±.3	.92± .06	.94± .05 14.21± .36		
033506.5-550535 267.80 -49.36 238.57 -35.19 033350.0-551530	 ESO　155- 54 PGC 13259	.L?.... 	.94± .07 .27± .05 .49± .06 .90	176 .00 .00	15.31 ±.19 15.05 ±.14 14.99	1.03± .04 .91	1.05± .03 13.25± .22 14.23± .40		10057± 60 9901 9999
033514.8-202222 211.72 -52.08 283.14 -42.46 033301.1-203218	NGC　1370 ESO　548- 48 IRAS03330-2032 PGC 13265	.E+..*. SE -3.5± .6	1.19± .04 .19± .03 1.14	50 .04 .00	 13.44 ±.14 12.97 13.39				1073± 39 988 941
033516.6-351559 236.43 -54.29 262.87 -41.47 033321.0-352554	NGC　1375 ESO　358- 24 PGC 13266	.LX.0*/ RCS -2.0± .5	1.35± .03 .42± .03 .74± .01 1.28	91 .00 .00	13.18 ±.13 13.02 ±.14 13.09	.78± .02 .32± .04 .73 .29	.82± .01 .33± .02 12.37± .04 13.73± .20		751± 20 630 647
033516.7-351335 236.36 -54.29 262.92 -41.48 033321.0-352330	NGC　1374 ESO　358- 23 PGC 13267	.E..... RBCS -4.5± .4	1.39± .04 .03± .04 .91± .02 1.38	 .00 .00	12.00M±.08 12.00 ±.11 11.98	.92± .01 .46± .02 .91 .47	.94± .01 .50± .01 12.04± .05 13.85± .22		1351± 15 1230 1246
033520.3-323605 231.87 -54.27 266.43 -41.92 033321.0-324600	 ESO　358- 22 PGC 13269	PLAR0.. Sr -1.8± .5	1.22± .04 .08± .03 1.21	99 .00 .00	 13.85 ±.14 13.75				6562± 29 6447 6452
0335.4 +8006 129.80 19.49 18.58 9.21 0327.8 +7956	 UGC　2767 PGC 13274	.S..9?. U 9.0±1.6	1.19± .12 .05± .12 1.25	 .65 .05 .02					
033527.8-211259 213.05 -52.29 281.99 -42.54 033315.0-212254	 ESO　548- 49 PGC 13275	.S?.... 	1.02± .05 .22± .04 1.02	117 .03 .32 .11	 15.35 ±.14 14.99				1510± 60 1422 1380
033530.8-342648 235.03 -54.26 263.95 -41.66 033334.0-343642	IC　　335 ESO　358- 26 PGC 13277	.L..../ S -2.0±1.1	1.41± .04 .59± .03 1.32	84 .00 .00	 12.90 ±.14 12.88				1663± 17 1543 1557
033531.7+050335 180.14 -38.95 315.56 -36.45 033253.2+045340	IC　1956 UGC　2795 IRAS03329+0454 PGC 13279	.SB.4*. U 4.0± .9	1.20± .04 .73± .06 1.26	34 .69 1.07 .36	 15.28 ±.13 12.81 13.47			15.59±.1 402±　8 1.75	6401±　7 6389 6244
0335.5 +4124 153.11 -11.65 349.82 -16.63 0332.2 +4115	 UGC　2791 PGC 13280	.S..8*. U 8.0±1.1	1.14± .13 .03± .12 1.26	 1.24 .04 .01					5679 5773 5539
033533.4-322754 231.64 -54.22 266.61 -41.98 033334.0-323748	 ESO　358- 25 PGC 13281	.LA.-*. S -3.0± .5	1.29± .04 .29± .03 .85± .07 1.24	60 .00 .00	13.79M±.09 13.90 ±.14 13.80		12.81± .23 14.37± .24		1459± 26 1344 1349
033537.7-211730 213.19 -52.28 281.89 -42.59 033325.0-212724	IC　1962 ESO　548- 50 PGC 13283	.SBS8.. S　　(1) 8.0± .8 7.8± .9	1.43± .03 .70± .04 1.43	2 .03 .86 .35	 14.67 ±.14 13.78				1803 1788± 60 1700 1658

R.A. 2000 DEC.	Names	Type	$\log D_{25}$	p.a.	B_T	$(B-V)_T$	$(B-V)_e$	m_{21}	V_{21}
l b		S_T n_L	$\log R_{25}$	A_g	m_B	$(U-B)_T$	$(U-B)_e$	W_{20}	V_{opt}
SGL SGB		T	$\log A_e$	A_i	m_{FIR}	$(B-V)_T^o$	m'_e	W_{50}	V_{GSR}
R.A. 1950 DEC.	PGC	L	$\log D_o$	A_{21}	B_T^o	$(U-B)_T^o$	m'_{25}	HI	V_{3K}
033550.3+475805				1.73				14.98±.3	4784± 9
149.22 -6.34								181± 11	
354.94 -12.41								166± 8	4894
033217.1+474809	PGC 13289								4654
033603.3-352626	NGC 1379	.E.....	1.38± .03		11.80M±.10	.89± .01	.91± .01		
236.72 -54.13	ESO 358- 27	RBCS	.02± .03	.00	12.02 ±.11	.37± .03	.41± .01		1376± 19
262.60 -41.59		-5.0± .4	1.14± .02	.00		.88	12.87± .08		1254
033408.0-353618	PGC 13299		1.38		11.88	.38	13.67± .19		1273
0336.2 +6733		.SB?...	1.08± .08	13				14.52±.1	3135± 11
137.70 9.55	UGC 2789		.07± .06	3.18					
9.41 .72	IRAS03315+6723			.10	11.28			358± 8	3284
0331.5 +6723	PGC 13301		1.38	.03					3042
033621.4-064250		.SBS9..	1.20± .04	165				14.99±.3	3081± 11
193.06 -45.89	MCG -1-10- 9	UE (1)	.21± .05	.11				180± 16	
301.40 -40.58		9.0± .6		.21				129± 12	3033
033354.0-065243	PGC 13308	7.5± .8	1.21	.11					2933
0336.4 +4830		.S?....	1.16± .13					15.81±.1	5823± 11
148.98 -5.84	UGC 2794		.00± .12	1.88					
355.41 -12.13	IRAS03328+4820			.00	13.27			156± 8	5934
0332.8 +4820	PGC 13315		1.34	.00					5694
033626.9-345833	NGC 1380	.LA....	1.68± .03	7	10.87M±.10	.94± .01	.96± .01		
235.93 -54.06	ESO 358- 28	R	.32± .03	.00	11.04 ±.11	.45± .01	.50± .01		1841± 15
263.20 -41.76	IRAS03345-3508	-2.0± .6	1.12± .01	.00	12.87	.89	11.95± .05		1720
033431.0-350824	PGC 13318		1.63		10.92	.43	13.33± .19		1737
033631.5-351739	NGC 1381	.LA..*/	1.43± .03	139	12.44M±.10	.94± .01	.96± .01		
236.47 -54.04	ESO 358- 29	RBCS	.56± .03	.00	12.60 ±.10	.46± .01	.51± .02		1776± 20
262.77 -41.72		-1.6± .5	.58± .01	.00		.87	10.91± .04		1654
033436.1-352730	PGC 13321		1.34		12.50	.42	13.04± .17		1672
033631.6-435728	IC 1970	.SA.3*/	1.50± .03	75					
250.94 -52.94	ESO 249- 7	PBS	.63± .04	.00	12.86 ±.14				1112± 46
251.58 -39.56	IRAS03348-4407	2.6± .5		.87	11.55				972
033450.1-440718	PGC 13322		1.50	.31	11.99				1027
033639.1-205403	NGC 1377	.L..0..	1.25± .05	92					
212.70 -51.93	ESO 548- 51	S	.30± .04	.04	13.38 ±.14				1792± 60
282.44 -42.81	IRAS03344-2103	-2.0± .8		.00	11.27				1705
033426.0-210354	PGC 13324		1.21		13.32				1663
033645.0-361522	NGC 1369	.SBT0..	1.19± .06	12					
238.10 -53.96	ESO 358- 34	Sr	.05± .06	.00	13.77 ±.14				1414± 17
261.48 -41.57		.0± .5		.03					1289
033451.0-362512	PGC 13330		1.19		13.71				1313
033646.4-355958	NGC 1386	.LBS+..	1.53± .03	25	12.09M±.10	.86± .01	.90± .01		
237.66 -53.97	ESO 358- 35	PBS	.42± .03	.00	12.21 ±.10	.33± .02	.39± .01		864± 15
261.82 -41.63	IRAS03348-3609	-.6± .5	.88± .01	.00	11.25	.79	11.98± .04		741
033452.0-360948	PGC 13333		1.47		12.14	.28	13.55± .20		763
0336.7 +1324		.S..4..	1.14± .13	149					9076
172.91 -32.97	UGC 2796	U	.57± .12	.96					
324.81 -33.00		4.0± .9		.84					9089
0334.0 +1315	PGC 13334		1.23	.28					8919
033647.2-344422	NGC 1380A	.L..0*/	1.38± .04	179	13.31 ±.13	.90± .01	.88± .01		
235.52 -53.99	ESO 358- 33	CS	.54± .03	.00	13.36 ±.14	.32± .02	.33± .02		1508± 23
263.49 -41.87		-2.0± .8	.83± .02	.00		.84	12.95± .05		1387
033451.0-345412	PGC 13335		1.30		13.31	.28	13.75± .27		1404
033654.0-352228		.LA.-*.	1.04± .11		14.81M±.09		.88± .05		
236.60 -53.96	MCG -6- 9- 8	S	.00± .08	.00			.34± .10		1823± 29
262.64 -41.78		-3.0±1.2	.64± .06	.00			12.68± .21		1700
033458.7-353217	PGC 13343		1.04		14.78		14.87± .58		1720
033657.1-353023	NGC 1387	.LXS-..	1.45± .03		11.68M±.10	.99± .01	1.01± .01		
236.82 -53.95	ESO 358- 36	RBCS	.00± .03	.00	11.73 ±.11	.50± .01	.51± .01		1328± 24
262.46 -41.76	IRAS03350-3540	-3.0± .3	.85± .02	.00	12.06	.98	11.55± .06		1205
033502.0-354012	PGC 13344		1.45		11.68	.51	13.80± .21		1225
033659.0-575824	IC 1980	.S.3?/P	1.04± .05	19					
271.48 -47.79	ESO 117- 2	S	.38± .04	.00	14.95 ±.14				
235.28 -34.13		3.0±1.9		.52					
033552.0-580812	PGC 13345		1.04		.19				
033705.8-500906	IC 1978	.S..2?/	1.05± .05	7					
260.53 -51.05	ESO 200- 52	S	.79± .04	.00	15.69 ±.14				10534± 88
244.05 -37.49		2.0±2.0		.97					10384
033537.0-501854	PGC 13350		1.05		.39 14.62				10465

R.A. 2000 DEC. l　b SGL　SGB R.A. 1950 DEC.	Names PGC	Type S_T　n_L T L	$\log D_{25}$ $\log R_{25}$ $\log A_e$ $\log D_o$	p.a. A_g A_l A_{21}	B_T m_B m_{FIR} B_T^o	$(B-V)_T$ $(U-B)_T$ $(B-V)_T^o$ $(U-B)_T^o$	$(B-V)_e$ $(U-B)_e$ m'_e m'_{25}	m_{21} W_{20} W_{50} HI	V_{21} V_{opt} V_{GSR} V_{3K}
033705.8-050236 191.22 -44.82 303.59 -40.30 033436.9-051226	NGC 1376 MCG -1-10- 11 IRAS03346-0512 PGC 13352	.SAS6.. R　　(3) 6.0± .4 2.9± .5	1.30± .03 .06± .03 1.31	95 .10 .09 .03	 12.8　±.2 12.33 12.59			14.75±.1 175± 6 162± 11 2.13	4159± 9 4441± 66 4120 4015
033708.5-351141 236.29 -53.92 262.87 -41.86 033513.0-352130	NGC 1382 ESO 358- 37 PGC 13354	.LXS-*. BS -2.7± .6 	1.18± .05 .06± .03 .71± .01 1.17	 .00 .00 	12.92V±.13 13.79 ±.14 13.81	.95± .02 .27± .03 .93 .28	.93± .01 .37± .03 14.50± .27		 1802± 25 1680 1700
033711.8-354442 237.23 -53.89 262.13 -41.76 033517.1-355430	NGC 1389 ESO 358- 38 PGC 13360	.LXS-*. RBCS -3.3± .4 	1.36± .04 .22± .03 .70± .02 1.33	30 .00 .00 	12.42 ±.13 12.48 ±.11 12.44	.92± .01 .38± .01 .90 .38	.93± .01 .44± .01 11.41± .06 13.54± .24		 995± 22 871 893
033712.9-315959 230.87 -53.84 267.17 -42.39 033513.0-320948	 ESO 418- 14 IRAS03352-3209 PGC 13362	.S?.... 	1.00± .06 .45± .05 1.00	138 .00 .62 .22	15.05 ±.14 13.54 14.34				11873± 29 11758 11764
033722.9-430825 249.56 -52.97 252.54 -39.95 033540.0-431812	 ESO 249- 9 PGC 13365	.L?.... 	.82± .07 .09± .05 .81	 .00 .00 	16.11 ±.17 15.94 ±.14 15.76	.88± .05 .72 	 14.89± .39		16349±108 16211 16264
033728.0-243012 218.46 -52.71 277.46 -43.04 033519.0-244000	NGC 1385 ESO 482- 16 IRAS03353-2439 PGC 13368	.SBS6.. R　　(2) 6.0± .3 3.5± .7	1.53± .02 .23± .02 .99± .01 1.53	165 .01 .34 .11	11.45M±.10 11.58 ±.12 10.03 11.15	.51± .01 -.17± .02 .46 -.21	.46± .01 -.19± .02 12.01± .03 13.36± .15	13.61±.1 213± 8 172± 11 2.35	1493± 6 1979± 58 1402 1376
033730.1-511501 262.12 -50.60 242.74 -37.12 033604.0-512448	 ESO 200- 53 PGC 13369	.S..0*. S .0±1.2 	1.06± .05 .30± .05 1.04	172 .00 .22 	14.67 ±.14				
033734.2-511120 262.02 -50.61 242.80 -37.15 033608.0-512106	 ESO 200- 54 FAIR 753 PGC 13372	.SAS5?. S　　(1) 5.0± .9 3.3± .8	1.04± .05 .11± .05 .72± .04 1.04	 .00 .16 .05	14.12 ±.19 14.23 ±.14 13.08 14.02	.46± .03 .42 	.64± .03 13.21± .09 13.89± .33		 2304± 88 2152 2238
033737.9+030701 182.49 -39.81 313.58 -37.70 033501.3+025713	IC 338 MCG 0-10- 7 PGC 13373	.S?.... 	.94± .11 .00± .07 .98	 .48 .00 .00	14.70 ±.13 14.18				5773 5754 5620
033738.4-154245 205.13 -49.88 289.58 -42.61 033519.9-155233	 MCG -3-10- 14 PGC 13375	.SXS8*. E　　(1) 8.0±1.2 8.7± .8	1.11± .06 .04± .05 1.12	105 .15 .05 .02					
033738.8-182019 208.94 -50.87 285.98 -42.89 033523.0-183006	NGC 1383 ESO 548- 53 PGC 13377	.LXS0.. SE -2.0± .6 	1.28± .03 .34± .03 .66± .03 1.24	91 .16 .00 	13.45 ±.14 13.45 ±.14 13.26	.98± .01 .36± .02 .89 .30	1.00± .01 .46± .02 12.24± .10 13.84± .21		 1978± 39 1897 1846
033742.6-722329 287.85 -39.50 220.97 -26.61 033801.0-723312	 ESO 31- 18 PGC 13379	.SAS5.. S　　(1) 5.0± .8 3.3± .9	1.14± .04 .33± .04 1.16	95 .16 .50 .17	14.99 ±.14				
033743.8-225437 215.95 -52.26 279.67 -43.11 033533.0-230424	 ESO 482- 17 PGC 13381	.L?.... 	1.12± .05 .34± .03 1.07	132 .01 .00 	15.11 ±.14 15.07				1455 1362 1330
033749.8+723420 134.74 13.63 13.14 4.03 033225.2+722428	NGC 1343 UGC 2792 7ZW 8 PGC 13384	.SXS3*P P 3.0±1.0 	1.41± .04 .20± .05 1.54	80 1.39 .27 .10	13.5 ±.3 11.54 11.78			14.17±.1 144± 6 116± 4 2.29	2215± 7 2180± 35 2369 2133
0337.8 +0457 180.72 -38.58 315.74 -37.02 0335.2 +0448	 MCG 1-10- 3 PGC 13385	.S?.... 	.74± .14 .00± .07 .80	 .65 .00 .00	15.47 ±.12 14.76				9723± 57 9709 9569
033752.0-190032 209.97 -51.05 285.07 -42.99 033537.0-191018	NGC 1390 ESO 548- 54 PGC 13386	.SB.1P* SE 1.0± .6 	1.15± .03 .41± .03 1.17	19 .14 .42 .20	14.63 ±.14 14.34				1215± 60 1132 1084
033755.6-710336 286.50 -40.33 222.18 -27.38 033801.0-711318	 ESO 54- 15 IRAS03380-7113 PGC 13387	 	.80± .07 .26± .06 .52± .02 .81	 .11 	15.44 ±.14 15.74 ±.14 12.95 	.86± .07 .11± .13 	.90± .07 .16± .13 13.53± .07 		14554± 60 14382 14540

R.A. 2000 DEC. l b SGL SGB R.A. 1950 DEC.	Names PGC	Type S_T n_L T L	$\log D_{25}$ $\log R_{25}$ $\log A_e$ $\log D_o$	p.a. A_g A_i A_{21}	B_T m_B m_{FIR} B_T^o	$(B-V)_T$ $(U-B)_T$ $(B-V)_T^o$ $(U-B)_T^o$	$(B-V)_e$ $(U-B)_e$ m' m'_{25}	m_{21} W_{20} W_{50} HI	V_{21} V_{opt} V_{GSR} V_{3K}
0337.9 +0456 180.76 -38.57 315.74 -37.05 0335.3 +0447	 UGC 2802 PGC 13389	.S?.... 1.11	1.05± .08 .31± .06 	90 .65 .47 .16	 15.20 ±.14 14.03				 9624± 57 9610 9471
033758.9-061613 192.84 -45.32 302.11 -40.85 033531.2-062600	 MCG -1-10- 12 PGC 13394	.SXR6.. E (1) 6.0± .9 5.3± .8	1.13± .06 .19± .05 1.14	125 .12 .29 .10					
0338.0 +1213 174.17 -33.59 323.75 -33.84 0335.3 +1204	 UGC 2803 PGC 13396	.SB.3*. U 3.0±1.3 1.06	.96± .17 .15± .12 	90 1.05 .20 .07					6417 6425 6262
0338.1 +0155 183.79 -40.48 312.24 -38.25 0335.5 +0146	 UGC 2804 PGC 13397	.S..6*. U 6.0±1.4 1.03	1.00± .08 .84± .06 	173 .34 1.24 .42					
033806.3-352625 236.70 -53.71 262.49 -42.00 033611.2-353610	NGC 1396 PGC 13398	.LX.-*. S -3.0±1.2 .99	1.00± .11 .07± .08 	 .00 .00 	14.80S±.10 14.79		 14.47± .61 		 894± 29 771 793
033809.4-343103 235.14 -53.71 263.72 -42.19 033613.0-344048	 ESO 358- 42 PGC 13399	.LB.0*. S -2.0± .8 1.03	1.08± .05 .37± .03 .88± .13 	135 .00 .00 	15.40S±.10 15.74 ±.14 15.50		 14.41± .44 14.77± .27		 1041± 29 920 938
033809.9+405858 153.77 -11.70 349.80 -17.29 033449.2+404911	 UGC 2798 IRAS03348+4049 PGC 13400	.SX.4.. U 4.0± .8 1.40	1.29± .04 .43± .05 	70 1.19 .63 .21	 14.0 ±.3 12.55 12.11		15.82±.1 405± 8 3.49		4941± 11 5032 4803
033810.2-194409 211.10 -51.22 284.07 -43.11 033556.0-195354	 ESO 548- 57 PGC 13401	.SB?... 1.17	1.16± .04 .49± .05 	43 .10 .60 .25	 15.25 ±.14 14.51				 4150± 60 4065 4021
033813.3-330739 232.80 -53.68 265.60 -42.43 033615.1-331724	 ESO 358- 43 PGC 13404	.L..-.. S -3.0± .9 1.03	1.09± .05 .42± .03 	20 .00 .00 	 15.26 ±.14 15.24				 1371± 29 1253 1265
033815.0+011027 184.60 -40.92 311.36 -38.57 033540.2+010041	 CGCG 391- 21 KUG 0335+010B PGC 13406	.S..3P? E 3.0±1.8 .86	.84± .06 .30± .05 	70 .19 .41 .15	 15.00 ±.14 				
0338.2 -0540 192.18 -44.93 302.91 -40.76 0335.8 -0550	 PGC 13407	.SBT6.. E (1) 6.0± .9 5.3± .8	1.07± .09 .18± .08 1.09	10 .13 .26 .09					
033817.2-232509 216.81 -52.27 278.96 -43.24 033607.0-233454	 ESO 482- 18 PGC 13408	.S?.... .94	.94± .06 .01± .04 	 .02 .01 .00	 14.72 ±.14 14.68				 1687± 60 1592 1564
0338.2 +4117 153.60 -11.44 350.06 -17.11 0334.9 +4107	 UGC 2801 IRAS03349+4107 PGC 13410	.S..6*. U 6.0±1.4 1.28	1.14± .07 .90± .06 	107 1.48 1.33 .45	 12.95 				
0338.3 +0738 178.31 -36.71 318.82 -36.01 0335.6 +0728	 UGC 2806 IRAS03356+0728 PGC 13413	.S?.... 1.06	.96± .09 .51± .06 	102 1.01 .76 .25					8895 8889 8741
033828.9-042041 190.68 -44.15 304.61 -40.43 033559.4-043025	 MCG -1-10- 13 PGC 13417	.SBS9*. UE (1) 9.0± .5 8.7± .8	1.25± .05 .20± .05 1.26	15 .10 .21 .10					4075 4033 3928
033829.0-352658 236.71 -53.64 262.46 -42.08 033634.0-353642	NGC 1399 ESO 358- 45 PGC 13418	.E.1.P. R -5.0± .3 1.83	1.84± .02 .03± .02 1.13± .02 	 .00 .00 	10.55M±.10 9.85 ±.11 10.22	.96± .01 .50± .03 .95 .51	.98± .01 .56± .01 11.72± .08 14.66± .16		 1447± 12 1323 1346
033829.6-230140 216.21 -52.12 279.51 -43.28 033619.1-231124	NGC 1395 ESO 482- 19 PGC 13419	.E.2... R -5.0± .3 1.73	1.77± .03 .12± .03 1.21± .04 	 .02 .00 	10.55M±.06 10.63 ±.11 10.52	.96± .01 .58± .03 .94 .58	.98± .01 .61± .02 12.43± .14 14.07± .16		 1699± 19 1605 1575

R.A. 2000 DEC. / l b / SGL SGB / R.A. 1950 DEC.	Names / PGC	Type / S_T n_L / T / L	logD_25 / logR_25 / logA_e / logD_o	p.a. / A_g / A_i / A_21	B_T / m_B / m_FIR / B_T^o	(B-V)_T / (U-B)_T / (B-V)_T^o / (U-B)_T^o	(B-V)_e / (U-B)_e / m'_e / m'_25	m_21 / W_20 / W_50 / HI	V_21 / V_opt / V_GSR / V_3K
033833.1-052805		.SXR6*/	1.14± .06	.13					6148
192.00 -44.76	MCG -1-10- 14	E (1)	.04± .05	.06					
303.19 -40.77		6.0± .7		.06					6103
033604.6-053749	PGC 13421	5.3± .8	1.15	.02					6002
033838.6-182540	NGC 1393	.LAR0*.	1.29± .03	170	12.97 ±.13	.95± .02	.97± .01		
209.21 -50.68	ESO 548- 58	PSEr	.19± .03	.16	13.00 ±.14	.44± .03	.52± .02		2185± 60
285.89 -43.13		-1.9± .4	.85± .02	.00		.88	12.71± .07		2103
033623.1-183524	PGC 13425		1.28		12.80	.40	13.81± .20		2055
033839.1-052054		.S..7?.	1.14± .06	102					4965
191.88 -44.67	MCG -1-10- 15	E (1)	.54± .05	.12					4920
303.35 -40.76		7.0±1.8		.74					4819
033610.5-053038	PGC 13426	5.3±1.6	1.15	.27					
033845.0-440600	NGC 1411	.LAR-*.	1.36± .04	6	12.23 ±.13	.97± .01	1.00± .01		
251.02 -52.52	ESO 249- 11	RBCS	.14± .04	.00	12.04 ±.11	.38± .02	.40± .02		970± 31
251.24 -39.89		-3.0± .4	.66± .02	.00		.95	11.02± .05		830
033704.0-441542	PGC 13429		1.34		12.10	.38	13.55± .27		889
033851.7-353536	NGC 1404	.E.1...	1.52± .03		10.97 ±.13	.97± .01	.99± .01		
236.95 -53.56	ESO 358- 46	R	.05± .03	.00	10.87 ±.11	.56± .01	.57± .01		1929± 14
262.24 -42.12		-5.0± .3	.90± .02	.00		.95	10.96± .06		1805
033657.0-354518	PGC 13433		1.51		10.88	.57	13.44± .20		1829
033851.8-262011	NGC 1398	PSBR2..	1.85± .02	100	10.57M±.10	.90± .01	.96± .01	13.13±.1	1407± 6
221.53 -52.79	ESO 482- 22	R (2)	.12± .02	.00	10.44 ±.11	.43± .02	.53± .01	497± 5	1491± 58
274.91 -43.29	IRAS03367-2629	2.0± .3	1.24± .02	.15	12.23	.87	12.30± .07	447± 12	1305
033645.0-262954	PGC 13434	1.1± .5	1.85	.06	10.35	.41	14.36± .14	2.72	1290
033851.9+405924		.SX.5..	1.04± .06	40					
153.88 -11.61	UGC 2805	U	.19± .05	1.19	14.8 ±.3				
349.90 -17.38	IRAS03355+4049	5.0± .9		.29	13.45				
033531.1+404939	PGC 13435		1.15	.10					
033853.7-182117	NGC 1391	.LBS0..	1.06± .05	65	14.37 ±.13	1.03± .02			
209.13 -50.59	ESO 548- 59	SE	.35± .03	.16	14.25 ±.14	.42± .04			
285.99 -43.18		-1.5± .6		.00					
033638.1-183100	PGC 13436		1.03				13.67± .27		
033855.7+301533		.S..9*.	1.16± .13	140				16.04±.3	4158± 10
160.79 -20.04	UGC 2807	U	.25± .12	.80					4219
341.06 -24.13		9.0±1.2		.25				124± 7	4010
033550.4+300549	PGC 13438		1.23	.12					
033902.7-033740		.LAR+?.	1.22± .07	75					
189.97 -43.62	MCG -1-10- 16	E	.28± .08	.13					
305.57 -40.35		-1.0±1.2		.00					
033632.5-034722	PGC 13443		1.20						
033906.7-181736	NGC 1394	.L..0*/	1.13± .04	5	13.82 ±.13	1.01± .02			
209.07 -50.52	ESO 548- 60	PSE	.49± .03	.16	13.66 ±.14	.46± .03			
286.08 -43.23		-2.0± .8		.00					
033651.1-182718	PGC 13444		1.08				13.13± .24		
033911.3-222318	NGC 1403	.LXS0..	1.13± .05		13.65 ±.13	.91± .01	.92± .01		
215.28 -51.80	ESO 482- 25	S	.11± .04	.02	13.86 ±.14				4293± 14
280.40 -43.44		-2.5± .6	.66± .02	.00		.86	12.44± .08		4200
033700.0-223300	PGC 13445		1.12		13.66		13.90± .29		4169
033913.1+154920	NGC 1384	.S?....	.89± .11						9813
171.43 -30.83	MCG 3-10- 3		.24± .07	.81	15.36 ±.12				
327.63 -32.30				.33					9831
033624.1+153937	PGC 13448		.96	.12	14.14				9660
033913.6-352219		DLXS0*.	1.12± .09		15.30S±.10				
236.58 -53.49		S	.10± .08	.00					952± 29
262.52 -42.24		-2.0± .8		.00					828
033718.6-353200	PGC 13449		1.11		15.29		15.54± .52		852
033919.6-354326		DE.0.*.	1.06± .10		16.10S±.10				
237.17 -53.46		S	.03± .08	.00					
262.04 -42.19		-5.0±1.2		.00					
033725.2-355307	PGC 13452		1.05				16.32± .57		
033921.9-224325	NGC 1401	.LBS0*/	1.38± .04	130	13.07 ±.14	.82± .01	.91± .01		
215.82 -51.85	ESO 482- 26	PS	.57± .04	.01	13.32 ±.14	.37± .02	.42± .02		1547± 39
279.93 -43.48		-2.0± .5	.75± .03	.00		.75	12.31± .11		1454
033711.0-225306	PGC 13457		1.30		13.16	.33	13.41± .26		1425
033922.6-311919	NGC 1406	.SBS4*/	1.58± .02	15	12.40M±.11	.60± .06	.68± .03	13.47±.2	1068± 6
229.79 -53.34	ESO 418- 15	PSU (1)	.70± .03	.00	12.56 ±.11	.00± .06	-.02± .03	358± 7	957± 41
268.01 -42.94	IRAS03373-3129	4.0± .5	1.03± .03	1.02	10.35	.46	13.04± .08	317± 5	951
033722.0-312900	PGC 13458	3.4± .9	1.58	.35	11.46	-.10	13.43± .17	1.66	958

R.A. 2000 DEC. / l b / SGL SGB / R.A. 1950 DEC.	Names / PGC	Type / S_T n_L / T / L	$\log D_{25}$ / $\log R_{25}$ / $\log A_e$ / $\log D_o$	p.a. / A_g / A_i / A_{21}	B_T / m_B / m_{FIR} / B_T^o	$(B-V)_T$ / $(U-B)_T$ / $(B-V)_T^o$ / $(U-B)_T^o$	$(B-V)_e$ / $(U-B)_e$ / m' / m'_{25}	m_{21} / W_{20} / W_{50} / HI	V_{21} / V_{opt} / V_{GSR} / V_{3K}
033929.1-130657	IC 340	DL..0*.	1.17± .05	90					4159
201.82 -48.37	MCG -2-10- 5	E	.41± .05	.10					
293.19 -42.68		-2.0±1.2		.00					4091
033708.1-131638	PGC 13464		1.12						4023
0339.4 +1943		.I..9?.	1.13± .07					15.42±.1	1282± 10
168.42 -27.94	UGC 2809	U	.14± .06	.93				131± 6	
331.53 -30.30		10.0±1.6		.11				127± 5	1312
0336.6 +1934	PGC 13465		1.22	.07					1130
033930.4-183137	NGC 1402	.LB.0P*	.91± .05	88	14.35 ±.13	.79± .02			
209.47 -50.52	ESO 548- 61	SE	.15± .03	.15	13.96 ±.14	.23± .03			
285.77 -43.34	IRAS03372-1841	-2.0± .8		.00	12.09				
033715.0-184118	PGC 13467		.90				13.37± .30		
033931.3-184119	NGC 1400	.LA.-..	1.36± .02	40	11.92 ±.13	.96± .01	1.02± .01		
209.71 -50.57	ESO 548- 62	R	.06± .02	.14	12.09 ±.11	.56± .02	.59± .01		558± 14
285.55 -43.36	IRAS03372-1850	-3.0± .3	.99± .01	.00	13.11	.92	12.36± .03		475
033716.0-185100	PGC 13470		1.37		11.87	.53	13.46± .18		429
033934.9-200102		.S?....	1.12± .05	38					
211.70 -51.00	ESO 548- 63		.67± .04	.11	15.21 ±.14				1965± 60
283.70 -43.45				.98					1878
033721.0-201042	PGC 13471		1.13	.33	14.10				1838
033937.2+384124		.SA.6..	.98± .07	160					
155.43 -13.35	UGC 2808	U	.16± .05	.97	14.7 ±.3				
348.16 -18.98		6.0± .9		.23					
033620.1+383142	PGC 13474		1.07	.08					
033942.2-143411		.S..5*/	1.16± .05	150					
203.84 -48.95	MCG -3-10- 24	E	.85± .05	.07					10497
291.22 -42.96		5.0±1.3		1.28					10425
033722.7-144351	PGC 13479		1.17	.43					10363
033946.9-044017	NGC 1397	RSBR0..	1.21± .04	125					
191.31 -44.06	MCG -1-10- 17	E (1)	.10± .04	.10					
304.32 -40.84		.0± .8		.07					
033717.7-044957	PGC 13485	9.8± .8	1.22						
034002.5-192203		.SB.1?/	1.19± .04	38					
210.78 -50.69	ESO 548- 65	SE	.68± .03	.13	15.46 ±.14				1218± 60
284.62 -43.52		.7±1.0		.69					1132
033748.0-193142	PGC 13491		1.20	.34	14.63				1091
034005.3-374740		.S..3?P	.99± .07						
240.62 -53.20	ESO 301- 22A	S	.25± .06	.00	15.06 ±.14				
259.25 -41.89		3.0±1.7		.34					
033814.0-375718	PGC 13493		.99	.12					
034007.5-182640	IC 343	.LBT+*.	1.21± .03	118	14.10 ±.14	.91± .03	.98± .03		
209.43 -50.35	ESO 548- 66	PSE	.32± .03	.17	14.17 ±.14		.49± .04		1869± 60
285.90 -43.48		-1.0± .5	.87± .02	.00		.81	13.94± .05		1786
033752.0-183618	PGC 13495		1.17		13.93		14.19± .21		1741
034008.1+393630		.SAT5..	1.04± .08					16.13±.1	6500± 11
154.93 -12.56	UGC 2810	U	.04± .06	.99	14.3 ±.3				6586
348.97 -18.46		5.0± .8		.06				277± 8	6363
033649.4+392650	PGC 13497		1.13	.02	13.21			2.90	
0340.1 +7123		.I..9?.	1.36± .05	100				13.74±.1	1176± 7
135.62 12.81	UGC 2800	U	.32± .06	1.47				228± 6	
12.41 3.09		10.0±1.5		.24				208± 4	1329
0334.9 +7114	PGC 13498		1.50	.16					1094
034009.6-353716	NGC 1427A	.IBS9..	1.37± .04	76				13.86±.2	2024± 6
236.99 -53.29	ESO 358- 49	PS (1)	.20± .04	.00	13.44 ±.14			115± 7	2123± 29
262.13 -42.38		10.0± .6		.15				73± 8	1902
033815.0-354654	PGC 13500	7.8± .8	1.37	.10	13.29			.47	1929
0340.1 +6643		.I..9?.	1.08?					12.70±.1	1501± 11
138.52 9.11	UGCA 81	U	.23?	1.58				277± 16	
9.07 -.13		10.0±1.8		.18				264± 12	1647
0335.5 +6634	PGC 13501		1.23	.12					1409
034012.4-183452	NGC 1407	.E.0...	1.66± .02	35	10.7 ±.2	1.03± .01	1.04± .01		
209.64 -50.38	ESO 548- 67	R	.03± .02	.17	10.72 ±.11		.65± .02		1762± 14
285.71 -43.51		-5.0± .3	1.37± .04	.00		.97	13.08± .14		1678
033757.0-184430	PGC 13505		1.68		10.51		13.90± .23		1634
034019.0-185552		.LBS-*/	1.16± .04	133					
210.17 -50.48	ESO 548- 68	SE	.31± .03	.17	14.05 ±.14				1817± 60
285.23 -43.56		-2.5± .8		.00					1732
033804.0-190530	PGC 13511		1.13		13.85				1690

R.A. 2000 DEC. l b SGL SGB R.A. 1950 DEC.	Names PGC	Type S_T n_L T L	$\log D_{25}$ $\log R_{25}$ $\log A_e$ $\log D_o$	p.a. A_g A_i A_{21}	B_T m_B m_{FIR} B_T^o	$(B-V)_T$ $(U-B)_T$ $(B-V)_T^o$ $(U-B)_T^o$	$(B-V)_e$ $(U-B)_e$ m'_e m'_{25}	m_{21} W_{20} W_{50} HI	V_{21} V_{opt} V_{GSR} V_{3K}
034019.0-153153 205.27 -49.21 289.93 -43.23 033800.5-154131	NGC 1405 MCG -3-10- 28 PGC 13512	.L...*/ E -2.0±1.3	1.19± .07 .45± .05 1.14	150 .14 .00					
0340.3 +4336 152.48 -9.36 352.12 -15.87 0336.9 +4327	 UGC 2811 PGC 13513	.S..8*. U 8.0±1.2	1.04± .15 .13± .12 1.15	60 1.16 .16 .06					
034022.8-351636 236.41 -53.25 262.59 -42.49 033827.9-352613	 PGC 13515	DE.0... S -5.0± .8	1.08± .10 .09± .08 1.05	.00 .00	16.20S±.10 16.38± .55				
034029.0-265141 222.49 -52.53 274.15 -43.62 033823.0-270118	NGC 1412 ESO 482- 29 PGC 13520	.LX.0*/ S -2.0± .7	1.27± .04 .39± .04 1.21	131 .00 .00	 13.54 ±.14 13.51				1790± 52 1685 1675
034041.1-264712 222.39 -52.47 274.25 -43.67 033835.1-265648	 ESO 482- 32 PGC 13529	.IBS9P* S (1) 9.5± .8 6.7±1.2	1.18± .04 .38± .05 1.18	67 .00 .28 .19	 15.26 ±.14 14.98				1765 1660 1650
034041.3-221712 215.28 -51.43 280.55 -43.78 033830.1-222648	 ESO 548- 70 PGC 13531	.SB.7?/ S (1) 7.0±1.3 6.7±1.3	1.21± .04 .74± .03 1.21	65 .03 1.03 .37	 15.62 ±.14 14.56				1737 1643 1615
034042.7-373042 240.13 -53.10 259.58 -42.07 033851.0-374018	NGC 1419 ESO 301- 23 PGC 13534	.E...P* BS -5.4± .5	1.06± .03 .03± .03 .52± .01 1.05	 .00 .00	13.48M±.08 13.56 ±.14 13.48	.89± .01 .32± .03 .88 .33	.90± .01 .36± .03 11.55± .04 13.70± .17		1528± 23 1398 1433
034043.1-062456 193.53 -44.82 302.17 -41.54 033815.6-063433	 MCG -1-10- 19 PGC 13535	.SXR6.. E (1) 6.0± .8 4.2± .8	1.27± .05 .10± .05 1.28	165 .13 .14 .05					5251 5201 5109
0340.7 +1744 170.21 -29.19 329.77 -31.61 0337.8 +1734	IC 1977 UGC 2815 IRAS03378+1734 PGC 13536	.SB.3.. U 3.0± .8	1.15± .05 .26± .05 1.20	177 .57 .36 .13	 14.29 ±.13 12.62 13.29				9936 9959 9785
034056.9-214248 214.42 -51.21 281.35 -43.83 033845.0-215224	NGC 1414 ESO 548- 71 PGC 13543	.SBS4?/ S (1) 3.8± .8 5.2± .8	1.16± .04 .66± .03 1.16	172 .04 .97 .33	 14.65 ±.14 13.63				1555 1463 1433
034057.0-223349 215.73 -51.45 280.16 -43.85 033846.0-224324	NGC 1415 ESO 482- 33 IRAS03387-2243 PGC 13544	RSXS0.. R .0± .3	1.54± .02 .29± .03 .97± .03 1.52	148 .01 .22	12.77M±.10 12.48 ±.11 11.21 12.38	.92± .01 .36± .02 .86 .32	.94± .01 .33± .01 12.77± .08 14.58± .17	15.11±.2 349± 4 322± 5	1585± 7 1508± 56 1489 1463
034102.8-224307 215.98 -51.47 279.94 -43.87 033852.1-225242	NGC 1416 ESO 482- 34 PGC 13548	.E.1.*. R -5.0± .6	1.12± .05 .02± .03 1.12	 .01 .00	* 13.95 ±.14 13.90				2167± 60 2072 2046
034104.0-334643 233.91 -53.10 264.57 -42.91 033907.0-335618	 ESO 358- 50 PGC 13550	.LA.0?. S -2.0± .8	1.22± .05 .38± .03 1.17	173 .00 .00	* 13.86 ±.14 13.84				1255± 23 1134 1154
034110.5-011810 187.82 -41.84 308.68 -40.12 033838.1-012745	NGC 1409 MCG 0-10- 11 PGC 13553	.LX..*P R -2.0± .5	1.02± .14 .11± .07 1.03	130 .27 .00	*				7510± 42 7475 7365
034110.8-011757 187.81 -41.84 308.69 -40.12 033838.4-012732	NGC 1410 MCG 0-10- 12 PGC 13556	.E...P* RE -5.0± .9	1.07? .00? 1.11	120 .27 .00	*				7527± 41 7492 7381
034111.3+240042 165.56 -24.48 335.82 -28.23 033813.2+235106	 UGC 2816 IRAS03382+2350 PGC 13557	.S?.... 	1.08± .06 .52± .05 1.15	142 .66 .72 .26	* 15.46 ±.13 14.05			17.04±.3 460± 10 2.73	7675± 10 7717 7526
034112.2-770808 292.20 -36.21 216.62 -23.96 034241.0-771736	 ESO 31- 20 PGC 13559	PSBR4P. S (1) 4.3± .4 4.7± .6	1.13± .04 .21± .04 1.17	175 .38 .31 .11	* 14.88 ±.14 14.16				4834 4659 4837

R.A. 2000 DEC. l b SGL SGB R.A. 1950 DEC.	Names PGC	Type S_T n_L T L	$\log D_{25}$ $\log R_{25}$ $\log A_e$ $\log D_o$	p.a. A_g A_i A_{21}	B_T m_B m_{FIR} B_T^o	$(B-V)_T$ $(U-B)_T$ $(B-V)_T^o$ $(U-B)_T^o$	$(B-V)_e$ $(U-B)_e$ m'_e m'_{25}	m_{21} W_{20} W_{50} HI	V_{21} V_{opt} V_{GSR} V_{3K}
034114.5-235014 217.74 -51.72 278.37 -43.91 033905.0-235948	 ESO 482- 35 IRAS03390-2359 PGC 13561	.SBT2.. Sr 2.1± .6 	1.28± .03 .17± .03 1.28	1 .02 .21 .09	* 13.69 ±.14 13.44				1877 1779 1758
0341.2 +0537 180.79 -37.50 316.94 -37.52 0338.6 +0528	 UGC 2820 PGC 13562	.SB?... 	.95± .09 .07± .06 1.02	 .68 .10 .03	* 				6283 6269 6134
034123.2-174526 208.59 -49.83 286.89 -43.72 033907.0-175500	 ESO 548- 75 PGC 13563	.SBT5.. SE (2) 5.0± .6 3.0± .6	1.16± .04 .07± .03 1.18	16 .19 .10 .03	* 14.05 ±.14 13.72				7166± 60 7084 7038
034130.9-214057 214.44 -51.08 281.40 -43.97 033919.1-215030	NGC 1422 ESO 548- 77 IRAS03393-2150 PGC 13569	.SB.2P/ PS 2.4± .6 	1.35± .03 .61± .03 1.35	65 .04 .75 .31	* 13.99 ±.14 13.84 13.18				1619 1526 1497
034131.9-195421 211.76 -50.53 283.89 -43.90 033918.0-200354	 ESO 548- 76 PGC 13570	.L?.... 	1.07± .06 .06± .03 1.07	 .11 .00 	* 14.69 ±.14 14.56				1471± 60 1383 1347
034132.5-345315 235.76 -53.02 263.05 -42.80 033937.0-350248	 ESO 358- 51 IRAS03396-3502 PGC 13571	.S..0*. S .0±1.1 	1.19± .04 .34± .04 1.17	0 .00 .26 	* 13.95 ±.14 13.71 13.67				1708± 29 1584 1610
0341.6 +1600 171.76 -30.30 328.17 -32.70 0338.8 +1551	 UGC 2823 PGC 13572	.S..3*. U (1) 3.0±1.3 5.5±1.3	1.16± .07 .71± .06 1.23	8 .78 .97 .35	* 				9117 9134 8967
0341.6 +3712 156.69 -14.26 347.24 -20.24 0338.4 +3703	 UGC 2817 PGC 13574	.S?.... 	1.32± .10 .45± .12 1.15 1.43	132 .46 .23	* 				5755 5834 5617
034144.6-181603 209.38 -49.93 286.19 -43.85 033929.0-182536	IC 346 ESO 548- 78 PGC 13575	.LBT+.. PSEr -.8± .4 	1.31± .04 .21± .03 .97± .02 1.29	78 .19 .00 	13.62 ±.14 13.53 ±.14 13.36	.99± .02 .55± .03 .90 .48	.94± .02 .53± .02 13.96± .07 14.50± .24		1897± 60 1813 1771
0341.9 +0809 178.56 -35.71 319.88 -36.56 0339.2 +0800	 UGC 2829 PGC 13580	.L..... U -2.0± .9 	1.01± .08 .51± .03 1.02	162 .77 .00 	14.88 ±.11 				
034154.9-505729 261.30 -50.06 242.72 -37.87 034029.0-510700	IC 1989 ESO 200- 55 PGC 13581	.LA.-.. S -3.0± .6 	1.03± .06 .17± .04 1.00	132 .00 .00 	14.60 ±.14 14.43				11142± 60 10988 11079
034155.9-185340 210.32 -50.11 285.32 -43.94 033941.0-190312	 ESO 548- 79 PGC 13582	.S?.... 	1.03± .06 .05± .04 1.05	 .16 .05 .02	14.22 ±.14 13.98				2053± 60 1968 1928
0341.9 -5316 264.63 -49.21 240.06 -36.88 0340.6 -5326	 PGC 13583	 	 	 .00 	16.82 ±.15 	.78± .06 			
034157.3-044221 191.77 -43.63 304.48 -41.37 033928.1-045153	NGC 1417 MCG -1-10- 21 IRAS03394-0451 PGC 13584	.SXT3.. R (3) 3.0± .3 1.9± .5	1.43± .03 .21± .03 .93± .03 1.44	175 .15 .29 .11	12.8 ±.2 12.72 ±.20 12.21 12.29	.70± .04 .14± .07 .60 .06	.83± .01 .27± .03 12.94± .09 14.27± .26	14.22±.3 418± 34 395± 25 1.83	4120± 17 4087± 14 4055 3959
034201.3-471317 255.69 -51.20 247.14 -39.35 034027.1-472248	NGC 1433 ESO 249- 14 IRAS03404-4722 PGC 13586	PSBR2.. 4 (2) 2.0± .3 2.7± .5	1.81± .02 .04± .02 1.40± .02 1.81	 .00 .05 .02	10.70M±.04 10.78 ±.12 11.37 10.65	.79± .01 .21± .01 .77 .20	.85± .01 .32± .01 13.23± .04 14.48± .11	13.54±.2 197± 5 175± 7 2.88	1075± 6 972± 22 920 997
0342.0 +1559 171.86 -30.24 328.21 -32.79 0339.2 +1550	 UGC 2830 PGC 13587	.L...*. U -2.0±1.2 	1.04± .09 .18± .04 1.10	85 .78 .00 	15.16 ±.13 				
034203.4-182904 209.74 -49.94 285.89 -43.94 033948.0-183836	 ESO 548- 80 IRAS03398-1838 PGC 13589	.S?.... 	1.11± .05 .37± .04 1.13	4 .18 .55 .18	14.85 ±.14 14.05				12235± 60 12151 12109

R.A. 2000 DEC. l b SGL SGB R.A. 1950 DEC.	Names PGC	Type S_T n_L T L	$\log D_{25}$ $\log R_{25}$ $\log A_e$ $\log D_o$	p.a. A_g A_i A_{21}	B_T m_B m_{FIR} B_T^o	$(B-V)_T$ $(U-B)_T$ $(B-V)_T^o$ $(U-B)_T^o$	$(B-V)_e$ $(U-B)_e$ m'_e m'_{25}	m_{21} W_{20} W_{50} HI	V_{21} V_{opt} V_{GSR} V_{3K}
034203.4-211440 213.83 -50.83 282.02 -44.08 033951.0-212412	 ESO 548- 81 IRAS03398-2124 PGC 13590	.SBT1P? S 1.0±1.1 	1.22± .04 .14± .03 1.23	 .04 .14 .07	 12.84 ±.14 13.38 12.61				4341± 60 4249 4219
0342.0 +4108 154.27 -11.12 350.42 -17.74 0338.7 +4059	 UGC 2822 PGC 13591	.S..4.. U 4.0± .9 	1.00± .08 .43± .06 1.15	121 1.58 .63 .22					
034206.3-835302 298.07 -31.54 210.98 -19.73 034818.0-840218	 ESO 4- 4 PGC 13595	.SBS9.. S (1) 9.0± .8 8.9±1.2	1.05± .05 .26± .04 1.10	102 .57 .26 .13	 15.68 ±.14 				
0342.1 +0311 183.34 -38.90 314.22 -38.70 0339.5 +0302	 UGC 2831 PGC 13598	.SBS8*. UE (1) 7.7± .7 7.5±1.2	1.13± .04 .25± .04 1.18	60 .55 .31 .13					4374 4352 4227
034210.3-064556 194.22 -44.70 301.84 -41.98 033943.1-065528	 MCG -1-10- 23 PGC 13600	PSXT0*. E .0± .8 	1.25± .09 .11± .07 1.26	70 .15 .08 					
034210.7-275159 224.22 -52.33 272.70 -43.92 034006.0-280130	 ESO 419- 3 VV 610 PGC 13601	PSXS5P* Sr (1) 4.5± .5 3.3± .7	1.22± .04 .22± .03 1.22	138 .00 .33 .11	 13.60 ±.14 12.69 13.25				4109± 41 4001 3998
034211.5-295340 227.52 -52.60 269.87 -43.72 034009.5-300311	NGC 1425 ESO 419- 4 IRAS03401-3002 PGC 13602	.SAS3.. R (3) 3.0± .3 3.2± .4	1.76± .02 .35± .03 1.32± .04 1.76	129 .00 .49 .18	11.29M±.11 11.61 ±.11 12.49 10.95	.68± .02 .11± .04 .60 .05	.74± .01 .22± .02 13.33± .11 14.05± .15	13.26±.1 367± 5 358± 5 2.14	1507± 4 1635± 36 1396 1402
034216.1-044351 191.85 -43.58 304.48 -41.45 033946.9-045323	NGC 1418 MCG -1-10- 22 PGC 13606	.SBS3*. R (1) 3.0± .6 4.2± .8	1.11± .04 .17± .04 1.13	165 .15 .24 .09	14.31 ±.13 13.89	.71± .02 .19± .03 .62 .12	 14.30± .26	15.72±.3 422± 34 308± 25 1.74	4218± 17 3723± 63 4139 4043
034218.7-224517 216.16 -51.20 279.89 -44.16 034008.1-225448	 ESO 482- 36 PGC 13608	.SXS9.. S (1) 9.0± .5 10.0± .5	1.23± .04 .16± .04 1.23	5 .01 .16 .08	 14.97 ±.14 14.77				21430 21334 21311
034219.6-352336 236.60 -52.85 262.32 -42.86 034025.0-353306	NGC 1427 ESO 358- 52 PGC 13609	.E+.... RBCS -4.1± .4 	1.56± .03 .17± .04 1.01± .03 1.51	76 .00 .00 	11.77M±.10 11.85 ±.11 11.78	.91± .01 .42± .02 .90 .43	.92± .01 .43± .01 12.35± .09 14.14± .21		 1413± 21 1287 1316
034223.0-350912 236.20 -52.84 262.64 -42.92 034028.0-351842	NGC 1428 ESO 358- 53 PGC 13611	.LX.-P* RBCS -3.3± .7 	1.21± .04 .32± .03 1.16	118 .00 .00 	 13.77 ±.14 13.75				1630± 21 1505 1533
034224.1+391439 155.52 -12.58 348.97 -19.03 033905.7+390507	 UGC 2828 IRAS03391+3905 PGC 13613	.SBT4.. U 4.0± .8 	1.10± .05 .08± .05 1.20	 1.07 .11 .04	 14.5 ±.3 13.17 				
034229.5-132918 202.80 -47.88 292.84 -43.47 034009.0-133848	NGC 1421 MCG -2-10- 8 IRAS03401-1338 PGC 13620	.SXT4*. R (2) 4.0± .4 3.8± .9	1.55± .02 .61± .03 1.17± .02 1.56	179 .11 .90 .31	11.95 ±.14 12.29 ±.17 10.71 11.06	.53± .02 -.05± .02 .38 -.16	.64± .01 .04± .01 13.29± .04 13.03± .19	13.37±.1 381± 7 338± 11 2.00	2090± 5 2081± 24 2018 1958
034232.5-041757 191.41 -43.28 305.06 -41.39 034003.0-042727	IC 347 MCG -1-10- 24 IRAS03399-0427 PGC 13622	.LXR0*. E -2.0± .9 	1.10± .11 .09± .07 1.11	40 .16 .00 	 13.70 				
0342.7 +7118 135.84 12.87 12.49 2.88 0337.5 +7109	 UGC 2813 PGC 13633	.I..9*. U 10.0±1.1 	1.32± .10 .35± .12 1.46	35 1.47 .26 .18				14.88±.1 156± 6 129± 4 	1392± 7 1545 1311
034249.3-220637 215.23 -50.91 280.80 -44.28 034038.0-221606	NGC 1426 ESO 549- 1 PGC 13638	.E.4... R -5.0± .3 	1.42± .02 .20± .02 .92± .01 1.36	111 .02 .00 	12.29M±.05 12.28 ±.11 12.24	.90± .01 .43± .02 .88 .43	.92± .01 .47± .02 12.39± .04 13.87± .14		1422± 16 1328 1303
0342.8 +1559 172.02 -30.11 328.33 -32.96 0340.0 +1550	 UGC 2833 PGC 13639	.S..6?. U 6.0±1.9 	.96± .09 .51± .06 1.04	10 .86 .75 .25					9233 9250 9085

R.A. 2000 DEC. l b SGL SGB R.A. 1950 DEC.	Names PGC	Type S_T n_L T L	$\log D_{25}$ $\log R_{25}$ $\log A_o$ $\log D_o$	p.a. A_g A_l A_{21}	B_T m_B m_{FIR} B_T^o	$(B-V)_T$ $(U-B)_T$ $(B-V)_T^o$ $(U-B)_T^o$	$(B-V)_o$ $(U-B)_o$ m' m'_{25}	m_{21} W_{20} W_{50} HI	V_{21} V_{opt} V_{GSR} V_{3K}
034254.9-725600 288.08 -38.85 220.23 -26.60 034322.0-730524	 ESO 31- 21 FAIR 234 PGC 13645	.S?.... 	1.04± .06 .72± .05 1.07	110 .34 1.07 .36	 15.79 ±.14 14.29				14690± 63 14516 14683
034256.2-125459 202.11 -47.53 293.65 -43.48 034035.2-130427	 MCG -2-10- 9 IRAS03405-1304 PGC 13646	.S..5*/ E 5.0±1.1 	1.49± .04 .82± .05 1.50	32 .11 1.23 .41	 13.37 				2166 2096 2034
034257.7-190120 210.63 -49.92 285.16 -44.19 034043.0-191048	 ESO 549- 2 PGC 13648	.IBS9P. SE (2) 10.0± .6 8.5± .7	1.13± .03 .08± .03 1.14	25 .14 .06 .04	 14.78 ±.14 14.57				1108 1134± 60 1048 1010
0343.0 -5335 264.97 -48.94 239.62 -36.89 0341.7 -5344	 PGC 13654		 	 .00 	16.20 ±.14 	.98± .05 			17842± 38 17685 17787
034302.2-361621 238.05 -52.69 261.09 -42.81 034109.0-362548	 ESO 358- 54 PGC 13655	.SXS8.. S (1) 8.0± .8 6.7± .8	1.27± .04 .13± .05 1.27	 .00 .15 .06	 13.92 ±.14 13.76			15.45±.2 100± 16 67± 12 1.62	895± 11 767 801
034308.2-171108 208.01 -49.23 287.74 -44.09 034051.6-172036	 MCG -3-10- 40 PGC 13659	.SBS8*/ ES (1) 8.0±1.1 5.6±1.3	1.19± .04 .67± .04 1.21	105 .19 .83 .34					
034313.8-044352 192.04 -43.38 304.57 -41.68 034044.7-045320	NGC 1424 MCG -1-10- 26 IRAS03407-0453 PGC 13664	.SXT3*. R (1) 3.0± .6 3.1± .8	1.22± .03 .42± .05 .70± .02 1.23	170 .15 .58 .21	14.33 ±.13 13.32 13.55	.53± .03 -.13± .04 .38 -.24	.61± .03 -.04± .04 13.32± .03 14.22± .23		5640± 63 5594 5500
0343.3 -5339 265.04 -48.87 239.52 -36.90 0342.0 -5348	 PGC 13668		 	 .00 	16.8 ±.2 	1.08± .10 			16946± 49 16788 16891
034322.6-335622 234.20 -52.63 264.23 -43.35 034126.1-340548	 ESO 358- 56 PGC 13671	.L..0*/ S -2.0±1.3 	1.07± .06 .59± .03 .98	5 .00 .00 	15.58 ±.14 15.56 				1189± 29 1066 1091
034330.4-534131 265.08 -48.83 239.46 -36.91 034211.9-535056	 PGC 13679		 .46± .02 	 .00 	15.55 ±.14 	1.11± .03 	 13.34± .04 		18716± 49 18558 18661
0343.5 +4104 154.54 -11.00 350.56 -18.00 0340.2 +4055	 UGC 2835 PGC 13683	.S..3.. U (1) 3.0±1.0 4.5±1.3	1.10± .05 .83± .05 1.25	123 1.62 1.14 .41					
034335.6-160054 206.43 -48.68 289.39 -44.07 034117.7-161020	 MCG -3-10- 41 PGC 13684	.SBS8.. E (1) 8.0± .8 6.4±1.2	1.30± .05 .27± .05 1.31	168 .12 .33 .13					1216 1137 1089
034337.8-355111 237.35 -52.58 261.62 -43.02 034144.1-360036	NGC 1437 ESO 358- 58 IRAS03417-3600 PGC 13687	PSXT2.. RBCSr (2) 2.5± .3 3.8± .6	1.47± .03 .17± .04 1.16± .02 1.47	150 .00 .23 .08	12.41S±.15 12.52 ±.12 13.18 12.24	.75± .02 .18± .03 .71 .15	.78± .01 .23± .02 12.81± .05 14.19± .24		1185± 41 1058 1091
034339.3-211416 214.00 -50.47 282.04 -44.45 034127.0-212342	 ESO 549- 6 PGC 13689	.IBS9.. S (1) 10.0± .6 9.4± .6	1.17± .04 .34± .03 1.18	21 .10 .26 .17	 15.29 ±.14 				
034345.6-800620 294.78 -34.12 214.03 -22.20 034633.0-801536	 ESO 15- 1 IRAS03464-8015 PGC 13695	.IBS9.. S (1) 9.5± .6 8.9± .6	1.28± .04 .31± .05 1.31	105 .33 .23 .15	 14.55 ±.14 13.99				1641 1466 1653
0343.7 +2402 166.03 -24.07 336.23 -28.69 0340.8 +2353	 UGC 2838 PGC 13696	.S..6*. U 6.0±1.4 	1.13± .05 .92± .05 1.20	116 .74 1.36 .46					5974 6014 5829
0343.8 +1419 173.58 -31.13 326.77 -34.02 0341.0 +1410	 UGC 2839 PGC 13698	.S..8*. U 8.0±1.4 	.96± .09 .51± .06 1.05	78 1.00 .63 .25					6405 6416 6258

R.A. 2000 DEC. / l b / SGL SGB / R.A. 1950 DEC.	Names / PGC	Type / S_T n_L / T / L	logD25 / logR25 / logAo / logDo	p.a. / Ag / Ai / A21	BT / mB / mFIR / BTo	(B-V)T / (U-B)T / (B-V)To / (U-B)To	(B-V)e / (U-B)e / m'o / m'25	m21 / W20 / W50 / HI	V21 / Vopt / VGSR / V3K
0343.8 +0036		.S..6*.	.95± .07						12169
186.32 -40.16	UGC 2842	U	.05± .05	.34	14.94 ±.13				
311.34 -40.08		6.0±1.2		.08					12138
0341.3 +0027	PGC 13702		.98	.03	14.47				12026
034354.5+400056		.SBT3..	1.18± .05	155				16.03±.1	4921± 11
155.26 -11.79	UGC 2837	U	.27± .05	1.28	14.2 ±.3				
349.78 -18.75	IRAS03405+3951	3.0± .8		.37	12.40			149± 8	5007
034034.7+395129	PGC 13704		1.30	.13	12.55			3.35	4788
0343.9 +2238		.I..9?.	1.04± .09						3650± 10
167.09 -25.09	UGC 2840	U	.08± .06	.74					
334.96 -29.53		10.0±1.6		.06					3686
0341.0 +2229	PGC 13706		1.11	.04					3505
034357.0+391742	A 0340+39	.L..-*.	.99± .12	0					4963± 43
155.72 -12.35	UGC 2836	U	.03± .05	1.12	13.4 ±.3				
349.22 -19.22	MK 1405	-3.0±1.2		.00	11.44				5046
034038.2+390816	PGC 13707		1.13		12.17				4829
0343.9 +6919		.SB?...	1.04± .15	60				15.88±.1	1523± 11
137.19 11.38	UGC 2827		.08± .12	2.63					
11.16 1.42				.12				146± 8	1672
0339.0 +6910	PGC 13712		1.29	.04					1438
034400.3-040137		.SAR7*.	1.15± .06	75					
191.38 -42.83	MCG -1-10- 27	E (1)	.08± .05	.19					
305.55 -41.65		7.0± .8		.11					
034130.5-041102	PGC 13714	6.4± .8	1.17	.04					
034401.4-142136		.SXS4*.	1.31± .04	138					1595
204.22 -47.91	MCG -3-10- 42	E (1)	.47± .04	.10					
291.72 -43.96	IRAS03417-1430	4.0± .7		.69	12.87				1521
034141.8-143100	PGC 13716	3.6± .6	1.32	.23					1466
0344.3 +3920		.L...*.	1.04± .09	12					
155.75 -12.27	UGC 2841	U	.53± .04	1.12					
349.30 -19.25		-2.0±1.3		.00					
0341.0 +3911	PGC 13723		1.09						
034432.0-443838	NGC 1448	.SA.6*/	1.88± .02	41	11.40M±.13	.72± .03	.80± .03	12.07±.1	1164± 5
251.52 -51.39	ESO 249- 16	R (3)	.65± .02	.00	11.47 ±.11	.01± .07	.10± .07	405± 5	1198± 23
250.10 -40.69		6.0± .6	1.34± .04	.95	10.31	.59	13.67± .10	385± 6	1021
034253.0-444800	PGC 13727	4.4± .5	1.88	.32	10.48	-.10	14.05± .17	1.27	1091
034440.9+025004	NGC 1431	.L...P*	1.02± .08	160					
184.21 -38.63	UGC 2845	UE	.13± .04	.58	15.00 ±.12				
314.12 -39.43		-2.0± .8		.00					
034204.4+024041	PGC 13732		1.07						
034450.4-215521	NGC 1439	.E.1...	1.39± .03		12.27M±.05	.88± .01	.92± .01		
215.15 -50.41	ESO 549- 9	R	.03± .02	.07	12.29 ±.11	.39± .04	.44± .02		1664± 17
281.08 -44.74		-5.0± .3	1.05± .02	.00		.85	13.03± .06		1568
034239.0-220442	PGC 13738		1.39		12.18	.38	14.11± .16		1547
034451.7-590817	IC 1997	.S..1?P	1.08± .05	73	13.94 ±.16	.55± .03	.58± .02		
272.26 -46.33	ESO 117- 7	S	.31± .04	.00	14.54 ±.14				
233.44 -34.46	IRAS03437-5918	1.0±1.8	.62± .04	.32	13.16		12.53± .13		
034351.0-591736	PGC 13740		1.08	.16			13.39± .32		
0344.9 +0554		.S..8*.	1.19± .06	120					6100± 10
181.27 -36.62	UGC 2852	U	.91± .06	.63					
317.76 -38.23		8.0±1.3		1.12					6084
0342.3 +0545	PGC 13744		1.25	.46					5956
034458.5+455804		.E...*.	.99± .12						
151.68 -6.99	UGC 2844	U	.03± .05	1.63	14.84 ±.11				
354.49 -14.91		-6.0±1.2		.00					
034127.7+454841	PGC 13746		1.27						
0345.0 -0412		.S..5*/	1.07± .09	130					
191.78 -42.73		E	.71± .08	.16					
305.42 -41.94		5.0±1.3		1.06					
0342.5 -0422	PGC 13747		1.09	.35					
034459.7-620459		.S..3*/	1.08± .05	110					
275.93 -44.89	ESO 117- 8	S (1)	.68± .04	.05	15.99 ±.14				
230.40 -32.94		3.0±1.4		.94					
034411.0-621418	PGC 13748	5.6±1.4	1.09	.34					
034500.1-361913		DE.4...	1.10± .10		15.70S±.10				
238.11 -52.30		S	.17± .08	.00					
260.90 -43.18		-5.0± .9		.00					
034307.1-362833	PGC 13749		1.05				15.77± .54		

R.A. 2000 DEC. / l b / SGL SGB / R.A. 1950 DEC.	Names / PGC	Type / S_T n_L / T / L	logD25 / logR25 / logA_e / logD_o	p.a. / A_g / A_i / A_21	B_T / m_B / m_FIR / B_T^o	(B-V)_T / (U-B)_T / (B-V)_T^o / (U-B)_T^o	(B-V)_e / (U-B)_e / m'_e / m'_25	m_21 / W_20 / W_50 / HI	V_21 / V_opt / V_GSR / V_3K
034503.4-181603	NGC 1440	PLBT0*.	1.33± .03	28	12.56 ±.15	1.03± .01	1.05± .01		
209.81 -49.19	ESO 549- 10	PSEr	.12± .03	.20	12.60 ±.11	.61± .02	.63± .02		1504± 66
286.27 -44.63		-1.9± .3	.90± .04	.00		.96	12.55± .13		1418
034248.0-182524	PGC 13752		1.34		12.36	.56	13.78± .22		1382
034503.4-355822		.LX.-..	.98± .04	155	13.99 ±.13	.91± .01	.86± .01		
237.54 -52.29	ESO 358- 59	S	.08± .03	.00	14.08 ±.14		.27± .06		1007± 18
261.37 -43.27		-3.0± .8	.62± .02	.00		.90	12.58± .07		879
034310.0-360742	PGC 13753		.97		14.02		13.55± .24		915
0345.1 +2043		.I..9*.	1.00± .08	120					5425
168.75 -26.31	UGC 2850	U	.41± .06	.71					
333.32 -30.85		10.0±1.3		.31					5455
0342.2 +2034	PGC 13754		1.07	.20					5281
034512.0-353410		.IBS9*/	1.24± .04	102					805
236.88 -52.27	ESO 358- 60	S (1)	.63± .04	.00	15.86 ±.14				
261.91 -43.39		10.0±1.2		.47					678
034318.0-354330	PGC 13756	10.0± .9	1.24	.31	15.39				713
0345.2 -3655		.S?....	.70?		15.00S±.10				
239.10 -52.22	MCG -6- 9- 28		.00?	.00					
260.07 -43.09				.00					
0343.4 -3705	PGC 13758		.70				13.34±1.58		
034517.1+763813	IC 334	.S?....	1.40± .03		12.45 ±.13	1.12± .02	1.20± .02	15.05±.1	2536± 7
132.51 17.12	UGC 2824		.12± .07	.87	13.00 ±.19	.69± .03	.74± .03	248± 16	
16.40 6.48			1.07± .03	.17		.89	13.29± .09	244± 12	2696
033852.9+762846	PGC 13759		1.49	.06	11.56	.45	14.01± .28	3.42	2468
034517.2-230010	NGC 1438	.SBR0*.	1.30± .03	69	13.22 ±.15	.79± .02	.87± .02		
216.83 -50.61	ESO 482- 41	S	.36± .03	.00	13.28 ±.14	.23± .03	.32± .02		1524
279.54 -44.85		.0± .6	.91± .02	.27		.72	13.26± .05		1426
034307.0-230930	PGC 13760		1.28		12.95	.18	13.67± .24		1409
034522.0-183804	NGC 1452	PSBR0..	1.35± .03	113	12.76 ±.13	.99± .01	1.01± .01	15.16±.3	1737± 10
210.37 -49.25	ESO 549- 12	PSEr	.18± .03	.14	12.96 ±.11	.59± .02	.61± .01	227± 6	1874± 66
285.76 -44.73	IRAS03430-1847	.4± .3	.91± .02	.14		.91	12.80± .06	177± 5	1653
034307.0-184724	PGC 13765		1.36		12.57	.53	13.92± .21		1619
034534.6-543137		.SBS3?P	1.16± .05	10	15.06 ±.19	.59± .07	.79± .04		
266.06 -48.23	ESO 156- 11	S	.52± .04	.00	15.08 ±.14				10398± 88
238.35 -36.79	IRAS03443-5439	3.0± .9	.68± .04	.72		.43	13.95± .10		10238
034419.0-544054	PGC 13773		1.16	.26	14.28		14.40± .33		10347
0345.5 +4452		.S..6*.	1.26± .06	73				14.88±.1	8138± 11
152.45 -7.79	UGC 2849	U	.22± .06	1.46					
353.73 -15.73		6.0±1.1		.32				578± 8	8235
0342.1 +4443	PGC 13774		1.40	.11					8013
034540.8-641802		.S?....	1.14± .06	35					
278.53 -43.67	ESO 83- 10		.26± .06	.13	15.00 ±.14				5545
228.14 -31.80				.39					5375
034503.0-642718	PGC 13778		1.15	.13	14.45				5518
034543.1-040532	NGC 1441	.SBS3..	1.20± .04	95	13.9 ±.3	.97± .06			
191.78 -42.51	MCG -1-10- 29	R	.39± .04	.25					4227± 43
305.64 -42.08		3.0± .5		.54		.81			4180
034313.4-041451	PGC 13782		1.23	.20	13.07		13.78± .38		4090
034543.8+464024		.S..3*.	1.26± .04	125				14.85±.1	5477± 11
151.35 -6.36	UGC 2851	U (1)	.44± .05	1.80	14.97 ±.17				
355.10 -14.52		3.0±1.1		.61				441± 8	5579
034211.5+463104	PGC 13783		1.43	.22	12.51			2.12	5354
0345.9 -0310		.S?....	1.04± .09						
190.79 -41.96	MCG -1-10- 31		.09± .07	.30					4122
306.84 -41.85				.07					4078
0343.4 -0320	PGC 13793		1.06						3985
034554.8-362131		.I.0.*.	1.42± .04	1					1493± 10
238.16 -52.11	ESO 358- 61	S	.49± .04	.00	13.98 ±.14				1500± 60
260.79 -43.35	IRAS03440-3630	90.0		.60				122± 15	1364
034402.0-363048	PGC 13794		1.42	.24	13.36			99± 11	1403
034603.1-040819	NGC 1449	.L.....	.86± .07	160	14.51 ±.13	1.05± .02	1.13± .02		
191.89 -42.47	MCG -1-10- 32	R	.18± .04	.25		.56± .04	.57± .04		4111± 43
305.62 -42.17		-2.0± .5	.34± .03	.00		.94	11.70± .09		4064
034333.4-041737	PGC 13798		.86		14.19	.50	13.23± .39		3974
034607.2-040411	NGC 1451	.L..-?.	.84± .10	135	14.37 ±.13	1.03± .01			
191.83 -42.41	MCG -1-10- 33	RE	.20± .06	.25		.55± .03			3927± 56
305.71 -42.17		-3.3± .9		.00		.93			3881
034337.5-041328	PGC 13801		.85		14.06	.50	12.94± .55		3791

R.A. 2000 DEC. l b SGL SGB R.A. 1950 DEC.	Names PGC	Type S_T n_L T L	$\log D_{25}$ $\log R_{25}$ $\log A_e$ $\log D_o$	p.a. A_g A_i A_{21}	B_T m_B m_{FIR} B_T^o	$(B-V)_T$ $(U-B)_T$ $(B-V)_T^o$ $(U-B)_T^o$	$(B-V)_e$ $(U-B)_e$ m'_e m'_{25}	m_{21} W_{20} W_{50} HI	V_{21} V_{opt} V_{GSR} V_{3K}
034614.2-364144 238.71 -52.04 260.31 -43.34 034422.0-365100	NGC 1460 ESO 358- 62 PGC 13805	.LBT0.. BSr -1.9± .4 	1.22± .05 .07± .04 1.21	 .00 .00 	13.52 ±.14 13.50				1343± 25 1212 1253
034615.8-594834 273.00 -45.86 232.63 -34.26 034518.0-595748	NGC 1463 ESO 117- 9 IRAS03453-5957 PGC 13807	RSAT1?. Sr 1.0± .6 	1.14± .05 .05± .05 .48± .07 1.14	45 .02 .05 .02	14.3 ±.2 14.15 ±.14 13.13 	.79± .02 	.75± .02 12.18± .21 14.70± .34		
034618.8-345633 235.86 -52.04 262.69 -43.75 034424.0-350548	 ESO 358- 63 IRAS03444-3505 PGC 13809	.I.0.?. S 90.0 	1.67± .03 .57± .05 1.10± .05 1.67	133 .00 .59 .29	12.6 ±.2 12.57 ±.14 11.66 11.97	.82± .04 -.13± .05 .70 -.23	.78± .03 -.09± .05 13.63± .17 14.35± .28	13.77±.2 281± 11 268± 11 1.51	1932± 7 1935± 28 1805 1839
0346.3 +4153 154.44 -10.03 351.56 -17.85 0343.0 +4144	 UGC 2853 PGC 13811	.I..9*. U 10.0±1.2 	1.00± .16 .00± .12 1.14	 1.53 .00 .00					
0346.4 +1526 173.18 -29.89 328.30 -33.99 0343.6 +1517	 UGC 2856 PGC 13813	.S..4.. U 4.0±1.0 	.96± .09 .51± .06 1.07	23 1.12 .75 .25					8733 8746 8590
034627.3-035810 191.78 -42.29 305.88 -42.22 034357.4-040726	NGC 1453 MCG -1-10- 34 PGC 13814	.E.2+.. R -5.0± .4 	1.38± .05 .09± .05 .92± .03 1.39	 .25 .00 	12.58 ±.13 12.41 ±.18 12.21	1.05± .01 .61± .02 .96 .57	1.07± .01 .64± .02 12.67± .10 14.24± .29		3933± 33 3887 3798
0346.5 +3644 157.78 -14.02 347.53 -21.29 0343.3 +3635	 UGC 2854 PGC 13817	.S..6*. U 6.0±1.3 	1.04± .08 .47± .06 1.16	176 1.30 .69 .24					
034634.9-403852 245.08 -51.68 255.04 -42.34 034449.0-404806	 ESO 302- 6 PGC 13818	.LAT-*. S -3.0± .8 	1.16± .05 .22± .04 1.13	133 .01 .00 	14.34 ±.14 14.31				1402 1264 1321
034636.0-042710 192.35 -42.53 305.27 -42.40 034406.6-043625	 MCG -1-10- 35 IRAS03441-0436 PGC 13820	.S..5./ E 5.0± .9 	1.37± .04 .85± .05 1.39	153 .22 1.28 .43	 12.59 				3796 3748 3661
034638.1-163304 207.60 -48.21 288.75 -44.86 034420.9-164219	 MCG -3-10- 45 IRAS03443-1642 PGC 13821	.IB.9P* E (1) 10.0±1.3 5.3±1.2	1.11± .06 .35± .05 1.12	40 .13 .26 .18	 11.45 				
034649.7+680545 138.17 10.58 10.48 .37 034158.6+675626	IC 342 UGC 2847 IRAS03419+6756 PGC 13826	.SXT6.. R (1) 6.0± .3 2.0± .7	2.33± .02 .01± .03 2.65	 3.5 .01 .01	9.10S±.14 9.1 ±.3 6.95 5.58		 15.55± .18	8.25±.1 175± 7 151± 6 2.66	34± 4 -4± 17 178 -53
034658.0-032744 191.31 -41.90 306.59 -42.18 034427.7-033658	 MCG -1-10- 36 PGC 13830	.S..4*/ E 4.0±1.2 	1.21± .05 .57± .05 1.24	50 .36 .84 .29					3988 3943 3853
0346.9 -1148 201.31 -46.16 295.41 -44.26 0344.6 -1158	 PGC 13831	.S..6./ E 6.0± .9 	1.17± .08 .73± .08 1.18	55 .11 1.07 .37					
034658.1-253123 220.84 -50.84 275.92 -45.15 034451.1-254036	NGC 1459 ESO 482- 43 IRAS03448-2540 PGC 13832	.SBS4?. S (1) 4.0± .8 2.2± .8	1.24± .03 .18± .03 1.24	167 .00 .27 .09	13.62 ±.14 13.21 13.32				4210± 52 4104 4101
034658.3+383804 156.62 -12.50 349.10 -20.10 034340.2+382849	 UGC 2857 IRAS03436+3828 PGC 13833	.SB.4.. U 4.0± .8 	1.10± .05 .14± .05 1.23	35 1.36 .20 .07	 14.3 ±.3 13.35 12.67				6444 6524 6313
034706.1-372242 239.80 -51.84 259.34 -43.34 034515.1-373154	 ESO 302- 7 PGC 13837	.SXS5*. S (1) 5.0± .5 4.1± .5	1.13± .04 .17± .04 1.13	10 .00 .26 .09	15.03 ±.14 14.74				5757± 26 5624 5670
034709.0-361941 238.10 -51.86 260.75 -43.60 034516.4-362853	 PGC 13839	DE.2... S -5.0± .9 	1.06± .10 .09± .08 1.03	 .00 .00 					

R.A. 2000 DEC. / l b / SGL SGB / R.A. 1950 DEC.	Names / PGC	Type / S$_T$ n$_L$ / T / L	logD$_{25}$ / logR$_{25}$ / logA$_e$ / logD$_o$	p.a. / A$_g$ / A$_i$ / A$_{21}$	B$_T$ / m$_B$ / m$_{FIR}$ / B$_T^o$	(B-V)$_T$ / (U-B)$_T$ / (B-V)$_T^o$ / (U-B)$_T^o$	(B-V)$_e$ / (U-B)$_e$ / m'$_e$ / m'$_{25}$	m$_{21}$ / W$_{20}$ / W$_{50}$ / HI	V$_{21}$ / V$_{opt}$ / V$_{GSR}$ / V$_{3K}$
034710.2-334231	IC 1993	PSXT3..	1.39± .04		12.35 ±.14	.75± .02	.79± .01		971± 24
233.86 -51.83	ESO 358- 65	BSr (1)	.06± .05	.00	12.50 ±.14	.22± .03	.30± .02		847
264.34 -44.17	IRAS03451-3351	3.1± .4	1.18± .02	.08	13.09	.73	13.74± .04		877
034513.8-335144	PGC 13840	4.1± .6	1.39	.03	12.33	.21	14.00± .28		
034710.4-265641		.S?....	.89± .06	57					4138
223.07 -51.07	ESO 482- 44		.22± .04	.00	15.47 ±.14				4029
273.88 -45.10				.33					4032
034505.0-270554	PGC 13841		.89	.11	15.11				
034711.1-543513		.S..3?.	.97± .05	25					10478± 88
266.01 -47.99	ESO 156- 13	S (1)	.20± .04	.00	15.13 ±.14				10317
238.15 -36.97	IRAS03459-5444	3.0±1.7		.27					10428
034556.1-544424	PGC 13842	3.3±1.2	.97	.10	14.78				
034713.5-385806		.LX.0..	1.12± .05	72	15.1 ±.2	.75± .07	.70± .03		1038
242.37 -51.72	ESO 302- 8	S	.30± .04	.00	15.06 ±.14	.32± .10	.41± .05		903
257.20 -42.94		-2.0± .8	.82± .08	.00		.71	14.65± .26		954
034525.0-390718	PGC 13843		1.07		15.06	.30	14.81± .34		
0347.3 +4051		.S..6*.	1.06± .06					15.03±.1	5493± 11
155.24 -10.72	UGC 2859	U	.00± .05	1.91					5579
350.89 -18.67	IRAS03439+4042	6.0±1.2		.00	13.13			199± 8	5365
0343.9 +4042	PGC 13846		1.24	.00					
034722.4+455811		.S..8..	1.13± .07	160				15.01±.3	5054± 9
152.01 -6.73	UGC 2858	U	.14± .06	1.71				270± 11	5153
354.78 -15.21	IRAS03438+4549	8.0± .8		.18	13.16			249± 8	4932
034351.2+454857	PGC 13850		1.29	.07					
034726.3-681315	NGC 1473	.IBS9*.	1.17± .05	36	13.4 ±.2	.52± .05			1359
282.84 -41.37	ESO 54- 19	S (1)	.27± .05	.22	13.76 ±.14	-.29± .07			1186
224.27 -29.70	IRAS03472-6822	10.0± .9		.20	12.82	.39			1342
034713.0-682224	PGC 13853	6.7± .9	1.19	.14	13.21	-.38	13.43± .33		
034733.1-383437		.SBS8?.	1.37± .04	49					993± 10
241.73 -51.68	ESO 302- 9	S (1)	.56± .04	.00	14.34 ±.14			152± 15	
257.70 -43.11		8.0± .8		.69				139± 11	858
034544.0-384348	PGC 13854	7.8±1.2	1.37	.28	13.65				909
034735.1-462038			.68± .07	130	16.0 ±.3	.87± .07	.83± .05		
253.96 -50.50	ESO 249- 19		.29± .06	.00	16.31 ±.14				
247.74 -40.56	IRAS03460-4629		.71± .11		13.23		14.99± .36		
034600.1-462948	PGC 13855		.68						
0347.6 +7240		.S..8*.	1.04± .08	140				15.46±.1	4302± 11
135.26 14.17	UGC 2848	U	.37± .06	1.23					4456
13.71 3.57	IRAS03421+7230	8.0±1.3		.46				185± 8	4226
0342.1 +7230	PGC 13857		1.16	.19					
0347.6 +1315		PSXS1*.	1.16± .05	35					6644± 85
175.23 -31.21	UGC 2862	U	.14± .05	.95	15.0 ±.2				6649
326.22 -35.37	IRAS03448+1305	1.0±1.2		.15					6502
0344.8 +1305	PGC 13858		1.25	.07	13.78				
0347.6 +3921		.SAR3..	1.17± .07	50				14.83±.1	6368± 11
156.26 -11.86	UGC 2861	U	.35± .06	1.46					6450
349.76 -19.73	IRAS03443+3911	3.0± .8		.48	13.39			492± 8	6239
0344.3 +3911	PGC 13859		1.30	.17					
034739.6-042112		.SBS8*.	1.25± .05	160					4214
192.43 -42.25	MCG -1-10- 37	E (1)	.16± .05	.22					
305.50 -42.62		8.0± .8		.20					4166
034510.1-043023	PGC 13860	7.5± .8	1.27	.08					4080
034752.4-362820		DLBS-*.	1.08± .05	42	15.10S±.10				1856± 29
238.33 -51.71	ESO 358- 66	S	.31± .04	.00	15.65 ±.14				1725
260.51 -43.71		-3.5± .8		.00					1768
034600.1-363730	PGC 13864		.99		15.26		14.59± .30		
034807.0-364203		.S?....	.91± .06		14.6 ±.2	.46± .06	.42± .03		5237± 28
238.70 -51.66	ESO 358- 67		.08± .05	.00	14.68 ±.14	-.28± .07	-.21± .04		5106
260.18 -43.70	IRAS03462-3651		.59± .06	.12	14.08	.41	13.02± .16		5150
034615.0-365112	PGC 13870		.91	.04	14.51	-.31	13.79± .37		
034814.7-212827		.SXT5..	1.41± .03	17					1548
214.84 -49.52	ESO 549- 18	S (1)	.22± .03	.11	13.60 ±.14				1452
281.74 -45.53		5.0± .7		.33					1435
034603.0-213736	PGC 13871	4.4± .8	1.42	.11	13.16				
034817.7-245315			.74± .06						4212±104
219.97 -50.40	ESO 482- 45		.19± .05	.02	15.31 ±.14				4107
276.81 -45.48									4104
034610.0-250224	PGC 13874		.74						

R.A. 2000 DEC.	Names	Type	logD₂₅	p.a.	B_T	(B-V)_T	(B-V)_●	m₂₁	V₂₁
l　b		S_T　n_L	logR₂₅	A_g	m_B	(U-B)_T	(U-B)_●	W₂₀	V_opt
SGL　SGB		T	logA_●	A_i	m_FIR	(B-V)°_T	m'_●	W₅₀	V_GSR
R.A. 1950 DEC.	PGC	L	logD_o	A₂₁	B°_T	(U-B)°_T	m'₂₅	HI	V₃K

0348.3　+4155		.SB.3..	1.03± .06						
154.70　-9.79	UGC　2863	U	.00± .05	1.73					
351.83　-18.09		3.0± .8		.00					
0344.9　+4146	PGC 13875		1.19	.00					

034825.0-063735		PSBT3P*	1.14± .06	80					
195.19　-43.31	MCG -1-10- 38	E	.13± .05	.21					
302.58　-43.44	IRAS03459-0646	3.0±1.2		.18	13.13				
034557.9-064644	PGC 13879		1.16	.07					

0348.4　+7009	A　0343+70	.SX.5..	1.64± .03	112				12.42±.1	1207±　5
136.96　12.27	UGC　2855	U	.34± .05	1.81	13.5　±.3			449±　6	1200± 42
12.01　1.74		5.0± .7		.51				416±　7	1357
0343.3　+7000	PGC 13880		1.81	.17	11.15			1.10	1126

034827.1-162330	NGC　1461	.LAR0..	1.48± .03	155	12.81 ±.13	1.04± .03	1.05± .03		
207.63　-47.75	MCG -3-10- 47	R	.51± .04	.09	12.71 ±.14	.57± .03	.59± .03		1413± 37
289.04　-45.27		-2.0± .4	.67± .03	.00		.96	11.65± .08		1330
034609.9-163238	PGC 13881		1.41		12.66	.50	13.80± .21		1293

0348.6　+3508		.S..4..	1.22± .06	140					5429
159.17　-14.98	UGC　2868	U	.98± .06	.89					
346.53　-22.65		4.0±1.0		1.44					5499
0345.4　+3459	PGC 13884		1.30	.49					5297

0348.7　+4152		.S..0..	1.11± .07	97					
154.80　-9.78	UGC　2867	U	.60± .06	1.86					
351.85　-18.18		.0± .9		.45					
0345.3　+4143	PGC 13885		1.25						

034844.0-532712		.SAR4*.	.98± .05						
264.31　-48.20	ESO　156- 15	S　　(1)	.10± .05	.00	15.28 ±.14				
239.28　-37.70		4.0± .9		.15					
034726.1-533618	PGC 13887	3.3± .9	.98	.05					

0348.7　+1308		.S..3..	1.03± .06	47					6805± 85
175.55　-31.08	UGC　2871	U　　(1)	.11± .05	.97	15.03 ±.16				6809
326.28　-35.66	IRAS03459+1259	3.0± .8		.15	13.27				6665
0346.0　+1259	PGC 13888	3.5±1.1	1.12	.06	13.86				

034851.4-185841		.SAT3*.	1.15± .03	36					4295± 60
211.30　-48.60	ESO　549- 22	SE　　(1)	.35± .03	.12	14.92 ±.14				4205
285.34　-45.58		2.7± .7		.49					4179
034637.0-190748	PGC 13889	2.2± .9	1.16	.18	14.28				

0348.9　+4221		.S..2..	1.07± .14						
154.52　-9.38	UGC　2869	U	.79± .12	1.20					
352.24　-17.88		2.0±1.0		.98					
0345.5　+4212	PGC 13891		1.18	.40					

034857.9-220754		PSB.1..	1.02± .05	152					
215.88　-49.55	ESO　549- 23	r	.24± .04	.09	13.82 ±.14				
280.79　-45.70	IRAS03467-2216	1.0± .9		.25	11.76				
034647.0-221700	PGC 13894		1.02	.12					

0349.0　+4220		.L.....	1.11± .08						
154.54　-9.38	UGC　2870	U	.31± .04	1.21					
352.24　-17.91		-2.0± .9		.00					
0345.6　+4211	PGC 13896		1.20						

034903.2-355236		.S?....	.95± .06	55					5393± 29
237.37　-51.48	ESO　358- 69		.41± .05	.00	15.12 ±.14				5263
261.24　-44.08				.56					5305
034710.0-360142	PGC 13900		.95	.20	14.52				

0349.0　+0111		.SXS8..	.99± .08	155					4135± 10
186.75　-38.77	UGC　2872	U	.07± .06	.57					
312.68　-41.08		8.0± .9		.09					4102
0346.5　+0102	PGC 13903		1.04	.04					3999

034905.6-030340		.S..3./	1.17± .05	150					
191.26　-41.24	MCG -1-10- 39	E	.66± .05	.34					
307.33　-42.56		3.0± .9		.91					
034634.9-031246	PGC 13905		1.20	.33					

034907.4-485131	IC　2000	.SBS6*/	1.61± .03	83		1.01± .05	1.09± .05		
257.65　-49.61	ESO　201- 3	S　　(1)	.70± .05	.00	12.90 ±.14				894± 88
244.56　-39.80	IRAS03476-4900	5.5± .5	.52± .05	1.03	13.09	.88	14.16± .11		741
034738.1-490036	PGC 13912	4.4± .6	1.61	.35	11.86				833

034937.2-835742		.SAS9*.	1.08± .05						
297.98　-31.34	ESO　4- 6	S　　(1)	.11± .05	.57	15.51 ±.14				
210.78　-19.82		8.5± .6		.11					
035607.0-840630	PGC 13925	8.9± .6	1.13	.06					

R.A. 2000 DEC. l b SGL SGB R.A. 1950 DEC.	Names PGC	Type S_T n_L T L	$\log D_{25}$ $\log R_{25}$ $\log A_e$ $\log D_o$	p.a. A_g A_i A_{21}	B_T m_B m_{FIR} B_T^o	$(B-V)_T$ $(U-B)_T$ $(B-V)_T^o$ $(U-B)_T^o$	$(B-V)_e$ $(U-B)_e$ m'_e m'_{25}	m_{21} W_{20} W_{50} HI	V_{21} V_{opt} V_{GSR} V_{3K}
034942.1-265939 223.31 -50.53 273.74 -45.66 034737.0-270842	 ESO 482- 46 IRAS03476-2708 PGC 13926	.S..5*/ SU 5.3± .6 	1.53± .02 .83± .03 1.53	69 .00 1.25 .42	 13.98 ±.14 12.72			14.58±.2 234± 11 220± 8 1.44	1525± 7 1415 1423
034946.8-554904 267.48 -47.17 236.57 -36.70 034836.0-555806	 ESO 156- 17 PGC 13929	.S?.... 	.96± .06 .35± .05 .52± .05 .96	13 .00 .48 .17	15.38 ±.18 15.45 ±.14 14.84	1.01± .04 .86	.96± .04 13.47± .13 14.15± .36		 13316± 88 13153 13271
0349.8 -3556 237.48 -51.33 261.09 -44.22 0347.9 -3605	 PGC 13930		 	 .00 	15.40S±.10 				1326± 29 1195 1239
034950.3-713807 286.34 -39.23 221.03 -27.81 035007.1-714706	 ESO 54- 21 PGC 13931	.SXS8.. Sr (1) 7.9± .3 7.6± .5	1.66± .03 .28± .05 1.68	93 .25 .35 .14	 12.98 ±.14 12.38			13.04±.3 204± 7 .51	1428± 9 1253 1421
0349.9 +1750 171.93 -27.60 331.26 -33.41 0347.1 +1741	 UGC 2873 PGC 13933	.S..3.. U (1) 3.0± .9 4.5±1.2	1.02± .08 .56± .06 1.10	40 .90 .77 .28					8008 8026 7870
0350.1 +1741 172.08 -27.67 331.14 -33.53 0347.3 +1732	 UGC 2874 PGC 13938	.S..0.. U .0±1.0 	1.00± .08 .73± .06 1.05	171 .90 .54 	 15.60 ±.12 14.01				10057 10074 9919
035022.8-501806 259.68 -49.00 242.74 -39.36 034857.0-502706	 ESO 201- 4 PGC 13944	.SBT2*P S 2.0±1.2 	.99± .07 .32± .06 .99	 .00 .40 .16	 15.03 ±.14 				
035023.5+065809 181.35 -34.90 319.73 -38.98 034742.9+064907	NGC 1462 MCG 1-10- 10 PGC 13945	.S?.... 	.94± .11 .28± .07 .99	 .59 .43 .14	 15.10 ±.12 14.04				8223 8207 8087
035030.5-392236 242.94 -51.05 256.41 -43.43 034843.0-393136	 ESO 302- 12 PGC 13947	.SBS7.. S (1) 7.0± .8 6.7±1.2	1.07± .04 .05± .04 1.07	 .06 .07 .03	 15.13 ±.14 				
0350.5 +3632 158.55 -13.66 347.93 -22.03 0347.3 +3623	 UGC 2875 PGC 13948	.S?.... 	1.00± .08 .55± .06 1.10	129 1.11 .82 .27					5767 5840 5638
0350.5 +7301 135.20 14.58 14.12 3.68 0344.9 +7252	 UGC 2865 IRAS03449+7252 PGC 13949	.SB?... 	1.11± .05 .35± .05 1.20	20 .96 .53 .18	 14.8 ±.3 11.25 13.27				4312 4466 4238
035036.9-355436 237.42 -51.17 261.09 -44.38 034844.0-360336	 ESO 359- 2 PGC 13950	.LB.-?. S -3.7± .7 	1.13± .05 .23± .04 .92± .16 1.06	45 .00 .00 	14.19M±.10 14.48 ±.14 14.27		 13.40± .57 14.14± .30		1460± 29 1329 1374
035040.1-251648 220.75 -49.97 276.20 -46. 034833.0-252548	 ESO 482- 47 IRAS03485-2525 PGC 13952	.SXT5.. S (1) 4.5± .6 1.1± .6	1.18± .04 .04± .03 1.18	 .00 .06 .02	* 13.83 ±.14 14.02 13.75				4083 3976 3979
0350.8 +7407 134.47 15.43 14.89 4.46 0345.0 +7358	 UGC 2864 IRAS03450+7358 PGC 13957	.S..1?. U 1.0±1.9 	.96± .17 .39± .12 1.06	92 1.01 .40 .20	* 15.1 ±.3 13.28 				
0350.9 -5158 262.04 -48.39 240.75 -38.69 0349.6 -5207	 PGC 13960		 .98± .17 	 .00 	14.0 ±.5 	.73± .06 	.69± .03 14.42± .51 		5475± 88 5316 5423
0351.0 +7819 131.60 18.60 17.80 7.45 0344.0 +7810	 UGC 2860 PGC 13965	.S..6*. U 6.0±1.3 	1.07± .14 .50± .12 1.13	88 .63 .73 .25					
035107.4-030721 191.70 -40.85 307.47 -43.06 034836.8-031620	 MCG -1-10- 43 PGC 13967	PSBS0?. E .0± .9 	1.13± .08 .12± .05 1.16	50 .40 .09 					

R.A. 2000 DEC. l b SGL SGB R.A. 1950 DEC.	Names PGC	Type S_T n_L T L	$\log D_{25}$ $\log R_{25}$ $\log A_e$ $\log D_o$	p.a. A_g A_i A_{21}	B_T m_B m_{FIR} B_T^o	$(B-V)_T$ $(U-B)_T$ $(B-V)_T^o$ $(U-B)_T^o$	$(B-V)_e$ $(U-B)_e$ m'_e m'_{25}	m_{21} W_{20} W_{50} HI	V_{21} V_{opt} V_{GSR} V_{3K}
035109.6+034739 184.53 -36.76 316.12 -40.52 034832.2+033840	 CGCG 417- 8 PGC 13968	.S?.... 	.84± .10 .19± .10 .91	 .74 .28 .09	 15.24 ±.12 14.18				7015 6989 6881
0351.1 +3654 158.41 -13.30 348.31 -21.87 0347.9 +3645	 UGC 2877 PGC 13969	.SB.8*. U 8.0±1.2 	1.16± .13 .25± .12 1.26	25 1.11 .30 .12					3818 3891 3690
0351.4 -0740 196.94 -43.20 301.44 -44.44 0349.0 -0749	 MCG -1-10- 44 PGC 13977	.SB.9*/ E (1) 9.0±1.3 5.3± .8	1.24± .05 .78± .05 1.25	159 .17 .79 .39					4065 4005 3940
035140.8-382704 241.45 -50.88 257.55 -43.92 034952.0-383600	 ESO 302- 14 PGC 13985	.IBS9.. S (1) 10.0± .8 10.0± .8	1.23± .04 .05± .04 1.23	 .00 .04 .02	 14.84 ±.14 14.80			14.53±.3 74± 7 -.30	881± 9 744 801
035146.3+325830 161.15 -16.20 345.22 -24.55 034836.0+324932	 UGC 2879 PGC 13987	.I..9?. U 10.0±1.6 	1.14± .13 .10± .12 1.21	 .75 .07 .05				16.07±.3 162± 7	3938± 10 4000 3808
035152.8-085018 198.42 -43.69 299.88 -44.83 034927.9-085914	NGC 1467 MCG -2-10- 15 PGC 13991	PLAT+?. E -1.0±1.3 	1.04± .07 .21± .05 1.02	115 .13 .00					
0351.9 +2454 166.87 -22.18 338.28 -29.66 0348.9 +2446	 UGC 2880 PGC 13992	.S..6*. U 6.0±1.4 	1.10± .05 .83± .05 1.13	146 .39 1.21 .41					5983 6021 5849
0351.9 +4248 154.65 -8.67 352.99 -17.97 0348.5 +4240	 UGC 2878 PGC 13994	.S..6*. U 6.0±1.3 	1.04± .08 .59± .06 1.17	150 1.33 .86 .29				16.28±.1 276± 8	5390± 11 5479 5269
035158.8-595549 272.69 -45.17 232.04 -34.81 035103.0-600442	IC 2010 ESO 117- 11 IRAS03510-6004 PGC 13995	.S..1?. S 1.0±1.8 	1.06± .05 .37± .05 .61± .06 1.06	71 .00 .38 .18	14.35 ±.19 14.52 ±.14 12.91 	.67± .03 	.68± .03 12.89± .17 13.58± .35		
035200.5-551524 266.54 -47.09 236.99 -37.25 035048.6-552417	 PGC 13996	.LBS0.. S -2.0± .9 	1.02± .11 .10± .08 1.00	 .00 .00					13317± 88 13154 13273
035200.8-390254 242.39 -50.78 256.72 -43.81 035013.0-391148	 ESO 302- 15 PGC 13997	DLX.0.. S -2.0± .9 	1.00± .05 .09± .04 .99	122 .02 .00	 15.21 ±.14 				
035201.3-332811 233.56 -50.81 264.39 -45.21 035005.0-333706	 ESO 359- 3 PGC 13998	.S..2*/ S 1.5± .8 	1.26± .04 .41± .04 1.26	132 .00 .50 .20	 14.25 ±.14 13.73			161± 15 166± 11	1596± 10 1636± 29 1474 1512
035202.5-834957 297.85 -31.38 210.84 -19.96 035824.0-835836	IC 2051 ESO 4- 7 IRAS03583-8358 PGC 13999	.SBR4*. Sr (1) 4.2± .4 3.3± .5	1.42± .03 .21± .04 1.47	67 .57 .31 .11	 12.32 ±.14 11.29 11.43		13.63±.3 329± 12 298± 9 2.10	1735± 10 1675± 34 1555 1753
035209.0-443154 250.93 -50.09 249.59 -41.99 035031.0-444048	NGC 1476 ESO 249- 24 IRAS03504-4440 PGC 14001	.S..1?P S 1.0±1.7 	1.14± .04 .38± .04 1.14	86 .00 .38 .19	 13.98 ±.14 13.64 				
035209.9-085959 198.66 -43.70 299.68 -44.93 034945.1-090854	NGC 1470 MCG -2-10- 16 IRAS03497-0908 PGC 14002	.S..2./ E 2.0± .9 	1.12± .06 .67± .05 1.13	169 .13 .82 .34	 13.54 				6464 6399 6341
035212.6-062058 195.53 -42.36 303.30 -44.28 034945.2-062953	NGC 1468 MCG -1-10- 45 PGC 14004	.L..-.. E -3.0± .9 	1.07± .09 .18± .08 1.08	45 .25 .00					
035213.1-545313 266.01 -47.20 237.37 -37.46 035100.0-550206	 ESO 156- 18 PGC 14005	.LXS0*P S -2.0± .7 	1.33± .04 .30± .04 .88± .09 1.29	21 .00 .00	14.5 ±.2 14.17 ±.14 14.08	1.19± .04 .51± .07 1.04 .55	1.14± .02 .48± .03 14.35± .30 15.28± .32		13315 13152 13270

R.A. 2000 DEC.	Names	Type	logD$_{25}$	p.a.	B$_T$	(B-V)$_T$	(B-V)$_e$	m$_{21}$	V$_{21}$
l b		S$_T$ n$_L$	logR$_{25}$	A$_g$	m$_B$	(U-B)$_T$	(U-B)$_e$	W$_{20}$	V$_{opt}$
SGL SGB		T	logA$_e$	A$_i$	m$_{FIR}$	(B-V)$_T^o$	m'$_e$	W$_{50}$	V$_{GSR}$
R.A. 1950 DEC.	PGC	L	logD$_o$	A$_{21}$	B$_T^o$	(U-B)$_T^o$	m'$_{25}$	HI	V$_{3K}$
0352.3 +0222	UGC 2884	.S..7..	1.00± .08	17					
186.15 -37.41		U	1.02± .06	.66					
314.56 -41.36		7.0±1.0		1.38					
0349.7 +0214	PGC 14007		1.06	.50					
0352.3 +3614	UGC 2881	RLX.+..	1.19± .06	148					
159.03 -13.66		U	.17± .03	1.11	14.3 ±.3				
347.95 -22.48	IRAS03490+3605	-1.0± .8		.00					
0349.0 +3605	PGC 14008		1.27						
035221.5-420819		DE.6.*.	1.13± .05	94					
247.22 -50.42	ESO 302- 16	S	.46± .04	.00	15.64 ±.14				
252.62 -42.88		-5.0± .9		.00					
035039.1-421712	PGC 14010		.99						
0352.7 +3440	UGC 2882	.S..8*.	1.22± .06	94					4355± 10
160.14 -14.78		U	.40± .06	.92					
346.76 -23.58		8.0±1.2		.49					4421
0349.5 +3432	PGC 14017		1.31	.20					4228
035247.8-472839	NGC 1483	.SBS4*.	1.21± .03	125	13.11 ±.13	.43± .01	.48± .01	14.69±.3	1143± 9
255.35 -49.37	ESO 201- 7	RS (1)	.08± .04	.00	13.23 ±.14	-.19± .03	-.20± .03	136± 10	1081± 31
245.88 -40.93	IRAS03512-4737	4.2± .4	.90± .01	.12	13.13	.41	13.10± .02		986
035116.0-473730	PGC 14022	5.6± .8	1.21	.04	13.04	-.21	13.83± .22	1.61	1078
035249.1-432351			.68± .06						1475±108
249.15 -50.16	ESO 249- 25		.15± .05	.00	14.85 ±.14				1329
250.96 -42.52									1406
035109.0-433242	PGC 14023		.68						
035251.8+070316		.IB?...	.82± .13						5737
181.75 -34.38	MCG 1-10- 12		.08± .07	.56	15.34 ±.12				
320.19 -39.49				.06					5720
035011.1+065423	PGC 14024		.87	.04	14.72				5604
035255.3-534330		.LXS-*P	1.04± .11						
264.36 -47.52		S	.21± .08	.00					
238.60 -38.12		-3.0± .9		.00					
035139.0-535220	PGC 14026		1.01						
0353.0 -5623					15.36 ±.15	.67± .04			13449± 88
268.00 -46.52				.02					13284
235.66 -36.79									13408
0351.9 -5632	PGC 14029								
035302.8+353526		.SAT5..	1.59± .04	40				13.64±.1	5802± 4
159.58 -14.05	UGC 2885	U	.30± .06	1.12	13.5 ±.3			579± 4	5785± 27
347.54 -23.03	IRAS03497+3526	5.0± .7		.45	12.80			550± 4	5870
034948.7+352633	PGC 14030		1.70	.15	11.89			1.60	5675
0353.1 +3457	UGC 2886	.S..9*.	1.19± .12						
160.01 -14.51		U	.12± .12	.92					5490
347.04 -23.46		9.0±1.1		.12					5557
0349.9 +3449	PGC 14031		1.28	.06					5363
0353.2 +3456	UGC 2887	.I..9?.	1.00± .16						5385
160.04 -14.51		U	.04± .12	.92					
347.05 -23.48		10.0±1.7		.03					5452
0350.0 +3448	PGC 14032		1.09	.02					5258
035319.9+321952		.S?....	1.02± .06	20				15.06±.3	4039± 10
161.85 -16.47	UGC 2888		.20± .05	.61	15.0 ±.3				
344.92 -25.22				.30				215± 7	4098
035010.4+321100	PGC 14034		1.07	.10	14.03			.93	3911
0353.5 +3228	NGC 1465	.S..0..	1.24± .05	165					2283
161.78 -16.32	UGC 2891	U	.57± .05	.61	14.7 ±.3				
345.08 -25.16		.0± .9		.43					2342
0350.4 +3220	PGC 14039		1.27		13.60				2155
035334.5-660102	NGC 1490	.E.1...	1.11± .06	142					
279.93 -42.07	ESO 83- 11	S	.07± .04	.20	13.42 ±.14				5397± 56
225.93 -31.49		-5.0± .7		.00					5224
035309.0-660948	PGC 14040		1.12		13.14				5379
035335.3-485918	IC 2009	.IXS9..	1.21± .05	72					
257.54 -48.86	ESO 201- 8	S (1)	.17± .05	.00	14.61 ±.14				1584
243.99 -40.41		10.0± .8		.12					1428
035207.0-490806	PGC 14041	8.9± .8	1.21	.08	14.48				1528
0353.5 +1905	UGC 2892	.SBT4..	1.13± .05					16.01±.3	7963± 11
171.61 -26.10		U	.01± .05	.79				464± 7	
333.08 -33.41		4.0± .8		.02				446± 7	7982
0350.7 +1857	PGC 14042		1.20	.01					7830

R.A. 2000 DEC. / l b / SGL SGB / R.A. 1950 DEC.	Names / PGC	Type / S_T n_L / T / L	$\log D_{25}$ / $\log R_{25}$ / $\log A_e$ / $\log D_o$	p.a. / A_g / A_i / A_{21}	B_T / m_B / m_{FIR} / B_T^o	$(B-V)_T$ / $(U-B)_T$ / $(B-V)_T^o$ / $(U-B)_T^o$	$(B-V)_e$ / $(U-B)_e$ / m'_e / m'_{25}	m_{21} / W_{20} / W_{50} / HI	V_{21} / V_{opt} / V_{GSR} / V_{3K}
035336.9-092722		.LA.0*.	1.08± .07	177					
199.45 -43.61	MCG -2-10- 19	E	.41± .05	.12					
299.16 -45.38		-2.0± .9		.00					
035112.7-093611	PGC 14044		1.03						
035337.6+371550		.SX.4*.	1.19± .05	100					5580
158.55 -12.71	UGC 2889	U	.28± .05	1.45	14.8 ±.3				
348.95 -21.99		4.0±1.2		.41					5653
035020.8+370659	PGC 14045		1.32	.14	12.92				5456
0354.1 +1559		.S..6*.	1.00± .06						6583± 57
174.21 -28.18	UGC 2894	U	.13± .05	1.14	15.14 ±.15				6592
330.07 -35.26	IRAS03512+1550	6.0±1.2		.18					6450
0351.3 +1550	PGC 14063		1.11	.06	13.79				
035413.2-434520		.S?....	1.02± .06	167					1078±108
249.64 -49.85	ESO 249- 26		.40± .05	.00	15.22 ±.14				930
250.38 -42.63				.56					1011
035234.0-435406	PGC 14066		1.02	.20	14.65				
035413.6-084431		.SXS7..	1.11± .09	95					
198.69 -43.13	MCG -2-10- 21	E (1)	.06± .05	.15					
300.20 -45.37		7.0± .9		.08					
035148.6-085318	PGC 14067	6.4± .8	1.12	.03					
0354.2 +3630		.SB.7..	1.04± .08						5833
159.15 -13.19	UGC 2893	U	.08± .06	1.33					
348.45 -22.59		7.0± .9		.11					5904
0351.0 +3622	PGC 14068		1.16	.04					5709
035415.8+155542								17.05±.3	6662± 9
174.28 -28.20	CGCG 465- 12			1.14	15.2 ±.3			220± 10	6675± 57
330.03 -35.32	IRAS03514+1546								6671
035125.9+154654	PGC 14069								6530
035417.9-365814	NGC 1484	.SBS3?.	1.40± .04	80 *					1022± 26
239.09 -50.42	ESO 359- 6	BS (1)	.64± .04	.00	13.67 ±.14				887
259.36 -44.83	IRAS03524-3706	3.0± .8		.89	14.24				943
035227.0-370700	PGC 14071	4.4± .8	1.40	.32	12.77				
0354.3 +1735		.S?....	1.07± .07	97 *					10082
172.94 -27.04	UGC 2895		.25± .06	1.02					
331.71 -34.41	IRAS03514+1726			.38	13.30				10096
0351.4 +1726	PGC 14072		1.17	.13					9950
0354.4 +0635		.S..8*.	1.13± .05	103 *					3471± 10
182.49 -34.36	UGC 2899	U	.86± .05	.56					
319.89 -40.05		8.0±1.4		1.06					3451
0351.8 +0627	PGC 14076		1.18	.43					3341
035428.4-355802	IC 2006	.E.....	1.32± .03		12.21M±.10	.92± .01	.95± .01		1356± 20
237.51 -50.39	ESO 359- 7	RBC	.07± .03	.00	12.44 ±.11	.39± .02	.44± .01		1223
260.73 -45.12		-4.5± .3	.98± .02	.00		.91	12.57± .08		1275
035236.1-360648	PGC 14077		1.30		12.29	.40	13.62± .19		
035429.2-444509		.IBS9*/	1.15± .05	92					1272±108
251.15 -49.64	ESO 249- 27	S (1)	.46± .05	.00	15.47 ±.14				1123
249.09 -42.29		10.0±1.3		.35					1208
035252.0-445354	PGC 14078	7.8± .9	1.15	.23	15.12				
035429.5-202538	NGC 1481	.LA.-*.	1.01± .05	133					
214.00 -47.81	ESO 549- 32	SE	.18± .03	.07	14.44 ±.14				
283.31 -46.96		-3.3± .6		.00					
035217.0-203424	PGC 14079		.99						
035431.0+104214	IC 2002	.SB.1..	1.06± .05	0				14.93±.1	5227± 7
178.77 -31.69	UGC 2898	U	.05± .04	.66	14.66 ±.15			335± 8	
324.54 -38.10		1.0± .8		.06					5220
035146.5+103327	PGC 14080		1.12	.03	13.87			1.03	5096
035439.4-203009	NGC 1482	.LA.+P/	1.39± .03	103	13.10 ±.15	.95± .02	.99± .01		
214.12 -47.80	ESO 549- 33	PSE (1)	.25± .02	.07	13.14 ±.14	.05± .03	.15± .02		1765± 39
283.20 -47.00	IRAS03524-2038	-.8± .5	.98± .03	.00	9.48	.88	13.49± .09		1668
035227.0-203854	PGC 14084	5.0±1.5	1.36		13.03	.01	14.30± .22		1659
035447.2+061546			1.00?						7821
182.86 -34.51	CGCG 417- 10		.22?	.63	15.13 ±.13				
319.55 -40.27									7800
035207.3+060700	PGC 14088		1.06						7691
0354.8 +1822		.S?....	1.00± .16	117					7758
172.42 -26.39	UGC 2900		.55± .12	.87					
332.58 -34.07				.82					7774
0352.0 +1814	PGC 14091		1.08	.27					7626

R.A. 2000 DEC. / l b / SGL SGB / R.A. 1950 DEC.	Names / PGC	Type / S_T n_L / T / L	logD25 / logR25 / logAe / logDo	p.a. / Ag / Ai / A21	BT / mB / mFIR / BT°	(B-V)T / (U-B)T / (B-V)T° / (U-B)T°	(B-V)e / (U-B)e / m'e / m'25	m21 / W20 / W50 / HI	V21 / Vopt / VGSR / V3K
035456.5-655637		.S..3?.	1.25± .05	42					
279.75 -41.99	ESO 83- 12	S	.91± .04	.18	14.78 ±.14				
225.90 -31.64	IRAS03545-6605	3.0±1.9		1.26	13.42				
035431.0-660518	PGC 14093		1.27	.46					
0355.0 -4936		DE.1.P*	1.11± .10						
258.35 -48.47	ESO 201- 10	S	.27± .08	.00					
243.12 -40.34		-5.0±1.2		.00					
0353.6 -4945	PGC 14098		1.03						
035504.8-061320		.SXT8*.	1.14± .05	65					
195.89 -41.68	MCG -1-10- 47	E (1)	.20± .04	.24					
303.75 -44.93		8.3± .7		.24					
035237.3-062204	PGC 14100	7.5± .6	1.17	.10					
035508.8-172804		.SBT5*.	1.07± .04	3					
209.98 -46.67	ESO 549- 36	SE (2)	.16± .03	.04	14.46 ±.14				8488± 20
287.70 -46.97		5.0± .6		.23					8398
035253.0-173648	PGC 14102	4.9± .6	1.07	.08	14.14				8379
035521.5-564443	IC 2014	.SBS5*.	1.05± .05	17					
268.28 -46.09	ESO 156- 20	S (1)	.32± .04	.00	15.52 ±.14				10319± 88
235.07 -36.89		5.0± .6		.49					10153
035415.1-565324	PGC 14108	3.9± .6	1.05	.16	14.98				10281
035522.2-280936	IC 2007	.SBT4?.	1.13± .04	52					
225.44 -49.50	ESO 419- 11	Sr	.24± .04	.00	13.70 ±.14				
271.85 -46.79	IRAS03533-2818	4.1±1.0		.35	12.82				
035319.0-281818	PGC 14110		1.13	.12					
035538.8-513026		.S?....	1.01± .06	18					
261.04 -47.85	ESO 201- 11		.34± .05	.00	15.57 ±.14				13820± 88
240.85 -39.55				.52					13660
035417.0-513906	PGC 14116		1.01	.17	14.98				13771
035546.6-422202	NGC 1487	.P.....	1.52± .02	55	12.34M±.07	.44± .02	.40± .02	13.05±.3	856± 9
247.45 -49.76	ESO 249- 31	R	.19± .02	.00	11.98 ±.12	-.30± .04	-.33± .03	219± 10	733± 17
252.02 -43.40	VV 78	99.0	.91± .02	.15	11.79	.39	12.29± .04		684
035405.0-423042	PGC 14117		1.52	.10	12.11	-.34	14.33± .13	.85	762
0355.9 +7254	A 0350+72	.S..8P*	1.45± .03	33					
135.58 14.75	UGC 2890	PU	.72± .05	.90	14.3 ±.3			14.61±.1	1165± 11
14.32 3.33		7.5± .8		.89				149± 8	1318
0350.3 +7246	PGC 14123		1.54	.36	12.50			1.76	1093
035608.6-523235		.LBT+P.	1.11± .10						
262.47 -47.46		S	.17± .08	.00					11269± 88
239.62 -39.12		-1.0± .9		.00					11108
035449.6-524113	PGC 14127		1.09						11223
035613.3-554852		.LX.-*.	1.02± .11		15.08 ±.19	1.02± .03	1.03± .03		
266.97 -46.33		S	.14± .08	.00					
235.99 -37.48		-3.0± .9	.52± .08	.00			13.17± .26		
035504.0-555730	PGC 14131		1.00				14.68± .62		
035618.8-214915	NGC 1486	.S..4*P	.94± .05	2					
216.17 -47.83	ESO 549- 37	S	.22± .03	.09	15.21 ±.14				
281.24 -47.40		4.0±1.3		.32					
035408.0-215754	PGC 14132		.95	.11					
035624.0-535243		.E.2.*.	1.11± .10						
264.31 -46.99		S	.14± .08	.00					20490± 88
238.10 -38.49		-5.0±1.2		.00					20327
035508.9-540120	PGC 14135		1.07						20447
035633.3-660225	NGC 1503	.LBT+?.	.93± .06	140					
279.75 -41.80	ESO 83- 13	S	.06± .04	.18	14.40 ±.14				
225.69 -31.72	IRAS03561-6611	-1.0± .7		.00	12.47				
035609.0-661100	PGC 14137		.94						
035639.6-592342	IC 2017	.LA.0*P	1.03± .05	20					
271.64 -44.87	ESO 117- 15	S	.22± .03	.00	14.90 ±.14				
232.19 -35.60	IRAS03557-5932	-2.0± .8		.00	12.77				
035543.0-593218	PGC 14140		1.00						
0356.7 +3451		.S?....	1.02± .06						8452
160.66 -14.10	UGC 2901		.00± .05	1.03					
347.49 -24.07				.00					8517
0353.5 +3443	PGC 14142		1.11	.00					8329
035647.5-602537		.S?....	1.20± .05	2					
272.94 -44.41	ESO 117- 16		.50± .05	.00	15.15 ±.14				9328
231.13 -35.03				.76					9158
035555.0-603412	PGC 14143		1.20	.25	14.34				9299

R.A. 2000 DEC. l　b SGL　SGB R.A. 1950 DEC.	Names PGC	Type S_T　n_L T L	logD₂₅ logR₂₅ logA_e logD_o	p.a. A_g A_i A₂₁	B_T m_B m_FIR B°_T	(B-V)_T (U-B)_T (B-V)°_T (U-B)°_T	(B-V)_e (U-B)_e m'_e m'₂₅	m₂₁ W₂₀ W₅₀ HI	V₂₁ V_opt V_GSR V_3K
0356.8　+0805 181.54 -32.96 321.97 -39.87 0354.1　+0757	UGC　2903 PGC 14145	.S..0.. U .0± .9 	1.00± .08 .33± .06 1.02	120 .43 .25 	 15.20 ±.12 				
0357.0　+1628 174.35 -27.34 331.04 -35.57 0354.2　+1620	UGC　2904 PGC 14148	.S..3*. U　(1) 3.0±1.1 4.5±1.1	1.29± .06 .09± .06 1.39	 1.06 .13 .05	 14.8　±.3 13.60				 7925± 67 7934 7796
0357.0　+1630 174.33 -27.31 331.07 -35.55 0354.2　+1622	UGC　2905 PGC 14149	.I..9*. U 10.0±1.3 	.96± .07 .19± .05 1.06	153 1.06 .14 .09	 15.31 ±.14 14.10				 7886± 67 7895 7757
035709.1+340939 161.21 -14.57 347.00 -24.61 035356.7+340101	UGC　2902 PGC 14152	.L..... U -2.0± .8 	1.13± .10 .08± .05 1.22	15 .90 .00 	 14.9　±.3 				
035710.1-285236 226.63 -49.23 270.72 -47.09 035508.1-290112	ESO　419- 12 PGC 14154	.S..1*/ S 1.0±1.2 	1.26± .03 .79± .03 1.26	9 .00 .81 .40	 14.18 ±.14 13.32				 4114± 52 3995 4024
035710.3-184642 212.01 -46.68 285.80 -47.53 035456.1-185518	ESO　549- 40 IRAS03549-1855 PGC 14155	.S?.... 	1.15± .04 .42± .04 1.15	142 .02 .52 .21	 14.23 ±.14 12.84 13.62				 7593± 22 7499 7489
035713.5-252355 221.44 -48.55 275.88 -47.47 035507.0-253230	ESO　483- 2 PGC 14157	.SBS5.. S　(1) 5.0± .9 3.3± .9	1.13± .04 .32± .03 1.13	158 .00 .48 .16	 14.85 ±.14 				
035719.9-524126 262.60 -47.24 239.34 -39.20 035601.7-525000	 PGC 14158	.SBS5.. S　(1) 5.0± .8 5.6± .8	1.22± .07 .18± .08 1.22	 .00 .27 .09					
035720.4-521533 261.99 -47.37 239.83 -39.42 035600.9-522407	 PGC 14159	.S..1P. S 1.0± .8 	1.17± .08 .03± .08 1.17	 .00 .03 .01					 10001± 88 9839 9955
035727.8-461238 253.20 -48.86 246.99 -42.18 035554.1-462112	NGC　1493 ESO　249- 33 IRAS03558-4621 PGC 14163	.SBR6.. R　(2) 6.0± .7 4.8± .4	1.54± .03 .03± .04 1.24± .03 1.54	 .00 .05 .02	11.78 ±.16 11.79 ±.12 11.89 11.73	.51± .02 .50 	.54± .01 .18± .04 13.47± .07 14.27± .25	13.38±.2 136± 6 105± 6 1.63	1054± 7 1014± 41 900 995
035738.7-191302 212.67 -46.72 285.15 -47.66 035525.0-192136	NGC　1489 ESO　549- 42 PGC 14165	.SBR3.. SEr　(2) 3.4± .5 3.3± .7	1.14± .03 .39± .03 .76± .02 1.14	12 .04 .54 .20	14.61 ±.14 14.65 ±.14 13.96	.81± .04 .03± .06 .65 -.10	.89± .04 .13± .06 13.90± .06 14.18± .21		 11453± 60 11357 11350
035742.8-485427 257.16 -48.23 243.69 -41.05 035615.0-490300	NGC　1494 ESO　201- 12 IRAS03562-4902 PGC 14169	.SXS7*. RS　(2) 6.5± .5 4.0± .6	1.50± .03 .22± .04 1.50	179 .00 .30 .11	 12.27 ±.12 12.58 11.96			13.91±.3 203± 10 1.83	1123± 9 1094± 41 964 1069
035753.9-524652 262.68 -47.13 239.18 -39.23 035636.0-525524	IC　2018 ESO　156- 21 PGC 14173	RSXT4*. S　(1) 4.0± .9 4.4± .9	.90± .05 .16± .04 .90	140 .00 .23 .08	15.62 ±.17 15.76 ±.14 15.40	.84± .04 .74 	 14.59± .32		 10157± 88 9995 10113
035810.8-223404 217.42 -47.63 280.11 -47.82 035601.0-224236	ESO　483- 3 PGC 14180	.SB.7?. S 7.0±1.8 	.97± .05 .23± .03 .97	92 .05 .32 .11	 15.37 ±.14 				
0358.1　+1834 172.87 -25.69 333.32 -34.60 0355.3　+1826	NGC　1488 CGCG 466- 3 PGC 14181			 1.07 	 15.4　±.3 			15.91±.3 250± 7 	7011± 9 7026 6884
035813.8-352646 236.72 -49.61 261.18 -45.99 035621.1-353518	NGC　1492 ESO　359- 12 IRAS03563-3535 PGC 14186	.S..1?. S 1.0±1.6 	.98± .05 .12± .04 .98	10 .00 .12 .06	 14.29 ±.14 13.21 14.12				 4573± 29 4439 4495
035814.1-521942 262.03 -47.22 239.66 -39.50 035655.0-522812	NGC　1500 ESO　201- 13 PGC 14187	.LAR-*. S -4.0± .8 	1.04± .05 .07± .04 1.02	 .00 .00 	14.87 ±.18 14.44 ±.14 14.45	1.05± .03 .96 	 14.89± .34		 9976± 88 9814 9931

R.A. 2000 DEC. / l b / SGL SGB / R.A. 1950 DEC.	Names / PGC	Type / S_T n_L / T / L	$\log D_{25}$ / $\log R_{25}$ / $\log A_e$ / $\log D_o$	p.a. / A_g / A_i / A_{21}	B_T / m_B / m_{FIR} / B_T^o	$(B-V)_T$ / $(U-B)_T$ / $(B-V)_T^o$ / $(U-B)_T^o$	$(B-V)_e$ / $(U-B)_e$ / m'_e / m'_{25}	m_{21} / W_{20} / W_{50} / HI	V_{21} / V_{opt} / V_{GSR} / V_{3K}
035821.2-442759 / 250.55 -49.01 / 249.08 -43.04 / 035644.1-443630	NGC 1495 / ESO 249- 34 / IRAS03567-4436 / PGC 14190	.S..5?/ / BS / 5.0± .8	1.48± .04 / .74± .05 / .93± .04 / 1.48	105 / .00 / 1.12 / .37	13.33 ±.16 / 13.28 ±.14 / 12.43 / 12.18	.69± .02 / -.13± .03 / .54 / -.25	.64± .02 / -.04± .03 / 13.47± .11 / 13.72± .27		1245± 34 / 1094 / 1184
0358.3 +0643 / 183.11 -33.53 / 320.63 -40.85 / 0355.7 +0635	UGC 2910 / PGC 14191	.S..4.. / U / 4.0±1.0	.96± .09 / .69± .06 / 1.00	134 / .47 / 1.01 / .34					5508 / 5486 / 5383
035828.0-401342 / 244.09 -49.46 / 254.59 -44.61 / 035643.1-402212	ESO 302- 23A / PGC 14196	.L...P. / S / -2.0±1.2	1.00± .07 / .20± .06 / .97	.00 / .00	15.44 ±.14				
035831.4-613238 / 274.21 -43.74 / 229.86 -34.56 / 035744.2-614106	PGC 14199	.L..../ / S / -2.0±1.2	1.13± .09 / .17± .08 / 1.11	.03 / .00					
035831.5-810346 / 295.17 -33.03 / 212.83 -22.01 / 040209.0-811206	ESO 15- 5 / IRAS04021-8112 / PGC 14200	PSBR3*. / Sr (1) / 3.4± .5 / 3.3± .6	1.25± .05 / .08± .05 / 1.28	75 / .34 / .12 / .04	13.38 ±.14 / 12.65 / 12.89				4819± 19 / 4643 / 4837
035844.1+274608 / 165.98 -19.05 / 341.90 -29.07 / 035540.2+273737	MCG 5-10- 4 / PGC 14205	.S?....	.64± .17 / .28± .07 / .68	.45 / .43 / .14	15.79 ±.16 / 14.88			16.38±.3 / 328± 7 / 1.36	6761± 10 / 6804 / 6637
035856.3-455132 / 252.61 -48.67 / 247.28 -42.56 / 035722.0-460000	ESO 249- 35 / PGC 14212	.SB.6*/ / S / 6.3± .8	1.17± .05 / .95± .04 / 1.17	98 / .00 / 1.40 / .48	16.32 ±.14				
035854.5+102600 / 179.84 -31.05 / 324.93 -39.17 / 035610.2+101730	A 0356+10 / PGC 14213		.27± .03	.57	15.93 ±.14	1.17± .02 / .50± .05	1.18± .02 / .55± .05 / 12.77± .11		9130± 42 / 9120 / 9005
035858.1-590346 / 271.04 -44.74 / 232.32 -36.03 / 035800.7-591213	PGC 14214	PLXS0.. / S / -2.0± .9	1.03± .11 / .21± .08 / 1.00	.00 / .00					
0358.9 +4320 / 155.30 -7.44 / 354.32 -18.52 / 0355.5 +4311	UGC 2908 / IRAS03555+4311 / PGC 14215	.S?....	1.21± .05 / .32± .05 / 1.37	115 / 1.63 / .48 / .16				16.94±.1 / 200± 12	4646± 11 / 4733 / 4533
035859.4+793318 / 131.02 19.76 / 18.93 8.09 / 035117.9+792439	UGC 2896 / KUG 0351+794 / PGC 14216	.S..6*. / U / 6.0±1.3	1.10± .05 / .66± .06 / 1.15	.48 / .97 / .33	15.36 ±.18 / 13.90			15.37±.1 / 200± 8 / 1.14	2224± 11 / 2386 / 2166
035902.9-604026 / 273.09 -44.06 / 230.68 -35.12 / 035812.0-604853	PGC 14220	PLAS-*. / S / -3.0± .9	1.05± .11 / .06± .08 / 1.04	.00 / .00					14457± 15 / 14286 / 14430
0359.0 +0121 / 188.44 -36.66 / 314.20 -43.32 / 0356.4 +0113	UGC 2913 / IRAS03564+0113 / PGC 14221	.SBS1*. / U / 1.0±1.2	.95± .07 / .00± .05 / 1.01	.66 / .00 / .00				16.40±.3 / 147± 7 / 132± 7	3912± 11 / 3873 / 3791
035906.5-411625 / 245.68 -49.26 / 253.14 -44.37 / 035723.4-412453	PGC 14222	.LX.-*. / S / -3.0± .9	1.13± .09 / .26± .08 / 1.10	.00 / .00					
035914.3-421621 / 247.19 -49.14 / 251.82 -44.03 / 035733.1-422448	ESO 302- 24 / PGC 14224	.IBS9P* / S (1) / 9.5± .6 / 6.7± .6	1.03± .05 / .11± .05 / 1.03	78 / .00 / .08 / .06	14.70 ±.14 / 14.61				4485 / 4338 / 4421
035915.2-455215 / 252.61 -48.61 / 247.23 -42.61 / 035741.0-460042	ESO 249- 36 / PGC 14225	.IBS9.. / S (1) / 9.5± .5 / 10.0± .6	1.34± .05 / .04± .05 / 1.34	.00 / .03 / .02	15.19 ±.14 / 15.16			14.07±.3 / 104± 7 / -1.10	901± 9 / 748 / 844
0359.3 +0641 / 183.32 -33.36 / 320.74 -41.09 / 0356.7 +0633	UGC 2914 / PGC 14228	.LB.... / U / -2.0± .8	1.07± .07 / .15± .03 / 1.10	15 / .50 / .00	14.43 ±.10 / 13.85				5702 / 5679 / 5579

R.A. 2000 DEC. / l b / SGL SGB / R.A. 1950 DEC.	Names / PGC	Type / S_T n_L / T / L	logD_25 / logR_25 / logA_o / logD_o	p.a. / A_g / A_i / A_21	B_T / m_B / m_FIR / B_T^o	(B-V)_T / (U-B)_T / (B-V)_T^o / (U-B)_T^o	(B-V)_o / (U-B)_o / m'_o / m'_25	m_21 / W_20 / W_50 / HI	V_21 / V_opt / V_GSR / V_3K
0359.4 +4318 / 155.39 -7.40 / 354.36 -18.60 / 0355.9 +4310	/ UGC 2911 / IRAS03559+4310 / PGC 14232	.SB?...	1.00± .08 / .55± .06 / / 1.15	86 / 1.65 / .82 / .27	/ / 12.84 /			16.81±.1 / / 170± 8 /	5117± 11 / / 5204 / 5005
035930.2+214749 / 170.58 -23.21 / 336.63 -32.94 / 035633.5+213920	/ CGCG 487- 4 / / PGC 14235			/ / .63 /				17.10±.2 / / 104± 5 /	7510± 6 / / 7534 / 7385
035935.8-673807 / 281.37 -40.73 / 224.01 -30.97 / 035923.0-674630	NGC 1511 / ESO 55- 4 / IRAS03594-6746 / PGC 14236	.SA.1P* / RS / .7± .5 /	1.54± .03 / .46± .03 / .97± .02 / 1.56	125 / .21 / .47 / .23	11.88 ±.15 / 12.07 ±.10 / 9.75 / 11.31	.57± .02 / -.05± .03 / .42 / -.14	.59± .01 / -.04± .01 / 12.22± .07 / 13.29± .22	12.82±.3 / 273± 8 / / 1.28	1351± 7 / 1334± 48 / 1175 / 1339
0359.6 +4237 / 155.87 -7.90 / 353.88 -19.12 / 0356.2 +4229	/ UGC 2912 / / PGC 14237	.S..0.. / U / .0± .8 /	1.04± .08 / .04± .06 / / 1.18	/ 1.54 / .03 /				
035941.7-364222 / 238.67 -49.34 / 259.30 -45.94 / 035751.0-365048	/ ESO 359- 13 / / PGC 14239	.S?....	1.08± .05 / .34± .05 / / 1.08	150 / .00 / .35 / .17	/ 14.88 ±.14 / / 14.52				1437± 26 / 1300 / 1364 /
035948.2-840539 / 297.88 -31.06 / 210.49 -19.93 / 040645.1-841348	/ ESO 4- 10 / IRAS04068-8413 / PGC 14240	.LAR-P? / S / -3.7± .7 /	1.14± .05 / .27± .04 / .56± .04 / 1.14	135 / .57 / .00 /	14.07 ±.15 / 14.12 ±.14 / 13.65 / 13.45	1.01± .01 / .51± .02 / .84 / .41	1.02± .01 / .55± .02 / 12.36± .15 / 13.95± .32		4936 / 4760 / 4960 /
0359.8 +6707 / 139.77 10.64 / 10.71 -1.18 / 0355.0 +6659	/ UGCA 86 / / PGC 14241	.I..9?. / U / 10.0±1.8 /	.90± .22 / .06± .18 / / 1.21	/ 3.27 / .05 / .03				10.58±.0 / 180± 6 / 99± 8 /	67± 4 / / 209 / -12
035954.6+323640 / 162.72 -15.33 / 346.16 -26.08 / 035644.1+322813	/ UGC 2915 / / PGC 14246	.S..6*. / U / 6.0±1.4 /	1.04± .08 / .76± .06 / / 1.10	23 / .62 / 1.12 / .38				15.73±.3 / / 294± 7 /	5302± 10 / 5358 / 5182 /
0400.0 +0542 / 184.38 -33.84 / 319.68 -41.69 / 0357.4 +0534	/ UGC 2919 / / PGC 14248	.S..6*. / U / 6.0±1.2 /	1.00± .06 / .19± .05 / / 1.06	160 / .62 / .29 / .10					5388 / / 5362 / 5266
040003.9-532225 / 263.36 -46.64 / 238.30 -39.22 / 035848.0-533048	IC 2024 / ESO 156- 26 / IRAS03588-5330 / PGC 14249	.SBS6?/ / S (1) / 6.0± .9 / 4.4±1.3	1.05± .05 / .51± .04 / / 1.05	29 / .00 / .76 / .26	/ 15.54 ±.14 / 13.13 /				
040007.9-611702 / 273.77 -43.68 / 229.98 -34.87 / 035920.0-612524	/ ESO 117- 18 / IRAS03593-6125 / PGC 14251	.SBS4.. / S (1) / 4.0± .8 / 3.3± .9	1.13± .04 / .19± .04 / / 1.13	115 / .00 / .29 / .10	/ 14.17 ±.14 / 13.26 / 13.85				5783 / 5611 / 5758 /
0400.1 +0045 / 189.25 -36.78 / 313.58 -43.80 / 0357.6 +0037	/ UGC 2921 / / PGC 14252	.SXS8*. / UE (1) / 8.0± .7 / 8.7± .8	1.14± .05 / .10± .05 / / 1.20	95 / .69 / .12 / .05					3544 / 3503 / 3425 /
0400.2 +2253 / 169.88 -22.31 / 337.78 -32.41 / 0357.3 +2245	/ UGC 2918 / / PGC 14253	.S..8.. / U / 8.0± .8 /	1.10± .05 / .06± .05 / / 1.17	/ .66 / .07 / .03	/ 15.1 ±.2 / / 14.39			15.60±.3 / 228± 7 / 199± 7 / 1.18	5998± 8 / / 6025 / 5875
040017.1-613218 / 274.08 -43.55 / 229.72 -34.74 / 035930.3-614040	/ / / PGC 14254	.LA.-.. / S / -3.0± .9 /	1.03± .11 / .27± .08 / / .99	/ .00 / .00 /					
040019.4-674828 / 281.52 -40.58 / 223.80 -30.91 / 040008.0-675648	NGC 1511A / ESO 55- 5 / IRAS04001-6756 / PGC 14255	.SB.1?/ / RS / 1.3± .6 /	1.24± .04 / .67± .04 / / 1.27	110 / .27 / .68 / .33	/ 14.27 ±.14 / / 13.30				1358± 69 / 1182 / 1347 /
040022.0-523426 / 262.23 -46.84 / 239.17 -39.66 / 035904.0-524248	NGC 1506 / ESO 156- 27 / / PGC 14256	.LA.-.. / S / -3.0± .7 /	1.08± .06 / .15± .04 / / 1.06	80 / .00 / .00 /	/ 14.55 ±.14 / / 14.15				11367± 88 / 11204 / 11324 /
040023.7-530356 / 262.91 -46.68 / 238.62 -39.41 / 035907.1-531218	IC 2025 / ESO 156- 28 / / PGC 14257	.S..3*/ / S / 3.0±1.4 /	1.03± .05 / .74± .04 / / 1.03	120 / .00 / 1.02 / .37	16.16 ±.15 / 16.00 ±.14 / / 14.96	.78± .04 / / .56 /	/ / / 14.32± .30		12998± 88 / 12834 / 12956 /

R.A. 2000 DEC. / l b / SGL SGB / R.A. 1950 DEC.	Names / PGC	Type / S_T n_L / T / L	$\log D_{25}$ / $\log R_{25}$ / $\log A_e$ / $\log D_o$	p.a. / A_g / A_i / A_{21}	B_T / m_B / m_{FIR} / B_T^o	$(B-V)_T$ / $(U-B)_T$ / $(B-V)_T^o$ / $(U-B)_T^o$	$(B-V)_e$ / $(U-B)_e$ / m' / m'_{25}	m_{21} / W_{20} / W_{50} / HI	V_{21} / V_{opt} / V_{GSR} / V_{3K}
040024.4-613959 / 274.23 -43.48 / 229.58 -34.67 / 035938.2-614820	PGC 14258	.SBT4.. / S (1) / 4.0± .9 / 5.6±1.2	1.03± .09 / .14± .08 / / 1.03	/ .03 / .21 / .07					
040026.5-251055 / 221.37 -47.79 / 276.12 -48.21 / 035820.0-251918	ESO 483- 6 / IRAS03583-2519 / PGC 14259	PSBS3*/ / Sr / 3.2± .6	1.42± .03 / .76± .03 / / 1.42	139 / .00 / 1.05 / .38	14.13 ±.14 / / / 13.05				4154± 52 / 4041 / / 4063
040027.8+683440 / 138.83 11.76 / 11.71 -.14 / 035527.9+682611	NGC 1469 / UGC 2909 / PGC 14261	.LA.-*. / PU / -2.7± .7	1.28± .05 / .36± .03 / / 1.50	153 / 2.12 / .00	13.7 ±.3				
040029.0-490144 / 257.18 -47.76 / 243.27 -41.40 / 035902.0-491006	ESO 201- 14 / IRAS03590-4910 / PGC 14262	.L..0*. / S / -2.0±1.2	1.20± .05 / .42± .04 / / 1.14	163 / .00 / .00	13.71 ±.14 / / / 13.02				
0400.5 -5456 / 265.47 -46.06 / 236.53 -38.46 / 0359.4 -5504	PGC 14266			/ .00	16.32 ±.16	.31± .05			
040037.5-842201 / 298.11 -30.87 / 210.27 -19.75 / 040804.0-843006	ESO 4- 11 / PGC 14267	.SAS8*. / S (1) / 7.8± .8 / 7.2± .9	1.12± .04 / .29± .04 / / 1.17	70 / .57 / .36 / .15	15.00 ±.14 / / / 14.06				4860 / 4684 / / 4885
040041.5-524403 / 262.43 -46.74 / 238.96 -39.62 / 035924.0-525224	ESO 156- 29 / FAIR 762 / PGC 14270	.SAT6P* / S (1) / 5.7± .7 / 5.6± .8	1.09± .05 / .16± .05 / / 1.09	165 / .00 / .24 / .08	15.29 ±.14 / / / 15.03				3900±190 / 3736 / / 3858
040042.2-304956 / 229.74 -48.75 / 267.66 -47.53 / 035843.0-305818	ESO 419- 13 / PGC 14271	.L..+?P / S / -1.0±1.7	1.13± .04 / .40± .03 / / 1.07	2 / .00 / .00	14.28 ±.14				
040042.9-465056 / 253.99 -48.18 / 245.88 -42.42 / 035911.0-465918	ESO 250- 1 / PGC 14272	.SX.7?P / S (1) / 7.0±1.3 / 7.8±1.3	1.02± .05 / .43± .04 / / 1.02	108 / .00 / .59 / .21	15.70 ±.14				
0400.7 +1734 / 174.14 -25.93 / 332.76 -35.67 / 0357.8 +1726	UGC 2922 / IRAS03578+1726 / PGC 14274	.S..3.. / U (1) / 3.0± .9 / 3.5±1.2	1.02± .06 / .20± .05 / / 1.12	62 / 1.12 / .27 / .10	/ / / 13.79				5215 / 5225 / / 5092
040047.9+350044 / 161.19 -13.43 / 348.23 -24.59 / 035733.8+345220	UGC 2920 / PGC 14276	.S..6*. / U / 6.0±1.2	1.36± .04 / .82± .05 / / 1.46	63 / 1.13 / 1.20 / .41	14.7 ±.3 / / / 12.37				4158 / / 4221 / 4040
040053.2-524940 / 262.55 -46.68 / 238.83 -39.60 / 035936.0-525800	PGC 14278	RSBS3?P / S / 3.0± .9	1.10± .08 / .17± .08 / / 1.10	/ .00 / .23 / .08					
040054.6-673642 / 281.26 -40.63 / 223.94 -31.08 / 040042.1-674500	NGC 1511B / ESO 55- 6 / PGC 14279	.SB.7?/ / RS (1) / 7.1± .7 / 7.8±1.4	1.24± .04 / .80± .04 / / 1.26	98 / .21 / 1.10 / .40	15.17 ±.14				
0400.9 -1014 / 201.57 -42.38 / 298.61 -47.33 / 0358.6 -1023	MCG -2-11- 5 / PGC 14282	.SB?... / (1) / 1.8± .9	.99± .10 / .09± .07 / / 1.00	/ .13 / .14 / .05				15.62±.1 / 275± 15	11060± 10 / 10985± 59 / 10984 / 10949
040100.8-525910 / 262.76 -46.62 / 238.64 -39.54 / 035944.1-530730	ESO 156- 31 / PGC 14285	PS..0*. / S / .0±1.3	1.02± .05 / .31± .05 / / 1.00	137 / .00 / .23	15.13 ±.14 / / / 14.74				10467± 88 / 10303 / / 10426
040101.6+740458 / 135.05 15.86 / 15.36 3.95 / 035506.9+735630	UGC 2906 / IRAS03551+7356 / PGC 14286	.S..3*. / U / 3.0±1.0	1.44± .03 / .20± .05 / / 1.53	5 / .88 / .28 / .10	13.8 ±.3 / / 13.10 / 12.60			13.80±.1 / / 446± 8 / 1.11	2494± 11 / / 2647 / 2426
040102.3-611321 / 273.63 -43.60 / 229.96 -34.99 / 040014.3-612140	PGC 14287	PSX.1*P / S / 1.0±1.2	1.13± .08 / .24± .08 / / 1.13	/ .00 / .24 / .12					

R.A. 2000 DEC. / l b / SGL SGB / R.A. 1950 DEC.	Names / / / PGC	Type / S_T n_L / T / • L	$\log D_{25}$ / $\log R_{25}$ / $\log A_o$ / $\log D_o$	p.a. / A_g / A_i / A_{21}	B_T / m_B / m_{FIR} / B_T^o	$(B-V)_T$ / $(U-B)_T$ / $(B-V)_T^o$ / $(U-B)_T^o$	$(B-V)_o$ / $(U-B)_o$ / m'_o / m'_{25}	m_{21} / W_{20} / W_{50} / HI	V_{21} / V_{opt} / V_{GSR} / V_{3K}
0401.0 -0042		.SBT8?.	1.07± .05	145					4267
190.92 -37.44	UGC 2925	UE (1)	.08± .04	.65					
311.80 -44.56		8.3±1.0		.09					4221
0358.5 -0051	PGC 14288	8.7±1.2	1.13	.04					4150
0401.1 +0533		.SB.3..	.97± .07	125					9858
184.73 -33.71	UGC 2926	U	.16± .05	.65	15.13 ±.10				
319.67 -42.01	IRAS03585+0524	3.0± .9		.22	13.33				9831
0358.5 +0524	PGC 14293		1.03	.08	14.18				9738
040118.5-524229	IC 2028	.S..6*.	.91± .05	55	15.06 ±.15	.86± .03			
262.35 -46.66	ESO 156- 32	S (1)	.14± .04	.00	14.94 ±.14				9767± 88
238.92 -39.72		6.0±1.3		.21		.77			9603
040001.1-525048	PGC 14299	5.6± .9	.91	.07	14.75		14.09± .32		9725
040120.9+264932		.S?....	.91± .07	150				16.85±.3	9405± 10
167.12 -19.33	UGC 2924		.22± .05	.40					
341.50 -30.12				.33				333± 7	9443
035818.0+264110	PGC 14301		.95	.11					9284
040126.9-544529		.LA.-*.	1.02± .11						
265.16 -46.00		S	.27± .08	.00					
236.64 -38.67		-3.0±1.3		.00					
040015.4-545347	PGC 14303		.98						
040127.0+231454		.S?....	.89± .11					15.99±.3	7243± 10
169.81 -21.87	MCG 4-10- 4		.29± .07	.69	15.34 ±.12				7270
338.31 -32.40				.44				220± 7	
035828.4+230633	PGC 14304		.95	.15	14.16			1.68	7121
040139.9-590903		.S..2*/	1.09± .09						
270.96 -44.38		S	.45± .08	.00					
231.99 -36.26		2.0±1.3		.55					
040043.6-591720	PGC 14313		1.09	.23					
0401.6 +2306		RSBS1..	1.27± .05	80					6257
169.96 -21.93	UGC 2927	U	.19± .05	.69	14.6 ±.2				
338.21 -32.53		1.0± .8		.19					6284
0358.7 +2258	PGC 14314		1.33	.09	13.68				6135
0401.6 +2312		PSXS1..	.97± .07	140					7521
169.89 -21.86	UGC 2928	U	.19± .05	.69	15.24 ±.13				
338.31 -32.47		1.0± .9		.19					7548
0358.7 +2304	PGC 14315		1.04	.09	14.26				7400
040151.5-415631		.S?....	1.12± .06	25					4427
246.62 -48.69	ESO 302- 27		.45± .06	.00	15.13 ±.14				4279
252.00 -44.60				.68					4365
040010.0-420448	PGC 14319		1.12	.23	14.42				
0402.0 -1625		.S?....	1.08± .09						7211
209.47 -44.74	MCG -3-11- 7		.70± .07	.01					7119
289.46 -48.53				.86					7110
0359.8 -1634	PGC 14330		1.08	.35					
040207.1+230759	NGC 1497	.L.....	1.26± .07	60					6170±141
170.02 -21.84	UGC 2929	U	.17± .04	.69	14.07 ±.13				6196
338.31 -32.59		-2.0± .8		.00					6049
035908.6+225941	PGC 14331		1.31		13.29				
0402.3 +2549		.SAS5..	1.00± .08	15					5747
168.04 -19.89	UGC 2931	U	.04± .06	.47	14.58 ±.13				
340.79 -30.94	IRAS03593+2540	5.0± .8		.06					5782
0359.3 +2540	PGC 14333		1.05	.02	14.02				5627
040226.5+264930								17.62±.3	5640± 10
167.30 -19.16	CGCG 487- 12			.39	15.6 ±.4				5678
341.68 -30.31								84± 7	
035923.6+264113	PGC 14335								5521
040232.6-621853		.SB.4?/	1.29± .04	78					5335
274.88 -42.97	ESO 117- 19	S (1)	.77± .04	.04	14.62 ±.14				5162
228.77 -34.49	IRAS04018-6227	4.3± .5		1.13	13.26				5314
040150.1-622706	PGC 14337	3.3±1.3	1.29	.38	13.41				
0402.6 +7142		.S..2..	1.24± .11					13.87±.1	4517± 11
136.81 14.20	UGC 2916	U	.02± .12	1.02	14.2 ±.3				
13.90 2.09	IRAS03572+7134	2.0± .8		.02	12.53			303± 8	4666
0357.2 +7134	PGC 14341		1.34	.01	13.12			.74	4446
0402.7 +7816		.S?....	1.32± .06					15.03±.1	2171± 11
132.09 18.97	UGC 2907		.23± .07	.62	14.7 ±.3				2151± 39
18.21 7.02	7ZW 10			.34				97± 8	2329
0355.5 +7808	PGC 14343		1.38	.11	13.70			1.21	2110

R.A. 2000 DEC. l b SGL SGB R.A. 1950 DEC.	Names PGC	Type S_T n_L T L	$\log D_{25}$ $\log R_{25}$ $\log A_e$ $\log D_o$	p.a. A_g A_i A_{21}	B_T m_B m_{FIR} B_T^o	$(B-V)_T$ $(U-B)_T$ $(B-V)_T^o$ $(U-B)_T^o$	$(B-V)_e$ $(U-B)_e$ m'_e m'_{25}	m_{21} W_{20} W_{50} HI	V_{21} V_{opt} V_{GSR} V_{3K}
040248.1+015748 188.51 -35.55 315.48 -43.93 040012.4+014933	 UGC 2936 IRAS04002+0149 PGC 14345	.SBS7.. UE (1) 7.0± .6 6.4± .8	1.40± .03 .57± .04 1.47	30 .69 .79 .29	15.0 ±.2 11.27 13.50			14.77±.1 492± 8 .98	3813± 7 3812± 76 3774 3697
0402.9 +3352 162.32 -13.98 347.65 -25.68 0359.7 +3344	 UGC 2932 PGC 14348	.S..8*. U 8.0±1.2	1.00± .16 .04± .12 1.10	 1.07 .05 .02					4886 4944 4770
040255.7+212848 171.43 -22.86 336.91 -33.76 035959.1+212033	 MCG 4-10- 14 PGC 14349	.S?.... 	.94± .11 .57± .07 1.02	 .85 .84 .28					5954 5975 5834
040301.1+310323 164.33 -16.02 345.37 -27.61 035952.4+305508	 UGC 2934 PGC 14353	.S?.... 	1.14± .13 .69± .12 1.19	38 .56 1.03 .34				15.45±.3 312± 7 	5200± 10 5250 5083
040303.3+043239 186.03 -33.96 318.73 -42.88 040025.0+042424	 UGC 2938 PGC 14355	.SB.2.. U 2.0± .9	1.06± .06 .31± .05 1.11	175 .60 .38 .15	15.26 ±.15 14.23			15.93±.3 393± 7 1.54	5512± 10 5480 5395
040303.7-524428 262.29 -46.40 238.71 -39.93 040146.6-525240	 PGC 14357	.LA.-*. S -3.0±1.3	1.02± .11 .17± .08 .99	 .00 .00 					10615± 88 10451 10575
040304.2-360823 237.82 -48.65 259.81 -46.75 040113.0-361636	 ESO 359- 14 PGC 14359	.LAR0*P S -2.0± .9	.97± .06 .20± .04 .94	172 .00 .00 	15.24 ±.14				
040306.5-531848 263.08 -46.22 238.07 -39.63 040151.1-532700	 PGC 14360	.LAS0*. S -2.0± .9	1.02± .11 .27± .08 .98	 .00 .00 					18972± 88 18807 18933
040320.3+270104 167.31 -18.88 342.90 -30.33 040017.0+265250	 CGCG 487- 14 PGC 14369	 	 	 .39 	15.7 ±.4			16.43±.3 356± 7 	14602± 10 14640 14484
0403.3 +7143 136.84 14.25 13.95 2.06 0357.9 +7135	 CGCG 327- 13 PGC 14370	 	.91± .09 .64± .06 1.01	 1.02 	15.6 ±.2			14.35±.1 169± 12 	4509± 11 4658 4438
040326.5-514943 260.99 -46.59 239.70 -40.44 040207.0-515754	 ESO 201- 16 PGC 14371	.SBS5?. S (1) 5.0± .9 1.1±1.3	1.04± .05 .24± .05 1.04	30 .00 .36 .12	15.08 ±.14 14.65				11100 10936 11059
040330.5+262137 167.83 -19.33 341.45 -30.79 040028.1+261323	 MCG 4-10- 15 PGC 14374	.SB?... 	.92± .11 .32± .07 .96	 .45 .47 .16	14.94 ±.13 13.97			16.24±.3 406± 7 2.11	7064± 10 7100 6946
040332.6-432401 248.76 -48.23 249.93 -44.32 040154.1-433212	NGC 1510 ESO 250- 3 IRAS04019-4332 PGC 14375	.LA.0P? BS -2.3± .7	1.12± .03 .26± .02 .48± .01 1.08	90 .00 .00 	13.47M±.11 13.47 ±.14 13.42 13.46	.45± .01 -.19± .02 .42 -.20	.32± .01 -.24± .02 11.31± .04 13.29± .18		989± 23 838 932
040334.6-171201 210.67 -44.70 288.33 -48.95 040118.9-172013	 PGC 14377	.IBS9.. SE (2) 10.0± .6 9.9± .6	1.24± .05 .12± .06 1.24	155 .00 .09 .06					1890 1796 1793
0403.7 +1953 172.82 -23.84 335.53 -34.88 0400.8 +1945	IC 358 UGC 2940 PGC 14382	.L..... U -2.0± .9	1.01± .08 .45± .03 1.05	64 .99 .00 	15.07 ±.10 13.98				6770 6786 6651
040344.5+220940 171.05 -22.25 337.69 -33.49 040047.1+220127	IC 357 UGC 2941 IRAS04007+2201 PGC 14384	.SBS2.. U 2.0± .8	1.09± .06 .09± .05 1.17	175 .80 .12 .05	14.13 ±.12 11.44 13.15			15.36±.2 156± 7 145± 5 2.17	6261± 6 6240±141 6284 6142
0403.7 +4643 153.70 -4.34 357.38 -16.68 0400.2 +4635	 UGC 2937 PGC 14386	.S..0.. U .0± .9	1.11± .05 .65± .05 1.35	175 2.92 .49 					

R.A. 2000 DEC. l b SGL SGB R.A. 1950 DEC.	Names PGC	Type S_T n_L T L	logD_25 logR_25 logA_e logD_o	p.a. A_g A_i A_21	B_T m_B m_FIR B_T^o	(B-V)_T (U-B)_T (B-V)_T^o (U-B)_T^o	(B-V)_e (U-B)_e m'_e m'_25	m_21 W_20 W_50 HI	V_21 V_opt V_GSR V_3K
040351.0-540657 264.13 -45.87 237.10 -39.30 040238.1-541506	NGC 1515A ESO 156- 34 FAIR 397 PGC 14388	.SBR3$. R (1) 3.0± .5 3.3± .6	1.01± .04 .08± .04 1.01	124 .00 .11 .04	 15.44 ±.14 15.23				13265± 69 13099 13229
040352.6+242345 169.36 -20.66 339.76 -32.11 040052.5+241533	 MCG 4-10- 17 PGC 14390	.S?.... 	.74± .14 .09± .07 .79	 .53 .14 .05	 15.55 ±.12 14.85			16.11±.3 555± 7 1.22	6032± 10 6061 5914
040354.6-432103 248.67 -48.16 249.95 -44.40 040216.1-432912	NGC 1512 ESO 250- 4 IRAS04022-4329 PGC 14391	.SBR1.. R (1) 1.0± .5 1.1± .7	1.95± .02 .20± .03 1.38± .01 1.95	90 .00 .20 .10	11.13M±.10 10.97 ±.11 11.56 10.84	.81± .01 .17± .02 .77 .14	.87± .01 .27± .01 13.44± .03 15.23± .15	11.68±.1 244± 7 234± 7 .74	901± 6 735± 22 740 834
040359.8+461334 154.06 -4.68 357.06 -17.07 040025.4+460521	 UGC 2939 IRAS04004+4605 PGC 14395	.S..8*. U 8.0±1.2 	1.11± .14 .15± .12 1.36	90 2.71 .18 .07	 12.01 			15.29±.1 279± 11 199± 6 	4436± 6 4529 4332
040403.0-540610 264.10 -45.85 237.10 -39.34 040250.0-541418	NGC 1515 ESO 156- 36 IRAS04028-5414 PGC 14397	.SXS4.. R (1) 4.0± .3 3.4± .5	1.72± .02 .67± .02 1.00± .01 1.72	18 .00 .99 .34	12.05M±.12 11.98 ±.11 11.80 11.02	.85± .01 .30± .02 .72 .18	.93± .01 .35± .02 12.55± .12 13.84± .16	14.40±.3 366± 12 352± 9 3.04	1169± 10 1216± 48 1004 1134
040410.1+220921 171.13 -22.19 337.76 -33.57 040112.7+220110	 UGC 2942 PGC 14398	.S?.... 	1.12± .08 .08± .07 1.19	 .80 .11 .04				14.99±.3 353± 7 	6361± 10 6383 6243
040410.6-822028 296.20 -32.07 211.70 -21.26 040858.0-822824	 ESO 15- 6 PGC 14399	PSAR1*. Sr .7± .5 	1.14± .04 .35± .04 1.18	73 .40 .36 .18	 14.95 ±.14 14.13				4893 4716 4914
040421.9-475659 255.41 -47.36 244.15 -42.47 040253.0-480506	 ESO 201- 17 PGC 14404	PSXR1P* Sr 1.2± .5 	1.10± .05 .31± .04 1.10	4 .00 .32 .16	 14.19 ±.14 				
040425.0-361052 237.89 -48.38 259.63 -47.00 040234.0-361900	 ESO 359- 16 PGC 14407	.IB.9?/ S (1) 10.0±1.8 5.6± .9	1.18± .04 .60± .04 1.18	54 .00 .45 .30	 15.00 ±.14 14.55				1408 1270 1340
040427.3-021117 193.06 -37.56 310.28 -45.89 040155.9-021926	NGC 1507 UGC 2947 MK 1080 PGC 14409	.SBS9P$ R (1) 9.0± .5 5.3±1.2	1.56± .02 .63± .02 1.01± .02 1.59	11 .40 .64 .31	12.89 ±.15 12.87 ±.10 12.55 11.84	.57± .02 -.07± .03 .34 -.24	.59± .01 -.03± .02 13.43± .06 13.95± .18	13.40±.0 191± 6 167± 12 1.25	856± 4 863± 28 803 745
040429.9-175422 211.71 -44.75 287.27 -49.22 040215.0-180230	 ESO 550- 2 PGC 14412	.S?.... 	1.14± .04 .72± .04 1.14	45 .00 1.00 .36	 15.34 ±.14 14.28				8115± 60 8018 8020
040431.8-475829 255.44 -47.33 244.11 -42.48 040303.0-480636	 ESO 201- 18 PGC 14413	.SXS8.. S (1) 8.0± .6 7.8± .6	1.00± .05 .05± .05 1.00	125 .00 .07 .03	 15.42 ±.14 				
040433.6+255815 168.30 -19.44 341.28 -31.22 040131.6+255006	 UGC 2946 PGC 14415	.SB.8*. U 8.0±1.3 	1.00± .08 .43± .06 1.04	3 .47 .53 .22				16.59±.3 192± 7 	7080± 10 7114 6963
040435.1-460235 252.62 -47.67 246.47 -43.37 040302.1-461042	 ESO 250- 5 PGC 14416	.LBS0?. S -2.0± .9 	1.11± .06 .18± .04 1.08	85 .00 .00 	 14.30 ±.14 14.28				1289± 34 1133 1238
040437.1+331716 163.00 -14.16 347.45 -26.34 040125.2+330907	 UGC 2944 PGC 14418	.S?.... 	1.19± .12 .15± .12 1.28	105 1.01 .22 .07				15.38±.3 283± 7 	5593± 10 5649 5479
040439.5+334829 162.64 -13.78 347.87 -25.99 040126.8+334020	 UGC 2945 IRAS04014+3340 PGC 14420	.S..0.. U .0± .8 	.97± .09 .09± .06 1.07	60 1.10 .07 	 14.7 ±.3 12.60 				
040442.4-620101 274.37 -42.87 228.87 -34.87 040359.0-620906	 ESO 117- 21 PGC 14423	.SBS5.. S (1) 5.0± .9 5.6± .9	.98± .05 .10± .05 .98	 .03 .16 .05	 15.49 ±.14 				

R.A. 2000 DEC.	Names	Type	$\log D_{25}$	p.a.	B_T	$(B-V)_T$	$(B-V)_e$	m_{21}	V_{21}
l b		S_T n_L	$\log R_{25}$	A_g	m_B	$(U-B)_T$	$(U-B)_e$	W_{20}	V_{opt}
SGL SGB		T	$\log A_e$	A_i	m_{FIR}	$(B-V)_T^0$	m'_e	W_{50}	V_{GSR}
R.A. 1950 DEC.	PGC	L	$\log D_0$	A_{21}	B_T^0	$(U-B)_T^0$	m'_{25}	HI	V_{3K}
040455.4+204948		.S..2..	1.11± .07	167				16.76±.3	5204± 10
172.29 -22.98	UGC 2948	U	.54± .06	.87	15.34 ±.13				5222
336.64 -34.53		2.0± .9		.66				436± 7	5087
040159.5+204140	PGC 14427		1.19	.27	13.76			2.73	
040502.5+251550		.SB.2*.	1.00± .06	119				15.76±.3	7166± 9
168.91 -19.86	UGC 2949	U	.26± .05	.50	14.57 ±.12				7197
340.74 -31.76	IRAS04020+2507	2.0± .9		.32	12.67			383± 5	7050
040201.2+250743	PGC 14431		1.05	.13	13.69			1.95	
040504.5+705952	NGC 1485	.SA.3$/	1.33± .04	22				14.79±.1	1083± 11
137.46 13.81	UGC 2933	P	.49± .05	.99	13.4 ±.3				1231
13.58 1.42	IRAS03598+7051	3.0± .8		.67	13.46			297± 8	1012
035942.4+705140	PGC 14432		1.42	.24	11.71			2.84	
040504.7-350025		.LB.0?/	1.11± .05	13					
236.14 -48.20	ESO 359- 18	S	.59± .03	.00	15.88 ±.14				
261.25 -47.46		-2.0±1.7		.00					
040312.0-350830	PGC 14433		1.02						
040506.2+310138		.S?....	1.22± .12	167				14.34±.2	4954± 6
164.68 -15.73	UGC 2950		.76± .12	.51					5003
345.69 -27.96				1.15				447± 5	4840
040157.4+305330	PGC 14435		1.27	.38					
0405.1 +7951		.S..9*.	1.00± .16					15.60±.1	2202± 7
131.00 20.17	UGC 2917	U	.09± .12	.45				226± 6	
19.33 8.13		9.0±1.2		.09				159± 4	2363
0357.2 +7943	PGC 14436		1.04	.04					2146
040511.8-655028	NGC 1526	.S..4..	.92± .05	36					
278.94 -41.12	ESO 84- 3	S (1)	.18± .04	.13	14.59 ±.14				
225.23 -32.53	IRAS04048-6558	4.0± .8		.26	13.20				
040449.0-655830	PGC 14437	2.2±1.3	.93	.09					
040530.4+042443		.S?....	.82± .13						5325± 37
186.60 -33.55	MCG 1-11- 10		.00± .07	.59	15.15 ±.12				5291
318.94 -43.49	MK 1081			.00	13.11				5212
040252.2+041638	PGC 14448		.87		14.47				
040531.0+042649		.S..6*.	.95± .07	15				15.84±.3	5345± 10
186.57 -33.53	UGC 2954	U	.39± .05	.56	15.23 ±.12				5214± 57
318.98 -43.47		6.0±1.3		.58				275± 7	5308
040252.7+041844	PGC 14449		1.00	.20	14.06			1.58	5228
040542.6+252959								16.93±.3	6880± 10
168.84 -19.59	CGCG 487- 20			.44	15.3 ±.4				6912
341.06 -31.72								193± 7	6764
040241.0+252154	PGC 14451								
0405.7 +7839		.SXS3..	1.00± .08	64					
131.92 19.35	UGC 2923	U	.33± .06	.51					
18.57 7.21	IRAS03584+7831	3.0± .9		.46	13.77				
0358.4 +7831	PGC 14453		1.05	.17					
040559.9-174633		.SBS9./	1.36± .03	58				14.72±.1	1886± 6
211.71 -44.36	ESO 550- 5	SUE (1)	.55± .03	.00	14.73 ±.14			149± 8	
287.51 -49.57		9.1± .4		.56				144± 12	1788
040345.0-175436	PGC 14458	8.9± .6	1.36	.28	14.16			.29	1792
040602.3-223752		.SBS9..	1.01± .05	46					
218.24 -45.90	ESO 483- 8	S (1)	.40± .05	.06	16.08 ±.14				
279.93 -49.64		9.0± .9		.41					
040353.0-224554	PGC 14460	8.9± .9	1.02	.20					
040607.7-524012	NGC 1522	PL..0*P	1.08± .03	42	13.93 ±.13	.37± .01	.32± .01		
262.00 -45.97	ESO 156- 38	S	.19± .04	.00	14.08 ±.14	-.34± .03			1012± 91
238.47 -40.36	FAIR 301	-2.3± .7	.48± .02	.00	13.30	.34	11.82± .07		846
040451.0-524812	PGC 14462		1.05		13.98	-.35	13.72± .23		975
040615.5-083812		.IBS9P.	1.18± .05	80					
200.49 -40.48	MCG -1-11- 2	E (1)	.24± .05	.13					
301.39 -48.23		10.0± .6		.18					
040350.7-084613	PGC 14464	7.0± .8	1.19	.12					
040634.1+311655		.S..6*.	1.14± .07	124				16.28±.3	4548± 10
164.74 -15.33	UGC 2956	U	.49± .06	.58	15.3 ±.3				4597
346.13 -28.02		6.0±1.2		.72				323± 7	4436
040324.9+310853	PGC 14468		1.19	.25	13.95			2.08	
040636.7-575744	IC 2034	.S..5*/	1.08± .05	118					
269.08 -44.23	ESO 117- 22	S	.78± .04	.00	15.89 ±.14				
232.74 -37.47		5.0±1.1		1.17					
040537.0-580542	PGC 14469		1.08	.39					

R.A. 2000 DEC. l b SGL SGB R.A. 1950 DEC.	Names PGC	Type S_T n_L T L	$\log D_{25}$ $\log R_{25}$ $\log A_\bullet$ $\log D_\circ$	p.a. A_g A_i A_{21}	B_T m_B m_{FIR} B_T°	$(B-V)_T$ $(U-B)_T$ $(B-V)_T^\circ$ $(U-B)_T^\circ$	$(B-V)_\bullet$ $(U-B)_\bullet$ $m'_\bullet$ m'_{25}	m_{21} W_{20} W_{50} HI	V_{21} V_{opt} V_{GSR} V_{3K}
040650.0-211043 216.33 -45.30 282.20 -49.85 040439.0-211842	NGC 1518 ESO 550- 7 IRAS04046-2118 PGC 14475	.SBS8.. R (2) 8.0± .4 5.6± .6	1.48± .03 .35± .03 1.48	35 .04 .43 .17	12.28 ±.13 12.17 ±.12 12.05 11.76	.47± .02 -.27± .02 .39 -.33	 13.66± .20	12.83±.2 181± 8 137± 25 .90	927± 7 983± 42 822 840
040653.1+225147 171.05 -21.24 338.88 -33.62 040354.7+224347	UGC 2958 PGC 14478	.S..3.. U 3.0± .9	1.11± .05 .65± .05 1.18	20 .77 .90 .33	 14.98 ±.12 13.26			15.31±.3 480± 7 1.72	6221± 10 6244 6107
040703.6-551933 265.56 -45.06 235.46 -39.03 040555.0-552730	IC 2032 ESO 156- 42 PGC 14481	.IXS9P* S (1) 10.0± .6 8.1± .7	1.13± .05 .21± .04 1.13	78 .00 .16 .11	 14.78 ±.14 14.62				1070 901 1039
040703.8-623824 274.98 -42.36 228.06 -34.70 040624.1-624620	PGC 14482	.SBS3*. S 3.0± .9	1.05± .09 .03± .08 1.05	 .03 .04 .01					
040707.1+342603 162.57 -12.97 348.75 -25.92 040353.3+341803	UGC 2959 PGC 14483	.S?.... 	1.04± .06 .04± .05 1.17	 1.45 .06 .02				15.81±.3 78± 7	5546± 10 5604 5436
040713.0-171216 211.12 -43.89 288.43 -49.82 040457.5-172014	MCG -3-11- 12 PGC 14487	.SXS8.. E (1) 8.0± .8 8.9± .8	1.28± .03 .00± .04 1.29	 .01 .01 .00				15.00±.1 91± 7 82± 12	1859± 6 1763 1767
040713.3-623258 274.86 -42.38 228.14 -34.77 040633.1-624053	PGC 14488	.L..-*. S -3.0±1.3	1.01± .11 .23± .08 .98	 .03 .00					
040714.6-534058 263.32 -45.53 237.23 -39.96 040601.1-534854	IC 2033 ESO 156- 43 PGC 14491	.SXT1?. RS 1.1± .6	1.07± .05 .29± .04 1.07	129 .00 .30 .15	 15.01 ±.14 14.54				14067± 88 13900 14033
040718.1+243649 169.78 -19.95 340.54 -32.57 040417.6+242850	UGC 2960 PGC 14493	.I..9?. U 10.0±1.7	1.00± .08 .13± .06 1.06	5 .57 .10 .07				16.73±.3 234± 7	5343± 10 5371 5229
040718.5+264223 168.21 -18.48 342.40 -31.21 040415.4+263424	UGC 2961 PGC 14494	.I..9*. U 10.0±1.1	1.14± .13 .03± .12 1.17	 .33 .02 .01				16.96±.3 105± 7	3909± 10 3943 3796
040719.4-625400 275.28 -42.23 227.79 -34.56 040641.0-630154	NGC 1529 ESO 84- 4 PGC 14495	.L..0*/ S -2.0± .8	1.08± .05 .66± .03 .98	164 .04 .00	 14.39 ±.14				
040736.7-212546 216.75 -45.21 281.80 -50.03 040526.1-213342	ESO 550- 8 PGC 14500	.E?.... 	.88± .06 .15± .03 .84	77 .05 .00	14.70 ±.14 14.70 ±.14	.52± .03	 13.70± .34		
040738.9+035810 187.41 -33.39 318.72 -44.16 040501.1+035013	UGC 2963 IRAS04050+0350 PGC 14502	.S..7.. U 7.0± .8	1.30± .05 .51± .06 1.36	145 .61 .71 .26	 14.78 ±.17 12.28 13.45			14.38±.3 421± 7 .68	5298± 7 5262 5188
040742.1-224252 218.51 -45.56 279.78 -50.02 040533.0-225048	ESO 483- 9 PGC 14503	.LX.0P. S -2.0± .8	1.19± .04 .42± .03 1.13	147 .06 .00	 14.82 ±.14				
040742.6+254634 168.97 -19.07 341.65 -31.89 040440.5+253837	UGC 2962 PGC 14504	.I..9?. U 10.0±2.0	1.14± .13 1.16± .06 1.18	19 .41 .75 .50				15.79±.3 313± 7	5415± 10 5446 5302
040745.9-295135 228.62 -47.10 268.68 -49.20 040546.0-295930	ESO 420- 3 IRAS04057-2959 PGC 14505	.SAT5.. Sr (1) 4.7± .6 2.2± .8	1.30± .03 .17± .03 1.30	146 .00 .25 .08	 13.52 ±.14 13.20 13.25				4107± 52 3979 4032
040746.7-450224 251.03 -47.27 247.38 -44.31 040612.0-451018	ESO 250- 6 PGC 14506	.S?.... 	.99± .06 .71± .05 .58± .06 .99	142 .00 1.07 .36	15.9 ±.2 16.03 ±.14	.68± .05	.76± .05 14.33± .17 13.96± .38		

R.A. 2000 DEC. / l b / SGL SGB / R.A. 1950 DEC.	Names / PGC	Type / S_T n_L / T / L	$\log D_{25}$ / $\log R_{25}$ / $\log A_e$ / $\log D_o$	p.a. / A_g / A_i / A_{21}	B_T / m_B / m_{FIR} / B_T^o	$(B-V)_T$ / $(U-B)_T$ / $(B-V)_T^o$ / $(U-B)_T^o$	$(B-V)_e$ / $(U-B)_e$ / m'_e / m'_{25}	m_{21} / W_{20} / W_{50} / HI	V_{21} / V_{opt} / V_{GSR} / V_{3K}
040747.1+694847 / 138.46 13.11 / 13.00 .37 / 040234.6+694046	IC 356 / UGC 2953 / ARP 213 / PGC 14508	.SAS2P. / R / 2.0± .3 /	1.72± .04 / .13± .04 / 1.67± .08 / 1.83	90 / 1.19 / .16 / .07	11.39M±.12 / 12.4 ±.3 / 10.95 / 10.13		1.40± .02 / .86± .03 / 15.12± .23 / 14.52± .25	12.39±.1 / 492± 7 / 462± 8 / 2.19	888± 6 / / 1033 / 817
040803.6+231746 / 170.92 -20.75 / 339.48 -33.55 / 040504.6+230950	/ MCG 4-10- 24 / / PGC 14511	.E?.... / / /	.90± .17 / .07± .07 / / .96	/ .56 / .00 /	/ 14.70 ±.10 / / 14.04			17.31±.3 / / 249± 7 /	6278± 10 / / 6301 / 6165
040807.5-171136 / 211.21 -43.68 / 288.47 -50.03 / 040552.0-171930	NGC 1519 / ESO 550- 9 / IRAS04058-1719 / PGC 14514	.SBR3?. / RSEr (2) / 3.4± .9 / 4.4±1.1	1.33± .02 / .59± .03 / .91± .02 / 1.33	107 / .01 / .81 / .29	13.57 ±.14 / 13.73 ±.14 / 13.06 / 12.81	.63± .03 / -.03± .04 / .50 / -.12	.61± .02 / -.02± .03 / 13.61± .06 / 13.60± .20		1842 / / 1745 / 1751
040819.0-210306 / 216.31 -44.93 / 282.39 -50.20 / 040608.0-211100	NGC 1521 / ESO 550- 11 / / PGC 14520	.E.3.*. / R / -5.0± .5 /	1.44± .03 / .22± .03 / .93± .02 / 1.38	10 / .01 / .00 /	12.39M±.06 / 12.38 ±.11 / / 12.31	.97± .01 / .48± .02 / .93 / .50	.98± .01 / .52± .02 / 12.73± .06 / 14.05± .18		4174± 21 / 4066 / 4088
040819.4-584503 / 270.00 -43.74 / 231.77 -37.19 / 040723.1-585254	IC 2037 / ESO 118- 1 / IRAS04073-5852 / PGC 14521	.S..3./ / S / 3.0± .9 /	1.20± .05 / .75± .04 / .91± .02 / 1.20	92 / .00 / 1.03 / .37	/ 14.76 ±.14 / /				/ / /
040820.3-540308 / 263.76 -45.26 / 236.71 -39.89 / 040708.0-541100	/ ESO 156- 44 / / PGC 14523	.S?.... / / /	.88± .06 / .00± .05 / / .88	/ .00 / .00 / .00	/ 15.20 ±.14 / / 15.04				12467± 88 / 12299 / 12434
040824.4-475350 / 255.15 -46.71 / 243.78 -43.09 / 040656.1-480142	NGC 1527 / ESO 201- 20 / / PGC 14526	.LXR-*. / RS / -2.7± .4 /	1.57± .03 / .43± .03 / .88± .02 / 1.50	78 / .00 / .00 /	11.74 ±.13 / 11.80 ±.10 / / 11.76	.95± .01 / .52± .02 / .92 / .50	.96± .01 / .53± .01 / 11.63± .07 / 13.35± .22		1001± 38 / 841 / 957
0408.4 +6940 / 138.61 13.05 / 12.95 .23 / 0403.2 +6932	/ UGC 2955 / IRAS04032+6932 / PGC 14531	.S?.... / / /	1.07± .14 / .50± .12 / / 1.20	98 / 1.36 / .75 / .25	/ / 12.92 /			14.20±.1 / / 203± 8 /	1007± 11 / 1151 / 936
040829.8-604928 / 272.63 -42.94 / 229.69 -35.96 / 040742.0-605718	/ ESO 118- 6 / / PGC 14533	.E+4... / S / -4.0± .9 /	1.06± .05 / .38± .03 / / .94	4 / .00 / .00 /	/ 15.29 ±.14 / /				/ / /
040834.7+030720 / 188.40 -33.70 / 317.80 -44.74 / 040557.8+025927	/ UGC 2965 / / PGC 14537	.I..9*. / U / 10.0±1.2 /	.96± .09 / .00± .06 / / 1.02	/ .69 / .00 / .00	/ / /			15.46±.3 / / 224± 7 /	7319± 10 / 7280 / 7211
040840.4-493457 / 257.53 -46.33 / 241.74 -42.30 / 040716.0-494248	/ ESO 201- 21 / IRAS04072-4942 / PGC 14543	.S..2.. / S / 2.0± .9 /	1.17± .05 / .39± .05 / .64± .05 / 1.17	57 / .00 / .48 / .20	14.42 ±.19 / 14.37 ±.14 / 13.59 / 13.85	.76± .04 / / .65 /	.86± .03 / / 13.11± .16 / 14.12± .33		5421± 88 / 5258 / 5380
040845.6-624747 / 275.06 -42.12 / 227.77 -34.76 / 040807.0-625536	NGC 1534 / ESO 84- 6 / IRAS04081-6255 / PGC 14547	.SATO*/ / S / .1± .5 /	1.22± .05 / .31± .05 / / 1.21	76 / .03 / .23 /	13.76 ±.14 / 13.52 / 13.42				5333 / 5157 / 5317
040847.0+083056 / 183.36 -30.40 / 324.38 -42.24 / 040604.5+082303	/ UGC 2966 / / PGC 14551	.S..6*. / U / 6.0±1.3 /	1.00± .16 / .33± .12 / / 1.05	8 / .56 / .49 / .17	/ / /			15.14±.3 / / 171± 7 /	3591± 7 / 3568 / 3481
040854.1-555941 / 266.34 -44.60 / 234.56 -38.87 / 040748.0-560730	IC 2038 / ESO 157- 1 / / PGC 14553	.S..7P* / RS (1) / 7.0±1.2 / 6.7± .9	1.24± .04 / .62± .04 / / 1.24	151 / .00 / .86 / .31	15.50 ±.16 / 14.82 ±.14 / / 14.25	.74± .03 / / .62 /	/ / 15.01± .27 /		712 / 542 / 683
040857.2+271140 / 168.12 -17.88 / 343.10 -31.16 / 040553.3+270347	/ UGC 2964 / / PGC 14554	.SA.6.. / U / 6.0± .8 /	1.03± .05 / .06± .04 / / 1.06	/ .36 / .09 / .03	/ 15.09 ±.16 / / 14.60				8970±125 / 9005 / 8859
040900.4-484335 / 256.30 -46.45 / 242.71 -42.77 / 040734.0-485124	/ ESO 201- 22 / IRAS04075-4851 / PGC 14557	.S..5*/ / S / 5.0±1.2 /	1.40± .03 / .87± .05 / / 1.40	59 / .00 / 1.31 / .44	/ 14.73 ±.14 / / 13.40			14.52±.3 / 352± 12 / 342± 9 / .68	4069± 10 / 4014± 34 / 3903 / 4023
040901.5-453104 / 251.69 -46.99 / 246.64 -44.30 / 040728.0-453854	IC 2035 / ESO 250- 7 / / PGC 14558	.L...P? / RS / -2.0±1.0 /	1.07± .04 / .10± .03 / / 1.05	86 / .00 / .00 /	12.50 ±.13 / 12.72 ±.11 / / 12.61	.73± .01 / .29± .04 / .71 / .29	/ / 12.46± .24 /		1443± 43 / 1286 / 1396

R.A. 2000 DEC. / l b / SGL SGB / R.A. 1950 DEC.	Names / PGC	Type / S_T n_L / T / L	$\log D_{25}$ / $\log R_{25}$ / $\log A_e$ / $\log D_o$	p.a. / A_g / A_i / A_{21}	B_T / m_B / m_{FIR} / B_T^o	$(B-V)_T$ / $(U-B)_T$ / $(B-V)_T^o$ / $(U-B)_T^o$	$(B-V)_e$ / $(U-B)_e$ / m'_e / m'_{25}	m_{21} / W_{20} / W_{50} / HI	V_{21} / V_{opt} / V_{GSR} / V_{3K}
040901.6-010937 192.77 -36.04 312.26 -46.58 040629.1-011728	 UGC 2969 IRAS04064-0117 PGC 14559	.SBS4*. UE (1) 4.3± .7 3.1± .8	1.10± .05 .47± .04 1.13	127 .38 .69 .24	 14.64 ±.12 13.20 				
040902.0-560047 266.36 -44.58 234.53 -38.87 040756.1-560836	IC 2039 ESO 157- 2 PGC 14560	.L..0*P S -2.5± .9 	.98± .05 .15± .03 .69± .10 .95	121 .00 .00 	14.9 ±.2 14.97 ±.14 	.73± .04 	.75± .03 13.82± .34 14.26± .35		
0409.1 -0837 200.91 -39.85 301.68 -48.91 0406.7 -0845	 PGC 14562	.SBS9.. E (1) 9.0± .9 9.8± .8	1.11± .08 .19± .08 1.12	75 .11 .19 .09					
0409.1 +1705 176.03 -24.78 333.73 -37.58 0406.3 +1658	 UGC 2968 PGC 14563	.S..3*. U (1) 3.0±1.2 5.5±1.1	1.17± .07 .57± .06 1.31	22 1.54 .78 .28					7282± 10 7285 7171
040911.7+083852 183.32 -30.24 324.60 -42.27 040629.1+083101	NGC 1517 UGC 2970 IRAS04064+0831 PGC 14564	.S..6*. U 6.0±1.2 	1.04± .08 .04± .06 1.09	 .56 .06 .02	 14.07 ±.12 11.74 13.43			14.67±.3 181± 7 1.22	3482± 7 3644± 46 3463 3376
040914.0-302458 229.50 -46.87 267.74 -49.41 040715.0-303248	 ESO 420- 5 IRAS04072-3032 PGC 14566	.S?.... 	1.09± .05 .13± .04 1.09	90 .00 .19 .07	 13.93 ±.14 13.83 13.73				1097 967 1025
040932.1-214623 217.40 -44.88 281.24 -50.47 040722.0-215412	 ESO 550- 14 PGC 14573	.SBS7.. S (1) 7.0± .8 5.6± .8	1.22± .03 .21± .03 1.22	100 .04 .29 .10	 15.10 ±.14 14.76				4438 4329 4355
040951.4-560714 266.45 -44.43 234.33 -38.90 040846.0-561500	NGC 1533 ESO 157- 3 IRAS04088-5614 PGC 14582	.LB.-.. R -3.0± .7 	1.44± .03 .07± .03 1.00± .02 1.43	151 .00 .00 	11.70M±.12 11.83 ±.11 11.76	.98± .01 .49± .02 .97 .49	.99± .01 .54± .01 12.21± .08 13.60± .20	12.53±.3 311± 14 	790± 10 668± 41 612 756
040955.6-394119 243.12 -47.30 254.20 -46.86 040811.0-394906	IC 2036 ESO 303- 1 PGC 14586	PSBS4.. r 4.5± .9 	.98± .06 .10± .05 .98	92 .00 .15 .05	 14.47 ±.14 				
0409.9 -0122 193.15 -35.97 312.09 -46.87 0407.4 -0130	 UGC 2973 IRAS04073-0130 PGC 14587	.S...2.. U 2.0± .9 	.98± .07 .48± .05 1.01	74 .38 .59 .24	 15.17 ±.12 12.87 				
041001.8-311531 230.75 -46.82 266.40 -49.40 040804.0-312318	 ESO 420- 6 PGC 14590	.IBS9.. SU (1) 9.5± .6 10.0± .9	1.19± .04 .34± .04 1.19	134 .00 .25 .17	 15.23 ±.14 14.98				1383 1252 1314
041022.8-233703 219.99 -45.21 278.30 -50.58 040815.0-234448	 ESO 483- 12 IRAS04082-2344 PGC 14596	.S..0*/ S .0±1.2 	1.28± .04 .57± .04 1.26	18 .07 .42 	 14.38 ±.14 12.22 13.82				4229 4114 4149
041033.3-071002 199.49 -38.84 303.96 -48.88 040807.0-071747	 MCG -1-11- 4 PGC 14600	.S..7*/ E 7.0±1.2 	1.29± .05 .88± .05 1.31	112 .21 1.22 .44					
041050.1-093748 202.33 -39.95 300.31 -49.56 040826.5-094532	 MCG -2-11- 23 PGC 14606	.SBS6*. E (1) 6.0±1.2 6.4± .8	1.25± .05 .36± .05 1.27	115 .20 .53 .18					
041050.6-295535 228.87 -46.45 268.38 -49.85 040851.0-300318	 ESO 420- 8 PGC 14607	.SBS4.. S (1) 4.0± .8 3.3± .8	1.09± .04 .12± .03 1.09	129 .00 .17 .06	 14.39 ±.14 14.18				5618± 52 5488 5547
041100.5-425706 247.89 -46.92 249.70 -45.74 040922.0-430448	 ESO 250- 10 PGC 14616	.S?.... 	.97± .06 .41± .05 .53± .05 .97	60 .00 .56 .20	15.38 ±.19 15.37 ±.14 	.86± .04 	.92± .04 13.52± .16 14.06± .37		
041100.5-312429 231.01 -46.63 266.10 -49.57 040903.0-313212	 ESO 420- 9 PGC 14617	.SBS5.. S (1) 5.0± .8 3.3± .8	1.22± .04 .13± .03 1.22	122 .00 .20 .07	 13.75 ±.14 13.55				1320± 23 1187 1251

R.A. 2000 DEC. l b SGL SGB R.A. 1950 DEC.	Names PGC	Type S_T n_L T L	$\log D_{25}$ $\log R_{25}$ $\log A_e$ $\log D_o$	p.a. A_g A_i A_{21}	B_T m_B m_{FIR} B_T^o	$(B-V)_T$ $(U-B)_T$ $(B-V)_T^o$ $(U-B)_T^o$	$(B-V)_e$ $(U-B)_e$ m'_e m'_{25}	m_{21} W_{20} W_{50} HI	V_{21} V_{opt} V_{GSR} V_{3K}
0411.0 +8328 128.35 22.82 21.85 10.74 0400.0 +8320	 UGC 2935 PGC 14618	.S..7.. U 7.0± .8 	1.00± .08 .00± .06 1.03	 .34 .00 .00				16.75±.1 63± 8 	4326± 11 4491 4278
041100.9+263651 168.89 -17.96 342.95 -31.89 040757.5+262906	 UGC 2975 PGC 14619	.I..9?. U 10.0±1.6 	1.11± .14 .04± .12 1.15	 .38 .03 .02				16.34±.3 290± 7 	19937± 10 19969 19829
041101.0-562913 266.87 -44.17 233.82 -38.82 040957.0-563654	NGC 1536 ESO 157- 5 IRAS04099-5636 PGC 14620	.SBS5P* RS (1) 5.0± .5 6.1± .6	1.30± .03 .14± .03 .97± .02 1.30	155 .00 .21 .07	13.15 ±.14 13.24 ±.12 13.53 12.98	.63± .02 -.04± .03 .59 -.07	.66± .01 .03± .03 13.49± .04 14.17± .21		1565±104 1394 1539
041101.6+260931 169.24 -18.27 342.56 -32.19 040758.9+260147	 UGC 2974 PGC 14621	.I..9?. U 10.0±1.7 	1.00± .16 .00± .12 1.04	 .37 .00 .00				16.55±.3 213± 7 	5875± 10 5906 5767
041109.9-534113 263.10 -44.97 236.80 -40.44 040957.0-534854	IC 2043 ESO 157- 4 IRAS04098-5348 PGC 14623	.S..3*/ S 3.0± .9 	1.14± .05 .80± .04 1.14	15 .00 1.11 .40	15.51 ±.16 15.29 ±.14 14.19	.94± .04 .72 	 14.07± .29		11767± 88 11598 11736
041119.7-161348 210.37 -42.61 290.09 -50.72 040903.3-162130	 MCG -3-11- 18 PGC 14626	.SBS9*/ E (1) 9.0± .9 7.5±1.2	1.15± .06 .59± .05 1.16	102 .01 .60 .29					1880 1783 1792
0411.3 +2652 168.75 -17.72 343.24 -31.77 0408.3 +2645	 UGC 2976 PGC 14627	.S..6*. U 6.0±1.2 	1.17± .05 .34± .05 1.21	149 .38 .49 .17	 14.40 ±.12 13.52				1447± 10 1480 1339
041149.0+062721 185.79 -31.07 322.45 -43.92 040908.6+061940	 UGC 2979 PGC 14634	.S?.... 	1.11± .14 .15± .12 1.16	30 .53 .22 .07				15.79±.3 286± 5 	8187± 9 8156 8082
041159.3-325057 233.13 -46.60 263.84 -49.42 041004.0-325836	NGC 1531 ESO 359- 26 PGC 14635	.L..-P* PBS -3.2± .5 	1.13± .07 .16± .04 1.11	122 .00 .00 	13.15 ±.11 12.62				1190± 24 1053 1125
041203.3-583323 269.51 -43.35 231.60 -37.70 041107.0-584100	IC 2049 ESO 118- 9 PGC 14636	.SXS7?. S (1) 7.0±1.7 5.6±1.2	1.01± .05 .04± .05 1.01	3 .00 .05 .02	14.56 ±.14 14.50				1086 912 1065
0412.0 +2756 168.07 -16.86 344.28 -31.17 0409.0 +2749	 UGC 2977 PGC 14637	.S..2.. U 2.0± .9 	1.00± .06 .25± .05 1.05	85 .48 .31 .13					
041205.5-325228 233.17 -46.58 263.80 -49.43 041010.2-330006	NGC 1532 ESO 359- 27 IRAS04102-3259 PGC 14638	.SBS3P/ PBS (2) 2.7± .3 1.9± .7	2.10± .02 .58± .03 1.51± .02 2.10	33 .00 .80 .29	10.65M±.10 10.73 ±.11 11.45 9.88	.80± .02 .15± .05 .69 .05	.88± .01 .24± .01 13.65± .07 14.55± .16	11.41±.2 541± 9 513± 12 1.23	1196± 7 923± 40 1052 1124
041210.8-482459 255.72 -46.00 242.72 -43.38 041044.2-483236	 PGC 14643	.SXT5?. S (1) 5.0±1.8 5.6± .9	1.07± .09 .39± .08 1.07	 .00 .59 .20					
041211.1-501547 258.32 -45.64 240.55 -42.43 041049.1-502324	 ESO 201- 25 PGC 14644	 	.86± .07 .13± .06 .86	55 .00 	 15.15 ±.14 				13564± 88 13399 13528
041222.7+053251 186.74 -31.50 321.43 -44.48 040943.2+052512	 UGC 2982 IRAS04097+0525 PGC 14651	.SB?... 	.95± .05 .34± .04 1.00	111 .58 .51 .17	 15.27 ±.12 10.80 14.14			14.89±.2 406± 8 350± 7 .57	5305± 5 5444± 42 5273 5204
0412.4 +2742 168.31 -16.96 344.15 -31.39 0409.4 +2735	IC 359 UGC 2980 PGC 14653	.L..-*. U -3.0±1.1 	1.03± .11 .00± .05 1.10	 .50 .00 	 14.90 ±.14 				
041234.2-541136 263.71 -44.63 236.10 -40.33 041123.1-541911	 FAIR 764 PGC 14655	.L..-P. S -3.0± .9 	1.02± .11 .14± .08 .77± .08 1.00	 .00 .00 	14.7 ±.2 14.46	.96± .04 .83 	1.04± .03 13.99± .26 14.25± .62		12964± 79 12795 12936

R.A. 2000 DEC. / l b / SGL SGB / R.A. 1950 DEC.	Names / PGC	Type / S_T n_L / T / L	$\log D_{25}$ / $\log R_{25}$ / $\log A_e$ / $\log D_o$	p.a. / A_g / A_i / A_{21}	B_T / m_B / m_{FIR} / B_T^o	$(B-V)_T$ / $(U-B)_T$ / $(B-V)_T^o$ / $(U-B)_T^o$	$(B-V)_e$ / $(U-B)_e$ / m'_e / m'_{25}	m_{21} / W_{20} / W_{50} / HI	V_{21} / V_{opt} / V_{GSR} / V_{3K}
041234.3-324859 / 233.10 -46.47 / 263.84 -49.54 / 041039.0-325636	IC 2041 / ESO 359- 28 / / PGC 14656	.LXS0*. / S / -2.0± .5 /	1.03± .05 / .23± .03 / / 1.00	132 / .00 / .00 /	14.65 ±.14 / / / 14.63				1260± 29 / 1123 / 1196 /
041241.2-230936 / 219.57 -44.58 / 278.97 -51.14 / 041033.0-231712	/ ESO 483- 13 / / PGC 14658	.LA.-*. / S / -3.0± .6 /	1.21± .05 / .30± .04 / / 1.17	136 / .07 / .00 /	14.09 ±.14 / / / 14.01				910± 60 / 795 / 833 /
041243.2-574414 / 268.41 -43.54 / 232.36 -38.26 / 041144.0-575148	NGC 1543 / ESO 118- 10 / IRAS04117-5751 / PGC 14659	RLBS0.. / R / -2.0± .6 /	1.69± .02 / .24± .03 / 1.09± .02 / 1.65	93 / .00 / .00 /	11.46M±.12 / 10.99 ±.11 / / 11.18	.95± .01 / .55± .03 / .92 / .53	.97± .01 / .56± .01 / 12.33± .08 / 14.17± .18		1094± 32 / 921 / 1072 /
041251.0-330006 / 233.38 -46.44 / 263.54 -49.55 / 041056.0-330742	/ ESO 359- 29 / / PGC 14664	.IBS9.. / S (1) / 10.0± .6 / 8.9± .6	1.23± .04 / .17± .05 / / 1.23	20 / .00 / .13 / .09	14.77 ±.14 / / / 14.64				873 / 736 / 810 /
041252.6+022210 / 189.88 -33.26 / 317.50 -46.04 / 041016.5+021433	UGC 2983 / IRAS04102+0214 / / PGC 14665	.SBS3*. / UE (2) / 3.3± .7 / 3.9± .7	1.30± .04 / .47± .04 / / 1.36	135 / .62 / .65 / .23	14.76 ±.18 / 13.28 / / 13.46				5001± 10 / 4956 / 4900 /
041258.5-434850 / 249.10 -46.48 / 248.36 -45.69 / 041122.0-435624	ESO 250- 13 / FAIR 404 / / PGC 14668	.L?.... / / /	.88± .07 / .46± .03 / / .81	14 / .00 / .00 /	15.33 ±.18 / 15.33 ±.14 / / 15.12	1.01± .03 / / .84 /	/ / / 13.45± .39		13850±190 / 13693 / 13804 /
0412.9 -5359 / 263.41 -44.63 / 236.27 -40.49 / 0411.7 -5407	/ FAIR 765 / / PGC 14669			.00	15.46 ±.05	.57± .03			12669± 88 / 12499 / 12640 /
041259.7-323307 / 232.73 -46.36 / 264.20 -49.70 / 041104.0-324042	IC 2040 / ESO 359- 30 / IRAS04110-3240 / PGC 14670	.L..0*P / S / -1.5± .8 /	1.15± .05 / .27± .03 / / 1.11	65 / .00 / .00 /	13.59 ±.14 / 13.21 / / 13.57				1381± 29 / 1245 / 1317 /
0413.1 -5329 / 262.72 -44.74 / 236.80 -40.79 / 0411.9 -5336	/ / / PGC 14672			.00	15.98 ±.15	1.13± .04			12669± 88 (?)
0413.1 +7500 / 134.97 17.09 / 16.57 4.14 / 0406.9 +7453	CGCG 327- 16 / / / PGC 14673	.L?.... / / /	.84± .10 / .21± .10 / / .88	/ .65 / .00 /	15.07 ±.17 / / / 14.39			17.07±.1 / / 99± 8 /	1711± 11 / 1864 / 1650 /
041309.9-342538 / 235.45 -46.50 / 261.38 -49.22 / 041117.0-343312	/ ESO 359- 31 / / PGC 14674	.SBS8.. / S (1) / 8.0± .6 / 7.2± .6	1.07± .05 / .04± .04 / / 1.07	/ .00 / .05 / .02	14.56 ±.14 / / / 14.50				1486 / 1346 / 1425 /
041312.8+132515 / 179.82 -26.45 / 330.64 -40.52 / 041024.9+131740	UGC 2984 / IRAS04104+1317 / / PGC 14678	.SB.8*. / U / 8.0±1.1 /	1.22± .12 / .08± .12 / / 1.32	/ 1.03 / .10 / .04				14.05±.3 / / 119± 7 /	1538± 7 / 1528 / 1433 /
041330.0-615041 / 273.59 -41.99 / 228.24 -35.78 / 041248.0-615812	ESO 118- 12 / IRAS04128-6158 / / PGC 14686	.L..0*. / S / -2.0± .9 /	1.07± .06 / .30± .04 / / 1.03	51 / .00 / .00 /	14.49 ±.14 / 11.83 / /				
0413.5 +2732 / 168.61 -16.91 / 344.19 -31.68 / 0410.4 +2724	UGC 2985 / IRAS04104+2724 / / PGC 14687	.S..6*. / U / 6.0±1.3 /	1.06± .06 / .55± .05 / / 1.10	48 / .48 / .81 / .28	12.07				4004± 10 / 4037 / 3899 /
0413.5 +2735 / 168.57 -16.87 / 344.24 -31.65 / 0410.5 +2728	UGC 2986 / / / PGC 14688	.I..9*. / U / 10.0±1.3 /	.95± .07 / .39± .05 / / 1.00	70 / .51 / .30 / .20					4041 / 4075 / 3937 /
041338.6+252851 / 170.18 -18.32 / 342.43 -33.08 / 041036.6+252117	UGC 2988 / / / PGC 14693	.S..3*. / U / 3.0±1.1 /	1.45± .03 / .67± .05 / / 1.50	5 / .55 / .92 / .33	14.9 ±.2			13.76±.1 / / 474± 7 /	3825± 5 / 8052± 76 /
041341.0-313846 / 231.46 -46.10 / 265.53 -50.06 / 041144.0-314618	NGC 1537 / ESO 420- 12 / / PGC 14695	.LX.-P? / RS / -2.5± .4 /	1.59± .04 / .18± .04 / .93± .02 / 1.56	98 / .00 / .00 /	11.47M±.10 / 11.53 ±.11 / / 11.48	.89± .01 / .42± .01 / .87 / .42	.93± .01 / .48± .01 / 11.69± .06 / 13.83± .24		1362± 21 / 1227 / 1297 /

R.A. 2000 DEC. l　　b SGL　　SGB R.A. 1950 DEC.	Names PGC	Type S_T　n_L T L	$\log D_{25}$ $\log R_{25}$ $\log A_e$ $\log D_o$	p.a. A_g A_i A_{21}	B_T m_B m_{FIR} B_T^o	$(B-V)_T$ $(U-B)_T$ $(B-V)_T^o$ $(U-B)_T^o$	$(B-V)_e$ $(U-B)_e$ m'_e m'_{25}	m_{21} W_{20} W_{50} HI	V_{21} V_{opt} V_{GSR} V_{3K}
041350.4-320022 231.98 -46.11 264.96 -50.01 041154.0-320754	 ESO　420- 13 IRAS04118-3207 PGC 14702	.LAR+P? Sr -1.0± .8	1.00± .05 .05± .03 .99	 .00 .00	 13.31 ±.14 10.36				
0413.8　+3127 165.75 -14.11 347.49 -29.02 0410.7　+3120	 UGC 2990 PGC 14703	.S..2.. U 2.0± .9	1.07± .14 .40± .12 1.15	0 .88 .49 .20					
041356.0-532830 262.66 -44.63 236.73 -40.90 041243.0-533600	IC　2050 ESO　157- 11 FAIR 405 PGC 14704	.SAT3P. S 3.0± .9	1.03± .05 .10± .04 .83± .12 1.03	15 .00 .13 .05	14.4 ±.3 14.78 ±.14 14.49	.71± .05	.79± .03 14.03± .33 14.14± .39		12368± 79 12199 12340
041356.7+290913 167.47 -15.72 345.62 -30.64 041049.8+290140	A　0410+29 UGC 2989 5ZW 372 PGC 14705	.SB..$. R	1.12± .05 .29± .05 1.17	30 .48 .43 .14			16.85±.2	429± 12 434± 10	5639± 7 5298± 49 5671 5529
041358.1-253617 223.00 -44.92 275.00 -51.21 041153.1-254348	 ESO　483- 14 PGC 14706	.S?....	.91± .07 .20± .06 .92	 .03 .20 .10	 15.49 ±.14 15.10				12857± 52 12735 12784
041358.9-380553 240.79 -46.52 255.98 -48.17 041212.0-381324	 ESO　303- 5 PGC 14707	.E+1.P. S -4.0± .8	.90± .07 .09± .04 .88	 .00 .00	 15.19 ±.14 14.97				14981 14834 14927
0413.9　+3650 161.89 -10.27 351.70 -25.19 0410.6　+3643	 UGC 2991 IRAS04106+3643 PGC 14709	.SBR3.. U 3.0± .8	1.27± .04 .12± .05 1.46	113 2.00 .16 .06	 12.44				6032 6093 5931
0414.0　-1324 207.25 -40.90 294.69 -51.03 0411.7　-1332	 MCG -2-11- 26 PGC 14710	.L?....	1.12± .12 .09± .07 1.11	 .07 .00					9337 9246 9250
041414.6+243932 170.91 -18.79 341.81 -33.73 041113.5+243200	 MCG 4-10- 28 PGC 14714	.S?....	.74± .14 .09± .07 .80	 .67 .14 .05	 15.47 ±.12 14.63			17.25±.3 218± 7 2.58	6595± 10 6619 6491
041423.0-624845 274.73 -41.53 227.25 -35.24 041346.0-625612	 ESO　84- 9 PGC 14717	.SBS5.. S　　(1) 5.0± .9 2.2±1.3	1.02± .05 .15± .04 .84± .02 1.02	52 .01 .23 .08	14.18 ±.15 15.04 ±.14 14.38	.65± .03	.67± .02 13.87± .06 13.77± .31		5080± 40 4903 5068
0414.4　+0242 189.81 -32.75 318.18 -46.23 0411.8　+0235	 UGC 2994 PGC 14718	.SA.8*. UE　　(1) 7.5± .8 7.5± .8	1.18± .04 .27± .04 1.24	35 .58 .33 .13	 15.1 ±.2 14.16				3318 3274 3219
041431.2-504738 258.95 -45.16 239.66 -42.46 041310.9-505506	 PGC 14720	.LA.-*. S -3.0± .9	1.11± .10 .17± .08 1.09	 .00 .00					
041431.6-063824 199.52 -37.72 305.16 -49.68 041204.9-064554	 MCG -1-11- 8 PGC 14721	.SBT4.. E　　(1) 4.0± .9 4.2± .8	1.06± .06 .11± .05 1.07	170 .12 .17 .06					
041436.7-560339 266.09 -43.82 233.89 -39.47 041332.1-561106	NGC 1546 ESO　157- 12 IRAS04134-5611 PGC 14723	.LA.+?. RS -1.3± .6	1.48± .03 .26± .03 1.27± .03 1.44	147 .00 .00	11.8 ±.1 12.30 ±.11 10.71 12.02	.88± .03 .35± .04 .83 .32	.91± .01 .33± .01 13.65± .12 13.42± .21		1160± 66 988 1136
0414.7　+3510 163.20 -11.35 350.56 -26.50 0411.5　+3503	 UGC 2993 PGC 14725	.S..6?. U 6.0±1.7	1.04± .15 .04± .12 1.18	 1.44 .06 .02				16.43±.3 163± 7 142± 7	6232± 11 6288 6132
041458.2-542016 263.77 -44.25 235.67 -40.53 041348.1-542742	IC　2052 ESO　157- 13 PGC 14729	.S?....	.96± .06 .40± .05 .96	165 .00 .55 .20	 15.38 ±.14 14.74				11353± 88 11182 11327
041510.3-282858 227.07 -45.27 270.35 -51.05 041309.1-283624	NGC 1540A ESO　420- 14A PGC 14734	.S..1?P S 1.0±1.8	.88± .07 .19± .06 .88	 .00 .19 .09	 14.46 ±.14				

R.A. 2000 DEC. / l b / SGL SGB / R.A. 1950 DEC.	Names / PGC	Type / S_T n_L / T / L	$\log D_{25}$ / $\log R_{25}$ / $\log A_\bullet$ / $\log D_o$	p.a. / A_g / A_i / A_{21}	B_T / m_B / m_{FIR} / B_T^o	$(B-V)_T$ / $(U-B)_T$ / $(B-V)_T^o$ / $(U-B)_T^o$	$(B-V)_\bullet$ / $(U-B)_\bullet$ / $m'_\bullet$ / m'_{25}	m_{21} / W_{20} / W_{50} / HI	V_{21} / V_{opt} / V_{GSR} / V_{3K}
041517.8-505647			.74± .07		15.13 ±.15	-.09± .07	.13± .07		
259.12 -45.01	ESO 201- 26		.12± .06	.00	15.19 ±.14	-1.05± .10	-.70± .10		3783± 63
239.40 -42.48	IRAS04139-5104		.59± .04				13.57± .15		3616
041358.0-510412	PGC 14740		.74						3751
0415.3 +0057		.SBS9..	1.10± .08	25					
191.70 -33.54		E (1)	.19± .08	.47					
316.03 -47.20		9.0± .9		.19					
0412.8 +0050	PGC 14744	9.8± .8	1.15	.09					
041532.2-352035		.SB?...	.97± .06	58					
236.82 -46.09	ESO 360- 2		.06± .06	.00	15.03 ±.14				4293
259.80 -49.40				.09					4150
041341.0-352800	PGC 14749		.97	.03	14.92				4236
0415.5 +0029		.S..6*.	1.00± .08	6					
192.19 -33.76	UGC 2995	U	.73± .06	.45					
315.43 -47.44		6.0±1.4		1.07					
0413.0 +0022	PGC 14751		1.04	.36					
0415.5 +0109		.S...0..	.97± .07	124					
191.53 -33.39	UGC 2996	U	.56± .05	.47	15.23 ±.13				
316.33 -47.17		.0±1.0		.42					
0413.0 +0102	PGC 14752		.99						
041537.2-523613			.74± .07						
261.39 -44.59	ESO 157- 14		.21± .06	.00	15.35 ±.14				13228± 88
237.49 -41.60									13059
041422.0-524336	PGC 14753		.74						13199
041545.1-553531	NGC 1549	.E.0+..	1.69± .02	135	10.72M±.08	.93± .01	.94± .01		1214± 21
265.41 -43.80	ESO 157- 16	R	.08± .02	.00	10.49 ±.11	.50± .02	.51± .01		1042
234.25 -39.88		-5.0± .6	1.18± .01	.00		.92	12.00± .05		1191
041439.1-554254	PGC 14757		1.67		10.62	.51	13.97± .14		
041555.9-573850		PSBR3*.	1.03± .04	133					13705
268.10 -43.17	ESO 118- 15	S (1)	.08± .04	.00	14.91 ±.14				13531
232.11 -38.66	IRAS04149-5746	3.0±1.2		.11	14.00				13685
041457.0-574612	PGC 14761	3.3±1.2	1.03	.04	14.69				
0416.0 +0810		.S..0..	1.00± .06	178				16.53±.3	1592± 9
184.95 -29.18	UGC 2997	U	.29± .05	.81	14.79 ±.12				1555± 42
325.25 -43.97	IRAS04133+0803	.0± .9		.22	11.64			120± 7	1562
0413.3 +0803	PGC 14762		1.07		13.74				1491
0416.1 -5101					15.9 ±.2	.90± .06	.98± .05		20066± 62
259.19 -44.87				.00					19898
239.22 -42.55			.63± .06				14.50± .18		20034
0414.8 -5108	PGC 14764								
041610.4-554651	NGC 1553	.LAR0..	1.65± .02	150	10.28M±.08	.88± .01	.93± .01	13.61±.7	1080± 11
265.63 -43.69	ESO 157- 17	R	.20± .02	.00	10.18 ±.11	.48± .05	.52± .01	123± 15	1239± 18
234.01 -39.82	IRAS04150-5554	-2.0± .3	1.34± .02	.00	13.92	.85	12.13± .07		946
041505.0-555412	PGC 14765		1.62		10.23	.47	12.91± .15		1096
041612.7-164502		.SBS8P*	1.21± .04	168					1960
211.62 -41.72	MCG -3-11- 19	SE (1)	.48± .04	.05					
289.40 -51.93		7.7± .7		.59					1859
041357.0-165225	PGC 14768	6.7± .9	1.22	.24					1880
0416.3 -1507		.SBS9..	1.04± .09						
209.64 -41.06		E (1)	.00± .08	.07					
292.05 -51.82		9.0± .9		.00					
0414.1 -1515	PGC 14772	9.8± .8	1.05	.00					
041623.6-601222	IC 2056	RSXR4*.	1.27± .03	8	12.48 ±.13	.56± .01	.57± .01		
271.36 -42.27	ESO 118- 16	Sr (1)	.08± .04	.00	12.50 ±.12				1099± 66
229.53 -37.10	IRAS04155-6019	4.4± .6	.61± .02	.11	11.09	.54	11.02± .05		923
041535.0-601942	PGC 14773	3.4± .8	1.27	.04	12.37		13.50± .21		1084
041627.5-354403		.S?....	.96± .06	128					4428± 87
237.40 -45.93	ESO 360- 4		.23± .05	.00	15.03 ±.14				4284
259.13 -49.45				.34					4373
041437.0-355124	PGC 14774		.96	.11	14.67				
041630.3-475058		.LXR0*P	1.08± .05	65					9987± 34
254.75 -45.38	ESO 202- 1	S	.20± .04	.00	14.76 ±.14				9823
242.89 -44.29		-2.0± .6		.00					9951
041503.0-475818	PGC 14775		1.05		14.61				
041634.4+024534		.SBT3..	1.18± .04					16.48±.2	3349± 6
190.13 -32.28	UGC 2998	UE (1)	.05± .04	.50	14.35 ±.18			114± 7	
318.59 -46.70	IRAS04139+0238	3.0± .6		.07	13.21			85± 4	3303
041357.8+023811	PGC 14779	4.2± .8	1.23	.03	13.75			2.70	3253

R.A. 2000 DEC.	Names	Type	logD$_{25}$	p.a.	B$_T$	(B-V)$_T$	(B-V)$_e$	m$_{21}$	V$_{21}$
l b	S$_T$ n$_L$	logR$_{25}$	A$_g$	m$_B$	(U-B)$_T$	(U-B)$_e$	W$_{20}$	V$_{opt}$	
SGL SGB		T	logA$_e$	A$_i$	m$_{FIR}$	(B-V)^o_T	m'	W$_{50}$	V$_{GSR}$	
R.A. 1950 DEC.	PGC	L	logD$_o$	A$_{21}$	B^o_T	(U-B)^o_T	m'$_{25}$	HI	V$_{3K}$

041638.4-122356		.SBT3..	1.23± .05	130					
206.39 -39.90	MCG -2-11- 30	E (1)	.38± .05	.12					
296.46 -51.50		3.0± .8		.52					
041417.9-123118	PGC 14780	4.2± .8	1.24	.19					
041642.6-121202	IC 362	.E+..?.	1.24± .09	10					
206.17 -39.80	MCG -2-11- 31	E	.19± .07	.12					
296.78 -51.48		-4.0± .8		.00					
041421.9-121923	PGC 14782		1.20						
041654.5+030549		.SB.6*.	1.05± .04	25					3563
189.85 -32.02	UGC 3002	U	.16± .04	.52					
319.09 -46.62	KUG 0414+029	6.0±1.2		.23					3518
041417.6+025828	PGC 14790		1.10	.08					3468
0416.9 -0727		.S..5?/	1.22± .07	77					
200.79 -37.60		E	.78± .08	.16					
304.19 -50.48		5.0±1.8		1.17					
0414.5 -0735	PGC 14791		1.24	.39					
041700.3+005005	NGC 1541	.L..+*.	1.10± .05	77	14.56 ±.13	1.04± .03			
192.09 -33.27	UGC 3001	UE	.41± .04	.40	14.66 ±.10	.59± .04			
316.11 -47.63		-.7± .7		.00					
041425.7+004244	PGC 14792		1.07				13.89± .28		
041712.4-175123	NGC 1547	.SBT4P*	1.10± .04	133					
213.11 -41.90	ESO 550- 18	SE (2)	.35± .03	.04	14.41 ±.14				9601± 60
287.61 -52.23	IRAS04149-1758	3.8± .5		.51	12.32				9496
041458.0-175842	PGC 14799	3.9± .7	1.11	.17	13.80				9524
041712.8+044700	NGC 1542	.S..2..	1.10± .04	128				17.35±.3	3714± 10
188.29 -30.98	UGC 3003	U	.43± .04	.58	14.83 ±.12			416± 13	3537±113
321.29 -45.90	IRAS04145+0439	2.0± .9		.53	13.64				3673
041434.1+043940	PGC 14800		1.16	.22	13.68			3.45	3617
0417.2 +1027								16.20±.3	7500± 9
183.10 -27.56				1.13					7566± 33
328.09 -42.99	IRAS04144+1020				11.95			141± 7	7482
0414.4 +1020	PGC 14801								7406
041715.5-700128		PSXR0*.	.95± .06	150					
283.06 -38.17	ESO 55- 18	r	.31± .05	.39	15.09 ±.14				
220.70 -30.56		-.1± .9		.23					
041728.0-700842	PGC 14802		.97						
041719.1+022600		.SB?...	1.07± .06	155				15.67±.3	3218± 9
190.57 -32.31	UGC 3004	(1)	.18± .05	.48					
318.29 -47.01				.27				184± 7	3171
041442.8+021841	PGC 14803	4.2± .8	1.12	.09					3123
0417.3 +0226		.SBS7*.	1.03± .08	3					3245
190.57 -32.31	UGC 3005	UE	.79± .06	.48					
318.29 -47.01	IRAS04147+0218	7.0±1.0		1.10	12.58				3198
0414.7 +0218	PGC 14804		1.07	.40					3150
041725.3+022215		.LA.0*.	1.25± .05	157					
190.65 -32.33	UGC 3006	UE	.39± .03	.48	14.51 ±.12				
318.22 -47.06		-2.0± .7		.00					
041449.2+021456	PGC 14807		1.24						
041735.5-255631		.L?....	.79± .07	165					
223.73 -44.21	ESO 484- 3		.17± .05	.03	15.45 ±.14				9908± 52
274.29 -51.98				.00					9784
041531.0-260348	PGC 14810		.76		15.27				9841
0417.6 +3617		.S..0..	1.04± .06	17					
162.82 -10.16	UGC 3000	U	.50± .05	2.14					
351.86 -26.09	IRAS04143+3609	.0± .9		.38	12.89				
0414.3 +3609	PGC 14812		1.22						
041737.4-624704	NGC 1559	.SBS6..	1.54± .03	64	11.0 S±.3	.35± .03		12.98±.2	1292± 6
274.51 -41.20	ESO 84- 10	R (2)	.24± .03	.00	11.21 ±.12	-.08± .05		296± 5	1333± 38
226.98 -35.54		6.0± .7	1.20± .08	.36		.30	12.06± .17	226± 7	1115
041701.0-625418	PGC 14814	4.8± .5	1.54	.12	10.83	-.12	12.95± .34	2.03	1283
041745.5-500951	NGC 1556	.S..2P?	1.24± .03	167	13.47 ±.13	.41± .01	.42± .01		
257.93 -44.78	ESO 202- 4	S	.51± .04	.00	13.53 ±.14	-.27± .02	-.17± .02		990± 87
240.00 -43.24	IRAS04163-5017	2.0±1.3	.67± .01	.62	12.95	.31	12.31± .04		823
041624.0-501706	PGC 14818		1.24	.25	12.86	-.34	13.25± .22		959
0417.7 +0132			.74± .14					16.05±.3	4927± 9
191.52 -32.72	MCG 0-11- 47		.09± .07	.34	14.93 ±.14				
317.18 -47.50	2ZW 7							73± 7	4877
0415.2 +0125	PGC 14819		.77						4834

R.A. 2000 DEC. l b SGL SGB R.A. 1950 DEC.	Names PGC	Type S_T n_L T L	$\log D_{25}$ $\log R_{25}$ $\log A_e$ $\log D_o$	p.a. A_g A_i A_{21}	B_T m_B m_{FIR} B_T^o	$(B-V)_T$ $(U-B)_T$ $(B-V)_T^o$ $(U-B)_T^o$	$(B-V)_e$ $(U-B)_e$ m'_e m'_{25}	m_{21} W_{20} W_{50} HI	V_{21} V_{opt} V_{GSR} V_{3K}
041752.8-634027 275.58 -40.83 226.14 -34.97 041721.1-634740	 PGC 14822	.LA.0*. S -2.0± .9 	1.10± .10 .14± .08 1.08	 .03 .00 	 				
041754.2-563704 266.64 -43.22 232.95 -39.50 041652.0-564418	IC 2060 ESO 157- 19 PGC 14823	.L..-*P S -3.0±1.2 	1.02± .05 .21± .03 .91± .22 .99	157 .00 .00 	14.9 ±.4 14.32 ±.14 14.28	.94± .08 .87 	1.02± .04 14.89± .78 14.35± .49		6686± 88 6512 6666
041754.5-555604 265.74 -43.41 233.66 -39.92 041650.1-560318	IC 2058 ESO 157- 18 IRAS04168-5603 PGC 14824	.S..7?/ RS 6.6± .7 	1.47± .03 .88± .03 1.47	18 .00 1.21 .44	 13.90 ±.14 13.55 12.68			14.27±.3 255± 12 1.14	1359± 9 1268± 88 1185 1337
0418.4 -5040 258.62 -44.58 239.32 -43.04 0417.1 -5048	 PGC 14832		 .99± .19 	 .00 	14.8 ±.6 		 15.23± .52 		14553± 88 14385 14524
041828.7-604016 271.83 -41.86 228.88 -37.00 041742.6-604727	 PGC 14833	.E+1... S -5.0± .8 	1.31± .07 .17± .08 1.26	 .00 .00 					
041830.2+334636 164.77 -11.79 350.10 -28.05 041516.4+333920	 UGC 3007 PGC 14834	.SXS7.. U 7.0± .8 	1.07± .07 .07± .06 1.20	 1.35 .09 .03				15.51±.3 82± 7 	5631± 7 5681 5535
0418.5 +7948 131.47 20.57 19.76 7.72 0410.5 +7941	 UGC 2987 PGC 14836	.SB.8*. U 8.0±1.2 	1.16± .07 .41± .06 1.20	113 .40 .51 .21				16.28±.1 214± 8 	5424± 11 5584 5372
041840.2+023338 190.67 -31.96 318.68 -47.25 041603.8+022624	 UGC 3008 PGC 14839	RLBT+*. UE -1.0± .6 	1.13± .04 .28± .04 1.14	75 .48 .00 	 14.45 ±.11 				
041857.5+052640 187.96 -30.24 322.43 -45.96 041618.0+051927	 UGC 3010 PGC 14849	.SB.6*. U 6.0±1.3 	1.02± .06 .20± .05 1.07	132 .59 .29 .10	 15.02 ±.14 14.12			15.19±.1 223± 8 .97	3870± 7 3831 3776
041905.8+261039 170.53 -16.95 344.02 -33.51 041602.6+260326	 UGC 3009 PGC 14853	.S..6*. U 6.0±1.3 	1.19± .05 .70± .05 1.25	107 .58 1.03 .35				15.19±.3 260± 7 	3754± 10 3780 3657
041910.5+555253 149.21 4.00 5.31 -11.29 041508.1+554538	 PGC 14858			 4.01 				14.82±.3 406± 11 382± 8 	5213± 9 5324 5132
041914.2+032053 190.00 -31.40 319.80 -47.02 041637.0+031341	IC 365 MCG 1-11- 17 IRAS04166+0313 PGC 14860	PLBT+*. E -1.0± .7 	1.00± .06 .22± .04 1.03	30 .61 .00 					
041918.7-114207 205.94 -39.01 297.76 -52.02 041657.5-114918	 HICK 27B PGC 14863	.S?.... 		 .09 	16.44S±.15 				
041921.0+023550 190.75 -31.80 318.84 -47.39 041644.5+022838	 UGC 3011 PGC 14865	.LAT0?. E -2.0± .9 	1.06± .04 .29± .04 1.08	124 .51 .00 	 15.05 ±.08 				
041922.2-264744 225.01 -44.01 272.80 -52.25 041719.0-265454	 ESO 484- 5 PGC 14867	.S..5./ S 5.0± .7 	1.20± .04 .99± .03 1.21	20 .09 1.49 .50	 16.02 ±.14 				
041922.6-173656 213.06 -41.33 288.06 -52.74 041708.1-174406	 ESO 550- 20 IRAS04171-1744 PGC 14868	.I?.... 	.98± .05 .08± .04 .99	 .05 .06 .04	 14.63 ±.14 13.66 14.51				8861± 60 8756 8786
041925.3-505352 258.87 -44.38 238.96 -43.05 041806.0-510100	 ESO 202- 7 PGC 14871	.L?.... 	1.04± .07 .12± .05 .84± .06 1.02	65 .00 .00 	14.29 ±.19 14.43 ±.14 14.21	1.02± .04 .90 	1.04± .02 13.98± .22 14.04± .40		11160± 88 10991 11132

R.A. 2000 DEC. / l b / SGL SGB / R.A. 1950 DEC.	Names / PGC	Type / S_T n_L / T / L	$\log D_{25}$ / $\log R_{25}$ / $\log A_e$ / $\log D_o$	p.a. / A_g / A_i / A_{21}	B_T / m_B / m_{FIR} / B_T^o	$(B-V)_T$ / $(U-B)_T$ / $(B-V)_T^o$ / $(U-B)_T^o$	$(B-V)_e$ / $(U-B)_e$ / m'_e / m'_{25}	m_{21} / W_{20} / W_{50} / HI	V_{21} / V_{opt} / V_{GSR} / V_{3K}
041927.3-264356 224.93 -43.98 272.90 -52.28 041724.0-265106	 ESO 484- 6 PGC 14874	.L..-*. S -3.0±1.3 	.98± .05 .32± .04 .94	70 .10 .00 	 15.30 ±.14 				
041938.3+022435 190.98 -31.85 318.64 -47.54 041702.0+021725	NGC 1550 UGC 3012 PGC 14880	.LAS-*. UE -3.2± .5 	1.35± .04 .06± .02 .93± .04 1.41	30 .56 .00 	13.07 ±.15 13.40 ±.14 12.64	1.08± .02 .57± .04 .92 .45	1.10± .02 .61± .03 13.21± .15 14.53± .28		3689± 50 3640 3598
041953.7+020543 191.33 -31.97 318.26 -47.73 041717.8+015834	 UGC 3014 ARP 20 PGC 14892	.SB?... 	1.07± .06 .21± .05 1.11	45 .47 .31 .10	 14.44 ±.12 13.17 13.63			15.23±.2 184± 5 1.49	4214± 6 4164 4123
042000.4-545618 264.31 -43.39 234.46 -40.76 041853.0-550324	NGC 1566 ESO 157- 20 IRAS04189-5503 PGC 14897	.SXS4.. R (2) 4.0± .6 1.7± .5	1.92± .02 .10± .02 1.34± .03 1.92	60 .00 .15 .05	10.33M±.03 10.10 ±.12 10.06 10.15	.60± .01 -.04± .04 .57 -.06	.68± .01 .05± .02 12.44± .07 14.52± .11	11.68±.2 233± 5 206± 7 1.47	1496± 5 1449± 19 1320 1472
042008.7-364111 238.82 -45.24 257.33 -49.82 041820.0-364818	 ESO 360- 7 PGC 14903	.SBS9.. S (1) 9.0± .8 10.0± .8	1.27± .06 .32± .05 1.27	113 .00 .33 .16	 16.62 ±.14 16.29				1172 1024 1123
042016.1-450154 250.67 -45.08 245.91 -46.27 041843.0-450900	NGC 1558 ESO 250- 17 IRAS04187-4508 PGC 14906	PSXR4*. RSr (1) 3.7± .6 3.3± .8	1.39± .04 .38± .05 1.39	72 .00 .55 .19	 13.28 ±.14 13.52 12.70				4541± 34 4379 4504
042017.6-004135 194.16 -33.42 314.52 -49.00 041744.6-004842	NGC 1552 UGC 3015 PGC 14907	.LXR+?. UE -1.0± .6 	1.26± .07 .17± .05 1.25	110 .24 .00 	 13.93 ±.12 13.62				4924± 50 4865 4836
042019.5-223805 219.55 -42.74 279.65 -52.93 041811.1-224512	 ESO 484- 7 PGC 14909	.L?.... 	.84± .07 .61± .03 .75	32 .02 .00 	16.64 ±.13 16.32 ±.14 	1.22± .03 .35± .06 	 14.17± .36		
042026.3-314330 231.85 -44.69 264.82 -51.43 041830.0-315036	IC 2059 ESO 420- 17 PGC 14910	.LXR0*. Sr -1.6± .6 	1.11± .04 .46± .03 1.04	172 .04 .00 	 13.87 ±.14 13.79				2810± 39 2671 2754
042028.3-625215 274.46 -40.86 226.63 -35.72 041953.0-625918	 ESO 84- 13 PGC 14913	.IX.9.. S (1) 10.0± .8 10.0± .8	1.07± .05 .46± .06 1.07	32 .00 .34 .23	 16.46 ±.14 				
0420.5 -0202 195.58 -34.09 312.64 -49.58 0418.0 -0210	 MCG 0-12- 10 PGC 14914	.S?.... 	.82± .13 .23± .07 .85	 .31 .34 .11	 14.91 ±.14 14.23				4736±113 4673 4649
042040.9-144658 209.76 -39.98 292.81 -52.81 041823.2-145404	IC 367 MCG -2-12- 1 PGC 14917	PLA.-*. E -3.0± .7 	1.17± .07 .32± .05 1.13	140 .07 .00 					
042054.9+054022 188.08 -29.71 323.06 -46.27 041815.2+053317	 UGC 3017 KUG 0418+055 PGC 14920	.S..8*. U 8.0±1.4 	.95± .07 .65± .08 1.01	42 .63 .80 .32					3877 3837 3786
042055.9+023759 190.97 -31.45 319.15 -47.72 041819.4+023054	 MCG 0-12- 11 PGC 14923	PLBT+?. E -1.0± .8 	1.12± .06 .06± .05 1.16	 .56 .00 	 14.64 ±.15 				
0420.9 -0443 198.45 -35.38 308.78 -50.63 0418.5 -0451	 MCG -1-12- 1 PGC 14926	.S?.... 	1.08± .09 .61± .07 1.08	 .05 .89 .30					8945 8874 8861
042060.0-450421 250.71 -44.94 245.77 -46.36 041927.1-451124	 ESO 250- 18 PGC 14927	.S..3?/ S 3.0±1.9 	1.11± .05 .69± .04 1.11	25 .00 .95 .34	 15.11 ±.14 				
0421.0 -0217 195.91 -34.12 312.35 -49.79 0418.5 -0225	 UGC 3020 PGC 14929	.S..7*. U 7.0±1.3 	1.00± .06 .34± .05 1.02	158 .15 .47 .17					

R.A. 2000 DEC.	Names	Type	$\log D_{25}$	p.a.	B_T	$(B-V)_T$	$(B-V)_e$	m_{21}	V_{21}
l b		S_T n_L	$\log R_{25}$	A_g	m_B	$(U-B)_T$	$(U-B)_e$	W_{20}	V_{opt}
SGL SGB		T	$\log A_e$	A_i	m_{FIR}	$(B-V)_T^o$	m'_e	W_{50}	V_{GSR}
R.A. 1950 DEC.	PGC	L	$\log D_o$	A_{21}	B_T^o	$(U-B)_T^o$	m'_{25}	HI	V_{3K}
042103.6-481822		.SBS5P.	1.10± .05	15					
255.23 -44.56	ESO 202- 9	S	.48± .06	.00	15.32 ±.14				
241.78 -44.70		5.0±1.0		.71					
041938.0-482524	PGC 14931		1.10	.24					
042108.7-481516	NGC 1567	.E.0.P*	1.13± .06		12.17V±.15		.94± .02		
255.16 -44.55	ESO 202- 10	S	.02± .04	.00	13.36 ±.14		.41± .04		4546± 22
241.83 -44.74		-5.0±1.2	.83± .04	.00					4380
041943.0-482218	PGC 14934		1.12		13.29				4515
042113.4-215039		.SBS7*.	1.75± .02	132	12.67 ±.18	.61± .04		13.08±.2	906± 11
218.61 -42.31	ESO 550- 24	SU (1)	.42± .04	.03	12.75 ±.14	-.16± .10		184± 16	
280.96 -53.18		6.7± .4		.58		.52		175± 12	789
041904.0-215742	PGC 14936	6.7± .5	1.76	.21	12.11	-.23	15.24± .22	.76	839
0421.3 +3646		.S..6*.	1.16± .13	91					
163.00 -9.29	UGC 3016	U	.71± .12	1.43					
352.82 -26.24		6.0±1.3		1.04					
0418.0 +3639	PGC 14937		1.29	.35					
042127.9-555600	IC 2065	.S..1?.	1.07± .05	38					
265.55 -42.94	ESO 157- 21	S	.45± .04		14.60 ±.14				
233.26 -40.31	IRAS04204-5602	1.0±1.8		.46					
042024.0-560300	PGC 14943		1.07	.23					
042144.7-641326		.SXR4..	1.04± .07	90					
276.04 -40.23	ESO 84- 14	r	.08± .06	.03	14.20 ±.14				5300±190
225.30 -34.91	FAIR 767	4.5± .9		.13	13.72				5120
042117.0-642024	PGC 14953		1.04	.04	14.02				5296
042147.1-532355		.E+3...	1.10± .10						
262.17 -43.51		S	.27± .08	.00					
235.90 -41.89		-4.0± .9		.00					
042035.2-533054	PGC 14955		1.02						
042151.3-614650		RLBR+*.	1.06± .05	96					
273.03 -41.10	ESO 118- 22	S	.30± .04	.00	14.92 ±.14				5339
227.50 -36.57		-1.0± .9		.00					5160
042111.0-615348	PGC 14958		1.01		14.84				5331
0421.9 +3607		.E.....	1.04± .13	35					
163.56 -9.66	UGC 3021	U	.06± .05	2.02					
352.44 -26.80		-5.0± .8		.00					
0418.6 +3600	PGC 14959		1.35						
0421.9 +0403	IC 2057	.S?....	.84± .10						
189.78 -30.44	CGCG 419- 2		.09± .10	.68	15.02 ±.12				7434±113
321.17 -47.29				.13					7389
0419.3 +0356	PGC 14962		.90	.04	14.17				7346
042157.8+015021		.LBS.*.	1.21± .08	20					
191.92 -31.69	UGC 3023	UE	.17± .05	.48	14.48 ±.11				13200± 64
318.26 -48.31	ARAK 106	-2.0± .7		.00					13148
041922.1+014320	PGC 14964		1.24		13.80				13113
042159.1-565826	NGC 1574	.LAS-*.	1.53± .03	35	11.36 ±.14	.93± .01	.96± .01		
266.89 -42.58	ESO 157- 22	RS	.04± .04	.00	11.33 ±.11	.51± .02	.55± .01		925± 66
232.14 -39.72		-2.7± .4	.85± .03	.00		.92	11.10± .12		749
042059.0-570524	PGC 14965		1.52		11.33	.51	13.77± .23		909
042201.2-101020		.S..4*.	1.18± .05	65					
204.56 -37.74	MCG -2-12- 5	E (1)	.61± .05	.14					
300.44 -52.37		4.0±1.3		.90					
041938.4-101721	PGC 14969	4.2± .8	1.19	.31					
042208.8-433744	NGC 1571	.E+..P.	1.30± .05	172	13.23 ±.13	.94± .01	.97± .01		
248.65 -44.85	ESO 250- 19	BS	.12± .04	.00	13.22 ±.14	.47± .02	.51± .01		4397± 29
247.47 -47.24		-4.0± .6	.64± .02	.00		.90	11.92± .07		4236
042033.0-434442	PGC 14971		1.27		13.16	.49	14.43± .29		4360
042212.0-633640		.SBS9*.	1.24± .05	85					
275.27 -40.41	ESO 84- 15	S (1)	.52± .04	.03	15.33 ±.14				1279
225.80 -35.36		9.0±1.2		.53					1099
042141.0-634336	PGC 14974	8.9± .8	1.25	.26	14.77				1274
042223.7-561334		PSBS4..	1.01± .06	138					
265.89 -42.73	ESO 157- 23	r	.22± .05	.00	15.02 ±.14				13050±190
232.85 -40.23	FAIR 302	4.5± .9		.33					12875
042121.0-562030	PGC 14980		1.01	.11	14.62				13033
042228.3-562010		.L...P.	.97± .07						
266.03 -42.69	ESO 157- 24A	S	.19± .06	.00	15.09 ±.14				13007± 88
232.73 -40.17		-2.0±1.2		.00					12832
042126.0-562706	PGC 14984		.94		14.89				12990

R.A. 2000 DEC. / l b / SGL SGB / R.A. 1950 DEC.	Names / PGC	Type / ST nL / T / L	logD25 / logR25 / logAe / logDo	p.a. / Ag / Ai / A21	BT / mB / mFIR / BT°	(B-V)T / (U-B)T / (B-V)T° / (U-B)T°	(B-V)e / (U-B)e / m'e / m'25	m21 / W20 / W50 / HI	V21 / Vopt / VGSR / V3K
0422.4 +2718		.L..-?.	1.08± .17	150					
170.21 -15.63	UGC 3024	U	.12± .08	.96					
345.60 -33.28		-3.0±1.7		.00					
0419.4 +2711	PGC 14986		1.18						
0422.6 +0312									3878
190.69 -30.78	CGCG 419- 4			.62	15.3 ±.3				3830
320.21 -47.83									3791
0420.0 +0306	PGC 14992								
042242.6-403558	NGC 1572	PSBS1*.	1.39± .03	0	13.26 ±.16	.82± .02	.90± .02		
244.38 -44.84	ESO 303- 14	BS	.30± .04	.00	13.63 ±.14	.24± .03	.26± .03		6012± 34
251.46 -48.73	IRAS04210-4042	.7± .6	.93± .04	.30	10.85	.71	13.40± .12		5856
042101.1-404254	PGC 14993		1.39	.15	13.09	.20	14.31± .26		5972
042312.3+053429		.S..8*.	1.05± .05	73					4986± 7
188.55 -29.31	UGC 3025	U	.51± .05	.59					
323.35 -46.82	KUG 0420+054	8.0±1.3		.62					4945
042032.7+052733	PGC 15009		1.11	.25					4899
042328.5+751750	NGC 1530	.SBT3..	1.66± .02		12.25 ±.16	.80± .03	.88± .02	13.50±.0	2461± 4
135.22 17.77	UGC 3013	R	.28± .04	.58	12.9 ±.3	.12± .05	.20± .04	334± 5	2506± 63
17.26 3.96	7ZW 12	3.0± .3	1.23± .03	.38	10.59	.60	13.89± .09	316± 5	2613
041704.9+751048	PGC 15018		1.72	.14	11.41	-.04	14.72± .23	1.96	2404
042332.5-544402	IC 2066	.SB?...	.89± .06	129					
263.87 -42.95	ESO 157- 25		.12± .05	.00	15.26 ±.14				11200±190
234.26 -41.29	FAIR 770			.18					11026
042225.0-545054	PGC 15019		.89	.06	15.01				11182
042346.8-513557	NGC 1578	.SAS1P*	1.09± .05	177	13.92 ±.16	.79± .03	.80± .02		
259.65 -43.58	ESO 202- 14	S	.06± .05	.00	13.83 ±.14				6108± 88
237.63 -43.20	FAIR 771	1.0± .8	.61± .03	.06	13.01	.73	12.46± .07		5937
042230.0-514248	PGC 15025		1.09	.03	13.73		14.08± .32		6085
0423.9 -0050		.S..3..	1.03± .06	160					
194.89 -32.73	UGC 3029	U (1)	.17± .05	.27	15.07 ±.15				
314.88 -49.89		3.0± .9		.24					
0421.4 -0057	PGC 15027	4.5±1.1	1.05	.09					
042401.9-092345	IC 370	.SBT5..	1.13± .06	140					9776
203.97 -36.95	MCG -2-12- 11	E (1)	.08± .05	.28					
301.88 -52.67		5.0± .8		.12					9690
042138.2-093037	PGC 15029	1.9± .8	1.15	.04					9700
0424.2 +3054		.S..6*.	1.04± .15	60					
167.74 -12.89	UGC 3027	U	1.06± .06	.92					
348.87 -30.96	IRAS04210+3048	6.0±1.5		1.47					
0421.0 +3048	PGC 15031		1.13	.50					
042418.4-275639		.S..3*/	1.19± .04	128					
226.89 -43.19	ESO 420- 18	S	.92± .04	.01	15.89 ±.14				
270.59 -53.12		3.0±1.3		1.27					
042217.0-280330	PGC 15033		1.19	.46					
042421.0-473129		PLAT+P*	1.29± .05						
254.05 -44.11	ESO 202- 15	S	.07± .04	.00	13.62 ±.14				4320± 14
242.29 -45.58		-.7± .4		.00					4153
042254.0-473818	PGC 15035		1.28		13.56				4291
042426.2-233004			.71± .06						
221.04 -42.08	ESO 484- 14		.12± .05	.09	15.47 ±.14				8376±104
278.06 -53.80	IRAS04223-2336								8253
042219.0-233654	PGC 15040		.72						8315
042425.4-004447		.S?....	1.03± .06	135					
194.87 -32.58	UGC 3032		.12± .05	.27	14.58 ±.13				4734± 97
315.08 -49.97				.09					4673
042152.5-005138	PGC 15042		1.05		14.15				4652
042430.7+105318		.SB?...	1.00± .16					16.37±.3	10605± 10
183.95 -25.88	UGC 3033		.04± .12	1.16					10579
329.96 -44.23	IRAS04217+1046			.06	13.65			524± 7	
042145.3+104627	PGC 15044		1.11	.02					10517
042434.5+335232		.S?....	1.17± .05	143				14.52±.3	5479± 7
165.59 -10.81	UGC 3028		.35± .05	1.68					5526
351.22 -28.83	IRAS04213+3345			.53	12.02			435± 7	5390
042119.9+334540	PGC 15047		1.33	.18					
042435.0-575849	IC 2070	.SXS5*.	1.15± .05	85					7046
268.07 -41.97	ESO 118- 23	Sr (1)	.20± .05	.00	14.49 ±.14				6869
230.84 -39.34	IRAS04236-5805	4.8± .5		.30					7034
042339.1-580536	PGC 15048	1.1± .6	1.15	.10	14.15				

R.A. 2000 DEC.	Names	Type	$\log D_{25}$	p.a.	B_T	$(B-V)_T$	$(B-V)_e$	m_{21}	V_{21}
l b		S_T n_L	$\log R_{25}$	A_g	m_B	$(U-B)_T$	$(U-B)_e$	W_{20}	V_{opt}
SGL SGB		T	$\log A_e$	A_i	m_{FIR}	$(B-V)_T^o$	m'_e	W_{50}	V_{GSR}
R.A. 1950 DEC.	PGC	L	$\log D_o$	A_{21}	B_T^o	$(U-B)_T^o$	m'_{25}	HI	V_{3K}
042436.3+094134		.I..9?.	1.00± .16	.90				16.25±.3	7619± 7
185.03 -26.59	UGC 3034	U	.14± .12	.10					7590
328.60 -44.93		10.0±1.7		.07				404± 7	7532
042152.2+093443	PGC 15049		1.08						
042438.7-004534	A 0422+00	CI..9??	.10?	110					4630± 45
194.92 -32.54	MK 615	E	.06± .08	.27					4569
315.10 -50.02		10.0±1.8		.05	13.61				4549
042205.8-005224	PGC 15051		.13	.03					
0424.6 -2110		.I..9..	1.30± .06						
218.08 -41.34	UGCA 91	U	.31± .09	.03					
282.05 -54.01		10.0± .8		.23					
0422.5 -2117	PGC 15053		1.30	.16					
0424.6 +0712		.LB....	1.23± .14	20					
187.28 -28.05	UGC 3035	U	.18± .08	.53	14.25 ±.13				
325.66 -46.29		-2.0± .8		.00					
0422.0 +0706	PGC 15054		1.26						
042445.6-545631	NGC 1581	.L..-..	1.26± .05	80	13.61 ±.15	.76± .02	.71± .02		
264.09 -42.73	ESO 157- 26	S	.40± .04	.00	13.36 ±.14				1527± 88
233.90 -41.30	IRAS04236-5503	-3.0± .9	.55± .05	.00	13.11	.73	11.85± .16		1352
042339.1-550318	PGC 15055		1.20		13.45		13.76± .31		1510
0424.8 -0102		.S..4..	1.11± .07	99					
195.23 -32.64	UGC 3037	U	.54± .06	.35	15.34 ±.13				
314.74 -50.18		4.0± .9		.79					
0422.3 -0109	PGC 15057		1.14	.27					
042500.6-291406		.SBR1*P	.95± .05						
228.67 -43.30	ESO 420- 20	S	.11± .04	.00	15.13 ±.14				
268.41 -53.00		1.0±1.2		.12					
042301.0-292054	PGC 15058		.95	.06					
042501.8+071023		.S..3*.	1.09± .06	35				16.63±.3	8089± 10
187.38 -28.01	UGC 3038	U (1)	.63± .05	.53	15.27 ±.12				8052
325.67 -46.38		3.0±1.3		.87				485± 7	8004
042220.4+070334	PGC 15059	4.5±1.3	1.14	.31	13.81			2.50	
0425.1 +0301		.L.....	1.04± .09	33					
191.28 -30.38	UGC 3039	U	.74± .04	.68	15.51 ±.10				
320.38 -48.47	IRAS04224+0254	-2.0±1.0		.00					
0422.4 +0254	PGC 15062		1.01						
042507.0-264201		.SBS7*P	1.06± .06	138					5308
225.29 -42.74	ESO 484- 15	S (1)	.35± .03	.02	17.28 ±.14				
272.61 -53.53		7.0±1.2		.49					5178
042304.0-264848	PGC 15064	6.7±1.2	1.07	.18	16.75				5252
042548.2-674849		.SAS3*/	1.18± .04	152					5514
280.10 -38.44	ESO 55- 23	S (1)	.64± .04	.14	14.38 ±.14				5332
221.88 -32.67		3.0± .6		.88					5518
042545.0-675530	PGC 15078	3.3± .9	1.19	.32	13.32				
042555.8-083409		.LB.0P*	1.10± .08	25					
203.32 -36.15	MCG -1-12- 5	E	.16± .08	.29					
303.40 -52.92	IRAS04235-0840	-2.0±1.2		.00					
042331.3-084053	PGC 15081		1.11						
042618.8-033715	NGC 1576	.L..-*.	1.04± .07	55					
198.10 -33.67	MCG -1-12- 7	E	.24± .05	.08					
311.17 -51.50		-3.0±1.3		.00					
042349.0-034358	PGC 15089		1.02						
042619.6-491355		.S?....	.96± .06	159	15.35 ±.17	.94± .04	.89± .03		
256.33 -43.57	ESO 202- 18		.20± .05	.00	15.21 ±.14				
240.00 -44.91	IRAS04249-4920		.54± .03	.28	14.05		13.54± .08		
042457.0-492036	PGC 15091		.96	.10			14.49± .35		
042625.1-403913		.S?....	1.00± .05	160					6133
244.45 -44.14	ESO 303- 16		.18± .05	.00	15.34 ±.14				5975
250.91 -49.33				.18					6097
042444.0-404554	PGC 15095		1.00	.09	15.08				
042630.7-422625		RSXS4*.	1.01± .05	90					
246.94 -44.10	ESO 250- 21	S (1)	.07± .04	.00	14.97 ±.14				
248.47 -48.51		4.0± .9		.10					
042453.0-423306	PGC 15098	4.4± .8	1.01	.03					
042633.8-531114	IC 2073	.SBS6?P	1.07± .03	49	14.55 ±.15	.77± .03	.72± .03		
261.67 -42.86	ESO 157- 29	S	.38± .04	.00	14.39 ±.14				3956± 88
235.54 -42.59	IRAS04253-5317	6.0±1.3	.49± .02	.56	13.30	.67	12.49± .06		3782
042522.1-531754	PGC 15102		1.07	.19	13.89		13.82± .23		3938

R.A. 2000 DEC.	Names	Type	logD₂₅	p.a.	B_T	(B-V)_T	(B-V)_e	m₂₁	V₂₁
l b		S_T n_L	logR₂₅	A_g	m_B	(U-B)_T	(U-B)_e	W₂₀	V_opt
SGL SGB		T	logA_o	A_i	m_FIR	(B-V)_T^o	m'_e	W₅₀	V_GSR
R.A. 1950 DEC.	PGC	L	logD_o	A₂₁	B_T^o	(U-B)_T^o	m'₂₅	HI	V_3K

R.A. 2000 DEC. / coords	Names / PGC	Type	logD₂₅	p.a.	B_T	(B-V)_T	(B-V)_e	m₂₁	V₂₁
0426.5 +6619		.I..9?.	.96± .17	50				15.79±.1	3757± 11
142.26 11.92	UGC 3030	U	.29± .12	1.94					
12.26 -3.45		10.0±1.8		.22				190± 8	3890
0421.7 +6613	PGC 15103		1.14	.15					3692
042637.4-420531	IC 2068	PSAT0*.	1.09± .05	153					
246.46 -44.09	ESO 303- 17	BSr	.19± .04	.00	14.29 ±.14				4750±190
248.92 -48.70	FAIR 408	.3± .5		.14					4590
042459.1-421212	PGC 15106		1.08		14.07				4716
042637.5-343349		.LAT-?.	.96± .05	13					
236.01 -43.76	ESO 360- 9	S	.17± .03	.00	15.45 ±.14				
259.77 -51.80		-3.0± .8		.00					
042446.1-344030	PGC 15107		.94						
0426.7 +2024		.S?....	1.11± .07	47					1387± 10
176.29 -19.48	UGC 3045		.66± .06	1.76					
340.27 -38.71	IRAS04237+2017			.91					1390
0423.7 +2017	PGC 15109		1.27	.33					1300
042650.5+402825								16.11±.3	6787± 9
161.07 -5.93				2.45				269± 11	
356.35 -24.13								265± 8	6853
042324.6+402142	PGC 15111								6703
0426.8 +2957		.I..9?.	1.04± .15						5228± 10
168.85 -13.13	UGC 3044	U	.18± .12	1.15					
348.59 -32.04		10.0±1.7		.13					5261
0423.7 +2951	PGC 15112		1.15	.09					5142
0426.9 +0141		.S..4..	1.00± .08	50					
192.85 -30.74	UGC 3049	U	.33± .06	.44	15.13 ±.12				
318.90 -49.48		4.0± .9		.49					
0424.3 +0135	PGC 15113		1.04	.17					
042704.9-060955		.SBT5P*	1.12± .06	135					
200.89 -34.76	MCG -1-12- 8	E (1)	.11± .05	.05					
307.36 -52.52		5.0± .8		.16					
042437.8-061635	PGC 15123	4.2± .8	1.13	.05					
042715.8+323728		.SB.6*.	1.16± .13	30				15.63±.3	6445± 7
166.91 -11.26	UGC 3047	U	.05± .12	1.49					
350.74 -30.14		6.0±1.1		.07				204± 7	6487
042403.0+323047	PGC 15134		1.30	.02					6360
042718.3-101905		.E?....			16.50S±.15				
205.45 -36.64				.17					
300.68 -53.67	HICK 28C								
042455.7-102544	PGC 15135								
042718.6-101822		.S?....			16.17S±.15				
205.44 -36.63				.17					11294± 41
300.70 -53.67	HICK 28A								11203
042456.1-102502	PGC 15136								11224
042718.7+022236		.S..6*.	.98± .07					16.07±.2	6933± 7
192.26 -30.28	UGC 3051	U	.02± .05	.60				201± 7	
319.91 -49.26		6.0±1.2		.03				194± 5	6880
042442.5+021557	PGC 15137		1.03	.01					6854
042720.1-101935		.S?....			15.92S±.15				
205.46 -36.63				.17					11402± 41
300.67 -53.68	HICK 28B								11311
042457.5-102614	PGC 15141								11332
042730.0+305650		.S?....	.96± .17					14.85±.3	2147± 7
168.20 -12.36	UGC 3050		.10± .12	1.30					
349.49 -31.41	IRAS04243+3049			.15				133± 7	2183
042419.6+305010	PGC 15147		1.08	.05					2061
042730.2-833304		.SBS5*P	1.08± .05	108					
296.86 -30.79	ESO 4- 13	S (1)	.33± .04	.63	14.78 ±.14				
210.28 -20.83	IRAS04340-8339	5.0±1.2		.49	13.25				
043408.1-833924	PGC 15148	4.4±1.2	1.14	.16					
042732.5-541148		.E.4.*.	1.12± .05	42	14.66 ±.14	.70± .02			
262.98 -42.51	ESO 157- 30	S	.38± .04	.00	14.30 ±.14				1341± 51
234.34 -42.08		-5.0±1.2		.00		.69			1166
042624.0-541824	PGC 15149		1.01		14.46		14.32± .31		1326
042733.1-420953	NGC 1585	.SA.5*.	1.06± .05	175					
246.55 -43.91	ESO 303- 18	S (1)	.22± .04	.00	14.23 ±.14				4642
248.69 -48.82	IRAS04259-4216	5.0± .6		.33	12.73				4481
042555.0-421630	PGC 15150	3.3± .5	1.06	.11	13.87				4609

R.A. 2000 DEC. l b SGL SGB R.A. 1950 DEC.	Names PGC	Type S_T n_L T L	$\log D_{25}$ $\log R_{25}$ $\log A_e$ $\log D_o$	p.a. A_g A_i A_{21}	B_T m_B m_{FIR} B_T^o	$(B-V)_T$ $(U-B)_T$ $(B-V)_T^o$ $(U-B)_T^o$	$(B-V)_e$ $(U-B)_e$ m'_e m'_{25}	m_{21} W_{20} W_{50} HI	V_{21} V_{opt} V_{GSR} V_{3K}
042737.8-550137 264.07 -42.31 233.46 -41.56 042632.0-550812	NGC 1596 ESO 157- 31 PGC 15153	.LA..*/ R -2.0± .5	1.57± .02 .59± .02 .73± .01 1.48	20 .00 .00	12.10 ±.13 12.13 ±.10 12.09	.94± .01 .46± .02 .87 .41	.98± .01 .51± .02 11.24± .05 13.32± .18	14.40±.3 247± 9	1510± 8 1465± 24 1329 1491
0427.7 +7053 138.86 15.07 14.96 .21 0422.2 +7047	 UGC 3036 PGC 15157	.S?.... 	1.08± .07 .81± .06 1.14	102 .64 1.22 .41				15.82±.1 424± 8	7453± 11 7596 7393
042743.9-333718 234.75 -43.42 261.11 -52.34 042551.0-334354	 ESO 360- 11 PGC 15160	.S?.... 	.91± .06 .19± .06 .90	30 .01 .14	 15.31 ±.14 14.95				13838± 39 13691 13794
042746.0-623532 273.73 -40.18 226.17 -36.52 042711.0-624206	 ESO 84- 18 PGC 15163	.SAT1*. S .5± .6	1.18± .05 .10± .05 1.18	 .00 .10 .05	 14.04 ±.14 13.87				5677 5496 5674
042753.7-550326 264.10 -42.27 233.40 -41.57 042648.1-551000	NGC 1602 ESO 157- 32 IRAS04267-5510 PGC 15168	.IBS9P* RS (1) 9.5± .6 7.8± .8	1.29± .04 .26± .05 1.06± .03 1.29	83 .00 .19 .13	13.33 ±.15 13.79 ±.14 13.02 13.38	.35± .04 -.23± .05 .28 -.28	.27± .01 -.35± .03 14.12± .08 13.98± .29	14.77±.3 87± 9 1.26	1568± 8 1731± 31 1402 1564
042754.9-293244 229.25 -42.74 267.65 -53.54 042556.0-293920	 PGC 15169	.SXS9.. S (1) 9.0± .9 10.0± .9	1.04± .09 .06± .08 1.04	 .00 .06 .03					
042759.7-475437 254.48 -43.46 241.34 -45.88 042634.0-480112	 ESO 202- 23 IRAS04265-4801 PGC 15172	.P..... S 99.0	1.33± .04 .07± .04 1.33	 .00 .09 .03	 13.50 ±.14 12.43 13.39				 4950± 60 4781 4926
042759.7-425019 247.49 -43.82 247.73 -48.56 042623.0-425654	 ESO 251- 2 PGC 15173	.LBS0*P S -2.4± .5	1.12± .05 .22± .04 1.09	9 .00 .00	 14.05 ±.14				
042804.8-144549 210.64 -38.32 293.19 -54.59 042547.2-145225	 MCG -2-12- 22 PGC 15175	.SXS9*. E (1) 8.7± .7 9.0± .7	1.10± .05 .21± .04 1.11	20 .06 .21 .10					3396 3293 3331
0428.1 +2139 175.52 -18.41 341.73 -38.12 0425.2 +2132	 UGC 3053 IRAS04251+2132 PGC 15179	.S..6*. U 6.0±1.3	.93± .07 .06± .05 1.09	170 1.64 .08 .03	 14.75 ±.12 12.27 13.02				2407± 10 2413 2322
042810.3-173124 213.93 -39.34 288.39 -54.83 042556.0-173800	NGC 1584 ESO 551- 6 PGC 15180	.LAS-P* SE -2.7± .7	.91± .06 .06± .04 .91	 .04 .00	 14.88 ±.14				
0428.1 +0102 193.69 -30.82 318.23 -50.06 0425.6 +0056	 UGC 3054 PGC 15181	.S..8*. U 8.0±1.4	1.16± .07 1.00± .06 1.19	30 .31 1.23 .50					3924 3866 3847
042818.3-402814 244.20 -43.78 250.91 -49.73 042637.0-403448	 ESO 303- 20 PGC 15188	.S?.... 	1.06± .05 .05± .05 1.06	 .00 .07 .02	 14.96 ±.14 14.84				8993 8835 8959
042821.8-474857 254.34 -43.41 241.40 -45.99 042656.0-475530	NGC 1595 ESO 202- 25 PGC 15195	.E.3... S -5.0± .5	1.13± .03 .17± .03 .65± .01 1.08	17 .00 .00	13.69 ±.13 13.75 ±.14 13.65	.96± .01 .44± .02 .92 .47	.97± .01 .45± .02 12.43± .04 13.93± .20		 4702± 22 4533 4678
042826.0+393728 161.90 -6.30 356.02 -24.98 042501.5+393052	 UGC 3052 IRAS04250+3930 PGC 15197	.SB?... 	1.16± .07 .16± .06 1.38	10 2.32 .22 .08				14.98±.1 301± 11 280± 6	4224± 6 4286 4141
042830.8-534410 262.33 -42.46 234.71 -42.48 042721.0-535042	IC 2079 ESO 157- 33 PGC 15200	.SBT2.. S 2.0± .9	1.09± .05 .48± .04 1.09	125 .00 .60 .24	 14.81 ±.14 14.11				 10834± 41 10659 10819
0428.5 +1103 184.47 -24.99 330.96 -44.94 0425.8 +1057	 UGC 3055 PGC 15203	.L...?. U -2.0±1.7	1.15± .08 .28± .04 1.23	60 1.08 .00					

R.A. 2000 DEC. / l b / SGL SGB / R.A. 1950 DEC.	Names / PGC	Type / S_T n_L / T / L	$logD_{25}$ / $logR_{25}$ / $logA_e$ / $logD_o$	p.a. / A_g / A_i / A_{21}	B_T / m_B / m_{FIR} / B_T^o	$(B-V)_T$ / $(U-B)_T$ / $(B-V)_T^o$ / $(U-B)_T^o$	$(B-V)_e$ / $(U-B)_e$ / m'_e / m'_{25}	m_{21} / W_{20} / W_{50} / HI	V_{21} / V_{opt} / V_{GSR} / V_{3K}
042833.9-474652 254.29 -43.38 241.42 -46.03 042708.0-475324	NGC 1598 ESO 202- 26 IRAS04271-4753 PGC 15204	PSBS5P* Sr (1) 4.9± .4 2.2± .5	1.16± .03 .24± .04 .67± .01 1.16	123 .00 .36 .12	13.81 ±.13 13.91 ±.14 12.49 13.47	.52 .01 .44 	.58± .01 12.65± .04 13.84± .22		5108± 23 4939 5084
042837.4-532435 261.89 -42.51 235.04 -42.69 042726.6-533107	 PGC 15205		 	 .00 	15.77 ±.07	.77± .04			13176± 51 13001 13161
0428.7 +6935 139.92 14.26 14.29 -.90 0423.4 +6929	A 0423+69 CGCG 328- 3 7ZW 14 PGC 15211	 	.56± .15 .12± .06 .65	 .91 	 15.7 ±.2 				10280± 60 10420 10219
0428.7 +7020 139.34 14.77 14.72 -.29 0423.3 +7014	A 0423+70 UGC 3042 IRAS04233+7014 PGC 15212	.SXS4.. U (1) 4.0± .8 1.6± .8	1.33± .04 .45± .05 .88± .04 1.39	70 .69 .66 .23	14.00 ±.18 14.4 ±.3 12.05 12.72	.78± .03 .52 	.80± .03 13.89± .08 14.36± .29	14.40±.1 299± 11 278± 7 1.46	3059± 6 2947± 59 3199 2998
042845.3-123032 208.12 -37.26 297.11 -54.44 042625.2-123705	 MCG -2-12- 24 PGC 15214	.SBS9.. E (1) 9.0± .8 7.5± .8	1.17± .05 .05± .05 1.18	30 .05 .06 .03					1803 1705 1737
042849.5-445147 250.27 -43.58 244.99 -47.66 042717.0-445818	 ESO 251- 4 PGC 15219	.E.5... S -5.0± .9 	1.11± .06 .30± .04 1.02	24 .00 .00 	 14.17 ±.14 				
042849.7-465753 253.16 -43.42 242.37 -46.53 042722.0-470424	 ESO 251- 5 FAIR 410 PGC 15220	.S?.... 	.83± .06 .06± .05 .83	 .00 .09 .03	 14.60 ±.14 13.47 14.48				4398± 63 4230 4373
0428.9 +7055 138.90 15.16 15.06 .18 0423.4 +7049	 UGC 3043 PGC 15225	.S..6*. U 6.0±1.5 	1.07± .14 1.09± .06 1.14	75 .73 1.47 .50				15.43±.1 381± 8 	7405± 7 7547 7345
042901.1-533648 262.14 -42.41 234.77 -42.61 042751.1-534318	IC 2081 ESO 157- 34 PGC 15231	.LB.-?. S -3.0± .9 	1.08± .05 .15± .04 1.06	90 .00 .00 	 14.46 ±.14 14.27				12738± 41 12563 12723
042906.5-372847 240.08 -43.50 255.08 -51.17 042720.0-373518	 ESO 303- 21 IRAS04273-3735 PGC 15237	.S?.... 	.90± .06 .18± .05 .90	170 .02 .27 .09	 14.65 ±.14 12.20 14.31				8706 8552 8669
042907.4-534936 262.42 -42.35 234.53 -42.49 042758.0-535606	IC 2082 ESO 157- 35 PGC 15239	.L...P. RS -2.4± .3 	1.13± .05 .23± .04 .77± .03 1.10	 .00 .00 	13.84 ±.13 14.95 ±.14 14.18	1.11± .02 .58± .05 .98 .62	1.04± .01 .56± .04 13.18± .10 13.80± .29		11785± 28 11609 11770
042918.5-532930 261.97 -42.39 234.86 -42.72 042808.1-533559	 PGC 15250		 	 .00 	15.60 ±.16	1.03± .04			12881± 37 12706 12866
0429.3 +6931 140.01 14.26 14.30 -.99 0424.0 +6925	 UGC 3046 IRAS04240+6925 PGC 15254	.S..3.. U 3.0±1.0 	1.17± .05 .10± .05 1.26	 .91 .13 .05	 14.7 ±.3 12.61 13.58			15.25±.1 240± 8 1.62	4699± 11 4839 4638
0429.3 -0445 199.74 -33.58 309.86 -52.61 0426.9 -0452	 PGC 15259	.S..6*/ E 6.0±1.3 	1.18± .08 .73± .08 1.19	85 .06 1.07 .37					
042923.1-625321 274.02 -39.90 225.75 -36.45 042850.1-625948	 ESO 84- 20 IRAS04288-6259 PGC 15261	.S..3*/ S (1) 3.0±1.3 5.6±1.3	1.06± .05 .49± .04 1.06	171 .00 .68 .25	 14.88 ±.14 12.91 				
0429.4 +7025 139.32 14.86 14.81 -.26 0423.9 +7018	 UGC 3048 IRAS04239+7018 PGC 15263	.SXT4*. U 4.0±1.2 	1.03± .06 .20± .05 1.10	125 .69 .29 .10	 14.9 ±.3 12.15 13.95			15.14±.1 247± 8 1.10	3049± 11 3190 2989
0429.4 -5336 262.12 -42.35 234.72 -42.66 0428.3 -5343	 PGC 15272	 	 .89± .16	 .00 	14.2 ±.4	.95± .06	1.03± .03 14.13± .55		

R.A. 2000 DEC. l b SGL SGB R.A. 1950 DEC.	Names PGC	Type S_T n_L T L	$\log D_{25}$ $\log R_{25}$ $\log A_\bullet$ $\log D_\circ$	p.a. A_g A_i A_{21}	B_T m_B m_{FIR} B_T°	$(B-V)_T$ $(U-B)_T$ $(B-V)_T^\circ$ $(U-B)_T^\circ$	$(B-V)_\bullet$ $(U-B)_\bullet$ m' m'_{25}	m_{21} W_{20} W_{50} HI	V_{21} V_{opt} V_{GSR} V_{3K}
042928.8-364225 239.03 -43.38 256.18 -51.55 042741.0-364854	ESO 360- 13 IRAS04276-3648 PGC 15273	.S.5?P/ S 5.0±1.8	1.12± .05 .75± .04 1.12	.35 .00 1.13 .38	15.32 ±.14 13.11				
042930.7-264248 225.61 -41.79 272.29 -54.49 042728.0-264918	NGC 1591 ESO 484- 25 IRAS04274-2649 PGC 15276	.SBR2P. S 2.0± .9	1.08± .04 .19± .03 1.08	30 .05 .23 .09	13.77 ±.14 12.24 13.45				4127± 52 3994 4077
042932.7-414631 246.01 -43.55 248.95 -49.33 042754.0-415300	ESO 303- 22 PGC 15278	.SBS6.. S (1) 6.0± .9 5.6± .9	1.08± .05 .33± .04 1.08	40 .00 .48 .16	15.47 ±.14 14.97				4485 4324 4454
0429.5 -5344 262.28 -42.31 234.57 -42.60 0428.4 -5350	PGC 15279		.61± .05	.00	14.87 ±.18	.87± .04	.89± .03 13.41± .13		
042935.0+004609 194.18 -30.68 318.08 -50.49 042700.4+003938	UGC 3058 PGC 15283	.SBS9.. E (1) 9.0± .9 7.5± .8	1.10± .05 .28± .04 1.13	82 .25 .29 .14	15.11 ±.16				
042940.8-272431 226.53 -41.92 271.08 -54.40 042739.0-273100	NGC 1592 ESO 421- 2 VV 647 PGC 15292	.P..... S 99.0	1.15± .04 .33± .03 1.14	96 .10 .25	14.49 ±.14 14.13				927± 37 792 878
042942.3+034050 191.39 -29.05 322.09 -49.16 042704.6+033420	UGC 3059 PGC 15294	.SA.8.. U 8.0± .8	1.42± .03 .50± .05 1.48	40 .64 .61 .25	14.7 ±.2 13.44			14.85±.1 384± 8 1.16	4811± 9 4760 4735
042949.4-534851 262.38 -42.26 234.46 -42.58 042840.1-535518	ESO 157- 36 PGC 15301	.S?.... 	.99± .06 .69± .05 .99	117 .00 1.04 .35	15.07 ±.14 13.96				13029± 33 12854 13015
042959.5-265008 225.80 -41.72 272.04 -54.58 042757.0-265636	ESO 484- 28 PGC 15311	.E?.... 	.90± .04 .17± .03 .86	27 .05 .00	14.38 ±.13 14.54 ±.14 14.35	.91± .01 .86	13.46± .24		4038± 37 3904 3989
0430.0 -5140 259.52 -42.61 236.73 -43.94 0428.8 -5147	PGC 15315		.52± .06	.00	14.59 ±.17	.68± .03	.76± .03 12.68± .20		
043009.8-424052 247.26 -43.42 247.64 -48.98 042833.0-424718	ESO 251- 6 PGC 15318	.S.3?P/ S 3.0±1.9	1.16± .05 .94± .04 1.16	0 .00 1.29 .47	16.35 ±.14				
043010.7+065620 188.43 -27.10 326.38 -47.59 042729.5+064952	UGC 3061 VV 555 PGC 15319	.S?.... 	1.33± .05 .02± .05 1.38	 .51 .03 .01	14.6 ±.4 13.70 14.04				4333± 32 4291 4255
043021.2-363834 238.96 -43.20 256.16 -51.74 042833.4-364500	PGC 15321	PLA.0?. S -2.0± .8	1.23± .08 .06± .08 1.22	.00 .00					
043037.0-582743 268.40 -41.07 229.69 -39.61 042944.1-583406	ESO 118- 28 PGC 15330	PSBS3.. r 3.3± .9	.93± .06 .05± .03 .93	80 .00 .06 .02	15.06 ±.14				
043038.1-001816 195.40 -31.03 316.73 -51.20 042804.7-002442	NGC 1586 UGC 3062 PGC 15331	.SBS4.. UE (2) 4.0± .6 3.5± .7	1.24± .04 .28± .04 1.25	155 .14 .41 .14	13.97 ±.13 13.40				3535± 50 3471 3462
043039.6+003943 194.45 -30.51 318.12 -50.78 042805.2+003317	NGC 1587 UGC 3063 PGC 15332	.E...P. PUE -5.0± .5	1.23± .06 .06± .05 .89± .03 1.25	144 .21 .00	12.70 ±.13 12.98 ±.10 12.61	1.02± .01 .47± .02 .94 .44	1.05± .01 .57± .01 12.64± .09 13.70± .36	15.90±.1 498± 11	3690± 9 3871± 34 3641 3628
043042.8-045218 200.05 -33.34 309.87 -52.96 042814.4-045844	IC 373 MCG -1-12- 13 PGC 15335	RLBT+.. E -1.0± .8	1.13± .06 .20± .05 1.10	10 .07 .00					

R.A. 2000 DEC. l　　b SGL　　SGB R.A. 1950 DEC.	Names PGC	Type S_T　n_L T L	$\log D_{25}$ $\log R_{25}$ $\log A_e$ $\log D_o$	p.a. A_g A_I A_{21}	B_T m_B m_{FIR} B_T^o	$(B-V)_T$ $(U-B)_T$ $(B-V)_T^o$ $(U-B)_T^o$	$(B-V)_e$ $(U-B)_e$ m'_e m'_{25}	m_{21} W_{20} W_{50} HI	V_{21} V_{opt} V_{GSR} V_{3K}
043043.7-535849 262.56 -42.09 234.17 -42.57 042935.0-540512	IC　2083· ESO 157- 37 FAIR 774 PGC 15339	.SBS0P. S .0± .9	1.03± .05 .26± .04 1.02	93 .00 .19	 14.92 ±.14 14.53				12963± 41 12786 12950
043043.8+003955 194.46 -30.49 318.14 -50.80 042809.4+003329	NGC 1588 UGC 3064 MK　616 PGC 15340	.E...P? PE -4.6± .7	1.14± .06 .26± .03 .63± .03 1.10	175 .21 .00	13.89 ±.13 13.81 ±.10 13.58	.98± .01 .46± .02 .90 .43	.99± .01 .51± .02 12.53± .09 13.95± .32		3458± 25 3397 3385
043045.9+005150 194.27 -30.38 318.42 -50.71 042811.2+004525	NGC 1589 UGC 3065 PGC 15342	.S..2./ PUE　(1) 2.2± .5 4.5±1.1	1.50± .02 .49± .04 1.01± .02 1.53	160 .25 .61 .25	12.80 ±.14 13.40 ±.14 12.21	1.04± .05 .51± .03 .87 .35	1.12± .05 .59± .03 13.34± .07 13.93± .21		3795± 50 3735 3722
043046.8-433930 248.60 -43.28 246.27 -48.58 042912.1-434554	 ESO 251- 7 PGC 15343	.S..3?/ S 3.0±1.9	1.08± .05 .62± .04 1.08	172 .00 .85 .31	 14.84 ±.14				
043050.3+645047 143.68　11.24 11.77 -4.92 042605.8+644418	NGC 1569 UGC 3056 7ZW　16 PGC 15345	.IB.9.. R　(2) 10.0± .4 7.5± .7	1.56± .02 .31± .02 .94± .01 1.76	120 2.18 .23 .16	11.86M±.09 11.9 ±.2 9.16 9.45	.83± .01 -.14± .02	.79± .01 -.17± .01 11.96± .04 13.71± .14	12.43±.1 84± 5 74± 8 2.83	-89± 5 -74± 17 40 -152
043051.7-054755 201.05 -33.76 308.43 -53.30 042824.3-055420	NGC 1594 MCG -1-12- 14 IRAS04284-0554 PGC 15348	.SBT4.. PE　(1) 4.0± .6 2.2± .8	1.25± .05 .14± .05 1.25	100 .09 .20 .07	 13.43			15.33±.1 297± 11 291± 11	4328± 10 4315± 59 4247 4259
043052.2-161824 212.78 -38.29 290.60 -55.40 042836.5-162448	 MCG -3-12- 11 PGC 15349	.SXR1P* E 1.0±1.2	1.07± .06 .11± .05 1.07	155 .01 .11 .05					
043057.9+053225 189.85 -27.75 324.76 -48.49 042818.2+052600	 UGC 3066 IRAS04282+0526 PGC 15355	.SXR7*. U 7.0±1.1	1.17± .07 .09± .06 1.23	110 .68 .12 .04	 14.46 ±.18 12.81 13.64			14.67±.3 327± 7 .98	4640± 7 4594 4565
043059.7-020012 197.15 -31.84 314.30 -51.97 042828.2-020637	 UGC 3070 PGC 15356	.SXS3P* E　(1) 3.0± .9 4.2±1.6	1.14± .04 .23± .04 1.15	160 .16 .32 .12	 14.51 ±.14				
0431.0　+0805 187.51 -26.27 327.96 -47.13 0428.3　+0759	 UGC 3067 PGC 15358	.E...?. U -5.0±1.6	1.04± .09 .00± .04 1.13	 .58 .00	 14.68 ±.13				
0431.1　-0324 198.60 -32.53 312.19 -52.54 0428.6　-0331	 PGC 15364	.LBR+?. E -1.0±1.3	1.05± .09 .42± .08 1.00	92 .13 .00					
043108.5-631922 274.46 -39.57 225.19 -36.28 043038.1-632542	 ESO 84- 21 FAIR 775 PGC 15367	.SBR3.. S　(1) 3.0± .9 3.3± .9	1.08± .05 .35± .04 1.08	124 .02 .48 .17	 14.49 ±.14 13.33 13.95				5100±190 4918 5101
043110.6+073757 187.96 -26.50 327.43 -47.42 042828.6+073133	NGC 1590 UGC 3071 2ZW　13 PGC 15368	.P..... R 99.0	.93± .05 .07± .04 .99	 .56 .11 .04	14.49 ±.14 14.45 ±.12 12.04 13.77	.80± .05 .08± .04 .63 -.05	 13.82± .31	15.67±.2 293± 5 1.86	3897± 7 3806± 61 3856 3820
0431.1　+0050 194.36 -30.30 318.46 -50.82 0428.6　+0044	 UGC 3072 PGC 15369	.I..9*. U 10.0±1.2	1.11± .14 .25± .12 1.13	 .25 .18 .12					3606 3545 3534
043112.9-612710 272.15 -40.15 226.84 -37.61 043033.1-613330	 ESO 118- 30 PGC 15373	.E+4... S -5.0± .8	.96± .07 .26± .04 .97± .11 .88	 .00 .00	14.2 ±.3 15.30 ±.14 14.83	.96± .06 .79	1.04± .02 14.57± .39 13.35± .46		17962± 26 17781 17960
0431.2　-6130 272.21 -40.13 226.79 -37.58 0430.6　-6136	 PGC 15379			 .00	15.4 ±.2	.95± .04			15846± 47 15664 15844
043122.4+063823 188.90 -27.03 326.24 -48.00 042841.5+063200	 UGC 3074 KUG 0428+065 PGC 15386	.S..7*. U 7.0±1.2	.96± .06 .05± .05 1.01	 .51 .08 .03	 15.15 ±.15 14.56			16.53±.3 115± 7 1.95	4431± 7 4388 4355

R.A. 2000 DEC. / l b / SGL SGB / R.A. 1950 DEC.	Names / / / PGC	Type / ST nL / T / L	logD25 / logR25 / logA• / logD°	p.a. / Ag / AI / A21	BT / mB / mFIR / BT°	(B-V)T / (U-B)T / (B-V)T° / (U-B)T°	(B-V)• / (U-B)• / m'• / m'25	m21 / W20 / W50 / HI	V21 / Vopt / VGSR / V3K
043123.8-035532 / 199.17 -32.73 / 311.44 -52.79 / 042854.4-040154	/ MCG -1-12- 15 / IRAS04289-0401 / PGC 15387	.SBR5*. / E (1) / 5.0± .8 / 4.2± .8	1.14± .06 / .07± .05 / / 1.15	95 / .10 / .11 / .04	/ / 13.95 /				4752 / 4677 / / 4683
043124.2-542504 / 263.11 -41.91 / 233.62 -42.37 / 043017.0-543124	IC 2085 / ESO 157- 38 / IRAS04303-5430 / PGC 15388	.L..0P/ / S / -2.0± .8 /	1.38± .04 / .67± .03 / 1.00± .05 / 1.28	101 / .00 / .00 /	13.89 ±.17 / 14.26 ±.14 / / 14.10	.77± .03 / .26± .05 / .70 / .21	.85± .02 / .31± .04 / 14.38± .18 / 13.98± .27		/ 1010± 30 / 833 / 998
0431.4 +0809 / 187.52 -26.14 / 328.13 -47.18 / 0428.7 +0803	/ UGC 3075 / IRAS04287+0803 / PGC 15389	.S?.... / / /	1.00± .08 / .55± .06 / / 1.05	108 / .58 / .82 / .27	/ / 12.66 /				4021 / / 3983 / 3945
043130.4-520204 / 259.95 -42.33 / 236.15 -43.89 / 043016.0-520824	/ ESO 202- 31 / / PGC 15390	.L..../ / S / -2.0± .9 /	1.01± .05 / .40± .03 / / .95	44 / .00 / .00 /	/ 15.00 ±.14 / / 14.82				11840± 88 / 11665 / 11825 /
043138.8-043521 / 199.90 -33.00 / 310.45 -53.08 / 042910.1-044143	NGC 1599 / MCG -1-12- 16 / IRAS04292-0441 / PGC 15403	.SBS5P* / PE (1) / 5.0± .9 / 4.2±1.6	.93± .08 / .04± .05 / / .94	/ .12 / .06 / .02	14.06 ±.14 / / 13.07 /	.37± .05 / -.25± .04 / /	/ / / 13.46± .43		
043139.6-543605 / 263.34 -41.83 / 233.40 -42.28 / 043033.1-544224	NGC 1617 / ESO 157- 41 / IRAS04305-5442 / PGC 15405	.SBS1.. / R / 1.0± .7 /	1.63± .03 / .31± .03 / 1.03± .02 / 1.63	107 / .00 / .31 / .15	11.38M±.12 / 11.44 ±.11 / 12.87 / 11.09	.94± .01 / .50± .01 / .88 / .44	.98± .01 / .53± .01 / 12.02± .06 / 13.61± .20		/ 1063± 21 / 886 / 1052
043139.9-050516 / 200.42 -33.24 / 309.66 -53.26 / 042911.8-051138	NGC 1600 / MCG -1-12- 17 / / PGC 15406	.E.3... / R / -5.0± .3 /	1.39± .07 / .17± .07 / 1.18± .02 / 1.36	170 / .08 / .00 /	11.93M±.08 / 12.37 ±.18 / / 11.85	1.01± .01 / .54± .03 / .95 / .55	1.03± .01 / .59± .01 / 13.34± .08 / 13.47± .41		/ 4718± 23 / 4639 / 4650
043141.9-050343 / 200.40 -33.22 / 309.71 -53.26 / 042913.7-051005	NGC 1601 / MCG -1-12- 18 / / PGC 15413	.L...*/ / R / -2.0± .7 /	.78± .13 / .27± .05 / .46± .04 / .75	95 / .07 / .00 /	14.80 ±.15 / / / 14.65	.99± .02 / .47± .09 / .90 / .45	1.00± .02 / / 12.59± .13 / 12.92± .69		/ 4997± 56 / 4918 / 4929
0431.7 +0826 / 187.31 -25.92 / 328.52 -47.08 / 0429.0 +0820	/ UGC 3076 / / PGC 15414	.S..3.. / U (1) / 3.0±1.0 / 3.5±1.3	1.00± .08 / .73± .06 / / 1.06	109 / .62 / 1.00 / .36					
0431.7 -1239 / 208.67 -36.66 / 297.05 -55.18 / 0429.4 -1246	/ MCG -2-12- 35 / / PGC 15417	.S?.... / / /	1.04± .09 / .78± .07 / / 1.04	/ .03 / 1.17 / .39					9985± 70 / 9885 / 9924 /
043150.1-050545 / 200.45 -33.21 / 309.68 -53.30 / 042921.9-051206	NGC 1603 / MCG -1-12- 19 / / PGC 15424	.E?.... / / /	.89± .15 / .16± .07 / .61± .08 / .86	115 / .08 / .00 /	14.7 ±.2 / / / 14.53	.92± .05 / / .85 /	.94± .02 / .22± .08 / 13.23± .28 / 13.76± .80		5117± 60 / 5038 / 5050 /
043156.8+011147 / 194.13 -29.95 / 319.10 -50.83 / 042921.8+010527	A 0429+01 / UGC 3080 / / PGC 15429	.SXT5.. / UE (2) / 5.0± .5 / 3.6± .7	1.27± .04 / .06± .04 / / 1.29	140 / .30 / .09 / .03	14.1 ±.4 / 14.2 ±.2 / / 13.81	.75± .06 / / .65 /	/ / / 15.12± .45	14.56±.1 / 170± 5 / 145± 4 / .72	3541± 5 / 3482± 59 / 3481 / 3469
0432.0 +0222 / 193.00 -29.30 / 320.75 -50.30 / 0429.4 +0216	/ UGC 3081 / / PGC 15436	.SBR1.. / U / 1.0± .8 /	.99± .09 / .15± .06 / / 1.03	10 / .48 / .16 / .08	/ 15.25 ±.15 / /				
0432.0 +6336 / 144.71 10.51 / 11.18 -6.01 / 0427.4 +6330	/ UGCA 92 / / PGC 15439	.I..9?. / U / 10.0±1.8 /	1.30? / .30? / / 1.53	/ 2.43 / .23 / .15				13.00±.1 / 61± 7 / 77± 12 /	-99± 5 / / 26 / -163
043203.3-050158 / 200.42 -33.13 / 309.81 -53.33 / 042935.1-050818	NGC 1606 / MCG -1-12- 22 / / PGC 15443	.LXR+*. / RE / -1.1± .5 /	.63± .22 / .00± .07 / / .64	/ .08 / .00 /					
043206.3+003401 / 194.77 -30.25 / 318.24 -51.15 / 042932.0+002740	NGC 1608 / UGC 3082 / / PGC 15447	.L..0*. / UE / -2.0± .7 /	1.21± .05 / .43± .03 / / 1.17	130 / .19 / .00 /	/ 14.43 ±.11 / /				
043211.6-285141 / 228.59 -41.69 / 268.40 -54.62 / 043012.0-285800	/ ESO 421- 8 / / PGC 15451	.S?.... / / /	1.03± .06 / .40± .06 / / 1.03	94 / .02 / .60 / .20	/ 15.97 ±.14 / / 15.29				10482± 52 / 10343 / 10438 /

R.A. 2000 DEC.	Names	Type	logD$_{25}$	p.a.	B$_T$	(B-V)$_T$	(B-V)$_e$	m$_{21}$	V$_{21}$
l b		S$_T$ n$_L$	logR$_{25}$	A$_g$	m$_B$	(U-B)$_T$	(U-B)$_e$	W$_{20}$	V$_{opt}$
SGL SGB		T	logA$_e$	A$_i$	m$_{FIR}$	(B-V)$_T^o$	m'$_e$	W$_{50}$	V$_{GSR}$
R.A. 1950 DEC.	PGC	L	logD$_o$	A$_{21}$	B$_T^o$	(U-B)$_T^o$	m'$_{25}$	HI	V$_{3K}$

043215.7-494031		.S..3*/	1.42± .04	133					1871± 32
256.76 -42.55	ESO 202- 35	S	.86± .04	.00	13.38 ±.14				1698
238.68 -45.43	IRAS04309-4946	3.0±1.3		1.18	12.54				1853
043055.1-494648	PGC 15455		1.42	.43	12.18				
0432.3 +0143		.S..6*.	1.11± .07	168					
193.68 -29.59	UGC 3083	U	.83± .06	.40					
319.90 -50.66		6.0±1.4		1.22					
0429.7 +0137	PGC 15458		1.15	.42					
043222.3+331306		.S..4..	1.16± .07	9				14.93±.3	5420± 7
167.20 -10.07	UGC 3078	U	.59± .06	2.08					5461
352.11 -30.40	IRAS04291+3306	4.0± .9		.86	12.51			423± 7	5341
042908.1+330646	PGC 15461		1.36	.29					
043234.2-625146		RLA.0*P	1.13± .09						
273.83 -39.57		S	.28± .08	.00					
225.45 -36.72		-2.0±1.2		.00					
043201.7-625800	PGC 15475		1.09						
043238.9-332449		.SXR4P*	1.00± .05						9780± 20
234.66 -42.38	ESO 360- 14	Sr (1)	.08± .04	.00	14.93 ±.14				9632
260.86 -53.37	FAIR 1126	4.3± .7		.12					9742
043046.0-333106	PGC 15477	3.3± .8	1.00	.04	14.75				
043241.5-434256	NGC 1616	.SXT4P*	1.26± .03	36	13.30 ±.13	.69± .01	.70± .01		
248.66 -42.94	ESO 251- 10	BS	.29± .04	.00	13.65 ±.14				4472± 40
245.92 -48.85	IRAS04311-4349	3.5± .6	.74± .02	.42	12.35	.61	12.49± .06		4307
043107.0-434912	PGC 15479		1.26	.14	13.01		13.73± .22		4447
043245.1-042226	NGC 1609	PL..+P*	1.04± .07	170					
199.84 -32.66	MCG -1-12- 25	E	.19± .05	.12					
310.94 -53.27		-1.0±1.3		.00					
043016.2-042843	PGC 15480		1.02						
043248.5+102324		.S..6*.	1.24± .06	158				15.07±.3	4065± 7
185.75 -24.56	UGC 3084	U	.67± .06	.87					4032
331.06 -46.18	IRAS04300+1017	6.0±1.2		.98				338± 7	3990
043003.4+101706	PGC 15481		1.32	.33					
0432.8 -1245		.SB?...	1.04± .09						10660± 70
208.92 -36.46	MCG -2-12- 37		.28± .07	.04					10559
296.95 -55.46				.35					10600
0430.5 -1252	PGC 15484		1.04	.14					
043249.8-504621			.70± .07		13.96 ±.13	.84± .02			
258.22 -42.32	ESO 202- 37		.03± .05	.00	13.94 ±.14				
237.36 -44.83									
043132.1-505236	PGC 15485		.70						
043250.0-753227	IC 2089	.SXR9..	1.03± .07						
288.48 -34.60	ESO 32- 15	r	.09± .06	.44	15.12 ±.14				
215.52 -27.24		9.1± .8		.09					
043418.1-753836	PGC 15487		1.07	.05					
043250.0+715252	NGC 1560	.SAS7./	1.99± .01	23	12.16 ±.14	.72± .02	.73± .02	11.26±.1	-36± 5
138.37 16.02	UGC 3060	R	.76± .03	.67	11.8 ±.3	.00± .03	.03± .02	157± 5	-194± 73
15.85 .79		7.0± .3	1.41± .02	1.04	12.16	.43	14.70± .05	125± 6	106
042708.2+714629	PGC 15488		2.06	.38	10.39	-.21	15.09± .17	.48	-93
0432.8 +0032		.S..9*.	.96± .09					16.02±.3	5478± 11
194.92 -30.11	UGC 3086	U	.00± .06	.19				176± 7	
318.33 -51.33		9.0±1.2		.00				136± 7	5415
0430.3 +0026	PGC 15491		.98	.00					5408
043254.0-041203		.S..7*/	1.20± .05	74					4905
199.68 -32.54	MCG -1-12- 28	E	.88± .05	.11					
311.24 -53.24		7.0±1.3		1.21					4828
043024.9-041819	PGC 15495		1.21	.44					4839
043306.1-041753	NGC 1611	PLBT+P?	1.27± .07	77					
199.81 -32.54	MCG -1-12- 29	E	.52± .08	.12					
311.11 -53.32	IRAS04306-0424	-1.0±1.2		.00	12.95				
043037.1-042409	PGC 15501		1.21						
043310.8+052119	A 0430+05	.L...*.	.90± .09	130					
190.37 -27.40	UGC 3087	U	.13± .03	.61					9910± 19
324.96 -49.06	2ZW 14	-2.0±1.3		.00					9862
043031.3+051503	PGC 15504		.95						9838
043313.2-041023	NGC 1612	.SBR0..	1.09± .06	40					
199.70 -32.46	MCG -1-12- 30	E	.11± .05	.13					
311.33 -53.30	IRAS04307-0416	.0± .9		.08	13.45				
043044.0-041639	PGC 15507		1.09						

R.A. 2000 DEC. / l b / SGL SGB / R.A. 1950 DEC.	Names / PGC	Type / S_T n_L / T / L	$\log D_{25}$ / $\log R_{25}$ / $\log A_\bullet$ / $\log D_o$	p.a. / A_g / A_i / A_{21}	B_T / m_B / m_{FIR} / B_T^o	$(B-V)_T$ / $(U-B)_T$ / $(B-V)_T^o$ / $(U-B)_T^o$	$(B-V)_\bullet$ / $(U-B)_\bullet$ / $m'_\bullet$ / m'_{25}	m_{21} / W_{20} / W_{50} / HI	V_{21} / V_{opt} / V_{GSR} / V_{3K}
0433.2 +7633		.SXS3..	1.02± .10					14.98±.1	7886± 11
134.61 19.04	UGC 3057	U	.27± .07	.63	15.00 ±.18			325± 8	8038
18.47 4.65	IRAS04263+7627	3.0± .8		.37	12.29				7834
0426.3 +7627	PGC 15509		1.08	.13	13.94			.90	7834
0433.2 +0740		.I..9?.	1.07± .14	160					8106
188.25 -26.05	UGC 3088	U	.16± .12	.65					
327.92 -47.83	IRAS04305+0734	10.0±1.7		.12					8065
0430.5 +0734	PGC 15512		1.13	.08					8033
043322.8-184040		.SB?...	1.11± .04	16					6372± 60
215.88 -38.59	ESO 551- 13		.27± .04	.04	14.50 ±.14				6256
286.39 -56.10	IRAS04311-1846			.41					6319
043110.0-184654	PGC 15517		1.12	.14	14.01				
043325.4-041558	NGC 1613	.LXT+?.	1.02± .09	135					
199.82 -32.46	MCG -1-12- 31	E	.10± .05	.12					
311.21 -53.39		-1.0± .9		.00					
043056.3-042213	PGC 15518		1.02						
043337.9-131547		.E+..?.	1.32± .04	5					9796± 33
209.59 -36.49	MCG -2-12- 39	E	.10± .05	.09					9694
296.13 -55.73		-4.0± .8		.00					9739
043118.8-132200	PGC 15524		1.30						
043347.1+010558		.SXS7*.	1.11± .05	125					5559
194.51 -29.61	UGC 3091	UE (1)	.03± .04	.27	14.80 ±.20				5497
319.30 -51.28		6.7± .7		.04					5490
043112.2+005945	PGC 15531	6.4± .8	1.14	.02	14.46				
0433.8 +1654		.S..4..	1.04± .06						
180.31 -20.39	UGC 3089	U	.08± .05	1.23	15.07 ±.17				
338.34 -42.30	IRAS04309+1648	4.0± .9		.12					
0430.9 +1648	PGC 15533		1.15	.04					
043352.2-114217		PSBT2*.	1.50± .05	45					5143
207.86 -35.79	MCG -2-12- 41	E	.50± .05	.05					
298.88 -55.52		2.0± .8		.61					5044
043131.3-114830	PGC 15534		1.50	.25					5084
043353.5-654212		.IBS9..	1.08± .09						
277.22 -38.49		S (1)	.11± .08	.06					
222.92 -34.76		10.0± .9		.08					
043337.7-654820	PGC 15535	10.0± .8	1.09	.06					
043359.9-083430	NGC 1614	.SBS5P.	1.12± .03	85	13.63 ±.13	.69± .01	.67± .01	16.00±.3	4778± 10
204.45 -34.38	MCG -1-12- 32	R	.06± .04	.23		-.01± .03	-.06± .02		4723± 28
304.27 -54.85	MK 617	5.0± .4	.56± .01	.09	9.56	.59	11.92± .03		4681
043135.5-084042	PGC 15538		1.14	.03	13.28	-.08	13.92± .23	2.69	4710
0434.3 +7311		.L..-*.	1.11± .16	55					
137.40 16.95	UGC 3069	U	.25± .08	.46	14.80 ±.17				
16.66 1.81		-3.0±1.2		.00					
0428.4 +7305	PGC 15548		1.13						
0434.6 +0809		.S?....		157					
188.03 -25.51	MCG 1-12- 10			.65	15.00 ±.12				7581± 46
328.77 -47.84									7540
0431.9 +0803	PGC 15556								7509
043443.6-303242		.S?....			15.10S±.15				13328± 41
230.95 -41.48				.04					13184
265.31 -54.70	HICK 29A								13289
043246.6-303850	PGC 15559								
043502.3+731548	NGC 1573	.E.....	1.28± .07	35	12.81 ±.15	1.08± .02	1.11± .01		
137.37 17.04	UGC 3077	U	.16± .04	.48	13.14 ±.15				4221± 14
16.75 1.84	7ZW 18	-5.0± .8	1.04± .04	.00		.93	13.50± .13		4367
042903.2+730933	PGC 15570		1.31		12.43		13.80± .38		4166
0435.0 -1413		.SBS9..	1.12± .06	95					
210.86 -36.56	MCG -2-12- 42	E (1)	.19± .05	.12					
294.49 -56.20	IRAS04327-1419	9.0± .9		.19					
0432.7 -1419	PGC 15573	8.7±1.6	1.13	.09					
043505.1+075919		.S..6*.	1.00± .05	118					
188.26 -25.51	UGC 3093	U	.68± .04	.68	15.50 ±.12				
328.67 -48.02	IRAS04324+0753	6.0±1.4		1.01					
043222.6+075311	PGC 15574		1.06	.34					
0435.1 -1313		.L?....	1.02± .14						10519± 55
209.74 -36.15	MCG -2-12- 43		.27± .07	.10					10416
296.28 -56.08				.00					10463
0432.8 -1320	PGC 15576		.99						

R.A. 2000 DEC. / l b / SGL SGB / R.A. 1950 DEC.	Names / PGC	Type / S_T n_L / T / L	$\log D_{25}$ / $\log R_{25}$ / $\log A_e$ / $\log D_0$	p.a. / A_g / A_i / A_{21}	B_T / m_B / m_{FIR} / B_T^0	$(B-V)_T$ / $(U-B)_T$ / $(B-V)_T^0$ / $(U-B)_T^0$	$(B-V)_e$ / $(U-B)_e$ / m'_e / m'_{25}	m_{21} / W_{20} / W_{50} / HI	V_{21} / V_{opt} / V_{GSR} / V_{3K}
043512.3-541219		PSXR2..	1.12± .04	12					
262.69 -41.40	ESO 157- 42	Sr	.31± .04	.00	14.86 ±.14				13000±190
233.35 -42.93	FAIR 776	1.5± .6		.38					12822
043405.1-541824	PGC 15578		1.12	.15	14.35				12991
0435.2 -0143	A 0432-01	CE...P?	.54± .19	65					
197.52 -30.79		E	.05± .08	.16					9730± 72
315.41 -52.83	2ZW 17	-5.0±1.8		.00					9659
0432.7 -0150	PGC 15579		.55						9665
0435.5 +1910		.S?....	1.04± .08	0					
178.70 -18.66	UGC 3094		.47± .06	1.58					7407
340.92 -41.08	IRAS04326+1904			.71	11.07				7401
0432.6 +1904	PGC 15592		1.19	.24					7333
043540.6-215919		.SB4?P/	1.21± .04	4					
220.13 -39.16	ESO 551- 16	S	.56± .03	.04	14.81 ±.14				1774
280.32 -56.52		4.0±1.0		.82					1648
043332.0-220524	PGC 15597		1.21	.28	13.93				1727
0435.7 +8844		.S..6*.	1.04± .08	112					
124.10 26.42	UGC 2886A	U	.76± .06	.47					
25.44 14.68		6.0±1.4		1.12					
0350.0 +8837	PGC 15599		1.08	.38					
0435.8 +0215		.L.....	1.08± .09	115					
193.70 -28.56	UGC 3097	U	.29± .04	.53	15.10 ±.12				
321.31 -51.19	IRAS04332+0209	-2.0± .9		.00	11.93				
0433.2 +0209	PGC 15600		1.10						
0435.9 +7256		.S..6*.	1.11± .07	62					
137.68 16.88	UGC 3085	U	.66± .06	.56					
16.62 1.54		6.0±1.3		.97					
0430.0 +7250	PGC 15604		1.16	.33					
043601.8+195703	NGC 1615	.LA.-*.	1.09± .10	115	*		.69± .04		
178.14 -18.08	UGC 3096	PU	.22± .05	1.86	14.57 ±.11				
341.76 -40.63		-3.0± .7	1.08± .17	.00			13.25± .39		
043305.5+195058	PGC 15608		1.30						
043606.7-030857	NGC 1618	.SBR3*.	1.37± .04	150	13.5 ±.2	.79± .03			
199.08 -31.32	MCG -1-12- 34	R (1)	.46± .05	.17		-.05± .06			
313.37 -53.60	IRAS04336-0314	3.0± .9		.63	12.44				
043336.4-031501	PGC 15611	3.1±1.2	1.38	.23			14.03± .32		
043618.7-024955		PLBT+*.	1.07± .05	40	13.65S±.15				
198.78 -31.12	MCG 0-12- 51	E	.17± .04	.16	14.46 ±.10				4697± 41
313.90 -53.52	HICK 30A	-.5± .6		.00					4622
043348.1-025558	PGC 15620		1.06		13.97			13.44± .32	4635
043620.8-650832		.LXR+*/	1.18± .05	101	14.26 ±.13	1.03± .01	1.04± .01		
276.44 -38.44	ESO 84- 26	Sr	.52± .04	.06	14.21 ±.14	.62± .02	.66± .02		5476
223.15 -35.34		-1.1± .9	.41± .02	.00		.89	11.80± .07		5292
043602.1-651430	PGC 15622		1.10		14.09	.55	13.70± .31		5483
043622.0-102238	A 0434-10	.SBS3P.	.94± .08	85					
206.72 -34.66	MCG -2-12- 45	E	.12± .05	.13					10825± 45
301.43 -55.85	MK 618	3.0± .9		.17	12.13				10729
043359.7-102840	PGC 15623		.95	.06					10769
043623.3-024759		.S?....			15.73S±.15				
198.76 -31.08				.16					4508± 41
313.96 -53.53	HICK 30C								4433
043352.7-025402	PGC 15624								4446
043624.6-093051		.IBS9*/	1.17± .05	170					2415
205.79 -34.27	MCG -2-12- 46	E (1)	.44± .05	.24					
302.93 -55.67		10.0± .9		.33					2321
043401.4-093653	PGC 15625	7.5±1.2	1.19	.22					2358
043625.0-045916	NGC 1621	.E+....	1.12± .08	95					
201.00 -32.16	MCG -1-12- 35	E	.24± .08	.21					
310.51 -54.34		-4.0± .9		.00					
043356.8-050518	PGC 15626		1.08						
0436.4 +1420		.S?....	1.11± .14	170					4425± 10
182.89 -21.48	UGC 3102		.29± .12	1.05					
336.21 -44.45	IRAS04336+1414			.44	12.79				4403
0433.6 +1414	PGC 15627		1.21	.15					4354
043630.4-025200		PLXT+*.	1.05± .09	160	14.19S±.15				
198.85 -31.09	MCG 0-12- 54	E	.10± .08	.16	14.44 ±.11				4625± 41
313.88 -53.58	HICK 30B	-1.0±1.3		.00					4550
043359.8-025803	PGC 15631		1.05		14.13			14.08± .52	4563

R.A. 2000 DEC. / l b / SGL SGB / R.A. 1950 DEC.	Names / PGC	Type / S_T n_L / T / L	$\log D_{25}$ / $\log R_{25}$ / $\log A_o$ / $\log D_o$	p.a. / A_g / A_i / A_{21}	B_T / m_B / m_{FIR} / B_T^o	$(B-V)_T$ / $(U-B)_T$ / $(B-V)_T^o$ / $(U-B)_T^o$	$(B-V)_o$ / $(U-B)_o$ / m'_o / m'_{25}	m_{21} / W_{20} / W_{50} / HI	V_{21} / V_{opt} / V_{GSR} / V_{3K}
043636.7-031122	NGC 1622	.SXR2*.	1.56± .03	145	13.4 ±.2	.89± .03			
199.19 -31.23	MCG -1-12- 36	R	.70± .05	.16		.36± .06			
313.39 -53.73		2.0± .5		.87					
043406.5-031723	PGC 15635		1.57	.35			14.27± .29		
0436.6 -0217		.S..1..	.98± .07	157					
198.29 -30.78	UGC 3104	U	.30± .05	.14	15.13 ±.12				
314.77 -53.38	ARP 61	1.0± .9		.31					
0434.1 -0224	PGC 15637		1.00	.15					
043637.5-000840	NGC 1620	.SXT4..	1.46± .03	25	13.08 ±.13	.79± .01	.87± .01	13.88±.1	3511± 5
196.16 -29.67	UGC 3103	PUE (1)	.46± .04	.22	13.28 ±.13	.26± .02	.31± .02	438± 8	3497± 10
318.02 -52.48	IRAS04340-0014	4.3± .4	1.00± .01	.67	12.42	.63	13.57± .03	428± 11	3441
043403.9-001442	PGC 15638	3.1±1.2	1.48	.23	12.27	.13	14.08± .22	1.38	3445
0436.6 +7133		.I..9*.	1.16± .13	30					2916± 11
138.83 16.04	UGC 3090	U	.12± .12	.67				14.93±.1	
15.92 .35		10.0±1.1		.09				148± 16	3058
0431.0 +7127	PGC 15639		1.22	.06				138± 12	2861
043652.0-483250		.S?....	.89± .07		15.57 ±.15	.54± .03			
255.14 -41.93	ESO 202- 40		.02± .06	.00	14.95 ±.14				6190
239.33 -46.71				.03		.50			6017
043529.1-483848	PGC 15647		.89	.01	15.16		14.82± .39		6176
043657.3-521032		.SBS9..	1.09± .05	162					
259.96 -41.48	ESO 202- 41	S (1)	.13± .05	.00	14.99 ±.14				1642
235.25 -44.45		9.0± .6		.13					1465
043544.0-521630	PGC 15650	8.9± .8	1.09	.07	14.85				1632
0437.0 +4355		.S..4..	.96± .09	2				14.76±.1	3909± 7
159.83 -2.19	UGC 3098	U	.39± .06					283± 9	4226±113
.24 -22.56	IRAS04335+4349	4.0± .9		.58	10.88			232± 5	3982
0433.5 +4349	PGC 15652			.20					3840
043706.4-031816	NGC 1625	.SBT3*.	1.32± .04	50	12.97 ±.15	.69± .03	.77± .02		
199.38 -31.18	MCG -1-12- 38	R (1)	.62± .05	.11	13.53 ±.16	-.09± .05	.00± .03		3033± 50
313.29 -53.89	IRAS04346-0324	3.0± .6	1.12± .03	.85	12.72	.53	14.06± .08		2956
043436.3-032415	PGC 15654	4.2± .8	1.33	.31	12.24	-.21	12.88± .30		2972
043707.8-021813		.L..-..	1.39± .06	75					
198.37 -30.67	UGC 3105	UE	.27± .05	.12	13.86 ±.15				8862± 50
314.85 -53.50		-3.0± .6		.00					8788
043436.6-022413	PGC 15655		1.36		13.61				8801
043711.3-015111		.S..1*/	1.14± .05	55					
197.93 -30.43	UGC 3106	UE	.57± .05	.11	15.25 ±.13				
315.55 -53.33		1.4± .8		.59					
043439.6-015710	PGC 15656		1.15	.29					
043715.4-185408	NGC 1630	.LB.+P?	.85± .05	140					
216.55 -37.81	ESO 551- 19	SE	.15± .03	.00	15.14 ±.14				
285.99 -57.01		-.5±1.3		.00					
043503.0-190006	PGC 15659		.83						
043718.1-623459		.L..-./	1.21± .05	122					
273.27 -39.14	ESO 84- 28	S	.50± .04	.00	14.14 ±.14				6417
225.20 -37.30		-3.0± .5		.00					6233
043645.1-624054	PGC 15660		1.13		14.04				6421
043718.3-555522		.IBS9..	1.08± .05	73					
264.85 -40.78	ESO 157- 44	S (1)	.32± .04	.00	14.92 ±.14				
231.34 -42.00		10.0± .9		.24					
043617.0-560118	PGC 15661	6.7± .9	1.08	.16					
043720.7-691213		.SXS7*.	1.15± .04	78					
281.20 -36.93	ESO 55- 29	S (1)	.24± .04	.41	13.92 ±.14				5529
219.85 -32.35	IRAS04375-6918	7.0± .9		.33	12.83				5344
043732.0-691806	PGC 15665	5.6± .9	1.19	.12	13.15				5541
043720.7-481409		.S?....	1.11± .06		15.0 ±.2	.60± .04	.56± .03		
254.71 -41.87	ESO 202- 42		.03± .06	.00	14.65 ±.14				
239.62 -46.96			.59± .05	.05			13.43± .13		
043557.0-482006	PGC 15666		1.11	.02			15.30± .40		
043722.2+093246		.S..4..	1.00± .08	72				15.66±.3	8368± 10
187.22 -24.14	UGC 3107	U	.55± .06	.63					8330
331.03 -47.59	IRAS04346+0926	4.0± .9		.81	13.35			469± 7	8300
043438.0+092647	PGC 15668		1.06	.27					
0437.5 -0018		.S..8*.	1.01± .09	10					3720± 10
196.45 -29.55	UGC 3109	U	.10± .06	.16					
317.95 -52.75		8.0±1.2		.12					3652
0435.0 -0024	PGC 15673		1.02	.05					3658

R.A. 2000 DEC. l b SGL SGB R.A. 1950 DEC.	Names PGC	Type S_T n_L T L	$\log D_{25}$ $\log R_{25}$ $\log A_e$ $\log D_o$	p.a. A_g A_i A_{21}	B_T m_B m_{FIR} B_T^o	$(B-V)_T$ $(U-B)_T$ $(B-V)_T^o$ $(U-B)_T^o$	$(B-V)_e$ $(U-B)_e$ m'_e m'_{25}	m_{21} W_{20} W_{50} HI	V_{21} V_{opt} V_{GSR} V_{3K}
043736.3-044302 200.89 -31.77 311.12 -54.52 043507.8-044859	NGC 1628 MCG -1-12- 39 IRAS04351-0448 PGC 15674	.S..3P/ E 3.0±1.7 	1.25± .05 .62± .05 1.27	171 .17 .86 .31	 13.54 				3695 3614 3636
043738.0-045321 201.07 -31.85 310.85 -54.59 043509.6-045918	NGC 1627 MCG -1-12- 40 IRAS04351-0459 PGC 15675	.SAR5P. E (1) 5.0± .8 3.1± .8	1.20± .05 .01± .05 1.21	 .18 .02 .01	 13.01 				3866 3784 3807
043746.1-471511 253.39 -41.88 240.73 -47.61 043620.1-472106	 ESO 251- 14 PGC 15684	.S?.... 	.94± .07 .22± .06 .94	152 .00 .23 .11	15.16 ±.14 14.81				10000± 69 9828 9985
043747.3-512524 258.94 -41.46 235.95 -45.03 043632.0-513118	 ESO 202- 43 PGC 15686	.E+4... S -4.0± .8 1.12	1.19± .05 .24± .04 .76± .09 	135 .00 .00 	14.5 ±.2 14.19 ±.14 14.09	1.04± .04 .93 	1.06± .03 13.75± .32 14.80± .37		11108 10931 11098
043755.7-093109 205.99 -33.94 303.08 -56.03 043532.4-093705	IC 382 MCG -2-12- 49 IRAS04355-0937 PGC 15691	.SXT5*. E (1) 5.0± .8 .8± .8	1.36± .04 .22± .05 1.37	0 .18 .32 .11	 12.44 				4998 4903 4943
0438.0 +7219 138.28 16.61 16.42 .93 0432.2 +7213	 UGC 3092 PGC 15693	.S..7.. U 7.0±1.0 	1.16± .07 1.18± .06 1.22	162 .59 1.38 .50					
0438.0 +0852 187.92 -24.40 330.38 -48.11 0435.3 +0847	 UGC 3111 PGC 15695	.S..1.. U 1.0± .9 	1.07± .07 .50± .06 1.12	148 .57 .51 .25					
0438.0 -0057 197.16 -29.79 317.07 -53.15 0435.5 -0103	 UGC 3113 PGC 15697	.S..6*. U 6.0±1.3 	1.00± .16 .43± .12 1.01	41 .12 .63 .22					
0438.2 +4841 156.42 1.15 3.34 -18.80 0434.5 +4835	 WEIN 28 PGC 15704	 	.30? .30? 		 10.64 				5935±113 6020 5867
043824.2-203854 218.76 -38.14 282.72 -57.24 043614.0-204448	NGC 1631 ESO 551- 21 IRAS04362-2045 PGC 15705	PSXR0*. SEr .4± .5 	1.16± .04 .19± .03 1.15	44 .04 .14 	 14.30 ±.14 13.98				9404± 60 9280 9360
0438.4 +6519 143.86 12.16 12.69 -4.97 0433.6 +6513	 UGC 3100 IRAS04336+6513 PGC 15707	.S..6*. 6.0±1.2 	1.01± .06 .06± .05 1.17	 1.71 .09 .03	 13.49 			15.83±.1 125± 8 	3882± 11 4010 3823
0438.4 +4402 159.92 -1.94 .53 -22.62 0434.8 +4356	 UGC 3108 WEIN 35 PGC 15708	.S?.... 	1.26± .06 .00± .06 	 .00 .00	 12.51 			14.47±.1 285± 8 	3959± 7 4030 3890
043838.1-615110 272.31 -39.19 225.69 -37.94 043801.7-615700	 PGC 15713	.LXS-.. S -3.0± .9 	1.12± .09 .26± .08 1.08	 .00 .00 					
0438.6 +0010 196.14 -29.06 318.87 -52.79 0436.1 +0005	 CGCG 393- 54 MK 1083 PGC 15714	 		 .17 	 14.85 ±.12 				8106± 47 8038 8045
0438.6 +1114 185.92 -22.89 333.30 -46.81 0435.9 +1109	A 0435+11 2ZW 18 PGC 15715	 		 .89 					4431± 72 4398 4365
043850.2+025041 193.61 -27.61 322.71 -51.56 043613.4+024448	 UGC 3117 IRAS04362+0244 PGC 15719	.SXS5*. E (1) 5.0±1.2 5.3±1.2	1.23± .06 .06± .05 1.29	145 .62 .09 .03	 14.4 ±.2 13.33 13.70		14.22±.3 275± 7 .49		4625± 10 4566 4563
043850.5-524104 260.57 -41.12 234.45 -44.34 043739.0-524654	 ESO 157- 45 PGC 15720	.SBS8*/ S (1) 7.7±1.1 7.0± .7	1.05± .05 .56± .04 1.05	120 .00 .69 .28	 15.78 ±.14 				

R.A. 2000 DEC.	Names	Type	logD$_{25}$	p.a.	B$_T$	(B-V)$_T$	(B-V)$_\bullet$	m$_{21}$	V$_{21}$
l b		S$_T$ n$_L$	logR$_{25}$	A$_g$	m$_B$	(U-B)$_T$	(U-B)$_\bullet$	W$_{20}$	V$_{opt}$
SGL SGB		T	logA$_\bullet$	A$_i$	m$_{FIR}$	(B-V)^o_T	m'$_\bullet$	W$_{50}$	V$_{GSR}$
R.A. 1950 DEC.	PGC	L	logD$_o$	A$_{21}$	B^o_T	(U-B)^o_T	m'$_{25}$	HI	V$_{3K}$

043853.9-641230		.S..3*.	1.22± .04	11					8700
275.19 -38.48	ESO 84- 33	S (1)	.50± .05	.03	14.20 ±.14				8515
223.67 -36.22	IRAS04385-6418	3.0±1.2		.69	12.68				8707
043830.1-641818	PGC 15722	3.3±1.3	1.22	.25	13.41				
0438.9 +1850		.L.....	1.18± .06	15					3290
179.50 -18.25	UGC 3115	U	.27± .03	1.43	15.05 ±.16				3281
341.32 -41.89	IRAS04359+1844	-2.0± .8		.00	12.20				3221
0436.0 +1844	PGC 15723		1.30		13.57				
043855.0-525146		PSBS1*P	.99± .05	42					
260.81 -41.08	ESO 157- 46	S	.23± .04	.00	15.41 ±.14				
234.25 -44.23		1.0± .6		.23					
043744.0-525736	PGC 15724		.99	.11					
0438.9 +0536		.I..9*.	1.07± .14						8310
191.04 -26.06	UGC 3118	U	.00± .12	.47					
326.47 -50.14		10.0±1.2		.00					8259
0436.3 +0531	PGC 15726		1.11	.00					8247
043900.4-630224		.SXS5..	1.06± .05						
273.76 -38.82	ESO 84- 32	S (1)	.01± .05	.00	14.71 ±.14				
224.63 -37.09		5.0± .8		.01					
043830.1-630812	PGC 15727	4.4± .8	1.06	.00					
0439.1 +1131		.S..4..	1.04± .08	101					
185.75 -22.63	UGC 3119	U	.47± .06	.89					
333.72 -46.73	IRAS04363+1125	4.0± .9		.69	13.22				
0436.3 +1125	PGC 15731		1.12	.24					
043918.7-541236		.S.0?P/	1.07± .05	75					
262.56 -40.81	ESO 157- 47	S	.61± .05	.00	15.52 ±.14				
232.80 -43.37		.0±1.7		.46					
043812.1-541824	PGC 15736		1.04						
043922.1-520754		.S..5*/	1.14± .04	18					
259.83 -41.12	ESO 202- 47	S	.64± .04	.00	15.41 ±.14				
234.96 -44.76	IRAS04381-5213	4.7±1.1		.96					
043809.0-521342	PGC 15739		1.14	.32					
043931.5-070553	IC 385	.LAT0P?	1.07± .09	80					
203.63 -32.48		E	.38± .08	.27					
307.45 -55.75		-2.0± .9		.00					
043705.6-071142	PGC 15746		1.05						
043937.4-530043		.S?....	1.29± .05	28					1729
260.98 -40.96	ESO 157- 49		.65± .05	.00	14.37 ±.14				1550
233.99 -44.21	IRAS04384-5306			.90	12.17				1723
043827.0-530630	PGC 15749		1.29	.32	13.46				
043940.3-131501		.SAS6*.	1.20± .05	100					
210.31 -35.14	MCG -2-12- 53	E (1)	.12± .05	.22					
296.54 -57.18		6.0±1.2		.18					
043721.4-132050	PGC 15754	6.4±1.2	1.22	.06					
0439.7 +0300		.S?....	1.11± .07	104					4517± 38
193.58 -27.33	UGC 3121		.54± .06	.62	15.27 ±.13				4458
323.13 -51.66				.81					4456
0437.1 +0255	PGC 15755		1.17	.27	13.81				
043945.9-411613		.LA.0*.	1.24± .08						
245.36 -41.63	MCG -7-10- 15	S	.17± .08	.00					
248.13 -51.24		-2.0±1.1		.00					
043807.2-412200	PGC 15757		1.21						
043948.0-765014	IC 2103	.S..5*/	1.26± .04	89					4263
289.66 -33.65	ESO 32- 18	S (1)	.74± .04	.39	14.66 ±.14				4080
214.26 -26.47		5.0± .7		1.11					4286
044144.0-765554	PGC 15758	5.6±1.3	1.30	.37	13.14				
043951.2+070315		.SXT5..	1.34± .04	25				15.07±.3	4693± 7
189.85 -25.07	UGC 3122	U	.23± .05	.49	14.2 ±.2				4646
328.51 -49.53		5.0± .8		.34				341± 7	4630
043709.7+065726	PGC 15760		1.38	.11	13.33			1.62	
043958.5-085502		.SAT4P*	1.03± .07	0					
205.62 -33.22	MCG -2-12- 54	E (1)	.15± .05	.19					
304.36 -56.37	IRAS04375-0900	4.0± .9		.22	13.48				
043734.7-090050	PGC 15767	3.1±1.2	1.05	.08					
044008.2-003252	NGC 1635	RSBR0..	1.15± .04	5					
197.08 -29.13	UGC 3126	PUE	.03± .04	.11	13.33 ±.12				
318.06 -53.44	IRAS04375-0038	.3± .5		.03	13.64				
043735.1-003840	PGC 15773		1.16						

R.A. 2000 DEC. / l b / SGL SGB / R.A. 1950 DEC.	Names / PGC	Type / S_T n_L / T / L	$\log D_{25}$ / $\log R_{25}$ / $\log A_e$ / $\log D_o$	p.a. / A_g / A_i / A_{21}	B_T / m_B / m_{FIR} / B_T^o	$(B-V)_T$ / $(U-B)_T$ / $(B-V)_T^o$ / $(U-B)_T^o$	$(B-V)_e$ / $(U-B)_e$ / m'_e / m'_{25}	m_{21} / W_{20} / W_{50} / HI	V_{21} / V_{opt} / V_{GSR} / V_{3K}
044009.6+072108	NGC 1633	.SXS2..	1.00± .06					16.32±.3	4989± 10
189.63 -24.84	UGC 3125	U	.07± .05	.49	14.36 ±.12				4870± 46
328.95 -49.42		2.0± .8		.09				167± 7	4938
043727.8+071520	PGC 15774		1.05	.04	13.73			2.55	4922
044009.9+072018	NGC 1634		.64± .17	109					
189.64 -24.84	MCG 1-12- 15		.11± .07	.49	15.14 ±.18				7500± 64
328.94 -49.43	ARAK 109								7454
043728.1+071431	PGC 15775		.68						7438
044014.4-241908		.SB?...	1.07± .05						4422
223.39 -38.84	ESO 485- 6		.03± .04	.04	14.52 ±.14				4288
275.81 -57.28	IRAS04381-2424			.05					4384
043809.0-242454	PGC 15780		1.07	.02	14.41				
044017.2-584447		.L..0P*	.99± .05		13.52 ±.13	.46± .02	.28± .01		1171± 28
268.35 -39.78	ESO 118- 34	S	.03± .03	.00	13.50 ±.14	-.36± .02	-.49± .02		989
228.28 -40.32	IRAS04394-5850	-2.0±1.2	.32± .01	.00	12.41	.45	10.61± .02		1173
043927.0-585030	PGC 15782		.98		13.49	-.36	13.26± .30		
044023.0+401523		.SXS5..	1.00± .08	115				15.92±.3	5625± 9
162.98 -4.18	UGC 3120	U	.19± .06	3.97				248± 11	
358.43 -25.88		5.0± .9		.28				240± 8	5684
043656.3+400935	PGC 15788		1.37	.09					5557
044025.5-020130		.SBS8*/	1.30± .03	24					
198.58 -29.82	UGC 3127	UE (1)	.69± .04	.14	14.75 ±.13				
315.85 -54.14		8.0± .7		.85					
043754.0-020716	PGC 15789	7.5± .8	1.31	.34					
044026.7-630624		.S..3./	1.27± .05	6					
273.78 -38.64	ESO 84- 34	S (1)	.97± .04	.00	15.48 ±.14				
224.43 -37.15	IRAS04399-6312	3.0± .9		1.34	13.57				
043957.0-631206	PGC 15790	2.2±1.3	1.27	.49					
044026.7-443758		.LX.-P.	1.22± .05	24					9932
249.85 -41.53	ESO 251- 21	S	.25± .04	.00	13.93 ±.14				9762
243.56 -49.51		-3.0± .8		.00					9917
043855.0-444342	PGC 15791		1.18		13.78				
0440.5 +6638		.S..6*.	1.26± .06	72				14.13±.1	3725± 11
142.97 13.16	UGC 3114	U	.30± .06	1.44	14.7 ±.3				
13.57 -3.98	IRAS04355+6632	6.0±1.1		.44	13.09			363± 8	3856
0435.5 +6632	PGC 15793		1.40	.15	12.80			1.17	3668
044033.5+041144	A 0437+04	.LX.+..	1.01± .12		14.8 ±.4	.88± .07			
192.60 -26.51	UGC 3128	CU	.05± .05	.56	14.84 ±.12				4600± 56
324.92 -51.24		-1.0± .8		.00		.70			4544
043755.3+040558	PGC 15795		1.05		14.21		14.59± .72		4540
044038.8+170827		.I..9*.	1.36± .09	45				15.24±.3	4671± 10
181.18 -18.97	UGC 3129	U	.28± .12	1.34					4656
340.03 -43.35		10.0±1.1		.21				259± 7	4605
043745.8+170241	PGC 15799		1.49	.14					
044040.3-083628	NGC 1636	PSBT2*.	1.08± .05	0					
205.38 -32.92	MCG -1-12- 42	E	.16± .04	.27					
304.99 -56.45	IRAS04382-0842	1.5± .6		.20	12.84				
043816.1-084213	PGC 15800		1.10	.08					
044042.2-344540		.S..3*/	1.07± .05	176					
236.74 -40.91	ESO 361- 5	S	.82± .04	.00	16.23 ±.14				
257.64 -54.41		3.0±1.3		1.13					
043852.1-345124	PGC 15801		1.07	.41					
044044.1-524524		.S.4*P/	1.22± .05	64					
260.61 -40.83	ESO 157- 50	S	.72± .04	.00	14.83 ±.14				
234.10 -44.50		3.8± .8		1.06					
043933.1-525106	PGC 15802		1.22	.36					
044058.2-084235		.S..7*/	1.15± .05	145					4143
205.53 -32.90	MCG -2-12- 57	E (1)	.77± .04	.24					
304.85 -56.55		7.0± .9		1.06					4048
043834.1-084819	PGC 15808	6.4±1.2	1.17	.38					4092
0441.1 +7340		.SB.6*.	1.33± .04	115				15.25±.1	4486± 11
137.33 17.63	UGC 3110	U	.39± .05	.54	14.6 ±.3				
17.33 1.95	IRAS04350+7334	6.0±1.1		.58	12.46			406± 8	4632
0435.0 +7334	PGC 15810		1.38	.20	13.48			1.58	4435
044128.2-025130	NGC 1637	.SXT5..	1.60± .02	15	11.47 ±.13	.64± .02	.73± .01	12.71±.1	717± 4
199.56 -30.02	MCG 0-12- 68	R (2)	.09± .02	.13	11.47 ±.11	.05± .04	.09± .02	192± 5	710± 31
314.72 -54.72	IRAS04389-0257	5.0± .3	1.30± .01	.13	11.12	.59	13.46± .02	181± 5	639
043857.7-025712	PGC 15821	3.0± .5	1.61	.04	11.21	.01	14.12± .16	1.45	663

R.A. 2000 DEC. l b SGL SGB R.A. 1950 DEC.	Names PGC	Type S_T n_L T L	$\log D_{25}$ $\log R_{25}$ $\log A_e$ $\log D_o$	p.a. A_g A_i A_{21}	B_T m_B m_{FIR} B_T^o	$(B-V)_T$ $(U-B)_T$ $(B-V)_T^o$ $(U-B)_T^o$	$(B-V)_e$ $(U-B)_e$ m'_e m'_{25}	m_{21} W_{20} W_{50} HI	V_{21} V_{opt} V_{GSR} V_{3K}
044136.2-014832 198.53 -29.46 316.40 -54.32 043904.4-015413	NGC 1638 UGC 3133 PGC 15824	.LXT0?. PUE -2.3± .5 	1.30± .05 .13± .05 .93± .03 1.30	70 .10 .00 	12.91 ±.13 13.15 ±.09 12.93	.87± .01 .31± .03 .80 .30	.93± .01 .34± .02 13.05± .09 13.98± .30	15.72±.3 312± 34 312± 25 	3320± 17 3276± 66 3242 3263
044142.2-084217 205.62 -32.74 304.94 -56.72 043918.1-084758	 MCG -1-12- 43 PGC 15828	.SBT7?. E (1) 7.0±1.2 5.3±1.2	1.14± .06 .16± .05 1.17	10 .24 .22 .08				
044144.3-490140 255.68 -41.09 238.05 -47.04 044023.0-490718	 ESO 202- 52 PGC 15830	.S..3*/ S 3.0±1.4 	1.10± .05 .70± .04 1.10	177 .00 .97 .35	15.50 ±.14 			
044144.2-070513 203.92 -31.99 307.77 -56.27 043918.4-071054	IC 387 MCG -1-12- 44 IRAS04393-0710 PGC 15831	.SXT5.. PE (2) 4.5± .6 2.3± .6	1.20± .05 .13± .05 .96± .07 1.22	105 .25 .20 .07	13.7 ±.3 13.35 13.26	.86± .04 .75 	.94± .03 14.02± .18 14.26± .39	15.09±.1 412± 11 408± 11 1.76	4530± 10 4663± 59 4444 4483
044147.9-011801 198.06 -29.16 317.22 -54.14 043915.6-012342	 UGC 3134 IRAS04392-0123 PGC 15833	.SXS5.. UE (1) 4.5± .6 3.1±1.2	1.13± .04 .10± .04 1.13	65 .10 .16 .05	14.08 ±.13 13.22 			
0441.8 +0103 195.77 -27.92 320.78 -53.09 0439.3 +0058	 UGC 3136 PGC 15836	.S..3.. U (1) 3.0± .9 3.5±1.2	1.00± .16 .25± .12 1.04	30 .40 .35 .13	15.00 ±.12 14.18			 	10221± 79 10154 10165
0441.9 +7958 132.02 21.48 20.70 7.27 0433.5 +7953	 UGC 3101 PGC 15838	.SA.8*. U 8.0± .8 	1.04± .09 .14± .07 1.07	 .40 .18 .07				15.77±.1 188± 8 	4276± 11 4433 4230
044159.7-071843 204.19 -32.04 307.41 -56.40 043934.0-072422	IC 389 MCG -1-12- 45 PGC 15840	.LA.-*. E -3.0± .9 	1.15± .10 .13± .07 1.17	145 .26 .00 				
044202.7-443446 249.78 -41.24 243.37 -49.77 044031.0-444024	 ESO 251- 23 PGC 15842	.S?.... 	1.06± .07 .19± .06 1.06	77 .00 .28 .09	14.22 ±.14 13.88			 	10561 10390 10548
044203.9-071225 204.09 -31.98 307.60 -56.38 043938.2-071804	IC 390 MCG -1-12- 46 PGC 15844	.LBS+?/ E -1.0±1.3 	1.03± .07 .41± .05 .99	140 .26 .00 				
0442.1 +2002 179.00 -16.90 343.17 -41.59 0439.2 +1957	 UGC 3135 PGC 15846	.SB.6*. U 6.0±1.1 	1.16± .13 .12± .12 1.32	65 1.74 .17 .06				 	7400± 10 7393 7336
044211.4-625449 273.48 -38.51 224.40 -37.43 044141.0-630024	 ESO 84- 35 PGC 15849	.SA.3*. S (1) 3.0± .9 3.3±1.3	1.11± .05 .42± .04 1.11	1 .00 .58 .21	14.94 ±.14 			
044214.4-202607 218.87 -37.22 283.04 -58.15 044004.1-203145	NGC 1640 ESO 551- 27 IRAS04400-2031 PGC 15850	.SBR3.. R (3) 3.0± .3 2.9± .5	1.42± .02 .11± .02 .98± .02 1.42	45 .00 .16 .06	12.42 ±.13 12.48 ±.12 12.74 12.29	.75± .01 .13± .02 .72 .10	.85± .01 .29± .02 12.81± .05 14.07± .17	14.72±.2 158± 14 155± 11 2.38	1602± 9 1643± 58 1478 1565
044227.0-172717 215.41 -36.14 288.75 -58.23 044013.0-173254	 ESO 551- 30 PGC 15857	.SBS7.. SE (2) 6.6± .5 5.6± .5	1.13± .03 .09± .03 1.13	135 .08 .12 .04	14.94 ±.14 14.72			 	3230 3190± 60 3071 3149
044228.7-214129 220.39 -37.57 280.63 -58.11 044020.0-214706	 ESO 551- 31 PGC 15858	.SBS7*/ S (1) 7.0±1.2 5.6±1.2	1.17± .04 .46± .03 1.17	173 .01 .64 .23	15.10 ±.14 14.45			 	1805 1676 1768
0442.5 +6912 141.04 14.95 15.09 -1.89 0437.2 +6907	 UGC 3124 IRAS04372+6907 PGC 15862	.SB.2.. U 2.0± .9 	.96± .07 .24± .05 1.05	20 .93 .29 .12	15.1 ±.3 12.45 			
044254.5+003711 196.35 -27.93 320.33 -53.53 044020.1+003135	NGC 1642 UGC 3140 IRAS04403+0031 PGC 15867	.SAT5*. PUE (1) 5.0± .4 1.9± .6	1.26± .02 .06± .03 .84± .01 1.29	175 .29 .09 .03	13.30 ±.13 13.28 ±.13 12.30 12.88	.71± .02 .05± .02 .60 -.03	.79± .01 .14± .02 12.99± .04 14.30± .19	14.69±.2 145± 9 125± 7 1.78	4633± 6 4564 4578

R.A. 2000 DEC. / l b / SGL SGB / R.A. 1950 DEC.	Names / PGC	Type / S_T n_L / T / L	$\log D_{25}$ / $\log R_{25}$ / $\log A_e$ / $\log D_o$	p.a. / A_g / A_i / A_{21}	B_T / m_B / m_{FIR} / B_T^o	$(B-V)_T$ / $(U-B)_T$ / $(B-V)_T^o$ / $(U-B)_T^o$	$(B-V)_e$ / $(U-B)_e$ / m' / m'_{25}	m_{21} / W_{20} / W_{50} / HI	V_{21} / V_{opt} / V_{GSR} / V_{3K}
044257.3-080528		.SXS8..	1.27± .03	145				14.43±.1	2522± 11
205.14 -32.19	MCG -1-12- 47	UE (1)	.14± .04	.25				202± 16	
306.18 -56.85		8.3± .5		.17				181± 12	2428
044032.5-081103	PGC 15869	8.1± .6	1.30	.07					2474
044258.9-124616		.SBR3..	1.18± .04	35					
210.18 -34.21	MCG -2-12- 58	E (1)	.41± .04	.36					
297.67 -57.90		2.5± .6		.56					
044039.5-125151	PGC 15870	3.1± .6	1.21	.20					
044301.0-654853	NGC 1669	.S..1?.	.87± .06	97					
276.97 -37.57	ESO 84- 38	S	.31± .04	.07	14.88 ±.14				
221.97 -35.29	IRAS04428-6553	.5±1.4		.32					
044248.0-655424	PGC 15871		.88	.16					
044303.8-675536		.S..9?.	1.09± .05	40					
279.48 -36.89	ESO 55- 35	S	.13± .05	.17	13.75 ±.14				
220.34 -33.67	IRAS04430-6801	8.5±1.2		.13	13.35				
044306.0-680106	PGC 15872		1.10	.07					
044333.2-541041		.E+3...	1.09± .05	96	*		1.07± .03		
262.38 -40.21	ESO 158- 2	S	.17± .04	.00	15.04 ±.14				
232.24 -43.85		-4.0± .8	.87± .18	.00			14.59± .64		
044227.0-541612	PGC 15882		1.04						
044336.7-445829		.S?....	1.09± .05	57					13600±190
250.30 -40.96	ESO 251- 28		.55± .05	.00	14.74 ±.14				
242.62 -49.77	FAIR 421			.76					13428
044206.0-450400	PGC 15887		1.09	.28	13.87				13589
044344.0-051910	NGC 1643	.SBR4P?	1.04± .09	30	*				
202.37 -30.72	MCG -1-13- 1	E	.00± .07	.19					
311.07 -56.16	IRAS04412-0524	3.5± .8		.00	12.11				
044116.2-052442	PGC 15891		1.05	.00					
0443.7 +4007		.S?....	1.14± .13	75				16.45±.1	6337± 7
163.51 -3.79	UGC 3139		.14± .12	4.55					
358.91 -26.37	IRAS04402+4001			.21				256± 8	6393
0440.3 +4001	PGC 15892		1.57	.07					6273
0443.9 +2859		RS..0..	1.00± .08	165	*				
172.10 -10.96	UGC 3142	U	.09± .06	1.69					
351.14 -35.23		.0± .9		.07					
0440.8 +2854	PGC 15897		1.15						
044401.0-412748	NGC 1658	.S..4*.	1.16± .04	124	*				
245.65 -40.84	ESO 304- 16	S (1)	.48± .04	.00	14.38 ±.14				5190
247.18 -51.81		4.0±1.3		.71					5023
044223.0-413318	PGC 15899	2.2±1.3	1.16	.24	13.64				5175
044406.2-052759	NGC 1645	PLBT+P.	1.36± .03	95	*				4802
202.57 -30.71	MCG -1-13- 2	E	.35± .04	.20					
310.88 -56.30		-1.0± .6		.00					4715
044138.6-053330	PGC 15903		1.33						4754
044410.9-412949	NGC 1660	.S..1?P	1.02± .05	32	*				
245.70 -40.81	ESO 304- 18	S	.30± .04	.00	14.86 ±.14				
247.10 -51.82		1.0±1.8		.31					
044233.0-413518	PGC 15908		1.02	.15					
044412.7+002103		PSBT5*.	1.17± .04	80	*				
196.80 -27.79	UGC 3145	UE (1)	.10± .04	.28	14.8 ±.2				
320.19 -53.94		4.6± .7		.16					
044138.6+001532	PGC 15910	5.3±1.2	1.20	.05					
044429.0+753826	IC 381	.SXT4..	1.38± .06		13.08 ±.14	.77± .02	.85± .02	13.90±.1	2480± 5
135.82 19.00	UGC 3130	PU	.25± .07	.62	14.1 ±.3	.23± .04	.21± .03	288± 6	
18.54 3.51	IRAS04378+7532	3.5± .5	1.00± .02	.36	11.95	.57	13.57± .05	270± 6	2629
043750.4+753248	PGC 15917		1.44	.12	12.29	.08	14.20± .37	1.49	2432
0444.5 +7248		.S?....	1.07± .06	77				14.87±.1	4754± 11
138.20 17.30	UGC 3131		.28± .05	.49					
17.09 1.09	IRAS04386+7243			.42				309± 8	4897
0438.6 +7243	PGC 15919		1.11	.14					4704
0444.7 -0250		.IBS9..	1.12± .08	150					
200.01 -29.31		E (1)	.15± .08	.08					
315.32 -55.45		10.0± .9		.11					
0442.2 -0256	PGC 15921	7.5± .8	1.13						
0444.9 -0808		.SBS7..	1.14± .08	50					
205.45 -31.78		E (1)	.19± .08	.21					
306.34 -57.33		7.0± .9		.26					
0442.5 -0814	PGC 15923	6.4± .8	1.16	.09					

R.A. 2000 DEC. / l b / SGL SGB / R.A. 1950 DEC.	Names / PGC	Type: S_T n_L / T / L	$\log D_{25}$ / $\log R_{25}$ / $\log A_o$ / $\log D_o$	p.a. / A_g / A_i / A_{21}	B_T / m_B / m_{FIR} / B_T^o	$(B-V)_T$ / $(U-B)_T$ / $(B-V)_T^o$ / $(U-B)_T^o$	$(B-V)_o$ / $(U-B)_o$ / m'_o / m'_{25}	m_{21} / W_{20} / W_{50} / HI	V_{21} / V_{opt} / V_{GSR} / V_{3K}
044507.6-413441		.LXS0*.	.99± .06	148					
245.81 -40.63	ESO 304- 19	S	.47± .03	.00	14.71 ±.14				
246.84 -51.92		-2.0±1.2		.00					
044330.0-414006	PGC 15929		.92						
044511.7-155220	NGC 1650	.E+....	1.36± .08	170					
213.90 -34.94	MCG -3-13- 1	E	.24± .07	.11					
291.87 -58.80		-4.0± .8		.00					
044256.0-155746	PGC 15931		1.31						
0445.2 -0249		.SXT7*.	1.18± .08	135					
200.06 -29.19		E (1)	.14± .08	.08					
315.44 -55.56		6.5± .9		.19					
0442.7 -0255	PGC 15932	8.7±1.2	1.19	.07					
044517.4+404928				5.77				16.11±.3	5419± 9
163.17 -3.10								197± 11	
359.62 -25.97								176± 8	5477
044149.2+404400	PGC 15935								5357
044542.2-591457	NGC 1672	.SBS3..	1.82± .02	170	10.28M±.08	.60± .01	.67± .01	11.69±.3	1350± 7
268.79 -38.99	ESO 118- 43	R (3)	.08± .02	.00	10.57 ±.12	.01± .04	.03± .02	276± 6	1282± 16
227.16 -40.44	VV 826	3.0± .6	1.39± .02	.11	9.25	.58	12.75± .06	199± 7	1155
044455.0-592018	PGC 15941	3.1± .4	1.82	.04	10.25	-.01	14.04± .14	1.40	1346
044547.6-022329	NGC 1653	.E+..*.	1.17± .08		12.93 ±.13	.94± .01	1.00± .01		
199.71 -28.85	UGC 3153	UE	.00± .05	.13	12.74 ±.11	.48± .02	.51± .02		4339± 14
316.24 -55.51		-4.0± .5	.83± .02	.00		.87	12.57± .09		4260
044316.5-022853	PGC 15942		1.19		12.62	.47	13.77± .46		4291
044549.0-020458	NGC 1654	PSBR1..	.87± .08						
199.41 -28.69	UGC 3154	RE	.00± .05	.14	14.33 ±.13				4577± 63
316.74 -55.39	IRAS04433-0210	1.0± .9		.00	13.13				4499
044317.6-021022	PGC 15943		.89	.00	14.13				4529
044553.4-050813	NGC 1656	.L..+P*	1.17± .10	55					
202.48 -30.16	MCG -1-13- 5	E	.17± .07	.13					
311.72 -56.60		-1.0±1.2		.00					
044325.4-051337	PGC 15949		1.16						
044553.8-171646		.SBT4P*	1.26± .04	95					3483
215.57 -35.31	MCG -3-13- 4	SE (1)	.15± .04	.06					
289.14 -59.04	IRAS04436-1722	4.0± .5		.22	13.00				3363
044339.7-172209	PGC 15950	3.0± .6	1.27	.08					3447
044604.7-222156		.SAT0..	1.12± .04	90					8527± 60
221.52 -36.98	ESO 552- 3	S	.10± .03	.04	14.78 ±.14				8395
279.17 -58.88		-.2± .5		.08					8496
044357.0-222718	PGC 15956		1.12		14.54				
044606.0-444357	NGC 1668	.LX.-*.	1.20± .05	107					10073
249.97 -40.52	ESO 251- 30	BS	.25± .04	.00	13.76 ±.14				9900
242.52 -50.27		-3.0± .7		.00					10064
044435.0-444918	PGC 15957		1.16		13.61				
044607.4-020436	NGC 1657	.SXT4..	1.09± .05	150					
199.45 -28.62	UGC 3156	UE (1)	.19± .04	.14	14.63 ±.13				
316.81 -55.45		4.0± .6		.28					
044336.0-020959	PGC 15958	4.2± .8	1.10	.09					
0446.2 +0022		.S?....	.99± .10						6918± 54
197.08 -27.34	MCG 0-13- 5		.16± .07	.34	14.88 ±.13				6847
320.63 -54.38				.12					6869
0443.7 +0017	PGC 15965		1.01		14.32				
044616.8-572035		.SBS9P.	1.20± .05						1206
266.36 -39.30	ESO 158- 3	S (1)	.02± .05	.00	13.87 ±.14				1022
228.81 -41.88		9.3± .4		.02					1211
044522.0-572554	PGC 15966	6.4± .5	1.20	.01	13.85				
0446.3 +7625		.S?....	1.55± .05					13.29±.1	993± 6
135.22 19.56	UGC 3137	(1)	.98± .07	.58	15.1 ±.3			240± 8	
19.03 4.13				1.35				222± 12	1144
0439.4 +7620	PGC 15967	5.5±1.1	1.60	.49	13.14			-.34	946
044621.6-132100		.S..2./	1.13± .06	100					
211.22 -33.69	MCG -2-13- 2	E	.68± .05	.31					
296.81 -58.80		2.0± .9		.84					
044403.0-132621	PGC 15970		1.16	.34					
0446.3 -0351		.SBS8..	1.30± .06	100					
201.26 -29.44		E (1)	.15± .08	.07					
313.95 -56.24		8.0± .9		.19					
0443.9 -0357	PGC 15971	8.7± .8	1.31	.08					

R.A. 2000 DEC.	Names	Type	logD$_{25}$	p.a.	B$_T$	(B-V)$_T$	(B-V)$_e$	m$_{21}$	V$_{21}$
l b		S$_T$ n$_L$	logR$_{25}$	A$_g$	m$_B$	(U-B)$_T$	(U-B)$_e$	W$_{20}$	V$_{opt}$
SGL SGB		T	logA$_e$	A$_i$	m$_{FIR}$	(B-V)$_T^o$	m'$_e$	W$_{50}$	V$_{GSR}$
R.A. 1950 DEC.	PGC	L	logD$_o$	A$_{21}$	B$_T^o$	(U-B)$_T^o$	m'$_{25}$	HI	V$_{3K}$
0446.4 +0330	IC 392	.L.....	1.21± .06	170	13.3 ±.3		.61± .03		
194.13 -25.66	UGC 3158	U	.13± .03	.37	14.24 ±.14		-.21± .06		4265± 96
325.22 -52.84	VV 665	-2.0± .8	.94± .10	.00	13.26		13.49± .25		4203
0443.8 +0324	PGC 15973		1.23		13.65		13.90± .43		4214
0446.4 +1827		.SB?...	1.15± .05						4615± 10
180.97 -17.06	UGC 3157		.08± .05	1.45	15.0 ±.2				
342.64 -43.43	IRAS04435+1822			.12	11.62				4601
0443.5 +1822	PGC 15975		1.28	.04	13.41				4557
044629.5-563636		.LB?...	.74± .07						
265.42 -39.40	ESO 158- 4		.16± .06	.00	15.32 ±.14				12418
229.47 -42.43	IRAS04455-5641			.00	13.79				12234
044532.0-564154	PGC 15976		.72		15.13				12422
044630.0-044721	NGC 1659	.SAR4P.	1.21± .04	40	13.14 ±.14	.66± .03		15.63±.3	4584± 17
202.21 -29.86	MCG -1-13- 6	R (3)	.15± .04	.13	13.1 ±.2	-.03± .05		301± 34	4537± 50
312.41 -56.61	IRAS04440-0452	4.0± .4		.22	12.06	.57		153± 25	4492
044401.6-045242	PGC 15977	4.5± .6	1.23	.08	12.76	-.09	13.68± .27	2.80	4534
044632.5-071408		.S..6*.	1.17± .06	65					5163
204.72 -31.01	MCG -1-13- 7	E (1)	.56± .05	.28					
308.18 -57.45		6.0±1.3		.83					5069
044406.9-071929	PGC 15978	5.3±1.2	1.19	.28					5120
044637.3+003721		.S..6*.	1.07± .07	68				15.52±.3	8779± 10
196.89 -27.14	UGC 3161	U	.91± .06	.34					
321.08 -54.34		6.0±1.4		1.34				516± 7	8708
044402.9+003200	PGC 15982	1.10		.46					8730
044643.0-622756		.SXS6..	1.09± .04	152					
272.75 -38.12	ESO 85- 1	S (1)	.11± .04	.00	14.71 ±.14				7962
224.28 -38.11		6.0± .5		.16					7776
044611.1-623312	PGC 15984	4.8± .5	1.09	.05	14.51				7972
044646.0-630150		.SBS7P*	1.17± .04	87					
273.44 -37.97	ESO 85- 2	S (1)	.61± .04	.00	14.96 ±.14				
223.81 -37.68	IRAS04462-6307	7.0± .7		.84	13.19				
044617.1-630706	PGC 15985	6.3± .7	1.17	.30					
0446.7 +7007		.S..6*.	1.24± .11	135				15.34±.1	4560± 11
140.53 15.80	UGC 3143	U	.96± .12	.59					
15.87 -1.30		6.0±1.3		1.41				326± 8	4697
0441.3 +7002	PGC 15986	1.30		.48					4509
0446.8 -0614		.S..7?/	1.12± .08	110					
203.74 -30.48		E	.58± .08	.24					
309.97 -57.20		7.0±1.8		.80					
0444.4 -0620	PGC 15989	1.15		.29					
0446.9 +0818		.S..6?.	1.08± .06	140					4644± 10
189.80 -22.92	UGC 3162	U	.22± .05	.52					
331.65 -50.22		6.0±1.7		.33					4597
0444.2 +0813	PGC 15992	1.12		.11					4591
044658.8-355500		.IBS9..	1.22± .04	40					1344
238.47 -39.79	ESO 361- 9	S (1)	.18± .05	.00	15.29 ±.14				
254.85 -55.08		10.0± .8		.14					1184
044511.1-360018	PGC 15996	10.0± .8	1.22	.09	15.15				1328
0447.0 +0004		.SXS9..	1.16± .05						
197.49 -27.33	UGC 3164	UE (1)	.03± .05	.26	14.9 ±.3				4477± 38
320.34 -54.70		9.0± .6		.03					4404
0444.5 -0001	PGC 15998	8.7± .8	1.19	.02	14.58				4429
044706.0-500302		.SBR4..	1.05± .06	5					
256.94 -40.14	ESO 203- 2	r	.18± .06	.00	14.80 ±.14				5419
236.06 -47.04		4.5± .9		.27					5240
044548.1-500818	PGC 15999	1.05		.09	14.50				5417
044708.0-020313	NGC 1661	.SAS4P*	1.14± .05	35					
199.57 -28.39	UGC 3166	UE (1)	.21± .05	.13	14.04 ±.12				
317.03 -55.67		3.5± .6		.31					
044436.5-020832	PGC 16000	4.2± .8	1.16	.11					
0447.1 +6603		.S?....	1.22± .06	110				15.04±.1	4698± 11
143.86 13.31	UGC 3149		.25± .06	1.26	14.4 ±.3				
13.83 -4.82	IRAS04422+6557			.38	13.19			409± 8	4826
0442.2 +6557	PGC 16001		1.34	.13	12.78			2.13	4645
044720.8-101419		.E+..P.	1.05± .07	100					
207.98 -32.17	MCG -2-13- 6	E	.13± .05	.24					
302.81 -58.45		-4.0± .9		.00					
044458.5-101936	PGC 16006	1.04							

R.A. 2000 DEC. / l b / SGL SGB / R.A. 1950 DEC.	Names / PGC	Type / S_T n_L / T / L	$\log D_{25}$ / $\log R_{25}$ / $\log A_e$ / $\log D_o$	p.a. / A_g / A_l / A_{21}	B_T / m_B / m_{FIR} / B_T^o	$(B-V)_T$ / $(U-B)_T$ / $(B-V)_T^o$ / $(U-B)_T^o$	$(B-V)_e$ / $(U-B)_e$ / m'_e / m'_{25}	m_{21} / W_{20} / W_{50} / HI	V_{21} / V_{opt} / V_{GSR} / V_{3K}
0447.4 +2358		.IA.9..	1.36± .09	135					3780± 7
176.59 -13.50	UGC 3165	U	.22± .12	2.24					
347.84 -39.55	IRAS04444+2353	10.0± .8		.16	12.15				3783
0444.4 +2353	PGC 16009		1.57	.11					3722
044730.6-173550		.LAR0..	1.07± .05	11					9007± 20
216.11 -35.07	ESO 552- 4	r	.08± .03	.07	14.62 ±.14				8886
288.53 -59.44		-2.4± .9		.00					8974
044517.0-174106	PGC 16011		1.07		14.41				
0447.5 -0218		.S..7..	.99± .06						
199.88 -28.43	UGC 3168	U	.06± .05	.12	14.89 ±.14				
316.69 -55.87		7.0± .9		.09					
0445.0 -0224	PGC 16012		1.01	.03					
044732.3-622241		.SA3?P/	1.10± .05	13					
272.61 -38.05	ESO 119- 2	S	.43± .04	.00	15.37 ±.14				
224.26 -38.24	IRAS04470-6227	3.0± .9		.59	13.63				
044700.0-622754	PGC 16013		1.10	.21					
044735.1-504513		.LA.0*.	1.03± .11						
257.84 -40.00		S	.40± .08	.00					
235.22 -46.63		-2.0± .9		.00					
044619.1-505027	PGC 16015		.97						
0447.6 -0109		.S..1..	.97± .07	117					
198.77 -27.83	UGC 3170	U	.51± .05	.18	15.35 ±.12				
318.55 -55.39		1.0±1.0		.53					
0445.1 -0115	PGC 16017		.99	.26					
044744.9+014903		.SB.6*.	1.03± .05					15.62±.3	4553± 10
195.91 -26.27	UGC 3171	U	.09± .04	.38	14.78 ±.14			226± 13	4588±125
323.09 -54.00		6.0±1.2		.13					4485
044509.2+014347	PGC 16021		1.07	.05	14.24			1.33	4506
044744.9+093550		.S..6*.	1.04± .08	73				15.91±.3	8788± 10
188.77 -22.02	UGC 3169	U	.76± .06	.70					8744
333.42 -49.59		6.0±1.4		1.12				359± 7	8736
044500.5+093033	PGC 16022		1.11	.38					
044750.0-623642		.SBT3P*	1.19± .04	26					
272.89 -37.96	ESO 85- 4	S (1)	.68± .04	.00	14.90 ±.14				
224.04 -38.08		3.0± .6		.93					
044719.0-624154	PGC 16026	4.4±1.3	1.19	.34					
0447.8 +7251		.S?....	1.10± .05	42					2948
138.31 17.52	UGC 3147		.29± .05	.51	14.69 ±.18				3091
17.32 1.01	IRAS04418+7246			.44	11.81				2899
0441.8 +7246	PGC 16027		1.14	.15	13.73				
0447.9 +7455	A 0441+74	.IB.9..	1.29± .04	117	15.0 ±.2	.64± .06	.65± .04	13.89±.1	1636± 6
136.57 18.75	UGC 3144	PU (1)	.37± .05	.60	15.2 ±.3	-.09± .10	-.12± .07	156± 8	
18.36 2.78	DDO 33	10.0± .6	.88± .05	.28		.41	14.91± .12	142± 12	1783
0441.4 +7450	PGC 16030	8.0±1.5	1.34	.18	14.17	-.26	15.37± .31	-.47	1588
0448.0 +0846		.S..7..	1.19± .06	138					
189.55 -22.43	UGC 3172	U	.91± .06	.60					
332.49 -50.15		7.0± .9		1.26					
0445.3 +0841	PGC 16033		1.25	.46					
044802.8-251346		.S..5*/	1.24± .03	102					4537± 52
225.13 -37.39	ESO 485- 12	S	.74± .03	.00	14.97 ±.14				4397
273.49 -58.87	IRAS04460-2519	5.0±1.2		1.11	13.88				4511
044559.0-251900	PGC 16036		1.24	.37	13.83				
044812.7-134003		.E+..?.	1.17± .05	135					
211.79 -33.41	MCG -2-13- 9	E	.04± .05	.28					
296.32 -59.29		-4.0±1.1		.00					
044554.5-134517	PGC 16040		1.20						
044817.2-052540	NGC 1665	.LAS+P?	1.24± .05	130					2736
203.10 -29.78	MCG -1-13- 9	E	.19± .05	.16					2646
311.61 -57.26		-1.0±1.2		.00					2694
044549.5-053054	PGC 16044		1.22						
044819.1-814101		.S?....	1.04± .07						4704
294.58 -31.14	ESO 15- 15		.08± .06	.47	14.63 ±.14				4523
210.87 -22.73				.12					4734
045304.0-814600	PGC 16046		1.08	.04	14.01				
0448.3 +0013		.E?....	.78± .13						8533± 38
197.53 -26.97	CGCG 394- 14		.08± .04	.33	15.35 ±.07				8460
320.83 -54.92				.00					8487
0445.8 +0008	PGC 16049		.81		14.89				

R.A. 2000 DEC. l b SGL SGB R.A. 1950 DEC.	Names PGC	Type S_T n_L T L	$\log D_{25}$ $\log R_{25}$ $\log A_e$ $\log D_o$	p.a. A_g A_i A_{21}	B_T m_B m_{FIR} B_T^o	$(B-V)_T$ $(U-B)_T$ $(B-V)_T^o$ $(U-B)_T^o$	$(B-V)_e$ $(U-B)_e$ m' m'_{25}	m_{21} W_{20} W_{50} HI	V_{21} V_{opt} V_{GSR} V_{3K}
044823.4-594802 269.39 -38.54 226.34 -40.25 044739.0-595312	NGC 1688 ESO 119- 6 IRAS04476-5953 PGC 16050	.SBT7.. RSr (2) 6.7± .4 5.3± .5	1.38± .03 .11± .03 1.38	177 .00 .15 .06	12.56 ±.12 11.93 12.02			14.20±.3 170± 10	1230± 9 1223± 34 1044 1239
0448.4 -5452 263.15 -39.41 230.86 -43.86 0447.3 -5457	PGC 16051	.LAR-?. S -3.0± .9	1.06± .10 .20± .08 1.03	.00 .00					
0448.4 +7328 137.82 17.92 17.66 1.51 0442.3 +7322	UGC 3150 IRAS04423+7322 PGC 16052	.SX.4.. U 4.0± .8	1.21± .05 .10± .05 1.26	165 .54 .15 .05	14.3 ±.2 13.10 13.56			16.16±.1 191± 8 2.54	4501± 11 4645 4453
044829.7-013218 199.26 -27.84 318.12 -55.75 044557.7-013731	MCG 0-13- 13 MK 1086 PGC 16054	.SB?... .05± .07 .95	.94± .11 .18 .06 .02		14.73 ±.12 13.35 14.40				8995± 61 8917 8951
0448.5 +0803 190.27 -22.73 331.71 -50.68 0445.8 +0758	UGC 3173 PGC 16055	.S..7*. U 7.0±1.3	.96± .17 .10± .12 1.00	.44 .14 .05					4648 4599 4598
044832.9-063414 204.30 -30.26 309.65 -57.71 044606.5-063926	NGC 1666 MCG -1-13- 10 PGC 16057	.LBR+.. R -1.0± .4	1.14± .03 .08± .05 .68± .02 1.15	35 .23 .00	13.61 ±.13	.97± .02 .50± .04	.98± .02 .48± .04 12.50 .05 14.00± .24		
044833.6-474902 254.00 -40.03 238.36 -48.69 044710.0-475412	NGC 1680 ESO 203- 4 IRAS04471-4754 PGC 16058	RSBS3?. Sr (1) 3.3± .5 3.3±1.3	1.08± .05 .42± .04 1.08	102 .00 .58 .21	14.45 ±.14 13.98				
0448.5 +0014 197.54 -26.92 320.91 -54.95 0446.0 +0009	A 0446+00 UGC 3174 DDO 34 PGC 16059	.IXS9*. UE (2) 10.0± .7 8.8± .7	1.22± .05 .18± .05 1.18± .06 1.25	85 .33 .14 .09	14.4 ±.2 13.94	.47± .10 -.15± .12 .34 -.24	.49± .05 -.17± .07 15.80± .13 14.90± .35	14.27±.1 118± 8 108± 12 .24	670± 6 596 624
044834.4-035202 201.57 -28.97 314.32 -56.74 044605.0-035714	MCG -1-13- 11 IRAS04460-0357 PGC 16060	.L..+P/ E -1.0±1.8	1.10± .06 .54± .05 1.03	24 .07 .00					2768 2683 2725
044837.0-061913 204.06 -30.12 310.11 -57.64 044610.4-062426	NGC 1667 MCG -1-13- 13 IRAS04461-0624 PGC 16062	.SXR5.. R (3) 5.0± .4 2.7± .6	1.25± .03 .11± .04 .79± .01 1.27	20 .24 .16 .05	12.77 ±.13 13.0 ±.2 11.06 12.41	.70± .02 .03± .02 .60 -.05	.78± .02 .10± .02 12.21± .04 13.59± .23	15.38±.3 324± 34 296± 25 2.92	4547± 17 4587± 46 4459 4511
044842.7-045839 202.70 -29.47 312.46 -57.20 044614.6-050351	MCG -1-13- 12 PGC 16065	.SBS7.. E (1) 7.0± .9 6.4± .8	1.13± .06 .37± .05 1.14	160 .11 .51 .18					4823 4734 4781
044845.9-050732 202.86 -29.53 312.22 -57.26 044617.9-051244	IC 2095 MCG -1-13- 14 PGC 16067	.S..6P/ E 6.0±1.4	1.14± .06 .92± .05 1.15	125 .16 1.36 .46					
044852.6-544242 262.93 -39.36 230.96 -44.03 044748.8-544750	PGC 16068	.LA.-*. S -3.0±1.2	1.11± .10 .00± .08 1.12	.03 .00					
044855.8-234344 223.40 -36.77 276.33 -59.34 044650.1-234854	ESO 485- 16 PGC 16071	.SXT2*. S 2.0±1.2	1.12± .04 .39± .03 1.12	70 .02 .48 .19	15.31 ±.14 14.73				8129± 22 7992 8103
044857.2-573934 266.68 -38.89 228.17 -41.89 044804.0-574442	ESO 119- 8 FAIR 239 PGC 16072	RSBR1*. Sr 1.3± .5	1.03± .05 .01± .05 1.03	.00 .01 .01	14.82 ±.14 14.72				7040± 63 6855 7047
044857.8-320732 233.71 -38.78 260.78 -57.03 044704.0-321242	ESO 421- 18 PGC 16073	.SBR6*. r 5.6±1.2	1.01± .06 .41± .06 1.01	153 .00 .61 .21	15.87 ±.14 15.20				12614 12460 12596
044912.2-291227 230.08 -38.13 265.91 -58.09 044714.0-291736	A 0447-29 ESO 421- 19 DDO 228 PGC 16084	.SXS9*. SU (2) 9.0± .5 8.9± .5	1.43± .03 .05± .04 1.24± .05 1.43	55 .00 .05 .03	13.1 ±.2 13.43 ±.14 13.27	.58± .04 -.35± .05 .56 -.36	.54± .03 -.28± .04 14.75± .10 14.96± .27	14.28±.2 148± 16 143± 12 .99	1471± 11 1322 1450

R.A. 2000 DEC. l b SGL SGB R.A. 1950 DEC.	Names PGC	Type S_T n_L T L	$\log D_{25}$ $\log R_{25}$ $\log A_e$ $\log D_0$	p.a. A_g A_i A_{21}	B_T m_B m_{FIR} B_T^0	$(B-V)_T$ $(U-B)_T$ $(B-V)_T^0$ $(U-B)_T^0$	$(B-V)_{\bullet}$ $(U-B)_{\bullet}$ $m'_{\bullet}$ m'_{25}	m_{21} W_{20} W_{50} HI	V_{21} V_{opt} V_{GSR} V_{3K}
044918.7+564203 151.43 7.64 9.26 −12.98 044508.5+563650	 PGC 16089	 	 	 1.64 	 	 	 	15.19±.3 259± 11 248± 8 	5257± 9 5360 5202
0449.3 −1244 210.92 −32.79 298.20 −59.42 0447.0 −1250	 MCG −2-13- 11 PGC 16090	.SBS9*. E (1) 9.0±1.2 8.7±1.2	1.20± .05 .16± .05 1.24	25 .38 .16 .08	 	 	 	 	2201 2091 2166
0449.4 +6355 145.73 12.18 12.97 −6.78 0444.7 +6350	 UGC 3167 IRAS04447+6350 PGC 16092	.S..4.. U 4.0± .9 	1.16± .13 .78± .12 1.32	167 1.69 1.15 .39	 12.17 	 	 	15.19±.1 441± 8 	4735± 7 4857 4683
044933.8+001514 197.67 −26.70 321.13 −55.16 044659.8+001006	IC 395 UGC 3178 PGC 16095	.L...*. U −2.0±1.2 	1.04± .09 .09± .04 1.06	130 .33 .00 	13.9 ±.3 13.75 ±.10 	1.03± .08 .79± .05 	 13.76± .56 	 	
0449.5 +6929 141.21 15.60 15.75 −1.98 0444.2 +6924	 UGC 3163 PGC 16096	.S..1.. U 1.0± .9 	1.17± .05 .64± .05 1.25	86 .81 .65 .32	 15.1 ±.3 	 	 	 	
044936.7−075139 205.77 −30.61 307.49 −58.36 044711.7−075647	 MCG −1-13- 15 PGC 16097	PSAT0*. E .0± .9 	1.03± .07 .18± .05 1.03	165 .13 .14 	 	 	 	 	
044937.3−535443 261.88 −39.36 231.64 −44.67 044831.0−535948	 ESO 158- 7 FAIR 783 PGC 16098	.SXR3*. S (1) 3.0± .8 3.3±1.2	1.15± .04 .14± .04 1.16	50 .09 .19 .07	 14.64 ±.14 14.25	 	 	 	15100±190 14917 15104
044939.7−172916 216.21 −34.55 288.78 −59.95 044726.0−173424	 ESO 552- 9 PGC 16101	.SB?... 	.95± .05 .11± .04 .95	92 .08 .13 .05	 15.48 ±.14 15.17	 	 	 	 9273± 60 9150 9243
044940.3−420342 246.50 −39.81 245.40 −52.36 044804.0−420848	 ESO 304- 21 IRAS04480−4209 PGC 16102	.SXT4.. r 4.5± .9 	1.03± .06 .22± .05 1.03	32 .00 .32 .11	 14.93 ±.14 14.58	 	 	 	5810 5639 5802
0449.6 +8239 129.82 23.24 22.34 9.41 0439.0 +8234	 UGC 3132 PGC 16105	.S..0.. U .0± .9 	1.16± .07 .41± .06 1.17	162 .37 .31 	 15.1 ±.2 	 	 	 	
044942.6−024539 200.63 −28.18 316.37 −56.55 044712.0−025046	NGC 1670 MCG 0-13- 16 PGC 16107	.LA.0*. E −2.3± .7 	1.32± .06 .33± .05 1.29	112 .15 .00 	13.70 ±.11 	 	 	 	
044942.7−104226 208.77 −31.85 302.18 −59.13 044721.0−104733	 MCG −2-13- 12 PGC 16108	.SBS1P? E 1.0±1.8 	1.03± .07 .08± .05 1.05	100 .19 .08 .04	 	 	 	 	
044944.6+032003 194.78 −25.05 325.71 −53.63 044707.2+031455	A 0447+03 UGC 3179 MK 1087 PGC 16109	.L...P* R −2.0± .7 	1.02± .06 .11± .05 1.05	10 .33 .00 	14.46 ±.10 11.88 14.00	 	 	15.51±.2 273± 5 241± 7 	8337± 5 8306± 21 8271 8290
0449.8 +7258 138.31 17.70 17.50 1.03 0443.8 +7253	 UGC 3159 PGC 16110	.S..4.. U 4.0± .9 	.96± .09 .29± .06 1.01	5 .51 .43 .15	 15.46 ±.18 	 	 	 	
044949.4+214101 178.80 −14.48 346.38 −41.63 044650.5+213553	 UGC 3177 PGC 16111	.I..9*. U 10.0±1.2 	1.11± .14 .00± .12 1.30	 2.06 .00 .00	 	 	 	15.05±.1 187± 6 155± 4 	3917± 7 3911 3863
044951.2−525937 260.70 −39.43 232.52 −45.35 044842.0−530442	 ESO 158- 8 PGC 16115	.S?.... 	1.00± .06 .57± .05 .55± .04 1.00	173 .00 .86 .29	15.40 ±.18 16.08 ±.14 	.55± .04 	.63± .04 13.64± .11 13.80± .36	 	
044951.3−360612 238.81 −39.24 254.08 −55.51 044804.0−361118	 ESO 361- 12 PGC 16116	.S..4?/ S 4.0±1.9 	1.18± .05 .89± .04 1.18	138 .00 1.31 .45	 15.19 ±.14 	 	 	 	

R.A. 2000 DEC.	Names	Type	logD$_{25}$	p.a.	B$_T$	(B-V)$_T$	(B-V)$_e$	m$_{21}$	V$_{21}$
l b		S$_T$ n$_L$	logR$_{25}$	A$_g$	m$_B$	(U-B)$_T$	(U-B)$_e$	W$_{20}$	V$_{opt}$
SGL SGB		T	logA$_e$	A$_i$	m$_{FIR}$	(B-V)^o_T	m'	W$_{50}$	V$_{GSR}$
R.A. 1950 DEC.	PGC	L	logD$_o$	A$_{21}$	B^o_T	(U-B)^o_T	m'$_{25}$	HI	V$_{3K}$

044953.5-415637		PS.R1..	.91± .06	0					
246.35 -39.76	ESO 304- 24	r	.23± .05	.00	15.40 ±.14				
245.52 -52.46		1.0± .9		.23					
044817.0-420142	PGC 16117		.91	.11					
044953.9-104518		.SBS4..	1.06± .06	40					
208.85 -31.83	MCG -2-13- 13	E (1)	.26± .05	.16					
302.11 -59.18		4.0± .9		.38					
044732.3-105025	PGC 16118	4.2± .8	1.07	.13					
044955.0-315800	NGC 1679	.SBS9..	1.43± .03	150				13.35±.2	1059± 11
233.56 -38.55	ESO 422- 1	SU (1)	.13± .04	.00	12.00 ±.14			151± 16	
260.91 -57.28	IRAS04480-3203	9.5± .4		.09	11.91			137± 12	904
044801.0-320306	PGC 16120	6.7± .4	1.43	.06	11.91			1.38	1042
044957.9-180506		.L?....	1.08± .05	162					10144± 60
216.92 -34.70	ESO 552- 11		.58± .03	.07	15.08 ±.14				10020
287.57 -60.03				.00					10115
044745.0-181012	PGC 16121		1.00		14.86				
045010.1-394743		.IXS9..	1.08± .09						
243.57 -39.57		S (1)	.28± .08	.00					
248.47 -53.69		10.0± .9		.21					
044829.4-395247	PGC 16126	10.0± .9	1.08	.14					
045013.7-171600		.SBR4?/	1.30± .04	93					
216.02 -34.34	MCG -3-13- 16	SE (1)	.59± .04	.08					
289.23 -60.08		4.0± .8		.86					
044759.8-172105	PGC 16128	2.2±1.2	1.31	.29					
045018.4-611453		.S..5*P	1.02± .05	164					
271.12 -38.00	ESO 119- 12	S (1)	.12± .04	.00	14.65 ±.14				5941
224.88 -39.31	IRAS04496-6119	5.0±1.2		.19	13.36				5754
044941.0-611954	PGC 16130	5.6± .9	1.02	.06	14.43				5953
045023.5-050449	IC 2097	.IB.9*/	1.05± .09	65					
203.04 -29.15		E (1)	.49± .08	.11					
312.57 -57.62		10.0±1.3		.37					
044755.4-050954	PGC 16134	6.4±1.2	1.06	.25					
045027.0-612041		.SBS5..	1.13± .04	70	13.64 ±.15	.77± .03	.72± .01		
271.24 -37.96	ESO 119- 13	Sr (1)	.07± .04	.00	14.14 ±.14				5903± 40
224.78 -39.25	IRAS04499-6125	4.5± .5	.99± .03	.11		.72	14.08± .08		5716
044950.1-612542	PGC 16136	5.6± .9	1.13	.04	13.76		13.96± .28		5915
045029.1-033115		.LBT0*.	1.11± .06	20					
201.49 -28.38	MCG -1-13- 16	E	.39± .05	.13					
315.25 -57.04		-2.0± .9		.00					
044759.3-033619	PGC 16138		1.07						
045037.6+060034		.SB.3..	1.34± .04	97				13.96±.3	4613± 7
192.43 -23.42	UGC 3181	U	.34± .05	.31	14.03 ±.16				4556
329.58 -52.32	IRAS04479+0555	3.0± .8		.46	13.08			300± 7	4567
044757.2+055530	PGC 16141		1.37	.17	13.22			.58	
045038.7-030830		.SAS1P?	1.01± .07	140					
201.14 -28.16	MCG -1-13- 17	E	.38± .05	.10					
315.92 -56.92		1.0± .9		.39					
044808.5-031334	PGC 16142		1.02	.19					
045039.9-612118		.S..7*P	1.07± .05	155					
271.24 -37.93	ESO 119- 14	S (1)	.29± .04	.00	15.15 ±.14				
224.74 -39.26		7.0±1.2		.40					
045003.0-612618	PGC 16143	1.1± .8	1.07	.15					
045044.6-052508	IC 2098	.S..6*/	1.32± .04	85					2826
203.43 -29.24	MCG -1-13- 18	E	.88± .05	.11					
312.04 -57.83	IRAS04482-0530	6.0±1.2		1.30	12.95				2735
044816.9-053012	PGC 16144		1.33	.44					2788
045052.0-045333	NGC 1677	.L..0P/	1.02± .07	40					
202.91 -28.96	MCG -1-13- 19	E	.47± .05	.11					
312.98 -57.66	IRAS04484-0458	-2.0±1.3		.00	13.05				
044823.8-045836	PGC 16146		.96						
045053.0-445823		.S?....	.80± .07						
250.29 -39.67	ESO 251- 36		.08± .06	.00	15.26 ±.14				10250
241.40 -50.80	IRAS04493-4502			.09					10075
044923.1-450324	PGC 16147		.80	.04	15.05				10247
045057.1-553154		.LA.-..	1.08± .10						
263.92 -38.95		S	.16± .08	.00					
229.87 -43.64		-3.0± .9		.00					
044956.4-553654	PGC 16151		1.06						

R.A. 2000 DEC.	Names	Type	logD$_{25}$	p.a.	B$_T$	(B-V)$_T$	(B-V)$_e$	m$_{21}$	V$_{21}$
I　　b		S$_T$　n$_L$	logR$_{25}$	A$_g$	m$_B$	(U-B)$_T$	(U-B)$_e$	W$_{20}$	V$_{opt}$
SGL　　SGB		T	logA$_e$	A$_i$	m$_{FIR}$	(B-V)^o_T	m'$_e$	W$_{50}$	V$_{GSR}$
R.A. 1950 DEC.	PGC	L	logD$_o$	A$_{21}$	B^o_T	(U-B)^o_T	m'$_{25}$	HI	V$_{3K}$

0451.1　+0222		.S?....	.94± .11						8936± 38
195.88 -25.27	MCG　0-13- 17		.00± .07	.29	15.22 ±.16				8868
324.63 -54.44				.00					8893
0448.5　+0217	PGC 16154		.96	.00	14.88				
045106.5-395136		PSBR2?.	1.02± .05	157					6222
243.67 -39.40	ESO　304- 26	Sr	.21± .04	.00	15.21 ±.14				6054
248.21 -53.81		1.6± .6		.26					6214
044926.0-395636	PGC 16155		1.02	.11	14.89				
0451.2　-1350		.S?....	1.12± .08						5460
212.34 -32.82	MCG -2-13- 15		.65± .07	.21					
296.15 -60.04				.97					5346
0448.9　-1356	PGC 16159		1.14	.32					5429
045118.3-055743		.S..6*/	1.15± .06	155					
204.05 -29.37	MCG -1-13- 21	E	.68± .05	.11					
311.17 -58.15		6.0±1.3		1.00					
044851.3-060244	PGC 16163		1.16	.34					
045119.5+231229		.S..3*.	.96± .09	120				16.05±.3	4409± 10
177.77 -13.28	UGC　3183	U	.10± .06	2.29					4407
348.05 -40.72		3.0±1.3		.14				405± 7	4357
044818.5+230727	PGC 16164		1.18	.05					
045120.6-173012	A　0449-17	.LAS-P*	1.06± .04	5	14.4 ±.3	.97± .05			9530± 61
216.40 -34.18	ESO　552- 14	SE	.20± .03	.11	14.58 ±.14	.50± .09			9406
288.76 -60.35		-2.7± .7		.00		.84			9502
044907.0-173512	PGC 16165		1.05		14.29	.52	14.10± .38		
045121.8-335619	NGC　1687	.SXR2*.	1.10± .05	40					
236.11 -38.61	ESO　361- 13	BSr	.38± .04	.00	14.73 ±.14				
257.34 -56.76		2.1± .5		.47					
044931.0-340118	PGC 16166		1.10	.19					
045123.4-591403		.L..0*.	1.05± .05	55					16165± 63
268.58 -38.28	ESO　119- 15	S	.19± .04	.00	15.01 ±.14				15978
226.45 -40.93		-2.0±1.1		.00					16176
045037.0-591900	PGC 16167		1.02		14.77				
045126.6-030720		.S..3*/	1.22± .05	80					4582
201.23 -27.98	MCG -1-13- 22	E　　(1)	.59± .05	.10					
316.10 -57.10		3.0±1.2		.82					4497
044856.4-031221	PGC 16168		1.23	.30					4543
045129.3-613903		.IBS9..	1.36± .04	25					980
271.58 -37.77	ESO　119- 16	S　　(1)	.36± .05	.02	14.88 ±.14				793
224.40 -39.09		9.5± .6		.27					993
045054.0-614400	PGC 16172		9.4± .6	1.36	.18	14.59			
0451.5　+0543									9207± 38
192.84 -23.39	CGCG 420- 7			.32	15.3 ±.3				9149
329.41 -52.68									9163
0448.9　+0538	PGC 16176								
0451.5　+6712		.L..-*.	1.15± .15						
143.21　14.37	UGC　3176	U	.07± .08	1.11	14.2 ±.3				
14.78　-4.05		-3.0±1.1		.00					
0446.5　+6707	PGC 16178		1.28						
045135.5-023724	NGC　1678	.LA.0P?	1.03± .07	110					
200.75 -27.70	MCG　0-13- 19	E	.15± .06	.13	14.18 ±.10				
316.96 -56.92		-2.0± .9		.00					
044904.7-024224	PGC 16179		1.02						
045136.4-023337		.S?....	.74± .14						2100± 49
200.69 -27.67	MCG　0-13- 18		.38± .07	.16	15.26 ±.18				2017
317.07 -56.90	ARAK　113			.47					2061
044905.5-023837	PGC 16180		.75	.19	14.63				
045141.5-034835		.LXS+P*	1.14± .08	145					
201.94 -28.26	MCG -1-13- 25	E	.26± .08	.10					
314.99 -57.43		-1.0± .9		.00					
044912.0-035334	PGC 16185		1.11						
045141.9+053011		.S?....	1.14± .13	13				14.73±.3	4561± 7
193.06 -23.47	UGC　3184		.47± .12	.31					4502
329.16 -52.83				.71				392± 7	4517
044902.0+052511	PGC 16186		1.17	.24					
045142.1-061349	IC　2101	.SBS5P?	1.21± .05	37					4525
204.38 -29.40	MCG -1-13- 24	E　　(1)	.64± .05	.12					
310.76 -58.33	IRAS04492-0618	5.0±1.3		.96	12.07				4431
044915.4-061848	PGC 16187		3.1±1.6	1.22	.32				4489

R.A. 2000 DEC. / l b / SGL SGB / R.A. 1950 DEC.	Names / PGC	Type / S_T n_L / T / L	$\log D_{25}$ / $\log R_{25}$ / $\log A_e$ / $\log D_0$	p.a. / A_g / A_i / A_{21}	B_T / m_B / m_{FIR} / B_T^0	$(B-V)_T$ / $(U-B)_T$ / $(B-V)_T^0$ / $(U-B)_T^0$	$(B-V)_e$ / $(U-B)_e$ / m'_e / m'_{25}	m_{21} / W_{20} / W_{50} / HI	V_{21} / V_{opt} / V_{GSR} / V_{3K}
045145.1+065127 / 191.83 -22.73 / 330.96 -52.05 / 044903.7+064627	UGC 3187 / PGC 16189	.S..7.. / U / 7.0± .8 /	1.14± .07 / .23± .06 / / 1.18	0 / .39 / .31 / .11	/ 15.11 ±.19 / / 14.39			17.02±.3 / / 242± 7 / 2.52	4730± 7 / / 4675 / 4685
045146.7+034012 / 194.76 -24.44 / 326.65 -53.88 / 044908.9+033513	UGC 3186 / PGC 16191	.S..6*. / U / 6.0±1.3 /	1.19± .06 / .81± .06 / / 1.22	48 / .28 / 1.20 / .41				14.31±.3 / / 250± 7 /	4578± 7 / / 4513 / 4535
045147.8-570846 / 265.94 -38.59 / 228.25 -42.53 / 045053.0-571342	ESO 158- 9 / PGC 16192	.IBS9*/ / S (1) / 9.7± .8 / 8.3± .7	1.06± .05 / .71± .04 / / 1.06	172 / .00 / .53 / .36	/ 16.49 ±.14 / /				
045148.8+085156 / 190.03 -21.60 / 333.51 -50.83 / 044905.1+084657	UGC 3188 / IRAS04492+0846 / PGC 16193	.S?.... / / /	1.00± .05 / .25± .04 / / 1.06	85 / .65 / .38 / .13	/ 14.78 ±.12 / 13.53 / 13.72			16.87±.3 / / 235± 5 / 3.02	3522± 9 / / 3474 / 3477
045150.1-481703 / 254.58 -39.46 / 237.28 -48.80 / 045028.0-482200	ESO 203- 9 / PGC 16194	.SX.9*. / S (1) / 9.0±1.3 / 10.0±1.3	.92± .05 / .26± .05 / / .92	38 / .00 / .26 / .13	/ 14.74 ±.14 / /				
045150.2-054813 / 203.96 -29.18 / 311.54 -58.22 / 044922.9-055312	NGC 1681 / MCG -1-13- 26 / IRAS04493-0553 / PGC 16195	.S..3P* / E / 3.0±1.2 /	1.11± .06 / .06± .05 / / 1.12	40 / .11 / .08 / .03	/ / / 12.12				
045154.9-045710 / 203.11 -28.76 / 313.05 -57.93 / 044926.7-050209	IC 2102 / MCG -1-13- 27 / PGC 16197	.SBT6.. / E (1) / 6.0± .8 / 4.2± .8	1.17± .06 / .00± .05 / / 1.17	/ .10 / .00 / .00					3645 / / 3555 / 3608
045158.0-331039 / 235.18 -38.35 / 258.51 -57.19 / 045006.0-331536	ESO 361- 15 / IRAS04500-3315 / PGC 16199	.SBS7?/ / S (1) / 7.0±1.1 / 5.6±1.2	1.47± .03 / .83± .04 / / 1.47	94 / .00 / 1.15 / .42	/ 13.71 ±.14 / 13.52 / 12.56				1186± 52 / / 1028 / 1173
045206.0-283533 / 229.51 -37.37 / 266.67 -58.89 / 045007.1-284030	ESO 422- 5 / PGC 16201	.I..9*. / U / 10.0±1.2 /	1.09± .05 / .17± .04 / / 1.09	2 / .00 / .13 / .08	/ 14.88 ±.14 / / 14.75			14.55±.3 / 138± 16 / 132± 12 / -.29	1460± 11 / / 1311 / 1443
045218.9+493239 / 157.28 3.45 / 5.84 -19.34 / 044830.5+492740	PGC 16210			/ 3.80				16.37±.3 / 228± 11 / 154± 8 /	6108± 9 / / 6190 / 6054
045225.1-033357 / 201.80 -27.98 / 315.54 -57.50 / 044955.4-033853	MCG -1-13- 30 / PGC 16215	.LAS0P? / E / -2.0±1.7 /	1.21± .09 / .04± .07 / / 1.22	100 / .09 / .00 /					
0452.4 +7429 / 137.12 18.74 / 18.40 2.26 / 0446.0 +7424	UGC 3175 / IRAS04460+7424 / PGC 16216	.S..4.. / U / 4.0± .9 /	1.16± .07 / .78± .06 / / 1.21	173 / .57 / 1.15 / .39	/ 15.22 ±.18 / 13.40 /				
045231.1-030623 / 201.36 -27.74 / 316.33 -57.33 / 045000.9-031119	NGC 1684 / MCG -1-13- 31 / IRAS04500-0311 / PGC 16219	.E+..P* / E / -4.0± .8 /	1.39± .07 / .17± .07 / / 1.36	95 / .14 / .00 /	/ 13.10 / /				
045231.2-625908 / 273.18 -37.35 / 223.21 -38.13 / 045203.0-630400	NGC 1706 / ESO 85- 7 / FAIR 240 / PGC 16220	.SAT2.. / S (1) / 2.3± .5 / 1.1± .7	1.14± .04 / .15± .04 / .94± .03 / 1.14	124 / .01 / .19 / .08	13.53 ±.15 / 13.97 ±.14 / / 13.52	.93± .02 / / .86 /	.89± .01 / / 13.72± .09 / 13.71± .29		4889± 26 / / 4701 / 4904
045234.2-025658 / 201.21 -27.65 / 316.61 -57.28 / 045003.8-030154	NGC 1685 / MCG -1-13- 32 / IRAS04500-0301 / PGC 16222	.SBR0.. / E / .0± .9 /	1.10± .06 / .16± .05 / / 1.10	45 / .14 / .12 /	/ 13.11 / /				4527 / / 4442 / 4490
045237.5-595208 / 269.33 -38.01 / 225.75 -40.55 / 045154.0-595700	ESO 119- 18 / PGC 16224	PSBT2?. / Sr / 1.5± .9 /	1.01± .05 / .29± .04 / / 1.01	60 / .00 / .35 / .14	/ 15.63 ±.14 / /				
0452.7 +0423 / 194.23 -23.85 / 327.90 -53.67 / 0450.1 +0419	UGC 3191 / PGC 16227	.S..3*. / U (1) / 3.0±1.2 / 5.5±1.1	1.10± .05 / .22± .05 / / 1.13	75 / .29 / .30 / .11	/ 14.85 ±.15 / /				

R.A. 2000 DEC. / l b / SGL SGB / R.A. 1950 DEC.	Names / S_T n_L / / PGC	Type / T / L	$\log D_{25}$ / $\log R_{25}$ / $\log A_e$ / $\log D_o$	p.a. / A_g / A_i / A_{21}	B_T / m_B / m_{FIR} / B_T^o	$(B-V)_T$ / $(U-B)_T$ / $(B-V)_T^o$ / $(U-B)_T^o$	$(B-V)_e$ / $(U-B)_e$ / m'_e / m'_{25}	m_{21} / W_{20} / W_{50} / HI	V_{21} / V_{opt} / V_{GSR} / V_{3K}
045249.4+011533		.L..-*.	.86± .15	55	14.28 ±.15	1.08± .02	1.09± .02		
197.18 -25.48	UGC 3192	U	.08± .05	.32	14.48 ±.11				17700± 49
323.37 -55.37	ARAK 114	-3.0±1.2	.76± .05	.00		.84	13.57± .17		17627
045014.2+011038	PGC 16232		.89		13.82		13.24± .80		17661
045252.0-594433	NGC 1703	.SBR3..	1.47± .04		11.9 ±.2	.56± .01	.64± .01	14.25±.3	1526± 8
269.17 -38.00	ESO 119- 19	R (1)	.05± .04	.00	12.31 ±.14	-.12± .03	-.03± .02	72± 9	1526± 40
225.83 -40.67	IRAS04521-5949	3.0± .7	1.19± .03	.07	11.94	.54	13.32± .07		1339
045208.0-594924	PGC 16234	2.2± .7	1.47	.03	12.09	-.13	13.95± .29	2.14	1538
045252.6-251448	A 0450-25	.SBS9..	1.59± .03	110	*			13.47±.2	1374± 11
225.53 -36.34	ESO 485- 21	SU (2)	.08± .04	.00	13.27 ±.14			91± 16	
273.00 -59.93	DDO 229	8.5± .5		.08				71± 12	1231
045049.0-251942	PGC 16236	7.9± .5	1.59	.04	13.19			.24	1355
045252.7+030324		.SBT2*.	1.18± .04	170	*			15.85±.3	4454± 10
195.49 -24.53	UGC 3193	UE	.42± .04	.23	14.43 ±.12				
326.03 -54.44	IRAS04502+0258	2.3± .7		.52	12.03			399± 7	4387
045015.6+025830	PGC 16237		1.20	.21	13.64			2.00	4414
045254.4-152052	NGC 1686	.S...4./	1.22± .05	27	*				
214.17 -33.03	MCG -3-13- 19	E	.70± .05	.11					
293.21 -60.61		4.0± .9		1.03					
045038.3-152546	PGC 16239		1.23	.35					
0452.9 -0532		.SBS8*.	1.22± .07	10	*				
203.84 -28.80		E (1)	.18± .08	.13					
312.21 -58.38		8.0±1.3		.22					
0450.5 -0537	PGC 16241	8.7±1.2	1.24	.09					
045258.9+011951		.SBS3P*	1.20± .04	5	*				
197.13 -25.41	UGC 3194	UE	.37± .04	.32	14.91 ±.16				
323.51 -55.37	IRAS04503+0114	3.0± .8		.52	13.39				
045023.7+011457	PGC 16242		1.23	.19					
0453.1 +0422		.S..6*.	1.07± .08	110	*				
194.30 -23.77	UGC 3195	U	.20± .06	.29	15.10 ±.16				
327.97 -53.77		6.0±1.2		.29					
0450.5 +0418	PGC 16248		1.10	.10					
045310.2-564328			.91± .06		16.79 ±.14	.72± .06			
265.37 -38.47	ESO 158- 11		.19± .06	.00	15.73 ±.14				
228.45 -42.97									
045214.1-564818	PGC 16250		.91						
045317.5-030035		.SBS6..	1.13± .06	65					4601
201.37 -27.52	MCG -1-13- 33	E (1)	.19± .05	.13					
316.65 -57.47		6.0± .9		.28					4515
045047.2-030528	PGC 16256	4.2± .8	1.14	.09					4565
045318.6-703555			1.33± .06	50					273
282.23 -35.16	ESO 56- 19		.32± .06	.48	14.08 ±.14				85
217.56 -32.08	IRAS04538-7040				12.53				295
045347.1-704042	PGC 16258		1.37						
045334.0-175521		.S?....	1.02± .05	71					11963± 60
217.11 -33.84	ESO 552- 19		.15± .04	.06	15.28 ±.14				11837
287.91 -60.88				.18					11939
045121.0-180012	PGC 16265		1.03	.07	14.92				
0453.7 +6428		.S..6*.	1.07± .14	120				15.17±.1	4680± 11
145.57 12.90	UGC 3185	U	.11± .12	1.35					
13.65 -6.53		6.0±1.2		.16				276± 8	4802
0449.0 +6424	PGC 16271		1.20	.05					4631
045349.1-322623		PSBR2*.	1.05± .04						5490± 34
234.35 -37.83	ESO 361- 16	Sr	.09± .03	.00	14.65 ±.14				5333
259.48 -57.85		1.6± .6		.11					5479
045156.1-323112	PGC 16273		1.05	.05	14.48				
045407.1+721928		.L.....	1.15± .15						
139.05 17.58	UGC 3182	U	.00± .08	.52	14.17 ±.18				
17.46 .31		-2.0± .8		.00					
044813.6+721432	PGC 16280		1.21						
045413.6-532144	NGC 1705	.LA.-P*	1.28± .03	50	12.77 ±.13	.38± .01	.22± .01		
261.08 -38.74	ESO 158- 13	RS	.13± .04	.19	12.80 ±.11	-.45± .01	-.58± .01		673± 56
231.49 -45.54	IRAS04531-5326	-3.0± .6	.59± .01	.00	13.16	.32	11.21± .03		489
045306.0-532630	PGC 16282		1.28		12.58	-.48	13.70± .22		681
045414.1-435225		.LA.+?/	1.16± .05	105					
248.89 -39.05	ESO 251- 41	S	.84± .03	.00	14.76 ±.14				
242.19 -51.95		-1.0±2.0		.00					
045242.0-435712	PGC 16283		1.04						

R.A. 2000 DEC. / l b / SGL SGB / R.A. 1950 DEC.	Names / PGC	Type S_T n_L / T / L	logD25 / logR25 / logAe / logDo	p.a. / A_g / A_i / A_{21}	B_T / m_B / m_{FIR} / B_T^0	$(B-V)_T$ / $(U-B)_T$ / $(B-V)_T^0$ / $(U-B)_T^0$	$(B-V)_e$ / $(U-B)_e$ / m'_e / m'_{25}	m_{21} / W_{20} / W_{50} / HI	V_{21} / V_{opt} / V_{GSR} / V_{3K}
0454.2 -0537 / 204.10 -28.56 / 312.28 -58.71 / 0451.8 -0542	PGC 16284	.SBS7*. / E (1) / 7.0±1.3 / 8.7± .8	1.30± .06 / .23± .08 / / 1.32	10 / .13 / .31 / .11					
045415.6-114624 / 210.46 -31.29 / 300.55 -60.44 / 045155.3-115113	MCG -2-13- 18 / PGC 16285	.S..7?. / E (1) / 7.0±1.8 / 6.4±1.6	1.12± .06 / .40± .05 / / 1.15	90 / .35 / .55 / .20					4750 / / 4639 / 4722
0454.3 +0207 / 196.57 -24.71 / 325.01 -55.24 / 0451.7 +0203	UGC 3200 / PGC 16286	.SBT3.. / U / 3.0± .9 /	1.04± .08 / .14± .06 / / 1.07	110 / .29 / .19 / .07	/ / / 14.81 ±.14 / 14.25				9409± 38 / 9338 / 9371
045419.4+013824 / 197.03 -24.96 / 324.28 -55.50 / 045143.8+013335	NGC 1690 / UGC 3199 / PGC 16289	.E...*. / UE / -5.0± .7 /	.99± .05 / .53± .05 / / .89	116 / .31 / .00 /	14.84 ±.11 / / / 14.50				2100± 49 / 2028 / 2063
045419.6+013825 / 197.03 -24.96 / 324.28 -55.50 / 045144.0+013336	UGC 3198 / ARAK 115 / PGC 16290	.E?.... / / /	.99± .09 / .00± .03 / / 1.04	/ .31 / .00 /	14.89 ±.13 / / / 14.54				2100 / 2028 / 2063
0454.5 -0122 / 199.94 -26.44 / 319.62 -57.02 / 0452.0 -0127	MCG 0-13- 29 / PGC 16296	.S?.... / / /	.64± .17 / .00± .07 / / .66	/ .20 / .00 / .00	15.45 ±.13 / / / 15.20				8610± 38 / 8528 / 8575
045438.4+031606 / 195.55 -24.05 / 326.75 -54.70 / 045201.0+031119	NGC 1691 / UGC 3201 / MK 1088 / PGC 16300	RSBS0*. / UE / .0± .6 /	1.22± .04 / .03± .04 / / 1.24	85 / .23 / .02 /	13.01 ±.12 / 11.17 / 12.69			16.24±.3 / 357± 14 / /	4581± 10 / 4590± 61 / 4514 / 4543
0454.6 +6943 / 141.28 16.10 / 16.25 -1.98 / 0449.2 +6939	UGC 3189 / PGC 16301	.S..7.. / U / 7.0± .8 /	1.28± .04 / .54± .05 / / 1.33	40 / .61 / .74 / .27	15.0 ±.3 / / 13.66		/ / / .76	14.70±.1 / / 334± 8 /	4560± 11 / / 4695 / 4513
045439.4-121213 / 210.97 -31.38 / 299.73 -60.61 / 045219.5-121700	MCG -2-13- 19 / IRAS04523-1217 / PGC 16302	.SXS5*. / E (1) / 5.0±1.2 / 3.1±1.2	1.29± .05 / .44± .05 / / 1.33	120 / .37 / .67 / .22					4927 / / 4815 / 4900
045440.7-283926 / 229.76 -36.84 / 266.20 -59.40 / 045242.0-284412	ESO 422- 9 / PGC 16305	PSBS9*. / r / 9.1±1.2 /	1.02± .05 / .03± .04 / / 1.02	/ .00 / .03 / .02	15.36 ±.14				
045440.9-562956 / 265.05 -38.30 / 228.44 -43.28 / 045344.1-563440	PGC 16306	.LA.-*P / S / -3.0± .8 /	1.27± .08 / .28± .08 / / 1.23	/ .00 / .00 /					
045442.9-624759 / 272.89 -37.15 / 223.11 -38.43 / 045414.1-625242	ESO 85- 14 / IRAS04542-6252 / PGC 16309	.SBS9.. / S (1) / 9.4± .5 / 7.2± .6	1.45± .04 / .41± .05 / / 1.46	75 / .00 / .41 / .20	13.42 ±.14 / 13.30 / 13.00				1356 / 1167 / 1372
045451.8-180650 / 217.45 -33.62 / 287.51 -61.19 / 045239.0-181136	ESO 552- 20 / PGC 16315	.E+.... / SE / -4.0± .5 /	1.37± .03 / .25± .03 / / 1.31	147 / .11 / .00 /	* / 13.34 ±.14 / / 13.09				9415± 60 / 9288 / 9393
045452.9-371915 / 240.52 -38.40 / 251.30 -55.78 / 045308.1-372400	ESO 361- 19 / IRAS04531-3723 / PGC 16317	.SB.1?P / S / 1.0±1.1 /	1.25± .04 / .43± .04 / / 1.25	105 / .00 / .44 / .21	* / 14.21 ±.14 / 13.63 / 13.74				2366± 39 / 2200 / 2360
045455.7-613352 / 271.36 -37.39 / 224.06 -39.42 / 045420.5-613834	PGC 16318	.IBS9*. / S (1) / 10.0±1.2 / 11.1± .8	1.23± .07 / .20± .08 / / 1.23	/ .02 / .15 / .10	*				
045459.6-371536 / 240.45 -38.37 / 251.37 -55.83 / 045314.6-372020	PGC 16320	PLA.+?P / S / -1.0±1.8 /	1.06± .10 / .40± .08 / / 1.00	/ .00 / .00 /	*				
045503.3-040605 / 202.69 -27.66 / 315.12 -58.32 / 045234.2-041051	MCG -1-13- 35 / IRAS04525-0410 / PGC 16322	.S..3./ / E / 3.0± .9 /	1.18± .05 / .77± .05 / / 1.20	166 / .13 / 1.06 / .38	* / / 12.57 /				

R.A. 2000 DEC. / l b / SGL SGB / R.A. 1950 DEC.	Names / PGC	Type S_T n_L / T / L	$\log D_{25}$ / $\log R_{25}$ / $\log A_e$ / $\log D_o$	p.a. / A_g / A_i / A_{21}	B_T / m_B / m_{FIR} / B_T^o	$(B-V)_T$ / $(U-B)_T$ / $(B-V)_T^o$ / $(U-B)_T^o$	$(B-V)_e$ / $(U-B)_e$ / m' / m'_{25}	m_{21} / W_{20} / W_{50} / HI	V_{21} / V_{opt} / V_{GSR} / V_{3K}
045505.7-104127		.SAS3*.	1.46± .04	55	*				4530
209.42 -30.65	MCG -2-13- 21	E (1)	.64± .05	.17					
302.80 -60.41	IRAS04527-1046	3.0± .8		.88	13.16				4422
045244.0-104612	PGC 16324	3.1± .8	1.47	.32					4502
045506.8-120537		.LAR0*.	1.24± .09	80	*				
210.90 -31.23	MCG -2-13- 22	E	.11± .07	.34					
299.99 -60.70	IRAS04527-1210	-2.0± .8		.00	13.36				
045246.8-121022	PGC 16327		1.26						
0455.1 -1209		.SXT5*.	1.18± .05	160	*				4867
210.97 -31.25	MCG -2-13- 23	E (1)	.04± .05	.34					4755
299.87 -60.72		5.0± .8		.06					
0452.8 -1214	PGC 16329	4.2±1.6	1.22	.02					4841
0455.2 -0111		.SB.2..	1.04± .06		*				8484± 38
199.86 -26.20	UGC 3202	U	.04± .05	.20	14.67 ±.15				8403
320.06 -57.10		2.0± .8		.05					
0452.7 -0116	PGC 16333		1.06	.02	14.34				8450
045516.9-224340		.S?....	1.03± .05	165	*				6803
222.76 -35.07	ESO 485- 24		.11± .04	.06	15.09 ±.14				6665
277.88 -60.93				.13					
045310.0-224824	PGC 16334		1.04	.05	14.83				6785
045523.7-203417	NGC 1692	.LAS0*.	1.12± .03	5	13.99 ±.15	1.01± .03	1.02± .02		
220.29 -34.36	ESO 552- 21	SE	.03± .03	.04	14.21 ±.14				10381± 55
282.35 -61.21		-2.3± .7	.96± .03	.00		.90	14.28± .08		10248
045314.0-203900	PGC 16336		1.12		13.92		14.38± .22		10362
045530.0-314842		.SXT3*.	1.15± .04	110					5683± 34
233.66 -37.36	ESO 422- 10	Sr	.35± .03	.00	14.44 ±.14				5526
260.30 -58.43		2.7± .6		.49					5673
045336.0-315324	PGC 16338		1.15	.18	13.91				
045540.7-561408		.P.....	1.13± .05	70					1638
264.69 -38.20	ESO 158- 15	S	.25± .05	.00	15.34 ±.14				1451
228.54 -43.57		99.0		.18					1650
045443.0-561848	PGC 16344		1.13	.12	15.15				
045543.2-145951		.SXT7..	1.03± .07	25					
214.10 -32.27	MCG -3-13- 31	E (1)	.03± .05	.14					
294.06 -61.25		7.0± .9		.04					
045326.7-150434	PGC 16347	6.4±1.2	1.05	.01					
045545.5-601227		RLBS0..	1.14± .05	63					5930
269.66 -37.56	ESO 119- 21	S	.30± .03	.00	14.62 ±.14				5742
225.07 -40.54		-2.0± .8		.00					5945
045504.0-601706	PGC 16348		1.09		14.53				
0455.7 +6744		.S..6*.	.99± .06	163				15.59±.1	3693± 11
143.00 15.01	UGC 3196	U	.52± .05	.80					3823
15.39 -3.77		6.0±1.3		.76				194± 8	3646
0450.6 +6740	PGC 16349		1.07	.26					
045550.9-295301	NGC 1701	RSAR3*.	1.09± .04	137					5808± 52
231.32 -36.88	ESO 422- 11	Sr (1)	.14± .03	.00	13.63 ±.14				5654
263.74 -59.23	IRAS04539-2957	3.2± .7		.20	13.14				5797
045354.0-295742	PGC 16352	4.4±1.2	1.09	.07	13.39				
045558.1+025615		.SXS4..	1.09± .05	120				15.74±.2	4450± 7
196.05 -23.94	UGC 3206	UE (2)	.29± .04	.25	14.90 ±.14			263± 14	4381
326.58 -55.15		3.5± .6		.42				259± 10	4415
045321.1+025133	PGC 16355	3.6± .7	1.12	.14	14.20			1.39	
045558.3+021405		.S..1?.	.96± .08	60					4500± 49
196.71 -24.30	MCG 0-13- 35	E	.21± .05	.25	14.57 ±.12				4428
325.55 -55.54	ARAK 116	1.0±1.9		.21					4465
045322.1+020923	PGC 16356		.98	.10	14.05				
045610.3+020921		.SXT3P*	1.37± .03	124					4550± 38
196.81 -24.30	UGC 3207	UE (1)	.51± .04	.27	13.94 ±.13				4478
325.48 -55.62	IRAS04535+0204	2.5± .6		.70	13.12				4515
045334.1+020440	PGC 16359	4.5±1.1	1.40	.25	12.94				
045615.1+300306		.S..2..	1.23± .06	53				14.34±.3	3589± 10
172.96 -8.21	UGC 3205	U	.42± .06	1.98					3607
354.50 -36.03		2.0± .8		.52				427± 7	3542
045304.3+295824	PGC 16360		1.42	.21					
0456.2 +0135		.S..5..	1.04± .06					16.75±.3	8500± 11
197.35 -24.56	UGC 3208	U	.02± .05	.30	15.03 ±.19			100± 7	8681± 38
324.67 -55.94		5.0± .8		.03				87± 7	8441
0453.7 +0131	PGC 16364		1.07	.01	14.66			2.08	8480

R.A. 2000 DEC. l b SGL SGB R.A. 1950 DEC.	Names PGC	Type S_T n_L T L	$\log D_{25}$ $\log R_{25}$ $\log A_o$ $\log D_o$	p.a. A_g A_i A_{21}	B_T m_B m_{FIR} B_T^o	$(B-V)_T$ $(U-B)_T$ $(B-V)_T^o$ $(U-B)_T^o$	$(B-V)_o$ $(U-B)_o$ m'_o m'_{25}	m_{21} W_{20} W_{50} HI	V_{21} V_{opt} V_{GSR} V_{3K}
045619.2-154755 215.04 -32.44 292.41 -61.46 045403.6-155235	IC 2104 MCG -3-13- 34 IRAS04540-1552 PGC 16367	PSBS4.. E (1) 4.0± .8 1.9±1.2	1.27± .05 .20± .05 1.28	103 .14 .30 .10	 12.21 				5496 5374 5474
0456.3 +0042 198.20 -25.00 323.33 -56.42 0453.8 +0038	 MCG 0-13- 40 PGC 16368	 	.74± .14 .00± .07 .77	 .40 	 15.17 ±.12 				4546± 38 4470 4513
045626.5+030155 196.03 -23.79 326.84 -55.20 045349.4+025715	 UGC 3209 IRAS04538+0257 PGC 16369	.SBS4P. UE (1) 3.5± .6 3.6± .8	1.14± .04 .31± .04 1.17	32 .25 .45 .15	 14.86 ±.15 14.10				8372± 38 8303 8337
045629.3-174003 217.12 -33.10 288.46 -61.58 045416.0-174442	 ESO 552- 27 PGC 16371	.L?.... 	.97± .06 .05± .04 .98	 .12 .00 	 14.94 ±.14 14.68				9143± 60 9016 9123
045633.9-283010 229.70 -36.40 266.23 -59.85 045435.0-283448	IC 2106 ESO 422- 12 IRAS04545-2834 PGC 16373	PSBT3.. Sr (1) 3.1± .6 5.6±1.2	1.22± .04 .26± .03 1.22	157 .00 .36 .13	 13.78 ±.14 13.42 13.39				4955± 52 4804 4944
045635.8-185157 218.47 -33.51 285.91 -61.59 045424.1-185636	 ESO 552- 28 PGC 16374	.S?.... 	1.00± .05 .21± .04 1.00	172 .04 .29 .10	 15.44 ±.14 15.03				11288± 60 11158 11269
045637.5-101312 209.11 -30.10 303.91 -60.67 045415.3-101751	 MCG -2-13- 24 PGC 16375	.LXS0.. E -2.0± .9 	1.10± .06 .32± .05 1.07	10 .16 .00 					
045646.0-103537 209.52 -30.23 303.18 -60.79 045424.3-104015	 MCG -2-13- 25 IRAS04544-1040 PGC 16379	.SBT4.. E (1) 4.0± .8 3.1± .8	1.34± .04 .32± .05 1.35	20 .17 .47 .16	 13.59 				4112 4003 4087
0456.8 +0248 196.29 -23.82 326.60 -55.40 0454.2 +0244	 MCG 0-13- 42 PGC 16381	.S?.... 	.74± .14 .00± .07 .76	 .24 .00 .00	 15.32 ±.12 15.03				9044± 38 8974 9010
0456.8 -0035 199.51 -25.55 321.38 -57.17 0454.3 -0040	 MCG 0-13- 43 PGC 16382	 	.89± .11 .35± .07 .91	 .28 	 15.20 ±.12 				4493± 38 4412 4461
045656.2-045203 203.70 -27.62 314.12 -59.05 045428.0-045641	NGC 1700 MCG -1-13- 38 PGC 16386	.E.4... R -5.0± .3 	1.52± .03 .20± .03 .79± .02 1.48	65 .12 .00 	12.20M±.05 11.90 ±.17 12.00	.97± .01 .50± .02 .91 .49	.98± .01 .55± .01 11.68± .07 14.29± .20		3915± 23 3822 3886
045659.2-424801 247.54 -38.50 243.08 -53.00 045525.0-425236	 ESO 252- 1 PGC 16389	 	1.12± .07 .51± .06 1.12	92 .00 	14.4 ±.2 15.28 ±.14 	.30± .08 -.36± .09 			3501± 63 3326 3502
045659.6-044528 203.60 -27.55 314.32 -59.02 045431.2-045006	NGC 1699 MCG -1-13- 39 IRAS04545-0449 PGC 16390	.SAT3.. R (1) 3.0± .5 4.2± .8	.93± .05 .22± .04 .94	20 .15 .31 .11					
0457.1 -0050 199.79 -25.61 321.04 -57.35 0454.6 -0055	 CGCG 394- 46 PGC 16392	 	 	 .28 	 15.1 ±.3 				6129± 38 6048 6098
045716.1-590715 268.27 -37.56 225.78 -41.50 045630.1-591148	 ESO 119- 23 PGC 16395	.IB.9?. S (1) 10.0±1.7 12.2± .8	1.16± .07 .22± .05 1.16	 .00 .16 .11	 18.26 ±.14 				
045716.9-151726 214.59 -32.03 293.52 -61.66 045500.7-152201	NGC 1710 MCG -3-13- 37 PGC 16396	.LA.-*. E -3.0±1.2 	1.11± .06 .12± .05 1.11	15 .16 .00 					5246 5124 5225
045717.5-752520 287.60 -33.25 214.19 -28.23 045852.0-752948	 ESO 33- 4 PGC 16397	.SBT4.. Sr (1) 3.8± .4 2.6± .5	1.16± .04 .17± .04 1.21	12 .47 .25 .09	 14.18 ±.14 13.43				5192± 34 5006 5220

R.A. 2000 DEC.	Names	Type	logD₂₅	p.a.	B_T	(B-V)_T	(B-V)_o	m₂₁	V₂₁
l　　b		S_T　n_L	logR₂₅	A_g	m_B	(U-B)_T	(U-B)_o	W₂₀	V_opt
SGL　SGB		T	logA_e	A_i	m_FIR	(B-V)°_T	m'_e	W₅₀	V_GSR
R.A. 1950 DEC.	PGC	L	logD_o	A₂₁	B°_T	(U-B)°_T	m'₂₅	HI	V_3K

045722.7+015335		.SB.6*/	1.13± .04	170					4183± 38
197.23 -24.18	UGC　3211	UE　　(1)	.62± .04	.30	15.07 ±.12				4110
325.37 -56.02		5.5± .9		.91					4150
045446.8+014859	PGC 16400	5.3±1.2	1.16	.31	13.84				

045723.1+781121	IC　　391	.SAS5..	1.06± .04		13.0　±.2	.31± .07		14.65±.1	1556± 8
134.04 21.06	UGC　3190	RC	.02± .04	.45	13.1　±.2			164± 9	1574± 58
20.43　5.36	IRAS04497+7806	5.0± .4		.03	11.29	.19		117± 12	1709
044944.3+780635	PGC 16402		1.11	.01	12.54		13.12± .30	2.10	1514

045726.2-512245		.L?....	1.00± .06	161					10500±190
258.50 -38.41	ESO　203- 12		.69± .03	.00	15.42 ±.14				10316
232.98 -47.31	FAIR　785			.00					10509
045613.0-512718	PGC 16403		.90		15.26				

045748.5-731357		.E.4.?.	1.06± .06	20					
285.12 -33.97	ESO　33-　3	S	.25± .04	.43	13.72 ±.14				
215.51 -30.09		-5.0±1.8		.00					
045849.0-731824	PGC 16415		1.06						

0457.9　+0255		.S..1*.	.95± .07	48					
196.34 -23.53	UGC　3213	U	.44± .05	.24	15.49 ±.12				
327.04 -55.57		1.0±1.3		.45					
0455.3　+0251	PGC 16419		.97	.22					

045756.7-000736		.S..3./	1.51± .03	57					
199.22 -25.08	UGC　3214	UE　　(1)	.69± .04	.30	13.96 ±.15				
322.37 -57.18	IRAS04553-0012	3.0± .6		.95	13.26				
045523.1-001209	PGC 16420	4.5±1.1	1.54	.34					

0457.9　+6459		.S..6*.	1.09± .06	99					
145.43 13.56	UGC　3204	U	.59± .05	1.18	15.3　±.3				
14.28　-6.29		6.0±1.3		.87					
0453.1　+6455	PGC 16421		1.20	.30					

045758.7+681917	IC　　396	.S?....	1.32± .05	85				15.62±.1	880± 11
142.65 15.52	UGC　3203		.16± .06	.67	13.0　±.3				825± 57
15.85　-3.36	IRAS04527+6814			.24	12.11			220± 8	1009
045243.9+681438	PGC 16423		1.39	.08	12.13			3.41	833

045809.2-213704		.S?....	1.04± .05						6691
221.74 -34.09	ESO　552- 32		.08± .04	.02	14.39 ±.14				6554
280.00 -61.74				.10					6676
045601.0-214136	PGC 16431		1.04	.04	14.20				

045812.6-074653	IC　　398	.SBS5$.	1.11± .05	158					
206.80 -28.68	MCG -1-13- 40	R	.47± .04	.17					
308.91 -60.37	IRAS04558-0751	5.0±1.0		.70	11.32				
045547.7-075125	PGC 16433		1.12	.23					

045812.9-202152	NGC　1716	.SXS4P.	1.14± .03	20					6818± 60
220.32 -33.66	ESO　552- 34	SEr　　(1)	.08± .03	.04	13.93 ±.14				6683
282.67 -61.89	IRAS04560-2026	4.2± .5		.12	12.10				6803
045603.1-202624	PGC 16434	2.2± .8	1.15	.04	13.73				

045815.6-094734		.S..6./	1.15± .06	156					
208.87 -29.56	MCG -2-13- 26	E	.61± .05	.19					
304.97 -60.95	IRAS04558-0952	6.0± .9		.89	12.87				
045553.0-095206	PGC 16436		1.17	.30					

045817.6-334835		.SBS6..	1.08± .04						
236.27 -37.17	ESO　361- 23	S　　(1)	.10± .04	.00	15.94 ±.14				
256.35 -58.10		6.0± .8		.15					
045627.0-335306	PGC 16438	5.6± .8	1.08	.05					

045819.4+004431		.L..0P*	1.11± .06	90					
198.44 -24.56	UGC　3215	UE	.05± .03	.31	14.32 ±.12				
323.83 -56.82		-2.3± .7		.00					
045544.9+003959	PGC 16439		1.14						

045824.3-620139	NGC　1765	.LBT-*.	1.09± .05	150					
271.83 -36.90	ESO　119- 24	S	.08± .04	.01	13.94 ±.14				
223.28 -39.31		-3.5± .8		.00					
045752.0-620606	PGC 16444		1.07						

0458.4　-0033									4807± 38
199.70 -25.19	MCG　0-13- 53				.31 15.28 ±.12				4726
321.79 -57.50									4778
0455.9　-0038	PGC 16448								

0458.5　-0127		.I..9*.	1.04± .15	15					
200.58 -25.62	UGC　3216	U	.37± .12	.30					
320.34 -57.95		10.0±1.3		.28					
0456.0　-0132	PGC 16452		1.07	.19					

R.A. 2000 DEC. l b SGL SGB R.A. 1950 DEC.	Names PGC	Type S_T n_L T L	$\log D_{25}$ $\log R_{25}$ $\log A_e$ $\log D_o$	p.a. A_g A_i A_{21}	B_T m_B m_{FIR} B_T^o	$(B-V)_T$ $(U-B)_T$ $(B-V)_T^o$ $(U-B)_T^o$	$(B-V)_e$ $(U-B)_e$ m'_e m'_{25}	m_{21} W_{20} W_{50} HI	V_{21} V_{opt} V_{GSR} V_{3K}
045842.6-555333 264.20 -37.82 228.41 -44.10 045744.0-555800	ESO 158- 16 PGC 16461	.SAS8*. S (1) 8.0± .8 7.8± .8	1.02± .05 .11± .05 1.02	55 .00 .14 .06	15.47 ±.14				
045844.1-002842 199.66 -25.09 321.99 -57.52 045610.9-003313	NGC 1709 MCG 0-13- 54 PGC 16462	.LB.0*. E -2.0±1.3	.94± .08 .11± .03 .96	10 .31 .00	15.17 ±.08 14.78				4731± 38 4650 4702
0458.7 -0053 200.06 -25.29 321.32 -57.73 0456.2 -0058	UGC 3221 PGC 16464	.S..6*. U 6.0±1.4	.96± .07 .54± .05 .99	162 .33 .80 .27	15.38 ±.12 14.23				4367± 38 4284 4338
045847.3-213407 221.74 -33.94 280.07 -61.90 045639.0-213836	ESO 552- 40 PGC 16465	.SBS2*P S 2.0± .8	1.09± .05 .22± .04 1.09	50 .02 .27 .11	14.30 ±.14 13.94				6818± 60 6680 6804
045848.9+070703 192.62 -21.11 333.05 -53.27 045607.2+070233	UGC 3220 PGC 16468	.S..7.. U 7.0± .9	1.02± .06 .27± .05 1.05	110 .34 .37 .13	15.29 ±.14 14.55			15.85±.2 274± 13 244± 6 1.17	8512± 7 8454 8478
045850.8+065900 192.75 -21.18 332.88 -53.36 045609.2+065430	UGC 3219 PGC 16469	.S..7.. U 7.0± .9	1.08± .06 .42± .05 1.12	100 .34 .58 .21	15.30 ±.14 14.35			16.16±.2 308± 13 297± 6 1.60	8481± 7 8422 8447
045854.5-002928 199.70 -25.06 322.01 -57.57 045621.3-003357	NGC 1713 UGC 3222 PGC 16471	.E+..*. UE -4.3± .5	1.14± .03 .07± .05 .65± .03 1.17	45 .31 .00	13.91 ±.14 13.57 ±.10 13.30	1.17± .02 1.06	1.13± .02 12.65± .08 14.44± .26		4514± 38 4433 4485
045855.5-631753 273.36 -36.58 222.24 -38.33 045830.0-632218	NGC 1771 ESO 85- 27 IRAS04585-6322 PGC 16472	.SXR5*/ RSr (1) 5.2± .5 2.2± .9	1.27± .04 .55± .04 .83± .03 1.27	136 .00 .82 .27	14.16 ±.16 14.32 ±.14 13.23 13.40	.74± .03 .06± .06 .61 -.05	.82± .02 .16± .04 13.80± .10 13.99± .27		5026 4836 5046
0458.9 -0129 200.66 -25.55 320.37 -58.05 0456.4 -0134	MCG 0-13- 57 PGC 16473		.74± .14 .21± .07 .77	.30	15.45 ±.13				4323± 38 4239 4295
045901.7-564007 265.16 -37.69 227.66 -43.54 045806.0-564433	PGC 16478	.SBS9.. S (1) 9.0± .9 10.0± .9	1.08± .09 .18± .08 1.08	.00 .19 .09					
045908.8-583917 267.64 -37.39 225.92 -42.02 045821.1-584342	ESO 119- 25 PGC 16480	RSBR1.. Sr 1.0± .6	1.09± .05 .33± .04 1.09	115 .00 .34 .17	14.28 ±.14 13.87				5916± 34 5727 5933
045909.4+045831 194.62 -22.19 330.26 -54.64 045630.1+045403	A 0456+04 UGC 3223 PGC 16482	.SB.1.. U 1.0± .8	1.15± .05 .25± .05 1.19	80 .43 .26 .13	13.83 ±.12 13.09				4723± 32 4659 4692
045917.4-110708 210.37 -29.90 302.40 -61.51 045656.4-111135	NGC 1721 MCG -2-13- 27 IRAS04569-1111 PGC 16484	PLXS0P. E -2.0± .8	1.32± .05 .31± .05 1.31	120 .29 .00	14569± 39 4457 4548				4569± 39 4457 4548
045920.8-075135 207.02 -28.46 308.94 -60.66 045655.9-075602	NGC 1720 MCG -1-13- 41 IRAS04569-0756 PGC 16485	.SBS2.. R 2.0± .4	1.21± .04 .19± .04 1.23	95 .21 .23 .09	11.02				4180± 46 4077 4157
045921.3+053703 194.06 -21.81 331.20 -54.30 045641.2+053235	A 0456+05 UGC 3224 PGC 16486	.S..3.. U (2) 3.0± .8 3.8± .6	1.19± .05 .10± .05 .92± .02 1.22	15 .38 .13 .05	13.66 ±.15 13.99 ±.14 13.28	.63± .02 .49	.71± .02 13.75± .05 14.21± .31	14.84±.1 261± 11 254± 6 1.51	4691± 7 4734± 59 4628 4660
045923.9-010929 200.41 -25.28 321.03 -58.00 045651.5-011356	MCG 0-13- 59 PGC 16490	.SBS4?/ E 4.0±1.3	1.06± .06 .63± .05 1.09	45 .34 .93 .32	15.39 ±.12 14.09				4400± 38 4316 4372
045926.0-105852 210.25 -29.81 302.70 -61.51 045704.8-110318	NGC 1723 MCG -2-13- 29 IRAS04571-1103 PGC 16493	.SBR1P. E 1.0± .7	1.51± .03 .21± .05 1.53	40 .28 .21 .10	13.18				

R.A. 2000 DEC.	Names	Type	logD₂₅	p.a.	B_T	(B-V)_T	(B-V)_e	m₂₁	V₂₁
l b		S_T n_L	logR₂₅	A_g	m_B	(U-B)_T	(U-B)_e	W₂₀	V_opt
SGL SGB		T	logA_e	A_i	m_FIR	(B-V)°_T	m'_e	W₅₀	V_GSR
R.A. 1950 DEC.	PGC	L	logD_o	A₂₁	B°_T	(U-B)°_T	m'₂₅	HI	V_3K

045927.2-163338		.SBS5..	1.11± .06	75					3443
216.21 -32.03	MCG -3-13- 42	E (1)	.08± .05	.16					
290.87 -62.26	IRAS04572-1638	5.0± .8		.11	13.22				3317
045712.6-163805	PGC 16494	4.2±1.2	1.12	.04					3426
045927.8-110722	NGC 1728	.S..1P/	1.30± .05	177					
210.40 -29.86	MCG -2-13- 30	E	.46± .05	.29					4507± 39
302.42 -61.55		1.0±1.6		.47					4395
045706.8-111149	PGC 16495		1.32	.23					4486
045931.7-154924	NGC 1730	.SBR1..	1.35± .04	94					3964
215.42 -31.74	MCG -3-13- 43	E	.34± .05	.15					
292.46 -62.24	IRAS04573-1553	1.0± .8		.35					3840
045716.3-155350	PGC 16499		1.36	.17					3947
045934.2-001540	NGC 1719	.S..1*/	1.05± .05	102					
199.57 -24.80	UGC 3226	UE	.57± .04	.31	14.53 ±.14				4165± 38
322.53 -57.60		1.0± .8		.58					4084
045700.8-002007	PGC 16501		1.08	.28	13.59				4137
045934.8-224158		.SXS9..	1.22± .04	38					1764
223.10 -34.12	ESO 486- 3	S (1)	.34± .05	.03	15.65 ±.14				
277.61 -61.92		9.0± .5		.34					1624
045728.0-224624	PGC 16502	9.7± .5	1.22	.17	15.28				1753
045935.3-191158		.SXT6*.	1.20± .03	140					
219.14 -32.96	ESO 552- 43	SE (2)	.36± .03	.09	14.83 ±.14				6747± 60
285.14 -62.28		5.5± .6		.52					6614
045724.1-191624	PGC 16504	3.6± .7	1.21	.18	14.18				6733
045940.0-455830		.LA.0*.	1.08± .05	59					
251.61 -38.15	ESO 252- 4	S	.28± .04	.00	14.70 ±.14				13488
238.58 -51.33		-2.0±1.2		.00					13308
045813.1-460254	PGC 16506		1.04		14.50				13495
045941.4-111617		.SBS5?/	1.24± .05	160					
210.58 -29.87	MCG -2-13- 31	E	.68± .05	.29					
302.13 -61.64	IRAS04573-1120	5.0±1.8		1.02	13.51				
045720.6-112043	PGC 16507		1.26	.34					
045942.0-074520	NGC 1726	.LAS0*.	1.17± .04	0	12.66 ±.13	.98± .01	1.02± .01		
206.96 -28.34	MCG -1-13- 42	R	.12± .04	.17	13.18 ±.18	.52± .03	.56± .02		3992± 37
309.20 -60.71		-2.0± .6	.90± .02	.00		.89	12.65± .05		3889
045717.0-074946	PGC 16508		1.17		12.61	.49	13.06± .27		3970
0459.7 +8009		.E...?.	1.12± .12						
132.32 22.21	UGC 3197	U	.09± .07	.30	15.0 ±.2				16040± 59
21.45 7.05		-5.0±1.6		.00					16196
0451.0 +8005	PGC 16509		1.14		14.41				15999
045950.2-235535		.SXT1..	1.07± .04	12					
224.54 -34.44	ESO 486- 4	S	.35± .03	.04	15.39 ±.14				11357± 60
275.00 -61.75		1.0± .6		.35					11214
045745.0-240000	PGC 16511		1.07	.17	14.86				11347
045953.1-120213		.LX.-?.	1.10± .06	85					
211.41 -30.15	MCG -2-13- 32	E	.03± .05	.30					
300.56 -61.84		-3.0± .8		.00					
045733.1-120638	PGC 16513		1.13						
045958.3-260136	NGC 1744	.SBS7..	1.91± .02	168	11.6 ±.3	.48± .03	.53± .02	12.02±.1	748± 6
227.00 -35.02	ESO 486- 5	R (3)	.27± .02	.00	11.86 ±.12	-.13± .07	-.12± .04	212± 7	643± 58
270.65 -61.30	IRAS04579-2605	7.0± .3	1.45± .07	.37	13.11	.42	14.31± .16	198± 10	599
045756.1-260600	PGC 16517	5.6± .4	1.91	.14	11.43	-.17	15.32± .28	.46	738
045959.5-172736		RSBR1..	1.15± .04	82					
217.25 -32.25	ESO 552- 45	r	.31± .04	.19	14.34 ±.14				6560± 22
288.92 -62.41	IRAS04577-1731	1.0± .9		.31					6432
045746.1-173200	PGC 16518		1.17	.15	13.75				6545
0500.0 +5349		.S?....	1.11± .14	160				14.91±.1	4074± 7
154.66 7.04	UGC 3217		.36± .12	1.72					4165
9.14 -16.24				.50				504± 8	4028
0456.0 +5345	PGC 16521		1.27	.18					
050010.7-745414			.85± .06	37	15.1 ±.2	.49± .10			
286.94 -33.26	ESO 33- 9		.29± .05	.45	15.11 ±.14	.39± .11			5955± 87
214.33 -28.77	IRAS05015-7458				13.49				5768
050137.0-745830	PGC 16526		.89						5984
050015.7-032110	NGC 1729	.SAS5..	1.21± .05	150					3644± 10
202.65 -26.16	MCG -1-13- 43	E (1)	.08± .05	.18					
317.47 -59.19	IRAS04577-0325	5.0± .8		.11	12.38				3553
045745.8-032533	PGC 16529	1.9± .8	1.23	.04					3619

R.A. 2000 DEC. / l b / SGL SGB / R.A. 1950 DEC.	Names / PGC	Type / S_T n_L / T / L	$\log D_{25}$ / $\log R_{25}$ / $\log A_e$ / $\log D_o$	p.a. / A_g / A_i / A_{21}	B_T / m_B / m_{FIR} / B_T^o	$(B-V)_T$ / $(U-B)_T$ / $(B-V)_T^o$ / $(U-B)_T^o$	$(B-V)_e$ / $(U-B)_e$ / m' / m'_{25}	m_{21} / W_{20} / W_{50} / HI	V_{21} / V_{opt} / V_{GSR} / V_{3K}
050018.7-124129 / 212.15 -30.32 / 299.22 -62.06 / 045759.5-124552	MCG -2-13- 34 / PGC 16530	.SA.7*/ / E / 7.0±1.2 /	1.29± .06 / .52± .05 / / 1.32	50 / .25 / .72 / .26					3759 / / 3642 / 3741
0500.5 +0423 / 195.35 -22.20 / 329.80 -55.27 / 0457.9 +0419	IC 2112 / CGCG 420- 27 / PGC 16534			.34	15.2 ±.3				10710 / 10642 / 10681
0500.6 +6214 / 147.88 12.17 / 13.25 -8.86 / 0456.0 +6210	UGC 3218 / PGC 16537	.SA.3.. / U / 3.0± .8 /	1.22± .05 / .43± .05 / / 1.38	145 / 1.73 / .59 / .21	/ 14.3 ±.2 / / 11.97			14.06±.1 / / 490± 8 / 1.87	5226± 7 / / 5341 / 5181
050040.5-494447 / 256.40 -37.97 / 234.14 -48.84 / 045923.0-494906	ESO 203- 15 / FAIR 786 / PGC 16539	.SXT4.. / Sr (1) / 3.7± .6 / 3.3±1.2	1.08± .05 / .35± .04 / / 1.08	135 / .00 / .51 / .17	/ 14.87 ±.14 / / 14.29				10800±190 / 10616 / 10811
050041.5-250433 / 225.94 -34.60 / 272.51 -61.69 / 045838.0-250854	ESO 486- 7 / PGC 16541	.SAR3*. / r / 3.3±1.2 /	1.24± .05 / .35± .04 / / 1.25	168 / .03 / .48 / .18	/ 15.00 ±.14 / / 14.40				11330± 52 / 11184 / 11322
0500.7 +7112 / 140.32 17.37 / 17.39 -.92 / 0455.0 +7108	UGC 3212 / PGC 16542	.S..9*. / U / 9.0±1.3 /	1.04± .08 / .47± .06 / / 1.09	13 / .53 / .48 / .24				16.80±.1 / / 49± 8 /	1225± 11 / 1362 / 1182
050049.1-384029 / 242.41 -37.40 / 248.11 -56.03 / 045907.1-384448	NGC 1759 / ESO 305- 1 / PGC 16547	.E+..*. / BS / -3.9± .5 /	1.16± .05 / .04± .04 / / 1.14	/ .00 / .00 /	/ 14.11 ±.14 / /				
050101.8-085727 / 208.35 -28.58 / 307.03 -61.38 / 045838.2-090147	MCG -2-13- 36 / IRAS04586-0901 / PGC 16553	.SBT8*. / E (1) / 8.0±1.2 / 5.3± .8	1.11± .06 / .13± .05 / / 1.14	95 / .25 / .16 / .07	/ / 12.25 /				
0501.4 +0346 / 196.05 -22.34 / 329.14 -55.81 / 0458.8 +0342	CGCG 420- 29 / PGC 16564			.35	15.4 ±.3				8695 / 8625 / 8667
050130.1-631734 / 273.28 -36.29 / 221.94 -38.50 / 050105.0-632148	ESO 85- 30 / IRAS05010-6321 / PGC 16567	.L..+?P / S / -1.0±1.7 /	1.20± .04 / .33± .03 / / 1.15	147 / .00 / .00 /	/ 13.78 ±.14 / 13.10 / 13.76				1308 / 1118 / 1330
050132.6-161000 / 216.01 -31.42 / 291.78 -62.74 / 045917.6-161417	MCG -3-13- 51 / IRAS04592-1614 / PGC 16568	PSBS3*. / E / 3.0±1.3 /	1.19± .05 / .55± .05 / / 1.20	85 / .16 / .76 / .28	/ / 12.87 /				
050135.5-041551 / 203.72 -26.31 / 316.12 -59.87 / 045906.6-042009	HICK 31B / PGC 16570	.S?....		.25	15.21S±.15				4117± 32 / 4023 / 4095
050135.5-041524 / 203.71 -26.30 / 316.13 -59.87 / 045906.6-041942	HICK 31D / PGC 16571	.S?....		.25					
0501.6 +0334 / 196.26 -22.40 / 328.91 -55.97 / 0459.0 +0330	A 0459+03 / 2ZW 28 / PGC 16572	.RING.. / R /		.31	15.5 ±.2	.44± .04 / -.41± .07			8583± 18 / 8512 / 8556
050137.9-041528 / 203.72 -26.30 / 316.14 -59.88 / 045909.0-041945	HICK 31C / PGC 16573	.S?....		.25					4068± 41 / 3974 / 4046
050138.4-041525 / 203.72 -26.29 / 316.14 -59.88 / 045909.4-041943	NGC 1741 / MCG -1-13- 45 / MK 1089 / PGC 16574	.P..... / R / 99.0 /	1.15± .06 / .30± .05 / / 1.18	130 / .25 / .31 / .15	13.30S±.15 / / 11.73 / 14.91		15.34± .34 / -1.04	14.03±.1 / 218± 10 / 115± 11 /	4058± 9 / 4004± 21 / 3956 / 4028
0501.6 -0030 / 200.10 -24.47 / 322.61 -58.17 / 0459.1 -0035	MCG 0-13- 62 / PGC 16575	.S?....	.74± .14 / .21± .07 / / .76	.29 / .31 / .10	15.30 ±.14 / / 14.67				4751± 38 / 4668 / 4727

R.A. 2000 DEC. / l b / SGL SGB / R.A. 1950 DEC.	Names / PGC	Type / S_T n_L / T / L	$\log D_{25}$ / $\log R_{25}$ / $\log A_e$ / $\log D_o$	p.a. / A_g / A_i / A_{21}	B_T / m_B / m_{FIR} / B_T^o	$(B-V)_T$ / $(U-B)_T$ / $(B-V)_T^o$ / $(U-B)_T^o$	$(B-V)_e$ / $(U-B)_e$ / m'_e / m'_{25}	m_{21} / W_{20} / W_{50} / HI	V_{21} / V_{opt} / V_{GSR} / V_{3K}
050139.2+433838 162.90 1.02 4.08 -25.23 045804.1+433418	 WEIN 91 PGC 16576	 	.90? .90? 	 	 	 	 	15.45±.1 576± 12 	7181± 11 7241 7138
050140.2-152354 215.19 -31.10 293.47 -62.72 045924.3-152812	 HICK 32B PGC 16578	.S?.... 	 	 .16 	15.90S±.15 	 	 	 	 12125± 41 12001 12111
050142.1-340156 236.70 -36.51 255.33 -58.62 045952.0-340612	 ESO 361- 25 IRAS04598-3406 PGC 16579	.S...P. S 	1.14± .06 .39± .06 1.14	164 .00 .58 .19	 14.26 ±.14 12.24 13.65	 	 	 	 5276± 39 5112 5276
050144.0-041720 203.76 -26.29 316.10 -59.92 045915.2-042137	IC 399 MK 1090 PGC 16582	.IXS9P* RE 10.0± .5 	.76± .10 .28± .06 .79	80 .25 .21 .14	 	 	 	 	 3991± 38 3897 3969
050145.2-152656 215.25 -31.10 293.36 -62.74 045929.4-153113	 MCG -3-13- 53 HICK 32A PGC 16583	.L?.... 	 	 .16 	14.40S±.15 	 	 	 	 12547± 41 12422 12533
050146.6-180931 218.21 -32.11 287.38 -62.83 045934.1-181348	NGC 1738 ESO 552- 49 IRAS04595-1813 PGC 16585	.SBS4P* SE (1) 3.8± .7 2.2± .8	1.11± .04 .26± .04 1.12	38 .10 .38 .13	 13.70 ±.14 11.53 13.19	 	 	 	 3978± 60 3847 3966
050147.6-181001 218.22 -32.11 287.36 -62.84 045935.0-181418	NGC 1739 ESO 552- 50 PGC 16586	.SBS4P* SE (1) 4.3± .7 3.3± .8	1.14± .04 .29± .04 1.15	105 .10 .42 .14	 14.24 ±.14 13.69	 	 	 	 3892± 60 3761 3880
050154.8-031746 202.81 -25.78 317.92 -59.54 045924.8-032202	NGC 1740 MCG -1-13- 46 PGC 16589	.L..-*. E -3.0±1.2 	1.19± .10 .12± .07 1.21	125 .25 .00 	 	 	 	 	
050200.3-483846 255.00 -37.77 235.09 -49.77 050040.0-484300	 ESO 203- 16 FAIR 787 PGC 16594	RLBR0.. Sr -1.7± .6 	1.05± .05 .08± .04 1.04	 .00 .00 	 14.31 ±.14 14.25	 	 	 	 4000±190 3817 4012
050200.8-210814 221.55 -33.09 280.81 -62.69 045952.0-211230	 ESO 552- 52 PGC 16595	.E?.... 	.99± .04 .15± .03 .70± .02 .95	 .03 .00 	13.92 ±.13 14.06 ±.14 13.89	.90± .01 .85 	.92± .01 12.91± .07 13.50± .25	 	 4584± 32 4446 4575
050205.8-693408 280.76 -34.75 217.51 -33.35 050226.0-693818	NGC 1809 IRAS05023-6937 PGC 16599	.S..5*. S 5.0±1.1 	1.50± .04 .59± .07 1.15± .06 1.54	143 .40 .88 .29	13.0 ±.2 15.38 ±.14 12.15 13.31	.86± .03 .17± .05 .65 -.01	.81± .02 .14± .03 14.20± .16 13.88± .34	 	 1301± 49 1111 1327
050209.6-081429 207.76 -28.01 308.64 -61.44 045945.3-081844	NGC 1752 MCG -1-13- 47 IRAS04597-0818 PGC 16600	.SBR5*. PE (1) 4.5± .6 3.1± .8	1.42± .03 .51± .05 1.00± .06 1.45	110 .29 .76 .25	13.27 ±.19 12.50 12.20	.88± .06 .26± .04 .70 .10	.97± .05 .29± .03 13.76± .20 13.96± .28	 	3583 3477 3565
0502.2 +0738 192.64 -20.12 334.62 -53.60 0459.5 +0734	 UGC 3229 IRAS04595+0734 PGC 16602	.SX.5.. U 5.0± .8 	1.07± .07 .00± .06 1.11	 .45 .00 .00	 15.1 ±.2 13.63 14.54	 	 	 	11359 11301 11330
050217.7-302716 232.41 -35.65 261.62 -60.31 050022.0-303130	 ESO 422- 23 PGC 16603	.SBT3P. r 3.3± .8 	1.15± .04 .25± .04 1.15	19 .00 .34 .12	 14.18 ±.14 13.79	 	 	 	 6340± 34 6182 6338
050218.2-200004 220.30 -32.64 283.29 -62.87 050008.0-200418	 ESO 552- 53 PGC 16604	.SBR3.. r 3.3± .8 	1.12± .05 .10± .04 1.12	 .04 .13 .05	 14.25 ±.14 14.02	 	 	 	 7239± 60 7103 7229
050219.4-102123 209.95 -28.90 304.35 -62.06 045957.5-102538	 MCG -2-13- 38 IRAS04599-1025 PGC 16605	PSAT1*. E 1.0± .8 	1.31± .05 .36± .05 1.34	105 .28 .36 .18	 13.12 	 	 	 	3986 3874 3969
0502.4 -0822 207.92 -28.00 308.43 -61.54 0500.0 -0826	 MCG -1-13- 49 IRAS05000-0826 PGC 16607	.SBT7.. E (1) 7.0± .8 6.4±1.2	1.09± .06 .00± .05 1.11	 .23 .00 .00	 13.00 	 	 	 	

R.A. 2000 DEC. / l b / SGL SGB / R.A. 1950 DEC.	Names / PGC	Type / ST nL / T / L	logD25 / logR25 / logAe / logD₀	p.a. / Ag / Ai / A21	BT / mB / mFIR / B°T	(B-V)T / (U-B)T / (B-V)°T / (U-B)°T	(B-V)₀ / (U-B)₀ / m' / m'25	m21 / W20 / W50 / HI	V21 / Vopt / VGSR / V3K
050227.5+861312 / 126.65 25.35 / 24.36 12.37 / 044355.5+860824	NGC 1544 / UGC 3160 / / PGC 16608	.S?.... / / /	1.10± .04 / .17± .04 / / 1.14	130 / .40 / .25 / .08	/ 14.16 ±.18 / / 13.49				4517±125 / 4683 / 4479
050232.3-032039 / 202.94 -25.66 / 317.97 -59.70 / 050002.4-032453	NGC 1753 / MCG -1-13- 48 / IRAS05000-0324 / PGC 16610	PSB.1P? / E / 1.0±1.2 /	1.14± .06 / .38± .05 / / 1.16	15 / .25 / .39 / .19					
0502.5 -1531 / 215.43 -30.95 / 293.22 -62.94 / 0500.3 -1536	/ MCG -3-13- 58 / / PGC 16611	/ / /	1.19± .07 / .35± .07 / / 1.21	/ .21 / /					4303 / 4178 / 4290
050234.7+001443 / 199.50 -23.90 / 324.06 -57.99 / 050000.7+001029	/ UGC 3231 / / PGC 16613	.SBS7*. / UE (1) / 7.3± .7 / 6.4± .8	1.18± .04 / .34± .04 / / 1.21	138 / .33 / .46 / .17	/ 14.53 ±.13 / / 13.72			14.77±.3 / / 259± 7 / .88	4838± 7 / 4756 / 4814
050241.3-221336 / 222.84 -33.29 / 278.38 -62.70 / 050034.0-221748	/ ESO 552- 55 / / PGC 16616	.S?.... / / /	1.06± .05 / .37± .05 / / 1.06	105 / .02 / .37 / .18	/ 15.47 ±.14 / / 14.90				14406 / 14265 / 14398
050243.2-610821 / 270.62 -36.55 / 223.44 -40.32 / 050207.0-611230	NGC 1796 / ESO 119- 30 / IRAS05021-6112 / PGC 16617	.SBT5P* / RSr (2) / 5.1± .7 / 5.2± .7	1.27± .04 / .27± .04 / .85± .01 / 1.27	102 / .00 / .41 / .14	12.86 ±.13 / 13.07 ±.12 / 12.06 / 12.57	.54± .01 / -.10± .02 / .48 / -.14	.57± .01 / -.06± .01 / 12.60± .02 / 13.37± .24		966± 56 / 776 / 988
050244.9-253212 / 226.64 -34.28 / 271.31 -62.03 / 050042.1-253624	/ ESO 486- 14 / / PGC 16618	.S?.... / / /	1.01± .06 / .11± .04 / / 1.01	/ .00 / .11 / .05	/ 14.80 ±.14 / / 14.61				6958± 52 / 6810 / 6953
050300.7-431750 / 248.26 -37.43 / 241.26 -53.54 / 050128.0-432200	/ ESO 252- 7 / / PGC 16628	.SB?... / / /	1.07± .05 / .25± .05 / / 1.07	47 / .00 / .37 / .12	/ 14.64 ±.14 / / 14.25				2956 / 2778 / 2965
0503.0 +0015 / 199.56 -23.79 / 324.19 -58.09 / 0500.5 +0011	/ UGC 3233 / / PGC 16631	.S..6*. / U / 6.0±1.2 /	1.03± .06 / .09± .05 / / 1.06	25 / .33 / .13 / .04	/ 14.86 ±.15 / / 14.39				4663± 38 / 4581 / 4640
050311.5-224849 / 223.55 -33.37 / 277.06 -62.72 / 050105.0-225300	/ ESO 486- 17 / / PGC 16635	.L..0*/ / S / -2.5± .9 /	1.05± .05 / .61± .03 / / .96	139 / .01 / .00 /	13.80V±.13 / 14.73 ±.14 / / 14.68	.99± .03 / / /	/ .91 / / 13.39± .28		4760± 62 / 4617 / 4754
050316.5-224956 / 223.58 -33.36 / 277.01 -62.74 / 050110.0-225406	/ ESO 486- 19 / / PGC 16638	.L..-*/ / S / -3.5± .6 /	1.16± .04 / .47± .03 / / 1.02	158 / .01 / .00 /	12.45V±.13 / 13.62 ±.14 / / 13.38	.88± .02 / / /	/ .83 / / 12.82± .26		4630± 62 / 4487 / 4624
050317.2-025612 / 202.65 -25.30 / 318.86 -59.68 / 050046.8-030022	/ MCG -1-13- 50 / IRAS05007-0300 / PGC 16639	.S..3P/ / E / 3.0± .7 /	1.35± .03 / .76± .04 / / 1.37	95 / .24 / 1.05 / .38	/ / 11.90 /				4322 / / 4231 / 4302
050318.2-634506 / 273.79 -36.01 / 221.40 -38.24 / 050256.0-634912	/ ESO 85- 34 / IRAS05029-6349 / PGC 16640	.S..1?. / S / 1.0±1.7 /	1.12± .05 / .26± .05 / / 1.12	86 / .01 / .27 / .13	/ 13.76 ±.14 / 13.73 /				
050318.6-254150 / 226.87 -34.21 / 270.90 -62.11 / 050116.0-254600	/ ESO 486- 20 / / PGC 16641	.S?.... / / /	1.02± .05 / .01± .05 / / 1.02	110 / .00 / .01 / .00	/ 14.66 ±.14 / / 14.61				6846 / 6697 / 6842
050318.6+182718 / 183.42 -13.85 / 346.72 -46.15 / 050023.3+182306	/ UGC 3232 / IRAS05003+1822 / PGC 16642	.S?.... / / /	1.00± .16 / .33± .12 / / 1.18	175 / 1.88 / .50 / .17				14.42±.3 / / 391± 7 /	5009± 7 / 4985 / 4976
050320.0-252526 / 226.55 -34.13 / 271.47 -62.18 / 050117.1-252936	/ ESO 486- 21 / / PGC 16643	.S?.... / / /	1.08± .05 / .16± .04 / / 1.08	107 / .00 / .20 / .08	/ 14.41 ±.14 / / 14.21				865 / 717 / 861
050324.7+162417 / 185.17 -15.01 / 344.79 -47.70 / 050031.9+162006	A 0500+16 / UGC 3234 / DDO 35 / PGC 16644	.I..9*. / U (1) / 10.0±1.1 / 9.0±1.4	1.26± .06 / .00± .06 / .96± .02 / 1.41	/ 1.64 / .00 / .00	15.01 ±.15 / / / 13.37	.67± .05 / .01± .06 / .29 / -.26	.75± .04 / -.02± .06 / 15.30± .05 / 16.16± .36	13.92±.1 / 142± 8 / 124± 5 / .55	1402± 5 / / 1372 / 1371

R.A. 2000 DEC.	Names	Type	logD$_{25}$	p.a.	B$_T$	(B-V)$_T$	(B-V)$_e$	m$_{21}$	V$_{21}$
l b		S$_T$ n$_L$	logR$_{25}$	A$_g$	m$_B$	(U-B)$_T$	(U-B)$_e$	W$_{20}$	V$_{opt}$
SGL SGB		T	logA$_e$	A$_i$	m$_{FIR}$	(B-V)^o_T	m'$_e$	W$_{50}$	V$_{GSR}$
R.A. 1950 DEC.	PGC	L	logD$_o$	A$_{21}$	B^o_T	(U-B)^o_T	m'$_{25}$	HI	V$_{3K}$

050327.8+003933		.S..6*/	1.15± .04	112				14.51±.3	4196± 7
199.23 -23.50	UGC 3237	UE	.70± .04	.35	15.02 ±.12				
324.94 -57.96	IRAS05008+0035	6.0± .7		1.03	13.18			294± 7	4116
050053.3+003523	PGC 16646		1.18	.35	13.62			.54	4174
0503.5 +0440									8908
195.52 -21.42	CGCG 420- 32			.31	14.9 ±.3				8840
330.98 -55.71									8883
0500.9 +0436	PGC 16649								
050333.1+063933		.S..4..	1.02± .06	18				15.59±.2	8511± 7
193.71 -20.37	UGC 3236	U	.46± .05	.44	15.33 ±.12			420± 13	
333.69 -54.48		4.0± .9		.67				406± 6	8448
050051.8+063523	PGC 16650		1.06	.23	14.16			1.20	8485
0503.5 +6548	A 0458+65								9960± 43
145.10 14.51				.73					10083
15.17 -5.84	7ZW 23								9917
0458.6 +6544	PGC 16651								
050336.6+013425	NGC 1762	.SAT5*.	1.23± .03	175	13.35 ±.13	.75± .01	.77± .01		4633± 10
198.39 -23.00	UGC 3238	UE (1)	.19± .04	.36	13.32 ±.12				4555
326.41 -57.50	IRAS05010+0130	4.7± .7	.77± .02	.28	12.55	.60	12.69± .05		4610
050101.1+013015	PGC 16654	3.1± .8	1.27	.09	12.67		13.89± .22		
050343.4-271440		.S?....	1.05± .06						11426± 52
228.70 -34.55	ESO 486- 22		.10± .05	.00	14.67 ±.14				11274
267.66 -61.75				.15					11424
050143.0-271848	PGC 16655		1.05	.05	14.46				
050400.9-235947		.LAR-*.	1.12± .05						12440± 60
224.97 -33.56	ESO 486- 23	S	.02± .04	.05	14.16 ±.14				12294
274.42 -62.67		-3.0± .8		.00					12436
050156.0-240354	PGC 16659		1.12		13.93				
0504.1 -0108		.SB?...	.94± .11						7045± 38
201.04 -24.25	MCG 0-13- 68		.28± .07	.29	15.25 ±.12				6959
322.14 -59.03				.35					7025
0501.6 -0113	PGC 16662		.96	.14	14.59				
0504.2 -0024		.S..2..	1.04± .08	154					7031± 38
200.36 -23.86	UGC 3239	U	.47± .06	.29	15.39 ±.13				6947
323.39 -58.69		2.0± .9		.58					7011
0501.7 -0029	PGC 16666		1.07	.24	14.45				
0504.2 +0439		.S..4..	1.00± .08	0				17.65±.3	10695± 10
195.64 -21.27	UGC 3240	U	.43± .06	.38	15.41 ±.12			162± 13	
331.15 -55.86	IRAS05016+0434	4.0± .9		.63	12.71			116± 10	10626
0501.6 +0434	PGC 16668		1.04	.22	14.33			3.10	10671
050419.1-633452		PSBT3*.	1.17± .04	90	13.49 ±.14	.76± .03	.69± .01		4844± 40
273.56 -35.93	ESO 85- 38	Sr (1)	.17± .04	.01	13.87 ±.14				4653
221.40 -38.45	IRAS05039-6338	2.5± .6	.93± .02	.24	13.06	.69	13.63± .05		4868
050356.1-633854	PGC 16670	2.2± .9	1.17	.09	13.39		13.76± .28		
050419.8-100434	IC 401	.SBT3?.	1.21± .05	55					3286± 81
209.90 -28.33	MCG -2-13- 40	E (1)	.45± .05	.29					3174
305.20 -62.47	MK 1092	3.0± .9		.63	13.46				3272
050157.6-100840	PGC 16672	3.6± .8	1.24	.23					
050429.8-870145		.IBS9*.	1.08± .05	59					
299.91 -28.43	ESO 4- 17	S (1)	.18± .05	.68	15.97 ±.14				
207.61 -18.34		10.0± .4		.14					
052310.0-870506	PGC 16673	10.0± .4	1.14	.09					
0504.5 -1635		.SB?...	1.12± .08						3340
216.77 -30.92	MCG -3-13- 63		.00± .07	.20					3211
290.92 -63.47	IRAS05022-1639			.00	13.93				3331
0502.2 -1639	PGC 16675		1.14	.00					
0504.5 +7652		.S..4..	1.04± .09					16.16±.1	7774± 11
135.46 20.68	UGC 3227	U	.57± .07	.50					
20.17 4.01		4.0± .9		.84				334± 8	7923
0457.4 +7648	PGC 16677		1.08	.28					7734
050434.2-170533		.S..5*/	1.15± .06	60					
217.33 -31.10	MCG -3-13- 64	E	.82± .05	.16					
289.77 -63.50		5.0±1.3		1.22					
050220.4-170938	PGC 16678		1.17	.41					
050435.2-553128		.S..2..	1.05± .05	70					
263.63 -37.04	ESO 158- 17	S	.14± .04	.00	14.43 ±.14				
227.84 -44.92		2.0± .8		.17					
050336.0-553530	PGC 16680		1.05	.07					

R.A. 2000 DEC.	Names	Type	$\log D_{25}$	p.a.	B_T	$(B-V)_T$	$(B-V)_e$	m_{21}	V_{21}
l b		S_T n_L	$\log R_{25}$	A_g	m_B	$(U-B)_T$	$(U-B)_e$	W_{20}	V_{opt}
SGL SGB		T	$\log A_e$	A_i	m_{FIR}	$(B-V)_T^o$	m'_e	W_{50}	V_{GSR}
R.A. 1950 DEC.	PGC	L	$\log D_0$	A_{21}	B_T^o	$(U-B)_T^o$	m'_{25}	HI	V_{3K}
050437.3-873420		.IBS9*.	1.12± .07		17.89 ±.14				
300.47 -28.20	ESO 1- 2	S	.20± .05	.65					
207.35 -17.86		10.0± .8		.15					
052809.1-873730	PGC 16681		1.18	.10					
0504.8 +0429		.S..3..	1.11± .07	123	15.42 ±.12				
195.87 -21.24	UGC 3242	U (1)	.66± .06	.38					
331.06 -56.07		3.0± .9		.91					
0502.2 +0425	PGC 16687	5.5±1.3	1.15	.33					
050500.4-811841		.LA.-*.	1.15± .05	2	14.26 ±.18	1.03± .02	1.05± .02		
293.89 -30.73	ESO 15- 18	S	.33± .04	.35	14.31 ±.14	.41± .03	.45± .03		4903
210.50 -23.37		-4.0± .8	.48± .07	.00		.90	12.15± .23		4720
050934.0-812230	PGC 16696		1.10		13.86	.35	14.18± .33		4936
050503.2-612901	NGC 1796A	PSAT2*.	1.23± .05	150					8700
270.98 -36.22	ESO 119- 35	RS	.55± .05	.00	14.87 ±.14				8509
222.87 -40.21		2.0± .9		.67					8723
050429.0-613300	PGC 16698		1.23	.27	14.11				
050506.9-733910		RSBT1..	1.00± .05	67					
285.40 -33.34	ESO 33- 11	Sr	.06± .04	.47	14.09 ±.14				
214.75 -30.01		1.0± .5		.06					
050615.1-734306	PGC 16700		1.04	.03					
050507.5-314704		.SXR5..	1.15± .04	9					
234.16 -35.36	ESO 422- 27	S (1)	.19± .03	.00	14.25 ±.14				3957± 52
258.64 -60.30		5.0± .8		.28					3795
050314.0-315106	PGC 16702	4.4± .8	1.15	.09	13.95				3960
050508.5-185916		.S?....	1.04± .05	54					
219.45 -31.66	ESO 552- 65		.37± .04	.08	15.47 ±.14				7801
285.47 -63.60				.55					7666
050257.1-190318	PGC 16703		1.05	.19	14.79				7795
0505.1 +2100		.SXS6*.	.96± .09						5315± 10
181.55 -12.01	UGC 3243	U	.05± .06	2.09					
349.48 -44.46		6.0±1.2		.07					5299
0502.2 +2056	PGC 16706		1.16	.02					5284
050514.2-182328		.IBS9..	1.07± .04	130					1704
218.81 -31.42	ESO 552- 66	SE (1)	.23± .04	.10	16.25 ±.14				
286.82 -63.65		10.0± .5		.17					1571
050302.1-182730	PGC 16708	9.8± .6	1.08	.11	15.98				1698
050515.1-375847	NGC 1792	.SAT4..	1.72± .02	137	10.87M±.10	.68± .01	.67± .01	13.36±.2	1225± 7
241.69 -36.45	ESO 305- 6	R (2)	.30± .03	.04	10.69 ±.11	.08± .02	.02± .02	316± 9	1218± 18
248.24 -57.14	IRAS05035-3802	4.0± .3		.45	9.38	.61		277± 12	1052
050332.0-380248	PGC 16709	4.0± .5	1.72	.15	10.30	.02	13.55± .17	2.90	1232
050517.9-581319		.SB.9*/	1.06± .05	60					
266.97 -36.66	ESO 119- 34	S (1)	.63± .05	.00	16.85 ±.14				
225.43 -42.85		9.0±1.3		.64					
050429.0-581718	PGC 16712	10.0±1.3	1.06	.31					
050518.1-090851	NGC 1779	PSXR0?.	1.37± .04	105	13.02 ±.16	.93± .06	1.00± .06		
209.07 -27.71	MCG -2-13- 41	PE	.27± .05	.25		.40± .04	.49± .04		3561
307.28 -62.45	IRAS05029-0912	.0± .5	.82± .05	.20	13.04	.80	12.61± .16		3451
050254.8-091253	PGC 16713		1.39		12.51	.32	14.07± .28		3548
050526.6-493401	NGC 1803	.SBS4*.	1.11± .03	62	13.38 ±.13	.53± .01	.56± .01		
256.15 -37.20	ESO 203- 18	S (1)	.21± .04	.00	13.51 ±.14				4127± 33
233.47 -49.50	IRAS05041-4938	4.0± .9	.62± .01	.31	11.83	.46	11.97± .01		3942
050409.0-493800	PGC 16715	2.2± .9	1.11	.11	13.10		13.25± .23		4143
050527.2-115218	NGC 1784	.SBR5..	1.60± .02	105	12.44 ±.14	.76± .03	.91± .01	12.87±.1	2316± 4
211.89 -28.84	MCG -2-13- 42	R (3)	.20± .03	.43	12.2 ±.2	.22± .04	.34± .02	342± 5	2255± 30
301.49 -63.14	IRAS05030-1156	5.0± .3	1.12± .02	.31	11.77	.61	13.53± .05	324± 5	2197
050307.1-115619	PGC 16716	2.7± .5	1.64	.10	11.64	.10	14.80± .18	1.13	2304
050534.5-493549		.LXR+..	.87± .06						4389± 59
256.19 -37.18	ESO 203- 19	r	.06± .05	.00	14.07 ±.14				
233.41 -49.49	FAIR 305	-1.3± .9		.00					4204
050417.0-493948	PGC 16720		.86		14.00				4405
0505.5 +3401		.S?....	1.11± .14						6692± 10
171.01 -4.22	UGC 3244		.07± .12	4.82					
359.16 -33.86				.10					6719
0502.3 +3357	PGC 16721		1.56	.03					6656
0505.6 +6701		.SXS6*.	1.36± .09	150				15.64±.1	3958± 11
144.19 15.37	UGC 3235	U	.39± .12	.59					
15.89 -4.85		6.0±1.1		.58				218± 8	4084
0500.5 +6657	PGC 16722		1.42	.20					3917

R.A. 2000 DEC.	Names	Type S_T n_L	$\log D_{25}$	p.a.	B_T	$(B-V)_T$	$(B-V)_e$	m_{21}	V_{21}
l b			$\log R_{25}$	A_g	m_B	$(U-B)_T$	$(U-B)_e$	W_{20}	V_{opt}
SGL SGB		T	$\log A_e$	A_i	m_{FIR}	$(B-V)_T^o$	m'_e	W_{50}	V_{GSR}
R.A. 1950 DEC.	PGC	L	$\log D_o$	A_{21}	B_T^o	$(U-B)_T^o$	m'_{25}	HI	V_{3K}
0505.6 +7601		.S..4..	1.02± .10						
136.27 20.29	UGC 3228	U	.65± .07	.55					
19.85 3.22	IRAS04587+7556	4.0±1.0		.95	12.71				
0458.7 +7556	PGC 16723		1.07	.32					
050549.4-283519		.L?....	.99± .06						
230.42 -34.45	ESO 422- 28		.07± .04	.00	14.62 ±.14				11422± 52
264.62 -61.73				.00					11266
050351.1-283918	PGC 16728		.98		14.45				11424
050550.2+413640								16.25±.3	5312± 9
164.98 .40								294± 11	
3.69 -27.37								226± 8	5363
050218.9+413238	PGC 16729								5274
0505.9 +0032		.S?....	.95± .07						
199.68 -23.01	UGC 3246		.05± .05	.34	14.71 ±.12				8168± 38
325.40 -58.55				.07					8086
0503.4 +0029	PGC 16734		.98	.02	14.25				8150
0506.0 -1644		.S?....	1.12± .08						3389
217.09 -30.64	MCG -3-13- 69		.28± .07	.11					
290.61 -63.84				.35					3259
0503.8 -1648	PGC 16735		1.13	.14					3382
050608.8-553652		.SXS7*.	1.03± .05		14.35 ±.13	.44± .02	.50 ± .02		
263.71 -36.81	ESO 158- 18	S (1)	.04± .04	.00	14.52 ±.14				3791± 40
227.52 -44.98		7.0± .9	.81± .02	.06		.41	13.89± .04		3601
050510.0-554048	PGC 16737	5.6± .9	1.03	.02	14.36		14.25± .29		3811
050609.7-450251		.SXR2..	1.20± .06	173					9821± 34
250.49 -36.98	ESO 252- 10	r	.10± .06	.00	14.20 ±.14				9640
238.44 -52.79		2.2± .7		.12					9835
050441.0-450648	PGC 16738		1.20	.05	13.98				
050614.8-090628	IC 402	.SXT6..	1.36± .03	140				14.33±.1	3339± 11
209.15 -27.48	MCG -2-13- 43	UE (1)	.18± .04	.28				244± 16	
307.51 -62.66	IRAS05038-0910	6.3± .4		.27	13.92			222± 12	3229
050351.5-091026	PGC 16742	5.3± .6	1.39	.09					3327
050620.8-192803	NGC 1780	PLBR0*.	.96± .04	84					
220.10 -31.56	ESO 553- 1	SE	.22± .02	.05	14.73 ±.14				
284.34 -63.86		-1.7± .5		.00					
050410.0-193200	PGC 16743		.93						
0506.4 -3157		.LB?...	1.02± .14						686± 63
234.44 -35.12	MCG -5-13- 5		.18± .07	.00					523
258.08 -60.47				.00					691
0504.5 -3201	PGC 16744		.99						
050626.2-315716	NGC 1800	.IBS9..	1.30± .03	113	13.13 ±.13	.54± .03	.42± .01	14.82±.2	803± 8
234.44 -35.12	ESO 422- 30	S (1)	.26± .03	.00	13.02 ±.12	-.17± .02	-.19± .01	84± 34	744± 34
258.07 -60.47	IRAS05045-3201	10.0± .6	.69± .01	.20	13.30	.47	12.07± .03	60± 8	637
050433.0-320112	PGC 16745	7.8± .6	1.30	.13	12.87	-.22	13.85± .21	1.82	804
050633.2-173516		.LAR0*.	1.18± .04	37	13.38 ±.13	1.03± .01	1.04± .01		
218.07 -30.84	ESO 553- 2	SE	.11± .03	.13	13.45 ±.14				4782± 37
288.65 -63.97		-2.3± .5	.88± .02	.00		.94	13.27± .08		4650
050420.0-173912	PGC 16748		1.18		13.21		13.88± .25		4777
050636.1+753542		.S..0..	1.00?						
136.68 20.11	UGC 3230	U							
19.71 2.81		.0± .8		.58	14.11 ±.18				
045950.9+753136	PGC 16750								
050637.2-173322	A 0504-17	.SXT5..	1.17± .03	136	14.1 ±.3	.71± .06	.79± .04	15.59±.1	4514± 10
218.04 -30.81	ESO 553- 3	PSE (3)	.22± .03	.13	14.14 ±.14			357± 15	4491± 59
288.72 -63.99	IRAS05044-1737	5.0± .4	.85± .08	.33	13.35	.61	13.81± .20		4381
050424.0-173718	PGC 16751	1.1± .4	1.18	.11	13.64		14.19± .35	1.84	4508
050638.2+084024		.S?....	.96± .09					15.44±.2	3371± 7
192.34 -18.64	UGC 3247		.00± .06	.60	14.89 ±.14			198± 7	
337.12 -53.74				.00				152± 5	3313
050354.5+083627	PGC 16752		1.02	.00	14.27			1.17	3348
050644.1-481136		.SB?...	.95± .06	175					3700±190
254.43 -36.98	ESO 203- 20		.16± .05	.00	15.17 ±.14				3515
234.70 -50.65	FAIR 789			.16					3716
050523.0-481530	PGC 16755		.95	.08	14.96				
050649.7-273929		.IBS9P*	1.13± .04	133					1270
229.41 -33.99	ESO 422- 33	S (1)	.33± .03	.00	14.45 ±.14				
266.33 -62.26		10.0± .6		.25					1116
050450.1-274324	PGC 16758	7.4± .7	1.13	.16	14.20				1273

R.A. 2000 DEC. l b SGL SGB R.A. 1950 DEC.	Names PGC	Type S_T n_L T L	$logD_{25}$ $logR_{25}$ $logA_e$ $logD_0$	p.a. A_g A_i A_{21}	B_T m_B m_{FIR} B_T^0	$(B-V)_T$ $(U-B)_T$ $(B-V)_T^0$ $(U-B)_T^0$	$(B-V)_e$ $(U-B)_e$ m'_e m'_{25}	m_{21} W_{20} W_{50} HI	V_{21} V_{opt} V_{GSR} V_{3K}
050650.7-202047		RSBR3..	1.05± .05	50					
221.12 -31.76	ESO 553- 5	r	.20± .04	.04	14.68 ±.14				
282.30 -63.90		3.3± .9		.28					
050441.0-202442	PGC 16759		1.06`	.10					
050656.3-594326	NGC 1824	.SBS9..	1.51± .04	160	13.0 ±.2	.44± .03	.45± .02		1263± 34
268.78 -36.25	ESO 119- 36	S (1)	.58± .05	.01	13.17 ±.14	-.10± .05	-.12± .03		1072
223.99 -41.77	IRAS05061-5947	9.0± .8	1.13± .05	.59	13.83	.30	14.15± .12		1286
050614.0-594718	PGC 16761	6.7± .8	1.51	.29	12.50	-.20	13.95± .29		
050706.5+035849		.S..3..	1.23± .05	135				15.27±.2	8929± 7
196.65 -21.02	UGC 3248	U (1)	.34± .05	.44	14.55 ±.15			449± 13	8856
330.95 -56.83		3.0± .8		.47				432± 6	8910
050428.2+035454	PGC 16764	3.5±1.1	1.27	.17	13.57			1.54	
0507.1 +3049		.I..9?.	1.40± .11						
173.76 -5.87	UGCA 100	U	.00± .18	3.20					
357.41 -36.72		10.0±1.4		.00					
0503.9 +3046	PGC 16765		1.70	.00					
0507.5 -1138		.S..7/.	1.27± .05	0					
211.88 -28.28		E	.85± .06	.41					
302.24 -63.60		7.0± .6		1.17					
0505.2 -1142	PGC 16772		1.31	.43					
050742.8-373051 NGC 1808		RSXS1..	1.81± .02	133	10.76M±.10	.82± .01	.90± .01	12.92±.2	1005± 9
241.21 -35.90	ESO 305- 8	R	.22± .03	.07	10.77 ±.11	.29± .02	.27± .01	286± 13	1014± 17
248.43 -57.81	IRAS05059-3734	1.0± .3	1.29± .01	.22	8.28	.76	12.73± .03	203± 25	835
050559.0-373442	PGC 16779	1.81	.11	10.47	.24	14.10± .16	2.34	1017
050744.0-625924		.SBS9..	1.22± .04						
272.75 -35.66	ESO 85- 47	S (1)	.07± .05	.00	14.57 ±.14				1469
221.43 -39.15		9.0± .5		.07					1277
050718.1-630312	PGC 16780	10.0± .6	1.22	.04	14.49				1495
050744.1-080107 NGC 1797		PSBT1P.	1.05± .05	90					
208.23 -26.67	MCG -1-14- 2	E	.14± .04	.22					4441± 49
310.02 -62.68	MK 1093	1.0± .6		.15	10.84				4332
050519.5-080459	PGC 16781		1.07	.07					4431
050744.6-075810 NGC 1799		.LB.0P?	1.04± .05	65					
208.18 -26.65	MCG -1-14- 1	E	.25± .04	.20					
310.12 -62.67		-2.0± .9		.00					
050520.0-080202	PGC 16783		1.02						
0507.7 -1617	A 0505-16	.SBS9..	1.37± .04	60	13.57 ±.17	.50± .06	.58± .03	14.19±.1	2039± 6
216.81 -30.09	MCG -3-14- 1	PE (2)	.23± .05	.18		-.19± .06	-.10± .04	50± 6	
291.65 -64.24	DDO 36	8.5± .5	1.20± .04	.23		.40	15.06± .09	36± 12	1909
0505.5 -1621	PGC 16784	8.4± .7	1.38	.11	13.15	-.26	14.68± .30	.93	2035
050754.3-611125 NGC 1796B		.SXT4*.	1.04± .05						
270.56 -35.93	ESO 119- 37	Sr (1)	.56± .05	.00	15.76 ±.14				5758
222.73 -40.65	IRAS05073-6115	4.0± .8		.83	13.37				5566
050719.0-611512	PGC 16787	3.3±1.3	1.04	.28	14.89				5783
050755.4-181127 NGC 1794		RLBS0P*	1.10± .04	45					
218.86 -30.76	ESO 553- 7	SE	.06± .03	.07	13.71 ±.14				4990
287.24 -64.29	IRAS05057-1815	-1.7± .5		.00	12.83				4855
050543.0-181518	PGC 16788		1.10		13.56				4987
050756.4-611135			.93± .07						
270.56 -35.92	ESO 119- 37A		.27± .06	.00	15.83 ±.14				
222.72 -40.65									
050721.1-611522	PGC 16789		.93						
050808.3-381841		.SBS8..	1.55± .03	63				12.92±.2	1025± 7
242.20 -35.94	ESO 305- 9	S (1)	.11± .05	.06	13.08 ±.14			127± 16	
247.12 -57.41		8.0± .7		.14				114± 6	851
050626.0-382230	PGC 16790	8.9± .7	1.55	.06	12.88			-.01	1036
050813.8-762055		.S.3*P/	1.02± .05	71					
288.37 -32.31	ESO 33- 16	S	.78± .04	.46	16.07 ±.14				
213.01 -27.78		3.0±1.4		1.07					
051009.0-762436	PGC 16793		1.06	.39					
050818.3-195005		.L?....	1.00± .06	115					
220.69 -31.26	ESO 553- 10		.33± .03	.05	14.97 ±.14				8964
283.41 -64.29				.00					8825
050608.0-195354	PGC 16796		.95		14.78				8963
050821.5-462049		PSBS4..	1.08± .06	172					
252.14 -36.66	ESO 252- 12	r	.39± .05	.00	15.06 ±.14				
236.46 -52.16	IRAS05069-4624	4.5± .9		.58	13.17				
050656.1-462436	PGC 16797		1.08	.19					

R.A. 2000 DEC. l b SGL SGB R.A. 1950 DEC.	Names PGC	Type S_T n_L T L	$\log D_{25}$ $\log R_{25}$ $\log A_e$ $\log D_o$	p.a. A_g A_i A_{21}	B_T m_B m_{FIR} B_T^o	$(B-V)_T$ $(U-B)_T$ $(B-V)_T^o$ $(U-B)_T^o$	$(B-V)_e$ $(U-B)_e$ m'_e m'_{25}	m_{21} W_{20} W_{50} HI	V_{21} V_{opt} V_{GSR} V_{3K}
050827.4-612203 270.76 -35.84 222.53 -40.54 050753.1-612548	 ESO 119- 38 IRAS05078-6125 PGC 16803	.SXS5P. S (1) 5.0± .8 3.3± .8	1.13± .05 .21± .04 1.13	103 .00 .31 .10	 14.71 ±.14 14.35				8895± 34 8703 8921
0508.6 -1656 217.58 -30.14 290.17 -64.47 0506.4 -1700	 MCG -3-14- 4 PGC 16809	.S?.... 	1.12± .08 .08± .07 1.13	 .12 .11 .04					2028 1896 2025
050843.3-291637 231.42 -34.01 262.73 -62.05 050646.0-292024	NGC 1811 ESO 422- 37 PGC 16811	.S.1*P/ S 1.0±1.2 	1.23± .04 .60± .03 1.23	60 .00 .61 .30	 14.46 ±.14 13.80				3949± 52 3790 3955
0508.7 +7028 141.34 17.51 17.65 -1.86 0503.1 +7024	A 0503+70 UGC 3245 IRAS05031+7024 PGC 16813	.SXS5.. U 5.0± .8 	1.17± .05 .00± .05 1.22	 .47 .00 .00	 14.6 ±.3 14.07			15.31±.1 386± 8 1.24	4845± 7 4979 4806
050847.7+131538 188.62 -15.70 343.02 -50.90 050558.6+131150	 UGC 3251 PGC 16814	.S?.... 	1.00± .16 .25± .12 1.19	100 2.05 .35 .13				15.37±.3 340± 7 	5330± 10 5286 5308
0508.8 +7525 136.91 20.15 19.77 2.61 0502.1 +7522	 UGC 3241 PGC 16816	.L..... U -2.0± .8 	1.00? 	 .61 	 14.43 ±.15 				
050853.3-291508 231.40 -33.97 262.75 -62.10 050656.1-291854	NGC 1812 ESO 422- 39 IRAS05068-2918 PGC 16819	.S..1P. S 1.0± .8 	1.09± .04 .14± .03 1.09	8 .00 .15 .07	 13.64 ±.14 13.54 13.44				3915± 52 3756 3921
050901.7-201520 221.22 -31.24 282.39 -64.42 050652.0-201906	 ESO 553- 12 PGC 16822	.S?.... 	1.06± .06 .18± .04 1.06	95 .07 .22 .09	 14.13 ±.14 13.79				4535 4395 4535
050909.3-091539 209.65 -26.91 307.64 -63.39 050646.2-091925	 MCG -2-14- 1 PGC 16826	.IBS9.. EU (1) 9.5± .6 8.7± .8	1.19± .05 .31± .05 1.22	145 .27 .23 .15				14.94±.1 189± 8 196± 12 	2706± 6 2594 2699
0509.6 -0016 200.95 -22.61 325.00 -59.77 0507.1 -0020	 MCG 0-14- 3 PGC 16837	.S?.... 	.82± .13 .17± .07 .84	 .35 .13 	 15.40 ±.12 14.82				6961± 38 6874 6949
050940.7-521137 259.42 -36.49 230.07 -47.97 050831.0-521518	 ESO 203- 22 PGC 16839	.S..5./ S 5.0±1.0 	1.15± .05 1.22± .05 1.15	163 .00 1.50 .50	 16.39 ±.14 				
050948.5-253612 227.27 -32.78 270.20 -63.54 050746.0-253954	 ESO 486- 32 IRAS05077-2539 PGC 16842	.SBT4.. S (1) 4.0± .8 2.2± .8	1.20± .04 .13± .04 1.20	124 .00 .18 .06	 14.56 ±.14 14.12 14.31				10774± 52 10622 10779
0509.8 +0728 193.85 -18.61 336.51 -55.13 0507.1 +0725	 UGC 3255 IRAS05071+0725 PGC 16843	.SB.3?. U 3.0±1.3 	1.09± .06 .63± .05 1.14	17 .53 .87 .31	 15.35 ±.12 12.89 13.90				5689± 59 5626 5672
050957.0-221800 223.56 -31.73 277.58 -64.35 050750.1-222142	 ESO 553- 15 PGC 16847	.S?.... 	1.08± .05 .11± .04 1.08	61 .03 .13 .05	 14.88 ±.14 14.62				9682 9537 9685
051003.7-365732 240.63 -35.35 248.78 -58.51 050819.0-370112	NGC 1827 ESO 362- 6 IRAS05083-3701 PGC 16849	.SXS6*/ BS (1) 6.0± .6 6.7±1.2	1.48± .04 .66± .05 1.48	120 .04 .98 .33	 13.26 ±.14 13.14 12.24			14.02±.2 193± 16 180± 12 1.45	1049± 11 1088± 52 878 1063
051004.4-145648 215.62 -29.05 294.89 -64.70 050748.1-150029	 MCG -2-14- 2 PGC 16850	.SBS8*. E (1) 8.0± .9 6.4± .7	1.14± .05 .42± .04 1.17	155 .29 .51 .21					1979 1851 1977
0510.1 -0043 201.44 -22.73 324.36 -60.11 0507.6 -0047	A 0507+00 MCG 0-14- 8 2ZW 32 PGC 16852	.E...*. E -5.0±1.3 	.80± .11 .04± .05 .87	100 .50 .00 	 15.31 ±.10 14.69				8170± 67 8081 8159

R.A. 2000 DEC. / l b / SGL SGB / R.A. 1950 DEC.	Names / PGC	Type / S_T n_L / T / L	$\log D_{25}$ / $\log R_{25}$ / $\log A_e$ / $\log D_o$	p.a. / A_g / A_i / A_{21}	B_T / m_B / m_{FIR} / B_T^o	$(B-V)_T$ / $(U-B)_T$ / $(B-V)_T^o$ / $(U-B)_T^o$	$(B-V)_e$ / $(U-B)_e$ / m'_e / m'_{25}	m_{21} / W_{20} / W_{50} / HI	V_{21} / V_{opt} / V_{GSR} / V_{3K}
0510.2 +0054 / 199.91 -21.91 / 327.09 -59.24 / 0507.6 +0050	UGC 3256 / IRAS05076+0050 / PGC 16853	.SX.3.. / U / 3.0± .8 /	1.00± .06 / .02± .05 / / 1.04	/ .34 / .03 / .01	/ 14.33 ±.12 / 13.42 / 13.88			15.51±.3 / 251± 7 / 217± 7 / 1.61	8697± 11 / / 8613 / 8685
051031.6+630958 / 147.78 13.63 / 14.68 -8.54 / 050549.4+630613	UGC 3250 / IRAS05058+6306 / PGC 16858	.SB.3.. / U / 3.0± .8 /	1.28± .04 / .13± .05 / / 1.40	100 / 1.36 / .18 / .06	/ 13.9 ±.3 / 12.83 / 12.31			14.96±.1 / / 430± 8 / 2.59	4530± 7 / / 4645 / 4492
051037.3-613119 / 270.91 -35.56 / 222.13 -40.56 / 051004.0-613454	ESO 119- 43 / / PGC 16859	.L..0./ / S / -2.0± .8 /	1.14± .05 / .44± .03 / / 1.07	13 / .00 / .00 /	/ 14.81 ±.14 / /				
051042.9+002429 / 200.45 -22.05 / 326.41 -59.62 / 050808.7+002050	UGC 3258 / IRAS05081+0020 / PGC 16861	.SBR2P. / E / 2.0± .9 /	.88± .06 / .06± .04 / / .91	/ .41 / .07 / .03	/ 13.98 ±.16 / 13.13 / 13.48			17.10±.3 / 365± 13 / / 3.60	2821± 10 / / 2735 / 2810
051043.8-572100 / 265.79 -36.03 / 225.34 -43.99 / 050952.0-572436	ESO 158- 20 / / PGC 16862	.SBS3?. / S (1) / 3.0±1.1 / 4.4± .9	1.04± .05 / .50± .05 / / 1.04	/ .00 / .69 / .25	/ 15.43 ±.14 / /				
051046.7-313552 / 234.27 -34.14 / 257.85 -61.45 / 050853.1-313930	A 0508-31 / ESO 422- 41 / DDO 230 / PGC 16864	.SXT8*. / PSUr (2) / 8.5± .4 / 7.8± .5	1.47± .03 / .08± .04 / 1.15± .03 / 1.47	/ / .00 / .09 .04	13.13 ±.15 / 13.04 ±.14 / 13.84 / 12.99	.43± .04 / -.20± .06 / / .41 .22	.39± .03 / -.23± .04 / 14.37± .06 / 15.12± .23	13.61±.2 / 138± 16 / 130± 6 / .57	981± 7 / 953± 52 / 817 / 991
051048.1-024054 / 203.39 -23.53 / 321.07 -61.23 / 050817.4-024433	A 0508-02 / MCG 0-14- 10 / MK 1094 / PGC 16868	.I.0.P? / E / 90.0 /	.84± .09 / .07± .06 / / .88	/ .40 / .08 / .03	/ 14.10 ±.16 / / 13.59			15.48±.1 / 163± 4 / 132± 5 / 1.85	2831± 6 / 2848± 23 / 2737 / 2823
051106.0-182541 / 219.43 -30.14 / 286.63 -65.04 / 050854.0-182918	ESO 553- 16 / / PGC 16874	.SBS9.. / SE (2) / 8.5± .6 / 8.2± .8	1.25± .03 / .52± .03 / / 1.25	118 / .06 / .53 / .26	/ 15.33 ±.14 / / 14.74				3566 / / 3429 / 3569
051107.9-092324 / 210.02 -26.52 / 307.67 -63.90 / 050845.0-092701	MCG -2-14- 3 / IRAS05087-0926 / PGC 16875	.SBS5*/ / E (1) / 5.0±1.2 / 3.1±1.6	1.41± .04 / .68± .05 / / 1.44	91 / .33 / 1.02 / .34	/ / 12.08 /				2670 / / 2556 / 2666
051109.1-342336 / 237.60 -34.67 / 252.75 -60.12 / 050920.0-342712	ESO 362- 8 / IRAS05093-3427 / PGC 16877	.L?.... / / /	1.09± .06 / .34± .03 / / 1.04	167 / .02 / .00 /	/ 13.62 ±.14 / 13.77 / 13.53				4785± 24 / / 4616 / 4798
051123.1-221449 / 223.63 -31.40 / 277.57 -64.68 / 050916.0-221824	ESO 553- 18 / IRAS05092-2218 / PGC 16885	.S?.... / / /	.97± .05 / .14± .04 / / .97	160 / .03 / .17 / .07	/ 14.64 ±.14 / 13.75 / 14.34				9912± 22 / / 9766 / 9917
051132.6+170324 / 185.75 -13.03 / 347.61 -48.48 / 050838.8+165947	UGC 3261 / IRAS05086+1659 / PGC 16887	.S..8*. / U / 8.0±1.2 /	1.07± .07 / .21± .06 / / 1.22	55 / 1.57 / .25 / .10	/ 14.7 ±.3 / 13.22 / 12.87			15.98±.3 / / 264± 7 / 3.01	5176± 7 / / 5143 / 5157
0511.5 -0034 / 201.48 -22.35 / 324.99 -60.33 / 0509.0 -0038	UGC 3264 / / PGC 16889	.S..0.. / U / .0± .9 /	1.00± .06 / .29± .05 / / 1.04	110 / .50 / .22 /	/ 14.57 ±.12 / /				
051138.5-202538 / 221.66 -30.73 / 281.83 -65.01 / 050929.1-202912	ESO 553- 20 / IRAS05094-2029 / PGC 16892	PSBT2P* / SE (1) / 2.3± .7 / 4.2±1.2	1.28± .03 / .35± .03 / / 1.29	110 / .04 / .43 / .18	/ 13.57 ±.14 / 11.70 / 13.05				3997 / / 3855 / 4001
051141.6-030530 / 203.90 -23.53 / 320.54 -61.62 / 050911.4-030905	MCG -1-14- 3 / IRAS05091-0309 / PGC 16893	.S..3?/ / E / 3.0±1.9 /	1.09± .06 / .63± .05 / / 1.13	0 / .42 / .87 / .31	/ / 12.66 /				
0511.7 -1447 / 215.63 -28.63 / 295.34 -65.08 / 0509.4 -1451	A 0509-14 / MCG -2-14- 4 / / PGC 16894	.SXT6.. / PUE (1) / 5.8± .4 / 4.7± .6	1.43± .03 / .12± .04 / 1.15± .02 / 1.46	100 / .33 / .18 / .06	13.01 ±.14 / 12.7 ±.2 / / 12.42	.56± .04 / .05± .04 / .45 / -.03	.64± .02 / .06± .03 / 14.25± .04 / 14.70± .22	14.10±.1 / 229± 14 / 210± 11 / 1.62	1979± 9 / 2110± 66 / 1853 / 1982
051146.0+672918 / 144.12 16.12 / 16.63 -4.68 / 050635.2+672538	UGC 3252 / IRAS05065+6725 / PGC 16897	.SXS5.. / U / 5.0± .8 /	1.38± .04 / .08± .05 / / 1.44	30 / .57 / .13 / .04	/ 13.6 ±.3 / 12.33 / 12.91			14.24±.1 / / 310± 8 / 1.29	6115± 11 / / 6241 / 6078

R.A. 2000 DEC. / l b / SGL SGB / R.A. 1950 DEC.	Names / PGC	Type / S_T n_L / T / L	$\log D_{25}$ / $\log R_{25}$ / $\log A_e$ / $\log D_o$	p.a. / A_g / A_i / A_{21}	B_T / m_B / m_{FIR} / B_T^o	$(B-V)_T$ / $(U-B)_T$ / $(B-V)_T^o$ / $(U-B)_T^o$	$(B-V)_e$ / $(U-B)_e$ / m'_e / m'_{25}	m_{21} / W_{20} / W_{50} / HI	V_{21} / V_{opt} / V_{GSR} / V_{3K}
051146.1-150803	NGC 1821	.IBS9..	1.05± .06	50					
216.00 -28.75	MCG -3-14- 7	E (1)	.16± .05	.22					
294.53 -65.13	IRAS05095-1511	10.0± .9		.12	12.13				
050930.0-151138	PGC 16898	5.3± .8	1.07	.08					
051146.3+051201	NGC 1819	.LB....	1.22± .06	120				16.05±.2	4470± 5
196.17 -19.40	UGC 3265	U	.15± .03	.66	13.41 ±.10			323± 6	4826± 97
334.00 -56.98	MK 1194	-2.0± .8		.00	11.05				4400
050906.6+050826	PGC 16899		1.27		12.68				4459
051159.4-325822	A 0510-33	.SXS9..	1.55± .03					12.63±.2	933± 7
235.96 -34.20	ESO 362- 9	SU (2)	.07± .04	.00	12.88 ±.14			91± 16	
255.06 -61.01	DDO 231	9.3± .3		.07				81± 6	765
051008.1-330154	PGC 16904	9.2± .4	1.55	.03	12.81			-.21	945
051202.8+135231		.S?....	1.00± .08	20				16.94±.3	6072± 9
188.54 -14.70	UGC 3266		.19± .06	1.68					
344.59 -50.98				.28				412± 5	6028
050912.9+134857	PGC 16905		1.16	.09					6055
051203.4-154116	NGC 1832	.SBR4..	1.41± .03	10	11.96 ±.13	.63± .01	.71± .01	13.85±.1	1937± 7
216.61 -28.90	MCG -3-14- 10	R (3)	.18± .03	.19	12.5 ±.2	-.01± .01	.08± .01	289± 11	1994± 40
293.21 -65.24	IRAS05098-1544	4.0± .3	1.03± .01	.27	10.91	.54	12.60± .03	265± 11	1808
050948.0-154449	PGC 16906	1.9± .5	1.43	.09	11.65	-.08	13.39± .21	2.11	1940
0512.0 +6645		.SB?...	1.14± .07	45				15.92±.1	6149± 11
144.77 15.75	UGC 3254		.32± .06	.48					
16.35 -5.36				.49				160± 8	6273
0507.0 +6642	PGC 16907		1.18	.16					6112
0512.1 +0527		.SB?...	.82± .13						4415
195.99 -19.18	MCG 1-14- 3		.04± .07	.62	15.17 ±.12				4344
334.47 -56.88	IRAS05094+0524			.06	13.11				4403
0509.4 +0524	PGC 16909		.88	.02	14.46				
051216.5-572355	NGC 1853	.SBS7?.	1.30± .04	43					1402
265.83 -35.82	ESO 158- 22	S (1)	.46± .05	.00	13.58 ±.14				1210
225.07 -44.07	IRAS05114-5727	6.7± .7		.63	13.04				1429
051125.0-572724	PGC 16911	5.6± .7	1.30	.23	12.94				
0512.2 +7554		.S..8..	1.11± .07	75				16.72±.1	4605± 11
136.58 20.58	UGC 3249	U	.15± .06	.59	15.2 ±.3				
20.16 2.95	IRAS05053+7551	8.0± .8		.18				150± 8	4751
0505.4 +7551	PGC 16912		1.17	.07	14.40			2.25	4567
051223.1-142137		.IB.9P/	1.11± .06	90					
215.26 -28.31	MCG -2-14- 5	E (1)	.52± .05	.34					
296.41 -65.20		10.0±1.8		.39					
051006.1-142509	PGC 16917	6.4± .8	1.14	.26					
0512.4 +5118		.L...?.	1.11± .16	116					
157.82 7.06	UGC 3260	U	.60± .08	1.62	15.08 ±.15				
9.67 -19.34		-2.0±1.8		.00					
0508.5 +5115	PGC 16918		1.20						
051234.2-395136		.SBS5..	1.12± .05	108					4876
244.24 -35.30	ESO 305- 14	S (1)	.27± .05	.11	14.13 ±.14				4698
243.86 -57.13		5.0± .9		.40					4894
051055.0-395506	PGC 16920	2.2± .9	1.13	.13	13.59				
0512.8 +0243		.S?....	.77± .11						8347
198.58 -20.44	CGCG 395- 16		.02± .10	.44	14.61 ±.14				8267
330.67 -58.72				.03					8338
0510.2 +0240	PGC 16922		.81	.01	14.10				
051250.6-413802		.S..5..	1.18± .05	157					9950± 52
246.40 -35.48	ESO 305- 15	S (1)	.55± .04	.01	14.55 ±.14				9770
241.35 -56.00	IRAS05112-4141	5.0± .9		.82	13.44				9969
051115.0-414130	PGC 16924	2.2±1.3	1.18	.27	13.66				
0512.9 -0911		.S?....	1.12± .08						2590
210.04 -26.02	MCG -2-14- 6		.74± .07	.35					
308.41 -64.28				1.11					2476
0510.6 -0915	PGC 16926		1.15	.37					2589
051318.1-221333		.S?....	1.19± .05	64					9103
223.78 -30.97	ESO 553- 26		.49± .05	.04	15.15 ±.14				8956
277.42 -65.12				.68					9111
051111.1-221700	PGC 16934		1.19	.25	14.36				
051328.9-613313		.SAS5*.	1.08± .05	150					
270.89 -35.22	ESO 119- 44	S (1)	.07± .05	.04	14.98 ±.14				
221.73 -40.72		5.0± .8		.10					
051256.1-613636	PGC 16937	4.4± .8	1.08	.03					

R.A. 2000 DEC. l b SGL SGB R.A. 1950 DEC.	Names PGC	Type S_T n_L T L	$\log D_{25}$ $\log R_{25}$ $\log A_e$ $\log D_o$	p.a. A_g A_i A_{21}	B_T m_B m_{FIR} B_T^o	$(B-V)_T$ $(U-B)_T$ $(B-V)_T^o$ $(U-B)_T^o$	$(B-V)_e$ $(U-B)_e$ m'_e m'_{25}	m_{21} W_{20} W_{50} HI	V_{21} V_{opt} V_{GSR} V_{3K}
051343.3-124538 213.75 -27.37 300.33 -65.29 051124.4-124904	 MCG -2-14- 7 PGC 16943	.SBT7.. E (1) 7.0± .8 6.4± .8	1.17± .06 .16± .05 1.21	100 .41 .22 .08					
051406.2-103738 211.62 -26.39 305.36 -64.93 051144.7-104102	NGC 1843 MCG -2-14- 8 IRAS05117-1041 PGC 16949	.SXS6*. UE (1) 6.2± .7 4.2± .8	1.31± .04 .10± .05 1.36	110 .48 .14 .05	 11.98 			13.75±.1 208± 16 187± 12 	2611± 11 2492 2613
051407.6+063112 195.30 -18.21 336.50 -56.56 051126.4+062747	 UGC 3269 PGC 16951	.S..4.. U 4.0± .9 	.91± .04 .15± .03 .96	175 .51 .22 .07	14.91S±.15 14.74 ±.12 14.02		 13.96± .28	17.11±.2 314± 13 133± 5 3.02	8980± 6 8912 8971
0514.1 -0608 207.17 -24.41 315.15 -63.49 0511.7 -0612	 PGC 16954	.S..5?/ E 5.0±1.8 	1.10± .08 .63± .08 1.16	45 .60 .95 .32					
0514.1 +7219 139.93 18.86 18.81 -.37 0508.2 +7216	 UGC 3259 PGC 16955	.SX.7.. U 7.0± .8 	1.25± .04 .12± .05 1.29	120 .50 .17 .06	 15.0 ±.4 14.31			15.62±.1 222± 8 1.25	4939± 11 5076 4902
0514.2 +6234 148.53 13.66 14.82 -9.26 0509.6 +6231	 UGCA 105 PGC 16957	.I..9?. U 10.0±1.3 	1.74± .07 .20± .18 1.88	 1.48 .10	 13.9 ±.3 12.32			11.67±.1 131± 7 118± 6 -.76	111± 5 223 76
051421.6-604728 269.94 -35.21 222.17 -41.41 051345.1-605048	 ESO 119- 45 IRAS05137-6050 PGC 16960	.SBS4.. S (1) 4.0± .8 3.3± .8	1.01± .05 .06± .04 1.02	18 .02 .09 .03	 14.91 ±.14 				
051429.6-621011 271.61 -35.02 221.16 -40.26 051400.0-621330	 ESO 119- 46 IRAS05140-6213 PGC 16964	.S.3?P/ S 3.0±1.9 	1.06± .05 .50± .04 1.06	57 .01 .69 .25	 15.04 ±.14 11.84 				
051429.7-235356 225.73 -31.25 273.36 -65.03 051225.0-235718	 ESO 486- 37 PGC 16966	.LA.-.. S -3.0± .8 	1.08± .05 .15± .04 1.06	 .02 .00 	 14.68 ±.14 				
051431.3-621341 271.68 -35.00 221.12 -40.22 051402.0-621700	 ESO 119- 47 FAIR 243 PGC 16968	.SB?... 	1.10± .06 .27± .06 1.10	52 .01 .37 .14	 14.07 ±.14 12.43 13.65				5100± 63 4906 5130
051436.1-612854 270.78 -35.09 221.64 -40.85 051403.0-613212	 ESO 119- 48 PGC 16971	PLAS+.P S -1.0± .8 	1.29± .05 .22± .04 1.26	55 .04 .00 	 13.60 ±.14 				
051443.6-611124 270.42 -35.11 221.83 -41.10 051409.0-611442	 ESO 119- 49 PGC 16973	.LA.0*/ S -2.0± .8 	1.12± .05 .37± .04 1.07	88 .03 .00 	 14.79 ±.14 				
051445.4-545653 262.79 -35.63 226.71 -46.26 051345.1-550012	 ESO 159- 1 PGC 16974	.L...?/ S -2.0±1.3 	1.08± .05 .74± .03 1.01	135 .36 .00 	 15.58 ±.14 				
051501.0-412329 246.17 -35.04 241.18 -56.46 051325.0-412648	 ESO 305- 17 PGC 16976	.IBS9.. S (1) 10.0± .8 7.8± .8	1.26± .05 .41± .05 1.26	64 .01 .31 .20	 14.61 ±.14 14.29				1084± 60 903 1105
051507.4-534418 261.30 -35.62 227.69 -47.26 051403.0-534736	 ESO 159- 2 IRAS05140-5347 PGC 16979	.SBS4*. S (1) 4.0± .6 2.8± .8	1.30± .04 .15± .05 1.34	60 .36 .22 .07	 13.77 ±.14 13.13 13.16				4289± 34 4097 4316
051516.1-303141 233.30 -32.95 258.94 -62.80 051321.1-303500	 ESO 423- 2 IRAS05133-3035 PGC 16983	.SBS7*/ SU (1) 7.0± .6 5.6± .8	1.45± .03 .55± .04 1.45	19 .00 .76 .27	 13.10 ±.14 13.01 12.34			14.02±.2 238± 16 235± 12 1.41	1481± 11 1509± 52 1318 1498
051519.4+062830 195.51 -17.98 336.80 -56.81 051238.2+062510	 UGC 3270 PGC 16984	.S?.... 	.95± .07 .39± .05 1.00	177 .53 .59 .20	 15.30 ±.12 14.14			16.20±.2 368± 13 355± 6 1.86	8636± 7 8567 8628

R.A. 2000 DEC. l　　b SGL　SGB R.A. 1950 DEC.	Names PGC	Type S_T　n_L T L	$\log D_{25}$ $\log R_{25}$ $\log A_e$ $\log D_o$	p.a. A_g A_i A_{21}	B_T m_B m_{FIR} B_T^o	$(B-V)_T$ $(U-B)_T$ $(B-V)_T^o$ $(U-B)_T^o$	$(B-V)_e$ $(U-B)_e$ m'_o m'_{25}	m_{21} W_{20} W_{50} HI	V_{21} V_{opt} V_{GSR} V_{3K}
0515.5　+0711 194.89 -17.57 337.80 -56.36 0512.8　+0708	 MCG　1-14-　9 PGC 16989	.S?.... 	.96± .05 .36± .03 1.01	 .53 .53 .18	15.43S±.15 15.20 ±.12	.72± .04 .10± .07	 14.17± .29		
051539.3-224231 224.52 -30.61 276.01 -65.56 051333.0-224548	 ESO　486- 40 IRAS05135-2245 PGC 16990	.SB?... 	1.07± .05 .36± .04 1.07	23 .03 .54 .18	 14.76 ±.14 13.96 14.16				5779 5630 5791
0515.8　+0710 194.95 -17.52 337.87 -56.43 0513.1　+0707	 MCG　1-14- 10 PGC 16995	.S?.... 	.82± .13 .08± .07 .87	 .53 .11 .04	 15.27 ±.12 14.58			16.32±.3 335± 13 300± 10 1.70	8923± 10 8856 8916
051548.3-232831 225.38 -30.83 274.17 -65.42 051343.0-233148	 ESO　486- 41 IRAS05137-2331 PGC 16996	.SAR3P. S　　(1) 3.0± .8 3.3±1.2	1.18± .04 .36± .03 1.19	78 .06 .49 .18	 14.86 ±.14 13.18				
0515.8　+7128 140.77 18.53 18.58　-1.20 0510.0　+7125	 UGC　3267 PGC 16997	.S..6*. U 6.0±1.2 	1.19± .05 .15± .05 1.23	15 .46 .22 .08	 14.8　±.3 14.08			14.89±.1 348± 8 .73	7036± 11 7171 7000
0515.9　-0656 208.17 -24.38 313.86 -64.21 0513.5　-0700	 PGC 17003	.SBS8*. E　　(1) 8.0±1.3 7.5±1.6	1.07± .09 .06± .08 1.13	140 .60 .08 .03					
0516.1　+6815 143.67 16.88 17.31　-4.15 0510.8　+6812	 UGC　3268 PGC 17011	.S?.... 	.98± .09 .17± .06 1.03	160 .50 .26 .09				15.00±.1 231± 8 	3934± 11 4061 3899
051609.1-540617 261.75 -35.46 227.19 -47.06 051506.1-540930	 ESO　159-　3 PGC 17012	.LB.0*P S -2.0± .8 	1.32± .04 .20± .04 .59± .04 1.33	72 .36 .00 	14.01 ±.15 13.97 ±.14 13.57	.79± .02 .23± .03 .65 .17	.85± .01 .32± .02 12.45± .13 15.00± .28		3902± 34 3710 3931
051611.4-000858 201.69 -21.13 326.99 -61.07 051337.9-001214	 UGC　3271 MK　1095 PGC 17013	.S?.... 	1.04± .06 .13± .05 1.08	175 .43 .18 .07	14.1 v±.2 14.35 ±.12 13.56 13.64	.47± .03 -.75± .03 .28 -.89	.43± .03 -.78± .03 9.20± .15 13.81± .40		9929± 28 9839 9927
0516.2　+0606 195.96 -17.98 336.58 -57.23 0513.5　+0603	 UGC　3272 IRAS05135+0603 PGC 17014	.S?.... 	1.00± .08 .55± .06 1.05	121 .55 .82 .27	 13.22				9285 9214 9279
0516.3　-1328 214.77 -27.07 298.86 -66.04 0514.0　-1331	 MCG -2-14- 10 IRAS05140-1331 PGC 17015	.S?.... 	1.08± .09 .61± .07 1.11	 .32 .91 .30	 12.61				3475 3347 3482
051628.7-451230 250.84 -35.17 236.07 -53.96 051501.1-451542	 ESO　252- 15 PGC 17021	.S?.... 	.94± .06 .10± .05 .72± .02 .94	0 .02 .15 .05	15.04 ±.15 15.28 ±.14 14.94	.61± .02 .52 	.56± .02 14.13± .05 14.36± .35		10391± 40 10205 10416
051635.8+062605 195.71 -17.73 337.13 -57.08 051354.7+062251	A　0513+06 UGC　3274 VV　161 PGC 17025	.S?.... 	1.08± .09 .32± .07 1.11	 .55 .24 					8100± 59 8030 8094
051638.9-234717 225.79 -30.75 273.32 -65.53 051434.1-235030	 ESO　486- 44 IRAS05145-2350 PGC 17026	.SB?... 	.95± .06 .27± .05 .95	19 .02 .28 .14	 15.29 ±.14 12.46 14.88				9264± 52 9112 9278
051639.2-370600 241.08 -34.08 247.02 -59.48 051455.0-370912	 ESO　362- 11 IRAS05149-3709 PGC 17027	.S..4*/ S 4.0± .8 	1.65± .03 .78± .05 1.66	76 .10 1.14 .39	 13.05 ±.14 11.74 11.80			12.77±.2 287± 16 281± 12 .58	1348± 11 1367± 52 1173 1370
051646.5+063720 195.57 -17.59 337.44 -56.98 051405.1+063406	 UGC　3275 PGC 17031	.S?.... 	1.24± .05 .75± .05 1.29	37 .53 1.12 .37				15.51±.2 563± 13 516± 6 	7972± 7 7902 7966
051708.1-645741 274.92 -34.32 218.97 -38.03 051655.0-650048	NGC　1892 ESO　85- 61 IRAS05169-6500 PGC 17042	.S..6*. S 6.0±1.2 	1.46± .03 .55± .02 .95± .03 1.47	74 .13 .81 .27	12.83 ±.15 12.72 ±.14 12.20 11.83	.59± .02 -.14± .03 .45 -.24	.61± .01 -.09± .02 13.07± .07 13.60± .25	230± 9 	1362± 8 1168 1395

R.A. 2000 DEC.	Names	Type	logD$_{25}$	p.a.	B$_T$	(B-V)$_T$	(B-V)$_e$	m$_{21}$	V$_{21}$
l b		S$_T$ n$_L$	logR$_{25}$	A$_g$	m$_B$	(U-B)$_T$	(U-B)$_e$	W$_{20}$	V$_{opt}$
SGL SGB		T	logA$_e$	A$_i$	m$_{FIR}$	(B-V)$_T^o$	m'	W$_{50}$	V$_{GSR}$
R.A. 1950 DEC.	PGC	L	logD$_o$	A$_{21}$	B$_T^o$	(U-B)$_T^o$	m'$_{25}$	HI	V$_{3K}$
051708.1-540403		RLBR+*.	1.01± .05	60					10300±190
261.70 -35.32	ESO 159- 4	Sr	.13± .03	.36	14.51 ±.14				10108
227.05 -47.17	FAIR 791	-1.1± .5		.00					10329
051605.0-540712	PGC 17043		1.03		13.99				
051709.9+065545		.S..3..	1.11± .05	56				16.63±.3	8386± 9
195.35 -17.35	UGC 3279	U (1)	.74± .05	.67	15.16 ±.12				8317
337.97 -56.84	IRAS05144+0652	3.0± .9		1.02				627± 5	8381
051428.3+065233	PGC 17044	.5±1.3	1.18	.37	13.41			2.84	
051726.9-234444		.S?....	1.17± .05	100					9366
225.81 -30.56	ESO 486- 49		.55± .05	.01	14.67 ±.14				9214
273.30 -65.72	IRAS05153-2347			.67	13.13				9381
051522.0-234754	PGC 17049		1.17	.27	13.90				
051737.5+064800		.SB.6?.	.98± .04	25	14.51S±.15			15.72±.2	8257± 7
195.53 -17.32	UGC 3282	U	.12± .03	.67	15.24 ±.15			322± 13	8187
337.94 -57.01		6.0±1.7		.17				286± 6	8253
051455.9+064450	PGC 17053		1.05	.06	13.99		13.99± .26	1.67	
051742.6-153123	IC 407	.S..5*.	1.16± .05	165					2828
217.04 -27.58	MCG -3-14- 13	E (1)	.68± .05	.25					2695
293.83 -66.58	IRAS05154-1534	5.0±1.3		1.01	13.28				2838
051527.1-153432	PGC 17056	4.2±1.2	1.18	.34					
0517.7 +5333		.S..9..	1.29± .06	45				13.07±.1	616± 6
156.43 9.01	UGC 3273	U	.36± .06	1.22	14.8 ±.3			196± 8	
11.44 -17.66		9.0± .8		.37				185± 12	701
0513.7 +5330	PGC 17057		1.40	.18	13.17			-.27	587
0517.7 -0111		.S..6*.	1.16± .07	105					
202.87 -21.29	UGC 3283	U	.88± .06	.74					
325.62 -61.97	IRAS05152-0114	6.0±1.3		1.30					
0515.2 -0114	PGC 17058		1.23	.44					
0517.7 +0013		.L...?.	.59?	40					9015± 67
201.56 -20.61		E	.06?	.52					8925
328.07 -61.19		-2.0±1.8		.00					9015
0515.2 +0009	PGC 17059		.64						
0517.7 +0013		.L...?.	.59?	40					
201.55 -20.61		E	.06?	.52					
328.08 -61.19		-2.0±1.8		.00					
0515.2 +0010	PGC 17060		.64						
051748.0+302747		.L?....	1.00± .10	160				16.52±.3	6268± 7
175.41 -4.22	UGC 3280		.66± .04	5.29					6277
359.60 -38.24				.00				199± 7	6250
051435.7+302437	PGC 17062		1.41						
051800.1-642809		.S..5*/	1.18± .05	176					
274.31 -34.30	ESO 85- 65	S	.72± .04	.06	15.78 ±.14				
219.18 -38.51		5.0±1.3		1.08					
051744.1-643112	PGC 17065		1.18	.36					
051800.6-335447		.LA.-*.	1.18± .04	108	15.0 ±.2	1.03± .03	1.01± .02		
237.39 -33.18	ESO 362- 13	S	.65± .03	.00	14.96 ±.14	.33± .04	.31± .04		11060
251.98 -61.58		-3.0± .8	.58± .08	.00		.90	13.39± .28		10889
051611.0-335754	PGC 17066		1.09		14.81	.36	14.16± .32		11081
051850.0-862309		.SAT6..	1.11± .04	175					
299.15 -28.50	ESO 4- 19	S (1)	.12± .04	.72	15.16 ±.14				
207.72 -19.00	IRAS05331-8625	6.3± .5		.18					
053349.0-862536	PGC 17077	7.2± .6	1.18	.06					
051854.7+191031		.E?....	1.08± .17					15.38±.3	5967± 7
184.95 -10.40	UGC 3285		.12± .08	2.24					5937
351.58 -47.86				.00				417± 7	5957
051558.2+190726	PGC 17079		1.37						
051901.1-370516	IC 2122	.LXS-*.	1.18± .05						4663± 39
241.16 -33.61	ESO 362- 14	BS	.06± .03	.10	13.83 ±.14				4487
246.45 -59.86		-2.7± .4		.00					4687
051717.1-370818	PGC 17081		1.18		13.66				
051901.8-213239	A 0516-21	.IBS9..	1.16± .04	15	15.6 ±.2	.86± .33		14.70±.2	1841± 11
223.55 -29.49	ESO 553- 33	SU (2)	.16± .04	.11	15.10 ±.14	-.88± .31		136± 16	
278.48 -66.55	DDO 37	10.0± .6		.12		.79		132± 12	1693
051654.1-213542	PGC 17082	9.8± .7	1.17	.08	15.03	-.95	15.87± .30	-.41	1857
0519.1 +6528		.S..6*.	1.28± .04	156				15.91±.1	5116± 7
146.31 15.67	UGC 3277	U	.46± .05	.73	14.9 ±.3				5235
16.48 -6.83		6.0±1.1		.67				347± 8	
0514.2 +6525	PGC 17084		1.35	.23	13.52			2.16	5084

R.A. 2000 DEC. l b SGL SGB R.A. 1950 DEC.	Names PGC	Type S_T n_L T L	$\log D_{25}$ $\log R_{25}$ $\log A_e$ $\log D_o$	p.a. A_g A_i A_{21}	B_T m_B m_{FIR} B_T^o	$(B-V)_T$ $(U-B)_T$ $(B-V)_T^o$ $(U-B)_T^o$	$(B-V)_e$ $(U-B)_e$ m'_e m'_{25}	m_{21} W_{20} W_{50} HI	V_{21} V_{opt} V_{GSR} V_{3K}
051918.7-613944 270.92 -34.51 220.88 -40.99 051847.0-614242	 ESO 119- 52 IRAS05188-6142 PGC 17092	.SBS4*. S (1) 4.0± .6 4.1± .7	1.13± .04 .24± .04 1.14	84 .08 .35 .12	 14.58 ±.14 				
051919.1+012006 200.72 -19.72 330.40 -60.86 051643.8+011703	 UGC 3287 IRAS05167+0116 PGC 17094	.SBS6P* UE 6.0± .9 1.00	.94± .07 .09± .05 .05	145 .63 .14 	 12.82 			15.46±.3 146± 10 	8186± 10 8131± 46 8096 8186
051921.4+165230 186.96 -11.58 349.65 -49.78 051627.7+164927	 UGC 3286 IRAS05164+1649 PGC 17095	.SB.4.. U 4.0± .9 1.21	1.04± .08 .47± .06 .24	57 1.82 .69 	 13.17 			15.37±.3 375± 7 	6959± 10 6921 6951
051930.9+040728 198.19 -18.29 334.79 -59.15 051652.4+040426	 UGC 3288 PGC 17100	.S..8*. U 8.0±1.3 1.05	1.00± .08 .19± .06 .09	 .52 .23 				15.97±.3 154± 7 	3061± 7 2982 3061
051934.4-774351 289.70 -31.29 211.67 -26.85 052204.1-774642	NGC 1956 ESO 16- 2 PGC 17102	.SAS1.. S .6± .4 1.34	1.28± .04 .36± .05 .18	68 .58 .36 	 14.05 ±.14 13.04				4844± 34 4656 4881
051935.6-323930 236.02 -32.57 253.86 -62.54 051744.1-324230	 ESO 362- 18 IRAS05177-3242 PGC 17103	.SBS0*P S -.3± .7 1.07	1.08± .05 .19± .04 	160 .00 .14 	 13.81 ±.14 12.83 13.61				3773± 20 3603 3796
051936.4+840312 128.96 24.68 23.75 10.26 050637.8+835948	 UGC 3253 IRAS05066+8359 PGC 17104	.SBR3.. U 3.0± .8 1.25	1.22± .05 .20± .05 .10	93 .31 .28 	 13.21 ±.19 13.38 12.59			15.13±.1 335± 8 2.44	4130± 7 4292 4094
051938.3-245336 227.26 -30.45 270.24 -65.87 051735.1-245636	 ESO 486- 52 PGC 17106	.SAR1.. r 1.0± .9 .95	.95± .06 .08± .05 .04	 .00 .08 	 14.99 ±.14 				
051939.3-614421 271.00 -34.47 220.78 -40.95 051908.1-614718	 ESO 119- 53 FAIR 244 PGC 17108	PSBT1*. Sr .7± .4 1.14	1.14± .04 .12± .04 .06	5 .08 .12 	 14.37 ±.14 14.11				4680± 63 4485 4714
051945.0-250354 227.46 -30.48 269.81 -65.84 051742.0-250654	IC 2121 ESO 486- 53 PGC 17110	.LXS0*P S -2.3± .6 1.24	1.27± .05 .22± .04 	160 .00 .00 	 13.81 ±.14 13.65				10344± 52 10188 10363
051948.4-320831 235.44 -32.41 254.77 -62.85 051756.0-321130	NGC 1879 ESO 423- 6 DDO 232 PGC 17113	.SBS9.. SU (2) 8.7± .4 7.6± .5	1.39± .02 .16± .04 1.05± .01 1.39	60 .00 .16 .08	13.16 ±.13 13.14 ±.14 13.87 12.99	.40± .03 -.14± .04 .36 -.17	.48± .02 -.15± .03 13.90± .03 14.59± .20	14.50±.2 138± 16 126± 12 1.43	1247± 11 1078 1270
051950.2-454650 251.60 -34.63 234.67 -53.93 051824.1-454948	A 0518-45 ESO 252- 18A PGC 17116	.L?.... 	.91± .09 .11± .04 .91	 .10 .00 	16.2 ±.4 15.95	.85± .08 -.39± .18 .72 -.36	 15.35± .61		10495± 42 10308 10524
052000.4-535616 261.53 -34.90 226.66 -47.52 051857.0-535912	 ESO 159- 6 FAIR 792 PGC 17122	PSBR1.. Sr 1.0± .6 1.03	1.00± .05 .12± .05 .06	116 .37 .13 	 15.19 ±.14 14.56				11200±190 11007 11232
052006.7+063456 196.05 -16.91 338.43 -57.61 051725.4+063157	 UGC 3289 IRAS05174+0631 PGC 17125	.SB.3.. U 3.0± .9 1.06	.99± .06 .50± .05 .25	50 .65 .69 	 15.39 ±.12 13.98			16.19±.3 308± 7 1.96	8893± 10 8821 8893
0520.1 +0549 196.74 -17.28 337.43 -58.14 0517.5 +0547	 CGCG 421- 30 PGC 17126	 	.92± .05 .35± .04 .99	 .73 	15.27S±.15 15.46 ±.12 	.78± .03 .16± .04 		15.64±.3 464± 13 414± 10 	8570± 10 8496 8570
0520.1 +0554 196.66 -17.23 337.54 -58.08 0517.5 +0552	 CGCG 421- 31 PGC 17127	 			 .73 	15.1 ±.3 			8644 8570 8644
052017.6-251927 227.79 -30.44 269.11 -65.88 051815.0-252224	IC 411 ESO 486- 56 PGC 17130	.L?.... 1.05	1.07± .06 .14± .04 	141 .00 .00 	 14.12 ±.14 13.98				9549± 52 9392 9569

R.A. 2000 DEC. l b SGL SGB R.A. 1950 DEC.	Names PGC	Type S_T n_L T L	$\log D_{25}$ $\log R_{25}$ $\log A_e$ $\log D_o$	p.a. A_g A_i A_{21}	B_T m_B m_{FIR} B_T^o	$(B-V)_T$ $(U-B)_T$ $(B-V)_T^o$ $(U-B)_T^o$	$(B-V)_e$ $(U-B)_e$ m'_e m'_{25}	m_{21} W_{20} W_{50} HI	V_{21} V_{opt} V_{GSR} V_{3K}
052019.6-611536 270.42 -34.44 221.02 -41.40 051946.0-611830	 ESO 119- 54 IRAS05197-6118 PGC 17131	PSBROP. Sr -.4± .5 	1.13± .05 .21± .05 1.13	173 .07 .16 	 14.33 ±.14 14.03				 5157 4962 5191
052019.9+174322 186.37 -10.92 350.71 -49.24 051725.2+174023	 UGC 3290 PGC 17132	.S..6*. U 6.0±1.3 	1.04± .08 .37± .06 1.22	23 1.93 .55 .19				 15.70±.3 282± 7 	 6287± 7 6252 6280
052021.5-611748 270.46 -34.43 220.99 -41.37 051948.0-612042	 ESO 119- 55 PGC 17134	.SAS1*P S 1.0± .6 	1.26± .04 .25± .05 1.27	175 .07 .25 .12	 13.72 ±.14 13.34				 4784 4589 4818
052026.2+063417 196.11 -16.84 338.51 -57.68 051744.9+063119	 UGC 3291 PGC 17136	.SB.6*. U 6.0±1.3 	.99± .04 .39± .03 1.06	94 .77 .58 .20	15.79S±.15 15.18 ±.12 14.03	.63± .05 -.06± .08 .32 -.28	 14.59± .26	17.29±.2 141± 13 71± 6 3.07	8878± 7 8806 8878
0520.6 +7232 140.00 19.38 19.33 -.37 0514.6 +7229	 UGC 3281 PGC 17140	.L..... U -2.0± .8 	1.20± .07 .19± .04 1.22	150 .45 .00 	 14.53 ±.20 				
0520.6 +0848 194.15 -15.63 341.46 -56.14 0517.9 +0845	 UGC 3293 IRAS05179+0845 PGC 17143	.S..6*. U 6.0±1.2 	1.22± .06 .30± .06 1.37	140 1.63 .45 .15					4689± 10 4624 4688
0520.8 +6614 145.71 16.21 16.94 -6.18 0515.7 +6611	 CGCG 307- 10 7ZW 35 PGC 17146	 	 	 .54 	 14.8 ±.3 12.97 				13294± 82 13414 13262
0520.8 +0314 199.16 -18.44 333.89 -59.97 0518.2 +0312	 MCG 1-14- 27 PGC 17147	.L?.... 	.72± .22 .00± .07 .78	 .53 .00 	 15.05 ±.11 14.41				 8270± 79 8187 8273
0520.9 +0314 199.18 -18.43 333.91 -59.99 0518.3 +0311	 MCG 1-14- 28 IRAS05183+0311 PGC 17152	.S?.... 	.74± .14 .21± .07 .79	95 .53 .26 .10	 15.30 ±.14 14.50				 8205± 79 8122 8208
052103.7+040023 198.50 -18.01 335.10 -59.52 051825.3+035728	 UGC 3294 (1) IRAS05184+0357 PGC 17156	.SAT3.. U 3.0± .7 3.5±1.0	1.47± .03 .27± .05 1.53	133 .56 .38 .14	 13.8 ±.2 13.19 12.85			14.03±.1 399± 4 374± 4 1.04	4145± 4 4065 4148
052104.2-365725 241.10 -33.19 246.13 -60.26 051920.0-370018	 ESO 362- 19 (1) IRAS05193-3659 PGC 17157	.SBS9.. S 8.5± .6 8.3± .6	1.35± .04 .52± .04 1.36	3 .10 .53 .26	 14.13 ±.14 13.49			14.80±.2 141± 16 135± 12 1.04	1303± 11 1126 1330
0521.1 +7621 136.45 21.25 20.82 3.14 0514.1 +7618	 UGC 3276 (1) PGC 17159	.S..3.. U 3.0± .9 3.5±1.2	1.15± .08 .77± .07 1.20	 .52 1.06 .38	 15.60 ±.18 14.00			14.47±.1 288± 8 .09	2503± 11 2648 2469
052115.6-610334 270.16 -34.34 221.03 -41.63 052041.0-610624	 ESO 119- 58 PGC 17161	.SBROP* S .0± .7 	1.18± .04 .42± .04 1.17	116 .06 .32 	 14.50 ±.14 14.05				 5177 4982 5212
0521.2 +7243 139.86 19.51 19.44 -.22 0515.2 +7240	 UGC 3284 PGC 17162	.SXS5.. U 5.0± .8 	1.19± .06 .19± .06 1.23	125 .45 .28 .09				15.54±.1 199± 8 	4700± 7 4837 4666
052122.1+045312 197.75 -17.50 336.48 -58.99 051842.7+045018	 UGC 3296 IRAS05187+0450 PGC 17164	.S..2.. U 2.0± .8 	1.18± .05 .13± .05 1.24	140 .58 .16 .07	 13.95 ±.13 12.48 13.17			14.82±.2 343± 7 318± 5 1.59	4266± 6 4287± 38 4189 4270
052124.5+151431 188.64 -12.05 348.72 -51.37 051832.8+151137	 UGC 3295 PGC 17165	.S..6?. U 6.0±1.8 	1.00± .08 .33± .06 1.21	107 2.20 .49 .17				14.92±.3 389± 7 	5603± 7 5559 5599
052124.8-165233 218.83 -27.28 290.42 -67.52 051911.0-165526	 PGC 17166	.IBS9*. S (1) 10.0± .9 6.4± .8	1.22± .05 .46± .06 1.24	160 .25 .35 .23					3282 3144 3299

R.A. 2000 DEC. / l b / SGL SGB / R.A. 1950 DEC.	Names / PGC	Type / S_T n_L / T / L	$\log D_{25}$ / $\log R_{25}$ / $\log A_e$ / $\log D_o$	p.a. / A_g / A_i / A_{21}	B_T / m_B / m_{FIR} / B_T^o	$(B-V)_T$ / $(U-B)_T$ / $(B-V)_T^o$ / $(U-B)_T^o$	$(B-V)_e$ / $(U-B)_e$ / m'_e / m'_{25}	m_{21} / W_{20} / W_{50} / HI	V_{21} / V_{opt} / V_{GSR} / V_{3K}
0521.6 +8428 / 128.56 24.91 / 23.97 10.63 / 0507.8 +8425	A 0508+84 / UGC 3257 / IRAS05078+8425 / PGC 17170	.SB.1.. / U / 1.0± .8 /	1.15± .05 / .32± .05 / / 1.18	157 / .25 / .33 / .16	/ 14.72 ±.19 / 13.44 /				
052146.0+064120 / 196.18 -16.50 / 339.10 -57.83 / 051904.5+063828	NGC 1875 / MCG 1-14-31 / HICK 34A / PGC 17171	.L?.... / / /	1.20± .11 / .55± .07 / / 1.21	/ .77 / .00 /	14.57S±.15 / / 13.66 /		/ / / 14.06± .59		/ 8997± 41 / 8925 / 8999
052148.7-234845 / 226.26 -29.63 / 272.47 -66.66 / 051944.0-235136	NGC 1886 / ESO 487- 2 / IRAS05197-2351 / PGC 17174	.S..4./ / S / 3.5± .6 /	1.49± .03 / .85± .03 / / 1.49	60 / .04 / 1.25 / .42	13.62 ±.14 / 12.13 /				
052150.0+064036 / 196.20 -16.49 / 339.10 -57.85 / 051908.6+063744	/ MCG 1-14-32 / HICK 34B / PGC 17176	.LA.-*. / R / -3.0± .7 /		/ .84 / /					/ 9514± 35 / 9441 / 9516
052151.2+524953 / 157.39 9.12 / 11.71 -18.57 / 051750.8+524658	/ UGC 3292 / / PGC 17178	.S?.... / / /	1.00± .08 / .73± .06 / / 1.11	130 / 1.18 / 1.09 / .36				15.88±.1 / / 459± 8 /	10092± 11 / 10173 / 10067 /
052156.4+032909 / 199.09 -18.09 / 334.61 -60.03 / 051918.6+032618	IC 412 / UGC 3298 / VV 225 / PGC 17180	.S?.... / / /	1.00± .08 / .16± .06 / / 1.04	/ .51 / .12 /	/ 14.56 ±.12 / 13.87 /				/ 4311± 46 / 4229 / 4316
052158.7+032856 / 199.10 -18.08 / 334.61 -60.04 / 051920.9+032605	IC 413 / UGC 3299 / / PGC 17181	.S?.... / / /	.96± .09 / .09± .06 / / 1.00	/ .51 / .10 / .05	/ 14.66 ±.12 / / 13.99				/ 4333± 46 / 4250 / 4337
052234.2-795107 / 292.02 -30.54 / 210.52 -24.99 / 052610.0-795342	NGC 2012 / ESO 16- 5 / / PGC 17194	.LA.-*. / S / -3.0± .7 /	1.05± .04 / .24± .03 / .86± .03 / 1.08	117 / .48 / .00 /	13.98 ±.14 / 14.49 ±.14 / / 13.68	1.05± .02 / / .89 /	1.13± .02 / / 13.77± .11 / 13.52± .24		/ 4862± 14 / 4676 / 4900
052234.7-112958 / 213.47 -24.87 / 304.49 -67.16 / 052014.4-113246	NGC 1888 / MCG -2-14- 13 / IRAS05202-1132 / PGC 17195	.SBS5P. / R / 5.0± .4 /	1.48± .02 / .57± .04 / .95± .01 / 1.53	145 / .48 / .85 / .28	12.83 ±.13 / / 11.43 / 11.49	.92± .02 / .30± .03 / .69 / .09	1.00± .01 / .44± .03 / 13.07± .04 / 13.67± .19	14.69±.1 / 470± 9 / 456± 7 / 2.92	2432± 5 / 2547± 51 / 2308 / 2449
052235.4-112949 / 213.47 -24.87 / 304.50 -67.16 / 052015.0-113237	NGC 1889 / MCG -2-14- 14 / / PGC 17196	.E+..P* / RE / -4.0± .5 /	.85± .08 / .14± .06 / / .88	165 / .48 / .00 /					/ 2482± 26 / 2356 / 2497
0522.7 +0341 / 199.01 -17.81 / 335.18 -60.04 / 0520.1 +0339	/ CGCG 421- 43 / / PGC 17203			/ .51 / /	15.2 ±.3 /				11025 / 10943 / 11031 /
0522.9 -0008 / 202.57 -19.64 / 329.03 -62.46 / 0520.4 -0011	/ UGC 3301 / IRAS05204-0011 / PGC 17208	.SB.0.. / U / .0± .8 /	.95± .09 / .31± .06 / / 1.01	20 / .81 / .23 /	15.24 ±.12 / 13.84 /				
052314.9-112528 / 213.47 -24.69 / 304.78 -67.30 / 052054.5-112813	/ MCG -2-14- 15 / / PGC 17217	.L..0*/ / E / -2.0±1.3 /	1.20± .05 / .82± .05 / / 1.13	51 / .48 / .00 /					2496 / / 2370 / 2512
0523.4 +4333 / 165.29 4.17 / 7.89 -27.14 / 0519.8 +4330	/ UGC 3300 / IRAS05198+4330 / PGC 17221	.S..6?. / U / 6.0±1.7 /	1.00± .16 / .09± .12 / / 1.24	/ 2.59 / .13 / .04	/ / 12.68 /			14.93±.1 / / 291± 8 /	6318± 7 / 6368 / 6299 /
052334.7-694522 / 280.47 -32.89 / 215.46 -34.12 / 052400.0-694800	LMC / ESO 56-115 / / PGC 17223	.SBS9.. / R (2) / 9.0± .3 / 5.8± .5	3.81± .05 / .07± .04 / / 3.84	170 / .27 / .07 / .04	.91S±.05 / / .74 / .57	.51± .03 / .00± .05 / .43 / -.06	/ / / 14.64± .27	2.75±.2 / / / 2.15	324± 10 / 277± 19 / 119 / 351
052353.4-495420 / 256.64 -34.23 / 229.53 -51.15 / 052238.0-495700	/ ESO 204- 4 / FAIR 306 / PGC 17227	.S?.... / / /	.92± .06 / .16± .05 / / .92	160 / .00 / .16 / .08	/ 15.34 ±.14 / 13.29 / 15.05				/ 10050±190 / 9858 / 10085
052356.5-171539 / 219.49 -26.87 / 289.39 -68.12 / 052143.2-171821	IC 416 / MCG -3-14- 14 / IRAS05217-1718 / PGC 17229	.SBS5P* / SE (2) / 5.0± .6 / 3.6± .7	1.15± .05 / .29± .04 / / 1.16	70 / .16 / .43 / .14	/ / 13.06 /				

R.A. 2000 DEC.　l　b　SGL　SGB　R.A. 1950 DEC.	Names　PGC	Type　S_T　n_L　T　L	$\log D_{25}$　$\log R_{25}$　$\log A_e$　$\log D_o$	p.a.　A_g　A_i　A_{21}	B_T　m_B　m_{FIR}　B_T^o	$(B-V)_T$　$(U-B)_T$　$(B-V)_T^o$　$(U-B)_T^o$	$(B-V)_e$　$(U-B)_e$　m'_e　m'_{25}	m_{21}　W_{20}　W_{50}　HI	V_{21}　V_{opt}　V_{GSR}　V_{3K}
0524.0 -0519 / 207.59 -21.84 / 319.29 -65.38 / 0521.6 -0522	PGC 17232	.SBT7*. / E (1) / 7.0±1.3 / 5.3±1.6	1.07± .09 / .03± .08 / / 1.16	20 / .87 / .03 / .01					
052420.6-464446 / 252.84 -33.94 / 232.60 -53.70 / 052257.0-464724	ESO 253- 1 / PGC 17237	.SBT6.. / Sr (1) / 6.3± .6 / 6.7± .8	1.12± .05 / .11± .05 / / 1.13	38 / .10 / .16 / .06	14.65 ±.14				
052431.8-612154 / 270.49 -33.92 / 220.37 -41.57 / 052359.0-612430	ESO 120- 1 / PGC 17239	.SBR6?. / r / 5.6±1.8	.95± .07 / .18± .05 / / .95	12 / .09 / .26 / .09	15.46 ±.14				
052436.3-460241 / 252.01 -33.83 / 233.30 -54.27 / 052311.0-460518	ESO 253- 2 / PGC 17241	.SBS9.. / S (1) / 9.0± .8 / 10.0± .8	1.07± .05 / .17± .05 / / 1.08	/ .10 / .17 / .08	15.53 ±.14				
052450.2-482718 / 254.91 -33.99 / 230.74 -52.40 / 052331.0-482954	ESO 204- 6 / FAIR 794 / PGC 17246	PSBR1*. / Sr / 1.0± .6	1.03± .05 / .35± .04 / / 1.03	65 / .07 / .35 / .17	15.41 ±.14 / / / 14.81				13600±190 / 13408 / 13635
052454.7-124122 / 214.92 -24.85 / 301.74 -67.97 / 052235.8-124400	MCG -2-14- 16 / PGC 17248	.S..2./ / E / 2.0± .9	1.23± .05 / .75± .05 / / 1.27	68 / .44 / .92 / .38					
0524.9 +0429 / 198.57 -16.93 / 337.09 -59.92 / 0522.3 +0427	UGC 3303 / PGC 17250	.I..9?. / U / 10.0±1.3	1.56± .04 / .12± .06 / / 1.62	/ .62 / .09 / .06				13.59±.1 / 171± 5 / 163± 4	522± 5 / / 441 / 530
052505.9-483513 / 255.07 -33.96 / 230.55 -52.32 / 052347.1-483748	ESO 204- 7 / FAIR 795 / PGC 17254	.SBS3?/ / S / 3.0±1.3	1.13± .05 / .84± .04 / / 1.14	94 / .07 / 1.16 / .42	15.91 ±.14 / / / 14.62				7400±190 / 7208 / 7435
0525.2 +0024 / 202.35 -18.88 / 330.72 -62.60 / 0522.7 +0022	UGC 3306 / PGC 17259	.L..-*. / U / -3.0±1.2	1.08± .17 / .29± .08 / / 1.15	52 / .85 / .00	14.63 ±.10				
052526.3-513839 / 258.76 -34.05 / 227.62 -49.87 / 052416.1-514112	ESO 204- 8 / PGC 17264	.SBS3.. / S / 3.0± .9	1.03± .05 / .38± .04 / / 1.03	55 / .00 / .53 / .19	15.94 ±.14				
052529.0+215117 / 183.55 -7.63 / 355.61 -46.50 / 052228.9+214840	UGC 3304 / PGC 17266	.S?....	1.07± .14 / .00± .12 / / 1.30	2.40 / .00 / .00				14.95±.3 / 293± 7	5623± 7 / 5599 / 5621
052547.5-395427 / 244.76 -32.80 / 240.54 -58.97 / 052409.0-395700	ESO 305- 25 / IRAS05241-3956 / PGC 17274	.SAS4*. / Sr (1) / 3.7± .6 / 4.4± .6	1.07± .05 / .20± .04 / / 1.07	87 / .00 / .30 / .10	14.27 ±.14 / 11.77 / 13.94				4553 / 4370 / 4587
052556.6-464347 / 252.85 -33.67 / 232.26 -53.88 / 052433.1-464618	NGC 1930 / ESO 253- 4 / PGC 17276	.LXS+*. / S / -1.5± .6	1.27± .05 / .18± .04 / .73± .01 / 1.26	32 / .11 / .00	13.37 ±.13 / 13.39 ±.14 / / 13.21	1.00± .01 / .44± .04 / .92 / .42	1.01± .01 / .48± .02 / 12.51± .04 / 14.15± .29		4260± 23 / 4069 / 4296
0526.0 +0857 / 194.73 -14.42 / 343.39 -56.93 / 0523.3 +0855	UGC 3308 / PGC 17281	.S..6*. / U / 6.0±1.2	.96± .09 / .00± .06 / / 1.11	1.60 / .00 / .00					8517± 10 / / 8450 / 8524
052627.0-211711 / 223.95 -27.78 / 278.30 -68.30 / 052419.1-211942	ESO 553- 43 / PGC 17287	.L...+P / S / -1.0± .8	1.03± .05 / .08± .03 / / 1.03	/ .14 / .00	14.84 ±.14				
052634.7-315036 / 235.52 -30.94 / 253.58 -64.22 / 052442.0-315306	ESO 423- 16 / PGC 17290	RSBS0.. / r / -.1± .9	.97± .07 / .06± .06 / / .97	/ .00 / .04	14.87 ±.14 / / / 14.65				11753± 52 / 11581 / 11784
052644.7-191242 / 221.80 -26.97 / 283.93 -68.68 / 052434.0-191512	ESO 553- 44 / IRAS05245-1915 / PGC 17294	.S..5*/ / SE / 5.0± .9 / 2.2±1.3	1.19± .03 / .75± .03 / / 1.20	61 / .16 / 1.13 / .38	14.81 ±.14 / 13.89 / 13.47				8336 / 8189 / 8362

R.A. 2000 DEC. l　　b SGL　SGB R.A. 1950 DEC.	Names PGC	Type S_T　n_L T L	$\log D_{25}$ $\log R_{25}$ $\log A_e$ $\log D_o$	p.a. A_g A_i A_{21}	B_T m_B m_{FIR} B_T^o	$(B-V)_T$ $(U-B)_T$ $(B-V)_T^o$ $(U-B)_T^o$	$(B-V)_e$ $(U-B)_e$ m'_e m'_{25}	m_{21} W_{20} W_{50} HI	V_{21} V_{opt} V_{GSR} V_{3K}
052647.5-634541 273.32 -33.44 218.55 -39.59 052628.0-634806	NGC 1947 ESO 85- 87 IRAS05264-6347 PGC 17296	.L..-P. R -3.0± .3	1.48± .03 .07± .02 1.03± .01 1.49	119 .16 .00	11.65M±.07 11.76 ±.11 12.59 11.50	1.01± .02 .50± .03 .96 .46	1.04± .01 .51± .01 12.47± .04 13.75± .16		1157± 25 961 1197
052705.8+454036 163.88 5.88 9.50 -25.47 052324.5+453805	 MCG 8-10- 3 PGC 17303	.I?.... 	.64± .17 .00± .07 .81	 1.83 .00 .00				15.50±.3 124± 11 91± 8	6087± 9 6143 6071
052709.5-631430 272.70 -33.45 218.82 -40.07 052647.0-631654	 ESO 85- 88 PGC 17305	.IBS9.. S　(1) 10.0± .6 10.0± .6	1.20± .07 .29± .05 1.21	130 .16 .22 .14	16.90 ±.14				
0527.9 +6352 148.22 15.65 16.74 -8.66 0523.1 +6350	 UGC 3307 PGC 17317	.E..... U -5.0± .9	.96± .12 .12± .05 1.06	85 .85 .00	14.6 ±.3				
052802.2-051839 208.07 -20.95 320.41 -66.26 052534.6-052103	NGC 1924 MCG -1-14- 11 IRAS05255-0521 PGC 17319	.SBR4.. E　(1) 4.0± .8 3.6± .8	1.20± .03 .12± .05 .80± .01 1.27	50 .74 .18 .06	13.25 ±.13 11.81 12.31	.73± .01 .52	.81± .01 12.74± .03 13.79± .25		2538 2426 2558
0528.2 +7639 136.36 21.76 21.31 3.27 0521.0 +7637	A 0521+76 UGC 3302 PGC 17322	.SXR4*. U　(1) 4.0± .8 2.7± .8	1.02± .10 .00± .07 1.00± .03 1.06	 .49 .00 .00	13.59 ±.17 14.68 ±.20 13.54	.73± .04 .59	.81± .02 14.08± .08 13.52± .53	15.71±.1 227± 11 219± 7 2.18	4174± 7 4173± 59 4319 4142
0528.2 -1607 218.76 -25.49 292.56 -69.14 0525.9 -1609	A 0526-16 MCG -3-14- 17 IRAS05259-1609 PGC 17323	.SBS6.. E　(2) 6.0± .9 3.5± .6	1.27± .05 .00± .05 1.29	 .24 .00 .00	13.5 ±.4 12.87 13.25	.49± .05 .42	 14.71± .48	14.31±.1 95± 15 1.06	2174± 10 2181± 59 2034 2200
052836.2-412928 246.72 -32.54 237.68 -58.19 052701.0-413148	 ESO 306- 2 PGC 17329	.S..3*/ S 3.0± .9	1.13± .04 .74± .04 1.13	38 .01 1.02 .37	15.81 ±.14				
052842.1-565605 265.13 -33.62 222.79 -45.67 052750.0-565824	 ESO 159- 12 PGC 17331	RLXR0*. Sr -1.5± .7	.99± .05 .14± .03 .99	175 .16 .00	14.71 ±.14				
052907.8-195605 222.78 -26.71 281.72 -69.14 052658.0-195824	 ESO 554- 2 PGC 17340	.SBS6.. SE　(1) 6.3± .4 6.1± .6	1.17± .03 .12± .03 1.18	165 .10 .17 .06	14.12 ±.14				
052908.4-392518 244.33 -32.08 240.31 -59.77 052729.1-392736	 ESO 306- 3 IRAS05274-3927 PGC 17341	RSBS4?. S　(1) 4.0± .6 3.3±1.2	1.09± .05 .31± .04 1.09	15 .00 .46 .16	14.44 ±.14 13.74				
052913.7-421236 247.58 -32.53 236.61 -57.73 052740.0-421454	 ESO 306- 4 VV 599 PGC 17343	.E+4.P. S -4.0± .6	1.12± .05 .31± .04 1.02	18 .00 .00 14.41	14.76 ±.14				23075 22888 23113
0529.2 +6722 145.12 17.50 18.12 -5.44 0524.0 +6719	 UGC 3309 IRAS05240+6719 PGC 17344	.S?.... 	1.12± .05 .29± .05 1.16	10 .48 .43 .15	 15.0 ±.2 13.29 14.10			15.90±.1 355± 8 1.65	6051± 11 6173 6024
052944.4-534516 261.31 -33.46 225.02 -48.47 052841.0-534730	 ESO 159- 13· PGC 17353	.S..1?P S 1.0±1.8	1.03± .05 .43± .05 1.06	40 .37 .44 .21	15.25 ±.14				
0530.0 +5555 155.39 11.79 14.09 -16.17 0525.9 +5553	 UGC 3314 PGC 17359	.S?.... 	1.22± .12 .94± .12 1.39	120 1.80 1.38 .47				15.16±.1 193± 8	2183± 7 2271 2164
053018.2-565213 265.05 -33.40 222.56 -45.83 052926.1-565424	 ESO 159- 16 PGC 17365	.SB.5?P S 5.0±1.7	1.09± .08 .23± .07 1.10	 .16 .35 .12	18.52 ±.14				
053029.1-245235 228.13 -28.11 268.17 -68.20 052826.1-245448	 ESO 487- 17 PGC 17373	.IBS9.. S　(1) 10.0± .8 11.1± .8	1.30± .06 .12± .05 1.31	40 .07 .09 .06	16.24 ±.14 16.08				1845 1684 1879

R.A. 2000 DEC. / l b / SGL SGB / R.A. 1950 DEC.	Names / PGC	Type / S_T n_L / T / L	$\log D_{25}$ / $\log R_{25}$ / $\log A_e$ / $\log D_o$	p.a. / A_g / A_i / A_{21}	B_T / m_B / m_{FIR} / B_T^o	$(B-V)_T$ / $(U-B)_T$ / $(B-V)_T^o$ / $(U-B)_T^o$	$(B-V)_e$ / $(U-B)_e$ / m'_e / m'_{25}	m_{21} / W_{20} / W_{50} / HI	V_{21} / V_{opt} / V_{GSR} / V_{3K}
053034.3-450706		.L?....	1.05± .06	178					
251.04 -32.69	ESO 253- 8		.38± .03	.09	15.22 ±.14				10604± 24
232.88 -55.64	IRAS05291-4509			.00	13.64				10413
052907.1-450918	PGC 17374	1.00			14.97				10644
053040.1-332318		.SXT3P.	.90± .06						
237.51 -30.49	ESO 363- 3	r	.09± .05	.00	14.88 ±.14				
249.52 -64.02	IRAS05288-3325	3.3± .9		.12	13.36				
052850.1-332530	PGC 17375		.90	.05					
0530.9 +5551		.I..9?.	1.14± .13	100				14.60±.1	2201± 7
155.52 11.86	UGC 3316	U	.18± .12	1.74				204± 6	
14.18 -16.28		10.0±1.7		.13				185± 4	2289
0526.8 +5549	PGC 17378		1.30	.09					2182
053059.9-535251		.L...?/	1.13± .06	135					
261.47 -33.28	ESO 159- 17	S	.80± .03	.37	15.55 ±.14				
224.68 -48.46		-2.0±1.9		.00					
052957.0-535500	PGC 17381		1.05						
053140.2-102336		.SBR2?/	1.22± .04	155					
213.40 -22.39	MCG -2-15- 1	E	.51± .04	.61					
308.93 -69.03	IRAS05293-1025	2.0±1.2		.62	11.68				
052918.6-102545	PGC 17395		1.27	.25					
053140.7-420947		PSAR2P*	1.13± .04						
247.61 -32.08	ESO 306- 9	Sr	.06± .04	.02	14.06 ±.14				14508± 17
236.02 -58.06	FAIR 1135	1.5± .6		.08	13.46				14319
053007.0-421154	PGC 17396		1.13	.03	13.82				14549
053142.0-734459		.S..7./	1.31± .04	170					
284.96 -31.51	ESO 33- 22	S	1.19± .04	.46	15.56 ±.14				
212.78 -30.78		7.0±1.0		1.38					
053256.0-734700	PGC 17397		1.35	.50					
053150.4-230841 IC 2130		.SBS8..	1.26± .03	103					1829
226.40 -27.24	ESO 487- 19	S (1)	.31± .03	.09	13.84 ±.14				
272.44 -69.06		8.0± .8		.39					1671
052945.0-231048	PGC 17402	6.7± .8	1.27	.16	13.36				1864
053214.9-075501 IC 421		.SXT4..	1.51± .03	80				13.30±.1	3557± 11
211.06 -21.18	MCG -1-15- 1	UE (1)	.07± .05	.68				316± 16	
315.47 -68.32	IRAS05297-0757	4.0± .5		.10				282± 12	3436
052950.3-075707	PGC 17407	1.9± .8	1.58	.03					3585
053216.1-503420		.L..0P/	1.12± .05	5					
257.54 -32.93	ESO 204- 13	S	.43± .04	.02	15.55 ±.14				
227.16 -51.36		-2.0± .8		.00					
053103.0-503624	PGC 17408		1.05						
053221.3-455556		.S?....	1.18± .06	76					
252.05 -32.48	ESO 253- 12		.54± .05	.10	15.20 ±.14				3996
231.57 -55.18				.75					3804
053056.0-455800	PGC 17410		1.19	.27	14.32				4039
053228.6-135538 IC 2132		.S..1P*	1.17± .05	175					
216.98 -23.67	MCG -2-15- 2	E	.30± .05	.43					
299.13 -69.98	IRAS05301-1357	1.0±1.2		.31	11.85				
053011.4-135743	PGC 17415		1.21	.15					
053232.0-495409		.S..3?.	1.05± .05	75					
256.74 -32.84	ESO 204- 14	S (1)	.36± .04	.02	14.92 ±.14				
227.69 -51.94	IRAS05313-4956	3.0±1.8		.49					
053117.1-495612	PGC 17416	5.6± .9	1.05	.18					
053240.8-325739		.LBR+..	.94± .06	26					
237.15 -29.98	ESO 363- 6	r	.20± .05	.00	15.18 ±.14				
249.70 -64.61		-1.3± .9		.00					
053050.1-325942	PGC 17420		.90						
053248.2-140350 NGC 1954		.SAT4P*	1.62± .03	155	12.44 ±.15	.63± .03	.78± .02	13.35±.1	3130± 5
217.16 -23.66	MCG -2-15- 3	PE (2)	.28± .04	.43		.05± .04	.08± .03	477± 6	3033± 63
298.76 -70.08	IRAS05305-1405	4.3± .6	1.29± .03	.41	12.44	.46	14.38± .06	426± 5	2992
053031.1-140554	PGC 17422	3.7± .7	1.66	.14	11.58	-.07	14.70± .23	1.63	3162
053252.9-074636		.L..+P*	1.05± .07	145					
211.00 -20.98	MCG -1-15- 2	E	.26± .05	.67					
315.99 -68.41	VV 848	-1.0±1.3		.00					
053028.2-074840	PGC 17425		1.07						
0532.9 +7923		.S..8*.	1.07± .14	64				16.63±.1	4543± 11
133.81 23.15	UGC 3311	U	.62± .12	.34					
22.48 5.74		8.0±1.3		.76				130± 8	4694
0524.5 +7921	PGC 17426		1.10	.31					4511

R.A. 2000 DEC.	Names	Type	logD$_{25}$	p.a.	B$_T$	(B-V)$_T$	(B-V)$_\bullet$	m$_{21}$	V$_{21}$
l b		S$_T$ n$_L$	logR$_{25}$	A$_g$	m$_B$	(U-B)$_T$	(U-B)$_\bullet$	W$_{20}$	V$_{opt}$
SGL SGB		T	logA$_\bullet$	A$_i$	m$_{FIR}$	(B-V)^o_T	m'$_\bullet$	W$_{50}$	V$_{GSR}$
R.A. 1950 DEC.	PGC	L	logD$_o$	A$_{21}$	B^o_T	(U-B)^o_T	m'$_{25}$	HI	V$_{3K}$
053308.6+492200				1.19				15.80±.3	7245± 9
161.31 8.74								160± 11	
12.04 -22.45								122± 8	7311
052917.6+491954	PGC 17429								7233
053311.8-523831		.S..0P/	1.18± .05	9	14.2 ±.2	.94± .05	.97± .02		
260.01 -32.90	ESO 159- 19	S	.29± .04	.07	14.71 ±.14				
225.23 -49.69	IRAS05320-5240	-.5± .6	.87± .07	.00	12.47		14.06± .22		
053205.0-524030	PGC 17432		1.14				14.24± .34		
053312.8-362359	NGC 1963	.S..6*/	1.44± .04	109				13.72±.2	1324± 11
241.05 -30.69	ESO 363- 7	BS	.69± .05	.00	13.27 ±.14			243± 16	
243.63 -62.46	IRAS05314-3626	5.5± .8		1.02	11.86			241± 12	1142
053128.0-362600	PGC 17433		1.44	.35	12.24			1.13	1366
053321.1-215651	NGC 1964	.SXS3..	1.75± .01	32	11.58 ±.13	.77± .01	.87± .01	12.90±.1	1663± 5
225.28 -26.51	ESO 554- 10	R (3)	.42± .02	.03	11.51 ±.11	.21± .03	.36± .02	429± 6	1699± 43
275.48 -69.72	IRAS05312-2158	3.0± .3	.97± .01	.58	10.68	.67	11.92± .04	407± 6	1507
053114.1-215852	PGC 17436	2.4± .4	1.76	.21	10.92	.13	14.14± .16	1.77	1700
053335.1+061634		.S..2..	1.01± .06	147				16.57±.3	8067± 10
198.09 -14.18	UGC 3321	U	.42± .05	1.51					
342.61 -60.17		2.0± .9		.51				509± 7	7987
053054.0+061433	PGC 17444		1.15	.21					8088
0533.6 +7343	A 0527+73	.I...9..	1.17± .07		15.2 ±.2	.75± .12	.73± .05	14.40±.1	1239± 6
139.35 20.77	UGC 3317	U (1)	.21± .06	.49		-.14± .14	-.13± .07	121± 8	
20.62 .40	DDO 38	10.0± .7	1.18± .06	.16		.58	16.58± .13	107± 12	1377
0527.2 +7341	PGC 17445	9.0±1.0	1.21	.11	14.54	-.27	15.34± .44	-.24	1211
0533.9 +7913		.S..9*.	1.04± .15	160				15.86±.1	4688± 11
133.99 23.12	UGC 3313	U	.18± .12	.34					
22.47 5.57		9.0±1.2		.18				76± 8	4838
0525.6 +7911	PGC 17450		1.07	.09					4657
053401.2-231832	NGC 1979	.LA.*..	1.34± .05						
226.77 -26.83	ESO 487- 24	S	.09± .04	.02	12.85 ±.14				
271.56 -69.48		-3.0±1.0		.00					
053156.0-232030	PGC 17452		1.33						
053406.8-282757		.SXS5*.	1.25± .05						3804± 52
232.29 -28.45	ESO 423- 20	S (1)	.09± .04	.04	13.53 ±.14				3634
258.55 -67.39	IRAS05321-2830	5.0± .7		.14	13.69				3844
053209.0-282954	PGC 17455	3.3± .6	1.26	.05	13.32				
0534.2 +7011		.S..6*.	1.22± .06	127				15.53±.1	4216± 11
142.73 19.23	UGC 3319	U	.76± .06	.42					
19.50 -2.95	IRAS05285+7009	6.0±1.3		1.12	12.74			330± 8	4344
0528.5 +7009	PGC 17456		1.26	.38					4190
053417.7-555254		.SXT7..	1.14± .05						
263.87 -32.85	ESO 159- 20	Sr (1)	.06± .05	.19	14.85 ±.14				
222.57 -46.97		7.4± .6		.08					
053322.0-555448	PGC 17458	7.8± .8	1.16	.03					
053420.2+064721		.S..6?.	1.00± .08	25				15.07±.3	7874± 7
197.73 -13.76	UGC 3322	U	.25± .06	1.59					
343.56 -59.91		6.0±1.8		.37				446± 7	7796
053138.5+064523	PGC 17462		1.15	.13					7896
053421.9-233158	IC 2138	.SXR2..	1.08± .05	88					
227.03 -26.83	ESO 487- 27	r	.19± .04	.03	13.91 ±.14				
270.87 -69.49		2.2± .9		.23					
053217.1-233354	PGC 17463		1.08	.10					
053423.2-304804	NGC 1989	.LAS-*.	1.16± .05	106					10782± 52
234.87 -29.06	ESO 423- 21	S	.12± .04	.00	14.13 ±.14				10608
253.41 -66.19		-3.0± .6		.00					10823
053229.0-305000	PGC 17464		1.14		13.97				
053432.1-305347	NGC 1992	.SAT0?.	1.02± .05	45					10576± 52
234.98 -29.05	ESO 423- 23	S	.18± .04	.00	14.69 ±.14				10402
253.17 -66.16		.0± .6		.14					10618
053238.1-305542	PGC 17466		1.01		14.39				
053439.7-341048		.IXS9..	1.10± .06	173					868
238.63 -29.88	ESO 363- 8	S (1)	.18± .04	.00	16.63 ±.14				
246.89 -64.16		10.0± .9		.13					689
053251.0-341242	PGC 17467	10.0± .9	1.10	.09	16.49				911
053441.6-291359		.LA.0?P	1.25± .05						3968± 52
233.17 -28.55	ESO 423- 24	S	.02± .04	.00	13.13 ±.14				3797
256.65 -67.10		-2.0± .7		.00					4009
053245.0-291554	PGC 17469		1.24		13.07				

R.A. 2000 DEC.	Names	Type	logD$_{25}$	p.a.	B$_T$	(B-V)$_T$	(B-V)$_e$	m$_{21}$	V$_{21}$
l　　b		S$_T$　n$_L$	logR$_{25}$	A$_g$	m$_B$	(U-B)$_T$	(U-B)$_e$	W$_{20}$	V$_{opt}$
SGL　SGB		T	logA$_e$	A$_i$	m$_{FIR}$	(B-V)$_T^o$	m'$_e$	W$_{50}$	V$_{GSR}$
R.A. 1950 DEC.	PGC	L	logD$_o$	A$_{21}$	B$_T^o$	(U-B)$_T^o$	m'$_{25}$	HI	V$_{3K}$
053450.3-100137		.S..5./	1.08± .06	50					
213.40 -21.52	MCG -2-15- 6	E	.87± .05	.77					
310.59 -69.66		5.0±1.0		1.30					
053228.4-100332	PGC 17475		1.15	.43					
053459.0-505520	NGC 2007	.SBT5*.	1.24± .04	83	13.91V±.14		.51± .05		
257.99 -32.53	ESO 204- 19	Sr　　(1)	.47± .04	.02	14.84 ±.14				4523± 62
226.28 -51.30		5.0± .7	.82± .02	.71					4326
053347.0-505712	PGC 17478	3.3±1.2	1.24	.24	14.08				4569
053503.8-505802	NGC 2008	.S..5..	1.17± .05	93	13.80V±.13	.65± .03	.72± .03		
258.04 -32.52	ESO 204- 20	S　　(1)	.30± .05	.06	14.64 ±.14				10341± 62
226.22 -51.26	IRAS05338-5059	5.0± .8	.64± .01	.45	13.57	.51			10144
053352.0-505954	PGC 17480	1.1± .8	1.17	.15	13.98		14.40± .29		10387
0535.0 +7648		.S..9*.	1.19± .12	45				15.53±.1	4468± 11
136.41 22.18	UGC 3318	U	.08± .12	.49					4613
21.73 3.27		9.0±1.1		.08				216± 8	
0527.8 +7646	PGC 17481		1.24	.04					4438
053507.2-484635		.L..-*.	1.32± .07						
255.46 -32.32		S	.50± .08	.16					
228.13 -53.12		-3.0±1.2		.00					
053349.2-484827	PGC 17482		1.27						
0535.3 +4054		.S..4..	1.15± .06	30				15.25±.1	6689± 7
168.71 4.55	UGC 3325	U	.16± .05	2.24					6726
8.96 -30.44		4.0± .8		.24				329± 8	6685
0531.8 +4053	PGC 17483		1.36	.08					
053525.4-174856	NGC 1993	.LAT-*.	1.17± .05	80	*				
221.21 -24.55	ESO 554- 14	SE	.02± .04	.13	13.41 ±.14				
287.51 -70.84		-2.5± .5		.00					
053313.0-175048	PGC 17487		1.19						
053534.4+494603		.S..7*.	1.00± .16	172	*			15.71±.1	6243± 7
161.17 9.28	UGC 3323	U	.19± .12	1.10				176± 11	6309
12.59 -22.22		7.0±1.3		.26				166± 6	6233
053142.2+494408	PGC 17489		1.10	.09					
053534.8-290909		.E+4.P*	1.03± .05	153	*				17831± 52
233.15 -28.34	ESO 424- 1	S	.24± .04	.00	15.26 ±.14				17660
256.57 -67.31		-4.0± .6		.00					17874
053338.1-291100	PGC 17490		.96		14.99				
0535.6 -1526		.S..7?/	1.13± .08	175	*				
218.83 -23.57		E	.87± .08	.22					
294.84 -70.89		7.0±1.8		1.20					
0533.4 -1528	PGC 17491		1.15	.43					
053552.0-532912		.S..4*.	1.01± .05	58	*				
261.03 -32.54	ESO 159- 21	S　　(1)	.20± .04	.41	15.36 ±.14				
224.04 -49.16		4.0±1.2		.29					
053448.0-533100	PGC 17495	3.3±1.2	1.04	.10					
053613.8-164016		.S..6*/	1.14± .06	80	*				7755
220.13 -23.93	MCG -3-15- 5	E　　(1)	.70± .05	.18					
291.03 -71.07		6.0±1.3		1.03					7609
053400.0-164205	PGC 17502	5.3±1.6	1.15	.35					7794
053617.4+071929		.S..3..	1.02± .06	169	*			16.22±.3	3857± 10
197.50 -13.07	UGC 3328	U	.49± .05	1.89	15.0 ±.3				3779
344.98 -59.83		3.0± .9		.68				380± 7	3881
053335.2+071740	PGC 17504		1.19	.25	12.43			3.54	
053626.1-521102		.SBS9*P	1.11± .05	17	*				1279
259.50 -32.39	ESO 204- 22	S　　(1)	.09± .05	.10	15.50 ±.14				1081
224.94 -50.33		9.0± .6		.09					1326
053518.0-521248	PGC 17507	10.0± .7	1.12	.04	15.30				
0536.4 +6335		.S..6*.	1.04± .08	40	*				
148.96 16.34	UGC 3324	U	1.06± .06	.66					
17.54 -9.25		6.0±1.5		1.47					
0531.7 +6334	PGC 17508		1.10	.50					
053633.0+163830		.S..4..	1.11± .07	10	*			16.18±.3	5253± 10
189.40 -8.23	UGC 3329	U	.36± .06	2.15					5206
354.78 -52.30		4.0± .9		.53				475± 7	5271
053339.4+163641	PGC 17509		1.31	.18					
053646.4-222420		.SAS3..	1.15± .04	7	*				
226.06 -25.92	ESO 554- 19	S　　(1)	.33± .03	.06	14.37 ±.14				
273.55 -70.36	IRAS05346-2225	3.0± .8		.45	13.72				
053440.1-222606	PGC 17511	3.3± .8	1.16	.16					

R.A. 2000 DEC. / l b / SGL SGB / R.A. 1950 DEC.	Names / PGC	Type / S_T n_L / T / L	$\log D_{25}$ / $\log R_{25}$ / $\log A_\bullet$ / $\log D_\bullet$	p.a. / A_g / A_i / A_{21}	B_T / m_B / m_{FIR} / B_T^o	$(B-V)_T$ / $(U-B)_T$ / $(B-V)_T^o$ / $(U-B)_T^o$	$(B-V)_\bullet$ / $(U-B)_\bullet$ / $m'_\bullet$ / m'_{25}	m_{21} / W_{20} / W_{50} / HI	V_{21} / V_{opt} / V_{GSR} / V_{3K}
053647.3+142518 / 191.35 -9.34 / 352.88 -54.20 / 053356.5+142330	UGC 3330 / IRAS05338+1423 / PGC 17512	.S?....	1.04± .15 / .04± .12 / / 1.21	/ 1.77 / .06 / .02				15.56±.2 / 198± 7 / 173± 5	5214± 6 / / 5159 / 5234
053653.2-151215 / 218.73 -23.21 / 295.62 -71.17 / 053437.5-151401	MCG -3-15- 6 / / PGC 17515	RSBR0*. / E / .0± .9	1.07± .06 / .11± .05 / / 1.09	45 / .26 / .09	*				
053715.9-491525 / 256.06 -32.02 / 227.21 -52.90 / 053559.3-491707	/ / PGC 17524	.LX.-.. / S / -3.0± .8	1.11± .10 / .14± .08 / / 1.10	/ .05 / .00	*				
053718.7-262553 / 230.35 -27.15 / 262.50 -68.98 / 053518.0-262736	ESO 487- 30 / / PGC 17525	.S..7?/ / S (1) / 7.0±1.7 / 6.7±1.2	1.23± .05 / .79± .05 / / 1.23	155 / .00 / 1.10 / .40	* / 15.85 ±.14 / / 14.74				1499 / / 1332 / 1543
053719.0-422435 / 248.10 -31.09 / 234.18 -58.53 / 053546.0-422618	ESO 306- 12 / IRAS05356-4226 / PGC 17526	.S?....	1.10± .05 / .28± .05 / / 1.10	155 / .04 / .39 / .14	* / 14.98 ±.14 / / 14.47				10946 / / 10755 / 10994
0537.8 +0006 / 204.21 -16.26 / 334.65 -65.24 / 0535.3 +0005	UGC 3331 / / PGC 17535	.S..3.. / U / 3.0± .8	1.02± .06 / .02± .05 / / 1.17	/ 1.57 / .03 / .01	*				
053813.0-675112 / 278.01 -31.85 / 214.98 -36.40 / 053822.7-675247	/ / PGC 17536	PSA.3?. / S / 3.0±1.0	1.09± .09 / .42± .08 / / 1.12	/ .28 / .58 / .21	*				
0538.3 +7935 / 133.72 23.46 / 22.78 5.85 / 0529.8 +7933	UGC 3320 / IRAS05298+7933 / PGC 17540	.SBS3.. / U / 3.0± .9	.99± .10 / .45± .07 / / 1.03	/ .43 / .62 / .22	* / 15.30 ±.18 / 11.90 / 14.22			15.57±.1 / / 234± 8 / 1.12	4739± 11 / / 4890 / 4709
053836.2-342328 / 239.10 -29.15 / 245.26 -64.63 / 053648.0-342506	ESO 363- 12 / / PGC 17544	.SBR3P. / r / 3.3± .9	.96± .06 / .04± .03 / / .96	/ .00 / .06 / .02	* / 15.41 ±.14 / / 15.27				10709± 52 / / 10528 / 10757
053849.2+410738 / 168.88 5.22 / 9.76 -30.49 / 053517.5+410558	/ / PGC 17547			1.94				16.45±.3 / 55± 11 / 45± 8	6117± 9 / / 6153 / 6118
053858.4-414413 / 247.39 -30.67 / 234.51 -59.25 / 053724.0-414548	ESO 306- 13 / IRAS05374-4145 / PGC 17552	.SB?...	1.02± .07 / .08± .06 / / 1.03	137 / .05 / .12 / .04	* / 14.03 ±.14 / 13.45 / 13.86				1021 / / 830 / 1070
053903.1+153438 / 190.64 -8.27 / 354.69 -53.52 / 053610.8+153300	UGC 3332 / IRAS05361+1532 / PGC 17554	.S?....	1.04± .15 / .59± .12 / / 1.24	30 / 2.13 / .88 / .29	/ / 12.67			14.48±.3 / / 450± 7	5817± 7 / / 5765 / 5840
0539.3 -1701 / 220.80 -23.38 / 289.88 -71.80 / 0537.1 -1703	MCG -3-15- 7 / / PGC 17556	.SB?...	1.12± .08 / .46± .07 / / 1.14	/ .22 / .56 / .23	*				4302 / / 4154 / 4345
0539.6 +7718 / 136.02 22.62 / 22.13 3.67 / 0532.1 +7716	UGC 3326 / IRAS05321+7716 / PGC 17561	.S..6*. / U / 6.0±1.2 / 1.59	1.55± .03 / 1.26± .06	/ .44 / 1.47 / .50	* / 15.3 ±.2 / 12.83 / 13.40			14.64±.1 / / 528± 8 / .74	4085± 11 / 4087± 76 / 4230 / 4057
053952.9-403041 / 246.03 -30.27 / 235.75 -60.32 / 053816.0-403212	ESO 306- 16 / / PGC 17566	.S?....	.99± .07 / .21± .06 / / .99	147 / .05 / .22 / .11	* / 14.97 ±.14 / / 14.56				11192 / / 11002 / 11242
053958.8-583507 / 267.08 -32.10 / 219.80 -44.92 / 053914.1-583636	ESO 120- 12 / IRAS05392-5836 / PGC 17567	PSBT7.. / Sr (1) / 6.8± .5 / 7.8± .8	1.29± .04 / .15± .05 / / 1.31	96 / .20 / .21 / .08	* / 14.09 ±.14 / / 13.68				1273 / / 1073 / 1323
054003.1-815856 / 294.17 -29.31 / 208.95 -23.28 / 054523.1-820012	ESO 16- 9 / IRAS05455-8200 / PGC 17568	.SXS8*P / S (1) / 8.0± .9 / 7.8±1.3	1.01± .05 / .29± .04 / / 1.09	28 / .84 / .35 / .14	* / 15.63 ±.14				

R.A. 2000 DEC. / l b / SGL SGB / R.A. 1950 DEC.	Names / PGC	Type / S_T n_L / T / L	$\log D_{25}$ / $\log R_{25}$ / $\log A_e$ / $\log D_o$	p.a. / A_g / A_i / A_{21}	B_T / m_B / m_{FIR} / B_T^o	$(B-V)_T$ / $(U-B)_T$ / $(B-V)_T^o$ / $(U-B)_T^o$	$(B-V)_e$ / $(U-B)_e$ / m'_e / m'_{25}	m_{21} / W_{20} / W_{50} / HI	V_{21} / V_{opt} / V_{GSR} / V_{3K}
054006.3-405012 / 246.41 -30.29 / 235.28 -60.09 / 053830.1-405142	ESO 306- 17 / PGC 17570	.E+3... / S / -4.0± .8	1.39± .04 / .21± .04 / / 1.33	177 / .03 / .00	* / 13.35 ±.14 / / 13.16				10734± 34 / 10544 / 10785
054009.8-553219 / 263.49 -32.01 / 221.74 -47.66 / 053913.0-553348	ESO 159- 23 / IRAS05392-5533 / PGC 17571	.S..3.. / S (1) / 3.0± .8 / 3.3± .8	1.23± .05 / .32± .05 / / 1.26	118 / .25 / .44 / .16	* / 14.42 ±.14 / 13.48 / 13.68				7183 / 6983 / 7233
054011.9-220011 / 225.95 -25.04 / 274.04 -71.23 / 053805.0-220142	ESO 554- 23 / IRAS05380-2201 / PGC 17572	.S?....	1.03± .05 / .14± .04 / / 1.04	130 / .10 / .21 / .07	14.68 ±.14 / 12.49 / 14.31				8963± 69 / 8803 / 9010
054039.5+162738 / 190.08 -7.48 / 355.99 -52.97 / 053746.1+162607	UGC 3338 / IRAS05377+1626 / PGC 17587	.S?....	1.00± .16 / .09± .12 / / 1.23	2.47 / .13 / .04				15.30±.3 / / 306± 10	4864± 7 / 4814 / 4888
054051.0-134856 / 217.77 -21.77 / 300.35 -71.98 / 053833.7-135025	MCG -2-15- 9 / PGC 17589	.SBS6.. / E (1) / 6.0± .9 / 4.2± .8	1.06± .06 / .05± .05 / / 1.12	130 / .59 / .08 / .03	*				
054057.6-820713 / 294.32 -29.25 / 208.87 -23.16 / 054626.0-820824	NGC 2144 / ESO 16- 10 / IRAS05464-8208 / PGC 17592	PSAT1*. / S / .8± .4	1.16± .05 / .10± .05 / / 1.24	/ .84 / .10 / .05	* / 13.93 ±.14 / 12.74				
054100.9-354227 / 240.70 -28.99 / 242.26 -64.07 / 053915.0-354354	ESO 363- 15 / IRAS05393-3543 / PGC 17595	.SAS7.. / S (1) / 7.0± .8 / 5.6± .8	1.39± .04 / .17± .05 / / 1.39	2 / .00 / .23 / .08	* / 13.92 ±.14 / 14.12 / 13.68			14.50±.2 / 133± 16 / 126± 12 / .74	1276± 7 / 1092 / 1327
054119.8-181640 / 222.26 -23.42 / 285.69 -72.20 / 053908.0-181806	ESO 554- 24 / PGC 17597	.SBT6P* / SE (2) / 5.5± .6 / 4.2± .7	1.18± .04 / .23± .03 / / 1.20	56 / .24 / .33 / .11	* / 14.29 ±.14				
054151.3-641804 / 273.81 -31.75 / 216.32 -39.82 / 054136.0-641924	NGC 2082 / ESO 86- 21 / IRAS05415-6419 / PGC 17609	.SBR3.. / R / 3.0± .4 / 4.6± .9	1.26± .03 / .03± .03 / .96± .01 / 1.28	/ .15 / .04 / .02	12.62 ±.13 / 12.79 ±.12 / 11.83 / 12.52	.56± .02 / -.10± .03 / .51 / -.13	.64± .01 / -.05± .02 / 12.91± .02 / 13.71± .20		1241± 49 / 1041 / 1291
054156.0+182928 / 188.49 -6.17 / 358.04 -51.36 / 053900.0+182803	UGC 3341 / IRAS05389+1828 / PGC 17616	.SB.2.. / U / 2.0± .8	1.26± .06 / .00± .06 / / 1.57	3.31 / .00 / .00	12.01			13.99±.2 / 152± 7 / 143± 5	4569± 7 / 4525 / 4593
054200.6-225643 / 227.09 -24.98 / 270.81 -71.34 / 053955.1-225806	ESO 487- 35 / IRAS05399-2258 / PGC 17619	.SBS8*/ / SU (1) / 7.8± .6 / 4.8± .7	1.44± .03 / .65± .04 / / 1.45	104 / .07 / .80 / .32	13.41 ±.14 / 13.15 / 12.54			14.02±.2 / 165± 16 / 159± 12 / 1.15	1731± 11 / 1568 / 1781
0542.0 -1233 / 216.66 -21.00 / 304.53 -72.03 / 0539.7 -1235	PGC 17621	.I..9.. / E (1) / 10.0± .9 / 7.5± .8	1.21± .07 / .28± .08 / / 1.27	85 / .59 / .21 / .14					
054204.4+692246 / 143.82 19.47 / 19.88 -3.93 / 053634.0+692116	NGC 1961 / UGC 3334 / ARP 184 / PGC 17625	.SXT5.. / R (2) / 5.0± .3 / 2.8± .6	1.66± .02 / .19± .02 / 1.30± .02 / 1.70	85 / .42 / .29 / .10	11.73 ±.14 / 11.67 ±.15 / 10.77 / 10.98	.74± .04 / / .58	.82± .02 / .30± .07 / 13.72± .05 / 14.39± .17	12.69±.1 / 690± 5 / 621± 5 / 1.62	3930± 4 / 3983± 22 / 4057 / 3911
054218.6-475633 / 254.64 -31.04 / 227.17 -54.46 / 054058.7-475753	PGC 17629	.LXR0P* / S / -2.5± .6	1.20± .09 / .17± .08 / / 1.21	/ .27 / .00					14726± 15 / 14529 / 14779
054220.7-550548 / 262.99 -31.68 / 221.63 -48.19 / 054122.5-550707	PGC 17632	.SBT3*. / S (1) / 3.0± .7 / 3.9± .9	1.05± .09 / .25± .08 / / 1.09	/ .39 / .35 / .13					
054220.9-253232 / 229.82 -25.79 / 263.33 -70.39 / 054019.0-253354	ESO 487- 36 / PGC 17633	.LAS0P* / S / -2.0± .6	1.14± .05 / .17± .04 / / 1.11	32 / .00 / .00	14.46 ±.14 / 14.32				9030± 52 / 8862 / 9081
054233.4-493705 / 256.59 -31.20 / 225.67 -53.04 / 054118.0-493824	ESO 204- 30 / IRAS05412-4938 / PGC 17639		.72± .07 / .05± .06 / / .73	.10	15.44 ±.14 / 13.27				12239±104 / 12041 / 12292

R.A. 2000 DEC. l b SGL SGB R.A. 1950 DEC.	Names PGC	Type S_T n_L T L	$\log D_{25}$ $\log R_{25}$ $\log A_e$ $\log D_o$	p.a. A_g A_i A_{21}	B_T m_B m_{FIR} B_T^o	$(B-V)_T$ $(U-B)_T$ $(B-V)_T^o$ $(U-B)_T^o$	$(B-V)_e$ $(U-B)_e$ m'_e m'_{25}	m_{21} W_{20} W_{50} HI	V_{21} V_{opt} V_{GSR} V_{3K}
0542.7 +6953 143.36 19.75 20.10 -3.46 0537.2 +6952	 CGCG 329- 10 PGC 17645	.S?.... 	.77± .11 .18± .10 .81	 .43 .27 .09	15.25 ±.20 14.54				4016±125 4142 3995
054305.9-203117 224.70 -23.87 278.13 -72.26 054057.0-203236	 ESO 554- 27 IRAS05409-2032 PGC 17651	 	.92± .06 .29± .05 .92	106 .04 	14.13 ±.14 12.99 				3237±104 3079 3287
054306.3-524202 260.19 -31.41 223.16 -50.38 054200.0-524318	 ESO 159- 25 PGC 17652	.IBS9.. S (1) 10.0± .5 10.0± .5	1.11± .05 .15± .05 1.13	55 .17 .11 .07	 15.54 ±.14 15.26			14.59±.3 109± 7 -.75	1100± 9 900 1153
054311.8-343655 239.62 -28.29 243.32 -65.16 054124.0-343812	 ESO 363- 17 PGC 17654	.S..5.. S (1) 5.0± .8 3.3± .8	1.03± .05 .03± .04 1.03	 .00 .05 .02	 14.62 ±.14 				
054314.3+163010 190.36 -6.93 356.91 -53.24 054020.8+162850	 UGC 3348 PGC 17656	.S?.... 	1.11± .07 .44± .06 1.37	9 2.77 .66 .22				15.00±.3 312± 7 	5171± 7 5120 5199
054315.2-300443 234.70 -27.02 252.01 -68.16 054120.0-300600	NGC 2049 ESO 424- 11 PGC 17657	.SAS1?. S 1.0± .8 	1.31± .04 .33± .04 1.31	168 .00 .34 .16	 13.67 ±.14 				
054315.8-550652 263.02 -31.55 221.44 -48.24 054217.6-550807	 PGC 17658	.LA.-P. S -3.0± .5 	1.36± .07 .21± .08 1.37	 .39 .00 					
054328.5-302944 235.16 -27.10 251.04 -67.94 054134.0-303100	 ESO 424- 13 IRAS05415-3030 PGC 17662	.SBT8.. Sr (1) 8.0± .6 5.6± .8	1.26± .04 .10± .05 1.26	96 .00 .12 .05	 13.61 ±.14 13.48				1302 1125 1356
054352.5-191739 223.53 -23.25 282.06 -72.67 054142.0-191854	 ESO 554- 29 PGC 17668	.SBS6P* SUE (1) 6.1± .6 6.7± .9	1.31± .03 .43± .03 1.32	10 .16 .63 .22	 14.02 ±.14 13.22			13.93±.2 273± 16 247± 12 .50	2749± 11 2593 2800
0544.9 +6910 144.14 19.62 20.07 -4.21 0539.5 +6909	 UGC 3344 PGC 17675	.SX.4.. U (1) 4.0± .7 3.5±1.0	1.39± .09 .20± .12 1.43	25 .47 .29 .10	 14.2 ±.3 				
0544.0 +5112 160.61 11.17 14.45 -21.31 0540.1 +5111	 UGC 3346 PGC 17678	.S..8*. U 8.0±1.3 	1.12± .07 .66± .06 1.25	135 1.41 .82 .33	 14.8 ±.3 12.56			15.23±.1 378± 8 2.34	5949± 7 6018 5946
054404.6-495324 256.94 -30.99 225.10 -52.92 054250.0-495436	 ESO 204- 32 IRAS05428-4954 PGC 17680	.S?.... 	.89± .06 .06± .06 .90	 .10 .06 .03	 14.99 ±.14 13.41 14.71				9958 9759 10013
054415.6-553201 263.52 -31.43 220.97 -47.92 054319.0-553312	NGC 2087 ESO 159- 26 IRAS05433-5533 PGC 17684	.SBR1*P S 1.0±1.0 	.92± .05 .12± .04 .95	136 .32 .12 .06	 14.69 ±.14 12.51 				
0544.2 +7537 137.81 22.20 21.88 1.98 0537.4 +7536	 CGCG 347- 21 PGC 17685	.L?.... 	.60± .15 .04± .10 .65	 .50 .00 	15.49 ±.20 14.87				7639±125 7780 7614
0544.5 +6918 143.99 19.63 20.06 -4.07 0539.0 +6917	 UGC 3342 PGC 17692	.S..6*. U 6.0±1.3 	1.24± .04 .67± .05 1.28	42 .42 .99 .34	 15.20 ±.19 13.77			14.97±.4 224± 35 .86	3974± 9 4098 3954
054430.2-553944 263.67 -31.40 220.84 -47.82 054334.0-554054	 ESO 159- 27 PGC 17693	PSBT1*. Sr (1) 1.3± .6 3.3±1.3	.97± .05 .26± .04 1.00	59 .32 .27 .13	 15.49 ±.14 				
054441.5-515750 259.36 -31.10 223.36 -51.14 054333.0-515900	 ESO 204- 34 PGC 17704	.IBS9.. S (1) 10.0± .4 10.0± .4	1.10± .05 .10± .05 1.12	33 .18 .07 .05	 14.99 ±.14 				

R.A. 2000 DEC. / l b / SGL SGB / R.A. 1950 DEC.	Names / PGC	Type / S_T n_L / T / L	$\log D_{25}$ / $\log R_{25}$ / $\log A_e$ / $\log D_0$	p.a. / A_g / A_l / A_{21}	B_T / m_B / m_{FIR} / B_T^0	$(B-V)_T$ / $(U-B)_T$ / $(B-V)_T^0$ / $(U-B)_T^0$	$(B-V)_.$ / $(U-B)_.$ / $m'_.$ / m'_{25}	m_{21} / W_{20} / W_{50} / HI	V_{21} / V_{opt} / V_{GSR} / V_{3K}
054448.6+164557		.SB?...	1.07± .07	177				16.70±.3	5403± 7
190.33 -6.47	UGC 3352		.45± .06	3.07					
357.66 -53.20	IRAS05419+1645			.68	13.40			127± 7	5352
054154.8+164444	PGC 17707		1.36	.23					5433
054451.5-250549		.S?....	.98± .06	138					12126± 62
229.57 -25.10	ESO 488- 4		.25± .04	.01	14.62 ±.14				11958
263.78 -71.10				.31					12180
054249.0-250700	PGC 17708		.98	.12	14.18				
054500.3-522146		.SBS9..	1.17± .07	117					
259.82 -31.09	ESO 204- 36	S	.16± .05	.20	17.12 ±.14				
223.00 -50.81		9.0± .6		.16					
054353.1-522254	PGC 17714		1.19	.08					
054500.5-480516		.LXT-*.	.97± .05	17					
254.88 -30.61	ESO 204- 35	S	.12± .04	.30	15.45 ±.14				
226.37 -54.56		-2.7± .7		.00					
054341.0-480624	PGC 17715		.99						
054501.7+050341				2.20				12.68±.3	365± 8
200.62 -12.30								177± 8	
345.42 -62.95									275
054222.0+050230	PGC 17716								404
054502.0-261138		.S?....	.97± .06	161					12472± 62
230.72 -25.43	ESO 488- 5		.45± .05	.00	15.43 ±.14				12302
260.73 -70.62				.46					12527
054301.0-261248	PGC 17717		.97	.22	14.82				
0545.1 +7914		.L.....	1.08± .09	122					
134.19 23.61	UGC 3335	U	.36± .04	.32	15.16 ±.16				
22.97 5.43		-2.0± .9		.00					
0536.7 +7913	PGC 17725		1.06						
054522.6-254728		.LAT+..	.95± .06						11681± 62
230.33 -25.23	ESO 488- 6	r	.07± .04	.00	14.71 ±.14				11511
261.70 -70.88		-1.3± .9		.00					11736
054321.1-254836	PGC 17735		.94		14.53				
054525.0+722120		.S?....	1.32± .03	80				14.61±.1	1090± 8
141.07 20.97	UGC 3343		.61± .04	.42	13.93 ±.18			197± 16	1142± 76
21.01 -1.17	IRAS05393+7220			.92	12.93			174± 8	1222
053922.0+722003	PGC 17736		1.36	.31	12.59			1.72	1068
054529.3-255558		.E?....	1.22± .07						13341± 62
230.49 -25.25	ESO 488- 9		.14± .05	.04	14.11 ±.14				13171
261.28 -70.84				.00					13396
054328.0-255706	PGC 17746		1.18		13.87				
0545.5 +7654		.S..7..	1.00± .08	177					
136.56 22.76	UGC 3339	U	.63± .06	.42					
22.32 3.18		7.0±1.0		.87					
0538.2 +7653	PGC 17750		1.04	.31					
0545.6 +6903		.S?....	.89± .07					15.36±.1	4286± 11
144.28 19.62	MCG 12- 6- 13		.03± .05	.47	15.01 ±.13				4344± 82
20.09 -4.34	7ZW 45			.04				148± 12	4410
0540.2 +6902	PGC 17757		.93	.01	14.47			.87	4268
054540.9-253211		PSXR2..	.95± .05						12974± 62
230.09 -25.08	ESO 488- 10	r	.11± .04	.00	14.81 ±.14				12805
262.29 -71.06	IRAS05436-2533	2.2± .9		.13	13.49				13029
054339.0-253318	PGC 17758		.95	.05	14.55				
054551.9-392942		RSAR0..	1.01± .05	13					
245.16 -28.94	ESO 306- 25	S	.17± .04	.07	14.95 ±.14				
235.18 -61.82		.0± .9		.13					
054413.1-393048	PGC 17768		1.00						
054553.8-220000	NGC 2073	.LAT-*.	1.19± .05						
226.47 -23.80	ESO 554- 31	S	.05± .04	.00	13.44 ±.14				
272.78 -72.50		-3.4± .5		.00					
054347.0-220106	PGC 17772		1.19						
0546.0 +5841		.S..2..	1.20± .05	163					
154.02 15.04	UGC 3351	U	.71± .05	1.23					4736±113
17.11 -14.27		2.0± .9		.87					4828
0541.7 +5840	PGC 17776		1.32	.35					4728
054620.8-232826		.L.0*P/	1.21± .05	34					
228.02 -24.23	ESO 488- 12	S	.61± .03	.02	14.87 ±.14				
268.06 -72.08	IRAS05442-2329	-2.0±1.2		.00	13.43				
054416.0-232930	PGC 17791		1.12						

R.A. 2000 DEC. / l b / SGL SGB / R.A. 1950 DEC.	Names / PGC	Type / S_T n_L / T / L	logD_25 / logR_25 / logA_e / logD_o	p.a. / A_g / A_i / A_21	B_T / m_B / m_FIR / B_T^o	(B-V)_T / (U-B)_T / (B-V)_T^o / (U-B)_T^o	(B-V)_e / (U-B)_e / m'_e / m'_25	m_21 / W_20 / W_50 / HI	V_21 / V_opt / V_GSR / V_3K
054623.2-520522 259.53 -30.86 222.90 -51.15 054515.1-520624	NGC 2101 ESO 205- 1 IRAS05451-5206 PGC 17793	.IBS9P. S (1) 10.0± .5 5.6±1.1	1.29± .05 .19± .05 .73± .02 1.31	85 .21 .14 .10	13.69V±.14 13.68 ±.14 13.59 13.54	.42± .05 -.26± .07 .32 -.33	.22± .02 -.61± .05 14.93± .29	13.77±.3 99± 7 .13	1192± 9 1204± 62 991 1249
054632.2+690302 144.32 19.69 20.16 -4.36 054104.5+690151	 UGC 3349 PGC 17794	.S..2.. U 2.0± .9 	1.01± .06 .18± .05 1.06	85 .48 .22 .09	 14.42 ±.18 13.68			16.72±.1 208± 35 158± 8 2.94	4314± 9 4437 4296
054647.2-164656 221.31 -21.64 290.55 -73.59 054433.6-164759	NGC 2076 MCG -3-15- 12 IRAS05445-1648 PGC 17804	.L..+*/ SE -.6± .7 	1.34± .06 .21± .05 1.33	45 .21 .00 	14.0 ±.2 10.85 13.76	1.01± .02 .43± .04 .91 .36	 15.06± .38		2156 2005 2210
054649.2-323440 237.63 -27.01 245.64 -67.12 054458.0-323542	 ESO 363- 22 PGC 17805	.LB.0?/ S -2.0± .9 	1.03± .05 .37± .03 .98	150 .00 .00 	 15.46 ±.14 				
054653.2-184340 223.26 -22.38 283.64 -73.46 054442.0-184442	IC 2143 ESO 554- 34 IRAS05447-1844 PGC 17810	.SBT3?. SE (2) 3.2± .5 3.3± .5	1.29± .03 .38± .03 1.32	98 .28 .52 .19	 13.41 ±.14 12.19 12.58				3134 2978 3189
054655.8-253810 230.30 -24.84 261.60 -71.26 054454.0-253912	 ESO 488- 13 PGC 17813	.LAR+.. r -1.3± .9 	1.03± .07 .19± .05 1.00	64 .00 .00 	15.1 ±.4 14.80 ±.14 14.63	1.05± .03 .54± .09 .89 .58	 14.62± .51		13739± 62 13569 13796
054702.4-341505 239.46 -27.43 242.56 -65.97 054514.0-341606	NGC 2090 ESO 363- 23 IRAS05452-3416 PGC 17819	.SAT5.. R (2) 5.0± .3 4.1± .4	1.69± .02 .31± .03 .92± .01 1.69	13 .00 .47 .16	11.99 ±.13 11.77 ±.12 11.69 11.39	.79± .01 .18± .02 .73 .12	.85± .01 .27± .01 12.08± .03 14.50± .17	12.29±.1 298± 8 289± 11 .74	931± 6 746 989
054704.8-513307 258.92 -30.70 223.13 -51.68 054555.1-513406	NGC 2104 ESO 205- 2 IRAS05459-5133 PGC 17822	.SBS9P. S (1) 8.5± .6 5.6± .8	1.30± .04 .33± .05 .98± .01 1.32	160 .17 .34 .16	13.18 ±.13 13.52 ±.14 13.64 12.83	.48± .02 .36 	.54± .01 13.57± .03 13.72± .28		1181± 33 981 1239
0547.1 +1733 189.93 -5.58 359.10 -52.77 0544.2 +1732	 UGC 3356 IRAS05442+1732 PGC 17823	.S..4.. U 4.0± .9 	.96± .09 .15± .06 1.32	 3.84 .22 .07	 10.80 				5582± 10 5533 5615
054714.9+505215 161.16 11.45 14.85 -21.78 054319.2+505111	 UGC 3355 PGC 17825	.L...*. U -2.0±1.2 	1.09± .10 .00± .05 1.25	 1.38 .00 	 14.5 ±.3 				
0547.3 +5606 156.45 13.97 16.50 -16.78 0543.0 +5605	 UGC 3354 IRAS05430+5605 PGC 17831	.S..2*. U 2.0±1.2 	1.32± .04 .55± .05 1.43	164 1.21 .68 .28	 14.7 ±.2 11.66 12.76			14.52±.1 393± 8 1.48	3086± 7 3170 3081
054724.9-253425 230.27 -24.72 261.61 -71.39 054523.1-253524	 ESO 488- 16 PGC 17834	.S?.... 	.97± .07 .07± .04 .97	40 .00 .08 .03	15.1 ±.4 14.76 ±.14 14.59	1.04± .03 .50± .09 .93 .47	 14.65± .51		13273± 62 13103 13331
054725.3-251524 229.95 -24.61 262.49 -71.54 054523.1-251624	 ESO 488- 15 PGC 17835	.E?.... 	.87± .07 .11± .05 .84	39 .00 .00 	 14.77 ±.14 14.58				12652± 62 12482 12710
054726.3-251448 229.94 -24.61 262.51 -71.55 054524.0-251548	 ESO 488- 19 PGC 17837	.E?.... 	.97± .07 .13± .05 .93	39 .00 .00 	 14.83 ±.14 14.64				12760± 62 12590 12818
0547.4 +7938 133.83 23.85 23.18 5.77 0538.8 +7936	 UGC 3340 IRAS05388+7936 PGC 17839	.S..2.. U 2.0± .9 	1.15± .08 .44± .07 1.18	 .36 .54 .22	 14.64 ±.18 11.73 13.70				4476 4626 4448
054751.6-173608 222.23 -21.73 287.56 -73.81 054539.0-173706	NGC 2089 ESO 554- 36 IRAS05456-1738 PGC 17860	.LX.-*. SE -3.0± .5 	1.27± .04 .21± .04 1.28	39 .32 .00 	 12.93 ±.14 				
054752.5-250903 229.88 -24.48 262.64 -71.68 054550.1-251000	 ESO 488- 22 PGC 17861	.S?.... 	.91± .05 .13± .04 .91	135 .00 .18 .07	 15.12 ±.14 14.86				11108± 62 10938 11166

R.A. 2000 DEC. / l b / SGL SGB / R.A. 1950 DEC.	Names / PGC	Type / S_T n_L / T / L	$\log D_{25}$ / $\log R_{25}$ / $\log A_e$ / $\log D_o$	p.a. / A_g / A_i / A_{21}	B_T / m_B / m_{FIR} / B_T^o	$(B-V)_T$ / $(U-B)_T$ / $(B-V)_T^o$ / $(U-B)_T^o$	$(B-V)_e$ / $(U-B)_e$ / m'_e / m'_{25}	m_{21} / W_{20} / W_{50} / HI	V_{21} / V_{opt} / V_{GSR} / V_{3K}
054807.8+461512								16.14±.3	6048± 9
165.32 9.30				1.05				195± 11	
13.48 -26.23								166± 8	6099
054424.4+461412	PGC 17866								6054
054809.2-472417		.E+3...	1.01± .05	161					
254.18 -29.99	ESO 253- 27	S	.23± .04	.26	14.93 ±.14				
226.16 -55.42		-4.0± .9		.00					
054648.0-472512	PGC 17867		.98						
054816.1-193422		.S?....	1.11± .04	32					
224.24 -22.39	ESO 554- 37		.55± .03	.18	14.55 ±.14				13045
280.47 -73.64	IRAS05460-1935			.81					12887
054606.0-193518	PGC 17872		1.13	.28	13.48				13102
054819.3-251658		.S?....	1.02± .06	148					
230.05 -24.43	ESO 488- 24		.45± .05	.00	15.26 ±.14				11079± 62
262.11 -71.71				.62					10909
054617.1-251754	PGC 17874		1.02	.22	14.56				11138
054827.5-325841		.E+1.*.	1.03± .05						
238.17 -26.79	ESO 363- 27	S	.08± .04	.00	14.20 ±.14				
244.28 -67.09		-4.0±1.2		.00					
054637.0-325936	PGC 17881		1.01						
054828.3-474537		.S?....	1.12± .07	145	15.5 ±.2	.86± .04			
254.60 -29.99	ESO 205- 3		.20± .05	.34	15.60 ±.14	-.64± .06			15095± 63
225.77 -55.13	IRAS05471-4746			.20	11.94	.62			14896
054708.1-474630	PGC 17883		1.15	.10	14.83	-.71	15.46± .43		15155
054836.2-184017		PSBR0?.	1.29± .04	135					
223.37 -21.98	ESO 554- 38	SE	.44± .03	.34	13.63 ±.14				
283.63 -73.87		-.3± .7		.33					
054625.0-184112	PGC 17891		1.30						
054838.0-252842		.E+1...	1.12± .07						
230.28 -24.43	ESO 488- 27	S	.07± .05	.00	14.02 ±.14				11959± 36
261.45 -71.68		-4.0± .8		.00					11789
054636.1-252936	PGC 17893		1.09		13.84				12019
054840.6-254530		.L?....	.92± .07	63					
230.57 -24.52	ESO 488- 28		.43± .05	.00	15.46 ±.14				12186± 62
260.65 -71.55				.00					12015
054639.0-254624	PGC 17896		.86		15.28				12246
054844.6-390319		.SBR5*.	1.02± .05	79					
244.82 -28.29	ESO 306- 28	S (1)	.33± .05	.14	15.34 ±.14				
234.79 -62.51		5.0±1.3		.49					
054705.0-390412	PGC 17901	3.3± .9	1.03	.16					
0548.7 +7933		.SB.3*.	1.00± .08						
133.94 23.88	UGC 3347	U	.04± .06	.36					
23.22 5.68		3.0±1.2		.05					
0540.2 +7932	PGC 17904		1.03	.02					
054852.8-253755		.S?....	.99± .06						
230.46 -24.43	ESO 488- 31		.14± .04	.00	14.71 ±.14				11151± 62
260.93 -71.65				.21					10980
054651.0-253848	PGC 17912		.99	.07	14.44				11211
054903.8-471003		PSBR2..	.97± .07						
253.94 -29.80	ESO 253- 29	r	.08± .06	.22	14.91 ±.14				
226.13 -55.70	IRAS05477-4710	2.2± .8		.10	12.94				
054742.0-471054	PGC 17920		.99	.04					
054904.0-252620		.S?....	.91± .07						
230.27 -24.32	ESO 488- 32		.05± .06	.00	15.34 ±.14				12773± 62
261.41 -71.78				.08					12602
054702.0-252712	PGC 17921		.91	.03	15.19				12833
054921.2+175027		.S?....	1.07± .07	147				15.88±.2	4559± 7
189.96 -4.99	UGC 3360		.40± .06	4.56				230± 14	
.06 -52.75	IRAS05464+1749			.60	12.16			219± 10	4510
054626.0+174934	PGC 17928		1.50	.20					4595
054922.2-252051		.E?....	1.01± .07						
230.20 -24.23	ESO 488- 33		.10± .05	.00	14.55 ±.14				12054± 62
261.56 -71.89				.00					11883
054720.0-252142	PGC 17929		.98		14.37				12115
054929.3-330346		RSA.0..	.78± .06	178					
238.33 -26.61	ESO 363- 31	r	.12± .03	.00	15.37 ±.14				
243.72 -67.18		-.1± .9		.09					
054739.0-330436	PGC 17937		.77						

R.A. 2000 DEC. / l b / SGL SGB / R.A. 1950 DEC.	Names / PGC	Type S_T n_L / T / L	$\log D_{25}$ / $\log R_{25}$ / $\log A_e$ / $\log D_o$	p.a. / A_g / A_i / A_{21}	B_T / m_B / m_{FIR} / B^o_T	$(B-V)_T$ / $(U-B)_T$ / $(B-V)^o_T$ / $(U-B)^o_T$	$(B-V)_e$ / $(U-B)_e$ / m'_e / m'_{25}	m_{21} / W_{20} / W_{50} / HI	V_{21} / V_{opt} / V_{GSR} / V_{3K}
054931.9-193305		PSXR2*.	1.08± .05	25					
224.34 -22.11	ESO 555- 1	S	.23± .04	.25	14.54 ±.14				
280.33 -73.94		2.0± .7		.29					
054721.8-193355	PGC 17941		1.10	.12					
054952.6-432400		.L?....	1.06± .07	148					4479± 52
249.71 -29.01	ESO 253- 30		.12± .05	.16	14.44 ±.14				4283
229.43 -59.03				.00					4541
054822.0-432448	PGC 17949		1.06		14.21				
054954.5-242323		RLA.+..	.94± .07						
229.27 -23.78	ESO 488- 37	r	.06± .05	.00	15.44 ±.14				
264.17 -72.44		-1.3± .9		.00					
054751.0-242412	PGC 17951		.93						
054957.6+510531		.SB.6*.	1.14± .07	162					
161.17 11.93	UGC 3359	U	.86± .06	1.20	15.2 ±.3				
15.36 -21.70	IRAS05460+5104	6.0±1.4		1.27					
054601.2+510439	PGC 17954		1.25	.43					
055004.2-243818		.SBR0..	1.00± .06	174					
229.53 -23.84	ESO 488- 38	r	.14± .04	.00	15.00 ±.14				
263.37 -72.36		-.1± .9		.11					
054801.0-243906	PGC 17960		.99						
0550.2 -1018		.SBR3?.	1.36± .06	45					
215.40 -18.21		E	.13± .08	1.47					
313.76 -73.33		3.0±1.8		.17					
0547.9 -1019	PGC 17965		1.50	.06					
055024.6+494246		.SB.8*.	.96± .09	110				16.30±.1	3304± 11
162.44 11.33	UGC 3361	U	.08± .06	1.30					3366
15.00 -23.04		8.0±1.2		.09				157± 8	3309
054632.2+494156	PGC 17968		1.08	.04					
055026.8-194337		.S..4*/	1.33± .03	105					
224.60 -21.98	ESO 555- 2	SE	.80± .03	.22	14.47 ±.14				
279.54 -74.12	IRAS05482-1944	4.4± .7		1.18	12.65				
054817.0-194424	PGC 17969		1.35	.40					
055028.2-334432		PSXR2..	.97± .06	58					
239.13 -26.61	ESO 364- 2	r	.18± .05	.00	15.06 ±.14				
242.13 -66.83	IRAS05486-3344	2.2± .9		.22					
054839.0-334518	PGC 17970		.97	.09					
055046.4-213403	NGC 2106	.LBS0..	1.43± .04	100					
226.48 -22.59	ESO 555- 3	S	.30± .04	.10	13.12 ±.14				
272.98 -73.71		-2.0± .5		.00					
054839.0-213448	PGC 17975		1.40						
055046.8-181016		.L..0P*	1.25± .06	125					2941± 15
223.08 -21.31	MCG -3-15- 21	SE	.08± .04	.16					2785
285.23 -74.45		-2.0± .6		.00					3002
054834.9-181101	PGC 17976		1.25						
055049.5-314422	A 0548-31	.LA.-*P	.97± .05	178					12000± 61
237.01 -25.96	ESO 424- 27	S	.26± .03	.00	14.99 ±.14				11818
245.70 -68.31		-3.0±1.3		.00					12063
054857.0-314506	PGC 17977		.93		14.81				
055054.3-144646		.SBS7..	1.47± .03	160				13.83±.1	904± 6
219.77 -19.94	MCG -2-15- 11	UE (1)	.12± .04	.55				131± 8	755
298.04 -74.51		7.2± .4		.16				105± 12	963
054838.2-144730	PGC 17978	5.3± .6	1.52	.06					
055056.9-385529		.L?....	1.05± .06	111					13625± 52
244.79 -27.85	ESO 306- 30		.12± .03	.11	14.85 ±.14				13433
234.20 -62.87				.00					13689
054917.0-385612	PGC 17979		1.05		14.54				
0551.0 -1451		.SBR3..	1.04± .07	85					
219.86 -19.93	MCG -2-15- 12	E (1)	.05± .05	.55					
297.77 -74.56		3.0± .9		.08					
0548.8 -1452	PGC 17985	4.2± .8	1.09	.03					
055115.3-533431		.S..3?/	1.24± .05	85					
261.33 -30.27	ESO 160- 2	S	.62± .04	.44	13.88 ±.14				4549
220.84 -50.13	IRAS05501-5335	2.5±1.3		.86	11.89				4346
055012.1-533512	PGC 17993		1.28	.31	12.54				4610
055119.0-381912	IC 2150	.SBR5*.	1.42± .03	84					
244.14 -27.63	ESO 306- 32	BSr (1)	.51± .04	.10	13.60 ±.14				
234.86 -63.40	IRAS05496-3819	4.7± .5		.76	13.12				
054938.0-381954	PGC 18000	1.1± .6	1.43	.25					

R.A. 2000 DEC. / l b / SGL SGB / R.A. 1950 DEC.	Names / PGC	Type / S_T n_L / T / L	$\log D_{25}$ / $\log R_{25}$ / $\log A_e$ / $\log D_o$	p.a. / A_g / A_l / A_{21}	B_T / m_B / m_{FIR} / B_T^o	$(B-V)_T$ / $(U-B)_T$ / $(B-V)_T^o$ / $(U-B)_T^o$	$(B-V)_e$ / $(U-B)_e$ / m'_e / m'_{25}	m_{21} / W_{20} / W_{50} / HI	V_{21} / V_{opt} / V_{GSR} / V_{3K}
0551.3 +7941 / 133.84 24.04 / 23.37 5.79 / 0542.7 +7940	UGC 3353 / IRAS05427+7940 / PGC 18004	.LB.... / U / -2.0±.8	1.00?	/ / .36 /	14.58 ±.16				
0551.4 +7415 / 139.36 22.11 / 21.95 .55 / 0544.9 +7415	UGC 3357 / / PGC 18005	.S...0.. / U / .0± .9	1.15± .05 / .48± .05 / / 1.18	58 / .61 / .36 /	14.90 ±.18				
0551.4 +4850 / 163.30 11.05 / 14.90 -23.92 / 0547.6 +4850	UGC 3366 / / PGC 18007	.S..6*. / U / 6.0±1.2	1.07± .14 / .11± .12 / / 1.21	50 / 1.44 / .16 / .05				15.28±.3 / 285± 11 / 261± 8 /	5856± 9 / / 5914 / 5863
0551.5 -1119 / 216.51 -18.36 / 310.70 -73.97 / 0549.2 -1120	/ / PGC 18010	.SXT4?. / E / 4.0±1.8	1.16± .08 / .11± .08 / / 1.29	170 / 1.30 / .17 / .06					
055135.4-590246 / 267.67 -30.61 / 217.60 -45.11 / 055053.0-590324	ESO 120- 16 / IRAS05508-5903 / PGC 18011	.S..3./ / S / 3.0± .8	1.35± .04 / .75± .04 / / 1.38	1 / .25 / 1.03 / .37	14.57 ±.14 / 12.11				
055136.1+394949 / 171.26 6.64 / 11.90 -32.53 / 054806.7+394905	UGC 3368 / / PGC 18013	.S..6*. / U / 6.0±1.3	1.11± .05 / .58± .05 / / 1.25	141 / 1.46 / .85 / .29				14.79±.3 / 319± 11 / 303± 8 /	5172± 7 / / 5199 / 5189
055140.0-180125 / 223.03 -21.06 / 285.69 -74.67 / 054928.0-180206	ESO 555- 5 / / PGC 18015	.SBS8.. / SE (2) / 8.0± .6 / 7.5± .6	1.10± .04 / .11± .04 / / 1.13	91 / .24 / .14 / .06	15.49 ±.14 / 15.10				3046 / / 2890 / 3108
055200.0-831752 / 295.55 -28.65 / 208.12 -22.15 / 055855.1-831812	ESO 4- 23 / / PGC 18024	.SB.5*P / S / 5.0±1.3	1.02± .05 / .41± .04 / / 1.10	20 / .82 / .62 / .21	15.51 ±.14				
0552.1 -1107 / 216.38 -18.14 / 311.56 -74.04 / 0549.8 -1108	/ / PGC 18027	.SXS5?. / E / 5.0±1.8	1.09± .09 / .09± .08 / / 1.22	100 / 1.30 / .13 / .04					
055211.5-072724 / 212.93 -16.55 / 323.23 -72.56 / 054946.5-072803	NGC 2110 / MCG -1-15- 4 / PGC 18030	.LX.-.. / E / -3.0± .8	1.23± .09 / .13± .07 / / 1.41	20 / 1.56 / .00 /					2284± 32 / 2153 / 2342
0552.2 -1406 / 219.26 -19.38 / 300.70 -74.75 / 0549.9 -1407	MCG -2-15- 13 / IRAS05499-1407 / PGC 18031	.SXR5.. / E (1) / 5.0± .8 / 1.9± .8	1.13± .06 / .09± .05 / / 1.19	80 / .73 / .13 / .04	13.10				
055212.6+414707 / 169.61 7.71 / 12.73 -30.70 / 054839.2+414625	UGC 3369 / / PGC 18033	.S..8*. / U / 8.0±1.2	1.19± .12 / .37± .12 / / 1.31	165 / 1.25 / .46 / .19					
055219.1-342052 / 239.89 -26.41 / 240.38 -66.63 / 055031.0-342130	ESO 364- 7 / IRAS05505-3421 / PGC 18034	.SBR5*P / S (1) / 5.0±1.2 / 2.2± .9	1.13± .06 / .07± .06 / / 1.14	/ .05 / .10 / .03	14.14 ±.14 / / / 13.98				3183± 87 / 2996 / 3248
0552.5 +6647 / 146.74 19.27 / 20.09 -6.69 / 0547.4 +6647	UGC 3365 / / PGC 18039	.S..1*. / U / 1.0±1.3	1.34± .05 / .88± .06 / / 1.39	154 / .50 / .90 / .44	14.91 ±.18 / / / 13.45			15.82±.1 / / 537± 8 / 1.94	5151± 11 / / 5266 / 5139
055236.3-174717 / 222.89 -20.76 / 286.51 -74.92 / 055024.0-174754	IC 2151 / ESO 555- 8 / IRAS05504-1747 / PGC 18040	.SBS4*. / SE (2) / 3.7± .7 / 4.3± .6	1.17± .03 / .22± .03 / / 1.19	99 / .25 / .32 / .11	14.16 ±.14 / 13.15 / / 13.51				13155 / 12999 / 13218
055300.3-175237 / 223.01 -20.71 / 286.12 -75.01 / 055048.1-175312	IC 438 / ESO 555- 9 / IRAS05508-1753 / PGC 18047	.SAT5.. / SUE (2) / 5.3± .4 / 1.6± .4	1.45± .02 / .12± .03 / / 1.47	55 / .24 / .18 / .06	12.75 ±.14 / 12.13 / / 12.32			13.64±.1 / 323± 9 / 311± 12 / 1.26	3120± 6 / / 2963 / 3184
055314.3-590359 / 267.70 -30.40 / 217.31 -45.17 / 055232.0-590430	ESO 120- 21 / / PGC 18051	.IB.9P. / S / 10.0± .8	1.08± .05 / .29± .05 / / 1.11	112 / .25 / .21 / .14	16.10 ±.14				

R.A. 2000 DEC. l b SGL SGB R.A. 1950 DEC.	Names PGC	Type S_T n_L T L	$\log D_{25}$ $\log R_{25}$ $\log A_e$ $\log D_o$	p.a. A_g A_i A_{21}	B_T m_B m_{FIR} B_T^o	$(B-V)_T$ $(U-B)_T$ $(B-V)_T^o$ $(U-B)_T^o$	$(B-V)_e$ $(U-B)_e$ m'_e m'_{25}	m_{2i} W_{20} W_{50} HI	V_{21} V_{opt} V_{GSR} V_{3K}
055349.8-324623 238.31 -25.66 242.51 -68.02 055159.0-324654	 ESO 364- 11 PGC 18058	.LA.+.. r -1.3± .9 	1.01± .06 .25± .03 .98	112 .13 .00 	 14.83 ±.14 				
0554.0 +7641 136.96 23.14 22.73 2.85 0546.8 +7641	 UGC 3364 PGC 18063	.SA.7.. U 7.0± .8 	1.02± .10 .00± .07 1.05	 .40 .00 .00				15.95±.1 138± 8 	4430± 11 4572 4407
055414.4-515832 259.55 -29.64 221.23 -51.77 055306.0-515900	 ESO 205- 7 FAIR 799 PGC 18066	.SBR3.. S (1) 3.0± .9 4.4± .9	.94± .05 .05± .04 .96	 .24 .06 .02	 15.26 ±.14 14.94				2000±190 1797 2064
055442.2+151137 192.90 -5.22 360.00 -55.70 055150.3+151107	 UGC 3376 IRAS05518+1509 PGC 18070	.SB?... 	1.00± .16 .14± .12 1.39	170 4.20 .21 .07	 12.27 			15.39±.3 105± 7 	3945± 7 3884 3992
055445.2-150804 220.52 -19.23 296.97 -75.47 055229.5-150832	 MCG -3-15- 27 IRAS05524-1508 PGC 18073	.SBR0?. E .0± .9 	1.07± .07 .47± .05 1.13	160 .87 .35 	 13.39 				
055453.9+462625 165.73 10.41 14.77 -26.39 055109.9+462555	 UGC 3374 IRAS05511+4625 PGC 18078	.SB?... 	1.32± .05 .15± .06 1.48	90 1.66 .22 .07	15.0 v±.2 12.08 13.09	.88± .01 -.29± .03 .44 -.60	.68± .02 -.69± .04 10.13± .25 16.08± .37	15.15±.1 417± 16 1.99	6141± 7 6189± 46 6191 6155
0555.1 +8555 127.34 26.07 25.10 11.76 0537.0 +8554	 UGC 3336 PGC 18084	.SBS3.. U 3.0± .8 	1.20± .05 .19± .05 1.23	45 .30 .26 .09	 15.1 ±.3 14.49			15.19±.1 229± 8 .60	5511± 11 5675 5478
0555.4 +5154 160.84 13.06 16.46 -21.15 0551.4 +5154	 UGC 3375 IRAS05514+5154 PGC 18089	.SAT5.. U 5.0± .8 	1.29± .04 .29± .05 1.39	45 1.07 .43 .14	 13.9 ±.3 12.70 12.40			14.38±.1 505± 8 1.84	5783± 7 5850 5790
055530.1-765515 288.31 -29.53 209.93 -28.31 055744.0-765530	IC 2160 ESO 33- 32 IRAS05576-7655 PGC 18092	PSBS5P* Sr (1) 4.7± .4 1.7±1.2	1.31± .03 .40± .04 .75± .01 1.36	107 .50 .60 .20	13.86 ±.13 13.76 ±.14 13.13 12.70	.74± .01 .52 	.77± .01 13.10± .03 14.28± .21		4730± 40 4537 4779
0555.6 -1540 221.13 -19.25 294.83 -75.71 0553.4 -1541	 PGC 18095	.SXS9*. E (1) 9.0± .9 8.7± .8	1.07± .09 .13± .08 1.13	85 .65 .13 .06					
0555.7 +0323 203.43 -10.77 347.84 -65.92 0553.0 +0323	A 0553+03 UGCA 116 2ZW 40 PGC 18096	CI...P. R 11.0± .5 	 .32± .03 	 2.44 	15.48 ±.13 	.82± .03 .07± .06 	.80± .01 .04± .02 12.57± .10 	14.34±.1 168± 4 143± 5 	789± 4 808± 63 689 847
055546.5-693336 279.86 -30.13 212.42 -35.36 055612.0-693354	NGC 2150 ESO 57- 55 IRAS05562-6933 PGC 18097	RSXR2*. Sr 2.1± .7 	1.05± .06 .09± .05 1.08	143 .32 .11 .04	13.6 ±.4 14.01 ±.14 11.52 13.49	.60± .04 -.01± .07 .48 -.08	 13.48± .50		4440± 51 4240 4495
055553.2-612411 270.42 -30.17 215.75 -43.10 055522.1-612430	 ESO 120- 23 PGC 18100	.LA.-.. S -3.0± .8 	1.06± .05 .19± .04 1.05	7 .15 .00 	 15.33 ±.14 				
0555.9 +8551 127.40 26.07 25.11 11.71 0538.0 +8551	 MCG 14- 3- 13 PGC 18101	.S?.... 	.85± .12 .48± .07 .88	 .30 .73 .24				16.68±.1 151± 8 	5756± 11 5920 5724
055608.3-380915 244.23 -26.68 233.32 -64.08 055427.0-380936	 ESO 307- 5 IRAS05544-3809 PGC 18105	.SBR1P* Sr 1.0± .7 	1.09± .05 .33± .05 1.10	102 .10 .34 .17	 15.05 ±.14 12.77 				
055616.6+483237 163.96 11.61 15.65 -24.43 055227.2+483212	 MCG 8-11- 12 VV 601 PGC 18109	.S?.... 	.99± .10 .00± .07 1.12	 1.38 .00 .00	 14.8 ±.3 13.35			15.03±.3 313± 11 277± 8 1.68	5757± 9 5780± 39 5814 5770
055626.5-340605 239.90 -25.52 239.11 -67.37 055438.0-340624	 ESO 364- 17 IRAS05546-3406 PGC 18117	.SXR6.. r 5.6± .9 	1.04± .06 .08± .05 1.04	160 .00 .12 .04	 14.12 ±.14 12.68 				

R.A. 2000 DEC. l b SGL SGB R.A. 1950 DEC.	Names PGC	Type S_T n_L T L	$\log D_{25}$ $\log R_{25}$ $\log A_e$ $\log D_o$	p.a. A_g A_i A_{21}	B_T m_B m_{FIR} B_T^o	$(B-V)_T$ $(U-B)_T$ $(B-V)_T^o$ $(U-B)_T^o$	$(B-V)_e$ $(U-B)_e$ m'_e m'_{25}	m_{21} W_{20} W_{50} HI	V_{21} V_{opt} V_{GSR} V_{3K}
0556.6 +7518 138.43 22.81 22.54 1.47 0549.8 +7518	A 0549+75 UGC 3371 DDO 39 PGC 18121	.I..9*. PU (1) 10.0± .5 9.0± .9	1.66± .04 .10± .06 1.14± .05 1.71	.51 .07 .05	14.7 ±.2 14.15	.46± .07 .01± .09 .31 -.10	.41± .06 .10± .07 15.92± .14 17.64± .31	13.62±.1 140± 5 129± 12 -.57	816± 6 954 796
0556.8 +7831 135.14 23.90 23.32 4.59 0548.8 +7831	 UGC 3370 PGC 18123	.S..4.. U 4.0±1.0	1.11± .14 .83± .12 1.14	134 .37 1.22 .42					
0557.1 +7631 137.20 23.25 22.86 2.64 0549.9 +7631	 UGC 3372 PGC 18128	.S..3.. U (1) 3.0± .9 4.5±1.2	1.04± .09 .66± .07 1.08	 .44 .92 .33					
055712.6-372838 243.56 -26.30 233.79 -64.76 055530.1-372854	 ESO 364- 18 PGC 18130	.E+2... S -4.0± .9	1.01± .05 .23± .03 .96	136 .14 .00	15.09 ±.14				
055717.1-522209 260.07 -29.23 220.29 -51.59 055610.0-522224	 ESO 205- 9 PGC 18133	.S..4./ S 3.7± .7	1.27± .05 .86± .04 1.29	106 .20 1.26 .43	15.51 ±.14				
0557.4 +1158 196.04 -6.25 358.39 -58.86 0554.6 +1158	NGC 2119 UGC 3380 PGC 18136	.E..... U -5.0± .8	1.08± .17 .08± .08 1.56	145 3.12 .00					
055734.1-520311 259.72 -29.15 220.42 -51.90 055626.1-520324	 ESO 205- 10 IRAS05564-5203 PGC 18139	.S..3?/ S 3.0±2.0	1.06± .05 .77± .04 1.08	168 .22 1.07 .39	16.03 ±.14 13.71				
055738.3-183533 224.17 -19.97 282.65 -76.00 055527.1-183548	 ESO 555- 14 IRAS05554-1835 PGC 18140	.S..3P* SE 3.0± .6	.97± .06 .23± .04 .99	143 .17 .32 .12	15.28 ±.14 13.59				
055741.4-515805 259.62 -29.12 220.44 -51.99 055633.0-515818	 ESO 205- 11 FAIR 800 PGC 18142	RSBR3*. S (1) 2.5± .6 3.3± .9	1.04± .05 .08± .05 1.06	21 .22 .12 .04	15.03 ±.14 14.63				8600±190 8396 8667
0557.8 +6444 148.96 18.93 20.10 -8.81 0552.9 +6444	 UGC 3377 PGC 18144	.S..6*. U 6.0±1.3	1.04± .06 .16± .05 1.09	110 .53 .24 .08					
055751.0-455130 252.76 -28.09 224.80 -57.56 055626.1-455142	 ESO 254- 6 PGC 18146	RLBR+.. r -1.3± .9	.92± .07 .04± .06 .93	15 .19 .00	14.60 ±.14				
055752.4-200504 225.66 -20.50 276.57 -75.73 055543.0-200518	NGC 2124 ESO 555- 16 IRAS05557-2005 PGC 18147	.SAS3?. SE (2) 3.0± .6 3.7± .7	1.43± .03 .49± .03 1.44	2 .12 .68 .25	13.44 ±.14 13.61				
055753.2-231052 228.75 -21.65 265.12 -74.61 055548.0-231106	IC 2152 ESO 488- 47 IRAS05559-2311 PGC 18148	RSBR1*. Sr 1.2± .5	1.23± .05 .16± .04 1.24	 .04 .16 .08	13.44 ±.14 13.83				
055804.0-595130 268.65 -29.83 216.11 -44.64 055725.4-595140	 PGC 18154	.S..7*. S (1) 7.0±1.2 6.7±1.2	1.11± .08 .17± .08 1.13	 .17 .23 .08					
055826.0+682739 145.35 20.44 21.05 -5.21 055303.1+682720	A 0553+68 UGC 3379 PGC 18161	PSX.3.. U 3.0± .8	1.25± .04 .18± .05 1.29	115 .42 .25 .09	13.68 ±.18 12.97			14.19±.1 372± 8 1.13	4114± 11 4233 4103
055840.1-252456 231.06 -22.28 257.63 -73.64 055638.1-252506	 ESO 488- 49 PGC 18169	.SBS8.. S (1) 8.0± .8 6.7± .9	1.22± .04 .40± .04 1.23	136 .09 .49 .20	15.01 ±.14 14.43				1798 1624 1872
055846.0-590735 267.82 -29.70 216.34 -45.36 055804.1-590742	NGC 2148 ESO 120- 24 IRAS05581-5907 PGC 18171	.SAT3P* S 2.5± .9 5.6±1.2	1.04± .05 .13± .05 1.06	150 .20 .18 .07	14.59 ±.14 13.63				

R.A. 2000 DEC. / l b / SGL SGB / R.A. 1950 DEC.	Names / PGC	Type / S_T n_L / T / L	$\log D_{25}$ / $\log R_{25}$ / $\log A_e$ / $\log D_o$	p.a. / A_g / A_i / A_{21}	B_T / m_B / m_{FIR} / B_T^o	$(B-V)_T$ / $(U-B)_T$ / $(B-V)_T^o$ / $(U-B)_T^o$	$(B-V)_e$ / $(U-B)_e$ / m'_e / m'_{25}	m_{21} / W_{20} / W_{50} / HI	V_{21} / V_{opt} / V_{GSR} / V_{3K}
055847.3-263908	NGC 2131	.IBS9*.	1.06± .05	118					1672
232.33 -22.69	ESO 488- 50	S (1)	.39± .04	.00	14.59 ±.14				
254.03 -72.95		10.0±1.3		.29					1495
055647.0-263918	PGC 18172	5.6± .9	1.06	.19	14.30				1746
055853.0-232021		.SXR1..	1.26± .04	130					
228.99 -21.49	ESO 488- 51	Sr	.35± .04	.04	14.24 ±.14				
264.18 -74.75		1.0± .6		.36					
055648.1-232030	PGC 18175		1.26	.18					
055858.7-481923		.LXT0P.	1.00± .05	8					
255.55 -28.35	ESO 205- 12	S	.21± .04	.24	15.01 ±.14				
222.58 -55.41		-2.0± .8		.00					
055740.0-481930	PGC 18178		1.00						
055901.9-522833		.S..2*.	1.02± .10						8900±190
260.23 -28.98		S	.17± .08	.20					8695
219.82 -51.60	FAIR 801	2.0±1.3		.21					8968
055755.2-522840	PGC 18180		1.04	.08					
0559.0 +7830	A 0551+78	.SX.5..	1.15± .08		13.81 ±.17	.63± .04	.72± .03	14.96±.1	4758± 10
135.20 23.99	UGC 3373	U (1)	.22± .07	.37	14.7 ±.2			287± 11	4821± 59
23.42 4.54		5.0± .8	1.00± .03	.33		.47	14.30± .07	266± 11	4906
0551.0 +7830	PGC 18181	3.3± .8	1.18	.11	13.45		13.85± .46	1.41	4737
055902.4-522706		.L..0/.	1.02± .11		14.6 ±.2	1.01± .03			
260.20 -28.98		S	.10± .08	.20		.38± .07			
219.84 -51.62		-2.0± .9		.00					
055755.6-522713	PGC 18182		1.02				14.31± .62		
055908.9-512806		.S..1*.	1.25± .04	64.					5300±190
259.10 -28.82	ESO 205- 13	S	.30± .04	.18	14.22 ±.14				5096
220.42 -52.54	FAIR 802	1.0±1.1		.30	13.31				5369
055759.1-512812	PGC 18187		1.27	.15	13.67				
055936.3-551953		.LAS0*.	1.13± .09						11000± 15
263.49 -29.25		S	.23± .08	.25					10795
218.08 -48.97		-2.0± .9		.00					11067
055839.2-551957	PGC 18196		1.13						
055937.7-600421		.SXT5P.	1.17± .04	4					
268.91 -29.65	ESO 120- 26	S (1)	.11± .04	.17	14.63 ±.14				
215.76 -44.50		5.0± .6		.16					
055900.0-600424	PGC 18197	2.6± .7	1.19	.05					
0559.7 +6208		.SBT1..	1.06± .06					15.69±.1	4496± 7
151.56 18.07	UGC 3382	U	.02± .05	.62	14.6 ±.2			188± 8	4596
19.69 -11.39		1.0± .8		.02				1.81	4494
0555.1 +6208	PGC 18200		1.11	.01	13.88				
055950.0-522427		.SBT2?.	1.20± .05	115					
260.17 -28.85	ESO 205- 14	S	.61± .04	.20	14.89 ±.14				
219.68 -51.71		2.0± .7		.75					
055843.0-522430	PGC 18202		1.21	.31					
055950.9-390759		.LXT0*P	1.23± .08						
245.49 -26.22	MCG -7-13- 1	S	.14± .08	.11					
230.73 -63.64		-2.0± .8		.00					
055811.6-390803	PGC 18204		1.22						
060004.8-335508	IC 2153	.P.....	1.02± .07	69					2836± 87
239.95 -24.74	ESO 364- 22	S	.11± .06	.07	14.16 ±.14				2647
237.84 -67.98	IRAS05582-3355	99.0			12.47				2911
055816.0-335512	PGC 18212		1.02						
060021.3-213957		.I?....	1.00± .05	48					1825
227.46 -20.56	ESO 555- 19	(1)	.21± .05	.14	15.68 ±.14				
269.65 -75.76				.16					1657
055814.1-214000	PGC 18221	5.6± .8	1.01	.11	15.38				1901
060023.9-194233		.SX.3?.	1.05± .04	175					
225.54 -19.81	ESO 555- 18	E (1)	.29± .03	.19	14.32 ±.14				
277.44 -76.40	IRAS05582-1942	3.0±1.3		.40	12.58				
055814.0-194236	PGC 18223	4.2±1.2	1.07	.14					
060027.3-600919		.SBR3..	1.14± .05	46					
269.02 -29.55	ESO 120- 27	S (1)	.60± .04	.17	15.68 ±.14				
215.58 -44.46		3.0± .6		.83					
055950.0-600918	PGC 18227	4.1± .7	1.15	.30					
060034.8-285934	A 0558-28	.SBS7*.	1.41± .03	153	14.1 ±.2	.40± .06	.42± .04	14.04±.2	1393± 7
234.87 -23.10	ESO 425- 2	SU (2)	.25± .04	.02	13.84 ±.14	.21± .08	.31± .05	182± 16	
247.24 -71.73	DDO 233	7.0± .4	1.21± .07	.35		.34	15.62± .21	177± 6	1211
055838.0-285936	PGC 18232	7.1± .5	1.41	.13	13.55	.17	15.36± .26	.36	1469

R.A. 2000 DEC. l b SGL SGB R.A. 1950 DEC.	Names PGC	Type S_T n_L T L	$\log D_{25}$ $\log R_{25}$ $\log A_e$ $\log D_o$	p.a. A_g A_i A_{21}	B_T m_B m_{FIR} B_T^o	$(B-V)_T$ $(U-B)_T$ $(B-V)_T^o$ $(U-B)_T^o$	$(B-V)_e$ $(U-B)_e$ m'_e m'_{25}	m_{21} W_{20} W_{50} HI	V_{21} V_{opt} V_{GSR} V_{3K}
060041.6-400241 246.52 -26.30 229.38 -62.94 055904.0-400242	 ESO 307- 13 PGC 18236	.E+3... S -4.0± .8 	1.06± .05 .21± .04 1.01	62 .14 .00 	 14.66 ±.14 14.31				13814± 62 13617 13889
060050.4-583536 267.23 -29.39 216.23 -45.95 060006.1-583534	 PGC 18246	.LBS0P. S -2.0± .8 	1.22± .08 .11± .08 1.22	 .16 .00 				
060055.0-504419 258.32 -28.44 220.45 -53.32 055943.0-504418	NGC 2152 ESO 205- 15 PGC 18249	PSBR1*P S 1.0±1.2 	1.04± .05 .16± .04 1.06	69 .12 .16 .08	 14.75 ±.14 				
060108.1-234033 229.53 -21.14 262.07 -75.04 055903.6-234033	NGC 2139 ESO 488- 54 IRAS05590-2340 PGC 18258	.SXT6.. (2) 6.0± .3 3.8± .7	1.42± .02 .13± .03 1.00± .01 1.43	140 .04 .19 .06	11.99 ±.13 12.17 ±.12 11.02 11.85	.36± .01 -.32± .02 .32 -.35	.44± .01 -.23± .02 12.48± .03 13.63± .18	13.00±.1 249± 8 201± 11 1.08	1843± 5 1828± 31 1670 1919
060108.2-214412 227.60 -20.42 269.09 -75.90 055901.0-214412	A 0559-21 ESO 555- 22 IRAS05590-2144 PGC 18259	.SBS4?. S (1) 4.0±1.1 3.3±1.2	1.37± .03 .53± .03 4.0±1.1 1.39	59 .15 .78 .27	 13.49 ±.14 12.39 				
0601.7 +7307 140.79 22.38 22.37 -.74 0555.5 +7307	 UGC 3384 PGC 18277	.S..9*. U 9.0±1.1 	1.24± .11 .00± .12 1.29	 .57 .00 .00				14.18±.1 86± 5 78± 5 	1086± 6 1218 1071
060206.3-583637 267.27 -29.23 216.00 -45.99 060122.1-583630	 ESO 121- 2 PGC 18293	PSBR3.. Sr (1) 3.1± .5 3.3± .7	1.04± .05 .10± .05 1.06	89 .16 .13 .05	 15.21 ±.14 				
0602.1 +3606 175.54 6.63 12.99 -36.69 0558.8 +3607	 UGC 3390 PGC 18299	.SX.8.. U 8.0± .8 	1.28± .06 .12± .06 1.42	35 1.41 .15 .06				14.08±.1 218± 5 202± 5 	1517± 4 1527 1551
060226.2-445349 251.89 -27.10 224.21 -58.78 060059.0-445342	 ESO 254- 9 PGC 18305	RSXR1.. r 1.0± .9 	1.08± .06 .52± .05 1.11	175 .27 .53 .26	 15.25 ±.14 				
060237.9+652217 148.56 19.64 20.74 -8.31 055738.5+652218	 UGC 3386 IRAS05576+6522 PGC 18312	.S..1*. U 1.0±1.2 	1.09± .06 .23± .05 1.14	43 .49 .24 .12	 14.36 ±.18 12.87 				
060242.8-123001 218.84 -16.38 309.33 -76.92 060023.8-122954	IC 441 MCG -2-16- 1 PGC 18315	.SBT5.. E (1) 5.0± .8 4.2±1.2	1.15± .06 .09± .05 1.27	40 1.29 .13 .04					2218 2070 2294
060248.2-634554 273.17 -29.46 213.74 -41.10 060230.1-634542	NGC 2178 ESO 86- 53 PGC 18322	.E.1... S -5.0± .9 	1.05± .06 .05± .04 1.07	 .17 .00 	 13.64 ±.14 				
0602.9 +6449 149.10 19.46 20.65 -8.85 0558.0 +6450	 UGC 3387 PGC 18327	.S..6*. U 6.0±1.3 	1.14± .05 .52± .05 1.19	155 .52 .77 .26					
060311.7+074938 200.39 -7.02 356.98 -63.19 060028.7+074946	A 0600+07 UGC 3393 2ZW 42 PGC 18336	.E?.... 	.54± .19 .00± .04 .94	 2.50 .00 				18.13±.3 	5263± 10 5386± 33 5185 5339
060336.6-203917 226.77 -19.47 272.50 -76.83 060128.0-203906	 ESO 555- 27 IRAS06014-2039 PGC 18349	.SBS7P. SUE (2) 7.0± .4 5.4± .6	1.36± .03 .09± .03 1.38	50 .21 .13 .05	 13.55 ±.14 13.34 13.21			13.83±.2 168± 16 147± 12 .57	1982± 11 1814 2062
060339.2-263954 232.75 -21.67 251.61 -73.79 060139.0-263942	 ESO 488- 59 PGC 18350	.LA.-*. S -3.5± .8 	1.13± .05 .22± .04 1.06	103 .00 .00 	 14.42 ±.14 				
060339.8-320848 238.35 -23.49 239.23 -69.84 060148.0-320836	 ESO 425- 4 PGC 18351	.SBR1.. r 1.0± .9 	1.10± .06 .18± .04 1.10	10 .00 .18 .09	 14.20 ±.14 				

R.A. 2000 DEC. l b SGL SGB R.A. 1950 DEC.	Names PGC	Type S_T n_L T L	$\log D_{25}$ $\log R_{25}$ $\log A_e$ $\log D_o$	p.a. A_g A_i A_{21}	B_T m_B m_{FIR} B_T^o	$(B-V)_T$ $(U-B)_T$ $(B-V)_T^o$ $(U-B)_T^o$	$(B-V)_e$ $(U-B)_e$ m'_e m'_{25}	m_{21} W_{20} W_{50} HI	V_{21} V_{opt} V_{GSR} V_{3K}
0603.7 +5730 156.19 16.61 19.11 -16.01 0559.4 +5731	UGC 3391 PGC 18352	.SA.6*. U 6.0±1.2	1.06± .06 .12± .05 1.14	.89 .18 .06					
060348.3-693459 279.86 -29.43 211.59 -35.54 060414.0-693442	NGC 2187A ESO 57- 68A PGC 18355	.SAS1?. S 1.0±1.0	1.39± .06 .34± .05 1.43	.37 13.06 ±.14 .35 .17				
060359.6-634159 273.11 -29.33 213.60 -41.20 060341.1-634142	ESO 86- 56 IRAS06036-6341 PGC 18359	.SXR1.. r 1.0± .9	1.13± .05 .33± .05 1.14	164 .17 14.47 ±.14 .34 .17				
0604.0 +0839 199.75 -6.43 358.23 -62.55 0601.3 +0840	UGC 3395 PGC 18360	.SXS3.. U 3.0± .8	1.04± .15 .00± .12 1.31	2.86 .00 .00				15.27±.3 224± 7 212± 7	5388± 8 5301 5454
060427.0-202115 226.56 -19.18 273.48 -77.12 060218.0-202100	ESO 555- 29 PGC 18369	.S..7?/ E 7.0±1.8	1.22± .04 .85± .03 1.25	14 .36 15.35 ±.14 1.18 .43				
060428.0-193721 225.85 -18.89 276.67 -77.34 060218.0-193706	ESO 555- 28 PGC 18370	.IBS9.. SE (2) 9.5± .6 9.9± .6	.97± .05 .36± .04 1.01	155 .41 17.15 ±.14 .27 .18 16.47				882 717 963
060430.8-530547 261.06 -28.25 218.23 -51.31 060326.0-530530	ESO 160- 11 PGC 18371	PSAT1P* S .7±1.0	1.11± .04 .25± .04 1.12	72 .15 14.88 ±.14 .26 .13				
060434.7-123730 219.15 -16.02 309.34 -77.39 060215.9-123715	MCG -2-16- 2 IRAS06022-1237 PGC 18373	.S..3?/ E 3.0±1.7	1.40± .04 .73± .05 1.54	105 1.44 1.00 12.65 .36				2231 2082 2310
060434.7+573740 156.13 16.76 19.25 -15.92 060015.6+573751	NGC 2128 UGC 3392 IRAS06002+5737 PGC 18374	.L..-*. U -3.0±1.1	1.18± .15 .13± .08 1.27	60 .89 13.60 ±.15 .00 13.24 12.66				3142±155 3226 3150
060443.0-260722 232.29 -21.26 252.58 -74.32 060242.0-260706	ESO 488- 60 PGC 18377	.SBT6.. SU (1) 6.3± .4 5.0± .6	1.45± .03 .37± .04 1.45	154 .00 13.63 ±.14 .54 .18 13.08			13.90±.2 253± 16 229± 12 .64	1814± 11 1635 1896
060444.7-732405 284.23 -29.24 210.34 -31.86 060555.0-732342	NGC 2199 ESO 34- 3 FAIR 247 PGC 18379	PSAR1*. Sr .8± .4	1.29± .05 .41± .05 1.33	37 .39 13.76 ±.14 .42 .21 12.89				4470± 63 4272 4526
0604.8 +5608 157.55 16.17 18.95 -17.37 0600.6 +5609	UGC 3394 PGC 18380	.SB?... 	1.26± .06 .00± .06 1.35	.91 14.7 ±.3 .00 .00 13.76			15.20±.1 135± 8 1.44	1821± 7 1900 1831
0604.9 +8007 133.58 24.74 24.05 6.07 0556.0 +8008	UGC 3385 PGC 18384	.L..... U -2.0± .9	1.15± .08 .43± .04 1.12	168 .34 14.89 ±.16 .00				
060517.5-275125 234.09 -21.74 247.58 -73.27 060319.1-275106	IC 2158 ESO 425- 7 IRAS06033-2751 PGC 18388	.SBR2*. Sr (1) 1.9± .5 4.4± .8	1.23± .04 .11± .04 1.23	.00 12.90 ±.14 .13 12.98 .05				
060522.0-454932 253.03 -26.80 222.59 -58.14 060357.0-454912	ESO 254- 12 PGC 18391	.SXR6.. r 5.6± .8	.95± .06 .13± .06 .97	50 .20 15.18 ±.14 .20 .07				
060539.9-863754 299.20 -27.75 206.93 -19.00 062157.0-863654	ESO 5- 4 IRAS06220-8636 PGC 18394	.S..3*/ S 3.3± .6	1.58± .03 .66± .05 1.64	93 .68 13.55 ±.14 .91 10.74 .33				
060545.3-330451 239.47 -23.36 236.60 -69.36 060355.0-330430	ESO 364- 29 PGC 18396	.IBS9.. S (1) 10.0± .4 10.0± .5	1.42± .04 .15± .05 1.42	52 .04 13.60 ±.14 .11 .08 13.45				790± 63 600 873

R.A. 2000 DEC. l b SGL SGB R.A. 1950 DEC.	Names PGC	Type S_T n_L T L	$\log D_{25}$ $\log R_{25}$ $\log A_e$ $\log D_o$	p.a. A_g A_i A_{21}	B_T m_B m_{FIR} B^o_T	$(B-V)_T$ $(U-B)_T$ $(B-V)^o_T$ $(U-B)^o_T$	$(B-V)_e$ $(U-B)_e$ m'_e m'_{25}	m_{21} W_{20} W_{50} HI	V_{21} V_{opt} V_{GSR} V_{3K}
060552.9-490205 256.58 -27.36 220.25 -55.20 060436.0-490142	 ESO 205- 19 FAIR 803 PGC 18400	.S?.... 	.94± .06 .11± .05 .95	 .13 .11 .05	 15.37 ±.14 14.98				 12300±190 12095 12377
060557.4-355746 242.48 -24.19 232.20 -66.98 060412.1-355724	 ESO 364- 30 PGC 18403	PSBS2.. r 2.2± .9 	1.06± .06 .52± .05 1.08	77 .20 .64 .26	 15.32 ±.14 				
060627.0-395149 246.64 -25.18 227.41 -63.63 060449.0-395124	 ESO 307- 17 IRAS06048-3951 PGC 18407	.S.1?P/ S 1.0±1.7 	1.23± .05 .65± .04 1.25	107 .19 .66 .32	 15.34 ±.14 13.02 				
0606.5 +6034 153.41 18.18 20.14 -13.09 0602.0 +6035	 UGC 3398 PGC 18409	.S..6*. U 6.0±1.2 	1.14± .05 .35± .05 1.20	112 .59 .52 .18				15.68±.1 283± 8 	7045± 11 7138 7050
060635.9-472956 254.91 -26.94 221.05 -56.67 060515.0-472930	 ESO 254- 17 PGC 18413	.E.2.?P S -5.0±1.7 	1.05± .05 .14± .04 1.03	125 .19 .00 	 14.63 ±.14 				
060636.4-275243 234.22 -21.48 246.81 -73.46 060438.0-275218	 ESO 425- 8 PGC 18414	.SB.9*. S (1) 9.0±1.2 11.1± .9	1.01± .05 .70± .06 1.01	80 .00 .72 .35	 16.57 ±.14 15.85				1478 1296 1563
060642.9-434350 250.81 -26.10 223.79 -60.16 060513.0-434324	 ESO 254- 16 PGC 18417	PSBS3.. r 3.3± .9 	.97± .06 .31± .05 1.01	152 .35 .42 .15	 15.46 ±.14 				
060651.1-752158 286.47 -28.99 209.64 -29.99 060833.0-752124	IC 2164 ESO 34- 5 FAIR 248 PGC 18424	PSAT2*. Sr (1) 2.1± .5 3.3± .9	1.03± .05 .07± .04 1.08	108 .49 .09 .04	 14.56 ±.14 14.14 13.87				 10790± 63 10594 10845
060730.2-614826 270.97 -28.82 213.78 -43.13 060701.0-614754	 ESO 121- 6 IRAS06070-6147 PGC 18437	.S.5P*/ S (1) 5.2± .5 5.0± .9	1.58± .03 .74± .05 1.60	41 .15 1.11 .37	10.5 ±.2 13.45 ±.14 11.08 	.81± .02 .38± .07 	 11.41± .29		
060741.6-195453 226.44 -18.30 274.28 -77.98 060532.0-195424	 ESO 555- 36 IRAS06055-1954 PGC 18444	.SBS5?/ S 5.0±1.3 	1.30± .05 1.01± .04 1.34	146 .47 1.50 .50	 15.60 ±.14 13.49 				
060744.7-232917 229.94 -19.66 259.51 -76.43 060540.0-232848	 ESO 489- 6 IRAS06056-2328 PGC 18445	.SBS7*. S (1) 7.3± .7 5.6± .6	1.27± .04 .21± .04 1.29	57 .19 .29 .11	 13.43 ±.14 13.16 12.94				2220 2045 2307
060745.1-472501 254.87 -26.73 220.77 -56.82 060624.0-472430	 ESO 254- 22 FAIR 805 PGC 18446	.SAT5.. S (1) 5.0± .6 2.8± .6	1.11± .05 .29± .04 1.13	135 .19 .43 .14	 14.61 ±.14 13.92				 12400±190 12195 12480
060802.2-214448 228.26 -18.94 266.11 -77.35 060555.0-214418	NGC 2179 ESO 555- 38 IRAS06059-2144 PGC 18453	.SAS0.. R .0± .3 1.25	1.23± .03 .16± .03 .83± .05 	170 .28 .12 	13.22 ±.17 13.27 ±.11 12.81	.90± .02 .35± .04 .78 .28	.98± .01 .42± .02 12.86± .16 13.83± .25	16.55±.2 416± 34 363± 25 	2798± 9 2731± 66 2625 2883
060823.8-523040 260.12 -27.58 217.65 -52.06 060717.0-523006	NGC 2191 ESO 160- 14 PGC 18464	.LBR0*. S -2.0± .6 	1.24± .05 .31± .04 1.22	118 .16 .00 	12.33V±.13 13.26 ±.14 13.03	.92± .02 .48± .04 .81 .44	 13.58± .30		 4514± 38 4307 4591
060835.5-484953 256.45 -26.88 219.64 -55.54 060718.1-484918	 ESO 205- 23 PGC 18472	.S..1?/ S 1.0±1.8 	1.17± .05 .61± .04 1.18	18 .17 .62 .30	 15.56 ±.14 				
060845.9-335504 240.54 -23.03 233.80 -69.02 060657.0-335430	 ESO 364- 33 IRAS06069-3354 PGC 18477	.S?.... 	.93± .06 .06± .05 .93	 .04 .06 .03	 14.51 ±.14 11.75 14.27				 11433± 62 11240 11519
060852.7-654351 275.45 -28.90 212.24 -39.38 060847.1-654312	 ESO 86- 62 PGC 18482	.E+4.*. S -4.0±1.2 	1.28± .05 .19± .04 1.25	173 .16 .00 	 13.57 ±.14 				

R.A. 2000 DEC.	Names	Type	$\log D_{25}$	p.a.	B_T	$(B-V)_T$	$(B-V)_e$	m_{21}	V_{21}
l　　b		S_T　n_L	$\log R_{25}$	A_g	m_B	$(U-B)_T$	$(U-B)_e$	W_{20}	V_{opt}
SGL　　SGB		T	$\log A_e$	A_i	m_{FIR}	$(B-V)_T^o$	m'_e	W_{50}	V_{GSR}
R.A. 1950 DEC.	PGC	L	$\log D_o$	A_{21}	B_T^o	$(U-B)_T^o$	m'_{25}	HI	V_{3K}

060853.5-250840		.SBS7..	1.06± .04	142					5904
231.68 -20.03	ESO 489- 8	S　　(1)	.33± .03	.00	14.66 ±.14				
253.24 -75.67	IRAS06068-2508	7.0± .9		.46	12.78				5726
060651.0-250806	PGC 18484	5.6±1.2	1.06	.17	14.18				5992
060857.6-274817		.SBS8..	1.13± .05	176					3030
234.34 -20.96	ESO 425- 10	S　　(1)	.23± .04	.00	13.72 ±.14				
245.67 -73.88	IRAS06069-2747	8.0± .9		.28	13.52				2847
060659.0-274742	PGC 18488	5.6± .9	1.13	.12	13.43				3118
060901.2-214323		.IBS9..	1.17± .04	30					1738
228.33 -18.72	ESO 555- 39	S　　(1)	.21± .04	.30	15.88 ±.14				
265.73 -77.57		9.5± .6		.15					1566
060654.0-214248	PGC 18490	10.0± .6	1.20	.10	15.43				1826
060908.1+420507		.S..1..	1.09± .06	18				15.86±.1	3604± 11
170.86　10.63	UGC 3407	U	.15± .05	1.53	13.7 ±.3				
16.32 -31.24	IRAS06055+4205	1.0± .9		.15	12.86			102± 12	3633
060534.1+420539	PGC 18494		1.23	.08	12.02			3.77	3639
060910.5-542426		.LB?...	.93± .07						7300±190
262.65 -27.76	ESO 160- 16		.16± .06	.25	14.90 ±.14				
216.54 -50.29	FAIR 806			.00					7092
060810.0-542348	PGC 18496		.93		14.54				7377
060935.7-620641		.SXR4?/	1.14± .05	82					
271.34 -28.59	ESO 121- 9	S　　(1)	.67± .05	.16	15.28 ±.14				
213.35 -42.90		4.3± .8		.98					
060908.0-620600	PGC 18510	2.6± .8	1.15	.33					
060936.4-200619		.S.6?P/	1.15± .04	39					
226.81 -17.97	ESO 555- 40	S	.74± .04	.45	16.20 ±.14				
272.66 -78.34		5.5± .9		1.09					
060727.0-200542	PGC 18511		1.19	.37					
060939.6-473721		.LA.-P.	.75?	15					
255.17 -26.46	ESO 205- 27	S	.08± .08	.15	14.73 ±.14				
220.09 -56.74		-3.0± .6		.00					
060819.0-473642	PGC 18514		.75						
060947.8-194344		PLX.-*.	1.16± .05	161					
226.47 -17.78	ESO 556- 1	S	.25± .04	.40	14.48 ±.14				
274.37 -78.51		-2.7± .6		.00					
060738.0-194306	PGC 18518		1.17						
0609.8　+6758		.S..7..	1.01± .06	135					
146.25　21.25	UGC 3402	U	.05± .05	.40					
21.97　-5.91		7.0± .9		.07					
0604.5　+6759	PGC 18520		1.04	.03					
060955.4-333903		.S?....	1.16± .06	102					8894± 50
240.35 -22.71	ESO 364- 35		.26± .05	.01	14.59 ±.14				
233.63 -69.38				.38					8701
060806.0-333824	PGC 18527		1.16	.13	14.13				8982
061000.4-333822		.S?....	1.05± .06						8663± 50
240.35 -22.69	ESO 364- 36		.19± .05	.01	14.39 ±.14				
233.61 -69.39	IRAS06081-3338			.28	13.44				8471
060811.0-333742	PGC 18532		1.06	.09	14.05				8751
061002.0-223733		.SXS8..	1.15± .05	128					
229.30 -18.85	ESO 489- 11	S　　(1)	.14± .05	.34	14.94 ±.14				
261.52 -77.33		8.0± .5		.17					
060756.0-223654	PGC 18533	8.1± .5	1.18	.07					
0610.1　+7955		.E.....							
133.87　24.90	UGC 3396	U			.35	15.1 ±.3			
24.23　5.82		-5.0± .8							
0601.3　+7956	PGC 18535								
061009.6-340622	NGC 2188	.SBS9./	1.64± .02	175	12.14 ±.13	.47± .01	.43± .01	13.38±.1	749± 6
240.84 -22.81	ESO 364- 37	R　　(2)	.59± .03	.02	12.20 ±.12	-.27± .02	-.25± .01	132± 8	727± 17
232.83 -69.01		9.0± .3	1.13± .01	.60		.34	13.28± .03	126± 11	553
060821.0-340542	PGC 18536	5.7± .7	1.65	.29	11.56	-.37	13.75± .19	1.53	835
061033.4-623221	NGC 2205	PLXT-*.	1.13± .06	80					8385± 15
271.83 -28.52	ESO 86- 63	S	.18± .04	.16	13.71 ±.14				
213.06 -42.52		-3.4± .6		.00					8178
061008.1-623136	PGC 18551		1.12		13.43				8456
061033.5-473943		.SBT6P*	1.07± .05	50					
255.24 -26.32	ESO 205- 29	Sr　　(1)	.24± .05	.15	15.30 ±.14				
219.81 -56.75		5.9± .5		.35					
060913.0-473900	PGC 18552	4.4± .6	1.08	.12					

R.A. 2000 DEC. l b SGL SGB R.A. 1950 DEC.	Names PGC	Type S_T n_L T L	$\log D_{25}$ $\log R_{25}$ $\log A_e$ $\log D_0$	p.a. A_g A_i A_{21}	B_T m_B m_{FIR} B_T^o	$(B-V)_T$ $(U-B)_T$ $(B-V)_T^o$ $(U-B)_T^o$	$(B-V)_e$ $(U-B)_e$ m' m'_{25}	m_{21} W_{20} W_{50} HI	V_{21} V_{opt} V_{GSR} V_{3K}
0610.5 +7952 133.93 24.90 24.24 5.76 0601.7 +7952	 UGC 3397 IRAS06017+7952 PGC 18553	.S..2.. U 2.0± .9 	.99± .10 .24± .07 1.02	 .35 .29 .12	 15.39 ±.20 13.76 				
0610.6 +7123 142.80 22.45 22.67 -2.58 0604.7 +7123	 UGC 3403 IRAS06047+7123 PGC 18557	.SB.6?. U 6.0±1.1 	1.36± .04 .52± .05 1.41	27 .52 .76 .26	 14.3 ±.2 12.72 13.07			14.62±.1 230± 8 1.29	1264± 7 1390 1255
0610.7 +6435 149.67 20.14 21.42 -9.25 0605.8 +6436	 UGC 3409 PGC 18558	.S..9*. U 9.0±1.3 	1.17± .05 .56± .05 1.21	10 .40 .57 .28				15.45±.1 123± 8 	1362± 11 1467 1364
061044.4-333807 240.39 -22.55 233.25 -69.48 060855.0-333724	 ESO 364- 38 PGC 18559	.S?.... 	1.05± .07 .83± .05 1.05	85 .01 1.24 .41	 16.46 ±.14 15.12				14872± 62 14679 14961
061051.0-335302 240.66 -22.61 232.82 -69.28 060902.1-335218	 ESO 364- 39 PGC 18566	.SAT2*P S 2.0± .6 	1.23± .05 .10± .05 1.24	7 .02 .12 .05	 13.81 ±.14 13.58				8890± 62 8697 8979
0610.8 +6200 152.23 19.21 20.95 -11.79 0606.2 +6201	 UGC 3411 PGC 18568	.S..6?. U 6.0±1.6 	1.23± .06 .19± .06 1.28	35 .60 .28 .09	 15.1 ±.4 14.23			15.27±.1 242± 8 .94	6798± 11 6895 6804
061116.4-213557 228.42 -18.19 265.10 -78.09 060909.0-213512	 ESO 556- 2 PGC 18583	.IAS9.. SU (1) 9.7± .4 10.0± .5	1.30± .04 .07± .04 1.34	 .42 .05 .03	 14.28 ±.14 13.81			13.40±.2 108± 16 91± 12 -.44	854± 11 681 945
0611.8 +8037 133.14 25.15 24.43 6.49 0602.5 +8038	 UGC 3400 PGC 18594	.S..8*. U 8.0±1.4 	1.04± .08 .76± .06 1.08	48 .39 .94 .38					
061208.7-233218 230.39 -18.74 256.79 -77.23 061004.0-233130	 ESO 489- 15 PGC 18601	 	.86± .07 .15± .06 .89	38 .29 	 14.63 ±.14 				1760± 35 1583 1852
061210.1-214824 228.71 -18.08 263.72 -78.17 061003.1-214736	NGC 2196 ESO 556- 4 IRAS06100-2147 PGC 18602	PSAS1.. R (3) 1.0± .3 1.8± .5	1.45± .02 .11± .03 1.16± .01 1.49	35 .45 .11 .06	11.82 ±.13 12.11 ±.11 12.57 11.39	.81± .01 .20± .04 .67 .10	.92± .01 .33± .01 13.11± .04 13.63± .19	13.90±.1 392± 5 391± 11 2.45	2321± 6 2301± 40 2147 2412
061216.6+641608 150.05 20.19 21.53 -9.60 060724.3+641651	 UGC 3414 7ZW 61 PGC 18607	.S?.... 	.83± .08 .00± .05 .87	 .48 .00 .00	 14.2 ±.2 12.47 13.66				4495± 61 4599 4499
0612.4 +5131 162.40 15.25 19.10 -22.13 0608.4 +5132	 UGC 3417 IRAS06084+5132 PGC 18608	.S..3.. U (1) 3.0±1.0 5.5±1.3	1.07± .14 .79± .12 1.17	165 1.04 1.10 .40	 15.6 ±.3 12.85 				
061226.1+442620 168.99 12.22 17.57 -29.07 060846.9+442706	 UGC 3418 IRAS06087+4427 PGC 18611	.SBT3.. U 3.0± .9 	1.02± .06 .02± .05 1.13	 1.22 .03 .01	 14.3 ±.3 12.75 13.00			16.42±.1 57± 8 3.41	6743± 7 6779 6777
061226.9-561610 264.81 -27.57 215.05 -48.64 061133.0-561518	 ESO 160- 18 IRAS06114-5615 PGC 18612	.L..+*/ S -1.0±1.4 	1.06± .05 .69± .03 .97	117 .14 .00	 15.39 ±.14 13.45 				
061232.9-215417 228.84 -18.03 263.08 -78.20 061025.9-215327	 PGC 18617	.IB.9*. S (1) 10.0± .9 10.0±1.2	1.10± .08 .31± .08 1.14	 .45 .23 .16					
061232.9-450428 252.53 -25.40 220.87 -59.30 061106.0-450336	 ESO 254- 37 PGC 18618	 	.96± .07 .10± .06 .99	117 .25 	 15.02 ±.14 				4437± 87 4232 4523
061253.7-332610 240.34 -22.06 232.43 -69.88 061104.0-332518	 ESO 364- 43 PGC 18636	.S?.... 	1.02± .06 .41± .05 1.03	75 .15 .61 .20	 15.09 ±.14 14.28				8866± 62 8672 8958

R.A. 2000 DEC. l　　b SGL　SGB R.A. 1950 DEC.	Names PGC	Type S_T　n_L T L	$\log D_{25}$ $\log R_{25}$ $\log A_o$ $\log D_o$	p.a. A_g A_i A_{21}	B_T m_B m_{FIR} B_T^o	$(B-V)_T$ $(U-B)_T$ $(B-V)_T^o$ $(U-B)_T^o$	$(B-V)_o$ $(U-B)_o$ m'_o m'_{25}	m_{21} W_{20} W_{50} HI	V_{21} V_{opt} V_{GSR} V_{3K}
061302.7-274353 234.61 -20.09 243.38 -74.54 061104.0-274300	 ESO　425- 14 PGC 18641	.LA.-*. S -2.7± .5	1.25± .05 .47± .05 .42± .01 1.18	108 .00 .00	13.83 ±.13 13.62 ±.14	.95± .01 .44± .02	1.09± .01 .52± .01 11.42± .03 13.77± .28		
0613.1　+7014 144.05　22.28 22.67　-3.74 0607.5　+7015	 UGC　3415 PGC 18646	.I..9?. U 10.0±1.9	1.19± .06 .81± .06 1.24	154 .52 .61 .41				15.40±.1 172±　8	3907± 11 4029 3901
061317.1-433949 251.05 -24.93 221.61 -60.67 061147.0-433854	NGC　2200 ESO　254- 39 PGC 18652	.SBR5.. BSr　(1) 5.0± .5 4.4± .8	1.00± .05 .05± .05 1.03	171 .24 .08 .03	14.89 ±.14				
061318.5-511902 259.34 -26.62 217.06 -53.43 061208.0-511806	 ESO　205- 34 IRAS06121-5117 PGC 18653	.SXS9.. S　　(1) 9.0± .6 7.2± .6	1.12± .05 .06± .04 1.13	 .15 .06 .03	13.89 ±.14 13.40				
061332.1-434220 251.11 -24.90 221.50 -60.65 061202.1-434124	NGC　2201 ESO　254- 40 PGC 18658	.SXR3*. Sr 2.9± .7	1.16± .05 .17± .05 1.18	113 .26 .23 .08	14.25 ±.14				
0613.5　+5304 161.01　16.05 19.59 -20.64 0609.5　+5305	 UGC　3424 PGC 18660	.I..9?. U 10.0±1.7	1.14± .13 .18± .12 1.24	145 1.02 .13 .09					
0613.5　+6944 144.58　22.15 22.62　-4.24 0608.0　+6945	A　0608+69 UGC　3416 PGC 18662	.S..6*. U 6.0±1.1	1.33± .03 .29± .04 1.38	90 .47 .43 .15	14.7　±.3 13.78			15.07±.1 286±　8 1.15	4002± 11 4062± 76 4124 3999
061335.5-531403 261.46 -26.93 216.09 -51.60 061231.0-531306	 ESO　160- 19 PGC 18663	.SXS4.. S　　(1) 4.0± .8 5.6±1.1	1.11± .05 .17± .05 1.13	15 .20 .25 .08	15.04 ±.14				
0613.6　+8104 132.68　25.33 24.58　6.92 0604.0　+8105	 UGC　3401 PGC 18664	.S..3.. U　　(1) 3.0± .9 3.5±1.3	1.25± .04 .90± .05 1.29	20 .40 1.25 .45	15.56 ±.18				
0613.8　+8000 133.83　25.08 24.40　5.87 0605.0　+8001	 UGC　3404 PGC 18672	.L..... U -2.0± .8		 .35	15.3　±.3				
061403.9-425122 250.23 -24.59 221.94 -61.47 061232.1-425024	 ESO　254- 43 PGC 18675	PSAR3.. r 3.3± .9	1.05± .05 .25± .05 1.06	57 .20 .34 .12	15.10 ±.14				
0614.2　+8028 133.33　25.20 24.49　6.33 0605.0　+8029	 UGC　3405 PGC 18682	.S..4.. U 4.0± .9	1.19± .07 .64± .07 1.22	49 .34 .93 .32	15.32 ±.19 14.02			14.39±.1 348±　8 .05	3791± 11 3872± 57 3944 3772
061430.2-535631 262.27 -26.92 215.58 -50.96 061328.0-535530	 ESO　160- 20 PGC 18689	RSBR0.. r -.1± .9	1.04± .06 .23± .05 1.05	6 .28 .18	14.94 ±.14				
061434.6-254747 232.82 -19.07 247.66 -76.19 061233.0-254648	 ESO　489- 20 IRAS06125-2546 PGC 18692	.S..1?P S 1.0±1.6	.96± .07 .03± .04 .98	 .16 .03 .02	14.17 ±.14 13.09				
0614.7　+6634 147.84　21.22 22.19　-7.38 0609.6　+6634	 UGC　3425 IRAS06096+6634 PGC 18699	.S..3.. U　　(1) 3.0± .9 4.5±1.2	1.40± .04 .86± .05 1.44	90 .40 1.19 .43	15.0　±.2 12.86 13.42			15.28±.1 419±　8 1.43	4057± 11 4168 4058
061505.3+001123 208.59　-7.99 353.81 -71.29 061231.4+001223	 UGC　3433 PGC 18705	.SXS4*. E　　(1) 4.0±1.2 3.1±1.6	1.31± .03 .51± .04 1.48	92 1.83 .75 .26					2339±　6 2220 2428
061507.3-192501 226.69 -16.50 273.73 -79.80 061257.0-192400	 ESO　556-　5 PGC 18708	.SBS4?. SE　　(1) 4.0± .7 3.3±1.2	1.15± .03 .28± .03 1.19	9 .44 .42 .14	14.57 ±.14				

R.A. 2000 DEC.	Names	Type	logD25	p.a.	BT	(B-V)T	(B-V)e	m21	V21
l b		ST nL	logR25	Ag	mB	(U-B)T	(U-B)e	W20	Vopt
SGL SGB		T	logAe	Ai	mFIR	(B-V)T^o	m'e	W50	VGSR
R.A. 1950 DEC.	PGC	L	logDo	A21	BT^o	(U-B)T^o	m'25	HI	V3K
061507.9+710812	A 0609+71A	.SXT3..	1.31± .04	43				14.34±.1	4059± 5
143.18 22.71	UGC 3422	U (1)	.10± .05	.49	14.1 ±.3			424± 5	
22.98 -2.88		3.0± .8		.14				404± 7	4183
060918.8+710905	PGC 18709	2.2± .8	1.36	.05	13.42			.87	4053
0615.1 +8027		.S..3..	1.24± .07	60				14.30±.1	3887± 11
133.36 25.24	UGC 3410	U	.68± .07	.34	14.99 ±.18				4026± 57
24.53 6.30		3.0± .8		.94				310± 12	4042
0606.0 +8028	PGC 18711		1.27	.34	13.67			.28	3870
061519.4-263432	A 0613-26	.IXS9..	1.24± .04		15.01 ±.18	.36± .10	.32± .05	15.27±.2	1799± 11
233.66 -19.20	ESO 489- 22	PSU (2)	.16± .04	.21	14.62 ±.14	-.24± .12	-.15± .07	97± 16	
244.92 -75.74	DDO 234	10.0± .5	1.08± .04	.12		.26	15.90± .08	69± 12	1615
061319.0-263330	PGC 18715	9.6± .7	1.26	.08	14.44	-.31	15.66± .30	.75	1895
0615.3 +6744		.S..8*.	1.15± .05	77				16.01±.1	3878± 11
146.68 21.66	UGC 3428	U	.54± .05	.43					
22.44 -6.24		8.0±1.3		.67				198± 8	3992
0610.1 +6745	PGC 18716		1.20	.27					3878
061529.0-223603		.SBT7..	1.09± .04	153					2053
229.79 -17.67	ESO 489- 23	S (1)	.15± .04	.29	14.88 ±.14				
258.33 -78.40		7.0± .6		.21					1877
061323.0-223500	PGC 18718	6.7± .6	1.12	.08	14.37				2150
0615.5 +7943		.S..6*.	1.02± .10					15.03±.1	4003± 11
134.15 25.08	UGC 3412	U	.36± .07	.43					
24.42 5.58		6.0±1.2		.53				259± 8	4151
0606.8 +7944	PGC 18719		1.06	.18					3983
061536.0+710204	A 0609+71B	.L...*.	1.26± .13		14.03M±.08	1.06± .01	1.14± .01	15.37±.1	3998± 6
143.30 22.72	UGC 3426	U	.06± .08	.49	13.55 ±.17	.18± .03	.16± .02	424± 16	4030± 24
23.01 -2.99	MK 3	-2.0±1.1	.23± .03	.00	11.92	.91	10.60± .12	258± 15	4124
060948.1+710300	PGC 18722		1.31		13.39	.08	15.06± .69		3994
0615.7 -7053		.SBS8..	1.30± .06					14.72±.3	1294± 9
281.36 -28.43		S (1)	.42± .08	.41					
210.03 -34.49		8.0± .8		.52				76± 7	1092
0616.4 -7052	PGC 18727	10.0± .9	1.34	.21					1358
0615.8 +6650		.S?....	1.04± .09						3926±113
147.60 21.41	MCG 11- 8- 25		.38± .07	.40	15.20 ±.18				4037
22.33 -7.13	IRAS06106+6651			.28	11.76				3928
0610.6 +6651	PGC 18729		1.06		14.45				
061551.3-695002		.E+3...	1.11± .10						
280.15 -28.39		S	.17± .08	.44					
210.28 -35.53		-4.0± .9		.00					
061618.9-694852	PGC 18730		1.12						
061559.2-264554	NGC 2206	.SXT4*.	1.38± .02	138	12.91 ±.13	.69± .01	.80± .01		6279± 10
233.90 -19.13	ESO 489- 26	SU (1)	.28± .04	.10	12.91 ±.14	.03± .07			6258± 45
243.95 -75.70	IRAS06140-2644	4.3± .6	.92± .02	.41	12.65	.57	13.00± .04		6094
061359.1-264448	PGC 18736	3.3±1.1	1.39	.14	12.36	-.06	13.96± .20		6375
061601.2+755613		.S..3..	1.41± .04					14.35±.1	5100± 7
138.19 24.14	UGC 3420	U (1)	.49± .06	.39	14.1 ±.2			509± 9	5110± 76
23.83 1.84		3.0± .8		.68				475± 12	5238
060901.1+755708	PGC 18739	3.5±1.1	1.45	.24	13.02			1.09	5087
061613.0-515026		.E+5...	1.10± .05	2					
260.02 -26.28	ESO 206- 1	S	.34± .04	.16	15.27 ±.14				
216.09 -53.05		-4.0± .6		.00					
061504.0-514918	PGC 18745		1.02						
0616.2 +5702		.S..6*.	1.22± .06	136				15.46±.1	4998± 11
157.34 17.98	UGC 3432	U	.65± .06	.63	15.28 ±.20				
20.73 -16.79		6.0±1.3		.95				292± 8	5077
0612.0 +5704	PGC 18747		1.28	.32	13.68			1.45	5016
061622.1-212221	NGC 2207	.SXT4P.	1.63± .02	141	12.2 ±.2	.68± .01	.76± .01	13.04±.2	2746± 7
228.68 -17.01	ESO 556- 8	R (3)	.19± .02	.53	11.59 ±.12		.23± .07	274± 8	2728± 26
263.17 -79.24	IRAS06142-2121	4.0± .3	.97± .05	.28	10.47	.50	12.54± .10	286± 25	2571
061414.4-212114	PGC 18749	2.3± .5	1.68	.10	10.90		14.72± .24	2.04	2843
061627.7-212231	IC 2163	.SBT5P.	1.48± .03	98	11.6 ±.3	.63± .03	.69± .01	13.43±.3	2688± 17
228.70 -16.99	ESO 556- 9	R	.39± .03	.53	12.55 ±.14			308± 34	2798± 63
263.10 -79.25		5.0± .5		.58		.42			2521
061420.0-212124	PGC 18751		1.53	.19	11.26		12.88± .34	1.98	2794
061632.0+321348		.SB?...	.96± .09					14.93±.3	7625± 7
180.41 7.41	UGC 3434		.00± .06	1.20					
15.44 -41.18	IRAS06132+3214			.00				121± 7	7615
061316.2+321453	PGC 18753		1.07	.00					7682

R.A. 2000 DEC. l b SGL SGB R.A. 1950 DEC.	Names PGC	Type S_T n_L T L	$\log D_{25}$ $\log R_{25}$ $\log A_e$ $\log D_o$	p.a. A_g A_i A_{21}	B_T m_B m_{FIR} B_T^o	$(B-V)_T$ $(U-B)_T$ $(B-V)_T^o$ $(U-B)_T^o$	$(B-V)_e$ $(U-B)_e$ m'_e m'_{25}	m_{21} W_{20} W_{50} HI	V_{21} V_{opt} V_{GSR} V_{3K}
0616.7 -1453 222.57 -14.29 300.58 -80.74 0614.5 -1452	 PGC 18759	.SX.5*. E 5.0±1.3	1.09± .09 .16± .08 .25 1.23	150 1.45 .08					
061701.8+810821 132.64 25.47 24.72 6.97 060719.9+810914	 UGC 3413 IRAS06072+8109 PGC 18764	.SBR4.. U 4.0± .7	1.33± .04 .12± .05 1.37	85 .40 .17 .06	 13.5 ±.2 12.80 12.94			14.36±.1 268± 8 1.36	4211± 7 4363 4189
061705.3-272310 234.61 -19.13 241.58 -75.38 061506.1-272200	 ESO 489- 29 IRAS06150-2722 PGC 18765	.S..4./ SU (1) 3.7± .5 4.4± .8	1.47± .03 .81± .03 1.48	2 .13 1.18 .40	 13.35 ±.14 12.56 12.02			14.90±.2 307± 16 306± 12 2.48	1696± 11 1510 1795
061716.7-553301 264.13 -26.79 214.33 -49.50 061620.0-553148	 ESO 160- 22 FAIR 811 PGC 18773	PSBR3*. Sr (1) 3.2± .6 2.2± .9	1.12± .05 .21± .05 1.14	33 .22 .30 .11	 14.10 ±.14 13.93 13.53				7400±190 7190 7483
061731.4-230424 230.43 -17.42 255.01 -78.50 061526.0-230312	 ESO 489- 31 PGC 18775	.SBS9.. S (1) 8.5± .6 9.4± .6	1.11± .05 .34± .05 1.14	42 .37 .34 .17	 16.12 ±.14				
0617.6 +7851 135.11 24.97 24.38 4.70 0609.4 +7852	 UGC 3423 PGC 18778	.S..8*. U 8.0±1.3	1.04± .09 .84± .07 1.07	 .40 1.04 .42				15.40±.1 222± 8	4273± 11 4419 4255
061742.8-563639 265.31 -26.91 213.86 -48.49 061650.1-563524	 ESO 160- 23 FAIR 812 PGC 18780	.LB?... 	.98± .07 .14± .05 .98	0 .20 .00	 14.87 ±.14 14.52				10100±190 9890 10182
061749.1-210337 228.52 -16.57 263.74 -79.68 061541.0-210224	 ESO 556- 12 IRAS06157-2101 PGC 18781	.SBS9.. SE (2) 8.7± .5 8.3± .5	1.21± .03 .25± .04 1.26	126 .58 .25 .12	 14.96 ±.14				
061820.1-660102 275.85 -27.95 210.94 -39.32 061816.0-655942	 ESO 87- 3 IRAS06182-6559 PGC 18791	PSBS4*. Sr (1) 3.7± .5 3.3±1.0	1.21± .05 .57± .04 1.23	112 .20 .83 .28	 14.65 ±.14				
061825.8-245106 232.24 -17.91 247.89 -77.47 061622.8-244950	 MCG -4-15- 22 PGC 18792	.E.0.*. S -5.0±1.2	1.02± .11 .03± .08 1.04	 .19 .00					
061830.5-183216 226.19 -15.41 277.26 -80.83 061619.0-183100	NGC 2211 ESO 556- 13 PGC 18794	.LBR0*. SE -2.3± .6	1.14± .04 .29± .03 1.16	22 .61 .00	 13.70 ±.14				
061835.5-183110 226.18 -15.39 277.33 -80.86 061624.0-182954	NGC 2212 ESO 556- 14 PGC 18796	.LBT+P* SE -1.0± .6	1.17± .04 .26± .02 1.19	136 .61 .00	 14.49 ±.14				
061840.1+782119 135.66 24.90 24.36 4.20 061042.7+782223	NGC 2146 UGC 3429 IRAS06106+7822 PGC 18797	.SBS2P. R (1) 2.0± .3 3.4± .7	1.78± .01 .25± .03 1.31± .01 1.81	56 .33 .30 .12	11.38 ±.13 11.06 ±.13 7.89 10.58	.79± .02 .29± .04 .66 .19	.87± .01 .35± .03 13.42± .03 14.51± .16	12.75±.0 499± 6 194± 11 2.05	893± 4 873± 12 1035 874
0618.7 +6326 151.14 20.57 22.10 -10.53 0614.0 +6328	 UGC 3436 PGC 18800	.SXS5.. U 5.0± .8	1.15± .07 .10± .06 1.19	 .41 .15 .05				15.05±.1 202± 8	4297± 7 4397 4306
061859.3-243748 232.08 -17.71 248.20 -77.72 061656.0-243630	 ESO 489- 35 PGC 18804	.LX.-.. S -3.0± .9	1.15± .05 .40± .04 1.12	138 .21 .00	 13.55 ±.14				
061904.2-165651 224.74 -14.64 287.23 -81.29 061650.8-165533	 MCG -3-16- 23 IRAS06168-1655 PGC 18805	.SB.4?. SE (1) 4.0±1.0 4.4±1.2	1.17± .04 .63± .04 1.27	5 1.08 .92 .31	 13.33				
061911.9+800411 133.82 25.32 24.64 5.90 061017.1+800515	IC 440 UGC 3427 IRAS06102+8005 PGC 18807	.SAR2.. U (1) 2.0± .8 3.5±1.1	1.22± .07 .28± .07 1.26	 .44 .35 .14	 14.16 ±.18 13.59 13.33			16.44±.1 339± 8 2.96	4344± 11 4305± 46 4491 4322

R.A. 2000 DEC. l b SGL SGB R.A. 1950 DEC.	Names PGC	Type S_T n_L T L	$\log D_{25}$ $\log R_{25}$ $\log A_e$ $\log D_o$	p.a. A_g A_i A_{21}	B_T m_B m_{FIR} B_T^o	$(B-V)_T$ $(U-B)_T$ $(B-V)_T^o$ $(U-B)_T^o$	$(B-V)_e$ $(U-B)_e$ m' m'_{25}	m_{21} W_{20} W_{50} HI	V_{21} V_{opt} V_{GSR} V_{3K}
061917.5-242756 231.94 -17.58 248.53 -77.88 061714.0-242636	 ESO 489- 37 PGC 18808	.LBR0.. S -2.0± .8 	1.22± .05 .44± .04 1.18	55 .21 .00 	13.56 ±.14 				
061919.2+423711 171.22 12.55 18.58 -31.10 061544.2+423828	 CGCG 203- 6 PGC 18809	 	 1.14 	 	14.8 ±.3 				7597± 57 7624 7642
061922.2+764846 137.31 24.56 24.16 2.67 061203.9+764955	 UGC 3431 PGC 18812	.E..... U -5.0± .8 	1.26± .13 .14± .08 1.28	173 .39 .00 	13.38 ±.15 				
0619.5 +6419 150.28 20.95 22.32 -9.67 0614.7 +6421	 UGC 3437 PGC 18815	.S..6*. U 6.0±1.2 	1.14± .05 .20± .05 1.18	35 .37 .29 .10					
0619.9 +5122 163.03 16.27 20.31 -22.48 0616.0 +5124	 UGC 3443 PGC 18825	.I..9?. U 10.0±2.0 	1.00± .16 .73± .12 1.08	166 .87 .54 .36					
062000.3-371142 A 0618-37 244.68 -21.88 224.30 -67.16 061817.0-371018	 ESO 365- 6 PGC 18828	.E?.... 	1.10± .09 .24± .06 1.07	92 .24 .00 	15.38 ±.14 15.00				9664± 55 9463 9763
062015.0-573438 NGC 2221 266.44 -26.72 ESO 121- 24 213.03 -47.63 IRAS06194-5733 061926.0-573312 PGC 18833		.S.1?P/ S 1.0±1.3 	1.29± .05 .64± .05 1.31	0 .17 .66 .32	13.86 ±.14 11.42 13.00 				2445± 54 2234 2527
062016.2-573151 NGC 2222 266.39 -26.71 ESO 121- 25 213.05 -47.67 061927.1-573024 PGC 18835		.SB1?P/ S 1.0±1.3 	1.08± .05 .57± .04 1.10	150 .19 .58 .29	14.21 ±.14 13.40 				2602± 76 2392 2685
0620.3 +6635 148.01 21.75 UGC 3438 22.74 -7.45 IRAS06151+6636 0615.2 +6636 PGC 18837		.SXS3.. U 3.0± .9 	.97± .07 .24± .05 1.01	101 .42 .34 .12	14.89 ±.18 				
062029.4-572946 266.36 -26.68 ESO 161- 1 213.01 -47.71 061940.1-572818 PGC 18839		.SBS8P* S (1) 8.3± .5 6.7± .6	1.07± .05 .28± .05 1.09	150 .19 .34 .14	14.81 ±.14 				
0620.5 +2758 184.61 6.22 UGC 3447 15.39 -45.52 0617.4 +2800 PGC 18841		.E...*. U -5.0±1.2 	1.08± .17 .29± .08 1.22	90 1.41 .00 	14.92 ±.18 				
0620.7 -1603 A 0618-16 224.08 -13.91 MCG -3-17- 1 293.11 -81.75 0618.5 -1602 PGC 18848		.SAT6*. PE (2) 6.0± .6 3.3± .7	1.19± .05 .17± .05 1.33	130 1.51 .25 .08				14.96±.1 216± 11 197± 11 	2852± 10 2827± 59 2687 2955
062047.5-275502 235.44 -18.56 ESO 425- 18 237.68 -75.45 061849.0-275336 PGC 18851		.SXS7.. S (1) 7.0± .8 6.7± .8	1.12± .05 .09± .05 1.14	 .16 .12 .04	14.36 ±.14 				
0620.9 -0827 217.07 -10.60 UGCA 127 336.74 -78.95 0618.5 -0826 PGC 18855		.S..6?. U (1) 6.0±1.5 3.1±1.6	1.59± .04 .54± .06 1.82	70 2.45 .80 .27				11.59±.1 315± 9 294± 12 	734± 7 588 836
0621.0 -0551 214.72 -9.41 345.48 -77.10 0618.6 -0550 PGC 18857		.S..7?/ E 7.0±1.8 	1.12± .08 .66± .08 1.32	46 2.13 .90 .33					
062105.5-200251 A 0618-20 227.88 -15.47 ESO 556- 15 266.90 -80.84 IRAS06189-2001 061856.1-200124 PGC 18858		.SBS1P. RSE 1.0± .6 	1.42± .02 .14± .03 1.50	141 .77 .14 .07	12.63 ±.14 11.24 11.69 			13.05±.2 352± 16 317± 12 1.29	1981± 11 1807 2086
062112.8+275131 184.78 6.29 UGC 3450 15.55 -45.67 061803.9+275257 PGC 18860		.SB.3.. U 3.0± .9 	.96± .09 .29± .06 1.09	162 1.40 .41 .15	15.30 ±.18 13.45 			16.19±.3 437± 7 2.59	6296± 10 6269 6366

R.A. 2000 DEC.	Names	Type	logD₂₅	p.a.	B_T	(B-V)_T	(B-V)_•	m₂₁	V₂₁
l b		S_T n_L	logR₂₅	A_g	m_B	(U-B)_T	(U-B)_•	W₂₀	V_opt
SGL SGB		T	logA_•	A_i	m_FIR	(B-V)_T^o	m'_•	W₅₀	V_GSR
R.A. 1950 DEC.	PGC	L	logD_o	A₂₁	B_T^o	(U-B)_T^o	m'_₂₅	HI	V_3K
062115.6-642732 NGC 2228		.LA.0*.	.88± .06						
274.12 -27.50 ESO 87- 7		S	.04± .04	.18	14.60 ±.14				7260± 63
210.94 -40.91 FAIR 249		-2.3± .8		.00					7052
062101.1-642600	PGC 18862		.90		14.31				7336
062123.5-645721 NGC 2229		.LXS-?/	1.16± .05	133					
274.68 -27.54 ESO 87- 8		S	.59± .03	.17	14.40 ±.14				
210.80 -40.42		-.7± .8		.00					
062112.0-645548	PGC 18867		1.09						
062126.2-280653		.LA.-*.	1.20± .05	106					
235.69 -18.50 ESO 425- 19		S	.13± .04	.13	13.38 ±.14				
236.77 -75.36		-3.0± .5		.00					
061928.0-280524	PGC 18871		1.20						
0621.4 +5106		.S..6*.	1.11± .05	146					
163.38 16.38 UGC 3449		U	.82± .05	.74	15.6 ±.3				
20.51 -22.78		6.0±1.4		1.21					
0617.5 +5108	PGC 18872		1.18		.41				
062127.2-645933 NGC 2230		PLA.-?.	1.06± .06						
274.72 -27.54 ESO 87- 9		S	.09± .04	.17	14.08 ±.14				
210.79 -40.39		-4.0± .6		.00					
062116.0-645800	PGC 18873		1.05						
062130.5-652451		.S?....	.96± .07						
275.20 -27.57 ESO 87- 10			.05± .06	.21	14.46 ±.14				4900
210.68 -39.97 IRAS06213-6523				.06	13.48				4693
062122.0-652318	PGC 18876		.98		.03 14.14				4975
062130.8-220511 NGC 2216		.SXR2*.	1.14± .04	20					
229.86 -16.19 ESO 556- 17		S	.10± .03	.48	13.70 ±.14				
256.17 -79.83 IRAS06194-2203		1.7± .7		.12	12.58				
061924.0-220342	PGC 18877		1.18		.05				
062133.2+590741 A 0617+59A		.S..0..	1.15± .07	101					
155.57 19.39 UGC 3445		U	.52± .06	.49	14.25 ±.20				3119± 58
21.78 -14.85		.0± .9		.39					3204
061708.1+590905	PGC 18878		1.17		13.32				3138
062135.0-444231		.SAR2..	1.04± .06	98					
252.58 -23.76 ESO 255- 5		r	.22± .05	.24	14.55 ±.14				
218.09 -60.15		2.2± .9		.26					
062007.1-444100	PGC 18879		1.06		.11				
062138.8-594421		.SBT4..	1.51± .04	115					
268.86 -26.87 ESO 121- 26		Sr (1)	.20± .05	.14	12.58 ±.14				
212.13 -45.55 IRAS06210-5942		4.2± .5		.30	12.52				
062059.0-594248	PGC 18880		3.3± .8 1.53		.10				
062138.7+590733 A 0617+59B		.L...*.	1.12± .10	150					
155.58 19.40 UGC 3446		U	.15± .05	.49	13.86 ±.15				3116± 58
21.79 -14.86		-2.0±1.2		.00					3201
061713.6+590858	PGC 18881		1.15		13.32				3135
062139.0-650158 NGC 2233		.L..-*/	.95± .06	45					
274.77 -27.52 ESO 87- 11		S	.59± .03	.17	14.86 ±.14				
210.75 -40.35		-3.5±1.0		.00					
062128.0-650024	PGC 18882		.80						
062139.6-271400 NGC 2217		RLBT+..	1.65± .02		11.71 ±.15	1.00± .01	1.04± .01	13.69±.1	1619± 5
234.85 -18.13 ESO 489- 42		R	.03± .02	.15	11.27 ±.11	.54± .02	.63± .01	299± 4	1609± 18
238.65 -76.12 IRAS06196-2712		-1.0± .3	1.00± .04	.00	12.55	.95	12.20± .14	242± 5	1431
061940.1-271230	PGC 18883		1.66		11.26	.51	14.78± .19		1723
062143.1-273336		.SXS9*.	1.25± .04	43					1807
235.17 -18.24 ESO 426- 1		S (1)	.43± .05	.23	15.05 ±.14				
237.83 -75.86		9.4± .5		.44					1620
061944.0-273206	PGC 18886		8.9± .5 1.27		.22 14.38				1912
0621.7 +7113		.S..6..	1.02± .06	130					
143.26 23.25 UGC 3440		U	.17± .05	.45	15.4 ±.2				
23.52 -2.88		6.0± .9		.25					
0615.9 +7115	PGC 18888		1.06		.08				
0621.8 +0021		.S..0..	1.03± .06	65					2728± 10
209.22 -6.41 UGC 3457		U	.11± .05	2.47					
358.70 -71.96 IRAS06192+0023		.0± .9		.08	11.47				2607
0619.2 +0023	PGC 18893		1.25						2827
062150.3-201331		.SXS9..	1.16± .04	100					1868
228.12 -15.38 ESO 556- 19		SE (2)	.05± .03	.72	14.88 ±.14				
265.38 -80.91		8.5± .5		.05					1694
061941.0-201200	PGC 18894		8.7± .6 1.23		.02 14.10				1974

R.A. 2000 DEC. l b SGL SGB R.A. 1950 DEC.	Names PGC	Type S_T n_L T L	$\log D_{25}$ $\log R_{25}$ $\log A_e$ $\log D_o$	p.a. A_g A_i A_{21}	B_T m_B m_{FIR} B_T^o	$(B-V)_T$ $(U-B)_T$ $(B-V)_T^o$ $(U-B)_T^o$	$(B-V)_e$ $(U-B)_e$ m'_e m'_{25}	m_{21} W_{20} W_{50} HI	V_{21} V_{opt} V_{GSR} V_{3K}
062157.5-363320 244.16 -21.30 223.97 -67.90 062013.0-363148	ESO 365- 10 IRAS06201-3632 PGC 18895	.SAS3*P S 3.0±1.2	1.29± .04 .53± .05 1.32	107 .22 .73 .26	14.74 ±.14				
062214.0-644343 274.44 -27.43 210.74 -40.66 062201.0-644206	ESO 87- 12 FAIR 250 PGC 18903	.L?....	.96± .06 .08± .05 .97	.18 .00	15.20 ±.14 14.91				7430± 63 7222 7506
062221.7-645601 274.67 -27.44 210.68 -40.46 062210.0-645424	NGC 2235 ESO 87- 13 PGC 18906	.E.2.*. S -4.7± .7	1.12± .06 .14± .04 1.11	68 .17 .00	14.00 ±.14				
062228.2+663443 148.09 21.95 22.95 -7.49 061720.7+663610	UGC 3448 PGC 18909	.S..0.. U .0± .9	1.11± .05 .64± .05 1.11	45 .42 .48	14.7 ±.2				
062234.7+515434 162.69 16.85 20.83 -22.01 061836.0+515604	NGC 2208 UGC 3452 PGC 18911	.L...*. U -2.0±1.2	1.22± .08 .20± .05 1.27	110 .73 .00	13.79 ±.15 12.98				5625±155 5685 5658
062239.3-243652 232.40 -16.93 245.32 -78.30 062036.0-243517	 PGC 18912	.LA.-.. S -3.0± .9	1.06± .10 .09± .08 1.09	.30 .00					
0622.8 +7342 140.67 23.99 23.94 -.43 0616.5 +7344	UGC 3444 PGC 18915	.I..9*. U 10.0±1.2	1.14± .13 .27± .12 1.17	60 .37 .21 .14				15.84±.1 163± 8	3348± 11 3479 3341
062259.6-625803 272.48 -27.14 211.06 -42.41 062236.1-625624	ESO 87- 14 PGC 18920		.86± .06 .35± .05 .88	10 .16	15.37 ±.14				5090± 87 4881 5169
062309.1-684400 278.95 -27.69 209.74 -36.73 062326.0-684218	ESO 57- 80 IRAS06234-6842 PGC 18923	.S..3?/ S 3.0±1.8	1.21± .04 .75± .04 1.23	117 .27 1.04 .38	15.03 ±.14 13.47				
062325.5-160942 224.45 -13.36 292.20 -82.39 062111.0-160805	MCG -3-17- 4 IRAS06211-1608 PGC 18930	.SXS8.. E (1) 8.0± .8 6.4± .8	1.15± .06 .20± .05 1.30	10 1.59 .25 .10					2893 2727 3001
0623.5 -1012 218.97 -10.78 331.47 -80.57 0621.2 -1011	 PGC 18942	.IBS9.. E (1) 10.0± .9 6.4±1.6	1.11± .08 .24± .08 1.38	160 2.85 .18 .12					
062345.9-321258 239.93 -19.50 227.99 -72.03 062154.0-321118	ESO 426- 2 IRAS06219-3211 PGC 18948	PSBR0.. Sr -.1± .5	1.19± .05 .03± .05 1.20	.11 .02	14.32 ±.14				
062346.1-605836 270.28 -26.79 211.43 -44.38 062312.1-605654	ESO 121- 28 PGC 18950		.94± .07 .10± .06 .96	11 .15	15.21 ±.14				12271± 54 12061 12353
062351.3-231140 231.14 -16.13 249.44 -79.52 062146.0-231000	ESO 489- 47 IRAS06217-2309 PGC 18953	.SAR4*. S (1) 4.0± .9 3.3± .9	1.10± .04 .13± .04 1.14	86 .40 .18 .06	14.40 ±.14 13.24				
062356.3+783151 135.53 25.19 24.64 4.34 061554.3+783318	NGC 2146A UGC 3439 IRAS06155+7833 PGC 18960	.SXS5*. PU 5.3± .6	1.48± .03 .42± .05 1.03± .03 1.52	30 .41 .62 .21	13.5 ±.2 13.8 ±.2 13.50 12.59	.63± .03 .02± .05 .45 -.11	.71± .02 .11± .04 14.14± .09 14.71± .28	13.62±.0 235± 5 224± 5 .83	1494± 4 1534± 49 1639 1479
062400.2+044238 205.60 -3.93 5.18 -68.19 062121.0+044417	UGC 3459 PGC 18964	.S..6?. U 6.0±1.7	1.04± .15 .13± .12 1.56	135 5.49 .19 .06				15.78±.3 299± 5	2874± 7 2765 2973
062419.8-373042 245.30 -21.16 221.94 -67.17 062237.0-372900	ESO 365- 15 PGC 18971	PSBS2.. r 2.2±1.0	1.04± .06 .61± .05 1.07	10 .33 .76 .31	15.38 ±.14				

R.A. 2000 DEC.	Names	Type	logD$_{25}$	p.a.	B$_T$	(B-V)$_T$	(B-V)$_{\bullet}$	m$_{21}$	V$_{21}$
l b		S$_T$ n$_L$	logR$_{25}$	A$_g$	m$_B$	(U-B)$_T$	(U-B)$_{\bullet}$	W$_{20}$	V$_{opt}$
SGL SGB		T	logA$_{\bullet}$	A$_i$	m$_{FIR}$	(B-V)^o_T	m'$_{\bullet}$	W$_{50}$	V$_{GSR}$
R.A. 1950 DEC.	PGC	L	logD$_o$	A$_{21}$	B^o_T	(U-B)^o_T	m'$_{25}$	HI	V$_{3K}$

0624.5 -5540		.SXS6..	1.19± .07						
264.47 -25.81		S	.45± .08	.34					
212.76 -49.60		6.0± .9		.66					
0623.6 -5539	PGC 18976		1.22	.23					
062435.8-225019	NGC 2223	.SXR3..	1.51± .02	175	*			13.87±.1	2722± 6
230.87 -15.83	ESO 489- 49	R (3)	.07± .03	.37 12.39 ±.12				338± 8	2660± 27
250.23 -79.89	IRAS06224-2248	3.0± .3		.10 12.94				300± 6	2539
062230.0-224836	PGC 18978	2.8± .4	1.54	.04 11.90				1.94	2829
062438.4-234356		.LBR0P.	1.19± .09		*				
231.73 -16.17		S	.26± .08	.40					
246.68 -79.26		-2.0± .8		.00					
062233.9-234213	PGC 18981		1.19						
062439.2-223549		.L...*/	1.15± .06	87	*				
230.65 -15.73	ESO 489- 50	S	.89± .03	.42 15.85 ±.14					
251.20 -80.06		-2.0±2.0		.00					
062233.1-223406	PGC 18983		1.07						
0624.8 +7205		.I..9*.	1.01± .06		*			15.77±.1	4156± 11
142.42 23.71	UGC 3454	U	.05± .05	.39					4282
23.87 -2.05		10.0±1.2		.03				77± 8	4153
0618.8 +7207	PGC 18986		1.05	.02					
0624.8 +7306		.SB?...	1.01± .05	50	*			15.46±.1	1044± 11
141.35 23.97	UGC 3453		.22± .04	.33 14.60 ±.18					1010±125
24.00 -1.05				.33				154± 8	1173
0618.6 +7308	PGC 18987		1.04	.11 13.94				1.41	1039
0624.9 +8219		.S?....	1.03± .06	122	*				4310
131.43 25.99	UGC 3435		.43± .05	.26 14.72 ±.19					4464
25.17 8.09	IRAS06140+8220			.60 11.59					4287
0614.1 +8220	PGC 18991		1.05	.22 13.83					
062455.3+493034		.S?....	.89± .11						5935±113
165.14 16.26	MCG 8-12- 24		.14± .07	.65 14.9 ±.3					5986
20.86 -24.44	IRAS06210+4932			.18 11.70					5974
062103.6+493214	PGC 18992		.95	.07 13.98					
062502.7+401740									6524± 38
173.84 12.53	CGCG 203- 9			1.36 15.0 ±.3					6541
19.38 -33.56									6579
062132.6+401922	PGC 18996								
062509.5+283310		.S..6*.	1.04± .08	0	*			15.97±.3	9299± 9
184.56 7.38	UGC 3462	U	.13± .06	1.18					9273
16.94 -45.15		6.0±1.2		.19				252± 5	9373
062159.6+283453	PGC 18998		1.15	.06					
062510.1-372022		RSBR2..	1.13± .05	148	*				
245.18 -20.95	ESO 365- 16	r	.39± .05	.35 15.10 ±.14					
221.68 -67.38		2.2± .9		.48					
062327.0-371836	PGC 19000		1.17	.20					
062517.3-374805		.LBT+..	1.00± .07	9	*				
245.66 -21.07	ESO 308- 5	r	.30± .05	.38 15.13 ±.14					
221.24 -66.95		-1.3±1.0		.00					
062335.0-374618	PGC 19003		1.00						
062535.7-225629		.S..4..	1.05± .04		*				
231.06 -15.66	ESO 489- 53	S (1)	.17± .03	.33 14.47 ±.14					
248.88 -79.98	IRAS06234-2254	4.0± .9		.25					
062330.1-225442	PGC 19014	4.4± .9	1.08	.08					
062552.5-275913		.S..4*/	1.19± .05	86	*				
235.95 -17.55	ESO 426- 8	S	.91± .04	.17 15.94 ±.14					
233.68 -75.99		4.3± .8		1.35					
062354.0-275724	PGC 19024		1.21	.46					
0625.9 +6444		.S..3..	1.38± .04	122	*				4292
150.10 21.72	UGC 3458	U (1)	.60± .05	.38 14.7 ±.2					4395
23.06 -9.36	IRAS06209+6445	3.0± .8		.83 13.01					4304
0620.9 +6445	PGC 19026	4.5±1.1	1.42	.30 13.51					
062558.0-220019	NGC 2227	.SBT5..	1.33± .04	19 13.20 ±.15	.71± .02	.76± .01		2261± 10
230.21 -15.21	ESO 556- 23	PS (2)	.27± .04	.37 13.40 ±.14	.18± .08			2221± 59
252.64 -80.68	IRAS06238-2158	5.0± .4	1.02± .02	.40 12.96	.56	13.79± .05		2080
062351.0-215830	PGC 19030	3.5± .4	1.36	.13 12.52	.07	14.03± .25		2371
062601.0-265955		.SXT4..	1.04± .05		*				
235.00 -17.14	ESO 489- 54	r	.08± .05	.29 14.83 ±.14					
235.76 -76.85		4.5± .8		.12					
062401.1-265806	PGC 19031		1.07	.04					

R.A. 2000 DEC. / l b / SGL SGB / R.A. 1950 DEC.	Names / PGC	Type / S_T n_L / T / L	$\log D_{25}$ / $\log R_{25}$ / $\log A_e$ / $\log D_o$	p.a. / A_g / A_i / A_{21}	B_T / m_B / m_{FIR} / B_T^o	$(B-V)_T$ / $(U-B)_T$ / $(B-V)_T^o$ / $(U-B)_T^o$	$(B-V)_e$ / $(U-B)_e$ / m' / m'_{25}	m_{21} / W_{20} / W_{50} / HI	V_{21} / V_{opt} / V_{GSR} / V_{3K}
062612.2-634512		.L?....	.96± .06						5414
273.41 -26.89	ESO 87- 20		.11± .05	.18	14.88 ±.14				5205
210.40 -41.70				.00					5494
062553.0-634318	PGC 19039		.96		14.62				
062622.7-314733		.SBR4..	1.05± .06	48					
239.72 -18.83	ESO 426- 9	r	.12± .04	.10	14.70 ±.14				
226.91 -72.65		4.5± .9		.18					
062430.0-314542	PGC 19047		1.06	.06					
0626.5 +7724		.S..0..	1.00± .08	160					
136.78 25.08	UGC 3456	U	.23± .06	.43	15.22 ±.19				
24.63 3.21		.0± .8		.17					
0619.0 +7726	PGC 19050		1.03						
0626.9 +5226		PSBS2..	.96± .09	5					
162.43 17.67	UGC 3465	U	.15± .06	.56	15.18 ±.19				
21.61 -21.57	IRAS06228+5228	2.0± .9		.18					
0622.9 +5228	PGC 19062		1.01	.07					
062655.2+590441	IC 2166	.SXS4..	1.48± .03	115				13.89±.1	2693± 5
155.87 20.02	UGC 3463	U	.15± .05	.43	13.2 ±.2			334± 8	2671± 76
22.47 -14.99		4.0± .7		.21				304± 11	2776
062230.5+590629	PGC 19064		1.52	.07	12.49			1.33	2716
0627.2 +1848		.S..8*.	1.11± .07	96				15.21±.3	2473± 7
193.48 3.35	UGC 3468	U	.54± .06	2.41				208± 7	
14.83 -54.81	IRAS06243+1850	8.0±1.3		.66				172± 7	2411
0624.3 +1850	PGC 19072		1.34	.27					2563
062718.1+493649		.S..2..	1.22± .06	20					
165.20 16.66	UGC 3467	U	.18± .06	.80	14.3 ±.2				
21.29 -24.39		2.0± .8		.22					
062326.2+493840	PGC 19073		1.30	.09					
062723.2-471045			1.07± .07	155	14.66 ±.14	.66± .05	.50± .04		11630± 19
255.47 -23.44	ESO 255- 7		.37± .06	.14	14.48 ±.14	-.15± .07	-.20± .06		11419
215.00 -58.01	IRAS06259-4708		.55± .03		10.85		12.90± .09		11730
062601.0-470848	PGC 19078		1.09						
062736.0-542658		.E+5...	1.15± .05	124	16.87 ±.14	1.16± .06			14520± 60
263.23 -25.13	ESO 161- 8	S	.38± .04	.28	15.02 ±.14				14307
212.45 -50.89		-4.0± .9		.00		.96			14613
062635.1-542500	PGC 19085		1.08		15.45		16.65± .31		
062757.9+741805	A 0621+74	.SBS8*.	1.25± .03	12				16.27±.3	5292± 17
140.13 24.48	UGC 3460	U	.31± .04	.36	14.7 ±.2			336± 34	4800± 63
24.36 .11	MK 4	8.0±1.2		.38				155± 8	5391
062127.7+741953	PGC 19094		1.28	.15	13.90			2.21	5253
0628.0 +8318		RSX.1?.	.88± .10	55					
130.36 26.25	UGC 3441	U	.30± .06	.25	15.34 ±.18				
25.38 9.06		1.0± .9		.31					
0616.0 +8320	PGC 19095		.90	.15					
062813.9-485148			.74± .07		15.40 ±.15	1.07± .05	1.09± .05		
257.29 -23.74	ESO 206- 12		.03± .05	.17	15.56 ±.14	.30± .08	.34± .08		
214.09 -56.40			.49± .04				13.34± .12		
062656.0-484948	PGC 19098		.76						
0628.4 -1706		.SBS1?.	1.09± .09	125					
225.86 -12.68		E	.13± .08	1.36					
283.23 -83.47		1.0±1.8		.13					
0626.2 -1705	PGC 19105		1.22	.06					
062828.2-484543		.SXS5..	1.10± .05						15000±190
257.19 -23.68	ESO 206- 14	S (1)	.08± .05	.17	14.92 ±.14				14788
214.06 -56.50	FAIR 815	5.0± .9		.12					15100
062710.1-484342	PGC 19106	5.6±1.2	1.12	.04	14.54				
0628.7 +5602		.S..6*.	1.07± .07	10					
159.00 19.20	UGC 3469	U	1.09± .06	.58					
22.34 -18.04		6.0±1.5		1.47					
0624.5 +5604	PGC 19112		1.12	.50					
062851.3-443957		.SBR1..	1.07± .06	89					
252.92 -22.51	ESO 255- 11	r	.42± .05	.25	14.65 ±.14				
215.59 -60.52	IRAS06273-4438	1.0± .9		.43					
062723.0-443754	PGC 19117		1.10	.21					
062913.8-262945		.SAR0..	1.10± .05						
234.80 -16.29	ESO 490- 6	r	.05± .03	.56	13.88 ±.14				
234.16 -77.66		-.1± .8		.04					
062713.0-262742	PGC 19126		1.15						

R.A. 2000 DEC.	Names	Type	logD$_{25}$	p.a.	B$_T$	(B-V)$_T$	(B-V)$_\bullet$	m$_{21}$	V$_{21}$
l b		S$_T$ n$_L$	logR$_{25}$	A$_g$	m$_B$	(U-B)$_T$	(U-B)$_\bullet$	W$_{20}$	V$_{opt}$
SGL SGB		T	logA$_\bullet$	A$_i$	m$_{FIR}$	(B-V)^o_T	m'$_\bullet$	W$_{50}$	V$_{GSR}$
R.A. 1950 DEC.	PGC	L	logD$_o$	A$_{21}$	B^o_T	(U-B)^o_T	m'$_{25}$	HI	V$_{3K}$

062917.6-172133		.LX.-*/	1.27± .06	85					
226.17 -12.59	MCG -3-17- 5	ES	.36± .05	1.27					
280.69 -83.62		-3.3± .6		.00					
062704.6-171930	PGC 19127		1.38						
0629.3 +7429		.S..1?.	1.15± .05	46					
139.95 24.62	UGC 3464	U	.56± .05	.37	15.32 ±.19				
24.48 .30	IRAS06228+7431	1.0±1.8		.57	13.66				
0622.8 +7431	PGC 19128		1.18	.28					
062934.3-263758		.LX.0*.	1.07± .09						
234.97 -16.27	MCG -4-16- 6	S	.02± .05	.61					
233.53 -77.58		-2.0± .7		.00					
062733.7-263553	PGC 19133		1.14						
062950.3-575256			.89± .06	45					9975± 63
267.02 -25.51	ESO 121- 34		.16± .06	.38	15.34 ±.14				9762
211.10 -47.56									10065
062902.0-575048	PGC 19143		.92						
063020.9-321644		PSBS3..	.92± .06						
240.52 -18.22	ESO 426- 18	r	.00± .05	.24	14.94 ±.14				
223.69 -72.51	IRAS06284-3214	3.3± .9		.00	13.04				
062829.0-321436	PGC 19155		.94	.00					
063029.0+393014		.S..9*.	1.36± .05	85				13.96±.1	486± 6
175.03 13.14	UGC 3475	U	.28± .06	1.25	14.5 ±.3			177± 5	
20.48 -34.49		9.0±1.1		.29				167± 4	498
062700.6+393219	PGC 19161		1.48	.14	12.99			.83	549
063029.3+331807		.I..9*.	1.00± .08	63					469
180.77 10.51	UGC 3476	U	.55± .06	2.05	14.9 ±.3				
19.47 -40.64		10.0±1.3		.41					458
062711.9+332013	PGC 19162		1.19	.27	12.45				542
0630.9 +5936		.S..5..	1.09± .06	43					
155.52 20.67	UGC 3473	U	.91± .05	.32					
23.05 -14.53	IRAS06264+5938	5.0±1.0		1.37					
0626.4 +5938	PGC 19170		1.12	.46					
063057.7-234346		.SBS9..	1.20± .04						2840
232.32 -14.84	ESO 490- 7	S (1)	.21± .04	.48	14.54 ±.14				
240.40 -80.19		9.0± .8		.22					2656
062853.1-234136	PGC 19173	8.9± .8	1.24	.11	13.84				2959
063102.1-713005		.S..6*/	1.16± .05	6					
282.11 -27.23	ESO 58- 3	S	.98± .04	.45	16.03 ±.14				
208.45 -34.09		6.0±1.4		1.43					
063146.0-712748	PGC 19175		1.20	.49					
0631.0 +8455		.S..0..	1.04± .08	110					
128.59 26.56	UGC 3442	U	.50± .06	.26	14.90 ±.19				
25.64 10.66		.0± .9		.38					
0616.0 +8457	PGC 19177		1.04						
063109.8-522507		.IXS9..	1.13± .07					15.05±.3	1190± 9
261.19 -24.16	ESO 206- 16	S (1)	.08± .05	.19	17.14 ±.14				977
212.18 -52.99		10.0± .8		.06				69± 7	
063002.0-522254	PGC 19180	10.0± .8	1.15	.04	16.89			-1.87	1288
063122.1+500537		.S..7..	1.28± .06	0				14.67±.1	6492± 7
164.99 17.45	UGC 3477	U	.71± .06	.73	15.4 ±.2				6543
22.06 -23.99		7.0± .9		.97				397± 8	6536
062729.1+500745	PGC 19186		1.35	.35	13.63			.69	
063135.8+742553	A 0625+74	.SAS5..	1.22± .03		13.58 ±.18	.77± .04	.78± .02	15.03±.1	5578± 10
140.05 24.75	UGC 3471	U (1)	.01± .03	.39	14.4 ±.3			248± 11	5568± 59
24.62 .22	IRAS06248+7428	5.0± .8	.99± .04	.01	12.83	.64	14.02± .11	232± 11	5710
062503.9+742757	PGC 19191	1.8± .8	1.26	.00	13.37		14.52± .27	1.66	5573
063153.5-201002		.L..0P/	1.29± .04	85					
229.06 -13.20	ESO 557- 3	SE	.39± .03	1.52	14.61 ±.14				
256.40 -82.92	IRAS06297-2007	-1.8± .7		.00	13.17				
062944.1-200748	PGC 19198		1.40						
063157.4-264609		.SBT7?P	1.33± .03	33					1824
235.31 -15.83	ESO 490- 10	S (1)	.24± .03	.56	13.69 ±.14				
231.03 -77.71		7.0±1.5		.33					1634
062957.0-264354	PGC 19201	6.7± .8	1.38	.12	12.80				1943
063220.1-673851		.SBT4P.	.97± .06						
277.82 -26.74	ESO 58- 4	r	.09± .05	.26	14.97 ±.14				
208.88 -37.93		4.5± .9		.14					
063227.0-673630	PGC 19211		1.00	.05					

R.A. 2000 DEC. / l b / SGL SGB / R.A. 1950 DEC.	Names / PGC	Type / S_T n_L / T / L	$\log D_{25}$ / $\log R_{25}$ / $\log A_e$ / $\log D_o$	p.a. / A_g / A_i / A_{21}	B_T / m_B / m_{FIR} / B_T^o	$(B-V)_T$ / $(U-B)_T$ / $(B-V)_T^o$ / $(U-B)_T^o$	$(B-V)_e$ / $(U-B)_e$ / m'_e / m'_{25}	m_{21} / W_{20} / W_{50} / HI	V_{21} / V_{opt} / V_{GSR} / V_{3K}
063221.1-675345 / 278.10 -26.76 / 208.85 -37.68 / 063230.1-675124	ESO 58- 5 / FAIR 253 / PGC 19212	RLX.0.. / r / -2.4± .9 /	1.02± .07 / .13± .05 / / 1.03	130 / .26 / .00 /	/ 14.39 ±.14 / / 14.07				4020± 63 / 3813 / 4097
063223.5+351124 / 179.20 11.68 / 20.29 -38.82 / 062903.0+351338	UGC 3479 / PGC 19215	.S..2.. / U / 2.0± .9 /	1.04± .08 / .59± .06 / / 1.17	179 / 1.39 / .72 / .29	/ 14.9 ±.3 / / 12.66			15.43±.3 / / 432± 7 / 2.48	7238± 10 / / 7234 / 7311
063234.8+401230 / 174.54 13.80 / 21.06 -33.84 / 062905.0+401444	UGC 3481 / IRAS06290+4014 / PGC 19221	.I?.... / / /	1.19± .12 / .74± .12 / / 1.29	96 / 1.07 / .55 / .37	/ 15.2 ±.3 / / 13.62				6378± 38 / 6392 / 6442
0632.6 +7133 / 143.15 24.17 / 24.41 -2.65 / 0626.7 +7135	UGC 3474 / IRAS06267+7135 / PGC 19222	.S..6*. / U / 6.0±1.3 /	1.35± .03 / .98± .04 / / 1.39	158 / .33 / 1.44 / .49	/ 15.40 ±.19 / 13.35 / 13.61			14.38±.1 / / 360± 8 / .28	3634± 11 / 3502± 76 / 3755 / 3633
063247.3+634026 / 151.43 22.12 / 23.69 -10.50 / 062759.7+634238	UGC 3478 / IRAS06280+6342 / PGC 19228	.S..3.. / U (1) / 3.0± .9 / 2.5±1.2	1.21± .05 / .49± .05 / / 1.24	42 / .37 / .68 / .25	/ 13.6 ±.2 / 12.64 / 12.48			14.89±.1 / 389± 34 / 361± 8 / 2.17	3830± 9 / / 3928 / 3848
063249.0-340743 / 242.53 -18.40 / 220.28 -70.91 / 063100.1-340524	ESO 365- 27 / / PGC 19229	.SBR5.. / Sr (1) / 4.8± .5 / 2.2± .6	1.18± .04 / .18± .04 / / 1.22	126 / .44 / .27 / .09	/ 14.62 ±.14 / /				
0632.8 +7552 / 138.52 25.13 / 24.83 1.64 / 0625.9 +7554	UGC 3472 / IRAS06259+7554 / PGC 19233	.L..... / U / -2.0± .8 /	1.18± .05 / .34± .03 / / 1.18	60 / .45 / .00 /	/ 15.1 ±.2 / / 14.57				5203±125 / 5339 / 5195
063252.9-254319 / 234.39 -15.23 / 232.51 -78.73 / 063051.0-254100	ESO 490- 12 / IRAS06308-2541 / PGC 19234	.LBR+*. / Sr / -1.2± .7 /	1.09± .05 / .22± .03 / / 1.10	118 / .44 / .00 /	/ 14.23 ±.14 / 13.78 /				
0633.0 +5332 / 161.70 18.93 / 22.72 -20.58 / 0629.0 +5335	UGC 3480 / PGC 19237	.SXT5.. / U / 5.0± .8 /	1.04± .06 / .00± .05 / / 1.10	/ .60 / .00 / .00	/ 14.9 ±.2 / /				
063313.6-280133 / 236.63 -16.06 / 227.45 -76.69 / 063115.0-275912	ESO 426- 22 / / PGC 19238	PSBR2.. / r / 2.2± .9 /	1.02± .05 / .09± .03 / / 1.05	/ .39 / .11 / .05	/ 14.84 ±.14 / /				
0633.2 +5120 / 163.88 18.18 / 22.52 -22.78 / 0629.3 +5123	UGC 3483 / PGC 19239	.S..4.. / U / 4.0±1.0 /	1.04± .08 / .76± .06 / / 1.09	97 / .56 / 1.12 / .38					
063319.2-625936 / 272.69 -25.99 / 209.49 -42.56 / 063255.0-625712	ESO 87- 28 / / PGC 19242	.L..OP. / S / -2.0± .9 /	1.06± .05 / .20± .04 / .87± .05 / 1.05	123 / .20 / .00 /	14.5 ±.2 / 14.52 ±.14 / / 14.18	.98± .04 / .84 / /	1.06± .03 / / 14.37± .17 / 14.17± .35		8444± 15 / 8233 / 8530
0633.4 +8252 / 130.85 26.35 / 25.50 8.62 / 0622.0 +8255	UGC 3461 / PGC 19245	.S..9*. / U / 9.0±1.2 /	1.04± .15 / .04± .12 / / 1.06	/ .26 / .04 / .02					
0633.5 +2102 / 192.18 5.68 / 17.93 -52.88 / 0630.5 +2104	UGC 3489 / IRAS06305+2104 / PGC 19249	.S..4.. / U / 4.0± .9 /	1.30± .10 / .84± .12 / / 1.44	123 / 1.50 / 1.24 / .42	/ / 12.26 /			14.55±.3 / 497± 7 / 475± 7 /	5456± 7 / / 5399 / 5551
063335.8-341541 / 242.72 -18.30 / 219.69 -70.83 / 063147.0-341318	ESO 365- 28 / / PGC 19250	.SXS5*P / S / 5.0±1.1 /	1.21± .04 / .27± .04 / . / 1.24	50 / .38 / .40 / .13	/ 14.48 ±.14 / /				
063339.9+404117 / 174.17 14.18 / 21.37 -33.39 / 063009.3+404336	UGC 3487 / IRAS06301+4042 / PGC 19252	.S..7.. / U / 7.0± .8 /	1.09± .08 / .18± .06 / / 1.18	110 / 1.01 / .24 / .09	/ 14.8 ±.3 / / 13.49			15.29±.1 / / 316± 12 / 1.71	5211± 7 / / 5227 / 5276
063343.0-245904 / 233.77 -14.77 / 233.52 -79.47 / 063140.0-245642	ESO 490- 14 / IRAS06316-2456 / PGC 19255	.SBS4*. / S (1) / 4.0±1.1 / 4.4± .8	1.22± .04 / .15± .04 / / 1.27	16 / .56 / .22 / .07	/ 13.68 ±.14 / 13.17 /				

R.A. 2000 DEC. / l b / SGL SGB / R.A. 1950 DEC.	Names / PGC	Type / S_T n_L / T / L	$\log D_{25}$ / $\log R_{25}$ / $\log A_e$ / $\log D_o$	p.a. / A_g / A_i / A_{21}	B_T / m_B / m_{FIR} / B_T^o	$(B-V)_T$ / $(U-B)_T$ / $(B-V)_T^o$ / $(U-B)_T^o$	$(B-V)_e$ / $(U-B)_e$ / m'_e / m'_{25}	m_{21} / W_{20} / W_{50} / HI	V_{21} / V_{opt} / V_{GSR} / V_{3K}
063358.9-344848	NGC 2255	.SXR5*.	1.18± .04	152	14.16 ±.14				
243.30 -18.42	ESO 365- 31	Sr (1)	.32± .04	.28	12.84				
219.02 -70.32	IRAS06321-3446	4.5± .7		.47					
063211.1-344624	PGC 19260	3.3± .9	1.20	.16					
0634.0 +5851		.S..4..	1.23± .05	72					
156.41 20.82	UGC 3484	U	.88± .05	.31	15.50 ±.18				
23.39 -15.31		4.0± .9		1.30					
0629.7 +5854	PGC 19261		1.26	.44					
0634.1 +5208		.S..0..	1.01± .06	50					
163.15 18.60	UGC 3488	U	.18± .05	.51	14.91 ±.18				
22.76 -22.00		.0± .9		.13					
0630.2 +5211	PGC 19267		1.05						
063449.6-745329		.SXR5..	1.07± .05	90					
285.94 -27.20	ESO 34- 9	Sr (1)	.08± .05	.49	14.76 ±.14				
207.69 -30.75		4.7± .6		.13					
063620.1-745054	PGC 19276	4.4± .9	1.11	.04					
063500.4-214610		.SBS9..	1.20± .04						1944
230.86 -13.20	ESO 557- 6	S (1)	.05± .04	.99	14.41 ±.14				
243.41 -82.31		9.0± .8		.05					1762
063253.0-214342	PGC 19279	7.8± .8	1.29	.02	13.37				2068
0635.2 +8521		.S..5..	1.00± .08	23					
128.12 26.70	UGC 3455	U	1.02± .06	.28					
25.77 11.08		5.0±1.0		1.50					
0619.0 +8523	PGC 19285		1.03	.50					
063514.7-352918		PSBS1*.	.88± .07	135					
244.06 -18.41	ESO 365- 33	r	.14± .06	.37	15.07 ±.14				
217.80 -69.73	IRAS06334-3526	1.0± .9		.14	12.77				
063328.0-352648	PGC 19287		.92	.07					
0635.3 +8331		.S..4..	1.00± .06	133					
130.15 26.49	UGC 3466	U	.42± .05	.25					
25.62 9.26		4.0± .9		.61					
0623.0 +8334	PGC 19290		1.03	.21					
0635.3 +1458		.S?....	1.04± .08	114					3790± 7
197.78 3.30	UGC 3498		.59± .06	2.41					
17.02 -58.93	IRAS06324+1501			.88	12.46				3711
0632.4 +1501	PGC 19292		1.27	.29					3896
063527.1+485038		.S?....	1.21± .06	100				15.56±.1	5899± 11
166.47 17.61	UGC 3493		.26± .06	.62	14.5 ±.2				5945
22.64 -25.30				.40				600± 8	5950
063137.7+485304	PGC 19294		1.27	.13	13.44			1.98	
063527.4-542720		.S?....	1.01± .06	54					12800±190
263.52 -24.02	ESO 161- 17		.27± .05	.29	15.27 ±.14				12586
210.67 -51.07	FAIR 818			.28					12899
063426.0-542448	PGC 19295		1.04	.14	14.55				
063538.9-391532		.SXS5..	1.15± .05						
247.83 -19.63	ESO 308- 16	S (1)	.09± .05	.31	14.18 ±.14				
215.39 -66.07	IRAS06339-3912	5.0± .8		.13	13.09				
063359.0-391300	PGC 19300	1.1±1.1	1.18	.04					
063608.2+392452		.SB?...	.74± .14						6967± 38
175.57 14.10	MCG 7-14- 6		.00± .07	.98	15.0 ±.3				6977
21.77 -34.70				.00					7037
063240.1+392722	PGC 19304		.83	.00	13.95				
063613.7+825803	IC 442	.S?....	1.03± .06					15.82±.1	4264± 11
130.77 26.44	UGC 3470		.00± .05	.25	13.77 ±.18				4474± 46
25.60 8.70				.00				512± 8	4430
062441.9+830016	PGC 19306		1.05	.00	13.47			2.35	4253
0636.2 +7420		.S..2..	1.13± .07	130					
140.22 25.03	UGC 3486	U	.74± .06	.37	15.46 ±.18				
24.92 .10	IRAS06297+7423	2.0± .9		.91	12.06				
0629.7 +7423	PGC 19307		1.17	.37					
063641.0-663146		PSBS4..	1.17± .04	16					7811
276.65 -26.17	ESO 87- 32	Sr (1)	.24± .04	.23	14.28 ±.14				7602
208.51 -39.09	IRAS06366-6629	3.7± .6		.35	11.66				7893
063639.0-662906	PGC 19315	2.2±1.3	1.19	.12	13.65				
063645.5-350418		PSB.0..	1.03± .06	140					
243.76 -17.98	ESO 365- 35	r	.11± .03	.32	14.82 ±.14				
217.80 -70.21		-.1± .9		.08					
063458.0-350142	PGC 19317		1.06						

R.A. 2000 DEC. l b SGL SGB R.A. 1950 DEC.	Names PGC	Type S_T n_L T L	logD_25 logR_25 logA_e logD_o	p.a. A_g A_i A_21	B_T m_B m_FIR B_T^o	(B-V)_T (U-B)_T (B-V)_T^o (U-B)_T^o	(B-V)_e (U-B)_e m'_e m'_25	m_21 W_20 W_50 HI	V_21 V_opt V_GSR V_3K
063717.4-552140 264.55 -23.98 210.08 -50.20 063619.0-551900	 ESO 161- 19 FAIR 819 PGC 19326	PSBR3.. r 3.3± .8 	1.04± .05 .03± .05 1.08	170 .43 .04 .01	 14.81 ±.14 14.23				15400±190 15185 15500
063721.0+675134 147.19 23.69 24.51 -6.37 063204.8+675405	IC 445 UGC 3497 PGC 19328	.L...?. U -2.0±1.7 	1.08± .17 .12± .08 1.09	25 .25 .00	 14.19 ±.15 13.86				5210± 57 5321 5222
063729.2-593900 269.15 -24.89 209.33 -45.94 063648.0-593618	 ESO 121- 41 PGC 19332	.IBS9.. S (1) 10.0± .9 10.0± .9	1.06± .04 .41± .04 1.08	8 .30 .30 .20	 16.14 ±.14				
063756.7-255959 235.12 -14.30 226.56 -78.97 063555.0-255718	 ESO 490- 17 PGC 19337	.IXS9?. S (1) 10.0±1.5 10.0±1.1	1.22± .04 .09± .04 1.29	155 .67 .06 .04	 13.71 ±.14 12.98				498 307 625
063800.2-241159 233.43 -13.56 230.62 -80.62 063556.1-240918	 ESO 490- 18 PGC 19340	.SBR1.. r 1.0± .9 	.97± .06 .18± .05 1.06	56 .96 .18 .09	 14.96 ±.14				
0638.0 +2238 191.20 7.32 19.93 -51.43 0635.0 +2241	 UGC 3503 PGC 19341	.S..8*. U 8.0±1.3 	1.22± .12 .65± .12 1.33	115 1.16 .80 .32	 15.1 ±.2 13.14				1392± 7 1339 1491
0638.0 -1501 224.94 -9.69 305.68 -85.86 0635.8 -1459	 PGC 19343	.SXS8?. E (1) 8.0± .9 8.7± .8	1.35± .04 .31± .06 1.58	160 2.48 .38 .15					
063819.6-515702 260.99 -22.98 210.48 -53.61 063710.0-515418	 ESO 206- 17 PGC 19349	.S..6*/ S 6.3± .8 	1.22± .05 .80± .04 1.23	0 .13 1.17 .40	 15.55 ±.14				
0638.4 +4914 166.26 18.21 23.21 -24.95 0634.6 +4917	 UGC 3501 PGC 19352	.I..9*. U 10.0±1.3 	1.04± .06 .55± .05 1.09	101 .56 .41 .28				16.00±.1 78± 5 55± 5	449± 6 495 502
063828.3-245049 234.09 -13.73 228.42 -80.07 063625.1-244806	NGC 2263 ESO 490- 19 IRAS06364-2448 PGC 19355	PSBR2.. Sr 2.1± .5 	1.42± .03 .23± .04 1.51	143 .89 .29 .12	 12.86 ±.14 12.29				
063840.1-801450 291.97 -27.24 206.93 -25.44 064228.0-801154	 ESO 16- 16 IRAS06423-8011 PGC 19358	.SB.7*P S 7.0±1.2 	.93± .06 .18± .05 1.00	150 .74 .25 .09	 15.44 ±.14				
063842.5-201708 229.85 -11.80 245.85 -84.00 063633.0-201424	 ESO 557- 9 IRAS06365-2014 PGC 19360	.SBS5.. E (1) 5.0± .8 4.4±1.2	1.28± .03 .35± .03 1.47	98 2.05 .53 .18	 14.35 ±.14 13.50				
063845.2-351533 244.10 -17.66 215.92 -70.11 063658.0-351248	 ESO 366- 4 PGC 19363	.SBR7./ r 6.8± .9 	1.21± .05 .69± .05 1.24	79 .31 .95 .35	 15.29 ±.14				
063851.9-250945 234.42 -13.77 227.24 -79.82 063649.0-250700	 ESO 490- 20 PGC 19366	.IBS9.. S (1) 10.0± .8 10.0± .8	1.18± .05 .06± .05 1.26	 .83 .05 .03	 14.82 ±.14 13.94				2792 2602 2921
0639.0 +7649 137.56 25.66 25.27 2.56 0631.8 +7652	 UGC 3495 PGC 19370	.S..2.. U 2.0±1.0 	1.00± .08 .69± .06 1.04	127 .41 .84 .34					
063936.2+204440 193.08 6.81 20.18 -53.35 063637.4+204726	 UGC 3505 PGC 19385	.S..6*. U 6.0±1.3 	1.14± .07 .69± .06 1.26	55 1.24 1.01 .34				15.52±.3 341± 7	5374± 10 5314 5478
0639.7 +6527 149.79 23.33 24.58 -8.78 0634.8 +6530	 UGC 3502 PGC 19390	.SB.3*. U 3.0± .9 	1.13± .05 .44± .05 1.15	119 .24 .61 .22	 14.75 ±.18				

R.A. 2000 DEC. l　b SGL　SGB R.A. 1950 DEC.	Names PGC	Type S_T　n_L T L	$\log D_{25}$ $\log R_{25}$ $\log A_o$ $\log D_o$	p.a. A_g A_i A_{21}	B_T m_B m_{FIR} B_T^o	$(B-V)_T$ $(U-B)_T$ $(B-V)_T^o$ $(U-B)_T^o$	$(B-V)_o$ $(U-B)_o$ m'_o m'_{25}	m_{21} W_{20} W_{50} HI	V_{21} V_{opt} V_{GSR} V_{3K}
063949.9-381443 247.12 -18.52 213.84 -67.22 063808.1-381154	 ESO　308- 23 IRAS06381-3811 PGC 19391	.SBR3*. S　　(1) 3.0±1.2 4.4±1.2	1.21± .04 .42± .04 1.25	95 .44 .58 .21	14.52 ±.14 14.08 				
064006.9+600458 155.41 21.92 24.27 -14.15 063538.6+600743	 UGC　3504 IRAS06356+6007 PGC 19397	.SXS6.. U 6.0± .7 	1.43± .04 .09± .05 1.47	135 .35 .13 .04	 13.0　±.2 12.59 12.53			14.24±.1 222± 8 215 12 1.67	2104± 6 2188 2135
064015.0-722721 283.25 -26.61 207.49 -33.21 064109.0-722424	 ESO　58- 9 PGC 19400	.SAS7*. S 7.0± .9 	1.12± .05 .35± .06 1.17	97 .48 .49 .18	15.08 ±.14 				
064032.2+500618 165.53 18.83 23.65 -24.11 063639.6+500906	 UGC　3506 IRAS06366+5008 PGC 19409	.L..-*. U -3.0±1.3 	.90± .11 .05± .04 .96	 .54 .00 	14.37 ±.16 13.64 13.75 				5668±155 5717 5721
064043.4-583132 268.04 -24.24 208.89 -47.10 063957.0-582836	 ESO　122- 1 IRAS06399-5828 PGC 19413	.SBS3P. S　　(1) 3.0± .5 2.2± .8	1.43± .04 .31± .05 .89± .03 1.47	3 .38 .43 .16	13.10 ±.15 13.14 ±.14 11.34 12.29	.79± .04 -.09± .05 .63 -.22	.87± .02 .01± .03 13.04± .09 14.33± .27	307± 9 	2598± 8 2649± 57 2384 2697
0640.7　+5314 162.41 19.90 23.90 -20.99 0636.7　+5317	A　0636+53 CGCG 260- 24 PGC 19414			 .37 	15.9　±.2 				10275± 40 10335 10322
064046.7-582814 267.99 -24.22 208.89 -47.16 064000.1-582518	 ESO　122- 2 PGC 19415		.87± .07 .08± .06 .91	0 .38 	14.88 ±.14 				2729± 57 2515 2827
064051.9-322854 241.57 -16.23 216.33 -72.92 063900.0-322600	NGC　2267 ESO　426- 29 PGC 19417	.LBR0.. Sr -1.5± .4 	1.22± .05 .11± .04 1.26	36 .52 .00 	13.24 ±.14 				
0641.1　+5316 162.40 19.97 23.97 -20.96 0637.1　+5319	A　0637+53 MCG　9-11- 23 PGC 19427	.S?....	.69± .15 .33± .07 .72	 .37 .50 .17					11130± 40 11190 11177
0641.3　+5344 161.94 20.15 24.03 -20.49 0637.3　+5347	 UGC　3507 PGC 19430	RSA.3?. U　　(1) 3.0± .9 4.5±1.1	1.00± .08 .09± .06 1.03	120 .37 .12 .04	15.1　±.2 				
064132.3+401011 175.27 15.35 23.09 -34.03 063802.9+401304	 UGC　3510 IRAS06380+4013 PGC 19439	.S?....	1.14± .07 .13± .06 1.22	100 .90 .19 .06	 12.87 				6392± 38 6404 6466
064136.8-505758 260.11 -22.22 209.80 -54.64 064024.0-505500	Carina ESO　206- 20A PGC 19441	.E?....	2.37± .02 .18± .06 2.33	70 .11 .00 					229± 60 13 339
064213.1-610414 270.82 -24.61 208.33 -44.58 064138.1-610112	 ESO　122- 4 PGC 19456	.SBR2.. Sr　　(1) 2.0± .6 5.6±1.2	1.03± .05 .06± .05 1.07	 .42 .07 .03	15.22 ±.14 				
064215.5+753725 138.91 25.65 25.39　1.35 063524.4+754014	A　0635+75 UGCA　130 MK　5 PGC 19459	.I?....	.84± .07 .17± .07 .88	 .36 .13 .08	15.6　±.2 15.12	.55± .05 -.31± .07 .42 -.40	 14.24± .44 	16.39±.1 80± 4 63± 4 1.19	792± 5 870± 63 927 789
064222.5-265336 236.38 -13.75 220.30 -78.44 064022.1-265036	 ESO　490- 31 PGC 19461	.SBT5*. S　　(1) 5.3± .6 6.2± .5	1.18± .04 .40± .04 1.27	55 .97 .61 .20	15.22 ±.14 				
064241.7-272731 236.94 -13.91 219.27 -77.91 064042.0-272430	NGC　2272 ESO　490- 33 PGC 19466	.LXS-.. PS -3.0± .4 	1.39± .04 .18± .04 1.45	123 .71 .00 	12.74 ±.14 				
0642.7　+4225 173.21 16.42 23.53 -31.81 0639.2　+4228	 UGC　3512 PGC 19469	.I..9.. U 10.0± .9 	1.16± .13 .34± .12 1.23	120 .74 .26 .17					

R.A. 2000 DEC.	Names	Type	logD$_{25}$	p.a.	B$_T$	(B-V)$_T$	(B-V)$_e$	m$_{21}$	V$_{21}$
I b		S$_T$ n$_L$	logR$_{25}$	A$_g$	m$_B$	(U-B)$_T$	(U-B)$_e$	W$_{20}$	V$_{opt}$
SGL SGB		T	logA$_e$	A$_i$	m$_{FIR}$	(B-V)^o_T	m'$_e$	W$_{50}$	V$_{GSR}$
R.A. 1950 DEC.	PGC	L	logD$_o$	A$_{21}$	B^o_T	(U-B)^o_T	m'$_{25}$	HI	V$_{3K}$

064249.9-353414		PSXS5*.	1.20± .05	39					
244.72 -17.00	ESO 366- 9	Sr (1)	.23± .05	.40	14.20 ±.14				
213.37 -69.95	IRAS06411-3531	5.1± .7		.35	13.68				
064103.1-353112	PGC 19472	4.4± .8	1.24	.12					
064251.9+412515		.S..6*.	1.00± .08	82					7316± 38
174.17 16.06	UGC 3513	U	.43± .06	.90	15.4 ±.3				7332
23.48 -32.80		6.0±1.3		.63					7389
063920.1+412814	PGC 19475		1.08	.22	13.86				
064252.3-232832	NGC 2271	.LX.-..	1.32± .05	71	*		1.02± .01		
233.22 -12.25	ESO 490- 34	S	.16± .04	1.02	13.17 ±.14				2596± 37
225.83 -81.72		-3.0± .8	1.03± .08	.00			13.31± .21		2408
064047.0-232530	PGC 19476		1.43		12.11				2730
064302.1-330151		.SAR2..	.95± .06	174	*				
242.28 -16.02	ESO 366- 10	r	.24± .05	.43	15.48 ±.14				
214.49 -72.47		2.2± .9		.29					
064111.1-325848	PGC 19479		.99	.12					
064306.8-741416		.I?....	1.20± .07		15.3 ±.2	1.13± .08	1.09± .00		
285.26 -26.59	ESO 34- 11A		.27± .08	.62	16.37 ±.14	.30± .11	.39± .00		6416± 19
207.11 -31.45			.58± .14	.20		.89	13.68± .48		6215
064425.0-741106	PGC 19480		1.25	.13	15.20	.05	15.46± .47		6487
064306.8-741416 A	0644-74	.E?....	1.18± .06	5	13.82 ±.17	.88± .04	1.06± .02		
285.26 -26.59	ESO 34- 11		.24± .04	.62	13.60 ±.14	.19± .06	.44± .04		6505± 19
207.11 -31.45	IRAS06443-7411		.72± .03	.00	12.51	.68	12.91± .08		6304
064425.0-741106	PGC 19481		1.21		12.98	.11	14.13± .36		6576
064308.3+225225		.S..8*.	1.11± .07	6				15.99±.3	1298± 7
191.52 8.48	UGC 3516	U	.44± .06	.98					1244
21.83 -51.31		8.0±1.3		.54				124± 7	1404
064006.8+225526	PGC 19483		1.20	.22					
064312.2+392734		.E?....							12997± 38
176.08 15.37	MCG 7-14- 11			.80	15.2 ±.3				13005
23.42 -34.76									13075
063944.3+393034	PGC 19487								
064316.0-492923		.LBR+..	.94± .06	120					
258.67 -21.55	ESO 206- 21	r	.31± .05	.16	15.31 ±.14				
209.57 -56.13		-1.3± .9		.00					
064159.0-492618	PGC 19490		.91						
064319.6-350929		.SBR0*.	.98± .07	160					
244.36 -16.76	ESO 366- 11	r	.36± .05	.28	14.89 ±.14				
213.26 -70.38		-.1±1.3		.27					
064132.0-350624	PGC 19492		.99						
064331.9-723541		.SBR5..	1.25± .04	137	13.88 ±.15	.71± .02	.79± .01		
283.44 -26.38	ESO 34- 12	Sr (1)	.28± .04	.52	14.02 ±.14				5547± 40
207.19 -33.09	FAIR 257	5.1± .4	.82± .02	.42	12.56	.50	13.47± .05		5344
064427.0-723230	PGC 19498	3.7± .5	1.30	.14	12.98		14.27± .27		5621
0643.6 +2826		.S..0..	1.00± .08	164					
186.46 10.99	UGC 3518	U	.55± .06	.85	14.9 ±.3				
22.62 -45.76		.0± .9		.41					
0640.5 +2830	PGC 19499		1.05						
0643.7 +5737		.S..2..	1.08± .06	12					
158.08 21.66	UGC 3514	U	.65± .05	.33	15.39 ±.18				
24.60 -16.62		2.0± .9		.80					
0639.4 +5741	PGC 19500		1.11	.33					
064342.0+651223 A	0638+65	.S..6*.	1.17± .05	135				14.65±.1	3563± 5
150.16 23.66	UGC 3511	U	.13± .05	.24	13.15 ±.20			327± 16	3665
24.98 -9.06		6.0±1.1		.19				324± 8	3585
063845.8+651522	PGC 19501		1.19	.06	12.70			1.88	
064348.3-734031		.SAT4P.	1.13± .05	45					
284.64 -26.48	ESO 34- 13	r	.45± .05	.59	14.69 ±.14				4073
207.09 -32.01	IRAS06449-7337	4.5± .9		.67	13.00				3871
064458.0-733718	PGC 19504		1.18	.22	13.40				4145
064348.8-405901		.SBR1..	1.02± .07	129					
250.11 -18.75	ESO 308- 26	r	.40± .06	.40	15.38 ±.14				
211.08 -64.61	IRAS06421-4055	1.0±1.0		.41	13.85				
064212.0-405554	PGC 19506		1.06	.20					
0644.0 +1224		.S?....	1.10± .05	10					3997± 10
201.04 4.00	UGC 3524		.35± .05	2.04					
20.53 -61.76	IRAS06412+1227			.52					3906
0641.2 +1227	PGC 19512		1.30	.17					4118

R.A. 2000 DEC. / l b / SGL SGB / R.A. 1950 DEC.	Names / / / PGC	Type / S_T n_L / T / L	$\log D_{25}$ / $\log R_{25}$ / $\log A_e$ / $\log D_o$	p.a. / A_g / A_i / A_{21}	B_T / m_B / m_{FIR} / B_T^o	$(B-V)_T$ / $(U-B)_T$ / $(B-V)_T^o$ / $(U-B)_T^o$	$(B-V)_e$ / $(U-B)_e$ / m'_e / m'_{25}	m_{21} / W_{20} / W_{50} / HI	V_{21} / V_{opt} / V_{GSR} / V_{3K}
064408.1-564122			1.00± .07						10196± 87
266.21 -23.37	ESO 161- 24		.22± .06	.28	14.56 ±.14				9980
208.43 -48.97	IRAS06431-5637				12.49				10299
064314.0-563812	PGC 19517		1.02						
064408.1-271026		.S..6*/	1.24± .04	28					
236.80 -13.50	ESO 490- 36	S	.72± .03	.81	14.97 ±.14				
218.09 -78.26	IRAS06421-2707	5.7± .7		1.05	13.18				
064208.0-270718	PGC 19518		1.31	.36					
064422.7-260633		PSAR0*.	1.32± .04	168					
235.83 -13.02	ESO 490- 37	Sr	.20± .04	.89	12.96 ±.14				
219.09 -79.31	IRAS06423-2603	.3± .6		.15					
064221.0-260324	PGC 19522		1.40						
064425.2-634300	NGC 2297	.SXT4..	1.15± .05		13.37 ±.14	.69± .02	.77± .01		
273.73 -24.89	ESO 87- 40	Sr	.05± .05	.26	13.75 ±.14				3395± 33
207.74 -41.96	FAIR 256	3.7± .6	.94± .02	.07	12.17	.60	13.56± .07		3183
064404.0-633948	PGC 19524		1.17	.02	13.21		13.85± .30		3487
064426.6-175556	IC 2171	.IBS9*.	1.22± .03	93					772
228.27 -9.57	ESO 557- 12	SE (2)	.55± .03	2.53	15.18 ±.14				
254.59 -86.63	IRAS06422-1752	10.0± .7		.42					595
064214.0-175248	PGC 19526	8.3± .8	1.46	.28	12.23				910
064440.7-712727		.S..3*/	1.35± .04	6					
282.19 -26.15	ESO 58- 13	S	.81± .05	.53	14.70 ±.14				
207.16 -34.23	IRAS06453-7124	3.0± .9		1.12	12.32				
064522.0-712412	PGC 19528		1.40	.40					
064448.5-273820	NGC 2280	.SAS6..	1.80± .02	163	10.9 ±.2	.60± .03	.68± .01	11.94±.1	1906± 6
237.30 -13.55	ESO 427- 2	R (3)	.31± .03	.82	11.37 ±.12	.15± .03	.13± .02	403± 7	1963± 46
216.91 -77.84	IRAS06428-2735	6.0± .3	1.57± .06	.46	11.08	.34	14.19± .17	379± 11	1710
064249.0-273510	PGC 19531	2.2± .3	1.88	.16	9.97	-.04	13.91± .25	1.81	2041
064503.5+402452		.E?....	1.18± .15	70					
175.31 16.07	UGC 3525		.17± .08	1.08	14.4 ±.3				7272± 38
23.91 -33.83				.00					7284
064133.8+402800	PGC 19536		1.30		13.21				7350
064503.9-462655		.RING..	1.07± .06	20					
255.67 -20.34	ESO 255- 18	S	.12± .05	.26	14.52 ±.14				
209.50 -59.19		10.0± .6		.09					
064339.0-462342	PGC 19537		1.10	.06					
064525.9-263937		.SXT0*.	1.20± .04	7					
236.44 -13.03	ESO 490- 38	S	.23± .04	1.05	14.26 ±.14				
217.23 -78.82		.0± .8		.17					
064325.0-263624	PGC 19542		1.29						
064526.1+342910		.S?....	.95± .07	165				15.01±.3	5427± 10
180.99 13.83	UGC 3529		.39± .05	.97	14.7 ±.3				5416
23.62 -39.75	IRAS06421+3432			.59	13.76			396± 7	5517
064207.3+343220	PGC 19544		1.04	.20	13.14			1.68	
0645.4 +6232		.SB.8*.	1.08± .06	58					
153.02 23.20	UGC 3520	U	.26± .05	.38					
25.06 -11.72		8.0±1.2		.32					
0640.8 +6236	PGC 19545		1.12	.13					
064530.4+254942		.SX.4..	1.11± .07	150				15.62±.3	11989± 9
189.05 10.25	UGC 3531	U	.20± .06	1.08	14.0 ±.3				11946
22.97 -48.40	IRAS06424+2552	4.0± .8		.29	13.66			403± 5	12094
064224.8+255253	PGC 19547		1.21	.10	12.51			3.01	
0645.6 +2224		.S?....	1.16± .13						4468± 10
192.19 8.80	UGC 3534		.41± .12	.94					
22.69 -51.81				.62					4412
0642.6 +2228	PGC 19549		1.25	.21					4578
0645.6 +5327		.S..3..	1.11± .05	54					
162.43 20.67	UGC 3526	U (1)	.69± .05	.31	15.55 ±.18				
24.69 -20.80		3.0± .9		.95					
0641.6 +5331	PGC 19550	5.5±1.3	1.14	.35					
064542.4+712037	IC 449	.E...*.	1.23± .14						
143.62 25.14	UGC 3515	PU	.11± .08	.38	13.48 ±.15				
25.43 -2.93		-5.0± .6		.00					
063954.0+712343	PGC 19554		1.26						
064548.3-473152		.SAS9..	1.24± .04					13.80±.3	1063± 9
256.80 -20.56	ESO 255- 19	S (1)	.11± .05	.25	14.20 ±.14				
209.10 -58.12		9.0± .4		.11				139± 7	847
064426.1-472836	PGC 19559	9.4± .4	1.26	.06	13.83			-.08	1181

R.A. 2000 DEC.	Names	Type	logD$_{25}$	p.a.	B$_T$	(B-V)$_T$	(B-V)$_e$	m$_{21}$	V$_{21}$
l b		S$_T$ n$_L$	logR$_{25}$	A$_g$	m$_B$	(U-B)$_T$	(U-B)$_e$	W$_{20}$	V$_{opt}$
SGL SGB		T	logA$_e$	A$_i$	m$_{FIR}$	(B-V)$_T^o$	m'$_e$	W$_{50}$	V$_{GSR}$
R.A. 1950 DEC.	PGC	L	logD$_o$	A$_{21}$	B$_T^o$	(U-B)$_T^o$	m'$_{25}$	HI	V$_{3K}$

064553.3-181239 NGC 2283	.SBS6..	1.56± .02	2				12.30±.3	822± 11
228.67 -9.38 ESO 557- 13	SE (2)	.12± .03	1.06	12.94 ±.14			187± 16	
246.98 -86.68	5.5± .5		.18				174± 12	643
064341.0-180924 PGC 19562	2.7± .5	1.66	.06	11.70			.54	961

064603.7+292053	.L.....	1.04± .11	142					4689±155
185.85 11.85 UGC 3536	U	.28± .05	.87	14.0 ±.3				4659
23.44 -44.89	-2.0± .9		.00					4789
064253.0+292406 PGC 19568	1.10		13.02				

064612.5+434732	.SBS3..	1.15± .07	100				15.58±.1	6287± 7
172.11 17.51 UGC 3532	U	.37± .06	.63	15.0 ±.2				6209± 46
24.34 -30.46 IRAS06425+4350	3.0± .8		.51	12.79			366± 8	6309
064235.8+435045 PGC 19571	1.21		.19	13.81			1.58	6358

064621.7-260629	PSXS0P*	1.46± .04	165					
236.01 -12.61 ESO 490- 41	Sr	.44± .04	1.02	13.47 ±.14				
216.74 -79.41 IRAS06443-2602	-.4± .4		.33	12.86				
064420.0-260312 PGC 19574	1.53						

064624.1+434929	.S?....	1.10± .07	100					6329± 34
172.09 17.56 UGC 3535		.13± .06	.63	14.47 ±.19				6353
24.38 -30.43			.16					6402
064247.4+435243 PGC 19576	1.16		.06	13.62			

064631.5+602032 NGC 2273B	.SBT6*.	1.43± .04	55	13.10 ±.19	.57± .05	.58± .03	14.22±.1	2101± 4
155.38 22.76 UGC 3530	PU	.26± .05	.34	13.7 ±.3	-.25± .06	-.16± .04	240± 4	
25.10 -13.93 IRAS06421+6023	6.0± .5	1.16± .04	.39	13.33	.43	14.39± .09	196± 5	2185
064202.5+602345 PGC 19579	1.46	.13	12.59	-.35	14.44± .28	1.50	2136

064636.2-703653	.L?....	1.04± .07	100					6902± 87
281.29 -25.87 ESO 58- 14		.28± .05	.52	14.70 ±.14				6696
207.02 -35.08			.00					6982
064708.1-703330 PGC 19581	1.05		14.08				

064645.5+333709	.SB.6*.	.97± .07					16.82±.2	5184± 7
181.92 13.73 UGC 3537	U	.03± .05	.93	14.7 ±.3			241± 7	
23.92 -40.63	6.0±1.2		.04				197± 5	5170
064328.1+334025 PGC 19586	1.06		.01	13.70			3.11	5277

064647.2-262831	.SBS7..	1.40± .03	90					1708
236.40 -12.68 ESO 490- 45	S (1)	.45± .04	1.12	14.04 ±.14				
215.87 -79.06 IRAS06447-2625	7.0± .6		.61					1514
064446.0-262512 PGC 19589	6.7± .6	1.50	.22	12.30				1847

064713.3+422038								5984± 38
173.59 17.16 CGCG 204- 17			.65	15.3 ±.3				6002
24.49 -31.92								6060
064339.8+422355 PGC 19601							

064713.9+741411 NGC 2256 ☾	.LX.-?.	1.36± .11						
140.48 25.75 UGC 3519	PU	.06± .08	.38	13.5 ±.2				
25.65 -.05	-3.4± .6		.00					
064047.5+741722 PGC 19602	1.40						

064717.3+333401 NGC 2274	.E.....	1.22± .08	169					5067± 34
182.01 13.81 UGC 3541	U	.00± .05	.93	13.1 ±.3				5052
24.07 -40.69	-5.0± .8		.00					5160
064400.0+333719 PGC 19603	1.37		12.05				

064717.9+333555 NGC 2275	.S?....	1.10± .05	20					4858± 32
181.98 13.82 UGC 3542		.12± .05	.93	14.1 ±.3				4843
24.07 -40.66			.18					4952
064400.6+333913 PGC 19605	1.18		.06	12.94			

064723.9-264409 NGC 2295	.S..2*/	1.32± .03	46					
236.70 -12.66 ESO 490- 47	S	.54± .03	1.21	13.58 ±.14				
214.94 -78.83	2.0±1.1		.67					
064523.1-264048 PGC 19607	1.44		.27				

064732.8+473917	.SBS1..	1.12± .04	0					
168.38 19.08 UGC 3538	U	.27± .04	.43	14.68 ±.19				
24.78 -26.61	1.0± .9		.28					
064347.1+474235 PGC 19612	1.16		.14				

064735.2-654156	.S?....	.97± .06	85					11005
275.94 -24.94 ESO 87- 42		.04± .05	.25	14.98 ±.14				10794
207.16 -39.99 IRAS06474-6538			.06					11095
064726.0-653830 PGC 19614	.99		.02	14.60			

064739.9-264447 NGC 2292	.LX.0P.	1.61± .06	1	*				2321± 37
236.73 -12.61 ESO 490- 48	S	.05± .05	1.21	11.82 ±.14				2125
214.63 -78.83	-2.0± .7		.00					2460
064539.0-264124 PGC 19617	1.74		10.57				

R.A. 2000 DEC. l b SGL SGB R.A. 1950 DEC.	Names PGC	Type S_T n_L T L	$\log D_{25}$ $\log R_{25}$ $\log A_e$ $\log D_o$	p.a. A_g A_i A_{21}	B_T m_B m_{FIR} B_T^o	$(B-V)_T$ $(U-B)_T$ $(B-V)_T^o$ $(U-B)_T^o$	$(B-V)_e$ $(U-B)_e$ m'_e m'_{25}	m_{21} W_{20} W_{50} HI	V_{21} V_{opt} V_{GSR} V_{3K}
064742.9-264511 236.74 -12.60 214.56 -78.83 064542.1-264148	NGC 2293 ESO 490- 49 PGC 19619	.LXS+P. S -1.0± .7	1.62± .06 .10± .05 .98± .04 1.72	125 1.21 .00	12.28 ±.15 11.70 ±.14 10.73	1.07± .01 .77	1.09± .01 12.67± .12 15.00± .34		1978± 32 1782 2117
064746.4+742855 140.22 25.82 25.69 .20 064116.2+743209	NGC 2258 UGC 3523 PGC 19622	.LAR0.. PU -2.0± .5	1.36± .11 .17± .08 .88± .10 1.38	150 .39 .00	13.0 ±.2 13.04 ±.15	1.14± .02	1.16± .01 12.93± .33 14.24± .63		
0647.9 +8409 129.48 26.89 25.99 9.87 0634.6 +8412	 UGC 3500 IRAS06346+8412 PGC 19627	.S?....	1.23± .05 .70± .05 1.25	3 .24 1.05 .35	 14.70 ±.18 13.33 13.39			14.71±.1 210± 8 .97	4388± 11 4546 4365
064808.1+333431 182.08 13.97 24.30 -40.69 064450.9+333753	 UGC 3544 PGC 19632	.S..6*. U 6.0±1.4	1.04± .08 .76± .06 1.12	155 .84 1.12 .38				15.31±.3 362± 7	7353± 10 7338 7448
064837.4-641624 274.43 -24.55 207.09 -41.42 064819.1-641254	NGC 2305 ESO 87- 44 PGC 19641	.E.2.*P S -5.0±1.1	1.31± .05 .13± .04 .82± .02 1.32	142 .31 .00	12.74 ±.13 12.82 ±.14 12.42	1.02± .01 .92	1.03± .01 .51± .03 12.33± .09 13.96± .30		3499± 20 3287 3593
064837.5+433035 172.55 17.82 24.83 -30.76 064501.6+433358	 CGCG 204- 18 PGC 19642			 .59	 15.4 ±.3				6771± 38 6793 6846
064851.1-642001 274.50 -24.54 207.06 -41.36 064833.1-641630	NGC 2307 ESO 87- 45 IRAS06485-6416 PGC 19648	.SBT3.. Sr (1) 2.6± .6 4.4± .8	1.24± .05 .03± .05 1.06± .03 1.26	 .29 .05 .02	12.92 ±.15 13.38 ±.14 12.79	.88± .02 .78	.96± .01 13.71± .09 13.87± .30		4558± 40 4345 4651
0648.9 +6614 149.19 24.42 25.56 -8.04 0643.9 +6618	A 0643+66 UGC 3539 PGC 19652	.SB.4?. U 4.0± .9	1.39± .04 1.04± .05 1.41	116 .20 1.47 .50	 15.24 ±.19 13.54			15.07±.1 312± 8 1.03	3305± 11 3409 3327
064904.0-472024 256.80 -19.97 208.08 -58.34 064741.0-471654	 ESO 256- 2 PGC 19656	PSBS2*. r 2.2±1.3	1.01± .06 .49± .05 1.03	177 .22 .60 .24	 16.06 ±.14				
064933.8+442552 171.71 18.30 25.06 -29.85 064556.0+442919	 MCG 7-14- 16 IRAS06459+4429 PGC 19665	.S?....	.58± .18 .23± .07 .63	 .57 .28 .11	 15.3 ±.4 13.45 14.34				6339± 38 6365 6413
064939.4-571516 267.01 -22.78 207.26 -48.44 064847.0-571142	 ESO 161- 26 FAIR 821 PGC 19668	.SXT1*. S 1.0± .9	1.08± .05 .37± .05 1.11	87 .31 .38 .19	 15.17 ±.14 14.34				11200±190 10983 11307
0649.7 +8550 127.61 27.02 26.07 11.54 0632.0 +8553	 UGC 3496 PGC 19670	.I..9*. U 10.0±1.1	1.32± .10 .13± .12 1.35	 .30 .10 .06				15.74±.1 70± 7 60± 12	1586± 6 1748 1559
064944.0+200454 194.73 8.65 24.07 -54.19 064646.4+200823	 UGC 3553 PGC 19671	.S..6*. U 6.0±1.3	1.00± .08 .33± .06 1.09	158 .95 .49 .17	 15.25 ±.19 13.78			16.27±.3 255± 7 2.32	5229± 10 5163 5348
064945.9+293131 186.03 12.65 24.58 -44.75 064635.2+293500	 UGC 3551 PGC 19674	.S..7*. U 7.0±1.5	1.06± .06 1.02± .05 1.13	38 .81 1.38 .50	 15.6 ±.3 13.43			15.88±.3 374± 7 1.95	4821± 10 4790 4925
064947.4-640723 274.30 -24.40 206.93 -41.58 064928.0-640348	 ESO 87- 46 PGC 19677	.SB9*P/ S 9.0±1.3	1.14± .05 .50± .05 1.17	48 .31 .51 .25	 15.86 ±.14				
064951.2+282217 187.11 12.20 24.55 -45.90 064642.1+282547	 UGC 3552 PGC 19679	.SB.6*. U 6.0±1.3	1.02± .06 .27± .05 1.09	75 .76 .39 .13	 14.5 ±.3 13.37			15.86±.3 261± 7 2.36	4865± 10 4830 4971
0649.8 +6136 154.15 23.47 25.55 -12.68 0645.3 +6140	 UGC 3545 PGC 19680	.SB?...	1.05± .08 .00± .06 1.08	 .35 .00 .00					3527 3615 3561

R.A. 2000 DEC.	Names	Type	logD$_{25}$	p.a.	B$_T$	(B-V)$_T$	(B-V)$_e$	m$_{21}$	V$_{21}$
l b	S$_T$ n$_L$	logR$_{25}$	A$_g$	m$_B$	(U-B)$_T$	(U-B)$_e$	W$_{20}$	V$_{opt}$
SGL SGB		T	logA$_e$	A$_i$	m$_{FIR}$	(B-V)$_T^o$	m'	W$_{50}$	V$_{GSR}$
R.A. 1950 DEC.	PGC	L	logD$_o$	A$_{21}$	B$_T^o$	(U-B)$_T^o$	m'$_{25}$	HI	V$_{3K}$

064959.5+253758		.SXT4..	1.08± .06					15.25±.2	4645± 7
189.67 11.08	UGC 3555	U	.05± .05	.73				199± 7	5077± 79
24.47 -48.64	IRAS06468+2541	4.0± .8		.08	12.97			176± 5	4603
064654.4+254128	PGC 19683		1.14	.03					4759

065008.6+605045	NGC 2273	.SBR1*.	1.51± .03	50	12.55 ±.14	.89± .01	.94± .01	14.81±.1	1840± 4
154.97 23.31	UGC 3546	PU	.12± .05	.31	12.34 ±.19	.38± .02	.41± .01	363± 5	1903± 36
25.57 -13.44	MK 620	.5± .5	.94± .03	.12	11.19	.78	12.74± .08	354± 7	1926
064537.6+605413	PGC 19688		1.54	.06	12.02	.30	14.65± .24	2.73	1877

0650.2 +2008		.E.....	1.00± .19	130					
194.73 8.78	UGC 3558	U	.05± .08	.94	14.72 ±.18				
24.28 -54.13		-5.0± .8		.00					
0647.3 +2012	PGC 19692		1.14						

065025.8+430312		.SB.3?.	1.06± .06	148					5640± 38
173.12 17.96	UGC 3554	U	.27± .05	.55	14.61 ±.18				5660
25.20 -31.23	IRAS06468+4306	3.0± .9		.37	13.15				5718
064650.9+430643	PGC 19697		1.11	.13	13.65				

065029.7+093957		.S..6*.	.96± .09	164				15.28±.3	7424± 10
204.21 4.18	UGC 3565	U	.69± .06	1.96					7321
23.59 -64.60		6.0±1.4		1.01				328± 7	7557
064744.8+094330	PGC 19700		1.14	.34					

065035.7+162107		.S..8*.	1.00± .08					15.26±.3	2547± 7
198.21 7.19	UGC 3564	U	.00± .06	1.16					2468
24.19 -57.92		8.0±1.2		.00				80± 7	2672
064742.7+162440	PGC 19701		1.11	.00					

065040.0-520825		.SBS9..	1.35± .04	6				14.08±.3	1085± 9
261.76 -21.21	ESO 207- 7	S (1)	.18± .05	.16	14.09 ±.14				867
207.29 -53.56		9.0± .4		.18				165± 7	1201
064930.0-520448	PGC 19705	8.7± .4	1.37	.09	13.74			.24	

065041.4+340227		.S..6*.	1.00± .08	65				15.99±.3	5416± 10
181.85 14.65	UGC 3559	U	.55± .06	.78					5402
25.01 -40.24		6.0±1.3		.81				250± 7	5513
064723.5+340600	PGC 19707		1.07	.27					

065052.1+332745	NGC 2288				.22?		15.3 ±.2	.95± .05	
182.42 14.45	MCG 6-15- 11		.00± .07	.77	15.9 ±.2	.51± .08			
25.04 -40.82									
064735.1+333119	PGC 19714		.29						

065053.6+332844	NGC 2289	.L.....	1.03± .11	125	14.23 ±.15	1.06± .03	1.08± .02		
182.40 14.46	UGC 3560	U	.21± .05	.77	14.3 ±.3	.56± .05	.54± .03		
25.05 -40.80		-2.0± .8	.66± .04	.00			13.02± .13		
064736.7+333218	PGC 19716		1.08				13.69± .59		

065056.9+332615	NGC 2290	RSA.1*.	1.12± .04	50	14.16 ±.15	.98± .02	1.00± .02	17.06±.3	5043± 9
182.45 14.46	UGC 3562	PU	.25± .04	.77	14.3 ±.3	.42± .04	.47± .04		5026
25.06 -40.84		.5± .6	.73± .03	.26		.72	13.30± .10	438± 5	5141
064740.1+332949	PGC 19718		1.19	.13	13.09	.23	13.97± .28	3.85	

065058.6+333131	NGC 2291	.LA.0*.	1.02± .07		14.2 ±.2	1.03± .04	1.05± .03		
182.37 14.50	MCG 6-15- 13	P	.00± .06	.77	14.7 ±.3	.49± .06	.54± .05		
25.07 -40.76		-2.0± .8	1.04± .08	.00			14.85± .26		
064741.5+333505	PGC 19719		1.11				14.17± .43		

0651.0 +5301		.S..4..	1.11± .05	75					
163.14 21.30	UGC 3556	U	.57± .05	.32					
25.53 -21.26		4.0± .9		.84					
0647.0 +5305	PGC 19720		1.14	.29					

065105.1-563434		.SXS6*.	1.02± .05	118					
266.36 -22.42	ESO 161- 29	S (1)	.25± .05	.28	15.50 ±.14				
206.99 -49.13		6.0±1.3		.37					
065010.0-563054	PGC 19722	5.6±1.3	1.04	.13					

065105.7+394033					.71	15.3 ±.3			9152± 38
176.47 16.86	CGCG 204- 22								9159
25.26 -34.61									9238
064737.7+394407	PGC 19723								

065106.6+125519	IC 454	.SB.2..	1.24± .05	140				14.02±.2	3945± 6
201.36 5.78	UGC 3570	U	.28± .05	1.44				438± 7	
24.20 -61.35		2.0± .8		.34				412± 5	3853
064817.8+125855	PGC 19725		1.37	.14					4075

065110.5-302456		.SAT3..	1.00± .06	33					
240.49 -13.40	ESO 427- 13	r	.24± .05	.63	15.00 ±.14				
209.41 -75.27	IRAS06492-3021	3.3± .9		.33	13.46				
064915.0-302118	PGC 19728		1.06	.12					

R.A. 2000 DEC. / l b / SGL SGB / R.A. 1950 DEC.	Names / PGC	Type / S_T n_L / T / L	$\log D_{25}$ / $\log R_{25}$ / $\log A_o$ / $\log D_o$	p.a. / A_g / A_i / A_{21}	B_T / m_B / m_{FIR} / B_T^o	$(B-V)_T$ / $(U-B)_T$ / $(B-V)_T^o$ / $(U-B)_T^o$	$(B-V)_o$ / $(U-B)_o$ / m'_o / m'_{25}	m_{21} / W_{20} / W_{50} / HI	V_{21} / V_{opt} / V_{GSR} / V_{3K}
065111.3+333138	NGC 2294	.E.6.$.	1.22± .06		14.9 ±.2	1.06± .04			
182.38 14.54	MCG 6-15- 14 P		.44± .05	.77	14.5 ±.3	.45± .06			
25.13 -40.76	IRAS06478+3335	-5.0±1.9		.00	12.20				
064754.3+333513	PGC 19729		1.21				14.90± .40		
0651.3 +2645	A 0648+26	.L?....	.72± .22						
188.75 11.82	MCG 4-16- 7		.00± .07	.63	14.8 ±.3				4740± 40
24.96 -47.52				.00					4698
0648.2 +2649	PGC 19731		.80		14.12				4851
065120.3+495311		.S?....	.87± .08	68					
166.36 20.39	UGC 3561		.27± .05	.45	14.7 ±.2				5501±155
25.53 -24.40	IRAS06474+4956			.40	13.27				5547
064729.0+495645	PGC 19732		.91	.13	13.81				5565
065120.8-293533		PSBS2?.	1.08± .05	98					
239.74 -13.03	ESO 427- 14	r	.46± .04	.81	15.54 ±.14				
209.47 -76.09		2.2±1.8		.56					
064924.0-293154	PGC 19733		1.15	.23					
065140.1+290420		.S..6*.	1.11± .05	105				15.47±.3	10780± 9
186.63 12.85	UGC 3571	U	.19± .05	.71	14.2 ±.3				10747
25.14 -45.21		6.0±1.2		.28				555± 5	10887
064830.0+290757	PGC 19740		1.18	.10	13.15			2.22	
065148.0+272906		.S..3..	1.29± .04	140				15.40±.3	4828± 9
188.12 12.22	UGC 3573	U	.86± .05	.69	14.6 ±.3				4789
25.14 -46.80		3.0± .9		1.18				466± 5	4789
064840.3+273244	PGC 19743		1.35	.43	12.65			2.32	4938
065148.0+482944		.L.....	1.00± .10	100					
167.79 20.02	UGC 3567	U	.09± .04	.39	14.15 ±.15				6045±155
25.59 -25.79		-2.0± .9		.00					6085
064800.4+483321	PGC 19744		1.03		13.67				6112
065154.5+405226									
175.36 17.44	CGCG 204- 23			.61	15.1 ±.3				4898± 38
25.47 -33.41									4909
064824.2+405603	PGC 19745								4982
0651.9 +2728	A 0648+27	.L?....	.72± .22						
188.15 12.24	MCG 5-16- 10		.11± .07	.69	14.9 ±.3				12260±101
25.18 -46.81				.00					12221
0648.8 +2732	PGC 19747		.79		13.98				12370
065158.8+502354		.S..4..	1.16± .05	16					
165.87 20.65	UGC 3568	U	.84± .05	.39	15.59 ±.18				
25.65 -23.89		4.0±1.0		1.23					
064806.1+502731	PGC 19749		1.20	.42					
065211.9+742535	IC 450	.LX.+*.	.88± .06	130	15.0 ±.2	1.06± .02	1.10± .02	19.45±.1	5850± 10
140.33 26.11	UGC 3547	PU	.20± .05	.36	14.88 ±.16	.12± .04	.10± .04		5501± 32
25.99 .14	MK 6	-.5± .6	.25± .07	.00	13.33	.90	11.73± .20		5947
064543.5+742907	PGC 19756		.89		14.47	.03	13.76± .40		5820
065215.1+391201									
177.02 16.90	CGCG 204- 25			.56	15.4 ±.3				9169± 38
25.52 -35.09									9174
064848.1+391540	PGC 19759								9257
0652.2 +1514		.L?....	.95± .10						
199.39 7.06	CGCG 85- 11		.33± .04	1.18	15.12 ±.11				4675± 46
24.89 -59.04				.00					4591
0649.4 +1518	PGC 19760		1.06		13.88				4804
065220.8+151509		.SBS2..	1.20± .05	20				14.83±.2	4531± 6
199.39 7.09	UGC 3578	U	.28± .05	1.17	13.93 ±.18			435± 7	4578± 46
24.93 -59.03	IRAS06494+1518	2.0± .8		.34	13.46			407± 5	4447
064929.3+151850	PGC 19763		1.31	.14	12.37			2.33	4661
0652.3 +4943		.S..6*.	1.00± .08	150					
166.58 20.50	UGC 3572	U	.19± .06	.38					
25.70 -24.57		6.0±1.3		.27					
0648.5 +4947	PGC 19764		1.04	.09					
0652.4 +5710		.S..8..	1.11± .05						
158.91 22.66	UGC 3569	U	.05± .05	.22	15.2 ±.3				5110± 57
25.81 -17.12		8.0± .8		.06					5182
0648.2 +5714	PGC 19767		1.13	.03	14.90				5157
065252.0+742853	IC 451	.SXR3*.	1.13± .05	155					
140.27 26.16	UGC 3550	PU (1)	.07± .05	.36	14.6 ±.2				
26.03 .19		2.5± .6		.10					
064622.8+743228	PGC 19775	4.5±1.1	1.16	.04					

R.A. 2000 DEC. / l b / SGL SGB / R.A. 1950 DEC.	Names / PGC	Type / S_T n_L / T / L	$\log D_{25}$ / $\log R_{25}$ / $\log A_e$ / $\log D_o$	p.a. / A_g / A_i / A_{21}	B_T / m_B / m_{FIR} / B_T^o	$(B-V)_T$ / $(U-B)_T$ / $(B-V)_T^o$ / $(U-B)_T^o$	$(B-V)_e$ / $(U-B)_e$ / m'_e / m'_{25}	m_{21} / W_{20} / W_{50} / HI	V_{21} / V_{opt} / V_{GSR} / V_{3K}
0652.9 +7130 / 143.55 25.74 / 26.01 -2.78 / 0647.1 +7134	UGC 3557 / / / PGC 19776	.S..7*. / U / 7.0±1.2 /	1.13± .07 / .21± .03 / / 1.16	5 / .35 / .29 / .11					
065256.8-714545 / 282.64 -25.55 / 206.37 -33.95 / 065340.0-714154	ESO 58- 19 / / / PGC 19778	.LX.-. / S / -1.0± .8 /	1.22± .05 / .16± .04 / / 1.25	104 / .62 / .00 /	13.59 ±.14 / / /				
065300.1+121113 / 202.22 5.87 / 25.13 -62.10 / 065012.2+121457	UGC 3582 / / / PGC 19781	.SBS3.. / U / 3.0± .9 /	1.00± .08 / .04± .06 / / 1.13	/ 1.42 / .05 / .02				16.08±.3 / / 155± 7 /	8390± 10 / / 8295 / 8524
065302.6-391610 / 249.09 -16.49 / 207.12 -66.44 / 065122.0-391224	ESO 309- 5 / IRAS06513-3912 / / PGC 19785	.SBT3?. / Sr / 3.1± .7 /	1.16± .05 / .27± .05 / / 1.21	155 / .53 / .37 / .13	14.39 ±.14 / / /				
065304.0-645501 / 275.23 -24.22 / 206.44 -40.79 / 065249.0-645112	ESO 87- 49 / FAIR 261 / / PGC 19787	.S?.... / / /	.90± .06 / .09± .05 / / .93	/ .26 / .13 / .05	14.82 ±.14 / 13.15 / / 14.36				10204± 51 / 9992 / 10299 /
065307.0+500202 / 166.30 20.71 / 25.84 -24.26 / 064915.4+500544	UGC 3576 / / / PGC 19788	.SBS3.. / U / 3.0± .8 /	1.20± .05 / .26± .05 / / 1.24	128 / .39 / .36 / .13	14.35 ±.20 / / / 13.56			15.68±.1 / / 372± 8 / 1.99	5948± 7 / 5994 / 6013 /
065310.6+571045 / 158.93 22.76 / 25.91 -17.11 / 064855.7+571426	UGC 3574 / IRAS06488+5714 / / PGC 19789	.SAS6.. / U / 6.0± .6 /	1.62± .04 / .06± .06 / / 1.64	/ .22 / .08 / .03	13.2 ±.5 / 13.23 / / 12.91			13.27±.1 / 159± 6 / 146± 6 / .33	1442± 5 / 1418± 57 / 1514 / 1489
065312.1+270453 / 188.63 12.34 / 25.58 -47.21 / 065005.0+270837	UGC 3584 / / / PGC 19792	.S..4.. / U / 4.0± .9 /	1.19± .06 / .81± .06 / / 1.25	14 / .66 / 1.20 / .41	14.9 ±.3 / / / 12.98			15.35±.3 / / 297± 7 / 1.96	4438± 10 / 4397 / 4550 /
065313.8+471005 / 169.21 19.82 / 25.83 -27.12 / 064929.8+471348	UGC 3579 / IRAS06495+4713 / / PGC 19794	.S..2.. / U / 2.0±1.0 /	1.09± .06 / .77± .05 / / 1.14	60 / .47 / .94 / .38	15.63 ±.18 / 13.49 / /				
065327.3+772436 / 137.03 26.53 / 26.09 3.12 / 064600.8+772812	UGC 3548 / MK 701 / / PGC 19803	.S..1.. / U / 1.0± .9 /	1.04± .04 / .10± .04 / / 1.07	/ .34 / .10 / .05	14.64 ±.19 / 12.87 / / 14.14			16.18±.1 / / 192± 8 / 1.99	5047± 11 / 4963± 61 / 5183 / 5040
065329.7+144347 / 199.98 7.10 / 25.45 -59.56 / 065038.8+144733	UGC 3586 / / / PGC 19804	.S..7*. / U / 7.0±1.3 /	1.04± .15 / .37± .12 / / 1.15	40 / 1.17 / .51 / .19				16.01±.3 / / 193± 7 /	2405± 7 / 2319 / 2536 /
0653.6 +5458 / 161.25 22.22 / 25.96 -19.32 / 0649.5 +5502	CGCG 261- 8 / / / PGC 19808	.E?.... / / /	.60± .15 / .03± .10 / / .64	/ .29 / .00 /	15.45 ±.19 / / / 14.99				10868±125 / 10932 / 10921 /
0653.6 +2719 / 188.45 12.52 / 25.72 -46.97 / 0650.5 +2723	UGC 3585 / / / PGC 19809	.S..6*. / U / 6.0±1.2 /	.98± .07 / .17± .05 / / 1.04	0 / .66 / .25 / .09	14.7 ±.3 / / /				
065353.6-405144 / 250.70 -16.92 / 206.67 -64.85 / 065216.0-404754	NGC 2310 / ESO 309- 7 / / PGC 19811	.L..../ / R / -2.0± .8 /	1.64± .03 / .72± .04 / .85± .02 / 1.58	47 / .46 / .00 /	12.74 ±.13 / 12.56 ±.09 / / 12.15	.98± .01 / .46± .01 / .80 / .30	.96± .01 / .44± .01 / 12.48± .06 / 14.02± .21		/ 1187± 66 / 973 / 1322
065354.4-322837 / 242.68 -13.70 / 206.98 -73.23 / 065202.0-322448	ESO 427- 17 / / / PGC 19812	.SBT6.. / r / 5.6± .9 /	.93± .07 / .10± .06 / / 1.01	/ .82 / .14 / .05	15.46 ±.14 / / /				
065355.1+191759 / 195.88 9.20 / 25.72 -54.99 / 065058.5+192146	UGC 3587 / IRAS06509+1921 / / PGC 19813	.S?.... / / /	1.45± .04 / .61± .06 / / 1.53	107 / .89 / .92 / .31	13.84 ±.20 / 13.68 / / 12.03			13.29±.1 / 229± 5 / 214± 4 / .95	1264± 4 / / 1194 / 1390
065359.2-631305 / 273.44 -23.74 / 206.33 -42.49 / 065334.0-630912	ESO 87- 50 / / / PGC 19816	.S..5?/ / S / 5.0±1.9 /	1.17± .05 / .81± .04 / / 1.20	121 / .37 / 1.21 / .40	15.84 ±.14 / / /				

R.A. 2000 DEC. / l b / SGL SGB / R.A. 1950 DEC.	Names / PGC	S_T n_L / T / L	$\log D_{25}$ / $\log R_{25}$ / $\log A_o$ / $\log D_o$	p.a. / A_g / A_i / A_{21}	B_T / m_B / m_{FIR} / B_T^o	$(B-V)_T$ / $(U-B)_T$ / $(B-V)_T^o$ / $(U-B)_T^o$	$(B-V)_o$ / $(U-B)_o$ / m'_o / m'_{25}	m_{21} / W_{20} / W_{50} / HI	V_{21} / V_{opt} / V_{GSR} / V_{3K}
0654.0 +8402		.S?....	1.05± .08	75	15.2 ±.2			15.03±.1	4424± 11
129.62 27.04	UGC 3521		.25± .06	.24					
26.15 9.75				.37				308± 8	4581
0641.0 +8406	PGC 19817		1.07	.12	14.62			.29	4403
0654.2 +6512		.SBR3..	1.08± .07	90	15.0 ±.2			15.90±.1	4591± 11
150.43 24.74	UGC 3577	U	.19± .06	.23					4691
26.09 -9.09		3.0± .8		.26				281± 8	
0649.3 +6516	PGC 19820		1.10	.09	14.50			1.31	4619
065431.4+300421		.SB.2..	.96± .09	45	15.0 ±.3			16.39±.3	5223± 9
185.94 13.82	UGC 3590	U	.39± .06	.68					5193
26.02 -44.22		2.0± .9		.48				256± 5	
065119.9+300810	PGC 19827		1.02	.20	13.82			2.38	5332
065432.8+454215		.L.....	1.25± .06	30	14.60 ±.16				
170.75 19.56	UGC 3588	U	.49± .03	.44					
26.08 -28.59		-2.0± .9		.00					
065052.3+454603	PGC 19829		1.23						
065435.1+502112	A 0650+50				.37 15.2 ±.3				5872± 40
166.05 21.03	CGCG 234- 26								5918
26.10 -23.94	MK 373								5937
065042.7+502500	PGC 19831								
0654.7 +7045		.S..6*.	1.24± .04	166				15.97±.1	3902± 11
144.40 25.77	UGC 3575	U	1.10± .05	.31					4020
26.15 -3.54		6.0±1.4		1.47				269±' 8	
0649.1 +7049	PGC 19838		1.27	.50					3916
065449.1+233006		.SXS3..	1.03± .06	137	14.3 ±.3			15.16±.3	4547± 10
192.11 11.18	UGC 3594	U	.27± .05	.79					4492
26.10 -50.79		3.0± .9		.37				262± 7	
065147.1+233357	PGC 19840		1.10	.13	13.13			1.90	4668
0654.8 +8229		.S..6*.	1.19± .05	128	15.50 ±.19			16.00±.1	1941± 11
131.36 26.99	UGC 3540	U	.70± .05	.29					2094
26.17 8.20		6.0±1.3		1.03				130± 8	
0644.0 +8233	PGC 19841		1.22	.35	14.18			1.48	1924
065500.3+805800		.E.....	.78± .13		14.49 ±.20				
133.07 26.90	UGC 3549	U	.00± .04	.30					
26.17 6.68		-5.0± .9		.00					
064535.1+810138	PGC 19847		.83						
065501.3-530814		.IBS9..	1.04± .08	100					
262.99 -20.88	ESO 162- 1	S (1)	.29± .07	.19					
206.17 -52.57		10.0± .9		.22					
065354.0-530418	PGC 19848	7.8± .9	1.05	.15					
065511.3+241351		.S?....	.96± .09		13.9 ±.3			17.45±.3	4670± 9
191.47 11.56	UGC 3599		.02± .06	.69					7271±155
26.23 -50.06	IRAS06521+2417			.03	13.01			165± 5	
065208.2+241743	PGC 19854		1.02	.01					
065513.8+402017		RSBS1..	.93± .07						13160± 38
176.12 17.84	UGC 3592	U	.12± .05	.47	14.97 ±.18				13169
26.22 -33.95	IRAS06517+4024	1.0± .9		.13	12.83				13249
065144.8+402409	PGC 19855		.98	.06	14.21				
065514.4+405736		.SX.5*.	1.22± .05	40	14.4 ±.2			15.90±.1	6694± 11
175.51 18.06	UGC 3593	U	.30± .05	.48					6705
26.22 -33.33		5.0± .8		.44				380± 8	
065144.1+410128	PGC 19856		1.26	.15	13.47			2.29	6782
065521.0-470645		PSBR3..	1.04± .05						
256.94 -18.89	ESO 256- 7	Sr (1)	.10± .04	.19	14.76 ±.14				
206.06 -58.60	IRAS06539-4702	2.6± .6		.14	13.76				
065357.1-470248	PGC 19858	3.3±1.2	1.06	.05					
065525.3-263626		.SXS8..	1.24± .04	98					2056
237.34 -10.98	ESO 491- 9	S (1)	.49± .04	1.94	14.84 ±.14				
205.70 -79.10	IRAS06534-2632	8.0± .9		.60	13.03				1859
065324.0-263230	PGC 19861	7.8± .9	1.42	.24	12.29				2206
065526.1-264532		.IBS9..	1.12± .05	49					
237.48 -11.04	ESO 491- 10	S (1)	.25± .05	1.94	15.62 ±.14				
205.69 -78.95		10.0± .8		.19					
065325.1-264136	PGC 19862	10.0± .8	1.30	.13					
065527.0+155606		.SA.8*.	1.21± .09		13.9 ±.3			14.68±.3	2074± 7
199.10 8.06	UGC 3602	U	.00± .05	1.02					1991
26.37 -58.36		8.0±1.1		.00				119± 7	
065234.7+160000	PGC 19863		1.31	.00	12.89			1.79	2206

R.A. 2000 DEC. l b SGL SGB R.A. 1950 DEC.	Names PGC	Type S_T n_L T L	$\log D_{25}$ $\log R_{25}$ $\log A_o$ $\log D_o$	p.a. A_g A_i A_{21}	B_T m_B m_{FIR} B_T^o	$(B-V)_T$ $(U-B)_T$ $(B-V)_T^o$ $(U-B)_T^o$	$(B-V)_o$ $(U-B)_o$ m'_o m'_{25}	m_{21} W_{20} W_{50} HI	V_{21} V_{opt} V_{GSR} V_{3K}
065528.7-652948 275.91 -24.11 206.11 -40.21 065517.0-652548	 ESO 87- 54 PGC 19865	PSAT3.. Sr (1) 3.0± .5 3.3± .9	1.07± .05 .20± .05 1.09	134 .28 .27 .10	 14.58 ±.14 				
065530.6+693352 145.72 25.65 26.22 -4.73 065002.1+693741	A 0650+69 UGC 3580 IRAS06500+6937 PGC 19867	.SAS1P* PU 1.0± .5 	1.53± .03 .27± .04 1.55	3 .21 .28 .14	12.71 ±.19 12.50 12.21			13.30±.1 237± 7 218± 12 .95	1201± 6 1283± 76 1316 1219
065536.1+394557 176.71 17.70 26.31 -34.53 065208.2+394950	 UGC 3596 IRAS06521+3949 PGC 19869	.L...?. U -2.0±1.7 	1.03± .08 .06± .03 1.08	 .47 .00 	13.7 ±.2 13.56 ±.17 13.31 13.60± .44	.90± .02 .36± .03 		14.77±.1 533± 11 	1340± 9 5274± 24
0655.6 +3904 177.40 17.46 26.32 -35.22 0652.2 +3908	 UGC 3600 PGC 19871	.I..9*. U 10.0±1.3 	1.04± .15 .37± .12 1.10	35 .62 .28 .19				15.79±.1 98± 6 84± 4 	398± 7 402 491
065549.7+400001 176.50 17.83 26.36 -34.29 065221.4+400355	 UGC 3601 PGC 19875	.S?.... 	.74± .12 .13± .06 .78	5 .46 .18 .06	14.8 ±.3 14.12				5163± 37 5170 5253
065550.8+404138 175.82 18.08 26.36 -33.60 065221.1+404532	 MCG 7-15- 2 PGC 19876	.S?.... 	.82± .13 .00± .07 .86	 .44 .00 .00	15.04 ±.18 14.59				6250± 38 6260 6339
065551.3+135420 200.98 7.25 26.58 -60.38 065301.4+135816	 UGC 3605 PGC 19877	.S..6*. U 6.0±1.5 	1.04± .08 1.06± .06 1.15	142 1.14 1.47 .50				16.16±.3 470± 7 	7971± 10 7881 8107
0655.8 +8454 128.65 27.11 26.19 10.62 0641.0 +8458	 UGC 3522 7ZW 92 PGC 19878	.S?.... 	1.30± .05 .50± .06 1.33	132 .24 .69 .25	15.2 ±.3 14.29			14.75±.1 201± 16 202± 12 .21	2137± 11 2296 2114
0656.0 +5524 160.90 22.67 26.32 -18.89 0651.9 +5528	 UGC 3595 PGC 19884	RSBR3*. U 3.0± .8 	1.10± .06 .09± .05 1.12	 .20 .12 .04	14.8 ±.2 				
065604.4+840450 129.58 27.09 26.20 9.79 064258.3+840825	 UGC 3528 IRAS06421+8407 PGC 19886	.SB.2.. U 2.0± .8 	1.14± .05 .24± .05 1.17	40 .24 .30 .12	14.38 ±.18 				
065605.0-671839 277.87 -24.44 206.04 -38.40 065606.0-671436	 ESO 87- 57 PGC 19887	.S?.... 	1.02± .06 .27± .06 1.07	159 .48 .41 .14	15.17 ±.14 14.24				7418± 69 7207 7510
065618.0+452937 171.07 19.78 26.43 -28.80 065238.1+453333	NGC 2303 UGC 3603 PGC 19891	.E..... U -5.0± .8 	1.18± .15 .00± .08 1.25	 .44 .00 	13.57 ±.16 13.05				6256±155 6284 6335
065619.1+061605 207.90 3.93 27.03 -68.02 065338.2+062003	 UGC 3607 IRAS06536+0620 PGC 19892	.SXS5.. U 5.0± .9 	1.00± .08 .04± .06 1.19	 2.07 .06 .02				14.95±.2 264± 7 222± 5 	6766± 6 6650 6910
065621.6-625039 273.12 -23.39 205.97 -42.86 065554.1-624636	 ESO 87- 56 FAIR 263 PGC 19895	.L?.... 	1.06± .07 .37± .05 1.06	168 .43 .00 	15.44 ±.14 14.88				9020± 63 8805 9121
0656.5 +6039 155.37 24.03 26.36 -13.64 0652.0 +6043	 UGC 3598 IRAS06520+6043 PGC 19899	.IB.9.. U 10.0± .8 	1.31± .04 .22± .05 1.34	25 .34 .16 .11	14.9 ±.4 14.41			14.93±.1 102± 7 72± 12 .41	1995± 6 2078 2036
065707.8-453252 255.51 -18.06 205.43 -60.16 065540.0-452848	 ESO 256- 10 PGC 19912	PSXS1.. r 1.0± .9 	.92± .06 .16± .05 .95	110 .25 .16 .08	15.43 ±.14 				
0657.2 +2026 195.17 10.38 27.03 -53.85 0654.2 +2030	 UGC 3611 IRAS06542+2030 PGC 19913	.S..0.. U .0± .9 	.96± .09 .29± .06 1.00	65 .62 .22 	14.3 ±.3 12.17 13.42				5020 4953 5148

R.A. 2000 DEC.	Names	Type	logD$_{25}$	p.a.	B$_T$	(B-V)$_T$	(B-V)$_e$	m$_{21}$	V$_{21}$
l b		S$_T$ n$_L$	logR$_{25}$	A$_g$	m$_B$	(U-B)$_T$	(U-B)$_e$	W$_{20}$	V$_{opt}$
SGL SGB		T	logA$_e$	A$_i$	m$_{FIR}$	(B-V)$_T^o$	m'$_e$	W$_{50}$	V$_{GSR}$
R.A. 1950 DEC.	PGC	L	logD$_o$	A$_{21}$	B$_T^o$	(U-B)$_T^o$	m'$_{25}$	HI	V$_{3K}$
065715.3+133222		.SB.8..	1.16± .13	85				15.37±.3	3997± 7
201.47 7.39	UGC 3613	U	.49± .12	1.11					
27.28 -60.75		8.0± .9		.60				217± 7	3905
065425.9+133624	PGC 19914		1.26	.25					4135
065717.0-265403		.IBS9..	1.19± .07						2125
237.79 -10.72		S (1)	.11± .08	1.76					
203.59 -78.80		10.0± .8		.08					1927
065516.0-265000	PGC 19917	11.1± .8	1.35	.06					2277
065725.0-244334		.S..5?/	1.28± .05	166					
235.81 -9.77	ESO 491- 12	S	.66± .05	1.88	14.80 ±.14				
202.73 -80.97	IRAS06553-2439	5.0±1.8		.99	13.28				
065521.0-243930	PGC 19920		1.46	.33					
065734.5+462413		.S?....	1.27± .06	20					6435± 37
170.23 20.28	UGC 3608		.29± .06	.33	13.61 ±.18				6466
26.67 -27.88	IRAS06538+4628			.44	10.95				6513
065352.6+462814	PGC 19924		1.30	.15	12.81				
065735.3-454836		.LAS0*.	1.29± .05	142	14.0 ±.2	1.04± .04	1.12± .02		
255.80 -18.08	ESO 256- 11	S	.35± .04	.18	13.56 ±.14	.60± .05	.57± .03		
205.29 -59.89		-2.0± .8	.83± .07	.00			13.65± .25		
065608.1-454430	PGC 19925		1.25				14.42± .33		
065740.3+390514		.SBS3..	.92± .07	25					5197± 38
177.53 17.84	UGC 3612	U	.14± .05	.49	14.3 ±.2				5200
26.80 -35.20	IRAS06541+3909	3.0± .9		.19	12.48				5291
065413.8+390916	PGC 19927		.97	.07	13.60				
065745.3+133429		.SB?...	.96± .09					15.40±.3	3988± 10
201.49 7.52	UGC 3617		.00± .06	1.09					
27.52 -60.71				.00				151± 7	3896
065455.8+133833	PGC 19928		1.06	.00					4127
0657.8 +6341		.S..8..	1.19± .05	12					3993
152.16 24.82	UGC 3606	U	.26± .05	.22	14.7 ±.2				4087
26.49 -10.61		8.0± .8		.32					4027
0653.1 +6345	PGC 19931		1.21	.13	14.16				
065752.2+225153		.S..7*.	1.14± .07	170				15.84±.3	5578± 10
193.00 11.55	UGC 3616	U	.19± .06	.53					
27.22 -51.42		7.0±1.2		.26				161± 7	5520
065451.1+225557	PGC 19932		1.19	.09					5704
065759.2+354403		.S..2..	1.16± .05	31					
180.82 16.68	UGC 3615	U	.70± .05	.60	15.27 ±.18				
26.94 -38.55	IRAS06546+3548	2.0± .9		.86	12.45				
065438.7+354807	PGC 19933		1.22	.35					
065816.8+450322		.SB.2..	1.11± .05	60					
171.63 19.97	UGC 3614	U	.06± .05	.48	14.4 ±.2				
26.83 -29.23		2.0± .8		.08					
065438.1+450726	PGC 19941		1.16	.03					
0658.3 +5758		.SBS3*.	1.00± .16	90					
158.27 23.62	UGC 3610	U	.25± .12	.22					
26.62 -16.31		3.0±1.3		.35					
0654.0 +5803	PGC 19942		1.02	.13					
065827.8+341242		.SB.5..	.93± .07						
182.33 16.20	UGC 3619	U	.03± .05	.61	15.0 ±.3				
27.10 -40.07		5.0± .9		.04					
065509.9+341648	PGC 19944		.99	.01					
065837.9+451240	NGC 2308	.S..2..	1.26± .04	170					5853± 38
171.50 20.08	UGC 3618	U	.18± .05	.42	14.1 ±.2				5879
26.90 -29.07		2.0± .8		.22					5935
065458.8+451646	PGC 19949		1.30	.09	13.38				
065904.8+141738		.I..9*.	1.18± .06					14.65±.1	2336± 6
200.98 8.12	UGC 3621	U	.05± .06	1.01				171± 5	2247
28.13 -59.98		10.0±1.1		.04				156± 3	2476
065614.5+142147	PGC 19963		1.27	.03					
065904.8+800013	A 0650+80	.SXT5*.	1.13± .04	100	13.74 ±.15	.85± .03	.77± .02	15.86±.1	4955± 7
134.16 27.01	UGC 3581	U (1)	.09± .04	.34	14.23 ±.19			283± 11	4986± 59
26.35 5.71		5.0± .8	.94± .02	.14		.72	13.93± .06	256± 11	5101
065021.6+800410	PGC 19964	2.2± .9	1.16	.05	13.43		14.01± .28	2.39	4946
0659.3 +6916		.S..3..	1.00± .08	130					
146.08 25.93	UGC 3609	U (1)	.19± .06	.16	15.3 ±.2				
26.55 -5.01		3.0± .9		.26					
0653.9 +6921	PGC 19969	4.5±1.2	1.01	.09					

R.A. 2000 DEC. l b SGL SGB R.A. 1950 DEC.	Names PGC	Type S_T n_L T L	$\log D_{25}$ $\log R_{25}$ $\log A_e$ $\log D_o$	p.a. A_g A_i A_{21}	B_T m_B m_{FIR} B_T^o	$(B-V)_T$ $(U-B)_T$ $(B-V)_T^o$ $(U-B)_T^o$	$(B-V)_e$ $(U-B)_e$ m' m'_{25}	m_{21} W_{20} W_{50} HI	V_{21} V_{opt} V_{GSR} V_{3K}
065921.4-255400 237.07 -9.88 200.72 -79.76 065719.0-254948	 ESO 491- 13 PGC 19970	RSXR0*. Sr -.1± .6 	1.25± .04 .35± .05 1.40	10 1.83 .26 	 14.42 ±.14 				
065931.5-531115 263.27 -20.26 205.07 -52.51 065824.0-530700	 ESO 162- 5 PGC 19973	.SBS9.. S (1) 9.0± .9 8.9± .9	1.07± .08 .32± .07 1.09	11 .21 .32 .16					
065934.2+330127 183.57 15.96 27.43 -41.25 065618.4+330538	 UGC 3622 IRAS06562+3305 PGC 19974	.S?.... 1.18	1.12± .05 .65± .05 	67 .71 .98 .33	 15.5 ±.3 13.48 13.76			15.42±.3 301± 7 1.34	4991± 10 4970 5100
0659.8 +2733 188.79 13.88 27.72 -46.71 0656.7 +2738	 UGC 3624 PGC 19979	.S..3.. U (1) 3.0±1.0 5.5±1.3	1.04± .08 .76± .06 1.09	146 .52 1.05 .38					
065950.2+352730 181.23 16.92 27.42 -38.82 065630.3+353141	IC 2175 UGC 3623 PGC 19981	.SB.5*. U 5.0± .8 1.28	1.24± .04 .33± .05 	66 .44 .50 .17	 14.6 ±.2 				
070004.5-323716 243.35 -12.57 202.55 -73.05 065812.0-323300	 ESO 366- 30 PGC 19985	.IBS9?P S (1) 9.8± .8 8.6± .7	1.11± .05 .54± .05 1.18	50 .83 .41 .27	 16.06 ±.14 14.82				2754 2547 2905
070018.3-300947 241.09 -11.50 201.64 -75.50 065822.0-300530	IC 456 ESO 427- 24 PGC 19993	.LBT0.. Sr -1.6± .6 1.38	1.33± .04 .21± .04 	110 .73 .00 	 12.96 ±.14 				
070042.4-272206 238.55 -10.23 200.00 -78.27 065842.0-271748	 ESO 491- 15 IRAS06586-2717 PGC 19996	.SX.5*/ S (1) 4.5± .9 4.4± .9	1.31± .04 .69± .03 1.45	64 1.45 1.03 .34	 14.34 ±.14 12.54 				
070042.7-211448 232.97 -7.57 192.81 -84.30 065834.0-211030	 ESO 558- 5 IRAS06585-2110 PGC 19997	.SAR2?. S 2.0±1.7 1.28	1.15± .06 .25± .05 	84 1.37 .31 .13					
0701.0 +5116 165.43 22.28 27.19 -23.01 0657.1 +5120	 UGC 3627 IRAS06571+5120 PGC 20007	.S..7.. U 7.0± .9 1.10	1.07± .06 .18± .05 	80 .34 .25 .09	 14.54 ±.18 				
070103.4+015437 212.33 3.01 31.12 -72.32 065827.5+015855	 UGC 3630 IRAS06584+0158 PGC 20008	.SXR3?. E 3.0±1.7 1.48	1.23± .04 .28± .04 	5 2.63 .39 .14	 10.48 				1774± 7 1643 1929
0701.6 +1407 201.41 8.61 29.37 -60.12 0658.8 +1412	 UGC 3634 PGC 20020	.SBR1.. U 1.0± .9 1.10	1.02± .06 .08± .05 	135 .94 .08 .04					
070140.9+171056 198.62 9.94 29.08 -57.07 065847.1+171516	 UGC 3635 PGC 20021	.S..8*. U 8.0±1.3 1.03	.96± .09 .22± .06 	115 .73 .27 .11	 15.10 ±.18 14.10			16.64±.3 139± 7 2.43	4941± 10 4861 5080
0701.8 +0455 209.73 4.55 30.96 -69.30 0659.1 +0459	 UGC 3637 IRAS06591+0459 PGC 20027	.S..8*. U 8.0±1.3 1.21	1.04± .15 .47± .12 	125 1.81 .58 .24	 13.24 				3550± 7 3428 3703
0701.8 +8634 126.78 27.23 26.28 12.28 0641.0 +8638	 UGC 3528A PGC 20028	.S?.... .92	.89± .11 .14± .07 	 .34 .18 .07	 15.46 ±.18 14.89			15.83±.1 297± 8 .86	4544± 11 4786± 46 4720 4529
070206.9-282730 239.69 -10.42 199.20 -77.15 070008.0-282306	 ESO 427- 26 IRAS07001-2823 PGC 20037	.L..+./ S -1.0± .9 1.21	1.16± .04 .52± .03 	64 1.31 .00 	 14.70 ±.14 12.91 				
070221.5+111413 204.11 7.49 30.11 -62.99 065934.9+111836	 UGC 3641 PGC 20039	.S..3.. U 3.0± .8 1.16	1.06± .06 .00± .05 	 1.09 .00 .00				17.00±.3 271± 7 	10314± 10 10213 10462

R.A. 2000 DEC.	Names	Type	$\log D_{25}$	p.a.	B_T	$(B-V)_T$	$(B-V)_\bullet$	m_{21}	V_{21}
l b		S_T n_L	$\log R_{25}$	A_g	m_B	$(U-B)_T$	$(U-B)_\bullet$	W_{20}	V_{opt}
SGL SGB		T	$\log A_\bullet$	A_i	m_{FIR}	$(B-V)_T^o$	$m'_\bullet$	W_{50}	V_{GSR}
R.A. 1950 DEC.	PGC	L	$\log D_0$	A_{21}	B_T^o	$(U-B)_T^o$	m'_{25}	HI	V_{3K}
070226.3+195807		.S?....	.96± .09	93				16.49±.3	7658± 10
196.14 11.29	UGC 3639		.59± .06	.57	15.0 ±.3				7588
29.14 −54.28				.88				402± 7	
065929.1+200230	PGC 20043		1.01	.29	13.52			2.68	7794
070232.6+503529	NGC 2315	.S..0..	1.13± .05	116					
166.20 22.31	UGC 3633	U	.58± .05	.32	14.57 ±.19				6159±155
27.47 −23.68		.0± .9		.44					6205
065840.3+503951	PGC 20045		1.13		13.72				6231
070236.7−420409	NGC 2328	PLX.-?.	1.21± .05	115	12.75V±.13	.61± .02			
252.51 −15.86	ESO 309- 16	BS	.11± .04	.47	13.04 ±.14	.17± .04			1159± 57
203.03 −63.59	IRAS07010-4159	-2.9± .5		.00	12.26	.49			942
070101.0−415942	PGC 20046		1.25		12.73	.09	13.99± .30		1301
070240.5−284150	NGC 2325	.E.4...	1.52± .03	6	*		1.04± .01		
239.96 −10.41	ESO 427- 28	R	.24± .04	1.20	12.20 ±.11		.65± .07		2159± 30
198.81 −76.89		-5.0± .3	1.36± .05	.00			14.09± .19		1956
070042.0−283724	PGC 20047		1.64		10.96				2316
0702.6 +7057		.SB.4?.	1.21± .05	155	*			14.82±.1	3277± 11
144.27 26.44	UGC 3626	U	.56± .05	.25	15.09 ±.20				3395
26.79 −3.32		4.0± .9		.83				305± 8	3294
0657.0 +7102	PGC 20048		1.24		13.99			.55	
070245.5−292545		.L..0*P	1.29± .04	151	*				
240.64 −10.71	ESO 427- 29	S	.22± .04	.62	13.40 ±.14				
199.17 −76.16		-2.0±1.1		.00					
070048.0−292118	PGC 20049		1.33						
070248.2+370716		.S..0..	1.00± .08	102	*				
179.84 18.08	UGC 3640	U	.55± .06	.55	15.16 ±.19				
28.11 −37.14		.0± .9		.41					
065925.6+371140	PGC 20050		1.02						
0702.8 +7635		.S..2..	1.00± .08	40	*				
137.99 26.99	UGC 3620	U	.94± .06	.38					
26.63 2.31		2.0±1.0		1.16					
0655.8 +7640	PGC 20055		1.04	.47					
0702.9 +6245		.SB?...	.92± .11						4548± 54
153.28 25.21	MCG 10-10- 21		.17± .07	.24	14.92 ±.18				4638
27.09 −11.52				.24					4588
0658.3 +6250	PGC 20058		.94	.09	14.41				
070303.0+492532		.SB.2*.	1.06± .06	48	*				
167.43 22.07	UGC 3638	U	.18± .05	.34	14.33 ±.18				5567± 38
27.60 −24.84	IRAS06592+4929	2.0± .9		.22	13.37				5608
065914.0+492956	PGC 20062		1.09	.09	13.71				5643
070303.8+391426									
177.76 18.87	CGCG 205- 12			.45	15.1 ±.3				6591± 60
28.06 −35.02	MK 1196				13.02				6594
065937.3+391851	PGC 20063								6691
070320.8+863329		.E?....	.97± .15						
126.80 27.25	UGC 3536A		.00± .07	.34	14.56 ±.15				4742± 46
26.30 12.27	ARAK 124			.00					4906
064235.8+863719	PGC 20066		1.02		14.14				4715
070351.4+222211		.S..6*.	1.31± .04	115	*			15.01±.3	4544± 7
194.05 12.60	UGC 3652	U	.51± .05	.44	13.9 ±.3				4482
29.46 −51.86	IRAS07008+2226	6.0±1.2		.74	13.39			369± 7	4679
070051.1+222640	PGC 20083		1.35	.25	12.67			2.09	
0704.0 +2914		.S..1..	.93± .07	63	*				5121± 79
187.57 15.38	UGC 3649	U	.27± .05	.55	14.7 ±.3				5085
28.92 −44.99	IRAS07008+2919	1.0± .9		.28	13.58				5243
0700.8 +2919	PGC 20087		.99	.13	13.85				
070405.2−415251		.S..3?/	1.24± .04	18	*				
252.43 −15.54	ESO 309- 17	S	.82± .04	.45	14.64 ±.14				
202.38 −63.76	IRAS07024-4148	3.0±1.9		1.13	13.38				
070229.0−414818	PGC 20094		1.28	.41					
0704.2 +5413		.E.....	1.08± .17	55	*				
162.47 23.52	UGC 3646	U	.08± .08	.22	14.57 ±.17				
27.60 −20.04		-5.0± .8		.00					
0700.2 +5418	PGC 20097		1.09						
0704.3 +1832		.S..0..	1.02± .06	174	*				
197.65 11.09	UGC 3656	U	.53± .05	.47	14.9 ±.3				
30.06 −55.67		.0± .9		.40					
0701.4 +1837	PGC 20099		1.03						

R.A. 2000 DEC. / l b / SGL SGB / R.A. 1950 DEC.	Names / PGC	Type / S_T n_L / T / L	$\log D_{25}$ / $\log R_{25}$ / $\log A_e$ / $\log D_o$	p.a. / A_g / A_i / A_{21}	B_T / m_B / m_{FIR} / B_T^o	$(B-V)_T$ / $(U-B)_T$ / $(B-V)_T^o$ / $(U-B)_T^o$	$(B-V)_e$ / $(U-B)_e$ / m'_e / m'_{25}	m_{21} / W_{20} / W_{50} / HI	V_{21} / V_{opt} / V_{GSR} / V_{3K}
070420.9+640115		.LA.0..	1.18± .06	30	13.3 ±.2	.93± .03	1.01± .01	14.01±.1	4498± 6
151.94 25.59	UGC 3642	U	.12± .03	.20	13.46 ±.15	.38± .05	.47± .02	462± 6	4501± 27
27.20 -10.25		-2.0± .8	.94± .06	.00		.83	13.53± .22	435± 8	4592
065934.8+640543	PGC 20103		1.18		13.14	.35	13.75± .38		4535
0704.7 +1736		.I..9*.	1.26± .11	65				15.22±.1	1183± 6
198.55 10.78	UGC 3658	U	.39± .12	.53				147± 5	
30.34 -56.60		10.0±1.2		.29				133± 4	1104
0701.8 +1741	PGC 20112		1.31	.20					1326
0704.8 +5631	A 0700+56	.IB.9..	1.14± .07	40	14.54 ±.15	.42± .04	.50± .04	14.64±.1	1384± 6
160.05 24.14	UGC 3647	U (1)	.13± .06	.16		-.22± .06	-.25± .05	72± 6	
27.58 -17.74	DDO 40	10.0± .8	1.04± .03	.10		.34	15.23± .06	50± 7	1451
0700.6 +5635	PGC 20116	9.0±1.1	1.15	.07	14.29	-.27	14.75± .40	.29	1442
070458.3+460828		.S?....	.80± .08	9	14.7 ±.2				6170±155
170.92 21.42	UGC 3655		.14± .05	.39	12.91				6198
28.11 -28.11	IRAS07012+4612			.21					6256
070117.5+461300	PGC 20121		.84	.07	14.06				
070517.5-583114		.SBS9..	1.54± .05						
268.95 -21.15		S (1)	.32± .08	.37					
204.22 -47.15		9.0± .7		.33					
070428.4-582634	PGC 20125	10.0± .8	1.57	.16					
0705.4 +6406		.S..4..	1.13± .05	104					
151.87 25.72	UGC 3648	U	.71± .05	.20	15.37 ±.18				
27.32 -10.16		4.0± .9		1.04					
0700.7 +6411	PGC 20127		1.14	.35					
070533.7+334940		.S..6*.	1.02± .06	80					
183.27 17.42	UGC 3664	U	.19± .05	.40	15.1 ±.2				
29.03 -40.40		6.0±1.3		.28					
070216.9+335416	PGC 20131		1.06	.09					
0705.5 +7103		.SXS7..	1.34± .04	140				14.58±.1	3242± 11
144.19 26.69	UGC 3644	U	.25± .05	.27	14.4 ±.3			238± 16	
27.03 -3.21		7.0± .8		.35				234± 12	3360
0659.9 +7108	PGC 20133		1.36	.13	13.82			.64	3260
070541.5+503449	NGC 2320	.E.....	1.15± .15	140	12.9 ±.2	.99± .02	1.07± .01		
166.36 22.79	UGC 3659	U	.24± .08	.32	13.76 ±.15				5725± 60
28.01 -23.67		-5.0± .8	1.09± .06	.00		.86	13.83± .19		5770
070149.5+503924	PGC 20136		1.13		13.03		13.05± .82		5800
070559.1+504521	NGC 2321	.SB.1..	1.15± .07	135					
166.19 22.88	UGC 3663	U	.11± .06	.31	14.5 ±.2				
28.05 -23.49		1.0± .8		.11					
070206.7+504957	PGC 20141		1.17	.06					
070600.3+503036	NGC 2322	.SB.1*.	1.06± .06	136					
166.45 22.82	UGC 3662	U	.45± .05	.32	14.65 ±.19				
28.07 -23.74		1.0±1.3		.46					
070208.5+503513	PGC 20142		1.09	.22					
0706.0 +7526		.S..2..	1.01± .06	0					
139.30 27.10	UGC 3636	U	.52± .05	.22	15.01 ±.19				
26.86 1.17	IRAS06593+7531	2.0± .9		.64	12.68				
0659.3 +7531	PGC 20143		1.03	.26					
070601.5+281751		.S?....	.82± .13						4881± 97
188.65 15.43	MCG 5-17- 10		.23± .07	.51	15.1 ±.3				4841
29.63 -45.91	MK 1197			.33	12.94				5007
070253.1+282229	PGC 20144		.86	.11	14.26				
0706.1 +7753		.S..6*.	1.00± .16	170					
136.55 27.25	UGC 3632	U	.14± .12	.22					
26.75 3.61		6.0±1.2		.20					
0658.6 +7758	PGC 20149		1.02	.07					
070627.8+301926		.I..9..	1.07± .14	150				14.58±.3	990± 7
186.75 16.29	UGC 3672	U	.21± .12	.53					
29.58 -43.88		10.0± .9		.15				99± 7	957
070316.4+302406	PGC 20154		1.12	.10					1112
070634.5+635059		.SX.1*.	1.22± .05	110				17.01±.1	4262± 11
152.17 25.80	UGC 3660	U	.27± .05	.19	13.62 ±.18				4281± 52
27.46 -10.41	IRAS07018+6355	1.0±1.2		.27				118± 8	4356
070149.8+635536	PGC 20158		1.24	.13	13.11			3.77	4302
070640.2+235335		.S..3..	1.02± .06	120				15.66±.3	6777± 10
192.89 13.81	UGC 3676	U (1)	.17± .05	.37	14.7 ±.3				
30.32 -50.30		3.0± .9		.23				347± 7	6720
070338.0+235816	PGC 20161	4.5±1.2	1.05	.08	14.01			1.56	6912

R.A. 2000 DEC. l　　b SGL　　SGB R.A. 1950 DEC.	Names PGC	Type S_T　n_L T L	$\log D_{25}$ $\log R_{25}$ $\log A_e$ $\log D_o$	p.a. A_g A_i A_{21}	B_T m_B m_{FIR} B_T^0	$(B-V)_T$ $(U-B)_T$ $(B-V)_T^0$ $(U-B)_T^0$	$(B-V)_e$ $(U-B)_e$ m'_e m'_{25}	m_{21} W_{20} W_{50} HI	V_{21} V_{opt} V_{GSR} V_{3K}
070645.8+252706 191.43　14.46 30.17　-48.74 070341.4+253147	 UGC　3674 PGC 20165	.S..6*. U 6.0±1.3	1.00± .08 .33± .06 1.04	110 .43 .49 .17	 15.0　±.3 14.07			16.15±.3 351±　5 1.92	8831±　9 8780 8964
070650.8-375914 248.96　-13.50 200.11　-67.57 070507.1-375430	 ESO　309- 19 IRAS07051-3754 PGC 20167	.S..4*. S　　(1) 4.0±1.3 5.6±1.3	1.15± .04 .86± .05 1.23	129 .85 1.27 .43	 15.81 ±.14				
070656.6-610917 271.72　-21.74 204.17　-44.50 070618.9-610430	 PGC 20170	.LXS-.. S -3.0± .9	1.09± .10 .18± .08 1.13	 .49 .00					
0706.9　-2201 234.33　-6.62 182.82　-83.08 0704.8　-2157	 ESO　558- 11 PGC 20171	.IBS9.. S　　(1) 9.5± .6 8.6± .7	1.26± .07 .24± .08 1.41	 1.62 .18 .12					737 544 904
070700.8+445100 172.36　21.37 28.60　-29.38 070323.1+445541	 UGC　3673 PGC 20172	.SX.6.. U 6.0± .9	.98± .07 .10± .05 1.02	35 .44 .15 .05	 15.4　±.2				
070712.6+141036 201.96　9.85 32.05　-59.96 070422.6+141520	 UGC　3682 PGC 20176	.S..7*. U 7.0±1.5	.96± .09 .98± .06 1.02	71 .66 1.35 .49				16.58±.3 362±　7	8111± 10 8019 8262
0707.4　+7111 144.06　26.85 27.17　-3.08 0701.7　+7116	 UGC　3657 PGC 20184	.S..6*. U 6.0±1.4	1.06± .06 .98± .05 1.08	148 .27 1.44 .49					
070725.5-281334 239.99　-9.27 193.90　-77.16 070526.1-280848	 ESO　427- 34 IRAS07054-2808 PGC 20187	.SBS8.. S　　(1) 8.0± .5 7.2± .6	1.28± .04 .21± .05 1.37	20 1.02 .26 .11	 14.46 ±.14 13.17				2387 2184 2550
070728.3+444724 172.44　21.43 28.69　-29.43 070350.8+445207	 UGC　3679 IRAS07038+4452 PGC 20190	.S?....	1.01± .09 .15± .06 1.05	145 .44 .22 .07	 15.2　±.2 12.12 14.50			15.88±.1 398±　8 1.31	5831±　7 5854 5922
070732.6+674055 147.96　26.46 27.36　-6.58 070222.3+674535	A　　0702+67 MK　　375 PGC 20194		.32± .18 .00± .12 .34	 .20					3600±110 3706 3629
070755.4+130523 203.03　9.53 32.63　-61.03 070506.7+131010	 UGC　3688 PGC 20204	.S..3.. U　　(1) 3.0± .9 4.5±1.2	1.00± .08 .43± .06 1.05	43 .56 .60 .22	 15.25 ±.18 14.04			14.97±.3 456±　7 .72	8225± 10 8128 8378
0707.9　+7133 143.66　26.93 27.19　-2.71 0702.2　+7138	 UGC　3665 PGC 20207	.S..1.. U 1.0± .9	1.04± .06 .54± .05 1.06	105 .30 .55 .27	 15.30 ±.18				
0707.9　+5255 164.00　23.73 28.25　-21.31 0704.0　+5300	 UGC　3680 PGC 20208	.I..9*. U 10.0±1.3	1.11± .14 .66± .12 1.14	 .31 .49 .33				16.64±.1 66±　6 64±　5	602± 10 655 673
070759.6+483958 168.47　22.63 28.53　-25.56 070412.9+484443	 MCG　8-13- 61 PGC 20209	.S?....	1.04± .09 .09± .07 .55± .04 1.07	 .36 .07	14.47 ±.15 14.45 ±.18 13.95	.99± .02 .45± .06 .84 .38	1.02± .01 .54± .05 12.71± .16 14.27± .52		5350± 31 5387 5432
070801.7+151046 201.13　10.46 32.23　-58.94 070510.5+151533	 UGC　3691 PGC 20214	.SA.6.. U 6.0± .8	1.34± .04 .31± .05 1.40	65 .67 .46 .16	 12.6　±.3 11.43			14.22±.1 250±　5 221±　6 2.63	2203±　5 2182± 27 2113 2353
070810.8+504054 166.37　23.20 28.43　-23.55 070418.7+504540	NGC　2326 UGC　3681 IRAS07043+5045 PGC 20218	.SBT3.. U 3.0± .8	1.27± .04 .02± .05 1.30	 .33 .03 .01	 13.9　±.3 13.49				5985± 10 6030 6062
070814.2+460659 171.12　21.95 28.75　-28.10 070433.8+461145	 UGC　3683 PGC 20220	.L..... U -2.0± .8	1.30± .12 .18± .08 1.32	50 .40 .00	 13.76 ±.18 13.27				5623±155 5651 5712

R.A. 2000 DEC. / l b / SGL SGB / R.A. 1950 DEC.	Names / PGC	Type / S_T n_L / T / L	$\log D_{25}$ / $\log R_{25}$ / $\log A_e$ / $\log D_o$	p.a. / A_g / A_i / A_{21}	B_T / m_B / m_{FIR} / B_T^o	$(B-V)_T$ / $(U-B)_T$ / $(B-V)_T^o$ / $(U-B)_T^o$	$(B-V)_e$ / $(U-B)_e$ / m'_e / m'_{25}	m_{21} / W_{20} / W_{50} / HI	V_{21} / V_{opt} / V_{GSR} / V_{3K}
0708.3 +3459				.64	16.66 ±.15	1.24± .03	1.25± .02		23089± 56
182.35 18.37	MCG 6-16- 21								23074
29.66 -39.21			.33± .04				13.80± .15		23204
0705.0 +3504	PGC 20221								
070820.7+184654	NGC 2339	.SXT4..	1.43± .02	175	12.51 ±.13	.72± .02	.87± .01	13.69±.1	2206± 5
197.83 12.06	UGC 3693	R (2)	.12± .03	.61	12.29 ±.18	.10± .02	.19± .02	340± 10	2361± 56
31.69 -55.35	IRAS07054+1851	4.0± .3	1.02± .01	.18	9.98	.54	13.10± .03	332± 4	2131
070525.2+185142	PGC 20222	3.7± .6	1.49	.06	11.63	-.03	14.23± .19	2.00	2353
070821.4+351007	NGC 2333	.S..1..	.98± .09	35				15.76±.3	4731± 10
182.18 18.44	UGC 3689	U	.14± .06	.64	14.17 ±.19				4274±155
29.66 -39.02	IRAS07050+3515	1.0± .9		.14	13.49			374± 7	4715
070502.5+351455	PGC 20223		1.04	.07	13.33			2.36	4844
0708.3 +5114		PSBS3..	1.02± .08	170					
165.79 23.37	UGC 3684	U	.44± .06	.28	14.83 ±.18				
28.43 -22.99	IRAS07044+5118	3.0± .9		.61	13.49				
0704.4 +5118	PGC 20225		1.05	.22					
0708.4 +2953		.L...?.	1.18± .15	13					
187.34 16.53	UGC 3692	U	.38± .08	.49	14.9 ±.3				
30.22 -44.29		-2.0±1.7		.00					
0705.3 +2958	PGC 20232		1.18						
070830.4-491246		.E+0.*.	1.11± .10						
259.83 -17.58		S	.09± .08	.21					
202.22 -56.38		-4.0±1.2		.00					
070710.6-490754	PGC 20233		1.12						
070834.4+503755	NGC 2326A	.SAS9*.	1.00± .06	15					
166.44 23.25	UGC 3687	P	.21± .05	.33	15.25 ±.20				
28.50 -23.59		9.0± .9		.21					
070442.5+504242	PGC 20237		1.03	.10					
070837.5+324051		.S..7..	1.17± .05	95				15.12±.3	4772± 10
184.64 17.60	UGC 3694	U	.29± .05	.34	14.6 ±.2				4748
29.97 -41.50		7.0± .8		.40				277± 7	
070522.6+324540	PGC 20239		1.20	.15	13.87			1.11	4893
070902.5+065548		.S..2..	.96± .09	45				15.31±.3	5967± 10
208.73 7.05	UGC 3707	U	.10± .06	1.16	14.9 ±.2				5849
35.10 -67.10	IRAS07063+0700	2.0± .9		.12	12.59			244± 7	
070620.9+070040	PGC 20244		1.07	.05	13.60			1.66	6129
070905.4+613541	A 0704+61	.SBT3..	1.52± .03					13.13±.1	1797± 4
154.69 25.70	UGC 3685	U	.08± .05	.33	12.8 ±.2			103± 5	
27.88 -12.65		3.0± .7		.12				79± 5	1882
070433.1+614029	PGC 20250		1.55	.04	12.30			.79	1844
070905.7+752111	IC 2174	PSBR1*.	1.01± .06		14.3 ±.3	.84± .07			
139.41 27.29	UGC 3666	PU	.06± .04	.22	14.81 ±.19	.36± .08			
27.06 1.09		.5± .6		.07					
070226.1+752555	PGC 20252		1.03	.03			14.02± .44		
0709.1 +5326	A 0705+53	.I..9..	1.15± .07		15.65 ±.15	.52± .06	.61± .05	15.88±.1	3144± 6
163.48 24.02	UGC 3690	U (1)	.11± .06	.34		-.29± .10	-.20± .07	68± 7	
28.40 -20.78	DDO 41	10.0± .8	.91± .03	.08		.40	15.69± .06	45± 12	3200
0705.1 +5331	PGC 20253	9.0±1.5	1.19	.05	15.23	-.38	16.00± .40	.59	3215
070908.2+483658	NGC 2329	.L..-*.	1.11± .08	175	13.48 ±.14	1.03± .02	1.05± .01		
168.57 22.80	UGC 3695	U	.07± .04	.36	13.60 ±.15	.52± .07	.55± .07		5729± 18
28.74 -25.60		-3.0±1.1	.83± .03	.00		.89	13.12± .10		5766
070521.7+484148	PGC 20254		1.15		13.09	.46	13.73± .45		5812
0709.1 +2842		.S..7..	1.00± .08						
188.54 16.22	UGC 3702	U	.00± .06	.57	14.9 ±.3				
30.56 -45.46		7.0± .8		.00					
0706.0 +2847	PGC 20256		1.05	.00					
070910.7-512801		.IAS9..	1.08± .05						
262.08 -18.29	ESO 207- 22	S (1)	.06± .05	.27	15.37 ±.14				
202.44 -54.13		10.0± .8		.04					
070757.0-512306	PGC 20257	10.0± .8	1.11	.03					
070912.1+203606	NGC 2341	.P.....	.92± .07	136	13.84 ±.14	.62± .04		15.39±.2	5227± 6
196.22 13.00	UGC 3708	R	.02± .05	.48	13.8 ±.3	.06± .03		324± 6	5226± 32
31.73 -53.52	IRAS07063+2043	99.0		.03	11.17	.47		208± 4	5157
070614.2+204058	PGC 20259		.97	.01	13.29	-.05	13.25± .39	2.09	5371
0709.3 +4423		.I..9*.	1.00± .16	0				14.87±.1	426± 10
172.96 21.62	UGC 3698	U	.19± .12	.44				105± 6	
29.09 -29.82		10.0±1.3		.14				50± 5	447
0705.7 +4428	PGC 20264		1.04	.09					520

R.A. 2000 DEC. / l b / SGL SGB / R.A. 1950 DEC.	Names / / / PGC	Type / S_T n_L / T / L	$\log D_{25}$ / $\log R_{25}$ / $\log A_o$ / $\log D_o$	p.a. / A_g / A_l / A_{21}	B_T / m_B / m_{FIR} / B_T^o	$(B-V)_T$ / $(U-B)_T$ / $(B-V)_T^o$ / $(U-B)_T^o$	$(B-V)_o$ / $(U-B)_o$ / m'_o / m'_{25}	m_{21} / W_{20} / W_{50} / HI	V_{21} / V_{opt} / V_{GSR} / V_{3K}
070918.6+203811 196.20 13.04 31.76 -53.49 070620.7+204303	NGC 2342 UGC 3709 PGC 20265	.S...P. R 	1.14± .07 .03± .06 1.18	126 .48 .05 .02	13.1 ±.2 12.7 ±.3 12.44	.54± .04 -.05± .03 .39 -.16	 13.56± .42	14.71±.2 398± 7 359± 6 2.25	5276± 7 5209± 28 5202 5416
070921.6+361702 181.15 19.02 29.81 -37.90 070600.8+362154	 UGC 3703 PGC 20267	.SAS5.. U 5.0± .8 	1.08± .06 .00± .05 1.11	 .33 .00 .00	 14.9 ±.3 14.57				7264 7253 7378
070923.4+483807 168.57 22.85 28.79 -25.58 070536.9+484258	 UGC 3696 PGC 20268	.E?.... 	1.00± .19 .14± .08 1.02	77 .36 .00 	 13.81 ±.19 13.36				6164± 28 6201 6248
0709.5 -0525 219.83 1.54 45.65 -79.12 0707.1 -0521	 PGC 20274	.IXS9*. E (1) 10.0±1.3 7.5±1.6	1.32± .06 .14± .08 1.75	170 4.56 .10 .07					
070933.8+501056 166.95 23.28 28.70 -24.03 070543.3+501548	NGC 2332 UGC 3699 PGC 20276	.L...*. U -2.0±1.1 	1.18± .15 .17± .08 1.19	60 .30 .00 	 13.82 ±.15 13.44				5806± 19 5849 5885
070945.0+484125 168.53 22.92 28.85 -25.52 070558.5+484617	 MCG 8-13- 82 PGC 20283	.L?.... 	.90± .17 .00± .07 .93	 .29 .00 	 14.95 ±.15 14.57				5750± 31 5787 5834
070947.5-273414 239.62 -8.51 190.83 -77.65 070747.0-272918	 ESO 491- 20 IRAS07077-2729 PGC 20285	.SBT3*P S 2.5± .6 	1.10± .04 .14± .04 1.21	65 1.15 .19 .07	13.82 ±.14 10.24 				
070949.5-273432 239.63 -8.50 190.80 -77.65 070749.0-272936	 ESO 491- 21 PGC 20287	.SBR2?P S 1.7± .7 	1.17± .04 .49± .03 1.28	20 1.15 .60 .24	13.64 ±.14 				
0710.0 +7453 139.93 27.33 27.15 .63 0703.5 +7458	 UGC 3675 IRAS07035+7458 PGC 20293	.S..6*. U 6.0±1.3 	1.19± .05 .91± .05 1.21	18 .22 1.34 .46	 15.49 ±.18 				
071006.5-631543 274.01 -22.00 203.89 -42.37 070939.0-631042	 ESO 88- 4 IRAS07096-6310 PGC 20294	 	.88± .07 .12± .06 .93	99 .52 	 13.94 ±.14 12.53 				2357±104 2140 2466
071013.6+442725 172.94 21.80 29.28 -29.73 070637.2+443220	NGC 2337 UGC 3711 IRAS07066+4432 PGC 20298	.IB.9.. U 10.0± .7 	1.35± .05 .13± .06 1.39	120 .40 .10 .06	 12.95 ±.18 12.53 12.45			13.42±.1 164± 7 144± 12 .91	434± 5 455 529
0710.4 +3944 177.77 20.38 29.75 -34.44 0707.0 +3949	 UGC 3716 PGC 20302	.S..6*. U 6.0±1.3 	1.03± .06 .24± .05 1.05	55 .28 .36 .12					
071030.7+614709 154.51 25.90 28.03 -12.44 070557.7+615203	 UGC 3704 PGC 20304	.S..1.. U 1.0± .9 	1.10± .05 .64± .05 1.12	50 .22 .65 .32	 14.7 ±.2 				
071032.4+751938 139.45 27.38 27.15 1.07 070353.7+752428	NGC 2314 UGC 3677 PGC 20305	.E.3... R -5.0± .4 	1.24± .03 .09± .05 .66± .01 1.25	25 .22 .00 	13.18 ±.13 12.91 ±.12 12.75	.99± .01 .54± .03 .90 .51	1.05± .01 .63± .03 11.97± .04 14.17± .25		3848± 25 3980 3856
071033.7+500709 167.07 23.42 28.88 -24.08 070643.5+501205	IC 458 UGC 3713 PGC 20306	.L..-*. U -3.0±1.3 	.95± .21 .33± .08 .94	170 .30 .00 	 14.49 ±.19 14.09				6620± 31 6662 6700
071039.2+394327 177.79 20.41 29.79 -34.45 070712.4+394824	 UGC 3718 PGC 20313	.L..... U -2.0± .9 	1.05± .08 .40± .03 1.02	118 .28 .00 	 15.04 ±.15 				
071040.6-381425 249.52 -12.92 198.28 -67.21 070857.1-380924	 ESO 310- 1 PGC 20315	.SBR7*. S (1) 6.7±1.0 5.6± .7	1.04± .05 .50± .04 1.13	119 .96 .69 .25	 15.83 ±.14 				

R.A. 2000 DEC.	Names	Type	logD$_{25}$	p.a.	B$_T$	(B-V)$_T$	(B-V)$_e$	m$_{21}$	V$_{21}$
l b		S$_T$ n$_L$	logR$_{25}$	A$_g$	m$_B$	(U-B)$_T$	(U-B)$_e$	W$_{20}$	V$_{opt}$
SGL SGB		T	logA$_e$	A$_i$	m$_{FIR}$	(B-V)^o_T	m'$_e$	W$_{50}$	V$_{GSR}$
R.A. 1950 DEC.	PGC	L	logD$_o$	A$_{21}$	B^o_T	(U-B)^o_T	m'$_{25}$	HI	V$_{3K}$

071042.6+342521		.LB....	1.24± .06	3					4918±155
183.09 18.63	UGC 3723	U	.28± .03	.61 14.12 ±.16					4900
30.35 -39.73	IRAS07073+3430	-2.0± .8		.00 12.69					5037
070725.0+343018	PGC 20316		1.26	13.44					
071043.5-733036			.88± .06						3080± 87
284.81 -24.58	ESO 35- 1		.11± .05	.96 14.32 ±.14					2875
204.87 -32.16	IRAS07117-7325			12.28					3164
071144.0-732530	PGC 20317		.97						
071044.4+501207									6310± 31
166.99 23.47	MCG 8-13- 89			.30 15.33 ±.18					6353
28.91 -24.00									6390
070653.9+501703	PGC 20318								
071052.9-514821		RLBS+*.	1.12± .05	170					
262.51 -18.16	ESO 207- 25	Sr	.26± .03	.36 14.69 ±.14					
202.06 -53.76		-1.2± .7		.00					
070940.0-514318	PGC 20326		1.12						
071100.8+484112									5600± 31
168.59 23.12	CGCG 234- 89			.29 15.4 ±.3					5637
29.08 -25.51									5685
070714.4+484609	PGC 20330								
071101.3+483047		.S..2..	1.00± .06	160					5820± 31
168.77 23.07	UGC 3719	U	.61± .05	.31					5856
29.09 -25.68		2.0±1.0		.75					5905
070715.3+483545	PGC 20331		1.03	.30					
071104.9+500811	IC 464	.E?....	.90± .11						4910± 31
167.07 23.51	CGCG 234- 87		.29± .04	.30 14.77 ±.12					4952
28.97 -24.06				.00					4991
070714.6+501309	PGC 20334		.87	14.40					
071107.0+255458		.S?....	1.02± .06	110					7565± 97
191.39 15.54	UGC 3726		.41± .05	.37 14.8 ±.3					7514
31.57 -48.19	MK 1198			.60 12.96					7702
070802.2+255957	PGC 20335		1.05	.20 13.81					
071110.7+501027	NGC 2340	.E.....	1.26± .13	8 12.7 ±.2	1.01± .03	1.06± .01			5949± 24
167.03 23.53	UGC 3720	U	.17± .08	.30 13.58 ±.15					5992
28.99 -24.02		-5.0± .8	1.32± .06	.00	.89	14.78± .20			6030
070720.4+501525	PGC 20338		1.26	12.86		13.56± .71			
0711.2 +2623	IC 2180	.S..2..	1.04± .08	0					
190.96 15.76	UGC 3727	U	.13± .06	.45 14.5 ±.3					
31.56 -47.72		2.0± .9		.16					
0708.2 +2628	PGC 20344		1.08	.06					
071121.3+715006	A 0705+71	.S..7*P	1.52± .03	76				14.15±.1	3137± 5
143.37 27.22	UGC 3697	V	1.23± .05	.16 13.5 ±.3				304± 5	3143± 17
27.44 -2.41	IRAS07055+7155	7.0±1.2		1.38 12.20				262± 5	3257
070532.5+715501	PGC 20348		1.54	.50 11.92				1.73	3156
0711.4 +3010		.SB.3*.	1.00± .08	140					
187.31 17.23	UGC 3728	U	.25± .06	.48 15.00 ±.18					
31.07 -43.94	IRAS07082+3015	3.0±1.3		.35 13.47					
0708.2 +3015	PGC 20351		1.05	.13					
071127.8+481423		.SB.3..	1.14± .07	55				14.94±.1	5921± 7
169.08 23.07	UGC 3724	U	.13± .06	.34 14.27 ±.19					5956
29.20 -25.95		3.0± .8		.17				391± 8	6007
070742.5+481922	PGC 20353		1.17	.06 13.71				1.17	
071133.6+501453	IC 465	.S?....	.94± .11						6270± 31
166.97 23.61	MCG 8-13- 98		.11± .07	.30 14.59 ±.18					6313
29.05 -23.94				.09					6351
070743.1+501953	PGC 20357		.96	14.11					
071138.4+290958		.S..8*.	1.10± .05	158				15.76±.3	4902± 10
188.31 16.90	UGC 3731	U	.43± .05	.39 14.73 ±.18					4863
31.27 -44.95	IRAS07084+2915	8.0±1.3		.53 12.74				407± 7	5033
070829.0+291500	PGC 20361		1.13	.22 13.80				1.74	
0711.6 +7210		.SAT6*.	1.26± .06					15.02±.1	2915± 11
143.00 27.27	UGC 3701	U	.00± .06	.17 14.6 ±.4					3036
27.44 -2.08		6.0± .8		.00				132± 8	2932
0705.8 +7215	PGC 20362		1.28	.00 14.42				.60	
071140.8-264216		.SXT3P.	1.33± .04	143					
239.03 -7.75	ESO 492- 2	Sr	.16± .05	1.30 12.98 ±.14					
187.63 -78.34	IRAS07096-2637	3.1± .4		.23 10.89					
070939.0-263712	PGC 20363		1.46	.08					

R.A. 2000 DEC.	Names	Type	$\log D_{25}$	p.a.	B_T	$(B-V)_T$	$(B-V)_e$	m_{21}	V_{21}
l　　b		S_T　n_L	$\log R_{25}$	A_g	m_B	$(U-B)_T$	$(U-B)_e$	W_{20}	V_{opt}
SGL　SGB		T	$\log A_e$	A_i	m_{FIR}	$(B-V)_T^o$	m'_e	W_{50}	V_{GSR}
R.A. 1950 DEC.	PGC	L	$\log D_o$	A_{21}	B_T^o	$(U-B)_T^o$	m'_{25}	HI	V_{3K}

071141.7+495145		.L..-*.	1.15± .15	140					5985±155
167.39　23.53	UGC　3725	U	.24± .08	.27	14.03 ±.15				6026
29.10 -24.33		-3.0±1.2		.00					6067
070752.3+495645	PGC 20364		1.15		13.67				
071144.1+393411		.S..6*.	1.00± .06	78					
178.02　20.56	UGC　3729	U	.58± .05	.26	15.48 ±.18				
30.06 -34.58		6.0±1.4		.85					
070817.6+393912	PGC 20366		1.02	.29					
071157.0+330517		.S..7..	1.00± .06	55				15.72±.3	7366± 9
184.50　18.40	UGC　3735	U	.15± .05	.61					7342
30.85 -41.04		7.0± .9		.20				231± 5	7490
070841.7+331020	PGC 20372		1.05	.07					
071203.6-603032		.LB.0?P	1.03± .05	155					
271.26 -20.95	ESO　122- 16	S	.11± .04	.56	14.62 ±.14				
203.23 -45.09	FAIR　26	-2.0±1.7		.00					
071122.0-602524	PGC 20376		1.08						
071209.0+414657		.S..2..	1.07± .07	175					
175.80　21.33	UGC　3732	U	.79± .06	.30	15.30 ±.20				
29.93 -32.37		2.0±1.0		.98					
070838.3+415200	PGC 20380		1.10	.40					
071211.3+732810		.S?....	1.14± .07					14.76±.1	2685± 11
141.54　27.39	UGC　3705		.03± .06	.21	14.2 ±.2				2810
27.38　-.78				.04				126± 8	2699
070602.1+733308	PGC 20383		1.16	.01	13.90			.84	
0712.3　+8059		.S..6*.	1.07± .06	160					
133.07　27.58	UGC　3668	U	.18± .05	.20					
26.85　6.72		6.0±1.2		.27					
0703.0　+8104	PGC 20387		1.09	.09					
0712.3　+2834									4851± 46
188.93　16.82	CGCG 146- 38			.44	15.4 ±.3				4810
31.57 -45.51									4984
0709.2　+2840	PGC 20390								
0712.4　+2343			1.00± .08						4450± 10
193.59　14.95	UGC　3737		.14± .06	.34					4494± 97
32.38 -50.33									4391
0709.4　+2349	PGC 20393		1.03						4593
0712.4　+2835									4896± 46
188.93　16.84	CGCG 146- 39			.44	15.1 ±.3				4855
31.60 -45.49									5029
0709.3　+2841	PGC 20394								
071228.3+471001	NGC　2344	.SAT5*.	1.23± .04		12.81 ±.15	.81± .03	.88± .02	14.12±.1	974± 6
170.25　22.94	UGC　3734	PU　　(1)	.01± .05	.42	13.10 ±.18	.18± .04	.27± .02	170± 14	914± 50
29.48 -27.00		4.5± .5	1.05± .03	.02		.70	13.55± .07	147± 8	1004
070845.8+471505	PGC 20395	2.5±1.3	1.27	.01	12.49	.09	13.80± .29	1.63	1063
0712.5　+5149		.S..6*.	1.00± .08	157					
165.33　24.14	UGC　3733	U	.73± .06	.31					
29.07 -22.35		6.0±1.4		1.07					
0708.6　+5155	PGC 20397		1.03	.36					
071233.8+714456	A　0706+71	.S...$P.	1.26± .11	35				16.06±.1	3064± 11
143.48　27.30	UGC　3714	R	.07± .12	.15	12.77 ±.18				2889± 38
27.54　-2.49				.10				150± 8	3170
070646.3+714956	PGC 20398		1.27	.03	12.51			3.52	3070
071312.9+121601	NGC　2350	.S..0..	1.13± .05	110					1877± 38
204.35　10.33	UGC　3747	U	.28± .05	.74	13.3 ±.3				1776
35.52 -61.66	IRAS07104+1221	.0± .8		.21	11.31				2038
071025.2+122110	PGC 20416		1.19		12.28				
0713.2　+1859	IC　2181	.S..2..	.97± .07	140					
198.13　13.20	UGC　3744	U	.27± .05	.50	14.5 ±.3				
33.65 -55.01		2.0± .9		.33					
0710.3　+1905	PGC 20417		1.02	.13					
071314.3+735036		.S..4..	1.32± .04	103					
141.13　27.48	UGC　3717	U	.44± .05	.21	13.83 ±.18				
27.43　-.40	IRAS07069+7355	4.0± .8		.64	12.80				
070700.1+735538	PGC 20418		1.34	.22					
071323.4+273051		.SB.3*.	1.13± .05					15.76±.3	7825± 10
190.06　16.63	UGC　3745	U	.13± .05	.33	14.3 ±.2				7780
32.06 -46.55		3.0±1.2		.17				340± 7	7962
071016.5+273600	PGC 20426		1.16	.06	13.76			1.93	

R.A. 2000 DEC. / l b / SGL SGB / R.A. 1950 DEC.	Names / PGC	Type / S_T n_L / T / L	$\log D_{25}$ / $\log R_{25}$ / $\log A_e$ / $\log D_o$	p.a. / A_g / A_i / A_{21}	B_T / m_B / m_{FIR} / B_T^o	$(B-V)_T$ / $(U-B)_T$ / $(B-V)_T^o$ / $(U-B)_T^o$	$(B-V)_e$ / $(U-B)_e$ / m'_e / m'_{25}	m_{21} / W_{20} / W_{50} / HI	V_{21} / V_{opt} / V_{GSR} / V_{3K}
071325.3-682134 / 279.41 -23.10 / 204.06 -37.27 / 071331.1-681618	ESO 58- 25 / IRAS07135-6816 / PGC 20427	.SBT5?. / S / 5.0± .9	1.16± .04 / .37± .05 / / 1.22	111 / .65 / .55 / .18	14.89 ±.14 / 13.67				
071328.4+350552 / 182.63 19.40 / 31.00 -39.00 / 071009.9+351101	UGC 3742 / IRAS07101+3511 / PGC 20429	.S..8.. / U / 8.0± .9	1.23± .03 / .30± .04 / / 1.25	73 / .25 / .37 / .15	14.11 ±.18 / 13.48			15.40±.2 / 246± 14 / 157± 10 / 1.77	3801± 7 / 3785 / 3922
071331.3-361418 / 247.90 -11.57 / 195.78 -69.06 / 071144.0-360906	ESO 367- 5 / PGC 20431	.SXR1?/ / S / 1.0±1.2	1.09± .05 / .40± .04 / / 1.20	6 / 1.14 / .40 / .20	15.32 ±.14				
0713.5 +3538 / 182.09 19.59 / 30.94 -38.46 / 0710.2 +3544	UGC 3743 / PGC 20432	.SXS6.. / U / 6.0± .8	1.13± .05 / .06± .05 / / 1.15	/ .24 / .09 / .03				15.72±.3 / 133± 14	5087± 10 / 5073 / 5207
071335.0+733503 / 141.42 27.49 / 27.47 -.66 / 070724.6+734006	A 0708+73B / MCG 12- 7- 34 / KUG 0707+736 / PGC 20434	.I?....	.86± .07 / .27± .06 / / .88	/ .21 / .21 / .14					
071347.4+501525 / 167.06 23.96 / 29.43 -23.90 / 070957.1+502034	UGC 3741 / PGC 20441	.S..6*. / U / 6.0±1.3	1.02± .06 / .35± .05 / / 1.04	60 / .30 / .52 / .18	15.31 ±.18 / 14.47			16.09±.1 / 323± 8 / 1.44	5301± 11 / 5343 / 5384
071348.2+122002 / 204.35 10.49 / 35.79 -61.57 / 071100.5+122513	UGC 3754 / PGC 20443	.SX.6*. / U / 6.0±1.2	1.00± .08 / .14± .06 / / 1.07	135 / .74 / .20 / .07				15.77±.3 / 219± 7	8262± 10 / 8161 / 8424
071351.9+103106 / 206.02 9.71 / 36.49 -63.36 / 071106.3+103618	UGC 3755 / PGC 20445	.I..9.. / U / 10.0± .8	1.22± .05 / .24± .05 / / 1.29	160 / .71 / .18 / .12	14.1 ±.2 / 13.23			14.87±.3 / 57± 7 / 1.53	323± 7 / 216 / 488
071353.8+230449 / 194.35 15.00 / 33.03 -50.94 / 071052.9+231000	UGC 3751 / IRAS07108+2309 / PGC 20446	.S?....	1.11± .05 / .55± .05 / / 1.15	30 / .35 / .83 / .28	14.7 ±.3 / 13.25 / 13.52			14.45±.3 / 262± 7 / .65	2295± 10 / 2233 / 2441
071355.6+231422 / 194.20 15.07 / 33.01 -50.78 / 071054.6+231933	UGC 3753 / PGC 20449	.S..1*. / U / 1.0±1.2	1.09± .06 / .35± .05 / / 1.12	35 / .35 / .36 / .18	14.7 ±.3 / 13.90			15.91±.3 / 412± 7 / 1.83	5433± 10 / 5371 / 5578
071404.1+351649 / 182.49 19.57 / 31.13 -38.81 / 071045.3+352200	UGC 3752 / IRAS07107+3521 / PGC 20450	.S?....	.67± .10 / .07± .05 / / .70	/ .25 / .09 / .03	14.8 ±.3 / 12.37 / 14.43			17.00±.3 / 128± 5 / 2.53	4705± 9 / 4657±155 / 4689 / 4826
071414.7+454156 / 171.88 22.84 / 29.96 -28.44 / 071035.8+454707	MK 376 / PGC 20457			/ .40	14.91 ±.15	.58± .03 / -.60± .05			16759± 36 / 16784 / 16854
071415.7+842250 / 129.24 27.55 / 26.65 10.11 / 070047.6+842741	NGC 2268 / UGC 3653 / IRAS07006+8427 / PGC 20458	.SXR4.. / R (1) / 4.0± .3 / 3.4± .8	1.51± .02 / .21± .03 / 1.03± .02 / 1.53	63 / .22 / .31 / .11	12.24 ±.13 / 12.09 ±.13 / 11.18 / 11.62	.72± .02 / .05± .04 / .62 / -.03	.78± .02 / .14± .03 / 12.88± .04 / 14.12± .18	13.84±.1 / 401± 5 / 366± 6 / 2.11	2222± 7 / 2304± 58 / 2380 / 2203
071420.8+732851 / 141.54 27.54 / 27.54 -.75 / 070812.1+733358	A 0708+73A / UGC 3654 / ARP 141 / PGC 20460	.RING.. / R / 10.0± .5	1.45± .06 / .28± .05 / 1.13± .11 / 1.47	165 / .21 / .21 / .14	13.2 ±.2 / 13.4 ±.2 / 12.88	.68± .06 / .14± .07 / .55 / .04	.76± .02 / .21± .03 / 14.38± .37 / 14.60± .39	14.84±.1 / 192± 11 / 135± 12 / 1.82	2709± 11 / 2736± 31 / 2837 / 2726
071421.9+352355 / 182.40 19.67 / 31.19 -38.69 / 071103.0+352907	UGC 3756 / IRAS07110+3529 / PGC 20462	.S..3*. / U (1) / 3.0±1.4 / 4.5±1.3	1.04± .08 / .76± .06 / / 1.06	149 / .25 / 1.05 / .38	15.54 ±.18 / 13.91				
0714.3 +5508 / 161.85 25.15 / 29.08 -19.03 / 0710.3 +5514	UGC 3746 / PGC 20464	.S..3.. / U (1) / 3.0± .9 / 3.5±1.1	.98± .07 / .21± .05 /	165 / / .28 / .10	15.26 ±.19				
071443.4-364817 / 248.53 -11.59 / 195.48 -68.46 / 071257.0-364300	ESO 367- 6 / PGC 20474	PSBR1.. / S / 1.0± .9	1.04± .05 / .15± .05 / / 1.14	/ 1.02 / .16 / .08	15.02 ±.14				

R.A. 2000 DEC. / l b / SGL SGB / R.A. 1950 DEC.	Names / PGC	Type / S_T n_L / T / L	logD_25 / logR_25 / logA_o / logD_o	p.a. / A_g / A_i / A_21	B_T / m_B / m_FIR / B_T^o	(B-V)_T / (U-B)_T / (B-V)_T^o / (U-B)_T^o	(B-V)_o / (U-B)_o / m'_o / m'_25	m_21 / W_20 / W_50 / HI	V_21 / V_opt / V_GSR / V_3K
071447.9+064647 / 209.51 8.27 / 38.76 -66.98 / 071206.6+065203	UGC 3767 / PGC 20478	.S..8.. / U / 8.0± .8 /	1.11± .14 / .20± .12 / / 1.20	140 / .97 / .24 / .10				15.54±.3 / 249± 7 /	5812± 7 / 5692 / 5982
071450.4+165845 / 200.17 12.70 / 34.84 -56.95 / 071157.2+170400	UGC 3766 / IRAS07119+1704 / PGC 20479	.S..6*. / U / 6.0±1.3 /	1.04± .08 / .37± .06 / / 1.09	145 / .50 / .55 / .19	14.8 ± .3 / 12.87 / 13.71			15.13±.3 / 331± 7 / 1.23	4910± 10 / 4825 / 5067
071456.5+004533 / 214.94 5.57 / 43.19 -72.81 / 071221.9+005050	UGC 3769 / PGC 20484	.S?.... / / / 1.10	.96± .17 / .39± .12 / /	115 / 1.48 / .59 / .20	15.35 ±.18 / 13.24			15.58±.3 / 442± 7 / 2.14	8248± 10 / 8109 / 8423
0714.9 +4843 / 168.73 23.76 / 29.79 -25.41 / 0711.2 +4849	UGC 3757 / PGC 20486	.S..6*. / U / 6.0±1.4 /	1.11± .07 / 1.13± .06 / / 1.14	166 / .36 / 1.47 / .50					
071458.9+344845 / 183.03 19.59 / 31.43 -39.26 / 071141.0+345400	UGC 3763 / PGC 20487	.S..6*. / U / 6.0±1.4 /	.96± .09 / .69± .06 / / .98	38 / .22 / 1.01 / .34				16.15±.3 / 307± 5 /	7158± 9 / 7140 / 7281
0715.1 +3807 / 179.70 20.71 / 31.02 -35.96 / 0711.7 +3813	UGC 3761 / PGC 20492	.S..6*. / U / 6.0±1.4 /	1.07± .07 / .79± .06 / / 1.09	161 / .26 / 1.17 / .40					
071512.9-503414 / 261.57 -17.08 / 200.64 -54.90 / 071356.0-502854	ESO 207- 31 / PGC 20500	PSBR1.. / r / 1.0± .9 /	.92± .07 / .04± .06 / / .95	145 / .32 / .04 / .02	15.35 ±.14				
0715.3 +6525 / 150.57 26.95 / 28.30 -8.77 / 0710.5 +6531	UGC 3748 / PGC 20509	.I..9*. / U / 10.0±1.3 /	1.06± .06 / .51± .05 / / 1.08	83 / .16 / .38 / .26					
071523.3-550434 / 265.98 -18.71 / 201.63 -50.44 / 071420.0-545912	ESO 162- 15 / PGC 20510	.SBS5.. / S (1) / 5.0± .8 / 5.6±1.2	1.17± .07 / .16± .07 / / 1.21	120 / .42 / .24 / .08					
071529.4+232542 / 194.17 15.48 / 33.53 -50.54 / 071228.2+233100	UGC 3770 / IRAS07123+2330 / PGC 20513	.I..9.. / U / 10.0± .9 /	.98± .09 / .17± .06 / / 1.02	27 / .34 / .12 / .08	14.9 ± .3 / 14.42			15.25±.3 / 327± 5 / .74	6379± 7 / 6317 / 6526
071531.2-292132 / 241.81 -8.18 / 188.16 -75.56 / 071333.0-291612	ESO 428- 11 / PGC 20514	.LA.-*. / S / -3.0±1.1 /	1.16± .05 / .08± .04 / .91± .07 / 1.30	/ 1.22 / .00 /	12.9 ± .1 / 13.31 ±.14	1.05± .02 / / 12.93± .25 / 13.35± .31	1.10± .01		
071532.7+645532 / 151.14 26.91 / 28.36 -9.27 / 071043.2+650046	IC 2179 / UGC 3750 / PGC 20516	.E.1+.. / R / -5.0± .5 /	1.05± .07 / .02± .04 / .77± .04 / 1.07	/ .16 / .00 /	13.45 ±.14 / 13.42 ±.17 / 13.21	1.01± .02 / .54± .04 / .93 / .52	1.07± .01 / .64± .03 / 12.79± .13 / 13.64± .41		4336± 49 / 4432 / 4377
071535.6+150841 / 201.95 12.09 / 35.72 -58.73 / 071244.6+151400	UGC 3772 / PGC 20518	.S?.... / / / 1.16	1.11± .14 / .20± .12 / /	40 / .49 / .27 / .10	14.2 ± .3 / 13.36			16.67±.3 / 246± 7 / 3.21	4735± 10 / 4643 / 4896
0715.7 +6758 / 147.73 27.26 / 28.11 -6.23 / 0710.6 +6804	UGC 3749 / PGC 20526	.S..6*. / U / 6.0±1.3 /	1.25± .04 / .92± .05 / / 1.27	21 / .17 / 1.35 / .46	15.19 ±.18				
071552.3+120653 / 204.78 10.85 / 36.92 -61.70 / 071305.0+121213	UGC 3775 / PGC 20530	.S..9*. / U / 9.0±1.1 /	1.19± .12 / .05± .12 / / 1.26	/ .79 / .05 / .02				15.91±.1 / 93± 5 / 75± 4	2134± 6 / 2031 / 2299
071554.3-572036 / 268.25 -19.43 / 201.98 -48.18 / 071459.0-571512	ESO 162- 17 / IRAS07149-5715 / PGC 20531	.S.3?P/ / S / 2.5±1.2 /	1.31± .06 / .43± .07 / / 1.35	62 / .50 / .60 / .22	12.20				1130± 19 / 909 / 1256
071603.0+564905 / 160.08 25.71 / 29.16 -17.34 / 071152.2+565423	UGC 3765 / PGC 20536	.L..... / U / -2.0± .9 /	.90± .11 / .00± .04 / / .93	/ .23 / .00 /	14.29 ±.16 / 14.02				3220± 46 / 3286 / 3286

R.A. 2000 DEC.	Names	Type / S_T n_L / T / L	$\log D_{25}$ / $\log R_{25}$ / $\log A_e$ / $\log D_o$	p.a. / A_g / A_i / A_{21}	B_T / m_B / m_{FIR} / B_T^o	$(B-V)_T$ / $(U-B)_T$ / $(B-V)_T^o$ / $(U-B)_T^o$	$(B-V)_e$ / $(U-B)_e$ / m'_e / m'_{25}	m_{21} / W_{20} / W_{50} / HI	V_{21} / V_{opt} / V_{GSR} / V_{3K}
071604.3+644237 151.38 26.94 28.43 -9.48 071116.2+644753	NGC 2347 UGC 3759 IRAS07112+6447 PGC 20539	PSAR3*. R (2) 3.0± .5 2.4± .7	1.25± .03 .15± .03 .85± .05 1.27	175 .16 .21 .08	13.21 ±.15 13.08 ±.13 12.14 12.74	.76± .02 .17± .05 .67 .10	.87± .01 .18± .04 12.95± .16 13.95± .21	14.80±.1 442± 8 425± 8 1.99	4422± 6 4483± 48 4518 4465
071614.0+325614 184.99 19.18 32.04 -41.09 071259.2+330134	UGC 3774 IRAS07129+3301 PGC 20542	.S..2.. U 2.0± .9	1.18± .05 .68± .05 1.20	60 .31 .83 .34	15.17 ±.18 13.08 13.95			16.76±.3 442± 5 2.46	7242± 9 7217 7371
0716.3 +3945 178.13 21.46 31.10 -34.31 0712.9 +3951	UGC 3773 PGC 20545	.I..9*. U 10.0±1.3	1.07± .07 .50± .06 1.10	58 .28 .37 .25					
071621.7-353006 247.48 -10.73 193.72 -69.64 071433.0-352442	ESO 367- 7 PGC 20546	.LBR+.. Sr -.6± .6	1.26± .04 .23± .05 1.30	.75 .00	14.38 ±.14				
071625.9-293718 242.14 -8.12 187.79 -75.24 071428.0-293154	ESO 428- 13 PGC 20548	.S?....	1.14± .07 .57± .06 1.26	125 1.24 .58 .28	15.04 ±.14 13.12				7993 7785 8166
071629.8-520514 263.12 -17.46 200.69 -53.37 071517.0-515948	ESO 208- 1 PGC 20550	.S?....	.84± .06 .25± .05 .90	48 .64 .37 .12	15.29 ±.14 14.26				2505± 19 2282 2643
071631.3-291924 241.88 -7.96 187.30 -75.51 071433.0-291400	ESO 428- 14 PGC 20551	.LXR0P. Sr -1.6± .6	1.27± .04 .22± .04 1.38	135 1.27 .00	13.28 ±.14 11.99				1630 1422 1803
0716.5 +7545 138.97 27.77 27.49 1.53 0709.8 +7551	UGC 3739 PGC 20552	.I..9*. U 10.0±1.2	.99± .08 .11± .06 1.00	.13 .08 .05				14.92±.1 97± 6 87± 5 1.00	1121± 10 1253 1129
071635.6-382919 250.25 -11.97 195.53 -66.73 071452.0-382354	ESO 310- 6 PGC 20555	.SXR2.. S 2.0± .8	1.06± .05 .10± .05 1.16	130 1.07 .12 .05	14.88 ±.14				
071638.5-622040 273.31 -21.01 202.76 -43.21 071605.0-621512	NGC 2369 ESO 122- 18 IRAS07160-6215 PGC 20556	.SBS1.. R (2) 1.0± .4 1.5± .6	1.55± .03 .52± .03 .89± .02 1.60	177 .55 .53 .26	13.21 ±.13 13.07 ±.10 9.86 12.01	.92± .01 .42± .03 .67 .23	1.00± .01 .43± .02 13.15± .04 14.53± .21		3275± 36 3056 3390
071638.9+335914 183.98 19.63 31.99 -40.04 071322.5+340436	UGC 3776 IRAS07133+3404 PGC 20559	.S?....	1.23± .03 .61± .04 1.25	66 .21 .92 .31	14.30 ±.18 13.21 13.15			15.39±.3 302± 7 1.93	3883± 10 3913±155 3862 4010
071639.0-352225 247.38 -10.62 193.46 -69.75 071450.0-351700	ESO 367- 8 PGC 20560	.LBR0.. Sr -1.7± .5	1.38± .04 .24± .04 1.00± .06 1.43	65 .75 .00	13.4 ±.2 13.66 ±.14 12.78	1.26± .03 .79± .04 1.04 .58	1.27± .01 .82± .02 13.92± .21 14.57± .32		2919± 57 2704 3085
071643.1+295113 188.07 18.18 32.69 -44.14 071332.9+295636	UGC 3777 IRAS07135+2956 PGC 20562	.S..6*. U 6.0±1.3	1.24± .04 .71± .05 1.31	152 .69 1.04 .36	14.57 ±.18 12.84 12.82			14.26±.2 292± 34 273± 7 1.09	3213± 8 3176 3349
071653.3+670644 148.71 27.29 28.30 -7.08 071149.2+671203	UGC 3764 IRAS07118+6711 PGC 20567	.SB?...	1.07± .04 .19± .04 1.09	120 .18 .29 .10	14.14 ±.18 13.25 13.65			15.88±.1 230± 8 2.14	4130± 11 4103±125 4233 4165
071654.9+283146 189.38 17.73 32.99 -45.45 071346.7+283710	UGC 3778 PGC 20568	.S..9*. U 9.0±1.2	1.00± .16 .00± .12 1.03	.31 .00 .00				15.83±.3 136± 7	4746± 7 4704 4885
071656.9-575958 268.96 -19.53 201.91 -47.51 071604.0-575430	ESO 122- 17 PGC 20569	.S..3./ S 3.0± .9	1.00± .05 .42± .04 1.05	33 .56 .58 .21	15.13 ±.14				
071701.3-354727 247.80 -10.73 193.57 -69.33 071513.0-354200	ESO 367- 9 IRAS07152-3541 PGC 20571	.SXT4.. S (1) 4.0± .9 3.3±1.2	.97± .05 .02± .04 1.09	1.33 .02 .01	15.29 ±.14				

R.A. 2000 DEC. / I b / SGL SGB / R.A. 1950 DEC.	Names / PGC	Type S_T n_L / T / L	$\log D_{25}$ / $\log R_{25}$ / $\log A_o$ / $\log D_o$	p.a. / A_g / A_i / A_{21}	B_T / m_B / m_{FIR} / B_T^o	$(B-V)_T$ / $(U-B)_T$ / $(B-V)_T^o$ / $(U-B)_T^o$	$(B-V)_o$ / $(U-B)_o$ / m'_o / m'_{25}	m_{21} / W_{20} / W_{50} / HI	V_{21} / V_{opt} / V_{GSR} / V_{3K}
071708.6-362201 / 248.34 -10.96 / 193.91 -68.76 / 071521.2-361634	/ PGC 20574	.LA.-.. / S / -3.0± .8	1.25± .08 / .14± .08 / / 1.38	1.18 / .00 / /					
071728.7+340443 / 183.95 19.82 / 32.20 -39.93 / 071412.1+341008	UGC 3780 / IRAS07141+3410 / PGC 20585	.S?....	1.06± .04 / .48± .04 / / 1.08	59 / .21 / .72 / .24	14.89 ±.18 / 11.74 / 13.93			15.50±.3 / 309± 7 / 1.33	3960± 10 / 3939 / 4088
071731.9+335830 / 184.05 19.80 / 32.23 -40.03 / 071415.5+340356	UGC 3779 / KUG 0714+340 / PGC 20586	.S?....	1.15± .04 / .82± .04 / / 1.17	88 / .21 / 1.23 / .41	15.44 ±.18 / / 13.98			15.20±.3 / 276± 7 / .81	3849± 10 / 3827 / 3977
0717.6 +2321 / 194.44 15.91 / 34.32 -50.54 / 0714.6 +2326	NGC 2357 / UGC 3782 / IRAS07146+2326 / PGC 20592	.S..6*. / U / 6.0±1.2	1.55± .03 / .87± .05 / / 1.59	122 / .45 / 1.28 / .43	14.0 ±.3 / 12.46 / 12.25			13.75±.1 / 361± 6 / 336± 6 / 1.06	2269± 5 / 2207 / 2419
071741.6-555831 / 266.99 -18.72 / 201.33 -49.50 / 071641.1-555300	ESO 162- 18 / PGC 20593	.S..4*. / S (1) / 4.0±1.3 / 4.4±1.3	1.12± .06 / .60± .05 / / 1.16	135 / .47 / .88 / .30					
0717.8 +0757 / 208.79 9.45 / 39.91 -65.68 / 0715.1 +0802	UGC 3785 / IRAS07150+0802 / PGC 20595	.S..1*. / U / 1.0±1.2	1.08± .06 / .37± .05 / / 1.17	12 / .97 / .38 / .18	/ / 12.88				5490± 56 / 5373 / 5663
071751.7+632912 A 0713+63 / 152.77 26.98 / 28.75 -10.68 / 071311.0+633436	MCG 11- 9- 41 / MK 379 / PGC 20599	.S?....	/ / / .28± .05	.20	14.79 ±.16 / 14.9 ±.3	1.01± .03 / .41± .07	.92± .02 / .32± .07 / 11.68± .18		4756 / 4847 / 4803
0717.9 +0940 / 207.23 10.25 / 39.08 -63.99 / 0715.2 +0946	UGC 3787 / PGC 20602	.L..-?. / U / -3.0±1.7	1.08± .17 / .29± .08 / / 1.15	0 / .86 / .00					
0717.9 +2444 / 193.14 16.51 / 34.10 -49.16 / 0714.9 +2450	UGC 3783 / PGC 20603	.S..6*. / U / 6.0±1.4	1.11± .07 / 1.05± .06 / / 1.14	24 / .33 / 1.47 / .50	15.9 ±.3 / / 14.04				6183 / 6126 / 6331
0717.9 +7031 / 144.88 27.66 / 28.07 -3.67 / 0712.4 +7037	UGC 3771 / PGC 20604	.SBS3.. / U / 3.0± .9	1.03± .06 / .09± .05 / / 1.04	35 / .15 / .13 / .05	14.80 ±.19				
0717.9 +2638 / 191.31 17.24 / 33.70 -47.28 / 0714.9 +2644	UGC 3784 / PGC 20608	.S..6*. / U / 6.0±1.4	1.00± .08 / .94± .06 / / 1.04	158 / .40 / 1.38 / .47	16.71 ±.18				
071816.2+304231 / 187.35 18.80 / 33.00 -43.25 / 071504.9+304800	UGC 3786 / PGC 20620	.S..8*. / U / 8.0±1.3	1.07± .07 / .70± .06 / / 1.10	27 / .35 / .86 / .35	15.53 ±.18 / / 14.31			15.59±.3 / 183± 7 / .93	3401± 10 / 3367 / 3537
0718.5 +3122 / 186.72 19.08 / 32.94 -42.58 / 0715.3 +3128	MCG 5-18- 3 / PGC 20629	.SB?...	.89± .11 / .35± .07 / / .92	/ .34 / .43 / .18					3438± 10 / 3406 / 3573
071832.2+270923 / 190.86 17.55 / 33.77 -46.76 / 071526.0+271453	UGC 3791 / IRAS07154+2714 / PGC 20631	.S..8*. / U / 8.0±1.4	1.09± .04 / .73± .04 / / 1.11	15 / .28 / .90 / .37	15.38 ±.18 / / 14.19			15.43±.1 / 355± 8 / .87	5090± 7 / 5042 / 5234
071835.6-572442 / 268.46 -19.12 / 201.46 -48.06 / 071740.1-571907	/ PGC 20635	.IBS9.. / S (1) / 10.0± .6 / 11.1± .6	1.17± .08 / .16± .08 / / 1.21	/ .47 / .12 / .08					
071843.9-625607 / 274.00 -20.97 / 202.54 -42.59 / 071813.0-625030	NGC 2369A / ESO 88- 8 / IRAS07182-6250 / PGC 20640	.SXT4.. / RSr (1) / 3.6± .4 / 4.4± .5	1.27± .04 / .18± .04 / / 1.33	33 / .58 / .26 / .09	13.68 ±.14 / 13.73 / 12.82				3213± 40 / 2995 / 3328
071918.3+511732 / 166.18 25.05 / 30.26 -22.79 / 071525.7+512304	UGC 3792 / PGC 20668	.SA.0.. / U / .0± .8	1.26± .13 / .14± .08 / / 1.28	65 / .28 / .11	13.75 ±.20 / / 13.27				5973±155 / 6018 / 6058

R.A. 2000 DEC.	Names	Type	logD$_{25}$	p.a.	B$_T$	(B-V)$_T$	(B-V)$_e$	m$_{21}$	V$_{21}$
l b		S$_T$ n$_L$	logR$_{25}$	A$_g$	m$_B$	(U-B)$_T$	(U-B)$_e$	W$_{20}$	V$_{opt}$
SGL SGB		T	logA$_e$	A$_i$	m$_{FIR}$	(B-V)$_T^o$	m'$_e$	W$_{50}$	V$_{GSR}$
R.A. 1950 DEC.	PGC	L	logD$_o$	A$_{21}$	B$_T^o$	(U-B)$_T^o$	m'$_{25}$	HI	V$_{3K}$

071926.7-353925		PSBT2*.	1.38± .03	72					
247.90 -10.23	ESO 367- 17	Sr	.39± .04	1.24	13.88 ±.14				
192.14 -69.32	IRAS07176-3533	1.5± .5		.48	11.73				
071738.1-353348	PGC 20676		1.50	.20					
071931.6+592121		RSAR2..	1.20± .05	170					
157.39 26.60	UGC 3789	U	.06± .05	.18	13.30 ±.18				3243±155
29.38 -14.77	7ZW 140	2.0± .8		.08	12.53				3318
071511.2+592653	PGC 20679		1.21	.03	13.01				3304
071947.1+305459		.S..1*.	1.08± .06	168				16.09±.3	4748± 9
187.27 19.18	UGC 3802	U	.75± .05	.35	15.34 ±.19				
33.40 -43.00		1.0±1.4		.77				414± 5	4714
071635.5+310034	PGC 20685		1.11	.38	14.16			1.55	4885
0719.8 -7243		.IBS9..	1.30± .06					14.80±.3	1501± 9
284.15 -23.75	S (1)	.23± .08	.80					
204.00 -32.87		10.0± .8		.17				175± 7	1294
0720.6 -7238	PGC 20690	10.0± .8	1.38	.11					1590
071957.4-630400	NGC 2381	PSBT1*.	1.20± .05						
274.18 -20.88	ESO 88- 10	Sr	.05± .05	.58	13.44 ±.14				3060± 63
202.38 -42.44	FAIR 266	1.2± .4		.05					2842
071927.1-625818	PGC 20694		1.25	.02	12.78				3175
071958.9+220534		.S..1*.	.91± .07					16.79±.3	10149± 9
195.87 15.90	UGC 3803	U	.03± .05	.33	14.8 ±.3				
35.47 -51.71		1.0±1.2		.03				291± 5	10082
071659.5+221110	PGC 20695		.95	.01	14.29			2.48	10304
072000.7+175649		.S?....	1.09± .06	147				15.81±.3	8290± 10
199.80 14.23	UGC 3805		.48± .05	.39	14.7 ±.3				
36.73 -55.79	IRAS07171+1802			.72	13.55			450± 7	8207
071706.5+180226	PGC 20699		1.13	.24	13.50			2.07	8452
0720.0 +5300		.SXS7..	1.02± .06					15.71±.1	5931± 7
164.35 25.52	UGC 3799	U	.05± .05	.30	14.67 ±.19				
30.18 -21.07		7.0± .9		.07				205± 8	5983
0716.1 +5306	PGC 20700		1.04	.03	14.29			1.40	6011
0720.2 +6015		.S..4..	1.01± .06	50					
156.40 26.82	UGC 3796	U	.77± .05	.21					
29.36 -13.86		4.0±1.0		1.13					
0715.8 +6021	PGC 20703		1.03	.38					
072022.5+225402		.SB.6*.	.95± .07					16.54±.3	5484± 9
195.13 16.30	UGC 3806	U	.03± .05	.36	14.8 ±.3				
35.40 -50.90		6.0±1.2		.04				203± 5	5420
071722.1+225940	PGC 20713		.98	.01	14.39			2.14	5638
072029.7-620313	NGC 2369B	PSBR4*.	1.18± .04						
273.18 -20.49	ESO 123- 5	RSr (1)	.02± .04	.66	14.18 ±.14				
202.10 -43.44	IRAS07199-6157	3.9± .4		.03	13.66				
071954.1-615730	PGC 20717	4.8± .4	1.24	.01					
072031.2-580349		.S..1?/	1.13± .05	21	14.9 ±.2	.39± .10			
269.20 -19.11	ESO 123- 4	S	.62± .04	.53	14.77 ±.14	-.08± .13			1088± 63
201.24 -47.38		1.0±1.1		.63		.14			866
071938.0-575806	PGC 20718		1.18	.31	13.64	-.25	13.85± .34		1216
072034.1-674239		.SBS7..	1.11± .05	46					
278.94 -22.28	ESO 58- 28	S (1)	.63± .04	.67	15.69 ±.14				
203.13 -37.83		7.0± .9		.87					
072033.0-673654	PGC 20720	7.8±1.0	1.18	.31					
072048.3-340712 A	0718-34	.E?....	1.37± .07						8900±128
246.63 -9.30			.26± .08	1.06					8685
190.00 -70.70				.00					9073
071857.0-340130	PGC 20731		1.46						
072100.5-752308		.SXR5*.	1.12± .04						
286.99 -24.35	ESO 35- 5	S (1)	.15± .04	1.01	15.04 ±.14				
204.29 -30.22	IRAS07224-7517	5.0± .8		.23	13.21				
072226.1-751718	PGC 20742	5.6± .9	1.21	.08					
072101.6+251039		.SBS7*.	1.23± .06	50				14.58±.1	2385± 5
193.00 17.32	UGC 3808	U	.27± .06	.33	14.6 ±.3			207± 8	
35.04 -48.63		7.0±1.1		.38				194± 6	2329
071758.2+251620	PGC 20744		1.26	.14	13.83			.61	2536
0721.0 +6338		.S..6*.	1.11± .05	79					
152.65 27.36	UGC 3801	U	.78± .05	.18					
29.09 -10.49		6.0±1.4		1.15					
0716.4 +6344	PGC 20749		1.12	.39					

R.A. 2000 DEC. l b SGL SGB R.A. 1950 DEC.	Names PGC	Type S_T n_L T L	$\log D_{25}$ $\log R_{25}$ $\log A_.$ $\log D_.$	p.a. A_g A_i A_{21}	B_T m_B m_{FIR} B_T^o	$(B-V)_T$ $(U-B)_T$ $(B-V)_T^o$ $(U-B)_T^o$	$(B-V)_.$ $(U-B)_.$ $m'_.$ m'_{25}	m_{21} W_{20} W_{50} HI	V_{21} V_{opt} V_{GSR} V_{3K}
072108.0-690700 280.41 -22.65 203.31 -36.43 072118.1-690112	NGC 2397A ESO 58- 29 IRAS07212-6901 PGC 20754	.SAT6*. RS 6.2± .5	1.06± .04 .07± .04 1.13	 .82 .10 .03	14.84 ±.14				
072108.9-342338 246.91 -9.36 190.07 -70.41 071918.1-341754	 ESO 367- 18 IRAS07193-3418 PGC 20755	.SXS4?. S (1) 4.0± .8 3.3±1.2	1.09± .04 .24± .04 1.19	36 1.06 .35 .12	15.23 ±.14 13.32				
072121.0-690007 280.30 -22.59 203.26 -36.54 072130.0-685418	NGC 2397 ESO 58- 30 IRAS07214-6854 PGC 20766	.SBS3*. RSr (3) 2.8± .4 4.0± .4	1.39± .03 .32± .04 .89± .01 1.47	123 .82 .45 .16	12.68 ±.13 12.86 ±.11 10.73 11.51	.85± .01 .10± .02 .60 -.10	.89± .01 .17± .01 12.62± .02 13.69± .23		 1311± 31 1099 1411
072126.9+131543 204.32 12.57 39.17 -60.30 071838.3+132126	 UGC 3813 PGC 20774	.I..9*. U 10.0±1.2	1.11± .14 .15± .12 1.16	90 .57 .11 .07	14.3 ±.3 13.61			15.56±.3 230± 7 1.88	4095± 10 3995 4266
072127.2-504528 262.15 -16.24 199.02 -54.55 072010.0-503942	 ESO 208- 3 PGC 20775	.SAT7*P S (1) 7.0±1.2 10.0±1.2	1.19± .05 .08± .05 1.23	 .47 .11 .04	14.79 ±.14				
072142.9+354415 182.59 21.19 33.02 -38.17 071823.9+354958	 UGC 3811 PGC 20799	.S..7.. U 7.0± .8	1.14± .07 .12± .06 1.16	50 .27 .16 .06	14.9 ±.3 14.48			15.70±.3 371± 5 1.17	8044± 9 8028 8172
0721.7 +4650 171.05 24.39 31.28 -27.16 0718.1 +4656	 UGC 3810 PGC 20805	.S..6*. U 6.0±1.2	1.00± .06 .09± .05 1.03	 .34 .13 .04	15.0 ±.2				
072156.5-685045 280.15 -22.50 203.17 -36.69 072204.1-684454	NGC 2397B ESO 58- 31 PGC 20813	.IBS9P. S (1) 10.0± .8 7.8± .8	1.00± .06 .23± .05 1.07	103 .80 .17 .11	14.98 ±.14				
072200.8+050848 211.80 9.14 44.39 -68.10 071921.3+051433	 UGC 3819 PGC 20817	.I..9*. U 10.0±1.3	1.00± .08 .55± .06 1.08	130 .86 .41 .27	15.45 ±.18 14.18			15.95±.3 393± 5 1.50	10022± 9 9895 10203
0722.1 +5552 161.29 26.36 30.14 -18.19 0718.0 +5558	 CGCG 261- 59 PGC 20823			 .16	15.22 ±.18				12008± 40 12071 12082
072209.8-291405 242.36 -6.83 182.77 -75.10 072011.0-290818	 ESO 428- 23 IRAS07202-2908 PGC 20825	PSBT2*. Sr (1) 2.4± .6 5.6± .8	1.30± .04 .25± .04 1.44	25 1.57 .31 .12	13.43 ±.14 10.66				
072210.9-055546 221.73 4.09 61.16 -78.17 071943.6-055000	 MCG -1-19- 1 IRAS07196-0549 PGC 20827	.S..3*/ E 3.0±1.3	1.16± .06 .77± .05 1.35	80 2.03 1.07 .39	 12.83				1610 1450 1798
072219.1+491737 168.45 25.06 31.04 -24.72 071832.3+492322	 UGC 3812 IRAS07185+4922 PGC 20833	.E..... U -5.0± .9	.85± .24 .00± .08 .91	 .35 .00	14.36 ±.17 13.92				6380±155 6417 6473
072220.5+171714 200.65 14.46 37.93 -56.32 071927.2+172300	 UGC 3820 IRAS07194+1723 PGC 20835	.S..6*. U 6.0±1.3	1.19± .05 .76± .05 1.22	100 .33 1.12 .38	14.9 ±.3 12.79 13.44			14.83±.3 303± 7 1.01	2527± 10 2441 2693
072222.4+220501 196.10 16.40 36.35 -51.62 071923.1+221047	NGC 2365 UGC 3821 PGC 20838	.SX.1.. U 1.0± .7	1.38± .04 .28± .05 1.41	170 .29 .28 .14	13.3 ±.3 12.72			15.70±.2 435± 14 367± 7 2.84	2278± 7 2451±155 2210 2436
072235.3+713554 143.69 28.08 28.33 -2.57 071652.3+714135	 UGC 3804 IRAS07168+7141 PGC 20844	.S..6*. U 6.0±1.1	1.25± .04 .22± .05 1.26	13 .08 .32 .11	13.10 ±.19 12.96 12.69			14.55±.3 333± 34 322± 25 1.75	2887± 9 3005 2911
0722.7 +4506 172.94 24.11 31.71 -28.86 0719.1 +4512	 UGC 3817 PGC 20852	.I..9*. U 10.0±1.1	1.26± .06 .29± .06 1.29	 .36 .22 .15				14.85±.1 53± 4 43± 4	438± 5 459 543

R.A. 2000 DEC. / l b / SGL SGB / R.A. 1950 DEC.	Names / PGC	Type / S_T n_L / T / L	$\log D_{25}$ / $\log R_{25}$ / $\log A_e$ / $\log D_o$	p.a. / A_g / A_i / A_{21}	B_T / m_B / m_{FIR} / B_T^o	$(B-V)_T$ / $(U-B)_T$ / $(B-V)_T^o$ / $(U-B)_T^o$	$(B-V)_e$ / $(U-B)_e$ / m'_e / m'_{25}	m_{21} / W_{20} / W_{50} / HI	V_{21} / V_{opt} / V_{GSR} / V_{3K}
072247.4+185539 199.15 15.23 37.52 -54.70 071952.1+190127	 UGC 3823 PGC 20860	.S..3.. U (1) 3.0± .9 3.5±1.2	1.02± .06 .20± .05 1.04	103 .25 .27 .10	 14.7 ±.3 14.08			15.61±.3 416± 7 1.43	8513± 10 8433 8677
0722.8 +7749 136.64 28.12 27.63 3.62 0715.4 +7755	 UGC 3794 PGC 20863	.S..9*. U 9.0±1.1	1.17± .07 .04± .06 1.18	120 .13 .05 .02				15.07±.1 172± 6 152± 5	2648± 10 2786 2651
072251.1+223520 195.66 16.70 36.38 -51.10 071951.2+224108	 UGC 3824 PGC 20864	.L..... U -2.0± .9	.95± .21 .05± .08 .98	 .30 .00	 14.39 ±.15 14.01				5267±155 5201 5425
072251.3-620141 273.25 -20.22 201.72 -43.42 072215.0-615548	 ESO 123- 9 IRAS07222-6155 PGC 20865	.SAT5.. Sr (1) 5.1± .4 5.2± .5	1.32± .04 .34± .04 1.38	144 .66 .50 .17	 14.42 ±.14 13.46				
072307.5+221231 196.05 16.61 36.59 -51.46 072008.1+221820	 UGC 3827 IRAS07201+2218 PGC 20881	.SB?... 	.99± .06 .03± .05 1.02	 .29 .04 .01	 14.48 ±.18 13.33 14.12			15.34±.2 229± 6 184± 4 1.21	5367± 6 5300 5526
072312.4+580357 158.90 26.87 30.02 -16.00 071857.9+580944	 UGC 3816 PGC 20884	.L..... U -2.0± .9	1.04± .08 .15± .03 1.03	112 .17 .00	13.81 ±.13 13.58 ±.20 13.52	.98± .02 .52± .03 .90 .48	 13.50± .42		 3342± 24 3412 3409
072312.9+372734 180.93 22.02 33.08 -36.42 071951.1+373323	 UGC 3822 PGC 20886	.SBT4.. U 4.0± .8	1.12± .05 .21± .05 1.15	15 .32 .32 .11	 14.7 ±.2				
072316.1+322945 185.97 20.42 34.06 -41.32 072002.4+323534	IC 2185 MCG 5-18- 8 ARAK 129 PGC 20889	.S?.... 	.78± .08 .22± .06 .81	 .27 .33 .11	 15.3 ±.2 13.27 14.69				4529± 82 4500 4667
072330.9+023653 214.26 8.33 47.74 -70.37 072054.2+024245	 UGC 3830 IRAS07208+0242 PGC 20894	.SB.7?. E (1) 7.0±1.3 6.4±1.2	1.07± .05 .45± .04 1.16	116 .95 .62 .23	 14.86 ±.12 12.27				
072333.3+412607 176.84 23.26 32.47 -32.48 072004.3+413157	 UGC 3825 PGC 20900	.SXS4.. U 4.0± .8	1.08± .05 .04± .04 1.10	 .26 .06 .02	 14.9 ±.3 14.51			15.95±.1 169± 8 1.42	8281± 11 8287 8397
072338.7-300300 243.24 -6.92 183.04 -74.23 072141.0-295706	 ESO 428- 28 IRAS07216-2957 PGC 20903	.S..5*/ S 5.0±1.2	1.37± .04 .70± .04 1.51	57 1.55 1.05 .35	 14.55 ±.14 11.07				
072339.3-364012 249.20 -9.92 190.78 -68.09 072152.0-363418	 ESO 367- 22 PGC 20904	PSBS3P. S 3.0± .8	1.11± .05 .11± .05 1.23	55 1.24 .16 .06	 15.01 ±.14				
072342.3-293906 242.89 -6.72 182.34 -74.58 072144.0-293312	 ESO 428- 29 IRAS07217-2933 PGC 20908	.SXT3*. Sr (1) 3.1± .6 5.6±1.1	1.21± .04 .10± .04 1.37	 1.60 .14 .05	 13.44 ±.14 11.69				
072345.4+332659 185.05 20.84 34.00 -40.37 072030.2+333250	 UGC 3829 MK 1199 PGC 20911	.S?.... 	1.01± .05 .12± .06 1.03	 .19 .15 .06	 13.8 ±.2 11.16 13.44				4028± 32 4003 4164
072354.7-355319 248.52 -9.53 189.96 -68.82 072206.0-354724	 ESO 367- 23 PGC 20915	PSXT3*. S (1) 3.0±1.3 5.6±1.3	1.12± .04 .39± .04 1.25	125 1.32 .54 .20	 15.23 ±.14				
072355.3-273143 241.02 -5.70 178.12 -76.41 072154.0-272548	NGC 2380 ESO 492- 12 PGC 20916	.LX.0*. S -1.7± .6	1.31± .05 .03± .04 1.02± .02 1.53	 1.98 .00	12.27 ±.13 12.52 ±.14 10.38	1.05± .01 .60	1.13± .01 12.86± .07 13.61± .30		 1782± 37 1575 1965
0724.4 +6142 154.87 27.52 29.71 -12.37 0719.9 +6147	 UGC 3826 IRAS07199+6147 PGC 20927	.SXS7.. U 7.0± .7	1.54± .02 .06± .04 1.56	85 .21 .08 .03	 14.1 ±.7 13.85			13.81±.1 65± 6 53± 7 -.06	1733± 6 1727± 76 1816 1790

R.A. 2000 DEC.	Names	Type	logD$_{25}$	p.a.	B$_T$	(B-V)$_T$	(B-V)$_e$	m$_{21}$	V$_{21}$
l　　b		S$_T$　n$_L$	logR$_{25}$	A$_g$	m$_B$	(U-B)$_T$	(U-B)$_e$	W$_{20}$	V$_{opt}$
SGL　　SGB		T	logA$_e$	A$_i$	m$_{FIR}$	(B-V)^o_T	m'$_e$	W$_{50}$	V$_{GSR}$
R.A. 1950 DEC.	PGC	L	logD$_o$	A$_{21}$	B^o_T	(U-B)^o_T	m'$_{25}$	HI	V$_{3K}$

072435.4+575808	A　　0720+58	.SXT3..	1.23± .05	0				16.11±.1	3510± 10
159.04　27.04	UGC　3828	U	.26± .05	.17	12.9　±.2				3580
30.22 −16.07		3.0± .8		.36				249±　8	3580
072021.5+580401	PGC 20933		1.24	.13	12.38			3.60	3579

072447.4+324815		.S?....	.98± .09					15.85±.3	4695± 10
185.78　20.83	UGC　3833		.07± .05	.20	15.0　±.2				4667
34.42 −40.97	KUG 0721+329			.10				170±　7	4834
072133.4+325411	PGC 20938		1.00	.03	14.70			1.11	

072457.0−093936	NGC　2377	.SAS5*.	1.23± .03	170	13.54 ±.13	.85± .02	.93± .01	13.18±.1	2457±　5
225.35　2.94	UGCA　132	PE　　(1)	.12± .07	2.79		.08± .03	.17± .02	193±　6	2332± 39
77.54 −80.52	IRAS07225-0933	4.5± .6	.90± .01	.19	10.72		13.53± .03	175±　6	2284
072233.8−093338	PGC 20948	3.1±1.6	1.49	.06	10.55		14.23± .27	2.56	2648

072500.4+492933		.SBS4..	1.08± .06	85					5949± 10
168.34　25.54	UGC　3831	U	.08± .05	.37	13.64 ±.18				5986
31.48 −24.46	IRAS07212+4935	4.0± .9		.11	12.20				6044
072113.4+493528	PGC 20953		1.11	.04	13.12				

072501.0+234659	NGC　2370	.SB?...	.93± .07	43				14.87±.3	5500±　9
194.71　17.62	UGC　3835		.23± .05	.24	14.6　±.2				5564±155
36.79 −49.83	IRAS07220+2352			.34	12.44			437±　5	5438
072159.7+235256	PGC 20955		.95	.11	14.04			.72	5658

072501.1−322948		.S..6*/	1.28± .04	78					
245.57　−7.78	ESO　428- 31	S　　　(1)	.71± .04	1.35	14.50 ±.14				
185.64 −71.89	IRAS07231-3223	6.0± .7		1.05	12.35				
072307.0−322348	PGC 20956	4.4±1.3	1.41	.36					

072502.0+271930		.S?....	.94± .11						7713± 97
191.25　18.95	MCG　5-18- 11		.00± .07	.36	14.59 ±.18				7664
35.80 −46.35	MK　1200			.00	12.71				7864
072156.0+272527	PGC 20957		.97	.00	14.15				

072506.9−633203		.S?....	.90± .06	150					10190± 63
274.86 −20.49	ESO　88- 13		.10± .05	.75	15.48 ±.14				9972
201.71 −41.89	FAIR　267			.15					10307
072438.0−632600	PGC 20961		.97	.05	14.52				

072509.9+093102		.SB.3..	1.02± .06					16.20±.2	5267±　7
208.17　11.78	UGC　3839	U	.06± .05	.44	14.1　±.3			205±　7	5153
43.07 −63.69		3.0± .9		.09				199±　5	5447
072225.6+093116	PGC 20964		1.06	.03	13.50			2.67	

072512.1−265900		.SBS5?.	1.13± .05	74					
240.67　−5.19	ESO　492- 14	S	.44± .05	2.24	15.48 ±.14				
175.86 −76.72		5.0±1.8		.66					
072310.0−265300	PGC 20965		1.35	.22					

0725.2　+5326		.SB.2..	1.07± .06	107					
164.06　26.37	UGC　3832	U	.30± .05	.22	14.99 ±.19				
30.94 −20.55		2.0± .9		.37					
0721.3　+5332	PGC 20971		1.09	.15					

072520.7+191036		.E.....	1.04± .18						8468± 38
199.17　15.88	UGC　3840	U	.00± .08	.31	14.0　±.3				8388
38.45 −54.33		−5.0± .8		.00					8635
072225.2+191635	PGC 20973		1.09		13.56				

072535.8−751050			.89± .06	50					4441± 19
286.86 −24.02	ESO　35- 7		.21± .05	1.00	15.05 ±.14				4238
203.93 −30.38	IRAS07269-7504				12.75				4525
072656.0−750442	PGC 20979		.99						

0725.6　+1907		.L.....	1.07± .07	120					8351± 38
199.25　15.92	UGC　3842	U	.15± .03	.25	14.4　±.3				8271
38.58 −54.37		−2.0± .8		.00					8518
0722.7　+1913	PGC 20980		1.07		14.05				

072541.3−302420		.SBS5P.	1.27± .04						
243.76　−6.69	ESO　428- 32	S　　　(1)	.09± .04	1.63	13.30 ±.14				
182.20 −73.71	IRAS07237-3018	5.0± .8		.13	11.90				
072344.1−301818	PGC 20983	4.4± .8	1.42	.04					

072544.4+295704	A　　0722+30	.S...*.	.60?						5528± 60
188.71　20.04	UGC　3841	R	.00± .07	.28					5489
35.34 −43.74	MK　1201			.00					5674
072234.7+300304	PGC 20988		.63						

072544.4+200607		.S..6*.	.96± .09	147				16.58±.3	5273±　9
198.32　16.34	UGC　3843	U	.69± .06	.22	15.7　±.3				5197
38.26 −53.40		6.0±1.4		1.01				202±　5	5439
072247.8+201207	PGC 20989		.98	.34	14.47			1.77	

R.A. 2000 DEC.	Names	Type	logD$_{25}$	p.a.	B$_T$	(B-V)$_T$	(B-V)$_e$	m$_{21}$	V$_{21}$
l b		S$_T$ n$_L$	logR$_{25}$	A$_g$	m$_B$	(U-B)$_T$	(U-B)$_e$	W$_{20}$	V$_{opt}$
SGL SGB		T	logA$_e$	A$_i$	m$_{FIR}$	(B-V)$_T^o$	m'$_e$	W$_{50}$	V$_{GSR}$
R.A. 1950 DEC.	PGC	L	logD$_o$	A$_{21}$	B$_T^o$	(U-B)$_T^o$	m'$_{25}$	HI	V$_{3K}$
0726.1 +0158		.LA.0*.	1.08± .09	95	14.75 ±.17				
215.13 8.61	CGCG 1- 6	E	.30± .08	.92					
50.26 -70.71		-2.0±1.3		.00					
0723.5 +0205	PGC 21005		1.14						
072618.8-843113		.S..3P?	1.02± .05	58	14.48 ±.14				
296.89 -26.17	ESO 5- 6	S	.28± .04	.57					
205.38 -21.14	IRAS07344-8424	3.0±1.7		.38	12.28				
073422.0-842448	PGC 21010		1.08	.14					
072635.1+431740		.E...?.	1.14± .06	5	14.02 ±.15				3128± 37
175.05 24.31	UGC 3844	U	.17± .05	.26					3141
32.80 -30.55		-5.0±1.6		.00					3241
072302.6+432342	PGC 21014		1.14		13.72				
072637.3+334927	NGC 2373	.S?....	.79± .06	0	14.7 ±.2			16.46±.3	7786± 9
184.88 21.52	UGC 3848		.11± .04	.24					7523± 82
34.68 -39.90	ARAK 131			.16	13.11			169± 5	7759
072321.7+335530	PGC 21016		.82	.05	14.24			2.16	7921
072643.6+470540		.SBS4..	1.19± .05	176	13.56 ±.18				3032± 10
171.00 25.27	UGC 3845	U	.16± .05	.36					
32.17 -26.79	IRAS07230+4711	4.0± .8		.24	13.45				3060
072302.7+471143	PGC 21020		1.22	.08	12.94				3135
072645.1+644832		.SBT6..	1.09± .05		15.1 ±.3				
151.40 28.08	UGC 3836	U	.04± .05	.18					
29.56 -9.25	KUG 0721+649	6.0± .8		.06					
072158.5+645432	PGC 21021		1.11	.02					
072657.4+202208		.S..6*.	.99± .06	90				15.92±.3	8505± 10
198.19 16.71	UGC 3856	U	.07± .05	.27					8429
38.63 -53.08		6.0±1.2		.10				185± 7	8672
072400.5+202813	PGC 21029		1.02	.04					
072704.5+801041	NGC 2336	.SXR4..	1.85± .01	178	11.05M±.13	.62± .02	.72± .02	12.72±.0	2200± 4
133.96 28.22	UGC 3809	R (2)	.26± .02	.13	11.28 ±.14	.06± .04	.15± .03	464± 4	2205± 23
27.53 5.98	IRAS07184+8016	4.0± .3	1.60± .03	.39	12.10	.53	14.44± .08	440± 5	2345
071828.0+801635	PGC 21033	1.1± .5	1.87	.13	10.62	-.01	14.51± .16	1.97	2196
072709.5+334956	NGC 2375	.SBS3..	1.13± .05	170	14.44 ±.20			15.54±.2	7860± 7
184.91 21.63	UGC 3854	U	.14± .05	.24					
34.82 -39.87	IRAS07238+3356	3.0± .8		.19	13.64			263± 7	7835
072353.9+335601	PGC 21035		1.15	.07	13.95			1.52	7998
072709.5+334953	NGC 2379	.LA..*.	.89± .11		14.5 ±.2	.98± .03			4030± 56
184.91 21.63	UGC 3857	PU	.00± .05	.24	14.75 ±.15	.56± .07			4006
34.82 -39.87	ARAK 132	-2.0± .7		.00		.89			4169
072354.0+335558	PGC 21036		.92		14.36	.52	13.84± .61		
072712.2-512215		.SBR1*.	1.04± .05	19	14.62 ±.14				
263.11 -15.66	ESO 208- 15	Sr	.09± .04	.53					
197.75 -53.76		1.4± .7		.09					
072556.0-511606	PGC 21038		1.09	.04					
072712.6+854520	NGC 2276	.SXT5..	1.45± .02	20	11.93 ±.13	.52± .03	.60± .02	13.99±.0	2417± 4
127.67 27.71	UGC 3740	R (2)	.02± .03	.23	12.17 ±.14	-.09± .06	.00± .03	187± 4	2372± 34
26.77 11.51	7ZW 134	5.0± .3	1.21± .02	.04	10.31	.45	13.47± .03	108± 4	2578
071022.1+855058	PGC 21039	3.6± .7	1.47	.01	11.76	-.14	13.96± .18	2.22	2394
0727.3 +1938		.S?....	.82± .13		14.8 ±.3				9907± 58
198.93 16.51	MCG 3-19- 13	.	.00± .07	.22					9828
39.07 -53.77	IRAS07244+1944			.00	12.68				10075
0724.4 +1944	PGC 21044		.84	.00	14.52				
0727.4 +1936		.S?....	.82± .13		15.0 ±.3				9813± 58
198.95 16.51	MCG 3-19- 14		.08± .07	.22					9735
39.10 -53.79				.10					9982
0724.5 +1943	PGC 21045		.84	.04	14.64				
072733.3-454105		.SBS9..	1.10± .04		16.89 ±.14				1007± 79
257.80 -13.22	ESO 257- 17	S (1)	.08± .05	.82					783
195.12 -59.26		9.0± .8		.08					1168
072603.1-453454	PGC 21050	10.0± .8	1.18	.04	15.98				
072748.0+202345		.S?....	1.20± .06	13	15.1 ±.2			15.22±.3	8151± 10
198.24 16.90	UGC 3862		.60± .06	.23					8075
38.94 -53.01				.90				432± 7	8319
072451.1+202954	PGC 21056		1.22	.30	13.90			1.02	
072754.0-673427	IC 2202	.SAS4*.	1.31± .04	165	13.63 ±.14				
279.04 -21.57	ESO 88- 16	S (1)	.47± .04	.71					
202.24 -37.86	IRAS07278-6728	4.3± .4		.69	12.49				
072750.0-672812	PGC 21057	3.0± .5	1.37	.24					

R.A. 2000 DEC. l　　b SGL　SGB R.A. 1950 DEC.	Names PGC	Type S_T　n_L T L	$\log D_{25}$ $\log R_{25}$ $\log A_e$ $\log D_o$	p.a. A_g A_i A_{21}	B_T m_B m_{FIR} B_T^o	$(B-V)_T$ $(U-B)_T$ $(B-V)_T^o$ $(U-B)_T^o$	$(B-V)_e$ $(U-B)_e$ m'_e m'_{25}	m_{21} W_{20} W_{50} HI	V_{21} V_{opt} V_{GSR} V_{3K}
072806.7-622145 273.81 -19.77 200.98 -42.98 072731.0-621530	IC　　2200A ESO　123- 11 PGC 21062	.LBS-P? RS -2.7± .7 	1.12± .05 .18± .04 1.19	35 .71 .00 	 13.74 ±.14 12.99				3248± 27 3028 3370
072811.3+723420 142.60 28.53 28.63 -1.55 072218.7+724024	A　　0722+72 UGC　3838 MK　　　7 PGC 21065	.P..... R 99.0 	.92± .06 .37± .05 .93	17 .08 .28 .19	 14.3 ±.3 13.93			15.06±.1 232± 6 243± 25 .94	3060± 9 3085± 19 3186 3087
072813.3+583024 158.51 27.59 30.63 -15.47 072358.0+583632	 UGC　3855 PGC 21067	.S..2.. U 2.0± .9 	1.30± .04 .65± .05 1.31	52 .16 .80 .32	 14.22 ±.18 13.23				3089± 38 3160 3159
072815.0+631521 153.17 28.12 29.94 -10.77 072337.7+632128	 UGC　3850 PGC 21071	PSXS1.. U 1.0± .8 	1.21± .05 .00± .05 1.23	 .19 .00 .00	 13.69 ±.20 13.45			15.79±.1 292± 8 2.35	4709± 11 4798 4763
072817.3+404614 177.81 23.93 33.63 -33.00 072449.9+405223	A　　0724+40 UGC　3860 DDO　43 PGC 21073	.I..9.. U　　　(1) 10.0± .8 9.0±1.5	1.12± .05 .17± .05 .90± .02 1.14	 .23 .13 .08	15.07 ±.14 15.1 ±.3 14.71	.46± .05 -.23± .07 .36 -.30	.42± .03 -.14± .05 15.06± .09 15.08± .31	14.72±.1 51± 4 38± 6 -.08	354± 4 356 476
072817.8-622104 273.81 -19.74 200.95 -42.99 072742.1-621448	IC　　2200 ESO　123- 12 IRAS07276-6214 PGC 21075	RSX.3P. RSr　　(1) 2.7± .5 3.3±1.3	1.11± .05 .26± .04 .57± .02 1.18	58 .71 .36 .13	14.05 ±.14 13.82 ±.14 12.32 12.84	.84± .02 .22± .03 .61 .04	.94± .02 .34± .03 12.39± .06 13.79± .29		3123± 43 2903 3246
072819.1-625340 274.35 -19.93 201.08 -42.45 072746.1-624724	 ESO　88- 15 PGC 21076	.S..6*/ S　　　(1) 5.7±1.2 5.6±1.4	1.21± .05 1.00± .04 1.28	71 .70 1.47 .50	 15.82 ±.14 				
072837.6+305548 187.96 20.96 35.93 -42.67 072526.6+310200	 UGC　3866 PGC 21086	.S?.... 	1.00± .08 .63± .06 1.03	178 .28 .94 .31	 15.52 ±.18 14.26			15.77±.3 330± 7 1.20	6972± 10 6936 7119
072841.1-750315 286.78 -23.79 203.68 -30.48 072958.0-745654	 ESO　35- 9 PGC 21088	.IBS9.. S　　　(1) 10.0± .9 10.0± .9	1.19± .05 .56± .05 1.29	17 .98 .42 .28	 16.03 ±.14 				
072849.7+353255 183.27 22.50 34.87 -38.12 072531.6+353907	 MCG　6-17- 9 ARAK　134 PGC 21094	.S?.... 	.82± .07 .17± .06 .85	 .25 .26 .09	 14.99 ±.20 13.55 14.47				3915± 82 3897 4052
072853.5+334907 185.05 21.97 35.29 -39.82 072538.2+335519	NGC　2388 UGC　3870 IRAS07256+3355 PGC 21099	.S?.... 	1.00± .06 .24± .05 1.02	65 .24 .34 .12	 14.67 ±.18 10.22 14.06			15.07±.2 455± 13 285± 7 .89	4134± 5 4109 4274
072854.0+203528 198.16 17.21 39.28 -52.76 072556.9+204141	 UGC　3873 IRAS07259+2041 PGC 21100	.S..6*. U 6.0±1.2 	1.28± .04 .66± .05 1.30	144 .23 .97 .33	 14.47 ±.18 12.41 13.25			14.37±.3 341± 7 .79	4464± 10 4389 4633
072854.0+490819 168.88 26.08 32.22 -24.72 072508.4+491430	 UGC　3863 IRAS07250+4914 PGC 21101	PSB.1.. U 1.0± .9 	1.09± .04 .34± .04 1.12	76 .30 .35 .17	 14.01 ±.20 13.81 13.28				5887± 76 5922 5986
072854.4+691252 146.42 28.54 29.16 -4.86 072336.5+691900	NGC　2366 UGC　3851 DDO　42 PGC 21102	.IBS9.. R　　　(2) 10.0± .3 8.7± .5	1.91± .02 .39± .02 1.63± .02 1.92	25 .18 .29 .19	11.43M±.10 11.67 ±.16 11.95 11.03	.58± .07 .45 	.35± .02 -.33± .02 14.94± .05 14.84± .14	11.25±.0 113± 2 96± 2 .03	100± 3 87± 11 209 134
0728.9　+4011 178.45 23.89 33.89 -33.54 0725.5　+4018	 UGC　3868 PGC 21104	.I..9.. U 10.0± .8 	1.07± .08 .03± .06 1.09	155 .22 .03 .02					
072900.1-665349 278.39 -21.25 201.96 -38.51 072851.0-664730	 ESO　88- 17 IRAS07288-6647 PGC 21107	.SXR4P* Sr 4.0± .7 	1.26± .05 .48± .05 1.33	63 .74 .71 .24	 13.69 ±.14 11.91 12.20				5181± 69 4966 5291
072904.9+335137 185.02 22.02 35.33 -39.77 072549.4+335750	NGC　2389 UGC　3872 KUG 0725+339 PGC 21109	.SXT5.. R 5.0± .4 	1.30± .02 .15± .03 .80± .01 1.32	83 .24 .22 .07	13.40 ±.13 13.34 ±.18 12.90	.47± .02 -.15± .03 .36 -.23	.61± .01 12.89± .04 14.36± .18	14.06±.1 393± 5 343± 4 1.09	3957± 4 3783± 58 3931 4097

R.A. 2000 DEC.	Names	Type S_T n_L	logD25 logR25	p.a. A_g	B_T m_B	(B-V)_T (U-B)_T	(B-V)_e (U-B)_e	m_21 W_20	V_21 V_opt
l b		T	logA_e	A_i	m_FIR	(B-V)_T^o	m'_e	W_50	V_GSR
SGL SGB									
R.A. 1950 DEC.	PGC	L	logD_o	A_21	B_T^o	(U-B)_T^o	m'_25	HI	V_3K

072905.1+265510		.S?....	1.14± .05					16.71±.3	7949± 10
192.00 19.65	UGC 3874		.17± .05	.30	14.5 ±.2				
37.19 -46.58				.26				368 7	7898
072559.9+270123	PGC 21110		1.16	.09	13.91			2.72	8106

072906.1-665444		.E.5.P.	.84± .06	114					
278.41 -21.25	ESO 88- 18	S	.52± .03	.76	14.91 ±.14				5146± 69
201.95 -38.49		-5.0±1.0		.00					4931
072857.0-664824	PGC 21111		.80		14.07				5256

0729.1 +5101		.S..4..	1.05± .08	40					
166.82 26.50	UGC 3865	U	.44± .06	.26					
31.95 -22.84		4.0±1.0		.65					
0725.3 +5108	PGC 21114		1.08	.22					

0729.2 +1424		.L.....	1.00± .10	48					
204.07 14.77	UGC 3877	U	.14± .04	.31	14.5 ±.3				
42.21 -58.71		-2.0± .9		.00					
0726.4 +1431	PGC 21116		1.01						

072916.7+421646		.SXS5..	1.02± .06	45					
176.27 24.52	UGC 3871	U	.14± .05	.22	14.73 ±.18				
33.55 -31.48		5.0± .9		.21					
072546.4+422259	PGC 21119		1.04	.07					

072917.2+275404		.SAS7..	1.35± .04	2				14.37±.1	860± 6
191.04 20.04	UGC 3876	U	.24± .05	.21	13.7 ±.2			208± 8	637±125
36.96 -45.61		7.0± .8		.33					812
072610.5+280018	PGC 21120		1.37	.12	13.12			1.12	1014

072924.4+071034		.I..9*.	1.11± .14	5				15.96±.3	3953± 10
210.79 11.69	UGC 3883	U	.44± .12	.45					
47.13 -65.58		10.0±1.3		.33				176± 7	3831
072642.8+071649	PGC 21122		1.15	.22					4142

072925.4+720741	IC 2184	.I?....	.92± .07					15.10±.1	3605± 9
143.10 28.62	UGC 3852		.11± .12	.05	14.0 ±.2			245± 6	3557± 16
28.79 -1.97	MK 8			.08	12.24			226± 25	3712
072338.5+721350	PGC 21123		.93	.05	13.91			1.13	3618

072943.9+334124		.S..8*.	1.36± .03	103				14.83±.2	4797± 5
185.24 22.09	UGC 3879	U	.84± .04	.26	15.04 ±.20			271± 7	
35.54 -39.91	KUG 0726+337	8.0±1.3		1.03				250± 5	4772
072628.7+334740	PGC 21136		1.38	.42	13.73			.68	4939

072944.9+752954		.SA.8*.	1.13± .05	57				16.07±.1	2464± 11
139.26 28.59	UGC 3846	U	.41± .08	.12					
28.32 1.37	KUG 0723+756	8.0± .9		.50				140± 8	2594
072310.2+753603	PGC 21138		1.14	.20					2477

0729.8 +5247		.SXT7..	1.06± .05	50				16.28±.1	5225± 11
164.90 26.93	UGC 3875	U	.06± .05	.24	14.9 ±.2				
31.76 -21.08		7.0± .8		.08				193± 8	5274
0725.9 +5254	PGC 21142		1.09	.03	14.58			1.67	5314

072954.5+372708	IC 2190	.SB.4*.	.95± .07	21					
181.37 23.29	UGC 3880	U	.25± .05	.22	14.81 ±.18				
34.70 -36.21		4.0± .9		.37					
072633.2+373324	PGC 21144		.97	.12					

072955.6-840226		.SBR8..	1.02± .05						
296.39 -25.98	ESO 5- 9	S (1)	.10± .04	.57	15.52 ±.14				
205.21 -21.60		8.0± .8		.12					
073704.1-835548	PGC 21145	7.8± .9	1.07	.05					

073005.0+340141	NGC 2393	.S..5..	1.09± .04	103				15.68±.2	4886± 5
184.92 22.27	UGC 3884	U	.20± .04	.24	14.68 ±.19			296± 7	
35.55 -39.57	IRAS07267+3407	5.0± .9		.30	13.53			281± 5	4862
072649.3+340758	PGC 21154		1.11	.10	14.11			1.46	5028

073011.7-621505	NGC 2417	.SAT4..	1.44± .04	81	12.84 ±.14	.81± .02	.89± .01	13.61±.2	3184± 6
273.80 -19.50	ESO 123- 15	RSr (2)	.16± .04	.71	13.09 ±.14	.16± .04	.25± .03	304± 8	3182± 33
200.63 -43.04	IRAS07295-6208	4.4± .4	1.14± .02	.24	11.95	.60	14.03± .05	274± 9	2964
072935.0-620842	PGC 21155	2.2± .4	1.51	.08	12.00	.00	14.49± .25	1.52	3308

073017.8-313557		.SAS3*.	1.20± .04	75					
245.29 -6.37	ESO 428- 37	S (1)	.36± .04	1.76	14.20 ±.14				
181.15 -72.19	IRAS07283-3129	3.0± .6		.49	12.37				
072822.0-312936	PGC 21161	3.3± .9	1.37	.18					

073018.2+493110		.SA.8..	1.07± .08	115					
168.51 26.38	UGC 3881	U	.29± .06	.28	15.3 ±.2				
32.41 -24.31		8.0± .9		.36					
072631.8+493727	PGC 21162		1.10	.15					

R.A. 2000 DEC. l　　b SGL　SGB R.A. 1950 DEC.	Names PGC	Type S_T　n_L T L	$\log D_{25}$ $\log R_{25}$ $\log A_e$ $\log D_o$	p.a. A_g A_i A_{21}	B_T m_B m_{FIR} B_T^o	$(B-V)_T$ $(U-B)_T$ $(B-V)_T^o$ $(U-B)_T^o$	$(B-V)_e$ $(U-B)_e$ m'_e m'_{25}	m_{21} W_{20} W_{50} HI	V_{21} V_{opt} V_{GSR} V_{3K}
073019.7+795222 134.29　28.38 27.72　　5.70 072156.3+795830	IC　　467 UGC　3834 IRAS07218+7958 PGC 21164	.SXS5*. R 5.0± .5 	1.51± .02 .40± .03 1.10± .04 1.53	80 .15 .60 .20	13.21M±.15 12.66 ±.18 12.82 12.22	.62± .04 .01± .06 .50 -.08	.57± .03 .10± .04 14.10± .09 14.62± .20	14.07±.1 300± 6 281± 6 1.65	2042± 5 2057± 24 2187 2041
073024.0-614724 273.35 -19.31 200.47 -43.49 072945.0-614100	 ESO　123- 16 IRAS07297-6141 PGC 21167	.SBS8.. S　　　(1) 8.0± .9 6.7± .9	1.26± .04 .50± .04 1.33	109 .71 .61 .25	 14.68 ±.14 13.15 				
073024.8+360647 182.80　22.98 35.13 -37.51 072705.8+361305	 UGC　3887 KUG 0727+362 PGC 21168	.SBS6.. U 6.0± .8 	1.06± .05 .07± .05 1.08	140 .23 .10 .03	 14.9　±.2 				
0730.6　+5533 161.85　27.50 31.43 -18.33 0726.6　+5540	 UGC　3882 PGC 21174	.SAR6*. U 6.0± .8 	1.16± .05 .37± .05 1.18	170 .16 .54 .18					
073040.9-620131 273.60 -19.36 200.49 -43.25 073003.0-615506	 ESO　123- 17 PGC 21175	.LX.0*. S -2.0± .9 	1.10± .05 .38± .04 1.12	58 .71 .00 	 14.63 ±.14 				
073044.4-722252 284.05 -22.87 202.98 -33.09 073124.0-721624	 ESO　　59- 2 PGC 21177	.SX.3.. r 3.3± .9 	1.02± .07 .49± .05 1.09	110 .73 .67 .24	 15.89 ±.14 				
073048.6+734223 141.30　28.71 28.66　　-.40 072442.2+734837	 UGC　3859 ARAK　133 PGC 21181	.S..1.. U 1.0± .8 	1.01± .06 .11± .05 1.02	120 .10 .11 .05	 13.6　±.2 13.28			15.85±.1 390± 8 2.52	5379± 11 5503 5399
073054.7+390111 179.80　23.93 34.59 -34.63 072730.7+390731	 UGC　3888 PGC 21187	.SB.3.. U 3.0± .9 	1.11± .05 .29± .05 1.14	115 .27 .40 .15	 15.2　±.2 				
073056.6+723103 142.66　28.74 28.84　-1.57 072505.6+723719	 UGC　3864 VV　　141 PGC 21189	.S?.... 	1.09± .04 .21± .04 1.10	 .08 .31 .11	 14.80 ±.19 14.40				2569± 27 2689 2593
073059.4-515101 263.81 -15.32 196.99 -53.15 072944.1-514436	 ESO　208- 18 IRAS07296-5144 PGC 21191	 	.87± .07 .24± .06 .94	115 .80 	 15.49 ±.14 				12037± 87 11812 12188
073107.4+592857 157.47　28.08 30.86 -14.45 072648.5+593516	 UGC　3885 IRAS07268+5935 PGC 21195	.S?.... 	1.02± .06 .07± .05 1.04	 .19 .11 .04	 13.98 ±.18 13.00 13.66			16.35±.1 162± 8 2.66	3809± 7 3884 3878
073110.3+130338 205.54　14.63 43.85 -59.87 072822.2+131000	 UGC　3892 PGC 21197	.S..6?. U 6.0±1.8 	1.15± .05 .53± .05 1.17	80 .21 .78 .27	 14.6　±.3 13.58			16.64±.3 439± 7 2.80	8103± 10 8000 8287
073118.3-681117 279.77 -21.48 201.99 -37.20 073118.0-680448	 ESO　　59- 1 PGC 21199	.IBS9.. S　　　(1) 10.0± .4 10.0± .5	1.33± .04 .10± .05 1.39	 .67 .07 .05	 13.74 ±.14 				
073120.7+314359 187.36　21.78 36.49 -41.76 072808.7+315022	 MCG　5-18- 16 MK　　1407 PGC 21200	.S?.... 	.99± .10 .45± .07 .99	 .25 .33 	 15.33 ±.18 14.66				4827± 63 4794 4976
073123.1+000316 217.47　　8.90 56.63 -71.86 072849.2+000940	 UGC　3895 PGC 21201	.I..9*. U 10.0±1.2 	.96± .09 .00± .06 1.04	 .88 .00 .00				16.35±.1 38± 4 33± 4 	1456± 5 1311 1653
0731.4　+6228 154.09　28.41 30.42 -11.49 0726.9　+6235	 UGC　3886 PGC 21207	.SAS5.. U 5.0± .8 	1.08± .06 .02± .05 1.09	 .19 .03 .01				15.42±.1 114± 8 	4883± 11 4968 4942
073143.4-691905 280.93 -21.82 202.21 -36.08 073152.2-691234	 PGC 21212	.S..9/. S 9.0± .9 	1.15± .08 .31± .08 1.23	 .82 .32 .16					

R.A. 2000 DEC. l b SGL SGB R.A. 1950 DEC.	Names PGC	Type S_T n_L T L	$\log D_{25}$ $\log R_{25}$ $\log A_e$ $\log D_o$	p.a. A_g A_i A_{21}	B_T m_B m_{FIR} B_T^o	$(B-V)_T$ $(U-B)_T$ $(B-V)_T^o$ $(U-B)_T^o$	$(B-V)_e$ $(U-B)_e$ m' m'_{25}	m_{21} W_{20} W_{50} HI	V_{21} V_{opt} V_{GSR} V_{3K}
073143.6+631432 153.23 28.50 30.34 -10.73 072707.0+632053	A 0727+63 MCG 11- 9- 53 MK 73 PGC 21213	.S?.... 	.98± .07 .42± .12 .19± .04 .99	 .18 .62 .21	15.2 ±.2 15.00 ±.19 14.27	.72± .04 -.13± .08 .57 -.25	.74± .02 11.66± .13 13.88± .50	 	 4447± 60 4535 4503
073147.7-532922 265.41 -15.89 197.46 -51.54 073037.0-532254	 ESO 163- 6 PGC 21217	.LXR+*P S -1.0± .9 	1.04± .08 .31± .04 1.05	51 .61 .00 	 	 	 	 	
0731.9 +1820 200.61 16.97 41.39 -54.76 0729.0 +1827	NGC 2407 UGC 3896 PGC 21220	.L..-*. U -3.0±1.2 	1.04± .11 .04± .05 1.06	75 .13 .00 	 14.38 ±.17 	 	 	 	
0732.3 +7255 142.19 28.83 28.88 -1.15 0726.4 +7302	 UGC 3878 PGC 21230	.S..8.. U 8.0± .9 	1.11± .07 .25± .06 1.12	102 .09 .30 .12	 	 	 	16.08±.1 155± 8 	3037± 11 3159 3060
073219.6+854232 127.71 27.81 26.88 11.47 071545.1+854831	NGC 2300 UGC 3798 -2.0± .3 PGC 21231	.LA.0.. R 	1.45± .03 .14± .03 1.02± .02 1.46	 .23 .00 	12.07 ±.13 12.10 ±.12 11.82	1.08± .01 .70± .02 1.00 .64	1.09± .01 .71± .02 12.66± .06 13.86± .21	 	 1963± 19 2124 1941
073227.2+105412 207.69 13.99 45.87 -61.82 072941.5+110040	 UGC 3900 PGC 21237	.SB.3.. U 3.0± .9 	1.00± .06 .10± .05 1.01	150 .17 .13 .05	 14.3 ±.3 13.95	 	 	16.90±.3 328± 7 2.90	8535± 10 8424 8723
073232.0+551148 162.31 27.71 31.77 -18.65 072829.3+551813	A 0728+55 MK 75 PGC 21240	 	 	 .17 	 	 	 	 	8904± 38 8962 8988
073236.5-840424 296.45 -25.92 205.14 -21.55 073945.0-835736	 ESO 5- 10 PGC 21243	.E.4... S -5.0± .7 	1.19± .05 .21± .04 .68± .07 1.21	35 .55 .00 	13.79 ±.19 13.87 ±.14 	1.10± .02 .50± .03 	1.13± .01 .55± .02 12.68± .25 14.18± .33	 	
073237.9+353653 183.46 23.26 35.81 -37.91 072919.9+354320	 UGC 3899 KUG 0729+357 PGC 21244	.S..2.. U 2.0± .9 	1.06± .04 .57± .04 1.08	46 .25 .70 .29	 15.33 ±.18 14.34	 	 	15.34±.2 225± 9 .72	3884± 7 3866 4024
0733.0 +6052 155.92 28.46 30.87 -13.04 0728.6 +6059	A 0728+60 7ZW 162 PGC 21254	 	 	 .19 	 	 	 	 	9031± 43 9110 9096
0733.0 +6500 151.23 28.77 30.20 -8.96 0728.3 +6507	 UGC 3893 PGC 21257	.SAR3*. U (1) 3.0±1.2 4.5±1.1	.97± .07 .02± .05 .99	 .18 .03 .01	 14.61 ±.18 	 	 	 	
073304.6+650440 151.15 28.77 30.19 -8.89 072817.9+651106	 UGC 3894 PGC 21258	.E..... U -5.0± .8 	1.06± .11 .04± .05 1.07	 .14 .00 	14.24 ±.15 14.00	 	 	 	6755± 38 6850 6806
073309.4+191157 199.91 17.58 41.48 -53.85 073014.2+191827	 UGC 3903 PGC 21261	PSXT3.. U 3.0± .9 	1.20± .05 .45± .05 1.22	174 .18 .62 .22	 14.44 ±.18 13.58	 	 	16.76±.3 462± 7 2.96	8040± 10 7959 8217
073320.7+593730 157.34 28.37 31.12 -14.26 072901.6+594358	 UGC 3897 PGC 21273	.LBR+.. U -1.0± .8 	1.11± .07 .10± .03 1.11	45 .17 .00 	 14.09 ±.15 	 	 	 	
073323.5+312956 187.75 22.11 37.12 -41.90 073012.0+313627	IC 2193 UGC 3902 IRAS07301+3135 PGC 21276	.S..3.. U (1) 3.0± .8 2.5±1.1	1.17± .05 .20± .05 1.19	90 .20 .28 .10	 14.2 ±.2 13.37 13.69	 	 	16.08±.3 430± 7 2.29	5021± 10 4986 5172
0733.4 +3033 188.71 21.81 37.42 -42.82 0730.3 +3040	 UGC 3904 PGC 21280	RSBR0.. U .0± .8 	1.13± .05 .00± .05 1.15	 .24 .00 	 14.0 ±.2 13.71	 	 	16.90±.3 171± 14 	4734± 10 4696 4888
0733.8 +6527 150.72 28.87 30.20 -8.50 0729.0 +6534	 UGC 3898 PGC 21288	.SXT3.. U (1) 3.0± .9 2.5±1.2	.92± .07 .24± .05 .94	18 .16 .33 .12	 15.25 ±.18 14.72	 	 	 	 6662±125 6758 6712

R.A. 2000 DEC.	Names	Type	logD₂₅	p.a.	B_T	(B-V)_T	(B-V)₀	m₂₁	V₂₁
l b		S_T n_L	logR₂₅	A_g	m_B	(U-B)_T	(U-B)₀	W₂₀	V_opt
SGL SGB		T	logA₀	A_i	m_FIR	(B-V)°_T	m'₀	W₅₀	V_GSR
R.A. 1950 DEC.	PGC	L	logD₀	A₂₁	B°_T	(U-B)°_T	m'₂₅	HI	V_3K

0733.8 +7338	A 0727+73	.S..2..	1.07± .07	112					
141.37 28.93	UGC 3889	U	.79± .06	.11	15.3 ±.2				
28.88 -.43		2.0±1.0		.98					
0727.8 +7345	PGC 21289		1.08	.40					

073353.4+370134		.S..6*.	1.11± .05	80					
182.07 23.92	UGC 3907	U	.29± .05	.21	15.3 ±.2				
35.77 -36.48		6.0±1.2		.43					
073033.2+370806	PGC 21291		1.13	.15					

073356.6-502631		.LX.-..	1.47± .04	110	12.19 ±.13	.99± .01	1.06± .01		1037± 31
262.69 -14.30	ESO 208- 21	S	.15± .04	.65	12.01 ±.14	.51± .03	.59± .01		812
195.61 -54.39		-3.0± .7	.93± .02	.00		.83	12.33± .08		1194
073237.0-501954	PGC 21293		1.54		11.45	.37	14.06± .27		

0734.1 +2306		.L.....	1.04± .09	160					
196.20 19.30	UGC 3911	U	.13± .04	.19	14.67 ±.17				
40.17 -50.02		-2.0± .8		.00					
0731.1 +2313	PGC 21296		1.04						

073410.4+312424	IC 2196	.E.....	1.15± .15	150					4742±155
187.90 22.24	UGC 3910	U	.10± .08	.26	13.72 ±.15				4707
37.37 -41.96		-5.0± .8		.00					4894
073059.0+313058	PGC 21300		1.16		13.38				

0734.2 +6653	A 0729+66	.I..9*.	1.48± .10						
149.09 28.96	UGCA 133	PU (1)	.18± .18	.18					
30.01 -7.09	DDO 44	10.0± .6		.14					
0729.2 +6659	PGC 21302	9.0±1.0	1.49	.09					

073412.6+043247		.IB.9*.	1.25± .06	120				14.37±.1	1231± 5
213.74 11.58	UGC 3912	U	.20± .06	.47	13.3 ±.3			191± 8	
52.48 -67.54	IRAS07315+0439	10.0± .8		.15	13.46			162± 6	1098
073133.9+043922	PGC 21303		1.29	.10	12.66			1.62	1428

073413.7-673446	.	.SA.6?.	1.03± .05						
279.26 -21.01	ESO 88- 22	S (1)	.08± .04	.75	14.45 ±.14				
201.50 -37.74	IRAS07341-6728	6.0± .9		.11	13.18				
073408.1-672806	PGC 21305	5.6±1.2	1.10	.04					

073448.5+223436		.S..7..	1.06± .06	85				15.32±.3	4452± 10
196.79 19.25	UGC 3916	U	.64± .05	.35	15.54 ±.18				4383
40.63 -50.49		7.0± .9		.88				245± 7	4624
073149.2+224113	PGC 21321		1.09	.32	14.29			.71	

073450.4+623244		.S..2..	.92± .05	80					
154.05 28.81	UGC 3905	U	.12± .04	.19	14.51 ±.19				
30.81 -11.35	IRAS07302+6239	2.0± .9		.14	13.36				
073018.1+623918	PGC 21322		.94	.06					

073451.2-694649		.IXS9?.	1.13± .04	113					
281.50 -21.71	ESO 59- 6	S (1)	.30± .04	.79	15.18 ±.14				
201.99 -35.57		9.7± .7		.22					
073503.0-694006	PGC 21323	9.4± .7	1.21	.15					

073451.2+333220		.S?....	1.22± .06	160				14.73±.3	4010± 10
185.76 23.06	UGC 3913		.76± .06	.19	15.46 ±.19				3983
36.94 -39.84				1.14				256± 7	4158
073136.7+333856	PGC 21324		1.24	.38	14.11			.24	

073451.5-691701	NGC 2434	.E.0+..	1.39± .04		12.33 ±.13	1.07± .01	1.09± .01		1390± 27
281.00 -21.54	ESO 59- 5	R	.03± .04	.76	12.44 ±.11	.49± .04	.57± .02		1178
201.87 -36.06		-5.0± .7	1.02± .02	.00		.89	12.92± .06		1496
073459.1-691018	PGC 21325		1.50		11.61	.32	14.19± .26		

073455.6+311633	IC 2199	.SB?...	1.04± .06	25				15.04±.3	4680± 10
188.09 22.35	UGC 3915		.26± .05	.17	14.1 ±.2				4577± 63
37.61 -42.05	IRAS07317+3123			.39	12.06			323± 7	4642
073144.5+312310	PGC 21328		1.05	.13	13.48			1.44	4831

073459.6+853210	IC 455	.L.....	1.04± .09	82					1887± 61
127.89 27.88	UGC 3815	U	.18± .04	.23	14.27 ±.15				2047
26.96 11.31		-2.0± .9		.00					1866
071901.9+853821	PGC 21334		1.04		14.01				

073502.2+324926	NGC 2410	.SB.3?.	1.39± .05	31				14.81±.3	4678± 10
186.51 22.87	UGC 3917	U	.54± .06	.16	13.76 ±.18				4460±155
37.19 -40.53	IRAS07318+3255	3.0± .8		.74	12.52			490± 7	4647
073148.8+325603	PGC 21336		1.40	.27	12.83			1.72	4827

073507.3+650033		.S?....	.82± .13						6316
151.24 28.98	MCG 11-10- 5		.28± .07	.20	15.3 ±.2				6410
30.41 -8.92	MK 76			.21					6368
073021.5+650707	PGC 21337		.82		14.75				

R.A. 2000 DEC. / l b / SGL SGB / R.A. 1950 DEC.	Names / PGC	Type / S_T n_L / T / L	$\log D_{25}$ / $\log R_{25}$ / $\log A_e$ / $\log D_o$	p.a. / A_g / A_i / A_{21}	B_T / m_B / m_{FIR} / B_T^o	$(B-V)_T$ / $(U-B)_T$ / $(B-V)_T^o$ / $(U-B)_T^o$	$(B-V)_e$ / $(U-B)_e$ / m' / m'_{25}	m_{21} / W_{20} / W_{50} / HI	V_{21} / V_{opt} / V_{GSR} / V_{3K}
073507.4-465529 / 259.53 -12.59 / 193.44 -57.69 / 073339.0-464848	ESO 257- 19 / IRAS07336-4648 / PGC 21338	.SBS6?/ / S / 6.2± .6	1.39± .04 / .62± .05 / 1.46	130 / .74 / .92 / .31	14.48 ±.14 / 12.20				
073508.6-551712 / 267.32 -16.19 / 197.43 -49.67 / 073403.0-551030	ESO 163- 7 / PGC 21339	.IBS9.. / S (1) / 10.0± .9 / 8.9± .9	1.08± .06 / .21± .08 / 1.13	.50 / .16 / .11					
073517.4-765500 / 288.86 -23.94 / 203.61 -28.57 / 073705.0-764812	ESO 35- 11 / PGC 21340	.SBR2P* / S (1) / 2.0± .6 / 4.4± .9	1.07± .05 / .10± .04 / 1.16	.96 / .12 / .05	14.63 ±.14				
073519.3+190249 / 200.27 17.99 / 42.39 -53.86 / 073224.3+190928	UGC 3920 / IRAS07323+1909 / PGC 21341	.SX.3.. / U (1) / 3.0± .8 / 3.5±1.1	1.12± .07 / .01± .06 / 1.13	.15 / .01 / .00	14.0 ±.2 / 13.59 / 13.79			16.65±.3 / 165± 7 / 142± 7 / 2.86	8520± 11 / 8553± 38 / 8440 / 8702
073520.9-500230 / 262.42 -13.93 / 195.05 -54.71 / 073400.1-495548	ESO 208- 26 / IRAS07339-4955 / PGC 21343	RLB.+.. / r / -1.3± .9	1.07± .07 / .33± .05 / 1.08	106 / .65 / .00	14.84 ±.14 / 12.68 / 14.14				3003± 19 / 2777 / 3161
073524.6-662115 / 278.08 -20.46 / 201.03 -38.91 / 073510.0-661430	ESO 88- 23 / FAIR 269 / PGC 21345	.SBR5.. / S (1) / 4.5± .6 / 3.9± .6	1.03± .05 / .02± .05 / 1.09	.60 / .03 / .01	14.97 ±.14 / 14.15 / 14.29				7790± 63 / 7574 / 7905
073533.3-542907 / 266.59 -15.80 / 197.02 -50.43 / 073425.0-542224	ESO 163- 8 / PGC 21351	.SBS9.. / S (1) / 9.0± .8 / 10.0± .8	1.13± .07 / .21± .07 / 1.18	16 / .58 / .21 / .10					
073538.5+113113 / 207.46 14.96 / 46.96 -60.97 / 073252.2+113753	UGC 3924 / PGC 21356	.S..9*. / U / 9.0±1.2	1.00± .16 / .04± .12 / 1.01	.12 / .04 / .02	14.7 ±.3 / 14.51			16.43±.2 / 113± 7 / 103± 5 / 1.89	5162± 6 / 5053 / 5354
0735.6 +6624 / 149.63 29.09 / 30.23 -7.53 / 0730.7 +6631	UGC 3908 / IRAS07307+6631 / PGC 21357	.S..3.. / U (1) / 3.0±1.0 / 4.5±1.3	1.08± .06 / .74± .05 / 1.09	50 / .19 / 1.03 / .37	13.58				
073542.0+113649 / 207.38 15.02 / 46.92 -60.88 / 073255.7+114330	NGC 2416 / UGC 3925 / IRAS07329+1143 / PGC 21358	.S..6*. / U / 6.0±1.3	1.02± .06 / .20± .05 / 1.03	110 / .12 / .29 / .10	14.1 ±.3 / 13.71			15.58±.3 / 206± 7 / 1.77	5101± 10 / 4992 / 5292
073612.5-694743 / 281.56 -21.61 / 201.86 -35.54 / 073624.0-694054	ESO 59- 7 / PGC 21369	PSBT0*. / Sr / -.3± .8	1.09± .05 / .32± .03 / 1.14	104 / .71 / .24	15.07 ±.14				
073613.1-704237 / 282.49 -21.92 / 202.09 -34.64 / 073633.0-703548	ESO 59- 9 / PGC 21370	PSAS8?. / Sr (1) / 7.7± .5 / 7.8± .6	1.18± .05 / .02± .05 / 1.25	.73 / .03 / .01	14.28 ±.14				
073616.8+330722 / 186.29 23.21 / 37.43 -40.18 / 073303.0+331404	IC 2201 / UGC 3926 / PGC 21372	.S..1.. / U / 1.0± .9	1.13± .05 / .63± .05 / 1.15	67 / .19 / .64 / .31	14.89 ±.18				
073623.9-693150 / 281.30 -21.50 / 201.77 -35.79 / 073633.0-692500	NGC 2442 / ESO 59- 8 / IRAS07367-6924 / PGC 21373	.SXS4P. / RS (3) / 3.7± .3 / 2.5± .3	1.74± .02 / .05± .03 / 1.45± .03 / 1.81	104 / .75 / .07 / .02	11.24M±.12 / 11.13 ±.12 / 13.99± .06 / 10.36	.82± .02 / .23± .07 / .63 / .09	.90± .01 / .32± .03 / 13.99± .06 / 14.70± .18	12.81±.3 / 499± 10 / 2.44	1449± 7 / 1399± 66 / 1236 / 1554
073628.0-473805 / 260.29 -12.70 / 193.45 -56.95 / 073501.0-473118	NGC 2427 / ESO 208- 27 / PGC 21375	.SXS8.. / R (2) / 8.0± .3 / 7.0± .4	1.72± .02 / .38± .02 / 1.79	122 / .79 / .47 / .19	12.33S±.10 / 12.00 ±.12 / 10.93	.84± .04 / -.15± .07 / .59 / -.35	14.82± .16	12.76±.2 / 269± 6 / 226± 7 / 1.64	972± 6 / 746 / 1137
0736.6 +5503 / 162.56 28.27 / 32.40 -18.69 / 0732.6 +5510	UGC 3922 / PGC 21380	.S..6*. / U / 6.0±1.3	1.20± .05 / .88± .05 / 1.22	164 / .17 / 1.29 / .44				15.90±.1 / 473± 8	8785± 7 / 8842 / 8871
073637.5+742647 / 140.44 29.08 / 28.93 .40 / 073022.6+743325	UGC 3906 / VV 539 / PGC 21381	.96± .06 / .08± .06 / .97		.10	14.73 ±.18				3704± 26 / 3830 / 3723

R.A. 2000 DEC. l b SGL SGB R.A. 1950 DEC.	Names PGC	Type S_T n_L T L	$\log D_{25}$ $\log R_{25}$ $\log A_o$ $\log D_o$	p.a. A_g A_l A_{21}	B_T m_B m_{FIR} B_T^o	$(B-V)_T$ $(U-B)_T$ $(B-V)_T^o$ $(U-B)_T^o$	$(B-V)_o$ $(U-B)_o$ m'_o m'_{25}	m_{21} W_{20} W_{50} HI	V_{21} V_{opt} V_{GSR} V_{3K}
073637.9+175306 201.52 17.82 43.48 -54.88 073344.4+175950	NGC 2418 UGC 3931 ARP 165 PGC 21382	.E..... U -5.0± .8 	1.26± .13 .00± .08 1.28	 .12 .00 	 13.16 ±.16 12.97				5057± 33 4970 5240
0736.7 +5518 162.28 28.32 32.37 -18.44 0732.7 +5525	 UGC 3923 PGC 21388	.S..3.. U (1) 3.0± .9 3.5±1.2	1.00± .06 .21± .05 1.01	153 .18 .29 .11	 14.80 ±.18 				
073654.5+653558 150.57 29.19 30.50 -8.31 073205.5+654240	NGC 2403 UGC 3918 IRAS07321+6543 PGC 21396	.SXS6.. R (2) 6.0± .3 5.4± .5	2.34± .01 .25± .02 1.68± .02 2.36	127 .17 .36 .12	8.93M.07 9.00 ±.16 8.63 8.41	.47± .03 .38 	.55± .01 -.14± .03 12.84± .04 14.88± .10	9.58±.1 244± 5 231± 5 1.05	131± 3 107± 34 226 181
0736.9 +6433 151.77 29.15 30.68 -9.34 0732.2 +6440	 UGC 3919 PGC 21397	.S..6*. U 6.0±1.4 	1.19± .05 1.06± .05 1.20	81 .19 1.47 .50					
073656.5+734248 141.28 29.14 29.08 -.32 073052.3+734927	A 0730+73 UGC 3909 KUG 0730+738 PGC 21398	.SB.6*. U 6.0±1.2 	1.42± .03 .74± .04 1.43	82 .10 1.08 .37	 14.81 ±.17 13.62			14.29±.1 172± 8 .30	945± 11 1069 967
073656.7+351433 184.14 23.99 37.00 -38.09 073339.7+352118	NGC 2415 UGC 3930 ARAK 136 PGC 21399	.I..9$. P 10.0±1.7 	.96± .06 .00± .04 .97	 .17 .00 .00	12.78 ±.13 12.9 ±.3 10.87 12.62	.42± .01 -.20± .03 .35 -.25	 12.41± .35	14.98±.1 239± 5 109± 8 2.36	3784± 5 3810± 28 3764 3930
073656.9+584617 158.36 28.75 31.74 -15.02 073242.0+585300	A 0732+58 CGCG 286- 36 MK 9 PGC 21400	.L...P$ P -2.0±1.8 	.72± .10 .09± .12 .72	 .14 .00 	 15.29 ±.17 13.34 14.97				11954± 48 12025 12029
073703.3+133600 205.66 16.16 46.18 -58.90 073414.7+134246	 UGC 3936 IRAS07342+1342 PGC 21401	.SB.4.. U 4.0± .9 	1.08± .06 .08± .05 1.11	100 .33 .11 .04	 13.8 ±.3 13.36 13.36			16.83±.3 346± 7 3.43	4725± 10 4623 4915
073703.4+222101 197.22 19.64 41.51 -50.57 073404.4+222747	 UGC 3932 PGC 21402	.S..2.. U 2.0± .9 	1.14± .05 .72± .05 1.16	150 .16 .89 .36	 15.03 ±.18 13.93			15.13±.3 382± 7 .84	4578± 10 4508 4753
073708.0+095443 209.13 14.60 48.92 -62.34 073423.5+100130	 UGC 3938 PGC 21404	.S..5.. U 5.0± .9 	1.02± .06 .09± .05 1.03	 .19 .14 .05	 14.4 ±.3 14.00			16.10±.3 196± 7 2.05	8863± 10 8748 9058
073712.1-692005 281.13 -21.37 201.63 -35.97 073719.4-691312	 PGC 21406	.IB.9*. S (1) 10.0±1.1 11.1± .8	1.21± .07 .04± .08 1.28	 .75 .03 .02					
0737.3 +1958 199.57 18.79 42.70 -52.84 0734.4 +2005	 UGC 3939 PGC 21414	.L..... U -2.0± .8 	.93± .13 .04± .05 .94	25 .15 .00 	 14.67 ±.15 				
073723.4+140452 205.24 16.43 46.01 -58.42 073434.3+141140	 UGC 3941 PGC 21416	.SB.6*. U 6.0±1.3 	1.16± .07 .59± .06 1.18	74 .24 .86 .29	 14.9 ±.3 13.81			15.23±.3 232± 7 1.12	4767± 10 4666 4957
0737.4 +5942 157.31 28.89 31.63 -14.10 0733.1 +5949	 UGC 3927 PGC 21417	.L..... U -2.0± .8 	1.14± .09 .38± .05 1.10	15 .17 .00 	 15.17 ±.18 				
073727.8+353926 183.74 24.21 37.01 -37.66 073410.1+354613	 UGC 3934 PGC 21419	.S..9?. U 9.0±1.7 	1.00± .16 .09± .12 1.02	 .22 .09 .04				15.75±.3 102± 5 	4087± 7 4068 4232
073736.3+353623 183.80 24.22 37.06 -37.70 073418.7+354310	 UGC 3937 IRAS07343+3543 PGC 21425	.SB?... 	1.30± .03 .57± .04 1.32	151 .22 .86 .29	 14.12 ±.18 12.33 13.02			14.74±.1 321± 7 1.43	3994± 7 3975 4140
073736.4-694512 281.56 -21.48 201.70 -35.55 073747.0-693818	 ESO 59- 10 PGC 21426	.SAT6.. S (1) 6.0± .9 6.7± .9	1.01± .05 .27± .04 1.07	136 .71 .40 .13	 15.26 ±.14 				

R.A. 2000 DEC. / l b / SGL SGB / R.A. 1950 DEC.	Names / PGC	Type / S_T n_L / T / L	$logD_{25}$ / $logR_{25}$ / $logA_e$ / $logD_o$	p.a. / A_g / A_i / A_{21}	B_T / m_B / m_{FIR} / B_T^o	$(B-V)_T$ / $(U-B)_T$ / $(B-V)_T^o$ / $(U-B)_T^o$	$(B-V)_e$ / $(U-B)_e$ / m'_e / m'_{25}	m_{21} / W_{20} / W_{50} / HI	V_{21} / V_{opt} / V_{GSR} / V_{3K}
073737.1-521816 / 264.68 -14.59 / 195.58 -52.44 / 073622.0-521124	ESO 208- 31 / PGC 21429	.S..5./ / S (1) / 5.0± .8 / 2.2±1.3	1.21 .04 / .93± .04 / / 1.29	167 / .91 / 1.40 / .47	15.48 ±.14				
073737.4+415651 / 177.06 25.93 / 35.40 -31.51 / 073408.7+420338	A 0734+42 / UGC 3933 / PGC 21431	.SXS4.. / U (1) / 4.0± .8 / 2.7± .9	1.09± .06 / .14± .05 / .84± .02 / 1.11	0 / .19 / .21 / .07	14.19 ±.14 / 14.34 ±.18 / / 13.80	.71± .03 / / .60 /	.77± .03 / / 13.88± .04 / 14.14± .33	15.27±.1 / 360± 11 / 353± 7 / 1.39	5897± 6 / 5856± 59 / 5903 / 6025
0737.6 +2702 / 192.61 21.48 / 39.84 -46.02 / 0734.6 +2709	UGC 3942 / PGC 21437	.S..3.. / U (1) / 3.0±1.0 / 4.5±1.3	1.04± .08 / .76± .06 / / 1.06	151 / .19 / 1.05 / .38	15.54 ±.19				
073749.4+462351 / 172.22 26.97 / 34.40 -27.15 / 073411.4+463039	UGC 3935 / PGC 21443	.SX.5.. / U / 5.0± .9 /	.99± .06 / .09± .05 / / 1.02	15 / .34 / .13 / .04	15.3 ±.2				
073753.4+633116 / 152.96 29.21 / 30.98 -10.33 / 073316.8+633802	A 0733+63A / CGCG 310- 5 / ARAK 135 / PGC 21444	RSXS0.. / P / .0± .9 /		.18	15.8 ±.3				
073753.5-551059 / 267.40 -15.79 / 196.81 -49.67 / 073647.1-550406	ESO 163- 10 / PGC 21445	.SBT3P. / S (1) / 3.0± .8 / 2.2±1.2	1.10± .08 / .05± .07 / / 1.14	/ .50 / .07 / .02					
073759.1+031838 / 215.30 11.86 / 56.22 -68.21 / 073521.7+032528	UGC 3946 / PGC 21450	.I..9.. / U / 10.0± .8 /	1.11± .05 / .23± .05 / / 1.15	15 / .44 / .18 / .12	13.6 ±.3 / / / 13.03			14.51±.1 / 117± 7 / 85± 6 / 1.37	1197± 5 / / 1060 / 1401
0738.0 +7052 / 144.53 29.32 / 29.66 -3.10 / 0732.5 +7059	UGC 3921 / PGC 21451	.S..6*. / U / 6.0±1.4 /	1.04± .08 / .98± .06 / / 1.05	53 / .06 / 1.44 / .49					
073805.4-551124 / 267.42 -15.76 / 196.77 -49.65 / 073659.0-550430	ESO 163- 11 / IRAS07369-5504 / PGC 21453	.SBS3?/ / S / 3.0± .8 /	1.42± .05 / .61± .07 / / 1.47	3 / .50 / .85 / .31	11.21				
073812.0-692827 / 281.30 -21.33 / 201.57 -35.81 / 073820.1-692130	ESO 59- 11 / PGC 21457	.SBS0P* / S / -.5± .6 /	1.28± .05 / .26± .04 / / 1.31	163 / .75 / .00 /	13.52 ±.14				
073822.5-504525 / 263.30 -13.81 / 194.63 -53.88 / 073703.0-503830	ESO 208- 33 / PGC 21466	.SBS9.. / S (1) / 9.0± .8 / 10.0± .8	1.33± .04 / .20± .05 / / 1.39	153 / .70 / .21 / .10	14.59 ±.14				
0738.5 +6337 / 152.85 29.28 / 31.03 -10.22 / 0733.8 +6344	A 0733+63B / MCG 11-10- 11 / PGC 21471	.E.0.P$ / P / -5.0±1.1 /	.52± .20 / .17± .07 / / .49	/ .18 / .00 /	15.9 ±.2				
073831.9-684616 / 280.61 -21.06 / 201.34 -36.49 / 073834.0-683918	ESO 59- 12 / IRAS07385-6839 / PGC 21472	PSXS1*. / Sr (1) / 1.4± .5 / 7.8±1.2	1.29± .04 / .28± .05 / / 1.35	17 / .66 / .29 / .14	14.16 ±.14				
073836.9+373809 / 181.72 24.99 / 36.74 -35.68 / 073516.1+374500	UGC 3944 / IRAS07352+3744 / PGC 21475	.S..6*. / U / 6.0±1.2 /	1.25± .04 / .35± .04 / / 1.28	130 / .27 / .51 / .17	14.6 ±.2 / / / 13.84			15.31±.1 / 303± 8 / / 1.29	3895± 7 / / 3884 / 4036
0738.7 +0213 / 216.38 11.52 / 58.14 -69.07 / 0736.1 +0220	UGC 3950 / PGC 21479	.S..0.. / U / .0±1.0 /	.98± .07 / .56± .05 / / .99	176 / .45 / .42 /	14.5 ±.3				
073902.3+335501 / 185.66 24.00 / 37.92 -39.28 / 073547.5+340154	UGC 3947 / PGC 21491	.S..9*. / U / 9.0±1.3 /	.96± .17 / .15± .12 / / .97	10 / .14 / .15 / .07				16.62±.3 / / 124± 7 /	3913± 10 / / 3887 / 4064
0739.1 -0045 / 219.12 10.24 / 63.29 -71.51 / 0736.6 -0039	PGC 21494	.SB.9*/ / E (1) / 9.0±1.3 / 7.5± .8	1.05± .09 / .38± .08 / / 1.12	55 / .66 / .39 / .19					

R.A. 2000 DEC.	Names	Type	logD$_{25}$	p.a.	B$_T$	(B-V)$_T$	(B-V)$_\bullet$	m$_{21}$	V$_{21}$
l b		S$_T$ n$_L$	logR$_{25}$	A$_g$	m$_B$	(U-B)$_T$	(U-B)$_\bullet$	W$_{20}$	V$_{opt}$
SGL SGB		T	logA$_\bullet$	A$_i$	m$_{FIR}$	(B-V)^o_T	m'$_\bullet$	W$_{50}$	V$_{GSR}$
R.A. 1950 DEC.	PGC	L	logD$_\bullet$	A$_{21}$	B^o_T	(U-B)^o_T	m'$_{25}$	HI	V$_{3K}$

073909.8+192956		.S..4..	1.00± .08	58				16.06±.3	5104± 9
200.21 19.00	UGC 3952	U	.73± .06	.16	15.64 ±.19				5023
43.62 −53.16		4.0±1.0		1.07				266± 5	5287
073614.4+193650	PGC 21495		1.02	.36	14.38			1.32	

073911.1+590941		.SBS5..	1.23± .03	108				14.76±.1	3527± 11
157.95 29.07	UGC 3943	U	.39± .04	.26	14.7 ±.2				3599
31.96 −14.59	IRAS07349+5916	5.0± .9		.59	13.59			267± 8	3602
073455.1+591633	PGC 21496		1.25	.20	13.79			.77	

073925.2+085358		.S..7*.	1.16± .07	71				15.67±.3	5106± 10
210.32 14.67	UGC 3955	U	.88± .06	.20	15.1 ±.3				4987
50.92 −63.05		7.0±1.3		1.22				337± 7	5306
073641.8+090054	PGC 21503		1.18	.44	13.67			1.56	

073934.9+492121			.64± .17						
169.02 27.83	MCG 8-14- 30		.00± .07	.25	15.1 ±.2				6250± 82
34.06 −24.19	ARAK 138				13.45				6285
073550.1+492815	PGC 21506		.66						6358

073945.3+484432	A 0736+48	.SBS7..	1.04± .06		14.53 ±.15	.72± .04	.67± .03	15.94±.1	6382± 10
169.71 27.75	UGC 3949	U (1)	.00± .05	.26	14.6 ±.2		.62	369± 15	6358± 59
34.22 −24.79		7.0± .8	.80± .02	.00			14.02± .05	350± 11	6413
073602.1+485127	PGC 21513	3.8± .9	1.06	.00	14.28		14.57± .35	1.66	6491

0739.7 +5419		.SBR6..	.99± .08	155					
163.46 28.62	UGC 3948	U	.11± .06	.16	15.3 ±.2				
33.02 −19.32		6.0± .9		.16					
0735.8 +5426	PGC 21514		1.01	.05					

073948.0+242816		.S..7*.	1.00± .08	22				16.56±.3	7356± 9
195.37 21.01	UGC 3956	U	.55± .06	.15					7293
41.53 −48.36		7.0±1.3		.76				233± 5	7530
073646.6+243513	PGC 21515		1.01	.27					

0739.8 +3901		.I..9..	1.00± .16						
180.32 25.60	UGC 3954	U	.14± .12	.28					
36.65 −34.28		10.0± .9		.10					
0736.5 +3908	PGC 21517		1.03	.07					

0739.9 +7110		.I..9..	1.14± .13					15.89±.1	2462± 10
144.18 29.47	UGC 3940	U	.00± .12	.10				128± 6	
29.76 −2.78		10.0± .8		.00				116± 5	2577
0734.4 +7117	PGC 21520		1.15	.00					2494

074015.1-703829		.SBS7?.	1.06± .05	172					
282.54 −21.58	ESO 59- 15	S (1)	.24± .05	.70	15.15 ±.14				
201.68 −34.63		7.0±1.7		.33					
074033.0-703124	PGC 21533	7.8±1.2	1.13	.12					

074017.4-013426		.SXS8*.	1.22± .05	125				14.72±.1	1461± 11
219.98 10.12	UGC 3964	UE (1)	.19± .05	.75	13.6 ±.3			171± 16	
65.58 −71.99		8.0± .5		.23				169± 12	1309
073745.3-012726	PGC 21535	6.4± .8	1.29	.10	12.60			2.03	1671

0740.3 +5754		.SXS3..	1.02± .06	2					
159.40 29.11	UGC 3953	U	.18± .05	.17	15.3 ±.2				
32.37 −15.79		3.0± .9		.25					
0736.2 +5801	PGC 21540		1.03	.09					

0740.4 +2316		.E.....	1.11± .16	50					
196.62 20.70	UGC 3960	U	.07± .08	.12	14.27 ±.19				
42.25 −49.48		-5.0± .8		.00					
0737.4 +2323	PGC 21542		1.11						

0740.4 +6904		.S..3..	1.07± .07	2					
146.60 29.56	UGC 3945	U (1)	.33± .06	.15	15.4 ±.2				
30.19 −4.84		3.0± .9		.45					
0735.2 +6911	PGC 21545	3.5±1.2	1.09	.16					

0740.4 +8347		.S?....	1.03± .05	135					
129.83 28.28	UGC 3890		.62± .04	.17	15.05 ±.20				2034±125
27.40 9.62	IRAS07284+8354			.93	13.08				2189
0728.4 +8354	PGC 21547		1.05	.31	13.94				2020

074029.4+135227		.SX.3..	1.33± .04	133				15.05±.3	8715± 10
205.76 17.03	UGC 3962	U	.45± .05	.25	14.0 ±.2				8613
47.49 −58.35	IRAS07377+1358	3.0± .8		.62				549± 7	8909
073740.6+135927	PGC 21552		1.35	.22	13.08			1.75	

074033.9+341349	IC 2203	.SB.6..	1.09± .06	160				15.44±.3	4525± 10
185.44 24.40	UGC 3958	U	.09± .05	.18	14.30 ±.19				4569±155
38.22 −38.90	IRAS07373+3420	6.0± .9		.13				191± 7	4500
073718.8+342048	PGC 21555		1.11	.04	13.97			1.42	4677

R.A. 2000 DEC. l b SGL SGB R.A. 1950 DEC.	Names PGC	Type S_T n_L T L	$\log D_{25}$ $\log R_{25}$ $\log A_e$ $\log D_o$	p.a. A_g A_i A_{21}	B_T m_B m_{FIR} B_T^o	$(B-V)_T$ $(U-B)_T$ $(B-V)_T^o$ $(U-B)_T^o$	$(B-V)_e$ $(U-B)_e$ m'_e m'_{25}	m_{21} W_{20} W_{50} HI	V_{21} V_{opt} V_{GSR} V_{3K}
074039.8+391359 180.14 25.81 36.77 -34.03 073716.4+392058	NGC 2424 UGC 3959 IRAS07372+3920 PGC 21558	.SBR3*/ R (1) 3.0± .5 3.5±1.1	1.58± .03 .81± .04 .88± .05 1.61	81 .30 1.12 .41	13.59 ±.15 13.69 ±.19 13.14 12.19	.98± .03 .41± .06 .74 .17	1.01± .02 .45± .05 13.48± .16 14.32± .22	14.10±.3 435± 9 425± 7 1.51	3109± 8 3252± 30 3114 3258
074058.4+552539 162.22 28.93 32.96 -18.20 073656.4+553238	 UGC 3957 PGC 21568	.E..... U -5.0± .8	1.08± .17 .00± .08 .79± .10 1.11	 .16 .00	14.0 ±.2 13.69	1.09± .03 .96	1.10± .02 13.40± .37 14.39± .90	 	10212± 37 10270 10301
0741.0 +6653 149.11 29.63 30.67 -6.97 0736.1 +6700	 UGC 3951 PGC 21573	.S..3.. U (1) 3.0±1.0 4.5±1.4	1.00± .08 .94± .06 1.01	20 .15 1.30 .47					
074112.9+474019 170.94 27.79 34.73 -25.78 073732.5+474720	 UGC 3961 PGC 21578	.SB.7.. U 7.0± .9	1.00± .08 .09± .06 1.03	140 .27 .12 .04	15.2 ±.2				
074114.4+273658 192.32 22.42 40.71 -45.25 073808.8+274400	 UGC 3969 IRAS07381+2743 PGC 21580	.S..6*. U 6.0±1.4	.99± .06 .75± .05 1.01	135 .20 1.10 .37	 15.68 ±.20 13.49 14.35			16.02±.3 464± 7 1.30	8130± 10 8079 8299
074118.2+341356 185.49 24.54 38.41 -38.86 073803.1+342058	IC 2204 UGC 3965 IRAS07380+3420 PGC 21581	RSBR2.. U 2.0± .8	1.02± .06 .00± .05 1.03	18 .18 .00 .00	 15.9 ±.4 15.69			15.33±.2 92± 7 79± 5 -.36	4664± 7 4639 4818
074120.1-551049 267.61 -15.34 196.09 -49.53 074013.0-550342	 ESO 163- 13 PGC 21582	.SBS9.. S (1) 9.0± .9 10.0±1.3	1.06± .08 .35± .07 1.11	173 .58 .36 .17					
0741.4 +4006 179.23 26.17 36.70 -33.14 0738.0 +4013	A 0738+40 UGC 3966 DDO 46 PGC 21585	.I..9.. U (1) 10.0± .8 9.0±1.0	1.24± .06 .01± .06 1.34± .06 1.26	 .22 .01 .00	13.9 ±.3 13.68	.41± .08 -.19± .09 .35 -.23	.38± .05 -.15± .06 16.10± .14 14.96± .42	14.16±.1 86± 6 71± 8 .48	361± 5 359 498
0741.5 +5141 166.45 28.52 33.86 -21.83 0737.7 +5149	 UGC 3963 PGC 21589	.S..8*. U 8.0±1.3	1.13± .07 .73± .06 1.15	54 .21 .90 .37					
0741.9 +1648 203.10 18.54 46.19 -55.49 0739.0 +1655	A 0739+16 UGC 3974 DDO 47 PGC 21600	.IBS9.. PU (1) 10.0± .4 9.0±1.2	1.49± .04 .02± .06 1.46± .06 1.50	 .10 .02 .01	13.6 ±.3 13.5 ±.7 13.50	.48± .19 -.10± .16 .45 -.12	.43± .05 -.12± .07 16.41± .14 15.88± .36	12.89±.1 79± 4 56± 6 -.62	270± 4 179 462
074206.4-852519 297.96 -26.04 205.19 -20.19 075148.0-851748	 ESO 5- 11 PGC 21606	.L..-./ S -2.7± .7	1.03± .05 .39± .04 1.05	115 .62 .00	14.84 ±.14				
074229.7+625648 153.64 29.70 31.60 -10.80 073757.1+630352	 UGC 3967 KUG 0737+630 PGC 21616	.SB.7?. U 7.0± .8	1.04± .05 .08± .05 1.06	 .16 .11 .04				15.59±.1 140± 8	5876± 11 5962 5939
074232.4+494841 168.60 28.38 34.46 -23.64 073846.9+495547	A 0738+49 UGC 3973 MK 79 PGC 21618	.SB.3.. U 3.0± .8	1.08± .06 .00± .05 .33± .07 1.11	 .24 .00 .00	13.9 v±.2 13.36 ±.19 12.79 13.32	.59± .01 -.52± .02 .49 -.59	.50± .01 -.73± .02 11.04± .26 14.17± .36	15.92±.1 218± 21 172± 7 2.59	6652± 8 6570± 30 6682 6754
0742.6 +1115 208.46 16.41 50.43 -60.56 0739.9 +1123	 CGCG 58- 28 PGC 21623			.08	15.4 ±.6				8801± 57 8689 9002
074240.8+651038 151.07 29.78 31.17 -8.61 073755.9+651743	A 0737+65 MK 78 PGC 21624		.62± .11 .26± .12 .63	 .13	 13.21				11194± 29 11288 11250
074244.3+181931 201.70 19.32 45.60 -53.99 073950.4+182640	 UGC 3980 IRAS07398+1826 PGC 21628	.S?.... 	1.11± .07 .74± .06 1.12	18 .12 1.10 .37	 15.28 ±.18 13.64 14.01			15.56±.3 495± 7 1.18	8402± 10 8315 8592
0742.8 +6614 149.84 29.81 30.97 -7.56 0738.0 +6622	 UGC 3968 PGC 21636	.SBR5.. U 5.0± .8	1.15± .05 .07± .05 1.17	 .14 .10 .03	 14.8 ±.3 14.50			15.54±.1 191± 8 1.00	6780± 11 6906±125 6879 6833

R.A. 2000 DEC. l　　b SGL　SGB R.A. 1950 DEC.	Names PGC	Type S_T　n_L T L	$\log D_{25}$ $\log R_{25}$ $\log A_\circ$ $\log D_\circ$	p.a. A_g A_i A_{21}	B_T m_B m_{FIR} B_T°	$(B-V)_T$ $(U-B)_T$ $(B-V)_T^\circ$ $(U-B)_T^\circ$	$(B-V)_\bullet$ $(U-B)_\bullet$ $m'_\bullet$ m'_{25}	m_{21} W_{20} W_{50} HI	V_{21} V_{opt} V_{GSR} V_{3K}
0742.9　+6009 156.84　29.62 32.23　-13.51 0738.6　+6017	 UGC　3971 PGC 21638	.S..0.. U .0± .9 	.98± .07 .25± .05 .99	103 .19 .19 	 15.41 ±.19 				
074318.7+521905 NGC 2426 165.80　28.88 UGC　3977 34.01　-21.17 073926.6+522614 PGC 21648		.E..... U -5.0± .8 	1.04± .09 .00± .04 1.07	 .18 .00 	 14.13 ±.15 13.86				 5858± 57 5904 5959
074330.4+225552 197.24　21.24 UGC　3987 43.46　-49.58 IRAS07405+2303 074031.0+230303 PGC 21654		.S?.... 	.95± .07 .71± .05 .96	49 .12 1.06 .35	 15.7 ±.2 13.12 14.47			 361± 5 1.11	15.94±.3 7261± 9 7191 7443
074331.8-514057 264.53　-13.50 ESO　208- 34 193.86　-52.75 IRAS07422-5133 074214.0-513342 PGC 21656		.SBT2P* Sr 2.1± .7 	1.02± .05 .42± .04 1.11	174 .93 .52 .21	 15.41 ±.14 12.30 13.93				 2578± 87 2351 2739
074332.6+313205 188.46　24.17 UGC　3986 39.91　-41.34 074021.7+313916 PGC 21657		.S..6*. U 6.0±1.5 	1.00± .08 1.02± .06 1.02	 .17 1.47 .50	 16.2 ±.2 14.50			 164± 5 1.26	16.26±.3 3736± 9 3700 3898
074336.6+494000 IC　471 168.79　28.53 UGC　3982 34.68　-23.75 073951.6+494710 PGC 21659		.E..... U -5.0± .9 	.78± .13 .00± .04 .82	 .24 .00 	 14.3 ±.2 13.98				 5519± 37 5554 5629
074343.2-564558 269.25　-15.73 ESO　163- 14 196.33　-47.92 IRAS07427-5638 074241.0-563842 PGC 21660		.SAS2*P S 2.0±1.2 	1.20± .07 .50± .05 1.27	73 .76 .62 .25	 12.22 				
0743.7　+5221 NGC 2429A 165.76　28.95 UGC　3983 34.07　-21.11 0739.9　+5229 PGC 21664		.S?.... 	1.17± .05 .58± .05 1.19	145 .22 .87 .29					 5451± 57 5497 5552
074350.0+493649 IC　472 168.86　28.56 UGC　3985 34.73　-23.79 074005.2+494400 PGC 21665		.SXT3.. U 3.0± .8 	1.21± .05 .20± .05 1.23	167 .24 .28 .10	 14.2 ±.2 13.67			 449± 8 2.29	16.06±.1 5682± 11 5667± 38 5716 5791
074352.9+565915 160.49　29.50 UGC　3981 33.04　-16.59 MK　81 073946.2+570626 PGC 21666		.L..... U -2.0± .8 	1.19± .06 .23± .03 1.17	73 .18 .00 	 14.12 ±.15 13.89				 3288 3351 3373
074408.4+291452 A　0741+29 190.88　23.56 UGC　3995 40.93　-43.51 074100.9+292205 PGC 21673		.S...P. R 	1.40± .04 .35± .05 1.40	85 .09 .52 .17	 13.28 ±.18 12.64			 453± 7 440± 6 2.35	15.16±.1 4750± 5 4769± 25 4705 4919
074413.5+313906 NGC 2435 188.39　24.35 UGC　3996 40.05　-41.19 074102.5+314620 PGC 21676		.S..1.. U 1.0± .8 	1.33± .04 .63± .05 1.35	36 .15 .65 .32	 13.74 ±.20 12.89				 4189±155 4153 4352
0744.3　+4744 170.97　28.33 UGC　3988 35.29　-25.59 0740.7　+4752 PGC 21680		.S..6*. U 6.0±1.5 	1.00± .08 1.02± .06 1.02	36 .25 1.47 .50					
0744.5　+6716 148.66　29.97 UGC　3979 30.92　-6.52 IRAS07395+6723 0739.5　+6723 PGC 21684		.S..4.. U 4.0± .9 	1.27± .03 .77± .04 1.29	153 .18 1.14 .39	 14.78 ±.18 12.91 13.43			 374± 8 1.20	15.02±.1 4061± 11 4032± 76 4162 4109
0744.6　-1307 230.73　5.50 106.08　-77.71 0742.3　-1300 PGC 21687		.SAS5*. E 5.0±1.3 	1.19± .07 .19± .08 1.34	20 1.53 .28 .09					
074437.8+402158 179.13　26.83 UGC　3997 37.33　-32.75 KUG 0741+404 074112.8+402913 PGC 21688		.I..9?. U 10.0±1.6 	1.07± .05 .00± .05 1.09	 .24 .00 .00					
074438.4-580914 270.60　-16.22 ESO　123- 23 196.77　-46.56 IRAS07436-5801 074341.0-580154 PGC 21690		.SBS6?/ S　　(1) 5.7±1.0 4.4± .9	1.34± .04 .83± .04 1.43	107 .96 1.22 .41	 14.97 ±.14 				

R.A. 2000 DEC.	Names	Type	$\log D_{25}$	p.a.	B_T	$(B-V)_T$	$(B-V)_e$	m_{21}	V_{21}
l b		S_T n_L	$\log R_{25}$	A_g	m_B	$(U-B)_T$	$(U-B)_e$	W_{20}	V_{opt}
SGL SGB		T	$\log A_e$	A_i	m_{FIR}	$(B-V)_T^o$	m'_e	W_{50}	V_{GSR}
R.A. 1950 DEC.	PGC	L	$\log D_o$	A_{21}	B_T^o	$(U-B)_T^o$	m'_{25}	HI	V_{3K}

0744.6 +5350		.S..9*.	1.00± .16						
164.09 29.27	UGC 3989	U	.00± .12	.19					
33.86 −19.64		9.0±1.2		.00					
0740.7 +5358	PGC 21691		1.02	.00					
074441.6+734916		.SB?...	1.04± .04	160					5100± 60
141.12 29.67	UGC 3972		.24± .04	.10	14.65 ±.18				5224
29.59 −.11	IRAS07387+7356			.35					5124
073839.2+735626	PGC 21693		1.05	.12	14.17				
0744.8 +7247		.SB.8..	1.15± .07	160				14.88±.1	2480± 6
142.29 29.76	UGC 3975	U	.07± .06	.10	14.7 ±.3			82± 7	
29.80 −1.11		8.0± .8		.09				59± 12	2600
0739.0 +7255	PGC 21698		1.16	.04	14.51			.33	2508
074456.2−414650		.SAS7..	1.16± .05						
255.72 −8.64	ESO 311− 7	S	.04± .05	1.35	14.22 ±.14				
186.43 −61.80		7.0± .8		.06					
074316.0−413930	PGC 21703		1.28	.02					
074511.3+075558		.S?....	1.28± .04	15				14.38±.3	5044± 10
211.87 15.53	UGC 4005		.72± .05	.13	14.6 ±.3				4970± 35
54.71 −63.29	IRAS07424+0803			1.09	13.15			469± 7	4915
074229.1+080317	PGC 21710		1.29	.36	13.36			.66	5247
074513.9+530429	NGC 2431	PSBS1*.	.97± .07	35					5679±155
164.98 29.27	UGC 3999	U	.03± .05	.22	14.26 ±.18				5727
34.13 −20.37		1.0± .9		.03					5779
074119.9+531145	PGC 21711		.99	.01	13.94				
074513.8+700159	A 0739+70	.SBS3..	1.30± .03	55	*		.73± .03	14.51±.1	3882± 10
145.47 29.95	UGC 3984	U (1)	.37± .04	.11	14.04 ±.19			339± 11	3906± 47
30.41 −3.81	IRAS07398+7009	3.0± .8	1.28± .14	.52	13.45		14.59± .36	325± 11	3994
073953.8+700913	PGC 21712	1.6± .9	1.31	.19	13.39			.94	3922
074514.2+110344		.S?....	.95± .07	165				16.06±.3	4897± 10
208.94 16.90	UGC 4006		.13± .05	.10	14.60 ±.18				4784
51.77 −60.48	IRAS07424+1110			.20	13.75			220± 7	5101
074228.6+111103	PGC 21713		.96	.07	14.28			1.71	
074515.4−712431	NGC 2466	.SAS5*.	1.19± .03		13.54 ±.13	.58± .01	.70± .01	14.49±.3	5364± 9
283.47 −21.48	ESO 59− 18	RS (1)	.05± .04	.74	13.42 ±.14	−.27± .10		172± 12	5284± 35
201.43 −33.79	IRAS07456−7117	5.3± .5	.82± .01	.08		.36	13.13± .03		5148
074539.0−711706	PGC 21714	1.7± .6	1.26	.03	12.64	−.42	14.23± .22	1.83	5462
0745.2 +5610		.SB.9..	1.05± .08	75				15.50±.1	988± 7
161.44 29.61	UGC 3998	U	.28± .06	.16					1048
33.41 −17.34		9.0± .9		.28				74± 8	1077
0741.2 +5618	PGC 21715		1.06	.14					
074528.5−674801		.SAR5..	1.05± .05						5100± 63
279.90 −20.10	ESO 59− 17	S (1)	.03± .05	.64	14.72 ±.14				4884
200.27 −37.27	FAIR 270	5.0± .8		.05					5216
074521.0−674036	PGC 21717	4.4± .8	1.11	.02	14.00				
0745.4 +0755		.S?....	.89± .11	134					4819± 35
211.90 15.60	MCG 1−20− 5		.24± .07	.13	14.8 ±.3				4695
54.86 −63.26	IRAS07427+0803			.36	13.35				5027
0742.7 +0803	PGC 21718		.90	.12	14.31				
074545.1+045852		.S..1?.	1.14± .13	90				14.76±.3	2768± 10
214.67 14.34	UGC 4010	U	.57± .12	.16	14.6 ±.3				2635
58.39 −65.80	IRAS07430+0506	1.0±1.8		.58	13.52			230± 7	2980
074306.0+050613	PGC 21730		1.15	.28	13.87			.61	
0745.9 +5059		.S..9..	1.09± .06	50					
167.36 29.09	UGC 4002	U	.47± .05	.21					
34.74 −22.37		9.0± .9		.48					
0742.1 +5107	PGC 21733		1.11	.24					
074621.9+481749		.SB.1..	1.06± .06	129					
170.42 28.75	UGC 4007	U	.49± .05	.24	14.67 ±.13				
35.50 −24.98	IRAS07427+4827	1.0± .9		.50	12.97				
074240.8+482510	PGC 21755		1.08	.24					
0746.3 +6227		.E.....	1.02± .11	135					
154.22 30.13	UGC 4001	U	.24± .05	.16	14.50 ±.15				
32.15 −11.18		−5.0± .9		.00					
0741.9 +6235	PGC 21756		.97						
0746.4 +5857		.S..1*.	1.15± .05	87					
158.26 29.98	UGC 4003	U	.31± .05	.17	14.64 ±.19				
32.94 −14.59		1.0±1.2		.31					
0742.2 +5905	PGC 21758		1.16	.15					

R.A. 2000 DEC. / l b / SGL SGB / R.A. 1950 DEC.	Names / PGC	Type / S_T n_L / T / L	$\log D_{25}$ / $\log R_{25}$ / $\log A_e$ / $\log D_o$	p.a. / A_g / A_i / A_{21}	B_T / m_B / m_{FIR} / B_T^o	$(B-V)_T$ / $(U-B)_T$ / $(B-V)_T^o$ / $(U-B)_T^o$	$(B-V)_e$ / $(U-B)_e$ / m'_e / m'_{25}	m_{21} / W_{20} / W_{50} / HI	V_{21} / V_{opt} / V_{GSR} / V_{3K}
074625.8-183254 235.67 3.18 131.66 -77.40 074412.1-182530	 ESO 560- 12 IRAS07442-1825 PGC 21759	.SBS6?/ SE (1) 5.7±1.0 6.7±1.3	1.25± .05 .56± .05 1.51	145 2.67 .83 .28	 12.26 				3304 3108 3522
074637.9+444728 174.35 28.16 36.49 -28.37 074304.5+445450	 UGC 4008 IRAS07430+4454 PGC 21767	.S..0.. U .0± .9 	1.12± .08 .22± .07 1.12	 .18 .16 	 13.35 				9219± 34 9235 9347
0746.7 -0547 224.53 9.54 79.73 -73.89 0744.3 -0540	 PGC 21768	.SBS7?. E (1) 7.0±1.8 6.4±1.6	1.22± .07 .08± .08 1.29	145 .66 .10 .04					
074653.0+390200 180.69 26.92 38.24 -33.93 074330.6+390924	NGC 2444 UGC 4016 PGC 21774	.RING.A R -2.0± .4 	1.07± .04 .16± .07 .46± .03 1.07	 .23 .00 	14.20 ±.14 13.91	1.01± .01 .53± .01 .91 .48	1.04± .01 .54± .01 11.99± .12 14.01± .29	14.80±.3 433± 34 341± 25 	4048± 17 3974± 32 4025 4177
074654.7+390101 180.71 26.92 38.25 -33.94 074332.3+390825	NGC 2445 UGC 4017 IRAS07435+3908 PGC 21776	.RING.B R 10.0± .4 	1.15± .03 .11± .07 .60± .02 1.17	 .23 .09 .06	13.9 ±.2 11.80 13.58	.61± .05 -.02± .08 .50 -.10	 12.41± .05 14.21± .31	14.54±.1 488± 9 330± 12 .90	4002± 8 3903± 36 3990 4143
074656.0+071744 212.66 15.64 56.25 -63.65 074414.5+072510	 UGC 4025 IRAS07442+0725 PGC 21779	.SB.3*. U 3.0± .9 	1.04± .06 .43± .05 1.05	37 .13 .60 .22	 14.6 ±.3 13.21 13.87			16.18±.3 335± 7 2.09	5068± 10 4942 4942 5279
0746.9 +5215 165.95 29.42 34.60 -21.10 0743.1 +5223	 UGC 4011 PGC 21782	.S..6*. U 6.0±1.5 	1.04± .06 .88± .05 1.06	143 .21 1.30 .44					
074702.1+413212 177.96 27.54 37.51 -31.50 074335.2+413936	 UGC 4018 PGC 21786	.E...*. U -5.0±1.2 	1.00± .19 .09± .08 1.01	154 .21 .00 					8489± 56 8491 8627
0747.1 +2756 192.46 23.75 42.32 -44.56 0744.0 +2804	A 0744+28 MCG 5-19- 4 PGC 21789	.S..4.. P (1) 4.0± .8 1.8± .9	.99± .10 .09± .07 1.00	 .09 .14 .05	14.1 ±.4 14.54 ±.18 14.19	.84± .04 .75 	.92± .04 13.65± .69	15.53±.1 379± 21 347± 15 1.29	8254± 8 8209± 59 8202 8428
074706.9+622256 154.31 30.22 32.25 -11.24 074238.3+623019	A 0742+62 MK 82 PGC 21790	 	.50± .14 .00± .12 .51	 .16 					5664± 60 5748 5733
074710.1+302921 189.82 24.58 41.30 -42.13 074401.0+303646	IC 475 MCG 5-19- 5 ARAK 140 PGC 21795	.S?.... 	.74± .14 .21± .07 .74	 .14 .16 	 15.4 ±.2 15.04				4246± 82 4205 4415
074720.2+265554 193.52 23.46 42.83 -45.51 074416.0+270320	NGC 2449 UGC 4026 PGC 21802	.S..2.. U 2.0± .9 	1.13± .05 .32± .05 1.14	137 .13 .39 .16	 14.26 ±.18 13.69			16.67±.3 230± 7 2.82	4778± 10 4809± 38 4725 4957
074725.9-780901 290.36 -23.71 203.20 -27.23 074934.0-780124	 ESO 17- 6 PGC 21804	.SBR3.. r 3.3± .8 	1.09± .09 .07± .07 1.16	 .68 .10 .04					
074729.1-694233 281.86 -20.67 200.68 -35.39 074736.0-693500	 ESO 59- 19 IRAS07476-6934 PGC 21809	.SBS3?. S (1) 3.0±1.3 3.3±1.3	1.09± .05 .50± .04 1.15	64 .65 .69 .25	 14.66 ±.14 13.24 				
074729.1+605559 155.99 30.21 32.62 -12.64 074307.4+610323	A 0743+61 UGC 4013 MK 10 PGC 21810	.S..3.. U (1) 3.0± .8 3.5±1.1	1.25± .03 .38± .04 .99± .07 1.27	130 .19 .53 .19	13.34 ±.19 13.95 ±.18 13.14 12.89	.68± .04 -.42± .09 .51 -.55	.66± .02 -.44± .04 13.78± .25 13.49± .27	14.70±.1 508± 15 1.62	8753± 12 8720± 29 8826 8822
074734.2-412706 255.68 -8.05 185.21 -61.87 074553.0-411936	 ESO 311- 12 PGC 21815	.S..0?/ S .0±1.7 	1.54± .04 .91± .05 1.63	14 1.49 .68 	 12.83 ±.14 				
074744.4+481322 170.55 28.96 35.77 -25.00 074403.5+482049	 UGC 4022 PGC 21819	.S..6*. U 6.0±1.3 	1.02± .06 .41± .05 1.05	133 .31 .60 .20	 15.35 ±.18 				

R.A. 2000 DEC.	Names	Type	logD$_{25}$	p.a.	B$_T$	(B-V)$_T$	(B-V)$_e$	m$_{21}$	V$_{21}$
l b		S$_T$ n$_L$	logR$_{25}$	A$_g$	m$_B$	(U-B)$_T$	(U-B)$_e$	W$_{20}$	V$_{opt}$
SGL SGB		T	logA$_e$	A$_i$	m$_{FIR}$	(B-V)$_T^o$	m'$_e$	W$_{50}$	V$_{GSR}$
R.A. 1950 DEC.	PGC	L	logD$_o$	A$_{21}$	B$_T^o$	(U-B)$_T^o$	m'$_{25}$	HI	V$_{3K}$

074748.8+621932		.SBS6..	1.06± .04					15.77±.1	5709± 11
154.38 30.30	UGC 4015	U	.11± .04	.16	15.1 ±.2				5792
32.34 -11.27	KUG 0743+624	6.0± .8		.15				197± 8	5778
074320.6+622657	PGC 21821		1.07	.05	14.73			.99	
074751.6-184454		.SBS4?.	1.43± .03	141					
236.01 3.37	ESO 560- 13	SE (1)	.79± .04	2.53					
132.25 -77.03	IRAS07456-1837	4.0±1.2		1.16	12.31				
074538.1-183724	PGC 21822	4.4±1.2	1.67	.39					
0748.0 +5900	A 0743+59	.SX.3..	1.31± .04	18	14.15 ±.15	.77± .03	.81± .03	14.66±.1	6494± 10
158.22 30.19	UGC 4020	U (1)	.31± .05	.13	14.4 ±.2			377± 11	6508± 59
33.13 -14.50	SBS 0743+591C	3.0± .8	.83± .02	.43		.64	13.79± .05	368± 11	6565
0743.8 +5908	PGC 21832	2.2± .8	1.32	.16	13.60		14.76± .28	.91	6575
074813.8-183819		.IB.9..	1.19± .07	5					921
235.96 3.50	ESO 560- 14	S (1)	.44± .07	2.44					
131.69 -76.96		10.0± .9		.33					725
074600.1-183048	PGC 21844	10.0±1.3	1.42	.22					1141
074819.3+341956		.SB.4*.	1.36± .04	63				14.34±.3	4415± 10
185.84 25.96	UGC 4029	U	.68± .05	.17	14.42 ±.18				4389
40.16 -38.38	IRAS07450+3427	4.0± .8		1.00	12.80			350± 7	4575
074504.7+342726	PGC 21847		1.37	.34	13.22			.78	
074820.1+231417		.S..6*.	.97± .05	165				15.83±.3	6852± 10
197.36 22.39	UGC 4031	U	.13± .04	.22	14.93 ±.19			261± 13	
44.91 -48.93		6.0±1.3		.20					6783
074520.6+232147	PGC 21849		.99	.07	14.48			1.28	7039
0748.5 +6520		.SB.2..	1.03± .06	50					
150.88 30.40	UGC 4021	U	.11± .05	.10	14.89 ±.20				
31.74 -8.32		2.0± .9		.14					
0743.8 +6528	PGC 21853		1.04	.06					
0748.5 +6211		.S..6*.	1.19± .05	115				15.62±.1	1716± 11
154.54 30.38	UGC 4024	U	.63± .05	.17	14.78 ±.18				1799
32.46 -11.39		6.0±1.3		.92				208± 8	1786
0744.1 +6219	PGC 21854		1.20	.31	13.69			1.62	
074833.4-662942		.SBR4*.	1.03± .08						
278.76 -19.31	ESO 89- 3	S (1)	.00± .07	.71					
199.46 -38.45		4.0±1.2		.00					
074816.0-662206	PGC 21855	4.4± .9	1.10	.00					
074835.1+300909		.S..6*.	1.00± .08	122				15.80±.3	8122± 9
190.28 24.77	UGC 4032	U	.84± .06	.15	15.9 ±.2				8079
41.83 -42.36		6.0±1.4		1.24				358± 5	8293
074526.7+301640	PGC 21857		1.01	.42	14.46			.92	
0748.6 +5549		.E?....							10670± 60
161.90 30.05	MCG 9-13- 57			.17	15.3 ±.3				10729
33.98 -17.58									10763
0744.6 +5557	PGC 21859								
074839.9+543638	NGC 2446	.S..3..	1.29± .04	130				14.60±.1	5672± 7
163.29 29.94	UGC 4027	U (1)	.29± .05	.15	13.73 ±.18				5575±155
34.28 -18.76	IRAS07446+5444	3.0± .8		.39	12.53			468± 8	5726
074441.9+544408	PGC 21860	1.5±1.1	1.30	.14	13.15			1.31	5769
074844.7+661154		.L.....	1.16± .06	136					
149.89 30.40	UGC 4023	U	.28± .03	.11	14.11 ±.15				
31.56 -7.48		-2.0± .8		.00					
074355.1+661922	PGC 21864		1.13						
074918.7-542814		.IBS9P.	1.39± .05	162				14.02±.3	1047± 9
267.49 -14.00	ESO 163- 19	S (1)	.22± .07	.70					820
194.08 -49.83		10.0± .8		.16				51± 7	1205
074808.0-542036	PGC 21889	7.8± .8	1.46	.11					
074919.2+305406		.S..7..	1.04± .08	34				15.94±.3	3971± 10
189.55 25.15	UGC 4034	U	.59± .06	.14	15.52 ±.18				3931
41.73 -41.60		7.0± .9		.81				199± 7	4141
074609.7+310140	PGC 21891		1.05	.29	14.56			1.09	
074924.6+742001	A 0743+74	.L..-*.	.92± .06	122	14.4 ±.2	.80± .03	.81± .02		
140.49 29.95	UGC 4014	U	.28± .04	.10	14.54 ±.19	.27± .05	.36± .03		3907± 31
29.79 .46	MK 11	-3.0±1.3	.36± .07	.00		.73	11.71± .24		4032
074317.0+742730	PGC 21896		.89		14.31	.26	13.18± .38		3931
074936.1+265428		.S?....	.98± .07	83				16.00±.3	7065± 10
193.73 23.93	UGC 4038		.42± .05	.09	15.17 ±.18				7010
43.53 -45.37				.63				423± 7	7245
074632.1+270203	PGC 21900		.98	.21	14.41			1.38	

R.A. 2000 DEC.	Names	Type	logD25	p.a.	BT	(B-V)T	(B-V)e	m21	V21
l b		ST nL	logR25	Ag	mB	(U-B)T	(U-B)e	W20	Vopt
SGL SGB		T	logAe	Ai	mFIR	(B-V)To	m'e	W50	VGSR
R.A. 1950 DEC.	PGC	L	logDo	A21	BTo	(U-B)To	m'25	HI	V3K
074941.3+295632		.S..3*.	1.11± .05	140				15.43±.3	3706± 10
190.58 24.93	UGC 4039	U (1)	.79± .05	.18	15.33 ±.18				
42.22 -42.49		3.0±1.3		1.08				297± 7	3662
074633.1+300407	PGC 21909	3.5±1.3	1.13	.39	14.05			.99	3879
0749.7 +6120		.SXR2..	.95± .07					16.28±.1	5935± 11
155.53 30.51	UGC 4033	U (1)	.09± .05	.19	14.82 ±.18				
32.80 -12.18		2.0± .9		.11				251± 8	6014
0745.4 +6128	PGC 21914	4.5±1.1	.97	.04	14.46			1.77	6009
074950.8+335743	IC 2207	.S..6*.	1.31± .04	124				14.84±.3	4794± 7
186.34 26.16	UGC 4040	U	.83± .05	.16	15.17 ±.19				
40.68 -38.64	IRAS07466+3405	6.0±1.3		1.22	13.27			446± 7	4766
074636.9+340519	PGC 21918		1.32	.42	13.77			.65	4956
074951.3+184947		.S..6*.	1.14± .05	147				15.39±.3	4634± 10
201.91 21.07	UGC 4044	U	.31± .05	.16	14.6 ±.2				
47.94 -52.92		6.0±1.2		.45				199± 7	4548
074657.2+185723	PGC 21919		1.16	.15	13.98			1.26	4831
074958.7+555527	A 0745+56A	.S?....	.64± .17		15.7 ±.2	.85± .06			
161.80 30.24	MCG 9-13- 66		.28± .07	.16		.13± .11			10563± 41
34.14 -17.44	ARAK 141			.43	13.40	.69			10622
074556.7+560302	PGC 21924		.65	.14	15.06	.00	13.02± .87		10657
074959.3+300130		RSBR3*.	.95± .07	25				16.45±.3	8292± 10
190.52 25.02	UGC 4042	U	.08± .05	.18	14.75 ±.18				
42.27 -42.39		3.0± .9		.11				394± 7	8249
074651.1+300906	PGC 21927		.97	.04	14.41			2.00	8465
075008.8+552303		.E.....	.90± .22						5778± 38
162.42 30.22	UGC 4035	U	.00± .08	.15	14.07 ±.17				5835
34.30 -17.96		-5.0± .9		.00					5873
074608.6+553038	PGC 21939		.92		13.84				
075009.4+304358	A 0747+30	.SB.3..	1.22± .05	120	13.86 ±.13	.57± .02	.65± .02	15.38±.1	4282± 6
189.78 25.27	UGC 4047	U (1)	.20± .05	.13	14.0 ±.2	-.01± .04	-.02± .04	324± 9	4363± 59
42.02 -41.70	IRAS07469+3051	3.0± .8	.70± .02	.28	12.98	.47	12.85± .05	265± 5	4242
074700.2+305135	PGC 21940	1.1± .8	1.23	.10	13.45	-.08	14.30± .29	1.83	4454
075010.1+304103		.SB?...	.96± .09					16.17±.3	4434± 7
189.84 25.26	UGC 4046		.22± .06	.13					4393
42.05 -41.75				.32				82± 7	4605
074701.0+304840	PGC 21942		.97	.11					
075011.4+342454	A 0746+34	.S?....	.96± .17						8520± 40
185.87 26.35	UGC 4045		.00± .12	.17	15.2 ±.2				8494
40.60 -38.19				.00					8682
074656.8+343231	PGC 21944		.98	.00	14.93				
0750.3 +5524		.S?....	.94± .11						5975± 38
162.40 30.25	MCG 9-13- 69		.47± .07	.15	15.0 ±.2				6032
34.32 -17.93				.71					6070
0746.3 +5532	PGC 21950		.95	.24	14.14				
075018.5-725221		.SBS4..	1.07± .05	132					
285.08 -21.66	ESO 35- 17	S (1)	.25± .05	.82	15.53 ±.14				
201.44 -32.28	IRAS07509-7244	4.0± .8		.37	13.21				
075056.1-724436	PGC 21951	3.3± .9	1.15	.13					
0750.5 +5554	A 0745+56B	.S?....	.94± .11		15.3 ±.2	.63± .09			
161.83 30.32	MCG 9-13- 71		.00± .07	.16	14.89 ±.19	-.09± .13			
34.23 -17.44				.00					
0746.5 +5602	PGC 21961		.95	.00			14.83± .60		
0750.7 +5421		.S..7..	1.32± .04	5				15.19±.1	3402± 7
163.61 30.22	UGC 4043	U	.95± .05	.13	14.79 ±.18				
34.65 -18.93	IRAS07468+5429	7.0± .9		1.31	12.79			419± 8	3454
0746.8 +5429	PGC 21970		1.33	.47	13.33			1.38	3501
075048.3+742132	A 0744+74	.SXS5$.	1.05± .04	10	13.1 ±.3	.43± .04		14.46±.1	3952± 6
140.45 30.04	UGC 4028	R	.10± .04	.10	13.1 ±.3	-.39± .06		259± 7	4030± 20
29.87 .51	MK 12	5.0± .6		.15		.36		213± 6	4083
074441.0+742906	PGC 21971		1.06	.05	12.81	-.44	12.95± .38	1.60	3982
075056.0+235333		.S..3..	1.46± .03	175				15.04±.1	2121± 7
196.93 23.18	UGC 4054	U (1)	.67± .04	.29	14.3 ±.3			252± 8	2156± 76
45.42 -48.11		3.0± .9		.92					2054
074755.9+240113	PGC 21976	3.5±1.2	1.49	.33	13.05			1.65	2310
0750.9 +1758		.S..8..	1.00± .16					15.99±.3	8491± 11
202.87 20.99	UGC 4058	U	.00± .12	.19				226± 7	
48.91 -53.61		8.0± .8		.00				183± 7	8402
0748.1 +1806	PGC 21978		1.02	.00					8691

R.A. 2000 DEC. l b SGL SGB R.A. 1950 DEC.	Names PGC	Type S_T n_L T L	$\log D_{25}$ $\log R_{25}$ $\log A_o$ $\log D_o$	p.a. A_g A_l A_{21}	B_T m_B m_{FIR} B_T^o	$(B-V)_T$ $(U-B)_T$ $(B-V)_T^o$ $(U-B)_T^o$	$(B-V)_\bullet$ $(U-B)_\bullet$ m' m'_{25}	m_{21} W_{20} W_{50} HI	V_{21} V_{opt} V_{GSR} V_{3K}
075108.1+340323 186.32 26.44 40.97 -38.47 074754.1+341104	 UGC 4055 PGC 21983	.S..7.. U 7.0± .9 	1.06± .06 .09± .05 1.07	135 .16 .12 .04	 15.2 ±.3 14.87			15.97±.3 188± 7 1.05	4748± 7 4720 4912
075117.8+501047 168.42 29.83 35.85 -22.96 074732.7+501827	 UGC 4051 ARAK 142 PGC 21990	.E?.... 	.72± .12 .14± .03 .71	 .21 .00 	 14.6 ±.3 14.34				6109±155 6145 6223
075126.2+140111 206.77 19.51 51.96 -57.16 074837.6+140854	 UGC 4060 IRAS07486+1409 PGC 22002	.S?.... 	1.08± .06 .52± .05 1.08	104 .09 .78 .26	 14.68 ±.19 13.22 13.78			15.24±.3 410± 7 1.20	4678± 10 4574 4885
075135.6+425249 176.68 28.65 38.04 -29.99 074806.7+430031	 UGC 4056 KUG 0748+430 PGC 22008	.SXS5.. U 5.0± .9 	1.04± .04 .20± .04 1.06	30 .20 .30 .10	 14.81 ±.18 				
0751.7 +7102 144.27 30.43 30.71 -2.70 0746.3 +7110	 CGCG 331- 18 PGC 22020	 	.84± .10 .06± .10 .85	 .10 	 15.47 ±.18 				6519±125 6633 6557
075154.7+730058 141.99 30.26 30.26 -.78 074605.7+730837	NGC 2441 UGC 4036 IRAS07460+7308 PGC 22031	.SXR3*. R (2) 3.0± .5 2.5± .7	1.30± .03 .06± .03 1.31	 .09 .08 .03	13.0 ±.2 12.70 ±.14 12.85 12.60	.81± .07 .76 	 14.22± .27	15.08±.3 156± 34 128± 25 2.45	3470± 17 3590± 58 3600 3510
075155.2+271803 193.51 24.55 44.03 -44.83 074850.8+272547	 UGC 4061 PGC 22032	.S..3.. U (1) 3.0±1.0 5.5±1.4	1.07± .14 1.01± .12 1.08	161 .12 1.38 .50	 15.97 ±.20 14.41			15.54±.3 503± 7 .63	7903± 10 7849 8085
075213.2+302138 190.33 25.58 42.75 -41.92 074904.7+302923	 UGC 4064 PGC 22042	.I..9*. U 10.0±1.2 	1.14± .13 .39± .12 1.15	154 .13 .29 .20				16.14±.3 152± 7 	4306± 10 4264 4480
0752.3 +7335 141.32 30.23 30.16 -.22 0746.4 +7343	 UGC 4037 PGC 22044	.S..6*. U 6.0±1.4 	1.04± .08 .76± .06 1.05	157 .10 1.12 .38					
075238.9+733009 141.42 30.26 30.20 -.30 074644.2+733751	 UGC 4041 IRAS07467+7337 PGC 22050	.E?.... 	.95± .10 .39± .04 .85	132 .10 .00 	 13.9 ±.3 12.04 13.77				3449 3571 3477
0752.6 +5024 168.20 30.07 36.02 -22.68 0748.9 +5032	 UGC 4062 PGC 22052	.S..4.. U 4.0±1.0 	1.07± .07 .79± .06 1.09	135 .22 1.17 .40					
0752.6 +6232 154.14 30.86 32.84 -10.94 0748.2 +6240	 UGC 4059 PGC 22053	.S..6*. U 6.0±1.2 	1.06± .08 .26± .06 1.07	50 .15 .38 .13	 15.2 ±.2 14.61			15.32±.1 248± 8 .58	6572± 11 6655 6643
0752.9 +4009 179.77 28.31 39.20 -32.54 0749.5 +4017	 UGC 4068 PGC 22063	.S..6*. U 6.0±1.3 	1.16± .07 .88± .06 1.18	14 .24 1.30 .44					
0753.0 +6805 147.68 30.75 31.53 -5.54 0748.0 +6813	 MCG 11-10- 35 PGC 22068	.S?.... 	.80± .08 .17± .06 .82	 .16 .26 .09	 15.13 ±.19 14.69				4011±125 4115 4061
0753.1 +5514 162.63 30.63 34.77 -17.99 0749.1 +5522	 UGC 4065 IRAS07491+5522 PGC 22072	.S..7.. U 7.0±1.0 	1.20± .05 .87± .05 1.21	159 .14 1.20 .44	 15.41 ±.18 12.56 14.05			14.96±.1 510± 8 .48	7482± 11 7538 7580
075326.3+501343 168.42 30.17 36.20 -22.82 074941.5+502132	 UGC 4070 PGC 22093	.SXR1*. U 1.0± .9 	1.00± .06 .10± .05 1.01	135 .18 .10 .05	 14.73 ±.18 				
075329.1+143643 206.41 20.20 52.31 -56.41 075040.0+144434	 UGC 4077 PGC 22096	.SBT7*. U 7.0± .9 	.97± .09 .02± .06 .98	 .12 .03 .01	 14.56 ±.19 14.40			16.03±.3 165± 7 1.62	4665± 10 4563 4874

R.A. 2000 DEC.	Names	Type	$\log D_{25}$	p.a.	B_T	$(B-V)_T$	$(B-V)_e$	m_{21}	V_{21}
l b		S_T n_L	$\log R_{25}$	A_g	m_B	$(U-B)_T$	$(U-B)_e$	W_{20}	V_{opt}
SGL SGB		T	$\log A_e$	A_i	m_{FIR}	$(B-V)^o_T$	m'_e	W_{50}	V_{GSR}
R.A. 1950 DEC.	PGC	L	$\log D_o$	A_{21}	B^o_T	$(U-B)^o_T$	m'_{25}	HI	V_{3K}

0753.6 +5252		.SBS3..	1.07± .06	17					
165.37 30.51	UGC 4072	U	.35± .05	.13	14.57 ±.18				
35.49 -20.25	IRAS07498+5300	3.0± .9		.49	13.06				
0749.8 +5300	PGC 22110		1.08	.18					
075351.5-721253		.S..4..	1.10± .05	103					
284.54 -21.17	ESO 59- 22	S (1)	.66± .04	.83	15.42 ±.14				
200.92 -32.83		4.0± .9		.97					
075420.0-720454	PGC 22119	3.3±1.0	1.17	.33					
075405.5+742312	A 0748+74	.S..1..	1.40± .04	55					
140.39 30.25	UGC 4057	U	.52± .05	.10	13.40 ±.18				
30.08 .59		1.0± .8		.53					
074759.4+743059	PGC 22127		1.41	.26					
075410.7+552939	NGC 2456	.E.....	1.04± .18	30					7572± 38
162.35 30.80	UGC 4073	U	.13± .08	.12	14.13 ±.15				7629
34.85 -17.71		-5.0± .8		.00					7670
075011.0+553730	PGC 22129		1.02		13.90				
075418.5+541519		.SAT6..	.99± .06						7186± 38
163.79 30.73	UGC 4074	U	.00± .05	.10	13.84 ±.19				7238
35.21 -18.90	IRAS07504+5423	6.0± .8		.00					7288
075022.7+542311	PGC 22134		1.00	.00	13.71				
075420.5+042734	NGC 2470	.S..2..	1.29± .03	128	13.64 ±.14	.94± .05	.98± .05		4114± 38
216.15 16.02	UGC 4091	U	.52± .05	.05	13.9 ±.3	.44± .03	.45± .03		3977
63.41 -65.05	IRAS07517+0435	2.0± .8	.68± .03	.64	13.44	.80	12.53± .08		4337
075142.1+043528	PGC 22137		1.30	.26	12.94	.32	13.66± .23		
0754.4 +1344		.S..1..	1.04± .08	35					
207.36 20.05	UGC 4089	U	.67± .06	.10	15.30 ±.18				
53.39 -57.10		1.0±1.0		.68					
0751.6 +1352	PGC 22140		1.05	.33					
075425.0+392217		.S?....	.79± .08						5864±125
180.71 28.42	MCG 7-16- 23		.06± .06	.21	14.88 ±.19				5857
39.80 -33.20				.08					6016
075102.8+393010	PGC 22141		.81	.03	14.55				
075427.0+161249		.I..9*.	.96± .09					17.06±.3	4754± 7
204.95 21.06	UGC 4086	U	.00± .06	.10					4658
51.45 -54.87		10.0±1.2		.00				84± 7	4962
075136.0+162043	PGC 22143		.97	.00					
0754.4 +3839		.I..9..	1.00± .08	58					
181.51 28.26	UGC 4081	U	.43± .06	.20					
40.07 -33.89		10.0± .9		.32					
0751.1 +3847	PGC 22144		1.02	.22					
075427.7+581615		.S?....	.89± .11						6086± 63
159.12 30.99	MCG 10-12- 11		.35± .07	.14	15.37 ±.18				6153
34.15 -15.01	MK 1411			.43					6174
075018.4+582407	PGC 22145		.90	.18	14.73				
075433.3+294236		.I..9*.	1.04± .08	150				15.66±.3	4995± 10
191.19 25.86	UGC 4084	U	.59± .06	.11	15.47 ±.18				4950
43.68 -42.37		10.0±1.3		.44				226± 7	5174
075125.9+295030	PGC 22151		1.05	.29	14.92			.45	
0754.5 +6011		.S..6*.	1.04± .06	160					
156.89 31.06	UGC 4075	U	.96± .05	.16					
33.67 -13.16		6.0±1.5		1.41					
0750.3 +6019	PGC 22153		1.05	.48					
075450.2+500214		.E...*.	1.04± .18						6743± 38
168.67 30.37	UGC 4082	U	.04± .08	.16	14.22 ±.15				6778
36.48 -22.94		-5.0±1.2		.00					6861
075106.0+501008	PGC 22162		1.06		13.96				
0754.9 +7237		.SB.6*.	1.12± .05	165				15.24±.1	3462± 11
142.42 30.52	UGC 4067	U	.25± .05	.09	15.1 ±.2				
30.57 -1.11		6.0±1.2		.37				211± 8	3581
0749.2 +7245	PGC 22167		1.12	.13	14.62			.50	3495
0754.9 +5407		.S?....	.74± .14						7585± 38
163.96 30.80	MCG 9-13- 88		.21± .07	.10	15.3 ±.2				7636
35.34 -19.01				.31					7688
0751.0 +5415	PGC 22169		.75	.10	14.82				
075504.1-762441		.SAS5*/	1.51± .04	135					
288.75 -22.70	ESO 35- 18	S (1)	.66± .05	.92	14.08 ±.14				
202.22 -28.79	IRAS07564-7616	5.3± .7		.99	13.06				
075631.0-761636	PGC 22174	5.0± .6	1.59	.33					

R.A. 2000 DEC. / l b / SGL SGB / R.A. 1950 DEC.	Names / PGC	Type / S_T n_L / T / L	$\log D_{25}$ / $\log R_{25}$ / $\log A_e$ / $\log D_o$	p.a. / A_g / A_i / A_{21}	B_T / m_B / m_{FIR} / B_T^o	$(B-V)_T$ / $(U-B)_T$ / $(B-V)_T^o$ / $(U-B)_T^o$	$(B-V)_e$ / $(U-B)_e$ / m' / m'_{25}	m_{21} / W_{20} / W_{50} / HI	V_{21} / V_{opt} / V_{GSR} / V_{3K}
075505.9+554213 162.12 30.95 34.93 -17.47 075105.7+555007	A 0751+55 UGC 4079 MK 84 PGC 22175	.S...P. R 	1.00± .06 .29± .05 1.01	3 .12 .44 .15	14.2 ±.2 13.9 ±.3 12.77 13.49	.47± .06 -.18± .08 .35 -.27	 13.31± .39	 	 6138± 60 6195 6235
075506.0+432914 176.14 29.40 38.55 -29.22 075136.3+433710	 UGC 4087 PGC 22176	.S..7.. U 7.0± .9 	1.02± .08 .19± .06 1.04	100 .24 .27 .10	 15.2 ±.2 				
075512.4-280959 244.97 .02 161.04 -71.33 075310.0-280200	 ESO 430- 1 IRAS07531-2802 PGC 22177	.SBR3?. S 3.3± .9 	1.27± .06 .01± .07 	 .02 .01	 12.06 				
0755.3 +5613 161.52 31.01 34.82 -16.97 0751.3 +5621	 UGC 4083 PGC 22185	.S..6*. U 6.0±1.5 	1.00± .16 1.02± .06 1.01	 .13 1.47 .50					
075520.3+531949 164.88 30.80 35.62 -19.75 075127.4+532745	 UGC 4085 IRAS07514+5327 PGC 22186	.S?.... 	1.02± .06 .19± .05 1.03	65 .10 .28 .09	 14.38 ±.18 12.58 13.96				 7327± 37 7376 7434
075522.9+264435 194.37 25.09 45.33 -45.09 075219.5+265233	IC 480 UGC 4096 IRAS07523+2652 PGC 22188	.S..4.. U 4.0± .9 	1.22± .06 .76± .06 1.23	168 .16 1.12 .38	 15.04 ±.18 12.12 13.73			15.41±.2 353± 15 334± 7 1.30	4626± 7 4569 4813
075525.0+391110 180.97 28.57 40.09 -33.32 075203.3+391907	A 0752+39 MCG 7-17- 1 MK 382 PGC 22190	.S?.... 	.88± .05 .06± .05 .90	 .21 .09 .03	 15.30 ±.18 14.96				10137± 32 10129 10290
075525.9-212029 239.15 3.58 140.56 -74.62 075315.0-211230	 ESO 561- 2 IRAS07532-2112 PGC 22192	.SBS5*. SE (1) 5.0± .5 9.8±1.6	1.34± .04 .18± .04 1.57	140 2.40 .28 .09	 13.43 			13.25±.3 175± 16 148± 12 	922± 11 719 1147
0755.5 +5612 161.54 31.04 34.85 -16.97 0751.5 +5620	 UGC 4088 PGC 22198	.S..6*. U 6.0±1.5 	1.00± .08 1.02± .06 1.01	 .13 1.47 .50					
075541.2+845538 128.48 28.42 27.51 10.81 074149.6+850316	 UGC 3993 PGC 22202	.L...?. U -2.0±1.6 	1.20± .14 .11± .08 1.21	35 .23 .00 	13.81 ±.16 13.51			15.49±.1 194± 11 186± 8 	4365± 9 4523 4348
075548.8+244221 196.52 24.50 46.52 -46.95 075248.0+245020	 UGC 4099 PGC 22205	.SBR3.. U 3.0± .8 	1.21± .05 .05± .05 1.22	100 .18 .06 .02	 13.9 ±.3 13.65			15.03±.2 319± 15 307± 7 1.36	4673± 7 4608 4865
075552.2-521827 266.00 -12.11 191.39 -51.46 075434.0-521024	NGC 2502 ESO 209- 8 PGC 22210	.LXS0.. S -2.0± .5 	1.31± .05 .20± .04 .78± .02 1.40	126 1.02 .00 	13.19 ±.13 13.09 ±.14 12.10	1.23± .01 .73± .02 .98 .46	1.26± .01 .77± .01 12.58± .08 14.12± .29		 1095± 15 867 1264
075556.4+850933 128.22 28.37 27.46 11.03 074131.7+851711	IC 469 UGC 3994 IRAS07419+8516 PGC 22213	.SXT2*. PU 1.5± .6 	1.34± .04 .34± .05 1.36	90 .22 .41 .17	 13.52 ±.18 12.87				2080± 39 2239 2063
075601.3+582552 158.94 31.20 34.31 -14.81 075151.7+583350	 UGC 4092 IRAS07517+5833 PGC 22219	.S..2.. U 2.0± .9 	1.04± .06 .38± .05 1.05	56 .14 .47 .19	 14.45 ±.20 13.78				5662±155 5730 5750
075606.6-681642 280.80 -19.38 199.26 -36.53 075559.1-680836	 ESO 59- 23 IRAS07560-6808 PGC 22224	.SBT5.. S (1) 5.0± .8 2.2± .9	1.16± .04 .31± .04 1.23	2 .65 .47 .16	 14.84 ±.14 13.79 				
075613.8+601819 156.75 31.27 33.84 -12.99 075156.9+602617	IC 2209 UGC 4093 MK 13 PGC 22232	.SBT3*. R 3.0± .5 	1.03± .03 .07± .04 .72± .03 1.05	145 .16 .09 .03	14.28 ±.15 14.41 ±.18 14.07	.57± .05 -.04± .07 .51 -.08	.52± .05 13.37± .09 14.12± .24	13.67±.1 420± 6 337± 5 -.43	1427± 10 1413± 25 1500 1506
0756.2 +1139 209.57 19.60 55.99 -58.71 0753.5 +1148	 UGC 4109 PGC 22235	.SBR3.. U 3.0± .8 	1.04± .15 .04± .12 1.04	 .05 .05 .02	 14.5 ±.2 				

R.A. 2000 DEC. l b SGL SGB R.A. 1950 DEC.	Names PGC	Type S_T n_L T L	$\log D_{25}$ $\log R_{25}$ $\log A_e$ $\log D_o$	p.a. A_g A_i A_{21}	B_T m_B m_{FIR} B_T^o	$(B-V)_T$ $(U-B)_T$ $(B-V)_T^o$ $(U-B)_T^o$	$(B-V)_e$ $(U-B)_e$ m'_e m'_{25}	m_{21} W_{20} W_{50} HI	V_{21} V_{opt} V_{GSR} V_{3K}
075618.3+780052 136.22 29.85 29.31 4.13 074905.6+780845	 UGC 4066 PGC 22238	.S..6*. U 6.0±1.1 	1.24± .05 .01± .05 1.25	 .14 .02 .01	 13.9 ±.2 13.73			14.91±.1 115± 6 93± 8 1.18	2298± 6 2435 2310
075624.2-784200 291.10 -23.48 202.87 -26.58 075840.1-783348	 ESO 17- 9 IRAS07586-7833 PGC 22244	.SA.2?. S 2.0± .9 	1.09± .06 .54± .05 1.13	42 .44 .66 .27	 12.83 				
075625.7+270048 194.17 25.40 45.50 -44.75 075322.0+270850	 UGC 4105 PGC 22246	.S?.... 	1.06± .06 .23± .05 1.07	85 .11 .34 .11	 14.67 ±.18 14.19			16.11±.3 394± 7 1.81	6565± 10 6509 6752
0756.6 +6336 152.90 31.30 33.01 -9.80 0752.1 +6344	 UGC 4094 PGC 22252	.SX.2.. U 2.0± .8 	1.01± .06 .05± .05 1.02	110 .11 .06 .03	 14.73 ±.19 14.49				6959±125 7046 7028
075645.6+395544 180.21 28.98 40.12 -32.53 075322.7+400346	NGC 2476 UGC 4106 PGC 22260	.E...?. U -5.0±1.7 	1.15± .08 .24± .04 .59± .03 1.12	135 .24 .00 	13.58 ±.13 13.39 ±.18 13.21	.94± .01 .47± .03 .85 .44	.99± .01 .54± .03 12.02± .10 13.73± .43		 3729± 24 3724 3882
075648.7+072839 213.62 17.92 60.67 -62.22 075407.2+073643	NGC 2485 UGC 4112 IRAS07541+0736 PGC 22266	.S..1.. U 1.0± .8 	1.20± .05 .00± .05 1.20	 .03 .00 .00	 13.07 ±.18 12.01 12.99			17.06±.2 275± 7 255± 6 4.07	4610± 7 4483 4832
075650.2+734716 141.05 30.52 30.41 .05 075053.6+735513	 UGC 4080 PGC 22268	.S..1.. U 1.0± .9 	1.02± .06 .18± .05 1.03	60 .10 .19 .09	 14.05 ±.20 				
075652.8+602100 156.70 31.35 33.91 -12.93 075235.8+602901	NGC 2460 UGC 4097 IRAS07525+6028 PGC 22270	.SAS1.. R (1) 1.0± .3 5.0±1.3	1.39± .03 .12± .04 .79± .02 1.41	40 .14 .12 .06	12.72 ±.13 12.52 ±.13 12.38 12.34	.91± .01 .33± .02 .84 .28	.99± .01 .42± .02 12.16± .06 14.24± .22	13.27±.1 372± 4 331± 5 .87	1451± 6 1442± 56 1525 1532
075653.6+663638 149.38 31.20 32.26 -6.88 075203.8+664438	 UGC 4095 IRAS07520+6644 PGC 22271	.S?.... 	.91± .07 .34± .05 .93	51 .15 .47 .17	 14.7 ±.2 12.38 14.00			15.00±.1 329± 11 311± 8 .82	4080± 9 4178 4137
075654.7-245429 242.38 2.03 151.64 -72.84 075448.0-244624	 ESO 494- 7 IRAS07547-2446 PGC 22272	.S..4*/ S 4.0±1.2 	1.44± .04 .66± .05 1.81	51 3.96 .98 .33	 14.09 ±.14 13.46 				
075701.8+493404 169.26 30.67 36.99 -23.30 075319.1+494206	 UGC 4107 IRAS07531+4942 PGC 22279	.SAT5.. U 5.0± .8 	1.15± .04 .01± .04 1.17	 .17 .02 .01	 13.73 ±.18 13.21 13.52			15.88±.1 145± 8 2.36	3504± 8 3539± 76 3537 3625
075702.5+142313 207.01 20.90 53.88 -56.23 075413.6+143117	 UGC 4115 PGC 22280	.IA.9.. U 10.0± .8 	1.26± .11 .25± .12 1.27	145 .11 .19 .13				14.06±.1 112± 6 83± 5 	338± 5 235 552
075710.7+234656 197.59 24.48 47.46 -47.69 075411.1+235500	NGC 2480 UGC 4116 IRAS07541+2354 PGC 22289	.SB?... 	1.13± .07 .31± .06 1.15	160 .19 .46 .15	 14.7 ±.2 13.50 14.03			16.85±.3 96± 7 2.67	2326± 7 2339± 79 2257 2521
075712.2+564036 161.00 31.29 34.95 -16.46 075309.2+564838	NGC 2463 MCG 10-12- 31 ARAK 145 PGC 22291	 	.64± .17 .00± .07 .65	 .14 	 15.1 ±.2 				8283± 60 8344 8379
075713.3+234601 197.61 24.48 47.48 -47.70 075413.7+235406	NGC 2481 UGC 4118 PGC 22292	.S?.... 	1.14± .05 .44± .05 1.16	18 .19 .65 .22	 13.9 ±.2 13.06			17.00±.3 149± 7 3.72	2332± 10 2182± 34 2252 2516
0757.2 +5831 158.83 31.37 34.45 -14.66 0753.1 +5840	 UGC 4104 PGC 22295	.S..1.. U 1.0±1.0 	1.13± .05 .81± .05 1.14	25 .10 .82 .40	 15.50 ±.18 				
075717.3+312808 189.51 26.95 43.62 -40.51 075407.6+313613	 UGC 4113 PGC 22297	.S..6*. U 6.0±1.5 	.96± .09 .98± .06 .97	66 .13 1.44 .49				16.15±.3 214± 5 	5274± 9 5235 5451

R.A. 2000 DEC. / l b / SGL SGB / R.A. 1950 DEC.	Names / PGC	Type / S_T n_L / T / L	logD_25 / logR_25 / logA_e / logD_o	p.a. / A_g / A_i / A_21	B_T / m_B / m_FIR / B_T^o	(B-V)_T / (U-B)_T / (B-V)_T^o / (U-B)_T^o	(B-V)_e / (U-B)_e / m'_e / m'_25	m_21 / W_20 / W_50 / HI	V_21 / V_opt / V_GSR / V_3K
075721.0+662616		.S?....	.96± .07	105				14.97±.1	4913± 9
149.57 31.26	UGC 4098		.23± .05	.14	14.60 ±.19			337± 11	
32.35 -7.04	IRAS07525+6634			.35	13.25			320± 8	5010
075232.5+663418	PGC 22301		.98	.12	14.08			.77	4971
075726.0+355621		.IB.9..	1.08± .06	136				15.58±.3	774± 9
184.66 28.18	UGC 4117	U	.31± .05	.21	15.2 ±.2				
41.78 -36.28		10.0± .9		.23				60± 5	753
075409.8+360426	PGC 22305		1.10	.15	14.76			.67	939
075727.5-191437		.SBS6P/	1.23± .04	105					
237.59 5.07	ESO 561- 3	SE	.75± .04	1.69					
132.43 -74.70	IRAS07551-1906	6.0±1.0		1.10	12.42				
075514.1-190630	PGC 22306		1.39	.37					
075746.3+323332	IC 2211	.S..1..	.90± .07	140				15.72±.3	5360± 10
188.36 27.35	UGC 4119	U	.25± .05	.21	14.6 ±.2				5340± 36
43.27 -39.45	IRAS07546+3241	1.0±1.0		.25	13.49			358± 7	5324
075435.2+324138	PGC 22314		.92	.12	14.12			1.48	5533
075756.3+250941	NGC 2486	.S..1..	1.22± .03	100	14.16 ±.15	.85± .04	.93± .03	16.08±.2	4648± 7
196.22 25.11	UGC 4123	U	.26± .05	.14	13.95 ±.19	.29± .08	.39± .05	430± 10	4603± 58
46.93 -46.35		1.0± .8	.80± .03	.26		.73	13.65± .10	413± 10	4584
075455.1+251748	PGC 22317		1.24	.13	13.62	.23	14.49± .24	2.33	4841
075759.0-141705		.SBS4..	1.12± .06	15					2287
233.37 7.72	MCG -2-21- 1	E (1)	.30± .05	1.05					
113.44 -74.73		4.0± .9		.44					2098
075540.0-140856	PGC 22319	3.1± .8	1.21	.15					2520
075759.4+525125	NGC 2474	.E.0...	.79?						5629± 34
165.47 31.15	MCG 9-13- 96	P	.00?	.12					5675
36.16 -20.10		-5.0±1.0		.00					5739
075408.4+525931	PGC 22321		.81						
075800.5+525144	NGC 2475	.E.1...	.90?		14.0 ±.2	.95± .03	1.03± .02		5583± 46
165.46 31.16	MCG 9-13- 97	P	.00?	.12		.54± .06	.58± .03		5629
36.17 -20.09		-5.0± .9	.68± .07	.00		.87	12.86± .24		5692
075409.4+525950	PGC 22322		.92		13.80	.54	13.49±1.59		
075802.5+562134	NGC 2468	.LA..*.	1.11± .11						
161.38 31.39	UGC 4110	P	.35± .07	.11					
35.16 -16.73		-2.0± .9		.00					
075400.7+562940	PGC 22325		1.07						
075804.4+564052	NGC 2469	.S..4P*	1.05± .06	160					3217± 82
161.00 31.41	UGC 4111	P	.17± .04	.11	13.5 ±.2				3278
35.07 -16.42	ARAK 147	4.0± .9		.25	11.81				3313
075401.6+564858	PGC 22327		1.06	.08	13.10				
075814.9-495106		.SBS6*/	1.79± .03	152	12.68 ±.14				1119± 8
264.01 -10.58	ESO 209- 9	S	.86± .05	.73	10.55			331± 9	
189.15 -53.53	IRAS07568-4942	6.0± .7		1.26					890
075650.0-494254	PGC 22338		1.86	.43	10.68				1296
075816.8+395002		.S..6*.	1.01± .05	12					
180.38 29.25	UGC 4120	U	.85± .04	.26	15.71 ±.20				
40.48 -32.53	KUG 0754+399	6.0±1.5		1.25					
075454.3+395810	PGC 22340		1.04	.43					
075820.2+250859	NGC 2487	.SB.3..	1.42± .04	115	13.23 ±.13	.73± .02	.82± .01	15.17±.2	4833± 4
196.27 25.19	UGC 4126	U	.09± .05	.14	13.1 ±.3	.13± .02	.24± .02	305± 8	4986± 54
47.05 -46.33	IRAS07553+2517	3.0± .7	1.01± .02	.13	13.46	.65	13.77± .04	273± 4	4771
075519.0+251708	PGC 22343		1.44	.05	12.90	.07	14.96± .25	2.22	5028
075828.1+374713	NGC 2484	.L...*.	.97± .13	145	14.1 ±.2	1.05± .05	1.07± .02		
182.68 28.83	UGC 4125	U	.09± .05	.19	14.71 ±.15	.61± .07	.70± .03		12990± 43
41.29 -34.47	ARAK 148	-2.0±1.2	.82± .06	.00		.88	13.70± .20		12976
075509.0+375522	PGC 22350		.98		14.11	.63	13.60± .70		13150
075830.0-142117	NGC 2501	.LXR0?.	1.12± .06	120					
233.50 7.80	MCG -2-21- 2	E	.15± .05	1.04					
113.76 -74.62	IRAS07561-1413	-2.0± .9		.00	13.21				
075611.1-141306	PGC 22354		1.21						
0758.6 +0801	NGC 2496	.E.....	1.15± .09	2	*				
213.30 18.56	UGC 4127	U	.08± .05	.00	13.95 ±.19				
60.81 -61.51		-5.0± .8		.00					
0755.9 +0810	PGC 22359		1.13						
0758.9 +5802	A 0754+58	.S..9*.	1.37± .05	10	15.20 ±.15	.51± .05	.59± .04	14.53±.1	1092± 6
159.41 31.57	UGC 4121	U (1)	.50± .06	.11		-.01± .07	-.04± .06	165± 7	
34.80 -15.08	DDO 48	9.0±1.2	1.00± .03	.51		.37	15.69± .06	145± 11	1158
0754.8 +5810	PGC 22369	8.0±1.5	1.38	.25	14.58	-.11	15.66± .33	-.30	1184

R.A. 2000 DEC. / l b / SGL SGB / R.A. 1950 DEC.	Names / PGC	Type / S_T n_L / T / L	$\log D_{25}$ / $\log R_{25}$ / $\log A_o$ / $\log D_o$	p.a. / A_g / A_i / A_{21}	B_T / m_B / m_{FIR} / B_T^o	$(B-V)_T$ / $(U-B)_T$ / $(B-V)_T^o$ / $(U-B)_T^o$	$(B-V)_e$ / $(U-B)_e$ / m'_e / m'_{25}	m_{21} / W_{20} / W_{50} / HI	V_{21} / V_{opt} / V_{GSR} / V_{3K}
0758.9 +5906 158.15 31.60 34.51 -14.04 0754.8 +5915	 UGC 4122 PGC 22373	.E..... U -5.0± .8 	1.20± .07 .15± .04 1.17	3 .12 .00 	 14.29 ±.18 14.08				6009± 79 6079 6096
0759.1 +5908 158.11 31.62 34.51 -14.01 0754.9 +5917	 UGC 4124 SBS 0754+592 PGC 22376	.L..... U -2.0± .8 	.98± .12 .21± .05 .96	150 .12 .00 	 15.19 ±.16 14.98				5958± 79 6028 6045
075907.2-003817 221.41 14.71 74.16 -68.16 075634.0-003004	NGC 2494 UGC 4141 IRAS07565-0030 PGC 22377	PSBT0.. UE .0± .6 	.95± .06 .10± .04 .96	95 .12 .07 	 14.1 ±.3 11.95 				
0759.1 +3147 189.28 27.42 43.97 -40.07 0756.0 +3156	 UGC 4131 PGC 22378	.SX.5.. U 5.0± .8 	1.07± .06 .13± .05 1.09	140 .17 .20 .07	 14.9 ±.3 				
075913.3+325454 188.07 27.75 43.48 -39.01 075601.8+330306	A 0756+33 UGC 4132 IRAS07560+3302 PGC 22381	.S..4.. U 4.0± .9 	1.22± .06 .58± .06 1.25	28 .25 .85 .29	 13.7 ±.2 11.58 12.56			14.91±.3 505± 7 2.06	5219± 10 5210±141 5185 5394
075917.3+242707 197.08 25.16 47.74 -46.89 075617.0+243520	 UGC 4135 PGC 22383	.S?.... 	1.00± .06 .62± .05 1.01	105 .15 .93 .31	 15.57 ±.18 14.46			15.32±.3 314± 7 .55	5837± 10 5771 6033
075922.2+180647 203.56 22.88 51.84 -52.65 075629.3+181500	 UGC 4140 PGC 22389	.S..4.. U 4.0± .9 	1.18± .05 .79± .05 1.18	144 .07 1.17 .40	 15.04 ±.18 13.77			15.54±.3 256± 7 1.37	4709± 10 4619 4918
075924.3+162514 205.25 22.23 53.13 -54.16 075633.3+163327	A 0756+16 UGC 4139 PGC 22391	.SAS5.. U (1) 5.0± .8 2.9± .8	1.13± .05 .05± .05 1.04± .05 1.13	 .07 .08 .03	13.8 ±.2 14.3 ±.2 13.79	.63± .04 .57 	.61± .03 14.45± .12 14.11± .35	15.42±.1 204± 6 180± 5 1.60	4889± 6 4848± 59 4792 5101
075925.6+551746 162.63 31.53 35.66 -17.70 075527.6+552557	 MCG 9-13- 99 MK 1412 PGC 22393	.S?.... 	.89± .11 .43± .07 .90	 .09 .64 .21	 15.1 ±.2 14.30				6266± 63 6321 6368
075929.7+270143 194.40 26.05 46.39 -44.49 075626.2+270956	NGC 2492 UGC 4138 PGC 22397	.L..-*. U -3.0±1.2 	1.01± .11 .02± .05 .78± .04 1.02	95 .08 .00 	13.70 ±.14 14.19 ±.15 13.75	1.00± .02 .92 	1.05± .01 13.09± .13 13.57± .60		 6663± 37 6606 6853
075932.9-013906 222.38 14.32 76.39 -68.77 075700.8-013051	 UGC 4149 PGC 22400	.SBS9*. UE (1) 9.0± .7 8.7± .8	1.11± .05 .27± .05 1.12	105 .14 .28 .14					
075938.7+245858 196.55 25.41 47.54 -46.37 075637.8+250712	NGC 2498 UGC 4142 IRAS07566+2507 PGC 22403	.SB.1*. U 1.0±1.2 	1.04± .06 .14± .05 1.05	113 .14 .14 .07	 14.32 ±.18 11.44 13.97			16.94±.3 273± 7 2.90	4720± 10 4738±155 4656 4915
075940.2+152316 206.30 21.88 54.05 -55.05 075650.4+153130	 UGC 4145 IRAS07568+1531 PGC 22404	.S..1.. U 1.0± .9 	1.15± .05 .33± .05 1.16	140 .10 .34 .17	 14.06 ±.18 11.79 13.57			16.66±.3 365± 7 2.93	4623± 10 4523 4838
0759.7 +1824 203.30 23.09 51.77 -52.34 0756.9 +1833	 UGC 4147 PGC 22408	.S..2.. U 2.0±1.0 	1.08± .06 .75± .05 1.08	66 .05 .92 .37	 15.28 ±.20 				
075952.6+053630 215.73 17.77 64.41 -63.31 075713.0+054446	NGC 2504 UGC 4152 IRAS07572+0544 PGC 22414	.S?.... 	.67± .10 .10± .05 .68	 .04 .14 .05	 14.9 ±.4 13.12 14.70			15.60±.3 111± 7 .85	2618± 10 2588± 38 2482 2844
075953.8+331723 187.71 27.98 43.49 -38.61 075641.8+332538	IC 2214 UGC 4143 IRAS07566+3325 PGC 22417	.SB.2.. U 2.0± .9 	.84± .08 .03± .05 .87	 .25 .03 .01	 14.50 ±.20 12.39 14.16				5888± 38 5856 6063
075954.5+472446 171.82 30.86 38.17 -25.22 075617.3+473300	 UGC 4136 PGC 22418	.S..1.. U 1.0± .9 	1.14± .05 .65± .05 1.16	142 .16 .66 .32	 14.96 ±.18 				

R.A. 2000 DEC. / l b / SGL SGB / R.A. 1950 DEC.	Names / PGC	Type / S_T n_L / T / L	$\log D_{25}$ / $\log R_{25}$ / $\log A_e$ / $\log D_o$	p.a. / A_g / A_i / A_{21}	B_T / m_B / m_{FIR} / B_T^o	$(B-V)_T$ / $(U-B)_T$ / $(B-V)_T^o$ / $(U-B)_T^o$	$(B-V)_e$ / $(U-B)_e$ / m'_e / m'_{25}	m_{21} / W_{20} / W_{50} / HI	V_{21} / V_{opt} / V_{GSR} / V_{3K}
0800.0 +5622 / 161.37 31.67 / 35.43 -16.63 / 0756.0 +5631	UGC 4133 / PGC 22428	.S..6*. / U / 6.0±1.4	1.28± .04 / 1.16± .05 / 1.29	164 / .11 / 1.47 / .50					8898± 79 / 8958 / 8996
080006.3+130904 / 208.54 21.07 / 56.15 -56.96 / 075718.9+131720	UGC 4154 / PGC 22433	.S..7.. / U / 7.0± .9	1.15± .05 / .36± .05 / 1.15	147 / .06 / .49 / .18	14.55 ±.20 / / / 13.98			15.78±.3 / 261± 7 / 1.63	4622± 10 / 4514 / 4841
080012.2+601719 / 156.77 31.76 / 34.33 -12.87 / 075556.4+602533	UGC 4128 / IRAS07559+6025 / PGC 22438	.S..6*. / U / 6.0±1.1	1.19± .05 / .18± .05 / 1.21	45 / .14 / .26 / .09	13.82 ±.18 / 12.95 / / 13.40			15.49±.1 / 313± 8 / 2.00	5985± 7 / 6060 / 6069
0800.2 +5621 / 161.39 31.69 / 35.46 -16.64 / 0756.2 +5630	UGC 4134 / PGC 22440	.SB?...	1.02± .08 / .38± .06 / 1.03	25 / .11 / .57 / .19	15.37 ±.18 / / / 14.63				8968± 79 / 9027 / 9067
080019.9+264210 / 194.81 26.12 / 46.80 -44.72 / 075716.8+265026	IC 485 / UGC 4156 / IRAS07572+2650 / PGC 22443	.S..1*. / U / 1.0±1.3	1.07± .14 / .62± .12 / 1.07	153 / .05 / .63 / .31	15.38 ±.18 / / / 14.60			16.34±.3 / 505± 7 / 1.44	8338± 10 / 8280 / 8530
080020.4+263657 / 194.90 26.10 / 46.85 -44.80 / 075717.5+264513	IC 486 / UGC 4155 / IRAS07572+2645 / PGC 22445	.SB.1.. / U / 1.0± .9	.97± .07 / .11± .05 / .98	145 / .05 / .12 / .06	14.60 ±.18 / / / 14.33			15.81±.3 / 376± 7 / 1.42	8057± 10 / 7999 / 8249
080023.7+421137 / 177.81 30.11 / 40.06 -30.17 / 075657.2+421953	UGC 4148 / KUG 0756+423 / PGC 22446	.S..7*. / U / 7.0±1.2	1.40± .03 / .87± .04 / 1.41	10 / .14 / 1.19 / .43	15.2 ±.2 / / / 13.82			14.68±.1 / 141± 16 / 135± 12 / .43	737± 11 / / 740 / 885
080023.8+394949 / 180.49 29.64 / 40.94 -32.41 / 075701.5+395805	NGC 2493 / UGC 4150 / PGC 22447	.LB.... / U / -2.0± .7	1.29± .03 / .00± .05 / .87± .04 / 1.31	.19 / .00	13.00 ±.15 / 12.91 ±.15 / / 12.70	1.03± .05 / .48± .03 / .95 / .45	1.05± .05 / .52± .03 / 12.84± .13 / 14.31± .26		3934± 38 / 3928 / 4090
080029.9+222356 / 199.31 24.71 / 49.34 -48.66 / 075732.2+223213	NGC 2503 / UGC 4158 / PGC 22453	.SXT4.. / U / 4.0± .8	1.02± .05 / .02± .04 / 1.05	.25 / .02 / .01	14.5 ±.2 / / / 14.19			16.41±.2 / 232± 6 / 221± 7 / 2.20	5506± 7 / 5442±125 / 5431 / 5708
080033.0+395016 / 180.49 29.68 / 40.97 -32.39 / 075710.7+395833	NGC 2495 / MCG 7-17- 8 / MK 383 / PGC 22457	.I?....	.60± .10 / .24± .06 / .61	.19 / .18 / .12	15.7 ±.2 / 15.8 ±.2 / / 15.38	.53± .09 / .57± .09 / .37 / .46	/ / 12.92± .55		8395± 48 / 8389 / 8551
0800.6 +1540 / 206.11 22.21 / 54.17 -54.68 / 0757.8 +1549	MCG 3-21- 7 / PGC 22460	.S?....	.89± .11 / .35± .07 / .90	.10 / .53 / .18	15.26 ±.20 / / / 14.60				4722± 38 / 4623 / 4937
080038.6-611558 / 274.44 -15.82 / 195.58 -42.95 / 075950.0-610736	ESO 124- 7 / PGC 22461	.SAR3.. / r / 3.3± .9	1.00± .06 / .18± .05 / 1.08	149 / .81 / .24 / .09	15.26 ±.14				
080044.4+371223 / 183.44 29.13 / 42.04 -34.87 / 075726.5+372041	UGC 4157 / PGC 22468	.I..9*. / U / 10.0±1.3	.96± .09 / .10± .06 / .98	120 / .25 / .07 / .05	15.2 ±.2 / / / 14.89				3955 / 3938 / 4119
080045.6+163239 / 205.26 22.58 / 53.53 -53.90 / 075754.5+164057	UGC 4162 / PGC 22469	.S?....	.94± .07 / .63± .05 / .95	28 / .06 / .94 / .31	15.4 ±.2 / / / 14.38			16.54±.3 / 367± 7 / 1.84	4369± 10 / 4273 / 4583
0800.7 +7921 / 134.65 29.81 / 29.16 5.49 / 0753.0 +7930	UGC 4103 / PGC 22471	.S..7.. / U / 7.0± .9	1.07± .06 / .51± .05 / 1.08	105 / .11 / .70 / .25				15.64±.1 / 155± 8	2133± 11 / 2274 / 2140
080049.8+273000 / 194.00 26.48 / 46.52 -43.94 / 075745.8+273818	IC 2217 / UGC 4160 / ARAK 149 / PGC 22476	.S?....	.78± .09 / .15± .05 / .78	80 / .07 / .23 / .08	14.9 ±.3 / 12.13 / / 14.55			16.64±.3 / 179± 7 / 2.01	5206± 10 / 5090±155 / 5151 / 5396
0800.9 +5907 / 158.14 31.86 / 34.76 -13.96 / 0756.8 +5916	UGC 4146 / PGC 22482	.S..6*. / U / 6.0±1.5	1.03± .06 / 1.00± .05 / 1.05	56 / .12 / 1.47 / .50					

R.A. 2000 DEC. l　b SGL　SGB R.A. 1950 DEC.	Names PGC	Type S_T　n_L T L	$\log D_{25}$ $\log R_{25}$ $\log A_e$ $\log D_o$	p.a. A_g A_i A_{21}	B_T m_B m_{FIR} B_T^o	$(B-V)_T$ $(U-B)_T$ $(B-V)_T^o$ $(U-B)_T^o$	$(B-V)_e$ $(U-B)_e$ m'_e m'_{25}	m_{21} W_{20} W_{50} HI	V_{21} V_{opt} V_{GSR} V_{3K}
080118.9+251723 196.37 25.87 47.87 -45.94 075817.7+252543	 UGC 4167 PGC 22495	.S..7.. U 7.0± .9 	.98± .07 .16± .05 .99	0 .12 .21 .08	 	 	 	15.48±.3 145± 10 	4553± 10 4489 4749
080123.1+152219 206.49 22.26 54.70 -54.86 075833.4+153040	 UGC 4170 PGC 22501	.E...*. U -5.0±1.2 	1.04± .18 .09± .08 1.03	140 .07 .00 	 14.22 ±.15 14.07	 	 	 	 4637± 38 4536 4854
080131.8+094228 212.03 19.94 60.16 -59.72 075848.1+095050	 UGC 4171 PGC 22506	.S..3.. U　(1) 3.0± .9 4.5±1.2	1.36± .04 .86± .05 1.36	113 .07 1.18 .43	 14.40 ±.18 13.11	 	 	15.84±.3 503± 7 2.30	4879± 10 4759 5104
080137.5+154236 206.18 22.44 54.51 -54.54 075847.4+155058	NGC 2507 UGC 4172 IRAS07587+1550 PGC 22510	.S..0P. P .0± .7 	1.39± .04 .13± .05 1.39	 .05 .10 	 13.2 ±.3 13.00	 	 	 	 4562± 38 4463 4778
080146.6+563316 161.17 31.92 35.62 -16.40 075745.0+564136	NGC 2488 UGC 4161 PGC 22520	.L..-*. U -3.0±1.2 	1.15± .15 .24± .08 1.01± .06 1.13	100 .11 .00 	13.40 ±.17 14.07 ±.15 13.54	1.00± .03 .88 	1.04± .02 13.94± .20 13.43± .81	 	 8598±155 8658 8697
080151.9+612448 155.45 31.95 34.20 -11.74 075731.6+613308	 UGC 4159 IRAS07573+6133 PGC 22524	.S?.... 	.98± .05 .25± .05 1.00	90 .18 .37 .12	 14.1 ±.2 13.57	 	 	 	 1591 1670 1671
080153.6+504418 168.00 31.57 37.43 -21.95 075808.9+505239	NGC 2500 UGC 4165 IRAS07581+5052 PGC 22525	.SBT7.. R　(2) 7.0± .3 6.0± .6	1.46± .02 .04± .03 1.47	 .14 .06 .02	12.20 ±.13 12.20 ±.14 12.14 12.00	.58± .02 -.22± .07 .54 -.25	 14.23± .18	13.57±.0 114± 4 98± 5 1.55	516± 5 437± 58 553 636
080157.6+083310 213.19 19.53 61.64 -60.62 075915.1+084133	NGC 2508 UGC 4174 PGC 22528	.E...?. U -5.0±1.5 	1.16± .10 .13± .05 1.12	130 .00 .00 	 13.71 ±.15 13.64	 	 	 	 4378± 38 4254 4605
0802.0　+1502 206.88 22.27 55.22 -55.08 0759.2　+1511	 UGC 4175 PGC 22531	.LB..?. U -2.0± .8 	1.00± .19 .05± .08 1.00	 .08 .00 	 14.9 ±.2 	 	 	 	
0802.0　+5638 161.07 31.95 35.62 -16.30 0758.0　+5647	 UGC 4164 PGC 22533	.S..0.. U .0± .9 	1.13± .05 .53± .05 1.11	140 .10 .40 	 15.11 ±.18 	 	 	 	
0802.0　+7301 141.87 30.99 30.97 -.58 0756.3　+7310	 UGC 4137 PGC 22534	.S..8.. U 8.0± .8 	1.14± .07 .20± .06 1.15	145 .09 .24 .10	 15.3 ±.3 	 	 	 	
080206.0+004830 220.45 16.05 72.86 -66.61 075931.3+005654	 UGC 4179 PGC 22539	.SB.3.. U 3.0± .8 	1.07± .06 .08± .05 1.10	 .24 .10 .04	 	 	 	15.37±.3 161± 7 	5563± 7 5413 5798
0802.1　+0928 212.32 19.97 60.66 -59.83 0759.4　+0937	NGC 2510 UGC 4178 PGC 22541	.L...*. U -2.0±1.2 	1.00± .19 .14± .08 .99	120 .06 .00 	 14.41 ±.15 	 	 	 	
080208.1+074026 214.05 19.19 62.75 -61.31 075926.5+074850	 UGC 4177 PGC 22542	.S..2.. U 2.0± .9 	1.04± .08 .47± .06 1.05	90 .09 .58 .24	 15.05 ±.18 14.29	 	 	16.30±.3 454± 7 1.78	9999± 10 9872 10228
080211.6+565633 160.72 31.98 35.56 -16.01 075808.8+570455	NGC 2497 UGC 4168 PGC 22547	.E?.... 	1.15± .15 .07± .08 1.15	 .10 .00 	 14.15 ±.17 13.92	 	 	 	 8173±155 8234 8271
080215.3+092342 212.42 19.97 60.80 -59.88 075931.9+093206	NGC 2511 MCG 2-21- 8 MK 1207 PGC 22549	.S?.... 	.94± .11 .40± .07 .95	 .11 .55 .20	 15.07 ±.19 14.38	 	 	 	 4467± 63 4345 4694
080222.3+065242 214.83 18.89 63.84 -61.92 075941.5+070107	 UGC 4183 PGC 22554	.SBR2.. U 2.0± .8 	1.23± .05 .18± .05 1.24	73 .13 .23 .09	 14.0 ±.2 13.52	 	 	15.76±.3 468± 7 2.15	9280± 10 9150 9510

R.A. 2000 DEC. l b SGL SGB R.A. 1950 DEC.	Names PGC	Type S_T n_L T L	$\log D_{25}$ $\log R_{25}$ $\log A_e$ $\log D_o$	p.a. A_g A_i A_{21}	B_T m_B m_{FIR} B_T^o	$(B-V)_T$ $(U-B)_T$ $(B-V)_T^o$ $(U-B)_T^o$	$(B-V)_e$ $(U-B)_e$ m'_e m'_{25}	m_{21} W_{20} W_{50} HI	V_{21} V_{opt} V_{GSR} V_{3K}
080225.0+092446 212.42 20.01 60.85 -59.84 075941.6+093311	NGC 2513 UGC 4184 PGC 22555	.E..... U -5.0± .7 	1.40± .11 .09± .08 1.04± .03 1.39	170 .11 .00 	12.59 ±.13 12.81 ±.19 12.48	.99± .01 .92 	1.00± .01 13.28± .10 14.34± .58	 	 4662± 60 4540 4889
080228.1-722525 285.03 -20.65 200.12 -32.43 080255.0-721654	 ESO 59- 24 PGC 22558	.SBS5*/ S (1) 5.0± .7 2.6± .8	1.14± .05 .69± .04 1.22	143 .81 1.04 .35	 15.48 ±.14 	 	 	 	
080232.3+612328 155.47 32.03 34.29 -11.74 075812.3+613150	A 0758+61 UGC 4169 IRAS07581+6131 PGC 22561	.S..6.. U 6.0± .8 	1.17± .04 .30± .04 1.18	140 .18 .44 .15	 13.63 ±.19 13.20 13.00	 	 	13.85±.1 217± 8 .70	1589± 7 1667 1669
080236.2+272615 194.21 26.84 47.06 -43.85 075932.4+273440	IC 2219 UGC 4180 IRAS07595+2734 PGC 22565	.S..5.. U 5.0± .9 	1.12± .05 .32± .05 1.12	175 .08 .48 .16	 14.39 ±.18 12.88 13.81	 	 	15.40±.2 352± 6 332± 7 1.44	5228± 7 5163±155 5172 5420
080243.3+404044 179.63 30.25 41.11 -31.46 075919.8+404909	 UGC 4176 KUG 0759+408 PGC 22575	.SB.7.. U 7.0± .9 	1.25± .03 .53± .04 1.26	112 .15 .74 .27	 14.9 ±.2 14.01	 	 	15.44±.1 202± 8 1.16	3086± 7 3084 3242
080247.2-121907 232.27 9.73 107.14 -73.22 080025.9-121040	NGC 2517 MCG -2-21- 3 PGC 22578	.LXT0*. PE -1.7± .8 	1.17± .08 .14± .05 .95± .06 1.21	70 .61 .00 	12.7 ±.2 	.93± .01 .47± .04 	.95± .01 .51± .02 12.96± .21 13.05± .46	 	
080249.8+154829 206.21 22.75 54.87 -54.31 075959.7+155655	NGC 2514 UGC 4189 IRAS08000+1556 PGC 22581	.SBS4.. PU (1) 4.0± .6 1.1± .8	1.11± .03 .02± .05 .78± .01 1.12	 .05 .02 .01	13.99 ±.13 13.97 ±.19 13.40 13.88	.56± .02 -.08± .03 .51 -.11	.64± .01 -.01± .03 13.38± .02 14.36± .23	15.54±.1 136± 5 119± 4 1.65	4856± 5 4893± 43 4757 5074
0802.9 +1618 205.72 22.97 54.50 -53.86 0800.1 +1627	 UGC 4190 PGC 22585	.L...*. U -2.0±1.3 	1.13± .10 .40± .05 1.08	32 .07 .00 	14.52 ±.15 	 	 	 	
080307.2+232333 198.51 25.62 49.55 -47.50 080008.4+233200	NGC 2512 UGC 4191 MK 384 PGC 22596	.SB.3.. U 3.0± .8 	1.14± .03 .18± .05 .63± .01 1.15	113 .17 .24 .09	13.85 ±.13 14.01 ±.18 11.67 13.45	.75± .01 .18± .02 .65 .10	.78± .01 .13± .02 12.49± .02 13.95± .23	16.64±.2 349± 8 324± 11 3.10	4702± 6 4649± 30 4630 4903
0803.2 +3047 190.65 27.97 45.51 -40.69 0800.1 +3056	 MCG 5-19- 36 PGC 22603	.S?.... 	.89± .11 .14± .07 .90	 .16 .21 .07	 14.95 ±.18 14.50	 	 	 	12160± 38 12118 12345
080324.0+415455 178.25 30.61 40.77 -30.25 075958.4+420322	 UGC 4188 PGC 22609	.L..... U -2.0± .8 	1.12± .10 .32± .05 1.09	115 .15 .00 	14.52 ±.15 	 	 	 	
080325.9+100259 211.92 20.51 60.56 -59.17 080041.8+101128	 UGC 4197 PGC 22611	.S..3*. U (1) 3.0±1.3 4.5±1.2	1.27± .04 .72± .05 1.27	133 .03 .99 .36	14.40 ±.18 	 	 	 	
080327.6-702158 283.09 -19.70 199.35 -34.34 080334.0-701324	 ESO 59- 25 IRAS08035-7013 PGC 22614	.SBS5*/ S (1) 5.0± .6 3.3± .9	1.19± .04 .65± .04 1.26	55 .73 .98 .33	15.18 ±.14 13.26 	 	 	 	
080328.1+250606 196.75 26.27 48.62 -45.91 080027.3+251434	A 0800+25 CGCG 118- 56 MK 385 PGC 22615	 	 	 .06 	15.16 ±.18 13.16 	 	 	18.50±.3 165± 6 	8300± 10 8100±110 8234 8497
080329.0+332744 187.75 28.75 44.30 -38.19 080017.1+333612	 MCG 6-18- 9 ARAK 151 PGC 22616	.L?.... 	.72± .22 .00± .07 .75	 .24 .00 	14.80 ±.19 14.38	 	 	 	11735± 82 11703 11913
080338.1+432036 176.62 30.90 40.29 -28.89 080009.9+432904	 UGC 4192 KUG 0800+434 PGC 22618	.S..7.. U 7.0±1.0 	1.08± .05 .84± .05 1.09	174 .16 1.16 .42	 	 	 	 	
0803.7 +0957 212.04 20.54 60.78 -59.21 0801.0 +1006	 UGC 4198 PGC 22621	.L...?. U -2.0±1.6 	1.18± .15 .13± .08 1.16	155 .03 .00 	14.2 ±.2 	 	 	 	

R.A. 2000 DEC. l　b SGL　SGB R.A. 1950 DEC.	Names PGC	Type S_T　n_L T L	$logD_{25}$ $logR_{25}$ $logA_o$ $logD_o$	p.a. A_g A_i A_{21}	B_T m_B m_{FIR} B^o_T	$(B-V)_T$ $(U-B)_T$ $(B-V)^o_T$ $(U-B)^o_T$	$(B-V)_o$ $(U-B)_o$ m'_o m'_{25}	m_{21} W_{20} W_{50} HI	V_{21} V_{opt} V_{GSR} V_{3K}
080356.3+084159 213.27　20.04 62.28　-60.22 080113.7+085030	 CGCG　59- 32 MK　　1208 PGC 22634			 .07 	 14.84 ±.18 13.02 				 4932± 97 4808 5161
080400.1+100035 212.02　20.62 60.83　-59.13 080116.1+100906	 CGCG　59- 34 MK　　1209 PGC 22638		.74± .12 .08± .06 .74	 .03 	 15.27 ±.13 				 10106± 97 9986 10334
0804.0　+8438 128.74　28.68 27.78　10.59 0751.0　+8446	 UGC　4078 IRAS07510+8446 PGC 22640	.S..4.. U 4.0± .9 	1.32± .04 .69± .05 1.34	82 .21 1.02 .35	 15.1　±.2 13.52 13.88			16.40±.1 213± 8 2.17	1860± 11 2017 1846
080406.0+050651 216.70　18.48 66.94　-63.06 080127.1+051522	 UGC　4203 MK　　1210 PGC 22641	.S?.... 	.91± .18 .00± .12 .92	 .07 .00 .00	 14.34 ±.18 12.76 14.24			16.69±.3 62± 7 2.45	4046± 10 4035± 60 3910 4279
080407.3+533257 164.73　32.12 36.88　-19.17 080015.4+534126	NGC　2505 UGC　4193 IRAS08001+5341 PGC 22644	.SB.1.. U 1.0± .9 	1.09± .06 .34± .05 1.10	0 .09 .35 .17	 14.09 ±.19 				
0804.1　+3010 191.38　27.98 46.05　-41.19 0801.0　+3019	 CGCG 148-107 PGC 22645			 .14 	 15.3　±.3 				 4148 4103 4335
080409.4+740256 140.66　30.98 30.83　　.44 075813.0+741121	NGC　2523A UGC　4166 KUG 0758+741 PGC 22649	.SBS5*. PU 5.0± .6 	1.02± .04 .17± .04 1.03	95 .10 .25 .08	 14.49 ±.18 14.12			15.63±.1 217± 8 1.43	3804± 11 3928 3834
0804.2　+5523 162.56　32.22 36.32　-17.41 0800.3　+5532	 MCG　9-13-116 PGC 22657	.SB?... 	.94± .11 .19± .07 .95	 .13 .26 .09	 14.93 ±.18 14.47				 9714± 54 9769 9819
080420.0+774902 136.36　30.30 29.76　　4.06 075717.4+775726	 UGC　4151 IRAS07572+7757 PGC 22660	.S..8*. U 8.0±1.1 	1.14± .05 .02± .05 1.15	 .16 .02 .01	 13.19 ±.19 13.10 13.01			15.88±.1 151± 7 124± 8 2.86	2286± 6 2422 2300
080421.2+613252 155.28　32.25 34.45　-11.52 080001.0+614121	 UGC　4196 IRAS08002+6141 PGC 22661	.S..7.. U 7.0± .9 	1.01± .06 .12± .04 1.02	165 .19 .17 .06	 14.92 ±.19 				
080431.1+401222 180.25　30.49 41.67　-31.79 080108.7+402054	 UGC　4200 KUG 0801+403 PGC 22670	PSBS4.. U　　　(1) 4.0± .9 1.1± .8	.96± .05 .09± .04 .97	5 .14 .13 .04	 14.54 ±.18 14.19				 12202± 59 12197 12360
0804.6　+6646 149.12　31.95 32.95　-6.51 0759.8　+6655	 UGC　4187 PGC 22674	.S..6*. U 6.0±1.3 	1.04± .06 .67± .05 1.06	75 .18 .98 .33					
080442.3+245217 197.10　26.46 49.13　-46.01 080141.9+250050	 UGC　4208 PGC 22677	.S..7.. U 7.0± .9 	.96± .17 .22± .12 .97	160 .07 .30 .11				16.56±.3 172± 7 	4971± 10 4905 5172
080442.8+352354 185.67　29.48 43.73　-36.29 080128.1+353226	 UGC　4201 PGC 22678	.S..6*. U 6.0±1.4 	1.04± .15 .88± .12 1.06	164 .25 1.30 .44				16.31±.3 373± 5 	8725± 9 8701 8898
0804.7　+1046 211.37　21.12 60.32　-58.39 0802.0　+1055	 UGC　4211 IRAS08020+1055 PGC 22680	.S..8*. U 8.0±1.2 	1.08± .06 .30± .05 1.08	33 .00 .37 .15	 14.53 ±.18 13.77 14.13				 10283± 60 10166 10511
080448.3+204126 201.46　25.03 51.85　-49.77 080152.8+205000	 UGC　4207 PGC 22681	.S..4?. U 4.0±1.6 	1.21± .04 .27± .04 1.22	125 .15 .39 .13	 14.2　±.2 13.64			15.35±.2 522± 13 521± 7 1.57	9360± 7 9279 9571
0805.0　+2942 191.95　28.03 46.53　-41.55 0801.9　+2951	 CGCG 148-109 PGC 22688			 .17 	 15.4　±.3 				 6500 6453 6689

R.A. 2000 DEC.	Names	Type	logD$_{25}$	p.a.	B$_T$	(B-V)$_T$	(B-V)$_e$	m$_{21}$	V$_{21}$
l b		S$_T$ n$_L$	logR$_{25}$	A$_g$	m$_B$	(U-B)$_T$	(U-B)$_e$	W$_{20}$	V$_{opt}$
SGL SGB		T	logA$_e$	A$_i$	m$_{FIR}$	(B-V)^o_T	m'	W$_{50}$	V$_{GSR}$
R.A. 1950 DEC.	PGC	L	logD$_o$	A$_{21}$	B^o_T	(U-B)^o_T	m'$_{25}$	HI	V$_{3K}$
0805.0 +8548		.SB.9..	1.04± .15						
127.45 28.38	UGC 4063	U	.08± .12	.24					
27.45 11.71		9.0± .9		.08					
0749.0 +8557	PGC 22691		1.06	.04					
080505.4+250346		.S..6*.	.99± .06	35				15.62±.3	5017± 10
196.93 26.60	UGC 4210	U	.80± .05	.07	15.7 ±.2				4952
49.13 -45.80		6.0±1.4		1.18				312± 7	
080204.7+251220	PGC 22693		1.00	.40	14.43			.79	5218
0805.1 +6646		.SBR3..	1.25± .06	20				15.38±.1	4888± 11
149.10 32.00	UGC 4195	U	.30± .06	.18	14.4 ±.2				4986
32.99 -6.49	IRAS08002+6655	3.0± .8		.41				372± 8	
0800.2 +6655	PGC 22695		1.27	.15	13.73			1.50	4948
080509.8-673510		.LXS0..	1.19± .08	119					
280.53 -18.32	ESO 89- 9	S	.15± .06	.71					
197.98 -36.89		-2.0± .8		.00					
080454.1-672630	PGC 22697		1.24						
080510.2+470305		.S..3..	1.14± .04	122					
172.37 31.70	UGC 4205	U (1)	.67± .04	.15	15.20 ±.18				
39.23 -25.29	KUG 0801+471	3.0± .9		.92					
080134.7+471139	PGC 22698	4.5±1.3	1.15	.33					
080526.6+464227		.L..-*.	1.08± .17	80					
172.78 31.71	UGC 4209	U	.17± .08	.12					
39.40 -25.60		-3.0±1.2		.00					
080151.9+465102	PGC 22707		1.07						
0805.4 +5558		.I..9*.	1.04± .15	130					
161.87 32.41	UGC 4204	U	.13± .12	.13					
36.31 -16.80		10.0±1.2		.10					
0801.5 +5607	PGC 22710		1.05	.06					
080529.6-485051		RSBS0..	.99± .06						
263.74 -9.05	ESO 209- 16	r	.08± .03	1.48	15.17 ±.14				
186.63 -53.87	IRAS08040-4842	-.1± .9		.06					
080401.1-484212	PGC 22711		1.12						
080532.4-244850		.S..1./	1.35± .04	47					
243.34 3.74	ESO 494- 22	S	.46± .05	2.32	14.02 ±.14				
148.46 -71.15		1.0± .8		.47					
080325.0-244012	PGC 22716		1.57	.23					
080534.0+102310		.S?....	1.04± .06	30				16.94±.3	10198± 10
211.83 21.13	UGC 4215		.05± .05	.04	14.4 ±.2				10079
61.05 -58.60				.07				327± 7	
080249.6+103147	PGC 22717		1.04	.02	14.27			2.65	10427
080538.1-112541	NGC 2525	.SBS5..	1.46± .02	75	12.26 ±.13	.62± .03		14.40±.1	1581± 5
231.85 10.79	MCG -2-21- 4	R (3)	.18± .03	.40	12.1 ±.2	.05± .07		223± 6	1802± 42
104.81 -72.32	IRAS08032-1117	5.0± .3	.99± .02	.27	10.97	.48	12.70± .04	211± 5	1401
080315.8-111703	PGC 22721	3.2± .5	1.50	.09	11.55	-.05	13.98± .17	2.76	1827
080538.9+261003	IC 492	.SBS4*.	1.04± .06					16.07±.2	5156± 6
195.80 27.08	UGC 4212	PU	.02± .05	.05	14.27 ±.18			253± 6	
48.64 -44.74		3.5± .6		.03				247± 4	5095
080236.9+261840	PGC 22724		1.04	.01	14.15			1.90	5355
0805.7 +3013									
191.44 28.32	CGCG 148-111			.16	15.4 ±.3				2356
46.45 -41.01									2311
0802.6 +3022	PGC 22726								2545
0805.8 +1228		.S..2..	.96± .09	150					
209.83 22.07	UGC 4216	U	.22± .06	.04	14.93 ±.18				
59.01 -56.81		2.0± .9		.27					
0803.1 +1237	PGC 22733		.96	.11					
0805.9 +7231		.S..8*.	1.01± .06	15					
142.40 31.35	UGC 4194	U	.39± .05	.08					
31.39 -.98	IRAS08002+7239	8.0±1.3		.48					
0800.3 +7239	PGC 22734		1.01	.20					
080559.3-272346		.SAT0?.	1.29± .04	150					
245.57 2.45	ESO 494- 25	Sr	.23± .05	3.40	13.45 ±.14				
155.06 -69.82	IRAS08039-2715	-.2± .6		.17	12.99				
080355.0-271506	PGC 22736		1.60						
080611.1-273140		.SXS3P*	1.68± .03	155					
245.71 2.41	ESO 494- 26	S (1)	.17± .05	3.43	12.47 ±.14				
155.32 -69.71	IRAS08041-2723	3.0± .5		.24	11.01				
080407.0-272300	PGC 22746	3.3±1.0	2.01	.09					

R.A. 2000 DEC. / l b / SGL SGB / R.A. 1950 DEC.	Names / PGC	Type ST nL / T / L	logD25 / logR25 / logA● / logD0	p.a. / Ag / Ai / A21	BT / mB / mFIR / BT°	(B-V)T / (U-B)T / (B-V)T° / (U-B)T°	(B-V)● / (U-B)● / m'● / m'25	m21 / W20 / W50 / HI	V21 / Vopt / VGSR / V3K
080611.3+123241	IC 2226	.SX.2..	1.10± .03		14.20 ±.15	.84± .03	.92± .03	17.20±.2	10845± 8
209.79 22.17	UGC 4220	U (2)	.05± .05	.04	14.2 ±.2			297± 21	10906± 59
59.06 -56.71		2.0± .8	.88± .02	.06		.74	14.09± .06	275± 6	10734
080324.7+124120	PGC 22747	2.8± .7	1.10	.02	13.99		14.43± .25	3.20	11072
080613.6+174226	NGC 2522	.S..0..	.99± .06	32					
204.64 24.23	UGC 4218	U	.47± .05	.12	14.8 ±.2				
54.53 -52.26		.0± .9		.35					
080321.5+175105	PGC 22749		.98						
0806.2 +8446		.IB.9..	1.17± .07	135				15.27±.1	1008± 11
128.57 28.69	UGC 4100	U	.18± .06	.23					
27.79 10.73		10.0± .8		.14				88± 8	1166
0753.0 +8455	PGC 22751		1.19	.09					993
0806.3 +5058		.SAS5..	.99± .06	118					
167.80 32.29	UGC 4213	U	.29± .05	.16	15.30 ±.19				
38.07 -21.51		5.0± .9		.44					
0802.6 +5107	PGC 22752		1.01	.15					
0806.3 +0537		.SB.3*.	.96± .09	40					
216.49 19.21	UGC 4222	U	.39± .06	.15	15.20 ±.18				
67.19 -62.31		3.0±1.3		.54					
0803.7 +0546	PGC 22753		.97	.20					
080624.4+010210	IC 494	.LA.0*.	1.12± .07	50					
220.77 17.11	UGC 4224	UE	.31± .03	.13	14.12 ±.15				
74.36 -65.68		-2.0± .6		.00					
080349.5+011050	PGC 22755		1.09						
0806.4 +2653				.08	15.4 ±.3				7647
195.09 27.48	CGCG 148-113								7589
48.45 -44.01									7845
0803.4 +2702	PGC 22756								
0806.5 +6703		.S..4..	1.21± .05	132					
148.76 32.11	UGC 4206	U	.96± .05	.16					
33.04 -6.19		4.0±1.0		1.41					
0801.7 +6712	PGC 22758		1.23	.48					
0806.6 +5508		.S..3..	1.08± .06	157					
162.87 32.55	UGC 4214	U (1)	.62± .05	.12	15.32 ±.18				
36.73 -17.54		3.0± .9		.86					
0802.7 +5517	PGC 22762	4.5±1.3	1.09	.31					
0806.6 +7320		.S?....	1.04± .06	99					3590± 57
141.44 31.27	UGC 4199		.48± .05	.10	14.73 ±.19				3711
31.20 -.18	IRAS08008+7329			.72	13.62				3624
0800.8 +7329	PGC 22763		1.05	.24	13.90				
080642.8+390525		.SAT3..	1.33± .05	150				15.57±.1	12433± 11
181.61 30.69	UGC 4219	U (1)	.19± .06	.16	14.8 ±.4				
42.59 -32.70		3.0± .7		.26				380± 8	12423
080322.5+391405	PGC 22766	2.5±1.0	1.34	.09	14.30			1.17	12597
080647.9+051829		.L.....	1.21± .09	145					
216.83 19.17	UGC 4228	U	.09± .05	.08	13.43 ±.16				
67.81 -62.48		-2.0± .8		.00					
080408.8+052710	PGC 22767		1.20						
080653.8+225035		.S..2..	1.05± .06	172				16.17±.3	6754± 10
199.42 26.24	UGC 4225	U	.44± .05	.10	15.12 ±.18				
51.06 -47.63		2.0± .9		.54				368± 7	6680
080355.9+225916	PGC 22772		1.06	.22	14.40			1.55	6962
080658.5+080021	NGC 2526	.S?....	.95± .07	140				15.47±.3	4603± 9
214.29 20.41	UGC 4231		.29± .05	.09	14.70 ±.19				4669± 35
64.35 -60.34	IRAS08042+0808			.30	13.39			335± 5	4480
080416.7+080903	PGC 22778		.96	.15	14.26			1.07	4840
0807.0 +8006		.I..9*.	1.29± .06	136				13.70±.1	862± 6
133.74 29.92	UGC 4173	U	.50± .06	.10	15.3 ±.3			80± 7	
29.22 6.28		10.0±1.1		.37					1005
0759.0 +8015	PGC 22783		1.30	.25	14.78			-1.33	867
0807.1 +0802	IC 2228		.64± .17						4635± 35
214.27 20.45	MCG 1-21- 13		.00± .07	.09	15.5 ±.2				4509
64.35 -60.29									4869
0804.4 +0811	PGC 22786		.65						
080708.5-280308		.SXS7?.	1.36± .05	110					1023
246.27 2.31	ESO 430- 20	S (1)	.48± .07	3.56					
156.25 -69.25	IRAS08050-2754	7.0±1.6		.67	12.23				808
080505.0-275424	PGC 22788	5.6±1.2	1.69	.24					1254

R.A. 2000 DEC. l b SGL SGB R.A. 1950 DEC.	Names PGC	Type S_T n_L T L	$\log D_{25}$ $\log R_{25}$ $\log A_e$ $\log D_o$	p.a. A_g A_i A_{21}	B_T m_B m_{FIR} B_T^o	$(B-V)_T$ $(U-B)_T$ $(B-V)_T^o$ $(U-B)_T^o$	$(B-V)_e$ $(U-B)_e$ m'_e m'_{25}	m_{21} W_{20} W_{50} HI	V_{21} V_{opt} V_{GSR} V_{3K}
080720.5-614311 275.25 -15.33 194.82 -42.22 080632.0-613424	 ESO 124- 11 IRAS08065-6134 PGC 22799	.L..../ S -2.0±1.3 	1.13± .05 .65± .03 .59± .04 1.11	114 .75 .00 	14.92 ±.15 15.05 ±.14 13.58 	.96± .03 .29± .04 	1.03± .02 .39± .03 13.36± .13 13.80± .32	 	
080720.5+510755 167.62 32.45 38.18 -21.31 080335.8+511636	NGC 2518 UGC 4221 PGC 22800	.L..-*. U -3.0±1.2 	1.08± .17 .08± .08 1.09	35 .16 .00 	 14.02 ±.15 13.78	 	 	 	 5266±155 5304 5389
080721.4+402354 180.15 31.06 42.19 -31.43 080359.0+403236	 UGC 4226 PGC 22802	.SAR6.. U 6.0± .8 	1.22± .06 .08± .06 1.24	35 .19 .12 .04	 14.7 ±.3 14.32	 	 	15.61±.1 364± 8 1.25	7911± 7 7906 8071
0807.3 +1745 204.71 24.50 54.89 -52.09 0804.5 +1754	 UGC 4232 PGC 22803	.S..4.. U 4.0±1.0 	1.00± .06 .53± .05 1.01	57 .12 .79 .27	 15.26 ±.19 	 	 	 	
080725.5+391143 181.53 30.84 42.70 -32.55 080405.1+392025	NGC 2528 UGC 4227 IRAS08040+3920 PGC 22805	.SXT3.. U (1) 3.0± .8 .5±1.1	1.19± .04 .01± .04 1.21	 .20 .01 .00	 13.38 ±.18 12.64 13.14	 	 	16.07±.1 112± 8 2.93	3928± 11 3945±155 3919 4092
080730.3-273032 245.85 2.67 154.83 -69.47 080526.1-272147	 PGC 22808	.IB.9*. S (1) 10.0±1.2 10.0±1.2	1.26± .07 .34± .08 1.56	3.16 .25 .17 	 	 	 	 	
0807.5 +0430 217.68 18.97 69.27 -62.97 0804.9 +0439	A 0804+04 MCG 1-21- 14 PGC 22810	.L?.... 	.72± .22 .00± .07 .73	 .08 .00 	 15.03 ±.18 14.82	 	 	 	 9125± 40 8986 9363
080741.3+390018 181.75 30.85 42.84 -32.71 080421.3+390901	A 0804+39 UGC 4229 MK 622 PGC 22816	.S?.... 	.78± .07 .03± .05 .80	 .20 .04 .02	 14.6 ±.2 12.99 14.31	 	 	16.32±.1 368± 16 1.99	6964± 11 6954± 25 6952 7127
080750.4-614601 275.32 -15.31 194.78 -42.15 080702.0-613712	 ESO 124- 12 PGC 22822	 	.82± .07 .24± .06 .89	 .75 	15.8 ±.2 15.80 ±.14 	.03± .09 -.34± .12 	 	 	8108± 87 7884 8255
080755.9+174902 204.70 24.65 55.03 -51.97 080503.8+175747	NGC 2529 UGC 4237 IRAS08050+1757 PGC 22827	.SBS7.. U 7.0± .8 	1.15± .05 .17± .05 1.16	170 .12 .24 .09	 14.24 ±.20 13.13 13.86	 	 	15.23±.3 159± 7 1.28	5029± 10 4936 5248
080757.4+260145 196.13 27.53 49.39 -44.65 080455.7+261030	 UGC 4236 PGC 22830	.SA.8.. U 8.0± .9 	.96± .09 .26± .06 .96	145 .09 .32 .13	 15.32 ±.19 14.90	 	 	16.75±.3 148± 7 1.72	4204± 10 4142 4406
0808.0 +6246 153.80 32.62 34.50 -10.22 0803.6 +6255	 UGC 4223 IRAS08036+6255 PGC 22834	.S..6*. U 6.0±1.4 	1.04± .08 .98± .06 1.05	129 .14 1.44 .49	 	 	 	 	
080806.0+145014 207.73 23.53 57.61 -54.52 080517.1+145900	 UGC 4240 IRAS08052+1458 PGC 22835	.S?.... 	1.18± .05 .71± .05 1.18	177 .04 1.06 .35	 15.18 ±.18 14.03	 	 	15.76±.3 535± 15 1.38	8563± 10 8460 8788
080809.9+390930 181.60 30.98 42.88 -32.53 080449.7+391815	NGC 2524 UGC 4234 PGC 22838	.S..0.. U .0± .8	1.14± .13 .14± .12 .69± .04 1.15	125 .20 .10 	13.64 ±.14 13.67 ±.18 13.29	.94± .02 .83 	.95± .01 12.58± .12 13.85± .73	 	 4030±155 4020 4195
0808.4 +1110 211.39 22.10 61.32 -57.55 0805.7 +1119	 UGC 4244 PGC 22846	.S..2.. U 2.0± .9 	.96± .09 .15± .06 .96	55 .04 .18 .07	 14.75 ±.18 	 	 	 	
0808.6 +2804 193.99 28.31 48.40 -42.73 0805.6 +2813	 CGCG 148-116 PGC 22855	 	 	 .13 	 15.3 ±.3 	 	 	 	 5608 5554 5805
080846.3+181138 204.40 24.98 55.01 -51.54 080553.9+182027	 UGC 4245 IRAS08058+1820 PGC 22860	.SB.3.. U 3.0± .9 	1.19± .05 .57± .05 1.20	110 .13 .79 .29	 14.74 ±.18 13.53 13.78	 	 	14.78±.3 357± 7 .71	5217± 10 5126 5437

R.A. 2000 DEC. / I b / SGL SGB / R.A. 1950 DEC.	Names / PGC	Type / S_T n_L / T / L	$\log D_{25}$ / $\log R_{25}$ / $\log A_e$ / $\log D_o$	p.a. / A_g / A_i / A_{21}	B_T / m_B / m_{FIR} / B_T^o	$(B-V)_T$ / $(U-B)_T$ / $(B-V)_T^o$ / $(U-B)_T^o$	$(B-V)_e$ / $(U-B)_e$ / m'_e / m'_{25}	m_{21} / W_{20} / W_{50} / HI	V_{21} / V_{opt} / V_{GSR} / V_{3K}
080850.0+574612	NGC 2521	.LA.-P?	1.08± .17	45	13.8 ±.3	.96± .06	1.04± .06		
159.75 32.89	UGC 4235	PU	.22± .08	.10		.52± .12	.56± .12		5283±155
36.18 -14.95	7ZW 212	-2.6± .8	.88± .15	.00		.88	13.66± .53		5347
080445.8+575458	PGC 22866		1.06		13.62	.51	13.52± .93		5382
080851.3+001829		RSXT1..	1.08± .05	85					
221.75 17.30	UGC 4248	UE	.09± .04	.12	14.3 ±.2				
76.69 -65.73		1.0± .6		.09					
080617.2+002718	PGC 22867		1.09	.04					
080903.1+164034		.SB.7?.	1.04± .08	18				15.26±.3	2838± 10
205.97 24.46	UGC 4247	U	1.06± .06	.12	16.0 ±.2				
56.34 -52.82		7.0±1.0		1.38				191± 7	2741
080612.3+164923	PGC 22873		1.05	.50	14.50			.26	3061
0809.1 +2800									
194.10 28.39	CGCG 148-117			.10	14.2 ±.3				11176
48.58 -42.74									11122
0806.1 +2809	PGC 22877								11374
080912.3-613936		.L..-..	1.18± .05	133					
275.31 -15.11	ESO 124- 14	S	.34± .04	.75	14.09 ±.14				
194.52 -42.19		-3.0± .8		.00					
080823.0-613042	PGC 22879		1.23						
080913.3+165907		.S?....	.97± .07	52				16.00±.3	4839± 10
205.68 24.62	UGC 4249		.66± .05	.11	15.4 ±.2				
56.14 -52.54				.99				312± 7	4743
080622.1+170757	PGC 22880		.98	.33	14.26			1.41	5061
0809.2 +0017		.L..-*.	1.05± .08	145					
221.82 17.37	UGC 4251	U	.17± .03	.12	14.32 ±.15				
76.89 -65.68	IRAS08066+0025	-3.0±1.2		.00	13.43				
0806.6 +0025	PGC 22881		1.04						
080915.6-002205		.S..4*.	1.17± .04	98					
222.42 17.07	UGC 4253 (1)	UE	.59± .04	.12	14.48 ±.18				
78.11 -66.11	IRAS08067-0013	4.3± .8		.86	13.63				
080642.1-001314	PGC 22883	4.2±1.2	1.18	.29					
0809.3 +5745		.S..2..	1.01± .06	146					
159.77 32.96	UGC 4241	U	.45± .05	.10	14.90 ±.19				
36.26 -14.94		2.0± .9		.55					
0805.3 +5754	PGC 22890		1.02	.22					
080924.0+003633		.SBT9P?	1.13± .04	92				15.09±.3	1807± 10
221.54 17.56	UGC 4254 (1)	UE	.27± .04	.14	14.12 ±.18				
76.37 -65.43	IRAS08068+0045	8.5± .8		.28				161± 7	1656
080649.6+004525	PGC 22894	6.4± .8	1.15	.14	13.70			1.26	2051
080927.1+435605		.S..2..	1.12± .04	75					
176.13 32.03	UGC 4246	U	.61± .04	.21	15.35 ±.18				
41.19 -27.98		2.0± .9		.75					
080558.6+440455	PGC 22896		1.13	.31					
080936.0+413542		.I?....	.83± .07						750± 33
178.85 31.69	MCG 7-17- 19		.25± .06	.21	15.48 ±.18				
42.16 -30.16	KUG 0806+417			.19					750
080611.8+414433	PGC 22900		.85	.12	15.08				908
080955.6-743042		.IBS9..	1.17± .07	131					
287.29 -21.08	ESO 35- 21	S (1)	.47± .05	.74	17.16 ±.14				
200.52 -30.29		9.5± .6		.35					
081044.0-742142	PGC 22908	10.0± .7	1.24	.23					
081000.0+400611		.S..6*.	1.23± .03	17					
180.59 31.50	UGC 4252	U	.48± .04	.21	15.0 ±.2				
42.86 -31.52	KUG 0806+402	6.0±1.2		.71					
080638.5+401503	PGC 22909		1.25	.24					
081000.6-645610		.SBR4?/	1.36± .05	100					
278.32 -16.64	ESO 89- 12	S (1)	.79± .05	.67					
196.15 -39.16	IRAS08094-6446	4.0±1.2		1.16	13.60				
080927.0-644712	PGC 22910	3.3±1.3	1.43	.39					
0810.0 +2251									
199.69 26.93	CGCG 119- 3			.11	15.4 ±.3				12134± 22
52.02 -47.29									12060
0807.1 +2300	PGC 22913								12345
0810.0 +5751		.S?....	.85± .12						7795± 60
159.65 33.05	MCG 10-12- 82		.10± .07	.10	15.11 ±.18				
36.31 -14.82	SBS 0806+579A			.13					7859
0806.0 +5800	PGC 22914		.86	.05	14.80				7894

R.A. 2000 DEC. / l b / SGL SGB / R.A. 1950 DEC.	Names / PGC	Type / S_T n_L / T / L	$\log D_{25}$ / $\log R_{25}$ / $\log A_e$ / $\log D_o$	p.a. / A_g / A_i / A_{21}	B_T / m_B / m_{FIR} / B_T^o	$(B-V)_T$ / $(U-B)_T$ / $(B-V)_T^o$ / $(U-B)_T^o$	$(B-V)_e$ / $(U-B)_e$ / m'_e / m'_{25}	m_{21} / W_{20} / W_{50} / HI	V_{21} / V_{opt} / V_{GSR} / V_{3K}
081005.4+461134		.S..6*.	1.08± .06		14.9 ±.3				
173.50 32.44	UGC 4250	U	.00± .05	.16					
40.43 -25.82		6.0±1.2		.00					
080632.5+462026	PGC 22915		1.09	.00					
0810.1 +2455	IC 497	.S?....	.89± .11		15.36 ±.19				4208± 22
197.50 27.63	MCG 4-20- 1		.35± .07	.13	13.38				4142
50.69 -45.44	IRAS08070+2503			.48					4415
0807.1 +2503	PGC 22918		.90	.18	14.71				
081010.1+245333		.S..6*.	1.33± .04	33				14.39±.3	4164± 10
197.53 27.64	UGC 4257	U	1.05± .05	.13	15.23 ±.18			261± 15	
50.72 -45.46		6.0±1.3		1.47					4098
080710.0+250227	PGC 22921		1.34	.50	13.61			.28	4371
081015.1+335727	NGC 2532	.SXT5..	1.34± .03	10	13.01 ±.14	.60± .04		14.70±.1	5260± 4
187.60 30.24	UGC 4256	R	.08± .03	.16	12.79 ±.18	-.06± .04		207± 6	5153± 56
45.72 -37.20	IRAS08070+3406	5.0± .4		.11	11.52	.52		188± 6	5228
080703.2+340620	PGC 22922		1.35	.04	12.63	-.12	14.35± .20	2.04	5442
081025.4+671436		.S..2..	1.10± .05	158				15.99±.1	4949± 11
148.50 32.47	UGC 4243	U	.23± .05	.18	14.42 ±.18				4998±128
33.35 -5.90	IRAS08056+6723	2.0± .8		.29	13.53			341± 8	5049
080536.1+672327	PGC 22930		1.12	.12	13.90			1.97	5010
081047.6+465444		.S..7..	1.20± .05	122					
172.67 32.64	UGC 4258	U	.58± .05	.16	15.4 ±.2				
40.28 -25.10		7.0± .9		.80					
080713.3+470339	PGC 22941		1.22	.29					
081056.4+364942		.S?....	.96± .09					16.34±.3	6421± 9
184.39 31.03	UGC 4261		.34± .06	.18	14.78 ±.19				5466± 35
44.52 -34.50	IRAS08076+3658			.47	13.06			175± 5	6347
080740.2+365838	PGC 22945		.98	.17	14.07			2.10	6541
081059.2+724741	A 0805+72	.S?....	.73± .09	165	14.9 ±.2	.41± .03			3150± 63
142.01 31.68	UGC 4242		.05± .08	.11	14.7 ±.3	-.38± .05			3269
31.66 -.61	MK 14			.07	13.15	.35			3188
080521.7+725633	PGC 22947		.74	.02	14.63	-.42	13.26± .56		
0811.0 +0505	IC 2231	.E...*.	1.15± .09						
217.56 20.01	UGC 4265	U	.00± .05	.11					
69.86 -61.96		-5.0±1.1		.00					
0808.4 +0514	PGC 22950		1.17						
0811.1 +5701		.S?....	.82± .13						8514± 60
160.64 33.20	MCG 10-12- 85		.36± .07	.10	15.4 ±.2				8575
36.73 -15.56	SBS 0807+571			.53					8617
0807.1 +5710	PGC 22954		.83	.18	14.70				
081109.7+462754	A 0807+46	.I..9*.	1.21± .03		14.13 ±.18	.40± .07	.36± .04	14.72±.1	2254± 11
173.20 32.65	UGC 4260	U (1)	.04± .04	.15	14.0 ±.2	-.25± .06	-.27± .03	147± 16	2273
40.52 -25.50	DDO 49	10.0±1.1	.96± .04	.03		.34	14.42± .10	142± 12	
080736.4+463650	PGC 22955	8.0±1.4	1.22	.02	13.88	-.29	14.92± .27	.82	2397
081113.2+251222	NGC 2535	.SAR5P.	1.39± .02		13.31 ±.13	.54± .02	.62± .01	14.13±.1	4099± 3
197.28 27.96	UGC 4264	R (1)	.31± .03	.15	13.19 ±.18	-.13± .02	-.06± .02	164± 5	4079± 18
50.83 -45.07	IRAS08082+2521	5.0± .3	.85± .01	.46	12.10	.42	13.05± .03	130± 4	4033
080812.8+252119	PGC 22957	1.0±1.4	1.41	.15	12.64	-.22	14.35± .17	1.34	4306
081116.1+251048	NGC 2536	.SBT5P.	.95± .05	112	14.7 ±.2	.55± .05			
197.31 27.96	MCG 4-20- 5	R	.15± .05	.15	14.75 ±.18	.05± .12			4142± 26
50.86 -45.09	KUG 0808+253B	5.0± .5		.23		.46			4076
080815.8+251946	PGC 22958		.97	.08	14.32	-.01	13.93± .35		4349
081122.8+033803	NGC 2538	PSB.1..	1.16± .05	25	13.49 ±.18				
218.96 19.42	UGC 4266	U	.07± .05	.07	11.77				3944
72.14 -62.96	IRAS08087+0347	1.0± .8		.07					3802
080845.4+034702	PGC 22962		1.17	.03	13.30				4187
081138.7+762517	A 0805+76	.SB.7..	1.38± .03	83				13.75±.1	1544± 7
137.84 30.99	UGC 4238	U	.24± .04	.13	13.33 ±.19			180± 16	
30.57 2.85	IRAS08054+7633	7.0± .8		.33	13.91			167± 12	1675
080509.4+763409	PGC 22969		1.39	.12	12.87			.76	1566
081140.1-693911		.SXS4..	1.09± .05	125					
282.76 -18.75	ESO 59- 27	S (1)	.27± .04	.70	14.84 ±.14				
198.26 -34.75		4.0± .8		.40					
081137.0-693006	PGC 22973	5.6± .9	1.16	.14					
081151.3-181801		.S..4*/	1.31± .04	91					
238.59 8.47	ESO 561- 23	SE	.81± .04	.96					
127.58 -71.46	IRAS08096-1809	3.7± .7		1.20	13.66				
080936.1-180900	PGC 22980		1.40	.41					

R.A. 2000 DEC. l b SGL SGB R.A. 1950 DEC.	Names PGC	Type S_T n_L T L	$\log D_{25}$ $\log R_{25}$ $\log A_o$ $\log D_o$	p.a. A_g A_i A_{21}	B_T m_B m_{FIR} B_T^o	$(B-V)_T$ $(U-B)_T$ $(B-V)_T^o$ $(U-B)_T^o$	$(B-V)_e$ $(U-B)_e$ m'_e m'_{25}	m_{21} W_{20} W_{50} HI	V_{21} V_{opt} V_{GSR} V_{3K}
081201.0+192152 203.51 26.13 55.16 -50.14 080907.4+193053	 UGC 4269 IRAS08091+1930 PGC 22990	.S?.... 	1.06± .05 .31± .04 1.07	90 .07 .47 .16	 14.57 ±.18 12.85 13.99			16.44±.3 640± 13 2.30	8533± 10 8446 8754
081233.0-571440 271.61 -12.51 191.34 -45.97 081126.1-570534	 PGC 23009	.E+2... S -4.0± .8 	1.09± .10 .09± .08 1.28	 1.45 .00 					
081243.8-795307 292.61 -23.24 202.50 -25.21 081520.0-794354	 ESO 18- 1 IRAS08153-7943 PGC 23013	.SBS3.. S (1) 3.0± .8 4.4± .9	1.02± .07 .07± .05 1.07	 .47 .09 .03	 13.30 				
081245.7+262142 196.16 28.65 50.54 -43.88 080944.0+263045	NGC 2540 UGC 4275 IRAS08097+2630 PGC 23017	.SBT6*. U 6.0±1.2 	1.12± .04 .17± .04 1.13	125 .15 .25 .09	 14.21 ±.18 12.94 13.78			15.98±.3 342± 15 2.11	6301± 10 6240 6507
081246.8+092309 213.63 22.30 64.95 -58.37 081003.7+093213	 UGC 4276 PGC 23018	PSB.1*. U 1.0± .9 	.96± .05 .28± .04 .97	147 .02 .28 .14	 14.90 ±.18 				
081250.6-273317 246.53 3.64 153.29 -68.43 081046.0-272412	 ESO 494- 35 PGC 23020	.L..0*/ S -2.0±1.1 	1.20± .05 .30± .04 1.43	23 2.40 .00 	 13.32 ±.14 				
0812.8 +5755 159.56 33.42 36.66 -14.63 0808.8 +5804	 MCG 10-12- 91 IRAS08088+5804 PGC 23022	.L?.... 	.93± .16 .30± .07 .89	 .10 .00 	 15.29 ±.15 15.07				8034± 60 8098 8135
081253.1+554024 162.25 33.45 37.43 -16.75 080856.8+554926	NGC 2534 UGC 4268 MK 85 PGC 23024	.E.1.\$P P -5.0±1.6 	1.14± .07 .05± .03 .83± .19 1.15	 .16 .00 	13.7 ±.4 13.58 ±.15 13.38	.77± .05 .26± .07 .70 .25	.85± .02 .30± .04 13.33± .66 14.26± .53		 3676± 31 3732 3785
081256.4+733349 141.09 31.67 31.55 .17 080710.8+734247	NGC 2523B UGC 4259 IRAS08072+7342 PGC 23025	.SAS3*/ R (1) 3.0± .5 4.5±1.2	1.33± .04 .83± .04 1.33	92 .10 1.14 .41	 14.74 ±.18 13.74 13.47			15.42±.1 397± 8 1.54	3833± 11 3955 3868
081257.0+545811 163.09 33.45 37.69 -17.41 080902.8+550713	 UGC 4267 IRAS08089+5507 PGC 23026	.S?.... 	.96± .09 .29± .06 .97	35 .10 .36 .15	 14.4 ±.2 13.90				2617±155 2670 2729
081258.4+361515 185.15 31.31 45.24 -34.87 080943.4+362418	NGC 2543 UGC 4273 IRAS08096+3624 PGC 23028	.SBS3.. R 3.0± .4 	1.37± .03 .24± .03 1.39	45 .21 .33 .12	 12.70 ±.18 11.94 12.14			13.72±.3 294± 10 1.46	2471± 10 2467± 44 2449 2649
081315.1+455929 173.81 32.96 41.08 -25.82 080943.1+460833	NGC 2537 UGC 4274 MK 86 PGC 23040	.SBS9P. V (1) 9.0± .4 5.7± .9	1.24± .02 .07± .03 .90± .01 1.26	 .16 .07 .04	12.32 ±.13 12.11 ±.19 12.02	.63± .01 -.14± .02 .57 -.18	.59± .01 -.13± .02 12.31± .02 13.19± .19	14.15±.0 110± 4 96± 4 2.09	447± 4 444± 17 464 593
0813.3 +5750 159.64 33.49 36.74 -14.68 0809.3 +5800	 UGC 4270 PGC 23047	.SXT4.. U 4.0± .8 	1.21± .05 .16± .05 1.22	130 .09 .23 .08	 14.5 ±.2 14.20			15.09±.1 334± 8 .81	2479± 11 2543 2580
081341.0+455941 173.82 33.03 41.15 -25.79 081009.0+460846	NGC 2537A MCG 8-15- 51 PGC 23057	.SBT5.. R 5.0± .5 	.83± .09 .04± .05 .50± .02 .84	 .16 .05 .02	16.02 ±.15 	.67± .05 -.06± .10 	.69± .05 .03± .10 14.01± .05 14.91± .51		
0813.7 +0038 222.05 18.54 78.04 -64.59 0811.2 +0048	 UGC 4285 PGC 23064	.S..6*. U 6.0±1.2 	1.12± .05 .19± .05 1.12	175 .08 .27 .09	 14.5 ±.2 				
0813.8 +5238 165.88 33.54 38.65 -19.55 0810.1 +5248	 UGC 4277 PGC 23069	.S..6*. U 6.0±1.2 	1.59± .02 .98± .05 1.60	110 .11 1.44 .49	 14.9 ±.3 13.33			14.22±.1 575± 8 .40	5459± 11 5654± 76 5507 5585
081359.0+454443 174.12 33.06 41.31 -26.00 081027.6+455350	IC 2233 UGC 4278 PGC 23071	.SBS7*/ R 7.0± .4 	1.67± .02 1.00± .03 1.24± .03 1.68	172 .16 1.38 .50	13.07 ±.15 13.48 ±.18 11.70	.44± .04 -.20± .06 .21 -.37	.46± .03 -.17± .04 14.76± .09 13.76± .19	13.13±.1 196± 14 180± 11 .92	563± 9 554± 18 577 709

R.A. 2000 DEC. l b SGL SGB R.A. 1950 DEC.	Names PGC	Type S_T n_L T L	$\log D_{25}$ $\log R_{25}$ $\log A_e$ $\log D_o$	p.a. A_g A_i A_{21}	B_T m_B m_{FIR} B_T^o	$(B-V)_T$ $(U-B)_T$ $(B-V)_T^o$ $(U-B)_T^o$	$(B-V)_e$ $(U-B)_e$ m'_e m'_{25}	m_{21} W_{20} W_{50} HI	V_{21} V_{opt} V_{GSR} V_{3K}
0814.0 +2352 198.96 28.14 52.53 -45.96 0811.1 +2401	IC 2239 MCG 4-20- 6 IRAS08111+2401 PGC 23078	.L?.... 	1.12± .12 .09± .07 1.13	.20 .00 	14.5 ±.2 11.97 14.17				6048± 19 5978 6261
081414.4+212125 201.64 27.33 54.35 -48.15 081118.7+213033	NGC 2545 UGC 4287 IRAS08113+2130 PGC 23086	RSBR2.. R (1) 2.0± .4 2.3± .9	1.30± .02 .24± .03 .86± .02 1.31	170 .12 .29 .12	13.16 ±.13 13.14 ±.13 12.98 12.70	.76± .02 .19± .02 .66 .13	.86± .01 .29± .02 12.95± .06 13.93± .20	15.86±.2 446± 12 420± 6 3.03	3373± 5 3358± 49 3293 3592
081416.0+182638 204.68 26.28 56.61 -50.67 081123.5+183547	 UGC 4286 PGC 23089	.S?.... 	1.19± .06 .36± .06 1.20	45 .13 .54 .18	 14.32 ±.18 13.61			15.07±.3 355± 7 1.28	5143± 10 5116± 46 5051 5367
081418.4-181722 238.90 8.97 127.39 -70.88 081203.0-180812	 ESO 561- 30 PGC 23090	.IBS9.. SE (1) 10.0± .5 9.8±1.2	1.26± .05 .40± .05 1.35	98 .90 .30 .20					1624 1425 1873
081422.1+391505 181.76 32.17 44.15 -32.00 081102.5+392413	 UGC 4283 PGC 23093	.SBS3.. U 3.0± .9 	1.12± .04 .17± .04 1.14	130 .16 .23 .08	14.8 ±.2				
081433.6+544804 163.30 33.68 37.97 -17.49 081040.3+545712	 UGC 4280 IRAS08106+5457 PGC 23103	.S..1.. U 1.0± .9 	1.16± .05 .66± .05 1.17	3 .11 .68 .33	 14.50 ±.20 13.49 13.68				3127±155 3179 3241
081440.3+490344 170.18 33.48 40.12 -22.86 081101.9+491253	NGC 2541 UGC 4284 PGC 23110	.SAS6.. R (2) 6.0± .3 6.2± .6	1.80± .02 .30± .03 1.30± .02 1.82	165 .18 .44 .15	12.26 ±.14 12.06 ±.18 11.56	.46± .02 -.23± .04 .36 -.30	.52± .02 -.13± .03 14.25± .04 15.38± .17	12.07±.0 207± 3 186± 4 .36	556± 4 628± 41 586 692
0814.6 +5813 159.18 33.65 36.78 -14.26 0810.6 +5823	 UGC 4281 PGC 23111	.S..2.. U 2.0±1.0 	1.06± .06 .60± .05 1.07	81 .10 .74 .30					
081443.3-663021 280.00 -16.98 196.40 -37.52 081417.1-662106	 ESO 89- 13 PGC 23113	.SBS7?. S (1) 7.0±1.8 7.8±1.3	1.14± .07 .29± .07 1.22	105 .87 .40 .14					
0814.7 +0132 221.34 19.19 76.86 -63.80 0812.2 +0142	 UGC 4291 PGC 23117	.S..4.. U 4.0±1.0 	1.00± .08 .84± .06 1.01	115 .09 1.24 .42	15.8 ±.2				
0814.8 +5800 159.44 33.68 36.88 -14.46 0810.8 +5810	MCG 10-12-100 SBS 0810+581 PGC 23119	.SB?... 	.99± .10 .24± .07 1.00	 .10 .35 .12	 15.17 ±.18 14.67				7885± 60 7950 7986
081459.2+733449 141.04 31.81 31.68 .23 080914.5+734355	NGC 2523 UGC 4271 ARP 9 PGC 23128	.SBR4.. R (2) 4.0± .3 1.1± .6	1.47± .03 .21± .03 1.13± .03 1.48	57 .10 .31 .11	12.63 ±.16 12.43 ±.13 13.08 12.07	.74± .06 .20± .09 .65 .13	.88± .03 13.77± .08 14.30± .22		3415± 58 3537 3450
0815.3 +2844 193.75 29.92 49.81 -41.49 0812.3 +2854	 CGCG 149- 6 PGC 23142			.13	15.2 ±.3				5924 5872 6126
0815.4 +2133 201.53 27.66 54.56 -47.84 0812.5 +2143	 CGCG 119- 19 PGC 23146			.13	15.6 ±.3			17.30±.3 242± 13	4277± 10 4197 4497
0815.5 +0820 214.95 22.46 67.22 -58.76 0812.8 +0830	 UGC 4296 PGC 23147	.S..6*. U 6.0±1.5 	1.00± .08 1.02± .06 1.00	112 .01 1.47 .50	 14.36	15.9 ±.3			9006 8879 9249
081532.6-205239 241.25 7.81 134.97 -70.22 081320.0-204324	 ESO 561- 33 PGC 23149	.SXT5*. SE (1) 5.2± .6 5.3±1.6	1.19± .05 .26± .05 1.29	84 1.07 .39 .13					
0815.6 +0515 217.94 21.11 71.39 -61.05 0813.0 +0525	 UGC 4298 PGC 23152	.S..9*. U 9.0±1.2 	.96± .09 .00± .06 .96	 .04 .00 .00				16.14±.3 113± 7 96± 7	4181± 8 4044 4428

R.A. 2000 DEC. l　　b SGL　SGB R.A. 1950 DEC.	Names PGC	Type S_T　n_L T L	$\log D_{25}$ $\log R_{25}$ $\log A_{\bullet}$ $\log D_o$	p.a. A_g A_i A_{21}	B_T m_B m_{FIR} B_T^o	$(B-V)_T$ $(U-B)_T$ $(B-V)_T^o$ $(U-B)_T^o$	$(B-V)_{\bullet}$ $(U-B)_{\bullet}$ $m'_{\bullet}$ m'_{25}	m_{21} W_{20} W_{50} HI	V_{21} V_{opt} V_{GSR} V_{3K}
081543.3-285116 247.97　3.45 155.41 -67.23 081340.1-284200	 PGC 23156	.SBS9.. S　　(1) 9.0± .8 10.0± .8	1.15± .08 .10± .08 1.39	 2.52 .10 .05					1700 1483 1938
0815.7　+5819 159.05　33.79 36.88 -14.12 0811.7　+5829	A　0811+58 UGC　4289 PGC 23160	.E..... U -5.0± .9	1.00± .19 .09± .08 .78± .07 .99	105 .14 .00	13.09V±.11 14.43 ±.15 14.17				7880± 59 7946 7981
081559.7+231145 199.84　28.33 53.55 -46.35 081302.0+232100	 UGC　4299 IRAS08130+2321 PGC 23169	.S..4.. U 4.0± .9	1.27± .03 .79± .03 1.28	150 .16 1.17 .40	13.99S±.15 14.96 ±.18 12.67 13.03		13.20± .24	15.92±.2 405± 10 286± 10 2.49	4286± 7 4213 4503
081601.3+270438 195.63　29.57 51.00 -42.91 081259.1+271353	UGC　4300 PGC 23170	.S..6*. U 6.0±1.1	1.13± .07 .02± .06 1.14	 .10 .03 .01	14.1 ±.2 13.92			15.71±.3 318± 15 1.78	7665± 7 7607 7872
081602.5+283725 193.93　30.02 50.06 -41.53 081258.3+284640	UGC　4301 PGC 23173	.S..4.. U 4.0± .9	1.07± .06 .35± .05 1.08	146 .12 .51 .17	14.58 ±.18 13.92			15.80±.3 362± 7 1.71	5968± 10 5916 6171
0816.1　+7340 140.90　31.87 31.73　.35 0810.4　+7350	UGC　4279 PGC 23181	.SB.8.. U 8.0± .9	1.00± .08 .14± .06 1.01	100 .11 .17 .07					
081617.1+255828 196.86　29.29 51.77 -43.87 081316.1+260744	UGC　4303 MK　　623 PGC 23184	.S?....	.84± .08 .12± .05 .85	120 .15 .18 .06	15.44 ±.18 13.16 15.04				12635± 45 12572 12845
0816.4　+2348 199.22　28.64 53.26 -45.75 0813.5　+2358	UGC　4304 PGC 23193	.L..... U -2.0± .9	1.23± .06 .68± .03 1.14	168 .18 .00	15.09 ±.17 14.84				4143± 22 4072 4359
081633.8-715135 285.01 -19.44 198.84 -32.56 081648.0-714212	ESO　60- 3 PGC 23200		.78± .06 .04± .05 .84	 .69	15.25 ±.14				1419± 51 1207 1532
0816.6　+2123 201.81　27.86 55.05 -47.84 0813.7　+2133	IC　2253 MCG　4-20- 11 PGC 23204	.E?....		.14	14.9 ±.3				4741± 22 4661 4962
0816.6　+2446 198.19　29.00 52.67 -44.88 0813.7　+2456	IC　2254 CGCG 119- 25 PGC 23206			.16	15.3 ±.3				7472± 22 7405 7686
081654.5+241038 198.86　28.85 53.14 -45.39 081355.7+241956	IC　2256 MCG　4-20- 12 KUG 0813+243 PGC 23214	.I?....	.85± .07 .39± .06 .87	.17 .29 .20	15.5 ±.2 15.03				2071± 22 2001 2286
0816.9　+2030 202.78　27.61 55.80 -48.57 0814.0　+2040	CGCG 119- 27 PGC 23215			.13	15.4 ±.3				4318± 22 4234 4542
081706.1-272727 246.97　4.48 151.87 -67.64 081501.1-271806	NGC　2559 ESO　494- 41 VV　　475 PGC 23222	.SBS4P* SU　　(1) 3.9± .4 3.3± .6	1.57± .02 .34± .04 1.19± .10 1.75	6 1.97 .50 .17	* 11.71 ±.14 9.23		.86± .01 12.88± .23	13.37±.2 391± 8 371± 6 3.98	1561± 7 1542± 37 1344 1801
0817.1　+6430 151.63　33.47 34.89　-8.25 0812.6　+6440	UGC　4295 PGC 23223	.SB.7*. U 7.0± .9	1.01± .06 .36± .05 1.02	53 .12 .50 .18	*				
081715.7+011227 221.97　19.58 78.37 -63.56 081440.8+012147	UGC　4310 PGC 23225	.SAS9.. UE　　(1) 8.5± .6 8.7± .8	1.09± .05 .03± .04 1.10	35 .10 .03 .02	* 14.4 ±.3 14.25			15.42±.3 157± 7 1.16	4359± 7 4209 4611
0817.4　+2109 202.14　27.95 55.46 -47.95 0814.5　+2119	CGCG 119- 28 PGC 23231			.14	15.4 ±.3				2156± 22 2075 2379

R.A. 2000 DEC. l b SGL SGB R.A. 1950 DEC.	Names PGC	Type ST nL T L	logD25 logR25 logA. logD.	p.a. Ag Ai A21	BT mB mFIR BT°	(B-V)T (U-B)T (B-V)T° (U-B)T°	(B-V). (U-B). m'. m'25	m21 W20 W50 HI	V21 Vopt VGSR V3K
081725.8+214100 201.58 28.13 55.07 -47.50 081429.9+215020	A 0814+21 UGC 4308 PGC 23232	.SBT5.. U (1) 5.0± .7 1.0±1.0	1.34± .04 .08± .05 1.35	110 .14 .12 .04	* 13.5 ±.3 13.27			14.51±.1 263± 4 244± 4 1.20	3565± 4 3518± 48 3486 3787
081727.5-244104 244.70 6.09 145.02 -68.71 081519.0-243142	 ESO 494- 42 PGC 23234	.L..0*. S -2.0± .8 	1.23± .05 .36± .04 .82± .07 1.34	55 1.43 .00 	14.0 ±.2 14.19 ±.14 	.86± .04 .27± .05 	.94± .02 .28± .03 13.58± .24 14.14± .32		
0817.5 +6433 151.57 33.50 34.91 -8.19 0813.0 +6443	 UGC 4302 PGC 23235	.SB.3*. U 3.0± .8 	1.04± .08 .06± .06 1.05	40 .12 .08 .03	 15.1 ±.2 14.86			16.42±.1 223± 8 1.54	11412± 11 10965±106 11497 11483
081736.7+352644 186.32 32.05 46.69 -35.23 081423.4+353604	 UGC 4306 IRAS08143+3536 PGC 23239	.S?.... 	.96± .09 .31± .06 .98	135 .21 .47 .16	 15.06 ±.18 11.73 14.37			15.66±.3 183± 7 1.13	2400± 10 2375 2585
0817.6 +2052 202.45 27.90 55.74 -48.17 0814.7 +2102	NGC 2553 MCG 4-20- 14 PGC 23240	.S?.... 	.94± .11 .11± .07 .95	 .16 .09 	 14.83 ±.18 14.52				 4721± 22 4639 4945
081738.3-294353 248.93 3.31 156.75 -66.41 081536.0-293430	 ESO 431- 1 IRAS08155-2934 PGC 23242	.SAT5*. Sr (1) 5.0± .6 5.6±1.1	1.25± .06 .12± .07 1.50	 2.62 .18 .06	 12.57 				
0817.6 +7730 136.52 31.06 30.54 3.98 0810.9 +7739	 UGC 4282 IRAS08109+7739 PGC 23244	.S..4.. U 4.0± .9 	1.01± .06 .25± .05 1.02	147 .13 .37 .13	 15.00 ±.18 13.26 				
081741.4+043628 218.82 21.26 73.10 -61.17 081503.1+044550	 UGC 4316 PGC 23245	.SX.8.. U 8.0± .8 	1.14± .05 .01± .05 1.15	 .05 .01 .01	 14.3 ±.3 14.20			15.55±.1 56± 5 46± 5 1.35	4222± 4 4083 4472
081742.8-300753 249.27 3.10 157.56 -66.19 081541.0-295830	 ESO 431- 2 IRAS08156-2958 PGC 23246	.SBT5*. SUr (1) 5.4± .4 3.9± .5	1.50± .04 .13± .06 1.76	5 2.77 .20 .07	 11.37 			12.74±.2 231± 8 192± 6 	1651± 7 1432 1889
081743.1+731909 141.29 32.05 31.95 .05 081203.0+732825	NGC 2523C UGC 4290 PGC 23247	.E...?. PU -5.0±1.0 	1.18± .08 .27± .04 1.11	95 .08 .00 	 13.93 ±.15 				
081753.0+125351 210.72 24.92 62.92 -54.81 081506.5+130313	 UGC 4317 PGC 23255	.SBS3.. U 3.0± .9 	.96± .09 .39± .06 .97	23 .10 .54 .20	 15.05 ±.18 14.35			16.43±.3 366± 7 1.89	9703± 10 9591 9942
081753.6+232816 199.71 28.84 53.91 -45.89 081455.8+233738	NGC 2554 UGC 4312 IRAS08149+2337 PGC 23256	.S..0.. U .0± .7 	1.50± .08 .13± .12 1.51	 .17 .10 	12.9 ±.2 12.7 ±.3 13.39 12.49	.93± .01 .54± .02 .83 .50	 14.93± .53	16.39±.3 433± 14 	4158± 10 4126± 21 4079 4370
0817.9 +0322 220.02 20.74 75.06 -61.99 0815.3 +0331	 UGC 4318 IRAS08153+0331 PGC 23257	.SB?... 	.96± .09 .39± .06 .97	 .06 .59 .20	 15.15 ±.18 13.57 14.46				 8849± 49 8706 9100
081756.2+004440 222.49 19.50 79.43 -63.73 081521.6+005403	NGC 2555 UGC 4319 IRAS08153+0054 PGC 23259	.SBT2.. UE 1.5± .5 	1.28± .04 .14± .05 1.29	115 .09 .17 .07	 13.14 ±.18 12.83				4401 4249 4654
081801.6+244416 198.35 29.27 53.07 -44.77 081502.2+245338	IC 2267 UGC 4315 KUG 0815+248 PGC 23266	.SB.6?. U 6.0±1.2 	1.32± .03 .81± .05 1.33	153 .17 1.19 .41	 14.81 ±.18 13.44			13.90±.3 222± 15 .06	2062± 10 2048± 22 1992 2274
0818.0 +3723 184.08 32.54 45.82 -33.42 0814.8 +3733	 UGC 4311 PGC 23271	.S..3*. U (1) 3.0±1.4 3.5±1.3	1.00± .08 .73± .06 1.01	135 .16 1.00 .36					
081808.8-673440 281.15 -17.22 196.58 -36.41 081748.0-672512	 ESO 89- 15 PGC 23277	.SXS6*. S (1) 6.0± .9 6.7± .9	1.05± .08 .19± .07 1.10	167 .61 .28 .09					

R.A. 2000 DEC. l b SGL SGB R.A. 1950 DEC.	Names PGC	Type ST nL T L	logD25 logR25 logAo logDo	p.a. Ag Al A21	BT mB mFIR BoT	(B-V)T (U-B)T (B-V)oT (U-B)oT	(B-V)o (U-B)o m'o m'25	m21 W20 W50 HI	V21 Vopt VGSR V3K
0818.2 +0218		.I..9..	1.28± .07	50					
221.06 20.31		E (1)	.23± .08	.12					
76.87 -62.66		10.0± .9		.17					
0815.6 +0228	PGC 23279	9.8± .8	1.30	.11					
0818.3 +1137		.S..2..	1.00± .08	171					
212.05 24.50	UGC 4321	U	.73± .06	.13	15.67 ±.18				
64.43 -55.76		2.0±1.0		.89					
0815.6 +1147	PGC 23285		1.01	.36					
081829.6+204536	A 0815+20	.S..2*/	.99± .09	27	14.9 ±.4	.82± .04		17.18±.2	4797± 6
202.66 28.05	UGC 4324	P	.46± .06	.15	15.29 ±.18	.17± .07		322± 13	4823± 65
56.09 -48.17	IRAS08155+2055	2.0±1.2		.57	13.49	.66		306± 5	4715
081534.8+205500	PGC 23289		1.00	.23	14.46	.06	13.53± .62	2.48	5022
081829.8-214901	NGC 2564	.L..-..	1.09± .09	60					
242.43 7.87	ESO 562- 1	S	.17± .06	1.06					
137.18 -69.34		-3.0± .9		.00					
081618.0-213936	PGC 23290		1.20						
081845.6-252951	NGC 2566	PSBT2P*	1.53± .02	110				13.87±.2	1649± 11
245.55 5.88	ESO 495- 3	SUr (1)	.17± .04	1.48	11.81 ±.14			216± 16	
146.76 -68.14	IRAS08166-2520	2.5± .4		.24	9.78			177± 12	1436
081638.0-252024	PGC 23303	3.3±1.1	1.67	.09	10.07			3.71	1895
081845.7-252214	IC 2311	.E.0.*.	1.32± .05		12.50 ±.13	1.01± .01	1.04± .01		
245.44 5.95	ESO 495- 2	S	.04± .04	1.47	12.49 ±.14	.48± .03	.52± .01		1844± 37
146.45 -68.19		-5.0± .8	.88± .02	.00		.67	12.39± .08		1631
081638.0-251248	PGC 23304		1.54		11.00	.23	13.97± .29		2090
0818.7 +2431	IC 501								
198.64 29.37	CGCG 119- 42			.19	15.4 ±.3				7530± 22
53.43 -44.87									7461
0815.8 +2441	PGC 23305								7746
0818.9 +2206								16.79±.3	3500± 10
201.26 28.61	CGCG 119- 44			.15	15.6 ±.3			148± 13	
55.20 -46.96								135± 10	3422
0816.0 +2216	PGC 23309								3722
081858.4+574811	NGC 2549	.LAR0./	1.59± .02	177	12.19 ±.13	.97± .01	.99± .01		
159.66 34.24	UGC 4313	R	.48± .03	.13	12.18 ±.11	.49± .02	.56± .01		1070± 23
37.48 -14.46		-2.0± .3	.76± .02	.00		.89	11.48± .06		1133
081456.7+575735	PGC 23313		1.54		12.04	.42	13.82± .18		1174
081858.7-673006		.LXS-*.	1.14± .08	64					
281.12 -17.12	ESO 89- 17	S	.27± .06	.61					
196.45 -36.45		-3.5± .6		.00					
081837.0-672036	PGC 23314		1.15						
081902.2+211114		.SAR6..	1.30± .03	125	14.15M±.10	.58± .02	.66± .02	15.05±.1	4096± 4
202.26 28.31	UGC 4329	U	.12± .03	.16	13.9 ±.3	-.08± .04	.01± .03	245± 5	
55.93 -47.74	IRAS08161+2120	6.0± .8	.87± .02	.18	13.59	.49	13.96± .06	224± 4	4015
081607.0+212040	PGC 23319		1.32	.06	13.77	-.14	15.21± .19	1.22	4320
081903.7+831559		.S..3..	1.33± .04	155				14.59±.1	5696± 11
130.11 29.46	UGC 4262	U (1)	.18± .05	.17	13.9 ±.2				
28.60 9.41	IRAS08078+8325	3.0± .7		.24				397± 8	5849
080832.5+832511	PGC 23321	4.5±1.0	1.35	.09	13.42			1.08	5689
081906.0+704251	Holmberg II	.I..9..	1.90± .02	15	11.10 ±.15	.44± .07	.40± .02	11.21±.0	158± 3
144.28 32.69	UGC 4305	R (1)	.10± .03	.10	10.9 ±.2		-.30± .04	72± 2	157± 18
32.94 -2.36	DDO 50	10.0± .3	1.69± .03	.08	12.84	.39	15.04± .06	66± 2	269
081353.5+705213	PGC 23324	8.0± .8	1.91	.05	10.89		15.19± .18	.27	207
0819.1 +2055	NGC 2556								
202.54 28.24	CGCG 119- 45			.15	15.4 ±.3				4635± 22
56.15 -47.95									4553
0816.2 +2105	PGC 23325								4860
0819.1 +2147		.S?....	1.04± .09					15.58±.3	4510± 10
201.62 28.54	MCG 4-20- 19		.09± .07	.13	14.7 ±.2			274± 13	
55.50 -47.21				.07				251± 10	4431
0816.2 +2157	PGC 23326		1.04		14.40				4733
0819.2 +2126	NGC 2557	.LB....	1.07± .07	55					
202.00 28.44	UGC 4330	U	.05± .03	.13	14.20 ±.16				4865± 21
55.79 -47.50		-2.0± .8		.00					4785
0816.3 +2136	PGC 23329		1.07		14.00				5089
081914.1-784148		.SAT4..	1.35± .04	150					
291.59 -22.45	ESO 18- 2	S (1)	.31± .05	.37					
201.71 -26.21	IRAS08212-7832	3.5± .6		.45					
082112.0-783212	PGC 23330	2.8± .8	1.39	.15					

R.A. 2000 DEC. / l b / SGL SGB / R.A. 1950 DEC.	Names / PGC	Type / ST nL / T / L	logD25 / logR25 / logAe / logDo	p.a. / Ag / Ai / A21	BT / mB / mFIR / BT°	(B-V)T / (U-B)T / (B-V)T° / (U-B)T°	(B-V)e / (U-B)e / m' / m'25	m21 / W20 / W50 / HI	V21 / Vopt / VGSR / V3K
081915.0-251122		.SXR1*.	1.12± .04						
245.35 6.14	ESO 495- 5	S	.03± .04	1.42	13.47 ±.14				
145.89 -68.15	IRAS08171-2501	1.0±1.1		.03	12.11				
081707.0-250154	PGC 23332		1.25	.01					
081915.5+244733	IC 2283	.S?....	1.00± .06					15.52±.3	4657± 10
198.39 29.56	MCG 4-20- 20		.10± .06	.12	14.8 ±.2			139± 13	
53.38 -44.59	KUG 0816+249A			.14				135± 10	4589
081616.2+245700	PGC 23333		1.01	.05	14.34				4873
0819.2 +1918	IC 2290	.L?....	.60± .15					17.10±.3	5681± 10
204.27 27.70	CGCG 88- 59		.11± .10	.16	15.8 ±.2			208± 13	
57.49 -49.31				.00				197± 10	5593
0816.4 +1928	PGC 23334		.60		15.51				5910
0819.3 +2030	NGC 2558	.SXT2..	1.24± .05	160	13.81 ±.15	.86± .02	.94± .01	14.91±.2	4990± 7
203.00 28.14	UGC 4331	U	.11± .05	.16	13.9 ±.2	.30± .03	.49± .02	400± 11	
56.53 -48.29		2.0± .8	.80± .03	.14		.77	13.30± .08	395± 10	4906
0816.4 +2040	PGC 23337		1.26	.06	13.48	.24	14.60± .30	1.37	5216
0819.3 +2045								17.11±.3	5031± 10
202.74 28.23	CGCG 119- 51			.15	14.1 ±.4			237± 13	
56.34 -48.07								216± 10	4948
0816.4 +2055	PGC 23338								5257
081920.3+500027	NGC 2552	.SAS9$.	1.54± .02	45	12.56 ±.15	.44± .09	.52± .03	13.67±.0	519± 4
169.10 34.29	UGC 4325	R (2)	.18± .03	.18	12.59 ±.19	-.16± .06	-.17± .03	141± 3	431± 46
40.51 -21.71	IRAS08156+5009	9.0± .3	1.36± .02	.18		.36	14.85± .05	134± 6	551
081540.6+500953	PGC 23340	8.0± .6	1.56	.09	12.21	-.22	14.67± .21	1.37	654
081922.1+350249		.S?....	.74± .14						
186.87 32.32	MCG 6-19- 1		.21± .07	.19	14.9 ±.3				5100
47.30 -35.44	ARAK 157			.31					5073
081609.6+351216	PGC 23341		.76	.10	14.36				5288
081922.4+234450	IC 2288	.S?....	.76± .07					16.17±.3	4603± 10
199.54 29.25	MCG 4-20- 23		.33± .05	.20	15.8 ±.2			263± 13	
54.14 -45.49	ARAK 158			.49				218± 10	4531
081624.4+235417	PGC 23342		.78	.17	15.09			.91	4822
081923.5-850841		.SBS9P*	1.06± .05						
297.96 -25.22	ESO 6- 1	S (1)	.07± .04	.65	15.16 ±.14				
204.33 -20.23		9.0± .7		.07					
082717.0-845854	PGC 23344	9.2± .7	1.12	.03					
081924.5-244711		.LB.0?P	1.13± .05	98					
245.04 6.40	ESO 495- 6	S	.31± .04	1.35	14.28 ±.14				
144.85 -68.26	IRAS08172-2437	-2.0± .8		.00	12.84				
081716.1-243742	PGC 23345		1.23						
0819.4 +2103					15.47 ±.14	.59± .02		16.82±.3	4852± 11
202.43 28.35	CGCG 119- 53			.16	15.4 ±.3	-.14± .04		212± 17	4719± 94
56.14 -47.80									4768
0816.5 +2113	PGC 23347								5075
081933.9-070223		.SBS7?.	1.03± .07	75					
229.79 15.99	MCG -1-22- 1	E (1)	.15± .05	.38					
96.19 -67.56		7.0±1.3		.21					
081706.8-065254	PGC 23350	5.3±1.6	1.06	.08					
081937.0+043924	NGC 2561	.SB?...	1.04± .06	138					
219.01 21.71	UGC 4336		.26± .05	.03	14.14 ±.20				4073
73.74 -60.79	IRAS08169+0448			.39	12.09				3933
081658.7+044853	PGC 23351		1.04	.13	13.70				4325
0819.6 +2122	IC 2293	.S?....	.94± .11						
202.11 28.51	MCG 4-20- 24		.11± .07	.13	14.93 ±.18				4039± 22
55.96 -47.51				.16					3958
0816.7 +2132	PGC 23352		.95	.06	14.62				4264
081938.5+210655	A 0816+21	.S?....	1.12± .04	53	14.82M±.10	.96± .02	1.00± .02	16.64±.2	5488± 7
202.39 28.42	UGC 4332		.30± .03	.16	14.9 ±.2	.34± .03	.39± .03		5505± 94
56.16 -47.73	IRAS08166+2116		.71± .02	.41	13.15	.83	13.87± .04	458± 7	5407
081643.4+211623	PGC 23355		1.14	.15	14.21	.22	14.54± .22	2.28	5714
0819.7 +1954		.S?....	.83± .08						
203.68 28.01	MCG 3-21- 28		.01± .06	.16	15.22 ±.18				4487± 22
57.13 -48.75				.02					4401
0816.8 +2004	PGC 23358		.84	.01	15.02				4715
081943.6+791409		.S..0..	.98± .07	90					
134.54 30.69	UGC 4292	U	.48± .05	.10	14.8 ±.2				
30.03 5.64	ARAK 155	.0± .9		.36					
081216.5+792329	PGC 23360		.97						

R.A. 2000 DEC. l　　b SGL　SGB R.A. 1950 DEC.	Names PGC	Type ST　nL T L	logD25 logR25 logAe logDo	p.a. Ag Ai A21	BT mB mFIR BT°	(B-V)T (U-B)T (B-V)T° (U-B)T°	(B-V)e (U-B)e m'e m'25	m21 W20 W50 HI	V21 Vopt VGSR V3K
081948.2+220138 201.43　28.77 55.52　-46.93 081652.2+221107	NGC　2565 UGC　4334 MK　386 PGC 23362	PSB.4*. PU　(1) 3.5± .6 3.0±1.5	1.28± .02 .35± .02 .71± .01 1.29	167 .12 .51 .17	13.40 ±.13 13.67 ±.18 13.24 12.84	.81± .01 .27± .01 .69 .17	.89± .01 .36± .01 12.44± .04 13.77± .18	14.57±.2 437± 10 412± 8 1.56	3591± 7 3675± 46 3514 3816
0819.9　+2058 202.56　28.43 56.35　-47.81 0817.0　+2108	NGC　2560 UGC　4337 PGC 23367	.S..0.. U .0± .9	1.16± .05 .63± .05 1.14	93 .15 .47	14.3 ±.4 14.87 ±.18 14.08	1.02± .02 .60± .04 .84 .48	 13.38± .49	 	 4925± 65 4843 5151
0819.9　+6249 153.60　33.97 35.78　-9.73 0815.6　+6259	 UGC　4322 PGC 23371	.E..... U -5.0± .8	1.04± .11 .14± .05 1.02	45 .13 .00	 14.70 ±.16				
081958.8-070434 229.88　16.06 96.39　-67.48 081731.8-065504	 MCG -1-22- 2 PGC 23373	.LA.-.. E -3.0± .9	1.16± .06 .15± .05 1.18	130 .34 .00					
0820.0　+2735 195.37　30.57 51.75　-42.05 0817.0　+2745	 CGCG 149- 16 PGC 23377			 .09	 15.02 ±.18				6075 6018 6285
082006.2-102917 232.92　14.31 105.08　-68.65 081742.6-101947	 MCG -2-22- 1 PGC 23378	.SBT7*. E　(1) 7.0±1.2 6.4± .8	1.05± .07 .11± .05 1.07	 .28 .15 .05					
0820.1　+2102 202.51　28.50 56.36　-47.73 0817.2　+2112	 MCG 4-20- 29 PGC 23379	.S?....	.82± .13 .08± .07 .83	 .15 .11 .04	 15.34 ±.18 15.04				 5180± 94 5098 5406
082007.1+172123 206.40　27.17 59.41　-50.84 081716.1+173053	 UGC　4343 PGC 23380	.I..9.. U 10.0± .9	1.00± .16 .14± .12 1.01	145 .12 .10 .07				15.65±.3 184± 7	4648± 10 4552 4881
082010.2+270530 195.93　30.45 52.10　-42.47 081708.3+271500	 UGC　4341 PGC 23383	.S..0.. U .0± .9	1.03± .05 .24± .04 1.03	43 .09 .18	 14.76 ±.18 14.40				6004 5945 6215
082012.5+260121 197.12　30.14 52.81　-43.41 081711.8+261051	 UGC　4340 KUG 0817+261 PGC 23385	.SX.7*. U 7.0± .8	1.08± .05 .19± .05 1.09	120 .06 .26 .09	 14.8 ±.2 14.47			16.24±.3 202± 15 1.68	4674± 10 4611 4888
0820.3　+2052 202.71　28.49 56.55　-47.85 0817.4　+2102	A　0817+21 UGC　4344 PGC 23391	.SA.8.. U 8.0± .8	1.19± .06 .05± .06 1.20	 .15 .06 .02	14.8 ±.4 14.4 ±.4 14.34	.64± .05 .03± .07 .56 -.02	 15.47± .53	15.41±.2 123± 6 104± 7 1.04	5037± 8 4954 5263
082020.5+665854 148.64　33.46 34.34　-5.82 081536.7+670822	 UGC　4323 PGC 23393	.E...?. U -5.0±1.6	1.20± .09 .18± .05 1.17	50 .12 .00	 14.08 ±.16 13.90				 3943±106 4041 4009
082023.7+210756 202.44　28.59 56.37　-47.62 081728.7+211727	NGC　2562 UGC　4345 ARAK 159 PGC 23395	.S..0*. PU -.3± .7	1.00± .04 .18± .03 1.01	3 .16 .13	13.88 ±.14 14.26 ±.20 13.63	1.02± .02 .60± .05 .91 .56	 13.30± .27		 4999± 48 4917 5225
082035.2-845838 297.80　-25.13 204.24　-20.38 082806.0-844848	 ESO　6- 2 IRAS08279-8449 PGC 23402	.SXR3*. S　(1) 3.0± .6 3.3± .7	1.17± .04 .40± .04 1.23	114 .65 .55 .20	 14.39 ±.14 13.60				
082035.7+210409 202.53　28.61 56.48　-47.65 081740.7+211340	NGC　2563 UGC　4347 PGC 23404	.L..0*. PU -2.0± .6	1.32± .08 .14± .06 .81± .03 1.32	80 .15 .00	13.24 ±.13 13.22 ±.17 13.01	1.03± .01 .94	1.05± .01 .59± .03 12.78± .09 14.38± .46		 4674± 48 4592 4900
082035.9+683603 146.73　33.21 33.80　-4.30 081541.2+684531	 UGC　4326 IRAS08157+6845 PGC 23405	.S..4.. U 4.0± .9	1.17± .04 .44± .04 1.18	152 .13 .65 .22	 13.95 ±.20 12.62 13.15				 4727± 76 4831 4786
082037.5+401643 180.80　33.53 44.97　-30.58 081717.1+402614	 UGC　4342 PGC 23407	.S..0.. U .0± .9	1.07± .07 .50± .06 1.06	43 .16 .37	 15.19 ±.18				

R.A. 2000 DEC. / l b / SGL SGB / R.A. 1950 DEC.	Names / PGC	Type / S_T n_L / T / L	$\log D_{25}$ / $\log R_{25}$ / $\log A_e$ / $\log D_o$	p.a. / A_g / A_i / A_{21}	B_T / m_B / m_{FIR} / B_T^o	$(B-V)_T$ / $(U-B)_T$ / $(B-V)_T^o$ / $(U-B)_T^o$	$(B-V)_e$ / $(U-B)_e$ / m'_e / m'_{25}	m_{21} / W_{20} / W_{50} / HI	V_{21} / V_{opt} / V_{GSR} / V_{3K}
082040.8+255419		.SBS2..	.93± .05					17.25±.2	5871± 7
197.28 30.21	UGC 4346	U	.08± .04	.06	14.82 ±.18			157± 7	
53.02 -43.46	KUG 0817+260	2.0± .9		.10				123± 5	5807
081740.4+260351	PGC 23409		.94	.04	14.59			2.61	6085
082041.8-012134		.SBS5*.	1.10± .05	97					
224.79 19.09	UGC 4349	UE (1)	.65± .04	.28	15.03 ±.18				
84.27 -64.47		4.5± .7		.98					
081809.3-011202	PGC 23410	3.1±1.6	1.13	.33					
0820.7 +5406		RSBR1*.	.98± .09	125					
164.13 34.58	UGC 4335	U	.12± .06	.13	14.69 ±.18				
39.11 -17.82		1.0± .8		.13					
0816.9 +5416	PGC 23412		.99	.06					
082045.3+192146	IC 2308	.S?....	.52± .20					16.44±.3	5732± 10
204.35 28.05	MCG 3-22- 1		.28± .07	.16				341± 13	5708± 88
57.90 -49.08	ARAK 160			.43	12.92			305± 10	5643
081752.2+193118	PGC 23415		.53	.14					5962
082048.2-085507	NGC 2574	.SBT2*.	1.35± .04	30					2869
231.63 15.28	MCG -1-22- 3	E	.28± .05	.25					
101.15 -67.99	IRAS08183-0845	2.0±1.1		.34					2690
081823.0-084534	PGC 23418		1.37	.14					3128
082049.4+223928		.S?....	.97± .04		14.52S±.15	.54± .03	.62± .03	15.89±.3	4138± 10
200.84 29.20	MCG 4-20- 34		.19± .03	.19	14.80 ±.18	-.19± .06	-.10± .06	220± 13	4062
55.34 -46.26	KUG 0817+228			.28		.43		214± 10	
081752.6+224900	PGC 23420		.99	.10	14.15	-.27	13.75± .26	1.65	4361
0820.8 +1638	A 0818+16	PSBS3..	1.00± .06	115					
207.21 27.05	UGC 4350	U	.32± .05	.08					
60.28 -51.34		3.0± .9		.44					
0818.0 +1648	PGC 23421		1.00	.16					
0820.8 +5515		.SB.1..	.96± .09	120					
162.74 34.59	UGC 4338	U	.10± .06	.15	15.3 ±.2				
38.69 -16.74		1.0± .9		.10					
0817.0 +5525	PGC 23422		.97	.05					
082117.5-254623		.S..3*/	1.27± .05	34					
246.10 6.20	ESO 495- 9	S	.68± .04	1.41	15.05 ±.14				
146.86 -67.50	IRAS08192-2536	3.0±1.2		.94	13.55				
081910.0-253648	PGC 23437		1.41	.34					
082124.5-131907	NGC 2578	.SBR0P.	1.31± .04	80	13.5 ±.2	.92± .03	1.00± .02		
235.56 13.08	MCG -2-22- 2	R	.22± .05	.27		.41± .04	.50± .03		
112.99 -68.96	IRAS08191-1310	.0± .4	.82± .03	.16	13.62		13.07± .10		
081903.6-130932	PGC 23440		1.33				14.37± .32		
082124.6+190852	NGC 2572	.S..1?.	1.12± .06	133					
204.64 28.11	UGC 4355	U	.44± .05	.18	14.69 ±.18				
58.27 -49.18	ARAK 161	1.0±1.8		.44					
081831.7+191827	PGC 23441		1.13	.22					
0821.4 +2051	NGC 2569	.E?....	.78± .13						5104± 22
202.83 28.72	MCG 4-20- 35		.08± .04	.16	15.26 ±.11				5021
56.89 -47.73				.00					5331
0818.5 +2101	PGC 23442		.78		15.02				
0821.4 +2055	NGC 2570	.S..2..	1.04± .06	75	15.1 ±.3	.58± .05		16.55±.3	6541± 14
202.76 28.74	UGC 4354	U	.25± .05	.15	14.94 ±.20	-.06± .07		277± 25	
56.84 -47.68		2.0± .9		.30		.45			6458
0818.5 +2105	PGC 23443		1.05	.12	14.47	-.13	14.53± .44	1.96	6768
082128.1+031010	IC 2327	.S..1?.	1.13± .05	168				14.79±.1	2684± 7
220.65 21.43	UGC 4356	U	.49± .05	.07	14.2 ±.2			315± 11	2463± 49
76.67 -61.48	IRAS08188+0319	1.0±1.8		.50	13.39			295± 8	2535
081851.2+031945	PGC 23447		1.14	.25	13.54			1.00	2935
0821.5 +2107									4952± 22
202.55 28.84	CGCG 119- 69			.15	15.34 ±.18				4870
56.71 -47.49									5179
0818.6 +2117	PGC 23448								
082141.3+735923	NGC 2544	.SBS1*.	1.04± .04	70	13.8 ±.2	.95± .05	.91± .03		
140.44 32.17	UGC 4327	PU	.15± .04	.04	13.6 ±.2	.23± .08	.21± .04		2828± 25
31.98 .77	MK 87	.5± .6	.81± .08	.15	12.65	.89	13.37± .24		2951
081555.2+740853	PGC 23453		1.05	.07	13.49	.20	13.53± .33		2863
0821.7 +5617		.SB?...	.89± .11						9102± 54
161.48 34.68	MCG 9-14- 34		.00± .07	.18	15.3 ±.2				9160
38.41 -15.74				.00					9214
0817.8 +5627	PGC 23454		.90		15.02				

R.A. 2000 DEC. / l b / SGL SGB / R.A. 1950 DEC.	Names / PGC	Type / S_T n_L / T / L	$\log D_{25}$ / $\log R_{25}$ / $\log A_o$ / $\log D_o$	p.a. / A_g / A_i / A_{21}	B_T / m_B / m_{FIR} / B_T^o	$(B-V)_T$ / $(U-B)_T$ / $(B-V)_T^o$ / $(U-B)_T^o$	$(B-V)_o$ / $(U-B)_o$ / m'_o / m'_{25}	m_{21} / W_{20} / W_{50} / HI	V_{21} / V_{opt} / V_{GSR} / V_{3K}
082154.1+223828		.S..6*.	1.08± .04	60	15.7 ±.4	.52± .06		15.83±.2	3741± 6
200.95 29.43	UGC 4361	U	.43± .04	.19	15.18 ±.19	-.24± .11		195± 13	
55.67 -46.15	KUG 0818+228	6.0±1.3		.64		.37		189± 5	3665
081857.4+224804	PGC 23465		1.10	.22	14.44	-.35	14.86± .46	1.17	3965
0822.1 +2119		.SB?...	.74± .14					16.92±.3	6469± 10
202.39 29.04	MCG 4-20- 38		.09± .07	.14	15.66 ±.18			208± 13	
56.73 -47.25				.14				184± 10	6388
0819.2 +2129	PGC 23470		.75	.05	15.34			1.54	6696
082207.2-080947		.SBS6..	1.15± .06	85					
231.13 15.95	MCG -1-22- 4	E (1)	.16± .05	.25					
99.57 -67.42		6.0± .8		.23					
081941.2-080009	PGC 23471	4.2±1.6	1.18	.08					
082210.4+031605	IC 503	.SB.1..	1.06± .06					14.64±.1	4131± 9
220.65 21.63	UGC 4366	U	.07± .05	.07	13.86 ±.18			311± 11	
76.76 -61.28	IRAS08195+0325	1.0± .9		.08	12.66			264± 8	3987
081933.4+032543	PGC 23474		1.07	.04	13.66			.94	4387
0822.3 +2105									6731± 22
202.66 29.00	CGCG 119- 72			.15	15.4 ±.3				6649
56.97 -47.42									6959
0819.4 +2115	PGC 23477								
082224.4+253035		.SX.2*.	1.04± .08					16.26±.2	8339± 8
197.86 30.46	UGC 4364	U	.04± .06	.10	14.7 ±.2			239± 6	
53.76 -43.61		2.0±1.1		.04				223± 7	8274
081924.5+254013	PGC 23480		1.05	.02	14.45			1.80	8556
082226.0+192502	IC 2329	.S..8*.	1.32± .04	117				14.84±.3	2082± 10
204.46 28.44	UGC 4365	U	.70± .05	.15	14.62 ±.19				
58.36 -48.82		8.0±1.2		.86				206± 7	1994
081933.0+193440	PGC 23483		1.33	.35	13.60			.89	2314
082228.6-010245		.SBT7?.	1.07± .05	135					
224.73 19.63	UGC 4370	UE (1)	.25± .04	.14	14.49 ±.18				
84.27 -63.92		6.5±1.2		.34					
081955.8-005306	PGC 23485	5.3± .8	1.09	.12					
082229.5-112506		.S..4*/	1.31± .04	100					4614
234.05 14.32	MCG -2-22- 5	E	.67± .05	.31					4429
107.98 -68.33	IRAS08200-1115	4.0±1.2		.99	12.41				4875
082006.6-111526	PGC 23486		1.34	.34					
082240.8+041548	IC 504	.L.....	1.09± .10	140					
219.76 22.21	UGC 4372	U	.17± .05	.10	14.02 ±.15				
75.41 -60.51		-2.0± .8		.00					
082002.9+042528	PGC 23495		1.08						
082241.1-694626		.SBT3P.	1.10± .04	85					
283.36 -17.97	ESO 60- 4	S (1)	.11± .04	.56	14.60 ±.14				
197.27 -34.25	IRAS08225-6936	3.0± .9		.16	11.56				
082234.0-693642	PGC 23496	4.4± .9	1.15	.06					
082243.1+223311	NGC 2577	.L..-*.	1.26± .13	105	13.4 ±.2	1.01± .03			
201.12 29.58	UGC 4367	U	.21± .08	.16	13.50 ±.15	.60± .05			2145± 65
55.97 -46.13		-3.0±1.1		.00		.94			2068
081946.6+224250	PGC 23498		1.25		13.28	.56	14.05± .71		2370
0822.7 +5618		RLAR0*.	1.08± .17						
161.45 34.81	UGC 4357	U	.08± .08	.18	14.61 ±.17				
38.54 -15.67		-2.0± .8		.00					
0818.8 +5628	PGC 23499		1.09						
082244.7+241753	NGC 2575	.SAT6*.	1.36± .03	145				14.49±.2	3870± 7
199.22 30.15	UGC 4368	U	.07± .04	.08	13.4 ±.3			248± 13	
54.70 -44.63	IRAS08198+2427	6.0± .7		.11	13.61			243± 6	3800
081946.2+242732	PGC 23501		1.37	.04	13.17			1.28	4091
082253.8+033420		.SBS6..	1.19± .05	55				15.14±.3	3983± 10
220.45 21.93	UGC 4374	U	.13± .05	.09	14.0 ±.2				
76.54 -60.94	IRAS08202+0343	6.0± .8		.19	13.48			199± 7	3840
082016.6+034400	PGC 23504		1.20	.06	13.66			1.42	4239
0822.9 +6656		.S..6*.	1.08± .06	28					4304±106
148.65 33.71	UGC 4353	U	.39± .05	.12	15.15 ±.18				4402
34.59 -5.77		6.0±1.3		.57					4372
0818.2 +6706	PGC 23506		1.09	.19	14.44				
0822.9 +2105									4473± 22
202.72 29.13	CGCG 119- 77			.15	15.3 ±.3				4391
57.15 -47.35									4701
0820.0 +2115	PGC 23507								

R.A. 2000 DEC. l b SGL SGB R.A. 1950 DEC.	Names PGC	Type S_T n_L T L	$\log D_{25}$ $\log R_{25}$ $\log A_e$ $\log D_o$	p.a. A_g A_i A_{21}	B_T m_B m_{FIR} B_T^o	$(B-V)_T$ $(U-B)_T$ $(B-V)_T^o$ $(U-B)_T^o$	$(B-V)_e$ $(U-B)_e$ m'_e m'_{25}	m_{21} W_{20} W_{50} HI	V_{21} V_{opt} V_{GSR} V_{3K}
082257.7+254427 NGC 2576 197.65 30.65 UGC 4371 53.75 -43.35 081957.5+255407 PGC 23512		.S..3.. U (1) 3.0± .9 4.5±1.2	1.23± .05 .73± .05 1.24	41 .09 1.01 .37	15.07 ±.18 13.91			15.28±.3 545± 7 1.00	8353± 10 8289 8570
082258.5+274226 195.45 31.23 UGC 4373 52.44 -41.63 081956.0+275206 PGC 23513		.SB.6*. U 6.0±1.2	1.04± .06 .25± .05 1.05	55 .07 .36 .12	15.0 ±.2 14.53			15.89±.3 232± 7 1.23	5945± 7 5888 6157
082307.9-050008 NGC 2583 228.44 17.79 MCG -1-22- 8 92.51 -65.86 082038.9-045026 PGC 23516		.E...*. PE -5.0± .8	.89± .08 .00± .05 .91	 .16 .00	14.4 ±.2 14.16	.96± .03 .48± .05 .87 .48	13.82± .48		5910± 60 5741 6172
082308.3-775107 290.88 -21.89 ESO 18- 7 201.14 -26.90 082444.0-774118 PGC 23517		.SBT4P* Sr (1) 4.0± .5 3.9± .6	1.12± .06 .30± .05 1.16	37 .41 .44 .15					5291± 87 5090 5382
082313.1-005156 224.66 19.88 UGC 4381 84.18 -63.67 IRAS08206-0042 082040.1-004214 PGC 23519		.S..5*/ UE (1) 4.7± .7 4.2±1.2	1.13± .04 .63± .04 1.15	33 .16 .95 .32	14.77 ±.18 12.92				
0823.2 +7102 143.82 32.96 33.14 -1.94 0818.0 +7112 PGC 23521		.I?....	 	 .09				15.83±.1 33± 6 26± 12	114± 6 227 163
0823.2 +2239 A 0820+22 201.05 29.73 UGC 4375 56.03 -45.98 IRAS08202+2249 0820.2 +2249 PGC 23522		.SX.5*. U 5.0± .7	1.39± .03 .19± .03 1.40	0 .16 .28 .09	12.79S±.15 13.6 ±.3 12.49		14.11± .22	15.47±.3 298± 13 282± 10 2.89	2059± 10 1983 2284
082315.3-045811 NGC 2584 228.42 17.84 MCG -1-22- 9 92.48 -65.82 082046.2-044829 PGC 23523		.SBS4?. E (1) 4.0±1.3 3.1± .8	1.06± .06 .28± .05 1.07	2 .16 .41 .14					
082318.2-261149 246.71 6.33 ESO 495- 11 147.42 -66.93 IRAS08211-2602 082111.0-260206 PGC 23527		.SBS9.. S (1) 9.0± .8 5.6± .8	1.13± .04 .04± .03 1.26	 1.38 .04 .02	13.79 ±.14 12.14 12.37				1705 1491 1955
0823.3 +7910 134.55 30.87 UGC 4328 30.22 5.65 0816.0 +7920 PGC 23530		.SB.8*. U 8.0±1.2	1.16± .07 .37± .06 1.17	10 .08 .46 .19	15.2 ±.2 14.64			15.93±.1 186± 8 1.10	4000± 11 4140 4013
082326.2-045453 NGC 2585 228.40 17.90 MCG -1-22- 10 92.41 -65.76 IRAS08209-0445 082057.0-044510 PGC 23537		.SBS3P. E 3.0± .8	1.25± .05 .33± .05 1.26	95 .16 .46 .17	 12.57				
082329.8+270809 196.13 31.17 MCG 5-20- 11 52.95 -42.08 KUG 0820+272 082028.0+271751 PGC 23539		.S?....	.94± .06 .17± .06 .95	 .07 .25 .09	15.10 ±.18 14.74				7133 7074 7347
0823.5 +2120 IC 2339 202.50 29.36 MCG 4-20- 45 57.14 -47.05 IRAS08206+2130 0820.6 +2130 PGC 23542		.SBS5P. R 5.0± .5	1.04± .09 .28± .07 1.05	 .15 .43 .14	 12.61				5323± 38 5241 5551
082336.5-150211 237.34 12.60 MCG -2-22- 8 117.93 -68.64 082117.3-145228 PGC 23545		.LA.+*. E -1.0±1.2	1.11± .06 .03± .05 1.13	90 .28 .00					
0823.6 +2120 IC 2338 202.51 29.37 MCG 4-20- 44 57.16 -47.05 0820.7 +2130 PGC 23546		.SXS6P. R 6.0± .6	.74± .14 .09± .07 .75	 .15 .14 .05				15.10±.1 324± 11 164± 12	5400± 11 5358± 38 5316 5626
082340.9-605234 275.55 -13.22 ESO 124- 15 191.98 -42.10 IRAS08227-6042 082244.0-604248 PGC 23550		.SBS4.. S (1) 4.0± .8 4.4± .9	1.24± .05 .31± .05 1.33	42 .90 .46 .16	14.34 ±.14 13.32				
0823.7 +2126 IC 2341 202.42 29.43 UGC 4384 57.11 -46.95 0820.8 +2136 PGC 23552		.L..-*. U -3.0±1.2	1.11± .16 .31± .08 1.08	1 .14 .00	14.56 ±.15 14.34				4846± 71 4765 5075

R.A. 2000 DEC. I b SGL SGB R.A. 1950 DEC.	Names PGC	Type S_T n_L T L	$\log D_{25}$ $\log R_{25}$ $\log A_\bullet$ $\log D_o$	p.a. A_g A_i A_{21}	B_T m_B m_{FIR} B_T^o	$(B-V)_T$ $(U-B)_T$ $(B-V)_T^o$ $(U-B)_T^o$	$(B-V)_\bullet$ $(U-B)_\bullet$ $m'_\bullet$ m'_{25}	m_{21} W_{20} W_{50} HI	V_{21} V_{opt} V_{GSR} V_{3K}
082351.6-255015 246.48 6.64 146.46 -66.94 082144.0-254030	 ESO 495- 12 PGC 23558	.SBS3?. S (1) 3.0±1.7 4.4±1.3	1.27± .04 .61± .04 1.39	87 1.31 .85 .31	14.88 ±.14 				
082352.1+144511 209.48 26.99 63.03 -52.46 082103.9+145454	UGC 4385 MK 1214 PGC 23559	.I?.... 	.94± .06 .09± .04 .95	 .11 .07 .04	 14.51 ±.18 14.33			15.25±.3 169± 13 .87	1969± 10 1863 2211
082358.8-065342 230.25 17.00 97.01 -66.50 082131.5-064357	MCG -1-22- 11 PGC 23564	.S..3P* E (1) 3.0±1.3 3.1±1.6	1.13± .06 .50± .05 1.14	120 .19 .69 .25					
082401.7+210133 202.89 29.35 57.53 -47.26 082107.0+211117	UGC 4386 IRAS08211+2111 PGC 23567	.S..3. U (1) 3.0± .9 4.5±1.2	1.27± .03 .56± .03 1.28	21 .16 .77 .28	14.17M±.10 14.47 ±.18 13.15 13.28	.97± .02 .32± .03 .80 .15	 13.96± .21	15.91±.2 448± 25 469± 10 2.35	4640± 8 4557 4869
0824.0 +1130 212.82 25.71 66.50 -54.99 0821.3 +1140	CGCG 60- 8 PGC 23570			.09	15.1 ±.6				9423± 57 9306 9671
0824.1 -1429 236.94 13.00 116.48 -68.47 0821.8 -1420	MCG -2-22- 9 PGC 23574	.S?.... 	1.12± .08 1.02± .07 1.15	 .31 1.50 .50					4780 4587 5041
082410.2-184639 240.61 10.66 128.23 -68.51 082154.8-183653	IC 2367 ESO 562- 5 IRAS08219-1836 PGC 23579	.SBR3.. SE (2) 3.0± .5 3.9± .7	1.38± .03 .15± .04 1.41	55 .25 .21 .07	 12.58 				
0824.2 -0744 231.05 16.62 99.08 -66.78 0821.8 -0735	 PGC 23586	PLXR0?. E -2.0±1.3 	1.14± .08 .00± .08 1.16	 .19 .00 					
0824.3 +2031 203.46 29.24 58.01 -47.65 0821.4 +2041	IC 2348 MCG 4-20- 49 PGC 23589	.S?.... 	.85± .05 .28± .03 .86	 .15 .35 .14	15.73S±.15 15.57 ±.18 15.11	.78± .05 .07± .07 .65 -.01	 14.10± .29	16.84±.3 348± 13 314± 10 1.59	5973± 10 5889 6204
082420.7+254039 A 197.83 30.92 54.16 -43.25 082120.7+255024	0821+25 MK 624 PGC 23591	 	.32± .18 .00± .12 .33	 .07 					8440± 45 8375 8659
0824.4 +5450 163.23 35.11 39.35 -16.94 0820.6 +5500	UGC 4380 PGC 23598	.S..6*. U 6.0±1.2 	1.06± .06 .00± .05 1.07	 .12 .00 .00	 14.7 ±.2 14.59			16.19±.1 121± 8 1.60	7485± 7 7537 7605
082430.3+183550 205.53 28.60 59.68 -49.23 082138.2+184536	NGC 2581 UGC 4388 IRAS08216+1845 PGC 23599	.SB?... 	1.04± .06 .12± .05 1.05	10 .13 .17 .06	 14.29 ±.18 13.95			15.34±.3 254± 7 1.33	5912± 10 5820 6147
082432.5-045316 228.52 18.15 92.67 -65.50 082203.3-044329	NGC 2586 MCG -1-22- 12 PGC 23603	PS..3?. E 3.0±1.3 	1.06± .06 .24± .05 1.07	3 .10 .33 .12					
082434.4+740044 140.36 32.36 32.16 .87 081849.7+741025	NGC 2550 UGC 4359 IRAS08188+7410 PGC 23604	.S..3*. P 3.0±1.3 	1.00± .08 .43± .06 1.01	103 .10 .60 .22	 13.6 ±.4 12.65 				
082449.3+732446 141.04 32.53 32.39 .32 081912.0+733428	NGC 2551 UGC 4362 ARAK 162 PGC 23608	.SAS0.. RC .2± .4 	1.22± .04 .17± .04 1.22	55 .07 .12 	13.1 ±.2 13.11 ±.13 12.87	.99± .04 .93 	 13.65± .29		 2263± 58 2384 2302
0824.8 +4654 172.93 35.03 42.73 -24.21 0821.3 +4704	UGC 4387 PGC 23609	.S..6*. U 6.0±1.4 	1.13± .05 .93± .05 1.14	97 .13 1.37 .47					
0824.8 +6652 148.69 33.91 34.80 -5.77 0820.1 +6701	UGC 4376 IRAS08201+6701 PGC 23611	.S?.... 	1.00± .16 .14± .12 1.01	150 .12 .21 .07	 14.94 ±.19 13.05 14.59				4127± 82 4225 4196

R.A. 2000 DEC. l b SGL SGB R.A. 1950 DEC.	Names PGC	Type S_T n_L T L	$\log D_{25}$ $\log R_{25}$ $\log A_o$ $\log D_o$	p.a. A_g A_i A_{21}	B_T m_B m_{FIR} B_T^o	$(B-V)_T$ $(U-B)_T$ $(B-V)_T^o$ $(U-B)_T^o$	$(B-V)_o$ $(U-B)_o$ m' m'_{25}	m_{21} W_{20} W_{50} HI	V_{21} V_{opt} V_{GSR} V_{3K}
082501.8-003530 224.64 20.41 84.27 -63.14 082228.5-002542	NGC 2590 UGC 4392 IRAS08224-0025 PGC 23616	.SAS4*. UE (2) 3.5± .6 2.2± .8	1.35± .03 .49± .04 .79± .02 1.36	77 .10 .72 .24	13.94 ±.13 13.66 ±.18 12.13 12.99	.89± .02 .74 	.97± .02 13.38± .05 14.30± .21	 	4998± 10 4990± 19 4840 5258
082502.0+671900 148.16 33.85 34.65 -5.35 082017.9+672844	 UGC 4377 KUG 0820+674 PGC 23617	.S..6*. U 6.0±1.2 	1.04± .05 .36± .05 1.05	5 .12 .53 .18	 	 	 	 	
082502.4+742600 139.87 32.28 32.03 1.27 081912.5+743542	A 0819+74 UGC 4363 DDO 51 PGC 23618	PSBS7*. U (1) 7.0±1.1 8.0±1.4	1.16± .04 .01± .04 1.16	 .07 .02 .01	14.8 ±.4 14.5 ±.3 14.50	.28± .10 -.06± .09 .24 -.09	 15.40± .45 	14.90±.1 108± 16 60± 12 .39	3522± 11 3646 3557
082512.1+202007 203.74 29.37 58.43 -47.70 082218.2+202955	NGC 2582 UGC 4391 PGC 23630	PSXS2.. U (1) 2.0± .8 1.1± .8	1.09± .03 .00± .01 .79± .01 1.10	 .16 .00 .00	13.90 ±.13 13.97 ±.18 13.71	.87± .02 .24± .04 .80 .20	.90± .01 .32± .04 13.34± .03 14.19± .24	17.39±.2 245± 6 244± 4 3.67	4439± 5 4461± 45 4354 4671
082530.3-680706 282.01 -16.90 196.11 -35.62 082510.0-675712	NGC 2601 ESO 60- 5 IRAS08251-6757 PGC 23637	.SXR1P? RS .7± .5 	1.20± .05 .17± .04 1.25	120 .53 .18 .09	 13.42 ±.14 11.70 12.67	 	 	 	3244± 74 3027 3376
082541.1+280710 195.19 31.92 52.87 -40.98 082238.4+281700	 UGC 4395 PGC 23643	.S..6*. U 6.0±1.3 	1.24± .04 .92± .05 1.25	154 .07 1.35 .46	 15.33 ±.18 13.90	 	 	15.78±.3 213± 7 1.42	2193± 10 2138 2407
082544.4+275227 195.47 31.86 53.05 -41.19 082242.0+280217	IC 2361 UGC 4394 IRAS08227+2802 PGC 23646	.S?.... 	1.06± .06 .39± .05 1.07	78 .09 .59 .20	 14.83 ±.18 14.14	 	 	17.32±.3 212± 5 2.99	2087± 9 2031 2301
082545.5+192659 204.74 29.18 59.33 -48.36 082252.6+193649	IC 2363 MCG 3-22- 11 IRAS08228+1936 PGC 23650	.SBS4*. P (1) 4.0± .9 2.0±1.0	.94± .11 .05± .07 .95	 .18 .07 .02	14.75 ±.18 14.46	 	 	 	7616± 73 7527 7851
0825.8 +7814 135.56 31.27 30.67 4.82 0818.9 +7824	 UGC 4360 PGC 23652	.S..8*. U 8.0±1.3 	1.00± .08 .55± .06 1.01	80 .11 .68 .27	 	 	 	 	
082552.7-133042 236.33 13.89 113.98 -67.92 082331.8-132051	 MCG -2-22- 11 PGC 23654	.L..+P* E -1.0±1.3 	1.01± .07 .10± .05 1.02	0 .25 .00 	 	 	 	 	
082558.4-114645 234.84 14.84 109.45 -67.58 082335.8-113653	 MCG -2-22- 12 PGC 23658	.S..7?/ E 7.0±1.8 	1.19± .05 .82± .05 1.21	75 .21 1.13 .41	 	 	 	 	2786 2599 3050
082604.7+455806 174.09 35.18 43.36 -24.98 082234.9+460756	 UGC 4393 IRAS08225+4607 PGC 23660	.SB?... 	1.35± .03 .15± .04 1.36	45 .15 .23 .08	 13.35 ±.20 13.25 12.96	 	 	14.08±.1 136± 7 113± 12 1.04	2125± 6 2152± 43 2142 2282
0826.1 +2140 202.38 30.03 57.62 -46.46 0823.2 +2150	 UGC 4400 PGC 23661	.S..6*. U 6.0±1.4 	1.13± .04 .96± .03 1.15	5 .14 1.41 .48	16.19S±.15 15.75 ±.18 14.44	.54± .06 -.12± .10 .30 -.29	.60± .06 -.07± .10 14.32± .26	15.39±.2 234± 10 222± 10 .47	4392± 7 4312 4622
082607.8+212725 202.62 29.96 57.79 -46.64 082312.7+213716	A 0823+21 UGC 4399 PGC 23662	.S..8*. U 8.0±1.3 	.98± .07 .45± .05 .59± .03 .99	30 .13 .56 .23	15.2 ±.2 15.41 ±.18 14.62	.53± .03 -.02± .03 .39 -.12	.62± .03 -.04± .03 13.63± .04 13.80± .40	16.39±.2 239± 10 235± 6 1.54	4487± 6 4406 4718
082612.9-540200 269.93 -9.15 186.58 -47.76 082453.0-535206	 ESO 164- 10 PGC 23666	.SX.9*. S (1) 9.0±1.0 10.0± .8	1.19± .08 	 1.65 	 	 	 	 	1054± 76 824 1238
082617.8+113001 213.08 26.21 67.25 -54.64 082333.0+113953	 UGC 4403 PGC 23668	.S?.... 	.98± .06 .47± .05 .99	79 .11 .71 .24	 15.26 ±.18 14.39	 	 	16.03±.3 361± 13 1.40	9530± 10 9527±125 9413 9780
082619.8-131811 236.21 14.09 113.47 -67.78 082358.7-130818	IC 2375 MCG -2-22- 14 PGC 23672	.SBS3P* E 3.0±1.2 	1.27± .05 .65± .05 1.29	83 .21 .90 .32	 	 	 	 	5892 5702 6156

R.A. 2000 DEC. / l b / SGL SGB / R.A. 1950 DEC.	Names / PGC	Type / S_T n_L / T / L	$\log D_{25}$ / $\log R_{25}$ / $\log A_e$ / $\log D_o$	p.a. / A_g / A_i / A_{21}	B_T / m_B / m_{FIR} / B_T^o	$(B-V)_T$ / $(U-B)_T$ / $(B-V)_T^o$ / $(U-B)_T^o$	$(B-V)_e$ / $(U-B)_e$ / m'_e / m'_{25}	m_{21} / W_{20} / W_{50} / HI	V_{21} / V_{opt} / V_{GSR} / V_{3K}
0826.3 +2750	IC 2365	.L.....	1.06± .11	45					6078
195.56 31.98	UGC 4402	U	.12± .05	.11	14.37 ±.15				6022
53.23 -41.16		-2.0± .8		.00					6293
0823.3 +2800	PGC 23673		1.05		14.17				
082621.0+225359		.S?....	.95± .06					15.85±.3	5379± 10
201.06 30.49	CGCG 119- 95		.78± .05	.18	15.92 ±.15			420± 13	5303
56.73 -45.39	KUG 0823+230B			1.14				344± 10	
082324.4+230351	PGC 23676		.97	.39	14.58			.88	5607
0826.4 +2016	IC 2369								7598± 22
203.93 29.61	CGCG 89- 27			.13	15.1 ±.3				7513
58.83 -47.60									7832
0823.5 +2026	PGC 23678								
082626.0-131824	IC 2379	.SBR1P?	1.01± .07	150					
236.23 14.11	MCG -2-22- 16	E	.21± .05	.21					
113.49 -67.76		1.0± .9		.21					
082404.9-130830	PGC 23681		1.02	.11					
082629.0+221543	UGC 4404	.S..8*.	1.04± .05	151				16.42±.2	8496± 7
201.77 30.31		U	.80± .05	.17	15.69 ±.19			429± 14	8498± 22
57.26 -45.92	IRAS08235+2225	8.0±1.4		.99	13.75			359± 7	8418
082333.1+222535	PGC 23684		1.06	.40	14.51			1.50	8725
0826.5 +2311		.S..0..	1.12± .05	28	14.7 ±.4	1.00± .03		15.91±.3	5652± 10
200.76 30.62	UGC 4405	U	.55± .05	.10	15.03 ±.18	.42± .05		517± 13	5578
56.57 -45.13		.0± .9		.41		.84		460± 10	5879
0823.6 +2321	PGC 23685		1.10		14.38	.33	13.75± .49		
0826.6 +2225									5374± 22
201.61 30.39	CGCG 119- 98			.17	15.34 ±.18				5297
57.18 -45.76									5603
0823.7 +2235	PGC 23687								
0826.6 +2257		RLB.0..	1.13± .07	125				16.68±.3	5898± 10
201.03 30.57	UGC 4406	U	.10± .03	.18	14.6 ±.2			251± 13	5823
56.77 -45.31		-2.0± .8		.00				240± 10	6126
0823.7 +2307	PGC 23688		1.13		14.30				
0826.7 +4848		.I..9*.	1.12± .07						
170.63 35.45	UGC 4401	U	.02± .06	.15					
42.22 -22.34		10.0±1.2		.01					
0823.2 +4858	PGC 23691		1.13	.01					
0826.8 +1722	NGC 2593	.S..0..	.96± .07	172					
207.07 28.66	UGC 4408	U	.29± .05	.10	14.95 ±.18				
61.48 -49.92		.0± .9		.22					
0824.0 +1732	PGC 23692		.96						
0826.9 +2022	IC 2373	.S..6?.	1.03± .08		15.1 ±.2	.44± .05	.52± .03	15.50±.3	7507± 14
203.86 29.75	UGC 4409	U	.00± .06	.13	14.7 ±.3	-.26± .08	-.23± .04	119± 25	7422
58.89 -47.45		6.0±1.6	.72± .05	.00		.36	14.22± .12		7741
0824.0 +2032	PGC 23695		1.04	.00	14.79	-.32	15.10± .47	.71	
0827.0 -1550		.SBS3P.	1.24± .07	10					
238.50 12.84		E	.76± .08	.18					
120.28 -67.88		3.0± .9		1.04					
0824.7 -1541	PGC 23697		1.26	.38					
0827.1 +2139		PSBS0..	1.04± .06		14.1 ±.2	.93± .04	1.01± .02	16.70±.3	7561± 10
202.49 30.24	UGC 4414	U	.04± .05	.13	14.5 ±.2	.38± .05	.50± .03	134± 14	7584± 22
57.92 -46.35		.0± .9	.89± .06	.03		.82	14.04± .22		7485
0824.2 +2149	PGC 23700		1.05		14.02	.39	14.03± .38		7796
082708.0+255815	NGC 2592	.E.....	1.23± .14	45					1989± 22
197.72 31.61	UGC 4411	U	.08± .08	.13	13.28 ±.15				1925
54.70 -42.68		-5.0± .8		.00					2210
082407.9+260810	PGC 23701		1.23		13.12				
082716.7+225240		.SBS3..	1.35± .03	165	13.75S±.15			14.61±.2	5530± 7
201.16 30.68	UGC 4416	U	.37± .03	.18	13.9 ±.2			399± 13	5454
57.01 -45.30	IRAS08243+2302	3.0± .8		.51	12.55			385± 6	5758
082420.2+230235	PGC 23705		1.37	.18	13.06		14.45± .23	1.37	
0827.3 +2312		.S?....	.82± .13					16.84±.3	5285± 10
200.81 30.80	MCG 4-20- 61		.57± .07	.12	16.0 ±.2			358± 13	5211
56.78 -45.02				.85				320± 10	5513
0824.4 +2322	PGC 23709		.83	.28	15.00			1.55	
082724.9+015037		.SXS3..	1.02± .06	0				15.80±.3	9412± 10
222.67 22.11	UGC 4421	U	.14± .05	.12	14.7 ±.2			260± 13	
80.82 -61.20		3.0± .9		.19					9263
082449.3+020033	PGC 23711		1.03	.07	14.31			1.42	9675

R.A. 2000 DEC.	Names	Type	logD₂₅	p.a.	B_T	(B-V)_T	(B-V)_e	m₂₁	V₂₁
l b		S_T n_L	logR₂₅	A_g	m_B	(U-B)_T	(U-B)_e	W₂₀	V_opt
SGL SGB		T	logA_e	A_i	m_FIR	(B-V)_T^o	m'_e	W₅₀	V_GSR
R.A. 1950 DEC.	PGC	L	logD_o	A₂₁	B_T^o	(U-B)_T^o	m'₂₅	HI	V_3K

R.A. 2000 DEC. / l b / SGL SGB / R.A.1950	Names / PGC	Type / S_T n_L / T / L	logD₂₅ / logR₂₅ / logA_e / logD_o	p.a. / A_g / A_i / A₂₁	B_T / m_B / m_FIR / B_T^o	(B-V)_T / (U-B)_T / (B-V)_T^o / (U-B)_T^o	(B-V)_e / (U-B)_e / m'_e / m'₂₅	m₂₁ / W₂₀ / W₅₀ / HI	V₂₁ / V_opt / V_GSR / V_3K
082726.7+171702 / 207.22 28.76 / 61.74 −49.91 / 082436.2+172658	NGC 2596 / UGC 4419 / IRAS08245+1726 / PGC 23714	.S..3.. / U (1) / 3.0± .9 / .5±1.2	1.17± .04 / .41± .05 / .75± .03 / 1.18	65 / .10 / .56 / .20	14.2 ±.2 / 14.23 ±.18 / 13.14 / 13.51	.73± .02 / .07± .04 / .59 / −.04	.81± .02 / .15± .03 / 13.41± .07 / 13.86± .29	15.03±.2 / 436± 13 / 432± 6 / 1.32	5934± 7 / / 5837 / 6175
082727.8+254325 / 198.02 31.61 / 54.96 −42.85 / 082428.1+255321	/ UGC 4418 / KUG 0824+258 / PGC 23717	.S..7.. / U / 7.0± .9 /	.96± .05 / .37± .05 / / .97	/ .13 / .51 / .19	/ 15.46 ±.18 / / 14.80			15.98±.3 / / 150± 7 / .99	2245± 10 / / 2180 / 2467
0827.5 +6413 / 151.80 34.62 / 36.05 −8.12 / 0823.1 +6423	/ UGC 4398 / / PGC 23719	.SAT4.. / U / 4.0± .8 /	1.05± .05 / .14± .05 / / 1.07	170 / .11 / .20 / .07	/ 14.9 ±.2 / / 14.49			16.03±.1 / / 339± 8 / 1.48	11059± 11 / 11199± 81 / 11149 / 11143
082734.0−124525 / 235.91 14.64 / 112.18 −67.39 / 082512.3−123528	/ MCG −2−22− 17 / IRAS08252−1235 / PGC 23723	.SAT6.. / E (1) / 6.0± .8 / 3.1± .8	1.36± .04 / .18± .05 / / 1.38	25 / .21 / .27 / .09	/ / 13.46 /				2804 / / 2615 / 3070
082741.7+212843 / 202.73 30.31 / 58.22 −46.42 / 082446.7+213840	NGC 2595 / UGC 4422 / 3ZW 59 / PGC 23725	.SXT5.. / PU / 4.5± .5 /	1.50± .03 / .12± .04 / 1.11± .02 / 1.51	45 / .12 / .19 / .06	12.93 ±.14 / 12.9 ±.3 / 12.85 / 12.59	.66± .03 / .08± .04 / .58 / .02	.77± .01 / .20± .02 / 13.97± .04 / 14.95± .22	14.56±.1 / 376± 4 / 338± 4 / 1.90	4330± 4 / 4469± 42 / 4250 / 4564
0827.7 −1446 / 237.68 13.57 / 117.48 −67.63 / 0825.4 −1437	/ / / PGC 23727	.SBR3?. / E (1) / 3.0±1.8 / 5.3± .8	1.21± .07 / .31± .08 / / 1.23	60 / .21 / .43 / .16					
0827.8 +7331 / 140.86 32.71 / 32.55 .49 / 0822.2 +7341	/ UGC 4390 / / PGC 23731	.SB.7.. / U / 7.0± .7 /	1.28± .06 / .07± .06 / / 1.28	/ .07 / .09 / .03	/ 14.9 ±.5 / / 14.76			14.46±.1 / 188± 8 / 162± 12 / −.33	2169± 6 / / 2290 / 2209
0827.8 +6316 / 152.93 34.80 / 36.45 −8.98 / 0823.5 +6326	/ MCG 11−11− 5 / / PGC 23737	.L?.... / / /	.93± .16 / .19± .07 / / .91	/ .12 / .00 /	/ 15.07 ±.15 / / 14.85				6825±106 / 6909 / 6911
0828.0 +5426 / 163.70 35.64 / 40.01 −17.11 / 0824.2 +5436	/ UGC 4415 / / PGC 23741	.SBS7.. / U / 7.0± .8 /	1.05± .08 / .17± .06 / / 1.06	155 / .12 / .24 / .09	/ 15.0 ±.2 / / 14.64			15.04±.1 / / 149± 8 / .32	3610± 7 / / 3660 / 3734
082806.2+550556 / 162.89 35.62 / 39.74 −16.49 / 082415.3+551553	/ MCG 9−14− 41 / ARAK 163 / PGC 23746	.S?.... / / /	.52± .20 / .00± .07 / / .53	/ .15 / .00 / .00	/ 15.2 ±.3 / 13.12 / 14.93				11329± 82 / 11382 / 11450
082808.1+201532 / 204.10 29.99 / 59.34 −47.38 / 082514.5+202530	/ UGC 4424 / / PGC 23748	.S?.... / / /	1.15± .04 / .19± .03 / / 1.16	0 / .13 / .28 / .09				15.36±.2 / 232± 13 / 214± 6 /	4442± 7 / / 4356 / 4678
0828.1 −0131 / 225.93 20.62 / 86.99 −63.02 / 0825.6 −0122	/ UGC 4430 / / PGC 23749	.S..0.. / U / .0± .9 /	1.00± .08 / .25± .06 / / 1.00	142 / .09 / .19 /	/ 14.88 ±.18 / /				
082810.7+554237 / 162.14 35.60 / 39.50 −15.93 / 082418.0+555234	A 0824+55 / UGC 4417 / MK 88 / PGC 23750	.S?.... / / /	.26± .24 / .00± .06 / / .27	/ .15 / .00 / .00	15.2 ±.2 / 15.1 ±.8 / / 14.98	.60± .04 / −.08± .08 / .51 / −.15	/ / 11.35±1.23 /		9180± 30 / 9235 / 9298
082813.0+000131 / 224.49 21.40 / 84.18 −62.13 / 082539.1+001130	/ UGC 4431 / / PGC 23752	.S..4.. / U / 4.0± .9 /	1.06± .06 / .08± .05 / / 1.07	145 / .16 / .12 / .04	/ 14.4 ±.2 / / 14.09			15.81±.3 / / 231± 7 / 1.68	9061± 10 / / 8906 / 9326
082814.4+280325 / 195.44 32.44 / 53.56 −40.76 / 082512.0+281323	/ UGC 4425 / / PGC 23753	.SB?... / / /	1.14± .05 / .43± .05 / / 1.15	130 / .11 / .65 / .22	/ 14.82 ±.19 / / 14.03			16.05±.3 / / 269± 7 / 1.81	5897± 10 / / 5841 / 6113
082815.3+010014 / 223.57 21.89 / 82.50 −61.55 / 082540.5+011013	/ UGC 4432 / / PGC 23755	.SA.4*. / UE (1) / 4.3± .7 / 5.3± .8	1.18± .03 / .18± .04 / / 1.19	135 / .12 / .26 / .09	/ 14.3 ±.3 / /				
082822.1−120829 / 235.48 15.14 / 110.70 −67.08 / 082559.8−115829	/ MCG −2−22− 18 / / PGC 23761	.SBS9.. / E (1) / 9.0± .8 / 7.5± .8	1.11± .06 / .07± .05 / / 1.14	85 / .25 / .07 / .04					

R.A. 2000 DEC. / l b / SGL SGB / R.A. 1950 DEC.	Names / PGC	Type / S_T n_L / T / L	$\log D_{25}$ / $\log R_{25}$ / $\log A_e$ / $\log D_o$	p.a. / A_g / A_i / A_{21}	B_T / m_B / m_{FIR} / B_T^o	$(B-V)_T$ / $(U-B)_T$ / $(B-V)_T^o$ / $(U-B)_T^o$	$(B-V)_e$ / $(U-B)_e$ / m'_e / m'_{25}	m_{21} / W_{20} / W_{50} / HI	V_{21} / V_{opt} / V_{GSR} / V_{3K}
082822.5+250729	IC 508	.S?....	.85± .07						2093± 33
198.76 31.62	MCG 4-20- 63		.09± .06	.13	14.93 ±.18				2026
55.63 -43.26	KUG 0825+252			.13					2317
082523.5+251728	PGC 23762		.86	.04	14.67				
0828.4 +4151	A 0825+42	.I..9*.	1.30± .04		15.0 ±.2	.43± .10	.39± .05	14.87±.1	393± 5
179.15 35.21	UGC 4426	PU (1)	.29± .05	.13		.33± .12	.30± .07	101± 6	
45.75 -28.53	DDO 52	10.0± .7	1.15± .05	.22		.33	16.21± .12	83± 8	393
0825.1 +4201	PGC 23769	9.0±1.0	1.31	.15	14.63	.26	15.58± .32	.10	566
082828.9+853632		.S..1..	1.17± .05	83					
127.50 28.86	UGC 4297	U	.46± .05	.23	14.47 ±.18				
27.94 11.67		1.0± .8		.47					
081351.9+854611	PGC 23770		1.20	.23					
0828.4 +3026	IC 2378	.S?....	.89± .11						14830± 28
192.73 33.11	MCG 5-20- 18		.00± .07	.13					14784
52.09 -38.67				.00					15040
0825.4 +3036	PGC 23771		.90						
082831.4+172758	A 0825+17	.S...P.	1.02± .06	15					
207.13 29.07	UGC 4433	R	.12± .05	.16	14.29 ±.18				
61.90 -49.61				.18					
082540.7+173758	PGC 23774		1.03	.06					
082838.0+734458	NGC 2550A	.S..5..	1.21± .03	0					3641± 10
140.58 32.70	UGC 4397	CU	.07± .04	.06	13.44 ±.18				
32.52 .73	IRAS08230+7354	5.0± .7		.11	12.96				3763
082259.0+735453	PGC 23781		1.21	.04	13.26				3680
082841.6-214414		PSBT2?.	1.38± .04	104					
243.70 9.88	ESO 562- 7	S	.48± .04	.40					
135.61 -67.04		2.0±1.2		.60					
082629.0-213412	PGC 23784		1.42	.24					
082846.8+403957		.S..6*.	1.02± .06	18					
180.60 35.12	UGC 4429	U	.33± .05	.16	15.30 ±.19				
46.40 -29.57		6.0±1.3		.49					
082526.9+404957	PGC 23788		1.03	.17					
082854.5+343905		.S?....	1.00± .05	105				16.45±.3	6286± 9
187.82 34.15	UGC 4434		.44± .05	.15	14.71 ±.20				6256
49.66 -34.92	IRAS08256+3449			.64				266± 5	6483
082543.6+344905	PGC 23798		1.01	.22	13.89			2.34	
082902.2-605805		.S..2*/	1.03± .05	133					
275.99 -12.72	ESO 124- 16	S	.65± .04	.89	16.04 ±.14				
191.29 -41.69	IRAS08280-6048	2.0±1.4		.80					
082804.1-604800	PGC 23804		1.11	.33					
0829.0 +6320		.L.....	1.20± .14	120					6918±106
152.83 34.92	UGC 4420	U	.28± .08	.10	14.36 ±.16				7002
36.55 -8.87		-2.0± .8		.00					7004
0824.7 +6330	PGC 23805		1.17		14.16				
0829.1 +5530		.S..4..	1.04± .06	7				15.28±.1	7758± 11
162.38 35.75	UGC 4427	U	.34± .05	.14	14.82 ±.18				7740± 46
39.72 -16.07		4.0± .9		.49				446± 8	7811
0825.3 +5540	PGC 23812		1.05	.17	14.14			.97	7877
082919.3+023729		.SBS7*.	1.13± .05	7					
222.17 22.90	UGC 4439	UE (1)	.35± .05	.10	14.9 ±.2				
80.18 -60.34		6.5± .8		.49					
082643.0+024732	PGC 23816	5.3± .8	1.14	.18					
0829.3 +5530		.S?....	.74± .12	129					
162.36 35.78	MCG 9-14- 44		.29± .06	.14	15.2 ±.3				7777± 46
39.74 -16.04	IRAS08255+5540			.22	13.44				7831
0825.5 +5540	PGC 23820		.74		14.69				7897
0829.4 +3139		.S?....	.82± .13						
191.35 33.60	MCG 5-20- 20		.28± .07	.13	15.3 ±.2				5468
51.54 -37.49				.21					5426
0826.3 +3150	PGC 23823		.81		14.87				5675
082938.9+520431	A 0825+52	.L..0P$	.70± .15	115	15.1 ±.2	.54± .06	.56± .03	15.95±.1	1732± 9
166.60 35.94	UGCA 140	P	.31± .09	.14		-.35± .10	-.37± .05	171± 6	1500± 77
41.24 -19.17	MK 89	-2.0±1.5	.26± .07	.00		.46	11.88± .24	134± 5	1769
082556.4+521433	PGC 23834		.67		14.93	-.39	12.68± .81		1863
0829.7 +6257		.S..6*.	1.11± .07	152					
153.26 35.05	UGC 4428	U	.83± .06	.10					
36.77 -9.18		6.0±1.4		1.22					
0825.4 +6308	PGC 23838		1.12	.42					

R.A. 2000 DEC. l b SGL SGB R.A. 1950 DEC.	Names PGC	Type S_T n_L T L	$\log D_{25}$ $\log R_{25}$ $\log A_o$ $\log D_o$	p.a. A_g A_i A_{21}	B_T m_B m_{FIR} B_T^o	$(B-V)_T$ $(U-B)_T$ $(B-V)_T^o$ $(U-B)_T^o$	$(B-V)_e$ $(U-B)_e$ m' m'_{25}	m_{21} W_{20} W_{50} HI	V_{21} V_{opt} V_{GSR} V_{3K}
0829.7 +7351 140.43 32.75 32.55 .85 0824.1 +7401	 UGC 4413 PGC 23840	.S..6*. U 6.0±1.3 	1.19± .06 .68± .06 1.20	105 .06 .99 .34	 15.42 ±.19 				
082950.1+484651 170.67 35.95 42.72 -22.16 082615.3+485654	 UGC 4436 IRAS08262+4856 PGC 23843	.S..4.. U 4.0± .9 	1.11± .05 .64± .05 1.12	42 .15 .94 .32	 15.38 ±.18 13.11 				
0829.9 +5157 166.74 35.98 41.33 -19.25 0826.2 +5208	 UGC 4437 PGC 23845	.S..6*. U 6.0±1.3 	1.06± .06 .60± .05 1.07	33 .14 .88 .30					
0829.9 -0732 231.63 17.93 99.90 -65.39 0827.5 -0722	 MCG -1-22- 14 PGC 23847	.S?.... 	1.08± .09 .70± .07 1.09	 .15 1.05 .35					1534 1358 1803
082959.7+524150 A 0826+52 165.83 35.98 41.02 -18.58 MK 90 082615.7+525153	 UGC 4438 PGC 23850	.S.R.P* R 	.87± .08 .06± .05 .89	15 .15 .08 .03	 14.1 ±.2 13.11 13.85				4279± 38 4322 4411
083002.2+171528 207.51 29.32 62.54 -49.56 082711.8+172533	 UGC 4444 PGC 23852	.SBS6?. U 6.0±1.7 	1.17± .05 .24± .05 1.19	125 .16 .36 .12	 14.11 ±.18 13.58			16.12±.2 189± 13 179± 6 2.42	2080± 7 1983 2324
083002.7+212921 202.94 30.83 58.87 -46.11 082707.9+213926	NGC 2598 UGC 4443 PGC 23855	.SB.1$. P 1.0± .9 	1.06± .05 .43± .04 1.07	3 .10 .44 .22	14.6 ±.3 14.98 ±.18 14.28	1.01± .04 .03± .07 .87 -.07	 13.66± .40	18.29±.3 307± 14 3.79	4609± 10 4584± 22 4524 4839
0830.0 +6700 148.40 34.39 35.21 -5.44 0825.4 +6711	 CGCG 311- 7 PGC 23856	 		 .16 	 15.39 ±.18 				4127±106 4225 4197
0830.2 -0254 227.47 20.38 90.21 -63.30 0827.7 -0244	 CGCG 4- 40 PGC 23859	 		 .08 	 15.3 ±.6 				12292± 60 12128 12560
083014.0+395318 181.59 35.29 47.09 -30.14 082655.6+400323	 UGC 4441 PGC 23860	.S..3.. U (1) 3.0±1.0 4.5±1.3	1.03± .06 .76± .05 1.04	38 .11 1.04 .38	 15.63 ±.18 				
0830.6 +2036 203.96 30.66 59.78 -46.76 0827.7 +2046	 UGC 4446 IRAS08277+2046 PGC 23878	.S..6*. U 6.0±1.5 	1.04± .08 1.06± .06 1.05	135 .13 1.47 .50	 16.1 ±.2 13.54 14.49			16.27±.3 339± 13 336± 10 1.28	6002± 10 5917 6239
083041.4+521757 A 0827+52 166.32 36.10 41.30 -18.90 082658.6+522803	 UGC 4442 PGC 23880	.SXS4.. U (1) 4.0± .8 2.7± .8	1.04± .06 .07± .05 1.05	110 .14 .10 .03	14.5 ±.2 14.36 ±.18 14.15	.53± .05 .45 	 14.36± .38	15.92±.1 178± 11 171± 11 1.74	5086± 10 5090± 59 5128 5220
0830.7 +2436 199.53 31.99 56.64 -43.40 0827.8 +2447	 CGCG 119-119 PGC 23881	.S?.... 	.95± .09 .09± .10 .96	 .09 .13 .04	 15.1 ±.2 14.83				12959± 22 12890 13187
083051.8+195035 204.81 30.44 60.48 -47.35 082758.8+200043	 UGC 4447 PGC 23885	.S..8*. U 8.0±1.3 	1.00± .16 .43± .12 1.01	30 .08 .53 .22				16.70±.3 163± 7 	4654± 7 4566 4893
083052.7+744116 139.45 32.59 32.29 1.65 082503.1+745119	 UGC 4423 KUG 0825+748 PGC 23886	.S..7.. U 7.0± .9 	1.02± .05 .19± .05 1.03	100 .06 .26 .09	 15.2 ±.2 				
083103.2-041146 228.77 19.90 93.01 -63.74 082833.3-040137	 MCG -1-22- 16 PGC 23894	.LA.-*. E -3.0± .9 	1.13± .11 .08± .07 1.14	35 .13 .00 					
0831.2 +0936 215.56 26.50 71.04 -55.24 0828.5 +0947	 UGC 4452 PGC 23900	.L...?. U -2.0±1.6 	1.08± .17 .04± .08 1.09	35 .15 .00 	 14.3 ±.2 				

R.A. 2000 DEC.	Names	Type	logD$_{25}$	p.a.	B$_T$	(B-V)$_T$	(B-V)$_\bullet$	m$_{21}$	V$_{21}$
l b		S$_T$ n$_L$	logR$_{25}$	A$_g$	m$_B$	(U-B)$_T$	(U-B)$_\bullet$	W$_{20}$	V$_{opt}$
SGL SGB		T	logA$_\bullet$	A$_I$	m$_{FIR}$	(B-V)$_T^o$	m'$_\bullet$	W$_{50}$	V$_{GSR}$
R.A. 1950 DEC.	PGC	L	logD$_o$	A$_{21}$	B$_T^o$	(U-B)$_T^o$	m'$_{25}$	HI	V$_{3K}$

083124.4-040212		PLX.OP?	1.11± .06	55					
228.67 20.06	MCG -1-22- 17	E	.30± .05	.13					
92.79 -63.59		-2.0±1.3		.00					
082854.3-035202	PGC 23908		1.08						
083124.8+273453		.S..6*.	1.04± .08	132				15.39±.3	5885± 10
196.21 32.99	UGC 4450	U	.88± .06	.10	15.7 ±.2				
54.69 -40.81		6.0±1.4		1.30				236± 7	5827
082823.3+274503	PGC 23910		1.05	.44	14.26			.69	6106
0831.4 +6059		.SAS5..	1.16± .05					15.34±.1	6330± 7
155.61 35.54	UGC 4445	U	.00± .05	.13	14.8 ±.3				
37.75 -10.91		5.0± .8		.00				47± 8	6405
0827.3 +6110	PGC 23913		1.18	.00	14.63			.70	6428
0831.5 +2201		.L.....	1.04± .18	156					4692± 22
202.47 31.33	UGC 4453	U	.53± .08	.10	15.27 ±.15				4613
58.84 -45.47		-2.0± .9		.00					4927
0828.6 +2212	PGC 23917		.97		15.10				
0831.5 -0112		.SBR1..	.97± .05	20					1347± 54
226.08 21.52	UGC 4455	U	.08± .04	.09	14.60 ±.18				1188
87.39 -62.13		1.0± .9		.09					1616
0829.0 -0102	PGC 23918		.98	.04	14.41				
083139.4-594708		.SBR4*P	1.21± .05	20					6366± 51
275.18 -11.79	ESO 124- 18	S (1)	.18± .05	.95	14.45 ±.14				6139
190.10 -42.53	FAIR 275	4.3± .7		.27					6533
083036.0-593654	PGC 23924	2.2± .8	1.30	.09	13.19				
083146.5-040712		.E+....	1.22± .07	45	13.7 ±.3	.96± .04	1.04± .02		
228.80 20.09	MCG -1-22- 18	E	.13± .08	.13					
93.05 -63.55		-4.0± .8	1.18± .11	.00			15.06± .39		
082916.5-035700	PGC 23929		1.21				14.49± .51		
083153.5-594657		.S?....	.88± .06						6381± 87
275.19 -11.77	ESO 124- 19		.06± .05	.95	14.88 ±.14				6154
190.06 -42.52	IRAS08308-5936			.10					6548
083050.1-593642	PGC 23930		.97	.03	13.80				
083158.1+191248	A 0829+19A	.SXT5P.	1.25± .06	125				15.43±.3	11159± 10
205.60 30.46	UGC 4457	R	.27± .06	.08	14.2 ±.2				11140± 48
61.34 -47.71	ARP 58	5.0± .4		.40				283± 7	11068
082905.8+192300	PGC 23935		1.26	.13	13.65			1.64	11400
083203.6+240028	IC 509	.SAT5..	1.26± .04		13.57 ±.15	.60± .04	.63± .02	15.39±.1	5488± 4
200.32 32.08	UGC 4456	U (1)	.05± .04	.09	13.8 ±.3		.00± .07	110± 5	5534± 47
57.44 -43.75	IRAS08290+2411	5.0± .8	1.01± .02	.08		.53	14.11± .05	82± 5	5416
082906.2+241040	PGC 23936	2.2± .8	1.27	.03	13.42		14.57± .25	1.94	5718
0832.0 +1911	A 0829+19B	CE.0.P$							
205.63 30.48		R		.08					
61.38 -47.71		-6.0±1.0							
0829.2 +1922	PGC 23937								
083210.9-020946	IC 510	.SBT4P.	1.03± .05	150					
227.05 21.18	UGC 4460	E	.16± .04	.09					
89.35 -62.50		4.0± .9		.24					
082939.1-015933	PGC 23940		1.04	.08					
083211.2+223347	NGC 2599	.SA.1..	1.27± .03		13.08M±.10	.84± .01	.86± .01	14.98±.2	4750± 5
201.94 31.65	UGC 4458	U	.05± .03	.10	13.12 ±.18	.35± .01	.36± .01	288± 6	4690± 71
58.60 -44.94	MK 389	1.0± .7	.75± .02	.05	13.18	.77	12.34± .06	257± 5	4672
082915.4+224400	PGC 23941		1.28	.02	12.88	.33	14.16± .19	2.08	4983
0832.2 +5550		.S?....	1.04± .09		14.56 ±.19	1.09± .04	1.24± .02		
161.92 36.17	MCG 9-14- 50		.38± .07	.16		.49± .06	.73± .03		
39.99 -15.57	1ZW 15		.54± .07	.57			12.75± .26		
0828.4 +5601	PGC 23943		1.05	.19			13.65± .53		
0832.2 +3031									9233
192.86 33.93	CGCG 149- 43			.12	15.3 ±.3				9187
52.94 -38.18									9446
0829.2 +3042	PGC 23945								
083228.1+523622	A 0828+52	.S?....	.94± .11						5101± 38
165.93 36.36	MCG 9-14- 53		.00± .07	.15	14.59 ±.18				5144
41.42 -18.51	MK 91			.00	11.50				5235
082844.9+524634	PGC 23955		.95	.00	14.41				
083246.9+001339		.S..7P?	1.13± .05	170					
224.90 22.50	UGC 4467	E (1)	.36± .05	.12	14.8 ±.2				
85.22 -61.08	IRAS08302+0023	7.0±1.7		.50					
083012.9+002354	PGC 23973	6.4±1.6	1.14	.18					

R.A. 2000 DEC. / l b / SGL SGB / R.A. 1950 DEC.	Names / PGC	Type / S_T n_L / T / L	$\log D_{25}$ / $\log R_{25}$ / $\log A_e$ / $\log D_o$	p.a. / A_g / A_i / A_{21}	B_T / m_B / m_{FIR} / B_T^o	$(B-V)_T$ / $(U-B)_T$ / $(B-V)_T^o$ / $(U-B)_T^o$	$(B-V)_e$ / $(U-B)_e$ / m' / m'_{25}	m_{21} / W_{20} / W_{50} / HI	V_{21} / V_{opt} / V_{GSR} / V_{3K}
0832.9 -0102									
226.11 21.91	CGCG 4- 50			.11	15.3 ±.6				11661
87.50 -61.75									11503
0830.4 -0052	PGC 23976								11931
083300.2+260049		.SB.6?.	1.14± .04	110				15.73±.3	5260± 10
198.12 32.88	UGC 4464	U	.72± .04	.11	15.33 ±.18				
56.20 -41.95	KUG 0830+261	6.0±1.3		1.06				295± 7	5196
083000.6+261104	PGC 23978		1.15	.36	14.13			1.24	5486
083305.3-122119	IC 513	.LBT0?.	1.03± .07	40					
236.32 15.99	MCG -2-22- 19	E	.20± .05	.30					
111.80 -65.99		-2.0± .9		.00					
083043.0-121103	PGC 23983		1.04						
083310.8-205353		.SXS5..	1.19± .04	52					
243.61 11.21	ESO 562- 13	SE (1)	.18± .04	.16					
133.08 -66.16		5.0± .6		.27					
083057.0-204336	PGC 23986	3.3± .8	1.21	.09					
083317.8-175723		.SB.3P/	1.15± .05	162					
241.15 12.91	ESO 562- 14	E	.70± .04	.22					
125.77 -66.39	IRAS08310-1747	3.0±1.8		.96	12.44				
083101.0-174706	PGC 23992		1.17	.35					
083317.9+411535		.S..1..	1.08± .06	160					
180.00 36.04	UGC 4465	U	.17± .05	.14	14.75 ±.20				
46.97 -28.66	IRAS08300+4125	1.0± .9		.18	13.64				
082957.7+412551	PGC 23993		1.09	.09					
083322.3+523150		.S..4..	1.23± .05	43					
166.02 36.50	UGC 4461	U	.52± .05	.15	14.32 ±.18				
41.59 -18.52	IRAS08296+5242	4.0± .9		.77	13.10				
082939.6+524205	PGC 23996		1.24	.26					
083322.6-225824	NGC 2613	.SAS3..	1.86± .01	113	11.16 ±.13	.91± .01	.99± .01	12.61±.1	1678± 5
245.36 10.05	ESO 495- 18	R (3)	.61± .02	.37	11.21 ±.11	.38± .02	.47± .02	614± 6	1619± 40
138.06 -65.73	IRAS08311-2248	3.0± .3	1.35± .02	.85	10.80	.70	13.40± .05	599± 6	1467
083111.0-224806	PGC 23997	3.0± .4	1.90	.31	9.96	.19	13.79± .16	2.34	1940
083322.9+293217	NGC 2604	.SBT6..	1.32± .05					14.33±.2	2094± 7
194.09 33.92	UGC 4469	U	.00± .06	.11	13.0 ±.2			151± 14	2012±155
53.85 -38.91	IRAS08303+2942	6.0± .8		.01	12.76			143± 7	2043
083019.2+294233	PGC 23998		1.33	.00	12.85			1.48	2310
0833.4 +8557		.SB?...	1.19± .05					16.37±.1	1913± 11
127.10 28.81	UGC 4348		.07± .05	.24	15.1 ±.4				2074
27.88 12.02				.10				190± 8	
0818.0 +8607	PGC 24001		1.22	.03	14.76			1.58	1895
083330.8+413131		.L.....	1.18± .15	165					
179.68 36.11	UGC 4468	U	.22± .08	.12	14.38 ±.17				
46.87 -28.40		-2.0± .8		.00					
083010.3+414147	PGC 24002		1.16						
083341.4+742420		.E.....	.90± .22	120					
139.71 32.85	UGC 4448	U	.19± .08	.06	14.39 ±.19				
32.58 1.47		-5.0± .9		.00					
082757.2+743433	PGC 24015		.85						
083350.1-131029	NGC 2612	.L..-./	1.43± .04	115					
237.13 15.69	MCG -2-22- 20	E	.68± .05	.27					
113.88 -65.96		-3.0± .8		.00					
083128.6-130010	PGC 24028		1.36						
083356.6+265821	NGC 2607	.S?....	.96± .06					15.17±.3	3528± 7
197.10 33.36	UGC 4473		.04± .05	.04	14.63 ±.18				3468
55.75 -41.03	KUG 0830+271			.06				67± 7	3753
083056.0+270839	PGC 24038		.97	.02	14.51			.64	
083356.8-215308		.SBS8*/	1.29± .04	68				14.53±.2	1777± 11
244.53 10.79	ESO 562- 19	SU (1)	.67± .04	.31				217± 16	
135.40 -65.82		7.5± .6		.82				210± 12	1569
083144.1-214248	PGC 24039	7.2± .9	1.32	.33					2042
083402.5+573354		.S?....	.71± .10	129					
159.77 36.26	MCG 10-12-146		.09± .06	.22	15.3 ±.2				5387± 57
39.49 -13.91	7ZW 239			.14					5449
083005.7+574411	PGC 24047		.73	.05	14.89				5501
0834.1 +6610	A 0829+66	.I..9..	1.19± .05	120	14.48 ±.17	.35± .06	.31± .04		19± 10
149.30 34.95	UGC 4459	U (1)	.06± .05	.10	14.9 ±.3	-.40± .08	-.42± .05		6±106
35.92 -6.05	DDO 53	10.0± .8	1.05± .03	.04		.31	15.22± .07		114
0829.5 +6621	PGC 24050	9.0±1.0	1.20	.03	14.41	-.43	15.15± .32		95

R.A. 2000 DEC. l b SGL SGB R.A. 1950 DEC.	Names PGC	Type S_T n_L T L	$\log D_{25}$ $\log R_{25}$ $\log A_o$ $\log D_o$	p.a. A_g A_i A_{21}	B_T m_B m_{FIR} B_T^o	$(B-V)_T$ $(U-B)_T$ $(B-V)_T^o$ $(U-B)_T^o$	$(B-V)_o$ $(U-B)_o$ m'_o m'_{25}	m_{21} W_{20} W_{50} HI	V_{21} V_{opt} V_{GSR} V_{3K}
083424.4-314922 252.73 5.05 156.64 -62.25 083223.0-313900	 ESO 431- 17 FAIR 1145 PGC 24063	.SAT2.. r 2.2± .9 	1.08± .09 .42± .07 1.24	68 1.77 .52 .21	 12.15 				
083433.2-023253 227.73 21.49 90.73 -62.18 083201.7-022232	NGC 2615 UGC 4481 IRAS08320-0222 PGC 24071	.SBT3.. UE 2.5± .5 	1.28± .04 .26± .05 1.28	40 .10 .35 .13	 13.30 ±.18 13.78 				
083442.0+750820 138.85 32.69 32.34 2.17 082848.5+751836	A 0828+75 MCG 13- 6- 25 MK 15 PGC 24079	.L?.... 	.64± .10 .27± .06 .61	 .06 .00 					6511± 38 6638 6545
0834.7 +5243 165.77 36.70 41.69 -18.25 0831.0 +5254	NGC 2600 UGC 4475 PGC 24082	.S..3.. U (1) 3.0± .9 3.5±1.2	1.09± .06 .48± .05 1.11	78 .15 .66 .24	 15.00 ±.18 				
083447.6-573848 273.63 -10.23 188.04 -44.09 083336.0-572824	 ESO 165- 1 PGC 24085	PSBR2*/ Sr 2.5± .6 	1.22± .06 .62± .05 1.31	106 .99 .85 .31					
0835.0 +2941 194.01 34.31 54.15 -38.59 0832.0 +2952	 MCG 5-20- 26 PGC 24098	.S?.... 	.94± .11 .00± .07 .95	 .10 .00 .00	 14.9 ±.2 14.77 				14456 14406 14674
083506.1-030546 228.31 21.33 91.92 -62.33 083235.1-025523	 MCG 0-22- 20 ARAK 170 PGC 24100	 	.64± .17 .00± .07 .64	 .08 	 15.3 ±.2 				6659± 63 6495 6932
083513.4+284511 195.13 34.11 54.82 -39.37 083210.8+285534	 UGC 4482 PGC 24104	.S..8*. U 8.0±1.4 	1.00± .08 .73± .06 1.01	125 .11 .89 .36	 15.74 ±.18 14.73 			16.85±.3 160± 7 1.76	2045± 10 1992 2266
0835.2 -0822 233.11 18.60 102.86 -64.45 0832.8 -0812	 PGC 24106	.L...P. E -2.0± .9 	1.18± .08 .16± .08 1.17	125 .09 .00 					
0835.2 +5548 161.93 36.59 40.40 -15.43 0831.4 +5559	 UGC 4478 PGC 24108	RSBS1.. U 1.0± .9 	.93± .10 .32± .06 .94	50 .16 .33 .16	 15.18 ±.18 				
083517.0+282827 195.45 34.05 55.03 -39.60 083214.8+283850	NGC 2608 UGC 4484 ARP 12 PGC 24111	.SBS3*. R (2) 3.0± .5 3.3± .7	1.36± .02 .22± .03 .93± .01 1.37	60 .12 .30 .11	13.01 ±.13 12.94 ±.13 12.06 12.55	.71± .01 .09± .02 .63 .03	.77± .01 .13± .02 13.15± .02 14.14± .19	15.71±.2 264± 11 210± 8 3.05	2135± 8 2120± 20 2078 2354
0835.3 +2333 201.10 32.66 58.66 -43.71 0832.4 +2344	 MCG 4-20- 69 PGC 24114	.S?.... 	.97± .10 .05± .07 .98	 .09 .07 .02	 14.53 ±.18 14.34 				5135± 22 5061 5370
083532.9+303156 193.06 34.61 53.70 -37.82 083228.2+304220	A 0832+30 MCG 5-20- 28 MK 390 PGC 24127	.S?.... 	.74± .14 .21± .07 .74	 .11 .16 	 15.5 ±.2 15.07 			16.32±.3 292± 6 	7640± 10 7200±110 7590 7852
083534.1-015058 227.22 22.07 89.71 -61.61 083301.9-014033	NGC 2616 UGC 4489 PGC 24129	.LXT0?. E -2.0± .8 	1.20± .08 .09± .05 1.20	145 .11 .00 	13.5 ±.2 14.1 ±.3 13.49	.99± .04 .44± .07 .87 .45	 14.14± .47		9000±190 8839 9273
083536.1-320850 253.14 5.06 156.96 -61.87 083335.0-315824	 ESO 431- 18 IRAS08335-3158 PGC 24131	.SBS9*. SU (1) 9.0± .5 7.1± .5	1.46± .04 .61± .05 1.62	169 1.77 .63 .31	 12.70 			14.15±.2 241± 16 234± 12 	1552± 11 1329 1802
083538.9-040521 229.30 20.94 93.98 -62.68 083308.8-035456	NGC 2617 MCG -1-22- 27 PGC 24136	.S..0P* E .0±1.2 	1.04± .07 .13± .05 1.05	 .13 .10 					
083539.9+462930 173.56 36.86 44.76 -23.80 083210.9+463953	A 0832+46 MK 92 PGC 24140	.S...$. R 	.80± .09 .26± .12 .28± .05 .81	 .13 	14.6 ±.2 	.47± .05 -.28± .09 	.53± .02 -.30± .04 11.52± .11 		4413± 46 4431 4574

R.A. 2000 DEC. / l b / SGL SGB / R.A. 1950 DEC.	Names / PGC	Type / S_T n_L / T / L	logD25 / logR25 / logA_o / logD_o	p.a. / Ag / Al / A21	BT / mB / mFIR / B_T^o	$(B-V)_T$ / $(U-B)_T$ / $(B-V)_T^o$ / $(U-B)_T^o$	$(B-V)_o$ / $(U-B)_o$ / m'_o / m'25	m21 / W20 / W50 / HI	V21 / Vopt / VGSR / V3K
083548.5+014315	UGC 4491	PSBT1*.	1.38± .03	60				15.42±.3	4102± 9
223.89 23.89		UE	.45± .04	.10	13.48 ±.18				
83.62 -59.60	IRAS08332+0153	1.0± .6		.46	13.73			412± 5	3952
083313.2+015341	PGC 24152		1.39	.22	12.88			2.32	4373
083553.3+004226	NGC 2618	PSAT2..	1.38± .04	140					
224.86 23.41	UGC 4492	UE	.10± .05	.13	13.0 ±.3				
85.31 -60.17		1.5± .5		.12					
083318.8+005252	PGC 24156		1.40	.05					
083605.5+015338	UGC 4494	PSXS1*.	1.11± .05	70					3800
223.76 24.03		UE	.32± .05	.10	14.65 ±.19				
83.42 -59.44		1.3± .7		.33					3650
083329.9+020404	PGC 24166		1.12	.16	14.17				4071
083615.0-262434	ESO 495- 21	.I.0.?P	1.24± .04						896± 22
248.57 8.58		S	.11± .04	.98	12.45 ±.14				
145.40 -64.14		90.0							680
083407.0-261406	PGC 24175		1.33						1157
083629.9-114952	MCG -2-22- 22	.S..7*/	1.28± .05	70					5914
236.34 16.98		E	.96± .05	.25					
110.95 -65.07		7.0±1.3		1.32					5727
083407.0-113923	PGC 24189		1.30	.48					6189
083635.5-202817	ESO 562- 23	.LBR+?/	1.34± .04	170					
243.72 12.11		SE	.64± .03	.34					
131.78 -65.42		-.5± .6		.00					
083421.0-201748	PGC 24195		1.28						
0836.6 +7749	UGC 4466	.S..9*.	1.14± .13	30				16.72±.1	1416± 11
135.80 31.93		U	.10± .12	.04					
31.35 4.66		9.0±1.2		.10				93± 8	1551
0830.0 +7800	PGC 24200		1.14	.05					1438
083639.4-832753	ESO 6- 3	.SXS6?.	1.21± .04	8					
296.52 -24.09		S (1)	.82± .05	.48	15.41 ±.14				
203.19 -21.59	IRAS08415-8317	6.0± .9		1.21	13.45				
084128.0-831712	PGC 24203	5.6± .9	1.25	.41					
083642.9+661355	A 0832+66	.S..1*.	1.00± .08	77	15.9 ±.2	.68± .05			
149.17 35.19	UGC 4490	P	.55± .06	.10	15.65 ±.18	-.27± .09			5271± 32
36.13 -5.89	MK 93	1.0±1.3		.56	12.95	.51			5366
083210.1+662420	PGC 24206		1.01	.27	15.03	-.36	14.38± .48		5348
083703.1+694650	UGC 4483	.I..9*.	1.05± .04	162				14.59±.1	156± 5
144.96 34.38		U	.17± .04	.15	15.0 ±.2			56± 4	
34.69 -2.65	KUG 0832+699	10.0±1.2		.13				49± 4	264
083207.0+695716	PGC 24213		1.06	.08	14.77			-.27	216
083710.6-221507	ESO 563- 2	.SBS5?/	1.25± .05	95					
245.28 11.19		S (1)	.66± .05	.13					
135.93 -65.02		5.3± .7		1.00					
083458.0-220436	PGC 24219	7.8± .7	1.26	.33					
083719.0-205625	ESO 563- 3	.SBR0P/	1.20± .04	88					
244.21 11.98		SE	.66± .04	.34					
132.84 -65.20	IRAS08350-2045	-.4± .8		.49	12.53				
083505.0-204554	PGC 24225		1.20						
083724.7-550727	NGC 2640	.LX.-..	1.35± .06	104					1051± 57
271.78 -8.46	ESO 165- 2	S	.06± .06	1.93					
185.58 -45.93	IRAS08360-5456	-3.0± .6		.00	11.40				821
083605.0-545654	PGC 24229		1.59						1238
083726.7+400208	UGC 4498	.SB.1..	.98± .07	10	*				
181.63 36.68		U	.16± .05	.12	14.71 ±.18				
48.42 -29.37		1.0± .9		.16					
083409.1+401238	PGC 24230		.99	.08					
083726.9+780134	NGC 2591	.S..6*/	1.48± .03	32	*			14.40±.1	1323± 5
135.56 31.90	UGC 4472	PU	.69± .04	.04	12.95 ±.20			277± 6	
31.30 4.86	IRAS08307+7811	5.5± .8		1.01	12.30			257± 6	1459
083044.9+781158	PGC 24231		1.49	.34	11.90			2.15	1344
083728.0+245630	NGC 2620	.S?....	1.31± .04	93				15.48±.3	7838± 10
199.70 33.54	UGC 4501		.59± .05	.08	14.40 ±.19				
58.14 -42.29				.88				654± 7	7770
083430.0+250700	PGC 24233		1.32	.29	13.39			1.79	8071
083732.6+284222	NGC 2619	.S..4..	1.36± .03	35	*			14.06±.3	3474± 7
195.34 34.59	UGC 4503	U	.21± .03	.07	13.19 ±.19				
55.41 -39.14	IRAS08345+2852	4.0± .7		.31	13.03			394± 7	3420
083430.3+285253	PGC 24235		1.37	.11	12.78			1.17	3697

R.A. 2000 DEC.	Names	Type	logD$_{25}$	p.a.	B$_T$	(B-V)$_T$	(B-V)$_e$	m$_{21}$	V$_{21}$
l b		S$_T$ n$_L$	logR$_{25}$	A$_g$	m$_B$	(U-B)$_T$	(U-B)$_e$	W$_{20}$	V$_{opt}$
SGL SGB		T	logA$_e$	A$_i$	m$_{FIR}$	(B-V)^o_T	m'$_e$	W$_{50}$	V$_{GSR}$
R.A. 1950 DEC.	PGC	L	logD$_o$	A$_{21}$	B^o_T	(U-B)^o_T	m'$_{25}$	HI	V$_{3K}$
083734.1-165556		.SB.1P?	1.09± .06	75	*				
240.88 14.33	MCG -3-22- 10	E	.35± .05	.20					
123.23 -65.38	IRAS08352-1645	1.0±1.8		.36	13.47				
083516.0-164524	PGC 24236		1.11	.18					
083741.6+513908	A 0834+51A	.SX.8..	1.42± .04	140	*			13.71±.1	692± 5
167.09 37.20	UGC 4499	U	.13± .05	.13	13.5 ±.3			145± 5	660± 48
42.62 -19.02		8.0± .7		.16				126± 4	730
083402.0+514938	PGC 24242		1.44	.06	13.25			.40	833
083743.0+203019		.S..7..	1.14± .04	30	*			15.79±.3	4703± 7
204.73 32.18	UGC 4504	U	.27± .04	.10	14.7 ±.2			239± 13	
61.81 -45.87		7.0± .8		.37					4617
083449.7+204050	PGC 24244		1.15	.13	14.17			1.48	4947
083756.1-022747	A 0835-02	CE...?.	.63± .09		*				1917± 34
228.12 22.26	UGC 4508	E	.00± .04	.08	15.0 ±.3				1754
91.46 -61.41	ARAK 173	-5.0±1.9		.00					2193
083524.5-021714	PGC 24253		.65		14.90				
083801.4-094912		.SXR4..	1.05± .06	110	*				
234.79 18.39	MCG -2-22- 23	E (1)	.32± .05	.13					
106.56 -64.23	IRAS08356-0938	4.0± .9		.47					
083536.6-093839	PGC 24259	3.1±1.2	1.06	.16					
083811.0+245344	NGC 2622	.S?....	.89± .11						8554± 20
199.81 33.68	MCG 4-21- 8		.24± .07	.06	15.01 ±.19				8485
58.35 -42.24	MK 1218			.29					8787
083513.1+250417	PGC 24269		.89	.12	14.58				
0838.2 -0738		.SBR2P?	1.12± .08	5	*				
232.89 19.62		E	.58± .08	.07					
101.85 -63.50		2.0± .9		.71					
0835.8 -0728	PGC 24273		1.13	.29					
083819.5-733233		.LAS0*.	1.16± .05	125	*				5279± 15
287.42 -18.90	ESO 36- 5	S	.09± .03	.71	13.71 ±.14				5070
198.02 -30.34		-2.0± .8		.00					5392
083839.0-732154	PGC 24280		1.23		12.92				
0838.3 +6902		.S..6*.	.98± .05	70	*				6666±125
145.79 34.68	UGC 4495	U	.18± .04	.14	15.07 ±.18				6771
35.10 -3.27		6.0±1.3		.26					6730
0833.5 +6913	PGC 24281		.99	.09	14.64				
083823.1+194259	NGC 2625		.50± .17						4510± 45
205.67 32.06	CGCG 89- 57		.07± .06	.05	15.9 ±.2				4422
62.69 -46.41	MK 625								4757
083530.7+195333	PGC 24285		.51						
083823.9+173753		.L?....	.72± .22						7988± 57
207.96 31.32	MCG 3-22- 20		.00± .07	.06	15.15 ±.18				7892
64.60 -48.04	ARAK 174			.00					8239
083533.6+174827	PGC 24286		.72		14.97				
083824.2+254501	NGC 2623	.P.....	1.38± .06		13.99 ±.15	.63± .03	.64± .02	17.05±.3	5535± 8
198.84 33.97	UGC 4509	R	.53± .07	.09		.11± .06	.20± .05		5472± 27
57.76 -41.51	ARP 243	99.0		.65	9.85	.47		126± 16	5464
083525.3+255535	PGC 24288		1.39	.27	13.19	.02	14.40± .37	3.59	5761
083833.7+304756	IC 2387	.S..6*.	1.06± .06	18				15.26±.3	7679± 10
192.92 35.31	UGC 4511	U	.35± .05	.12	14.73 ±.18				7634
54.23 -37.25	IRAS08354+3058	6.0±1.3		.52	13.28			370± 7	7897
083529.1+305830	PGC 24299		1.07	.18	14.06			1.02	
083837.0-642032		.IBS9..	1.20± .07	92					
279.49 -13.73	ESO 90- 4	S (1)	.11± .07	.71					
192.37 -38.24		10.0± .8		.08					
083750.1-640954	PGC 24303	8.9± .9	1.26	.06					
083844.6+433252		.S..3..	1.17± .04	43					
177.28 37.25	UGC 4507	U	.60± .04	.10	14.91 ±.18				
46.78 -26.16	KUG 0835+437	3.0± .9		.83					
083521.5+434326	PGC 24309		1.18	.30					
083847.2-750923		.IXT9..	1.50± .04	165				13.61±.3	1135± 9
288.88 -19.75	ESO 36- 6	Sr (1)	.36± .05	.63	14.19 ±.14				929
198.88 -28.91		9.6± .5		.27				144± 7	1241
083926.0-745842	PGC 24312	8.9± .8	1.56	.18	13.29			.14	
083904.7-144425		.SAS5*.	1.09± .06	115					
239.22 15.87	MCG -2-22- 25	E (1)	.28± .05	.30					
118.04 -64.89	IRAS08367-1433	5.0±1.2		.42					
083644.4-143348	PGC 24328	4.2±1.2	1.11	.14					

R.A. 2000 DEC.	Names	Type S_T n_L	$\log D_{25}$ $\log R_{25}$	p.a. A_g	B_T m_B	$(B-V)_T$ $(U-B)_T$	$(B-V)_\bullet$ $(U-B)_\bullet$	m_{21} W_{20}	V_{21} V_{opt}
l b		T	$\log A_\bullet$	A_i	m_{FIR}	$(B-V)_T^o$	$m'_\bullet$	W_{50}	V_{GSR}
SGL SGB									
R.A. 1950 DEC.	PGC	L	$\log D_o$	A_{21}	B_T^o	$(U-B)_T^o$	m'_{25}	HI	V_{3K}

083907.9-085144		.S..0*.	1.09± .06	60					
234.10 19.14	MCG -1-22- 31	E	.53± .05	.15					
104.63 -63.69		.0±1.3		.40					
083642.1-084107	PGC 24331		1.07						
0839.2 +7142		.SXS6..	1.17± .04	45					
142.66 34.03	UGC 4500	U	.14± .04	.07	15.0 ±.3				
34.05 -.82		6.0± .8		.20					
0834.1 +7153	PGC 24341		1.18	.07					
0839.6 +6057		.S..6*.	1.07± .05	110				15.96±.1	7912± 11
155.48 36.52	UGC 4512	U	.30± .04	.13	15.10 ±.19				
38.68 -10.53		6.0±1.2		.44				382± 8	7987
0835.5 +6108	PGC 24348		1.08	.15	14.50			1.32	8014
0839.6 +5327	A 0835+53	.SB.6?.	1.33± .05	70				14.15±.1	694± 6
164.81 37.40	UGC 4514	U	.36± .06	.10	13.87 ±.19			167± 8	
42.06 -17.28		6.0±1.6		.53				147± 12	739
0835.9 +5338	PGC 24351		1.34	.18	13.23			.74	829
083954.3+734515		.SB.6*.	1.21± .03	75				15.22±.1	3285± 11
140.30 33.45	UGC 4502	U	.75± .04	.03	15.44 ±.19				
33.23 1.05	KUG 0834+739	6.0±1.3		1.10				205± 8	3407
083422.1+735550	PGC 24360		1.21	.37	14.30			.55	3327
084004.8-342429		.SBS9*.	1.17± .05	165					1929
255.52 4.45	ESO 371- 3	S (1)	.18± .05	2.01	14.75 ±.14				
159.94 -59.91		9.0±1.2		.18					1704
083806.1-341348	PGC 24370	7.8±1.2	1.36	.09	12.56				2179
084009.5+522725		.SBR3*.	1.17± .05	175				15.74±.1	4974± 7
166.06 37.54	UGC 4515	U	.38± .05	.10	14.38 ±.18				
42.60 -18.14		3.0± .8		.52				376± 8	5015
083628.6+523803	PGC 24372		1.18	.19	13.72			1.83	5113
084014.4+053802		.S..7..	1.14± .05	49				15.03±.3	1940± 10
220.66 26.72	UGC 4524	U	.81± .05	.10	15.17 ±.18				
79.01 -56.33		7.0±1.0		1.11				159± 7	1802
083735.5+054843	PGC 24374		1.15	.40	13.95			.68	2212
084022.6+233226	NGC 2628	.SXR5?.	1.06± .06					15.20±.2	3622± 7
201.55 33.75	UGC 4519	U	.02± .05	.07	13.97 ±.18			268± 7	
59.98 -43.06	IRAS08374+2342	5.0±1.7		.03	13.48			259± 5	3548
083726.4+234306	PGC 24381		1.06	.01	13.84			1.35	3861
084044.5-040719	NGC 2642	.SBR4..	1.31± .02	140	13.35 ±.15	.77± .05	.94± .04	15.20±.3	4342± 17
230.04 22.01	MCG -1-22- 33	R (3)	.03± .03	.07	12.6 ±.2	.05± .05		150± 34	4406± 58
95.27 -61.55	IRAS08382-0356	4.0± .3	.94± .03	.04	12.94	.72	13.54± .09	116± 25	4179
083814.4-035636	PGC 24395	1.8± .5	1.32	.01	12.99	.01	14.68± .21	2.20	4627
084048.7+732913	IC 511	.S..0*.	1.20± .06	143					
140.58 33.60	UGC 4510	RU	.44± .03	.04	14.44 ±.18				
33.40 .84		-.4± .6		.33					
083520.2+733950	PGC 24397		1.18						
084050.8-320244		.SBS5?/	1.33± .05	129					
253.73 6.01	ESO 432- 2	S (1)	1.01± .05	1.49					
155.67 -60.95	IRAS08387-3151	5.0± .7		1.50	13.17				
083849.0-315200	PGC 24398	3.3±1.3	1.47	.50					
084052.7+425015		.S..4..	1.07± .04	112					
178.20 37.59	UGC 4521	U	.56± .04	.15	15.37 ±.18				
47.54 -26.60	KUG 0837+430	4.0± .9		.82					
083731.2+430056	PGC 24399		1.09	.28					
084054.3+192118		.S..2..	1.15± .05	59				15.61±.3	4379± 10
206.32 32.49	UGC 4526	U	.61± .05	.07	14.84 ±.18				
63.69 -46.33		2.0± .9		.76				340± 7	4289
083802.4+193200	PGC 24400		1.15	.31	13.97			1.33	4629
084058.8+161101		.S..8..	1.04± .06					16.25±.3	4290± 7
209.80 31.35	UGC 4528	U	.03± .05	.07					
66.73 -48.76		8.0± .9		.03				92± 7	4189
083810.1+162144	PGC 24403		1.05	.01					4547
0841.1 +6651		.S..8*.	1.21± .05	123					3864±106
148.30 35.48	UGC 4516	U	.75± .05	.10	15.13 ±.18				3961
36.27 -5.14		8.0±1.3		.92					3939
0836.6 +6702	PGC 24410		1.22	.37	14.11				
084118.4-044251		PSBR1..	1.18± .05	140					4407
230.66 21.82	MCG -1-22- 34	E	.16± .05	.06					
96.53 -61.68		1.0± .8		.16					4238
083848.8-043207	PGC 24414		1.18	.08					4687

R.A. 2000 DEC.	Names	Type	logD₂₅	p.a.	B_T	(B-V)_T	(B-V)_e	m₂₁	V₂₁
l b		S_T n_L	logR₂₅	A_g	m_B	(U-B)_T	(U-B)_e	W₂₀	V_opt
SGL SGB		T	logA_e	A_i	m_FIR	(B-V)°_T	m'_e	W₅₀	V_GSR
R.A. 1950 DEC.	PGC	L	logD_o	A₂₁	B°_T	(U-B)°_T	m'_25	HI	V_3K
0841.5 +5115		.SBS3..	1.03± .06	88					
167.56 37.81	UGC 4525	U	.17± .05	.12	14.88 ±.19				
43.37 -19.11		3.0± .9		.23					
0837.9 +5126	PGC 24423		1.04	.08					
0841.5 +6626									11603±106
148.79 35.62	CGCG 311- 12			.12	15.50 ±.18				11699
36.49 -5.50									11680
0837.0 +6637	PGC 24424								
084132.8+045858	NGC 2644	.S?....	1.33± .03	14				15.21±.2	1939± 7
221.47 26.70	UGC 4533		.41± .04	.06	13.31 ±.18			257± 13	1990± 76
80.31 -56.49	IRAS08389+0509			.61	12.55			201± 7	1800
083854.5+050943	PGC 24425		1.33	.20	12.62			2.39	2214
084135.0-201858		.SBT4*.	1.21± .03	82					
244.29 13.15	ESO 563- 11	SE (2)	.31± .03	.47					
131.11 -64.28	IRAS08393-2008	4.2± .7		.46					
083920.1-200812	PGC 24427	4.7± .7	1.26	.16					
084137.8+464735		.S..4..	1.03± .05	123					
173.21 37.89	UGC 4529	U	.68± .05	.10					
45.60 -23.06	KUG 0838+469	4.0±1.0		1.00					
083809.3+465819	PGC 24428		1.04	.34					
084138.5-204434		.SBS7P*	1.23± .03	69					1890
244.65 12.91	ESO 563- 12	SE (2)	.33± .03	.46					
132.08 -64.22		7.3± .6		.46					1684
083924.1-203348	PGC 24429	6.0± .6	1.28	.17					2164
084141.2+185225		.S..4..	1.19± .05	34				15.22±.3	4619± 10
206.93 32.49	UGC 4532	U	.96± .05	.07	15.47 ±.18				4527
64.34 -46.59		4.0±1.0		1.41				222± 7	4871
083849.9+190310	PGC 24431		1.19	.48	13.96			.78	
084152.9+325208		.SB.3..	1.05± .05	40				16.38±.3	7728± 10
190.61 36.44	UGC 4531	U	.41± .04	.16	14.56 ±.19			466± 13	
53.61 -35.12		3.0± .9		.57					7691
083846.0+330253	PGC 24438		1.06	.21	13.77			2.40	7942
0842.0 +3049									8045
193.11 36.04	CGCG 150- 17			.12	15.4 ±.3				7999
55.01 -36.83									8266
0839.0 +3100	PGC 24440								
0842.1 +5735		.SX.4*.	1.02± .06	37					
159.59 37.33	UGC 4530	U	.18± .05	.20					
40.48 -13.42		4.0± .9		.27					
0838.2 +5746	PGC 24442		1.03	.09					
084225.9+371313	NGC 2638	.S..0..	1.23± .05	72					3730±155
185.28 37.29	UGC 4534	U	.44± .05	.12	13.75 ±.19				3710
51.02 -31.34		.0± .8		.33					3930
083913.3+372400	PGC 24453		1.22		13.25				
084232.5-195225		.SBT5*.	1.18± .03	131					1623
244.05 13.59	ESO 563- 13	SE (1)	.62± .03	.42					
130.05 -64.10	IRAS08402-1941	4.7± .8		.93					1418
084017.0-194136	PGC 24454	6.7± .9	1.22	.31					1899
0842.5 +7058		.S..6*.	1.00± .06						
143.42 34.49	UGC 4522	U	.03± .05	.11	14.80 ±.19				
34.60 -1.37		6.0±1.2		.04					
0837.5 +7109	PGC 24455		1.01	.01					
084235.6+103512		.SB.8..	1.19± .05					14.92±.3	2047± 10
215.92 29.45	UGC 4540	U	.02± .05	.21	13.8 ±.2				1926
73.29 -52.58		8.0± .8		.02				137± 7	2315
083952.2+104600	PGC 24457		1.21	.01	13.53			1.38	
084240.1+141710	NGC 2648	.S..1..	1.51± .03	148				15.68±.2	2060± 10
212.03 30.99	UGC 4541	U	.47± .05	.12	12.74 ±.18				1918± 57
69.15 -49.91	ARP 89	1.0± .8		.48				416± 10	1947
083953.3+142758	PGC 24464		1.52	.24	12.11			3.33	2318
0842.7 +2717									7673
197.37 35.32	CGCG 150- 18			.09	15.3 ±.3				7614
57.67 -39.69									7904
0839.7 +2728	PGC 24467								
0842.7 +1416		.S?....	.94± .11	79					2126± 57
212.05 31.00	MCG 2-22- 6		.68± .07	.12	15.4 ±.2				2018
69.19 -49.91	IRAS08399+1427			1.02	13.64				2388
0839.9 +1427	PGC 24469		.95	.34	14.27				

R.A. 2000 DEC.	Names	Type	logD$_{25}$	p.a.	B$_T$	(B-V)$_T$	(B-V)$_\bullet$	m$_{21}$	V$_{21}$
l b		S$_T$ n$_L$	logR$_{25}$	A$_g$	m$_B$	(U-B)$_T$	(U-B)$_\bullet$	W$_{20}$	V$_{opt}$
SGL SGB		T	logA$_\bullet$	A$_I$	m$_{FIR}$	(B-V)$_T^o$	m'	W$_{50}$	V$_{GSR}$
R.A. 1950 DEC.	PGC	L	logD$_o$	A$_{21}$	B$_T^o$	(U-B)$_T^o$	m'$_{25}$	HI	V$_{3K}$
084245.5+354535		.S..6*.	.96± .09	149				15.81±.3	2914± 9
187.10 37.13	UGC 4537	U	.98± .06	.10					
51.97 -32.56		6.0±1.5		1.44				176± 5	2888
083934.9+355623	PGC 24470		.97	.49					3119
084248.4+725843	NGC 2614	.SAR5*.	1.39± .03	150				15.23±.1	3457± 6
141.10 33.90	UGC 4523	R	.08± .04	.07	13.6 ±.3			259± 8	
33.75 .44	IRAS08373+7309	5.0± .3		.12	13.76			252± 6	3576
083726.7+730927	PGC 24473		1.40	.04	13.39			1.80	3504
084252.7+250404		.SA.9..	1.05± .05					15.96±.1	5185± 6
199.99 34.74	UGC 4542	U	.02± .05	.05	14.7 ±.3			165± 5	
59.40 -41.49		9.0± .8		.02				153± 7	5116
083955.1+251453	PGC 24475		1.06	.01	14.60			1.36	5422
084255.8+133829		.SB?...	1.06± .06					15.38±.2	5029± 7
212.75 30.79	UGC 4545		.08± .05	.10	14.12 ±.18			159± 14	
69.91 -50.34				.12				117± 7	4918
084009.6+134918	PGC 24476		1.07	.04	13.88			1.46	5292
084257.3-200308		.SB.7?/	1.36± .03	83					
244.26 13.57	ESO 563- 14	SE	.68± .03	.45					
130.43 -63.98	IRAS08407-1952	6.7± .7		.93	11.89				
084042.0-195218	PGC 24479		1.40	.34					
084304.0-112822		.SXR5*.	1.11± .06	135					11524
236.96 18.51	MCG -2-22- 27	E (1)	.11± .05	.23					
110.92 -63.42		5.0± .6		.16					11337
084040.5-111732	PGC 24482	1.9± .6	1.13	.05					11806
084307.7+181320		.S?....	1.04± .06	50				15.80±.3	6273± 10
207.79 32.58	UGC 4548		.75± .05	.06	15.4 ±.2				
65.34 -46.87	IRAS08403+1824			1.13	13.48			367± 7	6179
084017.1+182410	PGC 24485		1.04	.38	14.21			1.22	6527
0843.2 +2649									5157
197.96 35.30	CGCG 150- 20			.09	15.2 ±.3				5096
58.13 -40.02									
0840.2 +2700	PGC 24486								5390
084315.7-203945	A 0841-20	.SXS9..	1.36± .03	92					1732
244.81 13.27	ESO 563- 16	SE (2)	.38± .03	.43					
131.80 -63.85	IRAS08410-2028	9.0± .5		.38	13.33				1526
084101.0-202854	PGC 24489	6.5± .5	1.40	.19					2007
084316.2+130510		.S..3..	1.43± .04	4				13.85±.3	2068± 10
213.38 30.64	UGC 4550	U (1)	.81± .05	.24	14.26 ±.19				
70.62 -50.69	IRAS08405+1315	3.0± .9		1.11	13.16			264± 7	1955
084030.5+131600	PGC 24490	5.5±1.2	1.45	.40	12.89			.56	2332
0843.3 +1044		.S..3..	1.02± .06	40					
215.86 29.67	UGC 4552	U (1)	.20± .05	.18	14.62 ±.18				
73.32 -52.34		3.0± .9		.28					
0840.6 +1055	PGC 24492	4.5±1.2	1.03	.10					
084321.6+454410		.SA.8..	1.52± .08	160				13.80±.1	1960± 6
174.56 38.17	UGC 4543	U	.25± .12	.11	14.3 ±.5			130± 8	
46.43 -23.85		8.0± .7		.31				108± 12	1974
083955.6+455459	PGC 24493		1.53	.13	13.89			-.22	2129
084322.7-172319		.IBS9..	1.31± .04	10					2019
242.09 15.20		E (1)	.36± .05	.31					
124.33 -64.00		10.0± .9		.27					1819
084104.7-171227	PGC 24494	9.8± .8	1.34	.18					2298
0843.4 +8438		.S..3..	1.12± .05	70					
128.38 29.54	UGC 4474	U (1)	.77± .05	.20	15.64 ±.18				
28.64 10.92	IRAS08315+8448	3.0±1.0		1.07	13.62				
0831.5 +8448	PGC 24497	4.5±1.3	1.14	.39					
084326.0+033653		.SB?...	1.09± .06	30				15.19±.3	8136± 10
223.07 26.47	UGC 4553		.34± .05	.07	14.51 ±.18				
82.84 -56.95	IRAS08408+0347			.51	13.27			388± 7	7991
084048.9+034744	PGC 24499		1.10	.17	13.89			1.13	8413
0843.5 +6415		.S..3..	1.07± .06	130					
151.35 36.32	UGC 4535	U (1)	.40± .05	.18	15.5 ±.2				11601±106
37.63 -7.37		3.0± .9		.55					11688
0839.2 +6426	PGC 24501	2.5±1.2	1.09	.20	14.65				11690
084338.0+501224	NGC 2639	RSAR1*$	1.26± .05	140	12.56 ±.13	.91± .03			3336± 11
168.87 38.19	UGC 4544	R	.22± .05	.11	12.72 ±.14	.34± .06			3198± 13
44.20 -19.89	IRAS08400+5023	1.0± .6		.23	12.03	.82			3315
084003.1+502314	PGC 24506		1.27	.11	12.26	.28	13.16± .29		3434

R.A. 2000 DEC. l　　b SGL　SGB R.A. 1950 DEC.	Names PGC	Type S_T　n_L T L	$\log D_{25}$ $\log R_{25}$ $\log A_o$ $\log D_o$	p.a. A_g A_i A_{21}	B_T m_B m_{FIR} B_T^o	$(B-V)_T$ $(U-B)_T$ $(B-V)_T^o$ $(U-B)_T^o$	$(B-V)_\bullet$ $(U-B)_\bullet$ $m'_\bullet$ m'_{25}	m_{21} W_{20} W_{50} HI	V_{21} V_{opt} V_{GSR} V_{3K}
084340.6+220532 203.49　34.01 62.02　−43.77 084046.1+221623	 UGC　4554 PGC 24509	.S..6*. U 6.0±1.3 	1.11± .05 .55± .05 1.12	9 .10 .81 .28	 15.14 ±.18 14.22			15.82±.3 256± 15 1.32	3710± 10 3630 3956
0843.7　+6511 150.22　36.13 37.24　−6.53 0839.3　+6522	 UGC　4536 PGC 24510	.SBR5.. U 5.0± .7 	1.21± .06 .00± .06 1.23	 .20 .01 .00	 14.9 ±.4 14.65				7383±106 7474 7467
084348.4−785654 292.46　−21.54 200.65　−25.46 084530.0−784554	 ESO　18− 13 IRAS08455−7845 PGC 24516	.SBS5.. S　　　(1) 5.0± .8 3.3±1.2	1.28± .05 .45± .05 1.33	121 .45 .68 .23	 13.42 				
0843.8　+5200 166.59　38.13 43.35　−18.28 0840.2　+5211	 UGC　4546 PGC 24517	.S...1.. U 1.0±1.0 	1.13± .05 .81± .05 1.13	25 .09 .82 .40	 15.01 ±.20 14.03				5181 5221 5325
0843.8　+6503 150.38　36.18 37.31　−6.64 0839.5　+6514	 UGC　4538 PGC 24520	.SXS5.. U 5.0± .8 	1.06± .08 .02± .06 1.08	140 .19 .03 .01				16.22±.1 190± 8 	6982± 11 7072 7067
084402.1−125137 238.31　17.93 114.09　−63.45 084139.8−124043	 MCG −2−23− 1 PGC 24525	.SBR3*. E　　　(1) 3.0± .6 3.1± .7	1.21± .04 .28± .04 1.23	85 .25 .38 .14					5713 5523 5995
0844.0　+6457 150.49　36.22 37.38　−6.72 0839.7　+6508	 MCG 11−11− 26 PGC 24526	.E?.... 	.78± .13 .00± .04 .81	 .19 .00 	 15.24 ±.11 14.95				6948±106 7038 7034
084406.7+494738 169.39　38.28 44.48　−20.22 084032.7+495830	 UGC　4551 PGC 24528	.L...?. U −2.0±1.7 	1.31± .05 .48± .03 1.25	113 .10 .00 	13.4 ±.2 13.18 ±.18 13.15	.94± .01 .47± .02 .86 .41	 13.63± .34 		1745± 31 1776 1898
084407.3+665745 148.08　35.74 36.49　−4.91 083933.6+670835	 UGC　4539 IRAS08395+6708 PGC 24529	.S..6*. U 6.0±1.3 	.96± .07 .13± .05 .97	18 .11 .20 .07	 14.53 ±.18 13.44 14.21			15.65±.1 130± 8 1.37	3660± 11 3528±128 3757 3735
084408.0+300707 194.08　36.32 55.96　−37.17 084104.7+301800	 UGC　4559 IRAS08410+3018 PGC 24530	.S...2.. U 2.0± .8 	1.51± .03 .82± .05 1.53	50 .18 1.01 .41	 14.1 ±.2 12.93			14.42±.1 360± 9 348± 6 1.08	2085± 6 2037 2310
084408.1+344304 188.45　37.24 52.91　−33.30 084059.1+345356	NGC　2649 UGC　4555 IRAS08409+3453 PGC 24531	.SXT4*. PU 3.5± .5 	1.20± .03 .02± .04 1.21	 .10 .03 .01	 13.07 ±.18 13.28 12.92			15.75±.2 251± 6 247± 7 2.82	4244± 6 4075± 76 4213 4453
084409.0+333101 189.93　37.03 53.68　−34.32 084101.5+334153	 UGC　4558 PGC 24532	.S...4.. U 4.0± .9 	1.22± .05 .72± .05 1.23	17 .11 1.06 .36	 15.07 ±.18 13.84			15.31±.3 471± 5 1.11	7673± 9 7638 7887
0844.2　+7655 136.63　32.64 32.11　4.02 0838.0　+7706	 UGC　4527 PGC 24539	.I..9*. U 10.0±1.2 	1.14± .13 .18± .12 1.15	40 .06 .13 .09				15.92±.1 85± 6 79± 5 	721± 10 853 749
084414.9+414301 179.68　38.13 48.77　−27.28 084055.8+415353	 UGC　4556 IRAS08409+4153 PGC 24540	.I?.... 	.93± .05 .42± .04 .94	21 .11 .31 .21	 14.4 ±.3 13.98				1794± 33 1792 1979
084422.4+585025 157.98　37.45 40.17　−12.17 084024.5+590117	 UGC　4549 IRAS08403+5901 PGC 24545	.S..8?. U 8.0±1.6 	1.20± .05 .09± .05 1.22	10 .15 .11 .04	 13.98 ±.20 13.21 13.72			14.92±.1 134± 8 1.15	1288± 11 1355 1402
0844.4　+0932 217.23　29.40 75.08　−52.98 0841.7　+0943	 UGC　4565 PGC 24548	.S..6*. U 6.0±1.5 	.96± .09 .98± .06 .98	114 .25 1.44 .49					4067 3942 4338
084430.1−202101 244.73　13.69 131.03　−63.60 084215.1−201006	 ESO　563− 17 IRAS08422−2010 PGC 24558	.SAT1?. SE 1.3± .6 	1.29± .03 .27± .04 1.33	29 .46 .28 .13	 12.75 				

R.A. 2000 DEC. l b SGL SGB R.A. 1950 DEC.	Names PGC	Type S_T n_L T L	logD_25 logR_25 logA_e logD_o	p.a. A_g A_l A_21	B_T m_B m_FIR B_T^o	(B-V)_T (U-B)_T (B-V)_T^o (U-B)_T^o	(B-V)_e (U-B)_e m'_e m'_25	m_21 W_20 W_50 HI	V_21 V_opt V_GSR V_3K
084442.6+094759 217.00 29.58 74.84 -52.74 084200.0+095854	 UGC 4567 PGC 24566	.S?.... 	.96± .09 .69± .06 .98	167 .21 1.03 .34	 15.5 ±.2 14.26			15.55±.3 242± 7 .95	4083± 10 3959 4354
084443.2+102811 216.30 29.87 74.03 -52.28 084200.0+103906	 UGC 4568 IRAS08419+1039 PGC 24567	.I..9?. U 10.0±1.8 	1.20± .04 .64± .04 1.22	80 .20 .48 .32	 14.65 ±.18 13.65 13.97			15.21±.3 222± 13 .92	4095± 10 3973 4365
0844.8 +5229 165.96 38.25 43.25 -17.78 0841.2 +5240	 UGC 4560 PGC 24572	.S..2.. U 2.0± .9 	1.08± .06 .61± .05 1.09	122 .10 .75 .31	 15.55 ±.18 				
084452.0+010914 225.65 25.58 87.04 -58.05 084217.1+012010	 UGC 4571 PGC 24573	.S..6*. U 6.0±1.3 	.96± .09 .39± .06 .97	160 .11 .58 .20	 15.35 ±.18 14.65			15.66±.3 182± 7 .81	4007± 10 3855 4288
0844.9 +4744 172.02 38.45 45.64 -21.97 0841.4 +4755	 UGC 4562 PGC 24575	.SBS5.. U 5.0± .9 	1.07± .07 .16± .06 1.08	127 .10 .24 .08				16.13±.1 306± 8 	8667± 11 8690 8829
084506.2-312016 253.72 7.16 153.56 -60.46 084303.0-310918	 ESO 432- 8 PGC 24588	.LBR+.. r -1.3± .9 	1.12± .09 .14± .04 1.22	141 1.23 .00 					
084508.2-334740 255.67 5.65 157.87 -59.32 084308.0-333642	NGC 2663 ESO 371- 14 PGC 24590	.E..... PBS -4.6± .4 	1.54± .04 .16± .04 1.26± .06 1.74	110 1.59 .00 	* 11.85 ±.14 10.23		1.24± .01 .80± .01 13.66± .21 		2102± 26 1876 2356
084516.3+274924 196.91 36.01 57.88 -38.93 084215.8+280020	 UGC 4570 PGC 24594	.SX.8.. U 8.0± .9 	1.00± .16 .09± .12 1.01	 .12 .11 .04	*			16.22±.3 202± 7 	6440± 7 6382 6672
084516.4+093846 217.23 29.64 75.18 -52.75 084233.9+094943	NGC 2657 UGC 4573 IRAS08425+0949 PGC 24595	.SAT7*. U 7.0± .8 	1.13± .07 .01± .06 1.15	 .21 .02 .01	* 13.61 ±.18 13.37				4141± 10 4016 4413
084518.5+443140 176.11 38.48 47.41 -24.74 084154.9+444236	 HICK 35C PGC 24596	.L?.... 	 	 .11 	16.04S±.15				16357± 41 16367 16532
084520.6+443033 176.14 38.48 47.42 -24.75 084157.1+444129	 HICK 35B PGC 24597	.L?.... 	 	 .11 	15.48S±.15				16338± 41 16348 16513
084521.3+443114 176.12 38.48 47.42 -24.74 084157.7+444210	 MCG 8-16- 28 HICK 35A PGC 24601	.S?.... 	.52± .20 .46± .07 .53	 .11 .67 .23	16.07S±.15 15.18		12.37±1.01		15919± 41 15929 16094
084521.7+854427 127.22 29.09 28.17 11.93 083101.8+855505	IC 499 UGC 4463 PGC 24602	.S..1.. U 1.0± .8 	1.32± .04 .19± .05 1.34	80 .25 .19 .09	* 13.37 ±.18 12.90				1883± 61 2043 1867
084538.1+365603 185.75 37.88 51.83 -31.26 084226.3+370701	A 0842+37 UGC 4572 MK 626 PGC 24620	.S?.... 	.81± .06 .00± .04 .82	 .11 .00 .00	 14.1 ±.3 13.07 13.94				3920± 40 3899 4123
084540.2+235208 201.61 35.00 61.05 -42.08 084244.0+240306	 UGC 4575 PGC 24621	.S..3.. U (1) 3.0± .9 5.5±1.2	1.09± .06 .42± .05 1.10	118 .09 .58 .21	* 15.13 ±.20 14.37			15.98±.3 585± 5 1.40	12913± 9 12840 13156
084555.7+124655 214.01 31.11 71.69 -50.45 084310.5+125754	 UGC 4582 PGC 24629	.S?.... 	1.03± .06 .53± .05 1.04	150 .17 .79 .26	 15.01 ±.18 13.99			15.43±.3 270± 7 1.17	8987± 10 8976± 57 8873 9254
0845.9 +1246 214.01 31.12 71.70 -50.44 0843.2 +1257	 MCG 2-23- 5 IRAS08432+1257 PGC 24631	.S?.... 	.94± .11 .40± .07 .95	5 .17 .60 .20	 15.07 ±.19 13.43 14.25				8931± 57 8817 9198

R.A. 2000 DEC.	Names	Type S_T n_L	$\log D_{25}$ $\log R_{25}$	p.a. A_g	B_T m_B	$(B-V)_T$ $(U-B)_T$	$(B-V)_\bullet$ $(U-B)_\bullet$	m_{21} W_{20}	V_{21} V_{opt}
l b		T	$\log A_\bullet$	A_i	m_{FIR}	$(B-V)^o_T$	$m'_\bullet$	W_{50}	V_{GSR}
SGL SGB									
R.A. 1950 DEC.	PGC	L	$\log D_o$	A_{21}	B^o_T	$(U-B)^o_T$	m'_{25}	HI	V_{3K}

084559.8+123705	NGC 2661	.S..6*.	1.14± .05		*			15.22±.2	4111± 6
214.19 31.06	UGC 4584	U	.01± .05	.17	13.52 ±.18			113± 6	
71.89 -50.56	IRAS08432+1248	6.0±1.1		.02	12.54			82± 6	3997
084314.7+124804	PGC 24632		1.16	.01	13.31			1.90	4379
084601.3-191812	NGC 2665	RSBT1..	1.31± .03	144	*				
244.08 14.59	ESO 563- 19	SUE	.13± .03	.40					
128.63 -63.32	IRAS08437-1907	1.3± .4		.13	11.21				
084345.1-190712	PGC 24634		1.34	.06					
084612.2+354148		.S..6*.	.96± .09	81	*			16.36±.3	4047± 9
187.32 37.81	UGC 4579	U	.98± .06	.09					4021
52.72 -32.25		6.0±1.5		1.44				169± 5	4255
084302.2+355247	PGC 24640		.97	.49					
084614.1+413447		.S..3..	1.06± .06	65	*				
179.89 38.49	UGC 4578	U (1)	.20± .05	.09	14.86 ±.19				
49.21 -27.21		3.0± .9		.27					
084255.6+414546	PGC 24641	3.5±1.2	1.07	.10					
0846.2 +2721		.S?....	.94± .11		*				
197.55 36.10	MCG 5-21- 5		.57± .07	.13	15.5 ±.2				5602
58.46 -39.19	IRAS08432+2732			.85					5543
0843.2 +2732	PGC 24643		.95	.28	14.44				5836
084633.5+482550		.S..4..	.91± .09	7	*				
171.12 38.72	UGC 4580	U	.29± .06	.10	14.7 ±.2				7012±155
45.55 -21.22		4.0± .9		.43					7037
084302.9+483650	PGC 24654		.92	.15	14.08				7172
084634.6+362623	A 0843+36		.62± .11		*				
186.40 38.00	CGCG 180- 2		.26± .12	.07	15.8 ±.2				3215
52.33 -31.58	MK 627								3192
084323.6+363724	PGC 24655		.63						3421
084635.6+190109		.S..6*.	1.23± .05	159	*			15.95±.3	4265± 9
207.25 33.63	UGC 4588	U	.80± .05	.09	14.97 ±.18				4174
65.51 -45.74		6.0±1.3		1.17				274± 5	4521
084344.4+191210	PGC 24656		1.23	.40	13.69			1.86	
0846.7 +0657		.S..2..	.95± .07	70	*				
220.16 28.77	UGC 4594	U	.29± .05	.22	14.80 ±.18				
79.03 -54.23		2.0± .9		.36					
0844.1 +0709	PGC 24665		.97	.15					
0846.8 +2810	IC 2393	.E.....	1.11± .16	20	*				
196.59 36.43	UGC 4589	U	.15± .08	.17	14.18 ±.15				6319
57.97 -38.44		-5.0± .8		.00					6263
0843.8 +2822	PGC 24669		1.09		13.91				6551
084658.7+214251		.SB?...	1.19± .05	143	*			15.35±.3	3691± 10
204.22 34.62	UGC 4592		.44± .05	.09	14.9 ±.2				3610
63.19 -43.60				.66				132± 7	3941
084404.9+215353	PGC 24673		1.19	.22	14.10			1.03	
084658.9+281414		.S..6*.	1.15± .05	18	*			16.95±.3	6437± 9
196.53 36.48	UGC 4591	U	.89± .05	.17	15.58 ±.18				6381
57.97 -38.37	IRAS08439+2825	6.0±1.4		1.31	13.39			313± 7	6669
084358.2+282516	PGC 24674		1.16	.44	14.07			2.43	
084705.5-334546		PSBT0*.	1.52± .04		*				
255.90 5.99	ESO 371- 16	S	.08± .05	1.50	12.78 ±.14				
157.43 -58.98	IRAS08450-3334	.0±1.0		.06	12.11				
084505.0-333442	PGC 24676		1.66						
084706.8+281409	IC 2394	.SBS3..	1.17± .05	90	*			16.50±.3	6385± 10
196.54 36.50	UGC 4595	U	.32± .05	.17	14.6 ±.2			329± 15	
58.00 -38.36		3.0± .8		.44					6329
084406.0+282512	PGC 24678		1.19	.16	13.94			2.40	6618
0847.1 +1937		RLAR+..	1.11± .07	155	*				
206.61 33.97	UGC 4596	U	.06± .03	.06	14.32 ±.20				
65.09 -45.18		-1.0± .8		.00					
0844.3 +1949	PGC 24680		1.10						
084714.6+725907	NGC 2629	.LAR0*.	1.25± .09	105	13.33 ±.14	1.05± .01			
140.97 34.20	UGC 4569	P	.07± .07	.03	12.82 ±.16	.62± .03			3650± 31
34.04 .59		-2.0± .7		.00		1.00			3769
084155.6+731006	PGC 24682		1.24		13.02	.62	14.25± .49		3698
084716.7-200204		.SA.4*/	1.48± .02	64					
244.87 14.40	ESO 563- 21	SUE	.84± .03	.53					
130.20 -62.97	IRAS08450-1951	4.0± .6		1.24	12.56				
084501.1-195100	PGC 24685		1.53	.42					

R.A. 2000 DEC. / l b / SGL SGB / R.A. 1950 DEC.	Names / PGC	Type / S_T n_L / T / L	$\log D_{25}$ / $\log R_{25}$ / $\log A_e$ / $\log D_o$	p.a. / A_g / A_i / A_{21}	B_T / m_B / m_{FIR} / B_T^o	$(B-V)_T$ / $(U-B)_T$ / $(B-V)_T^o$ / $(U-B)_T^o$	$(B-V)_e$ / $(U-B)_e$ / m'_e / m'_{25}	m_{21} / W_{20} / W_{50} / HI	V_{21} / V_{opt} / V_{GSR} / V_{3K}
084723.0+493325 169.67 38.81 45.09 -20.17 084350.2+494428	A 0843+49 UGC 4587 PGC 24688	.L...?. U -2.0±1.7 	1.14± .09 .28± .05 1.11	8 .07 .00 	 13.78 ±.16 13.67				3060± 60 3090 3216
084739.6+255330 199.39 36.01 59.91 -40.19 084441.5+260434	 UGC 4597 IRAS08446+2604 PGC 24698	.S?.... 1.03	1.02± .06 .31± .05 	70 .11 .47 .16	 14.84 ±.18 13.43 14.22			15.76±.3 324± 7 1.39	6541± 10 6476 6781
0847.6 +1324 213.53 31.75 71.46 -49.70 0844.9 +1336	 UGC 4599 PGC 24699	RLA.0.. U -2.0± .7 	1.30± .05 .00± .03 1.32	 .10 .00 	 13.6 ±.3 				
0847.8 +5351 164.16 38.57 42.98 -16.36 0844.1 +5403	NGC 2656 MCG 9-15- 25 VV 703 PGC 24707	.LA.-P* P -3.0± .8 	1.12± .12 .00± .07 1.12	 .07 .00 	14.9 ±.2 14.63 	1.14± .04 .40± .06 1.00 .46	 15.35± .67		13557± 55 13604 13695
084756.1+254948 199.48 36.05 60.03 -40.20 084458.0+260053	 UGC 4602 PGC 24710	.S..2.. U 2.0± .9 	1.09± .06 .42± .05 1.10	130 .11 .52 .21	 14.89 ±.18 14.20			16.65±.3 393± 7 2.24	5653± 10 5588 5893
084756.3+733217 140.32 34.06 33.83 1.11 084231.8+734318	IC 2389 UGC 4576 IRAS08425+7343 PGC 24711	.SBS3$. R 3.0± .5 	1.20± .06 .66± .05 1.20	126 .04 .91 .33	14.0 ±.3 13.18 13.03	.64± .06 .49 13.18± .43		14.53±.1 173± 34 180± 8 1.17	2382± 9 2599± 58 2508 2432
084805.1+174210 208.88 33.49 67.15 -46.50 084515.3+175316	IC 2406 UGC 4606 IRAS08452+1753 PGC 24721	.S..0.. U .0± .8 	1.14± .05 .24± .05 1.14	173 .05 .18 	 14.18 ±.18 12.35 				
0848.1 +7253 141.04 34.29 34.13 .54 0842.8 +7305	NGC 2641 UGC 4577 PGC 24722	.L...*. U -2.0±1.2 	1.13± .07 .08± .03 1.12	5 .07 .00 	 14.63 ±.20 				
084806.6+740558 139.68 33.87 33.59 1.61 084235.8+741700	NGC 2633 UGC 4574 ARP 80 PGC 24723	.SBS3.. R (1) 3.0± .3 1.8± .8	1.39± .03 .20± .03 .74± .07 1.39	175 .05 .28 .10	12.9 ±.2 12.53 ±.13 10.16 12.31	.66± .04 .60 	.74± .04 12.11± .21 14.21± .27	14.17±.1 292± 14 254± 8 1.77	2160± 7 2169± 23 2284 2203
0848.1 +2929 195.08 37.02 57.31 -37.20 0845.1 +2941	IC 2404 CGCG 150- 34 PGC 24725			.12	15.4 ±.3				8017 7966 8247
084810.0+173651 208.99 33.48 67.26 -46.55 084520.3+174757	IC 2407 UGC 4607 PGC 24726	.S..4.. U 4.0±1.0 	1.07± .06 .70± .05 1.08	86 .05 1.03 .35	 15.28 ±.19 14.16			16.57±.3 416± 5 2.06	6175± 9 6079 6435
084810.4+374519 184.80 38.48 51.83 -30.29 084457.8+375625	IC 2401 UGC 4600 PGC 24728	.L...*. U -2.0±1.2 	.94± .13 .09± .05 .94	110 .06 .00 	 14.82 ±.15 				
0848.2 +7834 134.73 32.21 31.56 5.59 0841.5 +7845	 UGC 4563 PGC 24731	.I..9*. U 10.0±1.2 	1.00± .08 .14± .06 1.00	130 .00 .10 .07					
084818.3-025831 230.05 24.21 94.86 -59.35 084547.1-024724	 CGCG 5- 9 PGC 24736			.02	15.4 ±.6				3941± 57 3776 4227
084818.9-030115 230.10 24.19 94.94 -59.37 084547.7-025007	 MCG 0-23- 2 PGC 24737	.SAS5*. E (1) 5.0± .8 3.1±1.6	1.17± .05 .25± .05 1.17	127 .02 .38 .13	 14.16 ±.19 13.73				3899± 57 3734 4185
0848.3 +0103 226.22 26.30 88.07 -57.36 0845.8 +0115	 UGC 4610 PGC 24743	.S..4.. U 4.0± .9 	1.00± .08 .55± .06 1.01	145 .11 .81 .27	 15.45 ±.18 				
084824.4+471721 172.58 39.05 46.44 -22.07 084456.4+472827	 MCG 8-16- 30 ARAK 179 PGC 24745	.S?.... 	.64± .17 .11± .07 .64	 .08 .14 .06	 15.1 ±.3 13.27 14.81				8840± 60 8861 9007

R.A. 2000 DEC.	Names	Type	logD$_{25}$	p.a.	B$_T$	(B-V)$_T$	(B-V)$_\bullet$	m$_{21}$	V$_{21}$
l b		S$_T$ n$_L$	logR$_{25}$	A$_g$	m$_B$	(U-B)$_T$	(U-B)$_\bullet$	W$_{20}$	V$_{opt}$
SGL SGB		T	logA$_\bullet$	A$_i$	m$_{FIR}$	(B-V)$_T^o$	m'$_\bullet$	W$_{50}$	V$_{GSR}$
R.A. 1950 DEC.	PGC	L	logD$_o$	A$_{21}$	B$_T^o$	(U-B)$_T^o$	m'$_{25}$	HI	V$_{3K}$

084824.6+734012	NGC 2636	.E.0.*.	.80± .12		14.66 ±.13	.88± .01			
140.15 34.04	UGC 4583	R	.00± .04	.04	14.47 ±.19				1896± 60
33.80 1.24		-5.0± .8	.00		.85			2017
084258.9+735115	PGC 24747		.80		14.53		13.64± .63		1941
084825.1+181949	IC 2409	.SXS1..	1.00± .06	165				15.64±.3	6329± 10
208.21 33.79	UGC 4608	U	.12± .05	.05	14.40 ±.18				6235
66.64 -45.98	IRAS08455+1830	1.0± .9		.13	13.60			140± 7	6588
084534.8+183056	PGC 24748		1.00	.06	14.14			1.44	
084825.1+735803	NGC 2634	.E.1.*.	1.23± .07		12.91 ±.15	.93± .01	.95± .01		
139.82 33.94	UGC 4581	R	.02± .06	.03	12.59 ±.17	.47± .04	.52± .04		2268± 31
33.67 1.50		-5.0± .5	.92± .04	.00		.90	13.00± .13		2390
084256.1+740906	PGC 24749		1.23		12.71	.48	14.02± .42		2311
0848.5 +7832		.I..9*.	1.00± .16					15.35±.1	3752± 11
134.74 32.23	UGC 4566	U	.09± .12	.00					3890
31.58 5.58		10.0±1.2		.07				213± 8	3772
0841.8 +7844	PGC 24758		1.00	.04					
0848.5 +1248									6506
214.28 31.71	CGCG 61- 13			.08	15.4 ±.6				6392
72.37 -49.98									6776
0845.8 +1300	PGC 24759								
084837.3+735618	NGC 2634A	.SBS4$/	1.25± .05	73				14.22±.1	2086± 9
139.85 33.96	UGC 4585	P	.68± .05	.03	14.36 ±.18			228± 34	
33.69 1.48	IRAS08433+7406	4.0± .9		1.00	14.09			180± 25	2209
084308.7+740721	PGC 24760		1.26	.34	13.32			.56	2130
084840.2+010218		RSBS1*.	1.29± .04	70				15.50±.2	8643± 7
226.29 26.35	UGC 4613	UE	.11± .04	.11	13.8 ±.3			210± 7	8738± 57
88.19 -57.31		1.3± .6		.12				195± 5	8492
084605.5+011326	PGC 24762		1.30	.06	13.45			2.00	8929
084850.5+295212		.S..7..	1.26± .04	107				15.49±.3	5957± 10
194.67 37.26	UGC 4611	U	.98± .05	.13	15.42 ±.18				5908
57.20 -36.80		7.0± .9		1.35				403± 7	6186
084548.0+300320	PGC 24771		1.27	.49	13.91			1.09	
084859.4+700634	A 0844+70	.S?....	.81± .11					15.72±.1	3626± 9
144.22 35.27	UGC 4593		.20± .06	.04	13.9 ±.4			210± 11	3569± 28
35.47 -1.91	IRAS08441+7017			.27	12.63				3729
084407.0+701740	PGC 24775		.81	.10	13.62			2.00	3683
084859.9+461458	A 0845+46				15.3 ±.2	.46± .04	.54± .03		
173.92 39.15				.09		-.42± .07	-.44± .05		6600± 77
47.09 -22.92	MK 96		.19± .03		13.63		11.69± .07		6617
084534.0+462606	PGC 24777								6771
084900.5-074954		.SBS4..	1.17± .05	20					2912
234.60 21.75	MCG -1-23- 2	E (1)	.19± .05	.09					2734
104.14 -61.05		4.0± .8		.28					3200
084633.5-073844	PGC 24778	3.1± .8	1.18	.09					
084912.5+601313	NGC 2654	.SB.2*/	1.63± .02	63	12.74 ±.15	.98± .02	1.06± .02	13.64±.1	1341± 4
156.13 37.81	UGC 4605	PU (1)	.73± .04	.17	12.77 ±.11	.50± .05	.58± .05	427± 3	1360± 56
40.08 -10.66	IRAS08451+6024	2.0± .5	1.02± .03	.89	13.65	.80	13.33± .08	397± 4	1413
084511.4+602421	PGC 24784	4.0±1.0	1.65	.36	11.68	.32	13.92± .20	1.59	1451
0849.2 +7517		.S..6*.	1.00± .08	166					
138.30 33.51	UGC 4586	U	.84± .06	.02					
33.11 2.71		6.0±1.4		1.24					
0843.5 +7529	PGC 24787		1.00	.42					
084916.9+360713		.S?....	.94± .06					15.44±.3	7556± 9
186.90 38.49	UGC 4614		.14± .05	.08	14.38 ±.19				7731±155
53.07 -31.56	IRAS08461+3618			.21	13.11			148± 5	7532
084606.8+361822	PGC 24788		.94	.07	14.06			1.31	7766
084922.3+190430	NGC 2672	.E.1+..	1.47± .04		12.7 ±.5	1.01± .01	1.02± .01		
207.46 34.26	UGC 4619	R	.02± .04	.05	11.99 ±.19	.62± .03	.60± .02		4255± 46
66.18 -45.27		-5.0± .3	.85± .28	.00		.96	12.48± .97		4164
084631.3+191540	PGC 24790		1.47		11.96	.63	15.00± .55		4513
084922.6+364237	NGC 2668	.S..2..	1.08± .07	155				15.46±.3	7529± 9
186.16 38.59	UGC 4616	U	.27± .06	.07	14.73 ±.18				7507
52.72 -31.05	IRAS08460+3654	2.0± .8		.34				412± 5	7736
084611.6+365347	PGC 24791		1.09	.14	14.25			1.08	
084924.8+190426	NGC 2673	.E.0.P.	1.09± .09		*		.95± .01		
207.47 34.26	UGC 4620	R	.00± .06	.05			.71± .05		3849± 46
66.19 -45.26		-5.0± .4		.00					3758
084633.8+191536	PGC 24792		1.10						4107

R.A. 2000 DEC. / l b / SGL SGB / R.A. 1950 DEC.	Names / PGC	Type / S_T n_L / T / L	$\log D_{25}$ / $\log R_{25}$ / $\log A_e$ / $\log D_o$	p.a. / A_g / A_i / A_{21}	B_T / m_B / m_{FIR} / B_T^o	$(B-V)_T$ / $(U-B)_T$ / $(B-V)_T^o$ / $(U-B)_T^o$	$(B-V)_e$ / $(U-B)_e$ / m'_e / m'_{25}	m_{21} / W_{20} / W_{50} / HI	V_{21} / V_{opt} / V_{GSR} / V_{3K}
084926.0+700946		.S..3$.	.62± .11		15.6 ±.3	.45± .07	.46± .04		
144.14 35.28	CGCG 332- 17	P	.26± .12	.04		-.18± .12	-.20± .07		3794± 48
35.47 -1.84	MK 95	3.0±1.8	.37± .09	.37		.37	12.89± .24		3903
084433.4+702053	PGC 24795		.63		.13 15.17	-.24	12.90± .71		3856
084927.5+293117		.SA.7..	1.15± .07	57				15.41±.3	8176± 7
195.13 37.31	UGC 4617	U	.47± .06	.09 15.0 ±.2					8125
57.59 -37.01		7.0± .9		.64				379± 7	8407
084625.5+294227	PGC 24796		1.16		.23 14.23			.95	
084957.5+701759	NGC 2650	.SBT3*.	1.20± .05	82				15.92±.1	3826± 9
143.97 35.28	UGC 4603	PU	.13± .05	.06 14.08 ±.20				334± 11	
35.45 -1.70		2.5± .6		.17					3936
084504.1+702908	PGC 24817		1.21		.06 13.82			2.03	3888
085012.0+350435		.S?....	1.00± .05	140				15.81±.3	2306± 10
188.26 38.52	UGC 4621		.26± .04	.07 13.9 ±.3					2340±155
53.93 -32.33	ARAK 182			.38 12.78				253± 7	2277
084703.3+351547	PGC 24829		1.00		.13 13.39			2.29	2520
0850.2 +0328		.S..6*.	1.19± .06	124					8471
224.12 27.89	UGC 4625	U	1.03± .06	.13 15.43 ±.19					8326
84.84 -55.65		6.0±1.4		1.47					8755
0847.6 +0340	PGC 24830		1.20		.50 13.78				
085020.3+411722		.SA.7..	1.14± .07						
180.33 39.23	UGC 4622	U	.03± .06	.09 14.8 ±.3					
50.12 -27.07		7.0± .8		.04					
084702.9+412835	PGC 24833		1.15		.01				
085020.7-163450		.SBT4*.	1.11± .06	140					5503
242.42 17.03	MCG -3-23- 9	E (1)	.11± .05	.23					
122.62 -62.31		4.0± .9		.16					5304
084801.6-162335	PGC 24836	1.9± .8	1.13		.05				5788
085021.2-324045			.83± .07						14660± 20
255.48 7.20	ESO 371- 19		.09± .06	1.23 15.42 ±.14					14436
154.98 -58.87	IRAS08482-3229								14921
084819.1-322930	PGC 24837		.95						
085022.2+732742	NGC 2646	.LBR0*.	1.13± .05						3680± 27
140.33 34.25	UGC 4604	R	.03± .03	.04 13.09 ±.17					3800
34.02 1.12	ARAK 180	-2.0± .4		.00					3726
084500.1+733851	PGC 24838		1.13		13.00				
085023.0+255707		.S..2..	.99± .05	30				16.81±.3	8297± 10
199.52 36.62	UGC 4624	U	.36± .04	.12 15.21 ±.18				381± 13	
60.51 -39.77		2.0± .9		.44					8232
084725.1+260820	PGC 24839		1.00		.18 14.56			2.07	8539
085031.7-030557									4287± 63
230.49 24.62	CGCG 5- 15			.02 14.8 ±.6					4122
95.56 -58.91	MK 1414			12.18					4575
084800.5-025443	PGC 24844								
085042.2-050034		.E+....	1.07± .07						
232.28 23.64	MCG -1-23- 4	E	.01± .05	.03					
99.05 -59.66		-4.0± .9		.00					
084812.7-044919	PGC 24851		1.07						
085046.0-215740		.SB.1*P	1.28± .04	15					
246.96 13.89	ESO 563- 28	S	.51± .04	.65					
134.16 -61.96	IRAS08485-2146	1.0±1.1		.52 10.80					
084832.0-214624	PGC 24854		1.34		.25				
085056.0-343211		.SAS5*.	1.50± .04	60				13.60±.3	2366± 10
257.01 6.12	ESO 371- 20	S (1)	.47± .05	1.47 13.48 ±.14				397± 12	
158.00 -57.91	IRAS08489-3420	5.0±1.1		.71 12.47				391± 9	2140
084856.1-342054	PGC 24860	4.4± .8	1.64		.24 11.29			2.08	2624
085057.4+291209 A	0847+29								
195.62 37.55				.10					7900
58.16 -37.08	MK 628								7848
084755.9+292324	PGC 24862								8133
085057.8+653816 A	0846+65								6979± 38
149.44 36.75	CGCG 311- 19			.15 15.4 ±.3					7072
37.70 -5.78	MK 97			12.53					7064
084634.3+654929	PGC 24863								
0851.0 +2910		.SB?...	.89± .11						
195.65 37.56	MCG 5-21- 13		.00± .07	.10 15.07 ±.19					8065± 30
58.19 -37.09	VV 473			.00					8013
0848.0 +2922	PGC 24864		.90		.00 14.88				8298

R.A. 2000 DEC.	Names	Type	logD$_{25}$	p.a.	B$_T$	(B-V)$_T$	(B-V)$_\bullet$	m$_{21}$	V$_{21}$
l b		S$_T$ n$_L$	logR$_{25}$	A$_g$	m$_B$	(U-B)$_T$	(U-B)$_\bullet$	W$_{20}$	V$_{opt}$
SGL SGB		T	logA$_\bullet$	A$_i$	m$_{FIR}$	(B-V)$_T^o$	m'$_\bullet$	W$_{50}$	V$_{GSR}$
R.A. 1950 DEC.	PGC	L	logD$_o$	A$_{21}$	B$_T^o$	(U-B)$_T^o$	m'$_{25}$	HI	V$_{3K}$
085101.6+241908		.I..9*.	.96± .09					15.54±.1	2735± 6
201.52 36.30	UGC 4626	U	.00± .06	.11				123± 5	
61.98 -40.97		10.0±1.2		.00				107± 3	2664
084805.4+243023	PGC 24865		.97	.00					2981
085107.1-173350		.SAT7*.	1.55± .02	75					2001
243.36 16.60	MCG -3-23- 10	ESU (1)	.53± .04	.37					
124.75 -62.15	IRAS08488-1722	6.9± .5		.74					1800
084848.9-172233	PGC 24870	4.4± .8	1.58	.27					2286
0851.3 +1920		.L.....	1.12± .10						
207.35 34.80	UGC 4631	U	.02± .05	.04	14.4 ±.2				
66.43 -44.76		-2.0± .8		.00					
0848.5 +1932	PGC 24877		1.12						
0851.4 +4732	NGC 2676	.L...*.	1.08± .09						
172.24 39.56	UGC 4627	U	.04± .04	.10	14.09 ±.15				6010± 60
46.78 -21.59		-2.0±1.2		.00					6032
0848.0 +4744	PGC 24881		1.09		13.90				6177
085128.3+610118	A 0847+61	.S?....	.69± .15						
155.06 37.92	MCG 10-13- 25		.33± .07	.17					3750± 77
39.95 -9.82	MK 99			.25					3825
084725.1+611233	PGC 24882		.69						3857
085133.1+305156	NGC 2679	.LB..*.	1.26± .13						
193.60 38.04	UGC 4632	U	.00± .08	.07					1995± 46
57.08 -35.65		-2.0± .8		.00					1949
084829.9+310313	PGC 24884		1.27						2223
0851.5 +5107		.S..6*.	1.22± .06	74					
167.60 39.39	UGC 4628	U	.69± .06	.08	15.27 ±.19				
44.90 -18.47		6.0±1.3		1.01					
0848.0 +5119	PGC 24887		1.23	.34					
085135.9-071541		.SAR0?.	1.20± .05	35					
234.47 22.60	MCG -1-23- 5	E	.15± .05	.11					
103.45 -60.26		.0±1.2		.11					
084908.3-070422	PGC 24888		1.20						
085138.1-022115		.S?....	1.16± .06	56					
229.95 25.25	UGC 4638	(1)	.27± .05	.07					3333± 46
94.50 -58.33				.41					3170
084906.2-020957	PGC 24889	5.3±1.6	1.17	.14					3622
085139.3+570625	A 0847+57								
159.95 38.68				.21					6830± 63
41.88 -13.25	MK 17								6890
084749.0+571741	PGC 24891								6955
085144.0-020804		.SXT5*.	1.51± .03	3				13.95±.1	3316± 11
229.76 25.38	UGC 4640	UE (1)	.46± .04	.05				322± 16	
94.14 -58.21	IRAS08491-0156	5.3± .6		.69				321± 12	3154
084912.0-015646	PGC 24893	4.2± .8	1.51	.23					3605
085148.7+291632		.S?....	.96± .09	173				16.41±.3	8088± 10
195.58 37.75	UGC 4636		.80± .06	.09	15.9 ±.2				8036
58.29 -36.91				1.21				339± 7	
084847.2+292750	PGC 24895		.97	.40	14.60			1.41	8321
085156.9+165641		.L...?.	1.18± .09						
210.13 34.07	UGC 4639	U	.02± .05	.05	13.82 ±.19				
68.90 -46.44		-2.0±1.6		.00					
084908.1+170800	PGC 24902		1.18						
085157.3+714848	A 0846+72	.E?....	.80± .09						
142.16 34.93	MCG 12- 9- 23		.26± .12	.05					3389± 53
34.89 -.28	MK 98			.00					3504
084652.5+720003	PGC 24903		.73						3444
0852.0 +8417		.I..9..	1.00± .08	15					
128.62 29.88	UGC 4557	U	.33± .06	.19					
28.99 10.72		10.0± .9		.25					
0841.0 +8429	PGC 24907		1.02	.17					
0852.0 +5255		.SB.3..	1.04± .06	87					
165.27 39.30	UGC 4633	U	.33± .05	.07					
44.03 -16.87		3.0± .9		.45					
0848.4 +5307	PGC 24908		1.04	.16					
085205.1+533703	NGC 2675	.E.....	1.18± .15	80	14.3 ±.2	1.04± .03			
164.39 39.22	UGC 4629	U	.13± .08	.05	14.03 ±.16	.54± .04			9231± 31
43.68 -16.27		-5.0± .8		.00		.94			9277
084824.3+534820	PGC 24909		1.15		13.95	.58	14.86± .79		9372

R.A. 2000 DEC. / l b / SGL SGB / R.A. 1950 DEC.	Names / PGC	Type / S_T n_L / T / L	$\log D_{25}$ / $\log R_{25}$ / $\log A_e$ / $\log D_o$	p.a. / A_g / A_i / A_{21}	B_T / m_B / m_{FIR} / B_T^o	$(B-V)_T$ / $(U-B)_T$ / $(B-V)_T^o$ / $(U-B)_T^o$	$(B-V)_e$ / $(U-B)_e$ / m' / m'_{25}	m_{21} / W_{20} / W_{50} / HI	V_{21} / V_{opt} / V_{GSR} / V_{3K}
085218.2-174439		PLBS0*.	1.21± .05	133					
243.69 16.72	ESO 563- 31	SE	.08± .03	.48					
125.14 -61.87		-2.0± .5		.00					
085000.0-173318	PGC 24913		1.26						
085238.4-023609	NGC 2690	.S..2*/	1.27± .04	19					
230.33 25.33	UGC 4647	UE	.59± .04	.04	14.02 ±.18				
95.14 -58.22		2.0± .7		.73					
085006.8-022447	PGC 24926		1.28		.30				
085239.6-332752		.SXS9..	1.28± .04	5					2261
256.40 7.08	ESO 371- 24	S (1)	.25± .05	1.26	15.07 ±.14				
155.91 -58.09		9.0± .8		.26					2036
085038.0-331630	PGC 24928	10.0±1.1	1.39	.13	13.55				2523
085240.6+212526		.SAS4..	1.21± .06	90				15.02±.3	7699± 10
205.08 35.78	UGC 4643	U	.08± .06	.08	13.9 ±.2				
64.86 -42.98		4.0± .8		.12				317± 7	7617
084947.5+213647	PGC 24929		1.22		.04	13.66			1.32	7955
085241.0+332503	NGC 2683	.SAT3..	1.97± .01	44	10.64M±.07	.89± .01	.92± .01	12.78±.1	410± 5
190.46 38.76	UGC 4641	R (1)	.63± .02	.08	10.36 ±.13	.27± .02	.36± .01	434± 7	358± 17
55.54 -33.42		3.0± .3	1.27± .01	.87	10.39	.75	12.39± .04	425± 6	370
084934.8+333623	PGC 24930	4.0± .9	1.98	.31	9.63	.14	13.76± .10	2.84	626
085247.1-733714		.LA.*P.	1.06± .05	100					
288.10 -18.10	ESO 36- 10	S	.19± .04	.63	14.65 ±.14				
197.07 -29.75		-2.0±1.2		.00					
085258.1-732548	PGC 24935		1.10						
085256.1+422455		.S?....	.84± .06						
178.90 39.78	UGC 4642		.04± .04	.07	14.3 ±.2				7487±155
49.92 -25.85	ARAK 184			.05	12.83				7488
084937.4+423616	PGC 24940		.85		.02	14.12				7676
085256.8-251811		.SXT5..	1.12± .04						
249.98 12.24	ESO 496- 13	S (1)	.07± .04	.59	14.38 ±.14				
140.82 -60.85		5.0± .8		.10					
085046.0-250648	PGC 24941	3.3± .8	1.18		.03				
0853.0 -0102		.S..8*.	.96± .09						4511
228.91 26.24	UGC 4651	U	.00± .06	.05					
92.59 -57.41		8.0±1.2		.00					4352
0850.5 -0051	PGC 24944		.96		.00				4801
085311.0+762909	A 0847+76	.S..6*.	1.55± .03	60				14.48±.1	2885± 11
136.88 33.28	UGC 4623	U	.61± .05	.04	13.35 ±.18				
32.76 3.87		6.0±1.1		.89				367± 8	3016
084711.1+764026	PGC 24947		1.55		.30	12.41			1.76	2917
085311.3+090850	IC 523	.S..2..	1.21± .06					15.33±.3	8774± 9
218.74 31.17	UGC 4652	U	.11± .06	.14	14.0 ±.3				
77.93 -51.61		2.0± .8		.14				334± 5	8647
085029.5+092013	PGC 24948		1.22		.06	13.66			1.61	9054
085314.9+731121	A 0847+73		.85± .12		15.2 ±.2	.56± .03		16.07±.1	2316± 10
140.55 34.54	UGCA 146		.07± .09	.07	14.72 ±.19	-.03± .06		136± 6	2441± 38
34.33 .97	MK 16							50± 5	2444
084757.6+732240	PGC 24949		.85						2373
085328.0-734552			.82± .06	42					
288.25 -18.15	ESO 36- 13		.31± .05	.63	15.41 ±.14				7590± 87
197.11 -29.60	IRAS08536-7334				13.07				7381
085340.0-733424	PGC 24958		.88						7706
0853.5 +0446		.S..6*.	1.04± .08	156					6193
223.28 29.24	UGC 4655	U	1.06± .06	.17	15.9 ±.2				
83.81 -54.23		6.0±1.5		1.47					6052
0850.9 +0458	PGC 24960		1.06		.50	14.27				6479
085333.1+511853	NGC 2681	PSXT0..	1.56± .02		11.09M±.10	.80± .01	.81± .01		692± 11
167.32 39.68	UGC 4645	R	.04± .03	.10	10.93 ±.15	.31± .05	.35± .01		683± 12
45.08 -18.16	ARAK 185	.0± .3	1.10± .02	.03	11.07	.76	12.07± .07		725
084958.0+513016	PGC 24961		1.57		10.90	.29	13.66± .16		840
0853.5 +4519		.I..9*.	1.30± .06	135				15.13±.1	1881± 10
175.11 39.96	UGC 4648	U	.15± .06	.11				86± 6	
48.35 -23.31		10.0±1.2		.12				72± 5	1894
0850.2 +4531	PGC 24963		1.31		.08					2059
085337.5+390808		.S..2..	1.27± .04	90					
183.17 39.69	UGC 4650	U	.32± .05	.08	14.3 ±.2				
52.02 -28.56	IRAS08503+3919	2.0± .8		.39	13.66				
085023.9+391931	PGC 24964		1.27		.16				

R.A. 2000 DEC. l b SGL SGB R.A. 1950 DEC.	Names PGC	Type S_T n_L T L	$\log D_{25}$ $\log R_{25}$ $\log A_o$ $\log D_o$	p.a. A_g A_i A_{21}	B_T m_B m_{FIR} B_T^o	$(B-V)_T$ $(U-B)_T$ $(B-V)_T^o$ $(U-B)_T^o$	$(B-V)_o$ $(U-B)_o$ m'_o m'_{25}	m_{21} W_{20} W_{50} HI	V_{21} V_{opt} V_{GSR} V_{3K}
085341.2+732926 IC 520 140.20 34.46 UGC 4630 34.21 1.25 ARAK 183 084821.0+734046 PGC 24970		.SXT2\$. R 2.0± .7	1.29± .05 .10± .05 1.00± .03 1.30	0 .07 .13 .05	12.55 ±.15 12.54 ±.15 12.74 12.32	.84± .02 .28± .03 .78 .24	.92± .01 .39± .02 13.04± .09 13.61± .32		3528 3649 3575
0853.8 +5223 165.92 39.62 MCG 9-15- 42 44.55 -17.20 0850.2 +5235 PGC 24974		.E?....		.08	15.11 ±.18				9097 9138 9245
0853.8 +7332 140.13 34.45 UGC 4634 34.20 1.31 0848.5 +7344 PGC 24978		.I..9*. U 10.0±1.2	1.07± .14 .00± .12 1.08	.07 .00 .00					
085353.9+181045 208.94 34.95 UGC 4656 68.18 -45.22 085104.0+182210 PGC 24980		.S..3.. U (1) 3.0± .9 3.5±1.3	1.14± .13 .69± .12 1.14	162 .04 .95 .34	15.19 ±.18 14.15			15.43±.3 369± 7 .94	8442± 10 8348 8706
085354.5+350854 188.31 39.28 UGC 4653 54.63 -31.85 ARP 195 085046.3+352018 PGC 24981		.SBS3.. U 3.0± .8	1.27± .07 .22± .07 1.27	.06 .30 .11	14.5 ±.3 12.24 14.06				16560± 51 16531 16776
0853.9 +1840 208.38 35.14 UGC 4657 67.71 -44.85 0851.1 +1852 PGC 24982		.S..6?. U 6.0±2.0	1.14± .07 .98± .06 1.14	18 .04 1.44 .49	15.71 ±.18 14.21				4360 4267 4623
085404.8-071059 234.77 23.16 MCG -1-23- 8 103.71 -59.65 085137.1-065933 PGC 24988		.LA.0*. E -2.0±1.2	1.06± .07 .13± .05 1.05	135 .10 .00					
085421.6+324050 IC 2421 191.47 38.98 UGC 4658 56.40 -33.83 IRAS08512+3252 085116.5+325216 PGC 24996		.SAT5.. PU (1) 5.0± .5 1.1± .8	1.34± .03 .03± .04 1.02± .04 1.35	 .07 .05 .02	13.88 ±.18 14.3 ±.4 13.81	.61± .04 .56	.63± .03 14.47± .09 15.36± .25	14.45±.1 133± 4 110± 4 .63	4383± 4 4389± 41 4345 4607
085421.6+324046 191.47 38.98 MCG 6-20- 13A 56.40 -33.83 KUG 0851+328B 085116.5+325212 PGC 24997		.S?....	.80± .07 .45± .06 .81	144 .07 .66 .22	15.5 ±.2 14.74				4322± 57 4284 4547
0854.3 +2022 206.46 35.81 UGC 4663 66.22 -43.51 0851.5 +2034 PGC 24999		.I..9*. U 10.0±1.3	1.22± .12 .65± .12 1.23	0 .07 .49 .32					
085424.3+343324 189.09 39.29 UGC 4660 55.12 -32.28 085116.9+344450 PGC 25001		.S..9*. U 9.0±1.1	1.07± .08 .03± .06 1.07	 .06 .03 .01				15.25±.1 69± 4 61± 3	2203± 4 2173 2422
085427.1-030403 NGC 2695 231.03 25.47 MCG 0-23- 10 96.32 -58.01 085155.9-025236 PGC 25003		.LXS0?. E -2.0± .8	1.23± .09 .16± .07 .72± .03 1.21	175 .04 .00	12.83 ±.14 13.15 ±.15 12.91	.95± .01 .91	.97± .01 11.92± .12 13.48± .51		1825± 60 1660 2117
085432.4-325610 256.24 7.72 ESO 371- 26 154.70 -57.97 085230.0-324442 PGC 25006		.LXT0*/ S -1.7± .5	1.44± .04 .69± .04 1.47	67 1.15 .00	13.68 ±.14				
085435.8+571007 IC 522 159.79 39.06 UGC 4654 42.20 -13.00 085046.2+572132 PGC 25009		.L..... U -2.0± .9	1.00± .10 .09± .04 1.01	165 .22 .00	13.97 ±.17				
085440.6+470619 172.79 40.12 UGC 4659 47.53 -21.69 085114.3+471745 PGC 25012		.SA.8*. U 8.0± .8	1.27± .04 .33± .05 1.28	115 .10 .41 .17	15.0 ±.3 14.46			15.40±.1 180± 8 .77	1749± 7 1732± 76 1768 1919
085446.4+393213 NGC 2691 182.67 39.95 UGC 4664 51.98 -28.10 MK 391 085132.3+394340 PGC 25020		.S..1?. U 1.0±1.6	1.09± .03 .19± .04 .59± .02 1.10	165 .09 .19 .09	13.93 ±.13 13.92 ±.18 12.65 13.60	.79± .02 .16± .03 .70 .12	.75± .01 .14± .02 12.37± .09 13.76± .22	16.08±.1 353± 13 218± 25 2.39	3981± 7 3931± 40 3969 4180
085446.8+201316 IC 2423 206.68 35.85 UGC 4667 66.46 -43.57 085155.0+202444 PGC 25021		.SXS3.. U 3.0± .9	1.02± .06 .09± .05 1.02	100 .07 .13 .05	14.46 ±.18 14.19			16.55±.3 401± 7 2.31	9100± 10 9013 9360

R.A. 2000 DEC. / l b / SGL SGB / R.A. 1950 DEC.	Names / PGC	Type / S_T n_L / T / L	$\log D_{25}$ / $\log R_{25}$ / $\log A_e$ / $\log D_o$	p.a. / A_g / A_i / A_{21}	B_T / m_B / m_{FIR} / B_T^o	$(B-V)_T$ / $(U-B)_T$ / $(B-V)_T^o$ / $(U-B)_T^o$	$(B-V)_e$ / $(U-B)_e$ / m'_e / m'_{25}	m_{21} / W_{20} / W_{50} / HI	V_{21} / V_{opt} / V_{GSR} / V_{3K}
085453.5+490938 170.10 40.06 46.43 -19.91 085123.3+492105	NGC 2684 UGC 4662 IRAS08514+4921 PGC 25024	.S?.... 	.97± .07 .09± .05 .77± .07 .98	 .06 .13 .04	13.6 ±.3 13.6 ±.2 12.79 13.41	.69± .05 -.20± .08 .64 -.24	.77± .03 -.22± .05 12.94± .18 13.10± .43	16.89±.3 171± 13 3.44	2860± 10 2878± 48 2889 3023
0854.9 +2641 198.97 37.80 60.98 -38.56 0852.0 +2653	 CGCG 150- 45 PGC 25028			.12	15.4 ±.3				8156 8094 8399
085459.4-025918 231.03 25.63 96.29 -57.86 085228.1-024749	NGC 2697 MCG 0-23- 11 IRAS08524-0247 PGC 25029	.LAS+*. E -1.0± .8 	1.26± .09 .22± .07 1.23	120 .04 .00	 13.32 ±.15 13.18				
085506.8+185601 208.20 35.49 67.75 -44.47 085216.3+190730	 UGC 4669 PGC 25035	.S..8*. U 8.0±1.1 	1.13± .07 .01± .06 1.13	 .04 .02 .01	 14.5 ±.3 14.41			16.04±.3 161± 7 1.63	4105± 7 4013 4368
085516.6-320242 255.65 8.40 153.05 -58.19 085313.0-315112	 ESO 432- 12 IRAS08532-3150 PGC 25045	.LAS0?. S -2.0±1.7 	1.28± .05 .57± .04 1.31	98 1.04 .00					
085521.2-250531 250.16 12.80 140.15 -60.36 085310.0-245400	 ESO 496- 19 PGC 25053	.SBS6.. S (1) 6.0± .8 3.3± .9	1.15± .05 .25± .04 1.21	176 .59 .37 .13	 15.40 ±.14				
085535.2+584402 157.78 38.90 41.53 -11.57 085141.3+585530	NGC 2685 UGC 4666 ARP 336 PGC 25065	RLB.+P. V -1.0± .3 	1.65± .02 .28± .03 1.03± .02 1.63	38 .16 .00	12.12M±.10 11.93 ±.11 11.86	.86± .01 .36± .02 .78 .29	.93± .01 .42± .01 12.73± .06 14.55± .17	13.90±.1 352± 4 278± 6	883± 4 869± 25 949 1003
085536.2-672704 283.12 -14.19 192.75 -34.70 085458.0-671530	 ESO 90- 9 IRAS08549-6715 PGC 25066	.SXS5*. S (1) 4.5± .6 4.8± .8	1.25± .05 .54± .05 1.31	29 .59 .81 .27	 12.93				
085536.4-031105 231.31 25.65 96.76 -57.80 085305.3-025934	NGC 2698 MCG 0-23- 12 PGC 25067	.LA.+?. PE -1.0± .7 	1.16± .07 .40± .07 1.11	96 .03 .00	 13.6 ±.2				
085538.7+781328 134.92 32.69 32.05 5.46 084909.1+782453	NGC 2655 UGC 4637 ARP 225 PGC 25069	.SXS0.. R .0± .3 	1.69± .02 .08± .03 1.11± .02 1.69	 .03 .06	10.96 ±.13 10.95 ±.13 12.36 10.85	.86± .01 .43± .02 .83 .42	.91± .01 .46± .01 12.00± .05 14.07± .18	13.42±.1 356± 5 182± 5	1404± 5 1407± 40 1540 1427
0855.7 +5733 159.26 39.14 42.14 -12.59 0851.9 +5745	 MCG 10-13- 38 VV 761 PGC 25071		1.02± .10 .65± .07 1.03	 .16					 11875± 27 11936 12000
085548.8-030740 231.29 25.73 96.70 -57.73 085317.6-025608	NGC 2699 MCG 0-23- 14 ARAK 187 PGC 25075	.E...*. E -5.0±1.3 	1.02± .07 .03± .05 1.01	45 .04 .00	 13.63 ±.17 13.57				 1825± 60 1660 2118
085553.1+023125 225.85 28.65 87.65 -55.00 085317.1+024257	 UGC 4673 PGC 25081	.SBS7*. UE (1) 7.0± .6 7.5± .8	1.19± .05 .07± .05 1.20	20 .13 .10 .04	 14.0 ±.3 13.75			15.16±.3 134± 7 1.37	3818± 7 3670 4108
085555.8+131343 214.70 33.51 73.83 -48.39 085310.6+132515	 UGC 4670 PGC 25085	.L...*. U -2.0±1.1 	1.20± .09 .14± .05 1.19	 .10 .00	 13.67 ±.16				
0855.9 +5740 159.11 39.15 42.11 -12.47 0852.1 +5752	 UGC 4668 PGC 25086	.S..4.. U 4.0±1.0 	1.04± .06 .86± .05 1.05	84 .16 1.26 .43	 15.72 ±.19				
085608.0-032139 231.55 25.67 97.16 -57.75 085337.0-031007	NGC 2708 MCG 0-23- 15 IRAS08535-0309 PGC 25097	.SXS3P? PE 2.5± .5 	1.42± .04 .30± .05 .96± .03 1.42	20 .03 .42 .15	12.80 ±.15 13.04 ±.20 12.05 12.43	.79± .02 .20± .03 .71 .13	.91± .01 .33± .02 13.09± .08 13.99± .27	14.25±.1 481± 7 434± 8 1.67	2008± 6 2060± 63 1843 2302
085612.8-023350 230.82 26.11 95.82 -57.40 085341.1-022217	NGC 2706 UGC 4680 IRAS08536-0222 PGC 25102	.S..4?/ UE (1) 3.7±1.0 5.3±1.6	1.26± .04 .50± .04 1.26	167 .04 .74 .25	 13.80 ±.18 11.08				

R.A. 2000 DEC. / l b / SGL SGB / R.A. 1950 DEC.	Names / PGC	Type / S$_T$ n$_L$ / T / L	logD$_{25}$ / logR$_{25}$ / logA$_\bullet$ / logD$_0$	p.a. / A$_g$ / A$_i$ / A$_{21}$	B$_T$ / m$_B$ / m$_{FIR}$ / B^o_T	(B-V)$_T$ / (U-B)$_T$ / (B-V)^o_T / (U-B)^o_T	(B-V)$_\bullet$ / (U-B)$_\bullet$ / m'$_\bullet$ / m'$_{25}$	m$_{21}$ / W$_{20}$ / W$_{50}$ / HI	V$_{21}$ / V$_{opt}$ / V$_{GSR}$ / V$_{3K}$
085612.9-031439	NGC 2709	.LA.0P*	.92± .11	0					
231.46 25.75	MCG 0-23- 16	E	.14± .05	.03	14.65 ±.15				
96.98 -57.69		-2.0±1.3		.00					
085341.8-030306	PGC 25103		.91						
085624.2+131100		.S..6*.	1.13± .05	120				16.34±.3	4140± 7
214.81 33.60	UGC 4677	U	.15± .05	.13	14.3 ±.2				
74.00 -48.34		6.0±1.2		.22				211± 7	4027
085339.1+132233	PGC 25113		1.14	.08	13.96			2.30	4416
085627.8-315904		.SBS9*.	1.20± .07	129					2102
255.76 8.64	ESO 432- 13	S (1)	.34± .07	1.01					
152.76 -57.99		9.0±1.2		.35					1879
085424.1-314730	PGC 25116	7.8±1.2	1.29	.17					2370
085640.3-675214		PSBS3..	1.31± .04	120					6330± 63
283.53 -14.37	ESO 60- 18	Sr (1)	.17± .05	.45	14.16 ±.14				
192.95 -34.30	FAIR 278	3.5± .4		.26					6112
085604.0-674036	PGC 25127	5.6± .7	1.35	.09	13.41				6474
085640.6+002227		.SAT8*.	1.15± .04	175				14.91±.2	2522± 7
228.07 27.74	UGC 4684	UE (1)	.09± .04	.14	14.0 ±.2			183± 13	
91.12 -55.93		8.0± .6		.11				157± 7	2367
085406.4+003401	PGC 25128	6.4± .8	1.17	.05	13.73			1.14	2814
085642.4+520619		.S?....	1.11± .05	69					4098± 46
166.24 40.09	UGC 4671		.07± .05	.07	13.60 ±.18				
45.11 -17.23	IRAS08531+5217			.10	12.31				4138
085306.4+521751	PGC 25130		1.11	.03	13.41				4249
085648.2+392259	NGC 2704	.SBR2..	1.02± .06					16.23±.1	7116± 11
182.92 40.32	UGC 4678	U	.00± .05	.02	14.27 ±.18				7132± 50
52.45 -28.01		2.0± .8		.00				320± 8	7105
085334.7+393433	PGC 25134		1.02	.00	14.17			2.06	7320
0856.8 +7941		.SXS3..	1.06± .06						
133.32 32.11	UGC 4644	U	.02± .05	.03	15.0 ±.3				
31.38 6.77		3.0± .8		.03					
0849.8 +7953	PGC 25138		1.06	.01					
085658.1+520400	NGC 2692	.SB.2*.	1.10± .05	165				15.93±.1	4032± 11
166.28 40.13	UGC 4675	PU	.38± .04	.07	14.18 ±.19				3778± 49
45.16 -17.24		1.7± .7		.47				169± 8	4060
085322.3+521533	PGC 25142		1.10	.19	13.60			2.14	4171
085658.6+511949	NGC 2694	.E.1...	1.08± .11		15.4 ±.3	1.03± .05			
167.24 40.21	MCG 9-15- 56	P	.01± .10	.07	14.7 ±.2	.65± .10			5123± 56
45.56 -17.87		-5.0±1.1		.00		.96			5160
085324.4+513122	PGC 25143		1.09		14.75	.66	15.77± .68		5277
085659.4+512051	NGC 2693	.E.3.*.	1.42± .04	160	12.84 ±.15	.96± .01	1.04± .01		
167.22 40.21	UGC 4674	R	.16± .03	.07	12.62 ±.12	.61± .03	.58± .03		4886± 27
45.55 -17.86		-5.0± .5	.83± .04	.00		.90	12.48± .15		4923
085325.2+513224	PGC 25144		1.39		12.56	.62	14.54± .24		5041
085701.8+131153		.S?....	1.11± .05	150				15.41±.2	3978± 6
214.87 33.75	UGC 4685		.40± .05	.10	14.24 ±.19			329± 7	
74.14 -48.22	IRAS08542+1323			.60	12.77			304± 5	3865
085416.6+132328	PGC 25145		1.12	.20	13.51			1.70	4254
085701.8-244024	NGC 2717	.LXS-..	1.32± .05		13.43 ±.17	1.11± .02	1.12± .01		
250.07 13.36	ESO 496- 21	S	.13± .04	.83	13.21 ±.14	.60± .02	.64± .02		
139.17 -60.08		-3.0± .8	.63± .06	.00			12.07± .19		
085450.0-242848	PGC 25146		1.41				14.57± .30		
0857.0 +5148		.S..7..	1.06± .06	55					
166.61 40.18	UGC 4676	U	.44± .05	.08					
45.32 -17.46		7.0± .9		.61					
0853.5 +5200	PGC 25147		1.06	.22					
085706.1-084333		.S..4*.	1.22± .05	145					9407
236.61 22.91	MCG -1-23- 11	E (1)	.69± .05	.08					
107.09 -59.40		4.0±1.3		1.02					9227
085439.7-083157	PGC 25148	4.2±1.2	1.23	.35					9702
085706.8-244500		.SBT7..	1.22± .04						2377
250.14 13.32	ESO 496- 22	S (1)	.07± .04	.83	13.90 ±.14				
139.31 -60.05	IRAS08549-2433	7.0± .8		.10	13.00				2164
085455.1-243324	PGC 25152	4.4± .8	1.29	.04	12.96				2659
0857.1 +5128		.S..6*.	1.13± .05	91					
167.05 40.23	UGC 4679	U	.93± .05	.08					
45.51 -17.73		6.0±1.4		1.36					
0853.6 +5140	PGC 25154		1.13	.46					

R.A. 2000 DEC.	Names	Type	logD25	p.a.	BT	(B-V)T	(B-V)o	m21	V21
l b		ST nL	logR25	Ag	mB	(U-B)T	(U-B)o	W20	Vopt
SGL SGB		T	logAo	Ai	mFIR	(B-V)oT	m'	W50	VGSR
R.A. 1950 DEC.	PGC	L	logDo	A21	BoT	(U-B)oT	m'25	HI	V3K
085720.6+025521	NGC 2713	.SBT2..	1.56± .02	107	12.72 ±.13	.97± .01	1.05± .01	14.65±.2	3922± 6
225.66 29.17	UGC 4691	R (2)	.38± .03	.15	12.46 ±.13	.47± .01	.57± .01	619± 12	3972± 30
87.40 -54.48	IRAS08547+0306	2.0± .3	.86± .01	.47	12.81	.84	12.51± .03	604± 6	3777
085444.3+030657	PGC 25161	1.9± .6	1.58	.19	11.93	.36	14.42± .17	2.53	4215
085723.3+171714	NGC 2711	.SB?...	.95± .07	170				15.52±.3	6147± 10
210.33 35.41	UGC 4688		.15± .05	.04	14.64 ±.18				6049
69.92 -45.30	IRAS08545+1728			.22	13.82			134± 7	6416
085434.5+172850	PGC 25164		.95	.07	14.34			1.10	
085727.4-690340		.SBS7..	1.50± .04	157				13.01±.3	1445± 9
284.54 -15.05	ESO 60- 19	S (1)	.41± .05	.39	12.79 ±.14				1228
193.71 -33.30	IRAS08569-6851	7.0± .5		.57	12.03			217± 7	1584
085658.0-685200	PGC 25169	5.6± .6	1.53	.21	11.82			.98	
085736.2+030526	NGC 2716	RLBR+..	1.11± .04	30	12.7 ±.2	.87± .04	.88± .03		
225.53 29.31	UGC 4692	R	.10± .03	.13	13.60 ±.15	.37± .09			3537± 56
87.21 -54.33		-1.0± .4	1.17± .09	.00		.79	14.02± .31		3391
085459.7+031703	PGC 25172		1.11		13.11	.35	12.85± .32		3828
085741.7+430739		.S?....	.87± .08	90					9197± 50
178.01 40.67	UGC 4686		.22± .05	.09	14.7 ±.2				9201
50.32 -24.78	IRAS08543+4319			.33	13.31				9386
085422.8+431915	PGC 25175		.88	.11	14.19				
0857.8 +1229		.SB.6*.	.95± .07	55					
215.74 33.64	UGC 4694	U	.17± .05	.12	14.80 ±.18				
75.14 -48.55		6.0±1.3		.25					
0855.1 +1241	PGC 25181		.96	.09					
0857.9 +5904		.I..9..	1.12± .07	110				16.50±.1	920± 6
157.27 39.12	UGC 4683	U	.21± .06	.17				70± 7	
41.62 -11.13		10.0± .8		.15					988
0854.0 +5916	PGC 25185		1.14	.10					1040
0858.1 +5211		.S..2..	1.08± .06	120					9415
166.09 40.29	UGC 4690	U	.29± .05	.05	14.75 ±.18				9455
45.25 -17.05		2.0± .9		.36					9566
0854.5 +5223	PGC 25194		1.08	.15	14.25				
085812.1-061154		.SBR4*.	1.30± .05	175					4864
234.49 24.56	MCG -1-23- 13	E (1)	.41± .05	.04					
102.56 -58.35		4.0±1.2		.60					4690
085543.5-060015	PGC 25197	4.2±1.2	1.30	.20					5160
085819.3-652200		.SAS5..	1.15± .07	2					
281.62 -12.68	ESO 90- 11	S (1)	.16± .07	.62					
190.97 -36.20		5.0± .8		.24					
085729.1-651018	PGC 25200	3.3± .9	1.21	.08					
085822.5-664342		.L..../	1.31± .05	50					
282.71 -13.53	ESO 90- 12	S	.60± .04	.46					
191.97 -35.12	IRAS08576-6631	-2.0±1.2		.00	12.28				
085739.0-663200	PGC 25202		1.27						
0858.4 +0619	A 0855+06		.66?					16.34±.2	3700± 10
222.37 31.05	UGC 4703		.11?	.13				164± 34	3556± 67
82.90 -52.33								152± 25	3561
0855.8 +0631	PGC 25205		.67						3985
085832.7+281601		.SB.2..	1.20± .05	55				15.47±.3	7997± 9
197.27 38.96	UGC 4698	U	.30± .05	.09	14.20 ±.19				7941
60.54 -36.83		2.0± .8		.37				604± 5	8239
085533.1+282740	PGC 25210		1.21	.15	13.67			1.65	
085840.1-731933		.LB?...	1.00± .06						11510± 63
288.11 -17.59	ESO 36- 15		.05± .03	.46	15.00 ±.14				11300
196.48 -29.76	FAIR 279			.00					11630
085844.0-730748	PGC 25216		1.05		14.37				
085845.6+393036		.E?....	1.11± .08						8350± 56
182.79 40.71	UGC 4699		.03± .04	.02	14.28 ±.17				8339
52.73 -27.70				.00					8553
085532.2+394215	PGC 25220		1.11		14.13				
085846.2-034237	NGC 2722	.SAT4P*	1.30± .04	110					
232.27 26.04	MCG -1-23- 14	PE (1)	.19± .04	.03					
98.27 -57.30	IRAS08562-0330	3.8± .6		.27	12.41				
085615.4-033056	PGC 25221	4.5± .7	1.30	.09					
085850.2+413450		.S..3..	1.06± .06					16.37±.1	8481± 11
180.06 40.84	UGC 4700	U (1)	.02± .05	.07	14.38 ±.19				8479
51.45 -25.97		3.0± .8		.03				192± 12	8677
085533.9+414630	PGC 25224	3.5±1.1	1.06	.01	14.22			2.14	

R.A. 2000 DEC. l b SGL SGB R.A. 1950 DEC.	Names PGC	Type S_T n_L T L	$\log D_{25}$ $\log R_{25}$ $\log A_e$ $\log D_o$	p.a. A_g A_i A_{21}	B_T m_B m_{FIR} B_T^o	$(B-V)_T$ $(U-B)_T$ $(B-V)_T^o$ $(U-B)_T^o$	$(B-V)_e$ $(U-B)_e$ m'_e m'_{25}	m_{21} W_{20} W_{50} HI	V_{21} V_{opt} V_{GSR} V_{3K}
085850.6+061736 222.45 31.12 83.03 -52.27 085611.6+062917	NGC 2718 UGC 4707 MK 703 PGC 25225	PSXS2.. U 2.0± .7 	1.33± .05 .00± .06 1.35	 .13 .00 .00	 12.73 ±.20 11.65 12.56			14.36±.1 132± 6 115± 6 1.80	3843± 5 3785± 44 3706 4131
085851.4+384836 183.71 40.68 53.19 -28.27 085539.0+390016	 UGC 4702 PGC 25226	.L...?. U -2.0±1.6 	1.18± .15 .03± .08 1.18	 .05 .00 	 14.5 ±.2 				
085854.1+662809 148.15 37.28 38.01 -4.66 085429.8+663947	A 0854+66 UGC 4687 MK 100 PGC 25227	.P..... R 99.0 	.81± .11 .07± .06 .82	45 .14 .09 .03	 14.5 ±.2 14.20				3598± 72 3694 3682
0858.9 +2854 196.49 39.18 60.12 -36.28 0855.9 +2906	 CGCG 150- 54 PGC 25228			 .07 	 15.3 ±.3 				12727 12674 12967
085856.6-045407 233.41 25.42 100.37 -57.72 085626.8-044225	NGC 2721 MCG -1-23- 15 PGC 25231	.SBT4P. E (1) 4.0± .8 3.1±1.6	1.37± .04 .17± .05 1.38	30 .07 .25 .09					3723 3553 4020
085900.5+391234 183.19 40.74 52.96 -27.92 085547.7+392414	 UGC 4704 KUG 0855+394 PGC 25232	.S..8*. U 8.0±1.2 	1.61± .02 1.01± .04 1.61	115 .06 1.23 .50	 15.0 ±.3 13.70			14.02±.1 129± 8 129± 12 -.18	596± 6 584 801
0859.0 +5337 164.21 40.25 44.61 -15.75 0855.4 +5349	 UGC 4696 PGC 25235	.S..6*. U 6.0±1.5 	1.06± .06 .95± .05 1.06	136 .04 1.40 .47	 15.78 ±.19 				
085906.3+534610 164.02 40.23 44.54 -15.62 085527.1+535750	NGC 2701 UGC 4695 IRAS08554+5357 PGC 25237	.SXT5*. R (2) 5.0± .5 3.9± .7	1.34± .02 .13± .03 .98± .02 1.34	23 .06 .20 .07	12.73 ±.15 12.46 ±.14 12.15 12.32	.42± .02 -.14± .03 .37 -.18	.55± .01 -.05± .02 13.12± .05 13.94± .20	13.96±.1 263± 6 244± 6 1.58	2326± 5 2299± 66 2373 2471
085908.6+110859 217.36 33.36 77.01 -49.20 085625.3+112041	NGC 2720 UGC 4710 PGC 25238	.L..-*. U -3.0±1.2 	1.08± .17 .04± .08 1.09	 .11 .00 	 13.82 ±.15 				
0859.2 +8532 127.27 29.42 28.50 11.88 0846.0 +8544	 UGC 4612 PGC 25240	.S..9*. U 9.0±1.2 	1.14± .13 .18± .12 1.16	175 .25 .18 .09				15.37±.1 142± 8 	1604± 11 1763 1590
0859.3 +6613 148.43 37.40 38.17 -4.85 0855.0 +6625	 UGC 4693 PGC 25243	.S..8.. U 8.0± .9 	1.01± .06 .28± .05 1.03	110 .22 .34 .14					
085931.2+445456 175.64 41.00 49.56 -23.10 085609.7+450638	NGC 2712 UGC 4708 IRAS08561+4506 PGC 25248	.SBR3*. R (3) 3.0± .5 1.2± .5	1.46± .02 .26± .02 .97± .01 1.46	178 .04 .35 .13	12.75 ±.13 12.43 ±.13 12.16 12.19	.67± .01 .06± .02 .60 .01	.76± .01 .15± .02 13.09± .03 14.24± .17	14.03±.1 332± 10 312± 10 1.72	1818± 6 1833± 14 1832 2004
085935.2+455534 174.31 40.99 48.98 -22.25 085612.0+460716	 UGC 4709 ARAK 188 PGC 25251	.S?.... 	.72± .09 .09± .05 .72	10 .07 .14 .05	 14.8 ±.3 13.41 14.53				8393± 50 8408 8572
085948.7+554223 161.49 40.03 43.60 -13.90 085604.9+555405	NGC 2710 UGC 4705 IRAS08560+5554 PGC 25258	.SBT3.. U 3.0± .8 	1.30± .04 .31± .05 1.30	125 .07 .43 .16	 13.66 ±.18 13.52 13.14				2538± 10 2592 2674
085949.1+050340 223.85 30.75 84.94 -52.78 085711.0+051524	 MCG 1-23- 16 ARAK 189 PGC 25259	.L?.... 	.72± .22 .00± .07 .73	 .12 .00 	 14.98 ±.19 14.80				3778± 82 3638 4069
0859.9 -0724 235.87 24.24 105.04 -58.34 0857.5 -0713	 MCG -1-23- 16 PGC 25264	.IB?... 	1.19± .07 .14± .07 1.19	 .05 .11 .07					5806 5629 6104
090006.8+165531 211.03 35.88 70.95 -45.10 085718.5+170716	 UGC 4721 PGC 25269	.SB.6*. U 6.0±1.2 	1.00± .08 .04± .06 1.00	 .05 .06 .02	 14.8 ±.2 14.70			15.80±.3 156± 7 1.08	6256± 7 6157 6528

R.A. 2000 DEC. / l b / SGL SGB / R.A. 1950 DEC.	Names / PGC	Type / S_T n_L / T / L	$\log D_{25}$ / $\log R_{25}$ / $\log A_e$ / $\log D_o$	p.a. / A_g / A_i / A_{21}	B_T / m_B / m_{FIR} / B_T^o	$(B-V)_T$ / $(U-B)_T$ / $(B-V)_T^o$ / $(U-B)_T^o$	$(B-V)_e$ / $(U-B)_e$ / m'_e / m'_{25}	m_{21} / W_{20} / W_{50} / HI	V_{21} / V_{opt} / V_{GSR} / V_{3K}
0900.1 +3200 / 192.61 40.06 / 58.08 -33.65 / 0857.1 +3212	CGCG 150- 56 / PGC 25273			.01	14.81 ±.18				1881± 35 / 1840 / / 2112
0900.1 +7001 / 143.93 36.19 / 36.34 -1.53 / 0855.4 +7013	UGC 4697 / PGC 25275	.SB.6*. / U / 6.0±1.3	1.19± .05 / .77± .05 / / 1.20	57 / .09 / 1.13 / .38				15.75±.1 / / 188± 8	3731± 11 / / 3840 / 3797
090014.5+031042 / 225.82 29.92 / 87.70 -53.73 / 085738.1+032227	NGC 2723 / UGC 4723 / PGC 25280	.L..0*. / PU / -2.0± .7	.96± .12 / .00± .05 / / .98	/ .13 / .00 /	14.23 ±.15 / / / 14.04				3725± 56 / 3579 / 4018
090015.5+354343 / 187.79 40.64 / 55.48 -30.63 / 085707.4+355528	NGC 2719 / UGC 4718 / PGC 25281	.I..9P$ / R / 10.0± .5	1.13± .05 / .59± .04 / / 1.13	133 / .08 / .44 / .29				14.28±.3 / / 231± 7	3157± 10 / 3197± 31 / 3135 / 3379
090015.6+401748 / 181.77 41.05 / 52.50 -26.88 / 085701.5+402932	UGC 4716 / IRAS08570+4029 / PGC 25282	.SBS3.. / U / 3.0± .8	1.10± .05 / .17± .05 / / 1.10	/ .00 / .23 / .08	15.1 ±.2 / 13.45				/ /
0900.2 +6454 / 149.98 37.89 / 38.91 -5.95 / 0856.0 +6506	UGC 4706 / PGC 25283	.SBS3.. / U / 3.0± .8	1.13± .05 / .06± .05 / / 1.14	165 / .13 / .09 / .03	15.0 ±.3 / 14.74			16.16±.1 / / 150± 8 / 1.39	10802± 11 / / 10892 / 10894
090016.1+354316 / 187.80 40.64 / 55.49 -30.64 / 085708.1+355500	NGC 2719A / MCG 6-20- 18 / IRAS08571+3555 / PGC 25284	.I..9P* / RC / 9.9± .6	.93± .10 / .14± .06 / / .94	/ .08 / .11 / .07	/ 12.58 /				3081± 31 / 3055 / 3300
090018.2+344013 / 189.17 40.51 / 56.22 -31.49 / 085711.5+345158	UGC 4720 / PGC 25287	.S?....	.98± .07 / .37± .05 / / .98	178 / .04 / .55 / .18	15.37 ±.18 / / / 14.76			15.45±.3 / / 238± 7 / .50	3198± 10 / 3168 / 3420
0900.3 -3404 / 257.90 7.90 / 155.65 -56.39 / 0858.2 -3353	ESO 371- 30 / PGC 25288	.IXS9.. / S (1) / 10.0± .5 / 10.0± .8	1.22± .04 / .11± .05 / / 1.33	150 / 1.12 / .09 / .06	15.05 ±.14 / / / 13.84				1338 / / 1113 / 1604
0900.3 +2948 / 195.44 39.67 / 59.74 -35.38 / 0857.3 +3000	CGCG 150- 57 / PGC 25289			.06	15.4 ±.3				14474 / 14424 / / 14712
090020.7+522934 / 165.64 40.59 / 45.39 -16.61 / 085644.8+524118	UGC 4713 / IRAS08567+5241 / PGC 25290	.S..3.. / U (1) / 3.0± .8 / 3.5±1.1	1.20± .05 / .10± .05 / / 1.20	177 / .01 / .13 / .05	13.57 ±.18 / 13.02 / / 13.36			15.76±.1 / / 525± 8 / 2.34	9033± 11 / 9036± 50 / 9075 / 9184
090021.9-255210 / 251.51 13.19 / 141.16 -59.09 / 085811.0-254024	ESO 497- 1 / PGC 25291	.IBS9.. / S (1) / 9.5± .6 / 8.9± .6	1.30± .06 / .49± .05 / / 1.36	123 / .65 / .37 / .25	16.55 ±.14 / / / 15.53				1907 / / 1692 / 2190
090023.4+253639 / 200.70 38.70 / 63.07 -38.63 / 085726.8+254824	UGC 4722 / PGC 25292	.S..8*. / U / 8.0±1.3	1.18± .05 / .78± .05 / / 1.19	32 / .10 / .96 / .39	15.16 ±.18 / / / 14.10			14.58±.2 / 157± 13 / 139± 7 / .10	1794± 7 / / 1728 / 2045
090031.9+172237 / 210.56 36.14 / 70.60 -44.71 / 085743.2+173423	UGC 4724 / PGC 25301	.I..9*. / U / 10.0±1.4	1.04± .06 / .72± .05 / / 1.04	52 / .09 / .54 / .36	15.49 ±.19 / / / 14.86			16.13±.3 / / 156± 7 / .91	3875± 10 / / 3778 / 4146
0900.5 +5113 / 167.30 40.78 / 46.12 -17.68 / 0857.0 +5125	UGC 4717 / PGC 25305	.S..2.. / U / 2.0± .9	1.02± .06 / .35± .05 / / 1.02	40 / .04 / .43 / .18	15.01 ±.18 / / / 14.49				/ 4873 / 4910 / 5030
0900.6 +5040 / 168.01 40.85 / 46.43 -18.13 / 0857.1 +5052	UGC 4719 / IRAS08570+5052 / PGC 25308	.S..5.. / U / 5.0± .9	1.36± .04 / .92± .05 / / 1.36	95 / .05 / 1.38 / .46	14.87 ±.18 / 12.94 / / 13.41			15.37±.1 / / 542± 8 / 1.50	5116± 11 / / 5150 / 5275
090041.5+173714 / 210.29 36.26 / 70.39 -44.51 / 085752.6+174900	UGC 4729 / PGC 25309	.SBS6*. / U / 6.0±1.2	1.00± .06 / .07± .05 / / 1.00	/ .09 / .10 / .03	14.48 ±.18 / / / 14.27			15.82±.3 / / 111± 7 / 1.51	3900± 10 / / 3804 / 4171

R.A. 2000 DEC.	Names	Type	logD$_{25}$	p.a.	B$_T$	(B-V)$_T$	(B-V)$_e$	m$_{21}$	V$_{21}$
l b		S$_T$ n$_L$	logR$_{25}$	A$_g$	m$_B$	(U-B)$_T$	(U-B)$_e$	W$_{20}$	V$_{opt}$
SGL SGB		T	logA$_e$	A$_i$	m$_{FIR}$	(B-V)$_T^o$	m'$_e$	W$_{50}$	V$_{GSR}$
R.A. 1950 DEC.	PGC	L	logD$_o$	A$_{21}$	B$_T^o$	(U-B)$_T^o$	m'$_{25}$	HI	V$_{3K}$
0900.7 +1711		.S..2..	1.00± .08	55					
210.80 36.11	UGC 4728	U	.55± .06	.07					
70.83 -44.81		2.0± .9		.68					
0857.9 +1723	PGC 25311		1.01	.27					
0900.8 +3200		.S..6*.	1.14± .07	67					1994
192.65 40.21	UGC 4725	U	.98± .06	.01	15.76 ±.18				
58.22 -33.56		6.0±1.4		1.44					1953
0857.8 +3212	PGC 25318		1.14	.49	14.30				2226
0901.0 +1037		.S..0..	1.11± .05						
218.17 33.55	UGC 4731	U	.03± .05	.14	14.0 ±.2				
78.10 -49.19		.0± .8		.02					
0858.3 +1049	PGC 25328		1.12						
090101.8+354543	NGC 2724	.SXS5..	1.26± .06	2				14.34±.3	3220± 10
187.77 40.80	UGC 4726	U	.05± .06	.08	14.3 ±.3				
55.61 -30.52	IRAS08579+3557	5.0± .8		.08				225± 7	3194
085753.8+355730	PGC 25331		1.26	.03	14.12			.19	3439
090103.4+110549	NGC 2725	.S?....	.85± .08					15.71±.3	2074± 10
217.66 33.76	UGC 4732		.07± .05	.08	14.4 ±.2				
77.54 -48.87				.07				185± 7	1954
085820.3+111737	PGC 25332		.85	.03	14.26			1.41	2358
0901.2 +5312		.S?....	.94± .11						8983± 79
164.68 40.62	MCG 9-15- 77		.00± .07	.07	15.2 ±.2				9027
45.12 -15.93	IRAS08575+5324			.00	13.27				9131
0857.5 +5324	PGC 25339		.94	.00	15.04				
090114.4+040712		.S..6*.	1.07± .05	178				16.12±.3	8430± 10
225.01 30.60	UGC 4733	U	1.01± .05	.16	15.85 ±.20			347± 13	
86.58 -53.01		6.0±1.5		1.47					8287
085837.1+041900	PGC 25341		1.09	.50	14.18			1.44	8723
090127.6-261802		.SBT8..	1.28± .04	178				15.85±.2	1960± 11
252.01 13.10	ESO 497- 2	Sr (1)	.20± .04	.65	14.68 ±.14			178± 16	
141.86 -58.75		8.0± .4		.25				180± 12	1744
085917.1-260612	PGC 25350	6.7± .6	1.34	.10	13.77			1.98	2242
090128.9+034312	NGC 2729	.L...?.	.90± .09						
225.45 30.46	UGC 4737	U	.18± .03	.11	14.4 ±.2				
87.20 -53.18	ARAK 191	-2.0±1.8		.00					
085852.0+035501	PGC 25352		.88						
090136.6-641616		.S?....	1.12± .16						6606± 87
280.99 -11.71	ESO 90- 14		.59± .14	.67					6383
189.80 -36.85	IRAS09006-6404			.81	11.71				6768
090040.0-640424	PGC 25356		1.18	.29					
090138.1-244726		RSBS1..	.94± .06	128					
250.84 14.09	ESO 497- 3	r	.31± .03	.93	15.49 ±.14				
138.99 -59.03	IRAS08594-2435	1.0± .9		.31	12.89				
085926.1-243536	PGC 25359		1.03	.15					
090141.3+110447	NGC 2728	.S..3..	1.04± .06	60				15.59±.3	5748± 10
217.76 33.90	UGC 4738	U (1)	.13± .05	.08	14.44 ±.19				
77.72 -48.77		3.0± .9		.18				290± 7	5628
085858.2+111636	PGC 25360	5.5±1.1	1.05	.07	14.13			1.39	6032
090158.0+600905		.S..6*.	1.05± .06	81					
155.77 39.37	UGC 4727	U	.93± .05	.21					
41.52 -9.94		6.0±1.4		1.37					
085801.6+602053	PGC 25369		1.07	.47					
090158.7+600912	A 0858+60	.S?....	.94± .07	93					3235± 45
155.76 39.37	UGC 4730		.45± .05	.21	14.6 ±.3				3307
41.52 -9.94	MK 18			.66	12.42				3351
085802.3+602100	PGC 25370		.96	.22	13.73				
0902.0 +7817		.S..7*.	1.31± .04	110				15.37±.1	1408± 11
134.69 32.95	UGC 4701	U	.33± .05	.03	15.1 ±.4				
32.30 5.68		7.0±1.1		.46				164± 8	1545
0855.6 +7829	PGC 25371		1.31	.17	14.64			.57	1432
090206.1-645418		.SXR3*.	1.25± .06	144					1639± 87
281.52 -12.08	ESO 90- 15	Sr (1)	.66± .05	.68					1417
190.24 -36.32	IRAS09011-6442	3.2± .7		.91	12.19				1798
090112.0-644224	PGC 25373	5.6± .9	1.32	.33					
090206.4+232313		.S?....	1.06± .06	166				15.98±.3	31± 10
203.57 38.46	UGC 4740		.35± .05	.11	14.78 ±.18				
65.35 -40.07				.53				210± 7	-43
085912.2+233503	PGC 25374		1.07	.18	14.14			1.66	289

R.A. 2000 DEC.	Names	Type	$\log D_{25}$	p.a.	B_T	$(B-V)_T$	$(B-V)_e$	m_{21}	V_{21}
l b		S_T n_L	$\log R_{25}$	A_g	m_B	$(U-B)_T$	$(U-B)_e$	W_{20}	V_{opt}
SGL SGB		T	$\log A_e$	A_i	m_{FIR}	$(B-V)_T^o$	m'	W_{50}	V_{GSR}
R.A. 1950 DEC.	PGC	L	$\log D_o$	A_{21}	B_T^o	$(U-B)_T^o$	m'_{25}	HI	V_{3K}
090208.2+081800	NGC 2731	.S?....	.90± .07	70				16.46±.3	2583± 10
220.80 32.77	UGC 4741		.16± .05	.19	14.5 ±.2				2454
81.21 -50.41	IRAS08594+0829			.24	11.95			184± 7	
085927.5+082951	PGC 25376		.91	.08	14.03			2.35	2872
0902.2 +1432		.S..2..	1.07± .07	128					
214.00 35.45	UGC 4742	U	.62± .06	.11	15.38 ±.18				
73.97 -46.38		2.0± .9		.76					
0859.5 +1444	PGC 25383		1.08	.31					
090216.1+165020	NGC 2730	.SB.8*.	1.23± .05	80				15.30±.2	3830± 7
211.37 36.33	UGC 4743	U	.13± .05	.08	13.48 ±.18			196± 13	
71.55 -44.79	IRAS08594+1702	8.0±1.1		.16	13.21			187± 6	3731
085928.0+170211	PGC 25384		1.23	.06	13.23			2.00	4104
090223.7-681755			1.02± .07	153					
284.22 -14.22	ESO 60- 23		.65± .05	.43	15.53 ±.14				4070± 87
192.75 -33.64									3852
090147.0-680600	PGC 25389		1.06						4214
090225.0-735914			.83± .06	165					
288.83 -17.79	ESO 36- 16		.32± .05	.52	15.30 ±.14				5694± 19
196.67 -29.07	IRAS09025-7347				13.50				5486
090233.1-734718	PGC 25392		.88						5812
0902.6 +3116									
193.69 40.45	CGCG 151- 3			.05	15.37 ±.18				4118± 35
59.13 -33.92									4074
0859.6 +3128	PGC 25398								4353
090238.6+255606	NGC 2735	.SXT3$P	1.09± .03	94	14.13 ±.14	.86± .02	.94± .02	14.56±.2	2450± 5
200.47 39.27	UGC 4744	R	.45± .05	.11		.34± .03	.35± .03	416± 6	
63.30 -38.06	IRAS08597+2608	3.0± .5	.49± .02	.62	12.79	.74	12.07± .08	393± 5	2385
085941.9+260758	PGC 25399		1.10	.22	13.38	.23	13.32± .25	.96	2702
090240.3-681338		.S?....	1.45± .04	55					
284.18 -14.16	ESO 60- 24		.84± .04	.43	13.98 ±.14				4094± 87
192.67 -33.68	IRAS09020-6801			1.16	13.32				3876
090203.1-680142	PGC 25400		1.49	.42	12.36				4239
0902.6 +2556	NGC 2735A	.I..9*P	.34?						
200.46 39.28	MCG 4-22- 3	R	.00± .07	.11					
63.31 -38.05		10.0± .9		.00					
0859.7 +2608	PGC 25402		.35	.00					
090243.7+250528		.S..6*.	.96± .09	170				15.49±.3	6006± 10
201.52 39.07	UGC 4745	U	.80± .06	.10	16.0 ±.2				
64.03 -38.69		6.0±1.4		1.18				334± 7	5938
085947.8+251720	PGC 25405		.97	.40	14.68			.41	6260
090244.2+252521		.S..4..	1.15± .05	145				15.34±.3	6036± 10
201.11 39.16	UGC 4746	U	.35± .05	.12	14.8 ±.2				
63.75 -38.44		4.0± .9		.52				313± 7	5969
085948.1+253713	PGC 25406		1.16	.18	14.09			1.08	6289
090245.1-204254		.IB.9P*	.91± .06	155					
247.72 16.86	ESO 564- 10	SE	.08± .04	.73	14.77 ±.14				
131.04 -59.31		10.0±1.8		.06					
090029.0-203100	PGC 25407		.98	.04					
090314.4+303533	IC 2428	.S..6*.	1.27± .03	75				15.44±.2	4310± 7
194.59 40.45	UGC 4747	U	.62± .04	.04	14.49 ±.18			334± 13	4277± 76
59.75 -34.37	IRAS09002+3047	6.0±1.2		.91	13.52			330± 7	4264
090012.9+304727	PGC 25423		1.27	.31	13.51			1.62	4548
0903.3 +1001		.S..6*.	1.07± .07	10					
219.12 33.79	UGC 4748	U	.50± .06	.12					
79.37 -49.13		6.0±1.3		.73					
0900.6 +1013	PGC 25426		1.08	.25					
090318.2+783350		.SX.3..	1.10± .05	105				16.75±.1	1257± 11
134.36 32.89	UGC 4714 (1)	U	.11± .05	.02	13.75 ±.18				
32.21 5.95		3.0± .8		.15				113± 8	1395
085649.0+784538	PGC 25427	3.5±1.1	1.10	.05	13.57			3.13	1280
0903.4 +0322		.S?....	.94± .11	60					
226.08 30.70	MCG 1-23- 19		.57± .07	.12	15.4 ±.2				7935± 57
88.14 -52.96				.85					7790
0900.8 +0334	PGC 25436		.95	.28	14.37				8231
090325.2+204003									
206.98 37.91	CGCG 121- 7			.10	15.3 ±.3				9530± 60
68.10 -41.87	MK 1222								9445
090033.7+205158	PGC 25437								9796

R.A. 2000 DEC. / l b / SGL SGB / R.A. 1950 DEC.	Names / PGC	Type S_T n_L / T / L	logD_25 / logR_25 / logA_o / logD_o	p.a. / A_g / A_i / A_21	B_T / m_B / m_FIR / B_T^o	(B-V)_T / (U-B)_T / (B-V)_T^o / (U-B)_T^o	(B-V)_. / (U-B)_. / m'_. / m'_25	m_21 / W_20 / W_50 / HI	V_21 / V_opt / V_GSR / V_3K
090333.6+032228 / 226.10 30.74 / 88.16 -52.92 / 090057.0+033424	MCG 1-23- 20 / PGC 25441	.LB?...	.90± .17 / .00± .07 / / .91	/ .12 / .00 /	14.68 ±.15 / / / 14.51				3694± 57 / 3549 / 3990
090337.3-675759 / 284.03 -13.92 / 192.40 -33.83 / 090258.0-674600	NGC 2788B / ESO 60- 25 / / PGC 25443	.S..3P? / S / 3.0±1.9	1.17± .04 / .53± .04 / / 1.21	165 / .44 / .72 / .26	14.51 ±.14				
0903.7 +2918 / 196.28 40.29 / 60.84 -35.32 / 0900.7 +2930	IC 2429 / MCG 5-22- 3 / / PGC 25446	.S?....		/ .07 / /	15.1 ±.3				3011 / 2959 / 3254
0903.8 -0029 / 229.98 28.82 / 94.01 -54.77 / 0901.3 -0018	UGC 4754 / / / PGC 25450	.S..4.. / U / 4.0±1.0	1.00± .08 / .73± .06 / / 1.01	135 / .12 / 1.07 / .36					
090351.3+853005 / 127.26 29.52 / 28.60 11.89 / 085057.0+854146	IC 512 / UGC 4646 / IRAS08508+8541 / PGC 25451	.SXS6.. / U / 6.0± .7	1.46± .03 / .10± .05 / / 1.49	175 / .25 / .15 / .05	12.9 ±.2 / 12.85 / / 12.51			14.43±.1 / 142± 16 / / 1.87	1614± 11 / / 1773 / 1601
0903.9 +2154 / 205.55 38.42 / 67.09 -40.88 / 0901.1 +2206	NGC 2737 / UGC 4751 / / PGC 25453	.S..2.. / U / 2.0± .9	.96± .07 / .40± .05 / / .97	61 / .10 / .49 / .20	15.03 ±.19				
090400.2+215805 / 205.47 38.45 / 67.04 -40.83 / 090107.6+221001	NGC 2738 / UGC 4752 / VV 481 / PGC 25454	.S?....	1.16± .05 / .37± .05 / / 1.17	55 / .09 / .55 / .18	13.99 ±.19 / 11.96 / / 13.34			15.09±.1 / 288± 5 / 236± 5 / 1.57	3108± 4 / 3065± 47 / 3027 / 3371
090402.3-720325 / 287.33 -16.49 / 195.26 -30.56 / 090351.1-715124	ESO 60- 26 / IRAS09038-7151 / / PGC 25455	.RINGP. / S / -2.0±1.1	1.01± .06 / .11± .03 / / 1.05	/ .49 / .00 /	14.48 ±.14 / 12.92				
090403.8+033457 / 225.96 30.95 / 87.98 -52.70 / 090127.1+034654	MCG 1-23- 21 / / / PGC 25457	.SB?...	.82± .13 / .17± .07 / / .83	/ .12 / .24 / .09	15.04 ±.20 / / / 14.62				7964± 54 / 7819 / 8260
090408.4-720301 / 287.33 -16.48 / 195.25 -30.56 / 090357.0-715100	ESO 60- 27 / / / PGC 25460	.RINGP. / S / 10.0± .6	1.03± .07 / .18± .06 / / 1.07	128 / .49 / .13 / .09	14.75 ±.14				
0904.2 +7455 / 138.26 34.55 / 34.13 2.85 / 0858.8 +7507	UGC 4736 / / / PGC 25464	.SA.8.. / U / 8.0± .8	1.16± .07 / .09± .06 / / 1.16	160 / .02 / .11 / .04				16.14±.1 / / 193± 8 /	6350± 11 / / 6476 / 6392
090422.9+275711 / 198.04 40.14 / 62.04 -36.27 / 090124.3+280908	IC 2430 / UGC 4755 / IRAS09013+2809 / PGC 25467	.S..0.. / U / .0± .9	1.04± .06 / .28± .05 / / 1.03	43 / .11 / .21 /	14.50 ±.18 / 13.70 / / 14.13				2980± 50 / 2923 / 3227
0904.5 +1333 / 215.38 35.56 / 75.58 -46.62 / 0901.8 +1345	MCG 2-23- 26 / / / PGC 25471	.S?....	.82± .13 / .28± .07 / / .83	124 / .10 / .43 / .14	15.2 ±.2 / / / 14.61				8313± 57 / 8202 / 8596
090433.4+451726 / 175.12 41.88 / 50.14 -22.30 / 090112.3+452923	UGC 4753 / / / PGC 25472	.S..7.. / U / 7.0±1.0	1.16± .05 / .84± .05 / / 1.16	33 / .04 / 1.16 / .42	15.59 ±.18				
090433.6+513649 / 166.68 41.35 / 46.46 -17.01 / 090100.7+514846	A 0901+51 / UGC 4749 / MK 101 / PGC 25473	.S..... / R	.87± .08 / .02± .05 / .30± .05 / .87	/ .00 / .03 / .01	14.0 ±.2 / 13.86 ±.18 / 13.02 / 13.86	.55± .04 / .02± .07 / .52 / .00	.63± .02 / .05± .04 / 10.96± .12 / 13.17± .44	15.72±.3 / 112± 34 / 115± 25 / 1.85	4750± 17 / 4780± 38 / 4793 / 4912
090434.9+143542 / 214.20 35.99 / 74.45 -45.92 / 090148.9+144740	IC 2431 / UGC 4756 / MK 1224 / PGC 25476		.74± .14 / .09± .07 / / .75	/ .11 /	14.6 ±.2 / / 11.54	.57± .03 / -.11± .05			14951± 36 / 14843 / 15231
090438.8+182726 / 209.73 37.44 / 70.50 -43.25 / 090149.4+183925	NGC 2744 / UGC 4757 / VV 612 / PGC 25480	.SBS2*P / R / 2.0± .4	1.22± .05 / .19± .05 / / 1.23	/ .08 / .24 / .10	13.9 ±.2 / / 12.79 / 13.54	.44± .09 / / .36	/ / / 14.36± .33	14.99±.1 / 250± 7 / 240± 6 / 1.35	3428± 6 / 3397± 32 / 3333 / 3699

R.A. 2000 DEC. l b SGL SGB R.A. 1950 DEC.	Names PGC	Type S_T n_L T L	$\log D_{25}$ $\log R_{25}$ $\log A_e$ $\log D_o$	p.a. A_g A_i A_{21}	B_T m_B m_{FIR} B_T^o	$(B-V)_T$ $(U-B)_T$ $(B-V)_T^o$ $(U-B)_T^o$	$(B-V)_e$ $(U-B)_e$ m'_e m'_{25}	m_{21} W_{20} W_{50} HI	V_{21} V_{opt} V_{GSR} V_{3K}
0904.7 +2201 205.48 38.63 67.17 -40.67 0901.9 +2213	 UGC 4758 PGC 25489	.SB.3.. U 3.0±.9 	1.08± .07 .24± .06 1.09	15 .11 .34 .12	 14.8 ±.2 				
0904.8 +1334 215.40 35.64 75.64 -46.56 0902.1 +1346	 MCG 2-23- 27 PGC 25493	.S?.... 	.94± .11 .11± .07 .95	174 .10 .16 .06	 14.61 ±.18 14.29				8545± 28 8434 8828
090454.4+250017 201.80 39.52 64.58 -38.43 090158.8+251216	NGC 2743 UGC 4760 IRAS09019+2512 PGC 25496	.S..8*. U 8.0±1.2 	1.06± .06 .16± .05 1.07	105 .11 .19 .08	 14.24 ±.18 13.60 13.93			16.58±.3 181± 7 2.57	3001± 10 2933 3257
0904.9 +1727 210.95 37.14 71.56 -43.91 0902.1 +1739	 UGC 4761 PGC 25497	.E..... U -5.0± .8 	1.08± .10 .02± .05 .81± .05 1.09	 .08 .00 	14.25 ±.16 14.18 ±.17 	1.04± .03 	1.05± .02 13.79± .18 14.60± .55		
090457.4+595558 155.92 39.79 41.96 -9.93 090102.8+600756	NGC 2726 UGC 4750 IRAS09010+6007 PGC 25498	.S..1?. PU 1.0±1.0 	1.20± .04 .46± .04 1.21	87 .17 .46 .23	 13.4 ±.2 13.39 12.74			16.87±.1 370± 8 3.90	1518± 11 1413± 72 1587 1634
090510.8-190507 246.76 18.32 127.82 -58.81 090253.0-185306	NGC 2754 ESO 564- 16 PGC 25504	.L..0P. SE -2.0±1.2 	.91± .05 .18± .03 .94	130 .53 .00 	 15.23 ±.14 				
090521.6+181852 209.98 37.54 70.80 -43.23 090232.4+183053	NGC 2749 UGC 4763 PGC 25508	.E.3... R -5.0± .4 .	1.24± .05 .08± .05 1.05± .02 1.23	 .08 .00 	12.71 ±.13 13.03 ±.12 12.74	.93± .01 .87 	.99± .01 .51± .04 13.45± .08 13.72± .33		 4201± 24 4107 4475
0905.3 +2815 197.73 40.42 62.02 -35.90 0902.4 +2827	 CGCG 151- 10 PGC 25510			 .07 	 15.4 ±.3 				7995 7939 8242
090526.7+253259 201.16 39.78 64.24 -37.94 090230.7+254500	 UGC 4764 PGC 25512	.S?.... 	1.02± .06 .14± .05 1.03	25 .13 .21 .07	 14.9 ±.2 14.51			15.43±.2 152± 14 142± 7 .86	2938± 7 2872 3193
090530.9-190238 246.77 18.41 127.73 -58.73 090313.0-185036	NGC 2758 ESO 564- 20 IRAS09032-1850 PGC 25515	PSB.4P? SE (1) 4.0± .7 5.6±1.2	1.27± .03 .61± .03 1.32	19 .53 .89 .30	 13.99 ±.14 13.30 12.56				1957 1754 2253
090532.8-191220 246.91 18.31 128.05 -58.72 090315.0-190018	IC 2437 ESO 564- 21 PGC 25518	.LXT-*. SE -2.8± .5 	1.26± .04 .20± .03 .71± .05 1.30	123 .55 .00 	13.92 ±.15 13.68 ±.14 	1.02± .02 .55± .03 	1.08± .01 .56± .02 12.96± .17 14.58± .25		
0905.6 +4519 175.07 42.07 50.30 -22.17 0902.3 +4531	 UGC 4762 PGC 25521	.I..9.. U 10.0± .9 	1.00± .16 .33± .12 1.00	50 .04 .25 .17				16.43±.1 97± 6 94± 5 	2017± 10 2030 2202
090542.7+182020 209.99 37.63 70.86 -43.16 090253.5+183222	NGC 2752 UGC 4772 IRAS09028+1832 PGC 25523	.SB.3*/ PU (1) 3.0± .6 3.5±1.2	1.29± .04 .64± .04 1.30	58 .08 .89 .32	 14.51 ±.18 13.00 			15.01±.3 649± 7 	8875± 10 4022± 71
090545.7+362116 187.13 41.82 56.10 -29.47 090237.7+363317	 UGC 4767 PGC 25524	.L..... U -2.0± .8 	1.11± .16 .07± .08 .55± .09 1.11	25 .05 .00 	14.2 ±.2 13.85 ±.15 13.78	1.08± .03 .23± .05 .99 .25	1.17± .03 .28± .05 12.40± .32 14.41± .87		 7233± 50 7210 7453
090548.1+252609 201.33 39.83 64.41 -37.97 090252.2+253811	NGC 2750 UGC 4769 VV 541 PGC 25525	.SX.5.. U 5.0± .7 	1.34± .04 .05± .05 1.35	 .14 .07 .02	 12.60 ±.18 11.62 12.38			14.06±.1 191± 5 136± 5 1.66	2674± 4 2644± 18 2605 2928
0905.9 +5143 166.49 41.55 46.58 -16.80 0902.4 +5156	NGC 2740 MCG 9-15- 86 PGC 25531	.S?.... 	.89± .11 .00± .07 .89	 .02 .00 .00	 14.95 ±.18 14.87				 8903± 79 8942 9060
090559.8+352240 188.43 41.76 56.82 -30.23 090253.0+353442	NGC 2746 UGC 4770 IRAS09028+3535 PGC 25533	.SBT1.. U 1.0± .8 	1.21± .04 .03± .04 1.22	 .04 .03 .02	 14.0 ±.2 13.88			15.99±.2 264± 6 255± 7 2.09	7065± 7 7025± 50 7037 7288

R.A. 2000 DEC. l　　b SGL　SGB R.A. 1950 DEC.	Names PGC	Type S_T　n_L T L	$\log D_{25}$ $\log R_{25}$ $\log A_o$ $\log D_o$	p.a. A_g A_i A_{21}	B_T m_B m_{FIR} B_T^o	$(B-V)_T$ $(U-B)_T$ $(B-V)_T^o$ $(U-B)_T^o$	$(B-V)_o$ $(U-B)_o$ m'_o m'_{25}	m_{21} W_{20} W_{50} HI	V_{21} V_{opt} V_{GSR} V_{3K}
090600.3+184554 209.51　37.84 70.51　-42.81 090310.7+185757	 UGC　4773 IRAS09031+1858 PGC 25535	.S?.... 	1.18± .05 .50± .05 1.18	67 .09 .76 .25	 14.63 ±.18 13.76			16.01±.3 334± 7 2.00	3437± 10 3345 3710
090606.0-042123 234.00　27.21 100.66　-55.86 090335.6-040920	 MCG -1-23- 17 PGC 25539	.SAS5?. E　　(1) 5.0±1.2 4.2±1.2	1.21± .05 .32± .05 1.21	135 .07 .48 .16					6474 6306 6777
090606.5-151836 243.75　20.82 120.55　-58.45 090345.3-150632	 MCG -2-23- 9 PGC 25540	.SBS7?. E　　(1) 7.3± .7 7.2± .7	1.08± .05 .16± .04 1.09	110 .15 .22 .08					
0906.2　+2534 201.18　39.96 64.38　-37.80 0903.3　+2547	 UGC　4774 PGC 25545	.SB.8.. U 8.0± .9 	.98± .09 .08± .06 .99	 .13 .10 .04	15.0 ±.2 14.80			15.80±.3 144± 14 .96	2979± 7 2913 3235
090617.2+500522 168.66　41.81 47.57　-18.14 090247.9+501724	 UGC　4771 PGC 25547	PSBT3.. U 3.0± .9 	1.10± .05 .25± .05 1.10	30 .05 .34 .12	15.2 ±.2 				
090627.4-280129 254.10　12.84 144.49　-57.25 090418.1-274924	 ESO　433- 2 IRAS09042-2748 PGC 25551	.SBS8.. S　　(1) 8.0± .5 8.9± .6	1.28± .06 .05± .07 1.35	 .77 .07 .03					2024 1806 2308
090634.1-071429 236.73　25.69 105.68　-56.73 090406.1-070224	 MCG -1-23- 19 PGC 25555	.LAR-?. E -3.0± .7 	1.18± .07 .21± .05 1.17	80 .14 .00 					
090634.2+061812 223.51　32.82 84.82　-50.69 090355.3+063017	 UGC　4781 PGC 25556	.S..6*. U 6.0±1.2 	1.28± .03 .52± .04 1.30	127 .14 .77 .26	 14.32 ±.20 13.39			14.38±.2 158± 6 143± 5 .72	1443± 5 1307 1738
090634.9-754827 290.51　-18.71 197.62　-27.43 090701.1-753618	 ESO　　36- 19 IRAS09070-7537 PGC 25558	PSB.4P* Sr　(1) 3.9± .5 3.9± .6	1.30± .04 .42± .05 1.35	157 .56 .61 .21	13.76 ±.14 11.53 12.56				4558± 54 4353 4668
090639.3+192008 208.90　38.18 70.10　-42.29 090349.2+193213	 UGC　4780 PGC 25561	.SX.8.. U 8.0± .8 	1.24± .05 .10± .05 1.25	60 .11 .12 .05	14.1 ±.3 13.81			14.77±.3 129± 7 .91	3284± 7 3194 3557
090640.1+343709 189.47　41.80 57.49　-30.75 090334.4+344913	 UGC　4777 PGC 25562	.I..9*. U 10.0±1.2 	1.34± .04 .70± .05 1.34	140 .03 .53 .35	14.9 ±.2 14.35			15.14±.1 202± 7 180± 12 .44	2054± 6 2156± 76 2024 2281
0906.6　-0633 236.12　26.10 104.49　-56.48 0904.2　-0621	 MCG -1-23- 20 PGC 25563	.S..7*/ E 7.0±1.3 	1.24± .04 1.00± .04 1.24	137 .10 1.37 .50					3763 3589 4067
090649.1-152959 244.02　20.84 120.94　-58.30 090428.0-151754	NGC　2763 MCG -2-23- 10 IRAS09044-1517 PGC 25570	.SBR6P. R　　(3) 6.0± .3 3.4± .4	1.36± .02 .06± .03 1.01± .02 1.38	120 .17 .09 .03	12.64 ±.14 12.5 ±.2 12.12 12.35	.62± .03 -.07± .08 .56 -.11	.66± .03 -.02± .08 13.18± .04 14.14± .18	14.27±.1 215± 5 189± 8 1.89	1893± 7 1860± 27 1695 2191
090650.0+261630 200.35　40.27 63.93　-37.19 090353.4+262835	IC　2435 UGC　4782 ARAK　194 PGC 25571	.E..... U -5.0± .9 	.88± .14 .32± .05 .80	120 .09 .00 	 15.27 ±.16 				
090708.3+281901 197.76　40.81 62.33　-35.60 090409.6+283107	 UGC　4786 PGC 25596	.S..6*. U 6.0±1.5 	.96± .09 .98± .06 .97	81 .07 1.44 .49				17.75±.3 224± 5 	6546± 9 6491 6794
090711.5+504245 167.81　41.88 47.33　-17.54 090341.2+505450	 UGC　4778 IRAS09036+5054 PGC 25600	.S..3.. U　　(1) 3.0± .9 2.5±1.1	.91± .09 .00± .06 .91	 .04 .00 .00	 14.25 ±.19 14.13				11293± 50 11328 11456
0907.2　+2519 201.57　40.11 64.81　-37.84 0904.3　+2532	NGC　2753 MCG　4-22- 15 PGC 25603	 	.74± .14 .21± .07 .75	 .14 	 15.4 ±.2 				 2832± 40 2765 3089

R.A. 2000 DEC.	Names	Type	logD$_{25}$	p.a.	B$_T$	(B-V)$_T$	(B-V)$_e$	m$_{21}$	V$_{21}$
l b		S$_T$ n$_L$	logR$_{25}$	A$_g$	m$_B$	(U-B)$_T$	(U-B)$_e$	W$_{20}$	V$_{opt}$
SGL SGB		T	logA$_e$	A$_i$	m$_{FIR}$	(B-V)$_T^o$	m'$_e$	W$_{50}$	V$_{GSR}$
R.A. 1950 DEC.	PGC	L	logD$_o$	A$_{21}$	B$_T^o$	(U-B)$_T^o$	m'$_{25}$	HI	V$_{3K}$

090716.2+371254	IC 2434	.SB?...	1.19± .05	13					
186.01 42.20	UGC 4785		.36± .05	.02	14.35 ±.18				7158± 50
55.79 -28.60	IRAS09041+3725			.54	12.51				7138
090407.4+372500	PGC 25609		1.19	.18	13.75				7376

090734.1+602846	NGC 2742	.SAS5*.	1.48± .03	87	12.03 ±.17	.59± .02	.67± .01	13.93±.1	1288± 7
155.12 39.96	UGC 4779	R (2)	.29± .03	.20	12.26 ±.14	.08± .03	.06± .02	327± 5	1273± 22
41.94 -9.30	IRAS09036+6040	5.0± .5	1.18± .03	.44	11.59	.48	13.42± .08	296± 5	1360
090338.6+604052	PGC 25640	3.3± .6	1.50	.15	11.52	.00	13.56± .23	2.26	1404

090735.1+331630		.S..8..	1.33± .04	6				15.15±.1	552± 7
191.29 41.81	UGC 4787	U	.65± .05	.03	14.21 ±.18			141± 16	
58.63 -31.69		8.0± .9		.80				125± 6	516
090431.0+332837	PGC 25644		1.33	.33	13.38			1.45	784

090736.5+032339	NGC 2765	.L.....	1.32± .04	107					
226.66 31.62	UGC 4791	U	.28± .05	.10	13.08 ±.15				3827± 76
89.01 -52.05		-2.0± .8		.00					3682
090460.0+033547	PGC 25646		1.29		12.91				4126

090739.3+663430		.L..-*.	1.15± .15	0					
147.66 38.06	UGC 4775	U	.19± .08	.17	14.12 ±.15				
38.70 -4.13		-3.0±1.2		.00					
090318.6+664635	PGC 25649		1.14						

090741.8-233715	NGC 2772	.S.3*P/	1.19± .04	71					
250.82 15.90	ESO 497- 14	S	.23± .05	1.07	14.19 ±.14				
136.36 -57.86	IRAS09054-2325	3.0±1.1		.32	12.30				
090528.0-232506	PGC 25654		1.29	.12					

090758.4+414237	NGC 2755	.S?....	1.08± .06	130					
179.94 42.55	UGC 4789		.17± .05	.00	14.15 ±.18				7547± 50
52.93 -24.89	IRAS09047+4154			.26	12.95				7545
090443.5+415445	PGC 25670		1.08	.09	13.85				7748

0908.0 +2029		.S..4..	1.00± .08	44					
207.64 38.88	UGC 4792	U	1.02± .06	.10	16.1 ±.2				
69.31 -41.24		4.0±1.0		1.47					
0905.2 +2042	PGC 25673		1.01	.50					

090807.1+780511	NGC 2715	.SXT5..	1.69± .02	22	11.79 ±.14	.56± .02	.64± .01	12.56±.2	1323± 7
134.73 33.32	UGC 4759	R (2)	.47± .02	.02	11.88 ±.13	-.09± .03	.01± .02	431± 5	1275± 25
32.67 5.67	IRAS09018+7817	5.0± .3	1.32± .03	.70	11.87	.46	13.88± .10	299± 4	1456
090152.5+781715	PGC 25676	3.3± .6	1.70	.23	11.12	-.16	13.95± .17	1.21	1346

0908.1 +0555	A 0905+06	.S..9*.	1.28± .06		14.49 ±.18	.55± .06	.63± .03	15.33±.1	1308± 5
224.13 32.99	UGC 4797	U (1)	.04± .06	.12	14.1 ±.5	-.02± .08	.01± .05	87± 5	
85.66 -50.57	DDO 54	9.0±1.1	1.15± .04	.04		.51	15.73± .10	74± 6	1171
0905.5 +0607	PGC 25679	9.0±1.0	1.29	.02	14.29	-.05	15.65± .37	1.02	1605

0908.2 +5138		.SB.2*.	1.06± .06						
166.53 41.91	UGC 4790	U	.27± .05	.03	15.06 ±.19				
46.94 -16.67		2.0± .9		.33					
0904.7 +5151	PGC 25683		1.06	.13					

090817.5+212637	NGC 2764	.L...*.	1.19± .03	15	13.63 ±.13	.76± .01	.71± .01	15.56±.2	2718± 5
206.50 39.23	UGC 4794	P	.25± .06	.14	13.65 ±.11	.23± .03	.16± .03	339± 8	2597± 66
68.48 -40.52	IRAS09054+2138	-2.0±1.2	.57± .01	.00	11.67	.68	11.97± .05	306± 5	2636
090525.6+213846	PGC 25690		1.17		13.46	.20	13.83± .26		2986

090823.6-013646		.SXS5..	.98± .07						
231.74 29.19	UGC 4802	U	.10± .05	.06	14.75 ±.14				
96.62 -54.24		5.0± .9		.15					
090551.1-012436	PGC 25698		.98	.05					

0908.4 +8608		.S..7..	1.02± .06	67					
126.59 29.26	UGC 4682	U	.30± .05	.37					
28.33 12.47		7.0± .9		.41					
0854.0 +8620	PGC 25708		1.05	.15					

0908.5 -0937					14.3 ±.4	1.08± .07	1.09± .03		
239.19 24.70				.18					16441± 20
110.16 -56.91			.98± .23				14.68± .79		16259
0906.1 -0925	PGC 25714								16746

090837.5-014511		.SB.4..	1.00± .06						
231.91 29.16	UGC 4804	U	.07± .05	.08	14.61 ±.20				
96.88 -54.25		4.0± .9		.10					
090605.0-013300	PGC 25717		1.00	.03					

090837.5+373717	NGC 2759	.L..-*.	1.00± .19	50	14.0 ±.2	.96± .02			
185.49 42.50	UGC 4795	U	.14± .08	.01	14.16 ±.16	.49± .04			6944± 22
55.75 -28.12		-3.0±1.2		.00		.89			6926
090528.3+374927	PGC 25718		.98		13.98	.52	13.51±1.00		7161

R.A. 2000 DEC. / l b / SGL SGB / R.A. 1950 DEC.	Names / PGC	Type / S_T n_L / T / L	$\log D_{25}$ / $\log R_{25}$ / $\log A_\bullet$ / $\log D_0$	p.a. / A_g / A_i / A_{21}	B_T / m_B / m_{FIR} / B_T^0	$(B-V)_T$ / $(U-B)_T$ / $(B-V)_T^0$ / $(U-B)_T^0$	$(B-V)_\bullet$ / $(U-B)_\bullet$ / $m'_\bullet$ / m'_{25}	m_{21} / W_{20} / W_{50} / HI	V_{21} / V_{opt} / V_{GSR} / V_{3K}
090838.5+323536 / 192.24 41.93 / 59.34 -32.09 / 090535.3+324746	/ MCG 6-20- 32 / / PGC 25719	.L?.... / / /		.02 / / /	14.8 ±.3 / / /				4299± 13 / 4260 / / 4534
090842.5+444837 / 175.73 42.64 / 51.09 -22.28 / 090523.1+450047	/ UGC 4798 / / PGC 25726	.SA.6.. / U / 6.0± .8 /	1.04± .06 / .06± .05 / / 1.04	.02 / .08 / / .03	14.9 ±.2 / / / 14.74			15.60±.1 / / 272± 8 / .83	8023± 11 / 8034 / / 8212
090848.2+295149 / 195.85 41.49 / 61.45 -34.18 / 090548.1+300400	NGC 2766 / UGC 4801 / / PGC 25735	.S..2.. / U / 2.0± .9 /	1.13± .05 / .39± .05 / / 1.14	132 / .05 / .48 / .20	/ 14.50 ±.18 / / 13.93			15.91±.3 / / 401± 7 / 1.79	4187± 10 / 4178± 19 / 4136 / 4430
090901.6+535056 / 163.59 41.67 / 45.78 -14.77 / 090525.0+540306	NGC 2756 / UGC 4796 / IRAS09054+5402 / PGC 25757	.S..3.. / U (1) / 3.0± .8 / 2.5±1.1	1.24± .05 / .18± .05 / / 1.24	0 / .02 / .25 / .09	/ 13.20 ±.18 / 13.02 / 12.90				4019± 52 / 4066 / 4168 /
090903.3-675557 / 284.34 -13.51 / 191.90 -33.54 / 090821.0-674342	NGC 2788 / ESO 61- 2 / IRAS09083-6743 / PGC 25761	.S..2?/ / RS / 2.0± .9 /	1.25± .04 / .67± .04 / / 1.30	114 / .52 / .82 / .33	/ 13.26 ±.14 / 11.87 / 11.90				1576± 63 / 1358 / 1724 /
090920.4+204154 / 207.52 39.23 / 69.41 -40.88 / 090629.3+205407	/ UGC 4809 / / PGC 25780	.SB.6?. / U / 6.0±1.7 /	1.20± .06 / .44± .06 / / 1.21	103 / .11 / .65 / .22	/ 14.64 ±.20 / / 13.87			15.21±.3 / / 199± 7 / 1.12	3022± 10 / 2937 / 3293 /
0909.3 +5453 / 162.21 41.51 / 45.23 -13.87 / 0905.7 +5506	/ UGC 4800 / / PGC 25781	.SBS6?. / U / 6.0±1.8 /	1.21± .05 / .50± .05 / / 1.21	120 / .02 / .74 / .25	/ 14.64 ±.18 / / 13.87			15.08±.1 / / 238± 8 / .96	2433± 11 / 2484 / 2577 /
090922.6+154746 / 213.38 37.52 / 74.28 -44.26 / 090635.8+155959	IC 528 / UGC 4811 / MK 1225 / PGC 25783	.S..2.. / U (1) / 2.0± .8 / 4.5±1.1	1.17± .05 / .27± .05 / / 1.18	163 / .07 / .33 / .14	15.00S±.15 / 14.24 ±.19 / / 14.26		/ / / 15.04± .31	15.51±.3 / / 351± 7 / 1.12	3781± 10 / 3801± 31 / 3680 / 4065
090930.0-332044 / 258.62 9.82 / 153.12 -54.93 / 090726.1-330830	/ ESO 372- 7 / / PGC 25797	/ / /	.95± .21 / .21± .14 / / 1.07	173 / 1.24 / /				/ 39± 16 / /	1136± 11 / / 911 / 1410
090933.3+330725 / 191.57 42.20 / 59.13 -31.55 / 090629.8+331938	NGC 2770 / UGC 4806 / IRAS09065+3319 / PGC 25806	.SAS5*. / R / 5.0± .5 /	1.58± .02 / .52± .03 / 1.12± .02 / 1.59	148 / .02 / .78 / .26	12.77 ±.13 / 12.16 ±.19 / 12.28 / 11.76	.55± .03 / -.01± .05 / .44 / -.09	.63± .03 / .05± .05 / 13.86± .03 / 14.24± .18	13.59±.1 / 348± 8 / 332± 6 / 1.56	1951± 7 / 1955± 63 / 1915 / 2185
090941.9+373605 / 185.54 42.71 / 55.96 -28.00 / 090632.9+374818	IC 527 / UGC 4810 / / PGC 25821	.S?.... / / /	1.22± .06 / .05± .06 / / 1.22	/ .01 / .08 / .03	14.2 ±.3 / / / 14.06			15.44±.3 / 379± 7 / 364± 7 / 1.35	6889± 11 / 6871 / 7107 /
090944.4+071026 / 223.04 33.92 / 84.39 -49.55 / 090704.8+072240	NGC 2773 / UGC 4815 / IRAS09070+0722 / PGC 25825	.S?.... / / /	.86± .10 / .35± .06 / / .87	83 / .12 / .53 / .18	/ 15.0 ±.3 / 12.00 / 14.28				5497± 61 / 5364 / 5794 /
090946.6-230033 / 250.65 16.66 / 135.11 -57.47 / 090732.0-224818	A 0907-22 / ESO 497- 17 / DDO 56 / PGC 25827	.IXS9.. / PSU (2) / 10.0± .4 / 9.8± .5	1.11± .04 / .13± .04 / 1.00± .04 / 1.19	65 / .84 / .09 / .06	15.32 ±.19 / 15.64 ±.14 / / 14.59	.86± .08 / .55± .11 / .64 / .37	.94± .07 / .53± .10 / 15.81± .09 / 15.39± .29	15.90±.2 / 77± 16 / 76± 12 / 1.25	724± 11 / / 513 / 1018
090957.6+621445 / 152.81 39.72 / 41.23 -7.65 / 090556.9+622658	NGC 2742A / UGC 4803 / IRAS09059+6227 / PGC 25836	.SBS3P$ / P / 3.0± .8 /	1.19± .05 / .39± .05 / / 1.21	90 / .17 / .54 / .19	/ 14.03 ±.18 / 12.31 / 13.27				7602± 61 / 7682 / 7711 /
0909.9 +7921 / 133.34 32.79 / 32.06 6.80 / 0903.3 +7934	/ UGC 4776 / / PGC 25837	.S..9*. / U / 9.0±1.2 /	1.11± .14 / .15± .12 / / 1.11	/ .02 / .15 / .07				16.46±.1 / 66± 5 / 36± 5 /	2070± 6 / / 2211 / 2090
091004.3-330916 / 258.56 10.04 / 152.73 -54.89 / 090800.0-325700	/ ESO 372- 8 / IRAS09079-3256 / PGC 25842	.SBS9*. / SU (1) / 9.2± .5 / 8.3± .6	1.36± .04 / .26± .05 / / 1.47	156 / 1.17 / .26 / .13	/ / 13.24 /			14.72±.2 / 147± 16 / 138± 12 /	1541± 11 / / 1317 / 1816
091005.9+543449 / 162.59 41.68 / 45.50 -14.07 / 090627.9+544702	/ UGC 4807 / / PGC 25845	.SAT6*. / U / 6.0± .9 /	1.02± .06 / .02± .05 / / 1.02	/ .02 / .03 / .01	/ 14.36 ±.18 / / 14.29			15.74±.1 / / 164± 8 / 1.44	3960± 7 / 3923± 50 / 4009 / 4105

R.A. 2000 DEC.	Names	Type	logD$_{25}$	p.a.	B$_T$	(B-V)$_T$	(B-V)$_e$	m$_{21}$	V$_{21}$
l b		S$_T$ n$_L$	logR$_{25}$	A$_g$	m$_B$	(U-B)$_T$	(U-B)$_e$	W$_{20}$	V$_{opt}$
SGL SGB		T	logA$_e$	A$_i$	m$_{FIR}$	(B-V)^o_T	m'$_e$	W$_{50}$	V$_{GSR}$
R.A. 1950 DEC.	PGC	L	logD$_o$	A$_{21}$	B^o_T	(U-B)^o_T	m'$_{25}$	HI	V$_{3K}$
0910.1 +4436		.S..3..	1.00± .08						
175.98 42.89	UGC 4814	U (1)	.94± .06	.05					
51.43 -22.30		3.0±1.0		1.30					
0906.8 +4449	PGC 25847	4.5±1.4	1.00	.47					
0910.1 +5024	NGC 2767	.S?....	.64± .17						4944± 50
168.12 42.39	UGC 4813		.11± .07	.00	14.8 ±.3				4977
47.92 -17.53				.14					5109
0906.7 +5037	PGC 25852		.64	.06	14.65				
091011.6-791403		.SBR6*.	1.16± .07	80					
293.48 -20.69	ESO 18- 15	S (1)	.26± .07	.47					
199.67 -24.55		6.0± .9		.39					
091132.0-790142	PGC 25855	6.7± .9	1.20	.13					
0910.2 +5957		.S..6*.	1.08± .06	165					
155.65 40.42	UGC 4808	U	.28± .05	.17	15.4 ±.2				
42.51 -9.55		6.0±1.2		.41					
0906.4 +6010	PGC 25858		1.09	.14					
091020.5+070219	NGC 2775	.SAR2..	1.63± .02	155	11.03M±.10	.90± .01	.96± .01	15.36±.1	1354± 5
223.27 33.99	UGC 4820	R	.11± .03	.11	11.19 ±.13	.38± .01	.47± .01	435± 4	1340± 12
84.69 -49.50		2.0± .3	1.07± .01	.13		.85	12.05± .03	405± 4	1219
090741.1+071435	PGC 25861		1.64	.05	10.84	.33	13.76± .15	4.47	1650
091025.2-232929		.S..0*/	1.21± .04	88					
251.14 16.47	ESO 497- 18	S	.85± .03	.77	15.31 ±.14				
135.96 -57.26		.0±1.3		.64					
090811.0-231712	PGC 25867		1.24						
091032.3+502559	NGC 2769	.S..1..	1.24± .04	146					4820± 13
168.09 42.44	UGC 4816	U	.62± .05	.00	13.9 ±.2				4854
47.96 -17.48		1.0± .9		.63					4986
090703.4+503814	PGC 25870		1.24	.31	13.22				
0910.6 +1928		.SB.8..	1.00± .16	165					3129± 10
209.13 39.12	UGC 4822	U	.33± .12	.10					
70.86 -41.53		8.0± .9		.41					3040
0907.8 +1941	PGC 25874		1.01	.17					3404
091039.8+502244	NGC 2771	PSBR2..	1.35± .04						5053± 38
168.15 42.47	UGC 4817	U	.08± .05	.00	13.6 ±.3				5087
48.01 -17.51		2.0± .8		.10					5219
090711.0+503459	PGC 25875		1.36	.04	13.44				
091042.1+071224	NGC 2777	.S..2$.	.93± .06					15.35±.2	1488± 7
223.14 34.15	UGC 4823	P	.15± .04	.12	14.2 ±.2			173± 8	
84.56 -49.34	IRAS09080+0724	2.0±1.8		.19	13.48				1355
090802.6+072441	PGC 25876		.94	.08	13.91			1.37	1786
0910.8 -0854	A 0908-08	.SXT3*/	1.63± .03	33	11.91 ±.13	.65± .02	.61± .01	13.73±.3	1834± 8
238.92 25.59	MCG -1-24- 1	PUE	.62± .05	.15		.37± .03	.47± .03	551± 9	
109.15 -56.18		2.8± .4	1.14± .01	.85		.49	13.10± .02	510± 7	1654
0908.4 -0842	PGC 25886		1.64	.31	10.90	.25	13.35± .21	2.53	2141
091101.7+132442		.S?....	1.09± .04	80				15.44±.3	8629± 10
216.33 36.95	UGC 4827		.57± .04	.12	14.90 ±.18			454± 13	
77.23 -45.52	IRAS09083+1337			.85	13.50				8518
090817.1+133700	PGC 25892		1.10	.28	13.88			1.27	8917
0911.0 +3205									12701
193.01 42.35	CGCG 151- 18			.04	15.4 ±.3				12661
60.18 -32.15									12940
0908.0 +3218	PGC 25893								
0911.1 +1939		.S..6*.	1.16± .07	5					8982
208.96 39.29	UGC 4828	U	1.00± .06	.10	15.73 ±.18				
70.79 -41.31		6.0±1.4		1.47					8893
0908.3 +1952	PGC 25895		1.17	.50	14.13				9257
0911.3 +1917		.S..2..	1.05± .06	104					
209.43 39.21	UGC 4830	U	.55± .05	.10	15.26 ±.18				
71.19 -41.54		2.0± .9		.68					
0908.5 +1930	PGC 25902		1.06	.28					
0911.3 -1503	A 0908-14	.IBS9*.	1.31± .05	70	14.2 ±.3	.39± .08	.43± .05	14.57±.1	2049± 4
244.36 21.98	MCG -2-24- 1	PE (2)	.08± .05	.15		-.19± .09	-.21± .06	119± 6	
120.29 -57.17	DDO 57	9.5± .6	1.21± .07	.06		.32	15.78± .15	87± 12	1854
0908.9 -1450	PGC 25903	8.8± .6	1.33	.04	14.03	-.24	15.44± .38	.50	2353
091119.9-305220		.IBS9..	1.21± .05	64					
257.01 11.76	ESO 433- 7	S (1)	.32± .05	1.04					
148.84 -55.42	IRAS09092-3039	10.0± .8		.24	13.55				
090913.0-304000	PGC 25905	6.7± .8	1.31	.16					

R.A. 2000 DEC. / l b / SGL SGB / R.A. 1950 DEC.	Names / PGC	Type S_T n_L / T / L	$\log D_{25}$ / $\log R_{25}$ / $\log A_o$ / $\log D_o$	p.a. A_g / A_i / A_{21}	B_T / m_B / m_{FIR} / B_T^o	$(B-V)_T$ / $(U-B)_T$ / $(B-V)_T^o$ / $(U-B)_T^o$	$(B-V)_o$ / $(U-B)_o$ / m'_o / m'_{25}	m_{21} / W_{20} / W_{50} / HI	V_{21} / V_{opt} / V_{GSR} / V_{3K}
091127.6-144901	NGC 2781	.LXR+..	1.48± .02	75	12.53 ±.15	.91± .01	.92± .01	14.37±.1	2025± 7
244.18 22.15	MCG -2-24- 2	R	.30± .03	.14	12.46 ±.16	.38± .01	.43± .01	431± 6	2007± 49
119.86 -57.12	IRAS09091-1436	-1.0± .3	.68± .05	.00	13.42	.82	11.42± .16	365± 5	1830
090905.6-143641	PGC 25907		1.45		12.33	.32	14.06± .21		2329
0911.4 +2848	IC 2443	.S?....	.82± .13	.04	15.04 ±.18				10252
197.38 41.85	MCG 5-22- 11		.00± .07	.00					10199
62.82 -34.60			.82	.00	14.93				10502
0908.5 +2901	PGC 25908								
091134.3+511518		.SA.7..	1.39± .04	98				15.31±.1	2185± 7
166.95 42.48	UGC 4824	U	.59± .05	.00	14.25 ±.20				2198± 50
47.62 -16.71		7.0± .8		.82				246± 8	2223
090804.1+512736	PGC 25910		1.39	.30	13.42			1.59	2348
091135.5+325056		.SX.7..	1.02± .06						4311± 10
192.02 42.58	UGC 4831	U	.00± .05	.00	15.0 ±.2				
59.72 -31.50		7.0± .8		.00					4274
090832.5+330315	PGC 25911		1.02	.00	15.00				4548
091137.7+600222	NGC 2768	.E.6.*.	1.91± .02	95	10.84M±.10	.97± .01	.99± .01		1335± 23
155.49 40.56	UGC 4821	R	.28± .02	.17	10.76 ±.12	.46± .02	.53± .01		1406
42.61 -9.40		-5.0± .3	1.33± .03	.00		.92	13.10± .10		1455
090745.2+601440	PGC 25915		1.85		10.61	.43	14.68± .17		
091139.8+463814	A 0908+46	.S?....	.79± .09						4269± 28
173.20 43.04	UGC 4829		.03± .05	.03	14.5 ±.2				4287
50.41 -20.49	MK 102			.03					4452
090818.1+465033	PGC 25917		.79	.01	14.43				
0911.7 +3457		.SB.8..	.98± .09	90					2073± 10
189.18 42.89	UGC 4834	U	.26± .06	.02					
58.21 -29.83		8.0± .9		.33					2044
0908.7 +3510	PGC 25919		.99	.13					2302
091153.0+161643		.S..4..	1.10± .05	60				15.98±.3	9272± 10
213.10 38.26	UGC 4839	U	.78± .05	.08	15.26 ±.20				9171
74.34 -43.49		4.0±1.0		1.14				406± 7	9555
090906.1+162903	PGC 25923		1.10	.39	13.98			1.61	
091154.4-200703		.S..6*/	1.61± .02	168				13.47±.2	2178± 11
248.66 18.90	ESO 564- 27	SUE	1.09± .03	.56	14.39 ±.14			349± 16	
129.69 -57.20	IRAS09096-1954	6.3± .7		1.47	12.95			332± 12	1973
090937.0-195442	PGC 25926		1.67	.50	12.35			.62	2478
0912.0 +7145		.S..6*.	1.11± .07	175					
141.47 36.40	UGC 4819	U	.83± .06	.09					
36.27 .47		6.0±1.4		1.22					
0907.2 +7158	PGC 25937		1.12	.42					
091206.7-152553		.S..7*.	1.12± .06	91					2057
244.81 21.89	MCG -2-24- 3	E (1)	.38± .05	.14					
121.02 -57.02		7.0±1.2		.52					1861
090945.2-151331	PGC 25938		1.13	.19					2362
091209.6+353157	A 0909+35	.S..9?.	1.32± .05	160	14.91 ±.16	.39± .05	.45± .04	14.99±.1	1879± 5
188.41 43.02	UGC 4837	U (1)	.20± .06	.00	14.8 ±.4	-.06± .07	-.09± .06	128± 5	
57.86 -29.34	DDO 55	9.0±1.6	.95± .03	.21		.33	15.15± .07	108± 6	1853
090903.6+354417	PGC 25940	9.0±1.4	1.32	.10	14.69	-.10	15.86± .34	.20	2107
091213.0-305440		.L..0*/	1.30± .05	166					
257.17 11.88	ESO 433- 8	S	.79± .03	1.04					
148.80 -55.22		-2.0±1.2		.00					
091006.0-304218	PGC 25943		1.30						
091215.0+445720	NGC 2776	.SXT5..	1.48± .02		12.14 ±.13	.52± .01	.59± .01	13.56±.1	2626± 5
175.49 43.25	UGC 4838	R (2)	.05± .02	.04	11.98 ±.14	-.07± .04	-.06± .04	214± 6	2618± 42
51.55 -21.80	IRAS09089+4509	5.0± .3	1.05± .01	.07	11.62	.48	12.88± .03	188± 5	2638
090856.1+450940	PGC 25948	1.9± .8	1.48	.02	11.94	-.09	14.26± .17	1.59	2816
091218.7-241022	NGC 2784	.LAS0*.	1.74± .02	73	11.30 ±.13	1.14± .01	1.16± .01		691± 35
251.97 16.35	ESO 497- 23	R	.39± .02	.79	11.15 ±.10	.72± .01	.73± .01		478
137.09 -56.74		-2.0± .3	.95± .01	.00		.92	11.54± .04		985
091005.0-235800	PGC 25950		1.77		10.41	.50	13.90± .17		
091223.0+122955		.S..2..	1.04± .08	144				16.56±.3	8915± 10
217.54 36.87	UGC 4842	U	.59± .06	.08	15.15 ±.18				
78.56 -45.84		2.0± .9		.72				526± 7	8800
090939.2+124217	PGC 25954		1.05	.29	14.26			2.01	9206
091224.6+350139	NGC 2778	.E.....	1.15± .08	40	13.35 ±.13	.93± .01	.96± .01		2019± 26
189.10 43.02	UGC 4840	U	.14± .04	.02	13.10 ±.18	.50± .03	.52± .03		1991
58.27 -29.70		-5.0± .8	.72± .02	.00		.91	12.44± .08		2249
090919.2+351400	PGC 25955		1.11		13.21	.51	13.73± .43		

R.A. 2000 DEC.	Names	Type	logD25	p.a.	BT	(B-V)T	(B-V)o	m21	V21
l b		ST nL	logR25	Ag	mB	(U-B)T	(U-B)o	W20	Vopt
SGL SGB		T	logAe	Ai	mFIR	$(B-V)_T^o$	m'e	W50	VGSR
R.A. 1950 DEC.	PGC	L	logDo	A21	B_T^o	$(U-B)_T^o$	m'25	HI	V3K

091225.6+095721		.SB.7*.	1.20± .04	98				14.30±.3	2117± 10
220.39 35.78	UGC 4845	U	.42± .05	.21	14.5 ±.2			232± 13	
81.53 -47.39		7.0± .9		.58					1994
090943.9+100943	PGC 25956		1.22	.21	13.70			.39	2412
091237.1-585031			.98± .07						
277.74 -7.13	ESO 126- 1		.12± .06	2.87	14.41 ±.14				2844± 87
183.97 -39.99	IRAS09112-5838				12.06				2616
091117.0-583806	PGC 25964		1.25						3034
091244.6+345535	NGC 2780	.SB?...	.95± .07	150				16.97±.3	1979± 9
189.25 43.08	UGC 4843		.13± .05	.02	14.28 ±.19				2019± 49
58.41 -29.73				.20				283± 5	1951
090939.4+350757	PGC 25967		.95	.07	14.05			2.86	2210
0912.8 +3013	IC 2444	.S?....	.99± .10						
195.59 42.40	MCG 5-22- 12		.05± .07	.03	14.68 ±.19				6534
61.97 -33.34				.04					6486
0909.8 +3026	PGC 25969		.99		14.51				6780
0912.9 +3324		.S..9..	.96± .17	100					3423
191.32 42.94	UGC 4850	U	.15± .12	.03					
59.56 -30.88		9.0± .9		.15					3388
0909.9 +3337	PGC 25972		.96	.07					3659
091300.8+202204		.SB.6*.	1.06± .06	25				15.19±.3	2623± 7
208.28 39.94	UGC 4853	U	.08± .05	.12	14.6 ±.2				
70.53 -40.51		6.0±1.2		.12				78± 7	2537
091010.4+203427	PGC 25973		1.07	.04	14.39			.76	2898
091302.6+493817		.SX.4*.	1.15± .05	170				14.85±.1	3999± 11
169.08 42.95	UGC 4844	U	.10± .05	.01	13.94 ±.18				4017± 50
48.78 -17.90	IRAS09096+4950	4.0± .8		.15	13.52			220± 8	4030
090935.9+495040	PGC 25976		1.15	.05	13.75			1.05	4170
0913.0 +4301		.S..8*.	1.14± .13	40					
178.13 43.48	UGC 4849	U	.39± .12	.00					
52.91 -23.27		8.0±1.2		.48					
0909.8 +4314	PGC 25977		1.14	.20					
091312.2-192431		.IBS9..	1.30± .03	140				14.30±.2	765± 7
248.28 19.59	ESO 564- 30	SUE (2)	.20± .04	.32	14.91 ±.14			123± 16	
128.37 -56.91		9.7± .4		.15				118± 6	561
091054.1-191206	PGC 25983	9.9± .6	1.33	.10	14.44			-.24	1066
0913.2 +3120		.S?....	1.04± .09						12458
194.11 42.69	MCG 5-22- 14		.00± .07	.02	14.7 ±.3				12415
61.18 -32.43				.00					12701
0910.2 +3133	PGC 25984		1.04		14.48				
0913.2 +3147	IC 2445	.SB?...	.91± .07	10					1900
193.51 42.76	UGC 4854		.18± .05	.02	15.07 ±.18				1859
60.83 -32.09				.27					2141
0910.2 +3200	PGC 25985		.92	.09	14.77				
091313.8+031353		.SB.6*.	1.18± .05	63				15.65±.3	3793± 10
227.66 32.75	UGC 4857	U	.29± .05	.10	14.19 ±.19				3647
90.40 -50.92		6.0±1.2		.43				232± 7	3647
091037.5+032617	PGC 25986		1.19	.15	13.64			1.86	4097
091320.9+173828	A 0910+17		.44± .22						7710± 40
211.64 39.08	MCG 3-24- 11		.00± .07	.09	16.0 ±.3				7614
73.27 -42.32	ARAK 196								7991
091032.9+175052	PGC 25993		.45						
091322.3+192208	A 0910+19	.I..9..	1.19± .05		15.19 ±.15	.30± .07	.35± .06	15.38±.3	3045± 10
209.55 39.69	UGC 4858	U (1)	.04± .05	.10	14.4 ±.4	-.25± .08	-.16± .07		
71.57 -41.14	DDO 58	10.0± .8	.93± .03	.03		.25	15.33± .06	115± 7	2955
091032.7+193433	PGC 25996	9.0±1.0	1.20	.02	14.94	-.29	15.87± .31	.42	3322
091325.0+791114	NGC 2732	.L..../	1.32± .04	67	12.9 ±.3	.96± .03	.98± .02		
133.43 33.01	UGC 4818	R	.42± .03	.02	13.01 ±.12	.52± .06	.56± .06		1973± 23
32.29 6.74		-2.0± .4	.67± .12	.00		.90	11.69± .42		2113
090652.9+792333	PGC 25999		1.26		12.93	.48	13.28± .36		1995
091326.3+525857		.L..-*.	1.04± .08	145					7618± 31
164.58 42.47	UGC 4851	U	.15± .03	.00	14.11 ±.15				7662
46.85 -15.12		-3.0±1.3		.00					7773
090952.9+531120	PGC 26000		1.02		14.00				
091330.5-604722		PSBR3*.	1.06± .06						
279.26 -8.38	ESO 126- 2	r	.03± .05	2.25	13.60 ±.14				2940± 63
185.62 -38.53	FAIR 280	3.3± .9		.04	10.57				2714
091216.0-603454	PGC 26001		1.28	.01	11.29				3122

R.A. 2000 DEC. / l b / SGL SGB / R.A. 1950 DEC.	Names / PGC	Type (S_T n_L / T / L)	$\log D_{25}$ / $\log R_{25}$ / $\log A_o$ / $\log D_o$	p.a. / A_g / A_i / A_{21}	B_T / m_B / m_{FIR} / B_T^o	$(B-V)_T$ / $(U-B)_T$ / $(B-V)_T^o$ / $(U-B)_T^o$	$(B-V)_o$ / $(U-B)_o$ / m'_o / m'_{25}	m_{21} / W_{20} / W_{50} / HI	V_{21} / V_{opt} / V_{GSR} / V_{3K}
091330.6+285656	IC 2446	.S..1..	1.15± .05	148				15.53±.3	7934± 10
197.32 42.31	UGC 4855	U	.40± .05	.01	14.65 ±.19				7881
63.12 -34.20	IRAS09105+2909	1.0± .9		.41				617± 7	7881
091032.1+290920	PGC 26002		1.16	.20	14.13			1.20 \	8185
091332.5-633740		.SAT2..	1.34± .04	74					
281.37 -10.30	ESO 91- 3	Sr (1)	.17± .05	.76	13.02 ±.14				
188.06 -36.47	FAIR 281	2.2± .5		.21	13.25				
091228.0-632512	PGC 26003	5.0± .6	1.41	.08					
091332.7+295959		.S?....	.52± .20		15.37S±.15				
195.93 42.52	MCG 5-22- 20		.17± .07	.04					6741± 41
62.29 -33.41	IRAS09105+3012			.25	13.40				6692
091033.0+301224	PGC 26004		.52	.09	15.05		12.38±1.01		6988
091333.9+300052		.S?....			16.27S±.15				
195.91 42.53	MCG 5-22- 18			.04					6731± 41
62.28 -33.39	HICK 37D								6682
091034.2+301317	PGC 26005								6978
0913.6 +7035		.SBS2..	1.11± .05	57					
142.72 37.01	UGC 4836	U	.29± .05	.15					
37.00 -.45	7ZW 266	2.0± .9		.36					
0908.9 +7048	PGC 26007		1.13	.15					
091335.9+122626		.S..1?.	.95± .07	77					4905
217.76 37.11	UGC 4861	U	.21± .05	.08	14.2 ±.2				
78.89 -45.65	ARAK 197	1.0±1.8		.21	13.27				4790
091052.2+123851	PGC 26008		.96	.10	13.84				5197
091337.3+295959		.L?....			16.37S±.15				
195.93 42.54	MCG 5-22- 16			.04					7357± 41
62.31 -33.40	HICK 37C								7308
091037.6+301223	PGC 26009								7604
0913.6 +2032		.L?....	.90± .17						8410± 58
208.13 40.13	MCG 4-22- 28		.00± .07	.10	14.95 ±.16				8325
70.50 -40.28				.00					8685
0910.8 +2045	PGC 26011		.91		14.72				
091339.5+295933		.S..3..	1.27± .04	77					
195.94 42.54	UGC 4856	U (1)	.90± .05	.04	15.10 ±.18				6721± 32
62.32 -33.40	HICK 37A	3.0± .9		1.25					6673
091039.9+301158	PGC 26012	4.5±1.3	1.27	.45	13.77				6969
091339.9+295937	NGC 2783	.E.....	1.32± .12	168	13.60S±.15		.95± .01		
195.94 42.55	UGC 4859	U	.13± .08	.04	13.28 ±.17		.46± .02		6696± 21
62.32 -33.40		-5.0± .8	1.13± .10	.00			13.99± .36		6648
091040.2+301202	PGC 26013		1.29		13.32		14.86± .65		6944
0913.6 +6219		.SB.1..	.99± .06	77	*				
152.53 40.11	UGC 4846	U	.30± .05	.15	15.06 ±.18				
41.55 -7.35		1.0± .9		.31					
0909.7 +6232	PGC 26015		1.00	.15					
091344.7-692005	NGC 2836	.SAT4*.	1.42± .04	118	*				
285.72 -14.12	ESO 61- 3	RS	.14± .04	.38	12.64 ±.14				
192.57 -32.19	IRAS09131-6907	4.0± .5		.21	13.28				
091308.0-690736	PGC 26017		1.46	.07					
091344.7+762832	NGC 2748	.SA.4..	1.48± .03	38	12.4 ±.3	.73± .06			
136.25 34.36	UGC 4825	R (1)	.42± .03	.01	12.31 ±.15				1456± 58
33.80 4.48	IRAS09080+7640	4.0± .4		.62	10.82	.64			1587
090802.7+764053	PGC 26018	4.6± .8	1.48	.21	11.69		13.59± .33		1492
091345.4+345014	A 0910+35	.S?....	.64± .17						7200± 26
189.40 43.27	MCG 6-20- 48		.19± .07	.00	15.6 ±.2				7171
58.66 -29.67				.28					7431
091040.4+350239	PGC 26019		.64	.09	15.29				
091349.8-693841	NGC 2822	.E...$.	1.52± .06	90					
285.96 -14.32	ESO 61- 4	R	.17± .05	.41	11.64 ±.14				
192.80 -31.95	IRAS09132-6926			.00	12.92				
091315.0-692612	PGC 26026		1.53						
091353.7-350409		.SBT7*.	1.18± .05	130					2398
260.53 9.33	ESO 372- 9	S (1)	.34± .05	.94					2172
155.23 -53.43		7.0±1.2		.46					2671
091151.0-345142	PGC 26028	6.7± .9	1.26	.17					
091405.5+400652	NGC 2782	.SXT1P.	1.54± .02		12.30 ±.13	.67± .01	.65± .01	14.45±.1	2562± 5
182.15 43.68	UGC 4862	R	.13± .04	.00	12.07 ±.13	-.01± .03	-.03± .01	190± 4	2532± 14
55.01 -25.49	ARP 215	1.0± .3	.89± .02	.13	10.87	.62	12.24± .08	145± 4	2551
091054.1+401918	PGC 26034		1.54	.07	12.02	-.02	14.54± .19	2.36	2770

R.A. 2000 DEC.	Names	Type	logD$_{25}$	p.a.	B$_T$	(B-V)$_T$	(B-V)$_o$	m$_{21}$	V$_{21}$
l b		S$_T$ n$_L$	logR$_{25}$	A$_g$	m$_B$	(U-B)$_T$	(U-B)$_o$	W$_{20}$	V$_{opt}$
SGL SGB		T	logA$_o$	A$_i$	m$_{FIR}$	(B-V)^o_T	m'$_o$	W$_{50}$	V$_{GSR}$
R.A. 1950 DEC.	PGC	L	logD$_o$	A$_{21}$	B^o_T	(U-B)^o_T	m'$_{25}$	HI	V$_{3K}$

091412.8+164439	A 0911+16	.SAR2..	1.28± .04					16.09±.1	8368± 6
212.81 38.95	UGC 4864	U (1)	.00± .05	.08	13.7 ±.3			259± 6	8293± 41
74.38 -42.76		2.0± .7		.00				253± 5	8267
091125.6+165706	PGC 26037	4.0± .8	1.28	.00	13.53			2.56	8650

091419.7-024704		.S?....	1.12± .08						7989
233.80 29.79	MCG 0-24- 3		.08± .07	.09	14.6 ±.3				
99.49 -53.36				.09					7825
091148.1-023436	PGC 26043		1.13	.04	14.30				8298

0914.4 +4047		.S..4..	1.02± .06	68					
181.21 43.75	UGC 4863	U	.70± .05	.00					
54.61 -24.91		4.0±1.0		1.03					
0911.2 +4100	PGC 26048		1.02	.35					

091426.3+360645		.SB?...	.94± .11						6505± 46
187.68 43.54	MCG 6-20- 50		.00± .07	.00	14.82 ±.18				6480
57.86 -28.59				.00					6731
091119.9+361912	PGC 26049		.94		14.72				

091434.8+300825		.L...?.	1.38± .07	35					6909± 31
195.80 42.77	UGC 4869	U	.55± .05	.03	13.98 ±.17				6861
62.38 -33.15	IRAS09115+3020	-2.0±1.6		.00	13.26				7157
091135.1+302053	PGC 26055		1.30		13.85				

091437.0-215823		.SBS4?.	1.31± .04	15					
250.58 18.19	ESO 564- 32	S (1)	.75± .04	.52	15.21 ±.14				
133.01 -56.46		4.0±1.2		1.10					
091221.0-214554	PGC 26056	4.4±1.3	1.36	.37					

091437.1-193305		.SXS5..	1.29± .03	118					
248.63 19.75	ESO 564- 31	SE (1)	.50± .03	.36	14.39 ±.14				
128.62 -56.57	IRAS09123-1920	4.5± .6		.75	13.25				
091219.0-192036	PGC 26057	3.3± .9	1.33	.25					

091437.3+360613		.S..3*.	1.04± .06	51					7474± 57
187.69 43.57	UGC 4866	U	.67± .05	.00	15.48 ±.18				7450
57.90 -28.58	IRAS09114+3618	3.0±1.4		.92	12.75				7701
091131.0+361841	PGC 26058	3.5±1.3	1.04	.33	14.51				

091439.1-602601		.SBS4..	1.26± .05	135					
279.09 -8.03	ESO 126- 3	S (1)	.21± .05	2.40	13.38 ±.14				
185.18 -38.68	IRAS09133-6013	4.0± .6		.31	11.40				
091323.1-601330	PGC 26062	4.4± .6	1.49	.11					

091442.1-604443		.SBS7..	1.16± .05						
279.33 -8.24	ESO 126- 4	S (1)	.01± .05	2.31	14.67 ±.14				
185.45 -38.46		6.5± .6		.02					
091327.0-603212	PGC 26066	7.8± .7	1.38	.01					

091443.1+405248		.SXS7..	1.23± .05	85				14.81±.1	2496± 7
181.09 43.81	UGC 4867	U	.10± .05	.00	14.7 ±.3				2491
54.60 -24.81		7.0± .8		.14				199± 8	2705
091130.7+410516	PGC 26068		1.23	.05	14.51			.25	

091447.5+741355	Holmberg III	.SXS7..	1.45± .02	150	13.02S±.07			13.72±.1	1127± 7
138.63 35.48	UGC 4841	R	.09± .03	.02	13.8 ±.4			173± 6	
35.09 2.64	IRAS09093+7426	7.0± .3		.12				165± 12	1251
090934.5+742620	PGC 26071		1.45	.04	12.90		14.89± .15	.78	1176

0914.8 +5322									8686
164.00 42.61	CGCG 264- 86			.00	14.9 ±.3				8731
46.80 -14.68									8840
0911.3 +5335	PGC 26076								

091455.8+465415	A 0911+47	.S.....	1.02± .06	142					4222± 72
172.76 43.57	UGC 4870	R (1)	.35± .05	.01	14.2 ±.2				4242
50.73 -19.95				.53					4406
091134.4+470643	PGC 26082	3.0±1.3	1.02	.18	13.63				

0914.9 +3915	A 0911+39	.SBS9..	1.23± .05	20	15.40 ±.15	.23± .05	.31± .05		
183.34 43.82	UGC 4871	U (1)	.44± .05	.00		-.23± .07	-.26± .07		
55.74 -26.06	DDO 59	9.0± .9	.79± .02	.45			14.84± .05		
0911.7 +3928	PGC 26083	9.0±1.1	1.23	.22			15.29± .30		

0914.9 +4003		.SB.3..	1.29± .04	12					
182.23 43.85	UGC 4872	U	1.00± .05	.00					
55.20 -25.43		3.0± .9		1.38					
0911.8 +4016	PGC 26086		1.29	.50					

091460.0+294349	NGC 2789	.S..0..	1.29± .06						6303± 36
196.37 42.78	UGC 4875	U	.00± .06	.01	13.2 ±.2				6254
62.79 -33.40	IRAS09120+2956	.0± .8		.00	12.16				6553
091200.8+295618	PGC 26089		1.29		13.11				

R.A. 2000 DEC. / l b / SGL SGB / R.A. 1950 DEC.	Names / PGC	Type: S_T n_L / T / L	logD25 / logR25 / logA_e / logD_o	p.a. / Ag / Ai / A21	B_T / m_B / m_FIR / B_T^o	(B-V)_T / (U-B)_T / (B-V)_T^o / (U-B)_T^o	(B-V)_e / (U-B)_e / m'_e / m'_25	m21 / W20 / W50 / HI	V21 / Vopt / VGSR / V3K
091502.9+194150	NGC 2790		.64± .17		15.4 ±.3				7846± 43
209.31 40.17	MCG 3-24- 16		.11± .07	.10	13.52				7757
71.61 -40.62	MK 1228								8123
091213.2+195419	PGC 26092		.65						
091505.1-281542		.SXT5..	1.27± .04	98					
255.58 14.12	ESO 433- 10	Sr (1)	.12± .05	.86					
144.05 -55.35	IRAS09129-2803	4.8± .4		.18	13.21				
091255.0-280312	PGC 26093	5.0± .6	1.35	.06					
091515.7+405504	NGC 2785	.I..9?.	1.18± .05	120					
181.04 43.91	UGC 4876	U	.46± .05	.00	14.73 ±.18				2737
54.67 -24.71	IRAS09120+4107	10.0±1.7		.34	10.76				2732
091203.4+410733	PGC 26100		1.18	.23	14.38				2946
091517.2+115308	IC 530	.S..2..	1.26± .04	90				15.24±.2	4969± 7
218.61 37.25	UGC 4880	U	.46± .06	.09	13.98 ±.18			497± 13	4899± 76
79.89 -45.66	IRAS09125+1205	2.0± .8		.56				490± 7	4852
091234.1+120538	PGC 26101		1.27	.23	13.28			1.73	5263
0915.3 +4840		.SB.9..	1.04± .06						
170.33 43.44	UGC 4874	U	.09± .05	.01					
49.69 -18.48	VV 131	9.0± .9		.09					
0911.9 +4853	PGC 26104		1.04	.04					
091526.5+312356		.SB.9..	1.24± .06	25				15.27±.3	1877± 7
194.15 43.16	UGC 4878	U	.42± .06	.03					
61.56 -32.08		9.0± .8		.43				145± 7	1834
091225.6+313626	PGC 26108		1.24	.21					2121
091535.2-353850		PSXR0*.	1.24± .07	40					
261.20 9.19	ESO 372- 12	Sr	.24± .06	.96					
155.88 -52.89	IRAS09135-3526	.0± .6		.18	11.48				
091333.0-352618	PGC 26112		1.31						
091536.5-630416	NGC 2842	PSBT0*.	1.19± .05	120	13.41 ±.15	.89± .02	.93± .01		
281.12 -9.75	ESO 91- 4	Sr	.09± .05	.78	13.28 ±.14	.24± .02	.32± .02		2857± 40
187.38 -36.72	FAIR 282	.0± .4	.64± .03	.06	11.71	.67	12.10± .10		2633
091429.0-625142	PGC 26114		1.26		12.45	.10	14.01± .31		3029
091538.7-062711		.L..OP*	1.18± .05	20					6398
237.47 27.98	MCG -1-24- 3	E	.57± .05	.10					
105.58 -54.34	IRAS09131-0614	-2.0±1.3		.00	12.59				6224
091309.8-061439	PGC 26116		1.10						6709
091542.7-864601		.E.2.*.	.99± .05	178					
300.04 -25.20	ESO 6- 6	S	.12± .03	.67	14.53 ±.14				
204.23 -18.34		-5.0± .9		.00					
092523.0-863312	PGC 26121		1.06						
0915.8 +1008		.S..2..	1.00± .06	35					
220.65 36.61	UGC 4884	U	.21± .05	.23	14.54 ±.18				
82.05 -46.61	IRAS09131+1020	2.0± .9		.26					
0913.1 +1020	PGC 26127		1.02	.11					
091554.5+205543	IC 2453								8959± 50
207.87 40.76	CGCG 121- 46			.13	15.4 ±.3				8876
70.61 -39.63	MK 1229				12.73				9235
091303.7+210815	PGC 26131								
091557.0-651347		.SXT7?.	1.14± .05	178					
282.75 -11.20	ESO 91- 6	S (1)	.54± .04	.70	15.53 ±.14				
189.15 -35.12		7.0±1.1		.74					
091458.0-650112	PGC 26133	7.8± .9	1.21	.27					
091601.7+174919	IC 2454	.S..1?.	1.06± .06					17.59±.3	8622± 10
211.72 39.74	UGC 4886	U	.02± .05	.06	14.12 ±.18				
73.67 -41.72	IRAS09132+1801	1.0±1.7		.02	13.53			266± 7	8527
091313.7+180151	PGC 26139		1.06	.01	13.93			3.65	8905
091601.8+173528	NGC 2794	.SB?...	1.08± .04		13.99 ±.14	.84± .02	.93± .02		
212.00 39.66	UGC 4885		.01± .05	.07	13.86 ±.18				8760±141
73.91 -41.88			.76± .02	.01		.77	13.28± .07		8664
091313.9+174800	PGC 26140		1.08	.00	13.81		14.20± .28		9043
091603.6+525025		.IA.9..	1.22± .06	85					
164.66 42.89	UGC 4879	U	.11± .06	.01	13.78 ±.19				600± 50
47.27 -15.01	VV 124	10.0± .8		.08					643
091231.3+530256	PGC 26142		1.22	.06	13.69				757
091604.2+173746	NGC 2795	.E.....	1.15± .15	170					
211.95 39.68	UGC 4887	U	.14± .08	.07	13.81 ±.15				8561± 32
73.88 -41.84		-5.0± .8		.00					8465
091316.4+175018	PGC 26143		1.12		13.62				8844

R.A. 2000 DEC. / l b / SGL SGB / R.A. 1950 DEC.	Names / PGC	Type / S_T n_L / T / L	$\log D_{25}$ / $\log R_{25}$ / $\log A_e$ / $\log D_o$	p.a. / A_g / A_i / A_{21}	B_T / m_B / m_{FIR} / B_T^o	$(B-V)_T$ / $(U-B)_T$ / $(B-V)_T^o$ / $(U-B)_T^o$	$(B-V)_e$ / $(U-B)_e$ / m' / m'$_{25}$	m_{21} / W_{20} / W_{50} / HI	V_{21} / V_{opt} / V_{GSR} / V_{3K}
091604.4-090907		PSBT2P*	1.21± .05	160					1680
239.99 26.48	MCG -1-24- 4	E	.19± .05	.07					
110.16 -54.99		2.0±1.2		.24					1499
091337.7-085634	PGC 26144		1.22	.10					1991
091604.5+674529	A 0911+67	.S?....	.62± .11						9414± 38
145.88 38.38	CGCG 311- 34		.11± .12	.16 15.8 ±.2					9515
38.75 -2.69	MK 103			.16					9497
091141.9+675759	PGC 26145		.64	.05 15.41					
091611.1+061957		.SB.6*.	1.00± .06	55				15.21±.3	3678± 10
224.86 34.93	UGC 4890	U	.21± .05	.04 14.62 ±.18					3543
86.84 -48.68		6.0±1.2		.31				200± 7	3981
091332.3+063230	PGC 26150		1.01	.11 14.25				.86	
091611.3-161847	NGC 2811	.SBT1..	1.40± .03	20	12.23 ±.13	.96± .01	1.02± .01		
246.21 22.10	MCG -3-24- 3	R	.46± .03	.13 12.79 ±.16	.59± .02	.60± .01		2514± 56
122.77 -56.11	IRAS09138-1606	1.0± .4	.83± .02	.47 13.56	.83	11.87± .07		2317
091350.3-160614	PGC 26151		1.41	.23 11.83	.48	12.92± .20		2821
0916.2 +7916	A 0909+79	.S..6*.	1.00± .08	106					
133.26 33.08	UGC 4847	U	.94± .06	.02					
32.35 6.88		6.0±1.4		1.38					
0909.7 +7929	PGC 26154		1.00	.47					
091619.6-233804	NGC 2815	.SBR3*.	1.54± .02	10	12.81 ±.14	.93± .01	1.01± .01	14.25±.2	2540± 8
252.17 17.40	ESO 497- 32	R (2)	.49± .03	.64 12.73 ±.11	.47± .02	.56± .02	586± 9	2560± 66
135.91 -55.90	IRAS09140-2325	3.0± .5	.88± .02	.68 12.57	.68	12.70± .06	556± 7	2329
091405.0-232530	PGC 26157	3.0± .7	1.60	.25 11.42	.25	14.16± .19	2.58	2838
091631.3-173549		.SB.7?.	1.21± .05	110					5030
247.33 21.35	MCG -3-24- 4 (1)	E	.70± .05	.15					
125.09 -56.10		7.0±1.8		.96					4830
091411.4-172315	PGC 26168	5.3±1.6	1.22	.35					5336
0916.6 +0716		.S..4..	1.08± .06	122					
223.91 35.48	UGC 4900	U	.22± .05	.09 14.40 ±.18					
85.74 -48.07		4.0± .9		.33					
0914.0 +0729	PGC 26173		1.08	.11					
091641.1+185746	NGC 2802	.S?....	1.04± .09						8753± 34
210.39 40.29	UGC 4897		.28± .07	.09					8662
72.68 -40.84				.21					9033
091352.1+191020	PGC 26177		1.03						
0916.7 +3054	NGC 2796	.S..1?.	1.06± .08	80					6951± 19
194.87 43.35	UGC 4893	U	.21± .06	.04					6906
62.19 -32.28		1.0±1.7		.21					7198
0913.7 +3107	PGC 26178		1.06	.10					
091643.3+594620	A 0912+59	.S?....	.89± .11						4230± 63
155.58 41.25	MCG 10-13- 71		.52± .07	.15					4301
43.30 -9.26	MK 19			.78					4354
091253.5+595853	PGC 26180		.90	.26					
091643.5+185716	NGC 2803	.S?....	1.08± .09						8858± 34
210.40 40.29	UGC 4898		.00± .07	.09					8767
72.69 -40.84				.00					9139
091354.6+190950	PGC 26181		1.09						
0916.7 +1950									8622± 58
209.31 40.59	CGCG 91- 45			.09 15.24 ±.18					8534
71.83 -40.24									8901
0913.9 +2003	PGC 26182								
091644.3+195606	NGC 2801	.SAS5..	1.06± .06	35				16.36±.3	7720± 10
209.19 40.62	UGC 4899	U	.06± .05	.09 14.7 ±.2					7576± 47
71.74 -40.17		5.0± .9		.09				203± 7	7627
091354.5+200840	PGC 26183		1.06	.03 14.45				1.88	7992
091646.1+532634	A 0913+53	.P.....	.60± .17		14.6 ±.2	.44± .03	.45± .02	16.52±.1	2203± 10
163.83 42.87	UGCA 154	R	.21± .09	.04 15.1 ±.4	-.30± .06	-.32± .05	184± 6	2208± 35
47.00 -14.46	MK 104	99.0	.36± .03			11.94± .07	163± 5	2249
091312.8+533907	PGC 26188		.61					2358
091646.7+342554	NGC 2793	.SBS9P.	1.10± .03					15.09±.1	1681± 5
190.05 43.85	UGC 4894	R	.07± .03	.00 13.55 ±.15				129± 6	1677± 40
59.51 -29.59	IRAS09137+3438	9.0± .4		.07 13.27				112± 5	1650
091342.6+343828	PGC 26189		1.10	.04 13.48				1.57	1915
091647.7-264900	NGC 2821	.S..4*/	1.31± .03	100					
254.73 15.36	ESO 497- 34	S	.66± .03	.67 13.87 ±.14					
141.44 -55.29	IRAS09145-2636	4.0±1.2		.97 11.90					
091436.0-263624	PGC 26192		1.37	.33					

R.A. 2000 DEC.	Names	Type	logD$_{25}$	p.a.	B$_T$	(B-V)$_T$	(B-V)$_e$	m$_{21}$	V$_{21}$
l　b		S$_T$　n$_L$	logR$_{25}$	A$_g$	m$_B$	(U-B)$_T$	(U-B)$_e$	W$_{20}$	V$_{opt}$
SGL　SGB		T	logA$_e$	A$_i$	m$_{FIR}$	(B-V)$_T^o$	m'$_e$	W$_{50}$	V$_{GSR}$
R.A. 1950 DEC.	PGC	L	logD$_o$	A$_{21}$	B$_T^o$	(U-B)$_T^o$	m'$_{25}$	HI	V$_{3K}$
0916.8　+8002		.I..9..	1.10± .07	95					
132.46　32.71	UGC　4852	U	.33± .06	.03	15.09 ±.20				
31.94　7.53		10.0± .8	.79± .11	.25					
0910.0　+8015	PGC 26195		1.10	.17					
091649.8+201155	NGC　2804	.L.....	1.35± .07	60	13.9　±.3	1.05± .05	1.06± .02		
208.87　40.73	UGC　4901	U	.04± .05	.08	13.2　±.2	.45± .08	.49± .04		8117± 31
71.51　-39.98		-2.0± .8	.79± .11	.00		.95	13.29± .38		8030
091359.8+202430	PGC 26196		1.35		13.24	.47	15.35± .47		8394
091649.8+272926		.S..3..	1.07± .06	163				15.90±.3	7073±　9
199.47　42.72	UGC　4895	U　　(1)	.46± .05	.03	14.88 ±.18				7015
64.98　-34.80		3.0± .9		.64				444±　5	7015
091353.1+274200	PGC 26197	3.5±1.2	1.08	.23	14.16			1.51	7331
0916.8　+0034									
230.92　32.15	CGCG　　6- 25			.06	15.3　±.6				8503± 58
94.87　-51.38									8349
0914.3　+0047	PGC 26202								8812
091658.5+161812		.S..6*.	1.03± .08					16.22±.3	11913± 10
213.66　39.40	UGC　4907	U	.11± .06	.07	14.8　±.2				11812
75.43　-42.55		6.0±1.2		.16				185±　7	11812
091411.8+163047	PGC 26209		1.03	.05	14.57			1.60	12200
0917.0　+2645		.S?....							7333
200.45　42.60	MCG　5-22- 32			.05	15.4　±.3				7272
65.64　-35.30									7593
0914.1　+2658	PGC 26215								
091705.3+252546		.L...?.	1.20± .09	157					
202.20　42.29	UGC　4902	U	.27± .05	.09	13.84 ±.15				1531± 97
66.79　-36.25	MK　1230	-2.0±1.7		.00	13.10				1465
091410.5+253821	PGC 26218		1.17		13.73				1795
091706.6+200415	NGC　2809	.L.....	1.10± .04		13.97 ±.14	.96± .02	1.02± .01		
209.06　40.75	UGC　4910	U	.02± .05	.08	13.72 ±.15				7983
71.69　-40.01		-2.0± .8	.71± .03	.00		.86	13.01± .11		7896
091416.7+201650	PGC 26220		1.11		13.65		14.31± .26		8261
091710.6-044512	NGC　2817	.SXT5..	1.30± .05	140					
236.13　29.27	MCG -1-24- 6	E　　(1)	.07± .05	.11					
103.04　-53.42	IRAS09146-0432	5.0± .8		.11	13.50				
091440.5-043236	PGC 26223	3.1± .8	1.31	.04					
0917.2　+3657		.S..9..	1.17± .09	38					4098
186.57　44.16	UGC　4903	U	.57± .05	.00					
57.75　-27.59		9.0± .9		.58					4077
0914.1　+3710	PGC 26224		1.17	.28					4323
0917.2　+3200		.S..3..	.87± .10	25					
193.40　43.63	UGC　4908	U　　(1)	.11± .06	.01	15.35 ±.19				
61.43　-31.38		3.0± .9		.15					
0914.2　+3213	PGC 26225	3.5±1.2	.87	.06					
0917.2　+2008									
208.99　40.80	CGCG　91- 55			.08	14.9　±.3				9095± 58
71.65　-39.95									9009
0914.4　+2021	PGC 26226								9373
091722.9+420002	NGC　2798	.SBS1P.	1.41± .02	160	13.04 ±.16	.72± .05	.68± .03	14.92±.1	1739±　7
179.53　44.30	UGC　4905	R	.42± .03	.00	12.94 ±.12	-.01± .11	-.03± .11	316±　6	1734± 24
54.29　-23.61	IRAS09141+4212	1.0± .4	.60± .04	.43	9.89	.63	11.53± .13	251±　4	1739
091409.5+421237	PGC 26232		1.41	.21	12.53	-.07	13.91± .21	2.19	1945
091729.4-003719		.S..4..	1.02± .06	119					
232.20　31.64	UGC　4915	U	.64± .05	.11	15.2　±.2				
96.73　-51.76	IRAS09149-0024	4.0± .9		.95	13.23				
091456.0-002442	PGC 26234		1.03	.32					
091729.8+255757		.L?....	.83± .10					15.60±.3	6538±　9
201.53　42.51	UGC　4912		.17± .03	.05	14.8　±.2				6538± 50
66.41　-35.80	IRAS09145+2610			.00	13.32			385±　5	6474
091434.6+261033	PGC 26235		.81		14.69				6801
091731.0-625303		.SBS6..	1.27± .05	177					
281.13　-9.47	ESO　91- 7	S　　(1)	.26± .05	.78	13.85 ±.14				
187.03　-36.70	IRAS09163-6240	6.0± .5		.38	13.30				
091622.1-624024	PGC 26236	5.9± .5	1.34	.13					
091731.5+415939	NGC　2799	.SBS9$.	1.27± .03	125				14.80±.3	1755± 17
179.54　44.32	UGC　4909	R	.58± .03	.00	14.32 ±.18			330± 16	1863± 23
54.31　-23.60	KUG 0914+422B	9.0± .9		.60				340± 25	1792
091418.1+421215	PGC 26238		1.27	.29	13.72			.79	1999

R.A. 2000 DEC.	Names	Type	logD$_{25}$	p.a.	B$_T$	(B-V)$_T$	(B-V)$_\bullet$	m$_{21}$	V$_{21}$
l b		S$_T$ n$_L$	logR$_{25}$	A$_g$	m$_B$	(U-B)$_T$	(U-B)$_\bullet$	W$_{20}$	V$_{opt}$
SGL SGB		T	logA$_\bullet$	A$_i$	m$_{FIR}$	(B-V)^o_T	m'$_\bullet$	W$_{50}$	V$_{GSR}$
R.A. 1950 DEC.	PGC	L	logD$_o$	A$_{21}$	B^o_T	(U-B)^o_T	m'$_{25}$	HI	V$_{3K}$
0917.6 -0755		.S?....	1.15± .08						7545
239.15 27.52	MCG -1-24- 7		.88± .07	.12					
108.27 -54.29				1.30					7368
0915.2 -0743	PGC 26245		1.16	.44					7857
091740.3+525940	A 0914+53	.S...1..	1.30± .04	48					
164.39 43.09	UGC 4906	U	.59± .05	.01	13.5 ±.2				2322± 50
47.38 -14.75		1.0± .8		.60					2366
091408.1+531216	PGC 26246		1.30	.29	12.89				2479
0917.7 +1953	NGC 2813	.L.....	1.11± .10	145					
209.35 40.83	UGC 4916	U	.06± .05	.09	14.5 ±.2				
71.99 -40.03		-2.0± .9		.00					
0914.9 +2006	PGC 26252		1.12						
091751.2-001645	IC 531	PSBT2?.	1.24± .04	60					
231.92 31.90	UGC 4923	UE	.53± .04	.08	14.44 ±.19				
96.29 -51.53	IRAS09152-0004	1.7± .7		.65	13.34				
091517.5-000407	PGC 26258		1.24	.26					
091753.0-222120	NGC 2835	.SBT5..	1.82± .01	8	11.01 ±.17	.49± .02	.61± .01	11.98±.1	888± 5
251.41 18.51	ESO 564- 35	R (2)	.18± .02	.44	11.08 ±.12	-.12± .06	-.03± .03	207± 6	876± 58
133.58 -55.67	IRAS09156-2208	5.0± .3	1.44± .03	.27	11.44	.35	13.70± .07	192± 6	679
091537.1-220842	PGC 26259	1.8± .3	1.86	.09	10.34	-.22	14.51± .19	1.55	1189
0918.0 -1205	Hydra A	PLA.-*.	.87± .12		13.9 ±.2	1.03± .06	1.04± .02		
242.93 25.09	MCG -2-24- 7	PE	.00± .05	.19			.41± .10		16141± 34
115.41 -55.13		-2.5± .7	1.17± .07	.00		.83	15.24± .21		15953
0915.6 -1153	PGC 26269		.89		13.46		13.11± .63		16452
091809.5+161157	NGC 2819	.E.....	1.15± .15						
213.92 39.62	UGC 4924	U	.04± .08	.07	13.78 ±.16				
75.79 -42.41		-5.0± .8		.00					
091522.9+162435	PGC 26274		1.15						
091812.3-182633		.SBR5..	1.23± .03	68					
248.30 21.11	ESO 564- 36	SE (1)	.13± .03	.22	14.17 ±.14				
126.62 -55.73		4.6± .5		.19					
091553.0-181354	PGC 26276	3.9± .6	1.25	.06					
091813.2-322858		.S.4*P/	1.36± .05	148					
259.23 11.76	ESO 433- 12	S	.69± .05	1.05					
150.66 -53.52	IRAS09161-3216	3.7±1.0		1.02	13.18				
091607.0-321618	PGC 26278		1.46	.35					
091814.4+453912	A 0915+45	.SXT5..	1.13± .05	5	14.47 ±.15	.78± .04			
174.42 44.26	UGC 4919	U (1)	.13± .05	.00	14.20 ±.18				8096± 59
52.02 -20.62		5.0± .8		.19		.70			8111
091455.8+455150	PGC 26282	2.7± .8	1.13	.06	14.13		14.65± .32		8287
091816.1+475655		.S..6*.	.89± .08						
171.24 44.02	UGC 4917	U	.06± .05	.01	15.28 ±.19				
50.56 -18.77		6.0±1.3		.08					
091453.8+480933	PGC 26283		.89	.03					
0918.2 +7420	A 0913+74	.S.....	1.10± .05	27					
138.37 35.64	UGC 4883	R	.30± .05	.04	13.2 ±.3				
35.22 2.87				.45					
0913.1 +7433	PGC 26284		1.10	.15					
091820.1+174514		.S..7..	1.23± .05	78				14.90±.3	3011± 10
212.05 40.23	UGC 4925	U	.90± .05	.05	15.29 ±.18				2916
74.23 -41.36		7.0± .9		1.24				206± 7	
091532.3+175753	PGC 26287		1.23	.45	13.99			.46	3296
0918.3 +6948		.S..2..	1.25± .04	145					
143.40 37.71	UGC 4896	U	.62± .05	.12	14.71 ±.18				
37.77 -.88		2.0± .9		.76					
0913.8 +7001	PGC 26289		1.26	.31					
091826.1+161820		.S?....	.82± .13						
213.82 39.72	MCG 3-24- 43		.23± .07	.05	15.38 ±.19				8968± 45
75.74 -42.29	MK 704			.31					8867
091539.5+163059	PGC 26292		.82	.11	14.95				9256
0918.5 +4933		.S..6*.	1.06± .06	26					
169.02 43.84	UGC 4921	U	.78± .05	.01					
49.59 -17.45		6.0±1.4		1.15					
0915.1 +4946	PGC 26294		1.06	.39					
091832.2+734531	IC 529	.SAS5*.	1.56± .02	145	12.62S±.11			13.70±.1	2260± 5
138.99 35.93	UGC 4888	PU	.34± .04	.03	12.04 ±.18			325± 6	2091± 63
35.56 2.40	IRAS09134+7358	5.3± .6		.51	12.57			303± 5	2380
091327.0+735807	PGC 26295		1.57	.17	11.91		14.44± .19	1.62	2310

R.A. 2000 DEC. / l b / SGL SGB / R.A. 1950 DEC.	Names / PGC	Type / S_T n_L / T / L	logD25 / logR25 / logA_o / logD_o	p.a. / A_g / A_i / A_21	B_T / m_B / m_FIR / B_T^o	(B-V)_T / (U-B)_T / (B-V)_T^o / (U-B)_T^o	(B-V)_o / (U-B)_o / m'_o / m'_25	m_21 / W_20 / W_50 / HI	V_21 / V_opt / V_GSR / V_3K
091835.6+343310		.S?....	1.15± .05	154				15.14±.2	6375 7
189.94 44.23	UGC 4926	(1)	.68± .05	.00	15.27 ±.18			496± 13	6571± 68
59.75 -29.26	IRAS09156+3445		.94	.94				491± 7	6346
091531.7+344549	PGC 26300	4.5±1.3	1.15	.34	14.29			.51	6612
091835.9+523046	NGC 2800	.E.....	1.15± .15	15					
165.00 43.33	UGC 4920	U	.19± .08	.01	13.81 ±.15				7622± 31
47.79 -15.06		-5.0± .8		.00					7664
091505.0+524325	PGC 26302		1.09		13.68				7782
091836.5+475219		.SA.9..	1.50± .03	55	13.87S±.10			14.45±.1	1991± 6
171.34 44.09	UGC 4922	U	.33± .03	.00	14.4 ±.5			248± 5	
50.66 -18.80		9.0± .7		.34				237± 12	2015
091514.3+480458	PGC 26304		1.50	.16	13.55		15.40± .18	.74	2173
091836.9-380041	NGC 2845	.SAR0*.	1.30± .04	67	12.92 ±.15	1.23± .03	1.24± .02		
263.35 8.00	ESO 314- 10	BSr	.30± .05	1.12	13.67 ±.14	.78± .06	.82± .04		
158.96 -51.36	IRAS09166-3747	.3± .4	1.13± .04	.22			14.06± .14		
091637.0-374800	PGC 26306		1.39				13.53± .29		
0918.7 +1349		.S..0..	.96± .09	165					
216.82 38.82	UGC 4931	U	.12± .06	.10	14.67 ±.18				
78.46 -43.81		.0± .9		.09					
0916.0 +1402	PGC 26310		.96						
091852.1-384518		.SAT0..	1.06± .06	25					
263.93 7.52	ESO 314- 11	r	.26± .06	1.20	15.11 ±.14				
159.99 -51.00		-.1± .8		.19					
091653.0-383236	PGC 26318		1.16						
091854.6+500115		.L..-?.	.94± .13	135					
168.37 43.83	UGC 4927	U	.04± .05	.00	14.54 ±.15				
49.36 -17.04		-3.0±1.7		.00					
091528.7+501355	PGC 26323		.94						
0918.9 +5119		.SBS4..	1.04± .06						
166.59 43.61	UGC 4928	U	.06± .05	.00	14.9 ±.2				
48.56 -15.99		4.0± .9		.09					
0915.5 +5132	PGC 26329		1.04	.03					
091902.3+261610	NGC 2824	.L.....	.97± .04	160	14.23 ±.13	.90± .01	.94± .01		
201.24 42.92	UGC 4933	U	.17± .05	.07	14.44 ±.16	.41± .03	.45± .03		2759± 31
66.46 -35.34	MK 394	-2.0± .9	.34± .03	.00	13.06	.84	11.42± .08		2696
091607.0+262850	PGC 26330		.95		14.20	.39	13.53± .27		3022
091910.3+201442		.S..3..	1.09± .06	147				17.52±.3	9020 9
209.05 41.26	UGC 4938	U (1)	.63± .05	.08	15.29 ±.18				9000± 58
71.95 -39.54		3.0± .9	.87	.87				544± 5	8934
091620.4+202723	PGC 26335	4.5±1.3	1.10	.31	14.28			2.93	9299
091912.0+485805		.S..6*.	.91± .07	115					
169.81 44.04	UGC 4930	U	.18± .05	.01	15.26 ±.18				
50.05 -17.86		6.0±1.3		.26					
091548.1+491046	PGC 26338		.92	.09					
091917.4+340029	NGC 2823	.SB.1..	.96± .07	30					
190.72 44.32	UGC 4935	U	.25± .05	.00	15.46 ±.19				7092± 56
60.29 -29.58		1.0± .9		.25					7060
091614.3+341310	PGC 26340		.96	.12	15.12				7329
091918.9+691211	NGC 2787	.LBR+..	1.50± .04	117	11.82 ±.14	1.06± .02	1.10± .01	14.32±.2	696± 8
144.05 38.05	UGC 4914	R	.19± .04	.17	11.84 ±.12	.67± .02	.69± .02	371± 6	649± 25
38.18 -1.33	IRAS09148+6924	-1.0± .3	.91± .03	.00		.99	11.86± .08	358± 5	797
091449.7+692450	PGC 26341		1.48		11.65	.60	13.70± .27		768
091923.1+334427	NGC 2825	.S..1*/	.98± .06						
191.09 44.31	MCG 6-21- 10	P	.45± .05	.00	15.27 ±.18				7847± 68
60.51 -29.77		1.0±1.3		.46					7814
091620.2+335708	PGC 26345		.98	.22	14.72				8086
091925.1+333719	NGC 2826	.L..+*/	1.18± .05	143					
191.26 44.30	UGC 4939	PU	.73± .03	.00	14.65 ±.17				6263± 68
60.60 -29.85		-.7± .7		.00					6229
091622.3+335001	PGC 26346		1.07		14.56				6502
0919.4 +2655									8102
200.40 43.16	CGCG 151- 58			.05	15.3 ±.3				8042
65.98 -34.81									8363
0916.5 +2708	PGC 26347								
0919.4 +5106		.S..8*.	1.19± .05	140				15.46±.1	546± 11
166.87 43.73	UGC 4932	U	.60± .05	.00	15.17 ±.19				
48.76 -16.12		8.0±1.2		.73				117± 8	583
0916.0 +5119	PGC 26351		1.19	.30	14.43			.73	713

R.A. 2000 DEC. l b SGL SGB R.A. 1950 DEC.	Names PGC	Type S_T n_L T L	$\log D_{25}$ $\log R_{25}$ $\log A_e$ $\log D_o$	p.a. A_g A_I A_{21}	B_T m_B m_{FIR} B_T^o	$(B-V)_T$ $(U-B)_T$ $(B-V)_T^o$ $(U-B)_T^o$	$(B-V)_e$ $(U-B)_e$ m' m'_{25}	m_{21} W_{20} W_{50} HI	V_{21} V_{opt} V_{GSR} V_{3K}
0919.5 +0553 225.83 35.44 88.09 -48.21 0916.9 +0606	 UGC 4946 PGC 26357	.SBR3.. U 3.0± .8 	1.12± .07 .18± .06 1.13	30 .09 .25 .09	 14.7 ±.3 				
0919.5 +5052 167.18 43.78 48.92 -16.30 0916.1 +5105	 UGC 4934 PGC 26363	.S..3.. U (1) 3.0±1.0 3.5±1.4	1.02± .06 .88± .05 1.02	4 .00 1.22 .44	 15.79 ±.19 				
091936.0+332541 191.53 44.32 60.78 -29.97 091633.5+333823	 MCG 6-21- 12 PGC 26366	.S?.... 	.82± .13 .00± .07 .82	 .00 .00 	 15.42 ±.18 15.31				7089± 68 7054 7329
091937.2+272721 199.70 43.32 65.56 -34.40 091640.8+274003	 UGC 4940 PGC 26368	.SX.2.. U 2.0± .9 	1.06± .06 .24± .05 1.06	58 .05 .29 .12	 14.59 ±.18 14.17			16.98±.3 447± 5 2.69	7767± 9 7709 8027
091941.9+334412 191.11 44.37 60.57 -29.73 091639.1+335654	NGC 2830 UGC 4941 PGC 26371	.SB.0*/ R .0± .5 	1.13± .05 .64± .03 1.10	112 .00 .48 	 15.27 ±.18 14.71			17.14±.3 482± 13 470± 10 	6105± 10 6237± 68 6074 6347
0919.7 +7506 137.50 35.35 34.87 3.55 0914.4 +7519	 MCG 13- 7- 22 PGC 26373	.S?.... 	.97± .10 .37± .07 .95	 .01 .28 					
091946.1+334438 191.10 44.39 60.57 -29.71 091643.3+335720	NGC 2831 MCG 6-21- 13 PGC 26376	.E.0... R -5.0± .5 	1.14± .10 .00± .10 1.14	 .00 .00 					5116± 28 5083 5355
091946.5+334502 191.09 44.39 60.57 -29.71 091643.7+335745	NGC 2832 MCG 6-21- 15 PGC 26377	.E+2.*. R -4.0± .3 	1.37± .07 .10± .06 .93± .03 1.35	160 .00 .00 	12.87 ±.13 12.49 ±.18 12.64	1.00± .01 .51± .02 .93 .55	1.02± .01 .59± .01 13.01± .10 14.49± .38		6869± 20 6836 7108
0919.8 -1214 243.34 25.33 115.80 -54.75 0917.4 -1202	A 0917-12 MCG -2-24- 11 DDO 60 PGC 26378	.IBS9P. PE (2) 10.0± .6 8.8± .7	1.23± .05 .09± .05 1.24	50 .16 .07 .04				14.44±.1 67± 7 49± 12 	1946± 6 1758 2259
0919.8 +6709 146.39 38.95 39.39 -2.98 0915.6 +6722	 UGC 4929 PGC 26381	.S..4.. U 4.0±1.0 	1.07± .07 1.09± .06 1.08	138 .15 1.47 .50					
091954.2+325600 192.23 44.32 61.22 -30.31 091652.3+330843	 UGC 4947 IRAS09168+3308 PGC 26382	.SB?... 	1.04± .06 .33± .05 1.04	5 .00 .49 .16	 15.12 ±.18 12.21 14.55				14922± 60 14885 15164
091954.8-685435 285.77 -13.43 191.76 -32.14 091912.0-684148	 ESO 61- 8 IRAS09192-6841 PGC 26383	.SAT5*. S (1) 5.0± .9 4.4±1.0	1.26± .04 .50± .04 1.30	108 .40 .75 .25	 13.47 ±.14 13.46 				
091955.4+005642 231.02 32.99 94.88 -50.53 091720.8+010926	 CGCG 6- 29 MK 1232 PGC 26385	 	.81± .11 .24± .06 .81	 .07 	 15.40 ±.13 				5331± 34 5179 5642
091958.1+371127 186.28 44.72 58.06 -27.06 091651.4+372410	IC 2461 UGC 4943 IRAS09168+3724 PGC 26390	.S..3.. U (1) 3.0± .9 3.5±1.2	1.37± .04 .75± .05 1.37	143 .00 1.04 .38	 14.8 ±.2 12.72 13.77				2265± 19 2246 2491
0919.9 +0055 231.06 32.99 94.93 -50.53 0917.4 +0108	 CGCG 6- 30 PGC 26392	 	 	 .07 	 15.3 ±.6 				5169± 34 5017 5481
0920.0 +3305 192.02 44.37 61.12 -30.17 0917.0 +3318	 UGC 4949 PGC 26397	.S..4.. U 4.0±1.0 	1.04± .06 .78± .05 1.04	144 .00 1.15 .39					
092002.1+010217 230.95 33.07 94.77 -50.47 091727.5+011501	 UGC 4956 PGC 26398	.E...*. UE -4.7± .6 	1.28± .07 .17± .05 .97± .06 1.24	15 .07 .00 	13.5 ±.2 13.41 ±.17 13.30	.92± .03 .86 	.98± .01 13.84± .22 14.45± .43		5125± 34 4974 5437

R.A. 2000 DEC. / l b / SGL SGB / R.A. 1950 DEC.	Names / PGC	Type / S_T n_L / T / L	logD_25 / logR_25 / logA_e / logD_o	p.a. / A_g / A_i / A_21	B_T / m_B / m_FIR / B_T^o	(B-V)_T / (U-B)_T / (B-V)_T^o / (U-B)_T^o	(B-V)_e / (U-B)_e / m'_e / m'_25	m_21 / W_20 / W_50 / HI	V_21 / V_opt / V_GSR / V_3K
092008.7+390948		.S..1..	.95± .07	62					
183.51 44.83	UGC 4950	U	.27± .05	.00	15.04 ±.18				
56.69 −25.51		1.0± .9		.28					
091659.6+392231	PGC 26403		.95	.13					
092009.7−163134	NGC 2848	.SXS5*.	1.43± .03	30	12.35 ±.15	.53± .03	.61± .01	13.62±.1	2044± 11
247.04 22.70	MCG −3−24− 7	R (3)	.21± .03	.14	12.7 ±.2	−.11± .05	−.01± .03	201± 16	2157± 46
123.27 −55.18	IRAS09178−1618	5.0± .5	1.17± .03	.32	12.48	.44	13.69± .08	192± 12	1852
091748.6−161849	PGC 26404	4.3± .5	1.45	.11	11.98	−.17	13.84± .22	1.53	2360
092013.1−601354		.SBS9..	1.03± .09						
279.43 −7.39		S (1)	.17± .08	2.78					
184.39 −38.33		9.0± .9		.17					
091854.3−600107	PGC 26406	10.0± .9	1.29	.08					
092013.3+084731		.S..7*.	1.24± .11	152				15.44±.3	8475± 10
222.76 36.97	UGC 4957	U	.96± .12	.13	15.32 ±.18				8348
84.58 −46.50		7.0±1.3		1.33				387± 7	8778
091732.7+090016	PGC 26407		1.25	.48	13.83			1.13	
0920.3 +2517		.S..8..	.96± .17	100					6461± 10
202.64 42.96	UGC 4955	U	.29± .12	.11					6394
67.56 −35.85		8.0± .9		.36					6728
0917.4 +2530	PGC 26409		.97	.15					
092020.4+640612	NGC 2805	.SXT7..	1.80± .02	125	11.52 ±.15	.49± .06	.54± .02	12.39±.0	1734± 4
150.01 40.19	UGC 4936	R	.12± .03	.17	11.5 ±.3		−.14± .03	118± 4	1699± 32
41.17 −5.46		7.0± .3	1.63± .03	.17		.42	15.16± .07	97± 5	1821
091617.1+641855	PGC 26410		1.82	.06	11.18		15.06± .19	1.16	1837
092022.0−075254		.S..3?.	1.21± .05	155					3448
239.55 28.09	MCG −1−24− 10	E (1)	.31± .05	.10					3271
108.51 −53.63	IRAS09178−0740	3.0±1.7		.43	12.85				3762
091754.2−074008	PGC 26411	5.3±1.6	1.22	.16					
0920.4 +3614		.S..7*.	1.19± .05	122					2184
187.63 44.74	UGC 4953	U	.65± .05	.00					2161
58.83 −27.73		7.0±1.3		.90					2414
0917.3 +3627	PGC 26412		1.19	.33					
092026.5+712418	A 0915+71	.S...$.	.32± .18						3530± 38
141.50 37.15		R	.26± .12	.09					3644
37.01 .54	MK 105								3595
091543.1+713700	PGC 26416		.33						
0920.4 +0704		.S..4..	.91± .07					17.35±.3	5539± 11
224.69 36.21	UGC 4959	U	.03± .05	.13	14.66 ±.18			185± 7	5406
86.76 −47.39		4.0± .9		.04				162± 7	5844
0917.8 +0717	PGC 26418		.93	.01	14.46			2.88	
092030.4−162945	NGC 2851	.LA.0*.	1.09± .06	5					
247.07 22.78	MCG −3−24− 8	E	.40± .05	.14					
123.22 −55.09		−2.0±1.3		.00					
091809.3−161659	PGC 26422		1.05						
092036.3+333902	NGC 2839	.E?....							7942± 68
191.26 44.55	MCG 6−21− 23			.00	15.22 ±.18				7908
60.80 −29.67									8182
091733.8+335147	PGC 26425								
0920.6 −1234	A 0918−12	.IXS9..	1.34± .05	170	14.8 ±.2	.45± .05	.53± .04	14.76±.1	1904± 6
243.77 25.28	MCG −2−24− 12	PE (2)	.14± .05	.14		−.34± .07	−.26± .06	92± 7	
116.43 −54.60	DDO 61	9.5± .6	.95± .05	.10		.37	15.05± .12	79± 12	1715
0918.2 −1221	PGC 26429	9.5± .7	1.36	.07	14.56	−.40	16.03± .33	.13	2217
092040.9+150602		.SB.1..	1.11± .05					16.78±.2	8652± 7
215.54 39.76	UGC 4962	U	.02± .05	.08	14.4 ±.3			198± 7	
77.48 −42.64		1.0± .8		.02				187± 5	8547
091755.4+151847	PGC 26431		1.12	.01	14.17			2.60	8944
0920.7 +1816		.L.....	1.04± .18						
211.68 40.94	UGC 4963	U	.32± .08	.03					
74.21 −40.59		−2.0± .9		.00					
0917.9 +1829	PGC 26433		.99						
092046.4−080325		.SXT5*.	1.13± .06	155					5892
239.78 28.07	MCG −1−24− 12	E (1)	.25± .05	.10					5714
108.85 −53.59		5.0± .9		.37					6207
091818.6−075038	PGC 26440	3.1±1.2	1.14	.12					
0920.8 +2808		.SB?...	.82± .13						7605
198.86 43.72	MCG 5−22− 41		.12± .07	.03	15.35 ±.18				7549
65.23 −33.72				.18					7863
0917.9 +2821	PGC 26442		.82	.06	15.09				

R.A. 2000 DEC. / l b / SGL SGB / R.A. 1950 DEC.	Names / PGC	Type / S_T n_L / T / L	$\log D_{25}$ / $\log R_{25}$ / $\log A_o$ / $\log D_o$	p.a. / A_g / A_i / A_{21}	B_T / m_B / m_{FIR} / B_T^o	$(B-V)_T$ / $(U-B)_T$ / $(B-V)_T^o$ / $(U-B)_T^o$	$(B-V)_\bullet$ / $(U-B)_\bullet$ / m' / m'_{25}	m_{21} / W_{20} / W_{50} / HI	V_{21} / V_{opt} / V_{GSR} / V_{3K}
0920.8 +2852		.S..8*.	1.00± .16	173					6541± 57
197.86 43.87	UGC 4964	U	.73± .12	.04	15.79 ±.18				6488
64.63 -33.18		8.0±1.4		.89					6797
0917.9 +2905	PGC 26443		1.00	.36	14.84				
092052.6+352206	NGC 2840	.SBT4..	1.02± .06						7477± 19
188.86 44.78	UGC 4960	U	.06± .05	.01	14.61 ±.18				7450
59.55 -28.33	IRAS09178+3534	4.0± .9		.09	13.34				7711
091748.1+353452	PGC 26445		1.02	.03	14.45				
092100.8-331136		.SA.3?/	1.38± .04	72					
260.17 11.69	ESO 372- 16	S	.74± .05	1.08					
151.46 -52.73	IRAS09189-3258	3.2± .5		1.02	12.69				
091855.0-325848	PGC 26455		1.48	.37					
092105.7+241820		.S?....	1.00± .06	60				16.53±.3	8015± 10
203.99 42.88	UGC 4965		.60± .05	.10	15.39 ±.19			435± 13	8024±125
68.59 -36.41				.90					7945
091812.4+243107	PGC 26461		1.01	.30	14.35			1.88	8285
092108.2-315330		.SBS5*.	1.20± .05	9					
259.23 12.61	ESO 433- 15	S (1)	.12± .05	.83					
149.41 -53.11	IRAS09190-3140	5.0±1.1		.19	13.74				
091901.1-314042	PGC 26463	5.6± .8	1.28	.06					
092110.8+455319		.S?....	.67± .10						
174.03 44.74	MCG 8-17- 84		.00± .06	.00	15.0 ±.2				1860±125
52.30 -20.12				.00					1876
091752.4+460605	PGC 26465		.67	.00	14.99				2051
092112.8+641505	NGC 2814	.S..3*.	1.08± .04	179	14.3 ±.2	.56± .04		17.30±.4	1634± 10
149.79 40.22	UGC 4952	RC	.59± .03	.16	14.3 ±.3	-.17± .05			1673± 42
41.16 -5.28	IRAS09170+6428	3.0±1.8		.81	13.00	.40			1724
091709.2+642750	PGC 26469		1.09	.29	13.32	-.28	13.08± .29	3.68	1739
092115.6+030909	IC 534	.S..3..	1.24± .06	148				14.78±.2	3517± 6
228.98 34.44	UGC 4968	U (1)	.86± .06	.10	14.94 ±.18			271± 7	
92.05 -49.21		3.0± .9		1.19				253± 5	3372
091839.4+032157	PGC 26471	4.5±1.3	1.25	.43	13.63			.72	3827
0921.2 +6924		.S..6*.	1.15± .05	93					
143.72 38.12	UGC 4944	U	.87± .05	.17	15.67 ±.18				
38.21 -1.06		6.0±1.4		1.28					
0916.8 +6937	PGC 26472		1.16	.44					
0921.3 +1813		.SB?...	1.04± .09	122					8833± 79
211.81 41.06	MCG 3-24- 51		.28± .07	.05	14.75 ±.18				8740
74.38 -40.52				.35					9119
0918.5 +1826	PGC 26474		1.04	.14	14.33				
092124.3+193359		.S..3..	1.11± .07	93				16.98±.3	8600± 9
210.13 41.54	UGC 4969	U (1)	.95± .06	.09	15.80 ±.18				8512
73.07 -39.61		3.0±1.0		1.31				525± 5	8883
091835.2+194647	PGC 26482	4.5±1.3	1.12	.48	14.34			2.17	
092127.5-115435	NGC 2855	RSAT0..	1.39± .02	130	12.63 ±.13	.98± .01	1.01± .01		1910± 40
243.33 25.85	MCG -2-24- 15	R	.05± .04	.16	12.29 ±.19	.47± .02	.52± .02		1723
115.35 -54.29	IRAS09190-1141	.0± .3	.74± .01	.04	13.32	.92	11.82± .03		2224
091902.7-114146	PGC 26483		1.40		12.29	.43	14.28± .20		
092128.1-223006	A 0919-22	.IBS9./	1.40± .03	27				13.62±.2	849± 11
252.10 19.03	ESO 565- 1	PSU (2)	.63± .04	.28	14.87 ±.14			134± 16	
133.72 -54.83	DDO 62	9.8± .4		.47				120± 12	640
091912.0-221718	PGC 26484	9.3±1.0	1.43	.31	14.12			-.82	1152
092129.5+641411	IC 2458	.I.0.P*	.67± .11		15.4 ±.2	.40± .04		12.75±.1	1534± 8
149.79 40.26	MCG 11-12- 5	P	.36± .06	.16	15.5 ±.3	-.64± .06		379± 6	1448± 41
41.20 -5.28	7ZW 276	90.0		.27		.29		324± 5	1619
091726.1+642657	PGC 26485		.67		14.98	-.70	12.70± .61		1634
0921.6 +7508		.S..7*.	1.11± .05	165					
137.39 35.44	UGC 4937	U	.25± .05	.00	14.9 ±.2				
34.95 3.65		7.0±1.2		.34					
0916.3 +7521	PGC 26489		1.11	.12					
092143.2+713236	A 0917+71	.S?....	.98± .05	145	14.6 ±.3	.37± .05	.44± .02		
141.29 37.18	UGC 4951		.12± .05	.07	14.66 ±.18	-.35± .08	-.26± .04		3525± 29
37.01 .71	MK 20		.51± .07	.18		.31	12.59± .16		3639
091659.6+714522	PGC 26492		.99	.06	14.38	-.39	14.04± .42		3589
0921.7 +3932		.S..6*.	1.20± .05	105				15.95±.1	2408± 11
182.99 45.14	UGC 4970	U	.97± .05	.00					2398
56.69 -25.03		6.0±1.3		1.43				230± 8	
0918.6 +3945	PGC 26495		1.20	.49					2626

R.A. 2000 DEC.	Names	Type	logD$_{25}$	p.a.	B$_T$	(B-V)$_T$	(B-V)$_e$	m$_{21}$	V$_{21}$
l b		S$_T$ n$_L$	logR$_{25}$	A$_g$	m$_B$	(U-B)$_T$	(U-B)$_e$	W$_{20}$	V$_{opt}$
SGL SGB		T	logA$_e$	A$_i$	m$_{FIR}$	(B-V)^o_T	m'$_e$	W$_{50}$	V$_{GSR}$
R.A. 1950 DEC.	PGC	L	logD$_o$	A$_{21}$	B^o_T	(U-B)^o_T	m'$_{25}$	HI	V$_{3K}$
092147.1+641529	NGC 2820	.SBS5P/	1.61± .02	59	13.28S±.06	.49± .05		13.05±.1	1579± 5
149.75 40.28	UGC 4961	R	.92± .03	.16	13.16 ±.18			370± 7	1581± 42
41.21 -5.24	IRAS09177+6428	5.0± .5		1.37	11.58	.27		324± 6	1666
091743.7+642816	PGC 26498		1.63	.46	11.73		13.90± .14	.87	1682
092148.1+400907	NGC 2844	.SAR1*.	1.19± .03	13	13.75 ±.13	.80± .01		15.49±.1	1486± 10
182.12 45.15	UGC 4971	R	.31± .03	.00	13.55 ±.13			319± 6	1495± 13
56.27 -24.54	IRAS09186+4021	1.0± .6		.32	13.61	.73		310± 5	1482
091838.0+402155	PGC 26501		1.19	.16	13.31		13.74± .23	2.03	1705
092151.6+332407		.L...*.	1.08± .11	0	14.20 ±.16	.99± .03	1.02± .01		
191.65 44.78	UGC 4972	U	.11± .05	.00	14.37 ±.16				7160± 68
61.22 -29.68		-2.0±1.1	.66± .05	.00		.91	12.99± .15		7125
091849.4+333655	PGC 26504		1.06		14.18		14.18± .57		7401
0922.0 -1138		.I..9..	1.07± .06	135					
243.19 26.12	MCG -2-24- 16	E (1)	.13± .05	.17					
114.94 -54.11		10.0± .9		.10					
0919.6 -1126	PGC 26511	9.8± .8	1.08	.06					
092201.8+505831	NGC 2841	.SAR3*.	1.91± .01	147	10.09M±.10	.87± .01	.93± .01	12.05±.0	638± 3
166.94 44.15	UGC 4966	R (1)	.36± .02	.00	10.12 ±.14	.34± .02	.44± .01	604± 4	635± 20
49.18 -15.98		3.0± .3	1.40± .02	.50	10.87	.80	12.54± .06	596± 4	674
091834.9+511119	PGC 26512	.5± .9	1.91	.18	9.60	.27	13.57± .12	2.27	807
092204.5+715042	NGC 2810	.E.....	1.23± .14		13.24 ±.15	1.01± .02	1.06± .01		
140.94 37.06	UGC 4954	U	.00± .08	.12	13.18 ±.15				3562± 31
36.86 .97		-5.0± .8	.81± .05	.00		.95	12.78± .16		3677
091718.9+720328	PGC 26514		1.25		13.04		14.38± .73		3625
0922.1 +0353		.SA.7..	1.20± .06						4135± 10
228.34 35.00	UGC 4978	U	.13± .06	.13	14.4 ±.4				
91.22 -48.67		7.0± .8		.18					3992
0919.5 +0406	PGC 26517		1.22	.06	14.05				4445
092207.6-094506		.SAT7*.	1.11± .06	125					
241.54 27.31	MCG -2-24- 17	E (1)	.13± .05	.16					
111.78 -53.68		7.0±1.2		.18					
091941.1-093216	PGC 26518	6.4± .8	1.13	.07					
092210.5+335058		.L...*.	1.18± .15	0	14.26 ±.15	.99± .01	1.00± .01		
191.03 44.90	UGC 4974	U	.09± .08	.00	14.15 ±.18				7017± 28
60.93 -29.30	.	-2.0±1.1	.54± .04	.00		.91	12.45± .12		6984
091907.9+340347	PGC 26520		1.17		14.11		14.79± .78		7257
0922.2 +7547		.I..9..	1.10± .07					16.02±.1	659± 6
136.68 35.14	UGC 4945	U	.04± .07	.00				42± 5	
34.60 4.20		10.0± .8		.03				34± 5	788
0916.8 +7600	PGC 26522		1.10	.02					700
0922.2 -0747		.SBS8*.	1.14± .08	140					
239.80 28.52		E (1)	.19± .08	.10					
108.59 -53.16		8.0±1.3		.23					
0919.8 -0735	PGC 26525	7.5± .8	1.15	.09					
0922.3 +0442		.SB.3*.	1.00± .06	172					
227.51 35.45	UGC 4980	U	.39± .05	.12	14.81 ±.18				
90.18 -48.22		3.0± .9		.53					
0919.7 +0455	PGC 26528		1.01	.19					
092224.9+471437	A 0919+47	.L?....	.56± .12						9100± 26
172.10 44.80	MCG 8-17- 86		.04± .06	.00	15.77 ±.18				9121
51.60 -18.91	MK 109			.00					9286
091904.7+472727	PGC 26531		.55		15.63				
092225.2-610259		.SBR3*.	1.24± .05	145					
280.21 -7.78	ESO 126- 10	S (1)	.42± .05	2.59	13.87 ±.14				
184.90 -37.57	IRAS09211-6050	3.0±1.2		.59	12.00				
092108.0-605006	PGC 26532	3.3±1.3	1.48	.21					
092234.3+463927		.SX.7*.	.97± .07	120					
172.91 44.90	UGC 4976	U	.22± .05	.00	15.29 ±.19				
52.01 -19.36		7.0±1.3		.31					
091915.1+465217	PGC 26538		.97	.11					
0922.6 +3640		.I..9*.	1.00± .16	135					
187.06 45.21	UGC 4979	U	.14± .12	.00					
58.89 -27.12		10.0±1.2		.10					
0919.5 +3653	PGC 26540		1.00	.07					
0922.6 +5014		.S..6*.	1.08± .06	31					
167.94 44.38	UGC 4975	U	.87± .05	.00					
49.72 -16.51		6.0±1.4		1.28					
0919.2 +5027	PGC 26541		1.08	.44					

R.A. 2000 DEC. l b SGL SGB R.A. 1950 DEC.	Names PGC	Type S_T n_L T L	$\log D_{25}$ $\log R_{25}$ $\log A_e$ $\log D_o$	p.a. A_g A_i A_{21}	B_T m_B m_{FIR} B_T^o	$(B-V)_T$ $(U-B)_T$ $(B-V)_T^o$ $(U-B)_T^o$	$(B-V)_e$ $(U-B)_e$ m'_e m'_{25}	m_{21} W_{20} W_{50} HI	V_{21} V_{opt} V_{GSR} V_{3K}
0922.6 +2157 207.20 42.56 71.04 -37.79 0919.7 +2210	 UGC 4985 IRAS09197+2210 PGC 26542	.SXS3.. U 3.0± .9 	1.21± .05 .42± .05 1.22	177 .13 .58 .21	 14.54 ±.20 13.12 13.75			15.33±.3 549± 7 1.37	10180± 10 10101 10457
0922.6 +6052 153.88 41.60 43.27 -7.94 0918.8 +6105	 UGC 4973 PGC 26543	.S..4.. U 4.0± .9 	.95± .05 .03± .04 .96	 .12 .04 .01	 14.79 ±.18 				
0922.8 +2919 197.36 44.39 64.64 -32.55 0919.9 +2932	 UGC 4983 PGC 26553	.SXS5.. U 5.0± .9 	.93± .07 .11± .05 .93	 .00 .17 .06	 15.2 ±.2 				
092254.9+030925 229.23 34.79 92.35 -48.85 092018.8+032217	NGC 2858 UGC 4989 PGC 26556	.S..0.. U .0± .8 	1.23± .05 .30± .05 1.22	117 .09 .22	 13.62 ±.18 				
0922.9 +0125 231.03 33.89 94.75 -49.64 0920.4 +0138	 CGCG 6- 35 PGC 26560	 	 	 .09 	 15.2 ±.6 				5192± 58 5041 5505
092300.9-322659 259.92 12.51 150.09 -52.57 092054.0-321406	IC 2469 ESO 433- 17 IRAS09208-3214 PGC 26561	.SBT2.. SUr (1) 2.0± .3 4.4±1.1	1.67± .02 .66± .04 1.75	36 .83 .81 .33	 13.03 				
092303.0+443314 175.86 45.20 53.46 -20.97 091947.1+444606	 UGC 4982 IRAS09198+4446 PGC 26563	.S..8*. U 8.0±1.3 	1.06± .06 .53± .05 1.06	4 .04 .65 .27	 14.2 ±.2 13.40 13.55				2691± 50 2701 2889
092314.3+400953 182.10 45.43 56.49 -24.36 092004.4+402245	NGC 2852 UGC 4986 PGC 26571	.SXR1$. R 1.0± .4 	.93± .08 .02± .06 .93	 .00 .02 .01	 14.09 ±.19 14.05				1835± 49 1828 2052
092315.3+344404 189.81 45.21 60.45 -28.49 092011.9+345656	 UGC 4988 PGC 26575	.SX.9.. U 9.0± .9 	1.04± .06 .11± .05 1.04	70 .00 .11 .05	 15.3 ±.3 15.17			16.46±.3 142± 14 1.23	1571± 10 1607± 60 1543 1809
092316.1-004338 233.24 32.79 97.86 -50.49 092042.8-003045	 UGC 4996 PGC 26576	.SXS4*. UE (1) 4.0± .7 4.2± .8	1.16± .04 .48± .04 1.17	130 .09 .71 .24	 14.21 ±.18 				
0923.2 +7203 140.65 37.05 36.82 1.20 0918.5 +7216	 UGC 4967 PGC 26579	.S..1.. U 1.0± .9 	1.01± .06 .07± .05 1.02	 .08 .07 .03	 14.73 ±.19 				
092317.1+401200 182.05 45.44 56.48 -24.33 092007.2+402452	NGC 2853 UGC 4987 PGC 26580	.LB..*. R -2.0± .4 	1.23± .05 .28± .03 1.19	25 .00 .00	 14.29 ±.17 14.26				1776± 49 1769 1993
0923.3 +0134 230.94 34.06 94.62 -49.48 0920.8 +0147	 CGCG 6- 37 PGC 26589	 	 	 .09 	 15.4 ±.6 				7742± 58 7592 8055
092324.1-634844 282.27 -9.64 187.26 -35.58 092216.0-633548	NGC 2887 ESO 91- 9 PGC 26592	.LAS-?. S -3.0±1.1 1.42	1.32± .05 .14± .04 .87± .02 	78 .87 .00	12.77 ±.13 12.51 ±.14 11.73	1.10± .01 .87 	1.14± .01 .63± .04 12.61± .07 13.92± .30		2907± 27 2684 3079
0923.4 +5430 162.10 43.59 47.18 -13.02 0919.9 +5443	 UGC 4984 PGC 26599	.S..9?. U 9.0±1.7 	1.02± .06 .17± .05 1.02	140 .03 .17 .08	 				
092330.8-230948 252.96 18.94 134.79 -54.31 092115.1-225654	NGC 2865 ESO 498- 1 PGC 26601	.E.3+.. R -5.0± .3 1.40	1.39± .02 .13± .03 .62± .03 	 .29 .00	 12.57 ±.14 12.35 ±.11 12.10	.91± .01 .41± .03 .82 .36	.90± .01 .42± .02 11.16± .11 14.19± .20		2611± 13 2401 2915
092334.9+244550 203.59 43.55 68.67 -35.69 092041.4+245843	 UGC 4994 PGC 26606	.S..2.. U 2.0± .9 	1.02± .06 .39± .05 1.02	33 .08 .48 .19	 14.93 ±.18 14.29			15.63±.3 474± 7 1.14	7597± 10 7529 7867

R.A. 2000 DEC.	Names	Type	$logD_{25}$	p.a.	B_T	$(B-V)_T$	$(B-V)_\bullet$	m_{21}	V_{21}
l b		S_T n_L	$logR_{25}$	A_g	m_B	$(U-B)_T$	$(U-B)_\bullet$	W_{20}	V_{opt}
SGL SGB		T	$logA_\bullet$	A_i	m_{FIR}	$(B-V)_T^o$	$m'_\bullet$	W_{50}	V_{GSR}
R.A. 1950 DEC.	PGC	L	$logD_\bullet$	A_{21}	B_T^o	$(U-B)_T^o$	m'_{25}	HI	V_{3K}
092336.2+020813	NGC 2861	.SBR4..	1.18± .04		13.52 ±.19				5134± 10
230.40 34.41	UGC 4999	UE (1)	.04± .04	.08	13.15				
93.87 −49.17	IRAS09210+0220	3.5± .6		.06					4986
092100.8+022107	PGC 26607	1.9± .8	1.19	.02	13.35				5447
092336.4−265249		.S..3?/	1.17± .04	33					
255.84 16.44	ESO 498- 3	S	.59± .03	.47	14.17 ±.14				
141.05 −53.79	IRAS09213−2639	3.0±1.7		.81	12.19				
092124.0−263954	PGC 26608		1.22	.29					
092347.6−253813		.LAT0..	1.36± .05	144					
254.92 17.32	ESO 498- 4	S	.22± .04	.64	13.31 ±.14				
138.97 −53.95		−2.5± .5		.00					
092134.1−252518	PGC 26624		1.41						
092347.7+421104		.S..6*.	1.06± .06	0					
179.21 45.48	UGC 4992	U	1.01± .05	.02	15.7 ±.2				
55.18 −22.73		6.0±1.5		1.47					
092035.3+422357	PGC 26625		1.06	.50					
092348.2−383002		.L..+?/	1.02± .05	78					
264.43 8.39	ESO 315- 5	S	.56± .03	1.07	16.24 ±.14				
158.98 −50.23		−1.0±1.8		.00					
092148.0−381706	PGC 26626		1.04						
092402.9+491214	NGC 2854	.SBS3..	1.22± .04	50				14.84±.1	2741± 11
169.31 44.78	UGC 4995	U	.41± .05	.02	13.82 ±.18				2732± 50
50.57 −17.20	IRAS09206+4925	3.0± .9		.56				284± 8	2770
092039.8+492508	PGC 26631		1.22	.20	13.22			1.41	2919
0924.1 +6833		.S..3..	1.06± .06	146					
144.55 38.72	UGC 4981	U (1)	.47± .05	.21	14.70 ±.19				
38.90 −1.61	IRAS09197+6846	3.0± .9		.65	13.73				
0919.7 +6846	PGC 26639	2.5±1.2	1.08	.24					
0924.2 +6201		.SB.6*.	.99± .05	175					
152.36 41.37	UGC 4990	U	.31± .04	.13	15.14 ±.18				
42.74 −6.90		6.0±1.3		.46					
0920.3 +6214	PGC 26642		1.00	.16					
0924.2 +1107		.S..6*.	1.00± .08	30					
220.69 38.89	UGC 5003	U	.33± .06	.10	14.88 ±.18				
82.63 −44.36		6.0±1.3		.49					
0921.5 +1120	PGC 26643		1.01	.17					
092416.7+491453	NGC 2856	.S?....	1.05± .04	134					
169.24 44.81	UGC 4997		.36± .04	.02	14.1 ±.2				2638± 50
50.57 −17.14	IRAS09208+4927			.54	11.31				2667
092053.6+492748	PGC 26648		1.05	.18	13.50				2816
092418.9+343046	NGC 2859	RLBR+..	1.63± .02	85	11.83 ±.13	.93± .01	.96± .01	17.22±.3	1687± 8
190.16 45.40	UGC 5001	R	.05± .02	.00	11.54 ±.12	.47± .02	.53± .02		1659± 18
60.81 −28.51		−1.0± .3	1.02± .02	.00		.91	12.42± .07	167± 6	1652
092116.0+344341	PGC 26649		1.62		11.65	.47	14.72± .17		1921
092422.0+281734		.SB?...	.87± .06	170				16.56±.3	6518± 10
198.86 44.52	UGC 5002		.20± .04	.03	14.99 ±.19			281± 13	6486±125
65.78 −33.07	IRAS09214+2830			.29	13.53				6463
092125.4+283030	PGC 26650		.88	.10	14.64			1.82	6778
092425.7−412339		.L..+*/	1.01± .06	102					
266.57 6.42	ESO 315- 7	S	.34± .04	1.42	15.47 ±.14				
162.88 −48.86		−1.0±1.3		.00					
092229.0−411042	PGC 26653		1.10						
0924.4 +7632		.E?....		.07	15.2 ±.3				1544± 82
135.81 34.87	MCG 13- 7- 27								1675
34.28 4.88	7ZW 277								1582
0918.9 +7645	PGC 26654								
092430.1+221624		.I..9..	1.16± .13					15.37±.3	3839± 7
206.96 43.07	UGC 5005	U	.09± .12	.10					
71.12 −37.25		10.0± .8		.07				111± 7	3761
092139.0+222920	PGC 26659		1.17	.04					4117
092430.3−374510		.SAT0*.	1.08± .05						
263.99 9.01	ESO 315- 6	Sr	.03± .04	.99	14.54 ±.14				
157.84 −50.41		−.5± .5		.00					
092229.0−373212	PGC 26660		1.17						
092431.7−631441		.SXS4..	1.00± .05	35					
281.96 −9.15	ESO 91- 12	S (1)	.03± .04	2.02	15.19 ±.14				
186.66 −35.88		4.0± .9		.04					
092321.0−630142	PGC 26662	5.6±1.2	1.19	.02					

R.A. 2000 DEC. l b SGL SGB R.A. 1950 DEC.	Names PGC	Type S_T n_L T L	$\log D_{25}$ $\log R_{25}$ $\log A_e$ $\log D_o$	p.a. A_g A_i A_{21}	B_T m_B m_{FIR} B_T^o	$(B-V)_T$ $(U-B)_T$ $(B-V)_T^o$ $(U-B)_T^o$	$(B-V)_e$ $(U-B)_e$ m' m'_{25}	m_{21} W_{20} W_{50} HI	V_{21} V_{opt} V_{GSR} V_{3K}
092433.9+343937 189.95 45.47 60.74 -28.37 092130.9+345233	 UGC 5004 PGC 26663	.I..9*. U 10.0±1.2 	1.12± .07 .15± .06 1.12	130 .00 .12 .08	 	 	 	16.84±.2 151± 14 106± 7 	1843± 7 1890± 60 1815 2083
092438.0+492120 169.07 44.85 50.55 -17.02 092114.8+493416	NGC 2857 UGC 5000 ARP 1 PGC 26666	.SAS5.. R (1) 5.0± .3 1.8± .8	1.35± .04 .05± .05 1.19± .05 1.35	 .01 .07 .02	12.9 ±.2 13.8 ±.3 13.08	.63± .05 .59 	.72± .02 14.31± .12 14.37± .30	14.83±.1 158± 7 142± 6 1.73	4887± 7 4864± 59 4917 5065
092439.5+173940 212.88 41.61 75.64 -40.27 092152.2+175236	A 0921+17 MCG 3-24- 55 MK 398 PGC 26668	 	.64± .17 .28± .07 .65	 .10 	15.4 ±.2 15.7 ±.3 	.57± .06 -.18± .09 	 	16.81±.3 249± 21 238± 11 	4021± 13 4200±110 3928 4313
092440.1+250648 203.21 43.87 68.58 -35.27 092146.5+251944	IC 536 UGC 5006 PGC 26669	.S..1?. U 1.0±2.0 	1.08± .06 .71± .05 1.08	23 .08 .73 .36	 15.39 ±.18 14.48	 	 	17.13±.3 576± 5 2.29	8186± 9 8119 8456
092441.2-250534 254.65 17.84 138.00 -53.83 092227.0-245236	A 0922-24 ESO 498- 5 IRAS09224-2452 PGC 26671	.SXS4P. PS (2) 4.3± .5 2.0± .5	1.13± .04 .09± .03 .91± .02 1.17	155 .42 .14 .05	13.96 ±.14 14.23 ±.14 13.48 13.52	.87± .02 .74 	.82± .01 14.00± .05 14.22± .26	 	2413± 59 2200 2714
0924.7 +2001 209.90 42.42 73.31 -38.73 0921.9 +2014	 UGC 5009 PGC 26673	.S..6*. U 6.0±1.2 	1.04± .06 .11± .05 1.05	150 .12 .16 .05	 14.8 ±.2 14.54	 	 	 	4277± 10 4191 4561
092453.3+410336 180.81 45.73 56.13 -23.47 092142.6+411633	NGC 2860 UGC 5007 IRAS09216+4116 PGC 26685	.SB.1.. U 1.0± .9 	1.16± .05 .40± .05 1.16	108 .01 .41 .20	 14.64 ±.18 12.40 14.16	 	 	 	4247 4243 4461
092455.3+264638 200.97 44.32 67.17 -34.06 092200.2+265935	NGC 2862 UGC 5010 IRAS09219+2659 PGC 26690	.S?.... 	1.40± .03 .66± .04 1.40	114 .04 .99 .33	 13.77 ±.18 12.72	 	 	14.54±.2 601± 6 593± 5 1.49	4096± 5 4156± 76 4036 4362
092502.0-371011 263.65 9.50 156.94 -50.54 092300.0-365712	 ESO 372- 23 PGC 26699	.S..5*/ S 4.5± .9 	1.30± .05 .95± .05 1.39	51 .99 1.43 .48	 	 	 	 	
0925.0 +6432 149.23 40.50 41.34 -4.81 0921.0 +6445	 MCG 11-12- 9 PGC 26700	 	.69± .15 .16± .07 .70	116 .15 	 15.5 ±.2 	 	 	 	 5369± 57 5458 5472
0925.0 +6434 149.19 40.48 41.32 -4.79 0921.0 +6447	 MCG 11-12- 10 PGC 26701	 	.82± .13 .36± .07 .83	 .15 	 15.0 ±.3 	 	 	 	 5021± 57 5110 5123
0925.1 +6824 144.67 38.87 39.07 -1.68 0920.8 +6837	 UGC 4998 PGC 26705	.I..9*. U 10.0±1.1 	1.21± .06 .29± .06 1.23	80 .17 .22 .14	 15.0 ±.3 	 	 	 	
092512.3-064950 239.41 29.68 107.39 -52.19 092243.5-063651	IC 2471 MCG -1-24- 15 PGC 26707	.L..0P* E -2.0±1.3 	1.04± .07 .24± .05 1.02	45 .10 .00 	 	 	 	 	
092513.9-064302 239.31 29.75 107.21 -52.15 092245.0-063003	NGC 2876 MCG -1-24- 16 PGC 26710	PL..0P? E -2.0±1.2 	1.23± .07 .17± .05 1.21	95 .10 .00 	 	 	 	 	
0925.2 +1735 213.03 41.72 75.83 -40.21 0922.4 +1748	 UGC 5012 IRAS09224+1748 PGC 26711	.S..2.. U 2.0±1.0 	.98± .07 .56± .05 .99	145 .10 .69 .28	 15.1 ±.2 	 	 	 	
092518.5-340612 261.47 11.70 152.38 -51.59 092313.0-335312	NGC 2883 ESO 372- 24 IRAS09232-3353 PGC 26713	.IBS9P* BS 10.0± .7 	1.44± .04 .49± .07 1.52	176 .95 .37 .24	 12.74 	 	 	 	1182 958 1465
092518.7+340649 190.75 45.57 61.29 -28.67 092216.3+341947	 UGC 5011 PGC 26714	.SB.6*. U 6.0±1.2 	1.02± .06 .07± .05 1.02	165 .00 .11 .04	 	 	 	16.03±.2 156± 14 155± 7 	6753± 7 6722 6994

R.A. 2000 DEC.	Names	Type	$\log D_{25}$	p.a.	B_T	$(B-V)_T$	$(B-V)_\bullet$	m_{21}	V_{21}
l b		S_T n_L	$\log R_{25}$	A_g	m_B	$(U-B)_T$	$(U-B)_\bullet$	W_{20}	V_{opt}
SGL SGB		T	$\log A_\bullet$	A_i	m_{FIR}	$(B-V)_T^o$	$m'_\bullet$	W_{50}	V_{GSR}
R.A. 1950 DEC.	PGC	L	$\log D_o$	A_{21}	B_T^o	$(U-B)_T^o$	m'_{25}	HI	V_{3K}
092522.7-122332	IC 537	RS..0?.	1.07± .06	175					
244.41 26.30	MCG -2-24- 20	E	.04± .05	.17					
116.44 -53.43	IRAS09229-1210	.0±1.7		.03	12.84				
092258.1-121033	PGC 26717		1.08						
0925.5 +1104		.S..6*.	1.00± .08	140					
220.93 39.16	UGC 5017	U	.84± .06	.08	15.82 ±.19				
82.95 -44.13		6.0±1.4		1.24					
0922.8 +1117	PGC 26721		1.01	.42					
0925.6 +3451		.S..8*.	1.00± .08	38					4868
189.71 45.71	UGC 5014	U	1.02± .06	.00					
60.78 -28.08		8.0±1.5		1.23					4839
0922.6 +3504	PGC 26727		1.00	.50					5106
092543.0+112556	NGC 2872	.E.2...	1.32± .07	22	12.86 ±.13	.99± .01	1.01± .01		
220.53 39.36	UGC 5018	R	.06± .06	.09	12.61 ±.15	.52± .02	.60± .01		3014± 36
82.57 -43.88		-5.0± .4	.77± .02	.00		.94	12.20± .09		2896
092300.6+113856	PGC 26733		1.32		12.62	.52	14.32± .42		3317
092547.2+021344	NGC 2877	.P.....	.67± .11						
230.65 34.92	MCG 0-24- 15	E	.00± .05	.12	15.1 ±.2				6959± 63
94.12 -48.65	ARAK 201	99.0			12.83				6811
092311.7+022644	PGC 26738		.69						7274
092547.5+020522	NGC 2878	.S..1*.	.91± .06	174					
230.80 34.84	UGC 5022	UE	.44± .04	.07	15.1 ±.2				
94.32 -48.71		1.0± .8		.45					
092312.1+021822	PGC 26739		.92	.22					
092547.8+112531	NGC 2874	.SBR4..	1.38± .03	43	13.36 ±.15	.85± .02	.93± .01	16.00±.1	3775± 4
220.55 39.38	UGC 5021	R	.51± .04	.09	13.36 ±.18	.29± .03	.34± .02	404± 5	3659± 41
82.60 -43.87	IRAS09230+1138	4.0± .4	.76± .03	.75	11.98	.71	12.65± .09	397± 6	3656
092305.5+113831	PGC 26740		1.39	.25	12.50	.17	13.82± .25	3.24	4077
092548.0+341638		.SX.8..	1.27± .06	20				15.52±.2	1648± 7
190.53 45.69	UGC 5015	U	.01± .06	.00	14.9 ±.5			126± 14	1758± 60
61.25 -28.49		8.0± .8		.01				122± 7	1619
092245.6+342937	PGC 26741		1.27		14.92			.60	1890
0925.9 -1159	NGC 2881	.S?....	1.04± .09						
244.16 26.65	MCG -2-24- 21		.09± .07	.18					4221± 60
115.82 -53.23	ARP 275			.13	12.77				4034
0923.4 -1146	PGC 26747		1.05	.05					4538
092600.8+192303	A 0923+19	.S?....	.86± .06					16.34±.2	2522± 5
210.85 42.50	UGC 5023		.06± .04	.07	14.65 ±.20			202± 5	2473± 51
74.19 -38.91	MK 400			.08	12.87			171± 3	2433
092312.2+193603	PGC 26750		.87	.03	14.48			1.83	2808
092601.5+343913		.S..6*.	1.33± .04	79				15.43±.2	1621± 7
190.00 45.77	UGC 5020	U	.68± .05	.00	15.0 ±.2			192± 14	1682± 60
61.00 -28.17		6.0±1.2		1.00				195± 7	1592
092258.6+345213	PGC 26752		1.33	.34	13.97			1.12	1861
092603.4+124402		.L...?.	.83± .10	70	14.88 ±.15	.54± .02			
219.05 40.00	UGC 5025	U	.08± .03	.08	14.4 ±.2	-.55± .04			8658± 86
81.17 -43.05	MK 705	-2.0±1.9		.00		.43			8545
092320.1+125703	PGC 26753		.82		14.52	-.52	13.69± .55		8958
092608.0+345347	A 0923+35								
189.66 45.81				.00					4800±110
60.83 -27.98	MK 399				13.19				4772
092304.9+350647	PGC 26757								5039
092609.3+491837		.SB.1*.	1.02± .06	30					
169.08 45.10	UGC 5016	U	.35± .05	.01	15.24 ±.18				
50.78 -16.90		1.0±1.3		.36					
092246.6+493137	PGC 26759		1.02	.18					
092611.6-763735	NGC 2915	.I.0...	1.28± .03	129	13.25M±.12	.57± .01	.53± .01	12.54±.3	460± 5
291.97 -18.36	ESO 37- 3	V	.29± .04	.56	12.93 ±.14	-.12± .02	-.14± .02	153± 7	462± 33
197.17 -26.06	IRAS09265-7624	90.0	.72± .01	.22	13.25	.39	12.32± .03		257
092631.0-762430	PGC 26761		1.31		12.33	-.24	13.75± .21		569
0926.2 +0306		.S..4..	1.00± .08	173					
229.79 35.47	UGC 5027	U	.73± .06	.13	15.52 ±.19				
93.00 -48.15		4.0±1.0		1.07					
0923.6 +0320	PGC 26762		1.01	.36					
0926.2 +6122		.S..3..	1.20± .04	120					
153.06 41.83	UGC 5013	U (1)	.24± .04	.10	15.1 ±.3				
43.33 -7.29		3.0± .8		.33					
0922.4 +6135	PGC 26766		1.20	.12					

R.A. 2000 DEC. / l b / SGL SGB / R.A. 1950 DEC.	Names / PGC	Type / S_T n_L / T / L	$\log D_{25}$ / $\log R_{25}$ / $\log A_e$ / $\log D_o$	p.a. / A_g / A_i / A_{21}	B_T / m_B / m_{FIR} / B_T^o	$(B-V)_T$ / $(U-B)_T$ / $(B-V)_T^o$ / $(U-B)_T^o$	$(B-V)_e$ / $(U-B)_e$ / m' / m'_{25}	m_{21} / W_{20} / W_{50} / HI	V_{21} / V_{opt} / V_{GSR} / V_{3K}
092619.7-280208	NGC 2888	.E+..*.	1.14± .03	158	13.58 ±.13	.96± .01	.98± .01		2203± 66
257.15 16.09	ESO 434- 2	RS	.13± .04	.49	13.24 ±.11	.38± .05	.42± .05		1986
142.76 -52.98		-4.0± .3	.55± .01	.00		.83	11.82± .04		2500
092408.0-274906	PGC 26768		1.18		12.86	.29	13.97± .22		
092623.1-603657		.SBS9..	1.15± .05	63					
280.24 -7.13	ESO 126- 11	S (1)	.12± .05	3.01	14.35 ±.14				
184.10 -37.51		9.0± .8		.12					
092503.1-602354	PGC 26772	8.9± .8	1.43	.06					
092624.6-113321	NGC 2884	.S..0$.	1.31± .04	175					5394
243.86 27.02	MCG -2-24- 22	R	.28± .04	.17					
115.13 -53.03	IRAS09239-1120	.0± .7		.21					5208
092359.3-112019	PGC 26773		1.31						5712
092628.3-154223		.SBS9P*	1.17± .04	30					1977
247.42 24.39	MCG -2-24- 23	E (1)	.31± .04	.21					
122.07 -53.60		8.7± .7		.32					1782
092406.2-152920	PGC 26776	7.0± .6	1.19	.16					2292
092635.5+075713	NGC 2882	.S?....	1.19± .05	80				15.90±.2	2149± 6
224.62 37.97	UGC 5030		.31± .05	.12	13.58 ±.19			303± 7	
86.87 -45.64	IRAS09239+0810			.46	12.85			285± 5	2020
092355.8+081015	PGC 26781		1.20	.15	12.99			2.76	2458
092639.7+455050		.L.....	1.04± .06	15					
173.94 45.70	UGC 5026	U	.15± .04	.05	14.14 ±.15				4316± 31
53.13 -19.56		-2.0± .8		.00					4332
092322.7+460351	PGC 26785		1.02		14.02				4510
092644.3+010900		.S..2..	1.05± .05	122					
231.92 34.54	UGC 5032	U	.67± .05	.09	15.47 ±.18				
95.77 -48.91		2.0±1.0		.82					
092409.7+012203	PGC 26787		1.06	.33					
092656.7-244658	NGC 2891	.LA.-*.	1.18± .05		13.51 ±.13	.87± .01	.88± .01		
254.77 18.42	ESO 498- 8	S	.03± .04	.28	13.30 ±.14	.35± .01	.36± .01		
137.38 -53.36		-3.0±1.1	.40± .02	.00			11.00± .08		
092442.0-243354	PGC 26794		1.21				14.19± .31		
092659.3-120633	IC 2482	.E+....	1.36± .04	145					
244.44 26.78	MCG -2-24- 25	E	.16± .05	.17					
116.09 -52.99		-4.0± .8		.00					
092434.4-115329	PGC 26796		1.33						
0927.1 +2135		RSBS1..	1.04± .06						
208.09 43.45	UGC 5035	U	.03± .05	.10	14.8 ±.3				
72.28 -37.25		1.0± .9		.03					
0924.3 +2149	PGC 26803		1.05	.02					
092712.6-113837	NGC 2889	.SXT5..	1.34± .03	65	12.44 ±.13	.71± .01	.80± .01	15.09±.1	3337± 5
244.08 27.12	MCG -2-24- 26	R (3)	.06± .04	.16	12.4 ±.2	.10± .02	.19± .02	364± 6	3387± 66
115.34 -52.86	IRAS09247-1125	5.0± .4	.96± .01	.09	11.65	.64	12.73± .03	267± 25	3152
092447.3-112533	PGC 26806	2.5± .6	1.35	.03	12.17	.05	13.84± .20	2.89	3656
092722.9+302629	IC 2473	.SBR4..	1.24± .04					16.29±.2	8070± 7
196.03 45.54	UGC 5038	U	.05± .04	.05	13.8 ±.3			334± 13	8075± 13
64.56 -31.07	IRAS09244+3039	4.0± .8		.07	13.15			316± 7	8025
092424.5+303933	PGC 26817		1.25	.02	13.61			2.66	8326
092726.0-320035		.IBS9..	1.16± .04					15.02±.2	1082± 11
260.27 13.49	ESO 434- 5	SU (1)	.08± .04	.66	15.07 ±.14			91± 16	
148.97 -51.80		10.0± .5		.06				84± 12	860
092518.0-314730	PGC 26819	10.0± .6	1.22	.04	14.35			.63	1371
0927.4 +4049		.S..8*.	1.06± .06	5					
181.12 46.22	UGC 5036	U	.56± .05	.01					
56.70 -23.33		8.0±1.3		.69					
0924.3 +4103	PGC 26822		1.06	.28					
092728.5+035549	IC 2481	.S?....	.98± .07	160				16.15±.3	5329± 9
229.14 36.17	UGC 5040A		.23± .05	.13	14.52 ±.18			275± 7	
92.14 -47.48	IRAS09248+0408			.29	13.56			246± 7	5187
092451.8+040853	PGC 26826		.99	.12	14.04			1.99	5643
092734.7+121610		.S?....	.89± .11		15.99S±.15				
219.80 40.14	MCG 2-24- 12		.64± .07	.09	15.8 ±.2				8760± 41
81.99 -43.02	HICK 38A			.93					8646
092451.8+122914	PGC 26831		.90	.32	14.86		13.69± .62		9063
092736.1+284756		.I..9?.	1.36± .05					15.04±.3	4149± 10
198.35 45.31	UGC 5040	U	.02± .06	.02	14.0 ±.6				
65.96 -32.21		10.0±1.5		.01				63± 7	4097
092439.4+290100	PGC 26833		1.36	.01	13.97			1.06	4410

R.A. 2000 DEC. / l b / SGL SGB / R.A. 1950 DEC.	Names / PGC	Type / S_T n_L / T / L	$\log D_{25}$ / $\log R_{25}$ / $\log A_e$ / $\log D_o$	p.a. / A_g / A_i / A_{21}	B_T / m_B / m_{FIR} / B_T^o	$(B-V)_T$ / $(U-B)_T$ / $(B-V)_T^o$ / $(U-B)_T^o$	$(B-V)_e$ / $(U-B)_e$ / m'_e / m'_{25}	m_{21} / W_{20} / W_{50} / HI	V_{21} / V_{opt} / V_{GSR} / V_{3K}
0927.7 +7013		.S..6*.	1.00± .08	55					
142.46 38.23	UGC 5024	U	.19± .06	.15					
38.18 -.07		6.0±1.3		.27					
0923.2 +7027	PGC 26841		1.01	.09					
092743.6+121714		.S?....			15.29S±.15				
219.80 40.18	MCG 2-24- 13			.09					8692± 31
82.00 -42.98	HICK 38B								8577
092500.8+123019	PGC 26842								8994
092744.6+121717		.S?....	.60?		15.95S±.15				
219.81 40.18	MCG 2-24- 14		.30?	.09					8735± 31
82.00 -42.98	HICK 38C			.44					8621
092501.7+123022	PGC 26844		.61	.15	15.38		13.06±1.59		9038
092749.8+682440 A 0923+68		.SBS8P.	.87± .06	145					
144.52 39.09	UGC 5028	R	.24± .05	.21	14.3 ±.3				3698± 14
39.27 -1.52	MK 111	8.0± .6		.30	11.95				3801
092330.2+683743	PGC 26849		.89	.12	13.75				3780
092753.1+295916 IC 2476		.L..-*.	1.17± .09						
196.70 45.58	UGC 5043	U	.03± .05	.06	13.85 ±.18				8025± 30
65.03 -31.32		-3.0±1.1		.00					7977
092455.3+301221	PGC 26854		1.17		13.68				8282
092754.4+572239 NGC 2870		.S..4..	1.39± .03	123				14.52±.1	3214± 5
158.06 43.37	UGC 5034	U	.58± .04	.07	13.83 ±.18			359± 8	3221± 42
45.92 -10.35	IRAS09241+5735	4.0± .8		.85	13.39			331± 5	3276
092415.5+573543	PGC 26856		1.40	.29	12.88			1.35	3355
0928.0 -0550		PSBR3P?	1.22± .07	80					
238.96 30.83		E	.13± .08	.08					
106.18 -51.23		3.0±1.8		.17					
0925.5 -0537	PGC 26861		1.23	.06					
092802.5+682517		.SXS5..	1.21± .05	13					
144.49 39.10	UGC 5029	U	.21± .05	.18	14.13 ±.19				3874± 52
39.28 -1.50		5.0± .8		.32					3977
092342.9+683821	PGC 26864		1.22	.11	13.61				3957
092806.8+171150		.S?....	.94± .07	16					
213.86 42.21	UGC 5046		.43± .05	.13	15.0 ±.2				4215± 50
76.80 -39.93	IRAS09253+1724			.65	11.97				4118
092520.1+172456	PGC 26869		.96	.22	14.17				4508
0928.1 +6455		.S..6*.	1.19± .05	167				16.30±.1	5207± 11
148.58 40.64	UGC 5033	U	1.15± .05	.16					5298
41.37 -4.30		6.0±1.4		1.47				264± 8	5308
0924.1 +6509	PGC 26871		1.20	.50					
092810.0+443958		.SXR5..	1.08± .06						
175.58 46.09	UGC 5045	U	.10± .05	.04	14.18 ±.18				7706± 50
54.15 -20.31	IRAS09248+4453	5.0± .9		.15	13.60				7717
092455.2+445303	PGC 26873		1.08	.05	13.95				7906
0928.3 +2942 IC 2480									
197.11 45.63	CGCG 151- 94			.01	15.13 ±.18				8120
65.34 -31.44									8071
0925.4 +2956	PGC 26883								8378
092821.8-360956		.LXS-*.	1.05± .08						
263.41 10.68	ESO 373- 3	S	.06± .04	.96					
155.10 -50.30		-3.0± .8		.00					
092618.1-355648	PGC 26884		1.16						
092825.8-360744		.LBS0P.	1.10± .09	11					
263.39 10.72	ESO 373- 4	S	.23± .06	.99					
155.04 -50.30	IRAS09263-3554	-2.0± .9		.00	12.73				
092622.0-355436	PGC 26887		1.18						
092826.4-604803		.LA.0*.	1.25± .05	63	13.6 ±.2	1.16± .01	1.22± .01		
280.56 -7.08	ESO 126- 14	S	.23± .04	3.06	13.41 ±.14	.68± .02	.71± .02		
184.06 -37.20		-2.0± .8	.56± .06	.00			11.89± .21		
092706.0-603454	PGC 26888		1.56				14.11± .34		
092829.2-242202		PSXS2..	.98± .05	67					
254.71 18.96	ESO 498- 10	r	.39± .04	.16	15.24 ±.14				
136.62 -53.06		2.2± .9		.48					
092614.0-240854	PGC 26890		.99	.20					
092837.7+383220		.I..9*.	1.10± .04	11					
184.44 46.47	UGC 5048	U	.67± .04	.00	15.19 ±.18				
58.54 -24.92	KUG 0925+387A	10.0±1.3		.50					
092530.9+384527	PGC 26895		1.10	.34					

R.A. 2000 DEC. / l b / SGL SGB / R.A. 1950 DEC.	Names / PGC	Type / S_T n_L / T / L	$\log D_{25}$ / $\log R_{25}$ / $\log A_o$ / $\log D_o$	p.a. / A_g / A_l / A_{21}	B_T / m_B / m_{FIR} / B_T^o	$(B-V)_T$ / $(U-B)_T$ / $(B-V)_T^o$ / $(U-B)_T^o$	$(B-V)_o$ / $(U-B)_o$ / m' / m'_{25}	m_{21} / W_{20} / W_{50} / HI	V_{21} / V_{opt} / V_{GSR} / V_{3K}
0928.7 +5133		.S..8*.	1.14± .05	160					
165.83 45.05	UGC 5047	U	.87± .05	.00					
49.67 -14.87		8.0±1.3		1.07					
0925.3 +5147	PGC 26899		1.14	.44					
092844.4-605122		.S..4./	1.15± .05	142					
280.62 -7.10	ESO 126- 15	S	.64± .04	3.05	15.21 ±.14				
184.08 -37.14		4.0± .9		.94					
092724.1-603812	PGC 26900		1.44	.32					
092855.5+491419		.S..6*.	1.06± .06	158					
169.06 45.56	UGC 5049	U	.79± .05	.01	15.68 ±.18				
51.20 -16.68		6.0±1.4		1.16					
092533.7+492726	PGC 26903		1.06	.39					
0928.9 +6628		.SB.8..	1.06± .06						
146.69 40.05	UGC 5042	U	.00± .05	.17					
40.50 -3.01		8.0± .8		.00					
0924.8 +6642	PGC 26904		1.07	.00					
092858.8-144827		.SBS7..	1.15± .06						2025
247.10 25.43	MCG -2-24- 27	E (1)	.11± .05	.25					
120.67 -52.91	IRAS09265-1435	7.0± .8		.15	13.73				1832
092635.9-143518	PGC 26905	6.4± .8	1.18	.05					2343
092902.0-375028		.S..5./	1.29± .05	158					
264.70 9.58	ESO 315- 12	S	1.10± .04	.97	15.79 ±.14				
157.42 -49.55		5.2± .6		1.50					
092700.0-373718	PGC 26907		1.38	.50					
092908.5-023254	IC 539	.S..6*.	1.00± .06						
236.01 32.98	UGC 5054	U	.07± .05	.09	14.14 ±.18				
101.41 -49.85	IRAS09266-0219	6.0±1.2		.10	13.33				
092636.5-021945	PGC 26909		1.01	.04					
0929.2 +2626									
201.72 45.18	CGCG 152- 15			.04	15.4 ±.3				13245
68.27 -33.61									13184
0926.3 +2640	PGC 26912								13514
092916.6-202246		RSBROP.	1.20± .03	144					
251.72 21.80	ESO 565- 11	SE	.12± .03	.08	13.67 ±.14				
129.97 -53.12	IRAS09269-2009	.3± .4		.09	12.73				
092658.0-200936	PGC 26918		1.20						
092930.0+074304	NGC 2894	.S..1..	1.29± .04	27				14.57±.2	2146± 6
225.33 38.49	UGC 5056	U	.30± .05	.12	13.27 ±.18			409± 7	
87.70 -45.15	IRAS09268+0756	1.0± .8		.31				400± 5	2016
092650.6+075614	PGC 26932		1.30	.15	12.82			1.60	2457
092935.1+622925	NGC 2880	.LB.-..	1.31± .05	140	12.46 ±.17	.92± .04	.94± .01		
151.46 41.77	UGC 5051	R	.23± .04	.09	12.75 ±.12	.45± .06	.49± .03		1551± 27
42.97 -6.15		-3.0± .4	.86± .05	.00		.87	12.25± .19		1633
092541.9+624233	PGC 26939		1.28		12.53	.43	13.29± .30		1666
092946.4+020349	NGC 2898	.L..+P*	1.01± .13	125					
231.47 35.67	MCG 0-24- 18	E	.11± .07	.15	14.41 ±.15				
95.03 -47.83	IRAS09271+0217	-1.0±1.3		.00					
092711.1+021700	PGC 26950		1.01						
092957.8-621055		.L..P*/	1.29± .05	120					
281.65 -7.95	ESO 126- 17	S	.21± .04	2.54	13.45 ±.14				
185.19 -36.15	IRAS09286-6157	-2.0± .9		.00					
092841.0-615742	PGC 26956		1.55						
0930.0 +6008		.S..8*.	1.04± .08	67					
154.36 42.70	UGC 5053	U	.37± .06	.12					
44.44 -7.98		8.0±1.3		.46					
0926.3 +6022	PGC 26959		1.05	.19					
093009.1+200522	IC 2487	.S..3*/	1.26± .03	164				14.85±.2	4339± 6
210.37 43.65	UGC 5059	PU (1)	.61± .04	.13	14.25 ±.18			391± 6	4294± 50
74.31 -37.72	IRAS09273+2018	3.3± .7		.84	13.49			373± 7	4253
092720.2+201834	PGC 26966	4.5±1.2	1.27	.31	13.24			1.31	4626
093011.3+555109	A 0926+56	PSBS3..	1.19± .03	155				15.18±.1	7541± 11
159.94 44.14	UGC 5055	U	.08± .04	.04	14.2 ±.2				7455± 31
47.12 -11.37	MK 114	3.0± .8		.11	12.62			222± 8	7587
092636.8+560420	PGC 26970		1.19	.04	13.96			1.17	7680
093014.3+040828	NGC 2900	.SB.6*.	1.23± .04					15.25±.1	5346± 7
229.54 36.87	UGC 5065	U	.07± .04	.10	13.7 ±.2			122± 8	5354± 76
92.35 -46.77	IRAS09276+0421	6.0±1.1		.11	13.69				5205
092737.5+042140	PGC 26974		1.24	.04	13.51			1.70	5662

R.A. 2000 DEC. / l b / SGL SGB / R.A. 1950 DEC.	Names / PGC	Type S_T n_L / T / L	logD_25 / logR_25 / logA_o / logD_o	p.a. / A_g / A_i / A_21	B_T / m_B / m_FIR / B_T^o	(B-V)_T / (U-B)_T / (B-V)_T^o / (U-B)_T^o	(B-V)_o / (U-B)_o / m'_o / m'_25	m_21 / W_20 / W_50 / HI	V_21 / V_opt / V_GSR / V_3K
093016.6+293221 197.46 46.01 65.84 -31.27 092719.6+294533	NGC 2893 UGC 5060 MK 401 PGC 26979	RSB.0.. U .0± .8	1.05± .04 .04± .04 1.05	 .02 .03	13.91 ±.17 13.77 ±.19 12.21 13.76	.67± .02 -.06± .04 .64 -.06	 13.90± .29	15.69±.1 172± 5 81± 3	1699± 5 1672± 25 1649 1958
093016.7-303637 259.69 14.90 146.54 -51.57 092807.0-302324	 ESO 434- 7 PGC 26980	.LB..*P S -2.0±1.2	1.16± .04 .45± .03 1.17	79 .71 .00	 14.87 ±.14				
093016.9-302301 259.52 15.06 146.19 -51.63 092807.0-300948	NGC 2904 ESO 434- 6 PGC 26981	.LXS-?. S -3.0± .7	1.17± .05 .18± .04 .78± .02 1.23	90 .71 .00	13.43 ±.13 13.69 ±.14 12.80	1.06± .02 .55± .04 .87 .39	1.08± .01 .59± .02 12.82± .07 13.69± .31		2395± 37 2175 2689
0930.3 +2638 201.52 45.46 68.30 -33.29 0927.4 +2652	IC 2486 UGC 5062 PGC 26982	.SXT4.. U 4.0± .9	.93± .07 .03± .05 .94	 .03 .04 .01	14.79 ±.18 14.63				13728 13667 13997
0930.4 +1621 215.19 42.42 78.13 -40.01 0927.7 +1634	 MCG 3-24- 63 IRAS09277+1634 PGC 26990	.L?....	.97± .15 .23± .07 .94	179 .10 .00	15.21 ±.16 12.83 14.98				8653± 46 8554 8949
0930.6 +0606 227.28 37.95 89.90 -45.73 0928.0 +0620	 UGC 5069 PGC 26998	.L..... U -2.0± .8	1.04± .09 .09± .04 1.04	15 .13 .00					
093047.1+491532 168.95 45.85 51.44 -16.47 092725.7+492844	A 0927+49 UGC 5063 MK 115 PGC 27000	.P..... R 99.0	.94± .07 .20± .05 .94	50 .01 .27 .10	 15.24 ±.18 14.89			16.69±.1 359± 12 1.69	7695± 11 7778± 60 7727 7878
093053.0-144410 247.37 25.83 120.64 -52.44 092829.9-143056	NGC 2902 MCG -2-24- 30 PGC 27004	.LAS0*. R -2.0± .4	1.15± .05 .08± .04 .71± .03 1.17	35 .27 .00	12.21V±.15 13.24 ±.18 12.94	1.02± .02 .53± .04	14.68±.1 171± 12		1990± 9 2035± 66 1798 2310
093053.7-354115 263.44 11.38 154.13 -49.98 092849.0-352800	 ESO 373- 5 PGC 27006	.SAT5.. S (1) 5.0± .7 3.3± .7	1.46± .05 .03± .07 1.55	 .99 .05 .02					
093106.2-131054 246.09 26.88 118.12 -52.18 092841.9-125740	IC 542 MCG -2-24- 31 IRAS09286-1257 PGC 27012	PSBT0*. E .0± .9	1.07± .06 .49± .05 1.06	105 .21 .37	 13.54				
093107.8+462302 173.02 46.40 53.40 -18.65 092751.1+463616	 UGC 5066 PGC 27016	.SB.4.. U 4.0± .9	1.02± .05 .26± .04 1.03	70 .09 .39 .13	 15.25 ±.20				
0931.1 +5237 164.24 45.16 49.29 -13.81 0927.7 +5251	 CGCG 265- 23 PGC 27020			 .00	15.0 ±.3				7252 7295 7417
093113.1+300122 196.81 46.29 65.60 -30.78 092815.8+301436	 UGC 5070 IRAS09282+3014 PGC 27023	.S?....	1.11± .04 .37± .04 1.11	43 .00 .56 .19	 14.47 ±.18 13.51 13.89			15.59±.3 261± 5 1.51	4189± 9 4138± 50 4140 4446
0931.2 +7627 135.62 35.25 34.64 5.06 0925.8 +7640	 UGC 5050 ARP 207 PGC 27026	.S?....	1.06± .06 .68± .05 1.06	51 .00 1.02 .34	 15.44 ±.18 13.33 14.40				2288± 26 2420 2327
093114.8+734838 138.39 36.67 36.25 2.96 092619.0+740150	 UGC 5052 IRAS09262+7401 PGC 27027	.SBS1.. U 1.0± .8	1.19± .05 .18± .05 1.19	120 .04 .18 .09	 14.05 ±.18 12.89 13.79				3300± 60 3422 3354
0931.3 +6746 145.05 39.67 39.91 -1.83 0927.1 +6800	 UGC 5058 PGC 27029	.S..5.. U 5.0± .9	1.08± .06 .49± .05 1.09	124 .19 .73 .24	 15.29 ±.18				
0931.3 -1635 249.00 24.70 123.70 -52.50 0929.0 -1622	 MCG -3-25- 1 PGC 27031	.SB?...	1.12± .08 .22± .07 1.14	 .27 .30 .11					1924 1727 2242

R.A. 2000 DEC. l b SGL SGB R.A. 1950 DEC.	Names PGC	Type S_T n_L T L	$\log D_{25}$ $\log R_{25}$ $\log A_e$ $\log D_o$	p.a. A_g A_i A_{21}	B_T m_B m_{FIR} B_T^o	$(B-V)_T$ $(U-B)_T$ $(B-V)_T^o$ $(U-B)_T^o$	$(B-V)_e$ $(U-B)_e$ m'_e m'_{25}	m_{21} W_{20} W_{50} HI	V_{21} V_{opt} V_{GSR} V_{3K}
0931.5 +6736		.SB.1..	.99± .06	110					
145.23 39.76	UGC 5061	U	.25± .05	.17	15.17 ±.18				
40.02 -1.95		1.0± .9		.25					
0927.3 +6750	PGC 27041		1.01	.12					
093136.6+411901		.SB.3*.	.97± .05	45					
180.36 46.98	UGC 5072	U	.26± .04	.02	15.46 ±.19				
57.00 -22.45	KUG 0928+415	3.0±1.3		.35					
092826.9+413216	PGC 27047		.98	.13					
093136.6-164409	NGC 2907	.SAS1$/	1.26± .03	115	12.7 ±.2	1.05± .02	1.13± .01	15.42±.1	2090± 9
249.16 24.65	MCG -3-25- 2	R	.22± .04	.24	13.10 ±.18	.48± .04	.57± .04	504± 12	2035± 66
123.95 -52.46	IRAS09292-1630	1.0± .4	.87± .02	.22	13.89	.94	12.57± .07		1892
092915.0-163053	PGC 27048		1.29	.11	12.44	.38	13.33± .27	2.88	2406
0931.6 +0429		.S..0?.	1.03± .08						
229.20 37.34	UGC 5075	U	.00± .06	.09	14.15 ±.19				
92.13 -46.30		.0±1.7		.00					
0929.0 +0443	PGC 27049		1.04						
093141.2-160237		.S..6?/	1.23± .05	140				15.78±.1	5979± 9
248.61 25.12	MCG -3-25- 3	E	.82± .05	.28				276± 12	
122.82 -52.38	IRAS09292-1549	6.0±1.3		1.21					5783
092919.1-154921	PGC 27054		1.26	.41					6297
093145.8+034341		.S..6*.	.96± .09	170				15.36±.3	3219± 10
230.05 36.98	UGC 5078	U	.69± .06	.09					
93.15 -46.63		6.0±1.4		1.01				192± 7	3077
092909.3+035657	PGC 27059		.97	.34					3537
0931.8 +2947		.S..1?.	1.02± .06	45					
197.17 46.39	UGC 5074	U	.11± .05	.01	14.63 ±.19				12777
65.91 -30.85		1.0±1.8		.12					12729
0928.9 +3001	PGC 27064		1.02	.06	14.34				13037
0931.9 -1640		.S..7?/	1.12± .06	93					
249.17 24.75		E	.89± .06	.27					
123.85 -52.37		7.0±1.8		1.23					
0929.6 -1627	PGC 27066		1.15	.44					
093159.8-084359		.S..5?/	1.16± .06	26					
242.33 29.86	MCG -1-25- 1	E	.85± .05	.15					
111.10 -51.08		5.0±1.9		1.27					
092932.2-083042	PGC 27069		1.17	.42					
093206.5+082634	NGC 2906	.S..6*.	1.16± .05	75				15.74±.2	2140± 6
224.90 39.40	UGC 5081	U	.23± .05	.08	13.4 ±.2			361± 7	
87.31 -44.22	IRAS09294+0839	6.0±1.2		.34	12.38			344± 5	2013
092926.6+083951	PGC 27074		1.17	.12	12.92			2.71	2452
093209.7+213002	NGC 2903	.SXT4..	2.10± .01	17	9.68M±.10	.67± .01	.71± .01	11.45±.0	556± 3
208.71 44.54	UGC 5079	R (3)	.32± .02	.07	9.60 ±.13	.06± .02	.11± .01	382± 4	565± 16
73.34 -36.44		4.0± .3	1.47± .01	.47	8.62	.59	12.42± .04	371± 5	476
092919.9+214319	PGC 27077	2.3± .4	2.10	.16	9.10	.00	14.21± .12	2.19	841
0932.1 +6826		.S..9*.	1.00± .08						
144.24 39.43	UGC 5067	U	.02± .06	.17					
39.56 -1.25		9.0±1.2		.02					
0927.9 +6840	PGC 27079		1.02	.01					
0932.5 +5152		.I..9*.	1.00± .09						
165.20 45.55	UGC 5076	U	.11± .06	.00	15.2 ±.2				
49.94 -14.26		10.0±1.1		.09					
0929.1 +5206	PGC 27091		1.00	.06					
093245.5-331443		.IBS9..	1.11± .06	172					862
261.98 13.40	ESO 373- 7	S	.25± .05	.72					
150.34 -50.38		10.0± .9		.19					639
093038.1-330124	PGC 27104		1.17	.12					1151
0932.8 +5944		.SB.3..	1.23± .05	77					
154.70 43.17	UGC 5077	U	.42± .05	.11	14.9 ±.2				
44.97 -8.08	VV 464	3.0± .8		.58	13.26				
0929.0 +5958	PGC 27108		1.24	.21					
093253.4+673659	NGC 2892	.E+..P*	1.15± .15						
145.15 39.88	UGC 5073	PU	.00± .08	.17	14.05 ±.17				
40.12 -1.87		-3.5± .6		.00					
092840.8+675016	PGC 27111		1.17						
0932.9 +2730		.S?....	1.02± .06	15					
200.48 46.21	UGC 5084		.17± .05	.01	14.70 ±.18				10061
68.04 -32.27				.25					10004
0930.0 +2744	PGC 27114		1.02	.08	14.38				10329

R.A. 2000 DEC.	Names	Type	logD$_{25}$	p.a.	B$_T$	(B-V)$_T$	(B-V)$_\bullet$	m$_{21}$	V$_{21}$
l　　b		S$_T$　n$_L$	logR$_{25}$	A$_g$	m$_B$	(U-B)$_T$	(U-B)$_\bullet$	W$_{20}$	V$_{opt}$
SGL　SGB		T	logA$_\bullet$	A$_i$	m$_{FIR}$	(B-V)$_T^o$	m'$_\bullet$	W$_{50}$	V$_{GSR}$
R.A. 1950 DEC.	PGC	L	logD$_o$	A$_{21}$	B$_T^o$	(U-B)$_T^o$	m'$_{25}$	HI	V$_{3K}$
0932.9　+2129	UGC　5086	.I..9*.	.96± .09					14.64±.1	448± 10
208.79　44.71		U	.00± .06	.07				198± 16	
73.49　-36.31		10.0±1.2		.00				171± 5	368
0930.1　+2143	PGC　27115		.97	.00					734
093303.4+295541	IC　　2490	.S..4..	1.18± .05	175				15.47±.3	7348± 10
197.04　46.67	UGC　5087	U　　(1)	.19± .05	.00	14.2　±.2				7357± 19
66.01　-30.57		4.0± .8		.27				373± 7	7302
093006.3+300900	PGC 27121	4.5±1.1	1.18	.09	13.86			1.52	7610
0933.2　+2736		.S?....	.94± .11						9916
200.36　46.30	MCG　5-23- 11		.11± .07	.00	15.08 ±.19				9859
68.00　-32.16				.14					10184
0930.3　+2750	PGC 27127		.94	.06	14.85				
093315.4-164606		.SBR3*.	1.29± .04	35				13.93±.1	2120± 7
249.48　24.92	MCG -3-25- 4	UE　　(1)	.13± .05	.26				245± 10	
124.05　-52.07	IRAS09309-1632	2.7± .6		.18				238± 12	1923
093053.7-163246	PGC 27130	3.1±1.2	1.32	.07					2438
0933.2　+2308		.I?....	.89± .11						7800± 67
206.60　45.25	MCG　4-23- 10		.00± .07	.07	14.97 ±.18				7726
72.01　-35.17	VV　　553			.00	13.75				8081
0930.4　+2321	PGC 27131		.89	.00	14.90				
093320.8-330157		.S..6./	1.76± .02	89	12.8　±.2	.76± .05		12.72±.2	929± 11
261.92　13.64	ESO　373- 8	S	.82± .04	.72		.26± .10		236± 16	
149.97　-50.32	IRAS09312-3248	6.0± .5		1.21	11.67	.44		222± 12	706
093113.0-324836	PGC 27135		1.82	.41	10.87	.03	14.38± .25	1.44	1219
093333.5-254722		PSBR2..	1.00± .06	16					
256.63　18.81	ESO　498- 13	r	.35± .05	.20	15.05 ±.14				
138.71　-51.75		2.2± .9		.43					
093119.0-253400	PGC 27149		1.02	.17					
093341.9+395904		.SB.4..	1.01± .05	150					
182.30　47.43	UGC　5089	U	.18± .04	.01	15.2　±.2				
58.29　-23.19	KUG 0930+402	4.0± .9		.26					
093034.3+401225	PGC 27154		1.01	.09					
093343.5-111919		.SA.4?.	1.28± .05	142				14.86±.1	2660± 9
244.94　28.56	MCG -2-25- 2	E　　(1)	.56± .05	.13				327± 12	
115.32　-51.23	IRAS09312-1106	4.0±1.2		.83	13.22				2475
093117.7-110557	PGC 27158	4.2±1.2	1.29	.28					2983
093346.5+100909	NGC　2911	.LAS.*P	1.29± .05	140	12.50 ±.15	.96± .02	1.03± .01	15.72±.2	3183± 5
223.18　40.56	UGC　5092	R	.11± .05	.08	13.05 ±.12	.46± .03	.55± .02	519± 8	3167± 32
85.60　-42.96	ARP　　232	-2.0± .3	1.23± .03	.00		.90	14.14± .11	327± 6	3062
093105.5+102230	PGC 27159		1.28		12.71	.45	13.52± .31		3493
093351.0+441525		.S..8*.	1.22± .07						
176.01　47.14	UGC　5091	U	.55± .07	.07	14.89 ±.19				
55.26　-19.97		8.0±1.3		.68					
093037.9+442846	PGC 27166		1.22	.28					
0933.8　+1009	NGC　2912							18.25±.3	3440± 9
223.18　40.58				.08				176± 10	
85.60　-42.94								121± 7	3319
0931.1　+1022	PGC 27167								3751
093352.1+465149		.SX.5..	.87± .10						
172.23　46.79	UGC　5090	U	.00± .06	.04	15.35 ±.20				
53.46　-17.98		5.0± .9		.00					
093035.4+470510	PGC 27169		.87	.00					
093402.2+551425	A　　0930+55A		.47± .12		15.98V±.16	.10± .03		16.34±.1	756± 5
160.53　44.84			.15± .07	.02		-.65± .05		82± 4	770± 42
47.95　-11.51	KUG 0930+554							41± 4	810
093030.4+552746	PGC 27182		.47						909
0934.0　+5513	A　　0930+55B								914± 54
160.54　44.85				.02					968
47.96　-11.52									1068
0930.5　+5526	PGC 27183								
093403.0+092853	NGC　2913	.S?....	1.06± .06	140				16.25±.3	3058± 9
224.00　40.32	UGC　5095		.21± .05	.08	14.08 ±.18			232± 7	3071± 50
86.43　-43.26				.31				225± 7	2935
093122.4+094215	PGC 27184		1.06	.10	13.67			2.48	3370
093403.3+100636	NGC　2914	.SBS2..	1.02± .06	15	14.1　±.2	.95± .06		17.18±.3	3151± 17
223.27　40.61	UGC　5096	R	.19± .04	.07	14.0　±.2			238± 34	3301± 45
85.70　-42.92	ARP　　137	2.0± .5		.24		.87		225± 25	3049
093122.3+101958	PGC 27185		1.02	.10	13.69		13.55± .36	3.39	3480

R.A. 2000 DEC. l b SGL SGB R.A. 1950 DEC.	Names PGC	Type S_T n_L T L	$\log D_{25}$ $\log R_{25}$ $\log A_e$ $\log D_o$	p.a. A_g A_i A_{21}	B_T m_B m_{FIR} B_T^o	$(B-V)_T$ $(U-B)_T$ $(B-V)_T^o$ $(U-B)_T^o$	$(B-V)_e$ $(U-B)_e$ m'_o m'_{25}	m_{21} W_{20} W_{50} HI	V_{21} V_{opt} V_{GSR} V_{3K}
093407.0-151823 248.42 26.04 121.71 -51.73 093144.1-150500	 MCG -2-25- 3 PGC 27188	.SXS9*. E (1) 9.0±1.3 8.7±1.2	1.01± .07 .21± .05 1.03	100 .28 .21 .10					
093408.8+110140 222.20 41.04 84.66 -42.40 093127.1+111502	A 0931+11 CGCG 63- 11 MK 706 PGC 27189			 .04	 15.3 ±.2				2488± 33 2370 2798
0934.1 +2412 205.20 45.73 71.20 -34.31 0931.3 +2426	 UGC 5094 PGC 27190	.L..... U -2.0± .9	1.00± .10 .21± .04 .97	150 .05 .00	 15.02 ±.17				
093410.6+001433 234.09 35.60 98.21 -47.61 093136.6+002755	 UGC 5097 MK 1233 PGC 27192	.S?.... 	.76± .09 .16± .05 .76	55 .08 .22 .08	 12.04			15.67±.3 245± 5	4885± 9 4779± 30 4724 5199
093412.7-205135 252.94 22.33 130.73 -51.96 093154.1-203812	NGC 2920 ESO 565- 15 IRAS09318-2038 PGC 27197	.S..1P* SE 1.0±1.7	.94± .05 .16± .04 .95	129 .10 .16 .08	 13.93 ±.14 13.27				
093414.4-611649 281.40 -6.95 183.95 -36.36 093253.0-610324	 ESO 126- 19 IRAS09328-6103 PGC 27201	.IBS9P. S (1) 10.0± .8 10.0± .8	1.28± .04 .07± .05 1.58	 3.21 .06 .04	 14.02 ±.14 12.74				
093427.1-023012 236.89 34.09 102.09 -48.60 093155.0-021649	NGC 2917 UGC 5098 PGC 27207	.L..+.. UE -1.0± .6	1.10± .05 .53± .04 1.03	169 .10 .00 	 14.55 ±.16 14.40			14.95±.1 333± 12	3675± 9 3514 3999
093431.6-205518 253.05 22.34 130.83 -51.88 093213.0-204154	NGC 2921 ESO 565- 17 IRAS09321-2041 PGC 27214	PSXT1P. SE 1.0± .5	1.45± .03 .44± .03 1.46	83 .10 .45 .22	 12.95 ±.14 13.24				
093434.0+000523 234.31 35.59 98.48 -47.58 093200.1+001847	 UGC 5099 PGC 27216	.SBS2P* E 2.0±1.2	1.08± .05 .24± .04 1.08	90 .08 .29 .12	 14.37 ±.18 13.95			14.71±.3 264± 7 .65	4915± 10 4762 5238
093438.8+055026 228.22 38.68 90.95 -44.99 093200.9+060350	 UGC 5100 IRAS09319+0604 PGC 27219	.SBS3.. U 3.0± .8	1.08± .07 .22± .06 1.09	30 .12 .30 .11	 14.44 ±.19 13.08 13.98			15.68±.3 333± 7 1.59	5514± 10 5379 5831
093441.3+320406 194.00 47.31 64.54 -28.81 093142.4+321729	 MCG 5-23- 17 KUG 0931+322B PGC 27223	.S?.... 	.91± .06 .24± .06 .91	 .00 .36 .12	 15.06 ±.18 14.67				6669 6630 6923
093443.9-215542 253.87 21.68 132.46 -51.80 093226.0-214218	 ESO 565- 19 IRAS09324-2142 PGC 27227	.E?.... 	1.05± .05 .24± .03 1.00	171 .14 .00	13.66 ±.13 13.65 ±.14 12.06	.85± .01	 13.32± .31		
093447.9+101703 223.17 40.85 85.63 -42.67 093206.8+103027	NGC 2919 UGC 5102 IRAS09321+1030 PGC 27232	.SXR3*. R 3.0± .6	1.24± .04 .44± .04 1.25	159 .08 .60 .22	 13.65 ±.19 12.91 12.95			15.41±.2 337± 7 319± 7 2.24	2432± 8 2490± 50 2313 2744
093450.3-162306 249.44 25.46 123.48 -51.66 093228.2-160941	IC 546 MCG -3-25- 7 PGC 27234	.LBT+.. E -1.0± .9	1.03± .07 .21± .05 1.02	100 .25 .00					
093457.3+214221 208.71 45.22 73.67 -35.81 093207.6+215545	NGC 2916 UGC 5103 IRAS09321+2155 PGC 27244	.SAT3$. P 3.0±1.5	1.39± .03 .17± .04 .94± .02 1.40	20 .07 .23 .08	12.74 ±.14 12.42 ±.18 12.55 12.29	.69± .02 .07± .03 .62 .01	.80± .01 .21± .02 12.93± .04 14.12± .21	14.61±.3 398± 14 381± 7 2.23	3730± 7 3665± 40 3649 4015
093507.5+050708 229.09 38.41 91.94 -45.24 093230.1+052033	 UGC 5107 PGC 27248	.SB.7.. U 7.0± .8	1.25± .04 .52± .05 1.26	47 .08 .72 .26	 14.5 ±.2 13.73			14.75±.3 175± 7 .76	2008± 7 1871 2327
093510.9-162356 249.51 25.51 123.51 -51.58 093248.7-161030	NGC 2924 MCG -3-25- 8 PGC 27253	.E+..*. PE -4.0± .6	1.14± .07 .04± .05 1.17	150 .25 .00	* 13.16 ±.19 12.84				4585± 66 4389 4905

R.A. 2000 DEC. / l b / SGL SGB / R.A. 1950 DEC.	Names / PGC	Type S_T n_L / T / L	$\log D_{25}$ / $\log R_{25}$ / $\log A_e$ / $\log D_o$	p.a. / A_g / A_i / A_{21}	B_T / m_B / m_{FIR} / B_T^o	$(B-V)_T$ / $(U-B)_T$ / $(B-V)_T^o$ / $(U-B)_T^o$	$(B-V)_e$ / $(U-B)_e$ / m'_e / m'_{25}	m_{21} / W_{20} / W_{50} / HI	V_{21} / V_{opt} / V_{GSR} / V_{3K}
093514.3+344356	IC 2491	.L...?.	1.04± .09	75	*				
190.09 47.66	UGC 5104	U	.13± .04	.00	14.73 ±.16				
62.51 -26.83		-2.0±1.7		.00					
093212.8+345720	PGC 27254		1.02						
093518.8+302421	A 0932+30		.50± .14					17.13±.3	7416± 10
196.46 47.23			.00± .12	.00				206± 6	7363± 27
66.01 -29.89	MK 402								7364
093221.6+303746	PGC 27258		.50						7669
093520.7-084834		.S...P.	1.09± .06	25	*				1926
242.99 30.46	MCG -1-25- 3	E	.11± .05	.10					
111.54 -50.29				.16					1748
093253.0-083508	PGC 27260		1.10	.05					2251
0935.3 +3554		.I..9*.	.99± .06	140	*				1588
188.34 47.74	UGC 5105	U	.10± .05	.02					
61.61 -25.96		10.0±1.2		.08					1565
0932.3 +3608	PGC 27261		.99	.05					1828
093525.5+294934		.SB.2..	1.11± .05	143	*				
197.31 47.16	UGC 5108	U	.19± .04	.02	14.5 ±.2				8075
66.52 -30.27	IRAS09324+3002	2.0± .9		.24	12.93				8027
093228.9+300259	PGC 27266		1.11	.10	14.16				8337
0935.7 -0443		.E?....						16.54±.1	3691± 9
239.26 33.03	MCG -1-25- 4			.10				188± 12	
105.46 -49.06									3524
0933.2 -0430	PGC 27281								4016
093544.5+314217	NGC 2918	.E.....	1.15± .15	65	*				
194.57 47.49	UGC 5112	U	.14± .08	.00	13.56 ±.16				6789
65.02 -28.91		-5.0± .8		.00					6749
093246.2+315543	PGC 27282		1.11		13.46				7045
093549.9-620029		.SAS8..	1.01± .05	164	*				
282.04 -7.36	ESO 126- 20	S	.16± .05	2.93	15.30 ±.14				
184.48 -35.74		8.0± .9		.20					
093430.0-614700	PGC 27288		1.28	.08					
0935.8 +6121		.S?....	1.06± .06	87	*				
152.48 42.90	UGC 5101		.23± .04	.10	15.2 ±.2				11945± 35
44.25 -6.59	IRAS09320+6134			.31	10.47				12022
0932.0 +6134	PGC 27292		1.07	.11	14.72				12068
093556.7-074336		.L..+*/	1.32± .04	105	*				
242.11 31.25	MCG -1-25- 5	E	.49± .05	.10					
109.95 -49.88		-1.0±1.2		.00					
093328.2-073009	PGC 27298		1.26						
093605.6+245655	IC 545		.64± .17						
204.33 46.34	MCG 4-23- 13		.00± .07	.00	15.3 ±.2				6141± 42
70.88 -33.49	ARAK 205								6074
093313.3+251022	PGC 27307		.64						6418
093605.9-122614	NGC 2947	.SXR4..	1.17± .05	25	*				2815
246.33 28.29	MCG -2-25- 4	E (1)	.04± .05	.14					
117.24 -50.86	IRAS09336-1212	4.0± .8		.06	12.73				2628
093340.8-121246	PGC 27309	4.2± .8	1.18	.02					3139
093612.5-082605		.E+..?.	1.25± .09	150	*				
242.81 30.86	MCG -1-25- 6	E	.28± .07	.11					
111.05 -49.99		-4.0± .8		.00					
093344.5-081237	PGC 27316		1.19						
093613.7-105829		.SXT6*.	1.17± .05	95	*				
245.08 29.25	MCG -2-25- 5	E (1)	.37± .05	.13					
114.96 -50.56		6.0± .9		.54					
093347.5-104501	PGC 27317	6.4±1.2	1.18	.18					
093626.2-003417		.IXS9*.	.96± .06	75	*				
235.31 35.61	MCG 0-25- 4	E (1)	.11± .05	.08	15.0 ±.2				
99.67 -47.41		10.0±1.3		.08					
093352.8-002049	PGC 27331	8.7±1.2	.97	.05					
093627.9-372029		.LX.-*.	1.08± .07	172	*		1.04± .02		
265.43 10.95	ESO 373- 10	S	.31± .04	.94					
155.91 -48.37		-3.0± .7	1.24±.11	.00			15.14± .39		
093424.0-370700	PGC 27332		1.16						
093632.8-635643		.SXS9?.	1.10± .05	85	*				
283.42 -8.73	ESO 91- 15	S (1)	.25± .05	2.22	14.76 ±.14				
186.19 -34.42		9.0± .9		.26					
093519.0-634312	PGC 27341	8.9±1.3	1.31	.13					

R.A. 2000 DEC.	Names	Type	logD$_{25}$	p.a.	B$_T$	(B-V)$_T$	(B-V)$_\bullet$	m$_{21}$	V$_{21}$
l b		S$_T$ n$_L$	logR$_{25}$	A$_g$	m$_B$	(U-B)$_T$	(U-B)$_\bullet$	W$_{20}$	V$_{opt}$
SGL SGB		T	logA$_\bullet$	A$_i$	m$_{FIR}$	(B-V)^o_T	m'$_\bullet$	W$_{50}$	V$_{GSR}$
R.A. 1950 DEC.	PGC	L	logD$_o$	A$_{21}$	B^o_T	(U-B)^o_T	m'$_{25}$	HI	V$_{3K}$

093644.7-210742	NGC 2935	PSXS3..	1.56± .02	0	12.35 ±.15	.80± .01	.86± .01	13.16±.1	2277± 5
253.59 22.57	ESO 565- 23	R (3)	.11± .03	.12	11.88 ±.12	.34± .04	.38± .01	312± 6	2214± 66
131.15 -51.36	IRAS09344-2054	3.0± .3	.86± .03	.16	11.41	.74	12.14± .09	283± 6	2072
093426.0-205412	PGC 27351	2.1± .5	1.57	.06	11.76	.29	14.71± .19	1.35	2592

0936.7 +3104									8022
195.54 47.63	CGCG 152- 38			.01	15.3 ±.3				7979
65.71 -29.20									8281
0933.8 +3118	PGC 27352								

0936.8 +6647		.S..4..	1.20± .05	120					
145.86 40.60	UGC 5111	U	.81± .05	.13	15.42 ±.18				
40.93 -2.28	IRAS09327+6700	4.0± .9		1.20	13.21				
0932.7 +6700	PGC 27358		1.21	.41					

093652.7+374141	NGC 2922	.I..9?.	1.06± .04	103				15.26±.3	4369± 10
185.69 48.09	UGC 5118	U	.38± .04	.00	14.59 ±.18			288± 13	4356± 19
60.49 -24.45	IRAS09337+3755	10.0±1.8		.28	12.81				4350
093348.4+375510	PGC 27361		1.06	.19	14.31			.76	4600

0936.8 +6633		.S..2..	1.06± .06	43					
146.13 40.72	UGC 5113	U	.68± .05	.13	15.51 ±.18				
41.08 -2.46		2.0±1.0		.84					
0932.8 +6647	PGC 27362		1.07	.34					

093658.2+311952		.S?....	.90± .06						8043
195.54 47.70	MCG 5-23- 21		.16± .06	.00	15.24 ±.18				8001
65.54 -28.99	KUG 0934+315A			.23					8301
093400.3+313321	PGC 27367		.90	.08	14.97				

093700.6-605550		.SBS9..	1.13± .05	135					
281.41 -6.47	ESO 126- 22	S (1)	.35± .05	3.66	15.39 ±.14				
183.35 -36.32		9.0± .9		.35					
093537.0-604218	PGC 27369	7.8± .9	1.47	.17					

0937.0 +1353		.S..7..	1.14± .07	131					8583
219.13 42.91	UGC 5121	U	.98± .06	.05	15.64 ±.18				8475
82.00 -40.21		7.0±1.0		1.35					8889
0934.3 +1407	PGC 27371		1.14	.49	14.21				

093709.2+195010		.S?....	.81± .08	155				16.89±.3	8483± 10
211.46 45.13	UGC 5123		.16± .05	.09	14.8 ±.2				8461± 50
75.89 -36.61	IRAS09343+2003			.19	13.23			419± 7	8396
093421.2+200340	PGC 27379		.82	.08	14.48			2.33	8775

093712.7+380532		.L?....	.82± .11						5982± 38
185.10 48.15	UGC 5119		.00± .03	.00	14.57 ±.12				5968
60.25 -24.12	ARAK 206			.00					6215
093407.9+381902	PGC 27383		.82		14.48				

093715.3+233533	NGC 2927	.SXT3..	1.28± .04	155				15.44±.3	7547± 10
206.32 46.26	UGC 5122	U	.24± .05	.05	13.70 ±.19				7556± 50
72.33 -34.18		3.0± .8		.34				516± 7	7476
093424.3+234903	PGC 27385		1.28	.12	13.25			2.07	7830

0937.2 +7322		.IX.9..	1.13± .07	135				15.57±.1	2135± 11
138.56 37.27	UGC 5110	U	.03± .06	.04	15.1 ±.3				2256
36.85 2.88		10.0± .8		.02				122± 8	2193
0932.5 +7336	PGC 27386		1.13	.01	15.02			.53	

093729.5+230940	NGC 2929	.S?....	1.09± .06	144				15.14±.3	7509± 9
206.95 46.19	UGC 5126		.56± .05	.05	14.7 ±.2				7549± 50
72.78 -34.42	IRAS09346+2323			.84	12.70			451± 5	7437
093438.9+232311	PGC 27398		1.10	.28	13.78			1.08	7794

093731.3+325030	NGC 2926	.S?....	.92± .07	120				16.40±.3	4362± 10
192.95 47.98	UGC 5125		.03± .05	.03	14.40 ±.18				4417± 50
64.40 -27.84	IRAS09345+3304			.04	13.34			243± 7	4329
093432.1+330401	PGC 27400		.92	.01	14.31			2.08	4617

0937.5 +2311	NGC 2930	.S?....	.82± .13						6599± 60
206.91 46.21	MCG 4-23- 18		.17± .07	.05	15.1 ±.2				6526
72.76 -34.40				.26					6883
0934.7 +2325	PGC 27404		.82	.09	14.72				

0937.6 +4837	A 0934+48								10145± 72
169.52 47.08				.04					10173
52.76 -16.24	1ZW 19								10332
0934.3 +4851	PGC 27408								

093738.6-335508		.SAR1..	1.11± .16						
263.22 13.62	ESO 373- 11	r	.02± .14	.79					
150.90 -49.21		1.0± .8		.02					
093531.1-334136	PGC 27413		1.18	.01					

R.A. 2000 DEC. / I b / SGL SGB / R.A. 1950 DEC.	Names / PGC	Type / S_T n_L / T / L	$\log D_{25}$ / $\log R_{25}$ / $\log A_{\bullet}$ / $\log D_0$	p.a. / A_g / A_I / A_{21}	B_T / m_B / m_{FIR} / B_T^0	$(B-V)_T$ / $(U-B)_T$ / $(B-V)_T^0$ / $(U-B)_T^0$	$(B-V)_{\bullet}$ / $(U-B)_{\bullet}$ / $m'_{\bullet}$ / m'_{25}	m_{21} / W_{20} / W_{50} / HI	V_{21} / V_{opt} / V_{GSR} / V_{3K}
0937.6 +3704 / 186.62 48.24 / 61.09 -24.80 / 0934.6 +3718	UGC 5127 / PGC 27416	.SX.7*. / U / 7.0± .9	1.14± .05 / .48± .05 / / 1.14	147 / .00 / .66 / .24					
093741.0-220214 / 254.46 22.10 / 132.59 -51.12 / 093523.0-214842	NGC 2945 / ESO 565- 28 / IRAS09354-2148 / PGC 27418	.LA.-*. / S / -3.0± .5	1.21± .04 / .13± .03 / / 1.21	168 / .13 / .00	13.23 ±.14 / / / 13.03				4630± 15 / 4423 / / 4945
093743.3+024508 / 232.08 37.72 / 95.41 -45.74 / 093507.5+025840	NGC 2936 / UGC 5130 / / PGC 27422	.I?.... / / /	1.12± .08 / .08± .07 / .77± .03 / 1.13	/ .09 / .06 / .04	13.9 ±.2 / / / 13.75	.84± .04 / .02± .07 / .76 / -.05	.80± .02 / .11± .03 / 13.29± .07 / 14.15± .49		6989± 38 / 6845 / 7312
093745.0+024451 / 232.09 37.72 / 95.42 -45.74 / 093509.3+025823	NGC 2937 / UGC 5131 / / PGC 27423	.E..... / U / -5.0± .9	1.32± .09 / .46± .07 / .52± .10 / 1.19	/ .09 / .00	14.6 ±.2 / / / 14.41	.94± .04 / .51± .08 / .85 / .53	1.02± .02 / .50± .04 / 12.65± .35 / 15.03± .53		6839± 39 / 6694 / 7161
093746.8-105827 / 245.36 29.54 / 115.08 -50.18 / 093520.5-104455	/ MCG -2-25- 8 / / PGC 27425	.S..4*. / E (1) / 4.0±1.3 / 4.2±1.2	1.16± .06 / .55± .05 / / 1.17	175 / .13 / .81 / .28					5369 / / 5186 / 5695
093749.7-222420 / 254.78 21.87 / 133.17 -51.07 / 093532.0-221048	/ ESO 565- 29 / / PGC 27430	.SBS7*. / SU (1) / 7.3± .4 / 7.0± .5	1.30± .03 / .16± .04 / / 1.32	54 / .17 / .22 / .08	14.28 ±.14 / / / 13.89			14.60±.2 / 186± 16 / 187± 12 / .63	2412± 11 / / 2205 / 2726
093751.3-620904 / 282.31 -7.30 / 184.43 -35.47 / 093631.0-615530	/ ESO 126- 23 / IRAS09365-6155 / PGC 27431	.SAS5*/ / S (1) / 4.7±1.0 / 4.8± .7	1.20± .05 / .42± .05 / / 1.48	11 / 2.98 / .64 / .21	14.38 ±.14 / / 12.07				
093758.1+252943 / 203.70 46.88 / 70.73 -32.81 / 093505.5+254315	/ UGC 5129 / / PGC 27437	.S..1.. / U / 1.0± .8	1.21± .05 / .35± .05 / / 1.21	103 / .00 / .36 / .18	14.11 ±.18 / / / 13.70			15.35±.3 / / 320± 5 / 1.48	4063± 9 / 4035± 50 / 3998 / 4340
093801.4-202039 / 253.20 23.33 / 129.89 -51.07 / 093542.0-200706	/ ESO 565- 30 / / PGC 27441	.LAT-*. / S / -3.0± .6	1.22± .05 / .22± .04 / / 1.20	13 / .13 / .00	13.49 ±.14				
0938.0 -1139 / 246.01 29.15 / 116.16 -50.26 / 0935.6 -1126	/ MCG -2-25- 9 / / PGC 27442	.S?.... / /	1.04± .09 / .09± .07 / / 1.05	/ .16 / .12 / .05				15.91±.1 / 168± 12	1853± 9 / / 1668 / 2179
0938.0 +0936 / 224.46 41.25 / 87.01 -42.35 / 0935.4 +0950	NGC 2940 / MCG 2-25- 12 / / PGC 27448	.L?.... / /	.95± .10 / .10± .04 / / .95	/ .07 / .00	14.51 ±.11 / / / 14.31				8622± 56 / 8500 / 8936
093807.7-335540 / 263.30 13.68 / 150.87 -49.11 / 093600.0-334206	/ ESO 373- 12 / / PGC 27450	.SXS8.. / S (1) / 8.0± .8 / 7.8± .8	1.20± .07 / .16± .07 / / 1.27	151 / .79 / .19 / .08					2640 / / 2417 / 2931
093808.1+093121 / 224.57 41.23 / 87.12 -42.38 / 093527.8+094453	NGC 2939 / UGC 5134 / IRAS09354+0945 / PGC 27451	.S..4.. / U / 4.0± .8	1.39± .04 / .45± .05 / / 1.40	154 / .07 / .67 / .23	13.25 ±.18 / / 12.22 / 12.49			14.17±.3 / 364± 7 / 346± 7 / 1.45	3338± 9 / 3355± 49 / 3216 / 3653
0938.2 +2803 / 200.03 47.47 / 68.51 -31.05 / 0935.3 +2817	IC 2495 / MCG 5-23- 23 / / PGC 27455	.S?.... / /	.89± .11 / .07± .07 / / .89	/ .00 / .10 / .03	14.80 ±.18 / / / 14.64				10110 / 10056 / 10379
093813.9-763530 / 292.45 -17.83 / 196.58 -25.62 / 093821.2-762153	/ / / PGC 27456	.I..9?. / S (1) / 10.0±1.1 / 11.1± .6	1.59± .04 / .62± .08 / / 1.65	/ .61 / .47 / .31				
093814.2-632905 / 283.25 -8.26 / 185.63 -34.58 / 093658.1-631530	/ ESO 91- 16 / IRAS09369-6315 / PGC 27458	.S..3?/ / S / 3.0±1.8	1.19± .05 / .56± .04 / / 1.42	122 / 2.44 / .78 / .28	14.15 ±.14 / / 11.12				
093819.4-390028 / 266.85 9.98 / 158.02 -47.43 / 093617.0-384654	/ ESO 315- 17 / / PGC 27466	.SAR6.. / S (1) / 6.0± .8 / 4.4± .8	1.22± .05 / .05± .05 / / 1.31	/ 1.02 / .07 / .02	14.32 ±.14 / / / 13.22			14.95±.3 / / 99± 7 / 1.70	2470± 9 / / 2243 / 2746

R.A. 2000 DEC. l b SGL SGB R.A. 1950 DEC.	Names PGC	Type S_T n_L T L	$logD_{25}$ $logR_{25}$ $logA_o$ $logD_o$	p.a. A_g A_i A_{21}	B_T m_B m_{FIR} B_T^o	$(B-V)_T$ $(U-B)_T$ $(B-V)_T^o$ $(U-B)_T^o$	$(B-V)_o$ $(U-B)_o$ m' m'_{25}	m_{21} W_{20} W_{50} HI	V_{21} V_{opt} V_{GSR} V_{3K}
093820.7-335140 263.29 13.76 150.75 -49.09 093613.0-333806	 ESO 373- 13 PGC 27468	.S..0?/ S .0±1.9	1.15± .06 .91± .05 1.18	24 .79 .68 	 				
093823.3+433033 176.98 48.04 56.44 -19.98 093512.3+434406	 UGC 5133 PGC 27472	.L..... U -2.0± .9 	1.07± .07 .56± .03 .99	149 .02 .00 	14.73 ±.17 				
093824.0+761910 135.48 35.67 35.06 5.21 093307.4+763240	NGC 2938 UGC 5115 PGC 27473	.SBT6.. U 6.0± .8 	1.24± .06 .22± .06 1.24	105 .01 .33 .11	 14.2 ±.2 13.81			14.10±.1 204± 8 .17	2285± 11 2416 2326
093829.6-614947 282.15 -7.01 184.07 -35.62 093708.1-613612	 ESO 126- 24 IRAS09371-6136 PGC 27476	.SXS4.. S (1) 4.0± .9 3.3± .9	1.17± .05 .25± .05 1.47	150 3.20 .36 .12	14.02 ±.14 12.19 				
093830.6+023420 232.41 37.78 95.77 -45.65 093555.0+024754	 UGC 5138 PGC 27477	.S..3.. U 3.0±1.0 	1.04± .08 .76± .06 1.05	170 .07 1.05 .38	15.47 ±.19 				
093831.8-635600 283.58 -8.57 186.01 -34.27 093717.0-634224	 ESO 91- 17 PGC 27481	.SXT4.. S (1) 4.0± .8 4.4± .9	1.14± .05 .14± .05 1.35	145 2.30 .21 .07	14.34 ±.14 				
093833.1+170200 215.32 44.47 78.96 -38.07 093547.3+171533	NGC 2943 UGC 5136 PGC 27482	.E..... U -5.0± .8 	1.34± .12 .25± .08 1.27	130 .03 .00 	13.42 ±.17 13.26				8377± 27 8281 8677
093848.8+074600 226.72 40.53 89.31 -43.14 093609.6+075934	 UGC 5142 PGC 27504	.SA.7*. U 7.0±1.2 	.96± .17 .00± .12 .97	 .06 .00 .00	 			16.41±.3 141± 7 	6767± 10 6638 7084
093853.3-045134 239.96 33.58 106.04 -48.35 093622.7-043759	A 0936-04C MCG -1-25- 8 IRAS09364-0437 PGC 27508	.SBT3P/ RCE 3.1± .6 	1.04± .07 .55± .05 1.06	61 .13 .75 .27	15.68S±.15 14.75		14.38± .39		6853± 26 6686 7180
093853.6-045056 239.95 33.59 106.03 -48.34 093623.1-043721	A 0936-04A MCG -1-25- 9 HICK 40A PGC 27509	.E..... RC -5.0± .5 	1.13± .11 .11± .07 1.12	175 .13 .00 	13.77S±.15 13.55		14.15± .58		6622± 21 6455 6950
093855.1-045158 239.97 33.59 106.06 -48.34 093624.6-043823	A 0936-04B MCG -1-25- 10 HICK 40B PGC 27513	.LAR-*P R -3.0± .5 	1.04± .09 .20± .05 1.03	125 .13 .00 	15.02S±.15 14.79		14.60± .50		6821± 26 6654 7149
093855.5-045128 239.97 33.59 106.05 -48.34 093625.0-043753	A 0936-04E MCG -1-25- 11 HICK 40E PGC 27515	.SXS1P* RCE .8± .6 	.82± .09 .39± .05 .83	125 .13 .40 .19	 				6625± 41 6458 6952
093855.8-045014 239.95 33.61 106.02 -48.33 093625.3-043639	A 0936-04D MCG -1-25- 12 HICK 40D PGC 27516	.SBS0P* RCE .3± .5 	.93± .08 .30± .05 .93	80 .13 .23 	15.06S±.15 14.61		13.80± .46		6466± 26 6299 6794
093859.2+065725 227.67 40.17 90.32 -43.51 093620.7+071100	NGC 2948 UGC 5141 IRAS09363+0710 PGC 27518	.SB.4.. U 4.0± .8 	1.14± .05 .20± .05 1.15	7 .14 .29 .10	13.71 ±.18 12.67 13.25			15.48±.3 390± 7 380± 7 2.14	4983± 9 4959± 50 4851 5301
093901.0+170138 215.39 44.57 79.06 -37.98 093615.2+171513	NGC 2946 UGC 5143 IRAS09362+1715 PGC 27521	.SB?... 	1.08± .06 .52± .05 1.08	13 .06 .78 .26	 14.90 ±.18 12.46 14.00			16.54±.3 391± 7 2.27	8953± 10 8857 9253
093907.8+340016 191.25 48.41 63.73 -26.78 093607.8+341351	NGC 2942 UGC 5140 IRAS09361+3413 PGC 27527	.SAS5*. R (3) 5.0± .4 1.7± .5	1.35± .03 .10± .04 1.15± .02 1.35	165 .00 .14 .05	13.15 ±.14 13.15 ±.17 13.43 12.98	.57± .03 .52	.65± .02 .08± .03 14.39± .06 14.53± .21	14.58±.1 270± 6 232± 5 1.55	4423± 5 4585± 44 4395 4675
093912.9-210354 253.98 23.03 131.03 -50.78 093654.0-205018	 ESO 565- 33 PGC 27529	.SBS9.. SE (2) 9.0± .5 9.2± .5	1.23± .03 .45± .04 1.24	104 .12 .46 .23	15.61 ±.14 				

R.A. 2000 DEC. / l b / SGL SGB / R.A. 1950 DEC.	Names / PGC	Type / S_T n_L / T / L	$\log D_{25}$ / $\log R_{25}$ / $\log A_e$ / $\log D_o$	p.a. / A_g / A_i / A_{21}	B_T / m_B / m_{FIR} / B_T^o	$(B-V)_T$ / $(U-B)_T$ / $(B-V)_T^o$ / $(U-B)_T^o$	$(B-V)_e$ / $(U-B)_e$ / m'_e / m'_{25}	m_{21} / W_{20} / W_{50} / HI	V_{21} / V_{opt} / V_{GSR} / V_{3K}
093917.4+363343 / 187.40 48.56 / 61.75 -24.94 / 093614.7+364718	MCG 6-21- 68 / PGC 27530	.S?.... / / /	.74± .14 / .09± .07 / / .73	.00 / .07 /	15.35 ±.19 / / / 15.19				6021± 46 / 6001 / / 6261
093917.4-190606 / 252.44 24.40 / 127.92 -50.75 / 093657.0-185230	NGC 2956 / ESO 565- 34 / IRAS09369-1852 / PGC 27531	.SBS3?. / SE (2) / 3.0± .8 / 3.3± .7	.96± .05 / .45± .04 / / .97	55 / .13 / .62 / .22	15.29 ±.14 / / 12.56 /				
093918.0+321842 / 193.80 48.31 / 65.13 -27.95 / 093619.6+323217	NGC 2944 / UGC 5144 / IRAS09363+3232 / PGC 27533	.SBS5P$ / R / 5.0± .5 /	1.04± .05 / .43± .05 / / 1.04	.00 / .65 / .22	/ / 12.79 / 14.05			15.29±.2 / 372± 14 / / 1.03	6798± 8 / 6601± 29 / 6747 / 7040
0939.3 +0624 / 228.35 39.97 / 91.05 -43.70 / 0936.7 +0638	MCG 1-25- 8 / PGC 27535	.S?.... / / /	.94± .11 / .19± .07 / / .95	.10 / .28 / .09	14.81 ±.18 / / / 14.38				6583± 60 / 6450 / / 6903
093922.3+363428 / 187.38 48.57 / 61.75 -24.92 / 093619.6+364803	MCG 6-21- 69 / PGC 27539	.S?.... / / /	.74± .14 / .21± .07 / / .73	.00 / .16 /	15.50 ±.20 / / / 15.25				6007± 46 / 5986 / / 6246
093926.6+322200 / 193.72 48.34 / 65.11 -27.89 / 093628.2+323535	A 0936+32A / MCG 6-21- 72 / / PGC 27546	.SB?... / / /	.89± .11 / .24± .07 / / .89	.00 / .36 / .12				15.68±.3 / 282± 5	6862± 9 / 6836± 26 / 6822 / 7115
093926.6+322200 / 193.72 48.34 / 65.11 -27.89 / 093628.2+323535	A 0936+32B / MCG 6-21- 71 / / PGC 27547	.LB?... / / /	.72± .22 / .00± .07 / / .72	.00 / .00 /	16.1 ±.3 / / / 16.00	.93± .08 / .57± .13 / .87 / .61	/ / / 14.55±1.15		6394± 51 / 6357 / / 6649
093927.2+382547 / 184.58 48.59 / 60.34 -23.57 / 093622.5+383922	UGC 5147 / KUG 0936+386 / / PGC 27549	.S..6*. / U / 6.0±1.5 /	1.05± .05 / .98± .05 / / 1.05	147 / .00 / 1.44 / .49					
0939.5 +1130 / 222.41 42.44 / 85.09 -41.03 / 0936.8 +1144	UGC 5148 / IRAS09368+1144 / / PGC 27558	.S..6*. / U / 6.0±1.4 /	1.06± .06 / .78± .05 / / 1.07	54 / .07 / 1.15 / .39	15.2 ±.2 / / 13.58 /				
0939.7 -0349 / 239.12 34.37 / 104.67 -47.82 / 0937.2 -0336	/ PGC 27566	.IBS9P. / E (1) / 10.0± .9 / 7.5±1.2	1.12± .08 / .19± .08 / / 1.14	160 / .12 / .14 / .09					
0939.8 +2100 / 210.15 46.09 / 75.24 -35.39 / 0937.0 +2114	UGC 5149 / / / PGC 27570	.S..6*. / U / 6.0±1.5 /	.96± .09 / .98± .06 / / .97	141 / .07 / 1.44 / .49					4734 / / 4653 / 5025
094006.8-250332 / 257.19 20.37 / 137.30 -50.36 / 093751.0-244954	ESO 498- 20 / / / PGC 27584	.SXR3P. / r / 3.3± .8 /	.99± .05 / .00± .04 / / 1.00	/ .16 / .01 / .00	14.96 ±.14 / / /				
0940.3 +8207 / 129.82 32.25 / 31.41 9.74 / 0933.0 +8221	UGC 5114 / / / PGC 27597	.IB.9*. / U / 10.0± .8 /	1.24± .11 / .42± .12 / / 1.24	140 / .03 / .32 / .21				16.17±.1 / / 89± 8	1609± 11 / 1759 / / 1617
094024.7+145520 / 218.29 44.07 / 81.51 -38.95 / 093740.6+150858	NGC 2954 / UGC 5155 / / PGC 27600	.E..... / U / -5.0± .8 /	1.23± .06 / .18± .06 / .73± .04 / 1.19	160 / .09 / .00	13.30 ±.15 / 13.37 ±.15 / / 13.19	.91± .01 / .42± .02 / .85 / .42	.97± .01 / .49± .02 / 12.44± .13 / 13.98± .35		3821± 29 / 3718 / / 4127
094027.2+482015 / 169.79 47.60 / 53.33 -16.14 / 093709.9+483353	UGC 5151 / MK 1418 / / PGC 27602	.I?.... / / /	.81± .06 / .07± .04 / / .81	/ .03 / .05 / .03	13.9 ±.3 / 13.72 / / 13.81			14.86±.3 / 130± 16 / 103± 12 / 1.01	773± 11 / 802± 28 / 804 / 966
0940.4 +7111 / 140.73 38.66 / 38.42 1.33 / 0936.0 +7124	Holmberg I / UGC 5139 / DDO 63 / PGC 27605	.IXS9.. / PU (1) / 10.0± .5 / 9.0± .9	1.56± .03 / .09± .05 / 1.56± .05 / 1.57	/ .06 / .07 / .05	13.0 ±.2 / 14.5 ±.8 / / 13.01		.37± .04 / -.35± .06 / 16.33± .10 / 15.46± .28	13.28±.0 / 44± 3 / 26± 3 / .23	136± 3 / / 250 / 207
094029.1-325033 / 262.90 14.81 / 149.08 -48.92 / 093820.1-323654	ESO 373- 19 / / / PGC 27606	.S.1*P/ / S / 1.0±1.2 /	1.15± .06 / .54± .05 / / 1.21	114 / .73 / .55 / .27					

R.A. 2000 DEC. / l b / SGL SGB / R.A. 1950 DEC.	Names / PGC	Type / S_T n_L / T / L	$\log D_{25}$ / $\log R_{25}$ / $\log A_e$ / $\log D_o$	p.a. / A_g / A_i / A_{21}	B_T / m_B / m_{FIR} / B_T^o	$(B-V)_T$ / $(U-B)_T$ / $(B-V)_T^o$ / $(U-B)_T^o$	$(B-V)_e$ / $(U-B)_e$ / m' / m'_{25}	m_{21} / W_{20} / W_{50} / HI	V_{21} / V_{opt} / V_{GSR} / V_{3K}
0940.5 -0353 / 239.33 34.49 / 104.86 -47.65 / 0938.0 -0340	/ / / PGC 27612	.SBS9.. / E (1) / 9.0± .9 / 9.8± .8	1.12± .08 / .13± .08 / / 1.14	50 / .12 / .13 / .06					
094033.6+252922 / 203.91 47.45 / 71.20 -32.37 / 093741.4+254300	UGC 5156 / / / PGC 27615	.SB?... / / /	1.04± .08 / .53± .06 / / 1.04	130 / .00 / .79 / .26	15.30 ±.18 / / / 14.45			16.45±.3 / 432± 7 / / 1.74	9953± 10 / 9889 / / 10232
0940.5 -0856 / 244.06 31.37 / 112.23 -49.07 / 0938.1 -0843	A 0938-08 / MCG -1-25- 15 / / PGC 27616	.SBS6.. / E (1) / 6.0± .8 / 4.2± .8	1.32± .04 / .10± .05 / / 1.33	55 / .13 / .14 / .05				14.25±.1 / 211± 12	2069± 9 / / 1891 / 2397
094036.6+033438 / 231.69 38.77 / 94.80 -44.73 / 093800.4+034817	NGC 2960 / UGC 5159 / MK 1419 / PGC 27619	.S..1?. / U / 1.0±1.6 /	1.25± .04 / .16± .05 / / 1.26	40 / .07 / .16 / .08	13.29 ±.18 / 13.32 / / 13.00			16.70±.3 / 435± 13 / / 3.62	4932± 10 / 4722± 49 / 4782 / 5248
094041.8+115317 / 222.12 42.86 / 84.87 -40.59 / 093759.8+120656	NGC 2958 / UGC 5160 / IRAS09379+1206 / PGC 27620	.S.R4.. / U / 4.0± .9 /	1.02± .06 / .12± .05 / / 1.02	10 / .05 / .18 / .06	13.97 ±.19 / 12.69 / / 13.70			15.77±.3 / 410± 7 / 380± 7 / 2.01	6663± 9 / 6639± 50 / 6548 / 6974
0940.7 +1319 / 220.35 43.48 / 83.29 -39.79 / 0938.0 +1333	CGCG 63- 34 / / / PGC 27621			.09	14.9 ±.6				3739± 79 / 3630 / / 4048
094043.7-321346 / 262.51 15.29 / 148.16 -49.02 / 093834.0-320006	ESO 434- 19 / / / PGC 27623	.IX.9.. / S (1) / 10.0± .8 / 7.8± .9	1.13± .05 / .34± .05 / / 1.18	126 / .57 / .26 / .17	15.18 ±.14 / / / 14.35				1035 / / 814 / 1331
094045.3-052608 / 240.85 33.60 / 107.09 -48.08 / 093815.1-051229	IC 553 / MCG -1-25- 16 / / PGC 27625	PSBT1P* / E / 1.0±1.2 /	1.06± .06 / .12± .05 / / 1.07	60 / .11 / .13 / .06					
0940.8 +1133 / 222.55 42.74 / 85.27 -40.74 / 0938.1 +1146	UGC 5164 / 8ZW 53 / / PGC 27630	.S..8*. / U / 8.0±1.4 /	1.06± .06 / .80± .05 / / 1.06	22 / .06 / .98 / .40	13.62				6711 / 6596 / / 7024
094054.0+051000 / 230.00 39.67 / 92.84 -43.95 / 093816.7+052340	NGC 2962 / UGC 5167 / / PGC 27635	RLXT+.. / R / -1.0± .3 / 1.41	1.42± .02 / .13± .03 / .89± .02 / 1.41	3 / .10 / .00 /	12.96 ±.13 / 12.56 ±.12 / / 12.62	1.03± .01 / .54± .04 / .97 / .51	1.04± .01 / .56± .02 / 12.90± .06 / 14.60± .19	15.90±.2 / 407± 21 / 416± 4 /	1966± 6 / 1991± 49 / 1830 / 2289
0940.9 +2746 / 200.61 48.00 / 69.23 -30.80 / 0938.0 +2800	UGC 5162 / / / PGC 27636	.S?....	1.00± .16 / .34± .06 / / 1.00	160 / .00 / .50 / .17					9904 / 9849 / / 10176
0940.9 +2746 / 200.61 48.00 / 69.23 -30.80 / 0938.0 +2800	UGC 5161 / / / PGC 27637	.S?....	1.02± .08 / .29± .06 / / 1.02	150 / .00 / .43 / .14					10079 / 10024 / / 10351
094058.3+473711 / 170.81 47.83 / 53.89 -16.62 / 093742.2+475050	A 0937+47 / UGC 5157 / / PGC 27641	.S..... / R / / 1.22	1.22± .05 / .65± .05 / / 1.22	18 / .01 / .98 / .33	14.50 ±.18 / / / 13.48			15.22±.1 / 305± 8 / / 1.41	4836± 7 / 4826± 50 / 4860 / 5029
0940.9 +2111 / 210.01 46.40 / 75.27 -35.06 / 0938.1 +2125	UGC 5165 / IRAS09381+2125 / / PGC 27643	.S?....	1.06± .06 / .56± .05 / / 1.06	107 / .05 / .84 / .28	15.44 ±.18 / 12.82				
094100.1+065610 / 228.02 40.59 / 90.69 -43.08 / 093821.6+070950	IC 551 / UGC 5168 / / PGC 27645	.S?....	.88± .08 / .03± .05 / / .89	/ .10 / .02 /	14.51 ±.18 / / / 14.26			16.56±.3 / / 337± 7 /	8574± 10 / 8581± 50 / 8443 / 8894
094106.5-290617 / 260.34 17.62 / 143.47 -49.59 / 093854.0-285236	ESO 434- 20 / / / PGC 27652	PSBS1.. / r / 1.0±1.0 /	1.07± .06 / .83± .05 / / 1.12	3 / .47 / .85 / .42	16.23 ±.14				
0941.2 +6103 / 152.48 43.61 / 44.95 -6.41 / 0937.5 +6117	UGC 5153 / / / PGC 27662	.S..4.. / U / 4.0± .9 /	.97± .07 / .02± .05 / / .98	/ .06 / .03 / .01	15.3 ±.2				

R.A. 2000 DEC. / l b / SGL SGB / R.A. 1950 DEC.	Names / PGC	Type: S_T n_L / T / L	logD$_{25}$ / logR$_{25}$ / logA$_e$ / logD$_o$	p.a. / A$_g$ / A$_i$ / A$_{21}$	B$_T$ / m$_B$ / m$_{FIR}$ / B$_T^o$	(B-V)$_T$ / (U-B)$_T$ / (B-V)$_T^o$ / (U-B)$_T^o$	(B-V)$_e$ / (U-B)$_e$ / m'$_e$ / m'$_{25}$	m$_{21}$ / W$_{20}$ / W$_{50}$ / HI	V$_{21}$ / V$_{opt}$ / V$_{GSR}$ / V$_{3K}$
0941.2 +3209 / 194.10 48.70 / 65.59 -27.76 / 0938.3 +3223	CGCG 152- 48 / PGC 27663			.00	15.4 ±.3				6580 / 6542 / / 6837
094116.8+103843 / 223.72 42.43 / 86.38 -41.14 / 093835.7+105223	IC 552 / UGC 5171 / / PGC 27665	.L..... / U / -2.0± .9	1.01± .08 / .25± .03 / / .97	175 / .04 / .00	14.44 ±.16 / / / 14.31				5809± 31 / 5691 / / 6124
094117.1+355300 / 188.44 48.94 / 62.60 -25.14 / 093815.4+360640	NGC 2955 / UGC 5166 / IRAS09382+3606 / PGC 27666	PSAR3.. / R (1) / 3.0± .4 / 2.5±1.2	1.24± .03 / .29± .03 / / 1.24	162 / .00 / .40 / .14	13.61 ±.13 / 13.61 ±.13 / 12.41 / 13.16	.69± .01 / .15± .05 / .59 / .07	13.93± .22	14.90±.3 / 488± 13 / / 1.60	7013± 10 / 7026± 66 / 6990 / 7257
0941.3 +2806 / 200.14 48.15 / 69.01 -30.51 / 0938.4 +2820	IC 2498 / CGCG 152- 49 / / PGC 27668			.01	15.2 ±.3				9873 / 9819 / / 10144
094120.7-285741 / 260.27 17.76 / 143.24 -49.56 / 093908.0-284400	ESO 434- 21 / / PGC 27670	PSXS0*. / r / -.1±1.3	.96± .06 / .25± .03 / / .99	155 / .47 / .19	15.06 ±.14				
094133.3+112452 / 222.83 42.84 / 85.55 -40.67 / 093851.7+113833	UGC 5173 / IRAS09388+1138 / PGC 27681	.S..3.. / U (1) / 3.0± .9 / 4.5±1.2	1.36± .04 / .92± .05 / / 1.36	131 / .04 / 1.27 / .46	14.72 ±.18 / / 13.05 / 13.37			14.77±.2 / 503± 7 / 491± 5 / .95	6238± 6 / 6122 / / 6551
094143.4-281130 / 259.78 18.38 / 142.05 -49.61 / 093930.0-275748	ESO 434- 23 / / PGC 27690	.SXS7.. / S (1) / 7.0± .6 / 6.1± .8	1.25± .04 / .25± .04 / / 1.29	10 / .46 / .34 / .12	14.54 ±.14				
0941.7 -0719 / 242.81 32.63 / 109.95 -48.38 / 0939.3 -0706	MCG -1-25- 20 / / PGC 27696	.SXS8*. / E (1) / 8.0± .9 / 7.5± .8	1.11± .06 / .14± .05 / / 1.12	125 / .10 / .17 / .07					
0941.8 +2738 / 200.86 48.17 / 69.51 -30.74 / 0938.9 +2752	MCG 5-23- 26 / / PGC 27702	.S?....	1.04± .09 / .49± .07 / / 1.01	.02 / .37	15.21 ±.18 / / / 14.69				8543 / 8488 / / 8816
094152.3+484014 / 169.23 47.76 / 53.28 -15.74 / 093834.8+485355	UGC 5172 / / PGC 27709	.SAS9.. / U / 9.0± .8	1.19± .05 / .00± .05 / / 1.19	.02 / .00 / .00	14.9 ±.4 / / / 14.93			15.58±.1 / 74± 7 / 57± 12 / .65	2594± 6 / 2623 / / 2783
094154.6-083610 / 244.00 31.85 / 111.84 -48.67 / 093926.5-082228	NGC 2969 / MCG -1-25- 21 / MK 1235 / PGC 27714	.SAS5P*. / E (1) / 5.0±1.2 / 3.1±1.2	1.13± .06 / .04± .05 / / 1.14	40 / .13 / .07 / .02	/ / / 12.92			15.75±.1 / 192± 12	4960± 9 / 4783 / / 5289
094157.1+121743 / 221.80 43.32 / 84.64 -40.11 / 093914.9+123125	IC 555 / UGC 5178 / / PGC 27716	.L..... / U / -2.0± .9	1.11± .10 / .40± .05 / / 1.06	18 / .04 / .00	14.31 ±.15 / / / 14.16				6731± 31 / 6618 / / 7043
094203.8+002008 / 235.37 37.28 / 99.25 -45.76 / 093929.7+003351	NGC 2967 / UGC 5180 / IRAS09394+0033 / PGC 27723	.SAS5.. / R (3) / 5.0± .3 / 3.0± .5	1.48± .02 / .04± .03 / 1.07± .02 / 1.50	65 / .17 / .06 / .02	12.30 ±.14 / 11.93 ±.14 / 11.11 / 11.87	.67± .02 / .07± .04 / .61 / .03	.74± .02 / .11± .04 / 13.14± .04 / 14.47± .19	13.20±.2 / 145± 6 / 121± 4 / 1.31	1892± 5 / 2212± 58 / 1743 / 2222
094211.4+044024 / 230.76 39.68 / 93.67 -43.89 / 093934.5+045407	NGC 2966 / UGC 5181 / MK 708 / PGC 27734	.SB?...	1.35± .04 / .41± .05 / / 1.36	72 / .10 / .61 / .20	13.56 ±.18 / / 11.37 / 12.84			15.17±.3 / 252± 7 / 242± 7 / 2.13	2044± 7 / 1897± 40 / 1902 / 2364
094212.1-044249 / 240.43 34.33 / 106.22 -47.52 / 093941.4-042906	MCG -1-25- 22 / / PGC 27735	.S..4*. / E (1) / 4.0±1.2 / 4.2±1.2	1.24± .05 / .64± .05 / / 1.25	57 / .13 / .94 / .32					4471 / 4305 / / 4800
094223.3-061509 / 241.92 33.42 / 108.45 -47.93 / 093953.6-060126	MCG -1-25- 24 / / PGC 27747	.SB.7?. / E / 7.0±1.7	1.25± .05 / .48± .05 / / 1.25	26 / .09 / .66 / .24				15.48±.1 / 288± 12	1863± 9 / 1692 / / 2193
094225.0+041704 / 231.23 39.53 / 94.19 -44.01 / 093948.4+043048	UGC 5182 / / PGC 27753	.L..... / U / -2.0± .8	1.09± .07 / .12± .03 / / 1.08	135 / .09 / .00	13.85 ±.15				

R.A. 2000 DEC. / l b / SGL SGB / R.A. 1950 DEC.	Names / PGC	Type / ST nL / T / L	logD25 / logR25 / logAo / logDo	p.a. / Ag / Ai / A21	BT / mB / mFIR / BT o	(B-V)T / (U-B)T / (B-V)T o / (U-B)T o	(B-V)o / (U-B)o / m'o / m'25	m21 / W20 / W50 / HI	V21 / Vopt / VGSR / V3K
0942.4 -1658 / 251.30 26.41 / 124.65 -49.88 / 0940.1 -1645	MCG -3-25- 15 / PGC 27757	.SB?...	1.19± .07 / .07± .07 / / 1.21	.24 / .10 / .03					4177 / / 3981 / 4501
094233.3-034159 / 239.51 35.01 / 104.83 -47.11 / 094001.9-032815	NGC 2974 / MCG 0-25- 8 / IRAS09400-0328 / PGC 27762	.E.4... / R / -5.0± .3	1.54± .02 / .23± .03 / .91± .02 / 1.49	42 / .12 / .00	11.87 ±.13 / 11.98 ±.11 / 13.62 / 11.79	1.00± .01 / .57± .02 / .95 / .55	1.02± .01 / .59± .01 / 11.91± .06 / 13.98± .20	16.52±.3 / 232± 34 / 235± 25	2072± 17 / 2006± 31 / 1893 / 2386
094236.4+585108 / 155.19 44.67 / 46.52 -7.98 / 093859.2+590451	NGC 2950 / UGC 5176 / / PGC 27765	RLBR0.. / R / -2.0± .3	1.43± .03 / .18± .03 / .69± .03 / 1.41	145 / .03 / .00	11.84 ±.13 / 11.98 ±.12 / / 11.86	.90± .02 / .50± .03 / .86 / .48	.92± .01 / .51± .02 / 10.78± .09 / 13.41± .20		/ 1337± 17 / 1406 / 1475
094249.1-064426 / 242.46 33.20 / 109.20 -47.97 / 094019.7-063041	MCG -1-25- 26 / IRAS09403-0630 / PGC 27773	.S..3*/ / E / 3.0±1.2	1.34± .04 / .75± .05 / / 1.35	6 / .08 / 1.03 / .37	13.64				5939 / / 5767 / 6269
094249.6-160656 / 250.66 27.06 / 123.32 -49.72 / 094026.8-155311	MCG -3-25- 16 / PGC 27774	.SXS8.. / E (1) / 8.0± .9 / 7.5± .8	1.17± .08 / .40± .05 / / 1.19	110 / .23 / .49 / .20					3620 / / 3426 / 3945
094253.9+315051 / 194.62 49.02 / 66.11 -27.72 / 093956.5+320435	NGC 2964 / UGC 5183 / IRAS09399+3204 / PGC 27777	.SXR4*. / R (2) / 4.0± .3 / 3.7± .7	1.46± .02 / .26± .02 / .93± .01 / 1.47	97 / .02 / .38 / .13	11.99 ±.13 / 12.12 ±.14 / 10.41 / 11.64	.68± .01 / -.03± .04 / .62 / -.08	.75± .01 / .02± .03 / 12.13± .02 / 13.52± .16	14.07±.1 / 311± 3 / 263± 3 / 2.30	1321± 4 / 1311± 19 / 1282 / 1580
094254.8+285852 / 198.93 48.64 / 68.53 -29.66 / 093959.9+291236	UGC 5185 / IRAS09400+2912 / PGC 27780	.S..5.. / U / 5.0± .9	1.22± .03 / .45± .04 / / 1.23	168 / .00 / .67 / .22	14.28 ±.18 / 13.29 / 13.56			15.44±.3 / 549± 7 / 1.66	8511± 10 / 8540± 50 / 8462 / 8781
094258.4+092811 / 225.39 42.25 / 88.02 -41.39 / 094018.2+094156	UGC 5189 / IRAS09401+0943 / PGC 27784	.I..9?. / U / 10.0±1.5	1.24± .07 / .28± .07 / / 1.24	/ .05 / .21 / .14	13.39				/ 3180± 40 / 3058 / 3498
0942.9 +3314 / 192.50 49.16 / 64.98 -26.75 / 0940.0 +3328	UGC 5186 / PGC 27785	.I..9.. / U / 10.0± .9	1.11± .14 / .66± .12 / / 1.11	43 / .01 / .49 / .33					548 / / 515 / 802
094301.8+585824 / 155.00 44.67 / 46.48 -7.86 / 093924.5+591208	UGC 5179 / MK 1423 / PGC 27788	.S?....	.97± .09 / .26± .06 / / .98	150 / .03 / .39 / .13	15.05 ±.18 / / 14.63				/ 1229± 63 / 1298 / 1367
094302.1+374925 / 185.48 49.30 / 61.36 -23.52 / 093958.8+380309	UGC 5184 / IRAS09399+3803 / PGC 27789	.SB.3.. / U / 3.0± .9	.97± .05 / .22± .04 / / .97	70 / .00 / .31 / .11	14.84 ±.18 / 13.40 / 14.48				/ 6584±125 / 6569 / 6821
094302.7-021509 / 238.18 35.98 / 102.89 -46.50 / 094030.3-020124	MCG 0-25- 9 / ARAK 210 / PGC 27791		.64± .17 / .00± .07 / / .65	/ .09	15.2 ±.3 / 13.44			16.12±.1 / 228± 12	4778± 9 / 4559± 63 / 4614 / 5103
094306.1+410534 / 180.48 49.15 / 58.89 -21.16 / 093959.2+411918	UGC 5187 / KUG 0939+413B / PGC 27792	.SB?...	1.00± .05 / .11± .04 / / 1.00	135 / .02 / .17 / .06	14.13 ±.18 / / 13.94			14.62±.1 / 130± 8 / .63	1465± 11 / 1481± 50 / 1464 / 1689
094308.7-102300 / 245.83 30.93 / 114.60 -48.77 / 094041.8-100915	NGC 2979 / MCG -2-25- 12 / IRAS09407-1009 / PGC 27795	PSAR1?. / E / 1.0± .8	1.17± .05 / .20± .05 / / 1.18	15 / .13 / .20 / .10	12.51				
094308.8+211108 / 210.24 46.88 / 75.67 -34.67 / 094020.2+212453	UGC 5192 / IRAS09402+2124 / PGC 27796	.SB.3*. / U / 3.0± .8	1.24± .06 / .29± .06 / / 1.24	90 / .05 / .40 / .15	14.2 ±.2 / 13.42 / 13.72			14.97±.3 / 431± 7 / 1.10	4940± 10 / / 4860 / 5233
094312.0-093646 / 245.15 31.44 / 113.46 -48.59 / 094044.6-092301	NGC 2980 / MCG -1-25- 28 / IRAS09407-0923 / PGC 27799	.SXS5?. / PE (1) / 4.5± .6 / 2.5± .6	1.21± .04 / .28± .04 / / 1.22	160 / .14 / .42 / .14	13.6 ±.2 / / 11.91 / 13.01	.61± .05 / -.01± .05 / .49 / -.10	/ / / 13.81± .30	14.50±.1 / 505± 12 / 1.35	5720± 9 / / 5541 / 6050
094312.0+315541 / 194.50 49.09 / 66.10 -27.62 / 094014.5+320926	NGC 2968 / UGC 5190 / / PGC 27800	.I.0... / R / 90.0	1.36± .02 / .16± .02 / / 1.36	45 / .02 / .12	12.78 ±.13 / 12.81 ±.12 / / 12.63	1.05± .02 / .64± .03 / 1.01 /	1.13± .02 / .73± .03 / / 14.04± .19	16.01±.3 / 123± 10 / 70± 7	1435± 9 / 1580± 27 / 1410 / 1707

R.A. 2000 DEC. / l b / SGL SGB / R.A. 1950 DEC.	Names / PGC	Type / S_T n_L / T / L	$\log D_{25}$ / $\log R_{25}$ / $\log A_\bullet$ / $\log D_o$	p.a. / A_g / A_i / A_{21}	B_T / m_B / m_{FIR} / B_T^o	$(B-V)_T$ / $(U-B)_T$ / $(B-V)_T^o$ / $(U-B)_T^o$	$(B-V)_\bullet$ / $(U-B)_\bullet$ / $m'_\bullet$ / m'_{25}	m_{21} / W_{20} / W_{50} / HI	V_{21} / V_{opt} / V_{GSR} / V_{3K}
094312.1+002450 / 235.50 37.56 / 99.31 -45.47 / 094038.0+003836	UGC 5195 / IRAS09406+0038 / PGC 27803	.S..4.. / U / 4.0±1.0	1.00± .08 / .69± .06 / 1.01	68 / .14 / 1.01 / .34	15.47 ±.19 / 13.36				
094316.8-094445 / 245.29 31.37 / 113.66 -48.60 / 094049.5-093059	NGC 2978 / MCG -1-25- 29 / IRAS09408-0931 / PGC 27808	PSXT4?. / PE (1) / 3.7± .5 / 3.1± .7	1.02± .05 / .07± .04 / 1.04	85 / .14 / .10 / .04	12.26			15.08±.3 / 349± 15 / 250± 12	1802± 11 / 1622 / 2132
094317.2-195216 / 253.77 24.55 / 129.16 -49.83 / 094057.0-193830	ESO 566- 2 / IRAS09409-1938 / PGC 27809	.SB.0?/ / SE / -.2± .8	1.20± .04 / .65± .03 / 1.18	82 / .13 / .49	15.52 ±.14				
094318.1-095649 / 245.47 31.24 / 113.96 -48.64 / 094050.9-094303	MCG -2-25- 13 / IRAS09408-0942 / PGC 27810	.SXS7.. / E (1) / 7.0± .8 / 6.4± .8	1.42± .04 / .30± .05 / 1.43	150 / .16 / .42 / .15				14.58±.1 / 329± 12	2697± 9 / 2517 / 3027
094319.2+361453 / 187.90 49.36 / 62.63 -24.59 / 094017.6+362838	NGC 2965 / UGC 5191 / PGC 27813	.L..... / U / -2.0± .8	1.08± .17 / .12± .08 / 1.06	85 / .00 / .00	14.44 ±.16 / 14.34				6733± 13 / 6712 / 6976
0943.4 -0517 / 241.21 34.21 / 107.19 -47.41 / 0940.9 -0504	A 0940-05 / MCG -1-25- 31 / PGC 27817	.SBS8$/ / R / 8.0± .6	1.12± .05 / .75± .05 / 1.13	90 / .13 / .92 / .37				14.63±.1 / 182± 12	1867± 9 / 1864± 39 / 1699 / 2197
094328.5-052155 / 241.29 34.18 / 107.30 -47.42 / 094058.2-050808	MCG -1-25- 33 / PGC 27825	.SBS7P* / E (1) / 7.0±1.2 / 5.3±1.2	1.13± .06 / .26± .05 / 1.14	35 / .13 / .36 / .13					1893 / 1725 / 2223
0943.4 +3402 / 191.29 49.32 / 64.42 -26.11 / 0940.5 +3416	UGC 5194 / PGC 27826	.S..6*. / U / 6.0±1.4	1.00± .08 / .84± .06 / 1.00	49 / .00 / 1.24 / .42					
094330.3+315834 / 194.44 49.16 / 66.11 -27.54 / 094032.9+321220	NGC 2970 / MCG 5-23- 30 / MK 405 / PGC 27827	.E.1.*. / P / -5.0± .9	.81± .07 / .11± .05 / .61± .04 / .78	.02 / .00	14.38 ±.14 / 14.93 ±.18 / 14.53	.76± .03 / .15± .07 / .74 / .15	.77± .02 / .19± .06 / 12.92± .12 / 13.14± .38		1664± 44 / 1626 / 1923
0943.5 -0516 / 241.21 34.24 / 107.18 -47.38 / 0941.0 -0503	A 0941-05 / MCG -1-25- 32 / PGC 27828	.SBS9$/ / R / 9.0± .5	1.09± .05 / .51± .05 / 1.10	75 / .13 / .52 / .26					
094333.1+392455 / 183.04 49.35 / 60.22 -22.31 / 094028.1+393841	UGC 5193 / PGC 27830	.E...*. / U / -5.0±1.2	1.17± .09 / .00± .05 / 1.17	.00 / .00	14.8 ±.3				
0943.5 +8348 / 128.23 31.26 / 30.38 11.08 / 0935.0 +8402	UGC 5128 / PGC 27832	.S..6*. / U / 6.0±1.2	1.01± .05 / .09± .04 / 1.02	.16 / .13 / .04	15.1 ±.2				
094336.4-055445 / 241.83 33.87 / 108.09 -47.55 / 094106.5-054059	MCG -1-25- 34 / IRAS09411-0541 / PGC 27833	.SBS8.. / UE (1) / 7.5± .5 / 6.4± .8	1.34± .04 / .00± .05 / 1.35	10 / .12 / .01 / .00				14.96±.1 / 107± 10 / 85± 12	2021± 7 / 1852 / 2352
094337.6-324423 / 263.34 15.33 / 148.69 -48.31 / 094128.0-323036	ESO 373- 20 / PGC 27836	.IXS9*P / S (1) / 10.0± .7 / 9.2± .7	1.21± .06 / .13± .07 / 1.28	142 / .70 / .10 / .07					911 / 690 / 1208
094340.4+110344 / 223.58 43.14 / 86.32 -40.42 / 094059.1+111730	NGC 2984 / UGC 5200 / PGC 27838	.L?.... / /	.87± .10 / .00± .03 / .87	.03 / .00	14.39 ±.17 / 14.27				6200± 50 / 6083 / 6515
094340.8-202841 / 254.31 24.19 / 130.11 -49.74 / 094121.1-201454	NGC 2983 / ESO 566- 3 / PGC 27840	.LBT+.. / R / -1.0± .3	1.40± .02 / .23± .03 / .81± .03 / 1.38	95 / .14 / .00	12.76 ±.14 / 12.74 ±.11 / 12.58	.99± .02 / .54± .03 / .91 / .48	1.01± .01 / .58± .02 / 12.30± .10 / 14.04± .19		2015± 56 / 1812 / 2336
094346.2+361046 / 188.01 49.45 / 62.76 -24.58 / 094044.7+362432	NGC 2971 / UGC 5197 / PGC 27843	.SBR3.. / U / 3.0± .9	1.06± .06 / .18± .05 / 1.06	135 / .00 / .25 / .09	14.78 ±.19				

R.A. 2000 DEC. / l b / SGL SGB / R.A. 1950 DEC.	Names / PGC	Type / S_T n_L / T / L	$\log D_{25}$ / $\log R_{25}$ / $\log A_e$ / $\log D_o$	p.a. / A_g / A_i / A_{21}	B_T / m_B / m_{FIR} / B_T^o	$(B-V)_T$ / $(U-B)_T$ / $(B-V)_T^o$ / $(U-B)_T^o$	$(B-V)_e$ / $(U-B)_e$ / m'_e / m'_{25}	m_{21} / W_{20} / W_{50} / HI	V_{21} / V_{opt} / V_{GSR} / V_{3K}
094346.9+745143	NGC 2977	.S..3*.	1.26± .03	145	13.3 ±.2				3072± 76
136.71 36.79	UGC 5175	P	.36± .04	.04	13.0 ±.2				
36.25 4.31	IRAS09388+7505	3.0±1.2	.80± .03	.49	12.04		12.80± .09		3198
093851.6+750527	PGC 27845		1.26	.18	12.60		13.55± .25		3122
094354.0-343330		.SBS8*.	1.20± .05	27					2651
264.64 14.02	ESO 373- 21	S (1)	.65± .05	.84					
151.29 -47.79		8.0±1.3		.80					2428
094146.0-341942	PGC 27856	6.7± .9	1.28	.33					2943
094357.0+424021		.S..2..	1.10± .06	65					
178.05 49.15	UGC 5199	U	.40± .05	.03					
57.85 -19.91		2.0± .9		.49					
094048.3+425408	PGC 27859		1.10	.20					
094357.5+414114		.SB.1..	1.04± .06	150					
179.55 49.26	UGC 5198	U	.17± .05	.04	14.80 ±.19				
58.58 -20.62		1.0± .9		.18					
094050.0+415500	PGC 27860		1.04	.09					
094403.3+293619	A 0941+29		.62± .11					17.54±.3	5093± 13
198.05 48.98			.11± .12	.04					5028± 22
68.19 -29.06	MK 406							206± 16	5029
094108.1+295006	PGC 27867		.63						5345
094405.2-320954		.IXS9*.	1.03± .04	10				14.95±.2	1209± 11
263.01 15.82	ESO 434- 27	SU (1)	.07± .04	.59	15.22 ±.14			63± 16	
147.82 -48.35		10.0± .5		.05				54± 12	988
094155.0-315606	PGC 27869	10.0± .6	1.09	.03	14.57			.34	1507
094407.3-003935		.SB.9P?	1.02± .05	152				14.65±.1	1492± 9
236.77 37.13	UGC 5205	UE	.20± .04	.13	14.82 ±.20			162± 12	
100.87 -45.67		9.3±1.0		.20					1338
094133.9-002547	PGC 27875		1.03	.10	14.49			.07	1821
094409.1+655841	A 0940+66	.P.....	.62± .11					16.64±.1	3322± 11
146.34 41.64	UGC 5188	R	.09± .05	.16	14.6 ±.4				3273± 46
42.01 -2.44	MK 119	99.0		.12	12.27			151± 8	3415
094010.1+661227	PGC 27879		.64	.04	14.33			2.27	3420
094414.2-285055		.LA.-*.	1.19± .05	102	13.12 ±.16	.95± .02	1.02± .01		
260.68 18.28	ESO 434- 28	S	.11± .04	.55	13.65 ±.14				
142.91 -48.96		-3.0±1.1	1.01± .05	.00			13.66± .17		
094201.0-283706	PGC 27882		1.24				13.66± .30		
094416.3-211642	NGC 2986	.:E.2...	1.50± .03	105	11.72 ±.13	.97± .01	1.00± .01		
255.04 23.72	ESO 566- 5	R	.06± .03	.10	11.64 ±.11	.57± .03	.61± .01		2310± 31
131.34 -49.60		-5.0± .3	1.08± .02	.00		.93	12.61± .07		2106
094157.1-210254	PGC 27885		1.50		11.54	.56	14.07± .21		2630
094417.0+762105	A 0939+76	.P.....	.99± .08	40					2354± 38
135.20 35.94	MCG 13- 7- 36	R	.38± .06	.03	15.14 ±.15				
35.30 5.46	MK 118	99.0		.00					2486
093905.8+763450	PGC 27887		.93		15.07				2396
0944.4 +5546		.SXS5..	1.17± .04	45				15.90±.1	7627± 11
159.12 46.06	UGC 5201	U	.19± .04	.02	14.9 ±.3				
48.75 -10.15	SBS 0941+559	5.0± .8		.29				241± 8	7684
0941.0 +5600	PGC 27893		1.17	.10	14.53			1.28	7782
094434.7-314726	IC 2507	.IBS9P*	1.22± .04	43					1250
262.82 16.17	ESO 434- 31	S (1)	.31± .03	.59	13.33 ±.14				
147.24 -48.33		9.5± .6		.24					1030
094224.1-313336	PGC 27903	5.6±1.2	1.27	.16	12.50				1550
094437.9-291926		.S..9*.	1.17± .04	170					
261.08 17.99	ESO 434- 32	U	.15± .04	.53	14.74 ±.14				
143.60 -48.79		9.0±1.2		.15					
094225.1-290536	PGC 27904		1.22	.07					
094447.7-314932	A 0942-31	.SBS9P*	1.33± .03					13.18±.2	1256± 11
262.88 16.17	ESO 434- 33	SU (2)	.10± .04	.59	13.19 ±.14			157± 16	
147.28 -48.27	DDO 235	9.0± .4		.10				122± 12	1036
094237.1-313542	PGC 27918	7.9± .6	1.39	.05	12.49			.64	1555
094456.2+310558	NGC 2981	.SXT4*.	1.08± .06	95				15.37±.3	10406± 9
195.83 49.37	UGC 5208	U	.06± .05	.02	14.4 ±.2				
67.07 -27.91		4.0± .9		.09				367± 5	10364
094159.8+311947	PGC 27925		1.08	.03	14.24			1.10	10669
094457.1+164228		.SBT1..	1.08± .06					17.13±.3	5958± 11
216.56 45.77	UGC 5213	U	.04± .05	.07	14.04 ±.18			373± 7	5980± 50
80.45 -37.02		1.0± .8		.04				310± 7	5863
094212.0+165618	PGC 27926		1.08	.02	13.86			3.26	6264

R.A. 2000 DEC.	Names	Type	logD$_{25}$	p.a.	B$_T$	(B-V)$_T$	(B-V)$_\bullet$	m$_{21}$	V$_{21}$
l b		S$_T$ n$_L$	logR$_{25}$	A$_g$	m$_B$	(U-B)$_T$	(U-B)$_\bullet$	W$_{20}$	V$_{opt}$
SGL SGB		T	logA$_\bullet$	A$_i$	m$_{FIR}$	(B-V)$_T^o$	m'$_\bullet$	W$_{50}$	V$_{GSR}$
R.A. 1950 DEC.	PGC	L	logD$_o$	A$_{21}$	B$_T^o$	(U-B)$_T^o$	m'$_{25}$	HI	V$_{3K}$
094458.5-194332		.SB.3P?	.91± .05	90					
253.96 24.93	ESO 566- 7	SE	.19± .04	.15	15.36 ±.14				9734± 97
128.95 -49.43		3.0±1.8		.26					9533
094238.1-192942	PGC 27928		.92	.09	14.88				10056
0945.0 +2927	IC 558	.S?....	.99± .10						
198.33 49.17	MCG 5-23- 33		.00± .07	.04	14.52 ±.19				9334
68.48 -29.00				.00					9286
0942.1 +2941	PGC 27931		.99		14.34				9603
094504.4+321411		.I.9..	.96± .09					16.73±.3	547± 7
194.09 49.51	UGC 5209	U	.00± .06	.00					510
66.15 -27.12		10.0± .9		.00				49± 7	806
094207.0+322800	PGC 27935		.96	.00					
0945.0 +2716	IC 2505	.L?....	1.02± .14	143					
201.61 48.81	MCG 5-23- 34		.05± .07	.01	14.9 ±.2				9919± 46
70.40 -30.44				.00					9862
0942.2 +2730	PGC 27936		1.01		14.70				10195
094510.0+683549	NGC 2959	PSXT2P*	1.13± .05					16.86±.1	4429± 11
143.29 40.38	UGC 5202	PU	.01± .04	.19	13.65 ±.18				4524± 57
40.39 -.39	IRAS09409+6849	1.5± .6		.01	13.35			263± 8	4537
094059.9+684937	PGC 27939		1.15	.00	13.40			3.45	4518
0945.2 +5341		.SB.2..	1.00± .08	55					
161.91 46.87	UGC 5207	U	.25± .06	.01	14.90 ±.18				
50.24 -11.64		2.0± .9		.31					
0941.8 +5355	PGC 27944		1.00	.13					
094514.6+090643	A 0942+09	.S..4..	1.21± .05	23				14.89±.3	5484± 9
226.18 42.58	UGC 5215	U (1)	.28± .05	.06	13.64 ±.18			419± 7	5533± 41
88.82 -41.09		4.0± .8		.41				403± 7	5363
094234.7+092033	PGC 27946	5.0±1.3	1.22	.14	13.13			1.62	5806
094517.5-272421		.SBT0..	1.09± .05	86					
259.81 19.49	ESO 499- 1	r	.55± .04	.31	14.90 ±.14				
140.69 -48.94		-.1± .9		.42					
094303.0-271030	PGC 27951		1.09						
0945.3 +2305		.I..9*.	1.00± .16	65				17.00±.1	2132± 10
207.76 47.91	UGC 5214	U	.25± .12	.05				53± 6	2059
74.25 -33.09		10.0±1.3		.19				29± 5	2421
0942.5 +2319	PGC 27954		1.00	.13					
094521.8-270527		.SXS7..	1.12± .04						4379
259.60 19.73	ESO 499- 2	S (1)	.05± .03	.47	14.58 ±.14				
140.21 -48.96		7.0± .6		.07					4165
094307.0-265136	PGC 27957	5.6± .6	1.16	.03	14.02				4690
094522.0+683631	NGC 2961	.S..3*.	1.06± .08	44					
143.26 40.39	MCG 12- 9- 63	P	.60± .06	.19	15.52 ±.18				
40.40 -.37		3.0±1.4		.82					
094111.9+685020	PGC 27958		1.08	.30					
094522.5+094552		.SX.7*.	1.15± .07	165				15.25±.3	3279± 10
225.41 42.91	UGC 5216	U	.30± .06	.04	14.31 ±.19				
88.09 -40.73		7.0±1.2		.42				218± 7	3158
094242.2+095943	PGC 27959		1.15	.15	13.85			1.25	3598
094525.8-182236	NGC 2989	.SXS4*.	1.23± .03	38	13.58 ±.13	.54± .01	.62± .01	14.70±.1	4172± 6
252.97 25.95	ESO 566- 9	PSE (4)	.26± .03	.16	13.46 ±.12	-.08± .03	.00± .03	312± 34	4136± 66
126.88 -49.27	IRAS09430-1808	4.3± .5	.70± .01	.38	12.32	.43	12.57± .02	377± 25	3973
094304.3-180844	PGC 27962	2.0± .5	1.25	.13	12.95	-.16	13.95± .20	1.62	4496
094530.1-302034		.IB.9*/	1.43± .03	61				14.08±.2	1004± 11
261.96 17.37	ESO 434- 34	SU (1)	.73± .04	.51	14.64 ±.14			141± 16	
145.06 -48.43		10.0± .7		.55				133± 12	786
094318.1-300642	PGC 27966	7.8± .8	1.48	.36	13.58			.13	1307
094530.2+062249		.SX.7..	1.24± .04	125				14.73±.2	3087± 6
229.41 41.28	UGC 5218	U	.28± .05	.05	14.2 ±.2			191± 7	
92.10 -42.37		7.0± .8		.39				182± 5	2955
094252.2+063640	PGC 27968		1.25	.14	13.71			.87	3411
094539.4-311128	NGC 2997	.SXT5..	1.95± .01	110	10.06S±.10			11.50±.1	1087± 4
262.58 16.76	ESO 434- 35	R (3)	.12± .02	.54	10.03 ±.12			281± 4	1090± 17
146.29 -48.23		5.0± .3		.18	9.18			256± 5	868
094328.1-305736	PGC 27978	1.6± .3	2.00	.06	9.31		14.33± .13	2.13	1388
094541.8+045633	NGC 2987	.S..2..	1.18± .05	160				15.49±.3	3742± 9
231.06 40.57	UGC 5220	U	.34± .05	.10	13.85 ±.18			353± 7	3738± 50
93.88 -42.99	IRAS09430+0510	2.0± .9		.42	12.83			344± 7	3605
094304.7+051025	PGC 27981		1.19	.17	13.29			2.03	4067

R.A. 2000 DEC. l b SGL SGB R.A. 1950 DEC.	Names PGC	Type S_T n_L T L	$\log D_{25}$ $\log R_{25}$ $\log A_e$ $\log D_o$	p.a. A_g A_i A_{21}	B_T m_B m_{FIR} B_T^o	$(B-V)_T$ $(U-B)_T$ $(B-V)_T^o$ $(U-B)_T^o$	$(B-V)_e$ $(U-B)_e$ m'_e m'_{25}	m_{21} W_{20} W_{50} HI	V_{21} V_{opt} V_{GSR} V_{3K}
094542.1-141939 249.71 28.78 120.71 -48.83 094317.8-140547	NGC 2992 MCG -2-25- 14 IRAS09432-1405 PGC 27982	.S..1P. R 1.0± .3	1.55± .02 .51± .04 .76± .01 1.57	15 .27 .52 .25	13.14 ±.13 12.82 ±.15 10.98 12.19	.96± .02 .40± .03 .79 .25	1.01± .01 .38± .02 12.43± .04 14.44± .20	14.11±.1 433± 6 355± 5 1.67	2314± 6 2334± 24 2125 2644
094545.5+335213 191.59 49.78 64.92 -25.89 094246.6+340604	 UGC 5217 PGC 27987	.SBS6.. U 6.0± .9	1.00± .06 .26± .05 1.00	5 .00 .38 .13	15.28 ±.19				
094548.4-142208 249.76 28.77 120.77 -48.81 094324.1-140815	NGC 2993 MCG -2-25- 15 IRAS09434-1408 PGC 27991	.S..1P. R 1.0± .4	1.13± .03 .16± .04 .44± .02 1.15	95 .27 .17 .08	13.11 ±.13 13.42 ±.18 10.70 12.75	.47± .04 -.39± .06 .36 -.45	.26± .01 -.60± .01 10.80± .06 13.18± .23		2420± 9 2227± 51 2224 2742
094550.0+282821 199.85 49.19 69.47 -29.52 094255.9+284213	 UGC 5219 KUG 0942+287 PGC 27993	.SXS7.. U 7.0± .9	1.02± .05 .10± .05 1.02	170 .03 .14 .05					7432± 10 7380 7704
094550.7-173524 252.42 26.57 125.68 -49.12 094328.6-172132	 MCG -3-25- 21 PGC 27994	.SBS8*. E (1) 8.0±1.2 6.4± .8	1.14± .06 .31± .05 1.16	155 .20 .39 .16					4012 3815 4338
094552.3+344108 190.33 49.84 64.28 -25.31 094252.6+345459	 MCG 6-22- 9 IRAS09428+3454 PGC 27996	.S?....	.89± .11 .35± .07 .89	 .00 .53 .18	14.9 ±.2 12.86 14.31				6176± 13 6148 6426
094552.8+025832 233.26 39.55 96.37 -43.81 094317.0+031224	 UGC 5224 PGC 27997	.SB.8.. U 8.0± .8	1.23± .06 .38± .06 1.24	70 .12 .47 .19				14.79±.1 178± 8 172± 12	1933± 6 1790 2261
094553.5-001608 236.69 37.72 100.59 -45.11 094319.8-000216	IC 560 UGC 5223 PGC 27998	.L..+*. UE -.7± .7	1.12± .05 .33± .05 1.08	15 .14 .00	14.32 ±.15				
094603.7+042416 231.73 40.36 94.61 -43.15 094327.0+043808	 UGC 5226 PGC 28009	.L...*. U -2.0±1.2	1.00± .19 .05± .08 1.01	 .13 .00	13.94 ±.16 13.74				5020± 50 4881 5346
094603.9+014007 234.70 38.86 98.07 -44.31 094329.0+015400	A 0943+01 UGC 5228 PGC 28010	.SBS5*. UE (1) 5.0±1.0 4.2± .8	1.39± .03 .53± .04 1.41	127 .21 .80 .27	13.66 ±.18 12.64			14.01±.2 280± 7 262± 5 1.11	1872± 6 1725 2201
094604.1-035822 240.43 35.54 105.64 -46.37 094332.9-034429	IC 562 MCG -1-25- 36 IRAS09435-0344 PGC 28011	.S..4?/ PE 4.0±1.3	1.15± .05 .59± .05 1.16	26 .10 .87 .30	12.32			14.78±.1 509± 12	4801± 9 4637 5133
094607.5-210853 255.27 24.12 131.14 -49.17 094348.1-205500	 ESO 566- 10 PGC 28013	.L..0*P S -2.0±1.2	1.18± .05 .52± .03 1.12	75 .12 .00	14.99 ±.14				
094608.7-631630 283.78 -7.52 184.77 -34.03 094448.1-630236	 ESO 91- 18 IRAS09447-6302 PGC 28015	.S..5./ S 5.0± .6	1.25± .04 .80± .04 1.52	65 2.89 1.21 .40	15.82 ±.14 13.45				
0946.1 +6858 142.81 40.26 40.22 -.05 0942.0 +6912	 UGC 5210 PGC 28018	.S..6*. U 6.0±1.3	1.35± .04 1.06± .05 1.37	153 .20 1.47 .50				14.84±.1 304± 8	4441± 11 4547 4525
094616.3-120606 247.92 30.38 117.39 -48.35 094350.4-115212	 MCG -2-25- 16 PGC 28023	.SAT8*. E (1) 8.0± .9 7.5± .8	1.04± .07 .08± .05 1.05	135 .06 .09 .04					
094617.6-685455 287.53 -11.78 189.86 -30.44 094519.0-684100	 ESO 61- 15 IRAS09453-6840 PGC 28025	.SAS5.. S (1) 5.0± .8 3.3± .9	1.34± .04 .23± .05 1.41	16 .68 .35 .12	13.50 ±.14 13.17				
094617.6+054230 230.31 41.10 93.04 -42.51 094340.1+055623	NGC 2990 UGC 5229 ARAK 214 PGC 28026	.S..5*. R (1) 5.0± .7 3.4± .9	1.11± .04 .27± .04 1.11	85 .05 .40 .13	13.1 ±.2 13.49 ±.18 11.35 12.85	.39± .04 -.24± .05 .31 -.30	 12.83± .31	14.68±.1 309± 5 274± 5 1.70	3088± 5 3150± 34 2955 3414

R.A. 2000 DEC. / l b / SGL SGB / R.A. 1950 DEC.	Names / PGC	Type / S_T n_L / T / L	$\log D_{25}$ / $\log R_{25}$ / $\log A_o$ / $\log D_o$	p.a. / A_g / A_i / A_{21}	B_T / m_B / m_{FIR} / B_T^o	$(B-V)_T$ / $(U-B)_T$ / $(B-V)_T^o$ / $(U-B)_T^o$	$(B-V)_o$ / $(U-B)_o$ / m'_o / m'_{25}	m_{21} / W_{20} / W_{50} / HI	V_{21} / V_{opt} / V_{GSR} / V_{3K}
094619.1-302618 / 262.16 17.42 / 145.15 -48.24 / 094407.1-301224	NGC 3001 / ESO 434- 38 / / PGC 28027	.SXT4.. / PSUr (3) / 3.9± .3 / 1.8± .4	1.46± .02 / .17± .04 / .87± .02 / 1.51	6 / .51 / .25 / .08	12.72 ±.14 / 12.34 ±.12 / / 11.73	.78± .04 / .07± .04 / .62 / -.06	.86± .01 / .15± .02 / 12.56± .04 / 14.48± .20	13.58±.1 / 412± 8 / 385± 11 / 1.77	2486± 6 / 2220± 66 / 2265 / 2787
094619.2-290800 / 261.23 18.38 / 143.22 -48.46 / 094406.0-285406	/ ESO 434- 37 / / PGC 28028	.SBR1.. / r / 1.0± .9 /	1.07± .05 / .31± .04 / / 1.12	0 / .55 / .32 / .16	/ 14.86 ±.14 / /				
094620.4+030241 / 233.27 39.68 / 96.35 -43.68 / 094344.6+031634	IC 563 / MCG 1-25- 22 / / PGC 28032	.SBR2*P / R / 2.0± .5 /	.94± .11 / .28± .07 / / .95	/ .12 / .35 / .14	/ 14.78 ±.19 / / 14.25				6093± 58 / 5950 / 6421
094621.4+030415 / 233.24 39.70 / 96.32 -43.66 / 094345.6+031808	IC 564 / UGC 5230 / / PGC 28033	.SAS6$P / R / 6.0± .5 /	1.24± .05 / .59± .05 / / 1.25	68 / .13 / .87 / .30	/ 14.10 ±.18 / / 13.07				6026± 58 / 5883 / 6354
094622.7-463906 / 273.06 5.21 / 166.98 -42.95 / 094428.0-462512	/ ESO 262- 4 / / PGC 28036	.S..6?/ / S / 6.0±1.9 /	1.27± .06 / .78± .07 / / 1.43	3 / 1.73 / 1.14 / .39					
094628.2-743544 / 291.42 -16.03 / 194.62 -26.64 / 094607.0-742148	/ ESO 37- 5 / / PGC 28044	.SXS3P. / S / 3.0± .8 /	1.47± .04 / .57± .05 / / 1.55	95 / .86 / .79 / .29	/ 13.92 ±.14 / /				
094628.6+454510 / 173.29 49.12 / 55.95 -17.36 / 094316.5+455903	A 0943+46 / UGC 5225 / 1ZW 21 / PGC 28045	.S?.... / / / .95	.95± .07 / .00± .05 / .43± .04 / .95	/ .02 / .00 /	14.69 ±.15 / 14.73 ±.18 / / 14.61	.93± .01 / .51± .02 / .88 / .53	1.00± .01 / .55± .02 / 12.33± .14 / 14.27± .39		4906± 49 / 4923 / 5110
094630.2-213418 / 255.67 23.88 / 131.79 -49.08 / 094411.0-212024	NGC 2996 / ESO 566- 12 / / PGC 28049	.L..+P* / SE / -1.0± .6 /	1.19± .04 / .07± .03 / / 1.19	115 / .11 / .00 /	13.56 ±.14 / / / 13.32				8775± 15 / 8570 / 9096
094634.5-235506 / 257.46 22.21 / 135.37 -48.98 / 094417.1-234112	/ ESO 499- 4 / / PGC 28053	.SBT6.. / S (1) / 5.5± .6 / 4.4± .6	1.15± .04 / .46± .03 / / 1.17	82 / .15 / .68 / .23	/ 14.96 ±.14 / /				
094635.4+421132 / 178.70 49.69 / 58.58 -19.92 / 094327.8+422525	/ UGC 5227 / KUG 0943+424 / PGC 28054	.S..6*. / U / 6.0±1.3 /	1.05± .04 / .56± .04 / / 1.05	68 / .00 / .82 / .28					
094641.9-145145 / 250.34 28.59 / 121.56 -48.66 / 094417.9-143751	/ MCG -2-25- 17 / / PGC 28066	RSBR1?. / E / 1.0± .9 /	1.10± .06 / .10± .05 / / 1.13	65 / .25 / .10 / .05					
0946.7 +5426 / 160.77 46.83 / 49.90 -10.94 / 0943.3 +5440	A 0943+54A / MCG 9-16- 43 / / PGC 28068	.L?.... / / / .90	.90± .17 / .00± .07 / / .90	/ .00 / .00 /	15.2 ±.2 / 14.95 ±.15 / /	1.05± .07 / .66± .11 / /	/ / / 14.55± .88		
094645.6+134621 / 220.64 45.01 / 83.86 -38.32 / 094402.7+140015	/ UGC 5232 / ARAK 216 / PGC 28069	.S?.... / / / .90	.89± .08 / .22± .05 / / .90	/ .07 / .33 / .11	/ 14.7 ±.2 / / 14.27				7222± 50 / 7115 / 7534
094648.4-333619 / 264.45 15.13 / 149.70 -47.45 / 094439.1-332224	/ ESO 373- 26 / / PGC 28074	.SBS7.. / S (1) / 7.0± .8 / 6.7± .8	1.22± .05 / .12± .04 / / 1.29	/ .75 / .17 / .06					2465 / / 2244 / 2761
094649.8+220056 / 209.44 47.95 / 75.52 -33.48 / 094401.0+221450	NGC 2991 / UGC 5233 / / PGC 28079	.L..... / U / -2.0± .8 /	1.15± .15 / .10± .08 / .89± .04 / 1.14	/ .05 / .00 /	13.53 ±.15 / / / 13.37	.91± .03 / .41± .04 / .82 / .43	.96± .01 / .51± .03 / 13.47± .12 / 13.89± .81		7455± 29 / 7378 / 7747
0946.8 +5428 / 160.71 46.83 / 49.89 -10.90 / 0943.4 +5442	A 0943+54B / MCG 9-16- 44 / / PGC 28080	.E?....		/ .00 / /	15.2 ±.3 / / /				
094650.0+160236 / 217.68 45.93 / 81.47 -37.03 / 094405.5+161630	/ UGC 5234 / IRAS09440+1616 / PGC 28081	.S.R5.. / U / 5.0± .8 /	1.21± .05 / .16± .05 / / 1.22	115 / .10 / .23 / .08	14.2 ±.2 / 13.51 / / 13.79			14.99±.3 / / 349± 7 / 1.13	6017± 10 / 5918 / 6324

R.A. 2000 DEC. l b SGL SGB R.A. 1950 DEC.	Names PGC	Type S_T n_L T L	$\log D_{25}$ $\log R_{25}$ $\log A_e$ $\log D_o$	p.a. A_g A_i A_{21}	B_T m_B m_{FIR} B_T^o	$(B-V)_T$ $(U-B)_T$ $(B-V)_T^o$ $(U-B)_T^o$	$(B-V)_e$ $(U-B)_e$ m' m'_{25}	m_{21} W_{20} W_{50} HI	V_{21} V_{opt} V_{GSR} V_{3K}
094653.8+003023 236.07 38.37 99.70 -44.58 094419.6+004418	UGC 5238 (1) PGC 28087	.SBS8*. UE 7.5± .6 7.5± .8	1.34± .04 .44± .05 1.36	55 .25 .55 .22	 14.3 ±.3 13.48			14.09±.1 230± 6 219± 12 .39	1777± 5 1627 2107
094654.6+230126 208.00 48.24 74.58 -32.84 094405.0+231520	UGC 5235 PGC 28088	.S..6*. U 6.0±1.4 	1.11± .07 .95± .06 1.12	151 .06 1.40 .48				15.54±.3 380± 7 	7296± 10 7223 7586
0947.0 +2144 209.86 47.90 75.82 -33.62 0944.2 +2158	UGC 5236 PGC 28095	.S..9*. U 9.0±1.2 	1.11± .07 .01± .06 1.11	 .06 .01 .00				16.75±.1 76± 5 54± 4 	3793± 10 3716 4086
0947.0 +7948 131.77 33.92 33.14 8.16 0941.0 +8002	UGC 5203 PGC 28098	.S..7.. U 7.0± .9 	1.36± .04 1.10± .05 1.36	80 .04 1.38 .50				14.47±.1 190± 8 	1551± 11 1694 1573
094704.3+423117 178.18 49.74 58.40 -19.62 094356.5+424511	UGC 5231 ARAK 215 PGC 28099	.SB.2*. U 2.0± .9 	1.00± .06 .26± .05 1.00	40 .00 .32 .13	 15.20 ±.18 14.83				5394± 82 5398 5613
094705.8+005745 235.63 38.67 99.13 -44.36 094431.3+011140	UGC 5242 (1) PGC 28101	.SBS9*. UE 9.3± .7 8.7± .8	1.13± .04 .19± .04 1.15	10 .19 .20 .10				15.56±.1 103± 6 101± 4 	1856± 7 1707 2186
094711.2+560615 158.47 46.29 48.82 -9.65 094342.1+562009	A 0943+56 CGCG 265- 38 MK 123 PGC 28111	 	.62± .11 .26± .12 .62	 .00 	 15.5 ±.3 				7673± 38 7731 7826
094721.5+725913 138.45 38.09 37.66 3.05 094245.7+731306	NGC 2957A MCG 12-10- 1 PGC 28113	.E.1.*. P -5.0±1.4 	.90± .17 .35± .07 .80	 .04 .00 					6701± 25 6821 6763
094712.8-245020 258.26 21.65 136.75 -48.78 094456.1-243624	 ESO 499- 5 PGC 28117	.SXS5*. S (1) 4.5± .6 3.3± .6	1.40± .03 .56± .03 1.41	145 .14 .85 .28	 13.79 ±.14 				
094719.1+725919 138.45 38.08 37.65 3.05 094243.2+731312	NGC 2957 MCG 12-10- 2 MK 121 PGC 28119	.S?.... 	1.01± .09 .45± .06 .99	 .04 .34 					6852± 22 6973 6914
094715.6+675450 143.92 40.90 40.98 -.78 094310.0+680843	NGC 2976 UGC 5221 KUG 0943+681 PGC 28120	.SA.5P. R (1) 5.0± .3 6.8± .8	1.77± .01 .34± .02 1.34± .01 1.78	143 .11 .51 .17	10.82 ±.13 10.96 ±.14 10.27	.66± .01 .00± .02 .57 -.07	.70± .01 .01± .02 13.01± .02 13.69± .16	13.08±.1 97± 5 2.64	3± 5 11± 34 106 93
094716.3+220523 209.38 48.06 75.53 -33.36 094427.5+221918	NGC 2994 UGC 5239 IRAS09444+2219 PGC 28122	.L..... U -2.0± .8 	1.12± .04 .12± .08 .66± .03 1.11	125 .06 .00 	13.98 ±.14 14.07 ±.15 13.84 13.85	.91± .02 .40± .03 .81 .42	.98± .02 .41± .03 12.77± .09 14.13± .31		7386± 31 7310 7679
094720.1+463637 171.96 49.10 55.45 -16.63 094407.2+465032	UGC 5237 PGC 28126	.S..6*. U 6.0±1.2 	1.17± .05 .35± .05 1.17	155 .00 .52 .18	 14.74 ±.20 14.20			16.09±.1 246± 8 1.72	4687± 7 4708 4887
094721.5+254442 204.04 49.00 72.15 -31.04 094429.8+255837	UGC 5240 PGC 28128	.S..6*. U 6.0±1.4 	1.04± .08 .76± .06 1.04	109 .04 1.12 .38	 15.69 ±.18 14.49			15.22±.3 334± 7 .35	6929± 10 6867 7211
094734.4-020153 238.79 37.02 103.16 -45.36 094501.9-014757	UGC 5245 (1) PGC 28136	.SB.8?/ UE 8.0± .8 7.5±1.6	1.41± .03 .84± .04 1.42	160 .08 1.04 .42	 14.7 ±.2 13.57			14.78±.1 163± 16 154± 12 .79	1425± 11 1267 1757
094739.9-305657 262.74 17.23 145.81 -47.85 094528.1-304300	 ESO 434- 40 PGC 28144	.L?.... 	1.05± .05 .34± .04 .50± .21 1.05	50 .49 .00 	* 14.10 ±.14 13.57		.97± .02 .47± .04 11.84± .71 		2482± 42 2263 2785
094743.2-325015 264.07 15.83 148.54 -47.45 094533.1-323618	IC 2510 ESO 373- 29 PGC 28147	.SBT2*. Sr (1) 2.3± .5 5.6± .7	1.10± .05 .27± .04 1.17	148 .73 .33 .13	*				

R.A. 2000 DEC.	Names	Type	logD25	p.a.	BT	(B-V)T	(B-V)e	m21	V21
l b		ST nL	logR25	Ag	mB	(U-B)T	(U-B)e	W20	Vopt
SGL SGB		T	logAe	Ai	mFIR	(B-V)T°	m'e	W50	VGSR
R.A. 1950 DEC.	PGC	L	logDo	A21	BT°	(U-B)T°	m'25	HI	V3K

094745.3+023738		.SB.7*.	1.38± .04	15	*			14.02±.1	1882± 11
233.97 39.74	UGC 5249	U	.51± .05	.13	13.81 ±.20			240± 16	1738
97.09 -43.53		7.0± .8		.71				207± 12	2211
094509.8+025135	PGC 28148		1.39	.26	12.97			.79	

094745.4-313027		.IBS9..	1.25± .04	99	*				991
263.15 16.83	ESO 434- 41	S (1)	.37± .04	.54	14.69 ±.14				
146.62 -47.73		10.0± .6		.28					772
094534.1-311630	PGC 28149	10.0± .8	1.30	.18	13.87				1293

094745.6-062618	NGC 3007	.S..0./	1.10± .06	90	*				
243.10 34.34	MCG -1-25- 38	E	.41± .05	.12					
109.28 -46.71	IRAS09452-0612	.0± .9		.31	12.70				
094515.9-061221	PGC 28150		1.09						

094747.2+390504	A 0944+39		.50± .14					14.72±.1	1589± 10
183.48 50.19	CGCG 210- 34		.15± .12	.02	15.5 ±.4			224± 10	1625± 37
61.11 -21.97	MK 407							172± 5	1582
094443.5+391900	PGC 28153		.50						1825

094749.8+725756	NGC 2963	.SB.2..	1.09± .06	165	*				6538± 45
138.44 38.13	UGC 5222	CU	.32± .05	.04	14.34 ±.18				6658
37.70 3.06	MK 122	2.3± .7		.39	12.87				6600
094314.5+731150	PGC 28155		1.10	.16	13.84				

094750.8+155110	IC 565	.S..6?.	1.21± .05	50	*			14.30±.3	5852± 10
218.07 46.09	UGC 5248	U	.89± .05	.09					5753
81.84 -36.94		6.0±1.9		1.30				520± 7	6160
094506.5+160507	PGC 28159		1.22	.44					

094758.2+540054		.S?....	.96± .07						7402± 39
161.25 47.15	UGC 5241		.02± .05	.00	14.06 ±.19				7452
50.33 -11.13	MK 1425			.03	13.03				7567
094433.5+541450	PGC 28166		.96	.01	13.98				

094805.2+325257	A 0945+33							16.30±.3	1551± 10
193.17 50.20	CGCG 182- 20			.01	14.86 ±.18			173± 6	1462± 42
66.10 -26.21	MK 408							69± 5	1512
094507.6+330654	PGC 28169								1804

094830.2+575816	A 0944+58	.SB....	.93± .09		*				8433± 38
155.89 45.73	UGC 5243	R	.27± .06	.01	15.12 ±.18				8499
47.69 -8.14	MK 21			.41					8578
094457.6+581213	PGC 28182		.93	.14	14.65				

094836.0+332518	NGC 3003	.S..4$.	1.76± .01	79	12.33 ±.13	.43± .02	.51± .02	12.49±.0	1480± 4
192.34 50.34	UGC 5251	R (2)	.63± .02	.00	12.15 ±.12	-.16± .04	-.08± .03	289± 4	1476± 56
65.74 -25.77	IRAS09456+3339	4.0± .6	1.18± .02	.93	11.72	.30	13.72± .06	264± 4	1448
094537.9+333916	PGC 28186	5.8± .7	1.76	.32	11.30	-.25	14.42± .16	.88	1736

094843.7+440452	NGC 2998	.SXT5..	1.46± .02	53	13.11M±.10	.57± .02	.65± .02	13.65±.3	4777± 7
175.72 49.81	UGC 5250	R (2)	.33± .03	.00	12.93 ±.14			388± 9	4767± 12
57.47 -18.29	IRAS09455+4418	5.0± .3	.89± .02	.50	12.39	.48	13.05± .05	373± 7	4786
094534.3+441850	PGC 28196	1.9± .6	1.46	.17	12.53		14.44± .17	.96	4987

0948.7 +6411	A 0944+64	.S..6*.	1.21± .05	31				15.45±.1	3025± 11
148.12 42.94	UGC 5244	U	.83± .05	.16	15.19 ±.18				3114
43.55 -3.48		6.0±1.3		1.22				234± 8	3136
0944.9 +6425	PGC 28197		1.23	.42	13.79			1.25	

094853.7-080306	NGC 3029	.SXR5..	1.16± .06	130	14.5 ±.3	.51± .07		15.34±.1	6580± 10
244.82 33.52	MCG -1-25- 47	E (1)	.18± .05	.09				286± 11	6526± 59
111.68 -46.86		5.0± .8		.27		.41		272± 11	6404
094625.0-074907	PGC 28206	1.8± .8	1.16	.09	14.10		14.68± .43	1.15	6912

094901.9-070818		.LBT+P.	1.21± .07	95					
244.00 34.14	MCG -1-25- 41	E	.23± .05	.15					
110.39 -46.59		-1.0± .8		.00					
094632.6-065418	PGC 28217		1.19						

094903.0-070922		.E...*.	1.18± .08	95					
244.02 34.13	MCG -1-25- 42	E	.11± .05	.15					
110.42 -46.59		-5.0±1.2		.00					
094633.8-065522	PGC 28219		1.17						

094903.0-024920	NGC 3017	.E...*.	1.01± .13	90					
239.86 36.84	MCG 0-25- 19	E	.03± .07	.08	14.05 ±.15				
104.41 -45.29		-5.0±1.3		.00					
094631.0-023520	PGC 28220		1.02						

0949.2 +6925		.SA.8..	2.12± .02	155				15.42±.3	143± 11
142.13 40.24	UGC 5247	U	.34± .05	.17	14.0 ±**			98± 16	3500± 60
40.12 .47		8.0± .8		.42				69± 12	
0945.0 +6939	PGC 28225		2.13	.17					

R.A. 2000 DEC.	Names	Type	logD25	p.a.	BT	(B-V)T	(B-V)e	m21	V21
l b		ST nL	logR25	Ag	mB	(U-B)T	(U-B)e	W20	Vopt
SGL SGB		T	logAe	Ai	mFIR	(B-V)T0	m'	W50	VGSR
R.A. 1950 DEC.	PGC	L	logD0	A21	BT0	(U-B)T0	m'25	HI	V3K
094916.7-475519		.LAR+P.	1.15± .08	10					
274.26 4.55	ESO 213- 2	S	.07± .06	1.93					
168.21 -41.94	IRAS09473-4741	-1.5± .6		.00	12.74				
094723.0-474118	PGC 28234		1.36						
094922.5+010841	NGC 3015	.LX.0P?	.72± .08	95					
235.86 39.24	UGC 5261	E	.14± .04	.13	14.9 ±.3				7559± 63
99.22 -43.76	ARAK 218	-2.0±1.9		.00	12.26				7411
094647.9+012242	PGC 28240		.71		14.64				7890
094925.4-325031	IC 2511	PSXS1*/	1.46± .03	38					
264.35 16.07	ESO 374- 49	Sr	.70± .03	.74	13.01 ±.14				
148.43 -47.10		1.4± .4		.72					
094715.0-323630	PGC 28246		1.53	.35					
094926.1+143926		.S?....	1.16± .05	129				15.29±.3	5921± 10
219.87 45.96	UGC 5258		.98± .05	.08	15.55 ±.19				
83.37 -37.29				1.47				312± 7	5818
094642.8+145327	PGC 28248		1.17	.49	13.97			.83	6233
094928.4-214431	NGC 3025	.LA.0P*	1.18± .05	110					
256.34 24.24	ESO 566- 15	SE	.11± .03	.12	13.88 ±.14				
132.03 -48.39		-2.0± .6		.00					
094709.0-213030	PGC 28249		1.18						
094930.3+553446	A 0946+55		.59± .10		16.3 ±.2	.63± .03		17.18±.1	1592± 10
159.00 46.79	UGCA 184		.30± .07	.00	16.0 ±.3	-.41± .05		104± 6	1500± 63
49.42 -9.82	MK 22							69± 5	1646
094603.2+554846	PGC 28251		.59						1747
094939.2-051001	NGC 3022	PLX.0*.	1.21± .09						
242.26 35.51	MCG -1-25- 46	E	.00± .07	.13					
107.69 -45.89		-2.0±1.2		.00					
094708.7-045600	PGC 28257		1.23						
094941.4+003721	NGC 3018	.SBS3P?	1.07± .05	27				14.44±.2	1863± 6
236.47 39.01	UGC 5265	E	.24± .04	.18	14.13 ±.18			171± 6	1833± 32
99.93 -43.90		3.0±1.7		.33				116± 5	1713
094707.2+005122	PGC 28258		1.09	.12	13.61			.71	2194
094941.4+321252	NGC 3011	.L.....	.95± .06					17.89±.3	1527± 7
194.26 50.48	UGC 5259	U	.05± .03	.01	14.30 ±.16			198± 6	1441± 36
66.91 -26.41	MK 409	-2.0± .9		.00				166± 5	1487
094644.7+322653	PGC 28259		.94		14.26				1785
094951.1+124141	NGC 3016	.S..3..	1.08± .06	70				15.37±.2	8970± 5
222.48 45.22	UGC 5266	U (1)	.14± .05	.02	13.75 ±.19			499± 6	8869± 48
85.56 -38.27	IRAS09471+1255	3.0± .9		.19	12.84			458± 5	8858
094709.1+125543	PGC 28269	4.5±1.1	1.08	.07	13.47			1.83	9285
094952.1+344250	NGC 3012	.E...*.	1.03± .11						
190.33 50.66	UGC 5262	U	.02± .05		14.53 ±.16				
64.88 -24.69		-5.0±1.2		.00					
094653.1+345651	PGC 28270		1.02						
094952.6+003713	NGC 3023	.SXS5P*	1.46± .03	70				13.78±.1	1879± 7
236.50 39.04	UGC 5269	PE (1)	.30± .04	.18				149± 8	1866± 20
99.96 -43.85	IRAS09472+0051	5.0± .5		.46	12.14			127± 6	1728
094718.4+005115	PGC 28272	3.1±1.3	1.48	.15					2210
0949.8 +0905		.S..3..	1.20± .05	44					
226.95 43.57	UGC 5267	U (1)	.81± .05	.03	14.89 ±.18				
89.60 -40.11	IRAS09472+0919	3.0± .9		1.12	13.52				
0947.2 +0919	PGC 28274	4.5±1.3	1.20	.41					
094954.3-191108	NGC 3028	.S..3P*	1.12± .04	48					
254.44 26.14	ESO 566- 16	SE	.14± .04	.14	13.55 ±.14				
128.18 -48.25	IRAS09475-1856	2.6± .7		.19	12.35				
094733.1-185706	PGC 28276	3.1±1.6	1.13	.07					
0949.9 -1207		.I?....	.89± .11						2632
248.63 31.03	MCG -2-25- 19		.81± .07	.10					
117.65 -47.47				.60					2448
0947.5 -1153	PGC 28278		.90	.40					2965
0949.9 -3254	IC 2514	.SAS2*/	1.19± .07						
264.48 16.10	MCG -5-23- 19	S	.71± .07	.74					
148.47 -46.97		2.4± .5		.88					
0947.8 -3240	PGC 28283		1.26	.36					
094959.0-250032		.SBT6*.	1.30± .03	87					
258.88 21.96	ESO 499- 8	S (1)	.22± .03	.15	14.34 ±.14				
136.93 -48.14		6.0±1.1		.32					
094742.0-244630	PGC 28285	5.6± .8	1.32	.11					

R.A. 2000 DEC.	Names	Type	logD₂₅	p.a.	B_T	(B-V)_T	(B-V)_●	m₂₁	V₂₁
l b		S_T n_L	logR₂₅	A_g	m_B	(U-B)_T	(U-B)_●	W₂₀	V_opt
SGL SGB		T	logA_●	A_i	m_FIR	(B-V)°_T	m'_●	W₅₀	V_GSR
R.A. 1950 DEC.	PGC	L	logD_●	A₂₁	B°_T	(U-B)°_T	m'₂₅	HI	V_3K

0950.0 +3143					.03 14.92 ±.18				4959
195.02 50.51	CGCG 152- 70								4921
67.37 -26.68									5222
0947.1 +3158	PGC 28289								

095006.6+124854	NGC 3020	.SBR6*.	1.50± .03	105				13.31±.1	1440± 4
222.36 45.33	UGC 5271	PU	.30± .04	.02 12.63 ±.20				233± 6	1450± 48
85.47 -38.15	IRAS09474+1302	6.0± .5		.44 12.55				214± 5	1330
094724.5+130256	PGC 28296		1.51	.15 12.17				1.00	1756

095008.1-735517	NGC 3059	.SBT4..	1.56± .03	11.7 ±.2	.68± .04	.73± .02	12.94±.3	1260± 6
291.15 -15.36	ESO 37- 7	R (2)	.05± .03	.82 11.62 ±.12	-.06± .03	-.08± .02	145± 7	1292± 17
193.88 -26.91	IRAS09496-7341	4.0± .3	1.26± .04	.07 10.41	.47	13.50± .10		1056
094938.0-734112	PGC 28298	4.4± .4	1.64	.02 10.74	-.21	14.23± .25	2.18	1391

095010.6+441740	NGC 3009	.S?....	.89± .07				16.00±.1	4666± 11
175.32 50.03	UGC 5264		.03± .05	.01 14.52 ±.18				4604± 50
57.51 -17.96				.04			100± 12	4675
094701.3+443141	PGC 28303		.89	.01 14.44			1.54	4876

0950.1 +3424	A 0947+34				.00				6582± 72
190.80 50.72									6554
65.17 -24.85									6835
0947.2 +3439	PGC 28304								

095011.0+280046	A 0947+28	.E.1.$.	.62± .10				16.03±.1	1447± 7
200.80 50.06	MCG 5-23- 40	P	.08± .06	.04			109± 5	1503± 61
70.60 -29.10	KUG 0947+282	-5.0±1.9		.00				1394
094717.9+281448	PGC 28305		.60					1723

095014.0-120331		.S..7*/	1.18± .05	125				2720
248.64 31.13	MCG -2-25- 20	E	.92± .05	.10				
117.58 -47.39	IRAS09477-1149	7.0±1.3		1.27 12.87				2536
094747.9-114929	PGC 28308		1.19	.46				3053

095014.2+455731	A 0947+46				.00				7404± 42
172.79 49.73									7423
56.30 -16.76	MK 125								7609
094702.8+461133	PGC 28309								

0950.2 +1617		.S..6*.	1.08± .06					5908
217.80 46.79	UGC 5274	U	.00± .05	.09 14.3 ±.2				
81.79 -36.22		6.0±1.2		.00				5811
0947.5 +1632	PGC 28310		1.08	.00 14.18				6217

095018.5-230127		.SBR3*.	1.34± .04	11				
257.46 23.45	ESO 499- 9	Sr (1)	.31± .04	.11 13.55 ±.14				
133.96 -48.17		3.4± .4		.43				
094800.1-224724	PGC 28313	4.1± .7	1.35	.16				

095020.9+721645	NGC 2985	PSAT2..	1.66± .02	0 11.18S±.08	.74± .05	.83± .03	12.58±.1	1322± 11
139.01 38.68	UGC 5253	R (1)	.10± .03	.07 11.20 ±.13			320± 16	1299± 23
38.29 2.67	IRAS09459+7230	2.0± .3		.13 10.93	.70			1436
094552.6+723045	PGC 28316	1.0± .9	1.66	.05 10.98		14.06± .15	1.55	1384

095021.3+312915	A 0947+31	.I..9..	1.32± .03	115 14.46 ±.14	.40± .04	.35± .01	14.31±.1	520± 5
195.42 50.56	UGC 5272	U (1)	.42± .04	.04 14.1 ±.2	-.54± .05	-.45± .02	108± 5	
67.63 -26.79	DDO 64	10.0± .8	.86± .02	.31	.29	14.25± .04	82± 6	480
094725.3+314317	PGC 28317	8.0±1.5	1.33	.21 14.01	-.62	14.89± .23	.09	784

0950.4 +4345		.S?....	.89± .11					4809± 19
176.12 50.17	MCG 7-20- 68		.19± .07	.01 14.92 ±.18				4819
57.94 -18.30				.26				5024
0947.3 +4400	PGC 28322		.89	.09 14.62				

095027.5-214809		.SBS4..	1.12± .04	170				
256.57 24.36	ESO 566- 18	E (1)	.13± .03	.12 14.08 ±.14				
132.12 -48.16	IRAS09481-2134	4.0± .8		.19 12.90				
094808.0-213406	PGC 28323	1.9± .8	1.13	.06				

095027.5+124557	NGC 3024	.S..5*/	1.32± .03	125				13.96±.2	1415± 5
222.48 45.39	UGC 5275	P	.66± .04	.02 13.79 ±.19			257± 7	1532± 48
85.58 -38.11	IRAS09477+1259	5.0±1.2		.99 12.87			231± 5	1307
094745.5+130000	PGC 28324		1.32	.33 12.77			.86	1732

0950.5 +6429		.S..4..	1.06± .06	65				
147.61 42.96	UGC 5260	U	.39± .05	.17				
43.49 -3.12		4.0±1.0		.57				
0946.7 +6444	PGC 28328		1.07	.19				

095042.1+302941		.S?....	1.14± .07	125				15.33±.3	8759± 9
196.98 50.53	UGC 5276		.17± .06	.03 14.5 ±.2				8716
68.53 -27.39				.13			413± 5	
094747.0+304344	PGC 28341		1.13	14.16				9027

R.A. 2000 DEC. l b SGL SGB R.A. 1950 DEC.	Names PGC	Type S_T n_L T L	logD₂₅ logR₂₅ logA₀ logD₀	p.a. A_g A_i A₂₁	B_T m_B m_FIR B°_T	(B-V)_T (U-B)_T (B-V)°_T (U-B)°_T	(B-V)₀ (U-B)₀ m'_₀ m'₂₅	m₂₁ W₂₀ W₅₀ HI	V₂₁ V_opt V_GSR V_3K
095045.4+224513 208.75 49.02 75.49 −32.31 094756.5+225916	 UGC 5278 PGC 28343	.SBS3.. U 3.0± .9	.96± .07 .31± .05 .97	58 .08 .43 .16	 15.02 ±.18 14.44			16.64±.3 409± 5 2.04	8586± 9 8513 8879
095054.3+283257 NGC 3026 200.02 50.30 70.24 −28.63 094800.8+284701	UGC 5279 IRAS09480+2847 PGC 28351	.I..9.. U 10.0± .8	1.43± .02 .53± .05 1.44	82 .02 .40 .27	 13.52 ±.18 13.53 13.09			14.79±.0 220± 5 189± 7 1.43	1490± 4 1468± 76 1439 1765
0950.9 +0416 232.72 41.31 95.49 −42.12 0948.3 +0431	UGC 5284 PGC 28352	.S..8*. U 8.0±1.2	1.00± .16 .09± .12 1.00	 .05 .11 .04					5071 4933 5400
0950.9 +6210 150.37 44.13 45.08 −4.81 0947.2 +6225	UGC 5268 PGC 28353	.S..2.. U 2.0± .9	1.12± .04 .51± .04 1.13	134 .07 .62 .25	 15.09 ±.18				
095055.2−091935 246.36 33.07 113.68 −46.67 094827.3−090530	MCG −1-25- 48 IRAS09484-0904 PGC 28354	RSBTOP* E .0± .8	1.18± .05 .05± .05 1.19	 .13 .04	 14.09				5716 5539 6050
095056.5−045954 242.35 35.86 107.60 −45.53 094825.8−044550	MCG −1-25- 49 VV 110 PGC 28356	RSBR1.. E 1.0± .8	1.29± .05 .40± .05 1.30	155 .13 .41 .20	 12.96				
095057.3+333316 NGC 3021 192.18 50.84 66.00 −25.31 094759.5+334720	UGC 5280 IRAS09479+3347 PGC 28357	.SAT4*. R (1) 4.0± .4 3.4± .9	1.20± .04 .25± .03 1.20	110 .00 .37 .13	 13.04 ±.16 11.48 12.66			14.52±.1 302± 5 249± 5 1.73	1541± 4 1506± 40 1509 1797
095106.2+090031 227.26 43.79 89.89 −39.89 094826.6+091435	UGC 5286 IRAS09484+0914 PGC 28366	.SA.7.. U 7.0± .8	1.31± .04 .17± .05 1.31	35 .05 .23 .08	 13.6 ±.2 13.26				5200± 50 5077 5523
095108.5+154348 IC 568 218.69 46.77 82.53 −36.35 094824.6+155753	UGC 5285 IRAS09484+1557 PGC 28368	.SBT3.. U 3.0± .9	1.14± .07 .18± .06 1.15	 .08 .24 .09	 14.3 ±.2 12.79 13.92			16.46±.3 289± 7 2.45	8720± 10 8621 9031
0951.1 +3307 192.85 50.86 66.38 −25.57 0948.2 +3322	UGC 5282 PGC 28370	.S..9*. U 9.0±1.3	1.09± .06 .39± .05 1.09	40 .00 .40 .20					1557 1524 1816
095112.9−182829 254.12 26.86 127.14 −47.90 094851.1−181424	ESO 566- 19 PGC 28373	.SBS6.. E (1) 6.0± .8 4.4± .8	1.23± .03 .00± .03 1.24	 .16 .00 .00	 13.98 ±.14 13.79			15.04±.1 182± 9 157± 12 1.25	3701± 7 3503 4029
095115.9−324518 NGC 3038 264.59 16.39 148.18 −46.74 094905.1−323112	ESO 374- 2 PGC 28376	.SAT3.. PSr (2) 2.9± .3 3.4± .5	1.40± .03 .27± .03 .94± .03 1.45	130 .56 .38 .14	12.42 ±.15 12.62 ±.11 11.60	.85± .02 .23± .05 .66 .07	.93± .01 .31± .02 12.61± .11 13.59± .23		2686± 41 2466 2986
095117.3+074938 228.71 43.25 91.29 −40.42 094838.5+080343	UGC 5288 PGC 28378	.S..8*. U 8.0±1.2	1.10± .05 .19± .05 1.11	155 .02 .23 .10	 14.09 ±.18 13.83			14.03±.1 108± 4 93± 4 .10	557± 5 491± 36 429 880
095122.2−123728 249.34 30.95 118.48 −47.21 094856.4−122322	MCG −2-25- 23 PGC 28380	.SBS6.. E (1) 6.0± .9 5.3± .8	1.14± .06 .18± .05 1.15	50 .15 .27 .09					
095123.6−270036 NGC 3037 260.59 20.70 139.86 −47.64 094908.1−264630	ESO 499- 10 PGC 28381	.IBS9.. S (1) 10.0± .8 5.6± .8	1.09± .04 .06± .03 1.11	 .16 .05 .03	 13.66 ±.14 13.45				889 676 1203
095125.1+445526 174.30 50.13 57.21 −17.35 094815.3+450931	UGC 5283 PGC 28383	.SA.7.. U 7.0± .8	1.13± .05 .18± .05 1.13	145 .00 .25 .09	 15.1 ±.3 14.79				4658± 10 4672 4868
095127.4−050615 242.55 35.90 107.80 −45.44 094856.7−045209	MCG −1-25- 50 IRAS09489-0452 PGC 28388	.SBR2.. E 2.0± .8	1.19± .05 .29± .05 1.20	170 .13 .35 .14	 13.67				

R.A. 2000 DEC. l b SGL SGB R.A. 1950 DEC.	Names PGC	Type S_T n_L T L	$\log D_{25}$ $\log R_{25}$ $\log A_o$ $\log D_o$	p.a. A_g A_i A_{21}	B_T m_B m_{FIR} B_T^o	$(B-V)_T$ $(U-B)_T$ $(B-V)_T^o$ $(U-B)_T^o$	$(B-V)_e$ $(U-B)_e$ m'_e m'_{25}	m_{21} W_{20} W_{50} HI	V_{21} V_{opt} V_{GSR} V_{3K}
095128.0+325631 193.16 50.91 66.59 −25.65 094830.9+331036	UGC 5287 KUG 0948+331 PGC 28390	.SBS6.. U 6.0± .8	1.16± .04 .22± .04 1.16	15 .00 .32 .11	14.44 ±.20 14.12			15.62±.2 163± 10 139± 5 1.39	1469± 6 1474± 19 1436 1729
0951.6 +6529 146.37 42.54 42.91 −2.30 0947.7 +6543	UGC 5277 IRAS09477+6543 PGC 28401	.SBT4.. U 4.0± .8	1.17± .04 .07± .04 1.18	.16 .10 .03	14.8 ±.3 13.20				
095139.6−134244 250.32 30.26 120.09 −47.32 094914.5−132838	MCG −2−25− 25 · PGC 28403	.S..3P? E (1) 3.0±1.7 5.3± .8	1.17± .05 .25± .05 1.18	130 .13 .34 .12					4002 3814 4334
0951.7 +0126 235.97 39.90 99.15 −43.10 0949.1 +0141	A 0949+01 MCG 0−25− 24 DDO 65 PGC 28408	.IBS9*. PUE (2) 10.0± .6 9.5± .7	1.03± .06 .12± .05 1.04	55 .10 .09 .06				15.43±.1 90± 16 84± 12	1853± 11 1706 2185
0951.9 +1256 222.46 45.78 85.62 −37.71 0949.2 +1311	UGC 5291 PGC 28414	.L..−*. U −3.0±1.2	1.00± .19 .09± .08 .99	0 .04 .00	14.39 ±.15				
095155.0−064923 244.27 34.89 110.23 −45.82 094925.4−063517	NGC 3035 MCG −1−25− 52 IRAS09494−0635 PGC 28415	.SBT4.. E (1) 4.0± .8 3.1± .8	1.21± .05 .06± .05 1.23	25 .19 .09 .03	13.37				
095157.7−330431 264.93 16.24 148.59 −46.53 094947.1−325024	ESO 374− 3 PGC 28416	.SXT6*. Sr (1) 5.5± .4 4.1± .7	1.34± .03 .46± .03 1.40	154 .63 .67 .23	14.13 ±.14				
095200.0−251843 259.46 22.05 137.33 −47.66 094943.0−250436	ESO 499− 11 PGC 28418	.SBT7.. S (1) 7.0± .8 7.8± .8	1.08± .04 .05± .04 1.09	.19 .06 .02	14.94 ±.14 14.67				2636 2426 2954
095204.1+405131 180.59 50.86 60.36 −20.14 094859.3+410537	UGC 5290 KUG 0948+410 PGC 28420	.S..0?. U .0±1.8	.98± .05 .19± .04 .97	0 .00 .15	14.81 ±.18				
095208.0+291413 199.01 50.67 69.84 −27.98 094914.1+292820	NGC 3032 UGC 5292 IRAS09492+2928 PGC 28424	.LXR0.. R −2.0± .3	1.30± .04 .05± .04 .49± .03 1.29	95 .02 .00	13.18 ±.14 12.75 ±.12 12.39 12.88	.67± .01 .10± .02 .65 .10	.63± .01 .02± .01 11.12± .10 14.42± .26	17.10±.2 241± 10 157± 6	1533± 5 1568± 56 1485 1806
095219.8−292614 262.47 19.03 143.36 −47.12 095006.0−291206	ESO 435− 3 PGC 28439	.LAS+P* S −1.0± .9	1.18± .05 .30± .05 1.15	103 .22 .00	15.03 ±.14 14.64				11475± 15 11259 11784
095229.2+020917 235.35 40.46 98.36 −42.65 094954.1+022325	NGC 3039 UGC 5297 PGC 28452	.SXS3P* PUE 2.8± .6	1.11± .04 .32± .04 1.12	12 .09 .44 .16	14.24 ±.18 13.67				5036± 50 4891 5368
0952.6 −0012 237.89 39.13 101.39 −43.52 0950.1 +0002	UGC 5299 PGC 28463	.S..8*. U 8.0±1.4	.96± .17 .69± .12 .97	148 .09 .84 .34					2919 2767 3253
095254.1+425055 177.44 50.75 58.96 −18.63 094947.3+430503	A 0949+43 UGC 5295 IRAS09498+4304 PGC 28470	.SXS3.. U 3.0± .8	1.33± .04 .25± .05 1.33	150 .00 .35 .13	14.1 ±.2 13.46 13.73			14.21±.1 277± 34 278± 11 .36	4785± 9 4805± 39 4792 5005
095307.0+164044 217.67 47.57 81.88 −35.43 095022.5+165453	NGC 3041 UGC 5303 IRAS09503+1654 PGC 28485	.SXT5.. R (2) 5.0± .3 3.6± .7	1.57± .02 .19± .03 1.57	95 .08 .29 .10	12.3 ±.2 12.32 ±.16 12.20 11.94	.77± .03 .71	14.50± .24	14.05±.1 294± 6 279± 5 2.01	1414± 4 1309± 46 1317 1722
095309.9+374528 185.49 51.31 62.94 −22.13 095008.5+375937	A 0950+37 MK 410 PGC 28486		.50± .14 .15± .12 .50	.04					6946± 42 6932 7188
0953.1 +0752 228.98 43.67 91.53 −39.99 0950.5 +0806	UGC 5304 ARP 255 PGC 28487	.S?.... 1.04± .09	1.04± .09 .09± .07 1.04	.04 .14 .05	14.34 ±.19 13.42 14.09				12308± 48 12182 12634

R.A. 2000 DEC. l b SGL SGB R.A. 1950 DEC.	Names PGC	Type S_T n_L T L	$logD_{25}$ $logR_{25}$ $logA_e$ $logD_o$	p.a. A_g A_l A_{21}	B_T m_B m_{FIR} B_T^o	$(B-V)_T$ $(U-B)_T$ $(B-V)_T^o$ $(U-B)_T^o$	$(B-V)_e$ $(U-B)_e$ m'_e m'_{25}	m_{21} W_{20} W_{50} HI	V_{21} V_{opt} V_{GSR} V_{3K}
0953.2 +5828 154.85 46.08 47.81 -7.34 0949.7 +5843	UGC 5296 PGC 28489	.S..9*. U 9.0±1.2	1.01± .06 .06± .05 1.01	.03 .06 .03				16.22±.1 93± 8	1520± 7 1587 1663
095316.7-255540 260.14 21.78 138.21 -47.32 095100.0-254130	ESO 499- 13 PGC 28490	.LBS0.. S -2.0± .8	1.16± .05 .07± .03 1.18	.28 .00	13.94 ±.14				
095317.8-183844 254.65 27.09 127.43 -47.42 095055.9-182434	NGC 3045 MCG -3-25- 28 PGC 28492	.SAR3?. SE (2) 3.0± .8 3.6± .7	1.14± .05 .34± .04 1.15	110 .12 .47 .17					
095318.3+360507 188.16 51.38 64.30 -23.25 095018.6+361916	A 0950+36 MCG 6-22- 24 KUG 0950+363 PGC 28494	.S?.... 	.90± .06 .01± .06 .90	.00 .02 .01	14.80 ±.18 14.75				5482± 73 5461 5730
0953.3 +4250 177.42 50.83 59.02 -18.58 0950.2 +4305	UGC 5301 PGC 28495	.S..6*. U 6.0±1.4	1.11± .05 .91± .05 1.11	159 .00 1.34 .45					
095320.0+004158 237.07 39.79 100.32 -43.03 095045.8+005608	NGC 3042 UGC 5307 PGC 28498	.L..0*/ UE -2.0± .7	1.08± .06 .20± .03 1.05	111 .06 .00	13.85 ±.17				
0953.4 +0752 229.01 43.74 91.57 -39.92 0950.8 +0807	UGC 5308 PGC 28502	.S?.... 	1.00± .06 .00± .05 1.01	.04 .00 .00	14.4 ±.2 14.34			16.07±.3 77± 7 57± 7 1.74	5358± 8 5327± 56 5232 5684
095333.3-193459 255.44 26.46 128.83 -47.40 095112.0-192048	ESO 566- 24 IRAS09512-1920 PGC 28510	.SBR4.. SE (1) 3.7± .4 2.2± .8	1.20± .03 .09± .03 1.21	67 .13 .13 .05	13.60 ±.14 12.91				
095336.1-122858 249.65 31.44 118.41 -46.65 095110.1-121447	NGC 3058 MCG -2-25- 26 VV 741 PGC 28513	.S?.... 	1.12± .08 .28± .07 1.13	35 .12 .43 .14	11.52				7525 7340 7859
095339.8+013446 236.19 40.37 99.25 -42.61 095105.0+014857	NGC 3044 UGC 5311 IRAS09511+0148 PGC 28517	.SBS5$/ R (1) 5.0± .5 4.2±1.6	1.69± .02 .84± .02 1.14± .01 1.70	13 .05 1.26 .42	12.46 ±.13 12.53 ±.12 10.64 11.18	.53± .02 -.19± .03 .35 -.32	.55± .02 -.11± .03 13.65± .04 13.68± .17	13.13±.2 351± 9 324± 6 1.53	1292± 6 1344± 34 1147 1627
0953.6 +0222 235.33 40.83 98.24 -42.28 0951.1 +0237	UGC 5312 PGC 28520	.S..6*. U 6.0±1.4	1.02± .08 .78± .06 1.03	85 .09 1.15 .39	15.60 ±.19				
0953.8 +0852 227.88 44.32 90.48 -39.36 0951.2 +0907	UGC 5314 PGC 28531	.S..6*. U 6.0±1.5	1.00± .08 1.02± .06 1.01	168 .06 1.47 .50	16.0 ±.2 14.48				6406 6283 6731
095356.5+232259 208.12 49.90 75.43 -31.35 095107.4+233710	UGC 5313 IRAS09511+2337 PGC 28533	.S?.... 	.84± .08 .14± .05 .85	130 .06 .17 .07	14.9 ±.2 12.93 14.58			16.34±.3 229± 5 1.69	3962± 9 3957± 50 3892 4254
095358.5-311806 264.05 17.86 145.95 -46.46 095146.0-310354	ESO 435- 5 PGC 28534	.SBT5./ S (1) 5.0± .9 2.2±1.5	1.27± .04 .82± .04 1.30	85 .31 1.23 .41	15.12 ±.14				
095358.7-271706 261.24 20.88 140.17 -47.04 095143.0-270254	NGC 3051 ESO 499- 16 PGC 28536	PLBS-P* S -2.7± .6	1.32± .05 .03± .04 1.32	.08 .00	12.79 ±.14 12.67				2552± 15 2339 2866
0954.1 +5820 154.95 46.24 47.98 -7.36 0950.6 +5835	UGC 5306 PGC 28542	.SBS3.. U 3.0± .9	.96± .09 .18± .06 .96	30 .03 .25 .09	15.32 ±.19				
095414.2+371756 186.21 51.54 63.46 -22.29 095113.5+373207	UGC 5315 KUG 0951+375 PGC 28551	.S..6*. U 6.0±1.4	1.06± .04 .79± .04 1.06	59 .03 1.16 .39					

R.A. 2000 DEC.	Names	Type	logD₂₅	p.a.	B_T	(B-V)_T	(B-V)•	m₂₁	V₂₁
l b		S_T n_L	logR₂₅	A_g	m_B	(U-B)_T	(U-B)•	W₂₀	V_opt
SGL SGB		T	logA•	A_i	m_FIR	(B-V)°_T	m'•	W₅₀	V_GSR
R.A. 1950 DEC.	PGC	L	logD₀	A₂₁	B°_T	(U-B)°_T	m'₂₅	HI	V_3K

095418.3+231721		.SBR6..	1.25± .04	102				15.40±.3	4140± 9
208.29 49.95	UGC 5320	U	.37± .05	.06	14.17 ±.20				
75.58 -31.34	IRAS09515+2331	6.0± .8		.54	13.50			304± 5	4070
095129.3+233133	PGC 28557		1.26	.18	13.55			1.67	4433
0954.3 +6820		.SAS8..	1.29± .04					15.16±.1	4376± 11
142.97 41.23	UGC 5302	U	.06± .05	.14	14.5 ±.4				
41.19 -.02	IRAS09503+6834	8.0± .8		.08				244± 8	4480
0950.3 +6834	PGC 28563		1.30	.03	14.31			.82	4465
095423.0-330701		.S.3?P/	1.15± .04	130					
265.36 16.54	ESO 374- 8	S	.65± .04	.57	15.59 ±.14				
148.49 -46.02		3.0±1.8		.90					
095212.0-325248	PGC 28565		1.20	.32					
095427.0-065713	IC 574	.LA.-*.	1.14± .07	5					
244.88 35.28	MCG -1-25- 56	E	.08± .05	.16					
110.65 -45.25		-3.0± .6		.00					
095157.5-064300	PGC 28569		1.15						
095428.0-183822	NGC 3052	.SXR5*.	1.31± .02	102	12.78 ±.15	.58± .02	.66± .01	14.78±.2	3768± 8
254.87 27.28	ESO 566- 26	R (4)	.19± .03	.11	12.92 ±.12			314± 34	3586± 66
127.45 -47.14	IRAS09521-1824	5.0± .4	.97± .02	.28	12.02	.49	13.12± .05	276± 7	3568
095206.0-182409	PGC 28570	2.0± .4	1.32	.09	12.46		13.70± .20	2.24	4095
095429.0-254210	NGC 3054	.SXR3..	1.58± .02	118	*			14.78±.2	2433± 8
260.19 22.13	ESO 499- 18	R (3)	.21± .02	.28	12.23 ±.11			403± 9	2194± 50
137.84 -47.07		3.0± .3		.28				393± 25	2217
095212.0-252757	PGC 28571	1.5± .5	1.61	.10	11.66			3.02	2745
095433.0-065128	IC 575	.S..1P/	1.22± .04	50	*				
244.81 35.37	MCG -1-25- 58	E	.15± .04	.16					
110.53 -45.20	ARP 292	1.0± .8		.15					
095203.4-063715	PGC 28575		1.24	.07					
095433.0-281749	NGC 3056	RLAS+*.	1.26± .04	16	12.57 ±.14	.90± .01	.92± .01		
262.06 20.21	ESO 435- 7	R	.20± .03	.18	12.73 ±.11	.31± .02	.36± .02		1017± 66
141.62 -46.80		-1.0± .4	.75± .03	.00		.82	11.81± .08		803
095218.1-280336	PGC 28576		1.25		12.48	.25	13.25± .25		1329
095439.5+372432	IC 2515	.S..3..	1.04± .06	173					
186.03 51.62	UGC 5321	U	.72± .05	.03	15.2 ±.2				5858
63.44 -22.15	IRAS09516+3738	3.0±1.0		.99	13.39				5843
095138.7+373844	PGC 28581		1.04	.36	14.15				6102
095449.8+091619	NGC 3049	.SBT2..	1.34± .03	25				14.55±.1	1494± 4
227.57 44.72	UGC 5325	U	.18± .04	.04	13.04 ±.19			213± 5	1439± 27
90.18 -38.96	MK 710	2.0± .8		.22	12.12			199± 7	1372
095210.2+093032	PGC 28590		1.34	.09	12.77			1.70	1818
0955.1 +1417									
221.17 47.06	CGCG 92- 72			.04	15.0 ±.3				7194± 57
84.70 -36.33									7090
0952.4 +1432	PGC 28602								7510
095509.1-330815	IC 2522	.SBS5P.	1.44± .03	0				13.26±.1	3012± 5
265.50 16.63	ESO 374- 10	PSU (3)	.14± .03	.58	12.60 ±.12			318± 7	
148.47 -45.86	IRAS09529-3254	5.3± .4		.21	11.38			281± 6	2792
095258.1-325400	PGC 28606	2.8± .5	1.50	.07	11.80			1.39	3313
095510.0-331239	IC 2523	.SBS4P?	1.12± .04	25					
265.55 16.57	ESO 374- 11	PS (1)	.23± .03	.58	13.62 ±.14				
148.57 -45.84	IRAS09529-3258	3.7± .5		.34	12.34				
095259.0-325824	PGC 28607	4.4± .9	1.18	.12					
0955.2 +1418									
221.17 47.09	CGCG 92- 73			.04	15.3 ±.3				7182± 57
84.70 -36.30									7079
0952.5 +1433	PGC 28611								7498
095517.4+041617	NGC 3055	.SXS5..	1.32± .03	63	12.7 ±.2	.59± .06		14.75±.3	1832± 17
233.54 42.22	UGC 5328	R (2)	.21± .03	.10	12.61 ±.15			280± 34	1880± 58
96.14 -41.13	IRAS09526+0430	5.0± .4		.31	11.56	.52		257± 25	1698
095241.0+043031	PGC 28617	3.3± .7	1.33	.10	12.23		13.63± .25	2.42	2167
095524.7+331547		.I..9..	1.02± .06					15.88±.2	1412± 6
192.72 51.75	UGC 5326	U	.07± .05	.01	14.39 ±.18			120± 6	1415± 50
66.93 -24.81		10.0± .9		.05				100± 5	1380
095228.0+333001	PGC 28623		1.02	.04	14.32			1.52	1672
095529.5+082328	A 0952+08	.S..2$P	.95± .07	90					
228.57 44.43	MCG 2-25- 56	P	.35± .06	.08	15.14 ±.18				1283± 56
91.29 -39.23	VV 373	2.0±1.8		.43					1159
095250.5+083743	PGC 28627		.95	.17	14.62				1609

R.A. 2000 DEC.	Names	Type	logD$_{25}$	p.a.	B$_T$	(B-V)$_T$	(B-V)$_e$	m$_{21}$	V$_{21}$
l b	S$_T$ n$_L$	logR$_{25}$	A$_g$	m$_B$	(U-B)$_T$	(U-B)$_e$	W$_{20}$	V$_{opt}$
SGL SGB	T	logA$_e$	A$_l$	m$_{FIR}$	(B-V)^o_T	m'$_e$	W$_{50}$	V$_{GSR}$
R.A. 1950 DEC.	PGC	L	logD$_o$	A$_{21}$	B^o_T	(U-B)^o_T	m'$_{25}$	HI	V$_{3K}$
095533.5+690400 NGC 3031	.SAS2..	2.43± .01	157	7.89M±.03	.95± .01	.99± .01	9.99±.1	-34± 4
142.09 40.90 UGC 5318	R (2)	.28± .02	.16	7.92 ±.13			434± 5	-49± 10
40.77 .59 IRAS09514+6918	2.0± .3	1.82± .04	.34	8.59	.86	12.40± .12	422± 5	69
095127.7+691813 PGC 28630	2.2± .5	2.44	.14	7.40		14.19± .08	2.46	48
095533.8+162558 NGC 3053	.SB.1?.	1.26± .04	140				15.69±.3	3731± 10
218.34 48.02 UGC 5329	U	.31± .05	.09	13.63 ±.18			421± 13	
82.54 -35.08 IRAS09528+1640	1.0± .8		.31	12.98			405± 10	3635
095249.7+164013 PGC 28631		1.26	.15	13.18			2.36	4042
095539.5+721213 NGC 3027	.SBT7*.	1.63± .02	130	12.18 ±.15	.42± .05	.46± .02	12.46±.1	1061± 5
138.78 39.05 UGC 5316	R (1)	.33± .03	.04	12.17 ±.18	-.15± .06	-.12± .03	223± 6	1046± 58
38.64 2.89 VV 358	7.0± .4	1.43± .03	.45	12.94	.34	14.82± .10	205± 7	1179
095115.9+722626 PGC 28636	5.0± .9	1.64	.16	11.67	-.21	14.37± .19	.62	1128
095541.6-062149 NGC 3064	.S..4?/	1.05± .05	145					
244.58 35.90 MCG -1-26- 1	E	.54± .04	.15					
109.96 -44.79	4.0±1.3		.80					
095311.7-060734 PGC 28638		1.07	.27					
0955.7 +1624	.I..9*.	1.04± .15					16.80±.1	1105± 10
218.39 48.05 UGC 5332	U	.59± .12	.09				63± 6	
82.59 -35.06	10.0±1.3		.44				62± 5	1009
0953.0 +1639 PGC 28641		1.05	.29					1416
095546.6+322544	.S..6*.	1.00± .06	167					
194.07 51.78 UGC 5331	U	.43± .05	.03	15.46 ±.18				
67.69 -25.31	6.0±1.3		.64					
095250.6+323959 PGC 28645		1.00	.22					
095553.1-230322	.S.1?P/	.99± .06	54					
258.51 24.31 ESO 499- 21	S	.58± .05	.11	15.36 ±.14				
133.95 -46.88	1.0±1.6		.59					
095334.0-224906 PGC 28654		1.00	.29					
095554.0+694057 NGC 3034	.I.0../	2.05± .01	65	9.30M±.09	.89± .01	.90± .01	11.54±.2	203± 4
141.40 40.57 UGC 5322	R	.42± .02	.13	9.24 ±.12	.31± .03	.38± .02	214± 12	300± 10
40.38 1.06 ARP 337	90.0	1.45± .01	.31	5.58	.79	11.98± .04	146± 6	323
095145.3+695511 PGC 28655		2.05		8.83	.22	13.38± .12		296
095603.8+102954 IC 577	.S?....	.72± .09						
226.25 45.58 UGC 5334		.02± .05	.06	14.9 ±.3				9009± 40
88.99 -38.09			.03					8892
095323.5+104410 PGC 28662		.72	.01	14.71				9332
095605.0-374611	.SXT4..	1.11± .06	5					
268.71 13.19 ESO 316- 4	r	.14± .06	.77	14.74 ±.14				
154.74 -44.54	4.5± .8		.20					
095358.0-373154 PGC 28663		1.18	.07					
095607.9-215923	.SXT4P.	1.23± .05	16					
257.76 25.13 ESO 566- 30	S (1)	.24± .05	.09	14.67 ±.14				
132.39 -46.84	4.0± .8		.35					
095348.0-214506 PGC 28667	5.6± .8	1.24	.12					
095611.9+755154 NGC 3061	PSBT5..	1.22± .03		13.47S±.11			14.68±.1	2457± 8
135.12 36.80 UGC 5319	PU	.04± .04	.07	13.71 ±.19			225± 16	
36.16 5.58 IRAS09513+7606	5.0± .5		.06	13.78			208± 8	2587
095118.1+760608 PGC 28670		1.23	.02	13.38		14.32± .21	1.28	2503
095614.0+591825 NGC 3043	.S..3*/	1.24± .03	84				15.81±.1	2995± 6
153.54 46.05 UGC 5327	P	.48± .04	.02	13.60 ±.14			277± 16	2934± 50
47.52 -6.48 IRAS09527+5932	3.0±1.2		.67	12.65			233± 8	3065
095241.4+593240 PGC 28672		1.24	.24	12.89			2.68	3134
0956.2 +1029 IC 578	.SB.1*.	.98± .07	72					
226.29 45.61 UGC 5337	U	.34± .05	.06	14.75 ±.19				8944± 40
89.03 -38.06 IRAS09535+1043	1.0± .9		.35	12.49				8828
0953.5 +1043 PGC 28674		.98	.17	14.23				9268
095617.9-125817	.LB.-P?	1.02± .12	55					
250.60 31.59 MCG -2-26- 3	E	.11± .07	.16					
119.27 -46.09	-3.0±1.3		.00					
095352.1-124400 PGC 28678		1.03						
095619.7+164954 NGC 3060	.S..3..	1.35± .04	78				15.20±.3	3690± 10
217.89 48.34 UGC 5338	U (1)	.60± .05	.04	13.82 ±.18			503± 13	3745± 50
82.25 -34.71 IRAS09535+1704	3.0± .9		.83	12.92			481± 10	3598
095335.4+170411 PGC 28680	4.5±1.2	1.35	.30	12.93			1.97	4003
095620.3+271344 IC 2520	.S?....	.81± .08					15.42±.3	1238± 9
202.39 51.27 UGC 5335		.07± .05	.04	14.7 ±.2				1224± 36
72.29 -28.55 IRAS09534+2727			.10	11.77			169± 5	1182
095328.7+272800 PGC 28682		.82	.04	14.60			.78	1519

R.A. 2000 DEC. l b SGL SGB R.A. 1950 DEC.	Names PGC	Type S_T n_L T L	logD_25 logR_25 logA_o logD_o	p.a. A_g A_i A_21	B_T m_B m_FIR B_T^o	(B-V)_T (U-B)_T (B-V)_T^o (U-B)_T^o	(B-V)_o (U-B)_o m'_o m'_25	m_21 W_20 W_50 HI	V_21 V_opt V_GSR V_3K
095621.8-311759 264.46 18.20 145.82 -45.96 095409.0-310342	 ESO 435- 10 PGC 28685	.SBS6.. S (1) 6.0± .9 6.7±1.0	1.25± .03 .46± .03 1.27	4 .26 .68 .23	 14.69 ±.14 				
095625.9-260542 260.83 22.13 138.37 -46.60 095409.0-255124	 ESO 499- 23 PGC 28690	.LAT-.. S -3.0± .8 	1.32± .05 .23± .04 1.30	109 .13 .00 	 12.75 ±.14 				
095634.8+110947 225.49 46.00 88.33 -37.65 095354.1+112404	 MCG 2-26- 4 MK 1241 PGC 28698	.L?.... 	.90± .17 .00± .07 .90	 .06 .00 	 14.63 ±.15 14.39				12408± 81 12294 12731
095635.8+012542 236.92 40.88 99.83 -41.99 095401.1+013959	NGC 3062 CGCG 8- 2 PGC 28699	.S..3.. E 3.0±1.1 	.75± .14 .33± .08 .76	65 .04 .45 .16	 15.5 ±.2 				
095636.6+203853 212.45 49.71 78.49 -32.48 095349.7+205310	 UGC 5341 IRAS09537+2053 PGC 28700	.S..6*. U 6.0±1.3 	1.47± .03 1.13± .05 1.47 1.47	57 .07 1.47 .50	 15.03 ±.20 13.59 13.45			14.07±.3 607± 7 .12	7568± 10 7488 7869
095639.5-134630 251.35 31.09 120.44 -46.12 095414.1-133212	IC 579 MCG -2-26- 5 IRAS09542-1332 PGC 28702	.SBT2*. E 2.0± .9 	1.09± .06 .47± .05 1.10	127 .15 .58 .24	 12.47 				
0956.6 +7917 131.90 34.58 33.80 8.08 0951.0 +7932	 UGC 5310 PGC 28703	.SBS3.. U 3.0± .9 	1.04± .08 .13± .06 1.05	 .07 .18 .06	 15.3 ±.3 				
095642.6+153817 219.59 47.96 83.55 -35.29 095359.1+155234	 UGC 5342 MK 712 PGC 28707	.S?.... 	1.04± .06 .30± .05 1.04	22 .07 .44 .15	 14.47 ±.19 13.28 13.93				4560± 27 4461 4873
095643.1+462730 171.63 50.71 56.76 -15.61 095332.7+464147	A 0953+46 MCG 8-18- 44 MK 129 PGC 28708	.S?.... 	.82± .13 .17± .07 .81	 .01 .13 	 15.15 ±.19 14.94				4660± 38 4681 4865
0956.7 +2849 199.89 51.61 70.94 -27.47 0953.8 +2903	A 0953+29 UGC 5340 DDO 68 PGC 28714	.I..9P* PU 10.0± .7 	1.43± .05 .45± .06 .89± .02 1.43	0 .03 .34 .23	14.76 ±.14 14.1 ±.3 14.30	.28± .05 -.30± .06 .16 -.39	.32± .04 -.27± .06 14.70± .04 15.62± .31	13.73±.1 97± 5 78± 6 -.80	503± 5 454 779
095645.9+164832 217.98 48.43 82.35 -34.63 095401.6+170250	 UGC 5343 PGC 28716	.S..8.. U 8.0± .9 	1.16± .07 .36± .06 1.16	115 .04 .44 .18	 14.8 ±.2 14.35			15.74±.3 214± 7 1.21	3748± 10 3654 4059
095648.0-071047 245.56 35.58 111.18 -44.74 095418.5-065629	 MCG -1-26- 2 IRAS09542-0656 PGC 28718	PSBS2?. E 2.0±1.2 	1.27± .05 .61± .05 1.29	0 .16 .75 .30	 12.72 				
095650.4+600516 152.49 45.76 47.03 -5.86 095316.1+601933	A 0953+60A MCG 10-14- 53 MK 128 PGC 28719	.S?.... 	.63± .09 .35± .06 .63	 .02 .53 .18					9300±110 9374 9436
095700.6+595803 152.63 45.83 47.13 -5.93 095326.7+601220	A 0953+60B MCG 10-14- 54 MK 23 PGC 28729	.L?.... 	.52± .12 .04± .06 .52	 .02 .00 	 15.8 ±.2 15.63 				9140± 43 9214 9276
095703.2-321519 265.22 17.56 147.12 -45.64 095451.0-320100	IC 2526 ESO 435- 12 PGC 28732	.LXS0*. S -2.0± .6 	1.32± .04 .48± .03 1.30	55 .47 .00 	 13.65 ±.14 				
0957.1 +1533 219.76 48.02 83.69 -35.24 0954.4 +1548	 UGC 5344 PGC 28736	.SX.7.. U 7.0± .9 	1.04± .06 .02± .05 1.04	 .07 .03 .01	 14.5 ±.2 14.40			16.75±.2 51± 6 35± 5 2.34	4106± 7 4007 4420
0957.3 +0431 233.62 42.78 96.11 -40.57 0954.7 +0446	 UGC 5347 PGC 28741	.S..7.. U 7.0±1.0 	1.14± .05 .89± .05 1.15	18 .07 1.23 .45	 15.2 ±.2 13.86				2155 2019 2487

R.A. 2000 DEC. / l b / SGL SGB / R.A. 1950 DEC.	Names / PGC	Type / S_T n_L / T / L	$\log D_{25}$ / $\log R_{25}$ / $\log A_e$ / $\log D_o$	p.a. / A_g / A_i / A_{21}	B_T / m_B / m_{FIR} / B_T^o	$(B-V)_T$ / $(U-B)_T$ / $(B-V)_T^o$ / $(U-B)_T^o$	$(B-V)_\bullet$ / $(U-B)_\bullet$ / $m'_\bullet$ / m'_{25}	m_{21} / W_{20} / W_{50} / HI	V_{21} / V_{opt} / V_{GSR} / V_{3K}
095719.4-022501									
241.06 38.72	CGCG 8- 4			.08	14.4 ±.3				14290± 58
104.85 -43.21									14133
095447.1-021042	PGC 28743								14628
095721.1+071120		.S?....	.52± .20					16.58±.1	6480± 7
230.53 44.21	MCG 1-26- 4		.00± .07	.00	15.6 ±.3			229± 21	6659± 63
92.96 -39.39	ARAK 223			.00					6355
095442.9+072539	PGC 28745		.52		15.52				6812
095723.8-192120	NGC 3072	.S..0?/	1.27± .03	71					
255.99 27.25	ESO 566- 33	SE	.49± .03	.13	13.73 ±.14				
128.55 -46.49		-.3± .9		.37					
095502.0-190700	PGC 28749		1.26						
095727.9+451418		.S?....	.64± .17		16.49S±.15				
173.47 51.11	MCG 8-18- 46		.57± .07	.00					9717± 41
57.77 -16.38	HICK 41C			.84					9733
095419.3+452837	PGC 28753		.64	.28	15.58		13.10± .86		9928
0957.5 +6902	Holmberg IX	.I..9..	1.40± .05		14.3 ±.3	.20± .06	.26± .04	12.33±.2	46± 6
141.98 41.06	UGC 5336	U (1)	.09± .06	.16		-.40± .08	-.31± .05		119± 60
40.92 .69	DDO 66	10.0± .7	1.26± .06	.07		.14	16.05± .14	69± 5	154
0953.4 +6916	PGC 28757	9.0±1.0	1.41	.05	14.08	-.44	15.91± .41	-1.79	133
095732.7+333704	IC 2524	.S?....	.80± .07					15.91±.3	1487± 10
192.17 52.21	MCG 6-22- 39		.17± .06	.01	14.9 ±.2			202± 6	1486± 42
66.96 -24.25	MK 411			.25				124± 5	1457
095435.9+335123	PGC 28758		.81	.08	14.67			1.15	1747
095735.8+451345		.S?....	1.18± .05	65	14.56S±.15				
173.47 51.14	UGC 5345		.75± .05	.00	14.5 ±.2				3751± 32
57.79 -16.37	IRAS09544+4528			1.13					3767
095427.2+452804	PGC 28764		1.18	.38	13.40		13.41± .31		3962
095737.6-181044	NGC 3076	.S..2P*	.99± .04						
255.12 28.14	ESO 566- 34	SE	.04± .04	.11	14.03 ±.14				
126.84 -46.37		1.8± .7		.04					
095515.0-175624	PGC 28766		1.00	.02					
095740.8+451531		.S?....	.97± .07	28	15.30S±.15				
173.42 51.15	UGC 5346		.51± .05	.00	15.24 ±.19				7241± 41
57.78 -16.33	HICK 41B			.77					7258
095432.2+452950	PGC 28770		.97	.26	14.46		13.72± .38		7452
095748.3-283021		.S..5*/	1.41± .03	53					
262.78 20.52	ESO 435- 14	S	.92± .03	.18	14.44 ±.14				
141.79 -46.06		5.0±1.3		1.38					
095533.0-281600	PGC 28778		1.42	.46					
0957.9 +1026	NGC 3069	.L?....	.90± .17						5296
226.64 45.97	MCG 2-26- 5		.35± .07	.06	15.08 ±.17				5180
89.35 -37.71				.00					5620
0955.3 +1041	PGC 28788		.85		14.94				
0957.9 +4720		.S..8*.	1.06± .08	120					
170.19 50.69	UGC 5348	U	.75± .06	.01					
56.26 -14.84		8.0±1.3		.92					
0954.8 +4735	PGC 28789		1.06	.37					
095806.6+371735		.S..8*.	1.42± .03	38				14.43±.1	1381± 11
186.17 52.31	UGC 5349	U	.51± .04	.05	14.2 ±.2			213± 16	
64.03 -21.72	KUG 0955+375	8.0±1.1		.63				198± 12	1366
095506.7+373155	PGC 28795		1.43	.26	13.47			.70	1627
095807.1+102140	NGC 3070	.E.....	1.16± .04		13.25 ±.13	.96± .01	.97± .01		
226.77 45.96	UGC 5350	U	.00± .08	.06	13.03 ±.16	.41± .03			5392± 31
89.46 -37.72		-5.0± .8	.71± .02	.00		.90	12.29± .06		5275
095527.1+103601	PGC 28796		1.17		13.02	.43	14.05± .31		5717
0958.2 +1555	IC 581	.SB.2*.	1.02± .08	130					
219.41 48.41	UGC 5352	U	.38± .06	.09					
83.49 -34.82		2.0± .9		.47					
0955.5 +1610	PGC 28800		1.03	.19					
095820.9-251022		.SBS8*.	1.42± .03	22				15.09±.3	2384± 11
260.52 23.11	ESO 499- 26	SU (1)	.69± .03	.15	14.18 ±.14			177± 15	
136.99 -46.24		7.5± .8		.85				153± 12	2175
095603.1-245600	PGC 28803	6.7± .8	1.44	.34	13.18			1.57	2705
095822.0+322212	NGC 3067	.SXS2$.	1.39± .02	105	12.78 ±.13	.69± .01	.64± .01	15.33±.1	1476± 4
194.22 52.32	UGC 5351	R (2)	.42± .03	.07	12.89 ±.13	.08± .02	.18± .02	256± 5	1487± 28
68.13 -24.93	IRAS09554+3236	2.0± .8	.89± .01	.51	10.68	.59	12.72± .02	245± 6	1442
095526.5+323633	PGC 28805	5.4± .7	1.39	.21	12.25	.01	13.53± .19	2.88	1741

R.A. 2000 DEC.	Names	Type	$\log D_{25}$	p.a.	B_T	$(B-V)_T$	$(B-V)_e$	m_{21}	V_{21}
l b		S_T n_L	$\log R_{25}$	A_g	m_B	$(U-B)_T$	$(U-B)_e$	W_{20}	V_{opt}
SGL SGB		T	$\log A_e$	A_i	m_{FIR}	$(B-V)_T^o$	m'_e	W_{50}	V_{GSR}
R.A. 1950 DEC.	PGC	L	$\log D_o$	A_{21}	B_T^o	$(U-B)_T^o$	m'_{25}	HI	V_{3K}
095824.5-265534	NGC 3078	.E.2+..	1.40± .04	177	12.14 ±.13	1.01± .01	1.04± .01		2495± 25
261.78 21.80	ESO 499- 27	R	.08± .04	.18	12.03 ±.11	.59± .02	.63± .01		2283
139.51 -46.09		-5.0± .3	.88± .02	.00		.95	12.03± .05		2812
095608.0-264112	PGC 28806		1.40		11.86	.56	13.93± .24		
0958.4 +0103		.S..6?.	1.04± .08	108					
237.67 41.05	UGC 5355	U	.59± .06	.02					
100.53 -41.70		6.0±1.9		.86					
0955.9 +0118	PGC 28809		1.04	.29					
0958.5 +2852	NGC 3068	.L..-*.	.96± .12						6322± 27
199.90 52.02	UGC 5353	U	.08± .05	.05					6274
71.18 -27.13	ARP 174	-3.0±1.2		.00					6600
0955.7 +2907	PGC 28815		.96						
095836.5+131517			.64± .17						2759± 63
223.09 47.39	MCG 2-26- 8		.11± .07	.03	15.3 ±.3				2652
86.37 -36.16	MK 1242								3079
095554.6+132939	PGC 28817		.64						
0958.6 +0515		.S..6*.	1.00± .08	50					
233.02 43.46	UGC 5357	U	.55± .06	.05					
95.42 -39.96		6.0±1.3		.81					
0956.0 +0530	PGC 28818		1.00	.27					
0958.7 +1123		.SBS3..	1.19± .05	77				15.60±.3	2914± 9
225.56 46.59	UGC 5358	U	.21± .05	.08	14.2 ±.2			213± 7	
88.42 -37.07		3.0± .8		.29				200± 7	2801
0956.1 +1138	PGC 28821		1.20	.10	13.77			1.73	3237
095847.3-283717		.I.0.?P	1.22± .04						919± 76
263.04 20.57	ESO 435- 16	S	.17± .04	.20	13.43 ±.14				705
141.92 -45.83	VV 592	90.0		.25					1233
095632.0-282254	PGC 28822		1.24	.08	12.97				
0958.8 +3136	NGC 3071								6418
195.47 52.36	CGCG 153- 8			.07	15.3 ±.3				6380
68.85 -25.35									6686
0955.9 +3151	PGC 28825								
0958.8 +1912		.S..4..	1.09± .06	97					
214.82 49.75	UGC 5359	U	.45± .05	.03	14.87 ±.18				
80.26 -32.88		4.0± .9		.67					
0956.1 +1927	PGC 28828		1.09	.23					
095853.0-302123	NGC 3082	.LAS-*.	1.26± .04	26					
264.25 19.27	ESO 435- 18	S	.44± .03	.13	13.49 ±.14				
144.37 -45.58		-3.0± .9		.00					
095639.0-300700	PGC 28829		1.21						
095853.3+474411		.SB?...	1.33± .04	75				14.19±.1	1171± 7
169.53 50.74	UGC 5354		.30± .05	.00	14.1 ±.2				1156± 50
56.08 -14.46	VV 618			.45				170± 8	1197
095541.9+475832	PGC 28830		1.33	.15	13.61			.43	1370
095856.2+142516	NGC 3075	.S..5..	1.08± .06	135				14.47±.3	3582± 9
221.58 47.96	UGC 5360	U	.18± .05	.05	14.32 ±.18			298± 7	3566± 50
85.18 -35.48	IRAS09562+1439	5.0± .9		.27	13.27			269± 7	3479
095613.6+143938	PGC 28833		1.08	.09	13.98			.39	3899
095905.2+102138	IC 584		.52± .20						5459± 63
226.93 46.17	MCG 2-26- 10		.00± .07	.07	15.5 ±.3				5342
89.61 -37.51	ARAK 226								5784
095625.1+103601	PGC 28839		.52						
095906.1-301459		.SBS5?/	1.50± .03	153					
264.21 19.38	ESO 435- 19	S	.96± .03	.20	13.78 ±.14				
144.21 -45.55		5.0± .9		1.43					
095652.0-300036	PGC 28840		1.52	.48					
095906.5-270741	NGC 3084	PSBS2P.	1.23± .04	2					2464± 69
262.05 21.75	ESO 499- 29	Sr	.14± .03	.18	13.17 ±.14				2252
139.78 -45.92		1.7± .4		.18					2781
095650.0-265318	PGC 28841		1.25	.07	12.79				
095908.8-341330	NGC 3087	.E+..*.	1.39± .04		11.58V±.13	1.05± .01	1.06± .01		2673± 24
266.89 16.32	ESO 374- 15	PBS	.08± .04	.62	12.62 ±.11	.56± .01	.60± .01		2452
149.73 -44.82		-4.2± .5	.79± .03	.00		.89			2973
095658.0-335906	PGC 28845		1.46		11.96	.43	14.39± .26		
0959.1 +5100		.SBT4..	1.08± .06	58					
164.64 49.77	UGC 5356	U	.29± .05	.00	14.97 ±.19				
53.71 -12.12		4.0± .9		.43					
0955.9 +5115	PGC 28846		1.08	.15					

R.A. 2000 DEC. l b SGL SGB R.A. 1950 DEC.	Names PGC	Type S_T n_L T L	$\log D_{25}$ $\log R_{25}$ $\log A_e$ $\log D_o$	p.a. A_g A_i A_{21}	B_T m_B m_{FIR} B_T^o	$(B-V)_T$ $(U-B)_T$ $(B-V)_T^o$ $(U-B)_T^o$	$(B-V)_e$ $(U-B)_e$ m' m'_{25}	m_{21} W_{20} W_{50} HI	V_{21} V_{opt} V_{GSR} V_{3K}
0959.2 +0317 235.37 42.48 97.87 -40.66 0956.6 +0332	 UGC 5365 PGC 28852	.I..9*. U 10.0±1.2 	.96± .17 .05± .12 .96	 .02 .04 .02					3884 3744 4219
095916.0+314157 195.34 52.47 68.84 -25.22 095621.2+315620	 CGCG 153- 9 MK 413 PGC 28856			 .07 	15.48 ±.18 12.96				11394± 48 11356 11661
0959.2 +5215 162.82 49.35 52.82 -11.23 0955.9 +5229	 MCG 9-17- 2 1ZW 23 PGC 28858	.S?.... 	.82± .13 .17± .07 .81	 .00 .13 	15.00 ±.20 13.29 14.69				12226± 58 12270 12403
095920.8-280754 262.80 21.03 141.20 -45.77 095705.1-275330	 ESO 435- 20 TOL 2 PGC 28863		.92± .07 .15± .06 .94	100 .18 	14.42 ±.14				1246± 72 1033 1561
0959.3 +3044 196.90 52.41 69.68 -25.81 0956.4 +3059	Leo A UGC 5364 DDO 69 PGC 28868	.IB.9.. PU (1) 10.0± .5 9.0± .9	1.71± .02 .22± .05 1.52± .04 1.72	 .07 .16 .11	12.92 ±.18 12.8 ±.7 12.68	.33± .08 -.19± .08 .26 -.24	.31± .04 -.22± .05 16.01± .09 15.78± .25	12.88±.1 46± 5 33± 5 .09	20± 4 -19 292
095928.7-265148 261.93 22.01 139.39 -45.86 095712.0-263724	 ESO 499- 32 PGC 28874	.SBR1.. Sr 1.0± .5 	1.31± .03 .17± .03 1.33	148 .18 .18 .09	13.41 ±.14				
095928.9-192936 256.50 27.49 128.78 -46.01 095707.0-191512	NGC 3085 ESO 566- 38 PGC 28875	.L..0*/ SE -2.3± .7 	1.08± .04 .52± .03 1.01	119 .15 .00 	13.99 ±.14				
095929.7-224930 259.02 25.04 133.59 -46.06 095710.1-223506	NGC 3081 ESO 499- 31 PGC 28876	RSXR0.. 4 .0± .3 	1.32± .02 .11± .03 .94± .02 1.33	158 .14 .08 	12.85M±.05 12.95 ±.11 12.61	.88± .01 .33± .02 .81 .30	.93± .01 .34± .02 13.01± .06 14.04± .15	14.43±.2 266± 6 254± 5 	2367± 9 2391± 34 2164 2694
095929.7-671813 287.45 -9.76 187.53 -30.48 095816.0-670348	 ESO 92- 3 PGC 28877	.SB.8*/ S (1) 8.0± .8 7.8± .9	1.16± .04 .48± .04 1.25	96 .96 .59 .24	15.03 ±.14				
095936.7-281949 262.98 20.91 141.47 -45.69 095721.0-280524	NGC 3089 ESO 435- 24 PGC 28882	.SXT3.. R (1) 3.0± .4 3.4± .9	1.25± .03 .23± .03 1.27	139 .20 .32 .12	13.29 ±.11 12.76				2623± 66 2409 2938
095941.2+352332 189.27 52.68 65.81 -22.75 095643.3+353756	NGC 3074 UGC 5366 IRAS09567+3537 PGC 28888	.SXT5.. U (2) 5.0± .7 2.8± .6	1.37± .04 .05± .05 1.09± .02 1.37	145 .02 .07 .02	13.30 ±.15 14.2 ±.4 13.57 13.29	.64± .03 -.01± .07 .59 -.04	.72± .02 .08± .04 14.24± .05 14.86± .27	14.54±.1 152± 7 128± 5 1.23	5144± 5 5102± 46 5121 5397
095942.6-020900 241.27 39.35 104.77 -42.56 095710.1-015436	 CGCG 8- 10 PGC 28891			 .11 	15.3 ±.6				11267± 58 11111 11606
095945.5-673832 287.68 -10.01 187.83 -30.26 095833.0-672406	 ESO 92- 4 PGC 28895	.SXS9*. S (1) 8.5± .4 8.6± .7	1.21± .04 .47± .04 1.30	62 .96 .48 .24	14.83 ±.14				
0959.8 +1258 223.65 47.53 86.86 -36.05 0957.1 +1313	IC 585 UGC 5371 PGC 28897	.L..-*. U -3.0±1.2 	.94± .13 .00± .05 .95	 .06 .00 	14.42 ±.15				
095949.7-025241 242.02 38.92 105.73 -42.77 095717.5-023816	NGC 3083 MCG 0-26- 2 PGC 28900	.S..1?. E 1.0±1.9 	1.01± .07 .46± .05 1.02	50 .08 .47 .23	 14.6 ±.2 13.93				6342± 34 6184 6681
095953.3+451623 173.26 51.52 58.05 -16.04 095645.3+453047	 MCG 8-18- 52 PGC 28904	.S?.... 	1.04± .09 .49± .07 1.04	 .01 .61 .25					6924± 97 6941 7136
095955.7-293655 263.92 19.98 143.28 -45.46 095741.0-292230	IC 2531 ESO 435- 25 PGC 28909	.S..5*/ SU 5.3± .7 	1.84± .02 1.09± .03 1.86	75 .25 1.50 .50	12.9 ±.2 13.19 ±.14 11.33	.89± .08 -.57± .15 .62 -.83	 14.20± .23	13.08±.2 483± 16 477± 12 1.25	2477± 11 2262 2789

R.A. 2000 DEC.	Names	Type	logD$_{25}$	p.a.	B$_T$	(B-V)$_T$	(B-V)$_e$	m$_{21}$	V$_{21}$
l b		S$_T$ n$_L$	logR$_{25}$	A$_g$	m$_B$	(U-B)$_T$	(U-B)$_e$	W$_{20}$	V$_{opt}$
SGL SGB		T	logA$_e$	A$_i$	m$_{FIR}$	(B-V)^o_T	m'$_e$	W$_{50}$	V$_{GSR}$
R.A. 1950 DEC.	PGC	L	logD$_o$	A$_{21}$	B^o_T	(U-B)^o_T	m'$_{25}$	HI	V$_{3K}$

095956.1+130237 NGC 3080		.S..1..	.95± .07					16.56±.2	10632± 6
223.58 47.59 UGC 5372		U	.03± .05	.06	14.32 ±.18			182± 12	10580± 38
86.81 -35.99 MK 1243		1.0± .9		.03				162± 5	10523
095714.5+131702 PGC 28910			.96	.01	14.10			2.44	10951
095959.9+051957 Sextans B		.IBS9..	1.71± .02	110	11.85 ±.14	.52± .03	.48± .02	12.51±.1	301± 4
233.20 43.78 UGC 5373		PU (1)	.16± .04	.05	11.4 ±.2	-.13± .05	-.15± .03	58± 4	
95.53 -39.62 DDO 70		10.0± .5	1.48± .02	.12		.47	14.74± .04	41± 7	168
095722.9+053422 PGC 28913		8.0± .9	1.71	.08	11.56	-.17	14.84± .19	.86	634
100005.0-341356 IC 2532		PSBT1..	1.20± .05	38					2900± 51
267.05 16.44 ESO 374- 16		BSr	.09± .05	.62	13.91 ±.14				2679
149.68 -44.63 IRAS09579-3359		.6± .5		.09	12.70				3200
095754.0-335930 PGC 28915			1.26	.05	13.16				
100006.2-313308 NGC 3095		.SXT5..	1.54± .02	126	*			13.70±.2	2723± 6
265.27 18.52 ESO 435- 26		R (2)	.24± .03	.34	12.43 ±.12			352± 8	2849± 47
145.98 -45.13 IRAS09578-3118		5.0± .3		.36	10.80			340± 7	2507
095753.0-311842 PGC 28919		3.4± .5	1.57	.12	11.71			1.87	3032
100010.3-193720		.E?....	.60?		14.24S±.15				4005± 41
256.73 27.50 MCG -3-26- 6			.00± .07	.15					3806
128.98 -45.85 HICK 42C				.00			12.21±1.33		4336
095748.5-192254 PGC 28922			.62		14.03				
100011.0-025835 NGC 3086		.S..3..	1.05± .06	145	*				9766± 42
242.19 38.93 MCG 0-26- 3		E (1)	.50± .05	.08	14.7 ±.2				9607
105.90 -42.71		3.0± .9		.69					10105
095738.9-024409 PGC 28924		3.1±1.6	1.06	.25	13.82				
100013.0-194022		.L?....			16.21S±.15				4076± 41
256.78 27.47				.15					3877
129.05 -45.84 HICK 42D									4407
095751.2-192556 PGC 28926									
100013.9-193814 NGC 3091		.E.3.*.	1.47± .03	149	12.13M±.10	1.00± .01	1.03± .01		3878± 32
256.76 27.50 ESO 566- 41		R	.20± .03	.15	12.11 ±.11		.55± .06		3679
129.00 -45.84 HICK 42A		-5.0± .5	1.04± .02	.00		.93	12.90± .08		4209
095752.0-192348 PGC 28927			1.43		11.91		13.96± .21		
1000.2 +5432 A 0956+54		.S.....	1.20± .05	157					
159.52 48.61 UGC 5369		R	.77± .05	.00	14.89 ±.18				
51.28 -9.51				1.16					
0956.9 +5447 PGC 28928			1.20	.39					
100019.4-244808		.SBS9P.	1.05± .04						2282
260.62 23.68 ESO 499- 34		S (1)	.05± .04	.09	14.83 ±.14				2074
136.42 -45.81		9.0± .8		.05					2605
095801.1-243342 PGC 28932		8.9± .8	1.06	.02	14.69				
100027.0+032230		.S..8*.	1.29± .04	151				14.51±.3	2050± 9
235.52 42.79 UGC 5376		U	.46± .05	.00	14.02 ±.19			389± 7	2080± 50
97.94 -40.34 IRAS09578+0336		8.0±1.2		.56	11.30			371± 7	1912
095751.2+033656 PGC 28939			1.29	.23	13.45			.83	2386
100030.2-025819 NGC 3090		.E+....	1.23± .09	90	13.6 ±.2	1.03± .03	1.04± .01		6191± 34
242.25 38.99 MCG 0-26- 5		E	.07± .07	.08	13.46 ±.18				6033
105.93 -42.64 IRAS09579-0243		-4.0± .8	.82± .09	.00		.95	13.14± .31		6531
095758.1-024353 PGC 28945			1.22		13.35		14.59± .53		
100030.5-020940		.SBR2..	.99± .06	155				17.60±.1	6233± 9
241.44 39.50 UGC 5380		U	.12± .05	.11	14.23 ±.18			112± 12	5994± 40
104.88 -42.37		2.0± .9		.15					6064
095757.9-015514 PGC 28946			1.00	.06	13.91			3.64	6560
100031.0+031214		.SX.8*.	1.14± .05	75				15.52±.3	2139± 7
235.72 42.70 UGC 5377		U	.10± .05	.00	14.1 ±.2				1999
98.16 -40.39		8.0± .8		.12				158± 5	2474
095755.3+032640 PGC 28947			1.14	.05	13.95			1.52	
100031.6-311438 IC 2533		.LA.-*.	1.26± .05	1	12.95 ±.14	.97± .01	.98± .01		
265.14 18.81 ESO 435- 27		S	.14± .04	.25	12.89 ±.14	.42± .02	.48± .01		
145.53 -45.09		-3.0±1.2	.64± .03	.00			11.64± .12		
095818.1-310012 PGC 28948			1.27				13.76± .30		
100031.5+042425		.S..3..	1.15± .05	103				15.53±.3	4161± 10
234.37 43.38 UGC 5378		U (1)	.32± .05	.00	14.04 ±.18				4185± 50
96.71 -39.89 IRAS09579+0438		3.0± .8		.45	13.67			262± 7	4026
095755.1+043851 PGC 28949		3.5±1.1	1.15	.16	13.56			1.81	4496
100032.9-193945 NGC 3096		.LBT0..	1.02± .05	170	14.35S±.15				4198± 41
256.84 27.53 ESO 566- 42		SE	.13± .03	.14	14.08 ±.14				3999
129.04 -45.76 HICK 42B		-2.0± .5		.00					4529
095811.0-192518 PGC 28950			1.02		14.00		14.00± .30		

R.A. 2000 DEC. l　b SGL　SGB R.A. 1950 DEC.	Names PGC	Type S_T　n_L T L	$\log D_{25}$ $\log R_{25}$ $\log A_o$ $\log D_o$	p.a. A_g A_i A_{21}	B_T m_B m_{FIR} B_T^o	$(B-V)_T$ $(U-B)_T$ $(B-V)_T^o$ $(U-B)_T^o$	$(B-V)_o$ $(U-B)_o$ m' m'_{25}	m_{21} W_{20} W_{50} HI	V_{21} V_{opt} V_{GSR} V_{3K}
1000.6 -1458 253.13 30.93 122.35 -45.34 0958.2 -1444	A 0958-14 MCG -2-26- 12 PGC 28954	.SXT5.. PE (1) 5.0± .6 2.7± .8	1.24± .05 .24± .05 1.26	95 .18 .37 .12	 	 	 	15.43±.1 328± 8 303± 11 	9093± 7 9011± 59 8903 9428
100038.7-322157 265.91 17.96 147.08 -44.88 095826.1-320730	 ESO 435- 29 PGC 28956	.SBT6.. r 5.6± .9 	1.01± .06 .10± .05 1.05	 .48 .15 .05	 15.04 ±.14 	 	 	 	
100041.2-313945 265.45 18.51 146.11 -44.99 095828.1-312518	NGC 3100 ESO 435- 30 PGC 28960	.LXS0P. R -2.0± .4 	1.50± .04 .29± .04 1.10± .06 1.49	154 .34 .00 	12.0 ±.2 12.15 ±.14 	.89± .06 .41± .05 	.97± .06 .50± .05 12.96± .18 13.62± .28	 	
100045.3+044402 234.03 43.61 96.35 -39.71 095808.7+045829	 UGC 5383 MK 713 PGC 28964	.L..... U -2.0± .8 	1.09± .07 .16± .03 1.07	155 .01 .00 	 14.16 ±.15 14.04	 	 	 	6808± 81 6673 7142
100047.5-030046 242.35 39.02 106.01 -42.58 095815.4-024620	NGC 3092 MCG 0-26- 8 PGC 28967	.LBS+?. E -1.0±1.3 	1.09± .06 .30± .05 1.05	30 .08 .00 	 14.30 ±.15 14.14	 	 	 	5853± 58 5694 6193
100051.9+553713 157.99 48.25 50.57 -8.68 095728.8+555139	NGC 3073 UGC 5374 MK 131 PGC 28974	.LX.-.. PU -2.5± .6 	1.11± .06 .03± .04 .53± .04 1.10	 .00 .00 	14.07 ±.14 13.66 ±.15 13.86	.67± .02 .05± .03 .66 .06	.62± .02 .10± .03 12.21± .12 14.40± .34	15.96±.1 218± 6 200± 5 	1217± 10 1176± 60 1273 1376
100053.6-025821 242.33 39.07 105.97 -42.55 095821.5-024354	NGC 3093 MCG 0-26- 7 PGC 28977	.E..... E -5.0±1.0 	.84± .12 .32± .05 .76	50 .08 .00 	 15.18 ±.18 15.01	 	 	 	6099± 58 5941 6439
100103.1+363707 187.23 52.92 65.00 -21.73 095804.3+365134	 UGC 5382 PGC 28984	.S..2.. U 2.0± .9 	.97± .07 .19± .05 .97	65 .00 .24 .10	 14.67 ±.18 14.36	 	 	 	6627± 19 6610 6877
1001.0 -0930 248.54 34.81 114.73 -44.27 0958.6 -0916	 PGC 28987	.SBS9*. E (1) 9.0± .9 9.8± .8	1.18± .05 .30± .06 1.20	85 .24 .30 .15	 	 	 	 	
100104.8+165620 218.39 49.43 82.91 -33.70 095820.8+171047	 UGC 5385 PGC 28989	.S..6*. U 6.0±1.2 	1.12± .07 .09± .06 1.12	 .06 .14 .05	 14.5 ±.3 14.28	 	 	16.23±.3 273± 7 1.90	7850± 10 7757 8163
1001.0 +5544 157.80 48.22 50.51 -8.58 0957.7 +5559	 MCG 9-17- 9 PGC 28990	 	.89± .11 .35± .07 .89	 .00 	 14.8 ±.2 	 	 	 	3952± 61 4010 4112
100126.0+154617 220.09 49.06 84.16 -34.26 095842.8+160045	NGC 3094 UGC 5390 IRAS09586+1600 PGC 29009	.SBS1.. U 1.0± .8 	1.30± .05 .16± .06 1.31	75 .08 .17 .08	 13.23 ±.18 10.67 12.96	 	 	14.99±.2 261± 8 242± 10 1.96	2406± 6 2220±141 2309 2722
100130.3-340641 267.21 16.72 149.43 -44.37 095919.0-335212	IC 2534 ESO 374- 19 PGC 29016	PLAT+P* BSr -.6± .5 	1.34± .04 .08± .04 1.39	 .62 .00 	 13.38 ±.14 	 	 	 	
100131.1+371214 186.25 52.99 64.59 -21.28 095831.8+372642	IC 2530 MCG 6-22- 53 PGC 29019	.S?.... 	.64± .17 .00± .07 .64	 .00 .00 .00	 15.2 ±.2 15.13	 	 	 	6614± 56 6599 6861
100132.0-193229 256.94 27.78 128.88 -45.52 095910.0-191800	 ESO 567- 5 PGC 29021	.S..3?/ SE 3.0±1.2 	1.15± .04 .67± .03 1.16	71 .14 .93 .34	 15.37 ±.14 	 	 	 	
100132.5-202305 257.58 27.16 130.09 -45.56 095911.0-200836	 ESO 567- 6 PGC 29022	.S..7*/ E 7.0±1.3 	1.18± .05 .86± .04 1.19	7 .17 1.19 .43	 15.93 ±.14 	 	 	 	
100135.1+124554 224.22 47.82 87.36 -35.79 095853.7+130022	 UGC 5395 ARAK 228 PGC 29024	.L..-*. U -3.0±1.2 	.88± .14 .04± .05 .88	40 .04 .00 	 14.53 ±.16 	 	 	 	

R.A. 2000 DEC. l b SGL SGB R.A. 1950 DEC.	Names PGC	Type S_T n_L T L	logD25 logR25 logA. logD_o	p.a. Ag Ai A21	BT mB mFIR BT^o	(B-V)T (U-B)T (B-V)T^o (U-B)T^o	(B-V). (U-B). m'. m'25	m21 W20 W50 HI	V21 Vopt VGSR V3K
100135.6-025943 242.50 39.19 106.08 -42.39 095903.5-024514	NGC 3101 MCG 0-26- 11 PGC 29025	.S..1?/ E 1.0±1.9 	1.07± .09 .58± .05 1.07	150 .09 .60 .29	 15.28 ±.18 	 	 	 	
1001.6 +2136 211.57 51.11 78.36 -30.98 0958.8 +2150	 UGC 5392 IRAS09588+2150 PGC 29028	.S..6*. U 6.0±1.4 	1.04± .08 .98± .06 1.05	11 .07 1.44 .49	 13.46 	 	 	 	6211 6135 6512
100137.7+393737 182.25 52.81 62.66 -19.65 095836.2+395205	 UGC 5389 PGC 29030	.S..6*. U 6.0±1.4 	1.28± .04 1.09± .05 1.28	122 .01 1.47 .50	 15.60 ±.18 14.09	 	 	15.63±.1 352± 8 1.04	6980± 7 6975 7217
100140.2+104519 226.87 46.91 89.57 -36.76 095900.0+105947	 UGC 5396 PGC 29032	.S..6*. U 6.0±1.3 	1.19± .05 .55± .05 1.19	162 .06 .81 .27	 14.43 ±.18 13.54	 	 	15.33±.2 346± 7 337± 5 1.52	5400± 6 5286 5726
1001.6 -0815 247.54 35.77 113.08 -43.84 0959.2 -0801	 MCG -1-26- 11 PGC 29033	DLBS0*. E -2.0±1.3 	1.05± .06 .37± .05 1.02	90 .19 .00 	 	 	 	 	
100141.1+371452 186.18 53.02 64.58 -21.22 095841.9+372920	 UGC 5391 PGC 29034	.S..9.. U 9.0± .8 	1.35± .04 .43± .05 1.35	170 .00 .44 .21	 14.5 ±.2 14.06	 	 	14.38±.1 216± 16 187± 12 .11	1569± 11 1539± 56 1553 1815
100142.0+330814 193.02 53.06 67.99 -23.90 095846.3+332242	 UGC 5393 KUG 0958+333 PGC 29036	.SBR8*. U 8.0±1.1 	1.29± .04 .25± .05 1.29	 .01 .31 .13	 14.5 ±.3 14.13	 	 	14.65±.1 166± 16 144± 12 .39	1448± 11 1417 1711
100148.4+362959 187.42 53.08 65.21 -21.70 095849.8+364427	 UGC 5394 KUG 0958+367 PGC 29043	.S..6*. U 6.0±1.3 	1.15± .04 .85± .05 1.15	56 .00 1.25 .42	 	 	 	 	
100153.2+721013 138.44 39.45 39.01 3.19 095734.6+722440	NGC 3065 UGC 5375 7ZW 303 PGC 29046	.LAR0.. R -2.0± .4 	1.23± .04 .01± .04 .49± .08 1.24	 .10 .00 	13.5 ±.2 12.86 ±.12 12.87	.97± .01 .48± .02 .93 .47	.98± .01 .52± .02 11.39± .27 14.43± .31	15.15±.1 318± 8 	2000± 11 2001± 25 2118 2069
100158.2+554043 157.81 48.36 50.64 -8.53 095835.4+555511	NGC 3079 UGC 5387 IRAS09585+5555 PGC 29050	.SBS5./ R (1) 7.0± .4 3.0±1.3	1.90± .01 .74± .02 1.21± .02 1.90	165 .00 1.03 .37	11.54 ±.14 11.43 ±.13 9.00 10.45	.68± .02 .03± .03 .53 -.09	.80± .01 .18± .02 13.08± .04 14.04± .17	12.16±.1 473± 7 435± 5 1.34	1125± 6 1101± 29 1182 1285
100206.6+030324 236.20 42.95 98.55 -40.09 095931.0+031754	IC 588 UGC 5399 PGC 29057	RSBR1.. U 1.0± .9 	.93± .07 .06± .05 .93	 .00 .06 .03	 14.61 ±.18 14.45	 	 	 	7032± 54 6892 7368
100210.0+720725 138.47 39.49 39.06 3.17 095752.0+722153	NGC 3066 UGC 5379 MK 133 PGC 29059	PSXS4P. R 4.0± .4 	1.04± .03 .05± .04 .58± .02 1.05	 .10 .08 .03	13.55 ±.15 13.1 ±.3 11.92 13.25	.62± .02 -.07± .03 .57 -.10	.70± .02 .02± .03 11.94± .07 13.47± .22	15.29±.1 157± 16 95± 12 2.01	2049± 8 2010± 38 2165 2116
100213.8+134148 223.07 48.37 86.46 -35.18 095931.9+135618	 UGC 5400 PGC 29061	.L..... U -2.0± .9 	1.00± .19 .09± .08 .99	160 .07 .00 	 14.15 ±.15 13.98	 	 	 	6989± 31 6885 7310
100217.4+244237 206.81 52.07 75.53 -29.03 095928.3+245706	NGC 3098 UGC 5397 PGC 29067	.L..../ R -2.0± .4 	1.36± .03 .57± .03 .69± .05 1.28	90 .11 .00 	12.89 ±.17 13.22 ±.11 13.00	.90± .01 .81 	.92± .01 11.83± .17 13.12± .23	15.59±.3 274± 33 	1311± 17 1401± 28 1272 1628
100229.5-314037 265.77 18.74 146.04 -44.61 100016.0-312606	NGC 3108 ESO 435- 32 PGC 29076	.LAS+.. S -1.0± .8 	1.40± .04 .14± .04 .81± .03 1.41	110 .33 .00 	12.78 ±.15 12.48 ±.14 12.25	1.03± .02 .91 	1.08± .01 12.32± .11 14.29± .28	 	2673± 17 2455 2981
1002.5 +1900 215.58 50.50 81.05 -32.27 0959.8 +1915	 UGC 5401 PGC 29082	.I..9*. U 10.0±1.2 	1.09± .07 .62± .06 1.09	65 .06 .46 .31	 	 	 	15.84±.1 142± 6 140± 5 	2010± 10 1925 2319
1002.5 +1910 215.33 50.57 80.89 -32.17 0959.8 +1925	 UGC 5403 IRAS09598+1925 PGC 29085	.S..0.. U .0± .9 	1.11± .05 .34± .05 1.10	80 .06 .25 	 14.49 ±.18 12.12 14.15	 	 	 	2077 1993 2385

R.A. 2000 DEC.	Names	Type	logD₂₅	p.a.	B_T	(B-V)_T	(B-V)_•	m₂₁	V₂₁
l b		S_T n_L	logR₂₅	A_g	m_B	(U-B)_T	(U-B)_•	W₂₀	V_opt
SGL SGB		T	logA_•	A_i	m_FIR	(B-V)°_T	m'_•	W₅₀	V_GSR
R.A. 1950 DEC.	PGC	L	logD_o	A₂₁	B°_T	(U-B)°_T	m'₂₅	HI	V_3K

100236.3-060043		.S..7*/	1.45± .03	16				14.14±.1	662± 5
245.64 37.43	MCG -1-26- 12	EU (1)	1.02± .05	.06				137± 6	
110.14 -43.04		7.0± .8		1.38				130± 8	495
100006.0-054612	PGC 29086	7.5±1.2	1.46	.50					1003

100238.4-452956		PSBS2..	1.13± .16	22					
274.55 7.87	ESO 262- 15	r	.13± .14	1.16					
164.05 -40.85	IRAS10006-4515	2.2± .8		.15	12.43				
100038.1-451524	PGC 29089		1.24	.06					

1002.6 +7044		.S..3..	.98± .07	178					
139.84 40.39	UGC 5386	U (1)	.10± .05	.20	15.01 ±.19				
40.06 2.22	IRAS09585+7058	3.0± .9		.14	13.64				
0958.5 +7058	PGC 29092	2.5±1.2	1.00	.05					

100244.1-420526		.S..5./	1.38± .04	117					
272.48 10.59	ESO 316- 18	S	.87± .04	.93	14.85 ±.14				
159.82 -42.02	IRAS10006-4150	4.7± .5		1.30	12.66				
100040.0-415054	PGC 29096		1.46	.43					

100248.8+431110 A 0959+43									
176.39 52.48				.00					5391± 42
60.03 -17.09	MK 134								5400
095943.8+432540	PGC 29106								5614

100300.6-152142		.S..3?/	1.18± .05	102					
253.93 31.06	MCG -2-26- 24	E	.76± .05	.18					
123.00 -44.81		3.0±1.8		1.04					
100035.8-150710	PGC 29122		1.19	.38					

100304.7+104602		.S?....	1.09± .06	165				15.77±.3	3001± 10
227.09 47.22	UGC 5409		.55± .05	.06	14.87 ±.18				2887
89.76 -36.46				.82				81± 10	
100024.6+110034	PGC 29126		1.10	.27	13.97			1.52	3328

100304.8-022359	IC 587	.SXR4P?	1.12± .05	107					
242.20 39.85	UGC 5411	E	.36± .04	.10	14.22 ±.18				
105.47 -41.84	IRAS10005-0209	4.0±1.8		.53	13.91				
100032.3-020927	PGC 29127		1.13	.18					

100306.7-260932	NGC 3109	.SBS9./	2.28± .01	93	10.39S±.07		.52± .03	9.67±.1	404± 3
262.10 23.07	ESO 499- 36	R (3)	.71± .02	.16	10.39 ±.12		-.11± .08	120± 3	408± 58
138.30 -45.10	DDO 236	9.0± .3		.72	11.69			116± 3	194
100049.0-255500	PGC 29128	8.2± .5	2.30	.35	9.51		14.89± .11	-.19	725

100311.5-424909		PSXT3*.	1.13± .08	89	*				
272.99 10.06	ESO 262- 16	r	.21± .07	.94					
160.70 -41.70	IRAS10011-4234	3.3±1.2		.29	13.61				
100108.0-423436	PGC 29134		1.22	.10					

100316.9-145643		.SBR6..	1.23± .05	85	*				4860
253.64 31.40	MCG -2-26- 27	E (1)	.38± .05	.19					
122.42 -44.70		6.0± .8		.56					4671
100051.8-144211	PGC 29136	5.3± .8	1.24	.19					5197

100317.5-385003		.S?....	.93± .07						12597± 19
270.53 13.24	ESO 316- 20		.04± .06	.54	14.83 ±.14				12373
155.60 -42.89	IRAS10011-3835			.04	13.39				12885
100110.0-383530	PGC 29139		.99	.02	14.09				

100319.0-212550		.SBS6..	1.12± .04		*			16.20±.2	3069± 11
258.71 26.66	ESO 567- 10	E (1)	.04± .03	.23	14.43 ±.14			60± 16	
131.60 -45.17		6.0± .8		.06				40± 12	2867
100058.0-211118	PGC 29140	6.4± .8	1.14	.02	14.13			2.05	3398

100319.0-645804		.SBR3..	1.27± .05	40	*				
286.32 -7.67	ESO 92- 6	Sr (1)	.25± .05	2.89	13.45 ±.14				
185.08 -31.52	IRAS10019-6443	3.4± .5		.34	12.51				
100155.0-644330	PGC 29141	4.4± .6	1.54	.12					

1003.3 +5355		.S..0..	1.04± .06	60	*				
160.11 49.27	UGC 5405	U	.38± .05	.00	15.37 ±.19				
52.05 -9.63		.0± .9		.28					
1000.0 +5410	PGC 29142		1.02						

100321.1+684402	NGC 3077	.I.0.P.	1.73± .02	45	10.61 ±.13	.76± .01	.70± .01	11.06±.1	14± 4
141.90 41.66	UGC 5398	R	.08± .02	.23	10.79 ±.13	.14± .02	.09± .01	91± 6	-16± 32
41.51 .83	IRAS09592+6858	90.0	1.30± .01	.06	10.24	.69	12.60± .04	65± 4	120
095921.9+685833	PGC 29146		1.75		10.41	.09	13.89± .17		102

100323.5+482157		.S?....	.83± .08						7684±125
168.23 51.27	MCG 8-18- 60		.18± .06	.00	14.99 ±.19				7714
56.16 -13.48				.28					7882
100012.5+483628	PGC 29147		.83	.09	14.67				

R.A. 2000 DEC. / l b / SGL SGB / R.A. 1950 DEC.	Names / PGC	Type / S_T n_L / T / L	$\log D_{25}$ / $\log R_{25}$ / $\log A_\bullet$ / $\log D_o$	p.a. / A_g / A_i / A_{21}	B_T / m_B / m_{FIR} / B_T^o	$(B-V)_T$ / $(U-B)_T$ / $(B-V)_T^o$ / $(U-B)_T^o$	$(B-V)_\bullet$ / $(U-B)_\bullet$ / $m'_\bullet$ / m'_{25}	m_{21} / W_{20} / W_{50} / HI	V_{21} / V_{opt} / V_{GSR} / V_{3K}
100323.8-372339 / 269.64 14.39 / 153.71 -43.24 / 100115.0-370906	/ ESO 374- 25 / / PGC 29148	.S?.... / / /	.96± .06 / .08± .05 / .11 / 1.02	164 / .59 / / .04	/ 15.07 ±.14 / / 14.31				7006± 69 / 6783 / / 7298
1003.4 +1305 / 224.09 48.38 / 87.30 -35.22 / 1000.8 +1320	/ UGC 5413 / / PGC 29156	.S..6*. / U / 6.0±1.5 /	1.00± .08 / 1.02± .06 / / 1.01	15 / .06 / 1.47 / .50					
100329.8-335703 / 267.44 17.10 / 149.09 -43.99 / 100118.0-334230	IC 2536 / ESO 374- 26 / IRAS10012-3342 / PGC 29157	.SB.5P/ / BS / 5.3± .7 /	1.28± .04 / .64± .04 / / 1.33	45 / .54 / .96 / .32	/ 14.57 ±.14 / 13.33 / 13.04				5384± 19 / 5164 / / 5686
100333.6-672652 / 287.85 -9.63 / 187.40 -30.08 / 100218.0-671218	NGC 3136A / ESO 92- 7 / / PGC 29160	.IBS9?/ / RS (1) / 9.5± .6 / 9.4± .7	1.31± .04 / .78± .04 / / 1.40	83 / .98 / .59 / .39	/ 15.60 ±.14 / /				
100337.1-150654 / 253.85 31.34 / 122.68 -44.64 / 100112.1-145221	/ MCG -2-26- 28 / / PGC 29163	PLXR0?. / E / -2.0± .9 /	1.09± .06 / .27± .05 / / 1.07	92 / .17 / .00 /					
100342.2-270139 / 262.82 22.49 / 139.51 -44.91 / 100125.0-264706	/ ESO 499- 37 / / PGC 29166	.SXS7*. / SU (1) / 7.0± .6 / 5.6± .8	1.51± .03 / .38± .04 / / 1.53	53 / .17 / .52 / .19	/ 13.32 ±.14 / / 12.63			13.27±.2 / 196± 16 / 141± 12 / .45	961± 11 / / 750 / 1280
100345.3-802519 / 296.14 -19.84 / 198.67 -22.08 / 100421.0-801042	NGC 3149 / ESO 19- 1 / IRAS10043-8010 / PGC 29171	RSAT3*. / Sr (1) / 3.4± .4 / 5.0± .5	1.30± .06 / .02± .07 / / 1.34	/ .44 / .03 / .01	/ / 12.06 /				
1003.7 +6335 / 147.63 44.66 / 45.17 -2.78 / 1000.1 +6350	/ UGC 5406 / / PGC 29172	.S..6*. / U / 6.0±1.3 /	1.02± .06 / .38± .05 / / 1.03	62 / .05 / .56 / .19					
1003.8 +5021 / 165.20 50.69 / 54.72 -12.05 / 1000.6 +5036	/ UGC 5412 / / PGC 29174	.S..6*. / U / 6.0±1.3 /	1.06± .05 / .37± .04 / / 1.06	25 / .00 / .54 / .18					
100351.9+592611 / 152.70 46.85 / 48.13 -5.70 / 100022.2+594043	A 1000+59 / UGC 5408 / MK 25 / PGC 29177	.E?.... / / /	.74± .07 / .07± .04 / / .72	/ .00 / .00 /	14.84 ±.13 / 14.4 ±.3 / 13.03 / 14.71	.57± .02 / -.27± .04 / .54 / -.26	/ / / 13.38± .37	15.75±.3 / / /	2602± 10 / 2997± 35 / 2707 / 2776
100352.8-273416 / 263.23 22.10 / 140.27 -44.82 / 100136.1-271942	IC 2537 / ESO 499- 39 / / PGC 29179	.SXT5.. / R (3) / 5.0± .4 / 3.3± .4	1.41± .02 / .17± .03 / 1.08± .04 / 1.43	26 / .22 / .26 / .09	12.78 ±.19 / 12.81 ±.12 / / 12.31	.69± .05 / .08± .09 / .59 / .00	.77± .05 / .15± .09 / 13.67± .10 / 14.26± .23	13.98±.2 / 347± 10 / 331± 12 / 1.59	2786± 7 / / 2574 / 3104
100356.2-344828 / 268.07 16.49 / 150.23 -43.73 / 100145.0-343354	IC 2538 / ESO 374- 27 / IRAS10017-3434 / PGC 29181	.SAR5P. / BSr (1) / 5.3± .5 / 2.2± .9	1.18± .04 / .29± .04 / / 1.23	1 / .53 / .44 / .15	/ 14.62 ±.14 / 13.25 / 13.61				8412± 19 / 8191 / / 8712
100357.2-062950 / 246.38 37.36 / 110.91 -42.85 / 100127.1-061516	/ MCG -1-26- 13 / / PGC 29184	.L..+P. / E / -1.0± .9 /	1.09± .07 / .41± .05 / / 1.03	95 / .07 / .00 /					
100357.9+404528 / 180.29 53.10 / 62.08 -18.56 / 100055.7+410001	NGC 3104 / UGC 5414 / ARP 264 / PGC 29186	.IXS9.. / R / 10.0± .3 /	1.52± .03 / .17± .05 / / 1.52	35 / .05 / .13 / .08	/ 13.6 ±.4 / / 13.42			13.94±.1 / 114± 6 / 100± 11 / .44	612± 6 / / 612 / 845
100359.0+221635 / 210.79 51.83 / 78.09 -30.14 / 100111.8+223108	/ UGC 5420 / / PGC 29188	.L...*. / U / -2.0±1.2 /	1.03± .08 / .13± .03 / / 1.02	75 / .08 / .00 /					6139± 31 / 6066 / / 6439
1004.0 +1412 / 222.65 48.98 / 86.20 -34.54 / 1001.3 +1427	A 1001+14 / CGCG 64- 47 / / PGC 29190	.SXS5.. / P / 5.0± .9 /		/ .07 / /					8991± 60 / 8889 / 9311
100402.2-062832 / 246.38 37.39 / 110.89 -42.82 / 100132.1-061358	NGC 3110 / MCG -1-26- 14 / IRAS10015-0614 / PGC 29192	.SBT3P. / E / 3.0± .9 /	1.19± .05 / .32± .05 / / 1.19	5 / .07 / .45 / .16	/ / 10.50 /				5080 / / 4913 / 5422

R.A. 2000 DEC. / l b / SGL SGB / R.A. 1950 DEC.	Names / PGC	Type / S_T n_L / T / L	$\log D_{25}$ / $\log R_{25}$ / $\log A_e$ / $\log D_o$	p.a. / A_g / A_i / A_{21}	B_T / m_B / m_{FIR} / B_T^o	$(B-V)_T$ / $(U-B)_T$ / $(B-V)_T^o$ / $(U-B)_T^o$	$(B-V)_e$ / $(U-B)_e$ / m' / m'_{25}	m_{21} / W_{20} / W_{50} / HI	V_{21} / V_{opt} / V_{GSR} / V_{3K}
100404.0-272001		.IX.9..	1.30± .06						361
263.10 22.31		S (1)	.11± .08	.22					
139.93 -44.80		10.0± .5		.08					150
100147.0-270527	PGC 29194	11.1± .8	1.32	.06					680
100405.6+311110	NGC 3106	.L.....	1.25± .06		13.27 ±.15	.89± .02	.94± .01		
196.32 53.45	UGC 5419	U	.00± .03	.05	13.38 ±.18	.31± .03	.41± .02		6203± 31
70.01 -24.75		-2.0± .8	.96± .04	.00		.82	13.56± .13		6165
100111.9+312543	PGC 29196		1.26		13.17	.33	14.39± .33		6475
100408.8+063038					14.60 ±.18				
232.59 45.29	CGCG 36- 27			.03					1128± 61
94.73 -38.18	MK 714								999
100131.2+064512	PGC 29198								1462
100416.8-752843		.SXS5..	1.40± .04						
292.92 -15.96	ESO 37- 10	Sr (1)	.05± .05	.90	13.15 ±.14				
194.53 -25.20	IRAS10037-7514	4.8± .4		.08	12.04				
100347.0-751406	PGC 29202	4.4± .6	1.49	.03					
100417.0-312141	IC 2539	.SAS4*.	1.28± .03	25					
265.87 19.23	ESO 435- 34	S (1)	.60± .03	.46	14.04 ±.14				2850± 60
145.52 -44.28	IRAS10020-3107	4.0±1.3		.89	12.94				2633
100203.1-310706	PGC 29203	5.6± .9	1.32	.30	12.68				3159
100417.9+034114		.S..6*.	1.00± .08	79					
235.92 43.75	UGC 5424	U	.33± .06	.00	14.98 ±.18				
98.08 -39.33		6.0±1.3		.49					
100142.0+035548	PGC 29205		1.00	.17					
1004.3 +5323		.SBS3..	1.10± .05	135					
160.77 49.62	UGC 5417	U	.32± .05	.00	15.3 ±.2				
52.54 -9.89	SBS 1001+536A	3.0± .8		.44					
1001.0 +5338	PGC 29206		1.10	.16					
100418.5-745549		.LXSO..	1.11± .05	48					
292.57 -15.53	ESO 37- 9	S	.28± .04	.89	14.61 ±.14				
194.05 -25.54		-2.0± .8		.00					
100344.0-744112	PGC 29207		1.17						
100421.9+133717	NGC 3107	.S..4*.	.87± .06	140				16.05±.2	2791± 8
223.51 48.81	UGC 5425	P	.10± .04	.08	14.2 ±.3			295± 21	2743± 60
86.87 -34.77	ARAK 229	4.0±1.3		.15				199± 11	2686
100140.2+135151	PGC 29209		.87	.05	13.93			2.07	3112
1004.4 +1445		.S..6*.	1.14± .07	177					
221.95 49.30	UGC 5426	U	1.16± .06	.06	15.8 ±.2				
85.68 -34.18		6.0±1.4		1.47					
1001.7 +1500	PGC 29212		1.15	.50					
100424.5-022541		.S?....	.99± .10						
242.50 40.09	MCG 0-26- 14		.33± .07	.09	14.66 ±.19				5913± 58
105.66 -41.54				.46					5757
100152.0-021106	PGC 29213		1.00	.17	14.07				6254
100426.2-412459		.S..3?/	1.41± .04	19					
272.32 11.32	ESO 316- 21	S	.96± .05	.81	15.04 ±.14				
158.83 -41.93	IRAS10023-4110	3.0±1.9		1.32	12.25				
100221.0-411024	PGC 29214		1.49	.48					
100426.2-282635	NGC 3113	.SXS7*.	1.52± .02	87				13.49±.2	1091± 7
263.93 21.51	ESO 435- 35	PSU	.44± .03	.24	13.35 ±.14			240± 9	
141.47 -44.62	IRAS10021-2812	7.3± .4		.61				193± 12	878
100210.0-281200	PGC 29216		1.55	.22	12.50			.77	1408
100430.7+600626	NGC 3102	.L..-*.	.89± .07						
151.79 46.58	UGC 5418	U	.00± .03	.00	14.32 ±.16				3065± 31
47.71 -5.17		-3.0±1.2		.00					3140
100059.8+602100	PGC 29220		.89		14.27				3203
100431.9+380022	IC 2535	.S?....	.95± .06						
184.85 53.53	MCG 6-22- 65		.01± .06	.00	14.51 ±.18				6966± 13
64.36 -20.30	KUG 1001+382			.02					6955
100132.5+381456	PGC 29222		.95	.01	14.46				7211
100434.1-371959		.S?....	1.07± .06	116					
269.79 14.58	ESO 374- 28		.34± .05	.44	14.69 ±.14				5061± 19
153.55 -43.03	IRAS10024-3705			.25	12.77				4838
100225.1-370524	PGC 29224		1.09		13.93				5354
1004.6 +5519		.S..8*.	1.24± .04	172					1108
158.05 48.86	UGC 5421	U	.82± .05	.00					
51.17 -8.51	SBS 1001+555	8.0±1.3		1.00					1165
1001.3 +5534	PGC 29229		1.24	.41					1271

R.A. 2000 DEC.	Names	Type	logD25	p.a.	BT	(B-V)T	(B-V)e	m21	V21
l b		ST nL	logR25	Ag	mB	(U-B)T	(U-B)e	W20	Vopt
SGL SGB		T	logAe	Ai	mFIR	(B-V)T°	m'e	W50	VGSR
R.A. 1950 DEC.	PGC	L	logDo	A21	BT°	(U-B)T°	m'25	HI	V3K
100440.3+292159		.S..8*.	1.07± .06	120				16.27±.3	495± 7
199.38 53.40	UGC 5427	U	.15± .05	.04	14.6 ±.2			78± 5	450
71.69 -25.78		8.0±1.2		.18				1.78	773
100148.1+293634	PGC 29230		1.08	.07	14.42				
1004.8 +5735		.SBS3..	1.01± .06	170					
154.99 47.85	UGC 5422	U	.10± .05	.00	15.3 ±.2				
49.54 -6.91		3.0± .9		.14					
1001.4 +5750	PGC 29237		1.01	.05					
1004.9 +4442		.S..8*.	1.00± .08	0					
173.82 52.52	UGC 5429	U	.09± .06	.00					
59.12 -15.78		8.0±1.2		.11					
1001.8 +4457	PGC 29247		1.00	.04					
1004.9 +0503		.E...*.	1.04± .18						
234.46 44.65	UGC 5432	U	.00± .08	.00	14.09 ±.16				
96.53 -38.63		-5.0±1.2		.00					
1002.3 +0518	PGC 29249		1.04						
1004.9 +2130		.S..6*.	1.13± .05	48					
212.09 51.83	UGC 5431	U	.85± .05	.07					
78.99 -30.39		6.0±1.4		1.25					
1002.2 +2145	PGC 29253		1.14	.42					
1005.1 +6633	A 1001+66	.I..9*.	.96± .09		*		.80± .07		
144.13 43.10	UGC 5428	PU (1)	.00± .06	.13			.50± .10		
43.17 -.59	DDO 71	10.0± .7	1.36± .11	.00			16.96± .25		
1001.3 +6647	PGC 29257	9.0±1.1	.97	.00					
100508.5+220717		.I..9*.	1.11± .14	37	*			16.32±.3	3998± 7
211.15 52.05	UGC 5433	U	.54± .12	.08					
78.42 -30.01		10.0±1.3		.40				174± 7	3925
100221.4+222153	PGC 29258		1.12	.27					4300
1005.2 +4430		.S..7..	1.05± .06	40	*				
174.12 52.61	UGC 5430	U	1.00± .05	.00					
59.31 -15.88		7.0±1.0		1.38					
1002.1 +4445	PGC 29261		1.05	.50					
100514.1-074307	NGC 3115	.L..-./	1.86± .01	43	9.87M±.04	.97± .01	.99± .01	14.00±.3	658± 8
247.78 36.78	MCG -1-26- 18	R	.47± .02	.10	9.71 ±.15	.54± .01	.59± .01	254± 9	670± 12
112.65 -42.86		-3.0± .3	1.03± .01	.00		.92	10.77± .02		492
100244.7-072831	PGC 29265		1.80		9.75	.50	12.87± .09		1005
100514.2+212724		.SXS3..	1.11± .05	25				15.35±.3	5580± 10
212.19 51.88	UGC 5434	U	.12± .05	.07	14.3 ±.2				
79.07 -30.37	IRAS10024+2142	3.0± .8		.16				260± 7	5504
100227.6+214200	PGC 29266		1.12	.06	14.03			1.27	5883
1005.2 +7915		.S?....	1.07± .05	175					8980±125
131.60 34.90	UGC 5402		.17± .04	.01	14.65 ±.18				9121
34.12 8.33				.23					9007
0959.8 +7930	PGC 29271		1.07	.08	14.34				
100521.9+191709		.S?....	1.14± .13	130				14.98±.3	3767± 9
215.52 51.22	UGC 5436		.39± .12	.08	14.32 ±.18			307± 7	3751± 34
81.22 -31.56	IRAS10026+1931			.59	12.94			289± 7	3682
100236.8+193145	PGC 29277		1.15	.20	13.64			1.15	4075
100522.9-341313	NGC 3120	.SXS4*.	1.25± .03	1	13.52 ±.14	.74± .02	.82± .02		2670± 60
267.94 17.13	ESO 374- 29	BS (1)	.15± .05	.52	13.53 ±.14	.13± .06	.19± .05		2450
149.35 -43.56	IRAS10031-3358	3.7± .6	.79± .02	.23	12.65	.57	12.96± .04		2972
100311.0-335836	PGC 29278	4.4± .8	1.30	.08	12.76	.01	14.23± .23		
100525.5-174755		.S..3..	1.03± .04	120					
256.36 29.69	ESO 567- 12	E (1)	.38± .03	.13	14.74 ±.14				
126.50 -44.49		3.0± .9		.52					
100302.0-173318	PGC 29281	4.2±1.2	1.04	.19					
1005.5 +7022		.I..9..	.94± .09	140				15.95±.1	340± 5
140.03 40.80	UGC 5423	U	.18± .06	.20	15.19 ±.18			67± 7	514± 60
40.48 2.12		10.0± .8		.13				39± 7	454
1001.4 +7037	PGC 29284		.96	.09	14.86			1.00	421
100534.2+273106		.S..8*.	1.07± .07	112				15.93±.3	6348± 7
202.50 53.33	UGC 5437	U	.50± .06	.06					
73.47 -26.75		8.0±1.3		.61				198± 7	6295
100243.5+274543	PGC 29291		1.08	.25					6633
1005.6 +0416		..S..7..	1.07± .07	97					
235.51 44.36	UGC 5440	U	.40± .06	.00					
97.55 -38.80		7.0± .9		.55					
1003.0 +0431	PGC 29295		1.07	.20					

R.A. 2000 DEC. l b SGL SGB R.A. 1950 DEC.	Names PGC	Type S_T n_L T L	$\log D_{25}$ $\log R_{25}$ $\log A_e$ $\log D_o$	p.a. A_g A_i A_{21}	B_T m_B m_{FIR} B_T^o	$(B-V)_T$ $(U-B)_T$ $(B-V)_T^o$ $(U-B)_T^o$	$(B-V)_e$ $(U-B)_e$ m'_e m'_{25}	m_{21} W_{20} W_{50} HI	V_{21} V_{opt} V_{GSR} V_{3K}
1005.6 +8017 130.69 34.20 33.38 9.06 0959.8 +8031	NGC 3057 UGC 5404 DDO 67 PGC 29296	.SBS8.. U (1) 8.0± .8 7.0±1.0	1.35± .03 .23± .06 1.01± .02 1.35	5 .03 .29 .12	13.49 ±.14 13.9 ±.2 13.71 13.28	.50± .02 -.16± .04 .44 -.20	.51± .02 -.13± .03 14.03± .04 14.50± .26	14.85±.1 142± 16 132± 12 1.45	1529± 11 1674 1550
100539.5-370514 269.81 14.90 153.15 -42.88 100330.0-365036	 ESO 374- 30 IRAS10034-3650 PGC 29298	.S?.... 	1.04± .06 .34± .05 1.08	75 .44 .50 .17	 15.17 ±.14 13.86 14.20				4865± 19 4643 5159
1005.6 -0744 247.90 36.85 112.72 -42.75 1003.2 -0730	 UGCA 200 PGC 29299	.I..9*. U 10.0±1.3 	1.04± .18 .20± .18 1.05	 .11 .15 .10					
100541.7-075853 248.12 36.69 113.04 -42.81 100312.4-074416	 MCG -1-26- 21 PGC 29300	DLASOP* E -2.0±1.2 	1.24± .07 .08± .05 1.23± .15 1.25	0 .18 .00 	13.2 ±.3 13.01	.73± .05 .67 	.81± .02 14.87± .51 14.08± .47		715± 62 544 1057
100548.1-172606 256.15 30.01 126.00 -44.37 100324.4-171128	IC 2541 MCG -3-26- 17 PGC 29309	.SBR4?. E (1) 4.0± .9 3.1±1.2	1.10± .06 .38± .05 1.11	5 .13 .56 .19					4987 4794 5323
100548.2-672239 287.99 -9.45 187.20 -29.94 100431.6-670800	NGC 3136 ESO 92- 8 PGC 29311	.E...*. RS -5.0± .4 	1.49± .03 .16± .03 1.09± .03 1.60	30 .98 .00 	11.7 ±.1 11.74 ±.11 10.71	1.01± .01 .51± .03 .78 .32	1.03± .01 .56± .01 12.68± .11 13.73± .22		1713± 30 1497 1878
100548.2-181450 256.78 29.42 127.14 -44.44 100325.0-180012	 ESO 567- 13 PGC 29312	.S..8*. E (1) 8.0±1.2 7.5±1.6	1.06± .04 .18± .03 1.07	88 .13 .22 .09	 15.24 ±.14 14.87			15.29±.1 263± 12 .33	4941± 9 4746 5276
100552.6-441338 274.25 9.23 162.23 -40.77 100350.0-435900	 ESO 263- 3 IRAS10037-4358 PGC 29320	 	1.27± .06 .26± .06 .75± .04 1.36	 .98 	14.18 ±.15 14.07 ±.14 12.86 	.61± .08 -.15± .07 	.51± .03 -.24± .05 13.42± .12 		3350± 19 3123 3620
100556.2-230326 260.42 25.84 133.90 -44.57 100336.0-224848	 ESO 499- 41 PGC 29323	RLBR+*. r -1.3±1.0 	1.02± .06 .28± .03 .99	93 .13 .00 	14.56 ±.14 				
100601.4-160724 255.15 31.01 124.18 -44.18 100336.9-155245	NGC 3128 MCG -3-26- 20 PGC 29330	.SBS3?. E (1) 3.0± .9 1.9± .8	1.20± .05 .38± .05 1.21	175 .13 .53 .19					4635 4444 4973
100602.5-670352 287.81 -9.18 186.89 -30.10 100444.0-664912	 ESO 92- 9 PGC 29331	.S..9*/ S (1) 9.0±1.3 8.9±1.3	1.09± .05 .53± .04 1.29	107 2.14 .54 .27	 15.83 ±.14 				
100605.3-335309 267.84 17.49 148.86 -43.48 100353.0-333830	IC 2545 ESO 374- 32 IRAS10039-3338 PGC 29334	 	.81± .07 .20± .06 .86	 .49 	 15.27 ±.14 10.97 				10267± 19 10048 10570
100608.2+471547 169.70 52.04 57.31 -13.89 100259.4+473025	NGC 3111 UGC 5441 PGC 29338	.L..-*. U -3.0±1.2 	.95± .21 .05± .08 .94	 .00 .00 	 14.01 ±.17 				
100610.5+025440 237.17 43.69 99.25 -39.21 100335.0+030918	NGC 3117 UGC 5445 PGC 29340	.E...?. U -5.0±1.7 	.94± .13 .00± .05 .94	 .00 .00 	 14.29 ±.15 				
1006.2 -1220 252.08 33.76 118.97 -43.59 1003.8 -1206	 PGC 29345	.SBS9.. E (1) 9.0± .9 9.8± .8	1.26± .07 .34± .08 1.28	100 .22 .35 .17					
100615.2-160129 255.12 31.12 124.05 -44.12 100350.6-154651	 MCG -3-26- 21 PGC 29346	.SB.5?/ E 5.0±1.3 	1.22± .05 .89± .05 1.23	50 .13 1.33 .44					5007 4817 5345
1006.2 +2856 200.17 53.69 72.31 -25.77 1003.4 +2911	A 1003+29 MCG 5-24- 11 PGC 29347	.S?.... 	.72± .10 .02± .06 .72	 .06 .01 	14.5 ±.2 15.1 ±.2 14.65	.43± .04 -.13± .06 .40 -.13	 12.90± .57	17.39±.2 115± 5 120± 5 	1363± 7 1323± 30 1314 1641

R.A. 2000 DEC. I b SGL SGB R.A. 1950 DEC.	Names PGC	Type S_T n_L T L	$\log D_{25}$ $\log R_{25}$ $\log A_{\bullet}$ $\log D_o$	p.a. A_g A_I A_{21}	B_T m_B m_{FIR} B_T^o	$(B-V)_T$ $(U-B)_T$ $(B-V)_T^o$ $(U-B)_T^o$	$(B-V)_{\bullet}$ $(U-B)_{\bullet}$ $m'_{\bullet}$ m'_{25}	m_{21} W_{20} W_{50} HI	V_{21} V_{opt} V_{GSR} V_{3K}
100624.8-160736 255.24 31.07 124.20 -44.09 100400.2-155257	NGC 3127 MCG -3-26- 22 PGC 29357	.S..3*/ E 3.0±1.3	1.09± .06 .75± .05 1.10	55 .13 1.03 .37					
100631.0+325648 193.40 54.06 68.86 -23.24 100336.4+331127	 UGC 5446 KUG 1003+331 PGC 29365	.S..6*. U 6.0±1.3	1.17± .04 .66± .04 1.17	51 .03 .98 .33	 15.17 ±.18				
100634.5-295610 265.33 20.65 143.47 -43.99 100419.1-294130	NGC 3125 ESO 435- 41 IRAS10042-2941 PGC 29366	.E...?. PS -5.0±1.7	1.05± .05 .19± .03 .42± .01 1.04	114 .27 .00	13.50 ±.13 13.41 ±.11 11.54 13.16	.50± .01 -.47± .01 .43 -.51	.45± .01 -.52± .01 11.09± .01 13.29 .28		 1080± 47 866 1394
1006.6 +1426 222.74 49.65 86.35 -33.89 1003.9 +1441	 UGC 5448 PGC 29372	.S..3*. U (1) 3.0±1.2 4.5±1.2	1.02± .06 .13± .05 1.03	175 .10 .18 .06					
1006.6 +2810 201.47 53.67 73.05 -26.17 1003.8 +2825	 CGCG 153- 16 PGC 29376		 	.07	15.3 ±.3				4669 4619 4952
100639.7-191317 257.71 28.83 128.53 -44.30 100417.0-185838	NGC 3124 ESO 567- 17 PGC 29377	.SXT4.. R (4) 4.0± .3 1.0± .5	1.47± .02 .07± .03 1.11± .02 1.49	165 .18 .11 .04	12.86 ±.15 12.71 ±.12 12.46	.71± .02 .10± .07 .63 .04	.80± .01 .19± .07 13.90± .05 14.89± .19	13.91±.2 295± 10 246± 25 1.42	3562± 6 3351± 66 3364 3895
100645.9-474152 276.45 6.53 166.35 -39.37 100447.1-472712	 ESO 263- 4 PGC 29384	.SBR1.. r 1.0± .8	1.09± .06 .20± .06 1.22	106 1.40 .21 .10	 14.48 ±.14				
100652.2+142221 222.87 49.68 86.46 -33.86 100410.1+143701	NGC 3121 UGC 5450 PGC 29387	.E..... U -5.0± .8	1.24± .08 .09± .05 1.23	20 .10 .00	 13.57 ±.17 13.34				 9053± 22 8952 9374
1006.8 +6750 142.59 42.46 42.37 .44 1003.0 +6805	 UGC 5442 PGC 29388	.I..9*. U 10.0±1.1	1.26± .06 .29± .06 1.27	 .12 .22 .15					
1006.8 +3128 195.91 54.07 70.18 -24.10 1004.0 +3143	IC 2540 MCG 5-24- 14 PGC 29389	.E?.... 	1.02± .14 .11± .07 .99	 .07 .00	 14.66 ±.16 14.49				 6534 6497 6806
100656.0-450252 274.89 8.68 163.15 -40.31 100454.1-444812	 ESO 263- 5 IRAS10049-4448 PGC 29393	.SBS3P? Sr (1) 3.5± .8 4.4± .7	1.29± .05 .49± .05 1.39	64 1.05 .72 .24	 13.84 ±.14 12.33 12.05				 4040± 60 3813 4307
100701.1-430441 273.72 10.28 160.72 -40.95 100457.0-425000	 ESO 263- 7 IRAS10049-4249 PGC 29402	.S..3*/ S 3.0±1.3	1.15± .05 .60± .04 1.22	105 .74 .83 .30	 14.91 ±.14 12.66				
100701.1-430335 273.71 10.29 160.70 -40.96 100457.0-424854	 ESO 263- 6 FAIR 426 PGC 29403	.L?.... 	.99± .07 .42± .05 1.01	120 .74 .00	 15.08 ±.14 14.22				 7700±190 7474 7975
100705.4-055302 246.46 38.35 110.39 -41.93 100434.9-053822	 MCG -1-26- 24 PGC 29405	.S..3./ E 3.0± .9	1.13± .06 .70± .05 1.14	170 .11 .97 .35					
1007.1 +1559 220.62 50.40 84.81 -32.98 1004.4 +1614	 UGC 5453 PGC 29408	.I..9.. U 10.0± .9	1.03± .08 .09± .06 1.04	80 .09 .07 .05	 14.63 ±.20 14.48				840 745 1158
100710.5+123946 225.28 48.99 88.31 -34.66 100429.5+125426	 UGC 5454 PGC 29413	.SX.8*. U 8.0±1.2	1.04± .08 .18± .06 1.05	40 .09 .22 .09	 14.7 ±.2 14.37			15.81±.3 131± 7 1.35	2791± 10 2684 3116
100711.9+330147 193.26 54.21 68.88 -23.08 100417.3+331627	NGC 3118 UGC 5452 IRAS10042+3316 PGC 29415	.S..4.. U 4.0± .9	1.40± .03 .83± .04 1.41	41 .04 1.22 .41	 14.30 ±.18 13.75 13.04			14.05±.3 227± 7 200± 7 .60	1342± 9 1311 1608

R.A. 2000 DEC. / l b / SGL SGB / R.A. 1950 DEC.	Names / PGC	Type / S_T n_L / T / L	$logD_{25}$ / $logR_{25}$ / $logA_e$ / $logD_o$	p.a. / A_g / A_i / A_{21}	B_T / m_B / m_{FIR} / B_T^o	$(B-V)_T$ / $(U-B)_T$ / $(B-V)_T^o$ / $(U-B)_T^o$	$(B-V)_e$ / $(U-B)_e$ / m'_e / m'_{25}	m_{21} / W_{20} / W_{50} / HI	V_{21} / V_{opt} / V_{GSR} / V_{3K}
100717.0-262823 / 263.10 23.43 / 138.66 -44.15 / 100459.1-261342	ESO 500- 2 / PGC 29423	.L?....	.83± .07 / .11± .05 / .43± .03 / .84	153 / .20 / .00	15.46 ±.14 / 15.24 ±.14	1.07± .02 / .16± .03	.98± .02 / .26± .03 / 13.10± .10 / 14.21± .37		
100719.4+470023 / 170.00 52.31 / 57.65 -13.92 / 100411.2+471503	UGC 5451 / PGC 29427	.I..9.. / U / 10.0± .8	1.13± .05 / .35± .05 / / 1.13	103 / .00 / .26 / .17	/ 14.12 ±.18 / / 13.86				571± 50 / 596 / / 777
100719.7+102144 / 228.37 47.94 / 90.83 -35.73 / 100440.1+103625	A 1004+10 / UGC 5456 / PGC 29428	.I.0.$. / R / 90.0	1.21± .05 / .31± .05 / .76± .01 / 1.22	148 / .04 / .46 / .15	13.72 ±.13 / 13.52 ±.18 / / 13.15	.39± .01 / -.23± .02 / .32 / -.28	.41± .01 / -.25± .03 / 13.01± .02 / 13.86± .29		535 / 420 / / 864
100727.6+121634 / 225.85 48.88 / 88.77 -34.78 / 100446.9+123115	IC 591 / UGC 5458 / ARAK 231 / PGC 29435	.S?....	1.01± .06 / .19± .05 / / 1.02	170 / .10 / .29 / .10	/ 14.14 ±.20 / 13.09 / 13.73			16.79±.2 / 174± 13 / 178± 5 / 2.96	2841± 6 / 2759± 63 / 2731 / 3166
100746.0-412000 / 272.78 11.76 / 158.48 -41.36 / 100540.0-410518	ESO 316- 29 / IRAS10056-4105 / PGC 29450	PSBT2?. / Sr / 2.0± .6	1.24± .04 / .51± .04 / / 1.32	76 / .80 / .63 / .26	/ 14.65 ±.14 / 12.08				
1007.8 -1129 / 251.69 34.64 / 117.92 -43.05 / 1005.4 -1115	PGC 29455	.IBS9.. / E (1) / 10.0± .9 / 9.8± .8	1.05± .09 / .29± .08 / / 1.08	85 / .23 / .22 / .14					
100754.4+292735 / 199.36 54.11 / 72.09 -25.17 / 100502.6+294217	UGC 5461 / PGC 29459	.S..4.. / U / 4.0±1.0	.99± .06 / .73± .05 / / 1.00	1 / .08 / 1.07 / .36	/ 15.66 ±.20 / / 14.47			15.74±.3 / / 271± 5 / .90	4799± 9 / 4755 / / 5078
1007.9 +6821 / 141.96 42.22 / 42.07 .87 / 1004.0 +6836	UGC 5449 / PGC 29460	.SB.6?. / U / 6.0±1.3	.96± .09 / .15± .06 / / .97	70 / .10 / .22 / .07	/ 15.15 ±.18				
100755.5-351349 / 269.01 16.66 / 150.56 -42.85 / 100544.0-345906	IC 2548 / ESO 374- 37 / PGC 29461	PSBR3.. / BSr (1) / 2.8± .4 / 2.2± .8	1.29± .04 / .05± .04 / / 1.33	65 / .44 / .07 / .03	/ 13.90 ±.14 / / 13.36				4691± 18 / 4471 / / 4991
100759.0-022951 / 243.32 40.73 / 106.13 -40.72 / 100526.5-021509	IC 592 / UGC 5465 / PGC 29465	.S..4.. / U / 4.0± .9	.92± .07 / .06± .05 / / .93	/ .06 / .08 / .03	/ 14.2 ±.2 / / 14.00			16.31±.1 / 142± 12 / / 2.28	6050± 9 / 5976± 58 / 5893 / 6391
100807.5+293231 / 199.23 54.17 / 72.05 -25.08 / 100515.7+294713	A 1005+29 / UGC 5464 / DDO 72 / PGC 29468	.S..9.. / U (1) / 9.0± .8 / 9.0±1.5	1.14± .05 / .21± .05 / .81± .03 / 1.15	135 / .08 / .22 / .11	15.62 ±.17 / / / 15.32	.62± .08 / .05± .11 / .55 / .00	.58± .06 / .03± .08 / 15.16± .08 / 15.65± .33	16.01±.3 / / 106± 7 / .59	1011± 7 / 967 / / 1290
100809.2+515042 / 162.62 50.77 / 54.09 -10.54 / 100454.8+520524	A 1004+52 / UGC 5460 / PGC 29469	.SBT7.. / U / 7.0± .7	1.36± .04 / .02± .05 / / 1.36	/ .00 / .02 / .01	/ 13.5 ±.3 / / 13.45			14.56±.1 / 105± 7 / 90± 12 / 1.10	1093± 5 / 1137 / / 1276
100810.2+530455 / 160.84 50.28 / 53.17 -9.69 / 100454.1+531936	A 1004+53 / UGC 5459 / IRAS10049+5319 / PGC 29472	.SBS5?/ / PU / 5.3± .6	1.68± .03 / .81± .05 / 1.05± .03 / 1.68	132 / .00 / 1.22 / .41	13.19 ±.14 / 13.5 ±.2 / 12.16 / 12.06	.62± .03 / -.10± .04 / .46 / -.22	.70± .03 / -.01± .04 / 13.93± .09 / 14.39± .22	13.20±.1 / 278± 6 / 264± 6 / .73	1111± 5 / 1159 / / 1287
100811.5+422507 / 177.32 53.60 / 61.32 -16.87 / 100508.6+423949	UGC 5462 / PGC 29474	.S..1.. / U / 1.0±1.0	.96± .07 / .55± .05 / / .96	107 / .00 / .56 / .28	/ 15.1 ±.2 / / 14.40				6895± 13 / 6901 / / 7122
100812.5+095835 / 229.03 47.94 / 91.39 -35.72 / 100533.1+101317	NGC 3130 / UGC 5468 / PGC 29475	.S..0.. / U / .0± .9	1.00± .08 / .25± .06 / / .99	30 / .07 / .19	/ 14.38 ±.19 / / 14.00				8206± 50 / 8090 / / 8536
100813.2+184233 / 216.76 51.66 / 82.23 -31.31 / 100528.6+185715	UGC 5467 / IRAS10054+1857 / PGC 29478	.L...?. / U / -2.0±1.7	1.00± .08 / .00± .03 / / 1.01	/ .09 / .00	/ 14.16 ±.15 / 13.35 / 14.03				2894± 50 / 2809 / / 3206
1008.3 +1431 / 222.89 50.06 / 86.51 -33.49 / 1005.6 +1446	MCG 3-26- 32 / PGC 29480	.SB?...	.82± .13 / .08± .07 / / .83	/ .10 / .10 / .04	/ 15.29 ±.18 / / 15.02				8651± 54 / 8551 / / 8973

R.A. 2000 DEC.	Names	Type	logD$_{25}$	p.a.	B$_T$	(B-V)$_T$	(B-V)$_\bullet$	m$_{21}$	V$_{21}$
l b		S$_T$ n$_L$	logR$_{25}$	A$_g$	m$_B$	(U-B)$_T$	(U-B)$_\bullet$	W$_{20}$	V$_{opt}$
SGL SGB		T	logA$_\bullet$	A$_i$	m$_{FIR}$	(B-V)$_T^o$	m'$_\bullet$	W$_{50}$	V$_{GSR}$
R.A. 1950 DEC.	PGC	L	logD$_o$	A$_{21}$	B$_T^o$	(U-B)$_T^o$	m'$_{25}$	HI	V$_{3K}$
100818.5-023133	IC 593	.S?....	.87± .08					16.13±.1	6078± 9
243.42 40.78	UGC 5469		.13± .05	.06	14.5 ±.2			256± 12	5970± 58
106.20 -40.65				.19					5920
100546.0-021650	PGC 29482		.88	.06	14.21			1.85	6418
100820.9+315150	NGC 3126	.S..3..	1.45± .03	123				14.73±.1	5179± 7
195.28 54.40	UGC 5466	U (1)	.71± .04	.07	13.55 ±.18			606± 8	5144± 42
70.05 -23.62	IRAS10054+3206	3.0± .8		.98					5143
100527.4+320632	PGC 29484	4.5±1.2	1.46	.36	12.46			1.91	5449
100827.5+121827	Leo I	.E.....	1.99± .02	80	11.18 ±.15		.63± .03		
225.98 49.11	UGC 5470	PU	.12± .04	.10	10.1 ±.2		.15± .05		168± 60
88.88 -34.56	DDO 74	-5.0± .3	1.70± .03	.00			15.17± .06		60
100546.8+123310	PGC 29488		1.98		10.80		15.85± .21		494
100833.9+765423	A 1003+77								10073
133.55 36.65				.03					10207
35.94 6.83	MK 136								10115
100341.9+770904	PGC 29494								
100834.8-003923	IC 594	.SBR4..	1.02± .05	127				16.29±.1	6436± 7
241.54 42.01	UGC 5472	U	.29± .04	.09	14.57 ±.18			417± 9	6409± 58
103.89 -39.96		4.0± .9		.43					6286
100601.3-002440	PGC 29496		1.03	.15	14.01			2.14	6778
100836.3+181350	NGC 3131	.SB.3*.	1.38± .04	54				15.54±.3	5098± 9
217.54 51.58	UGC 5471	U	.54± .05	.12	13.76 ±.18			542± 7	5251± 50
82.76 -31.49	IRAS10058+1828	3.0± .8		.74	13.02			521± 7	5016
100552.0+182833	PGC 29499		1.39	.27	12.86			2.41	5416
100840.5-400538		RLA.+..	.99± .06	170					
272.16 12.86	ESO 316- 30	r	.27± .05	.80	15.07 ±.14				
156.84 -41.54		-1.3± .9		.00					
100633.1-395054	PGC 29505		1.03						
1008.6 +7038		.I..9..	1.26± .06					15.16±.1	1290± 6
139.53 40.84	UGC 5455	U	.00± .06	.19				71± 7	
40.48 2.50		10.0± .8		.00				57± 12	1404
1004.6 +7053	PGC 29506		1.28	.00					1369
100851.4-670139	IC 2554	.SBS4P*	1.49± .03	175	12.51 ±.19	.72± .03	.73± .01	13.73±.3	1378± 9
288.02 -8.99	ESO 92- 12	RS	.38± .04	2.22	12.43 ±.14	-.01± .07	-.03± .03	393± 12	1251± 18
186.68 -29.90	IRAS10075-6647	4.0± .8	1.02± .04	.56	10.10		13.10± .11		1136
100731.1-664654	PGC 29512		1.69	.19	9.67		13.84± .28	3.87	1519
100901.5+322929		.SXT6..	1.13± .04					15.54±.3	5893± 10
194.21 54.58	UGC 5474	U	.02± .04	.04	14.6 ±.3				5861
69.61 -23.11	KUG 1006+327	6.0± .8		.03				152± 7	
100607.6+324413	PGC 29526		1.14	.01	14.48			1.05	6162
100903.4-111358		.L..0..	1.13± .06	15					
251.72 35.03	MCG -2-26- 31	E	.35± .05	.17					
117.64 -42.71		-2.0± .9		.00					
100635.8-105914	PGC 29527		1.09						
100906.0-382433		RSXR0P.	1.24± .05	170					
271.20 14.26	ESO 316- 32	Sr	.25± .05	.45	13.47 ±.14				4845± 60
154.65 -41.90	IRAS10069-3809	-.1± .6		.19	12.74				4622
100657.0-380948	PGC 29529		1.27		12.76				5136
100907.4-290351	NGC 3137	.SAS6..	1.80± .02	1	12.1 ±.4	.60± .05		12.18±.2	1109± 11
265.21 21.67	ESO 435- 47	SU (1)	.45± .04	.23	12.21 ±.14	-.10± .11		262± 16	
142.18 -43.53	IRAS10068-2849	6.3± .4		.65	12.54	.46		255± 12	896
100651.0-284906	PGC 29530	5.6± .5	1.82	.22	11.31	-.20	14.82± .42	.65	1426
100908.1-382339		.E.4*P.	1.18± .06	140					
271.19 14.28	ESO 316- 33	S	.22± .04	.45	13.52 ±.14				4512± 60
154.63 -41.90		-5.0±1.3		.00					4289
100659.0-380854	PGC 29531		1.18		13.00				4803
1009.1 -1158	NGC 3138	.S..4..	1.08± .06	80					
252.38 34.52	MCG -2-26- 32	E	.51± .05	.22					
118.65 -42.82		4.0± .9		.75					
1006.7 -1144	PGC 29532		1.10	.25					
100912.7+150019		.E.....	1.21± .09						
222.35 50.46	UGC 5477	U	.00± .05	.10	13.69 ±.20				9212± 31
86.14 -33.05		-5.0± .9		.00					9114
100630.4+151503	PGC 29536		1.22		13.45				9533
1009.2 +4021		.SX.3..	1.13± .05						7031
180.72 54.16	UGC 5476	U	.03± .05	.05	14.3 ±.2				
63.11 -18.08	IRAS10062+4035	3.0± .8		.05	13.53				7030
1006.2 +4035	PGC 29539		1.13	.02	14.15				7268

R.A. 2000 DEC. / l b / SGL SGB / R.A. 1950 DEC.	Names / PGC	Type / S_T n_L / T / L	$\log D_{25}$ / $\log R_{25}$ / $\log A_0$ / $\log D_0$	p.a. / A_g / A_i / A_{21}	B_T / m_B / m_{FIR} / B_T^0	$(B-V)_T$ / $(U-B)_T$ / $(B-V)_T^0$ / $(U-B)_T^0$	$(B-V)_0$ / $(U-B)_0$ / m' / m'_{25}	m_{21} / W_{20} / W_{50} / HI	V_{21} / V_{opt} / V_{GSR} / V_{3K}
100927.9-163742	NGC 3140	.S..5..	.95± .09	20	14.8 ±.3	.81± .05			
256.26 31.20	MCG -3-26- 28	PE (1)	.22± .05	.15					8458± 59
125.01 -43.42		5.0± .8	.33			.68	13.83± .54		8267
100703.4-162257	PGC 29548	1.8± .9	.96	.11	14.27				8797
1009.5 +3009	A 1006+30	.I..9..	1.22± .06		14.69 ±.15	.48± .05	.50± .04	15.26±.1	1378± 5
198.25 54.53	UGC 5478	U (1)	.01± .06	.06	14.3 ±.4	-.17± .07	-.19± .06	67± 5	
71.72 -24.47	DDO 73	10.0± .8	.95± .02	.01		.45	14.93± .05	55± 6	1336
1006.6 +3023	PGC 29549	9.0±1.0	1.22	.01	14.57	-.19	15.59± .37	.69	1655
1009.5 +5827		.SBS5..	1.25± .04	30					
153.41 47.97	UGC 5475	U	.38± .05	.00	15.1 ±.3				
49.35 -5.86		5.0± .8	.56						
1006.1 +5842	PGC 29550	1.25	.19						
100938.8-395616		.LA.-P.	1.19± .05	66					
272.22 13.10	ESO 316- 34	S	.26± .04	.80	13.86 ±.14				
156.58 -41.40		-3.0± .9	.00						
100731.1-394130	PGC 29554	1.26							
100949.1-424847		.SXT3..	1.09± .05	14					
273.98 10.79	ESO 263- 13	S	.11± .05	.70	14.94 ±.14				9980±190
160.18 -40.55	FAIR 427	3.0± .8	.15						9755
100744.0-423400	PGC 29565	1.16	.06	14.01					10256
100951.8+301921		.S..1*.	1.19± .05	91				15.86±.3	6305± 10
197.96 54.62	UGC 5481	U	.52± .05	.05	14.76 ±.19				
71.61 -24.31	IRAS10069+3034	1.0±1.2	.53	13.34			544± 7	6264
100659.7+303407	PGC 29567	1.19	.26	14.11			1.49	6582
101000.5-022746		.S?....	.95± .07					16.37±.1	6227± 9
243.71 41.14	UGC 5483		.00± .05	.07	14.8 ±.2			99± 12	
106.30 -40.23			.00						6072
100728.0-021300	PGC 29576	.96	.00	14.68			1.69	6571
101002.8-241947		.SXR3..	1.04± .05	99					
262.13 25.48	ESO 500- 5	r	.16± .04	.15	14.74 ±.14				
135.65 -43.62		3.3± .9	.22						
100743.0-240500	PGC 29577	1.05	.08						
1010.0 -1238		.S..8..	1.20± .07	100					
253.14 34.19		E (1)	.43± .08	.24					
119.60 -42.73		8.0± .9	.53						
1007.6 -1224	PGC 29578	8.7± .8	1.23	.21					
101004.0-123453	NGC 3143	.SBS3..	.92± .07	105	14.9 ±.2	.62± .02			
253.09 34.24	MCG -2-26- 33	R (1)	.13± .05	.24					3536± 62
119.52 -42.71		3.0± .5	.17			.52			3354
100737.1-122007	PGC 29579	4.2± .8	.94	.06	14.46		14.03± .41		3878
101005.2-114644	NGC 3139	.L..0P*	1.16± .06	75					
252.41 34.82	MCG -2-26- 34	E	.08± .05	.22					
118.44 -42.56		-2.0±1.3	.00						
100737.8-113157	PGC 29583	1.18							
1010.1 +5429		.S?....	1.14± .05	100					
158.67 49.93	UGC 5479		.17± .05	.00					1105± 36
52.33 -8.52			.26						1160
1006.8 +5444	PGC 29584	1.14	.09						1274
1010.1 +1741									
218.56 51.72	CGCG 93- 63			.09	15.4 ±.3				10369± 57
83.54 -31.47									10281
1007.4 +1756	PGC 29589								10684
101008.5-380753		RLBT+P*	1.13± .05						
271.20 14.61	ESO 316- 38	S	.11± .04	.44	14.11 ±.14				4969± 19
154.23 -41.77		-1.3± .7	.00						4746
100759.0-375306	PGC 29590	1.15		13.60					5261
101010.0-122602	NGC 3145	.SBT4..	1.49± .03	20	12.54 ±.13	.80± .01	.88± .01	14.48±.1	3652± 6
252.98 34.37	MCG -2-26- 36	R (3)	.29± .03	.22	12.43 ±.20	.28± .03	.37± .03	457± 11	3656± 10
119.33 -42.66		4.0± .3	1.03± .02	.43		.67	13.18± .05	459± 25	3472
100743.0-121115	PGC 29591	1.0± .5	1.51	.15	11.83	.17	14.09± .20	2.51	3996
1010.2 +5829		.S..8*.	.97± .07	170					
153.30 48.03	UGC 5480	U	.19± .05	.00	15.38 ±.20				
49.39 -5.77		8.0±1.3	.23						
1006.8 +5844	PGC 29595	.97	.09						
101013.3-670018	NGC 3136B	.E+....	1.18± .05	30	12.7 ±.2	.98± .02	1.05± .01		
288.11 -8.89	ESO 92- 13	RS	.23± .03	2.27	13.32 ±.14				1780± 31
186.57 -29.80		-4.0± .6	1.18± .04	.00			14.05± .14		1564
100852.1-664530	PGC 29597	1.44		10.82			13.01± .31		1948

R.A. 2000 DEC. / l b / SGL SGB / R.A. 1950 DEC.	Names / PGC	Type / S_T n_L / T / L	$\log D_{25}$ / $\log R_{25}$ / $\log A_o$ / $\log D_o$	p.a. / A_g / A_i / A_{21}	B_T / m_B / m_{FIR} / B_T^o	$(B-V)_T$ / $(U-B)_T$ / $(B-V)_T^o$ / $(U-B)_T^o$	$(B-V)_e$ / $(U-B)_e$ / m'_e / m'_{25}	m_{21} / W_{20} / W_{50} / HI	V_{21} / V_{opt} / V_{GSR} / V_{3K}
101017.0+321638		.S..6*.	1.07± .07					15.08±.3	1466± 7
194.60 54.83	UGC 5482	U	.21± .06	.07					1433
69.97 -23.04		6.0±1.2		.30				106± 7	
100723.5+323124	PGC 29603		1.08	.10					1736
101021.6-380147		PS..1..	.99± .07	123					
271.17 14.71	ESO 316- 40	r	.26± .05	.44	14.64 ±.14				4137± 19
154.08 -41.75	IRAS10082-3747	1.0± .9		.27	13.12				3915
100812.1-374700	PGC 29608		1.03	.13	13.89				4430
101028.1+021343		.SB.4..	1.04± .06	70					
238.81 44.15	UGC 5487	U	.37± .05	.04	14.89 ±.18				
100.60 -38.48	VV 722	4.0± .9		.54					
100753.1+022830	PGC 29614		1.04	.18					
101028.2+275726	IC 2550	.SXT3*.	1.02± .06	135				16.26±.3	4788± 9
202.04 54.47	UGC 5484	U (1)	.09± .05	.10	14.41 ±.18				4738
73.81 -25.62	IRAS10076+2811	3.0±1.2		.13	14.13			258± 5	5074
100737.8+281213	PGC 29615	4.5±1.1	1.03	.05	14.15			2.07	
101029.7-390836		RSBR0..	1.03± .06	161	14.72 ±.17	1.10± .03	1.11± .02		
271.87 13.83	ESO 316- 42	r	.12± .05	.49	14.50 ±.14	.52± .05	.53± .03		5685± 19
155.51 -41.45		-.1± .8	.62± .05	.09		.92	13.31± .18		5462
100821.0-385348	PGC 29616		1.07		13.92	.41	14.46± .38		5974
101034.9-254924	A 1008-25	.SBS8*.	1.30± .03	140				15.03±.2	2518± 8
263.28 24.40	ESO 500- 6	SU (2)	.10± .04	.11	14.28 ±.14			72± 11	
137.71 -43.45	DDO 237	8.2± .7		.12				58± 8	2310
100816.0-253436	PGC 29623	7.6± .6	1.31	.05	14.04			.94	2842
101036.3-382918		.S?....	.95± .06						
271.49 14.37	ESO 316- 43		.06± .06	.44	15.40 ±.14				4802± 19
154.66 -41.60				.10					4579
100827.0-381430	PGC 29625		.99	.03	14.83				5093
101039.9+200415		.S..1..	1.32± .04	0				15.30±.2	3824± 6
214.99 52.66	UGC 5489	U	.29± .05	.07	13.61 ±.19			427± 13	3824± 50
81.25 -30.09		1.0± .8		.30				426± 5	3745
100754.7+201902	PGC 29631		1.32	.15	13.20			1.96	4134
101040.3+242453	IC 2551	.S?....	1.01± .06					16.27±.2	6351± 6
208.00 53.86	UGC 5488		.11± .05	.11	14.46 ±.18			315± 10	6335± 25
77.09 -27.66	MK 717			.16	11.87			221± 10	6286
100752.4+243940	PGC 29632		1.02	.06	14.14			2.07	6647
101046.4-345042	IC 2552	.LXS-*.	1.21± .05						3104± 18
269.26 17.32	ESO 374- 40	BS	.04± .04	.44	13.42 ±.14				2884
149.90 -42.35		-3.0± .6		.00					3406
100834.0-343554	PGC 29637		1.26		12.66				
101047.8-285406		.E?....	.98± .07	116	14.43 ±.13	.92± .02	.93± .01		4219± 37
265.41 22.03	ESO 435- 49		.13± .05	.21	14.83 ±.14				4007
141.91 -43.19			.91± .02	.00		.83	14.47± .08		4537
100831.0-283918	PGC 29639		.97		14.34		13.99± .38		
101052.9-663849		.LBT0?.	1.17± .05	116					
287.96 -8.57	ESO 92- 14	Sr	.36± .03	2.42	13.38 ±.14				
186.19 -29.94		-1.9± .8		.00					
100930.0-662400	PGC 29644		1.39						
101055.0+455658	NGC 3135	.S?....	.95± .07	90					7270± 50
171.38 53.21	UGC 5486		.20± .05	.00	14.40 ±.20				7291
58.89 -14.17				.31					7482
100749.0+461145	PGC 29646		.95	.10	14.05				
101059.9-450907		.SBS3?.	1.13± .05	106					5200±190
275.54 9.02	ESO 263- 14	S (1)	.23± .05	1.00	13.83 ±.14				4974
162.96 -39.60	FAIR 428	3.0± .9		.31	11.33				5468
100857.1-445418	PGC 29651	4.4±1.2	1.22	.11	12.48				
101101.3-044248	Sextans A	.IBS9..	1.77± .02	135	11.86M±.07	.39± .08	.35± .04	11.79±.1	324± 4
246.17 39.86	MCG -1-26- 30	E (2)	.08± .02	.07			-.32± .12	62± 4	374± 19
109.25 -40.67	DDO 75	10.0± .9	1.69± .04	.06		.35	15.43± .09	63± 4	164
100830.1-042800	PGC 29653	9.5± .6	1.77	.04	11.73		15.34± .12	.02	671
101102.3-394137		RS..0*P	1.11± .05	64					
272.29 13.45	ESO 316- 44	S	.27± .04	.51	14.03 ±.14				
156.17 -41.21		.0±1.3		.20					
100854.0-392648	PGC 29655		1.15						
101107.7-000233		.S?....			16.11S±.15					10074± 33
241.43 42.89	CGCG 8- 59				15.3 ±.6				9926
103.43 -39.15	HICK 43B			.01					10417
100833.9+001215	PGC 29657								

R.A. 2000 DEC. / l b / SGL SGB / R.A. 1950 DEC.	Names / PGC	Type / S_T n_L / T / L	$\log D_{25}$ / $\log R_{25}$ / $\log A_\bullet$ / $\log D_o$	p.a. / A_g / A_i / A_{21}	B_T / m_B / m_{FIR} / B_T^o	$(B-V)_T$ / $(U-B)_T$ / $(B-V)_T^o$ / $(U-B)_T^o$	$(B-V)_\bullet$ / $(U-B)_\bullet$ / m' / m'_{25}	m_{21} / W_{20} / W_{50} / HI	V_{21} / V_{opt} / V_{GSR} / V_{3K}
1011.1 -1346 / 254.31 33.56 / 121.19 -42.65 / 1008.7 -1332	A 1008-13 / MCG -2-26- 39 / DDO 76 / PGC 29661	.SBS9.. / PE (2) / 8.5± .6 / 7.9± .8	1.30± .05 / .37± .05 / / 1.32	30 / .25 / .37 / .18	15.2 ±.2 / / / 14.57	.70± .08 / -.01± .10 / .54 / -.13	.78± .08 / -.04± .10 / / 15.63± .33	15.17±.1 / 343± 12 / / .42	3628± 9 / / 3444 / 3970
101110.1-205207 / 259.85 28.29 / 130.88 -43.32 / 100848.0-203718	NGC 3146 / ESO 567- 23 / / PGC 29663	.SBR3?. / SE (1) / 3.3± .7 / 3.3±1.2	1.00± .04 / .07± .03 / / 1.02	100 / .25 / .10 / .04	13.97 ±.14				
1011.1 +0113 / 240.08 43.68 / 101.90 -38.69 / 1008.6 +0128	/ CGCG 8- 60 / / PGC 29664	.SBS4P. / E / 4.0± .9 /	1.35? / .32? / / 1.35	45 / .00 / .47 / .16	14.0 ±.3				
101112.7-000404 / 241.48 42.90 / 103.47 -39.14 / 100838.9+001045	/ CGCG 8- 61 / HICK 43C / PGC 29665	.L?....	/ / /	/ .01 / /	16.10S±.15 / 15.4 ±.6				9916± 41 / 9768 / 10259
101115.6-354138 / 269.87 16.70 / 150.99 -42.09 / 100903.9-352649	/ MCG -6-23- 9 / / PGC 29669	.LA.-P* / S / -3.0± .8 /	1.25± .08 / .14± .08 / / 1.30	/ .55 / .00 /					
101117.7+002638 / 240.95 43.23 / 102.85 -38.94 / 100843.6+004127	/ UGC 5493 / / PGC 29671	.SXT5.. / UE (1) / 4.5± .6 / 3.1± .8	1.20± .04 / .10± .04 / .94± .03 / 1.20	15 / .00 / .15 / .05	14.08 ±.15 / 13.55 ±.18 / / 13.69	.74± .04 / / .70 /	.69± .02 / / 14.27± .06 / 14.66± .27	15.41±.3 / 257± 7 / 231± 7 / 1.66	3644± 7 / / 3498 / 3987
101119.0-171217 / 257.10 31.07 / 125.86 -43.04 / 100854.7-165728	/ MCG -3-26- 30 / / PGC 29675	.LX.-P* / ES / -2.7± .7 /	1.25± .05 / .15± .04 / / 1.25	100 / .18 / .00 /					
101119.9-000123 / 241.45 42.95 / 103.43 -39.10 / 100846.1+001326	/ CGCG 8- 62 / HICK 43A / PGC 29677	.S?....	/ / /	/ .01 / /	15.80S±.15 / 15.2 ±.6				10150± 33 / 10002 / 10493
101130.4+304731 / 197.20 55.01 / 71.44 -23.74 / 100838.2+310220	/ UGC 5490 / / PGC 29683	.S..6*. / U / 6.0±1.2 /	1.11± .05 / .13± .05 / / 1.12	10 / .10 / .20 / .07	14.4 ±.2 / / / 14.10		15.85±.3 / / 212± 7 / 1.69		5115± 7 / 5076 / 5391
1011.5 +8449 / 126.81 31.04 / 30.15 12.29 / 1003.0 +8504	/ UGC 5438 / / PGC 29684	.S..4.. / U / 4.0± .9 /	1.01± .06 / .57± .05 / / 1.04	100 / .34 / .84 / .28					
101136.9-211514 / 260.22 28.07 / 131.42 -43.23 / 100915.0-210024	/ ESO 567- 25 / / PGC 29686	.S?....	.76± .06 / .01± .05 / / .78	/ .23 / .02 / .01	14.38 ±.14 / / / 14.08				9230±104 / 9030 / 9563
101141.0-375532 / 271.32 14.95 / 153.87 -41.52 / 100931.0-374042	/ ESO 316- 46 / / PGC 29690	RLXTO*. / Sr / -2.1± .4 /	1.10± .05 / .19± .03 / / 1.12	114 / .44 / .00 /	14.11 ±.14 / / / 13.60				4700± 19 / 4478 / 4994
101143.0-313832 / 267.38 19.99 / 145.59 -42.67 / 100928.0-312342	NGC 3157 / ESO 435- 51 / IRAS10094-3123 / PGC 29691	.SBS4*/ / S (1) / 4.5± .6 / 4.4±1.1	1.39± .03 / .67± .03 / / 1.43	38 / .40 / 1.01 / .34	13.94 ±.14 / 12.63 / / 12.51				2850± 60 / 2634 / 3162
101151.5+585323 / 152.62 48.01 / 49.24 -5.34 / 100826.5+590812	/ UGC 5491 / KUG 1008+591 / PGC 29696	.S..3.. / U / 3.0±1.1 /	.99± .06 / .28± .04 / / .99	72 / .00 / .38 / .14	15.9 ±.2 / 15.35 ±.18 / / 15.15	.40± .06 / -.36± .10 / .29 / -.44	/ / 14.98± .39 /		9100± 53 / 9171 / 9246
101151.5+585331 / 152.62 48.00 / 49.24 -5.34 / 100826.5+590820	A 1008+59 / MCG 10-15- 26 / MK 26 / PGC 29697	/ (1) / / 4.5±1.4	.52± .20 / .17± .07 / / .52	/ .00 / /	16.0 ±.3				9122± 35 / 9194 / 9268
1011.8 +1626 / 220.67 51.62 / 85.06 -31.77 / 1009.1 +1641	/ UGC 5495 / IRAS10091+1641 / PGC 29698	.S..6*. / U / 6.0±1.2 /	1.48± .03 / .88± .05 / / 1.49	101 / .12 / 1.30 / .44	14.30 ±.19 / 13.75				
101156.3+584404 / 152.81 48.09 / 49.36 -5.44 / 100831.6+585853	A 1008+58 / UGCA 206 / MK 27 / PGC 29702	.P..... / R / 99.0 /	.49± .11 / .26± .07 / / .49	/ .00 / /				16.95±.1 / 146± 6 / 62± 5 /	2096± 10 / 2171± 38 / 2172 / 2248

R.A. 2000 DEC. l　　b SGL　SGB R.A. 1950 DEC.	Names PGC	Type S_T　n_L T L	$\log D_{25}$ $\log R_{25}$ $\log A_o$ $\log D_o$	p.a. A_g A_i A_{21}	B_T m_B m_{FIR} B_T^o	$(B-V)_T$ $(U-B)_T$ $(B-V)_T^o$ $(U-B)_T^o$	$(B-V)_o$ $(U-B)_o$ m'_o m'_{25}	m_{21} W_{20} W_{50} HI	V_{21} V_{opt} V_{GSR} V_{3K}
1012.0 +5943 151.55 47.59 48.65 -4.76 1008.6 +5958	 UGC 5492 PGC 29706	.S..2.. U 2.0± .9 	1.08± .06 .41± .05 1.08	25 .00 .51 .21	 	 	 	 	
1012.0 +2305 210.32 53.85 78.55 -28.15 1009.3 +2319	 UGC 5498 IRAS10093+2319 PGC 29708	.S..1*. U 1.0±1.3 	1.17± .05 .79± .05 1.18	63 .07 .81 .40	 15.25 ±.18 13.69 	 	 	 	
101208.3+461738 170.72 53.31 58.77 -13.78 100902.3+463228	 UGC 5496 PGC 29711	.S..8*. U 8.0±1.3 	1.10± .05 .68± .05 1.10	152 .00 .83 .34	 15.41 ±.18 	 	 	 	
101209.3-385303 271.98 14.23 155.07 -41.21 101000.1-383812	 ESO 316- 47 IRAS10099-3838 PGC 29712	.SXR3.. r 3.3± .8 	1.07± .06 .04± .06 1.12	 .48 14.25 ±.14 .06 13.14 .02 13.67	 14.25 ±.14 13.14 13.67	 	 	 	5295± 19 5072 5586
101211.9+045525 236.07 46.08 97.65 -37.03 100935.4+051016	 UGC 5501 MK 718 PGC 29714	.S?.... 	.74± .09 .00± .05 .74	 .01 14.8 ±.2 .00 12.70 .00 14.76	14.8 ±.2 12.70 14.76 	 	 	 	8441± 39 8309 8780
101218.1+275143 202.31 54.86 74.16 -25.35 100928.0+280633	 UGC 5499 IRAS10094+2806 PGC 29715	.SB.3*. U 3.0±1.1 	1.44± .03 .68± .05 1.46	42 .14 13.92 ±.19 .93 13.37 .34 12.81	 13.92 ±.19 13.37 12.81	 	 	14.45±.3 429± 5 1.30	4752± 9 4757± 50 4702 5039
101219.9-471740 276.98 7.40 165.43 -38.64 101019.0-470248	 ESO 263- 15 IRAS10103-4703 PGC 29716	.S..6*/ S 6.0± .9 	1.46± .04 .94± .05 1.58	109 1.24 14.42 ±.14 1.38 12.19 .47 11.79	 14.42 ±.14 12.19 11.79	 	 	14.16±.3 348± 12 344± 9 1.90	2525± 10 2299 2785
101222.1-254609 263.58 24.70 137.61 -43.05 101003.1-253118	 ESO 500- 10 PGC 29719	.SBT3P* r 3.3±1.4 	1.04± .06 .62± .05 1.05	178 .10 15.51 ±.14 .85 .31	 15.51 ±.14 	 	 	 	
101223.9+582353 153.20 48.32 49.65 -5.63 100900.0+583843	A 1009+58 MK 28 PGC 29720	 	.50± .14 .00± .12 .50	 .00 	 	 	 	 	9101± 38 9171 9250
101232.2-451416 275.82 9.10 162.95 -39.31 101029.0-445924	 ESO 263- 16 IRAS10105-4459 PGC 29723	.SAS5P. S 5.0± .8 	1.23± .05 .13± .05 1.32	 .99 13.85 ±.14 .19 .06	 13.85 ±.14 	 	 	 	
101237.8-344346 269.50 17.64 149.65 -42.00 101025.0-342854	IC 2556 ESO 374- 42 (1) IRAS10104-3428 PGC 29727	PSBS7.. BSr 7.0± .4 5.6± .6	1.31± .04 .31± .04 1.34	108 .41 14.29 ±.14 .43 13.62 .16 13.43	 14.29 ±.14 13.62 13.43	 	 	 	2351± 19 2132 2655
101241.1+030750 238.26 45.13 99.80 -37.63 101005.6+032242	NGC 3156 UGC 5503 PGC 29730	.L...*. R -2.0± .6 	1.29± .03 .24± .03 .68± .01 1.26	47 13.07 ±.13 .04 12.97 ±.12 .00 12.96	13.07 ±.13 12.97 ±.12 12.96	.77± .01 .26± .02 .73 .24	.75± .01 .29± .01 11.96± .04 13.80± .21	17.08±.3 687± 13 	 1118± 41 980 1459
1012.7 +3035 197.60 55.26 71.80 -23.65 1009.9 +3050	 CGCG 153- 31 PGC 29738	 	 	 .11 15.2 ±.3 	 15.2 ±.3 	 	 	 	5208 5169 5485
101247.8-275022 265.07 23.14 140.42 -42.84 101030.1-273530	 ESO 436- 1 IRAS10104-2735 PGC 29743	.S..4*/ S 4.3± .7 	1.46± .03 .92± .03 1.47	137 .17 14.08 ±.14 1.36 12.11 .46 12.53	 14.08 ±.14 12.11 12.53	 	 	14.25±.3 357± 15 339± 12 1.25	2603± 11 2392 2924
101248.5+430847 175.81 54.27 61.33 -15.75 100946.1+432338	IC 598 UGC 5502 PGC 29745	.S..0.. U .0± .9 	1.15± .05 .48± .05 1.13	8 .01 13.9 ±.2 .36 13.54	13.9 ±.2 13.54	 	 	 	2268± 50 2278 2494
101248.7+070608 233.55 47.42 95.23 -35.98 101011.0+072100	 UGC 5504 PGC 29746	.S..7*. U 7.0±1.4 	1.14± .13 .86± .12 1.14	81 .02 15.44 ±.18 1.19 .43 14.23	 15.44 ±.18 14.23	 	 	15.68±.3 149± 7 1.02	1545± 10 1420 1882
101250.2+124008 226.25 50.23 89.13 -33.45 101009.6+125500	NGC 3153 UGC 5505 IRAS10101+1254 PGC 29747	.S..6*. U 6.0±1.1 	1.33± .04 .36± .05 1.34	170 .08 13.35 ±.18 .54 13.29 .18 12.72	 13.35 ±.18 13.29 12.72	 	 	14.56±.1 263± 6 246± 8 1.66	2808± 5 2827± 63 2703 3136

R.A. 2000 DEC. l b SGL SGB R.A. 1950 DEC.	Names PGC	Type S_T n_L T L	logD25 logR25 logAe logD0	p.a. Ag Ai A21	BT mB mFIR BT°	(B-V)T (U-B)T (B-V)T° (U-B)T°	(B-V)e (U-B)e m' m'25	m21 W20 W50 HI	V21 Vopt VGSR V3K
101251.3+044728 236.36 46.14 97.88 -36.93 101014.9+050220	 UGC 5506 PGC 29749	.S?.... 	1.06± .06 .17± .05 1.06	17 .01 .25 .08	 	 	 	16.28±.3 496± 7 	9576± 10 9444 9916
101305.5+672436 142.55 43.19 43.09 .57 100917.9+673927	A 1009+67 UGC 5494 MK 138 PGC 29764	.P...$. R 99.0 	1.00± .16 .73± .12 1.01	172 .13 .54 .36	 	 	 	 	 4500±110 4602 4598
101319.9-355859 270.40 16.71 151.26 -41.62 101108.0-354406	 ESO 374- 44 FAIR 1149 PGC 29778	PSBT1*. Sr 1.1± .6 	1.06± .05 .14± .04 1.10	112 .40 .15 .07	 15.05 ±.14 14.40	 	 	 	 8530± 14 8310 8830
101327.1-381154 271.78 14.94 154.11 -41.12 101117.1-375700	 ESO 317- 3 PGC 29790	.E.1... S -5.0± .7 	1.15± .05 .10± .04 1.18	 .41 .00 	 13.65 ±.14 13.17	 	 	 	 4819± 19 4597 5112
101330.0-434306 275.07 10.45 161.03 -39.63 101125.0-432812	 ESO 263- 18 IRAS10113-4327 PGC 29795	.S..4./ S 4.0± .9 	1.31± .05 .77± .04 1.37	128 .62 1.12 .38	 14.82 ±.14 12.84 	 	 	 	
101330.0+383701 183.49 55.21 65.08 -18.56 101031.9+385154	NGC 3151 MCG 7-21- 18 PGC 29796	.LA..*. P -2.0± .9 	1.09± .12 .24± .07 1.06	 .00 .00 	 14.80 ±.17 14.69	 	 	 	 7140± 79 7133 7386
101330.9-434906 275.13 10.37 161.15 -39.60 101126.1-433412	 ESO 263- 19 PGC 29797	.SXR4.. Sr (1) 4.2± .6 5.6±1.2	1.20± .05 .15± .05 1.26	38 .67 .23 .08	 13.86 ±.14 12.94	 	 	 	 4210± 60 3985 4484
101331.4+032232 238.15 45.45 99.62 -37.34 101055.8+033725	NGC 3165 UGC 5512 PGC 29798	.SAS8*. PU 8.3± .7 	1.20± .04 .29± .04 1.20	177 .01 .35 .14	14.5 ±.2 14.01 ±.19 13.88	.63± .09 .04± .11 .57 -.01	 14.66± .30	15.87±.2 146± 4 138± 7 1.85	1332± 5 1317± 50 1195 1673
101332.0+224423 211.04 54.08 79.09 -28.06 101045.5+225916	NGC 3162 UGC 5510 IRAS10107+2259 PGC 29800	.SXT4.. R (2) 4.0± .3 3.0± .7	1.48± .02 .08± .03 1.06± .01 1.48	 .07 .12 .04	12.21 ±.13 12.04 ±.14 12.01 11.94	.57± .01 .53 	.58± .01 13.00± .03 14.25± .17	13.70±.1 187± 5 171± 6 1.72	1298± 7 1456± 56 1231 1604
101332.3+201023 215.18 53.33 81.57 -29.47 101047.4+202516	 UGC 5509 PGC 29802	.S..3.. U (1) 3.0± .9 4.5±1.3	1.17± .05 .84± .05 1.18	66 .06 1.16 .42	 15.11 ±.19 13.83	 	 	15.14±.3 562± 7 .89	8351± 10 8272 8661
101334.8+385026 183.09 55.20 64.90 -18.41 101036.5+390519	NGC 3152 MCG 7-21- 18A PGC 29805	PLB.0*. P -2.0± .9 	1.01± .14 .24± .07 .98	 .04 .00 	 15.20 ±.16 15.07	 	 	 	 6471± 79 6464 6716
101338.4-005534 242.91 42.83 104.79 -38.87 101105.0-004040	 UGC 5515 PGC 29807	.E+..P* UE -4.3± .6 	1.18± .07 .07± .04 1.33± .23 1.17	90 .04 .00 	* 13.68 ±.17 13.44	 	1.09± .05 15.48± .82 	 	 13438± 88 13288 13783
101342.9-345130 269.77 17.67 149.77 -41.75 101130.1-343636	 ESO 374- 45 IRAS10115-3436 PGC 29811	.LB?... 	1.00± .07 .41± .05 .98	157 .39 .00 	 15.06 ±.14 14.60	 	 	 	 4390± 13 4171 4693
1013.7 +1807 218.42 52.67 83.64 -30.52 1011.0 +1822	 UGC 5514 PGC 29813	.S..6*. U 6.0±1.4 	1.12± .05 .93± .05 1.12	139 .08 1.36 .46	* 15.65 ±.19 14.19	 	 	 	 3646± 46 3559 3961
101345.0+032531 238.14 45.52 99.59 -37.27 101109.3+034025	NGC 3166 UGC 5516 IRAS10111+0340 PGC 29814	.SXT0.. R .0± .3 	1.68± .02 .31± .03 .92± .01 1.67	87 .01 .23 	11.32M±.10 11.30 ±.12 11.14 11.05	.93± .01 .40± .01 .87 .35	.95± .01 .44± .01 11.49± .04 13.81± .16	15.58±.2 139± 6 177± 5 	1345± 5 1326± 30 1208 1686
101347.4-005448 242.93 42.86 104.79 -38.83 101114.0-003954	 CGCG 8- 77 PGC 29820	 	 	 .04 	 15.3 ±.6 	 	 	 	 13947± 88 13797 14292
101350.8+384555 183.21 55.26 65.00 -18.41 101052.7+390048	NGC 3158 UGC 5511 PGC 29822	.E.3.*. R -5.0± .5 	1.30± .05 .05± .05 .90± .02 1.29	 .04 .00 	12.94 ±.13 12.84 ±.12 12.74	1.01± .01 .56± .03 .94 .59	1.02± .01 .61± .03 12.93± .09 14.29± .32	 	 6865± 27 6858 7110

R.A. 2000 DEC. l b SGL SGB R.A. 1950 DEC.	Names PGC	Type S_T n_L T L	$\log D_{25}$ $\log R_{25}$ $\log A_o$ $\log D_o$	p.a. A_g A_i A_{21}	B_T m_B m_{FIR} B_T^o	$(B-V)_T$ $(U-B)_T$ $(B-V)_T^o$ $(U-B)_T^o$	$(B-V)_e$ $(U-B)_e$ m'_e m'_{25}	m_{21} W_{20} W_{50} HI	V_{21} V_{opt} V_{GSR} V_{3K}
101352.3+003259 241.38 43.80 103.02 -38.30 101118.1+004753	 UGC 5521 PGC 29824	.SXS5.. U 5.0± .9 	.95± .05 .06± .04 .95	 .01 .08 .03	 14.61 ±.18 14.48			16.09±.3 176± 13 1.58	6232± 10 6087 6576
101353.2+383912 183.41 55.28 65.10 -18.48 101055.1+385405	NGC 3159 MCG 7-21- 21 PGC 29825	.E.2.P* P -5.0± .8 	1.32± .08 .00± .06 1.32	 .00 .00 	 14.61 ±.13 14.1 ±.3 14.44	.98± .02 .91 	 16.22± .45 		 6917± 58 6910 7163
101354.7-350655 269.97 17.48 150.09 -41.67 101142.0-345200	 ESO 374- 46A PGC 29829	.E?.... 	.60? .00? 1.03± .05 .68	 .47 .00 	 13.52 ±.17 15.35 ±.14 	1.12± .02 	1.13± .02 14.16± .15 11.52±1.59		
1013.9 +1807 218.45 52.72 83.66 -30.48 1011.2 +1822	 MCG 3-26- 43 PGC 29832	.S?.... 	.82± .13 .00± .07 .82	 .08 .00 	 15.14 ±.18 15.01				3562± 46 3476 3878
101359.1+070124 233.89 47.62 95.48 -35.76 101121.5+071618	 UGC 5522 PGC 29835	.S..7.. U 7.0± .7 	1.48± .03 .24± .05 1.48	145 .00 .34 .12	 13.2 ±.3 12.89			13.39±.1 221± 8 203± 6 .37	1221± 5 1096 1558
101359.5+383921 183.40 55.30 65.11 -18.46 101101.5+385415	NGC 3161 MCG 7-21- 22 ARAK 234 PGC 29837	.E.2... P -5.0± .9 	1.32± .09 .11± .06 1.28	 .00 .00 	 14.5 ±.3 14.43				6171± 58 6164 6417
101401.7-350825 270.00 17.48 150.12 -41.64 101149.0-345330	 ESO 374- 46 PGC 29840	.E?.... 	.91± .07 .12± .05 .95	12 .44 .00 	 14.39 ±.14 13.81				9090± 13 8870 9392
101403.7-215837 261.24 27.88 132.43 -42.69 101142.0-214342	 ESO 567- 26 PGC 29841	.S..4./ S (1) 4.0± .9 3.3±1.3	1.30± .04 .72± .03 1.32	168 .24 1.05 .36	 14.27 ±.14 				
101407.1+383910 183.40 55.33 65.13 -18.44 101109.1+385404	NGC 3163 UGC 5517 PGC 29846	.LA.-*. PU -3.0± .7 	1.17± .07 .00± .04 1.17	 .00 .00 	 14.33 ±.15 14.01 ±.18 14.11	1.02± .02 .96 	 15.04± .38 		 6199± 27 6192 6445
1014.1 +3928 181.98 55.23 64.47 -17.92 1011.2 +3943	 UGC 5518 PGC 29851	.I..9.. U 10.0± .8 	1.16± .07 .14± .06 1.16	60 .02 .10 .07				16.08±.1 79± 6 81± 4 	2085± 7 2081 2327
1014.2 +6907 140.67 42.18 41.91 1.80 1010.3 +6922	 UGC 5508 PGC 29852	.S..6*. U 6.0±1.2 	1.07± .06 .06± .05 1.09	 .13 .08 .03	 15.0 ±.3 14.76				10507 10615 10595
101414.4+032808 238.20 45.65 99.59 -37.14 101138.7+034303	NGC 3169 UGC 5525 PGC 29855	.SAS1P. R 1.0± .3 	1.64± .02 .20± .03 1.22± .02 1.64	45 .01 .20 .10	11.08M±.10 11.53 ±.13 11.02	.85± .02 .26± .09 .80 .23	.93± .01 .35± .01 12.77± .07 13.64± .17	12.69±.2 508± 5 158± 25 1.57	1233± 5 1261± 24 1098 1575
101422.1+220738 212.13 54.10 79.80 -28.24 101136.0+222233	 UGC 5524 PGC 29865	.S..6*. U 6.0±1.3 	1.24± .05 .81± .05 1.25	43 .07 1.19 .40	 15.29 ±.19 14.02			16.00±.3 181± 7 1.58	1644± 10 1573 1949
101428.1+774915 132.46 36.24 35.49 7.69 100932.3+780408	NGC 3197 UGC 5500 PGC 29870	.S..4.. U (1) 4.0± .8 3.5±1.1	1.11± .05 .12± .05 1.11	155 .01 .18 .06	 14.34 ±.19 14.09				8087 8224 8124
101435.1-274132 265.31 23.50 140.18 -42.45 101217.0-272636	NGC 3173 ESO 500- 16 IRAS10122-2726 PGC 29883	.SAS5.. S (1) 5.0± .4 3.7± .5	1.32± .04 .08± .03 1.35	7 .24 .12 .04	 13.56 ±.14 13.29 13.18			14.43±.3 212± 15 200± 12 1.21	2501± 11 2520± 60 2292 2824
101439.4-272438 265.14 23.73 139.80 -42.45 101221.0-270942	 ESO 500- 17 PGC 29888	PSBS3.. r 3.3± .9 	1.05± .06 .46± .05 1.07	30 .20 .63 .23	 14.98 ±.14 				
101439.7-004955 243.03 43.08 104.78 -38.59 101206.3-003500	 UGC 5528 PGC 29889	.SXR1P* E 1.0±1.2 	1.03± .05 .12± .04 1.04	150 .05 .12 .06	 14.37 ±.18 14.03				14630± 88 14481 14975

R.A. 2000 DEC. / l b / SGL SGB / R.A. 1950 DEC.	Names / PGC	Type / S_T n_L / T / L	$\log D_{25}$ / $\log R_{25}$ / $\log A_o$ / $\log D_o$	p.a. / A_g / A_i / A_{21}	B_T / m_B / m_{FIR} / B_T^o	$(B-V)_T$ / $(U-B)_T$ / $(B-V)_T^o$ / $(U-B)_T^o$	$(B-V)_o$ / $(U-B)_o$ / m' / m'_{25}	m_{21} / W_{20} / W_{50} / HI	V_{21} / V_{opt} / V_{GSR} / V_{3K}
101442.2-445102 / 275.91 9.64 / 162.33 -39.08 / 101238.1-443606	/ ESO 263- 21 / IRAS10126-4436 / PGC 29891	.IBS9.. / S (1) / 10.0± .8 / 6.7± .8	1.20± .05 / .08± .05 / / 1.26	19 / .60 / .06 / .04	/ 13.51 ±.14 / 12.62 /				
101442.4-285220 / 266.12 22.58 / 141.77 -42.34 / 101225.1-283724	NGC 3175 / ESO 436- 3 / VV 796 / PGC 29892	.SXS1$. / R (2) / 1.0± .7 / 5.3± .7	1.70± .02 / .57± .03 / 1.14± .02 / 1.72	56 / .26 / .58 / .28	12.13M±.10 / 12.04 ±.10 / 10.25 / 11.23	.90± .03 / .29± .06 / .73 / .14	.93± .02 / .28± .03 / 13.28± .05 / 14.04± .15		1095± 66 / 883 / 1415
101444.5-342020 / 269.63 18.21 / 149.04 -41.64 / 101231.1-340524	IC 2558 / ESO 375- 1 / IRAS10125-3405 / PGC 29895	.P...*. / S / 99.0 /	1.07± .05 / .24± .04 / / 1.11	13 / .40 / .36 / .12	14.3 ±.3 / 14.48 ±.14 / 13.16 / 13.67	.58± .05 / -.19± .07 / .42 / -.30	/ / / 13.89± .39		2573± 19 / 2355 / 2878
101445.7-340338 / 269.46 18.44 / 148.67 -41.68 / 101232.1-334842	IC 2559 / ESO 375- 2 / IRAS10125-3348 / PGC 29898	.SBS3?. / BS (1) / 3.3± .6 / 5.6±1.2	1.23± .04 / .38± .04 / / 1.26	18 / .40 / .52 / .19	14.39 ±.14 / 12.83 / 13.44				2899± 19 / 2681 / 3205
101445.7-002026 / 242.53 43.41 / 104.20 -38.40 / 101212.0-000530	CGCG 8- 81 / / / PGC 29899			.02	15.4 ±.6				9953± 88 / 9805 / 10298
101447.0-394826 / 272.96 13.78 / 156.08 -40.49 / 101238.1-393330	/ ESO 317- 5 / / PGC 29901	.LXR+.. / r / -1.3± .9 /	1.01± .06 / .00± .05 / / 1.08	/ .71 / .00 /	13.80 ±.14				
101449.4-381120 / 272.00 15.10 / 154.02 -40.86 / 101239.0-375624	/ ESO 317- 6 / IRAS10126-3756 / PGC 29905	.SBT3P? / S / 2.7±1.0 /	1.18± .04 / .44± .04 / / 1.22	120 / .39 / .61 / .22	14.43 ±.14 / 12.32 / 13.40				4529± 19 / 4307 / 4823
101452.0-200044 / 259.97 29.51 / 129.77 -42.42 / 101229.1-194548	/ ESO 567- 29 / / PGC 29907	/ / /	.80± .07 / .07± .06 / / .82	/ .23 / /	15.04 ±.14				3560 / 3363 / 3897
101454.2-230302 / 262.18 27.17 / 133.89 -42.51 / 101233.0-224806	/ ESO 500- 18 / IRAS10125-2248 / PGC 29911	.LXT0?. / S / -2.0± .8 /	1.20± .04 / .36± .03 / / 1.17	0 / .24 / .00 /	14.22 ±.14 / 13.17 /				
101457.5-433709 / 275.23 10.68 / 160.81 -39.41 / 101252.0-432212	/ ESO 263- 23 / IRAS10128-4322 / PGC 29915	.SA.0?P / S / .0± .8 /	1.34± .04 / .58± .05 / / 1.37	79 / .67 / .44 /	13.84 ±.14 / 11.52 / 12.69				3060± 60 / 2835 / 3335
101500.5+650821 / 144.86 44.75 / 44.89 -.83 / 101121.6+652316	UGC 5520 / / / PGC 29919	.S..6*. / U / 6.0±1.1 /	1.31± .04 / .23± .05 / / 1.32	100 / .11 / .34 / .11	14.1 ±.2 / 13.63			14.38±.1 / 267± 8 / 258± 8 / .64	3315± 9 / 3410 / 3427
101503.9+211024 / 213.77 53.98 / 80.82 -28.62 / 101218.5+212520	UGC 5529 / / / PGC 29924	.S..8*. / U / 8.0±1.3 /	1.10± .05 / .68± .05 / / 1.10	21 / .07 / .83 / .34	15.34 ±.18 / 14.42			15.52±.3 / / 304± 7 / .77	6197± 10 / 6122 / 6505
101512.1+564024 / 155.16 49.52 / 51.19 -6.51 / 101152.6+565520	NGC 3164 / UGC 5527 / / PGC 29928	.S?.... / / /	.94± .07 / .12± .05 / / .94	0 / .00 / .18 / .06	14.56 ±.18 / 14.34				7783± 50 / 7846 / 7942
1015.2 -1506 / 256.27 33.25 / 123.18 -41.86 / 1012.8 -1452	MCG -2-26- 41 / / / PGC 29929	.SXT8?. / E (1) / 8.0±1.1 / 7.5±1.2	1.17± .05 / .04± .05 / / 1.19	/ .22 / .05 / .02				14.77±.1 / 148± 12	3475± 9 / / 3289 / 3818
1015.2 +2106 / 213.91 54.00 / 80.92 -28.63 / 1012.5 +2121	A 1012+21 / CGCG 123- 30 / ZZW 44 / PGC 29934			.07	15.4 ±.3				6154± 49 / 6079 / 6463
1015.4 +1402 / 224.78 51.42 / 88.07 -32.22 / 1012.8 +1417	UGC 5533 / / / PGC 29943	.S..3.. / U (1) / 3.0±1.0 / 3.5±1.3	1.04± .08 / .67± .06 / / 1.05	98 / .14 / .92 / .33	15.35 ±.18				
101533.1-450758 / 276.20 9.49 / 162.61 -38.84 / 101329.0-445300	/ ESO 263- 24 / FAIR 429 / PGC 29947	.LBR0.. / Sr / -1.6± .6 /	1.18± .08 / .18± .06 / / 1.22	0 / .63 / .00 /	14.56 ±.14 / 12.70 / 13.91				1050±190 / 825 / 1320

R.A. 2000 DEC.	Names	Type	logD$_{25}$	p.a.	B$_T$	(B-V)$_T$	(B-V)$_e$	m$_{21}$	V$_{21}$
l　　b		S$_T$　n$_L$	logR$_{25}$	A$_g$	m$_B$	(U-B)$_T$	(U-B)$_e$	W$_{20}$	V$_{opt}$
SGL　　SGB		T	logA$_o$	A$_i$	m$_{FIR}$	(B-V)$_T^o$	m'$_e$	W$_{50}$	V$_{GSR}$
R.A. 1950 DEC.	PGC	L	logD$_o$	A$_{21}$	B$_T^o$	(U-B)$_T^o$	m'$_{25}$	HI	V$_{3K}$
101533.2+741318 NGC　3144		.SBS2P*	1.08± .05	0					6446
135.61　38.83 UGC　5519		PU	.22± .04	.04	14.29 ±.18				
38.21　　5.32		1.5± .6		.27					6572
101112.5+742813 PGC 29949			1.09	.11	13.92				6504
101536.7-203846 NGC　3171		.L..-P.	1.22± .05	176					
260.59　29.13 ESO　567- 31		SE	.18± .03	.25	13.85 ±.14				3470± 60
130.64 -42.28		-2.5± .5		.00					3272
101314.0-202348 PGC 29950			1.22		13.55				3806
1015.7　+5540 A　1012+55		.SBR3*.							7254± 60
156.46　50.08 CGCG 266- 31		P		.00	14.7 ±.3				7314
51.99　-7.13		3.0± .9							7418
1012.4　+5555 PGC 29953									
101542.5-342451		.LX.-*.	1.04± .11						8745± 15
269.84　18.27 MCG -6-23- 21		S	.00± .08	.37					8527
149.09 -41.43		-3.0± .9		.00					9051
101328.9-340953 PGC 29954			1.09						
1015.7　+0719		.S..7..	1.40± .04	147				14.71±.3	3756± 9
233.87　48.14 UGC　5537		U	1.15± .05	.03	14.91 ±.18			304± 7	
95.37 -35.24		7.0± .9		1.38				288± 7	3633
1013.1　+0734 PGC 29956			1.40	.50	13.49			.72	4094
1015.7　+1856		.S..0..	1.04± .06	127					
217.43　53.40 UGC　5535		U	.77± .05	.05	15.5 ±.2				
83.11 -29.69		.0±1.0		.58					
1013.0　+1911 PGC 29957			1.01						
101544.9-201740 ESO　567- 32		.SBR1P*	1.15± .04	133					
260.36　29.42		SE	.31± .03	.23	14.27 ±.14				
130.17 -42.23		1.0± .7		.32					
101322.0-200242 PGC 29959			1.17	.16					
101548.5+434713 A　1012+44		.S?....	.80± .07						5168± 45
174.51　54.64 MCG　7-21- 34			.10± .06	.00	14.8 ±.2				5181
61.18 -14.92 MK　139				.14					5392
101246.1+440210 PGC 29962			.80	.05	14.65				
101553.9-340652		.SBS9..	1.15± .05	73					3091
269.69　18.53 ESO　375- 3		S　　(1)	.46± .04	.40	16.24 ±.14				
148.69 -41.44		9.0± .8		.47					2874
101340.1-335154 PGC 29966		10.0± .8	1.19	.23	15.36				3398
101555.2+024108		.IB.9*.	1.11± .08					14.48±.1	1274± 7
239.45　45.51 UGC　5539		U	.07± .06	.10				166± 6	
100.72 -37.05		10.0± .8		.06				151± 4	1136
101320.0+025606 PGC 29969			1.12	.04					1618
101609.2-154730 NGC　3178		.SAT6P*	1.12± .05	70					3480
257.01　32.90 MCG -3-26- 34		E　　(1)	.27± .04	.18					
124.13 -41.73		5.5± .6		.39					3292
101343.7-153232 PGC 29980		5.3± .8	1.14	.13					3822
1016.1　+5825		.S..3..	1.15± .05	98					7650± 10
152.78　48.73 UGC　5534		U	.33± .05	.01	14.72 ±.19				
49.97　-5.25		3.0± .9		.46					7720
1012.8　+5840 PGC 29983			1.16	.17	14.19				7799
101619.3-333353 IC　2560		PSBR3*.	1.50± .03	45					2873± 13
269.42　19.03 ESO　375- 4		BSr　(1)	.20± .04	.31	12.53 ±.14				2656
147.95 -41.44 IRAS10140-3318		3.3± .4		.28	11.82				
101405.0-331854 PGC 29993		2.8± .5	1.53	.10	11.92				3182
1016.3　+0449		.S..5..	1.08± .06	167					13786± 60
237.05　46.86 UGC　5543		U	.26± .05	.01	14.33 ±.18				13655
98.29 -36.13 IRAS10137+0504		5.0± .9		.39	13.57				14127
1013.7　+0504 PGC 29995			1.08	.13	13.86				
101621.6+374649		.S..6?.	1.23± .03	111					1216± 19
184.83　55.87 UGC　5540		U	.72± .04	.00	14.60 ±.18				1206
66.14 -18.65 KUG 1013+380		6.0±1.8		1.05					1467
101324.8+380147 PGC 29997			1.23	.36	13.54				
1016.4　+6014 NGC　3168		.E.....	1.00± .10						
150.47　47.77 UGC　5536		U	.05± .04	.00	14.40 ±.15				9395± 49
48.64　-4.01		-5.0± .8		.00					9472
1013.0　+6029 PGC 30001			.99		14.26				9534
101626.2-421448		PSXT3..	1.05± .05	87					
274.66　11.96 ESO　317- 8		Sr　　(1)	.16± .05	.58	14.73 ±.14				
159.02 -39.55 IRAS10143-4159		3.2± .6		.22	13.61				
101419.0-415948 PGC 30003		4.4±1.2	1.10	.08					

R.A. 2000 DEC. / l b / SGL SGB / R.A. 1950 DEC.	Names / PGC	Type / S_T n_L / T / L	$\log D_{25}$ / $\log R_{25}$ / $\log A_e$ / $\log D_o$	p.a. / A_g / A_i / A_{21}	B_T / m_B / m_{FIR} / B_T^o	$(B-V)_T$ / $(U-B)_T$ / $(B-V)_T^o$ / $(U-B)_T^o$	$(B-V)_e$ / $(U-B)_e$ / m'_e / m'_{25}	m_{21} / W_{20} / W_{50} / HI	V_{21} / V_{opt} / V_{GSR} / V_{3K}
101628.4+451918 171.91 54.32 60.05 −13.84 101324.7+453416	A 1013+45 UGCA 208 MK 140 PGC 30005	 	.54± .11 .19± .07 .54	 .00 	 15.4 ±.4 	 	 	15.70±.1 188± 6 147± 5 	1661± 10 1607± 52 1678 1876
101634.5+210729 214.02 54.30 81.09 −28.35 101349.2+212228	NGC 3177 UGC 5544 IRAS10138+2122 PGC 30010	.SAT3.. R (2) 3.0± .4 3.7± .7	1.16± .03 .09± .03 1.17	135 .07 .13 .05	13.04 ±.13 13.14 ±.15 10.70 12.87	.64± .01 .07± .04 .60 .04	 13.45± .23	15.61±.2 293± 18 186± 7 2.69	1302± 5 1220± 56 1227 1610
1016.6 +1234 227.05 51.00 89.75 −32.68 1013.9 +1249	 UGC 5548 IRAS10139+1249 PGC 30013	.S?.... 	1.00± .08 .55± .06 1.01	67 .09 .82 .27	 15.15 ±.19 13.37 14.19	 	 	 	 9403± 57 9298 9732
1016.6 +1658 220.61 52.88 85.21 −30.52 1013.9 +1713	 UGC 5547 U PGC 30014	.L...*. −2.0±1.2 	1.06± .11 .21± .05 1.04	125 .10 .00 	 14.57 ±.15 	 	 	 	
1016.8 +5824 152.73 48.82 50.05 −5.19 1013.5 +5839	 UGC 5541 U PGC 30018	.I..9.. 10.0± .9 	1.14± .05 .37± .05 1.14	20 .01 .28 .19	 15.3 ±.2 	 	 	 	
101653.2+732404 136.29 39.46 38.88 4.84 101239.4+733902	NGC 3147 UGC 5532 IRAS10126+7339 PGC 30019	.SAT4.. R (2) 4.0± .3 2.7± .6	1.59± .02 .05± .03 1.07± .03 1.59	155 .05 .08 .03	11.43 ±.16 11.38 ±.14 10.69 11.25	.82± .04 .78 	.90± .04 12.27± .09 14.08± .20	14.38±.3 345± 34 345± 25 3.10	2820± 17 2721± 56 2935 2875
101654.3−485255 278.52 6.53 166.95 −37.36 101454.0−483754	 ESO 213− 11 IRAS10149−4837 PGC 30022	.SAS5.. S (1) 5.0± .5 2.2± .5	1.54± .04 .19± .07 1.67	7 1.40 .28 .09	 11.52 	 	 	13.00±.3 369± 7 	2742± 9 2516 2997
1016.9 +6017 150.36 47.79 48.64 −3.93 1013.5 +6032	 UGC 5542 U PGC 30027	.E...*. −5.0±1.2 	1.00± .10 .00± .04 1.00	 .00 .00 	 14.59 ±.16 14.45	 	 	 	 9230± 56 9307 9369
101658.6+493742 165.04 52.92 56.73 −10.98 101350.1+495241	 UGC 5545 IRAS10138+4952 PGC 30031	.S?.... 	.97± .07 .10± .05 .97	130 .00 .15 .05	 14.38 ±.18 13.21 14.17	 	 	 	12601± 50 12637 12797
1017.0 +5449 157.48 50.67 52.77 −7.55 1013.8 +5504	 UGC 5546 U PGC 30036	.SB.3.. 3.0± .9 	1.04± .06 .19± .05 1.04	150 .00 .26 .10	 14.96 ±.19 	 	 	 	
101707.9−332517 269.48 19.24 147.73 −41.29 101453.3−331016	 PGC 30038	.IBS9.. S (1) 10.0± .8 10.0±1.2	1.12± .08 .07± .08 1.15	 .31 .05 .04	 	 	 	 	2776 2560 3085
101711.0−032955 246.33 41.82 108.32 −38.84 101438.9−031455	IC 600 MCG 0−26− 34 VV 97 PGC 30041	.SBS8.. UE (1) 7.7± .5 6.4± .6	1.37± .03 .30± .04 1.38	25 .06 .36 .15	 13.00 ±.18 12.58	 	 	14.19±.1 182± 10 172± 12 1.47	1309± 7 1153 1656
101711.9+042010 237.81 46.75 98.95 −36.12 101435.8+043510	 UGC 5551 U PGC 30042	.I..9*. 10.0±1.3 	.96± .17 .10± .12 .96	 .02 .07 .05	 	 	 	15.77±.1 78± 5 57± 3 	1341± 5 1208 1683
1017.2 +5328 159.35 51.33 53.81 −8.43 1014.0 +5343	 UGC 5549 U (1) PGC 30044	.S..4.. 4.0± .9 3.5±1.1	1.04± .06 .07± .05 1.04	55 .00 .11 .04	 15.0 ±.2 14.76	 	 	 	13622± 60 13673 13798
101718.0−183949 259.47 30.91 128.01 −41.75 101454.0−182448	 ESO 567− 40 PGC 30047	.SB?... 	.89± .06 .00± .06 .92	 .26 .00 .00	 15.85 ±.14 15.56	 	 .52	16.09±.1 116± 12 	4015± 9 3821 4354
101720.1+152925 222.95 52.44 86.82 −31.12 101438.1+154425	A 1014+15 MK 629 PGC 30049	 	 	 .08 	 13.23 	 	 	 	9710± 45 9615 10033
1017.4 +1704 220.55 53.10 85.21 −30.30 1014.7 +1720	 UGC 5552 U PGC 30052	.S..4.. 4.0± .9 	1.19± .05 .65± .05 1.19	32 .09 .96 .33	 14.89 ±.18 	 	 	 	

R.A. 2000 DEC. / l b / SGL SGB / R.A. 1950 DEC.	Names / PGC	Type S_T n_L / T / L	$\log D_{25}$ / $\log R_{25}$ / $\log A_o$ / $\log D_o$	p.a. / A_g / A_i / A_{21}	B_T / m_B / m_{FIR} / B_T^o	$(B-V)_T$ / $(U-B)_T$ / $(B-V)_T^o$ / $(U-B)_T^o$	$(B-V)_o$ / $(U-B)_o$ / m'_o / m'_{25}	m_{21} / W_{20} / W_{50} / HI	V_{21} / V_{opt} / V_{GSR} / V_{3K}
101738.7+214119	NGC 3185	RSBR1..	1.37± .02	130	12.99S±.07	.82± .05		15.41±.1	1230± 4
213.22 54.70	UGC 5554	R	.17± .02	.10	12.73 ±.13			283± 4	1239± 29
80.70 -27.84	IRAS10148+2156	1.0± .4		.17	12.53	.76		261± 4	1159
101453.2+215620	PGC 30059		1.38	.08	12.65		14.26± .14	2.69	1539
101741.8+742055	NGC 3155	.S?....	1.16± .07	35					2944
135.37 38.84	UGC 5538		.16± .06	.05	13.85 ±.18				3070
38.21 5.52	IRAS10133+7436			.24	13.65				3002
101322.2+743555	PGC 30064		1.16	.08	13.55				
101747.6+215225	NGC 3187	.SBS5P.	1.47± .02		13.91M±.11	.47± .04	.48± .03	14.84±.1	1578± 3
212.93 54.79	UGC 5556	R	.37± .03	.09	13.2 ±.2	-.20± .06	-.13± .06	257± 3	1583± 32
80.54 -27.71	IRAS10150+2207	5.0± .3	.99± .03	.56	13.00	.37	14.11± .07	218± 6	1507
101502.0+220726	PGC 30068		1.48	.19	13.11	-.27	15.18± .16	1.55	1886
101753.0+600332	A 1014+60	.S?....	.62± .11						12792± 72
150.53 48.01			.26± .12	.00					12868
48.89 -3.99	MK 29			.39					12933
101428.3+601833	PGC 30075		.62	.13					
101757.1+410652	NGC 3179	.L.....	1.27± .06	48					7258± 31
178.90 55.64	UGC 5555	U	.53± .03	.00	14.09 ±.15				7261
63.60 -16.32		-2.0± .8		.00					7495
101457.7+412153	PGC 30078		1.19		13.98				
101805.7+214959	NGC 3190	.SAS1P/	1.64± .02	125	12.12M±.10	.97± .01	.98± .01	15.62±.2	1338± 6
213.03 54.85	UGC 5559	R	.45± .02	.10	11.96 ±.12	.48± .02	.50± .01	601± 6	1289± 18
80.62 -27.68	IRAS10153+2204	1.0± .3	.85± .02	.46	11.53	.85	11.74± .06	438± 11	1262
101520.2+220501	PGC 30083		1.65	.22	11.48	.37	14.04± .15	3.92	1641
1018.1 +6020		.S..2..	1.02± .06	78					
150.15 47.87	UGC 5553	U	.70± .05	.00					
48.69 -3.78		2.0±1.0		.86					
1014.7 +6036	PGC 30084		1.02	.35					
1018.2 +0701	IC 601	.S?....	.82± .13	141					3672± 41
234.74 48.50	MCG 1-26- 33		.46± .07	.02	15.4 ±.3				3548
96.01 -34.80				.68					4011
1015.6 +0717	PGC 30086		.82	.23	14.72				
101817.3+412526	NGC 3184	.SXT6..	1.87± .01	135	10.36M±.10	.58± .02	.66± .01	12.18±.0	593± 4
178.34 55.64	UGC 5557	R (2)	.03± .02	.00	10.45 ±.15	-.03± .03	.06± .02	147± 3	404± 40
63.39 -16.08	KUG 1015+416	6.0± .3	1.59± .02	.04		.57	13.74± .03	126± 4	596
101517.8+414028	PGC 30087	3.5± .5	1.87	.01	10.35	-.04	14.49± .12	1.83	826
101819.7+070259	IC 602	.S...*.	.92± .07	177				14.77±.2	3744± 6
234.74 48.53	UGC 5561	R	.20± .05	.02	14.1 ±.3			315± 7	3724± 35
96.01 -34.77	ARAK 237			.30	11.96			198± 6	3620
101542.3+071801	PGC 30090		.92	.10	13.72			.95	4083
1018.3 -1300					14.16 ±.16	1.14± .03	1.09± .02		
255.23 35.33				.37					
120.54 -40.79			.87± .05				14.00± .19		
1015.9 -1245	PGC 30092								
101822.8+455718		.S..1..	1.04± .06	12					
170.70 54.45	UGC 5558	U	.67± .05	.00	15.33 ±.18				
59.77 -13.18		1.0±1.0		.68					
101518.8+461220	PGC 30094		1.04	.33					
101823.7-130616		PLXS-*.	1.16± .05	85					
255.32 35.26	MCG -2-26- 42	E	.10± .06	.35					
120.68 -40.80		-3.0± .6		.00					
101556.6-125113	PGC 30095		1.19						
101825.0+215342	NGC 3193	.E.2...	1.48± .02		11.83M±.03	.95± .01	.96± .01		
212.97 54.93	UGC 5562	R	.05± .03	.09	12.01 ±.12	.46± .02	.50± .01		1379± 14
80.61 -27.58	HICK 44B	-5.0± .3	.95± .02	.00		.92	12.07± .07		1308
101539.5+220845	PGC 30099		1.48		11.73	.45	14.12± .14		1687
101826.0-334316		.IBS9..	1.08± .09						
269.90 19.16		S (1)	.06± .08	.36					
148.07 -40.98		10.0± .8		.04					
101611.4-332813	PGC 30100	10.0± .8	1.11	.03					
101836.5-175857	NGC 3200	.SXT5*.	1.62± .02	169	12.83M±.10	.79± .02	.87± .02	13.12±.1	3516± 6
259.23 31.63	ESO 567- 45	VSUE (4)	.51± .03	.30	12.80 ±.11	.25± .03	.34± .03	551± 9	3537± 66
127.13 -41.38		4.5± .4	1.07± .02	.76		.60	13.69± .05	541± 11	3325
101612.0-174354	PGC 30108	2.1± .4	1.64	.25	11.73	.10	14.49± .14	1.14	3857
101848.7+382813		.S?....	1.08± .04	3					
183.50 56.26	UGC 5563		.40± .04	.00	14.59 ±.18				6750± 13
65.89 -17.84	KUG 1015+387			.60					6742
101551.8+384316	PGC 30120		1.08	.20	13.95				6998

R.A. 2000 DEC. l b SGL SGB R.A. 1950 DEC.	Names PGC	Type S_T n_L T L	$\log D_{25}$ $\log R_{25}$ $\log A_\bullet$ $\log D_o$	p.a. A_g A_i A_{21}	B_T m_B m_{FIR} B_T^o	$(B-V)_T$ $(U-B)_T$ $(B-V)_T^o$ $(U-B)_T^o$	$(B-V)_\bullet$ $(U-B)_\bullet$ $m'_\bullet$ m'_{25}	m_{21} W_{20} W_{50} HI	V_{21} V_{opt} V_{GSR} V_{3K}
101851.5-323552 269.28 20.12 146.58 -41.05 101636.1-322048	IC 2563 ESO 436- 9 PGC 30125	RSB.7?P S 7.0±1.7 	1.04± .05 .37± .04 1.06	106 .26 .51 .18	 15.57 ±.14 14.80				1348 1133 1660
101901.7-374016 272.39 16.00 153.13 -40.16 101650.1-372512	 ESO 375- 7 IRAS10168-3725 PGC 30131	.L?.... 	.98± .06 .50± .03 .94	54 .27 .00 	 14.65 ±.14 14.31				 4833± 19 4612 5130
1019.0 +2116 214.06 54.90 81.29 -27.78 1016.3 +2132	 CGCG 124- 1 PGC 30133		 	 .07 	 15.2 ±.3 			16.83±.3 63± 10 	1079± 9 1085±115 1006 1389
101905.0+462716 169.82 54.40 59.45 -12.77 101600.7+464219	NGC 3191 UGC 5565 IRAS10160+4642 PGC 30136	.SBS4P. P 4.0±1.0 	.92± .05 .16± .04 .92	5 .00 .23 .08	 14.1 ±.2 12.46 13.82				9145± 60 9169 9357
101908.8+344029 190.33 56.68 69.14 -20.11 101615.0+345533	IC 2561 UGC 5567 PGC 30147	.S?.... 	.96± .07 .25± .05 .96	17 .00 .37 .12	 14.86 ±.18 14.47				4523 4501 4787
101912.0+635811 145.78 45.83 46.06 -1.28 101538.7+641314	A 1015+64 MCG 11-13- 18 MK 141 PGC 30151	.E.3.$. P -5.0±1.8 	 	 .00 	15.42 ±.15 13.32 	.67± .03 -.36± .06 .55 -.29			12265± 38 12355 12384
101913.8+590751 151.55 48.67 49.70 -4.48 101551.5+592254	 UGC 5564 HICK 45A PGC 30153	.S..3.. U (1) 3.0± .9 4.5±1.2	1.11± .05 .50± .05 1.11	75 .00 .68 .25	15.73S±.15 14.88		 14.91± .33		21811± 41 21884 21957
101925.1-053924 248.94 40.76 111.20 -38.92 101654.1-052419	IC 603 MCG -1-26- 41 PGC 30166	RSBR1*. E .5± .6 	1.09± .06 .13± .06 1.10	25 .13 .13 .07					
101933.3+581220 152.69 49.22 50.43 -5.06 101612.9+582724	NGC 3182 UGC 5568 PGC 30176	.SAR1$. R 1.0± .7 	1.26± .09 .07± .10 1.26	155 .02 .07 .03	 13.00 ±.18 12.88				2130± 50 2200 2281
101934.5-264153 265.62 24.96 138.77 -41.39 101715.0-262648	NGC 3203 ESO 500- 24 PGC 30177	.LAR+$/ R -1.0± .4 1.39	1.46± .03 .69± .02 .72± .03 	58 .34 .00 	13.1 ±.2 13.18 ±.09 12.79	.97± .02 .48± .04 .78 .32	1.00± .02 .52± .04 12.19± .09 13.56± .25		2394± 66 2187 2720
1019.6 +0619 235.91 48.40 96.98 -34.77 1017.0 +0635	 UGC 5573 PGC 30178	.S..6*. U 6.0±1.2 	1.18± .05 .27± .05 1.18	128 .01 .40 .14	 14.11 ±.19 13.66			15.58±.3 415± 14 379± 15 1.78	8567± 12 8441 8908
101938.2+572509 153.70 49.65 51.03 -5.57 101619.3+574013	NGC 3188A MCG 10-15- 64 MK 30 PGC 30179	.S?.... 	.58± .10 .21± .06 .58	 .00 .31 .10					7954± 38 8021 8110
101941.0-254848 265.04 25.68 137.60 -41.40 101721.0-253342	NGC 3208 ESO 500- 25 IRAS10173-2533 PGC 30180	.SXT4.. PSr (2) 4.5± .4 3.3± .6	1.26± .03 .06± .03 .99±.01 1.28	20 .28 .10 .03	13.42 ±.13 13.73 ±.14 13.22 13.18	.68± .02 .14± .04 .59 .07	.74± .01 .13± .03 13.86± .03 14.40± .20	14.47±.1 189± 9 167± 7 1.26	2896± 7 3007± 59 2692 3226
101941.7-174459 259.28 31.97 126.86 -41.10 101717.1-172954	 ESO 567- 48 PGC 30181	.S..9*/ SE (1) 9.0± .7 7.8±1.3	1.24± .03 .75± .04 1.26	51 .27 .76 .37	 16.06 ±.14 15.03		 -.05	15.35±.1 117± 12 	901± 9 710 1242
101943.0+222705 212.18 55.38 80.26 -27.03 101657.3+224210	 UGC 5574 PGC 30182	.S..6?. U 6.0±1.9 	1.11± .05 .69± .05 1.12	93 .05 1.02 .35				16.28±.3 129± 7 	1467± 10 1398 1773
101943.0+572516 153.69 49.66 51.04 -5.56 101624.0+574020	NGC 3188 UGC 5569 MK 31 PGC 30183	RSBR2.. U 2.0± .9 	.96± .05 .03± .04 .96	 .00 .04 .02	 14.62 ±.18 14.51				7769± 38 7836 7925
1019.7 +5204 161.06 52.30 55.13 -9.04 1016.6 +5220	 UGC 5571 PGC 30187	.S..9*. U 9.0±1.3 	1.07± .14 .50± .12 1.07	172 .00 .51 .25				15.15±.1 59± 5 48± 5 	662± 6 708 846

R.A. 2000 DEC. l　　b SGL　　SGB R.A. 1950 DEC.	Names PGC	Type S_T　n_L T L	$\log D_{25}$ $\log R_{25}$ $\log A_o$ $\log D_o$	p.a. A_g A_i A_{21}	B_T m_B m_{FIR} B_T^o	$(B-V)_T$ $(U-B)_T$ $(B-V)_T^o$ $(U-B)_T^o$	$(B-V)_o$ $(U-B)_o$ m'_o m'_{25}	m_{21} W_{20} W_{50} HI	V_{21} V_{opt} V_{GSR} V_{3K}
101947.0+223535 211.94　55.43 80.14　-26.94 101701.2+225040	 UGC　5575 PGC 30188	.I..9*. U 10.0±1.3	1.04± .08 .37± .06 1.04	55 .05 .28 .19				17.39±.3 129± 5	1466± 9 1398 1772
101954.9+453309 171.21　54.83 60.26　-13.23 101651.8+454814	NGC　3198 UGC　5572 PGC 30197	.SBT5.. R　　(2) 5.0± .3 2.6± .5	1.93± .01 .41± .02 1.49± .02 1.93	35 .00 .61 .20	10.87M±.10 10.78 ±.13 10.91 10.22	.54± .02 -.04± .03 .46 -.10	.62± .01 .02± .02 13.81± .04 14.37± .13	11.50±.1 319± 4 303± 4 1.08	663± 4 667± 14 684 880
102002.6-213106 262.14　29.11 131.87　-41.28 101740.0-211600	 ESO　567- 51 PGC 30204	.E?....	.94± .06 .01± .04 .98	 .26 .00	14.13 ±.13 14.31 ±.14 13.90	.96± .01 .87	 13.79± .36		3705± 37 3507 4041
102003.6+383701 183.17　56.48 65.93　-17.56 101706.9+385206	 UGC　5577 IRAS10171+3852 PGC 30206	.S?....	.96± .05 .06± .04 .96	100 .00 .09 .03	14.30 ±.18 13.77 14.19				2008± 50 2002 2256
102010.9+274904 202.82　56.58 75.33　-23.96 101722.0+280410	NGC　3204 UGC　5580 PGC 30214	.SXR3.. U 3.0± .8	1.12± .05 .16± .05 1.12	110 .07 .22 .08	 14.34 ±.20 14.01			15.83±.3 315± 7 1.74	4968± 10 4920 5258
102018.2+424916 175.77　55.68 62.50　-14.91 101717.8+430422	 UGC　5578 PGC 30217	.S..6*. U 6.0±1.4	1.13± .05 .81± .05 1.13	136 .00 1.18 .40	15.53 ±.18				
102032.1+430114 175.41　55.66 62.37　-14.75 101731.6+431620	NGC　3202 UGC　5581 PGC 30236	.SBR1.. U 1.0± .8	1.07± .06 .15± .05 1.07	20 .00 .15 .07	 14.14 ±.18 13.90				6715± 39 6726 6944
1020.5　+7316 136.16　39.73 39.15　　4.96 1016.4　+7332	 UGC　5570 PGC 30239	.S..4.. U 4.0±1.0	1.04± .06 1.01± .05 1.04	160 .09 1.47 .50					
1020.6　+5712 153.86　49.87 51.27　-5.60 1017.3　+5728	 UGC　5579 PGC 30240	.S..2.. U 2.0± .9	.94± .07 .36± .05 .94	122 .00 .45 .18	15.45 ±.18				
102038.5+253015 206.98　56.29 77.52　-25.18 101751.2+254522	NGC　3209 UGC　5584 PGC 30242	.E..... U -5.0± .8	1.11± .16 .07± .08 1.10	80 .07 .00	13.72 ±.15 13.56				6197± 50 6140 6494
102043.9+651020 144.29　45.20 45.27　　-.36 101708.1+652526	 UGC　5576 IRAS10171+6525 PGC 30247	.S?....	1.13± .05 .25± .05 1.13	0 .04 .38 .13	14.10 ±.18 13.52 13.67			15.79±.1 298± 11 2.00	3296± 9 3391 3408
102043.9-374644 A　1018-37 272.75　16.11 153.18　-39.81 101832.0-373136	 ESO　317- 16 PGC 30248	.SXS5P* PS　　(2) 4.5± .6 3.1± .6	1.08± .05 .06± .04 1.10	0 .25 .09 .03	 14.71 ±.14 14.33				7633± 59 7413 7930
102049.8+425818 175.47　55.73 62.44　-14.74 101749.4+431325	NGC　3205 UGC　5585 PGC 30254	.S?....	1.14± .05 .09± .05 1.14	 .00 .09 .05	 14.17 ±.19 13.98				7035± 50 7046 7264
102057.4+252156 207.25　56.34 77.69　-25.19 101810.1+253703	 UGC　5588 ARAK　238 PGC 30263	.S?....	.72± .09 .18± .05 .72	 .05 .24 .09	15.0 ±.4 13.01 14.66			16.04±.3 200± 7 1.29	1291± 10 1359± 63 1235 1591
102100.6+425908 175.43　55.76 62.45　-14.70 101800.3+431415	NGC　3207 UGC　5587 PGC 30267	.S?....	1.13± .10 .22± .05 1.12	73 .00 .16	 14.19 ±.18 13.92				6992± 50 7003 7221
102105.5-662927 288.71　-7.86 185.43　-29.16 101936.0-661418	 ESO　92- 21 PGC 30273	.LAS-*. S -3.0±1.2	1.10± .06 .16± .04 1.44	48 2.83 .00	13.89 ±.14				
102109.7-460038 277.52　9.32 163.27　-37.64 101905.1-454530	 ESO　263- 29 PGC 30277	RLBT+*. Sr -.6± .6	1.34± .05 .61± .04 1.34	1 .97 .00	 14.19 ±.14 13.18				2960± 60 2735 3228

R.A. 2000 DEC. l b SGL SGB R.A. 1950 DEC.	Names PGC	Type S_T n_L T L	$\log D_{25}$ $\log R_{25}$ $\log A_e$ $\log D_o$	p.a. A_g A_i A_{21}	B_T m_B m_{FIR} B_T^o	$(B-V)_T$ $(U-B)_T$ $(B-V)_T^o$ $(U-B)_T^o$	$(B-V)_e$ $(U-B)_e$ m'_e m'_{25}	m_{21} W_{20} W_{50} HI	V_{21} V_{opt} V_{GSR} V_{3K}
102117.7+193907 217.04 54.88 83.20 -28.19 101833.8+195415	NGC 3213 UGC 5590 PGC 30283	.S..4*. PU 4.3± .7 	1.05± .05 .10± .04 1.06	133 .06 .15 .05	 14.18 ±.18 13.96			 16.79±.3 147± 5 2.78	1347± 9 1412± 50 1270 1664
102118.4-394803 274.02 14.50 155.69 -39.27 101908.1-393254	 ESO 317- 17 PGC 30285	RSXR1.. r 1.0± .8 	1.14± .06 .18± .06 1.19	173 .55 .18 .09	 14.34 ±.14 13.57				2841± 19 2619 3131
1021.3 +2755 202.67 56.84 75.38 -23.69 1018.5 +2811	IC 2565 CGCG 154- 4 1ZW 24 PGC 30288			 .08 					15019± 72 14971 15309
1021.5 +2232 212.22 55.81 80.43 -26.61 1018.8 +2248	 UGC 5592 PGC 30305	.S..7.. U 7.0± .9 	1.02± .06 .17± .05 1.02	70 .05 .23 .08	 15.1 ±.2 				
102134.5-341517 270.79 19.10 148.63 -40.26 101919.8-340008	NGC 3223 ESO 375- 12 IRAS10193-3400 PGC 30308	.SAS3.. R (2) 3.0± .3 2.3± .5	1.61± .02 .22± .03 1.15± .02 1.66	135 .50 .31 .11	11.79 ±.14 11.81 ±.11 11.25 10.98	.82± .02 .26± .04 .65 .12	.90± .01 .36± .03 13.03± .04 14.15± .19	13.84±.1 419± 9 402± 7 2.75	2895± 7 2886± 14 2676 3201
102138.5+123431 227.98 52.08 90.45 -31.60 101858.4+124940	 UGC 5595 PGC 30310	.S..4.. U 4.0± .9 	1.08± .06 .05± .05 1.09	 .12 .07 .03	 14.4 ±.2 14.23			15.86±.3 136± 5 1.60	2905± 7 2801 3236
1021.6 +2354 209.86 56.18 79.15 -25.85 1018.9 +2410	NGC 3216 UGC 5593 PGC 30312	.E...*. U -5.0±1.1 	1.13± .10 .14± .05 1.09	0 .03 .00 	 14.36 ±.19 				
102141.5-344145 271.07 18.75 149.20 -40.17 101927.1-342636	NGC 3224 ESO 375- 13 PGC 30314	.E+.... BS -4.0± .5 	1.28± .05 .09± .04 .77± .04 1.33	133 .55 .00 	12.00V±.15 13.15 ±.14 12.48	.99± .02 .50± .04 .84 .39	1.00± .02 .54± .04 14.17± .30		3088± 18 2870 3394
102148.4+565550 154.09 50.16 51.60 -5.66 101831.1+571058	NGC 3206 UGC 5589 IRAS10184+5710 PGC 30322	.SBS6.. R 6.0± .3 	1.47± .02 .18± .03 1.47	0 .00 .26 .09	 12.57 ±.18 13.17 12.31			13.54±.1 191± 5 166± 8 1.15	1159± 5 1192± 50 1224 1318
102150.4+741041 135.26 39.16 38.53 5.61 101735.9+742548	NGC 3183 UGC 5582 IRAS10176+7425 PGC 30323	.SBS4*. PU 3.5± .5 	1.37± .03 .23± .03 1.38	170 .12 .33 .11	12.68S±.11 12.58 ±.18 11.62 12.18		 13.82± .19	14.08±.1 330± 8 308± 6 1.79	3088± 7 3214 3148
102152.5+235141 209.97 56.21 79.23 -25.84 101906.2+240650	 UGC 5597 PGC 30328	.S..2.. U 2.0±1.0 	1.07± .06 .76± .05 1.08	168 .03 .93 .38	 15.55 ±.18 14.52			15.92±.3 404± 7 1.02	6283± 10 6220 6586
102155.6+480152 167.00 54.29 58.52 -11.39 101850.5+481701	 UGC 5594 PGC 30334	.S..3.. U (1) 3.0± .9 4.5±1.3	1.04± .06 .58± .05 1.04	122 .00 .80 .29	 15.35 ±.18 				
102157.4-221604 263.07 28.80 132.89 -40.87 101935.1-220054	NGC 3233 ESO 568- 1 IRAS10195-2200 PGC 30336	PSBR0.. Sr .0± .4 	1.24± .04 .31± .04 1.25	140 .25 .24 	13.53 ±.14 12.39 12.98				3700± 60 3501 4036
102207.3+175017 220.09 54.42 85.11 -28.95 101924.4+180527	 MCG 3-27- 5 HICK 46A PGC 30347	.L?.... 	.72± .22 .11± .07 .71	 .10 .00 	 15.45 ±.17 15.23				8201± 41 8116 8521
102212.8+175135 220.07 54.45 85.10 -28.92 101929.9+180645	 MCG 3-27- 7 HICK 46C PGC 30349	.L?.... 		 .10 	16.49S±.15 				7906± 41 7821 8226
102219.6+363459 186.76 57.18 67.93 -18.44 101924.9+365009	IC 2566 MCG 6-23- 8 PGC 30357	.S?.... 	.89± .11 .19± .07 .89	 .00 .26 .09	 14.92 ±.18 14.60				7802± 46 7788 8059
102220.4+213409 213.98 55.71 81.48 -26.98 101935.5+214919	NGC 3221 UGC 5601 IRAS10195+2149 PGC 30358	.SBS6*/ PU (1) 6.0± .6 6.0±1.1	1.51± .03 .68± .04 1.51	167 .03 1.00 .34	 13.8 ±.2 10.83 12.72			13.71±.2 563± 5 516± 5 .66	4105± 5 4117± 36 4034 4415

R.A. 2000 DEC. l b SGL SGB R.A. 1950 DEC.	Names PGC	Type S_T n_L T L	$\log D_{25}$ $\log R_{25}$ $\log A_o$ $\log D_o$	p.a. A_g A_i A_{21}	B_T m_B m_{FIR} B_T^o	$(B-V)_T$ $(U-B)_T$ $(B-V)_T^o$ $(U-B)_T^o$	$(B-V)_e$ $(U-B)_e$ m'_e m'_{25}	m_{21} W_{20} W_{50} HI	V_{21} V_{opt} V_{GSR} V_{3K}
102224.3+011158 242.56 45.87 103.20 -36.08 101949.9+012708	IC 605 UGC 5606 IRAS10197+0127 PGC 30363	.S?.... 	.80± .07 .07± .04 .81	 .08 .11 .04	 14.8 ±.2 13.54 14.54			16.16±.3 256± 13 1.59	6492± 10 6483± 50 6350 6839
102224.4+035951 239.35 47.60 99.96 -35.05 101948.7+041501	 UGC 5607 PGC 30364	.S..3.. U (1) 3.0± .9 3.5±1.2	1.22± .05 .46± .05 1.22	62 .03 .64 .23	 14.04 ±.18 13.32			15.52±.2 459± 14 397± 6 1.96	6834± 8 6824± 50 6702 7179
102227.0-330804 270.26 20.12 147.15 -40.23 102011.3-325253	 PGC 30367	.IX.9*. S (1) 10.0± .9 12.2± .7	1.61± .04 .14± .08 1.63	 .22 .11 .07					
102230.2+363558 186.72 57.22 67.93 -18.41 101935.6+365108	IC 2568 UGC 5603 PGC 30371	.SB.1.. U 1.0± .9 	1.09± .06 .33± .05 1.09	98 .00 .34 .17					8052± 46 8038 8309
1022.5 +4351 173.80 55.78 61.92 -13.94 1019.5 +4407	 UGC 5599 PGC 30372	.S..7.. U 7.0± .9 	1.07± .08 .24± .06 1.07	175 .00 .33 .12					5263 5278 5488
102234.7+195314 216.83 55.25 83.15 -27.81 101950.7+200824	NGC 3222 UGC 5610 PGC 30377	.LB..*. PU -2.3± .7 	1.12± .05 .08± .03 1.11	 .07 .00 	13.7 ±.3 14.06 ±.16 13.82	.94± .06 .86 	 13.96± .40 		5585± 27 5507 5899
1022.6 +2721 203.78 57.05 76.08 -23.77 1019.8 +2737	 UGC 5608 PGC 30382	.S..3.. U (1) 3.0± .9 4.5±1.2	1.02± .06 .35± .05 1.02	70 .06 .49 .18					
102240.8+461419 169.81 55.06 60.03 -12.42 101937.7+462929	A 1019+46 UGC 5604 PGC 30386	.S..5.. U (1) 5.0± .8 2.2± .8	1.36± .04 .29± .05 1.16± .07 1.36	47 .00 .44 .15	13.7 ±.2 14.3 ±.3 13.44	.66± .07 .57 	.74± .03 15.03± .22 14.61± .30 	15.09±.1 381± 7 364± 6 1.51	5059± 7 4983± 59 5082 5272
1022.7 +2051 215.22 55.59 82.22 -27.27 1020.0 +2107	 CGCG 124- 18 PGC 30390			 .03 	15.1 ±.3 				7442±115 7368 7754
102251.3-242017 264.68 27.29 135.63 -40.70 102030.0-240506	 ESO 500- 32 PGC 30399	.SXS8*. S (1) 8.0± .8 7.8± .8	1.18± .04 .08± .04 1.21	 .30 .10 .04	 14.76 ±.14 14.35			15.88±.3 130± 15 115± 12 1.48	2365± 11 2162 2697
102301.8-421412 275.69 12.65 158.60 -38.37 102053.1-415900	 ESO 317- 20 IRAS10209-4159 PGC 30407	.SAT5*. S (1) 5.3± .6 4.4± .6	1.24± .04 .04± .05 1.29	 .51 .05 .02	 13.17 ±.14 12.59 12.59				2500± 60 2277 2782
102302.3-391000 273.94 15.21 154.80 -39.08 102051.1-385448	 ESO 317- 19 IRAS10208-3854 PGC 30409	PSXR1.. r 1.0± .8 	1.06± .06 .10± .06 1.09	63 .37 .10 .05	 14.50 ±.14 13.42 13.99				2833± 19 2612 3126
102307.3+415029 177.24 56.42 63.64 -15.11 102008.5+420540	 UGC 5611 PGC 30413	.SB.6*. U 6.0±1.3 	.98± .07 .23± .05 .98	45 .00 .33 .11	 15.43 ±.20 				
102308.0-393724 274.22 14.84 155.37 -38.97 102057.0-392212	 ESO 317- 21 PGC 30416	.LA.-.. Sr -2.7± .6 	1.12± .05 .17± .04 1.17	115 .58 .00 	 13.85 ±.14 13.23				2635± 19 2414 2927
102308.5+570216 153.80 50.26 51.63 -5.46 101951.6+571727	NGC 3214 MCG 10-15- 71 PGC 30419	.SA.0*. P .0± .9 	1.19± .06 .31± .05 1.17	 .00 .23 	 14.9 ±.2 				
102311.4+175807 220.05 54.71 85.13 -28.67 102028.5+181318	A 1020+18 MCG 3-27- 14 MK 630 PGC 30420			 .10 	14.52 ±.18 13.28 				3545± 45 3460 3865
102311.6-492812 279.71 6.60 167.19 -36.20 102110.1-491300	 ESO 214- 2 IRAS10211-4913 PGC 30421	PSBR1*. Sr .7± .7 	1.15± .07 .21± .07 1.28	146 1.38 .22 .11	 12.38 				

R.A. 2000 DEC. / l b / SGL SGB / R.A. 1950 DEC.	Names / PGC	Type / S_T n_L / T / L	$\log D_{25}$ / $\log R_{25}$ / $\log A_e$ / $\log D_o$	p.a. / A_g / A_i / A_{21}	B_T / m_B / m_{FIR} / B_T^o	$(B-V)_T$ / $(U-B)_T$ / $(B-V)_T^o$ / $(U-B)_T^o$	$(B-V)_e$ / $(U-B)_e$ / m' / m'_{25}	m_{21} / W_{20} / W_{50} / HI	V_{21} / V_{opt} / V_{GSR} / V_{3K}
102312.9-472018		.SBT5*.	1.22± .05	161					
278.55 8.40	ESO 263- 31	Sr (1)	.38± .05	1.09	14.95 ±.14				
164.71 -36.89		5.3± .6		.58					
102109.0-470506	PGC 30422	6.3± .7	1.33	.19					
1023.3 +0956		.S..6*.	1.13± .05	140					9725
231.99 51.14	UGC 5616	U	1.09± .05	.09	15.81 ±.19				
93.47 -32.41		6.0±1.5		1.47					9612
1020.7 +1012	PGC 30430		1.14	.50	14.20				10062
102327.4+195355	NGC 3226	.E.2.*P	1.50± .04	15	12.3 ±.2	.90± .05	.95± .01		
216.93 55.44	UGC 5617	R	.05± .04	.02	12.25 ±.14		.47± .03		1322± 19
83.26 -27.63		-5.0± .4	1.06± .06	.00		.88	13.07± .20		1244
102043.6+200907	PGC 30440		1.48		12.22		14.64± .29		1636
102330.2-352719	IC 2573	.SBS7*/	1.18± .05	2					4160
271.85 18.34	ESO 375- 17	BS (1)	.64± .05	.36	15.37 ±.14				
150.09 -39.68		6.5± .9		.88					3943
102116.0-351206	PGC 30442	5.6±1.3	1.21	.32	14.12				4465
1023.5 +1806		.S...0..	1.00± .06	85					
219.86 54.83	UGC 5619	U	.32± .05	.10	14.99 ±.18				
85.03 -28.53		.0± .9		.24					
1020.8 +1822	PGC 30444		.99						
102331.5+195148	NGC 3227	.SXS1P.	1.73± .02	155	11.1 ±.2	.82± .03	.83± .01	14.02±.1	1157± 3
217.00 55.45	UGC 5620	R (1)	.17± .03	.02	11.51 ±.15	.27± .04	.25± .02	419± 4	1145± 18
83.30 -27.63	IRAS10207+2007	1.0± .3	1.42± .06	.17	10.79	.77	13.71± .20	391± 5	1079
102047.6+200700	PGC 30445	3.5± .9	1.74	.08	11.16	.24	14.19± .24	2.78	1472
102332.7+105735	IC 606	.S?....	.64± .17						9465± 39
230.63 51.70	MCG 2-27- 6		.00± .07	.07	15.4 ±.2				9356
92.41 -31.92	MK 721			.00	13.53				9800
102053.5+111247	PGC 30448		.64	.00	15.30				
1023.5 +5220		.S..8*.	1.04± .06	4					9647
160.25 52.69	UGC 5613	U	.64± .05	.00	15.08 ±.19				9695
55.31 -8.43	IRAS10203+5235	8.0±1.4		.78	11.55				9831
1020.4 +5235	PGC 30449		1.04	.32	14.28				
102334.8-383919		.S?....	1.11± .16	155					2830± 60
273.74 15.69	ESO 317- 22		.06± .14	.37	14.93 ±.14				2610
154.13 -39.09				.09					3125
102123.1-382406	PGC 30451		1.15	.03	14.45				
102336.3+281845		.SXS4..	1.04± .06	90				16.79±.3	6558± 10
202.10 57.39	UGC 5621	U	.09± .05	.04	14.9 ±.3				6512
75.35 -23.06		4.0± .9		.13				198± 7	6847
102047.5+283357	PGC 30453		1.04	.04	14.69			2.06	
102341.8+334627		.SXS4..	1.07± .06					15.41±.3	9999± 10
191.93 57.65	UGC 5622	U	.17± .05	.01	15.1 ±.2				9975
70.51 -19.89		4.0± .9		.25				217± 7	10268
102049.5+340139	PGC 30459		1.07	.08	14.73			.60	
102345.2+570138	NGC 3220	.S..3*/	1.23± .04	97				15.32±.3	1192± 10
153.74 50.33	UGC 5614	P	.46± .04	.00	13.80 ±.19			145± 13	
51.70 -5.40		3.0±1.2		.63					1258
102028.6+571650	PGC 30462		1.23	.23	13.16			1.93	1351
1023.7 +1233	NGC 3230	.L.....	1.36± .06	115					
228.40 52.53	UGC 5624	U	.30± .04	.10	13.8 ±.2				
90.74 -31.14		-2.0± .8		.00					
1021.1 +1249	PGC 30463		1.33						
102349.1+334828		.S..6*.	1.00± .06	125				15.96±.3	10161± 10
191.86 57.68	UGC 5623	U	.21± .05	.01					
70.50 -19.85		6.0±1.3		.30				219± 7	10137
102056.7+340340	PGC 30468		1.00	.10					10430
102356.5-031100		.SBR3..	1.26± .05	150				15.18±.1	5670± 9
247.55 43.28	MCG 0-27- 5	E (1)	.22± .05	.11	13.23 ±.18			429± 12	
108.58 -37.14		3.0± .8		.30					5516
102124.2-025547	PGC 30473	1.9± .8	1.27	.11	12.78			2.30	6020
1023.9 +7851		.E...?.	1.11± .16						
131.15 35.80	UGC 5596	U	.00± .08	.01	14.31 ±.18				
35.01 8.73		-5.0±1.6		.00					
1019.0 +7907	PGC 30475		1.11						
1024.0 +7052	A 1020+71	.SBS8..	1.53± .03	165	12.6 ±.3	.48± .05	.55± .05	14.00±.1	1011± 11
138.16 41.62	UGC 5612	U (1)	.18± .05	.20	14.1 ±.5	-.01± .06	-.04± .06	165± 16	
41.15 3.60	DDO 77	8.0± .7	1.32± .07	.22		.39	14.73± .15	146± 12	1126
1020.1 +7107	PGC 30484	8.0± .9	1.55	.09	12.56	-.07	14.66± .36	1.35	1090

R.A. 2000 DEC. l b SGL SGB R.A. 1950 DEC.	Names PGC	Type S_T n_L T L	$logD_{25}$ $logR_{25}$ $logA_●$ $logD_o$	p.a. A_g A_i A_{21}	B_T m_B m_{FIR} B_T^o	$(B-V)_T$ $(U-B)_T$ $(B-V)_T^o$ $(U-B)_T^o$	$(B-V)_●$ $(U-B)_●$ $m'_●$ m'_{25}	m_{21} W_{20} W_{50} HI	V_{21} V_{opt} V_{GSR} V_{3K}
102407.3-053800 250.00 41.62 111.58 -37.79 102136.2-052247	MCG -1-27- 7 PGC 30487	.S..5?/ E 5.0±1.8 	1.22± .05 .81± .05 1.23	106 .10 1.21 .40	 				5237 5076 5587
102410.8+783737 131.33 35.98 35.20 8.59 101917.4+785248	UGC 5600 PGC 30491	.L...?. U -2.0±1.6 	1.15± .07 .15± .03 1.13	170 .01 .00 	 14.19 ±.16 14.13				2823± 46 2964 2856
102413.0+210300 215.09 55.98 82.24 -26.88 102128.6+211813	UGC 5629 PGC 30493	.S..9*. U 9.0±1.2 	1.11± .14 .11± .12 1.11	 .03 .11 .05				15.22±.3 115± 7 	1246± 7 1173 1558
102415.4+164437 222.17 54.47 86.50 -29.05 102133.4+165950	IC 607 UGC 5628 ARP 43 PGC 30496	.SBT4.. U 4.0± .8 	1.25± .05 .09± .05 1.26	105 .10 .13 .04	 14.0 ±.3 13.70			15.66±.3 336± 10 1.91	5575± 10 5486 5898
102417.5-322856 270.20 20.88 146.25 -39.93 102201.1-321342	NGC 3241 ESO 436- 16 IRAS10220-3213 PGC 30498	.SAR2*. R (2) 2.0± .5 4.0± .6	1.35± .03 .16± .03 1.38	123 .26 .20 .08	* 12.92 ±.11 12.69 12.43				2844± 66 2630 3158
1024.4 +2800 202.69 57.53 75.73 -23.08 1021.6 +2816	NGC 3232 MCG 5-25- 4 PGC 30508	.S?.... 	.82± .13 .00± .07 .83	* .10 .00 .00	 15.19 ±.18 15.03				6242 6195 6532
102424.8+783636 131.33 36.00 35.22 8.59 101932.0+785148	UGC 5609 PGC 30510	.S?.... 	1.10± .06 .20± .05 1.10	15 .01 .20 .10	 14.40 ±.18 14.16				2729± 42 2869 2761
1024.4 +5723 153.17 50.21 51.48 -5.09 1021.2 +5739	UGC 5626 PGC 30513	.I..9.. U 10.0± .8 	1.21± .05 .42± .05 1.21	53 .00 .31 .21	* 14.9 ±.2 				
102430.9-214726 263.27 29.54 132.29 -40.26 102208.0-213212	NGC 3240 ESO 568- 3 IRAS10221-2132 PGC 30515	.SAT6*. SE (2) 5.8± .6 4.4± .6	1.04± .04 .07± .03 1.07	85 .32 .10 .03	* 13.90 ±.14 13.10 13.46				3970± 60 3772 4307
102431.9-361620 272.51 17.78 151.09 -39.34 102218.1-360106	ESO 375- 20 IRAS10223-3601 PGC 30518	.S?.... 	.91± .06 .02± .05 .94	 .28 .03 .01	 14.86 ±.14 13.60 14.48				10497± 19 10279 10800
102431.9-233314 264.49 28.14 134.60 -40.31 102210.0-231800	ESO 500- 34 IRAS10221-2317 PGC 30519	RSBS0.. r -.1± .9 	1.04± .05 .28± .04 1.05	152 .28 .21 	* 14.45 ±.14 10.64 13.90				3950± 60 3749 4284
102433.4-365556 272.91 17.23 151.92 -39.22 102220.0-364042	ESO 375- 22 PGC 30522	PLA.0*P S -2.0±1.2 	1.00± .05 .08± .03 1.02	17 .22 .00 	* 14.97 ±.14 14.71				2749± 19 2530 3050
1024.6 +2006 216.73 55.78 83.21 -27.28 1021.9 +2022	UGC 5632 PGC 30526	PSXS3.. U 3.0± .9 	1.02± .06 .23± .05 1.02	163 .04 .31 .11	* 15.0 ±.2 				
1024.6 +1445 225.32 53.73 88.58 -29.92 1021.9 +1500	A 1021+15 UGC 5633 DDO 79 PGC 30531	.SB.8.. U (1) 8.0± .7 8.0±1.0	1.49± .04 .10± .06 1.06± .03 1.50	175 .13 .12 .05	14.35 ±.16 13.6 ±.7 14.06	.67± .05 -.11± .07 .61 -.15	.65± .03 -.09± .05 15.14± .06 16.38± .31	14.44±.1 177± 4 166± 6 .33	1383± 5 1287 1711
102442.6-391821 274.30 15.27 154.89 -38.74 102231.0-390306	ESO 317- 23 IRAS10225-3903 PGC 30534	PSBT1.. Sr 1.0± .6 	1.28± .04 .36± .05 1.32	14 .44 .37 .18	13.96 ±.14 10.32 13.12				2892± 19 2671 3185
1024.7 +0624 236.91 49.48 97.51 -33.58 1022.1 +0640	CGCG 37- 27 PGC 30535	 	 	 .05 	15.3 ±.6 				13094± 79 12970 13436
1024.7 +0625 236.89 49.49 97.50 -33.57 1022.1 +0641	CGCG 37- 28 PGC 30536	 	 	 .05 	14.49 ±.18 				13074± 79 12950 13416

R.A. 2000 DEC. l b SGL SGB R.A. 1950 DEC.	Names PGC	Type S_T n_L T L	$\log D_{25}$ $\log R_{25}$ $\log A_e$ $\log D_o$	p.a. A_g A_i A_{21}	B_T m_B m_{FIR} B_T^o	$(B-V)_T$ $(U-B)_T$ $(B-V)_T^o$ $(U-B)_T^o$	$(B-V)_e$ $(U-B)_e$ m'_e m'_{25}	m_{21} W_{20} W_{50} HI	V_{21} V_{opt} V_{GSR} V_{3K}
102447.6-435751 276.93 11.38 160.59 -37.61 102240.0-434236	ESO 263- 33 PGC 30544	PLAS-*. S -3.0± .8 	1.10± .05 .05± .05 1.17	 .62 .00 	14.03 ±.14 				
102448.2-264127 266.65 25.66 138.71 -40.23 102228.1-262612	ESO 500- 37 PGC 30545	RLB.+.. r -1.3± .9 	1.06± .05 .47± .03 1.01	160 .24 .00 	15.44 ±.14 				
1024.9 +2801 202.68 57.64 75.78 -22.98 1022.1 +2817	NGC 3235 UGC 5635 PGC 30553	.L..-*. U -3.0±1.2 	1.08± .17 .12± .08 1.07	85 .10 .00 	14.29 ±.16 14.10				6409± 61 6363 6699
102457.0-172602 260.19 33.02 126.61 -39.82 102231.8-171047	MCG -3-27- 10 PGC 30554	.SXT1P* E 1.0± .9 	1.07± .06 .11± .05 1.09	165 .27 .11 .05					
102505.6+170935 221.64 54.82 86.20 -28.67 102223.3+172450	NGC 3239 UGC 5637 ARP 263 PGC 30560	.IBS9P. R 10.0± .3 	1.70± .02 .18± .03 1.09± .01 1.71	 .11 .13 .09	11.73 ±.13 12.3 ±.4 11.78 11.53	.42± .01 -.34± .02 .35 -.39	.37± .01 -.30± .02 12.67± .02 14.65± .18	12.64±.0 189± 3 152± 4 1.02	753± 3 830± 47 666 1075
1025.1 +2805 202.57 57.69 75.75 -22.91 1022.3 +2820	IC 2572 UGC 5636 IRAS10222+2820 PGC 30562	.S..1.. U 1.0± .9 	.97± .07 .27± .05 .98	27 .10 .28 .13	15.24 ±.19 				
102510.4+580902 152.12 49.87 50.95 -4.53 102152.5+582416	NGC 3225 UGC 5631 IRAS10218+5824 PGC 30569	.S..6*. U 6.0±1.1 	1.31± .04 .29± .05 1.31	155 .00 .43 .15	13.30 ±.18 13.35 12.86			14.26±.1 259± 9 249± 12 1.25	2134± 7 2204 2287
1025.3 +5530 155.58 51.33 53.01 -6.20 1022.1 +5546	A 1022+55 MCG 9-17- 64 PGC 30579	.SXS5*. P 5.0± .8 	.99± .10 .05± .07 .99	 .00 .07 .02					7621± 60 7681 7788
1025.3 +2627 205.60 57.52 77.28 -23.76 1022.6 +2643	 UGC 5638 PGC 30584	.L..... U -2.0± .8 	1.11± .08 .25± .04 1.08	117 .05 .00 	14.79 ±.20 14.52				14430 14378 14726
1025.4 +1708 221.72 54.88 86.25 -28.62 1022.7 +1724	MCG 3-27- 27 PGC 30585	.SB?... 	.99± .10 .16± .07 1.00	 .12 .24 .08	14.87 ±.18 14.51				815± 79 728 1138
102526.5-152056 258.69 34.70 123.93 -39.45 102300.1-150540	MCG -2-27- 1 PGC 30591	.S..7*/ E 7.0±1.3 	1.17± .05 .90± .05 1.19	168 .28 1.24 .45					
102529.3-394940 274.73 14.92 155.50 -38.48 102318.0-393424	NGC 3244 ESO 317- 24 IRAS10232-3934 PGC 30594	.SAT6.. RBS (1) 6.0± .5 3.3± .8	1.31± .04 .13± .04 1.01± .03 1.37	170 .59 .20 .07	12.89 ±.15 13.09 ±.14 12.22 	.61± .03 .00± .08 	.69± .02 .05± .04 13.43± .06 13.97± .27		
1025.5 +1714 221.57 54.94 86.17 -28.55 1022.8 +1730	 UGC 5639 PGC 30595	.SX.5.. U 5.0± .9 	1.14± .05 .21± .05 1.15	125 .11 .32 .11	14.6 ±.2 				
102531.4+514025 161.00 53.27 56.03 -8.63 102223.1+515540	 MK 142 PGC 30597	 	 	 .00 	16.06 ±.15 	.45± .04 -.68± .08 			13474± 46 13520 13662
102535.5-021253 246.95 44.24 107.57 -36.45 102302.7-015737	IC 609 UGC 5641 ARP 44 PGC 30600	.SXT4P. UE (1) 3.5± .6 3.1±1.2	1.18± .04 .32± .04 1.19	10 .08 .47 .16	14.07 ±.18 			16.07±.1 393± 12 	5538± 9 12608± 65
102541.3+114414 229.96 52.54 91.86 -31.10 102301.8+115930	 UGC 5642 PGC 30604	.S..4.. U 4.0± .9 	1.27± .04 .84± .05 1.28	98 .03 1.24 .42	14.68 ±.18 13.40			15.22±.2 256± 6 243± 7 1.40	2350± 6 2342± 76 2244 2684
102543.9+393848 180.96 57.39 65.77 -16.06 102247.6+395404	NGC 3237 UGC 5640 PGC 30610	RLX.0.. U -2.0± .8 	1.11± .07 .00± .03 1.11	 .00 .00 	13.96 ±.16 13.85				7079± 31 7078 7324

R.A. 2000 DEC. l b SGL SGB R.A. 1950 DEC.	Names PGC	Type S_T n_L T L	logD_25 logR_25 logA_o logD_o	p.a. A_g A_i A_21	B_T m_B m_FIR B_T^o	(B-V)_T (U-B)_T (B-V)_T^o (U-B)_T^o	(B-V)_o (U-B)_o m'_o m'_25	m_21 W_20 W_50 HI	V_21 V_opt V_GSR V_3K
102546.4+134301 227.09 53.50 89.80 -30.18 102306.0+135817	A 1023+13 UGC 5644 HICK 47A PGC 30616	.SAR.*. R 	1.00± .06 .21± .05 1.01	15 .12 .32 .11	15.15S±.15 14.66		 14.45± .37 		9581± 41 9482 9911
1025.7 +2633 205.45 57.62 77.25 -23.63 1023.0 +2649	 MCG 5-25- 9 PGC 30617	.S?.... 	.94± .11 .19± .07 .94	 .04 .23 .09	 14.73 ±.18 14.40				5094 5042 5390
102548.7+134341 227.08 53.52 89.80 -30.16 102308.3+135857	 MCG 2-27- 13 HICK 47B PGC 30619	.L?.... 	.90± .17 .00± .07 .91	 .12 .00 	15.96S±.15 15.70		 15.31± .87 		9487± 41 9388 9817
102551.1-434453 276.98 11.67 160.27 -37.48 102343.0-432936	NGC 3256A ESO 263- 34 PGC 30626	.SBS9P* PS (1) 9.0± .8 8.9± .9	1.13± .04 .32± .04 1.19	85 .57 .33 .16	 15.05 ±.14 				
102553.4+142145 226.14 53.82 89.15 -29.85 102312.6+143701	A 1023+14 UGC 5646 PGC 30630	.S..... R 	1.44± .03 .56± .05 1.45	165 .12 .85 .28	 13.74 ±.20 12.76			14.67±.3 232± 7 221± 7 1.63	1370± 9 1350± 50 1273 1698
1025.9 +7111 137.72 41.50 41.00 3.92 1022.0 +7127	 UGC 5634 PGC 30631	.SX.3.. U 3.0± .9 	1.04± .06 .20± .05 1.06	152 .16 .28 .10	 15.3 ±.2 				
102559.4-325405 270.77 20.74 146.74 -39.52 102343.0-323848	IC 2576 ESO 375- 23 IRAS10236-3238 PGC 30634	.S?.... 	.90± .05 .08± .04 .93	 .28 .11 .04	 14.81 ±.14 13.50 14.35				9345± 19 9131 9658
102614.4-454412 278.12 10.02 162.62 -36.87 102408.0-452854	 ESO 263- 35 IRAS10241-4528 PGC 30646	.SAT3*. Sr (1) 3.1± .6 2.2± .8	1.25± .05 .25± .05 .83± .02 1.32	111 .77 .35 .13	13.75 ±.15 13.66 ±.14 12.25 12.55	.73± .01 .48 	.81± .01 13.39± .05 14.20± .30		4600± 40 4376 4870
102621.7-023719 247.55 44.10 108.13 -36.39 102349.1-022202	NGC 3243 UGC 5652 PGC 30655	.L..0P* UE -2.3± .6 	1.15± .09 .10± .05 1.14	125 .04 .00 	 13.65 ±.15 13.52				5540± 66 5388 5891
102622.1-345748 272.07 19.07 149.35 -39.18 102407.0-344230	NGC 3249 ESO 375- 24 IRAS10240-3442 PGC 30657	.SXT6*. BS (1) 6.0± .7 3.3± .8	1.20± .04 .09± .04 1.23	139 .35 .13 .04	 13.86 ±.14 13.16 13.37				3486± 18 3270 3794
102625.6+173036 221.29 55.25 86.03 -28.22 102343.3+174553	 UGC 5651 PGC 30659	.SX.6*. U 6.0±1.2 	1.17± .05 .13± .05 1.18	20 .11 .19 .06	 13.96 ±.20 13.63			14.98±.3 225± 7 212± 7 1.28	5568± 9 5528± 50 5481 5889
1026.4 +6739 141.14 44.00 43.75 1.69 1022.8 +6754	A 1022+67 DDO 78 PGC 30664	.I..9.. P (1) 10.0± .9 9.0±1.0	 	 .06 					
102629.1-200230 262.44 31.20 130.04 -39.70 102405.0-194712	 ESO 568- 9 PGC 30666	.S?.... 	1.06± .05 .10± .04 1.10	 .34 .15 .05	 14.28 ±.14 13.77			14.31±.3 221± 15 202± 12 .49	3108± 11 2914 3449
102631.6+152026 224.75 54.39 88.23 -29.25 102350.4+153543	 UGC 5654 PGC 30669	.S..4.. U 4.0± .9 	1.06± .06 .23± .05 1.07	128 .13 .33 .11	 14.9 ±.2 14.40			16.15±.3 287± 7 1.64	9831± 10 9738 10158
1026.5 +2013 216.79 56.23 83.35 -26.84 1023.8 +2029	IC 610 UGC 5653 PGC 30670	.S..4./ PU (1) 4.0± .6 2.5±1.2	1.27± .04 .76± .05 1.27	29 .01 1.12 .38	 14.72 ±.18 				
102632.4-395636 274.97 14.93 155.59 -38.26 102421.0-394118	NGC 3250 ESO 317- 26 PGC 30671	.E.4... R -5.0± .7 	1.45± .04 .14± .04 .91± .03 1.50	148 .57 .00 	12.18 ±.13 12.17 ±.11 11.56	1.05± .01 .65± .03 .90 .53	1.07± .01 .67± .02 12.22± .09 14.05± .26		 2824± 32 2603 3115
102634.7-024949 247.81 44.00 108.40 -36.40 102402.2-023431	 MCG 0-27- 13 PGC 30676	.S..7P? E 7.0±1.8 	1.12± .06 .48± .05 1.12	13 .05 .66 .24	 14.67 ±.18 13.94			16.22±.1 378± 12 2.04	5891± 9 5739 6242

R.A. 2000 DEC. / l b / SGL SGB / R.A. 1950 DEC.	Names / PGC	Type / S_T n_L / T / L	$\log D_{25}$ / $\log R_{25}$ / $\log A_o$ / $\log D_o$	p.a. / A_g / A_i / A_{21}	B_T / m_B / m_{FIR} / B_T^o	$(B-V)_T$ / $(U-B)_T$ / $(B-V)_T^o$ / $(U-B)_T^o$	$(B-V)_.$ / $(U-B)_.$ / $m'_.$ / m'_{25}	m_{21} / W_{20} / W_{50} / HI	V_{21} / V_{opt} / V_{GSR} / V_{3K}
1026.6 +4514 / 171.05 56.04 / 61.27 -12.51 / 1023.6 +4530	UGC 5650 / PGC 30680	.S..3.. / U (1) / 3.0±1.0 / 4.5±1.4	1.02± .06 / .81± .05 / / 1.02	52 / .00 / 1.12 / .40					
102640.6-190300 / 261.77 32.01 / 128.76 -39.57 / 102416.0-184742	ESO 568-11 / IRAS10242-1847 / PGC 30683	.SXS4*P / S (1) / 4.0±1.1 / 3.3±1.2	1.20± .04 / .17± .04 / / 1.22	126 / .23 / .25 / .08	14.60 ±.14 / 13.39				
102641.6+035138 / 240.48 48.36 / 100.61 -34.12 / 102406.0+040656	NGC 3246 / UGC 5661 / IRAS10240+0407 / PGC 30684	.SX.8.. / U / 8.0± .8	1.38± .04 / .25± .05 / / 1.39	100 / .05 / .31 / .13	13.2 ±.2 / 13.55 / 12.80			14.09±.1 / 259± 13 / 241± 6 / 1.16	2150± 6 / 2018 / 2496
102643.3+571339 / 153.13 50.56 / 51.80 -4.96 / 102327.6+572856	NGC 3238 / UGC 5649 / PGC 30686	.LAR0*. / PU / -2.3± .7	1.14± .08 / .01± .04 / / 1.14	/ .00 / .00 /	13.90 ±.16 / 13.78				7369± 31 / 7436 / 7527
102649.7+345512 / 189.71 58.24 / 69.93 -18.70 / 102357.1+351030	UGC 5656 / PGC 30694	.SXS4.. / U / 4.0± .8	1.12± .05 / .26± .05 / / 1.12	135 / .00 / .39 / .13	14.8 ±.2 / 14.40			15.66±.3 / 325± 7 / 1.13	6662± 10 / 6643 / 6927
102653.2+622009 / 146.85 47.52 / 47.87 -1.68 / 102327.9+623526	A 1023+62 / CGCG 313-18 / MK 143 / PGC 30701	.S?....	.61± .09 / .14± .08 / / .61	/ .00 / .21 / .07	15.8 ±.2 / 15.55				9586± 42 / 9671 / 9715
102653.8+440022 / 173.13 56.49 / 62.31 -13.23 / 102354.0+441540	A 1023+44 / MCG 7-22- 6 / MK 144 / PGC 30702	.S?....	.73± .09 / .04± .06 / / .73	/ .00 / .06 / .02	15.10 ±.14 / 13.55 / 15.01				8237± 38 / 8253 / 8462
102656.9-240525 / 265.36 28.04 / 135.32 -39.76 / 102435.0-235006	ESO 500-41 / IRAS10245-2350 / PGC 30708	RSAR2.. / r / 2.2± .9	1.03± .05 / .17± .03 / / 1.06	173 / .31 / .20 / .08	14.22 ±.14 / 12.80				
102701.2+283829 / 201.64 58.17 / 75.51 -22.25 / 102412.7+285347	NGC 3245A / UGC 5662 / PGC 30714	.SBS3./ / R / 3.0± .4	1.52± .03 / .99± .04 / / 1.53	150 / .05 / 1.37 / .50	14.7 ±.2 / 13.24			14.97±.2 / 198± 8 / 173± 6 / 1.24	1325± 7 / 1486± 63 / 1283 / 1615
102702.1+561609 / 154.35 51.13 / 52.57 -5.54 / 102348.0+563127	A 1023+56 / UGCA 211 / MK 32 / PGC 30715	.P..... / R / 99.0	.54± .19 / .25± .09 / / .54	/ .00 /	16.14 ±.15	.31± .04 / -.13± .06		16.19±.1 / 82± 5 / 57± 5	833± 5 / 867± 38 / 897 / 997
102702.3-361337 / 272.93 18.10 / 150.93 -38.85 / 102448.0-355818	ESO 375-26 / IRAS10248-3558 / PGC 30716	.S..4*/ / S / 4.0±1.2	1.28± .04 / .73± .04 / / 1.30	30 / .28 / 1.07 / .36	14.56 ±.14 / 12.17 / 13.19				3373± 19 / 3155 / 3677
102705.7-454001 / 278.21 10.16 / 162.49 -36.75 / 102459.0-452442	ESO 263-36 / FAIR 430 / PGC 30721		.90± .07 / .22± .06 / / .97	175 / .77	15.44 ±.14				1550±190 / 1326 / 1821
102710.3-031913 / 248.45 43.77 / 109.04 -36.41 / 102438.0-030354	CGCG 9-42 / PGC 30732			/ .07	14.60 ±.18				11510± 55 / 11357 / 11861
1027.2 +2026 / 216.52 56.46 / 83.24 -26.59 / 1024.5 +2042	A 1024+20 / PGC 30735			/ .02					5772± 72 / 5698 / 6087
1027.2 +7124 / 137.42 41.42 / 40.90 4.14 / 1023.3 +7140	UGC 5645 / IRAS10233+7140 / PGC 30737	.SBS3.. / U / 3.0± .9	1.09± .06 / .53± .05 / / 1.10	142 / .13 / .74 / .27	12.24				
102718.3+283029 / 201.90 58.22 / 75.67 -22.27 / 102430.0+284548	NGC 3245 / UGC 5663 / IRAS10245+2845 / PGC 30744	.LAR0*$ / R / -2.0± .5	1.51± .03 / .26± .04 / .95± .04 / 1.48	177 / .05 / .00	11.70 ±.15 / 11.86 ±.12 / 12.41 / 11.73	.91± .02 / .47± .03 / .86 / .44	.94± .01 / .51± .02 / 11.94± .14 / 13.50± .22		1348± 22 / 1304 / 1638
102721.1-351625 / 272.43 18.93 / 149.71 -38.93 / 102506.0-350106	ESO 375-28 / PGC 30750	.LX.0*. / S / -2.0±1.3	1.09± .05 / .50± .03 / / 1.04	61 / .27 / .00	15.15 ±.14				

R.A. 2000 DEC.	Names	Type	logD25	p.a.	B_T	(B-V)_T	(B-V)_o	m_21	V_21
I b		S_T n_L	logR25	A_g	m_B	(U-B)_T	(U-B)_o	W_20	V_opt
SGL SGB		T	logA_o	A_i	m_FIR	(B-V)_T^o	m'_o	W_50	V_GSR
R.A. 1950 DEC.	PGC	L	logD_o	A_21	B_T^o	(U-B)_T^o	m'_25	HI	V_3K
102723.0-335238	IC 2578	.S..5*/	1.17± .04	141					
271.61 20.09	ESO 375- 29	S	.72± .04	.31	14.99 ±.14				
147.94 -39.12	IRAS10250-3337	5.0±1.3		1.08	12.52				
102507.0-333718	PGC 30753		1.20	.36					
102724.0-430814		.SAT3..	1.02± .06						
276.88 12.33	ESO 263- 37	r	.09± .05	.55	14.47 ±.14				
159.44 -37.37	IRAS10252-4252	3.3± .9		.12	12.89				
102515.0-425254	PGC 30754		1.07	.04					
1027.5 +1656									10379
222.38 55.26	CGCG 94- 65			.10	15.2 ±.3				10292
86.74 -28.28									10703
1024.8 +1712	PGC 30763								
102742.7-400008	NGC 3250C	PSAT2*.	1.25± .04	56					
275.20 15.01	ESO 317- 28	PSr	.51± .04	.50	14.28 ±.14				
155.60 -38.03	IRAS10255-3944	1.6± .6		.63	12.76				
102531.0-394448	PGC 30774		1.30	.25					
102744.3-402608	NGC 3250B	.SB.1P?	1.36± .04	6	13.74 ±.19	1.06± .03	1.12± .01		
275.44 14.64	ESO 317- 29	PS	.59± .04	.48	13.83 ±.14	.46± .06	.55± .03		2520± 19
156.13 -37.93	IRAS10255-4010	1.0±1.2	.86± .05	.60	12.35	.82	13.53± .16		2299
102533.0-401048	PGC 30775		1.41	.29	12.69	.24	13.92± .29		2810
102745.6+225056	NGC 3248	.L.....	1.40± .11	135					
212.37 57.27	UGC 5669	U	.34± .08	.00	13.38 ±.17				
80.99 -25.24		-2.0± .8		.00					
102500.6+230615	PGC 30776		1.35						
1027.7 +2707		.S..4..	1.11± .07	115					
204.53 58.15	UGC 5670	U	.95± .06	.03					
76.99 -22.94		4.0±1.0		1.40					
1025.0 +2723	PGC 30780		1.11	.48					
102751.5-435420	NGC 3256	.P.....	1.58± .02	100	12.15M±.08	.64± .01	.62± .01		
277.37 11.73	ESO 263- 38	R	.25± .03	.59	11.85 ±.11	-.08± .02	-.19± .01		2781± 24
160.34 -37.09	VV 65	99.0	.80± .02	.35	8.34	.44	11.64± .08		2558
102543.0-433900	PGC 30785		1.63	.13	11.09	-.22	14.26± .16		3059
102753.6-400450	NGC 3250A	.S..3*/	1.09± .04	89					
275.27 14.96	ESO 317- 30	PS	.75± .04	.50	15.67 ±.14				
155.69 -37.98	IRAS10257-3949	2.6± .9		1.04	12.90				
102542.0-394930	PGC 30790		1.14	.38					
102755.5+192927	A 1025+19	.S?....	.52± .20						
218.24 56.30	MCG 3-27- 42		.08± .07	.03					12384± 72
84.26 -26.93	2ZW 47			.10					12306
102512.3+194447	PGC 30791		.52	.04					12701
102757.9-394857	NGC 3250D	.L...*/	1.24± .04	29					
275.13 15.19	ESO 317- 31	PS	.80± .03	.40	14.32 ±.14				
155.36 -38.02		-1.7± .8		.00					
102546.0-393336	PGC 30792		1.16						
1027.9 +4959		.S..6*.	1.07± .07	5					
163.23 54.39	UGC 5668	U	.21± .06	.00					
57.60 -9.38		6.0±1.2		.30					
1024.9 +5015	PGC 30795		1.07	.10					
102802.0-420639		PSBR3?.	1.18± .04	74					
276.42 13.26	ESO 317- 32	Sr (1)	.47± .04	.52	14.56 ±.14				
158.16 -37.50		3.4± .5		.65					
102552.0-415118	PGC 30798		1.22	.23					
102817.6+794929	NGC 3212	.SB?...	1.19± .06	107					
130.19 35.19	UGC 5643		.15± .06	.01	14.11 ±.19				9769± 42
34.38 9.49	IRAS10232+8004			.22	12.75				9913
102313.2+800447	PGC 30813		1.19	.07	13.82				9795
102818.7-313109	IC 2580	.SBT5..	1.27± .03		13.2 ±.2	.66± .06	.67± .02		3132± 10
270.37 22.15	ESO 436- 25	RSr (2)	.03± .03	.24	13.27 ±.14	.01± .05			3137± 59
144.90 -39.18	IRAS10260-3115	4.8± .5	1.09± .06	.05	13.34	.58	14.07± .15		2920
102601.0-311548	PGC 30814		1.29	.02	12.96	-.05	14.31± .26		3450
102819.1-352715	NGC 3258A	.LX.+*.	1.06± .05	169	14.1 ±.3	.92± .03	.98± .02		
272.71 18.88	ESO 375- 32	PS	.39± .03	.28	14.56 ±.14	.42± .06	.51± .03		2930± 60
149.90 -38.71		-1.3± .7	.59± .10	.00		.78	12.58± .34		2713
102604.0-351154	PGC 30815		1.03		14.16	.32	13.29± .40		3236
102820.8+223420		.S?....	1.26± .04	158				15.83±.3	531± 10
212.93 57.33	UGC 5672		.52± .05	.00	14.48 ±.19				465
81.34 -25.27				.78				86± 7	840
102536.0+224940	PGC 30818		1.26	.26	13.70			1.87	

R.A. 2000 DEC. l　　b SGL　SGB R.A. 1950 DEC.	Names PGC	Type S_T　n_L T L	$\log D_{25}$ $\log R_{25}$ $\log A_e$ $\log D_o$	p.a. A_g A_i A_{21}	B_T m_B m_{FIR} B_T^o	$(B-V)_T$ $(U-B)_T$ $(B-V)_T^o$ $(U-B)_T^o$	$(B-V)_e$ $(U-B)_e$ m'_e m'_{25}	m_{21} W_{20} W_{50} HI	V_{21} V_{opt} V_{GSR} V_{3K}
102822.5+682459 140.20　43.60 43.28　2.31 102441.3+684018	IC　2574 UGC　5666 DDO　81 PGC 30819	.SXS9.. R　(1) 9.0± .3 8.0± .8	2.12± .02 .39± .03 1.84± .04 2.13	50 .07 .40 .19	10.80 ±.19 10.8　±.3 11.68 10.33	.44± .08 .34	.47± .03 15.49± .09 15.28± .22	10.87±.1 123± 6 115± 3 .35	47± 3 -4± 58 154 141
102824.8-342821 272.15　19.71 148.66 -38.83 102609.0-341300	ESO　375- 33 PGC 30823	.SAT4.. S　(1) 4.0± .8 4.4± .8	1.02± .05 .10± .04 1.05	143 .36 .15 .05	15.13 ±.14				
1028.4　-0314 248.67　44.06 109.06 -36.08 1025.9　-0259	CGCG　9- 52 PGC 30828			.06	15.3　±.6				10519± 66 10366 10870
102827.5+124222 229.11　53.60 91.20 -30.06 102547.8+125743	NGC　3253 UGC　5674 IRAS10257+1257 PGC 30829	.SXT4.. U　(2) 4.0± .8 3.1± .6	1.08± .03 .02± .05 .76± .02 1.08	 .08 .03 .01	14.30 ±.14 13.96 ±.18 13.57 14.00	.72± .03 .63	.81± .03 13.59± .04 14.47± .24	15.58±.1 260± 6 231± 7 1.58	9689± 8 9711± 59 9587 10022
1028.5　+1934 218.18　56.46 84.26 -26.77 1025.8　+1950	UGC　5675 PGC 30831	.S..9*. U 9.0±1.1	1.27± .06 .19± .06 1.27	 .03 .20 .10				15.89±.1 77± 5 63± 5	1109± 5 1032 1426
102835.5+033339 241.27　48.55 101.17 -33.79 102600.0+034900	UGC　5677 PGC 30832	.S..8*. U 8.0±1.4	1.16± .07 1.00± .06 1.17	6 .06 1.23 .50				15.36±.3 109± 7	1153± 10 1021 1500
1028.7　+0340 241.16　48.64 101.05 -33.72 1026.1　+0356	UGC　5678 PGC 30839	.S..4.. U 4.0± .9	1.04± .06 .18± .05 1.04	50 .06 .27 .09	14.55 ±.18				
102841.8+794848 130.19　35.21 34.40　9.50 102338.2+800407	NGC　3215 UGC　5659 PGC 30840	.S?....	1.04± .06 .02± .05 1.04	130 .01 .03 .01	14.00 ±.18 13.90				9468± 46 9612 9493
102842.7+395016 180.40　57.91 65.97 -15.49 102546.9+400537	A　1025+40 CGCG 212- 11 MK　415 PGC 30842		.62± .11 .11± .12 .62	 .00	15.7　±.2				8769± 35 8769 9014
102847.1-353928 272.91　18.77 150.14 -38.59 102632.0-352406	NGC　3257 ESO　375- 36 PGC 30849	.LXS-*. RBS -2.7± .4	1.00± .03 .06± .03 .49± .02 1.03	0 .28 .00	14.11 ±.13 14.05 ±.14 13.76	.98± .02 .48± .05 .88 .43	1.00± .02 .52± .05 12.05± .08 13.84± .21		3172± 32 2955 3478
1028.8　+0412 240.55　48.99 100.46 -33.50 1026.2　+0428	CGCG　37- 47 8ZW　81 PGC 30852			.07	15.2　±.6				2234± 82 2104 2580
1028.8　+6649 141.77　44.73 44.53　1.33 1025.2　+6705	UGC　5671 PGC 30853	.I..9.. U 10.0± .9	1.14± .13 .39± .12 1.14	155 .03 .29 .20				15.46±.1 116± 6 108± 5	1120± 10 1222 1224
102853.6+194535 217.92　56.60 84.13 -26.60 102610.3+200057	UGC　5681 IRAS10261+2000 PGC 30855	.SB.3.. U 3.0±1.0	1.09± .06 .65± .05 1.09	165 .02 .90 .33	15.42 ±.18 14.44			18.73±.3 77± 5 3.97	8134± 9 8057 8451
1028.9　+2620 206.08　58.27 77.87 -23.17 1026.1　+2635	UGC　5679 IRAS10261+2635 PGC 30856	.S?....	1.21± .05 .46± .05 1.21	122 .00 .68 .23	14.8　±.2 13.76 14.06				6488± 10 6436 6785
102853.8-313634 270.54　22.15 145.00 -39.04 102636.0-312112	ESO　436- 27 IRAS10266-3121 PGC 30857	PLASOP. S -2.0± .7	1.37± .06 .30± .03 1.35	0 .23 .00	12.6　±.2 12.73 ±.14	.88± .07 .15± .13	 13.55± .36		
102854.2-353622 272.90　18.82 150.07 -38.57 102639.0-352100	NGC　3258 ESO　375- 37 PGC 30859	.E.1... R -5.0± .3	1.46± .03 .07± .03 1.00± .03 1.48	75 .28 .00	12.49 ±.13 12.44 ±.11 12.14	1.01± .01 .54± .04 .92 .49	1.05± .01 .58± .02 12.98± .09 14.61± .23		2792± 28 2576 3098
102857.5-354815 273.03　18.66 150.32 -38.53 102642.5-353253	PGC 30860	.IXS9.. S　(1) 10.0± .8 10.0± .8	1.15± .08 .14± .08 1.18	 .27 .11 .07					

R.A. 2000 DEC. l　　b SGL　SGB R.A. 1950 DEC.	Names PGC	Type S_T　n_L T L	$\log D_{25}$ $\log R_{25}$ $\log A_\bullet$ $\log D_\circ$	p.a. A_g A_i A_{21}	B_T m_B m_{FIR} B_T°	$(B-V)_T$ $(U-B)_T$ $(B-V)_T^\circ$ $(U-B)_T^\circ$	$(B-V)_\bullet$ $(U-B)_\bullet$ $m'_\bullet$ m'_{25}	m_{21} W_{20} W_{50} HI	V_{21} V_{opt} V_{GSR} V_{3K}
102900.9-400459 275.46　15.07 155.63 -37.77 102649.0-394936	NGC　3250E ESO　317- 34 IRAS10268-3949 PGC 30865	.SXS6*. PS　　(1) 6.0± .6 4.4± .8	1.32± .04 .17± .04 1.36	142 .48 .25 .08	 13.19 ±.14 12.11 12.45			13.92±.3 188± 10 1.39	2818±　9 2570± 19 2554 3066
102901.3-442411 277.82　11.42 160.87 -36.76 102653.1-440848	NGC　3256B ESO　263- 39 IRAS10268-4408 PGC 30867	.SBS4*. PS　　(1) 4.2± .6 5.6± .9	1.26± .04 .53± .04 1.33	135 .68 .78 .26	 13.78 ±.14 11.57 				
102902.1-443923 277.96　11.20 161.17 -36.70 102654.0-442400	NGC　3261 ESO　263- 40 IRAS10268-4423 PGC 30868	.SBT3.. R　　(2) 3.0± .3 1.5± .4	1.57± .03 .12± .03 1.63	85 .68 .16 .06	 12.00 ±.12 11.71 11.15			13.13±.2 367± 10 369± 12 1.93	2553±　6 2582± 66 2330 2828
1029.0　+5442 156.21　52.21 53.98　-6.30 1025.9　+5458	 UGC　5676 PGC 30871	.SB.8*. U 8.0±1.2 	1.13± .07 .18± .06 1.13	15 .00 .22 .09	 14.54 ±.20 14.31				1412 1470 1584
102905.9-435111 277.54　11.89 160.21 -36.89 102657.1-433548	NGC　3256C ESO　263- 41 FAIR　431 PGC 30873	.SBT7.. R　　(1) 7.0± .4 5.6± .6	1.19± .04 .13± .04 1.25	159 .59 .18 .07	 13.31 ±.14 12.03 12.53				2550±190 2328 2828
102906.2-353540 272.93　18.85 150.05 -38.53 102651.0-352018	NGC　3260 ESO　375- 40 PGC 30875	.E...P* RBS -4.7± .5 	1.09± .03 .11± .04 .42± .03 1.10	2 .28 .00 	13.71 ±.13 13.72 ±.14 13.40	1.06± .01 .97 	1.14± .01 11.30± .10 13.88± .22		2416± 32 2200 2722
102906.5-440941 277.71　11.63 160.57 -36.81 102658.0-435418	NGC　3262 ESO　263- 42 PGC 30876	.LXT+P. RBSr -1.5± .3 	1.05± .04 .09± .04 1.11	108 .59 .00 	 14.24 ±.14 13.57				2834± 87 2611 3111
102910.6-302040 269.82　23.23 143.38 -39.09 102652.0-300518	IC　2582 ESO　436- 28 IRAS10268-3005 PGC 30880	.SXR5.. Sr　　(1) 4.8± .6 2.2± .8	1.15± .04 .10± .03 1.16	 .14 .15 .05	 13.68 ±.14 12.56 13.37				4210± 60 4000 4531
102912.6+060738 238.28　50.23 98.37 -32.67 102636.0+062300	 UGC　5687 PGC 30885	.S..6*. U 6.0±1.3 	1.31± .04 .96± .05 1.31	111 .02 1.40 .48	 14.79 ±.18 13.36			14.71±.2 275± 7 258± 5 .88	3563±　6 3439 3907
102913.7-440617 277.69　11.69 160.50 -36.81 102705.1-435054	NGC　3263 ESO　263- 43 IRAS10270-4351 PGC 30887	.SBT6*/ R 6.0± .3 	1.71± .04 .56± .05 1.76	97 .59 .82 .28	12.5　±.2 11.46 ±.14 10.78 10.38	.61± .05 .31± .12 .35 .12	 14.50± .30	12.70±.3 586± 21 2.04	3015± 13 2792 3292
102916.9+260558 206.56　58.32 78.14 -23.22 102630.2+262120	NGC　3251 UGC　5684 IRAS10264+2621 PGC 30892	.SB?... 	1.30± .04 .70± .05 1.30	55 .04 1.05 .35	 14.27 ±.18 13.26 13.16			15.35±.3 471± 5 1.85	5087±　9 5131± 50 5035 5387
102920.0+292928 200.11　58.75 75.05 -21.36 102631.4+294450	NGC　3254 UGC　5685 IRAS10265+2944 PGC 30895	.SAS4.. R　　(2) 4.0± .3 3.3± .6	1.70± .02 .50± .03 1.14± .02 1.71	46 .04 .74 .25	12.41 ±.13 12.17 ±.13 11.49	.68± .03 .09± .04 .57 .00	.76± .02 .19± .03 13.60± .05 14.52± .17	13.29±.1 431± 5 399± 7 1.54	1365±　9 1223± 14 1286 1613
1029.4　+6017 148.97　49.03 49.64　-2.75 1026.1　+6033	 UGC　5680 PGC 30898	.S?.... 	.95± .05 .54± .04 .95	165 .00 .82 .27	 15.1　±.2 14.24				7024±125 7103 7166
1029.4　-0344 249.41　43.89 109.75 -35.99 1026.9　-0329	 CGCG　9- 57 PGC 30899			 .05 	 15.3　±.6 				8549± 66 8395 8901
102929.7+193715 218.25　56.69 84.35 -26.55 102646.5+195238	 UGC　5690 IRAS10267+1952 PGC 30903	.SB?... 	.97± .07 .08± .05 .97	 .03 .11 .04	 14.46 ±.18 13.44 14.27			16.12±.3 209± 5 1.80	8072±　9 8071± 50 7995 8390
102930.1-034541 249.45　43.89 109.78 -35.98 102658.0-033018	 MCG　0-27- 22 PGC 30904	.E?....		 .05 	 14.8　±.6 				9241± 66 9087 9593
102931.5-351535 272.81　19.18 149.61 -38.50 102716.0-350012	A　　1027-35A ESO　375- 41 PGC 30905	.L..-./ RS -2.8± .6 	1.11± .05 .61± .03 1.06	148 .28 .00 	 14.61 ±.14 14.31				1862± 66 1646 2169

R.A. 2000 DEC. l b SGL SGB R.A. 1950 DEC.	Names PGC	Type S_T n_L T L	$\log D_{25}$ $\log R_{25}$ $\log A_o$ $\log D_o$	p.a. A_g A_i A_{21}	B_T m_B m_{FIR} B_T^o	$(B-V)_T$ $(U-B)_T$ $(B-V)_T^o$ $(U-B)_T^o$	$(B-V)_o$ $(U-B)_o$ m'_o m'_{25}	m_{21} W_{20} W_{50} HI	V_{21} V_{opt} V_{GSR} V_{3K}
102932.2-395029 275.41 15.33 155.31 -37.72 102720.1-393506	 ESO 317- 36 IRAS10273-3935 PGC 30907	.SBS1P. S 1.0± .8 	1.15± .04 .21± .04 1.19	86 .37 .22 .11	 13.40 ±.14 11.70 			
102937.2-240647 265.93 28.39 135.36 -39.15 102715.1-235124	 ESO 501- 1 PGC 30915	.SXS7.. S (1) 7.0± .6 5.6± .6	1.29± .03 .23± .03 1.32	159 .31 .32 .12	 13.85 ±.14 13.21			14.17±.3 252± 15 235± 12 .85	3776± 11 3940± 60 3580 4116
102938.6-035046 249.57 43.85 109.89 -35.97 102706.6-033523	 MCG -1-27- 9 PGC 30917	.SAR5*. E (1) 5.0± .9 5.3± .8	1.06± .06 .14± .05 1.06	90 .05 .20 .07				
102946.4-382054 274.62 16.61 153.46 -37.96 102733.0-380530	 ESO 317- 38 IRAS10275-3805 PGC 30927	RSB.1P? Sr 1.0±1.0 	1.11± .05 .49± .04 1.13	73 .27 .50 .25	 14.55 ±.14 13.14 			
102946.8+130104 228.90 54.04 91.05 -29.63 102707.0+131627	 UGC 5695 PGC 30928	.S?.... 	1.10± .05 .43± .05 1.10	96 .08 .65 .22	 14.66 ±.18 13.92			15.74±.3 231± 7 1.61	2940± 10 2840 3273
102949.6+161059 224.01 55.46 87.81 -28.15 102708.2+162622	A 1027+16 CGCG 94- 75 MK 631 PGC 30932	 		 .06 	 15.1 ±.3 				3200± 45 3111 3526
102949.7-351924 272.91 19.16 149.68 -38.43 102734.3-350400	NGC 3267 ESO 375- 42 PGC 30934	.LXR0.. R -2.0± .4 	1.26± .04 .22± .03 1.26	148 .28 .00 	 13.50 ±.14 13.17				3745± 41 3528 4052
1029.8 +1950 217.92 56.84 84.18 -26.37 1027.1 +2005	 UGC 5696 IRAS10271+2005 PGC 30935	.S..3.. U (1) 3.0± .9 2.5±1.2	1.13± .05 .23± .05 1.13	37 .02 .31 .11	 14.40 ±.19 			
102951.8-345442 272.67 19.51 149.16 -38.47 102736.0-343918	IC 2584 ESO 375- 43 PGC 30938	.L...*/ PBS -2.0± .6 1.23	1.30± .04 .65± .03 .33± .05 1.23	133 .30 .00 	13.6 ±.2 13.62 ±.14 13.28	.93± .02 .47± .04 .78 .36	.97± .02 .46± .04 10.74± .15 13.33± .30		2549± 19 2333 2857
1029.8 -3522 272.95 19.12 149.75 -38.41 1027.6 -3507	A 1027-35B MCG 6-23- 0 PGC 30939	.E...*. RS -4.6± .4 		 .28 					1781± 66 1565 2088
102957.6-351330 272.87 19.26 149.55 -38.41 102742.1-345806	NGC 3269 ESO 375- 44 PGC 30945	.LAR+.. R -1.0± .4 	1.40± .03 .34± .03 1.38	8 .28 .00 	 13.24 ±.14 12.91				3799± 41 3583 4106
103000.6-351930 272.94 19.18 149.68 -38.39 102745.1-350406	NGC 3268 ESO 375- 45 PGC 30949	.E.2... R -5.0± .3 1.54	1.54± .03 .14± .03 1.08± .06 1.54	71 .28 .00 	12.5 ±.2 12.43 ±.11 12.13	1.05± .02 .55± .08 .96 .50	1.07± .01 .59± .03 13.41± .22 14.85± .27		2805± 22 2589 3112
103010.7-030951 249.01 44.42 109.13 -35.64 102738.3-025427	 MCG 0-27- 23 PGC 30960	.LXT+*. E -1.0± .8 1.31	1.31± .08 .09± .07 .93± .07 1.31	40 .06 .00 	13.75 ±.19 13.4 ±.2 13.39	1.02± .03 .55± .08 .89 	1.07± .02 13.89± .25 14.96± .49		11438± 36 11286 11790
103014.7-441836 277.96 11.61 160.69 -36.58 102806.0-440312	 ESO 263- 46 PGC 30966	.SBS7*. S (1) 7.3±1.1 6.7± .9	1.14± .05 .54± .04 1.19	13 .58 .74 .27	 15.32 ±.14 			
1030.2 +4407 172.57 57.03 62.60 -12.68 1027.3 +4423	 UGC 5698 PGC 30971	.S..4.. U 4.0± .9 	1.28± .04 .95± .05 1.28	127 .00 1.39 .47					8707± 10 8724 8933
103020.2-342413 272.47 19.99 148.51 -38.44 102804.1-340848	 ESO 375- 47 IRAS10280-3408 PGC 30976	.L..+*/P S -1.0±1.3 	1.11± .05 .46± .03 1.07	10 .31 .00 	 14.56 ±.14 13.38 14.21				3040± 19 2825 3350
1030.3 +2244 212.85 57.82 81.44 -24.79 1027.6 +2300	 UGC 5704 PGC 30978	.S..6*. U 6.0±1.4 	1.14± .07 1.16± .06 1.14	71 .00 1.47 .50				

R.A. 2000 DEC.	Names	Type	logD$_{25}$	p.a.	B$_T$	(B-V)$_T$	(B-V)$_e$	m$_{21}$	V$_{21}$
l b		S$_T$ n$_L$	logR$_{25}$	A$_g$	m$_B$	(U-B)$_T$	(U-B)$_e$	W$_{20}$	V$_{opt}$
SGL SGB		T	logA$_\bullet$	A$_i$	m$_{FIR}$	(B-V)$_T^o$	m'$_\bullet$	W$_{50}$	V$_{GSR}$
R.A. 1950 DEC.	PGC	L	logD$_\bullet$	A$_{21}$	B$_T^o$	(U-B)$_T^o$	m'$_{25}$	HI	V$_{3K}$
1030.3 +7003	A 1026+70A	.SB.9*.	1.60± .04	145	13.8 ±.2	.59± .07	.54± .03	14.45±.1	1921± 6
138.45 42.58	UGC 5688	U (1)	.31± .06	.10		-.25± .09	-.16± .04	84± 7	2055± 60
42.12 3.48	DDO 80	9.0±1.0	1.16± .05	.32		.49	15.12± .11	54± 12	2035
1026.6 +7018	PGC 30983	9.0± .9	1.61	.16	13.38	-.32	15.87± .31	.91	2007
103023.8-302342		.SXT5*.	1.19± .04					15.35±.3	4079± 10
270.09 23.34	ESO 436- 29	S (1)	.10± .04	.15	13.46 ±.14			93± 6	4107± 41
143.42 -38.82	IRAS10280-3008	5.0± .8		.15	12.64			68± 5	3870
102805.1-300818	PGC 30984	2.2± .9	1.21	.05	13.14			2.16	4402
103023.9-351512		.LXT0..	1.06± .10						
272.97 19.28		S	.27± .08	.28					
149.57 -38.32		-2.0± .9		.00					
102808.2-345947	PGC 30985		1.05						
103026.6-352131	NGC 3271	.LBR0..	1.49± .03	106	12.86 ±.13	1.09± .01	1.10± .01		
273.04 19.20	ESO 375- 48	R	.23± .03	.25	12.63 ±.11	.59± .02	.64± .02		3794± 66
149.70 -38.30		-2.0± .3	.73± .02	.00		.98	12.00± .08		3578
102811.0-350606	PGC 30988		1.48		12.42	.52	14.59± .22		4101
103029.2-353649	NGC 3273	.LAR0..	1.23± .03	97	13.55 ±.13	1.06± .02	1.07± .01		
273.19 18.99	ESO 375- 49	R	.35± .03	.25	13.50 ±.14	.54± .03	.60± .03		2429± 66
150.02 -38.25		-2.0± .4	.50± .02	.00		.95	11.54± .06		2213
102813.8-352124	PGC 30992		1.20		13.24	.45	13.67± .19		2735
103032.0-461613		.SXS9*.	1.08± .05	110					
279.05 9.97	ESO 263- 47	S (1)	.14± .05	.86	14.91 ±.14				
162.99 -36.01		9.0± .6		.15					
102825.0-460048	PGC 30994	10.0± .6	1.16	.07					
1030.5 -0312									11319± 66
249.15 44.46	CGCG 9- 66			.06	15.13 ±.18				11167
109.21 -35.57									11671
1028.0 -0257	PGC 30995								
1030.6 +7037	A 1026+70B	.S..9*.	1.51± .04	0	13.47 ±.19	.80± .06	.70± .03		
137.89 42.18	UGC 5692	U (1)	.26± .06	.07	14.4 ±.5	.07± .07	.01± .04		180± 60
41.69 3.85	DDO 82	9.0±1.0	1.14± .04	.27		.73	14.66± .09		295
1026.8 +7052	PGC 30997	9.0±1.3	1.52	.13	13.24	.00	15.22± .32		262
1030.6 +7413		.I..9..	1.08± .08	140					2805
134.66 39.53	UGC 5686	U	.04± .06	.16					
38.87 6.11		10.0± .8		.03					2932
1026.5 +7429	PGC 30998		1.09	.02					2865
1030.6 +5330		.S..3..	1.06± .06						
157.69 53.04	UGC 5703	U (1)	.12± .05	.00	15.2 ±.3				
55.07 -6.87		3.0± .9		.17					
1027.5 +5346	PGC 31003	4.5±1.1	1.06	.06					
1030.8 +7353		.S..7..	1.23± .05	154					
134.94 39.79	UGC 5689	U	1.02± .05	.13	15.43 ±.18				
39.15 5.90	IRAS10267+7408	7.0± .9		1.38	13.51				
1026.7 +7408	PGC 31011		1.24	.50					
103051.8-364425	NGC 3275	.SBR2..	1.45± .03		11.8 V±.2				
273.90 18.08	ESO 375- 50	R (1)	.12± .04	.23	12.52 ±.11				3211± 66
151.42 -38.01	IRAS10286-3628	2.0± .7	.85± .06	.15	12.63				2994
102837.1-362900	PGC 31014	1.1± .8	1.47	.06	12.11				3514
1030.9 -0347		.S..5P/	1.02± .09	15					
249.83 44.12		E	.76± .08	.06					
109.94 -35.65		5.0±1.8		1.13					
1028.4 -0332	PGC 31017		1.03	.38					
103100.2-343350	IC 2587	PLBS-..	1.31± .05	10	13.3 ±.2	.98± .03	1.00± .02		
272.68 19.93	ESO 375- 51	PBSr	.13± .04	.31	13.38 ±.14	.51± .05	.49± .05		2111± 18
148.69 -38.29	IRAS10287-3418	-2.9± .4	.84± .07	.00	13.66	.88	12.97± .25		1896
102844.0-341824	PGC 31020		1.33		13.02	.44	14.40± .32		2421
103100.6-365832		.SB.0*P	1.12± .05	0					
274.06 17.90	ESO 375- 52	S	.31± .04	.23	14.39 ±.14				9126± 19
151.70 -37.95		.0±1.2		.23					8909
102846.0-364306	PGC 31022		1.13		13.79				9428
103101.5-490244		.SBT7?.	1.18± .07	125					
280.58 7.64	ESO 214- 13	Sr (1)	.43± .07	1.20					
166.20 -35.12	IRAS10289-4847	6.8±1.0		.59	12.37				
102857.0-484718	PGC 31023	5.6±1.3	1.29	.21					
103102.8-284308	IC 2586	.E.4...	1.17± .03	79	13.54 ±.13	1.00± .01	1.01± .01		
269.19 24.80	ESO 436- 30	S	.13± .04	.16	13.68 ±.14				3663± 25
141.27 -38.77	IRAS10288-2824	-5.0± .8	.67± .03	.00		.93	12.38± .10		3455
102843.0-282742	PGC 31025		1.16		13.39		14.06± .23		3988

R.A. 2000 DEC. / l b / SGL SGB / R.A. 1950 DEC.	Names / PGC	Type / S_T n_L / T / L	$\log D_{25}$ / $\log R_{25}$ / $\log A_e$ / $\log D_o$	p.a. / A_g / A_i / A_{21}	B_T / m_B / m_{FIR} / B_T^o	$(B-V)_T$ / $(U-B)_T$ / $(B-V)_T^o$ / $(U-B)_T^o$	$(B-V)_e$ / $(U-B)_e$ / m'_e / m'_{25}	m_{21} / W_{20} / W_{50} / HI	V_{21} / V_{opt} / V_{GSR} / V_{3K}
1031.0 +7851 / 130.82 36.02 / 35.23 9.00 / 1026.3 +7907	UGC 5682 / PGC 31027	.SXT5.. / U / 5.0± .8 /	1.04± .06 / .03± .05 / / 1.04	.01 / .04 / .01	15.1 ±.3				
103107.1+284748 / 201.52 59.08 / 75.91 −21.42 / 102819.1+290313	NGC 3265 / UGC 5705 / IRAS10282+2903 / PGC 31029	.E...*. / PU / −5.0± .7 /	1.11± .06 / .11± .03 / / 1.09	73 / .06 / .00	13.87 ±.15 / 12.35 / 13.79			16.72±.3 / / 185± 11	1447± 8 / 1421± 25 / 1403 / 1735
103109.5−395644 / 275.74 15.41 / 155.36 −37.39 / 102857.0−394118	NGC 3276 / ESO 317- 40 / PGC 31031	.L...*. / S / −2.0±1.9 /	1.02± .06 / .27± .04 / / 1.02	74 / .40 / .00	14.44 ±.14				
1031.1 +3430 / 190.37 59.16 / 70.83 −18.21 / 1028.3 +3446	UGC 5706 / PGC 31032	.I..9.. / U / 10.0± .8 /	1.14± .13 / .10± .12 / / 1.14	.02 / .07 / .05				15.88±.1 / 44± 4 / 36± 4	1494± 5 / / 1474 / 1762
103111.3−461502 / 279.14 10.04 / 162.93 −35.91 / 102904.1−455936	ESO 263- 48 / IRAS10290-4559 / PGC 31035	.LAS0?. / S / −2.0± .6 /	1.42± .04 / .25± .04 / 1.04± .04 / 1.48	168 / .86 / .00	12.5 ±.2 / 12.70 ±.14 / 12.54 / 11.73	1.04± .03 / .56± .03 / .80 / .37	1.12± .01 / .63± .01 / 13.17± .13 / 13.87± .31		2889± 57 / 2666 / 3158
1031.2 +7206 / 136.48 41.13 / 40.56 4.82 / 1027.3 +7222	UGC 5700 / PGC 31036	.SB?... / / / 1.15	1.13± .04 / .07± .04 / / 1.15	.19 / .09 / .03	14.8 ±.3 / 14.48				6689±125 / 6809 / 6762
103112.8+042815 / 240.80 49.63 / 100.44 −32.84 / 102837.0+044340	UGC 5708 / PGC 31037	.SB.7*. / U / 7.0±1.1 /	1.54± .02 / .75± .04 / / 1.54	168 / .05 / 1.04 / .38	13.8 ±.2 / 12.71			13.46±.1 / 195± 7 / 174± 12 / .38	1175± 5 / 1307± 76 / 1047 / 1523
103114.4+430815 / 174.20 57.51 / 63.52 −13.13 / 102816.5+432340	UGC 5707 / KUG 1028+433 / PGC 31040	.SXS6.. / U / 6.0± .7 /	1.37± .03 / .14± .05 / / 1.37	155 / .00 / .21 / .07	14.1 ±.3 / 13.91			14.68±.1 / 123± 16 / 111± 12 / .70	2800± 11 / / 2814 / 3031
103116.1+192308 / 218.91 57.00 / 84.81 −26.30 / 102833.2+193833	UGC 5709 / PGC 31042	.S..7*. / U / 7.0±1.2 /	1.13± .05 / .20± .05 / / 1.13	115 / .04 / .27 / .10	14.8 ±.3 / 14.48			15.89±.3 / / 247± 5 / 1.31	6222± 7 / 6145 / 6541
103119.1−083528 / 254.43 40.73 / 115.77 −36.79 / 102849.2−082002	MCG -1-27- 11 / PGC 31047	PSBT1P? / E / 1.0±1.2 /	1.12± .06 / .12± .05 / / 1.13	65 / .10 / .12 / .06					
103122.9−420338 / 276.93 13.63 / 157.93 −36.91 / 102912.1-414812	ESO 317- 41 / IRAS10292-4148 / PGC 31051	.SBR4*P / S (1) / 4.0± .8 / 3.3± .9	1.17± .05 / .54± .04 / / 1.23	106 / .56 / .80 / .27	14.42 ±.14 / 11.68				
103124.9−351314 / 273.13 19.42 / 149.50 −38.12 / 102909.0-345748	NGC 3258C / ESO 375- 53 / PGC 31053	.SBR1.. / S / 1.0± .8 /	1.07± .04 / .13± .04 / / 1.10	48 / .31 / .14 / .07	14.61 ±.14 / 14.13				2597± 19 / 2381 / 2905
103125.2−295720 / 270.02 23.82 / 142.84 −38.63 / 102906.0-294154	ESO 436- 31 / PGC 31055	.SBS7.. / S (1) / 7.0± .8 / 6.7± .9	1.11± .05 / .10± .05 / / 1.12	.15 / .14 / .05	15.03 ±.14 / 14.72			15.39±.3 / 144± 6 / 107± 5 / .61	4061± 10 / 3852 / 4383
103129.5−324251 / 271.69 21.53 / 146.34 −38.39 / 102912.1-322724	ESO 436- 32 / IRAS10291-3227 / PGC 31058	.S..1*. / S / 1.0±1.2 /	1.01± .05 / .33± .05 / / 1.03	165 / .26 / .34 / .17	15.09 ±.14 / 13.38				
103130.2+245209 / 209.05 58.58 / 79.58 −23.45 / 102844.6+250735	NGC 3270 / UGC 5711 / IRAS10287+2507 / PGC 31059	.SXR3*. / U (1) / 3.0± .8 / 2.5±1.1	1.50± .03 / .58± .04 / / 1.50	10 / .00 / .80 / .29	13.9 ±.3 / 13.6 ±.2 / 13.29 / 12.82	.85± .05 / .41± .07 / .70 / .28	/ / / 14.80± .34	14.44±.1 / 544± 4 / 514± 5 / 1.33	6264± 4 / 6284± 42 / 6208 / 6567
1031.5 +3222 / 194.54 59.32 / 72.75 −19.35 / 1028.7 +3238	UGC 5712 / PGC 31063	.S..6*. / U / 6.0±1.4 /	1.00± .08 / .94± .06 / / 1.00	43 / .03 / 1.38 / .47					
103132.9−393333 / 275.60 15.77 / 154.87 −37.39 / 102920.0-391806	ESO 317- 42 / PGC 31064	.L..-.. / S / −3.0± .9 /	1.11± .05 / .23± .05 / / 1.10	62 / .26 / .00	14.04 ±.14				

R.A. 2000 DEC.	Names	Type	logD$_{25}$	p.a.	B$_T$	(B-V)$_T$	(B-V)$_\bullet$	m$_{21}$	V$_{21}$
l　　b		S$_T$　n$_L$	logR$_{25}$	A$_g$	m$_B$	(U-B)$_T$	(U-B)$_\bullet$	W$_{20}$	V$_{opt}$
SGL　SGB		T	logA$_\bullet$	A$_i$	m$_{FIR}$	(B-V)^o_T	m'$_\bullet$	W$_{50}$	V$_{GSR}$
R.A. 1950 DEC.	PGC	L	logD$_o$	A$_{21}$	B^o_T	(U-B)^o_T	m'$_{25}$	HI	V$_{3K}$
103134.8+002839		.S..4..	1.04± .06	155					8582± 50
245.52　47.15	UGC　5715	U	.18± .05	.14	14.20 ±.18				8441
105.01 -34.17		4.0± .9		.27					8933
102900.7+004405	PGC 31067		1.05	.09	13.74				
103135.6-395721	NGC　3278	.SAS5?.	1.12± .05	62					
275.82　15.44	ESO　317- 43	S	.15± .05	.40	13.02 ±.14				
155.35 -37.31	IRAS10293-3941	5.0±1.2		.22	10.96				
102923.1-394154	PGC 31068		1.16	.07					
1031.6　+2559		.S..4..	1.22± .05	5					
206.95　58.82	UGC　5713	U	.48± .05	.03	14.58 ±.19				
78.56 -22.84	IRAS10288+2614	4.0± .9		.70	13.40				
1028.8　+2614	PGC 31075		1.22	.24					
103140.1-452903		.SBT6P.	1.19± .05	164					
278.81　10.74	ESO　263- 51	S　　　(1)	.38± .04	.77	14.21 ±.14				
162.00 -36.03	IRAS10295-4513	5.7± .7		.56	12.61				
102932.0-451336	PGC 31076	5.6± .9	1.26	.19					
103143.4+251827		.S..9*.	1.11± .07					14.95±.1	1277±　7
208.24　58.71	UGC　5716	U	.20± .06	.00				120±　6	
79.20 -23.18		9.0±1.2		.20				118±　3	1222
102857.6+253353	PGC 31081		1.11	.10					1579
103145.1+464024		.SBS5..	.98± .07	85					
168.07　56.35	UGC　5714	U	.10± .05	.00	14.60 ±.18				
60.67 -10.93	IRAS10287+4655	5.0± .9		.15					
102844.2+465550	PGC 31083		.98	.05					
103148.2-263357		.LX.-..	1.18± .05						
267.99　26.67	ESO　501- 3	S	.07± .03	.24	13.77 ±.14				
138.51 -38.67		-3.0± .8		.00					
102927.0-261830	PGC 31085		1.20						
103148.4-360151		.SB.1?.	1.13± .05	119					
273.67　18.78	ESO　375- 54	S	.86± .05	.24	15.22 ±.14				
150.50 -37.93		1.0±1.8		.88					
102933.0-354624	PGC 31086		1.16	.43					
103150.0-302309	IC　2588	PSBR1*.	1.15± .04	150				15.02±.3	3520± 11
270.37　23.52	ESO　436- 33	Sr	.08± .04	.10	13.65 ±.14			368± 15	3541± 15
143.38 -38.51	IRAS10295-3007	1.0± .6		.08				325± 12	3317
102931.1-300742	PGC 31088		1.16	.04	13.42			1.56	3848
103152.2-345127	NGC　3281	.SAS2P*	1.52± .03	140	12.7　±.2	.98± .03	1.05± .02		
273.01　19.78	ESO　375- 55	RBCSr	.30± .03	.31	12.62 ±.11		.47± .07		3439± 41
149.03 -38.07	IRAS10295-3435	2.1± .3	1.05± .05	.37	11.25	.83	13.44± .15		3224
102936.0-343600	PGC 31090		1.55	.15	11.93		14.40± .26		3748
103155.8-352433	NGC　3258D	.SBS3*.	1.21± .04	5					2722± 19
273.34　19.32	ESO　375- 58	PS	.26± .04	.31	13.99 ±.14				2506
149.71 -37.99		3.0± .7		.35					3029
102940.0-350906	PGC 31094		1.24	.13	13.31				
1031.9　+7748		.I..9*.	1.12± .07	0					1621
131.61　36.86	UGC　5701	U	.35± .06	.02					
36.09　8.39		10.0±1.2		.26					1759
1027.4　+7804	PGC 31098		1.12	.17					1659
103159.0-351151		.L?....	1.09± .05	123					2289± 19
273.23　19.51	ESO　375- 59		.10± .03	.31	14.43 ±.14				2073
149.45 -38.01				.00					2597
102943.0-345624	PGC 31103		1.11		14.09				
103212.4-091422		RLA.-*.	1.13± .11	165					
255.22　40.39	MCG -1-27- 13	E	.11± .07	.10					
116.63 -36.72		-3.0± .9		.00					
102942.7-085855	PGC 31116		1.13						
103212.8-012936		PSXR1P*	.96± .17	40				15.47±.1	1171±　9
247.79　45.94	UGC　5723	E	.15± .12	.12				365± 12	
107.35 -34.65		1.0±1.3		.15					1024
102939.6-011409	PGC 31117		.97	.07					1523
103216.7+274013	NGC　3274	.SX.7?.	1.33± .03	100	13.21 ±.13	.39± .02		12.88±.1	537±　5
203.75　59.21	UGC　5721	PU　　　(1)	.32± .03	.06	13.17 ±.14	-.14± .04		174±　5	519± 46
77.09 -21.81	IRAS10294+2755	6.8± .7		.44	13.14	.31		157±　8	491
102929.6+275540	PGC 31122	8.0± .9	1.34	.16	12.69	-.20	13.92± .22	.03	831
103217.0+120324		.S?....	.82± .13						9418± 61
230.84　54.10	MCG　2-27- 24		.17± .07	.09	15.09 ±.19				9315
92.37 -29.51	MK　　722			.26					9754
102937.8+121851	PGC 31123		.83	.09	14.69				

R.A. 2000 DEC. l b SGL SGB R.A. 1950 DEC.	Names PGC	Type S_T n_L T L	$\log D_{25}$ $\log R_{25}$ $\log A_o$ $\log D_o$	p.a. A_g A_i A_{21}	B_T m_B m_{FIR} B_T^o	$(B-V)_T$ $(U-B)_T$ $(B-V)_T^o$ $(U-B)_T^o$	$(B-V)_o$ $(U-B)_o$ m' m'_{25}	m_{21} W_{20} W_{50} HI	V_{21} V_{opt} V_{GSR} V_{3K}
103220.3+560500 153.94 51.84 53.19 -5.08 102908.6+562027	NGC 3264 UGC 5719 IRAS10291+5620 PGC 31125	.SB.8*. 8.0±1.0 	1.46± .04 .37± .06 1.46	177 .00 .46 .19	12.5 ±.4 13.9 ±.3 13.00	.52± .04 -.07± .05 .44 -.12	 13.70± .48 	14.15±.1 175± 16 137± 12 .96	942± 11 1005 1107
103220.6-240216 266.46 28.81 135.28 -38.53 102958.0-234648	IC 2589 ESO 501- 4 IRAS10299-2346 PGC 31126	.S?.... 	.95± .06 .20± .05 .97	10 .25 .28 .10	 14.27 ±.14 12.72 13.71				3700± 60 3500 4036
103222.5-221804 265.30 30.22 133.06 -38.46 102959.0-220236	NGC 3282 ESO 568- 16 PGC 31129	RLBR0?. S -2.0± .9 	1.29± .04 .52± .03 1.24	82 .19 .00 	 13.99 ±.14 13.74				3880± 60 3683 4219
1032.3 +0233 243.35 48.66 102.71 -33.27 1029.8 +0249	 UGC 5726 PGC 31130	.I..9?. U 10.0±1.9 	1.00± .08 .43± .06 1.01	157 .07 .32 .22					6593 6459 6942
103225.2-345958 273.19 19.72 149.19 -37.94 103009.0-344430	NGC 3258E ESO 375- 60 PGC 31131	.S..3?/ S 3.0±1.7 	1.20± .04 .67± .04 1.23	27 .31 .92 .33	 15.42 ±.14 				
103228.2-363158 274.07 18.43 151.10 -37.72 103013.0-361630	 ESO 375- 61 PGC 31134	.SXS7*. S (1) 7.0±1.2 8.9± .8	1.05± .05 .16± .05 1.07	145 .25 .22 .08	 15.73 ±.14 				
103228.7-273152 268.74 25.96 139.74 -38.50 103008.0-271624	 ESO 501- 5 PGC 31135	 	.66± .07 .04± .05 .68	 .22 	 15.23 ±.14 				4027± 76 3821 4355
1032.4 +2709 204.76 59.19 77.58 -22.05 1029.7 +2725	 MCG 5-25- 21 PGC 31136	.S?.... 	.89± .11 .00± .07 .89	 .01 .00 .00	 15.07 ±.19 14.93				 12043 11995 12339
103231.5+542356 156.20 52.80 54.54 -6.10 102922.2+543923	A 1029+54 UGC 5720 MK 33 PGC 31141	.I..9P* R 10.0± .5 	1.00± .06 .03± .05 1.00	 .00 .02 .01	 13.4 ±.2 13.39			15.70±.1 181± 5 113± 5 2.29	1461± 7 1390± 42 1516 1633
103235.3+650229 143.26 46.23 46.17 .51 102906.8+651756	NGC 3259 UGC 5717 IRAS10291+6517 PGC 31145	.SXT4*. R (3) 4.0± .5 5.2± .6	1.34± .03 .26± .03 1.34	20 .00 .38 .13	 12.97 ±.13 12.91 12.57			13.65±.3 295± 34 259± 25 .95	1686± 17 1760± 38 1794 1813
103239.4-172526 261.91 34.17 126.86 -38.00 103013.4-170958	 PGC 31148	.IB.9.. S (1) 10.0± .8 9.8± .8	1.18± .05 .17± .06 1.21	10 .25 .12 .08					2645 2458 2992
1032.7 +1955 218.19 57.51 84.47 -25.74 1030.0 +2011	 UGC 5729 PGC 31151	.SB.2.. U 2.0± .9 	1.02± .06 .28± .05 1.02	5 .03 .35 .14	 14.94 ±.18 				5779 5689 6107
103244.2-283646 269.46 25.10 141.11 -38.41 103024.0-282118	 ESO 436- 34 IRAS10303-2821 PGC 31154	.S..3./ S 3.0± .9 	1.32± .04 .74± .03 1.34	60 .17 1.02 .37	 14.42 ±.14 13.05 13.20				3614± 76 3407 3940
1032.7 +1551 225.07 55.97 88.53 -27.68 1030.1 +1606	IC 616 UGC 5730 IRAS10301+1606 PGC 31159	.S..6*. U 6.0±1.2 	1.06± .06 .03± .05 1.07	 .08 .04 .01	 14.22 ±.19 14.07				5779 5689 6107
103248.7-342359 272.92 20.27 148.42 -37.94 103032.0-340830	 ESO 375- 62 PGC 31160	.L...*/ S -2.0±1.8 	1.18± .05 .84± .03 1.09	173 .30 .00 	 14.79 ±.14 14.44				3160± 60 2946 3471
103248.8-273123 268.80 26.01 139.73 -38.42 103028.1-271554	NGC 3285A ESO 501- 8 PGC 31161	.SBT6*. PS (1) 6.3± .5 3.9± .6	1.07± .04 .14± .03 1.09	171 .21 .21 .07	 14.52 ±.14 14.09			15.40±.3 214± 6 152± 5 1.24	4300± 10 4094 4629
103250.2-301611 270.49 23.73 143.22 -38.30 103031.0-300042	 ESO 436- 35 PGC 31164	.SXR8.. Sr (1) 8.0± .6 6.7± .9	1.08± .04 .10± .03 1.09	 .10 .12 .05	 14.60 ±.14 14.37			15.84±.3 167± 6 141± 5 1.42	3451± 10 3451± 19 3242 3773

R.A. 2000 DEC.	Names	Type	logD₂₅	p.a.	B_T	(B-V)_T	(B-V)_●	m₂₁	V₂₁
l b		S_T n_L	logR₂₅	A_g	m_B	(U-B)_T	(U-B)_●	W₂₀	V_opt
SGL SGB		T	logA_●	A_i	m_FIR	(B-V)_T^o	m'_●	W₅₀	V_GSR
R.A. 1950 DEC.	PGC	L	logD_o	A₂₁	B_T^o	(U-B)_T^o	m'₂₅	HI	V_3K

1032.9 -0629		.S?....	1.04± .09		14.60 ±.13	.95± .02			
252.91 42.54	MCG -1-27- 15		.28± .07	.10					4998± 62
113.34 -35.90				.39		.84			4837
1030.4 -0614	PGC 31165		1.05	.14	14.07			13.92± .51	5351
103255.4+283043	NGC 3277	.SAR2..	1.29± .03		12.50 ±.13	.82± .01	.83± .01	16.17±.1	1408± 7
202.15 59.45	UGC 5731	R (2)	.05± .03	.03	12.58 ±.14	.24± .03	.32± .04	381± 34	1460± 56
76.40 -21.24	IRAS10301+2846	2.0± .4	.88± .01	.06	13.26	.79	12.39± .04	253± 4	1366
103008.0+284611	PGC 31166	2.5± .7	1.29	.02	12.44	.22	13.67± .20	3.71	1700
103259.4-345311	NGC 3281C	.L..../	1.15± .05	160					
273.23 19.88	ESO 375- 63	RCS	.67± .03	.31	14.30 ±.14				2779± 19
149.03 -37.84		-2.0± .7		.00					2564
103043.0-343742	PGC 31173		1.09		13.95				3088
103302.6-385859		.SBS7*/	1.36± .04	11					
275.54 16.42	ESO 317- 46	S	.85± .04	.27	14.85 ±.14				
154.10 -37.21		7.0±1.3		1.17					
103049.0-384330	PGC 31178		1.38	.42					
103308.1-270547		.S?....	1.15± .05	172					
268.60 26.40	ESO 501- 10		.89± .04	.22	14.72 ±.14				4238± 76
139.18 -38.36				1.34					4033
103047.1-265018	PGC 31182		1.17	.45	13.14				4568
103313.6-072754		.LBT0..	1.15± .06	35					
253.88 41.88	MCG -1-27- 18	E	.08± .05	.13					
114.54 -36.07		-2.0± .8		.00					
103043.1-071225	PGC 31191		1.15						
1033.2 +6430		.S..4..	1.14± .05	29					
143.76 46.64	UGC 5727	U	.55± .05	.00	14.95 ±.18				
46.63 .24		4.0± .9		.81					
1029.8 +6446	PGC 31192		1.14	.28					
103317.7+644458	NGC 3266	.LX.0$.	1.19± .05	105					
143.49 46.48	UGC 5725	R	.07± .03	.01	13.42 ±.15				
46.44 .39		-2.0± .4		.00					
102950.3+650026	PGC 31198		1.18						
103330.2-265354		.L..0*.	1.14± .04	59					
268.55 26.61	ESO 501- 13	S	.39± .03	.26	13.94 ±.14				3594± 47
138.93 -38.28		-2.0±1.1		.00					3390
103109.1-263824	PGC 31212		1.11		13.63				3924
103335.9-272718	NGC 3285	.SBS1P.	1.41± .03	108	13.05 ±.13	1.02± .01	1.06± .01	14.88±.3	3378± 10
268.92 26.16	ESO 501- 15	R	.23± .03	.21	12.97 ±.11	.43± .02	.53± .02	594± 6	3502± 26
139.63 -38.25	IRAS10312-2711	1.0± .3	.87± .01	.23	13.51	.90	12.89± .04	556± 5	3190
103115.0-271148	PGC 31217		1.43	.11	12.52	.34	14.39± .21	2.25	3725
1033.8 +1252		.S..4..	.94± .09	0					
229.94 54.84	UGC 5735	U	.24± .06	.07	14.75 ±.18				
91.71 -28.80		4.0± .9		.35					
1031.2 +1308	PGC 31235		.95	.12					
103351.6-003342		.S..4..	1.04± .06	80					
247.21 46.88	UGC 5736	U	.09± .05	.13	14.7 ±.2				
106.43 -33.97		4.0± .9		.13					
103118.0-001812	PGC 31236		1.05	.04					
103353.8-274954		.S?....	.97± .06	74				16.45±.2	2694± 7
269.21 25.89	ESO 436- 38		.18± .04	.18	15.70 ±.14			198± 6	
140.11 -38.18				.27				68± 5	2488
103133.0-273424	PGC 31238		.99	.09	15.24			1.12	3022
1033.9 +1112	A 1031+11	.S..6*.	.96± .09	110	15.2 ±.2	.73± .09			
232.44 54.02	UGC 5737	U	.05± .06	.09	14.68 ±.19	-.25± .12			
93.46 -29.51		6.0±1.2		.07					
1031.3 +1128	PGC 31241		.97	.02			14.73± .50		
103359.5-301013		.S..4*/	1.22± .04	84					
270.66 23.95	ESO 436- 39	S	.76± .03	.15	14.92 ±.14				3365± 27
143.07 -38.06	IRAS10316-2954	4.0±1.3		1.12	12.88				3156
103140.0-295442	PGC 31242		1.24	.38	13.63				3688
103400.0-272012		.SBT2..	.92± .05						
268.92 26.31	ESO 501- 16	r	.12± .04	.21	14.78 ±.14				9683± 76
139.48 -38.16	IRAS10316-2704	2.2± .9		.14					9478
103139.0-270442	PGC 31243		.94	.06	14.33				10012
103401.3-351701		.SXT1P.	1.15± .05						
273.65 19.66	ESO 375- 64	S	.14± .05	.31	14.66 ±.14				2573± 19
149.49 -37.58	IRAS10317-3501	1.0± .8		.14	13.54				2358
103145.0-350130	PGC 31248		1.18	.07	14.17				2881

R.A. 2000 DEC. l b SGL SGB R.A. 1950 DEC.	Names PGC	Type S_T n_L T L	$\log D_{25}$ $\log R_{25}$ $\log A_e$ $\log D_o$	p.a. A_g A_i A_{21}	B_T m_B m_{FIR} B_T^o	$(B-V)_T$ $(U-B)_T$ $(B-V)_T^o$ $(U-B)_T^o$	$(B-V)_e$ $(U-B)_e$ m'_e m'_{25}	m_{21} W_{20} W_{50} HI	V_{21} V_{opt} V_{GSR} V_{3K}
103407.3-351925 273.69 19.63 149.54 -37.56 103151.0-350354	NGC 3289 ESO 375- 65 PGC 31253	.LBT+*/ S -1.0±1.1 	1.35± .04 .60± .03 .78± .04 1.29	153 .31 .00 	13.44 ±.15 13.68 ±.14 13.22	.93± .02 .46± .03 .75 .33	.94± .01 .47± .03 12.83± .12 13.54± .25	 	 2702± 19 2487 3010
103413.8-731426 293.33 -12.98 191.27 -24.79 103301.0-725854	IC 2596 ESO 38- 2 FAIR 283 PGC 31265	.SBT2.. Sr 2.0± .5 	1.08± .06 .17± .05 1.13	177 .56 .21 .09	 11.87 	 	 	 	 3390± 63 3184 3527
103415.0+525217 158.13 53.84 55.92 -6.83 103108.3+530747	 UGC 5734 PGC 31269	.S..0.. U .0± .9 	1.19± .05 .71± .05 1.15	156 .00 .54 	 14.2 ±.2 13.56	 	 	 	 7109 7161 7292
103419.0-342413 273.20 20.43 148.38 -37.63 103202.0-340842	NGC 3281D ESO 375- 68 PGC 31273	.SBS7*/ PS (1) 6.7± .7 5.6±1.3	1.31± .04 .70± .04 1.35	160 .35 .97 .35	 14.67 ±.14 13.33	 	 	 	 5366± 19 5152 5677
103420.1+134512 228.69 55.36 90.87 -28.31 103140.2+140043	 UGC 5739 IRAS10316+1400 PGC 31275	.I..9?. U 10.0±1.9 	1.02± .06 .53± .05 1.02	152 .07 .39 .26	 14.5 ±.3 12.09 14.02	 	 	16.73±.3 217± 5 2.45	2976± 9 3001± 50 2880 3310
103421.3-321113 271.93 22.31 145.61 -37.84 103203.1-315542	 ESO 436- 40 IRAS10320-3155 PGC 31276	PSBT2*P S 2.0± .9 	1.10± .04 .40± .03 1.12	48 .27 .49 .20	 14.73 ±.14 12.99 	 	 	 	
103423.0+734549 134.80 40.05 39.40 6.02 103022.9+740118	NGC 3252 UGC 5732 IRAS10303+7401 PGC 31278	.SB.7?/ PU 6.7± .7 	1.30± .04 .50± .04 1.31	35 .13 .69 .25	 14.12 ±.18 13.07 13.29	 	 	14.83±.1 268± 16 249± 12 1.29	1136± 11 1262 1199
103424.6-262931 268.48 27.06 138.41 -38.08 103203.1-261400	 ESO 501- 17 IRAS10320-2613 PGC 31280	.S?.... 	.93± .05 .50± .04 .93	15 .25 .37 	 15.02 ±.14 13.17 14.33	 	 	 	 4429± 27 4225 4760
103430.3+351527 188.79 59.77 70.59 -17.22 103138.9+353058	 UGC 5738 IRAS10316+3530 PGC 31285	.S?.... 	.96± .05 .20± .04 .96	30 .02 .30 .10	 14.2 ±.2 13.74 13.91	 	 	15.60±.1 155± 11 145± 8 1.59	1516± 9 1500 1782
1034.5 +7909 130.43 35.89 35.09 9.31 1029.8 +7925	 UGC 5728 PGC 31287	.S..9*. U 9.0±1.2 	1.14± .13 .18± .12 1.14	0 .01 .18 .09	 	 	 	 	2742 2884 2772
103437.0-273914 269.25 26.12 139.88 -38.02 103216.1-272342	NGC 3285B ESO 501- 18 IRAS10322-2723 PGC 31293	.SXR3*. RSr (1) 3.3± .3 5.0± .6	1.18± .03 .12± .03 1.20	43 .19 .16 .06	 13.89 ±.14 13.60 13.52	 	 	15.69±.3 165± 6 121± 5 2.12	2952± 10 3149± 76 2750 3284
103437.6-432644 278.20 12.76 159.42 -36.02 103227.0-431112	 ESO 264- 5 PGC 31295	.SBS9.. S (1) 9.0± .6 9.4± .6	1.05± .05 .05± .05 1.10	 .55 .05 .02	 15.13 ±.14 	 	 	 	
103438.4-283513 269.83 25.35 141.06 -37.99 103218.0-281942	 ESO 436- 42 IRAS10323-2819 PGC 31296	.E?.... 	.82± .07 .24± .05 .78	49 .18 .00 	 14.44 ±.14 12.14 14.21	 	 	 	 3451± 22 3244 3778
103442.7+111157 232.62 54.17 93.56 -29.35 103204.0+112728	NGC 3279 UGC 5741 IRAS10320+1127 PGC 31302	.S..7.. U 7.0± .9 	1.46± .03 .92± .05 1.47	152 .12 1.26 .46	 13.95 ±.18 12.17 12.56	 	 	15.60±.3 356± 7 325± 7 2.58	1392± 9 1422± 50 1287 1731
103445.1-015809 248.92 46.07 108.15 -34.20 103212.1-014237	 UGC 5745 PGC 31304	.LBT+.. UE -.5± .6 	1.03± .06 .11± .05 1.02	135 .11 .00 	 13.86 ±.16 13.74	 	 	 	 957 810 1310
1034.7 +2531 208.08 59.43 79.40 -22.48 1032.0 +2547	 UGC 5743 PGC 31306	.S..6*. U 6.0±1.3 	1.00± .08 .49± .06 1.00	147 .00 .72 .25	 	 	 	 	
1034.7 +5045 161.19 54.98 57.67 -8.06 1031.7 +5101	 UGC 5740 PGC 31307	.SX.9.. U 9.0± .8 	1.24± .05 .16± .05 1.24	140 .00 .17 .08	 15.1 ±.4 14.89	 	 	14.58±.1 163± 16 117± 12 -.39	651± 11 695 845

R.A. 2000 DEC. l b SGL SGB R.A. 1950 DEC.	Names PGC	Type S_T n_L T L	logD_25 logR_25 logA_o logD_o	p.a. A_g A_i A_21	B_T m_B m_FIR B_T^o	(B-V)_T (U-B)_T (B-V)_T^o (U-B)_T^o	(B-V)_o (U-B)_o m' m'_25	m_21 W_20 W_50 HI	V_21 V_opt V_GSR V_3K
103447.5-282956 269.81 25.44 140.95 -37.96 103227.0-281424	 ESO 436- 44 PGC 31310	.L?.... 	1.06± .06 .16± .04 1.06	105 .18 .00 	 13.90 ±.14 13.68				3177± 47 2971 3504
103447.7+213902 215.38 58.51 83.07 -24.46 103204.1+215433	NGC 3287 UGC 5742 IRAS10320+2154 PGC 31311	.SBS7.. R (2) 7.0± .4 6.0± .8	1.32± .03 .33± .03 1.32	20 .01 .46 .17	 13.03 ±.15 12.35 12.56			15.21±.1 198± 11 166± 6 2.49	1306± 6 1151± 49 1236 1617
103448.2-271250 269.01 26.51 139.32 -37.99 103227.0-265718	 ESO 501- 20 PGC 31312	.L?.... 	.96± .06 .11± .03 .75± .08 .97	37 .27 .00 	14.5 ±.2 14.70 ±.14 14.30	1.02± .04 .51± .08 .91 .46	1.04± .02 .55± .05 13.70± .26 13.89± .35		 4306± 76 4101 4636
103450.5-283502 269.87 25.38 141.06 -37.95 103230.1-281930	 ESO 436- 46 PGC 31316	.SBT4.. Sr (1) 3.7± .5 1.1± .8	1.34± .03 .12± .04 1.36	127 .18 .17 .06	 13.44 ±.14 13.07			14.95±.2 301± 6 230± 5 1.82	3438± 7 3790± 60 3237 3770
103450.5-283056 269.83 25.44 140.97 -37.95 103230.1-281524	 ESO 436- 45 PGC 31317	 	.77± .07 .02± .06 .79	 .18 	 15.10 ±.14 				3235± 76 3028 3562
103455.6-062816 253.39 42.89 113.48 -35.41 103224.6-061245	 MCG -1-27- 20 PGC 31326	.SBT5?. E (1) 5.0± .9 3.1±1.2	1.07± .06 .27± .05 1.08	75 .10 .41 .14					
103459.8-280450 269.59 25.82 140.42 -37.93 103239.0-274918	 ESO 437- 2 IRAS10326-2748 PGC 31330	.S?.... 	1.01± .05 .40± .04 1.02	154 .18 .60 .20	 14.67 ±.14 13.37 13.88				2311± 76 2105 2639
103504.6+463336 167.85 56.92 61.12 -10.54 103204.7+464907	A 1032+46 UGC 5744 MK 146 PGC 31331	.L?.... 	.77± .08 .05± .04 .76	110 .00 .00 	 14.3 ±.2 13.36 14.29				3330± 41 3358 3545
103508.5-434132 278.41 12.59 159.69 -35.87 103258.0-432600	NGC 3366 ESO 264- 7 IRAS10329-4325 PGC 31335	PSBR3*. BS (1) 3.3± .5 2.6± .6	1.34± .04 .31± .05 1.06± .02 1.40	37 .60 .43 .16	12.03 ±.14 12.77 ±.14 11.58 11.34	.75± .01 .25± .03 .54 .09	.83± .01 .30± .03 12.82± .05 12.80± .29		3000± 60 2779 3280
103508.9+450512 170.34 57.50 62.34 -11.41 103210.4+452043	 UGC 5746 PGC 31336	.SB.2.. U 2.0± .8 	1.16± .05 .20± .05 1.16	40 .00 .25 .10	 14.06 ±.18 13.74				7588± 50 7610 7810
103516.9-071448 254.19 42.38 114.43 -35.52 103246.2-065916	 MCG -1-27- 21 PGC 31345	.SBS1P? E 1.0±1.3 	1.08± .06 .27± .05 1.09	85 .12 .27 .13					
103517.6-171636 262.40 34.67 126.77 -37.35 103251.4-170104	NGC 3290 ARP 53 PGC 31346	.SXT4*P R (1) 4.0± .5 3.3±1.2	.99± .07 .28± .06 1.01	60 .22 .42 .14				16.20±.1 430± 12 	10576± 9 10616± 71 10391 10924
103518.5-365245 274.77 18.44 151.42 -37.11 103303.0-363712	 ESO 375- 69 IRAS10330-3637 PGC 31348	.S?.... 	.87± .06 .13± .05 .89	153 .18 .19 .06	 14.38 ±.14 13.20 13.99				3145± 19 2929 3449
103520.2-272145 269.22 26.46 139.51 -37.87 103259.0-270612	 ESO 501- 21 PGC 31353	.L?.... 	1.16± .05 .69± .03 1.07	162 .19 .00 	 14.59 ±.14 14.33				4554± 76 4349 4884
103523.6-244521 267.58 28.62 136.21 -37.85 103301.0-242948	A 1033-24 ESO 501- 23 DDO 238 PGC 31359	.SBS8.. SU (2) 8.0± .5 8.7± .6	1.54± .03 .10± .04 1.33± .08 1.55	14 .18 .12 .05	13.2 ±.3 13.15 ±.14 12.86	.63± .06 -.21± .08 .56 -.26	.59± .06 -.20± .08 15.34± .20 15.49± .34	13.55±.2 79± 16 62± 12 .65	1048± 11 848 1383
103523.7-281857 269.82 25.67 140.71 -37.83 103303.0-280324	 ESO 437- 4 IRAS10330-2803 PGC 31360	.SXR4P* Sr (1) 3.7± .5 2.6± .7	1.27± .03 .21± .03 1.28	153 .18 .31 .11	 13.96 ±.14 13.37 13.45			15.20±.2 324± 6 300± 5 1.64	3288± 7 3265± 47 3082 3615
103525.6-263927 268.79 27.05 138.62 -37.86 103304.0-262354	 ESO 501- 25 PGC 31366	.LX.0.. S -2.0± .8 	1.23± .04 .29± .03 1.21	155 .26 .00 	 14.15 ±.14 13.84				3834± 76 3630 4165

R.A. 2000 DEC. l b SGL . SGB R.A. 1950 DEC.	Names PGC	Type S_T n_L T L	$\log D_{25}$ $\log R_{25}$ $\log A_\bullet$ $\log D_o$	p.a. A_g A_i A_{21}	B_T m_B m_{FIR} B_T^o	$(B-V)_T$ $(U-B)_T$ $(B-V)_T^o$ $(U-B)_T^o$	$(B-V)_\bullet$ $(U-B)_\bullet$ $m'_\bullet$ m'_{25}	m_{21} W_{20} W_{50} HI	V_{21} V_{opt} V_{GSR} V_{3K}
103526.8-242303 267.35 28.93 135.74 -37.83 103304.0-240730	ESO 501- 24 IRAS10330-2407 PGC 31368	RLXR+*. r -1.3± .9	1.06± .06 .29± .03 1.05	102 .27 .00	14.37 ±.14 13.23				
103527.4-140749 260.05 37.17 122.85 -36.88 103259.8-135216	MCG -2-27- 9 IRAS10329-1352 PGC 31369	.LBT+P? E -1.0±1.2	1.28± .05 .60± .05 1.21	85 .19 .00	13.63				
103531.1-061106 253.27 43.20 113.18 -35.20 103259.9-055533	NGC 3292 MCG -1-27- 23 PGC 31370	.L..OP? E -2.0±1.8	1.05± .12 .10± .07 1.05	175 .10 .00					
103532.2-414427 277.44 14.31 157.34 -36.22 103320.0-412854	NGC 3318A ESO 317- 50 PGC 31373	.SXT5*. S 5.0± .7	1.16± .04 .53± .04 1.21	3 .47 .80 .27	15.83 ±.14				
103535.0+441852 171.64 57.86 63.03 -11.80 103237.3+443424	A 1032+44 UGC 5747 MK 148 PGC 31376	.P...$. R 99.0	.94± .06 .63± .05 .94	133 .00 .86 .31	15.51 ±.20 14.59				7200±110 7219 7426
103538.1-484833 281.12 8.22 165.65 -34.47 103332.0-483300	ESO 214- 14 PGC 31378	.S?....	1.01± .10 .80± .07 1.11	63 1.11 1.21 .40					5914± 34 5691 6174
103539.5+283355 202.16 60.05 76.70 -20.70 103252.4+284928	UGC 5749 ARAK 248 PGC 31379	.S?....	.65± .11 .05± .05 .65	 .04 .07 .02	15.1 ±.3 13.25 14.95			16.73±.3 157± 7 1.76	4413± 10 4413± 50 4371 4705
103545.4+205927 216.72 58.52 83.83 -24.59 103302.2+211500	UGC 5750 PGC 31386	.SB.8*. U 8.0±1.3	1.04± .08 .29± .06 1.04	 .02 .36 .15				16.78±.3 108± 7	7148± 10 7078 7464
103546.2+210259 216.61 58.54 83.78 -24.56 103303.0+211832	UGC 5751 IRAS10330+2118 PGC 31388	.S?....	1.31± .04 .65± .05 1.31	172 .02 .97 .32	14.53 ±.19 13.24 13.50			15.13±.3 388± 7 1.30	7041± 10 6971 7356
103547.5-322133 272.30 22.32 145.79 -37.52 103329.0-320600	NGC 3302 ESO 437- 7 PGC 31391	.LA.0.. S -2.0± .5	1.22± .04 .13± .03 1.23	118 .29 .00	13.51 ±.14 13.16				4075± 18 3863 4392
103549.7+633210 144.52 47.51 47.57 -.14 103226.5+634742	A 1032+63 CGCG 313- 27 MK 147 PGC 31395			 .00					7020± 38 7111 7144
103551.3-341610 273.41 20.72 148.17 -37.33 103334.0-340036	ESO 375- 70 PGC 31397	.S?....	.97± .06 .13± .05 1.00	30 .33 .18 .06	15.02 ±.14 14.48				3845± 19 3631 4157
103603.9-241922 267.44 29.06 135.67 -37.69 103341.1-240348	IC 2594 ESO 501- 28 PGC 31405	.LA.-.. S -3.0± .8	1.24± .04 .07± .03 1.27	 .27 .00	13.39 ±.14 13.06				3580± 60 3380 3916
1036.0 -2730 269.46 26.42 139.69 -37.70 1033.7 -2715	A 1033-27 PGC 31407	.E+..P* RS -4.5± .6		 .18					2381± 47 2176 2710
103609.5-371416 275.12 18.22 151.83 -36.89 103354.0-365842	ESO 375- 71 PGC 31414	.IBS9.. S (1) 10.0± .4 10.0± .5	1.47± .04 .08± .05 1.49	93 .16 .06 .04	13.10 ±.14 12.87			15.85±.2 60± 16 39± 6 2.94	957± 7 741 1260
103612.5-270946 269.27 26.73 139.25 -37.68 103351.1-265412	NGC 3305 ESO 501- 30 PGC 31421	.E.0... S -5.0± .8	1.05± .03 .01± .03 .56± .01 1.09	 .27 .00	13.77 ±.13 13.88 ±.14 13.49	1.02± .01 .58± .02 .92 .54	1.05± .01 .60± .02 12.06± .04 13.96± .22		3987± 19 3783 4317
103615.3-082005 255.41 41.74 115.81 -35.55 103345.1-080431	IC 624 MCG -1-27- 26 PGC 31426	.SX.1*. E 1.0± .8	1.43± .04 .55± .05 1.43	138 .07 .56 .27				14.68±.1 605± 12	5042± 9 4877 5396

R.A. 2000 DEC. l　　b SGL　SGB R.A. 1950 DEC.	Names PGC	Type S_T　n_L T L	logD25 logR25 logA_e logD_o	p.a. A_g A_i A_21	B_T m_B m_FIR B_T^o	(B-V)_T (U-B)_T (B-V)_T^o (U-B)_T^o	(B-V)_e (U-B)_e m'_e m'_25	m_21 W_20 W_50 HI	V_21 V_opt V_GSR V_3K
1036.2 +1326		.I..9..	.96± .17						2956
229.57 55.62	UGC 5758	U	.39± .12	.05					
91.43 -28.03		10.0± .9		.29					2859
1033.6 +1342	PGC 31427		.96	.20					3290
103616.1+371928	NGC 3294	.SAS5..	1.55± .02	122	12.2 ±.3	.41± .07		14.35±.1	1586± 6
184.62 59.84	UGC 5753	R (2)	.29± .02	.00	11.66 ±.14			399± 7	1436± 58
69.01 -15.75	IRAS10333+3734	5.0± .3		.44	10.95	.34		378± 6	1576
103323.7+373501	PGC 31428	1.5± .6	1.55	.15	11.31		14.04± .32	2.89	1842
1036.2 +2658	IC 2590	.L.....	1.04± .18						
205.35 60.00	UGC 5756	U	.00± .08	.04	14.25 ±.16				6363
78.24 -21.43		-2.0± .8		.00					6315
1033.5 +2714	PGC 31429		1.04		14.12				6660
103617.3-273146	NGC 3307	.SBROP?	.96± .04	28	14.49V±.13	.98± .03			
269.52 26.43	ESO 501- 31	PS	.43± .03	.18	15.28 ±.14	.35± .05			3897± 76
139.71 -37.65		.1± .6		.32		.83			3692
103356.1-271612	PGC 31430		.96		14.82	.26	14.06± .26		4226
1036.3 +5836	NGC 3286	.E?....	.90± .11						8128± 56
150.18 50.77	MCG 10-15-112		.19± .04	.00	14.61 ±.12				8201
51.52 -3.10				.00					8280
1033.1 +5852	PGC 31433		.84		14.49				
103621.5+134243		.S?....	1.15± .05	1				16.04±.3	3000± 9
229.17 55.77	UGC 5760		.60± .05	.07	14.4 ±.2				2997± 50
91.16 -27.89	IRAS10337+1358			.90	12.94			294± 5	2904
103341.8+135817	PGC 31435		1.15	.30	13.41			2.32	3333
103622.3-272616	NGC 3308	.LXS-*.	1.23± .04	32	12.94 ±.13	1.03± .01	1.04± .01		
269.48 26.52	ESO 501- 34	R	.12± .03	.27	13.29 ±.14	.59± .02	.61± .01		3610± 16
139.60 -37.64		-3.0± .4	1.00± .02	.00		.93	13.43± .06		3406
103401.1-271042	PGC 31438		1.25		12.78	.54	13.68± .25		3940
103622.4-252240		.SB?...	.97± .05						4080± 60
268.18 28.23	ESO 501- 32		.12± .04	.23	14.24 ±.14				3879
137.00 -37.64	IRAS10339-2506			.14	13.62				4414
103400.1-250706	PGC 31440		.99	.06	13.82				
103623.8+124224	NGC 3299	.SXS8..	1.34± .04	3				15.73±.1	641± 6
230.73 55.29	UGC 5761	R	.12± .04	.03	13.3 ±.2			131± 11	
92.20 -28.32		8.0± .4		.14				120± 8	541
103344.5+125758	PGC 31442		1.35	.06	13.07			2.59	977
103624.5-265958		.LBRO*.	1.18± .04	120					
269.21 26.89	ESO 501- 35	S	.46± .03	.27	14.21 ±.14				4158± 30
139.05 -37.63		-2.0± .8		.00					3954
103403.0-264424	PGC 31443		1.14		13.88				4489
103626.6+583327	NGC 3288	.SX.4*.	1.05± .07	175					
150.23 50.81	UGC 5752	PU	.12± .05	.00	14.79 ±.20				8164± 56
51.57 -3.12		3.7± .7		.17					8237
103312.9+584901	PGC 31446		1.05	.06	14.57				8316
103633.0-280353		.L?....	1.08± .05	154					
269.90 26.02	ESO 437- 8		.64± .03	.18	14.69 ±.14				4333± 76
140.39 -37.59				.00					4128
103412.0-274818	PGC 31456		1.00		14.45				4661
103633.1+382619		.L..-*.	1.08± .10	25					
182.42 59.69	UGC 5759	U	.04± .05	.00	14.22 ±.16				7712± 31
68.09 -15.07		-3.0±1.2		.00					7709
103340.0+384153	PGC 31457		1.08		14.10				7965
1036.5 -0115									
248.63 46.89	CGCG 9- 85			.13	15.1 ±.6				10984±108
107.49 -33.55									10839
1034.0 -0100	PGC 31458								11337
103634.2-443159		.SBS8P.	1.06± .05						
279.07 12.00	ESO 264- 11	S (1)	.10± .04	.62	15.00 ±.14				
160.62 -35.43		8.0± .9		.12					
103424.0-441624	PGC 31460	8.9± .8	1.12	.05					
103634.9-281259		.E?....	1.02± .06	15				16.95±.3	3460± 11
270.00 25.90	ESO 437- 9		.44± .03	.17	14.85 ±.14			154± 15	3811± 26
140.58 -37.57				.00				120± 12	3306
103414.0-275724	PGC 31462		.92		14.62				3839
103636.3-273105	NGC 3309	.E.3...	1.27± .05		12.60 ±.13	1.01± .01	1.04± .01		
269.57 26.48	ESO 501- 36	R	.06± .03	.18	12.20 ±.11	.62± .03	.63± .01		4100± 16
139.70 -37.58		-5.0± .4	.98± .02	.00		.93	12.99± .06		3895
103415.1-271530	PGC 31466		1.28		12.12	.60	13.78± .28		4429

R.A. 2000 DEC.	Names	Type	$logD_{25}$	p.a.	B_T	$(B-V)_T$	$(B-V)_\bullet$	m_{21}	V_{21}
l b		S_T n_L	$logR_{25}$	A_g	m_B	$(U-B)_T$	$(U-B)_\bullet$	W_{20}	V_{opt}
SGL SGB		T	$logA_\bullet$	A_i	m_{FIR}	$(B-V)_T^o$	m'	W_{50}	V_{GSR}
R.A. 1950 DEC.	PGC	L	$logD_o$	A_{21}	B_T^o	$(U-B)_T^o$	m'_{25}	HI	V_{3K}
103636.6-425111		RLXR+..	.87± .07	164	15.14 ±.14				4508± 87
278.20 13.45	ESO 264- 12	r	.03± .05	.56					4288
158.62 -35.80		-1.3± .9		.00					4791
103425.0-423536	PGC 31468		.92		14.51				
103638.5+141019	NGC 3300	.LXR0*$	1.28± .05	173	13.20 ±.11				3045± 40
228.50 56.05	UGC 5766	R	.28± .04	.08					2950
90.72 -27.62		-2.0± .4		.00					3377
103358.6+142553	PGC 31472		1.25		13.08				
103639.1-344529		.SBS9*/	1.17± .04	121	15.72 ±.14				
273.84 20.39	ESO 375- 72	S (1)	.62± .04	.32					
148.76 -37.11		9.0±1.2		.63					
103422.0-342954	PGC 31473	7.8± .9	1.20	.31					
103639.2+350309	IC 2591	.S?....	1.09± .04	13	14.41 ±.18			15.13±.3	6797± 10
189.10 60.23	UGC 5763		.24± .04	.04					6755± 50
71.03 -16.97	IRAS10337+3518			.37	13.52			351± 7	6779
103348.4+351843	PGC 31474		1.09	.12	13.96			1.05	7063
103641.3-273342	A 1034-27A	.L..-/.							4714± 50
269.62 26.46		RS		.18					4509
139.75 -37.56		-3.4± .6							5044
103420.0-271807	PGC 31476								
1036.7 +3132	A 1033+31	.IBS9*.	1.30± .04	60	15.21 ±.19	.53± .07	.48± .05	14.59±.1	586± 5
196.18 60.42	UGC 5764	R (1)	.24± .05	.04	14.4 ±.4	-.40± .08	-.31± .07	120± 5	
74.14 -18.90	DDO 83	10.0± .4	.95± .06	.18		.46	15.45± .20	101± 6	556
1033.9 +3148	PGC 31477	9.0±1.0	1.30	.12	14.83	-.45	15.93± .30	-.37	867
103643.3-273141	NGC 3311	.E+2...	1.54± .04		12.65 ±.13	1.00± .01	1.04± .01		3785± 20
269.60 26.49	ESO 501- 38	V	.08± .04	.18	11.93 ±.14	.57± .03	.61± .01		3581
139.71 -37.56		-4.0± .3	1.05± .02	.00		.92	13.39± .04		4115
103422.0-271606	PGC 31478		1.54		12.08	.55	15.14± .24		
103649.0-180623	A 1034-17	PSBR2P.	1.16± .03	141	14.28 ±.14				
263.36 34.22	ESO 568- 19	SE	.23± .03	.21					
127.86 -37.09	IRAS10343-1750	1.5± .6		.28	13.31				
103423.1-175048	PGC 31485		1.18	.11					
103650.1-275511		.L?....	1.05± .05	141	14.31 ±.13	1.03± .01	1.05± .01		4745± 76
269.87 26.17	ESO 437- 11		.27± .03	.18	14.33 ±.14	.56± .03	.60± .03		4540
140.20 -37.53			.44± .02	.00		.92	12.00± .09		5074
103429.0-273936	PGC 31488		1.02		14.07	.51	13.73± .31		
103652.7-322053		.SAS2*/	1.41± .03	86	13.48 ±.14				2840± 19
272.51 22.46	ESO 437- 14	S	.53± .03	.31					2629
145.75 -37.29	IRAS10345-3205	2.4± .5		.66	12.23				3157
103434.0-320518	PGC 31493		1.44	.27	12.49				
103653.6-270311		.SB?...	.89± .06	34	15.36 ±.14			16.41±.3	3533± 11
269.35 26.91	ESO 501- 41		.18± .04	.27				203± 15	3329
139.11 -37.53				.25				195± 12	3864
103432.1-264736	PGC 31494		.91	.09	14.82			1.50	
103654.1-275505		.L?....	1.00± .06	12	14.59 ±.13	1.02± .01			3610± 76
269.88 26.18	ESO 437- 13		.42± .03	.18	14.54 ±.14	.62± .04			3405
140.20 -37.51				.00		.91	13.40± .32		3939
103433.0-273930	PGC 31495		.96		14.34	.55			
103655.6+215258	NGC 3301	PSBT0..	1.55± .02	52	12.31 ±.13	.88± .01	.90± .01	18.05±.3	1321± 10
215.23 59.05	UGC 5767	R	.54± .03	.01	12.43 ±.11	.37± .04	.41± .04		1333± 56
83.12 -23.91	IRAS10341+2208	.0± .3	.83± .01	.40		.78	11.95± .05	298± 10	1255
103412.1+220833	PGC 31497		1.52		11.94	.29	13.56± .18		1635
103657.1-261141		.S?....	1.04± .05	126	14.53 ±.14				3280± 27
268.82 27.63	ESO 501- 42		.39± .04	.28					3078
138.03 -37.52	IRAS10345-2555			.48	12.94				3613
103435.1-255606	PGC 31500		1.07	.20	13.74				
103657.2-070128	IC 626	.L..-P.	1.15± .10	60					
254.40 42.83	MCG -1-27- 28	E	.00± .07	.12					
114.29 -35.06		-3.0± .8		.00					
103426.4-064553	PGC 31501		1.17						
103657.9+001349		.S?....	1.00± .06	88	14.58 ±.18				8846±108
247.12 47.98	UGC 5772		.24± .05	.16					8706
105.83 -32.98				.37					9198
103424.0+002924	PGC 31503		1.01	.12	14.01				
103658.0-281041		.LBS0*/	1.33± .04	29	13.55 ±.14				2746± 25
270.05 25.97	ESO 437- 15	S	.65± .03	.17					2540
140.53 -37.49		-1.7± .7		.00					3074
103437.1-275506	PGC 31504		1.25		13.34				

R.A. 2000 DEC. / l b / SGL SGB / R.A. 1950 DEC.	Names / PGC	Type S_T n_L / T / L	$\log D_{25}$ / $\log R_{25}$ / $\log A_o$ / $\log D_o$	p.a. / A_g / A_i / A_{21}	B_T / m_B / m_{FIR} / B_T^o	$(B-V)_T$ / $(U-B)_T$ / $(B-V)_T^o$ / $(U-B)_T^o$	$(B-V)_o$ / $(U-B)_o$ / m'_o / m'_{25}	m_{21} / W_{20} / W_{50} / HI	V_{21} / V_{opt} / V_{GSR} / V_{3K}
103659.6+180813	NGC 3303	.P.....	1.48± .04						6165± 17
221.99 57.81	UGC 5773	R	.15± .06	.03					6085
86.78 −25.72	ARP 192	99.0		.23					6489
103417.9+182348	PGC 31508		1.49	.08					
103702.4-273353	NGC 3312	.SAS3P$	1.52± .02	175				14.96±.3	2869± 11
269.69 26.50	ESO 501- 43	R	.42± .03	.27	12.72 ±.11			628± 15	2853± 18
139.76 −37.49	IRAS10346-2718	3.0± .4		.58	12.96			576± 12	2661
103441.0-271818	PGC 31513		1.55	.21	11.85			2.90	3195
103710.4+123913	NGC 3306	.SBS9$.	1.13± .04	141				15.01±.1	2887± 5
230.98 55.43	UGC 5774	R	.42± .04	.03	14.0 ±.2			292± 8	
92.35 −28.18	IRAS10345+1254	9.0±1.0		.43	11.99			270± 4	2787
103431.2+125448	PGC 31528		1.14	.21	13.49			1.31	3223
103712.8-264012		.L?....	.87± .06	95					4578± 76
269.17 27.26	ESO 501- 45		.18± .03	.28	15.16 ±.14				4375
138.63 −37.46	IRAS10348-2624			.00					4910
103451.0-262436	PGC 31530		.88		14.81				
103713.3-274106	NGC 3314A	.S..2*/	1.19± .07	143				15.60±.2	2872± 10
269.80 26.42		S	.35± .07	.18				210± 6	2916± 61
139.91 −37.44		1.5± .8		.43				184± 5	2669
103452.0-272530	PGC 31531		1.20	.18					3203
103713.3-274106	NGC 3314B	.SAS5*.	1.56?					15.26±.3	4426± 10
269.80 26.42	ESO 501- 46A	S	.08?	.18				182± 6	
139.91 −37.44		5.0± .8		.12				173± 5	4221
103452.0-272530	PGC 31532		1.58	.04					4755
103715.7-413742	NGC 3318	.SXT3..	1.38± .03	78	12.19 ±.14	.60± .01	.68± .01		2768± 79
277.67 14.57	ESO 317- 52	R (2)	.28± .03	.35	12.71 ±.11	−.02± .07	.04± .03		2548
157.12 −35.93	IRAS10350-4122	3.0± .4	1.08± .02	.38	11.23	.45	13.08± .05		3056
103503.0-412206	PGC 31533	4.2± .6	1.41	.14	11.75	−.13	13.23± .23		
103717.4-272806	A 1034-27B	.LB.0*.	1.00± .05	72	14.8 ±.4	.97± .04	.99± .04		4849± 28
269.68 26.61	ESO 501- 47	RS	.36± .03	.27	14.64 ±.14	.47± .08	.52± .08		4645
139.63 −37.43		−2.3± .5		.00		.83			5179
103456.0-271230	PGC 31537		.98		14.32	.40	13.77± .46		
103719.4+433516		.S..0..	1.19± .06	60					
172.72 58.40	UGC 5771	U	.15± .06	.00	14.3 ±.2				
63.82 −11.98		.0± .8		.11					
103422.7+435051	PGC 31539		1.18						
103719.6-271130	NGC 3315	.L..-?.	1.04± .03		14.42 ±.13	1.05± .02	1.06± .03		3840± 30
269.52 26.84	ESO 501- 48	S	.05± .03	.27	13.86 ±.14	.56± .04	.61± .04		3636
139.29 −37.43		−3.0±1.6	.62± .02	.00		.95	13.01± .08		4170
103458.0-265554	PGC 31540		1.07		13.83	.51	14.36± .21		
103720.4-273330		.L?....	.96± .06	170					4112± 27
269.75 26.54	ESO 501- 49		.51± .03	.27	15.20 ±.14				3908
139.75 −37.42				.00					4442
103459.1-271754	PGC 31542		.91		14.86				
1037.3 +3704		.S..6*.	1.12± .05	112					
185.03 60.10	UGC 5775	U	.84± .05	.00					
69.36 −15.71	IRAS10345+3720	6.0±1.4		1.24	12.85				
1034.5 +3720	PGC 31545		1.12	.42					
103725.6-251906	NGC 3313	PSBT2..	1.59± .03	55					3740± 60
268.37 28.41	ESO 501- 50	SUr (1)	.09± .04	.24	12.38 ±.14				3539
136.93 −37.40	IRAS10350-2503	2.4± .4		.11	12.44				4075
103503.0-250330	PGC 31551	3.3± .7	1.61	.04	12.00				
103729.1-261900		PSAT1*.	1.43± .03	117					3395± 26
269.01 27.59	ESO 501- 51	Sr	.24± .03	.28	13.02 ±.14				3193
138.19 −37.40		.5± .5		.24					3728
103507.1-260324	PGC 31557		1.46	.12	12.46				
1037.5 +0537		.S?....	1.02± .08	123				15.76±.3	8574± 12
240.88 51.57	UGC 5779		.75± .06	.07	15.4 ±.2			385± 14	8325± 57
99.86 −30.96				1.13				338± 15	8440
1034.9 +0553	PGC 31559		1.03	.38	14.11			1.27	8911
1037.5 +6849		.S..4..	1.24± .04	129					
138.99 43.89	UGC 5765	U	.86± .05	.05	15.09 ±.18				
43.48 3.22		4.0± .9		1.26					
1033.9 +6905	PGC 31560		1.25	.43					
103734.9-412754	NGC 3318B	.SBS5..	1.17± .04	110					2770± 60
277.64 14.74	ESO 317- 53	R (1)	.13± .04	.35	13.92 ±.14				2551
156.91 −35.90	IRAS10353-4112	5.0± .4		.20					3059
103522.0-411218	PGC 31565	5.6± .8	1.21	.07	13.35				

R.A. 2000 DEC. / l b / SGL SGB / R.A. 1950 DEC.	Names / PGC	Type / S_T n_L / T / L	$\log D_{25}$ / $\log R_{25}$ / $\log A_e$ / $\log D_o$	p.a. / A_g / A_i / A_{21}	B_T / m_B / m_{FIR} / B_T^o	$(B-V)_T$ / $(U-B)_T$ / $(B-V)_T^o$ / $(U-B)_T^o$	$(B-V)_e$ / $(U-B)_e$ / m'_e / m'_{25}	m_{21} / W_{20} / W_{50} / HI	V_{21} / V_{opt} / V_{GSR} / V_{3K}
1037.6 +0535 / 240.94 51.57 / 99.91 -30.95 / 1035.0 +0551	IC 628 / UGC 5780 / / PGC 31567	.SXS2.. / U / 2.0± .9 /	.95± .07 / .10± .05 / / .96	115 / .08 / .13 / .05	/ 14.57 ±.18 / /			15.88±.3 / 320± 14 / 274± 15 /	7170± 12 / 8718± 57 / /
103737.4-273536 / 269.83 26.54 / 139.79 -37.36 / 103516.1-272000	NGC 3316 / ESO 501- 54 / / PGC 31571	.LBT0*. / RS / -1.8± .3 /	1.10± .05 / .07± .03 / / 1.12	/ .27 / .00 /	/ 13.66 ±.14 / / 13.32			15.91±.3 / 235± 6 / 214± 5 /	4213± 10 / 3916± 25 / 3966 / 4500
103737.6+372722 / 184.27 60.08 / 69.06 -15.46 / 103445.5+374258	NGC 3304 / UGC 5777 / / PGC 31572	.SBS1$. / R / 1.0± .5 /	1.22± .04 / .46± .04 / / 1.22	158 / .00 / .46 / .23	/ 14.30 ±.18 / / 13.75				6896± 50 / 6889 / 7154 /
103738.1-261636 / 269.02 27.64 / 138.14 -37.36 / 103516.1-260100	/ ESO 501- 53 / IRAS10352-2600 / PGC 31574	.S?.... / / /	1.00± .05 / .56± .04 / / 1.03	130 / .28 / .84 / .28	/ 14.85 ±.14 / / 13.71				3814± 76 / 3612 / 4147 /
103740.6-270328 / 269.51 27.00 / 139.12 -37.35 / 103518.9-264751	/ / HICK 48C / PGC 31577	.S?.... / / /	/ / /	/ .27 / /	16.37S±.15 / / /				4203± 36 / 3999 / 4534 /
103742.9-283618 / 270.46 25.71 / 141.06 -37.32 / 103522.0-282042	/ ESO 437- 17 / / PGC 31583	.L?.... / / /	1.07± .06 / .62± .03 / / .99	138 / .16 / .00 /	/ 15.19 ±.14 / / 14.97				3502± 27 / 3296 / 3829 /
103744.9-263748 / 269.26 27.36 / 138.58 -37.34 / 103523.0-262212	/ ESO 501- 56 / / PGC 31585	.L..../ / S / -2.0±1.2 /	1.24± .04 / .57± .03 / / 1.18	72 / .28 / .00 /	/ 13.79 ±.14 / / 13.45				3456± 76 / 3253 / 3788 /
103747.7-270455 / 269.55 26.99 / 139.15 -37.32 / 103526.1-264918	IC 2597 / ESO 501- 58 / HICK 48A / PGC 31586	.E+4... / S / -4.0± .7 / 1.40	1.41± .04 / .15± .03 / .85± .04 /	4 / .27 / .00 /	12.84 ±.15 / 12.75 ±.14 / / 12.48	1.00± .01 / .60± .04 / .91 / .55	1.03± .01 / .61± .02 / 12.58± .15 / 14.47± .25		3007± 21 / 2803 / 3338 /
103749.2-240955 / 267.72 29.42 / 135.48 -37.28 / 103526.1-235418	/ ESO 501- 57 / / PGC 31587	PSBS1.. / r / 1.0± .9 /	.97± .06 / .17± .05 / / .99	167 / .21 / .17 / .08	/ 15.38 ±.14 / /				
103749.7-270719 / 269.58 26.96 / 139.20 -37.32 / 103528.0-265142	/ ESO 501- 59 / IRAS10354-2651 / PGC 31588	.S?.... / / /	.96± .05 / .06± .04 / / .99	97 / .27 / .09 / .03	14.97S±.15 / 14.21 ±.14 / 13.18 / 14.19		/ / 14.48± .32 /		2385± 36 / 2181 / 2716 /
103756.8-285431 / 270.69 25.48 / 141.44 -37.26 / 103536.0-283854	/ ESO 437- 19 / / PGC 31593	.SB?... / / /	1.05± .05 / .26± .04 / / 1.06	57 / .15 / .38 / .13	/ 14.22 ±.14 / / 13.66				4174± 76 / 3968 / 4501 /
1037.9 +1343 / 229.48 56.12 / 91.35 -27.53 / 1035.3 +1359	/ UGC 5781 / / PGC 31594	.I..9*. / U / 10.0±1.3 /	1.00± .08 / .33± .06 / / 1.01	110 / .06 / .25 / .17	/ / /				3027 / / 2931 / 3361
103801.5-810551 / 297.75 -19.60 / 198.39 -20.57 / 103755.0-805012	/ ESO 19- 3 / IRAS10379-8050 / PGC 31600	.SBT5*P / S (1) / 5.0±1.1 / 1.22	1.21± .05 / .24± .05 / /	137 / .09 / .36 / .12	/ 13.07 / /				
103802.4+641556 / 143.49 47.19 / 47.14 .50 / 103438.9+643132	A 1034+64 / UGC 5776 / MK 149 / PGC 31601	.S?.... / / / .66	.66± .13 / .00± .06 / /	/ .00 / .00 / .00	15.0 ±.2 / 14.8 ±.3 / / 14.91	.65± .03 / .23± .05 / .64 / .22	/ / 13.15± .71 / 1.87	16.78±.1 / 118± 7 / /	1700± 6 / 1619±101 / 1793 / 1819
103805.2+014441 / 245.71 49.21 / 104.23 -32.21 / 103530.7+020017	/ UGC 5783 / / PGC 31604	.SBT5?. / UE (1) / 4.5± .9 / 1.08	1.07± .05 / .44± .04 / /	54 / .07 / .66 / .22	/ 14.79 ±.18 / /				
103806.5+391036 / 180.86 59.82 / 67.64 -14.40 / 103513.3+392612	/ MCG 7-22- 35 / IRAS10352+3926 / PGC 31607	.S?.... / / / .91	.91± .06 / .27± .06 / /	/ .00 / .40 / .14	/ 14.94 ±.18 / / 14.50				9105± 13 / 9105 / 9355 /
103807.2+045336 / 241.95 51.24 / 100.73 -31.09 / 103531.3+050913	/ UGC 5784 / / PGC 31608	.S?.... / / / 1.09	1.08± .06 / .66± .05 / /	125 / .07 / .99 / .33	/ 15.03 ±.19 / / 13.94			15.31±.3 / 408± 7 / 1.04 /	6607± 10 / 6482 / 6955 /

R.A. 2000 DEC.	Names	Type	logD$_{25}$	p.a.	B$_T$	(B-V)$_T$	(B-V)$_e$	m$_{21}$	V$_{21}$
l　b		S$_T$　n$_L$	logR$_{25}$	A$_g$	m$_B$	(U-B)$_T$	(U-B)$_e$	W$_{20}$	V$_{opt}$
SGL　SGB		T	logA$_o$	A$_i$	m$_{FIR}$	(B-V)$_T^o$	m'$_e$	W$_{50}$	V$_{GSR}$
R.A. 1950 DEC.	PGC	L	logD$_o$	A$_{21}$	B$_T^o$	(U-B)$_T^o$	m'$_{25}$	HI	V$_{3K}$
103807.7-442713		.S?....	.94± .07						7250±190
279.27　12.21	ESO 264- 18		.06± .06	.59	14.82 ±.14				7029
160.44 -35.17	FAIR 433			.09					7528
103557.0-441136	PGC 31609		.99	.03	14.10				
103809.6-500931		.S..5*/	1.22± .06	160					
282.15　7.26	ESO 214- 16	S	.99± .05	1.25					
167.06 -33.69		5.0±1.3		1.48					
103604.0-495354	PGC 31613		1.34	.49					
103809.8-250619		.S?....	.90± .06	160					3597
268.39　28.68	ESO 501- 62		.15± .05	.24	15.28 ±.14				3397
136.67 -37.23				.20					3932
103547.0-245042	PGC 31614		.92	.07	14.81				
103810.9-284701		.L?....	1.09± .07						3956± 15
270.67　25.61	ESO 437- 21		.02± .05	.15	14.00 ±.14				3750
141.28 -37.21				.00					4283
103550.0-283124	PGC 31616		1.10		13.79				
1038.2　+7922		.S..1?.	1.32± .05	115					1957
130.11　35.83	UGC 5757	U	.40± .06	.03	14.6 ±.2				
35.02　9.57		1.0±1.6		.41					2100
1033.5　+7938	PGC 31621		1.32	.20	14.10				1986
103814.3-380601		.SBS6..	1.30± .04					14.15±.3	3050± 9
275.96　17.70	ESO 317- 54	S　(1)	.03± .05	.25	14.01 ±.14				
152.81 -36.36		6.1± .4		.04				81± 7	2833
103559.0-375024	PGC 31622	5.0± .4	1.32	.01	13.70			.43	3350
103817.8-285313		.SB.4*/	1.21± .04	165				15.40±.3	4356± 11
270.75　25.54	ESO 437- 22	S	.69± .03	.15	14.99 ±.14			293± 15	4277± 27
141.41 -37.18		4.0± .9		1.02				279± 12	4139
103557.0-283736	PGC 31626		1.22	.35	13.79			1.26	4672
1038.4　+3009		.S?....	.87± .10	5					6365± 10
199.03　60.75	UGC 5785		.20± .06	.00	15.42 ±.18				
75.60 -19.35				.30					6330
1035.6　+3025	PGC 31630		.87	.10	15.07				6652
103826.4-023411		.SB.4*.	1.29± .04	136				15.91±.1	8258± 9
250.47　46.31	UGC 5787	UE　(1)	.45± .04	.12	14.11 ±.20			415± 12	
109.18 -33.50		3.7± .7		.65					8110
103553.6-021834	PGC 31634	4.2± .8	1.30	.22	13.28			2.40	8612
1038.5　-0709	IC 630	.L...P?	1.30± .06	130					2159± 38
254.92　42.98	MCG -1-27- 29	E	.10± .08	.10					1998
114.57 -34.72	MK 1259	-2.0±1.6		.00					2513
1036.0　-0654	PGC 31636		1.30						
103833.5-274414		.SBS7*P	1.17± .04	94					4484± 47
270.11　26.54	ESO 501- 65	S	.17± .03	.28	13.73 ±.14				4280
139.97 -37.15	IRAS10361-2728	7.0±1.1		.23	12.69				4814
103612.0-272836	PGC 31638		1.20	.08	13.20				
103836.9+443116	A 1035+44		.63± .09						3688± 59
170.90　58.28	MCG 8-20- 2		.24± .06	.00	15.4 ±.3				3709
63.18 -11.24	MK 150								3914
103539.8+444653	PGC 31639		.63						
103840.1-283414		.S?....	1.08± .05	31					3459± 76
270.64　25.85	ESO 437- 25		.51± .04	.15	14.53 ±.14				3253
141.01 -37.11	IRAS10363-2818			.76	13.44				3787
103619.0-281836	PGC 31642		1.09	.25	13.60				
103840.4-461456		.SBR1..	1.01± .07						
280.27　10.70	ESO 264- 20	r	.16± .06	.80	15.22 ±.14				
162.52 -34.66		1.0± .9		.16					
103631.0-455918	PGC 31643		1.08	.08					
103842.9-284614		.L?....	1.02± .06	36					3867± 76
270.77　25.69	ESO 437- 27		.45± .03	.15	15.52 ±.14				3661
141.26 -37.09				.00					4194
103622.0-283036	PGC 31646		.96		15.31				
103846.1+533008	NGC 3310	.SXR4P.	1.49± .02		11.15M±.10	.35± .01	.32± .01	12.63±.1	980± 6
156.61　54.06	UGC 5786	R　(2)	.11± .03	.00	11.17 ±.16	-.43± .01	-.45± .02	204± 6	1018± 12
55.82 -5.91	ARP 217	4.0± .3	.61± .01	.16	9.48	.32	9.91± .02	174± 11	1043
103540.3+534545	PGC 31650	3.2± .7	1.49	.06	10.99	-.45	13.16± .15	1.58	1169
1038.8　+0541		.S..3..	1.03± .06	10					
241.11　51.87	UGC 5788	U　(1)	.20± .05	.05	14.40 ±.18				
99.93 -30.63		3.0± .9		.27					
1036.2　+0557	PGC 31651	4.5±1.2	1.03	.10					

R.A. 2000 DEC. l b SGL SGB R.A. 1950 DEC.	Names PGC	Type S_T n_L T L	$\log D_{25}$ $\log R_{25}$ $\log A_e$ $\log D_o$	p.a. A_g A_i A_{21}	B_T m_B m_{FIR} B_T^o	$(B-V)_T$ $(U-B)_T$ $(B-V)_T^o$ $(U-B)_T^o$	$(B-V)_e$ $(U-B)_e$ m'_e m'_{25}	m_{21} W_{20} W_{50} HI	V_{21} V_{opt} V_{GSR} V_{3K}
103850.7-113854 258.88 39.62 119.99 -35.62 103621.8-112316	NGC 3321 MCG -2-27- 10 IRAS10363-1123 PGC 31653	.SXR5*. UE (1) 5.3± .6 4.2±1.2	1.40± .04 .31± .05 1.41	30 .14 .46 .15	 13.66 			14.07±.1 270± 10 263± 12 	2487± 7 2315 2840
103855.1-263820 269.51 27.50 138.59 -37.08 103633.0-262242	 ESO 501- 66 PGC 31659	.L?.... 	1.07± .06 .52± .03 1.02	54 .29 .00 	 14.84 ±.14 14.50				 3142± 76 2939 3474
103857.1-763521 295.37 -15.70 194.24 -22.85 103800.0-761942	 ESO 38- 4 PGC 31661	.SBS7.. S (1) 6.5± .6 7.8± .7	1.26± .06 .50± .07 1.31	63 .61 .70 .25					
103904.0-265233 269.69 27.32 138.89 -37.04 103642.1-263654	 ESO 501- 67 IRAS10366-2636 PGC 31665	.SBT3.. S (1) 3.0± .8 3.3± .9	.85± .06 .23± .04 .87	127 .29 .32 .12	 15.36 ±.14 13.32 14.67				 10594± 27 10391 10926
103910.0+414118 175.98 59.34 65.62 -12.80 103615.2+415656	NGC 3319 UGC 5789 KUG 1036+419 PGC 31671	.SBT6.. R (2) 6.0± .3 3.8± .5	1.79± .01 .26± .02 1.63± .04 1.79	37 .00 .39 .13	11.48 ±.17 11.64 ±.16 11.18	.41± .06 .36 	.48± .03 -.08± .05 15.12± .10 14.64± .19	12.59±.0 223± 3 195± 4 1.28	742± 4 771± 10 755 985
103911.6-460827 280.30 10.84 162.37 -34.60 103702.0-455248	 ESO 264- 24 PGC 31672	.L..-*/ S -3.0±1.2 	1.26± .05 .52± .04 .41± .05 1.28	148 .80 .00 	14.24 ±.16 14.12 ±.14 	1.05± .01 .53± .02 	1.09± .01 .57± .02 11.78± .18 14.10± .32		
103911.7-002438 248.39 47.95 106.77 -32.66 103638.0-000900	IC 632 UGC 5792 PGC 31673	.S..1?. U 1.0±1.8 	.97± .07 .22± .05 .98	30 .13 .23 .11	 14.67 ±.18 14.24			14.96±.1 371± 12 .60	5620± 9 5697±108 5479 5974
103915.2-301757 271.78 24.46 143.16 -36.92 103655.0-300218	 ESO 437- 30 IRAS10368-3002 PGC 31677	.SBT4*/ Sr 3.9± .8 	1.49± .03 .72± .04 1.50	126 .15 1.05 .36	 13.69 ±.14 12.27 12.46				 3894± 18 3686 4217
103918.1-265021 269.72 27.38 138.84 -36.99 103656.1-263442	 ESO 501- 68 PGC 31683	.S?.... 	1.24± .04 .45± .04 1.26	13 .31 .66 .22	 14.29 ±.14 13.31			15.24±.2 321± 6 315± 5 1.71	3095± 10 2892 3427
103918.6-445957 279.74 11.84 161.03 -34.85 103708.0-444418	 ESO 264- 25 PGC 31685	.SXT5.. Sr (1) 5.2± .6 3.3± .9	1.20± .05 .35± .05 1.26	160 .67 .53 .18	 14.94 ±.14 13.71				 6320± 60 6099 6596
103920.3-001154 248.20 48.13 106.54 -32.56 103646.6+000345	NGC 3325 UGC 5795 PGC 31689	.E...*. UE -5.0± .7 	1.10± .09 .06± .05 1.10	55 .10 .00 	 13.67 ±.15 13.48				 5672±108 5531 6025
103922.6-293509 271.39 25.08 142.27 -36.92 103702.0-291930	 ESO 437- 31 PGC 31690	.SXT7.. Sr 7.5± .6 	1.12± .04 .23± .03 1.13	132 .12 .28 .12	 14.76 ±.14 14.35				3881 3675 4207
103924.4-002316 248.42 48.01 106.77 -32.60 103650.7-000737	IC 633 UGC 5796 ARAK 250 PGC 31691	.S?.... 	.76± .09 .34± .05 .77	102 .13 .52 .17	 15.2 ±.3 14.53				5547± 54 5405 5900
103925.5-275445 270.40 26.49 140.18 -36.95 103704.1-273906	A 1037-27 ESO 437- 32 PGC 31692	.SBT4P* S (1) 4.0± .6 3.0± .7	.95± .05 .08± .03 .97	55 .17 .12 .04	 14.69 ±.14 				
103925.5+014257 246.08 49.44 104.39 -31.91 103651.0+015836	 UGC 5797 PGC 31693	.I..9*. U 10.0±1.2 	1.02± .08 .05± .06 1.02	 .05 .04 .02	 14.28 ±.18 14.19				713 578 1065
103927.3+475650 164.96 56.99 60.42 -9.12 103627.4+481228	 UGC 5791 PGC 31697	.S?.... 	1.13± .07 .55± .06 1.13	43 .00 .68 .27	 14.47 ±.19 13.78				858± 10 872± 36 893 1068
103931.4+050632 242.03 51.65 100.65 -30.68 103655.5+052211	NGC 3326 UGC 5799 MK 1260 PGC 31701	.S..1.. U 1.0± .9 	.78± .09 .04± .05 .79	 .07 .04 .02	 14.6 ±.2 14.40				7859± 63 7735 8208

R.A. 2000 DEC. l b SGL SGB R.A. 1950 DEC.	Names PGC	Type S_T n_L T L	$\log D_{25}$ $\log R_{25}$ $\log A_\bullet$ $\log D_o$	p.a. A_g A_i A_{21}	B_T m_B m_{FIR} B_T^o	$(B-V)_T$ $(U-B)_T$ $(B-V)_T^o$ $(U-B)_T^o$	$(B-V)_\bullet$ $(U-B)_\bullet$ $m'_\bullet$ m'_{25}	m_{21} W_{20} W_{50} HI	V_{21} V_{opt} V_{GSR} V_{3K}
103934.5-235521 267.94 29.84 135.20 -36.88 103711.0-233942	NGC 3335 ESO 501- 71 PGC 31706	.LXR+*. Sr -1.2± .7 	1.06± .05 .12± .03 1.06	130 .20 .00 	 14.02 ±.14 13.76				 3730± 60 3532 4068
103936.7+472346 165.85 57.26 60.90 -9.42 103637.4+473925	NGC 3320 UGC 5794 IRAS10366+4739 PGC 31708	.S..6*. R (2) 6.0± .6 5.2± .7	1.34± .03 .35± .03 1.34	20 .00 13.08 ±.14 .52 12.34 .18 12.55				14.36±.1 304± 6 270± 11 1.64	2329± 9 2361 2541
103939.2+251919 208.89 60.47 80.21 -21.64 103654.3+253458	NGC 3323 UGC 5800 IRAS10368+2535 PGC 31712	.SB?... 	1.10± .05 .27± .05 1.10	 .01 14.28 ±.18 .41 12.61 .14 13.83					5164± 50 5111 5468
1039.6 +2644 206.04 60.71 78.89 -20.91 1036.9 +2700	IC 2598 CGCG 154- 37 PGC 31713			 .05 	15.0 ±.3				5830± 37 5782 6129
103943.6-462016 280.48 10.71 162.57 -34.46 103734.0-460436	 ESO 264- 26 PGC 31717	PSA.1.. r 1.0± .9 	.99± .06 .18± .05 1.07	109 .80 15.12 ±.14 .18 .09 14.06					6339± 87 6118 6610
103947.1+475553 164.94 57.04 60.47 -9.08 103647.4+481132	 UGC 5798 PGC 31720	.S?.... 	1.00± .06 .61± .05 1.00	45 .00 14.8 ±.3 .92 .31 13.82					1517± 10 1505± 40 1550 1726
103949.6-450710 279.88 11.78 161.15 -34.73 103739.0-445130	 ESO 264- 27 IRAS10376-4451 PGC 31722	.SBS4?. S (1) 4.0±1.9 5.6±1.4	1.10± .05 .60± .04 1.16	101 .67 15.44 ±.14 .88 12.91 .30					4104± 19 3889 4411
103950.0-360210 275.13 19.63 150.24 -36.32 103733.0-354630	NGC 3333 ESO 376- 2 IRAS10375-3546 PGC 31723	.SX.4P/ PBS 3.8± .5 	1.30± .03 .75± .05 .72± .02 1.33	160 13.93 ±.13 .25 13.91 ±.14 1.11 11.80 .38 12.54		.74± .02 .06± .04 .52 -.12	.82± .01 .12± .03 13.02± .04 13.41± .22		4104± 19 3889 4411
1039.8 +1536 226.82 57.40 89.67 -26.28 1037.2 +1552	 UGC 5802 PGC 31725	.S..6*. U 6.0±1.5 	.96± .09 .98± .06 .97	2 .08 1.44 .49					6660 6571 6990
1039.9 +1354 229.60 56.63 91.39 -27.02 1037.2 +1410	 PGC 31727			 .06 				17.95±.3 44± 13 27± 10	1010± 10 915 1344
103957.9+240531 211.35 60.29 81.40 -22.20 103713.7+242110	NGC 3327 UGC 5803 PGC 31729	.SAR3*. U (1) 3.0± .8 3.5±1.1	1.04± .06 .11± .05 1.04	85 .01 14.17 ±.18 .16 .06 13.96					6288± 50 6230 6595
103958.4-301140 271.87 24.63 143.02 -36.77 103738.0-295600	 ESO 437- 33 IRAS10376-2955 PGC 31730	PSXR1*. Sr 1.0± .6 	1.17± .04 .14± .03 1.19	4 .18 13.73 ±.14 .14 12.62 .07 13.36					3480± 60 3272 3804
104000.7+390720 180.80 60.19 67.90 -14.13 103708.0+392259	 MCG 7-22- 39 IRAS10371+3922 PGC 31733	.SB?... 	.96± .06 .17± .06 .96	 .00 14.67 ±.18 .25 13.23 .08 14.38					9268± 13 9268 9519
104004.4-301604 271.93 24.58 143.11 -36.74 103744.0-300024	 ESO 437- 35 IRAS10377-3000 PGC 31738	.SBS4*. S 4.0±1.3 	1.15± .04 .48± .03 1.17	16 .18 14.22 ±.14 .71 13.01 .24					
104005.8-460652 280.42 10.94 162.30 -34.45 103756.0-455112	 ESO 264- 28 PGC 31739	.LBR+.. r -1.3± .9 	1.09± .07 .25± .05 1.13	177 .81 14.87 ±.14 .00 					
104008.7-234916 268.00 30.00 135.08 -36.74 103745.1-233336	NGC 3331 ESO 501- 72 IRAS10377-2333 PGC 31743	.SBS5*. S (1) 5.0±1.1 3.3±1.1	1.07± .04 .10± .03 1.09	 .20 13.94 ±.14 .14 12.67 .05 13.57					3730± 60 3532 4068
104016.7-274634 270.49 26.71 140.01 -36.77 103755.0-273054	NGC 3336 ESO 437- 36 IRAS10379-2730 PGC 31754	.SXT5P. Sr (1) 4.9± .4 2.8± .6	1.29± .04 .10± .04 1.32	123 .28 13.01 ±.14 .15 12.38 .05 12.56				14.30±.3 320± 6 300± 5 1.69	4000± 10 3970± 43 3794 4328

R.A. 2000 DEC. l b SGL SGB R.A. 1950 DEC.	Names PGC	Type S_T n_L T L	logD_25 logR_25 logA_e logD_o	p.a. A_g A_i A_21	B_T m_B m_FIR B_T^o	(B-V)_T (U-B)_T (B-V)_T^o (U-B)_T^o	(B-V)_e (U-B)_e m'_e m'_25	m_21 W_20 W_50 HI	V_21 V_opt V_GSR V_3K
104016.7-461935 280.56 10.77 162.53 -34.37 103807.0-460354	 ESO 264- 30 PGC 31755	.L?.... 1.00	.92± .07 .12± .05 1.00	75 .87 .00 	14.61 ±.14 13.64				6164± 63 5943 6435
104017.8+382914 182.02 60.39 68.48 -14.44 103725.5+384454	 UGC 5804 PGC 31758	.SX.3.. U (1) 3.0± .8 3.5±1.1	1.17± .05 .17± .05 1.17	110 .00 .24 .09	14.8 ±.3				
104018.8-483411 281.67 8.82 165.12 -33.79 103811.0-481830	 ESO 214- 17 IRAS10381-4818 PGC 31760	.SBT7.. Sr (1) 6.9± .5 4.4±1.0	1.63± .04 .09± .07 1.73	60 1.03 .13 .05				12.46±.3 35± 7 	1052± 9 830 1313
104020.9-362447 275.43 19.36 150.68 -36.17 103804.1-360906	NGC 3347A ESO 376- 4 IRAS10380-3609 PGC 31761	.SBS6*/ PS (1) 5.8± .6 5.6± .6	1.30± .04 .43± .04 1.31	5 .21 .63 .21	13.47 ±.14 12.00 12.62				2899± 19 2684 3205
104024.3+213710 216.18 59.75 83.81 -23.34 103741.4+215250	 UGC 5805 PGC 31762	.SB.8.. U 8.0± .8 	1.14± .07 .18± .06 1.14	90 .01 .23 .09	14.6 ±.3 14.38			15.32±.3 136± 5 .85	1230± 7 1164 1545
104028.7+091100 236.83 54.27 96.37 -28.89 103751.1+092640	NGC 3332 UGC 5807 PGC 31768	RLA.-.. U -3.0± .8 	1.15± .15 .00± .08 1.16	 .08 .00 	13.34 ±.15 13.18				5727± 70 5616 6070
1040.5 +1217 232.26 55.96 93.12 -27.59 1037.8 +1233	 UGC 5808 IRAS10378+1233 PGC 31771	.SAS3.. U (1) 3.0± .9 3.5±1.1	1.08± .06 .05± .05 1.08	 .05 .06 .02	14.08 ±.19 13.96 				
104032.9-461129 280.53 10.91 162.36 -34.36 103823.0-455548	 ESO 264- 31 PGC 31777	.E?.... 1.17	1.08± .07 .17± .05 .88± .07 1.17	37 .86 .00 	13.9 ±.1 14.44 ±.14 13.16	1.07± .04 .81 	1.14± .02 13.82± .25 13.91± .37		6705± 37 6484 6976
104038.9-461859 280.61 10.81 162.50 -34.31 103829.0-460318	 ESO 264- 32 PGC 31781	PSBR1.. r 1.0± .8 	1.11± .05 .11± .05 1.19	 .87 .12 .06	14.68 ±.14 13.60				6938± 87 6717 7209
104038.9+371956 184.29 60.70 69.52 -15.03 103747.5+373536	 UGC 5806 KUG 1037+375 PGC 31782	.S..6*. U 6.0±1.2 	1.02± .05 .06± .04 1.02	 .00 .09 .03					
1040.7 +3948 179.39 60.17 67.40 -13.62 1037.9 +4004	 UGC 5810 PGC 31789	.S..3.. U (1) 3.0±1.0 5.5±1.3	1.00± .08 .84± .06 1.00	39 .00 1.16 .42					
1040.8 -0900 257.15 41.96 116.94 -34.60 1038.3 -0845	 PGC 31791	.S..5*/ E 5.0±1.3 	1.07± .09 .71± .08 1.08	56 .07 1.06 .35					
104050.7-275753 270.72 26.62 140.24 -36.64 103829.0-274212	 ESO 437- 38 PGC 31794	.L?.... 1.01	1.03± .06 .34± .03 1.01	84 .29 .00 	14.49 ±.14 14.14				4510± 76 4306 4840
104054.0-361718 275.47 19.53 150.51 -36.07 103837.0-360136	NGC 3347C ESO 376- 5 PGC 31797	.SBS7.. PS (1) 7.2± .7 5.6±1.1	1.18± .04 .10± .04 1.20	20 .21 .14 .05	14.84 ±.14				
104056.0+122829 232.07 56.14 92.99 -27.42 103817.1+124410	 UGC 5812 PGC 31801	.I..9.. U 10.0± .9 	1.02± .06 .35± .05 1.02	27 .04 .26 .17	15.10 ±.19 14.80			16.92±.2 68± 13 57± 5 1.95	1008± 6 909 1346
104059.1-270500 270.22 27.38 139.14 -36.61 103837.0-264918	 ESO 501- 75 IRAS10386-2649 PGC 31805	.SAS5*. RS (1) 5.0± .6 3.3± .6	1.34± .03 .27± .03 1.37	23 .31 .41 .14	13.50 ±.14 12.96 12.76				5190± 60 4987 5522
1041.0 -1730 263.92 35.30 127.27 -36.02 1038.6 -1714	 MCG -3-27- 24 IRAS10386-1714 PGC 31811	.S?.... 1.09	1.08± .09 .18± .07 1.09	 .15 .22 .09					6037 5852 6385

R.A. 2000 DEC. / I b / SGL SGB / R.A. 1950 DEC.	Names / PGC	Type / S_T n_L / T / L	$\log D_{25}$ / $\log R_{25}$ / $\log A_o$ / $\log D_o$	p.a. / A_g / A_i / A_{21}	B_T / m_B / m_{FIR} / B_T^o	$(B-V)_T$ / $(U-B)_T$ / $(B-V)_T^o$ / $(U-B)_T^o$	$(B-V)_o$ / $(U-B)_o$ / m'_o / m'_{25}	m_{21} / W_{20} / W_{50} / HI	V_{21} / V_{opt} / V_{GSR} / V_{3K}
104105.4-242948 / 268.64 29.56 / 135.93 -36.55 / 103842.0-241406	ESO 501- 76 / IRAS10386-2414 / PGC 31812	.SBR2*. / r / 2.2±1.3	.86± .06 / .09± .05 / / .88	63 / .21 / .11 / .05	/ 15.11 ±.14 / 13.48 /				
104108.8+362221 / 186.19 60.96 / 70.41 -15.48 / 103818.1+363802	UGC 5813 / / PGC 31817	.S..6*. / U / 6.0±1.2	1.19± .04 / .30± .04 / / 1.19	35 / .01 / .44 / .15	/ 14.9 ±.2 / / 14.40			15.57±.3 / 465± 13 / / 1.02	13277± 10 / / 13267 / 13540
104111.4-732207 / 293.84 -12.84 / 191.13 -24.29 / 103953.0-730624	ESO 38- 5 / / PGC 31820	.SXT3*. / r / 3.3±1.2	1.09± .09 / .14± .07 / / 1.14	32 / .55 / .20 / .07					
104111.6-370842 / 275.98 18.82 / 151.55 -35.91 / 103855.0-365300	ESO 376- 7 / / PGC 31821	PLX.0P* / S / -2.5± .6	1.19± .05 / .23± .04 / .49± .02 / 1.18	99 / .14 / .00 /	13.96 ±.13 / 13.80 ±.14 / / 13.68	.95± .01 / .49± .02 / .87 / .47	1.02± .01 / .50± .02 / 11.90± .07 / 14.24 ±.30		4302± 11 / 4086 / 4606
104120.7+062138 / 240.88 52.78 / 99.49 -29.80 / 103844.3+063720	UGC 5818 / / PGC 31834	.S..6*. / U / 6.0±1.2	1.07± .08 / .16± .06 / / 1.07	137 / .01 / .24 / .08	/ 14.6 ±.2 / / 14.36			15.79±.3 / / 268± 7 / 1.35	6255± 10 / 6188± 79 / 6134 / 6601
1041.4 +6942 / 137.82 43.47 / 42.98 4.02 / 1037.8 +6958	UGC 5809 / / PGC 31838	.S..6*. / U / 6.0±1.5	1.00± .08 / 1.02± .06 / / 1.00	93 / .03 / 1.47 / .50					
104126.0-232300 / 268.00 30.53 / 134.55 -36.43 / 103902.0-230718	A 1039-23 / ESO 501- 79 / DDO 85 / PGC 31840	.SXS9*. / SU (2) / 9.0± .4 / 9.8± .4	1.32± .03 / .15± .03 / 1.07± .03 / 1.34	48 / .20 / .16 / .08	14.02 ±.17 / 14.23 ±.14 / / 13.78	.48± .08 / -.09± .09 / .39 / -.15	.50± .04 / -.06± .05 / 14.86± .07 / 15.11 ±.25	14.15±.2 / 148± 16 / 132± 12 / .29	1199± 11 / 1003 / 1538
104127.8-314648 / 273.08 23.45 / 144.96 -36.37 / 103908.0-313106	ESO 437- 42 / / PGC 31841	.SBT5.. / S / 5.0± .8	1.25± .04 / .06± .05 / / 1.27	/ .28 / .09 / .03	/ 14.41 ±.14 / / 14.03				2610± 18 / 2401 / 2930
104131.4+371843 / 184.26 60.87 / 69.64 -14.90 / 103840.1+373425	NGC 3334 / UGC 5817 / / PGC 31845	.L...?. / U / -2.0±1.7	1.04± .09 / .04± .04 / .73± .05 / 1.03	/ .00 / .00 /	13.85 ±.16 / 13.99 ±.15 / / 13.82	1.01± .02 / .52± .03 / .94 / .56	1.02± .01 / .53± .03 / 12.99± .16 / 13.82 ±.50		7202± 31 / 7195 / 7461
104142.4-284648 / 271.39 26.03 / 141.25 -36.44 / 103921.0-283106	ESO 437- 44 / / PGC 31855	.SAS5.. / S / 5.0± .7	1.49± .04 / .04± .05 / / 1.51	/ .16 / .06 / .02	/ 14.11 ±.14 / / 13.86			13.94±.3 / 128± 15 / 97± 12 / .06	4396± 11 / 4409± 76 / 4191 / 4724
1041.7 +1538 / 227.14 57.82 / 89.86 -25.86 / 1039.1 +1554	IC 635 / UGC 5821 / IRAS10390+1554 / PGC 31858	.S?....	1.21± .06 / .67± .06 / / 1.21	5 / .05 / 1.00 / .33	/ 14.95 ±.18 / / 13.87				6600 / 6512 / 6931
104150.5+211842 / 216.97 59.97 / 84.29 -23.20 / 103907.8+213424	MCG 4-25- 40 / MK 725 / PGC 31862	.S?....	.89± .11 / .07± .07 / / .89	/ .01 / .10 / .03	/ 14.95 ±.18 / / 14.80				7538± 61 / 7471 / 7854
104150.6+384303 / 181.42 60.64 / 68.46 -14.06 / 103858.5+385845	UGC 5819 / IRAS10389+3859 / PGC 31863	.S..4.. / U / 4.0± .9	1.17± .05 / .64± .05 / / 1.17	120 / .00 / .94 / .32	/ / 13.66 /				10693± 50 / 10692 / 10946
104152.8+211508 / 217.09 59.96 / 84.35 -23.22 / 103910.2+213050	UGC 5822 / IRAS10391+2130 / PGC 31864	RSBR1.. / U / 1.0± .8	1.06± .06 / .07± .05 / / 1.06	/ .01 / .07 / .04	/ 14.7 ±.3 / / 14.55			15.82±.2 / 263± 7 / 248± 5 / 1.24	7447± 7 / 7379 / 7763
104153.2+004729 / 247.77 49.27 / 105.67 -31.63 / 103919.1+010312	UGC 5823 / VV 113 / / PGC 31865	.I..9*. / U / 10.0±1.3	1.00± .06 / .08± .05 / / 1.01	165 / .13 / .06 / .04	/ 14.47 ±.18 / / 14.28				5684±108 / 5547 / 6037
104159.4-284637 / 271.45 26.07 / 141.24 -36.37 / 103938.0-283054	ESO 437- 45 / / PGC 31874	.LA.0P. / S / -2.0± .8	1.14± .05 / .11± .04 / .54± .04 / 1.14	/ .16 / .00 /	14.33 ±.15 / 14.00 ±.14 / / 13.94	1.01± .02 / .44± .03 / .93 / .41	1.03± .02 / .48± .02 / 12.52± .12 / 14.62± .31		3786± 76 / 3581 / 4114
104159.9-365607 / 276.02 19.08 / 151.27 -35.78 / 103943.0-364024	NGC 3347B / ESO 376- 10 / IRAS10397-3640 / PGC 31875	.SAS8*/ / RCS (1) / 7.6± .4 / 6.1± .6	1.51± .03 / .63± .04 / / 1.52	95 / .09 / .78 / .32	13.71 ±.14 / 13.84 /				

R.A. 2000 DEC. / l b / SGL SGB / R.A. 1950 DEC.	Names / PGC	Type / S_T n_L / T / L	$\log D_{25}$ / $\log R_{25}$ / $\log A_e$ / $\log D_o$	p.a. / A_g / A_i / A_{21}	B_T / m_B / m_{FIR} / B_T^o	$(B-V)_T$ / $(U-B)_T$ / $(B-V)_T^o$ / $(U-B)_T^o$	$(B-V)_e$ / $(U-B)_e$ / m'_e / m'_{25}	m_{21} / W_{20} / W_{50} / HI	V_{21} / V_{opt} / V_{GSR} / V_{3K}
104202.0-331443 / 274.01 22.26 / 146.76 -36.14 / 103943.0-325900	ESO 376- 9 / PGC 31876	.LB.0?/ / S / -2.0±1.2	1.19± .05 / .56± .03 / / 1.14	127 / .29 / .00	12.83V±.13 / 13.67 ±.14 / / 13.40	.97± .02 / / .82	/ / / 13.22± .28		3055± 18 / 2844 / 3371
104206.5-403437 / 277.94 15.94 / 155.66 -35.22 / 103952.0-401854	ESO 318- 2 / PGC 31880	.SAS1*. / S / 1.0±1.0	1.14± .05 / .61± .04 / / 1.17	168 / .36 / .62 / .30	14.86 ±.14				
104207.5+134452 / 230.33 57.02 / 91.82 -26.61 / 103928.1+140035	NGC 3338 / UGC 5826 / PGC 31883	.SAS5.. / R (3) / 5.0± .3 / 2.3± .5	1.77± .02 / .21± .02 / 1.28± .02 / 1.78	100 / .06 / .32 / .11	11.64 ±.14 / 11.39 ±.16 / / 11.15	.59± .02 / -.01± .03 / .53 / -.05	.72± .01 / .10± .02 / 13.53± .05 / 14.82± .17	12.22±.1 / 349± 4 / 332± 4 / .96	1301± 4 / 1297± 58 / 1207 / 1636
104210.8-370925 / 276.17 18.91 / 151.53 -35.71 / 103954.0-365342	ESO 376- 11 / PGC 31885	.SBS9.. / S (1) / 8.5± .6 / 8.3± .6	1.15± .05 / .10± .05 / / 1.16	/ .09 / .10 / .05	14.75 ±.14				
104211.4+234500 / 212.26 60.71 / 82.00 -21.93 / 103927.6+240043	UGC 5825 / PGC 31887	.S..1?. / U / 1.0±1.8	1.06± .06 / .37± .05 / / 1.06	92 / .00 / .38 / .19	14.91 ±.18 / / 14.49			15.29±.3 / 305± 7 / .62	3485± 10 / 3427 / 3794
104214.2+474558 / 164.86 57.49 / 60.85 -8.85 / 103915.3+480140	A 1039+48 / MCG 8-20- 16 / MK 151 / PGC 31888	.S?....	.80± .09 / .39± .06 / / .78	165 / .00 / .29	15.3 ±.2 / / 14.99				1519± 60 / 1553 / 1729
104217.8-002242 / 249.16 48.54 / 107.03 -31.91 / 103944.1-000659	NGC 3340 / UGC 5827 / PGC 31892	.SBS4.. / E (1) / 4.0± .9 / 1.9± .8	1.00± .05 / .06± .04 / / 1.01	145 / .13 / .09 / .03	13.8 ±.2 / / 13.54			15.24±.1 / 292± 12 / 1.67	5566± 9 / 5646±108 / 5426 / 5921
104219.0-173857 / 264.32 35.36 / 127.49 -35.74 / 103952.3-172314	MCG -3-27- 26 / PGC 31895	.L...P/ / E / -1.5± .9	1.16± .05 / .58± .04 / / 1.09	65 / .13 / .00					
1042.3 +2800 / 203.60 61.46 / 78.05 -19.74 / 1039.6 +2816	MCG 5-25- 33 / PGC 31899	.L?....	1.02± .14 / .11± .07 / / 1.00	/ .03 / .00	14.84 ±.18 / / 14.64				11594 / 11552 / 11889
104222.4-361038 / 275.68 19.77 / 150.34 -35.79 / 104005.0-355454	ESO 376- 12 / PGC 31903	.SB?...	.97± .06 / .27± .05 / / .99	14 / .21 / .39 / .13	14.74 ±.14 / / 14.11				4634± 19 / 4420 / 4941
104225.5-445308 / 280.17 12.21 / 160.75 -34.33 / 104014.1-443724	ESO 264- 34 / IRAS10402-4437 / PGC 31906	.LXS0*. / S / -2.0±1.2	1.11± .05 / .33± .04 / / 1.14	106 / .65 / .00	15.01 ±.14				
1042.4 +1545 / 227.09 58.03 / 89.83 -25.65 / 1039.8 +1601	UGC 5828 / PGC 31908	.S?....	1.09± .06 / .84± .05 / / 1.10	28 / .05 / 1.26 / .42					14989± 57 / 14902 / 15320
1042.5 +1543 / 227.17 58.03 / 89.88 -25.65 / 1039.9 +1559	MCG 3-27- 75 / PGC 31913	.S?....	.94± .11 / .00± .07 / / .94	/ .05 / .00 / .00	14.69 ±.18 / / 14.56				14608± 57 / 14521 / 14939
1042.6 +2357 / 211.90 60.85 / 81.86 -21.74 / 1039.9 +2413	UGC 5830 / PGC 31917	.SXR3*. / U (1) / 3.0± .9 / 3.5±1.2	1.09± .06 / .34± .05 / / 1.09	30 / .01 / .47 / .17	14.83 ±.19				
104237.9-235608 / 268.62 30.22 / 135.25 -36.18 / 104014.1-234024	ESO 501- 80 / IRAS10402-2340 / PGC 31919	.S..5?/ / S / 5.0±1.6	1.37± .03 / .64± .03 / / 1.39	105 / .23 / .96 / .32	13.72 ±.14 / 13.76 / 12.52				1070± 60 / 873 / 1408
104244.4+342730 / 190.05 61.54 / 72.28 -16.26 / 103955.2+344313	A 1039+34 / UGC 5829 / DDO 84 / PGC 31923	.I..9.. / U (1) / 10.0± .6 / 9.0± .9	1.67± .02 / .05± .05 / 1.24± .04 / 1.67	/ .04 / .04 / .03	13.73 ±.18 / 14.0 ±.9 / / 13.67	.21± .06 / -.29± .08 / .18 / -.31	.29± .05 / -.19± .06 / 15.42± .08 / 16.78± .23	12.85±.1 / 93± 4 / 75± 4 / -.84	630± 4 / 599± 33 / 612 / 901
104246.4-362114 / 275.85 19.66 / 150.54 -35.69 / 104029.0-360530	NGC 3347 / ESO 376- 13 / PGC 31926	.SBT3.. / R (2) / 3.0± .3 / 1.7± .5	1.56± .02 / .24± .03 / .93± .02 / 1.58	173 / .19 / .33 / .12	12.17M±.08 / 12.25 ±.11 / / 11.66	.88± .01 / .30± .02 / .77 / .20	.90± .01 / .37± .01 / 12.42± .06 / 14.23± .15	13.32±.3 / 367± 9 / 1.55	3010± 8 / 2893± 66 / 2794 / 3315

R.A. 2000 DEC. / I b / SGL SGB / R.A. 1950 DEC.	Names / PGC	Type / S_T n_L / T / L	$\log D_{25}$ / $\log R_{25}$ / $\log A_o$ / $\log D_o$	p.a. / A_g / A_i / A_{21}	B_T / m_B / m_{FIR} / B_T^o	$(B-V)_T$ / $(U-B)_T$ / $(B-V)_T^o$ / $(U-B)_T^o$	$(B-V)_o$ / $(U-B)_o$ / m'_o / m'_{25}	m_{21} / W_{20} / W_{50} / HI	V_{21} / V_{opt} / V_{GSR} / V_{3K}
104248.7+132734		.SB?...	1.06± .06	95				15.57±.2	1216± 4
230.94 57.02	UGC 5832		.05± .05	.06	13.74 ±.18			138± 6	1178± 50
92.20 -26.59	ARP 291			.08	13.77			105± 4	1121
104009.6+134318	PGC 31930		1.06	.03	13.59			1.95	1552
104251.4-473656		.SBS7P*	1.19± .05						
281.58 9.86	ESO 264- 35	S (1)	.21± .05	.88	14.05 ±.14				770± 60
163.90 -33.63		7.0± .7		.29					549
104042.0-472112	PGC 31932 .	5.6± .9	1.28	.11	12.88				1036
104302.4-362151	NGC 3354	.S..*P.	.91± .03		13.72M±.12	.52± .01	.58± .01		
275.90 19.68	ESO 376- 14	S	.09± .03	.19	13.82 ±.14	-.09± .03	-.07± .03		2812± 19
150.54 -35.64	IRAS10407-3606		.55± .01	.13	12.88	.44	11.96± .02		2598
104045.0-360606	PGC 31941		.92	.04	13.43	-.15	12.89± .20		3119
104306.2+202509	A 1040+20	.L.....	1.18± .06	148	14.4 ±.4	.69± .05		16.99±.2	1323± 8
218.85 59.97	UGC 5833	U	.52± .03	.02	14.75 ±.16	-.09± .08		121± 6	1290± 42
85.31 -23.37	MK 416	-2.0± .9		.00		.63		82± 4	1251
104024.0+204053	PGC 31945		1.11		14.67	-.12	13.88± .51		1641
1043.1 +2139		.SXR4..	1.00± .06	170					
216.49 60.36	UGC 5836	U	.27± .05	.00	15.18 ±.19				
84.12 -22.77		4.0± .9		.39					
1040.4 +2155	PGC 31947	2.5±1.2	1.00	.13					
104308.9-300257		.SXS5..	1.20± .04					15.50±.3	3190± 10
272.43 25.12	ESO 437- 49	S (1)	.09± .03	.19	14.56 ±.14			154± 6	2880± 60
142.81 -36.09		5.0± .8		.14				141± 5	2974
104048.0-294712	PGC 31948	3.3± .9	1.21	.05	14.22			1.23	3506
104310.1+390218			.69± .15						6548± 19
180.66 60.81	MCG 7-22- 47		.05± .07	.00	15.2 ±.2				6548
68.33 -13.67									6548
104018.1+391802	PGC 31949		.69						6800
104311.9-261503		.SBT4*.	1.23± .04	66					
270.19 28.35	ESO 501- 82	S (1)	.37± .03	.30	14.21 ±.14				4540± 60
138.12 -36.11	IRAS10407-2558	4.0± .8		.54	13.35				4339
104049.1-255918	PGC 31951	3.3±1.2	1.26	.18	13.34				4874
104313.1-535918		.S..5?.	1.50± .05						
284.70 4.28		S	.17± .08	1.98					
171.07 -31.79		5.0±1.4		.25					
104109.8-534333	PGC 31952		1.69	.08					
104313.7-461245		.SB?...	1.03± .06	102					
280.95 11.12	ESO 264- 36		.18± .05	.91	14.34 ±.14				7000±190
162.26 -33.90	FAIR 436			.18	11.12				6780
104103.1-455700	PGC 31954		1.12	.09	13.16				7272
104325.1+404629		.SB.3..	1.06± .06	130					
177.23 60.38	UGC 5838	U	.35± .05	.00	14.1 ±.2				8993
66.87 -12.66	IRAS10405+4102	3.0± .9		.49	13.28				9000
104032.0+410213	PGC 31959		1.06	.18	13.58				9237
104328.1-255157		.LAS0..	1.15± .04	155					
270.01 28.70	ESO 501- 84	S	.11± .03	.30	14.12 ±.14				4560± 60
137.64 -36.05		-2.0± .8		.00					4360
104105.0-253612	PGC 31962		1.17		13.75				4895
104329.6+394116		.S..0..	1.04± .06	147					
179.34 60.70	UGC 5839	U	.48± .05	.00	14.84 ±.18				
67.81 -13.25		.0± .9		.36					
104037.2+395700	PGC 31965		1.01						
104330.6-304621		.SB?...	1.01± .05	173					
272.91 24.54	ESO 437- 50		.18± .04	.25	14.25 ±.14				3790± 60
143.69 -35.98	IRAS10411-3030			.27					3583
104110.0-303036	PGC 31966		1.03	.09	13.71				4113
104330.8+245525	NGC 3344	RSXR4..	1.85± .01		10.45 ±.13	.59± .02	.67± .01	11.84±.0	586± 4
210.04 61.25	UGC 5840	R (2)	.04± .02	.03	10.69 ±.16	-.07± .03	.03± .02	166± 5	575± 13
81.06 -21.09	IRAS10407+2511	4.0± .3	1.47± .02	.05	10.97	.57	13.29± .04	155± 7	531
104046.6+251110	PGC 31968	1.9± .5	1.85	.02	10.46	-.08	14.45± .16	1.36	891
1043.5 +1206									
233.24 56.49	CGCG 66- 1			.02	15.0 ±.6				7922± 57
93.68 -27.00									7822
1040.9 +1222	PGC 31971								8261
104333.5-362439	NGC 3358	RSXS0..	1.52± .02	141	12.29M±.08	.90± .03	.98± .01	13.33±.3	2988± 6
276.02 19.69	ESO 376- 17	R (1)	.25± .03	.18	12.51 ±.11	.37± .07	.47± .02	367± 9	2880± 66
150.59 -35.53	IRAS10412-3609	.0± .3	1.00± .02	.18		.79	12.91± .06		2773
104116.1-360854	PGC 31974	1.1± .8	1.52		11.96	.31	14.11± .15		3294

R.A. 2000 DEC. l b SGL SGB R.A. 1950 DEC.	Names PGC	Type S_T n_L T L	$\log D_{25}$ $\log R_{25}$ $\log A_e$ $\log D_o$	p.a. A_g A_i A_{21}	B_T m_B m_{FIR} B_T^0	$(B-V)_T$ $(U-B)_T$ $(B-V)_T^0$ $(U-B)_T^0$	$(B-V)_e$ $(U-B)_e$ m'_e m'_{25}	m_{21} W_{20} W_{50} HI	V_{21} V_{opt} V_{GSR} V_{3K}
104336.3-095124 258.58 41.75 118.13 -34.12 104106.4-093539	 MCG -2-28- 1 PGC 31979	.SBS7P* EU (1) 7.4± .4 5.3± .4	1.30± .03 .05± .03 1.31	170 .06 .07 .02				14.35±.1 165± 10 139± 12	2079± 7 1912 2434
1043.6 +1205 233.29 56.50 93.71 -26.99 1041.0 +1221	 CGCG 66- 2 PGC 31980			.04	15.2 ±.6				7893± 57 7793 8232
104338.7+145218 228.82 57.88 90.86 -25.79 104059.0+150803	NGC 3346 UGC 5842 IRAS10410+1507 PGC 31982	.SBT6.. R (2) 6.0± .3 3.5± .6	1.46± .02 .06± .03 1.46	 .07 .09 .03	 12.28 ±.15 12.36 12.12			14.31±.0 166± 5 159± 7 2.16	1260± 4 1110± 50 1169 1592
104347.8-242204 269.15 30.00 135.79 -35.93 104124.0-240618	 ESO 501- 86 IRAS10413-2406 PGC 31987	.SXS4.. S (1) 4.0± .8 3.3± .8	1.27± .03 .33± .03 1.28	19 .18 .48 .16	 13.87 ±.14 13.28				
104348.1+155329 227.13 58.37 89.86 -25.31 104108.0+160914	IC 638 CGCG 94-117 MK 632 PGC 31988			.07	15.3 ±.3 13.22				11937± 39 11851 12268
104348.9-023139 251.83 47.28 109.61 -32.20 104116.1-021554	 UGC 5847 PGC 31993	PSBS3*. U 3.0±1.3	1.06± .06 .27± .05 1.07	5 .16 .38 .14	 14.9 ±.2				
104350.4-381546 277.05 18.11 152.82 -35.24 104134.0-380000	 ESO 318- 4 IRAS10415-3800 PGC 31995	.SAS5?/ S (1) 5.0± .7 4.4±1.2	1.43± .03 .76± .04 1.44	61 .16 1.14 .38	 13.56 ±.14 12.57 12.24				3077± 19 2861 3378
1043.8 +2809 203.35 61.81 78.10 -19.38 1041.1 +2825	 UGC 5844 PGC 31996	.S..7.. U 7.0± .9	1.14± .05 .77± .05 1.15	121 .03 1.06 .38					
1043.8 +6942 137.59 43.61 43.10 4.19 1040.3 +6958	 UGC 5834 PGC 31997	.SB.3.. U 3.0± .9	1.08± .06 .75± .05 1.09	49 .03 1.04 .38					
104352.5-011736 250.56 48.17 108.21 -31.82 104119.2-010151	 UGC 5849 MK 1261 PGC 31998	.I?.... 	.97± .09 .16± .06 .99	 .16 .12 .08	 14.23 ±.14 13.95			15.08±.1 738± 21 1.05	7808± 12 7918± 54 7671 8168
1043.9 +8054 128.76 34.73 33.90 10.66 1039.0 +8110	 UGC 5820 PGC 32004	.S..4.. U 4.0± .8	1.13± .04 .45± .04 1.13	75 .06 .67 .23	 15.05 ±.18				
104358.0+114215 233.95 56.37 94.14 -27.08 104119.7+115800	NGC 3351 UGC 5850 IRAS10413+1158 PGC 32007	.SBR3.. R (2) 3.0± .3 3.3± .5	1.87± .01 .17± .02 1.33± .01 1.88	13 .05 .24 .09	10.53M±.10 10.65 ±.14 9.99 10.28	.80± .01 .18± .02 .75 .14	.85± .01 .24± .01 12.68± .04 14.30± .13	12.96±.1 277± 4 268± 5 2.60	778± 4 771± 24 677 1117
104407.1-162812 263.90 36.57 126.11 -35.16 104139.8-161226	 MCG -3-28- 1 VV 410 PGC 32017	.SXT5P? E (1) 5.0± .9 .8±1.6	1.03± .07 .11± .05 1.04	15 .10 .17 .06	 13.04				
104409.5-204810 266.93 33.02 131.41 -35.63 104144.0-203224	 ESO 569- 1 PGC 32018	.E?.... 	.79± .07 .19± .05 .76	83 .14 .00	15.20 ±.13 14.63 ±.14 14.74	.98± .03 .91	 13.68± .38		3654± 37 3463 3998
104412.4+064532 241.08 53.59 99.37 -28.99 104136.0+070118	NGC 3356 UGC 5852 VV 529 PGC 32021	.S..4.. U (1) 4.0± .8 1.5±1.1	1.23± .05 .32± .05 .70± .10 1.23	102 .06 .48 .16	13.8 ±.2 13.40 ±.19 13.02	.50± .06 -.10± .11 .39 -.18	.52± .03 -.03± .06 12.83± .34 13.98± .32	14.74±.1 357± 5 329± 6 1.56	6175± 5 5800± 67 6055 6520
104415.6+222217 215.24 60.82 83.57 -22.19 104132.8+223803	NGC 3352 UGC 5851 PGC 32025	.L..... U -2.0± .8	1.21± .08 .14± .05 .75± .03 1.19	0 .00 .00	13.53 ±.14 13.70 ±.16 13.52	.98± .02 .48± .03 .91 .50	.99± .01 .52± .02 12.77± .12 14.10± .47		5744± 31 5681 6058
104416.2-111435 259.91 40.77 119.82 -34.24 104146.8-105849	NGC 3360 MCG -2-28- 3 PGC 32026	.SAS5*. E (1) 5.0± .9 5.3±1.2	1.07± .03 .14± .05 .66± .01 1.08	55 .11 .20 .07	14.41 ±.13 14.05	.68± .02 .57	.75± .02 13.20± .04 14.28± .25		8441± 62 8271 8796

R.A. 2000 DEC.	Names	Type	$\log D_{25}$	p.a.	B_T	$(B-V)_T$	$(B-V)_\bullet$	m_{21}	V_{21}
l b	S_T n_L	$\log R_{25}$	A_g	m_B	$(U-B)_T$	$(U-B)_\bullet$	W_{20}	V_{opt}	
SGL SGB	T	$\log A_\bullet$	A_i	m_{FIR}	$(B-V)_T^o$	$m'_\bullet$	W_{50}	V_{GSR}	
R.A. 1950 DEC.	PGC	L	$\log D_o$	A_{21}	B_T^o	$(U-B)_T^o$	m'_{25}	HI	V_{3K}

104418.6-224934		PSXT2*.	1.19± .04	78					
268.29 31.36	ESO 501- 88	Sr (1)	.15± .03	.21	13.96 ±.14				
133.90 -35.74		2.0± .6		.19					
104154.0-223348	PGC 32030	4.4±1.1	1.21	.08					

104420.8+140507	NGC 3357	.E+..*.	1.16± .06	90					
230.27 57.65	UGC 5206	PU	.03± .04	.07	13.65 ±.16				9809± 50
91.74 -25.98		-4.3± .5		.00					9716
104141.4+142053	PGC 32032		1.16		13.43				10144

1044.3 +5246		.E?....	.60± .15						
156.83 55.17	MCG 9-18- 14		.09± .10	.00	14.9 ±.3				7568±125
56.92 -5.67				.00					7621
1041.3 +5302	PGC 32033		.57		14.77				7753

1044.3 +3043	NGC 3350								
197.92 62.06	CGCG 155- 2			.00	15.3 ±.3				10395
75.82 -17.95									10364
1041.6 +3059	PGC 32035								10681

104424.0-403828		.SBT1..	.87± .07	25					
278.37 16.09	ESO 318- 7	r	.17± .05	.38	15.07 ±.14				
155.65 -34.78		1.0± .9		.17					
104209.0-402242	PGC 32038		.91	.08					

104424.0-321235		.SBT4..	1.16± .04	173					
273.90 23.40	ESO 437- 56	S (1)	.13± .03	.31	14.09 ±.14				2850± 19
145.45 -35.72		4.0± .8		.19					2641
104204.0-315648	PGC 32039	4.4± .8	1.19	.06	13.57				3169

1044.4 +5625		.S..9*.	1.32± .05	115				15.18±.1	822± 7
151.84 52.98	UGC 5848	U	.33± .06	.00	14.8 ±.3			145± 6	826± 67
53.93 -3.54		9.0±1.1		.33				130± 4	889
1041.3 +5641	PGC 32041		1.32	.16	14.47			.55	987

104426.4-252234		.E?....	1.02± .06	178	13.67 ±.15	.88± .02	.94± .02		
269.92 29.23	ESO 501- 89		.12± .03	.29	14.25 ±.14				
137.04 -35.82			.93± .03	.00			13.81± .07		
104203.0-250648	PGC 32042		1.03				13.48± .34		

104429.1-111228	NGC 3361	.SXT5?.	1.30± .05	155	13.44 ±.13	.60± .02		14.66±.1	1926± 9
259.93 40.83	MCG -2-28- 4	E (1)	.23± .05	.11				279± 12	1924± 62
119.79 -34.18	IRAS10419-1056	5.0± .8		.34	12.88	.52			1757
104159.6-105642	PGC 32044	3.1±1.2	1.31	.11	12.98		14.24± .29	1.57	2281

1044.5 +6022	A 1041+60	.I..9..	1.22± .05		15.2 ±.2	.22± .16	.18± .07	14.32±.1	1019± 6
147.02 50.41	UGC 5846	U (1)	.21± .05	.00	14.9 ±.3	-.09± .19	.00± .10	63± 7	1100
50.72 -1.22	DDO 86	10.0± .8	1.05± .05	.16		.16	15.95± .10	44± 12	1163
1041.2 +6037	PGC 32048	9.0±1.0	1.22	.10	14.97	-.13	15.67± .33	-.75	

1044.6 +2610		.S?....	1.16± .05	97					6258
207.54 61.72	UGC 5855		.63± .05	.05	14.91 ±.18				
80.02 -20.24	IRAS10418+2626			.94	12.72				6210
1041.8 +2626	PGC 32055		1.17	.31	13.89				6560

104439.1+384534		.S?....	.94± .11						10723± 13
181.07 61.16	MCG 7-22- 53		.05± .07	.00	14.85 ±.18				10723
68.74 -13.58				.07					10976
104147.7+390120	PGC 32058		.94	.02	14.71				

104440.3+764838	NGC 3329	RSAR3*.	1.25± .03	140					
131.69 38.08	UGC 5837	R	.25± .03	.04	13.10 ±.14				1812± 49
37.32 8.36	IRAS10405+7704	3.0± .6		.34	13.07				1948
104031.1+770423	PGC 32059		1.26	.12	12.71				1857

1044.7 +5827		.S..6*.	1.10± .06	38					
149.25 51.71	UGC 5853	U	.52± .05	.00	15.35 ±.18				
52.30 -2.31		6.0±1.4		.76					
1041.6 +5843	PGC 32069		1.10	.26					

104448.9+381049		.S?....	.91± .07	97					
182.22 61.33	UGC 5856		.64± .05	.00	15.3 ±.2				7759
69.26 -13.87	IRAS10419+3826			.95	13.61				7757
104157.8+382636	PGC 32071		.91	.32	14.32				8015

104449.3+155706		.S..8..	1.13± .05	160				15.92±.3	6481± 10
227.23 58.62	UGC 5858	U	.06± .05	.07	14.6 ±.3			93± 13	
89.92 -25.06		8.0± .8		.08					6395
104209.3+161253	PGC 32072		1.13	.03	14.43			1.46	6812

1044.8 +2728									
204.85 61.95	CGCG 154- 45			.02	15.3 ±.3				13292
78.85 -19.54									13249
1042.1 +2744	PGC 32077								13590

R.A. 2000 DEC.	Names	Type	$\log D_{25}$	p.a.	B_T	$(B-V)_T$	$(B-V)_e$	m_{21}	V_{21}
l b		S_T n_L	$\log R_{25}$	A_g	m_B	$(U-B)_T$	$(U-B)_e$	W_{20}	V_{opt}
SGL SGB		T	$\log A_e$	A_i	m_{FIR}	$(B-V)_T^o$	m'_e	W_{50}	V_{GSR}
R.A. 1950 DEC.	PGC	L	$\log D_0$	A_{21}	B_T^o	$(U-B)_T^o$	m'_{25}	HI	V_{3K}

104451.6+063541	NGC 3362	.SX.5..	1.15± .05	90				15.91±.3	8318± 10
241.47 53.61	UGC 5857	U (1)	.11± .05	.05	13.48 ±.18				8322± 48
99.62 -28.90		5.0± .8		.16				187± 7	8200
104215.2+065128	PGC 32078	2.5±1.1	1.15	.05	13.23			2.63	8666
1045.0 +7640		.SXS5..	1.19± .05	130				14.63±.3	1766± 10
131.77 38.19	UGC 5841	U	.23± .05	.04	14.8 ±.2			93± 13	
37.44 8.30	IRAS10408+7656	5.0± .8		.34					1902
1040.8 +7656	PGC 32081		1.19	.11	14.37			.15	1812
104502.8+434216		.SB.1..	1.13± .05	25					
171.51 59.66	UGC 5859	U	.24± .05	.00	14.65 ±.20				
64.55 -10.76		1.0± .8		.24					
104208.0+435803	PGC 32084		1.13	.12					
104504.0+002601									7922±108
249.01 49.60	CGCG 10- 4			.06	14.9 ±.6				7785
106.38 -31.00									8276
104230.0+004148	PGC 32085								
104509.3+045633		.S..6?.	.98± .05						22837
243.69 52.63	UGC 5865	U	.03± .04	.04	14.9 ±.2				22714
101.43 -29.44		6.0±1.7		.05					23187
104233.6+051220	PGC 32086		.98	.02	14.67				
1045.1 +3859		.S..6*.	1.18± .05	151					
180.56 61.20	UGC 5861	U	.97± .05	.00					
68.60 -13.37		6.0±1.4		1.42					
1042.3 +3915	PGC 32087		1.18	.48					
1045.1 +0007		.SB.3..	1.00± .06						11780±108
249.38 49.40	UGC 5867	U	.00± .05	.07	14.42 ±.20				11642
106.74 -31.07		3.0± .9		.00					12134
1042.6 +0023	PGC 32088		1.00	.00	14.27				
104510.3+220453	NGC 3363	.S?....	1.12± .05	0				15.40±.3	5766± 10
215.94 60.94	UGC 5866		.20± .05	.00	14.34 ±.19				5702
83.96 -22.15	IRAS10424+2220			.30	12.96			401± 7	6081
104227.7+222040	PGC 32089		1.12	.10	14.01			1.29	
104510.8-100353		PSXT7..	1.17± .05						2523
259.17 41.83	MCG -2-28- 6	E (1)	.00± .05	.05					
118.48 -33.78		7.0± .8		.00					2357
104240.9-094806	PGC 32091	6.4± .8	1.17	.00					2879
104522.6+555733	NGC 3353	.S..3$P	1.13± .03		13.25 ±.13	.46± .01	.41± .01	14.75±.1	944± 5
152.31 53.38	UGC 5860	P	.15± .04	.00	13.26 ±.15	-.35± .02	-.37± .02	120± 4	935± 36
54.39 -3.70	MK 35	3.0±1.7	.51± .02	.20	11.49	.43	11.29± .05	96± 5	1009
104216.5+561320	PGC 32103		1.13	.07	13.04	-.37	13.39± .22	1.63	1112
1045.4 +1326								17.73±.3	3142± 10
231.55 57.57				.04				95± 13	
92.52 -26.02								83± 10	3047
1042.7 +1342	PGC 32107								3478
104543.1+112042		.SXT3*.	1.15± .05	95				15.38±.2	6571± 6
234.92 56.54	UGC 5869	U (1)	.23± .05	.05	14.3 ±.2			289± 10	
94.71 -26.84	MK 1262	3.0±1.2		.32				276± 5	6469
104305.0+113630	PGC 32119	5.5±1.2	1.15	.11	13.87			1.39	6911
1045.7 +7706		.S..8*.	.99± .06						
131.41 37.88	UGC 5854	U	.05± .05	.03	15.0 ±.2				
37.11 8.57		8.0±1.2		.06					
1041.6 +7722	PGC 32121		1.00	.03					
104547.2+371242		.S..3.	1.08± .06						
184.12 61.72	UGC 5868	U (1)	.04± .05	.00	14.7 ±.2				
70.21 -14.24		3.0± .8		.05					
104256.9+372830	PGC 32123	3.5±1.1	1.08	.02					
104549.8+273713		.L?....	.90± .17						13298± 61
204.58 62.19	MCG 5-26- 3		.00± .07	.02	14.75 ±.15				13255
78.83 -19.28	MK 726			.00	12.59				13595
104304.6+275301	PGC 32127		.90		14.53				
104558.7+345753		.L...?.	1.04± .09						2032± 50
188.82 62.14	UGC 5870	U	.00± .04	.00	14.12 ±.15				2017
72.21 -15.42		-2.0±1.6		.00					2302
104309.7+351341	PGC 32134		1.04		14.09				
1046.0 +7321	NGC 3343	.E.....	1.11± .16	55					
134.30 40.89	UGC 5863	U	.15± .08	.17	14.40 ±.16				
40.22 6.45		-5.0± .8		.00					
1042.3 +7337	PGC 32143		1.09						

R.A. 2000 DEC. l b SGL SGB R.A. 1950 DEC.	Names PGC	Type S_T n_L T L	$\log D_{25}$ $\log R_{25}$ $\log A_\bullet$ $\log D_o$	p.a. A_g A_i A_{21}	B_T m_B m_{FIR} B_T^o	$(B-V)_T$ $(U-B)_T$ $(B-V)_T^o$ $(U-B)_T^o$	$(B-V)_\bullet$ $(U-B)_\bullet$ $m'_\bullet$ m'_{25}	m_{21} W_{20} W_{50} HI	V_{21} V_{opt} V_{GSR} V_{3K}
104605.2-451937 280.97 12.13 161.10 -33.61 104353.0-450348	 ESO 264- 39 PGC 32144	PSBT1.. Sr .5± .6 	1.19± .05 .06± .05 1.25	 .72 .06 .03	 14.34 ±.14 13.49				5520± 60 5300 5796
104609.7+493238 161.41 57.17 59.75 -7.29 104310.5+494826	 UGC 5872 IRAS10431+4948 PGC 32149	.S..4.. U 4.0± .9 	1.03± .08 .32± .06 1.03	74 .00 .47 .16	 14.97 ±.18 				
104613.1+014846 247.75 50.76 104.96 -30.27 104338.6+020435	NGC 3365 UGC 5878 PGC 32153	.S..6*/ UE (1) 6.0± .8 5.3±1.2	1.65± .02 .76± .04 1.22± .04 1.66	159 .08 1.12 .38	13.17 ±.15 13.04 ±.20 11.92	.61± .04 -.01± .05 .44 -.13	.67± .03 .04± .04 14.76± .11 14.37± .21	13.10±.1 244± 16 231± 12 .80	986± 11 853 1333 1339
104634.5+134509 231.31 57.96 92.34 -25.64 104355.5+140058	NGC 3367 UGC 5880 IRAS10439+1400 PGC 32178	.SBT5.. R (2) 5.0± .3 2.4± .6	1.40± .02 .06± .03 1.06± .02 1.41	 .05 .09 .03	12.05 ±.14 12.07 ±.14 11.12 11.90	.55± .02 -.16± .04 .51 -.19	.62± .02 12.84± .04 13.76± .18	14.46±.1 247± 6 222± 5 2.53	3042± 4 2879± 56 2948 3377
1046.6 +5209 157.38 55.81 57.63 -5.74 1043.6 +5225	 UGC 5876 PGC 32182	.S..6*. U 6.0±1.2 	1.13± .05 .25± .05 1.13	103 .00 .36 .12	 15.3 ±.3 				
104637.7+631322 143.60 48.59 48.54 .65 104321.1+632911	NGC 3359 UGC 5873 IRAS10433+6329 PGC 32183	.SBT5.. R (2) 5.0± .3 3.0± .5	1.86± .01 .22± .03 1.46± .02 1.86	170 .00 .34 .11	11.03M±.05 11.03 ±.14 11.26 10.69	.46± .02 -.20± .03 .41 -.24	.52± .01 -.10± .02 13.75± .04 14.63± .11	11.69±.0 256± 6 242± 6 .89	1013± 3 1008± 14 1104 1140
1046.6 +1300 232.54 57.61 93.12 -25.93 1044.0 +1316	 PGC 32186			 .05 				18.66±.3 82± 13 76± 10 	6526± 10 6430 6863
1046.7 +2555 208.22 62.15 80.50 -19.96 1043.9 +2611	 UGC 5881 IRAS10439+2611 PGC 32188	.S..1.. U 1.0± .9 	1.05± .06 .36± .05 1.05	55 .04 .37 .18	 14.87 ±.18 12.64 				
104643.0-400056 278.46 16.86 154.82 -34.44 104427.0-394506	NGC 3378 ESO 318- 12 IRAS10444-3945 PGC 32189	.S..4*. BS (1) 4.3± .7 3.3± .8	1.18± .04 .03± .04 1.22	 .40 .05 .02	 13.48 ±.14 11.94 13.00				5186± 19 4970 5481
104644.8-251438 270.34 29.62 136.90 -35.29 104421.0-245848	NGC 3369 ESO 501- 95 PGC 32191	.LA.-?. S -3.0± .8 	1.15± .04 .24± .03 1.16	114 .35 .00 	 14.64 ±.14 				
104645.2+114916 234.43 57.01 94.34 -26.41 104406.9+120505	NGC 3368 UGC 5882 IRAS10441+1205 PGC 32192	.SXT2.. R (1) 2.0± .3 3.4± .7	1.88± .01 .16± .02 1.39± .02 1.88	5 .06 .19 .08	10.11 ±.13 9.94 ±.13 10.48 9.77	.86± .01 .31± .02 .81 .27	.94± .01 .41± .01 12.49± .06 13.95± .16	12.77±.1 354± 4 341± 4 2.92	897± 4 943± 33 798 1238
104648.8-861717 300.83 -23.94 202.90 -17.65 104922.0-860124	 ESO 7- 1 IRAS10494-8601 PGC 32195	.SB9P?/ S 9.0±1.8 	1.12± .06 .48± .05 1.17	36 .51 .49 .24					
104650.9-012326 251.46 48.62 108.58 -31.14 104417.6-010737	 UGC 5886 PGC 32196	.SB?... 	1.00± .06 .18± .05 1.01	115 .10 .27 .09	 14.33 ±.18 13.90				11487±108 11345 11843
1046.8 -1608 264.33 37.22 125.84 -34.46 1044.4 -1553	 MCG -3-28- 3 PGC 32197	PSXT0.. E .0± .7 	1.32± .04 .19± .05 1.33	45 .14 .14 					7997 7817 8348
1046.9 +5955 147.20 50.94 51.27 -1.23 1043.8 +6011	 UGC 5879 PGC 32204	.S..7.. U 7.0±1.0 	1.12± .05 .94± .05 1.12	28 .00 1.30 .47	 15.75 ±.18 				
104700.8-095630 259.54 42.21 118.45 -33.31 104430.7-094040	NGC 3375 MCG -2-28- 8 PGC 32205	RLAT0?. E -2.0± .8 	1.17± .05 .11± .06 .78± .05 	45 .06 .00 	13.43 ±.15 13.34	.82± .02 .77 	.88± .01 12.82± .16 13.91± .34		 2365± 60 2200 2721
104703.2+263235 206.95 62.32 79.97 -19.59 104418.8+264825	 UGC 5884 VV 727 PGC 32206	.SAS3*. U (1) 3.0±1.2 3.5±1.2	1.06± .06 .16± .05 1.06	97 .04 .22 .08	 14.29 ±.18 13.13 13.98				6287± 50 6241 6588

R.A. 2000 DEC.	Names	Type	logD$_{25}$	p.a.	B$_T$	(B-V)$_T$	(B-V)$_e$	m$_{21}$	V$_{21}$
l b		S$_T$ n$_L$	logR$_{25}$	A$_g$	m$_B$	(U-B)$_T$	(U-B)$_e$	W$_{20}$	V$_{opt}$
SGL SGB		T	logA$_e$	A$_i$	m$_{FIR}$	(B-V)^o_T	m'$_e$	W$_{50}$	V$_{GSR}$
R.A. 1950 DEC.	PGC	L	logD$_o$	A$_{21}$	B^o_T	(U-B)^o_T	m'$_{25}$	HI	V$_{3K}$

104703.7+171626	NGC 3370	.SAS5..	1.50± .03	148				13.63±.2	1280± 5
225.35 59.67	UGC 5887	R (2)	.25± .03	.04	12.29 ±.14			284± 9	1367± 58
88.87 -23.99	IRAS10444+1732	5.0± .3		.38	11.55			270± 5	1200
104423.2+173216	PGC 32207	3.4± .6	1.51	.13	11.87			1.64	1609
1047.0 +3003		.S..3..	1.16± .05	138					9589
199.38 62.63	UGC 5885	U (1)	.45± .05	.05	14.9 ±.2				9556
76.75 -17.81		3.0± .9		.62					9878
1044.3 +3019	PGC 32208	3.5±1.2	1.17	.23	14.20				
104710.0+725027	NGC 3348	.E.0...	1.31± .07		12.17 ±.13	1.02± .01	1.05± .01		
134.63 41.35	UGC 5875	R	.01± .06	.19	12.03 ±.13	.47± .03	.53± .02		2837± 27
40.69 6.22		-5.0± .3	.94± .03	.00		.95	12.36± .10		2960
104326.4+730616	PGC 32216		1.34		11.87	.44	13.68± .39		2906
1047.2 +5402		.I..9..	1.02± .06					15.52±.1	767± 10
154.60 54.77	UGC 5883	U	.04± .05	.00	15.2 ±.3			82± 6	
56.13 -4.59		10.0± .9		.03				59± 5	825
1044.2 +5418	PGC 32221		1.02	.02	15.16			.34	946
104719.2-242621	NGC 3383	.SBT4?.	1.15± .04	24					
269.98 30.37	ESO 501- 97	Sr (1)	.10± .03	.22	13.53 ±.14				3380± 60
135.92 -35.13	IRAS10449-2410	3.8± .7		.15	12.39				3183
104455.0-241030	PGC 32224	3.3± .8	1.17	.05	13.15				3718
104722.6+140417	NGC 3377A	.SXS9..	1.35± .05		14.22 ±.16	.61± .06	.58± .03	15.42±.1	572± 5
230.96 58.29	UGC 5889	R (1)	.02± .05	.04	13.6 ±.4	-.05± .08	.01± .05	55± 5	
92.11 -25.32	DDO 88	9.0± .4	1.07± .03	.02		.59	15.06± .07	44± 6	481
104443.5+142007	PGC 32226	9.0±1.0	1.35	.01	14.08	-.06	15.75± .30	1.33	908
104726.7+060248	NGC 3376	.S?....	.90± .06	167					5837± 50
242.87 53.77	UGC 5891		.32± .04	.06	14.8 ±.2				5718
100.48 -28.51				.49					6186
104450.6+061838	PGC 32231		.90	.16	14.20				
104729.6+071502		.SB.3..	.95± .07						8129± 50
241.27 54.53	UGC 5892	U	.03± .05	.07	14.32 ±.18				8014
99.20 -28.05		3.0± .9		.04					8477
104453.1+073053	PGC 32234		.96	.01	14.15				
104730.3-012932		.S..2..	1.13± .05	132					11399±108
251.75 48.67	UGC 5896	U	.46± .05	.08	14.77 ±.18				11257
108.76 -31.02		2.0± .9		.57					11755
104457.0-011342	PGC 32235		1.14	.23	14.00				
104739.8+261743		.SXS2..	1.19± .05	155				15.45±.3	6537± 10
207.52 62.42	UGC 5894	U	.31± .05	.04	14.31 ±.19				6490
80.28 -19.59		2.0± .9		.38				444± 7	6839
104455.5+263333	PGC 32243		1.19	.16	13.82			1.48	
104739.8+385600		.E...?.	1.09± .10		14.15 ±.17	1.02± .03	1.03± .02		10611± 31
180.41 61.68	UGC 5893	U	.02± .05	.00					10612
68.92 -13.00		-5.0±1.7	.70± .06	.00		.92	13.14± .21		10864
104448.9+391150	PGC 32244		1.09		13.99		14.56± .55		
1047.6 +2946		.SB.2?.	1.04± .06	73					9422
200.01 62.75	UGC 5895	U	.59± .05	.05	15.53 ±.18				
77.08 -17.84		2.0±1.0		.73					9388
1044.9 +3002	PGC 32245		1.04	.30	14.66				9712
1047.6 +5604		.I..9..	1.08± .07						1239± 10
151.82 53.56	UGC 5888	U	.00± .06	.00	14.8 ±.3				
54.48 -3.37		10.0± .8		.00					1305
1044.6 +5620	PGC 32248		1.08	.00	14.79				1407
104741.7+135900	NGC 3377	.E.5+..	1.72± .02	35	11.24M±.10	.86± .01	.89± .01		
231.18 58.31	UGC 5899	R	.24± .02	.07	10.79 ±.12	.31± .03	.40± .01		692± 13
92.24 -25.29		-5.0± .3	1.06± .02	.00		.84	11.98± .08		600
104502.6+141451	PGC 32249		1.66		10.98	.30	14.25± .17		1028
104741.9-385115		.SBS7*/	1.35± .04	75					
278.04 17.96	ESO 318- 13	S	.85± .04	.27	15.02 ±.14				17± 19
153.41 -34.42		7.0±1.3		1.17					-197
104525.0-383524	PGC 32250		1.38	.42	13.58				317
104742.1+110440	A 1045+11	.SA.5..	1.42± .04	75				14.46±.3	2718± 9
235.81 56.80	UGC 5897	U	.44± .05	.04	13.41 ±.20			310± 7	2731± 50
95.21 -26.50		5.0± .8		.66				296± 7	2616
104504.1+112031	PGC 32251		1.43	.22	12.69			1.55	3060
104749.9+123457	NGC 3379	.E.1...	1.73± .02		10.24M±.03	.96± .01	.98± .01		
233.49 57.63	UGC 5902	R	.05± .02	.05	10.10 ±.14	.53± .01	.57± .01		889± 12
93.68 -25.85		-5.0± .3	1.07± .01	.00		.94	11.31± .04		793
104511.4+125048	PGC 32256		1.72		10.17	.52	13.73± .10		1228

R.A. 2000 DEC. / l b / SGL SGB / R.A. 1950 DEC.	Names / PGC	Type / S_T n_L / T / L	$\log D_{25}$ / $\log R_{25}$ / $\log A_\bullet$ / $\log D_o$	p.a. / A_g / A_i / A_{21}	B_T / m_B / m_{FIR} / B_T^o	$(B-V)_T$ / $(U-B)_T$ / $(B-V)_T^o$ / $(U-B)_T^o$	$(B-V)_\bullet$ / $(U-B)_\bullet$ / $m'_\bullet$ / m'_{25}	m_{21} / W_{20} / W_{50} / HI	V_{21} / V_{opt} / V_{GSR} / V_{3K}
104750.4+334342		.S..8*.	1.13± .05	158				15.79±.3	1648± 7
191.37 62.66	UGC 5898	U	.56± .05	.00					1629
73.52 -15.75		8.0±1.3		.69				134± 7	1922
104502.5+335933	PGC 32259		1.13	.28					
1047.9 +2815		.SXS2..	1.06± .06	15					6294
203.32 62.71	UGC 5903	U	.18± .05	.00	14.66 ±.20				6254
78.50 -18.56		2.0± .9		.22					6589
1045.2 +2831	PGC 32264		1.06	.09	14.38				
104801.2+431110	NGC 3374	.SB.5..	1.10± .05	160					
172.05 60.36	UGC 5901	U	.13± .05	.00	14.40 ±.19				
65.30 -10.61	IRAS10451+4327	5.0± .8		.19	13.10				
104507.6+432701	PGC 32266		1.10	.06					
104803.9-205058	NGC 3450	.SBR3..	1.40± .03	140					
267.87 33.49	ESO 569- 6	PSEU (2)	.05± .03	.16	12.72 ±.14				3920± 60
131.57 -34.72	IRAS10456-2034	3.2± .3		.06	12.97				3730
104538.1-203506	PGC 32270	3.2± .5	1.42	.02	12.46				4265
104804.0-313157	NGC 3390	.S..3./	1.55± .03	177	12.85 ±.13	.96± .02	1.02± .01		
274.26 24.38	ESO 437- 62	PS	.79± .04	.31	13.23 ±.10	.38± .03	.44± .02		2820± 66
144.57 -34.97	IRAS10457-3116	3.0± .6	1.03± .02	1.09	12.81	.73	13.49± .08		2613
104543.0-311606	PGC 32271		1.58	.40	11.66	.15	13.46± .21		3142
104805.1-015533									
252.36 48.45	CGCG 10- 17			.08	15.2 ±.6				11297±108
109.30 -31.01									11154
104532.0-013942	PGC 32273								11653
104808.5+181124	IC 642	.E?....	1.15± .15						
223.88 60.28	UGC 5905		.04± .08	.04	13.63 ±.15				5928± 50
88.09 -23.35			1.14	.00					5851
104527.8+182715	PGC 32278				13.50				6254
104812.0+045546	NGC 3385	.L.....	1.17± .06	97					
244.51 53.20	UGC 5908	U	.23± .03	.09	13.61 ±.16				7818± 50
101.76 -28.74		-2.0± .8	1.14	.00					7696
104536.3+051137	PGC 32285				13.40				8169
104812.5+283608	NGC 3380	PSB.1$.	1.23± .04					16.41±.3	1606± 9
202.57 62.80	UGC 5906	P	.10± .04	.00	13.38 ±.18			135± 7	1615± 52
78.21 -18.33	IRAS10454+2851	1.0± .8		.10	13.35			119± 7	1568
104527.3+285159	PGC 32287		1.23	.05	13.26			3.09	1900
104817.2+123749	NGC 3384	.LBS-*.	1.74± .02	53	10.85M±.05	.93± .01	.95± .01		
233.52 57.75	UGC 5911	R	.34± .02	.07	10.72 ±.13	.44± .01	.54± .01		735± 26
93.68 -25.73		-3.0± .3	.92± .02	.00		.89	11.04± .05		639
104538.7+125341	PGC 32292		1.70		10.75	.41	13.60± .11		1074
104818.5+005027		.S?....	.87± .08	110					
249.43 50.46	UGC 5913		.39± .05	.07	14.9 ±.3				4800± 46
106.23 -30.10				.53					4665
104544.4+010619	PGC 32293		.88	.19	14.23				5154
104819.4+500123		.L?....	.97± .15						
160.31 57.21	MCG 8-20- 26		.41± .07	.00	14.92 ±.16				6780± 56
59.55 -6.73				.00					6824
104520.5+501714	PGC 32294		.90		14.81				6980
104824.1-250940	NGC 3393	PSBT1*.	1.34± .03						
270.66 29.89	ESO 501-100	Sr	.04± .04	.36	13.10 ±.14				3730± 60
136.82 -34.92		1.0± .4		.04					3532
104600.1-245348	PGC 32300		1.37	.02	12.66				4067
104825.1+344244	NGC 3381	.SB..P.	1.31± .03					14.34±.2	1629± 7
189.21 62.67	UGC 5909	R	.04± .04	.00	12.73 ±.18			118± 7	1506± 33
72.71 -15.13	IRAS10456+3458			.06	12.55			63± 6	1608
104536.7+345835	PGC 32302		1.31	.02	12.66			1.66	1894
104825.6-454128		.SAT5..	1.15± .04	53					
281.51 12.00	ESO 264- 41	S (1)	.08± .04	.86	14.68 ±.14				
161.42 -33.13		5.0± .8		.12					
104613.0-452536	PGC 32303	4.4± .8	1.23	.04					
104827.6+263508	A 1045+26	.SAS5..	1.13± .05		13.88 ±.14	.64± .03	.67± .02	15.08±.1	6295± 8
206.96 62.64	UGC 5912	U (1)	.02± .05	.01	14.0 ±.2			151± 6	6295± 59
80.10 -19.29		5.0± .8	.85± .02	.03		.59	13.62± .04	133± 6	6249
104543.4+265100	PGC 32305	2.9± .8	1.13	.01	13.84		14.33± .32	1.23	6596
104827.9+123201	NGC 3389	.SAS5..	1.44± .02	112	12.36M±.06	.46± .01		14.04±.2	1301± 6
233.72 57.74	UGC 5914	R (2)	.31± .02	.06	12.42 ±.15	-.16± .02		277± 9	1270± 31
93.80 -25.73		5.0± .4		.46		.38		240± 6	1203
104549.4+124753	PGC 32306	4.2± .7	1.44	.15	11.84	-.22	13.62± .12	2.04	1638

R.A. 2000 DEC.	Names	Type	logD$_{25}$	p.a.	B$_T$	(B-V)$_T$	(B-V)$_\bullet$	m$_{21}$	V$_{21}$
l b	S$_T$ n$_L$	logR$_{25}$	A$_g$	m$_B$	(U-B)$_T$	(U-B)$_\bullet$	W$_{20}$	V$_{opt}$
SGL SGB	T	logA$_\bullet$	A$_i$	m$_{FIR}$	(B-V)^o_T	m'	W$_{50}$	V$_{GSR}$
R.A. 1950 DEC.	PGC	L	logD$_o$	A$_{21}$	B^o_T	(U-B)^o_T	m'$_{25}$	HI	V$_{3K}$

104828.0+382350		.SBR3..	1.27± .06						7660
181.42 61.97	UGC 5910	U	.06± .06	.00 14.2 ±.3					
69.48 -13.15		3.0± .8		.08					7659
104537.6+383941	PGC 32307		1.27	.03 14.08					7916
104829.6-213804		.SXS7?.	1.20± .05	107					
268.48 32.89	ESO 569- 9	E (1)	.24± .03	.19 14.72 ±.14					
132.53 -34.69		7.0± .8		.33					
104604.0-212212	PGC 32311	6.4± .8	1.22	.12					
104831.3+722528	NGC 3364	.SXT5..	1.19± .04		13.46S±.11				
134.87 41.74	UGC 5890	PU	.01± .04	.17 13.65 ±.18					2713± 50
41.09 6.07	IRAS10448+7241	4.5± .5		.02 13.02					2836
104450.8+724119	PGC 32314		1.21	.01 13.30		14.25± .23			2785
104838.5+662144	A 1045+66	.S..3..	1.30± .05	152					
140.21 46.43	UGC 5904	U (1)	.79± .06	.01					6548± 34
46.09 2.63	7ZW 346	3.0± .9		1.09 12.44					6650
104516.6+663735	PGC 32321	3.5±1.2	1.30	.40					6656
1048.7 +2143		.SX.6*.	1.02± .06					15.36±.3	7368± 8
217.15 61.62	UGC 5916	U	.08± .05	.00				125± 7	
84.73 -21.60		6.0± .9		.12				112± 7	7305
1046.0 +2159	PGC 32325		1.02	.04					7685
1048.7 +1218								17.95±.3	888± 10
234.13 57.67				.06				85± 13	
94.06 -25.77								64± 10	791
1046.0 +1234	PGC 32327								1227
104843.9-452511		.S..3./	1.30± .04	1					
281.43 12.27	ESO 264- 43	S	.64± .04	.62 14.67 ±.14					
161.09 -33.13	IRAS10465-4509	3.0± .9		.88 12.55					
104631.1-450918	PGC 32328		1.35	.32					
104844.2+260314	A 1046+26		.44± .22					16.47±.3	7630± 10
208.12 62.62	MCG 4-26- 9		.00± .07	.00				279± 6	7660± 31
80.63 -19.50	MK 727			12.46					7585
104600.2+261906	PGC 32329		.44						7936
1048.7 +2646									6444
206.58 62.72	CGCG 155- 17			.01 15.3 ±.3					6399
79.97 -19.15									6745
1046.0 +2702	PGC 32330								
1048.8 +6611		.I..9*.	1.22± .12						
140.36 46.58	UGC 5907	U	.11± .12	.01					
46.25 2.54		10.0±1.1		.08					
1045.5 +6627	PGC 32337		1.22	.05					
104852.7+500221	A 1045+50	.SB?...	.89± .11						
160.20 57.28	MCG 8-20- 28		.35± .07	.00 15.0 ±.2					6896± 27
59.59 -6.65	MK 152			.43					6940
104554.0+501813	PGC 32341		.89	.18 14.52					7096
104853.0+480315		.S..2..	.96± .07	25					
163.37 58.32	UGC 5915	U	.23± .05	.00 15.30 ±.13					
61.26 -7.77		2.0± .9		.28					
104556.0+481907	PGC 32342		.96	.11					
104854.0+464314		.I..9..	1.04± .06	163					
165.60 58.98	UGC 5917	U	.26± .05	.00 14.85 ±.18					
62.38 -8.51		10.0± .9		.19					
104558.0+465906	PGC 32343		1.04	.13					
1048.9 +1212								15.69±.3	1325± 10
234.36 57.66	CGCG 66- 25			.08				143± 13	
94.20 -25.76	MK 1263							124± 10	1227
1046.3 +1228	PGC 32346								1665
104856.5+141318	NGC 3391	.S?....	1.00± .06	35				15.12±.2	2959± 6
231.07 58.69	UGC 5920		.17± .05	.07 13.9 ±.2				265± 13	
92.14 -24.92	IRAS10462+1429			.26 13.44				218± 5	2868
104617.4+142910	PGC 32347		1.00	.09 13.55				1.49	3294
104901.6-003823		.SB?...	1.02± .06	10				15.60±.1	1846± 9
251.26 49.54	UGC 5922		.20± .05	.04 14.47 ±.18				188± 12	1897±108
107.94 -30.39				.29					1707
104628.0-002230	PGC 32351		1.02	.10 14.12				1.39	2202
1049.0 -0445		.SXS5*.	1.12± .08	60					
255.44 46.51		E (1)	.09± .08	.06					
112.59 -31.58		5.0± .9		.13					
1046.5 -0430	PGC 32352	1.9± .8	1.13	.04					

R.A. 2000 DEC. / l b / SGL SGB / R.A. 1950 DEC.	Names / PGC	Type / S_T n_L / T / L	$\log D_{25}$ / $\log R_{25}$ / $\log A_o$ / $\log D_o$	p.a. / A_g / A_i / A_{21}	B_T / m_B / m_{FIR} / B_T^o	$(B-V)_T$ / $(U-B)_T$ / $(B-V)_T^o$ / $(U-B)_T^o$	$(B-V)_.$ / $(U-B)_.$ / $m'_.$ / m'_{25}	m_{21} / W_{20} / W_{50} / HI	V_{21} / V_{opt} / V_{GSR} / V_{3K}
104904.6+521958 156.73 56.01 57.69 -5.33 104603.8+523550	A 1046+52 MCG 9-18- 32 MK 153 PGC 32356	.S?.... 	.91± .08 .25± .06 .91	 .00 .38 .13	15.0 ±.2 14.7 ±.2 14.47	.18± .03 -.64± .05 .12 -.69	 13.75± .47	 	2447± 53 2500 2635
104907.7+065507 242.16 54.64 99.73 -27.80 104631.3+071100	 UGC 5923 MK 1264 PGC 32364	.S..0?. U .0±1.9 	.97± .07 .36± .05 .96	173 .04 .27 	 14.5 ±.2 14.20				722± 50 607 1070
104912.2+220107 216.62 61.82 84.50 -21.36 104630.0+221700	 UGC 5924 U PGC 32367	.S..1.. U 1.0± .9 	1.17± .05 .55± .05 1.17	52 .00 .56 .28	 14.75 ±.18 14.09			15.70±.3 592± 7 1.34	7636± 10 7573 7952
104912.4+275530 204.09 62.96 78.95 -18.48 104627.7+281123	 UGC 5921 U PGC 32368	.S..8*. U 8.0±1.3 	1.20± .05 .64± .05 1.20	161 .00 .78 .32	 14.98 ±.18 14.19			15.44±.3 152± 7 .93	1407± 10 1366 1704
104913.3-311817 274.37 24.70 144.28 -34.74 104652.0-310224	 ESO 437- 65 IRAS10468-3102 PGC 32369	.SBT4P. Sr 4.2± .6 	1.24± .04 .34± .04 1.26	32 .27 .50 .17	14.24 ±.14 13.21 13.45 				3140± 18 2933 3463
1049.2 +1225 234.09 57.84 94.01 -25.61 1046.6 +1241	 CGCG 66- 29 PGC 32371			 .05 	15.4 ±.6 			16.39±.3 93± 13 74± 10 	1383± 10 1286 1722
104916.5-193811 267.35 34.66 130.13 -34.32 104650.1-192218	 ESO 569- 12 IRAS10468-1922 PGC 32374	.LX.0P* SE -1.5± .5 	1.25± .04 .17± .03 1.24	105 .11 .00 	13.53 ±.14 11.86 				
104917.0+001949 250.27 50.28 106.88 -30.03 104643.0+003542	 MCG 0-28- 14 PGC 32375	.SB?... 	.94± .11 .11± .07 .94	 .07 .17 .06	 14.81 ±.18 14.51				12297±108 12161 12652
1049.2 +1222 234.17 57.83 94.06 -25.62 1046.6 +1238	 PGC 32376			 .05 				17.78±.3 70± 13 44± 10 	1350± 10 1253 1689
104921.2-004007 251.38 49.58 108.00 -30.32 104647.7-002414	 MCG 0-28- 13 E PGC 32383	.LA.-P* E -3.0± .9 	1.02± .07 .10± .05 1.01	 .04 .00 	14.40 ±.16 14.18 				11669±108 11530 12025
104923.5+431830 171.63 60.54 65.33 -10.33 104630.2+433423	 UGC 5925 KUG 1046+435 PGC 32384	.S..7.. U 7.0±1.0 	1.12± .04 .88± .04 1.12	157 .00 1.21 .44	15.59 ±.18 				
1049.4 +6444 141.71 47.70 47.48 1.76 1046.1 +6500	 UGCA 220 U PGC 32385	.I..9*. U 10.0±1.1 	1.23± .14 .16± .18 1.23	 .00 .12 .08					
104925.7+324629 IC 2604 193.39 63.07 74.56 -15.97 104638.6+330222	 UGC 5927 VV 538 PGC 32390	.SBS9P? PU 9.3± .7 	1.10± .04 .14± .04 1.10	40 .00 .14 .07	14.7 ±.2 14.56 				1680± 39 1658 1959
1049.4 +1613 227.73 59.74 90.20 -23.94 1046.8 +1629	NGC 3399 MCG 3-28- 12 PGC 32395	.L?.... 	1.23± .10 .00± .07 1.23	 .04 .00 	13.8 ±.2 13.64 				6826 6742 7157
1049.5 +0448 245.03 53.36 102.02 -28.48 1046.9 +0504	 UGC 5929 U PGC 32396	.S..4.. U 4.0± .9 	1.04± .08 .37± .06 1.05	68 .10 .55 .19	14.83 ±.18 				
104930.6+225754 214.72 62.14 83.64 -20.85 104648.0+231347	A 1046+23 MK 417 PGC 32398			 .00 					9820± 31 9761 10132
1049.6 +6531 140.91 47.12 46.83 2.23 1046.2 +6547	A 1046+65 UGC 5918 DDO 87 PGC 32405	.I..9*. PU (1) 10.0± .6 9.0± .9	1.38± .04 .00± .05 1.14± .05 1.38	 .00 .00 .00	15.1 ±.2 15.10	.54± .09 .05± .11 .54 .05	.49± .05 14.30±.1 .03± .07 73± 6 16.29± .11 62± 8 16.83± .30 -.80		338± 5 438 452

R.A. 2000 DEC. l b SGL SGB R.A. 1950 DEC.	Names PGC	Type S_T n_L T L	$logD_{25}$ $logR_{25}$ $logA_o$ $logD_o$	p.a. A_g A_i A_{21}	B_T m_B m_{FIR} B_T^o	$(B-V)_T$ $(U-B)_T$ $(B-V)_T^o$ $(U-B)_T^o$	$(B-V)_e$ $(U-B)_e$ m'_e m'_{25}	m_{21} W_{20} W_{50} HI	V_{21} V_{opt} V_{GSR} V_{3K}
104941.0+002148 250.34 50.37 106.89 -29.93 104707.0+003742	 MCG 0-28- 15 PGC 32410	.E?.... 	 	 .07 	 14.7 ±.6 				11618± 49 11482 11973
1049.8 +5154 157.24 56.35 58.12 -5.48 1046.8 +5210	 UGC 5928 PGC 32423	.L..-*. U -3.0±1.2 	1.00± .10 .00± .04 1.00	 .00 .00 	 14.67 ±.17 				
104949.5+325851 192.93 63.14 74.42 -15.79 104702.3+331445	NGC 3395 UGC 5931 ARAK 257 PGC 32424	.SXT6P* R (1) 6.0± .6 4.6± .9	1.32± .03 .23± .03 1.32	50 .00 .34 .12	12.4 ±.2 12.46 ±.16 12.09	.33± .02 -.24± .04 .27 -.28	.38± .02 -.19± .04 13.26± .24	13.57±.1 216± 4 162± 6 1.37	1620± 4 1634± 17 1599 1899
1049.8 +0113 249.44 51.01 105.96 -29.61 1047.3 +0129	 CGCG 10- 28 PGC 32429	 	 	 .09 	 15.3 ±.6 				11691±108 11558 12046
104956.1+325922 192.90 63.16 74.43 -15.77 104709.0+331516	NGC 3396 UGC 5935 PGC 32434	.IB.9P. R 10.0± .4 	1.49± .02 .42± .03 1.49	100 .00 .31 .21	 12.61 ±.14 12.29			13.57±.3 160± 10 1.07	1625± 10 1667± 20 1612 1911
104958.4+315419 195.30 63.24 75.41 -16.32 104711.8+321013	 UGC 5934 PGC 32437	.SBS8.. U 8.0± .8 	1.11± .07 .05± .06 1.11	 .02 .07 .03	 14.7 ±.3 14.59			14.81±.3 90± 7 .19	1608± 7 1583 1891
104959.0+001924 250.47 50.40 106.96 -29.87 104725.1+003518	 MCG 0-28- 17 PGC 32439	.S?.... 	.82± .13 .08± .07 .82	 .07 .09 .04	 14.87 ±.19 14.59				11652± 49 11516 12007
1050.0 +0022 250.44 50.44 106.91 -29.83 1047.5 +0038	 CGCG 10- 31 PGC 32447	 	 	 .07 	 15.4 ±.6 				11115±108 10979 11470
1050.1 +0519 244.53 53.81 101.53 -28.16 1047.5 +0535	 UGC 5940 PGC 32449	.SB.3.. U 3.0± .9 	1.04± .08 .59± .06 1.05	113 .10 .81 .29	 15.15 ±.18 				
105007.3+362031 185.58 62.75 71.46 -13.98 104718.4+363625	 UGC 5936 PGC 32452	RLA.+*. U -1.0± .8 	1.11± .07 .12± .03 1.09	83 .00 .00 	 14.21 ±.15 14.10				7229± 31 7221 7494
105009.7-171434 265.93 36.77 127.28 -33.83 104742.3-165839	NGC 3420 MCG -3-28- 11 PGC 32453	RSBT1*. SE .7± .7 	1.12± .05 .08± .04 1.13	30 .11 .08 .04	 				
1050.2 -0115 252.26 49.31 108.75 -30.29 1047.7 -0100	A 1047-01 UGC 5943 PGC 32463	.SXR5.. UE (1) 4.5± .6 3.3± .9	1.05± .05 .11± .04 .98± .03 1.05	155 .07 .16 .05	14.01 ±.16 14.28 ±.18 13.88	.74± .04 .68 	.70± .02 14.40± .07 13.83± .31	16.06±.1 123± 11 108± 11 2.13	4544± 10 4531± 59 4403 4900
105017.6+375723 182.14 62.43 70.06 -13.09 104727.9+381317	IC 2606 MCG 6-24- 21 PGC 32465	.S?.... 	.89± .11 .35± .07 .89	 .00 .43 .18	 14.9 ±.2 14.44				7725± 19 7723 7983
105018.1-120631 262.15 40.98 121.19 -32.96 104748.7-115036	NGC 3404 MCG -2-28- 11 IRAS10477-1150 PGC 32466	.SB.2?/ PE 1.8± .7 	1.33± .04 .64± .05 1.34	80 .11 .79 .32	 13.39 				
105020.3-170238 265.83 36.96 127.05 -33.76 104752.8-164643	NGC 3409 MCG -3-28- 12 PGC 32470	.SB.5?/ E (1) 5.0±1.3 4.2±1.6	1.08± .06 .69± .05 1.09	10 .11 1.03 .34	 				
1050.3 +1316 232.98 58.51 93.27 -25.01 1047.7 +1332	 UGC 5944 PGC 32471	.I..9.. U 10.0± .8 	1.00± .16 .00± .12 1.01	 .06 .00 .00	 14.8 ±.3 				
105021.3+412812 175.01 61.39 67.01 -11.19 104729.4+414406	 UGC 5941 PGC 32472	.S?.... 	1.04± .09 .09± .07 1.04	 .00 .13 .05	 				7144± 50 7156 7386

R.A. 2000 DEC. l b SGL SGB R.A. 1950 DEC.	Names PGC	Type S_T n_L T L	$\log D_{25}$ $\log R_{25}$ $\log A_o$ $\log D_o$	p.a. A_g A_i A_{21}	B_T m_B m_{FIR} B_T^o	$(B-V)_T$ $(U-B)_T$ $(B-V)_T^o$ $(U-B)_T^o$	$(B-V)_o$ $(U-B)_o$ m'_o m'_{25}	m_{21} W_{20} W_{50} HI	V_{21} V_{opt} V_{GSR} V_{3K}
105021.5-385107 278.52 18.21 153.32 -33.90 104804.0-383512	 ESO 318- 19 PGC 32473	PSXS2.. r 2.2± .9 	1.05± .06 .34± .05 1.07	68 .28 .42 .17	 14.97 ±.14 14.22				4765± 19 4551 5065
1050.3 +1734 225.47 60.51 88.97 -23.15 1047.7 +1750	 UGC 5945 PGC 32474	.IB.9.. U 10.0± .8 	1.22± .06 .29± .06 1.23	95 .04 .22 .15	 14.4 ±.2 14.13				1132 1054 1460
105026.2-125044 262.76 40.40 122.07 -33.07 104757.0-123449	NGC 3411 MCG -2-28- 12 PGC 32479	.E+.... E -4.0± .9 	1.32± .08 .00± .07 .97± .06 1.34	 .13 .00 	12.9 ±.2 12.70	1.03± .02 .96 	1.05± .01 13.28± .21 14.50± .48		4417± 60 4245 4772
1050.4 +6446 141.55 47.75 47.51 1.88 1047.1 +6502	 UGC 5932 IRAS10471+6502 PGC 32484	.S..4.. U 4.0± .8 	1.16± .05 .41± .05 1.16	52 .01 .60 .20	 14.98 ±.20 				
1050.5 +1938 221.55 61.34 86.95 -22.19 1047.8 +1954	A 1047+19 UGC 5947 DDO 89 PGC 32486	.I..9P* PU 10.0± .7 	1.12± .07 .40± .06 .88± .02 1.12	25 .06 .30 .20	14.92 ±.14 14.75 ±.18 14.50	.42± .05 -.21± .07 .30 -.29	.50± .04 -.23± .06 14.81± .04 14.34± .41	15.27±.1 91± 16 75± 12 .57	1253± 11 1182 1576
105039.0+654339 140.60 47.04 46.73 2.43 104719.8+655933	NGC 3394 UGC 5937 IRAS10473+6559 PGC 32495	.SAT5.. U 5.0± .8 	1.28± .04 .13± .05 1.28	35 .00 .19 .06	 13.09 ±.18 13.24 12.88				3403± 56 3504 3516
1050.6 +1544 228.84 59.78 90.82 -23.89 1048.0 +1600	 UGC 5948 PGC 32496	.I..9?. U 10.0±1.8 	1.11± .07 .54± .06 1.11	35 .04 .40 .27					1120 1035 1452
1050.7 +1519 229.60 59.60 91.25 -24.05 1048.1 +1535	 UGC 5950 PGC 32498	.I..9?. U 10.0±1.6 	1.26± .06 .29± .06 1.26	 .04 .22 .15					6400 6314 6733
105045.6+282806 202.96 63.35 78.63 -17.91 104800.8+284401	NGC 3400 UGC 5949 PGC 32499	.SBS1*. PU .5± .6 	1.13± .05 .21± .04 1.13	100 .00 .21 .10	 14.11 ±.18 13.88				1408± 36 1370 1703
105053.2+132446 232.87 58.70 93.18 -24.83 104814.6+134041	NGC 3412 UGC 5952 PGC 32508	.LBS0.. R -2.0± .3 	1.56± .02 .25± .03 .94± .03 1.53	155 .06 .00 	11.45 ±.13 11.39 ±.13 11.35	.91± .01 .39± .02 .86 .36	.92± .01 .43± .02 11.64± .09 13.49± .19		865± 27 772 1202
1050.9 +6546 140.53 47.03 46.71 2.48 1047.6 +6602	NGC 3392 MCG 11-13- 42 PGC 32512	.E?.... 	.90± .11 .12± .04 .87	 .00 .00 	 14.71 ±.11 14.66				3275± 56 3376 3387
105057.8-122654 262.59 40.80 121.63 -32.87 104828.4-121059	NGC 3421 MCG -2-28- 13 PGC 32514	PSBT1P. E 1.0± .8 	1.30± .05 .10± .05 1.31	175 .13 .10 .05					
105058.0-233931 270.32 31.48 135.03 -34.26 104833.1-232336	 ESO 501-102 PGC 32515	.SBS5*P S (1) 4.5± .6 3.3± .7	1.16± .04 .10± .04 1.18	 .25 .14 .05	 13.34 ±.14 12.92			14.95±.3 154± 15 120± 12 1.99	3979± 11 4090± 60 3788 4323
105058.7-020857 253.39 48.78 109.80 -30.38 104825.6-015302	IC 651 UGC 5956 ARAK 258 PGC 32517	.SBS9P* RE (1) 9.0±1.3 5.3± .8	.89± .05 .00± .04 .90	 .08 .00 .00	 13.6 ±.3 11.40 13.56			15.58±.1 192± 12 2.02	4469± 9 4259± 63 4322 4821
105100.9+361137 185.83 62.95 71.69 -13.90 104812.2+362732	 UGC 5951 KUG 1048+364A PGC 32519	.S..7.. U 7.0±1.0 	.91± .06 .89± .05 .91	161 .00 1.23 .44					7135 7126 7400
105113.3+055031 244.16 54.36 101.09 -27.71 104837.4+060627	NGC 3423 UGC 5962 PGC 32529	.SAS6.. R (2) 6.0± .3 3.9± .6	1.58± .02 .07± .02 1.08± .02 1.59	10 .09 .10 .04	11.59M±.10 11.58 ±.15 11.39	.45± .04 .41 	 12.50± .05 14.17± .14	13.24±.1 177± 6 154± 8 1.82	1011± 7 835± 66 890 1359
105115.0-170029 266.03 37.11 127.05 -33.54 104847.4-164433	NGC 3431 MCG -3-28- 14 IRAS10487-1644 PGC 32531	.SX.3?. E (1) 3.0±1.3 1.9±1.2	1.11± .06 .62± .05 1.12	130 .11 .85 .31	 13.38 				5376 5195 5727

R.A. 2000 DEC.	Names	Type	logD₂₅	p.a.	B_T	(B-V)_T	(B-V)_e	m₂₁	V₂₁
l b		S_T n_L	logR₂₅	A_g	m_B	(U-B)_T	(U-B)_e	W₂₀	V_opt
SGL SGB		T	logA_e	A_i	m_FIR	(B-V)_T^o	m'_e	W₅₀	V_GSR
R.A. 1950 DEC.	PGC	L	logD_o	A₂₁	B_T^o	(U-B)_T^o	m'₂₅	HI	V_3K

105115.8+275102 A 1048+28	.S..4..	1.16± .07	179				16.22±.2	1182± 6
204.35 63.41 UGC 5958	U	.71± .06	.00	15.32 ±.18			196± 10	
79.26 −18.13	4.0± .9		1.04				184± 5	1142
104831.3+280657 PGC 32532		1.16	.35	14.27			1.59	1479

105116.3+275833 NGC 3414	.L...P.	1.55± .04		11.96 ±.18	.97± .03	.99± .02	17.40±.3	1414± 9
204.08 63.42 UGC 5959	R	.14± .04	.00	11.82 ±.12	.55± .05	.59± .05	371± 21	1434± 27
79.15 −18.06 ARP 162	−2.0± .3	.84± .06	.00		.94	11.65± .20	321± 7	1376
104831.8+281428 PGC 32533		1.53		11.84	.54	14.24± .28		1713

105117.5−122409 NGC 3422	.L..+?/	1.10± .06	54					
262.64 40.89 MCG −2-28- 15	E	.62± .05	.13					
121.59 −32.78	−1.0±1.8		.00					
104848.1−120814 PGC 32534		1.02						

105117.6+135643 NGC 3419	RLXR+..	1.09± .05	115				16.42±.2	3035± 7
232.08 59.05 UGC 5964	R	.07± .03	.05	13.46 ±.16			251± 21	3028± 40
92.69 −24.52 ARAK 259	−1.0± .5		.00				237± 7	2944
104838.8+141238 PGC 32535		1.09		13.36				3371

105117.7+443412 A 1048+44	.P.....	.79± .06	85					
169.01 60.33 UGC 5953	R	.27± .04	.00	13.8 ±.5				1800±110
64.45 −9.35 MK 155	99.0		.37					1824
104824.1+445007 PGC 32536		.79	.13	13.43				2028

105119.9+140127 NGC 3419A	.SBS3*/	1.25± .04	137				14.45±.2	3074± 6
231.95 59.10 UGC 5965	R	.92± .04	.05	14.89 ±.18			271± 7	3048± 42
92.61 −24.48	3.3± .6		1.27				255± 5	2983
104841.0+141723 PGC 32540		1.25	.46	13.56			.44	3410

105121.1−342544	.IB.9?/	1.32± .04	49					
276.46 22.20 ESO 376- 22	S	.62± .04	.27	14.06 ±.14				1364± 53
148.02 −34.13 TOL 51	10.0±1.2		.46					1154
104901.1−340948 PGC 32542		1.35	.31	13.32				1678

105121.1+324604 NGC 3413	.L..../	1.34± .03	178				14.56±.2	645± 6
193.34 63.48 UGC 5960	R	.38± .03	.00	13.08 ±.16			186± 10	655± 52
74.79 −15.63 IRAS10485+3301	−2.0± .4		.00	13.09			156± 6	624
104834.4+330200 PGC 32543		1.28		13.07				924

105123.1−100803	.L...*/	1.19± .05	150					2486
260.85 42.72 MCG −2-28- 16	E	1.02± .05	.05					
118.95 −32.30	−2.0±1.3		.00					2321
104852.9−095207 PGC 32548		1.04						2843

105124.3+280645 NGC 3418	.SXS0*.	1.14± .05	75					1251± 39
203.78 63.46 UGC 5963	PU	.10± .04	.00	14.07 ±.19				1212
79.03 −17.97	.2± .7		.08					1548
104839.7+282240 PGC 32549		1.13		13.98				

105124.6−195320	.SBS7*.	1.52± .02	151				14.06±.2	3108± 11
268.03 34.72 ESO 569- 14 (1)	E	.85± .03	.13	13.83 ±.14			245± 11	
130.50 −33.85 IRAS10489−1937	7.0±1.2		1.18	13.61			233± 8	2921
104858.0−193724 PGC 32550	3.3±1.3	1.53	.43	12.52			1.11	3455

1051.4 +0917 NGC 3428	.SXS3..	1.19± .05	170					7953± 50
239.44 56.53 UGC 5968	U	.36± .05	.04	13.95 ±.18				7846
97.48 −26.37	3.0± .9		.49					8298
1048.8 +0933 PGC 32552		1.19	.18	13.35				

105125.3−034345	.S..3*.	1.10± .06	17					6644
255.10 47.68 MCG 0-28- 21	E (1)	.41± .05	.05	14.43 ±.18				
111.61 −30.72	3.0±1.3		.57					6497
104852.8−032749 PGC 32553	3.1±1.6	1.11	.21	13.76				7001

105125.8+083404 NGC 3425	.L.....	1.01± .08						6627± 31
240.48 56.11 UGC 5967	U	.00± .03	.02	14.12 ±.15				6518
98.23 −26.65	−2.0± .8		.00					6973
104848.9+085000 PGC 32555		1.01		14.00				

1051.4 +4757	.S..2..	1.14± .07	58					
163.12 58.74 UGC 5961	U	.69± .06	.00					
61.58 −7.47	2.0± .9		.84					
1048.5 +4813 PGC 32557		1.14	.34					

105126.7+081752 NGC 3427	.S..0..	1.05± .08	77					
240.87 55.95 UGC 5966	U	.32± .06	.03	14.2 ±.2				6263
98.52 −26.74	.0± .9		.24					6153
104849.9+083348 PGC 32559		1.03		13.84				6610

105131.8+552325 IC 644	.S?....	.98± .09	78					2867± 50
152.11 54.41 UGC 5954		.45± .06	.00	14.6 ±.2				2931
55.36 −3.31			.46					3039
104828.8+553920 PGC 32564		.98	.22	14.10				

R.A. 2000 DEC.	Names	Type	logD$_{25}$	p.a.	B$_T$	(B-V)$_T$	(B-V)$_e$	m$_{21}$	V$_{21}$
l b		S$_T$ n$_L$	logR$_{25}$	A$_g$	m$_B$	(U-B)$_T$	(U-B)$_e$	W$_{20}$	V$_{opt}$
SGL SGB		T	logA$_o$	A$_i$	m$_{FIR}$	(B-V)^o_T	m'$_e$	W$_{50}$	V$_{GSR}$
R.A. 1950 DEC.	PGC	L	logD$_o$	A$_{21}$	B^o_T	(U-B)^o_T	m'$_{25}$	HI	V$_{3K}$

105133.6-352826		.S..5*/	1.25± .04	107					
277.04 21.30	ESO 376- 23	S	.95± .04	.20	15.41 ±.14				
149.27 -34.01	IRAS10492-3512	5.0±1.3		1.42					
104914.0-351230	PGC 32565		1.27	.47					
1051.5 +0016		.SB.2?.	1.00± .08	112					
250.97 50.64	UGC 5973	U	.43± .06	.09					
107.16 -29.51		2.0± .9		.53					
1049.0 +0032	PGC 32567		1.01	.22					
105134.9+552755	NGC 3398	.S..1$.	1.10± .07						
152.01 54.36	MCG 9-18- 39	P	.31± .06	.00	15.3 ±.2				
55.30 -3.26		1.0±1.8		.31					
104831.9+554351	PGC 32568		1.10	.15					
105135.1+043457		.S..6*.	1.31± .04	130				14.25±.3	1041± 10
245.88 53.61	UGC 5974	U	.40± .05	.12	14.0 ±.2				919
102.47 -28.07		6.0±1.2		.59				155± 7	
104859.7+045053	PGC 32570		1.32	.20	13.29			.76	1393
105136.9-170716		RSBT2P?	1.34± .04	40					
266.20 37.06		SE (1)	.39± .06	.13					
127.19 -33.47		1.7± .6		.48					
104909.2-165120	PGC 32573	5.6± .8	1.35	.19					
105142.0+182852	NGC 3426	.S?....	1.03± .09						
224.01 61.17	UGC 5975		.13± .06	.05	14.08 ±.19				6109± 50
88.23 -22.46	ARAK 262			.18					6034
104901.5+184448	PGC 32577		1.03	.07	13.80				6435
105142.7+434244	NGC 3415	.LA.+*.	1.32± .05	10	13.45 ±.17	.80± .02		16.68±.1	3303± 9
170.53 60.76	UGC 5969	PU	.20± .03	.00	13.05 ±.11	.21± .03		136± 11	3177± 66
65.22 -9.76	IRAS10488+4358	-.5± .5		.00	12.61	.74			3321
104849.7+435840	PGC 32579		1.29		13.12	.20	14.41± .30		3532
105146.6+325402	NGC 3424	.SBS3*$	1.45± .02	112				14.55±.3	1501± 10
193.03 63.56	UGC 5972	R	.57± .03	.00	13.18 ±.18				1421± 43
74.71 -15.48	IRAS10489+3309	3.0± .8		.79	10.73			351± 10	1476
104859.8+330958	PGC 32584		1.45	.29	12.38			1.88	1776
105148.5+434550	NGC 3416	.S?....	.76± .08						3276± 82
170.42 60.76	MCG 7-22- 73		.40± .06	.00	15.4 ±.2				3297
65.19 -9.71	ARAK 260			.59					3508
104855.6+440146	PGC 32588		.76	.20	14.83				
1051.8 +7734		.I?....	1.04± .08	167					10644± 97
130.73 37.69	UGC 5942		.67± .06	.04					10783
36.91 9.11	IRAS10477+7749			.50	13.22				10685
1047.7 +7749	PGC 32589		1.04	.33					
105153.9+510024	NGC 3410	.S?....	.64± .17						7105± 56
158.22 57.13	MCG 9-18- 42		.11± .07	.00	15.1 ±.3				7153
59.05 -5.71	ARAK 261			.14					7300
104855.3+511620	PGC 32594		.64	.06	14.95				
105157.7+034729	NGC 3434	.SAR3..	1.33± .04	5				14.25±.2	3632± 6
246.97 53.15	UGC 5980	U (1)	.05± .05	.09	12.85 ±.19			281± 6	
103.36 -28.26		3.0± .7		.06				251± 7	3507
104922.6+040326	PGC 32595	.5±1.0	1.34	.02	12.67			1.55	3985
105203.0+553600		.S..6*.	1.10± .05	30					1195
151.76 54.33	UGC 5976	U	.06± .05	.00	14.9 ±.3				
55.22 -3.13		6.0±1.2		.08					1260
104860.0+555156	PGC 32604		1.10	.03	14.78				1366
105203.5+100900	NGC 3433	.SAS5..	1.55± .03	50				13.81±.1	2719± 4
238.32 57.16	UGC 5981	R (2)	.05± .03	.04	12.30 ±.19			273± 4	2591± 66
96.65 -25.89		5.0± .3		.08				255± 4	2615
104926.2+102456	PGC 32605	1.6± .6	1.55	.03	12.17			1.62	3062
105205.6+714629		.E.....	1.08± .09						1181± 50
135.10 42.42	UGC 5955	U	.00± .04	.10	14.12 ±.16				1302
41.78 5.93		-5.0± .8		.00					1257
104830.6+720225	PGC 32610		1.10		14.00				
105206.6-003339	IC 653	.S..0*.	1.27± .04	55				15.98±.1	5538± 9
252.03 50.13	UGC 5985	UE	.34± .04	.11	13.75 ±.18			514± 12	
108.13 -29.64		.0± .6		.25					5400
104933.0-001742	PGC 32611		1.27		13.31				5894
1052.1 +6646		.S..2..	1.00± .08	7					
139.44 46.35	UGC 5971	U	.49± .06	.00	15.39 ±.18				
45.96 3.14		2.0± .9		.60					
1048.8 +6702	PGC 32613		1.00	.25					

R.A. 2000 DEC.	Names	Type	logD₂₅	p.a.	B_T	(B-V)_T	(B-V)_e	m₂₁	V₂₁
l b		S_T n_L	logR₂₅	A_g	m_B	(U-B)_T	(U-B)_e	W₂₀	V_opt
SGL SGB		T	logA_e	A_i	m_FIR	(B-V)°_T	m'	W₅₀	V_GSR
R.A. 1950 DEC.	PGC	L	logD_o	A₂₁	B°_T	(U-B)°_T	m'₂₅	HI	V₃K

```
105210.9+325709  NGC  3430   .SXT5..  1.60± .02   30 12.2  ±.2    .65± .02              13.18±.1  1585±  4
192.90  63.64    UGC  5982    R    (2) .25± .02   .00 12.05 ±.14   .11± .05              344±  5   1570± 25
 74.71 -15.38    IRAS10494+3312 5.0± .3            .38 11.76       .59                   337±  4   1564
104924.2+331306  PGC 32614    2.5± .7  1.60        .13 11.70       .07     14.42± .23    1.35      1864

105212.2+582618  NGC  3408   .S..5*.  .99± .06   175                                               9526± 39
148.20  52.44    UGC  5977    P        .06± .04   .00 14.15 ±.18                                   9601
 52.87  -1.52    IRAS10490+5842 5.0±1.2            .09 13.27                                       9681
104905.9+584214  PGC 32616             .99        .03 14.01

1052.2  +3004               .S?....  1.27± .07                                                     10423± 37
199.41  63.75    MCG  5-26- 24         .18± .07   .03                                              10392
 77.34 -16.83                          .27                                                         10713
1049.5  +3020    PGC 32620             1.27        .09

105216.2-324015             PSBR2..  1.31± .03                                                     3165± 18
275.72  23.83    ESO  437- 67 Sr       .06± .03   .27 13.51 ±.14                                   2957
145.90 -34.04    IRAS10499-3224 2.1± .4            .07 13.01                                       3484
104955.1-322418  PGC 32625             1.34        .03 13.14

1052.2  +6123    NGC  3407   .L..-*.  1.15± .08   15                                               4994± 60
144.84  50.37    UGC  5978    U        .28± .04   .00 14.55 ±.16                                   5080
 50.43   .14                  -3.0±1.2            .00                                              5132
1049.1  +6139    PGC 32626             1.11            14.48

105225.2-462204             PSBS4P.  1.17± .05                                                     5874± 87
282.46  11.72    ESO  264- 46 Sr       .16± .05   .69 13.91 ±.14                                   5656
162.03 -32.32    IRAS10501-4606 3.7± .6            .24 11.91                                       6146
105012.1-460606  PGC 32635             1.24        .08 12.95

1052.4  +5941    A  1049+59  .SB.5$.  .79± .09                                                     8417± 60
146.71  51.60    MCG 10-16- 18 P       .16± .06   .00 15.1  ±.2                                    8497
 51.85   -.80                  5.0±1.5            .24                                              8565
1049.3  +5957    PGC 32637             .79         .08 14.78

105226.1+103249  NGC  3438   .S?....  .92± .07                                                     6486± 50
237.82  57.47    UGC  5988             .03± .05   .01 14.30 ±.18                                   6384
 96.29 -25.65                          .04                                                         6829
104948.7+104846  PGC 32638             .92         .01 14.21

105231.1+071328  NGC  3441   .S?....  .87± .08    5                                                6659± 63
242.66  55.49    UGC  5993             .27± .05   .05 14.5  ±.3                                    6546
 99.76 -26.90    ARAK  263             .40                                                         7008
104954.7+072925  PGC 32642             .88         .13 14.03

105231.3+363708  NGC  3432   .SBS9./  1.83± .01   38 11.67M±.11   .42± .02    .46± .01 12.14±.1   616±  4
184.77  63.16    UGC  5986    R    (1) .66± .02   .00 11.77 ±.13  -.36± .03  -.29± .02 259±  4    619± 42
 71.48 -13.42    ARP   206    9.0± .3  1.32± .02   .67            .28        13.73± .05 232±  5    609
104942.7+365305  PGC 32643    4.6±1.1  1.83        .33 11.04      -.46       14.02± .14  .77       880

1052.5  +1001               .S...6*.  1.19± .06  142                                    15.41±.3  6389±  9
238.64  57.18    UGC  5994    U        1.21± .06   .00 15.75 ±.20                        351±  7
 96.84 -25.84                  6.0±1.4            1.47                                   343±  7   6285
1049.9  +1017    PGC 32644             1.19         .50 14.20                            .71       6733

105232.2+194729             .I..9?.  1.18± .05  127                                     15.09±.3  1126±  9
221.61  61.84    UGC  5989    U        .50± .05   .06 14.33 ±.18                         159±  7   1114± 50
 87.04 -21.70    IRAS10498+2003 10.0±1.8          .38 13.74                             141±  7   1056
104951.2+200326  PGC 32645             1.18         .25 13.89                            .95       1448

105235.0+225604  NGC  3437   .SXT5*.  1.40± .02  122                                    14.20±.1  1283±  4
215.18  62.82    UGC  5995    R    (2) .49± .03   .00 12.95 ±.14                         330±  7   1153± 50
 84.03 -20.24    IRAS10498+2312 5.0± .6            .74 10.49                             321±  8   1224
104952.9+231201  PGC 32648    5.0± .7  1.40         .25 12.20                            1.75      1596

1052.5  +6758               .I..9..  1.23± .06                                          15.02±.1  1121± 11
138.30  45.45    UGC  5979    U        .22± .06   .03 15.1  ±.3                          104± 16
 44.98   3.85                  10.0± .8           .16                                   97± 12    1230
1049.2  +6814    PGC 32649             1.24         .11 14.88                            .03       1220

105238.1-450952             PSXR1..  1.03± .05   65
281.94  12.81    ESO  264- 47 Sr       .13± .05   .62 14.90 ±.14
160.64 -32.51                  .5± .6             .13
105024.0-445354  PGC 32650             1.09         .06

105238.4+342900             .S..2..  1.13± .04   14                                     15.94±.3  1569± 10
189.45  63.56    UGC  5990    U        .60± .04   .00 15.03 ±.18
 73.39 -14.51    KUG 1049+347  2.0± .9            .74                                   185±  7   1554
104951.0+344457  PGC 32652             1.13         .30 14.28                            1.36      1842

105247.7+493645             .L..-..  1.04± .18  155
160.22  58.04    UGC  5991    U        .18± .08   .00 14.51 ±.15
 60.31  -6.37                  -3.0± .9           .00
104950.6+495242  PGC 32659             1.01
```

R.A. 2000 DEC.	Names	Type	logD$_{25}$	p.a.	B$_T$	(B-V)$_T$	(B-V)$_{\bullet}$	m$_{21}$	V$_{21}$
l b		S$_T$ n$_L$	logR$_{25}$	A$_g$	m$_B$	(U-B)$_T$	(U-B)$_{\bullet}$	W$_{20}$	V$_{opt}$
SGL SGB		T	logA$_{\bullet}$	A$_i$	m$_{FIR}$	(B-V)^o_T	m'$_{\bullet}$	W$_{50}$	V$_{GSR}$
R.A. 1950 DEC.	PGC	L	logD$_o$	A$_{21}$	B^o_T	(U-B)^o_T	m'$_{25}$	HI	V$_{3K}$

105247.8-454040		.SXS5*.	1.31± .04	72					
282.20 12.36	ESO 264- 48	S (1)	.51± .05	.70	14.45 ±.14				
161.23 -32.39	IRAS10505-4524	5.0±1.2	.66± .04	.76	13.52				
105034.1-452442	PGC 32660	4.4±1.2	1.38	.25					
105249.5+073715	A 1050+07	.I..9..	1.16± .13					17.10±.3	3394± 7
242.19 55.80	UGC 5999	U (1)	.05± .12	.05					
99.37 -26.68	DDO 90	10.0± .8		.04				60± 7	3282
105013.0+075313	PGC 32661	9.0±1.0	1.16	.02					3742
105253.1-325534	NGC 3449	.SAS2*.	1.52± .03	148	12.19V±.15				
275.98 23.67	ESO 376- 25	PS (1)	.54± .03	.27	13.00 ±.11				3237± 66
146.20 -33.89	IRAS10505-3240	2.0± .4	.66± .04	.67	12.93				3029
105032.0-323936	PGC 32666	2.8± .6	1.55	.27	12.03				3556
105259.5+101240	NGC 3444	.S..4*/	1.06± .07	19					
238.47 57.39	UGC 6004	PU	.80± .05	.04	15.46 ±.19				
96.69 -25.65		3.7± .8		1.18					
105022.1+102838	PGC 32670		1.07	.40					
105300.5+173430	NGC 3443	.SA.7..	1.45± .03	145				14.83±.1	1132± 11
225.99 61.09	UGC 6000	U	.31± .05	.05	13.7 ±.3			176± 16	
89.27 -22.59		7.0± .8		.42				175± 12	1054
105020.4+175028	PGC 32671		1.45	.15	13.17			1.50	1460
105303.7+043743		.E?....	.67± .13						
246.24 53.92	UGC 6003		.00± .03	.13	14.7 ±.3				5780± 39
102.57 -27.71	MK 1267			.00					5658
105028.3+045341	PGC 32672		.70		14.49				6132
105305.3-401946		.E?....	1.18± .07	135	13.62 ±.13	1.04± .01	1.05± .01		
279.73 17.15	ESO 318- 21		.19± .05	.42	13.91 ±.14	.56± .03	.57± .02		4831± 31
154.99 -33.20			.80± .02	.00		.90	13.11± .06		4616
105048.0-400348	PGC 32673		1.19		13.26	.49	14.02± .37		5126
105308.1+501702	A 1050+50	.P...$.	.97± .07	110				15.72±.1	1385± 10
159.11 57.71	UGC 5998	R	.44± .05	.00	14.7 ±.2			189± 6	1394± 46
59.77 -5.95	MK 156	99.0		.66				173± 5	1431
105010.6+503300	PGC 32678		.97	.22	14.00			1.49	1585
105308.2+335436	NGC 3442	.S..1$.	.79± .07	30	13.8 ±.2	.41± .04		16.21±.2	1733± 5
190.70 63.74	UGC 6001	P	.12± .05	.00	13.7 ±.4	-.28± .07		179± 5	1725± 37
73.96 -14.72	MK 418	1.0±1.9		.12	12.53	.37		126± 4	1716
105021.2+341034	PGC 32679		.79	.06	13.65	-.29	12.30± .42	2.50	2008
105309.9+372528		.SBS3..	1.06± .06	110					
182.97 63.11	UGC 6002	U	.08± .05	.00	15.1 ±.3				
70.84 -12.89		3.0± .8		.11					
105021.1+374126	PGC 32680		1.06	.04					
105312.6-072546		.SAT5*.	1.11± .06	85					2469
259.04 45.13	MCG -1-28- 4	E (1)	.17± .05	.04					
115.95 -31.25		5.0± .9		.26					2312
105041.3-070948	PGC 32681	5.3±1.2	1.12	.09					2827
105316.9-221905		.LBT+*.	1.12± .05	179					
270.03 32.90	ESO 569- 16	S	.21± .03	.25	14.78 ±.14				
133.46 -33.64		-1.0±1.1		.00					
105051.1-220306	PGC 32685		1.12						
1053.3 +3149									10370
195.43 63.96	CGCG 155- 33			.09	15.13 ±.18				10345
75.87 -15.74									10653
1050.6 +3205	PGC 32693								
105323.4+164624	NGC 3447	.SXS9P.	1.57± .03	0				13.40±.1	1069± 5
227.56 60.83	UGC 6006	PU	.24± .03	.02	13.1 ±.4			154± 7	1048± 34
90.10 -22.85	IRAS10507+1702	8.5± .5		.25				108± 11	988
105043.7+170222	PGC 32694		1.57	.12	12.79			.48	1399
105323.4 -0035		.S..8*.	1.14± .09	143					5547
252.46 50.34	UGC 6011	U	.56± .05	.10					
108.29 -29.33		8.0±1.3		.69					5409
1050.9 -0020	PGC 32697		1.15	.28					5904
105329.0+164707	NGC 3447A	.IBS9P.	1.19± .05	110				14.30±.3	1082± 7
227.56 60.85	UGC 6007	PU	.29± .05	.02				116± 7	1014± 79
90.10 -22.83		10.0± .6		.22				84± 7	1001
105049.4+170305	PGC 32700		1.19	.14					1412
1053.5 +5046		.SB.3..	1.25± .04	154					
158.29 57.49	UGC 6008	U	.83± .05	.00	15.32 ±.18				
59.40 -5.62	IRAS10506+5102	3.0± .9		1.15	12.69				
1050.6 +5102	PGC 32705		1.25	.42					

R.A. 2000 DEC. l b SGL SGB R.A. 1950 DEC.	Names PGC	Type S_T n_L T L	logD_25 logR_25 logA_e logD_o	p.a. A_g A_i A_21	B_T m_B m_FIR B_T^o	(B-V)_T (U-B)_T (B-V)_T^o (U-B)_T^o	(B-V)_e (U-B)_e m'_e m'_25	m_21 W_20 W_50 HI	V_21 V_opt V_GSR V_3K
105341.1-214729 269.80 33.40 132.84 -33.50 105115.1-213130	NGC 3453 ESO 569- 17 IRAS10512-2131 PGC 32707	.SBS3*. SE (2) 3.0± .6 4.4± .9	1.03± .04 .25± .03 1.05	4 .26 .35 .13	 13.87 ±.14 11.59 13.23				3950± 60 3760 4294
105342.5+094341 239.39 57.25 97.27 -25.68 105105.3+095940	 UGC 6014 PGC 32708	.SX.8*. U 8.0±1.2 	1.16± .13 .25± .12 1.16	70 .05 .30 .12	 14.4 ±.2 14.06			16.37±.3 93± 5 2.18	1135± 7 1030 1480
1053.7 +2654 206.62 63.84 80.43 -18.11 1051.0 +2710	 UGC 6012 PGC 32709	.S..4.. U 4.0±1.0 	1.08± .06 .92± .05 1.08	157 .00 1.35 .46					6340 6297 6641
105350.2+570713 149.55 53.50 54.09 -2.08 105046.2+572311	NGC 3440 UGC 6009 IRAS10508+5722 PGC 32714	.SB.3$/ P 3.0±1.7 	1.32± .04 .62± .04 1.32	48 .00 .86 .31	 14.02 ±.18 13.65 13.15				1911± 10 1920± 50 1982 2074
105350.5-114334 262.79 41.80 120.95 -32.04 105120.7-112735	IC 654 MCG -2-28- 18 PGC 32716	PLB.+?. E -1.0±1.8 	1.07± .06 .27± .05 1.03	135 .07 .00					
1053.8 +6231 143.42 49.68 49.59 .93 1050.7 +6247	 UGC 6010 PGC 32718	.S..9*. U 9.0±1.2 	1.00± .16 .09± .12 1.00	175 .00 .09 .04					
105355.3+734120 133.44 40.95 40.25 7.10 105014.4+735718	NGC 3403 UGC 5997 IRAS10502+7357 PGC 32719	.SA.4*. R (2) 4.0± .4 5.3± .8	1.48± .02 .41± .03 1.49	73 .13 .60 .21	 13.02 ±.13 12.59 12.28			13.29±.1 297± 16 .80	1262± 11 1211± 58 1387 1325
1053.9 -1608 266.11 38.18 126.14 -32.77 1051.5 -1553	 MCG -3-28- 17 PGC 32724	 	.74± .14 .00± .07 .75	 .18				16.70±.1 113± 12	4370± 9 4192 4722
105400.0+493938 159.94 58.18 60.37 -6.18 105103.4+495537	 UGC 6013 IRAS10510+4955 PGC 32726	.L..... U -2.0± .9 	.96± .09 .07± .03 .94	45 .00 .00	 13.97 ±.18 13.48 13.87				6577± 50 6620 6780
105402.6+460140 165.96 60.09 63.47 -8.15 105108.6+461739	 UGC 6015 PGC 32729	.S..6*. U 6.0±1.3 	.95± .07 .14± .05 .95	0 .00 .20 .07	 14.81 ±.18				
105403.4-160141 266.05 38.30 126.00 -32.73 105135.1-154541	NGC 3456 MCG -3-28- 18 IRAS10515-1545 PGC 32730	.SBT5*. PE (1) 4.5± .6 3.1± .8	1.27± .04 .17± .04 1.29	85 .17 .25 .08	 12.23			14.43±.1 276± 9 271± 12	4193± 7 4015 4545
105405.1-454900 282.47 12.34 161.34 -32.14 105151.0-453300	 ESO 264- 49 PGC 32731	.LXR+*. Sr -.8± .7 	1.10± .05 .29± .05 1.11	162 .60 .00	 14.49 ±.14				
105406.1+203838 220.17 62.48 86.40 -20.98 105125.1+205437	 UGC 6018 PGC 32734	.I..9.. U 10.0± .8 	1.04± .08 .00± .06 1.04	 .00 .00 .00				16.75±.3 57± 5	1291± 7 1225 1611
105407.3-330706 276.33 23.62 146.41 -33.63 105146.0-325106	 ESO 376- 26 IRAS10518-3251 PGC 32736	PLXT+*. S -1.0± .6 	1.32± .04 .31± .04 1.30	100 .28 .00	 13.38 ±.14 13.05				3620± 60 3412 3938
1054.1 +4437 168.47 60.77 64.69 -8.89 1051.3 +4453	 UGC 6017 PGC 32738	.S..6*. U 6.0±1.4 	1.11± .05 .98± .05 1.11	80 .00 1.44 .49					
1054.2 +5418 153.13 55.40 56.48 -3.59 1051.2 +5434	 UGC 6016 PGC 32740	.I..9.. U 10.0± .8 	1.33± .05 .25± .06 1.33	30 .00 .19 .12				14.35±.2 118± 17 136± 25	1493± 10 1554 1671
105414.2-112420 262.64 42.12 120.60 -31.88 105144.3-110820	NGC 3452 MCG -2-28- 19 PGC 32742	.S..1*/ E 1.0±1.3 	1.02± .07 .60± .05 1.03	65 .02 .61 .30					

R.A. 2000 DEC.	Names	Type	logD25	p.a.	BT	(B-V)T	(B-V).	m21	V21
l　b		ST　nL	logR25	Ag	mB	(U-B)T	(U-B).	W20	Vopt
SGL　SGB		T	logA.	Ai	mFIR	(B-V)T^o	m'.	W50	VGSR
R.A. 1950 DEC.	PGC	L	logD.	A21	BT^o	(U-B)T^o	m'25	HI	V3K
105414.4+210807		.S..3..	1.11± .05	55				15.64±.3	9764± 10
219.18　62.67	UGC　6020	U　(1)	.84± .05	.00					
85.94　-20.73		3.0±1.0		1.16				604±　7	9700
105133.1+212406	PGC 32743	4.5±1.3	1.11	.42					10083
1054.2　+1748		.I..9*	1.07± .14	10					970
225.82　61.45	UGC　6022	U	.25± .12	.03					
89.19　-22.22		10.0±1.2		.19					894
1051.6　+1804	PGC 32747		1.07	.13					1298
105420.5-181154		.I?....	1.00± .05					15.66±.1	4134±　9
267.62　36.52	ESO　569- 20		.09± .04	.13	15.39 ±.14			66± 12	
128.58　-32.97				.07					3952
105153.0-175554	PGC 32752		1.01	.05	15.19			.43	4484
105420.7-042049		.S..2*.	1.10± .06	140					
256.52　47.69	MCG -1-28- 5	E	.49± .05	.03					
112.54　-30.19		2.0±1.3		.61					
105148.3-040449	PGC 32753		1.11	.25					
105420.8+271424		.S..7..	1.23± .05	50				15.09±.3	1334± 10
205.89　64.02	UGC　6023	U	.32± .05	.00	13.61 ±.18				
80.18　-17.83	IRAS10516+2730	7.0± .8		.44	12.56			233±　7	1292
105137.1+273023	PGC 32754		1.23	.16	13.16			1.77	1634
105428.9-461242		.SBR2..	1.01± .06	176					
282.71　12.02	ESO　264- 50	r	.27± .05	.73	14.93 ±.14				
161.77　-32.00	IRAS10522-4556	2.2± .9		.33	13.61				
105215.0-455642	PGC 32762		1.07	.13					
105429.1+172043		.SBS5$/	1.32± .04	116				15.35±.2	1108±　5
226.73　61.31	UGC　6026	R	.71± .04	.01	14.18 ±.18			224±　5	1153± 33
89.66　-22.37		5.0± .9		1.06				212± 10	1031
105149.3+173642	PGC 32763		1.32	.35	13.10			1.89	1438
1054.5　+5902		.S..1..	1.01± .06	55					
147.16　52.24	UGC　6019	U	.32± .05	.00	15.42 ±.19				
52.54　-.94		1.0± .9		.33					
1051.4　+5918	PGC 32765		1.01	.16					
105431.4+171708		PSXT3..	1.39± .03	80				14.10±.1	1107±　4
226.85　61.29	UGC　6028	R　(2)	.21± .03	.02	12.89 ±.13			215±　4	1113± 33
89.72　-22.39	IRAS10518+1733	3.0± .3		.29				206±　5	1029
105151.6+173308	PGC 32767	4.0± .8	1.40	.11	12.58			1.42	1437
1054.5　+5558		.S?....	.94± .11						14520± 60
150.89　54.35	MCG 9-18- 53		.11± .07	.00	15.4 ±.2				14587
55.11　-2.64				.14					14689
1051.5　+5614	PGC 32770		.94	.06	15.13				
105436.5+565928		.SXS9..	1.21± .03		12.9 ±.2	.35± .04			14.28±.1	2023±　7
149.59　53.67	UGC　6021	R　(1)	.04± .03	.00	12.86 ±.15	-.29± .08		161± 11	1990± 34
54.26　-2.06	IRAS10515+5715	9.0± .4		.04	12.23	.33		102± 11	2092
105133.0+571527	PGC 32772	5.7± .9	1.21	.02	12.83	-.31	13.69± .26	1.43	2184
105439.1+541824		.I.0...	1.75± .03	65	12.48 ±.13	.43± .01	.44± .02	12.91±.1	1350±　5
153.05　55.45	UGC　6024	R	.50± .04	.00	12.04 ±.11	-.19± .02	-.17± .02	328±　7	1388± 13
56.51　-3.54	ARP　205	90.0	.89± .01	.38	11.22	.33	12.42± .02	239±　8	1417
105138.5+543423	PGC 32774		1.73		11.84	-.24	14.84± .21		1534
105440.5-210354		.SBT5..	1.41± .02	112	13.0 ±.3	.44± .07		14.30±.3	3742± 17
269.57　34.13	ESO　569- 22	PSUE　(4)	.19± .03	.19	13.14 ±.12	-.05± .07		347± 34	3806± 66
132.00　-33.20	IRAS10521-2047	5.0± .4		.28	12.26	.34		337± 25	3558
105214.1-204754	PGC 32778	2.5± .4	1.43	.09	12.64	-.12	14.45± .33	1.57	4092
105444.4-170232		PSB.2*.	1.20± .05	155				15.44±.1	2663±　9
266.93　37.54	MCG -3-28- 22	E	.48± .05	.11				379± 12	
127.22　-32.72	IRAS10522-1646	2.0±1.2		.59	12.17				2483
105216.5-164632	PGC 32782		1.21	.24					3014
105448.6+611725		.S..3..	1.27± .04	35				15.38±.1	5158±　9
144.59　50.65	UGC　6025	U　(1)	.18± .05	.00	14.0 ±.2			362± 11	5118± 40
50.68　.34	IRAS10517+6132	3.0± .8		.25					5242
105139.6+613325	PGC 32786	2.5±1.1	1.27	.09	13.67			1.62	5295
105448.7+173718		.S?....	.96± .09					17.07±.3	1158±　9
226.27　61.50	UGC　6030		.00± .06	.01	13.6 ±.3				1156± 50
89.43　-22.18				.00				202±　5	1081
105208.8+175318	PGC 32787		.96	.00	13.61			3.46	1486
1055.0　+6401		.I..9..	1.11± .14					16.31±.1	1702±　6
141.75　48.64	UGC　6027	U	.25± .12	.02				97±　7	
48.41　1.86		10.0± .9		.18				72± 12	1797
1051.8　+6417	PGC 32799		1.11	.12					1825

R.A. 2000 DEC.	Names	Type	logD$_{25}$	p.a.	B$_T$	(B-V)$_T$	(B-V)$_•$	m$_{21}$	V$_{21}$
l b		S$_T$ n$_L$	logR$_{25}$	A$_g$	m$_B$	(U-B)$_T$	(U-B)$_•$	W$_{20}$	V$_{opt}$
SGL SGB		T	logA$_•$	A$_i$	m$_{FIR}$	(B-V)$_T^o$	m'	W$_{50}$	V$_{GSR}$
R.A. 1950 DEC.	PGC	L	logD$_o$	A$_{21}$	B$_T^o$	(U-B)$_T^o$	m'$_{25}$	HI	V$_{3K}$
105501.9+494334	A 1052+49	.P.....	1.00± .06	35				14.72±.1	1363± 7
159.65 58.28	UGC 6029	R	.06± .05	.00	14.03 ±.18			201± 5	1427± 74
60.41 -6.00	MK 157	99.0		.08	13.73			156± 4	1408
105205.5+495934	PGC 32800		1.00	.03	13.94			.75	1566
105513.5-260931	NGC 3463	.S..3..	1.19± .04	77					3810± 60
272.77 29.82	ESO 502- 2	S (1)	.37± .03	.34	13.78 ±.14				3613
138.09 -33.42	IRAS10528-2552	3.0± .8		.51	13.03				4146
105249.0-255330	PGC 32813	4.4± .9	1.22	.19	12.90				
105521.0+074144	NGC 3462	.L...*.	1.23± .14	60					
242.79 56.33	UGC 6034	U	.14± .08	.05	13.19 ±.15				
99.56 -26.07		-2.0±1.1		.00					
105244.7+075745	PGC 32822		1.21						
1055.4 +1707		.IB.9..	1.16± .05					15.32±.1	1072± 4
227.33 61.43	UGC 6035	U	.05± .05	.04	14.3 ±.3			48± 3	
89.98 -22.25		10.0± .8		.04				36± 3	994
1052.8 +1724	PGC 32826		1.16	.03	14.23			1.06	1402
105533.1+421759		.S..3..	1.03± .06	135					
172.62 61.98	UGC 6033	U (1)	.20± .05	.00	15.2 ±.2				
66.83 -9.92		3.0± .9		.27					
105242.0+423400	PGC 32831	3.5±1.2	1.03	.10					
105549.2-095137		.SAS5?.	1.10± .05	105					
261.82 43.59	MCG -2-28- 21	E (1)	.13± .04	.02					
118.91 -31.18	MK 1270	4.5± .6		.19	13.20				
105318.7-093535	PGC 32846	2.3± .7	1.10	.06					
105555.3+365140		.S..2..	1.20± .05	101					6502± 19
183.91 63.77	UGC 6036	U	.67± .05	.00	14.60 ±.18				6497
71.63 -12.71		2.0± .9		.83					6765
105307.3+370741	PGC 32850		1.20	.34	13.71				
1056.0 +7253		.S..7..	1.15± .05	78					
133.89 41.68	UGC 6032	U	.34± .05	.18					
41.00 6.79		7.0± .8		.46					
1052.4 +7310	PGC 32851		1.17	.17					
105601.9+570707	NGC 3458	.LX..*.	1.14± .05	5					1818± 40
149.21 53.72	UGC 6037	R	.20± .03	.00	13.33 ±.12				1889
54.26 -1.83		-2.0± .4		.00					1980
105258.9+572308	PGC 32854		1.11		13.31				
105609.7+472332		.S..3..	1.04± .06						
163.25 59.71	UGC 6038	U (1)	.02± .05	.00	15.1 ±.3				
62.50 -7.11		3.0± .8		.03					
105315.4+473934	PGC 32863	2.5±1.1	1.04	.01					
1056.2 +6040		.E?....			15.8 ±.2	1.07± .03			
145.05 51.21	MCG 10-16- 28			.00	15.71 ±.18	.55± .05			
51.28 .15	7ZW 352								
1053.1 +6057	PGC 32867								
1056.2 +4136		.S..8*.	1.00± .08	64					
173.87 62.36	UGC 6041	U	.63± .06	.00					
67.49 -10.18		8.0±1.4		.77					
1053.4 +4153	PGC 32870		1.00	.31					
1056.2 +1512		.S..6*.	1.04± .06	73					8151
231.04 60.72	UGC 6043	U	.29± .05	.00	15.04 ±.20				
91.97 -22.90		6.0±1.3		.42					8066
1053.6 +1529	PGC 32871		1.04	.14	14.58				8485
105615.7+094518	NGC 3466	.SBS3*.	1.07± .05	55					
240.04 57.77	UGC 6042	PU	.25± .04	.02	14.36 ±.18				
97.51 -25.09		2.5± .6		.34					
105338.6+100120	PGC 32872		1.07	.12					
1056.3 +5645		.S..7*.	1.15± .05	25					
149.60 53.99	UGC 6039	U	.48± .05	.00					
54.58 -1.99		7.0±1.2		.66					
1053.3 +5702	PGC 32876		1.15	.24					
105621.3-503333		PLXR+*.	1.18± .05	85					
284.94 8.25	ESO 215- 7	Sr	.36± .04	1.10	14.03 ±.14				
166.63 -30.78	IRAS10541-5017	-1.1± .7		.00	12.87				
105410.1-501730	PGC 32877		1.23						
105641.6+671107		.S?....			16.18S±.15				9939± 41
138.56 46.31	CGCG 314- 1			.00	15.5 ±.3				10045
45.85 3.74	HICK 49A								10043
105323.2+672709	PGC 32899								

R.A. 2000 DEC. l b SGL SGB R.A. 1950 DEC.	Names PGC	Type S_T n_L T L	$\log D_{25}$ $\log R_{25}$ $\log A_e$ $\log D_o$	p.a. A_g A_i A_{21}	B_T m_B m_{FIR} B_T^o	$(B-V)_T$ $(U-B)_T$ $(B-V)_T^o$ $(U-B)_T^o$	$(B-V)_e$ $(U-B)_e$ m'_e m'_{25}	m_{21} W_{20} W_{50} HI	V_{21} V_{opt} V_{GSR} V_{3K}
105644.3+094530 240.17 57.87 97.56 -24.98 105407.4+100133	NGC 3467 UGC 6045 PGC 32903	.L..... U -2.0± .9	.97± .04 .05± .08 .51± .04 .97	.02 .00	14.43 ±.15 14.11 ±.16 14.11	.99± .02 .51± .04 .89 .55	1.00± .02 .55± .04 12.47± .12 14.05± .32		9477± 31 9374 9822
1056.8 +0653 244.30 56.11 100.54 -26.03 1054.2 +0710	 UGC 6046 VV 149 PGC 32907	.SB.6*. U 6.0±1.3	.97± .07 .15± .05 .98	60 .05 .23 .08	14.63 ±.18				
105649.6-473927 283.73 10.90 163.33 -31.32 105436.1-472324	 ESO 264- 51 PGC 32909	PSBS1.. r 1.0± .9	1.02± .06 .48± .05 1.10	155 .79 .49 .24	15.46 ±.14				
105657.7-141803 265.56 40.12 124.11 -31.76 105428.6-140200	NGC 3469 MCG -2-28- 24 PGC 32912	PSBR2.. E 2.0± .8	1.23± .05 .16± .05 1.24	115 .12 .20 .08					4646 4473 5001
1057.2 +2832 203.00 64.77 79.30 -16.64 1054.5 +2849	 CGCG 155- 40 PGC 32926			.00	15.2 ±.3				13703 13667 13999
1057.5 +0541 246.12 55.45 101.87 -26.30 1054.9 +0558	 UGC 6049 PGC 32937	.S..2.. U 2.0± .9	.96± .09 .22± .06 .97	67 .07 .27 .11	14.88 ±.18				
105730.3-452046 282.81 13.03 160.67 -31.64 105515.0-450442	 ESO 264- 54 PGC 32938	.SBR1*. r 1.0±1.2	.99± .07 .09± .06 1.06	60 .65 .09 .04	15.09 ±.14				
105731.5+405648 175.02 62.84 68.20 -10.32 105441.8+411251	NGC 3468 UGC 6048 PGC 32940	.L..... U -2.0± .8	1.20± .14 .19± .08 1.17	8 .00 .00	13.96 ±.16			17.05±.3	2468± 10 7536± 31
105737.1-252540 272.89 30.72 137.25 -32.85 105512.0-250936	 ESO 502- 5 PGC 32944	.L?.... 	1.08± .06 .20± .03 1.09	110 .32 .00	13.85 ±.14 13.47				3790± 60 3595 4128
105747.8-475037 283.96 10.80 163.51 -31.13 105534.0-473433	 PGC 32955	.LA.0.. S -2.5± .6	1.07± .10 .10± .08 1.17	 .91 .00					
105751.2-200004 269.67 35.43 130.83 -32.36 105524.0-194400	 ESO 569- 27 PGC 32961	.LB.0P* E -2.0±1.2	1.11± .04 .21± .04 1.09	154 .13 .00	14.09 ±.14				
105751.9-392622 280.17 18.36 153.82 -32.39 105533.0-391018	 ESO 318- 24 PGC 32962	.SBS9P. S 9.0± .8	1.37± .04 .23± .05 1.41	153 .37 .24 .12	14.00 ±.14				
105753.6-045418 258.06 47.83 113.44 -29.49 105521.3-043814	IC 657 MCG -1-28- 9 PGC 32966	.SBT1?. E 1.0± .9	1.07± .06 .38± .05 1.07	12 .02 .39 .19					
105801.3+202859 221.18 63.29 87.00 -20.24 105520.7+204503	 MK 634 PGC 32975			.00	15.90 ±.15	.68± .03 -.44± .05			19859± 30 19793 20180
1058.0 +4011 176.48 63.19 68.91 -10.63 1055.2 +4028	 UGC 6050 PGC 32976	.S..6*. U 6.0±1.4	1.03± .06 .86± .05 1.03	46 .00 1.27 .43					
1058.0 +1706 227.94 61.98 90.29 -21.70 1055.4 +1723	NGC 3473 UGC 6052 PGC 32978	.SB.3*. U 3.0±1.2	1.06± .06 .05± .05 1.06	40 .00 .07 .03	14.35 ±.20 14.21				9115± 56 9037 9445
105803.9-061538 259.36 46.79 114.97 -29.80 105532.1-055934	IC 659 MCG -1-28- 10 PGC 32979	.E+.... E -4.0± .9	1.15± .10 .13± .07 1.11	60 .00 .00					

R.A. 2000 DEC. / l b / SGL SGB / R.A. 1950 DEC.	Names / PGC	Type / S_T n_L / T / L	$\log D_{25}$ / $\log R_{25}$ / $\log A_\bullet$ / $\log D_o$	p.a. / A_g / A_i / A_{21}	B_T / m_B / m_{FIR} / B_T^o	$(B-V)_T$ / $(U-B)_T$ / $(B-V)_T^o$ / $(U-B)_T^o$	$(B-V)_\bullet$ / $(U-B)_\bullet$ / $m'_\bullet$ / m'_{25}	m_{21} / W_{20} / W_{50} / HI	V_{21} / V_{opt} / V_{GSR} / V_{3K}
105805.7-044535		.S..5./	1.14± .06	150					
257.98 47.97	MCG -1-28- 11	E	.83± .05	.02					
113.29 -29.40		5.0± .9		1.24					
105533.5-042930	PGC 32985		1.14	.41					
1058.1 +0601		.SAS5*.	1.02± .06					16.66±.3	7871± 11
245.85 55.78	UGC 6053	UF (1)	.02± .05	.09	14.5 ±.2			81± 7	
101.58 -26.04		5.3± .7		.03				60± 7	7756
1055.5 +0618	PGC 32986	2.6±1.0	1.02	.01	14.28			2.37	8222
1058.1 +0916	NGC 3476	.E.....							
241.27 57.85	MCG 2-28- 32	F		.04	14.8 ±.6				
98.19 -24.85		-5.0± .9							
1055.5 +0933	PGC 32987								
1058.1 +1705	NGC 3474	.S?....	.89± .11						
227.99 61.99	MCG 3-28- 42		.07± .07	.00	14.85 ±.18				9115± 56
90.32 -21.68				.09					9037
1055.5 +1722	PGC 32989		.89	.03	14.69				9445
105809.5-693905		.SBS5?.	1.11± .07	150					
293.32 -8.92	ESO 63- 1	S (1)	.50± .05	2.34					
186.86 -24.66	IRAS10564-6923	5.0± .9		.75	12.77				
105624.0-692300	PGC 32990	3.3±1.4	1.33	.25					
105809.6-004634		.S..5?/							
254.01 51.01	CGCG 10- 46	F		.10	15.0 ±.6				
108.89 -28.26		5.0±1.8							
105536.0-003030	PGC 32991								
1058.1 +2004		.S..6*.	1.04± .08	12					4171
222.04 63.18	UGC 6054	U	1.06± .06	.00	16.1 ±.2				
87.41 -20.38		6.0±1.5		1.47					4104
1055.5 +2021	PGC 32992		1.04	.50	14.57				4493
105810.7-773753		.S..3?/	1.19± .07	115					
296.86 -16.12	ESO 38- 6	S	.74± .05	.50					
194.69 -21.41		3.0±1.8		1.03					
105658.0-772148	PGC 32994		1.24	.37					
1058.2 +0733		.LBR0*.							
243.78 56.80	CGCG 38- 68	F		.05	15.0 ±.6				
99.99 -25.46		-2.0±1.3							
1055.6 +0750	PGC 32995								
105812.9-261822		PSBS0..	.88± .06	155					
273.53 30.02	ESO 502- 7	r	.15± .05	.31	14.80 ±.14				
138.30 -32.75	IRAS10558-2602	-.1± .9		.12	12.62				
105548.1-260218	PGC 32998		.90						
1058.2 +0434		RSX.0*.							
247.80 54.85	CGCG 38- 69	F		.53	14.9 ±.6				
103.13 -26.50		.0±1.3							
1055.7 +0451	PGC 33002								
1058.3 +0814	IC 658	.E+..P.							
242.84 57.25	MCG 2-28- 33	F		.03	14.4 ±.6				
99.28 -25.19		-4.0± .9							
1055.7 +0831	PGC 33004								
105825.5+241340	NGC 3475	.S..1..	1.23± .05	65				15.70±.3	6431± 10
213.07 64.43	UGC 6058	U	.18± .05	.00	14.0 ±.2				6350± 53
83.47 -18.45	IRAS10556+2427	1.0±1.1		.18	13.03			527± 7	6377
105543.5+242944	PGC 33012		1.23	.09	13.73			1.88	6739
1058.4 +0437		.S..2P/							
247.80 54.92	CGCG 38- 72	F		.13	15.1 ±.6				
103.10 -26.44		2.0±1.8							
1055.9 +0454	PGC 33019								
105833.8-463505	NGC 3482	PSBT1..	1.28± .04	14					
283.53 11.99	ESO 264- 56	Sr	.14± .05	.72	13.51 ±.14				2870± 60
162.05 -31.24		.8± .4		.14					2653
105619.0-461900	PGC 33025		1.35	.07	12.61				3142
105837.6+090305		.LXR...	1.08± .10	25					
241.76 57.81	UGC 6062	UF	.12± .05	.04	13.70 ±.16				2607
98.49 -24.82		-1.5± .6		.00					2502
105600.9+091910	PGC 33030		1.06		13.61				2954
1058.6 -1531		.S?....	1.30± .06						7634
266.89 39.33	MCG -2-28- 26		.91± .07	.16					
125.62 -31.56	IRAS10562-1515			1.34	12.28				7459
1056.2 -1515	PGC 33032		1.31	.46					7988

R.A. 2000 DEC.	Names	Type	$\log D_{25}$	p.a.	B_T	$(B-V)_T$	$(B-V)_\bullet$	m_{21}	V_{21}
l b		S_T n_L	$\log R_{25}$	A_g	m_B	$(U-B)_T$	$(U-B)_\bullet$	W_{20}	V_{opt}
SGL SGB		T	$\log A_\bullet$	A_i	m_{FIR}	$(B-V)_T^o$	$m'_\bullet$	W_{50}	V_{GSR}
R.A. 1950 DEC.	PGC	L	$\log D_o$	A_{21}	B_T^o	$(U-B)_T^o$	m'_{25}	HI	V_{3K}

1058.7 +5535		.SXS4..	1.24± .05	15					
150.68 55.03	UGC 6059	U	.44± .05	.00	15.2 ±.3				
55.74 −2.35		4.0± .8		.65					
1055.7 +5552	PGC 33033		1.24	.22					
105843.2-501929		.S..4./	1.24± .05	179					
285.18 8.62	ESO 215- 12	S	.73± .04	1.05	15.43 ±.14				
166.27 −30.47	IRAS10564-5003	4.0± .9		1.07					
105631.0-500324	PGC 33034		1.34	.36					
105845.6+593037	NGC 3470	.SAR2*.	1.16± .05	170					
145.98 52.28	UGC 6060	PU	.07± .04	.00	14.10 ±.19				6651± 42
52.44 −.23		2.0± .6		.08					6731
105541.0+594641	PGC 33040		1.16	.03	13.95				6800
1058.7 +2508		.S?....	1.17± .04	65				15.79±.3	6052± 10
211.03 64.69	UGC 6063		.82± .04	.00	15.34 ±.18			445± 13	
82.65 −17.96	IRAS10560+2524			1.23	13.78				6004
1056.0 +2524	PGC 33041		1.17	.41	14.08			1.30	6359
1058.8 +0530		.L..+*/							
246.74 55.57	CGCG 38- 73	F		.07	15.4 ±.6				
102.19 −26.06		−1.0±1.3							
1056.2 +0547	PGC 33044								
1058.8 +5515		.SB.6*.	1.02± .06	12					
151.09 55.27	UGC 6061	U	.53± .05	.00					
56.03 −2.51		6.0±1.3		.77					
1055.8 +5532	PGC 33045		1.02	.26					
1058.8 +7237									8129± 79
133.86 42.02	CGCG 333- 64			.18	14.7 ±.3				8253
41.34 6.83									8200
1055.3 +7254	PGC 33047								
105853.5-495824		.IBS9..	1.22± .05	174					2726± 87
285.05 8.95	ESO 215- 13	S (1)	.16± .05	1.01	14.19 ±.14				2508
165.87 −30.52	IRAS10566-4942	10.0± .8		.12	12.82				2984
105641.0-494218	PGC 33052	6.7± .8	1.32	.08	13.06				
105855.5-145742	NGC 3479	.SXR4..	1.24± .05	175				15.65±.1	4545± 9
266.55 39.83	MCG −2-28- 27	E (1)	.15± .05	.17				330± 12	
124.97 −31.41	IRAS10563-1441	4.0± .8		.21					4371
105626.5-144137	PGC 33053	4.2± .8	1.26	.07					4899
105857.0+723835 A	1055+72								8105± 35
133.84 42.02				.18					8230
41.34 6.84	MK 159				13.15				8177
105524.9+725439	PGC 33056								
1059.0 +0630		.SA.2*/	1.13± .07	38					
245.47 56.27	UGC 6066	UF	.73± .06	.05	15.00 ±.18				
101.16 −25.66		1.8± .8		.90					
1056.4 +0647	PGC 33058		1.13	.37					
105900.1-282835	NGC 3483	PLAR+*.	1.26± .04	105					
274.90 28.20	ESO 438- 1	Sr	.16± .04	.25	13.07 ±.14				3730± 60
140.87 −32.64		−.7± .5		.00					3530
105636.0-281230	PGC 33060		1.26		12.76				4061
105900.9-432630		.SAT6*.	.96± .06	143					
282.21 14.86	ESO 264- 57	r	.09± .05	.83	15.02 ±.14				
158.43 −31.67	IRAS10567-4310	5.6±1.3		.13	11.20				
105644.0-431024	PGC 33062		1.04	.04					
1059.0 +0126		.SBT3?.							
251.82 52.79	CGCG 10- 51	F		.07	14.9 ±.6				
106.56 −27.36		3.0± .9							
1056.5 +0143	PGC 33065								
1059.0 +0459		.SXR3*.							10868
247.50 55.28	CGCG 38- 75	F (1)		.13	15.3 ±.6				10750
102.77 −26.17		3.0±1.3							11220
1056.5 +0516	PGC 33067	2.6±1.0							
105906.2+011055		.SA.2..							
252.13 52.60	CGCG 10- 52	F		.08	14.9 ±.6				
106.85 −27.43		2.0± .9							
105632.0+012700	PGC 33069								
105909.3+613151	NGC 3471	.S..1..	1.24± .05	14	13.23 ±.16	.71± .02	.72± .02	15.82±.1	2129± 10
143.73 50.82	UGC 6064	U	.32± .05	.00	13.2 ±.2	.17± .04	.14± .04	235± 6	2070± 52
50.76 .90	MK 158	1.0± .8	.76± .04	.32	10.91	.63	12.52± .13	209± 5	2214
105602.2+614756	PGC 33074		1.24	.16	12.86	.13	13.49± .30	2.80	2265

R.A. 2000 DEC. / l b / SGL SGB / R.A. 1950 DEC.	Names / PGC	Type / S_T n_L / T / L	$\log D_{25}$ / $\log R_{25}$ / $\log A_e$ / $\log D_o$	p.a. / A_g / A_i / A_{21}	B_T / m_B / m_{FIR} / B_T^o	$(B-V)_T$ / $(U-B)_T$ / $(B-V)_T^o$ / $(U-B)_T^o$	$(B-V)_e$ / $(U-B)_e$ / m' / m'_{25}	m_{21} / W_{20} / W_{50} / HI	V_{21} / V_{opt} / V_{GSR} / V_{3K}
1059.1 +0516 / 247.17 55.49 / 102.48 -26.05 / 1056.6 +0533	UGC 6068 / PGC 33078	.LA.+P? / UF / -1.3±1.0 / 1.07	1.11± .08 / .31± .04 / /	.5 / .08 / .00	14.72 ±.18				
105915.7-094901 / 262.74 44.13 / 119.07 -30.34 / 105645.0-093255	MCG -2-28- 30 / PGC 33079	.S..2P/ / E / 2.0±1.3 / 1.06	1.05± .05 / .47± .06 / /	80 / .02 / .58 / .24					
105916.2-094740 / 262.72 44.15 / 119.05 -30.33 / 105645.5-093134	MCG -2-28- 29 / IRAS10567-0931 / PGC 33080	.SXT3P* / EF (1) / 2.7± .5 / 2.5± .6 / 1.14	1.14± .06 / .15± .05 / /	30 / .02 / .21 / .07	12.99				
1059.2 +0148 / 251.46 53.09 / 106.18 -27.19 / 1056.7 +0205	CGCG 10- 54 / PGC 33081	RSBR0.. / F / .0± .9		.07	15.4 ±.6				
105917.5+243237 / 212.45 64.69 / 83.27 -18.13 / 105635.5+244843	IRAS10565+2448 / PGC 33083			.00	10.60				12912± 37 / 12862 / 13221
1059.3 +0110 / 252.21 52.65 / 106.87 -27.37 / 1056.8 +0127	CGCG 10- 55 / PGC 33090	CE..... / F / -5.0± .9		.08	15.16 ±.18				
1059.3 +0135 / 251.74 52.95 / 106.42 -27.24 / 1056.8 +0152	IC 662 / CGCG 10- 56 / PGC 33091	.E..... / F / -5.0± .9		.09	15.4 ±.6				
105926.6-662000 / 292.01 -5.87 / 183.44 -25.77 / 105733.0-660354	ESO 93- 3 / FAIR 286 / PGC 33098	.SXR0?. / Sr / .3±1.0 / 1.63	1.17± .07 / .22± .07 / /	140 / 5.09 / .17	10.60				1470± 63 / 1259 / 1647
1059.4 +7511 / 131.93 39.92 / 39.18 8.21 / 1055.7 +7527	NGC 3465 / UGC 6056 / IRAS10557+7527 / PGC 33099	.S..2.. / U / 2.0± .8 / 1.09	1.08± .09 / .09± .07 / /	/ .15 / .11 / .04	14.44 ±.19 / / / 14.11				7170 / / 7302 / 7226
105927.8+460721 / 164.84 60.85 / 63.90 -7.31 / 105635.4+462327	NGC 3478 / UGC 6069 / IRAS10565+4623 / PGC 33101	.SBT4.. / PU (1) / 3.5± .5 / 2.0±1.0 / 1.42	1.42± .03 / .35± .04 / /	132 / .00 / .51 / .17	13.57 ±.14 / 13.27 ±.14 / 13.48 / 12.87	.71± .03 / .18± .04 / .60 / .10	/ / / 14.67± .24		6667± 10 / 6658± 66 / 6698 / 6888
1059.5 +0740 / 243.99 57.12 / 100.00 -25.12 / 1056.9 +0757	CGCG 38- 78 / PGC 33104	.LAR0*. / F / -2.0±1.3		.04	14.3 ±.3				
1059.5 +0855 / 242.19 57.91 / 98.70 -24.66 / 1056.9 +0912	CGCG 66- 78 / PGC 33105	.L..... / F / -2.0± .9		.05	15.3 ±.6				
105937.3-153138 / 267.14 39.45 / 125.66 -31.33 / 105708.5-151532	MCG -2-28- 31 / PGC 33108	.S..6*/ / E (1) / 6.0±1.2 / 6.4±1.6 / 1.30	1.28± .05 / .81± .05 / /	85 / .16 / 1.19 / .41				15.55±.1 / 297± 12	3041± 9 / / 2866 / 3395
105944.4+100417 / 240.53 58.65 / 97.55 -24.18 / 105707.4+102023	UGC 6072 / PGC 33114	PSAR2.. / UF / 1.5± .6 / 1.00	1.00± .06 / .03± .05 / /	/ .02 / .03 / .01	14.41 ±.19 / / / 14.25			16.06±.3 / 212± 5 / 1.80	10647± 9 / / 10546 / 10992
105946.6+332332 / 191.50 65.17 / 75.15 -13.79 / 105701.0+333938	UGC 6070 / IRAS10569+3339 / PGC 33118	.S?....	.83± .07 / .11± .05 / / .83	/ .00 / .16 / .05	13.7 ±.4 / 12.76 / 13.58			15.40±.3 / 140± 5 / 1.77	1849± 9 / / 1832 / 2127
1059.7 +0358 / 249.00 54.71 / 103.91 -26.36 / 1057.2 +0415	CGCG 38- 79 / PGC 33120	.SB.1?. / F / 1.0±1.8		.13	15.0 ±.6				
105951.4-252943 / 273.45 30.91 / 137.36 -32.34 / 105726.0-251336	ESO 502- 8 / IRAS10574-2513 / PGC 33124	.S?....	.88± .05 / .11± .04 / / .90	96 / .23 / .16 / .05	14.43 ±.14 / 13.01 / 14.01				3810± 60 / 3615 / 4148

R.A. 2000 DEC. l b SGL SGB R.A. 1950 DEC.	Names PGC	Type S_T n_L T L	$\log D_{25}$ $\log R_{25}$ $\log A_o$ $\log D_o$	p.a. A_g A_i A_{21}	B_T m_B m_{FIR} B_T^o	$(B-V)_T$ $(U-B)_T$ $(B-V)_T^o$ $(U-B)_T^o$	$(B-V)_o$ $(U-B)_o$ m'_o m'_{25}	m_{21} W_{20} W_{50} HI	V_{21} V_{opt} V_{GSR} V_{3K}
105953.2+500054 158.33 58.74 60.59 −5.19 105658.2+501700	MCG 8-20- 61 PGC 33126	.S?.... 	.89± .11 .00± .07 .89	.00 .00 .00	14.73 ±.18 14.69				7564± 40 7610 7766
1059.9 +0921 241.66 58.25 98.30 −24.40 1057.3 +0938	NGC 3490 MCG 2-28- 36 PGC 33128	.E..... F −5.0± .9 		.04	14.8 ±.6				
105958.5+505411 156.96 58.22 59.84 −4.70 105702.9+511017	 UGC 6074 IRAS10570+5110 PGC 33136	.S?.... 	1.10± .06 .14± .05 1.10	.00 .21 .07	14.17 ±.18 11.92 13.95				2911± 50 2960 3108
105959.9+500322 158.25 58.73 60.56 −5.15 105704.9+501928	 UGC 6071 PGC 33138	.E..... U −5.0± .8 	1.15± .15 .24± .08 1.08	135 .00 .00	14.14 ±.15 14.03				7152± 25 7198 7353
110002.4+145037 232.64 61.34 92.76 −22.22 105724.0+150643	NGC 3485 UGC 6077 PGC 33140	.SBR3*. R (2) 3.0± .5 3.3± .7	1.36± .03 .06± .03 1.36	.00 .09 .03	12.62 ±.14 12.51			12.68±.3 142± 7 130± 6 .14	1434± 7 1483± 66 1350 1771
1100.0 +0146 251.73 53.20 106.29 −27.01 1057.5 +0203	MCG 0-28- 26 PGC 33144	.LAR-.. F −3.0± .9 		.09	14.3 ±.3				
1100.0 −1606 267.65 39.02 126.35 −31.31 1057.6 −1550	 PGC 33145	.L..-?. E −3.0±1.8 	1.14± .08 .55± .08 1.08	125 .14 .00					
1100.1 +1213 237.19 59.97 95.39 −23.25 1057.5 +1230	 UGC 6078 PGC 33147	.S..4.. U 4.0± .9 	1.04± .06 .33± .05 1.04	126 .01 .49 .17	14.78 ±.18				
1100.1 +7655 130.68 38.47 37.70 9.16 1056.3 +7712	 UGC 6065 PGC 33149	.S..3.. U 3.0± .9 	1.19± .07 .58± .07 1.19	.05 .80 .29	15.4 ±.2				
110009.3+455504 165.08 61.06 64.14 −7.31 105717.3+461110	 UGC 6076 PGC 33150	.E...?. U −5.0±1.8 		.00	14.5 ±.3				6410± 31 6441 6632
110010.1+102214 240.19 58.91 97.29 −23.96 105733.0+103820	CGCG 66- 81 MK 1275 PGC 33152	 		.02	15.3 ±.6				10993± 19 10893 11338
110011.5+454417 165.39 61.16 64.30 −7.40 105719.6+460023	 UGC 6075 PGC 33153	.S..3.. U (1) 3.0± .8 3.5±1.1	1.21± .05 .35± .05 1.21	107 .00 .49 .18	14.25 ±.18 13.71				6411± 50 6441 6634
1100.3 +0612 246.26 56.32 101.61 −25.47 1057.7 +0629	CGCG 38- 81 PGC 33158	.L..../ F −2.0± .9 		.04	15.1 ±.6				
1100.3 +0734 244.37 57.21 100.18 −24.97 1057.7 +0751	CGCG 38- 82 PGC 33159	.L..+*/ F −1.0±1.3 		.03	15.4 ±.6				
110018.2+135408 234.38 60.91 93.73 −22.54 105740.0+141015	NGC 3489 UGC 6082 PGC 33160	.LXT+.. R −1.0± .3 1.51	1.55± .02 .24± .02 .83± .02 	70 .02 .00	11.12 ±.13 11.30 ±.13 11.18	.83± .01 .34± .02 .79 .31	.84± .01 .37± .01 10.76± .08 13.14± .17	17.45±.3	708± 10 690± 13 613 1039
1100.3 +1002 240.73 58.75 97.64 −24.05 1057.7 +1019	 UGC 6081 PGC 33161	.S..5P. F 5.0± .9 1.03	1.03± .08 .39± .06 	.04 .59 .20					
1100.3 +1640 229.29 62.28 90.97 −21.39 1057.7 +1657	 UGC 6083 PGC 33163	.S..4.. U 4.0±1.0 1.10	1.10± .05 .93± .05 	143 .00 1.37 .46	15.5 ±.2				

R.A. 2000 DEC. l b SGL SGB R.A. 1950 DEC.	Names PGC	Type S_T n_L T L	$\log D_{25}$ $\log R_{25}$ $\log A_e$ $\log D_o$	p.a. A_g A_i A_{21}	B_T m_B m_{FIR} B_T^o	$(B-V)_T$ $(U-B)_T$ $(B-V)_T^o$ $(U-B)_T^o$	$(B-V)_e$ $(U-B)_e$ m'_e m'_{25}	m_{21} W_{20} W_{50} HI	V_{21} V_{opt} V_{GSR} V_{3K}
110023.6+285833 NGC 3486 202.07 65.49 UGC 6079 79.25 -15.82 105740.0+291440 PGC 33166		.SXR5.. R (2) 5.0± .3 2.6± .5	1.85± .01 .13± .02 1.32± .02 1.85	80 .01 .20 .07	11.05M±.10 10.68 ±.15 10.72	.52± .01 -.16± .02 .49 -.18	.60± .01 -.07± .02 13.15± .05 14.80± .14	12.04±.1 232± 4 202± 4 1.25	682± 4 665± 37 648 976
110023.7-095903 263.19 44.15 MCG -2-28- 32 119.33 -30.10 IRAS10579-0943 105753.0-094257 PGC 33167		.SXT4P* EF (1) 4.3± .6 3.1±1.2	1.18± .04 .37± .04 1.18	155 .00 .55 .19	 13.66				8189 8028 8547
1100.4 +0623 246.04 56.46 CGCG 38- 84 101.42 -25.38 1057.8 +0640 PGC 33168		.SBR4.. F (1) 4.0± .9 1.6±1.4		 .04	15.2 ±.6				
1100.5 +2941 200.34 65.54 UGC 6084 78.61 -15.45 IRAS10578+2957 1057.8 +2957 PGC 33175		.S..2.. U 2.0± .9	.90± .07 .29± .05 .90	175 .04 .36 .15	 15.25 ±.18 13.60 14.75				10352 10321 10644
110035.7+120946 NGC 3491 237.43 60.03 UGC 6088 95.51 -23.17 105758.1+122553 PGC 33180		.L..-*. U -3.0±1.2	.97± .04 .00± .04 .48± .03 .97	 .01 .00	14.27 ±.13 13.99 ±.16 14.05	1.03± .02 .57± .03 .97 .60	1.04± .02 .61± .03 12.16± .09 13.99± .25		6386± 31 6292 6727
1100.6 -0253 256.92 49.82 CGCG 10- 60 111.41 -28.28 1058.1 -0237 PGC 33186		PSBS2.. F 2.0± .9		 .06	15.2 ±.6				
1100.7 +6119 143.72 51.09 UGC 6080 51.03 .95 1057.6 +6136 PGC 33188		.S..7.. U 7.0± .9	1.33± .04 1.11± .05 1.33	129 .00 1.38 .50	 15.52 ±.18 14.13				2180 2267 2319
1100.7 +0952 241.10 58.73 UGC 6091 97.85 -24.03 1058.1 +1009 PGC 33190		.SBR4./ UF (1) 3.5± .7 1.6±1.0	1.00± .06 .42± .05 1.00	2 .02 .62 .21	 15.00 ±.18				
1100.7 +1032 IC 664 240.07 59.12 MCG 2-28- 42 97.17 -23.77 1058.1 +1049 PGC 33191		.L?.... 	1.12± .12 .00± .07 1.12	 .02 .00	13.98 ±.19 13.81				10127± 18 10028 10472
1100.8 +1043 239.81 59.25 UGC 6093 96.99 -23.68 1058.2 +1100 PGC 33198		.SXT4.. U (1) 4.0± .9 2.9± .8	1.12± .05 .07± .05 .66± .02 1.12	 .01 .11 .04	14.78 ±.13 14.1 ±.2 14.41	.80± .03 .71	.81± .03 13.57± .04 15.03± .32	16.71±.1 157± 15 2.26	10805± 10 10793± 59 10706 11149
110052.6+380617 180.57 64.39 UGC 6089 71.05 -11.23 105805.1+382224 PGC 33203		.S..6*. U 6.0±1.2	.99± .06 .07± .05 .99	 .00 .10 .04	 15.1 ±.2				
110057.4+103023 NGC 3492 240.20 59.15 UGC 6094 97.23 -23.73 8ZW 116 105820.4+104630 PGC 33207		.S?.... 	1.06± .04 .19± .06 .62± .02 1.06	100 .02 .14	14.17 ±.13 13.99 ±.18 13.79	1.02± .02 .88	1.09± .02 12.76± .04 13.87± .27		10874± 44 10774 11218
1100.9 +1904 224.65 63.42 UGC 6095 88.69 -20.23 1058.3 +1921 PGC 33208		.I..9?. U 10.0±1.8	1.14± .13 .47± .12 1.14	5 .00 .35 .24					1115 1045 1441
110059.9-140221 266.44 40.88 MCG -2-28- 37 124.00 -30.75 105830.4-134613 PGC 33211		PLBS+P? E -1.0± .8	1.19± .07 .11± .05 1.19	85 .15 .00					
110101.8+110249 239.37 59.48 CGCG 66- 94 96.69 -23.51 MK 728 105824.7+111856 PGC 33214				 .01	14.9 ±.6			18.14±.3 386± 12	10686± 9 10698± 18 10591 11032
110102.4-434038 282.65 14.80 ESO 265- 2 158.64 -31.27 105845.1-432430 PGC 33216		.SB.7?/ S 7.0± .8	1.12± .05 .77± .04 1.19	133 .72 1.06 .38	 16.18 ±.14				
1101.0 +0437 248.57 55.39 CGCG 38- 86 103.35 -25.83 1058.5 +0454 PGC 33220		.SB.6?. F 6.0±1.8		 .13	15.3 ±.6				

R.A. 2000 DEC. / l b / SGL SGB / R.A. 1950 DEC.	Names / PGC	Type / S_T n_L / T / L	$\log D_{25}$ / $\log R_{25}$ / $\log A_o$ / $\log D_o$	p.a. / A_g / A_i / A_{21}	B_T / m_B / m_{FIR} / B_T^o	$(B-V)_T$ / $(U-B)_T$ / $(B-V)_T^o$ / $(U-B)_T^o$	$(B-V)_e$ / $(U-B)_e$ / m'_e / m'_{25}	m_{21} / W_{20} / W_{50} / HI	V_{21} / V_{opt} / V_{GSR} / V_{3K}
110109.2-440226		.SXS6..	1.43± .04	110					
282.83 14.48	ESO 265- 3	S (1)	.36± .05	.81	14.07 ±.14				11240± 60
159.05 -31.20	IRAS10589-4346	6.3± .4		.53	13.34				11025
105852.0-434618	PGC 33225	6.3± .5	1.51	.18	12.68				11522
110114.4-495420		.SBT0*/	1.16± .05	82					
285.37 9.17	ESO 215- 15	S	.64± .04	.98	15.62 ±.14				
165.70 -30.16		.0±1.3		.48					
105901.0-493812	PGC 33230		1.22						
110115.8+033735	NGC 3495	.S..7*.	1.69± .02	20	12.41S±.15			13.43±.1	1136± 6
249.89 54.72	UGC 6098	R (2)	.61± .03	.86	12.49 ±.14			311± 7	1086± 24
104.43 -26.13	IRAS10586+0353	7.0± .5		.84	12.17			297± 12	1012
105840.9+035343	PGC 33234	5.0± .7	1.77	.30	10.75		14.17± .19	2.38	1488
1101.3 +2947		.S..0..	1.06± .06	157					
200.10 65.71	UGC 6097	U	.56± .05	.04	15.39 ±.18				
78.60 -15.25		.0± .9		.42					
1058.6 +3004	PGC 33238		1.03						
110120.7+030222		.SXR4..							
250.63 54.32	CGCG 38- 89	F (1)		.12	14.9 ±.6				
105.06 -26.30		4.0± .9							
105846.0+031830	PGC 33240	2.6±1.0							
110124.3+574037	NGC 3488	.SBS5*.	1.27± .04	175					2994± 10
147.68 53.84	UGC 6096	PU	.17± .04	.00	13.58 ±.18				2971± 50
54.18 -.92	IRAS10583+5756	5.3± .6		.25	13.45				3067
105823.1+575645	PGC 33242		1.27	.08	13.31				3153
1101.4 +1013									
240.76 59.07	CGCG 66- 98			.02	15.4 ±.6				10414± 88
97.56 -23.73									10314
1058.8 +1030	PGC 33244								10759
1101.4 -0212		.SBR2?.							
256.47 50.48	CGCG 10- 62	F		.11	15.1 ±.6				
110.73 -27.89		2.0±1.8							
1058.9 -0156	PGC 33248								
1101.4 +2743	NGC 3493	.S?....	1.06± .06	84					
205.14 65.64	UGC 6099		.52± .05	.03	15.19 ±.18				8963
80.53 -16.21	IRAS10587+2759			.79	13.45				8925
1058.7 +2759	PGC 33249		1.06	.26	14.33				9262
110134.2+453917		.S..1?.	.92± .07	10				15.91±.1	8718± 11
165.28 61.41	UGC 6100	U	.18± .05	.00	14.3 ±.2			363± 16	8778± 50
64.50 -7.24	IRAS10587+4555	1.0±1.8		.18	13.47				8751
105842.8+455525	PGC 33257		.92	.09	13.98			1.84	8944
110137.7-705709		.LAR+?.	1.13± .08	31					
294.14 -9.98	ESO 63- 3	S	.48± .04	.65					
188.04 -23.90		-1.0± .9		.00					
105953.1-704100	PGC 33260		1.12						
1101.7 +2841		.I..9..	1.00± .16	140				14.96±.1	699± 10
202.78 65.76	UGC 6102	U	.09± .12	.01	14.9 ±.2			117± 6	
79.66 -15.70		10.0± .9		.07				92± 5	665
1059.0 +2858	PGC 33264		1.00	.04	14.87			.06	995
1101.7 +4706		.S..4..	1.15± .07	64					
162.69 60.66	UGC 6101	U	1.04± .06	.00					
63.25 -6.45		4.0±1.0		1.47					
1058.9 +4723	PGC 33269		1.15	.50					
110151.4+163624		.S..4..	1.18± .05	50				15.08±.2	2946± 6
229.79 62.57	UGC 6104	U	.56± .05	.00	14.50 ±.18			257± 7	2936± 50
91.21 -21.09		4.0± .9		.82				246± 5	2868
105912.5+165233	PGC 33276		1.18	.28	13.65			1.15	3278
1101.8 +7511		.S..2..	1.12± .08						
131.76 40.00	UGC 6090	U	.32± .07	.15	14.69 ±.18				
39.25 8.35	IRAS10582+7528	2.0± .9		.40	13.25				
1058.2 +7528	PGC 33277		1.13	.16					
1101.9 +1017									
240.80 59.21	CGCG 66-101			.02	15.3 ±.6				10251± 19
97.55 -23.60									10151
1059.3 +1034	PGC 33278								10596
110158.4+451339	A 1059+45	.P.....	.89± .05		14.0 ±.2	.51± .04	.55± .02	15.47±.3	5990± 17
165.98 61.69	UGC 6103	R	.20± .04	.00	13.8 ±.3	-.17± .07	-.13± .04	151± 34	5947± 60
64.91 -7.40	MK 161	99.0	.52± .04	.29		.43	12.08± .09	128± 25	6016
105907.3+452947	PGC 33280		.89	.10	13.63	-.22	12.79± .35	1.74	6212

R.A. 2000 DEC. l b SGL SGB R.A. 1950 DEC.	Names PGC	Type S_T n_L T L	$\log D_{25}$ $\log R_{25}$ $\log A_o$ $\log D_o$	p.a. A_g A_i A_{21}	B_T m_B m_{FIR} B_T^o	$(B-V)_T$ $(U-B)_T$ $(B-V)_T^o$ $(U-B)_T^o$	$(B-V)_o$ $(U-B)_o$ m' m'_{25}	m_{21} W_{20} W_{50} HI	V_{21} V_{opt} V_{GSR} V_{3K}
1101.9 -1444 267.21 40.42 124.85 -30.64 1059.5 -1428	 PGC 33284	.SBS8.. E (1) 8.0± .9 8.7± .8	1.12± .08 .19± .08 1.14	150 .19 .23 .09				
110202.2-403603 281.46 17.66 155.07 -31.47 105943.0-401954	 ESO 318- 31 PGC 33288	PSBR1.. r 1.0± .9 	.90± .06 .05± .05 .96	 .55 15.08 ±.14 .05 .02				
110211.3+455324 164.75 61.38 64.35 -7.03 105919.8+460932	 UGC 6106 IRAS10593+4609 PGC 33294	.SB.3?. U 3.0± .9 	1.16± .05 .41± .05 1.16	160 .00 14.41 ±.18 .56 13.76 .20 13.80				6550± 50 6581 6772
1102.3 +1013 241.02 59.25 97.66 -23.53 1059.7 +1030	 CGCG 66-103 PGC 33303			 .01 15.3 ±.6				10572± 88 10472 10917
110220.3-140813 266.88 40.97 124.18 -30.45 105950.8-135203	NGC 3502 MCG -2-28- 41 PGC 33306	.SBT1P. E 1.0± .8 	1.07± .06 .00± .05 1.08	 .12 .00 .00				
110221.9-464433 284.20 12.12 162.10 -30.57 110006.1-462824	 ESO 265- 5 PGC 33307	.SAS8.. S (1) 7.5± .6 8.9± .6	1.02± .05 .14± .05 1.09	5 .70 15.65 ±.14 .18 .07				
1102.4 +0603 247.09 56.60 101.97 -25.03 1059.8 +0620	 MCG 1-28- 28 PGC 33310	.SBS3*. F (1) 3.0± .9 	.82± .13 .17± .07 .82	 .07 15.09 ±.19 .24 .09				
1102.4 +0828 243.70 58.18 99.46 -24.16 1059.8 +0845	 CGCG 66-104 PGC 33312	RLX.-.. F -3.0± .9 		 .05 15.0 ±.6				
1102.4 +0237 251.46 54.23 105.60 -26.17 1059.9 +0254	 MCG 1-28- 29 VV 463 PGC 33319	.SBS4P. F 4.0± .9 	.82± .13 .17± .07 .83	100 .14 15.14 ±.19 .25 .09 14.66				11942± 46 11817 12297
1102.4 +0305 250.90 54.56 105.10 -26.02 1059.9 +0322	 UGC 6111 PGC 33320	.SA.2?/ F 2.0±1.9 	.96± .09 .65± .06 .97	2 .12 15.4 ±.2 .80 .32				
110229.8-033921 258.22 49.52 112.41 -28.05 105957.0-032312	 MCG 0-28- 28 PGC 33323	.SAS1*. F 1.0±1.2 	.94± .11 .05± .07 .94	 .05 14.83 ±.20 .05 .02				
1102.5 +5035 156.95 58.72 60.31 -4.52 1059.6 +5052	 UGC 6109 PGC 33325	.SXR5.. U 5.0± .8 	1.08± .06 .00± .05 1.08	 .00 14.9 ±.3 .00 .00				
110231.0-533834 287.11 5.85 169.80 -29.15 110020.0-532224	 ESO 169- 5 PGC 33328	.LBT+?. S -1.0±1.2 	1.09± .09 .16± .06 1.21	30 1.52 .00				
110231.5-261003 274.43 30.61 138.18 -31.78 110006.1-255354	 ESO 502- 11 IRAS11000-2553 PGC 33329	.P..... S 99.0 	1.15± .04 .37± .03 1.17	88 .18 14.81 ±.14 .56 .19 14.05				3958± 37 3763 4294
110233.6+023621 251.52 54.23 105.63 -26.16 105959.1+025230	 MCG 1-28- 30 PGC 33330	.SBS4P* F 4.0±1.3 	.64± .17 .00± .07 .65	 .14 15.3 ±.2 .00 .00 15.10				11988± 46 11863 12343
1102.5 +3831 179.39 64.58 70.85 -10.73 1059.8 +3848	IC 2620 MCG 7-23- 14 PGC 33332	.S?.... 	.82± .13 .46± .07 .47± .07 .79	 15.1 ±.2 .00 .34	.87± .05	.95± .03 12.90± .21 12.90± .69		
110236.0+164402 229.72 62.78 91.16 -20.87 105957.2+170011	 UGC 6112 PGC 33333	.S..7?. U 7.0±1.6 	1.40± .04 .49± .05 1.40	123 .00 13.9 ±.2 .68 .25 13.23			14.30±.1 185± 7 175± 7 .83	1036± 7 959 1368

R.A. 2000 DEC.	Names	Type	logD$_{25}$	p.a.	B$_T$	(B-V)$_T$	(B-V)$_\bullet$	m$_{21}$	V$_{21}$
l b		S$_T$ n$_L$	logR$_{25}$	A$_g$	m$_B$	(U-B)$_T$	(U-B)$_\bullet$	W$_{20}$	V$_{opt}$
SGL SGB		T	logA$_\bullet$	A$_l$	m$_{FIR}$	(B-V)$_T^o$	m'$_\bullet$	W$_{50}$	V$_{GSR}$
R.A. 1950 DEC.	PGC	L	logD$_o$	A$_{21}$	B$_T^o$	(U-B)$_T^o$	m'$_{25}$	HI	V$_{3K}$
1102.6 +0522	CGCG 38- 96	.E.....							
248.09 56.19		F		.07	14.92 ±.18				
102.71 -25.20		-5.0± .9							
1100.1 +0539	PGC 33339								
110246.9+175933	NGC 3501	.S..6*.	1.59± .03	27				14.31±.1	1134± 5
227.26 63.37	UGC 6116	U	.88± .05	.00	13.57 ±.18			314± 7	1083± 57
89.95 -20.31		6.0±1.1		1.29				304± 6	1061
110007.6+181542	PGC 33343		1.59	.44	12.27			1.60	1462
1102.8 +5209		.I..9*.	.96± .09					15.92±.1	948± 6
154.59 57.77	UGC 6113	U	.00± .06	.00				56± 5	
58.99 -3.66		10.0±1.2		.00				48± 5	1002
1059.9 +5226	PGC 33346		.96	.00					1138
110250.5-233528		PSXT2..	1.18± .04	30				14.99±.3	3596± 11
273.06 32.91	ESO 502- 12	Sr	.29± .03	.25	14.22 ±.14			268± 15	
135.17 -31.54	IRAS11003-2319	2.1± .6		.36	13.49			247± 12	3406
110024.0-231918	PGC 33349		1.20	.15	13.58			1.26	3938
110259.8-161721	NGC 3508	.SAR3P?	1.03± .07	15					3889
268.55 39.24	MCG -3-28- 31	E	.09± .05	.14					
126.69 -30.65	IRAS11005-1601	3.0±1.2		.13	10.97				3714
110030.9-160111	PGC 33362		1.05	.05					4242
1103.0 +0327	CGCG 38- 98	.SBR5..							
250.64 54.93		F (1)		.14	15.1 ±.6				
104.77 -25.75		5.0± .9							
1100.5 +0344	PGC 33364	2.6±1.0							
110307.5+750704	NGC 3523	.S..4..	1.19± .07						7100
131.72 40.11	UGC 6105	U	.00± .07	.15	13.65 ±.18				7232
39.36 8.38	IRAS10594+7523	4.0± .8		.00	13.37				7156
105929.4+752313	PGC 33367		1.20	.00	13.45				
110310.0+501223	UGC 6117	.SB.1..	1.00± .06	143					7276± 50
157.42 59.05		U	.10± .05	.00	14.43 ±.18				7324
60.71 -4.64		1.0± .9		.11					7476
110015.9+502833	PGC 33370		1.00	.05	14.24				
110310.8+275825	NGC 3504	RSXS2..	1.43± .02		11.67M±.13	.75± .02	.69± .01	15.66±.1	1539± 5
204.60 66.04	UGC 6118	R (1)	.11± .02	.02	11.82 ±.14		.01± .02	197± 6	1518± 23
80.49 -15.76	IRAS11004+2814	2.0± .3	.90± .03	.13	9.96	.71	11.65± .08	172± 8	1502
110028.1+281435	PGC 33371	2.3± .8	1.43	.05	11.57		13.39± .16	4.04	1837
110311.5+561318	NGC 3499	.I.0.$.	.91± .07						1559± 50
149.13 55.05	UGC 6115	P	.05± .04	.00	14.40 ±.19				1628
55.55 -1.48		90.0		.04					1727
110012.5+562928	PGC 33375		.90		14.34				
110312.3+110438	NGC 3506	.S..5*.	1.08± .04					15.41±.2	6403± 6
239.94 59.93	UGC 6120	R (1)	.04± .04	.02	13.37 ±.16			207± 7	6373± 18
96.88 -23.00	ARAK 273	5.0± .6		.07				188± 4	6304
110035.3+112048	PGC 33379	2.3± .9	1.08	.02	13.25			2.14	6744
110314.3+032033	UGC 6119	.L..0..	.70± .10	17					7536± 49
250.84 54.87	ARAK 272	F	.39± .05	2.45	15.3 ±.4				7414
104.91 -25.76		-2.0±1.0		.00					7890
110039.5+033643	PGC 33380		.91		12.77				
1103.2 -1645	MCG -3-28- 32	.S?....	1.12± .08						7991
268.94 38.87			.85± .07	.15					
127.24 -30.66	IRAS11007-1629			.64					7815
1100.8 -1629	PGC 33381		1.09						8344
110323.7-230511	NGC 3511	.SAS5..	1.76± .01	76	11.53M±.10	.57± .02	.59± .01	12.63±.1	1106± 4
272.90 33.41	ESO 502- 13	R (3)	.45± .02	.22	11.64 ±.11	-.09± .06	-.05± .06	311± 6	1144± 31
134.59 -31.37		5.0± .3	1.27± .01	.67		.43	13.35± .04	295± 5	917
110057.0-224900	PGC 33385	4.0± .4	1.78	.22	10.68	-.19	14.04± .13	1.73	1450
1103.4 +0649	CGCG 38-103	.LXR-*.							
246.35 57.30		F		.04	14.92 ±.18				
101.27 -24.53		-3.0± .9							
1100.8 +0706	PGC 33386								
1103.4 +0654	CGCG 38-104	.SAR1*.							
246.23 57.36		F		.02	15.0 ±.6				
101.18 -24.54		1.0±1.3							
1100.8 +0711	PGC 33387								
110325.4+391518		.S?....	1.17± .05	97					6475±125
177.67 64.50	UGC 6121		.58± .05	.00	15.26 ±.19				6482
70.29 -10.22				.88					6728
110037.9+393128	PGC 33388		1.17	.29	14.35				

R.A. 2000 DEC.	Names	Type	logD$_{25}$	p.a.	B$_T$	(B-V)$_T$	(B-V)$_\bullet$	m$_{21}$	V$_{21}$
l b		S$_T$ n$_L$	logR$_{25}$	A$_g$	m$_B$	(U-B)$_T$	(U-B)$_\bullet$	W$_{20}$	V$_{opt}$
SGL SGB		T	logA$_\bullet$	A$_l$	m$_{FIR}$	(B-V)$_T^o$	m'$_\bullet$	W$_{50}$	V$_{GSR}$
R.A. 1950 DEC.	PGC	L	logD$_o$	A$_{21}$	B$_T^o$	(U-B)$_T^o$	m'$_{25}$	HI	V$_{3K}$

110325.6+180815	NGC 3507	.SBS3..	1.53± .03	110	11.73S±.15			14.09±.1	979± 4
227.11 63.57	UGC 6123	U	.07± .05	.00				154± 6	940± 57
89.88 -20.11		3.0± .7		.10				141± 5	906
110046.3+182425	PGC 33390		1.53	.03	11.63		14.07± .24	2.43	1307
1103.5 +0651		.L..-..							
246.33 57.34	CGCG 38-105	F		.04	14.81 ±.18				
101.24 -24.49		-3.0± .9							
1100.9 +0708	PGC 33394								
110332.4+110707		.S..9*.	1.07± .14					16.25±.3	6394± 7
239.97 60.02	UGC 6122	U	.07± .12	.02					6297
96.88 -22.91		9.0±1.2		.07				78± 7	6738
110055.3+112317	PGC 33396		1.07	.03					
1103.5 -0130		.LAR-?.							
256.39 51.35	CGCG 10- 64	F		.14	14.81 ±.18				
110.14 -27.19		-3.0± .9							
1101.0 -0114	PGC 33398								
1103.5 -1728		.S?....	1.19± .07						8080
269.49 38.29	MCG -3-28- 33		.92± .07	.19					
128.09 -30.69	IRAS11011-1712			1.27	13.04				7902
1101.1 -1712	PGC 33401		1.20	.46					8432
110338.5+451049		PSBS2..	1.11± .05	33					
165.75 61.96	UGC 6125	U	.35± .05	.00					
65.10 -7.17		2.0± .9		.43					
110047.9+452659	PGC 33404		1.11	.18					
110343.9+285309	NGC 3510	.SBS9./	1.60± .02	163	*			13.18±.0	705± 4
202.37 66.21	UGC 6126	R	.70± .02	.00	12.99 ±.14			204± 4	711± 26
79.70 -15.22	IRAS11010+2909	9.0± .4		.71	13.52			186± 4	673
110101.0+290919	PGC 33408		1.60	.35	12.28			.56	1001
110345.7-231441	NGC 3513	.SBT5..	1.45± .02	75	11.93M±.10	.43± .02	.51± .01	13.65±.2	1194± 7
273.08 33.31	ESO 502- 14	R (3)	.10± .02	.24	12.10 ±.12			103± 11	1190± 30
134.79 -31.30	IRAS11013-2258	5.0± .3	1.17± .02	.15	11.85	.35	13.19± .04	90± 8	1004
110119.0-225830	PGC 33410	4.0± .4	1.48	.05	11.60		13.79± .15	2.00	1537
110350.8-200535	IC 2623	.E?....	.89± .04	70	14.46 ±.14	.97± .02			
271.22 36.07	ESO 569- 33		.16± .03	.16	14.37 ±.14				3762± 37
131.12 -30.97				.00		.90			3579
110123.1-194924	PGC 33418		.86		14.20		13.49± .25		4110
1103.8 -0012		.S..2?.							
255.11 52.38	CGCG 10- 66	F		.07	15.4 ±.6				
108.76 -26.73		2.0±1.8							
1101.3 +0004	PGC 33421								
110354.2+405100	A 1101+41	.RING..	.80± .06						10357± 34
174.19 63.99	MCG 7-23- 19	R	.11± .05	.00					10369
68.92 -9.34	ARP 148			.08					10603
110106.0+410711	PGC 33423		.80	.06					
110400.2-184653	NGC 3514	.SXS5?.	1.04± .05	115					
270.44 37.22	ESO 570- 1	SE (2)	.07± .04	.14					
129.61 -30.77	IRAS11015-1830	5.0± .5		.11	12.96				
110132.0-183042	PGC 33430	2.5± .6	1.05	.04					
110402.4+280219	NGC 3512	.SXT5..	1.21± .02		12.98M±.15	.71± .04	.79± .04	14.96±.1	1376± 5
204.47 66.23	UGC 6128	R (2)	.03± .02	.01	12.99 ±.14			238± 4	1469± 58
80.52 -15.56	IRAS11013+2818	5.0± .4	.97± .04	.05	12.53	.69	13.33± .10	202± 4	1341
110119.7+281830	PGC 33432	3.4± .7	1.21	.02	12.92		13.77± .20	2.03	1674
110403.4+494917		.S..6*.	.91± .07	140					
157.84 59.40	UGC 6127	U	.06± .05	.00	14.43 ±.18				7194± 50
61.11 -4.72	IRAS11011+5005	6.0±1.3		.08	13.38				7240
110109.9+500528	PGC 33433		.91	.03	14.32				7396
1104.1 +0821		.SXT6..	1.02± .06						8319
244.38 58.43	UGC 6130	UF (1)	.04± .05	.06	14.42 ±.20				
99.75 -23.81		5.5± .6		.06					8213
1101.5 +0838	PGC 33436	3.6±1.0	1.02	.02	14.27				8667
110418.3+500204		.S..4..	.87± .10						
157.46 59.30	UGC 6129	U	.01± .06	.00	15.06 ±.18				
60.95 -4.57		4.0± .9		.02					
110124.7+501815	PGC 33444		.87	.01					
110423.7+044950	NGC 3509	.SAS4P.	1.33± .04	40				14.85±.2	7704± 7
249.33 56.12	UGC 6134	R	.35± .05	.19	13.53 ±.19			469± 12	7636± 26
103.45 -24.99	ARP 335	4.0± .4		.51	12.83			432± 6	7583
110148.6+050601	PGC 33446		1.34	.17	12.78			1.90	8053

R.A. 2000 DEC.	Names	Type	logD$_{25}$	p.a.	B$_T$	(B-V)$_T$	(B-V)$_\bullet$	m$_{21}$	V$_{21}$
l b		S$_T$ n$_L$	logR$_{25}$	A$_g$	m$_B$	(U-B)$_T$	(U-B)$_\bullet$	W$_{20}$	V$_{opt}$
SGL SGB		T	logA$_\bullet$	A$_i$	m$_{FIR}$	(B-V)$_T^o$	m'$_\bullet$	W$_{50}$	V$_{GSR}$
R.A. 1950 DEC.	PGC	L	logD$_o$	A$_{21}$	B$_T^o$	(U-B)$_T^o$	m'$_{25}$	HI	V$_{3K}$
110427.7+381232		.S?....	.90± .06		13.3 v±.2	.52± .03	.46± .04		
179.83 65.03	UGC 6132		.12± .05	.00	13.5 ±.3	-.50± .03	-.60± .04		9000± 24
71.32 -10.57	MK 421		.29± .07	.17		.44	10.21± .25		9003
110140.9+382843	PGC 33452		.90	.06	13.13	-.56	12.37± .38		9258
110428.1+440201		.L.....	1.17± .09	95					
167.72 62.67	UGC 6131	U	.15± .05	.00	14.35 ±.18				
66.18 -7.63		-2.0± .8		.00					
110138.4+441812	PGC 33454		1.15						
1104.4 +0239		.L..-./							
252.04 54.60	CGCG 38-110	F		.17	15.1 ±.6				
105.75 -25.69		-3.0± .9							
1101.9 +0256	PGC 33455								
110428.8-094729		.LBR0?/	1.19± .06	80					
264.19 44.89	MCG -2-28- 43	EF	.57± .04	.06					
119.37 -29.08		-1.5± .6		.00					
110157.9-093118	PGC 33456		1.11						
1104.5 -0043		.L.....							
255.88 52.11	CGCG 10- 68	F		.12	15.03 ±.18				
109.37 -26.72		-2.0± .9							
1102.0 -0027	PGC 33459								
110434.8+160343		.L...?.	1.14± .07	105					
231.51 62.89	UGC 6137	U	.20± .03	.00	14.01 ±.15				6360± 26
92.04 -20.72		-2.0±1.7		.00					6281
110156.2+161955	PGC 33460		1.11		13.91				6694
1104.5 +0511		.SXR3*.	1.04± .08	13					
248.91 56.41	UGC 6141	UF (1)	.18± .06	.13	14.64 ±.20				
103.09 -24.82		3.0± .6		.24					
1102.0 +0528	PGC 33461	1.6±1.0	1.05	.09					
1104.6 +2743		.S..9*.	1.32± .04						2575± 10
205.26 66.33	UGC 6138	U	.31± .05	.04	14.5 ±.4				
80.87 -15.59		9.0±1.1		.31					2538
1101.9 +2800	PGC 33463		1.32	.15	14.11				2874
110436.9+450735		.S?....	.99± .05						
165.66 62.14	UGC 6135		.02± .04	.00	13.3 ±.3				6512
65.24 -7.05	IRAS11017+4523			.03	12.30				6541
110146.6+452346	PGC 33465		.99	.01	13.20				6738
110437.3+281337	NGC 3515	.SA.5*.	.99± .06	55					
204.03 66.37	UGC 6139	PU	.10± .04	.00	14.60 ±.18				8797
80.41 -15.36	IRAS11018+2829	4.5± .6		.16	13.53				8762
110154.7+282948	PGC 33467		.99	.05	14.39				9095
1104.6 -1516		.S..7?/	1.12± .08	155					
268.30 40.32		E (1)	.58± .08	.12					
125.60 -30.09		7.0±1.8		.80					
1102.2 -1500	PGC 33468	6.4±1.6	1.14	.29					
110441.4-013230		.E.....	.52± .20						
256.78 51.50	MCG 0-28- 29	F	.00± .07	.16	15.4 ±.3				
110.27 -26.93	ARAK 275	-5.0±1.0		.00					
110208.1-011619	PGC 33469		.54						
110441.8+041745		.S?....	.84± .08	150					
250.11 55.80	UGC 6142	(1)	.24± .05	.17	14.9 ±.2				7559± 50
104.04 -25.10				.36					7440
110206.8+043357	PGC 33470	1.6±1.0	.86	.12	14.29				7913
1104.7 +0416		PSBR3P.	.72± .22						
250.15 55.81	UGC 6142A	F	.00± .07	.17	15.22 ±.19				
104.07 -25.08		3.0± .9		.00					
1102.2 +0433	PGC 33477		.73	.00					
1104.8 +6359		.I..9*.	1.04± .10						
140.50 49.34	UGC 6133	U	.00± .06	.02					
49.01 2.76		10.0±1.1		.00					
1101.7 +6416	PGC 33479		1.04	.00					
1104.9 +0417		.SBS2*.	.82± .13						
250.19 55.85	MCG 1-28- 36	F (1)	.00± .07	.17	14.92 ±.18				
104.07 -25.04	IRAS11023+0433	2.0± .9		.00	12.73				
1102.3 +0433	PGC 33485	1.6±1.4	.83	.00					
110458.5+290822 A	1102+29		.47± .15		15.7 ±.2	.28± .03		16.93±.1	646± 5
201.76 66.49	MCG 5-26- 46		.13± .06	.00		-.66± .05		92± 4	672± 32
79.60 -14.86	MK 36							44± 5	615
110215.6+292434	PGC 33486		.47						941

R.A. 2000 DEC. l b SGL SGB R.A. 1950 DEC.	Names PGC	Type S_T n_L T L	$\log D_{25}$ $\log R_{25}$ $\log A_e$ $\log D_o$	p.a. A_g A_i A_{21}	B_T m_B m_{FIR} B_T^o	$(B-V)_T$ $(U-B)_T$ $(B-V)_T^o$ $(U-B)_T^o$	$(B-V)_e$ $(U-B)_e$ m'_e m'_{25}	m_{21} W_{20} W_{50} HI	V_{21} V_{opt} V_{GSR} V_{3K}
110504.6+352157 186.35 65.91 73.93 −11.86 110219.3+353809	 UGC 6143 IRAS11023+3538 PGC 33495	.S..3*. U (1) 3.0±1.3 4.5±1.3	1.15± .05 .62± .05 1.15	12 .04 .86 .31	 15.17 ±.18 13.70 14.21				7432 7424 7702
110508.0+444448 166.26 62.41 65.62 −7.17 110218.0+450100	A 1102+45 MCG 8-20- 83 MK 162 PGC 33498	 	.78± .08 .35± .06 .78	 .00 	 14.9 ±.3 				6372± 74 6399 6600
110513.8−263730 275.30 30.49 138.75 −31.20 110248.0−262118	 ESO 502- 16 PGC 33505	.SBS9.. S (1) 9.0± .8 7.8± .9	1.21± .04 .47± .03 1.24	82 .27 .48 .23	 14.72 ±.14 13.97				1512 1318 1849
110514.7−235042 273.78 32.95 135.52 −31.01 110248.0−233430	 ESO 502- 15 PGC 33509	 	1.01± .06 .29± .06 1.03	152 .25 	 14.78 ±.14 				3779± 37 3590 4121
1105.2 −0047 256.16 52.17 109.51 −26.57 1102.7 −0031	 CGCG 10- 71 PGC 33510	.SBR0*/ F .0±1.3 	 	 .13 	 14.9 ±.6 				
1105.5 −0152 257.38 51.39 110.70 −26.81 1103.0 −0136	 PGC 33525	.IBS9.. E (1) 10.0± .9 9.8± .8	1.16± .08 .28± .08 1.18	115 .16 .21 .14					740 603 1099
1105.5 +0409 250.55 55.87 104.26 −24.93 1103.0 +0426	 UGC 6145 PGC 33528	.I..9.. U 10.0± .8 	1.14± .13 .07± .12 1.16	 .18 .05 .03					
1105.5 +0412 250.49 55.90 104.21 −24.92 1103.0 +0429	 UGC 6146 PGC 33529	.SXS9*. UF (1) 9.3± .8 6.6±1.0	1.07± .07 .62± .06 1.09	140 .20 .63 .31					6348 6230 6702
110537.6+563134 148.34 55.07 55.47 −1.04 110239.3+564746	NGC 3517 UGC 6144 IRAS11026+5647 PGC 33532	.SA.3*. PU 2.5± .6 	1.03± .05 .07± .04 1.03	120 .00 .09 .03	 13.50 				8286± 79 8357 8453
110539.0−095526 264.63 44.94 119.59 −28.83 110308.0−093914	 MCG −2-28- 45 PGC 33535	PL..-P* EF −3.3± .6 	1.25± .06 .05± .04 1.26	110 .10 .00 					
110542.4+350705 186.88 66.09 74.22 −11.87 110257.2+352317	 MCG 6-25- 1 MK 1279 PGC 33540	.S?.... 	.80± .07 .17± .06 .80	 .00 .25 .08	 15.46 ±.18 15.18				8673± 52 8665 8944
1105.7 −0129 257.05 51.71 110.30 −26.66 1103.2 −0113	 CGCG 10- 73 PGC 33546	.L...?/ F −2.0±1.8 	 	 .13 	 15.4 ±.6 				
110548.9−000215 255.54 52.83 108.75 −26.21 110315.1+001358	NGC 3521 UGC 6150 PGC 33550	.SXT4.. R (3) 4.0± .3 3.6± .4	2.04± .01 .33± .02 1.26± .01 2.04	163 .06 .48 .16	9.83M±.10 9.77 ±.13 8.81 9.26	.81± .01 .23± .02 .73 .16	.88± .01 .31± .01 11.80± .03 14.04± .13	11.26±.0 460± 4 439± 5 1.84	805± 4 782± 31 673 1162
110549.8−204731 272.15 35.69 131.99 −30.59 110322.0−203118	 ESO 570- 2 PGC 33552	.SBT4.. S (1) 4.0± .8 4.4± .9	1.19± .04 .34± .04 1.21	58 .15 .50 .17					
110550.8−095532 264.69 44.96 119.60 −28.78 110319.8−093919	 MCG −2-28- 48 PGC 33554	.LBS+P* E −1.3±1.1 	1.14± .05 .34± .04 1.10	100 .10 .00 					
110551.7−094211 264.52 45.15 119.36 −28.73 110320.6−092558	 MCG −1-28- 24 PGC 33555	.SXT4*. EF (1) 4.0± .6 5.3± .8	1.17± .05 .19± .04 1.17	5 .06 .28 .10					
1105.8 +0239 252.48 54.84 105.87 −25.36 1103.3 +0256	 CGCG 38-121 PGC 33558	.S..2?/ F 2.0±1.8 	 	 .47 	 15.2 ±.6 				

R.A. 2000 DEC. / l b / SGL SGB / R.A. 1950 DEC.	Names / PGC	Type / S_T n_L / T / L	$\log D_{25}$ / $\log R_{25}$ / $\log A_o$ / $\log D_o$	p.a. / A_g / A_i / A_{21}	B_T / m_B / m_{FIR} / B_T^o	$(B-V)_T$ / $(U-B)_T$ / $(B-V)_T^o$ / $(U-B)_T^o$	$(B-V)_o$ / $(U-B)_o$ / m'_o / m'_{25}	m_{21} / W_{20} / W_{50} / HI	V_{21} / V_{opt} / V_{GSR} / V_{3K}
110555.2-253519 / 274.91 31.48 / 137.56 -30.98 / 110329.0-251906	ESO 502- 17 / PGC 33560	.E?.... / / / .76	.83± .06 / .36± .03 / /	95 / .22 / .00	15.09 ±.13 / 14.62 ±.14	.62± .03	13.31± .35		
1105.9 +1949 / 224.09 64.78 / 88.51 -18.85 / 1103.2 +2005	A 1103+20 / UGC 6151 / DDO 91 / PGC 33562	.S..9*. / U (1) / 9.0±1.1 / 9.0±1.0	1.28± .06 / .15± .04 / 1.06± .03 / 1.28	/ .00 / .15 / .07	14.8 ±.2 / 14.3 ±.2 / / 14.42	.41± .08 / -.28± .10 / .37 / -.31	.49± .05 / -.19± .06 / 15.55± .07 / 15.67± .36	15.41±.1 / 38± 5 / 26± 7 / .92	1331± 6 / / 1266 / 1655
110558.0+362426 / 183.77 65.84 / 73.08 -11.20 / 110312.4+364039	UGC 6148 / PGC 33566	.S..3.. / U (1) / 3.0±1.0 / 3.5±1.3	1.13± .05 / .79± .05 / / 1.13	176 / .04 / 1.09 / .39	15.29 ±.18				
1106.0 +0821 / 244.96 58.79 / 99.94 -23.38 / 1103.4 +0838	MCG 2-28- 49 / PGC 33567	PSBR1*/ / F / 1.0±1.3	.94± .11 / .40± .07 / / .94	/ .07 / .41 / .20	15.07 ±.19				
110602.2+042542 / 250.36 56.13 / 104.03 -24.74 / 110327.2+044155	UGC 6155 / PGC 33569	.SXT5.. / UF (1) / 4.5± .6 / 3.6±1.0	1.13± .05 / .07± .05 / / 1.15	145 / .20 / .10 / .03	/ 13.88 ±.19 / / 13.54			16.33±.3 / / 194± 7 / 2.75	6425± 10 / 6412± 50 / 6307 / 6778
110604.7+031905 / 251.76 55.35 / 105.20 -25.10 / 110330.0+033518	CGCG 38-122 / PGC 33572	.E..... / F / -5.0± .9		.15	14.71 ±.18				
110606.1+295607 / 199.76 66.75 / 78.99 -14.28 / 110323.1+301220	UGC 6152 / IRAS11033+3012 / PGC 33573	.SB.3.. / U / 3.0± .9	1.07± .06 / .39± .05 / / 1.07	154 / .00 / .54 / .19	14.99 ±.18 / / / 14.39			17.28±.3 / 417± 15 / / 2.70	8932± 11 / / 8904 / 9224
1106.1 +0318 / 251.79 55.36 / 105.21 -25.07 / 1103.6 +0335	CGCG 38-124 / PGC 33577	.LB.0P. / F / -2.0± .9		.15	14.9 ±.6				
1106.1 +0419 / 250.53 56.09 / 104.14 -24.74 / 1103.6 +0436	MCG 1-28- 38 / PGC 33578	.LA.-.. / F / -3.0± .9	1.04± .09 / .21± .07 / / 1.03	/ .19 / .00	14.39 ±.15				
110615.6-181531 / 270.69 37.94 / 129.09 -30.17 / 110347.0-175918	ESO 570- 3 / PGC 33581	.S?.... / / / 1.07	1.06± .09 / .74± .07 / / 1.07	135 / .19 / 1.03 / .37				15.78±.1 / 277± 12	3899± 9 / / 3720 / 4250
1106.3 +1729 / 229.10 63.92 / 90.82 -19.74 / 1103.7 +1746	UGC 6157 / PGC 33584	.SAS8.. / U / 8.0± .8	1.28± .04 / .03± .05 / / 1.28	/ .00 / .04 / .02	13.9 ±.4 / / / 13.83			14.68±.1 / 153± 16 / 133± 12 / .83	2959± 11 / / 2886 / 3289
110631.4+483905 / 159.20 60.43 / 62.34 -4.97 / 110339.5+485518	A 1103+48 / UGC 6156 / IRAS11036+4855 / PGC 33600	.S?.... / / / 1.06	1.06± .06 / .56± .05 / / 1.06	3 / .00 / .85 / .28	15.09 ±.18 / 13.37 / 14.21				7500±110 / 7543 / 7709
110631.7-373908 / 280.97 20.70 / 151.57 -30.85 / 110410.0-372254	ESO 377- 10 / IRAS11041-3723 / PGC 33601	.SBT3*. / S (1) / 2.7± .5 / 3.3±1.2	1.41± .03 / .59± .04 / / 1.45	150 / .40 / .81 / .29	13.64 ±.14 / 13.31 / 12.41				2870± 60 / 2661 / 3176
110632.2+112305 / 240.41 60.77 / 96.91 -22.13 / 110355.2+113918	NGC 3524 / UGC 6158 / PGC 33604	.S..0.. / U / .0± .9	1.21± .05 / .54± .05 / / 1.18	14 / .00 / .41	13.8 ±.2 / / 13.34				1321± 57 / 1226 / 1664
110632.3-480226 / 285.40 11.23 / 163.42 -29.65 / 110416.0-474612	ESO 215- 21 / IRAS11042-4746 / PGC 33606	PSXR1*. / r / 1.0±1.0	1.10± .06 / .61± .05 / / 1.16	168 / .63 / .62 / .31	14.98 ±.14 / 12.14				
1106.5 +0017 / 255.41 53.20 / 108.45 -25.93 / 1104.0 +0034	CGCG 10- 76 / PGC 33608	.LA.+?/ / F / -1.0±1.8		.10	15.4 ±.6				
110640.5+200508 / 223.67 65.04 / 88.34 -18.58 / 110400.9+202122	NGC 3522 / UGC 6159 / PGC 33615	.E..... / U / -5.0± .9	1.08± .09 / .22± .04 / / 1.01	117 / .00 / .00	14.13 ±.15 / / 14.11			17.25±.3 / / 242± 11	1221± 10 / 1254± 50 / 1158 / 1546

R.A. 2000 DEC. / l b / SGL SGB / R.A. 1950 DEC.	Names / PGC	Type / S_T n_L / T / L	$\log D_{25}$ / $\log R_{25}$ / $\log A_o$ / $\log D_o$	p.a. / A_g / A_i / A_{21}	B_T / m_B / m_{FIR} / B_T^o	$(B-V)_T$ / $(U-B)_T$ / $(B-V)_T^o$ / $(U-B)_T^o$	$(B-V)_o$ / $(U-B)_o$ / m'_o / m'_{25}	m_{21} / W_{20} / W_{50} / HI	V_{21} / V_{opt} / V_{GSR} / V_{3K}
1106.7 +2301 / 216.95 65.99 / 85.52 −17.28 / 1104.1 +2318	UGC 6163 / PGC 33620	.S..1.. / U / / 1.0± .9	1.06± .06 / .20± .05 / / 1.06	90 / .00 / .20 / .10	/ 14.59 ±.18 / /				
1106.7 +5741 / 146.76 54.32 / 54.54 −.30 / 1103.8 +5758	A 1103+57 / MCG 10-16- 61 / PGC 33622	.SBS5P. / P / / 5.0± .9	1.02± .10 / .19± .07 / / 1.02	/ .00 / .28 / .09	/ 14.88 ±.18 / / 14.54				9820± 60 / 9895 / 9980
110647.3+723412 / 133.24 42.40 / 41.70 7.31 / 110322.7+725025	NGC 3516 / UGC 6153 / IRAS11033+7250 / PGC 33623	RLBS0*. / R / −2.0± .4 /	1.24± .03 / .11± .04 / .58± .07 / 1.24	/ .10 / .00 /	12.5 v±.2 / 12.46 ±.13 / 12.69 / 12.34	.81± .03 / −.06± .05 / .75 / −.08	.79± .03 / −.08± .05 / 10.92± .25 / 13.33± .27		2624± 21 / 2749 / 2696
110649.6+434322 / 167.86 63.19 / 66.67 −7.42 / 110400.7+435936	UGC 6161 / PGC 33625	.SB.8.. / U / / 8.0± .7	1.41± .05 / .32± .06 / / 1.41	40 / .00 / .40 / .16	/ 14.0 ±.3 / / 13.58			13.94±.1 / 131± 8 / 108± 12 / .20	758± 6 / / 782 / 991
1106.8 −0230 / 258.42 51.10 / 111.49 −26.68 / 1104.3 −0214	CGCG 10- 77 / PGC 33628	.SBR1?. / F / / 1.0±1.8		.19	/ 14.4 ±.3 / /				
110654.4+511215 / 155.20 58.87 / 60.15 −3.62 / 110400.9+512829	UGC 6162 / IRAS11041+5127 / PGC 33633	.S..7.. / U / 7.0± .8 /	1.40± .03 / .32± .04 / / 1.40	88 / .00 / .44 / .16	/ 13.8 ±.2 / / 13.39			14.30±.1 / 216± 16 / 211± 12 / .75	2203± 11 / 2097± 76 / 2253 / 2396
110656.3+071026 / 246.97 58.18 / 101.26 −23.58 / 110420.5+072640	NGC 3526 / UGC 6167 / IRAS11043+0726 / PGC 33635	.SA.5P/ / UF / 4.7±1.0 /	1.28± .03 / .64± .04 / / 1.29	55 / .09 / .97 / .32	/ 13.86 ±.20 / 13.23 / 12.79			15.00±.1 / 213± 7 / / 1.89	1420± 6 / 1271± 42 / 1309 / 1768
1107.0 +2835 / 203.17 66.92 / 80.32 −14.72 / 1104.3 +2852	UGC 6166 / PGC 33640	.S..4.. / U / / 4.0± .9	.99± .06 / .07± .05 / / .99	115 / .00 / .10 / .04	/ 14.74 ±.20 / / 14.57			15.74±.3 / 399± 13 / 368± 10 / 1.14	10174± 10 / / 10141 / 10470
110703.7+120340 / 239.44 61.27 / 96.29 −21.76 / 110426.5+121954	UGC 6169 / PGC 33642	.S..3*. / U (1) / 3.0±1.3 / 4.5±1.2	1.27± .04 / .77± .05 / / 1.27	0 / .00 / 1.06 / .38	/ 14.51 ±.18 / / 13.43			14.65±.3 / / 101± 10 / .83	1557± 10 / / 1465 / 1899
110704.3+454920 / 163.91 62.13 / 64.85 −6.33 / 110414.3+460534	UGC 6165 / IRAS11042+4605 / PGC 33643	.S..2.. / U / 2.0± .9 /	1.19± .05 / .68± .05 / / 1.19	59 / .00 / .84 / .34	/ 15.16 ±.18 / 13.33 /				
1107.1 +0747 / 246.13 58.63 / 100.63 −23.33 / 1104.5 +0804	UGC 6168 / PGC 33645	PSBR4*. / UF (1) / 3.7± .7 / 4.5±1.0	1.11± .14 / .54± .12 / / 1.12	128 / .08 / .79 / .27	/ 14.86 ±.18 / /				
110707.0-371021 / 280.86 21.19 / 151.01 −30.77 / 110445.1-365406	NGC 3533 / ESO 377- 11 / IRAS11047-3654 / PGC 33647	PSXR2*/ / PBSr / 1.8± .5 /	1.45± .04 / .64± .04 / / 1.48	65 / .39 / .79 / .32	/ 13.79 ±.14 / 13.58 / 12.57				3370± 60 / 3162 / 3677
110708.8-180126 / 270.78 38.25 / 128.85 −29.93 / 110440.1-174512	ESO 570- 4 / PGC 33648	.E+4.P. / S / −4.0± .8 /	1.14± .07 / .11± .04 / / 1.14	/ .19 / .00 /					
110714.5-111420 / 266.11 44.06 / 121.16 −28.72 / 110443.8-105806	MCG -2-28- 49 / PGC 33659	.SBS8*/ / E F (1) / 8.0± .7 / 7.5± .6	1.10± .06 / .39± .05 / / 1.11	150 / .11 / .48 / .19					
1107.2 +1833 / 227.09 64.58 / 89.87 −19.10 / 1104.6 +1850	UGC 6171 / PGC 33660	.IB.9*. / U / / 10.0± .8	1.40± .04 / .63± .05 / / 1.40	68 / .00 / .47 / .32	/ 14.4 ±.2 / / 13.91			14.47±.1 / 171± 16 / 144± 12 / .24	1200± 11 / / 1131 / 1527
110716.4+061809 / 248.28 57.66 / 102.19 −23.81 / 110440.9+063423	IC 669 / UGC 6174 / PGC 33662	.L..-.. / UF / −2.5± .6 /	1.09± .06 / .24± .05 / / 1.07	165 / .11 / .00 /	/ 14.10 ±.15 / / 13.86				8801± 50 / 8690 / 9153
1107.3 +7641 / 130.39 38.89 / 38.11 9.39 / 1103.6 +7657	UGC 6154 / 7ZW 367 / PGC 33665	.SB.1.. / U / 1.0± .9 /	.99± .10 / .20± .07 / / .99	/ .06 / .20 / .10	/ 14.77 ±.18 / 12.68 /				

R.A. 2000 DEC.	Names	Type	logD$_{25}$	p.a.	B$_T$	(B-V)$_T$	(B-V)$_\bullet$	m$_{21}$	V$_{21}$
l b		S$_T$ n$_L$	logR$_{25}$	A$_g$	m$_B$	(U-B)$_T$	(U-B)$_\bullet$	W$_{20}$	V$_{opt}$
SGL SGB		T	logA$_\bullet$	A$_i$	m$_{FIR}$	(B-V)$_T^o$	m'$_\bullet$	W$_{50}$	V$_{GSR}$
R.A. 1950 DEC.	PGC	L	logD$_o$	A$_{21}$	B$_T^o$	(U-B)$_T^o$	m'$_{25}$	HI	V$_{3K}$

110718.3-192821 NGC 3497		.LAS0*.	1.41± .03	59					
271.72 37.01 ESO 570- 6		SE	.25± .03	.14					3707± 52
130.52 -30.08		-1.7± .4		.00					3526
110450.1-191206 PGC 33667			1.39						4057
110718.4-253426		.IBS9..	1.15± .07	27					
275.22 31.63 ESO 502- 18		S (1)	.23± .05	.24 17.18 ±.14					
137.56 -30.67		10.0± .8		.17					
110452.0-251812 PGC 33668		11.1± .8	1.17	.11					
110718.4+283136 NGC 3527		RSBR2*.	.99± .06					17.16±.3	10093± 9
203.36 66.98 UGC 6170		U	.02± .05	.01 14.63 ±.20					10060
80.42 -14.70		2.0± .8		.03				137± 7	10390
110436.1+284750 PGC 33669			.99	.01 14.50				2.66	
110719.2+182543 A 1104+18B		.LB?...	1.12± .12	3					
227.39 64.54 MCG 3-28- 63			.27± .07	.00					7959± 35
90.01 -19.14				.00					7890
110440.2+184157 PGC 33670			1.08						8287
110719.3-193321 NGC 3529		.SBT3P*	1.01± .04	55					
271.78 36.94 ESO 570- 7		SE	.11± .03	.14					
130.62 -30.09 IRAS11048-1917		3.0± .6		.16 12.89					
110451.0-191706 PGC 33671			1.02	.06					
110719.8+232859		.S?....	1.00± .06	160					
215.94 66.23 UGC 6173			.13± .05	.00 14.41 ±.18					6599± 50
85.15 -16.97 IRAS11046+2345				.20 12.49					6548
110439.3+234513 PGC 33675			1.00	.07 14.18					6912
110720.5+182553 A 1104+18A		.LB?...	.42?	129					
227.39 64.54 MCG 3-28- 62			.11± .07	.00					8231± 35
90.01 -19.14				.00					8161
110441.5+184207 PGC 33677			.40						8559
1107.3 -0232		.E.....							
258.61 51.15 CGCG 10- 78		F		.19 15.03 ±.18					
111.57 -26.57		-5.0± .9							
1104.8 -0216 PGC 33678									
1107.4 +0642 IC 670		.E+....	1.01± .08						
247.75 57.96 UGC 6178		UF	.10± .03	.09 14.31 ±.15					
101.78 -23.64		-3.5± .6		.00					
1104.8 +0659 PGC 33680			.99						
110724.9+213927		.LB....	1.13± .07	20					
220.26 65.74 UGC 6176		U	.41± .03	.00 14.60 ±.15					
86.90 -17.75 MK 1282		-2.0± .9		.00					
110444.9+215541 PGC 33682			1.06						
110728.9+034758		PSB.1..							
251.61 55.94 CGCG 38-135		F		.17 15.1 ±.6					
104.82 -24.61		1.0± .9							
110454.1+040412 PGC 33686									
110731.7+004700 IC 671		.SXR4*.	1.10± .04	100					
255.17 53.73 UGC 6180		UE F (1)	.03± .04	.11 14.1 ±.2					
108.01 -25.56		4.2± .5		.05					
110457.7+010314 PGC 33689		4.6± .6	1.11	.02					
110745.7+193258		.I..9*.	1.02± .06					15.76±.1	1169± 5
225.08 65.08 UGC 6181		U	.01± .05	.00 14.7 ±.3				74± 4	
88.97 -18.58		10.0±1.2		.01				70± 5	1104
110506.4+194913 PGC 33699			1.02	.00 14.73				1.03	1494
110748.2-200121 NGC 3565		.LB.-P?	1.00± .08	129					
272.18 36.59 ESO 570- 8		SE	.24± .04	.13					
131.17 -30.04		-2.7± .7		.00					
110520.0-194506 PGC 33701			.98						
110749.5+022127		.SBT3*.							
253.46 54.95 CGCG 39- 3		F (1)		.13 14.9 ±.6					
106.37 -24.99		3.0±1.3							
110515.0+023742 PGC 33704		1.6±1.4							
110749.6-463127		.SBS6..	1.56± .04	141					
284.99 12.71 ESO 265- 7		S (1)	.48± .05	.68 12.38 ±.14					1640± 60
161.68 -29.68 IRAS11055-4615		5.5± .5		.71 10.94					1425
110532.0-461512 PGC 33705		4.4± .8	1.63	.24 10.98					1913
1107.9 +0759		.SXT8..	1.19± .05	160				15.79±.3	3327± 12
246.09 58.91 UGC 6185		UF (1)	.31± .05	.06 14.13 ±.19				202± 14	
100.50 -23.07		8.0± .6		.38				198± 15	3222
1105.3 +0816 PGC 33712		4.6±1.0	1.19	.15 13.68				1.96	3676

R.A. 2000 DEC. l b SGL SGB R.A. 1950 DEC.	Names PGC	Type S_T n_L T L	$\log D_{25}$ $\log R_{25}$ $\log A_e$ $\log D_o$	p.a. A_g A_i A_{21}	B_T m_B m_{FIR} B_T^o	$(B-V)_T$ $(U-B)_T$ $(B-V)_T^o$ $(U-B)_T^o$	$(B-V)_e$ $(U-B)_e$ m' m'_{25}	m_{21} W_{20} W_{50} HI	V_{21} V_{opt} V_{GSR} V_{3K}
1107.9 +0200 253.92 54.72 106.75 -25.07 1105.4 +0217	CGCG 11- 2 PGC 33716	.SBS5*. F (1) 5.0± .9 5.6±1.4	 	 	.16 14.0 ±.3 				
110758.9+352747 185.79 66.47 74.14 -11.30 110514.1+354402	 UGC 6183 PGC 33719	.SX.3.. U 3.0± .8 1.12	1.12± .05 .26± .05 	5 .00 .36 .13	 14.68 ±.19 14.25				8674± 19 8667 8944
110759.5+365221 182.40 66.10 72.87 -10.62 110514.1+370836	 MCG 6-25- 6 PGC 33720	.SB?... .94	.94± .11 .05± .07 	 .00 .07 .02	 14.55 ±.18 14.42				8041± 19 8040 8305
110803.2-122905 267.26 43.11 122.61 -28.78 110532.8-121250	 MCG -2-29- 1 PGC 33725	.SBS3P? E 3.0±1.9 1.07	1.05± .06 .54± .05 	175 .13 .75 .27					
110803.5+533701 151.59 57.38 58.15 -2.23 110508.7+535316	 UGC 6182 PGC 33726	.S?.... .97	.97± .07 .03± .05 	 .00 .04 .01	 14.12 ±.18 14.07				1255± 50 1316 1438
110808.3+455906 163.40 62.20 64.80 -6.09 110518.5+461521	 UGC 6184 PGC 33729	.S..6*. U 6.0±1.3 1.02	1.02± .06 .49± .05 	2 .00 .72 .25	 15.17 ±.18 				
110820.1-373722 281.30 20.88 151.52 -30.50 110558.1-372106	 ESO 377- 12 PGC 33744	.SBT2.. Sr (1) 2.1± .5 5.6± .9	1.14± .04 .16± .04 1.19	145 .52 .19 .08	 13.95 ±.14 13.20				3690± 60 3482 3996
110822.0-475546 285.64 11.45 163.24 -29.37 110605.0-473930	 ESO 215- 27 PGC 33745	.S..3./ S 3.4± .6 1.28	1.22± .04 .83± .04 	100 .62 1.14 .41	 15.09 ±.14 				
110823.7+032950 252.28 55.88 105.22 -24.49 110549.0+034606	 MCG 1-29- 3 IRAS11058+0346 PGC 33749	.SBS3*. F (1) 3.0± .9 1.6±1.0	.82± .13 .36± .07 .83	 .15 .50 .18	 15.2 ±.3 12.59 				
1108.4 +2949 200.02 67.25 79.33 -13.88 1105.7 +3006	CGCG 155- 71 PGC 33750	 	 	 .00 	 15.3 ±.3 				10718 10690 11010
1108.4 -1016 265.71 45.03 120.15 -28.24 1105.9 -1000	 PGC 33752	.SXR2.. F 2.0± .8 1.17	1.16± .08 .05± .08 	60 .09 .06 .03					
110830.0+450755 164.87 62.72 65.58 -6.46 110540.9+452410	 UGC 6187 IRAS11056+4524 PGC 33755	.S?.... .81	.81± .11 .07± .06 	 .00 .10 .03	 14.3 ±.3 13.01 14.13				6171± 50 6201 6397
110832.3-102931 265.91 44.86 120.40 -28.26 110601.3-101315	 MCG -2-29- 3 PGC 33759	PSXR7?. E (1) 7.0±1.6 5.3± .8	1.17± .05 .04± .05 1.18	80 .12 .06 .02				15.21±.1 471± 12 	6274± 9 6114 6633
110834.1+044956 250.65 56.87 103.84 -24.01 110559.0+050611	NGC 3535 UGC 6189 IRAS11059+0505 PGC 33760	.SAS1P* UF .8± .8 1.05	1.04± .08 .38± .06 	178 .17 .39 .19	 14.44 ±.20 13.41 13.80				6962± 50 6847 7315
110840.3+571345 146.98 54.84 55.07 -.32 110542.8+573000	NGC 3530 UGC 6188 ARAK 278 PGC 33766	.S?.... .81	.81± .08 .41± .05 	99 .00 .56 .20	 14.8 ±.3 14.24				1938± 50 2012 2101
110851.7+282831 203.53 67.32 80.63 -14.42 110609.6+284447	NGC 3536 UGC 6191 PGC 33779	.SB?... 1.05	1.05± .08 .13± .06 	155 .00 .20 .07	 14.7 ±.2 14.41			16.39±.3 493± 7 1.91	10925± 9 10893 11222
1108.8 +2636 208.33 67.16 82.37 -15.26 1106.2 +2653	 UGC 6193 PGC 33782	.S..3.. U (1) 3.0± .9 4.5±1.3	.95± .07 .34± .05 .95	168 .00 .47 .17	 15.38 ±.18 				

R.A. 2000 DEC. l b SGL SGB R.A. 1950 DEC.	Names PGC	Type S_T n_L T L	$\log D_{25}$ $\log R_{25}$ $\log A_.$ $\log D_0$	p.a. A_g A_i A_{21}	B_T m_B m_{FIR} B_T^0	$(B-V)_T$ $(U-B)_T$ $(B-V)_T^0$ $(U-B)_T^0$	$(B-V)_e$ $(U-B)_e$ $m'_.$ m'_{25}	m_{21} W_{20} W_{50} HI	V_{21} V_{opt} V_{GSR} V_{3K}
1108.8 +0618 248.78 57.96 102.33 -23.43 1106.3 +0635	CGCG 39- 14 PGC 33783	.SXS5P* F (1) 5.0±1.3 3.6±1.0		.12	15.0 ±.6				13233± 67 13122 13585
110855.8+263635 208.34 67.17 82.38 -15.25 110614.4+265251	UGC 6190 IRAS11062+2652 PGC 33786	.S?.... 	1.11± .05 .62± .05 1.11	92 .00 .94 .31	 15.23 ±.18 12.89 14.26			15.26±.3 561± 5 .68	6576± 9 6537 6879
110858.7-282216 277.07 29.30 140.83 -30.45 110633.0-280600	ESO 438- 5 PGC 33788	.SBS9*/ SU (1) 9.1± .6 8.9± .8	1.45± .04 .73± .04 1.48	62 .24 .74 .36	 14.87 ±.14 13.89			13.86±.1 165± 11 150± 8 -.39	1507± 6 1311 1839
1108.9 +0620 248.77 58.00 102.31 -23.40 1106.4 +0637	CGCG 39- 16 PGC 33792	.SBS4P* F 4.0±1.3 		.12	14.9 ±.6				13157± 67 13047 13509
110901.6+225524 217.54 66.47 85.87 -16.87 110621.4+231140	UGC 6194 PGC 33794	.S?.... 	1.18± .04 .00± .04 1.18	 .00 .01 .00	 13.9 ±.3 13.92			15.25±.2 127± 6 109± 7 1.33	2643± 7 2590 2958
1109.1 +2840 203.01 67.38 80.46 -14.28 1106.4 +2857	NGC 3539 MCG 5-26- 65 PGC 33799	.S?.... 	1.04± .09 .66± .07 1.00	 .00 .50 	 15.47 ±.18 14.83				9735 9703 10031
1109.1 +0042 255.78 53.94 108.23 -25.19 1106.6 +0059	CGCG 11- 7 PGC 33803	PSBS3?. F 3.0±1.8 		.12	15.4 ±.6				
110916.1+360116 184.27 66.58 73.76 -10.80 110631.4+361732	NGC 3540 UGC 6196 PGC 33806	.LB.... U -2.0± .8 	1.14± .06 .06± .05 1.13	 .00 .00 	 14.26 ±.18 14.16				6387± 19 6383 6655
1109.3 +2933 200.71 67.44 79.67 -13.83 1109.6 +2950	UGC 6198 PGC 33809	.L...?. U -2.0±1.7 	1.01± .08 .12± .03 .99	130 .00 .00 	 14.64 ±.16 14.48				10427 10399 10720
1109.3 -0051 257.53 52.78 109.92 -25.61 1106.8 -0035	CGCG 11- 8 PGC 33812	RSBS1.. F 1.0± .9 		.13	15.1 ±.6				
110922.5-225235 274.24 34.25 134.49 -29.98 110655.0-223618	ESO 502- 20 PGC 33813	.SXS5*. S (1) 5.0± .6 6.1± .6	1.15± .04 .08± .03 1.17	 .23 .11 .04	 14.50 ±.14 14.15			14.17±.3 134± 15 121± 12 -.01	1377± 11 1190 1721
1109.4 +1050 242.17 61.01 97.75 -21.68 1106.8 +1107	MCG 2-29- 5 PGC 33816	 	.82± .13 .00± .07 .82	 .04 	 14.87 ±.18 				1538± 54 1443 1883
110925.0-000553 256.74 53.37 109.11 -25.37 110651.2+001023	IC 673 UGC 6200 IRAS11068+0010 PGC 33817	PSXT5.. F (1) 5.0± .8 1.6±1.0	1.22± .05 .36± .05 1.23	165 .10 .54 .18	 14.06 ±.18 13.39 13.40			14.28±.3 356± 7 325± 7 .70	3859± 9 3729 4217
110932.4-372053 281.41 21.23 151.19 -30.27 110710.0-370436	NGC 3557B ESO 377- 15 PGC 33824	.E.5.*. R -5.0± .6 	1.31± .04 .53± .03 1.22	110 .44 .00 	13.6 ±.2 13.30 ±.14 12.92	.99± .03 .54± .06 .86 .46	 13.80± .29		2855± 66 2647 3162
110934.0-464453 285.36 12.62 161.88 -29.35 110716.0-462836	ESO 265- 9 PGC 33826	.SAR0*. S -2.0± .8 	1.06± .05 .04± .04 1.14	 .68 .00 	 14.60 ±.14 13.82				6370± 60 6156 6642
110935.1-301853 278.17 27.61 143.08 -30.37 110710.0-300236	ESO 438- 6 PGC 33829	.SBS9*. S 9.0± .8 	1.04± .05 .10± .04 1.06	 .31 .11 .05	 15.71 ±.14 15.29				2200 2001 2528
1109.6 -0049 257.59 52.85 109.91 -25.53 1107.1 -0033	CGCG 11- 11 PGC 33835	.SB.3P/ F 3.0±1.3 		.13	15.4 ±.6				

R.A. 2000 DEC.	Names	Type	logD$_{25}$	p.a.	B$_T$	(B-V)$_T$	(B-V)$_o$	m$_{21}$	V$_{21}$
l b		S$_T$ n$_L$	logR$_{25}$	A$_g$	m$_B$	(U-B)$_T$	(U-B)$_o$	W$_{20}$	V$_{opt}$
SGL SGB		T	logA$_o$	A$_i$	m$_{FIR}$	(B-V)$_T^o$	m'	W$_{50}$	V$_{GSR}$
R.A. 1950 DEC.	PGC	L	logD$_o$	A$_{21}$	B$_T^o$	(U-B)$_T^o$	m'$_{25}$	HI	V$_{3K}$
1109.6 +2145 220.42 66.27 87.04 -17.23 1107.0 +2202	NGC 3555 UGC 6203 PGC 33836	.E...?. U -5.0±1.6	1.26± .08 .03± .05 1.25	30 .00 .00	13.8 ±.2				
110943.6+464852 161.61 61.95 64.22 -5.44 110653.8+470508	 UGC 6201 PGC 33840	.S?.... 	.85± .08 .19± .05 .85	110 .00 .26 .09	14.5 ±.2 14.21				7497± 50 7534 7715
1109.7 +0048 255.86 54.12 108.18 -25.01 1107.2 +0105	 CGCG 11- 13 PGC 33844	RSAROP? F .0±1.8			 .11 15.4 ±.6				
110946.4-132259 268.39 42.57 123.71 -28.54 110716.1-130642	NGC 3546 MCG -2-29- 7 PGC 33846	.LA.-P* EF -2.8± .8	1.10± .06 .27± .05 1.08	100 .11 .00					
1109.8 +6217 141.42 51.03 50.76 2.37 1106.7 +6234	 UGC 6199 IRAS11067+6234 PGC 33850	.S?.... 	1.13± .05 .63± .05 1.13	157 .00 .94 .31	14.94 ±.18 13.97				4797 4889 4931
110951.9+241548 214.37 66.98 84.68 -16.11 110711.4+243205	A 1107+24A UGC 6204 PGC 33855	.S..3$P R 3.0±1.0	.90± .07 .21± .05 .90	177 .00 .29 .10	14.8 ±.2 14.46			15.30±.3 501± 5 .73	6173± 9 6009± 46 6120 6479
110953.3-234335 274.84 33.55 135.48 -29.94 110726.1-232718	IC 2627 ESO 502- 21 IRAS11074-2327 PGC 33860	.SAS4*. R (3) 4.0± .4 2.4± .5	1.43± .02 .06± .03 1.03± .02 1.45	 .30 12.61 ±.12 .09 11.64 .03 12.21	12.62 ±.14	.61± .04 .01± .05 .52 -.06	.69± .02 -.01± .03 13.26± .04 14.45± .19	13.94±.2 49± 7 37± 8 1.70	2081± 6 2071± 29 1892 2423
110954.3+241525 214.39 66.99 84.69 -16.10 110713.8+243142	A 1107+24B UGC 6207 PGC 33862	.S.3$P/ R 3.0±1.1	1.16± .05 .84± .05 1.16	64 .00 1.16 .42 13.78	15.0 ±.2				6288± 39 6240 6599
1109.9 +0930 244.45 60.27 99.15 -22.06 1107.3 +0947	 CGCG 67- 18 PGC 33863	.LAR-.. F -3.0± .9			 .04 14.9 ±.6				
110955.4+104323 242.53 61.03 97.92 -21.61 110718.8+105940	NGC 3547 UGC 6209 PGC 33866	.S..3*. R (1) 3.0± .6 5.7± .9	1.28± .03 .31± .04 1.28	7 .03 13.10 ±.14 .43 .16 12.63	13.2 ±.4	.42± .03 -.17± .04 .34 -.22	 13.68± .44	14.73±.3 195± 10 1.94	1614± 10 1513± 66 1516 1957
110955.6+365649 181.95 66.45 72.99 -10.24 110710.6+371306	NGC 3542 MCG 6-25- 13 IRAS11071+3712 PGC 33868	.S?.... 	.89± .11 .35± .07 .89	 .00 15.10 ±.20 .53 13.23 .18 14.52					9098± 13 9098 9362
1109.9 +0714 247.84 58.78 101.48 -22.86 1107.3 +0730	 UGC 6208 IRAS11073+0730 PGC 33869	.S..3.. U (1) 3.0±1.0 4.5±1.3	.96± .09 .69± .06 .97	107 .10 15.4 ±.2 .95 13.52 .34 14.33				15.84±.3 388± 14 384± 15 1.17	6334± 12 6227 6684
110957.5-373217 281.58 21.09 151.40 -30.18 110735.0-371600	NGC 3557 ESO 377- 16 PGC 33871	.E.3... R -5.0± .3	1.61± .03 .14± .04 1.00± .02 1.66	30 .59 11.48 ±.11 .00 10.79	11.41M±.06	1.03± .01 .61± .01 .87 .49	1.05± .01 .62± .01 12.13± .06 14.10± .21		3021± 28 2813 3327
110959.1+460545 162.81 62.40 64.87 -5.75 110709.9+462202	 UGC 6205 PGC 33875	.S..9*. U 9.0±1.1	1.26± .04 .12± .05 1.26	50 .00 14.9 ±.4 .12 .06 14.82				15.98±.1 100± 6 87± 5 1.10	1391± 10 1425 1613
1109.9 +0518 250.48 57.45 103.47 -23.51 1107.4 +0535	 UGC 6210 PGC 33876	.SXS4.. UF (1) 3.5± .6 4.0± .8	1.16± .05 .38± .05 1.18	15 .15 14.33 ±.18 .56 .19					
1110.2 +0441 251.38 57.07 104.14 -23.65 1107.7 +0458	 CGCG 39- 31 PGC 33896	.LB.0*/ F -2.0± .9			 .17 15.0 ±.6				
1110.3 +0358 252.32 56.57 104.90 -23.87 1107.8 +0415	 MCG 1-29- 7 PGC 33901	.SBR1.. F 1.0± .9	.99± .10 .33± .07 1.00	 .14 14.86 ±.18 .34 .17					

R.A. 2000 DEC. l b SGL SGB R.A. 1950 DEC.	Names PGC	Type S_T n_L T L	$\log D_{25}$ $\log R_{25}$ $\log A_e$ $\log D_o$	p.a. A_g A_i A_{21}	B_T m_B m_{FIR} B_T^o	$(B-V)_T$ $(U-B)_T$ $(B-V)_T^o$ $(U-B)_T^o$	$(B-V)_e$ $(U-B)_e$ m'_e m'_{25}	m_{21} W_{20} W_{50} HI	V_{21} V_{opt} V_{GSR} V_{3K}
1110.3 +0435 251.54 57.01 104.25 -23.66 1107.8 +0452	CGCG 39- 33 PGC 33902	.E+..*. F -4.0±1.3		.16	14.60 ±.18				
1110.4 +1007 243.64 60.76 98.57 -21.72 1107.8 +1024	CGCG 67- 22 PGC 33905	.SA.9.. F (1) 9.0± .9 7.6±1.4		.03	15.2 ±.6				
1110.4 +0315 253.24 56.06 105.65 -24.08 1107.9 +0332	CGCG 39- 37 PGC 33910	.E..... F -5.0± .9		.10	15.0 ±.6				
1110.4 +0403 252.25 56.65 104.82 -23.81 1107.9 +0420	CGCG 39- 38 PGC 33911	.SXR0*. F .0±1.3		.14	15.1 ±.6				
1110.5 +5512 149.07 56.49 56.95 -1.12 1107.6 +5529	UGC 6211 PGC 33914	.SX.4.. U 4.0± .9	1.04± .06 .10± .05 1.04	125 .00 .15 .05					
1110.5 +0345 252.66 56.45 105.14 -23.89 1108.0 +0402	CGCG 39- 39 PGC 33917	.L..... F -2.0± .9		.12	14.95 ±.18				
111035.0-490612 286.45 10.51 164.49 -28.82 110818.0-484954	ESO 215- 31 PGC 33919	PSBT3.. Sr (1) 2.6± .5 4.4± .8	1.37± .04 .13± .05 .64± .03 1.44	130 .82 .18 .07	13.64 ±.15 13.44 ±.14 12.51	.84± .06 .19± .09 .61 .02	.88± .06 .18± .09 12.33± .09 14.99± .28		2735± 51 2520 2997
1110.6 +2819 203.96 67.69 80.95 -14.15 1107.9 +2836	A 1107+28B PGC 33922			.00					
111035.8-373300 281.71 21.13 151.40 -30.06 110813.2-371642	NGC 3564 ESO 377- 18 PGC 33923	.L...*/ RBS -2.0± .3	1.25± .04 .37± .03 .71± .03 1.26	15 .54 .00	13.14 ±.14 13.31 ±.14 12.64	1.02± .02 .60± .03 .84 .46	1.03± .01 .61± .03 12.18± .11 13.35± .24		2830± 41 2623 3137
1110.6 +1227 239.81 62.21 96.24 -20.80 1108.0 +1244	CGCG 67- 23 PGC 33925			.00	15.2 ±.6			16.63±.3 244± 14 227± 15	6250± 12 6161 6592
111038.4+284604 202.81 67.72 80.54 -13.94 110756.5+290222	NGC 3550 UGC 6214 PGC 33927	.P..... R 99.0	1.02± .06 .02± .05 1.02	.00 .03 .01	14.12 ±.18 13.99				10325± 19 10294 10621
1110.6 +0430 251.75 57.01 104.37 -23.62 1108.1 +0447	MCG 1-29- 9 PGC 33929	RSBR0.. F .0± .8	1.04± .09 .05± .07 1.05	.16 .04	14.4 ±.2				
1110.6 +0450 251.32 57.25 104.02 -23.51 1108.1 +0507	UGC 6216 PGC 33930	.SB.1?/ UF 1.0±1.4	1.09± .06 .82± .05 1.11	173 .18 .84 .41	15.2 ±.2 14.08			15.37±.3 584± 14 449± 15 .89	5810± 12 5695 6164
1110.7 +2818 204.01 67.71 80.97 -14.13 1108.0 +2835	A 1107+28A MCG 5-27- 1 PGC 33931	.L?.... 	1.07± .13 .00± .07 1.07	.00 .00	14.37 ±.19 14.22				10386± 84 10354 10684
1110.7 +1006 243.76 60.80 98.62 -21.66 1108.1 +1023	CGCG 67- 24 PGC 33934	.LA.0.. F -2.0± .9		.03	15.4 ±.6				
111044.3-302042 278.44 27.69 143.12 -30.12 110819.0-300424	ESO 438- 8 IRAS11083-3004 PGC 33937	.S?.... 	1.01± .06 .27± .05 1.04	120 .35 .37 .13	14.37 ±.14 12.23 13.58				8907± 69 8708 9235
1110.7 +5737 146.16 54.72 54.87 .13 1107.8 +5754	MCG 10-16- 72 7ZW 377 PGC 33938	.S?.... 	.78± .09 .05± .06 .78	.00 .07 .02	15.23 ±.18 15.11				9439± 68 9515 9600

R.A. 2000 DEC. / l b / SGL SGB / R.A. 1950 DEC.	Names / PGC	Type / S_T n_L / T / L	$\log D_{25}$ / $\log R_{25}$ / $\log A_o$ / $\log D_o$	p.a. / A_g / A_i / A_{21}	B_T / m_B / m_{FIR} / B_T^o	$(B-V)_T$ / $(U-B)_T$ / $(B-V)_T^o$ / $(U-B)_T^o$	$(B-V)_o$ / $(U-B)_o$ / m' / m'_{25}	m_{21} / W_{20} / W_{50} / HI	V_{21} / V_{opt} / V_{GSR} / V_{3K}
111045.3+120057 / 240.63 61.98 / 96.71 -20.94 / 110808.4+121715	NGC 3559 / UGC 6217 / PGC 33940	.S...P. / R / / 1.13	1.13± .05 / .19± .05 / /	55 / .00 / .29 / .10	13.68 ±.18 / 13.38			15.58±.2 / 207± 13 / 203± 6 / 2.11	3242± 6 / 3311± 50 / 3152 / 3585
1110.7 +0113 / 255.72 54.60 / 107.82 -24.65 / 1108.2 +0130	CGCG 11- 17 / PGC 33942	.LA.0./ / F / -2.0± .9 /		.11	15.2 ±.6				
111046.5-154425 / 270.28 40.66 / 126.41 -28.72 / 110816.8-152807	MCG -2-29- 8 / PGC 33943	.SBT5*. / E (1) / 5.0±1.2 / 5.3±1.2	1.07± .06 / .19± .05 / / 1.09	150 / .22 / .28 / .09					
111047.9-283000 / 277.55 29.36 / 140.99 -30.05 / 110822.0-281342	ESO 438- 9 / IRAS11083-2813 / PGC 33949	PSBR2P. / r / 2.2± .9 /	.99± .06 / .15± .05 / / 1.02	/ .34 / .19 / .08	14.26 ±.14 / 11.99 / 13.66				7350 / 7154 / 7682
1110.8 +1005 / 243.82 60.81 / 98.65 -21.64 / 1108.2 +1022	CGCG 67- 26 / PGC 33951	.LA.-*. / F / -3.0±1.3 /		.03	15.4 ±.6				
111048.7-372642 / 281.70 21.24 / 151.28 -30.02 / 110826.0-371024	NGC 3568 / ESO 377- 20 / IRAS11084-3710 / PGC 33952	.SBS5*. / R / 5.0± .6 / 1.45	1.40± .03 / .50± .03 / .64± .15 /	7 / .54 / .75 / .25	* / 13.01 ±.14 / 10.91 / 11.71	1.02± .01 / .54± .02 / 11.26± .33		13.78±.2 / 320± 13 / 213± 25 / 1.82	2445± 9 / 2446± 66 / 2237 / 2752
111050.9-215830 / 274.10 35.20 / 133.50 -29.55 / 110823.0-214212	ESO 570- 10 / PGC 33956	.S?.... / / / 1.00	.99± .07 / .44± .05 / /	53 / .14 / .66 / .22				14.81±.3 / 299± 15 / 205± 12 /	3560± 11 / 3375 / 3906
111052.1-275348 / 277.26 29.91 / 140.30 -30.01 / 110826.0-273730	ESO 438- 10 / PGC 33957	.SXS9.. / S (1) / 8.7± .7 / 7.8± .6	1.15± .05 / .31± .05 / / 1.18	10 / .28 / .32 / .15	* / 14.95 ±.14 / / 14.35			15.79±.3 / 147± 15 / 135± 12 / 1.28	1487± 11 / 1292 / 1821
111055.1+010530 / 255.92 54.52 / 107.98 -24.65 / 110821.0+012148	CGCG 11- 18 / PGC 33959	.I..9?. / F / 10.0±1.8 /		.11	* / 15.3 ±.6				
111055.8+283235 / 203.41 67.78 / 80.78 -13.98 / 110814.0+284853	NGC 3558 / MCG 5-27- 8 / MK 422 / PGC 33960	.S?.... / / /	.94± .11 / .05± .07 / / .93	/ .00 / .04 /	14.65 ±.18 / 14.47				9515± 81 / 9483 / 9812
111056.3-355854 / 281.08 22.59 / 149.60 -30.05 / 110833.0-354236	ESO 377- 21 / IRAS11085-3542 / PGC 33962	.SXT3P? / Sr / 3.1± .7 / 1.34	1.31± .04 / .42± .04 / /	129 / .32 / .58 / .21	* / 13.82 ±.14 / 13.14 / 12.89				2890± 60 / 2684 / 3201
111056.6+532316 / 151.34 57.84 / 58.57 -1.98 / 110803.1+533933	NGC 3549 / UGC 6215 / IRAS11080+5339 / PGC 33964	.SAS5*. / R (2) / 5.0± .5 / 1.51	1.51± .03 / .44± .03 / /	38 / .00 / .67 / .22	12.78 ±.13 / 12.74 ±.13 / 12.49 / 12.08	.65± .03 / .22± .05 / .55 / .14	14.05± .20	13.99±.1 / 427± 8 / 421± 7 / 1.68	2863± 6 / 2817± 66 / 2923 / 3046
111056.8+270549 / 207.21 67.68 / 82.13 -14.63 / 110815.6+272207	MCG 5-27- 9 / KUG 1108+273 / PGC 33965	.S?.... / / /	.85± .07 / .27± .06 / / .85	/ .00 / .39 / .13	15.49 ±.18 / 15.05				10720 / 10683 / 11022
1110.9 +1910 / 226.61 65.63 / 89.66 -18.05 / 1108.3 +1927	UGC 6219 / PGC 33966	.S?.... / / /	1.19± .05 / .71± .05 / / 1.19	83 / .04 / 1.06 / .35	14.91 ±.18 / 13.77				6233± 57 / 6167 / 6559
1110.9 +0451 / 251.40 57.31 / 104.03 -23.43 / 1108.4 +0508	CGCG 39- 45 / PGC 33968	.SBR5.. / F (1) / 5.0± .9 / 5.6±1.4		.18	14.9 ±.6				
1111.0 -0334 / 260.77 50.89 / 112.97 -25.97 / 1108.5 -0318	CGCG 11- 19 / PGC 33973	.SA.5*/ / F / 5.0±1.3 /		.13	15.3 ±.6				
111104.4+343408 / 187.65 67.28 / 75.26 -11.17 / 110820.5+345026	UGC 6222 / PGC 33976	.I..9*. / U / 10.0±1.1 / 1.14	1.14± .13 / .00± .12 / /	.00 / .00 / .00				15.87±.3 / 82± 7 /	1960± 10 / 1951 / 2234

R.A. 2000 DEC. / l b / SGL SGB / R.A. 1950 DEC.	Names / PGC	Type / S_T n_L / T / L	$\log D_{25}$ / $\log R_{25}$ / $\log A_o$ / $\log D_o$	p.a. / A_g / A_i / A_{21}	B_T / m_B / m_{FIR} / B_T^o	$(B-V)_T$ / $(U-B)_T$ / $(B-V)_T^o$ / $(U-B)_T^o$	$(B-V)_o$ / $(U-B)_o$ / m'_o / m'_{25}	m_{21} / W_{20} / W_{50} / HI	V_{21} / V_{opt} / V_{GSR} / V_{3K}
1111.0 +0344 / 252.84 56.52 / 105.20 -23.78 / 1108.5 +0401	CGCG 39- 46 / PGC 33977	.E...?. / F / -5.0±1.8		.13	15.13 ±.18				
111106.4+433802 / 167.16 63.90 / 67.14 -6.80 / 110818.7+435420	IC 674 / UGC 6221 / PGC 33982	.S..2.. / U / 2.0± .8	1.22± .05 / .36± .05 / / 1.22	120 / .00 / .45 / .18	14.32 ±.18 / / / 13.79				7552± 50 / 7577 / 7786
1111.1 +0356 / 252.63 56.68 / 105.00 -23.69 / 1108.6 +0413	CGCG 39- 49 / PGC 33988	.L..... / F / -2.0± .9		.13	15.1 ±.6				
1111.2 +2841 / 203.01 67.84 / 80.67 -13.86 / 1108.5 +2858	NGC 3561 / MCG 5-27- 10 / PGC 33991	.S?....	.82± .13 / .00± .07 / / .82	.00 / .00					8549± 25 / 8518 / 8845
1111.2 +2842 / 202.97 67.84 / 80.65 -13.85 / 1108.5 +2859	MCG 5-27- 11 / IRAS11085+2859 / PGC 33992	.S?....	.94± .11 / .00± .07 / / .94	.00 / .00 / .00	11.62				8811± 35 / 8780 / 9107
1111.2 +0311 / 253.58 56.15 / 105.80 -23.91 / 1108.7 +0328	CGCG 39- 50 / PGC 33998	.E+.... / F / -4.0± .9		.11	14.81 ±.18				
111118.8+055014 / 250.22 58.06 / 103.05 -23.03 / 110843.5+060632	NGC 3567 / UGC 6230 / PGC 34004	.L...P. / F / -2.0± .9	.96± .13 / .05± .05 / / .97	67 / .19 / .00	14.25 ±.15 / / / 13.97				6338± 43 / 6227 / 6690
111119.0-365225 / 281.55 21.81 / 150.62 -29.94 / 110856.0-363606	NGC 3573 / ESO 377- 22 / IRAS11089-3636 / PGC 34005	.SA.0P/ / BS / .0± .6	1.56± .03 / .57± .05 / / 1.57	4 / .40 / .42	13.3 ±.2 / 13.18 ±.14 / 13.02 / 12.35	.99± .03 / .40± .05 / .79 / .23	/ / 14.54± .29		2366± 57 / 2159 / 2674
1111.3 -0958 / 266.33 45.67 / 119.99 -27.48 / 1108.8 -0942	A 1108-09 / MCG -2-29- 9 / PGC 34006	.SXT5.. / UE F (1) / 4.8± .4 / 2.2± .8	1.30± .03 / .09± .04 / / 1.31	105 / .11 / .14 / .05	14.1 ±.4 / / / 13.80	.71± .06 / / .62	/ / / 15.19± .44	15.15±.1 / 179± 8 / 163± 11 / 1.30	7782± 6 / 7772± 59 / 7624 / 8140
1111.3 +0316 / 253.52 56.23 / 105.72 -23.86 / 1108.8 +0333	CGCG 39- 52 / PGC 34008	.L..0.. / F / -2.0± .9		.11	15.2 ±.6				
1111.3 +0318 / 253.48 56.25 / 105.68 -23.85 / 1108.8 +0335	CGCG 39- 53 / PGC 34009	.L..0.. / F / -2.0± .9		.10	15.2 ±.6				
111122.9-480107 / 286.15 11.57 / 163.25 -28.86 / 110905.0-474448	ESO 215- 32 / PGC 34010	.LX.0?. / S / -2.0± .7	1.05± .06 / .22± .04 / / 1.08	45 / .61 / .00	14.20 ±.14 / / / 13.52				4311± 87 / 4097 / 4578
1111.3 +0549 / 250.26 58.07 / 103.07 -23.01 / 1108.8 +0606	MCG 1-29- 12 / PGC 34013	.L?....	.95± .10 / .10± .04 / / .96	159 / .19 / .00	14.66 ±.11 / / / 14.38				6256± 79 / 6145 / 6608
1111.4 +0654 / 248.77 58.82 / 101.95 -22.64 / 1108.8 +0711	UGC 6233 / PGC 34015	.SAR0*. / UF / .0± .7	.98± .07 / .45± .05 / / .96	3 / .11 / .34	14.8 ±.2				
1111.4 +6211 / 141.28 51.22 / 50.94 2.49 / 1108.4 +6228	UGC 6223 / PGC 34018	.S..6?. / U / 6.0±1.7	.99± .09 / .23± .06 / / .99	150 / .00 / .33 / .11	15.38 ±.20				
111126.1+435427 / 166.56 63.81 / 66.93 -6.61 / 110838.3+441045	UGC 6232 / PGC 34019	.S..1.. / U / 1.0± .9	1.19± .05 / .65± .05 / / 1.19	146 / .00 / .66 / .32	15.06 ±.18				
111127.7+470213 / 160.85 62.06 / 64.17 -5.07 / 110838.3+471831	UGC 6227 / PGC 34021	.L..-*. / U / -3.0±1.3	.90± .09 / .05± .03 / / .89	.00 / .00	14.31 ±.17 / / / 14.19				7580± 50 / 7618 / 7797

R.A. 2000 DEC.	Names	Type	logD$_{25}$	p.a.	B$_T$	(B-V)$_T$	(B-V)$_\bullet$	m$_{21}$	V$_{21}$
l b		S$_T$ n$_L$	logR$_{25}$	A$_g$	m$_B$	(U-B)$_T$	(U-B)$_\bullet$	W$_{20}$	V$_{opt}$
SGL SGB		T	logA$_\bullet$	A$_l$	m$_{FIR}$	(B-V)^o_T	m'$_\bullet$	W$_{50}$	V$_{GSR}$
R.A. 1950 DEC.	PGC	L	logD$_o$	A$_{21}$	B^o_T	(U-B)^o_T	m'$_{25}$	HI	V$_{3K}$

111128.2-310619		.S?....	1.01± .06	125					9983± 34
278.96 27.07	ESO 438- 11		.25± .05	.34	15.23 ±.14				9783
144.00 -29.98				.34					10309
110903.0-305000	PGC 34023		1.04	.12	14.48				

1111.4 +0257		RLBR0..							
253.93 56.01	CGCG 39- 58	F		.11	15.4 ±.6				
106.06 -23.94		-2.0± .9							
1108.9 +0314	PGC 34024								

111124.4+265748	NGC 3563	.LB..*.	1.03± .08	15					9921± 52
207.59 67.76	UGC 6234	U	.13± .03	.00					9884
82.30 -14.60		-2.0± .8		.00					10223
110843.2+271406	PGC 34025		1.01						

1111.4 +0626		.E.....							
249.45 58.51	CGCG 39- 59	F		.13	14.92 ±.18				
102.44 -22.78		-5.0± .9							
1108.9 +0643	PGC 34026								

111130.1-181719	NGC 3544	PSXT1*.	1.48± .02	94				15.84±.1	3614± 9
272.09 38.53	ESO 570- 11	PSE	.46± .03	.15	12.99 ±.16			296± 12	3571± 41
129.32 -28.94		.7± .4		.47					3435
110901.1-180100	PGC 34028		1.50	.23	12.32			3.28	3963

1111.5 +5632		.SB.6?.	1.09± .06						3054± 10
147.27 55.61	UGC 6228	U	.06± .05	.00	15.0 ±.3				
55.86 -.33		6.0±1.7		.08					3126
1108.6 +5649	PGC 34029		1.09	.03	14.91				3221

111131.8+554015	NGC 3556	.SBS6./	1.94± .01	80	10.69M±.10	.66± .02	.73± .01	11.79±.0	695± 3
148.32 56.26	UGC 6225	R (1)	.59± .02	.00	10.72 ±.13	.07± .03	.09± .02	326± 4	682± 13
56.63 -.76		6.0± .3	1.52± .01	.86	9.21	.55	13.71± .03	304± 4	763
110836.8+555633	PGC 34030	5.7± .7	1.94	.29	9.84	-.02	13.78± .13	1.66	866

1111.6 +6220		.I..9*.	.96± .09						4718
141.11 51.12	UGC 6229	U	.04± .06	.00					
50.83 2.58		10.0±1.2		.03					4810
1108.6 +6237	PGC 34039		.96	.02					4852

111143.1-022702			.64± .17						5459± 63
259.90 51.89	MCG 0-29- 4		.00± .07	.12	15.2 ±.3				5323
111.82 -25.50	ARAK 281								5818
110909.9-021043	PGC 34047		.65						

111154.1-475507			.85± .07	62					4312± 87
286.19 11.69	ESO 215- 33		.00± .06	.61	15.00 ±.14				4098
163.13 -28.79	IRAS11096-4738				13.38				4579
110936.1-473848	PGC 34058		.90						

111158.6+030759		.E.....	.96± .12	70					
253.89 56.22	UGC 6239	UF	.04± .05	.11	14.28 ±.15				
105.92 -23.76		-5.0± .6		.00					
110924.1+032418	PGC 34062		.97						

111204.0+282910		.S?....	.96± .05	42					1554
203.58 68.02	UGC 6241		.62± .04	.00	15.4 ±.2				1523
80.95 -13.79	KUG 1109+287			.93					1851
110922.4+284529	PGC 34068		.96	.31	14.42				

1112.0 +2734	NGC 3570	.L.....	1.00± .19						10511
206.00 67.97	UGC 6240	U	.00± .08	.00	14.51 ±.17				10476
81.80 -14.19		-2.0± .8		.00					10811
1109.4 +2751	PGC 34071		1.00		14.35				

111208.4+352709	NGC 3569	.LA.0..	1.04± .18						7467± 22
185.42 67.29	UGC 6238	U	.04± .08	.00	14.29 ±.15				7462
74.56 -10.56		-2.0± .8		.00					7737
110924.4+354328	PGC 34075		1.03		14.18				

1112.1 +2735	NGC 3574	.S?....	.64± .17						
205.97 68.00	MCG 5-27- 22		.00± .07	.00	15.79 ±.18				
81.79 -14.16				.00					
1109.5 +2752	PGC 34080		.64						

111213.5-241407		.IBS9..	1.08± .05	79					1464
275.68 33.33	ESO 502- 23	S (1)	.37± .05	.31	16.02 ±.14				
136.12 -29.46		10.0± .6		.28					1276
110946.1-235748	PGC 34084	8.3± .6	1.11	.18	15.44				1806

111215.2-054514		PSBS3*.	1.13± .06	95					
263.15 49.30	MCG -1-29- 5	E (1)	.42± .05	.09					
115.43 -26.25		3.0±1.2		.57					
110942.9-052855	PGC 34085	3.1± .8	1.13	.21					

R.A. 2000 DEC. / l b / SGL SGB / R.A. 1950 DEC.	Names / PGC	Type / S_T n_L / T / L	$\log D_{25}$ / $\log R_{25}$ / $\log A_\bullet$ / $\log D_\circ$	p.a. / A_g / A_i / A_{21}	B_T / m_B / m_{FIR} / B_T°	$(B-V)_T$ / $(U-B)_T$ / $(B-V)_T^\circ$ / $(U-B)_T^\circ$	$(B-V)_\bullet$ / $(U-B)_\bullet$ / $m'_\bullet$ / m'_{25}	m_{21} / W_{20} / W_{50} / HI	V_{21} / V_{opt} / V_{GSR} / V_{3K}
111217.3-280001 / 277.64 29.96 / 140.44 -29.70 / 110951.0-274342	ESO 438- 12 / PGC 34087	.S?.... / / /	1.00± .06 / .16± .04 / / 1.03	37 / .31 / .25 / .08	/ 14.29 ±.14 / / 13.72				/ / / 1322± 34 / 1127 / 1655
1112.3 +0345 / 253.26 56.75 / 105.30 -23.47 / 1109.8 +0402	CGCG 39- 66 / PGC 34090	.SXS3.. / F / 3.0± .9 /			/ .12 15.1 ±.6 / /				
111226.2+513804 / 153.45 59.22 / 60.22 -2.65 / 110934.5+515423	A 1109+51 / UGCA 229 / MK 164 / PGC 34094	.S...$. / R / /	.54± .19 / .48± .09 / / .54	18 / .00 / .73 / .24					3000±110 / 3055 / 3193
111233.5-362532 / 281.60 22.32 / 150.10 -29.71 / 111010.0-360912	ESO 377- 24 / FAIR 1150 / PGC 34101	.SAT5*P / S (1) / 5.0±1.1 / 4.4±1.1	1.14± .04 / .07± .04 / / 1.18	101 / .34 / .11 / .04	/ 13.54 ±.14 / 12.90 / 13.08				2930± 19 / 2724 / 3240
111239.8+090325 / 246.04 60.49 / 99.88 -21.59 / 111003.8+091944	IC 676 / UGC 6245 / MK 731 / PGC 34107	RLBR+.. / UF / -1.0± .5 /	1.39± .07 / .13± .05 / / 1.38	10 / .07 / .00 /	/ 12.79 ±.17 / 12.00 / 12.70				1414± 24 / 1314 / 1762
1112.6 +7848 / 128.77 37.17 / 36.37 10.66 / 1108.9 +7905	UGC 6235 / PGC 34110	.S..4.. / U / 4.0±1.0 /	1.04± .08 / .73± .06 / / 1.05	82 / .08 / 1.07 / .36					
111250.7+231524 / 217.31 67.40 / 85.95 -15.93 / 111010.9+233144	UGC 6246 / PGC 34121	.S..6*. / U / 6.0±1.4 /	1.16± .05 / .84± .05 / / 1.16	45 / .00 / 1.23 / .42	/ 15.53 ±.18 / / 14.27			15.25±.3 / / 282± 7 / .56	6340± 10 / 6290 / 6290 / 6654
111252.6+101203 / 244.31 61.27 / 98.74 -21.13 / 111016.3+102823	UGC 6248 / PGC 34124	.IXS9.. / UF (1) / 9.5± .6 / 9.6±1.0	1.16± .13 / .09± .12 / / 1.16	/ .03 / .07 / .04				16.21±.1 / 87± 6 / 38± 7 /	1289± 7 / / 1193 / 1635
111253.1+272607 / 206.42 68.14 / 82.02 -14.09 / 111012.0+274227	UGC 6247 / PGC 34125	.S..2.. / U / 2.0± .9 /	1.19± .05 / .31± .05 / / 1.19	125 / .00 / .38 / .15	/ 14.5 ±.2 / / 14.06			16.00±.3 / / 385± 7 / 1.79	6830± 10 / / 6795 / 7131
1112.9 +0009 / 257.62 54.13 / 109.13 -24.45 / 1110.4 +0026	MCG 0-29- 6 / PGC 34132	PSBS5*. / F (1) / 5.0±1.3 / 4.6±2.0	.82± .13 / .08± .07 / / .83	/ .11 / .11 / .04	/ 15.27 ±.18 / /				
111259.3+725244 / 132.49 42.38 / 41.66 7.86 / 110939.4+730903	NGC 3562 / UGC 6242 / PGC 34134	.E..... / U / -5.0± .8 /	/ / .88± .02 /	/ .10 / /	13.17 ±.13 / 13.2 ±.3 / .88 / .54	.97± .02 / .53± .04 / 13.06± .07 /	1.02± .01 / .56± .04 / /		6748± 31 / 6875 / 6818
111304.8+000004 / 257.83 54.03 / 109.31 -24.47 / 111031.0+001624	CGCG 11- 27 / PGC 34138	.E+.... / F / -4.0± .9 /			/ .11 14.71 ±.18 / /				
1113.0 +0503 / 251.84 57.82 / 104.02 -22.87 / 1110.5 +0520	MCG 1-29- 16 / PGC 34139	.SXS5.. / F (1) / 5.0± .9 / 3.6±1.0	.94± .11 / .19± .07 / / .96	/ .20 / .28 / .09	/ 14.76 ±.18 / /				
1113.0 +0513 / 251.62 57.94 / 103.84 -22.82 / 1110.5 +0530	CGCG 39- 70 / PGC 34140	PLBS+?. / F / -1.0± .9 /			/ .20 15.3 ±.6 / /				
111308.5-691556 / 294.38 -8.04 / 185.97 -23.56 / 111111.1-685936	ESO 63- 11 / IRAS11111-6859 / PGC 34147	.SBS8*/ / S / 8.0±1.2 /	1.51± .04 / .97± .07 / / 1.77	173 / 2.82 / 1.19 / .48	/ / 12.65 /				
1113.1 +2748 / 205.43 68.23 / 81.70 -13.87 / 1110.5 +2805	UGC 6250 / PGC 34152	.E?.... / / /	1.08± .17 / .08± .08 / / 1.06	/ .00 / .00 /	/ 14.20 ±.16 / / 14.06				9440 / 9407 / 9740
1113.2 +5954 / 143.26 53.17 / 53.05 1.55 / 1110.3 +6011	UGC 6249 / PGC 34156	.SA.6.. / U / 6.0± .8 /	1.19± .05 / .01± .05 / / 1.19	/ .00 / .02 / .01	/ 14.2 ±.2 / / 14.22				1058 / / 1142 / 1206

R.A. 2000 DEC. / l b / SGL SGB / R.A. 1950 DEC.	Names / PGC	Type / S_T n_L / T / L	$\log D_{25}$ / $\log R_{25}$ / $\log A_\bullet$ / $\log D_o$	p.a. / A_g / A_l / A_{21}	B_T / m_B / m_{FIR} / B_T^o	$(B-V)_T$ / $(U-B)_T$ / $(B-V)_T^o$ / $(U-B)_T^o$	$(B-V)_\bullet$ / $(U-B)_\bullet$ / $m'_\bullet$ / m'_{25}	m_{21} / W_{20} / W_{50} / HI	V_{21} / V_{opt} / V_{GSR} / V_{3K}
1113.2 +2551 / 210.64 68.03 / 83.53 -14.71 / 1110.6 +2608	UGC 6252 / PGC 34157	.S..6*. / U / / 6.0±1.3	.97± .07 / .06± .05 / / .97	175 / .00 / .08 / .03	/ 14.9 ±.2 / / 14.76				6442± 10 / / 6401 / 6748
1113.2 +0339 / 253.68 56.83 / 105.49 -23.29 / 1110.7 +0356	NGC 3580 / MCG 1-29- 18 / PGC 34159	.LA.0./ / F / / -2.0± .9	.94± .11 / .47± .07 / / .88	/ .12 / .00	14.99 ±.20				
111316.9-264520 / 277.25 31.17 / 139.02 -29.41 / 111050.0-262900	NGC 3585 / ESO 502- 25 / PGC 34160	.E.7... / R / -5.0± .3	1.67± .02 / .26± .02 / 1.08± .02 / 1.64	107 / .30 / .00	10.88 ±.13 / 10.77 ±.11 / / 10.50	.97± .01 / .52± .02 / .89 / .46	.99± .01 / .56± .01 / 11.77± .06 / 13.58± .18		1399± 27 / 1206 / 1735
1113.4 +3035 / 197.91 68.30 / 79.13 -12.57 / 1110.7 +3052	MCG 5-27- 27A / PGC 34169	.S?.... / / /	.82± .13 / .08± .07 / / .82	/ .00 / .11 / .04	15.14 ±.18 / / / 14.96				10243 / 10220 / 10532
1113.4 +5335 / 150.57 57.95 / 58.57 -1.56 / 1110.5 +5352	A 1110+53 / UGC 6251 / DDO 92 / PGC 34170	.SX.9*. / U (1) / 9.0± .9 / 8.0±1.0	1.26± .06 / .03± .06 / .95± .03 / 1.26	60 / .00 / .03 / .02	14.93 ±.15 / 15.0 ±.5 / / 14.90	.24± .05 / -.34± .07 / .23 / -.35	.32± .04 / -.36± .06 / 15.17± .06 / 16.00± .36	14.74±.1 / 57± 6 / 44± 7 / -.17	927± 6 / / 990 / 1110
1113.4 -0615 / 263.95 49.06 / 116.05 -26.10 / 1110.9 -0559	/ PGC 34171	.SBS7P. / E (1) / 7.0± .9 / 6.4± .8	1.20± .07 / .19± .08 / / 1.21	110 / .09 / .26 / .09					
111329.3+220912 / 220.17 67.23 / 87.07 -16.27 / 111049.8+222532	Leo II / UGC 6253 / DDO 93 / PGC 34176	.E.0.P. / R / -5.0± .3	2.08± .02 / .04± .04 / 1.74± .07 / 2.07	/ .00 / .00	12.6 ±.3 / / / 12.59	.59± .33 / / .59	.67± .05 / / 16.78± .16 / 17.87± .30		90± 60 / 36 / 408
1113.5 +1029 / 244.05 61.58 / 98.51 -20.88 / 1110.9 +1046	IC 2634 / MCG 2-29- 10 / PGC 34178	.SXS3.. / F (1) / 3.0±1.0 / 3.6±1.0	.64± .17 / .00± .07 / / .64	/ .03 / .00 / .00	15.3 ±.2				
111336.6-440103 / 284.94 15.41 / 158.71 -28.98 / 111116.1-434442	ESO 265- 16 / IRAS11113-4345 / PGC 34183	.SA.5*/ / S (1) / 4.5± .7 / 3.3±1.3	1.23± .05 / .71± .04 / / 1.28	50 / .56 / 1.07 / .36	15.55 ±.14 / 13.66				
111340.2-023638 / 260.69 52.06 / 112.14 -25.08 / 111107.0-022018	MCG 0-29- 7 / PGC 34189	.SBR3P? / F (1) / 3.0± .9 / 3.6±1.0	.94± .11 / .00± .07 / / .95	/ .12 / .00 / .00	14.52 ±.18				
111341.2+473436 / 159.45 62.03 / 63.88 -4.47 / 111052.2+475056	UGC 6255 / ARAK 283 / PGC 34192	.S?.... / / /	.87± .08 / .20± .05 / / .87	35 / .30 / .10	14.0 ±.3 / 12.56 / / 13.64				5351± 82 / 5391 / 5565
111345.3+481623 / 158.29 61.60 / 63.27 -4.12 / 111056.0+483243	NGC 3577 / UGC 6257 / PGC 34195	.SBR1.. / R / 1.0± .4	1.15± .04 / .01± .04 / / 1.15	/ .00 / .01 / .00	14.3 ±.2 / / / 14.25				5336± 10 / 5321± 49 / 5378 / 5546
111349.5+213106 / 221.78 67.11 / 87.71 -16.46 / 111110.3+214727	UGC 6258 / PGC 34198	.I..9.. / U / 10.0± .9	1.30± .03 / .61± .04 / / 1.30	175 / .00 / .46 / .31	14.9 ±.2 / / / 14.39			14.95±.1 / 189± 7 / 195± 12 / .25	1454± 6 / / 1398 / 1774
111349.6+093513 / 245.60 61.06 / 99.45 -21.13 / 111113.5+095133	IC 2637 / UGC 6259 / MK 732 / PGC 34199	.E+..P. / F / -4.0± .9	.92± .07 / .03± .05 / / .92	70 / .02 / .00	14.04 ±.18 / 12.43 / 13.89				8768± 29 / 8670 / 9115
1113.8 +6510 / 138.24 48.95 / 48.47 4.19 / 1110.8 +6527	A 1110+65 / UGC 6256 / PGC 34203	.S..6*. / U / / 6.0±1.4	1.11± .07 / .87± .06 / / 1.11	89 / .02 / 1.28 / .44					
1113.8 +0417 / 253.09 57.39 / 104.88 -22.94 / 1111.2 +0433	UGC 6260 / IRAS11112+0433 / PGC 34204	.S..4*/ / UF (1) / 4.0± .9 / 4.5±1.2	1.09± .06 / .53± .05 / / 1.10	43 / .12 / .78 / .26	14.80 ±.18 / 12.60				
1113.9 +1033 / 244.07 61.70 / 98.48 -20.76 / 1111.3 +1050	IC 2638 / UGC 6261 / PGC 34205	RLBR+.. / UF / -1.0± .6	1.01± .08 / .29± .03 / / .97	170 / .03 / .00	14.80 ±.15				

R.A. 2000 DEC.	Names	Type	logD$_{25}$	p.a.	B$_T$	(B-V)$_T$	(B-V)$_e$	m$_{21}$	V$_{21}$
l b		S$_T$ n$_L$	logR$_{25}$	A$_g$	m$_B$	(U-B)$_T$	(U-B)$_e$	W$_{20}$	V$_{opt}$
SGL SGB		T	logA$_e$	A$_i$	m$_{FIR}$	(B-V)$_T^o$	m'$_e$	W$_{50}$	V$_{GSR}$
R.A. 1950 DEC.	PGC	L	logD$_o$	A$_{21}$	B$_T^o$	(U-B)$_T^o$	m'$_{25}$	HI	V$_{3K}$
111356.7+121804	IC 677	.S..4..	1.19± .05	45				15.62±.2	3247± 6
241.13 62.77	UGC 6262	U (1)	.40± .05	.00	13.75 ±.20			269± 13	3272± 50
96.74 -20.11	IRAS11113+1234	4.0± .9		.59	13.01			277± 5	3159
111119.9+123425	PGC 34211	3.5±1.2	1.19	.20	13.14			2.28	3590
111403.0-140518	NGC 3591	PLBR+?.	1.09± .06	150					
270.09 42.49	MCG -2-29- 12	E	.16± .05	.22					
124.72 -27.65		-1.0± .9		.00					
111132.6-134857	PGC 34220		1.08						
1114.0 +0634	IC 678	.E.....							
250.14 59.07	MCG 1-29- 21	F		.14	14.9 ±.6				968± 50
102.55 -22.13		-5.0± .9							860
1111.5 +0651	PGC 34222								1319
1114.1 +5647	A 1111+57	.SBR4*.	.85± .12	.00	15.29 ±.18				10015± 60
146.50 55.65	MCG 10-16- 89	P	.15± .07	.22					10089
55.82 .11		4.0±1.4		.08	15.00				10180
1111.2 +5704	PGC 34223		.85						
111407.6-144216		.L...P.	1.12± .08	175					
270.53 41.96	MCG -2-29- 13	E	.33± .05	.20					
125.41 -27.74	IRAS11116-1426	-2.0± .9		.00	13.34				
111137.4-142555	PGC 34225		1.09						
1114.1 +2713	CGCG 156- 29			.00	15.4 ±.3				8085
207.05 68.40									8050
82.34 -13.93									8387
1111.5 +2730	PGC 34230								
111411.5+481912	NGC 3583	.SBS3..	1.45± .03	125				13.91±.1	2136± 6
158.11 61.63	UGC 6263	R (2)	.19± .04	.00	11.96 ±.14			356± 14	2098± 31
63.27 -4.04	IRAS11113+4835	3.0± .3		.26	10.86			327± 11	2178
111122.3+483533	PGC 34232	3.3± .6	1.45	.09	11.69			2.13	2345
1114.2 +2833		.L?....	.90± .17						10306
203.43 68.51	MCG 5-27- 30		.00± .07	.00	14.90 ±.16				10276
81.11 -13.32				.00					10603
1111.6 +2850	PGC 34237		.90		14.74				
1114.4 +2228		.S..6*.	1.04± .08	125					
219.53 67.54	UGC 6268	U	.98± .06	.00					
86.86 -15.93		6.0±1.4		1.44					
1111.8 +2245	PGC 34247		1.04	.49					
111428.4+171532	NGC 3592	.S..5?/	1.25± .04	120				16.32±.3	1298± 10
231.67 65.54	UGC 6267	PU	.46± .05	.00	14.38 ±.19				1227
91.89 -18.07		5.0±1.2		.69				183± 7	1227
111150.4+173153	PGC 34248		1.25	.23	13.69			2.40	1629
1114.4 +4313		.S..9..	1.10± .07	55					2204± 10
167.23 64.64	UGC 6266	U	.16± .06	.00					2229
67.80 -6.46		9.0± .8		.16					
1111.7 +4330	PGC 34249		1.10	.08					2440
1114.5 +2932		.SB.0..	1.00± .16	165					
200.75 68.59	UGC 6270	U	.43± .12	.00					
80.23 -12.82		.0± .9		.32					
1111.9 +2949	PGC 34254		.98						
111436.1+124907	NGC 3593	.SAS0*.	1.72± .02	92	11.86M±.08	.94± .01	.98± .01	14.69±.1	628± 4
240.42 63.21	UGC 6272	R	.43± .03	.00	11.78 ±.11	.32± .01	.30± .01	238± 3	624± 13
96.29 -19.77	IRAS11119+1305	.0± .4	1.05± .03	.32	9.92	.87	12.62± .10	194± 3	541
111159.3+130528	PGC 34257	1.69			11.50	.25	14.21± .13		968
111437.3+301853		.S...1..	1.10± .04	52					2022
198.64 68.58	UGC 6271	U	.44± .04	.00	14.64 ±.18				1998
79.52 -12.46	IRAS11119+3035	1.0± .9		.45	13.45				2313
111155.6+303514	PGC 34260		1.10	.22	14.17				
111438.8-335421		.L..../	1.18± .04	128	12.68V±.13	.98± .02			2666± 56
280.92 24.80	ESO 377- 29	S	.40± .03	.37	13.47 ±.14				2464
147.21 -29.33		-2.0±1.2		.00		.84			2984
111214.0-333800	PGC 34262		1.16		13.17		13.42± .26		
111441.9-234346	NGC 3597	.L..+*P	1.28± .04						3513± 39
276.02 34.04	ESO 503- 3	S	.11± .04	.22	13.62 ±.14				3326
135.60 -28.85	IRAS11122-2327	-1.3± .8		.00	10.47				3856
111214.0-232724	PGC 34266		1.28		13.34				
1114.7 +5634	A 1111+56	.SBR3$.	.82± .13	.00	15.57 ±.18				10369± 60
146.65 55.87	MCG 10-16- 93	P	.17± .07	.24					10442
56.05 .07		3.0±1.4		.09	15.25				10535
1111.8 +5651	PGC 34268		.82						

R.A. 2000 DEC.	Names	Type	logD$_{25}$	p.a.	B$_T$	(B-V)$_T$	(B-V)$_e$	m$_{21}$	V$_{21}$
l b		S$_T$ n$_L$	logR$_{25}$	A$_g$	m$_B$	(U-B)$_T$	(U-B)$_e$	W$_{20}$	V$_{opt}$
SGL SGB		T	logA$_e$	A$_i$	m$_{FIR}$	(B-V)$_T^o$	m'$_e$	W$_{50}$	V$_{GSR}$
R.A. 1950 DEC.	PGC	L	logD$_o$	A$_{21}$	B$_T^o$	(U-B)$_T^o$	m'$_{25}$	HI	V$_{3K}$
1114.8 -1041		.S..5P?	1.12± .08	60					
267.92 45.51		E (1)	.33± .08	.14					
120.99 -26.79		5.0±1.8		.49					
1112.3 -1025	PGC 34275	5.3±1.2	1.14	.16					
111450.2-025040		RLBT+*.	1.16± .12						8246
261.30 52.04	MCG 0-29- 8	F	.08± .07	.13	14.1 ±.2				
112.48 -24.86		-1.0± .8		.00					8110
111217.0-023418	PGC 34276		1.16		13.81				8605
111451.3+353014		.S..2..	1.20± .05	29					
184.79 67.81	UGC 6273	U	.66± .05	.02					6447± 50
74.78 -10.05		2.0± .9		.81					6443
111207.9+354635	PGC 34278		1.20	.33					6717
1114.9 +3130									10292
195.36 68.55	CGCG 156- 34			.00	15.2 ±.3				10273
78.44 -11.87									
1112.2 +3147	PGC 34280								10578
111456.8+334933		.S..6*.	.95± .07	145				16.11±.3	11073± 9
189.14 68.22	UGC 6274	U	.06± .05	.06	14.61 ±.18				
76.32 -10.81		6.0±1.3		.08				267± 5	11063
111214.0+340554	PGC 34284		.96	.03	14.42			1.66	11350
111457.3-472822		.IBS9P.	1.08± .06						
286.51 12.30	ESO 265- 19	S (1)	.15± .05	.63	15.76 ±.14				
162.55 -28.34		10.0± .6		.11					
111238.1-471200	PGC 34285	8.9± .7	1.14	.08					
111503.6-282328	A 1112-28	.SBT4..	1.30± .03	35				14.91±.3	3353± 11
278.47 29.86	ESO 438- 15	Sr (1)	.35± .03	.35	13.83 ±.14			320± 15	3385± 34
140.92 -29.11	IRAS11126-2807	4.1± .4		.51	13.22			302± 12	3161
111237.1-280706	PGC 34292	2.6± .5	1.34	.17	12.95			1.78	3688
111505.3+144717	NGC 3596	.SXT5..	1.60± .02		11.95S±.15			13.61±.0	1191± 4
236.89 64.42	UGC 6277	R (2)	.02± .03	.00	11.50 ±.14			134± 4	1176± 66
94.38 -18.90		5.0± .3		.03				116± 4	1111
111227.9+150338	PGC 34298	3.5± .6	1.60	.01	11.67		14.74± .20	1.92	1528
111510.1+310207		.L.....	1.18± .15						8202± 31
196.64 68.65	UGC 6276	U	.07± .08	.00	13.84 ±.17				8181
78.91 -12.04		-2.0± .8		.00					8490
111228.2+311828	PGC 34303		1.17		13.71				
111511.6+171549	NGC 3598	.L..-*.	1.27± .08	35					6097± 36
231.86 65.69	UGC 6278	U	.14± .05	.00	13.25 ±.15				6026
91.96 -17.91		-3.0±1.1		.00					6428
111233.6+173211	PGC 34306		1.25		13.16				
111513.8+604204	NGC 3589	.S..7*.	1.17± .05	48					1969± 10
142.14 52.69	UGC 6275	U	.26± .05	.00	14.35 ±.18				1966± 50
52.47 2.14		7.0±1.2		.36					2056
111216.0+605825	PGC 34308		1.17	.13	13.98				2112
111515.9-273940		.S?....	.89± .06						
278.16 30.54	ESO 503- 5		.10± .06	.25	14.57 ±.14				3961± 34
140.09 -29.03				.15					3768
111249.0-272318	PGC 34310		.91	.05	14.14				4295
1115.2 +2723									8183
206.66 68.66	CGCG 156- 37			.00	15.1 ±.3				8149
82.30 -13.64									
1112.6 +2740	PGC 34312								8484
111516.8-134358		.L.R0P?	1.12± .08	95					
270.21 42.95	MCG -2-29- 14	EF	.19± .05	.18					
124.38 -27.29	IRAS11127-1327	-2.0± .8		.00	12.54				
111246.2-132736	PGC 34313		1.11						
111519.0-034622		.E+....							
262.33 51.36	CGCG 11- 33	F		.10	15.0 ±.6				
113.51 -25.00		-4.0± .9							
111246.1-033000	PGC 34318								
111521.9+353008		.S..1..	1.10± .05	7					
184.72 67.91	UGC 6279	U	.35± .05	.02	14.68 ±.18				6399± 19
74.83 -9.96		1.0± .9		.35					6395
111238.6+354630	PGC 34320		1.10	.17	14.23				6669
111526.0+472649	NGC 3595	.E?....	1.20± .07	176					
159.27 62.34	UGC 6280		.33± .04	.00	13.1 ±.2				2248
64.14 -4.28				.00					2288
111237.7+474311	PGC 34325		1.10		13.05				2463

R.A. 2000 DEC.	Names	Type	logD$_{25}$	p.a.	B$_T$	(B-V)$_T$	(B-V)$_o$	m$_{21}$	V$_{21}$
l b		S$_T$ n$_L$	logR$_{25}$	A$_g$	m$_B$	(U-B)$_T$	(U-B)$_o$	W$_{20}$	V$_{opt}$
SGL SGB		T	logA$_o$	A$_i$	m$_{FIR}$	(B-V)$_T^o$	m'$_o$	W$_{50}$	V$_{GSR}$
R.A. 1950 DEC.	PGC	L	logD$_o$	A$_{21}$	B$_T^o$	(U-B)$_T^o$	m'$_{25}$	HI	V$_{3K}$

111527.3+180646	NGC 3599	.LA..*.	1.43± .09		12.82 ±.14	.87± .01	.85± .01		781± 52
230.08 66.14	UGC 6281	R	.10± .07	.00	12.57 ±.16	.32± .04	.33± .02		713
91.16 -17.51		-2.0± .3	.90± .03	.00		.85	12.81± .11		1110
111249.1+182308	PGC 34326		1.41		12.70	.32	14.57± .49		
111532.2-141019	IC 2668	.SBT1P.	1.10± .06	140					
270.58 42.59	MCG -2-29- 15	E	.23± .05	.20					
124.88 -27.31	IRAS11130-1353	1.0± .9		.24	12.79				
111301.7-135357	PGC 34333		1.12	.12					
111533.1+050700	NGC 3601	.SBR2?.	.74± .09						7859± 63
252.60 58.28	UGC 6282	F	.09± .05	.16	14.7 ±.3				7747
104.18 -22.27	ARAK 284	2.0±1.9		.11	12.53				8212
111258.1+052322	PGC 34335		.75	.04	14.38				
111535.8-484535		.SAR5..	1.45± .04						5476± 87
287.10 11.14	ESO 215- 37	Sr (1)	.08± .05	.53	13.53 ±.14				5262
163.96 -28.06		4.8± .4		.11					5740
111317.0-482912	PGC 34338	4.4± .6	1.50	.04	12.85				
111549.1-270640		.LB.-?/	1.14± .04	36					1517± 34
278.02 31.09	ESO 503- 7	S	.46± .03	.34	14.76 ±.14				1325
139.47 -28.87		-3.0± .8		.00					1853
111322.0-265018	PGC 34349		1.11		14.40				
111552.1+413527	NGC 3600	.S..1?.	1.61± .02	3				13.15±.1	719± 11
170.29 65.68	UGC 6283	U	.68± .04	.00	12.60 ±.18			214± 16	1443± 81
69.39 -7.01	MK 1443	1.0±1.5		.70	12.81			197± 12	750
111306.6+415149	PGC 34353		1.61	.34	11.90			.91	975
1115.9 -0020		.SBR0?.							
259.15 54.21	CGCG 11- 36	F			.16 15.3 ±.6				
109.91 -23.88		.0±1.8							
1113.4 -0004	PGC 34359								
111601.0-335759	A 1113-33	.SXS4*.	1.31± .04	148	14.05 ±.15	.76± .02	.82± .02	15.87±.3	2971± 17
281.24 24.85	ESO 377- 31	PS (2)	.37± .04	.34	14.29 ±.14			249± 34	3005± 59
147.28 -29.05		3.8± .7	.94± .02	.54		.59	14.24± .05	227± 25	2772
111336.0-334136	PGC 34362	2.4± .6	1.34	.18	13.28		14.52± .26	2.41	3291
111604.5-761253	NGC 3620	PSBS2P*	1.44± .05	78					
297.22 -14.41	ESO 38- 10	Sr	.38± .07	.40					
192.89 -21.03	IRAS11143-7556	1.7± .4		.46	9.03				
111423.0-755630	PGC 34366		1.47	.19					
111605.6+100931	IC 2672	.SXT5*.	1.00± .05	35				16.46±.3	5895± 10
245.45 61.85	UGC 6288	UF (1)	.08± .04	.05	14.7 ±.2			231± 13	
99.09 -20.41		4.7± .7		.12					5800
111329.5+102553	PGC 34368	4.6±1.0	1.00	.04	14.48			1.94	6241
111605.8+235450		.I..9..	1.04± .15					16.23±.3	6258± 7
216.11 68.29	UGC 6287	U	.13± .12	.00					6211
85.66 -14.97		10.0± .9		.10				77± 7	6570
111326.2+241113	PGC 34369		1.04	.06					
111609.0+003419		.SXT5*.	.89± .11						
258.24 54.95	MCG 0-29- 9	F (1)	.24± .07	.15	15.24 ±.18				
108.96 -23.57		5.0±1.3		.36					
111335.0+005042	PGC 34371	5.6±2.0	.90	.12					
1116.1 -0341		.SAS3?.							
262.53 51.55	CGCG 11- 37	F			.12 15.4 ±.6				
113.48 -24.78		3.0±1.3							
1113.6 -0325	PGC 34372								
111609.6+110230	IC 2674	.S..6*.	1.03± .07	20				16.38±.3	5883± 10
244.02 62.43	UGC 6290	U	.25± .08	.01	14.81 ±.19			214± 13	
98.21 -20.07		6.0±1.2		.37					5791
111333.2+111853	PGC 34373		1.03	.13	14.41			1.85	6227
1116.1 +5541	NGC 3594	.LB..*.	1.11± .10	10					6257
147.40 56.67	UGC 6286	U	.03± .05	.00	14.7 ±.2				6327
56.93 -.18		-2.0±1.1		.00					6428
1113.3 +5558	PGC 34374		1.11		14.62				
1116.0 +2926		.E.3.*.			13.98V±.05	1.07± .03			8334± 88
201.02 68.92	CGCG 156- 39	P		.00	15.3 ±.3				8308
80.47 -12.57		-5.0±1.3				.99			8628
1113.4 +2943	PGC 34375								
111611.2+292325	A 1113+29A	.E?....	.90± .17		13.98V±.14	1.07± .03			8757± 49
201.16 68.94	MCG 5-27- 35		.16± .07	.00	15.07 ±.15				8730
80.53 -12.57				.00		.99			9051
111330.0+293948	PGC 34376		.85		14.93		14.11± .87		

R.A. 2000 DEC.	Names	Type	$\log D_{25}$	p.a.	B_T	$(B-V)_T$	$(B-V)_e$	m_{21}	V_{21}
l b		S_T n_L	$\log R_{25}$	A_g	m_B	$(U-B)_T$	$(U-B)_e$	W_{20}	V_{opt}
SGL SGB		T	$\log A_e$	A_i	m_{FIR}	$(B-V)_T^o$	m'_e	W_{50}	V_{GSR}
R.A. 1950 DEC.	PGC	L	$\log D_o$	A_{21}	B_T^o	$(U-B)_T^o$	m'_{25}	HI	V_{3K}

111616.1-334935 NGC 3606		.E.0.*.	1.17± .05						
281.23 25.00 ESO 377- 32		S	.04± .04	.34	13.42 ±.14				3013± 21
147.12 -28.99		-5.0±1.1		.00					2812
111351.0-333312 PGC 34378			1.22		13.03				3331
111617.0+025317		.SXS4..	.92± .07						
255.64 56.75 UGC 6289		F (1)	.07± .05	.12	14.44 ±.18				8956± 50
106.55 -22.82 IRAS11137+0309		4.0± .9		.10	13.43				8837
111342.5+030940 PGC 34379		2.6±1.0	.93	.03	14.16				9312
1116.2 +2913		.S?....	.99± .10						
201.62 68.96 MCG 5-27- 36			.16± .07	.00	15.0 ±.2				13012± 29
80.69 -12.63				.12					12985
1113.6 +2930 PGC 34380			.98		14.70				13307
1116.3 +2915		.S?....	.99± .10						
201.52 68.98 MCG 5-27- 37			.09± .07	.00	14.79 ±.19				13581± 29
80.67 -12.59				.07					13554
1113.7 +2932 PGC 34385			.98		14.51				13875
1116.4 +0948 IC 2680		.SAR0?.							
246.11 61.68 CGCG 67- 47		F		.07	15.4 ±.6				
99.47 -20.46		.0±1.3							
1113.8 +1005 PGC 34387									
111628.0+291936		.SBS3..	1.23± .03						
201.34 69.00 UGC 6292		U	.25± .04	.00	14.4 ±.3				13895± 88
80.62 -12.55 IRAS11137+2935		3.0± .8		.34	13.27				13868
111346.9+293559 PGC 34393			1.23	.12	13.91				14189
111629.2+410442		.S..6*.	.94± .11						
171.25 66.03 UGC 6293		U	.19± .07	.00	15.16 ±.18				
69.90 -7.15		6.0±1.2		.28					
111344.0+412105 PGC 34399			.94	.09					
111632.4+284637 A 1113+29B									
202.87 69.01 CGCG 156- 45				.00					6950± 45
81.14 -12.78 MK 37									6921
111351.4+290300 PGC 34400									7246
1116.6 +2621		.S..4..	1.04± .08	104					
209.59 68.86 UGC 6294		U	.88± .06	.00					
83.41 -13.81		4.0±1.0		1.30					
1114.0 +2638 PGC 34411			1.04	.44					
1116.7 +2632									
209.09 68.91 CGCG 156- 47				.00	15.3 ±.3				10055
83.25 -13.71									10018
1114.1 +2649 PGC 34414									10359
111646.7+180104 NGC 3605		.E.4+..	1.19± .03	17	13.13 ±.13	.86± .02	.87± .01		
230.64 66.38 UGC 6295		R	.18± .02	.00	13.09 ±.13	.46± .06	.49± .04		649± 24
91.39 -17.26		-5.0± .4	.85± .03	.00		.85	12.87± .09		581
111408.7+181727 PGC 34415			1.14		13.10	.46	13.61± .22		978
111651.1+174804		.S?....	1.09± .06	166				16.64±.2	976± 6
231.14 66.30 UGC 6296			.44± .05	.00	14.51 ±.19			204± 6	
91.61 -17.33				.66				181± 7	908
111413.1+180427 PGC 34419			1.09	.22	13.84			2.58	1306
111654.1+180312 NGC 3607		.LAS0*.	1.69± .03	120	10.82M±.10	.93± .01	.96± .01		
230.59 66.42 UGC 6297		R	.30± .03	.00	11.00 ±.13	.48± .03	.53± .02		935± 22
91.37 -17.22		-2.0± .3	1.16± .03	.00		.89	12.15± .09		867
111416.1+181935 PGC 34426			1.64		10.87	.46	13.38± .20		1264
111658.7+180857 NGC 3608		.E.2...	1.50± .03	75	11.70M±.10	.94± .02	.96± .01	18.51±.3	1108± 10
230.40 66.48 UGC 6299		R	.09± .03	.00	11.77 ±.12	.39± .03	.48± .01		1205± 24
91.28 -17.17		-5.0± .3	1.05± .04	.00		.93	12.48± .15	138± 11	1056
111420.7+182520 PGC 34433			1.48		11.71	.40	14.00± .21		1452
111703.5+360829		.S?....	.95± .07	75					
182.82 68.06 UGC 6298			.26± .05	.00	14.97 ±.18				7613± 19
74.41 -9.36				.40					7612
111420.3+362452 PGC 34440			.95	.13	14.53				7880
111704.0-491206		.SXS5..	1.15± .04	21	13.57 ±.15	.54± .02	.63± .01		
287.49 10.82 ESO 215- 39		S (1)	.15± .04	.59	14.08 ±.14				4282± 40
164.42 -27.75		5.0± .6	.95± .02	.22		.35	13.81± .05		4069
111445.0-485542 PGC 34443		2.8± .6	1.21	.07	13.01		13.81± .28		4544
111704.9-345718		.SBS4*.	1.34± .04	160					
281.89 24.03 ESO 377- 34		S	.55± .04	.30	14.35 ±.14				
148.40 -28.82		4.0±1.2		.81					
111440.0-344054 PGC 34445		3.3± .9	1.37	.27					

R.A. 2000 DEC. l b SGL SGB R.A. 1950 DEC.	Names PGC	Type S_T n_L T L	$\log D_{25}$ $\log R_{25}$ $\log A_{\bullet}$ $\log D_o$	p.a. A_g A_i A_{21}	B_T m_B m_{FIR} B_T^o	$(B-V)_T$ $(U-B)_T$ $(B-V)_T^o$ $(U-B)_T^o$	$(B-V)_{\bullet}$ $(U-B)_{\bullet}$ $m'_{\bullet}$ m'_{25}	m_{21} W_{20} W_{50} HI	V_{21} V_{opt} V_{GSR} V_{3K}
1117.2 +1619 234.42 65.66 93.08 −17.83 1114.6 +1636	 UGC 6300 PGC 34461	.E..... U −5.0± .9 	1.09± .06 .21± .05 1.03	80 .00 .00 	 14.8 ±.2 				
111717.5−482630 287.25 11.54 163.57 −27.82 111458.0−481006	 ESO 215−40 PGC 34465	.LXR0?. S −2.0± .6 	1.17± .05 .34± .04 1.18	92 .54 .00 	 14.53 ±.14 				
111718.1−274918 278.71 30.59 140.30 −28.58 111451.0−273254	 ESO 438−17 PGC 34466	.SBS5.. S (1) 5.0± .6 5.6± .6	1.26± .03 .37± .03 1.30	9 .39 .56 .19	 14.19 ±.14 13.23			15.29±.3 162± 16 144± 12 1.87	1230± 11 1037 1564
1117.3 +2220 220.40 68.14 87.28 −15.37 1114.7 +2237	 UGC 6301 PGC 34468	.S..6*. U 6.0±1.3 	1.20± .06 .63± .06 1.20	139 .00 .93 .32	 15.2 ±.2 				
111727.3+043623 253.92 58.23 104.88 −21.99 111452.5+045247	 UGC 6306 PGC 34476	.I..9*. U 10.0±1.3 	1.07± .14 .40± .12 1.08	 .14 .30 .20	 			15.72±.3 103± 7 	1751± 10 1638 2105
111729.5+043317 254.00 58.20 104.93 −22.00 111454.7+044941	NGC 3611 UGC 6305 IRAS11149+0449 PGC 34478	.SAS1P. R 1.0± .3 1.34	1.32± .04 .09± .04 .78± .03 	 .14 .09 .04	12.77 ±.13 12.48 ±.13 11.43 12.37	.62± .01 .00± .02 .56 −.03	.65± .01 .06± .02 12.16± .10 14.02± .25	14.38±.1 337± 5 299± 4 1.96	1585± 4 1754± 56 1473 1940
111731.5+030154 255.90 57.06 106.51 −22.48 111457.1+031818	 CGCG 39−102 PGC 34481	PSBR1*. F 1.0± .9 	 	 .11 	 14.9 ±.6 				
111733.4−303118 280.03 28.14 143.37 −28.66 111507.0−301454	 ESO 438−19 PGC 34484	PSAR0P? r −.1± .9 	.92± .05 .13± .04 .94	152 .36 .10 	 15.11 ±.14 14.50				9839± 34 9642 10166
111733.6+360349 182.94 68.18 74.53 −9.30 111450.5+362013	 UGC 6303 PGC 34485	.S..6*. U 6.0±1.2 	1.07± .06 .04± .05 1.07	 .00 .06 .02	 14.7 ±.2 				
1117.5 −0011 259.54 54.57 109.88 −23.45 1115.0 +0005	 CGCG 11−43 PGC 34486	.L..+*/ F −1.0±1.3 	 	 .17 	 15.4 ±.6 				
111734.1−275342 278.81 30.54 140.39 −28.53 111507.0−273718	 ESO 438−18 PGC 34487	.SB?... 1.08	1.04± .05 .21± .06 1.08	9 .39 .31 .10	 15.14 ±.14 14.40				7300± 60 7107 7634
1117.5 +0717 250.33 60.19 102.14 −21.06 1115.0 +0734	 CGCG 39−104 PGC 34489	.E..... F −5.0± .9 	 	 .12 	 14.9 ±.6 				
111737.1−031012 262.53 52.18 113.04 −24.28 111504.1−025348	 MCG 0−29−10 PGC 34492	.SXR2.. F (1) 2.0± .9 1.6±1.0	1.08± .09 .42± .07 1.09	 .12 .51 .21	 14.64 ±.18 				
111739.0+270524 207.63 69.16 82.82 −13.29 111458.7+272148	 UGC 6308 KUG 1114+273 PGC 34495	.S..4.. U 4.0±1.0 	1.04± .05 .71± .05 1.04	94 .00 1.04 .36	 				
1117.6 −0118 260.72 53.70 111.07 −23.75 1115.1 −0102	 CGCG 11−45 PGC 34496	.SB.4*/ F 4.0±1.3 	 	 .13 	 14.9 ±.6 				
111740.3+380307 177.96 67.52 72.74 −8.37 111456.6+381931	 UGC 6307 PGC 34497	.S..8*. U 8.0±1.3 	1.15± .05 .60± .05 1.15	160 .00 .74 .30	 14.78 ±.18 				
1117.6 +0231 256.55 56.70 107.05 −22.60 1115.1 +0248	 CGCG 39−105 PGC 34498	.L..... F −2.0± .9 	 	 .12 	 15.13 ±.18 				

R.A. 2000 DEC.	Names	Type	logD$_{25}$	p.a.	B$_T$	(B-V)$_T$	(B-V)$_e$	m$_{21}$	V$_{21}$
l b		S$_T$ n$_L$	logR$_{25}$	A$_g$	m$_B$	(U-B)$_T$	(U-B)$_e$	W$_{20}$	V$_{opt}$
SGL SGB		T	logA$_e$	A$_i$	m$_{FIR}$	(B-V)^o_T	m'	W$_{50}$	V$_{GSR}$
R.A. 1950 DEC.	PGC	L	logD$_o$	A$_{21}$	B^o_T	(U-B)^o_T	m'$_{25}$	HI	V$_{3K}$

1117.7 +5821		.S..9*.	1.14± .07						
144.12 54.74	UGC 6304	U	.11± .06	.00					
54.68 1.29		9.0±1.2		.11					
1114.8 +5838	PGC 34502		1.14	.05					
111744.7-260224		.SXS6..	1.12± .04	90					
277.95 32.25	ESO 503- 11	S (1)	.25± .04	.26	15.49 ±.14				
138.29 -28.36		6.0± .8		.37					
111517.0-254600	PGC 34504	6.7± .8	1.14	.12					
111746.8+512837		.SB?...	1.14± .05	125					
152.53 59.93	UGC 6309		.24± .05	.00	13.73 ±.18				2843± 50
60.76 -2.01	IRAS11149+5144			.36	12.23				2899
111457.0+514501	PGC 34508		1.14	.12	13.36				3037
111750.0+263729	NGC 3609	.S..2..	1.08± .06	50					
208.95 69.15	UGC 6310	U	.10± .05	.00	14.02 ±.18				8089± 56
83.28 -13.46		2.0± .9		.12					8053
111509.9+265353	PGC 34511		1.08	.05	13.82				8393
111750.7-260806	NGC 3617	.E+....	1.25± .04	147	13.69 ±.16	.90± .01	.92± .01		2199± 27
278.02 32.17	ESO 503- 12	SU	.15± .03	.26	13.45 ±.14	.35± .02	.39± .02		2010
138.40 -28.35		-4.0± .5	.55± .05	.00		.82	11.93± .19		2537
111523.1-255142	PGC 34513		1.25		13.26	.30	14.58± .28		
111753.1-403536		.SBR6*.	1.43± .05	115					3050± 60
284.36 18.88	ESO 319- 11	S (1)	.21± .07	.52					2842
154.78 -28.48	IRAS11155-4019	6.0± .7		.30	13.18				3346
111530.0-401912	PGC 34519	2.2± .8	1.48	.10					
111754.4-015654	IC 680	PSBT1?.	.82± .13						
261.45 53.22	MCG 0-29- 12	F	.17± .07	.10	14.9 ±.2				
111.76 -23.88	IRAS11153-0140	1.0±1.3		.17	13.60				
111521.0-014030	PGC 34520		.83	.09					
111755.4-020531	A 1115-01	.SXS5P.	1.14± .04	145	13.9 ±.2	.61± .04	.68± .03	15.61±.1	7414± 7
261.59 53.10	UGC 6311	UEF (2)	.05± .04	.10	13.78 ±.19			236± 8	7446± 59
111.92 -23.91		5.0± .5	.87± .06	.07		.53	13.75± .15	229± 11	7281
111522.0-014907	PGC 34521	1.0± .6	1.15	.02	13.62		14.33± .34	1.97	7774
1118.0 +0750		.SB.1P/	1.04± .06	45					
249.67 60.65	UGC 6312	UF	.33± .05	.10	14.68 ±.18				
101.61 -20.78		.5± .6		.34					
1115.4 +0807	PGC 34527		1.05	.17					
1118.0 +0437		.I..9..	1.00± .08	40					2326
254.12 58.35	UGC 6317	U	.25± .06	.16					
104.91 -21.84		10.0± .9		.19					2214
1115.5 +0454	PGC 34530		1.02	.13					2680
111806.6-184331		.IBS9..	1.02± .05	55					
274.13 38.87	ESO 570- 16	E (1)	.04± .04	.14					
130.07 -27.45		10.0± .9		.03					
111537.0-182706	PGC 34534	9.8± .8	1.04	.02					
111806.8+232349	NGC 3615	.E.....	1.15± .15	40					
217.78 68.61	UGC 6313	U	.19± .08	.00	13.82 ±.15				6684± 50
86.35 -14.77		-5.0± .8		.00					6636
111527.6+234013	PGC 34535		1.09		13.72				6998
111811.9+302342		.S?....	.89± .08	45				16.26±.3	7880± 11
198.30 69.34	UGC 6314		.08± .05	.00				332± 15	
79.80 -11.74				.13					7858
111530.7+304006	PGC 34544		.89	.04					8170
111813.9+263710	NGC 3612	.S..8*.	1.00± .06	160				15.95±.3	8352± 9
208.99 69.24	UGC 6321	U	.10± .05	.00	14.67 ±.19				8340± 56
83.32 -13.38		8.0±1.3		.12				267± 5	8316
111533.9+265334	PGC 34546		1.00	.05	14.53			1.38	8655
111815.7-484449		PLBT+..	1.22± .05	92					4654± 87
287.51 11.31	ESO 216- 3	Sr	.04± .05	.50	14.04 ±.14				4441
163.88 -27.62		-.7± .5		.00					4918
111556.0-482824	PGC 34548		1.26		13.47				
1118.2 +2812		.SB?...	.89± .11						9533± 39
204.50 69.38	MCG 5-27- 52		.14± .07	.00	14.90 ±.18				9503
81.84 -12.68				.21					9831
1115.6 +2829	PGC 34551		.89	.07	14.63				
1118.2 +2816	A 1115+28	.LA.+*.	1.14± .07		14.89 ±.15	.52± .02			
204.32 69.38	UGC 6322	U	.00± .03	.00	14.02 ±.20	-.21± .04			9829± 44
81.78 -12.65		-1.0±1.1		.00		.43			9799
1115.6 +2833	PGC 34552	3.0±1.3	1.14		14.42	-.16	15.47± .37		10127

R.A. 2000 DEC.	Names	Type	logD25	p.a.	BT	(B-V)T	(B-V)●	m21	V21
l b		ST nL	logR25	Ag	mB	(U-B)T	(U-B)●	W20	Vopt
SGL SGB		T	logA●	Al	mFIR	(B-V)oT	m'●	W50	VGSR
R.A. 1950 DEC.	PGC	L	logD●	A21	BoT	(U-B)oT	m'25	HI	V3K

111816.8-324849	NGC 3621	.SAS7..	2.09± .01	159	10.18S±.15	.62± .01	.70± .01	10.20±.1	727± 5
281.21 26.10	ESO 377- 37	R (3)	.24± .02	.42	9.56 ±.12	-.08± .03	.01± .02	280± 5	623± 47
145.97 -28.57	IRAS11159-3235	7.0± .3	1.34± .08	.33	9.19	.47	12.55± .18	266± 6	526
111551.0-323224	PGC 34554	5.8± .3	2.13	.12	9.04	-.19	14.90± .18	1.04	1047

111817.8+185054		.S?....	.95± .07					15.91±.2	1123± 6
229.16 67.07	UGC 6320	.02± .05		.00	14.0 ±.2			106± 13	1123± 37
90.74 -16.60	ARAK 286			.03				78± 5	1059
111539.8+190718	PGC 34556		.95	.01	13.96			1.94	1450

111819.0+650140		.S?....	.85± .12						9857
137.77 49.33	UGC 6316	.66± .07		.00	14.9 ±.4				9960
48.83 4.53	IRAS11153+6518			.92	11.99				9975
111518.2+651804	PGC 34557		.85	.33	13.95				

1118.3 +1843		.L.....	1.24± .06	167					1083± 56
229.45 67.02	UGC 6324	U	.43± .03	.00	14.39 ±.17				1018
90.86 -16.64		-2.0± .9		.00					1410
1115.7 +1900	PGC 34558		1.18		14.37				

111820.9+222610									9608± 61
220.34 68.39	CGCG 126- 23			.00	15.4 ±.3				9557
87.29 -15.13	MK 733								9925
111542.0+224234	PGC 34560								

111821.0+454449	NGC 3614	.SXR5..	1.66± .02	80				13.59±.1	2333± 7
161.55 63.78	UGC 6318	R (2)	.24± .03	.00	12.27 ±.16			303± 8	2263± 66
65.89 -4.65	IRAS11155+4601	5.0± .3		.35	12.92			293± 6	2367
111534.3+460113	PGC 34561	1.9± .6	1.66	.12	11.91			1.56	2556

111821.9+454452	NGC 3614A	.SBS9..	.91± .09						
161.55 63.79	MCG 8-21- 14	R	.00± .06	.00					
65.89 -4.65		9.0± .5		.00					
111535.3+460116	PGC 34562		.91	.00					

111825.9+584714	NGC 3610	.E.5.*.	1.43± .03		11.70 ±.13	.86± .01	.88± .01		
143.54 54.46	UGC 6319	R	.07± .03	.00	11.51 ±.13	.46± .04			1787± 48
54.35 1.58		-5.0± .4	.71± .02	.00		.84	10.74± .06		1869
111531.4+590338	PGC 34566		1.41		11.58	.47	13.66± .20		1941

1118.4 +2518		.S..6?.	1.19± .05	10					7543
212.65 69.10	UGC 6325	U	.50± .05	.00	14.9 ±.2				
84.58 -13.89		6.0±1.8		.73					7502
1115.8 +2535	PGC 34568		1.19	.25	14.18				7851

111832.7+232809	NGC 3618	.SX.3*.	.95± .07	175				16.13±.3	6807± 9
217.66 68.72	UGC 6327	U (1)	.06± .05	.00	14.43 ±.18			306± 7	6788± 43
86.33 -14.65	MK 1288	3.0± .9		.08				283± 7	6759
111553.5+234433	PGC 34575	4.5±1.2	.95	.03	14.30			1.80	7120

1118.5 +0025		CE.....							
259.22 55.21	CGCG 11- 49	F			.68 15.3 ±.2				
109.31 -23.03		-5.0± .9							
1116.0 +0042	PGC 34578								

111834.3-020649		.E.....							
261.83 53.18	CGCG 11- 48	F			.12 14.81 ±.18				
111.99 -23.76		-5.0± .9							
111601.0-015024	PGC 34579								

111835.6+631642	A 1115+63								3237±101
139.22 50.81	CGCG 314- 19				.00 14.92 ±.18				3334
50.40 3.73	MK 165								3365
111537.0+633306	PGC 34582								

111836.2+580005	NGC 3613	.E.6...	1.59± .03	102	11.82 ±.13	.93± .01	.95± .01		
144.34 55.10	UGC 6323	R	.32± .04	.00	11.52 ±.12				2054± 56
55.06 1.22		-5.0± .3	.96± .02	.00		.91	12.11± .08		2133
111542.4+581629	PGC 34583		1.50		11.62		13.96± .21		2213

1118.6 +2831									9977
203.61 69.48	CGCG 156- 59				.00 15.13 ±.18				9948
81.58 -12.47									10274
1116.0 +2848	PGC 34588								

1118.6 +0729		.L...P.							
250.43 60.53	CGCG 39-113	F			.12 15.0 ±.6				
102.03 -20.74		-2.0± .9							
1116.1 +0746	PGC 34591								

111842.8-464043		PSBT1..	1.16± .05						5850±190
286.82 13.27	ESO 265- 22	Sr	.15± .05		.62 14.96 ±.14				5638
161.58 -27.80	FAIR 443	1.0± .5		.15					6122
111622.1-462418	PGC 34593		1.22	.08	14.11				

R.A. 2000 DEC. / l b / SGL SGB / R.A. 1950 DEC.	Names / PGC	S_T n_L / T / L	$\log D_{25}$ / $\log R_{25}$ / $\log A_\bullet$ / $\log D_o$	p.a. / A_g / A_i / A_{21}	B_T / m_B / m_{FIR} / B_T^o	$(B-V)_T$ / $(U-B)_T$ / $(B-V)_T^o$ / $(U-B)_T^o$	$(B-V)_\bullet$ / $(U-B)_\bullet$ / m' / m'_{25}	m_{21} / W_{20} / W_{50} / HI	V_{21} / V_{opt} / V_{GSR} / V_{3K}
1118.7 -0021		.SBS3..							
260.12 54.62	CGCG 11- 50	F (1)		.17	15.2 ±.6				
110.15 -23.22		3.0± .9							
1116.2 -0005	PGC 34598	3.6±1.4							
1118.7 +0731	NGC 3624	.SBR3..	.94± .11						
250.42 60.57	MCG 1-29- 29	F (1)	.19± .07	.12	14.66 ±.18				
102.01 -20.71		3.0± .9		.26					
1116.2 +0748	PGC 34599	1.6±1.0	.95	.09					
111853.9-292531		.S?....	1.02± .05	149					9064± 24
279.84 29.26	ESO 438- 20		.18± .04	.38	14.80 ±.14				8869
142.14 -28.32	IRAS11164-2908			.22	12.76				9394
111627.1-290906	PGC 34608		1.06	.09	14.10				
111855.3+130535	NGC 3623	.SXT1..	1.99± .01	174	10.25M±.05	.92± .02	.96± .01	14.20±.1	807± 3
241.33 64.22	UGC 6328	R (2)	.53± .02	.02	10.10 ±.13	.45± .04	.53± .01	502± 3	775± 25
96.43 -18.69	IRAS11163+1322	1.0± .3	1.45± .04	.54	11.34	.81	12.77± .12	473± 3	723
111618.6+132200	PGC 34612	3.3± .6	1.99	.27	9.65	.35	13.69± .09	4.28	1147
111856.9+001023		.SXT5*.	1.04± .06					15.67±.3	7475± 11
259.63 55.07	UGC 6329	UF (1)	.02± .05	.24	14.5 ±.2			221± 7	
109.61 -23.02		5.3± .7		.03				200± 7	7349
111623.0+002648	PGC 34613	4.6±2.0	1.06	.01	14.23			1.42	7833
111859.9-361443		PSBS3..	.93± .06	98					
282.82 22.99	ESO 377- 38	r	.09± .05	.37	15.03 ±.14				
149.86 -28.42	IRAS11166-3558	3.3± .9		.12	13.57				
111635.0-355818	PGC 34620		.97	.04					
111907.6+031355		.S..8?/	1.07± .07	57				14.94±.3	
256.22 57.48	UGC 6331	UF	.62± .06	.13	15.06 ±.18				6031± 10
106.44 -22.04	IRAS11165+0330	8.0±1.1		.76	12.49			378± 7	5915
111633.1+033020	PGC 34623		1.08	.31	14.16			.47	6386
1119.2 -0950		.SBS8..	1.15± .05	115					
268.63 46.80	MCG -2-29- 18	EF (1)	.15± .04	.10					
120.33 -25.55		8.0± .5		.18					
1116.7 -0934	PGC 34625	7.5± .6	1.16	.07					
111916.6+204848		RSB.1..	1.09± .06	130					
224.67 68.06	UGC 6332	U	.06± .05	.00	14.13 ±.19				6245± 50
88.94 -15.60		1.0± .8		.06					6188
111638.1+210513	PGC 34630		1.09	.03	13.99				6567
111917.2+225255		.S..8*.	1.06± .06	120				15.85±.3	6283± 10
219.35 68.73	UGC 6333	U	.31± .05	.00	14.98 ±.19				
86.96 -14.74		8.0±1.3		.38				218± 7	6234
111638.3+230920	PGC 34632		1.06	.15	14.58			1.11	6599
111921.8+574535	NGC 3619	RLAS+*.	1.43± .04		12.5 ±.3			15.58±.3	1542± 10
144.46 55.35	UGC 6330	R	.06± .04	.00	12.43 ±.12			351± 6	1649± 56
55.32 1.19	IRAS11164+5802	-1.0± .3	.93± .11	.00	13.61		12.67± .37	308± 5	1624
111628.6+580200	PGC 34641		1.42		12.42		14.38± .36		1705
1119.3 +0311		PSBR1*.							
256.36 57.49	CGCG 39-119	F		.13	15.2 ±.6				
106.50 -21.99		1.0± .9							
1116.8 +0328	PGC 34642								
111925.9+622901	A 1116+62				15.6 ±.2	.39± .04	.35± .04		
139.80 51.53				.00		-.06± .07	.03± .07		3600± 77
51.14 3.44	MK 166						11.79± .06		3694
111628.6+624526	PGC 34649		.15± .02						3733
111925.4-030413	A 1116-02	.E...P*	.58± .18	95					
263.04 52.52		E	.10± .08	.13					
113.07 -23.82		-5.0±1.8		.00					
111652.3-024748	PGC 34651		.57						
1119.4 +0137		.E.....							
258.22 56.29	CGCG 11- 54	F		.11	15.13 ±.18				
108.14 -22.46		-5.0± .9							
1116.9 +0154	PGC 34652								
111929.2-522220		PSXT1*.	1.17± .05	101					
289.00 7.99	ESO 216- 5	Sr	.33± .05	1.13	14.83 ±.14				
167.84 -26.90		1.3± .5		.33					
111711.1-520554	PGC 34654		1.28	.16					
111930.2+283908		RS..0*.	1.42± .04	82					6256± 50
203.27 69.66	UGC 6334	U	.32± .05	.00	13.6 ±.3				
81.55 -12.25		.0± .8		.24					6228
111649.7+285533	PGC 34656		1.40		13.26				6553

R.A. 2000 DEC. l b SGL SGB R.A. 1950 DEC.	Names PGC	Type S_T n_L T L	$\log D_{25}$ $\log R_{25}$ $\log A_\bullet$ $\log D_\circ$	p.a. A_g A_i A_{21}	B_T m_B m_{FIR} B_T°	$(B-V)_T$ $(U-B)_T$ $(B-V)_T^\circ$ $(U-B)_T^\circ$	$(B-V)_\bullet$ $(U-B)_\bullet$ $m'_\bullet$ m'_{25}	m_{21} W_{20} W_{50} HI	V_{21} V_{opt} V_{GSR} V_{3K}
1119.5 +5129 152.12 60.11 60.88 -1.76 1116.7 +5146	A 1116+51 PGC 34658			.00					1351± 38 1407 1545
1119.5 -0229 262.53 53.02 112.46 -23.63 1117.0 -0213	MCG 0-29-15 PGC 34659	.SAS5.. F (1) 5.0± .9 5.6±1.0	.99± .10 .09± .07 1.00	.11 .14 .05	14.9 ±.2				
1119.6 +2456 213.81 69.30 85.04 -13.80 1117.0 +2513	UGC 6336 PGC 34663	.S..2.. U 2.0±1.0	1.10± .05 .89± .05 1.10	158 .00 1.10 .45	15.71 ±.18				
1119.7 +5917 142.80 54.15 53.98 1.96 1116.8 +5934	UGC 6335 PGC 34666	.SAS6.. U 6.0± .8	1.20± .05 .03± .05 1.20	.00 .04 .01	14.6 ±.3 14.57				2927± 10 3011 3078
1119.8 +5428 148.14 57.94 58.26 -.30 1117.0 +5445	MCG 9-19-42 PGC 34670	.S..6*. U 6.0±1.2	1.00± .07 .00± .06 1.00	.00 .00 .00	15.0 ±.2				
111954.4+330526 190.58 69.38 77.48 -10.22 111712.8+332151	UGC 6337 IRAS11171+3321 PGC 34674	.S?.... 	.96± .17 .05± .12 .97	.09 .07 .02	14.97 ±.19 14.72				12791 12779 13071
111955.2-005250 261.06 54.38 110.79 -23.09 111721.6-003624	UGC 6340 IRAS11173-0036 PGC 34675	.SXT4*. UEF (2) 4.0± .5 3.8± .6	1.21± .04 .32± .04 1.23	155 .13 .48 .16	13.84 ±.18 13.52				
1119.9 -0237 262.79 52.97 112.64 -23.57 1117.4 -0221	CGCG 11-57 PGC 34677	.SA.3*/ F 3.0±1.3		.11	15.4 ±.6				
1119.9 -0128 261.67 53.90 111.42 -23.25 1117.4 -0112	CGCG 11-58 PGC 34678	.L..../ F -2.0± .9		.13	15.4 ±.6				
112000.7+360602 182.43 68.64 74.73 -8.84 111718.1+362228	UGC 6338 PGC 34681	.S..6*. U 6.0±1.3	1.06± .06 .36± .05 1.06	150 .03 .54 .18	14.89 ±.18				
112001.5-014108 261.91 53.74 111.65 -23.30 111728.0-012442	CGCG 11-60 PGC 34682	.LA.+.. F -1.0± .9		.13	15.1 ±.6				
1120.0 +1815 230.98 67.18 91.48 -16.46 1117.4 +1832	UGC 6341 PGC 34683	.S..8*. U 8.0±1.4	.99± .06 .80± .05 .99	152 .00 .98 .40					1641 1575 1970
112003.7+182130 230.76 67.23 91.39 -16.41 111725.9+183756	NGC 3626 UGC 6343 PGC 34684	RLAT+.. R -1.0± .3	1.43± .03 .14± .03 .94± .02 1.41	157 .00 .00	11.78 ±.13 11.77 ±.14 11.75	.82± .01 .29± .01 .79 .28	.84± .01 .33± .01 11.97± .05 13.46± .20	15.53±.1 357± 5 326± 4	1493± 4 1438± 38 1427 1821
1120.0 +3029 197.95 69.74 79.89 -11.33 1117.4 +3046	UGC 6342 PGC 34685	.L..-*. U -3.0±1.2	1.00± .19 .09± .08 .99	155 .00 .00	14.71 ±.16 14.60				7099 7078 7389
1120.1 -0346 263.93 52.05 113.87 -23.84 1117.6 -0330	CGCG 11-61 PGC 34687	.LBR0*. F -2.0±1.3		.13	15.4 ±.6				
112010.0-080622 267.60 48.40 118.51 -24.93 111737.9-074956	NGC 3638 MCG -1-29-7 IRAS11176-0749 PGC 34688	.SA.3*/ EF 3.0± .8	1.34± .04 .49± .05 1.35	44 .07 .67 .24	 13.53				
112010.2-030326 263.28 52.64 113.11 -23.64 111737.0-024700	A 1117-02 MCG 0-29-17 PGC 34689	.SAS5.. PEF (1) 4.7± .5 2.6±1.0	1.17± .05 .18± .05 .89± .07 1.18	30 .13 .27 .09	14.1 ±.3 13.91 ±.19 13.53	.52± .04 .41	.59± .04 14.02± .15 14.33± .40	15.59±.1 283± 12 1.97	7721± 9 7692± 59 7585 8080

R.A. 2000 DEC. l b SGL SGB R.A. 1950 DEC.	Names PGC	Type ST nL T L	logD25 logR25 logA. logDo	p.a. Ag Ai A21	BT mB mFIR BoT	(B-V)T (U-B)T (B-V)oT (U-B)oT	(B-V). (U-B). m'. m'25	m21 W20 W50 HI	V21 Vopt VGSR V3K
112012.2-212814	ESO 570- 19	PSA.5*P	1.29± .05	74					1270± 60
276.22 36.62		S (1)	.24± .04	.12					1090
133.22 -27.34		5.0±1.1		.36					1617
111743.0-211148	PGC 34691	5.6± .8	1.30	.12					
112012.8+671427	NGC 3622	.S?....	1.07± .07	7	13.65 ±.19	.49± .03	.57± .02	14.54±.3	1306± 17
135.77 47.55	UGC 6339		.40± .06	.00	14.0 ±.2		-.37± .04	92± 33	1328± 50
46.96 5.74	IRAS11171+6730		.58± .05	.60	12.88	.41	12.04± .14		1418
111710.3+673053	PGC 34692		1.07	.20	13.16		12.85±.44	1.18	1413
112014.5+125942	NGC 3627	.SXS3..	1.96± .01	173	9.65 ±.13	.73± .01	.81± .01	13.42±.0	727± 3
241.95 64.42	UGC 6346	R (3)	.34± .02	.01	9.56 ±.15	.20± .05	.28± .01	374± 4	703± 27
96.66 -18.42	ARAK 288	3.0± .3	1.35± .01	.47	8.59	.66	11.95± .04	339± 4	643
111737.9+131608	PGC 34695	3.0± .5	1.97	.17	9.13	.14	13.47± .15	4.12	1067
112014.8+023132	A 1117+02	.IBS9..	1.37± .04	75	13.8 ±.2	.36± .05	.45± .03	13.94±.1	1596± 6
257.46 57.11	UGC 6345	UF (1)	.26± .05	.11	13.5 ±.3	-.27± .07	-.25± .05	137± 8	
107.27 -21.99	DDO 94	9.5± .5	1.20± .05	.20		.26	15.28± .10	99± 12	1478
111740.5+024758	PGC 34696	8.6±1.0	1.38	.13	13.39	-.34	14.83± .30	.43	1952
112016.3+133522	NGC 3628	.S..3P/	2.17± .01	104	10.28S±.05	.80± .05		11.39±.0	847± 3
240.85 64.78	UGC 6350	R (1)	.70± .02	.02	10.51 ±.13			476± 3	809± 58
96.07 -18.20	IRAS11176+1351	4.5± .9		.96	8.77	.66		449± 3	765
111739.6+135148	PGC 34697		2.17	.35	9.32		14.21± .09	1.72	1186
112016.5+025757	NGC 3630	.L..../	1.66± .03	37					1509± 49
256.95 57.46	UGC 6349	R	.19± .02	.12	11.95 ±.13				1392
106.82 -21.85		-2.0± .4		.00					1864
111742.1+031423	PGC 34698		1.64		11.82				
1120.2 +0419		.E.....							
255.28 58.49	CGCG 39-125	F							
105.41 -21.42		-5.0± .9		.12	14.60 ±.18				
1117.7 +0436	PGC 34699								
112020.2-025502		RSBR0..	.89± .11						7800± 54
263.20 52.78	MCG 0-29- 19	F	.24± .07	.13	14.94 ±.19				7666
112.97 -23.56		.0± .9		.18					8159
111747.0-023836	PGC 34701		.89		14.52				
112020.8+340547		.S..2..	.98± .07	85					
187.74 69.25	UGC 6347	U	.20± .05	.06	15.11 ±.18				
76.59 -9.68		2.0± .9		.24					
111739.0+342213	PGC 34702		.98	.10					
1120.3 +5744		.S..9*.	1.04± .15						1951
144.29 55.45	UGC 6344	U	.08± .12	.00					
55.40 1.31		9.0±1.2		.08					2029
1117.5 +5801	PGC 34704		1.04	.04					2111
112025.3-101655	NGC 3636	.E.0...	1.13± .06		13.34 ±.13	.94± .01			1715± 62
269.32 46.56	MCG -2-29- 19	R	.00± .05	.13					1561
120.88 -25.36		-5.0± .5		.00		.90			2074
111753.6-100028	PGC 34709		1.15		13.19		14.00± .33		
112026.0+033512	NGC 3633	.SA.1*/	1.08± .05	72				16.17±.3	2598± 9
256.25 57.96	UGC 6351	PUF	.48± .04	.12	14.45 ±.20			344± 7	2595± 39
106.19 -21.62	IRAS11178+0351	1.3± .6		.49	11.95			324± 7	2483
111751.4+035138	PGC 34711		1.10	.24	13.81			2.12	2953
1120.4 +0407		.LAR-..							
255.60 58.37	CGCG 39-127	F							
105.64 -21.44		-3.0± .9		.12	15.1 ±.6				
1117.9 +0424	PGC 34712								
112031.4-090049	NGC 3635	.SXT4P*	1.10± .07	165					
268.41 47.67	MCG -1-29- 9	EF (1)	.13± .05	.09					
119.51 -25.05		3.6± .8		.18					
111759.5-084423	PGC 34717	2.6±1.0	1.10	.06					
112031.7+574655	NGC 3625	.SXS3*.	1.30± .03	148					1966± 50
144.22 55.43	UGC 6348	R	.49± .03	.00	13.88 ±.18				2045
55.37 1.34		3.0± .6		.68					2126
111739.0+580321	PGC 34718		1.30	.25	13.19				
112031.8+265747	NGC 3629	.SAS6*.	1.36± .03	30				14.00±.1	1517± 6
208.17 69.79	UGC 6352	R (2)	.15± .03	.00	12.77 ±.14			234± 6	1558± 66
83.23 -12.77		6.0± .4		.22				178± 6	1483
111751.9+271413	PGC 34719	4.7± .7	1.36	.07	12.55			1.37	1820
112035.5-012932		.LX..P.	1.04± .09	75					7264± 56
261.91 53.98	UGC 6359	UF	.18± .04	.13					7134
111.49 -23.10	IRAS11181-0112	-2.0± .6		.00					7623
111802.1-011306	PGC 34724		1.03						

R.A. 2000 DEC.	Names	Type S_T n_L	$\log D_{25}$ $\log R_{25}$	p.a. A_g	B_T m_B	$(B-V)_T$ $(U-B)_T$	$(B-V)_\bullet$ $(U-B)_\bullet$	m_{21} W_{20}	V_{21} V_{opt}
l b		T	$\log A_\bullet$	A_l	m_{FIR}	$(B-V)_T^o$	$m'_\bullet$	W_{50}	V_{GSR}
SGL SGB		L	$\log D_o$	A_{21}	B_T^o	$(U-B)_T^o$	m'_{25}	HI	V_{3K}
R.A. 1950 DEC.	PGC								

1120.6 -0718 267.12 49.15 117.68 -24.62 1118.1 -0702	 PGC 34728	.LA.-P* E -3.0± .9 	1.22± .07 .15± .08 1.21	30 .10 .00 	 	 	 	 	
112039.1-033244 263.89 52.31 113.66 -23.66 111806.0-031618	 CGCG 11- 65 PGC 34730	.SBS3.. F (1) 3.0± .9 1.6±1.0	 	 .13 	 15.3 ±.6 	 	 	 	
112039.6-101527 269.38 46.62 120.87 -25.30 111808.0-095901	NGC 3637 MCG -2-29- 20 PGC 34731	RLBR0.. R -2.0± .4 	1.20± .04 .01± .04 .85± .02 1.21	 .13 .00 	13.61 ±.13 12.84 ±.19 13.21	.89± .02 .84 	.98± .02 13.35± .06 14.45± .27	17.24±.1 205± 12 	1846± 9 1802± 45 1690 2203
112041.2+311320 195.82 69.81 79.28 -10.89 111800.2+312946	 UGC 6355 KUG 1118+314 PGC 34733	.S..7.. U 7.0± .9 	1.30± .03 .77± .04 1.30	102 .01 1.06 .39	 14.64 ±.18 13.56	 	 	 	2168± 56 2150 2455
1120.6 +3306 190.44 69.53 77.53 -10.06 1118.0 +3323	 UGC 6357 PGC 34735	.S..6*. U 6.0±1.5 	.96± .09 .98± .06 .97	18 .06 1.44 .49	 	 	 	 	10539 10528 10819
1120.7 +5642 145.34 56.30 56.34 .86 1117.9 +5659	 UGC 6353 PGC 34741	.S..3*. U (1) 3.0±1.3 2.5±1.2	1.03± .06 .24± .05 1.03	5 .00 .33 .12	 15.19 ±.20 	 	 	 	
1120.7 +0034 259.83 55.67 109.34 -22.46 1118.2 +0051	 CGCG 11- 68 PGC 34742	RSXT2?. F 2.0±1.3 	 	 .13 	 15.2 ±.6 	 	 	 	
112051.1-033850 264.05 52.25 113.79 -23.64 111818.0-032224	 MCG 0-29- 21 PGC 34749	.LA.+*. F -1.0±1.3 	.99± .10 .45± .07 .93	 .13 .00 	 14.89 ±.17 	 	 	 	
112053.2-292409 280.28 29.46 142.14 -27.89 111826.0-290742	 ESO 438- 23 PGC 34755	.L?.... 	1.05± .06 .21± .04 1.06	41 .34 .00 	 14.26 ±.14 13.79	 	 	 	8917± 34 8723 9247
112054.8-070045 266.97 49.43 117.38 -24.48 111822.4-064418	 MCG -1-29- 11 PGC 34756	.SXT6P* EF (1) 6.3± .7 5.8± .7	1.09± .06 .27± .05 1.09	85 .07 .40 .14	 	 	 	 	
112059.9+003203 259.96 55.67 109.40 -22.42 111826.0+004830	 CGCG 11- 75 PGC 34762	.L...P? F -2.0±1.8 	 	 .13 	 14.9 ±.6 	 	 	 	
112100.0+212014 223.72 68.62 88.61 -15.02 111821.6+213640	 UGC 6363 PGC 34763	.S..6?. U 6.0±1.9 	1.04± .06 .62± .05 1.04	57 .00 .91 .31	 15.32 ±.18 14.38	 	 	15.37±.3 310± 7 .68	6306± 10 6244± 57 6250 6624
112102.7+531017 149.53 59.04 59.50 -.76 111813.3+532643	NGC 3631 UGC 6360 ARP 27 PGC 34767	.SAS5.. R (2) 5.0± .3 1.8± .6	1.70± .02 .02± .07 1.70	 .00 .04 .01	11.01S±.08 10.97 ±.14 10.69 10.96	.58± .05 .57 	 14.32± .22	12.93±.1 125± 5 106± 5 1.96	1158± 4 1143± 31 1221 1343
1121.0 +3115 195.71 69.88 79.28 -10.81 1118.3 +3131	 UGC 6367 IRAS11183+3131 PGC 34768	.SBS2.. U 2.0± .9 	1.08± .06 .21± .05 1.08	170 .01 .26 .11	 14.61 ±.19 13.51 14.27	 	 	 	7092± 56 7074 7379
112103.9+342039 186.96 69.34 76.44 -9.44 111822.1+343705	IC 2735 UGC 6364 IRAS11184+3436 PGC 34772	.S..2.. U 2.0±1.0 	1.02± .06 .68± .05 1.02	100 .05 .84 .34	 15.44 ±.18 14.44	 	 	 	10970± 31 10963 11245
112105.5+212116 223.70 68.65 88.60 -14.99 111827.2+213743	 UGC 6366 PGC 34777	.S..4.. U 4.0±1.0 	1.06± .06 .74± .05 1.06	18 .00 1.08 .37	 15.67 ±.18 14.54	 	 	15.05±.3 231± 10 .14	6306± 10 6156± 57 6248 6622
112106.8+031408 256.92 57.80 106.61 -21.57 111832.3+033035	NGC 3640 UGC 6368 PGC 34778	.E.3... R -5.0± .3 	1.60± .04 .10± .04 1.03± .02 1.59	100 .10 .00 	11.36 ±.13 11.26 ±.12 11.18	.92± .01 .53± .02 .88 .51	.95± .01 .51± .01 12.00± .07 14.11± .27	 	 1314± 27 1199 1670

R.A. 2000 DEC.	Names	Type	logD25	p.a.	BT	(B-V)T	(B-V).	m21	V21
l b		ST nL	logR25	Ag	mB	(U-B)T	(U-B).	W20	Vopt
SGL SGB		T	logA.	Ai	mFIR	(B-V)T°	m'	W50	VGSR
R.A. 1950 DEC.	PGC	L	logD0	A21	BT°	(U-B)T°	m'25	HI	V3K
112108.7+031143	NGC 3641	.E...P.	1.03± .08		14.1 ±.2	.90± .02	.98± .01		
256.98 57.77	UGC 6370	PUF	.00± .05	.10					1780± 39
106.65 -21.57		-5.3± .4	.58± .04	.00		.86	12.44± .15		1664
111834.2+032810	PGC 34780		1.05		13.97		14.26± .46		2136
1121.1 +0521		.E.....	1.02± .14						
254.27 59.41	MCG 1-29- 35	F	.00± .07	.13	14.25 ±.16				
104.43 -20.87	VV 510	-5.0± .9		.00					
1118.6 +0538	PGC 34783		1.04						
112112.2-025903		.SXS3..	.89± .11						
263.56 52.85	MCG 0-29- 23	F	.07± .07	.13	14.78 ±.18				
113.11 -23.37	IRAS11186-0242	3.0± .9		.09	11.33				
111839.0-024236	PGC 34786		.90	.03					
1121.2 +0340		.E.....							
256.45 58.16	CGCG 39-133	F							
106.17 -21.39		-5.0± .9		.12	14.92 ±.18				
1118.7 +0357	PGC 34788								
112123.1+342125	IC 2738	.S?....	.74± .14						
186.88 69.40	MCG 6-25- 49		.00± .07	.05	15.29 ±.18				10450± 24
76.45 -9.38				.00					10444
111841.3+343752	PGC 34797		.74		15.08				10725
1121.4 +3916		.S..7..	1.00± .08	97					
174.24 67.69	UGC 6372	U	.55± .06	.00					
71.97 -7.15		7.0± .9		.76					
1118.7 +3933	PGC 34801		1.00	.27					
112125.5+030051	NGC 3643	.LBR+*/	.90± .17						
257.30 57.68	MCG 1-29- 36	PF	.44± .07	.09	15.11 ±.20				
106.86 -21.56		-1.0± .7		.00					
111851.1+031718	PGC 34802		.84						
1121.4 +0131		.SBR2*.							
259.04 56.53	CGCG 11- 77	F							
108.41 -22.01		2.0±1.3		.10	15.4 ±.6				
1118.9 +0148	PGC 34806								
1121.4 +0244	IC 683E	.E.....	.86± .10						
257.64 57.48	CGCG 39-138	F	.12± .06	.11	14.99 ±.11				
107.15 -21.64		-5.0± .9		.00					
1118.9 +0301	PGC 34807		.84						
112132.4+024839	NGC 3644	PSB.1P*	1.19± .05	63					
257.58 57.54	UGC 6373	UF	.36± .05	.11	14.6 ±.2				
107.08 -21.60		1.0± .7		.37					
111858.0+030506	PGC 34814		1.20	.18					
112135.6+182733	NGC 3639	.S?....	.80± .11					15.42±.2	5441± 6
230.96 67.60	UGC 6374		.08± .06	.00	14.6 ±.3			368± 13	5459± 36
91.44 -16.04	ARAK 289			.12				330± 5	5377
111857.9+184400	PGC 34819		.80	.04	14.48			.90	5769
1121.6 +0130		.E+..*.							
259.13 56.54	CGCG 11- 78	F							
108.44 -21.97		-4.0±1.3		.10	15.4 ±.6				
1119.1 +0147	PGC 34826								
1121.6 +0225		.SX.9?.	.96± .09						2583
258.08 57.26	UGC 6379	UF (1)	.00± .06	.12					
107.49 -21.69		9.3± .7		.00					2465
1119.1 +0242	PGC 34827		9.6±1.0	.97	.00				2939
1121.7 +0834	IC 2749	.SXS5P.							
249.90 61.81	CGCG 67- 60	F (1)							
101.21 -19.67		5.0± .9		.09	15.4 ±.6				
1119.1 +0851	PGC 34829		2.6±1.0						
112142.5+342146	IC 2744	.L?....	.72± .22						
186.82 69.46	MCG 6-25- 52		.00± .07	.05	15.40 ±.15				10610± 31
76.48 -9.31				.00					10604
111900.8+343813	PGC 34833		.72		15.19				10885
112143.2+201016	NGC 3646	.RING..	1.59± .02	50	11.78 ±.13	.65± .02	.74± .01	13.68±.2	4248± 5
226.85 68.35	UGC 6376	R (2)	.24± .02	.00	11.62 ±.14	-.02± .04	.07± .04	535± 5	4261± 26
89.80 -15.33		4.0± .3	1.26± .02	.35		.58	13.57± .04	510± 5	4190
111905.2+202643	PGC 34836	2.4± .6	1.59	.12	11.33	-.07	14.01± .16	2.23	4571
112143.7+461234		.S..2..	1.08± .06	97					
159.90 63.96	UGC 6375	U	.63± .05	.00	15.11 ±.18				
65.76 -3.92	IRAS11189+4629	2.0±1.0		.77					
111857.9+462901	PGC 34837		1.08	.31					

R.A. 2000 DEC.	Names	Type	logD$_{25}$	p.a.	B$_T$	(B-V)$_T$	(B-V)$_\bullet$	m$_{21}$	V$_{21}$
l b		S$_T$ n$_L$	logR$_{25}$	A$_g$	m$_B$	(U-B)$_T$	(U-B)$_\bullet$	W$_{20}$	V$_{opt}$
SGL SGB		T	logA$_\bullet$	A$_i$	m$_{FIR}$	(B-V)$_T^o$	m'$_\bullet$	W$_{50}$	V$_{GSR}$
R.A. 1950 DEC.	PGC	L	logD$_o$	A$_{21}$	B$_T^o$	(U-B)$_T^o$	m'$_{25}$	HI	V$_{3K}$
1121.7 +0407		.SA.2?/							
256.07 58.59	CGCG 39-146	F		.11	15.4 ±.6				
105.75 -21.13		2.0±1.8							
1119.2 +0424	PGC 34842								
112147.1+114420					15.07 ±.15	.37± .03		18.20±.2	15050± 14
244.75 63.94	MK 734			.07		-.68± .04		318± 18	14768± 86
98.05 -18.53								353± 19	14955
111910.9+120047	PGC 34843								15386
112148.5+292644			1.02± .06						9954
200.96 70.16	CGCG 156- 67		.35± .12	.08	15.17 ±.18				9930
81.03 -11.45	KUG 1119+297								10248
111908.2+294311	PGC 34845		1.03						
112154.4-123230		.SXT5*.	1.10± .06	10					
271.36 44.77	MCG -2-29- 21	EF (1)	.16± .05	.15					
123.42 -25.48		5.0± .6		.24					
111923.2-121603	PGC 34852	5.7± .7	1.11	.08					
1121.9 +0823	IC 2757	.SBS1..							
250.29 61.74	CGCG 67- 62	F (1)		.12	14.9 ±.6				
101.43 -19.66		1.0± .9							
1119.4 +0840	PGC 34858	1.6±1.0							
1122.0 +5036		.S..0..	1.15± .05	58					
152.81 61.01	UGC 6380	U	.30± .05	.00	14.51 ±.18				
61.85 -1.83		.0± .8		.23					
1119.2 +5053	PGC 34859		1.14						
1122.0 +3542		.S..6*.	1.00± .08	132					
183.11 69.14	UGC 6382	U	1.02± .06	.05					
75.27 -8.66		6.0±1.5		1.47					
1119.3 +3559	PGC 34861		1.00	.50					
112203.0+424913		.S..7..	1.08± .04	59					
166.27 66.02	UGC 6383	U	.76± .04	.02	15.34 ±.18				3158±125
68.83 -5.43		7.0±1.0		1.05					3183
111918.6+430540	PGC 34864		1.08	.38	14.26				3396
1122.0 +6938	A 1119+69	.S..7..	1.44± .05						1309
133.82 45.55	UGC 6378	U	1.04± .07	.00	14.66 ±.18				1427
44.89 6.99		7.0± .9		1.38					1399
1119.0 +6955	PGC 34869		1.44	.50	13.27				
1122.1 +3456		.S..6*.	1.11± .05	143					
185.17 69.39	UGC 6384	U	1.07± .05	.03					
75.98 -8.98		6.0±1.5		1.47					
1119.4 +3513	PGC 34870		1.11	.50					
112208.0-380352		.SBS5P*	1.20± .05	27					
284.18 21.53	ESO 319- 16	Sr (1)	.28± .05	.56					2950± 60
151.89 -27.76		4.9± .5		.42					2745
111943.0-374724	PGC 34874	5.6± .6	1.25	.14					3255
112213.2+241903		.L?....	.42?		15.09S±.15				7700± 41
215.90 69.74	MCG 4-27- 27		.00± .07	.00	16.3 ±.3				7657
85.89 -13.54	ARAK 290			.00					8011
111934.3+243530	PGC 34881		.42		15.27		12.04±1.73		
112214.5+241758	IC 2759	.S?....	.89± .11		15.55S±.15				
215.96 69.74	MCG 4-27- 26		.14± .07	.00	15.25 ±.18				8183± 41
85.91 -13.54	HICK 51B			.21					8140
111935.6+243425	PGC 34882		.89	.07	15.18		14.48± .62		8494
112214.8+201234	NGC 3649	.SBS1..	1.09± .04	140				15.47±.3	4979± 17
226.92 68.48	UGC 6386	R	.33± .04	.00	14.57 ±.18			576± 34	4442± 33
89.81 -15.20		1.0± .5		.34				418± 25	4810
111936.8+202901	PGC 34883		1.09	.17	14.17			1.13	5191
1122.2 +6907		.S..9*.	1.19± .12	5					
134.15 46.01	UGC 6381	U	.12± .12	.00					
45.36 6.77		9.0±1.1		.12					
1119.2 +6924	PGC 34886		1.19	.06					
112218.3+130353	IC 2763	.S..6?.	1.13± .05	99				16.22±.3	1574± 9
242.53 64.86	UGC 6387	U	.81± .05	.04					1491
96.79 -17.93		6.0±1.9		1.19				134± 5	1914
111941.8+132020	PGC 34887		1.13	.41					
112218.4+590434	NGC 3642	.SAR4*.	1.73± .02	105	11.65 ±.15	.49± .06		13.25±.3	1588± 9
142.55 54.53	UGC 6385	R (2)	.08± .03	.00	11.56 ±.16			65± 34	1623± 56
54.33 2.15	IRAS11194+5920	4.0± .3	1.22± .02	.12	12.41		13.24± .07	75± 25	1672
111925.6+592101	PGC 34889	1.1± .5	1.73	.04	11.47		14.96± .18	1.73	1741

R.A. 2000 DEC.	Names	Type	logD$_{25}$	p.a.	B$_T$	(B-V)$_T$	(B-V)$_e$	m$_{21}$	V$_{21}$
l b		S$_T$ n$_L$	logR$_{25}$	A$_g$	m$_B$	(U-B)$_T$	(U-B)$_e$	W$_{20}$	V$_{opt}$
SGL SGB		T	logA$_e$	A$_i$	m$_{FIR}$	(B-V)$_T^o$	m'$_e$	W$_{50}$	V$_{GSR}$
R.A. 1950 DEC.	PGC	L	logD$_o$	A$_{21}$	B$_T^o$	(U-B)$_T^o$	m'$_{25}$	HI	V$_{3K}$
112226.5+241756 215.99 69.78 85.93 -13.50 111947.6+243423	NGC 3651 UGC 6388 HICK 51A PGC 34898	.E..... U -5.0± .8	1.04± .18 .00± .08 1.04	.00 .00	14.17S±.15 14.16 ±.16 14.05		14.36± .94		7696± 41 7653 8007
112226.7+241737 216.00 69.78 85.93 -13.50 111947.8+243404	HICK 51F PGC 34899	.L?....		.00	15.23S±.15				7532± 41 7489 7843
112226.7-073911 267.98 49.08 118.17 -24.27 111954.4-072244	MCG -1-29- 13 PGC 34900	.LB.-*/ EF -3.0± .6	1.12± .08 .28± .08 1.09	70 .09 .00					
112228.4+241742 216.00 69.79 85.94 -13.49 111949.5+243409	HICK 51G PGC 34901			.00	16.20S±.15				7532± 41 7489 7843
112230.3+241645 216.05 69.79 85.95 -13.49 111951.5+243312	NGC 3653 MCG 4-27- 29 HICK 51C PGC 34905	.L?.... 	.95± .10 .17± .04 .93	.00 .00	14.59S±.15 14.81 ±.11 14.61		13.81± .55		8902± 41 8859 9213
112230.7+041450 256.19 58.80 105.69 -20.92 111956.0+043118	MCG 1-29- 38 IRAS11199+0431 PGC 34906	.SAT4P* F 4.0±1.3	.82± .13 .08± .07 .83	.11 .11 .04	14.87 ±.19 13.16 14.57				11419± 42 11307 11774
112230.9+241758 216.00 69.80 85.94 -13.48 111952.0+243426	MCG 4-27- 30 HICK 51D PGC 34907	.S?.... 	.44± .22 .00± .07 .44	.00 .00 .00	15.67S±.15 15.64		12.70±1.14		7529± 41 7486 7840
112231.9+395235 172.58 67.61 71.53 -6.69 111948.7+400903	NGC 3648 UGC 6389 PGC 34908	.L..... U -2.0± .8	1.11± .16 .20± .08 1.08	75 .00 .00	13.55 ±.17 13.52				2111 2125 2362
1122.5 -0050 261.94 54.80 110.96 -22.45 1120.0 -0034	CGCG 11- 80 PGC 34911	.LA.0.. F -2.0± .9		.12	15.4 ±.6				
112235.5+204212 225.72 68.74 89.37 -14.93 111957.4+205840	NGC 3650 UGC 6391 PGC 34913	.SA.3./ RCU (1) 3.0± .4 4.5±1.2	1.23± .04 .75± .04 1.23	54 .00 1.03 .37	14.69 ±.18				
112239.5+374600 177.65 68.54 73.46 -7.62 111957.0+380228	NGC 3652 UGC 6392 ARAK 291 PGC 34917	.S..6?. U 6.0±1.6 	1.30± .06 .48± .06 1.30	150 .00 .70 .24	12.9 ±.3 12.04 12.15				2096± 82 2103 2356
112243.8-074035 268.09 49.10 118.21 -24.21 112011.5-072407	MCG -1-29- 15 IRAS11201-0724 PGC 34925	.LASOP* EF -2.0± .6	1.37± .08 .16± .07 1.36	125 .09 .00	13.23				
1122.7 +6404 137.93 50.39 49.90 4.51 1119.8 +6421	UGC 6390 PGC 34929	.S..7.. U 7.0± .9	1.32± .06 .84± .07 1.32	.01 1.15 .42	14.68 ±.18 13.51				1008± 10 1108 1131
1122.8 -0254 264.05 53.14 113.15 -22.96 1120.3 -0238	CGCG 11- 82 PGC 34931	.LBR0?. F -2.0±1.3		.12	15.1 ±.6				
112253.6+342027 186.70 69.71 76.61 -9.10 112012.1+343655	UGC 6393 PGC 34933	.SB.1.. U 1.0± .9	.97± .07 .12± .05 .98	40 .05 .13 .06	14.78 ±.18 14.47				10160± 31 10154 10435
1122.9 +1326 242.03 65.21 96.47 -17.66 1120.3 +1343	IC 2782 UGC 6395 PGC 34934	.S..8*. U 8.0±1.2	1.03± .08 .06± .03 1.04	.05 .08 .03	14.5 ±.2				
112254.7+163527 235.58 66.97 93.38 -16.47 112017.6+165155	NGC 3655 UGC 6396 IRAS11202+1651 PGC 34935	.SAS5*. R (1) 5.0± .4 5.7± .9	1.19± .04 .18± .04 1.19	30 .00 .27 .09	12.30 ±.13 12.78 ±.19 10.79 12.18	.65± .02 .06± .03 .61 .03	14.70±.1 12.64± .24	310± 11 309± 12 2.43	1473± 6 1457± 58 1403 1806

R.A. 2000 DEC.	Names	Type	logD₂₅	p.a.	B_T	(B-V)_T	(B-V)_o	m₂₁	V₂₁

R.A. 2000 DEC.	Names	Type S_T n_L	$\log D_{25}$	p.a.	B_T	$(B-V)_T$	$(B-V)_e$	m_{21}	V_{21}
l b		T	$\log R_{25}$	A_g	m_B	$(U-B)_T$	$(U-B)_e$	W_{20}	V_{opt}
SGL SGB			$\log A_e$	A_i	m_{FIR}	$(B-V)_T^o$	m'_e	W_{50}	V_{GSR}
R.A. 1950 DEC.	PGC	L	$\log D_o$	A_{21}	B_T^o	$(U-B)_T^o$	m'_{25}	HI	V_{3K}

112256.5+340642		.E...*.	1.04± .18						12900± 13
187.33 69.77	UGC 6394	U	.04± .08	.05	14.30 ±.15				12893
76.83 -9.20		-5.0±1.2		.00					13175
112015.2+342310	PGC 34936		1.04		14.06				

1122.9 -0256		.SAS4*.							
264.12 53.13	CGCG 11- 83	F		.12	15.1 ±.6				
113.20 -22.94		4.0±1.3							
1120.4 -0240	PGC 34937								

112302.2+342952		.S..2..	1.22± .05	0				17.07±.3	6314± 10
186.24 69.69	UGC 6397	U	.77± .05	.05	14.94 ±.18			336± 13	6279± 16
76.48 -9.01	IRAS11203+3446	2.0± .9		.94	12.86				6300
112020.7+344620	PGC 34946		1.22	.38	13.89			2.80	6579

1123.0 +0248		.LXR0..							
258.14 57.78	CGCG 39-156	F		.09	15.3 ±.6				
107.21 -21.24		-2.0± .9							
1120.5 +0305	PGC 34948								

1123.0 +0534		PSBR2..							
254.68 59.89	CGCG 39-157	F (1)		.13	15.4 ±.6				
104.38 -20.35		2.0± .9							
1120.5 +0551	PGC 34949		1.6±1.0						

1123.1 +0247		.LA.0*.							
258.20 57.78	CGCG 39-160	F		.09	15.3 ±.6				
107.24 -21.22		-2.0± .9							
1120.6 +0304	PGC 34956								

112318.3-005523		.S..8?/	.95± .07	93					
262.29 54.84	UGC 6402	F	.53± .05	.12	14.9 ±.3				
111.10 -22.29	IRAS11207-0038	8.0±1.9		.65	12.81				
112044.7-003855	PGC 34967		.96	.27					

1123.3 +1338	IC 2787	.S..6?.	.96± .17						
241.78 65.40	UGC 6401	U	.00± .12	.07	14.8 ±.2				
96.31 -17.49		6.0±1.7		.00					
1120.7 +1355	PGC 34969		.97	.00					

1123.3 +5055		.S..9*.	1.44± .03	142	14.19S±.15			14.90±.1	805± 8
152.03 60.94	UGC 6399	U	.55± .05	.00	14.5 ±.3			166± 16	
61.67 -1.49		9.0±1.1		.56				156± 8	860
1120.6 +5112	PGC 34971		1.44	.27	13.69		14.89± .26	.93	1002

1123.4 +5341		.S..4..	1.02± .06	98					
148.35 58.88	UGC 6400	U	.41± .05	.00					
59.20 -.21		4.0± .9		.60					
1120.6 +5358	PGC 34975		1.02	.20					

112332.3-083929	NGC 3660	.SBR4..	1.43± .03	115				14.07±.1	3678± 9
269.10 48.36	MCG -1-29- 16	UEF (1)	.09± .05	.07				292± 12	
119.32 -24.24	MK 1291	4.0± .4		.13	12.29				3529
112100.1-082301	PGC 34980		1.44	.04					4037

112335.6-013359		.SBS3..							
263.03 54.35	CGCG 11- 85	F (1)		.20	15.1 ±.6				
111.80 -22.41		3.0± .9							
112102.1-011730	PGC 34982		1.6±1.0						

112338.4-134952	NGC 3661	.SAR0*/	1.20± .05	137					6700
272.72 43.83	MCG -2-29- 22	EF	.42± .05	.11					
124.92 -25.32		.3± .7		.32					6538
112107.3-133324	PGC 34986		1.18						7056

112339.4+535040	NGC 3656	PI.0.*P	1.21± .05	7					2905± 49
148.11 58.78	UGC 6403	R	.01± .05	.00	13.28 ±.18				2970
59.08 -.11	IRAS11208+5406	90.0		.01	12.08				3086
112050.5+540708	PGC 34989		1.21		13.23				

112345.2+174859	NGC 3659	.SBS9$.	1.32± .03	60				13.98±.1	1283± 7
233.08 67.75	UGC 6405	R (1)	.28± .03	.00	12.91 ±.15			256± 16	1234± 46
92.27 -15.82	IRAS11211+1805	9.0± .8		.29	12.47			220± 8	1216
112107.9+180528	PGC 34995		5.7± .9	1.32	.14	12.62		1.22	1611

112346.9-010616	NGC 3662	.SXR4P.	1.15± .04	25					
262.64 54.76	UGC 6408	PEF (1)	.22± .04	.12	13.71 ±.13				
111.33 -22.23	IRAS11211-0049	3.7± .5		.33	12.69				
112113.3-004947	PGC 34996		1.7± .7	1.17	.11				

1123.8 +0305		RLA.+..							
258.10 58.13	CGCG 39-164	F		.09	15.1 ±.6				
106.99 -20.96		-1.0± .9							
1121.3 +0322	PGC 35001								

R.A. 2000 DEC. l b SGL SGB R.A. 1950 DEC.	Names PGC	Type S_T n_L T L	$\log D_{25}$ $\log R_{25}$ $\log A_o$ $\log D_o$	p.a. A_g A_i A_{21}	B_T m_B m_{FIR} B_T^o	$(B-V)_T$ $(U-B)_T$ $(B-V)_T^o$ $(U-B)_T^o$	$(B-V)_o$ $(U-B)_o$ m'_o m'_{25}	m_{21} W_{20} W_{50} HI	V_{21} V_{opt} V_{GSR} V_{3K}
112354.5+525515 149.22 59.51 59.92 -.50 112106.2+531143	NGC 3657 UGC 6406 IRAS11212+5310 PGC 35002	.SXT5P. P 5.0± .8 	1.16± .05 .01± .05 1.16	 .02 .01	.00 13.12 ±.18 13.09			13.52±.1 208± 16 196± 12 .42	1215± 11 1227± 51 1278 1402
112358.2+383348 175.38 68.45 72.85 -7.03 112115.7+385016	NGC 3658 UGC 6409 PGC 35003	.LAR0*. PU -2.3± .6 	1.21± .07 .04± .06 1.20	 .00 .00	13.1 ±.2 13.18 ±.15 13.12	.94± .02 .44± .03 .92 .45	 13.90± .42		 2043± 31 2053 2300
112359.4+024131 258.61 57.83 107.41 -21.05 112125.1+025800	 UGC 6413 PGC 35005	.SXT4*. UF (1) 3.7± .7 3.6±1.0	1.02± .06 .00± .05 1.02	 .08 .00 .00	 14.27 ±.19 14.15			16.15±.3 84± 7 50± 7 2.00	6887± 11 6771 7243
112360.0-121745 271.83 45.23 123.27 -24.93 112128.5-120116	NGC 3663 MCG -2-29- 23 IRAS11214-1200 PGC 35006	.SAT4P. EF (1) 3.5± .6 3.9± .6	1.28± .05 .16± .05 1.29	85 .11 .23 .08	 13.61 			15.09±.1 364± 12 	5040± 9 4882 5398
1124.0 +0613 254.18 60.54 103.81 -19.90 1121.5 +0630	 CGCG 39-167 PGC 35013	RSBR0.. F .0± .9 	 	 .08 	14.9 ±.6 				
112405.9+454839 160.01 64.53 66.31 -3.73 112121.0+460507	 UGC 6410 PGC 35014	.SX.5.. U 5.0± .9 	1.04± .06 .11± .05 1.04	5 .00 .17 .06	14.61 ±.18 				
112406.5+484146 155.17 62.60 63.72 -2.41 112120.3+485814	 UGC 6411 PGC 35015	.S..4.. U 4.0±1.0 	1.02± .06 .75± .05 1.02	79 .00 1.10 .37	15.71 ±.18 				
112408.6-010929 262.82 54.77 111.42 -22.16 112135.0-005300	 MCG 0-29- 27 PGC 35016	.LXR0.. F -2.0±1.0 	.74± .14 .28± .07 .71	 .12 .00	15.4 ±.2 				
1124.1 +2437 215.30 70.23 85.79 -13.01 1121.5 +2454	 UGC 6414 PGC 35017	.S..4.. U 4.0±1.0 	1.22± .06 1.06± .06 1.22	51 .00 1.47 .50					
112408.9+004201 260.90 56.28 109.48 -21.62 112135.0+005830	 MCG 0-29- 28 PGC 35018	.S?.... 	.94± .11 .28± .07 .95	 .09 .35 .14	14.83 ±.19 14.37				7837± 56 7715 8194
1124.1 +2701 208.22 70.60 83.52 -12.01 1121.5 +2718	 UGC 6415 PGC 35019	.S..2.. U 2.0± .9 	1.32± .04 .69± .05 1.32	20 .00 .85 .35	14.8 ±.2 13.81				9999 9967 10301
112411.8+692445 133.74 45.84 45.18 7.05 112109.0+694113	NGC 3654 UGC 6407 IRAS11211+6941 PGC 35025	.S?.... 	1.08± .09 .28± .07 1.08	 .01 .43 .14	13.7 ±.2 13.20 13.22				1579± 50 1696 1670
112413.6-770936 298.03 -15.13 193.66 -20.28 112226.1-765306	 ESO 38- 12 IRAS11224-7652 PGC 35026	.SXT4*. Sr (1) 4.5± .6 3.3± .6	1.11± .07 .26± .07 1.16	161 .47 .38 .13					
112416.9-135128 272.93 43.88 124.98 -25.17 112145.8-133459	NGC 3667 MCG -2-29- 25 PGC 35028	PSAT2*. E (1) 2.0± .8 1.6±1.0	1.17± .05 .15± .05 1.18	85 .11 .18 .07				15.50±.1 557± 12 	5351± 9 5189 5707
112419.0+003837 261.02 56.26 109.55 -21.60 112145.1+005506	 MCG 0-29- 29 PGC 35030	.SBS5.. F (1) 5.0± .9 3.6±1.0	.89± .11 .00± .07 .89	 .08 .00 .00	14.55 ±.18 14.42				7906± 56 7784 8264
112421.4-135123 272.95 43.89 124.99 -25.15 112150.3-133454	NGC 3667A MCG -2-29- 26 PGC 35034	.SAT3P. F 3.0± .9 	1.15± .08 .39± .05 1.16	40 .11 .54 .19					
1124.3 +0308 258.22 58.24 106.98 -20.82 1121.8 +0325	 UGC 6417 PGC 35037	.S..6*/ UF 6.0±1.0 	.96± .09 .98± .06 .97	144 .09 1.44 .49	16.0 ±.3 14.46				10768 10653 11123

R.A. 2000 DEC.	Names	Type S$_T$ n$_L$	logD$_{25}$ logR$_{25}$	p.a. A$_g$	B$_T$ m$_B$	(B-V)$_T$ (U-B)$_T$	(B-V)$_o$ (U-B)$_o$	m$_{21}$ W$_{20}$	V$_{21}$ V$_{opt}$
l b / SGL SGB / R.A. 1950 DEC.	PGC	T L	logA$_o$ logD$_o$	A$_l$ A$_{21}$	m$_{FIR}$ B$_T^o$	(B-V)$_T^o$ (U-B)$_T^o$	m'$_o$ m'$_{25}$	W$_{50}$ HI	V$_{GSR}$ V$_{3K}$
1124.4 +3914		.I..9*.	1.07± .14						1932± 10
173.63 68.22	UGC 6416	U	.16± .12	.00					
72.27 -6.65		10.0±1.2		.12					1945
1121.7 +3931	PGC 35039		1.07	.08					2186
112424.8+031939	NGC 3664	.SBS9P.	1.31± .03		13.20 ±.13	.39± .03	.35± .02	14.01±.1	1382± 7
258.02 58.39	UGC 6419	R (2)	.03± .04	.12	12.86 ±.16	-.28± .04	-.30± .03	108± 8	1362± 29
106.79 -20.76	DDO 95	9.0± .4	1.07± .02	.03	13.19	.35	14.04± .03	117± 12	1267
112150.4+033608	PGC 35041	7.6± .6	1.32	.01	12.92		-.31	14.52± .21	1.07	1736
112425.1+031321	NGC 3664A	.SB.9P$	.92± .06					15.40±.3	1325± 9
258.14 58.31	UGC 6418	R (1)	.03± .05	.12	14.86 ±.19			55± 10	
106.90 -20.79		9.0± .5	1.07± .02	.03					1211
112150.7+032950	PGC 35042	8.6±1.0	.93	.02	14.71			.67	1680
112425.7+112034	NGC 3666	.SAT5*.	1.64± .02	100	12.70S±.15			13.15±.1	1062± 6
246.40 64.18	UGC 6420	R (2)	.56± .03	.13	12.35 ±.13			271± 8	1047± 66
98.70 -18.07	IRAS11218+1137	5.0± .3	1.07± .02	.84	11.73			256± 6	974
112149.7+113703	PGC 35043	4.7± .6	1.65	.28	11.53		14.36± .20	1.34	1405
112425.9+272725		.S..8*.	1.05± .08	100				15.08±.3	1503± 7
206.93 70.69	UGC 6421	U	.17± .06	.00					1472
83.15 -11.78		8.0±1.2		.21				164± 7	
112146.6+274354	PGC 35044		1.05	.09					1804
1124.6 +2336		.SA.4..	1.09± .06						
218.29 70.11	UGC 6425	U (1)	.05± .05	.00	14.6 ±.3				
86.80 -13.32		4.0± .8		.07					
1122.0 +2353	PGC 35056	4.5±1.1	1.09	.02					
112440.4+145657		.S..8*.	1.09± .06	121				15.87±.3	4155± 9
239.65 66.43	UGC 6424	U	.77± .05	.05	15.52 ±.18				4080
95.16 -16.70		8.0±1.4		.95				265± 5	
112203.7+151326	PGC 35061		1.10	.39	14.51			.98	4491
112443.3+384547	NGC 3665	.LAS0..	1.39± .04	30	11.77 ±.17	.95± .03	.96± .01		
174.71 68.49	UGC 6426	R	.08± .04	.00	11.94 ±.13	.47± .05	.51± .05		2061± 27
72.73 -6.81	IRAS11220+3902	-2.0± .3	.98± .04	.00	12.10	.92	12.16± .11		2072
112201.0+390216	PGC 35064		1.38		11.84	.47	13.39± .26		2317
1124.7 +2356	NGC 3670	.SB.0..	1.06± .06	35					
217.36 70.21	UGC 6427	U	.24± .05	.00	14.49 ±.18				
86.49 -13.17		.0± .9		.18					
1122.1 +2413	PGC 35067		1.05						
1124.8 -1333	A 1122-13	.SBT3P*	1.23± .05	115	13.5 ±.3	.45± .05	.52± .00	14.93±.1	5390± 7
272.90 44.21	MCG -2-29- 27	E (1)	.20± .05	.07				267± 8	5392± 59
124.69 -24.98		3.0±1.3	1.12± .13	.27		.36	14.55± .35	256± 11	5229
1122.3 -1317	PGC 35073	1.6± .8	1.23	.10	13.08		13.95± .41	1.75	5746
1124.9 +3754		.S..8*.	1.14± .13	62					2010
176.80 68.89	UGC 6428	U	.69± .12	.00					
73.53 -7.16		8.0±1.3		.84					2018
1122.2 +3811	PGC 35080		1.14	.34					2270
1125.0 +1705								19.23±.3	1209± 10
235.15 67.66				.00				40± 13	
93.10 -15.82								28± 10	1141
1122.4 +1721	PGC 35087								1540
112502.4-094743	NGC 3672	.SAS5..	1.62± .02	8	12.09 ±.15	.70± .01	.79± .01	12.92±.1	1862± 4
270.42 47.55	MCG -2-29- 28	R (3)	.33± .02	.07	11.7 ±.2			409± 4	1857± 18
120.63 -24.14	IRAS11225-0931	5.0± .3	1.08± .02	.49	10.76	.61	12.98± .05	377± 4	1711
112230.4-093113	PGC 35088	2.3± .4	1.63	.16	11.38		14.22± .18	1.38	2221
1125.1 +1653								17.89±.3	1019± 10
235.65 67.59				.00				34± 13	
93.31 -15.86								29± 10	950
1122.5 +1709	PGC 35096								1351
112512.6-264412	NGC 3673	.SBT3..	1.56± .02	70				14.45±.1	1940± 6
280.09 32.28	ESO 503- 16	R (3)	.18± .03	.27	12.34 ±.11			325± 7	2113± 41
139.23 -26.75	IRAS11227-2627	3.0± .3		.25	12.37			313± 5	1755
112244.0-262742	PGC 35097	3.3± .5	1.58	.09	11.80			2.56	2280
112517.9+002100		.SAT4*.	.94± .09	105					
261.69 56.16	UGC 6432	F (1)	.12± .06	.08	14.81 ±.18				
109.94 -21.45		4.0± .9		.18					
112244.0+003730	PGC 35102	2.6±1.0	.95	.06					
112520.3+634342	A 1122+64	.SAT5..	1.32± .06		13.77 ±.15	.43± .04	.57± .02	14.82±.1	3726± 5
137.83 50.84	UGC 6429	U (1)	.08± .07	.00	13.8 ±.2			69± 13	3724± 59
50.34 4.60		5.0± .7	.94± .03	.11		.39	13.96± .07	55± 5	3826
112225.2+640011	PGC 35105	1.8± .7	1.32	.04	13.63		15.02± .39	1.15	3852

R.A. 2000 DEC. l b SGL SGB R.A. 1950 DEC.	Names PGC	Type S_T n_L T L	$\log D_{25}$ $\log R_{25}$ $\log A_e$ $\log D_o$	p.a. A_g A_i A_{21}	B_T m_B m_{FIR} B_T^o	$(B-V)_T$ $(U-B)_T$ $(B-V)_T^o$ $(U-B)_T^o$	$(B-V)_e$ $(U-B)_e$ m'_e m'_{25}	m_{21} W_{20} W_{50} HI	V_{21} V_{opt} V_{GSR} V_{3K}
112527.3+574322 143.34 55.87 55.73 1.90 112237.0+575951	NGC 3669 UGC 6431 IRAS11226+5759 PGC 35113	.SB.6*/ P 6.0± .8 	1.35± .05 .61± .05 1.35	153 .00 .89 .30	 13.1 ±.2 12.73 12.24			 	
112527.3+574322 (cont)								14.62±.1 403± 16 405± 12 2.08	1940± 11 2019 2100
112531.4+632641 138.04 51.09 50.61 4.50 112236.6+634311	NGC 3668 UGC 6430 IRAS11225+6343 PGC 35123	.S..4.. U 4.0± .8 	1.24± .07 .11± .07 1.24	 .00 .17 .06	 13.13 ±.18 12.59 12.94				3522± 39 3620 3649
112532.1+380341 176.27 68.94 73.44 -6.98 112250.1+382011	A 1122+38 UGC 6433 VV 87 PGC 35124	.P..... R 99.0 	1.06± .05 .21± .05 1.06	79 .00 .16 .11	 14.57 ±.18 14.41				2102 2111 2361
1125.5 +2248 220.72 70.09 87.65 -13.46 1122.9 +2305	A 1122+23 MCG 4-27- 34 PGC 35125	.SBS3. P (1) 3.0± .8 1.6± .9	.94± .11 .00± .07 .68± .02 .94	 .00 .00 .00	14.49 ±.13 14.49 ±.18 14.44	.56± .03 .52 	.64± .03 13.38± .04 14.02± .58	15.38±.3 178± 21 175± 15 .94	6471± 12 6563± 59 6428 6791
112534.4-004605 262.94 55.29 111.12 -21.70 112300.7-002935	 UGC 6435 PGC 35126	.L..0*. UF -2.2± .7 	1.04± .09 .00± .04 1.05	 .12 .00 	 13.88 ±.15 13.65				7620± 35 7494 7978
112536.4+542256 147.03 58.55 58.73 .39 112248.0+543926	A 1122+54 MCG 9-19- 73 MK 40 PGC 35129	.L...P. R -2.0± .6 	1.12± .12 .91± .07 .98	 .00 .00 				15.65±.3 	6323± 10 6198± 19 6361 6472
1125.7 +1855 231.02 68.68 91.39 -14.96 1123.1 +1912	 UGC 6437 PGC 35137	.S..4.. U 4.0± .9 	1.07± .06 .34± .05 1.07	35 .00 .49 .17	 14.92 ±.19 				
112545.4-480830 288.49 12.31 163.05 -26.46 112323.0-475200	 ESO 216- 8 IRAS11233-4752 PGC 35140	.SXS5*/ Sr (1) 4.8± .5 3.3± .9	1.26± .05 .55± .04 1.31	90 .51 .82 .27	 14.62 ±.14 12.80 				
112546.3+144023 240.60 66.49 95.53 -16.56 112309.8+145653	IC 2810A UGC 6436 IRAS11231+1456 PGC 35142	.SB?... 	1.10± .07 .31± .06 1.10	30 .06 .46 .15	 14.9 ±.2 11.18 14.30			16.80±.3 411± 10 2.35	10243± 9 10167 10580
112552.5-352342 283.89 24.29 148.90 -27.03 112326.0-350712	A 1123-35 ESO 377- 46 PGC 35150	RLXT0.. Sr -2.0± .5 	1.07± .07 .06± .05 1.10	 .31 .00 	 14.35 ±.14 13.89				9733± 43 9532 10046
112553.6+095913 249.23 63.54 100.18 -18.21 112318.0+101543	IC 692 UGC 6438 ARAK 292 PGC 35151	.S?.... 	.84± .08 .14± .05 .85	125 .10 .11 	 14.5 ±.2 14.30			16.37±.3 67± 5 	1163± 9 1157± 50 1071 1509
1125.9 -1146 272.08 45.91 122.82 -24.36 1123.4 -1130	 PGC 35152	.SBR7P? E (1) 7.0±1.8 7.5±1.2	1.18± .08 .10± .08 1.19	100 .09 .14 .05					
112601.2+015900 260.18 57.58 108.31 -20.79 112327.0+021530	 UGC 6440 IRAS11234+0215 PGC 35158	.SBT4.. UF (1) 3.5± .6 2.6± .8	1.04± .06 .18± .05 1.05	43 .10 .27 .09	 14.40 ±.18 14.03 				
112603.4-524649 290.09 7.95 168.12 -25.85 112343.0-523018	 ESO 170- 2 IRAS11237-5230 PGC 35160	.SAR6*. S (1) 6.0±1.2 4.4± .8	1.10± .08 .04± .07 1.21	105 1.14 .06 .02	 13.08 				
112607.8+433506 163.67 66.19 68.48 -4.42 112324.2+435136	NGC 3675 UGC 6439 PGC 35164	.SAS3.. R (2) 3.0± .3 3.3± .6	1.77± .02 .28± .02 1.77	178 .00 .39 .14	11.00S±.15 10.90 ±.14 10.55		 14.01± .19	13.04±.1 426± 4 407± 4 2.35	767± 5 724± 36 796 1000
112608.7-541407 290.58 6.58 169.70 -25.62 112349.0-535736	 ESO 170- 3 PGC 35166	RSXR2?. Sr 1.5± .9 	1.13± .07 .39± .07 1.26	127 1.36 .48 .19					
1126.1 +2645 209.16 71.01 83.97 -11.72 1123.5 +2702	 CGCG 156- 74 PGC 35167	 		 .00 	15.4 ±.3 				9948 9915 10251

R.A. 2000 DEC. / l b / SGL SGB / R.A. 1950 DEC.	Names / PGC	Type / S_T n_L / T / L	$\log D_{25}$ / $\log R_{25}$ / $\log A_\bullet$ / $\log D_0$	p.a. / A_g / A_i / A_{21}	B_T / m_B / m_{FIR} / B_T^0	$(B-V)_T$ / $(U-B)_T$ / $(B-V)_T^0$ / $(U-B)_T^0$	$(B-V)_\bullet$ / $(U-B)_\bullet$ / $m'_\bullet$ / m'_{25}	m_{21} / W_{20} / W_{50} / HI	V_{21} / V_{opt} / V_{GSR} / V_{3K}
112609.4+032954	A 1123+03	.SXS3*.							
258.46 58.80	CGCG 39-180	F		.13	15.0 ±.6				
106.76 −20.29		3.0± .9							
112335.0+034624	PGC 35168								
1126.1 −0150		.SB.3?/							
264.21 54.49	CGCG 11- 95	F		.16	15.0 ±.6				
112.28 −21.87		3.0±1.8							
1123.6 −0134	PGC 35170								
112609.8−723704		.I.9..	1.22± .07		.87				
296.58 −10.81		S (1)	.08± .08		.06				
189.04 −21.54		10.0± .8							
112408.1−722033	PGC 35171	10.0± .8	1.30		.04				
1126.1 +0750		.S..7..	1.14± .07	176					6310
252.68 62.08	UGC 6442	U	1.16± .06	.19	15.8 ±.2				6211
102.36 −18.87		7.0±1.0		1.38					
1123.6 +0807	PGC 35174		1.16	.50	14.17				6659
112615.5+275159	NGC 3678	.S..4..	.91± .05					16.56±.3	7210± 11
205.76 71.12	UGC 6443	U	.03± .04	.00	14.40 ±.19			100± 7	7185± 50
82.94 −11.24	KUG 1123+281	4.0± .9		.05				72± 7	7180
112336.3+280829	PGC 35177		.91	.02	14.31			2.24	7508
112617.5+465826	NGC 3677	PSAR0*.	1.29± .07	130					
157.41 64.05	UGC 6441	PU	.07± .05	.00	13.30 ±.19				7475± 57
65.43 −2.87	IRAS11235+4714	−.3± .6		.05	12.92				7517
112332.8+471456	PGC 35181		1.29		13.13				7692
112619.1+210546		.SB?...	.82± .13		15.40S±.15				12979± 41
225.60 69.69	MCG 4-27- 36		.28± .07	.00					12926
89.36 −13.98	HICK 52A			.42					13300
112341.4+212216	PGC 35183		.82	.14	14.92		13.62± .68		13300
1126.3 +0255		.SXR4..	.84± .10						
259.23 58.38	CGCG 39-182	F (1)	.21± .10	.09	15.27 ±.18				
107.37 −20.42		4.0±1.0		.30					
1123.8 +0312	PGC 35185	2.6±1.0	.85	.10					
112625.4−361537			.78± .07	168	14.4 ±.4	1.13± .03			
284.34 23.52	ESO 377- 48		.16± .06	.34	15.37 ±.14	.69± .08			
149.87 −26.92									
112359.0−355906	PGC 35188		.81						
112626.8+570258	NGC 3674	.L..../	1.29± .05	33					1885
143.85 56.49	UGC 6444	RU	.48± .04	.00	13.24 ±.18				1962
56.40 1.71		−2.0± .4		.00					2049
112337.3+571928	PGC 35191		1.22		13.22				
112629.5+165151	NGC 3681	.SXR4..	1.40± .03		11.9 S±.3	.71± .05		13.50±.1	1239± 4
236.13 67.84	UGC 6445	R (1)	.10± .03	.00	12.39 ±.14			184± 4	1299± 53
93.46 −15.58	IRAS11238+1708	4.0± .3		.14	13.19	.68		157± 4	1171
112352.6+170822	PGC 35193	2.3± .8	1.40	.05	12.14		13.51± .30	1.31	1571
112634.5+112624	IC 2822	.SB.4..	1.19± .06	130				14.97±.2	3204± 6
247.03 64.64	UGC 6449	U	.32± .06	.13	14.4 ±.2			236± 6	3117
98.80 −17.54		4.0± .8		.48				220± 5	3547
112358.6+114254	PGC 35196		1.20	.16	13.82			.98	
112635.5−430331		.L...*.	1.08± .05	128					
286.87 17.15	ESO 265- 33	S	.49± .03	.38	14.61 ±.14				
157.42 −26.71		−2.0± .9		.00					
112411.1−424700	PGC 35198		1.05						
112640.6+534458		.SA.7..	1.55± .03	10	13.54S±.15		.48± .07	13.36±.1	645± 7
147.56 59.14	UGC 6446	U	.19± .05	.00	13.8 ±.5		−.29± .12	143± 6	
59.37 .25		7.0± .7		.26				135± 12	710
112353.1+540128	PGC 35202		1.55	.09	13.31		15.67± .24	−.04	826
112640.7−014138		CE.....			*				
264.25 54.68	CGCG 11- 96	F		.16	14.84 ±.18				
112.17 −21.70		−5.0± .9							
112407.2−012507	PGC 35203								
112640.9−541519		.S.5?P/	1.36± .05	86	*				
290.66 6.58	ESO 170- 4	S	.73± .07	1.36					
169.71 −25.54		5.0±1.7		1.09					
112421.0−535848	PGC 35204		1.49	.36					
112643.7+590917	IC 691	.I?....	.81± .08	150	14.5 ±.2	.56± .03		15.47±.1	1199± 7
141.67 54.78	UGC 6447		.18± .05	.00	14.5 ±.3	−.29± .05		158± 5	1225± 38
54.52 2.69	MK 169			.14	11.88	.51		141± 5	1284
112353.0+592547	PGC 35206		.81	.09	14.38	−.33	12.94± .48	1.01	1351

R.A. 2000 DEC.	Names	Type	logD25	p.a.	BT	(B-V)T	(B-V)o	m21	V21
l　b		ST　nL	logR25	Ag	mB	(U-B)T	(U-B)o	W20	Vopt
SGL　SGB		T	logA•	Ai	mFIR	(B-V)oT	m'•	W50	VGSR
R.A. 1950 DEC.	PGC	L	logDo	A21	BoT	(U-B)oT	m'25	HI	V3K
112646.1+391559		.S?....	.94± .11						
172.99　68.62	MCG　7-24-　6		.05± .07	.00	14.85 ±.18				6222± 61
72.45　-6.23				.07					6236
112404.1+393229	PGC 35207		.94	.02	14.73				6476
112648.5+351447	A　　1124+35	.L?....	.97± .15						
183.50　70.22	MCG　6-25- 72		.23± .07	.00	14.78 ±.15				9652± 11
76.14　-7.99	MK　　423			.00	12.84				9650
112407.6+353117	PGC 35210		.93		14.63				9923
112650.4+640816	A　　1123+64	.P.....	1.02± .10					15.73±.1	991　 6
137.28　50.57	UGC　6448	R	.36± .07	.00	14.91 ±.18			94± 5	991± 60
50.05　4.94	MK　　170	99.0		.54				60± 5	1092
112355.8+642446	PGC 35213		1.02	.18	14.36			1.18	1113
112655.6-011919		.SAT4..	.82± .13						
263.98　55.03	MCG　0-29- 33	F　　(1)	.08± .07	.11	15.12 ±.18				
111.80 -21.54		4.0± .9		.11					
112422.0-010248	PGC 35217	3.6±1.4	.83	.04					
112656.5+632531		.E?....	.72± .22						
137.84　51.19	MCG 11-14- 25A		.05± .07	.00	14.8　±.2				3304± 60
50.70　4.63	ARAK　293			.00					3403
112402.6+634201	PGC 35219		.70		14.78				3431
112705.3-285843	IC　　2764	RSAT0..	1.21± .04						
281.54　30.37	ESO　439-　8	Sr	.06± .04	.28	13.15 ±.14				1635± 30
141.76 -26.51		.4± .6		.05					1444
112437.0-284212	PGC 35222		1.24		12.80				1966
112711.2+170149	NGC　3684	.SAT4..	1.49± .02	130	12.0 S±.3	.62± .08		13.39±.1	1163± 4
235.98　68.07	UGC　6453	R　　(2)	.16± .02	.00	12.18 ±.14			241± 4	1394± 53
93.36 -15.36	IRAS11245+1718	4.0± .3		.24	11.94	.58		209± 4	1097
112434.4+171820	PGC 35224	4.1± .6	1.49	.08	11.90		13.90± .29	1.41	1496
1127.1　+0843	IC　　2828	.L?....	1.12± .12						
251.73　62.88	MCG　2-29- 28		.36± .07	.14	14.38 ±.15				1020±150
101.56 -18.34				.00					925
1124.6　+0900	PGC 35225		1.08		14.23				1368
112712.6-005943		.I..9*.	.98± .09					15.44±.1	976± 9
263.76　55.34	UGC　6457	UF	.01± .06	.07	14.46 ±.19			126± 12	
111.48 -21.38		10.0±1.2		.01					850
112439.0-004312	PGC 35227		.99	.00	14.39			1.05	1334
1127.2　-0608		.LAS+P*	1.20± .07	80					
268.37　50.99		E	.34± .08	.09					
116.88 -22.74		-1.0±1.3		.00					
1124.7　-0552	PGC 35230		1.16						
112717.5+034523									
258.58　59.18	CGCG　39-184			.08	15.0　±.6				10561
106.59 -19.94									10450
112443.0+040154	PGC 35234								10916
112718.9+383951		.S..6*.	1.23± .03	34					
174.33　68.99	UGC　6454	U	1.05± .04	.00					
73.05　-6.40	KUG 1124+389	6.0±1.4		1.47					
112437.2+385622	PGC 35235		1.23	.50					
112719.7+593740		.S?....	1.02± .06	30					
141.11　54.43	UGC　6452		.31± .05	.00	14.2　±.2				5169± 50
54.13　2.97	IRAS11244+5954			.47	13.19				5255
112428.9+595411	PGC 35236		1.02	.16	13.71				5318
112723.0-105709		.SBS3..	1.05± .06	140					
271.98　46.81	MCG　-2-29- 31	E　　(1)	.17± .05	.08					
122.01 -23.83		3.0± .9		.24					
112451.2-104038	PGC 35239	3.1± .8	1.06	.09					
112723.3-291531		RSBR1P*	1.21± .04	99					
281.73　30.13	ESO　439-　9	Sr	.63± .03	.27	14.76 ±.14				7454± 99
142.08 -26.47	IRAS11249-2859	1.0±1.8		.64					7263
112455.1-285900	PGC 35241		1.23	.31	13.75				7784
112728.9+195113									
229.18　69.46	CGCG　96- 48			.00	15.02 ±.18				6568± 61
90.66 -14.21	MK　　735								6511
112451.5+200744	PGC 35247								6892
112732.1+565243	NGC　3683	.SBS5$.	1.27± .04	128					
143.81　56.72	UGC　6458	R　　(1)	.43± .04	.00	13.33 ±.15				1656± 66
56.62　1.77	IRAS11247+5709	5.0± .9		.65	10.20				1733
112443.1+570914	PGC 35249	5.7± .9	1.27	.22	12.67				1821

R.A. 2000 DEC.	Names	Type	logD$_{25}$	p.a.	B$_T$	(B-V)$_T$	(B-V)$_e$	m$_{21}$	V$_{21}$
I b		S$_T$ n$_L$	logR$_{25}$	A$_g$	m$_B$	(U-B)$_T$	(U-B)$_e$	W$_{20}$	V$_{opt}$
SGL SGB		T	logA$_e$	A$_i$	m$_{FIR}$	(B-V)$_T^o$	m'$_e$	W$_{50}$	V$_{GSR}$
R.A. 1950 DEC.	PGC	L	logD$_o$	A$_{21}$	B$_T^o$	(U-B)$_T^o$	m'$_{25}$	HI	V$_{3K}$

112732.4-291037		.P..../	1.09± .05	147					
281.73 30.22	ESO 439- 10	S	.39± .05	.27 15.73 ±.14				7164± 34
141.99 -26.43		99.0		.59				6973
112504.0-285406	PGC 35250		1.12	.20 14.83				7494

112733.0+360416		.L?....	.82± .19	123					
181.06 70.08	MCG 6-25- 75		.09± .07	.00 15.48 ±.15				10262± 79
75.45 -7.49	IRAS11248+3620			.00 13.35				10264
112452.0+362047	PGC 35252		.80	15.32				10529

112733.8+360339		.S?....	.94± .11						
181.09 70.09	MCG 6-25- 74		.47± .07	.00 15.1 ±.2				10066± 79
75.46 -7.49				.71				10068
112452.9+362010	PGC 35254		.94	.24 14.34				10334

112736.9+002341		.SBS3..							
262.50 56.54	CGCG 11-100	F (1)		.07 15.1 ±.6				
110.07 -20.88		3.0± .9						
112503.0+004012	PGC 35259	1.6±1.0						

112742.7+663525	NGC 3682	.SAS0*$	1.22± .04	95 13.3 ±.2	.76± .03	.74± .02		
135.30 48.48	UGC 6459	R	.18± .05	.00 13.42 ±.18	.04± .05	.02± .03		1543± 42
47.88 6.10	IRAS11247+6651	.0± .5	.52± .06	.13 11.67	.72	11.37± .16		1652
112446.2+665156	PGC 35266		1.21	13.21	.02	13.78± .32		1652

1127.7 +2122		.S..2..	1.04± .08	2					
225.19 70.10	UGC 6461	U	.47± .06	.00				
89.23 -13.57		2.0± .9		.58				
1125.1 +2139	PGC 35267		1.04	.24				

112744.1+171325	NGC 3686	.SBS4..	1.51± .02	15 11.89M±.11	.57± .07	.61± .07	14.52±.0	1156± 4
235.72 68.28	UGC 6460	PU (2)	.11± .02	.02 11.84 ±.14		.00± .07	202± 4	1033± 53
93.23 -15.16	IRAS11251+1729	4.0± .5	1.10± .04	.17 11.47	.54	13.04± .08	189± 4	1089
112507.3+172956	PGC 35268	2.6± .6	1.51	.06 11.67		13.99± .16	2.79	1487

112744.5-090956	NGC 3688	PSBT3..	1.11± .06	170				
270.84 48.42	MCG -1-29- 24	EF (1)	.14± .05	.06				
120.12 -23.35		2.5± .6		.20				
112512.3-085325	PGC 35269	2.6±1.0	1.11	.07				

1127.7 -0454		.SB.7..	1.15± .08					971
267.52 52.11	MCG -1-29- 23	F	.60± .07	.12				
115.61 -22.31		7.0± .9		.83				834
1125.2 -0438	PGC 35271		1.16	.30				1331

1127.7 +0608		.SAT3..	.83± .08					
255.70 61.09	MCG 1-29- 42	F (1)	.03± .06	.11 14.97 ±.18				
104.21 -19.06		3.0±1.0		.04				
1125.2 +0625	PGC 35272	5.6±1.0	.84	.01				

112747.4+075939		.S?....	1.19± .12	3				16.21±.2	6356± 8
253.07 62.46	UGC 6462		.81± .12	.19 15.01 ±.18			244± 14	6259
102.35 -18.45				1.22			244± 8	6259
112512.3+081610	PGC 35273		1.21	.41 13.57			2.24	6705

1127.8 -0113		.LAR0P.						
264.22 55.24	CGCG 11-101	F		.10 15.1 ±.6				12451± 79
111.77 -21.29		-2.0± .9						12325
1125.3 -0057	PGC 35276							12809

112754.2-413650		PSBR1..	1.02± .19	45				
286.60 18.59	ESO 319- 22	r	.22± .14	.43				
155.81 -26.54	IRAS11254-4120	1.0± .9		.23 11.07				
112529.0-412018	PGC 35278		1.06	.11				

112757.6-011244		.LAR+P.						
264.25 55.26	CGCG 11-103	F		.10 15.2 ±.6				12932± 79
111.77 -21.26		-1.0± .9						12806
112524.0-005612	PGC 35281							13290

112800.6+293041	NGC 3687	PSXR4?.	1.28± .03					14.91±.1	2507± 6
200.63 71.51	UGC 6463	PU (2)	.00± .04	.00 12.86 ±.14			187± 4	2377± 66
81.56 -10.20	MK 736	3.5± .5		.01			178± 4	2484
112521.3+294712	PGC 35285	2.2± .7	1.28	.00 12.83			2.07	2800

1128.0 +7859	A 1124+79	.P.....	1.16± .07					14.43±.1	-93± 5
127.84 37.33	UGC 6456	R	.24± .06	.10 14.50 ±.19			58± 5	-49± 32
36.54 11.40	7ZW 403	99.0		.30			49± 3	53
1124.6 +7916	PGC 35286		1.17	.12 14.10			.21	-60

112804.7-363232	A 1125-36	.SXS5*.	1.34± .04	178 13.79 ±.15	.66± .03	.77± .02		3023± 10
284.79 23.38	ESO 378- 3	PS (2)	.20± .05	.38 13.85 ±.14	.01± .10			2976± 59
150.18 -26.59		5.2± .5	1.03± .02	.30	.52	14.43± .05		2821
112538.0-361600	PGC 35288	3.5± .5	1.37	.10 13.13	-.09	14.83± .28		3331

R.A. 2000 DEC. / l b / SGL SGB / R.A. 1950 DEC.	Names / PGC	Type / S_T n_L / T / L	$\log D_{25}$ / $\log R_{25}$ / $\log A_e$ / $\log D_o$	p.a. / A_g / A_i / A_{21}	B_T / m_B / m_{FIR} / B_T^o	$(B-V)_T$ / $(U-B)_T$ / $(B-V)_T^o$ / $(U-B)_T^o$	$(B-V)_e$ / $(U-B)_e$ / m'_e / m'_{25}	m_{21} / W_{20} / W_{50} / HI	V_{21} / V_{opt} / V_{GSR} / V_{3K}
112808.6+165515 236.55 68.21 93.56 −15.19 112531.8+171146	NGC 3691 UGC 6464 ARAK 294 PGC 35292	.SB.3$. P 3.0± .9	1.13± .03 .13± .03 1.13	15 .04 .18 .07	12.6 S±.3 13.43 ±.15 13.31 13.02		 12.77± .32	15.78±.1 184± 5 130± 5 2.69	1085± 4 997±181 1018 1417
112810.3+253944 212.71 71.32 85.19 −11.76 112531.9+255615	NGC 3689 UGC 6467 PGC 35294	.SXT5.. R (2) 5.0± .4 3.3± .8	1.22± .03 .17± .03 1.22	97 .00 .26 .09	13.03 ±.13 13.13 ±.15 12.80	.70± .02 −.01± .04 .65 −.05	 13.54± .22	15.75±.2 260± 34 346± 25 2.87	2739± 8 2722± 61 2703 3045
1128.1 +0419 258.22 59.76 106.09 −19.55 1125.6 +0436	 UGC 6466 PGC 35295	.SBT4P* UF (1) 4.3± .7 3.6±1.0	1.03± .06 .46± .05 1.04	5 .10 .67 .23	 15.17 ±.18 14.35				8658± 79 8549 9012
112811.7−131143 273.70 44.90 124.47 −24.11 112540.1−125511	NGC 3693 MCG −2−29−32 IRAS11256−1255 PGC 35299	PSAR3*. EF (1) 2.7± .6 2.6±1.0	1.51± .03 .68± .05 1.52	85 .06 .93 .34	 13.43			14.20±.1 521± 12	4953± 9 4794 5310
112812.3+215952 223.57 70.42 88.68 −13.22 112534.6+221623	 UGC 6465 VV 594 PGC 35300	.SB?... 	1.02± .06 .07± .05 1.02	120 .00 .11 .04	 14.58 ±.19 14.43			15.58±.3 198± 7 1.11	6307± 10 6200± 55 6255 6622
1128.2 +0903 251.61 63.30 101.31 −17.99 1125.6 +0920	IC 2850 MCG 2−29−30 IRAS11256+0920 PGC 35301	.S?.... 	.82± .13 .46± .07 .83	 .11 .68 .23	 15.3 ±.3 14.51			16.10±.3 288± 20 1.36	6298± 12 6204 6645
1128.2 +0908 251.48 63.37 101.23 −17.96 1125.6 +0925	IC 2853 UGC 6470 IRAS11256+0925 PGC 35302	.SB.2.. U 2.0± .9 	1.00± .06 .29± .05 1.01	15 .11 .35 .14	 14.58 ±.18 13.03 14.05			15.73±.2 398± 8 380± 7 1.53	6310± 6 6216 6657
112817.0+005328 262.22 57.04 109.61 −20.57 112543.0+011000	 MCG 0−29−36 PGC 35306	.SXS4.. F (1) 4.0± .9 3.6±1.0	.74± .14 .00± .07 .74	 .07 .00 .00	 15.17 ±.18 				
112817.1+023916 260.26 58.46 107.80 −20.04 112542.8+025548	 UGC 6469 IRAS11257+0255 PGC 35307	.SA.4*. F 4.0±1.3 	.92± .07 .40± .05 .93	123 .07 .59 .20	 14.7 ±.3 12.61 13.96			15.82±.3 374± 7 346± 7 1.66	6841± 9 6824± 50 6726 7196
112824.3+092423 251.14 63.58 100.99 −17.83 112548.9+094055	NGC 3692 UGC 6474 IRAS11258+0940 PGC 35314	.S..3.. U (1) 3.0± .8 1.5±1.1	1.50± .03 .65± .05 1.51	95 .12 .90 .33	12.94 ±.19 12.75 11.91			14.74±.2 404± 9 387± 6 2.51	1726± 6 1633 2073
1128.4 +0906 251.64 63.38 101.29 −17.91 1125.9 +0923	IC 2857 UGC 6475 PGC 35320	.S..6*. U 6.0±1.4 	1.27± .04 1.04± .05 1.28	161 .11 1.47 .50	 15.20 ±.18 13.59			15.09±.3 412± 20 1.00	6324± 12 6230 6671
112833.7+583351 141.90 55.41 55.16 2.64 112544.2+585023	NGC 3690 UGC 6472 IRAS11257+5850 PGC 35321	.IB.9P. R 9.0± .3 	1.31± .04 .12± .05 1.31	50 .00 .13 .06	12.1 ±.3 8.29 11.95				3033± 17 3116 3188
112829.5+583404 141.91 55.40 55.15 2.63 112540.0+585035	 MCG 10−17−2A PGC 35325			32 .00					
112831.3+583328 141.91 55.41 55.16 2.63 112541.9+585000	IC 694 UGC 6471 PGC 35326	.SB.9$P R 9.0± .5 	1.07? .08? 1.07	 .00 .08 .04					3132± 18 3215 3287
112834.5−013750 264.88 55.00 112.25 −21.23 112601.0−012118	IC 697 CGCG 11−106 PGC 35327	.L..... F −2.0± .9 		 .12	15.1 ±.6				
112836.4+232418 219.59 70.93 87.37 −12.57 112558.5+234050	 UGC 6476 PGC 35328	.S..4.. U (1) 4.0± .8 3.5±1.1	1.16± .05 .16± .05 .67± .01 1.16	115 .00 .23 .08	14.19 ±.13 14.03 ±.19 13.86	.57± .02 .05± .04 .49 .00	.65± .02 .03± .04 13.03± .03 14.45± .30	15.87±.2 318± 6 299± 6 1.94	7331± 6 7388± 50 7288 7645
1128.6 +7050 132.33 44.74 44.05 8.02 1125.6 +7107	 UGC 6468 PGC 35330	.S..7.. U 7.0± .9 	1.04± .08 .59± .06 1.04	153 .04 .81 .29					

R.A. 2000 DEC.	Names	Type	logD$_{25}$	p.a.	B$_T$	(B-V)$_T$	(B-V)$_e$	m$_{21}$	V$_{21}$
l b	S$_T$ n$_L$	logR$_{25}$	A$_g$	m$_B$	(U-B)$_T$	(U-B)$_e$	W$_{20}$	V$_{opt}$
SGL SGB		T	logA$_e$	A$_i$	m$_{FIR}$	(B-V)$_T^o$	m'$_e$	W$_{50}$	V$_{GSR}$
R.A. 1950 DEC.	PGC	L	logD$_o$	A$_{21}$	B$_T^o$	(U-B)$_T^o$	m'$_{25}$	HI	V$_{3K}$

112839.9+090555	IC 696	.SB.8*.	1.00± .06					15.75±.2	6282± 8
251.72 63.41	UGC 6477	U	.03± .05	.11	14.21 ±.18			206± 7	6311± 50
101.32 -17.87	IRAS11259+0922	8.0±1.2		.03	13.24			144± 7	6189
112604.6+092227	PGC 35332		1.01	.01	14.05			1.69	6630
112841.0+194348		.S?....	1.02± .06	24				16.05±.3	5831± 9
229.84 69.66	UGC 6478		.52± .05	.00	15.39 ±.18				
90.90 -14.00				.78				305± 5	5774
112603.8+200020	PGC 35335		1.02	.26	14.58			1.21	6155
112843.5-140717		RLBR+*.	1.15± .08	140					
274.44 44.13	MCG -2-29- 34	EF	.05± .05	.03					
125.50 -24.16		-.5± .6		.00					
112612.1-135045	PGC 35338		1.15						
112844.2-111658	NGC 3696	.SXS5P.	1.09± .09	90					
272.63 46.67	EF (1)	.09± .08	.05						
122.45 -23.58		4.5± .6		.13					
112612.3-110026	PGC 35340	3.2±1.1	1.10	.04					
112845.2-350208		.SAS6..	1.11± .04	80					
284.36 24.84	ESO 378- 5	S (1)	.09± .04	.31	14.82 ±.14				
148.51 -26.43		6.0± .8		.13					
112618.0-344536	PGC 35341	6.7± .8	1.14	.05					
112850.7+204744	NGC 3697	.SB?...	1.37± .03	93	13.77M±.08	.70± .02	.84± .02	14.52±.2	6263± 6
227.05 70.13	UGC 6479		.50± .03	.00	13.78 ±.18	-.02± .06	.21± .03	528± 6	6261± 32
89.89 -13.55	HICK 53A		.90± .01	.69		.57	13.76± .04	517± 6	6210
112613.3+210416	PGC 35347		1.37	.25	13.03	-.12	14.24± .17	1.24	6584
1128.8 +7302		.S..2..	1.08± .09						
130.98 42.77	UGC 6473	U	.28± .07	.08	15.01 ±.19				
42.05 8.97	IRAS11257+7318	2.0± .9		.35					
1125.7 +7319	PGC 35349		1.08	.14					
112854.7+352448	NGC 3694	.S?....	.85± .11	120					2254± 82
182.61 70.57	UGC 6480		.09± .06	.00	13.9 ±.3				2254
76.17 -7.53	ARAK 296			.13	13.84				2524
112614.2+354120	PGC 35352		.85	.05	13.70				
112855.4+032434		.SXS3..	.74± .14						
259.62 59.16	MCG 1-29- 46	F (1)	.09± .07	.06	15.1 ±.2				
107.08 -19.66		3.0± .9		.13					
112621.1+034106	PGC 35353	3.6±1.0	.74	.05					
112858.5+204500		.S?....	.85± .12		15.12S±.15				
227.21 70.14	MCG 4-27- 45		.15± .07	.00	15.26 ±.18				
89.94 -13.54	MK 1296			.22					
112621.2+210132	PGC 35355		.85	.08			13.83± .65		
112860.0-222856		RLBR+..	1.03± .09						8190± 52
279.05 36.54	ESO 571- 3	r	.08± .03	.14					8011
134.62 -25.44		-1.3± .9		.00					8535
112630.1-221224	PGC 35359		1.03						
112900.0+204422		.L?....	.72± .22		14.94S±.15				
227.24 70.14	MCG 4-27- 44		.11± .07	.00	15.50 ±.17				
89.96 -13.54	HICK 53B			.00					
112622.7+210054	PGC 35360		.70				13.12±1.12		
112902.6+171344		.S..6*.	1.35± .04	68				14.57±.3	3891± 10
236.14 68.55	UGC 6483	U	1.06± .05	.01	15.13 ±.18				
93.35 -14.87	IRAS11263+1730	6.0±1.3		1.47				324± 7	3825
112625.9+173016	PGC 35362		1.35	.50	13.63			.44	4222
112904.1+090648	IC 698	.L...?.	1.01± .08	147				16.48±.3	6353± 12
251.86 63.49	UGC 6482	U	.28± .03	.11	14.42 ±.16			217± 20	6189± 50
101.34 -17.78	IRAS11264+0923	-2.0±1.8		.00	11.69				6250
112628.8+092320	PGC 35364		.98		14.22				6691
1129.0 +0859	IC 699	.S..3*.	1.09± .06	12				16.55±.3	6219± 12
252.06 63.40	UGC 6485	U (1)	.56± .05	.14	14.69 ±.19			510± 20	
101.46 -17.81		3.0±1.3		.78					6125
1126.5 +0916	PGC 35365	4.5±1.2	1.10	.28	13.72			2.54	6567
1129.1 -0142		.E.....							
265.16 55.01	CGCG 11-107	F		.00	15.4 ±.6				
112.37 -21.11		-5.0± .9							
1126.6 -0126	PGC 35370								
112912.4+570757	NGC 3683A	.SBT5..	1.37± .03	75					2429± 10
143.21 56.64	UGC 6484	U	.14± .04	.00	12.60 ±.18				2446± 50
56.49 2.08	IRAS11263+5724	5.0± .7		.21	12.87				2508
112624.0+572429	PGC 35376		1.37	.07	12.38				2593

R.A. 2000 DEC. l b SGL SGB R.A. 1950 DEC.	Names PGC	Type S_T n_L T L	logD_25 logR_25 logA_e logD_o	p.a. A_g A_i A_21	B_T m_B m_FIR B_T^o	(B-V)_T (U-B)_T (B-V)_T^o (U-B)_T^o	(B-V)_e (U-B)_e m' m'_25	m_21 W_20 W_50 HI	V_21 V_opt V_GSR V_3K
112912.7+115155 247.28 65.40 98.62 -16.79 112636.9+120827	UGC 6486 PGC 35377	.I..9.. U 10.0± .8 	1.11± .14 .07± .12 1.12	.13 .05 .03				15.60±.3 114± 7	3230± 7 3146 3572
112913.1+194621 229.88 69.79 90.91 -13.87 112636.0+200253	MCG 3-29- 58 MK 1297 PGC 35379	.S?.... 	.94± .11 .11± .07 .93	.00 .09	14.78 ±.18 14.61				5703± 81 5646 6027
112914.1+203453 227.74 70.13 90.13 -13.55 112636.8+205126	HICK 54B PGC 35380	.S?.... 		.00	16.24S±.15				1412± 41 1358 1734
112915.2+203501 227.73 70.13 90.13 -13.55 112637.9+205133	IC 700 UGC 6487 VV 498 PGC 35382	.L?.... 	1.02± .14 .33± .07 .97	.00 .00	14.02S±.15 14.89 ±.15 14.43		13.15± .74	15.58±.3 105± 10	1418± 9 1383± 27 1361 1737
112917.6-014232 265.21 55.03 112.38 -21.08 112644.0-012600	CGCG 11-109 PGC 35390	RL...P. F -2.0± .9		.04	15.2 ±.6				
112924.1+345216 184.07 70.84 76.72 -7.67 112643.9+350848	UGC 6491 PGC 35396	.S..8.. U 8.0± .8	1.19± .06 .07± .06 1.19	.00 .09 .04	15.0 ±.4 14.90			15.30±.1 163± 16 144± 12 .36	2530± 11 2528 2802
112927.5-014450 265.31 55.02 112.43 -21.05 112654.0-012818	CGCG 11-110 PGC 35400	.E..... F -5.0± .9		.04	15.3 ±.6				
112928.4+214656 224.47 70.62 89.00 -13.03 112651.0+220328	A 1126+22 MK 172 PGC 35403			.00					10074± 45 10025 10392
112929.1+240538 217.69 71.30 86.80 -12.12 112651.2+242210	NGC 3701 UGC 6493 PGC 35405	.S..4.. U 4.0± .8	1.28± .04 .32± .05 .89± .02 1.28	145 .00 .47 .16	13.48 ±.13 13.80 ±.18 13.09	.53± .02 -.09± .03 .45 -.15	.56± .02 -.07± .03 13.42± .05 13.91± .28	14.06±.2 274± 6 261± 6 .81	2806± 6 2802± 47 2765 3118
1129.6 +2207 223.53 70.77 88.69 -12.87 1127.0 +2224	A 1127+22 UGC 6495 PGC 35412	.SXS3*. U (1) 3.0±1.2 1.1± .9	.97± .07 .03± .05 .77± .04 .97	.00 .04 .01	14.52 ±.15 14.59 ±.18 14.46	.62± .04 .57	.70± .03 13.86± .13 14.16± .38	16.04±.1 121± 10 113± 10 1.57	6501± 8 6547± 59 6454 6820
112938.6+353053 182.17 70.68 76.14 -7.35 112658.2+354725	NGC 3700 UGC 6494 PGC 35413	PSBR2.. U 2.0± .9	1.02± .06 .15± .05 1.02	130 .00 .18 .07	14.88 ±.19				
1129.6 +2815 204.61 71.88 82.89 -10.39 1127.0 +2832	MCG 5-27- 80 PGC 35414	.S?.... 	.94± .11 .11± .07 .94	.00 .14 .06	15.03 ±.19 14.80				9801 9775 10099
112943.2-315103 283.33 27.89 144.98 -26.12 112715.1-313430	ESO 439- 13 PGC 35416	.S?.... 	.91± .05 .18± .04 .94	15 .29 .25 .09	15.13 ±.14 14.52				8977± 34 8783 9300
112944.1-362333 285.07 23.63 150.02 -26.26 112717.0-360700	NGC 3706 ESO 378- 6 PGC 35417	.LAT-.. RBCS -3.0± .5	1.48± .04 .22± .04 .83± .02 1.50	78 .38 .00	12.38 ±.13 12.23 ±.11 11.87	1.04± .01 .57± .01 .92 .48	1.06± .01 .60± .01 12.02± .07 14.10± .25		2977± 14 2776 3286
1129.7 +0429 258.63 60.14 106.06 -19.12 1127.2 +0446	CGCG 39-194 PGC 35419	.SBS1?. F 1.0±1.8		.09	15.2 ±.6				
1129.7 +0548 256.93 61.16 104.72 -18.70 1127.2 +0605	CGCG 39-195 PGC 35420	.SBR1?. F 1.0±1.8		.08	15.4 ±.6				
112947.9-371245 285.39 22.86 150.93 -26.25 112721.0-365612	ESO 378- 7 PGC 35421	.IBT9*P S (1) 9.7±1.0 6.7±1.2	1.04± .05 .23± .04 1.08	4 .45 .18 .12	15.10 ±.14				

R.A. 2000 DEC. / I b / SGL SGB / R.A. 1950 DEC.	Names / PGC	Type / S_T n_L / T / L	$logD_{25}$ / $logR_{25}$ / $logA_o$ / $logD_o$	p.a. / A_g / A_i / A_{21}	B_T / m_B / m_{FIR} / B_T^o	$(B-V)_T$ / $(U-B)_T$ / $(B-V)_T^o$ / $(U-B)_T^o$	$(B-V)_o$ / $(U-B)_o$ / m'_o / m'_{25}	m_{21} / W_{20} / W_{50} / HI	V_{21} / V_{opt} / V_{GSR} / V_{3K}
112951.5+245614 / 215.16 71.56 / 86.04 -11.70 / 112713.4+251246	UGC 6496 / PGC 35424	.L?.... / / /	.78? / .17± .08 / / .75	/ .00 / .00 /	15.0 ±.2 / / / 14.94				6277± 50 / 6239 . / / 6586
1129.9 +1625 / 238.29 68.31 / 94.21 -14.98 / 1127.3 +1642	/ PGC 35426			.09				18.89±.3 / 43± 13 / 34± 10 /	1067± 10 / 999 / / 1400
112957.2-314603 / 283.35 27.99 / 144.89 -26.06 / 112729.1-312930	ESO 439- 14 / PGC 35427	.SAR0*. / r / -.1±1.3 / 1.00	.99± .06 / .35± .05 / / 1.00	13 / .29 / .26 /	15.41 ±.14 / / / 14.72				9208± 34 / 9014 / / 9531
113004.7-113247 / 273.23 46.58 / 122.81 -23.32 / 112732.8-111615	NGC 3704 / MCG -2-29- 37 / PGC 35435	.E+.... / E / -4.0± .8 / 1.20	1.21± .09 / .07± .07 / / 1.20	150 / .04 / .00 /					
1130.0 +3837 / 173.72 69.49 / 73.33 -5.93 / 1127.4 +3854	UGC 6497 / IRAS11274+3854 / PGC 35437	.S..7*. / U / 7.0±1.3 / 1.36	1.36± .05 / .96± .06 / / 1.36	86 / .00 / 1.32 / .48	13.83				6324 / 6336 / / 6580
113006.8+091636 / 252.02 63.78 / 101.27 -17.48 / 112731.5+093309	NGC 3705 / UGC 6498 / IRAS11275+0933 / PGC 35440	.SXR2.. / R (3) / 2.0± .3 / 3.4± .5	1.69± .02 / .38± .03 / 1.02± .01 / 1.70	122 / .12 / .47 / .19	11.86M±.10 / 11.65 ±.12 / 11.47 / 11.18	.79± .02 / .14± .03 / .68 / .05	.84± .01 / / 12.36± .04 / 14.18± .15	13.10±.1 / 354± 16 / 345± 12 / 1.73	1017± 11 / 1054± 66 / 926 / 1365
1130.1 +2348 / 218.67 71.37 / 87.13 -12.09 / 1127.5 +2405	A 1127+24 / PGC 35442			.00					
113012.1+014939 / 261.93 58.08 / 108.81 -19.84 / 112738.0+020612	MCG 0-29- 38 / PGC 35447	.SAS5.. / F (1) / 5.0± .9 / 2.6±1.0	.74± .14 / .09± .07 / / .74	/ .04 / .14 / .05	15.49 ±.18				
113013.4-085148 / 271.44 48.98 / 119.96 -22.68 / 112741.0-083515	NGC 3702 / MCG -1-29- 26 / PGC 35448	.SXR1P* / EF / .6± .7 / 1.12	1.11± .06 / .28± .05 / / 1.12	30 / .04 / .29 / .14					
1130.2 +3551 / 181.07 70.67 / 75.88 -7.09 / 1127.6 +3608	UGC 6499 / PGC 35450	.I..9*. / U / 10.0±1.2 / 1.19	1.19± .12 / .44± .12 / / 1.19	5 / .00 / .33 / .22					2221 / 2223 / / 2489
1130.3 +5807 / 141.99 55.90 / 55.66 2.65 / 1127.5 +5824	A 1127+58 / MCG 10-17- 7 / PGC 35451	.SAS9*. / R / 9.0± .5 / .97	.97± .10 / .05± .07 / / .97	/ .00 / .05 / .02	14.54 ±.14 / 14.85 ±.18	.47± .04 / -.11± .07	/ / 14.11± .56 /		
113020.0-410357 / 286.86 19.26 / 155.19 -26.10 / 112754.0-404724	ESO 319- 26 / PGC 35453	.S..5./ / S / 5.3± .8 / 1.21	1.17± .05 / 1.02± .05 / / 1.21	69 / .42 / 1.50 / .50					
1130.3 +0441 / 258.61 60.39 / 105.90 -18.92 / 1127.8 +0458	CGCG 39-197 / PGC 35458	.SBT3.. / F (1) / 3.0± .9 / 4.6±1.0		.09	15.0 ±.6				
113027.3+364407 / 178.59 70.37 / 75.09 -6.68 / 112746.8+370040	A 1127+37 / CGCG 185- 74 / MK 424 / PGC 35464	.S?....	.87± .08 / .32± .12 / / .87	/ .00 / .48 / .16	15.26 ±.19 / / / 14.77			16.58±.3 / 136± 6 / 96± 5 / 1.64	1977± 10 / 1994± 35 / 1984 / 2243
1130.4 +0029 / 263.46 57.02 / 110.20 -20.17 / 1127.9 +0046	CGCG 11-112 / PGC 35465	.LA.-*. / F / -3.0±1.3		.06	15.13 ±.18				
113028.4+480627 / 154.44 63.77 / 64.72 -1.73 / 112744.6+482300	A 1127+48 / MK 173 / PGC 35467			.00					8400±110 / 8447 / / 8611
113029.1-130530 / 274.34 45.24 / 124.48 -23.54 / 112757.4-124857	IC 2889 / MCG -2-29- 38 / PGC 35469	.SXS5P* / EF (1) / 4.6± .6 / 5.3± .7	1.10± .05 / .23± .04 / / 1.10	150 / .05 / .35 / .12					

R.A. 2000 DEC.	Names	Type	logD$_{25}$	p.a.	B$_T$	(B-V)$_T$	(B-V)$_e$	m$_{21}$	V$_{21}$
l b	S$_T$ n$_L$	logR$_{25}$	A$_g$	m$_B$	(U-B)$_T$	(U-B)$_e$	W$_{20}$	V$_{opt}$
SGL SGB	T	logA$_e$	A$_l$	m$_{FIR}$	(B-V)$_T^o$	m'$_e$	W$_{50}$	V$_{GSR}$
R.A. 1950 DEC.	PGC	L	logD$_o$	A$_{21}$	B$_T^o$	(U-B)$_T^o$	m'$_{25}$	HI	V$_{3K}$
113040.0+440931		.S..2..	1.01± .06	178					
161.28 66.48	UGC 6500	U	.43± .05	.00	15.22 ±.18				
68.32 -3.43		2.0± .9		.53					
112757.5+442604	PGC 35476		1.01	.21					
113044.8-151721		PSBR3..	1.23± .04	80					6569
275.74 43.28	MCG -2-29- 39	EF (1)	.11± .04	.07					
126.86 -23.90	IRAS11282-1500	3.3± .5		.15	13.06				6406
112813.5-150048	PGC 35480	3.6± .5	1.24	.05					6924
1130.8 +6030		.S..7*.	.92± .09						3195
139.67 53.92	UGC 6501	U	.17± .06	.00					
53.53 3.75		7.0±1.2		.23					3285
1128.0 +6047	PGC 35486		.92	.08					3339
113100.5+202814	IC 701	.SBT8P.	.98± .07	47	15.1 ±.4	.45± .04	.47± .04	15.53±.2	6143± 7
228.53 70.46	UGC 6503	R	.19± .05	.00	14.67 ±.18	-.31± .07	-.30± .07		6182± 48
90.40 -13.21	ARP 197	8.0± .6		.23		.38		178± 7	6090
112823.4+204447	PGC 35494		.98	.09	14.50	-.36	14.38± .51	.94	6465
113106.3+030751		.SB.3./							
260.80 59.26	CGCG 39-200	F		.05	14.9 ±.6				
107.55 -19.23		3.0± .9							
112832.1+032424	PGC 35499								
113106.9+224604	NGC 3710	.E.....	1.00± .19	105	14.06 ±.14	.99± .01	1.00± .01		
221.99 71.30	UGC 6504	U	.09± .08	.00	14.33 ±.15	.47± .03	.54± .02		6490± 31
88.22 -12.30		-5.0± .9	.59± .03	.00		.93	12.50± .11		6445
112829.4+230237	PGC 35502		.97		14.09	.51	13.81± .99		6805
1131.1 +5142		.SX.7*.	1.02± .06						
149.09 61.15	UGC 6505	U	.02± .05	.00					
61.51 -.05		7.0± .9		.03					
1128.4 +5159	PGC 35506		1.02	.01					
113109.2+283405	NGC 3712	.SB?...	1.22± .04	160					1583± 10
203.62 72.22	UGC 6506		.44± .05	.00	14.9 ±.3				
82.74 -9.96	ARP 203			.65					1559
112830.6+285038	PGC 35507		1.22	.22	14.23				1879
1131.2 +7716		.S..6*.	1.11± .14	43					
128.51 38.97	UGC 6502	U	1.13± .06	.08					
38.20 10.85		6.0±1.4		1.47					
1128.0 +7733	PGC 35510		1.12	.50					
1131.2 +2618		.S..3..	.95± .07						9882
210.99 72.09	UGC 6508	U (1)	.08± .05	.00	15.0 ±.2				
84.87 -10.86		3.0± .9		.11					9850
1128.6 +2635	PGC 35513	3.5±1.1	.95	.04	14.82				10186
113117.0+341220		.SBS5..	1.12± .05	175				15.43±.3	6309± 7
185.67 71.41	UGC 6507	U	.17± .05	.00	15.15 ±.19			256± 15	
77.50 -7.60		5.0± .8		.25					6305
112837.2+342853	PGC 35517		1.12	.08	14.86			.48	6584
1131.3 +2306		.S..7..	1.32± .04	79	15.5 ±.2	.68± .11		15.45±.3	2905± 10
221.02 71.45	UGC 6509	U	1.12± .05	.00		.15± .13		191± 13	
87.91 -12.12		7.0±1.0		1.38		.45		175± 10	2861
1128.7 +2323	PGC 35521		1.32	.50	14.11	-.03	14.15± .31	.84	3219
113127.2-372622		.SXR6..	.92± .06						
285.81 22.76	ESO 378- 9	r	.08± .05	.41	15.28 ±.14				
151.18 -25.92		5.6± .9		.12					
112900.0-370948	PGC 35531		.96	.04					
113128.6-015034		.LAR-..							
266.14 55.21	CGCG 11-113	F		.10	15.3 ±.6				
112.68 -20.59		-3.0± .9							
112855.1-013400	PGC 35533								
1131.4 +0604		PSXR2..	.82± .13						
257.25 61.63	MCG 1-29- 48	F	.08± .07	.08	14.92 ±.18				
104.60 -18.22		2.0± .9		.09					
1128.9 +0621	PGC 35534		.82	.04					
113131.8-021831		.SXT6..	1.28± .03	10	13.0 ±.2	.52± .04	.60± .02	14.47±.1	4742± 7
266.59 54.82	UGC 6510	UE F (1)	.03± .04	.09	13.2 ±.2			119± 12	
113.17 -20.71	IRAS11289-0202	5.5± .4	1.20± .06	.04	13.46	.46	14.45± .17		4614
112858.3-020157	PGC 35538	5.0± .5	1.29	.01	12.96		14.18± .27	1.50	5100
113131.8-301828	NGC 3717	.SA.3*/	1.78± .02	33	12.24 ±.16	1.04± .05	1.07± .05	13.16±.1	1733± 4
283.13 29.47	ESO 439- 15	R (1)	.73± .03	.23	12.23 ±.10	.45± .12	.50± .12	433± 4	1724± 30
143.30 -25.64	IRAS11290-3001	3.0± .5	1.06± .04	1.00	10.48	.85	13.03± .11	410± 4	1541
112903.0-300154	PGC 35539	5.0±1.0	1.80	.36	10.98	.24	14.17± .19	1.82	2060

R.A. 2000 DEC.	Names	Type	$\log D_{25}$	p.a.	B_T	$(B-V)_T$	$(B-V)_\bullet$	m_{21}	V_{21}
l　　b		S_T　n_L	$\log R_{25}$	A_g	m_B	$(U-B)_T$	$(U-B)_\bullet$	W_{20}	V_{opt}
SGL　SGB		T	$\log A_\bullet$	A_i	m_{FIR}	$(B-V)_T^o$	$m'_\bullet$	W_{50}	V_{GSR}
R.A. 1950 DEC.	PGC	L	$\log D_o$	A_{21}	B_T^o	$(U-B)_T^o$	m'_{25}	HI	V_{3K}

113132.3-141357	NGC 3715	PSBT4*.	1.13± .05	145					
275.36 44.32	MCG -2-29- 41	EF　　(1)	.18± .04	.05					
125.76 -23.51	IRAS11290-1357	4.0± .6		.27	11.98				
112900.7-135723	PGC 35540	2.4± .8	1.14	.09					
1131.6 -0337		.SAT4?.	.94± .11						
267.81 53.72	MCG 0-30- 2	F　　(1)	.40± .07	.12	15.42 ±.18				
114.55 -21.03		4.0±1.3		.59					
1129.1 -0321	PGC 35544	3.6±1.4	.95	.20					
113140.6+032923	NGC 3716	RLX.0?.	.87± .07	150				16.61±.3	6628± 10
260.60 59.64	UGC 6513	UF	.07± .03	.04	14.54 ±.17			597± 13	6636± 31
107.23 -18.98		-1.5±1.3		.00					6518
112906.3+034556	PGC 35545		.87		14.40				6983
113142.1+280914	NGC 3713	.L..-*.	1.08± .09	125					
204.98 72.33	UGC 6511	U	.17± .04	.00	14.20 ±.15				6989± 31
83.18 -10.02		-3.0±1.2		.00					6963
112903.7+282548	PGC 35546		1.05		14.10				7287
113144.7+342000		.S?....	1.06± .04	150				14.82±.3	1870± 10
185.20 71.46	UGC 6512		.16± .04	.00	15.0 ±.2				1867
77.42 -7.46	KUG 1129+346			.24				156± 7	1867
112905.0+343634	PGC 35549		1.06	.08	14.74			.00	2144
113150.8-302434	IC 2913	.LAR+*.	.90± .06						
283.24 29.40	ESO 439- 16	r	.01± .04	.33	13.75 ±.14				1770± 60
143.42 -25.58	IRAS11293-3008	-1.3±1.3		.00	12.59				1578
112922.0-300800	PGC 35554		.93		13.39				2097
113153.5+282126	NGC 3714	.I?....	.73± .08	68					
204.31 72.38	UGC 6516		.16± .05	.00	15.1 ±.3				6974± 39
83.00 -9.90	ARAK 297			.12					6950
112915.1+283800	PGC 35556		.73	.08	14.95				7271
113154.7-094332	IC 2910	PLBT+P*	1.22± .04	130					
272.60 48.41	MCG -1-30- 1	EF	.16± .05	.06					
120.98 -22.48		-.8± .5		.00					
112922.4-092659	PGC 35557		1.20						
1131.9 +2700									9849
208.76 72.32	CGCG 156- 97			.00	15.3 ±.3				9819
84.28 -10.44									10151
1129.3 +2717	PGC 35558								
113157.4-025522		.LA.+*/	.82± .13						
267.30 54.36	MCG 0-30- 3	F	.46± .07	.13	15.3 ±.3				
113.84 -20.77	IRAS11293-0238	-1.0±1.3		.00	13.44				
112924.0-023848	PGC 35560		.76						
113158.8-000258		.SAS6*.	.94± .11						
264.59 56.79	MCG 0-30- 4	F　　(1)	.28± .07	.06	15.23 ±.18				
110.87 -19.96		6.0±1.3		.42					
112925.0+001336	PGC 35563	5.6±1.4	.94	.14					
113202.5-320440		.S?....	1.02± .06	171					
283.94 27.85	ESO 439- 17		.46± .05	.29	16.09 ±.14				3849± 34
145.26 -25.64				.63					3655
112934.1-314806	PGC 35567		1.05	.23	15.14				4171
113203.1+364154		.S..4..	1.23± .06	40				15.01±.3	2491± 9
178.31 70.68	UGC 6517	U　　(1)	.25± .06	.00	13.86 ±.18				2477± 50
75.26 -6.40	IRAS11293+3658	4.0± .8		.37				244± 5	2497
112922.9+365828	PGC 35569	4.5±1.1	1.23	.13	13.47			1.41	2755
113205.5+704824		.E...P*							
131.99 44.90	MCG 12-11- 28B	RC		.02					15480± 60
44.20 8.26	HICK 55B	-5.0± .5							15603
112907.1+710457	PGC 35572								15563
113205.7+704840		.SB.1P*							
131.99 44.90	MCG 12-11- 28C	RC		.02					15571± 34
44.20 8.26	HICK 55C	1.0± .5							15694
112907.2+710514	PGC 35573								15654
113206.9+704917		.L...P*							
131.98 44.89	MCG 12-11- 28D	RC		.02					15891± 34
44.19 8.27	HICK 55D	-2.0± .7							16014
112908.4+710551	PGC 35574								15973
113207.0+704856		.LA..P*		15.93S±.15					
131.99 44.89	MCG 12-11- 28A	RC		.02					15649± 32
44.20 8.27	HICK 55A	-2.0± .5							15772
112908.5+710529	PGC 35575								15732

R.A. 2000 DEC. / l b / SGL SGB / R.A. 1950 DEC.	Names / PGC	Type / S_T n_L / T / L	logD$_{25}$ / logR$_{25}$ / logA$_\bullet$ / logD$_o$	p.a. / A$_g$ / A$_i$ / A$_{21}$	B$_T$ / m$_B$ / m$_{FIR}$ / B^o_T	(B-V)$_T$ / (U-B)$_T$ / (B-V)^o_T / (U-B)^o_T	(B-V)$_\bullet$ / (U-B)$_\bullet$ / m' / m'$_{25}$	m$_{21}$ / W$_{20}$ / W$_{50}$ / HI	V$_{21}$ / V$_{opt}$ / V$_{GSR}$ / V$_{3K}$
113207.6+704908		.S?....							
131.98 44.89	MCG 12-11- 28E			.02					36880± 60
44.19 8.27	HICK 55E								37003
112909.1+710542	PGC 35576								36963
113208.4-412540		.S..4*/	1.18± .04	29					
287.33 19.03	ESO 320- 2	S	.80± .04	.41	15.57 ±.14				
155.58 -25.75	IRAS11296-4109	4.0±1.3		1.18	13.29				
112942.0-410906	PGC 35577		1.22	.40					
113208.7-005634		.LX.0*.							
265.54 56.06	CGCG 12- 5	F		.06	15.13 ±.18				
111.81 -20.18		-2.0± .9							
112935.0-004000	PGC 35578								
1132.1 +0113		.S..7*/	1.07± .07	93					
263.34 57.87	UGC 6519	UF	.91± .06	.06	15.68 ±.19				
109.58 -19.55		7.0± .8		1.26					
1129.6 +0130	PGC 35579		1.08	.46					
113213.5+004914	NGC 3719	.SAT4P*	1.25± .03	15	13.64 ±.13	.60± .02	.68± .02		
263.79 57.54	UGC 6521	PUE F (2)	.14± .03	.08	13.38 ±.18	.01± .12	.11± .12		5889± 33
110.00 -19.65		3.6± .3	.84± .01	.21		.52	13.33± .02		5770
112939.6+010548	PGC 35581	3.1± .5	1.26	.07	13.22	-.05	14.37± .20		6246
1132.3 +2802		.L...?.	1.15± .15						
205.37 72.47	UGC 6522	U	.00± .08	.00	14.1 ±.2				6958
83.34 -9.94		-2.0±1.6		.00					6932
1129.7 +2819	PGC 35588		1.15		14.03				7256
113222.1+004817	NGC 3720	.SA.1*.	.98± .05	85	13.71 ±.13	.76± .02	.69± .01	15.22±.3	5818± 17
263.86 57.55	UGC 6523	PEF (2)	.04± .04	.08	13.73 ±.14	-.04± .03	.01± .03	393± 34	5958± 30
110.03 -19.62	IRAS11297+0104	.9± .7	.57± .02	.04	11.99	.68	12.05± .06	325± 25	5734
112948.2+010451	PGC 35594	2.1± .7	.99	.02	13.52	-.05	13.36± .29	1.68	6209
1132.3 +0510		.SA.2?/							
258.80 61.08	CGCG 40- 2	F		.05	15.1 ±.6				
105.58 -18.29		2.0±1.8							
1129.8 +0527	PGC 35598								
113227.5+623026	A 1129+62	.SB..$.	1.12± .05	165					
137.71 52.30	UGC 6520	R	.25± .05	.00	14.12 ±.18				3660± 72
51.80 4.79	MK 175			.38					3757
112937.5+624700	PGC 35601		1.12	.13	13.72				3792
1132.5 +2026		RSBR3*.	1.13± .05					15.68±.2	6052± 7
229.04 70.78	UGC 6525	U	.03± .05	.00	14.2 ±.2			115± 6	
90.57 -12.90		3.0± .8		.05				99± 6	5999
1129.9 +2043	PGC 35607		1.13	.02	14.11			1.54	6374
113232.3+743748	NGC 3752	.S..2..	1.23± .04	155					
129.78 41.43	UGC 6515	U	.37± .04	.15	13.75 ±.18				1917± 42
40.69 9.86	IRAS11293+7454	2.0± .8		.45	13.27				2051
112927.6+745422	PGC 35608		1.24	.18	13.13				1976
113232.8+525622		.L?....			16.37S±.15				
147.18 60.32				.00					7884± 97
60.49 .67	HICK 56E								7949
112947.9+531255	PGC 35609								8070
113235.4+525650		.L?....							
147.16 60.31				.00					8346± 41
60.49 .68	HICK 56D								8411
112950.5+531324	PGC 35615								8532
113235.7+530359	NGC 3718	.SBS1P.	1.91± .02	15	11.59M±.10	.81± .02	.92± .01	12.43±.0	994± 3
147.01 60.22	UGC 6524	R	.31± .03	.00	11.40 ±.14	.29± .03	.40± .02	470± 3	1031± 40
60.38 .74	ARP 214	1.0± .3	1.30± .02	.32	13.12	.74	13.61± .04	455± 3	1059
112950.8+532033	PGC 35616		1.91	.16	11.19	.24	15.19± .15	1.08	1180
113236.7+525651		.L?....			15.82M±.11				
147.15 60.32				.00					8110± 41
60.49 .69	HICK 56C								8175
112951.9+531325	PGC 35618								8296
113238.9+525653		.S?....	1.05± .05		16.22S±.15				
147.14 60.32	UGC 6527		.53± .06	.00				16.65±.1	8208± 12
60.49 .69	MK 176			.80				478± 21	7929± 17
112954.0+531327	PGC 35620		1.05	.27	14.17		13.80± .31	2.21	8180 / 8301
113239.1+351942		.S?....	1.20± .04	90					
182.06 71.31	UGC 6526		.71± .05	.00	14.3 ±.2			16.09±.3	1907± 9
76.58 -6.87	KUG 1129+356			1.05					1851± 50
112959.4+353616	PGC 35621		1.20	.36	13.20			148± 5 / 2.54	1906 / 2176

R.A. 2000 DEC. / l b / SGL SGB / R.A. 1950 DEC.	Names / PGC	Type / S_T n_L / T / L	logD₂₅ / logR₂₅ / logA_• / logD_•	p.a. / A_g / A_i / A₂₁	B_T / m_B / m_FIR / B°_T	(B-V)_T / (U-B)_T / (B-V)°_T / (U-B)°_T	(B-V)_• / (U-B)_• / m'_• / m'₂₅	m₂₁ / W₂₀ / W₅₀ / HI	V₂₁ / V_opt / V_GSR / V_3K
1132.7 +2001		.S?....	.85± .12					16.08±.3	6122± 10
230.25 70.64	MCG 3-30- 3		.20± .07	.01	15.42 ±.18			306± 13	
90.99 -13.01				.29				256± 10	6068
1130.1 +2018	PGC 35622		.85	.10	15.07			.91	6445
1132.7 +6225		.S?....	.92± .11						3650± 53
137.72 52.38	MCG 11-14- 29		.55± .07	.00	15.4 ±.2				
51.88 4.79	IRAS11298+6242			.83	13.32				3747
1129.8 +6242	PGC 35623		.92	.28	14.53				3783
113243.6-012558		RLBR+..							
266.22 55.73	CGCG 12-14	F		.07	15.1 ±.6				
112.35 -20.18		-1.0± .9							
113010.0-010924	PGC 35625								
113244.5+614940		.SAT5..	1.04± .06					16.19±.1	3250± 5
138.21 52.90	UGC 6528	U	.02± .05	.00	14.07 ±.18			49± 5	
52.44 4.53		5.0± .8		.03				30± 8	3344
112955.0+620614	PGC 35626		1.04	.01	14.02			2.16	3386
1132.7 -0211		.LX.-*.							
266.94 55.08	CGCG 12-13	F		.09	14.92 ±.18				
113.14 -20.38		-3.0± .9							
1130.2 -0155	PGC 35627								
113245.6+405034		.S..4..	1.01± .05	70					
167.64 68.78	UGC 6529	U	.12± .04	.00	15.1 ±.2				
71.52 -4.52	KUG 1130+411	4.0± .9		.18					
113004.7+410708	PGC 35629		1.01	.06					
113245.7-004428		.SBS9..	.94± .11						
265.57 56.31	MCG 0-30- 7	F (1)	.11± .07	.06	15.1 ±.2				
111.64 -19.97		9.0± .9		.12					
113012.0-002754	PGC 35630	7.6±1.0	.94	.06					
113246.7+525628		.S?....	1.04± .09		16.21S±.15				8245± 41
147.12 60.34	MCG 9-19-113		.78± .07	.00					8310
60.51 .71	HICK 56A			1.14					8431
113001.9+531302	PGC 35631		1.04	.39	15.03		14.30± .52		
113249.1+390506		.SBR8..	1.26± .04					15.27±.1	1565± 11
171.83 69.73	UGC 6531	U	.00± .05	.01	14.9 ±.5			90± 16	
73.13 -5.26		8.0± .8		.00				77± 12	1580
113008.5+392140	PGC 35634		1.26	.00	14.92			.35	1819
1132.8 +0543		.S..3..	.96± .09	178					
258.28 61.59	UGC 6533	U (1)	.69± .06	.09	15.5 ±.2				
105.07 -18.00	IRAS11302+0600	3.0±1.0		.95	12.83				
1130.3 +0600	PGC 35639	4.5±1.3	.97	.34					
113252.7+433054		.S?....	.74± .14						6864± 82
161.90 67.20	MCG 7-24- 21		.38± .07	.07	15.2 ±.3				6895
69.08 -3.35	ARAK 299			.57					7098
113011.1+434728	PGC 35641		.74	.19	14.56				
113253.1+014844		.SAT5..	.74± .14						
262.99 58.45	MCG 0-30- 8	F (1)	.00± .07	.07	15.22 ±.18				
109.04 -19.20		5.0± .9		.00					
113019.0+020518	PGC 35642	3.6±1.4	.74	.00					
1133.1 -1321	IC 706	.SAR0?/	1.12± .08						
275.33 45.27	MCG -2-30- 4	F	.74± .07	.05					
124.91 -22.96		.0±1.3		.56					
1130.6 -1305	PGC 35658		1.08						
1133.1 +7023		.SBR6*.	.97± .07						
132.13 45.31	UGC 6532	U	.08± .05	.01	15.09 ±.20				
44.62 8.17		6.0±1.2		.12					
1130.2 +7040	PGC 35661		.98	.04					
113310.5-101344		.SX.3*/	1.11± .06	175					
273.36 48.10	MCG -2-30- 3	EF (1)	.65± .05	.04					
121.58 -22.29	IRAS11306-0957	2.7± .8		.90					
113038.2-095709	PGC 35664	4.2±1.6	1.12	.32					
1133.2 +2426	NGC 3728	.S..3..	1.31± .04	25					6979± 10
217.24 72.21	UGC 6536	U (1)	.14± .05	.00	13.8 ±.3				
86.82 -11.19		3.0± .8		.19					6941
1130.6 +2443	PGC 35669	3.5±1.0	1.31	.07	13.55				7289
1133.2 +5018		.S..6*.	1.06± .06	67					
150.44 62.44	UGC 6535	U	.93± .05	.00					
62.93 -.37		6.0±1.4		1.36					
1130.5 +5035	PGC 35671		1.06	.46					

R.A. 2000 DEC. / l b / SGL SGB / R.A. 1950 DEC.	Names / PGC	Type / S_T n_L / T / L	$\log D_{25}$ / $\log R_{25}$ / $\log A_e$ / $\log D_o$	p.a. / A_g / A_i / A_{21}	B_T / m_B / m_{FIR} / B_T^o	$(B-V)_T$ / $(U-B)_T$ / $(B-V)_T^o$ / $(U-B)_T^o$	$(B-V)_e$ / $(U-B)_e$ / m'_e / m'_{25}	m_{21} / W_{20} / W_{50} / HI	V_{21} / V_{opt} / V_{GSR} / V_{3K}
113320.7+631649 136.95 51.67 51.14 5.21 113030.6+633323	A 1130+63 UGC 6534 IRAS11304+6333 PGC 35675	.S..7.. U 7.0± .8	1.40± .04 .61± .05 1.41	60 .07 .84 .30	 13.38 ±.18 13.37 12.47			14.58±.1 142± 16 137± 12 1.80	1273± 11 1373 1401
113320.9+470139 155.38 64.88 65.92 -1.77 113038.4+471813	NGC 3726 UGC 6537 IRAS11306+4718 PGC 35676	.SXR5.. R (2) 5.0± .3 2.2± .5	1.79± .02 .16± .02 1.79	10 .01 .24 .08	10.91S±.07 11.02 ±.14 11.34 10.67	.49± .05 .45	 14.30± .12	12.17±.1 284± 4 257± 4 1.41	849± 4 948± 56 894 1066
113322.9+550420 144.51 58.66 58.61 1.71 113037.4+552054	A 1130+55 UGCA 239 MK 177 PGC 35678	.P..... R 99.0	.51± .11 .15± .07 .51	 .00 .22 .08	15.8 ±.2 15.57	.46± .06 -.13± .10 .42 -.16	 12.84± .62		1800±110 1872 1974
113323.6+323613 190.27 72.25 79.17 -7.87 113044.6+325247	 UGC 6539 PGC 35679	.S..4.. U 4.0± .9	1.14± .05 .62± .05 1.14	25 .02 .92 .31	 15.25 ±.18 14.27			15.64±.3 345± 15 1.05	6260± 11 6253± 57 6251 6541
113325.5-022259 267.36 55.01 113.39 -20.27 113052.0-020624	 MCG 0-30- 9 PGC 35682	.SAT5*. F (1) 5.0± .9 5.6±1.0	.94± .11 .00± .07 .95	 .10 .00 .00	 14.8 ±.2				
113328.5+491408 151.91 63.28 63.92 -.80 113045.2+493043	A 1130+49 UGC 6541 MK 178 PGC 35684	.I..9.. U 10.0± .9 1.09	1.09± .06 .27± .05 .67± .02 1.09	133 .00 .20 .13	14.50 ±.14 13.97 ±.19 14.11	.35± .03 -.30± .04 .28 -.35	.26± .02 -.46± .03 13.34± .04 14.14± .33	16.27±.1 47± 4 31± 5 2.03	246± 5 309± 18 302 456
113329.8-265653 282.22 32.77 139.65 -24.93 113100.1-264018	 ESO 503- 22 PGC 35686	.IBS9*P S (1) 10.0± .8 5.6± .9	1.14± .04 .35± .03 1.16	125 .24 .26 .17	 14.81 ±.14 14.31				1911± 34 1726 2246
113330.0+341859 184.90 71.81 77.59 -7.14 113050.7+343533	IC 2928 UGC 6540 PGC 35687	.S..4.. U 4.0± .9	1.04± .06 .09± .05 1.04	155 .04 .14 .05	 14.52 ±.18 14.29			16.50±.3 200± 14 189± 15 2.16	8057± 12 8085± 19 8063 8339
1133.5 +0023 264.75 57.37 110.54 -19.46 1131.0 +0040	 CGCG 12- 21 PGC 35690	RLXR+.. F -1.0± .9		 .05	 15.3 ±.6				
113333.9-254541 281.74 33.88 138.35 -24.79 113104.0-252906	 ESO 503- 23 PGC 35691	.LXR0.. r -2.4± .9	.87± .06 .16± .05 .87	108 .19 .00	 15.27 ±.14				
113334.7-021649 267.33 55.12 113.29 -20.21 113101.2-020014	 MCG 0-30- 10 ARAK 300 PGC 35692	.S..2?. F 2.0±1.9	.64± .17 .11± .07 .65	 .10 .14 .06	 15.4 ±.3 12.86				
1133.6 +2324 220.57 72.04 87.84 -11.52 1131.0 +2341	 UGC 6544 PGC 35694	.S..4.. U 4.0± .9	1.04± .06 .29± .05 1.04	95 .00 .43 .15	 14.85 ±.18				
113340.8+615319 138.00 52.91 52.43 4.66 113051.8+620953	NGC 3725 UGC 6542 MK 179 PGC 35698	.SB.5.. U 5.0± .8	1.07± .08 .10± .06 1.07	145 .00 .15 .05	 13.71 ±.19 13.00 13.55			16.27±.2 163± 34 178± 25	3334± 9 3232± 55 3426 3467
113340.9-444959 288.73 15.88 159.31 -25.35 113115.0-443324	 ESO 266- 3 PGC 35699	.LA.-.. S -3.0± .9	1.09± .05 .19± .04 1.09	105 .25 .00	 14.44 ±.14				
1133.7 +1723 237.38 69.59 93.62 -13.78 1131.1 +1740	 UGC 6546 PGC 35701	.S..4.. U 4.0±1.0	1.05± .06 .89± .05 1.05	7 .02 1.30 .44	15.69 ±.15 15.77 ±.20 14.36	.76± .05 -.04± .08 .55 -.21	 13.56± .35	16.97±.3 322± 13 284± 10 2.17	5761± 10 5698 6091
1133.7 -1546 276.91 43.14 127.53 -23.29 1131.2 -1530	 PGC 35704	.S..5/. E 5.0± .9	1.12± .08 .71± .08 1.13	135 .09 1.06 .35					
113343.9-032602 268.40 54.14 114.50 -20.48 113110.6-030927	 MCG 0-30- 11 PGC 35705	.SXS8.. EF (1) 8.0± .6 7.0± .7	1.10± .06 .15± .05 1.12	55 .14 .18 .07	 14.6 ±.3 14.32				1616 1486 1975

R.A. 2000 DEC. l　　b SGL　SGB R.A. 1950 DEC.	Names PGC	Type S_T　n_L T L	$\log D_{25}$ $\log R_{25}$ $\log A_e$ $\log D_o$	p.a. A_g A_i A_{21}	B_T m_B m_{FIR} B_T^o	$(B-V)_T$ $(U-B)_T$ $(B-V)_T^o$ $(U-B)_T^o$	$(B-V)_e$ $(U-B)_e$ m'_e m'_{25}	m_{21} W_{20} W_{50} HI	V_{21} V_{opt} V_{GSR} V_{3K}
113344.6+323804 190.12　72.31 79.17　-7.79 113105.6+325438	 UGC　6545 PGC 35707	.S?.... 	.99± .06 .47± .05 1.00	133 .01 .71 .24	 14.7　±.2 13.93				2619± 50 2611 2900
113344.6+212250 226.72　71.41 89.78　-12.27 113107.7+213925	IC　　707 UGC　6543 ARAK　301 PGC 35708	.S?.... 	.74± .09 .09± .05 .74	 .00 .13 .04	14.3　±.3 15.0　±.3 13.16 14.53	.47± .02 -.12± .05 .41 -.16	 12.63± .56	16.59±.3 238± 13 2.01	6575± 10 6659± 63 6528 6896
1133.8　+2338 219.88　72.15 87.64　-11.38 1131.2　+2355	 UGC　6548 PGC 35710	.L..... U -2.0± .9 	1.01± .08 .51± .03 .93	40 .00 .00 					
113349.8+530737 146.64　60.28 60.41　　.93 113105.4+532411	NGC　3729 UGC　6547 IRAS11310+5324 PGC 35711	.SBR1P. R 1.0± .3 	1.45± .02 .17± .03 1.45	15 .00 .17 .09	12.03S±.07 12.43 ±.13 11.83 11.94	.60± .05 .56 	 13.69± .14	14.28±.3 214± 21 201± 15 2.26	1024±　7 1005± 29 1089 1208
113359.8+490340 152.02　63.46 64.11　　-.79 113116.8+492015	IC　　708 UGC　6549 PGC 35720	.E..... U -5.0± .8 	1.15± .15 .19± .08 1.09	95 .00 .00 	 13.96 ±.15 13.81				9591± 18 9643 9797
113407.1+364102 177.82　71.07 75.46　-6.03 113127.4+365737	 UGC　6551 KUG 1131+369 PGC 35725	.S..7.. U 7.0±1.0 	1.15± .04 .86± .04 1.15	113 .00 1.19 .43				15.46±.3 430± 14 273± 15 	6444± 12 6451 6708
113409.6-013541 266.92　55.78 112.63　-19.88 113136.1-011906	 CGCG　12- 24 PGC 35729	PSBR0.. F .0± .9 		 .08	 15.2　±.6 				
113411.6+123044 248.08　66.74 98.43　-15.42 113136.1+124719	NGC　3731 UGC　6553 PGC 35731	.E..... U -5.0± .8 	1.00± .19 .05± .08 1.00	50 .10 .00 					3212± 31 3132 3553
113412.8+341845 184.76　71.95 77.66　-7.00 113133.6+343520	IC　2933 UGC　6550 PGC 35732	.S..2.. U 2.0± .9 	1.10± .05 .54± .05 1.10	8 .04 .66 .27	 15.10 ±.18 				
113414.0-095045 273.46　48.56 121.24　-21.95 113141.6-093410	NGC　3732 MCG -2-30- 5 IRAS11316-0934 PGC 35734	.SXS0*. R .0± .6 	1.09± .04 .02± .04 .50± .01 1.09	85 .04 .02 	12.50V±.13 13.27 ±.19 11.44 13.06	.57± .01 -.17± .02 .54 -.17	 13.29± .25	16.22±.1 133±　6 	1692±　5 1682± 66 1545 2051
113414.5+490234 151.98　63.50 64.15　　-.77 113131.6+491909	IC　　709 MCG　8-21- 57 PGC 35736	.E?.... 		 .00 	 14.90 ±.18 				9549± 19 9600 9755
113420.2-062328 271.02　51.64 117.62　-21.10 113147.3-060653	 MCG -1-30- 2 IRAS11317-0606 PGC 35742	.SA.5*/ EF 5.0± .8 	1.18± .05 .76± .05 1.19	162 .08 1.14 .38	 13.32 				5012 4874 5371
1134.3　-0935 273.32　48.80 120.98　-21.86 1131.8　-0919	 MCG -1-30- 4 PGC 35744	.SB.8?/ F 8.0±1.3 	1.04± .09 .66± .07 1.04	 .04 .82 .33					
113423.5-023141 267.85　55.01 113.61　-20.08 113150.1-021506	 MCG　0-30- 13 PGC 35747	.SBR2.. F 2.0± .9 	.89± .11 .07± .07 .90	 .09 .08 .03	 14.88 ±.18 14.60				11591± 54 11464 11949
113427.6+331045 188.26　72.33 78.73　-7.43 113148.7+332720	 MCG　6-26- 5 PGC 35754	.L?.... 	.72± .22 .00± .07 .72	 .00 .00 	 15.02 ±.17 14.99				2583± 19 2577 2862
113428.7-093935 273.41　48.75 121.06　-21.85 113156.3-092300	NGC　3724 MCG -1-30- 7 IRAS11319-0922 PGC 35757	.L..+*/ F -1.0±1.3 	1.20± .06 .37± .05 1.14	124 .04 .00 	 12.82 				
113429.5-501223 290.54　10.80 165.16　-24.84 113205.0-495548	 ESO　216- 16 PGC 35759	.L2.... 	.89± .07 .35± .05 .90	0 .63 .00 	 15.42 ±.14 14.72				4481± 87 4272 4738

R.A. 2000 DEC.	Names	Type	logD$_{25}$	p.a.	B$_T$	(B-V)$_T$	(B-V)$_\bullet$	m$_{21}$	V$_{21}$
l b		S$_T$ n$_L$	logR$_{25}$	A$_g$	m$_B$	(U-B)$_T$	(U-B)$_\bullet$	W$_{20}$	V$_{opt}$
SGL SGB		T	logA$_\bullet$	A$_I$	m$_{FIR}$	(B-V)$_T^o$	m'$_\bullet$	W$_{50}$	V$_{GSR}$
R.A. 1950 DEC.	PGC	L	logD$_o$	A$_{21}$	B$_T^o$	(U-B)$_T^o$	m'$_{25}$	HI	V$_{3K}$

113433.9+713226		.S?....	1.06± .06	46					
131.31 44.32	UGC 6552		.36± .05	.04	14.41 ±.19				2807± 50
43.61 8.74				.55					2932
113136.6+714901	PGC 35767		1.06	.18	13.81				2885
113438.8+000731		.SAR0*.							
265.44 57.30	CGCG 12- 27	F		.07	14.1 ±.4				
110.90 -19.28		.0±1.3							
113205.0+002406	PGC 35772								
113440.7-140456	NGC 3734	PSXT4*.	1.13± .08	85					
276.24 44.78	MCG -2-30- 6	EF (1)	.11± .05	.04					
125.77 -22.74		3.5± .6		.16					
113208.9-134821	PGC 35773	2.0± .8	1.14	.06					
113443.8-381459		.SBS5*.	1.11± .04	53					
286.75 22.20	ESO 320- 4	S (1)	.45± .04	.46	14.67 ±.14				
152.08 -25.28	IRAS11323-3758	5.3±1.0		.68					
113216.1-375824	PGC 35775	6.7±1.3	1.15	.23					
113444.0-371253		.S..5*/	1.20± .04	160					
286.40 23.18	ESO 378-11	S (1)	.84± .04	.41	15.23 ±.14				
150.94 -25.27	IRAS11322-3656	5.0±1.3		1.26	13.55				
113216.1-365618	PGC 35776	5.6±1.4	1.24	.42					
113446.2+025513		.SBR3*.							
262.50 59.63	CGCG 40- 7	F (1)		.03	14.81 ±.18				
108.06 -18.42		3.0± .9							
113212.0+031148	PGC 35779	2.6±1.0							
113446.6+485723	IC 711	.E?....							
151.96 63.63	MCG 8-21- 62								9724± 19
64.26 -.72				.00	15.11 ±.18				9775
113203.9+491358	PGC 35780								9931
113446.9+001407		.E....							
265.38 57.41	CGCG 12- 29	F		.07	14.81 ±.18				
110.80 -19.21		-5.0± .9							
113213.1+003042	PGC 35781								
113449.4+490440	IC 712	.S?....	1.04± .09						
151.77 63.54	MCG 8-21- 63		.21± .07	.00	14.67 ±.18				10047± 18
64.16 -.66				.16					10099
113206.7+492115	PGC 35785		1.03		14.36				10253
113455.7-383106		.SXR0..	.96± .06	107					
286.89 21.96	ESO 320- 5	r	.18± .05	.46	14.65 ±.14				
152.37 -25.25		-.1± .9		.14					
113228.0-381430	PGC 35789		.99						
1135.0 +1606		.S..7..	1.22± .05	91					
240.90 69.14	UGC 6556	U	.88± .05	.06	15.38 ±.18				
94.99 -13.96		7.0±1.0		1.22					
1132.4 +1623	PGC 35792		1.23	.44					
113500.2+511308		.L..-*.	1.01± .12	90					
148.74 61.91	UGC 6555	U	.09± .05	.00	14.27 ±.15				8154± 31
62.22 .28		-3.0±1.2		.00					8213
113216.8+512943	PGC 35793		.99		14.14				8349
113501.8+545108	NGC 3733	.SXS6*.	1.68± .03	170	12.93 ±.15	.54± .07	.62± .03	12.98±.1	1185± 6
144.38 58.98	UGC 6554	PU	.34± .04	.00	12.8 ±.3	-.20± .09	-.15± .04	241± 7	
58.91 1.83	VV 459	6.0± .5	1.24± .02	.51		.47	14.62± .04	231± 6	1257
113217.1+550743	PGC 35797		1.68	.17	12.38	-.25	15.31± .22	.44	1360
113504.2+023257		.SBS8..	1.12± .05	123			16.04±.3		5230± 10
263.04 59.37	UGC 6558	UF (1)	.55± .05	.04	15.01 ±.18			259± 13	
108.46 -18.46		7.5± .7		.68					5118
113230.1+024933	PGC 35802	6.6±1.0	1.12	.27	14.28			1.49	5585
1135.1 +1557		.S..6*.	1.23± .05	176					5124
241.29 69.07	UGC 6559	U	1.19± .05	.06	15.81 ±.18				
95.14 -13.99		6.0±1.4		1.47					5056
1132.5 +1614	PGC 35803		1.23	.50	14.26				5457
113509.2-452318		.SB?...	1.11± .05						
289.16 15.43	ESO 266- 5	(1)	.36± .05	.42	15.47 ±.14				5025± 51
159.90 -25.06	IRAS11326-4506			.49	13.17				4818
113243.1-450642	PGC 35806	3.3± .9	1.15	.18	14.52				5302
1135.2 +5648		.I..9..	1.15± .15						
142.28 57.37	UGCA 240	U	.31± .18	.00					
57.14 2.68		10.0± .9		.23					
1132.5 +5705	PGC 35813		1.15	.16					

R.A. 2000 DEC. / l b / SGL SGB / R.A. 1950 DEC.	Names / PGC	Type / S_T n_L / T / L	$\log D_{25}$ / $\log R_{25}$ / $\log A_o$ / $\log D_o$	p.a. / A_g / A_i / A_{21}	B_T / m_B / m_{FIR} / B_T^o	$(B-V)_T$ / $(U-B)_T$ / $(B-V)_T^o$ / $(U-B)_T^o$	$(B-V)_\bullet$ / $(U-B)_\bullet$ / $m'_\bullet$ / m'_{25}	m_{21} / W_{20} / W_{50} / HI	V_{21} / V_{opt} / V_{GSR} / V_{3K}
1135.3 -0245		.SAS5?.							
268.42 54.93	CGCG 12- 32	F (1)		.07	14.3 ±.3				
113.92 -19.91		5.0±1.3							
1132.8 -0229	PGC 35819	5.6±2.0							
113521.8-214248		.S?....	1.06± .09					15.07±.3	3645± 11
280.40 37.81	ESO 571- 6		.12± .07	.10				230± 15	
134.00 -23.88				.18				215± 12	3470
113251.0-212612	PGC 35820		1.07	.06					3990
1135.4 +2030				.00	15.5 ±.3			16.27±.3	6679± 10
229.72 71.42	CGCG 126-104							300± 13	
90.78 -12.24								280± 10	6627
1132.8 +2047	PGC 35823								7000
113527.0+331810		.S..8*.	1.03± .04	139					
187.68 72.49	UGC 6561	U	.68± .04	.00	15.57 ±.18				1601± 60
78.70 -7.19	KUG 1132+335	8.0±1.4		.83					1596
113248.2+333446	PGC 35826		1.03	.34	14.74				1879
1135.4 +4328		.S..6*.	1.00± .08	150					
161.19 67.58	UGC 6562	U	.94± .06	.02					
69.33 -2.94		6.0±1.4		1.38					
1132.8 +4345	PGC 35829		1.00	.47					
113530.3-250848		RSA.0P.	.99± .06	37					
281.97 34.61	ESO 504- 1	r	.09± .03	.19	14.63 ±.14				
137.74 -24.29		-.1± .9		.07					
113300.1-245212	PGC 35830		1.01						
113533.0-375730	NGC 3742	PSXT2P.	1.38± .04	116	13.03 ±.15	.96± .02	1.04± .02		
286.82 22.53	ESO 320- 6	BSr	.15± .05	.45	12.91 ±.14	.55± .05	.57± .05		2790± 60
151.76 -25.12		1.5± .4	.91± .02	.19		.81	13.07± .07		2590
113305.0-374054	PGC 35833		1.42	.08	12.30	.42	14.38± .28		3094
113536.4+155828	A 1133+16		.32± .18						
241.44 69.18			.26± .12	.07					5230± 45
95.17 -13.87	MK 636								5163
113300.4+161504	PGC 35838		.33						5563
113536.5+000740		.SBS9..	.92± .06	2				16.86±.3	5955± 10
265.81 57.43	UGC 6568	F (1)	.28± .04	.06	14.7 ±.2			206± 13	5949± 50
110.97 -19.04	IRAS11330+0024	9.0± .9		.28	13.67				5836
113302.7+002416	PGC 35839	6.6±1.0	.93	.14	14.31			2.41	6312
113536.7+545657	NGC 3737	.LB....	1.02± .09						
144.14 58.95	UGC 6563	P	.02± .06	.00	13.85 ±.15				5580± 59
58.86 1.94		-2.0± .8		.00					5652
113252.2+551333	PGC 35840		1.02		13.77				5755
1135.6 +2503	NGC 3739	.S..4..	1.06± .06	17					
215.61 72.87	UGC 6564	U	.56± .05	.00	15.25 ±.18				
86.46 -10.45		4.0± .9		.83					
1133.0 +2520	PGC 35841		1.06	.28					
113544.9-382200		.S..0*.	1.09± .05	3					
287.00 22.15	ESO 320- 7	S	.29± .05	.46	14.25 ±.14				
152.21 -25.08	IRAS11332-3805	.0±1.2		.22	13.30				
113317.0-380524	PGC 35851		1.12						
1135.7 +5811		.SB.9..	1.03± .08	110					
140.84 56.23	UGC 6566	U	.14± .06	.00					1249
55.91 3.33		9.0± .8		.14					1332
1133.0 +5828	PGC 35852		1.03	.07					1406
113547.3-490148		.SXT5*.	1.14± .05	54	14.29 ±.14	.75± .02	.83± .02		
290.38 11.99	ESO 216- 21	Sr (1)	.16± .05	.62	14.74 ±.14	.57± .05			5565± 40
163.86 -24.73	IRAS11333-4845	4.8± .6	.77± .02	.24		.54	13.63± .06		5357
113322.0-484512	PGC 35853	3.3±1.2	1.20	.08	13.62		14.43± .30		5827
113548.8+543122	NGC 3738	.I..9..	1.40± .02	155	12.13 ±.13	.40± .02		13.70±.1	229± 4
144.56 59.32	UGC 6565	R (1)	.12± .03	.00	11.81 ±.17	-.19± .04		109± 5	218± 26
59.26 1.79	ARP 234	10.0± .3		.09	12.41	.37		78± 5	299
113304.5+544758	PGC 35856	5.7± .8	1.40	.06	11.93	-.21	13.72± .18	17.02±.3	406
113549.9+352008		.S..0..	1.04± .08	123				17.02±.3	1603± 10
181.30 71.92	UGC 6570	U	.29± .06	.04	14.50 ±.18			119± 15	1622± 37
76.85 -6.28	MK 1301	.0± .9		.22					1607
113310.8+353644	PGC 35859		1.03		14.21				1874
113553.0-375948	NGC 3749	.SAS1P/	1.50± .03	107					
286.90 22.51	ESO 320- 8	BS	.60± .05	.45	13.22 ±.14				2580± 60
151.80 -25.05	IRAS11333-3743	1.0± .5		.62	11.53				2380
113325.0-374312	PGC 35861		1.54	.30	12.72				2884

R.A. 2000 DEC. / l b / SGL SGB / R.A. 1950 DEC.	Names / PGC	Type / S_T n_L / T / L	$\log D_{25}$ / $\log R_{25}$ / $\log A_e$ / $\log D_o$	p.a. / A_g / A_i / A_{21}	B_T / m_B / m_{FIR} / B_T^o	$(B-V)_T$ / $(U-B)_T$ / $(B-V)_T^o$ / $(U-B)_T^o$	$(B-V)_e$ / $(U-B)_e$ / m'_e / m'_{25}	m_{21} / W_{20} / W_{50} / HI	V_{21} / V_{opt} / V_{GSR} / V_{3K}
113559.7+703207	NGC 3735	.SA.5*/	1.62± .02	131				13.65±.1	2696± 7
131.74 45.28	UGC 6567	R (2)	.70± .03	.00	12.57 ±.13			509± 8	2671± 46
44.59 8.44	IRAS11330+7048	5.0± .5		1.05	10.84			493± 6	2818
113304.8+704842	PGC 35869	4.2± .7	1.62	.35	11.51			1.80	2780
113604.4-030612		.SAS3P.	.94± .11						
268.99 54.72	MCG 0-30- 15	F	.28± .07	.09	14.93 ±.18				
114.32 -19.83		3.0± .9		.39					
113331.0-024936	PGC 35877		.95	.14					
113606.4+451707	NGC 3741	.I..9..	1.31± .05	5	14.3 ±.2	.30± .06	.19± .03		
157.57 66.45	UGC 6572	U	.26± .06	.00	13.9 ±.2	-.34± .09	-.42± .04		211± 50
67.71 -2.08		10.0± .8	.83± .04	.19		.24	13.98± .08		249
113325.2+453343	PGC 35878		1.31	.13	13.91	-.39	15.04± .37		436
113606.9-123409	NGC 3777	.SXS2*.	1.04± .07	35					
275.80 46.30	MCG -2-30- 8	EF	.24± .05	.06					
124.23 -22.09		1.7± .7		.29					
113334.8-121733	PGC 35879		1.05	.12					
1136.1 +1003	IC 2941	.S?....	.94± .11					16.51±.3	6216± 12
253.26 65.38	MCG 2-30- 3		.00± .07	.11	14.57 ±.18			260± 20	
101.03 -15.80				.00					6129
1133.6 +1020	PGC 35881		.95	.00	14.42			2.09	6561
1136.2 +2031	A 1133+20								
229.92 71.60	CGCG 126-108			.00	15.48 ±.18				6542± 72
90.83 -12.06									6491
1133.6 +2048	PGC 35882								6863
1136.2 +2946		.S..4..	1.00± .08	132					
199.41 73.28	UGC 6574	U	.73± .06	.00					
82.07 -8.46		4.0±1.0		1.07					
1133.6 +3003	PGC 35885		1.00	.36					
1136.2 -0251		.L.....							
268.85 54.96	CGCG 12- 36	F		.09	15.1 ±.6				
114.09 -19.72		-2.0± .9							
1133.7 -0235	PGC 35886								
113622.5-024930		.SX.1?.							
268.86 55.00	CGCG 12- 37	F		.09	15.0 ±.6				
114.06 -19.68		1.0±1.8							
113349.1-023254	PGC 35896								
113627.0+581128		.S..6*.	1.29± .04	9					
140.70 56.28	UGC 6575	U	.67± .05	.00	14.32 ±.18				1225± 50
55.95 3.41	SBS 1133+584	6.0±1.2		.99					1309
113341.4+582804	PGC 35900		1.29	.34	13.32				1382
1136.4 -0248		.SBS3*.							
268.88 55.03	CGCG 12- 38	F		.09	15.3 ±.6				
114.05 -19.66		3.0±1.3							
1133.9 -0232	PGC 35901								
113628.6-450342		.S.3*P/	1.18± .04	94					
289.29 15.82	ESO 266- 8	S	.75± .04	.23	15.03 ±.14				
159.53 -24.84		3.0±1.3		1.03					
113402.0-444706	PGC 35904		1.20	.37					
113629.2+213548	NGC 3758	.S?....	.74± .14					17.74±.3	8912± 9
226.83 72.08	MCG 4-27- 73		.05± .07	.00	15.2 ±.2			333± 12	8914± 29
89.83 -11.60	MK 739			.07	12.78				8865
113352.5+215224	PGC 35905		.74	.02	15.03			2.68	9230
113630.3-095048	NGC 3763	.SXT5P.	1.07± .05						
274.22 48.80	MCG -2-30- 9	EF (1)	.00± .04	.03					
121.38 -21.40	IRAS11339-0934	4.7± .5		.00	12.47				
113357.8-093411	PGC 35907	2.5± .6	1.07	.00					
113630.7+490753		.S?....	.74± .14						
151.22 63.68	MCG 8-21- 72		.00± .07	.00	14.99 ±.19				11223± 19
64.23 -.39				.00					11275
113348.5+492429	PGC 35908		.74	.00	14.91				11429
113630.8+474901		.LB....	1.14± .07						
153.20 64.67	UGC 6576	U	.03± .03	.02	15.0 ±.3				
65.43 -.94		-2.0± .8		.00					
113349.0+480537	PGC 35909		1.14						
113631.1+024500		.SXR7..							
263.40 59.74	CGCG 40- 9	F (1)		.03	15.1 ±.6				
108.37 -18.06		7.0± .9							
113357.0+030136	PGC 35910	5.6±1.0							

R.A. 2000 DEC.	Names	Type	logD$_{25}$	p.a.	B$_T$	(B-V)$_T$	(B-V)$_e$	m$_{21}$	V$_{21}$
l　b		S$_T$　n$_L$	logR$_{25}$	A$_g$	m$_B$	(U-B)$_T$	(U-B)$_e$	W$_{20}$	V$_{opt}$
SGL　SGB		T	logA$_e$	A$_i$	m$_{FIR}$	(B-V)$_T^o$	m'$_e$	W$_{50}$	V$_{GSR}$
R.A. 1950 DEC.	PGC	L	logD$_o$	A$_{21}$	B$_T^o$	(U-B)$_T^o$	m'$_{25}$	HI	V$_{3K}$
113633.4+362439 NGC　3755	.SXT5P.	1.50± .02	133				13.39±.1	1570±　5
177.95　71.62 UGC　6577	R	.36± .03	.00	13.5　±.2			281±　8	
75.91　-5.70		5.0± .3		.54			267±　6	1577
113354.3+364115 PGC 35913		1.50		.18	12.96			.25	1835
1136.5　-0254		.E.....							
269.00　54.95 CGCG　12- 39	F			.09	15.03　±.18				
114.16 -19.66		-5.0± .9							
1134.0　-0238	PGC 35914								
113636.7+490346		.S?....	.74± .14						9641± 19
151.29　63.74 MCG　8-21- 74		.00± .07	.00	15.44　±.18				9693
64.30　-.41				.00					
113354.7+492022 PGC 35920		.74			15.30				9847
1136.6　+0617		.LAR0P.	.87± .09						
259.10　62.61 MCG　1-30- 2	F	.06± .06	.08	14.69　±.15				
104.82 -16.92		-2.0±1.0		.00					
1134.1　+0634	PGC 35925		.87						
113642.4+545047 IC　2943		.S..1$.	.59± .09						5788± 35
143.99　59.12 CGCG 268- 62	P	.02± .05	.00	15.27　±.18				5861
59.02　2.04 MK　41				.03					
113358.4+550723 PGC 35926		.59		.01	15.21				5964
1136.7　-1259		.SA.6*/	1.21± .05	95					
276.26　45.97 MCG -2-30- 10	EF	.74± .05	.04					
124.72 -22.03		5.5± .9		1.09					
1134.2　-1243	PGC 35927		1.21	.37					
1136.8　+1752 NGC　3764		.S?....	1.04± .09						
237.32　70.47 MCG　3-30- 20		.09± .07	.00					
93.44 -12.92 2ZW　52				.07					
1134.2　+1809	PGC 35930		1.03						
113648.5+541746 NGC　3756		.SXT4..	1.62± .02	177	12.11S±.08	.62± .05		13.91±.1	1289±　5
144.57　59.58 UGC　6579	R　(2)	.30± .02	.00	12.04　±.13			304±　7	1071± 66
59.53　1.82 IRAS11340+5434	4.0± .3		.43	12.32	.56		286±　5	1358
113404.7+543422 PGC 35931	2.6± .6	1.62		.15	11.65		14.31± .13	2.12	1466
113648.5-021454		.SAS4*.	.94± .11						
268.52　55.56 MCG　0-30- 16	F　(1)	.05± .07	.06	14.73　±.19				
113.50 -19.42		4.0± .9		.07					
113415.0-015818 PGC 35932		5.6±1.4	.94	.02					
113648.9-083510		.SA.3*/	1.10± .06	107					
273.48　49.97 MCG -1-30- 11	FE	.60± .05	.06					
120.08 -21.04 IRAS11342-0818	2.8± .8		.82	12.87					
113416.2-081833 PGC 35934		1.11		.30					
113649.8-093404		.L..0*.	1.05± .07	130					
274.14　49.09 MCG -1-30- 12	E	.12± .05	.04					
121.11 -21.26		-2.0±1.2		.00					
113417.3-091728 PGC 35935		1.04							
113654.1+195814 A　1134+20A		.P.....	.86± .04	5	14.36S±.15			15.63±.1	6191±　4
231.72　71.51 UGC　6583	R	.22± .03	.00	14.6　±.3			374±　4	6229± 38
91.43 -12.12 MK　181	99.0		.30				304±　5	6138
113417.7+201450 PGC 35942		.86		.11	14.07		12.95± .27	1.45	6514
113654.3+195951		.S?....	.82± .13					17.38±.3	6630±　9
231.64　71.52 MCG　3-30- 21		.17± .07	.00	15.0　±.2			283± 10	
91.40 -12.11				.26					6577
113417.9+201627 PGC 35943		.82		.09	14.71			2.58	6953
113654.3+040718		.SBS8..							
261.96　60.92 CGCG　40- 11	F　(1)			.01	15.2　±.6			
107.02 -17.55		8.0± .9							
113420.0+042354 PGC 35944		7.6±1.0							
113654.4+544927 NGC　3759		.LX.0*.	1.04± .04		14.24　±.15	.98± .03	1.00± .03		
143.97　59.15 UGC　6581	PU	.00± .08	.00	14.12　±.15				5564± 27
59.05　2.06		-1.5± .6	.61± .04	.00		.93	12.78± .12	5637
113410.5+550603 PGC 35945		1.04			14.10		14.30± .32	5740
113658.1+550947 NGC　3759A		.SX.5P.	1.08± .06		14.28　±.19				5785± 50
143.59　58.88 UGC　6582	PU	.04± .05	.00					5858
58.75　2.21 IRAS11342+5526	4.5± .6		.06						5959
113414.0+552623 PGC 35948		1.08		.02	14.19				
113702.0+153414		.SX.4..	1.26± .03	175	14.39　±.13	.59± .03	.67± .03	14.75±.2	3962±　6
242.93　69.22 UGC　6586	U	.34± .04	.04	14.07　±.19	-.12± .04	-.02± .04	286±　6	3989± 50
95.69 -13.70 IRAS11344+1550	4.0± .8	.79± .02	.50	13.72	.49	13.83± .04	262±　6	3895
113426.3+155050 PGC 35952		1.26		.17	13.73	-.19	14.67± .24	.85	4297

R.A. 2000 DEC. / l b / SGL SGB / R.A. 1950 DEC.	Names / PGC	Type S_T n_L / T / L	$\log D_{25}$ / $\log R_{25}$ / $\log A_e$ / $\log D_o$	p.a. / A_g / A_i / A_{21}	B_T / m_B / m_{FIR} / B_T^o	$(B-V)_T$ / $(U-B)_T$ / $(B-V)_T^o$ / $(U-B)_T^o$	$(B-V)_e$ / $(U-B)_e$ / m'_e / m'_{25}	m_{21} / W_{20} / W_{50} / HI	V_{21} / V_{opt} / V_{GSR} / V_{3K}
113702.6-364855	ESO 378- 12	.SA.2*.	1.14± .04	43	14.33 ±.14				
286.74 23.71		S (1)	.34± .04	.45	12.90				
150.51 -24.80	IRAS11345-3632	2.0± .9		.42					
113434.0-363218	PGC 35954	4.4±1.3	1.18	.17					
113703.5+582450	NGC 3757	.L...?.	1.06± .07						
140.37 56.12	UGC 6584	U	.02± .03	.00	13.55 ±.16				1267± 50
55.78 3.58		-2.0±1.6		.00					1351
113418.2+584126	PGC 35955		1.06		13.53				1423
113704.4+240546	NGC 3765	.S?....	.89± .11					16.65±.3	9022± 9
219.05 72.98	MCG 4-28- 1		.10± .07	.00	15.00 ±.18			558± 10	
87.50 -10.52				.16					8984
113427.4+242222	PGC 35956		.89	.05	14.79			1.81	9333
1137.1 +0208	CGCG 12- 42	.SAS2*.							
264.34 59.33		F		.03	14.1 ±.4				
109.04 -18.08		2.0±1.3							
1134.6 +0225	PGC 35963								
1137.1 +0249		.SXS3P*	.95± .07	150					
263.58 59.89	UGC 6587	UF	.25± .05	.03	14.85 ±.18				8776± 46
108.35 -17.88		3.4± .8		.34					8666
1134.6 +0306	PGC 35964		.95	.12	14.41				9130
1137.2 +1525		.S..4..	1.11± .05	144					
243.33 69.16	UGC 6588	U	.41± .05	.04	14.71 ±.18				
95.85 -13.71		4.0± .9		.60					
1134.6 +1542	PGC 35966		1.12	.20					
113714.0+175024	NGC 3768	.L.....	1.26± .08	155					
237.56 70.54	UGC 6589	U	.19± .05	.00	13.43 ±.15				3354± 30
93.51 -12.83		-2.0± .8		.00					3294
113438.0+180700	PGC 35968		1.23		13.38				3683
113715.7+165234	NGC 3767	RLBR0?.	.98± .04		14.35 ±.15	.92± .02	.97± .02		
239.97 70.02	UGC 6590	U	.05± .03	.01	14.36 ±.15	.39± .04	.48± .04		6362± 31
94.44 -13.18	IRAS11346+1708	-2.0± .9	.50± .04	.00		.85	12.34± .12		6298
113439.8+170910	PGC 35969		.98		14.24	.42	14.01± .25		6693
1137.3 +2010		.S?....	.86± .05		15.39S±.15	.79± .05			
231.29 71.68	MCG 3-30- 26		.29± .03	.00	15.54 ±.18	.17± .08			
91.27 -11.96	IRAS11347+2026			.36					
1134.7 +2026	PGC 35973		.86	.15			13.84± .29		
113722.2+254428		.S?....	.89± .10						
213.52 73.38	UGC 6593		.23± .06	.00	14.9 ±.2				3285± 39
85.97 -9.82	MK 741			.32					3253
113445.1+260104	PGC 35977		.89	.12	14.51				3590
1137.4 +2223		.S?....	.94± .11		15.25 ±.13	.60± .03	.70± .03	16.42±.3	6864± 10
224.64 72.56	MCG 4-28- 3		.40± .07	.00	15.39 ±.18	-.06± .05	.03± .05	291± 13	
89.16 -11.10			.61± .02	.60		.48	13.79± .05	275± 10	6820
1134.8 +2240	PGC 35978		.94	.20	14.66	-.14	13.79± .58	1.55	7180
113724.7+614539	NGC 3762	.S..1..	1.27± .04	167					
137.46 53.23	UGC 6591	U	.58± .05	.00	13.5 ±.2				3463± 39
52.73 5.00		1.0± .9		.59					3558
113437.9+620215	PGC 35979		1.27	.29	12.90				3599
1137.4 +3122	IC 2947	.S?....	.82± .13						
193.77 73.33	MCG 5-28- 2		.08± .07	.00	14.9 ±.2				12730
80.68 -7.58				.11					12718
1134.8 +3139	PGC 35981		.82	.04	14.70				13015
113737.2+163320		.S..7..	1.41± .03	134					
240.88 69.91	UGC 6594	U	.84± .04	.01	14.45 ±.19				
94.79 -13.21	KUG 1135+168	7.0± .9		1.15					
113501.4+164957	PGC 35991		1.41	.42					
113743.6+220034	NGC 3746	.SBR3..	1.04± .06	127	15.01S±.15			17.61±.3	9022± 9
225.91 72.50	UGC 6597	R	.31± .05	.01	14.98 ±.18			558± 10	9074± 35
89.55 -11.18	HICK 57B	3.0± .5		.43					8980
113507.1+221710	PGC 35997		1.04	.16	14.49		14.28± .35	2.96	9342
113743.8+475340	NGC 3769	.SBR3*.	1.49± .03	152	12.55S±.15			13.30±.3	730± 8
152.72 64.75	UGC 6595	R (2)	.50± .03	.00	12.42 ±.15			272± 9	722± 36
65.44 -.73		3.0± .5		.69				228± 7	778
113502.4+481016	PGC 35999	3.8± .7	1.49	.25	11.78		13.58± .21	1.27	941
1137.7 +5608		.IB.9..	1.04± .06	140					2277
142.41 58.11	UGC 6596	U	.22± .05	.00	15.2 ±.2				
57.90 2.72	VV 148	10.0± .9		.16					2354
1135.0 +5625	PGC 36000		1.04	.11	15.04				2445

R.A. 2000 DEC. / l b / SGL SGB / R.A. 1950 DEC.	Names / PGC	Type S_T n_L / T / L	$\log D_{25}$ / $\log R_{25}$ / $\log A_o$ / $\log D_o$	p.a. / A_g / A_i / A_{21}	B_T / m_B / m_{FIR} / B_T^o	$(B-V)_T$ / $(U-B)_T$ / $(B-V)_T^o$ / $(U-B)_T^o$	$(B-V)_o$ / $(U-B)_o$ / m'_o / m'_{25}	m_{21} / W_{20} / W_{50} / HI	V_{21} / V_{opt} / V_{GSR} / V_{3K}
113744.6+220115	NGC 3745	.LBS-*.	.60?		16.18S±.15				
225.88 72.50	MCG 4-28- 4	R	.27± .07	.01					9413± 72
89.54 -11.17	HICK 57G	-3.0± .7		.00					9368
113508.1+221752	PGC 36001		.56		16.03		13.35±1.33		9730
113744.9-491043		RSXT5..	1.17± .05		13.31 ±.15	.87± .03	.83± .01		
290.74 11.94	ESO 216- 24	Sr (1)	.02± .05	.62	14.01 ±.14				5070± 40
164.00 -24.40	IRAS11353-4854	4.7± .6	.98± .03	.03	12.49	.69	13.70± .08		4862
113519.0-485406	PGC 36002	2.2± .8	1.23	.01	13.02		13.95± .31		5331
1137.7 -0716	A 1135-07	.SBT5..	1.14± .06		14.2 ±.3	.75± .06		15.57±.1	9460± 7
272.88 51.24	MCG -1-30- 13	PEF (1)	.00± .05	.12				282± 8	9320
118.77 -20.50		5.0± .5		.00		.66		259± 11	9818
1135.2 -0700	PGC 36003	2.9± .8	1.15	.00	14.03		14.75± .44	1.54	
113748.7+224133	NGC 3772	.SB.1..	1.06± .06	16					
223.78 72.75	UGC 6598	U	.27± .05	.00	14.44 ±.18				3478± 50
88.90 -10.90		1.0± .9		.28					3435
113512.1+225810	PGC 36005		1.06	.14	14.11				3793
1137.8 +2408		.IB.9..	1.07± .14					15.42±.1	1572± 10
219.04 73.16	UGC 6599	U	.21± .12	.00				149± 6	
87.53 -10.35		10.0± .9		.15				135± 5	1534
1135.2 +2425	PGC 36006		1.07	.10					1882
113749.2+220133	NGC 3748	.LB.0$/	.82± .19		15.77S±.15				
225.88 72.52	MCG 4-28- 7	R	.20± .07	.01	15.47 ±.15				8989± 72
89.54 -11.15	HICK 57E	-2.0± .9		.00					8944
113512.7+221810	PGC 36007		.79		15.47		14.22± .98		9306
113749.4+475302	NGC 3769A	.SB.9P*	1.03± .08	71					
152.71 64.77	MCG 8-21- 77	R (1)	.38± .06	.00	14.72 ±.18				761± 36
65.46 -.72	IRAS11350+4809	9.0± .6		.38	11.89				809
113508.1+480939	PGC 36008	4.5±1.3	1.03	.19	14.33				973
113751.8+215826	NGC 3750	.LX.-$.	.90± .17		14.91S±.15				
226.06 72.51	MCG 4-28- 8	R	.07± .07	.00	14.94 ±.15				9064± 35
89.59 -11.16	HICK 57C	-3.0± .7		.00					9019
113515.3+221503	PGC 36011		.89		14.79		14.08± .87		9381
1137.8 +0530		.SXS4..							
260.65 62.18	CGCG 40- 16	F (1)		.06	15.3 ±.6				
105.70 -16.89		4.0± .9							
1135.3 +0547	PGC 36012	5.6±1.4							
113753.8+215851	NGC 3753	.S.2$/P	1.23± .06	120	14.52S±.15				
226.05 72.52	UGC 6602	R	.56± .06	.00					8717± 35
89.59 -11.15	HICK 57A	2.0± .9		.69					8671
113517.3+221528	PGC 36016		1.23	.28	13.74		14.12± .37		9034
113754.1+215610	NGC 3751	.L..-P?	.90± .15	5	14.91S±.15				
226.19 72.51	UGC 6601	PU	.23± .05	.00	15.19 ±.15				9592± 72
89.63 -11.17	HICK 57F	-2.7±1.2		.00					9546
113517.6+221247	PGC 36017		.86		14.90		13.68± .77		9909
113755.2+215909	NGC 3754	.SB.3$P	.64± .17		15.06S±.15				
226.04 72.53	MCG 4-28- 11	R	.11± .07	.00					9012± 35
89.59 -11.14	IRAS11352+2216	3.0±1.3		.16	13.75				8967
113518.6+221545	PGC 36018		.64	.06	14.84		12.81± .86		9329
113759.3+593700	NGC 3770	.SB.1..	1.02± .06	107					
139.12 55.14	UGC 6600	U	.20± .05	.00	13.8 ±.2				3214± 50
54.73 4.19	IRAS11352+5953	1.0± .9		.20	13.08				3303
113513.9+595337	PGC 36025		1.02	.10	13.52				3363
113759.5-171404		.S..5*/	1.32± .04	7					6537
278.97 42.19	MCG -3-30- 3	SE (1)	.83± .04	.05					
129.29 -22.55	IRAS11354-1657	5.0± .9		1.25	12.94				6373
113527.8-165727	PGC 36026	3.1±1.6	1.33	.42					6889
113800.6-321931		.SXR4..	1.12± .04	105					
285.38 28.03	ESO 439- 18	Sr (1)	.13± .03	.30	14.09 ±.14				8924± 22
145.62 -24.39	IRAS11355-3203	3.7± .6		.19	13.69				8732
113531.1-320254	PGC 36028	2.2± .9	1.15	.06	13.55				9245
113802.3+351214		.S..6*.	1.34± .03	78					
181.14 72.38	UGC 6603	U	.67± .04	.00	14.7 ±.2				
77.16 -5.92	KUG 1135+354	6.0±1.2		.98					
113523.8+352851	PGC 36029		1.34	.33					
1138.0 -0536		.LA.+?.	1.12± .08	150					
271.76 52.76		E	.15± .08	.12					
117.06 -20.01		-1.0±1.3		.00					
1135.5 -0520	PGC 36031		1.11						

R.A. 2000 DEC.	Names	Type	logD₂₅	p.a.	B_T	(B-V)_T	(B-V)_•	m₂₁	V₂₁
l b		S_T n_L	logR₂₅	A_g	m_B	(U-B)_T	(U-B)_•	W₂₀	V_opt
SGL SGB		T	logA_•	A_i	m_FIR	(B-V)°_T	m'	W₅₀	V_GSR
R.A. 1950 DEC.	PGC	L	logD_o	A₂₁	B°_T	(U-B)°_T	m'₂₅	HI	V_3K

113809.1-093626	NGC 3789	.LXT+?.	1.11± .06	170					
274.62 49.19	MCG -1-30- 15	E	.31± .05	.03					
121.23 -20.95		-1.0±1.0		.00					
113536.5-091949	PGC 36036		1.06						
113809.3+584532		.E?....	1.00± .10						
139.84 55.90	UGC 6604		.00± .04	.00	13.92 ±.16				1323± 50
55.52 3.85				.00					1409
113524.3+590208	PGC 36037		1.00		13.90				1476
113813.0+120643	NGC 3773	.LA..*.	1.07± .04	165				16.42±.2	987± 6
250.54 67.19	UGC 6605	R	.07± .03	.10	13.34 ±.13			191± 34	973± 27
99.18 -14.64	MK 743	-2.0± .6		.00	12.85			99± 5	907
113537.8+122320	PGC 36043		1.07		13.23				1328
1138.2 +3139									
192.69 73.45	CGCG 157- 4			.00	15.4 ±.3				9782
80.48 -7.32									9772
1135.6 +3156	PGC 36044								10066
1138.2 -1303		.SXS6P*	1.14± .06	140					5135
276.78 46.06	MCG -2-30- 11	EF (1)	.08± .05	.04					
124.87 -21.68		5.6± .7		.11					4981
1135.7 -1247	PGC 36045	5.3± .8	1.15	.04					5491
113816.0-294343		.SXR4..	1.22± .04	100				14.56±.3	4090± 11
284.49 30.51	ESO 439- 20	Sr (1)	.44± .03	.25	14.26 ±.14			431± 15	4185± 34
142.79 -24.14		3.7± .6		.64				405± 12	3910
113546.0-292706	PGC 36047	4.4±1.0	1.24	.22	13.34			1.00	4426
113821.7-504301	NGC 3778	.LX.-*.	1.07± .05						
291.29 10.49	ESO 216- 26	S	.10± .04	.72	14.15 ±.14				4490± 60
165.65 -24.18		-3.0± .8		.00					4282
113556.1-502624	PGC 36051		1.15		13.36				4744
113826.7-103819	NGC 3775	.LXR+*.	1.05± .06	27					
275.38 48.29	MCG -2-30- 12	EF	.40± .05	.00					
122.33 -21.11		-.5± .6		.00					
113554.2-102142	PGC 36055		.99						
113830.2-085834	NGC 3774	.SBR3*.	1.01± .07	95					
274.32 49.80	MCG -1-30- 16	F	.34± .05	.01					
120.59 -20.72		3.0±1.3		.47					
113557.5-084157	PGC 36058		1.01	.17					
113830.4+204423		.S..7..	1.15± .05	75	15.01 ±.15	.50± .03	.58± .03	16.02±.2	3366± 6
230.00 72.18	UGC 6607	U	.27± .05	.00	14.6 ±.2	-.10± .05	-.13± .05	223± 8	
90.83 -11.49		7.0± .8	.72± .02	.38		.43	14.10± .06	114± 10	3316
113554.1+210100	PGC 36060		1.15	.14	14.49	-.15	14.95± .32	1.39	3686
113833.1-011107		.SXR2?.	1.07± .05	36					
268.23 56.69	UGC 6608	F	.19± .04	.04	14.23 ±.18				6200
112.53 -18.71	IRAS11359-0054	2.0± .9		.23	13.50				6078
113559.5-005430	PGC 36061		1.07	.09	13.90				6557
1138.5 +0657		.SA.5*.							
259.01 63.43	CGCG 40- 18	F (1)		.12	15.2 ±.6				
104.31 -16.26		5.0±1.3							
1136.0 +0714	PGC 36063	5.6±1.4							
1138.6 +2032		.E.....	1.15± .15	100					
230.63 72.11	UGC 6609	U	.24± .08	.00	14.33 ±.17				
91.03 -11.54		-5.0± .8		.00					
1136.0 +2049	PGC 36068		1.08						
1138.6 +0347		.IB.9..							
263.10 60.90	CGCG 40- 19	F (1)		.02	15.2 ±.6				
107.49 -17.23		10.0± .9							
1136.1 +0404	PGC 36073	7.6±1.0							
113844.2+334821		.S..6*.	1.32± .03	13					
185.39 73.00	UGC 6610	U	.73± .04	.00	15.0 ±.2				
78.52 -6.36	KUG 1136+340	6.0±1.2		1.07					
113606.0+340458	PGC 36079		1.32	.36					
113847.5-331008		.LA.+..	1.15± .09						
285.86 27.28		S	.14± .08	.30					
146.55 -24.28		-1.0± .8		.00					
113618.0-325330	PGC 36081		1.16						
113850.9-103502	NGC 3779	.SBS7P.	1.27± .05	85				15.15±.1	1700± 9
275.48 48.38	MCG -2-30- 13	EF (1)	.28± .05	.00				193± 12	
122.30 -21.01		6.5± .6		.38					1552
113618.3-101825	PGC 36084	5.7± .7	1.27	.14					2057

R.A. 2000 DEC. / l b / SGL SGB / R.A. 1950 DEC.	Names / PGC	Type / S_T n_L / T / L	$logD_{25}$ / $logR_{25}$ / $logA_o$ / $logD_o$	p.a. / A_g / A_i / A_{21}	B_T / m_B / m_{FIR} / B_T^o	$(B-V)_T$ / $(U-B)_T$ / $(B-V)_T^o$ / $(U-B)_T^o$	$(B-V)_o$ / $(U-B)_o$ / m'_o / m'_{25}	m_{21} / W_{20} / W_{50} / HI	V_{21} / V_{opt} / V_{GSR} / V_{3K}
113855.1+033452		PSBR3*.	.94± .11						
263.44 60.77	MCG 1-30- 5	F	.19± .07	.00	14.66 ±.18				
107.72 -17.23	MK 1302	3.0±1.3		.26	13.75				
113621.0+035129	PGC 36093		.94	.09					
113857.7-175807		.S?....	1.06± .06	131					
279.63 41.60	ESO 571- 12		.43± .05	.09					6597± 52
130.12 -22.45	IRAS11364-1741			.64	13.14				6432
113626.0-174130	PGC 36097		1.07	.21					6948
113901.8-374420	NGC 3783	PSBR2..	1.28± .04						
287.46 22.95	ESO 378- 14	RBCSr (2)	.05± .04	.50	12.57 ±.12				2926± 28
151.53 -24.43	IRAS11365-3727	1.5± .3		.07	11.90				2728
113633.0-372742	PGC 36101	1.7± .6	1.32	.03	11.97				3231
113903.2-001218	IC 716	.S..4P/	1.21± .03	132				14.96±.3	5429± 10
267.50 57.60	UGC 6612	UEF (1)	.77± .04	.04	14.80 ±.18			460± 13	
111.57 -18.31		3.6± .6		1.13					5311
113629.5+000419	PGC 36102	4.5±1.3	1.21	.38	13.60			.98	5786
113904.0+262139	NGC 3781	.S?....						17.40±.3	6798± 9
211.53 73.84	MCG 5-28- 4			.00	14.7 ±.3			243± 10	
85.53 -9.23									6769
113627.0+263816	PGC 36104								7101
113906.0-092053	NGC 3771	.E...*.	1.12± .08						
274.77 49.53		F	.00± .08	.01					
121.02 -20.66		-5.0±1.3		.00					
113633.2-090416	PGC 36107		1.13						
1139.1 -0920			.64± .17		13.57 ±.15	.98± .02	.82± .02		
274.79 49.54	MCG -1-30- 17		.28± .07	.01					
121.01 -20.65			.78± .04				12.96± .14		
1136.6 -0904	PGC 36113		.64						
113909.1-231814		.S?....	1.01± .05	67					
282.14 36.63	ESO 504- 8		.05± .04	.13	14.36 ±.14				8210± 52
135.86 -23.23	IRAS11366-2301			.06	13.57				8034
113638.0-230136	PGC 36115		1.02	.02	14.09				8552
113910.0-504008		PSBS3?.	1.05± .05						
291.40 10.58	ESO 216- 27	S (1)	.01± .05	.62	14.61 ±.14				
165.59 -24.06	IRAS11367-5023	3.0± .8		.01					
113644.0-502330	PGC 36117	3.3±1.2	1.11	.01					
113911.0+392002		.SB.1..	1.06± .04	57					
169.28 70.64	UGC 6613	U	.36± .04	.00	15.04 ±.18				
73.43 -4.03	IRAS11365+3936	1.0± .9		.36	12.82				
113631.9+393639	PGC 36118		1.06	.18					
113911.4-492456		PSXR4..	1.07± .05		13.68 ±.15	.63± .02	.75± .02		
291.04 11.78	ESO 216- 28	Sr (1)	.11± .05	.60	14.38 ±.14				5565± 40
164.24 -24.15	IRAS11367-4908	3.6± .6	.94± .02	.16	13.27	.43	13.87± .05		5358
113645.1-490818	PGC 36119	3.3±1.2	1.13	.06	13.26		13.60± .31		5825
1139.2 +1708		RSAR1?.	1.22± .04		14.37S±.15	.87± .03		15.01±.3	6351± 10
240.09 70.55	UGC 6614	U	.07± .03	.02	14.0 ±.3	.30± .05		262± 13	
94.36 -12.65		1.0±1.4		.07		.80		242± 10	6289
1136.6 +1725	PGC 36122		1.22	.03	14.11	.30	15.17± .25	.87	6681
113914.0+335552		.S?....	.78± .08						
184.87 73.06	MCG 6-26- 12		.04± .06	.00	15.33 ±.18				9834
78.44 -6.21	KUG 1136+342			.06					9832
113635.8+341229	PGC 36126		.78	.02	15.23				10109
113920.8+463051	NGC 3782	.SXS6*.	1.22± .03	0	13.10S±.15			13.69±.1	739± 6
154.44 65.96	UGC 6618	R (1)	.19± .03	.01	13.13 ±.14			132± 7	699± 63
66.82 -1.05	IRAS11366+4647	6.0± .6		.28	12.98			100± 5	782
113640.2+464728	PGC 36136	8.0± .9	1.22	.09	12.82		13.58± .24	.78	957
113921.1+581606		.SAS7*.	1.36± .05	80				14.38±.1	1154± 11
140.03 56.40	UGC 6616	U	.02± .06	.00	13.7 ±.3			112± 16	
56.04 3.79		7.0± .7		.03				109± 12	1238
113636.9+583243	PGC 36137		1.36	.01	13.71			.66	1310
113923.5+561615	NGC 3780	.SAS5*.	1.49± .03	90				14.07±.1	2399± 5
141.91 58.12	UGC 6615	R (2)	.10± .03	.00	12.17 ±.14			313± 6	2361± 66
57.87 2.98	IRAS11366+5632	5.0± .3		.15	11.97			289± 5	2476
113640.1+563252	PGC 36138	3.7± .6	1.49	.05	12.01			2.02	2566
113927.2+032811		.SAS4..	.82± .13						
263.80 60.75	MCG 1-30- 6	F (1)	.08± .07	.00	14.82 ±.19				
107.88 -17.14		4.0± .9		.11					
113653.0+034448	PGC 36141	5.6±1.0	.82	.04					

R.A. 2000 DEC. l b SGL SGB R.A. 1950 DEC.	Names PGC	Type S_T n_L T L	$\log D_{25}$ $\log R_{25}$ $\log A_\bullet$ $\log D_o$	p.a. A_g A_i A_{21}	B_T m_B m_{FIR} B_T^o	$(B-V)_T$ $(U-B)_T$ $(B-V)_T^o$ $(U-B)_T^o$	$(B-V)_\bullet$ $(U-B)_\bullet$ m' m'_{25}	m_{21} W_{20} W_{50} HI	V_{21} V_{opt} V_{GSR} V_{3K}
1139.4 +0746 258.24 64.20 103.58 −15.79 1136.8 +0802	MCG 1-30- 7 IRAS11368+0802 PGC 36142	.SBR7.. F (1) 7.0± .9 5.6±1.0		.07	14.8 ±.6 13.86				
113928.5−024143 269.93 55.50 114.15 −18.90 113655.0−022506	CGCG 12- 48 PGC 36145	.SBR1?. F 1.0±1.8		.08	15.2 ±.6				
1139.4 +6032 138.06 54.42 53.95 4.74 1136.7 +6049	UGC 6619 IRAS11367+6049 PGC 36146	.S?.... 	1.08± .06 .62± .05 1.09	8 .04 .93 .31	15.36 ±.18 13.29 14.37				3465 3557 3608
1139.5 +2618 211.77 73.93 85.62 −9.16 1136.9 +2635	NGC 3784 MCG 5-28- 6 PGC 36147	.S?.... 	.94± .11 .40± .07 .94	.00 .60 .20	15.29 ±.18 14.66				6854 6825 7157
1139.5 +2618 211.77 73.93 85.62 −9.16 1136.9 +2635	NGC 3785 UGC 6620 PGC 36148	.L..... U −2.0± .9	.99± .08 .39± .03 .93	25 .00 .00	15.16 ±.15 15.03				9157 9128 9460
1139.5 +0203 265.42 59.59 109.32 −17.54 1137.0 +0220	CGCG 12- 49 PGC 36149	.IX.9*. F (1) 10.0± .9 6.6±1.0		.03	14.1 ±.4				
1139.6 +0428 262.71 61.60 106.89 −16.78 1137.1 +0445	UGC 6622 PGC 36155	.SAS1.. F 1.0± .9	1.00± .08 .43± .06 1.00	141 .01 .44 .22	15.10 ±.18				
113941.7−092202 274.99 49.57 121.07 −20.53 113708.9−090524	NGC 3791 MCG −1-30- 20 PGC 36156	RLX..P* EF −2.0± .8	1.12± .08 .13± .08 1.11	50 .02 .00					
113942.4+315435 191.58 73.70 80.37 −6.93 113704.7+321113	NGC 3786 UGC 6621 MK 744 PGC 36158	.SXT1P. R 1.0± .4	1.34± .02 .23± .03 1.34	77 .00 .24 .12	13.24 ±.18 12.97			14.58±.1 451± 4 308± 5 1.49	2723± 4 2722± 24 2714 3006
113944.1+315558 191.50 73.70 80.35 −6.91 113706.4+321235	NGC 3788 UGC 6623 PGC 36160	.SXT2P. R 2.0± .4	1.33± .02 .51± .03 1.33	178 .00 .63 .26	13.46 ±.20 12.80			14.68±.2 542± 6 390± 7 1.63	2699± 6 2486± 29 2682 2973
113944.7+224108 224.29 73.17 89.08 −10.49 113708.3+225746	MCG 4-28- 16 PGC 36162	.S?.... 	.82± .13 .00± .07 .82	.00 .00 .00	14.99 ±.18 14.95				6935± 9 6893 7250
113947.6+195605 232.79 72.10 91.72 −11.50 113711.5+201242	UGC 6625 IRAS11371+2012 PGC 36166	.S?.... 	.87± .08 .06± .05 .87	55 .00 .08 .03	14.5 ±.2 12.71 14.38				10964± 50 10912 11287
113947.6+174249 238.88 70.99 93.86 −12.31 113711.8+175926	NGC 3790 UGC 6624 PGC 36167	.S..0.. U .0±1.0	1.04± .06 .54± .05 1.01	154 .01 .40	14.9 ±.2 14.40				3434± 50 3374 3762
113949.7−443338 289.73 16.47 158.96 −24.27 113722.0−441700	ESO 266- 12 IRAS11373−4416 PGC 36172	PSXS4.. r 4.5± .9	1.01± .06 .24± .05 1.05	139 .35 .35 .12	14.78 ±.14 12.69				
1139.8 +0852 256.77 65.11 102.52 −15.34 1137.3 +0909	IC 718 UGC 6626 PGC 36174	.I..9*. U 10.0±1.3	1.08± .06 .36± .05 1.08	178 .07 .27 .18	14.48 ±.18 14.13			15.55±.3 173± 20 1.24	1982± 12 1892 2328
1139.8 +1328 248.63 68.42 97.99 −13.79 1137.3 +1345	UGC 6627 PGC 36176	.S..8*. U 8.0±1.3	1.00± .08 .55± .06 1.01	41 .07 .68 .27	15.45 ±.18 14.70				3552 3478 3890
113953.4−033108 270.78 54.82 115.02 −19.02 113720.0−031430	CGCG 12- 51 PGC 36177	.E+.... F −4.0± .9		.09	15.13 ±.18				

R.A. 2000 DEC.	Names	Type	logD25	p.a.	BT	(B-V)T	(B-V)e	m21	V21
l b		ST nL	logR25	Ag	mB	(U-B)T	(U-B)e	W20	Vopt
SGL SGB		T	logAe	Ai	mFIR	(B-V)T°	m'e	W50	VGSR
R.A. 1950 DEC.	PGC	L	logDo	A21	BT°	(U-B)T°	m'25	HI	V3K
1139.9 -0052		.LX.0*.							
268.50 57.14	CGCG 12- 52	F		.04	15.3 ±.6				
112.33 -18.29		-2.0±1.3							
1137.4 -0036	PGC 36182								
114000.7-485744		.SA.2*/	1.31± .04	151					
291.04 12.25	ESO 216- 31	S	.64± .05	.60	15.24 ±.14				5470± 60
163.74 -24.04	IRAS11375-4841	2.0±1.2		.79	12.45				5263
113734.0-484106	PGC 36185		1.37	.32	13.80				5732
114005.3-542350		.SBS3*.	1.10± .08	168					
292.58 7.03	ESO 170- 10	S	.16± .07	1.28					
169.59 -23.59	IRAS11376-5406	3.0±1.2		.22	13.26				
113740.0-540712	PGC 36186		1.22	.08					
114005.7-005408		.SAS3..							
268.58 57.13	CGCG 12- 53	F (1)		.04	14.9 ±.6				
112.36 -18.26		3.0± .9							
113732.0-003730	PGC 36187	4.6±1.0							
114005.7+455632	A 1137+46	.SA.9..	1.46± .04		13.18S±.15			13.95±.1	850± 5
155.17 66.47	UGC 6628	U	.00± .06	.01	13.9 ±.5			52± 5	892
67.40 -1.17		9.0± .7		.00				40± 6	
113725.5+461309	PGC 36188		1.46	.00	13.23		15.32± .31	.72	1071
114006.7-005014		.E.....							
268.52 57.19	CGCG 12- 54	F		.04	15.0 ±.6				
112.30 -18.24		-5.0± .9							
113733.0-003336	PGC 36190								
114007.4+583644	NGC 3795	.S.....	1.33± .04	53					
139.57 56.15	UGC 6629	R	.59± .05	.00	14.06 ±.18				1091± 50
55.76 4.02	IRAS11373+5853			.88					1177
113723.4+585321	PGC 36192		1.33	.29	13.17				1245
114008.9+151939	NGC 3799	.SBS3*P	.90± .05					15.96±.3	3312± 9
244.79 69.66	UGC 6630	R	.20± .04	.06	14.7 ±.2			438± 7	3449± 31
96.20 -13.08	KUG 1137+156A	3.0± .5		.27				424± 7	3254
113733.4+153617	PGC 36193		.91	.10	14.38			1.49	3656
1140.2 +1727		.S?....	.94± .11					16.41±.3	3297± 9
239.70 70.93	MCG 3-30- 38		.11± .07	.02	14.63 ±.18			213± 10	3237
94.14 -12.31				.09				134± 7	3626
1137.6 +1744	PGC 36194		.93		14.48				
114012.0+171847		.S?....	.91± .07	117					
240.06 70.85	UGC 6631		.24± .05	.02	14.7 ±.2				3543± 50
94.28 -12.37	IRAS11376+1735			.30	13.48				3482
113736.3+173525	PGC 36195		.91	.12	14.37				3872
114013.0+152033	NGC 3800	.SXT3*P	1.31± .03	52				15.76±.2	3299± 5
244.78 69.69·	UGC 6634	R	.54± .05	.06	13.5 ±.2			429± 6	3457± 31
96.20 -13.06	IRAS11376+1537	3.0± .4		.75	11.29			414± 7	3236
113737.5+153711	PGC 36197		1.32	.27	12.65			2.84	3637
114013.9+244151	NGC 3798	.LB....	1.40± .06	60					
217.58 73.82	UGC 6632	U	.15± .05	.00	13.1 ±.2				3509± 39
87.21 -9.63	IRAS11376+2458	-2.0± .7		.00	13.01				3474
113737.3+245829	PGC 36199		1.38		13.08				3817
114016.3+174342	NGC 3801	.L...?.	1.54± .05	120	12.96 ±.13	.93± .02	1.01± .01		
239.03 71.09	UGC 6635	U	.22± .05	.00	12.57 ±.20	.39± .02	.48± .02		3388± 24
93.89 -12.20		-2.0±1.4	1.04± .02	.00		.88	13.65± .07		3329
113740.6+180020	PGC 36200		1.51		12.79	.39	15.00± .33		3717
1140.3 +1745	NGC 3802	.S.....	1.06± .06	85	14.25 ±.14	.94± .03	.98± .02	17.17±.3	3321± 9
238.94 71.11	UGC 6636	R	.64± .05	.00	15.0 ±.2	.40± .05	.49± .04	291± 10	3262
93.86 -12.18	IRAS11377+1802		.79± .02	.96	13.49	.80	13.69± .06	280± 7	3649
1137.7 +1802	PGC 36203		1.06	.32	13.51	.26	12.78± .34	3.34	
114018.6+090033	IC 719	.L...?.	1.10± .07	52				16.24±.2	1860± 8
256.75 65.29	UGC 6633	U	.51± .03	.07	14.1 ±.2			304± 20	1659± 63
102.42 -15.19	ARAK 308	-2.0±1.8		.00	13.02			248± 10	1768
113743.8+091711	PGC 36205		1.03		13.99				2203
114019.2+030004		PSBR3*.	.94± .11						
264.70 60.48	MCG 1-30- 8	F (1)	.11± .07	.01	14.61 ±.18				
108.42 -17.08		3.0± .9		.16					
113745.1+031642	PGC 36206	1.6±1.0	.94	.06					
114021.5-024214		.SA.1P?							
270.28 55.60	CGCG 12- 55	F		.07	15.3 ±.6				
114.22 -18.69		1.0±1.8							
113748.0-022536	PGC 36208								

R.A. 2000 DEC. / l b / SGL SGB / R.A. 1950 DEC.	Names / S_T n_L / / PGC	Type / / T / L	$\log D_{25}$ / $\log R_{25}$ / $\log A_e$ / $\log D_0$	p.a. / A_g / A_i / A_{21}	B_T / m_B / m_{FIR} / B_T^o	$(B-V)_T$ / $(U-B)_T$ / $(B-V)_T^o$ / $(U-B)_T^o$	$(B-V)_e$ / $(U-B)_e$ / m'_e / m'_{25}	m_{21} / W_{20} / W_{50} / HI	V_{21} / V_{opt} / V_{GSR} / V_{3K}
114025.1+282223 / 204.28 74.25 / 83.75 -8.17 / 113748.0+283901	A 1137+28 / UGC 6637 / MK 1507 / PGC 36211	.S?.... / / /	.86± .08 / .29± .05 / / .86	72 / .00 / .39 / .14	/ 15.0 ±.2 / / 14.59			16.82±.3 / 164± 6 / / 2.09	1836± 10 / 1893± 61 / 1816 / 2134
114030.3+601759 / 138.07 54.69 / 54.23 4.75 / 113745.8+603437	NGC 3796 / UGC 6638 / / PGC 36215	.S?.... / / /	1.11± .05 / .15± .05 / / 1.11	127 / .00 / .23 / .08	/ 13.53 ±.20 / / 13.30				1266± 50 / 1357 / 1411
114034.6-100510 / 275.76 49.01 / 121.88 -20.48 / 113801.9-094832	/ MCG -2-30- 14 / / PGC 36217	PSXT6.. / FUE (1) / 6.3± .4 / 5.0± .6	1.43± .03 / .09± .05 / / 1.43	5 / .03 / .13 / .04				14.02±.1 / 126± 5 / 109± 12 /	1738± 4 / / 1592 / 2095
114037.9-443132 / 289.86 16.54 / 158.92 -24.12 / 113810.0-441454	/ ESO 266- 13 / / PGC 36220	RLBR+.. / r / -1.3± .9 /	.91± .07 / .16± .05 / / .93	125 / .41 / .00 /	/ 15.24 ±.14 / /				
114042.0+202035 / 231.91 72.47 / 91.41 -11.15 / 113806.0+203713	NGC 3805 / UGC 6642 / / PGC 36224	.L..-*. / U / -3.0±1.1 /	1.15± .15 / .10± .08 / / 1.13	60 / .00 / .00 /	/ 13.65 ±.15 / / 13.55				6526± 45 / 6476 / 6847
114044.0+222540 / 225.40 73.29 / 89.42 -10.38 / 113807.7+224218	NGC 3808 / MCG 4-28- 21 / / PGC 36227	.SXT5*P / R / 5.0± .4 /	1.22± .07 / .27± .07 / / 1.22	123 / .00 / .40 / .13	14.1 ±.3 / / / 13.66	.65± .03 / .04± .04 / .55 / -.03	/ / 14.37± .50 /	15.95±.3 / 293± 9 / / 2.15	7078± 8 / 7050± 29 / 7033 / 7391
114044.8+222644 / 225.34 73.30 / 89.40 -10.37 / 113808.5+224322	NGC 3808A / MCG 4-28- 20 / IRAS11381+2243 / PGC 36228	.I.0.$P / R / 90.0 /	.74± .14 / .38± .07 / / .72	/ .00 / .28 /	/ 11.95 / /				7189± 29 / 7147 / 7505
1140.8 +1747 / 239.08 71.23 / 93.88 -12.06 / 1138.2 +1804	NGC 3806 / UGC 6641 / / PGC 36231	.SX.3.. / U / 3.0± .8 /	1.15± .03 / .02± .05 / .90± .02 / 1.15	/ .00 / .02 / .01	14.15 ±.14 / 14.0 ±.2 / / 14.05	.55± .04 / -.01± .05 / .52 / -.03	.63± .03 / -.03± .04 / 14.14± .04 / 14.70± .24	15.63±.2 / 160± 6 / 150± 5 / 1.57	3495± 7 / 3436 / 3823
114048.3+351232 / 180.40 72.90 / 77.38 -5.40 / 113810.3+352910	A 1138+35 / MCG 6-26- 16 / MK 426 / PGC 36232	.S?.... / / /	.78± .08 / .32± .06 / / .78	/ .00 / .47 / .16	14.8 ±.2 / 15.1 ±.2 / / 14.46	.47± .05 / -.39± .08 / .40 / -.44	/ / 12.76± .45 /	15.76±.3 / 163± 6 / 114± 5 / 1.14	1548± 10 / 1487± 42 / 1548 / 1814
1140.8 +0448 / 262.82 62.05 / 106.65 -16.40 / 1138.3 +0505	/ CGCG 40- 27 / / PGC 36237	.S..8?/ / F / 8.0±1.8 /		/ .03 / /	/ 14.9 ±.6 / /				
114054.8+561210 / 141.63 58.29 / 58.02 3.14 / 113812.1+562848	NGC 3804 / UGC 6640 / IRAS11381+5628 / PGC 36238	.SXS7.. / PU / 6.5± .5 /	1.35± .03 / .19± .04 / / 1.35	120 / .00 / .26 / .09	/ 13.5 ±.2 / 13.38 / 13.25			14.26±.1 / 187± 6 / 170± 5 / .91	1382± 5 / / 1460 / 1550
114056.0-442850 / 289.90 16.60 / 158.87 -24.07 / 113828.0-441212	/ ESO 266- 15 / IRAS11384-4412 / PGC 36240	.S..4.. / S (1) / 4.0± .8 / 2.2± .8	1.25± .05 / .25± .05 / .94± .02 / 1.29	140 / .41 / .36 / .12	13.00 ±.14 / 13.18 ±.14 / 11.58 / 12.29	.60± .01 / / .44 /	.63± .01 / / 13.19± .04 / 13.50± .30		3160± 40 / 2956 / 3440
114056.3+254651 / 213.82 74.18 / 86.25 -9.07 / 113819.6+260329	/ UGC 6645 / / PGC 36241	.SXR3.. / U (1) / 3.0± .8 / 4.5±1.1	1.13± .05 / .14± .05 / / 1.13	/ .00 / .19 / .07	14.3 ±.2 / / / 14.03			15.66±.3 / 303± 10 / / 1.56	6871± 9 / / 6840 / 7176
1140.9 +0503 / 262.55 62.27 / 106.41 -16.29 / 1138.4 +0520	/ UGC 6646 / / PGC 36242	.SA.5*/ / UF / 5.0± .9 /	1.00± .06 / .47± .05 / / 1.00	140 / .03 / .70 / .23	/ 15.11 ±.18 / / 14.34				5827 / / 5726 / 6179
114058.5+112817 / 252.94 67.22 / 100.05 -14.22 / 113823.5+114455	NGC 3810 / UGC 6644 / IRAS11383+1144 / PGC 36243	.SAT5.. / R (2) / 5.0± .3 / 2.4± .6	1.63± .02 / .15± .02 / 1.10± .01 / 1.64	15 / .15 / .23 / .08	11.35M±.10 / 11.32 ±.14 / 10.33 / 10.95	.58± .02 / -.07± .03 / .51 / -.12	.67± .02 / .02± .03 / 12.34± .03 / 13.97± .14	13.20±.1 / 266± 4 / 247± 3 / 2.17	994± 4 / 958± 36 / 913 / 1335
114059.4-222838 / 282.28 37.55 / 135.03 -22.70 / 113828.0-221200	/ ESO 571- 15 / IRAS11384-2212 / PGC 36245	.S..3./ / S / 3.0± .7 /	1.23± .06 / .84± .05 / / 1.24	37 / .13 / 1.16 / .42				
114104.0+014334 / 266.39 59.51 / 109.77 -17.28 / 113830.1+020012	/ CGCG 12- 57 / / PGC 36248	.LA.0?. / F / -2.0±1.8 /		/ .04 / /	/ 14.9 ±.6 / /				

R.A. 2000 DEC.	Names	Type	$\log D_{25}$	p.a.	B_T	$(B-V)_T$	$(B-V)_\bullet$	m_{21}	V_{21}
l b		S_T n_L	$\log R_{25}$	A_g	m_B	$(U-B)_T$	$(U-B)_\bullet$	W_{20}	V_{opt}
SGL SGB		T	$\log A_\bullet$	A_i	m_{FIR}	$(B-V)_T^o$	$m'_\bullet$	W_{50}	V_{GSR}
R.A. 1950 DEC.	PGC	L	$\log D_o$	A_{21}	B_T^o	$(U-B)_T^o$	m'_{25}	HI	V_{3K}

1141.0 +1013		.S..7*.	1.06± .08						
255.15 66.32	UGC 6647	U	.09± .06	.10	14.6 ±.3				
101.28 -14.61		7.0±1.2		.12					
1138.5 +1030	PGC 36250		1.07	.04					
114107.8+244916	NGC 3812	.E.....	1.23± .14						3632± 24
217.30 74.04	UGC 6648	U	.03± .08	.01	13.37 ±.16				3598
87.17 -9.39		-5.0± .8		.00					3940
113831.3+250554	PGC 36256		1.22		13.31				
1141.1 -0248		.SBS2?.							
270.68 55.60	CGCG 12- 58	F			.07 15.2 ±.6				
114.39 -18.53		2.0±1.8							
1138.6 -0232	PGC 36257								
114109.8-501938		PSBR1*.	1.01± .07	57					
291.61 10.99	ESO 216- 33	r	.30± .06	.55	15.46 ±.14				
165.20 -23.77		1.0±1.3		.31					
113843.0-500300	PGC 36258		1.06	.15					
1141.2 +1959		.S?....	.82± .13						6554±120
233.13 72.42	MCG 3-30- 43		.00± .07	.00	15.14 ±.18				6503
91.79 -11.18				.00					6876
1138.6 +2016	PGC 36262		.82		15.04				
114115.4+595312	NGC 3809	.L.....	1.01± .12						3443± 50
138.26 55.10	UGC 6649	U	.11± .05	.00	13.74 ±.18				3533
54.65 4.67	ARAK 310	-2.0± .8		.00					3590
113831.5+600950	PGC 36263		.99		13.69				
114116.2+474135	NGC 3811	.SBR6*.	1.34± .05	160	12.9 ±.4	.61± .05		14.19±.1	3105± 6
151.96 65.29	UGC 6650	U (1)	.12± .06	.02	12.88 ±.18	-.01± .07		286± 8	3037± 60
65.88 -.27	MK 185	6.0±1.1		.18	12.04	.56		270± 6	3153
113836.0+475813	PGC 36265	5.0±1.0	1.35	.06	12.68	-.04	14.17± .50	1.45	3317
114118.2+363249	NGC 3813	.SAT3*.	1.35± .03	87	12.23 ±.13	.58± .02		13.69±.1	1465± 6
176.19 72.42	UGC 6651	R (1)	.31± .03	.01	12.77 ±.14	-.08± .04		309± 5	1470± 40
76.18 -4.78		3.0± .4		.43		.51		294± 6	1474
113840.1+364927	PGC 36266	5.2± .9	1.35	.16	12.03	-.13	13.03± .21	1.50	1730
114123.1-062840		.SXS9P*	1.22± .04	65				15.53±.1	1715± 7
273.62 52.36	MCG -1-30- 22	FE (1)	.12± .05	.12				144± 10	2070± 39
118.18 -19.43	VV 544	9.3± .7		.12	13.73			77± 12	1591
113850.0-061202	PGC 36274	9.6±1.0	1.24	.06					2084
1141.6 +1704		.S?....	.84± .05		16.42S±.15	.65± .08			
241.24 70.99	MCG 3-30- 45		.49± .03	.02	15.7 ±.2	.02± .14			
94.64 -12.14				.73					
1139.0 +1721	PGC 36284		.84	.24			14.24± .29		
114138.0+375933	IC 2950		.80± .09						6610± 74
171.99 71.77	CGCG 186- 26		.07± .12	.00	14.88 ±.19				6624
74.86 -4.14	MK 638								6868
113859.7+381611	PGC 36287		.80						
114139.4+244800	NGC 3815	.S..2..	1.23± .05	72				15.13±.3	3711± 9
217.47 74.16	UGC 6654	U	.27± .05	.01	13.90 ±.18			368± 7	3725± 50
87.24 -9.29		2.0± .8		.33				338± 7	3678
113902.9+250438	PGC 36288		1.23	.14	13.52			1.48	4019
114139.6-021450		PSBS1P?							
270.40 56.15	CGCG 12- 60	F			.03 15.3 ±.6				
113.85 -18.26		1.0± .9							
113906.1-015812	PGC 36289								
114147.9+200617	NGC 3816	.L.....	1.28± .03	70	13.46 ±.13	.96± .02	.97± .01		5596± 45
233.00 72.59	UGC 6656	U	.23± .04	.00	13.39 ±.15	.44± .03	.52± .02		5546
91.74 -11.00		-2.0± .8	.79± .02	.00		.89	12.90± .08		5918
113912.0+202255	PGC 36292		1.24		13.35	.45	14.15± .24		
114150.6+155825		.SB.0?.	1.18± .05	11					3219
244.03 70.38	UGC 6653	U	.62± .05	.04	14.67 ±.18				3154
95.73 -12.47	IRAS11390+1614	.0± .9		.46	12.67				3551
113915.2+161503	PGC 36294		1.15		14.12				
114150.6+155825		.E?....	.67± .13	20				17.17±.3	745± 10
244.03 70.38	UGC 6655		.18± .03	.04	15.2 ±.3			106± 6	780± 49
95.73 -12.47	MK 747			.00				62± 5	682
113915.2+161503	PGC 36295		.63		15.19				1079
114152.1+023451		RSAS4..	.89± .11					16.46±.3	5912± 12
265.82 60.34	MCG 1-30- 10	F (1)	.07± .07	.03	14.93 ±.18			131± 14	
108.97 -16.83		4.0± .9		.10				117± 15	5803
113918.0+025130	PGC 36297	5.6±1.4	.89	.03	14.76			1.66	6266

R.A. 2000 DEC. / l b / SGL SGB / R.A. 1950 DEC.	Names / PGC	Type / S_T n_L / T / L	$\log D_{25}$ / $\log R_{25}$ / $\log A_e$ / $\log D_o$	p.a. / A_g / A_i / A_{21}	B_T / m_B / m_{FIR} / B_T^o	$(B-V)_T$ / $(U-B)_T$ / $(B-V)_T^o$ / $(U-B)_T^o$	$(B-V)_e$ / $(U-B)_e$ / m'_e / m'_{25}	m_{21} / W_{20} / W_{50} / HI	V_{21} / V_{opt} / V_{GSR} / V_{3K}
114153.1+101807	NGC 3817	.SB.0..	1.02± .06	140	14.29M±.10	.97± .02			6102± 32
255.38 66.52	UGC 6657	U	.08± .05	.10	14.17 ±.18				
101.27 -14.40	HICK 58C	.0± .9		.06		.88			6018
113918.3+103446	PGC 36299		1.02		14.01		14.02± .34		6446
114153.8+230237		.S?....	.82± .13					16.71±.3	10407± 9
223.65 73.75	MCG 4-28- 28		.00± .07	.00	15.09 ±.18			328± 10	
88.93 -9.90				.00					10367
113917.6+231915	PGC 36300		.82	.00	15.04			1.67	10720
114157.4-060921	NGC 3818	.E.5...	1.31± .04	103	12.67 ±.15	.96± .02	.97± .01		
273.59 52.71	MCG -1-30- 23	R	.22± .04	.12	12.57 ±.17				1498± 56
117.88 -19.21		-5.0± .4	.87± .05	.00		.92	12.51± .17		1364
113924.2-055243	PGC 36304		1.27		12.49		13.67± .29		1856
1142.0 +3200		.S?....	.97± .07	129					1481± 19
190.85 74.16	UGC 6658		.41± .05	.00	15.12 ±.18				1473
80.48 -6.44				.62					1763
1139.4 +3217	PGC 36305		.97	.21	14.49				
1142.0 +0739		RLX.0..							
259.57 64.52	CGCG 40- 32	F			15.03 ±.18				
103.91 -15.22		-2.0± .9		.04					
1139.5 +0756	PGC 36307								
114204.9+102302	NGC 3820	.S?....	.82± .13		15.45S±.15				
255.34 66.61	MCG 2-30- 14		.17± .07	.13	15.04 ±.20				6052± 41
101.21 -14.33	HICK 58E			.25					5968
113930.1+103941	PGC 36308		.83	.09	14.89		13.96± .68		6396
114206.0+102103	NGC 3819	.E?....	.90± .11		14.82S±.15				
255.40 66.59	MCG 2-30- 13		.06± .04	.10	14.55 ±.11				6270± 41
101.24 -14.33	HICK 58D			.00					6186
113931.2+103742	PGC 36311		.90		14.45		14.19± .59		6614
114208.9+201900	NGC 3821	RSXS2..	1.16± .05		13.7 ±.2	.82± .01			
232.49 72.76	UGC 6663	U	.04± .05	.00	13.58 ±.18	.23± .02			5552± 36
91.56 -10.85		2.0± .8		.05		.77			5502
113933.1+203538	PGC 36314		1.16	.02	13.53	.22	14.26± .34		5873
114210.0-181009		.SBS4*/	1.22± .04	108				15.95±.1	3637± 9
280.68 41.69	ESO 571- 16	SE (1)	.70± .04	.09				270± 12	3633± 52
130.48 -21.73	IRAS11396-1753	4.0± .8		1.02	12.64				3473
113938.0-175330	PGC 36315	2.2± .9	1.23	.35					3987
1142.1 +1256		.S..8*.	.96± .09	47					6143
250.73 68.46	UGC 6662	U	.39± .06	.07	15.20 ±.18				
98.71 -13.45		8.0±1.3		.48					6068
1139.6 +1313	PGC 36316		.97	.20	14.64				6482
114211.3+101640	NGC 3822	.L...?.	1.14± .07	178	14.14S±.15				6132± 32
255.56 66.55	UGC 6661	U	.25± .03	.13	13.70 ±.16				6048
101.32 -14.34	IRAS11395+1033	-2.0±1.7		.00	11.79				
113936.5+103318	PGC 36319		1.12		13.71		14.09± .37		6476
1142.2 +2002								17.91±.3	6102± 9
233.34 72.65	CGCG 97- 63			.00	15.6 ±.3			167± 10	
91.83 -10.94									6052
1139.6 +2019	PGC 36323								6424
114212.0-104628		PSAR2..	1.24± .05	95					6617
276.73 48.54	MCG -2-30- 16	EF (1)	.15± .05	.04					
122.69 -20.25		2.0± .6		.19					6470
113939.3-102950	PGC 36324	1.6±1.0	1.24	.08					6974
114212.3+002004		.S..3P*	.64± .14	15				15.87±.3	5488± 9
268.26 58.47	UGC 6665	EF	.04± .03	.06	14.7 ±.4			269± 10	5448± 49
111.26 -17.41	MK 1304	3.0±1.3		.06	11.85				5371
113938.5+003642	PGC 36325		.65	.02	14.55			1.30	5843
114214.0+200554		.S?....	.74± .14					17.53±.3	5968± 9
233.18 72.68	MCG 3-30- 48		.28± .07	.00	15.85 ±.20			369± 10	
91.78 -10.91				.39					5918
113938.2+202233	PGC 36328		.74	.14	15.41			1.97	6290
1142.2 +6210		.S..8*.	1.11± .05	176					
136.31 53.12	UGC 6660	U	.86± .05	.00					
52.58 5.69		8.0±1.4		1.06					
1139.5 +6227	PGC 36329		1.11	.43					
114214.8+195840								16.89±.2	7790± 7
233.54 72.63	CGCG 97- 62			.00	15.4 ±.3			279± 10	7752± 19
91.90 -10.95								201± 10	7735
113939.0+201518	PGC 36330								8107

R.A. 2000 DEC.	Names	Type	logD$_{25}$	p.a.	B$_T$	(B-V)$_T$	(B-V)$_e$	m$_{21}$	V$_{21}$
l b		S$_T$ n$_L$	logR$_{25}$	A$_g$	m$_B$	(U-B)$_T$	(U-B)$_e$	W$_{20}$	V$_{opt}$
SGL SGB		T	logA$_e$	A$_i$	m$_{FIR}$	(B-V)$_T^o$	m'$_e$	W$_{50}$	V$_{GSR}$
R.A. 1950 DEC.	PGC	L	logD$_o$	A$_{21}$	B$_T^o$	(U-B)$_T^o$	m'$_{25}$	HI	V$_{3K}$
114215.2-135200 NGC 3823	MCG -2-30- 17	.E+..P.	1.17± .05	100					
278.52 45.70		EF	.08± .05	.04					
125.94 -20.89		-3.5± .6		.00					
113942.8-133521	PGC 36331		1.15						
1142.2 +0846 IC 720	MCG 2-30- 16	.S?....	1.04± .09					17.15±.3	6371± 12
258.01 65.42			.28± .07	.03				178± 20	6587± 56
102.82 -14.81				.43					6292
1139.7 +0903	PGC 36333		1.04	.14					6727
114217.2+301348	UGC 6664	.S..6*.	1.17± .05	102				15.63±.3	9685± 9
197.29 74.54		U	.79± .05	.01	15.35 ±.18			421± 10	
82.17 -7.08		6.0±1.3		1.17					9671
113940.1+303026	PGC 36334		1.17	.40	14.12			1.11	9974
114219.1-083721	MCG -1-30- 24	.SXR0*.	1.11± .06	125					
275.42 50.52		EF	.50± .05	.08					
120.46 -19.72		.0± .6		.38					
113946.2-082043	PGC 36339		1.10						
114220.8+160043	UGC 6666	.S?....	1.07± .04	14				16.29±.3	3087± 9
244.16 70.50	IRAS11397+1617		.47± .04	.05	14.6 ±.2			206± 7	3107± 50
95.73 -12.35				.70	13.45			193± 7	3023
113945.4+161721	PGC 36342		1.07	.23	13.83			2.23	3420
1142.3 +5136	UGC 6667	.S..6*.	1.53± .03	89	14.18S±.15			14.78±.1	979± 6
146.28 62.28		U	.89± .05	.00	14.5 ±.2			189± 8	
62.34 1.48		6.0±1.2		1.31				176± 5	1041
1139.7 +5153	PGC 36343		1.53	.45	12.97		14.43± .24	1.37	1171
1142.3 +1458	UGC 6669	.I...9..	1.16± .13					15.41±.1	1016± 7
246.57 69.86		U	.29± .12	.10				89± 5	
96.75 -12.70		10.0± .8		.22				55± 4	948
1139.8 +1515	PGC 36344		1.17	.15					1351
114223.7+101552 NGC 3825	UGC 6668	.SB.1..	1.11± .05	160	13.96S±.15	.94± .03			
255.68 66.57	HICK 58B	U	.09± .05	.13	13.65 ±.18	.41± .07			6510± 32
101.36 -14.29		1.0± .8		.10		.84			6426
113948.9+103230	PGC 36348		1.13	.05	13.52	.38	14.14± .33		6853
114223.8+200701	MCG 3-30- 51	.S?....	1.05± .04		14.55M±.10	.71± .03	.69± .03	16.05±.2	5970± 7
233.18 72.72			.23± .03	.00	14.55 ±.18	.04± .04	.13± .04	355± 8	5972±120
91.78 -10.87			.63± .03	.34		.63	13.16± .08		5920
113948.0+202340	PGC 36349		1.05	.11	14.18	-.02	14.06± .26	1.76	6292
114228.9-082027 IC 721	MCG -1-30- 26	.SBS6?/	1.09± .06	76					6745
275.29 50.79	IRAS11399-0803	EF	.67± .05	.09					
120.18 -19.62		5.7±1.1		.99	13.38				6605
113955.9-080348	PGC 36354		1.10	.34					7103
114229.4+181956 A 1139+18	UGC 6670	.IB.9..	1.48± .03	153	13.34 ±.16	.42± .05	.48± .02	14.30±.1	921± 7
238.32 71.86	IRAS11398+1836	U	.54± .05	.00	13.7 ±.2	-.25± .06	-.15± .03	220± 10	
93.50 -11.49		10.0± .8	1.06± .03	.40	13.38	.29	14.13± .09	210± 8	864
113953.9+183635	PGC 36355		1.48	.27	13.04	-.35	14.25± .26	.99	1247
114230.6-021509	CGCG 12- 62	.LA.-..							
270.74 56.25		F		.04	15.13 ±.18				
113.91 -18.05		-3.0± .9							
113957.0-015830	PGC 36358								
114233.0+262922 NGC 3826	UGC 6671	.E.....	.95± .10	65					
211.39 74.63		U	.09± .04	.04	14.40 ±.16				9048± 47
85.72 -8.46		-5.0± .9		.00					9020
113956.5+264601	PGC 36359		.93		14.22				9350
114236.1+185041 NGC 3827	UGC 6673	.S?....	.97± .07	65	13.78 ±.13	.50± .01		15.96±.3	3132± 10
236.95 72.15	IRAS11400+1907		.10± .05	.00	14.0 ±.2	-.16± .03		235± 13	3273± 45
93.02 -11.28				.15	12.56	.46		229± 10	3084
114000.4+190720	PGC 36361		.97	.05	13.68	-.19	13.26± .37	2.23	3463
114239.5+244922	UGC 6674	.S..6*.	1.06± .06	153				15.97±.3	6300± 9
217.56 74.38		U	.18± .05	.05	14.7 ±.2			381± 10	
87.31 -9.07		6.0±1.2		.26					6267
114003.2+250600	PGC 36363		1.06	.09	14.38			1.50	6608
1142.6 +0858 IC 722	MCG 2-30- 19	.S?....	1.04± .09					16.55±.3	6535± 12
257.89 65.64			.21± .07	.00	14.53 ±.18			299± 20	
102.65 -14.65				.31					6447
1140.1 +0915	PGC 36365		1.04	.10	14.18			2.26	6881
114242.6-022139	CGCG 12- 63	.SBS3*.							
270.91 56.18		F		.03	15.4 ±.6				
114.04 -18.04		3.0±1.3							
114009.1-020500	PGC 36368								

R.A. 2000 DEC. / l b / SGL SGB / R.A. 1950 DEC.	Names / PGC	Type / S_T n_L / T / L	$logD_{25}$ / $logR_{25}$ / $logA_e$ / $logD_o$	p.a. / A_g / A_i / A_{21}	B_T / m_B / m_{FIR} / B_T^o	$(B-V)_T$ / $(U-B)_T$ / $(B-V)_T^o$ / $(U-B)_T^o$	$(B-V)_e$ / $(U-B)_e$ / m' / m'_{25}	m_{21} / W_{20} / W_{50} / HI	V_{21} / V_{opt} / V_{GSR} / V_{3K}
114245.2+524655	NGC 3824	.SAS1$/	1.13± .05	118					
144.79 61.33	UGC 6676	P	.29± .04	.00	14.46 ±.18				5693± 56
61.28 2.01	IRAS11400+5303	1.0± .8		.29	13.30				5760
114004.2+530333	PGC 36370		1.13	.14	14.09				5879
114246.8+200139		.S?....	.94± .11		14.92 ±.13	.80± .02	.81± .02	17.96±.3	6334± 10
233.59 72.76	MCG 3-30- 55		.28± .07	.00	15.05 ±.18	.11± .03	.20± .03	308± 15	6314±120
91.90 -10.82			.58± .02	.35		.71	13.31± .04		6284
114011.0+201818	PGC 36371		.94	.14	14.60	.03	13.74± .58	3.22	6656
1142.8 +2633		.S..2..	1.07± .06	142					
211.16 74.69	UGC 6677	U	.76± .05	.05	15.60 ±.18				
85.68 -8.38		2.0±1.0		.93					
1140.2 +2650	PGC 36372		1.08	.38					
1142.9 +2004									
233.50 72.81	CGCG 97- 74			.00	15.3 ±.3				6496
91.86 -10.77									6446
1140.3 +2021	PGC 36377								6818
114256.0+230743									8696± 9
223.63 74.00	CGCG 127- 34			.00	15.4 ±.3				
88.94 -9.64									8657
114019.9+232422	PGC 36380								9009
1142.9 +4050		.S..7..	1.09± .06	109					
164.37 70.33	UGC 6679	U	.89± .05	.00					
72.33 -2.76		7.0±1.0		1.22					
1140.3 +4107	PGC 36381		1.09	.44					
114256.5+195803								17.22±.2	7275± 6
233.83 72.77	CGCG 97- 73			.00	15.5 ±.3			266± 13	7282±120
91.97 -10.80								288± 10	7225
114020.8+201442	PGC 36382								7597
1142.9 +7723	NGC 3901	.S..6*.	1.25± .04	165					
127.68 39.10	UGC 6675	U	.37± .04	.13	14.39 ±.20				1708
38.34 11.49		6.0±1.2		.54					1851
1140.0 +7740	PGC 36386		1.26	.18	13.71				1750
1143.0 +1938		.S..3..	1.05± .06	110					
234.82 72.62	UGC 6680	U (1)	.30± .05	.00	15.04 ±.20				13125
92.29 -10.91		3.0± .9		.41					13074
1140.4 +1955	PGC 36388	4.5±1.2	1.05	.15	14.53				13448
114302.0+261531		.S..3*.	.95± .07	85				16.52±.3	9479± 9
212.30 74.71	UGC 6678	U (1)	.25± .05	.04	15.22 ±.18			308± 10	
85.98 -8.45		3.0±1.3		.34					9451
114025.6+263209	PGC 36392	5.5±1.2	.95	.12	14.76			1.63	9782
1143.0 -1251		.LB.+?/	1.20± .11						
278.22 46.70	MCG -2-30- 18	F	.88± .07	.05					
124.92 -20.50		-1.0± .9		.00					
1140.5 -1235	PGC 36393		1.07						
114303.5-232557		.SBS9*.	1.13± .04	14					1932
283.24 36.81	ESO 504- 10	S (1)	.48± .04	.16	15.54 ±.14				
136.13 -22.36		8.7± .7		.49					1757
114032.0-230918	PGC 36395	8.9± .9	1.14	.24	14.88				2274
1143.1 +5906	A 1140+59	.S..9*.	1.21± .06	100	14.5 ±.3		.57± .06	14.71±.1	1326± 5
138.54 55.90	UGC 6682	U (1)	.07± .06	.00			-.28± .08	82± 5	
55.46 4.58	DDO 96	9.0±1.1	1.27± .10	.07			16.30± .22	73± 12	1414
1140.4 +5923	PGC 36398	9.0±1.0	1.21	.03	14.43		15.23± .46	.24	1477
1143.2 +1922									
235.67 72.54	CGCG 97- 77			.00	15.13 ±.18				6531± 19
92.56 -10.96									6479
1140.6 +1939	PGC 36401								6855
1143.2 +1944		.S..0..	1.00± .06	71					
234.60 72.71	UGC 6683	U	.62± .05	.00	15.42 ±.19				7631± 42
92.21 -10.83		.0±1.0		.47					7580
1140.6 +2001	PGC 36402		.97		14.84				7954
114313.2+200018				.00				17.92±.3	7000± 9
233.82 72.84	CGCG 97- 79							251± 10	7016±120
91.96 -10.73									6950
114037.5+201657	PGC 36406								7322
114314.6-124559		.S..5*/	1.09± .06	40					4790
278.24 46.81	MCG -2-30- 21	EF	.77± .05	.05					
124.84 -20.43	IRAS11406-1229	5.0±1.1		1.16	13.60				4639
114042.1-122920	PGC 36409		1.10	.39					5145

R.A. 2000 DEC. l　　b SGL　SGB R.A. 1950 DEC.	Names PGC	Type S_T　n_L T L	$\log D_{25}$ $\log R_{25}$ $\log A_e$ $\log D_o$	p.a. A_g A_i A_{21}	B_T m_B m_{FIR} B_T^o	$(B-V)_T$ $(U-B)_T$ $(B-V)_T^o$ $(U-B)_T^o$	$(B-V)_e$ $(U-B)_e$ m'_e m'_{25}	m_{21} W_{20} W_{50} HI	V_{21} V_{opt} V_{GSR} V_{3K}
114318.6-125241 278.32　46.71 124.96　-20.44 114046.1-123602	NGC　3831 MCG -2-30- 23 PGC 36417	.LXR+P. EF -.5± .6 	1.43± .05 .68± .05 1.34	23 .05 .00 					
1143.3　+3128 192.55　74.55 81.09　-6.39 1140.7　+3145	 MCG　5-28- 23 PGC 36418	.SB?... 	1.04± .09 .28± .07 1.04	 .03 .43 .14	 15.05 ±.20 14.59				1216±115 1207 1500
1143.3　+3128 192.55　74.55 81.09　-6.39 1140.7　+3145	 UGC　6684 PGC 36419	.I..9?. U 10.0±1.8 	.96± .17 .15± .12 .96	160 .03 .11 .07	 15.07 ±.14 14.93			15.38±.1 147± 6 104± 5 .37	1789± 10 1784± 35 1780 2073
1143.3　+1629 243.47　70.99 95.36　-11.94 1140.8　+1646	 UGC　6686 PGC 36431	.S..3.. U　　(1) 3.0± .9 4.5±1.2	1.42± .03 .92± .03 .77± .02 1.42	51 .05 1.27 .46	14.99M±.10 14.69 ±.19 13.56	.79± .04 .18± .05 .57 -.01	 .27± .05 14.35± .07 14.63± .19	14.90±.3 412± 13 402± 10 .88	6546± 10 6484 6877
1143.3　+5530 141.74　59.07 58.81　3.18 1140.7　+5547	 UGC　6685 PGC 36432	.S..6*. U 6.0±1.4 	1.11± .05 .75± .05 1.11	6 .00 1.11 .38					
1143.3　+1811 239.08　71.96 93.72　-11.34 1140.8　+1828	 UGC　6687 PGC 36433	.S..8*. U 8.0±1.3 	1.19± .05 .52± .05 1.19	25 .00 .64 .26					3194 3138 3521
1143.4　+2139 228.71　73.59 90.39　-10.09 1140.8　+2156	 UGC　6689 PGC 36434	.S?.... 	.97± .07 .66± .05 .97	12 .00 .99 .33	 15.68 ±.19 14.67			15.74±.3 245± 10 .74	3524± 9 3480 3841
114324.6+194457 234.65　72.76 92.22　-10.78 114049.0+200136	IC　2951 UGC　6688 PGC 36436	.S..1.. U 1.0± .9 	1.15± .04 .31± .03 .73± .03 1.15	80 .00 .32 .16	14.46 ±.16 14.56 ±.20 14.10	.91± .03 .51± .04 .80 .46	.99± .02 .49± .03 13.60± .09 14.27± .25		6100± 10 6172± 19 6066 6439
114325.5+250019 217.03　74.59 87.20　-8.84 114049.3+251657	 MCG　4-28- 37 PGC 36437	.S?.... 	.82± .13 .17± .07 .82	 .05 .26 .09	 15.41 ±.18 15.07			16.54±.3 228± 10 1.38	6186± 9 6154 6493
114326.1+360638 176.84　73.00 76.76　-4.55 114048.5+362317	A　1140+36 CGCG 186- 27 MK　427 PGC 36438	 	.50± .14 .15± .12 .50	 .00 	 15.8 ±.3 				6378± 42 6386 6644
114327.4+524246 144.68　61.45 61.39　2.08 114046.8+525925	NGC　3829 UGC　6690 IRAS11407+5259 PGC 36439	.SBS3*. P 3.0± .9 	1.03± .04 .20± .04 1.03	95 .00 .28 .10	 14.82 ±.18 14.50				5620± 56 5687 5806
1143.4　+1009 256.37　66.67 101.55　-14.08 1140.9　+1026	NGC　3833 UGC　6692 PGC 36441	.S..5.. U 5.0± .9 	1.16± .05 .29± .05 1.17	27 .12 .43 .14	 14.22 ±.19 				
1143.5　+1906 236.56　72.46 92.84　-10.99 1140.9　+1923	NGC　3834 MCG　3-30- 65 PGC 36443	.S?.... 	1.15± .08 .15± .07 1.14	 .00 .11 	 14.4 ±.2 				
114329.7-014003 270.63　56.88 113.39　-17.66 114056.0-012324	IC　725 CGCG 12- 65 PGC 36444	.E...*. F -5.0±1.3 		 .04 	 14.9 ±.6 				
1143.5　-1647 280.42　43.09 129.09　-21.17 1140.9　-1631	NGC　3836 MCG -3-30- 10 VV　477 PGC 36445	.S?.... 	1.15± .08 .03± .07 1.15	 .07 .04 .01	 11.62 			14.61±.1 217± 12 	3660± 9 3674± 65 3500 4012
1143.5　+2243 225.18　74.00 89.38　-9.67 1140.9　+2300	NGC　3832 UGC　6693 PGC 36446	.SBT4.. U 4.0± .8 	1.29± .03 .08± .03 1.03± .02 1.29	 .00 .11 .04	13.63M±.10 13.4 ±.2 13.44	.67± .02 .01± .03 .61 -.03	.71± .02 .10± .03 14.07± .05 14.72± .20	14.74±.1 189± 4 171± 5 1.26	6909± 4 6906±120 6869 7223
1143.5　+6041 137.19　54.52 54.02　5.25 1140.8　+6058	 UGC　6691 PGC 36447	.S..8*. U 8.0±1.2 	1.08± .07 .32± .06 1.08	120 .06 .39 .16					

R.A. 2000 DEC.	Names	Type	logD$_{25}$	p.a.	B$_T$	(B-V)$_T$	(B-V)$_e$	m$_{21}$	V$_{21}$
l b		S$_T$ n$_L$	logR$_{25}$	A$_g$	m$_B$	(U-B)$_T$	(U-B)$_e$	W$_{20}$	V$_{opt}$
SGL SGB		T	logA$_.$	A$_i$	m$_{FIR}$	(B-V)$_T^0$	m'$_.$	W$_{50}$	V$_{GSR}$
R.A. 1950 DEC.	PGC	L	logD$_0$	A$_{21}$	B$_T^0$	(U-B)$_T^0$	m'$_{25}$	HI	V$_{3K}$
114334.7+085631	IC 724	.S..1..	1.37± .04	60				16.06±.3	5972± 12
258.35 65.75	UGC 6695	U	.40± .05	.03	13.40 ±.18			533± 20	5959± 14
102.76 -14.45	IRAS11409+0913	1.0± .8		.41					5879
114100.1+091310	PGC 36450		1.37	.20	12.88			2.97	6312
114339.8+332504		.SB.1..	.96± .09	102					
185.54 74.10	UGC 6696	U	.29± .06	.00	15.50 ±.19				
79.29 -5.57		1.0± .9		.30					
114102.7+334143	PGC 36455		.96	.15					
114346.5+550251		.SB?...	.87± .07						
142.10 59.49	MCG 9-19-165		.10± .06	.00	15.19 ±.18				6715± 63
59.25 3.05	MK 1452			.15					6790
114105.3+551930	PGC 36463		.87	.05	15.00				6889
1143.8 +1947		.L?....	.90± .17						
234.68 72.86	MCG 3-30- 67		.27± .07	.00	15.19 ±.15				5624± 19
92.21 -10.68				.00					5573
1141.2 +2004	PGC 36465		.86		15.11				5946
114348.7+195812		.I..9*.	1.27± .03	137	14.08M±.10	.49± .02	.50± .01	16.18±.2	6732± 6
234.14 72.95	UGC 6697	U	.76± .03	.00	14.47 ±.19	-.32± .03	-.28± .02	599± 9	6621± 32
92.04 -10.61	IRAS11412+2014	10.0±1.3	.76± .01	.57	12.62	.27	13.37± .03		6678
114113.0+201451	PGC 36466		1.27	.38	13.59	-.48	13.36± .21	2.21	7050
1143.9 +1959	NGC 3841	.S?....	.82± .13		14.59 ±.15	.99± .02	1.01± .01		
234.12 72.98	MCG 3-30- 73		.00± .07	.00	14.99 ±.18	.46± .05	.49± .04		6216± 19
92.03 -10.59			.62± .04	.00		.93	13.18± .14		6166
1141.3 +2016	PGC 36469		.82		14.67	.49	13.52± .68		6538
1143.9 +2000	NGC 3845	.S?....	.89± .11		15.01 ±.13	.98± .02			
234.07 72.98	MCG 3-30- 74		.35± .07	.00	15.31 ±.19	.40± .04			5643±120
92.01 -10.58				.43		.87			5593
1141.3 +2017	PGC 36470		.89	.18	14.62	.32	13.42± .61		5965
114354.1+075532	NGC 3843	.S..0..	.95± .07	42					
260.02 65.01	UGC 6699	U	.35± .05	.00	14.5 ±.2				5908± 50
103.79 -14.70		.0± .9		.26					5817
114119.7+081211	PGC 36471		.93		14.17				6255
114354.4+104659	NGC 3839	.S..8*.	1.00± .06	87				14.90±.3	5910± 12
255.51 67.21	UGC 6700	U	.29± .05	.21	14.1 ±.3			348± 20	5939± 50
100.97 -13.77	IRAS11413+1103	8.0±1.3		.36	11.59				5830
114119.7+110338	PGC 36475		1.02	.15	13.51			1.25	6254
114356.7+195342	NGC 3837	.E.....	.81± .16		14.25 ±.13	1.00± .01	1.01± .01		
234.42 72.94	UGC 6701	U	.02± .05	.00	14.6 ±.2	.55± .04	.59± .04		6280± 25
92.13 -10.61	ARAK 314	-5.0± .9	.53± .03	.00		.94	12.39± .10		6230
114121.1+201021	PGC 36476		.80		14.25	.58	13.22± .80		6602
1143.9 +2004	NGC 3840	.S..1..	1.03± .03	67	14.54M±.10	.70± .02	.73± .01	16.46±.2	7368± 7
233.86 73.03	UGC 6702	U	.15± .03	.00	14.50 ±.18	.01± .03	.07± .03	284± 11	7409± 42
91.95 -10.54	IRAS11413+2021	1.0± .9	.57± .02	.15		.61	12.91± .05	262± 10	7320
1141.3 +2021	PGC 36477		1.03	.07	14.29	.00	14.16± .19	2.10	7691
114358.2+201107								17.69±.3	6373± 9
233.55 73.08	CGCG 97- 92			.00				458± 10	6497±120
91.85 -10.50									6324
114122.6+202746	PGC 36478								6695
1144.0 +2001	NGC 3844	.S..0..	1.07± .03	28	14.85 ±.14	.98± .02	.99± .02		
234.06 73.01	UGC 6705	U	.68± .05	.00	15.18 ±.20	.46± .04	.50± .04		6829± 42
92.01 -10.55		.0± .9	.38± .04	.51		.81	12.24± .12		6780
1141.4 +2018	PGC 36481		1.03		14.34	.38	13.32± .24		7151
1144.0 +3003		.S?....	.99± .10						
197.77 74.93	MCG 5-28- 26		.33± .07	.01	15.18 ±.18				11767
82.48 -6.80				.50					11753
1141.4 +3020	PGC 36482		.99	.17	14.61				12056
114401.9+194657		.S?....	.94± .11						4857
234.79 72.91	MCG 3-30- 71		.47± .07	.00	15.54 ±.18				4937± 19
92.24 -10.63				.58					4886
114126.4+200336	PGC 36486		.94	.24	14.90				5259
114402.0+195701	NGC 3842	.E.....	1.15± .15	5	12.78 ±.13	.96± .01	.99± .01		
234.29 72.99	UGC 6704	U	.14± .08	.00	13.41 ±.17	.49± .03	.54± .02		6214± 17
92.08 -10.57		-5.0± .9	1.10± .02	.00		.90	13.77± .09		6164
114126.4+201340	PGC 36487		1.11		12.92	.52	13.16± .81		6535
114404.6-023327		.SAS5..	.99± .10						
271.62 56.16	MCG 0-30- 20	F (1)	.09± .07	.04	14.9 ±.2				
114.34 -17.76		5.0± .9		.14					
114131.0-021648	PGC 36489	5.6±1.0	.99	.05					

R.A. 2000 DEC. l　　b SGL　SGB R.A. 1950 DEC.	Names PGC	Type S$_T$　n$_L$ T L	logD$_{25}$ logR$_{25}$ logA$_\bullet$ logD$_o$	p.a. A$_g$ A$_i$ A$_{21}$	B$_T$ m$_B$ m$_{FIR}$ B$_T^o$	(B-V)$_T$ (U-B)$_T$ (B-V)$_T^o$ (U-B)$_T^o$	(B-V)$_\bullet$ (U-B)$_\bullet$ m'$_\bullet$ m'$_{25}$	m$_{21}$ W$_{20}$ W$_{50}$ HI	V$_{21}$ V$_{opt}$ V$_{GSR}$ V$_{3K}$
114405.3-523645 292.68　8.91 167.62　-23.15 114138.1-522006	 ESO　216- 35 PGC 36492	.SAR8*. S　　　(1) 7.6± .7 7.0± .7	1.02± .05 .10± .04 1.11	148 1.01 .12 .05	 15.62 ±.14 				
114405.6+600713 137.52　55.06 54.57　5.09 114123.0+602352	NGC　3835 UGC　6703 IRAS11413+6023 PGC 36493	.S..2*/ PU 2.2± .7	1.29± .04 .39± .04 1.29	60 .00 .49 .20	 13.2　±.2 13.27 12.66				2452 2544 2597
1144.1　+2953 198.39　74.97 82.64　-6.85 1141.5　+3010	 CGCG 157- 27 PGC 36495			 .02 	 15.2　±.3 				13876 13862 14166
114411.0+371108 173.40　72.62 75.82　-3.99 114133.4+372747	 CGCG 186- 30 MK　　428 PGC 36500	.S?.... 	.62± .11 .26± .12 .63	 .04 .39 .13	 15.3　±.3 14.81				12471± 42 12483 12732
114414.1+333054 185.06　74.18 79.25　-5.42 114137.0+334733	NGC　3847 UGC　6708 PGC 36504	.E..... U -5.0± .8 	1.04± .18 .00± .08 1.04	 .00 .00 	 14.27 ±.16 14.13				9542± 19 9541 9818
114414.4+575701 139.28　56.99 56.59　4.26 114132.6+581340	NGC　3838 UGC　6707 PGC 36505	.SA.0?. PU -.4± .8 	1.19± .04 .39± .04 1.17	141 .00 .29 	13.25 ±.13 13.1　±.3 12.90	.91± .01 .41± .02 .83 .35	 13.09± .28		 1299± 31 1384 1457
114415.1+550206 142.00　59.54 59.29　3.11 114134.2+551845	NGC　3846A UGC　6706 KUG 1141+553B PGC 36506	.SBS9*. PU 9.3± .6 	1.28± .03 .09± .04 1.28	40 .00 .09 .05	 13.8　±.2 13.75			15.19±.1 160± 8 145± 6 1.40	1443± 7 1518 1617
1144.3　+8328 125.18　33.36 32.53　13.64 1141.0　+8345	 UGC　6694 PGC 36514	.SA.9.. U 9.0± .9 	1.06± .06 .22± .05 1.10	150 .34 .22 .11					
1144.3　+0729 260.86　64.73 104.26　-14.73 1141.8　+0746	 CGCG　40- 35 PGC 36519	RLB.+.. F -1.0± .9 		 .00 	 15.4　±.6 				
1144.3　+0810 259.88　65.27 103.59　-14.51 1141.8　+0827	 MCG　1-30- 12 PGC 36520	.SB?... 	.94± .11 .57± .07 .94	 .00 .85 .28	 15.2　±.2 14.34			16.38±.3 257± 20 1.75	5890± 12 5800 6237
1144.4　+1943 235.11　72.95 92.33　-10.57 1141.8　+2000	 CGCG　97-109 PGC 36525			 .00 	 15.13 ±.18 				6823± 19 6772 7145
114425.1+485008 149.31　64.72 65.03　.67 114145.6+490647	 UGC　6713 PGC 36528	.S..9.. U 9.0± .8 	1.17± .07 .14± .06 1.17	 .01 .15 .07	 15.0　±.3 14.83			14.62±.1 104± 5 95± 4 -.29	899± 6 953 1105
114425.9+332119 185.56　74.27 79.42　-5.44 114148.9+333758	IC　2953 UGC　6709 PGC 36530	PSBR3.. U 3.0± .9 	1.08± .06 .12± .05 1.08	65 .00 .16 .06	 14.8　±.2 				
114426.2-282752 285.52　32.13 141.57　-22.68 114155.0-281112	 ESO　440- 1 PGC 36531	.S?.... 	1.06± .05 .34± .05 1.10	53 .35 .47 .17	 15.37 ±.14 14.48				8571± 34 8387 8901
114428.6+163331 243.78　71.24 95.39　-11.68 114153.4+165010	NGC　3853 UGC　6712 PGC 36535	.E..... U -5.0± .8 	1.22± .04 .22± .08 .70± .03 1.16	140 .07 .00 	13.41 ±.13 13.44 ±.16 13.31	.98± .01 .47± .03 .93 .47	.99± .01 .53± .03 12.40± .10 13.93± .31		3349± 31 3287 3679
1144.4　+1046 255.80　67.29 101.03　-13.64 1141.9　+1103	IC　727 UGC　6715 PGC 36536	.S..3.. U　　　(1) 3.0± .9 4.5±1.3	1.20± .05 .81± .05 1.21	161 .11 1.12 .41	 14.94 ±.18 13.66			15.73±.3 522± 20 1.66	6121± 12 6040 6463
114429.5+553904 141.33　59.02 58.73　3.38 114148.5+555543	NGC　3846 UGC　6710 IRAS11417+5556 PGC 36539	.SA.5*. RCU 4.5± .6 	1.03± .06 .14± .04 1.03	135 .00 .21 .07	 14.56 ±.18 13.44 14.34				1396± 50 1473 1566

R.A. 2000 DEC.	Names	Type	$\log D_{25}$	p.a.	B_T	$(B-V)_T$	$(B-V)_\bullet$	m_{21}	V_{21}
l b		S_T n_L	$\log R_{25}$	A_g	m_B	$(U-B)_T$	$(U-B)_\bullet$	W_{20}	V_{opt}
SGL SGB		T	$\log A_\bullet$	A_I	m_{FIR}	$(B-V)_T^o$	$m'_\bullet$	W_{50}	V_{GSR}
R.A. 1950 DEC.	PGC	L	$\log D_o$	A_{21}	B_T^o	$(U-B)_T^o$	m'_{25}	HI	V_{3K}
114429.9+694354		.S?....	.89± .07	132				15.15±.1	2702± 9
131.23 46.30	UGC 6711		.23± .05	.00	14.2 ±.3			164± 11	2545± 82
45.62 8.80	ARAK 317			.34	13.35			139± 8	2822
114142.7+700033	PGC 36542		.89	.12	13.81			1.22	2789
1144.5 +2006		.S?....	.94± .11						
234.01 73.16	MCG 3-30- 83		.57± .07	.00	15.4 ±.2				6761±120
91.98 -10.40	IRAS11419+2022			.85	13.13				6712
1141.9 +2022	PGC 36544		.94	.28	14.46				7082
1144.5 +1047		.I?....	.94± .11					16.35±.3	6200± 12
255.82 67.32	MCG 2-30- 26		.57± .07	.11	15.58 ±.18			255± 20	
101.02 -13.62				.43					6119
1142.0 +1104	PGC 36547		.95	.28	15.04			1.02	6542
1144.6 +1933	NGC 3857	.L?....	1.02± .14		15.1 ±.2	1.00± .02			
235.68 72.91	MCG 3-30- 84		.27± .07	.00	14.84 ±.15	.54± .04			6255± 19
92.51 -10.59				.00		.93			6204
1142.0 +1950	PGC 36548		.98		14.84	.56	14.37± .75		6578
114435.7-034805		.SBS8..	1.25± .04	107					1637
272.82 55.11	MCG -1-30- 27A	E F (1)	.42± .04	.06					
115.64 -17.97		7.7± .5		.52					1510
114202.2-033125	PGC 36551	6.6±1.0	1.26	.21					1994
1144.6 +6757		.S?....	.89± .07	148					
132.20 47.95	UGC 6714		.56± .05	.00	15.2 ±.2				2801±125
47.29 8.15	IRAS11418+6814			.83	12.25				2917
1141.8 +6814	PGC 36555		.89	.28	14.33				2900
1144.7 +1946									
235.08 73.04	CGCG 97-114			.00	15.3 ±.3				6419± 19
92.31 -10.49									6369
1142.1 +2003	PGC 36565								6741
1144.7 +1952									
234.78 73.09	CGCG 97-115			.00	15.4 ±.3				7792± 45
92.21 -10.45									7742
1142.1 +2009	PGC 36567								8114
114441.6+355805		.S..4..	1.23± .05						
176.87 73.29	UGC 6716	U (1)	.05± .05	.00	14.6 ±.3				
76.99 -4.38		4.0± .8		.08					
114204.3+361444	PGC 36568	3.5±1.0	1.23	.03					
1144.7 +0911		.SB.9*.	1.20± .06					15.69±.3	2869± 8
258.53 66.13	UGC 6717	U	.00± .06	.04				64± 20.	
102.62 -14.09		9.0± .8		.00					2783
1142.2 +0928	PGC 36571		1.21	.00					3214
114447.2+194623			.69± .15					17.48±.3	8293± 9
235.11 73.06	MCG 3-30- 87		.00± .07	.00				550± 10	
92.32 -10.47									8243
114211.7+200302	PGC 36573		.69						8615
114447.4+200731		.S?....	1.08± .06	30	14.44 ±.13	.85± .02	.90± .02	16.95±.3	6573± 8
234.04 73.23	UGC 6719		.19± .05	.00	14.39 ±.18	.28± .03	.34± .03	407± 9	6606± 42
91.98 -10.34			.65± .01	.29		.77	13.18± .03		6525
114211.8+202410	PGC 36574		1.08	.10	14.09	.22	14.19± .33	2.76	6895
114448.7+194739	NGC 3860	.S?....	.99± .06	38	14.22 ±.13	.80± .02	.82± .01	17.38±.3	5595± 8
235.06 73.07	UGC 6718		.25± .05	.00		.31± .03	.33± .03	473± 9	5461± 45
92.30 -10.45	IRAS11422+2003		.63± .02	.31	13.16	.71	12.86± .05		5540
114213.2+200418	PGC 36577		.99	.12	13.86	.26	13.41± .36	3.40	5912
114450.6-013603	IC 728	.SBT3..	1.09± .05	65					
271.11 57.10	UGC 6720	UEF (1)	.29± .04	.04	14.39 ±.18				
113.42 -17.31	IRAS11422-0119	3.3± .5		.40	13.61				
114216.9-011924	PGC 36580	2.4± .7	1.10	.15					
114452.0-091359	NGC 3865	.SXT3P*	1.31± .04	135				15.29±.1	5702± 9
276.71 50.22	MCG -1-30- 28	PEF (1)	.14± .04	.02	13.0 ±.2			636± 12	5684± 66
121.25 -19.25		3.0± .4		.20					5560
114219.0-085719	PGC 36581	3.1±1.2	1.32	.07	12.71			2.51	6059
114452.5+192721	NGC 3859	.S?....	1.07± .03	58	14.76M±.10	.62± .02	.70± .02	16.65±.3	5466± 10
236.09 72.92	UGC 6721		.55± .03	.00	14.98 ±.18	.14± .03	.12± .03	439± 15	5508±120
92.63 -10.56	IRAS11423+1943		.53± .02	.82	13.03	.48	12.86± .05		5415
114217.0+194400	PGC 36582		1.07	.27	13.96	.04	13.62± .19	2.41	5789
114453.1-503240		.SBR4P.	1.20± .05	172					
292.25 10.94	ESO 216- 37	S (1)	.20± .05	.50	14.48 ±.14				
165.39 -23.16	IRAS11423-5015	3.7± .7		.29	13.47				
114225.1-501600	PGC 36584	4.4± .8	1.25	.10					

R.A. 2000 DEC. l b SGL SGB R.A. 1950 DEC.	Names PGC	Type S_T n_L T L	$\log D_{25}$ $\log R_{25}$ $\log A_e$ $\log D_o$	p.a. A_g A_i A_{21}	B_T m_B m_{FIR} B_T^o	$(B-V)_T$ $(U-B)_T$ $(B-V)_T^o$ $(U-B)_T^o$	$(B-V)_e$ $(U-B)_e$ m'_e m'_{25}	m_{21} W_{20} W_{50} HI	V_{21} V_{opt} V_{GSR} V_{3K}
114454.2-440558 290.51 17.16 158.45 -23.37 114225.1-434918	 ESO 266- 20 PGC 36588	PSBR1.. r 1.0± .9 	.99± .07 .16± .06 1.04	115 .54 .16 .08	 15.04 ±.14 				
114454.2+194633 235.15 73.08 92.32 -10.44 114218.7+200312	 MCG 3-30- 92 PGC 36589	.S?.... 	.82± .13 .36± .07 .82	 .00 .44 .18	 15.74 ±.18 15.22			17.11±.3 468± 10 1.71	8288± 9 8248± 19 8231 8603
114455.8-515422 292.62 9.63 166.85 -23.07 114228.0-513742	 ESO 216- 38 PGC 36590	.SBR2.. Sr 2.1± .5 	1.16± .05 .17± .05 1.23	161 .79 .21 .08	 14.91 ±.14 				
1144.9 +0209 267.60 60.38 109.63 -16.22 1142.4 +0226	 PGC 36594	.I..9.. E (1) 10.0± .9 10.9± .8	1.10± .08 .15± .08 1.11	160 .04 .11 .08					
114458.6-022351 271.84 56.41 114.24 -17.50 114225.0-020712	 CGCG 12- 73 PGC 36597	.LA.0.. F -2.0± .9 	 	 .05 	 15.4 ±.6 				
114503.9+193714 235.68 73.04 92.49 -10.46 114228.5+195354	IC 2955 MCG 3-30- 96 ARAK 318 PGC 36603	 	.34? .11± .07 .41± .04 .34	 .00 	15.05 ±.14 16.4 ±.5 	1.03± .01 .59± .02 	1.04± .01 .64± .02 12.59± .13 		6345± 19 6294 6668
114504.0+195826 234.61 73.21 92.15 -10.33 114228.5+201505	NGC 3861 UGC 6724 IRAS11424+2015 PGC 36604	.S?.... (1) 4.5±1.1	1.36± .03 .26± .03 .88± .02 1.36	77 .00 .36 .13	13.47M±.10 13.62 13.10	.80± .02 .18± .02 .72 .11	.88± .01 .27± .02 13.41± .05 14.51± .17	15.31±.1 478± 5 469± 10 2.08	5082± 4 5068± 26 5033 5404
1145.0 +0729 261.19 64.84 104.32 -14.57 1142.5 +0746	 CGCG 40- 39 PGC 36605	 	 	 .00 	 15.5 ±.6 			15.68±.3 545± 14 414± 15 	5823± 12 5731 6171
114504.6+193626 235.72 73.04 92.50 -10.46 114229.1+195305	NGC 3862 UGC 6723 PGC 36606	.E..... U -5.0± .8 	1.17± .09 .00± .05 .77± .03 1.17	 .00 .00 	13.67 ±.13 13.52 ±.16 13.51	1.00± .01 .51± .04 .94 .55	1.03± .01 .52± .02 13.01± .10 14.53± .49		6469± 25 6419 6792
114505.4+082801 259.78 65.62 103.35 -14.25 114230.9+084441	NGC 3863 UGC 6722 IRAS11425+0844 PGC 36607	.S..4.. U 4.0± .8 	1.45± .03 .68± .05 1.45	75 .00 1.00 .34	 13.65 ±.18 13.01 12.62			14.55±.2 508± 7 491± 7 1.59	4492± 7 4572± 50 4405 4839
1145.1 +2125 230.03 73.86 90.76 -9.81 1142.5 +2142	 CGCG 127- 46 PGC 36608	 	 	 .00 	 15.5 ±.3 			17.55±.3 243± 13 199± 10 	7814± 10 7770 8132
114506.3+202614 233.20 73.44 91.71 -10.16 114230.7+204253	 UGC 6725 PGC 36609	.L..... U -2.0± .8 	1.21± .06 .11± .03 1.20	40 .00 .00 	 13.86 ±.19 13.75				6878± 50 6830 7198
114506.7-014010 271.28 57.07 113.51 -17.27 114233.0-012330	 CGCG 12- 74 PGC 36611	.LA.0.. F -2.0± .9 	 	 .04 	 15.3 ±.6 				
114509.3+494309 147.88 64.08 64.26 1.13 114230.0+495948	 UGC 6726 PGC 36613	.L..-*. U -3.0±1.3 	1.04± .11 .18± .05 1.02	65 .00 .00 	 14.57 ±.15 				
114509.7-014546 271.38 56.99 113.61 -17.28 114236.0-012906	 CGCG 12- 75 PGC 36614	.E+.... F -4.0± .9 	 	 .04 	 15.2 ±.6 				
114511.7+614226 136.12 53.69 53.15 5.83 114229.1+615905	 UGC 6727 IRAS11424+6159 PGC 36617	.S?.... 	.96± .07 .44± .05 .96	125 .01 .67 .22	 14.7 ±.2 12.46 14.00				10617± 50 10714 10753
114514.7-014210 271.36 57.06 113.55 -17.25 114241.0-012530	 CGCG 12- 76 PGC 36618	.LB.0?. F -2.0±1.3 	 	 .04 	 15.1 ±.6 				

R.A. 2000 DEC. l b SGL SGB R.A. 1950 DEC.	Names PGC	Type S_T n_L T L	$\log D_{25}$ $\log R_{25}$ $\log A_e$ $\log D_o$	p.a. A_g A_i A_{21}	B_T m_B m_{FIR} B_T^o	$(B-V)_T$ $(U-B)_T$ $(B-V)_T^o$ $(U-B)_T^o$	$(B-V)_e$ $(U-B)_e$ m'_e m'_{25}	m_{21} W_{20} W_{50} HI	V_{21} V_{opt} V_{GSR} V_{3K}
114515.0+195043 235.07 73.19 92.29 -10.34 114239.5+200722	 MCG 3-30- 98 ARAK 319 PGC 36619		.52± .20 .08± .07 .53± .04 .52	 .00 	14.84 ±.15 15.9 ±.3 	.92± .03 .36± .05 	1.00± .02 .46± .04 12.98± .14 		7646± 45 7596 7968
114516.5+192326 236.44 72.97 92.73 -10.50 114241.0+194006	NGC 3864 MCG 3-30- 97 PGC 36620	.S?....	.94± .11 .11± .07 .94	 .00 .16 .06	 15.08 ±.19 14.88				6997± 19 6946 7320
1145.2 -0920 276.92 50.16 121.39 -19.18 1142.7 -0903	NGC 3866 MCG -1-30- 29 PGC 36621	PSBT1.. FE 1.0± .6	1.14± .06 .24± .05 1.15	120 .03 .25 .12					
114517.0+200119 234.55 73.28 92.12 -10.27 114241.6+201759	 CGCG 97-133 PGC 36622			 .00 	 15.6 ±.3 			17.94±.3 216± 10 	5290± 9 5241 5611
114517.7+264602 210.56 75.27 85.69 -7.79 114241.6+270242	IC 2956 UGC 6729 PGC 36625	.SXS4.. U 4.0± .9	1.07± .06 .20± .05 1.08	65 .07 .29 .10	 14.55 ±.19 14.13			16.68±.3 421± 10 2.45	9045± 9 9020 9346
114522.7-015240 271.57 56.92 113.74 -17.26 114249.0-013600	 CGCG 12- 77 PGC 36634	.LB.-.. F -3.0± .9		 .04 	 15.3 ±.6 				
1145.3 +1928 236.25 73.03 92.66 -10.44 1142.8 +1945	NGC 3868 MCG 3-30-104 PGC 36638	.S?....	.89± .11 .43± .07 .87	 .00 .32 	 15.3 ±.2 14.83				6653±120 6602 6976
114523.8+201932 233.65 73.45 91.84 -10.14 114248.3+203612	 MCG 4-28- 47 ARAK 321 PGC 36639		.64± .17 .11± .07 .64	 .00 	 15.4 ±.3 				7243±120 7195 7564
114524.3-442528 290.69 16.87 158.80 -23.27 114255.0-440848	 ESO 266- 22 PGC 36640	.E?....	.92± .07 .04± .05 .85± .05 .98	 .40 .00 	14.43 ±.15 14.62 ±.14 	1.14± .04 	1.16± .03 14.17± .17 13.94± .39		
114525.8+000008 269.92 58.58 111.84 -16.73 114252.0+001648	 CGCG 12- 78 PGC 36642	.LA.0./ F -2.0± .9		 .05 	 15.1 ±.6 				
114526.2-100610 277.44 49.47 122.19 -19.32 114253.3-094930	 MCG -2-30- 27 PGC 36643	.SXS7.. EF (1) 6.5± .6 4.9± .7	1.44± .03 .55± .05 1.44	143 .00 .77 .28				14.26±.1 216± 6 197± 12 	1716± 5 1573 2073
114526.9+090943 258.89 66.22 102.70 -13.94 114252.4+092623	NGC 3876 UGC 6730 IRAS11428+0926 PGC 36644	.S..2*. U 2.0±1.2	1.06± .06 .20± .05 1.06	105 .03 .25 .10	 13.7 ±.2 12.85 13.42			15.78±.2 170± 7 147± 7 2.26	2897± 7 2811 3242
114527.2-503534 292.35 10.92 165.44 -23.07 114259.1-501854	 ESO 216- 39 PGC 36647	PSXR2.. r 2.2± .9	1.15± .05 .42± .05 1.20	157 .50 .51 .21	 15.20 ±.14 				
1145.4 +0943 258.02 66.65 102.15 -13.76 1142.9 +1000	 MCG 2-30- 30 PGC 36648	.L?....	1.02± .14 .11± .07 1.01	 .07 .00 	 14.62 ±.17 14.45			15.93±.3 401± 20 	6407± 12 6323 6751
114529.4+192350 236.51 73.02 92.74 -10.45 114254.0+194030	NGC 3867 UGC 6731 PGC 36649	.S?....	1.18± .03 .42± .05 .72± .01 1.18	173 .00 .64 .21	14.20 ±.13 14.42 ±.18 13.60	1.02± .02 .55± .03 .89 .43	1.00± .02 .56± .03 13.29± .03 13.89± .24		7457± 19 7406 7780
1145.5 +7941 126.58 36.97 36.19 12.41 1142.5 +7958	 UGC 6728 PGC 36651	.SB.0.. U .0± .9	.89± .10 .19± .06 .89	160 .19 .14 	 14.89 ±.19 				
114530.5-433428 290.47 17.70 157.88 -23.26 114301.1-431748	 ESO 266- 23 PGC 36652	.SBR3.. Sr (1) 2.6± .6 3.3± .8	1.05± .05 .02± .05 1.10	 .52 .03 .01	 14.79 ±.14 14.16				10510± 60 10308 10793

R.A. 2000 DEC. / l b / SGL SGB / R.A. 1950 DEC.	Names / PGC	Type / S_T n_L / T / L	$\log D_{25}$ / $\log R_{25}$ / $\log A_e$ / $\log D_o$	p.a. / A_g / A_i / A_{21}	B_T / m_B / m_{FIR} / B_T^o	$(B-V)_T$ / $(U-B)_T$ / $(B-V)_T^o$ / $(U-B)_T^o$	$(B-V)_e$ / $(U-B)_e$ / m'_e / m'_{25}	m_{21} / W_{20} / W_{50} / HI	V_{21} / V_{opt} / V_{GSR} / V_{3K}
114533.3+585843 138.14 56.16 55.70 4.82 114251.9+591522	A 1142+59 UGC 6732 7ZW 421 PGC 36655	.L?.... 	.94± .13 .08± .05 .92	 .00 .00 	13.87 ±.13 13.8 ±.2 13.80	.74± .01 .32± .01 .70 .33	 13.22± .66	 	2979± 43 3067 3131
114535.1+031350 266.72 61.38 108.60 -15.75 114301.0+033030	IC 730 MCG 1-30- 13 IRAS11430+0330 PGC 36658	.S..0P? F .0±1.8 	.90± .17 .16± .07 .89	 .05 .12 	 14.73 ±.19 11.86 	 	 	 	
1145.5 +0907 259.02 66.20 102.75 -13.92 1143.0 +0923	 UGC 6734 IRAS11430+0923 PGC 36659	.S..3.. U (1) 3.0± .9 4.5±1.3	1.16± .07 .71± .06 1.16	 .03 .97 .35	 15.10 ±.18 12.29 14.05	 	 	16.00±.3 379± 20 1.60	6257± 12 6171 6602
114536.5+555308 140.84 58.89 58.58 3.62 114255.9+560948	NGC 3850 UGC 6733 PGC 36660	.SBS5*. R 5.0± .5 	1.34± .04 .30± .04 1.34	130 .00 .46 .15	 14.0 ±.2 13.58	 	 	14.84±.1 171± 8 167± 6 1.10	1156± 7 1234 1324
114542.4-282204 285.80 32.31 141.50 -22.39 114311.0-280524	 ESO 440- 4 PGC 36664	.SBS8.. S (1) 8.0± .5 8.1± .5	1.40± .04 .39± .05 1.44	63 .40 .48 .19	 14.21 ±.14 13.32	 	 	13.75±.3 205± 15 178± 12 .23	1840± 11 1836± 34 1657 2170
1145.7 +1029 256.88 67.27 101.42 -13.45 1143.1 +1045	 UGC 6740 IRAS11431+1045 PGC 36666	.S?.... 	1.02± .06 .26± .05 1.03	80 .13 .39 .13	 14.60 ±.18 14.05	 	 	16.09±.3 394± 20 1.92	5483± 12 5401 5826
114545.6+104925 256.32 67.54 101.09 -13.33 114310.9+110605	NGC 3869 UGC 6737 PGC 36669	.S..1.. U 1.0± .9 	1.29± .04 .56± .05 1.30	135 .12 .58 .28	 13.65 ±.20 12.92	 	 	 	3026± 50 2946 3368
114545.6+194628 235.49 73.26 92.40 -10.25 114310.2+200308	NGC 3873 UGC 6735 PGC 36670	.E..... U -5.0± .9 	1.17± .09 .06± .05 .69± .05 1.15	95 .00 .00 	13.85 ±.15 14.01 ±.20 13.82	1.00± .03 .41± .10 .95 .44	1.02± .01 .46± .10 12.79± .17 14.55± .51	 	5507± 35 5457 5829
114546.1+030144 267.02 61.24 108.82 -15.77 114312.0+031824	 UGC 6736 PGC 36671	.SB.6?/ UF 5.5± .9 	1.16± .05 .63± .05 1.16	178 .03 .93 .32	 14.75 ±.18 13.77	 	 	16.07±.3 312± 14 307± 15 1.99	5986± 12 5880 6339
114546.5+200150 234.71 73.39 92.15 -10.16 114311.0+201830	 CGCG 97-138 PGC 36672	 	 	 .00 	 14.4 ±.3 	 	 	16.99±.2 131± 11 60± 10 	5313± 7 5255± 19 5256 5627
1145.7 +1933 236.16 73.16 92.61 -10.32 1143.2 +1950	 MCG 3-30-107 PGC 36673	.L?.... 	.72± .22 .00± .07 .72	 .00 .00 	 15.40 ±.16 15.32	 	 	 	5486± 19 5435 5809
1145.7 +1945 235.56 73.26 92.42 -10.25 1143.2 +2002	NGC 3875 UGC 6739 PGC 36675	.S..0.. U .0±1.0 	1.00± .06 .58± .05 .97	87 .00 .44 	 14.9 ±.3 14.33	 	 	 	6958± 18 6908 7280
114549.0+134558 250.77 69.67 98.22 -12.33 114314.1+140238	NGC 3872 UGC 6738 PGC 36678	.E.5... R -5.0± .4 	1.27± .05 .20± .04 .78± .03 1.23	 .12 .00 	12.74 ±.14 12.80 ±.12 12.61	1.00± .01 .94 	1.01± .01 12.13± .11 13.57± .29	 	3186± 27 3116 3523
1145.8 +2038 232.85 73.69 91.58 -9.91 1143.3 +2055	 CGCG 127- 49 PGC 36683	 	 	 .00 	 15.4 ±.3 	 	 	17.20±.3 297± 13 286± 10 	7061± 10 7014 7381
114555.6+210136 231.62 73.87 91.21 -9.77 114320.1+211816	 UGC 6743 PGC 36684	.SX.4.. U 4.0± .8 	1.20± .05 .01± .05 .84± .02 1.20	 .00 .02 .01	14.33 ±.14 14.0 ±.3 14.19	.65± .04 .19± .05 .60 .16	.81± .04 .22± .05 14.02± .04 15.13± .30	15.85±.2 197± 5 185± 7 1.66	6750± 6 6705 7069
114556.1+501203 147.02 63.75 63.87 1.44 114316.9+502843	NGC 3870 UGC 6742 MK 186 PGC 36686	.L...?. U -2.0±1.7 	1.02± .08 .09± .03 .41± .05 1.01	25 .00 .00 	13.50M±.11 13.4 ±.2 12.89 13.46	.46± .03 -.23± .04 .44 -.23	.41± .02 -.25± .03 10.99± .17 13.24± .42	15.51±.1 123± 6 98± 4 	756± 7 658± 52 813 953
114559.5+202623 233.52 73.62 91.78 -9.96 114324.0+204303	IC 732 MCG 4-28- 50 PGC 36688	.L?.... 	 	 .00 	 	 	 	17.65±.3 333± 10 	7288± 9 7241 7608

R.A. 2000 DEC. l b SGL SGB R.A. 1950 DEC.	Names PGC	Type S_T n_L T L	$\log D_{25}$ $\log R_{25}$ $\log A_e$ $\log D_0$	p.a. A_g A_i A_{21}	B_T m_B m_{FIR} B_T^o	$(B-V)_T$ $(U-B)_T$ $(B-V)_T^o$ $(U-B)_T^o$	$(B-V)_e$ $(U-B)_e$ m' m'_{25}	m_{21} W_{20} W_{50} HI	V_{21} V_{opt} V_{GSR} V_{3K}
114606.2-562328 293.94 5.34 171.63 -22.54 114339.0-560648	NGC 3882 ESO 170- 11 IRAS11436-5606 PGC 36697	.SBS4.. S (1) 4.0± .5 4.1± .6	1.37± .05 .24± .07 1.52	126 1.63 .35 .12	 10.00 			13.50±.3 270± 7 	1817± 9 1610 2044
114608.0+472939 150.72 65.96 66.39 .40 114329.4+474618	NGC 3877 UGC 6745 IRAS11434+4746 PGC 36699	.SAS5*. R (1) 5.0± .3 3.9± .8	1.74± .02 .63± .03 1.21± .02 1.74	35 .01 .94 .31	11.79M±.10 11.85 ±.13 10.99 10.85	.80± .02 .21± .03 .68 .10	.86± .01 .22± .02 13.40± .05 13.78± .15	13.94±.1 368± 7 341± 5 2.77	902± 6 838± 66 951 1115
1146.1 +0730 261.68 65.01 104.39 -14.30 1143.6 +0747	 CGCG 40- 42 PGC 36703	PSBR8*. F (1) 8.0± .9 5.6±1.0		.00	14.3 ±.3				
1146.1 +1036 256.90 67.44 101.34 -13.30 1143.6 +1053	 MCG 2-30- 35 PGC 36704	.I?.... 	.94± .11 .11± .07 .95	 .11 .09 .06	 14.51 ±.18 14.32			16.28±.3 136± 20 1.90	3038± 12 2957 3380
114612.4+202331 233.76 73.64 91.84 -9.93 114337.0+204011	NGC 3884 UGC 6746 PGC 36706	.SAR0.. U .0± .8 	1.32± .03 .19± .06 .90± .02 1.31	10 .00 .14 	13.50 ±.13 13.5 ±.2 13.24	.91± .02 .34± .02 .81 .35	.96± .01 .43± .02 13.49± .05 14.49± .25	15.81±.3 529± 13 505± 10 	6948± 10 6869± 45 6897 7265
114617.6-031046 273.00 55.86 115.13 -17.39 114344.1-025406	 CGCG 12- 79 PGC 36707	.S..7*/ F 7.0±1.3 		.06	15.3 ±.6				
114622.3+330944 185.75 74.72 79.75 -5.15 114345.6+332624	NGC 3880 MCG 6-26- 33 PGC 36712	.L?.... 	.82± .19 .00± .07 .82	 .00 .00 	 14.75 ±.16 14.60				10052± 19 10050 10329
1146.3 +1349 250.93 69.81 98.21 -12.19 1143.8 +1406	 UGC 6747 PGC 36713	.I..9*. U 10.0±1.4 	1.07± .07 .79± .06 1.08	70 .12 .60 .40	 15.58 ±.18 14.86				2693 2623 3029
114626.2+345107 179.89 74.10 78.18 -4.48 114349.5+350747	A 1143+35 CGCG 186- 45 MK 429 PGC 36716	.S?.... 	.50± .14 .15± .12 .50	 .00 .23 .08	 15.5 ±.4 15.23			16.51±.3 187± 6 176± 5 1.21	1382± 10 1308± 42 1382 1648
114632.3-300558 286.60 30.71 143.37 -22.39 114401.0-294918	 ESO 440- 6 PGC 36719	.L..+?P S -1.0±1.8 	1.21± .04 .53± .03 1.15	44 .19 .00 	 14.20 ±.14 13.99				1754± 34 1569 2080
114632.7+713733 130.04 44.60 43.89 9.65 114345.9+715413	A 1143+71 MK 187 PGC 36720			.01					9600±110 9727 9677
114635.5-035134 273.65 55.28 115.84 -17.50 114402.0-033454	A 1144-03A MCG -1-30- 32 PGC 36723	.SXT3$P R 3.0± .5 	1.08± .05 .28± .04 1.08	135 .05 .39 .14					5108± 56 4982 5465
114639.7-015946 272.19 56.96 113.95 -16.99 114406.0-014306	 UGC 6750 PGC 36726	.SBS4*/ F 4.0±1.3 	1.06± .06 .44± .05 1.06	13 .05 .65 .22	 14.83 ±.18 				
1146.7 +2358 221.54 75.07 88.46 -8.53 1144.1 +2415	 UGC 6751 PGC 36727	.S..4.. U 4.0± .9 	1.11± .05 .43± .05 1.12	97 .05 .63 .21	 14.99 ±.19 14.27			15.42±.3 326± 10 .94	6409± 9 6374 6719
114645.5-035052 273.71 55.31 115.84 -17.46 114412.1-033412	A 1144-03B MCG -1-30- 33 PGC 36733	.SBS3*P R 3.0± .4 	1.26± .04 .43± .04 1.27	85 .05 .59 .21				14.53±.1 519± 12 	5167± 9 5008± 56 5037 5520
114646.6-275522 285.91 32.80 141.05 -22.11 114415.1-273842	NGC 3885 ESO 440- 7 IRAS11442-2738 PGC 36737	.SAS0.. R (1) .0± .4 5.0±1.5	1.38± .03 .40± .03 .72± .02 1.40	123 .37 .30 	11.89V±.13 12.83 ±.10 10.62 12.13	.95± .01 .30± .02 .78 .16	.98± .01 .35± .02 13.60± .23	14.24±.3 562± 6 555± 5 	1802± 10 1918± 66 1623 2136
114646.9+204036 233.67 73.89 91.62 -9.71 114411.5+205716	NGC 3883 UGC 6754 PGC 36740	.SAT3.. U (1) 3.0± .7 2.5±1.0	1.47± .03 .09± .03 1.15± .02 1.47	 .00 .13 .05	13.40M±.10 13.1 ±.4 13.20	.75± .03 .17± .04 .69 .12	.89± .02 .38± .02 14.64± .06 15.35± .18	14.76±.1 224± 4 207± 5 1.51	7025± 4 7103± 46 6979 7345

R.A. 2000 DEC. l b SGL SGB R.A. 1950 DEC.	Names PGC	Type S_T n_L T L	$\log D_{25}$ $\log R_{25}$ $\log D_o$	p.a. A_g A_i A_{21}	B_T m_B m_{FIR} B_T^o	$(B-V)_T$ $(U-B)_T$ $(B-V)_T^o$ $(U-B)_T^o$	$(B-V)_\bullet$ $(U-B)_\bullet$ $m'_\bullet$ m'_{25}	m_{21} W_{20} W_{50} HI	V_{21} V_{opt} V_{GSR} V_{3K}
114649.5-034922 A 1144-03C 273.72 55.34 MCG -1-30- 34 115.82 -17.44 114416.0-033242 PGC 36742		.SBS5P. R 5.0± .5	.99± .06 .14± .04 1.00	25 .05 .22 .07					5396± 41 5270 5753
114649.6+692255 NGC 3879 131.14 46.70 UGC 6752 46.02 8.86 IRAS11441+6939 114404.5+693935 PGC 36743		.S..8*. U 8.0±1.2	1.42± .04 .75± .05 1.42	130 .00 13.59 ±.19 .92 13.76 .37 12.67				14.51±.1 215± 8 198± 8 1.46	1431± 9 1552 1522
1146.8 -1428 280.32 45.54 MCG -2-30- 29 126.82 -19.93 1144.3 -1412 PGC 36744		.SBR6.. F (1) 6.0± .8 4.6±1.4	1.12± .08 .22± .07 1.12	.03 .32 .11					4494 4340 4847
114700.7-001738 270.86 58.51 CGCG 12- 81 112.26 -16.43 114427.0-000058 PGC 36750		.LA.0P. F -2.0± .9		.05 14.9 ±.6					
114704.7-165116 NGC 3887 281.55 43.33 MCG -3-30- 12 129.32 -20.34 IRAS11445-1634 114432.1-163436 PGC 36754		.SBR4.. R (3) 4.0± .3 2.9± .5	1.52± .02 .12± .03 1.38± .11 1.52	20 11.41S±.15 .05 11.5 ±.2 .18 11.15 .06 11.21			13.27± .26 13.55± .20	13.18±.1 248± 9 246± 11 1.91	1209± 7 1050 1560
114705.2+195016 NGC 3886 235.83 73.56 UGC 6760 92.45 -9.94 114429.9+200656 PGC 36756		.L..-*. U -3.0±1.2	1.09± .07 .16± .03 1.07	132 * .00 14.11 ±.15 .00 14.03					5717± 46 5668 6038
114706.4+134225 251.52 69.86 UGC 6758 98.38 -12.06 IRAS11445+1359 114431.6+135905 PGC 36759		.S?.... 	1.16± .04 .02± .05 1.17	.10 13.7 ±.2 .04 13.64 .01 13.58				15.15±.3 191± 7 180± 7 1.55	3102± 11 3103± 50 3032 3438
114707.7+293440 199.34 75.65 UGC 6761 83.19 -6.36 114431.6+295120 PGC 36761		.S..0.. U .0± .9	1.00± .06 .46± .05 .99	97 * .04 15.19 ±.18 .34 14.71			17.68±.3 417± 10	6811± 9 6797 7102
114716.7-373304 289.10 23.59 ESO 378- 20 151.39 -22.79 114446.0-371624 PGC 36767		.LBT0?. Sr -2.1± .8	1.11± .05 .33± .03 1.10	29 12.76V±.13 .40 13.58 ±.14 .00 13.20	.93± .02 .39± .04 .78 .29		13.28± .29		3032± 57 2837 3336
1147.3 +5545 140.55 59.13 MCG 9-19-182 58.80 3.80 1144.7 +5602 PGC 36774		.E?.... 	 .45± .03	.00	15.45 ±.13	1.10± .02 .95	1.14± .02 13.19± .10		15459± 56 15537 15628
114722.7+601757 136.75 55.08 UGC 6762 54.56 5.53 IRAS11447+6034 114441.7+603437 PGC 36776		.S?.... 	1.02± .06 .00± .05 1.02	.04 13.93 ±.18 .00 .00 13.87					3574± 50 3667 3718
1147.3 +1958 235.54 73.69 MCG 3-30-114 92.35 -9.83 1144.8 +2015 PGC 36779		.S?.... 	.99± .04 .44± .03 .39± .01 .99	15.52M±.10 .00 15.35 ±.18 .61 .22 14.82	.91± .03 .35± .04 .79 .24	.99± .03 .33± .05 12.99± .03 14.24± .26	16.75±.3 433± 13 404± 10 1.70	6188± 10 6140 6509
1147.4 -0301 273.35 56.12 CGCG 12- 82 115.06 -17.07 1144.9 -0245 PGC 36784		.L...?. F -2.0±1.8		.06 15.4 ±.6					
1147.4 +0246 268.05 61.23 CGCG 40- 43 109.21 -15.44 1144.9 +0303 PGC 36786		.LX.0*. F -2.0± .9		.06 15.1 ±.6					
1147.5 +6907 131.20 46.97 UGC 6764 46.29 8.83 1144.8 +6924 PGC 36787		.S..8*. U 8.0±1.2	1.13± .07 .50± .06 1.13	70 .00 .61 .25					1458 1578 1550
114733.7+555817 NGC 3888 140.30 58.95 UGC 6765 58.61 3.91 MK 188 114453.9+561457 PGC 36789		.SXT5.. R (2) 5.0± .4 2.7± .7	1.24± .03 .13± .03 .83± .03 1.24	120 12.71 ±.15 .00 12.89 ±.15 .19 11.31 .06 12.60	.59± .02 -.08± .04 .55 -.11	.67± .02 .01± .03 12.35± .07 13.46± .22	14.35±.1 276± 16 270± 12 1.69	2408± 11 2404± 44 2487 2576
1147.6 +5430 141.68 60.24 UGC 6766 59.98 3.35 1145.0 +5447 PGC 36795		.S..4.. U 4.0±1.0	1.10± .06 .76± .05 1.10	135 .00 1.12 .38					

R.A. 2000 DEC. l b SGL SGB R.A. 1950 DEC.	Names PGC	Type S_T n_L T L	$\log D_{25}$ $\log R_{25}$ $\log A_e$ $\log D_o$	p.a. A_g A_i A_{21}	B_T m_B m_{FIR} B_T^o	$(B-V)_T$ $(U-B)_T$ $(B-V)_T^o$ $(U-B)_T^o$	$(B-V)_e$ $(U-B)_e$ m'_e m'_{25}	m_{21} W_{20} W_{50} HI	V_{21} V_{opt} V_{GSR} V_{3K}
114743.7+014935 269.14 60.45 110.18 -15.65 114509.8+020615	 UGC 6769 PGC 36800	.SBR3*. UEF (1) 3.3± .5 1.6± .7	1.09± .04 .36± .04 1.09	73 .06 .49 .18	 14.65 ±.18 14.03			15.60±.3 470± 13 1.39	8537± 10 8429 8891
1147.7 +5546 140.44 59.14 58.80 3.85 1145.1 +5603	 MCG 9-19-190 PGC 36805	.L?.... 	.97± .15 .14± .07 .54± .05 .95	 .00 .00 	15.41 ±.15 15.18	1.11± .02 .96 	1.13 ±.01 13.60± .16 14.77± .79		 15260± 59 15338 15429
114749.3+434453 156.47 69.01 69.99 -.79 114511.7+440133	 UGC 6768 PGC 36811	.SB.2*. U 2.0± .9 	.97± .07 .34± .05 .97	42 .00 .42 .17	 15.13 ±.18 				
114749.6+312042 192.23 75.51 81.58 -5.55 114513.4+313722	IC 2961 CGCG 157- 34 MK 748 PGC 36812	 		 .04 	 15.4 ±.3 				8253± 61 8245 8537
1147.8 +2002 235.53 73.82 92.33 -9.69 1145.3 +2019	 MCG 3-30-115 PGC 36816	.L?.... 	.97± .15 .14± .07 .95	 .00 .00 	 14.75 ±.15 14.64				7280±120 7232 7601
114754.1+040056 266.89 62.35 108.00 -14.97 114520.1+041736	 CGCG 40- 44 PGC 36817	.E..... F -5.0± .9 		 .04 	 15.0 ±.6 				
114759.1-505211 292.82 10.75 165.71 -22.66 114530.0-503530	 ESO 217- 1 PGC 36821	.SBS7.. S (1) 7.0± .6 7.2± .6	1.01± .05 .20± .05 1.09	95 .82 .28 .10	 16.25 ±.14 				
114800.1+042916 266.39 62.77 107.54 -14.80 114526.0+044556	 UGC 6771 PGC 36824	PSXR2.. UF (1) 1.5± .5 3.6±1.0	1.23± .04 .03± .05 1.23	 .01 .04 .01	 13.6 ±.2 13.47			16.72±.2 309± 10 287± 15 3.24	5964± 8 5981± 50 5864 6315
114800.5+494830 146.89 64.27 64.36 1.59 114522.1+500510	 UGC 6773 PGC 36825	.I..9.. U 10.0± .8 	1.21± .05 .29± .05 1.21	168 .00 .22 .15	 14.8 ±.2 14.59				925 983 1126
114801.1-105743 278.84 48.92 123.23 -18.90 114528.1-104103	NGC 3892 MCG -2-30- 30 PGC 36827	.LBT+.. R -1.0± .3 	1.47± .04 .12± .04 1.45	95 .01 .00 	 12.42 ±.18 12.38				1697± 66 1553 2052
114802.9-522511 293.22 9.25 167.37 -22.56 114534.1-520830	 ESO 217- 2 PGC 36830	RLB.0.. r -2.4± .9 	.95± .06 .20± .05 1.03	169 .97 .00 	 15.28 ±.14 				
114803.1+302134 196.09 75.75 82.53 -5.88 114527.0+303814	NGC 3891 UGC 6772 PGC 36832	.S..4.. U 4.0± .8 	1.31± .04 .08± .05 1.31	70 .04 .12 .04	 13.2 ±.2 13.01			14.76±.3 462± 10 1.71	6371± 9 6196± 50 6355 6654
1148.0 +5502 141.06 59.80 59.50 3.61 1145.4 +5519	A 1145+55 UGC 6774 PGC 36836	.S..6*. U 6.0±1.2 	1.31± .04 .91± .05 1.31	155 .00 1.33 .45					2420 2496 2593
114820.2+124258 254.13 69.36 99.45 -12.11 114545.6+125939	 MCG 2-30- 37 HICK 59B PGC 36853	.L?.... 	.72± .22 .00± .07 .73	 .07 .00 	15.60S±.15 15.23 ±.16 15.29		 14.05±1.12		3908± 41 3835 4246
1148.3 +2109 232.10 74.44 91.30 -9.19 1145.8 +2126	 CGCG 127- 56 PGC 36856	 		 .10 	 15.6 ±.3 			16.40±.3 450± 13 393± 10 	6834± 10 6790 7152
114827.5+124339 254.17 69.39 99.45 -12.08 114552.9+130019	IC 736 MCG 2-30- 39 HICK 59A PGC 36861	.E?.... 		 .07 	14.82S±.15 14.5 ±.6 				4114± 41 4041 4452
114830.8+124347 254.19 69.40 99.45 -12.06 114556.2+130028	IC 737 MCG 2-30- 40 IRAS11459+1300 PGC 36867	.S?.... 	.82± .13 .17± .07 .82	 .07 .25 .09	16.15S±.15 15.24 ±.18 11.98 15.43		 14.66± .68		4087± 41 4014 4425

R.A. 2000 DEC.　l b　SGL SGB　R.A. 1950 DEC.	Names　PGC	Type　S_T n_L　T　L	$\log D_{25}$　$\log R_{25}$　$\log A_o$　$\log D_o$	p.a.　A_g　A_i　A_{21}	B_T　m_B　m_{FIR}　B_T^o	$(B-V)_T$　$(U-B)_T$　$(B-V)_T^o$　$(U-B)_T^o$	$(B-V)_o$　$(U-B)_o$　m'_o　m'_{25}	m_{21}　W_{20}　W_{50}　HI	V_{21}　V_{opt}　V_{GSR}　V_{3K}
1148.5　+4343 156.26　69.11 70.06　-.68 1145.9　+4400	 UGC　6776 PGC 36868	.S..6*. U 6.0±1.4 	1.16± .05 .92± .05 1.16	33 .00 1.35 .46	 	 	 	 	
114832.5+124219 254.25　69.39 99.48　-12.07 114557.9+125900	 MCG　2-30- 41 HICK　59C PGC 36871	.S?.... 	.94± .11 .57± .07 .94	 .07 .84 .28	15.90S±.15 15.43 ±.20 	 	 14.01± .58 	 	
114835.0-255711 285.68　34.80 139.00　-21.46 114603.0-254030	 ESO　504- 16 PGC 36872	.L?.... 	.98± .06 .50± .03 .94	151 .31 .00 	 15.32 ±.14 14.96 	 	 	 	3155± 34 2978 3490
114837.8+323810 187.10　75.33 80.43　-4.91 114601.6+325450	 UGC　6777 PGC 36873	.E?.... 	.89± .08 .05± .05 .88	 .04 .00 	 15.17 ±.16 15.03 	 	 	16.57±.2 277± 9 	6999± 7 6996 7278
114839.1+484240 148.15　65.23 65.42　1.26 114601.1+485920	NGC　3893 UGC　6778 PGC 36875	.SXT5*. R　(2) 5.0± .4 1.4± .7	1.65± .02 .21± .03 1.65	165 .02 .32 .11	11.16S±.15 10.87 ±.15 10.67	 	 13.71± .19 	12.58±.1 302± 4 272± 4 1.81	973± 4 944± 23 1026 1178
114845.9-281741 A 1146-28 286.54　32.57 141.50　-21.71 114614.1-280100	 ESO　440- 11 DDO　239 PGC 36882	.SBS7*. SU　(2) 7.0± .4 6.0± .5	1.40± .03 .06± .04 1.43	 .37 .08 .03	 12.98 ±.14 12.53	 	 	13.85±.2 118± 16 105± 12 1.29	1934± 11 1753 2264
114849.0-020113 273.09　57.18 114.13　-16.47 114615.4-014432	 UGC　6780 PGC 36887	.SXS7*. UEF　(1) 6.7± .4 6.1± .6	1.51± .02 .52± .04 1.52	20 .04 .71 .26	 13.6 ±.3 12.87	 	 	13.58±.1 232± 16 224± 12 .45	1736± 11 1751± 76 1617 2092
114851.4+592501 137.12　55.95 55.46　5.37 114611.4+594141	NGC　3894 UGC　6779 PGC 36889	.E.4+.. R -5.0± .4 	1.45± .05 .21± .04 .93± .03 1.38	20 .00 .00 	12.63 ±.14 12.66 ±.15 12.60	1.00± .01 .59± .02 .97 .61	1.04± .01 .61± .02 12.77± .12 14.34± .29	 	 3223± 19 3314 3372
114852.9+010325 270.40　59.92 111.03　-15.60 114619.0+012006	 CGCG　12- 89 PGC 36893	.SAS5*. F　(1) 5.0± .9 5.6±1.0	 	 .04 	 15.0 ±.6 	 	 	 	
114856.4+235020 A 1146+24 222.60　75.53 88.78　-8.10 114621.0+240700	 UGC　6782 DDO　97 PGC 36896	.I..9.. U　(1) 10.0± .8 9.0±1.0	1.30± .04 .00± .05 1.12± .06 1.30	 .04 .00 .00	15.0 ±.2 14.99	.57± .15 -.15± .17 .56 -.16	.65± .07 -.17± .10 16.12± .13 16.36± .34	15.22±.1 90± 6 84± 6 .24	525± 6 491 834
114857.0+484030 148.10　65.29 65.48　1.30 114619.1+485710	NGC　3896 UGC　6781 PGC 36897	.SB.0*P P .0± .8 	1.15± .04 .15± .04 1.14	125 .02 .11 	 13.89 ±.18 13.75 	 	 	 	869± 57 923 1076
114857.6+311819 192.17　75.76 81.71　-5.34 114621.6+313500	 UGC　6783 PGC 36899	.I..9.. U 10.0± .9 	1.00± .08 .14± .06 1.00	 .05 .10 .07	 	 	 	16.30±.3 161± 15 	6507± 11 6500 6791
114859.5+350058 178.52　74.50 78.23　-3.94 114623.2+351738	NGC　3897 UGC　6784 PGC 36902	.S..4.. U　(1) 4.0± .7 1.1± .8	1.29± .04 .00± .05 .82± .03 1.29	 .00 .00 .00	13.56 ±.15 13.8 ±.3 13.57	.64± .02 .04± .06 .60 .01	.72± .02 13.15± .11 14.85± .28	14.69±.1 312± 6 292± 6 1.12	6411± 8 6434± 38 6417 6681
114901.7-010617 272.42　58.02 113.22　-16.17 114628.0-004936	 CGCG　12- 90 PGC 36903	.S..3*. F 3.0±1.3 	.86± .10 .29± .06 .87	 .04 .41 .15	 15.28 ±.13 	 	 	 	
114904.1-373059 289.47　23.72 151.38　-22.43 114633.0-371418	NGC　3903 ESO　378- 24 IRAS11465-3714 PGC 36906	.SXT5P* BS　(1) 4.7± .5 3.3± .6	1.05± .05 .05± .04 1.09	 .43 .08 .03	 13.47 ±.14 12.20 12.94	 	 	 	3300± 60 3105 3604
114904.6+592601 137.06　55.95 55.45　5.41 114624.7+594241	NGC　3895 UGC　6785 PGC 36907	.SBT1*. R 1.0± .4 	1.13± .04 .14± .04 1.13	125 .00 .14 .07	14.0 ±.2 13.96 ±.18 13.79	.95± .04 .32± .07 .90 .30	 14.15± .30	 	 3159± 27 3250 3308
114905.0-094346 278.51　50.16 122.02　-18.36 114631.8-092705	NGC　3905 MCG -1-30- 35 IRAS11465-0927 PGC 36909	.SBT5.. PE F　(2) 4.5± .4 1.3± .4	1.28± .04 .14± .04 1.01± .03 1.28	40 .04 .20 .07	13.38 ±.16 12.97 13.10	.57± .03 .50 	.65± .02 13.92± .06 14.27± .26	15.35±.1 266± 9 2.18	5774± 7 5731± 59 5633 6129

R.A. 2000 DEC. l b SGL SGB R.A. 1950 DEC.	Names PGC	Type S_T n_L T L	$\log D_{25}$ $\log R_{25}$ $\log A_\bullet$ $\log D_o$	p.a. A_g A_i A_{21}	B_T m_B m_{FIR} B_T^o	$(B-V)_T$ $(U-B)_T$ $(B-V)_T^o$ $(U-B)_T^o$	$(B-V)_\bullet$ $(U-B)_\bullet$ $m'_\bullet$ m'_{25}	m_{21} W_{20} W_{50} HI	V_{21} V_{opt} V_{GSR} V_{3K}
114908.9+270126	NGC 3900	.LAR+..	1.50± .03	2	12.2 ±.2	.85± .05	.88± .02	14.25±.1	1799± 7
209.80 76.14	UGC 6786	R	.27± .03	.07	12.37 ±.11			429± 5	1702± 56
85.77 -6.90	IRAS11465+2718	-1.0± .3	1.07± .07	.00		.78	13.04± .24	414± 5	1774
114633.3+271806	PGC 36914		1.46		12.24		13.90± .25		2096
114913.3-291635	NGC 3904	.E.2+*.	1.43± .03	8	11.83 ±.13	.98± .01	.99± .01	15.33±.3	1496± 17
286.98 31.66	ESO 440- 13	R	.14± .03	.19	11.79 ±.11	.53± .02	.58± .01	547± 34	1714± 31
142.56 -21.73		-5.0± .5	.91± .02	.00		.92	11.87± .07	468± 25	1363
114641.4-285954	PGC 36918		1.42		11.59	.49	13.63± .21		1874
114915.4+560502	NGC 3898	.SAS2..	1.64± .02	107	11.60M±.10	.90± .01	.94± .01	13.47±.1	1176± 5
139.79 58.96	UGC 6787	R (2)	.23± .02	.00	11.63 ±.13	.40± .02	.49± .02	485± 3	1163± 14
58.59 4.17	IRAS11465+5621	2.0± .3	1.00± .02	.28		.85	12.09± .09	467± 3	1255
114636.3+562142	PGC 36921	1.9± .6	1.64	.11	11.32	.36	14.07± .14	2.04	1342
114918.9+260722	NGC 3902	.SXS4*.	1.21± .04	85				14.90±.3	3601± 9
213.55 76.09	UGC 6790	PU	.11± .04	.07	13.64 ±.19			256± 7	3622± 50
86.64 -7.19	IRAS11467+2623	3.5± .6		.16	13.18			238± 7	3576
114643.4+262403	PGC 36923		1.22	.05	13.39			1.46	3904
114920.8+741809	NGC 3890	.S?....	.96± .07						
128.54 42.14	UGC 6788		.00± .05	.13	14.15 ±.18				6815± 50
41.42 10.78	IRAS11465+7434			.00	13.09				6950
114634.4+743449	PGC 36925		.97	.00	13.99				6875
114922.4+245621	NGC 3911	.SB?...	1.04± .06	110				15.62±.2	5956± 6
218.38 75.90	UGC 6795		.11± .05	.04	14.8 ±.2			213± 7	
87.77 -7.61				.17					5926
114647.0+251302	PGC 36926		1.04	.06	14.54			1.02	6262
1149.3 -0104		.S..3./	1.08± .05	76	14.01V±.13	.89± .04			
272.55 58.08	UGC 6793	UEF (1)	.56± .04	.04	14.81 ±.18				6406± 42
113.22 -16.08	IRAS11468-0048	2.8± .6		.77	13.27	.74			6290
1146.8 -0048	PGC 36928	4.5±1.3	1.08	.28	14.02		13.75± .29		6761
1149.3 +1638		.SB.8*.	1.15± .05	72	14.73 ±.13	.54± .03	.50± .03	15.48±.3	3451± 10
245.89 72.22	UGC 6794	U	.48± .05	.15	14.61 ±.18	-.15± .05	-.08± .05	257± 13	
95.73 -10.55		8.0±1.3	.72± .02	.58		.39	13.82± .04	230± 10	3392
1146.8 +1655	PGC 36929		1.16	.24	13.94	-.26	14.12± .31	1.30	3780
114923.3+394617		.S..6*.	1.44± .03	172					850
164.52 71.96	UGC 6792	U	.83± .04	.00	14.56 ±.19				
73.81 -2.05	KUG 1146+400	6.0±1.2		1.23					873
114646.6+400257	PGC 36930		1.44	.42	13.33				1099
114924.0+264425		.S..7..	1.29± .04	1				15.69±.3	1852± 9
211.00 76.18	UGC 6791	U	.85± .05	.05	14.99 ±.18			237± 10	
86.06 -6.95		7.0± .9		1.18					1828
114648.5+270105	PGC 36932		1.29	.43	13.75			1.52	2152
114924.7-050707	IC 2963	.L..+*/	1.15± .06	82					1680
275.68 54.44	MCG -1-30- 36	FE	.61± .05	.07					
117.32 -17.15	IRAS11468-0450	-.6±1.1		.00					1552
114651.2-045027	PGC 36933		1.07						2037
1149.4 -0327		PS..0P?	.64± .17						
274.49 55.95	MCG 0-30- 27	F	.00± .07	.04	15.3 ±.2				8142± 55
115.64 -16.71		.0±1.9		.00					8019
1146.9 -0311	PGC 36938		.64		15.11				8498
114929.6-333835		PSBS1..	.80± .06		*		.36± .06		
288.42 27.48	ESO 378- 25	r	.06± .05	.39	15.48 ±.14		-.09± .12		
147.23 -22.09	TOL 58	1.0±1.0		.06					
114658.1-332154	PGC 36940		.84	.03					
114929.7-010508	NGC 3907	.LBS-*.	1.07± .07	40	13.11V±.14	.97± .02	.99± .02		
272.60 58.09	UGC 6796	UE	.14± .03	.04	14.06 ±.15				6034± 42
113.24 -16.06		-2.5± .6	.76± .03	.00		.90			5918
114656.0-004827	PGC 36941		1.06		13.94		13.96± .39		6389
114932.6-032835		.E.....							
274.54 55.95	CGCG 12- 95	F		.05	15.0 ±.6				
115.66 -16.69		-5.0± .9							
114659.0-031154	PGC 36944								
1149.5 +7559		.SB.3..	1.01± .06	128					
127.80 40.55	UGC 6789	U	.20± .05	.11	14.82 ±.18				
39.81 11.37	IRAS11467+7616	3.0± .9		.28	13.74				
1146.7 +7616	PGC 36945		1.02	.10					
1149.6 -0331		.LX.+*.							
274.62 55.91	CGCG 12- 98	F		.05	14.81 ±.18				
115.72 -16.68		-1.0±1.3							
1147.1 -0315	PGC 36950								

R.A. 2000 DEC.	Names	Type	logD$_{25}$	p.a.	B$_T$	(B-V)$_T$	(B-V)$_\bullet$	m$_{21}$	V$_{21}$
l b		S$_T$ n$_L$	logR$_{25}$	A$_g$	m$_B$	(U-B)$_T$	(U-B)$_\bullet$	W$_{20}$	V$_{opt}$
SGL SGB		T	logA$_\bullet$	A$_i$	m$_{FIR}$	(B-V)$_T^o$	m'$_\bullet$	W$_{50}$	V$_{GSR}$
R.A. 1950 DEC.	PGC	L	logD$_o$	A$_{21}$	B$_T^o$	(U-B)$_T^o$	m'$_{25}$	HI	V$_{3K}$

1149.6 -0330		.LB.+?.							
274.61 55.93	CGCG 12- 99	F		.05	14.2 ±.3				
115.70 -16.67		-1.0±1.8							
1147.1 -0314	PGC 36951								
114940.2+482530	NGC 3906	.SBS7..	1.27± .04		13.49S±.15			15.83±.1	961± 6
148.21 65.56	UGC 6797	P	.05± .04	.03	13.8 ±.2			48± 6	959± 50
65.76 1.31	IRAS11469+4842	7.0± .8		.07	13.70			39± 7	1014
114702.6+484211	PGC 36953		1.28	.02	13.47		14.58± .26	2.33	1168
114949.6-033105		.SXR0?/							
274.68 55.94	CGCG 12-100	F		.05	14.9 ±.6				
115.72 -16.63		.0±1.3							
114716.1-031424	PGC 36962								
114950.1-384705		.S..3..	1.38± .04	163					2840± 60
289.98 22.53	ESO 320- 26	S (1)	.43± .04	.44	12.84 ±.14				2644
152.75 -22.34	IRAS11473-3830	3.0± .8		.60	11.51				3139
114719.0-383024	PGC 36964	2.2± .8	1.42	.22	11.79				
1149.8 +0640		.LAR0*.	.82± .13						
264.58 64.85	MCG 1-30- 16	F	.08± .07	.00	14.94 ±.16				
105.52 -13.69		-2.0± .9		.00					
1147.3 +0657	PGC 36966		.81						
114954.6-033947		.E...P*							
274.82 55.81	CGCG 12-101	F		.08	14.60 ±.18				
115.87 -16.65		-5.0±1.3							
114721.0-032306	PGC 36969								
114959.6+212000	NGC 3910	.L..-*.	1.20± .14	150					7833± 31
232.10 74.86	UGC 6800	U	.11± .08	.13	13.83 ±.17				7791
91.27 -8.77		-3.0±1.1		.00					8150
114724.5+213641	PGC 36971		1.20		13.59				
1150.0 +5152	A 1147+52	.S..6*.	1.33± .04	160				15.33±.3	1256± 8
143.81 62.70	UGC 6802	U	1.12± .05	.01				147± 9	
62.57 2.68		6.0±1.3		1.47				139± 7	1322
1147.4 +5209	PGC 36973		1.33	.50					1446
115002.8+150124			.62± .11					17.39±.3	754± 10
250.13 71.28	CGCG 97-159		.11± .12	.07	15.8 ±.2			80± 6	673± 61
97.35 -10.95	MK 750							60± 5	687
114728.1+151805	PGC 36976		.63						1085
115004.2+262847	NGC 3912	.SXS3$P	1.19± .03	5				15.30±.1	1791± 7
212.15 76.30	UGC 6801	P (1)	.26± .03	.05	13.44 ±.14			229± 5	1717± 32
86.37 -6.90	IRAS11474+2645	3.0± .8		.36	11.79			164± 4	1764
114728.8+264528	PGC 36979	5.0±1.5	1.20	.13	13.01			2.16	2089
115006.3+245515	NGC 3920	.S?....	1.00± .07						3609± 30
218.60 76.06	UGC 6803		.02± .05	.03	14.11 ±.18				3579
87.85 -7.46	ARAK 327			.02	13.24				3915
114731.0+251156	PGC 36981		1.00	.01	14.03				
115009.0-493705		.S?....	1.02± .06	107					3319± 87
292.85 12.05	ESO 217- 9		.40± .05	.47	15.09 ±.14				3115
164.36 -22.36				.61					3577
114739.0-492024	PGC 36987		1.06	.20	14.00				
1150.1 +0659		.S..7*/	1.22± .06	104					6002
264.31 65.16	UGC 6804	UF	1.16± .06	.00	15.46 ±.20				
105.23 -13.52		7.0± .8		1.38					5911
1147.6 +0716	PGC 36988		1.22	.50	14.06				6349
115011.8+420431		.E?....	.60± .08						1033± 50
158.91 70.51	UGC 6805		.12± .04	.00	14.6 ±.4				1065
71.71 -1.03	IRAS11475+4221			.00	13.85				1271
114735.1+422112	PGC 36990		.56		14.58				
115019.8+255742	A 1147+26	.S...P.	1.30± .05	45				15.71±.3	3760± 9
214.35 76.29	UGC 6806	R	.48± .06	.04	14.08 ±.18			261± 10	3757± 50
86.88 -7.04				.73					3734
114744.5+261423	PGC 36996		1.31	.24	13.30			2.17	4062
115020.5-024836		.SXS3P*	1.15± .06	130					8028
274.35 56.63	MCG 0-30- 30	EF (1)	.29± .05	.04	14.41 ±.20				7907
115.04 -16.32		2.5± .6		.40					8384
114746.9-023156	PGC 36998	2.6±1.0	1.15	.14	13.91				
115021.8-752223		.IXS9..	1.48± .04	40				13.76±.3	1830± 9
299.02 -12.97	ESO 39- 2	S (1)	.26± .07	.72					1635
191.43 -19.25	IRAS11480-7505	10.0± .7		.20				170± 7	1955
114800.1-750542	PGC 37000	8.9± .8	1.55	.13					

R.A. 2000 / l / SGL / R.A. 1950	DEC / b / SGB / DEC	Names / PGC	Type / S$_T$ n$_L$ / T / L	logD$_{25}$ / logR$_{25}$ / logA$_e$ / logD$_o$	p.a. / A$_g$ / A$_i$ / A$_{21}$	B$_T$ / m$_B$ / m$_{FIR}$ / B$_T^o$	(B-V)$_T$ / (U-B)$_T$ / (B-V)$_T^o$ / (U-B)$_T^o$	(B-V)$_e$ / (U-B)$_e$ / m' / m'$_{25}$	m$_{21}$ / W$_{20}$ / W$_{50}$ / HI	V$_{21}$ / V$_{opt}$ / V$_{GSR}$ / V$_{3K}$
1150.3	+2600		.I..9*.	1.02± .08						3712
214.18	76.31	UGC 6807	U	.01± .06	.04					
86.84	-7.01		10.0±1.2		.01					3686
1147.8	+2617	PGC 37002		1.02	.01					4014
1150.4	+3515		.S..6*.	1.14± .07	140					
177.22	74.65	UGC 6808	U	1.16± .06	.00					
78.11	-3.58		6.0±1.4		1.47					
1147.8	+3532	PGC 37004		1.14	.50					
115025.3	-383853		.IB.?P.	1.27± .05	45					
290.06	22.69	ESO 320- 27	S	.31± .05	.44	14.00 ±.14				
152.60	-22.22		10.0±1.7		.23					
114754.0	-382212	PGC 37005		1.31	.15					
115027.7	-183447		RSXR0?.	.97± .05	50					
283.35	41.96	ESO 572- 5	F	.30± .04	.07	15.37 ±.14				
131.29	-19.87		.0± .9		.23					
114755.0	-181806	PGC 37009		.97						
1150.4	+0441		.SXR3*.							
267.31	63.26	CGCG 40- 49	F (1)		.00	15.3 ±.6				
107.53	-14.15		3.0±1.3							
1147.9	+0458	PGC 37010	3.6±1.4							
115033.0	+063404	NGC 3914	PSBT3..	1.06± .06	40					
265.05	64.86	UGC 6809	F (1)	.30± .05	.00	14.0 ±.2				6141± 50
105.67	-13.56	IRAS11479+0650	3.0± .9		.41	12.73				6049
114758.8	+065045	PGC 37014	3.6±1.0	1.06	.15	13.58				6489
115033.9	-025431		PLASOP?	1.14± .06						
274.52	56.57	MCG 0-30- 31	EF	.05± .05	.04	14.00 ±.19				
115.15	-16.29		-2.3± .7		.00					
114800.3	-023751	PGC 37016		1.14						
115036.3	-003405		.P.....							
272.63	58.68	CGCG 12-105	F		.04	15.2 ±.6				1757± 60
112.80	-15.65		99.0							1643
114802.5	-001724	PGC 37019								2112
1150.6	+5033		.S..0..	1.04± .06	9					
145.16	63.87	UGC 6811	U	.43± .05	.00	14.81 ±.18				
63.83	2.26		.0± .9		.32					
1148.0	+5050	PGC 37022		1.02						
1150.6	+5036		.S..8*.	1.16± .13	54					
145.10	63.83	UGC 6812	U	.59± .12	.00					
63.78	2.28		8.0±1.3		.72					
1148.0	+5053	PGC 37023		1.16	.29					
115038.9	+552112	NGC 3913	PSAT7*.	1.42± .02		13.17M±.10	.57± .04	.61± .03	14.62±.0	954± 4
140.11	59.70	UGC 6813	PU (1)	.01± .03	.00	13.6 ±.3	-.05± .06	-.07± .04	54± 3	842± 72
59.35	4.07	IRAS11480+5537	7.0± .5	.97± .13	.02	13.63	.56	13.82± .30	41± 4	1032
114800.7	+553753	PGC 37024	5.0± .9	1.42	.01	13.18	-.06	15.07± .17	1.43	1125
1150.6	-1012		DLB.0?.	1.12± .08	45					
279.35	49.86		E	.33± .08	.03					
122.61	-18.09		-2.0±1.8		.00					
1148.1	-0956	PGC 37027		1.08						
1150.6	+6420		.S..8*.	1.04± .15	15					
133.44	51.53	UGC 6814	U	.37± .12	.04					
50.92	7.38		8.0±1.3		.46					
1148.0	+6437	PGC 37029		1.04	.19					
1150.6	+7749		.S..2..	1.06± .06	13					
127.00	38.82	UGC 6798	U	.68± .05	.14	15.29 ±.19				
38.07	12.03	IRAS11478+7806	2.0±1.0		.84					
1147.8	+7806	PGC 37031		1.07	.34					
115041.7	+200054	NGC 3919	.E.....	.95± .21						
236.79	74.39	UGC 6810	U	.00± .08	.14	14.34 ±.15				6166± 31
92.59	-9.08		-5.0± .9		.00					6119
114806.8	+201735	PGC 37032		.97		14.10				6486
115045.4	+514933	NGC 3917	.SA.6*.	1.71± .02	77	12.51M±.10	.72± .03	.78± .02	13.98±.1	970± 5
143.65	62.79	UGC 6815	R (1)	.61± .03	.01	12.43 ±.13	.08± .05	.12± .04	293± 5	
62.65	2.76	IRAS11481+5206	6.0± .4	1.26± .03	.90	13.10	.60	14.27± .10	279± 5	1036
114807.7	+520614	PGC 37036	5.7± .8	1.71	.31	11.57	-.02	14.36± .15	2.10	1160
1150.7	+5627	A 1148+56	.IB.9..	1.17± .05		14.12M±.11	.33± .06	.38± .03	14.60±.1	889± 6
139.11	58.73	UGC 6816	U (1)	.05± .05	.00	14.7 ±.3	-.25± .07	-.24± .04	131± 8	
58.32	4.50	DDO 98	10.0± .9	1.09± .03	.04		.31	15.02± .07	119± 12	970
1148.1	+5644	PGC 37037	8.0±1.4	1.17	.03	14.15	-.26	14.70± .29	.43	1054

R.A. 2000 DEC. / l b / SGL SGB / R.A. 1950 DEC.	Names / PGC	Type / S_T n_L / T / L	logD_25 / logR_25 / logA_. / logD_o	p.a. / A_g / A_i / A_21	B_T / m_B / m_FIR / B_T^o	(B-V)_T / (U-B)_T / (B-V)_T^o / (U-B)_T^o	(B-V)_. / (U-B)_. / m'_. / m'_25	m_21 / W_20 / W_50 / HI	V_21 / V_opt / V_GSR / V_3K
115046.6+454825		.SB?...	1.31± .04	78				14.46±.1	819± 6
151.76 67.78	UGC 6818		.33± .05	.00	14.3 ±.2			182± 7	
68.27 .49				.48				164± 5	864
114809.7+460506	PGC 37038		1.31	.16	13.77			.53	1039
1150.8 +3052	IC 2967	.E?....							6732
193.56 76.23	MCG 5-28- 38								6724
82.27 -5.14				.06	14.7 ±.3				7017
1148.2 +3109	PGC 37042								
115050.5+550844	NGC 3916	.SA.3*/	1.20± .04	45					
140.26 59.90	UGC 6819	PU (2)	.57± .04	.00	14.69 ±.18				
59.55 4.02	IRAS11481+5525	3.0± .6		.78	13.48				
114812.4+552525	PGC 37047	4.3± .8	1.20	.28					
1150.8 +2030		.S..0..	1.06± .06	138					
235.27 74.67	UGC 6820	U	.48± .05	.17	15.20 ±.18				
92.13 -8.87		.0± .9		.36					
1148.3 +2047	PGC 37049		1.05						
115053.1+385250	A 1148+39	.I..9..	1.61± .04	65	13.4 ±.3	.33± .07	.28± .07	13.43±.1	245± 5
166.20 72.75	UGC 6817	U (1)	.42± .06	.00	14.3 ±.5	-.20± .09	-.23± .09	61± 6	
74.75 -2.12	DDO 99	10.0± .7	1.34± .09	.31		.23	15.62± .22	37± 8	266
114816.9+390931	PGC 37050	9.0±1.4	1.61	.21	13.30	-.27	15.28± .38	-.08	498
115055.6+210844		.S?....	.69± .15					17.28±.3	6388± 9
233.12 74.97	MCG 4-28- 66		.16± .07	.15	15.7 ±.2			173± 10	6411± 52
91.52 -8.64	MK 1461			.24	13.53				6346
114820.7+212525	PGC 37051		.70	.08	15.30			1.90	6706
115055.9+202349		.SXT4*.	1.15± .05		14.44 ±.14	.64± .04	.72± .04	15.95±.3	6438± 12
235.65 74.62	UGC 6821	U	.06± .05	.17	14.0 ±.2	.18± .06	.17± .06	298± 21	
92.24 -8.90		4.0±1.2	.80± .02	.08		.55	13.93± .04		6393
114821.0+204030	PGC 37052		1.16	.03	14.03	.12	14.88± .31	1.90	6757
115059.0+204819	IC 742	.SB.2..	1.04± .03		14.74 ±.13	.87± .03	.88± .03	17.72±.2	6425± 7
234.31 74.83	UGC 6822	U	.02± .05	.16	14.5 ±.2	.22± .06		227± 10	
91.85 -8.74		2.0± .8	.75± .01	.02		.78	13.98± .03	217± 10	6381
114824.1+210500	PGC 37056		1.05	.01	14.44	.18	14.74± .24	3.28	6743
115102.1-284823	NGC 3923	.E.4+..	1.77± .02	50	10.8 ±.4	1.00± .02	1.02± .01		1668± 31
287.28 32.22	ESO 440- 17	R	.18± .02	.25	10.53 ±.11	.61± .03	.62± .03		1487
142.11 -21.28		-5.0± .3	1.22± .21	.00		.93	12.41± .74		1996
114830.0-283142	PGC 37061		1.76		10.27	.56	14.18± .42		
115105.8-343353		.SBR7*.	1.12± .05	95					3853
289.05 26.68	ESO 379- 1	S (1)	.30± .05	.43	15.62 ±.14				
148.24 -21.83		7.0±1.2		.42					3663
114834.0-341712	PGC 37062	6.7±1.3	1.16	.15	14.76				4166
115106.1+550439	NGC 3921	PSAS0P.	1.33± .03	20	13.06 ±.15	.68± .02	.75± .01	15.51±.1	5838± 9
140.25 59.98	UGC 6823	R	.21± .04	.00	13.25 ±.18	.25± .03	.35± .02	306± 11	5942± 22
59.63 4.03	MK 430	.0± .4	.86± .04	.16	13.29	.59	12.85± .13		5930
114828.0+552120	PGC 37063		1.32		12.90	.26	14.04± .23		6026
1151.2 +4305									
156.46 69.91				.00					
70.84 -.47									
1148.6 +4321	PGC 37069								
1151.2 +4305									6810± 60
156.45 69.91				.00					6846
70.84 -.47									7044
1148.6 +4322	PGC 37070								
115113.5+500918	NGC 3922	.S..0..	1.24± .05	38				15.27±.3	1003± 11
145.46 64.26	UGC 6824	U	.36± .05	.00	13.76 ±.18			96± 16	
64.24 2.20		.0± .8		.27				79± 12	1063
114836.2+502559	PGC 37072		1.22		13.47				1202
115113.8+520003	NGC 3931	.LA.-*.	1.06± .07	160				13.55±.1	837± 11
143.32 62.68	UGC 6825	R	.09± .04	.01	14.35 ±.15			113± 16	936± 36
62.52 2.90		-3.0± .4		.00					911
114836.2+521644	PGC 37073		1.05		14.33				1034
1151.2 -0605		.S..3*/	1.12± .08						
277.07 53.73	MCG -1-30- 39	F	.74± .07	.07					
118.44 -16.95		3.0±1.3		1.02					
1148.7 -0549	PGC 37076		1.12	.37					
1151.4 +3233		.S..6*.	1.07± .07	152					
186.68 75.91	UGC 6826	U	1.09± .06	.04					
80.73 -4.40		6.0±1.5		1.47					
1148.8 +3250	PGC 37088		1.07	.50					

R.A. 2000 DEC. / l b / SGL SGB / R.A. 1950 DEC.	Names / PGC	Type / S_T n_L / T / L	$\log D_{25}$ / $\log R_{25}$ / $\log A_e$ / $\log D_o$	p.a. / A_g / A_i / A_{21}	B_T / m_B / m_{FIR} / B_T^o	$(B-V)_T$ / $(U-B)_T$ / $(B-V)_T^o$ / $(U-B)_T^o$	$(B-V)_e$ / $(U-B)_e$ / m' / m'_{25}	m_{21} / W_{20} / W_{50} / HI	V_{21} / V_{opt} / V_{GSR} / V_{3K}
1151.4 +5326 / 141.75 61.44 / 61.18 3.46 / 1148.8 +5343	UGC 6828 / PGC 37091	.S..4.. / U (1) / 4.0± .8 / 3.5±1.1	1.13± .05 / .10± .05 / / 1.13	.00 / .15 / .05	15.0 ±.3				
115127.7+352552 / 176.25 74.76 / 78.03 -3.31 / 114851.9+354233	UGC 6827 / MK 431 / PGC 37093	.S...0.. / U / .0± .9	1.06± .06 / .37± .05 / / 1.04	127 / .00 / .28	14.37 ±.19 / / / 14.05				3084± 50 / 3092 / 3351
1151.4 +2352 / 223.15 76.10 / 88.96 -7.55 / 1148.9 +2409	IC 739 / UGC 6830 / PGC 37097	.SB.2*. / U / 2.0± .9	1.06± .06 / .17± .05 / / 1.06	150 / .07 / .20 / .08	14.65 ±.20				
115130.1-112527 / 280.31 48.80 / 123.91 -18.17 / 114856.9-110846	NGC 3942 / MCG -2-30- 35 / IRAS11489-1108 / PGC 37099	.SXT5P* / EF (1) / 5.3± .7 / 5.0± .7	1.15± .06 / .23± .05 / / 1.16	125 / .07 / .35 / .12	/ / 13.31			15.52±.1 / 264± 12	3696± 9 / / 3552 / 4050
115131.5-000303 / 272.57 59.24 / 112.34 -15.28 / 114857.8+001338	CGCG 12-107 / PGC 37100	.SBS1.. / F / 1.0± .9		.04	15.4 ±.6				
1151.5 -0005 / 272.63 59.21 / 112.39 -15.28 / 1149.0 +0011	CGCG 12-106 / PGC 37103	.LXR0?. / F / -2.0± .9		.04	14.81 ±.18				
1151.5 +0505 / 267.37 63.75 / 107.22 -13.77 / 1149.0 +0522	CGCG 40- 53 / PGC 37104	.E..... / F / -5.0± .9		.00	14.92 ±.18				
1151.5 +1528 / 249.87 71.86 / 97.04 -10.45 / 1149.0 +1545	UGC 6831 / PGC 37105	.L..... / U / -2.0± .8	1.05± .08 / .04± .03 / / 1.05	.08 / .00	14.39 ±.17				
115142.9+015931 / 270.76 61.08 / 110.31 -14.65 / 114909.0+021612	CGCG 12-108 / PGC 37125	.SBS4*. / F (1) / 4.0±1.3 / 2.6±1.4		.02	15.3 ±.6				
115143.0+210010 / 233.93 75.07 / 91.73 -8.51 / 114908.1+211651	NGC 3929 / UGC 6832 / PGC 37126	.E?....	.74± .12 / .14± .03 / / .73	80 / .18 / .00	15.0 ±.2 / / / 14.75				7113± 31 / 7070 / 7431
115144.8-465953 / 292.47 14.66 / 161.55 -22.15 / 114914.0-464312	ESO 266- 30 / PGC 37130	.SXT3.. / r / 3.3± .9	1.04± .06 / .14± .05 / / 1.08	/ .44 / .20 / .07	15.21 ±.14				
115146.2+380054 / 168.19 73.41 / 75.63 -2.29 / 114910.3+381735	NGC 3930 / UGC 6833 / KUG 1149+382 / PGC 37132	.SXS5.. / PU / 5.0± .5	1.50± .02 / .12± .04 / / 1.50	30 / .00 / .19 / .06	/ 13.1 ±.3 / / 12.87			13.55±.1 / 166± 8 / 150± 6 / .62	916± 6 / 915± 15 / 933 / 1172
115147.1+484053 / 147.17 65.54 / 65.65 1.73 / 114910.1+485734	NGC 3928 / UGC 6834 / MK 190 / PGC 37136	.SAS3$.	1.18± .03 / .00± .04 / .58± .02 / 1.18	/ .04 / .00	13.22 ±.13 / 12.98 ±.16 / 12.02 / 13.07	.67± .02 / -.07± .03 / .65 / -.07	.65± .01 / -.14± .02 / 11.61± .07 / 14.10± .24	15.21±.1 / 128± 6 / 90± 5	982± 10 / 957± 35 / 1035 / 1186
115148.2+363534 / 172.38 74.22 / 76.96 -2.81 / 114912.3+365215	UGC 6836 / KUG 1149+368 / PGC 37138	.S?....	.94± .05 / .25± .04 / / .94	90 / .00 / .37 / .12	15.42 ±.18 / / / 14.98				10951± 57 / 10964 / 11213
115152.1-312841 / 288.32 29.70 / 144.97 -21.39 / 114920.0-311200	ESO 440- 19 / PGC 37142	.SBR4.. / r / 4.5± .9	.98± .06 / .07± .05 / / 1.01	/ .28 / .10 / .03	15.06 ±.14 / / / 14.59				16055± 34 / 15870 / 16376
1151.8 +1832 / 241.90 73.83 / 94.11 -9.33 / 1149.3 +1849	UGC 6837 / PGC 37143	.S..6*. / U / 6.0±1.5	.96± .09 / .98± .06 / / .97	154 / .07 / 1.44 / .49	16.2 ±.2 / / / 14.67			16.58±.3 / 359± 13 / 344± 10 / 1.42	5975± 10 / / 5924 / 6299
1151.9 +4545 / 151.43 67.94 / 68.39 .65 / 1149.3 +4602	A 1149+46 / 1ZW 29 / PGC 37144			.00					5910± 60 / 5955 / 6131

R.A. 2000 DEC. l b SGL SGB R.A. 1950 DEC.	Names PGC	Type S_T n_L T L	$\log D_{25}$ $\log R_{25}$ $\log A_e$ $\log D_0$	p.a. A_g A_l A_{21}	B_T m_B m_{FIR} B_T^o	$(B-V)_T$ $(U-B)_T$ $(B-V)_T^o$ $(U-B)_T^o$	$(B-V)_e$ $(U-B)_e$ m'_e m'_{25}	m_{21} W_{20} W_{50} HI	V_{21} V_{opt} V_{GSR} V_{3K}
115155.7-023835 274.88 56.95 114.98 -15.89 114922.0-022154	 UGC 6838 PGC 37147	.SA.2.. F 2.0±1.0 	.95± .07 .26± .04 .96	113 .05 .32 .13	 15.06 ±.18 				
115200.3+210636 233.68 75.18 91.65 -8.41 114925.4+212317	 MCG 4-28- 77 PGC 37153	.S?.... 	.92± .05 .09± .03 .94	 .18 .13 .04	14.58S±.15 14.68 ±.18 14.28	.61± .02 .00± .04 .51 -.07	 13.83± .29	17.00±.2 251± 7 206± 10 2.68	6653± 6 6555±120 6610 6970
115201.3+164834 246.77 72.81 95.79 -9.89 114926.7+170515	NGC 3933 UGC 6839 PGC 37156	.S?.... 	1.06± .06 .29± .05 1.07	83 .12 .44 .15	14.26 ±.13 14.35 ±.19 13.71	.71± .02 .01± .04 .60 -.07	 13.67± .34	15.74±.3 342± 11 277± 10 1.88	3738± 10 3676± 56 3679 4064
115203.1-323459 288.69 28.64 146.15 -21.46 114931.0-321818	 ESO 440- 21 PGC 37159	PSBT3*. r 3.3± .9 	.89± .06 .16± .05 .92	174 .39 .22 .08	 15.19 ±.14 				
1152.0 +0624 266.01 64.93 105.96 -13.25 1149.5 +0641	 MCG 1-30- 18 PGC 37160	.LBR-*. F -3.0±1.3 	.74± .14 .00± .07 .74	 .00 .00 	 14.93 ±.18 				
1152.1 +5206 142.94 62.66 62.47 3.06 1149.5 +5223	A 1149+52 UGC 6840 DDO 100 PGC 37164	.SBT9.. PU (1) 9.0± .5 8.0± .9	1.27± .06 .50± .06 1.06± .05 1.27	 .00 .51 .25	14.3 ±.2 15.2 ±.3 14.11	.42± .09 -.03± .11 .31 -.11	.50± .04 -.05± .06 15.12± .11 14.30± .39	14.45±.1 149± 6 143± 7 .09	1046± 5 1113 1234
1152.2 +1650 246.77 72.88 95.77 -9.83 1149.6 +1707	NGC 3934 UGC 6841 IRAS11496+1707 PGC 37170	.S?.... 	1.03± .06 .04± .05 1.04	 .15 .05 .02	 14.5 ±.2 12.32 14.28				3779± 10 3626± 56 3717 4102
1152.2 +2103 233.98 75.21 91.72 -8.37 1149.7 +2120	 MCG 4-28- 78 PGC 37175	.S?.... 	.74± .14 .00± .07 .75	 .18 .00 .00	 15.24 ±.18 14.99			17.09±.3 306± 13 270± 10 2.09	6743± 10 6701 7060
115220.5-265423 286.99 34.13 140.13 -20.75 114948.0-263742	NGC 3936 ESO 504- 20 IRAS11497-2637 PGC 37178	.SBS4*/ PSU (3) 4.0± .6 3.1± .6	1.59± .02 .79± .03 1.63	63 .40 1.17 .40	 12.78 ±.10 12.43 11.20			14.29±.1 323± 7 304± 5 2.70	2022± 6 2062± 52 1845 2355
115224.5+322414 186.99 76.16 80.95 -4.25 114949.0+324055	NGC 3935 UGC 6843 KUG 1149+326 PGC 37183	.S?.... 	1.02± .04 .32± .04 1.03	114 .09 .48 .16	 14.2 ±.2 13.57				3067± 50 3065 3346
1152.4 +6953 130.22 46.40 45.72 9.50 1149.8 +7010	 UGC 6844 PGC 37193	.S..2.. U 2.0±1.0 	1.00± .08 .84± .06 1.00	35 .01 1.04 .42	 15.7 ±.2 				
115231.3-035220 276.04 55.89 116.26 -16.08 114957.7-033539	IC 2969 MCG -1-30- 40 IRAS11499-0335 PGC 37196	.SBR4?. E (1) 3.7± .7 4.6±1.4	1.09± .05 .20± .05 1.09	100 .04 .29 .10	 13.27 				
115233.2-123937 281.30 47.72 125.25 -18.19 115000.0-122256	 MCG -2-30- 36 PGC 37199	.SXT5P* EF (1) 5.0± .6 4.9± .7	1.13± .06 .11± .05 1.13	40 .08 .17 .06					
1152.5 -0339 275.90 56.09 116.05 -16.01 1150.0 -0323	 UGCA 249 PGC 37202	.IBS9.. UE F (1) 10.0± .4 7.5± .8	1.14± .05 .06± .05 1.15	10 .05 .05 .03	 14.3 ±.3 14.19			15.40±.1 115± 16 69± 12 1.18	1665± 11 1665 1543 2021
1152.5 +2336 224.53 76.26 89.31 -7.41 1150.0 +2353	 UGC 6846 PGC 37206	.E...?. U -5.0±1.7 	1.01± .11 .02± .05 .71± .12 1.01	 .04 .00 	14.0 ±.3 14.36 ±.16 14.14	1.00± .03 .93 	1.01± .02 13.05± .43 13.99± .66		 6842± 39 6809 7152
1152.6 +3522 176.02 74.99 78.17 -3.12 1150.0 +3539	 UGC 6848 PGC 37208	.S..8*. U 8.0±1.4 	.96± .09 .69± .06 .96	61 .00 .84 .34					6340 6349 6607
115237.5-022809 275.03 57.18 114.86 -15.68 115003.8-021128	 UGC 6850 MK 1307 PGC 37213	.P..... F 99.0 	.81± .11 .07± .06 .81	 .04 .09 .03	 14.5 ±.2 14.39				1012± 76 893 1367

R.A. 2000 DEC. l b SGL SGB R.A. 1950 DEC.	Names PGC	Type S_T n_L T L	logD$_{25}$ logR$_{25}$ logA$_\bullet$ logD$_0$	p.a. A$_g$ A$_i$ A$_{21}$	B$_T$ m$_B$ m$_{FIR}$ B^o_T	(B-V)$_T$ (U-B)$_T$ (B-V)^o_T (U-B)^o_T	(B-V)$_\bullet$ (U-B)$_\bullet$ m'$_\bullet$ m'$_{25}$	m$_{21}$ W$_{20}$ W$_{50}$ HI	V$_{21}$ V$_{opt}$ V$_{GSR}$ V$_{3K}$
115237.8+241819 221.69 76.47 88.65 -7.15 115002.9+243501	 UGC 6847 KUG 1150+245 PGC 37214	.S..4.. U 4.0± .9 	1.19± .03 .56± .04 1.19	151 .02 .82 .28	 14.97 ±.19 14.10			15.46±.1 266± 8 1.08	4942± 7 4911 5250
115239.2+500217 145.15 64.48 64.44 2.37 115002.4+501858	NGC 3924 UGC 6849 PGC 37217	.S..9*. U 9.0±1.1 	1.26± .06 .04± .06 1.26	 .01 .04 .02	 15.0 ±.5 14.91			15.65±.3 115± 9 90± 7 .72	995± 8 1055 1194
1152.6 +2920 199.80 76.88 83.87 -5.32 1150.1 +2937	 UGC 6853 PGC 37218	.L..... U -2.0± .9 	1.06± .07 .28± .03 1.02	35 .04 .00 	 14.65 ±.15 14.49				8639 8626 8929
115242.6+203757 235.63 75.10 92.17 -8.42 115007.9+205438	NGC 3937 UGC 6851 PGC 37219	.L..-*. U -3.0±1.1 	1.26± .13 .06± .08 .86± .04 1.27	15 .17 .00 	13.43 ±.15 13.41 ±.17 13.15	.97± .03 .87 	1.00± .01 13.22± .15 14.46± .70		6617± 39 6573 6935
115243.9+014422 271.47 60.97 110.64 -14.48 115010.0+020103	 UGC 6854 MK 752 PGC 37222	.SBT4P. F 4.0± .9 	1.00± .05 .04± .04 1.00	 .02 .06 .02	 14.18 ±.18 13.40 14.06			15.78±.2 159± 8 1.70	6128± 6 6094± 29 6020 6479
115246.6+205925 234.42 75.28 91.83 -8.29 115011.8+211606	NGC 3940 UGC 6852 PGC 37224	.E..... U -5.0± .8 	1.22± .08 .01± .05 .87± .06 1.25	 .18 .00 	13.8 ±.2 13.56 ±.19 13.40	.97± .04 .87 	1.02± .02 13.61± .21 14.88± .48		6461± 24 6419 6779
1152.7 +2328 225.12 76.26 89.45 -7.41 1150.2 +2345	 UGC 6855 PGC 37226	.L..... U -2.0± .9 	1.00± .19 .21± .08 .97	130 .00 .00 					
115249.8+440726 153.87 69.32 69.98 .19 115013.6+442407	NGC 3938 UGC 6856 IRAS11502+4423 PGC 37229	.SAS5.. R (2) 5.0± .3 1.1± .5	1.73± .02 .04± .02 1.38± .01 1.73	 .00 .06 .02	10.90M±.10 10.89 ±.14 10.90 10.83	.52± .01 -.10± .02 .51 -.11	.62± .01 .00± .01 13.22± .03 14.30± .14	12.46±.1 106± 4 90± 4 1.61	809± 4 771± 31 848 1036
115255.1+365914 170.72 74.19 76.68 -2.46 115019.4+371555	NGC 3941 UGC 6857 PGC 37235	.LBS0.. R -2.0± .3 	1.54± .03 .18± .04 .88± .04 1.52	10 .00 .00 	11.25M±.10 11.34 ±.12 11.27	.91± .03 .44± .03 .88 .43	.93± .01 .50± .02 11.16± .13 13.38± .19	15.58±.2 258± 4 232± 3 	922± 5 934± 40 937 1183
1152.9 +6113 135.02 54.51 53.95 6.51 1150.3 +6130	 UGC 6858 PGC 37236	.S..6*. U 6.0±1.2 	1.15± .05 .23± .05 1.16	15 .05 .34 .12	 14.8 ±.2 				
115256.7+202844 236.25 75.08 92.33 -8.43 115022.0+204525	NGC 3943 MCG 4-28- 84 PGC 37237	.S?.... 	1.04± .09 .00± .07 1.05	 .17 .00 .00	 14.29 ±.19 14.09			18.21±.3 249± 10 4.12	6538± 9 6623± 39 6498 6861
1152.9 -0424 276.60 55.44 116.84 -16.11 1150.4 -0408	A 1150-04 MCG -1-30- 43 VV 457 PGC 37238	.SXS8.. FE (1) 7.9± .6 6.4± .8	1.30± .05 .24± .05 1.31	88 .04 .30 .12				13.76±.1 167± 12 	1489± 9 1480± 30 1364 1844
1153.0 -1344 282.00 46.74 126.39 -18.30 1150.5 -1328	 PGC 37242	DLA.-*. E -3.0±1.3 	1.21± .07 .00± .08 1.22	 .04 .00 					
115304.0-363818 290.07 24.77 150.49 -21.58 115032.0-362136	 ESO 379- 6 IRAS11505-3621 PGC 37243	.S..5*/ S (1) 4.7± .7 3.3±1.3	1.40± .04 .81± .04 1.44	68 .37 1.22 .41	 14.16 ±.14 12.67 				
115305.5+261228 213.66 76.94 86.87 -6.37 115030.6+262909	NGC 3944 UGC 6859 PGC 37244	.L..-*. U -3.0±1.1 	1.15± .15 .10± .08 1.14	25 .03 .00 	 13.90 ±.16 13.81				3638± 31 3614 3939
115306.0-182235 284.07 42.35 131.20 -19.22 115033.0-180554	 ESO 572- 8 IRAS11505-1805 PGC 37245	.S?.... 	.88± .07 .04± .06 .88	 .04 .05 .02	 14.59 ±.14 13.80 14.49			15.95±.3 174± 15 124± 12 1.45	1766± 11 1606 2113
115308.3-323359 288.94 28.72 146.15 -21.23 115036.0-321718	 ESO 440- 25 PGC 37247	.L?.... 	1.01± .06 .07± .04 1.05	0 .39 .00 	 14.09 ±.14 13.58				8187± 30 8001 8505

R.A. 2000 DEC. / l b / SGL SGB / R.A. 1950 DEC.	Names / PGC	Type / S_T n_L / T / L	$\log D_{25}$ / $\log R_{25}$ / $\log A_\bullet$ / $\log D_0$	p.a. / A_g / A_i / A_{21}	B_T / m_B / m_{FIR} / B_T^0	$(B-V)_T$ / $(U-B)_T$ / $(B-V)_T^0$ / $(U-B)_T^0$	$(B-V)_\bullet$ / $(U-B)_\bullet$ / $m'_\bullet$ / m'_{25}	m_{21} / W_{20} / W_{50} / HI	V_{21} / V_{opt} / V_{GSR} / V_{3K}
115309.2+702600	A 1150+70								9600±110
129.89 45.90	CGCG 334- 54			.01	15.4 ±.3				9725
45.22 9.74	MK 191								9684
115028.7+704241	PGC 37248								
115311.9-390748		PSXR1*.	1.35± .04	121					
290.76 22.36	ESO 320- 30	Sr	.24± .05	.47	13.30 ±.14				3232± 19
153.15 -21.71	FAIR 1151	1.0± .5		.24	9.44				3037
115040.0-385106	PGC 37254		1.39	.12	12.55				3530
115314.7+604036	NGC 3945	.LBT+..	1.72± .03	15	11.8 ±.3	.95± .02	.97± .01		
135.33 55.03	UGC 6860	R	.18± .04	.08	11.40 ±.12	.55± .02	.58± .01		1220± 56
54.48 6.35	IRAS11506+6056	-1.0± .3	.88± .10	.00		.90	11.64± .33		1316
115036.7+605717	PGC 37258		1.70		11.36	.51	14.79± .31		1361
1153.2 +1139		.S..7..	1.14± .07	104					
258.69 69.36	UGC 6862	U	.98± .06	.01					
100.90 -11.32		7.0±1.0		1.35					
1150.7 +1156	PGC 37259		1.14	.49					
115317.0+232750		.S?....	.90± .06					16.71±.3	7428± 9
225.31 76.36	MCG 4-28- 87		.22± .06	.00	15.05 ±.13			328± 10	
89.50 -7.31	KUG 1150+237			.32					7394
115042.2+234431	PGC 37260		.90	.11	14.70			1.90	7738
115319.1-493506		.SXS5*.	1.21± .05	135					
293.35 12.20	ESO 217- 12	S (1)	.14± .05	.40	14.32 ±.14				3640± 60
164.31 -21.85	IRAS11507-4918	5.0± .8		.21	13.34				3437
115048.1-491824	PGC 37263	2.2±1.2	1.24	.07	13.70				3897
115320.1+204512	NGC 3947	RSBT3..	1.15± .03		13.88M±.10	.73± .02	.81± .02	15.53±.2	6197± 5
235.49 75.29	UGC 6863	U (1)	.06± .03	.23	13.84 ±.19	.09± .03	.18± .02	413± 5	6256± 46
92.10 -8.25	IRAS11507+2101	3.0± .8	.80± .01	.09	13.14	.63	13.40± .03	390± 7	6155
115045.5+210153	PGC 37264	3.0±1.0	1.17	.03	13.51	.01	14.31± .18	1.98	6515
115320.3-323900		.SBR1..	1.00± .06	93					
289.01 28.65	ESO 440- 26	r	.14± .06	.39	14.86 ±.14				8187± 34
146.25 -21.20		1.0± .8		.15					8000
115048.1-322218	PGC 37265		1.03	.07	14.22				8505
115321.9-131541	IC 743	.SB.5?/	1.06± .06	140					
281.88 47.22	MCG -2-30- 37	FE	.50± .05	.03					
125.91 -18.13	IRAS11508-1259	5.4±1.1		.75	13.51				
115048.6-125900	PGC 37267		1.06	.25					
115323.1-180959		.IBS9..	1.21± .06					15.63±.1	1737± 9
284.07 42.57	ESO 572- 9	S (1)	.07± .05	.04	17.30 ±.14			88± 12	
131.00 -19.11		10.0± .8		.06					1578
115050.0-175318	PGC 37270	11.1± .8	1.21	.04	17.20			-1.62	2085
115323.6-283312		.S..7*/	1.64± .02	78				13.01±.3	1702± 11
287.79 32.61	ESO 440- 27	SU (1)	.84± .03	.28	13.19 ±.14			280± 16	
141.90 -20.73	IRAS11508-2816	7.0± .8		1.16	11.90			267± 12	1522
115051.1-281630	PGC 37271	4.4± .8	1.67	.42	11.75			.84	2030
1153.5 +1052		.S..4..	1.04± .08	110					
260.17 68.80	UGC 6864	U	.88± .06	.00	15.6 ±.2				
101.67 -11.51	IRAS11509+1109	4.0±1.0		1.30					
1150.9 +1109	PGC 37276		1.04	.44					
115338.7-265942		.SBS9..	1.12± .05						
287.35 34.12	ESO 504- 24	S (1)	.12± .05	.40	15.19 ±.14				
140.27 -20.48		9.0± .8		.12					
115106.0-264300	PGC 37280	11.1± .8	1.15	.06					
115339.2+432729		.S?....	1.08± .09	35					
154.77 69.92	UGC 6865		.42± .07	.00	14.68 ±.18				5809± 69
70.66 .08	2ZW 55			.63					5847
115103.4+434410	PGC 37282		1.08	.21	14.02				6040
115340.6-035946	NGC 3952	.IB.9*/	1.20± .03	85	13.51 ±.14	.38± .03	.34± .03	14.56±.1	1576± 7
276.59 55.89	MCG -1-30- 44	EF (1)	.37± .04	.04	13.0 ±.2	-.24± .04	-.26± .04	187± 12	1595± 66
116.47 -15.83	IRAS11510-0342	9.8± .8	.62± .02	.28	12.79	.27	12.10± .05		1453
115107.0-034304	PGC 37285	5.6± .8	1.20	.19	13.06	-.32	13.42± .22	1.31	1932
115341.4+232255	NGC 3951	.S?....	1.03± .04	172	14.30M±.10	.67± .01		16.22±.2	6455± 6
225.76 76.43	UGC 6867		.30± .03	.00	14.58 ±.18	.07± .03		428± 8	6471± 31
89.62 -7.25	IRAS11511+2339			.45	12.82	.57		391± 10	6422
115106.6+233936	PGC 37288		1.03	.15	13.88	.00	13.57± .23	2.19	6766
115341.4+461239	A 1151+46	.SB?...	.74± .14						
150.05 67.76	MCG 8-22- 28		.09± .07	.00	15.28 ±.19				7200± 63
68.08 1.11	MK 42			.14					7247
115105.3+462920	PGC 37289		.74	.05	15.09				7418

R.A. 2000 DEC. / l b / SGL SGB / R.A. 1950 DEC.	Names / PGC	Type / S_T n_L / T / L	$\log D_{25}$ / $\log R_{25}$ / $\log A_e$ / $\log D_o$	p.a. / A_g / A_i / A_{21}	B_T / m_B / m_{FIR} / B_T^o	$(B-V)_T$ / $(U-B)_T$ / $(B-V)_T^o$ / $(U-B)_T^o$	$(B-V)_e$ / $(U-B)_e$ / m'_e / m'_{25}	m_{21} / W_{20} / W_{50} / HI	V_{21} / V_{opt} / V_{GSR} / V_{3K}
115341.5+475135	NGC 3949	.SAS4*.	1.46± .02	120	11.54S±.15	.45± .03		13.38±.1	798± 6
147.63 66.40	UGC 6869	R (1)	.24± .03	.03	11.64 ±.17			276± 7	681± 56
66.54 1.72	IRAS11510+4808	4.0± .3		.36	10.45	.39		260± 5	850
115105.2+480816	PGC 37290	5.7± .8	1.46	.12	11.19		13.09± .20	2.06	1007
115341.7+205301	NGC 3954	.E?....	.81± .16						6902± 31
235.19 75.42	UGC 6866		.02± .05	.23	14.65 ±.19				6860
92.01 -8.12	ARAK 331			.00					7220
115107.1+210943	PGC 37291		.85		14.32				
115342.7-725248		.SXT2..	1.07± .06	16					5160± 63
298.65 -10.49	ESO 39- 4	S	.04± .05	.92					4963
188.82 -19.49	FAIR 288	2.0± .8		.05					5299
115116.0-723606	PGC 37295		1.16	.02					
115342.8-540624		.S..4..	1.13± .05	164					
294.45 7.81	ESO 171- 1	S (1)	.48± .05	1.15	15.34 ±.14				
169.12 -21.62	IRAS11512-5349	4.0± .9		.71	13.36				
115112.0-534942	PGC 37296	2.2±1.3	1.24	.24					
1153.7 +1024		.LB....	1.05± .08	143					
261.11 68.46	UGC 6871	U	.12± .03	.00	14.22 ±.15				
102.16 -11.60	IRAS11512+1040	-2.0± .9		.00	13.52				
1151.2 +1040	PGC 37298		1.03						
115348.8-051008	IC 2974	.SAS5?/	1.34± .04	81					5747
277.45 54.83	MCG -1-30- 45	EF (1)	.68± .05	.12					5621
117.67 -16.10	IRAS11512-0453	5.0±1.0		1.02	13.20				6103
115115.2-045326	PGC 37304	3.9±1.1	1.35	.34					
115349.5+521939	NGC 3953	.SBR4..	1.84± .01	13	10.84M±.10	.77± .02	.86± .01	13.69±.1	1055± 5
142.21 62.59	UGC 6870	R (2)	.30± .02	.01	10.83 ±.13	.20± .03	.29± .02	425± 7	987± 31
62.36 3.39		4.0± .3	1.45± .02	.44		.71	13.47± .05	411± 5	1122
115112.9+523620	PGC 37306	1.8± .5	1.84	.15	10.38	.14	14.14± .14	3.17	1241
115350.7-272100		.SBS9..	1.26± .04	13					
287.52 33.79	ESO 504- 25	S (1)	.13± .04	.36	14.34 ±.14				
140.65 -20.48		9.0± .6		.14					
115118.1-270418	PGC 37307	10.0± .6	1.29	.07					
115350.8+332154	IC 2973	.SBS7*.	1.14± .04	125				15.32±.2	3206± 7
182.79 76.10	UGC 6872	U	.24± .04	.00	14.34 ±.18			211± 6	3199± 50
80.16 -3.62	KUG 1151+336	7.0±1.2		.32				182± 7	3208
115115.6+333835	PGC 37308		1.14	.12	14.00			1.21	3481
115357.9-230954	NGC 3955	.S..0P.	1.46± .03	165	12.64 ±.15	.69± .01	-.65± .01	15.21±.2	1491± 9
286.14 37.83	ESO 504- 26	RS	.49± .03	.16	12.56 ±.10	-.03± .03	-.05± .02	230± 14	1515± 35
136.25 -19.85	IRAS11514-2253	.0± .6	.81± .03	.36	10.79	.56	12.18± .08	190± 11	1323
115125.1-225312	PGC 37320		1.45		12.04	-.11	13.58± .21		1832
1153.9 +0620		.SAR5./	1.04± .15	105					
267.04 65.12	UGC 6875	F (1)	.37± .12	.00					
106.17 -12.82		5.0± .9		.56					
1151.4 +0637	PGC 37321	5.6±1.0	1.04	.19					
1153.9 +0937	A 1151+09								
262.46 67.86	CGCG 68- 87			.00	15.3 ±.6				
102.94 -11.80									
1151.4 +0954	PGC 37322								
115359.6+203412		.S..2..	.98± .04		14.77M±.10	.95± .02		16.63±.2	6848± 6
236.41 75.33	UGC 6876	U	.12± .03	.16	14.68 ±.18	.28± .04		482± 8	6903± 39
92.33 -8.16		2.0± .9		.15		.84		441± 10	6806
115125.1+205053	PGC 37324		1.00	.06	14.37	.21	14.23± .24	2.19	7168
115401.0-203400	NGC 3956	.SAS5*.	1.53± .02	58				13.73±.1	1653± 5
285.21 40.32	ESO 572- 13	R (4)	.53± .03	.08	12.87 ±.11			284± 6	1630± 52
133.53 -19.40	IRAS11514-2017	5.0± .4		.79	12.47			254± 6	1489
115128.0-201718	PGC 37325	4.3± .4	1.54	.26	11.99			1.47	1997
115401.1-193406	NGC 3957	.LA.+*/	1.49± .03	173					1703± 41
284.82 41.28	ESO 572- 14	R	.66± .02	.06	12.94 ±.10				1541
132.49 -19.22	IRAS11515-1917	-1.0± .4		.00	13.48				2048
115128.0-191724	PGC 37326		1.39		12.85				
115406.1-395148		.SBS5?/	1.42± .03	143					
291.13 21.69	ESO 320- 31	S (1)	.94± .04	.49	14.20 ±.14				
153.94 -21.57		4.7± .7		1.40					
115134.1-393506	PGC 37334	3.3±1.3	1.46	.47					
115407.6-034053		.S..2?.							
276.56 56.23	CGCG 12-113	F		.04	15.4 ±.6				
116.18 -15.64		2.0±1.8							
115134.0-032412	PGC 37336								

R.A. 2000 DEC.	Names	Type	logD$_{25}$	p.a.	B$_T$	(B-V)$_T$	(B-V)$_•$	m$_{21}$	V$_{21}$
l b		S$_T$ n$_L$	logR$_{25}$	A$_g$	m$_B$	(U-B)$_T$	(U-B)$_•$	W$_{20}$	V$_{opt}$
SGL SGB		T	logA$_•$	A$_i$	m$_{FIR}$	(B-V)^o_T	m'$_•$	W$_{50}$	V$_{GSR}$
R.A. 1950 DEC.	PGC	L	logD$_o$	A$_{21}$	B^o_T	(U-B)^o_T	m'$_{25}$	HI	V$_{3K}$
115412.3+000811 IC 745		.L.....	.82± .08						
273.59 59.71 UGC 6877		F	.03± .05	.03	14.2 ±.2				1087± 42
112.35 -14.58 MK 1308		-2.0± .9		.00					976
115138.5+002453 PGC 37339			.82		14.17				1440
1154.2 +2001								17.30±.3	6187± 10
238.40 75.11 CGCG 98- 2				.13	15.5 ±.3			298± 13	
92.88 -8.29								263± 10	6142
1151.7 +2018 PGC 37345									6507
115421.6-122843		.IBS9P.	1.27± .03	160					1483
281.83 48.04 MCG -2-30- 39		FUE (1)	.22± .04	.06					
125.16 -17.72		9.8± .4		.16					1338
115148.3-121201 PGC 37348		8.2± .5	1.28	.11					1836
115425.1-021906		.SXR7?/	1.23± .03	168				15.64±.1	2904± 9
275.68 57.50 UGC 6879		UEF (1)	.46± .04	.04	14.11 ±.18			246± 8	
114.83 -15.21		6.8± .5		.63					2786
115151.5-020225 PGC 37352		7.0± .8	1.24	.23	13.42			1.99	3259
115434.5+582202 NGC 3958		.SBS1..	1.17± .04	28	13.8 S±.3	.79± .05		16.50±.1	3367± 7
136.69 57.22 UGC 6880		RU	.35± .03	.02	13.4 ±.2			360± 11	3322± 50
56.72 5.68		1.0± .4		.36		.69			3455
115157.5+583843 PGC 37358			1.17	.18	13.15		13.63± .34	3.17	3520
115437.9-074525 NGC 3959		PSBR1..	1.04± .05	140					
279.40 52.50 MCG -1-30- 46		E F	.09± .04	.13					
120.35 -16.56		.7± .5		.10					
115204.4-072843 PGC 37363			1.05	.05					
115439.1-405554		PSBR3..	.98± .06	0					
291.51 20.68 ESO 320- 32		r	.23± .05	.50	14.80 ±.14				
155.08 -21.51		3.3± .9		.31					
115207.0-403912 PGC 37364			1.03	.11					
115439.8-135823 NGC 3962		.E.1...	1.41± .04	15	11.62 ±.13	.95± .01	.98± .01	14.95±.3	1815± 17
282.65 46.65 MCG -2-30- 40		R	.06± .04	.04	11.92 ±.18	.50± .02	.57± .01	336± 34	1818± 40
126.72 -17.97		-5.0± .3	1.07± .02	.00		.92	12.46± .06	253± 25	1667
115206.6-134141 PGC 37366			1.39		11.66	.50	13.49± .25		2167
1154.6 +2005		.I..9*.	1.16± .13	125				16.34±.1	618± 10
238.37 75.23 UGC 6881		U	.34± .12	.13				88± 6	
92.85 -8.18		10.0±1.2		.26				82± 5	573
1152.1 +2022 PGC 37367			1.17	.17					937
115449.7-165150		.SXS6*.	1.32± .03	3				14.32±.1	1813± 7
283.98 43.91 MCG -3-30- 19		UE F (1)	.26± .04	.04				205± 10	
129.71 -18.52		6.3± .5		.38				198± 12	1657
115216.5-163509 PGC 37373		5.5± .5	1.33	.13					2162
115450.9+025731		.SAS4P*	.87± .08					16.48±.3	6036± 12
271.26 62.30 MCG 1-30- 19		F	.03± .06	.08	14.88 ±.18			112± 14	
109.59 -13.62		4.0±1.3		.05				101± 15	5934
115217.0+031412 PGC 37374			.88	.02	14.71			1.75	6387
1154.8 +2816 NGC 3964		.S?....	.94± .11						8570
204.44 77.44 MCG 5-28- 43			.11± .07	.00	14.93 ±.18				8554
85.06 -5.26				.09					8864
1152.3 +2833 PGC 37375			.93		14.72				
115454.9-271500		.SXT7*.	1.16± .04						
287.76 33.95 ESO 504- 28		Sr (1)	.04± .03	.34	13.69 ±.14				
140.58 -20.23		6.7± .6		.06					
115222.0-265818 PGC 37377		6.2± .5	1.20	.02					
1154.9 +0439		.SXR0?.							
269.52 63.80 CGCG 41- 2		F		.00	14.9 ±.6				
107.91 -13.09		.0±1.8							
1152.4 +0456 PGC 37380									
115458.7+261208		.SXS6*.	1.10± .05					15.56±.2	5150± 4
213.93 77.36 UGC 6883		U	.02± .05	.02	14.6 ±.2			52± 4	
87.03 -5.98		6.0±1.2		.02				32± 5	5127
115224.1+262849 PGC 37382			1.10	.01	14.55			1.00	5451
1154.9 +2718		.S?....	.94± .11						6597
208.89 77.45 MCG 5-28- 44			.40± .07	.01	15.24 ±.18				6578
85.98 -5.59				.55					6894
1152.4 +2735 PGC 37383			.94	.20	14.63				
115459.3+582937 NGC 3963		.SXT4..	1.44± .03	12.5	S±.3	.59± .03		13.99±.1	3186± 7
136.51 57.12 UGC 6884		R (2)	.04± .03	.02	12.24 ±.14	-.01± .04		165± 7	3204± 66
56.62 5.77 IRAS11523+5846		4.0± .3		.06	12.17	.56		124± 7	3275
115222.5+584618 PGC 37386		2.2± .6	1.44	.02	12.19	-.03	14.44± .31	1.77	3340

R.A. 2000 DEC. / l b / SGL SGB / R.A. 1950 DEC.	Names / PGC	Type / S_T n_L / T / L	$\log D_{25}$ / $\log R_{25}$ / $\log A_e$ / $\log D_o$	p.a. / A_g / A_i / A_{21}	B_T / m_B / m_{FIR} / B_T^o	$(B-V)_T$ / $(U-B)_T$ / $(B-V)_T^o$ / $(U-B)_T^o$	$(B-V)_e$ / $(U-B)_e$ / m'_e / m'_{25}	m_{21} / W_{20} / W_{50} / HI	V_{21} / V_{opt} / V_{GSR} / V_{3K}
1155.0 +6919 / 130.18 47.01 / 46.34 9.52 / 1152.4 +6936	NGC 3961 / UGC 6885 / / PGC 37390	RSBR1*. / U / 1.0± .8 /	1.11± .07 / .00± .06 / / 1.11	/ .00 / .00 / .00	14.4 ±.2 / / / 14.33				6720 / / 6842 / 6810
115509.2-185536 / 284.91 41.97 / 131.87 -18.84 / 115236.1-183854	NGC 3969 / ESO 572- 17 / / PGC 37396	.LAR+*. / SE F / -.5± .4 /	1.14± .04 / .21± .03 / / 1.12	64 / .07 / .00 /	13.99 ±.14 / / / 13.82				/ 6689± 52 / 6529 / 7035
115510.5-075038 / 279.65 52.47 / 120.47 -16.45 / 115237.0-073357	NGC 3967 / MCG -1-30- 47 / / PGC 37398	PL..-*. / E F / -3.0± .5 /	1.19± .05 / .13± .04 / / 1.18	65 / .13 / .00 /					
1155.1 +7925 / 126.15 37.35 / 36.59 12.74 / 1152.4 +7942	/ MCG 13- 9- 6 / IRAS11524+7942 / PGC 37399	.S?.... / / /	.85± .12 / .10± .07 / / .87	/ .19 / .16 / .05	15.26 ±.18 / 13.34 / / 14.85				11527± 79 / 11676 / 11556
115511.4+061004 / 267.87 65.14 / 106.43 -12.59 / 115237.5+062646	A 1152+06 / UGC 6886 / / PGC 37400	.SBT4P. / UF (1) / 3.5± .6 / 2.2± .8	1.09± .06 / .16± .05 / .77± .03 / 1.09	80 / .00 / .24 / .08	14.13 ±.16 / 14.21 ±.18 / / 13.88	.72± .02 / .11± .04 / .64 / .05	.80± .02 / .20± .04 / 13.47± .07 / 14.03± .34	15.38±.2 / 371± 6 / 351± 6 / 1.42	6977± 6 / / 6886 / 7324
115514.4-374142 / 290.81 23.85 / 151.65 -21.22 / 115242.0-372500	IC 2977 / ESO 379- 9 / / PGC 37405	.I.0.?. / S / 90.0 /	1.32± .04 / .34± .05 / .88± .02 / 1.35	122 / .40 / .25 / .17	13.19 ±.13 / 13.28 ±.14 / / 12.58	.93± .02 / .32± .02 / .74 / .15	.97± .01 / .40± .01 / 13.08± .07 / 13.78± .28		3010± 15 / 2817 / 3312
115514.5+224137 / 229.06 76.52 / 90.40 -7.16 / 115240.0+225818	/ UGC 6887 / IRAS11526+2258 / PGC 37406	.SB?... / / /	1.05± .08 / .26± .06 / / 1.06	157 / .04 / .40 / .13	15.0 ±.2 / 12.98 / / 14.51			16.27±.2 / 335± 12 / 332± 10 / 1.63	6812± 8 / / 6777 / 7124
1155.2 +0029 / 273.76 60.14 / 112.08 -14.23 / 1152.7 +0046	/ UGC 6890 / / PGC 37407	.IBS9.. / UEF (1) / 10.0± .5 / 9.7± .6	1.13± .07 / .36± .07 / / 1.13	5 / .02 / .27 / .18					3214 / / 3105 / 3567
115517.4+172919 / 246.66 73.85 / 95.40 -8.93 / 115243.0+174600	/ UGC 6891 / KUG 1152+177 / PGC 37409	.S..2.. / U / 2.0± .9 /	1.20± .03 / .62± .03 / / 1.21	113 / .12 / .77 / .31	15.40S±.15 / 15.15 ±.19 / / 14.35	.83± .05 / .21± .07 / .64 / .07	/ / / 14.69± .22	15.64±.2 / 387± 12 / 366± 10 / .98	6781± 14 / / 6727 / 7107
115519.7+510739 / 143.05 63.75 / 63.57 3.17 / 115243.7+512420	A 1152+51 / / MK 192 / PGC 37413			/ .01 / /					3600±110 / 3665 / 3793
1155.3 -0628 / 278.91 53.76 / 119.09 -16.06 / 1152.8 -0612	/ MCG -1-30- 48 / / PGC 37415	.S..6?/ / FE / 6.9± .7 /	1.23± .04 / .98± .04 / / 1.23	130 / .09 / 1.36 / .49					2492 / / 2363 / 2847
115524.4+543926 / 139.52 60.63 / 60.26 4.45 / 115248.2+545607	/ UGC 6894 / KUG 1152+549 / PGC 37418	.S..6*. / U / 6.0±1.4 /	1.15± .04 / .78± .04 / / 1.15	89 / .00 / 1.15 / .39	15.49 ±.18 / / / 14.33			15.65±.3 / 155± 9 / 152± 7 / .92	767± 8 / / 844 / 941
115524.7+391325 / 163.37 73.20 / 74.76 -1.18 / 115249.5+393006	/ UGC 6893 / / PGC 37419	.S..7.. / U / 7.0± .8 /	1.02± .06 / .04± .05 / / 1.02	/ .00 / .06 / .02	15.3 ±.3 / / / 15.18				6183 / / 6206 / 6433
115524.8-501800 / 293.85 11.58 / 165.06 -21.50 / 115253.1-500118	/ ESO 217- 14 / / PGC 37420	.IBS9*. / S (1) / 10.0±1.2 / 5.6± .8	1.18± .04 / .44± .04 / / 1.22	18 / .46 / .33 / .22	14.3 ±.4 / 14.45 ±.14 / /	/ / /	.99± .05 / .37± .08 / / 13.95± .47		
115524.8+330735 / 183.18 76.50 / 80.51 -3.40 / 115250.0+332417	/ UGC 6892 / / PGC 37421	.S?.... / / /	.99± .06 / .58± .05 / / .99	82 / .00 / .87 / .29	15.33 ±.18 / / / 14.45				3149± 57 / 3151 / 3425
115527.9-120337 / 282.01 48.52 / 124.79 -17.36 / 115254.5-114655	NGC 3970 / MCG -2-30- 41 / / PGC 37425	.LAR0*/ / EF / -2.2± .6 /	1.06± .05 / .39± .04 / / 1.01	98 / .07 / .00 /					
115528.6+573945 / 137.02 57.91 / 57.43 5.54 / 115252.1+575626	A 1152+57 / / MK 193 / PGC 37427			/ .00 / /					5282± 74 / 5368 / 5439

R.A. 2000 DEC.	Names	Type	logD$_{25}$	p.a.	B$_T$	(B-V)$_T$	(B-V)$_\bullet$	m$_{21}$	V$_{21}$
l　　b		S$_T$　n$_L$	logR$_{25}$	A$_g$	m$_B$	(U-B)$_T$	(U-B)$_\bullet$	W$_{20}$	V$_{opt}$
SGL　SGB		T	logA$_\bullet$	A$_I$	m$_{FIR}$	(B-V)$_T^o$	m'$_\bullet$	W$_{50}$	V$_{GSR}$
R.A. 1950 DEC.	PGC	L	logD$_o$	A$_{21}$	B$_T^o$	(U-B)$_T^o$	m'$_{25}$	HI	V$_{3K}$

115528.9+115813	NGC 3968	.SXT4..	1.43± .04	10				13.90±.2	6393± 6
259.32 69.95	UGC 6895	U	.15± .05	.00	12.6 ±.2			516± 9	
100.77 -10.70	IRAS11529+1214	4.0± .7		.21	12.93			491± 7	6321
115254.7+121455	PGC 37429		1.43	.07	12.37			1.46	6731

1155.5 +5452	A 1152+55B	.S..0$.	.74± .14		16.3 ±.2	1.09± .10			
139.30 60.44	MCG 9-20- 30	P	.28± .07	.00		.04± .16			
60.06 4.54		.0±2.0		.21					
1152.9 +5509	PGC 37430		.72					14.12± .76	

115530.7-011542		RLX.+..							
275.33 58.58	CGCG 13- 2	F		.04	15.1 ±.6				
113.85 -14.66		-1.0± .9							
115256.9-005900	PGC 37434								

1155.5 +0947		.SB.6*.	1.04± .15	8					6524± 10
263.06 68.22	UGC 6897	U	.37± .12	.00					
102.91 -11.38		6.0±1.3		.55					6445
1153.0 +1004	PGC 37438		1.04	.19					6866

115534.6+255319	IC 746	.S?....	1.04± .04	169				15.20±.2	5028± 5
215.44 77.45	UGC 6898		.52± .04	.00	14.8 ±.2			298± 5	5000± 50
87.38 -5.96	IRAS11530+2609			.78				270± 7	5004
115300.0+261000	PGC 37440		1.04	.26	14.02			.92	5329

1155.6 +5454	A 1152+55A	.LA..P*	.82± .19		15.6 ±.2	1.16± .06			
139.25 60.42	MCG 9-20- 31	P	.27± .07	.00		.38± .09			
60.03 4.57		-2.0±1.0		.00					
1153.0 +5511	PGC 37442		.78					13.87± .98	

115536.5+295946	NGC 3971	.L.....	1.15± .15						6880± 65
196.48 77.42	UGC 6899	U	.07± .08	.00	13.66 ±.15				6871
83.48 -4.49		-2.0± .8		.00					7167
115301.8+301628	PGC 37443		1.14		13.56				

115536.5+011404		.SBS6..	1.42± .02	150				14.01±.1	1892± 5
273.26 60.85	UGC 6903	UE F (1)	.05± .03	.01	13.0 ±.3			187± 7	1957± 76
111.36 -13.94	IRAS11530+0130	5.8± .4		.08				180± 12	1786
115302.7+013046	PGC 37444	5.5± .5	1.42	.03	12.89			1.09	2245

115538.4+430243		.S?....	1.08± .06	32					7108± 50
154.74 70.47	UGC 6901		.29± .05	.00	14.36 ±.18				7145
71.18 .26	IRAS11530+4319			.44	12.21				7341
115303.1+431925	PGC 37448		1.08	.15	13.89				

1155.6 +3130	A 1153+31	.I..9*.	1.32± .05	115	14.80 ±.19	.67± .06	.75± .05		590
189.76 77.08	UGC 6900	PU (1)	.21± .06	.02	14.4 ±.5	-.03± .09	.06± .07		
82.05 -3.94	DDO 101	10.0± .6	1.02± .04	.16		.61	15.39± .09		586
1153.0 +3147	PGC 37449	9.0±1.0	1.32	.11	14.56	-.07	15.72± .36		872

115540.2-120135	NGC 3974	PSBR0*.	1.05± .05	10					
282.07 48.57	MCG -2-31- 1	E F	.02± .04	.07					
124.77 -17.31		.0± .5		.01					
115306.7-114453	PGC 37452		1.06						

1155.6 +8013			1.00± .08						
125.85 36.59	UGC 6896		1.02± .06	.16					12560± 67
35.82 13.00	IRAS11529+8030				11.67				12711
1152.9 +8030	PGC 37459		1.02						12584

115542.6+321124	NGC 3966		.62± .11						3433± 13
186.89 76.89	MCG 5-28- 48		.11± .12	.04	15.5 ±.3				3431
81.41 -3.68	KUG 1153+324								3712
115307.8+322805	PGC 37462		.63						

115544.7+175319		.S?....	.96± .06					16.36±.3	6346± 10
245.78 74.18	MCG 3-31- 1		.36± .06	.10	15.32 ±.18			336± 13	
95.06 -8.69	KUG 1153+181			.53				318± 10	6294
115310.4+181000	PGC 37463		.97	.18	14.66			1.52	6671

115546.1+551907	NGC 3972	.SAS4..	1.59± .02	120	12.98S±.15	.66± .04		14.54±.1	842± 7
138.85 60.06	UGC 6904	R	.56± .03	.00	12.77 ±.18	-.04± .05		263± 8	609± 63
59.65 4.74	IRAS11531+5535	4.0± .3		.82	12.71	.55		251± 6	918
115310.0+553548	PGC 37466		1.59	.28	12.07	-.12	14.41± .21	2.19	1010

115546.2+295632		.S..3..	1.04± .04	41				16.39±.3	6785± 11
196.70 77.46	UGC 6905	U (1)	.74± .04	.00				389± 15	
83.55 -4.48	KUG 1153+302	3.0±1.0		1.01					6775
115311.5+301313	PGC 37467	4.5±1.3	1.04	.37					7073

115549.5+250753								18.07±.3	6458± 9
218.90 77.37	CGCG 127-107			.00	15.6 ±.3			82± 10	
88.12 -6.18									6431
115315.0+252435	PGC 37474								6762

R.A. 2000 DEC. / l b / SGL SGB / R.A. 1950 DEC.	Names / PGC	Type / S_T n_L / T / L	$\log D_{25}$ / $\log R_{25}$ / $\log A_e$ / $\log D_o$	p.a. / A_g / A_i / A_{21}	B_T / m_B / m_{FIR} / B_T^o	$(B-V)_T$ / $(U-B)_T$ / $(B-V)_T^o$ / $(U-B)_T^o$	$(B-V)_e$ / $(U-B)_e$ / m'_e / m'_{25}	m_{21} / W_{20} / W_{50} / HI	V_{21} / V_{opt} / V_{GSR} / V_{3K}
115551.3-181142 / 284.84 42.72 / 131.14 -18.54 / 115318.0-175500	ESO 572- 18 / / / PGC 37476	.SBR5P* / SE F (1) / 5.0± .6 / 5.5± .8	1.20± .03 / .30± .03 / / 1.20	35 / .05 / .46 / .15	13.95 ±.14 / / / 13.43			15.13±.3 / 191± 15 / 180± 12 / 1.55	1584± 11 / 1587± 52 / 1426 / 1931
115553.8+603146 / 134.90 55.29 / 54.73 6.60 / 115317.2+604827	NGC 3975 / MCG 10-17-103 / / PGC 37480	.SA.3$. / P / 3.0±1.0 / 1.01	1.00± .07 / .21± .05 / / 1.01	/ .04 / .29 / .11					
115556.3-185836 / 285.17 41.97 / 131.96 -18.67 / 115323.0-184154	ESO 572- 19 / / / PGC 37482	.S..4*. / F / 4.0±1.3 / 1.04	1.03± .05 / .36± .04 / / 1.04	91 / .06 / .53 / .18	15.40 ±.14				
115557.0+064458 / 267.53 65.73 / 105.92 -12.23 / 115323.0+070140	NGC 3976 / UGC 6906 / IRAS11533+0701 / PGC 37483	.SXS3.. / R (3) / 3.0± .3 / 2.4± .5	1.58± .02 / .50± .03 / / 1.58	53 / .00 / .68 / .25	12.37 ±.12 / 12.16 / / 11.67			12.97±.1 / 440± 7 / 421± 5 / 1.06	2498± 5 / 2491± 66 / 2409 / 2845
115601.2-024316 / 276.66 57.30 / 115.35 -14.93 / 115327.5-022634	NGC 3979 / UGC 6907 / / PGC 37488	.LB.0P* / UE F / -2.0± .5 / 1.14	1.15± .06 / .08± .03 / / 1.14	110 / .04 / .00					
115607.3-195342 / 285.58 41.11 / 132.92 -18.80 / 115334.0-193700	NGC 3981 / ESO 572- 20 / ARP 289 / PGC 37496	.SAT4.. / R (3) / 4.0± .3 / 2.3± .4	1.72± .02 / .35± .03 / / 1.72	15 / .08 / .51 / .17	11.87 ±.11 / 10.88 / / 11.27			12.74±.1 / 312± 6 / 276± 6 / 1.30	1723± 5 / 1715± 35 / 1561 / 2068
115608.3+552332 / 138.69 60.01 / 59.60 4.81 / 115332.4+554013	NGC 3977 / UGC 6909 / / PGC 37497	RSAT2*. / R / 2.0± .5 / 1.23	1.23± .04 / .04± .04 / / 1.23	/ .00 / .05 / .02	14.3 ±.3 / / / 14.15			·	5722± 72 / 5801 / 5892
1156.1 +1507 / 253.20 72.41 / 97.76 -9.52 / 1153.6 +1524	/ UGC 6911 / / PGC 37501	.S..4.. / U / 4.0± .9 / 1.16	1.15± .05 / .58± .05 / / 1.16	18 / .11 / .85 / .29	15.04 ±.18				
115610.6+603121 / 134.85 55.31 / 54.75 6.63 / 115334.2+604802	NGC 3978 / UGC 6910 / IRAS11535+6047 / PGC 37502	.SX.4*. / PU (1) / 4.0± .6 / 3.5±1.0	1.21± .04 / .04± .04 / / 1.21	/ .00 / .06 / .02	13.3 ±.3 / 13.20 ±.18 / 12.00 / 13.10	.60± .04 / .02± .05 / .53 / -.03	/ / / 14.09± .37		9981± 34 / 10077 / 10123
1156.2 +5812 / 136.45 57.45 / 56.95 5.82 / 1153.6 +5829	/ UGC 6912 / 7ZW 430 / PGC 37504	.S?.... / / / 1.32	1.32± .05 / .30± .06 / / 1.32	/ .01 / .41 / .15	14.6 ±.3 / / / 14.21			14.34±.3 / 127± 16 / 96± 12 / -.01	1357± 11 / / 1445 / 1512
1156.2 -0338 / 277.42 56.47 / 116.29 -15.12 / 1153.7 -0322	/ CGCG 13- 7 / / PGC 37506	RLB.+*. / F / -1.0±1.3		/ .04	15.4 ±.6				
1156.2 +1701 / 248.48 73.73 / 95.93 -8.87 / 1153.7 +1718	/ UGC 6913 / / PGC 37507	.S..3.. / U / 3.0± .8 / 1.20	1.19± .05 / .28± .05 / / 1.20	0 / .11 / .39 / .14	14.9 ±.2 / / / 14.32				6800± 10 / / 6745 / 7127
115620.5+252230 / 217.91 77.53 / 87.93 -5.98 / 115346.0+253912	CGCG 127-109 / / / PGC 37511			/ .00	15.6 ±.3			16.39±.3 / / 283± 7 /	4731± 9 / / 4706 / 5034
115622.3-193306 / 285.52 41.45 / 132.58 -18.68 / 115349.0-191624	ESO 572- 22 / / / PGC 37513	RSB.7P? / E (1) / 6.7±1.0 / 7.5± .8	1.13± .04 / .58± .03 / / 1.14	44 / .08 / .80 / .29	15.00 ±.14 / / / 14.12			15.56±.3 / 192± 15 / 168± 12 / 1.15	1915± 11 / / 1754 / 2260
1156.3 +2352 / 224.57 77.16 / 89.37 -6.50 / 1153.8 +2409	NGC 3983 / UGC 6914 / / PGC 37514	.S..0.. / U / .0± .9 / 1.03	1.06± .06 / .56± .05 / / 1.03	114 / .00 / .42	15.00 ±.19				
115622.7+320112 / 187.40 77.08 / 81.63 -3.61 / 115348.0+321753	IC 2978 / UGC 6915 / KUG 1153+323 / PGC 37515	.S..7.. / U / 7.0± .9 / 1.04	1.04± .04 / .41± .04 / / 1.04	126 / .04 / .57 / .21	15.15 ±.18 / / / 14.53			16.47±.2 / 266± 8 / 199± 7 / 1.74	3225± 7 / / 3223 / 3505
115628.1+550729 / 138.83 60.27 / 59.87 4.76 / 115352.3+552410	NGC 3982 / UGC 6918 / IRAS11538+5524 / PGC 37520	.SXR3*. / R (1) / 3.0± .5 / 4.6± .8	1.37± .02 / .06± .03 / / 1.37	/ .00 / .09 / .03	11.86 ±.15 / 10.93 / / 11.77			14.14±.2 / 234± 7 / 214± 5 / 2.34	1109± 6 / 924± 44 / 1184 / 1277

R.A. 2000 DEC. l b SGL SGB R.A. 1950 DEC.	Names PGC	Type S_T n_L T L	$\log D_{25}$ $\log R_{25}$ $\log A_.$ $\log D_o$	p.a. A_g A_l A_{21}	B_T m_B m_{FIR} B^o_T	$(B-V)_T$ $(U-B)_T$ $(B-V)^o_T$ $(U-B)^o_T$	$(B-V)_.$ $(U-B)_.$ $m'_.$ m'_{25}	m_{21} W_{20} W_{50} HI	V_{21} V_{opt} V_{GSR} V_{3K}
115628.3+394430 161.58 72.99 74.35 -.80 115353.3+400112	 UGC 6916 KUG 1153+400B PGC 37521	.SB.4*. U 4.0± .9	1.22± .03 .54± .04 1.22	34 .00 .80 .27	 14.58 ±.18 				
115629.9+502542 143.46 64.45 64.30 3.08 115354.5+504223	 UGC 6917 PGC 37525	.SB.9.. U 9.0± .7	1.55± .04 .24± .06 1.55	130 .03 .25 .12	13.13S±.15 13.5 ±.3 12.91		 15.10± .29	13.76±.1 195± 7 183± 4 .73	910± 5 973 1107
1156.6 +5537 138.38 59.83 59.41 4.96 1154.0 +5554	A 1154+55 UGC 6919 PGC 37532	.S..8*. U 8.0±1.3	1.16± .05 .54± .05 1.16	87 .00 .67 .27	 15.17 ±.19 				
115641.9-322306 289.72 29.08 146.04 -20.47 115409.0-320624	 ESO 440- 32 PGC 37541	.SBR0.. r -.1± .9	1.14± .07 .33± .03 1.15	86 .34 .25 	 14.09 ±.14 13.38				8190± 30 8005 8508
115642.0+482006 145.94 66.27 66.28 2.36 115406.7+483648	NGC 3985 UGC 6921 ARAK 334 PGC 37542	.SBS9*. R 9.0± .5	1.11± .04 .19± .03 1.12	73 .05 .19 .09	13.12S±.15 13.11 ±.17 12.58 12.87		 13.08± .26	14.52±.1 165± 11 111± 12 1.56	957± 8 925± 28 1010 1161
115643.5+320121 187.28 77.15 81.65 -3.54 115408.9+321802	NGC 3986 UGC 6920 IRAS11541+3217 PGC 37544	.L..../ R -2.0± .4	1.49± .03 .65± .03 1.39	110 .02 .00 	 13.60 ±.16 13.56 13.53			15.85±.3 567± 10 537± 7 	3263± 9 3242± 50 3261 3542
115645.8-011736 275.90 58.68 113.97 -14.37 115412.0-010054	 CGCG 13- 8 PGC 37548	.SBR3P* F 3.0±1.3		 .05 	 15.2 ±.6 				
115646.8-381130 291.26 23.44 152.19 -20.95 115414.1-375448	 ESO 320- 35 IRAS11542-3754 PGC 37549	.SXS5P* Sr (1) 4.9± .4 3.3± .5	1.34± .04 .16± .05 1.39	7 .51 .25 .08	 13.57 ±.14 13.05 				
1156.7 +5050 142.91 64.12 63.93 3.28 1154.2 +5107	 UGC 6922 PGC 37550	.S?.... 	1.19± .03 .09± .03 1.19	 .04 .13 .04	14.14S±.10 14.9 ±.3 14.02		 14.70± .21	15.03±.1 144± 5 127± 5 .96	877± 5 942 1072
115649.9+530941 140.51 62.06 61.74 4.11 115414.4+532623	 UGC 6923 KUG 1154+534 PGC 37553	.I..9*. U 10.0±1.2	1.30± .03 .38± .04 .73± .02 1.30	175 .00 .28 .19	13.90M±.10 13.92 ±.18 13.62	.52± .03 -.09± .04 .42 -.16	.47± .03 -.12± .04 13.15± .04 14.31± .21	14.92±.1 173± 4 160± 6 1.11	1066± 4 1138 1248
1156.8 +1246 258.59 70.78 100.10 -10.13 1154.3 +1303	 UGC 6924 PGC 37554	.S..6*. U 6.0±1.3	1.00± .16 .55± .12 1.00	104 .00 .81 .27	 15.39 ±.18 				
1156.9 +5730 136.81 58.13 57.64 5.66 1154.3 +5747	 UGC 6926 PGC 37556	.S..8*. U 8.0±1.2	1.08± .06 .34± .05 1.08	98 .00 .42 .17					
115654.5+320936 186.64 77.14 81.53 -3.46 115420.0+322617	IC 2979 UGC 6925 PGC 37559	.LB.... U -2.0± .9	.95± .09 .06± .03 .94	0 .04 .00 	 14.47 ±.15 14.38				3122± 31 3121 3401
115658.3-195112 285.82 41.20 132.92 -18.59 115425.0-193430	 ESO 572- 23 PGC 37565	.LXS0*. SE -2.4± .5	1.17± .04 .03± .03 .42± .05 1.17	 .05 .00 	13.89 ±.15 13.52 ±.14 13.62	.94± .02 .42± .03 .91 .41	.95± .01 .46± .02 11.48± .16 14.54± .27		1728± 52 1567 2072
115659.4-195906 285.87 41.08 133.05 -18.61 115426.1-194224	 ESO 572- 24 PGC 37566	.SB.7?. S (1) 7.0±1.8 7.8±1.3	1.12± .04 .59± .04 1.12	115 .05 .82 .30	 16.20 ±.14 				
115708.8+302333 194.38 77.67 83.23 -4.04 115434.4+304015	 UGC 6927 PGC 37574	.L..... U -2.0± .9	1.02± .08 .12± .03 1.00	 .00 .00 	 14.35 ±.15 14.31				3309± 40 3302 3595
115710.1-141629 283.63 46.55 127.16 -17.44 115436.7-135947	 MCG -2-31- 3 PGC 37575	.SXT7*. EF (1) 6.5± .6 5.3± .9	1.10± .06 .31± .05 1.11	135 .06 .43 .16					8336 8187 8687

R.A. 2000 DEC. l b SGL SGB R.A. 1950 DEC.	Names PGC	Type S_T n_L T L	$\log D_{25}$ $\log R_{25}$ $\log A_e$ $\log D_o$	p.a. A_g A_i A_{21}	B_T m_B m_{FIR} B_T^o	$(B-V)_T$ $(U-B)_T$ $(B-V)_T^o$ $(U-B)_T^o$	$(B-V)_e$ $(U-B)_e$ m' m'_{25}	m_{21} W_{20} W_{50} HI	V_{21} V_{opt} V_{GSR} V_{3K}
115715.2-274200		.SBR8*.	1.04± .06						1794± 30
288.51 33.65	ESO 504- 30	r	.08± .05	.37	14.70 ±.14				1617
141.12 -19.78		7.9±1.2		.10					2123
115442.0-272518	PGC 37580		1.08	.04	14.22				
115717.2+491708	A 1154+49	.SXS7..	1.64± .03		12.65S±.15			13.41±.1	778± 7
144.54 65.51	UGC 6930	U	.20± .06	.06	13.5 ±.5			133± 8	838
65.42 2.79		7.0± .7		.28				119± 6	981
115442.0+493350	PGC 37584		1.65	.10	12.38		15.21± .28	.93	
115719.8+362457		.S..6*.	1.02± .06	60					
170.58 75.24	UGC 6929	U	.09± .05	.00	15.0 ±.2				
77.54 -1.85		6.0±1.2		.13					
115445.1+364139	PGC 37589		1.02	.05					
115721.3+251141	NGC 3987	.S..3..	1.35± .04	58	13.89 ±.13	.97± .03	1.03± .02	15.10±.2	4498± 6
218.51 77.72	UGC 6928	CU (1)	.73± .05	.00	14.27 ±.18	.31± .05	.35± .04	575± 7	4539± 24
88.18 -5.83	IRAS11547+2528	3.0± .6	.95± .02	1.01	11.14	.81	14.13± .06		4475
115447.0+252823	PGC 37591	4.5±1.2	1.35	.37	12.98	.14	13.67± .26	1.75	4804
115725.9+575548		.SB.9*.	1.16± .07	80					1175± 50
136.37 57.77	UGC 6931	U	.17± .06	.00	14.31 ±.19				1263
57.26 5.87	VV 241	9.0± .8		.18					1331
115450.3+581229	PGC 37598		1.16	.09	14.13				
115726.3+251358	NGC 3989	.S?....	.80± .07					16.63±.3	4713± 9
218.79 77.74	MCG 4-28-100		.28± .06	.00	15.72 ±.18			370± 10	
88.15 -5.80	KUG 1154+255			.41					4687
115452.0+253040	PGC 37599		.80	.14	15.29			1.20	5016
115728.2+310505		.I..9*.	.99± .06	170				16.41±.3	6882± 11
191.17 77.58	UGC 6932	U	.51± .05	.00	15.44 ±.18			170± 15	6886± 34
82.60 -3.73		10.0±1.3		.38					6878
115453.7+312147	PGC 37608		.99	.25	15.05			1.11	7165
1157.4 +2753	NGC 3988								6562± 65
206.19 78.02	MCG 5-28- 57								6546
85.63 -4.86				.00	14.27 ±.18				6856
1154.9 +2810	PGC 37609								
115730.1-734106	IC 2980	.E.3...	1.26± .07	42					8352± 63
299.10 -11.22	ESO 39- 5	S	.19± .06	.77					8156
189.60 -19.09		-5.0± .8		.00					8486
115500.1-732424	PGC 37612		1.33						
115730.7+322016	NGC 3991	.I..9P/	1.16± .03	33	13.5 ±.2	.39± .03		14.25±.2	3192± 5
185.69 77.20	UGC 6933	R	.55± .03	.02	14.0 ±.2	-.33± .06		245± 5	3204± 14
81.41 -3.27	VV 523	10.0± .5		.41	12.10	.24		157± 10	3194
115456.3+323658	PGC 37613		1.16	.27	13.29	-.44	12.77± .27	.69	3472
115731.9-011512		.SAR6*/	1.26± .04	141				14.99±.1	5536± 9
276.21 58.79	UGC 6934	UEF	.83± .04	.04	14.91 ±.18			455± 12	
113.98 -14.17		5.8± .7		1.22					5423
115458.1-005830	PGC 37614		1.26	.41	13.63			.95	5889
115736.0+321644	NGC 3994	.SAR5P$	1.02± .03	10	13.3 ±.2	.61± .03			3096± 27
185.91 77.24	UGC 6936	R	.25± .03	.02	13.9 ±.2	-.14± .05			3095
81.47 -3.28	ARAK 337	5.0± .5		.38		.54			3374
115501.5+323326	PGC 37616		1.02	.13	13.15	-.19	12.60± .28		
115736.2+532231	NGC 3992	.SBT4..	1.88± .01	68	10.60 ±.13	.77± .01	.85± .01	12.75±.0	1048± 4
140.09 61.92	UGC 6937	R (2)	.21± .02	.00	10.55 ±.14	.20± .02	.30± .01	475± 3	1059± 56
61.58 4.30	IRAS11549+5339	4.0± .3	1.50± .01	.31	12.15	.72	13.59± .03	459± 4	1121
115501.0+533913	PGC 37617	1.1± .5	1.89	.11	10.25	.16	14.34± .16	2.39	1229
115736.3+552733	NGC 3990	.L..-*/	1.15± .03	40	13.43M±.10	.88± .03	.91± .03		
138.25 60.04	UGC 6938	R	.24± .03	.00	13.58 ±.16	.38± .05	.46± .05		696± 40
59.61 5.03		-3.0± .6	.46± .04	.00		.86	11.23± .13		776
115501.0+554415	PGC 37618		1.12		13.46	.37	13.46± .19		866
115737.7+251432	NGC 3993	.S..3*/	1.26± .03	141				15.73±.3	4828± 9
218.79 77.79	UGC 6935	P	.57± .03	.00	14.50 ±.18			368± 10	4824± 58
88.16 -5.76	KUG 1155+255A	3.0±1.2		.79					4803
115503.4+253114	PGC 37619		1.26	.29	13.68			1.77	5131
1157.6 +5315		.S?....	1.00± .06	136				16.32±.3	1108± 8
140.18 62.04	UGC 6940		.79± .05	.00				109± 6	
61.70 4.27				1.19					1181
1155.1 +5332	PGC 37621		1.00	.40					1290
115744.4+321738	NGC 3995	.SA.9P.	1.44± .02	33	12.7 ±.2	.27± .02	.35± .02	13.79±.2	3254± 5
185.80 77.26	UGC 6944	R (1)	.42± .03	.02	12.85 ±.14	-.41± .04	-.32± .03	210± 5	3339± 28
81.47 -3.25	IRAS11550+3234	9.0± .4		.43	11.64	.15		133± 10	3257
115509.9+323420	PGC 37624	5.7± .9	1.45	.21	12.34	-.49	13.71± .24	1.24	3536

R.A. 2000 DEC. / l b / SGL SGB / R.A. 1950 DEC.	Names / PGC	Type / S_T n_L / T / L	$\log D_{25}$ / $\log R_{25}$ / $\log A_e$ / $\log D_o$	p.a. / A_g / A_i / A_{21}	B_T / m_B / m_{FIR} / B_T^o	$(B-V)_T$ / $(U-B)_T$ / $(B-V)_T^o$ / $(U-B)_T^o$	$(B-V)_e$ / $(U-B)_e$ / m'_e / m'_{25}	m_{21} / W_{20} / W_{50} / HI	V_{21} / V_{opt} / V_{GSR} / V_{3K}
1157.7 -1010		.IBS9..	1.20± .07	10					
281.89 50.49		E (1)	.28± .08	.07					
123.00 -16.39	MK 1309	10.0± .9		.21	13.05				
1155.2 -0953	PGC 37625	5.3± .8	1.21	.14					
115746.4+141749	NGC 3996	.S?....	.97± .07	50					
255.98 72.08	UGC 6941		.12± .05	.04	14.36 ±.18				6989± 50
98.69 -9.42	IRAS11551+1434			.18	13.23				6926
115512.4+143431	PGC 37628		.98	.06	14.11				7322
115747.3+251618	NGC 3997	.SB.3P.	1.22± .03		14.02 ±.14	.57± .02	.52± .01	15.07±.2	4768± 6
218.69 77.83	UGC 6942	P	.43± .04	.00	14.17 ±.18	-.08± .03	-.10± .03	288± 7	4742± 38
88.15 -5.71	IRAS11552+2532	3.0± .8	.61± .02	.59	12.93	.46	12.56± .06		4742
115513.0+253300	PGC 37629		1.22	.21	13.45	-.16	13.91± .22	1.40	5070
115750.4+290234	NGC 3984	.SBT3..	1.08± .06					15.64±.2	6408± 8
200.61 78.03	UGC 6943	U	.02± .05	.00	14.3 ±.2			79± 6	6396
84.56 -4.38		3.0± .8		.03				60± 7	6698
115516.0+291916	PGC 37632		1.08	.01	14.19			1.43	
1157.8 +2311	NGC 4002	.L?....	.90± .17	75					
228.00 77.26	MCG 4-28-104		.27± .07	.04	14.96 ±.18				6572± 56
90.14 -6.42				.00					6539
1155.3 +2328	PGC 37635		.86		14.81				6882
115755.2+362332		.S?....	1.08± .09						10436± 33
170.39 75.36	UGC 6945		.18± .07	.00					10451
77.61 -1.74	ARP 194			.25					10698
115520.7+364014	PGC 37639		1.08	.09					
115756.6+552715	NGC 3998	.LARO$.	1.43± .03	140	11.61 ±.13	.95± .01	.99± .01	15.39±.4	1040± 13
138.17 60.06	UGC 6946	R	.08± .04	.00	11.50 ±.14	.52± .01	.57± .01	633± 24	1066± 19
59.63 5.08		-2.0± .3	.76± .02	.00		.93	10.90± .06	574± 18	1129
115521.4+554357	PGC 37642		1.41		11.54	.52	13.42± .21		1219
115757.3+250842	NGC 4000	.S?....	1.02± .06	3				16.44±.2	4557± 6
219.32 77.84	UGC 6949		.70± .05	.00	15.47 ±.20			311± 7	
88.28 -5.72	IRAS11554+2524			1.05	13.42				4531
115523.0+252524	PGC 37643		1.02	.35	14.39			1.70	4860
1157.9 +2307	NGC 4003	.LB....	1.17± .09	10					
228.32 77.26	UGC 6948	U	.23± .05	.04	14.28 ±.17				6509± 32
90.21 -6.42		-2.0± .8		.00					6476
1155.4 +2324	PGC 37646		1.14		14.14				6818
115802.6-515300		.LB?...	.88± .07						11500±190
294.61 10.12	ESO 217- 15		.04± .05	.74	14.84 ±.14				11297
166.73 -21.05	FAIR 446			.00					11747
115530.1-513618	PGC 37650		.94		13.93				
115802.7-034436		PSB.3..	.61± .11						
278.23 56.55	MCG 0-31- 5	F (1)	.01± .06	.05	15.3 ±.2				
116.51 -14.72	IRAS11554-0327	3.0±1.0		.01	13.00				
115529.0-032754	PGC 37651	1.6±1.0	.62	.00					
115805.4+275238	NGC 4004	.P.....	1.26± .03	8	14.1 ±.2	.44± .03		14.28±.1	3377± 6
206.24 78.15	UGC 6950	R	.49± .04	.00	14.00 ±.18	-.27± .05		316± 6	3377± 23
85.69 -4.74	MK 432	99.0		.37	11.73	.30			3361
115531.1+280920	PGC 37654		1.26	.25	13.67	-.37	14.02± .28	.36	3671
115806.0-020715	NGC 4006	.E...P*	1.22± .08	20					
277.11 58.05	UGC 6951	UEF	.13± .05	.04	13.55 ±.16				
114.89 -14.27		-4.7± .6		.00					
115532.2-015033	PGC 37655		1.19						
115810.3+250718	NGC 4005	.S?....	1.08± .03	92	13.89 ±.14	.76± .02	.84± .02	16.56±.2	4464± 6
219.47 77.88	UGC 6952		.23± .05	.00	14.16 ±.18			432± 7	4425± 27
88.32 -5.68			.59± .02	.34		.69	12.33± .06		4436
115536.1+252400	PGC 37661		1.08	.11	13.63		13.59± .25	2.82	4766
115814.8-015336		.LARO..							
277.01 58.28	CGCG 13- 16	F		.04	15.2 ±.6				
114.67 -14.17		-2.0± .9							
115541.1-013654	PGC 37665								
115817.2+281134	NGC 4008	.E.5...	1.39± .05	167	12.96 ±.15	.94± .02	.99± .01		
204.98 78.19	UGC 6953	R	.28± .04	.00	12.73 ±.11	.24± .08			3562± 46
85.40 -4.59		-5.0± .4	.76± .04	.00		.91	12.25± .15		3548
115542.9+282816	PGC 37666		1.30		12.76	.26	14.19± .29		3855
1158.3 -1444	A 1155-14A	.IBS9P*	1.06± .06	115	15.8 ±.2				1979
284.20 46.20	MCG -2-31- 7	UEF (2)	.45± .07	.07		.22± .11	.19± .08		
127.69 -17.27	DDO 104	9.9± .6	.80± .04	.34			15.30± .08		1830
1155.7 -1427	PGC 37667	9.6± .8	1.07	.22	15.39	.12	14.86± .39		2329

R.A. 2000 DEC. l b SGL SGB R.A. 1950 DEC.	Names PGC	Type S_T n_L T L	$\log D_{25}$ $\log R_{25}$ $\log A_e$ $\log D_o$	p.a. A_g A_i A_{21}	B_T m_B m_{FIR} B_T^o	$(B-V)_T$ $(U-B)_T$ $(B-V)_T^o$ $(U-B)_T^o$	$(B-V)_e$ $(U-B)_e$ m'_e m'_{25}	m_{21} W_{20} W_{50} HI	V_{21} V_{opt} V_{GSR} V_{3K}
115817.7-510806 294.49 10.86 165.94 -21.03 115545.0-505124	ESO 217- 16 IRAS11557-5051 PGC 37668	.SXR3P* Sr (1) 2.6± .6 3.3±1.3	1.10± .05 .30± .04 1.16	157 .64 .41 .15	14.74 ±.14 13.19 13.65				5082± 87 4880 5332
115819.1-393242 291.90 22.19 153.65 -20.74 115546.0-391600	ESO 321- 1 IRAS11557-3916 PGC 37669	.SAT6.. Sr (1) 5.9± .5 5.0± .6	1.28± .05 .22± .05 1.32	140 .47 .33 .11	13.55 ±.14 13.21				
115820.8-505824 294.46 11.02 165.77 -21.02 115548.0-504142	ESO 217- 17 IRAS11557-5041 PGC 37670	.S..5*. S (1) 5.0±1.3 4.4± .9	1.18± .04 .48± .04 1.24	165 .55 .72 .24	14.95 ±.14 13.37				
1158.3 -0156 277.10 58.24 114.73 -14.16 1155.8 -0140	CGCG 13- 17 PGC 37673	.L..... F -2.0± .9		.04	15.3 ±.6				
115824.0-021638 277.35 57.94 115.07 -14.24 115550.2-015956	UGC 6958 IRAS11558-0200 PGC 37678	.SXT4.. UEF (2) 3.7± .5 4.1± .6	1.20± .04 .08± .04 1.21	65 .04 .12 .04	13.8 ±.2				
1158.4 -1431 284.15 46.40 127.49 -17.20 1155.8 -1414	A 1155-14B MCG -2-31- 6 DDO 103 PGC 37680	.I..9P. F (1) 10.0± .9 7.5±1.6	1.05± .05 .06± .05 .79± .02 1.06	10 .09 .04 .03	14.60 ±.14	.36± .05 -.22± .07	.44± .05 -.13± .07 14.04± .06 14.55± .32		
115825.4-222624 287.15 38.80 135.67 -18.71 115552.0-220942	A 1155-22 ESO 572- 30 DDO 106 PGC 37681	.SBS9*. RSU (1) 9.2± .4 9.1± .4	1.41± .03 .19± .04 1.30± .08 1.43	103 .15 .19 .10	13.6 ±.3 14.38 ±.14 13.89	.31± .07 -.02± .09 .22 -.08	.39± .07 -.05± .09 15.55± .18 15.04± .35	14.45±.2 RSU ± 16 135± 12 .46	1784± 11 1618 2124
1158.4 +5054 142.31 64.17 63.95 3.54 1155.8 +5111	A 1155+51 UGC 6956 DDO 102 PGC 37682	.SBS9.. PU (1) 9.0± .5 9.0±1.0	1.35± .05 .03± .03 1.17± .08 1.36	 .04 .03 .01	14.92S±.10 14.85	.41± .12 .07± .16 .39 .05	.49± .06 .05± .08 16.13± .21 16.47± .28	14.82±.1 61± 5 52± 5 -.04	917± 4 982 1110
1158.4 +1001 264.29 68.82 102.91 -10.63 1155.9 +1018	NGC 4012 UGC 6960 PGC 37686	.S..3.. U (1) 3.0± .9 4.5±1.2	1.29± .04 .54± .05 1.29	153 .00 .75 .27	14.25 ±.18				
115829.3+273144 207.94 78.24 86.05 -4.78 115555.1+274826	NGC 4016 UGC 6954 KUG 1155+278 PGC 37687	.SB.8*. U 8.0±1.2	1.19± .04 .29± .04 1.19	175 .00 .36 .15	13.8 S±.3 14.23 ±.19 13.72		13.82± .34	15.28±.2 126± 21 81± 16 1.42	3447± 7 3494± 65 3431 3743
115829.9+380433 165.03 74.41 76.06 -1.03 115555.5+382115	A 1155+38 UGC 6955 DDO 105 PGC 37689	.IBS9*. PU (1) 9.5± .5 9.0± .9	1.70± .03 .29± .05 1.23± .05 1.70	70 .00 .22 .15	13.8 ±.2 14.2 ±.8 13.57	.54± .06 -.04± .08 .47 -.09	.49± .03 -.06± .05 15.40± .10 16.39± .27	13.60±.1 157± 6 144± 12 -.12	909± 7 929 1163
115831.5-182042 285.73 42.75 131.43 -17.95 115558.0-180400	NGC 4024 ESO 572- 31 PGC 37690	.LX.-.. R -3.0± .4	1.28± .03 .11± .03 .85± .04 1.27	125 .06 .00	12.65 ±.15 12.84 ±.11 12.69	.94± .01 .91	.96± .01 12.39± .14 13.66± .22		1653± 41 1495 1999
115831.7+435648 151.86 70.09 70.53 1.08 115557.2+441330	NGC 4013 UGC 6963 IRAS11559+4413 PGC 37691	.S..3./ R (1) 3.0± .3 4.5±1.0	1.72± .02 .71± .03 1.27± .04 1.72	66 .00 .98 .35	12.19M±.11 12.39 ±.12 10.90 11.30	.96± .03 .32± .04 .83 .18	1.04± .02 .34± .03 14.06± .13 13.86± .16	13.56±.1 407± 8 392± 5 1.90	839± 6 694± 46 878 1065
115833.6+424410 154.08 71.05 71.67 .65 115559.1+430052	IC 749 UGC 6962 PGC 37692	.SXT6.. PU (2) 5.5± .5 4.7± .7	1.37± .04 .09± .04 1.37	150 .00 .13 .04	12.92S±.15 13.04 ±.15 12.84	.56± .03 -.04± .05 .54 -.06	.63± .02 -.06± .04 14.39± .25	14.08±.1 138± 12 1.19	784± 11 809± 40 823 1020
115833.8-020812 277.32 58.08 114.94 -14.16 115600.0-015130	MCG 0-31- 10 PGC 37693	.LBR+*. F -1.0±1.2	1.02± .14 .00± .07 1.02	.04 .00	14.21 ±.16				
115835.5+161040 252.03 73.57 96.94 -8.62 115601.5+162722	NGC 4014 UGC 6961 IRAS11560+1627 PGC 37695	.S..0.. U .0± .8	1.35± .05 .22± .06 1.35	120 .10 .17	 13.29 ±.18 12.00 12.97			14.71±.2 437± 10 403± 8	3760± 7 3775± 50 3704 4088
115837.8+471538 146.68 67.36 67.41 2.28 115603.2+473220	NGC 4010 UGC 6964 PGC 37697	.SBS7*/ PU 6.7± .6	1.63± .02 .73± .03 1.63	66 .00 1.01 .37	13.17S±.15 12.99 ±.18 12.08		14.32± .20	13.47±.1 269± 7 257± 5 1.03	907± 6 959 1119

R.A. 2000 DEC. / l b / SGL SGB / R.A. 1950 DEC.	Names / PGC	Type / S_T n_L / T / L	$\log D_{25}$ / $\log R_{25}$ / $\log A_o$ / $\log D_o$	p.a. / A_g / A_l / A_{21}	B_T / m_B / m_{FIR} / B_T^o	$(B-V)_T$ / $(U-B)_T$ / $(B-V)_T^o$ / $(U-B)_T^o$	$(B-V)_\bullet$ / $(U-B)_\bullet$ / $m'_\bullet$ / m'_{25}	m_{21} / W_{20} / W_{50} / HI	V_{21} / V_{opt} / V_{GSR} / V_{3K}
115840.2+251904	NGC 4018	.S..2..	1.23± .03	163				15.12±.2	4481± 6
218.67 78.03	UGC 6966	U	.74± .04		14.73 ±.18			364± 7	4463± 35
88.17 -5.51	IRAS11561+2535	2.0± .9		.91	11.97				4455
115606.0+253546	PGC 37699		1.23	.37	13.76			.99	4783
1158.6 +5325	A 1156+53	.I..9..	1.21± .05	152				15.31±.3	1113± 8
139.74 61.95	UGC 6969	U	.50± .05	.00	15.2 ±.2			155± 6	
61.59 4.46		10.0± .9		.38					1186
1156.1 +5342	PGC 37700		1.21	.25	14.78			.28	1293
115843.2+250235		.S?....	.94± .11					16.46±.3	4347± 9
219.97 77.99	MCG 4-28-110		.68± .07	.00				811± 10	4347± 43
88.44 -5.59				1.02					4321
115609.1+251917	PGC 37702		.94	.34					4651
115843.7+281725		.S?....	1.45± .03	85				15.01±.3	8232± 11
204.19 78.28	UGC 6968		.44± .05	.00	14.1 ±.4			541± 15	
85.35 -4.46				.67					8218
115609.5+283407	PGC 37704		1.45	.22	13.35			1.44	8525
115845.3+272715	NGC 4017	.SX.4..	1.25± .03		13.0 S±.3			14.14±.2	3454± 5
208.31 78.30	UGC 6967	U	.11± .04	.01	13.28 ±.18			285± 5	3401± 26
86.14 -4.75	ARP 305	4.0± .8		.15	12.20			256± 7	3434
115611.1+274357	PGC 37705		1.26	.05	13.01		13.86± .33	1.09	3747
115845.5-195036		.SBS7P.	1.29± .06						1638
286.35 41.33	ESO 572- 32	S (1)	.11± .05	.08	17.03 ±.14				
132.99 -18.18		7.0± .8		.15					1478
115612.0-193354	PGC 37707	7.8±1.2	1.30	.06	16.79				1982
1158.7 -0237		.SB.5?.							
277.76 57.65	CGCG 13- 21	F		.06	15.1 ±.6				
115.44 -14.25		5.0±1.8							
1156.2 -0221	PGC 37708								
1158.7 -0126		.SBS9..	1.19± .05	75				15.81±.1	1484± 5
276.90 58.74	UGC 6970	UEF (1)	.28± .05	.05				164± 7	1520± 46
114.26 -13.93		9.2± .5		.29				128± 7	1372
1156.2 -0110	PGC 37710	6.6± .6	1.19	.14					1838
115851.8+424320	IC 750	.S..2*/	1.42± .03	43	12.94S±.15	1.00± .02	1.03± .02		
153.97 71.10	UGC 6973	P	.35± .04	.00	12.85 ±.13	.36± .05	.46± .05		703± 28
71.70 .70		2.0±1.1		.43		.93			740
115617.3+430002	PGC 37719		1.42	.18	12.44	.29	13.99± .25		937
115852.4+423415	IC 751	.S..3$/	1.08± .04	30					
154.26 71.22	UGC 6972	P	.57± .04	.00	15.07 ±.18				
71.85 .64	KUG 1156+428A	3.0±1.8		.79					
115618.0+425057	PGC 37721		1.08	.29					
1158.8 +4543		.I..9..	1.04± .18					15.47±.1	1154± 11
148.81 68.68	UGCA 259	U	.09± .18	.00				343± 16	
68.87 1.77		10.0± .9		.07					1201
1156.3 +4600	PGC 37722		1.04	.05					1374
115855.6+302448	NGC 4020	.SB.7?/	1.32± .03	15				14.53±.3	757± 11
193.90 78.04	UGC 6971	PU	.35± .04	.00	13.28 ±.18			181± 16	814± 65
83.34 -3.67	IRAS11563+3041	7.3± .7		.48	13.01			172± 12	752
115621.4+304130	PGC 37723		1.32	.17	12.79			1.56	1044
115858.6-190136		.IB.9..	1.02± .05	24					
286.12 42.13	ESO 572- 34	EF (1)	.32± .03	.09	14.19 ±.14				1075± 28
132.15 -17.97	IRAS11564-1844	10.0± .9		.24	13.40				916
115625.0-184454	PGC 37727	6.4±1.2	1.02	.16	13.87				1420
115858.8+541412		.S?....	.80± .07						
138.93 61.23	MCG 9-20- 49		.04± .06	.00	14.78 ±.19				3564± 81
60.84 4.79	MK 433			.06					3640
115624.0+543054	PGC 37728		.80	.02	14.71				3740
115901.1+251319	NGC 4022	.LX.0*.	1.08± .10						
219.20 78.09	UGC 6975	U	.01± .05	.00	14.01 ±.16				4340± 27
88.29 -5.47		-2.0± .8		.00					4315
115627.0+253001	PGC 37729		1.08		13.94				4643
1159.0 +2149		.L...*.	1.15± .15						
234.11 76.97	UGC 6976	U	.00± .08	.06	14.1 ±.2				
91.55 -6.62		-2.0±1.1		.00					
1156.5 +2206	PGC 37731		1.16						
115905.1+245919	NGC 4023	.S?....	.96± .07	25				16.97±.3	4408± 9
220.32 78.06	UGC 6977		.14± .05	.00	14.60 ±.18			217± 10	4800± 52
88.52 -5.54				.17					4393
115631.1+251601	PGC 37732		.96	.07	14.39			2.52	4722

R.A. 2000 DEC. l b SGL SGB R.A. 1950 DEC.	Names PGC	Type S_T n_L T L	$\log D_{25}$ $\log R_{25}$ $\log A_e$ $\log D_o$	p.a. A_g A_i A_{21}	B_T m_B m_{FIR} B_T^o	$(B-V)_T$ $(U-B)_T$ $(B-V)_T^o$ $(U-B)_T^o$	$(B-V)_e$ $(U-B)_e$ m' m'_{25}	m_{21} W_{20} W_{50} HI	V_{21} V_{opt} V_{GSR} V_{3K}
115908.6+524226 A 1156+52 140.27 62.62 UGC 6983 62.29 4.28 115634.0+525908 PGC 37735		.SBT6.. U (1) 6.0± .7 5.0±1.0	1.54± .03 .16± .03 1.54	85 .01 .24 .08	13.12S±.08 13.8 ±.5 12.89		.54± .06 -.14± .10 15.28± .17	13.43±.1 194± 5 176± 5 .47	1080± 4 1060±122 1151 1264
1159.1 -0234 277.90 57.73 UGC 6978 115.42 -14.14 1156.6 -0218 PGC 37737		RSBS9*. UEF (1) 8.8± .5 7.6± .7	1.11± .07 .17± .07 1.11	130 .04 .17 .08	*			15.58±.1 141± 12	1536± 9 1420 1890
115909.9+374734 NGC 4025 165.52 74.69 UGC 6982 76.38 -1.01 DDO 107 115635.6+380416 PGC 37738		.SBS6.. PU (1) 6.0± .5 7.0±1.3	1.44± .02 .23± .04 .99± .02 1.44	40 .00 .33 .11	14.03 ±.14 14.3 ±.4 13.71	.48± .03 -.15± .05 .42 -.20	.55± .03 -.05± .04 14.47± .04 15.49± .20		3215 3235 3471
115910.0-532436 295.10 8.66 ESO 171- 4 168.35 -20.85 IRAS11566-5307 115637.0-530754 PGC 37739		.LXR0?. S -2.0± .7 	1.25± .05 .21± .04 1.34	62 1.04 .00 	13.58 ±.14 12.84				
1159.1 +2428 222.75 77.95 UGC 6980 89.02 -5.70 1156.6 +2445 PGC 37740		.I..9.. U 10.0± .9 	1.11± .14 .36± .12 1.11	153 .00 .27 .18					3349± 10 3322 3654
115912.3+304358 IC 2985 192.33 78.03 UGC 6981 83.06 -3.51 KUG 1156+310 115638.1+310040 PGC 37744		.SB?... 	.98± .05 .32± .04 .98	142 .00 .48 .16	15.09 ±.18 14.59			16.45±.3 158± 15 1.70	3322± 11 3317 3606
115912.6-003131 IC 753 276.39 59.63 UGC 6979 113.37 -13.57 ARAK 340 115638.8-001449 PGC 37745		.LBR0.. F -2.0±1.0 	.62± .14 .15± .03 .60	30 .04 .00 	15.2 ±.3				
115917.5-285418 289.38 32.59 ESO 440- 37 142.45 -19.50 115644.0-283736 PGC 37752		PLAT0.. S -2.0± .8 	1.01± .06 .04± .04 1.04	 .29 .00 	14.14 ±.14 13.82				2073± 30 1894 2399
115923.6-013909 IC 754 277.33 58.61 UGC 6984 114.51 -13.83 115649.8-012227 PGC 37757		.E+.... UEF -4.3± .4 	1.06± .07 .08± .03 1.04	40 .05 .00 	14.05 ±.15				
115925.6+505743 NGC 4026 141.94 64.20 UGC 6985 63.96 3.71 115651.1+511425 PGC 37760		.L..../ R -2.0± .3 	1.72± .03 .61± .03 .74± .06 1.63	178 .04 .00 	11.67M±.12 11.65 ±.11 11.60	.92± .04 .85 	.98± .01 .53± .02 10.90± .20 13.60± .19		930± 35 995 1123
115926.5+174518 248.24 74.78 UGC 6986 95.48 -7.91 KUG 1156+180 115652.6+180200 PGC 37761		.S?.... 	1.05± .04 .51± .04 1.06	129 .08 .76 .25	15.18 ±.18 14.30			15.90±.3 324± 13 1.34	6449± 10 6398 6773
1159.4 -0107 276.97 59.10 CGCG 13- 27 113.99 -13.67 1156.9 -0051 PGC 37764		.SB.7?/ F 7.0±1.8 		 .02 	15.4 ±.6				
115928.7+345338 174.81 76.45 MCG 6-26- 67 79.14 -1.98 MK 434 115654.5+351020 PGC 37769		.S?.... 	1.04± .09 .44± .07 1.04	 .00 .64 .22	15.34 ±.18 14.63				9823± 43 9833 10091
115929.6-191954 NGC 4027A 286.39 41.87 ESO 572- 36 132.49 -17.91 115656.0-190312 PGC 37772		.IBS9*. EF (1) 10.0± .6 9.6±1.4	.96± .03 .16± .03 .97	159 .06 .12 .08	14.94M±.04 15.03 ±.14 14.76	.45± .05 -.12± .10 .39 -.17	 14.19± .18		1747± 73 1588 2091
115930.1-191605 NGC 4027 286.37 41.93 ESO 572- 37 132.43 -17.90 8ZW 158 115656.5-185923 PGC 37773		.SBS8.. R (3) 8.0± .3 6.4± .5	1.50± .01 .12± .02 1.08± .01 1.50	167 .06 .15 .06	11.66M±.04 11.66 ±.12 10.40 11.44	.54± .01 -.04± .01 .49 -.07	.52± .01 -.05± .01 12.45± .02 13.68± .10	13.05±.2 205± 8 139± 11 1.55	1671± 6 1639± 18 1509 2012
115931.4-364306 291.47 25.00 ESO 379- 19 150.68 -20.30 IRAS11569-3626 115658.0-362624 PGC 37774		PSXR2.. r 2.2± .9 	.93± .06 .12± .05 .96	45 .33 .14 .06	14.83 ±.14 13.56				
115931.5+300928 194.98 78.22 UGC 6987 83.63 -3.64 KUG 1156+304 115657.4+302610 PGC 37775		.SBS3.. U 3.0± .9 	1.03± .04 .23± .04 1.03	10 .00 .32 .12	14.89 ±.19 14.50			16.34±.3 355± 15 1.72	8775± 11 8768 9061

R.A. 2000 DEC. / l b / SGL SGB / R.A. 1950 DEC.	Names / PGC	Type / S_T n_L / T / L	$\log D_{25}$ / $\log R_{25}$ / $\log A_o$ / $\log D_o$	p.a. / A_g / A_i / A_{21}	B_T / m_B / m_{FIR} / B_T^o	$(B-V)_T$ / $(U-B)_T$ / $(B-V)_T^o$ / $(U-B)_T^o$	$(B-V)_\bullet$ / $(U-B)_\bullet$ / $m'_\bullet$ / m'_{25}	m_{21} / W_{20} / W_{50} / HI	V_{21} / V_{opt} / V_{GSR} / V_{3K}
1159.5 −0321 / 278.61 57.04 / 116.23 −14.25 / 1157.0 −0305	CGCG 13- 28 / PGC 37778	.E...*. / F / −5.0±1.3		.05	15.4 ±.6				
1159.5 +1353 / 257.92 72.06 / 99.23 −9.14 / 1157.0 +1410	MCG 2-31- 7 / PGC 37779	.S?....	.94± .11 / .28± .07 / / .94	.04 / .43 / .14	14.98 ±.18 / / / 14.51			15.81±.3 / 77± 13 / 58± 10 / 1.16	1450± 10 / / 1386 / 1783
1159.6 −0339 / 278.86 56.77 / 116.54 −14.30 / 1157.1 −0323	CGCG 13- 29 / PGC 37784	.L..0.. / F / −2.0± .9		.06	15.0 ±.6				
115940.2+263247 / 212.87 78.44 / 87.08 −4.87 / 115706.2+264929	MCG 5-28- 71 / KUG 1157+268 / PGC 37786	.S?....	.98± .06 / .02± .06 / / .98	.00 / .03 / .01	15.0 ±.2 / / / 14.90			16.69±.3 / 146± 14 / 109± 15 / 1.77	6694± 12 / / 6674 / 6992
1159.7 −0322 / 278.71 57.04 / 116.27 −14.21 / 1157.2 −0306	MCG 0-31- 14 / PGC 37791	.SXS5.. / F (1) / 5.0± .9 / 1.6±1.0	1.04± .09 / .66± .07 / / 1.04	.06 / 1.00 / .33	15.50 ±.18				
1159.7 +2618 / 214.08 78.44 / 87.32 −4.94 / 1157.2 +2635	CGCG 157- 77 / PGC 37794			.00	15.3 ±.3				4100 / 4079 / 4399
1159.7 +3052 / 191.53 78.11 / 82.97 −3.34 / 1157.2 +3109	IC 2986 / MCG 5-28- 72 / PGC 37795	.L?.... / / / .90	.90± .17 / .00± .07	.00 / .00	14.75 ±.15 / / / 14.70				3139± 19 / 3134 / 3422
1159.8 +5542 / 137.47 59.94 / 59.49 5.42 / 1157.3 +5559	UGC 6988 / PGC 37809	.I..9*. / U / 10.0±1.3	1.11± .05 / .54± .05 / / 1.11	134 / .00 / .41 / .27					
115955.8−003254 / 276.74 59.68 / 113.45 −13.40 / 115722.0−001612	MCG 0-31- 15 / PGC 37810	.SA.4*. / F (1) / 4.0±1.3 / 5.6±1.4	.89± .11 / .07± .07 / / .89	.04 / .10 / .03	15.03 ±.19				
1200.0 +0811 / 267.85 67.49 / 104.83 −10.82 / 1157.5 +0828	NGC 4029 / UGC 6990 / PGC 37816	.S..3*. / U / 3.0±1.2	1.08± .06 / .20± .05 / / 1.08	150 / .00 / .28 / .10	14.29 ±.18 / / / 13.97			16.65±.3 / 431± 13 / 425± 10 / 2.58	6137± 10 / 6144± 50 / 6055 / 6480
1200.2 +3148 / 187.09 77.93 / 82.13 −2.92 / 1157.7 +3205	CGCG 157- 81 / PGC 37831			.00	15.2 ±.3				7723 / 7722 / 8003
1200.2 +5040 / 141.97 64.52 / 64.28 3.73 / 1157.7 +5057	UGC 6992 / PGC 37832	.S..8*. / U / 8.0±1.2	1.13± .05 / .34± .05 / / 1.13	63 / .02 / .42 / .17	14.71 ±.18				
120016.4+311331 / 189.74 78.11 / 82.68 −3.12 / 115742.4+313013	MCG 5-28- 74 / KUG 1157+315 / PGC 37838	.S?....	.91± .06 / .36± .06 / / .91	.00 / .53 / .18	15.1 ±.2 / / / 14.57				714± 19 / 711 / 996
120017.8+001736 / 276.24 60.49 / 112.64 −13.08 / 115744.0+003418	CGCG 13- 34 / PGC 37839	PSB.1P? / F / 1.0±1.3		.07	15.1 ±.6				
120023.8+002924 / 276.12 60.68 / 112.45 −13.00 / 115750.0+004606	CGCG 13- 35 / PGC 37844	.LX.0.. / F / −2.0± .9		.07	15.1 ±.6				
120024.1−010604 / 277.37 59.21 / 114.03 −13.44 / 115750.3−004922	NGC 4030 / UGC 6993 / IRAS11578-0049 / PGC 37845	.SAS4.. / R (4) / 4.0± .3 / 1.6± .4	1.62± .02 / .14± .02 / / 1.63	27 / .03 / .21 / .07	11.41 ±.15 / 9.88 / / 11.16			13.06±.1 / 347± 8 / 329± 7 / 1.83	1460± 5 / 1449± 19 / 1348 / 1812
120029.3−155654 / 285.43 45.18 / 129.06 −17.01 / 115755.6−154012	NGC 4035 / MCG -3-31- 10 / IRAS11579-1540 / PGC 37853	PSXT4P. / PEF (1) / 4.3± .5 / 3.4± .7	1.09± .06 / .06± .05 / / 1.10	0 / .10 / .09 / .03	/ / 13.48 /			14.66±.3 / 195± 15 / 146± 12 /	1567± 11 / 1604± 31 / 1420 / 1919